Greenfield's
Neuropathology

Greenfield's Neuropathology

NINTH EDITION

VOLUME 1

Edited by

Seth Love MBBCh PhD FRCP FRCPath
Professor of Neuropathology
Institute of Clinical Neurosciences
School of Clinical Sciences
University of Bristol
Southmead Hospital
Bristol, UK

Herbert Budka MD
Professor of Neuropathology
Consultant, Institute of Neuropathology
University Hospital Zurich
Zurich, Switzerland

James W Ironside CBE BMSc FRCPath FRCPEdin FMedSci FRSE
Professor of Clinical Neuropathology and Honorary
Consultant in Neuropathology
National Creutzfeldt–Jakob Disease Surveillance Unit
University of Edinburgh
Western General Hospital
Edinburgh, UK

Arie Perry MD
Professor of Pathology and Neurological Surgery
Director of Neuropathology
Director of Neuropathology Fellowship Training Program
University of California, San Francisco
San Francisco, CA, USA

CRC Press
Taylor & Francis Group
Boca Raton London New York

CRC Press is an imprint of the
Taylor & Francis Group, an **informa** business

CRC Press
Taylor & Francis Group
6000 Broken Sound Parkway NW, Suite 300
Boca Raton, FL 33487-2742

© 2015 by Taylor & Francis Group, LLC
CRC Press is an imprint of Taylor & Francis Group, an Informa business

No claim to original U.S. Government works

Printed on acid-free paper
Version Date: 20141203

Printed and bound in India by Replika Press Pvt. Ltd.

International Standard Book Number-13: (Volume I) 978-1-4441-6693-4 (Hardback)
(Two-volume set) 978-1-4987-2128-8 (Hardback)

Visit the Taylor & Francis Web site at
http://www.taylorandfrancis.com

and the CRC Press Web site at
http://www.crcpress.com

Contents

Preface

Greenfield's Neuropathology holds a special place in the heart of most neuropathologists. It has long been a standard-bearer of our specialty. In 1921, Joseph Godwin Greenfield and Edward Farquhar Buzzard published *Pathology of the Nervous System*, which had a key role in defining neuropathology as a distinct specialty. The authors set out to 'describe clearly the anatomical changes which are associated with disorders of nervous function, to discuss briefly questions of pathogenesis, and to indicate in a few words, where it is possible, the relationship between structural alterations and clinical signs and symptoms.' In 1958, a book entitled simply *Neuropathology*, by Greenfield, William Blackwood, William McMenemy, Alfred Meyer and Ronald Norman, updated and greatly expanded on most of the content of *Pathology of the Nervous System*. Unlike *Pathology of the Nervous System*, however, *Neuropathology* did not cover neoplastic diseases (dealt with instead in a companion book, Russell and Rubinstein's *Pathology of Tumours of the Nervous System*). However, tumours of the nervous system have been included in *Greenfield's Neuropathology* since the seventh edition in 1997.

Readers of a succession of editions over many decades have dipped into this venerable reference book seeking definitive advice and instruction on all matters neuropathological. Producing a new edition of *Greenfield's Neuropathology* has therefore been both a huge privilege and a massive responsibility. It has also been a balancing act, in which we have had to reconcile the tension between the physical constraints of a two-volume book and the ever-expanding amount of information encompassed within our field. Indeed, this may be the last edition of *Greenfield's Neuropathology* that can be produced in hardcover printed format. Accommodating the additional information has largely involved a combination of reorganisation and restraint, together with considerably increased use of photographs and diagrams.

The reorganisation has involved the merging of vascular disease, hypoxia and related conditions into a single chapter; the subdivision of movement disorders into separate chapters on extrapyramidal disorders, ataxias and motor neuron diseases; the inclusion of separate chapters on ageing and dementia, the latter encompassing an expanded section on vascular dementia; and the further subdivision of the tumour section from two chapters in the previous edition to twenty-one in the present one, which we hope will make this part of the book easier to navigate. The total number of chapters in the book has increased from twenty-four to forty-six. Restraint has been applied in relation to the inclusion of references and of some very

detailed molecular genetic and phenotypic information that is readily accessible through online resources such as OMIM, the database of Genotypes and Phenotypes (dbGaP), AlzGene and PDGene. We expect readers to look to *Greenfield's Neuropathology* for guidance and perspective rather than as a substitute for bibliographic databases and search engines.

The changes have involved a great deal of work on the part of our authors, who have shown unfailing courtesy and forbearance in responding to requests to condense prose, reorganise chapters and be selective in the inclusion of references. We are in their debt. Throughout, our objectives, much like those of Greenfield and Buzzard, have been to describe clearly the neuropathological changes that underlie neurological diseases, to discuss briefly their pathogenesis, and to try to relate molecular genetic, structural and biochemical alterations to clinical and neuroradiological manifestations.

Once a full account has been taken of the clinical and neuroradiological manifestations of neurological disease in a particular patient, a detailed visual examination of the diseased tissue is the starting point for almost all neuropathological investigations. Much of the excitement of neuropathology comes from discovering visual clues to disease, macroscopic or microscopic, whether in a section stained simply with haematoxylin and eosin, a series of confocal laser scanning images or a transmission electron micrograph. Neuropathology remains a highly visual specialty and most of us neuropathologists obtain immense aesthetic gratification from our work. Not surprisingly, therefore, we have placed a strong emphasis on visual aspects of this reference book, which includes over one thousand completely new photographs and drawings. It also incorporates new design elements such as the alternate colour coding of chapters that is intended to allow their easier navigation. To this same end, both volumes now include full indexes to the whole book. There are also improved search, annotation and bookmarking facilities in the bundled bonus e-book version of this edition. The e-book frees users from most of the physical limitations (not least of which are the size and weight) of the printed version and can be downloaded to a wide range of mobile and electronic devices, so that it is not necessary to be online to have full access to *Greenfield's Neuropathology*.

Publication of this ninth edition of *Greenfield's Neuropathology* would not have been possible without the support of many people, initially at Hodder Arnold and subsequently at Taylor and Francis. At Hodder Arnold, Joanna Koster, Editorial Director; Caroline

Makepeace, Head of Postgraduate and Professional Publishing; Mischa Barrett, Project Editor; and Miriam Trent, Editorial Assistant, were closely involved in the early stages. At Taylor and Francis, Barbara Norwitz, Executive Editor; Amy Blalock, Supervisor, Editorial Project Development; Rachael Russell, Senior Editorial Assistant; and Linda Van Pelt, Senior Project Manager, Medical, all worked on different stages of the title, and one person who merits special thanks is Sue Hodgson for her invaluable help as Executive Editor. Glenys Norquay provided freelance support and Jayne Jones designed the cover and interior pages.

We are pleased to present the ninth edition of *Greenfield's Neuropathology*. We hope you obtain as much satisfaction from reading this book as we have from editing it.

S Love
H Budka
J W Ironside
A Perry

November 2014

Contributors

Knarik Arkun, MD
Director
Neuropathology and Autopsy Service
Assistant Professor
Department of Pathology
Tufts Medical Center
Boston, MA, USA

Sylvia L Asa, MD, PhD
Medical Director
Laboratory Medicine Program
University Health Network
Lakeridge Health & Women's College Hospital
Senior Scientist
Ontario Cancer Institute
Professor
Department of Laboratory Medicine and Pathobiology
University of Toronto
Toronto, ON, Canada

Juan M Bilbao, FRCP (Canada)
Professor Emeritus of Neuropathology
St Michael's and Sunnybrook Hospitals
University of Toronto
Toronto, ON, Canada

Daniel J Brat, MD, PhD
Professor and Vice Chair
Translational Programs
Department of Pathology and Laboratory Medicine
Emory University School of Medicine
Georgia Research Alliance Distinguished Cancer Scientist
Atlanta, GA, USA

Susan C Brown, PhD
Reader in Translational Medicine
Comparative Biomedical Sciences
Royal Veterinary College
London, UK

Herbert Budka, MD
Professor of Neuropathology
Consultant, Institute of Neuropathology
University Hospital Zurich
Zurich, Switzerland

Steven A Chance, DPhil
Associate Professor in Clinical Neurosciences
Department of Neuropathology
University of Oxford
Oxford, UK

Leila Chimelli, MD, PhD
Professor of Pathology
Federal University of Rio de Janeiro
Rio de Janeiro, Brazil

Patrick F Chinnery,
BMedSci, MBBS, PhD, FRCP, FRCPath, FMedSci
Professor of Neurogenetics
Newcastle University
Newcastle upon Tyne, UK

H Brent Clark, MD, PhD
Director of Neuropathology
Professor of Laboratory Medicine and Pathology,
    Neurology, and Neurosurgery
University of Minnesota Medical School
Minneapolis, MN, USA

Tim J Crow, MBBS, PhD, FRCP, FRCPsych, FMedSci
SANE POWIC
University Department of Psychiatry
Warneford Hospital
Oxford, UK

Matthew D Cykowski, MD
Neuropathology Fellowship Program
Houston Methodist Hospital/MD Anderson Cancer Center
Houston, TX, USA

Jean Debarros, PhD, MBPsS
Research Clinical Psychologist
Counselling Service
University of Oxford
Oxford, UK

Martina Deckert, MD
Professor
Department of Neuropathology
University Hospital of Cologne
Cologne, Germany

Marc R Del Bigio, MD, PhD, FRCPC
Canada Research Chair in Developmental
    Neuropathology
Professor
Department of Pathology (Neuropathology)
University of Manitoba
Winnipeg, MB, Canada

Salvatore DiMauro, MD
Lucy G Moses Professor
Department of Neurology
Columbia University Medical Center
New York, NY, USA

Ann-Christine Duhaime, MD
Director
Pediatric Neurosurgery
Massachusetts General Hospital
Nicholas T Zervas Professor of Neurosurgery
Harvard Medical School
Boston, MA, USA

Charles Eberhart, MD, PhD
Professor of Pathology, Ophthalmology and Oncology
Director of Neuropathology and Ophthalmic Pathology
Johns Hopkins University School of Medicine
Baltimore, MD, USA

Margaret M Esiri, DM, FRCPath
Neuropathology Department
John Radcliffe Hospital
Emeritus Professor of Neuropathology
Nuffield Department of Clinical Neurosciences
University of Oxford
Oxford, UK

Phyllis L Faust, MD, PhD
Associate Professor of Clinical Pathology and Cell Biology
Department of Pathology and Cell Biology
Columbia University
New York, NY, USA

Isidro Ferrer, MD, PhD
Professor
Institute of Neuropathology
Bellvitge University Hospital and University of Barcelona
Hospitalet de Llobregat
Barcelona, Spain

Rebecca D Folkerth, MD
Director of Neuropathology
Department of Pathology
Brigham and Women's Hospital
Consultant in Neuropathology
Boston Children's Hospital
Associate Professor of Pathology
Harvard Medical School
Boston, MA, USA

Christine E Fuller, MD
Professor
Pathology and Neurology
Director
Neuropathology and Autopsy Pathology
Department of Pathology
Virginia Commonwealth University
Richmond, VA, USA

Gregory N Fuller, MD, PhD
Professor and Chief
Section of Neuropathology
The University of Texas
MD Anderson Cancer Center
Department of Pathology
Houston, TX, USA

Bernardino Ghetti, MD, FANA, FAAAS
Distinguished Professor
Indiana University
Chancellor's Professor
Indiana University-Purdue University Indianapolis
Department of Pathology and Laboratory Medicine
Division of Neuropathology
Indiana University School of Medicine
Indianapolis, IN, USA

Jeffrey A Golden, MD
Chair
Department of Pathology
Brigham and Women's Hospital
Ramzi S Cotran Professor of Pathology
Harvard Medical School
Boston, MA, USA

Glenda M Halliday, PhD
Professor of Neuroscience and NHMRC Senior Principal
    Research Fellow
School of Medical Sciences and Neuroscience Research
    Australia
University of New South Wales
Sydney, NSW, Australia

Brian N Harding, MA, DPhil, BM, BCh, FRCPath
Department of Pathology and
    Laboratory Medicine
Children's Hospital of Philadelphia
Perelman School of Medicine
University of Pennsylvania
Philadelphia, PA, USA

John B Harris, PhD, BPharm, FSoCBiol, MRPharmSoc
Emeritus Professor of Experimental
    Neurology
Medical Toxicology Centre
Newcastle University
Newcastle upon Tyne, UK

Jason F Harrison, MD, PhD
Neurosurgery Resident
Department of Neurosurgery
Virginia Commonwealth University
Richmond, VA, USA

Mark W Head, BSc, PhD
Reader
University of Edinburgh
Deputy Director
National CJD Research & Surveillance Unit
Edinburgh, Scotland

J Robin Highley, DPhil, FRCPath
Senior Clinical Lecturer in Neuropathology
Department of Neuroscience
Sheffield Institute of Translational Neuroscience
University of Sheffield
Sheffield, UK

Janice L Holton, BSc, MBChB, PhD, FRCPath
Professor of Neuropathology
Department of Molecular Neuroscience
University College London Institute of Neurology
London, UK

Henry H Houlden, MD, PhD
Professor of Neurology and Neurogenetics
Department of Molecular Neuroscience
University College London Institute of Neurology
London, UK

Paul G Ince, MBBS, MD, FRCPath
Professor of Neuropathology
Head of Department of Neuroscience
University of Sheffield
Sheffield, UK

James W Ironside, CBE, BMSc, FRCPath,
    FRCPEdin, FMedSci, FRSE
Professor of Clinical Neuropathology and Honorary
    Consultant in Neuropathology
National Creutzfeldt–Jakob Disease Surveillance Unit
University of Edinburgh
Western General Hospital
Edinburgh, UK

Thomas S Jacques, PhD, MRCP, FRCPath
Higher Education Funding Council for England
    Clinical Senior Lecturer
Honorary Consultant
Paediatric Neuropathologist
University College London Institute of Child Health and
    Great Ormond Street Hospital
Department of Histopathology
Great Ormond Street Hospital for Children NHS
    Foundation Trust
London, UK

Evelyn Jaros, PhD
Clinical Scientist in Neuropathology
Neuropathology/Cellular Pathology
Newcastle upon Tyne Hospitals NHS Foundation Trust
Honorary Senior Research Associate
Institute of Neuroscience and Institute for Ageing
Newcastle University
Campus for Ageing and Vitality
Newcastle upon Tyne, UK

Martin Jeffrey, BVMS, DVM, Dip ECVP,
    MRCVS, FRCPath
Consultant Pathologist
Pathology Department
Animal Health and Veterinary Laboratories Agency
    (AHVLA-Lasswade)
Penicuik, UK

Anne Jouvet, MD, PhD
Associate Professor of Pathology
Centre de Pathologie et Neuropathologie Est
Centre de Biologie et Pathologie Est
Groupement Hospitalier Est
Hospices Civils de Lyon
Lyon, France

Raj Kalaria, PhD, FRCPath
Professor of Cerebrovascular Pathology
    (Neuropathology)
Institute of Neuroscience
Newcastle University
National Institute for Health Research Biomedical
    Research Building
Campus for Ageing and Vitality
Newcastle upon Tyne, UK

B K Kleinschmidt-DeMasters, MD
Professor of Pathology, Neurology, and
    Neurosurgery
Department of Pathology
University of Colorado School of Medicine
Aurora, CO, USA

Jillian Kril, PhD, FFSc (RCPA)
Professor of Neuropathology
Sydney Medical School
The University of Sydney
Sydney, NSW, Australia

Nichola Z Lax, PhD
Research Associate
Wellcome Trust Centre for Mitochondrial
    Research
Institute of Neuroscience
Newcastle University
Newcastle upon Tyne, UK

David N Louis, MD
Pathologist-in-Chief
Massachusetts General Hospital
Benjamin Castleman Professor of
    Pathology
Harvard Medical School
James Homer Wright Pathology
    Laboratories
Massachusetts General Hospital
Boston, MA, USA

Seth Love, MBBCH, PhD, FRCP, FRCPath
Professor of Neuropathology
Institute of Clinical Neurosciences
School of Clinical Sciences
University of Bristol
Southmead Hospital
Bristol, UK

James Lowe, DM, FRCPath
Professor of Neuropathology
University of Nottingham
Honorary Consultant in Neuropathology to
    the Nottingham University Hospitals
    NHS Trust
School of Medicine
Faculty of Medicine and Health Sciences
University of Nottingham
Nottingham, UK

Sebastian Lucas, FRCP, FRCPath
Emeritus Professor of Histopathology
Department of Histopathology
St Thomas' Hospital
London, UK

Susan S Margulies, PhD
George H Stephenson Professor
Department of Bioengineering
University of Pennsylvania
Philadelphia, PA, USA

Michelle Fevre Montange, PhD
Centre de Recherche en Neuroscience de Lyon
  INSERM U1028
CNRS UMR 5292
Equipe Neuro-oncologie et Neuro-inflammation
Université de Lyon
Lyon, France

G R Wayne Moore, BSC, MD, CM, FRCPC, FRCPath
Clinical Professor
Department of Pathology and Laboratory Medicine
International Collaboration on Repair Discoveries (ICORD)
University of British Columbia
Vancouver General Hospital
Vancouver, BC, Canada

Christopher M Morris, PhD
Senior Lecturer
Medical Toxicology Centre
National Institutes of Health Research
Health Protection Research Unit in Chemical and
  Radiation Threats and Hazards
Institute of Neuroscience
Newcastle University
Newcastle upon Tyne, UK

Francesco Muntoni, FRCPCH, FMedSci
Director
Dubowitz Neuromuscular Centre
MRC Centre for Neuromuscular Diseases
University College London Institute of Child Health and
  Great Ormond Street Hospital for Children (GOSH)
London, UK

Arie Perry, MD
Professor of Pathology and Neurological Surgery
Director of Neuropathology
Director of Neuropathology Fellowship Training Program
University of California, San Francisco
San Francisco, CA, USA

Rahul Phadke, MBBS, MD, FRCPath
Consultant Neuropathologist
University College London Institute of Neurology
National Hospital for Neurology and Neurosurgery and
  Dubowitz Neuromuscular Centre
Great Ormond Street Hospital for Children
London, UK

Pedro Piccardo, MD
Senior Investigator
Chief Transmissible Spongiform Encephalopathy
  Pathogenesis Section
Laboratory of Bacterial and TSE Agents
Office of Blood Research and Review
Center for Biologics Evaluation and Research
U.S. Food and Drug Administration
Silver Spring, MD, USA
Professor
Neurobiology Division
The Roslin Institute
University of Edinburgh
Easter Bush, UK

James M Powers, MD
Professor Emeritus
Department of Pathology
University of Rochester School of Medicine
  and Dentistry
Rochester, NY, USA

Robin Reid, BSc, MBChB, FRCPath
Formerly Consultant Pathologist
Western Infirmary
Glasgow, UK

Guido Reifenberger, MD
Professor
Department of Neuropathology
Heinrich Heine University
Düsseldorf, Germany

Tamas Revesz, MD, FRCPath
Professor Emeritus in Neuropathology
UCL Institute of Neurology
University College London
London, UK

Hope T Richard, MD, PhD
Neuropathology Fellow
Department of Pathology
Virginia Commonwealth University
Richmond, VA, USA

Marc K Rosenblum, MD
Founder's Chair and Chief
Neuropathology and Autopsy Service
Memorial Sloan-Kettering Cancer Center
Professor of Pathology and Laboratory
  Medicine
Weill Medical College of Cornell
  University
New York, NY, USA

Robert E Schmidt, MD, PhD
Professor of Pathology and Immunology
Director
Division of Neuropathology
Medical Director
Electron Microscope Facility
Washington University School of Medicine
St Louis, MO, USA

Caroline A Sewry, PhD, FRCPath
Professor of Muscle Pathology
Dubowitz Neuromuscular Centre
Institute of Child Health and Great Ormond Street
  Hospital
London
Wolfson Centre for Inherited Neuromuscular
  Diseases
Robert Jones and Agnes Hunt Orthopaedic
  Hospital
Oswestry, UK

Sanjay Sisodiya, MA, PhD, FRCP, FRCPEdin
Professor of Neurology
Department of Clinical and Experimental Epilepsy
UCL Institute of Neurology
London
Consultant Neurologist
Epilepsy Society
National Hospital for Neurology and Neurosurgery
Chalfton St Peter, UK

Colin Smith, MD, FRCPath
Reader in Pathology
University of Edinburgh
Edinburgh, UK

Christine Stadelmann-Nessler, MD
Professor
Department of Neuropathology
University Medical Center Göttingen
Göttingen, Germany

Kinuko Suzuki, MD
Emeritus Professor of Pathology and Laboratory Medicine
University of North Carolina at Chapel Hill
Chapel Hill, NC, USA
Neuropathology
Tokyo Metropolitan Institute of Gerontology
Tokyo, Japan

Kunihiko Suzuki, MD
Director Emeritus
Neuroscience Center
University of North Carolina
Chapel Hill, NC, USA

Robert W Taylor, PhD, DSC, FRCPath
Professor of Mitochondrial Pathology
Wellcome Trust Centre for Mitochondrial Research
Institute of Neuroscience
Newcastle University
Newcastle upon Tyne, UK

Maria Thom, BSC, MBBS, MRCPath
Senior Lecturer
Institute of Neurology
University College London
London, UK

Douglas M Turnbull, MD, PhD, FRCP, FMedSc
Professor of Neurology
Director
Wellcome Trust Centre for Mitochondrial Research
Director
LLHW Centre for Ageing and Vitality
Newcastle University
Newcastle upon Tyne, UK

Alexandre Vasiljevic, MD
Associate Professor of Pathology
Centre de Pathologie et Neuropathologie Est
Centre de Biologie et Pathologie Est
Groupement Hospitalier Est
Hospices Civils de Lyon
Lyon, France

Harry V Vinters, MD
Professor of Pathology and Laboratory Medicine,
    Neurology
Chief of Neuropathology
Division of Neuropathology
Member of Brain Research Institute ACCESS
    Program
Department of Cellular and Molecular
    Pathology
University of California, Los Angeles
Los Angeles, CA, USA

Steven U Walkley, DVM, PhD
Director
Rose F Kennedy Intellectual and Developmental
    Disabilities Research Center
Head
Sidney Weisner Laboratory of Genetic
    Neurological Disease
Departments of Neuroscience, Pathology and
    Neurology
Albert Einstein College of Medicine
Bronx, NY, USA

Stephen B Wharton, BSC, MBBS, PhD, FRCPath
Professor and Honorary Consultant in Neuropathology
Department of Neuroscience
Sheffield Institute of Translational Neuroscience
University of Sheffield
Sheffield, UK

Clayton A Wiley, MD, PhD
Professor of Pathology
Director of Neuropathology
PERF Endowed Chair
Univeristy of Pittsburgh Medical Center
    Presbyterian Hospital
Pittsburgh, PA, USA

Robert G Will, MA, MD, FRCP
Professor of Clinical Neurology
National Creutzfeldt–Jakob Disease Surveillance Unit
University of Edinburgh
Western General Hospital
Edinburgh, UK

Abbreviations

AA	anaplastic astrocytoma
AACD	age-associated cognitive decline
AAMI	age-associated memory impairment
ABC	ATP-binding cassette
ABCA1	ATP-binding cassette transporter 1
ABRA	Aβ-related angiitis
ACA	anterior cerebral artery
ACC	adrenocortical carcinoma
ACCIS	Automated Childhood Cancer Information System
ACh	acetylcholine
AChR	acetylcholine receptor
ACTH	adrenocorticotropin
AD	Alzheimer disease
ADAMTS13	a disintegrin and metalloproteinase with a thrombospondin type 1 motif, member 13
ADC	apparent diffusion coefficient
ADCA	autosomal dominant cerebellar ataxia
ADEM	acute disseminated encephalomyelitis
ADK	adenosine kinase
ADNFLE	autosomal dominant nocturnal frontal lobe epilepsy
ADP	adenosine diphosphate
AFP	alpha-fetoprotein
AGA	aspartylglucosaminidase
AGE	advanced glycosylation end product
AGPS	alkylglycerone phosphate synthase
AGS	Aicardi-Goutières syndrome
AGU	aspartylglucosaminuria
AHLE	acute haemorrhagic leukoencephalitis
AHT	abusive head trauma
AIDP	acute inflammatory demyelinating polyneuropathy
AIDS	acquired immunodeficiency syndrome
AIP	aryl hydrocarbon receptor-interacting protein
AIS	axon initial segment
AISS	axonal index sector score
AL	amyloidosis
ALCL	anaplastic large cell lymphoma
ALD	adrenoleukodystrophy
ALK	anaplastic lymphoma kinase
ALL	acute lymphoblastic leukemia
ALS	amyotrophic lateral sclerosis
ALT	alternative lengthening of telomeres
AMAN	acute motor axonal neuropathy
AMN	adrenomyeloneuropathy
AMPA	α-amino-3-hydroxy-5-methyl-4-isoxazolepropionic acid
AMSAN	acute motor sensory axonal neuropathy
ANA	antinuclear antibody
ANCA	antineutrophil cytoplasmic autoantibody
ANCL	adult neuronal ceroid lipofuscinosis
Ang-1	angiopoietin-1
Ang-2	angiopoietin-2
ANI	asymptomatic neurocognitive impairment
AOA1	early-onset ataxia with oculomotor apraxia, type 1
APGBD	adult polyglucosan body disease
APLA	antiphospholipid antibody
ApoE	apolipoprotein E
APP	amyloid precursor protein
APrP	amyloid prion protein
APUD	amine precursor uptake and decarboxylation
AQP4	aquaporin-4
AR	androgen receptor
ARBD	alcohol-related brain damage
ARFGEF2	adenosine diphosphate (ADP)-ribosylation factor guanine exchange factor 2
ASA	arylsulfatase A
ASDH	acute subdural haematoma
ASE	acute schistosomal encephalopathy
ASL	arterial spin labelling
AT	ataxia telangiectasia
ATP	adenosine triphosphate
ATRT	atypical teratoid/rhabdoid tumour
ATTR	amyloid transthyretin
AVM	arteriovenous malformation
BA	Brodmann area
BACE	β-site APP-cleaving enzyme
BAC	bacterial artificial chromosome
BAV	Banna virus
BBB	blood-brain barrier
BDNF	brain-derived neurotrophic factor
BDV	Borna disease virus
BEAN	brain expressed protein associated with NEDD4
bFGF	basic fibroblast growth factor
BGC	basal ganglia calcification
BHC	benign hereditary chorea
BMAA	β-N-methylamino-L-alanine
BMD	Becker muscular dystrophy
BMP	bone morphogenetic protein
BOLD	blood oxygenation dependent
bp	base pair
BPAU	bromophenylacetylurea

BRC	brain reserve capacity
BRRS	Bannayan-Riley-Ruvalcaba syndrome
BSE	bovine spongiform encephalopathy
CAA	cerebral amyloid angiopathy
CADASIL	cerebral autosomal dominant arteriopathy with subcortical infarcts and leukoencephalopathy
CAE	childhood absence epilepsy
CAHS	chronic acquired hepatocerebral syndrome
CAMTA1	calmodulin-binding transcription activator 1
c-ANCA	cytoplasmic antineutrophil cytoplasmic antibody
CANOMAD	chronic ataxic neuropathy, ophthalmoplegia, M-protein agglutination, disialosyl antibodies
CAR	coxsackievirus and adenovirus receptor
CARASIL	cerebral autosomal recessive arteriopathy with subcortical infarcts and leukoencephalopathy
cART	combined antiretroviral therapy
CASK	calcium-dependent serine protein kinase
CBD	corticobasal degeneration
CBF	cerebral blood flow
CBS	corticobasal syndrome
CBTRUS	Central Brain Tumor Registry of the United States
CCM	cerebral cavernous malformation
CCSVI	chronic cerebrospinal venous insufficiency
CD	Cowden disease
CDE	common data elements
CDI	conformation dependent immunoassay
CDK5	cyclin-dependent kinase 5
CDKI	cyclin-dependent kinase inhibitor
CDKN2C	cyclin-dependent kinase inhibitor 2C
CDV	canine distemper virus
CEA	carcinoembryonic antigen
CESD	cholesteryl ester storage disease
CGH	comparative genomic hybridization
cGMP	cyclic guanosine monophosphate
CGRP	calcitonin gene-related peptide
CHD5	chromodomain helicase DNA binding domain 5
CHN	congenital hypomyelinating neuropathy
CHS	classical hippocampal sclerosis
CIM	critical illness myopathy
CIP	critical illness polyneuropathy
CIPD	chronic inflammatory demyelinating polyneuropathy
CIS	clinically isolated syndrome
CISP	chronic immune sensory polyradiculopathy
CK	cytokeratin; creatine kinase
CLA2	X-linked cerebellar ataxia
CLL	chronic lymphatic leukaemia
CM	cerebral malaria
CMD	congenital muscular dystrophy
CMRgl	cerebral metabolic rate for glucose
CMRO$_2$	cerebral metabolic rate for oxygen
CMROGl	cerebral metabolic rates of oxygen and glucose
CMT	Charcot–Marie–Tooth
CMV	cytomegalovirus
CN	cystic nephroma
CNC	Carney's complex
CNP	2′,3′-cyclic nucleotide 3′-phosphodiesterase
CNS	central nervous system
CNS PNET	central nervous system primitive neuroectodermal tumour
CNTF	ciliary neurotrophic factor
CO	carbon monoxide
COL4A1	collagen, type IV, alpha 1
COX	cytochrome *c* oxidase
COX-2	cyclooxygenase-2
CP	choroid plexus
CPCS	chronic post-concussion syndrome
CPM	central pontine myelinolysis
CPP	cerebral perfusion pressure; central precocious puberty
CPT	carnitine palmitoyltransferase
CR	cognitive reserve
CR3	complement receptor type 3
CRABP	cellular retinoic acid binding protein
CRBP	cytoplasmic retinol binding protein
CREB	cyclic adenine dinucleotide phosphate response element binding protein
CRH	corticotropin-releasing hormone
CRIMYNE	critical illness myopathy and neuropathy
CRMP-5	collapsing response mediator protein 5
CRV	cerebroretinal vasculopathy
CSDH	chronic subdural haematoma
CSF	cerebrospinal fluid
CSPα	cysteine string protein α
CT	computed tomography
CTD	connective tissue disease
CTE	chronic traumatic encephalopathy
CTF	Colorado tick fever
CTL	cytotoxic lymphocyte
CUP	cancer of unknown primary
CUTE	corticotropin upstream transcription-binding element
CuZnSOD	copper- and zinc-containing superoxide dismutase
CVD	cardiovascular disease
CVS	chorionic villus sampling
CVST	cerebral venous sinus thrombosis
CVT	cerebral venous thrombosis

CWD	chronic wasting disease
CX32	connexin 32
DAB	diaminobenzidine
DAG	dystrophin-associated glycoprotein
DAI	diffuse axonal injury
DAPAT	dihydroxyacetonephosphate acyltransferase
DASE	developmentally arrested structural elements
DAWM	diffusely abnormal white matter
DCX	doublecortin
DEHSI	diffuse excessive high-signal intensity
DFFB	DNA fragmentation factor subunit beta
DHA	docosahexaenoic acid
DHAP	dihydroxyacetone phosphate
DHPR	dihydropyridine receptor
DiI	dioctadecyl-tetramethylindocarbacyanine perchlorate
DILS	diffuse infiltrative lymphocytosis syndrome
DIR	double inversion recovery
DLB	dementia with Lewy bodies
DLBCL	diffuse large B cell lymphoma
DLK	dual leucine kinase
DM	dermatomyositis
DMD	Duchenne muscular dystrophy
DMNV	dorsal motor nucleus of the vagus
DMPK	dermatomyositis protein kinase
DNER	delta/notch-like epidermal growth factor-related receptor
DNL	disseminated necrotizing leukoencephalopathy
DNMT	DNA methyltransferase
DNT	dysembryoplastic neuroepithelial tumour
DPR	dipeptide repeat
DPX	di-n-butylphthalate-polystyrene-xylene
DRD	dopa-responsive dystonia
DRPLA	dentatorubropallidoluysian atrophy
DSD	Dejerine-Sottas disease
DSPN	diffuse sensory polyneuropathy
DTI	diffusion tensor imaging
DTICH	delayed traumatic intracerebral haemorrhage
DWI	diffusion weighted imaging
DXC	doublecortin
EA	episodic ataxia
EAAT	excitatory amino acid transporter
EAN	experimental allergic neuritis
EBP	elastin-binding protein
EBV	Epstein-Barr virus
EC	endothelial cell; entorhinal cortex
ECGF1	endothelial cell growth factor 1 (platelet-derived)
ECM	extracellular matrix
ECMO	extracorporeal membrane oxygenation

EDH	extradural haematoma
EEE	eastern equine encephalitis
EEG	electroencephalogram
EET	epoxyeicosatrienoic acid
EF HS	end folium hippocampal sclerosis
EGA	estimated gestational age
EGB	eosinophilic granular body
EGFR	epidermal growth factor receptor
EGL	external granule cell layer
EGR2	early growth response 2 gene
EIEE	early infantile epileptic encephalopathy
EL	encephalitis lethargica
ELBW	extreme low birth weight
ELISA	enzyme-linked immunosorbent assay
EM	electron microscopy
EMA	epithelial membrane antigen
EME	early myoclonic encephalopathy
EMG	electromyography
EMT	epithelial-mesenchymal transition
eNSC	embryonic neural stem cell
ENU	ethylnitrosourea
EPC	endothelial progenitor cell
EPMR	epilepsy with mental retardation
ER	endoplasmic reticulum
ERG	electroretinogram
ERK	extracellular signal-regulated kinase
ERM	ezrin, radixin and moesin
ESAM	endothelial cell-selective adhesion molecule
ESR	erythrocyte sedimentation rate
ETANTR	embryonal tumour with abundant neuropil and true rosettes
ETMR	embryonal tumor with multilayered rosettes
EVOH	ethylene-vinyl alcohol copolymer
FA	Friedreich's ataxia
FACS	fluorescence-activated cell sorting
FAD	familial Alzheimer's disease
FAF	familial amyloidosis of the Finnish type
FAK	focal adhesion kinase
FALS	familial amyotrophic lateral sclerosis
FAP	familial amyloid polyneuropathy; familial polyposis
FBD	familial British dementia
FBXO7	F-box only protein 7
FCD	focal cortical dysplasia; follicular dendritic cell
FCE	fibrocartilaginous embolism
fCJD	familial Creutzfeldt-Jakob disease
FDD	familial Danish dementia
FDF-2	fibroblast growth factor 2
FFI	fatal familial insomnia
FFPE	formalin-fixed paraffin-embedded tissue

FG	fast-twitch glycolytic		GIST	gastrointestinal stromal tumour
FGF	fibroblast growth factor		GLAST	glutamate/aspartate transporter
FHL1	four and a half LIM domains protein 1		GLB1	galactosidase, beta 1
FILIP	filamin-A-interacting protein		GLD	globoid cell leukodystrophy
FIPA	familial isolated pituitary adenoma		GLM	glial limiting membrane
FISH	fluorescence *in situ* hybridization		GM	gliomesodermal tissue
FKRP	fukutin-related protein		GOM	granular osmiophilic material
FLAIR	fluid-associated inversion recovery		GP	globus pallidus
FLNA	filamin A		GPI	glycosylphosphatidylinositol
FMD	fibromuscular dysplasia; Fukuyama muscular dystrophy		GROD	granular osmiophilic deposit
			GSC	glioma stem cell
fMRI	functional magnetic resonance imaging		GSD	glycogen storage disease
FOG	fast-twitch oxidative glycolytic		GSN	gelsolin
FPS	fasciitis-panniculitis syndrome		Gsp	G-protein oncogene
FR	fatigue resistant		GSS	Gerstmann-Sträussler-Scheinker disease
FS	febrile seizure			
FSH	follicle stimulating hormone		gTSE	genetic transmissible spongiform encephalopathy
FSHD	facioscapulohumeral muscular dystrophy			
FTBSI	focal traumatic brain stem injury		GU	genitourinary
FTD	frontotemporal dementia		GWAS	genome wide association studies
FTL	ferritin light		HAART	highly active antiretroviral therapy
FTLD	frontotemporal lobar degeneration		HACE	high altitude cerebral oedema
FUPB1	far-upstream element binding protein 1		HAD	HIV-associated dementia
FUS	fused-in-sarcoma protein		HAM	HTLV-1-associated myelopathy
FXTAS	fragile X tremor/ataxia syndrome		HAN	hereditary neuralgic amyotrophy
G-CIMP	glioma CpG island methylator phenotype		HANAC	hereditary angiopathy with nephropathy, aneurysms and muscle cramps
GABA	gamma-aminobutyric acid			
GAD	gracile axonal dystrophy; glutamic acid decarboxylase		HAS	high-altitude stupid
			HAT	human African trypanosomiasis
GAG	glycosaminoglycan		HB-EGF	heparin-binding epidermal growth factor
GALT	gut-associated lymphoid tissue			
GAP-43	growth-associated protein 43		HCG	human chorionic gonadotropin
GAT1	glutaric aciduria type 1		HCHWA-D	hereditary cerebral haemorrhage with amyloid angiopathy of the Dutch
Gb Ose3 Cer	globotriaosylceramide			
			HCHWA-F	hereditary cerebral haemorrhage with amyloid angiopathy of the Flemish
GBE	glycogen branching enzyme			
GBM	glioblastoma		HCHWA-I	hereditary cerebral haemorrhage with amyloid angiopathy of the Icelandic
GBS	Guillain–Barré syndrome			
GC	granule cell		HCMV	human cytomegalovirus
GCA	giant cell (or temporal) arteritis		HD	Huntington's disease
GCD	granule cell dispersion		HDL	high density lipoprotein
GCI	global cerebral ischaemia; glial cytoplasmic inclusion		HDL1	Huntington disease-like type 1
			HDL2	Huntington disease-like type 2
GCL	granule cell layer		HDL3	Huntington disease-like type 3
GCS	Glasgow Coma Scale		HE	hepatic encephalopathy
GDAP1	ganglioside-induced differentiation-associated protein 1		H&E	haematoxylin and eosin
			HERNS	hereditary endotheliopathy with retinopathy, nephropathy and stroke
GDNF	glial cell-derived neurotrophic factor			
GEMM	genetically engineered mouse model		HERV	human endogenous retrovirus
GFAP	glial fibrillary acidic protein		HES	hairy/enhancer of split
GFP	green fluorescent protein		HES-1	hairy/enhancer of split 1
GH	growth hormone		hGH	human growth hormone
GHR	GH receptor		HH	hypothalamic hamartoma
GI	gastrointestinal		HHV-8	human herpesvirus 8
			HIF	hypoxia inducible factor

HIHRATL	hereditary infantile hemiparesis, retinal arteriolar tortuosity and leukoencephalopathy
HIMAL	hippocampal malrotation
HIV	human immunodeficiency virus
HLA	human leukocyte antigen
HLH	helix-loop-helix
HMEG	hemimegalencephaly
HMERF	hereditary myopathy with early respiratory failure
HMG	high mobility group
HMGCR	3-hydroxy-3-methylglutaryl-coenzyme A reductase
H-MRS	proton magnetic resonance spectroscopy
HMSN	hereditary motor and sensory neuropathy
HNE	hydroxy-2-nonenal
HNPCC	hereditary nonpolyposis colorectal cancer
HNPP	hereditary neuropathy with liability to pressure palsy
H_2O_2	hydrogen peroxide
HPC	haemangiopericytoma
HPE	holoprosencephaly
HPF	high-power field
HPS	haematoxylin-phloxine-safranin
HPV	human papillomavirus
HRE	hypoxia response elements
HRP	horseradish peroxidase
HS	hippocampal sclerosis
HSA	hereditary systemic angiopathy
HSAN	hereditary sensory and autonomic neuropathy
HSP	heat-shock protein; hereditary spastic paraplegia
HSV	herpes simplex virus
5-HT	5-hydroxytryptamine
hTERT	human telomerase reverse transcriptase
HTLV-I	human T-cell lymphotropic virus I
HVR	hereditary vascular retinopathy
IBM	inclusion body myositis
ICA	internal carotid artery; internal cerebral artery
ICAM-1	intercellular adhesion molecule-1
ICD	I-cell disease; intracellular domain
ICE	interleukin-converting enzyme
ICH	intracerebral haematoma
iCJD	iatrogenic Creutzfeldt-Jakob disease
ICP	intracranial pressure
IDH	intradural haemorrhage
IENF	intra-epidermal nerve fibre
IFS	isolated familial somatotropinoma
IGF	insulin-related growth factor
IgM	immunoglobulin M
IHC	immunohistochemistry
IHI	incomplete hippocampal inversion

IIM	idiopathic inflammatory myopathy
IL-1β	interleukin-1 beta
ILAE	International League Against Epilepsy
ILOCA	idiopathic late-onset cerebellar ataxia
ILS	isolated lissencephaly sequence
IMAM	inflammatory myopathy with abundant macrophages
IMD	inherited myoclonus-dystonia
IML	inner molecular layer
IMNM	immune-mediated necrotizing myopathy
IMT	inflammatory myofibroblastic tumour
INAD	infantile neuroaxonal dystrophy
INCL	infantile neuronal ceroid lipofuscinosis
iNOS	inducible nitric oxide synthase
ION	inferior olivary nucleus
IPI	initial precipitating injury
iPSC	induced pluripotent stem cell
IPSP	inhibitory postsynaptic potential
IRD	infantile Refsum's disease
IRES	internal ribosomal entry site
IRIS	immune reconstitution inflammatory syndrome
IRS	insulin receptor substrate
ISF	interstitial fluid
ISPD	isoprenoid synthase domain-containing
ISSD	infantile sialic acid storage disease
ITPR-1	inositol triphosphate receptor type 1
IUGR	intrauterine growth restriction
IVH	intraventricular haemorrhage
JAK/STAT	Janus kinase and downstream signal transducer and activator of transcription
JAM	junctional adhesion molecule
JME	juvenile myoclonic epilepsy
JNCL	juvenile neuronal ceroid lipofuscinosis
JXG	juvenile xanthogranuloma
kb	kilobase
KO	knockout
KPS	Karnofsky performance status
KRS	Kufor Rakeb syndrome
KS	Korsakoff's syndrome
KSS	Kearns-Sayre syndrome
LA	lupus anticoagulant
LB	Lewy body
LCH	Langerhans cell histiocytosis
LCMV	lymphocytic choriomeningitis virus
LDD	Lhermitte-Duclos disease
LDL	low-density lipoprotein
LEAT	long-term epilepsy-associated tumour
LFB	Luxol fast blue
LFB-CV	Luxol fast blue-cresyl violet
LGI1	leucine-rich glioma-inactivated 1
LGMD	limb-girdle muscular dystrophy
LGN	lateral geniculate nucleus

LH	luteinizing hormone
LIF	leukaemia inhibitory factor
LINCL	late infantile neuronal ceroid lipofuscinosis
LMNA	lamin A/C
LNMP	last normal menstrual period
LOH	loss of heterozygosity
LPH	lipotropin
LRPN	lumbosacral radioplexus neuropathy
LSA	lenticulostriate artery
L-SS	Lewis-Sumner syndrome
LTD	long-term depression
LTP	long-term potentiation
MAG	myelin-associated glycoprotein
MAGE-A	melanoma-associated cancer-testis antigen
MAP	microtubule-associated protein
MAPK	mitogen-activated protein kinase
MATPase	myofibrillar adenosine triphosphatase
MBD	Marchiafava-Bignami disease
MBEN	medulloblastoma with extensive nodularity
MBP	myelin basic protein
MCA	middle cerebral artery
MCB	membranous cytoplasmic body
MCD	malformation of cortical development
MCI	mild cognitive impairment
MCM2	minichromosome maintenance 2
MCP-1	monocyte chemoattractant protein 1
MDC1A	merosin-deficient CMD
MELAS	mitochondrial encephalomyopathy, lactic acidosis and stroke-like episodes
MEN	multiple endocrine neoplasia
MEN2	multiple endocrine neoplasia type 2
MERRF	myoclonic epilepsy with ragged-red fibres
MFN2	mitofusin 2
MFS	Miller Fisher syndrome; mossy fibre spouting
MGUS	monoclonal gammopathy of unknown significance
MHC	myosin heavy chain
MHC-I	major histocompatibility complex class I
MHV	mouse hepatitis virus
MIBE	measles inclusion body encephalitis
MJD	Machado-Joseph disease
ML	mucolipidosis
MLC	myosin light chain
MLD	metachromatic leukodystrophy
MLI	mucolipidosis I
MM	methionine homozygosity
MMN	multifocal motor neuropathy
MMP	matrix metalloproteinase
MMR	mismatch repair; measles-mumps-rubella
MNCV	motor nerve conduction velocity
MND	motor neuron degeneration; mild neurocognitive disorder; motor neuron disease
MNGC	multinucleated giant cell
MNGIE	mitochondrial neuro-gastrointestinal encephalomyopathy
MnSOD	manganese-containing superoxide dismutase
MNU	methylnitrosourea
MOG	myelin-oligodendrocyte protein
MPNST	malignant peripheral nerve sheath tumour
MPO	myeloperoxidase
MPS	mucopolysaccharidosis
MPT	mitochondrial permeability transition
MPTP	N-methyl-4-phenyl-1,2,3,6-tetrahydropyridine
mPTS	membrane peroxisomal targeting sequence
MPZ	myelin protein zero
MR	magnetic resonance
MRC	Medical Research Council
MRI	magnetic resonance imaging
mRNA	messenger ribonucleic acid
MRS	magnetic resonance spectroscopy
MRT	malignant rhabdoid tumour
MS	multiple sclerosis
MSA	multiple system atrophy; myositis-specific autoantibody
MSA-C	cerebellar form of multiple system atrophy
MSB	Martius scarlet blue
MSD	multiple sulphatase deficiency
MSH	melanotropin
MSI	microsatellite instability
mtDNA	mitochondrial DNA
MTI	magnetization transfer imaging
MTLE	mesial temporal lobe epilepsy
MTMR2	myotubularin-related protein 2
mTOR	mammalian target of rapamycin
MTR	magnetization transfer ratio
MuSK	muscle-specific kinase
MV	valine heterozygous
MVE	Murray Valley encephalitis
NAA	N-acetylaspartate
NAD+	nicotinamide adenine dinucleotide
NADH-TR	nicotinamide adenine dinucleotide-tetrazolium reductase
NAHI	non-accidental head injury
NALD	neonatal adreno-leukodystrophy
NAM	necrotizing autoimmune myopathy
NARP	neuropathy, ataxia and retinitis pigmentosa
NAT	non-accidental trauma
NAWM	normal-appearing white matter

NBCCS	naevoid basal cell carcinoma syndrome		OML	outer molecular layer
NBIA	neurodegeneration with brain iron accumulation		OPC	oligdendrocyte precursor cell
NBIA1	neurodegeneration with brain iron accumulation, type 1		OPCA	olivopontocerebellar atrophy
			OPIDPN	organophosphate-induced delayed polyneuropathy
NBIA2	neurodegeneration with brain iron accumulation, type 2		ORF	open reading frame
NCAM	neural cell adhesion molecule		PACNS	primary angiitis of the central nervous system
NCI	neuronal cytoplasmic inclusion		PAFAH	platelet activating factor acetyl hydrolase
NCIPC	National Center for Injury Prevention and Control		PAMP	pathogen-associated molecular pattern
NCL	neuronal ceroid lipofuscinosis		PAN	polyarteritis nodosa; perchloric acid naphthoquinone
NCM	neurocutaneous melanosis		p-ANCA	perinuclear ANCA
NECD	notch extracellular domain		PARK1	Parkinson's disease and alpha-synuclein
NF	neurofilament protein		PAS	periodic acid–Schiff
NF1	neurofibromatosis type 1		PB	pineoblastoma
NF2	neurofibromatosis type 2		PBD	peroxisome biogenesis disorder
NFL	National Football League		PBH	parenchymal brain haemorrhage
NFP	neurofilament protein		PBP	progressive bulbar palsy
NFT	neurofibrillary tangle		PC	pineocytoma
NGF	nerve growth factor		PCD	Purkinje cell degeneration
NHNN	National Hospital for Neurology and Neurosurgery		PCNA	proliferating cell nuclear antigen
NIFID	neuronal intermediate filament inclusion disease		PCNSL	primary central nervous system lymphoma
NIID	neuronal intranuclear inclusion disease		PCP	planar cell polarity
NINDS	National Institute of Neurological Disorders and Stroke		PCR	polymerase chain reaction
			PCV	packed cell volume
NINDS-PSP	National Institute of Neurological Disorders and Stroke and the Society for Progressive Supranuclear Palsy		PD	Parkinson's disease; pars distalis
			PDC	parkinsonism/dementia complex
			PDCD	programmed cell death
NIRS	near-infrared spectroscopy		PDD	Parkinson's disease dementia
NK	natural killer		PDGF	platelet-derived growth factor
NMDA	*N*-methyl-D-aspartate		PDGFB	platelet-derived growth factor beta
NMDAR	*N*-methyl-D-aspartate receptor		PDH	pyruvate dehydrogenase
NMO	neuromyelitis optica		PECAM	platelet-endothelial cell adhesion molecule
nNOS	neuronal nitric oxide synthase		PEM	protein-energy malnutrition
NO	nitric oxide		PEO	progressive external ophthalmoplegia
NOS	not otherwise specified		PEP	postencephalitic parkinsonism
NOTCH3	notch homolog 3		PERM	progressive encephalomyelitis with rigidity and myoclonus
NPC	Niemann-Pick disease type C		PES	pseudotumoural encephalic schistosomiasis
NPH	normal pressure hydrocephalus			
NPY	neuropeptide Y		PET	paraffin-embedded tissue; positron emission tomography
NSAID	non-steroidal anti-inflammatory drug			
NSC	neural stem cell		PGNT	papillary glioneuronal tumour
NSE	neuron specific enolase		PGP	protein gene product
NTD	neural tube defect		PHF	paired helical filament
NTE	neuropathy target esterase		PHP	pseudo-Hurler polydystrophy
NTS	nucleus of the solitary tract		PhyH	phytanoyl-CoA hydroxylase
OCT	optimal cutting temperature; optical coherence tomography		PI	pars intermedia
			PiB	Pittsburgh compound B
OEF	oxygen extraction fraction		PICA	postero-inferior cerebellar artery
O-FucT-1	O-fucosetransferase 1		PKAN	pantothenate kinase-associated neurodegeneration
OH	hydroxyl radical			
OMIM	Online Mendelian Inheritance in Man			

PKC	protein kinase C
PLA2G6	phospholipase A2, group VI
PLAN	PLA2G6-associated neurodegeneration
PLP	proteolipid protein
PLS	primary lateral sclerosis
PMA	pilomyxoid astrocytoma; progressive muscular atrophy
PMCA	protein misfolding cyclic amplification
PMD	Pelizaeus-Merzbacher disease
PME	progressive myoclonic epilepsy
PML	progressive multifocal leukoencephalopathy
PMNS	post-malaria neurological syndrome
PMP	peroxisomal membrane protein
PMP2	peripheral myelin protein 2
PMS	psammomatous melanotic schwannoma
PN	pars nervosa
PNDC	progressive neuronal degeneration of childhood with liver disease
PNET	primitive neuroectodermal tumour
PNMA	paraneoplastic Ma antigen
PNS	peripheral nervous system
pO$_2$	partial pressure of oxygen
POEMS	polyneuropathy, organomegaly, endocrinopathy, M-protein, skin changes
POLG	polymerase γ
POMC	proopiomelanocortin
PPA	primary progressive aphasia
PPB	familial pleuropulmonary blastoma
PPCA	protective protein with cathepsin A-like activity
ppm	parts per million
pPNET	peripheral primitive neuroectodermal tumour
PPS	pentosan polysulphate; post-polio syndrome
PPT	pineal parenchymal tumour
PPTID	pineal parenchymal tumour of intermediate differentiation
PR	progesterone receptor
PRBC	parasitized red blood cell
PRES	posterior reversible encephalopathy syndrome
PRL	prolactin
PRNP	PrP gene
PROMM	proximal myotonic myopathy
PROP-1	prophet of Pit-1
ProtCa	activated protein C
ProtS	protein S
PrP	prion protein
PrP-CAA	PrP-cerebral amyloid angiopathy
PRR	pattern recognition receptor
PSAP	prosapson
PSD	post-stroke dementia
PSIR	phase-sensitive inversion recovery

PSP	progressive supranuclear palsy
PSP-CA	progressive supranuclear palsy with cerebellar ataxia
PSP-CST	atypical progressive supranuclear palsy with corticospinal tract degeneration
PSP-P	progressive supranuclear palsy with parkinsonism
PSP-PAGF	pure akinesia with gait freezing with subsequent development of typical signs of progressive supranuclear palsy
pSS	primary Sjögren's syndrome
PTAH	phosphotungstic acid haematoxylin
PTC	periodic triphasic complex
PTD	primary (idiopathic) torsion dystonia
ptd-FGFR4	pituitary tumour-derived FGFR4
PTLD	post-transplant lymphoproliferative disorder
PTPR	papillary tumour of the pineal region
PTRF	polymerase I and transcript release factor
PTS	peroxisomal targeting signal
Ptx2	pituitary homeobox factor 2
PVH/IVH	periventricular/intraventricular haemorrhage
PVL	periventricular leukomalacia
PWI	perfusion weighted imaging
PXA	pleomorphic xanthoastrocytoma
QuIC	quaking induced conversion
RALDH	retinaldehyde dehydrogenase
RANO	response assessment in neuro-oncology
RAR	retinoic acid receptor
RARE	retinoic acid response element
RC2	reaction centre type 2
RCA-1	*Ricinus communis* agglutinin 1
rCBF	regional cerebral blood flow
rCBV	regional cerebral blood volume; relative cerebral blood volume
RCC	renal cell carcinoma
RCDP	rhizomelic chondrodysplasia punctata
RDD	Rosai-Dorfman disease
RDP	rapid onset dystonia-parkinsonism
RE	Rasmussen encephalitis
REM	rapid eye movement
rhNGF	recombinant human nerve growth factor
RIG	radiation-induced glioma
RIM	radiation-induced meningioma
RING	Really Interesting New Gene
RIP1	receptor-interacting protein 1
RIS	radiologically isolated syndrome
RNI	reactive nitrogen intermediate
ROS	reactive oxygen species
RPLS	reversible posterior leukoencephalopathy syndrome
RPS	rhabdoid predisposition syndrome
Rpx	Rathke's pouch homeobox
RRF	ragged-red fibre

RRMS	relapsing-remitting form of multiple sclerosis
RSMD1	rigid spine muscular dystrophy type 1
RSV	Rous sarcoma virus
RTA	road traffic accident
RTK	receptor tyrosine kinase
RVCL	retinal vasculopathy with cerebral leukodystrophy
RXR	retinoid X receptor
SAH	subarachnoid haemorrhage
SANDO	sensory ataxic neuropathy, dysarthria and ophthalmoparesis
SAP	serum amyloid P
Sap-A	sapsosin-A
Sap-B	sapsosin-B
Sap-C	sapsosin-C
SAR	specific absorption rate
SBF2	set binding factor 2
SBMA	spinal and bulbar muscular atrophy
SBP	systemic blood pressure
SBS	shaken baby syndrome
SCA	spinocerebellar ataxia
SCAR1	spinocerebellar ataxia recessive type 1
SCD	subacute combined degeneration
SCI	spinal cord injury
sCJD	sporadic Creutzfeldt-Jakob disease
SCLC	small cell lung cancer
SCMAS	subunit c of mitochondrial ATP synthase
SCO	subcommissural organ
SCS	spinal cord schistosomiasis
SDF-1	stromal cell-derived factor 1
SDH	subdural haematoma; succinate dehydrogenase
SDS	Shy-Drager syndrome
SE	spin echo; status epilepticus
SEER	Surveillance, Epidemiology and End Results
SEGA	subependymal giant cell astrocytoma
SF-1	steroidogenic factor-1
sFI	sporadic fatal insomnia
SFT	solitary fibrous tumour
SFV	Semliki forest virus
Shh	Sonic hedgehog
SIADH	syndrome of inappropriate antidiuretic hormone secretion
SIS	second impact syndrome
SKL	serine-lysine-leucine
SLD	sudanophilic (orthochromatic) leukodystrophy
SLE	systemic lupus erythematosus; St. Louis encephalitis
Sm	Smith
SMA	spinal muscular atrophy
SMARD	spinal muscular atrophy with respiratory distress

SMC	smooth muscle cell
SMN	survival motor neuron
SMNA	sensorimotor neuropathy with ataxia
SMTM	sulcus medianus telecephali medii
SN	substantia nigra
SNAP	sensory nerve action potential
SNARE	soluble N-ethylmaleimide-sensitive factor attachment protein receptor complex
SND	striatonigral degeneration
SNP	single nucleotide polymorphism
SNPC	substantia nigra pars compacta
SNPR	substantia nigra pars reticulata
SO	slow-twitch oxidative
SOD	superoxide dismutase
SPECT	single photon emission computed tomography
SPLTLC1	serine-palmitoyltransferase 1
SPS	stiff-person syndrome
SRP	signal recognition protein
SSPE	subacute sclerosing pan-encephalitis
SUDEP	sudden unexpected death in epilepsy
SVD	small vessel disease
SVZ	subventricular zone
SWI	susceptibility-weighted imaging
SYN	synaptophysin
TACE	TNFα converting enzyme
TAI	traumatic axonal injury
TBI	traumatic brain injury
TBP	TATA box-binding protein
TCGA	The Cancer Genome Atlas
TCI	total contusion index
TCR	T-cell receptor
TEF	thyrotroph embryonic factor
TGA	transposition of the great arteries
TGF	transforming growth factor
TGM6	transglutaminase 6
THCA	trihydroxycholestanoic acid
TIA	transient ischaemic attack
TLE	temporal lobe epilepsy
TLR	Toll-like receptor
TME	transmissible mink encephalopathy
TMEV	Theiler's murine encephalomyelitis virus
TNF	tumour necrosis factor
TOCP	triorthocresylphosphate
Topo II alpha	topoisomerase II alpha
TPNH	triphosphopyridine nucleotide
TPP	thiamine pyrophosphate
TS	Tourette's syndrome; Turcot syndrome
tSAH	traumatic subarachnoid haemorrhage
TSC	tuberous sclerosis complex
TSE	transmissible spongiform encephalopathy
TSH	thyrotrophin

TSP	tropical spastic paraparesis
TTF-1	thyroid transcription factor 1
TTP	thrombotic thrombocytopenic purpura
TTR	transthyretin
UBO	unidentified bright object
UCH-L1	ubiquitin carboxy-terminal hydrolase
uPA	urokinase plasminogen activator
UPDRS	Unified Parkinson's Disease Rating Scale
UPR	unfolded protein response
UPS	ubiquitin-proteasome system
UV	ultraviolet
VaD	vascular dementia
VCAM-1	vascular cell adhesion molecule 1
VCI	vascular cognitive impairment
vCJD	variant Creutzfeldt-Jakob disease
VCP	vasolin-containing protein
VEE	Venezuelan equine encephalomyelitis
VEGF	vascular endothelial growth factor
VEP	visual evoked potential
VGKC	voltage-gated potassium channel
VHL	Von Hippel-Lindau
VLBW	very low birth weight
VLCFA	very-long-chain fatty acid
VLDL	very low density lipoprotein
VLM	visceral larva migrans
VM	vacuolar myelopathy
VMB	vascular malformation of the brain
VPF	vascular permeability factor
VPSPr	variably protease sensitive prionopathy
VSMC	vascular smooth muscle cell
VV	valine homozygous
vWF	von Willebrand factor
VZ	ventricular zone
WBC	white blood cell
WE	Wernicke's encephalopathy
WEE	western equine encephalitis
WHO	World Health Organization
WKS	Wernicke-Korsakoff syndrome
Wlds	wallerian degeneration slow
WM	white matter
WNV	West Nile virus
WSM	widely spaced myelin
XMEA	X-linked myopathy with excessive autophagy
YAC	yeast artificial chromosome
ZASP	Z-line alternatively spliced PDZ protein
ZPT	zinc pyridinethione
ZS	Zellweger syndrome

General Pathology of the Central Nervous System

Harry V Vinters and B K Kleinschmidt-DeMasters

NEURONS

The neuron is the excitable cell type responsible for the reception of stimuli and information, and conduction of electro-chemical impulses in the brain, spinal cord and ganglia. Neurons are 10–50 times less numerous than their supporting cells, the neuroglial astrocytes, oligodendrocytes and ependymal cells,[27] and are estimated to constitute only 5 per cent of the cells within the cerebral grey matter.[56] Yet they are responsible for the most critical and complex (arguably defining) cellular functions of the organ. They also undergo the greatest number of microscopic changes in response to acute and chronic cell injury and are the principal site of damage for several of the diseases associated with the highest morbidity and mortality in our society, i.e. cerebrovascular and neurodegenerative diseases.

The complex functions of the neuron are responsible for its high metabolic demand for glucose and oxygen/blood supply and are also reflected in its specialized morphological features. Neurons possess a nucleus, nucleolus, cytoplasm and many of the same cytoplasmic organelles found in other cells in the body. However, their extreme protein synthetic and energy requirements, the extraordinary length of their cell processes, and the need for a complex cytoskeletal architecture to support these long cell processes mandate the need for some of these subcellular structures to be better developed than in cells elsewhere in the body, or even in their neighbours, the neuroglial cells.

Under normal, non-injury conditions, usually only the nuclei and cell bodies of neurons are visible to the pathologist on routine histochemical stains used in daily practice, such as haematoxylin and eosin (H&E) or Luxol fast blue–H&E. Immunohistochemical stains commonly employed in routine neuropathology practice to identify proximal portions of neurons (the dendrites and/or soma) include primary antibodies to synaptophysin (a presynaptic vesicle protein), NeuN (a neuronal nuclear protein), microtubule-associated protein 2 (MAP-2), and some of the three polypeptide subunits of neurofilament, which constitutes the major cytoskeletal intermediate filament type for neurons. Low (NF-L, 68 kDa), medium (NF-M, 160kDa) and heavy (NF-H, 200 kDa) subunits exist within the neuron and selective antibodies have been developed over the past 20 years against each. Early work with antibodies directed against these various NF subunits showed no staining of neuronal perikarya and dendrites with antibodies directed against the heavy 200 kDa component.[47] It was subsequently recognized that the antibody directed against NF-L recognized a component in the central core of neurofilaments, and the NF-H antibody a component of the interneurofilamentous cross-bridges; because neurofilaments in mammalian axons were extensively cross-linked, it was not surprising that axons immunostained best with the antibody directed against NF-H.[97] Later work further showed that a lower ratio of NH-L to NH-M and NH-H was found in dendrites and that this proportion was essential for the shaping and growth of complex dendritic trees in motor neurons.[127] Antibodies were also raised to phosphorylated (SM131, NE14) and non-phosphorylated (SM132) NF subtypes. Phosphorylation of NFs was correlated with abundance of NFs and bundling and cross-linking between NF core filaments.[86] Anti-phosphorylated NF antibodies showed strongest immunostaining in axons where NFs are abundant and show this cross-linking, but not in dendrites and perikarya where NFs are sparse and are present singly.[86] Although there are variations among cell types and the distribution of NFs changes in disease states, a general principle is that antibodies directed against non-phosphory-lated subunits of NF best stain the dendrites and perikarya of neurons, whereas those against phosphorylated NFs are used to highlight axons. Neuron specific enolase (NSE), despite its name, is unfortunately not specific for neurons but does also highlight the neuronal cell body.[77] Among the many definitely non-neuronal entities that NSE stains, myeloma and lymphomas can be the most problematic for the diagnostic surgical neuropathologist.[168]

Specific subsets of neurons can be further identified by immunostaining for calretinin, galanin or any of the various specific neurotransmitters and neuromodulators that they produce (γ-aminobutyric acid (GABA), glutamine, dopa-mine, acetylcholine, neuropeptide Y, etc.), but these tech-niques are almost exclusively employed in research rather than routine daily practice.

Antibodies to markers of neuronal lineage have applica-tion both in the study of normal central nervous system (CNS) and peripheral nervous system (PNS) neurons and in assess-ing brain tumours of possible neuronal lineage/differentiation. Antibodies have been raised to α-synuclein, a presynaptic nerve terminal protein found in normal neurons, and immu-nostaining for this has found widest application in the study of inclusion bodies in neurodegenerative disorders. However, α-synuclein immunostaining has also been found in human brain tumours manifesting neuronal differentiation, such as ganglioglioma, medulloblastoma, neuroblastoma, primitive neuroectodermal tumours and central neurocytoma.[117] The proportion of tumours immunopositive for α-synuclein was reported to be lower than that labelled with more commonly used neuronal antibodies, including those to synaptophysin, MAP-2, NSE and tau, but higher than the proportion posi-tive for neurofilament or chromogranin A.[117] Other neuro-nal markers such as TrkA, TrkB, TrkC, the α1 subunit of the GABA receptor, N-methyl-D-aspartate receptor subunit 1, glutamate decarboxylase and embryonal neural cell adhesion molecule have also occasionally been utilized to detect putative neuronal lineage in human brain tumours.[270]

The full extent of the cell processes of neurons, termed neurites, cannot be discerned on H&E staining and is only fully appreciable with special stains. The neurites respon-sible for receiving synaptic information from other neurons and for afferent conduction of electrochemical impulses towards the cell body (soma) are termed dendrites. The full arborization pattern of dendrites is best visualized using Golgi staining techniques (a time-consuming process not usually available in non-research settings). The single elon-gate process responsible for efferent conduction of impulses away from the cell soma is the axon. Axons can be visualized using staining techniques widely available in most diagnos-tic laboratories, including the modified Bielschowsky and Bodian silver histochemical stains or immunohistochemical methods that target phosphorylated neurofilaments.

The number, length and position on the neuronal cell body of the branching dendrites determine the shape and morphological classification of the neuron. Unipolar neu-rons possess a single cell process that divides a short dis-tance from the cell body; an example is the dorsal root ganglion cell. Bipolar neurons have an elongate cell body

with two cell processes emerging at either end of the cell soma; examples include retinal bipolar cells and cells of the sensory cochlear and vestibular ganglia. The vast majority of neurons are, however, multipolar, with large numbers of dendrites arranged in a radiating pattern around the entire cell body (motor neurons of the spinal cord), at the apex of a triangular cell body (pyramidal cell of cerebral cortex) or near the top of a flask-shaped cell (Purkinje cell neuron of cerebellum). Multipolar neurons can be further subdivided based on the length of their efferent axonal process. Golgi type II neurons, with a short axon that terminates near the cell body, greatly outnumber Golgi type I neurons. Golgi type I neurons possess a long axon that may be up to sev-eral feet in length in the case of some motor neurons, or less lengthy in the case of pyramidal cells of the cerebral cortex or Purkinje cells of the cerebellar cortex.[224]

The cross-sectional diameter of the neuronal cell body, by contrast, is largely determined by the length of the axon. The size of neuronal cell bodies varies greatly, from the small, 5 μm-diameter, granule cell neurons of the cerebel-lum to the large, 135 μm-diameter, anterior horn cells of the spinal cord.[224] The volume of the neuronal soma parallels the length of the axon for which it is responsible: the lon-ger the axon, the larger the cell body must be – specifically, the larger the cytoplasmic volume and organelle machinery must be to sustain that axon. Hence, Golgi type I neurons have larger amounts of cell cytoplasm that are readily visible even on H&E preparations, whereas Golgi type II neurons have scant cytoplasm that may give the neuron a 'naked nucleus' appearance on routine stains. Examples of the latter include the 'lymphocyte-like' granule cell neurons of the cer-ebellar cortex, which have a densely basophilic nucleus but in routine preparations appear to possess no cytoplasm, or the small interneurons of the cerebral cortex, which because of the paucity of their cytoplasm may be difficult to distin-guish from neuroglial cells in H&E-stained sections.

The neuronal nucleus is the repository for the chromo-somes and in resting, non-mitotic conditions the chromatin is generally fairly evenly dispersed throughout the nucleus. The prominent large nucleolus seen especially in Golgi type I neurons is a reflection of the need for a high rate of pro-tein synthesis to maintain the numerous proteins within the large cytoplasmic volume, determined largely by the length of axon. The nuclear membrane is well defined on routine H&E staining, but the double-layering of the membrane and the presence of fine nuclear pores, through which sub-stances can diffuse into and out of the nucleus, is appre-ciable only on electron microscopy (EM). The nuclear pores are a conduit through which newly synthesized ribosomal subunits can pass from the nucleus into the cytoplasm. The cytoplasm contains both granular and non-granular endo-plasmic reticulum. The granular, RNA-containing, endo-plasmic reticulum extends throughout the cell body into the proximal parts of the dendrites; it is absent from the area of cytoplasm immediately adjacent to the axon, known as the axon hillock, and from the axon itself.

Subcellular organelles of the neuron can be variably appreciated on H&E staining. Components that contain appreciable amounts of DNA (nucleus) or RNA (nucleo-lus and abundant cytoplasmic rough endoplasmic reticu-lum arranged in parallel arrays known as Nissl substance) in the cell have affinity for the haematoxylin dye used in routine histochemical staining. Therefore, the nucleus, nucleolus and Nissl substance of large neurons manifest a distinct blue-purple colouration and are readily visible on staining with H&E. The DNA- and RNA-containing structures within the neuron can be further highlighted by other histochemical staining techniques such as the modified Nissl method, which originally used aniline but has been modified to use toluidine blue, cresyl violet or others. The monochromatic Nissl staining method is often employed by investigators interested in morphometric analyses of neuronal populations in normal or diseased states. The Nissl stain is often used to highlight neuronal loss in chronic neurodegenerative disorders.

The remaining, non-DNA or RNA containing organelles in the neuron, such as the mitochondria, Golgi complex, lysosomes, neurofilaments, microtubules and microfilaments, are individually unresolvable by H&E at the light microscopic level under normal conditions. These neuronal organelles blend together within the eosinophilic, pink cytoplasm in H&E-stained normal cells and can be appreciated only on EM. The complexity of the synapse is also appreciable only by EM. A number of antibodies directed to synaptic vesicle proteins have, however, been developed that can highlight the synapse and give an indication as to its function or dysfunction. The more common of these include synaptophysin, synaptobrevin (vesicle associated membrane protein, VAMP), synaptotagmin I and synaptic vesicle protein 2 (SV2).

In order to function normally, neurons require complex membrane pumps to exclude toxic calcium ions and to maintain the correct balance between internal (intracellular) and external (microenvironmental) electrolyte concentrations of sodium and potassium in order to transmit electrical signals.[56] With microenvironmental damage to neuronal membranes or with energy deprivation, these pumps fail, calcium ions flood the neuron and irreparable cell damage – known as necrosis – occurs. Local oxygen deprivation, such as that seen in stroke, may result in transient (recoverable) or permanent (irreparable, necrotic) injury to the neuron (see Chapter 2, Vascular Disease, Hypoxia and Related Conditions). This oxygen deprivation can affect cell energy requirements, membrane integrity, and/or the immediate surrounding microenvironment.[56] Thus calcium-channel blocking 'neuroprotectant' agents used in the treatment of stroke may work not just at the level of the neuron alone but also on the microvessels and supportive glial cells around them, the so-called 'neurovascular unit'.[56]

At the light microscopic level, acute sustained deprivation of energy (oxygen/blood supply or glucose), however, is best appreciated in the neuron itself. The irreparable cell damage can be visualized as the brightly eosinophilic 'red (dead) neuron' (Figure 1.1a). This change, seen most often with ischaemia, is manifested by cell shrinkage, nuclear pyknosis, loss of nucleolar detail and loss of basophilic cytoplasmic staining as a result of dissolution of granular endoplasmic reticulum. These result in a smaller, triangular cell, condensed nuclear chromatin, loss of the nucleolus and eosinophilia of the cytoplasm (Figure 1.1a). It should be emphasized that neurons may succumb within several minutes at normal body temperature to severe deprivation of oxygen. However, when body temperature is lowered, metabolism is slowed and considerably longer time periods

without oxygen may be endured by the human brain, with relatively lesser amounts of irretrievable neuronal loss. This explains the remarkable recovery of some people immersed for an hour or more at the bottom of a cold lake who, when retrieved and resuscitated, are able to survive in a relatively cognitively intact state! It should also be emphasized that at normal body temperature, neurons are actually irreparably damaged within minutes when subjected to complete lack of oxygen, but to fully appreciate the 'red cell change' in these same cells under the microscope, at least 8 hours and optimally 18–24 hours must elapse after the injury event before these changes can be confidently diagnosed. The corollary to this is that if a patient dies soon after a cardiac arrest and the family and treating physician of the deceased want to know exactly how widespread the ischaemic neuronal injury was in the patient's brain at autopsy, the pathologist reviewing the case will be unable to answer this question by using routine autopsy techniques. A spectrum of morphological changes ('necrophanerosis') evolves over variable time intervals prior to final ('definite') necrosis; these changes depend upon a variety of factors such as the rate and extent of blood (re-) perfusion, body temperature and others. Animal studies on ischaemic cell injury in neurons often avoid this problem by using rapid perfusion-fixation and EM to detect early, subtle organelle injury. During the acute phase, brain tissue surrounding a focus of ischaemic injury has an eosinophilic neuropil and exhibits significant vacuolation due to oedema (Figure 1.1a). This should not be mistaken for the spongiform change seen in transmissible spongiform encephalopathies. These changes are considered in detail in Chapter 2, Vascular Disease, Hypoxia and Related Conditions).

When neurons undergo cell death and necrosis, no effective neuronal mitosis or replenishment of neurons from stem cells is present within the adult human brain: neuron(s) and their function(s) are lost to the host. Irreversibly damaged neurons are removed over the next few days by phagocytosing microglial cells and macrophages. Astrocytes begin to proliferate in response to injury and may leave a distinctive, tell-tale indication of where the now-removed neurons formerly resided. The classic example of this is Bergmann astrocytosis in the layer of cerebellar cortex where the Purkinje cell neurons formerly resided (Figure 1.1b). Occasionally, morphological evidence of sublethal cell injury in neurons can be detected, best typified by peripheral (Figure 1.1c, bottom) and central (Figure 1.1c, top) chromatolysis. Chromatolysis refers to the response to injury usually seen in Golgi I motor neurons in the anterior horn of the spinal cord when their long axonal process is transected or severely injured. Chromatolysis can be thought of as reorganization by the cell soma and redistribution of Nissl substance in an attempt to reconstitute the axon; central and peripheral chromatolysis may be different phases of this process. If the axonal injury is too severe or the axonal transaction too proximate to the cell body, the efforts of the cell body and its chromatolytic response will be insufficient to produce a new healthy axon and the neuron will itself eventually disappear.

Sublethal injury to neurons may manifest not as eosinophilic change but by cell shrinkage and atrophy. This can occur in a variety of neurodegenerative disorders but is typified by neurons affected by trans-synaptic neuronal degeneration.

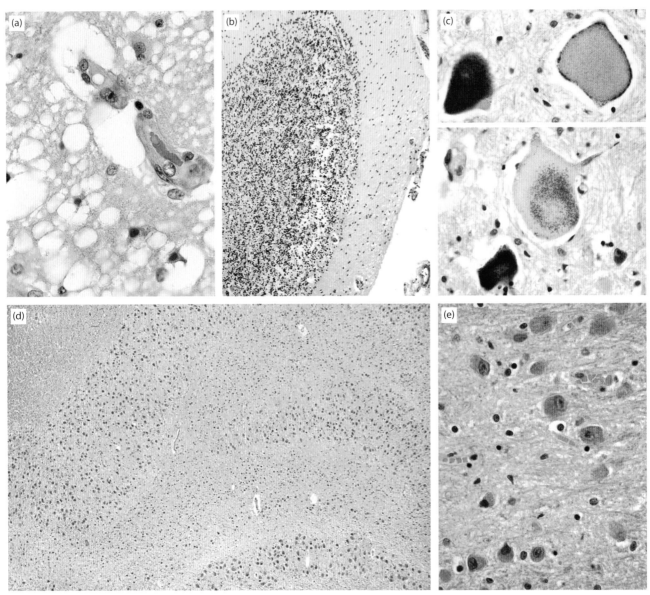

1.1 Neuronal abnormalities in diseases of the CNS. (a) Red, dead neurons with loss of nucleoli and Nissl substance after cerebral ischaemia. Note the vacuolated, oedematous background neuropil. **(b)** Absence of Purkinje cell neurons and gliosis, but good preservation of granule cell neurons as a result of chronic ischaemic cerebellar injury. **(c)** Central (bottom) and peripheral (top) chromatolysis. Nissl stain. **(d)** and **(e)** Trans-synaptic degeneration in lateral geniculate nuclei; see text for detailed description.

Trans-synaptic degeneration occurs when a neuron loses the major source of its axonal input from connecting (incoming) fibres, usually as a result of the loss of 'upstream' neurons that give rise to these axons. A good example of this process is seen following enucleation of one eye. Axons from retinal ganglion cells synapse on neurons in the lateral geniculate nuclei. The example of trans-synaptic degeneration illustrated in Figure 1.1d is from an autopsy performed on a female who underwent right eye removal for retinal melanoma 6 years prior to death, with subsequent wallerian degeneration of the ipsilateral optic nerve and trans-synaptic degeneration in the lateral geniculate nuclei. Because of the differing patterns of projection of axons from the ipsilateral and contralateral eye, the left lateral geniculate ganglion showed atrophy of neurons in layers 1, 4 and 6, whereas the right showed atrophy of neurons in layers 2, 3 and 5. Note the bands of preserved large cells alternating with the bands containing severely

shrunken neurons, making them nearly invisible at low magnification (Figure 1.1d). The atrophic neurons (lower left) are readily seen at higher magnification (Figure 1.1e) and contrast with the adjacent normal neurons from preserved layers (upper left).

A special variant of trans-synaptic neuronal degeneration occurs when axons emanating from the inferior olivary nucleus (ION) are disrupted and their synaptic connections lost, or input to the ION is interrupted. Lesions in the ipsilateral central tegmental tract or the contralateral dentate nucleus result in unilateral olivary hypertrophy. In these instances, the neurons individually enlarge to the degree that they collectively produce hypertrophy of the entire nucleus, visible grossly or at low magnification (Figure 1.1f), and even on high resolution neuroimaging studies carried out while a patient is alive. The illustrated example originates from a man with multiple small cavitary

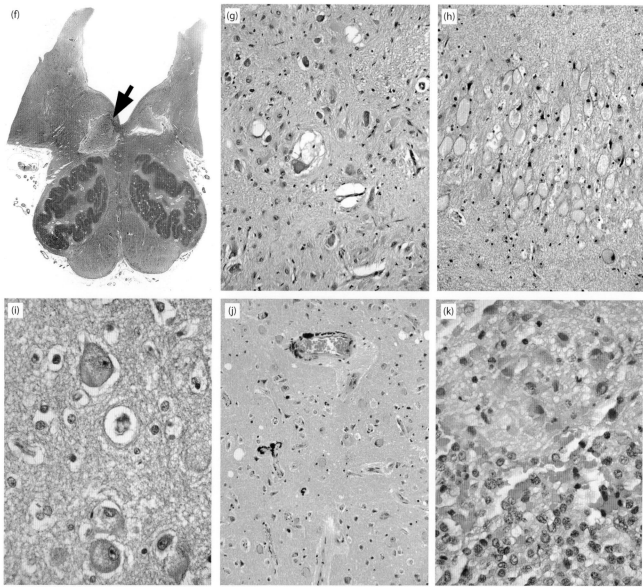

1.1 (*Continued*) Neuronal abnormalities in diseases of the CNS. (f) Bilateral olivary hypertrophy; this change on any given side is due to disruption of the ipsilateral tegmental tract or contralateral dentate nucleus. Note the area of remote infarction (arrow) in the medulla. Nissl stain. **(g)** Vacuolation/fenestration of neurons in inferior olivary nucleus in the example seen in panel **(f)**. **(h)** Neuronal storage diseases cause accumulation of abnormal cytoplasmic material, evidenced by cytoplasmic bloating. Tay–Sach's disease illustrated. **(i)** Neuronal alterations in some storage disorders manifest as fine vacuolation in the cytoplasm. Hunter's disease illustrated. **(j)** Neuronal enlargement and calcification of blood vessels may occur after cranial irradiation; the latter change is much more common than the former. **(k)** Rare pituitary adenomas (lower part of photomicrograph) manifest neuronal metaplasia (upper part), the so-called mixed pituitary adenoma-gangliocytoma.

remote infarcts that were present in the medulla (arrow) and elsewhere in the brain stem and cerebellum and that disrupted these tracts on both sides. Note the bilateral inferior olivary nuclear enlargement (Figure 1.1f). Microscopically, neurons showed characteristic vacuolation ('fenestration') and enlargement, accompanied by considerable astrocytosis (Figure 1.1g). The reason for this special microscopic response to trans-synaptic degeneration in the inferior olivary nucleus is unknown but involves fragmentation within the Golgi apparatus and trans-Golgi network[234] and redistribution of presynaptic vesicles, as manifested by an altered pattern of synaptophysin immunoreactivity.[116]

Less common reactions of neurons to injury include the accumulation of abnormal cytoplasmic storage material, such as in inherited, autosomal recessive storage disease disorders seen in childhood (see Chapter 6, Lysosomal Diseases). Two illustrated examples depict the neuronal changes seen in Tay–Sachs disease (Figure 1.1h) and Hunter's disease (Figure 1.1i). Neuronal enlargement and gigantism may occur in the brain tissue adjacent to a tumour after cranial radiation therapy for a nearby neoplasm, and may be accompanied by other manifestations of tissue injury such as calcification (Figure 1.1j). Unlike many epithelial cell types, neurons rarely undergo metaplasia. In rare pituitary adenomas, most often of growth hormone-secreting type, adenoma cells (Figure 1.1k, lower portion) transform focally into neurons (Figure 1.1k, top);[81] these cells, phenotypically identical to other neurons, may also express small amounts of pituitary hormones.

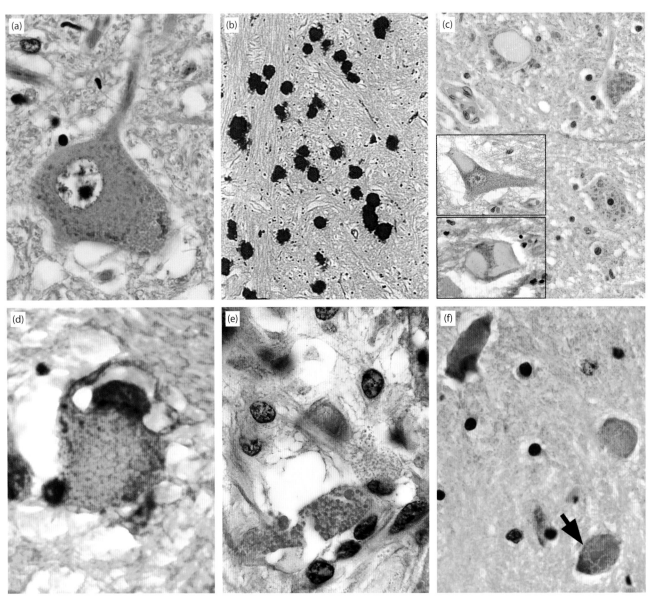

1.2 Inclusion bodies and abnormal deposits I. (a) Bunina body in anterior horn cell in a patient with amyotrophic lateral sclerosis (motor neuron disease). The significance of these structures is discussed in detail elsewhere. **(b)** Buscaino bodies (mucocytes, metachromatic bodies) in white matter can occur secondary to poor tissue fixation and post-mortem degeneration of myelin. On H&E staining, these are barely visible as pale blue bodies or almost clear vacuoles; the periodic acid–Schiff stain, used here, demonstrates these bodies strikingly. **(c)** Colloid bodies (hyaline inclusions) are pale eosinophilic areas within the cytoplasm of neurons and correspond on electron microscopy to dilated cisternae of endoplasmic reticulum. Although usually seen in the hypoglossal nucleus (large picture), they may also be found in the anterior horn cells of the spinal cord (top inset) and very rarely in other neurons, such as the nuclei of Clarke's column (bottom inset, lowest left). They are of no known pathological significance and should not be mistaken for pathological accumulations of proteins or chromatolytic change. **(d)** Cowdry A inclusion bodies are seen in herpetic viral infections of the nervous system (herpes simplex type I and II, cytomegalovirus infection, and varicella-zoster virus infection but not infections with Epstein–Barr virus). On electron microscopy, it can be appreciated that they are due to accumulations of virions within the nucleus of the host cell. Note the clearing of the host cell nuclear chromatin centrally, with margination of chromatin at the edge of the nuclear membrane and the 'owl's eye' appearance of the viral inclusion. In this case of cytomegalovirus infection, the cell cytoplasm is also enlarged (cytomegaly) and distended by viral particles. **(e)** Eosinophilic granular bodies (EGBs) are dot-like, refractile, proteinaceous deposits most commonly encountered in the background neuropil in or adjacent to certain types of low grade brain tumours, as here in a pleomorphic xanthoastrocytoma. They can be further highlighted by periodic acid–Schiff staining. **(f)** Eosinophilic crystalline inclusions can occasionally be seen in the cytoplasm of neurons of the inferior olivary nucleus, especially in aged individuals and are of no known pathological significance.

Although necrosis is the type of neuronal cell death that predominates in acute energy-deprivation states, neuronal apoptosis plays a critical role during embryonic development. Apoptosis or programmed cell death refers to a controlled, coordinated biochemical process leading to the death of affected cells and is a physiological part of normal development. In a wide variety of disparate organisms, apoptosis involves the triggering of a series of biochemical events in which caspases (cysteine aspartases) play a key role.[170] Although the morphological manifestations of apoptosis

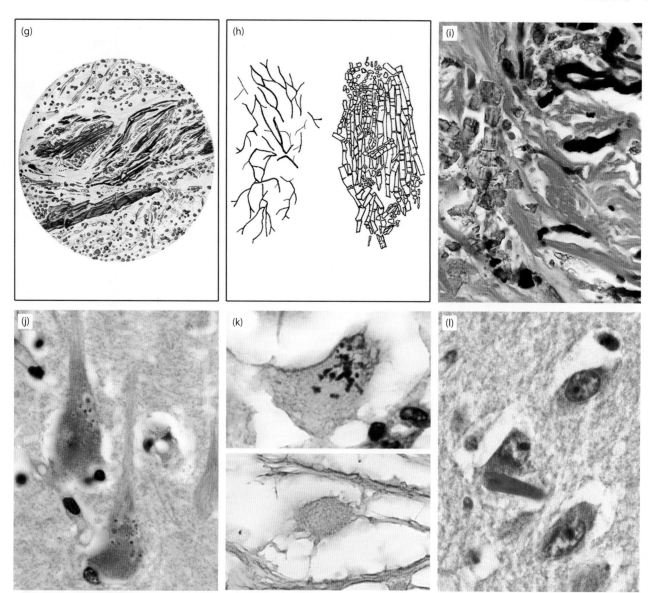

1.2 (*Continued*) Inclusion bodies and abnormal deposits I. (g) Gamna–Gandy bodies are foci containing linear, bamboo-like fibrous tissue and collagen fibres encrusted with iron pigments and calcium salts. They were originally described in the spleen in patients with congestive splenomegaly but can be seen around cavernous angiomas, cholesterol granulomas of temporal bone, pituitary adenomas, and a variety of other highly vascular primary and metastatic neoplasms and cysts in the nervous system that are subject to recurrent bouts of haemorrhage. However, when first described in the 1920s, the authors had to go to great lengths to exclude a fungal causation for these structures, which are illustrated in a colour drawing from a 1922 article. **(h)** Gamna–Gandy bodies, illustrated in black and white drawings of from a 1929 article by Hu *et al*.;[102] these authors showed that there was no morphological identity between the wavy encrusted fibres (left) or waxy septate, bamboo-like fibres (right) and true fungal mycelia. **(i)** Gamna–Gandy bodies in tissue from a region of recurrent brain haemorrhage. **(j)** Granulovacuolar degeneration (of Simchowicz, granulovacuolar bodies, GVBs) appear as tiny dots within clear vacuoles that can be seen particularly in the cytoplasm of pyramidal neurons of the hippocampal gyrus in normal ageing and, to a greater extent, in patients with Alzheimer's disease. These structures contain abnormal accumulations of several proteins including tubulin, neurofilament proteins and tau. **(k)** Granular mitoses (top) are clusters of chromatin often encountered in cells in highly mitotically active tissues. Although usually found, and illustrated, in the context of acute demyelinating lesions, this example comes from a case of cytomegalovirus ventriculitis and should not be mistaken for a micro-organism. Herring bodies (bottom) are spherical or ovoid eosinophilic structures with an apparent surrounding membrane that are normal findings in the posterior pituitary gland (neurohypophysis). They represent normal storage sites within axons for oxytocin and vasopressin. **(l)** Hirano bodies are elongate (when longitudinally sectioned) to oval (in cross-section), brightly eosinophilic neuronal inclusions that are encountered in pyramidal neurons of the hippocampal gyrus in normal ageing, and, to a greater extent, in patients with neurodegenerative diseases such as Alzheimer's disease. Although they often seem to be extraneuronal, by electron microscopy Hirano bodies can be seen to lie within the neuronal soma or cell processes. They are composed of actin and α-actinin.

(b) Reproduced with permission from Graeber MB, Blakemore WF, Kreutzberg GW. Cellular pathology of the central nervous system. In: Graham DI, Lantos PL (eds). Greenfield's Neuropathology, 7th edn. London: Arnold, 2002, pp. 123–192.

(h–j) From Kleinschmidt-Demasters, BK. Gamna–Gandy bodies in surgical neuropathology specimens: observations and a historical note. Journal of Neuropathology and Experimental Neurology 2004;63:106–12. Reproduced with permission from the Journal of Neuropathology and Experimental Neurology.

are classically described as 'cell shrinkage, membrane blebbing and nuclear DNA condensation and fragmentation',[146] these may not be seen in non-vertebrate systems.[216]

Neuronal apoptosis also occurs in pathological disease states and involves similar 'execution systems' and proteins.[170] At least 14 different mammalian caspases have been identified thus far, but these may have both death-related and death-unrelated functions in the cell.[170] Neuronal necrosis and apoptosis are not always mutually exclusive processes and the co-existence of both has been emphasized in some pathological conditions.[136] For instance, a shift from apoptotic to necrotic types of neuronal death may occur when energy levels are rapidly compromised.[136] The practical aspect of identifying a role for neuronal apoptosis in a disease process lies in the fact that small peptide caspase inhibitors have been developed and may have therapeutic utility. Caspase inhibitors may be useful in preserving sublethally injured neurons at the perimeter (penumbra) of an acute infarct that might be less severely affected by excitotoxic-ischaemic injury than is the necrotic core of the infarct.[170] They may also act to protect against the deleterious effects of oxygen radicals, cytokines and lipid peroxidation products that are generated in the necrotic core of the infarct and seep out to the penumbra.[139,146] Among human diseases, especially prominent neuronal apoptosis is seen in the (rare) perinatal disorder, pontosubicular necrosis. In neurodegenerative diseases, apoptosis may play differing roles at different time points during the disorder, explaining why caspase inhibitors may not be universally effective therapies. In addition, apoptosis can occur without involvement of the caspase system.[24] A further consideration is whether or not the preservation of neurons that would otherwise undergo apoptosis is desirable in neurodegenerative disorders such as Huntington's disease or Alzheimer's disease, especially if the preserved cells have aberrant function.[170] A role for neuronal apoptosis has been implicated in numerous disorders other than ischaemia and neurodegenerative disease; these include spinal cord trauma, head injury[194] and viral nervous system infections.[51] This complex topic has been the subject of several excellent reviews (e.g. Schulz and Nicotera,[215] Nicotera et al.,[170] Robertson et al.,[194] Paulson,[180] Mattson[146]).

Although individual cellular organelles in neurons are not distinguishable under normal, resting conditions on light microscopy using routine stains, in ageing or in disease processes, massive accumulations of some organelles can be discerned. These processes result in the development of 'inclusion bodies'.[27,76,77] Some of these have limited pathogenic implications (colloid bodies, Marinesco bodies), although others are almost exclusively seen in specific disease conditions (Lafora bodies). Yet more are seen in small numbers in 'normal' ageing but in significantly greater numbers in specific neurodegenerative disorders (neurofibrillary tangles, granulovacuolar degeneration/bodies, Pick bodies). Still other 'bodies' occur in the background tissues but are discussed here with neuronal inclusion bodies because their exact intracellular (Hirano bodies, Figure 1.2l) or extracellular (Gamna–Gandy bodies, Figure 1.2g,h) location may not be apparent in H&E-stained sections. A

pictorial, alphabetically arranged chronology of these 'bodies'–most of which develop in neurons–is depicted in Figures 1.2 and 1.3. Most of these are fully identifiable on H&E staining, including Bunina bodies, colloid bodies (Figure 1.2c), granulovacuolar bodies (Figure 1.2j), Lewy bodies (Figure 1.3c and d), neuroaxonal swellings, neurofibrillary tangles, Pick bodies (Figure 1.3j, insert) and Lafora bodies (Figure 1.3a). Special silver histochemical and immunohistochemical staining, however, can further delineate these normal and abnormal accumulations. Modified Bielschowsky or Bodian histochemical silver stains generally identify neurofilament-containing inclusions or structures in various diseases, such as globose or flame-shaped neurofibrillary tangles (Figure 1.3h,j), Pick bodies (Figure 1.3j) and neuroaxonal swellings, also known as 'spheroids' (Figure 1.3g). Identification of inclusions specific for certain neurodegenerative disorders can be achieved with immunohistochemical methods that identify tau (including its isoforms), ubiquitin, huntingtin or α-synuclein. It is an unresolved issue as to whether neuronal inclusions play a role in direct neuronal injury or represent a mechanism by which neurons protect themselves by sequestering abnormal proteins (reviewed by Paulson[180]).

Neurons are post-mitotic, fully differentiated cells that have little or no capacity to regenerate effectively and reconstitute functions lost when the cell is lost. Neuronal plasticity plays an important role in development and early childhood in overcoming major areas of brain tissue damage but this ability is lost in the adult brain, in which neurons cannot be innately regrown or replaced, even by the small numbers of neural stem cells that are known to be present (discussed later).

ASTROCYTES

Astrocytes are, together with oligodendrocytes/oligodendroglia, the two cell types in the nervous system often described as macroglia, to distinguish them from microglia (see later). These specialized glial cells outnumber neurons by over five-fold.[226] Generally considered to be, in part, the CNS counterpart of fibroblasts, with a significant role in producing scar tissue (described as 'astrocytic gliosis', 'astrogliosis' or simply 'gliosis') within the brain or spinal cord, astrocytes are now known to have myriad physiological and biochemical functions in both brain development and maintenance of homeostasis (especially with respect to the make-up of the interstitial fluid of the brain) and may even contribute to regeneration and repair after brain/spinal cord injury.[255] Many of these properties will be described in detail later. Based upon recent discoveries in molecular neurobiology, the function(s) of astrocytes within normal brain and their relationship to neurons are being so radically redefined that even the nomenclature defining these cells (in relation to neurons) has been called into question. Changes from astroglial to neuronal phenotype (in select cell populations) are now well documented, although brain parenchyma in some lesions (e.g. malformations of cortical development associated with epilepsy) contains cells that have features of both a neuronal and astrocytic phenotype.[254] As one expert in the field

1

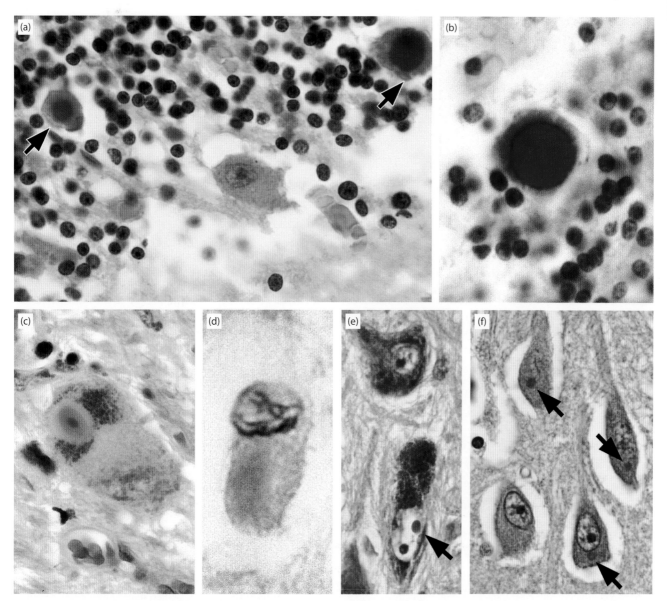

1.3 Inclusion bodies and abnormal deposits II. (a,b) Lafora bodies are basophilic inclusions that are composed of polyglucosans and occur in Lafora's disease, a neurodegenerative storage disease of children. These inclusions occur in many different types of cell and tissue, including neurons, choroid plexus, sweat glands, peripheral nerves, cardiac and striated muscle, and liver and skin. They closely resemble corpora amylacea and, like corpora amylacea, stain intensely with periodic acid–Schiff, but are usually surrounded by a corona of radiating filaments or spicules and are not restricted to the sites of predilection for corpora amylacea. In addition, corpora amylacea are infrequent in children. These figures illustrate Lafora bodies in the cerebellum. **(c)** Lewy bodies (brain stem type) are intracyoplasmic inclusions that represent abnormal proteinaceous accumulations consisting predominantly of α-synuclein. Like many proteinaceous deposits, they are readily visualized in H&E-stained sections. They are easiest to identify in pigmented neurons, such as this one from the substantia nigra compacta, where they displace the normal intracytoplasmic, brown neuromelanin pigment. Note the targetoid appearance; however, most are not so eye-catching. Lewy bodies can be encountered in the substantia nigra compacta and especially in the locus coeruleus in normal ageing, but even in this instance may represent preclinical disease. They are more numerous and more widely distributed in patients with idiopathic Parkinson's disease and related disorders. **(d)** Lewy bodies (intracortical type), when located in small neurons of the cerebral cortex, are far less well-defined in H&E-stained sections but can be highlighted by immunostaining for α-synuclein or ubiquitin. Cortical Lewy bodies are usually associated with disease, not normal ageing. **(e)** Marinesco bodies, sometimes referred to as 'maraschino cherry bodies' by residents trying to remember the names of all of the various bodies for board examinations, are intranuclear eosinophilic bodies (arrow), about the same size as the nucleolus. They are largely confined to the pigmented, neuromelanin-containing neurons of the substantia nigra compacta and are usually found in aged individuals. They are proteinaceous inclusions of no known pathological significance, but are very similar to the intranuclear bodies seen in large numbers of neurons in patients with the childhood degenerative disorder, neuronal intranuclear inclusion disease. **(f)** Negri bodies are a pathognomonic finding in rabies viral infection (rabies viral encephalitis) of the central nervous system. These well-circumscribed intracytoplasmic, red cell-like bodies are easily overlooked, particularly when the virus fails to elicit an inflammatory host reaction.

Continued

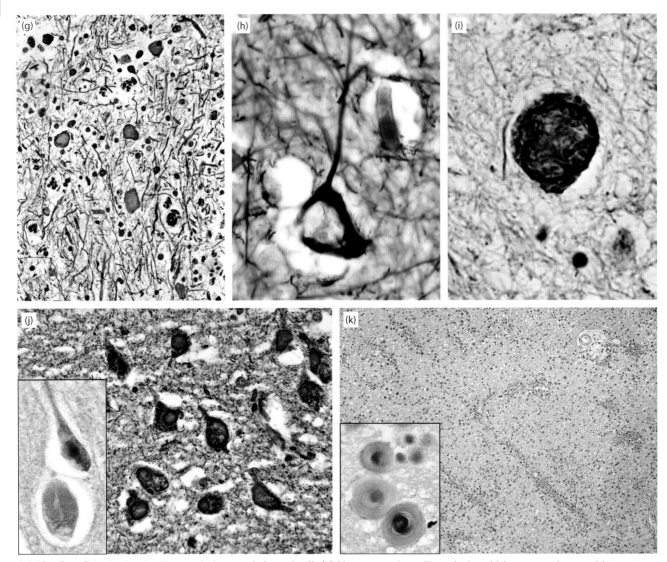

1.3 (*Continued*) Inclusion bodies and abnormal deposits II. (g) Neuroaxonal swellings (spheroids) are round or ovoid structures formed when transportation of intra-axonal neurofilaments is disrupted by axonal injury or transection. Although also discernible in sections stained with H&E they are better highlighted with silver stains, as here. They are illustrated here in the anterior horn of a patient with short-duration amyotrophic lateral sclerosis. **(h)** Neurofibrillary tangles (flame-shaped) are easily recognized by even novice pathologists by their flame-shaped profiles, demonstrated best with silver stains. The classical shape usually illustrated in textbooks is the one seen here in a pyramidal neuron of the cerebral cortex, and the intracytoplasmic location of the tangle, which loops around the (unstained) nucleus, is easily appreciated. Scattered tangles may be encountered in pyramidal neurons of the hippocampal gyrus in normal ageing, but they are seen in greater numbers and in a wider neocortical and brain stem distribution in patients with neurodegenerative diseases such as Alzheimer's disease. **(i)** Neurofibrillary tangles (globose) contain skein-like tangles of abnormal, hyperphosphorylated tau protein and may be seen in brain stem neurons in Alzheimer's disease or in progressive supranuclear palsy; the latter disease is illustrated here. The shape of the tangle is predicated on the shape of the neuronal cell body in which it resides. The coarse internal structure of the globose tangle distinguishes it from argentophilic Pick bodies seen in **(j)**. Pick body-like structure associated with neurodegenerative disease. Bodian silver stain. **(j)** Pick bodies are intracytoplasmic bodies found in the pyramidal neurons of the hippocampal gyrus, the granule cell neurons of the dentate gyrus, smaller cortical neurons especially in layer 2, and in brain stem neurons of patients with Pick's disease, a neurodegenerative disease associated with lobar atrophy of the frontal and temporal lobes. They have a relatively homogeneous appearance on both H&E (inset) and silver staining, in contrast to globose neurofibrillary tangles, but sometimes a degree of overlap exists. Unlike neurofibrillary tangles, Hirano bodies or granulovacuolar bodies (degeneration), Pick bodies are almost never encountered in normal aged individuals. **(k)** Polyglucosan bodies are histologically indistinguishable from the corpora amylacea but occur in very large numbers in individuals affected by adult polyglucosan body disease,[23] illustrated here in a section of white matter from a middle-aged patient with this disorder. The variably blue-grey bodies may have a concentric, targetoid appearance (inset). Corpora amylacea are a normal finding in aged individuals, but not in so great a number, and are usually more concentrated in (but not confined to) subpial, subependymal and perivascular locations and the spinal cord. In polyglucosan body disease, heart, skeletal muscle, liver, and dermal sweat glands in addition to peripheral nerves and brain may contain these bodies. They are composed largely of sulphated polysaccharides (polyglucosans) and stain deeply with haematoxylin, periodic acid–Schiff and methyl violet. By electron microscopy, polyglucosan bodies in the nervous system are seen to consist of densely packed 6–7 nm filaments that are not bounded by a unit membrane and lie within astrocytic processes, within axons and few within the neuropil, but not within the neuronal soma.

has boldly and bluntly stated, '…virtually every aspect of brain development and function involves a neuron–glial partnership. It is no longer tenable to consider glia as passive support cells'.[17]

By morphological criteria, astrocytes have been subclassified as protoplasmic (found mainly within the grey matter) or fibrous/fibrillary (located predominantly within the subcortical white matter).[226] The phenotype of astrocytes is defined by the location within their cytoplasm of the intermediate filament protein, glial fibrillary acidic protein (GFAP).[60,64,255] Though not all astrocytes express GFAP that is immunohistochemically detectable within the cytoplasm by light microscopy (and some non-CNS

cells do), the presence of this protein essentially remains, in daily diagnostic work, a defining feature of the cell type. GFAP is especially abundant within the cytoplasm of reactive or hypertrophic (and often neoplastic) astrocytes, though unfortunately the extent and robustness of GFAP immunoreactivity do not correlate well with the specific type or duration of CNS insult to which the astrocytes have reacted (Figures 1.4 and 1.5). GFAP immunohistochemistry has become the standard way to assess astrocytic gliosis (both qualitatively and quantitatively) in both animal studies and human CNS disease tissue examined at biopsy or autopsy. It has superseded older classic cytochemical stains such as the Holzer and phosphotungstic acid haematoxylin

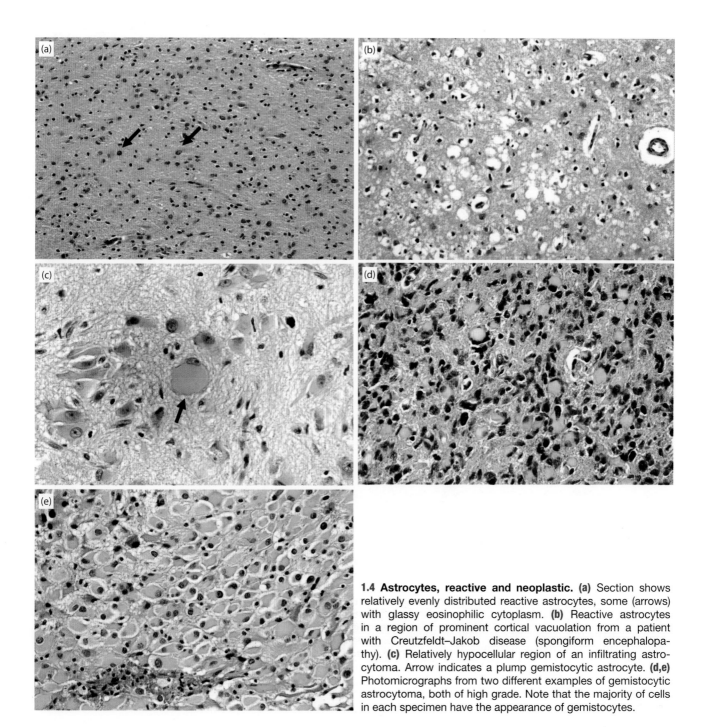

1.4 Astrocytes, reactive and neoplastic. (a) Section shows relatively evenly distributed reactive astrocytes, some (arrows) with glassy eosinophilic cytoplasm. **(b)** Reactive astrocytes in a region of prominent cortical vacuolation from a patient with Creutzfeldt–Jakob disease (spongiform encephalopathy). **(c)** Relatively hypocellular region of an infiltrating astrocytoma. Arrow indicates a plump gemistocytic astrocyte. **(d,e)** Photomicrographs from two different examples of gemistocytic astrocytoma, both of high grade. Note that the majority of cells in each specimen have the appearance of gemistocytes.

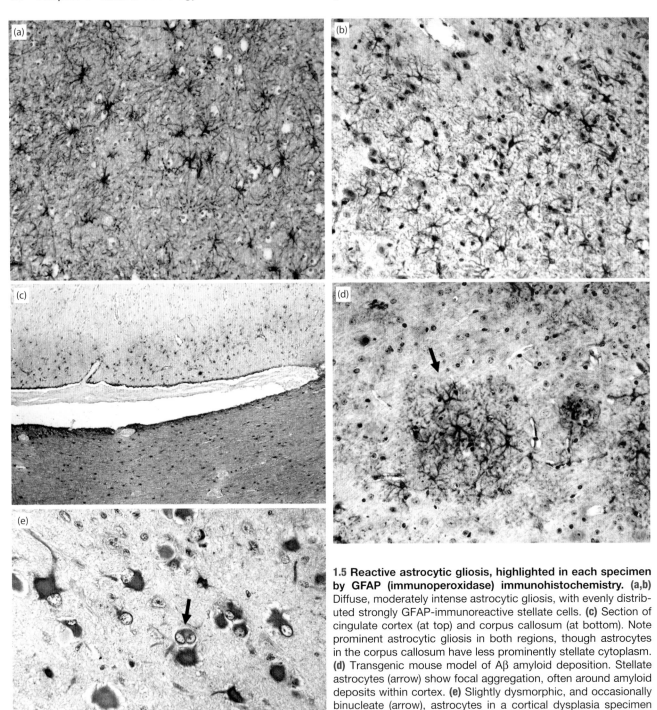

1.5 Reactive astrocytic gliosis, highlighted in each specimen by GFAP (immunoperoxidase) immunohistochemistry. (a,b) Diffuse, moderately intense astrocytic gliosis, with evenly distributed strongly GFAP-immunoreactive stellate cells. **(c)** Section of cingulate cortex (at top) and corpus callosum (at bottom). Note prominent astrocytic gliosis in both regions, though astrocytes in the corpus callosum have less prominently stellate cytoplasm. **(d)** Transgenic mouse model of Aβ amyloid deposition. Stellate astrocytes (arrow) show focal aggregation, often around amyloid deposits within cortex. **(e)** Slightly dysmorphic, and occasionally binucleate (arrow), astrocytes in a cortical dysplasia specimen (corticectomy for epilepsy in a child).

(PTAH) techniques, although the latter stains retain value in some settings. Vimentin and S100β are also prominent components of the astrocytic cytoplasm, though vimentin immunoreactivity in astroglial cells lacks specificity, as this epitope is expressed in many non-glial cell types. By electron microscopy, astrocytes contain abundant intermediate filaments, cytoplasmic dense bodies, gap junctions and multiple cellular processes.[64,255] Astrocytes may also express a variety of growth factor receptors, including those for epidermal growth factor and basic fibroblast growth factor.[60,93]

Prominent cytoplasmic GFAP immunoreactivity also characterizes neoplastic astrocytes within astrocytomas, especially gemistocytic astrocytomas, and other types of tumour-related astrocytes, e.g. the mini-gemistocytes commonly found in oligodendrogliomas (Figure 1.6) (see Chapter 26, Introduction to Tumours). The term 'gemistocyte/gemistocytic' used to describe an astrocyte does not, however, classify it as being malignant or reactive – gemistocytes are also common in brain tissue surrounding infarcts, vascular malformations, traumatic lesions, cerebritis/encephalitis and metastatic neoplasms, as well as in numerous other

1.6 (a–c) A predominantly oligodendroglial neoplasm (oligodendroglioma; micrographs are at various magnifications) contains numerous GFAP-immunoreactive astrocytic cells, including mini-gemistocytes. Note the different morphology of tumour cells (round, regular nuclei with clear cytoplasm) and the more characteristically stellate appearance of the astrocytic element. By contrast, note the widespread GFAP-immunoreactivity of tumour cells and their processes **(d)** in a predominantly astrocytic tumour (astrocytoma).

reactive settings. GFAP immunoreactive cells may even be encountered within the interstices of a metastatic neoplasm (Figure 1.7), leading to diagnostic difficulty in distinguishing an anaplastic primary glioma from a poorly differentiated metastasis.

The recent discovery of mutations in the active site of isocitrate dehydrogenase (IDH1) gene in >70 per cent of intermediate-grade diffuse gliomas, i.e. diffuse astrocytomas grade II, oligodendrogliomas grade II, mixed oligoastrocytomas grade II and anaplastic variants grade III,[15] and the finding that a high-fidelity antibody correlates well with a specific mutational (R132H) status[33,34] have provided the pathologist with a truly tumour-specific glial marker that can be used in daily diagnostic practice. Capper *et al.* demonstrated that the antibody can be used to distinguish diffuse glioma from non-neoplastic reactive gliosis associated with metastases, vascular malformations, abscesses, progressive multifocal leukoencephalopathy, and ischaemic or haemorrhagic lesions (Figure 1.8).[35] In addition, they showed that the IDH1 antibody was superior to p53 or Wilms Tumor 1 (WT1) antibodies in identifying neoplastic glial cells.[35]

Astrogliosis is sometimes subclassified (on purely morphologic grounds) as being isomorphic (when astrocytes arrange themselves along an anatomical structure such as a tract, e.g. the corticospinal tract, in association with wallerian degeneration) or anisomorphic (cells arranged more haphazardly, as at the edges of an infarct or cerebritis/abscess; see Figure 1.9).[64] Brisk reactive astrocytic gliosis can also be associated with the proliferation of Rosenthal fibres, which represent protein aggregates in astrocytic processes that also contain ubiquitin, αB-crystallin and heat shock protein HSP27 (Figure 1.10). Dominant missense mutations in the human GFAP gene are associated with a leukodystrophy (Alexander's disease) characterized by overwhelming proliferation of Rosenthal fibres within the diseased white matter.[137,159,245] The GFAP gene on chromosome 17 includes four α-helical segments within the central rod domain, joined by non-helical linkers. Of interest, GFAP-null mice show relatively subtle neuropathological abnormalities, although animals that overexpress GFAP 10–15-fold manifest a fatal encephalopathy associated with prominent astrocytic swelling.[159]

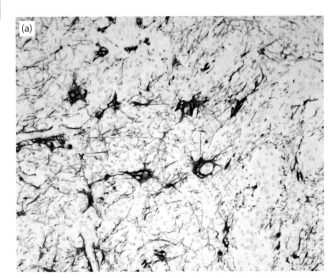 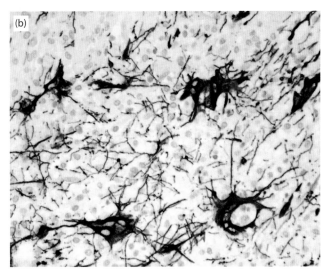

1.7 Both panels show GFAP-immunoreactive reactive astrocytes in an atypical teratoid rhabdoid tumour (AT/RT). Note ramified processes of the reactive astrocytes throughout the tumour.

Role of Astrocytes in CNS Development and Regeneration

Experimental evidence now suggests remarkable plasticity and regenerative potential for at least some populations of astrocytes and astrocyte precursors, a view that would have been somewhat heretical as recently as 20–30 years ago.[58] Since the late 1800s, radial glia have been recognized as key players in brain development. Their elongated fibres span the full width of the developing cerebral wall in most mammals. In the cerebellum, radial glia extend from the pia to the Purkinje cell layer, and are quite regularly and evenly spaced in the molecular layer (Figure 1.11). These cells, at least in the cerebrum, retain the capacity to divide. An increasingly complex understanding of their role in CNS development has coincided with more sophisticated ways to study this unique cell type.[187] In the late stages of cortical development, radial glia appear to divide asymmetrically in the ventricular zone to generate (more) radial glia and intermediate progenitor (IP) cells. IP cells then divide symmetrically in the subventricular zone to give rise to multiple neurons.[144] During development of the brain, radial glia (which provide the 'guidewires' by which neuroblasts in the germinal matrix find their way to the cerebral cortex) are thought to give rise to astrocytes.[211,257] Adult astrocytes may revert to their radial glial phenotype (in tissue culture) when exposed to embryonic brain extracts.[103] However, it is now clear that they can themselves also function as neural progenitor cells.[84] The molecular developmental and neurobiological events in this process are extraordinarily complex, and are well reviewed elsewhere.[8,89,129,158,162,226,241] Astrocytes, in addition to giving rise to new neurons in the adult hippocampus,[218] are now recognized as a major component of 'neurogenic niches', which have the potential to generate neuroepithelial cells from the subventricular zone during early brain development and possibly also at later time point.[8] Increased generation of neuronal progenitor cells after ischaemic stroke has even been demonstrated in human autopsy brain specimens originating from quite elderly individuals.[143] Astrocytes secrete molecules that may support neurogenesis (fibroblast growth factor/FGF,

insulin-like growth factor-1, glutamate, etc.) or inhibit it (astrocyte-derived bone morphogenetic protein).

In the developing mammalian brain, the subventricular zone (SVZ), a germinal region of the brain, contains abundant astrocytes and astrocyte precursors, together with migrating neuroblasts, undifferentiated immature precursors and ependymal elements. In experimental animals, it has been shown that SVZ astrocytes can divide to generate neuroblasts and immature neuronal precursors, and that such astrocytes placed into tissue culture can grow into multipotent neurospheres.[58] Many astrocytes may have a latent neurogenic potential that is suppressed by various inhibitory signals, or expressed only in certain well-defined anatomic locations, e.g. the subventricular zone surrounding the lateral ventricles. Experiments utilizing transgenic targeted cell fate mapping strategies have also shown that morphologically distinct GFAP-positive progenitor cells may represent the major source of cells that are key to constitutive adult neurogenesis in the adult mouse forebrain; in experimental systems, astrocytes appear to have important neuroprotective functions.[80,225] Whereas astrocytic gliosis has historically been thought to inhibit axonal regeneration, experiments in rats have shown that reactive astrocytes may in fact act as a permissive substrate for axon outgrowth from neurons sensitive to (implanted) nerve growth factor (NGF) within the brain.[115]

Sometimes contradictory experimental data continue to fuel the debate as to whether astrocytic scarring or some cellular components overexpressed during that process of scar formation inhibits CNS regeneration. Axonal sprouting in rats is increased in lesioned septohippocampal circuits in parallel with accentuated GFAP immunoreactivity, suggesting that astrocytes may produce trophic factors (e.g. nerve growth factor and related molecules) that facilitate this reparative response.[78] Yet in knockout mice that are rendered deficient in both GFAP and vimentin, improved anatomical regeneration, axonal plasticity and functional recovery have clearly been observed in lesioned spinal cords.[153] Organotypic slice culture experiments have demonstrated that astroglial-associated fibronectin may play a significant role in axonal regeneration within the white matter.[240]

1

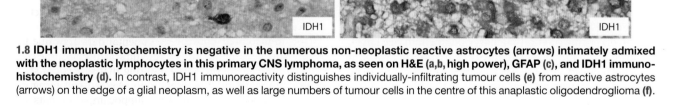

1.8 IDH1 immunohistochemistry is negative in the numerous non-neoplastic reactive astrocytes (arrows) intimately admixed with the neoplastic lymphocytes in this primary CNS lymphoma, as seen on H&E (a,b, high power), GFAP (c), and IDH1 immuno-histochemistry (d). In contrast, IDH1 immunoreactivity distinguishes individually-infiltrating tumour cells **(e)** from reactive astrocytes (arrows) on the edge of a glial neoplasm, as well as large numbers of tumour cells in the centre of this anaplastic oligodendroglioma **(f)**.

Trophic Effects and Influence of Astrocytes on Vascular Structure, Integrity and Physiology

The physical proximity of astrocytes and their processes to CNS microvasculature (Figure 1.12), together with the obvious neuroanatomical observation that cerebral blood vessels are 'swimming in' a sea of astrocytes, intuitively suggests that astrocytes and molecules released by them influence microvascular structure and function. Astrocytes may be instrumental in subdividing brain segments into microdomains, thus defining the functional architecture of the CNS through 'gliovascular units'.[167] The observation of this intimate neuroanatomical association between glia and blood vessels was first noted by Golgi over 120 years ago.[222] A modular organization has even been proposed to define the association of cerebral microvessels, neurons and astrocytes, which are now described as forming

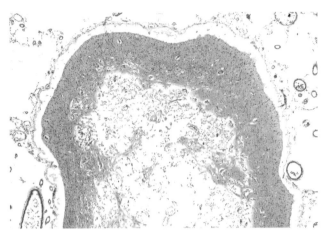

1.9 An example of anisomorphous/anisomorphic gliosis is seen in the lining of a cystic cavity that occupies most of this gyrus, sampled at necropsy from an infant with severe perinatal brain damage. The centre of the cavity contained irregular clumps of glial fibres admixed with numerous foamy macrophages.

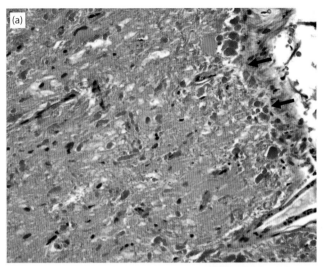

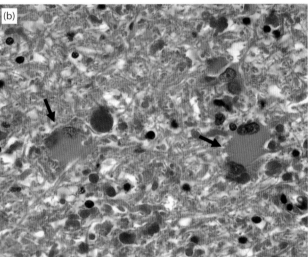

1.10 Rosenthal fibres and astroglia. This is an unusual corticectomy specimen from a child with intractable epilepsy. **(a)** Note numerous hyaline rod-like Rosenthal fibres aggregated at the pial surface (arrows) and in the underlying cortex. **(b)** Two gemistocyte-like cells (arrows), including a binucleated astrocyte (at right) are seen amid numerous Rosenthal fibres.

functional 'neurovascular units' (NVUs).[3] Other elements of the NVU include pericytes (in the case of capillaries) and medial vascular smooth muscle cells (SMCs) in the case of arterioles. Astrocytes are a key link in these units, because they communicate with both synapses and blood vessels, as well as with other astrocytes (via gap junctions and through the release of ATP).[126] They appear to act as crucial intermediaries in intercellular signalling in this putative neurovascular unit. The role of astrocytes in mediating many physiological and biochemical functions of the cerebral capillary endothelium, site of the blood–brain barrier (BBB, see later), has been established by elegant tissue grafting and transplantation, as well as cell culture experiments (for reviews, see Pardridge;[174] Nag;[163] Ballabh et al.[13]). Co-culture studies (first carried out in the mid-1980s, when cerebral capillary isolation techniques became routine) have been performed in which brain-derived capillary endothelial cells[252] are seeded on one side of a porous mesh separating two fluid-filled chambers, another cell type (astrocytes, pericytes, etc.) on its other side. Such protocols were used extensively to demonstrate the inductive effect of astrocytes on both structural and physiologic properties of the BBB, e.g. its well-known 'polarity' for transport of certain molecules.[19,163,174]

Though the morphological site of the BBB is widely accepted as being cerebral capillary endothelium (see later), its physiologic functions and integrity are affected by both adjacent pericytes and astrocytes in the NVU.[3,13,163] The tight junction proteins that mediate many BBB functions (see also discussion later) are expressed very early in human CNS development within the germinal matrix, cerebral cortex and subcortical white matter.[14] Proteins known to be crucial for calcium signalling between cells (purinergic receptors and gap junctions Cx43) are expressed mainly by perivascular astrocyte end-feet that are an invariable finding on the abluminal aspects of CNS blood vessels, both capillaries and larger arteries. Brain slice experiments show that electrical field stimulations cause an increase in astrocytic calcium, which is transmitted to perivascular end-feet, resulting in arteriolar smooth muscle cell oscillations and dilatation of these vessels.[126] Molecules that may mediate communications between astrocytic end-feet and vascular smooth muscle cells include prostaglandins, epoxyeicosatrienoic acids (EETs), potassium ions and arachidonic acid. Astrocyte-mediated control of cerebral blood flow also occurs through the action of calcium transients in astrocytic end-feet.[161,235] Increased blood flow that is coupled to neuronal activity (and is thus used as an indirect measure of brain activity by techniques such as functional magnetic resonance imaging [fMRI]) is modulated in part by cyclooxygenase-2 metabolites, EETs, adenosine and NO derived from neurons. Neuronal activation that results in increased astrocytic calcium is partly mediated by activation of metabotropic glutamate receptors (mGluRs). In tissue culture systems, calcium signalling may be influenced by adenosine and EETs that are produced by astrocytes. Astrocytes are even able to transmit signals to brain surface (pial) arterioles to ensure their continuous adequate supply of blood to parenchymal arterioles.

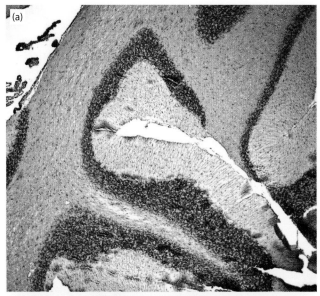

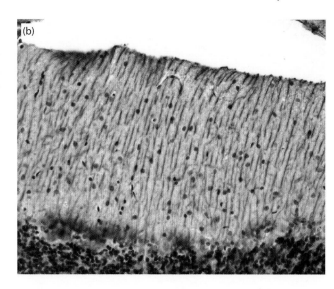

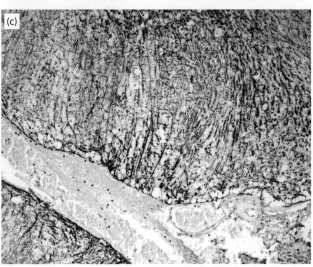

1.11 Radial glia in the cerebellum (all panels represent micrographs from GFAP-immunostained sections). (a,b) Rat cerebellum. Note GFAP-immunoreactive processes that extend from the Purkinje cell layer to the pial surface throughout the specimen, best seen at higher magnification in **(b)**. **(c)** Fragment of human cerebellum adjacent to a surgically resected lesion shows similar radial glia, though in a slightly more disorganized arrangement.

Astrocytes are also a key element in the regulation of water movement into and out of the brain through the BBB (also see later). An important molecule in this physiological regulation is aquaporin-4 (AQP4), the major water channel expressed within CNS perivascular astrocytic foot processes.[169] In normal circumstances, AQP4 activity is associated with osmotically induced water efflux, probably through functional linkage to ion/solute channels. In experimental animals with reduced AQP4, reduction of osmotic water efflux causes astrocytic foot processes to become swollen. In the setting of water influx to the CNS (with induced brain oedema), astrocytic foot processes swell more in wild-type animals than in AQP4-knockout (KO) mice.

Physiology, Metabolism and Neurochemistry of Astrocytes

Astrocytes play several roles in maintaining neurochemical homeostasis within the CNS.

One important way by which this occurs is through the regulation of glutamate levels in the extracellular fluid.[6] Astrocytes have been described as a 'ready source (for)

glutamate on demand'.[147] Glutamate functions as the major CNS excitatory neurotransmitter and can also act as a potent neurotoxin – so much so that glutamate toxicity has been implicated (with varying degrees of supporting evidence) as a key pathogenetic factor in diseases as different as ischaemic stroke, amyotrophic lateral sclerosis (motor neuron disease) and epilepsy. Brain extracellular glutamate is normally present at a concentration of approximately $2 \mu M$, whereas cytosolic concentrations are in the much higher range of $1-10 \text{mM}$, depending upon the cell type. Glutamate can be transported by a variety of CNS cell types, including neurons, astrocytes and even endothelia, but uptake of this neurotransmitter by astrocytes is considered quantitatively the most important. Glutamate uptake into astrocytes is mediated by both Na+-dependent and Na+-independent systems, the latter characteristically chloride-dependent glutamate/cystine antiporters, sensitive to quisqualate inhibition. The Na+-dependent glutamate transporters are termed EAAT1 and EAAT2. Because of the huge concentration gradient against which glutamate moves to gain access to the cytosol, significant brain energy is expended in moving glutamate from extracellular fluid into cells – this is estimated to be

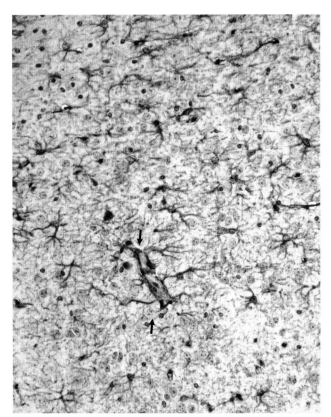

1.12 A close anatomic association frequently exists between astrocytes and capillaries. Arrows indicate a cerebral capillary that contains many astrocytic foot processes on its abluminal aspect. GFAP-immunostained section.

greater than 1 ATP per molecule of glutamate transported.[6] There is debate as to whether this ATP is generated by astrocytic glycolysis–tissue culture experiments suggest that this is not the case. When glutamate transporters analogous to EAAT1 and EAAT2 (GLAST and GLT-1) are 'knocked-out' *in vivo* (in rats) by use of antisense oligonucleotides, severe neurological abnormalities (e.g. paralysis) ensues in affected animals, probably the result of neurodegeneration related to excitotoxicity caused by glutamate.[198]

Intracellular (astrocytic) glutamate 'handling' may occur in several ways, though glutamine formation and its release into the extracellular space (where it may be taken up by neurons) or entry into the tricarboxylic acid cycle appear to be the most important of these. Glutamate uptake can be modulated by alterations in transporter activity and/or expression, and the transporter activity is in turn governed by both thermodynamic and kinetic factors, a detailed consideration of which is beyond the scope of this chapter. Glutamate uptake kinetics are influenced by signalling molecules (including other neurotransmitters).[82] Even amyloid precursor protein (APP), a molecule that is central to the pathogenesis of Alzheimer's disease[16] and often used as a marker of axonal injury,[10] can affect astrocytic glutamate levels. In tissue culture systems, APP increases glutamate uptake by a process that is sensitive to both protein kinase A and C inhibitors. Astrocytes may even participate in the determination of synaptic structure and function–synapses throughout the CNS show varying degrees of ensheathment by astrocytic processes. Astrocytic processes are more prominently distributed near synapses with a greater likelihood of

showing 'glutamate escape'. Glutamate released by synapses follows one (or more) of three subsequent pathways: it can (1) diffuse further between synapses, (2) be cleared by neuronal glutamate receptors, or (3) be cleared by transporters on astrocytes or their processes. The importance of each of these mechanisms varies in different regions of the CNS, determined to some extent by the degree of synaptic ensheathment by astrocytes. It appears that astrocytes are much more important synaptic glutamate sinks than are neurons.[6]

Of course, 'what comes into cells … may also go out'! Glutamate release from astrocytes occurs through the process of 'transport reversal', through anion channels activated by cell swelling or through gap junction hemichannels.[271] In the adverse circumstance of ATP depletion (e.g. during irreversible, severe cerebral ischaemia), the key membrane gradients keeping glutamate within astrocytes disappear, causing 'flooding' of the extracellular (including synaptic) spaces by this potentially neurotoxic molecule. Glutamate efflux may also occur in less dire circumstances, e.g. in response to certain signalling mediators and processes. Calcium-dependent glutamate release can occur in response to bradykinin, some prostaglandins and even extracellular ATP.[6] Glutamate release may also take place as a consequence of cytoplasmic swelling; the astrocyte responds to this by the 'expulsion' of chloride and glutamate (among other molecules) through volume-sensitive organic osmolyte-anion channels (VSOACs). The delicate balance of astrocytic modulation of glutamate levels has practical implications in our understanding of, for example, the neurobiology of traumatic brain injury (TBI). Glutamate neurotoxicity can greatly exacerbate the secondary CNS damage that occurs after TBI. Unfortunately, one of the consequences of TBI may be downregulation of the very receptors (EAAT1, EAAT2) that can potentially ameliorate glutamate-mediated injury. Clinical intervention (to minimize secondary injury after TBI) using glutamate receptor antagonists has also been largely unsuccessful as a preventative strategy.[272]

Tissue co-culture experiments utilizing retinal ganglion cells and neuroglia (including astrocytes) suggest that developing neurons *in vitro* form inefficient, largely inactive synapses that only become fully and vigorously functional when exposed to glial signals.[183] Glia may even control the number of mature synapses.[243] Neuronal stimulation may trigger electrophysiological and/or calcium responses in cultured astrocytes or experimental brain slices. In addition to glutamatergic pathways already described, NO-mediated signalling may also occur. Neuron-to-astrocyte signalling can activate subcellular compartments, the entire cell, or it can activate a multicellular astrocytic response in the form of a 'calcium wave'–a phenomenon that may also occur, somewhat surprisingly, in pure cultures of astrocytes.[20,210]

Water and ion homeostasis by astrocytes is achieved partly through hormonal mechanisms, viz. the varying effects of vasopressin (AVP), atrial natriuretic peptide (atriopeptin), angiotensinogen (AGT) and angiotensin (Ang) II on astroglial water and chloride uptake, which in turn is linked to their intrinsic osmoregulation.[221] Astrocytic swelling of a pronounced degree appears to be a key element in the cerebral oedema seen in patients who experience fulminant hepatic failure. This 'cellular oedema' has numerous negative consequences for the CNS, including a failure of astrocytes to take up neurotransmitters, reduction in size of the extracellular space leading to abnormally elevated extracellular ion concentrations, and even vascular compromise through compression of the microvasculature. A neuropathologic correlate of hepatic encephalopathy is the presence of characteristic Alzheimer type II astrocytes, identified predominantly within cortical grey matter and deep central grey structures (Figure 1.13).[64,171] This condition may be potentially severe enough to cause fatal internal herniation of brain tissue and results in severe metabolic encephalopathy.[172] Its proximate cause appears to be hyperammonaemia. Various lines of evidence suggest that the pathogenesis of ammonia-induced astrocytic swelling involves oxidative stress, induction of the mitochondrial permeability transition (MPT, associated with a sudden increase in permeability of the inner mitochondrial membrane to small molecules), and intracellular accumulation of glutamine, which then act as an intracellular osmolyte. It has been hypothesized that glutamine induces both oxidative stress and the MPT.

Further metabolic linkage between neurons and astrocytes may result from their utilization of specific energy substrates. Astrocytes have a prominent capacity for aerobic glycolysis and production of lactate even in the presence of normal oxygen levels. Glucose is the major energy substrate within adult CNS, but lactate and ketone bodies may serve as alternative energy substrates in prolonged starvation, diabetes mellitus or during hypoglycaemia.[181] In the course of normal brain function, approximately 90–95 per cent of brain energy consumption is attributed to neurons, only 5–10 per cent to glia (especially astrocytes). Recent data suggest that glial cells may function as 'nursing partners' for neurons, releasing a metabolic intermediate from glucose that can be taken up and oxidized by neurons. It has also been claimed that astrocytes 'sense' synaptic activity at glutamatergic synapses (see earlier) and metabolize glucose into lactate that can be passed to neurons. Energy transfer from astrocytes to axons may also occur (during aglycaemic conditions) through the degradation of astrocytic glycogen to lactate, the latter molecule then being

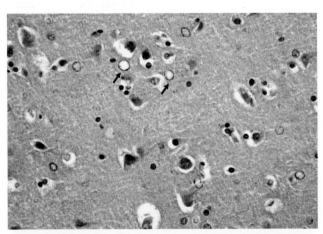

1.13 Alzheimer type II astrocytes. These are easily appreciated in H&E-stained sections, as in this section through the basal ganglia. Arrows indicate two such cells, characterized by enlarged clear nuclei, often with tiny eccentric nucleoli.

used by neurons as a supplementary energy source.[28] High-resolution imaging methods (e.g. two-photon fluorescence imaging of nicotinamide adenine dinucleotide) can now be used to study metabolic interactions between neurons and astrocytes at the single cell level.[112] For a thorough discussion of the physiological and biophysical aspects of neuronal and astrocytic metabolism in the course of afferent and efferent neural activity in the brain, see Gjedde *et al*.[82]

Pathological Reactions and Role in Neurologic Disease

Neurons and astrocytes (the latter vastly outnumbering the former),[226] once thought to act independently in CNS development and response to injury, are now known to be tightly linked in both processes, with well-defined cell-to-cell contacts between the two cellular elements in many regions of the CNS. Until the early 1990s, there was a widespread tendency to view reactive astroglia as generic cells with uniform biological properties regardless of their location within the CNS. This approach has been largely superseded by an appreciation of their significant functional and regional heterogeneity throughout the brain and spinal cord.[93] A second assumption was that, in the face of CNS injury, supportive astroglial cells become transformed into elements that actively inhibit axonal regrowth – the classic 'glial scar' that poses a major barrier to CNS regeneration after brain or spinal cord injury. A third major assumption about astrocytes was that brain or spinal cord injury led directly to glial proliferation ('astrocytic gliosis') and associated scarring. All of these assumptions have recently been questioned or re-evaluated. In experimental models, CNS injury of only certain types (e.g. that caused by physical tearing or laceration, as often occurs in traumatic brain and spinal cord injury) reliably induces gliosis. As well, the gliosis may be regionally accentuated; furthermore, there may be an increase in the GFAP content of individual astrocytes (astroglial hypertrophy) without an actual increase in their number (hyperplasia). Neurons may substantially regulate astroglial proliferation and differentiation. Experiments using cultured cerebellar granule cells (admittedly a highly specialized type of neuron but one fairly easy to maintain *in vitro*) and astroglia further highlight potential neuronal–glial interactions in the response to injury. When postnatal astroglia were grown in the absence of neurons, they expressed low levels of GFAP and grew rapidly. When granule cell neurons were added to the cultures (especially at a ratio of at least 4:1), DNA synthesis (in the astrocytes) decreased substantially and GFAP protein expression increased.[93] GFAP appears to be necessary but not sufficient for the formation of astroglial processes in the presence of neurons. As indicated earlier, astrocytes may support axonal growth; their degree of differentiation (rather than chronological age) may be a key determinant of glial support of axonal growth. Astroglia are also capable of expressing extracellular matrix molecules (such as laminin, heparan sulphate) and cell adhesion receptor systems – these and other molecules may play a role in axonal guidance following injury.

The function (and malfunction) of both microglia (see later) and astrocytes is inextricably linked to an understanding of cytokines, low molecular weight (MW) glycoproteins that may be secreted or function as membrane-bound complexes.[110,171] These molecules exert their actions in a paracrine or autocrine fashion. They interact with specific cell-surface receptors, most cytokines having the capability of acting as ligands for several different receptors. The biological properties of a given cytokine are determined by the receptor that is activated, more than the cytokine itself. Astrocytes both produce cytokines and are, under a variety of physiological and pathophysiological circumstances, highly responsive to them. Properties of cytokines and their receptors have been elucidated through the use of elegant (usually transgenic) animal models and tissue culture experiments. In these studies, genes specific for cytokines and/or their receptors are deleted or overexpressed, often in specific cell types or anatomically defined populations, allowing for a detailed dissection of their myriad effects on the CNS (for a detailed review, see John *et al.*[110]). Although many of these experimental studies looking at the effects of cytokines on neural cells are illuminating from the perspectives of cellular neurobiology, their relevance for understanding complex neurologic diseases is not always clear.

Reactive gliosis, a non-specific but highly characteristic response to almost any type of CNS injury, can be thought of as resulting from astrocytic proliferation (hyperplasia) and enlargement (hypertrophy), both associated with distinctive patterns of gene expression. As indicated earlier, reactive gliosis can be either a positive process that results in neuroglial survival or a negative one causing diminution in neuroglial growth, migration or both processes. Targeted ablation of reactive astrocytes can cause both increased neuronal degeneration and, simultaneously, an increase in neuritic outgrowth, together with accentuated chronic inflammation and a delay in post-injury re-establishment of BBB integrity.[32] Interleukin-1β (IL-1β), tumour necrosis factor-α (TNFα), IFNγ and transforming growth factor-β1 (TGFβ1) have all been implicated as players in the initiation or modulation of reactive gliosis. Furthermore, astrocytes possess receptors for all four of these cytokines, each apparently having a different role in the astrocytic response to injury, and each response in turn is mediated by specific gene expression patterns. IL-1β, especially, appears to have a pro-inflammatory role in CNS disease, but may also be involved in CNS regeneration,[110] some of these effects being modulated through ciliary neurotrophic factor (CNTF) or nerve growth factor (NGF). *In vitro* experiments suggest that IL-1β induces genes that are key elements in the acute or subacute immune response and include other cytokines, chemokines and several adhesion molecules. By contrast with IL-1β, IFNγ (produced in abundance by activated lymphocytes) appears to potentiate (rather than initiate) astrocytic gliosis, possibly through the induction of MHC class I and II molecules and chemokines, and the potentiation of IL-1β-induced expression of TNFα and nitric oxide synthase (NOS) (inducible NOS type II can be expressed by astrocytes). Complex interactions among infiltrating lymphocytes, microglia and astrocytes determine the microenvironment in which the CNS functions or malfunctions. NOS II may be downregulated in astrocytes via a complex cascade that involves both microglia and IL-4 (produced by TH2 lymphocytes) acting upon TH1 lymphocytes to modify their synthesis of IFNγ. Both IL-1β and IFNγ may thus be of importance in, for example, the pathogenesis and progression of multiple sclerosis (and its experimental animal model, experimental allergic encephalomyelitis, EAE), in which lymphocytic infiltration into spinal cord or brain is an integral part of the neuropathologic picture.[64,255]

TGFβ1, expressed in both astrocytes and microglia in the context of brain trauma, infarction or inflammation, is itself an apparent stimulant of astrocytic gliosis. Inhibiting TGFβ1 activity prevents formation of a glial membrane at the site of CNS injury and downregulates production of extracellular matrix molecules, e.g. fibronectin and laminin. Other effects of TGFβ1 include inhibition of astrocytic expression of MHC class II molecules, vascular cell adhesion molecule-1 (VCAM-1) and intercellular adhesion molecule-1 (ICAM-1) and TNFα, as well as the induction of numerous molecules important in CNS wound healing (fibronectin, tenascin, collagen, laminin, actin and actin depolymerizing factor among many others). TNFα itself is synthesized by astrocytes, may stimulate astrocytic gliosis but may also have a part to play in CNS repair after injury. It can also be cytotoxic to both oligodendroglia and neurons. These somewhat paradoxical effects are probably affected via the distinctive signalling pathways mediated by two TNF receptors, TNFRI/p55 and TNFRII/p75; the former appears to be linked to cell death, the latter to cell viability and growth. IL-6 activates many signalling cascades, and houses the signalling receptor gp130, which is activated through the Jak/Stat pathway. In the CNS, IL-6 promotes neuronal survival and neurite outgrowth, may impact cell-fate decisions (e.g. progression of stem cells to neurogenesis *versus* gliogenesis), and has an immunomodulatory function in the highly complex glial cytokine network—abnormalities of which may result (under some circumstances) in CNS inflammation and neurodegeneration. Complex and potentially confusing as all of these interactions may appear to be, they appear to become even more so with every passing week! Summaries of the interactions among microglial and astrocyte-secreted factors are to be found in recent reviews of these cell types.[226,268] Immune responses mediated through cytokines, chemokines and lymphokines, especially involving cell surface receptors, are especially important in understanding the evolution of viral infections of the CNS, especially those that impact on neurons.[39]

Astrocytic gliosis plays a part in virtually all neurological diseases and neuroanatomical lesions—whether they be degenerative, traumatic, metabolic, neoplastic, inflammatory or of any other aetiology. Astrocytes are commonly found in microscopic lesions as disparate as the neuritic plaques of Alzheimer disease,[205] the demyelinating plaques of multiple sclerosis, and foci of viral encephalitis that have little to do with plaques! Recently, it has been suggested that astrocytes play a major role in the generation of epileptic seizures through their modulation of glutamate and calcium signalling.[238] The potential roles of astrocytes in specific entities are further considered in the chapters dealing with these diseases. The molecular and cellular basis of astrogliosis itself has been the topic of substantial debate.[60] Reactive gliosis appears to vary quantitatively and qualitatively—the nature of the glial response being determined by both the nature of the lesion/injury and the microenvironment in which it occurs.[193] Astrogliosis is recognized by the apparent proliferation and hypertrophy of GFAP-expressing astrocytes. However, by definition this excludes a consideration (within a given lesion) of any reactive astrocytes that fail to express

GFAP. Furthermore, because of their apparently minimal proliferative and migratory potential, it has been suggested that reactive astrocytes (in regions of acute/subacute neural injury) may simply represent a change in the phenotype of locally residing astrocytes. The transition of resting astrocytes to activated cells is associated with the expression of new molecules not normally expressed (in the resting state), as well as the upregulation of molecules that are expressed at low levels in the resting state (for a catalogue of these molecules, see Eddleston and Mucke;[60] Ridet et al.[193]).

One of the myths to be banished in recent years is that astrocytic processes (in glial scars) are a major impediment to axonal regeneration, e.g. after traumatic spinal cord injury or stroke.[93] Axons appear to grow quite happily on astrocytic scars; it has therefore been postulated that, because scars are in fact a complex admixture of various cell types, extracellular matrix components and other elements, some combination of non-astrocytic components may actually impede axonal growth after a spinal cord injury or cerebral infarct. Despite this, strategies aimed at re-establishing spinal cord function are frequently aimed at bypassing the scarred and gliotic site of cord injury, which is still perceived by many investigators as a significant barrier to axonal regeneration.[30,220] Nevertheless, one strategy in the treatment of experimental spinal cord contusional or transection injury is to acutely transplant glial-restricted precursor cells (which have the potential to differentiate into oligodendroglia and astrocytes) into the lesion.[50,96] When reactive astrocytes were selectively ablated from a region of spinal cord injury in a novel transgenic mouse model, the absence of astrocytes caused failure of BBB repair, leukocyte infiltration, severe demyelination with local tissue disruption, and oligodendroglial/neuronal death, resulting in pronounced neurologic deficits in experimental animals.[67,225] One obvious conclusion from this work was that reactive astrocytes have important neuroprotective functions that could be harnessed in post-injury repair of neural tissue.

Given the central role of astrocytes in brain energetics, water and ion homeostasis, vascular regulation, and genesis of the neurovascular unit, discussed elsewhere in this chapter, astrocytes would appear to be an attractive cell population to target in new therapeutic approaches in neuroprotection (for review, see Nedergaard and Dirnagl[166]). As one example, sustained astrocytic expression of the glycoprotein clusterin significantly improved brain remodelling after ischaemia in mice.[106] Genetically modified astrocytes may be a way to deliver therapeutic factors into lesioned (whether artificially or by nature) portions of brain or spinal cord.[182] A comprehensive review of the complex biological and pathological features of astrocytes has recently been published.[226]

OLIGODENDROCYTES

Oligodendrocytes are neuroglial cells with small cell bodies, few (Greek: *oligos*, little, few) short cell processes, and no cytoplasmic filaments. They are found in grey matter, where they cluster around neuronal cell bodies (Figure 1.14a, arrow) and are seen in the pencil fibres of white matter that course through the putamen (Figure 1.14a, asterisks). In compact regions of white matter they are often arranged in rows between myelinated fibres (Figure 1.14b), and in

cortex often lie adjacent to neurons. The functions of oligodendrocytes are likely to be different in the grey and white matter. In the grey matter, the standard explanation for the role of oligodendrocytes that encircle neurons is that they play sustentative roles for neurons, analogous to the role of satellite or capsular Schwann cells in dorsal or peripheral sensory ganglia, or that they represent progenitor cells.[27]

In central nervous system white matter, oligodendrocytes are the cell type responsible for myelin formation and are thus analogous to Schwann cells in the peripheral nervous system. Oligodendrocytes must undergo a series of complex series of steps, from proliferation, migration, differentiation, to myelination before they finally are capable of producing an ensheathment of axons.[26] Oligodendrocytes are among the most vulnerable cells to injury in the CNS.[26] A large recent review addresses the development of oligodendrocytes, particularly in rodents[26] and several of these discoveries in rodents are likely analogous in human oligodendroglial development as well. New insights into oligodendroglial development include the fact that: (1) there appears to be a common progenitor cell origin for neurons and oligodendrocytes; (2) a ventral-to-dorsal progression occurs in oligodendroglial development; (3) there are multiple origins for oligodendrocytes and (4) there is an interrelationship between axonal signalling and myelination (reviewed in Bradl and Lassmann[26]).

Both CNS and PNS myelin can be histochemically stained with a number of different stains, the most commonly used of which is the Luxol fast blue (LFB) stain with which CNS myelin appears in paraffin sections as a slightly vacuolated robin's egg-blue substance surrounding the axon (Figure 1.14c). In autopsy tissues, LFB may be suboptimal or yield patchy, variably intense staining; therefore, immunohistochemistry incorporating primary antibodies to, for example, myelin basic protein may better demonstrate myelinated fibres. The tinctorial properties differ slightly for PNS myelin, which appears darker blue than CNS myelin on being stained with Luxol fast blue-periodic acid Schiff (LFB-PAS). This can be easily seen at a transition zone between oligodendrocyte-mediated CNS myelin and Schwann cell-mediated PNS myelin where cranial nerves or spinal nerves exit the CNS (Figure 1.14d, 5th cranial nerve illustrated). Additional differences exist between oligodendrocytes and Schwann cells. Schwann cells are surrounded by basement membrane and have a single cell process responsible for ensheathing only a single myelin segment lying between two nodes of Ranvier, whereas the oligodendrocyte is devoid of basement membrane and its several cell processes can each form several internodal segments of myelin. Schwann cells have a more limited responsibility, with one cell responsible for one internodal segment on one axon, which they spirally enwrap to produce myelin. Oligodendrocytes, in contrast, may form as many as 60 internodal segments.[224] In both the CNS and PNS, myelin serves similar functions, allowing increased speed of conduction of impulses, which propagate by saltation from node to node along myelinated axons; on loss of myelin, conduction slows or ceases.

Myelin integrity requires both its formation and subsequent maintenance. Myelin formation begins at about the 16th week of intrauterine life[224] and continues throughout childhood. Although myelin formation is most rapid during

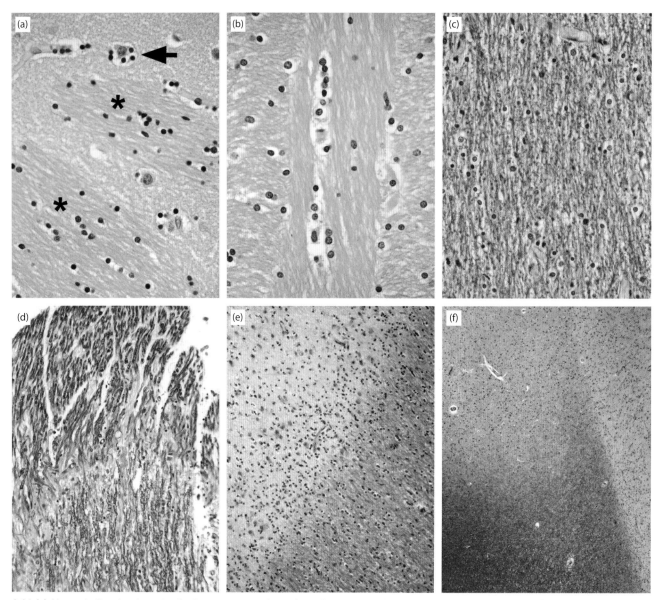

1.14 (a) Normal oligodendrocytes in the putamen, where they cluster around neuronal cell bodies (arrow) and are also seen in the white matter pencil fibres (asterisks). **(b)** In white matter, oligodendrocytes are arranged in rows between myelinated fibres. **(c)** Luxol fast blue-periodic acid Schiff (LFB-PAS) stain for myelin in normal white matter. **(d)** As demonstrated using the same histochemical LFB-PAS stain, the tinctoral properties differ for peripheral nervous system (PNS) myelin. PNS myelin appears darker blue, as seen in the illustrated transition zone between oligodendrocyte-mediated CNS myelin and Schwann cell-mediated PNS myelin, in the 5th cranial nerve root entry zone. **(e)** Edge of an active demyelinating plaque in multiple sclerosis (MS) shows oligodendrocyte proliferation, evidenced as a band of hypercellularity between the severely demyelinated zone (upper left) and less demyelinated edge of plaque (lower right). LFB-PAS. **(f)** Shadow plaque (upper left) represents an area of partial remyelination. LFB-PAS.

the first 2 years of life, diffusion tensor imaging studies suggest that myelination occurs well into the second decade.[123] Maintenance of myelin demands integrity of the oligodendrocyte cell body, high energy expenditure by the cell and, should the myelin be lost, effective remyelination by oligodendrocyte precursors. Loss of, or damage to, the cell body of the oligodendrocyte almost inevitably leads to loss of the myelin produced and supported by that cell's processes. Damage to axons in myelinated fibres leads to breakdown of the surrounding myelin sheath. Conversely, the oligodendrocyte is also important for axonal support and maintenance.[61]

In H&E-stained sections, the oligodendrocyte is seen as a round nucleus with evenly dispersed, dark chromatin, no nucleolus, and no visible cytoplasm (Figure 1.14a and b). In poorly-fixed tissues, most oligodendrocytes manifest an artefactual perinuclear clearing around the nucleus, the so-called 'perinuclear halo', leading to a 'fried-egg' appearance (Figure 1.14b). These perinuclear haloes are often not apparent in well-fixed small surgical biopsy specimens; hence identification of an oligodendrocyte–normal or neoplastic–cannot rest solely on the presence or absence of the 'halo'. The processes and cytoplasm of the oligodendrocyte cannot be discerned without special histochemical stains or EM. A number of antibodies have been used immunohistochemically to identify oligodendrocytes, such as Leu-7 (CD57), anti-myelin associated glycoprotein (MAG),

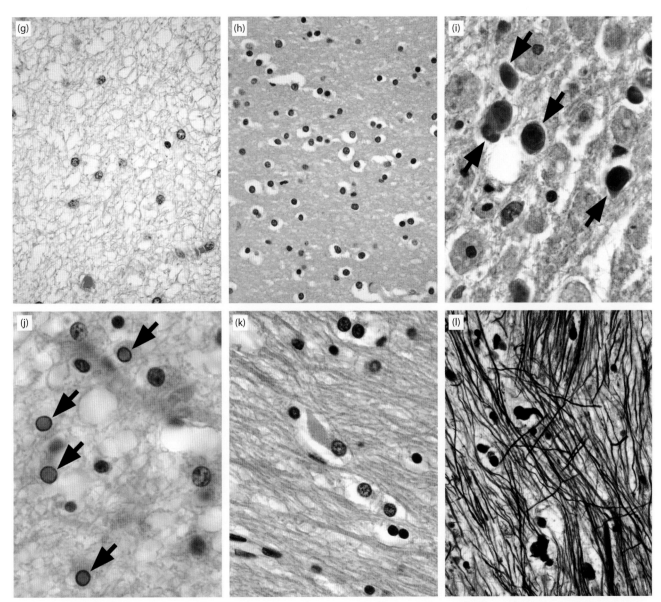

1.14 (*Continued*) (g) Centre of old, inactive MS plaque is hypocellular and contains almost no myelin or oligodendrocytes. **(h)** Photomicrograph illustrates normal myelin and axon content for comparison with panel (g). **(i)** Oligodendrocyte nuclei with characteristic glassy, violaceous inclusion bodies in progressive multifocal leukoencephalopathy. **(j)** Oligodendrocyte nuclei bearing viral inclusion bodies, in subacute sclerosing panencephalitis. **(k)** Oligodendrocytes in multiple system atrophy contain cytoplasmic, linear, pointed, densely eosinophilic, glial inclusions, in affected regions of brain. **(l)** The glial cytoplasmic inclusions in multiple system atrophy are better highlighted by modified Bielschowsky silver impregnation (note that this also stains the axons).

anti-myelin oligodendrocyte protein (MOG), anti-CD44 (a cell surface glycoprotein)[25] and anti-OLIG1 transcription factor.[11] Several of these markers have now been recognized to be present in oligodendrocytes in differing stages of development, as reviewed by Bradl and Lassmann.[26] On paraformaldehyde-fixed brain tissue, differentiated oligodendrocytes express carbonic anhydrase II, 2′:3′-cyclic nucleotide 3′-phosphodiesterase (CNP), galactosylceramide (GalC), Kir4.1 (inwardly rectifying K+ channel subunit), myelin basic protein (MBP), MAG, MOG and proteolipid protein (PLP), although myelinating oligodendrocytes express RIP and TPPP/p25 and oligodendrocyte precursor cells (OPC) express CNP, OLIG2, NG2 and O4.[26] Although OLIG2 especially is being used in daily practice, it is not specific for gliomas of oligodendroglial lineage, and other markers are not part of

the standard armamentarium of surgical neuropathologists. Hence the question, 'What is an oligodendrocyte?' or the corollary, 'What is an oligodendroglioma?'[31] remains a vexing one when we do not have a universally accepted immunohistochemical marker for these cells.

Compounding the problem have been *in vivo* studies that suggest neuron-like physiological properties in cells cultured from human oligodendroglial tumours,[178] as well as immunohistochemical studies demonstrating staining of oligodendroglial tumour cells for markers traditionally associated only with neurons, such as NF-H,[52] N-methyl-D-aspartate receptor subunit 1 or embryonal neural cell adhesion molecule.[270]

The repertoire of responses to injury available to the oligodendrocyte is limited. Proliferation of oligodendrocyte precursor cells has been documented in a number of different disease processes including radiation injury,[10] vanishing white matter disease[246] and multiple sclerosis (MS) but, at least in the first two conditions, may be counterbalanced by neuronal apoptosis.

Oligodendrocyte proliferation from precursor cells is most confidently identified by the pathologist at the edge of an active demyelinating plaque in multiple sclerosis (MS), as a band of hypercellularity (Figure 1.14e). This precursor cell proliferation may produce a region of partially effective remyelination, known as the 'shadow plaque'. This appears as an area of partial, hazy myelin staining (reflecting inadequately thin sheaths) that is intermediate in intensity between normal-appearing white matter with its intense robin's-egg blue colour on Luxol fast blue stain for myelin, and the unstained, totally demyelinated parts of a plaque (Figure 1.14f). Unfortunately, remyelination seems to be transient, because it disappears in older lesions.[74] Although remyelination initially occurs as a result of recruitment of oligodendrocyte precursor cells, these may progressively become depleted,[145] may become quiescent or may respond to axonal inhibitory signals and cease the myelination process.[142] Whether or not oligodendrocyte apoptosis occurs in MS is still debated.[16,74] In any case, the centre of an old, inactive MS plaque does not evince effective remyelination, is quite hypocellular and contains naked axons and chronic gliosis but virtually no myelin or oligodendrocyte nuclei (Figure 1.14g). The extent of this hypocellularity and oligodendrocyte loss can be best appreciated when the centre of the plaque is compared side by side with the oligodendrocyte content of normal subgyral white matter (Figure 1.14h).

In many other diseases in which oligodendrocytes and myelin are lost in the CNS, no phase of oligodendrocyte proliferation has been confidently identified. In these situations, oligodendrocytes are simply injured and destroyed. A number of different viruses can infect oligodendrocytes and cause lytic infections, with loss of cell bodies and their dependent myelin sheaths. Prior to cell lysis, if sufficiently large clusters of viral particles accumulate, they may be seen in the cell nucleus as viral inclusion bodies, even on routine H&E staining. The associated margination of the normal nuclear chromatin underlines the fact that the nuclear machinery has been 'commandeered' by the virus for its own purposes. The characteristic glassy, violaceous inclusion bodies in progressive multifocal leukoencephalopathy (Figure 1.14i) are composed of myriad virions that fill the entire nucleus; much less frequently these form a condensed, 'owl's eye' viral inclusion. Subacute sclerosing panencephalitis similarly produces relatively homogeneous viral inclusions that fill the nucleus and are associated with a margination of nuclear chromatin (Figure 1.14j).

Less dramatic oligodendrocyte changes accompany almost all other disorders that damage myelin. Indeed, by the time these disorders, termed 'leukoencephalopathies', are encountered by the pathologist, hypocellularity, and a reduction in the number of oligodendrocyte nuclei and amount of myelin with a corresponding increase in the white matter water-to-myelin ratio, are the non-specific findings. Leukoencephalopathy is the term usually applied to non-inherited white matter damage (usually maximal in cerebral hemispheric white matter) that is a result of toxic, metabolic or ischaemic processes, with the alternate term 'leukoaraiosis' also applied to chronic ischaemic white matter injury. Examples of toxic substances that cause oligodendrocyte and myelin loss include: chronic substance usage (ethyl alcohol, 'ecstasy', and toluene, i.e. 'glue sniffing'), ciclosporin (US spelling cyclosporine), therapeutic radiation, and chemotherapeutic agents (carmustine and methotrexate) (Figure 1.15).[70] The latter two categories of agents also affect blood vessels and produce white matter damage via ischaemic mechanisms.

Oligodendrocytes are the second most vulnerable cell, after neurons, to anoxic-ischaemic central nervous system injury. In any acute deprivation of regional blood supply to grey and white matter, i.e. stroke, oligodendrocytes are lost, along with the neurons and virtually any other tissue components in the epicentre of the lesion. Significant, widespread, acute or chronic damage specifically to white matter may also occur, because of the vulnerability of oligodendrocytes to ischaemic injury. Deprivation of oxygen/blood supply to the white matter is especially likely to affect the cerebral hemispheres. Acute hypoxic-ischaemic injury to white matter may be accompanied by a haemorrhagic component and is known as hypoxic-ischaemic leukoencephalopathy. Chronic hypoxic-ischaemic injury to white matter is often maximal in boundary zones (watershed territories) of arterial distribution in the cerebral white matter between the middle and posterior cerebral arteries and the middle and anterior cerebral arteries. It tends also to be maximal in the regions of white matter midway between the ventricular and pial surfaces, not in periventricular regions. Chronic hypoxic-ischaemic injury to white matter may result from either severe arteriolosclerosis of deep white matter blood vessels and hypoperfusion (leukoaraiosis, causing Binswanger's disease in the most severe form) or an inherited, autosomal dominant disorder, CADASIL (cerebral autosomal dominant arteriopathy with subcortical infarcts and leukoencephalopathy). This ischaemic injury to oligodendrocytes has usually been considered to be predominantly necrotic. However, recent studies of mild hypoxic/ischaemic insults occurring in the perinatal time period have suggested that O4+ oligodendrocyte precursor cells are particularly vulnerable to injury and die by apoptotic rather than necrotic mechanisms.[199] Ischaemic injury to the white matter is considered in more detail elsewhere in this text.

Inherited, usually autosomal recessive, storage diseases may cause oligodendrocyte injury and loss as a result of the accumulation of abnormal storage material within the cell cytoplasm. This type of white matter injury is designated a leukodystrophy (to reflect the intrinsic dysfunction in oligodendrocyte biology in these disorders, as opposed to the acquired injury to previously normal oligodendrocytes in leukoencephalopathies), and includes a number of rare disorders, the most well known of which are metachromatic leukodystrophy, Krabbe's disease, and adrenoleukodystrophy.

Finally, oligodendrocytes contain an extensive microtubular network and express tau, which is a microtubule-associated protein.[192] In neurodegenerative disorders, tau-positive inclusion bodies may form within oligodendrocytes, usually as 'coiled' bodies. These inclusion bodies can also be immunostained with antibodies against ubiquitin and heat shock proteins such as αB-crystallin.[192]

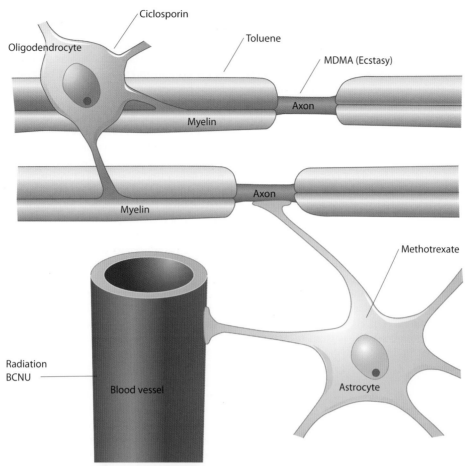

1.15 Examples of toxic substances that cause oligodendrocyte and myelin loss include: chronic substance usage (ethyl alcohol, 'ecstasy', and toluene, i.e. 'glue-sniffing'), ciclosporin (US spelling cyclosporine), therapeutic radiation, and chemotherapeutic agents (carmustine and methotrexate). The latter two categories of agents also affect blood vessels and produce white matter damage by ischaemic mechanisms.

Other inclusions are illustrated here in a case of multiple system atrophy, in which the cytoplasmic inclusions appear as linear, pointed, densely eosinophilic structures immediately adjacent to oligodendrocyte nuclei in select anatomic areas of the brain (Figure 1.14k). Although visible on careful inspection of H&E sections, they are better highlighted with silver stains as 'Papp–Lantos bodies' (Figure 1.14l) or by immunostaining for ubiquitin, alpha-synuclein or αB-crystallin. Ubiquitination is common to many very different types of neuronal and glial inclusions and makes antibodies to ubiquitin and p62 valuable and effective generic immunostains to have available in laboratories that study neurodegenerative disorders.

Reactions of the oligodendrocyte at the level of the myelin sheath include intramyelinic oedema, resulting in vacuolation of the neuropil, and myelin degeneration, resulting in myelin digestion by macrophages. Wallerian degeneration is the process that best illustrates the co-dependency of myelin sheaths and their axons. When axons are transected or otherwise severely injured, the axon distal to the transection will start to undergo dissolution. Following this, the myelin sheaths surrounding the axon start to break down into a string of myelin ovoids, separated by the oligodendrocytes that formed the myelin. These myelin debris-containing ovoids are quickly surrounded by astrocytic cell processes

and are then phagocytosed by microglia and macrophages. Entire tracts may undergo wallerian degeneration, such as the descending corticospinal tracts in the spinal cord when a severe injury has occurred at a more proximal point in the tract (e.g. the internal capsule). Extensive wallerian degeneration is appreciable on both pre-mortem neuroimaging studies and, should the patient succumb, on gross brain examination.

In the PNS, one of the more striking responses at the level of the myelin sheath of Schwann cells to injury follows chronic, repetitive myelin loss caused by various types of peripheral neuropathies. This results in exuberant proliferation of Schwann cells, usually seen as multiple concentric layers of cells around a thinly myelinated axon, evidence that the process is not fully successful. Although an analogous reaction does not affect oligodendrocytes in the CNS, in severe proximal peripheral nerve injury, onion bulb formation by Schwann cells may extend into spinal cord parenchyma.[186,214]

EPENDYMA

Ependymal cells constitute the lining of the ventricular system, including the aqueduct of Sylvius and foramina that connect the different ventricles. Highly specialized ependymal cells play a key role in fluid homeostasis between brain parenchyma and the cerebrospinal fluid, and ependymal cells are rich in the membrane water channel protein aquaporin-4.[173] This role in fluid homeostasis is relevant to both normal physiological conditions and disease states, especially ones that affect the ventricular lining (intraventricular haemorrhage, hydrocephalus, CNS infections resulting in ependymitis or ventriculitis). Ependymal cells have electrophysiological properties similar to those of astroglia.[203,204] They are also seen in the spinal cord 'central canal', even when the canal becomes vestigial and is identified only as a somewhat disorganized collection of cells that retain their phenotype. Ependymal cells resemble the relatively monotonous cuboidal and columnar epithelia that line the gastrointestinal and respiratory tracts (Figure 1.16), but do not have a well-defined basement membrane and are immunopositive for GFAP (see later). They show prominent cilia. Interspersed among the ependymal cells are tanycytes, which have radially directed basal processes that extend into the periventricular neuropil and enwrap blood vessels, or terminate on neurons, glia or the external glia limitans.[94] The innate immune response of ependymal cells may involve signalling through several pattern recognition receptors (PRRs) (see later under Identifying Microglia and Quantifying Microgliosis).

Ependymal cells arise from epithelium of the neural plate and the neural tube, which develops from the neural plate. Regions in the CNS destined to show an ependymal lining later in life, demonstrate (at 6 weeks of intrauterine development) only densely cellular pseudostratified columnar epithelium that is quite mitotically active. This mitotic activity essentially ends when the ventricular lining is fully developed, never to resume. However, the pseudostratified appearance of ependyma may occasionally be found in the ventricular lining of older patients, even adults (Figure 1.16). Immunohistochemical studies of human fetal ependyma show variably strong vimentin immunopositivity as early as 8 weeks into development in most ependymal regions (including spinal cord, the lining of all ventricles and cerebral aqueduct), though this diminishes (but does not disappear entirely) by 40 weeks of gestation.[203] Cytokeratin (CK-904) immunoreactivity is maximal in most regions, though surprisingly absent from the lining of the third ventricle, from 8 to 14 weeks of gestation but disappears thereafter. GFAP and S-100 antibodies show patterns of ependymal immunoreactivity that appear to be highly dependent on the precise locus within a given neuroanatomic structure being examined: as one example, GFAP immunoreactivity is prominent in the roof plate of the developing spinal cord but absent from its floor plate, although the opposite pattern of immunoreactivity is noted in the fourth ventricle and cerebral aqueduct.[203] By full term, fairly consistent and robust GFAP and S-100 immunoreactivity are noted only in the ependymal lining of the lateral and third ventricles and parts of the fourth ventricle.

Ependymal cells are connected to each other by gap junctions. Proteins known as connexins (Cx) are present at these junctions and contribute to 'intercellular communication, ion homeostasis, volume control and adherent connections between neighbouring cells'.[55] Connexin proteins Cx26, Cx30, Cx43, and Cx45 are mainly expressed at the apices of ependymal cells.[55] Aquaporins (AQPs) are well known to be expressed in the end feet of astrocytes, but the AQP4 channel that controls water movement in brain is also expressed in basal-lateral areas of ependymal cells, as are other members of the AQP family.[55]

Ependymal cells have apical cilia that beat in a coordinated fashion and this organized beating may, in part, be due to the gap junctions. Del Bigio notes that ependymal cilia 'may help to create concentration gradients of guidance molecules in cerebrospinal fluid that serve to direct neuroblast migration from the lateral ventricle wall into the olfactory bulb'.[55] Damage to ependymal cilia has, in rare human instances, been shown to produce hydrocephalus (or may indeed result from severe and/or prolonged hydrocephalus), although there are numerous examples of mice with mutations in ciliary proteins that have been proven to develop hydrocephalus, with or without occlusion of the cerebral aqueduct.[55]

Ependymal cells have a relatively circumscribed repertoire of responses to injury, and only limited regenerative capacity at all ages. However, subependymal zone (SEZ) cells in a thin layer surrounding the lateral ventricle have been found to show properties of neural stem cells.[206] In experimental models, injury to the cerebral cortex modestly increases metabolic activity in the SEZ (as measured by cytochrome oxidase activity), as well as its proliferative capacity. Cells in the SEZ may have the potential to repopulate lost neurons in the olfactory bulb and even regions of the cerebral cortex. Although spared in most degenerative and genetic diseases that afflict the nervous system, the ependyma is highly vulnerable in many other conditions, by virtue of its unique 'barrier' position and vulnerability to increases in ventricular size. It can undergo injury when stretched during the evolution of ventriculomegaly associated with hydrocephalus, a hematoma or infarct that involves the ventricular wall (e.g. germinal matrix haemorrhages that commonly extend into the ventricular cavities in distressed premature infants), and in infections or inflammatory processes that extend directly from the brain parenchyma or subarachnoid space.[204] Hydrocephalus in humans is often accompanied by neuroimaging abnormalities that include a subventricular band of 'transependymal oedema', which may be transient and has poorly characterized neuropathological correlates. In experimental models of hydrocephalus, discontinuities and gaps in the ependymal lining are filled by the processes of subependymal astrocytes, but residual ependymal cells do not become proliferative. Neither do subependymal cells undergo metaplasia to repopulate the ependymal ventricular lining.

The range of ependymal reactions to injury has been well summarized by Sarnat.[204] Atrophic ependymal cells, usually a response to ventriculomegaly, are characterized by flattening and loss of their cytoplasm. Ventricular enlargement, especially when rapidly evolving and progressive, can cause stretching and tearing of the ventricular lining. Sites of rupture are more likely to occur over the smooth ventricular surface than at the ventricular angles. Subventricular astrogliosis is characterized by glial cells that proliferate, often extending into the ventricular cavity. This phenomenon, thought to occur within 1–2 weeks subsequent to the ependymal injury, is often described (somewhat inaccurately, because true inflammation is almost never a histopathological feature) as 'granular ependymitis', and the protrusions of glial tissue as 'ependymal granulations'. The process is very patchy and multifocal throughout the ventricular

1

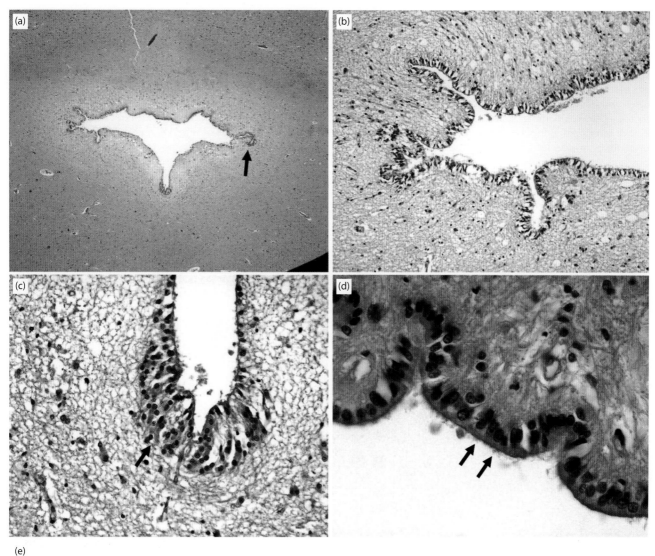

(e)

1.16 Histological features of normal ependyma. (a) Low-magnification view of the junction of the cerebral aqueduct (of Sylvius) with the fourth ventricle. Note small ependyma-lined 'outpouchings' (arrow), which are very common in any ependyma-lined structure. **(b,c)** Magnified views of the ependyma show a 'picket fence'-like structure with a slight tendency to stratification of ependymal cells in a few foci (arrow, [c]). **(d)** Relatively uniform ependymal nuclei, absence of a basement membrane deep to the ependyma, and wispy cilia on the ependymal surface (arrows). **(e)** Occasionally, as in this image from the wall of the third ventricle, the ependyma is arrayed in 'udder-like' folds – a finding of no pathological significance.

lining. These subventricular glial nodules may also represent the sequelae of a fairly indolent viral infection of the CNS (see later) and are commonly seen in the CNS of patients with acquired immunodeficiency syndrome (AIDS),[251] but are also commonly encountered as an incidental necropsy finding in individuals with no history of neurological disease.[204] When the ependyma has been injured by intraventricular haemorrhage (e.g. extending from a large germinal matrix bleed in a premature infant), macrophages, including

siderophages or hemosiderin-laden macrophages, may be seen at or near the locus of injury. Ependymal rosettes, characterized by tiny 'tubules' of ependymal cells in the periventricular region (and sometimes forming hemi-rosettes rather than complete rosettes) may represent abortive attempts at recapitulating the formation of the embryonic neural tube; they may also be the result of ependymal residua within brain parenchyma, in which luminal/ periventricular astrocytic overgrowth has occurred. Similar rosettes are seen, of course, in a variety of CNS tumours, most prominently ependymomas.

True ependymitis (to be distinguished from granular ependymitis, see earlier) may be suppurative/purulent, e.g. when encountered in association with a bacterial or fungal meningitis or brain abscess, polymorphonuclear leukocytes will then be present in abundance within and adjacent to the ependymal lining. Such an ependymitis may evolve into a ventriculitis; in its extreme form (for instance, when untreated) this can lead to filling of the ventricular cavity by pus and ventricular abscess formation.[204] Ependymitis may become so severe that it leads to fragments of ependyma being shed into the ventricular cavity and cerebrospinal fluid (CSF) pathways, rarely such cells are identified in samples obtained by lumbar puncture for evaluation of CSF cytology. Several viruses have a propensity to colonize ependyma and periventricular tissues. In the era of AIDS, this is most dramatically manifest with cytomegalovirus (CMV) infection, which can cause such a severe ependymitis/ventriculitis that a thick icing-like layer of exudative material is noted in cut sections through the fixed brain[251,253] (Figure 1.17c). This ependymal infection by CMV then commonly spreads in a 'ventriculofugal' direction, into the brain parenchyma. Much less commonly, adenovirus has been identified as causing a more subtle ependymitis/ventriculitis (Figure 1.17 b).[5] Mumps ependymitis, often with minimal inflammation, can lead to aqueductal stenosis, an important (though rare) cause of acquired hydrocephalus.[204] Ventriculomegaly becomes apparent weeks to months after a clinically apparent mumps virus infection, e.g. manifest as parotitis. Aqueductal stenosis has also been described after influenza and parainfluenza 2 infections. The only 'footprint' of many of these viral infections is the presence of microglial nodules in close proximity to the ependyma, and immunocytochemical evidence of viral infection within ependymal cells. In experimental animals, human respiratory syncytial virus (RSV) infection can also lead to viral antigen within ependymal cells and, eventually, aqueductal stenosis with hydrocephalus. Primary CNS neoplasms may extend to the ependymal lining (Figure 1.17a) and sometimes breach this barrier to gain access to the ventricular cavity, facilitating spread of such a tumour through the CSF pathways.

It is hardly surprising that the ependyma is vulnerable to infection by numerous pathogens, especially viruses, given its anatomically critical locus at the interface between CSF and brain parenchyma. Viruses may use specific receptors (e.g. CAR, JAM, CD46 and CD55) to target the ependymal and subependymal microenvironment. Choroid plexus

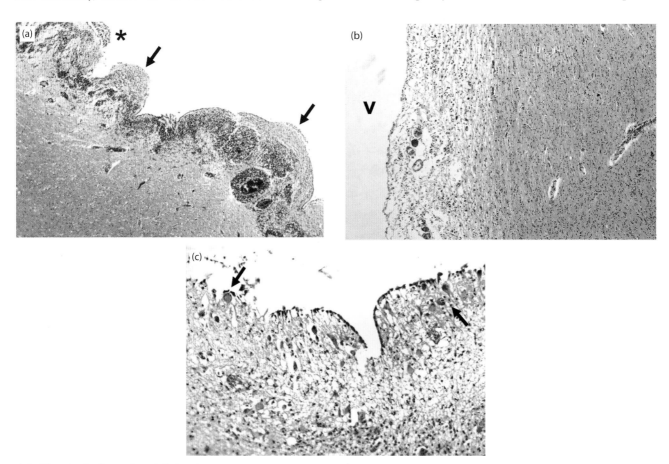

1.17 Ventricular/ependymal lining involved by neuropathologic lesions. (a) Widely infiltrating malignant primary brain tumour extends to the ependymal lining of the lateral ventricle. In one area, tumour appears (asterisk) to extend into the ventricular cavity. Other regions of the ependyma (arrows) show disruption. **(b)** Adenovirus encephalitis and ependymitis in a child with AIDS. Ventricular cavity is indicated by the 'V' (at left). Note almost complete loss of the ependyma, with spongy change and oedema in the periventricular region. **(c)** Cytomegalovirus ependymitis/ventriculitis affecting the fourth ventricle in a human immunodeficiency (HIV)-infected patient with AIDS. Note patchy denudation of the ependymal lining, with scattered cytomegalic cells (arrows) in or immediately adjacent to the ependyma. Sparse inflammatory cells are present in the periventricular neuropil.

epithelium, with anatomic similarities to ependyma, contains some cells that express pattern recognition receptors (PRRs) that in turn bind to pathogens opsonized with C3.[94] Thus both ependyma and choroid plexus epithelium, functionally linked by having crucial functions in maintaining ionic and fluid homeostasis in the brain, also play a direct role in preventing its colonization by micro-organisms, and maintaining the CSF in a sterile condition.

CHOROID PLEXUS

Choroid plexus is a villous, frond-like, convoluted structure located within ventricles and composed of epithelial cells, fenestrated blood vessels, and stroma; it is involved in the production of cerebrospinal fluid.[269] Four separate areas of brain contain choroid plexus: each lateral ventricle, the third ventricle, and the fourth ventricle. Choroid plexus cells are epithelial cells derived from neuroectoderm and thus constitute a subtype of macroglia. In early development, the epithelial cells are tall and pseudostratified, at intermediate stages of development they contain cytoplasmic glycogen, but by later stages of development they become cuboidal and lose glycogen.[269] Unlike ependymal cells that carry cilia, choroid plexus epithelial cells have frequent microvilli and cilia are rarely found. At their basal aspect, choroid plexus epithelial cells lie on a basal lamina and adjacent to highly fenestrated blood vessels that allow the choroid plexus (CP) to produce CSF from the blood. Wholesale diffusion of blood-borne substances, however, does not occur between the blood and CSF because of the presence of tight junctions between the CP epithelial cells. Tight junctions in CP cells contain the proteins occludin, claudin-3, claudin-5, and endothelial selective adhesion molecule (ESAM).[269]

A comprehensive review of the functions of choroid plexus in health and different disease states has recently been published.[269] Abnormalities in CP cells are few in number, although progressive calcification of the stroma of the choroid plexus occurs with ageing. Aβ amyloid and Biondi ring tangles accumulate in CP epithelial cells in Alzheimer's disease, with the latter also seen to accumulate as part of normal ageing.[269] The amyloid source for Aβ amyloid accumulation in CP cells has been suggested to be uptake from CSF. Biondi rings are biochemically and ultrastructurally different from either neurofibrillary tangles or Aβ amyloid, and the exact nature of these structures is still being explored. Interestingly, some workers have suggested that Biondi ring tangles are among the earliest manifestations of Alzheimer disease.[155]

MICROGLIA AND MACROPHAGES

Microglial cells have historically been considered the cells within the CNS that respond to invasion of the parenchyma by viral agents, have important phagocytic functions and constitute the neural component of the reticuloendothelial system. A variety of observations, many of them originating in neuropathological specimens from patients, have led to a reformulation of the putative role of microglia in various diseases and responses of brain and spinal cord to injury.[121,247] It has become frustratingly clear that some ailments in which microglial proliferation is a key element, e.g. Rasmussen encephalitis associated with intractable partial epilepsy and cerebral hemiatrophy in children, do not have an obvious viral aetiology and may in fact be autoimmune disorders.[1,18,66] Microglia may also play a significant role in the progression of lesions seen in diseases of the CNS that are not primarily inflammatory, e.g. Alzheimer's disease, and under some circumstances may even have trophic/nutritive functions.[121] Thus, our understanding of microglial function and potential has greatly expanded in recent years. Much of this has also come about as a consequence of the AIDS pandemic. It became manifestly clear early in the worldwide epidemic of infection by the human immunodeficiency virus (HIV) that understanding the CNS consequences of direct HIV infection of the brain required a sophisticated understanding of the role of its microglial cells, the major cell type that harbours this retrovirus and is productively infected by it.[49,185] For these and other reasons, the 'rediscovery' of the importance of microglia has been reflected (as pointed out in a recent mini-symposium on this cell type) by a massive increase in the literature pertinent to their biology – between 2001 and 2005, almost 3600 articles had been published on microglia, more than in the prior 15 years![57]

Identifying Microglia and Quantifying Microgliosis

For decades, the origin of microglia/macrophages in the CNS has been controversial – the question has been formulated, somewhat simplistically, as 'Do they originate in the brain or the bone marrow or in both sites?' The current view is that blood-derived monocytes move into the brain during early embryonic development, then differentiate into microglia that share many surface markers or antigens with their blood-borne and visceral counterparts, monocytes and macrophages.[121] Under normal circumstances, microglia are inconspicuous bystanders in the scaffolding of the brain and spinal cord. Unlike neurons or ependymal cells, they are not identifiable by their striking and distinctive morphological characteristics or anatomical location. They are estimated to comprise as many as 15 per cent of cells in some parts of the CNS.[94] Though cells with characteristic microglial morphology had been recognized previously – probably even by Nissl in the late 1800s – their discovery and confirmation as a distinctive cell type is widely attributed to the work of del Rio Hortega and Penfield in 1927. They and other investigators seeking to study microglial biology in the early to mid-1900s utilized the silver carbonate method to demonstrate their presence in histological specimens. These methods have been largely supplanted by immunohistochemical stains of microglia/macrophages with primary antibodies directed against macrophage/microglial epitopes and surface antigens or receptors involved with immune system activation, including integrins and the ligands for ICAM-1. Frequently used markers are CD45 (relatively non-specific), CD68 (Figure 1.18), HAM (human alveolar macrophage)-56, CD11b (Mac1), CD11c (LeuM5), CD64 (an immunoglobulin receptor), MHC Class I antigen, MHC Class II antigen (HLA-DR), *Ricinus communis* agglutinin I lectin (RCA),[121] and a newer marker of great utility, Iba1.[268] Some microglia express both MHC class I and II antigens and may interact biologically with both T-helper (T4) and T-cytotoxic

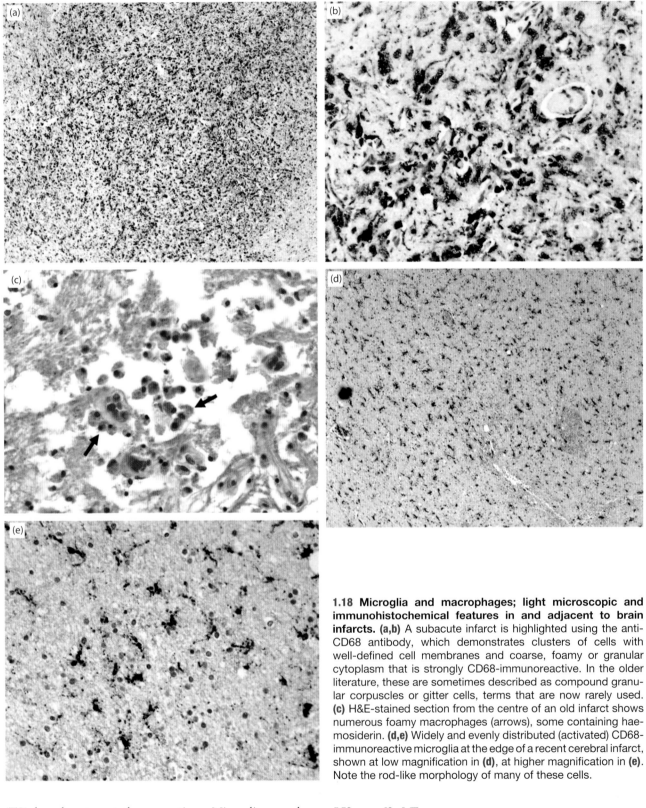

1.18 Microglia and macrophages; light microscopic and immunohistochemical features in and adjacent to brain infarcts. (a,b) A subacute infarct is highlighted using the anti-CD68 antibody, which demonstrates clusters of cells with well-defined cell membranes and coarse, foamy or granular cytoplasm that is strongly CD68-immunoreactive. In the older literature, these are sometimes described as compound granular corpuscles or gitter cells, terms that are now rarely used. **(c)** H&E-stained section from the centre of an old infarct shows numerous foamy macrophages (arrows), some containing haemosiderin. **(d,e)** Widely and evenly distributed (activated) CD68-immunoreactive microglia at the edge of a recent cerebral infarct, shown at low magnification in **(d)**, at higher magnification in **(e)**. Note the rod-like morphology of many of these cells.

(T8) lymphocytes at the same time. Microglia are also demonstrated histochemically (in the mouse) by staining for nucleotide diphosphatase (NDPase).[267] Quantitative estimates of microglial number in the mouse fascia dentata have been provided through the use of unbiased stereological cell counting techniques; estimated numbers of microglia in this structure were on the order of 12 000.[266]

Microglial Types

Microglia were subclassified using the older silver carbonate technique as being amoeboid, ramified or of intermediate form.[121] They are now more commonly described as resting, activated or amoeboid phagocytic microglia, but are known to modify their structure and repertoire of expressed cell surface antigens in response to their ambient microenvironment. Thus there are almost certainly many subsets of microglia, defined by the molecules they secrete and their immunophenotype. Some may be more harmful to the CNS when activated, although others may have significant protective functions.[217] Activated microglia usually take on a 'rod shape' within neural parenchyma. However, their definitive identification relies upon immunohistochemistry (see p. 29 and Figure 1.18). Perivascular cells surrounding brain capillary endothelium share phenotypic properties of both microglia and smooth muscle cells. Perivascular macrophages and macrophages in the subarachnoid space express higher levels of CD45, MHC class II antigen, and 'pattern recognition receptors' (e.g. CD14) than do their parenchymal counterparts, and turn over more quickly, with replenishment by bone marrow-derived elements.[110] When studied by electron microscopy, microglia have few microtubules and intermediate filaments (unlike astrocytes) but numerous cytoplasmic dense bodies that presumably contain molecules that facilitate their phagocytic potential. Their cell surface extends into pseudopodia and filopodia. Given the ability of microglia to sequester themselves unobtrusively within the microanatomy of the CNS, it is somewhat surprising that they can be coaxed from this setting and isolated in relatively pure tissue culture preparations – this has even been achieved from human autopsy brain tissue, obtained with a very short post-mortem autolysis time.[258] *In vitro*, they can be easily induced to reveal their phagocytic properties, e.g. by co-culturing the cells with latex beads, which they avidly take up into their cytoplasm. Primary cultures of microglia are difficult to obtain in pure form, but a number of different permanent/established cell lines have a microglial (immunohistochemical) phenotype.[121] Microglia *in vitro* can be stimulated to proliferate by molecules that originate from microglia themselves, e.g. some cytokines (e.g. IL-1β, IL-4, IFNγ) and appear to be tightly regulated by colony-stimulating factors, such as macrophage colony-stimulating factor (M-CSF, constitutively produced by astrocytes) and granulocyte/macrophage colony-stimulating factor (GM-CSF, expressed primarily during development). M-CSF is crucial to maintaining the normal population of ramified microglia within the brain, although GM-CSF promotes proinflammatory activity.

Functions of Microglia

As for astrocytes (see earlier), whether proliferation and activation of microglia within the CNS have predominantly deleterious consequences for the brain and spinal cord or have beneficial effects has been the subject of intense investigation and reconsideration. Microglia have numerous functions.[87,94,110,121,268] They are, in addition to being immunological sensors within the CNS and antigen presenting cells (APCs), certainly 'factories' for the production of a variety of cytokines and chemokines. The former are low molecular weight proteins including interleukins (ILs), tumour necrosis factors (TNFs), interferons (IFNs), transforming growth factors (TGFs), and colony stimulating factors (CSFs). Cytokines are important elements in not only the modulation of inflammation and immune responses but also the physiological processes vital to CNS growth and development. Chemokines, by

contrast, are small (8–10 kDa) inducible, secreted proinflammatory molecules crucial to various immune and inflammatory responses; they activate specific types of leukocytes and act as chemoattractants. Using reverse transcription polymerase chain reaction methodology, human microglia have also been shown to express transcripts for numerous cytokine receptors, as well as gp130 (an IL-6 receptor component), ciliary neurotrophic factor, and chemokine receptor CXCR4, an important co-receptor for HIV entry into (human) microglia. It remains controversial whether microglia are major producers of nitric oxide (NO). Microglia can also synthesize various neurotrophins and other molecules vital to neuronal survival, including nerve growth factor (NGF), neurotrophin 3 (NT3), brain-derived neurotrophic factor (BDNF), hepatocyte growth factor (HGF) and basic fibroblast growth factor (bFGF). This may explain the apparent neuroprotective function of microglia in some settings.[217]

Pathological Reactions of Microglia

Under normal circumstances, microglial cells represent uncommitted (undifferentiated, immature, resting) myeloid progenitors within the brain, capable of differentiating toward a dendritic morphology (under the influence of granulocyte/macrophage colony-stimulating factor [GM-CSF]) or a histiocytic morphology (in response to macrophage colony-stimulating factor (M-CSF) and/or cytokines).[110] Maintenance of the immature state of microglia is attributed, at least in part, to TGFβ and IL-10. The relatively 'immune privileged' status of the CNS is ascribed to the relative absence of cells that express MHC class I, as well as the constitutive microglial expression of macrophage migration inhibition factor (MIF, also an inhibitor of natural killer [NK] cells) and interleukin-1 receptor antagonist (IL-1Ra), which exerts its inhibitory effect by binding to the IL-1 receptor but not transmitting a signal as a result of this union. TGFβ is a growth factor with numerous effects on processes as diverse as immune homeostasis, angiogenesis, extracellular matrix remodelling, apoptosis and migration. TGFβ1 knockout mice develop a fatal multisystem autoimmune disease. IL-10 inhibits the production of numerous proinflammatory cytokines (e.g., IL-1, IL-6, IL-12, TNFα). Microglial activation occurs through either stimulation/activation of pattern recognition receptors (a major component of the innate immune response, see later) or the adaptive immune response itself. The innate immune response is the initial line of defense in response to a variety of pathogens, one that does not require prior exposure to foreign antigens in order to be 'triggered'.[120] Cellular elements of this innate immune response include macrophages, neutrophils, dendritic cells and natural killer cells, in addition to CNS microglia. Component cells of the innate immune system must be able to recognize antigens by virtue of a predetermined set of conserved receptors, ones expressed on many micro-organisms; these conserved structural motifs are known as pathogen-associated molecular patterns (PAMPs), although the cellular receptors that recognize them are described as pattern recognition receptors (PRRs). Toll-like receptors (TLRs) are a family of PRRs expressed on cells of the innate immune system. Microglia are known to express several TLRs, including TLR2, 3, 4 and 9.[120] Activated microglia secrete, in addition to proinflammatory chemokines and cytokines (e.g. TNFα), several toxic molecules that may damage structural CNS components, e.g. reactive nitrogen intermediates (RNIs), reactive oxygen species (ROS) and derivatives of arachidonic acid, and upregulate MHC class I and II expression.

Microglia in Disease/Injury States

The role of microglia in the pathogenesis and progression of various diseases and forms of brain/spinal cord injury and infections is considered in detail in other chapters of this book, but will briefly be introduced here. A common experimental paradigm has been to consider microglial responses to ischaemic, traumatic or seizure-related brain injury in experimental animals, often focusing on the hippocampus.[72] In elegant experiments using green fluorescent protein (GFP)-expressing (GFP+) bone marrow-derived cells and highly specific phenotypic markers, investigators using a modest entorhinal cortex lesion (that results in local axonal degeneration and microgliosis) have been able to demonstrate the time course of the microglial reaction, as well as the relative contributions of 'resident' and 'immigrant' microglia to the tissue response.[267] Whereas the immigrant (GFP+) microglial response reached a maximum at 7 days post-injury, the resident microglial reaction appeared to peak at about 3 days. Resident microglia, when undergoing early activation, express the stem cell antigen CD34.[132] Microglia (including resting and perivascular cells) play both beneficial and potentially harmful roles in the evolution of focal brain infarcts and global ischaemia.[48] Microglia–at least ones that are not rendered necrotic by a significant ischaemic insult–have been shown to become activated and increase in number within minutes to hours of ischaemic brain injury. Taking on various morphologies (amoeboid, round), these cells probably represent a combination of microglia that have undergone local proliferation and those that have migrated from the penumbra of the ischaemic lesion.

Microglial/macrophage responses have been studied in the spinal cords of patients who had experienced significant cerebral infarcts at time intervals ranging from 4 days to 4 months prior to death. An increase in the number of microglia/macrophages (labelled with antibodies to HLA-DR and CD68) was observed in the contralateral anterior horn of the spinal cord when the cerebral infarct had occurred less than 2 weeks prior to death, but was much more pronounced in the ipsilateral corticospinal tracts (probably a function of ongoing wallerian degeneration) weeks or months after the stroke.[212] Localized spinal cord injury may also be induced (in rats) by injection of zymosan, which causes axonal injury and focally pronounced demyelination in the adjacent spinal cord tissue, injury that appears to be mediated by activation of microglia and macrophages).[184] Microglial activation/proliferation need not be caused by a direct injury to the neural parenchyma. In rats, transection of the facial nerve generates an increased number of microglia in the facial nucleus (from which the nerve obviously originates)–this response is at a maximum 3 days after nerve transection.[165] Microglia subsequently isolated from this axotomized facial nucleus could be cultured and were shown to proliferate *in vitro* in an autocrine fashion.

The possible role of microglia (and other aspects of neuro-inflammation) in the exacerbation of developmental brain injuries has been reviewed.[38]

Subtle alterations in microglial structure (including increases in heterogeneous cytoplasmic inclusions) and immunophenotype (increased cell numbers expressing the ED1 macrophage marker, complement receptor 3/CR3) have been shown to occur with age in experimental animals.[41] Microglial changes (in both size and number) as a function of age are difficult to quantify in the human brain. Several factors are responsible for this. The brain undergoes physiological atrophy beginning in the sixth decade of life, meaning that assessments of microglial number (in autopsy brain specimens) must be undertaken using rigorous unbiased stereological techniques in order to provide meaningful data. Subtle morphological age-related changes are difficult to quantify in this extremely polymorphous cell population. However, human brain appears to show an increased number and density of MHC Class II cells as a function of age. Common neurodegenerative diseases (e.g. Alzheimer's disease [AD] and Parkinson's disease [PD]) are associated with activation and proliferation of microglia either focally or diffusely within the CNS. Although few would claim that activation of microglia is the key aetiological factor underlying these diseases, there is growing, compelling evidence for their probable role in the progression of neurodegeneration.[140,149] The implication of this scenario is that interference with microglial (and, for that matter, astroglial, see earlier) activation might stabilize or even somewhat ameliorate the disease while not curing it.[160] There is evidence from tissue culture systems, for instance, that the presence of microglia exacerbates the neurotoxicity of Aβ(1-42) peptide, which accumulates in the brains of patients with AD. Certainly activated microglia are easily demonstrated around Aβ plaques in the cerebral cortex in AD. In AD, CD68-immunoreactive microglia are more abundant within the subcortical white matter than in the overlying cortex.[256] In an inflammation-mediated rat model of PD, nigral dopaminergic neuronal degeneration clearly postdated a period of microglial activation, suggesting that the latter may have contributed to the former. Microglial activation and increase in number occurs in both the spinal cord and motor cortex of patients with motor neuron disease/amyotrophic lateral sclerosis (MND/ALS) but this could be argued to be as much a result of the neurodegeneration in these regions as a cause of it.[150] Nevertheless, evidence continues to accumulate that microglia and their secreted products are of some importance in this disorder. One study of MND/ALS spinal cord tissue from both sporadic and familial cases found immunohistochemical and molecular evidence (mRNA expression profiles reflecting, e.g. increases in the chemokine MCP-1) suggesting the involvement of immune and inflammatory–including microglial–responses in exacerbating neuronal degeneration in spinal cords of affected patients.[95] Microglial activation has even been suggested as one mechanism of progression in the trinucleotide repeat disease, Huntington disease.[202]

Although generally a helpful process in terms of fending off or at least containing potentially harmful pathogens (e.g. viruses that have gained entry to the CNS), microglial activation may also be harmful. As one example of the latter, radiation-induced microglial activation has been shown to interfere with hippocampal (granule cell) neurogenesis–this may be associated with clinical cognitive decline commonly seen in cancer patients undergoing CNS radiotherapy.[118] The putative molecular pathogenesis of this phenomenon is by no means clear; however, brain inflammation associated with radiation-induced microglial activation may lead to overproduction of IL-6, which in turn interferes with neuronal production (and facilitates the generation of astroglia) from pluripotent neuronal progenitor cells that reside in the hippocampus. This observation reflects a newly emerging role for microglia in brain repair and perhaps even regeneration.[87,268] Brain microglia and macrophages have been shown to express neurotrophins (neurotrophin-3, nerve growth factor) that can then selectively regulate microglial proliferation and even their function.[63] Microglia may also be involved in directing the migration and differentiation of neural precursor cells, even playing a part in influencing the differentiation of adult and embryonic neural precursor cells toward a well-defined neuronal phenotype.[2]

Microglia and macrophages may also be seen in substantial numbers in resected glioneuronal brain tumours that presented with intractable seizures. (Indeed, the more microglial/macrophage immunohistochemical markers are applied to brain tumour surgical material, the more obvious it has become that these cells are frequently a major component of both low- and high-grade glial tumours of all types!) In epilepsy-related neoplasms, the density of microglia was reported to correlate with preoperative epilepsy duration and seizure frequency.[9] This suggests either a role for microglia in causing the seizures or an inflammatory response to the neoplasm or the seizures themselves. As regards the relationship between infiltrating microglia within primary brain tumours, it is possible that diminution of the immune effector functions of these microglia enhances tumour growth, although microglia-derived cytokines and growth factors may facilitate tumour cell proliferation and invasion into normal brain parenchyma.[260] Microglia and other inflammatory cells are particularly abundant in brain tissues of patients with viral encephalitis. In this context, small collections of such cells are often described as inflammatory or microglial nodules (Figure 1.19). The former term is preferred, because microglia are often only one cellular component of a collection of inflammatory cells. Often, in viral encephalitis microglia accumulate around an infected neuron, as part of the process of neuronophagia (Figure 1.19b). Microglial activation is a prominent feature of Rasmussen encephalitis (RE)–a paediatric inflammatory brain disorder causing partial epilepsy and hemiatrophy, probably as the result of the actions of (autoimmune) cytotoxic T-lymphocytes or an (as yet) unidentified viral pathogen and is probably an important contributor to neurodegeneration in this uncommon disorder.[1,66] In RE, microglia are abundant among other inflammatory and reactive cells that infiltrate the CNS, including T-lymphocytes and astrocytes (Figure 1.20). Microglial cells and macrophages are the main types of cell that harbour HIV-1 in patients with HIV encephalitis/HIV-associated dementia; a marker for HIV-1 in the

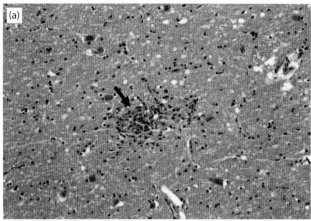

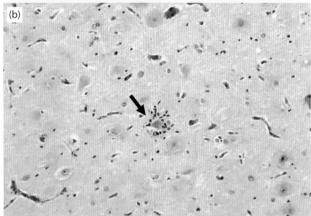

1.19 (a) Microglial nodule in the white matter. **(b)** A cluster of microglia around a 'dying' neuron (neuronophagia).

brain is the presence of multinucleated microglia/macrophages (Figure 1.21). A recent review on microglial cells has been published as part of a symposium on the biology and pathology of glial cells and particularly focuses on the role of microglia in neurodegenerative disorders, CNS trauma, and pain hypersensitivity syndromes following nerve injury (neuropathic pain due to aberrant excitability of dorsal horn neurons and mediated by microglial activation due to chemokine CCL2).[87] An extensive review of microglial function has also been published by Wirenfeldt *et al.*[268]

DISTINGUISHING PATHOLOGICAL ABNORMALITIES FROM ARTEFACTS AND INCIDENTALS

One of the greatest challenges for diagnostic autopsy or surgical pathologists, or researchers using human tissue in their studies, is to determine whether a new feature they are viewing grossly or microscopically in the tissue they are studying is real or relevant to their study. Central and peripheral nervous system anatomy is challenging enough in the first place, and the difficulties in tissue interpretation can be further compounded by the fact that these tissues are subject to a plethora of artefacts, incidental findings and variations of normal. Tangential cuts through complex anatomical areas can erroneously give the impression of an abnormality that

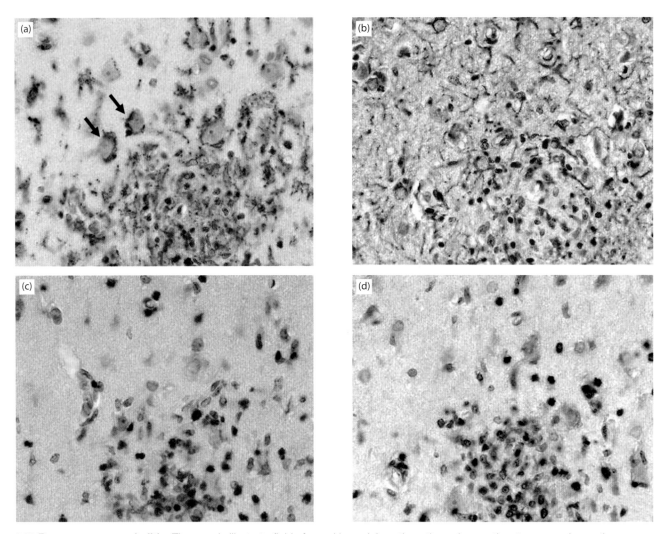

1.20 Rasmussen encephalitis. The panels illustrate fields from skip-serial sections through a corticectomy specimen; the surgery was performed for intractable seizures. **(a)** CD68-immunoreactive cells are found in clusters, and often closely apposed to neurons (arrows). **(b)** Other cell types present in this inflammatory infiltrate include astrocytes, **(c)** and lymphocytes immunoreactive for both ICAM-3, and **(d)** CD3 (T-lymphocytes).

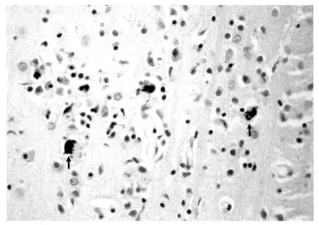

1.21 A case of HIV encephalitis. Multinucleated cells containing minimal cytoplasm (arrows) are characteristic of HIV-1 infection of the brain.

does not exist. Normal tissues can be compressed, distorted, or can predominate in what is otherwise thought to be a tissue block containing solely neoplasm. Oedema in the tissue,

and increased water content of CNS tissue in general, often lead to poor tissue fixation or preservation. Encountering patient tissues from extremes of the age spectrum may lead to misinterpretation of what is simply a normal, age-related change. Many 'bodies', cellular inclusions, pigments, crystals, clefts, unusual reactions to tissue injury, and surgical or treatment-related inert substances may be encompassed within human tissues of pathological interest.[27,76,77] These structures can be diagnostically challenging for autopsy and surgical pathologists, as well as for researchers using the tissues in their studies. Most of the inclusions occur within neurons and represent either accumulations of virions in infected cells or abnormal proteins or cytoskeletal components in neurodegenerative or storage diseases. We have tried to illustrate some of the more common of these changes, mostly as seen in sections stained by the routine haematoxylin and eosin (H&E) method most often used by diagnostic and research pathologists. Some of these inclusion bodies and abnormal deposits are described earlier, and alphabetically listed and illustrated in Figures 1.2 and 1.3. Further CNS and PNS artefacts, incidentals, pigments and miscellaneous bodies are illustrated in Figures 1.22 to 1.26.

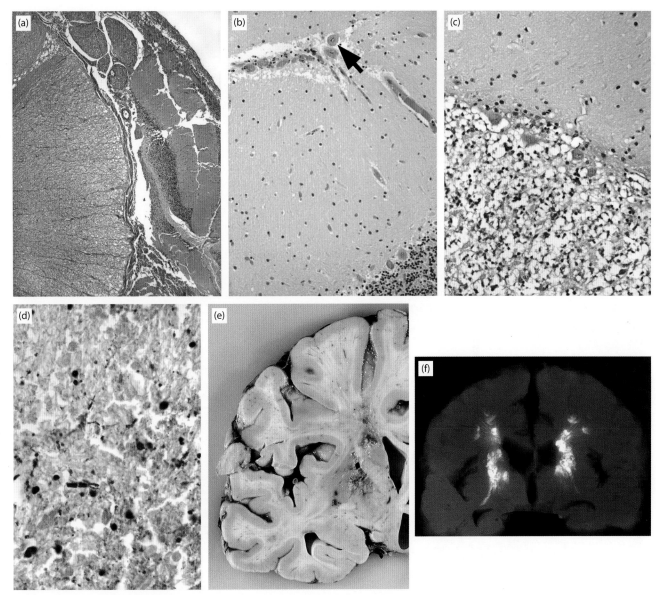

1.22 Artefacts and incidentals found at autopsy. (a) Displacement of cerebellar tissue from herniated cerebellar tonsils may occur when the necrotic, softened cerebellar tissue (right) is displaced from the posterior fossa and travels within the subarachnoid space to surround the spinal cord (left), where it is found at autopsy; this is a characteristic finding in patients who have suffered non-perfused ('respirator') brain change. **(b)** Artefactual displacement of individual neurons (arrow) into the leptomeninges can sometimes be encountered at autopsy, especially in the cerebellum. **(c)** Cerebellar granular cell layer autolysis ('état glacé') is a frequent finding at autopsy and is characterized histologically by the dissolution and smudgy appearance of this layer of neurons. The cerebellar granular cell layer presumably undergoes enzymatic alterations soon after death, especially if an elevated body temperature is not lowered by prompt refrigeration of the body in the morgue. This layer of cells is the 'pancreas of the brain' when it comes to autolytic change! Unlike anoxic injury to the cerebellum, the nearby Purkinje cells (which are considerably more vulnerable than the granule cells to hypoxia and ischaemia) are intact with normal basophilic cytoplasm. Contrast this autolytic change with the normal intact granule cell neurons seen in Figure 1.2a. **(d)** Ferrugination of axons secondary to dystrophic mineralization in a patient with a remote infarction of the brain stem, as illustrated here, can be easily mistaken for organisms, especially if the mineralization is not noticed before special stains are applied. Ferrugination of dead neurons ('tombstoning') can also occur in old infarcts and is more often illustrated in textbooks but less easily misinterpreted. **(e)** Calcification of the basal ganglia is commonly encountered at autopsy in older patients and may be especially prominent in individuals with chronic hypertension, but it is seldom as striking, or grossly apparent, as seen here. Note the yellow-white discolouration in both the globus pallidus and the nearby white matter. Microscopic calcification is also common in the endplate of the dentate gyrus and in the hilum of the dentate nucleus of the cerebellum in aged individuals. However, the degree of calcification in this 62-year-old male, who had an 'undefined progressive neurological disorder', is extreme, probably disease producing, and at the pathological end of the spectrum, so-called Fahr's disease. Coronal brain section at the level of the amygdala. Less extreme calcification is incidental finding. **(f)** Calcification of the basal ganglia provides an excellent opportunity to capitalize on your friendship with your local radiologist and obtain a post-mortem radiographic image, as was done here for the brain illustrated in panel **(e)**.

Continued

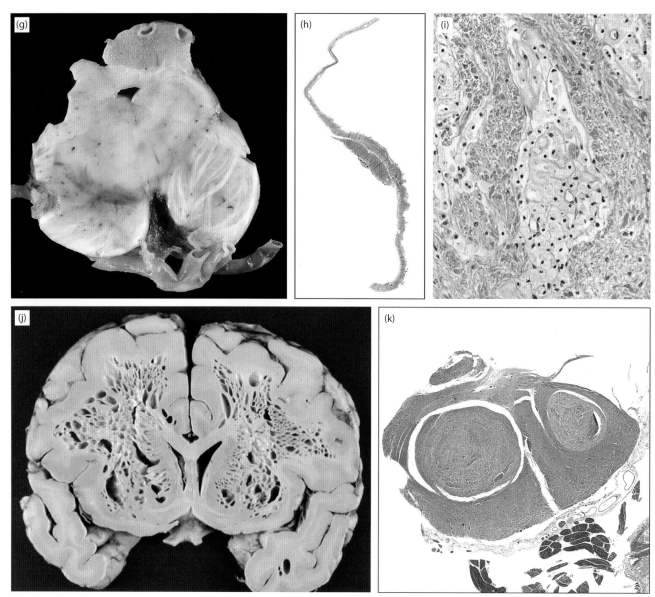

1.22 (*Continued*) Artefacts and incidentals found at autopsy. (g) Lipomas are one of the more common incidental mass lesions encountered at autopsy and can be located in a variety of usually midline sites, including the collicular plate of the midbrain, as illustrated here. The patient was asymptomatic from this lesion. **(h)** Meningiomas are one of the other frequent incidental mass lesions found at autopsy, especially in elderly patients. This tiny, flat, plaque-like, subdural-based lesion, seen on a whole mount section, was found in an older woman with multiple sclerosis and was too small to be responsible for any of her symptomatology. **(i)** Rosenthal fibres are brightly eosinophilic, sausage-like inclusions in astrocytic processes that are associated with long-standing gliosis. Although they are most commonly illustrated in the context of a childhood disorder, Alexander's disease, or in neoplasms such as pilocytic astrocytomas, pleomorphic xanthoastrocytomas or ganglion cell tumours, this example represents an incidental, massive accumulation of Rosenthal fibres in reaction to a macrophage-filled infarct in the cerebellum. This represented the only site of Rosenthal fibre deposition in the brain of this elderly woman. **(j)** Swiss cheese artefact develops when gas-forming bacteria, usually *Clostridia* sp. derived from the gastrointestinal tract, grow post-mortem within brain tissue. Note the concentration of these massive vacuoles in deeper brain areas where the formalin fixative used in autopsy preservation of tissues has not penetrated. The brain was immersed in formalin at the time of autopsy and the formalin penetrated a few centimetres into the tissues and killed the post-mortem overgrowth of micro-organisms in the more superficial areas. On microscopic inspection, no host inflammatory reaction is present around these vacuoles, but numerous bacteria can often be found. **(k)** Toothpaste artefact is seen when the spinal cord is removed in a less-than-gentle fashion at the time of autopsy and portions of the cord become 'intussuscepted' or squeezed into cord at lower or higher levels. This is illustrated here with Luxol fast blue–PAS staining, which highlights the pencil-like cores of displaced cord. This artefactual occurrence is particularly likely if the cord is already fragile and damaged by infarction or tumour.

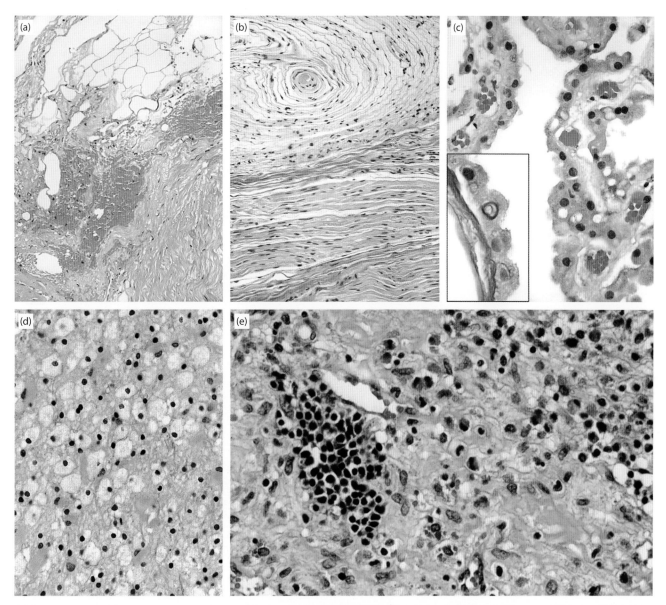

1.23 Artefacts, incidentals, interpretation problems and inert substances found in surgical pathology specimens. (a) Adipose tissue in dura mater (top) is more often observed on neuroimaging studies as a normal finding but when it is included in a surgical pathology specimen, such as in this patient who had falcine meningioma removed with the adjacent attached dura, can be misinterpreted. Also note the small amount of surgically induced haemorrhage within the specimen. **(b)** Pacinian corpuscles (top) are eye-catching, concentric, onion skin-like normal structures that occasionally show up in peripheral nerve (bottom) biopsy specimens. **(c)** Lipid vacuoles in choroid plexus cells can be seen in either surgical or autopsy pathology material and occur with normal ageing; the change is of no known pathological significance. Choroid plexus can also undergo a number of other age-related changes including cystic degeneration, xanthogranuloma of choroid plexus, calcification of the collagenous stroma, psammoma body calcification of the meningothelial cells entrapped in the normal choroid plexus, and formation of Biondi rings within the cytoplasm of choroid plexus cells, seen best with PAS (inset) or silver stains. The latter are also of unknown significance and are not associated with any specific disease. **(d)** Macrophages cause no problem in interpretation when they occur in cohesive sheets or lie within cavities of tissue damage in response to injury. However, in acute demyelinating lesions that prompt biopsy (as illustrated here) they diffusely permeate a neuropil that is largely intact except for the loss of myelin. Their admixture with reactive astrocytes further adds to the diagnostic difficulty; some examples are misdiagnosed as mixed oligoastrocytomas. **(e)** Chronic subdural hematomas can be resident to large numbers of eosinophils and even foci of extramedullary haematopoiesis, both of which are seen here. Although metastatic neoplasms, including hematopoietic neoplasms, often involve the dura, this granulation tissue response should not be interpreted as neoplastic, or even abnormal.

Continued

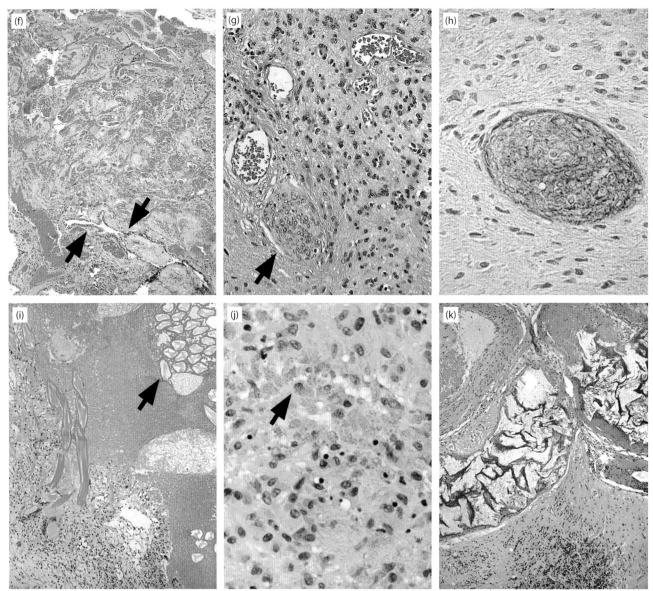

1.23 (*Continued*) Artefacts, incidentals, interpretation problems and inert substances found in surgical pathology specimens. **(f)** Artefactually compressed choroid plexus (top) within a surgical pathology specimen from a patient with colloid cyst of the third ventricle (arrows) can mimic a vascular malformation or even a choroid plexus papilloma unless the collapsed lumen and inconspicuous lining of the colloid cyst are recognized (arrows). **(g,h)** Entrapped normal meningothelial cell nests within a slowly growing, low-grade brain tumour (arrow) are a rare occurrence. The identity of the entrapped cells can be proven by immunostaining for epithelial membrane antigen **(h)**. More commonly, corpora amylacea can also become encompassed by an infiltrating glioma; this usually occurs in an older adult whose brain contains numerous corpora amylacea as part of normal ageing. **(i)** Textilomas (gossypibomas)[191] can occur in response to the surgical haemostatic packing material (arrow) that is employed intraoperatively, eliciting a brain mass that mimics recurrent brain tumour on neuroimaging.[191] The foreign body reaction may occur in response to gelatin sponge, oxidized cellulose and microfibrillar collagen (resorbable agents) or cotton or rayon-based hemostats (non-resorbable). Mass lesions usually follow abdominal surgery, not brain surgery, but CNS examples are well documented. The appearance of the inert substance differs, depending on the agent used. **(j)** Bioglue is an inert, bioadhesive surgical substance composed of bovine serum albumin and glutaraldehyde that is usually employed in cardiac surgical repair procedures; in this patient it was used in a neurosurgical operation. This 23-year-old woman had undergone a posterior cervical spinal cord untethering procedure several months previously and had had Bioglue placed over her dural suture as a sealant for a cerebrospinal fluid leak. She then developed a fluid collection in the posterior cervical tissue and aseptic meningitis, prompting wound revision and removal of the soft tissue and Bioglue. This specimen shows the granulomatous foreign-body type response and the macrophages containing tiny droplets of eosinophilic Bioglue (arrow); the large central inert blob of Bioglue was homogeneously densely eosinophilic, and visually uninteresting. **(k)** Embolization of arteriovenous malformations[134] with polyvinyl alcohol (PVA) prior to operative removal of the lesion can yield striking intravascular spicules of inert substance occluding many, but often not all, vascular channels of the malformation.[134] Note the patent channel in extreme upper left part of the panel. This 30-year-old woman presented with a spontaneous haemorrhage from her cerebellar arteriovenous malformation (AVM) and required three embolization procedures before the lesion could be removed. The material can cause vessel wall necrosis, neutrophilic infiltrates that simulate an acute infection, and a foreign body giant cell reaction.

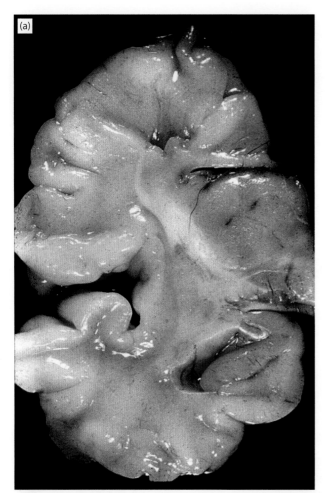

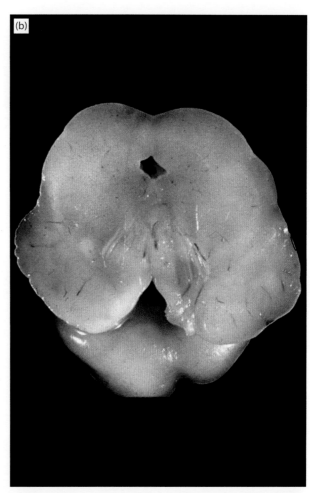

1.24 Colours, clefts and crystals in the nervous system. Bilirubin staining in the damaged central nervous system of immature infants (kernicterus) occurs when the CNS is injured and there is damage to the blood–brain barrier, allowing access of high levels of unconjugated bilirubin from the blood into the brain. Both the anoxia, sepsis or acidosis that was part of the original injury and the bilirubin itself may injure the brain. Free (unconjugated to albumin) bilirubin is reported to bind to phospholipids and gangliosides in cell membranes and to interfere with neuronal oxygen consumption and oxidative phosphorylation.[227] The morphological result is bright yellow discolouration of the subthalamic nucleus, hippocampus, thalamus, globus pallidus (all seen in [a]) and cranial nerve nuclei (3rd cranial nerve nucleus illustrated in [b]). The cerebellar vermis and dentate nuclei can also be affected. The pigmentation may be lost after formalin fixation of the autopsy brain.

STEM CELLS IN THE CENTRAL NERVOUS SYSTEM – PROMISE, POTENTIAL AND REALITY

The field of stem cell research is advancing so rapidly that any information on the subject that appears in a textbook is out-of-date before the book is in print.[233] Nevertheless, a comprehensive symposium on stem cells in the CNS appeared in 2006 in *Brain Pathology*[42,207,208,228,273] and served as an up-to-date reference for neuropathologists to that time. More recently, the potential role of embryonic (neural) stem cells (eNSCs) and induced pluripotent stem cells (iPSCs) in treating various neurologic disorders (especially traumatic brain and spinal cord injury) has been summarized in several reviews.[128,130] eNSC grafts may partially restore cord function after transection in a rat model.[100] Stroke therapy may eventually be revolutionized through the use of iPSCs, NSCs, and even mesenchymal stem cells, though several barriers to their effective use must be overcome (e.g. cell homing to appropriate tracts, survival and tracking, not to mention safety issues).[91] Multiple sclerosis (MS) patients may benefit from SC therapy, with an emphasis on cells that differentiate towards an oligodendroglial phenotype – the important caveat being that axonal injury and loss are now accepted as being mediators of long-term MS disability (see Chapter 23, Demyelinating Diseases).[105]

Stem cells are defined as cells with the potential for self-renewal and multilineage differentiation. Neural stem cells from the adult mammalian central nervous system were first confidently isolated in 1992.[190] It is now appreciated that neural stem cells, as well as glial progenitors, exist in multiple adult brain sites, including the subventricular zone, the lining of the lateral ventricles, the dentate gyrus, the hippocampus, and the subcortical white matter (reviewed by Sanai *et al.*[201]). In humans, the largest of these germinal zones is the subventricular zone. As noted earlier, although the presence of these stem cells is now undisputed, their innate ability effectively to replenish injured or dead neurons is highly limited. Nevertheless, overwhelming interest has been

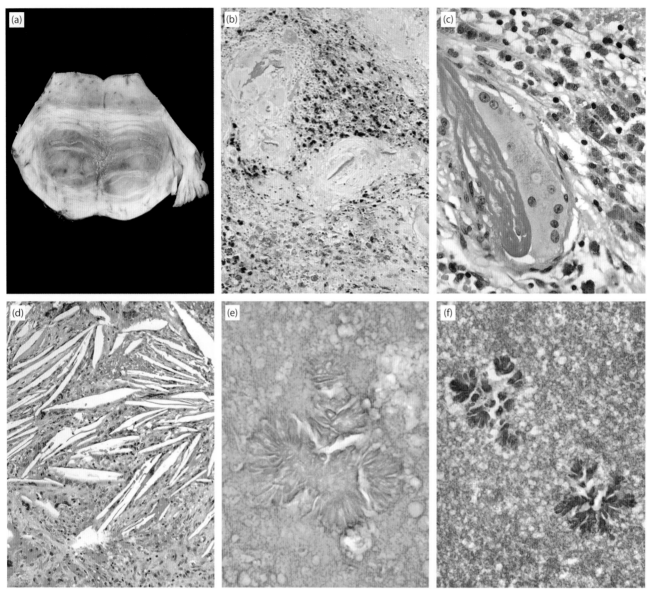

1.25 Pigments and crystals. (a) Bilirubin staining of damaged central nervous system tissue can occur when the blood–brain barrier is acutely or subacutely broken down and the patient has an elevated bilirubin level due to liver disease. This is an example of active central pontine myelinolysis, characterized by its typical triangular midline location in the basis pontis at the level of the 5th cranial nerve, occurring in a typical patient with severe concomitant liver failure and bilirubin elevation. Note how the greenish discolouration extends beyond the immediate nidus of the tissue injury. **(b)** Biliverdin staining can be seen around lesions with massive, recurrent haemorrhages and will occasionally elicit a multinucleated giant cell reaction by the host. This non-iron-containing pigment forms when erythrocytes are disrupted within the ingesting macrophage; the iron in haemoglobin is then oxidized to the trivalent state, forming methaemoglobin. The haem and globin then dissociate and the iron is liberated from hemin by the microsomal enzyme, haem oxygenase, yielding iron and biliverdin.[131] Biliverdin does not stain with the Perls' iron reaction. Note the more abundant robin's egg blue iron-containing material within macrophages also seen in the lesion. **(c)** Biliverdin pigment manifests a yellowish gold colour in contrast to the granular golden brown, refractile appearance of the haemosiderin within macrophages. **(d)** Cholesterol clefts also occur in areas of recurrent haemorrhage, including in the incidental xanthogranulomas often encountered at autopsy in choroid plexus located at the trigone of the lateral ventricle. Xanthogranulomas of the choroid plexus are incidental findings usually in aged individuals, but the cholesterol clefts illustrated here are seen in a pathological lesion, a cholesterol granuloma of temporal bone. **(e)** Crystals due to mucus can occur within the inspissated contents of colloid cysts or Rathke cleft cysts, manifest a spike-like faint yellow or non-stained colouration as seen here, and can be mistaken for the 'sulphur' granules of *Actinomyces* sp. infection. **(f)** Crystals due to mucus also stain with PAS. Same case as panel **(e)**.

generated in the possibility of therapeutically manipulating/expanding exogenous or endogenous stem cell populations and transplanting them into the CNS for the treatment of neurodegenerative disorders, spinal cord injury and multiple sclerosis. Any successful use of stem cells as therapy requires an understanding of how neurons differentiate and commit to lineage during normal embryology.[21,83,88,219] Identifying the factors responsible for proliferation, symmetrical versus asymmetrical cell division, and differentiation is critical. For example, proliferation and expansion of neural stem cells

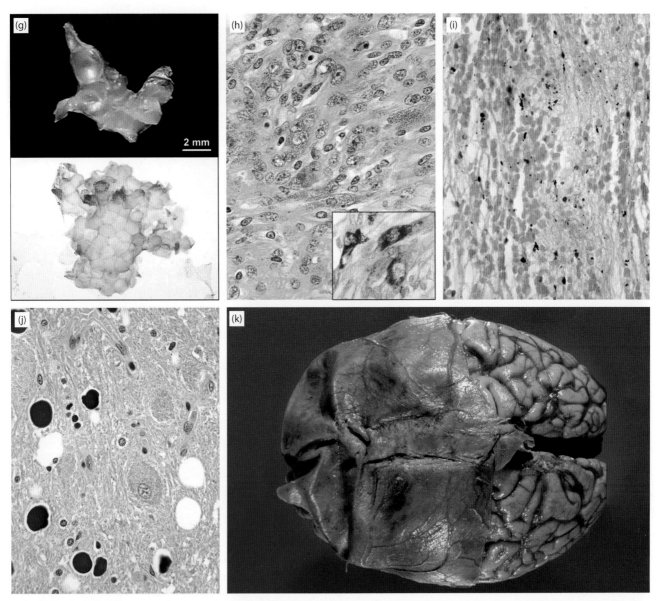

1.25 (*Continued*) Pigments and crystals. (g) Epidermoid cysts, as a gross specimen (top) and a touch preparation made at the time of intraoperative consultation (bottom) for this cerebellopontine angle mass (microscopic, H&E), can have a pearly white cyst content that can grossly mimic true crystals and can contain anucleate clusters of angular squamous cells. Pseudo-crystal. **(h)** Haemosiderin pigment (intracellular) is much finer, less refractile, and more likely to mimic melanin pigment when it is within tumour cells rather than macrophages (compare with Figure 1.4c) and may require a Perls' stain to confidently exclude melanin (inset). An anaplastic meningioma with intracellular iron pigment is illustrated. **(i)** Haematoidin pigment is an artefactual black, dot-like pigment that occurs in areas of fresh haemorrhage and represents a chemical reaction between haemoglobin in red blood cells and formalin fixative, not a true breakdown product of haemoglobin. This pigment appears birefringent when examined under polarized light. **(j)** Melanosis of the dentate nucleus (melanosis cerebelli) is a rare finding at autopsy and manifests with large, rounded globules of extracellular, homogeneous, non-refractile pigment that can be compared in size to the adjacent lipofuscin-containing neurons of the dentate nucleus. This material is Fontana-positive, bleaches with potassium permanganate and is thought to be composed of sulphur. It is probably artefactual, may represent melanization of lipofuscin and is almost never associated with cerebellar symptomatology. **(k)** Ochronosis of the dura mater occurs only rarely, in patients who have the autosomal recessive disorder alkaptonuria, characterized by increased urinary excretion of homogentistic acid. The black pigment is usually found in joints, the cardiovascular system, kidney and skin, but in this reportable example it was found in the dura mater. Gross photograph of lateral surface of formalin-fixed brain with attached, reflected dura mater.

(k) Reproduced with permission from Liu W, Prayson RA. Dura mater involvement in ochronosis (alkaptonuria). Arch Pathol Lab Med 2001;125:961–3.[141]

isolated from developing brain depend on the presence of basic fibroblastic growth factor or epidermal growth factor, and withdrawal of these factors will result in differentiation of the stem cells into neurons, astrocytes or oligodendrocytes (reviewed by Brüstle[29]). The localized microenvironment into which neural stem cells are transplanted has a strong influence on the stem cells and provides the signals that influence the cells to form the correct synapses and chemical phenotype

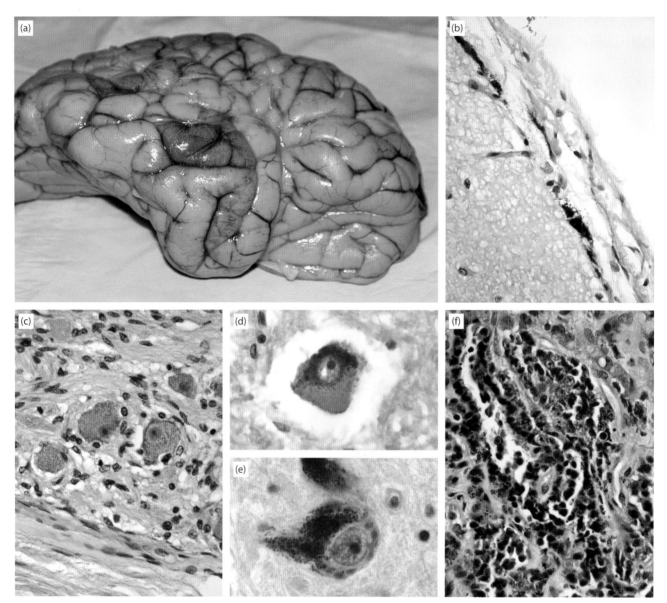

1.26 The many guises of melanin, lipofuscin and amyloid in the central nervous system (CNS) and peripheral nervous system (PNS). **(a)** Meningeal naevi represent both an uncommon incidental finding and a deposition of a colour/pigment in the brain. Although melanocytic melanin pigment is commonly found at the base of the brain at autopsy, particularly around the ventral medulla or hypothalamus, and especially in dark-skinned individuals, visible melanocytic accumulation elsewhere is rare. This example from the temporal lobe of a young Hispanic girl was mistaken by the referring coroner for an area of remote haemorrhage. **(b)** Meningeal naevi are much less exciting microscopically, where they appear as linear melanocytes within the leptomeninges. The elongate shape of the cells containing pigment, coupled with the fine black brown cytoplasmic content, distinguishes these cells from haemosiderin-containing macrophages (compare with Figure 1.4b). **(c)** Lipofuscin is seen in several locations within the neurons of older individuals but not normally in children. Although the coeliac ganglion is illustrated here, more commonly observed sites include neurons of the dorsal root ganglia, the dentate nucleus of the cerebellum and the inferior olivary nucleus. Note the satellite Schwann cells surrounding the neurons and the fact that the Nissl substance is arranged at the periphery in this type of neuron. **(d)** Lipofuscin also accumulates with age in large neurons, particularly the Golgi type I neurons of the anterior spinal cord, cerebral cortex and brain stem. Note the displacement of the basophilic Nissl substance in this neuron from the anterior horn cell region, as well as the artefactual perineuronal space. **(e)** Neuromelanin accumulates with age in neurons and represents a by-product of the metabolism of neurotransmitters in dopaminergic (substantia nigra) and noradrenergic (locus coeruleus) neurons. Young children do not start to manifest either gross or significant microscopic accumulation of neuromelanin pigment in these neurons until about the age of 7 years. **(f)** Melanin (melanocytic melanin) can occur in a variety of different types of central and peripheral nervous system neoplasms; a melanotic medulloblastoma is illustrated here.

appropriate for that anatomical area, but gene transfer or other techniques may aid in manipulating neuronal cell differentiation and commitment.[73] Reviews of numerous aspects of neural stem cell biology have been published (see also previous paragraph).[29,73,152,233,248]

Neural stem cells share many properties with the component cells in gliomas, such as high motility and proliferative potential, association with blood vessels and white-matter tracts, and immature antigenic profiles that reflect activation of developmental signalling pathways.[201]

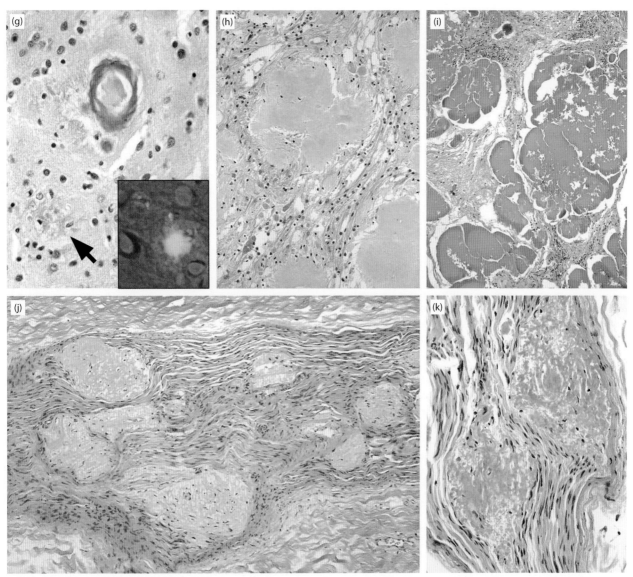

1.26 (*Continued*) The many guises of melanin, lipofuscin and amyloid in the central nervous system (CNS) and peripheral nervous system (PNS). (**g**) Amyloid (cerebral amyloid angiopathy) is most commonly encountered in the central nervous system in conjunction with Alzheimer's disease but can also occur as a primary vasculopathy, cerebral amyloid angiopathy. The amyloid in blood vessels of the cortical grey matter and leptomeninges is quite obvious on Congo red staining. The amyloid core in the centre of neuritic plaques can also be visualized with Congo red stain (arrow) but is better highlighted with thioflavin-S staining (inset). (**h**) Amyloid (primary cerebral amyloidoma) can also occur in large, tumour-like deposits in the cerebrum that cause mass effect and require neurosurgical intervention. These are not associated with cerebral amyloid angiopathy, Alzheimer's disease, inherited forms of cerebral amyloid angiopathy or systemic amyloidosis. The amyloid appears to be formed focally and locally for unknown reasons, but has the staining characteristics of all amyloids. (**i**) Pseudoamyloid deposits can also occur in brain. These appear to be large conglomerates of proteinaceous material and lack the affinity for Congo red dye or the thioflavin-S immunofluorescence of true amyloid. (**j**) Amyloid deposits due to systemic amyloidosis do not affect the central nervous system, because of the presence of the blood–brain barrier, but massive deposits are often found in the peripheral nervous system, as seen in this nerve from a patient with widespread systemic amyloidosis. (**k**) Amyloid deposits due to systemic amyloidosis show affinity for Congo red stain.

(f) Reproduced with permission from USCAP Case 2006-5 Illustrated. (h) Reproduced with permission from USCAP Case 2006-3 Illustrated. (i) Reproduced with permission from Case 1991-3, AANP Diagnostic Slide Session Illustrated.

Both neural stem cells and glioma tumour cells demonstrate hedgehog and Wnt pathway activity and expression of nestin, epidermal growth factor-receptor, telomerase and PTEN.[201] Nestin (named for 'neuroepithelial stem cell') is an intermediate filament strongly expressed in neuroepithelial cells but not in differentiated cells.[29] Self-renewing, multipotent cells in human tumours have been recognized by their expression of CD133 cell-surface markers.[79,223]

Neural stem cells are thought to be particularly vulnerable to oncogenic transformation. In animal models, highly proliferative stem cells within the subventricular zone show the highest degree of susceptibility to chemical or viral oncogenesis. In a tumour model in which avian sarcoma virus was injected into neonatal dog brains, gliomas initially developed in the periventricular regions, but as the tumours increased in size, their relationship with the subventricular zone diminished until they were found at day 10 deep within the white matter, unconnected to the subventricular zone.[249] This model may explain how spontaneous adult human gliomas can still originate in periventricular neural stem cells, yet be located in lobar areas by the time they are discovered clinically. Further evidence in support of the concept that human gliomas may arise from neural stem cells is the fact that neural stem cells share many properties with gliomas (see earlier). The strongest evidence implicating neural stem cells in brain tumour initiation, however, comes from the studies that showed that injection of as few as 100 CD133+ stem cells isolated from human brain tumours into NOD-SCID (non-obese diabetic, severe combined immunodeficient) mice produced a neoplasm, as many as 10^5 CD133- cells did not.[223]

The best-documented example of a human brain tumour that originates from progenitor cells is the medulloblastoma, a highly malignant posterior fossa brain tumour in children that is thought to arise from neural stem cells present in the external granule layer of the cerebellum.[231] The mechanisms of tumourigenesis are still incompletely understood, but studies demonstrate that abnormal co-expression of REST/NRSF (a transcriptional repressor of neuronal differentiation genes) and c-Myc oncogene in neural stem cells 'causes cerebellum-specific tumours by blocking neuronal differentiation and thus maintaining the "stemness" of these cells'.[231] This highly malignant tumour arises through mutations in the sonic hedgehog developmental signalling pathway that normally controls self-renewal in the cerebellar cells of the external granule cell layer. Various developmentally regulated genes that are important in normal brain development and the evolution of neoplasia, e.g. in medulloblastoma (SHH, PTCH, WNT, Notch, etc.) have even suggested molecular subclassifications of this tumour type that may have relevance for prognosis and appropriate treatment.[214]

In addition to their potential therapeutic roles in transplantation and reconstitution of neurons and axons irreparably lost in neurodegenerative disorders or trauma, neural stem cells have been suggested as possible delivery vectors for therapeutic genes.[73]

CONCEPT OF THE BLOOD–BRAIN BARRIER AND THE NEUROVASCULAR UNIT

The BBB is one of three well-defined barrier sites that separate blood from neural parenchyma and the fluid that bathes it, within the CNS; the others are the arachnoid epithelium, which forms the layer of the leptomeninges immediately 'deep' to the dura mater, and the choroid plexus epithelium, site of ongoing synthesis of the cerebrospinal fluid, thus site of the blood–CSF barrier. Barrier properties of these structures are determined by intercellular tight junctions that reduce intercellular (paracellular) permeability pathways for large molecules (for reviews, see Abbott et al.,[3] Ballabh et al.,[13] Najjar et al.,[164] Bicker et al.[22]).

BBB permeability changes have been implicated in disorders as diverse as Alzheimer's disease and neuropsychiatric diseases; successful delivery of drugs through the blood to the CNS is obviously dependent upon their making their way through the BBB (and surrounding cells, e.g. pericytes and astrocytes). Abnormalities of the NVU have been suggested to be of aetiologic importance in the pathogenesis of Alzheimer's disease and mixed dementias resulting from a combination of cerebrovascular disease and AD changes within the brain.[104,274] Virtually all CNS neurons are located within 8–20 μm of a brain capillary. The anatomical 'site' of the BBB is the cerebral capillary endothelium. A more encompassing (and recently popular) term, the 'neurovascular unit', is now used to describe the neuroanatomical cohesiveness and combined activity (especially important in cerebral blood flow and the brain's response to stroke) of brain microvessels together with their surrounding neurons and glia.[36] Current knowledge of the regulation of cerebrovascular tone by perivascular nerves is well summarized by Hamel.[90] Definition of the concept of the BBB, and identification of its anatomic site as being cerebral capillary endothelium, was achieved in the second half of the last century using elegant ultrastructural tracing techniques, e.g. injection of high molecular weight horseradish peroxidase (HRP) into animals prior to their sacrifice. Extensive characterization of BBB physiology continued through the 1980s and 1990s, when various BBB carrier/transport systems were identified, sometimes by examination of isolated cerebral capillaries and even cells cultured from them.[252] Recent comprehensive textbooks have summarized our current knowledge of BBB physiology, biochemistry and molecular pathology, including the novel and innovative techniques that have been used to examine the regulation of its unique biological properties.[163,174] Understanding BBB physiology has immense implications for the therapy of neurological diseases, because the vast majority of large molecules that might be of benefit in the treatment of such ailments do not readily cross the BBB. Novel strategies are being developed to overcome this problem, especially through the use of fusion proteins that can utilize specific BBB transporters to gain access to the CNS parenchyma.[176,177]

The BBB has to ensure the supply of the CNS with nutrients, the efflux of waste molecules from the brain, restriction of ionic and fluid movements between blood and brain or spinal cord, and protection of the CNS from significant fluctuations in blood ionic composition that might result from intense exercise or ingestion of a large meal. It has a high electrical resistance (up to $2000\,\Omega/cm^2$) and effectively regulates the brain/spinal cord microenvironment (i.e. its interstitial fluid) in part by separating neuroactive agents that act both centrally and peripherally.[3] Brain capillary endothelium is characterized by tight intercellular junctions, abundant cytoplasmic mitochondria, and a comparatively low pinocytotic rate, this last feature correlating with a lower rate of transcytosis/endocytosis than is observed in peripheral endothelia.

The structure of interendothelial junctions that define the BBB is itself highly complex. Two of the most important components of tight junctions are a group of proteins with four transmembrane domains and two extracellular loops, known as occludin and the claudins.[3,163] Adherens junctions, which are intertwined with tight junctions, contain vascular endothelial cadherin (VE-cadherin), although platelet-endothelial cell adhesion molecule (PECAM) functions as a mediator of homophilic adhesion. Catenins serve to provide linkage between adherens junctions and the endothelial cytoskeleton. Other junctional elements include members of the immunoglobulin superfamily, junctional adhesion molecules (JAMs) and endothelial cell-selective adhesion molecule (ESAM). The brain capillary endothelial cytoplasm contains a number of adaptor and regulatory/signalling proteins, whose function is to bind to the membranous proteins and modulate their interactions with the actin/vinculin-based cytoskeleton. Among these are zonula occludens 1, 2 and 3 (ZO-1,2,3), calcium-dependent serine protein kinase (CASK), cingulin and junction-associated coiled-coil protein (JACOP). Other proteins vital to signalling and regulation at the BBB include multi-PDZ-protein 1 (MUPP1), partitioning defective proteins (PAR3 and PAR6), membrane-associated guanylate kinase, ZO-1-associated nucleic acid-binding protein (ZONAB), afadin (AF6) and regulator of G-protein signaling 5 (RGS5). It is important to remember that certain periventricular structures within the CNS lack a BBB or contain a porous or 'leaky' one, allowing for rapid entry of blood-borne molecules into the local CNS parenchyma. These structures, known as the circumventricular organs (CVOs) and including the median eminence, neurohypophysis, subfornical organ and the area postrema in the floor above the obex of the fourth ventricle, are populated by neurons highly specialized for chemosensitivity and/or neurosecretion; the location of these neurons within the CVOs allows them to respond rapidly to blood- and CSF-borne substances, including potentially injurious ones, and initiate homeostatic neural and neurohumoral responses.

Although the neuroanatomic site of the BBB is cerebral capillary endothelium, this distinctive endothelium shows dynamic interactions with numerous other cell types. CNS capillary endothelial cells are usually surrounded by pericytes and/or are ensheathed by astrocytic foot processes; indeed astrocytes are often thought of as the cells that link the BBB to neural function (see also earlier section on interactions between astrocytes and the microvasculature). Bidirectional interactions between CNS capillary endothelial cells and their neighbouring cellular elements have been demonstrated using novel co-culture techniques.[19] Molecules hypothesized or proven to be of importance in the maintenance of BBB integrity include TGFβ, basic fibroblast growth factor (bFGF), glial cell derived neurotrophic factor (GDNF) and angiopoietin-1 (ANG1). Some properties of the BBB are influenced by the glycosaminoglycan-rich basement membrane (especially its components heparin sulphate proteoglycan and agrin) that envelops the basal (abluminal) aspect of cerebral capillary endothelial cells and separates them from adjacent pericytes or the foot processes/end feet of perivascular astrocytes. Several studies have shown that anatomical proximity and interaction between the perivascular astrocytes and the capillary endothelium plays a critical role in determining the properties of the BBB. Astrocytic processes or end feet that are in close proximity to cerebral capillary walls show distinctive anatomic and molecular features, including a high density of orthogonal arrays of particles, which contain aquaporin 4 and the Kir4.1 potassium channel. Axonal terminals synapse on the smooth muscle cells of arterioles, whose endothelium may also have BBB functions.[3,163,174] Various agents (e.g. ATP, histamine) can affect endothelial physiology and functions by ligand–receptor interactions, which in turn may be coupled to intracellular calcium release. Such ligand–receptor interactions can take place on either the luminal or abluminal aspect of CNS capillary endothelium, with resultant release of molecules into either the blood or the intercellular space of the CNS. There is indeed not only a blood–brain barrier, but also a brain–blood barrier.

How do molecules (including toxins) gain entry to the CNS across the BBB at sites where it is intact, i.e. most of the CNS? This depends very much on the size and biological properties of the molecules involved. Water-soluble molecules may pass through interendothelial junctions. Lipid-soluble substances (e.g. barbiturates, ethanol) and small gases such as oxygen and carbon dioxide cross the endothelial cell membranes (both luminal and abluminal) with relative ease. Specific transport proteins exist for several families of molecules, e.g. glucose transporter 1 (GLUT1) for glucose, LAT1 for large neutral amino acids, P-glycoprotein, excitatory amino acid transporters EAAT1-3, carriers for ciclosporin A, zidovudine, vinca, etc., a list that contains several commonly administered therapeutic agents. Some BBB transport systems are polarized, i.e. present or active on only the luminal or abluminal endothelial membrane, a property that may be induced, at least in part, by astrocytes.[19] Genes selectively expressed within cerebral capillary endothelium include those for carrier-mediated transporters, efflux transporters and receptor-mediated transporters.[175] Receptor-mediated transcytosis is responsible for the BBB transport of some large proteins, e.g. insulin and transferrin, although others (e.g. plasma proteins, including albumin) cross this barrier by adsorptive transcytosis.

The BBB is a highly dynamic structure, vulnerable to modulation by many factors and molecules. Impairment of the normal BBB function may well be one of the most common pathways by which neurological symptoms occur, in entities as aetiologically diverse as CNS neoplasms (vasogenic oedema), HIV and other viral infections, Alzheimer's disease and a variety of intoxications. Increasing BBB permeability through injury of the brain capillary endothelium (which may also cause severe derangement of normal CNS metabolism) occurs through the action of numerous agents representing various families of molecules, including bradykinin, histamine and glutamate; purine nucleotides; adenosine; phospholipase A2, arachidonic acid, prostaglandins and leukotrienes; interleukins; TNF and macrophage-inhibitory proteins; free radicals and NO.[3,163] With CNS inflammation, especially purulent meningoencephalitis, interendothelial junctions may open, leading to severe (sometimes fatal) cerebral oedema with a marked rise in intracranial pressure. Starvation and hypoxia may lead to GLUT1 transporter upregulation at the BBB. Factors that have the ability to cause the opposite effect, i.e. improved BBB function secondary to its 'tightening', include steroids, noradrenergic agents and increased intracellular cAMP.

Structural or functional alterations of the BBB in disease states are often inferred from neuroimaging studies but less often directly demonstrated by morphoanatomical investigations. Evidence of BBB 'leakiness' is usually surmised, in autopsy brain specimens, by looking for evidence of seepage of serum proteins (e.g. fibrinogen) into perivascular (especially pericapillary) brain parenchyma.[174] Such immunohistochemical investigations can be problematic because of post-mortem artefacts, and should therefore be carefully controlled by use of selected comparable 'normal' tissue specimens. Even so, when controls are chosen to match age and post-mortem autolysis interval of the 'disease' specimen, agonal factors that may influence BBB permeability cannot always be accurately assessed. The other approach to evaluating anatomical integrity of cerebral capillary endothelium–quantitative ultrastructural analysis of brain microvessels–relies on the availability of scarce brain biopsy material that has been appropriately harvested and processed, to ensure preservation of morphological details, which can be studied by electron microscopy.[229] Despite these limitations, many studies have clearly demonstrated BBB abnormalities in human disease states. Other mechanisms of disease-related BBB disruption have been inferred from experimental (including transgenic) animal models.[163] In acute brain injury, as with CNS trauma and ischaemic stroke, microvascular leakage may lead to fatal cerebral swelling secondary to vasogenic oedema (see later). One mechanism for this in animal studies appears to be hypoxia-induced expression of vascular endothelial growth factor (VEGF).[213] BBB abnormalities have been suggested to be a major cause of neurological morbidity in AIDS patients;[174] the molecular pathogenesis of this microvascular injury may have to do with direct effects of HIV proteins on cerebral capillary endothelium. HIV-1 Tat protein, for example, has been shown to alter tight junction protein expression and distribution in brain endothelial cells *in vitro* and can furthermore induce oxidative and inflammatory pathways in the endothelium.[7,239] The BBB may be transiently opened in epileptogenic foci that cause intractable seizures.[3]

BBB abnormalities have also been found in neurodegenerative diseases, e.g. Alzheimer's disease[230] and Parkinson's disease. In the latter, brain capillary dysfunction may result from reduced efficacy of P-glycoprotein.[3] Direct morphometric evaluation of brain biopsies from a small number of AD patients has shown several features suggestive of barrier 'leakiness', including diminished mitochondrial density within endothelial cytoplasm (perhaps a correlate of decreased BBB 'work capability'), an increased number of capillary profiles containing pericytes, and abnormalities of inter-endothelial tight junctions.[229,230] The specific role of BBB abnormalities in the progression of such chronic conditions remains to be ascertained; the abnormalities may be relatively non-specific, which does not exclude their playing a part in disease pathogenesis. Brain neoplasms are perhaps the situation in which subacute or chronic cerebral oedema secondary to BBB abnormalities is the greatest cause of direct morbidity and mortality. The molecular basis for this has been suggested to be underexpression of the tight junction proteins occludin, and claudins-1 and -5, as well as malfunction of aquaporin-4 (AQP4).[173] AQP4 expression is prominently upregulated around malignant brain tumours. Further discussion of brain tumour-related brain oedema appears elsewhere in this text, as well as the role of microvascular lesions in the pathogenesis of inflammatory/demyelinating conditions (e.g. multiple sclerosis).

DETERMINANTS OF INTRACRANIAL PRESSURE AND PRESSURE/VOLUME RELATIONSHIPS, AND CAUSES AND CONSEQUENCES OF RAISED INTRACRANIAL PRESSURE

The Pathology of Intracranial Expanding Lesions

An expanding mass lesion within a rigid cranial cavity will trigger a number of closely interrelated events, the first of which is distortion of the adjacent brain.[259] The major factor responsible for spatial compensation is a reduction in the volume of intracranial CSF. This is achieved by a reduction in the volume of the cerebral ventricles, subarachnoid space and extracerebral CSF cisterns. Compression of the major intracranial venous sinuses may also contribute to spatial compensation within the cranial cavity.

The basic sequence of events can therefore be summarized as local deformity, reduced volume of CSF, shift and distortion of the brain and eventually (in the intact skull) the appearance of internal hernias, i.e. displacement of brain tissue from one intracranial component into another, or into the spinal canal. These displacements result from the development of pressure gradients between intracranial compartments and lead to secondary vascular complications such as haemorrhage and ischaemia.[157,261] When the skull surrounding an expanding intracranial mass is not rigid, e.g. the unfused skull in infants or a displaced flap of bone resulting from skull fracture or surgery, displacement of the brain may occur through the bony defect as an external cerebral hernia.

Supratentorial Expanding Lesions

Expansion of an intrinsic lesion within a cerebral hemisphere–irrespective of the etiology of the 'mass'–results in compression of adjacent brain structures and overall expansion of the hemisphere (Figure 1.27). Sulci on the surface of the brain become narrowed and overlying gyri are flattened against the dura mater, obliterating the subarachnoid space. Reduced cerebral perfusion pressure in a patient with a high intracranial pressure is the major factor causing perisulcal infarcts.[107]

As the mass continues to expand, the lateral ventricles on the side of the lesion and the third ventricle become reduced in size and there is contralateral displacement of the midline structures: the pericallosal arteries, interventricular septum, thalamus, hypothalamus, third ventricle and midbrain. Clinical and neuroradiological studies have suggested that acute lateral displacement of the midbrain and hypothalamus may be fatal in the absence of established cerebral herniation.[156,195] Obliteration of the contralateral foramen of Monro may lead to enlargement of the contralateral lateral ventricle and a further increase in intracranial pressure. The sylvian fissure becomes narrowed and the lesser wing of the sphenoid bone may produce a groove on the inferior surface of the frontal lobe. The floor of the third ventricle is displaced towards the basal cisterns and the mammillary bodies become wedged into a narrowed interpeduncular fossa.

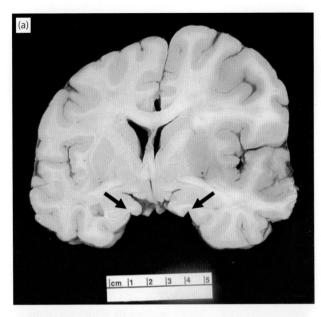

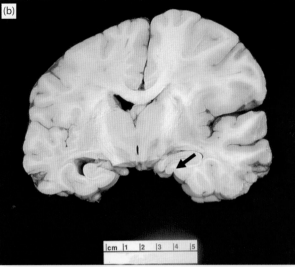

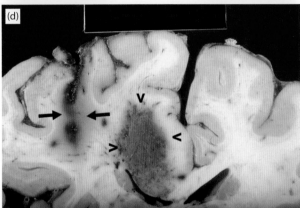

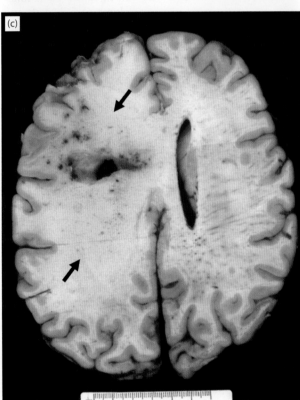

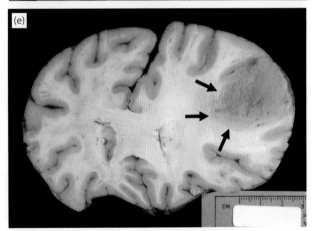

1.27 Cerebral oedema and herniation. (a,b) Coronal slices through the brain from a patient with a recent (24–36 hours) right middle cerebral artery territory infarct. Note the enlargement of the right cerebral hemisphere, with dusky discolouration and effacement of normal anatomical landmarks at the cortex–white matter junction. There is partial effacement of the right lateral ventricle. Arrows indicate bilateral uncal grooving from transtentorial herniation, which is wider on the right **(a)**. Only right uncal grooving is seen in **(b)**. **(c,d)** Cerebral oedema related to high-grade gliomas. **(c)** A large left cerebral hemispheric (predominantly frontal) expansile lesion (arrows) after partial surgical resection. There is diffuse left hemispheric oedema. **(d)** Coronal view of a large infiltrative necrotic and haemorrhagic mass in the left cingulate gyrus (patient died within a few days of biopsy), extending into the corpus callosum (arrowheads). Note subfalcine herniation of the cingulate gyrus, which also shows blurring of its cortex–white matter junction. Arrows indicate one of the biopsy needle tracks, at some distance from the neoplasm. **(e)** Large right frontal mass lesion (chloroma, indicated by arrows) with pronounced oedema of the right cerebral hemisphere and marked right-to-left shift of midline structures.

Although this sequence of events may occur with any expanding lesion within a cerebral hemisphere, certain displacements are selectively affected by the site of the lesion.

An expanding lesion in the frontal lobe will produce displacement of the free margin of the anterior part of the falx cerebri; the posterior part of the falx is rarely displaced laterally because it is firmly tethered at this level. A lesion in the temporal lobe will produce disproportionately severe shift of the third ventricle and will displace upwards the sylvian fissure and the adjacent branches of the middle cerebral artery.

As the lesion continues to expand, the next stage is the development of internal cranial hernias. The major sites of intracranial herniation are at the falx cerebri, tentorium cerebelli and foramen magnum.[45,135,154,209]

Supracallosal Subfalcine (or Cingulate) Hernia

Expansion of a mass in the frontal or parietal lobe will eventually result in herniation of the ipsilateral cingulate gyrus under the free edge of the falx to produce selective displacement of the pericallosal arteries away from this lesion and from the midline (Figure 1.27d). This may compromise circulation through the pericallosal arteries and result in infarction of the parietal parasagittal cortex, manifesting clinically as a weakness or sensory loss in one or both legs. A wedge of pressure necrosis may occur along the groove where the cingulate gyrus makes contact with the falx.[197] If the brain returns to its normal shape as a result of emergency treatment, this wedge of necrosis can be taken as a reliable marker of previous herniation at this site.

Tentorial (Uncal or Lateral Transtentorial) Hernia

Any supratentorial expanding hemispheric mass may produce herniation of the ipsilateral uncus and medial part of the parahippocampal gyrus medially and downward through the tentorial incisura; this occurs most frequently when the mass is located in the temporal lobe (Figure 1.28). The width of this hernia is influenced by variations in the capacity of the tentorial incisura,[43] as well as the size and location of the mass lesion. As the parahippocampal gyrus herniates, the midbrain is narrowed in its transverse axis and the cerebral aqueduct becomes compressed. The contralateral cerebral peduncle is pushed against the opposite free tentorial edge,[69] and the ipsilateral oculomotor nerve becomes compressed between the petroclinoid ligament or the free edge of the tentorium and the posterior cerebral artery. The oculomotor nerve is at first only flattened where it is compressed and angulated over the posterior cerebral artery, but later there is often haemorrhage into the nerve. The resulting paralysis of oculomotor nerve produces ptosis and dilatation of the pupil ipsilateral to the lesion, with loss of the direct response to light shone in the affected eye and of the consensual response to light shone in the opposite eye. There is loss of upward and medial movement of the eye and in its resting position it deviates laterally, because of unopposed action of the VIth cranial nerve. Dilatation of the pupil is the earliest consistent sign of tentorial herniation and may occur before there is any impairment of consciousness.[71,108,109,154,189,232,250]

As the tentorial hernia enlarges, a wedge of haemorrhagic necrosis appears along the lines of the groove in the parahippocampal gyrus. Compression of the contralateral cerebral peduncle against the free edge of the tentorium may lead to infarction, with or without haemorrhage in the dorsal part of the peduncle and adjacent tegmentum.[119] This

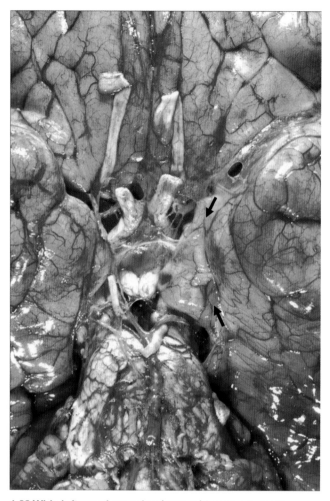

1.28 Wide left uncal grooving (arrows) in autopsy brain specimen from a patient with a large, inoperable brain tumour in the left temporal and parietal lobes.

lesion (Kernohan's notch) may produce weakness followed by extensor rigidity in the limbs ipsilateral to the expanding lesion. This phenomenon has been studied by magnetic resonance imaging (MRI), and is seen most often in older patients with brain atrophy and a pronounced degree of midbrain shift as a complication of chronic subdural hematoma.[40] It is much more common for tentorial herniation to be associated with contralateral limb weakness and eventual extensor rigidity, owing to compression of the cerebral peduncle on the side of the mass lesion by direct pressure from the herniating brain.

Any increase in the pressure gradient is commonly associated with abrupt worsening of the patient's neurological status, such as onset of decerebrate rigidity and loss of consciousness. Expansion of a supratentorial mass lesion may therefore be responsible for initiating tentorial herniation and establishing the beginnings of a transtentorial pressure gradient. Subsequently, any process that would normally induce a diffuse increase in intracranial pressure will increase the transtentorial pressure gradient and accentuate the process of herniation; major degrees of lateral midline shift may cause blockage of the foramen of Monro and narrowing of the cerebral aqueduct, resulting in hydrocephalus.

Other sequelae of raised intracranial pressure and tentorial herniation include the compression of arteries; occlusion of the anterior choroidal artery may lead to infarction in the medial part of the globus pallidus, in the internal capsule and in the optic tract. Compression of a posterior cerebral artery, the blood vessel most commonly affected, may lead to infarction in the thalamus, in the temporal lobe including the hippocampus, and of the medial and inferior cortex and subcortical white matter in the occipital lobe. Compression of a superior cerebellar artery may lead to cerebellar infarction. Infarction of the occipital cortex and cerebellum under these circumstances is often intensely haemorrhagic. These vascular effects usually occur on the same side as the tentorial hernia, but may also be bilateral and very occasionally contralateral.[236] Clinical and neuroradiological studies of survivors of transtentorial herniation have revealed a spectrum of complications, which range from a transient 'locked-in' syndrome[263] to more profound neurological deficits, the severity of which is generally related to the degree of herniation as assessed neuroradiologically.[244]

Central Transtentorial Herniation

This form of herniation occurs particularly in response to frontal and parietal lesions or to bilateral expanding lesions such as chronic subdural hematomas. It results from caudal displacement of the diencephalon and the rostral brain stem and may be preceded by a lateral transtentorial hernia. If intracranial pressure (ICP) increases rapidly in association with lesions of this type, both parahippocampal gyri may herniate through the tentorial incisure, leading to the formation of a circular or ring hernia that is most evident posteriorly and may compress the tectal plate. The clinical manifestations are bilateral ptosis and failure of upward gaze, followed by loss of the pupillary light reflex.

Although major degrees of 'diencephalic downthrust'[65] are readily identifiable on neuropathological examination, minor degrees of caudal displacement of the brain stem are less easy to identify, even in a properly fixed brain. The evidence for downward axial displacement of the brain stem in the herniation process has emerged from both experimental[237] and human post-mortem studies.[92] One study[196] obtained MR images of a patient showing the clinical manifestations of central transtentorial herniation and failed to establish downward axial displacement of the brain stem. Autopsy studies in patients in whom central herniation has been clinically established show backwards and downwards displacement of the mammillary bodies, compression of the pituitary stalk and caudal displacement of the posterior part of the floor of the third ventricle, which comes to lie below the level of the tentorial incisure. Focal infarction may occur in the mammillary bodies and in the anterior lobe of the pituitary gland, owing to impaired blood flow through the long hypothalamo-hypophysial portal vessels. The thalamus becomes distorted with elongation of individual neurons, and the oculomotor nerves become elongated and angulated. Infarction in territories supplied by the anterior choroidal, posterior cerebral and superior cerebellar arteries is also a frequent occurrence.

The clinical correlates of this state are loss of consciousness, decerebrate rigidity and bilateral dilatation of the pupils with loss of light response. The systemic blood pressure becomes elevated as a result of increased sympathetic activity and the heart rate slows. Important areas in the brain stem associated with arterial hypertension appear to be the floor of the fourth ventricle and the nucleus of the tractus solitarius, especially on the left side.[68,98] Alterations in respiration are also common.[101,237]

Haemorrhage and Infarction of the Midbrain and Pons

This is a common and often terminal event in patients with supratentorial expanding lesions, high ICP and tentorial herniation. Emphasis is usually placed on the occurrence of haemorrhage because this is obvious macroscopically, but microscopic examination shows infarction to be at least as frequent as haemorrhage. Both types of lesion occur adjacent to the midline in the tegmentum of the midbrain (Figure 1.29) and in the tegmental and basal parts of the pons. First described by Duret,[255] there has always been considerable debate about the pathogenesis of the haemorrhage and ischaemia. The most important factors are likely to be caudal displacement and anterior–posterior elongation of the rostral brain stem caused by side-to-side compression by the tentorial hernia, coupled with relative immobility of the basilar artery. With progressive displacement, the central perforating branches of the basilar artery that supply the rostral brain stem become stretched and narrowed,[92] leading to spasm, infarction or haemorrhage (Figure 1.30).[111] According to Klintworth,[124,125] brain stem haemorrhage is more likely when high ICP and axial brain stem shift have suddenly been reduced by surgical decompression, resulting in an increase in blood flow in the previously ischaemic brain stem.

Tonsillar Hernia (Foraminal Impaction, Cerebellar Cone)

Downward displacement of the cerebellar tonsil through the foramen magnum occurs as an early complication of

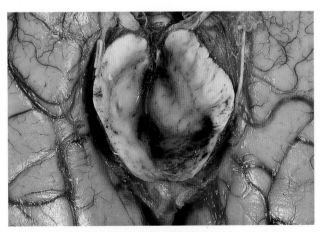

1.29 Transtentorial hernia: brain stem haemorrhage. A large centrally located haemorrhage is present in the midbrain as a consequence of transtentorial herniation. The cerebral peduncle on the right is compressed and distorted.

Reproduced with permission from Ironside JW, Pickard JD. Raised intracranial pressure, oedema and hydrocephalus In: Graham DI, Lantos PL (eds). Greenfield's Neuropathology, 7th edn. London: Arnold, 2002, pp. 193–233.

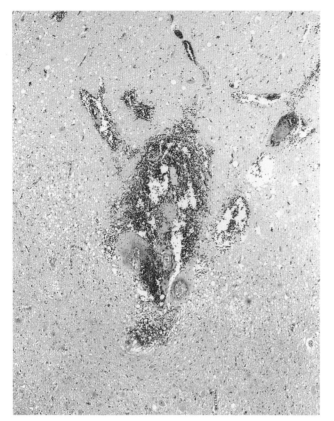

1.30 Transentorial hernia: brain stem haemorrhage.
Histological examination of the midbrain from the case illustrated in Figure 1.29 shows distortion of the white matter and multiple perivascular haemorrhages extending into the parenchyma as a consequence of transtentorial herniation.

Reproduced with permission from Ironside JW, Pickard JD. Raised intracranial pressure, oedema and hydrocephalus In: Graham DI, Lantos PL (eds). Greenfield's Neuropathology, 7th edn. London: Arnold, 2002, pp. 193–233.

expanding masses in the posterior cranial fossa, but may also occur in association with supratentorial space-occupying lesions.[113] The pathognomonic indication of this form of brain herniation is haemorrhagic necrosis at the tips of the cerebellar tonsils and a groove on the ventral surface of the medulla, where it is compressed against the anterior border of the foramen magnum. The accompanying distortion of the spinomedullary junction results in apnoea, which may occur at a stage when consciousness is still preserved. However, tonsillar herniation is usually the last in a sequence of intracranial events, at least one of which will already have been responsible for loss of consciousness. Most patients at this stage will also exhibit other abnormal neurological signs, such as decerebrate rigidity and impairment of brain stem reflexes. This latter situation is more likely if the source of raised ICP is a supratentorial expanding lesion; isolated apnoea is usually a sequel to an expanding lesion within the posterior cranial fossa.

Diffuse Brain Swelling

When intracranial pressure has become elevated as a result of diffuse brain swelling, the ventricles become small but remain symmetrical and there is no lateral shift of midline structures.[4] Nevertheless, bilateral tentorial hernias may occur, their size depending on the rate and severity of brain swelling and the dimensions of the tentorial incisura. Caudal displacement of the diencephalon and brain stem, and central transtentorial herniation are the major contributors to the neurological dysfunction and vegetative disturbance that may result in a fatal outcome in such patients.

Infratentorial Expanding Lesions

Hydrocephalus, with enlargement of both lateral ventricles and the third ventricle, is the most common abnormality associated with expanding lesions in the posterior cranial fossa, whether they be in the fourth ventricle, within or outside the cerebellum. When the lesion is not in the midline, the aqueduct and fourth ventricle are both compressed and displaced contralaterally. Tonsillar herniation occurs most rapidly with a supratentorial expanding lesion. Occasionally, the posterior inferior cerebellar arteries may be compressed, resulting in infarction in the inferior part of one or both cerebellar hemispheres. Herniation of the superior surface of the cerebellum may occur in an upward direction through the tentorial incisure, and is termed 'reversed tentorial hernia'. If the posterior cranial fossa lesion has been expanding very slowly, upward herniation of the superior vermis of the cerebellum can produce considerable distortion of the temporal lobes.

The clinical manifestations of upward tentorial herniation (Figure 1.31) are sudden appearance of bilateral extensor rigidity and loss of the pupillary light reflex. This is most likely to occur when sudden decompression by CSF drainage of enlarged lateral ventricles is carried out in the presence of an undecompressed expanding lesion in the posterior fossa.

External Cerebral Hernias

These occur as rare complications of rapidly expanding supratentorial masses when there is a displaceable defect in the skull, usually surgical or traumatic. This may amount to small protrusions of cortex through cranial burr holes, but if a larger cerebral decompression has been undertaken, major portions of the cerebral hemisphere may herniate through the calvarial defects. Haemorrhagic pressure necrosis occurs

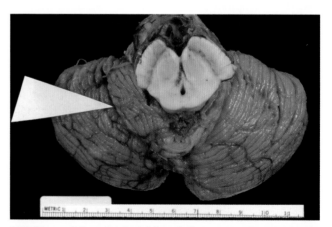

1.31 Upward transtentorial herniation, secondary to a cerebellar mass lesion. Note the notch, indicated by an arrow, most pronounced on the superior aspect of the left cerebellar hemisphere. A small Duret (midbrain) haemorrhage is also noted.

at the edge of the hernia, with swelling of brain tissue within the defect, because of venous obstruction and vasogenic oedema. Continuing herniation will result in extensive isch-aemic or haemorrhagic necrosis of the involved cortex and white matter. A recent study of acute intraoperative brain herniation during elective surgery found that most cases were due to extra-axial haemorrhage (subarachnoid or intraventricular), rather than the intraparenchymal haemor-rhages and acute brain oedema occurring in patients with severe head injury who undergo emergency neurosurgery.[262] Accordingly, the outcome for such elective surgical patients is better than for those with severe head injury.

HYDROCEPHALUS – PATHOPHYSIOLOGY, CAUSES AND CONSEQUENCES FOR THE CNS

The term 'hydrocephalus' defines a state or condition of the brain in which circulation of the cerebrospinal fluid is altered such that there is resultant expansion of fluid-filled intracranial compartments (Figure 1.32).[53,54] It can affect individuals at any age, and since the 1950s has been treatable, largely by the use of shunts that redirect excess CSF from the brain or intracranial cavity to another site, usually the peritoneal cavity; these shunts are often encountered at necropsy in the course of brain removal (Figure 1.32). Most of the CSF is produced by the choroid plexus (at a rate estimated to be 500 mL/day;[46] because CSF volume ranges between 90 and 150 mL, this is replaced or renewed 4–5 times per day), although brain extracel-lular fluid is responsible for a significant, though lesser,

amount. Hydrocephalus is a result of decreased clear-ance of CSF from the lateral and third ventricles through the aqueduct of Sylvius and thence the exit foramina of the fourth ventricle, decreased absorption of CSF into the venous system at the parasagittal arachnoid granulations over the cerebral convexities, or excessive production of CSF from a choroid plexus papilloma or (rarely) choroid plexus hyperplasia, the latter being much the least com-mon aetiology.[75] A diagnosis of hydrocephalus is not itself an aetiological diagnosis. Neoplastic, malformative, haem-orrhagic, degenerative and other causes of hydrocephalus are considered in their respective chapters. Here we shall deal with cellular/molecular mechanisms germane to the consequences of hydrocephalus for the brain and consider some aetiologies, as well as selected animal models that bear on the pathogenesis. An excellent monograph on the pathology of hydrocephalus, with unsurpassed descrip-tions of its causes and consequences, was published by Russell almost 60 years ago but remains topical in the twenty-first century.[200]

Hydrocephalus is not a benign condition and results in deleterious consequences for the brain. These are depen-dent upon the duration and magnitude of the ventricular enlargement (ventriculomegaly) as well as the age at onset and rate of progression.[53,54] Hydrocephalus may cause clini-cal symptoms that include motor, cognitive and endocrine disturbances. As ventriculomegaly progresses, the ependy-mal lining of the ventricles is compromised (ependymal cells are largely non-proliferative, see earlier), in particular over the periventricular white matter. Subependymal gliosis may result from movement of CSF into the interstitial space of the brain, with resultant oedema. Periventricular white matter may become rarefied, devoid of oligodendroglia and gliotic.

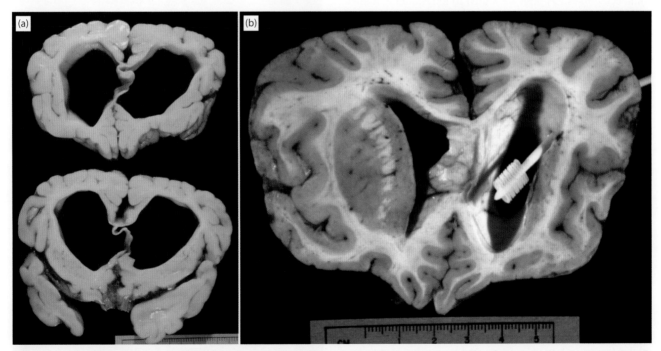

1.32 Hydrocephalus in a young child. (a) Note massive enlargement of the ventricular system (lateral ventricles are illustrated), with thinning leading almost to obliteration of the subcortical white matter. Despite this, the overlying neocortex appears relatively normal. **(b)** A young child who had hydrocephalus caused by a posterior fossa mass. The apparent loss of deep grey matter in the right cerebral hemisphere is an artefact of a slightly asymmetrical cut. Note large catheter tip in the right lateral ventricle; shunting was only partly successful in treating the hydrocephalus.

However, in the authors' experience, evidence of direct axonal injury in the form of neuroaxonal spheroids is observed only rarely, even in relatively severe hydrocephalus, when the process has progressed slowly over many years. In acute hydrocephalus, there may be petechial haemorrhages and evidence of injured axons, a change that can be highlighted using amyloid precursor protein (APP) immunohistochemistry.[59,188] Atrophy of the corpus callosum is commonly observed, as a consequence of both axonal stretching and pressure from above caused by an unyielding falx cerebri. Compression atrophy of the adjacent fimbria/fornix may effectively disconnect the hippocampal formation from the mammillary bodies (and produce trans-synaptic degeneration in the latter), causing memory storage and retrieval deficits. Animal studies suggest that myelin injury, possibly caused by oligodendroglial damage resulting from ventricular enlargement, may antedate axonal disruption.[53,54] Hypothalamic nuclei may be distorted or injured by the hydrocephalus (especially in children), possibly explaining the frequent neuroendocrine abnormalities observed in affected individuals. The brain stem and cerebellum are relatively spared in hydrocephalus.

The cerebral cortex is comparatively spared of injury except in cases of severe hydrocephalus, in the course of which cortical thinning can occur. This may be accompanied by polygyria, not to be confused with polymicrogyria. Atrophy of the basal ganglia and hypothalamic nuclei may occur. Neurodegenerative changes, when present (identifiable in the form of cytoplasmic shrinkage and vacuolation), are relatively non-specific and thought to most likely reflect retrograde change secondary to axonal injury. There may be loss of dendritic spines and a reduction of synaptic vesicle proteins. Neurofibrillary tangle formation in affected neuronal cell bodies has been attributed to longstanding hydrocephalus in adults, but may simply represent coincidental Alzheimer-type change. Effects of hydrocephalus on the choroid plexus itself are variable and include (subtle) epithelial atrophy, cytological changes, and 'stromal sclerosis', possibly reflecting diminished secretory activity.[53,54] Findings of unclear significance include alterations of various receptors and glycoproteins in the circumventricular zones of hydrocephalic brains.

From a neuroanatomical perspective, it seems logical that a disease process that causes stretching and distortion of axons (coincident with ventriculomegaly) leads to axonal and myelin abnormalities, whereas neuronal alterations may be secondary. Nevertheless, the mechanisms by which these alterations occur are incompletely understood. Molecular phenomena of likely importance in producing these changes include those of aetiologic significance in traumatic brain injury (TBI), another situation in which long axons are extremely vulnerable to stretching and tearing.[148] Damage to periventricular axons results in part from calcium-mediated activation of proteolytic calpains, molecules that injure cytoskeletal proteins.[53,54] Cerebral blood flow and oxidative metabolism may be reduced, especially in the deep white matter, in infants, children and adults with hydrocephalus; in adults, this process may be aggravated by concomitant hypertension and atherosclerotic cerebrovascular disease. The BBB (see earlier) may serve as an alternative route for regulation of brain water in hydrocephalics and manifest increased pinocytotic activity (which is quite low in normally functioning cerebral capillary endothelium). Elevated nitric oxide synthase activity and NO production may play protective roles in hydrocephalus, though the details are poorly understood.

Some aetiologies of hydrocephalus will be briefly considered. Acute intraventricular haemorrhage (IVH) in premature neonates is a major clinical problem, especially as increasingly young preterm infants are salvaged.[37] Even though the incidence of IVH appears to be declining, it occurs in one in five infants with a birth weight of less than 1000g. Many are destined to subsequently develop cerebral palsy. The abnormalities in CSF flow in surviving infants that result in hydrocephalus are the direct consequence of meningeal fibrosis, arachnoiditis and subependymal astrogliosis, all of which lead to impairment of its normal flow and absorption. In addition to compressive and ischaemic mechanisms contributing to hydrocephalus in these infants, there appears to be increased parenchymal and perivascular deposition of extracellular matrix proteins, probably as the result of upregulation of TGFβ. It is of note that astrocytic overexpression of TGFβ1 in transgenic mice leads to perivascular astrocytosis, increased perivascular deposition of extracellular matrix proteins such as laminin and fibronectin and striking ventriculomegaly.[44] Free radical mediated brain parenchymal injury may also contribute to morbidity after ventricular haemorrhage.

Abnormalities of the cerebral aqueduct (Figure 1.33), including agenesis, atresia, gliosis, forking and membranous occlusion, are further discussed elsewhere in this text. The terminology used to describe these aqueductal anomalies is inconsistent and often confusing.[12,138,151,242] It has been suggested that aqueductal stenosis can be the

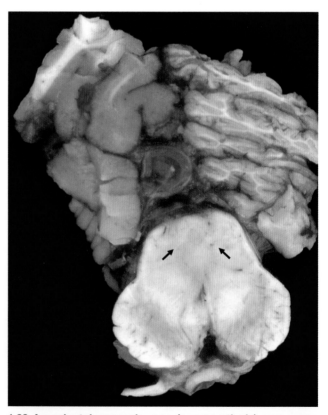

1.33 Aqueductal agenesis – one (comparatively) rare cause of congenital hydrocephalus. This condition may be X-linked. Arrows indicate region of the midbrain where the aqueduct would be expected.

result of hydrocephalus, rather than its cause.[264] Rare X-linked forms of aqueductal stenosis have been recognized for almost 60 years.[99,133]

Hydrocephalus is often discovered in adults, a population in whom it is almost certainly underestimated. The incidence of adult-onset chronic hydrocephalus has been estimated at 2.6 per 100 000;[62] in other words, adult-onset hydrocephalus may account for approximately half of the 80 000 cases of hydrocephalus diagnosed in the United States annually. It may represent decompensation of a compensated congenital hydrocephalus (e.g. secondary to aqueductal stenosis) that becomes symptomatic for the first time only in adulthood. Acquired neurological diseases that lead to meningeal fibrosis may result in hydrocephalus; the most common are subarachnoid haemorrhage and meningitis, especially suboptimally treated purulent or granulomatous meningitis.

Normal-pressure hydrocephalus (NPH), a type of communicating hydrocephalus, classically presents with the triad of gait disturbance, urinary incontinence and cognitive impairment, though it may manifest only one or two of these symptoms (most often gait disturbance) and can be entirely asymptomatic. The neuropsychological phenomena observed in affected patients include those common in subcortical frontal lobe disorders (inattention, forgetfulness, diminished intellectual agility), as well as apathy, emotional lability and disinhibition. First described in 1965 and sometimes known as Hakim–Adams syndrome, NPH is treated by ventricular shunting.[62] Though it is said to account for 5–10 per cent of dementia, NPH is encountered only infrequently in dementia autopsy series, including ours – perhaps because the patients respond well to shunting. At autopsy, NPH manifests as enlargement of the lateral and third ventricles, out of proportion to the degree of cerebral cortical atrophy (Figure 1.34) – a dissociation that is generally not found in AD brain specimens, notwithstanding the fact

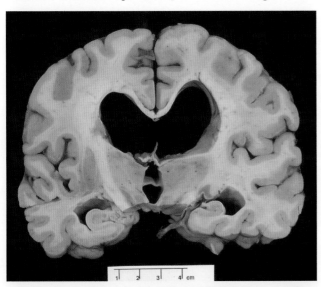

1.34 Normal pressure hydrocephalus (NPH). This condition, one potentially treatable cause of cognitive impairment, has caused ventriculomegaly (affecting the third and lateral ventricles) with macroscopic preservation of the cortical ribbon. Note that, unlike in patients with Alzheimer disease, the hippocampi are relatively normal in size, with no evidence of atrophy.

that not all AD brains are strikingly atrophic. However, a note of caution is in order: when brain biopsies from individuals undergoing shunt placement for idiopathic NPH are carefully examined, they often show evidence of AD-related lesions, in the form of Aβ plaques and, less commonly, amyloid angiopathy, neuropil threads and neurofibrillary tangles;[85] the severity of AD-related abnormalities in these study patients correlated with the severity of their dementia.

Treatment of Hydrocephalus, and Its Effects on CNS Structure/Function

The treatment of hydrocephalus usually includes therapy of the underlying causal condition, when this is possible. For example, treatment of ventriculomegaly secondary to a posterior fossa mass is surgical resection of this mass. Other modalities used to treat hydrocephalus over the decades have included attempts at ablation of the choroid plexus, third ventriculostomy and (most commonly) a shunt that redirects CSF from the brain to the peritoneal cavity or the cardiac chambers.[53,54] The introduction of the flexible, biocompatible shunt tubing now used in these procedures has dramatically altered the outlook for people with hydrocephalus, leading to pronounced symptomatic improvement in treated patients. However, although shunting is effective at relieving symptoms caused by hydrocephalus, it does not totally reverse resultant pathological abnormalities. The ventricular size in shunted patients does not always return to normal after shunting. Successful shunting leads to less severe atrophy of the corpus callosum and fornix than would be expected in untreated hydrocephalus, but there is residual periventricular gliosis. This may serve as an impediment to reconstitution of normal white matter in the periventricular region. As indicated earlier, the ependyma has very limited regenerative potential, thus regions of ependymal loss that occur in the evolution of hydrocephalus are not repaired. This may lead to persistent abnormalities of interstitial fluid content within the underlying brain parenchyma.

Clinically, shunt placement to treat hydrocephalus is one of the most commonly performed operations by paediatric neurosurgeons and the procedure has completely changed the natural history of hydrocephalus. Before shunting was available, hydrocephalus led to progressive neurological deterioration and early death, often before the third decade of life.[179] The procedure, however, is associated with significant complications, including shunt malfunction and shunt infection.[179] In a recently published, 20-year follow-up of patients who were younger than age 15 years at the time of their first shunt placement, 2.9 per cent died of shunt failure, 81 per cent required at least one shunt revision, and among those with revision, the mean number of revisions was 4.2.[179] Thus, a large National Institutes of Health workshop on hydrocephalus has called for additional research regarding better treatment options and further investigation of 'pathophysiological and recovery mechanisms of neuronal function'.[265]

Pharmacological therapy of hydrocephalus has included administration of carbonic anhydrase inhibitors (e.g. acetazolamide, which can reduce CSF production by over 60 per cent) and furosemide. Attempts to prevent hydrocephalus in infants following intraventricular haemorrhage (see earlier) by infusion of urokinase, streptokinase and tissue plasminogen activator in affected patients have met with limited success.[53,54] Given some of the neurobiological similarities between axonal injury associated with hydrocephalus and brain injury after trauma/stroke, neuroprotective approaches are also being tried.

ACKNOWLEDGEMENTS

This chapter would not have been possible without the photographic assistance of Ms Lisa Litzenberger and Carol Appleton and the manuscript preparation work of Ms Susan Peth.

REFERENCES

1. Aarli JA. Rasmussen's encephalitis: a challenge to neuroimmunology. *Curr Opin Neurol* 2000;**13**:297–9.
2. Aarum J, Sandberg K, Budd Haeberlein SL, Persson MAA. Migration and differentiation of neural precursor cells can be directed by microglia. *Proc Natl Acad Sci U S A* 2003;**100**:15983–8.
3. Abbott NJ, Ronnback L, Hansson E. Astrocyte–endothelial interactions at the blood–brain barrier. *Nat Rev Neurosci* 2006;**7**:41–53.
4. Aldrich EF, Eisenberg HM, Saydjari C, *et al*. Diffuse brain swelling in severely head injured children. A report from the NIH Traumatic Coma Data Bank. *J Neurosurg* 1992;**76**:450–54.
5. Anders KH, Park C-S, Cornford ME, Vinters HV. Adenovirus encephalitis and widespread ependymitis in a child with AIDS. *Pediatr Neurosurg* 1991;**16**:316–20.
6. Anderson CM, Swanson RA. Astrocyte glutamate transport: review of properties, regulation, and physiological functions. *Glia* 2000;**32**:1–14.
7. Andras IE, Pu H, Deli MA, *et al*. HIV-1 Tat protein alters tight junction protein expression and distribution in cultured brain endothelial cells. *J Neurosci Res* 2003;**74**:255–65.
8. Anthony TE, Klein C, Fishell G, Heintz N. Radial glia serve as neuronal progenitors in all regions of the central nervous system. *Neuron* 2004;**41**:881–90.
9. Aronica E, Gorter JA, Redeker S *et al*. Distribution, characterization and clinical significance of microglia in glioneuronal tumours from patients with chronic intractable epilepsy. *Neuropathol Appl Neurobiol* 2005;**31**:280–91.
10. Atkinson SL, Li YQ, Wong CS. Apoptosis and proliferation of oligodendrocyte progenitor cells in the irradiated rodent spinal cored. *Int J Radiat Oncol Biol Phys* 2005;**62**:535–44.
11. Azzarelli B, Miravalle L, Vidal R. Immunolocalization of the oligodendrocyte transcription factor 1 (Olig1) in brain tumors. *J Neuropathol Exp Neurol* 2004;**63**:170–79.
12. Baker DW, Vinters HV. Hydrocephalus with cerebral aqueductal dysgenesis and craniofacial anomalies. *Acta Neuropathol* 1984;**63**:170–73.
13. Ballabh P, Braun A, Nedergaard M. The blood–brain barrier: an overview. Structure, regulation, and clinical implications. *Neurobiol Dis* 2004;**16**:1–13.
14. Ballabh P, Hu F, Kumarasiri M, Braun A, Nedergaard M. Development of tight junction molecules in blood vessels of germinal matrix, cerebral cortex, and white matter. *Pediatr Res* 2005;**58**:791–8.
15. Balss J, Meyer J, Mueller W, *et al*. Analysis of the *IDH1* codon 132 mutation in brain tumors. *Acta Neuropathol* 2008;**116**:597–602.
16. Barnett MH, Prineas JW. Relapsing and remitting multiple sclerosis: pathology of the newly forming lesion. *Ann Neurol* 2004;**55**:458–68.
17. Barres BA. What is a glial cell? *Glia* 2003;**43**:4–5.
18. Bauer J, Bien CG, Lassmann H. Rasmussen's encephalitis: a role for autoimmune cytotoxic T lymphocytes. *Curr Opin Neurol* 2002;**15**:197–200.
19. Beck DW, Vinters HV, Hart MN, Cancilla PA. Glial cells influence polarity of the blood–brain barrier. *J Neuropathol Exp Neurol* 1984;**43**:219–24.
20. Berridge MJ, Bootman MD, Lipp P. Calcium: a life and death signal. *Nature* 1998;**395**:645–8.
21. Bertrand N, Castro D, Guillemot F. Proneural genes and the specification of neural cell types. *Nat Rev Neurosci* 2002;**3**:517–30.
22. Bicker J, Alves G, Fortuna A, Falcão. Blood–brain barrier models and their relevance for a successful development of CNS drug delivery systems: a review. *Eur J Pharm Biopharm* 2014;doi.org/10.1016/j.ejpb.2014.03.012.
23. Bigio EH, Weiner MF, Bonte FJ, White CL. Familial dementia due to adult polyglucosan body disease. *Clin Neuropathol* 1997;**16**:227–34.
24. Borner C, Monney L. Apoptosis without caspases: an inefficient molecular guillotine? *Cell Death Differ* 1999;**6**:497–507.
25. Bouvier-Labit C, Liprandi A, Monti G, et al. CD44H is expressed by cells of the oligodendrocyte lineage and by oligodendrogliomas in humans. *J Neurooncol* 2002;**60**:127–34.
26. Bradl M, Lassmann H. Oligodendrocytes: biology and pathology. *Acta Neuropathol* 2010;**119**:37–53.
27. Brat DJ. Overview of central nervous system anatomy and histology. In: Prayson R ed. *Neuropathology*. Philadelphia, PA: Churchill Livingstone, 2005:1–36.
28. Brown AM, Tekkok SB, Ransom BR. Energy transfer from astrocytes to axons: the role of CNS glycogen. *Neurochem Int* 2003;**45**:529–36.
29. Brüstle, O. Building brains: neural chimeras in the study of nervous system development and repair. *Brain Pathol* 1999;**9**:527–45.
30. Bunge MB. Bridging areas of injury in the spinal cord. *Neuroscientist* 2001;**7**:325–39.
31. Burger PC. What is an oligodendroglioma? *Brain Pathol* 2002;**12**:257–9.
32. Bush TG, Puvanachandra N, Horner CH, *et al*. Leukocyte infiltration, neuronal degeneration, and neurite outgrowth after ablation of scar-forming, reactive astrocytes in adult transgenic mice. *Neuron* 1999;**23**:297–308.
33. Capper D, Zentgraf H, Balss J, *et al*. Monoclonal antibody specific for *IDH1* R132H mutation. *Acta Neuropathol* 2009;**118**:599–601.
34. Capper D, Weißert S, Balss J, *et al*. Characterization of R132H mutation-specfiic IDH1 antibody binding in brain tumors. *Brain Pathol* 2010a;**20**:245–54.
35. Capper D, Sahm F, Hartmann C, *et al*. Application of mutant IDH1 antibody to differentiate diffuse gliomas from non-neoplastic central nervous system lesions and therapy-induced changes. *Am J Surg Pathol* 2010b;**34**:1199–1204.
36. Carmichael ST. Cellular and molecular mechanisms of neural repair after stroke: making waves. *Ann Neurol* 2006;**59**:735–42.
37. Cherian S, Whitelaw A, Thoresen M, Love S. The pathogenesis of neonatal post hemorrhagic hydrocephalus. *Brain Pathol* 2004;**14**:305–11.
38. Chew L-J, Takanohashi A, Bell M. Microglia and inflammation: impact on developmental brain injuries. *Ment Retard Dev Disabil Res Rev* 2006;**12**:105–12.
39. Chakraborty S, Nazmi A, Dutta K, Basu A. Neurons under viral attack: victims or warriors? *Neurochem Int* 2010;**56**:727–35.
40. Cohen AR, Wilson J. Magnetic resonance imaging of Kernohan's notch. *Neurosurgery* 1990;**27**:205–7.
41. Conde JR, Streit WJ. Microglia in the aging brain. *J Neuropathol Exp Neurol* 2006;**65**:199–203.
42. Conti L, Reitano E, Cattaneo E. Neural stem cell systems: diversities and properties after transplantation in animal models of diseases. *Brain Pathol* 2006;**16**:143–54.
43. Corsellis JAN. Individual variation in the size of the tentorial opening. *J Neurol Neurosurg Psychiatry* 1958;**21**:279–83.
44. Crews L, Wyss-Coray T, Masliah E. Insights into the pathogenesis of hydrocephalus from transgenic and experimental animal models. *Brain Pathol* 2004;**14**:312–16.
45. Cushing H. Some principles of cerebral surgery. *JAMA* 1909;**52**:184–95.
46. Cutler RWP, Page L, Galicich J, Watters GV. Formation and absorption of cerebrospinal fluid in man. *Brain* 1968;**91**:707–20.
47. Dahl D. Immunohistochemical differences between neurofilaments in perikarya, dendrites and axons. Immunofluorescence study with antisera raised to neurofilament polypeptides (200K, 150K, 70K) isolated by anion exchange chromatography. *Exp Cell Res* 1983;**149**:397–408.
48. Danton GH, Dietrich WD. Inflammatory mechanisms after ischemia and stroke. *J Neuropathol Exp Neurol* 2003;**62**:127–36.

49. D'Aversa TG, Eugenin EA, Berman JW. NeuroAIDS: contributions of the human immunodeficiency virus-1 proteins Tat and gp120 as well as CD40 to microglial activation. *J Neurosci Res* 2005;**81**:436–46.

50. Davies JE, Huang C, Proschel C, *et al.* Astrocytes derived from glial-restricted precursors promote spinal cord repair. *J Biol* 2006;**5**:7.1–21.

51. DeBiasi R, Kleinschmidt-DeMasters BK, Richardson-Burns S, Tyler KL. Central nervous system apoptosis in human herpes simplex virus and cytomegalovirus encephalitis. *J Infect Dis* 2002;**186**:1547–57.

52. Dehghani F, Maronde E, Schachenmayr W, Korf HW. Neurofilament H immunoreaction in oligodendrogliomas as demonstrated by a new polyclonal antibody. *Acta Neuropathol* 2000;**100**:122–30.

53. Del Bigio MR. Neuropathological changes caused by hydrocephalus. *Acta Neuropathol* 1993;**85**:573–85.

54. Del Bigio MR. Cellular damage and prevention in childhood hydrocephalus. *Brain Pathol* 2004;**14**:317–24.

55. Del Bigio MR. Ependymal cells: biology and pathology. *Acta Neuropathol* 2010;**119**:55–73.

56. Del Zoppo GJ. Stroke and neurovascular protection. *N Engl J Med* 2006;**354**:553–5.

57. de Vellis J, Kim SU. Foreword. *J Neurol Sci* 2005;**81**:301.

58. Doetsch F, Caille I, Lim DA, *et al.* Subventricular zone astrocytes are neural stem cells in the adult mammalian brain. *Cell* 1999;**97**:703–16.

59. Dolinak D, Reichard R. An overview of inflicted head injury in infants and young children, with a review of beta-amyloid precursor protein immunohistochemistry. *Arch Pathol Lab Med* 2006;**130**:712–17.

60. Eddleston M, Mucke L. Molecular profile of reactive astrocytes–implications for their role in neurologic disease. *Neuroscience* 1993;**54**:15–36.

61. Edgar JM, Garbern J. The myelinated axon is dependent on the myelinating cell for support and maintenance: molecules involved. *J Neurosci Res* 2004;**76**:593–8.

62. Edwards RJ, Dombrowski SM, Luciano MG, Pople IK. Chronic hydrocephalus in adults. *Brain Pathol* 2004;**14**:325–36.

63. Elkabes S, DiCicco-Bloom EM, Black IB. Brain microglia/macrophages express neurotrophins that selectively regulate microglial proliferation and function. *J Neurosci* 1996;**16**:2508–21.

64. Ellison D, Love S, Chimelli L, *et al.* Neuropathology. *A reference text of CNS pathology*, 3rd edn. Edinburgh: Mosby, 2013.

65. Esiri MM. *Oppenheimer's diagnostic neuropathology*, 2nd edn. Oxford: Blackwell Science, 1996:26–7.

66. Farrell MA, Droogan O, Secor DL, *et al.* Chronic encephalitis associated with epilepsy: immunohistochemical and ultrastructural studies. *Acta Neuropathol* 1995;**89**:313–21.

67. Faulkner JR, Herrmann JE, Woo MJ, *et al.* Reactive astrocytes protect tissue and preserve function after spinal cord injury. *J Neurosci* 2004;**24**:2143–55.

68. Fein JM. Hypertension and the central nervous system. *Clin Neurosurg* 1982;**29**:666–721.

69. Feldman E, Gandy SE, Becker R, *et al.* MRI demonstrates descending transtentorial herniation. *Neurology* 1988;**39**:622–7.

70. Filley CM, Kleinschmidt-DeMasters BK. Toxic leukoencephalopathy: an emerging medical and public health problem. *N Engl J Med* 2001;**345**:425–32.

71. Finney LA, Walker AE. *Transtentorial herniation*. Springfield, IL: Thomas, 1962.

72. Finsen BR, Jorgensen MB, Diemer NH, Zimmer J. Microglial MHC antigen expression after ischemic and kainic acid lesions of the adult rat hippocampus. *Glia* 1993;**7**:41–9.

73. Foster GA, Stringer BMJ. Genetic regulatory elements introduced into neural stem and progenitor cell populations. *Brain Pathol* 1999;**9**:547–67.

74. Frohman EM, Racke MK, Raine CS. Medical progress: multiple sclerosis – the plaque and its pathogenesis. *N Engl J Med* 2006;**354**:942–55.

75. Fujimura M, Onuma T, Kameyama M, *et al.* Hydrocephalus due to cerebrospinal fluid overproduction by bilateral choroid plexus papillomas. *Childs Nerv Syst* 2004;**20**:485–8.

76. Fuller GN, Burger PC. Central nervous system. In: Sternberg S ed. *Histology for pathologists*, 2nd edn. Philadelphia, PA: Lippincott-Raven, 1997:243–82.

77. Fuller GN, Goodman JC. Cells of the nervous system. In: Fuller GN, Goodman JC eds. *Practical review of neuropathology*. Philadelphia, PA: Lippincott Williams & Wilkins, 2001:7–73.

78. Gage FH, Olejniczak P, Armstrong DM. Astrocytes are important for sprouting in the septohippocampal circuit. *Exp Neurol* 1988;**102**:2–13.

79. Galli R, Binda E, Orfanelli U, *et al.* Isolation and characterization of tumorigenic, stem-like neural precursors from human glioblastoma. *Cancer Res* 2004;**64**:7011–21.

80. Garcia ADR, Doan NB, Imura T, *et al.* GFAP-expressing progenitors are the principal source of constitutive neurogenesis in adult mouse forebrain. *Nat Neurosci* 2004;**7**:1233–41.

81. Geddes JF, Jansen GH, Robinson SFD, *et al.* 'Gangliocytomas' of the pituitary. A heterogeneous group of lesions with differing histogenesis. *Am J Surg Pathol* 2000;**24**:607–13.

82. Gjedde A, Marrett S, Vafaee M. Oxidative and nonoxidative metabolism of excited neurons and astrocytes. *J Cereb Blood Flow Metab* 2002;**22**:1–14.

83. Gokhan S, Mehler MF. Basic and clinical neuroscience applications of embryonic stem cells. *The Anat Rec B New Anat* 2001;**265**:142–56.

84. Goldman S. Glia as neural progenitor cells. *Trends Neurosci* 2003;**26**:590–96.

85. Golomb J, Wisoff J, Miller DC, *et al.* Alzheimer's disease comorbidity in normal pressure hydrocephalus: prevalence and shunt response. *J Neurol Neurosurg Psychiatry* 2000;**68**:778–81.

86. Gotow T, Tanaka J. Phosphorylation of neurofilament H subunit as related to arrangement of neurofilaments. *J Neurosci Res* 1994;**37**:691–713.

87. Graeber MB, Streit WJ. Microglia: biology and pathology. *Acta Neuropathol* 2010;**119**:89–105.

88. Guillemot F. Cellular and molecular control of neurogenesis in the mammalian telencephalon. *Curr Opin Cell Biol* 2005;**17**:639–47.

89. Hagg T. Molecular regulation of adult CNS neurogenesis: an integrated view. *Trends Neurosci* 2005;**28**:589–95.

90. Hamel E. Perivascular nerves and the regulation of cerebrovascular tone. *J Appl Physiol* 2006;**100**:1059–64.

91. Hao L, Zou Z, Tian H, *et al.* Stem cell-based therapies for ischemic stroke. *Biomed Res Int* 2014;doi:10.1155/2014/468748.

92. Hassler O. Arterial pattern of human brain stem. Normal appearance and deformation in expanding supratentorial conditions. *Neurology* 1967;**17**:368–75.

93. Hatten ME, Liem RKH, Shelanski ML, Mason CA. Astroglia in CNS injury. *Glia* 1991;**4**:233–43.

94. Hauwel M, Furon E, Canova C, *et al.* Innate (inherent) control of brain infection, brain inflammation and brain repair: the role of microglia, astrocytes, "protective" glial stem cells and stromal ependymal cells. *Brain Res Rev* 2005;**48**:220–33.

95. Henkel JS, Engelhardt JI, Siklos L, *et al.* Presence of dendritic cells, MCP-1, and activated microglia/macrophages in amyotrophic lateral sclerosis spinal cord tissue. *Ann Neurol* 2004;**55**:221–35.

96. Hill CE, Proschel C, Noble M, *et al.* Acute transplantation of glial-restricted precursor cells into spinal cord contusion injuries: survival, differentiation, and effects on lesion environment and axonal regeneration. *Exp Neurol* 2004;**190**:289–310.

97. Hirokawa N, Glicksman MA, Willard MB. Organization of mammalian neurofilament polypeptides within the neuronal cytoskeleton. *J Cell Biol* 1984;**98**:1523–36.

98. Hoff JT, Reis DJ. Localization of regions mediating the Cushing response in the central nervous system of the cat. *Arch Neurol* 1970;**22**:228–40.

99. Holmes LB, Nash A, ZuRhein GM, *et al.* X-linked aqueductal stenosis: clinical and neuropathological findings in two families. *Pediatrics* 1973;**51**:697–704.

100. Hou S, Tom VJ, Graham L, *et al.* Partial restoration of cardiovascular function by embryonic neural stem cell grafts after complete spinal cord transection. *J Neurosci* 2013;**33**:17138–49.

101. Howell DA. Upper brain stem compression and foraminal impaction with intracranial space-occupying lesions and brain swelling. *Brain* 1959;**82**:525–50.

102. Hu CH, Reimann HA, Kurotchkin TG. Filaments in siderotic nodules of spleen in cases of splenomegaly of unknown origin. *Proc Soc Exp Biol Med* 1929;**6**:413–16.

103. Hunter KE, Hatten ME. Radial glial cell transformation to astrocytes is bidirectional: regulation by a diffusible factor in embryonic forebrain. *Proc Natl Acad Sci U S A* 1995;**92**:2061–5.

104. Iadecola C. The overlap between neurodegenerative and vascular factors in the pathogenesis of dementia. *Acta Neuropathol* 2010;**120**:287–96.

105. Iajimi AA, Hagh MF, Saki N, *et al.* Feasibility of cell therapy in multiple sclerosis: a systematic review of 83 studies. *Int J Hematol Oncol Stem Cell Res* 2013;**7**:15–33.

106. Imhof A, Charnay Y, Vallet PG, *et al.* Sustained astrocytic clusterin expression improves remodeling after brain ischemia. *Neurobiol Dis* 2006;**22**:274–83.

107. Janzer RC, Friede RL. Perisulcal infarcts: lesions caused by hypotension during increased intracranial pressure. *Ann Neurol* 1979;**6**:339–404.

108. Jefferson G. The tentorial pressure cone. *Arch Neurol Psychiatry* 1938;**40**:857–76.

109. Jennett WB, Stern WE. Tentorial herniation, the mid-brain and pupil. Experimental studies in brain compression. *J Neurosurg* 1960;**17**:598–608.

110. John GR, Lee SC, Brosnan CF. Cytokines: powerful regulators of glial cell activation. *Neuroscientist* 2003;**9**:10–22.

111. Johnson RT, Yates PO. Brain stem haemorrhages in expanding supratentorial conditions. *Acta Radiol (Stockh)* 1956;**46**:250–56.

112. Kasischke KA, Vishwasrao HD, Fisher PJ, *et al.* Neural activity triggers neuronal oxidative metabolism followed by astrocytic glycolysis. *Science* 2004;**305**:99–103.

113. Kaufmann GE, Clark K. Continuous simultaneous monitoring of intraventricular and cervical subarachnoid cerebrospinal fluid pressure to investigate the development of cerebral or tonsillar herniation. *J Neurosurg* 1970;**33**:145–50.

114. Ka-Wai K, Lau K-M, Ng H-K. Signaling pathway and molecular subgroups of medulloblastoma. *Int J Clin Exp Pathol* 2013;**6**:1211–22.

115. Kawaja MD, Gage FH. Reactive astrocytes are substrates for the growth of adult CNS axons in the presence of elevated levels of nerve growth factor. *Neuron* 1991;**7**:1019–30.

116. Kawanami T, Kato T, Llena JF, *et al.* Altered synaptophysin-immunoreactive pattern in human olivary hypertrophy. *Neurosci Lett* 1994;**176**:178–80.

117. Kawashima M, Suzuki SO, Doh-ura K, Iwaki T. alpha-Synuclein is expressed in a variety of brain tumors showing neuronal differentiation. *Acta Neuropathol* 2000;**99**:154–60.

118. Kempermann G, Neumann H. Microglia: the enemy within? *Science* 2003;**302**:1689–90.

119. Kernohan JW, Woltman HW. Incisura of the crus due to contralateral brain tumor. *Arch Neurol Psychiatry* 1929;**21**:274–87.

120. Kielian T. Toll-like receptors in central nervous system glial inflammation and homeostasis. *J Neurosci Res* 2006;**83**:711–30.

121. Kim SU, de Vellis J. Microglia in health and disease. *J Neurosci Res* 2005;**81**:302–13.

122. Kleinschmidt-DeMasters BK. Gamna–Gandy bodies in surgical neuropathology specimens: observations and a historical note. *J Neuropathol Exp Neurol* 2004;**63**:106–12.

123. Klingberg T, Vaidya CJ, Gabrieli JD, *et al.* Myelination and organization of the frontal white matter in children: a diffusion tensor MRI study. *Neuroreport* 1999;**10**:2817–21.

124. Klintworth GK. The pathogenesis of secondary brain stem hemorrhage as studied with an experimental model. *Am J Pathol* 1965;**47**:525–36.

125. Klintworth GK. Secondary brain stem haemorrhage. *J Neurol Neurosurg Psychiatry* 1966;**29**:423–5.

126. Koehler RC, Gebremedhin D, Harder DR. Role of astrocytes in cerebrovascular regulation. *J Appl Physiol* 2006;**100**:307–17.

127. Kong J, Tung VW, Aghajanian J, Xu Z. Antagonistic roles of neurofilament subunits NF-H and NF-M against NF-L in shaping dendritic arborization in spinal motor neurons. *J Cell Biol* 1998;**140**:1167–76.

128. Kramer AS, Harvey AR, Plant GW, *et al.* Systematic review of induced pluripotent stem cell technology as a potential clinical therapy for spinal cord injury. *Cell Transplant* 2013;**22**:571–617.

129. Kriegstein AR, Noctor SC. Patterns of neuronal migration in the embryonic cortex. *Trends Neurosci* 2004;**27**:392–9.

130. Kunkanjanawan T, Noisa P, Parnpai R. Modeling neurological disorders by human induced pluripotent stem cells. *J Biomed Biotechnol* 2011;doi:10.1155/2011/350131.

131. Kushner JP. Hypochromatic anemias. In: Wyngaarden JB, Smith LH Jr eds. *Cecil textbook of medicine*. Philadelphia, PA: WB Saunders, 1988:895.

132. Ladeby R, Wirenfeldt M, Dalmau I, *et al.* Proliferating resident microglia express the stem cell antigen CD34 in response to acute neural injury. *Glia* 2005;**50**:121–31.

133. Landrieu P, Ninane J, Ferriere G, Lyon G. Aqueductal stenosis in X-linked hydrocephalus: a secondary phenomenon? *Develop Med Child Neurol* 1979;**21**:637–52.

134. Lanman TH, Martin NA, Vinters HV. The pathology of encephalic arteriovenous malformations treated by prior embolotherapy. *Neuroradiology* 1988;**30**:1–10.

135. Le Beau J. *L'oedème du Cerveau*. Paris: Recht, 1938.

136. Leist M, Single B, Castoldi AF, *et al.* Intracellular ATP concentration: a switch deciding between apoptosis and necrosis. *J Exp Med* 1997;**185**:1481–6.

137. Li R, Johnson AB, Salomons G, *et al.* Glial fibrillary acidic protein mutations in infantile, juvenile, and adult forms of Alexander disease. *Ann Neurol* 2005;**57**:310–26.

138. Lichtenstein BW. Atresia and stenosis of the aqueduct of Sylvius. *J Neuropathol Exp Neurol* 1959;**18**:3–21.

139. Lipton SA, Rosenberg PA. Excitatory amino acids as a final common pathway for neurologic disorders. *N Engl J Med* 1994;**330**:613–22.

140. Liu B, Hong J-S. Role of microglia in inflammation-mediated neurodegenerative diseases: mechanisms and strategies for therapeutic intervention. *J Pharmacol Exp Ther* 2003;**304**:1–7.

141. Liu W, Prayson R. Dura matter involvement in ochronosis (alkaptonuria). *Arch Pathol Lab Med* 2001;**125**:961–3.

142. Lubetzki C, Williams A, Stankoff B. Promoting repair in multiple sclerosis: problems and prospects. *Curr Opin Neurol* 2005;**18**:237–44.

143. Macas J, Nern C, Plate KH, Momma S. Increased generation of neuronal progenitors after ischemic injury in the aged adult human forebrain. *J Neurosci* 2006;**26**:13114–21.

144. Martinez-Cerdeno V, Noctor SC, Kriegstein AR. The role of intermediate progenitor cells in the evolutionary expansion of the cerebral cortex. *Cereb Cortex* 2006;**16**:152–61.

145. Mason JL, Toews A, Hostettler JD, *et al.* Oligodendrocytes and progenitors become progressively depleted within chronically demyelinated lesions *Am J Pathol* 2004;**164**:1673–82.

146. Mattson M. Apoptotic and anti-apoptotic synaptic signaling mechanisms. *Brain Pathol* 2000;**10**:300–12.

147. Mazzanti M, Sul J-Y, Haydon PG. Glutamate on demand: astrocytes as a ready source. *Neuroscientist* 2001;**7**:396–405.

148. McArthur DL, Chute DJ, Villablanca P. Moderate and severe traumatic brain injury: epidemiologic, imaging and neuropathologic perpectives. *Brain Pathol* 2004;**14**:185–94.

149. McGeer PL, McGeer EG. Inflammation, autotoxicity and Alzheimer disease. *Neurobiol Aging* 2001;**22**:799–809.

150. McGeer PL, Kawamata T, Walker DA, *et al.* Microglia in degenerative neurological disease. *Glia* 1993;**7**:84–92.

151. McMillan JJ, Williams B. Aqueduct stenosis. Case review and discussion. *J Neurol Neurosurg Psychiatry* 1977;**40**:521–32.

152. Mehler MF, Gokhan S. Postnatal cerebral cortical multipotent progenitors: regulatory mechanisms and potential role in the development of novel neural regenerative strategies. *Brain Pathol* 1999;**9**:515–26.

153. Menet V, Prieto M, Privat A, Gimenez y Ribotta M. Axonal plasticity and functional recovery after spinal cord injury in mice deficient in both glial fibrillary acidic protein and vimentin genes. *Proc Natl Acad Sci U S A* 2003;**100**:8999–9004.

154. Meyer A. Herniation of the brain. *Arch Neurol Psychiatry* 1920;**4**:387–400.

155. Miklossy J, Kraftsik R, Pillevuit O, *et al.* Curly fiber and tangle-like inclusions in the ependyma and choroid plexus. A pathogenetic relationship with the cortical Alzheimer-type changes? *J Neuropathol Exp Neurol* 1998;**57**:1202–12.

156. Miller Fisher C. Brain herniation: a revision of classical concepts. *Can J Neurol Sci* 1995;**22**:83–91.

157. Moore MT, Stern K. Vascular lesions of the brain stem and occipital lobe occurring in association with brain tumours. *Brain* 1938;**61**:70–81.

158. Morest DK, Silver J. Precursors of neurons, neuroglia, and ependymal cells in the CNS: what are they? Where are they from? How do they get where they are going? *Glia* 2003;**43**:6–18.

159. Moser HW. Alexander disease: combined gene analysis and MRI clarify pathogenesis and extend phenotype. *Ann Neurol* 2005;**57**:307–8.

160. Mrak RE, Griffin WST. Glia and their cytokines in progression of neurodegeneration. *Neurobiol Aging* 2005;**26**:349–54.

161. Mulligan SJ, MacVicar BA. Calcium transients in astrocyte endfeet cause cerebrovascular constrictions. *Nature* 2004;**431**:195–9.

162. Nadarajah B. Radial glia and somal translocation of radial neurons in the developing cerebral cortex. *Glia* 2003;**43**:33–6.

163. Nag S ed. *The blood–brain barrier. Biology and research protocols.* Totowa, NJ: Humana Press, 2003.

164. Najjar S, Pearlman DM, Devinky O, *et al.* Neurovascular unit dysfunction with blood–brain barrier hyperpermeability contributes to major depressive disorder: a review of clinical and experimental evidence. *J Neuroinflammation* 2013;**10**:142.

165. Nakajima K, Graeber MB, Sonoda M, *et al. In vitro* proliferation of axotomized rat facial nucleus-derived activated microglia in an autocrine fashion. *J Neurosci Res* 2006;**84**:348–59.

166. Nedergaard M, Dirnagl U. Role of glial cells in cerebral ischemia. *Glia* 2005;**50**:281–6.

167. Nedergaard M, Ransom B, Goldman SA. New roles for astrocytes: redefining the functional architecture of the brain. *Trends Neurosci* 2003;**28**:523–30.

168. Nemeth J, Galain A, Mikol J, *et al.* Neuron-specific enolase and malignant

1

lymphomas. *Virchows Arch A Pathol Anat Histopathol* 1987;**412**:89–93.

169. Nicchia GP, Nico B, Camassa LMA, et al. The role of aqauporin-4 in the blood–brain barrier development and integrity: studies in animal and cell culture models. *Neuroscience* 2004;**129**:935–45.

170. Nicotera P, Leist M, Fava E, et al. Energy requirement for caspase activation and neuronal cell death. *Brain Pathol* 2000;**10**:276–82.

171. Norenberg MD. Astrocyte responses to CNS injury. *J Neuropathol Exp Neurol* 1994;**53**:213–20.

172. Norenberg MD, Rao KVR, Jayakumar AR. Mechanisms of ammonia-induced astrocyte swelling. *Metab Brain Dis* 2005;**20**:303–18.

173. Papadopoulos MC, Saadoun S, Binder DK, et al. Molecular mechanisms of brain tumor edema. *Neuroscience* 2004;**129**:1011–20.

174. Pardridge WM ed. *Introduction to the blood–brain barrier. Methodology, biology and pathology*. Cambridge, UK: Cambridge University Press, 1998.

175. Pardridge WM. Molecular biology of the blood–brain barrier. *Mol Biotechnol* 2005;**30**:57–70.

176. Pardridge WM. Molecular Trojan horses for blood–brain barrier drug delivery. *Curr Opin Pharmacol* 2006;**6**:494–500.

177. Pardridge WM. Blood–brain barrier delivery. *Drug Discov Today* 2007;**12**:54–61.

178. Patt S, Labrakakis C, Bernstein M, et al. Neuron-like physiological properties of cells from human oligodendroglial tumors. *Neuroscience* 1996;**71**:601–11.

179. Paulsen AH, Lundar T, Lindegaard K-L. Twenty-year outcome in young adults with childhood hydrocephalus: assessment of surgical outcome, work participation, and health-related quality of life. *J Neurosurg Pediatr* 2010;**6**:527–35.

180. Paulson HL. Toward an understanding of polyglutamine neurodegeneration. *Brain Pathol* 2000;**10**:293–9.

181. Pellerin L, Magistretti PJ. Neuroenergetics: calling upon astrocytes to satisfy hungry neurons. *Neuroscientist* 2004;**10**:53–62.

182. Pencalet P, Serguera C, Corti O, et al. Integration of genetically modified adult astrocytes into the lesioned rat spinal cord. *J Neurosci Res* 2006;**83**:61–7.

183. Pfrieger FW, Barres BA. Synaptic efficacy enhanced by glial cells in vitro. *Science* 1997;**277**:1684–7.

184. Popovich PG, Guan Z, McGaughy V, et al. The neuropathological and behavioral consequences of intraspinal microglial/macrophage activation. *J Neuropathol Exp Neurol* 2002;**61**:623–33.

185. Power C, Boisse L, Rourke S, Gill MJ. NeuroAIDS: an evolving epidemic. *Can J Neurol Sci* 2009;**36**:285–95.

186. Quan D, Kleinschmidt-DeMasters BK. A 71-year-old male with 4 decades of symptoms referable to both central and peripheral nervous system. *Brain Pathol* 2005;**15**:369–70.

187. Rakic P. Elusive radial glial cells: historical and evolutionary perspective. *Glia* 2003;**43**:19–32.

188. Reichard RR, White CL III, Hladik CL, Dolinak D. Beta-amyloid precursor protein staining in nonhomicidal pediatric medicolegal autopsies. *J Neuropathol Exp Neurol* 2003;**62**:237–47.

189. Reid WL, Cone WV. The mechanism of fixed dilation of the pupil resulting from ipsilateral cerebral compression. *JAMA* 1939;**112**:2030–34.

190. Reynolds BA, Weiss S. Generation of neurons and astrocytes from isolated cells of the adult mammalian central nervous system. *Science* 1992;**255**:1707–10.

191. Ribalta T, McCutcheon I, Neto A, et al. Textiloma (gossypiboma) mimicking recurrent intracranial tumor. *Arch Pathol Lab Med* 2004;**128**:749–58.

192. Richter-Landsberg C, Bauer NG. Tau-inclusion body formation in oligodendroglial: the role of stress proteins and proteasome inhibition. *Int J Dev Neurosci* 2004;**22**:443–51.

193. Ridet JL, Malhotra SK, Privat A, Gage FH. Reactive astrocytes: cellular and molecular cues to biological function. *Trends Neurosci* 1997;**20**:570–77.

194. Robertson G, Crocker S, Nicholson D, Schulz J. Neuroprotection by the inhibition of apoptosis. *Brain Pathol* 2000;**10**:283–92.

195. Ropper AH. Lateral displacement of the brain and level of consciousness in patients with an acute hemispheral mass. *N Engl J Med* 1986;**314**:953–8.

196. Ropper AH. Syndrome of transtentorial herniation: is vertical displacement necessary? *J Neurol Neurosurg Psychiatry* 1993;**56**:932–5.

197. Rothfus WE, Goldberg AL, Tabas JH, Deeb ZL. Callosomarginal infarction secondary to transfalial herniation. *Am J Neuroradiol* 1987;**8**:1073–6.

198. Rothstein JD, Dykes-Hoberg M, Pardo CA, et al. Knockout of glutamate transporters reveals a major role for astroglial transport in excitotoxicity and clearance of glutamate. *Neuron* 1996;**16**:675–86.

199. Rothstein RP, Levison SW. Gray matter oligodendrocyte progenitors and neurons die caspase-3 mediated deaths subsequent to mild perinatal hypoxic-ischemic insults. *Dev Neurosci* 2005;**27**:149–59.

200. Russell DS. *Observations on the pathology of hydrocephalus*. London, UK: HM Stationery Office, 1949.

201. Sanai N, Alvarez-Buylla A, Berger M. Mechanisms of disease: neural stem cells and the origin of gliomas. *N Engl J Med* 2005;**353**:811–22.

202. Sapp E, Kegel KB, Aronin N, et al. Early and progressive accumulation of reactive microglia in the Huntington disease brain. *J Neuropathol Exp Neurol* 2001;**60**:161–72.

203. Sarnat HB. Regional differentiation of the human fetal ependyma: immunocytochemical markers. *J Neuropathol Exp Neurol* 1992;**51**:58–75.

204. Sarnat HB. Ependymal reactions to injury. A review. *J Neuropathol Exp Neurol* 1995;**54**:1–15.

205. Sastre M, Klockgether T, Heneka MT. Contribution of inflammatory processes to Alzheimer's disease: molecular mechanisms. *Int J Dev Neurosci* 2006;**24**:167–76.

206. Schallert T, Leasure JL, Kolb B. Experience-associated structural events, subependymal cellular proliferative activity, and functional recovery after injury to the central nervous system. *J Cereb Blood Flow Metab* 2000;**20**:1513–28.

207. Scheffler B, Brüstle O. Symposium: stem cells in the central nervous system. *Brain Pathol* 2006;**16**:131.

208. Scheffler B, Edenhofer F, Brüstle O. Merging fields: stem cells in neurogenesis, transplantation, and disease modeling. *Brain Pathol* 2006;**16**:155–68.

209. Scheinker IM. Transtentorial herniation of the brain stem; a characteristic clinicopathologic syndrome: pathogenesis of hemorrhages in the brain stem. *Arch Neurol Psychiatry* 1945;**53**:289–98.

210. Schipke CG, Kettenmann H. Astrocyte responses to neuronal activity. *Glia* 2004;**47**:226–32.

211. Schmechel DE, Rakic P. A Golgi study of radial glial cells in developing monkey telencephalon: morphogenesis and transformation into astrocytes. *Anat Embryol* 1979;**156**:115–52.

212. Schmitt AB, Brook GA, Buss A, et al. Dynamics of microglial activation in the spinal cord after cerebral infarction are revealed by expression of MHC class II antigen. *Neuropathol Appl Neurobiol* 1998;**24**:167–76.

213. Schoch HJ, Fischer S, Marti HH. Hypoxia-induced vascular endothelial growth factor expression causes vascular leakage in the brain. *Brain* 2002;**125**:2549–57.

214. Shonka N, Brandes A, De Groot JR. Adult medulloblastoma, from spongioblastoma cerebelli to the present day: a review of treatment and the integration of molecular markers. *Oncology* 2012;**26**(11):1083–91.

215. Schulz J, Nicotera P. Targeted modulation of neuronal apoptosis: a double-edged sword? *Brain Pathol* 2000;**10**:273–5.

216. Schwartz LM, Smith SW, Jones MEE, Osborne BA. Do all programmed cell deaths occur via apoptosis? *Proc Natl Acad Sci U S A* 1993;**90**:980–84.

217. Schwartz M, Butovsky O, Bruck W, Hanisch U-K. Microglial phenotype: is the commitment reversible? *Trends Neurosci* 2006;**29**:68–74.

218. Seri B, Garcia-Verdugo J, McEwen BS, Alvarez-Buylla A. Astrocytes give rise to new neurons in the adult mammalian hippocampus. *J Neurosci* 2001;**21**:7153–60.

219. Shirasaki R, Pfaff S. Transcriptional codes and the control of neuronal identify. *Annu Rev Neurosci* 2002;**25**:251–81.

220. Silver J, Miller JH. Regeneration beyond the glial scar. *Nat Rev Neurosci* 2004;**5**:146–56.

221. Simard M, Nedergaard M. The neurobiology of glia in the context of water and ion homeostasis. *Neuroscience* 2004;**129**:877–96.

222. Simard M, Arcuino G, Takano T, et al. Signaling in the gliovascular interface. *J Neurosci* 2003;**23**:9254–62.

223. Singh SK, Hawkins C, Clarke ID, et al. Identification of human brain tumour initiating cells. *Nature* 2004;**432**:396–401.

224. Snell R. The neurobiology of the neuron and the neuroglia. In: Snell R ed. *Clinical neuroanatomy*, 6th edn. Philadelphia, PA: Lippincott Williams & Wilkins, 2005:31–67.

225. Sofroniew MV. Reactive astrocytes in neural repair and protection. *Neuroscientist* 2005;**11**:400–407.

226. Sofroniew MV, Vinters HV. Astrocytes: biology and pathology. *Acta Neuropathol* 2010;**119**:7–35.

227. Spranger M. Neurologic complications of metabolic diseases. In: Duckett S ed. *Pediatric neuropathology*. Baltimore, MD: Williams & Wilkins, 1995:756–66.

228. Steindler D. Redefining cellular phenotypy based on embryonic, adult, and cancer stem cell biology. *Brain Pathol* 2006;**16**:169–80.

229. Stewart PA, Magliocco M, Hayakawa K, et al. A quantitative analysis of blood–

brain barrier ultrastructure in the aging human. *Microvasc Res* 1987;**33**:270–82.

230. Stewart PA, Hayakawa K, Akers M-A, Vinters HV. A morphometric study of the blood–brain barrier in Alzheimer disease. *Lab Invest* 1992;**67**:734–42.

231. Su X, Gopalakrishnan V, Stearns D, *et al.* Abnormal expression of REST/NRSF and Myc in neural stem/progenitor cells causes cerebellar tumors by blocking neuronal differentiation. *Mol Cell Biol* 2006;**26**:1666–78.

232. Sunderland S. The tentorial notch and complications produced by herniation through that aperture. *Br J Surg* 1958;**45**:422–38.

233. Svendsen CN, Caldwell MA, Ostenfeld T. Human neural stem cells: isolation, expansion and transplantation. *Brain Pathol* 1999;**9**:499–513.

234. Takamine K, Okamoto K, Fujita Y, *et al.* The involvement of the neuronal Golgi apparatus and trans-Golgi network in the human olivary hypertrophy. *J Neurol Sci* 2000;**182**:45–50.

235. Takano T, Tian G-F, Peng W, *et al.* Astrocyte-mediated control of cerebral blood flow. *Nat Neurosci* 2006;**9**:260–67.

236. Teasdale E, Cardos E, Galbraith S, Teasdale G. CT scan in severe diffuse head injury: physiological and clinical correlations. *J Neurol Neurosurg Psychiatry* 1984;**47**:600–603.

237. Thompson RK, Malina S. Dynamic axial brain stem distortion as a mechanism explaining the cardiorespiratory changes in increased intracranial pressure. *J Neurosurg* 1959;**16**:664–75.

238. Tian G-F, Azmi H, Takano T, *et al.* An astrocytic basis of epilepsy. *Nat Med* 2005;**11**:973–81.

239. Toborek M, Lee YW, Pu H, *et al.* HIV-Tat protein induces oxidative and inflammatory pathways in brain endothelium. *J Neurochem* 2003;**84**:169–79.

240. Tom VJ, Doller CM, Malouf AT, Silver J. Astrocyte-associated fibronectin is critical for axonal regeneration in adult white matter. *J Neurosci* 2004;**24**:9282–90.

241. Tramontin AD, Garcia-Verdugo JM, Lim DA, Alvarez-Buylla A. Postnatal development of radial glia and the ventricular zone (VZ): a continuum of the neural stem cell compartment. *Cereb Cortex* 2003;**13**:580–87.

242. Turnbull IM, Drake CG. Membranous occlusion of the aqueduct of Sylvius. *J Neurosurg* 1966;**24**:24–33.

243. Ullian EM, Sapperstein SK, Christopherson KS, Barres BA. Control of synapse number by glia. *Science* 2001;**291**:657–61.

244. Uzan M, Yentur E, Hanci M, *et al.* Is it possible to recover from uncal herniation? Analysis of 71 head injured cases. *J Neurosurg Sci* 1998;**42**:89–94.

245. van der Knaap MS, Salomons GS, Li R, *et al.* Unusual variants of Alexander's disease. *Ann Neurol* 2005;**57**:327–38.

246. Van Haren K, van der Voorn JP, Peterson DR, *et al.* The life and death of oligodendrocytes in vanishing white matter disease. *J Neuropathol Exp Neurol* 2004;**63**:618–30.

247. van Rossum D, Hanisch UK. Microglia. *Metab Brain Dis* 2004;**19**:393–411.

248. Vescovi AL, Synder EY. Establishment and properties of neural stem cell clones: plasticity *in vitro* and *in vivo*. *Brain Pathol* 1999;**9**:569–78.

249. Vick NA, Lin MJ, Bigner DD. The role of the subependymal plate in glial tumorigenesis. *Acta Neuropathol* 1977;**40**:63–71.

250. Vincent C, David M, Thiebault F. Le cÔne de pression temporal dans les tumeurs des hémisphères cérébraux. Sa symptomatologie: sa gravité: les traitments qu'il convient de lui opposer. *Rev Neurol* 1936;**65**:536–45.

251. Vinters HV, Anders KH. *Neuropathology of AIDS.* Boca Raton, FL: CRC Press 1990.

252. Vinters HV, Reave S, Costello P, Girvin JP, Moore SA. Isolation and culture of cells derived from human cerebral microvessels. *Cell Tissue Res* 1987;**249**:657–67.

253. Vinters HV, Kwok MK, Ho HW, *et al.* Cytomegalovirus in the nervous system of patients with the acquired immune deficiency syndrome. *Brain* 1989;**112**:245–68.

254. Vinters HV, Fisher RS, Cornford ME, *et al.* Morphological substrates of infantile spasms: studies based on surgically resected cerebral tissue. *Childs Nerv Syst* 1992;**8**:8–17.

255. Vinters HV, Farrell MA, Mischel PS, Anders KH. *Diagnostic neuropathology.* New York: Marcel Dekker, 1998.

256. Vinters HV, Klement IA, Sung SH, Farag ES. Pathologic issues and new methodologies in the evaluation of non-Alzheimer dementias. *Clin Neurosci Res* 2004;**3**:413–26.

257. Voigt T. Development of glial cells in the cerebral wall of ferrets: direct tracing of their transformation from radial glia into astrocytes. *J Comp Neurol* 1989;**289**:74–88.

258. Walker DG, Lue L-F. Investigations with cultured human microglia on pathogenic mechanisms of Alzheimer's disease and other neurodegenerative diseases. *J Neurosci Res* 2005;**81**:412–25.

259. Walsh EK, Schettini A. Elastic behaviour of brain tissue *in vivo*. *Am J Physiol* 1976;**230**:1058–62.

260. Watters JJ, Schartner JM, Badie B. Microglia function in brain tumors. *J Neurosci Res* 2005;**81**:447–55.

261. Weinstein JD, Langfitt TW, Bruno L, *et al.* Experimental study of patterns of brain distortion and ischaemia produced by an intracranial mass. *J Neurosurg* 1968;**28**:513–21.

262. Whittle IR, Viswanathan R. Acute intraoperative brain herniation during elective neurosurgery: pathophysiology and management considerations. *J Neurol Neurosurg Psychiatry* 1996;**61**:584–90.

263. Wijdicks EF, Miller GM. Transient locked-in syndrome after uncal herniation. *Neurology* 1999;**12**:1296–7.

264. Williams B. Is aqueduct stenosis a result of hydrocephalus? *Brain* 1973;**96**:399–412.

265. Williams MA, McAllister JP, Walker ML, *et al.* Priorities for hydrocephalus research: report from a National Institutes of Health-sponsored workshop. *J Neurosurg* 2007;**107**(5 Suppl):345–57.

266. Wirenfeldt M, Dalmau I, Finsen B. Estimation of absolute microglial cell numbers in mouse fascia dentata using unbiased and efficient stereological cell counting principles. *Glia* 2003;**44**:129–39.

267. Wirenfeldt M, Babcock AA, Ladeby R, *et al.* Reactive microgliosis engages distinct responses by microglial subpopulations after minor central nervous system injury. *J Neurosci Res* 2005;**82**:507–14.

268. Wirenfeldt M, Babcock AA Vinters HV. Microglia - insights into immune system structure, function, and reactivity in the central nervous system. *Histol Histopathol* 2011;**26**:519–30.

269. Wolbur H, Paulus W. Choroid plexus: biology and pathology. *Acta Neuropathol* 2010;**119**:75–88.

270. Wolf HK, Buslei R, Blumcke I, *et al.* Neural antigens in oligodendrogliomas and dysembryoplastic neuroepithelial tumors. *Acta Neuropathol* 1997;**94**:436–43.

271. Ye Z-C, Wyeth MS, Baltan-Tekkok S, Ransom BR. Functional hemichannels in astrocytes: a novel mechanism of glutamate release. *J Neurosci* 2003;**23**:3588–96.

272. Yi J-H, Hazell AS. Excitotoxic mechanisms and the role of astrocytic glutamate transporters in traumatic brain injury. *Neurochem Int* 2006;**48**:394–403.

273. Zhang S. Neural subtype specification from embryonic stem cells. *Brain Pathol* 2006;**16**:132–42.

274. Zlokovic BV. Neurovascular mechanisms of Alzheimer's neurodegeneration. *Trends Neurosci* 2005;**28**:202–8.

Vascular Disease, Hypoxia and Related Conditions

Raj Kalaria, Isidro Ferrer and Seth Love

INTRODUCTION

The efficient functioning of the central nervous system (CNS) relies on an uninterrupted supply of oxygenated blood and nutrients, particularly glucose. The transportation of these fuels requires sufficient blood flow through a healthy cerebral vasculature with the capacity to respond appropriately to metabolic demands. If the oxygen or glucose content or the flow of blood falls below the level needed to maintain nervous tissue viability, this precipitates a series of acute and longer term changes within the brain parenchyma. The removal of metabolic wastes such as lactate by venous drainage also plays an important role in cerebral function.

This chapter describes the regulatory mechanisms that protect brain perfusion and oxygenation, the diseases that affect cerebral blood vessels and the damage to brain tissue that results from disturbances in brain oxygenation and cerebral blood flow (CBF).

TERMINOLOGY

Several purely clinical and radiological terms have come into general use even in pathology to describe specific syndromes and lesions. Table 2.1 provides the currently used definitions of key terms and concepts. Hypoxia describes a reduction in oxygen supply or content but is also sometimes applied to conditions in which the metabolic utilization of oxygen is impaired (histotoxic hypoxia). Ischaemia means a lack of blood supply but is

usually used to denote a reduction in blood supply below the level needed to maintain tissue function. The term hypoxia is often combined with ischaemia (hypoxia–ischaemia) although, as indicated later, the general or casual use of this expression is not recommended. Tissue hypoxia is a consequence of ischaemia; however, hypoxia is itself anti-ischaemic, as it stimulates an increase in CBF.[3,149] In most forms of hypoxia, CBF is increased, so that supply of nutrients and removal of potentially damaging end products continue unabated. During ischaemia, in addition to the decreased blood supply, the removal of damaging metabolites is also impaired. Brain ischaemia may accompany systemic hypoxia in specific circumstances, such as strangulation and cardiorespiratory arrest, in which cases the term describing the particular clinical event should be used unless the CBF or partial pressure of oxygen (pO_2) is known. The pO_2 is the pressure that the oxygen would exert in a liquid or gas if it alone occupied the total volume, regardless of other molecules that may be present and irrespective of the total pressure.

Hypoxia and ischaemia are different pathophysiological states. Ischaemia is always pathological but not so hypoxia. Hypoxia is graded as one ascends in altitude; it blends smoothly with physiology but not directly with pathology. Ischaemia interfaces more directly with tissue pathology. In itself, low oxygen tension in the blood is incapable of causing cerebral necrosis, but ischaemia of even 2 minutes[957] can result in necrosis within selectively vulnerable brain regions of the brain, e.g. the hippocampus.

The term hypoxia is often qualified to indicate whether it refers to the means of delivery or utilization of oxygen.

TABLE 2.1 Terminology: definitions of key terms and concepts

Term	Description	Type of damage in brain tissue	Comment
Anoxia	No oxygen	Not a specific or useful term by itself	Carries no specific meaning in the intact organism
Anoxaemia	No oxygen in blood	Impossible to assess in intact animal	An impossibility without cardiac bypass and removal of all blood O_2
Anaemic hypoxia	Low blood haemoglobin	No brain-damaging potential	Actually protective for stroke because of favourable rheology
Asphyxia	Inability to breathe	Can cause brain necrosis if ischaemia results	Includes suffocation, strangulation and some chemicals (cyanide, sulphide, azide) which paralyze breathing centres in medulla oblongata
Carbon monoxide (CO) toxicity	CO in blood, displacing O_2 from haemoglobin sites	Necrosis in pallido-reticularis, plus typical ischaemic distribution	Complex triad effected: anaemia (haemoglobin occupation by CO), histotoxic hypoxia (by binding to iron-rich globus pallidus), and global ischaemia due to heart failure
Haematoma	Localized bleeding (e.g. intracerebral, sub-arachnoid or sub-dural) from ruptured vessels or aneurysms	Haemorrhagic strokes result in tissue injury by causing compression of tissue from expanding bleeds	Not to be confused with hemiangioma
Hypoxia	Low oxygen, not further specified (tissue, blood, atmosphere)	See specific entities	Not a useful term without further qualification
Hypoxia/ischaemia	Combination of hypoxia and ischaemia	Hypoxia and ischaemia cause even greater necrosis	Occurs in strangulation and hanging; widely used incorrectly to describe pure ischaemia; cardiac arrest encephalopathy and global ischaemia are better terms, if that is what is meant
Hypoxaemia	Low oxygen in blood	Reversible synaptic alterations without neuronal necrosis	Seen in respiratory tract disease (larynx, trachea, bronchi, bronchioles), not in pure cardiovascular disease; tends to occur in younger patients; causes tissue hypoxia that is not necrotizing
Hypobaric hypoxia	Hypoxaemia accompanying decrease in ambient pO_2	Reversible synaptic alterations (at very high altitudes), but without neuronal necrosis	Temporary synaptic alterations produce 'high-altitude stupid' (HAS) syndrome; capillary leakage produces high altitude cerebral oedema (HACE), which is potentially lethal; both reverse on descent or on increasing inspired O_2
Histotoxic hypoxia	Tissue utilization of oxygen impaired	No brain-damaging potential without accompanying hypotension	Examples: poisoning by cyanide, sulphide and azide
Ischaemia	Cessation of blood flow to tissue; no perfusion	Variable cellular damage, neurons most vulnerable	Often also used (albeit imprecisely) to describe reduced blood flow — oligaemia
Oligaemia	Low blood flow, hypoperfusion	Selective vulnerability	Close to normal but still insufficient
Tissue hypoxia (global ischaemia)	Low tissue pO_2 due to global ischaemia	Necrosis (both pan-necrosis and selective neuronal necrosis) in brain regions of selective vulnerability	Decreased tissue pO_2 due to imbalance between delivery and utilization everywhere in brain
Tissue hypoxia (focal ischaemia)	Low tissue pO_2 due to focal ischaemia	Necrosis is usually pan-necrosis and does not spare glia	Decreased tissue pO_2 due to imbalance between delivery and utilization in focal arterial distribution
Watershed infarction	Localized to the border zones between territories of two major arteries (e.g. anterior cerebral artery [ACA] and middle cerebral artery [MCA] or MCA and posterior cerebral artery [PCA])	Ischaemic injury	Analogous to a lawn watered by multiple sprinklers: occlusion of the hose leads to a dry lawn in the territory centred on a sprinkler (or artery), but low pressure (hypotension) leads to a dry lawn between the sprinklers.

Hypoxic, anaemic and histotoxic hypoxia describe states in which, respectively, oxygen supply, blood oxygen transport and tissue utilization of oxygen are impaired (Table 2.1). Hypoxaemia is low blood oxygen content. Anoxaemia is zero blood oxygen (a physiological impossibility). Anoxia is a term often used, although there cannot be total absence of oxygen in the body. Ambient oxygen, however, can be zero, as for example on inhalation of pure nitrogen,[449] in drowning[939] and in an unscheduled space walk.

In clinical brain ischaemia, systemic hypoxaemia is usually absent. Conversely, hypoxic states are usually not accompanied by ischaemia. Pure hypoxaemia of the brain can result in a prolonged coma of 2 weeks, from which a complete and remarkable recovery is possible,[360,890] whereas prolonged coma after cardiac arrest or global ischaemia carries a very poor prognosis. Because hypoxia tends to occur in younger patients, recognition of a pure hypoxic insult, without accompanying ischaemia, is important in determining clinical prognosis. Hypoxia, thus, needs to be distinguished from ischaemia, while taking note that at tissue level, ischaemia always causes low tissue oxygenation (tissue hypoxia).

In ischaemic injury, the perfusion range between the threshold (in millilitres of blood per 100 grams of brain per minute, or mL/g/min) below which there is impairment of electrophysiological responses or function and the threshold below which irreversible damage occurs (typically around 15–18mL/100g/min) is termed the penumbra. The restitution of flow above the functional threshold can reverse the deficits without permanent damage. However, attempts to define precise ischaemic thresholds below which damage consistently takes place encounter difficulty because this depends on interacting factors including age, temperature, blood glucose concentration, and duration of ischaemia. Magnetic resonance (MR) imaging and MR spectroscopy are useful for assessing the effects of hypoxia and ischaemia on brain chemistry, structure and function.

The term stroke describes an acute disturbance or loss of brain function resulting from brain ischaemia or haemorrhage. The types of stroke and their pathological manifestations are described in detail later in this chapter.

DEVELOPMENT OF THE CNS VASCULAR SYSTEM

Normal and Abnormal Vasculogenesis and Angiogenesis

The formation of the brain vascular system is a tightly regulated developmental process. It involves an intricate interplay between the mesodermally derived vascular cells and the neuroectodermally derived CNS (Figure 2.1). Vasculogenesis is the differentiation of mesodermal precursors into endothelial cells whereas angiogenesis is the formation of new vessels from preexisting vessels or plexuses.

Embryonic blood vessels consist of endothelial cells and pericytes that organize and expand into highly branched conduits. This process is controlled by signalling systems involving a large number of specific receptors and their ligands, in addition to mediators of mitogenic, chemotactic, proteolytic and adhesive activities.[98,641,778] The regulation of embryonic or pathological angiogenesis when hypoxia- or tumour-induced is predominantly controlled by the same molecules (Table 2.1). There has been much therapeutic interest in tumour-induced angiogenesis, but the understanding of pro- and anti-angiogenic mechanisms has implications for cerebral ischaemia, vascular malformations, neurodegenerative disorders, CNS trauma, multiple sclerosis and diabetic retinopathy.[98,804,853] The development

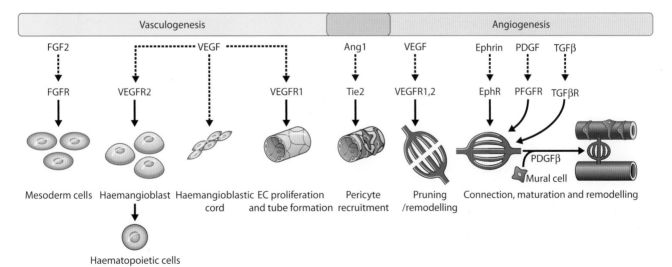

2.1 Processes of vasculogenesis and angiogenesis in the brain. Growth factors regulate differentiation of mesodermal cells into haemoangioblasts, which give rise to endothelial cells that proliferate to form cords and capillary tubes. Pericytes are recruited as support cells, with concomitant basal lamina production. Multiple growth factors activate specific receptors to model and prune branching vessels. Ang1, angiopoietin; FGF, fibroblast growth factor; PDGF, platelet-derived growth factor; PDGFR, PDGF receptor; TGFβ, transforming-growth factor β; TGFβR, TGFβ receptor; VEGF, vascular endothelial growth factor; VEGFR, VEGF receptor.

Adapted from Augustin et al.[76] and Patel–Hett and D'Amore.[778] Diagram courtesy of Y Yamamoto, Yamaguchi University Graduate School of Medicine, Japan.

of the vascular system (Box 2.1) is initiated around the third week of gestation, when splanchnopleuric mesodermal precursor cells at the periphery of blood islands in paraxial mesoderm differentiate into haemangioblasts. This occurs under the influence of mesoderm-inducing factors of the fibroblast growth factor family, which interact with vascular endothelial growth factor receptor 2 (VEGFR-2) on mesodermal precursor cells. The haemangioblasts differentiate into vessel-forming angioblasts and haematopoietic stem cells. Angioblasts cluster and acquire lumina, to form interconnecting tubes that constitute the primitive vascular plexus. Angioblasts from the splanchnopleuric region migrate into the head region to form a perineural vascular plexus around the developing brain (extracerebral vascularization). Later, this plexus develops into meningeal arteries and veins.

After development of the primitive perineural vascular plexus, brain blood vessels are formed (intracerebral vascularization) by capillary sprouts from the pre-existing vessels in this plexus. New blood vessels in adult organs are formed by similar angiogenesis. The capillary sprouts penetrate the developing CNS and extend into the subventricular zone, forming a new plexus there. Next, another capillary plexus is formed in the intermediate zone between the subventricular precursor cell zone and the cortical plate. Finally, the cortical plate is vascularized from the deep layers outwards.

Hypoxia is considered to be the major regulator of angiogenesis (Box 2.1). In the presence of oxygen, the α-subunit of hypoxia-inducible (transcription) factor 1 (HIF-1α) is hydroxylated by HIF-prolylhydroxylases, binds to a multimolecular complex that includes von Hippel–Lindau (VHL) tumour suppressor protein, and is rapidly degraded in proteasomes. Hypoxia inhibits the hydroxylation of HIF-1α and, as a result, constitutively expressed HIF-1β can bind to HIF-1α. Heterodimeric HIF-1 thus accumulates and is translocated to the nucleus where it binds to hypoxia-responsive elements to activate transcription of numerous hypoxia-inducible genes (estimated at 2–3 per cent of all genes), including genes of the pivotal vascular endothelial growth factor (VEGF) family (Figure 2.2).[174]

Perineural angioblasts differentiate into endothelial cells, which are attracted by different chemotactic factors. A most important chemotactic stimulus is VEGF, produced by neural progenitor cells in the periventricular matrix zone, towards which the VEGFR-2-expressing cells migrate.[188] VEGFR-2 is essential for the proliferation of endothelial cells, their migration and survival.[935] Endothelial and blood cells are not formed in mice lacking VEGFR-2. VEGFR-1 is expressed later than VEGFR-2; the former acts as a negative regulator of VEGF and seems necessary for the assembly of angioblasts into functional blood vessels.

The primary endothelial structures need to be stabilized by the formation of pericytes and smooth muscle cells, in which process angiopoietin-1 (Ang-1) ligand binding to TIE-2 receptors on endothelial cells is essential.[76,946] This stabilization through Ang-1/TIE-2 signalling also regulates

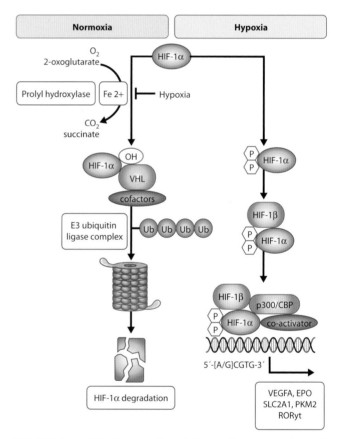

2.2 HIF-1 stabilization and activity under normoxic and hypoxic conditions. Under normoxia, hydroxylation at specific proline residues leads to binding of HIF-1α to VHL followed by HIF-1α destruction via the ubiquitin/proteasome pathway. During hypoxia, HIF-1α subunit is stabilized and dimerizes with the ubiquitously expressed HIF-1β subunit. Nuclear translocation is initiated, followed by binding of the HIF-1 heterodimer to hypoxia response elements (HREs) of enhancers and promoters of specific target genes. OH, hydroxyl group; VHL, von Hippel–Lindau tumour suppressor protein; P, phosphorylated subunit.

Diagram adapted from Trollman and Gassmann[1021] and redrawn courtesy of Y Yamamoto, Yamaguchi University Graduate School of Medicine, Japan.

BOX 2.1. Key events in the development of the cerebral vasculature

- Mesodermal precursor cells in paraxial mesoderm differentiate into haemangioblasts, induced by FGFs signalling via VEGFR-2 on precursor cells.
- Angioblasts form the perineural vascular plexus around the developing brain (vasculogenesis, extracerebral vascularization: leptomeningeal vascularization).
- Capillary sprouts emerge from the primitive plexus (angiogenesis) and penetrate into the brain, beginning from deeper layers upwards (intracerebral vascularization).
- Angiogenesis is downregulated postnatally and follows the metabolic needs of the growing CNS. In adult humans, fewer than 1 per cent of the endothelial cells proliferate.
- Angiogenesis may be re-upregulated, for example in ischaemia (the most important cause of the reactivation), upon metabolic demand and in neoplasia.
- Bone marrow derived endothelial progenitor cells (EPCs) also promote postnatal vasculogenesis and neovascularisation in ischaemic tissues. Their contribution is confirmed by the presence of various cytokines and other secreting proangiogenic factors in EPCs, such as VEGF and Ang-1.

the permeability of developing vessels. Sprouting and induction of further angiogenesis in mature vessels require destabilization of endothelium and pericyte contacts by angiopoetin-2 (Ang-2), an antagonist of Ang-1. Another set of ligands and receptors, ephrins, play a decisive role in determining the arterial or venous identity of the vessels (Figure 2.1; Table 2.2). Other signalling pathways with precise functions in endothelial-mural cell interactions during vascular development and maturation involve platelet-derived growth factor-B (PDGFB) and transforming growth factor-β (TGF-β) and their respective receptors.[323]

During sprouting, the extracellular matrix is degraded by proteolysis, enabling migration of the proliferating endothelial cells. The blood vessels penetrating the neuroectoderm form intracerebral branches of various sizes. This, together with the regression of supernumerary vessels, creates the vascular tree. The blood vessels mature,

through recruitment of pericytes and, in larger blood vessels, also smooth-muscle cells and fibroblasts and the formation of contacts with astrocytic processes to form a tight blood–brain barrier (BBB) (see Chapter 1).

The active phase of angiogenesis ceases soon after birth, after which the cerebral vasculature is expanded only to meet the needs of the growing brain, mainly by elongation of the pre-existing blood vessels. Experimental evidence suggests complete new 'loops' are formed from existing vascular networks.[406] In the adult brain, angiogenesis is normally minimal (fewer than 1 per cent of endothelial cells incorporate thymidine) but may be reactivated in pathological conditions,[395] e.g. hypoxia, ischaemia, trauma and neoplasia. The newly activated angiogenesis is mainly regulated by the same signalling molecules as during development (Figure 2.1; Table 2.2). Because VEGF and its receptors are hypoxia-inducible, their expression is upregulated in glial

TABLE 2.2 Key mediators in neovascularisation including vasculogenesis and angiogenesis

Parent molecule	Isoforms/ Receptors	Gene(s)	Function
Hypoxia-inducible (transcription) factor 1 (HIF-1)	HIF-1α; HIF-1β	*HIF1A; HIF1B*	In the presence of oxygen HIF-1 is degraded, but it persists under conditions of hypoxia. During hypoxic episodes, HIF-1 is translocated into the nucleus and binds to hypoxia-responsive elements (HREs) of several genes, including vascular endothelial growth factor (VEGF) family, to activate their transcription (Figure 2.2) and regulate angiogenesis.
Fibroblast growth factor 2 (FGF2)	Basic FGF	*FGF2*	FGF family members bind heparin and possess broad mitogenic/angiogenic activities. FGFs activate receptors (FGFRs) on endothelial cells or indirectly stimulate angiogenesis by inducing the release of angiogenic factors from other cell types.
Vascular endothelial growth factor A (VEGFA)	VEGFR-1 (FLT1); VEGFR-2 (KDR, Flk-1)	*VEGFA FLT1*	VEGFA is expressed in neuroectodermal cells and is the ligand for VEGFR-2 on endothelial cells. VEGF is an important chemotactic stimulus for endothelial cells (ECs). VEGFR-2 is essential for the proliferation, migration and survival of ECs.
Angiopoietin-1 (ANG1)	TIE-1; TIE-2	*ANGPT1 TIE1 TEK*	The ANG and TIE family binary switch mechanism allows vessels to maintain quiescence, while remaining able to respond to angiogenic stimuli. ANG-1 is the ligand binding to TIE-2 receptors on perivascular cells to induce formation of pericytes and smooth muscle cells around the vessels.
Ephrin	EPH receptor	*EPHRIN EPHB4*	EPH receptors and their ligands, the ephrins, regulate cell-contact-dependent patterning and can generate bidirectional signals. Ephrin-B2 and its receptor EPHB4 regulate vessel morphogenesis by several mechanisms.
Delta-like ligand (DLL4)	NOTCH	*DLL4 NOTCH*	NOTCH signalling is involved in vessel-branching; tip cells migrate and stalk cells proliferate. EGFR-2 upregulates DLL4 expression in tip cells and in neighbouring stalk cells, DLL4 activates NOTCH, which modulates VEGFR-2 and VEGFR-1.
Platelet-derived growth factor (PDGF)	PDGFR-β	*PDGFB PDGFRA PDGFRB*	The ligand/receptor pair PDGFB/PDGFR-β are involved in pericyte recruitment.
Transforming growth factor β (TGFβ)	TGFβR1, TGFβR2	*TGFB1 TGFB2 TGFBR1 TGFBR2*	TGFβ regulates proliferation, differentiation and survival of many cells. TGFβR binding activates multiple intracellular pathways resulting in phosphorylation of receptor-regulated SMAD (small 'mothers against' decapentaplegic) proteins. Both TGFβ1 and 2 stimulate expression of VEGF, plasminogen activator inhibitor and certain metalloproteinases involved in vascular remodelling, angiogenesis and degradation of the extracellular matrix.
Chemokine (C-X-C motif) ligand 12 (CXCL12)	Chemokine (C-X-C motif) receptor 4 (CXCR4)	*CXCL12 CXCR4*	CXC ligand 12 is a haemostatic chemokine, whose major function is to regulate haemopoietic-cell trafficking.

cells or endothelial cells at the periphery of infarcts.[98] VEGF expression in glioblastomas, especially in the perinecrotic palisading cells, is enhanced up to 50-fold and accompanied by a parallel increase in VEGFRs on the endothelium of the neoplastic blood vessels. Analogous upregulation occurs in the angiopoietin/TIE systems.[834]

Studies in mice challenge the previous notion of CNS angiogenesis as a passive process driven primarily by demands for oxygen and other nutrients by the growing neuronal populations. Angiogenesis in the mouse telencephalon progresses in an orderly, ventral-to-dorsal gradient regulated in a cell-autonomous manner by compartment-specific homeobox transcription factors.[1042] These same transcription factors, Nkx2.1, Dlx1, Dlx2 and Pax6, confer compartmental identities on telencephalic neurons and progenitor populations and therefore regulate the development of telencephalic neuronal networks and vascular networks, underscoring shared mechanisms in CNS vascular and neuronal development.[1042]

The identification of endothelial progenitor cells (EPCs) in peripheral blood as haematopoietic cells with the ability to differentiate into endothelial cells has also changed the dogma that vasculogenesis is only an embryogenic process.[59] CXCL12, a chemokine involved in both embryonic and tumour angiogenesis, may mediate signalling to promote the formation of new vessels by recruiting circulating EPCs or directly enhancing migration or growth of endothelial cells.[601,992]

Hypoxia-independent tumour angiogenesis occurs in haemangioblastomas in patients with VHL, because the loss of VHL tumour suppressor gene function hinders degradation of HIFs, which thus accumulate and upregulate hypoxia-regulated genes in the absence of hypoxia (Figure 2.2). It has also been suggested that some tumours may create vascular channels lined by tumour cells instead of endothelium, a phenomenon called vascular mimicry, which was first described in melanomas[305] and has been claimed to occur even in astrocytomas.[1116]

The orderly formation of the vasculature is vulnerable to disruption at several stages, resulting in vascular malformations of the brain (VMBs). The identification of gene mutations and risk factors associated with cerebral cavernous malformations (CCMs), sporadic brain arteriovenous malformations (AVMs), and the arteriovenous malformations of hereditary haemorrhagic telangiectasia has provided new insights and enabled the development of genetic testing and animal models for these diseases.[590,1065] Genes associated with angiogenesis and vascular remodelling are involved in VMBs (see Table 2.3). Studies suggest the angiogenic process most severely disrupted by VMB gene mutations is that of vascular stabilization subsequent to the formation of vessel tubes and recruitment of vascular smooth muscle cells.[590] Mutations in VMB genes also render vessels vulnerable to rupture when challenged by other genetic or environmental factors.

TABLE 2.3 Genes and protein products involved in vascular malformations and aneurysms

Classification	Gene	Protein product	Putative Function
CCM1	KRIT1	KRIT1, ankyrin repeat-containing protein	Binds β-catenin, stabilizes interendothelial junctions associated with actin stress fibres. Three CCM genes (CCM1, CCM2, and CCM3) have been identified to date. All 3 corresponding proteins are expressed in vascular endothelium and associated with cytoskeletal and interendothelial junction proteins and components of certain signal transduction pathways.
CCM2	CCM2	Malcavernin	Cellular responses to osmotic stress; modulates mitogen-activated protein (MAP) kinase and RhoA GTPase signalling and is part of a complex with MAP2K3, MAP3K3 and RAC1.
CCM3	PDCD10	Programmed cell death protein 10, TF-1 cell apoptosis-related protein 15	Cell proliferation and transformation (cancer cell lines); modulates extracellular signal-regulated kinase (ERK).
HHT1	ENG	Endoglin, CD105	TGFβ superfamily co-receptor; modulates signalling by TGFβ type II receptor, ALK-1 and ALK-5. Mutations in ENG or ALK1(ACVRL1) cause HHTs.
HHT2	ACVRL1	ALK-1, Activin A receptor type II-like 1, serine/threonine-protein kinase receptor R3	Type I receptor for TGF-β family ligands BMP9/GDF2 and BMP10. It is an important regulator of normal blood vessel development. On ligand binding, forms a receptor complex consisting of two type II and two type I transmembrane serine/threonine kinases.
HHT (a juvenile form)	SMAD4	Mothers against decapentaplegic homologue 4, deletion target in pancreatic carcinoma 4	Common downstream mediator of multiple TGFβ superfamily signalling pathways.

CCM, cerebral cavernous malformation; ECs, endothelial cells; HHT, hereditary hemorrhagic telangiectasia.

ANATOMY OF THE CNS VASCULATURE

Arterial Blood Supply

General Aspects

The metabolism of the brain is almost solely aerobic. The brain lacks significant energy reserves and requires a continuous supply of well-oxygenated blood. The exceptionally high demand for circulating blood and oxygen is reflected in the disproportionately high rate of CBF compared with flow to other parts of the body: 20 per cent of the cardiac output and 15 per cent of oxygen consumption in an adult at rest, although the brain makes up only 2 per cent of the body weight.

Large Arteries of the Brain

The blood to the brain is supplied by two pairs of large arteries. The anterior flow (approximately 70 per cent of CBF) enters the cranial cavity through the internal carotid arteries, and the posterior flow (approximately 30 per cent of CBF) through the vertebral arteries, which fuse to form the basilar artery. These two systems anastomose via the anterior and posterior communicating arteries at the base of the brain to form the circle of Willis (Figure 2.3).[299,402] There can be considerable variation in the relative size of the vertebral arteries and those forming the circle of Willis.[239] In the most common variants of the circle, there is hypoplasia of the precommunicating segment of the anterior cerebral artery (A1) or the posterior cerebral artery (P1).[275] Variations in the circle of Willis may affect the relative flow rates in proximal arteries with the circle.[984] Another important anastomotic pathway is between the external and internal carotid arteries via the ophthalmic arteries. When this anastomotic network is normal, then the occlusion of even one of the four main arteries will not necessarily lead to insufficient regional CBF. For example, collateral circulation is usually sufficient for slow occlusion of the distal segment of an internal carotid artery (ICA) by atherosclerosis or surgical clamping for treatment of a fusiform aneurysm, not to cause infarction.[389] Patients surviving with only one patent vertebral artery demonstrate the efficiency of the collateral circulation. The perfusion territories of the main arteries in the brain and brain stem are illustrated in Figures 2.4 to 2.7.

Leptomeningeal and Pial Arteries

The leptomeningeal anastomoses are in the subarachnoid space over the surface of the brain, where the territories of the distal branches of the anterior, middle and posterior cerebral arteries overlap in the border or watershed zones (Figure 2.4a–c). Corresponding border zones are also formed between the superior and inferior cerebellar arteries (Figure 2.4d). The leptomeningeal anastomoses are located at the periphery of the arterial trees and these zones tend to be the first to be deprived of sufficient blood flow in the event of arterial hypotension or a reduction in perfusion due to raised intracranial pressure.

Intraparenchymal Arteries

The branches of the arteries running in the subarachnoid space penetrate the brain parenchyma (Figure 2.8). These branches include both deep and superficial perforating

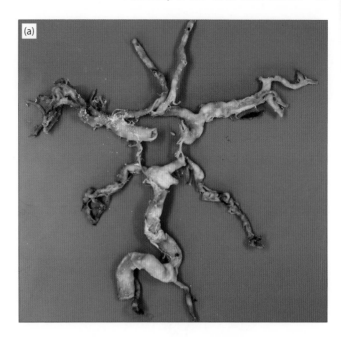

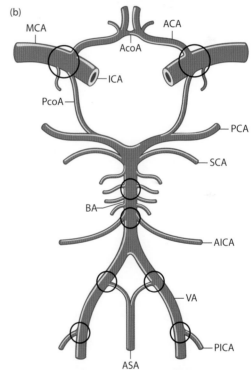

2.3 Main arteries of the circle of Willis. (a) Structure of the main arteries that anastomose to form a complete circle, in this case dissected from the base of the brain of a 70-year-old man. Some atherosclerosis can be seen. Note the small calibre of the posterior communicating arteries, which are often narrower in older people. **(b)** Schematic diagram of the circle of Willis showing the most common sites of atheroma (circles). ACA, anterior cerebral artery; AcoA, anterior communicating artery; AICA, anterior inferior cerebellar artery; ASA, anterior spinal artery; BA, basilar artery; ICA, internal carotid artery; MCA, middle cerebral artery; PCA, posterior cerebral artery; PCoA, posterior communicating artery; PICA, posterior inferior cerebellar artery; SCA, superior cerebellar artery; VA, vertebral artery.

Images kindly provided by T Polvikoski and A Oakley, Newcastle University, UK.

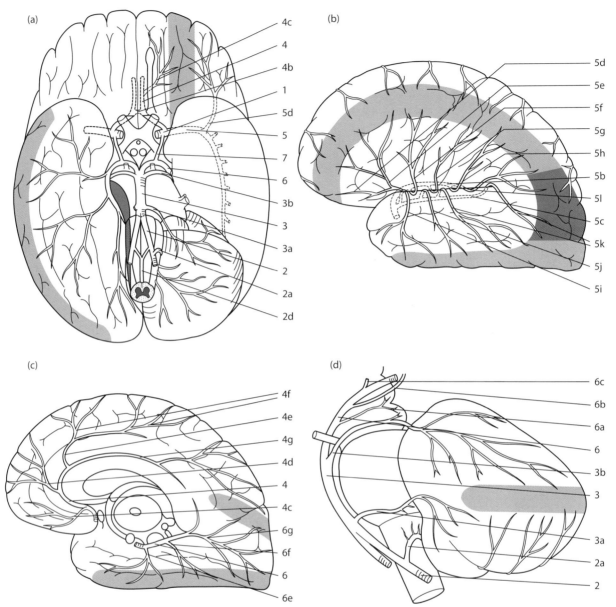

2.4 Perfusion territories of the cerebral and cerebellar arteries (see also Figure 2.6). (a) Basal view; **(b)** lateral view; **(c)** midline view; **(d)** cerebellum and brain stem. The border zones (watershed areas) between the territories are indicated by shading. 1, internal carotid artery; 1a, anterior choroidal artery; 2, vertebral artery; 2a, posterior interior cerebellar artery; 2b, paramedian branch; 2c, lateral bulbar branch; 2d, anterior spinal artery; 3, basilar artery; 3a, anterior inferior cerebellar artery; 3b, superior cerebellar artery; 3c, paramedian branch; 3d, short circumferential branch; 4, anterior cerebral artery; 4a, recurrent artery of Heubner; 4b, anterior communicating artery; 4c, medial orbitofrontal artery; 4d, frontopolar artery; 4e, callosomarginal artery; 4f, internal frontal branches; 4g, pericallosal artery; 5, middle cerebral artery; 5a, lenticulostriate artery; 5b, upper division; 5c, lower division; 5d, lateral orbitofrontal artery; 5e, ascending frontal (candelabra) branch; 5f, central (Rolandic) artery; 5g, anterior parietal artery; 5h, posterior parietal artery; 5i, temporal polar artery; 5j, anterior temporal artery; 5k, posterior temporal branches; 5l, angular artery; 6, posterior cerebral artery: 6a, quadrigeminal artery; 6b, posterior choroidal artery; 6c, thalamogeniculate artery; 6d, thalamo-perforating artery; 6e, anterior temporal artery; 6f, posterior temporal artery; 6g, calcarine artery; 6h, paramedian branch; 6i, short circumferential branch; 6j, long circumferential branch; 7, posterior communicating artery; 7a, hypothalamic artery.

Adapted from Romanul.[858] *With permission from Wolters Kluwer.*

arteries. The *deep perforators* branch off the main cerebral arteries at the base of the brain and consist of (i) the lenticulostriate arteries (LSAs), which emerge from the first segments of the anterior and middle cerebral artery (MCA) to supply the basal ganglia, and (ii) perforant branches, which leave the posterior cerebral and posterior communicating arteries and supply the thalamic nuclei. In the striatum, the distribution of microvascular territories from the lateral LSA, medial LSA and the recurrent artery of Heubner co-register with each of the functional corticostriatal zones, the sensorimotor, associative and limbic zones, with greater density of both large penetrating vessels and small pre-capillary arterioles and capillaries within the matrix compared to the striosomes.[282] The *superficial perforators* originate from the

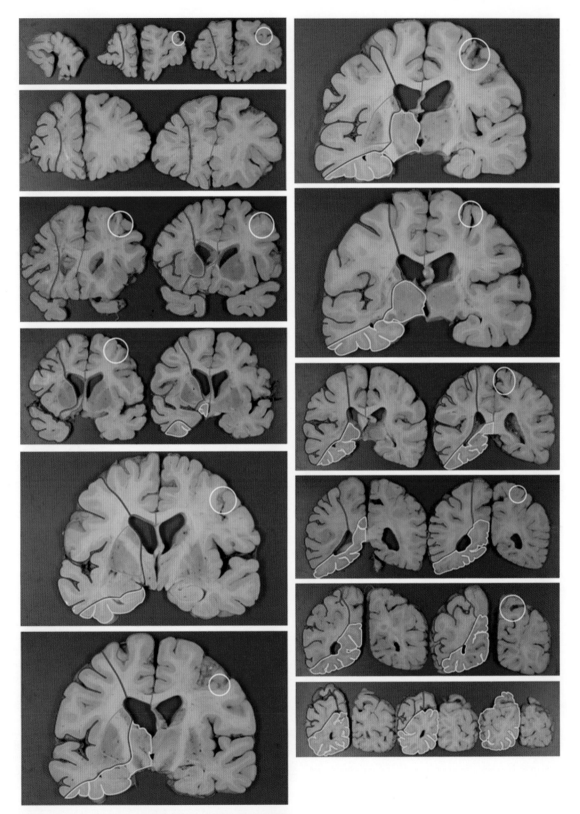

2.5 Arterial supply territories depicted in an adult brain sliced in the coronal plane. The extent of the three main arterial territories in the cerebrum along the rostro-caudal plane are shown: anterior cerebral (magenta), middle cerebral (blue) and posterior cerebral (yellow). The vascular supply to the striatum (delineated in green) includes the lateral lenticulostriate arteries, medial lenticulostriate arteries and the recurrent artery of Heubner (most medial), all of which branch off from the middle cerebral artery.[282] The hippocampal formation is perfused predominantly by the posterior cerebral artery with a minor contribution from the anterior choroidal artery. The slices include several small infarcts (white circles) in the right frontal lobe, in the territory of the middle cerebral artery.

Original material from the Newcastle Cognitive Function After Stroke study and kindly provided by T Polvikoski, Newcastle University, UK.

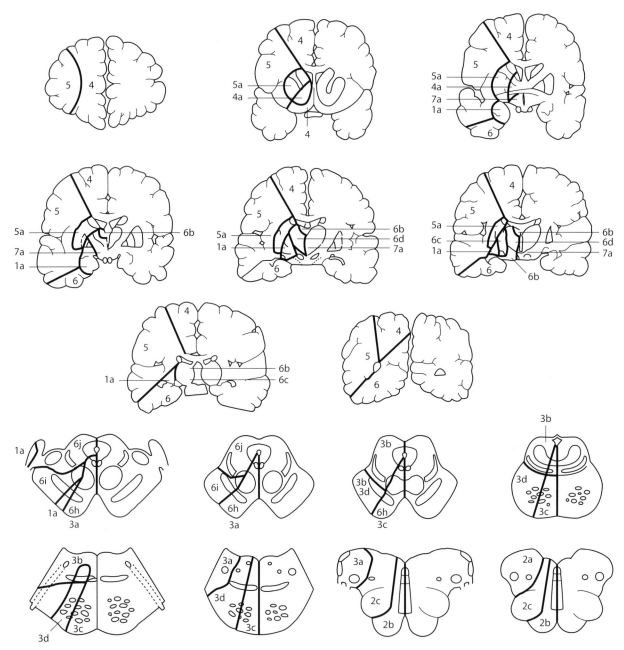

2.6 Arterial supply territories in coronal planes of the cerebrum and brain stem in greater detail. The key to the numbers is in the legend to Figure 2.4.

Adapted from Romanul.[858] With permission from Wolters Kluwer.

pial branches of the anterior, middle and posterior cerebral arteries over the surface of the brain and are of variable length: short ones supply the cortex and longer ones (the medullary arteries) the deep white matter. Short penetrators also exist in the brain stem as paramedian branches of the basilar artery.

The perforators are end arteries, i.e. they have very limited collateral connections with neighbouring blood vessels until they branch into capillaries. The capillaries do interconnect but their collateral flow is so local and restricted that the occlusion of a perforator usually results in a small region of ischaemic damage, described as a lacunar infarct (see later). The deep and superficial perforators do not

anastomose deep in the brain but meet in junctional zones (Figure 2.8a), where subcortical infarction may occur.[121] There is increasing interest in the magnetic resonance imaging of structural changes with high tensile magnets (e.g. 3 and 7 Tesla) that allow identification of small lacunar lesions *in vivo* (Figure 2.9).

Arteries of the Spinal Cord

The main arterial blood supply of the spinal cord is provided by one (occasionally two) anterior and two posterior spinal arteries (Figure 2.10). Branches from vertebral arteries join in the midline to form the (usually) single anterior

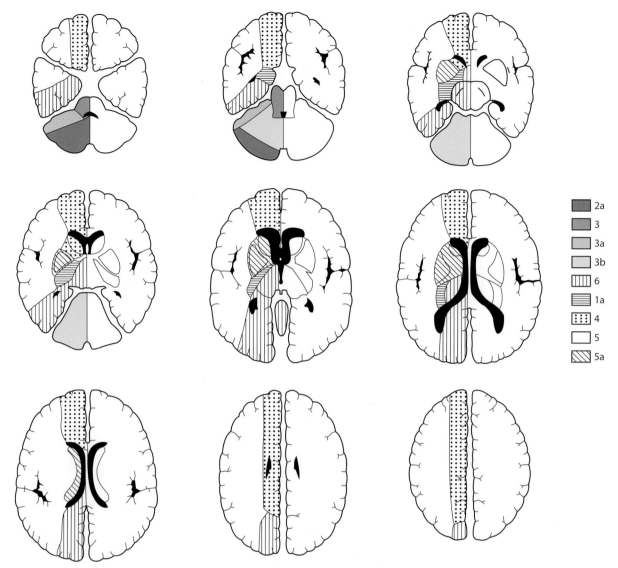

▨	2a
▨	3
▨	3a
▨	3b
▥	6
▤	1a
⠿	4
☐	5
▨	5a

2.7 Maps of arterial supply territories in horizontal planes through the cerebrum, cerebellum and brain stem. The key to the numbers is in the legend to Figure 2.4.

Adapted from Sacco[888] and Savoiado.[903] With kind permission from Springer Science and Business Media.

spinal artery, responsible for the greater part of the blood supply to the spinal cord. The anterior spinal artery may be of variable size or even discontinuous at different levels, depending on the pattern of replenishing tributary arteries along its passage downwards. Posterior spinal arteries are even more irregular, deriving from vertebral or posterior inferior cerebellar arteries.

Along the spinal canal, branches from larger arteries coursing along the spine enter the spinal canal through the intervertebral foramina and give off (i) a dural branch, which supplies the dural sleeve around the root, then (ii) a branch to serve the nerve roots and finally (iii) a medullary (tributary) branch to replenish blood to the anterior and posterior spinal arteries (Figure 2.10b). The term 'radicular artery' is commonly used to describe the combination of these three branches, not solely the branch supplying the nerve roots. The number of the tributary radicular arteries varies from 4 to 10. In the superior (cervicothoracic, C1 to

T1–2) territory, the tributary arteries emerge from vertebral arteries and costocervical or thyrocervical trunks (Figure 2.10a). In the intermediate (midthoracic T1–2 to T8) and inferior (thoracolumbar, T9 to conus medullaris) territories, the tributary arteries emerge from the intercostal or lumbar arteries as their dorsal rami (Figure 2.10b). The most important and largest tributary artery is arteria radicularis magna (of Adamkiewicz), which enters the spinal canal at a variable level, between T8 and L2, below which spinal artery blood flow is mainly downward. A border zone is created, usually at a lower thoracic level than the traditionally stated T4.[193,975]

The anterior spinal artery gives off sulcal (also called sulco-commissural or central) arteries. These supply the anterior central part of the cord, in an alternating pattern on the left and the right side (Figure 2.10c). This alternate pattern of distribution explains why occlusion of a sulcal artery can manifest as spinal hemisection, typically the

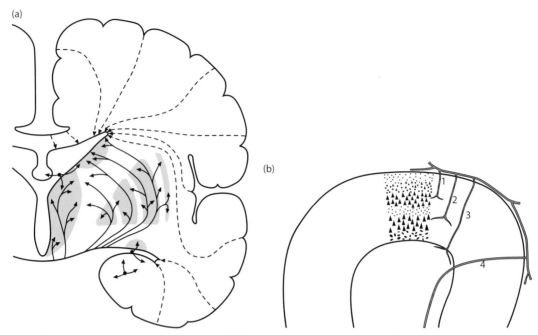

2.8 (a) General pattern of cerebral penetrating arteries. **(b)** The cortical penetrators reach three different depths (1–3). The longest penetrators (4) continue into the white matter.

Adapted from Romanul.[858]

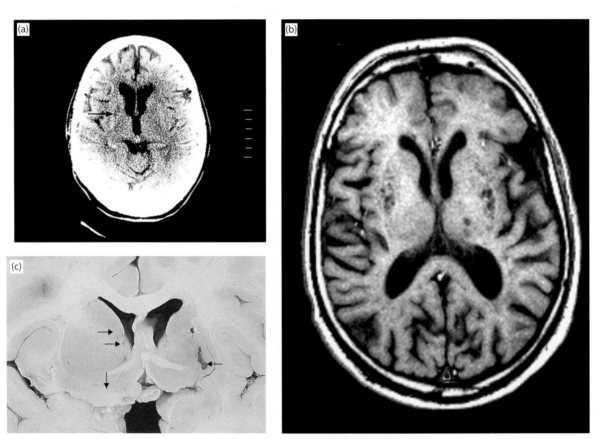

2.9 Sub-cortical infarction. (a) CT scan in an 82-year-old female with longstanding arterial hypertension, who had a small, clinically silent lacuna in the right basal ganglia (arrow). She died acutely of a large fresh atherothrombotic infarct in the territory of the left middle cerebral artery, which appears hypodense in this scan. **(b)** Several lacunar infarcts in lenticulostriate territories seen bilaterally on T2-weighted MRI in the axial plane of an elderly subject. **(c)** The lacunar change seen by CT scan was confirmed as a lacunar infarct (extreme right arrow) elongated in the direction of the lenticulostriate perforating arteries. There are also perivascular cavities (left arrows), best seen in the left caudate nucleus.

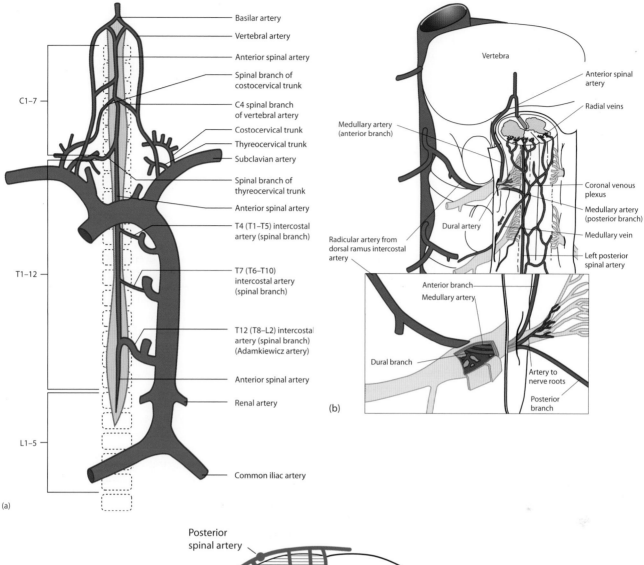

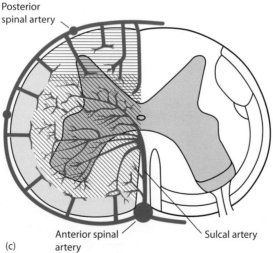

2.10 (a) Blood supply to the spinal cord. The main anterior (usually single) and posterior (usually paired) spinal arteries arise from the vertebral arteries, and receive tributaries from the intraosseous vertebral, intercostal, lumbar and other arteries that enter the spinal canal through the intervertebral formina at multiple levels. The levels at which the different tributaries enter the spinal canal vary considerably. **(b)** Blood supply of the spinal cord at a segmental level. **(c)** Blood supply within the spinal cord. The anterior spinal artery gives rise to sulcal arteries. In the depths of the anterior median fissure, alternate sulcal arteries deviate either left or right to supply the corresponding side of the cord.

(b) Adapted with permission from Rosenblum et al.[868]

(c) Reproduced with permission from DeGirolami and Kim.[238] © 2005 Wiley-Blackwell.

Brown–Séquard syndrome. The branches from the posterior spinal arteries supply the posterior horns and columns. The circumferential vascular plexus receiving blood from the radicular and spinal arteries supplies the superficial parts of the cord (Figure 2.10b,c).

Venous Sinus Systems

Venous Drainage of the Brain

The venous circulation of the brain (Figure 2.11) differs from the common, consistent antiparallel (i.e. running in parallel but in opposite directions) orientation of arteries and veins in most other organs. In addition, the cerebral venous drainage employs the dural sinuses as the final intracranial collecting blood vessels.[39,858]

The blood from most of the white matter and cortex of the cerebral hemispheres is drained by veins of various lengths, antiparallel to the pial penetrating arteries. In general, there are fewer veins than perforating arteries and the long veins also drain the cerebral cortex while coursing through it. After the veins of the superficial or cortical network exit the parenchyma and enter the subarachnoid space, they turn towards the dural sinuses. In the suprasylvian and paramedian regions, the frontal, parietal and occipital superior cerebral veins run upward to drain into the superior sagittal sinus. In the parasylvian region, the middle cerebral veins drain via the superficial sylvian vein

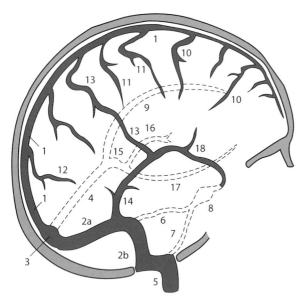

2.11 Anatomy of venous drainage pathways of the brain. In the case of paired blood vessels, only the right is shown in dark colour. The veins in black are on the surface of the brain; those depicted with dashed lines are within the parenchyma. 1, superior sagittal sinus; 2a, transverse portion of lateral sinus; 2b, sigmoid portion of lateral sinus; 3, confluence of sinuses; 4, straight sinus; 5, internal jugular vein; 6, superior petrosal vein; 7, inferior petrosal vein; 8, cavernous sinus; 9, inferior sagittal sinus; 10, frontal veins; 11, parietal vein; 12, occipital vein; 13, vein of Trolard; 14, vein of Labbé; 15, great vein of Galen; 16, internal cerebral vein; 17, basal vein; 18, superficial sylvian vein.

Adapted from Ameri and Bousser.[39] With permission from Elsevier.

to the cavernous sinus. On the posterior lateral and inferior surfaces of the temporal lobe, and on the lateral and inferior surfaces of the occipital lobe, the veins drain into the lateral sinuses. The middle cerebral veins connect superiorly, through the vein of Trolard, with the superior sagittal sinus, and inferiorly, through the vein of Labbé, with the lateral sinus. The number and location of the cortical veins vary considerably, which makes angiographic verification of their patency difficult. The superficial veins have thin walls, no tunica muscularis and no valves, permitting dilation and flow of venous blood in various directions. These features, together with numerous anastomoses, help to achieve efficient collateral flow in the case of venous thrombosis.

Within the parenchyma of the hemispheres, the veins of the superficial system anastomose extensively with the internal cerebral and basal veins of the deep network. The deep veins collect blood from the deep grey matter at the base of the brain and the choroid plexus of the lateral ventricles and drain into the centrally located great cerebral vein of Galen. The latter joins the straight sinus at the apex of the cerebellar tentorium. The inferior sagittal sinus, running along the lower edge of the falx, also joins the straight sinus. The straight sinus then merges with the superior sagittal and occipital sinuses at the confluence of sinuses (torcula herophili). The bulk of the venous blood flows bilaterally via the transverse and sigmoid sinuses (which together form the lateral sinus), through the jugular foramen into the jugular vein. The right lateral sinus is commonly larger than the left. In 14 per cent of cases, the transverse portion of the left sinus is not visualized on angiography, an anomaly that may be relevant when investigating venous thrombosis. Dural sinuses also receive blood from the diploë of the skull bones and are connected with the extracranial veins via the emissary veins, which traverse the cranium.

The posterior fossa veins drain the cerebellum and brain stem. At the surface, the veins form a subarachnoid plexus, from where blood drains in three directions: from the superior part it drains into the great cerebral vein of Galen, from the anterior part into the petrosal sinuses, and from the posterior and lateral parts into the adjacent straight, occipital and lateral sinuses.

Venous Drainage of the Cord

The venous drainage of the cord in general corresponds to the vascular architecture of the arterial supply of the cord (see earlier), but the number of veins within and around the cord, as well as exiting the spinal canal through the intervertebral foramina, is greater than that of arteries.[350] Within the cord, there are three main venous networks, anastomosing more freely than the arteries: the anterior part of the cord is drained by the anterior spinal vein, and the posterior part by a single posterior midline vein as opposed to paired posterior spinal arteries. From the periphery, radially oriented veins drain into the superficial plexus of veins around the cord. Radicular veins convey the blood into paravertebral and intervertebral plexuses, which drain into the azygous and pelvic veins. Because these veins do not have valves, there is a high potential for infections from the abdominal cavity to spread into the spinal cord.

Histology of Cerebral Vessels and Barrier of the Brain

The microscopic structure of the extracranial parts of the carotid and vertebral arteries is similar to that of other large arteries, whereas intracranial cerebral blood vessels have several distinct structural features for the specific functions and protection of the CNS (Figure 2.12). The endothelial cells of the intracranial blood vessels are joined by tight junctions and have no fenestrations. The muscle layer of the intracranial arteries is thinner than in extracranial arteries of a similar size, the external elastic lamina lacking and the adventitia leaner. To complement these features the brain is endowed with structurally unique protective systems. The intracellular organelles within endothelial cells have specific features related to BBB and transcellular transport functions.[40b] The cerebrospinal fluid (CSF)–blood barrier is controlled by specialized epithelial cells of the choroid plexus.[832] The exchange between the ependymal cells and ventricle walls is also a regulated surface. In addition to the perivascular drainage routes,[1082] the brain also has a lymphatic-like pathway, recently described as the glymphatic system.[457] This is thought to be an anatomically distinct drainage system or paravascular pathway for CSF and interstitial fluid (ISF) exchange that facilitates efficient clearance of solutes and waste from the brain.[1]

PHYSIOLOGY OF THE CEREBRAL CIRCULATION

The physiology and regulation of the CBF and the exchange of metabolites between blood and parenchyma across the BBB are described in Chapter 1. Autoregulation of CBF is the first order mechanism that ensures flow and the supply of oxygen, glucose and nutrients through the vascular beds. Cerebral resistance arteries dilate or constrict during changes in arterial pressure. Functional hyperaemia, or coupling of CBF to neural activity, is another vital mechanism whereby the neurovascular unit maintains the homeostasis of the cerebral microenvironment. Several different signals, including vasoactive peptides and nitric oxide, mediate the regional increases in CBF in tandem with waves of neural activity in different brain regions. Besides the BBB (see Chapter 1), the choroid plexus (CP)–CSF barrier system ensures not only CSF secretion but also constant scrutiny of nutrients and harmful substances to maintain homeostasis.[276] However, none of these protective mechanisms is immune to ischaemic insults or the effects of ageing and disease.

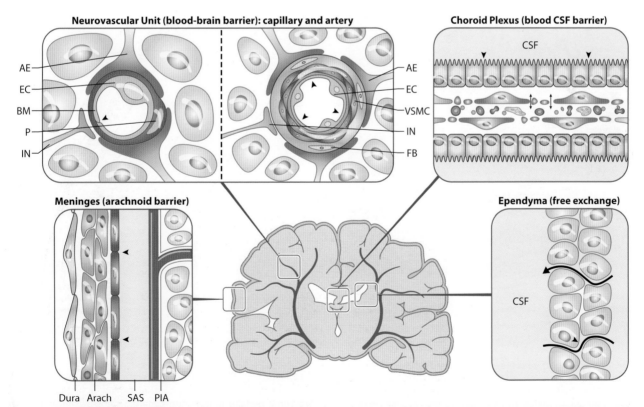

2.12 Schematic illustration of structural components of the perfused surfaces of the brain. AE, astrocytic endfeet; BM, basement membrane; EC, endothelial cell; IN, innervation; P, pericyte; VSMC, vascular smooth muscle cell; FB, fibroblast; CSF, cerebrospinal fluid; Arach, arachnoid; SAS, subarachnoid space.

Diagram kind courtesy of Y Yamamoto, Yamaguchi University Graduate School of Medicine, Japan.

Microcirculation and Neuronal Metabolism

As arterial perforators penetrate the brain parenchyma, they branch repeatedly, reducing in size until they end in capillaries that allow blood cells to pass only in single file. The capillaries vary in density throughout the brain, being more abundant in regions with high metabolic rates. Capillaries have long been known to be distributed more densely in grey than white matter (Figure 2.13).[221] Cerebral blood flow averages 50 mL/100 g of brain/min. This value for human CBF is an integrated average value for grey matter, where flow rates are greater than 80 mL/100 g/min, and white matter, where rates can be as low as 20–25 mL/100 g/min. Within grey or white matter, capillary density is richer where metabolic and, consequently, oxygen-delivery requirements are higher.[147] Within the white matter, capillary density is twice as high in the pyramidal tract as the fasciculus cuneatus.[221] Within the grey matter of the inferior colliculus, capillary density correlates with both the metabolic rate for glucose and the CBF.[375]

The rate of consumption of glucose and oxygen, per gram of brain tissue, is higher in species with small brains[1015] and in infants.[109] Small brains have a greater density of neurons (Figure 2.14), not glia, and this relationship has remained constant during phylogenesis. The metabolic correlate of this is that glycolytic rate remains constant across species,[1016] whereas the rate of oxidative phosphorylation (aerobic metabolism) increases in smaller brains[1015] because of increased neuronal density (Figure 2.14). The decrease in density of cortical neurons from the mouse (~142 500 neurons/mm³) to rat (~105 000 neurons/mm³), cat (~30 800 neurons/mm³), human (~10 500 neurons/mm³) and whale (~6800 neurons/mm³) is accompanied by a corresponding reduction in the rate of oxidative phosphorylation per unit volume of brain tissue. These phylogenetic and ontogenetic variations may explain the differential effects of hypoxia and ischaemia in these species (see later).

Cerebral blood flow is modulated by cerebral metabolism and, like the metabolic rate, is higher per gram of tissue in smaller brains, e.g. of rats or gerbils, than in humans. Glucose use per gram of a rat brain is roughly twice that of a human.[116] This is reflected anatomically in a higher capillary density (and reduced diffusion distance for oxygen) in species with small brains and in the developing brain. Capillary density also increases on adaptation to high altitude or reduced oxygen intake (see Hypoxic Neovascularization).

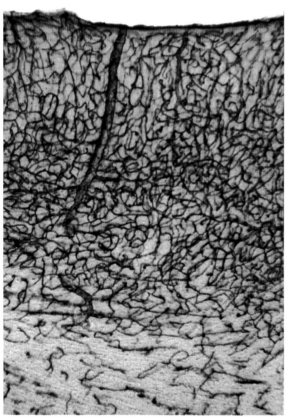

2.13 Cerebral blood vessels in cortex and white matter. Just to the left of centre, an arteriole penetrates about two-thirds of the cortex before ramifying. The vascular density is much lower in the white matter (bottom part of the photomicrograph) than in the cerebral cortex, reflecting the neuronal and synaptic energy requirements. Across other species as well, neuronal density determines blood vessel density. Capillary density also increases in adaptation to altitude (see text). Factor VIII immunocytochemistry.

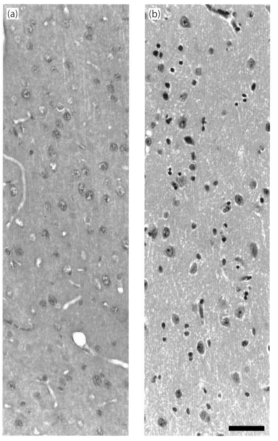

2.14 Neuronal density in rat **(a)** versus human **(b)**, in homologous areas of cortex. The neuronal density is higher in the rat. Neuronal density determines synaptic density, metabolic rate, capillary density, susceptibility to ischaemia and threshold of necrosis due to lactic acid production. Measured in moles of glucose used per gram of brain per minute, metabolic rate in the rat is twice that in man. Capillaries and venules are also more numerous per unit volume in rats (a) than in humans (b).[116]

When CBF drops, the critical points at which ischaemic injury occurs vary according to the level of metabolism. Ischaemic thresholds for infarction are lower in species with large brains, at approximately 12 mL/100 g/min in the larger non-human primate brain[64] compared to 45 mL/100 g/min in the small rodent brain.[472] The ratio of the normal blood flow to the threshold for infarction remains roughly constant across species, however, at 3:1 (in humans, flow drops from ~50 mL/100 g/min to 15–18 mL/100 g/min before infarction occurs) (Figure 2.15). Animals with small brains and greater neuronal densities have higher rates of basal blood flow to satisfy the higher rates of metabolism and, therefore, have higher absolute CBF thresholds for infarction. Thresholds for ischaemic damage are related to metabolic activity.[472] Neuropil necrosis can result not only from vascular occlusion but also from hypermetabolism and acidosis. This is exemplified by Wernicke's encephalopathy (see Chapter 9), in which acidosis-induced neuropil necrosis (sparing neurons) can be precipitated or exacerbated by a glucose load.

Cerebral Blood Flow Regulation

Mechanisms controlling CBF have been investigated intensively and are local rather than global: hypoxic increases in brain blood flow occur independent of sensorineural input from peripheral hypoxia sensors.[557,1019] Rather, local brain activity determines blood flow in any particular brain region, depending on oxygen and glucose consumption. In the intact brain, CBF, cerebral metabolic rate for oxygen (CMRO$_2$) and cerebral metabolic rate for glucose (CMRgl) are normally coupled, rising and falling in synchrony. This is termed local autoregulation or functional hyperaemia. From the ~50 mL of blood delivered to the brain every minute, 25 µmol of glucose is extracted, stoichiometrically requiring the simultaneous extraction of 150 µmol (3 mL) of O$_2$ for its oxidation: $C_6H_{12}O_6 \rightarrow 6CO_2 + 6H_2O$.

The normal coupling mechanism matching CBF to cerebral metabolism is still unknown. Candidates include pH (H$^+$), adenosine, nitric oxide and O$_2$-derived free radicals. Whatever the mechanism, coupling normally occurs between primary, activity-driven increases or decreases in brain glucose metabolism, and both oxygen utilization and blood flow. In contrast to most of the pathophysiological states, isolated hypoxaemia impairs neither CBF-metabolism coupling nor cerebral circulatory responses.[260,697]

The average human CBF value of 50 mL blood/100 g brain/min is not only an average for grey and white matter but also an average of regional brain blood flow coupled to local activity. In addition to the fine-tuning of local blood flow, the brain maintains or autoregulates constant overall CBF within the upper and lower limits of the autoregulatory range (usually cited as 50–160 mmHg) during fluctuations in systemic arterial pressure (Figure 2.16). If blood pressure falls to too low a level, perfusion drops. In man, this occurs in the range of 70–93 mmHg for different individuals. Human studies of the upper limit of autoregulation are ethically difficult because vascular rupture and haemorrhage can result, not only excessive perfusion, if blood pressure is too high. The upper limit of autoregulation is probably just over 160 mmHg for most individuals. Mathematical modelling suggests that the brain can maintain constant blood flow over a blood pressure range of approximately 69–153 mmHg.[326]

Certain conditions or age-related alterations in the systemic circulation, and degenerative changes in the extracerebral resistance arteries, can shift the lower and upper limits of the autoregulatory plateau to cause hypertensive encephalopathy or cerebral hypoperfusion, which may not necessarily cause ischaemic injury as obvious as in stroke, but an oligaemia that leads to disruption of the microcirculation, damage to the cerebral endothelium, breach of the BBB and oedema. Neurogenic responses via the brain stem autonomic nuclei may also disrupt CBF autoregulation. In addition, pathological changes of vascular smooth muscle and altered release of metabolic factors (e.g. vasoactive substances, cytokines) can influence autoregulatory responses.

Although the usual focus of attention in hypoxia and ischaemia is on the use and supply of brain oxygen and glucose, also important is the removal of metabolic wastes, such as lactate. There is good evidence that accumulation of H$^+$ contributes to cerebral necrosis.[558,720]

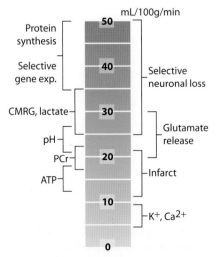

2.15 Thresholds for cerebral blood flow and critical events at which metabolic disturbances occur. PCr, phosphocreatine.

Adapted from Hossman et al.[439]

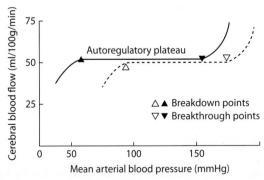

2.16 Autoregulatory thresholds for maintenance of cerebral blood flow (CBF). Constant CBF is maintained in the 50–150 mmHg range of mean arterial blood pressure. Below and above this range, the CBF varies with the arterial blood pressure. In hypertensive patients, the curve shifts to the right (dashed line).

Interstitial Fluid Drainage Pathways

The total volume of interstitial fluid (ISF) in the human adult brain is approximately 200–250 mL, 15–18 per cent of brain weight.[832] It is distributed in the narrow extracellular space between the parenchymal cells and their processes. Its precise origin has not been established. Some of the ISF may be derived from passage across the endothelium of CNS capillaries, some from CSF entering from the ventricles and subarachnoid space, and some from neuronal and glial metabolism. The pial–glial layer on the surface of the brain and the ependymal lining of the ventricles allow free exchange between ISF and CSF. About 10 per cent of ISF is formed by the metabolic activity of parenchymal cells.[1] ISF has an important role in the homeostasis of the extracellular environment. This homeostatic function necessitates that ISF is not static but flows within the extracellular space and is continuously renewed, as drainage via specific pathways[172,456,1082] provides space for fresh ISF.[1,1081] The motive force for ISF flow is assumed to be derived from systolic vascular pulsations.

In animal experiments, ISF has been shown to drain from the white matter into ventricular CSF, whereas from the cortex the flow of ISF occurs in the reverse direction to the blood flow in the penetrating arteries, i.e. towards the surface.[172] There are two pathways of ISF drainage: the bulk flow of particulate tracers such as Indian ink or fluospheres has been shown to occur in the expanded perivascular spaces between arterial tunica media and surrounding astrocytic end feet, whereas soluble and particulate tracers but not cells follow the basal lamina of capillaries and arteries.[172,1081]

In humans, the flow of ISF cannot be traced directly but the drainage pattern of ISF has been inferred from the distribution of amyloid β-peptide (Aβ) and development of cerebral amyloid angiopathy (CAA).[1082] Soluble Aβ is formed in the parenchyma, carried in the ISF and drained along the perivascular basement membrane of capillaries and arterioles. As cerebral blood vessels become more rigid with age, the drainage may slow down, increasing the polymerization and deposition of Aβ in the walls of the vessels along which it drains. This may further impair flow along the perivascular drainage pathways. Impaired perivascular drainage of ISF may contribute to other forms of CAA and dementias characterized by the deposition of aggregated proteins, such as amyloid familial British dementia (ABri), amyloid familial Danish dementia (ADan), cystatin C and transthyretin.[1082] In experimental animals, the perivascular drainage of antigens to cervical lymph nodes is important in the development of an immune response,[1083] and in humans these drainage routes may facilitate the spread of immune cells and inflammation from the primary site of tissue damage.[1] ISF pathways have also been implicated in the spread of malignant cells, in the pharmacokinetics of drugs within the CNS and in potential therapeutic approaches using stem cells.[2]

PATHOPHYSIOLOGY OF CELL DEATH IN ISCHAEMIA AND HYPOXIA

Neuronal Cell Death Mechanisms

Besides necrosis, several 'non-accidental' mechanisms of neuronal death have been described. They include delayed neuronal death, apoptosis, autophagy and necroptosis.[201,1041] These 'programmed' mechanisms involving specific signalling pathways have molecular signatures but cannot readily be distinguished on routine histopathological examination. The extent of necrotic tissue damage is conventionally classified into two categories, selective neuronal necrosis, which affects neurons but spares glia, and pan-necrosis, in which all tissue elements die — neurons as well as glia and blood vessels — in time progressing to cavitation. If the aetiology

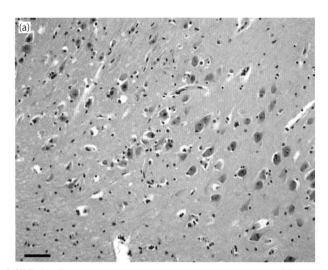
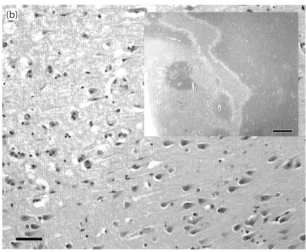

2.17 Selective neuronal necrosis versus pan-necrosis. (a) Selective neuronal necrosis in the hippocampus of a 78-year-old woman with septic shock. The acutely necrotic neurons toward the upper left part of the figure are abnormally acidophilic ('red' neurons). The selective neuronal death spares the glia and neuropil. **(b)** Pan-necrosis: left middle cerebral artery embolic infarct in a 51-year-old male. The necrosis (upper left part of the figure) involves all cellular elements and neuropil, producing a bubbly appearance surrounding acidophilic neurons. Inset: Sharp demarcation of the necrotic tissue, which at low power has a 'geographical' contour. The sharp, undulating border cuts across cell processes and crosses over grey and white matter boundaries, implying an abnormal microenvironment (likely to be acidosis) rather than a cellular mechanism. Bars = 50 μm; inset, bar = 500 μm.

of pan-necrosis is ischaemia, the term infarction is applied. Selective neuronal necrosis and pan-necrosis (Figure 2.17) can both affect virtually any brain region.

Selective neuronal necrosis should not be confused with selective vulnerability, which refers to the phenomenon whereby global brain insults cause focal lesions that predominate in certain brain regions (see later), the specific location depending on the type of insult. In ischaemic states, selectively vulnerable areas can show either pan-necrosis or selective neuronal necrosis.

Selective Neuronal Necrosis

Comparison of the patterns of neuronal death in ischaemia, hypoglycaemia and epilepsy[71] reveals that all three insults can cause selective neuronal necrosis (Table 2.4). This is because neurons are more vulnerable than astrocytes to death by overstimulation, not by virtue of their higher metabolic rates or larger size[705] but as a result of excessive release of excitatory amino acids.[24,198] In selective neuronal necrosis, recovery is abetted by the surviving neuropil and, despite the drop out of neurons, the surrounding cellular elements appear healthy.[701]

Pathophysiology: Excitotoxicity

It has long been known[755,756] that neurons can be damaged by overstimulation by glutamate or other excitatory amino acids such as aspartate. The neuroexcitatory activity of acidic amino acids coincides with their neurotoxic potential.[757]

Glutamate is a ubiquitous neurotransmitter,[1074] and acts via four glutamate receptors:[712] (i) N-methyl-D-aspartate (NMDA) receptors, (ii) kainate receptors, at which kainate is a favoured agonist (although the marine toxin domoic acid, associated with outbreaks of shellfish poisoning, appears to have even higher affinity), (iii) α-amino-3-hydroxy-5-methyl-4-isoxazole propionic acid (AMPA) receptors and (iv) metabotropic receptors, so named because they lead to activation of intracellular regulatory transduction events without necessarily generating transmembrane ion fluxes. The first three glutamate receptor subtypes cause ion fluxes accompanied by water movement across the dendritic cell membrane. They play a role in the morphological expression of excitotoxicity by causing dendritic swelling, but axons are spared.

When subjected to different patterns of stimulation, the NMDA receptors have the capability of causing long-term changes in synaptic efficiency. These may persist for hours to days or even longer and are likely to be involved in memory. The prolonged facilitation of synaptic transmission is known as long-term potentiation (LTP), and the prolonged reduction of synaptic transmission as long-term depression (LTD).[117] LTP is associated with postsynaptic alterations in dendrites and spines.[630] NMDA receptors are highly concentrated in the CA1 pyramidal neurons of the hippocampus. This may explain the selective vulnerability of Sommer's sector to the three insults of ischaemia, hypoglycaemia and epilepsy. Cell death in these conditions can be attributed to several common mechanisms.[842]

TABLE 2.4 Comparison of some clinical and metabolic aspects of ischaemia, hypoxia, hypoglycaemia and epilepsy

Feature	Global ischaemia	Hypoxaemia	Hypoglycaemia	Epilepsy
Patient age (usual)	Old	Young	Young	Young–middle aged
Spreading depolarisation	+	+	+	++
↓ ATP	+++	+	+	++
↓ Phosphocreatine	+++	+	+	+
↑ Lactate	+++	+	+	+
Fatty acid catabolism	+++	0	0	0
Protein catabolism	+++	0	0	+
Steady state attained	No	Yes	Yes	No
Gene activation	All genes studied affected	Some genes activated	Some genes activated	Many genes activated
Tissue pathology	Selective neuronal necrosis and infarction	Synaptic alterations only; no cell death	Selective cell death	Selective cell death
Local perfusion	Decreased	Decreased	Decreased/ increased	Increased
Acute-phase reactants in serum	All[619]	Erythropoietin only		
Clinical context	Cardiac arrhythmia or arrest, profound hypotension (older age group)	Anaphylaxis, asthma, bronchitis, bronchiolitis, epiglottitis, short anaesthetic accidents (younger age group)	Hypoglycaemic encephalopathy, coma	Myoclonus, seizures
Clinical decerebration	+++	+	++	+
Clinical recovery	±	+++	+++	++

ATP, adenosine triphosphate.

Pan-necrosis and Infarction

Pan-necrosis occurs in ischaemia but also in MELAS (mitochondrial encephalomyopathy, lactic acidosis, stroke-like episodes), to some extent in Wernicke's encephalopathy, and experimentally in the pars reticulata of the substantia nigra in epilepsy, non-ischaemic conditions all characterized by extreme local tissue acidosis and unimpaired or increased blood flow. Large regions of infarction leave only a fluid-filled cyst, within which and recovery of any nature at the tissue level is impossible. Multiple adjacent small infarct cavities can eventually close to form a glial scar.[701]

Pathophysiology: Acidosis and Pan-necrosis

Acidosis is a key alteration in pan-necrosis, damaging both neurons and glia[88,560,720] but sometimes sparing endothelial cells[63,425] perhaps because of metabolic acid washout by the circulation. In MELAS, the distribution of necrosis does not conform to arterial territories and is unaccompanied by ischaemia, although microvascular changes may develop.[588] Pan-necrosis in MELAS and some other mitochondrial diseases probably relates to local tissue acidosis (see later). Neuropil acidosis possibly accounts also for the neuronal soma-sparing necrosis seen in Wernicke's encephalopathy and, occasionally, ischaemia.

Pan-necrosis occurs when all cell types are overwhelmed by a drop in tissue pH.[721,942] Lactic acid is toxic and has a pK_a of 3.83. It tends to lower the tissue pH but the drop in pH is not due simply to equimolar H^+ and lactate production. It also results from protons released by the hydrolysis of ATP under anaerobic conditions.[777]

Conditions altering lactate production can leave pH_i unaltered,[710] and low tissue pH does not always accompany high lactate levels.[444,777] Intracellular buffering and compartmentation of lactate in glia[558] allow independent modulation of lactate levels and acidosis.[559] The dissociation of acid from lactate production has led to the terms lactosis and acidosis to describe accumulation of lactate and protons.[777] There is both hypermetabolism and excitation in pan-necrotic states. Wernicke's encephalopathy is a good example in which there is dendritic swelling[1072] and tissue acidosis.[388] Similarly, in ischaemic or epileptic pan-necrosis of the substantia nigra there is dendritic swelling,[463,464] which also accompanies a profound tissue acidosis.[465]

Acidosis of only a mild degree can protect against excitotoxic selective neuronal death,[1008] but severe acidosis overwhelms this minor protective action. Acidotic pan-necrosis can manifest with axonal swelling, sparing dendrites (Figure 2.18). This finding may be accounted for by the protective effect of mild acidosis on excitotoxicity.[1008] A clear demarcation within the tissue (Figure 2.17) between necrotic and non-necrotic brain suggests a threshold effect. With time the intervening neuropil is removed and pan-necrosis appears as a fluid-filled cyst surrounded by neuropil containing fibre-forming astrocyte.

Acidosis and Acidophilic 'Pink' Neurons

The swelling of dendrites is an early morphological feature in neuronal necrosis and is evident in ischaemia,[1102] hypoglycaemia[74] and epilepsy.[466] The dendritic location of excitatory receptors accounts for the dendritic swelling

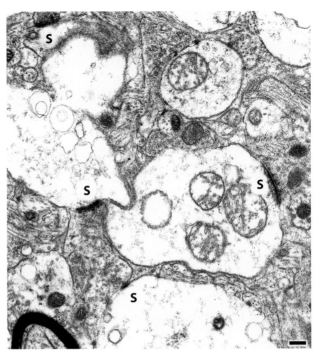

2.18 Selective swelling of dendrites in early selective neuronal necrosis. Synapses are marked S on the dendritic side of the synaptic membrane densities. Dendrites are subject to ion fluxes caused by excitatory activity, leading to transmembrane water fluxes and swelling. The dendritic mitochondria are also swollen, to several micrometres in diameter. The axon terminals that synapse with the swollen dendritic spines are not swollen and contain dark mitochondria. This axon-sparing dendritic lesion is a hallmark of the excitotoxic neuronal death, seen in ischaemia, hypoglycaemia and epilepsy. Bar = 1 μm.

(Figure 2.18). An early subsequent change is an increased affinity for acid dyes, i.e. acidophilia, that occurs whenever a cell or tissue dies. In nerve cells acidophilia is not due merely to the loss of the basophilic ribonucleic acid that constitutes Nissl substance but the nucleus and other cell structures show acidophilia.

Acidophilic neurons in pan-necrosis and selective neuronal necrosis appear similar (Figures 2.19 and 2.20). They will take up any acid dye of whatever colour, including safranin, which is yellow. In some centres haematoxylin–phloxine–safranin (HPS) staining is routinely used. The commonly used terms 'eosinophilic neuron', 'pink neuron' and 'red neuron' reflect the widespread histological use of eosin dye. Acidophilic neurons are important to distinguish from dark neurons (biopsy artefact), as the latter are not injured lethally but represent perturbed neurons at the time of fixation (see Experimental neuropathology, later in chapter).

Specificity of Acidophilic Neurons

Acidophilic neurons are seen in several acute, non-ischaemic causes of neuronal death, such as viral encephalitis or experimental seizure activity. The term 'ischaemic neuron' should thus be avoided as a morphological descriptor of neuronal acidophilia, as it is too specific with respect to cause. Necrotic neurons develop an affinity for acid dyes regardless of the cause of neuronal death.

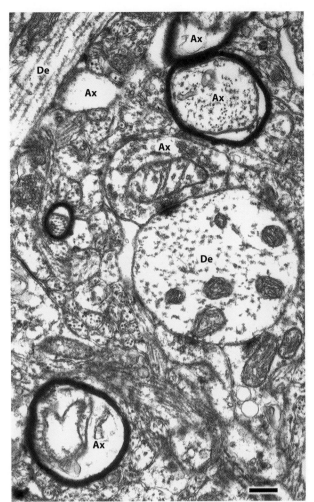

2.19 Selective swelling of axons in early pan-necrosis. Axons (Ax) are subject to transmembrane water fluxes and swelling. Dendrites (De) are not swollen and contain dark mitochondria. This selective axon lesion is almost a mirror image of the excitotoxic lesion in selective neuronal death and suggests an entirely different mechanism for pan-necrosis, not an exaggeration of the mechanism of selective neuronal necrosis. Pan-necrosis, often sparing of the nerve cell body, is seen in ischaemia, epilepsy with focal acidosis, and other acidotic, neuropil-cavitating conditions such as Wernicke's disease and mitochondrial encephalopathies. Neuropil cavitation, or pan-necrosis, does not result from pure hypoglycaemia, as acidosis cannot occur. Note that axonal mitochondria are swollen to several micrometres in diameter. Bar = 1 μm.

Electron Microscopy of the Acidophilic Neuron

The ultrastructural features of acidophilic neurons are those of cellular necrosis, not apoptosis. Acidophilic neurons have mitochondrial flocculent densities and large, confluent breaks in the nuclear and cell membrane (Figure 2.20). Cell membrane rupture precedes nuclear membrane abnormalities, the reverse order of apoptosis. Concomitant with the appearance of cytoplasmic acidophilia is the ultrastructural appearance of mitochondrial flocculent densities (Figure 2.19) related to irreversible protein alterations in the cell. Mitochondrial flocculent densities have been shown to be proteinaceous by their disappearance when the tissue is injected with trypsin or proteinase.[1022]

Delayed Neuronal Death

'Delayed neuronal death' or 'maturation phenomenon', was first applied to observations in the hippocampus of rodents subjected to global ischaemia[469,539] but may also occur after brief, focal ischaemia.[713] The fundamental observation is that neurons do not die immediately after the 5- to 10-minute period of global ischaemia but rather in the hours to days afterwards. In both experimental conditions and humans,[438] delayed neuronal death refers to the slow neuronal degeneration occurring even remotely days after an episode of complete ischaemia of short duration rather than that following chronic, partial incomplete ischaemia.[236,730]

Cell Death: Apoptosis versus Necrosis and Necroptosis

The death of cells in the CNS following injury may occur by combinations of apoptosis, necrosis, and hybrid forms along an apoptosis–necrosis continuum ('necroptosis'). Apoptosis is the best characterized form of programmed cell death in vascular diseases and the principal type of delayed neuronal death in ischaemia. Mild ischaemic injury preferentially induces cell death by an apoptotic-like mechanism rather than necrosis. Although some controversy prevails,[842] there is good evidence from experimental studies that apoptosis predominates in the hypoperfused regions of the brain, e.g. the penumbra. The triggers of apoptosis include oxygen free radicals, ionic imbalance, DNA damage, protease activation and death receptor ligand binding. There are two main types of apoptotic signalling cascade after cerebral ischaemia. Apoptosis can be initiated by internal events (the intrinsic pathway) involving disruption of mitochondria and the release of cytochrome C. The other principal pathway involves the binding of ligands to cell surface death receptors such as the Fas and tumour necrosis factor (TNF) receptors (extrinsic pathway).[143] Both pathways result in the activation of a series of downstream caspases, leading to DNA cleavage and cell death.[617]

Necroptosis

Until recently necrosis was considered an uncontrolled mode of cell death. It is now known that necrosis can be regulated by several molecular mediators.[201] ATP depletion is considered a key factor in initiating ischaemia-induced necrotic cell death. Serine/threonine kinase receptor-interacting protein 1 (RIP1) plays a crucial role in the initiation of necrosis by ligand-receptor interactions. Programmed necrosis (necroptosis) during ischaemic injury may be activated upon stimulation of death receptors by several ligands (TNFα, FasL and Trail) that can also activate apoptosis.

Autophagy

Autophagy (Greek 'self eating') is a regulated cell process that is activated for bulk removal of cellular proteins and organelles. Depending on the circumstances, the role of autophagy in cell death can be cytoprotective or cytotoxic. In autophagic degradation, proteins are targeted and transported to membrane-enclosed vesicles, which then fuse with lysosomes allowing their contents to be degraded by

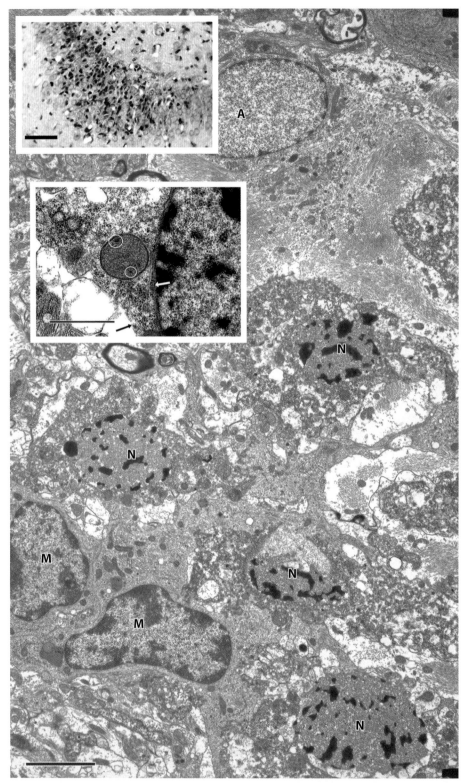

2.20 Electron-microscopic features of acidophilic neurons. Colour inset: Acidophilic neurons are seen in the dentate gyrus of the hippocampus. Electron microscopy of such a field shows three neurons (N) with the typical coarse tigroid nuclear pattern of necrosis, with ruptured nuclear and cytoplasmic membranes, and amorphous cytoplasm. An astrocyte (A) contains glial cytoplasmic fibrils. Two macrophages (M) are at the lower left. Greyscale inset: early features of necrosis are rupture of the cell membrane (black arrow) whereas the nuclear membrane (white arrow) is still intact. Two mitochondrial flocculent densities, also characteristic of necrosis, are in white circles. Bar = 1 μm; colour inset, bar = 100 μm.

lysosomal enzymes such as cathepsins B and D. Besides soluble cytosolic proteins, effete organelles such as mitochondria and peroxisomes or regions of the Golgi and endoplasmic reticulum are removed by autophagy. Autophagy can be activated after ischaemic injury,[819,825] in both neurons and glia.[609] There are three basic types of autophagic processes: macroautophagy, microautophagy and chaperone-mediated autophagy. The autophagy pathway is mediated by up to 50 autophagy-related genes and their encoded proteins such as ATG5, ATG12 and microtubule-associated protein light chain 3 (LC3).[825]

Selective Vulnerability

What is Selective Vulnerability?

Many of the insults that may be delivered to the entire brain produce only restricted regions of brain damage. Almost all diseases in neuropathology show selective vulnerability but little is known of the mechanisms. Because a pattern of selective vulnerability constitutes a type of 'signature' of a disease or a class of disease processes, it is diagnostically important. It may also provide insights into the pathogenesis of brain injury caused by a specific deficiency, genetic defect or toxin.

Selective Vulnerability in Hypoxic and Ischaemic States

Several factors contribute to selective vulnerability. Much progress has been made in understanding some forms of selective vulnerability. Experimental studies have shown that transient global cerebral ischaemia (GCI) causes accumulation of glutamate in the extracellular fluid of the brain,[106] leading to 'excitotoxic neuronal death': over-stimulation of NMDA receptors, excessive neuronal entry of Ca^{2+}, stimulation of a range of enzymes such as neuronal nitric oxide synthase, phospholipases and calpain, oxidative and enzymatic damage to macromolecules, and cell death. The NMDA receptor is abundant in the dendritic fields of vulnerable neurons[693] in the CA1 field of the hippocampus.[615,965] The synaptic connections that facilitate LTP make these hippocampal neurons particularly vulnerable. Such vulnerability is probably an exaggeration of the normal physiological mechanism that allows long-term changes in excitability. The selective vulnerability of CA1 over CA3 neurons may relate to higher concentrations of NMDA receptors on CA1 neurons, and their less robust and smaller perikarya, with less Nissl substance.

Other pathogenetic mechanisms may also be involved. For example, other excitatory receptors, such as AMPA receptors, may play a role in neuronal death after transient GCI[308,348] Neurotoxicity may be enhanced by zinc, known to be present in synaptic vesicles and co-released with transmitter:[309,951] zinc released into the synaptic cleft has been estimated to reach potentially neurotoxic concentrations.[63,441] Zinc in the hippocampus, however, is present primarily in *en passant* axonal dilatations of mossy fibres ending on CA3 neurons,[321] which are relatively resistant. There are numerous enzymatic differences between the vulnerable CA1 zone and the resistant CA3 zone.[315]

The dentate granule cells of the hippocampus are rich in NMDA receptors and yet are relatively resistant to ischaemia. In hypoglycaemic brain damage, where massive quantities of aspartate and also glutamate are released into the brain extracellular fluid,[899] the dentate gyrus is clearly vulnerable, in a pattern showing a relationship to CSF spaces.[74,75] It thus seems that access of an excitatory neurotransmitter to vulnerable neurons and their receptive fields must be considered, in addition to the intrinsic properties of the neuron, to give a comprehensive explanation of selective vulnerability.

In most brain regions, GCI causes selective loss of GABAergic neurons. Selective loss of GABAergic neurons occurs in the cerebral cortex,[950] thalamus (lateral reticular nucleus[314,872,957]) and striatum.[307] In contrast, in the hippocampus the CA1 zone shows preservation of GABAergic interneurons,[487] in spite of the stronger excitatory input that CA1 GABAergic neurons receive.[1110] The hilus of the hippocampus shows loss of somatostatinergic innervation of GABAergic neurons, rather than loss of GABAergic neurons.[488] Selective GABAergic neuron loss is one plausible explanation for epilepsy after ischaemic states.[950,958] The fundamental cause of selective destruction of neurons containing GABA is unclear but may be related to activation of excitatory receptors other than those of the NMDA subtype.[872]

Other examples of selective vulnerability include that of the cerebellar Purkinje cells to transient GCI. This was initially difficult to reconcile with the calcium hypothesis of neuronal cell death, because calcium was thought to enter neurons mainly by NMDA-gated channels or by voltage-sensitive Ca^{21} channels activated by depolarization. The use of cerebellar cultures enriched in Purkinje cells allowed the discovery of an AMPA subtype of glutamate receptor with direct Ca^{2+} permeability,[627] perhaps explaining the vulnerability of cerebellar Purkinje cells to degeneration in excitotoxic situations such as GCI.

Sometimes, selective vulnerability of a particular brain region is striking and easily correlated with a known pathogenetic factor. For example, the globus pallidus and pars reticularis of the substantia nigra are the two brain regions richest in iron and vulnerable to necrosis in CO toxicity (see Carbon Monoxide Toxicity, later in chapter). The high affinity of the CO molecule for the cytochrome haem iron in these regions far exceeds that of oxygen itself. This is a good example of histotoxic hypoxia. Similarly, necrosis of the putamen implicates methanol. Large neuronal size does not adequately explain selective vulnerability. For example, the largest neurons in the brain, the Betz cells of the motor cortex, are relatively resistant to a transient global ischaemic insult.[705]

CONSEQUENCES OF HYPOXIC INSULTS

As in the case of ischaemia, cerebral hypoxia results if the process or cascade of oxygen (O_2) delivery is interrupted from through the route from ambient air to the brain. The pO_2 decreases in steps from the ambient air, through the lungs and blood and finally to the brain. O_2 molecules always flow from an area of higher partial pressure to an area of lower partial pressure. Each step in this cascade allows O_2 molecules to 'flow down' a portion of the gradient, culminating in the delivery of O_2 to the tissue. At sea level, these steps give rise to an overall drop in pO_2 from

21 kPa (158 mmHg) in ambient air, to 4 kPa (30 mmHg) as the normal value for tissue pO_2. Values for focal partial pressures of O_2 in mitochondria are even lower.

Hypoxaemia

Hypoxaemia (Table 2.1) may result from restrictive or obstructive pulmonary parenchymal disease, upper airway obstruction or a low inspired oxygen concentration, such as during high-altitude mountaineering (Figure 2.21). At normal pO_2 values, over 98 per cent of the oxygen-carrying sites on the haemoglobin molecule are occupied by oxygen in arterial blood, with most molecules having all four sites occupied by O_2 and only a few having three O_2 molecules. In the superior sagittal sinus, roughly 70 per cent of the oxygen-carrying sites on haemoglobin molecules are occupied, i.e. three sites carry O_2 on most haemoglobin molecules and two sites carry O_2 on some. The oxygen extraction fraction (OEF) is thus 28 per cent (98 minus 70), implying that 28 per cent of the oxygen carried into the brain is extracted by passing through the brain. The OEF can increase in states of hypoxia, as part of the adaptation of the brain to hypoxia. In addition to O_2 carried on haemoglobin, a small amount of O_2 is dissolved in the blood (roughly 0.3 mL/dL), none of which is returned from the tissue to venous blood.[415] In hyperbaric oxygen treatment of profound anaemia, the dissolved oxygen alone is sufficient to supply tissue needs.[1003]

Brain Compensation in Hypoxaemia

Several mechanisms act together over acute and chronic adaptation to hypoxia (Box 2.2) to uphold CBF,[149] cerebral oxygenation, the cerebral metabolic rate and the cerebral rate of oxygen consumption ($CMRO_2$). Experimental work

shows a slight decline in phosphocreatine and ATP and an early increase in tissue lactate, followed by the achievement of a new, hypoxic steady state. Subsequent to these metabolic adjustments, brain pH and tissue pCO_2 remain stable during hypoxaemia,[432] indicating that tissue washout and removal of wastes are unimpaired.

Hypoxic Neovascularization

Formation of new vessels is stimulated in the brain by hypoxia,[120] mediated by a sequence of gene activation (Figure 2.2). HIF-1α activates VEGF, erythropoietin and glycolytic enzymes[927] and genes in many other classes.[829] HIF-1α is the key upstream mediator of hypoxia.[217,926] The primary oxygen-sensing mechanism that turns on HIF-1α is a reduced O_2-dependent degradation (see earlier). HIF-1 is a heterodimer of HIF-1α and HIF-1α, subunits and the increased HIF-1α allows heterodimerization of these subunits. This induces a conformational change that allows the complex to bind to hypoxia response elements. Induction of VEGF is the critical event in the growth of new capillaries, preceding brain vascularization itself.[650,1067] Similar upregulation and neovascularisation also occur in experimental models of intermittent hypoxaemia that simulate sleep apnoea.[89]

These changes all occur along a smooth spectrum of physiology, not pathophysiology, because HIF-1α levels vary over the physiological range of O_2 tension.[485] However, damage can occur when newly formed vessels leak, especially under the high CBF conditions of hypoxia. This leakage is the basis for brain water accumulation and high-altitude cerebral oedema, which can be fatal[384] or leave permanent neuropsychiatric impairment because of damage to the globus pallidus.[1032]

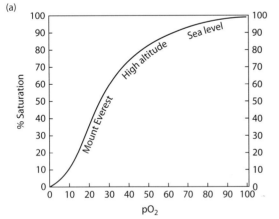

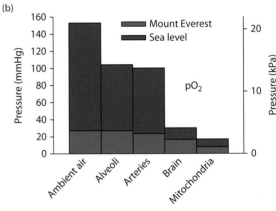

2.21 Oxygen availability to brain. (a) Oxygen–haemoglobin dissociation curve, showing sigmoid shape of saturation as a function of pO_2; 25 per cent, 50 per cent, 75 per cent and 100 per cent saturation correspond to, on average, one, two, three and four O_2 molecules per molecule of haemoglobin. At sea level, most oxygen delivered in arterial blood consists of the fourth O_2 molecule and often the third, leaving two or three O_2 molecules in venous blood haemoglobin molecules and an average venous saturation of 70 per cent. At 4 kPa (30 mmHg) on Mount Everest, however, haemoglobin is just over 50 per cent saturated; under such circumstances of extreme altitude, even arterial blood reaching the brain carries only two or three O_2 molecules per haemoglobin (as does venous blood at sea level). At the summit of Mount Everest, venous blood departs the brain with only one or two O_2 molecules per haemoglobin molecule, utilizing a different part of the oxygen–haemoglobin dissociation curve to effect oxygen delivery at altitude. **(b)** Stepwise drop in pO_2 at each successive step of oxygen delivery. Although the ambient pO_2 on top of Mount Everest is equal to the brain tissue pO_2 at sea level, a staircase of declining pO_2 at every step along the anatomical pathway of O_2 delivery nevertheless exists at all altitudes. The staircase is considerably flattened on Mount Everest and in the most hypoxic patients. The lowest recorded arterial pO_2 in a living human was 1 kPa (7.5 mmHg),[360] with a corresponding venous pO_2 of 0.27 kPa (2 mmHg).

- Hyperventilation: occurring almost immediately and accompanied by a respiratory alkalosis. Hypoxic hyperventilation is due to an increase in tidal volume.
- Oxygen extraction from blood: oxyhaemoglobin dissociation curve shifts to the right, decreasing affinity for O_2 with the ultimate consequence of decreasing brain tissue pO_2.
- Glycolysis: net effect is a relative preservation of ATP at the expense of phosphocreatine.
- CBF: hypoxic hyperaemia compensates for the decreased arterial blood O_2 by increasing flow through the tissue microcirculation even when brain tissue pO_2 is low.
- Erythropoietin: production increases circulating haemoglobin.
- Capillary density: hypoxic brain neovascularization is important for adaptation over time.

All above mechanisms act together during acute and chronic hypoxia to uphold cerebral oxygenation, the cerebral metabolic rate and even the cerebral rate of oxygen consumption ($CMRO_2$).
ATP, adenosine triphosphate; CBF, cerebral blood flow; pO_2, partial oxygen tension.

Tissue pO_2 Measurement in Hypoxia

There is a direct proportional relationship between brain tissue pO_2 and the FiO_2.[857] Measurement of brain tissue pO_2 shows that hypoxaemia leads to a further decrease in the tissue pO_2 from the normal. Tissue pO_2 levels can fall from 2.7–5.4 kPa (20–40 mmHg) to < 1 kPa (< 5 mmHg) during severe hypoxaemia[780] and can reach < 0.25 kPa (0–2 mmHg) when breathing 2–3.5 per cent O_2.[912] The effect is a flattening of the steps in the staircase of oxygen delivery (Figure 2.21b). In the non-injured brain subjected to iatrogenic ischaemia, e.g. during neurosurgical aneurysm clipping, a tissue pO_2 of 1.3 kPa (10 mmHg) can be tolerated before tissue hypoxia becomes critical and tissue pH decreases.[273,430] Brain tissue pO_2 monitoring appears to offer prognostic value in head injury, correlating with CBF better than either brain pCO_2 or pH.[261] Poor outcome in head injury was found to correlate with tissue pO_2 of <2.5 kPa (19 mmHg).[262]

High-Altitude Hypoxia and Brain Damage

When arterial oxygen levels drop acutely to 4 kPa (30 mmHg) or less, such as on rapid ascent to high altitudes, consciousness is dulled and memory impaired, unless considerable time elapses for adaptation and inspired oxygen is lowered gradually. Although neuronal necrosis is not produced,[688] synapses are affected by hypoxia[950] and synaptic dysfunction causes a syndrome that has been termed 'high-altitude stupid' (HAS). Even an experienced mountaineer may lose accrued climbing wisdom, show clear errors of judgement and make poor decisions at high altitudes. However, the syndrome reverses on return to low altitudes and permanent deficits are not detectable on neuropsychological testing.[479]

Adaptation occurs after several days at high altitude.[149,562] The increased CBF reaches a steady state after roughly 6 days[269,928] but the new capillaries formed at elevation in the remodelled brain vasculature may leak.[396,575] The resulting cerebral oedema is cured by return to lower altitudes,

where the hyperdynamic cerebral circulation subsides and fluid can be reabsorbed into the brain vasculature. High-altitude pulmonary oedema results from the grossly hyperdynamic pulmonary circulation at altitude. A paradoxical increase in O_2-derived free-radical generation may play a role in both cerebral and pulmonary oedema at high-altitude.[82] Thus, high-altitude hypoxia is accompanied by several distinct complications (HAS, cerebral and pulmonary oedema) but causes more change in tissue capillaries and brain synapses than in neuronal bodies. A plethora of other neurological conditions can occur at altitude,[95] not all necessarily due to hypoxia.

Severe Hypoxia Other Than at Altitude

Although very low arterial pO_2 levels of about 3 kPa (20 mmHg) are held widely to be incompatible with survival, survivors of such extreme hypoxia do not in fact have any permanent long-term physiological impairment.[360] A decerebrate state after pure hypoxia is also compatible with long-term recovery, free of neurological signs and symptoms.[360,890] For example, a 2-year-old boy with obstructive bronchitis remained in a coma for over 2 weeks without neurological residua after profound hypoxia.[890] Autopsies of subjects with pure hypoxia show no readily discernable pathological changes.[847] Thus, acute, severe, hypoxic hypoxia occurring without cardiac failure is not accompanied by brain damage. In contrast, when acute hypoxia is accompanied by cardiac arrest, e.g. as in near-drowning and pure nitrogen inhalation,[449] there is severe brain damage, seemingly worse than after cardiac arrest alone.[134,209] In this case hypoxia worsens the tissue damage from ischaemia.[688]

Experimental Hypoxia

The effects of uncomplicated acute hypoxic hypoxia on brain damage have been tested in controlled laboratory experiments (in contrast the widely known Levine model of unilateral carotid occlusion and immersion in nitrogen is relatively uncontrolled).[594] Neonatal rat pups with unilateral carotid artery ligation that were exposed to 7.7 per cent O_2 for 2 hours did not show brain damage in the hemisphere exposed only to hypoxia, opposite the occluded artery.[9] However, the neonatal brain differs from the adult brain in its response to global ischaemia in several respects, one of which is that glutamate is not released in newborn animals.[192] Younger mammals and infants also have lower basal metabolic rates,[109] and an adaptation to high altitude[427,428] and hibernation.[313,534]

In spontaneously breathing adult animals, hypoxia causes a precipitous fall in blood pressure, the animals sustaining damage in arterial boundary zones.[138] Profound hypoxic hypoxia by itself does not produce cerebral necrosis; concomitant hypotension, either spontaneous (i.e. due to the hypoxia) or induced, is necessary to produce cerebral necrosis.[234,688]

Hypoxia and Brain Synapses

There is no clear evidence that necrosis of cells or tissue results from uncomplicated hypoxic hypoxia but synaptic alterations can occur. These include expansion of presynaptic terminals containing multilamellar bodies, aggregated or clumped synaptic vesicles and various tubular arrays and profiles. Similar ultrastructural changes can be seen in other elements of the brain, including neuronal perikarya, glia and vascular cells.[1114,1115] Hamster brains exposed to hypoxia showed reversible swelling of mitochondria in neuronal perikarya.[386]

Infant non-human primates exposed to hypoxia of 4 per cent O_2 (3 kPa, 20–22 mmHg) showed selective synaptic degeneration of GABAergic terminals but no neuronal necrosis.[950] In mild global ischaemic insults, the lateral reticular nucleus of the thalamus, a solely GABAergic nucleus, is susceptible neuronal necrosis.[873,957] Loss of GABAergic synapses or neurons may explain post-hypoxic or post-ischaemic seizures.

The time-course of clinical recovery from pure hypoxic coma is up to 2 weeks.[360,890] The timing of recovery coincides with synaptic repair. Although ganglioside breakdown has been found in human infant brains exposed to hypoxia,[820] adult brains do not show change in free fatty acids but only small changes in prostaglandins and eicosanoids.[792] A combination of hypoxia with alcohol, to simulate sleep apnoea in alcoholics, did not produce evidence of brain damage, underscoring the innocuous nature of hypoxia, even in presence of alcohol.[219]

Anaemia

As in hypoxic hypoxia, compensatory mechanisms intervene to preserve brain oxygenation. When haemoglobin levels drop, CBF increases proportionally in an inverse relationship. Conversely, polycythaemia lowers CBF and increases stroke risk.[543,781] Patients with leukaemia can have very low haemoglobin levels but without signs of inadequate cerebral oxygenation unless leukostasis and very high white blood cell (WBC) counts (>300 000) supervene, when cerebral haemorrhage (not 'anaemic hypoxia') can result.[312] As in hypoxia, therefore, there is no evidence that uncomplicated anaemia causes brain damage.

Hypoxia versus Ischaemia

The mechanisms of tissue damage in hypoxia and ischaemia are quite different. In hypoxaemia, only delivery of O_2 is impaired, not removal of products of metabolism. Because CBF is maintained or increased in hypoxia,[3,207,343,796] other molecules continue to be delivered to the brain. Waste products such as CO_2 and H^+ also continue to be removed unabated, accounting for normal tissue pCO_2 and pH in pure hypoxaemia.[431] Hypoxia is thus a much simpler and less dire primary insult than ischaemia.[636] When hypoxia is added to ischaemia, necrosis occurs, and hypoxia then modulates the degree of damage.[688]

Neurotransmission Failure and Energy Failure

In the hypoxic brain, neurotransmission ceases before energy failure.[21,64,664] For example, there is no evidence of energy failure in the medulla at an appropriately early time that would account for hypoxic apnoea.[267] Lactate accumulation occurs without energy failure,[895] and so energy failure cannot be invoked to explain hypoxic lactate accumulation. Lactate begins to increase when the oxygen content of inspired air drops to 12 per cent, corresponding to a saturation of O_2 (S_aO_2) of 50 per cent.[381] When inspired oxygen is 7 per cent, corresponding to an S_aO_2 of 23–35 per cent, there is some hydrolysis of phosphocreatine but ATP levels are maintained.[381] Still lower inspired oxygen concentrations of 2–3 per cent cause electroencephalogram (EEG) changes and the phosphocreatine/creatine ratio decreases.[912]

Events in hypoxia do not progress beyond this reversible pathophysiological state of electrical failure and early energy failure. A steady state is reached,[381] unlike in ischaemia. Unlike in ischaemia the release of glutamate[106] is absent in pure hypoxaemia.[780] As a consequence, the stage of tissue necrosis is never reached in the brain during hypoxaemia.

Gene Expression in Hypoxia and Ischaemia

Analysis of gene activation in hypoxia and ischaemia shows further differences between these two kinds of insults (Table 2.4). A new steady state is achieved in hypoxia, with upregulation of mainly regulatory genes whereas in ischaemia there is mass activation of genes, in association with tissue destruction (see earlier).

Many classes of genes are differentially induced and regulated by hypoxia and ischaemia. These include genes for transcriptional regulators (*c-fos*, *c-jun*, *junB*, *TIS8* (*5zif-268*), *krox-20*), stress proteins (e.g. heat-shock proteins), glucose transporters, haem oxygenase, growth factors, interleukin-converting enzyme (ICE)-like proteases, bcl family (e.g. *ICE*, *Nedd2*, *Yama/CPP32*, *bcl-2*, *bcl-x*) and caspases involved in apoptosis. Some genes have adaptive value when stimulated,[233,525,1107] whereas others may be harmful.[60,226,277] Thresholds for activation of gene transcription vary. Even physiological stimuli and neurotransmitter (acetylcholine) release can activate genes such as the transcriptional regulator *c-fos*.[363,448,854,880] Stronger stimuli turn on heat-shock proteins, and still stronger stimuli switch on the genes for apoptosis. In hypoxaemia, one of the first genes to be upregulated is that for erythropoeitin.[659] Some genes stimulated are pro-apoptotic, such as *BNIP-3*.[150] Other genes, such as that for haem oxygenase, are turned on by ischaemia but not hypoxia alone.[108] The genes encoding heat-shock proteins such as hsp72,[538,742] tumour suppressor genes[199] and transcriptional regulators, such as *c-fos*, *c-jun*, *junA* and *junB*,[210,760] are all stimulated by ischaemia (Table 2.4).

Clinical Differences between Hypoxia and Ischaemia

A major reason to distinguish hypoxic and ischaemic insults[688,943] clinically is to predict the outcome of global insults that result in coma. Hypoxic coma, although rare, is reversible and tends to occur in young people. If a grave prognosis is spuriously ascribed to a coma accompanying a pure hypoxic insult, then treatment could be withdrawn from a patient who has not suffered widespread brain necrosis.

In medicolegal review, it is important to determine whether cardiorespiratory arrest or only respiratory arrest has occurred. The prognosis is very different if cardiac arrest has not occurred and blood pressure has been maintained. Total cerebral ischaemia of only 2 minutes can cause neuronal necrosis[957] whereas even profound arterial hypoxia, without cardiac arrest or hypotension, does not.[360,847,890]

Causes of pure hypoxia include allergic reactions and tracheobronchial infections (see Table 2.4). Mostly young people demonstrate this syndrome of respiratory failure without heart arrest. Hypoxic coma persists for up to 2 weeks, usually followed by a complete recovery.[360,847,890] The underlying brain-repair process is synaptic, accounting for this time course and eventual recovery. If only hypoxia has occurred, no neuropathological abnormalities will be detectable on conventional examination,[360,847,890] whereas GCI causes either extensive brain necrosis[209] or the changes of a non-perfused brain if capillary closure has occurred. The neuropathologist is often called upon to confirm a diagnosis of non-perfused ('respirator') brain. This pathophysiological and pathological distinction between hypoxia and ischaemia is thus important for the pathologist as well as the clinician.

Hyperoxic Brain Damage

Hyperoxia alone at normobaric pressures does not seem to damage the adult brain, but at hyperbaric pressures, such as those encountered during diving,[100,257,258] 100 per cent O_2 is toxic and causes brain necrosis.[86] Early changes are seen in cell processes and especially in mitochondria.[87] In the neonate, the situation is different, as even normobaric hyperoxia can cause widespread neuronal necrosis.[19]

The combination of hyperoxia and unilateral carotid ligation has the converse effect to that of hypoxia and carotid ligation. In hyperoxia alone, i.e. in the hemisphere contralateral to the ligation, necrosis is seen, but in the ipsilateral, carotid-ligated hemisphere, there is no necrosis.[86] Thus, the reduced flow protects the ligated ipsilateral hemisphere from hyperoxic damage.

Toxicity of hyperbaric gases is not discussed here. See review 237 on the narcotic and toxic properties of CO_2, O_2 and N_2 (nitrogen narcosis) at normal and high pressures.

Asphyxia

Hypoxic conditions arise from asphyxia, a term that denotes an inability to breathe. Forensic analysis distinguishes three causes: (i) suffocation, (ii) strangulation and (iii) chemical asphyxia.[746] Chemical asphyxia paralyzes breathing by causing hyperpolarization of the respiratory control neurons in the medulla oblongata. Although suffocation and strangulation are discussed here, more detailed analysis is available in textbooks of forensic neuropathology.[746]

Suffocation includes environmental suffocation e.g. lack of breathable O_2 in the ambient environment, smothering and gagging (blockade of external air passages), choking (bolus or aspiration, manual compression of trachea), drowning (H_2O contacting larynx, causing laryngospasm) and mechanical suffocation as in a snake encircling and progressively tightening around the victim's thorax.

Strangulation implies neck constriction in addition to ventilatory obstruction and includes band strangulation (a band is tightened around the neck), garrotting (a rope or a finer, neck-cutting ligature is tightened slowly, from behind), manual strangulation (occlusion of neck by fingers and hands) and mugging (forearm around the neck applied from the rear). In the special case of hanging, the weight of the individual is used to tighten a constricting band or rope around the neck.

'Birth asphyxia' refers to the mechanical constriction of the thorax in the birth canal during delivery and is physiological, ending with the first breath.

'Birth Asphyxia': Neonatal Hypoxia and Cerebral Palsy

Hypoxia is physiological *in utero*. The fetus has different, 'stickier' haemoglobin, facilitating O_2 transfer from mother to fetus. The temporary inability to breathe due to chest wall compression during a vertex delivery is normal if the umbilical cord is supplying oxygenated blood, as are physiological decelerations in fetal heart rate. Breech delivery, with the umbilical cord pressed against the pelvic rim until delivery of the aftercoming head, initially seems a plausible risk factor for cerebral ischaemia, cerebral necrosis and cerebral palsy. Large studies of breech deliveries, however, have shown no risk associated with vaginal delivery versus caesarean section for the development of subsequent neurological deficits.[225,767]

Converging evidence suggests that brain necrosis and cerebral palsy are not causally related to the birth process except in rare instances. The loose attribution of cerebral palsy to 'birth asphyxia' should be avoided. The generally litigious climate attributing cerebral palsy to obstetric or neonatal paediatric malpractice is based on spurious claims. Cerebral palsy has an incidence of roughly 1.5–3 per 1000 live births, generally unchanged over time.[795] Cerebral palsy is now known not to be linked to hypoxia of birth and rates have remained unchanged over the time period in which increased oxygen has been loaded successfully into newborn infants.[724] Hypoxia even in adult animals[688,780] and adult humans[360,847,890] causes no necrosis unless accompanied by ischaemia. The neonatal brain has a lower metabolic rate and is even less sensitive than in the adult, and so the neonatal hypoxic brain does not even activate some genes such as *HSP32* that are turned on by ischaemia.[108]

Epidemiological evidence over decades indicates a general success of the medical establishment in increasing newborn blood oxygen levels. The major effect has been not a decline in cerebral palsy but rather a decreased incidence in the diagnosis of 'birth asphyxia'.[1099] Neither the birth process itself nor isolated transient hypoxia has causal association with cerebral palsy, but several other disorders have been discovered to be associated with a risk. These include low birth weight, disorders of coagulation and intrauterine exposure to infection or inflammation, all of which show a positive association with cerebral palsy.[528,724]

Histotoxic Hypoxia

This describes the interference by toxins in the inability of the cell to use oxygen as an electron acceptor in the mitochondrial electron transport chain, preventing the oxidation of glucose to CO_2 and H_2O. Sulphide (S^2) cyanide (CN^2) and azide (NaN_3) can all inhibit mitochondrial cytochrome oxidases, sulphide being slightly more potent than cyanide and azide being relatively weak.[956] Sulphide,[8] cyanide[137] and azide[670] can also cause immediate apnoea and death, a phenomenon too rapid to be explained by inhibition of metabolism. This is due to their immediate effect on respiratory neurons of the medulla.

Sulphide, Cyanide and Azide

Exposure to sulphide is seen clinically in a number of circumstances. Hydrogen sulphide is formed in sewers and, together with occupational exposure to H_2S-containing gas ('sour gas') in the oil and gas industry,[156] accounts for most cases of histotoxic hypoxia the neuropathologist is likely to encounter. Cyanide exposure occurs industrially, and in suicide and homicide attempts, because the chemical is easily available. Granular crystallized prussic acid, Zyklon B, was used in Hitler's genocide in the Second World War in the gas chambers. The admixture with acid produced free cyanide gas, lethal at 300 parts per million (ppm). Although azide is an important industrial chemical, used also in rocket fuels and as a herbicide, insecticide and molluscicide, it is no longer used as a fumigating agent in buildings. Historically, there was widespread use of both cyanide and azide to rid ships, buildings, rooms and apparatuses of both infection and infestation by insects. Sometimes, the fumigator perished because of the action of these agents, in a manner similar to workers exposed to natural gas or sewers that contains H_2S. Gaseous sulphide smells of rotten eggs, and cyanide smells of apricot seeds or bitter almonds.

Exposure to any of these three agents causes brain damage, but heart failure always supervenes in sulphide-,[8] cyanide-[405] and azide-related[670] injury. Hence, the brain damage is attributable to the associated hypotension. Inhaled H_2S passes through the lungs into the blood and dissociates in an aqueous equilibrium: $H_2S \rightleftharpoons H^+ + HS^- \rightleftharpoons 2H^+ + S^{2-}$.

Exposure to low to moderate ambient concentrations of H_2S at 20–50 ppm causes eye and lung irritation.[614] Very low concentrations are sought by people for 'cures' at sulphur springs, where the associated air has the faint odour of rotten eggs, believed to be healthful. Higher concentrations of H_2S paralyse the olfactory nerves and sense of smell, making it impossible to recognize the signal rotten-egg odour.

The mechanism of immediate death is too fast to be accounted for by necrosis of cells due to cytochrome binding. Inhalation of 500 ppm sulphide causes immediate apnoea, related to hyperpolarization of neurons in the medulla oblongata that control breathing.[551,833] Together, the anosmia and apnoea obviate the possibility of life-saving self-removal in H_2S exposure. The immediate death often leads to a scenario in which a missing person exposed to H_2S is sought after by rescue workers, who are themselves then overcome by the gas. No neuropathological changes are usually evident in such cases.

If exposure is survived, however, brain necrosis can be evident. Whether necrosis is due to direct histotoxicity, cardiac hypotension or standstill is not clear in these physiologically uncontrolled human observations. Cardiac function and blood pressure remain unknown at the time of exposure, and GCI to the brain can thus never be ruled out. Experimental work suggests that the cerebral necrosis relates to the potent and immediate depression of blood pressure by cyanide or sulphide. Exposure to even very high (supra-lethal without ventilation) concentrations of these agents is incapable of producing cerebral necrosis unless hypotension supervenes.[84,137] Cyanide causes severe perturbations in cerebral energy metabolism, including reduction in cytochrome oxidase activity, depletion of ATP and production of lactate but not brain necrosis.[637] The situation is similar with sulphide exposure, in which necrosis does not occur even after several hours of EEG isoelectricity (flat EEG). In one series, a single ventilated animal that received a very high dose of sulphide (a supra-lethal dose in the unventilated animal) showed cerebral necrosis;[84] physiological monitoring of this animal had revealed persistent hypotension to <4.7 kPa (35 mmHg) for more than half an hour. Animals that do show necrosis in studies of both cyanide and sulphide encephalopathy do so in a distribution resembling that after global ischaemia.[84,637] Agents that produce histotoxic hypoxia produce profound hypotension; like hypoxic hypoxia and anaemic hypoxia, histotoxic hypoxia does not, by itself, damage the brain.[84,137,637]

Air Embolism and Decompression Sickness

When decompression occurs after there has been equilibrium at a higher pressure, as in divers surfacing after spending some time at considerable depth, previously dissolved nitrogen will leave the liquid phase of blood and spontaneously form intravascular bubbles. These bubbles vary in size, determining where they lodge in the circulation. The resulting decompression sickness has been termed 'the bends' and is a hazard in divers.

Target organs in decompression sickness include the spinal cord,[162] as well as the skin, bone, retina[807] and ear.[769] Bubbles can be demonstrated in the blood vessels of the spinal cord 20 minutes after return to atmospheric pressure in experimental decompression sickness.[768] Examination of the spinal cord of amateur and professional divers regularly reveals spinal cord microinfarcts, even in asymptomatic individuals.[769] Thus, the clinical neurological spectrum of spinal cord lesions due to air embolism extends from patients who are frankly paraplegic[162] to divers who were asymptomatic during life and totally unaware of their subclinical spinal cord neuropathology.

Air embolism also plays a role in the neurological damage that can be seen after cardiac bypass surgery.[744] The vernacular term 'pump head' has been used to describe the resulting neurological impairment. Air introduced into the cardiac chambers during open-heart surgery can embolize to the brain.[550] One study found, in addition to hippocampal necrosis suggestive of global ischaemia, microinfarcts in the white matter and in the cerebral cortex but no crystalline emboli, a combination suggesting micro-bubble air embolism.[1026] Because air requires a considerable time to dissolve in the blood (long enough to cause infarction), air embolism can be considered a form of transient ischaemia. Permanent neurological damage may result.[153,550]

DISORDERS OF CNS HYPOXIA

'Swiss Cheese Brain'

It is important to distinguish post-mortem alterations from true ante-mortem lesions. The 'Swiss cheese brain' is a common post-mortem alteration, the appearance resembling that of cheese with holes of various sizes (Figure 2.22). This is not equivalent to pan-necrosis or a diagnosis of cystic necrosis. The yellow colour of necrosis is absent and the walls of the cysts are smooth, unlike foci of necrosis. This post-mortem alteration is due to somatic death being non-instantaneous, with hypotensive shock redistributing blood flow away from an ischaemic bowel while preserving that to the heart and brain. Ischaemic bowel is a rich source of bacteria and anaerobes characteristically find their way into the blood stream just before death. When the heart fails, it pumps bacteria that seed the brain, normally a sterile organ, in a transient peri-mortem septicaemia before death and cardiac standstill. The post-mortem growth of bacilli can be revealed histologically in the walls of the cysts (Figure 2.22).

Dark Neurons

Dark neurons are commonly seen in both clinical and experimental neuropathology. They have plagued the interpretation of tissue sections since their early observation[165] and their profiles can occasionally be seen in normal tissue.[205]

Unlike acidophilic neurons, dark neurons (Figure 2.23) represent early, reversible alterations[72,73,961] and show only marked condensation ultrastructurally. If images of the cells can be lightened by photographic or computerized image analysis, the preservation of cellular substructure is evident. Recovery of dark neurons can be demonstrated through serial study over time, with reversal of the appearance of cytoplasmic condensation,[72] the organelles and cell membranes.[72,961]

Clinical Neuropathology

Although they are very dark, dark neurons have visibly intact cell membranes and internal organelles[72] and do not signify cell death. Dark neurons occur in the early stages of neuronal injury due to ischaemia,[227] hypoglycaemia[72] and epilepsy.[961] They can be shown to recover longitudinally over time after brain insults, including hypoglycaemia[72] and epilepsy.[961]

Mechanical stimulation of the cortex causes a wave of depolarization termed 'spreading depression'[589] and such mechanical handling of living neural tissue readily

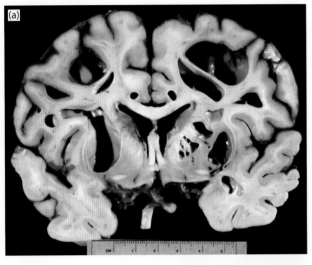

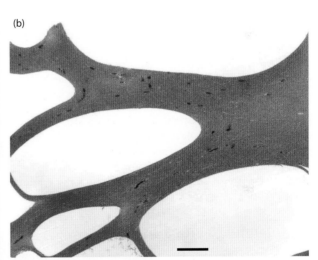

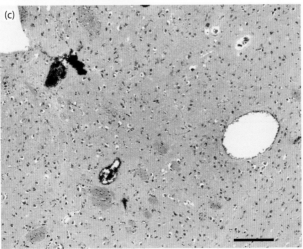

2.22 The 'Swiss cheese brain'. This alteration is easily mistaken by the uninitiated for cystic infarcts, especially if the cavities are few in number. The gross appearance **(a)** results from gas-forming anaerobic bacteria, which seed the brain peri-mortem. The cysts are smooth **(a, b)**, with no hint of yellow colour or ragged edge, as in necrosis. Gram-positive bacilli **(c)** can be demonstrated in the parenchyma and in the walls of the cysts.

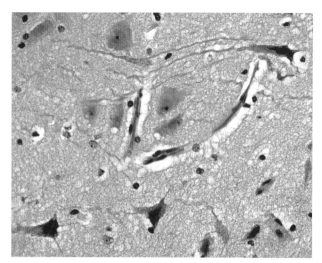

2.23 Dark neurons. Two capillaries cross the microscopic field, which includes both dark and normal neurons. Note the characteristic corkscrew dendrites of dark neurons. Dark neurons represent living neurons perturbed at the time of fixation. In life, the change is reversible.

produces dark neurons.[166] Spreading depression itself produces no permanent neuronal damage.[719] Other causes for dark neurons involve mild metabolic perturbation of the neuron: trans-synaptic stimulation and release of glutamate,[212,325,952,1039] or simply inhibition of the Na^+/K^+ ATPase pump induces dark neurons.[216,1017] A large variety of conditions can perturb the neuron into undergoing dark transformation.[166,324,529,821] Dark neurons are also a 'biopsy artefact.' Neurosurgically excised tissues while being retrieved and exposed to complete glucose deprivation *en route* to the laboratory may contain dark neurons, which are attributed to an undefined combination of spreading depression and hypoglycaemia.

Experimental Neuropathology

In experimental material, delayed fixation after death causes the tissue to be deprived of glucose, and dark neurons then appear.[166] A short post-mortem interval at autopsy in humans gives similar results to an experiment in the laboratory with delayed fixation. Neurons seem very sensitive to the perturbation that causes the cellular contraction at the time of fixation. They are occasionally seen in normal, aldehyde-fixed neural tissue.[205] Dark neurons should be distinguished from slow forms of neuronal degeneration, where the cytoplasm undergoes progressive darkening accompanied by systematic changes to organelles.[1028]

Carbon Monoxide Toxicity

Epidemiology

Exposure to even 0.04 per cent of carbon monoxide (CO) can cause collapse within a relatively short period. Unintentional exposure today causes about 40 deaths per year in the UK and 500 per year in the USA.[885,1076] This can complicate incomplete combustion of hydrocarbon fuels whilst cooking indoors or in tents, working in foundries and mines, working with running automobiles, using indoor heating during winters particularly in recreational vehicles and houseboats, water-skiing behind a boat, running unclean motors and building fires. The intentional exposure to CO occurs with suicidal or homicidal intent. The CO content of automobile exhausts has been reduced in recent years but this was long the commonest method of gaining exposure to CO with a view to suicide.

Carbon Monoxide Levels

Air pollution and cigarette consumption cause more widespread low-level exposure to CO than do the sources listed earlier. Non-smokers' normal carboxyhaemoglobin level is less than 0.5 per cent as a result of ambient air pollution and endogenous production of CO by haem oxygenase. Smokers have carboxyhaemoglobin levels that correlate with the amount smoked. Cigarette filters offer no protection as CO passes through them. Smoking one pack of cigarettes per day leads to carboxyhaemoglobin levels of 2–3 per cent. Smokers of more than two packs per day can have carboxyhaemoglobin levels of up to 6–7 per cent.

Low-dose CO exposure can be difficult to diagnose clinically. Signs and symptoms are diverse and may include headache, dizziness, nausea, vomiting, syncope, seizures, coma, dysrhythmias and cardiac ischaemia. The mechanisms of brain damage after CO poisoning are complex; CO poisoning involves several mechanisms and can cause prominent white matter damage.[814]

Brain Changes in Carbon Monoxide Poisoning

Changes in the brain in cases of CO poisoning are dependent upon the length of survival, and the severity of the initial insult. Acutely, the brain is pink because of the appearance of the bright-red carboxyhaemoglobin. Oedema may be seen, as in GCI, up to several days after the insult. Later, sharply demarcated necrosis appears in the globus pallidus.[814] If low-dose CO exposure has occurred, then incomplete necrosis of the globus pallidus will be evident. A granular appearance of the white matter may suggest Grinker's myelinopathy (see later), if survival has been long enough. Atrophy of the cerebral cortex and the hippocampus is seen and may be severe. Changes may also be seen in the substantia nigra. The lesions in CO poisoning follow the typical time course of selective neuronal necrosis or pan-necrosis in the CNS.

Mechanisms of Brain Damage in Carbon Monoxide Poisoning

Carbon Monoxide Binding to Haemoglobin

The displacement by the CO molecule of oxygen from its haem binding site on the haemoglobin reduces the effective concentration of oxygen-carrying haemoglobin. This is pathophysiologically equivalent to anaemic hypoxia. The affinity of the CO molecule for haemoglobin is almost 200 times that of oxygen itself. In survivors of CO poisoning, the reduced effective haemoglobin level is thus permanent until new red blood cells (which have a lifespan of about 120 days) are produced. To obviate this long-term CO-occupation of haemoglobin, patients are sometimes treated with hyperbaric O_2 in an attempt to displace CO by

O_2, although the hyperbaria may be unnecessary and ineffective,[152,351,497] possibly because it causes NMDA receptor activation. Compared with arterial hypoxaemia, CO hypoxia decreases cerebrovascular resistance even more,[1020] but then CO is a vasodilatory second messenger, analogous to nitric oxide (NO), acting via a cyclic guanosine monophosphate (cGMP)-dependent mechanism,[552] and has more protean effects than simply hypoxia. Paradoxically, smoking dilates the cerebral circulation and improves CBF.[127] Experimental evidence also suggests a beneficial effect under pathophysiological conditions such as organ transplantation, ischaemia/reperfusion, inflammation, sepsis, or shock states.[96]

Carbon Monoxide Binding to Brain

The second effect of CO is its direct binding to brain regions rich in haem iron. These include the globus pallidus and the pars reticulata of the substantia nigra (pallido-reticularis). The regional histotoxicity explains the selective vulnerability of the pallido-reticularis to CO toxicity (Figure 2.24). Neuroimaging techniques allow visualization of the pallidal lesion more easily than of the nigral lesion during life.

Unlike in the context of sulphide, cyanide or azide toxicity (see earlier), 'histotoxic hypoxia' can be applied here in a necrotizing sense, as CO causes histotoxicity leading to necrosis, through direct binding to brain. The binding of CO to the haem moiety of cytochromes reduces the ability of the brain to extract O_2, decreasing the oxygen extraction fraction and causing hyperaemia ('luxury perfusion') of brain.[545]

Cardiovascular Effects of Carbon Monoxide

Several factors probably influence the distribution of cerebral necrosis in CO toxicity. The binding of CO to brain

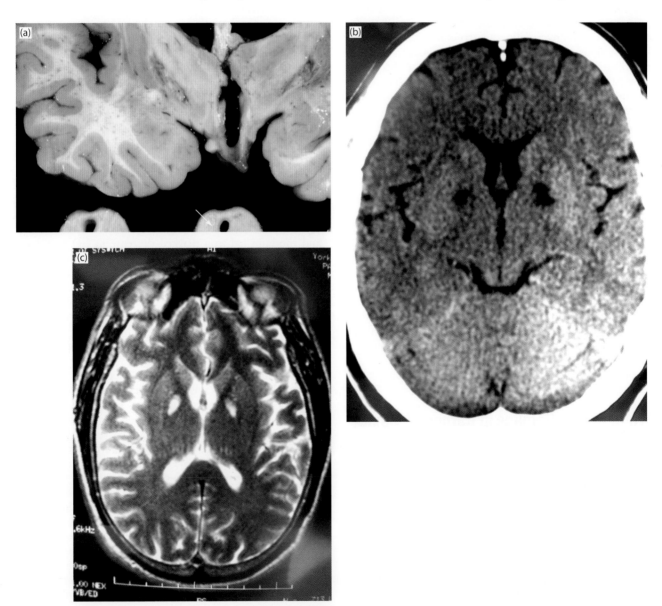

2.24 Carbon monoxide (CO) poisoning in a 5-year-old girl trapped in a house fire. (a) The characteristic necrosis of the globus pallidus and pars reticulata (arrows) of the substantia nigra is seen bilaterally. Low-dose CO exposure in a living patient in whom **(b)** CT scanning and **(c)** FLAIR MRI showed incomplete globus pallidus necrosis. The patient, who had suicidal ideation, and CO exposure from a late-model car in a garage, developed psychic akinesia with a preserved intelligence quotient.

regions rich in iron explains the necrosis in the globus pallidus and the pars reticulata of the substantia nigra. However, other brain regions are regularly affected in CO poisoning that are not especially iron-rich. Experimental evidence from laboratory animals[753] indicates that CO cardiotoxicity and hypotension are pathogenetic factors in CO encephalopathy. Comparison of outcomes in uncomplicated hypoxaemia and CO toxicity has shown that hypoxaemia causes hypertension whereas CO causes hypotension.[1020] In human CO poisoning, cardiac failure is associated with myocardial necrosis.[727] Anaemia causes high output failure, and effective anaemia is produced by CO in the blood (by the binding of CO to haemoglobin; see Anaemias, earlier in this chapter). The distribution of brain damage in the cerebral cortex, hippocampus and cerebellar Purkinje cells in human cases of CO poisoning[672,811,828] suggests a prominent role for hypotension,[753] in common with other forms of histotoxic hypoxia. Thus, clinical CO poisoning resembles GCI pathologically. Cardiac arrest can give rise to metabolic and structural lesions focally in the globus pallidus or the substantia nigra.[327,681] The pallido-reticular pattern of damage has also been seen in experimental epilepsy.[728] It is not known whether the post-ischaemic epileptiform activity seen after human GCI correlates with this pattern of vulnerability in the globus pallidus or pars reticulata of the substantia nigra.

A prominent role for hypotension in the pathogenesis of brain damage generally is further suggested by cyanide poisoning, another form of histotoxic hypoxia marked by cardiac failure. This can also produce selective lesions of the globus pallidus.[673] Conversely, the pallidum may be normal in CO intoxication.[580] These considerations indicate that the pallido-reticular pattern of brain damage, although associated with CO poisoning, is neither entirely sensitive nor entirely specific for CO, and that any condition leading to cardiac failure can give rise to lesions of the globus pallidus and the pars reticularis of the substantia nigra. When necrosis is isolated to the globus pallidus, however, CO is implicated. Like GCI, CO poisoning is exacerbated by high blood glucose levels.[788] In common with other forms of hypoxia is the paucity of permanent brain damage in survivors of CO poisoning.[331] It is unknown whether cases of documented CO poisoning without brain damage had less hypotension.

White Matter Damage in Carbon Monoxide Poisoning

CO toxicity can lead to demyelination and destruction of cerebral white matter, sometimes termed Grinker's myelinopathy.[373,727] The destruction of the white matter is often delayed after exposure to CO,[474] and accompanied by delayed neurological deterioration similar to that in GCI.[806,1059] White matter lesions can also be produced experimentally by cyanide.[137] Thus, as applies to the other features of CO poisoning, white matter lesions and delayed neurological deterioration are not specific for CO. The occurrence of white matter lesions in hypoxic states seems to be favoured by less severe insults.[594] The pathogenesis may involve lipid peroxidation and subsequent immune mediation of damage. In experimental CO poisoning in the rat, adduct formation between myelin basic protein (MBP)

and malonaldehyde caused immunological responses to the chemically modified MBP.[997]

Carbon Monoxide Exposure, Basal Ganglia and Psychic Akinesia

An organic psychiatric syndrome associated with exposure to low-dose CO has become a clearly defined nosological entity.[628,968] The classic clinical picture of severe rigidity and hypertonia does not develop when necrosis of the globus pallidus is incomplete. Instead, patients develop a syndrome of inactivity, lethargy and an amotivational state, termed 'psychic akinesia'. Neuroimaging reveals more subtle abnormalities in the globus pallidus than seen in classical pan-necrosis.

The association of partial necrosis of the pallidum with a psychiatric syndrome is in line with new concepts of basal ganglia function that extend beyond motor control. Basal ganglia neuronal loops are not limited to the motor cortex but also connect with the posterior parietal, premotor and prefrontal cortex,[513] and with the medial and lateral temporal lobes, hippocampus, anterior cingulate and orbitofrontal cortex.[513] This latter, limbic basal ganglia loop is known to be involved in emotion and in the motivational regulation of behaviour.

Carbon Monoxide in Automobile Exhaust and Incomplete Pallidal Necrosis

Carbon monoxide levels in automobile exhaust have decreased over the years from 12 per cent in the 1950s to 8 per cent in the 1960s and 1970s.[1058] In 1980, catalytic converters using the transition metals palladium, platinum and rhodium led to a marked decrease in CO in exhaust fumes. The use of computer chips that match O_2 delivery to fuel consumption has resulted in a further decline of CO levels to only about 0.2 per cent in modern cars. This has changed the clinical picture of automobile-related CO poisoning from complete globus pallidus necrosis to incomplete necrosis, with a concomitant shift in the clinical picture from death or a rigid pallidal motor syndrome to the syndrome of psychic akinesia. It is difficult to commit suicide with a modern, well-maintained car[999] but this is not widely known by the public, accounting for the increasing prevalence of incomplete globus pallidus necrosis and psychic akinesia.

Incomplete pallidal necrosis and psychic akinesia are illustrated by the case of a 52-year-old attorney, who presented with a long history of drug abuse. He required hospitalization and requested increasing doses of narcotics during his hospital stay. He also had suicidal thoughts. Neuroimaging showed incomplete globus pallidus necrosis. In a legal action, hypotension was alleged to have occurred but this could not be substantiated from his hospital course. The man's spouse reported that he spent all day in front of the television set and rarely moved. IQ testing revealed an above-average IQ of 126, and the cortex was preserved on imaging (Figure 2.24). He had been left alone with a new, late-model car in the garage and, in view of the suicidal ideation, the case was deemed one of probable attempted suicide by CO intoxication from automobile exhaust fumes.

DISEASES AFFECTING THE BLOOD VESSELS

Adequate flow of blood to the brain is normally delivered by cerebral arteries of sufficient luminal calibre, lined by non-activated but 'alert' endothelial cells on robust intima supported by structurally and functionally intact tunica media and adventitia. Many disease processes thicken the arterial wall (Box 2.3), particularly the intima, and narrow the lumen or weaken the wall, making it more susceptible to dilatation and rupture. Such disorders frequently damage the endothelium or produce factors that activate platelets and the clotting cascade and cause thrombosis.

Atherosclerosis

General Aspects

The most common vascular disease is atherosclerosis (Figure 2.25). It develops slowly and often begins in childhood, although a small minority escapes it until the ninth or tenth decade.[355,969] Atherosclerosis has many risk factors in common with stroke, including dyslipidaemia, hypertension, diabetes mellitus and cigarette smoking. Atherosclerosis is a disease of the whole arterial network, although marked regional variations exist. The carotid arteries are one of the most often affected extracranial sites and the progression of atheroma is similar to that in the aorta and coronary arteries. The origin and the intracavernous segment of the internal carotid artery, the first segment (M1) of the middle cerebral artery, the origin and the distal portion of the vertebral artery (VA1 and VA4), and mid-part of the basilar artery are particularly prone to atheroma.[191,748] In people of European ancestry, the internal carotid artery is most often involved, whereas intracranial arteries, especially the middle cerebral artery, are most commonly affected among African and Asian people. Atherosclerosis of the anterior

cerebral arteries can block the ostia of the small anterior perforating arteries.[856] Typical atheroma may also develop in the smaller distal segments of the main cerebral arteries in the oldest old.[506] The atherosclerotic lesions do not differ qualitatively between genders or ethnic groups, although there is some anatomical variation.[695] Compared to Afro-Caribbean people, in Caucasians atherosclerosis is usually more severe in extracranial arteries and is less common in intracranial vessels.[695] Quantitative differences between different ethnic groups may reflect exposure to different risk factors associated with social and economic circumstances. The characteristics of the fibrous caps in plaques seem to play a major role in the development of complications. Some generalizations about the risk factors have been made: high serum lipids and blood pressure are associated with a high prevalence of atherosclerosis in both the extracranial and intracranial arteries, whereas low or normal lipids and high blood pressure levels are associated mainly with intracranial and intracerebral arterial disease.

Pathogenesis of Atherosclerosis

The development of atherosclerotic lesions in the arterial walls is an increasingly complex process.[605,969] A key feature is accumulation of lipids in the arterial intima, which is initiated by dysfunction of the endothelium. This leads to a series of interdependent cellular and molecular processes (Figure 2.26), including modification of lipids, migration and proliferation of smooth muscle and inflammatory cells, production of pro-inflammatory mediators (Figure 2.27), and possibly also invasion by microorganisms.[556,605,874]

Lipid Metabolism

There is strong epidemiological and experimental evidence that increased dietary lipids, particularly cholesterol and saturated fats, are associated with the development of atherosclerotic lesions in the walls of large and medium-sized cerebral arteries (as in coronary arteries).[556,605] The normal metabolism of dietary cholesterol begins by its transport in serum chylomicrons together with triglycerides. The latter are removed by endothelial lipoprotein lipase in adipose tissue and skeletal muscle, and the remnants are taken up by hepatocytes, within which endogenous synthesis of cholesterol is adjusted according to the amount of exogenous cholesterol. Cholesterol, phospholipids and triglycerides are delivered from the liver to the body tissues by the apolipoproteins to ensure solubility of the lipids in the serum. At first, the very low-density lipoprotein (VLDL) particles are rich in triglycerides. Most of the triglycerides are released in adipose tissue and muscle, and VLDL particles are thereby transformed into low-density lipoprotein (LDL) particles with a diameter of about 20 nm, which provide extrahepatic cells, including those of the CNS, with cholesterol. To do this, LDL particles traverse the walls of capillaries and enter the extracellular fluid compartment. They bind to LDL receptors on the surface of cells, including smooth muscle cells in the tunica media, and are taken up by receptor-mediated endocytosis to be used in ordinary metabolic pathways. Cells that have acquired sufficient cholesterol cease to produce LDL receptors and, consequently,

---

BOX 2.3. Main types of vascular disorder affecting the CNS, and their consequences*

Structural vascular changes[†]
- Atherosclerosis and arteriolosclerosis
- Hypertensive disease – vascular coiling, kinking, twisting
- Cerebral amyloid or protein angiopathies, some also involving the spinal cord
- Arteritis** – giant cell arteritis, Takayasu disease
- Vasculitis** – primary angiitis, tuberculosis, other bacterial or fungal infection
- Aneurysms – sacular, berry, fusifom, cerebral
- Vascular malformations – cavernous, arteriovenous, venous, capillary telangiectasias, mixed malformations

Tissue changes
- Strokes – ischaemic or haemorrhagic; infarcts – large, lacunar, microinfarcts, watershed
- Ischaemic leukoencephalopathy – white matter changes including rarefaction, incomplete white matter infarcts, demyelination, axonal damage
- Various forms of global hypoxic and anoxic damage, often exacerbated by ischaemic damage

*Some of these have hereditary forms.
†Excluding haematological and neoplastic disorders.
**See Box 2.4 for more information on inflammatory vascular diseases.

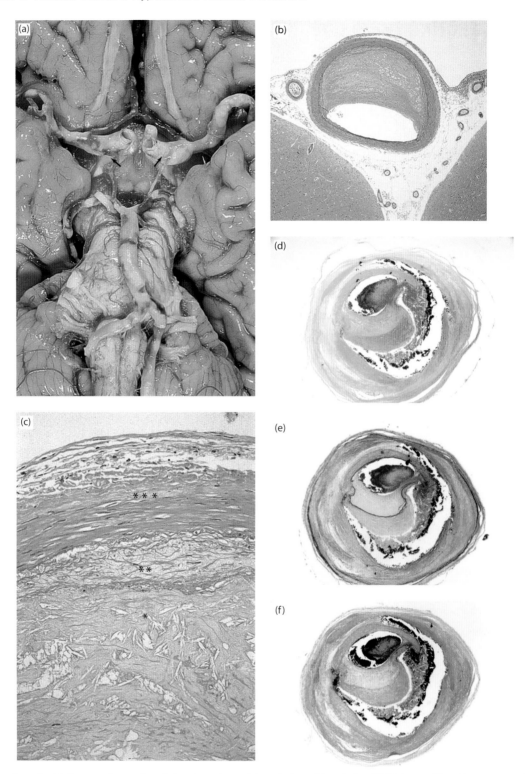

2.25 Atherosclerosis. (a) Severe atherosclerosis has transformed the arteries of the circle of Willis into stiffened, tortuous, yellowish, non-translucent tubes. In spite of the marked structural changes in the arterial walls, the atherosclerosis had not caused overt cerebrovascular symptoms. Note also the dilation (dolichoectasia) of the internal carotid arteries (arrows). **(b)** The presence of a fibrous atherosclerotic plaque in a small leptomeningeal branch of the middle cerebral artery only 1.8 mm in diameter demonstrates the aggravating effects of hypertension and diabetes mellitus, associated with extension of atherosclerotic changes into more distant arterial branches. The plaque impinges on the lumen, but luminal decreases of this magnitude should still allow adequate flow. **(c)** A higher magnification of the fibrous plaque shows the necrotic core with cholesterol crystals (*), foam cells (**) and fibrosis of the media with loss of smooth muscle cells (***). **(d)** Thrombotic occlusion of the left M1 from an 82-year-old man with diabetes mellitus and hypertension who developed right hemiparesis 81 days before death. He died of pneumonia. There is organizing thrombotic occlusion of the left M1 after plaque rupture at autopsy. **(e)** An elastin/van Gieson stain reveals patchy loss of elastic fibres from parts of the vessel wall but increased deposition of elastic fibres in other areas. **(f)** A trichrome stain shows extensive collagenous fibrosis.

Images (d–f) kindly provided by J Ogata, National Cerebral and Cardiovascular Disease Centre, Osaka, Japan.

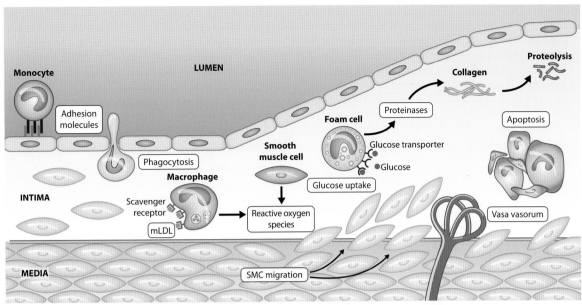

2.26 Schematic illustration of the pathogenesis of atherosclerotic lesions. Atherosclerosis begins by complex interaction of circulating factors and cells with cells in the vessel wall, including endothelial cells, lymphocytes, monocytes and smooth muscle cells (SMCs). The serial changes that occur within the intimal layer during progression of atherosclerosis are shown from left to right. The initial steps include adhesion of blood leukocytes to the activated endothelial monolayer, directed migration of the bound leukocytes into the intima, maturation of monocytes into macrophages and their uptake of lipid, yielding foam cells. SMCs migrate from the media to the intima, there is proliferation of resident intimal SMCs and media-derived SMCs and heightened synthesis of extracellular matrix macromolecules such as collagen, elastin and proteoglycans. Plaque macrophages and SMCs die in advancing lesions, some by apoptosis. Extracellular lipid derived from dead and dying cells accumulates in the central region of the plaque, often denoted the lipid or necrotic core. Advancing plaques also contain cholesterol crystals and microvessels. Thrombosis, the ultimate result of atherosclerosis, often involves a physical disruption of the atherosclerotic plaque, which develops a fibrous cap. This may induce blood coagulation components to come into contact with tissue factors in the interior of the plaque, triggering thrombosis that extends into the vessel lumen, where it can impede blood flow.

Modified from Libby et al.[605] and Kalaria et al.[506] Diagram redrawn courtesy of Y Yamamoto, Yamaguchi University Graduate School of Medicine, Japan.

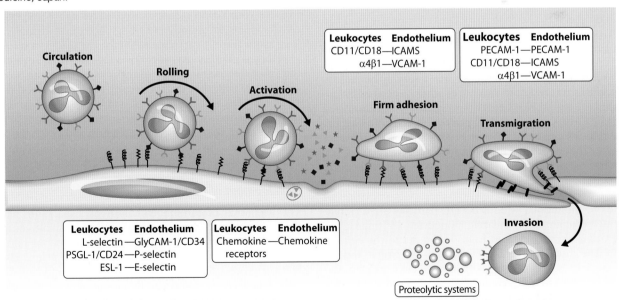

2.27 Schematic diagram showing extravasation of inflammatory cells. Initially, the circulating lymphocytes adhere to the endothelium only transiently and roll on it. This phase involves interaction between selectins and their ligands. During rolling, lymphocytes are activated by cytokines and/or chemokines, which leads to activation of their integrin receptors CD11/CD18 and α4β1. Shear stress and/or inflammatory mediators regulate endothelial cell expression of corresponding ligands, intercellular adhesion molecules (ICAMs) and vascular cell adhesion molecule 1 (VCAM-1). These interactions result in firm adhesion and extravasation of inflammatory cells. The molecules involved are indicated in the boxes.

Diagram courtesy of Y Yamamoto, Yamaguchi University Graduate School of Medicine, Japan.

to endocytose LDL particles. The surplus LDL particles enter lymphatics and return to the circulation. The efflux of excess cholesterol is aided by high-density lipoprotein (HDL) particles, which enter the tissue from the blood, incorporate the excess cholesterol from cells and return to the circulation. In blood, the HDL-borne enzyme lecithin-cholesterol acyltransferase esterifies cholesterol and the product is transferred to LDL. LDL-cholesterol returned to the circulation is ultimately transported back to the liver for excretion in the bile. This phenomenon, called reverse cholesterol transport, normally keeps an equilibrium between the influx and efflux of cholesterol in extrahepatic tissues, including the arterial intima.

Atherosclerotic Lesions

Initial Lesions

The lipid hypothesis posits that atherosclerosis commences when high plasma levels of LDL lead to the accumulation of excessive LDL-cholesterol in the arterial intima (Figure 2.26). LDL particles are retained much longer in the arterial intima than in other tissues because there are no lymphatic vessels in the intima, the nearest being in the media. Furthermore, the proteoglycan matrix slows down the movement of LDL particles across the intima, which is separated from the media by a poorly permeable internal elastic lamina. In consequence, concentrations of LDL in the intima increase some 10-fold compared to the extracellular fluid of other extrahepatic tissues. Intimal cells do not express LDL receptors like other extrahepatic tissues, which results in blockade of LDL uptake and metabolism.[556] LDL particles retained within the intima may be modified through oxidation and peroxidation, as well as proteolysis and aggregation, leading to accumulation of lipid droplets.[759] The cholesterol derived from the modified LDL particles and stored in macrophages can be returned to the circulation by HDL, but low HDL levels in atherosclerotic patients may impair this reverse cholesterol transport.

According to the response to injury/inflammatory hypothesis, a further prerequisite of atherogenesis is disruption of the endothelial barrier through injury to endothelial cells.[874] Causes may include hypertension, hyperlipidaemia, free radicals and toxic substances such as those in cigarette smoke. The endothelial injury allows excessive egress of lipids into the intima. There is strong evidence that chemical modification of LDL particles is essential for the initiation of the cascade that leads to pathological accumulation of lipids in the intima. Modified LDL particles are a potent chemoattractant for circulating monocytes and impede the migration of macrophages from the intima to the circulation. The monocytes collecting within the intima and, to a lesser extent, resident smooth muscle cells take up and deposit modified LDL-cholesterol much more effectively than native LDL, by exploiting specific scavenger receptors. These cells are transformed into intimal foam cells, clusters of which form the initial lesion visible to the naked eye as yellowish fatty streaks. Both the foam cells and extracellular lipids are stained strongly by common methods for demonstrating neutral lipid, such as oil red O and Sudan black B. The efflux of excess cholesterol from cells is mediated by the transmembrane protein adenosine triphosphate

(ATP)-binding cassette transporter A1 (ABCA1). Reduced expression of ABCA1 in carotid plaques appears a key factor in the pathogenesis of atherosclerotic lesions.[25]

Fatty streaks may appear in the carotid arteries during the first decade (in the vertebral and intracranial cerebral arteries they develop a decade later[355,695]) and are themselves innocuous. Their appearance does not always lead to more severe atherosclerosis.

Development of Atherosclerotic Plaques

In most cases, however, the fatty streaks do develop into more advanced lesions, as atherosclerotic plaques (atheroma), over a long period of time, even decades. The plaques are called fibrous plaques if the collagenous connective tissue component predominates over the lipids (Figures 2.25 and 2.26). The plaques are usually located at specific sites, such as on the outer aspects of the bifurcations of arteries, where the intima is thickened and laminar blood flow disturbed. Inflammatory cells enter the plaques from the circulation through interactions between adhesion molecules and their ligands, on endothelial cells and leukocytes (Figure 2.27). Inflammatory and endothelial cells secrete growth factors, such as platelet-derived, epidermal, fibroblast and transforming growth factors, and cytokines, such as interleukins (ILs), tumour necrosis factor (TNF) and leukotrienes, which recruit additional cells, including smooth muscle cells from the media, into these lesions (Figure 2.27). They also induce the transformation of smooth muscle cells from contractile cells into cells that actively synthesize extracellular proteins. The smooth muscle cells alone are responsible for the production of extracellular matrix components (mainly types I and III collagen) of the fibrous cap underneath the intact endothelial cell layer. Beneath the cap, clusters of smooth muscle cells, foam cells and lymphocytes collect, together with a central core of necrotic cell debris and extracellular lipids, frequently including cholesterol crystals. As long as the fibrous cap remains intact, the plaque is stable.

Fibrous plaques may form over 20–30 years. Although they may stenose the lumen of the affected artery to a considerable extent, they do not usually reduce the blood flow sufficiently to cause neurological symptoms. Even 5 per cent of the cross-sectional area of the lumen may allow blood flow sufficient for normal brain function. However, the turbulence caused by stenosis may cause endothelial damage and lead to the development of complicated plaques.

Development of Complicated Plaques

The likelihood of thromboembolism rises markedly when the atherosclerotic plaque converts into an unstable complicated plaque, in which the lipid core expands and the fibrous cap thins. Thinning of the fibrous caps leads to plaque instability, a process thought to be modulated by inflammatory cells, which are a consistent finding at sites of ulceration or rupture of an atheromatous plaque.[175,556] Macrophages, T-lymphocytes (mainly the Th1 subpopulation of CD4+ cells) and mast cells that accumulate in the intima secrete cytokines that affect the rate of synthesis and lysis of the matrix proteins. γ-interferon (γ-IFN) secreted by T-lymphocytes inhibits the proliferation of smooth muscle cells and, together with other cytokines, such as TNF-α and IL-1β, induces their apoptosis.[316,392,604,607] γ-IFN inhibits collagen synthesis by smooth muscle cells. Both vascular and inflammatory cells can produce matrix metalloproteinases (MMPs), including MMP-1, MMP-9 and MMP-3 (stromelysin), which, if not sufficiently counteracted by specific tissue inhibitors, degrade the extracellular matrix, proteoglycans and elastin, and weaken the fibrous cap. These events transform solid fibrous lesions with small lipid cores and thick caps into lesions with large lipid cores and thin caps. Plaques with thin caps become unstable, prone to ulceration and thrombosis.[607]

A critical event in atherosclerosis is damage to the endothelium. Examination of complicated plaques in cases of carotid artery occlusion reveals that the endothelial lining is always disrupted. This exposes the blood to tissue factors that activate coagulation, and thrombus forms over the plaque, causing narrowing of the lumen and predisposing to embolism. The high frequency of embolic strokes highlights the danger this poses. An embolus leads to abrupt occlusion of the vascular lumen, whereas a local thrombotic process is usually slow and may allow time for collateral channels to develop. Endothelial injury also occurs if the size of the plaque is increased abruptly by intramural bleeding from new blood vessels formed in the fibrous cap and at the margins of the plaque, although more frequently the blood originates from the circulation through defects caused by rupture of the plaque surface. Over time, haemosiderin may accumulate and, when calcium salts precipitate in the necrotic core, the plaque acquires the characteristic hard consistency.

Genetic Risk Factors

Familial hypercholesterolaemia, which is accompanied by a marked increase in the plasma level of LDL, is a major predisposing factor for atherosclerosis. Other genetic factors implicated in the pathogenesis of atherosclerosis include polymorphisms of the *ABCA1* gene and its promoter, and of the genes for MMP-3 and IL-6[164,447,1118] (see also Table 2.7, Genetics of Stroke).

Diseases of Small Arteries

Hypertensive Angiopathy

General Aspects

Hypertension is not itself a disease of blood vessels but because its deleterious effects are mediated by structural changes in blood vessels, particularly small arteries and arterioles, it is discussed here. Experimental studies have demonstrated that chronic hypertension shifts the autoregulatory limits to the right, towards higher pressure values (see Figure 2.16).[94] This is a protective response that allows maintenance of a constant CBF even at increased arterial pressure and prevents the ill effects of the excessive systemic pressure on delicate capillaries. Clinical experience shows that the same holds true for the human cerebral circulation. The upper limit of autoregulation may increase up to 4 kPa (30 mmHg). This shift to the right also raises the lower limit of autoregulation at which adequate CBF can be maintained (see Cerebral Blood Flow Regulation and Figure 2.16, earlier in this chapter). Symptoms of cerebral hypoperfusion develop when the mean arterial blood pressure falls to about 40 per cent of baseline levels. In hypertensive patients, such a reduction is reached at a correspondingly

higher level of arterial pressure than in normotensive people. One-third of asymptomatic hypertensive patients were found to have focal or diffuse cerebral hypoperfusion.[738] This may be exacerbated by excessive antihypertensive medication, to ischaemic levels severe enough to cause tissue damage, especially along the arterial border zones.[728] Similarly, in a hypertensive patient with stroke, decreases in blood pressure to levels tolerated by a normotensive subject may worsen the ischaemia.

Clinical Picture of Hypertensive Brain Disease

The symptoms of arterial hypertension depend on the rapidity of the rise of arterial pressure, as do the resulting vascular structural changes (see later) and the parenchymal damage. Sudden and severe malignant hypertension, precipitated for example by renal disease, release of catecholamines from a phaeochromocytoma, eclampsia, or as a rebound effect after discontinuation of antihypertensives, may cause acute hypertensive encephalopathy.[252,762] This syndrome is characterized by rapidly progressive diffuse cerebral dysfunction, severe headache, altered consciousness, nausea and vomiting, i.e. symptoms and signs of increased intracranial pressure. These may be followed by focal neurological abnormalities, such as visual disturbances and seizures, and a progressive decrease in the level of consciousness. Parenchymal brain haemorrhage (PBH) is a potentially fatal complication of acute hypertensive brain disease.[772,844]

A syndrome that shares many of the clinical manifestations of hypertensive encephalopathy but usually also includes cortical blindness and extensive bilateral white matter abnormalities on neuroimaging, suggestive of oedema in the posterior regions of the cerebral hemispheres, brain stem and cerebellum, has been termed reversible posterior leukoencephalopathy syndrome (RPLS) – also known as occipito-parietal encephalopathy or posterior reversible encephalopathy syndrome (PRES).[426] In patients with RPLS, acute hypertension is often present and associated diseases include renal disease, eclampsia and immunosuppressive therapy. RPLS is not always reversible or confined to posterior brain regions. The syndrome also occurs in children.[779] The findings on brain imaging suggest multifocal leakage across the BBB.[578]

In chronic arterial hypertension, the clinical symptoms are usually absent or mild, e.g. minor headaches and tinnitus. Patients are at risk of sudden unexpected PBH.

Pathogenesis and Pathology of Hypertensive Angiopathy

Acute Hypertension

Two different pathogenetic mechanisms have been proposed to explain acute hypertensive encephalopathy and also RPLS. The prevailing hypothesis is that the forced dilation of the resistance vessels by the high arterial pressure ('breakthrough' of autoregulation) is crucial to the development of the encephalopathy because it exposes smaller, distal blood vessels to excessive pressure and hyperperfusion.[584] The high arterial pressure disrupts the BBB and causes vasogenic oedema (Figure 2.28).[510,910,962] If the hypertension does not

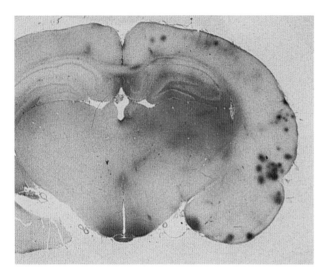

2.28 Experimental acute hypertension. This has caused multifocal disruption of the blood–brain barrier (BBB) and consequent vasogenic brain oedema. The acute rise in the arterial blood pressure was induced by clamping the abdominal aorta in a rat for 10 minutes followed by survival for 2 hours. The leakage is visible as small perivascular accumulations of albumin in the cortex. Diffuse spread of the oedema fluid to the surrounding parenchyma has already occurred in the deep grey matter, hippocampus and hypothalamus (anti-albumin antibody and haematoxylin counterstain).

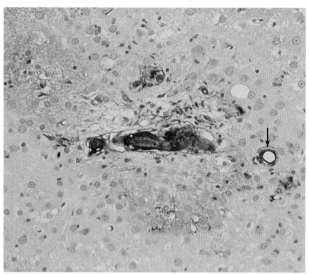

2.29 Chronic hypertension in a stroke-prone spontaneously hypertensive rat. A site of disruption of the blood–brain barrier (BBB) in the cortex is surrounded by leaked plasma constituents, which spread only focally in the cortex but diffusely in the underlying white matter. Note that the longstanding hypertension does not cause generalized leakage of the BBB, and the cortical disruption has not resulted in a major haemorrhage. The wall of the small artery (arrow) is thickened by the deposited plasma proteins. The lumen of the larger artery is filled by thrombus and there has been leakage of plasma proteins into the surrounding cortical parenchyma (anti-fibrinogen antibody and haematoxylin counterstain).

Reproduced by permission from Fredriksson et al.[310] With kind permission from Springer Science and Business Media.

persist, the BBB is rapidly reconstituted upon lowering of the blood pressure. However, over time, plasma proteins, including fibrin, are deposited in the walls of small arteries, a process also associated with destruction of smooth muscle cells (fibrinoid necrosis; see Figures 2.29 and 2.31).[311,870] Acute hypertension may secondarily lead to ischaemia as a consequence of injury to the endothelium, and thrombosis.[709]

An alternative view is that cerebral autoregulation attempts to protect the brain by exaggerated arteriolar vasoconstriction, which may focally be severe enough to result in ischaemic injury and cytotoxic brain oedema.[1009] In favour of this hypothesis is the observation of vasoconstriction of arteries[333] in the retina as well as the cerebrum.[470] The cerebral parenchyma is particularly vulnerable to damage because of marked regional variation in CBF.[253]

Chronic Hypertension

In humans, chronic hypertension has two main consequences. First, it aggravates atherosclerotic changes in extracranial and intracranial larger arteries. A critical level of circulating lipoproteins is a prerequisite for this effect.[205] In spontaneously hypertensive rats a special lipid-rich diet is needed to induce atherosclerosis.[437]

In addition, hypertension makes atherosclerosis extend more distally into the intracranial compartment, to affect blood vessels under 2 mm in diameter (see Figure 2.25b). The leptomeningeal arteries over the convexities are usually spared in normotensive atherosclerotic subjects, whereas in hypertensive patients they stand out as hardened, non-collapsed, yellowish blood vessels. Lesions similar to those in large vessel atherosclerosis may develop in arterioles down to 100 µm in diameter (see later).

Second, persistently high blood pressure leads to focal disruption of the BBB (Figure 2.29). The brightly

eosinophilic 'fibrinoid' material in thickened blood vessel walls is immunopositive for different plasma proteins (Figure 2.31) and electron microscopy reveals extracellular deposits of proteinaceous material. At these sites the basal lamina under the damaged or regenerated endothelial cells becomes thickened or reduplicated. Vascular smooth muscle cells degenerate and are replaced by collagenous connective tissue (Figure 2.30).[310,311]

Small Vessel Changes

In the CNS, this term applies to perforating arteries and arterioles, with diameters from 40 to 900 µm, which emerge from the leptomeningeal arteries, enter the brain parenchyma from the surface of the brain and extend a variable depth into the parenchyma (basal and pial penetrators; see earlier). Three common cerebrovascular disorders are associated with small vessel disease (SVD): (i) lacunar (small deep) infarcts; (ii) subcortical ischaemic vascular dementia, which may be caused by multiple lacunar infarcts or by more diffuse white matter ischaemic damage; and (iii) primary non-traumatic PBH. Since they were first reported[648] the pathogenesis of lacunar infarcts has been a topic of intensive research.[83] The risk factors and pathogenesis of PBHs are still much debated. Besides age, hypertension and diabetes mellitus are the main risk factors for SVD. The importance of these risk factors for clinical lacunar stroke has been questioned[473,915] but their contribution to the structural pathology of SVD is generally accepted.

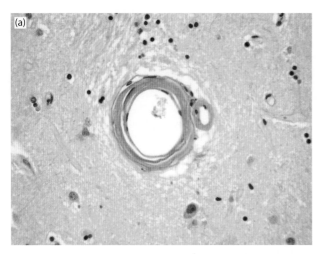

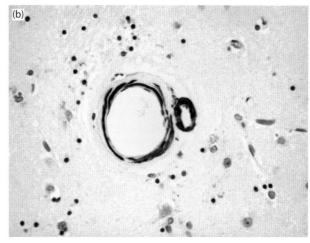

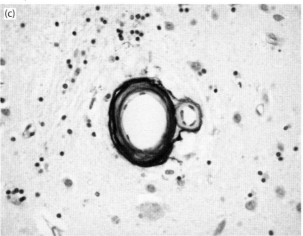

2.30 Two arterioles from a patient with hypertension. The walls of the arterioles are thickened **(a)**. The tunica media of the smaller arteriole shows hyperplasia, whereas in the larger arteriole there is an irregular degeneration of smooth muscle cells **(b)** and accumulation of collagen type I **(c)**. (a) Herovici's stain; (b) smooth muscle cell α-actin; (c) type I collagen.

The consequences of SVD, including lacunar infarcts and PBHs, are described later in this chapter or, in connection with vascular dementia, in Chapter 16.

The four major structural pathologies described in penetrating small arteries in SVD are atherosclerosis, fibrinoid necrosis (lipohyalinosis), arteriolosclerosis and micro-aneurysms. The original and most meticulous work on SVD was that of C Miller Fisher, who, performed serial-section analysis of small arteries entering 68 lacunar infarcts in 18 brains and laid the basis for current concepts of the pathogenesis of lacunar lesions.[298] He reported segmental arterial disorganization (later called lipohyalinosis) and small vessel atherosclerosis to be the two commonest causes of lacunar infarcts, and fibrinoid necrosis to be associated with PBHs. Small cerebral arteries and arterioles may be affected by many other diseases, such as hereditary angiopathies, inflammatory and infective vasculitides and toxic disorders (see later).

Atherosclerosis

Hypertension and diabetes, in addition to ageing, are considered to be the major risk factors for small vessel atherosclerosis, as for atherosclerosis in larger arteries. They are thought to cause atherosclerosis to extend more distally intracranially, including along the larger perforators (Figure 2.25).

Small vessel atherosclerosis affects vessels 200–900 μm in diameter.[506] Plaques may be found in proximal segments of penetrating arteries (micro-atheroma), at the junction of the branching and parent arteries (junctional atheroma) and in the parent vessel overlying the branch origin. The pathogenesis of atherosclerosis in the small vessels does not differ substantially from that in larger vessels. The atheroma in small arteries also contains CD68-positive macrophages.[506] Atheroma in small vessels causes complete occlusion at an earlier stage of plaque evolution.

Fibrinoid Necrosis and Lipohyalinosis

This type of SVD, affecting mainly arterioles 40–300 μm in diameter, was called lipohyalinosis by Miller Fisher,[296] although the term is poorly defined and variably applied.[870] It is closely linked to longstanding hypertension and is likely to increase the risk of rupture of small arteries with consequent PBH.[473,915]

In the early stages, the walls are thickened by eosinophilic fibrinoid material, composed mainly of plasma proteins, with abundant fibrin, formed by leakage of the BBB (see Figure 2.29) together with remnants of smooth muscle cells (Figure 2.30). At this fibrinoid-necrosis stage of SVD, the BBB is disrupted (Figure 2.29) and the affected vessels are probably most prone to rupture. With time, the fibrinoid material is replaced by collagen produced by fibroblasts and the arteriolar walls become more homogeneous and less structured (Figure 2.30c). It is likely that it was this stage of SVD that CM Fisher described as 'lipohyalinosis' (or segmental arterial disorganization).[296] Lipids are usually only a minor component in these lesions, and hyalinosis refers to acellular fibrosis. It has been recommended that the term 'lipohyalinosis' should be abandoned and replaced by descriptions based on appropriate stains: fibrinoid change if histological or immunocytochemical stains verify the presence of fibrin, and fibrosis if collagen is the main constituent of thickened arterial walls. The homogeneous eosinophilia in haematoxylin- and eosin (H&E)-stained sections may result from either fibrinoid change or collagenous fibrosis.[283] These two are probably consecutive changes and can be readily distinguished by use of special stains but appear deceptively similar with H&E.[870]

Arteriolosclerosis

Arteriolosclerosis describes non-fibrinoid hyaline thickening of 40–150 μm-diameter arteriolar vessels (Figure 2.31). There is clearly morphological overlap with the late stage of Fisher's lipohyalinosis. The frequency of arteriolosclerosis rises with age. The thickened walls narrow the lumen and increase the sclerotic index (SI = 1 − [internal diameter/external diameter]), a measure devised to indicate severity of degenerative fibrous thickening of the tunica media.[576] Healthy intracerebral arteries are relatively thin-walled with a wide lumen in relation to the wall thickness: their SI is below 0.4 and in young people even below 0.3. Arteriolosclerosis tends to be associated with ischaemic white matter disease and vascular dementia rather than lacunar infarcts. The thickened fibrotic arteries seldom rupture.

Differential Diagnosis

The three main arteriopathies that cause thickening of the penetrating small arteries are arteriolosclerosis, CAA (see Sporadic and Familial Cerebral Amyloid Angiopathy, later in chapter) and cerebral autosomal dominant arteriopathy with subcortical infarcts and leukoencephalopathy CADASIL (see CADASIL, later in chapter). These can be distinguished by location and staining. CAA is largely leptomeningeal and cortical; CADASIL predominantly affects leptomeningeal and subcortical vessels, sparing those in the cortex; SVD is most pronounced in subcortical arteries and arterioles. Vessels affected by arteriolosclerosis show tinctorial staining for collagen and are immunopositive for collagen I; arterioles with CAA are Congo red- and (in the most common form of CAA) Aβ-positive; in CADASIL, arteriolar walls show granular basophilia in sections stained with H&E and label with antibodies to the ectodomain of NOTCH3.[1105]

Micro-Aneurysms

According to the traditional view, Charcot–Bouchard or miliary micro-aneurysms arise in the context of hypertension, at weakened sites in vessel walls. They resemble small sacs, 0.3 to 2 mm across, arising from parent arteries/arterioles 100–300 μm in diameter.[297,1061,1062] The walls of the aneurysms consist of hyaline connective tissue, damaged smooth muscle cells and elastica interna.[297,981] Rupture of micro-aneurysms typically produces globular haemorrhages;[297] if 'healed' by thrombosis and fibrosis, these are transformed into fibrocollagenous balls[869] as demonstrated by intravascular injection of contrast or casting media and examination of serial microscopic sections.[297,869] Micro-aneurysms were previously identified both under the operating microscope and in serial paraffin sections in patients who had had lobar haemorrhages and negative angiography;[1062] half of these patients with a ruptured micro-aneurysm did not, however, have a history of hypertension.

The significance of micro-aneurysms in hypertensive haemorrhage has been questioned. In one Japanese study, only two aneurysms were found in association with 48 hypertensive PBHs.[981] Alkaline phosphatase histochemistry and high-resolution micro-radiography showed[183] the great majority of 'micro-aneurysms' to be complex tortuosities.[966] These were most common at the interface between the grey and white matter and their number increased with age, but hypertension had no effect on their prevalence (Figure 2.32). In surgical samples with PBH, true micro-aneurysms were virtually non-existent. Micro-aneurysms were also not found in relation to lacunar infarcts.[182] The discrepancies between these observations may be explained by the possibility that degenerative changes in the arterial walls in Japanese people with a high frequency of PBH are truly different from those in white populations because of genetic, dietary or environmental factors. However, definitive identification of micro-aneurysms in routine diagnostic analysis is very rare.

Venous Collagenosis

Venous collagenosis is mostly seen in older brains and increases in tandem with white matter disease.[147] There is gradual thickening of the walls of periventricular veins and venules. The walls of the veins are composed of collagen types I and III. The veins may be dilated or, in advanced cases, severely stenosed or occluded but lacking inflammatory cells (Figure 2.33). The functional significance is unclear.

Inflammatory Diseases of the Cerebral Vasculature

General Aspects

Inflammatory conditions of CNS blood vessels are a heterogeneous group with multiple aetiologies, including infectious and immunological processes. In some respects, atherosclerosis also has some features of an inflammatory disease (see earlier).[874] Vasculitides are challenging to the clinician, because of their highly variable symptomatology, difficulties in establishing the correct diagnosis, variable response to therapy, and frequently poor outcome. The American College of Rheumatology has published clinical diagnostic criteria for several vasculitides.[289] Vasculitides may present with a range of CNS symptoms and signs including headaches, meningism, encephalopathy, psychiatric syndromes, dementia, cranial nerve palsies, seizures or strokes, as well as neuropathy, muscle damage and even multiorgan involvement or non-specific systemic symptoms. Deciphering the

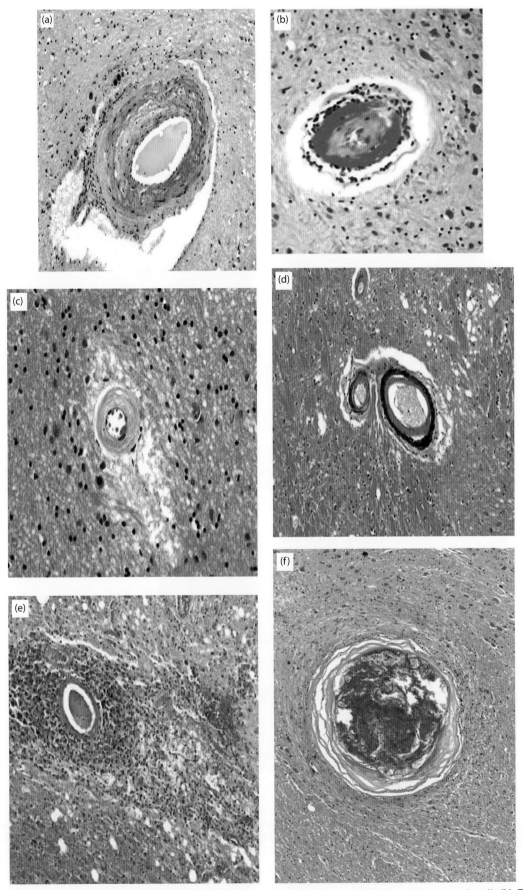

2.31 Progression and complications of arteriolosclerosis. (a) Collagenous thickening of the vessel wall. **(b)** Fibroid necrosis. **(c)** Lipohyalinosis. **(d)** Calcification (the basal ganglia are often affected). **(e)** Microhaemorrhage and microinfarction. **(f)** Microaneurysm.

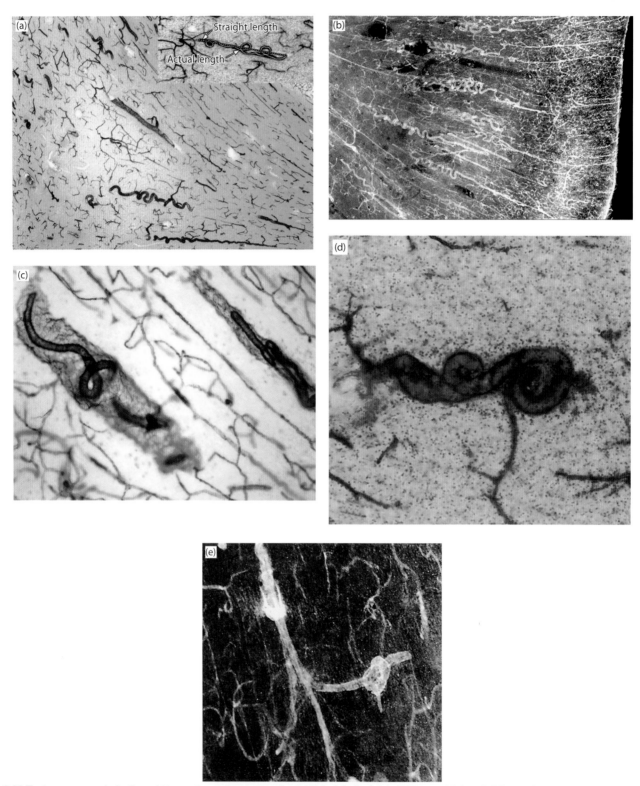

2.32 Tortuous vessels in the white matter in hypertension in the elderly. (a) A 1000-μm-thick celloidin section stained for alkaline phosphatase includes several twisting and tortuous arterioles, the long axes of which run along those of the nerve fibres. Inset shows the distinction between the actual length and straight distance between two points along the vessel. **(b)** Radiogram showing deep penetrating tortuous vessels. **(c)** This thick celloidin section immunolabelled for collagen IV shows tortuous arterioles in cavities and within a collagen bag or barrel. **(d)** Complex vascular coiling in a hypertensive patient who died from intracellular haemorrhage. The vessels in this celloidin section have been stained for alkaline phosphatase. **(e)** Higher power microradiograph of a 500-μm thick section shows an arteriolar knot-like structure.

(a–c) Images kindly provided by W Brown, Winston-Salem University, USA. (d) Reproduced by permission from Challa VR, et al.[183] With permission from Lippincott Williams & Wilkins/Wolters Kluwer Health. (e) Reproduced by permission from Spangler KM, et al.[966] With permission from Lippincott Williams & Wilkins/Wolters Kluwer Health.

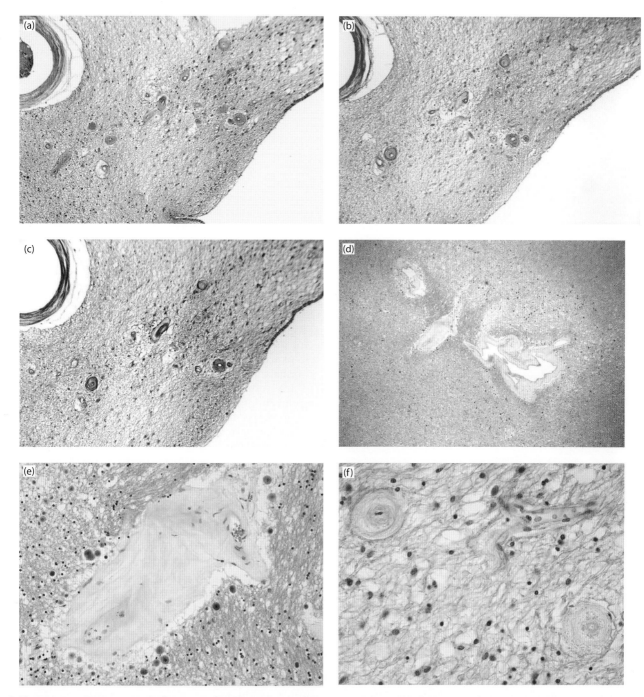

2.33 Venous collagenosis. (a–f) Venous collagenosis is most prominent in close to the lateral ventricles. Collagenous thickening varies in severity, causing partial to complete occlusion. Periventricular venous collagenosis is invariably present in older subjects, although the extent and severity vary. (b) Haematoxylin/van Gieson stain. (c) Masson trichrome.

Images kindly provided by T Polivikoski, Newcastle University, UK and J Ogata, National Cerebral and Cardiovascular Disease Centre, Osaka, Japan.

aetiology and pathogenesis of vasculitides affecting large, medium or small sized vessels, caused by various immunological processes and possibly triggered by infectious agents is also challenging for the pathologist. There is no single generally accepted classification of vasculitides, but they are often categorized according to the aetiology or presumed pathogenesis, and by the site and type of affected blood vessels (Box 2.4).

Non-infectious Vasculitides

Primary Cranial or Cerebral Vasculitides

These include four main types of vasculitis: Takayasu's arteritis, giant cell or temporal arteritis (GCA), primary angiitis of the central nervous system (PACNS) and Aβ-related angiitis (ABRA). All share the histopathological features of granulomatous angiitis.[679]

BOX 2.4. Inflammatory CNS vascular diseases

Non-infective vasculitides
- Primary cranial and/or cerebral:
  - Takayasu's arteritis; giant cell or temporal arteritis (GCA); primary angiitis of the central nervous system (PACNS); Aβ-related angiitis (ABRA); Kawasaki's disease
- Manifestations of systemic diseases:
  - Systemic lupus erythematosus (SLE); polyarteritis (or panarteritis) nodosa (PAN); antineutrophil cytoplasmic antibody (ANCA)-associated vasculitides
- Wegener's granulomatosis
- Eosinophilic granulomatosis with polyangiitis (EGPA) (Churg–Strauss syndrome)
- Vasculitis in Sjögren's syndrome; Behçet's syndrome; rheumatoid arthritis; malignancy-related

Infective vasculitides
- Bacterial:
  - Spirochaetal (e.g. luetic, due to borreliosis); purulent (e.g. streptococcal); granulomatous (e.g. tuberculous)
- Viral (e.g. due to varicella-zoster virus, Epstein–Barr virus)
- Other microbial (fungal, protozoal, mycoplasmal, rickettsial)

The following classification of vasculitides, adapted from Jennette *et al.*,[482] is based on proposed pathogenetic mechanisms:

Immunological injury
- Cell-mediated inflammation:
  - Takayasu's arteritis; GCA; PACNS; ABRA; Kawasaki's disease
- Immune complex-mediated inflammation:
  - SLE; PAN; Behçet's syndrome; infection-induced (e.g. group A streptococcus, hepatitis B and C virus); some malignancy-related vasculitides; some drug-induced vasculitides
- ANCA-mediated:
  - Wegener's granulomatosis; EGPA; some drug-induced vasculitides
- Mixed immunological disorders:
  - Sjögren's syndrome

Direct infection of blood vessels
- Bacterial (e.g. due to streptococci, mycobacteria, spirochaetes)
- Viral (e.g. due to varicella–zoster virus, Epstein–Barr virus)
- Other microbial (fungal, protozoal, mycoplasmal, rickettsial)

Takayasu's Arteritis

In Takayasu's arteritis, the aortic arch with its main arterial trunks and the descending aorta are the main sites of inflammation.[289] The involvement of the carotid and subclavian arteries is of particular neurological relevance. Patients are relatively young (15–45 years). Japanese females are most frequently affected. The incidence in Japan is 1 per 3000, whereas in the USA it is about 2.6 per million. About 20 per cent of Takayasu's arteritis cases are monophasic and self-limited. The overall 10-year survival is estimated to be 90 per cent.[289] Angiography reveals constant narrowing of the aorta and variable involvement of the subclavian, carotid, coronary and renal arteries. The inflammation affects primarily the tunica media, causing destruction of the elastic lamellae and inducing the formation of foreign body giant cells, a common finding wherever elastic tissue is destroyed by inflammation. Secondary fibrosis of all layers causes thickening and loss of compliance of the blood vessel walls, resulting in characteristic loss of carotid pulsations. The affected arteries are finally transformed into rigid, thick-walled tubes with severe narrowing, or occlusion by superimposed thrombosis. The arteritis may result in cerebral ischaemia, usually because of embolism. Rarely involvement of both carotids may cause severe haemodynamic changes. Revascularization of the obstructed carotid arteries can cause a marked transient hyperperfusion syndrome.

The histopathology of recent lesions is of intimal proliferation and granulomatous inflammation, with foreign body and Langhans' type giant cells in the media. There is multifocal destruction of elastic lamellae in the aorta and of smooth muscle cells in the carotid arteries. The vertebral arteries are seldom affected. In the chronic phase, affected vessels usually have thickened fibrotic walls. Occasionally the arteries become dilated; rarely, aortic dissection may ensue.

The pathogenesis of Takayasu's arteritis is poorly understood. There is early infiltration by dendritic cells, T-cells and macrophages, and production of IFN-γ and TNF-α.[603,925] It is hypothesized that an unknown stimulus triggers the expression of the 65-kDa heat-shock protein in aortic vascular cells and induces the cells to express major histocompatibility class I chain A antigens. γδ T-cells and NK cells expressing NKG2D receptors recognize these antigens on vascular smooth muscle cells and release perforin, resulting in cytolysis, and proinflammatory cytokines, causing recruitment of mononuclear inflammatory cells.[56] Vascular damage is probably the result of the release of perforin and MMPs, and of oxidative and nitrosative stress.[603,690,925,1086]

Giant Cell Arteritis

GCA is the most common of the three granulomatous vasculitides, with an incidence of about 20 per 100000 in individuals over 50 years.[289] Most patients are over 55 years. In rare fatal cases, it has been possible to analyze the topography of the inflammation in detail. The principal target of GCA is the extracranial arteries of the head and neck, particularly the superficial temporal arteries. It usually spares the intracranial blood vessels, but ophthalmic artery involvement is well recognized and the carotid and vertebral arteries may also rarely be affected, leading to brain infarction in a small percentage of cases (Figure 2.34). It may also affect more proximal vessels, i.e. the aorta and its branches and even the coronary arteries. The inflammation fades as the affected arteries perforate the dura, at which point the amount of elastic in the arterial wall is also markedly diminished. The key symptom is headache, and a serious sequel is blindness: transient amaurosis fugax in about 10–12 per cent of patients and permanent blindness in about 8 per cent. The blindness is usually due to extension of the disease into the ocular, most commonly, or the ophthalmic arteries or their branches but can also be caused by occipital infarction, probably as a result of emboli from thrombosed vertebral arteries. Stroke and TIA are, however, relatively rare manifestations, occurring in about 7 per cent of cases. GCA is commonly associated with polymyalgia rheumatica, and diagnostic clues include the very high erythrocyte sedimentation rate (ESR) as well as the prompt response to corticosteroid therapy in most cases. The aetiology of GCA is obscure. GCA is associated with HLA class II DR4.[739] The histopathological and immunological

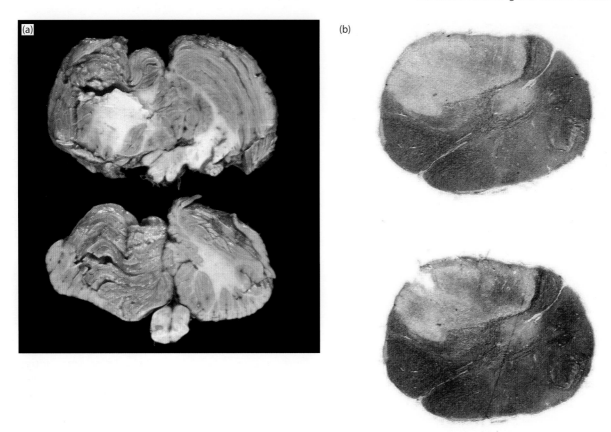

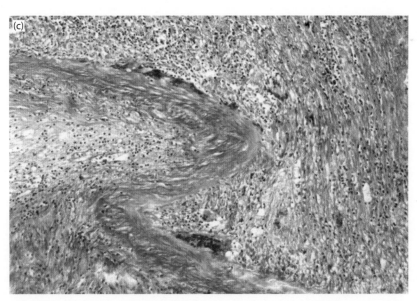

2.34 Giant cell arteritis. (a) Infarct of the cerebellar hemisphere and medulla oblongata. **(b)** Section of the medulla oblongata, showing the poorly stained region of infarction. **(c)** Typical lesions of giant cell arteritis in the vertebral artery. Masson trichrome.

characteristics indicate a cell-mediated immunological reaction to an unknown antigen.[1087,1088]

The affected blood vessel, usually the temporal artery, becomes tortuous, thickened and tender, with diminished pulsations.[679] The microscopic features of GCA vary to a considerable extent (Figure 2.35). The inflammatory

changes may extend along the length of the artery but are often focal. Therefore, a biopsy of 3–5 cm is recommended, and a negative biopsy cannot completely rule out GCA. The inflammation induces proliferation of the intima, which is infiltrated by varying numbers of CD4+ T-lymphocytes and lesser numbers of CD8+ T-lymphocytes, monocytes and

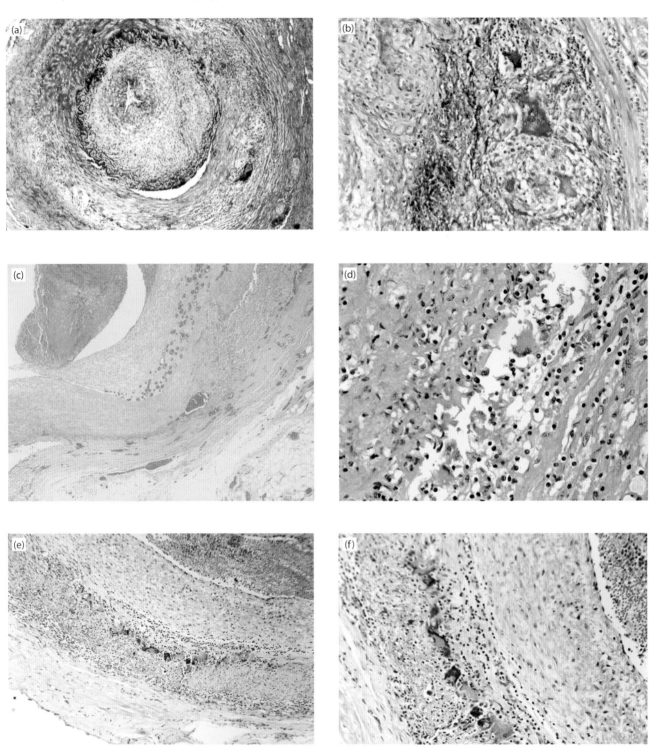

2.35 Giant cell arteritis. (a) Only fragments of internal elastic lamina remain. Marked proliferation of intimal cells has thickened the intima and severely narrowed the lumen. Lymphocytes are present in the adventitia and, to a lesser extent, the media. In the media, there are also histiocytes, including one large multinucleated giant cell of Langhans' type. Elastica/van Gieson. **(b)** Higher magnification shows several multinucleated giant cells and mild lymphocytic infiltration in the adventitia. **(c, d)** Arterial wall showing luminal thrombus, thick fibrotic intima and inflammatory infiltrate including multinucleated giant cells, in the outer media and neighbouring adventitia. **(e, f)** Inflammatory infiltrate with a row of multinucleated giant cells.

Images kindly provided by T Polvikoski, Newcastle University, UK (c,d) and J Ogata, National Cerebral and Cardiovascular Disease Centre, Osaka, Japan (e,f).

macrophages. Eosinophils may be present. Multinucleated giant cells are a characteristic feature, and are of either foreign body or Langhans' type, most frequently in the inner media next to an irregularly frayed internal elastic lamina, where histiocytic cells of epithelioid appearance also accumulate. This type of infiltrate is often seen in the adventitia as well. However, a lack of giant cells in temporal biopsies does not rule out a diagnosis of GCA. At later stages, the intima is markedly thickened and the internal elastic lamina and the media largely destroyed, and the media and adventitia may become fibrotic and thickened, with blurring of their interface. Foci of calcification may be present.[738a,739] The lumen of the artery is generally markedly reduced but the wall may become weakened to such an extent that an aneurysm is formed. The vasculitis may also induce local thrombosis; if located in an anatomically critical blood vessel, this may serve as a source of small emboli to the intracerebral arteries and be a rare cause of an infarct.

Primary Angiitis of the Central Nervous System (PACNS)

PACNS was originally defined in 1959.[223] It has also been called granulomatous angiitis of the CNS, isolated angiitis of the CNS and primary vasculitis of the CNS.[161,342,387] PACNS is rare (2.4 cases per million) and of unknown cause. Most patients are initially misdiagnosed as having another condition. The most common clinical features are headaches, confusion, memory impairment, hallucinations, seizures and multifocal neurological deficits.[387,922] The patients are normally young or middle-aged but paediatric cases have been reported.[104,105,654] Angiography typically demonstrates multiple narrowed segments in cerebral arteries but the findings are not specific. In a minority of patients, neuroimaging reveals infarcts, haemorrhages or white matter lesions. Functional brain imaging may reveal multiple small perfusion defects. In contrast to GCA, the ESR in PACNS is usually mildly elevated at most, and the CSF often has elevated protein and a moderate mononuclear leucocytosis. The outcome of PACNS varies, from spontaneous resolution to rapid decline and death.[680] Aggressive immunosuppressive therapy has proven effective in treating the disease in many patients.[113]

The diagnosis of PACNS is established by brain biopsy. A biopsy sample that includes leptomeninges and a wedge of cortex is best for diagnosis, particularly if it can safely be obtained from a region that shows abnormalities on imaging. Arteries are affected much more than veins.[342] PACNS may present in both granulomatous and non-granulomatous forms, either separately or coexisting in different parts of the brain of the same patient. The non-granulomatous form may manifest as polyarteritis-type necrotizing inflammation or as a simple lymphocytic vasculitis. The inflammatory infiltrate (Figures 2.36 and 2.37) is composed mainly of lymphocytes, macrophages and plasma cells.[342] Giant cells of Langhans or foreign body-type are usually but not always present, in the wall or immediately adjacent to affected vessels. A negative biopsy does not exclude PACNS. In one series,

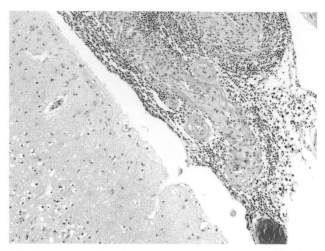

2.36 Primary angiitis of the central nervous system. The leptomeninges include a dense perivascular infiltrate of lymphocytes. There is also infiltration of the walls of several arterioles by lymphocytes and epithelioid macrophages.

PACNS was identified at biopsy in only four of seven patients.[268] The surrounding parenchyma may include infarcts or foci of intracerebral haemorrhage.[896] A small minority of PACNS cases has been associated with varicella-zoster virus infection.[346]

Aβ-Related Angiitis (ABRA)

Some patients with sporadic, Aβ-related cerebral amyloid angiopathy (CAA) develop a necrotising angiitis centred on Aβ-laden blood vessels. Aβ can often be identified within macrophages and giant cells in the inflammatory infiltrate. Because the inflammation is usually granulomatous, and generally affects blood vessels of similar size and distribution to those involved in PACNS, this concurrence of angiitis and CAA has been classified as a variant of PACNS. The neurological and angiographic findings are similar in PACNS and ABRA. However, ABRA has distinct clinical, pathological and pathogenetic features and should probably be regarded as a separate disease entity. It affects an older age group. Scolding et al.[922] reported that the mean age at presentation was 67.3 years in ABRA and 44.8 years in PACNS. Rapidly progressive dementia and hallucinations are more common than in PACNS. White matter oedema or discrete white matter lesions are present in about two-thirds of cases.

Close examination of the cerebral cortex in some cases of ABRA reveals scattered clusters of microglia that contain cytoplasmic Aβ.[922] Similar findings were reported at autopsy in patients who had participated in a trial of Aβ peptide immunization for the treatment of Alzheimer's disease.[161,435] The trial patients were also found to have more severe CAA than non-immunized Alzheimer's disease patients.[119] It was suggested that ABRA and Aβ-immunization-induced meningoencephalitis may share a common disease mechanism:[922] an immune response directed against Aβ and causing leptomeningeal and parenchymal inflammation, enhanced

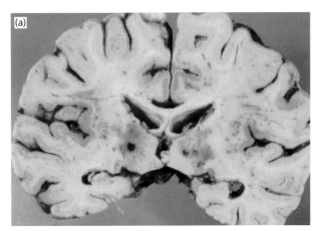
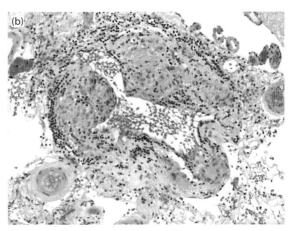

2.37 Primary angiitis of the central nervous system. (a) Small bilateral haemorrhagic infarcts in the thalamus. **(b)** Granulomatous vasculitis, with lymphocytes, epithelioid macrophages and multinucleated cells in the wall of a small leptomeningeal blood vessel.

microglial phagocytosis and clearance of parenchymal Aβ, and increased deposition of Aβ in cortical and leptomeningeal blood vessels. Further studies have provided some evidence in support of this hypothesis. In one patient with ABRA, Hermann et al.[416] found a significant increase in the proportion of circulating CD22+ memory cells directed against $Aβ_{1-42}$. However, some of the patient's B-cells produced antibodies that bound to Aβ in brain tissue. In another patient, with a diagnosis of probable ABRA, high levels of antibodies to $Aβ_{1-40}$ and $Aβ_{1-42}$ were detected in the CSF.[251] In addition, Melzer et al.[666] demonstrated an increased proportion of activated (CD69+) CD4+ cells in the CSF. Some patients with ABRA respond well to immunosuppressive treatment.

Manifestations of Systemic Diseases

Systemic Lupus Erythematosus (SLE)

SLE is a relapsing-remitting autoimmune disease with a preponderance in women and CNS involvement in up to 75 per cent of patients.[231] The average worldwide prevalence is estimated to be 25 per 100 000. SLE patients can develop a wide range of neurological and neuropsychiatric symptoms, including seizures, stroke, peripheral neuropathies, psychosis and depression.[810] Neuropsychiatric SLE is a complex entity comprising up to 19 different discrete syndromes. Damage to the cerebral microvasculature appears to be the main substrate of neuropsychiatric SLE. However, complications of SLE, such as uraemia, hypertension and infections, and the administration of steroid therapy, may contribute to some of the neuropsychiatric manifestations. Intracranial cerebrovascular pathology is a major cause of death, doubling mortality in SLE.[111]

Cerebral perfusion deficits have been detected by imaging methods in vivo.[977] The most common structural abnormality on neuroimaging is multiple foci of infarction.[445] Dysfunction of the blood–brain barrier and choroid plexus has been implicated in neurological disease in SLE.[2,271,841] Antinuclear antibodies are found in 95 per cent of cases; antibodies to dsDNA and Smith (Sm) antigen strongly support the clinical diagnosis. Two different pathogenetic

mechanisms have been proposed to explain neurological disease in SLE. Primary small vessel inflammatory vasculopathy is considered the main cause of infarction by some, whereas others emphasize thrombotic and thromboembolic mechanisms and ascribe minor significance to vasculitis. Proponents of the former view have reported fibrinoid necrosis, mononuclear inflammatory infiltrates and fibrotic thickening of the vessel walls, resembling changes to the renal vasculature, whereas those who support the latter view have found only limited vascular pathology. Deposition of immunoglobulins and complement in blood vessel walls can be demonstrated in biopsies from extracranial, more easily available tissues but studies on intracranial blood vessels are scarce. In recent years thrombosis and thromboembolism have gained favour as the more important contributors to CNS damage.[231,841]

The thrombotic hypothesis gained strong support from studies on circulating antiphospholipid antibodies (APLAs), which have a well-documented pro-coagulant effect and induce recurrent thromboses.[879] The frequency of APLAs in SLE ranges from 7–85 per cent. APLAs also occur in a wide variety of other conditions,[112] and in some healthy subjects.[990] In addition to vascular immune complexes and circulating APLAs, several autoantibodies have been demonstrated in neuropsychiatric lupus.[371] The pathogenesis of SLE as a complex multisystem autoimmune disorder remains far from clear.[231] There is a strong genetic component, as indicated by familial aggregation studies. These show a 10-fold higher disease concordance in monozygotic than dizygotic twins or siblings.[244] More than 30 different loci on several different chromosomes have been robustly associated with SLE. Recent genome-wide association studies suggest that B-cell receptor pathway and interferon signalling are important in SLE pathogenesis.

Polyarteritis Nodosa (PAN)

PAN is a transmural, necrotizing vasculitis.[793] It may start at any age, but the mean is around 40–50 years. The male:female ratio is about 2:1. PAN is a chronic disease with a fluctuating course and requires aggressive immunosuppressive therapy, which has improved its 5-year survival rate to 75–80 per cent from less than 15 per cent in untreated patients.[208,289] PAN causes necrotizing lesions predominantly in medium-sized and small arteries, with frequent distal spread into arterioles. There is sparing of capillaries, venules and veins. Its aetiology and pathogenesis have not been established definitively, but the mechanism most commonly implicated is an immune complex-mediated vasculitis. Various antigens are implicated,[793] including microorganisms, autoantigens and drugs. Many PAN patients are seropositive for hepatitis B or C virus (about 10–50 per cent and 20 per cent, respectively) but the precise pathogenetic relationship between PAN and the viruses is unclear.[176] The diagnosis of PAN is based on exclusion of other recognized causes of vasculitis.[208] Multiple small lesions on MR imaging support the diagnosis of neurological involvement but a normal arteriogram does not exclude it. Distinction should be made between limited versus systemic disease and idiopathic versus hepatitis B-related PAN because of differences in the pathogenetic mechanisms.

At least two different forms of idiopathic PAN are recognized. In systemic PAN, the visceral organs (heart, kidneys, gastrointestinal tract) are mainly affected. In the limited form of PAN, neurological manifestations are frequent: skeletal muscle is affected in 40–80 per cent of patients and peripheral nerves in 50–75 per cent. The reported frequencies of CNS involvement are highly variable, ranging from 4–53 per cent, but most studies seem to agree that it usually occurs in the late stages of the disease.[208] The variability probably reflects the difficulties in diagnosing brain involvement with certainty in what is a fairly uncommon generalized vasculitis. CNS disease is the second most common cause of death among patients with PAN.[793]

The necrotizing lesions are always very patchy, with a predilection for the branch points of the affected blood vessels. Most lesions are less than 1 mm in length and the vasculitis does not necessarily affect the whole circumference of the artery. Active lesions characteristically show fibrinoid necrosis, often with complete destruction of the blood vessel wall. CD8+ T-cells, macrophages and polymorphonuclear leukocytes predominate in the inflammatory infiltrate. Eosinophils and plasma cells may also be encountered in the adventitia. The necrotic wall occasionally undergoes aneurysmal dilation. In healed vessels, all layers of the artery show fibrous scarring and any residual inflammation is lymphocytic.[606]

Kawasaki's Disease

Kawasaki's disease (mucocutaneous lymph node syndrome) is an uncommon acute illness occurring mostly in children under 5 years.[979] The aetiology is not known. It may be initiated by certain infectious agents. Manifestations include fever, mucosal inflammation with ulceration, non-suppurative lymphadenopathy, oedema of hands and feet, a polymorphous rash and ischaemic cardiac symptoms. Kawasaki's disease is a necrotizing arteritis synchronously involving large, medium and small arteries.[979] It may also involve cerebral arteries, causing cerebral hypoperfusion,[454] and strokes or diffuse encephalopathy.[289]

Antineutrophil Cytoplasmic Antibody-Associated Vasculitides

Wegener's granulomatosis, microscopic polyangiitis and Eosinophilic granulomatosis with polyangiitis/Churg–Strauss syndrome are all usually associated with antineutrophil cytoplasmic antibodies (ANCAs).[97,110] In view of the paucity of immune deposits these are often referred to as pauci-immune systemic vasculitis. In ANCA-positive patients, polyneuropathies are of the multiplex type and associated with the histological detection of small vessel vasculitis.

Wegener's Granulomatosis

Wegener's granulomatosis is a rare arteritis characterized by (i) acute necrotizing granulomas of the upper respiratory tract, (ii) focal necrotizing or granulomatous vasculitis in the lungs and other sites, and (iii) focal or necrotizing glomerulonephritis. The vasculitis may also involve extracerebral and intracerebral cranial blood vessels, causing ischaemic stroke, haemorrhage or encephalopathy with or without seizures. The age of onset is variable but most patients are in their forties or fifties. Cytoplasmic ANCAs (c-ANCAs), directed against proteinase-3, are present in about 90 per cent of cases.[97,208] Neurological manifestations due to either vasculitis or granulomatous inflammation in the parenchyma are seen in about 30 per cent.[429] Parenchymal CNS disease usually represents extension from primary disease of the air sinuses, orbit or auditory canal.

Eosinophilic Granulomatosis with Polyangiitis (Churg–Strauss Syndrome)

The neurological abnormalities in eosinophilic granulomatosis with polyangiitis (EGPA) (previously known as Churg–Strauss syndrome) are similar to those in PAN.[694] However, the systemic symptoms, i.e. the presence of an allergic component manifesting as severe asthma and eosinophilia, and pulmonary involvement, are distinct and the pathogenetic mechanisms are considered to be different. PAN is an immune complex vasculitis, whereas in EGPA toxic proteins released from eosinophils have been implicated. Perinuclear ANCAs (p-ANCAs) directed at myeloperoxidase are present in about 80 per cent of cases.[97] Histopathologically, there is a necrotizing vasculitis of medium-sized and small arteries, and sometimes also of capillaries and venules, with an abundance of eosinophils among the inflammatory cells, as well as extravascular granulomas.[694] About two-thirds of patients develop peripheral neuropathies and involvement of the CNS, the latter including stroke, cerebral haemorrhage and ischaemic optic neuropathy.[110]

Sjögren's Syndrome

Sjögren's syndrome is an autoimmune disorder characterized by multiple abnormalities of cellular and humoral immunity.[28,702,921] An important diagnostic feature is the involvement of exocrine glands, including salivary and lacrimal glands, giving rise to sicca syndrome — dry eyes and mouth because of decreased secretory activity of the affected glands. Sjögren's syndrome can be a primary disease (pSS) or a 'secondary' disease in association with other connective tissue disorders. Peripheral nervous system involvement is well documented in pSS, but the frequency of involvement of the CNS is controversial. In one series, 20–25 per cent of pSS patients experienced either focal (hemiparesis, focal epilepsy, loss of vision, transverse myelitis or multiple sclerosis-like syndrome) or diffuse (aseptic meningoencephalitis, progressive encephalopathy or dementia) CNS disease.[28] Considerably lower numbers have been reported by others, the discrepancies being ascribed mainly to a lack of generally accepted diagnostic criteria and biased patterns of patient referral.[702] Rheumatoid factor, antinuclear antibody (ANA), anti-Ro/anti-La and anti-α-fodrin are often positive.[1073] Vasculitis has been reported in about 10–15 per cent of pSS patients. CNS pathology has been described as vasculitis or vasculopathy, which predominantly affects small venous vessels in the white matter. Necrotizing vasculitis of medium-sized vessels resembling PAN can occur.[921] The inflammatory cells are mainly lymphocytes, plasma cells and macrophages and also occur in the meninges.[28]

Behçet's Syndrome

This rare vasculitic multisystem disorder is characterized by recurrent ulcers of oral and genital mucous membranes, uveitis and arthritis.[110] The syndrome is most prevalent in the Middle East, Far East and the Mediterranean regions. In 30–40 per cent of patients, the CNS is affected, usually in the form of meningoencephalitis, which is often accentuated in the brain stem (rhombencephalitis). The diagnosis is based largely on the associated systemic findings. Increased concentrations of antibodies against phosphatidylserine and ribosomal phosphoproteins have been described but the pathogenesis is unclear. Both viral and immunological mechanisms have been proposed.[110] The vasculitis predominantly affects venules.[694] Cerebral lesions in Behçet's syndrome are often destructive. Perivascular infiltration by lymphocytes, macrophages and neutrophils is common, whereas involvement of the blood vessels is variable.[385,1113]

Rheumatoid Arthritis

Intracranial vasculitis, meningeal or parenchymatous, is a rare complication of rheumatoid arthritis and may be accompanied by white matter lesions, visible on MRI. Only a few cases have been reported, with variable outcome. Corticosteroid therapy may be beneficial but about 30 per cent of patients do not respond (Figure 2.38).

Vasculitis Associated with Hodgkin Disease

Vasculitis of the peripheral nervous system may occasionally complicate malignancies, but paraneoplastic vasculitis of the CNS is very rare except for that accompanying Hodgkin disease.[866]

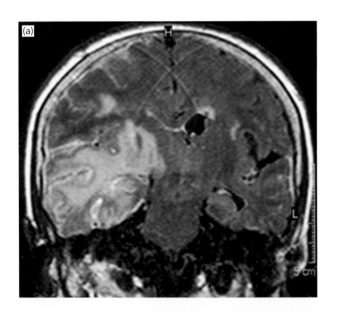

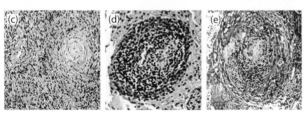

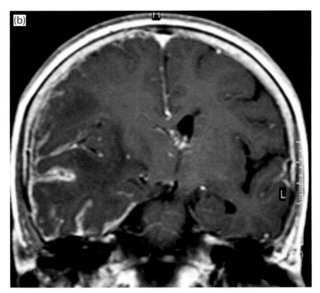

2.38 Rheumatoid arthritis with cerebral involvement. (a,b) Space-occupying lesion in the temporal lobe. **(c–e)** Inflammatory infiltrates composed of lymphocytes and mononuclear cells in the wall of small and medium-sized arteries and surrounding brain. Several blood vessels are occluded. Haematoxylin and eosin (H&E).

Drug-Induced Vasculitis

Vasculitis is one of the pathogenetic mechanisms purported to explain the increased incidence of stroke among drug abusers. There are, however, very few neuropathological studies to verify this.[158] In some cocaine abusers, definite vasculitis has been reported. The vasculitis seems not to be infective. The inflammatory infiltrate is mainly lymphocytic, and some patients respond to immunosuppressive therapy. There is a high frequency of intracerebral haemorrhage in users of cocaine, phenylpropanolamine and amphetamine,[949] suggesting that necrotizing vasculitis or arteritis may be more common than generally recognized (Figure 2.39), although drug-induced hypertension and vasospasm are also likely to contribute. The association between PAN and hepatitis B and C infection may account for the necrotizing vasculitis in some drug abusers.[176]

Infective Vasculitides

Many infectious agents are capable of invading blood vessel walls and inducing inflammation.[200] For some agents, blood vessel walls are even sites of predilection, and neurological complications including strokes and haemorrhages are common.[823]

Bacterial Vasculitis

Vasculitis is a fairly common complication in patients with bacterial (mainly pneumococcal) meningitis (Figure 2.40). Although small arteries are often affected, in rare cases involvement of large arteries (i.e. internal carotid artery) may produce large cerebral infarcts. Arterial vasculitis in tuberculous meningitis can cause brain infarcts of varying size (Figure 2.41) – see also Chapter 20.

Spirochaetal Vasculitis

A classical example of infective vasculitis is that resulting from the invasion of arterial walls by *Treponema pallidum* spirochaetes in syphilis.[341] This commonly affects extracranial arteries but can involve arteries in the CNS. The vasculitis in meningovascular neurosyphilis is known

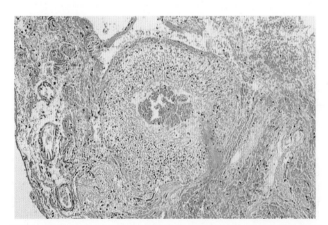

2.39 Intracerebral haematoma caused by drug-induced vasculitis. A necrotic blood vessel is seen next to an intracerebral lobar haematoma of clinically undetermined origin from a 27-year-old abuser of multiple hard drugs. Van Gieson.

as Heubner's arteritis. The intima and adventitia become thickened with a thinner media. The inflammatory infiltrate consists of lymphocytes and plasma cells. The end result of healed syphilitic vasculitis may be indistinguishable from an atherosclerotic fibrous plaque. Since the 1990s, neurosyphilis has become more prevalent, as a complication of human immunodeficiency virus (HIV) infection.[745]

Another spirochaete that has become increasingly important is *Borrelia burgdorferi*, the cause of Lyme neuroborreliosis. There is evidence that *B. burgdorferi* is able to penetrate the BBB and enter the brain parenchyma in the early stages of the disease. The spirochaete seems to bind to connective tissue molecules in the blood vessel wall and induces vasculitis. As a consequence, the BBB is disrupted, and multifocal enhancing lesions with mild oedematous tissue changes may develop (Figure 2.42). The sequestration of *B. burgdorferi* within vessel walls may impede access by antibiotics and lead to recurrent disease.

Viral Vasculitis

True cerebral vasculitis of the lymphocytic, necrotizing or granulomatous types may occur in viral infection,[964] particularly in immunosuppressed patients. Inflammatory alterations in the cerebral blood vessel walls have been reported, particularly in patients infected with the herpes viruses, notably varicella-zoster virus (see Chapter 19). In some patients with PAN, not only antibodies to hepatitis C virus but also viral nucleic acids have been demonstrated by polymerase chain reaction.[110]

Vasculitis in the Context of Human Immunodeficiency Virus Infection

Various types of vasculitis have associated with HIV infection.[154,160,338] These include PACNS- and PAN-like conditions, hypersensitivity vasculitis and lymphomatoid granulomatosis. Perivascular and transmural infiltration by CD8+ T-cells may occur in early HIV infection. The exact relationship between HIV infection and these various inflammatory vascular disorders remains unclear.[154] Vascular inflammation in HIV infection appears to be multifactorial and may result from HIV-induced immunological abnormalities and exposure to a variety of xenoantigens, such as HIV, other infectious agents and drugs.[338]

Vasculitis, with high numbers of CD8+ T-lymphocytes infiltrating blood vessel walls, has also been observed following combined antiretroviral treatment despite amelioration of other clinical parameters. This is thought to represent an overactive response of a reconstituted immune system to various infection-related antigens (see Chapter 19, Viral Infections).[682,685]

Fungal Vasculitis

The frequency of opportunistic infections has increased in parallel with the rising number of patients receiving immunosuppressive drugs or infected with HIV. Fungi of otherwise low pathogenicity are amongst the more common infective agents in such patients. Fungal emboli may lodge in the intracranial arteries, penetrate the wall and induce vasculitis (Figure 2.43), often complicated by thrombosis and infarcts

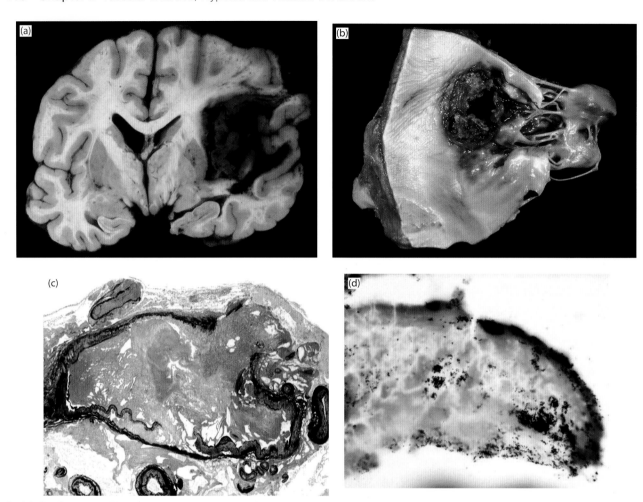

2.40 Infective vasculitis, secondary to subacute bacterial endocarditis. (a) Multiple foci of cerebral infarction and haemorrhage, in a patient found at autopsy to have **(b)** a large vegetation that had eroded through the mitral valve. **(c)** Histology showed a ruptured aneurysm within one of the foci of haemorrhage (elastin/van Gieson). **(d)** High-magnification view of a section through the blood clot stained by Gram's method revealed numerous Gram-positive cocci.

Courtesy of S Love, University of Bristol, UK.

or by vessel wall rupture and haemorrhage. Vasculitis is the most common cause of fungal brain necrosis (Figure 2.44). Fungi associated with infective cerebral vasculitis include *Aspergillus*, *Candida*, *Coccidioides* and *Mucor* species.[964]

INTRACRANIAL ANEURYSMS AND VASCULAR MALFORMATIONS

The term 'aneurysm' describes various forms of focal arterial dilatation. Intracranial aneurysms have been classified according to a range of features including aetiology (congenital, acquired, dissecting, infectious, tumourous), size, shape (fusiform, saccular) or association with a specific branch of a vessel. Two to five per cent of the population worldwide are estimated to develop intracranial aneurysms.[1054] Dissections of the arterial wall and arteriovenous fistulae are often considered together with aneurysms.

Saccular Aneurysms

Saccular (berry) aneurysms are the most common type (95 per cent versus 5 per cent fusiform) visualized by

angiography and MRI.[813] They are of clinical importance mainly because of their propensity to rupture. Ruptured intracranial aneurysms are responsible for about 85 per cent of cases of subarachnoid haemorrhage.[1038]

Epidemiology

The autopsy prevalence of intracranial saccular aneurysms in the general population is 2–5 per cent (range 0.2–9 per cent).[471,850] Aneurysms are rare in children: fewer than 5 per cent are encountered under 20 years, and about 60 per cent are diagnosed in patients aged 40–60 years. The overall prevalence of aneurysms is higher in women than men (ratio 3:2) but this female preponderance does not manifest until after the fifth decade; in younger age groups, males predominate. In about 12–15 per cent of patients with saccular aneurysms there is an autosomal dominant pattern of inheritance;[23] these patients tend to be younger (by about 5 years).[861] About 40–70 per cent of the aneurysms noted in various autopsy studies had ruptured before death.[502]

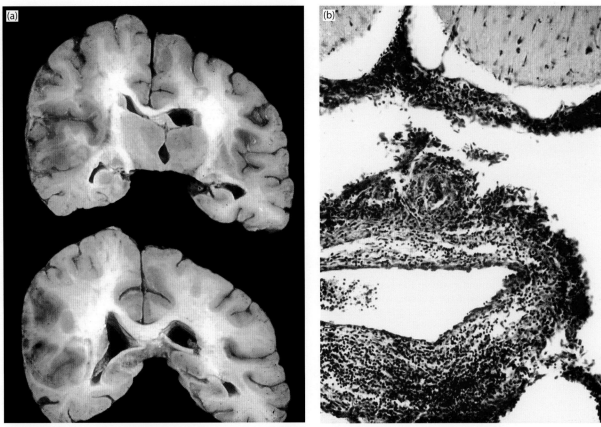

2.41 Tuberculous angiitis leading to cerebral infarction in the middle cerebral artery (MCA) territory. (a) Acute infarct, with marked ipsilateral oedema. **(b)** Vasculitis, with lymphocytes in the wall of the MCA. Typical granulomatous meningitis involved the cerebrum, cerebellum and brainstem.

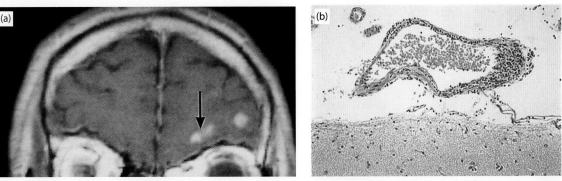

2.42 Infective vasculitis caused by *Borrelia burgdorferi*. **(a)** Gadolinium-enhanced T1-weighted magnetic resonance (MR) image of the brain of a 40-year-old, previously healthy man, who had two episodes of epileptic seizures. There are two focal enhancing lesions, which disappeared with antibiotic therapy, but recurrences similarly responsive to antibiotics occurred three times in different locations.**(b)** Histopathology of the surgically removed lesion (arrow in [a]) disclosed lymphocytic infiltration of the wall of a small leptomeningeal blood vessel. The underlying parenchyma was slightly oedematous, with minimal astrocytic reaction. Polymerase chain reaction analysis for *Borrelia burgdorferi* DNA was positive in three separate tissue samples.

Image (a) courtesy of J Oksi, Turku University Hospital, Turku, Finland.

Aetiology, Pathogenesis and Risk Factors

Both genetic and environmental factors are involved in the development of intracranial aneurysms. Several studies have indicated that hypertension, hypercholesterolemia and cigarette smoking are risk factors. Genetic factors are important contributors to early-onset familial saccular aneurysms and subarachnoid haemorrhage.[23] The age-adjusted prevalence of saccular aneurysms among first-degree relatives of affected families was reported to be four times higher than in the general population.[862] Familial aneurysms were more often multiple and more likely to rupture even if small.[861] The molecular pathways have still not been established for the majority of cases.[23] However, a number of gene loci have been linked to aneurysm development[30] and the risk is increased in a range of multisystem familial diseases (mostly affecting connective tissue), including

2.43 Fungal vasculitis. Fungal hyphae of phycomycosis (mucormycosis) attached to and invading the wall of an intracerebral artery of a 52-year-old male patient, who developed cerebro–rhino–ocular phycomycosis in the context of ketoacidotic diabetes mellitus. Methenamine silver.

autosomal dominant polycystic kidney disease, neurofibromatosis type I, Marfan syndrome, multiple endocrine neoplasia type I, pseudoxanthoma elasticum, hereditary haemorrhagic telangiectasia, and Ehlers–Danlos syndrome type II and IV.[171] See also Genetics and Genomics of Intracranial Aneurysms and Vascular Malformations (Table 2.3).

Several environmental factors have been implicated in the pathogenesis of saccular aneurysms. A Cochrane review,[203] identified that besides female gender, older age, posterior circulation aneurysms, larger aneurysms, previous symptoms, 'non-white' ethnicity, hypertension, low body mass index, cigarette smoking and alcohol consumption of more than 150 g per week were the strongest risk factors for intracranial aneurysms and subarachnoid haemorrhage. A risk locus associated with intracranial aneurysms was shown to be associated with elevated systolic blood pressure.[320]

Sites of Occurrence

Saccular aneurysms can be located in anterior or posterior circulation vessels (Figure 2.45b). A large majority (85–90 per cent) involve the terminal part of the ICA or the major branches of the anterior part of the circle of Willis (Figure 2.45b). The reported frequencies of multiple aneurysms in patients with subarachnoid haemorrhage vary from 8.6 to 31 per cent, and their most frequent location is the MCA.[852] In a review of over 300 cases, Pritz[813] subdivided the aneurysms into those not associated with a branch vessel, those associated with a side-branch vessel, and those at a bifurcation. Most were located at the bifurcation of major

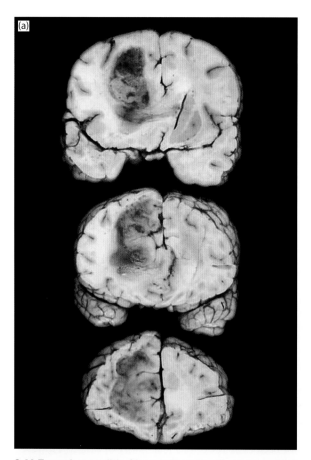

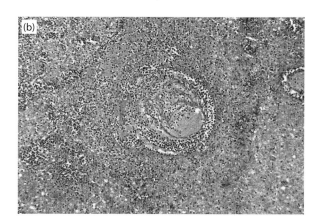

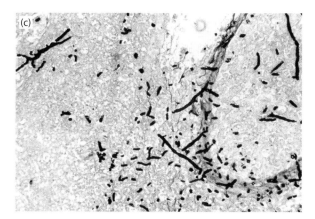

2.44 Fungal vasculitis (*Aspergillus*) leading to necrotizing cerebritis. (a) Large necrotic lesion in the frontal lobe. **(b)** Vasculitis, thrombosis and necrosis of the cerebral white matter. **(c)** Multiple hyphae in the blood vessel, the blood vessel wall and the brain parenchyma. Methenamine silver.

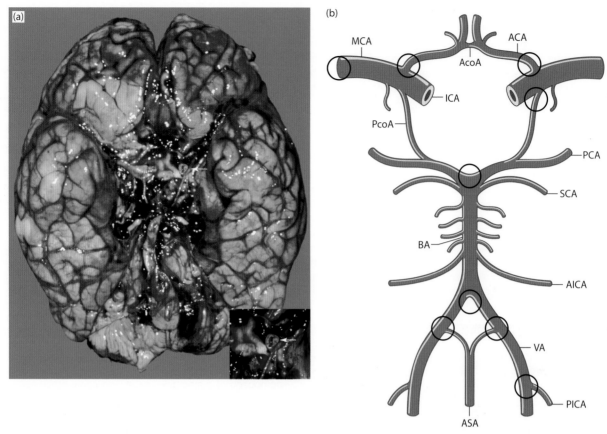

2.45 Aneurysms at the base of the brain. (a) A small berry aneurysm of the left middle cerebral artery (MCA) (arrow) had ruptured and caused fatal subarachnoid haemorrhage. Inset: higher magnification of the aneurysm. **(b)** Schematic diagram of the circle of Willis showing the most common sites of aneurysms formation (circles). Along the MCA, aneurysms occur at the branch points of M1 and M2 arteries. Aneurysms are also frequent at the bifurcation of the internal cerebral artery (ICA). ACA, anterior cerebral artery; ACoA, anterior communicating artery; AICA, anterior inferior cerebellar artery; ASA, anterior spinal artery; BA, basilar artery; ICA, internal cerebral artery; MCA, middle cerebral artery; PCA, posterior cerebral artery; PCoA, posterior communicating artery; PICA, posterior inferior cerebellar artery; SCA, superior cerebellar artery; VA, vertebral artery.

Images kindly provided by M Röyttä, Turku University Hospital, Turku, Finland and A Oakley, Newcastle University, UK.

conducting vessels (internal carotid or vertebrobasilar), primary vessels (anterior communicating artery region or middle cerebral artery bifurcation) or more distal vessels (distal anterior cerebral artery).

Macroscopic Appearance

Saccular aneurysms vary widely in size and shape (Figures 2.45 and 2.46). After post-mortem fixation, they may have shrunk by 30–60 per cent. Their diameters at autopsy generally range between a few and 25 mm; those over 25 mm are called giant aneurysms (Figure 2.46). A large proportion (30–40 per cent) of saccular aneurysms in children is of giant size.[502] At the basilar bifurcation, the average size was found to be twice that in other locations.[776]

Saccular aneurysms may be elongated or buckled and can have wide or narrow neck by which they communicate with the blood vessel lumen (Figures 2.47 and 2.48). The walls of some aneurysms are thin and translucent and tend to collapse at autopsy. Some aneurysms have opaque walls due to fibrous thickening or atherosclerosis, and protrude as rigid sacs from their site of origin. The walls may be

calcified, particularly in large specimens. Large aneurysms also often contain lamellated thrombus, which may seed thromboemboli into distal arteries. Aneurysms are sometimes completely fibrosed, with obliteration of the lumen. Others are buried in adjoining brain tissue instead of protruding free into the subarachnoid space (see Figure 2.121).

A ruptured aneurysm can be difficulty to recognize at autopsy. Small thin-walled aneurysms may be destroyed as they bleed, and only the torn edges may remain of larger aneurysms. Ruptured aneurysms are often obscured by large amounts of surrounding subarachnoid or intracerebral blood clot. Particularly if the location of the aneurysm has not been established previously by angiography or brain imaging, it is advisable to wash away the blood clot and to conduct a thorough search for the aneurysm before fixation of the brain. Dissection of fixed blood is tedious and the aneurysm may be difficult to identify within the clot. The distribution of the subarachnoid blood may suggest the site of the aneurysm. In the absence of recent bleeding, a site of previous haemorrhage is often marked by orange-brown discolouration of the pial surface and arachnoid. A more elaborate method for post-mortem visualization of ruptured aneurysms or malpositioned clips has been developed for use in forensic cases. After infusion of radio-opaque polymerizable rubber solution into the vasculature, the ruptured aneurysm; the patency of arteries and sometimes even associated arterial spasm can be assessed by X-ray (Figure 2.49) and later in brain sections, because the rubber cast can be cut.

Microscopic Appearance

The relationship of saccular aneurysms to arterial branching is seen particularly well in small aneurysms. The pads of intimal thickening can be seen by scanning electron microscopy proximal to normal branches or aneurysms (Figure 2.47). The pads appear as flattened areas that have lost the rugose pattern of the vessel wall. The most characteristic histological feature of the aneurysm wall is the absence of both the tunica media and the internal elastic lamina, both of which end abruptly at the neck of the aneurysm (Figure 2.48). Apart from the endothelium, the wall

of the aneurysmal pouch consists only of fibrous connective tissue of varying thickness. Even the endothelium may be incomplete, and blood clot may line the luminal surface. The aneurysm wall often includes atheroma, the severity of which tends to parallel the size of the aneurysm.[555] After

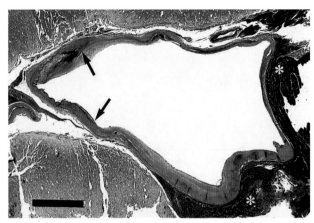

2.48 Saccular aneurysm that arose at the trifurcation of the right middle cerebral artery. The aneurysm had ruptured and caused both intracerebral haemorrhage (into the right temporal lobe) and subarachnoid haemorrhage. The elastic lamina (black) stops at the arrows, and the wall of the dilated aneurysm is composed of connective tissue with some atherosclerotic intimal thickening. The wall thins out towards the apex of the aneurysm on the right but the rupture itself is not visible. The apex is covered by the blood (*) in the subarachnoid space. Elastica/van Gieson stain.

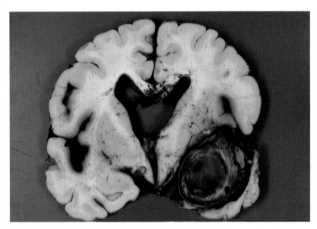

2.46 Giant saccular aneurysm of the middle cerebral artery (MCA).

2.47 Saccular aneurysm. Scanning electron microscopic picture of a small, thin-walled aneurysm (A) at the bifurcation of a middle cerebral artery. A pad of vessel wall thickening (P) is seen proximal to the bifurcation, and the rugose pattern of the normal vessel wall is disturbed. The direction of blood flow is indicated by the arrow.

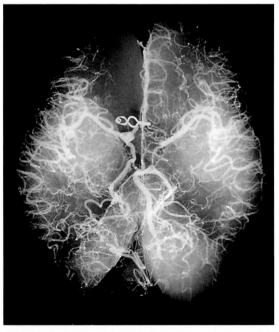

2.49 Cerebral rubber cast angiography. A silicone rubber angiographic X-ray of a brain after an accidental kinking of the right anterior cerebral artery against the clip after ligating an aneurysm of the anterior communicating artery. As a consequence, the territory of the right anterior cerebral artery has remained unfilled by the contrast medium.

recent rupture, the frayed remnants of the wall show necrosis and infiltration by inflammatory cells.[555]

Microscopic study of intact berry aneurysms discloses attenuation and disruption of the internal elastic lamina and focal or diffuse fibrosis of the vessel wall. This may be accompanied by deposition of glycosaminoglycans, accumulation of macrophages, calcification and haemosiderin deposits. Immunohistochemical studies have shown reduced smooth muscle actin, consistent with the atrophy of the media. Abnormalities of components of the extracellular matrix, including collagens and elastin, have also been demonstrated.[130,520]

Natural History and Pathogenesis

Small saccular aneurysms of the basal cerebral arteries do not usually cause neurological symptoms. In a minority of cases, patients may present with clinical manifestations of focal nerve or parenchymal compression. An aneurysm may, for example, compress the optic tract or the third, fourth, fifth (ophthalmic division) or sixth cranial nerves. Large aneurysms can cause epileptic seizures, particularly when the fundus is buried within the temporal lobe. Rarely, a giant aneurysm may expand to such an extent that the patient dies from complications of raised intracranial pressure.

The most common and serious complication of saccular aneurysms is rupture, with consequent subarachnoid haemorrhage, and possible arterial vasospasm and infarction or intracerebral haemorrhage.[502] If the fundus of the aneurysm is embedded in brain tissue, rupture may cause haemorrhage that is predominantly parenchymal rather than subarachnoid.

Most saccular aneurysms arise at or very close to bifurcations of the main intracranial arteries.[813] They rarely occur outside the cerebral circulation (Figures 2.47 and 2.48). Intracranial arteries differ from their extracranial counterparts by the thinness of their walls, absence of an external elastic lamina and lack of perivascular support. Histological studies of major cerebral arteries of normal individuals have demonstrated frequent gaps in the muscular tunica media at the bifurcations. The predilection of aneurysms for this location may be due the fact that arterial wall is often limited here to the intima, the internal elastic lamina, which may be fenestrated, and some adventitial connective tissue.

New saccular aneurysms have been detected during angiographic follow-up monitoring of patients with a previously diagnosed aneurysm.[501] These observations and experimental work indicate that most saccular aneurysms are acquired degenerative lesions made worse by haemodynamic stress. With increasing age, fibrous intimal pads are formed, particularly at the bifurcations of cerebral arteries, rendering their walls less elastic (Figure 2.49). The intimal arteriosclerotic thickening may alter the haemodynamic stress at sites of arterial bifurcation. In the presence of a gap in the muscular layer and fenestrated internal elastic lamina, this stress may result in the bulging outwards of the thin arterial wall. Alternatively, the elastic and muscular layers may degenerate without a pre-existing local defect, as the consequence of constant overstretching by haemodynamic stress at arterial branch points.[787]

The haemodynamic hypothesis is compatible with the well-documented association of saccular aneurysms with AVMs, which cause increased regional blood flow. About 3–9 per cent of patients with intracranial AVMs have saccular aneurysms in arteries haemodynamically related to the AVM. An increased frequency of saccular aneurysms has also been noted in coarctation of the aorta and in patients with an asymmetrical circle of Willis, particularly asymmetry of the proximal segments of the anterior cerebral arteries. Occlusion of one or more of the cerebral feeding blood vessels may enhance aneurysm formation through increased haemodynamic stress to the arteries involved in the collateral flow.

Experimental manipulation of haemodynamic factors can produce arterial mural atrophy and aneurysmal dilation. In rats and non-human primates with induced hypertension and unilateral carotid artery ligation, aneurysms arise on the anterior communicating artery, ipsilateral anterior cerebral artery and proximal segment of the posterior cerebral artery, whereas bilateral carotid ligation induces aneurysms in the posterior part of the circle of Willis, i.e. they develop in arteries with enhanced collateral blood flow. Under these experimental conditions, aneurysm formation was associated with apoptosis of smooth muscle cells in the tunica media.[553]

Loss of tensile strength of the arterial wall probably explain the association with certain connective tissue disorders, including Marfan's syndrome and Ehlers–Danlos syndrome types II and IV. Aneurysms in these conditions as well as in pseudo-xanthoma elasticum tend to be fusiform,[23] however, and to occur in relatively young patients. Saccular aneurysms sometimes complicate fibromuscular dysplasia (FMD)[204] and moyamoya disease.

Dolichoectasia and Fusiform Aneurysms

The term 'dolichoectasia' describes the elongation, widening and tortuosity of a blood vessel and is usually used in the context of the cerebral arteries. The basilar artery and the supraclinoid segment of the ICA are the two most commonly involved sites (Figure 2.50) but the process may extend to the adjoining segments of the vertebral and middle cerebral arteries. The dolichoectatic basilar artery is often S-shaped, with a luminal diameter exceeding 4.5 mm, the defined lower limit for diagnosis of ectasia. The dilated part of the artery may form a fusiform aneurysm.[813]

Dolichoectasia and fusiform aneurysms are commonly seen in patients with advanced cerebral atherosclerosis and have been associated with risk factors for atherosclerosis, such as hypertension, diabetes mellitus, hypercholesterolaemia and cigarette smoking.[490] Dolichoectasia and fusiform aneurysms of the basal artery have occasionally been reported in children and young non-atheromatous patients, sometimes in conditions such as Ehlers–Danlos syndrome type IV, Marfan's syndrome, pseudo-xanthoma elasticum, α1-antitrypsin deficiency[23,905,906] and various forms of arteritis. An increased risk of dolichoectasia was also reported for patients with autosomal dominant polycystic kidney disease.[907] Apart from the intimal atherosclerotic lesions, the affected arterial segments may have a thin internal elastic lamina and an atrophic and fibrotic tunica media. These lesions have been interpreted by some authors to represent

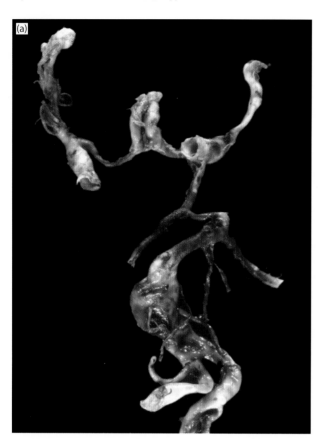
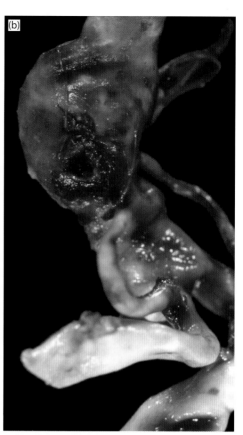

2.50 Fusiform aneurysm. (a) The basilar artery is dilated and tortuous, to such an extent that the lesion constitutes a fusiform aneurysm, which also compresses and distorts the brain stem. **(b)** A tear in the wall of the fusiform aneurysm caused fatal intracranial haemorrhage.

a unique form of arteriopathy, and in a review of 120 surgically treated giant fusiform aneurysms, 111 had neither atherosclerosis nor other known arteriopathy.[264] Studies have shown deficiencies in the internal elastic lamina and arterial media, and very rarely this complicates α-glucosidase deficiency.[169]

Progressive enlargement of the dolichoectatic arteries has been documented by serial CT, and rapid *de novo* development of giant fusiform aneurysms was demonstrated angiographically in children.[491] Dolichoectasia and fusiform aneurysms may, by mechanical compression, cause cranial nerve palsies, cranial nerve hyperexcitation syndromes (such as trigeminal or glossopharyngeal neuralgia) and hydrocephalus. Complications include TIAs, possibly due to emboli derived from thrombi or fragments of plaques in the walls of the enlarged arterial segment, and brain stem and cerebellar infarcts as a result of intraluminal thrombosis or atheromatous occlusion or distortion of the branches of the dilated basilar artery. Spontaneous haemorrhage is relatively rare. Dolichoectasia is commonly associated with small vessel disease, lacunar infarcts, status cribrosus and leukoaraiosis.[797]

Infective (Septic) Aneurysms

Infective, or septic, aneurysms, which account for about 2.5–4.5 per cent of intracerebral aneurysms, arise from microbe-carrying emboli, usually originating from an infected heart valve or pulmonary vein (Figure 2.51). About 3 per cent of patients with infective endocarditis have been claimed to develop such aneurysms. It is likely, however, that the majority (up to 80–90 per cent) of these patients have a pyogenic infection of the arterial wall but not a true infective aneurysm, which may be difficult to identify with certainty. The infected foci sometimes rupture, causing parenchymal brain haemorrhage, the risk of which has been estimated at 3–7 per cent for patients with endocarditis. Most frequently, infective aneurysms occur in the distal branches of the cerebral arteries, with particular predilection for the MCA. The aneurysms are multiple in about 20 per cent of cases.[692] The most common causative organisms are *Staphylococcus aureus* and *Streptococcus viridans*. Histologically, in the acute stage, an infected embolus may be seen adherent to an oedematous and necrotic arterial wall infiltrated by polymorphonuclear and other inflammatory cells. Experimental studies indicate that organisms spreading from the embolus may lead to weakening of the wall and aneurysmal dilatation within 24 hours.

Fungal aneurysms ('mycotic' in the proper sense; the term 'mycotic aneurysm' was previously used somewhat confusingly to describe all infective aneurysms) are caused by infected emboli from the heart, or fungal meningitis, the primary focus of which usually resides in the lungs or paranasal sinuses.[451,467] *Aspergillus* spp. are most often responsible. The hyphae may be seen spreading along the intima or through the wall of the affected artery, accompanied by necrosis and a variable inflammatory response (Figure 2.43).

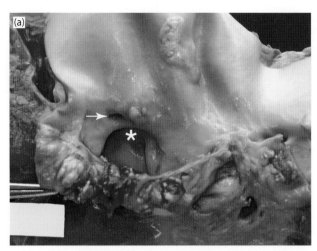 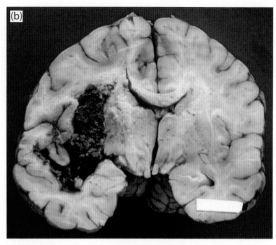

2.51 (a) The heart of a 23-year-old female with a Valsalva aneurysm (*) below the origin of right coronary artery (arrow). This had resulted in mechanical damage to the aortic valves with secondary streptococcal valvulitis. **(b)** A bacterial embolus from the valve caused an infective aneurysm in the left lenticulostriate artery, which ruptured, resulting in fatal intracerebral haemorrhage.

Miliary Aneurysms

Miliary aneurysms or microaneurysms of Charcot–Bouchard are controversial. They are considered to be caused by hypertension (discussed earlier in this chapter).

Miscellaneous Aneurysmal Lesions

Arterial Dissections and Dissecting Aneurysms of Intracranial Arteries

The term arterial dissection describes the extravasation of blood into the arterial wall, usually between the intima and media or, less frequently, between the media and adventitia. The intramural haematoma may extend between the planes of the artery and re-enter the original lumen, creating a parallel false lumen. The dissection does not necessarily result in actual arterial dilatation but if it does, the lesion is referred to as a dissecting aneurysm. It should be noted that, apart from cases with severe atherosclerosis and large aneurysms, the intracranial blood vessels have only sparse vasa vasorum, mainly supplying the adventitia. Thus blood in an intramural haematoma must originate from the lumen of the artery, through an intimal tear.

Dissections are most frequent in young and middle-aged adults but also occur in children. The vertebral, basilar, middle cerebral and internal carotid (above the clinoid processes), middle cerebral, vertebral and basilar arteries are particularly susceptible, and the lesions may be bilateral. Arterial dissection is, rarely, a cause of spontaneous subarachnoid haemorrhage. Many dissections are associated with blunt trauma to the head or neck or hyperextension injury, which stretches the arteries at the base of the brain sufficiently to rupture their intima. The trauma may seem trivial. Spontaneous dissection of intracranial arteries is less common than its extracranial counterpart but has been diagnosed increasingly *intra vitam* in recent years by advanced imaging methods.[44] It has been reported in association with many conditions, such as arteritis, atherosclerosis, hypertension, the use of oral contraceptives, cystic medial necrosis and mucoid degeneration, FMD and segmental mediolytic 'arteritis'.[280] Arterial dissection may be a complication in Marfan's syndrome, Ehlers–Danlos syndrome type IV, polycystic kidney disease type I and osteogenesis imperfecta. α1-Antitrypsin deficiency and inherited connective tissue disorders are more common in patients with dissections.[129,908] Abnormalities of the internal elastic lamina are often evident and include focal absence, splitting, fraying and reduplication. In approximately one-third of pathologically documented cases, however, no underlying vasculopathy has been noted. Most intracranial dissections present with an ischaemic stroke due to stenosis or occlusion of the original lumen of the affected artery by the intramural haematoma. Angiographic studies indicate that a benign course may be more frequent than previously recognized.[540] Recurrent dissections seem to be rare. Persistent fusiform aneurysmal dilation has been described as a sequel to dissection of the vertebral artery.[540,891]

Caroticocavernous Sinus Fistulae and Dural Arteriovenous Fistulae

Within the base of the skull, the ICA passes along the wall of the cavernous sinus, with arterial and venous blood being separated only by the arterial wall and a thin venous endothelium. Most fistulae between the carotid artery and the venous cavernous sinus occur in adult males and are the immediate or delayed consequence of head injury.[571] Spontaneous fistulae are more frequent in women and may be secondary to rupture of saccular aneurysms arising from the ICA and projecting into the cavernous sinus, or FMD, or connective tissue disorders such as Ehlers–Danlos syndrome type IV. Occasional cases have been reported in infancy and childhood.[356] A caroticocavernous sinus fistula usually produces a distinctive clinical picture, characterized by ipsilateral pulsating exophthalmos, with proptosis and a continuous bruit over the orbit. Large fistulae can divert sufficient blood from the brain to cause signs of cerebral ischaemia. Caroticocavernous sinus fistulae are usually treated by interventional neuroradiological techniques, although spontaneous cure by thrombosis, particularly in the indirect (dural) type of fistula (see later), does occur occasionally.

A dural arteriovenous fistula is an acquired arteriovenous shunt located within the dura. It has a variable natural history and symptomatology; very rarely this includes progressive cognitive decline, associated with venous hypertensive encephalopathy and potentially reversible.[450]

Aneurysms Associated with Neoplasms

These are very rare, usually caused by neoplastic emboli that destroy the internal elastica lamina and arterial tunica media. Germ cell tumours, lung carcinoma and cardiac myxoma are most often responsible.[206,1064]

Vascular Malformations

Vascular malformations of the brain are relatively common lesions that can cause serious neurological disability or death.[590] Most common are AVMs and cerebral cavernous malformations (CCMs), with detection rates of approximately 1.1 and 0.6 per 100 000 adults per year. Malformations have been classified on the basis of the calibre and configuration of the constituent vascular channels, their continuity with the normal cerebral vasculature, and the relation between the blood vessels and the intervening parenchyma. Such classifications include discrete arteriovenous, venous, cavernous, capillary and mixed types (Box 2.5). Vascular malformations are usually congenital lesions that arise as a result of disordered mesodermal differentiation between the third and eighth weeks of gestation.

BOX 2.5. Classification of brain vascular malformations

I	Congenital malformations in the brain parenchyma
	Arteriovenous malformation variant: vein of Galen malformation Cavernous haemangioma Venous angioma Capillary telangiectasia Mixed (or combined) • cavernous and venous • cavernous and capillary Other: haemangioma calcificans
II	Congenital malformations in meninges
	Arteriovenous malformation Venous angioma and varix (leptomeninges) Cavernous haemangioma (dura)
III	Vascular malformations as a part of CNS or generalized syndromes
	Phakomatoses (e.g. Sturge–Weber syndrome) Hereditary haemorrhagic telangiectasia (Rendu–Osler–Weber syndrome) Other, e.g. Wyburn–Mason syndrome, cerebro–hepato–renal cavernous angiomas
IV	Acquired vascular lesions simulating vascular malformations
	Radiation-induced lesions of the white matter Lesions secondary to venous sinus obstruction

CNS, central nervous system.
Adapted from Challa VR, *et al.*[184] With permission from Lippincott Williams & Wilkins/Wolters Kluwer Health.

Familial cases are rare.[40] The genetics of vascular malformations[590] have advanced in recent years with the discovery of several gene defects (Table 2.3). However, *de novo* malformations can occur, and radiation to the brain and dural venous sinus obstruction can cause acquired vascular lesions that are radiologically and pathologically identical to congenital vascular malformations. Vascular malformations may also occur as part of various syndromes, either generalized or limited to the CNS.[184]

Arteriovenous Malformations

AVMs, also known as arteriovenous aneurysms and angiomas, consist of tangled masses of tortuous arteries, veins and abnormal connecting channels. Although the AVM itself lacks a true capillary bed, dilated capillaries are usually present in the perinidal tissue. In view of their potential to bleed, arteriovenous malformations constitute the most clinically significant group of vascular malformations and the most frequent type in surgical specimens.

The prevalence of intracranial arteriovenous malformations is about 0.018 per cent, and the incidence about 1.2 per 100 000 per annum.[33,34] AVMs can become symptomatic at any age but the mean age of presentation is 40 years, with recurrent subarachnoid or intracerebral bleeding. Other common symptoms include seizures, headache and focal neurological deficits.[184] Some AVMs are discovered incidentally at autopsy. Hormonal changes during puberty and pregnancy may increase the risk of haemorrhage.[306]

AVMs range in size from the grossly invisible (cryptic vascular malformations) to large ones involving an entire hemisphere. A large majority of AVMs are supratentorial, occurring on the surfaces of the cerebral hemispheres (Figure 2.52) or deep in the basal ganglia or thalamus, but infratentorial lesions also occur (Figure 2.53). Occasionally, they are confined to the dura or leptomeninges.[184] The superficial cerebral examples are often wedge-shaped, with the apex extending inwards into the centrum semiovale and approaching the ventricular surface. The overlying leptomeninges are frequently thickened and brownish, owing to previous haemorrhage, and the surrounding brain parenchyma atrophic and discoloured. Microscopically, the constituent blood vessels correspond to arteries with muscular and elastic laminae, and veins dilated by the pressure to which they are exposed because of the shunting. The veins usually have thickened collagenous walls and appear 'arterialized', and yet the increased cellularity in their walls reflects proliferation of fibroblasts, not smooth muscle cells (Figure 2.52). Segmental saccular dilatations are common. The affected blood vessels are separated by brain parenchyma that often shows gliosis, haemosiderin pigmentation and foci of calcification.[184] The dilated perinidal capillaries are prone to leakage and may contribute to the accumulation of haemosiderin as well as to postoperative bleeding.[66,1025] AVMs are inherently angiogenic and overexpress vascular endothelial growth factor (VEGF), as well as the vasoconstricting peptide endothelin-1 (ET-1).[897] A grading scheme has been devised, based largely on the size and location of the malformation. The incidence of postoperative neurological complications correlates with the grade according to this scheme.[391]

2

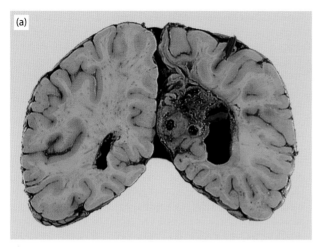

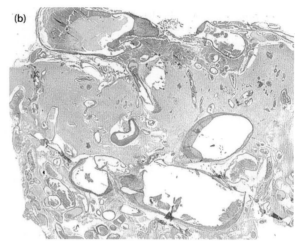

2.52 (a) Arteriovenous malformation (AVM) on the medial surface of the right hemisphere in the parieto-occipital region. **(b)** Microscopically, the arteriovenous malformation is composed of irregularly spaced blood vessels of variable sizes within, and separated by, brain parenchyma. There are arteries, veins and 'arterialized' veins with walls of uneven thickness. Van Gieson.

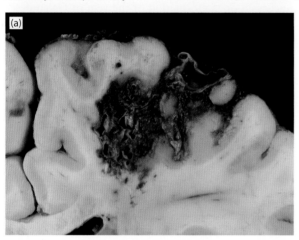

2.53 (a) This arteriovenous malformation (AVM) extends through the frontal white matter and cortex into the overlying subarachnoid space. Note the large calibre and irregular contour of many of the vascular channels. **(b)** Sections through an AVM composed of abnormally enlarged blood vessels partly separated by brain parenchyma. Elastica/van Gieson. Some of the vessels have abnormally thin walls but others are irregularly thickened. Most show marked collagenous fibrosis (red) but also include a discernible, if incomplete and disorganized, tunica media. The section includes two small arteries with a distinct internal elastic lamina (black). **(c)** The smooth muscle cells and internal elastic lamina are more obvious at higher magnification.

Courtesy of S Love, University of Bristol, UK.

AVMs may enlarge over time by recruitment of contiguous blood vessels, and cause shunting of blood from the arterial to the venous circulation. Large malformations can short-circuit ('steal') so much blood that the total CBF is increased, sometimes markedly, while tissue perfusion is reduced, resulting in chronic ischaemia.[610] In infants and children, excessive arteriovenous shunting may even lead to cardiac decompensation, particularly in the context of AVMs draining to and dilating the great vein of Galen ('vein of Galen aneurysms'). These lesions sometimes obstruct the aqueduct of Sylvius, causing hydrocephalus.[184] Occasionally, intracranial AVMs communicate with extracranial arteries, usually branches of the external carotid, forming cirsoid aneurysms.

Venous Malformations

Venous malformations, or venous angiomas, comprise conglomerates of varicose veins, separated from each other by nearly normal brain tissue. They may be the most common type of incidental vascular malformation found on neuro-imaging. Most are silent, but they rarely present with epileptic seizures or haemorrhage.[708] Angiography discloses a cluster of small veins converging into a 'Medusa head', from where blood is drained via a large central vein, either peripherally into the leptomeninges or centrally into the galenic system. Histology reveals clusters of dilated veins, the three-dimensional structure of a 'Medusa head' being difficult to appreciate in tissue sections.[184]

Cavernous Malformations

Cerebral cavernous malformations (CCMs) or cavernous haemangiomas, are compact, occasionally multiple lesions up to several centimetres in diameter that may occur anywhere in the brain or the leptomeninges.[95a,504] Their incidence is 0.56 per 100 000.[848] Although many remain asymptomatic, these malformations may present with seizures, headache, focal neurological deficits and, less commonly, haemorrhage. They can occur in families, with an autosomal inheritance pattern caused by mutations in at least three different genes (*CCM1*, *CCM2*, and *CCM3*) (Table 2.3). Distinctive hyperkeratotic cutaneous venous malformations may also occur in these patients.[220,572]

Cavernous haemangiomas may be mistaken macroscopically for fresh, demarcated brain haemorrhages. Histologically, they are composed of closely apposed dilated vascular channels, with little or no intervening brain parenchyma (Figure 2.54). The blood vessel walls are usually thin and consist of endothelium and a collagenous adventitia, but may include foci of calcification and even ossification.[184] Deposits of haemosiderin in the surrounding brain tissue are characteristic, contributing to the stereotyped appearance of these lesions on MRI. The vascular channels are often occluded and the lesions may not fill on angiography, but enhance in CT scans. Small, 'cryptic', angiographically covert CCMs or venous vascular malformations are occasionally identified in patients with brain haemorrhage but no evidence of hypertension, saccular aneurysms or arteriovenous malformations. CCMs are rarely associated with similar lesions in other organs, such as the kidney, liver, lung and skin.

Capillary Telangiectasias

Capillary telangiectasia describes the presence of dilated (ectatic) capillary-type blood vessels, separated by relatively

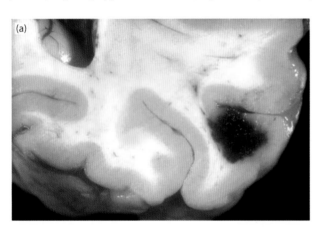

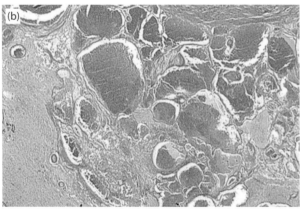

2.54 Cavernous haemangioma. (a) The temporal lobe contains a well-circumscribed lesion that resembles a haematoma. **(b)** The lesion is composed of closely apposed, dilated vascular channels, with fibrotic walls and little intervening brain parenchyma.

normal brain parenchyma (Figure 2.55). Haemorrhage from these lesions is very uncommon. They are usually found incidentally at autopsy as small areas of reddish blush, most frequently in the basis pontis and more rarely in other parts of the brain and spinal cord.[184]

Mixed Vascular Malformations

Some vascular malformations show features of more than one of the pathologically subtypes. In one series, 14 of 280 vascular malformations were found to be of mixed type.[77] Six were mixed cavernous and venous malformations, five were predominantly cavernous but with features of arteriovenous malformation or capillary telangiectasia in the same lesion, and three were mixed venous and arteriovenous malformations.

Malformation of the Vein of Galen

Aneurysm or malformation of the vein of Galen describes an arteriovenous malformation, with the vein of Galen acting as the main drainage system. Treatment is usually endovascular, although shunting may be needed to treat associated hydrocephalus.[582]

Genetics and Genomics of Intracranial Aneurysms and Vascular Malformations

Multiple genes and loci have been linked to intracranial aneurysms and vascular malformations (Table 2.3, and see also Reference 30). Familial cases of intracranial aneurysms are relatively common,[862,882] indicating the strong contribution of predisposing genetic factors. The occurrence of aneurysms in polycystic kidney disease and other genetic diseases is related to a combination of hypertension, and defects in proteins of the perivascular matrix or the cytoskeleton of the vessel wall. Autosomal dominant polycystic kidney disease is caused by mutations in the genes *PKD1* and *PKD2*, which encode proteins that interact with the extracellular matrix.[789,1043,1075] A defect in the extracellular matrix protein fibrillin-1 occurs in Marfan's syndrome, whereas type II collagen is abnormal in Ehlers–Danlos

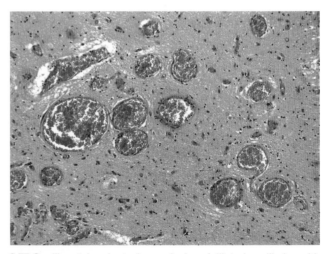

2.55 Capillary telangiectasia – a cluster of dilated capillaries with normal-looking intervening brain parenchyma.

syndrome type IV.[520,882] Abnormal neurofibromin may impair microtubule function and contribute to the increased risk of arterial aneurysms in neurofibromatosis 1 (NF1).[960]

Hereditary diseases causing arteriovenous malformations include hereditary haemorrhagic telangiectasia (Rendu–Osler–Weber syndrome), von Hippel–Lindau disease and Wyburn–Masson syndrome. However, Sturge–Weber–Dimitri disease (encephalotrigeminal angiomatosis) has no known genetic defect and no obvious family history. Hereditary haemorrhagic telangiectasia types 1 and 2 result from loss-of-function mutations in one copy of the endoglin (*ENG*) and activin A receptor type II-like 1 (also known as activin-like kinase receptor 1) (*ACVRL1*) genes. *ACVRL1* variants may also be associated with risk for sporadic arteriovenous malformations. Mutations in another gene, *SMAD4*, were described in some cases of combined juvenile polyposis and hereditary haemorrhagic telangiectasia syndrome. ALK-1, endoglin, and SMAD4 all are components of TGF-β superfamily signalling pathways (Table 2.3).

The mechanisms leading to malformations of cerebral blood vessels are gradually being unravelled in these disorders.[590] Most of the genes involved encode major protein components of the extracellular matrix surrounding blood vessels, such as collagens, elastin, glycoproteins fibrillin 1 and 2, proteoglycans, actin-related proteins, fibronectins and vascular growth promoting or maintenance factors, including cytokines and adhesion molecules.

Other proteins probably involved in the pathogenesis of berry aneurysms include MMP-9,[790] COL3A1 and COL1A2,[130,1111] angiotensin-converting enzyme,[530,982] heparan sulfate proteoglycan 2, versican, angiotensin-converting enzyme, interleukin 6 and endothelin receptor type A. There are reports of associations of polymorphisms in the corresponding genes or promoters, and at other loci and genes including the cyclin-dependent kinase inhibitor 2B antisense gene, with increased susceptibility to saccular aneurysms.[30]

Interventional Radiology

Endovascular neurosurgical procedures have been applied successfully to the treatment of a wide range of cerebrovascular pathologies, including atherosclerosis, acute stroke, vasospasm, saccular aneurysms, arteriovenous malformations and fistulae.[783] Angioplasty and stenting of the carotid artery are used in the treatment of atherosclerotic stenosis, as an alternative to carotid endarterectomy. Intracranial atherosclerotic disease can also be ameliorated by balloons and stents. Intra-arterial administration of thrombolytics, in combination with initial intravenous thrombolysis, is used for the treatment of acute stroke[459] as, increasingly, is endovascular thrombectomy.[932,959] Intra-arterial coils (Figure 2.56) are widely used to occlude saccular aneurysms,[85,783] although recanalization occurs in about 20 per cent of patients.[830] Impregnation of coils with healing promoters appears to reduce complications.[657,706,843] Clipping of saccular aneurysms remains the approach of choice in many cases. Coiling of aneurysms has been associated with a slightly lower mortality but a higher risk of recurrent bleeding.[687] The experience of the neurosurgeon and interventional neuroradiologist and the location and physical characteristics of the aneurysm are all important

considerations in deciding how best to manage an individual patient.

Arteriovenous malformations and fistulae can be embolized with various polymers, such as polyvinyl or ethylene vinyl alcohol particles and acrylic glues (Figure 2.57).[475,626,783]

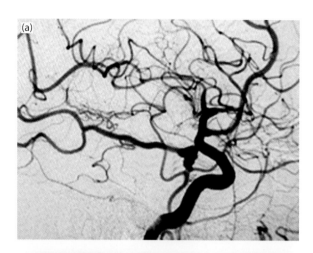

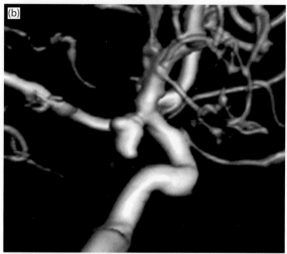

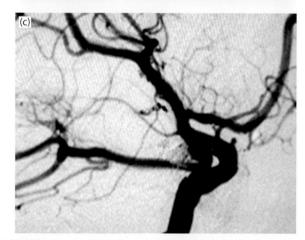

2.56 (a,b) Cerebral aneurysm at the origin of the posterior communicating artery. **(c)** Embolization with platinum detachable coils.

Images kindly provided by MA de Miquel, Hospital Universitari de Bellvitge, Barcelona, Spain.

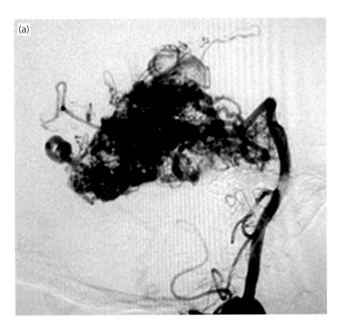
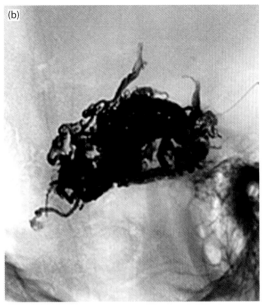
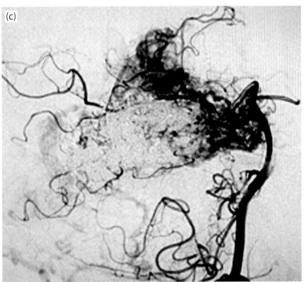

2.57 AVM in the territories of the middle and posterior cerebral arteries. (a) Conventional angiography; **(b)** embolization with ethylene–vinyl alcohol copolymer (EVOH) (onyx) cast of intravascular material; **(c)** partial embolization with onyx.

Images kindly provided by MA de Miquel, Hospital Universitari de Bellvitge, Barcelona, Spain.

This treatment may be helpful even though incomplete closure or recanalization often necessitates subsequent surgery or radiation therapy.

Miscellaneous Diseases of Blood Vessels

Moyamoya Syndrome

Moyamoya syndrome is a rare disorder resulting from stenosis of the major vessels of the circle of Willis and proximal segments of its principal branches at the base and over the convexity of the brain.[569,736] *Moyamoya* in Japanese describes the angiographic appearance of the vascular networks as 'something hazy' or like a puff of cigarette smoke drifting in the air (Figure 2.58).[974] Moyamoya syndrome covers both primary moyamoya disease and the angiographic moyamoya phenomenon, which is associated with various underlying disorders.[322] Since the first descriptions of moyamoya disease in Japan[569,736,974] it has been described in all major racial and ethnic groups.[151,196] Up to 15 per cent of moyamoya disease appears familial. About 50 per cent of the reported patients are children under 15 years,[974] with a slight preponderance in females. The most common clinical manifestation of moyamoya in children is alternating hemiparesis due to cerebral ischaemia. About 6 per cent of strokes in children are explained by moyamoya disease. The second peak of incidence occurs in adults in their 40s, often presenting with intracranial haemorrhage arising from thin-walled collateral vessels.[48]

The principal pathological alterations include bilateral stenosis or occlusion of the distal internal carotid arteries and the proximal parts of the anterior and middle cerebral arteries, combined with numerous dilated, thin-walled collateral arteries arising from the posterior parts of the circle of Willis.[389,1106] These collaterals form an irregular network of pial anastomoses that also penetrate the brain. The outer diameter of the stenosed or occluded arteries is often severely reduced, and their walls may be whitish and nodular. Histologically, their intima shows massive fibrous thickening, usually without atheromatous features (Figure 2.58b). The internal elastic lamina is generally preserved but often wavy and replicated, whereas the media becomes atrophic (Figure 2.58c). There is usually no inflammatory infiltration, but thrombosis, recanalization and aneurysm may occur. Electron microscopy shows that the intimal thickening is associated with proliferation of smooth muscle-like cells and accumulation of collagen fibrils and elastic tissue.[389]

Despite concerted efforts, the aetiology and pathogenesis of moyamoya disease remain largely unknown.[1078] Reduced blood flow in the major vessels of the anterior circulation of the brain leads to compensatory development of small collaterals near the apex of the carotid, on the cortical surface, leptomeninges, and branches of the external carotid artery.[923] Damage to the endothelium, with subsequent platelet aggregation and release of platelet-derived growth factor (PDGF), may be key to the intimal proliferation of smooth muscle-like cells and accumulation of collagen and elastic fibres. Local or systemic infections frequently precede the clinical manifestations of the moyamoya syndrome. An inflammation-related humoral factor is thought to induce repeated endothelial damage and intimal thickening.[389] Association of the moyamoya syndrome with a host of other acquired conditions, such as irradiation of the head and neck, suggests a multifactorial aetiology. The moyamoya phenomenon has also been reported in patients with a number of unrelated conditions, such as NF-1, tuberous

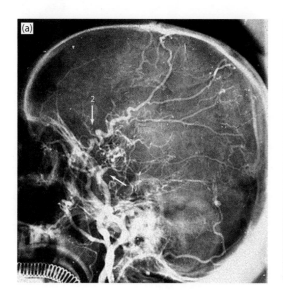
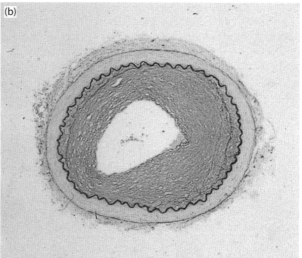
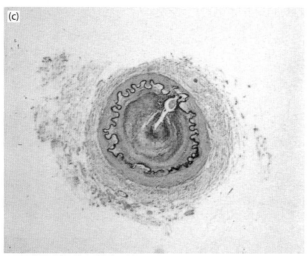

2.58 Moyamoya disease. (a) Angiography reveals occlusion of the left internal carotid artery (arrow 1). A network of small collateral blood vessels arise from the enlarged and meandering left middle meningeal artery (arrow 2). **(b)** A section through the basilar artery from another moyamoya patient shows marked thickening of the intima and narrowing of the lumen. Elastica/van Gieson. **(c)** The right posterior communicating artery from the same patient, with severe folding of the internal elastic lamina and probable thrombosis and recanalization of the lumen.

Reproduced from Haltia M, et al.[389] With permission from Dustri-Verlag.

sclerosis, Marfan's syndrome, Alpert's syndrome, sickle cell anaemia, Fanconi's anaemia, Graves' disease, Down's syndrome and Schimke's immuno-osseous dysplasia.

Currently, the major proteins believed to play an active role in the pathogenesis of moyamoya disease are VEGF, basic fibroblast growth factor, hepatocyte growth factor, TGFβ[1] and granulocyte colony-stimulating factor. Genetic studies show low penetrance autosomal dominant or polygenic inheritance patterns involving chromosomes 3, 6, 8, 12, and 17 in familial moyamoya disease.[1078] However, linkage and association studies in 20 families showed that heterozygous mutations in the α-actin (ACTA2) gene, which causes disordered proliferation of vascular smooth muscle cells, are likened to the syndrome of thoracic aortic aneurysms and dissections, premature coronary artery disease, ischaemic stroke, and moyamoya disease.[380]

Fibromuscular Dysplasia

Fibromuscular dysplasia (FMD) is a non-atherosclerotic, non-inflammatory segmental disease of the arterial wall, first described in the renal arteries and later in the cervicocephalic arterial tree.[213,923] Among patients undergoing carotid angiography, FMD has been diagnosed in approximately 0.25–3.2 per cent. FMD primarily affects women aged 20 to 60 years but may occur in all age groups including the elderly.[754] Cervicocephalic FMD involves the midportion of the cervical ICA, mostly with bilateral involvement (60–80 per cent of cases), and is considered the second most common stenosing lesion of the ICA after atheroma. The extradural segment of the vertebral arteries is the second most often affected part of the CNS vasculature, although involvement of the intracranial arteries is rare (7–20 per cent of cases).

Neurological symptoms of FMD are often trivial, such as headache, mental distress, vertigo, tinnitus or audible bruits. The two serious complications of FMD are cervicocephalic arterial dissection (about 10–15 per cent are caused by FMD) and intracranial, often multiple, aneurysms in about 7.5 per cent of patients.[204,230] In systemic FMD, involvement of the renal arteries may cause hypertension, which increases the risk of rupture of aneurysms. In addition, transient ischaemic attack (TIA) or infarction may occur, probably caused by thromboembolism or platelet emboli resulting from turbulent blood flow in the affected artery.[230,774] Imaging, either digital subtraction angiography (Figure 2.59a) or MR angiography, shows luminal narrowing alternating with aneurysmal dilation, giving a characteristic string-of-beads appearance (evident in about 80 per cent of FMD patients). The two other patterns, 'tubular stenosis' and 'semicircumferential lesions', are much less common.[764] Most FMD patients (85–90 per cent) can be treated pharmacologically, but in complicated cases endovascular stenting or bypass surgery may be needed.[274] Differential diagnoses include atherosclerotic stenosis and stenosis associated with vascular Ehlers–Danlos and Williams' syndromes and type 1 neurofibromatosis.

Three different types of FMD have been described – intimal, medial and periadventitial fibroplasia – the medial variant being the most common in cervicocephalic FMD. Histopathologically, affected vessels have stenotic segments with asymmetrical thickening of the wall, and dilated segments with a thinned wall (Figure 2.59b). In cases with

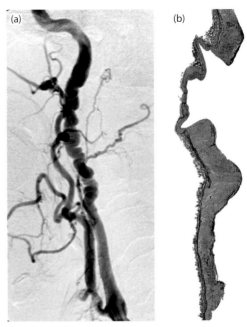

2.59 (a) Angiography of the common carotid artery in a patient with fibromuscular dysplasia. The lumen is irregularly widened and narrowed, giving the artery a 'string-of-beads' appearance. **(b)** At tissue level, the thickness of the wall (and the size of the lumen) varied correspondingly.

Reproduced with permission from Parisi.[774] © 2005 Wiley-Blackwell.

dissection, blood may be evident in the false lumen. In the narrowed segments, the thickened media is composed of fascicles of collagenous tissue with abundant fibroblasts and fewer smooth muscle cells. In the dilated regions, the arterial wall is thinned and fibrosed, with deficient media and disruption of the elastic lamina.[774]

The aetiology of fibromuscular dysplasia has not been established.[805] Local trauma, hormonal influences, ischaemia of the vessel wall and viral infections have been proposed.[774] The occasional familial cases suggest a genetic predisposition, probably because of a heritable connective tissue defect. Some FMD patients have α1-antitrypsin deficiency, which is also apparent in other aneurysmal diseases.[909]

Fabry Disease

Fabry disease is an X-linked disorder of glycosphingolipids that is caused by some 431 mutations in the α-galactosidase A gene on Xq22.1. These result in deficiency of α-galactosidase A, the manifestations of which include a systemic vasculopathy and small-fibre peripheral neuropathy. Patients are at high risk of premature stroke, cerebrovascular dolichoectasia, and white matter hyperintensities.[287,911] Cerebrovascular disease occurs at a mean age of 42 years in males and approximately 10 years later in females. Fabry disease was reported to be a relatively common cause of cryptogenic ischaemic stroke in young men, accounting for 5 per cent in Germany but lower in the multiracial US population. The Stroke Prevention in Young Men Study conducted in the Baltimore–Washington area implicated Fabry disease in 0.18 per cent of all strokes and

0.65 per cent of cryptogenic strokes.[669] The cerebrovascular abnormalities are associated with the generalized accumulation of neutral glycosphingolipid globotriaosylceramide in the walls of small blood vessels (as well as in neurons in selective brain stem and cortical regions, including the hippocampus; see Chapter 6, Lysosomal Disease).[669]

HAEMATOLOGICAL DISORDERS

The majority of the diseases of blood vessel walls do not themselves cause sufficient narrowing of the vascular lumen to decrease the blood flow to ischaemic levels. It is commonly the thrombus induced by the vascular injury or deformity that finally obstructs the lumen. Alternatively, emboli detach from the thrombus to obstruct arterial flow. The risk of thrombosis is increased in thrombophilic conditions or 'hypercoagulable states.'[159]

Haematological disorders or hypercoagulable states are uncommon primary causes of acute cerebrovascular disease (Box 2.6), accounting for only about 1 per cent of all strokes but a slightly higher proportion in young adults.[51] Several primary haematological or haemostatic disorders including essential thrombocythaemia, polycythaemia vera and thrombotic thrombocytopenic purpura, are associated with increased risk of ischaemic and haemorrhagic strokes.[653,990]

Thrombosis and Antithrombosis

Haemostasis is produced by the induction of two interdependent processes: aggregation of platelets and activation of the coagulation cascade. Endothelial damage induces aggregation of platelets, which release factors such as thromboxane A2 and adenosine diphosphate. These chemicals recruit further platelets to adhere to the platelet thrombus. Platelets also release factors that activate the intrinsic pathway of coagulation. Tissue factor (factor VII receptor), capable of activating the extrinsic pathway, is expressed on the surface of diseased endothelium, exposed intima or accumulated inflammatory cells. These thrombogenic processes also occur in complicated atherosclerotic plaques (Figure 2.25). The intrinsic and extrinsic pathways converge into the common pathway, as a result of which thrombin cleaves fibrinogen into fibrin (Figure 2.60). If the flow conditions are permissive, a fibrin thrombus is formed on the platelet matrix.

Under normal conditions, vascular patency and unhindered blood flow are ensured by an intact endothelium and the presence of antithrombotic molecules, many of which are secreted by the endothelial cells. These mediators prevent the aggregation of platelets and activation of the coagulation cascade (Figure 2.61). Antithrombin III prevents the action of thrombin, and this inhibitory effect is amplified by heparin. Thrombin bound to thrombomodulin on endothelial surfaces participates in the counteraction of its own pro-coagulant effect by activating endothelium-derived protein C, which, together with its circulating cofactor protein S, forms a proteolytic complex that destroys activated factors V and VIII of the coagulation cascade. Finally, the fibrin thrombi are degraded by fibrinolysis (Figure 2.62). Drug-induced fibrinolysis, is an established therapy for acute myocardial infarction and ischaemic stroke.[382,707,1068] Among the thrombolytic agents available for dissolving clots after stroke, recombinant tissue-type plasminogen activator is much more fibrin-specific than streptokinase and is approved for the early treatment of ischaemic strokes.

Thrombophilia

Defects in the antithrombotic pathways (see Figure 2.61), caused by deficiencies in various mediators, can lead to a hypercoagulable state and an increased risk of thrombotic brain infarction.[446,835] The five most commonly inherited disorders of coagulation are protein C deficiency, protein S deficiency, antithrombin deficiency, and factor V Leiden (G1691A) and prothrombin (G2020A) gene mutations. These inherited thrombophilias most often induce thrombi to form in the systemic venous circulation but are also risk factors for cerebral venous thrombosis.[835] Their role in arterial thrombosis is less clear. These antithrombotic deficiency disorders are probably risk factors for ischaemic strokes in children and young adults[743] but not necessarily in older adults.[698,835]

BOX 2.6. Haematological disorders associated with cerebrovascular disease

- Coagulopathies (disorders of coagulation)
  - Thrombosis and antithrombosis
  - Thrombophilia
  - Antiphospholipid antibodies
  - Polycythaemia and hyperviscosity
  - Disorders of platelets
- Anaemias
  - Several different types – nutrient deficiency or toxicity, and haemolytic
- Haemoglobinopathies
  - Sickle cell disease
  - β-Thalassemia major

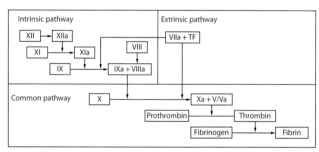

2.60 Blood clotting cascade. a, activated; TF, tissue factor.

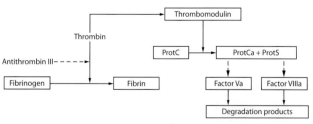

2.61 Physiological antithrombotic mechanisms. Inhibitory effects are marked with dashed lines. Activated protein C (ProtCa) and protein S (ProtS) inhibit the function of factors Va and VIIIa (which regulate formation of thrombin) by degrading these factors. a, activated.

The increased incidence of cerebral infarction in certain generalized conditions such as malignancy, infection, use of oral contraceptives, pregnancy and chemotherapy, may depend on acquired hypercoagulability because of deficient function of the antithrombotic molecules.[875]

In addition to the above factors, which increase the formation of thrombin, high plasma levels of fibrinogen and defective lysis of the incipient fibrin thrombi, e.g. due to deficiency of plasminogen (Figure 2.62), can lead to thrombotic episodes.[1040] Homocysteinaemia has been shown to lead to impaired fibrinolysis, thereby increasing the risk of cardiovascular disease, including ischaemic stroke.[566,1080] However, it is uncertain whether elevated serum fasting levels of homocysteine or the C677T mutation in thermolabile 5,10-methylenetetrahydrofolate reductase associated with hyperhomocysteinaemia increases the risk of stroke through this action on fibrinolysis.[398,566]

Antiphospholipid Antibodies

Antiphospholipid antibodies (aPLAs) are a heterogeneous family of autoantibodies associated with the antiphospholipid syndrome.[281] Their presence is associated with an increased risk of thrombotic events. Originally, aPLs were detected in patients with SLE. With improved testing methods and awareness, aPLs have been linked to many other diseases, including malignancies, HIV infection, drug ingestion and various autoimmune disorders, such as rheumatoid arthritis, immune thrombocytopenic purpura, Sjögren's syndrome and Behçet's syndrome. 'Secondary aPL syndrome' describes the complication of these conditions by aPLs and thrombotic disturbances. It is distinguished from the primary syndrome in which there is no underlying medical condition.[112] Population studies suggest that aPLs are present in as many as 2–5 per cent of healthy individuals.[990]

The major clinical manifestations of the aPL syndrome are arterial and/or venous circulatory disturbances, recurrent miscarriages and thrombocytopenia. The main neurological manifestation is ischaemic stroke, and aPLs have been shown to be an independent risk factor for first-time ischaemic stroke.[45,1010] The strokes often occur at an young age, with no other apparent causes.[131,986,1010] The prevalence of aPLs among young stroke patients has been estimated at 18–46 per cent, but aPLs are also detected in 10–18 per cent of older stroke patients.[726]

The thrombotic occlusions in patients with aPLs are most often venous. This increases the risk of paradoxical embolic strokes (i.e. strokes resulting from the passage of

embolic material of venous origin through a patent foramen ovale). When thromboses occur in the arterial circulation, the brain is the most common site.[131] It is quite likely that the pro-coagulant state induced by the aPLs promotes thrombosis in patients with other risk factors of stroke, such as atherosclerosis, hypertension and diabetes. Furthermore, stroke patients with aPLs have an eight-fold risk of recurrence.[597] Smaller arteries may also become occluded; for example, involvement of retinal arteries may give rise to manifestations of ocular ischaemia. Fibrin thrombi, obstruction by intimal proliferation and recanalization with persistent fibrous webs across arterial lumina have been described as typical features that suggest recurrent episodes of intravascular thrombosis and associated infarction.[128]

The two conventional assays for aPLs are: (i) demonstration of lupus anticoagulant (LA) phenomenon, the name of which derives from the fact that it was first detected in patients with SLE as a paradoxical prolongation of the *in vitro* coagulation time, even though an *in vivo* positive LA test indicates a thrombophilic state and a risk of ischaemic complications;[986] and (ii) detection of anticardiolipin antibodies in the patient's serum, typically by enzyme-linked immunosorbent assay (ELISA).[986] Knowledge of the antigenic specificities and pathogenetic mechanisms of aPLs changed during the 1990s. It was demonstrated that a 50-kDa protein in the plasma, $\beta 2$-glycoprotein I ($\beta 2$-GPI; also called apolipoprotein H), is a necessary cofactor for the binding of aPLs to the lipid antigens. Possibly, the combined $\beta 2$-GPI and phospholipid form neo-epitopes to which the aPLs bind.[879] Whether $\beta 2$-GPI acts as the primary antigen or needs priming factors to induce thrombosis is unclear; some of the differences probably reflect variation in the aPL spectrum in individual patients. In addition to $\beta 2$-GPI, aPLs recognize other phospholipid-binding plasma proteins, including prothrombin, protein C, protein S, thrombomodulin, annexin V and kininogens.[986]

Among the non-cardiolipin phospholipids targeted by aPLs are both negatively charged (phosphatidylserine, phosphatidylinositol, phosphatidylglycerol and phosphatidic acid) and neutral phospholipids (phosphatidylethanolamine and phosphatidylcholine).[1011] Of 77 young non-SLE patients with stroke, as many as 34 (44 per cent) had aPLs against one or more of the phospholipids mentioned earlier. Only 53 per cent had antibodies to cardiolipin, the remainder having aPLs with other phospholipid specificities, phosphatidylinositol being the most common (65 per cent). A single patient can have a variety of aPLs, i.e. to several different phospholipid antigens and of different immunoglobulin classes.[1011]

The precise thrombogenic mechanisms of aPLs have not been established. They may bind to phospholipids in the platelet membrane and cause increased platelet adhesion and aggregation. They may also bind to endothelial phospholipids in combination with $\beta 2$-GPI, to induce endothelial damage and initiation of thrombosis. Conversely, $\beta 2$-GPI has been proposed to be protective, binding to phospholipids in damaged endothelial plasma membranes and preventing the exposure of pro-coagulant factors in the endothelium to the circulation; this anticoagulant function may be impaired if $\beta 2$-GPI is sequestered by aPLs.[875,879]

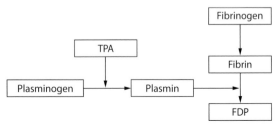

2.62 Fibrinolysis. FDP, fibrin degradation products; TPA, tissue plasminogen activator (natural or iatrogenic, such as recombinant TPA or streptokinase or urokinase).

Polycythaemia and Hyperviscosity

In polycythaemia, the number of red blood cells is increased, with a high haemoglobin concentration (>200 g/L) and haematocrit (>60 per cent). In patients with polycythaemia vera the risk of ischaemic stroke is increased up to five times, with a lesser increase in those with secondary polycythaemia.[195,400] Even an abnormally high haematocrit irrespective of high red blood cell count carries an increased risk of stroke, which may occur secondary to dehydration.[731] In the Framingham study, if the haematocrit was over 42–45 per cent the risk of stroke increased two-fold.[514] Similarly, the risk of cerebral infarction increased significantly when haematocrit values were above 45 per cent, especially if severe atherosclerosis was evident.[1005]

In patients with polycythaemia, stroke occurs because of hyperviscosity, which decreases CBF and gives rise to multifocal small infarcts.[291] Whether the sluggish flow results in the formation of thrombi or the stagnation of the blood itself causes ischaemia has not been established.

Increased numbers of white blood cells can also lead to infarction, most likely because of stasis. An excessive concentration of plasma proteins, e.g. in plasma cell dyscrasias, can lead to hyperviscosity, too. A hyperviscosity syndrome quite often complicates Waldenström's macroglobulinaemia, a neoplastic disease of B-lymphocytes in which there are high levels of plasma immunoglobulin M (IgM), but other paraproteinaemias can also give rise to this complication.[766]

Anaemias

The low blood haemoglobin of anaemia causes decreased oxygen-carrying capacity. Compensatory mechanisms during anaemia usually ensure adequate transport of oxygen to the brain. Low haemoglobin does not itself cause strokes.[514] The risk of strokes increases if additional factors affect blood flow.

Sickle Cell Disease

Sickle cell disease is one of the best known monogenic disorders. It belongs to a group of haemoglobinopathies in which the abnormal β-chains of the sickle haemoglobin S aggregate and polymerize to form rigid filamentous structures, 'tactoids', which deform erythrocytes into sickle cells.[125] Although the rheological properties of sickled erythrocytes suggest that microvascular occlusion might result primarily from intravascular aggregation, an additional major contributor is likely to be the extensive cranial vasculopathy that affects many patients. This is characterized by stenosis of the extracranial and intracranial segments of the internal carotid artery, and the anterior, middle and posterior cerebral arteries. The vasculopathy results from abnormal proliferation of fibroblasts and vascular smooth muscle cells in the vessel wall; contributory factors probably include increased blood flow as a result of anaemia, abnormal adherence of erythrocytes to the endothelium, haemolysis, endothelial activation, leukocyte adhesion, elevated production of endothelin-1 and scavenging of nitric oxide by cell-free haemoglobin dimers.[976] Intracranial haemorrhages and fat embolism may also occur.[884] Of the anaemias, sickle cell disease is the most commonly associated with silent infarction and stroke. Strokes occur most often in children under the age of 15 years. About 15 per cent of children with sickle cell disease experience cerebrovascular disorders. Cerebral infarcts occur in about 75 per cent and intracerebral haemorrhages in some 20 per cent, and these changes often occur bilaterally. In sickle cell disease, children with seem to have a greater risk of ischaemic stroke, and adults, intracranial haemorrhage.[749] Recent, pathway analyses based on genome-wide association studies have shed light on the importance of the TGF-β superfamily and oxidative stress in the pathogenesis of complex traits in sickle cell disease.[290]

Beta-Thalassaemia Major

The thalassaemias are another group of hereditary anaemias associated with strokes.[125,677,1096] In these, either the α- or the β-chain of the haemoglobin A molecule carries a genetic defect. Beta-thalassaemia major is the most clinically important of the thalassaemias. Patients may carry one of over 200 homozygous β-chain mutations, resulting in reduced or no β-globin synthesis and excessive α-globin, which precipitates within the red blood cells. The main feature of β-thalassaemia major is hypochromic, microcytic anaemia due to impaired production and haemolysis of erythrocytes. The disease usually manifests soon after birth. Patients have an increased risk of thrombotic stroke, to which the post-splenectomy thrombocytosis contributes. Cerebral haemorrhage has also been reported as an occasional complication of blood transfusion in β-thalassaemia.

Platelet Abnormalities

Both thrombocytosis and thrombocytopenia have been associated with TIAs, ischaemic stroke and intracerebral haemorrhage. The risk of micro-occlusion is increased if the platelet count is above 400 000 or if the platelets are abnormally adhesive.

Neurological complications are most common in thrombotic thrombocytopenic purpura (TTP), which is also known as thrombotic micro-angiopathy or Moschowitz disease.[766] It is a rare disorder that primarily affects women aged 20–50 years. In TTP, platelets form thrombi, which occlude mainly cerebral and renal micro-vessels. In parallel, platelets are consumed to such an extent that thrombocytopenia and petechial and purpuric haemorrhages occur.[809] TTP is caused by deficient activity of the metalloprotease ADAMTS13 (a disintegrin and metalloproteinase with a thrombospondin type 1 motif, member 13),[889,1023] which cleaves von Willebrand factor (vWF), a glycoprotein that is produced by endothelial cells and plays an important role in the formation of blood clots. In the absence of normal ADAMTS13 activity, large multimers of vWF accumulate, bind platelets and cause thrombosis, leading to the depletion of vWF multimers from the circulation. In idiopathic TTP, the impairment of ADAMTS13 activity is caused by inhibitory antibodies in the absence of underlying disease. In secondary TTP, deficient ADAMTS13 activity occurs in the context of other conditions, including neoplasia, pregnancy and HIV infection, or as an adverse reaction to several medications, such as ticlopidine and immunosuppressive drugs.

Rare familial forms of TTP are caused by mutations in the ADAMTS13 gene.[548,889] Several single nucleotide polymorphisms may also affect the expression and function of ADAMTS13.[1024]

Neurological symptoms in TTP,[988] are often dramatic and include seizures, stupor, coma and stroke. The changes on neuroimaging[526] and histopathological examination may be minimal, even in lethal cases.[7] The lumina of microvessels, predominantly in the grey matter, are occluded by hyaline, eosinophilic platelet thrombi (Figure 2.63). These can be highlighted by immunohistochemistry with antibody to CD61. In addition, the thrombi may contain fibrin and factor VIII. Endothelial hyperplasia may be prominent, and sometimes the blood vessel wall is necrotic, whereas the surrounding parenchyma may seem nearly normal. In severe cases, multiple small cerebral infarcts are present in the territory of the occluded microvessel. Rarely large vessel occlusion occurs.

Alpha2β1-Integrin (glycoprotein Ia–IIa) is one of the major collagen receptors on platelets via which platelets adhere to collagen exposed in damaged vessel walls and become activated. The density of this receptor molecule is regulated by two linked, silent polymorphisms (C807T and G873A) in the α2 gene coding sequence. Compared to individuals homozygous for C807, those homozygous or heterozygous for the T807 allele have higher α2β1-integrin density, enhancing adhesion to subendothelial collagen and promoting thrombus formation. The genotype T807 was shown to be an independent risk factor for stroke in young patients (<50 years).[173a]

CONSEQUENCES OF CEREBROVASCULAR DISORDERS AND IMPACT ON BRAIN TISSUES

Diseases affecting the cerebral blood vessels cause two basic types of sequelae: ischaemic damage, resulting from obstruction of the blood vessels, and haemorrhage,

produced by rupture of the vessel wall. Besides ischaemic and haemorrhagic stroke, other main categories include subarachnoid hemorrhage, cerebral venous thrombosis and spinal cord stroke.

Stroke Epidemiology and Risk Factors

Stroke is the third leading cause of death in developed countries (Table 2.5). It is an important cause of long-term disability in most industrialized populations[887,1071] and demands enormous resources from healthcare systems.[878,995] Stroke is defined as the abrupt onset of focal or global neurological symptoms caused by ischaemia or haemorrhage.[887] According to the previous definition from the World Health Organization (WHO), these symptoms must continue for more than 24 hours for a diagnosis of stroke, which is usually associated with permanent damage to brain. If the symptoms resolve within 24 hours, the episode is called a transient ischaemic attack (TIA). A proposed later qualification to this definition was that there should be no evidence of infarction on neuroimaging;[22] the authors suggested that a TIA should be defined as 'a brief episode of neurologic dysfunction caused by focal brain or retinal ischaemia, with clinical symptoms typically lasting less than one hour, and without evidence of acute infarction'.

The clinical diagnosis of stroke is usually accurate, diagnosis of the precise type of stroke often less so. Determination of the pathological type of stroke is best achieved by early brain imaging, or by autopsy in fatal cases.[62] The frequency of different types of stroke varies between racial and ethnic groups and with economic status.[284] In western countries, cerebral infarction accounts for approximately 60–80 per cent of first-time strokes, and parenchymal brain haemorrhage (PBH) for 5–11 per cent (Table 2.6). In Far Eastern countries, the figure for infarction is about 50–60 per cent and for PBH 16–44 per cent. Subarachnoid haemorrhage (SAH) has a smaller geographical variation, accounting for some 5–10 per cent of cases. The frequency of unspecified strokes varies within a wide range from 3 to 25 per cent, reflecting divergence in diagnostic resources and policies. The trend of change in the relative frequencies of different types of stroke has been similar in most countries, although quantitative differences exist. Since the 1970s, the proportion of strokes resulting from PBH has decreased, most

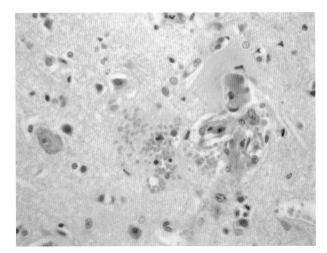

2.63 Idiopathic thrombocytopenic purpura. The cerebral cortex includes microthrombi and foci of microhaemorrhage.

Image courtesy of T Polvikoski, Newcastle University, UK.

TABLE 2.5 Annual risk of stroke (all subtypes combined)	
Age group (years)	**Approximate population risk**
0–14	1 in 100 000
15–24	1 in 20 000
25–34	1 in 10 000
35–44	1 in 5000
45–54	1 in 1000
55–64	1 in 300
65–74	1 in 100
75–85	1 in 50
>85	1 in 33

TABLE 2.6 Range of crude stroke incidence rates (per 100 000 person years) and pathological subtypes among all ages in different populations in high and low-to middle-income countries

Year	High-income countries (range low to high)				Low- to middle-income countries (range low to high)			
	Total	CI	ICH	SAH	Total	CI	ICH	SAH
1970–1979	125–460	n.r.	n.r.	n.r.	15–50	n.r.	n.r.	n.r.
1980–1989	156–466	n.r.	n.r.	n.r.	202–217	n.r.	n.r.	n.r.
1990–1999	131–451	137–264	24–48	4–9	167–281	n.r.	n.r.	n.r.
2000–2008	112–223	101–174	10–23	2–10	73–165	47–92	17–44	4–16

CI, cerebral infarction; ICH, intracerebral haemorrhage; SAH, subarachnoid haemorrhage; n.r., not reported.

Data from Feigin et al.[284] and Lovelock et al.[625] The 18 high-income countries included Australia, Barbados, Denmark, Estonia, Finland, France, Germany, Greece, Italy, Ireland, Japan, Norway, New Zealand, Portugal, Sweden, Netherlands, UK and USA. The 10 low to middle income countries were Brazil, Chile, French West Indies, Georgia, India, Nigeria, Mongolia, Sri Lanka, Russia and Ukraine.

significantly in Japan, whereas that of SAH has remained fairly constant.[993]

Stroke Incidence

Increasing age is the strongest risk factor for stroke (Table 2.5). The risk for a child under 15 years of age is 1 in 100 000, whereas it is 1 in 33 for people aged 85 years and over. The incidence of stroke varies greatly according to the age distribution of the population under study. It is higher in western countries and Japan because of the relatively high proportion of elderly people in these countries. Corresponding trends in stroke incidence (see later) are observed in younger (<75 year) and older (≥75 year) age groups, although the differences are far more pronounced in the older group. For meaningful comparison of incidence rates, these should be adjusted according to the age distribution of the population. The age-adjusted annual incidence of all first-time strokes in different countries has changed considerably over the last four decades (1970–2008).[284] Stroke incidence has decreased by 42 per cent in high-income countries but more than doubled in low- to middle-income countries (52 to 117 per 100 000 person-years). The decline in stroke has occurred in all age groups but has been greatest in the elderly. Of the pathological types, the lowest proportion with ischaemic stroke (73 per cent) was recorded in New Zealand, and the highest (90 per cent) in France in 2000–2008. The proportion with primary PBH ranged from 9 per cent in France to 13 per cent in Barbados and the proportion with SAH from 1 per cent in France to 6 per cent in New Zealand. From 1980 to 2008, the pooled age-adjusted incidence rates of ischaemic stroke in high-income countries fell by 11 per cent but there was no change in the incidence of PBH or SAH. Oxfordshire data reveal a fall in hypertension-associated PBH over the past 25 years, but not in the overall number of cases of PBH in older age-groups. The latter is probably explained by a rise in PBH associated with anticoagulant use and the increase in prevalence of amyloid angiopathy in the ageing population.[625]

In 2000–08, there was no difference in the pooled age-adjusted incidence of ischaemic stroke between high-income (70 per 100 000) and low- to middle-income countries (67 per 100 000), but the rates of PBH and SAH in low- to middle-income countries rose to almost twice those in high-income countries (PBH 22 per 100 000 and SAH 7 per 100 000) (Table 2.6; cf. Figure 2.118). Racial, ethnic and social backgrounds have a definite impact on stroke incidence. In general, white people have lower rates than non-white people. Differences may exist even within the same race: among Japanese males, the incidence of stroke was three times higher in Japan than in Hawaii.[983] However, the high and increasing incidence of stroke in low- to middle-income countries over the past four decades is likely to reflect health and demographic transitions in these countries.[214]

Stroke Mortality

Over the past few decades, deaths from stroke have declined in high-income countries and many middle- and low-income countries. The main cause of the decline is the reduced incidence of stroke[284] but the case fatality rates have also decreased as a result of lesser stroke severity or improved management.[124,770,877] In 2000–2008, early (21 days to 1 month) case fatality ranged from 17 per cent to 30 per cent (13–23 per cent for ischaemic stroke, 25–35 per cent for PBH, and 25–35 per cent for SAH) in high-income countries and from 18 to 35 per cent in low- to middle-income countries (13–19 per cent for ischaemic stroke, 30–48 per cent for primary PBH, and 40–48 per cent for SAH). Although early stroke case fatality has decreased in both high-income and low- to middle-income countries, over the past decade early stroke fatality has been 25 per cent greater in the latter than the former group.

Risk Factors and Consequences of Stroke

Besides age, gender and race-related differences, genetic, vascular and metabolic diseases and many potentially modifiable social habits markedly increase the risk of stroke.[1070] Hypertension increases the risk by up to four to five times in a dose-dependent manner, and is the most clinically important predisposing factor by virtue of its high prevalence.[1029] Atrial fibrillation increases the risk to the greatest extent (5.6–17.6-fold). A 2–4-fold increase is conferred by several conditions that share a propensity for embolization, including previous myocardial infarction, valvular heart disease and congestive heart disease. Carotid artery stenosis increases the risk of a completed stroke, especially if it is associated with a previous TIA. Hyperlipidaemia (up to 2-fold increase in risk) and diabetes mellitus (1.5–3-fold)[211] are the most common metabolic disorders associated with stroke. Certain medical treatments, such as open heart and coronary bypass surgery, increase the risk, as do older oral contraceptives with a high oestrogen content. Cigarette smoking increases the risk of stroke in a dose-dependent manner.[949] Smokers have an approximately three-fold increased risk of SAH and a two-fold increased risk of brain infarction, but the risk of PBH is not increased over that in non-smoking controls.[422,936] There seems to be a J-shaped relationship between alcohol and stroke, similar to that with myocardial disease. Alcohol in small amounts seems slightly to decrease the risk of stroke, whereas heavy drinking increases the risk by up to 2.5-fold.[335,840]

In recent years, multiple genetic risk factors have been recognized for ischaemic stroke and haemorrhage. Large scale GWAS and multicentre cohorts have enabled identification of at least 3 loci that associate with large artery disease and cardioembolic strokes (Table 2.7). Several smaller studies have reported mutations and polymorphisms in genes encoding a variety of proteins regulating haemostasis and vascular function. Strokes explained by gene defects occur especially in the younger age group, as do strokes induced by antiphospholipid autoantibodies (see Haematological Disorders, earlier in chapter).[541,743,986]

Migraine has also been reported as a risk factor (about 3.5-fold increase) in women under 45 years for ischaemic but not haemorrhagic stroke.[187,1030] A family history of migraine increases the risk irrespective of personal episodes of migraine, and is synergistic with other risk factors.[187,1030]

In addition, migraine is a common early symptom in CADASIL (see later).

As the risk of death from strokes has declined, the number of stroke survivors with cerebral comprise and cognitive dysfunction has increased.[31,549] Stroke patients are at increased risk of cognitive impairment. Meta-analysis of data from several studies yielded estimates of 1-in-10 patients being demented prior to a first stroke, 1-in-10 developing new dementia soon after a first stroke, and over 1-in-3 being demented after a recurrent stroke (see Chapter 16).[784,785]

Transient Ischaemic Attack

In its mildest form, impaired regional CBF or irregular perfusion causes a transient disturbance in neurological function without permanent tissue damage, i.e. a TIA (see earlier for clinical definition). The cause of TIA in most cases is small emboli from an extracranial source, either cardiac or an atherosclerotic plaque in the carotid or vertebrobasilar artery. Occasionally, TIA has a haemodynamic cause, e.g. a regional flow deficit beyond a stenosed artery, during transient hypotension.

Stroke and Infarction

When the ischaemia lasts long enough, permanent cell damage ensues. If it results in a clinically detectable functional deficit, this is a stroke (Figures 2.64–2.67). As noted earlier, ischaemia of moderate severity or short duration may cause selective necrosis of neurons only (incomplete infarction) rather than pan-necrosis of all tissue components (complete infarction). The preservation of non-neuronal tissue

TABLE 2.7 Genetics of stroke		
Type of stroke	**Type of study**	**Gene variant and risk**
Ischaemic stroke*		
Large artery (atherosclerotic)†	GWAS, multicentre and meta-analysis	• Two loci (*HDAC9* encoding histone deacetylase 9 and 9p21) identified although 9p21 locus did not reach genome-wide significance (OR 1.38). • Another susceptibility locus on 6p21.1 (rs556621; OR=1.62). • Variants in the ABO genes significantly associated with levels of vWF (rs505922, rs643434, rs8176743) and/or FVIII (rs505922, rs651007). All associations were subtype-specific.
Cardioembolic stroke	GWAS (deCODE) and meta-analysis	• Two gene regions (*PITX2* and *ZFHX3*) identified (OR 1.22 to 1.52) • ABO gene variants (see earlier)
Small vessel disease	GWAS, cohorts	• Locus on 17q25 encompassing 6 known genes including WBP2, TRIM65, TRIM47, MRPL38, FBF1 and ACOX1, as a novel genetic locus for cerebral white matter lesion burden
Cryptogenic stoke	Cohort	• Variants on 4q25 asscociate with stroke (OR 1.18); PITX2
Haemorrhagic stroke		
Intracerebral (deep)	Multicentre	• *APOE* ε4 (OR 2.0) and *APOE* ε2 (OR 1.8)
Lobar	Multicentre	• *APOE* ε4 (OR 2.0) and *APOE* ε2 (OR 1.8)
Sub-arachnoid‡	Multicentre	• Variants on 9p21 associate with abdominal aortic and intracranial aneurysms.

APOE genotypes showed a positive dose-response association with ischaemic strokes, low-density lipoprotein cholesterol and carotid intima-media thickness.[531]
†Other mutiple loci in coronary artery disease and myocardiac infarction, e.g. CDKN2A and CDKN2B also identified.
‡Strokes associated with intracranial aneurysms are considered in Table 2.3.
ECs, endothelial cells; GWAS, genome-wide association studies; OR, odds ratio.
Data from several references.[101, 372, 378, 379, 411, 434, 567, 916, 1018]

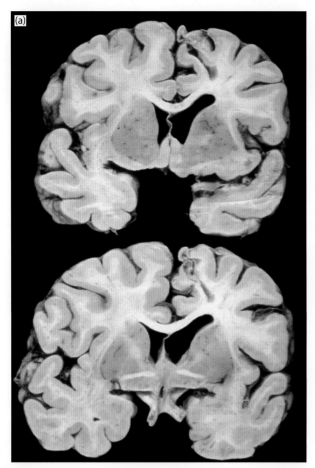
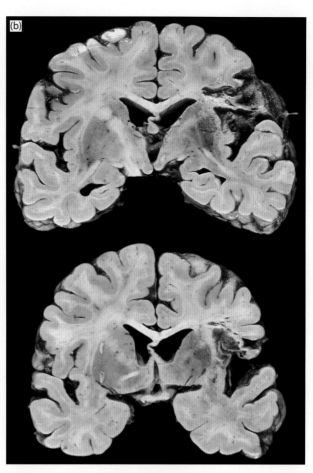

2.64 Chronic infarcts in territory of the middle cerebral artery. (a) Atrophy of the right caudate, putamen and internal capsule, with enlargement of the ipsilateral lateral ventricle. **(b)** Cystic infarct involving the right striatum, internal capsule, claustrum, extreme capsule and insular region.

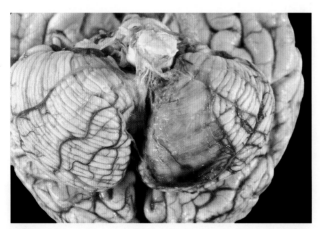

2.65 Recent infarct in territory of the postero-inferior cerebellar artery (PICA).

components may make radiological detection of incomplete infarction difficult, even with high-resolution imaging methods.[328,701]

Classification of Subtypes of Stroke

Based on different neuroimaging methods, the National Institute of Neurological Disease and Stroke (NINDS) Stroke Data Bank provided information on the relative incidence of the different ischaemic processes that lead to infarction.[887] The relative frequencies and subtypes of infarctions were classified as follows: (i) atherothrombotic, because of either (1A) large artery thrombosis or (1B) artery-to-artery embolism; (ii) cardioembolic; and (iii) lacunar. The limitations of even the most advanced imaging techniques were recognized by the inclusion of (iv) infarcts of undetermined cause (Table 2.8). In infarcts of known cause, the lumen of intracranial large to medium-sized arteries is most commonly occluded by an embolus. The frequency of locally formed thrombi in these arteries proved to be much lower than had been estimated previously. In contrast, small intraparenchymal penetrating arteries are most often occluded by a local process: thrombosis of a diseased small artery, micro-atheroma or occlusion of the origin of a penetrating artery by an atherosclerotic plaque. The presence of microemboli in retinal arteries has been interpreted as indirect evidence that microemboli may also enter small-calibre intracerebral penetrating arteries. More recent recommendations on stroke classification[37,38] indicate that aetiological classification of stroke should reflect the most likely cause (e.g. atherothrombotic, small vessel disease, cardioembolic) but without neglecting other vascular conditions that may be present (e.g. small vessel disease in the presence of severe large vessel obstruction).

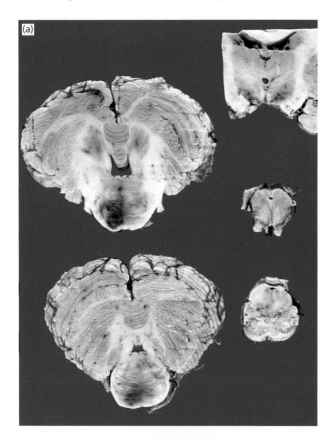

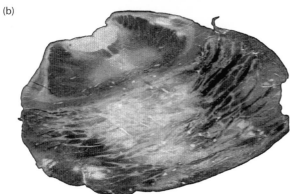

2.66 Acute infarct in the pons. (a) Macroscopic appearance. **(b)** Examination of a section stained with Luxol fast blue and cresyl violet reveals a large, poorly stained, central infarct and several small, more peripheral foci of infarction.

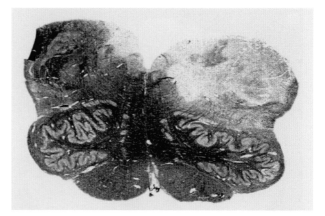

2.67 Acute infarct in the lateral part of the medulla oblongata. The infarct manifested clinically with Wallenberg syndrome.

either because the systemic circulation fails (e.g. in cardiac arrest) or because intracranial pressure rises sufficiently to impede blood flow. Depending on certain physical factors, GCI may involve transient (e.g. cardiac arrest encephalopathy) or permanent ('respirator brain') pathologies (see later under GCI). Irrespective of the type of ischaemia, two factors – the degree and duration of the insult – dictate the amount of permanent damage. Both types of ischaemic injury induce a complex set of cascades of cellular and molecular responses over hours to days to remove parenchymal tissue that is beyond repair or protect what is salvageable.[144]

Experimental models of ischaemia have been criticized as being poor representations of human stroke.[440,442] However, they have provided information on key factors involved in the evolution of ischaemic injury and may form the basis for rational therapeutic approaches.[699] In experimental studies, FCI is most commonly produced by occlusion of the MCA, with or without allowing recirculation. Experimental GCI is usually induced by reducing flow to the whole forebrain.[79,941]

In FCI, irreversible injury usually develops more slowly than after GCI. FCI produces a central core of densely ischaemic tissue surrounded by a zone of hypoperfusion in which the tissue may be perfused by surrounding collaterals but blood flow is critically reduced (penumbral zone). In GCI, similar gradients do not usually exist. The cells in the central core of FCI rapidly become irreversibly damaged, although cells in the penumbral zone may be salvageable by appropriate measures. There is a 'therapeutic window' during which this at-risk tissue within the penumbra may be rescued.[701]

Focal Cerebral Arterial Ischaemia

Focal Ischaemia in Clinical Practice

Neurons have the highest demand for oxygen and their function and viability are affected first, followed in declining order of vulnerability by oligodendrocytes, astrocytes and vascular cells. Cerebral ischaemic insults to the human brain primarily comprise two types. Focal cerebral ischaemia (FCI) is more frequently encountered in clinical practice than global cerebral ischaemia (GCI). In FCI, narrowing or occlusion of the lumen of an artery reduces CBF in a defined territory in which the function (and ultimately the viability) of neurons becomes impaired. In global cerebral ischaemia (GCI), blood flow to the whole brain becomes insufficient,

Atherothrombotic, Large Artery Thrombosis

Atherosclerosis is usually most severe in the extracranial carotid arteries but can involve large and, in severe cases, medium-sized intracranial arteries, particularly when the aggravating effect of hypertension is superimposed (Figure 2.25). The atherosclerotic plaques serve as sites of thrombosis when the endothelium has become ulcerated and/or stenosis has critically reduced the blood flow. Under circumstances of decreased cerebral perfusion pressure, stenosis exceeding 90–95 per cent may itself be sufficient to cause infarction on the basis of haemodynamic failure, a situation that is very difficult to verify *in vivo* and almost impossible post mortem.

TABLE 2.8 Relative frequencies of symptomatic stroke and infarcts[a] and incidence rates[b] according to subtypes of stroke

Subtype		Relative frequencies[a]		Incidence[b]
		All strokes (per cent)	Brain infarcts (per cent)	
Infarction		68.6		
	Atherothrombotic, large artery thrombosis	6.1	8.9	All 27, F 12, M 47
	Atherothrombotic, artery-to-artery embolism	3.8	5.4	
	Cardioembolic	13.2	19.3	All 40, F 37, M 42
	Lacunar	18.1	26.6	All 25, F 22, M 29
	Undetermined cause	27.4	39.9	All 52, F 50, M 51
	Total (infarcts)		100.0	All 147, F 124, M 173
Haemorrhage		28.6		
	Parenchymal/intracerebral	13.1		
	Subarachnoid	13.5		
Other		2.8		
	Total (all strokes)	100.0		

F, female; M, male.
[a]In Stroke Data Bank of National Institute of Neurological Disorders and Stroke (adapted from Sacco[888]).
[b]Age- and sex-adjusted incidence rates per 100 000 from Sacco.[888]

Thromboembolic and Related Strokes

The Stroke Data Bank has separated embolic strokes into two categories: strokes due to artery-to-artery embolism and strokes due to cardioembolic strokes. Even though the emboli to the intracerebral arteries from the extracranial arteries or heart often have identical pathological consequences, the clinical distinction is important as the optimum therapeutic strategies differ.[170]

Artery-to-Artery Embolism

Emboli breaking loose from thrombi formed on atherosclerotic, often ulcerated, lesions in the extracranial arteries may contain cholesterol and calcified particles in addition to coagulated blood and platelets. The thrombus formed in severely stenosed arteries with reduced flow is usually of the fibrin-dependent 'red' variety, whereas in arteries with brisker flow 'white' platelet–fibrin thrombi form on rough, sclerotic, often ulcerated surfaces. Artery-to-artery emboli involving the anterior cerebral circulation most commonly arise from the vicinity of the bifurcation of the common carotid artery. Arterial emboli that enter the posterior circulation originate in the vertebral arteries, either in the neck or within the cranial cavity.

Cardioembolic Strokes

The causes and composition of emboli arising from the heart are several-fold. The most common are fragments of thrombus formed because of atrial fibrillation or myocardial infarction (Table 2.8). The emboli are usually 'red clots'. Other common causes are emboli detaching from thrombi formed on damaged or prosthetic valves (10 per cent) or cardiomyopathy and ventricular aneurysm (7.5 per cent). Less common causes include emboli from marantic vegetations of non-bacterial thrombotic endocarditis, and paradoxical embolism via a patent foramen ovale.[1098] The emboli formed on valves contain platelets and fibrin, and fragments of emboli from calcified valves frequently contain calcium precipitates.

Paradoxical emboli enter the arterial side of the heart from the venous circulation through a patent foramen ovale, usually during a temporary rise in the right cardiac chamber pressure (e.g. Valsalva manoeuvre) in association with conditions that favour venous thrombosis, such as phlebitis, recent surgery or obstetric delivery. A patent foramen ovale is relatively common, as has been demonstrated by Doppler monitoring of the carotid arteries after intravenous injection of an agitated saline solution containing numerous micro-bubbles. The prevalence is 25–30 per cent in the general population, and a patent foramen ovale in a stroke patient is not necessarily related to the stroke.[1098] However, the frequency of a patent foramen was higher in patients with a stroke than in controls (40 per cent versus 10 per cent). In a French study paradoxical emboli were thought to account for at least 12.8 per cent of all embolic strokes whereas in an American study it was only 3.7 per cent.

The distribution of cardiogenic thromboemboli corresponds relatively closely to the volume of blood each major intracerebral artery receives from the heart. The largest proportion of these emboli, therefore, lodges in the middle cerebral arteries. The carotid artery, even its common segment, may be obstructed by a large embolus. Approximately one-fifth of internal carotid artery occlusions is embolic.[170] Cardiogenic thrombi may also find their way into the posterior circulation. They may account for up to 28.5 per cent and 55 per cent of infarcts in the territories of the posterior cerebral artery and superior cerebellar artery respectively.[518]

Other Embolic Strokes

Tumour Emboli

Emboli may be composed of tumour cells, which either detach from a neoplasm located within the cardiovascular system or proliferate free in the circulating blood. A common source of these is cardiac myxoma, most often in the left atrium. This tumour has a marked propensity to embolize. Embolic complications including neurological deficits are the presenting manifestations in about one-third of the patients. Cardiac myxoma occurs in two settings: sporadic tumour is most common among middle-aged female patients, whereas familial tumours occur most often in younger patients. Angiotropic lymphoma is another 'embolizing' neoplasm, in which clusters of malignant lymphoid cells proliferating in the circulating blood obstruct CNS blood vessels and result in multiple small infarcts or foci of haemorrhage (Figure 2.68) and spinal cord.

Fat Embolism

Destruction of fat-containing tissues, e.g. as a result of fracture of bones, pancreatitis, burns, or trauma to viscera or subcutaneous tissue, may lead to the formation of fat emboli. Symptomatic fat embolism ensues in up

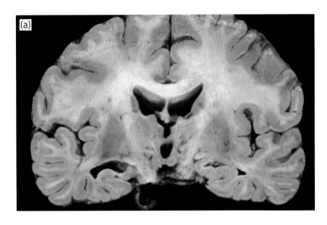

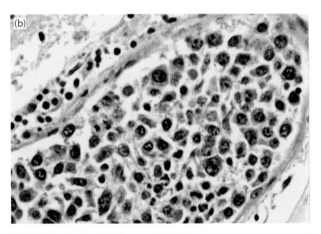

2.68 Intravascular lymphoma. (a) Multiple infarcts in the grey matter. **(b)** Malignant lymphoid cells within the lumen of a small blood vessel.

to 2.2 per cent of patients with fractures of long bones. After trauma, fat from the marrow of the fractured bone or from other traumatized adipose tissue enters the venous circulation, facilitated by increased tissue pressure at the site of trauma.[704,972] In non-traumatic fat embolism, factors that destabilize plasma lipids and increased levels of C-reactive protein, which agglutinates chylomicrons, promote coalescence of fat into larger globules. When the number of blood-borne fat globules exceeds the trapping capacity of the pulmonary capillaries, or if the globules can bypass the lungs via a patent foramen ovale or pulmonary arteriovenous shunts, they may enter the systemic circulation and be carried into the cerebral arteries. The blockade of micro-vessels in the brain causes multifocal ischaemia, which is aggravated by hypoxaemia due to pulmonary dysfunction. The BBB is disrupted, and vascular damage is often so extensive that petechial haemorrhages occur.[704,972]

The neurological symptoms include confusion, delirium, seizures and a decreased level of consciousness (even coma), which usually manifest 12–72 hours after the injury.[704,716,972] CT scans may show diffuse brain oedema and focal low-density areas. MRI is more sensitive than CT and reveals scattered 'spotty' areas of low and/or high intensity in T1-weighted images, and of high intensity in T2- and diffusion-weighted images. Although, complete recovery is possible, the reported mortality rates are high, varying from 13 to 87 per cent. The frequent association of fat embolism with underlying life-threatening disease makes it difficult to estimate the prognosis of fat embolism per se.[704]

Upon pathological examination, the brain appears swollen, with multifocal petechial perivascular haemorrhages, predominantly in the white matter (Figure 2.69).[512,972] These are characteristic findings in patients surviving for several days. A similar picture may occur in some other conditions, such as hypoxic-ischaemic encephalopathy, acute haemorrhagic leukoencephalitis, malaria, air embolism and carbon monoxide intoxication (see later). The diagnosis should be confirmed microscopically in frozen sections by demonstrating globules of neutral fat within microvessels surrounded by extravasated blood. Fat emboli may also cause perivascular anaemic microinfarcts, best seen with myelin stains. In long-term survivors, there may be white matter atrophy. The grey matter is usually spared, even though fat globules are more common in the blood vessels of the grey matter than the white matter. The greater anastomotic potential of the grey matter vasculature probably explains this discrepancy.

Air Embolism

As described earlier, small blood vessels may be occluded by bubbles of nitrogen or air in the context of decompression sickness in divers or as a complication of cardiac bypass surgery, resulting in foci of infarction in the spinal cord or brain (Figure 2.70).

Lacunae and Lacunar Infarcts

Lacunar infarcts are the most common type of infarct.[52] Although many develop subclinically, this type of infarct

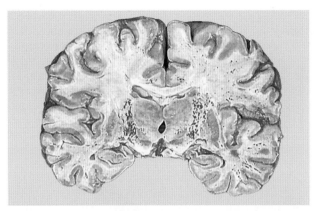

2.69 Brain of a patient with acute pancreatitis, who developed generalized convulsions and became unconscious, followed by cardiovascular collapse. He died within a couple of hours after the onset of the neurological symptoms. In the brain, there are numerous small petechial haemorrhages, predominantly in the white matter.

Image courtesy of H Aho, Turku University Hospital, Turku, Finland.

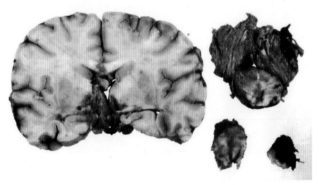

2.70 Marked brain oedema and haemorrhagic necrosis secondary to massive air embolism.

is also the most commonly identified cause of stroke (Table 2.8). A lacuna was defined in 1901 by Pierre Marie as 'a cavity as a result of a healed infarct resulting from obstruction or rupture of a small perforating artery, most commonly in the lenticular nucleus'. Lacunar infarcts were thoroughly analyzed by CM Fisher.[298,691] These studies gave rise to the concept that small lacunar infarcts (Figures 2.71 and 2.72) are responsible for specific types of stroke with focal neurological symptoms and signs – lacunar syndromes – such as pure motor or sensory stroke, dysarthria–clumsy hand syndrome and ataxia–hemiparesis. The pathological substrates of these syndromes were lacunae, small trabeculated cavities and remnants of small infarcts ranging in diameter from 0.5 to 15 mm. Lacunar infarcts are suggested to result from the occlusion of small arteries and arterioles as a result of degenerative changes that commonly occur in the context of longstanding hypertension (see Diseases of Small Arteries, earlier in chapter).[298,403,691]

High-resolution neuroimaging has facilitated the detection of these small cavities. MRI shows them to be sharply

defined lesions with a maximum diameter < 15 mm (a 10-mm cutoff has been used in some studies), corresponding to the territory of a single perforating artery (Figure 2.72).[671] The lacunar lesions visible on neuroimaging may, in addition to infarction, have a variety of other aetiologies, including haemorrhage, infection and neoplasia.[535] It is appropriate to use the general term 'lacunae' to describe all such lesions. Only when the clinical picture or the pathological findings allow determination of the cause of a lacuna should the designation be more specific, e.g. lacunar infarct or haemorrhage. Lacunae are broadly subdivided into three types: type I, lacunar infarcts; type II, lacunar haemorrhages; and type III, dilated perivascular spaces. Lammie *et al.*[577] suggested that an additional entity of incomplete lacunar infarct (type Ib) be added to this classification to describe small areas of non-cavitated rarefaction and neuronal loss.

Small perivascular cavities up to few millimetres in diameter are common in the basal ganglia and deep white matter of the elderly and in SVD. When these cavities are numerous, the condition is called *état lacunaire* in the grey matter and *état criblé* in the deep white matter.[403] It is impossible to exclude completely the possibility that a cavity surrounding a central thickened blood vessel might be the remnant of a tiny complete or incomplete lacunar infarct. However, perivascular cavities lack the structural features of an infarct[263] and are likely to arise from distortion (spiralling or kinking) of the small arteries and arterioles, and loss of parenchymal tissue.[1103] Degenerative changes in small arteries and arterioles (e.g. due to hypertension) may lead to leakage of the BBB[1069] and chronic local vasogenic oedema, which could contribute to the cavitation. Thus these perivascular empty spaces or type III lacunae (see earlier) are best termed 'perivascular cavities' (consistent with the original description of Durand–Fardel[403]) (Figure 2.73).

Lacunar infarcts[671,691] with a diameter up to 10–15 mm are thought to be caused by the occlusion of the perforating small arteries/arterioles with outer diameters of 100–200 μm (Figure 2.71). In contrast to the original proposal by Fisher that lacunar strokes are nearly exclusively hypertension-related, surveys suggest lacunar infarcts are associated with hypertension in 24–75 per cent of cases. The major risk factors for lacunar infarcts are hypertension and diabetes.[52] Small-artery disease alone explains fewer than 50 per cent of lacunar infarcts or 'infarcts limited to the territory of deep perforators'.[121] Lacunar infarcts are likely have two different underlying pathophysiologies. Many single symptomatic lacunar infarcts probably result from micro-emboli, or micro-atheromatosis[691] of the intracranial parent artery of the perforator supplying the infarct. In contrast, the presence of multiple, often asymptomatic, lacunar infarcts reflects underlying arteriolosclerosis (usually in the context of hypertension, diabetes or hyperinsulinism).[121,122] The extracranial arteries and heart have been implicated as potential sources of emboli in 17–23 per cent and 11–13 per cent, respectively, of lacunar infarcts. In the case of infarcts caused by micro-atheromatosis or arteriolosclerosis, the rapid onset of the stroke suggests that there is often an acute exacerbating factor, such as superimposed thrombus or intramural

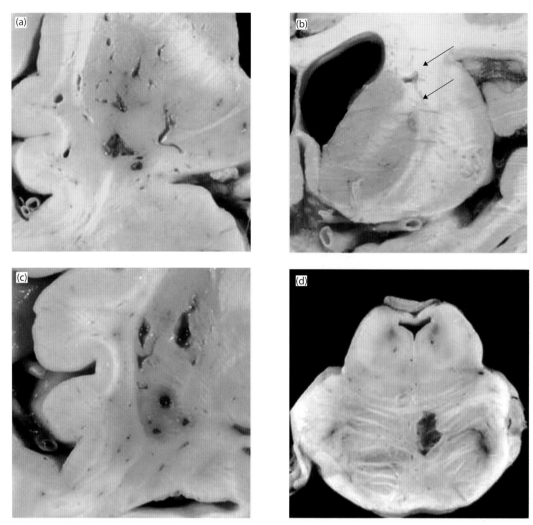

2.71 Lacunar infarcts. (a) Lacunar infarcts in the striatum. **(b)** Lacuna in the internal capsule (upper arrow) and non-cavitated ischaemic lesion in the putamen (lower arrow). **(c)** Lacunar infarcts in the caudate and putamen. **(d)** Lacunar infarct in the pons.

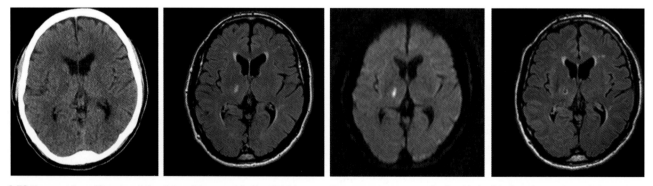

2.72 Progression of lacunar infarct in a 61-year-old. On CT 3 hours after symptom onset he had left side hemiparesis and dysarthria. On retrospective examination a faint hypodensity was seen at the lateral border.

Scans kindly provided by GJ Biessels, University Medical Center Utrecht, The Netherlands.

haemorrhage. Lacunar strokes are usually non-fatal, so that direct visualization of the underlying pathological process is rarely possible.

Because the perforators are end arteries, all of the tissue of their cylinder-shaped territory is usually damaged although some may be salvageable, e.g. through the upregulation of protective pathways by the cerebral endothelium.[353] The ischaemic parenchyma undergoes the same sequence of changes as in larger infarcts, but the end stage of cyst formation occurs rapidly.

Arterial Spasm

Focal ischaemia may develop in the territories of healthy intracerebral arteries, when hypercontraction of the smooth muscle cells reduces the arterial lumen to such a degree that the blood flow is affected. This is a common complication of SAH (see later).

Neuroimaging in Focal Ischaemia

Recent advances in neuroimaging have vastly improved our understanding of the pathophysiology of stroke-related brain injuries, particularly in terms of lesion progression with time.[699] CT scans during the early stages of brain infarction appear normal but, as time progresses, demarcation between normal and infarcted regions becomes apparent (Figure 2.74). Impaired demarcation between the grey and adjacent white matter within an infarct, and flattening of the ischaemic cerebral sulci are early changes. Over the following hours the infarcted tissue becomes more hypodense and better demarcated (Figure 2.75).

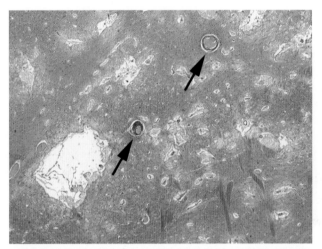

2.73 Multiple enlarged perivascular spaces in the putamen (*état lacunaire*). This patient had a history of hypertension and histology revealed widespread cerebral arteriolosclerosis (arrows) as well as enlargement of perivascular spaces in the cerebral white matter and basal ganglia.

Reduced density in the ischaemic region is usually accompanied by space-occupying effects that depend on the size of infarct. Haemorrhagic transformation is visualized as focally increased density in parts of the infarct. Injection of intravascular contrast medium increases the density at the sites of disruption of the BBB. In the subacute phase, 'fogging' occurs, as the infarcted tissue becomes more isodense. Older infarcts are characterized by focal atrophy and cyst formation, with enlargement of the corresponding ventricle (Figure 2.76).

MR imaging shows low T1 signal in the ischaemic region. On T2-weighted imaging, ischaemic zones appear as high-signal regions with a reduced diffusion coefficient. T2-weighted hyperintense areas are generally visible from the first 1–2 hours until 24–48 hours after the onset of the stroke (Figure 2.77). Modifications in brain structure can also be demonstrated by proton density MR imaging. Spin-density-weighted images and fluid attenuation inversion recovery (FLAIR) sequences show the limits of the cerebral oedema whereas T2* MR images show sites of haemorrhagic transformation (Figure 2.78). The sensitivity of T1 MR imaging for the detection of acute infarcts can be improved by contrast enhancement with gadolinium to detect BBB leakage.

The penumbra can be identified after infusion of paramagnetic contrast, which causes a reduction in the intensity of the T2-weighted signal. Differences in cerebral blood volume are evident as a mismatch between diffusion- and perfusion-weighted MR images, allowing the delineation of the penumbra (Figure 2.79). Normal MR imaging is also useful for the detection of wallerian degeneration. Reduced signal intensity in T2-weighted images is seen from the end of the first month. This is followed by isodensity and then hyperintensity on T2-weighted imaging within 2–3 months.

Single photon emission computed tomography (SPECT) and PET (Figures 2.79 and 2.80) can be used to study of regional blood flow, regional blood volume, regional cerebral metabolic rate of oxygen and regional cerebral metabolic rate of glucose in brain infarcts.[409] Functional MRI is increasingly used to measure blood flow and oxygen extraction and to analyze haemodynamic responses and neuronal connectivity in a range of neurological diseases, including stroke.[35,374,561,893]

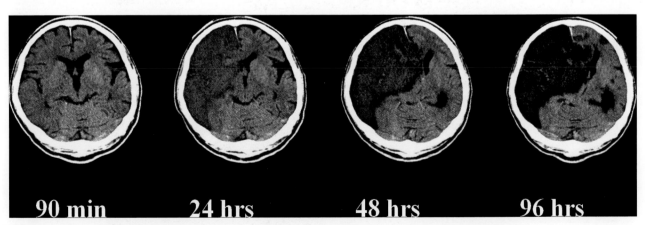

90 min   24 hrs   48 hrs   96 hrs

2.74 CT scans showing changes with time in the imaging characteristics of a major cerebral infarct.

Images kindly provided by K Nagata, Institute of Brain and Blood Vessels, Akita, Japan.

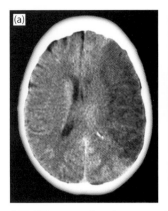
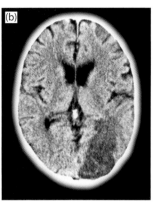
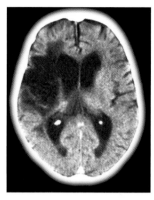
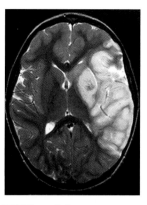

2.75 CT scans showing acute infarcts, in the territories of the left middle **(a)** and posterior **(b)** cerebral arteries.

Scans kindly provided by A Muntane–Sánchez, Hospital Universitari de Bellvitge, Barcelona, Spain.

2.76 CT scan of an old infarct in the territory of the right MCA.

Scan kindly provided by A Muntane–Sánchez, Hospital Universitari de Bellvitge, Barcelona, Spain.

2.77 Hyperintense T2-weighted MRI signal after infarction in the territory of the left MCA.

Scan kindly provided by A Muntane–Sánchez, Hospital Universitari de Bellvitge, Barcelona, Spain.

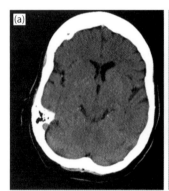
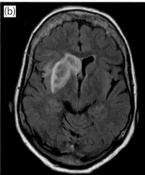
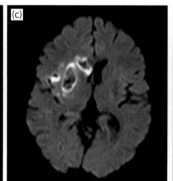
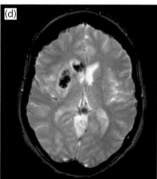

2.78 Haemorrhagic transformation of an infarct in MCA territory in a 75-year-old woman. (a) CT shows signs of early ischaemia in right MCA territory (signal attenuation caudate, basal ganglia fading). She woke up with left hemiparesis. An estimated 3-5 hour symptom duration at presentation. **(b)** MR FLAIR after 2 days shows extent of infarct. **(c)** MR DWI and **(d)** T2* (or gradient echo) show haemorrhagic transformation of the lesion.

Scans kindly provided by G J Biessels, University Medical Center Utrecht, The Netherlands.

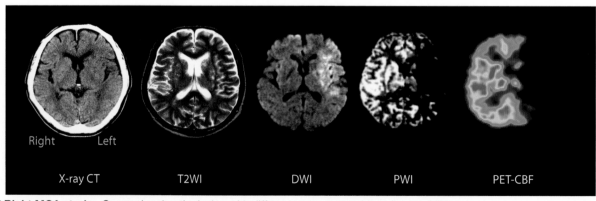

2.79 Right MCA stroke. Scans showing the lesion with different sequence and imaging modalities.

Images courtesy of K Nagata, Institute of Brain and Blood Vessels, Akita, Japan.

Pathophysiology of Focal Cerebral Ischaemia

General Aspects

Much of our understanding of the ischaemic cascade comes from experimental studies. The cascade consists of a complex series of events that are highly heterogeneous[144] and evolve over minutes to days and weeks after the initial hypoperfusive event (Figure 2.81). The main stages include energy failure due to disruption of blood flow, excitotoxicity, calcium overloading, oxidative stress, BBB dysfunction, microvascular injury, haemostatic activation, injury-related inflammation and immune responses, and cell death involving neurons, glia and endothelial cells.

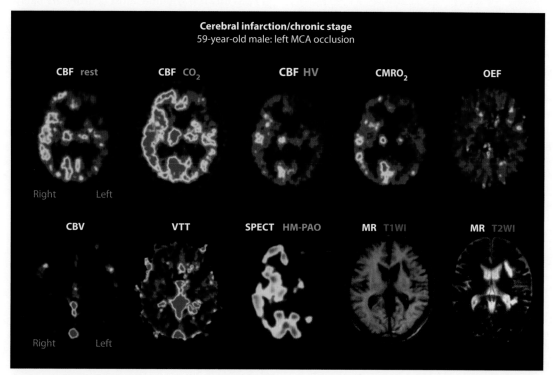

2.80 Use of neuroimaging techniques to demonstrate metabolic changes in the brain after a right MCA territory infarct. Abbreviations explained in Figure 2.83.

Images courtesy of K Nagata, Institute of Brain and Blood Vessels, Akita, Japan.

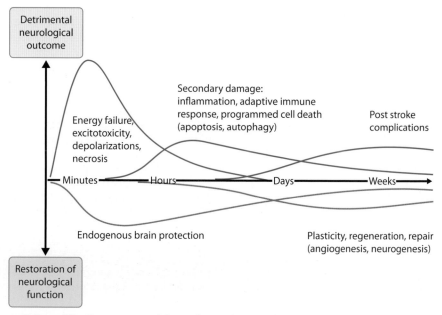

2.81 Schematic representation of the time course of damaging and reparative processes after focal ischaemic injury in the brain.

Kindly supplied by U Dirnagl, Universitätsmedizin Berlin, Berlin, Germany.

Microvascular damage and disruption of the BBB, which may occur days later, lead to vasogenic oedema and can cause haemorrhages. At the same time the tissue is undergoing a complex range of reparative and remodelling responses to limit damage and improve outcome.

Ischaemic Thresholds

Experimental studies provided initial data on the critical threshold values of CBF necessary for the maintenance of functional and structural integrity of the brain.[408] Approximately similar values were obtained in clinical studies using PET in stroke patients (Figure 2.82 and 2.15). Blood flow above about 40 per cent of the normal value (see Microcirculation and Neuronal Metabolism) ensures unimpaired spontaneous and evoked electrical activity of nerve cells. At flow of about 30–40 per cent of normal, increasing numbers of neurons are unable to produce sufficient energy to sustain neurotransmission. At about 30 per cent of normal blood flow, transmission ceases completely. Energy production in these electrically silent neurons can still maintain basic intracellular functions, e.g. the activity of ion pumps in the plasma membrane. Neurons may resume transmission if adequate CBF is restored. If regional CBF falls below about 15 per cent of normal, a sudden rise in extracellular K^+ indicates that the threshold of membrane failure is reached. At that stage, the cells are unable to generate sufficient energy to maintain transmembrane ion gradients and the efflux of K^+ is accompanied by influx of Na^+, Ca^{2+} and Cl^- ions, together with influx of water along the resulting osmotic gradient.[941] Membrane failure results in irreversible nerve cell injury unless adequate blood flow and energy production are restored quickly. The absolute flow values at these thresholds depend on the species, being higher in smaller animals, and influenced by physiological variables such as brain temperature. By and large, however, the threshold levels of blood flow seem to be proportional to the baseline blood flow in both animals and man.[13,408,941]

A fall in CBF affects not only electrophysiological function but also other aspects of neuronal metabolism. At around 50–60 per cent of normal blood flow protein synthesis ceases, at 40 per cent intracellular and extracellular

acidosis are induced, at 25 per cent brain oedema develops and below this general cell metabolism is disturbed (Figure 2.15). The development of irreversible injury depends not only on the severity of the ischaemic insult but also on its duration.[328] After MCA occlusion, neurons may tolerate flow rates of near zero, 10 or 15 mL/100 g/minute for 25, 40 or 80 minutes, respectively, and at flow rates above 17–18 mL/100 g/minute most neurons are likely to recover. The noticeable variation in ischaemic tolerance of individual types and groups of neurons indicates selective vulnerability. The duration and depth of ischaemia are not always the only decisive factors. Under certain conditions, neuronal death may occur even after short ischaemic episodes followed by reperfusion, sometimes long after recovery of many neuronal functions. Thus, the duration of the ischaemic insult after which neurons can still recover must be assessed after a sufficiently long recirculation period (up to several days) to provide assurance that the recovery is permanent and not simply the temporary restoration of specific cellular functions (e.g. energy production or electrical activity). Knowledge of the tolerance of human brain to focal ischaemia is important clinically. One review indicated that normotensive and normothermic patients who had undergone temporary iatrogenic occlusion of an intracranial artery during saccular aneurysm surgery under general anaesthesia, tolerated up to 14 minutes of ischaemia without evidence of infarction whereas 95 per cent of the patients tolerated 19 minutes' ischaemia. However, all those with occlusion for over 31 minutes had both clinical and radiological evidence of infarction.[898] The thresholds in man may be lowered by the longstanding effects of known cerebrovascular risk factors, such as hypertension and diabetes.

The Penumbra

The evolution of a focal brain infarct has been analyzed in detail in experimental models of permanent MCA occlusion (Figures 2.83 and 2.84). In FCI, the ischaemic or central core resulting from severely impaired blood flow rapidly infarcts, with dead cells killed by total energy failure, ion homeostatic deregulation, loss of membrane integrity, lipolysis and proteolysis. However, this is surrounded by a penumbra, a zone of tissue that, although electrically silent, has the capacity to recover if perfusion is restored.[64,701]

Tissue within the penumbra is functionally impaired but potentially salvageable by improving blood flow or intervening with appropriate therapy. Attempts have been made to determine its physiological characteristics at different time intervals after the ischaemic insult.[288] The penumbra was initially studied using electrodes inserted into the brain for measurement of the oxygen and ion concentrations. However, PET allows non-invasive quantitative *in situ* analysis of the main physiological parameters of brain energy metabolism including regional CBF (rCBF), blood volume, oxygen extraction fraction (OEF) and cerebral metabolic rates of oxygen ($CMRO_2$) and glucose (CMROglc).[93] PET studies of the penumbra in human stroke (Figure 2.80) have been complemented by a range of other imaging techniques, including functional MRI. The possible PET equivalent of the penumbra is the zone

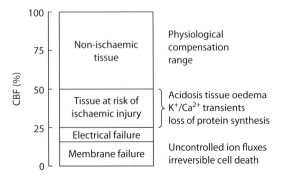

2.82 Ischaemic thresholds for neuronal function. CBF, cerebral blood flow. (See also Figure 2.15.)

Image courtesy of Dr P Lindsberg, Helsinki University Central Hospital, Helsinki, Finland.

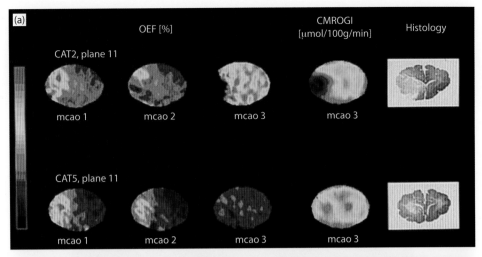

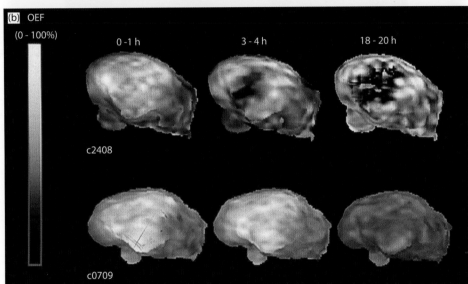

2.83 Focal experimental ischaemic stroke. (a) Sequential multi-tracer positron emission tomography (PET) scanning of the brains of two cats after middle cerebral artery (MCA) occlusion for approximately 1 hour (MCAO 1), 2–3 hours (MCAO 2) or 18–24 hours (MCAO 3). In both cats, in the core of ischaemic region the oxygen extraction fraction (OEF) has increased (misery perfusion) in the early phase (MCAO 1) because of low blood flow. It has already decreased markedly by 2–3 hours (MCAO 2) in cat 2, in which the evolved infarct (verified by histology) is finally demarcated as a region with a complete lack of oxygen and glucose utilization (MCAO 3). In contrast, in cat 5, the early misery perfusion reverses, as verified by normalization of OEF and cerebral metabolic rates of oxygen and glucose (CMROGI) and preservation of tissue integrity. **(b)** The same phenomenon is demonstrated on the lateral aspect of the brain. Upper panel: cat 2, with an evolving infarct; lower panel: cat 5, with reversal of the ischaemia.

(a) Reproduced by permission from APASS.[49] (b) Courtesy of Professor W–D Heiss, Max Planck Institut für Neurologische Forschung, Cologne, Germany.

of 'misery perfusion', characterized by reduced CBF with relatively preserved or even normal $CMRO_2$. A hallmark of misery perfusion is an increased OEF: as the flow of oxygenated blood decreases, more oxygen is extracted from the remaining flow in order to prevent a critical reduction in $CMRO_2$ (Figure 2.82).

PET studies[409,646] show that rCBF remains above 55 per cent of normal in patients with a TIA, i.e. in the absence of tissue destruction. Cortical tissue undergoes necrosis in regions with rCBF < 30 per cent or $CMRO_2$ < 40 per cent of normal (see Microcirculation and Neuronal Metabolism, earlier). In the peri-infarct zone, with rCBF 30–45 per cent

of normal during the first 2 days, metabolic functions are unstable and infarction occurs if low flow persists. Tissue recovery in such peri-infarct zones occurs only if misery perfusion is associated with $CMRO_2$ above a critical threshold level, i.e. if the peri-infarct zone represents penumbral tissue. Comparison of early CBF values (measured by PET within 3 hours of onset of the stroke) with the final infarct volume indicated that the initial penumbral tissue accounted for about 18 per cent of the final infarct volume and tissue that was initially sufficiently perfused for about 12 per cent,[410] indicating enlargement of the infarct and its incorporation of penumbral territory (Figure 2.80).

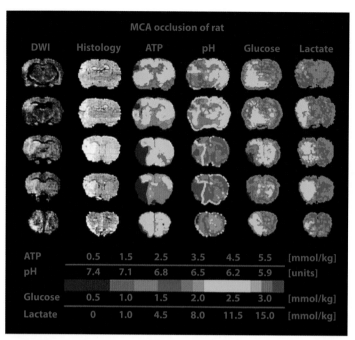

2.84 Focal experimental ischaemic stroke. Colour-coded presentation of an early evolving infarct 7 hours after middle cerebral artery (MCA) occlusion in a rat. Diffusion-weighted magnetic resonance image (DWI) reveals the infarct *in vivo* as a pale area. Metabolic analysis of frozen sections of the same brain demonstrates a decrease in adenosine triphosphate (ATP) and glucose concentration and lowering of pH parallel to the increase in lactate concentration in the infarct.

Reproduced from Back, et al.[80] With permission from Lippincott Williams & Wilkins/Wolters Kluwer Health.

The development of high-resolution PET has been useful to evaluate sequential physiological changes during the ischaemic episode (Figures 2.83 and 2.84).[410,1090] After MCA occlusion in cats[410] CBF in the core region fell immediately to critical levels. Around the core, in areas with reduced rCBF, CMRO$_2$ and CMROglc remained relatively preserved and OEF increased. As the core expanded, CMRO$_2$ and OEF declined in the region of reduced rCBF and the penumbral region of increased OEF moved outward towards the periphery of the ischaemic territory, indicating the dynamic state of the penumbra. In one cat, with only transient misery perfusion, as indicated by later normalization of OEF, CMRO$_2$ and CMROglc, the eventual infarct was small, demonstrating the viability of the penumbral tissue (Figure 2.83). These findings verified the sequence of events that had been deduced from single examinations in man at different post-stroke survival times.[93] The results also validated the concept of the penumbra as a dynamic zone around the core of an infarct[410] and have been validated using newer imaging techniques. In particular, diffusion- and perfusion-weighted MRI are used increasingly in clinical research and practice for early diagnosis of brain ischaemia (changes in the apparent diffusion coefficient of protons are evident as early as 10 minutes after the onset of ischaemia), to identify regions of reduced cerebral perfusion (by measuring the transit time of an exogenous tracer, gadolinium, or by magnetically spin-labelling water in the blood so that it can serve as an endogenous tracer), and to assess the mismatch between the two as an indication of the ischaemic penumbra (and hence of the potentially salvageable brain tissue).[270,336]

Penumbral regions are best defined in terms of function rather than structure (Figure 2.85). They are difficult to delineate in tissue preparations. In addition to close scrutiny of conventionally stained tissues, markers of several immediate early genes or DNA repair proteins, e.g. HSP70, haem oxygenase or poly(ADP-ribose) polymerase, may be of help in identifying tissue that, whilst not undergoing necrosis, shows evidence of ischaemic stress (Figure 2.86).[802] Multiple pathogenetic processes occur in the penumbral zone as it is incorporated into an evolving infarct.[941] These are considered to cause 'molecular' injury as opposed to the 'haemodynamic' injury in the infarct core.[533] The ischaemic core may induce spreading depression, as a result of repeated nerve cell depolarization caused by excessive release of K$^+$ and excitatory amino acids (glutamate and aspartate) from injured neurons. The release and accumulation of glutamate and, to a lesser extent, aspartate in the extracellular space causes overstimulation of α-amino-3-hydroxy-5-methyl-4-isoxazole propionic acid (AMPA), kainate and N-methyl-d-aspartic acid (NMDA) receptors on neighbouring neurons, with consequent influx of Na$^+$, Cl$^-$ and Ca^{2+} ions through their gated receptor channels. The depolarized neurons cause more calcium influx and glutamate release leading to local amplification of the initial ischaemic insult. The repeated depolarizations also exhaust the marginal energy supplies within the penumbra, which may transform from non-lethal to irreversible injury.[439] This interpretation is supported by studies in which agents that decrease presynaptic release of excitatory amino acids or that block glutamate receptors, reduced the frequency of perifocal depolarization, the volume of tissue showing ATP depletion, and the infarct

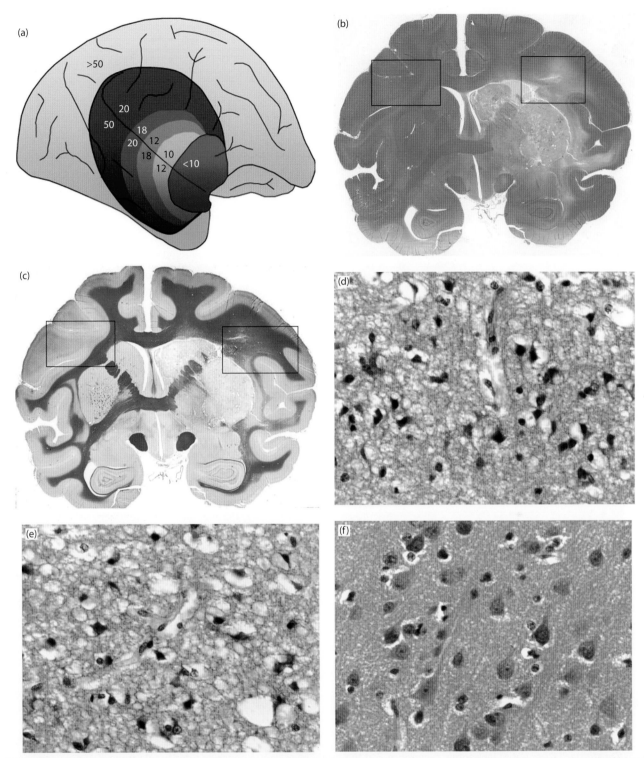

2.85 Concept of ischaemic penumbra and the gradient of tissue changes after focal stroke. (a) Spatial pattern of cerebral blood flow (CBF) reduction following middle cerebral artery (MCA) occlusion in a baboon. The demarcated areas demonstrate a gradient from the ischaemic core (red) through to the penumbra and oligaemic tissue (blue) to normally perfused tissue (grey). The penumbra is severely hypoperfused, non-functional, but still viable cortex surrounding the irreversibly damaged ischaemic core; with elapse of time, more of the penumbral region is recruited into the core until the tissue is reperfused. Coronal brain sections at the level of the anterior commissure, stained with haemoxylin and eosin (H&E), **(b)** and Luxol fast blue and cresyl violet (LFB/CV) **(c)** from an olive baboon subjected to MCA occlusion. The core and hypoperfused areas can be discerned. The images (from inset regions in b and c) show status of neurons and white matter within the ischaemic core **(d,g,j)**, poor penumbral region **(e,h,k)** and unaffected tissue **(f,i,l)**. Stains H&E, amyloid precursor protein immunohistochemistry and LFB–CV.

Image (a) modified from Moustafa and Baron.[701] With permission of John Wiley and Sons. © 2008 British Pharmacological Society.

Continued

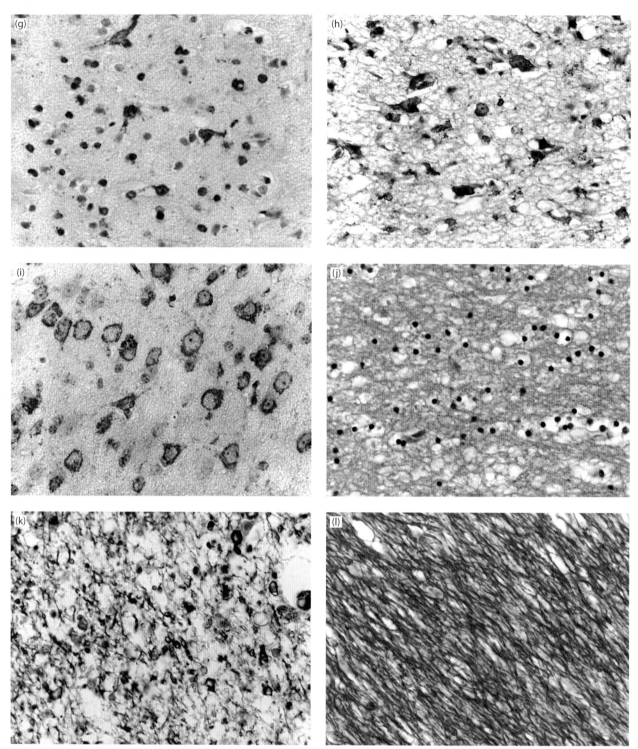

2.85 (Continued) Concept of ischaemic penumbra and the gradient of tissue changes after focal stroke. The core and hypo-perfused areas can be discerned. Insets show status of neurons and white matter within the ischaemic core **(d,g,j)**, poor penumbral region **(e,h,k)** and unaffected tissue **(f,i,l)**. Stains H&E, amyloid precursor protein immunohistochemistry and LFB–CV.

size.[678,953] These agents appeared to have real neuroprotective effects on penumbral neurons, because blood flow to the ischaemic region did not increase.[634,699]

Excitatory amino acid antagonists have shown benefit in experimental stroke but not in human trials.[703] Adenosine receptors modulate several neurotransmitter receptors; adenosine 2A receptors were beneficial in experimental stroke, probably by counteracting the effects of excitatory amino acids.[758] No evidence of significant advantage or harm was found, however, when using drugs that modulate the action of excitatory amino acids in human patients.[703] This may not be surprising though, as results from experimental focal ischaemia have demonstrated a therapeutic window of up to 3–4 hours and, in most human studies, drugs aimed at reducing

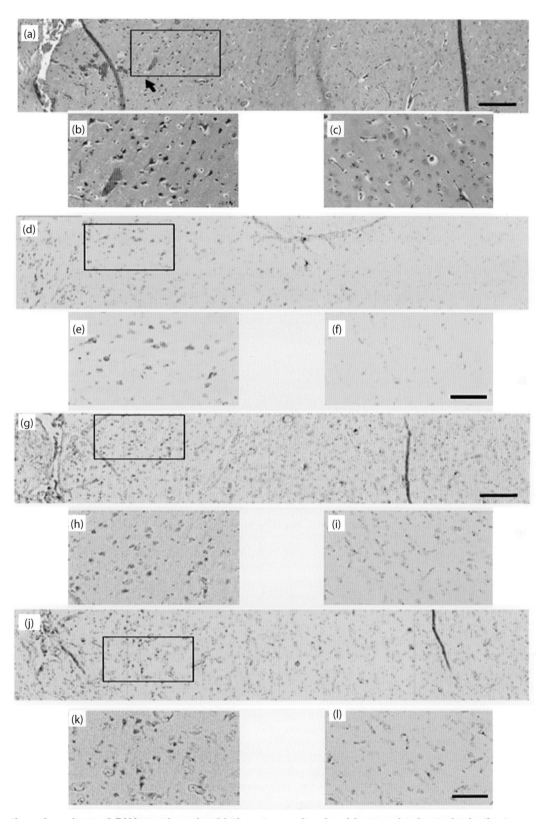

2.86 Activation of markers of DNA repair and oxidative stress after focal haemorrhagic stroke in the temporal cortex.
Differential spatial patterns of focal increase in markers towards the infarct core (I) through the hypoperfused region (P, penumbra) and normal (N) appearing cortex in a neurosurgical biopsy. **(a–c)** haemotoxylin and eosin (H&E), **(d–f)** poly(ADP-ribose) polymerase 1, **(g–i)** *trans*-4-hydroxy-2-nonenal (HNE) pyrrole adduct, **(j–l)** heme oxygenase-1 protein in peri-infarct (box) and normal-appearing region.

Images courtesy of A Oakley and J Sixsmith, Newcastle University, UK.

glutamate release or blocking glutamate receptors have been administered after longer intervals, of up to 24 hours.

The microcirculation in the penumbral zone may be restructured or become progressively impaired. Because the penumbra lies next to irreversibly injured tissue, reactive inflammatory changes may occur in the penumbral microvessels. These may cause occlusion of the hypoperfused vessels by leukocytes or platelet aggregates adhering to endothelium, which is stimulated by ischaemia to express adhesion molecules and additional pro-inflammatory factors (see later).[242,568]

Oxidative Stress Mechanisms

Oxidative stress is a significant consequence of ischaemic injury during both the initial insult and reperfusion.[190] Oxygen free radicals and their derivatives including superoxide anions, hydroxyl radicals, peroxynitrite and hydrogen peroxide and nitric oxide generated during cerebral ischaemia cause oxidative and nitrosative damage to a wide range of macromolecules including enzymes of respiration and nucleic acids.[616] Oxidative stress also causes mitochondrial dysfunction with release of cytochrome c and initiation of apoptotic cell death. Transgenic animals that overexpress or are deficient in antioxidant proteins such as superoxide dismutases or pro-oxidants such as NADPHoxidase[189] have been used to study the role of these processes in ischaemic brain damage and the potential for modulation of the damage by manipulation of oxidative stress.

Reperfusion Injury and Recanalization

Restoring natural perfusion is the best protective measure against tissue damage in FCI.[904] Apart from the persistence of the initial occlusive pathology, however, several other processes may interfere with restoration of blood flow. These include endothelial swelling, perivascular oedema, increased blood viscosity and intravascular changes that involve inflammatory cells, platelets and clotting factors.[771]

Polymorphonuclear leukocytes use surface adhesion molecules, particularly CD11b/CD18 of the integrin superfamily, to bind to endothelium. In experimental studies, antibody to CD18 was able to inhibit the adherence and improve reperfusion, most significantly in non-capillary micro-vessels.[696] Ischaemic injury stimulates endothelial expression of the ligands for leukocyte binding: intercellular adhesion molecule-1 (ICAM-1), the ligand of CD11b/CD18; and P-selectin, an adhesion glycoprotein of the selectin super-family.[617, 752] P-selectin is also involved in clot formation. It is expressed on platelets and promotes deposition of fibrin. Finally, the extrinsic coagulation system seems also to be involved; infusion of an antibody against a key molecule in the extrinsic pathway of the coagulation cascade, tissue factor, increased reflow after 3 hours of MCA occlusion.[486] Activation of the coagulation cascade may be more likely in human than experimental stroke, in that the artery is usually obstructed by a thrombotic mass instead of a suture or clip and the endothelium is often primarily injured, e.g. by atherosclerosis. Complement activation also seems to be an important component of reperfusion injury.[58]

Even successful reperfusion after FCI carries significant risks, especially if there is an increase in oxygen in excess of consumption, which may result in the formation of highly reactive oxygen free radicals, lipid peroxidation, protein oxidation and damage of DNA.[186,496] Toxic free radicals are probably produced even during the period of ischaemia with low tissue oxygen, e.g. in the penumbra with misery perfusion.[1109]

In the context of reperfusion after recanalization, knowledge of the fate of thrombi or thromboemboli that obstruct arteries in stroke patients is limited. Approximately 20 per cent of embolic (cardiac or artery-to-artery) MCA occlusions may recanalize spontaneously within 24 hours, and up to 80 per cent within 1 week.[1117] These data do not include TIAs, but the disappearance of symptoms within 24 hours suggests that rapid recanalization is achieved in the majority of TIAs caused by small thromboemboli.

Pre- and Post-Conditioning and the Induction of Ischaemic Tolerance

Prior sublethal transient ischaemic episodes cause brain cells to acquire tolerance to subsequent, otherwise detrimental, ischaemia.[344] However, even post-stroke conditioning, implemented by series of repetitive but brief interruptions of reperfusion subsequent to a prolonged ischaemic episode, may confer protection.[800] Experimental studies implicate several pathways in neuroprotection in ischaemic preconditioning.[255] These include activation of the N-methyl-D-aspartate (NMDA) and adenosine receptors; activation of intracellular signalling pathways such as mitogen activated protein kinases (MAPK) and protein kinase C (PKC); upregulation of Bcl-2 and heat shock proteins (HSPs); and activation of the ubiquitin-proteasome pathway and the autophagic-lysosomal pathway. Reprogramming of the Toll-like receptor signalling has also been implicated in neuroprotection after preconditioning.[971] In rodents, ischaemic tolerance also occurs after spreading depression, the silencing of electrical activity by slowly spreading waves of depolarisation that result from activation of NMDA channels, e.g. by elevated glutamate and K+ after stroke.[265] Depending on its intensity, the same type of insult may be either protective or injurious.[242] The understanding of ischaemic tolerance has potential for stroke prevention[256,585] and recurrent stroke injury. However, the clinical utility of ischaemic preconditioning remains to be demonstrated. Physical exercise has been suggested to be a promising preconditioning measure to induce ischaemic tolerance through the promotion of angiogenesis, mediation of inflammatory responses, inhibition of glutamate over-activation, protection of the BBB and inhibition of apoptosis.[1119]

Inflammatory and Immune Responses

Elucidation of the immunopathology of ischaemic injury may offer scope for improving clinical outcomes after stroke.[453,486] There is activation of both innate and adaptive immune cellular responses during the post-stroke period (Box 2.7). Experimental studies suggest that inflammation contributes to ischaemic brain injury,[1066] with greater morbidity after post-stroke inflammation. However, recent findings show that inflammatory processes may also have beneficial effects,[600] depending on the type and size of stroke injury, degree of vascularisation and pre-existing systemic infection or inflammatory disorders.[638] Molecular responses

BOX 2.7. Repertoire of inflammatory and immune responses in stroke

I Cells and molecules predominantly associated with inflammatory mechanisms and innate immune response

Microglia – derived from haematopoietic system and central nervous system (CNS) resident. Produce TNF, IL-1β, ROS and other pro-inflammatory mediators and IL-10, TGF-β and IGF-1 (tissue repair).

Perivascular macrophages – unlike microglia, continuously replenished from haematogenous precursors. M1 produce IL-1β, IL-12, IL-23 and TNF, chemokines, ROS, and NO; M2 produce anti-inflammatory cytokines (IL-10 and TGF-β), IL-1ra and arginase.

Mast cells (MC) – found in meninges and cerebral blood vessels. MC granules store vasoactive substances (histamine), cytokines (TNF), anticoagulants (heparin), and proteases (tryptase, chymase, MMP2, MMP9).

Blood monocytes – precursors of tissue macrophages. At least two different monocyte populations: 'classical' monocytes produce IL-10 and 'pro-inflammatory' monocytes produce TNFα.

Dendritic cells (DC) – specialized antigen-presenting cell (APC) of myeloid lineage. DCs associated with meninges, choroid plexus and the CSF. Cells expressing DC markers (CD11c+, MHC Class II+) appear in the brain parenchyma after focal ischaemia.

Neutrophils – secretory phagocytic cells containing granules with pro-inflammatory molecules including iNOS, NADPH oxidase, myeloperoxidase, MMP8, MMP9, elastase and cathepsins. Vesicle and granule exocytosis is induced by binding to E-selectin, and IL-8 stimulation.

II Cells and molecules predominantly associated with inflammatory events and adaptive immune response*

Lymphocytes - key cells in both innate and adaptive immune responses.

- B-lymphocytes (B-cells) – humoral immune responses through production of antibodies that attack and neutralize specific antigens.
- T-lymphocytes (T-cells) – involved in cellular immunity, in which extraneous antigens are suppressed by a cytotoxic cellular response. T-cells express CD4 or CD8.

Antigen presenting cells – APC are mainly DC and macrophages. They degrade foreign antigen into peptides and assemble MHC-class II-antigen complex, which is recognized by the T-cell receptor (TCR) of CD4+ T cells and interaction with co-stimulatory molecules (e.g. B7-1, B7-2 and CD28), resulting in T-cell activation.

Helper T cells – CD4+ T cells largely become helper T-cells (Th). Act as 'helpers' by coordinating and modulating immune responses. Th type 1 (Th1) cells secrete IFNγ and TNFα and Th2 cells secrete IL-4, IL-5, IL-9, IL-10, IL-13.

Cytotoxic T-cells – CD8+ T cells are cytotoxic. The TCR of CD8-expressing T-cells binds the antigen presented with MHC class I molecules and co-stimulatory molecules. Unlike MHC class II, MHC class I molecules are present on almost every cell.

Natural killer (NK) cells – lymphocytic cells with cytotoxic function also include natural killer T-cells (NKT). NK cells lack a TCR and do not require antigen presentation for their activation and cytotoxicity, which is triggered by interferons or cytokines (IFNγ, IL-2, and TNFα).

γΔT-cells – subset of effector lymphocyte with a TCR comprising γΔ chains. These cells have several different functions including cytolysis, antigen presentation, immunoregulation and production of growth factors.

*Unlike the innate immune cell repertoire, these cells develop memory against stimulating antigens.

Adapted from Macrez et al.[638] and Iadecola and Anrather.[453]

triggered almost immediately upon ischaemic injury involve the expression of cytokines including interleukin-1β, interleukin-6 and tumour necrosis factor-α, and chemokines such as MCP-1 and MIP-1α. These mediate a complex range of inflammatory processes, including the upregulation of leukocyte adhesion molecules (ICAM-1, E-selectin, P-selectin and Mac-1, LFA-1, L-selectin and PSGL-1) by cerebral endothelium and leukocytes. The microvasculature undergoes major changes including the opening of tight junctions and the BBB, luminal adhesion of circulating cells and transendothelial migration of circulating neurophils and monocytes.

Almost simultaneously, there is activation of resident cells including microglia and astrocytes. Perivascular macrophages, mast cells and dendritic cells, originating in vascular walls or the meninges, also drive the innate and adaptive immune processes.[453] Mediators released from the infiltrating cells may further activate or damage the endothelium through nitric oxide synthase and cyclooxygenase-2 pathways. The cytokine environment favours a Th1 helper T-cell response, stimulating macrophage and dendritic cell activity. The inflammatory responses are further amplified by the production and release of reactive oxygen species by infiltrating inflammatory cells, through NADPH oxidase and myeloperoxidase activity, and the production and activation of MMPs (especially MMP-9), facilitating inflammatory cell infiltration and leading to cell death further afield that may encroach on the penumbral region (Figure 2.85). Experimental studies have also shown that despite the early deleterious effects of inflammation, the inflammatory reaction and removal of cellular debris are important for later expression of neurotropic and protective factors, regeneration and neurovascular remodelling.

Mechanisms of Cell Death after Infarction

Neuronal death following ischaemic injury is largely attributed to necrosis (Figure 2.87). However, recent developments indicate that significant cell death occurs by apoptotic as well as hybrid mechanisms along an apoptosis–necrosis continuum (see earlier). Although the infarcted core is necrotic, within the penumbra caspase-mediated apoptosis is activated, although secondary necrosis may result from failure to implement the apoptotic signalling pathways fully, as they require energy. Cerebral ischaemia triggers two general pathways of apoptosis. The intrinsic pathway originates with mitochondrial release of cytochrome c and subsequent stimulation of caspase-3 whereas the extrinsic pathway is initiated by the activation of cell surface death receptors, which belong to the tumour necrosis factor superfamily, by Fas ligand, resulting in the stimulation of caspase-8. There is some evidence that ischaemic cell death may also be mediated by autophagy, which is activated during cerebral ischaemia for the bulk removal of damaged neuronal organelles and proteins.[547,826] Molecular markers of the autophagic process during ischaemic injury include increased production of microtubule-associated protein 1 light chain 3 (LC3-II), beclin-1, lysosome-associated membrane protein 2 and lysosomal cathepsin B. Oxidative and endoplasmic stresses in cerebral ischaemia are stimuli for autophagy in neurons.[10] There is also cross-talk between pathways involved in autophagy and apoptosis and both processes may occur in parallel, as suggested by upregulation of

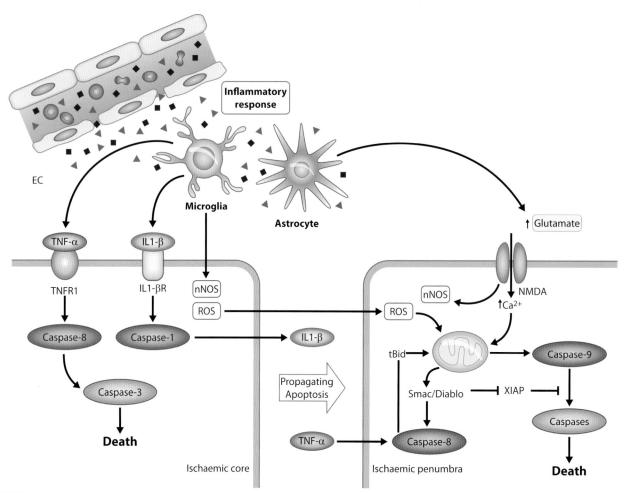

2.87 Mechanisms in focal ischaemic injury. Subsequent to stroke injury, the severity of depletion of energy and the disruption of the blood–brain barrier influence the inflammatory and excitotoxic responses from glia in the ischaemic core and the penumbra. The release of diffusible factors by damaged cells and the activation of caspases can induce an apoptotic cascade in cells surrounding necrotic core, increasing the extent of damage in response to the ischaemic event. IL-1βR, IL-1β receptor; NMDA, N-methyl-D-aspartate; nNOS, neuronal nitric oxide synthase; ROS, reactive oxygen species; TNF-α, tumour necrosis factor α; TNFR I, TNF-α receptor I.

Original schematic from Ribe, et al.,[842] kindly modified and produced by Y Yamamoto, Japan.

both autophagic and apoptotic markers in the ischaemic penumbra.[826] However, the role of autophagy in cell death after cerebral ischaemia is still unclear.[955]

Infarct Progression

Experimental studies in rodents and non-invasive imaging techniques in humans have shown that infarcts enlarge with time. In absence of early reperfusion, ongoing subacute injury to cells in the ischaemic penumbra resulting in expansion of the infarcted core over about 4–6 hours. Additional delayed cell death may occur over the subsequent hours or days.

Cell death during the subacute phase is associated mainly with peri-infarct spreading depression, which exacerbates the imbalance between metabolic demand and haemodynamic capacity in the penumbra. Other factors may include excitotoxicity, calcium toxicity, mitochondrial disturbances, oxidative damage, NO toxicity, endoplasmic reticulum stress and reduced protein synthesis. The subsequent delayed cell death is poorly understood. DNA damage leading to caspase activation probably contributes,[620,621]

particularly after reperfusion of previously ischaemic tissue, as does brain oedema.

Sequence of Microscopic Changes in Brain Infarcts, with Approximate Timing

1 hour: micro-vacuoles (swollen mitochondria) become visible within neurons. This is followed by perineuronal vacuolation (swelling of astrocyte processes).

4–12 hours: the neuronal cytoplasm becomes increasingly eosinophilic and Nissl bodies disappear. The nucleus initially appears pyknotic but later shows reduced basophilia and eventually disappears.

12–24 hours: neutrophil leukocytes infiltrate the tissue.

2 days: macrophages (foam cells) appear. In larger infarcts, these cells may remain present for months.

5 days: few or no neutrophils remain. Astrocytes start to proliferate around the core of the infarct.

Around 1 week: there is increased proliferation of astrocytes around the core. Vessels show endothelial hyperplasia, and neovascularization commences.

Zone of Transition

The transition from poorly to adequately perfused tissue is relatively abrupt. Similarly, the transition from infarcted to surviving tissue in histopathological sections is surprisingly sharp. Around the core of infarction after MCA occlusion, there is only a narrow zone of transition from complete loss of neurons to tissue with a normal number of neurons, i.e. the zone of selective neuronal necrosis and preservation of glial cells is narrower than would be expected on the basis of the difference between neuronal and glial vulnerability to ischaemia (Figures 2.83 and 2.85).

Pathological Changes in Brain Infarcts

Early Changes

Unlike with *in vivo* MR imaging methods, at post-mortem examination, infarction or ischaemic necrosis cannot be detected macroscopically until about 12 hours have elapsed. In fresh unfixed tissue, it may take 24 hours to become evident. In fixed tissue, the early infarct can be delineated by palpation, because the necrotic tissue does not become fixed and is velvety to touch, which contrasts with the firmer consistency of non-necrotic, fixed tissue. Visually, the loss of demarcation of the boundary between the grey and white matter is one of the earliest macroscopic changes (Figure 2.88 and 2.85).

The earliest morphological changes associated with infarction were documented in the classical studies of Brown and Brierley[145,146] but require electron microscopic examination of perfusion-fixed tissue for reliable assessment. Within minutes, neuronal mitochondria swell and loose their cristae, giving the cells a micro-vacuolated appearance. At the same time, the cytoplasm of the neurons begins to condense to become electron-dense. Over the following minutes and hours the nucleus becomes increasingly pyknotic with clumping of chromatin, and the nucleoli

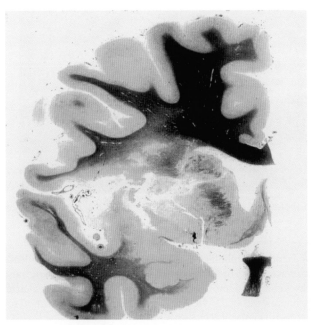

2.88 Acute infarct. Region of liquefactive necrosis in the territory of the MCA. Luxol fast blue and cresyl violet.

disappear. These neurons correspond to the acidophilic neurons that are first detectable light microscopically in conventionally fixed tissue at about 4 hours (Figure 2.89). They are detectable by conventional (H&E) staining, but with acid fuchsin stand out more prominently. Swelling of astrocytes results in perineuronal and perivascular vacuolation, which becomes more marked during the hours after onset of ischaemia. Astrocytes also show clumping of the normally finely stippled chromatin.

The inflammatory reaction begins with invasion of neutrophils, the earliest of which arrive by about 12 hours and which are present in larger numbers by 24 hours (Figure 2.89). Entry of blood-derived macrophages occurs at about 2 days, by which time there is pronounced axonal disruption (Figure 2.90). Reactive astrocytes begin to surround the necrotic tissue after about 5 days, when the increased density of capillaries also becomes apparent. Phagocytosed tissue debris is visible as pools of foamy material in macrophages ('gitter' cells), which may persist for months to years in large infarcts. The end result is a cavity lined by astrocytes and filled with clear fluid (Figure 2.89c).

The importance of reperfusion in the genesis of early ischaemic neuronal changes is highlighted by the difference between the structural changes in incomplete and/or temporary ischaemia (a condition that always prevails in focal ischaemia) and those in complete and/or permanent ischaemia. Acidophilic neurons with micro-vacuolation are seen only when they are in tissue that has received at least a trickle of blood flow during or after the ischaemic period. Complete permanent ischaemia gives rise to swollen neurons that stain only weakly with eosin and other acidic dyes. In FCI, the neurons in the central core of large infarcts undergo damage more akin to that in complete permanent ischaemia, presumably because the residual blood flow is so low.

Haemorrhagic Infarction

A substantial proportion of acute infarcts undergoes haemorrhagic transformation, particularly in the grey matter. In autopsy studies the proportion varies from 18 to 48 per cent. In patients with first-time infarcts, haemorrhagic transformation was found in 14.9 per cent of patients by imaging (CT or MR) within 7–10 days.[700] The majority of haemorrhagic infarcts are embolic arterial strokes: 51–71 per cent of these become haemorrhagic compared with only 2–21 per cent of non-embolic strokes.

Two generally accepted mechanisms by which an infarct becomes haemorrhagic are: (i) reperfusion of necrotic, leaking blood vessels and (ii) occlusion of venous drainage. Reperfusion occurs when the embolus fragments or is broken down by fibrinolytic enzymes, either naturally or through thrombolytic therapy.[1068] Haemorrhagic infarction has been considered to carry a higher risk of clinical deterioration, but small punctate (type I) haemorrhages do not result in additional neurological deficits. Non-anticoagulated 'stroke' patients with this type of CT-verified haemorrhagic transformation of their infarct are either stable or clinically improving in a large majority of cases. Larger intraparenchymal (type II) haemorrhages (Figure 2.91), which are associated with worsening of the patient's neurological status, are less frequent. They occur

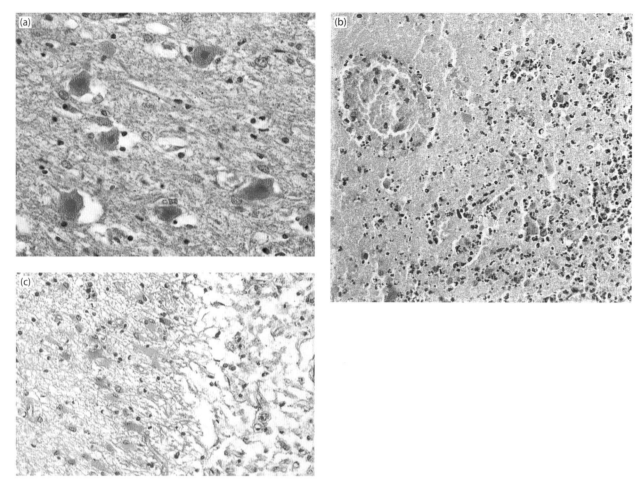

2.89 Ischaemic infarct. (a) In acute ischaemic nerve cell injury, the cytoplasm of the injured thalamic neurons stains bright red with eosin, and the nuclei are pyknotic and have lost their characteristic nucleoli. **(b)** On the second day after the insult, neutrophil leukocytes have invaded the infarct, where necrotic neurons appear as eosinophilic ghosts. **(c)** After 2 months, the necrotic tissue has been removed by macrophages and transformed into a cyst, in which some lipid-containing macrophages are still present (on the right). The cystic cavity is bounded by reactive astrocytes.

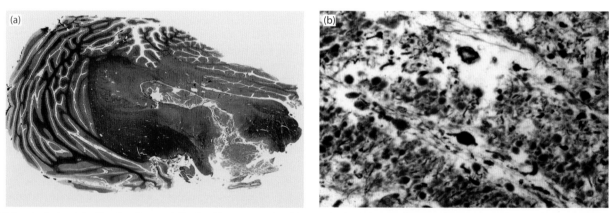

2.90 Acute infarct in the territory of the basilar artery, leading to locked-in syndrome. (a) Myelin staining showing extensive necrosis of the pons. **(b)** Silver impregnation reveals swollen axons in the vicinity of the infarct.

in about 8 per cent of all embolic infarcts and 6.5 per cent of all ischaemic hemispheric infarcts.[383]

Brain Oedema Following Focal Cerebral Ischaemia

Decreased blood flow to about 10 mL/100g/minute produces a breakdown of the ionic pumps and the initiation of cytotoxic (cellular) oedema. Later, BBB breakdown leads to vasogenic brain oedema. Perilesional cytotoxic and vasogenic oedema are maximal at 24–72 hours after ischaemic stroke.[867] Oedema progression may result in tissue necrosis (Figure 2.92).

Cytotoxic oedema is not truly a form of oedema in that the excess fluid is intracellular rather than interstitial.

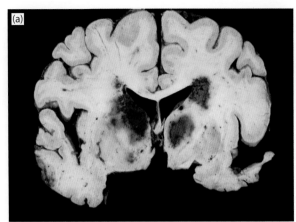

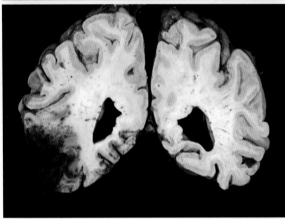

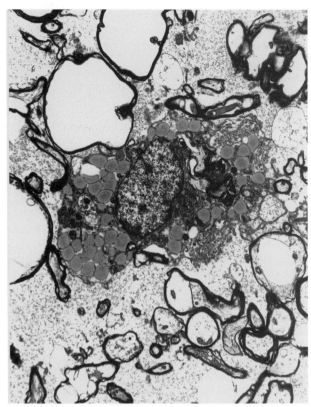

2.92 Electron micrograph showing a lipid-laden cell, probably a macrophage, in a region of marked oedema and early axonal degeneration in tissue surrounding a cerebral infarct.

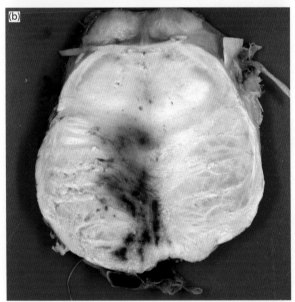

2.91 Haemorrhagic infarction. (a) Multiple foci of haemorrhagic infarction in the territories of the lenticulostriate arteries and left PCA. **(b)** Partly haemorrhagic infarct in the pons, caused by a thrombus in the basilar artery.

However, the term is now firmly embedded in common usage. This form of 'oedema' is characterized by intracellular accumulation of water, primarily in astrocytes but also in neurons. Dilated astrocytic cell bodies are seen on electron microscopy, and the extracellular (interstitial) space is decreased in cytotoxic oedema, in contrast to

vasogenic oedema (see later). Several mechanisms participate: (i) energy depletion and failure of the Na⁺/K⁺ ATP-ase pump, and voltage- and ligand-gated channels; (ii) increased influx of ions, cations in particular, through multiple different types of ion channel;[602] and (iii) sustained uptake of osmotically active solutes. Glutamate plays an important role. Activation of inotropic and metabotropic glutamate receptors in neurons, and impairment or reversal of glutamate transporter activity in neurons and astrocytes, leads to the influx of Ca^{2+}, Na^+, Cl^- and water, exacerbating cytotoxic oedema.[36,335] Aquaporins (AQPs) are also thought to be involved and may represent targets for therapeutic intervention. AQP4 deletion in mice reduces oedema after water intoxication and after FCI.[644,963] The expression of AQP4 mRNA is decreased in the core of the infarct but increases in the peri-infarct region, peaking on day 3 after FCI in the rat, in parallel with the development of oedema as monitored by MRI.[985] AQP9 is also increased in astrocytes after transient forebrain ischaemia in mice.[81] In the human brain, AQP4 immunoreactivity is decreased in the core of the infarct but is increased in astrocytes at its periphery.[46]

Vasogenic oedema results in the accumulation of fluid interstitially. It predominantly affects the white matter and is characterized microscopically by increased separation of myelinated fibres, especially around blood vessels, and reduced intensity of staining with dyes such as Luxol fast blue and solochrome cyanine. Myelinated fibres may appear swollen and beaded. Enlarged extracellular spaces are occupied by protein-rich fluid, and immunohistochemistry discloses exudation of albumin and other plasma proteins, including immunoglobulins. Astrocytes show elongated perivascular foot processes. With time, hypertrophy and hyperplasia of astrocytes and loss of oligodendrocytes and myelin occur.

Subacute and Chronic Infarcts

During the weeks after infarction, necrotic elements are phagocytosed mainly by haematogenous macrophages. Axonal swellings start to develop within hours in the vicinity of the infarct, become more prominent over the following days and persist for several weeks. Fibres degenerate distal to the infarct, a process that continues for months and even years, as macrophages filled with myelin debris mark the pathways of affected tracts. This can be visualized by MRI. New capillaries form around infarct and capillary buds extend into the necrotic tissue. Astrocytes increase in number and become enlarged, often gemistocytic (with abundant homogeneous-looking eosinophilic cytoplasm), and produce increased amounts of glial fibrillary acidic protein.

Chronic infarcts are characterized by local brain atrophy, glial proliferation and increased fibril formation (gliosis), and the presence of cystic cavities (Figure 2.93). Calcification and neuronal ferrugination may occur in the adjacent tissue. Chronic infarcts cause atrophy of affected fibre tracts, depending on the location of the infarct. Typical examples are the atrophy of the corticopontine and corticospinal (pyramidal) tracts after MCA occlusion involving the anterior part of the capsula interna (Figure 2.94). Infarcts of the hippocampus are followed by atrophy of the ipsilateral fimbria/fornix and mammillary body. More extensive cerebral atrophy after infarction can cause *ex vacuo* hydrocephalus.

White Matter Pathology

Constant perfusion of the white matter by deep penetrating arterioles is essential for functioning of axons and oligodendrocytes. White matter abnormalities depend on the severity and duration of ischaemic injury and can include swelling and breakdown of myelin, beading, swelling and degeneration of axons; beading and fragmentation of astrocyte processes (clasmatodendrosis), loss of oligodendroglia, activation of microglia and infiltration by macrophages. Chronic hypoperfusion leads to white matter rarefaction and astrocytic gliosis, degeneration of myelin, apoptosis of oligodendroglia, an increase in microglia, and loss of BBB function with extravasation of plasma proteins and enlargement of perivascular spaces.[945] Myelin proteins are affected differentially by chronic hypoperfusion, the adaxonal myelin-associated glycoprotein being particularly susceptible. This differential susceptibility has been used to quantify the severity of ante-mortem hypoperfusion, by comparison of the levels of myelin-associated glycoprotein and proteolipid protein in human post-mortem tissue.[91,92]

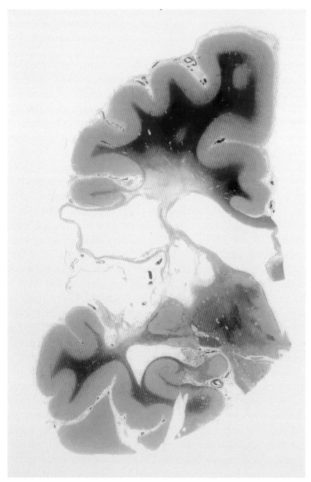

2.93 Chronic, cystic infarct in the territory of the middle cerebral artery (MCA). Woelcke.

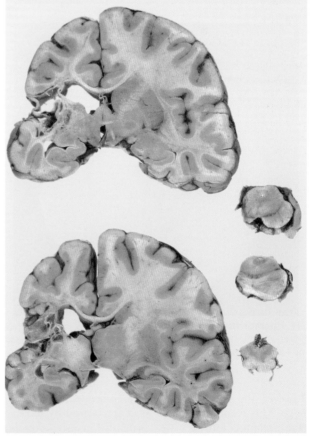

2.94 Chronic, cystic infarct in the territory of the left middle cerebral artery (MCA). The infarct has resulted in atrophy of the ipsilateral corticopontine and corticospinal tracts.

Leukoaraiosis and Binswanger's Disease

The term leukoaraiosis is used for non-specific signal change in the white matter on radiological imaging (decreased density on CT and increased signal intensity on T2/FLAIR MRI). It has multiple pathological counterparts, such as enlarged perivascular spaces (*état criblé*), white matter infarcts or foci of ischaemic damage short of infarction, and foci of demyelination. The term 'leukoaraiosis' should not be used as a synonym for white matter ischaemia.[121]

Patients with multiple small infarcts, commonly in the white matter and deep cerebral grey matter structures, often develop dementia. Originally termed atherosclerotic dementia, its description has evolved from multi-infarct dementia to vascular dementia. The infarcts usually result from SVD and repeated thrombotic events or from multiple emboli. Although in early studies of vascular dementia, the emphasis was on the number and cumulative volume of infarcts, diffuse ischaemic damage to the white matter is now thought to be a more significant contributor to dementia in most cases (see Chapter 16).

Binswanger's disease is generally described as dementia caused by subcortical ischaemic brain damage in people with longstanding hypertension (Figure 2.95).[121,611] There has been debate as to whether the symptoms, signs and structural alterations are sufficiently distinct to justify designation of this disease as a separate clinicopathological entity. The topic is covered in more detail in Chapter 16.

Pathophysiology of Global Cerebral Ischaemia

Transient Global Ischaemia

Global cerebral ischaemia (GCI), like focal ischaemia, can be either transient or permanent, depending on whether blood flow to the intracranial contents is restored. A critical variable is the duration of ischaemia. If flow is restored immediately some degree of global ischaemic damage may result. If flow is not restored in time, resistance increases in the cerebrovascular bed, precluding reperfusion, and permanent global ischaemia inevitably results. Cardiac arrest is an important cause of GCI. Cardiac arrest encephalopathy is the accurate designation for the brain consequences of resuscitated cardiac arrest, although non-perfused brain is another potential consequence if reflow of blood within brain tissue is not established.

Even in GCI, both selective neuronal necrosis and infarction can occur. Selective neuronal necrosis occurs if ischaemia is less severe or less prolonged, and infarction if ischaemia is more profound or prolonged. The consequences of transient GCI depend on the duration and completeness of the ischaemic insult. Some degree of residual perfusion (and resuscitative cardiac massage) gives a better outcome than total cardiac standstill and no blood pressure whatsoever during the insult.

A third factor affecting outcome is the brain temperature during the period of transient GCI. This is critical in determining the degree of resulting brain damage.[157] Clinical cases of GCI with hypothermia well illustrate this. A widely documented case is that of a 14-year-old girl who fell through ice into freezing water and suffered anoxia for 40 minutes. In spite of the prolonged submersion, there were ultimately no

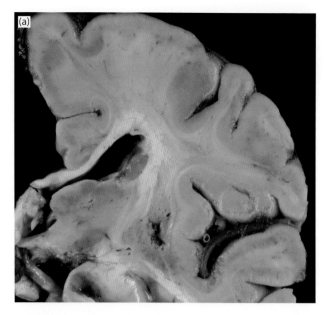

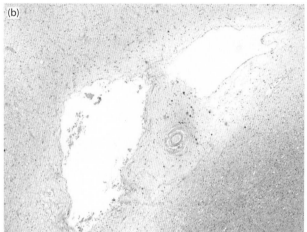

2.95 Binswanger's encephalopathy. (a) A coronal slice through the cerebrum of a patient with chronic, severe arterial hypertension shows much of the white matter to have a grey, pitted appearance. Note, too, the lacunar infarct in the putamen. **(b)** Histology shows the white matter to have a rarefied appearance, with multiple lacunae. There is also severe arteriolosclerosis. **(c)** Marked pallor of myelin staining, with relative preservation of the subcortical U fibres. Luxol fast blue–cresyl violet (LFB–CV).

long-term neurological deficits.[939] The powerful neuroprotective effect of hypothermia has given rise to a resurgence in hypothermia as a neuroprotective tool against ischaemia (see Pre- and Post-Conditioning and the Induction of Ischaemic Tolerance, earlier). Conversely, mild hyperthermia of 1–2°C augments ischaemic brain necrosis,[157] acidosis[394] and glutamate release.[686] Hypothermia has been shown to reduce H⁺ production in total circulatory arrest in humans.[431,433]

Enhanced tissue acidosis may also account for the aggravating effect of the fourth major factor in determining the outcome after transient GCI, i.e. the blood glucose level. The major role of blood glucose in determining outcome was first suggested by studies in which rabbits fed a carrot diet had markedly better survival after hypobaric hypoxia,[167] an effect probably mediated by caloric restriction.[222,723] Metabolic studies have shown that restricting substrate supply improves recovery of energy metabolism and reduces accumulation of lactate.[329] The ketotic or starved state improves tolerance to ischaemia[629] and reduces neuronal death.[647] High blood glucose increases neuronal death after ischaemia[817] and worsens outcome in both GCI and FCI. The chance of awakening after cardiac arrest is reduced by high blood glucose levels, and neurological deficits are worsened.[613] The same principle applies in near-drowning.[61] In FCI, poorer outcome with high blood glucose levels has been shown to occur in patients with or without diabetes,[818] indicating a direct glucose effect.

These are the four major factors determining the amount of brain damage seen after cardiac arrest: duration of ischaemia, degree of ischaemia, temperature during the period of circulatory stagnation and blood glucose level. When damage is severe, it is often clinically accompanied by epileptic activity,[944,1091] which independently augments necrosis.[1056]

Pathological Findings in GCI

GCI causes most damage in specific neuroanatomical locations: the cerebral cortex, striatum, hippocampus and cerebellar Purkinje cells.[209] In general, the damage within the affected regions is associated with similar neuronal, glial, vascular and inflammatory changes (Figure 2.96) to those in FCI.

Cerebral Cortex

In the cerebral hemispheres, in the mildest cases of hypotension or brief cardiac arrest, damage may be largely restricted to the triple watershed zone. Located occipitally, more posteriorly than generally realized (Figure 2.98), this is the intersection between the territories irrigated by the anterior, middle and posterior cerebral arteries. There is an inverse distribution of hypotensive and arterial occlusive ischaemia (compare Figures 2.97 and 2.98).

The parasagittal convexity is the junction between the anterior cerebral artery (ACA) and middle cerebral artery (MCA) territories and is affected in more severe GCI. In either triple or double watershed necrosis, damage can be unilateral because of a common asymmetry in the circle of Willis. Occlusion of a carotid artery, because it reduces flow to both anterior and middle cerebral arteries, simulates hypotension by reducing the pressure to both vessels and thus also gives this ACA–MCA boundary zone pattern.[358]

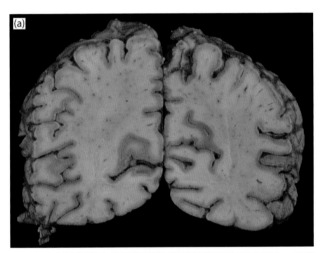

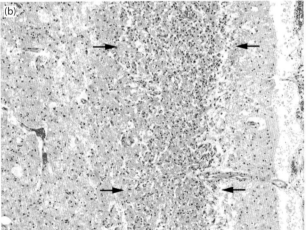

2.97 Ischaemic pan-necrosis in a patient who survived several weeks after prolonged cardiac arrest. (a) The cerebral cortex is abnormally thin and granular and has a yellowish-brown colour. **(b)** Histology shows almost full-thickness destruction of the cortex (arrows), which has been replaced by sheets of macrophages and newly-formed blood vessels.

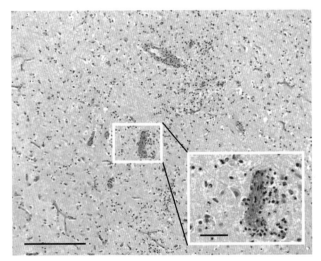

2.96 Infiltration of brain by inflammatory cells after global cerebral ischaemia (GCI) caused by cardiac arrest. Low magnification reveals dense inflammation, mimicking early abscess formation. Higher magnification (inset) shows the inflammatory cells to be mainly neutrophils and macrophages. The changes can be dated as being within a few days of GCI, after which the neutrophils vanish. Bar = 200 μm; inset, bar = 50 μm.

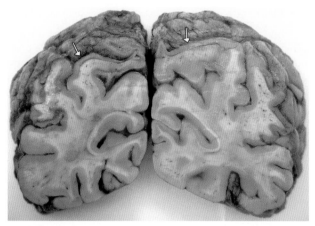

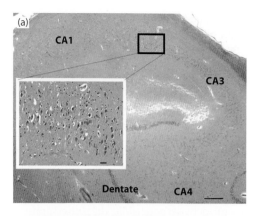

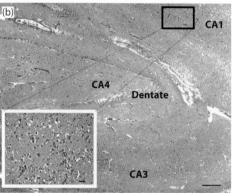

2.98 Watershed infarction. The superior convexity (arrows) and superomedial surface of the occipital lobes show thinning and cavitation in a distribution characteristic of cardiac arrest and hypotension. This is the watershed region between the perfusion territories of the anterior, middle and posterior cerebral arteries. The bilateral necrosis tapered anteriorly in this brain. The double watershed zone between the anterior and middle cerebral arteries was normal.

Although mild or less prolonged hypotension tends to affect only the triple watershed zone, severe hypotension or prolonged cardiac arrest causes the entire cerebral cortex to become necrotic, along with the rest of the brain.

Hippocampus

Within the hippocampus, the cells first affected by the mildest ischaemic insults are neurons in the CA1 area of Lorente de Nó,[615] also termed, in humans, the H$_1$ cells of Rose (Figure 2.99a).[865] Under experimental conditions these cells undergo necrosis after only 2 minutes of global ischaemia.[957]

What is usually seen is selective neuronal necrosis rather than pan-necrosis, although infarction of the entire hippocampus develops if the global ischaemic insult is severe (Figure 2.99b). Indeed, the entire brain becomes necrotic after insults of high magnitude (Figure 2.100). Thus, selective vulnerability of lesions is lost with severe insults, which can affect CA3 (H$_2$) field and even the dentate gyrus. The extent of neuronal death may relate not only to the four factors listed earlier but also to the survival period: more damage may be seen after longer survival times after cardiac arrest[438,790a] due to maturation of cell damage over days (see earlier, Delayed Neuronal Death). This increase in damage over time, however, applies only to selective neuronal necrosis, not infarction, which develops rapidly in minutes to hours.

Asymmetry in the circle of Willis can account for asymmetrical or entirely unilateral hippocampal damage. To a lesser extent, damage may also vary along the septo-temporal axis of the hippocampus. Sometimes, the hippocampus is entirely spared bilaterally, despite necrosis in the cerebral cortex, thalamus and cerebellum.[133]

White Matter

As in CO toxicity (see Carbon Monoxide Toxicity, earlier in chapter), GCI may sometimes selectively affect the white matter, especially when hypotension occurs during intensive care.[155]

2.99 Hippocampus in global ischaemia. Mild **(a)** and severe **(b)** damage. In (a), at the border between CA1 and CA3 (inset) there is selective neuronal necrosis in CA1. The inset reveals the hallmark of selective acute neuronal necrosis, with acidophilic neurons, sparing neuropil and showing little cellular reaction. In (b), both CA1 and CA3 show identical histology, with acute pan-necrosis rather than selective neuronal necrosis. The inset reveals abundant cellular reaction, including a mitosis (circled), probably in a macrophage. Magnification bars = 500 µm; inset, bar = 50 µm.

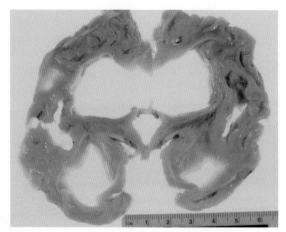

2.100 In this case global ischaemia of prolonged and severe nature because of hypotension during hip-replacement surgery resulted in near-total brain necrosis.

Thalamus

The thalamus is regularly damaged in GCI, by selective neuronal necrosis or infarction. A thalamic nuclear pattern can sometimes be discerned but the damage is often homogeneous. Coma after cardiac arrest is related not only to neocortical necrosis but also to thalamic damage.[537]

Basal Ganglia

When global ischaemia affects the basal ganglia, the corpus striatum is usually most vulnerable. In some cases, however, the globus pallidus can be affected,[327] which can simulate CO toxicity. This is important in medicolegal evaluation of causality. When globus pallidus necrosis occurs in GCI the remaining brain is also affected. In contrast, CO toxicity can solely affect the globus pallidus, sparing the remainder of the brain. However, if CO exposure is more severe, and is accompanied by hypotension, the entire brain becomes affected, as in GCI.

The pars reticulata of the substantia nigra is also sometimes affected selectively in GCI (Figure 2.101), and this does not necessarily imply CO toxicity. When disease elsewhere in the brain is absent and damage limited to the pallido-reticularis pathway, then CO is aetiologically implicated; if ischaemia is severe enough to damage the pallido-reticularis, other sensitive structures such as the hippocampal CA1 pyramidal neurons will be affected.

Necrosis of the pars reticulata of the substantia nigra[133,134] is noteworthy in view of necrosis in the identical midbrain region in pure primary insults of experimental status epilepticus[728] and the tendency for epileptiform activity to appear in the post-ischaemic brain.[178,944,1091,1112] It is not known whether those cases of ischaemia that show prominent pallidal necrosis have more overt or covert epileptic activity (e.g. periodic lateralized epileptiform discharges) than those not showing pallidal damage after GCI.

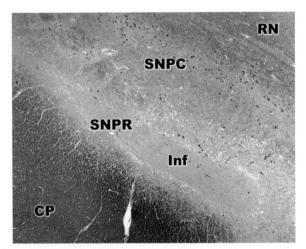

2.101 Midbrain in global ischaemia. From a 50-year-old man with a history of a seizure disorder, 120 kg obesity and severe coronary artery disease. Coronary artery bypass grafting was followed the next day by a 2- to 3-minute period of pulseless electrical activity. Death occurred 6 days later. An infarct (Inf) is sharply demarcated in the substantia nigra pars reticulata (SNPR), located between the cerebral peduncle (CP) and the substantia nigra pars compacta (SNPC). The section includes part of the red nucleus (RN). Similar bilateral damage was present in the globus pallidus. In addition to pallido-reticular necrosis, global ischaemia caused widespread pan-necrosis in the neocortex and cerebellum, as well as selective neuronal necrosis in CA1, CA3, CA4 and the dentate gyrus.

Brain Stem

Necrosis in the brain stem can also occur outside the substantia nigra in GCI, with a clinical picture of irreversible coma and absent brain stem reflexes[478,836] that can simulate the non-perfused brain (see later, under Permanent Global Ischaemia). Alternatively, the brain stem may be entirely spared, with the typical necrosis of forebrain structures only.[1100]

Cerebellum

The cerebellum, like the cerebrum, is supplied by three major arteries and in GCI can show widespread necrosis in severe cardiac arrest.[209] After shorter cardiac arrest, necrosis is limited to the boundary zones between the superior, posterior inferior and anterior inferior cerebellar arteries. Microscopically, Purkinje cells are affected first and show acidophilic necrosis after short survival (Figure 2.102a). Longer survival allows development of a characteristic isomorphic fibrillary gliosis in the cerebellar cortex,[209] Bergmann gliosis (Figure 2.102b), signifying previous loss of Purkinje neurons.

Spinal Cord

Unless the patient is paraplegic, the spinal cord is often neglected in neuropathology and in evaluation of global ischaemic changes. The spinal cord has a vertical watershed zone with thoracic vulnerability, traditionally described as centred around T4,[193] but series of ischaemic spinal cords have revealed lumbar[78,118,349] and low thoracic T7[458] and T9[193] predilection. This is surprising in view of the robust blood supply of the lumbar enlargement by the artery of Adamkiewicz.[948]

In addition, a horizontal watershed is sometimes seen in the spinal cord, between the anterior and posterior spinal arterial territories.[1095] Usually, the central grey matter is infarcted (Figure 2.103). Unlike in cerebral cortex, where large neurons are spared,[705] spinal ischaemia seems to have a predilection for motor neurons.[349]

In infants, spinal cord ischaemic damage may be seen after hypotension.[946] Rarely, the brain is spared but the spinal cord selectively affected in GCI caused by cardiac arrest.[483]

In addition to GCI, a variety of local vascular diseases can lead to spinal cord ischaemia and infarction, including aortic disease, which can occlude to mural ostia of aortic feeding vessels supplying the spinal cord.[332] Surgery on the aorta is a major cause of spinal ischaemia.[658]

Pituitary

The pituitary gland may undergo necrosis after GCI (Figure 2.104). The frequency of this is unknown. In the non-perfused brain the pituitary is regularly infarcted (see later, under Permanent Global Ischaemia).

Permanent Global Ischaemia (Non-Perfused or Respirator Brain)

If perfusion never returns to the intracranial contents after transient GCI,[295] then there is brain death, or non-perfused brain.[114] Blood flow stops at the level of the foramen lacerum in the carotid arteries. This was formerly termed 'respirator

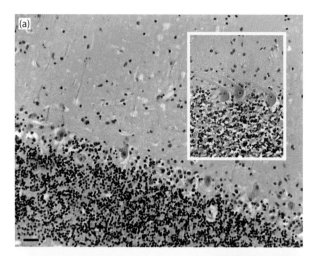

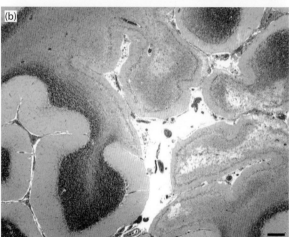

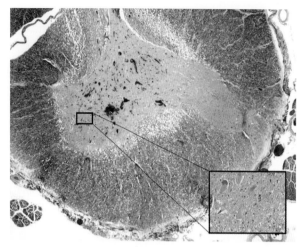

2.103 Spinal cord in global ischaemia. Watershed ischaemia in low thoracic spinal cord, seen as haemorrhagic necrosis of the grey matter. The inset shows acidophilic motor neurons from the anterior horn devoid of nuclear or cytoplasmic detail, showing only blue staining of lipofuscin with Luxol fast blue. Luxol fast blue and PAS.

2.102 Cerebellum in global ischaemia. (a) Selective neuronal necrosis in a 53-year-old woman with small bowel perforation, cardiac arrest and 5-day survival. Acidophilic Purkinje cells are bright red and lack nuclear detail. Normal Purkinje cells often have somewhat smudged nuclei and can be somewhat pink because of a paucity of Nissl substance, which is dispersed finely (inset). Selective neuronal necrosis was also evident in the CA1, CA3 and CA4 fields of the hippocampus, the neocortex, the lateral geniculate body and the dentate nucleus. No infarcts were seen. Bar = 100 μm. **(b)** Pan-necrosis in a 68-year-old man with atherosclerotic aneurysm of the aorta and old myocardial infarcts. Both watershed ischaemia and focal infarcts were seen in the cerebrum. The cerebellum does not cavitate after encephalomalacia as readily as the cerebrum. Cerebellar damage here is old, with shrunken folia and some cystic areas of pan-necrosis, together suggesting previous cardiac emboli. Watershed, hypotensive cerebellar damage would be less demarcated and involves Purkinje cells as the most vulnerable neurons in the cerebellum. A row of Bergmann glia replacing Purkinje cells is seen as a fine blue line at this magnification. Bar = 500 μm.

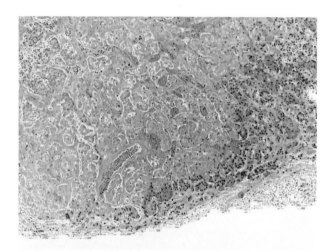

2.104 Pituitary infarct in global ischaemia in a 61-year-old male with adenocarcinoma, who suffered Gram-negative septic shock and abrupt loss of blood pressure when vasopressin line came out. The brain showed CA1 necrosis in the hippocampus.

brain' because of its frequent association with a clinical history of the patient's having been mechanically ventilated.[1014]

Brain perfusion requires the mean arterial blood pressure to exceed the intracranial pressure (ICP), to drive blood through the resistance of the cerebral vasculature. The cerebral perfusion pressure (CPP) equals the mean arterial blood pressure minus the ICP. When the CPP drops below about 6 kPa (45 mmHg), brain perfusion ceases permanently if flow is not restored immediately. Non-perfusion can result from an increase in ICP or a decrease in arterial blood pressure. Causes include ischaemic brain oedema (e.g. associated with recent infarction), sudden hypotension (e.g. during anaesthesia or due to cardiac arrhythmia) or traumatic brain injury with its rise in ICP.[649] Whatever the cause, perfusion stops and blood stagnates in the microcirculation throughout the brain.

Initially, series of cases were described[6,26,377] of patients who had been on a respirator for prolonged periods, and the neuropathological effects were thought to be related somehow to the respirator. It is now clear that the medulla oblongata can in some cases be perfused in 'respirator' brain, and such patients need not be on a respirator to develop the classical pathological changes, as they can breathe spontaneously. Thus, the term 'non-perfused brain' is more descriptive and pathophysiologically accurate. Non-perfused brain is extreme, irreversible GCI in which blood flow to the brain never returns and function cannot ever be restored. It is the pathological correlate of clinical brain death.

The non-perfused brain shows dusky brown discoloration of the cerebral cortex, blurring of the junction of the cortex and the white matter, and general friability of the brain upon gross examination (Figure 2.105a). With longer survival, there is increasing brown discolouration of the entire brain. The cerebral tissue is difficult to fix in formalin because of protein alterations, resulting in symmetrical patches of pink, unfixed white matter in the regions of the centrum semiovale furthest away from the ventricles.

The histology is better preserved than might be expected from the gross appearance. Although acidophilic neurons may be seen, there is usually poor staining of the tissue, including neurons (Figure 2.105b). Stagnant red blood corpuscles in the vasculature appear pale because of lysis of haemoglobin. Because of the lack of perfusion, there is no tissue reaction except possibly at the border of perfused and non-perfused tissue, usually at the level of the high cervical spinal cord or medulla oblongata. The pathological changes of non-perfused brain take about 12 hours to develop[114] but become more obvious by 24 hours, an important point if the neuropathologist is called upon to confirm brain death.

The spinal cord is usually well preserved pathologically, and clinically it may show autonomous neurophysiological activity (Lazarus sign).[421] Fragments of autolysed cerebellum[747] can be found around the cord. The dural venous sinuses are thrombosed.[775] The pituitary gland invariably undergoes infarction (Figure 2.106) because of the vulnerability of its portal venous blood supply from the hypothalamus to compression by increased ICP.

Even if ventilator support is continued, non-perfusion of the brain is generally followed after a variable time period by somatic death. This is due to the cardiac consequences of increased ICP[937] and also poorly understood brain–body relationships. Deterioration in haemodynamic stability[855]

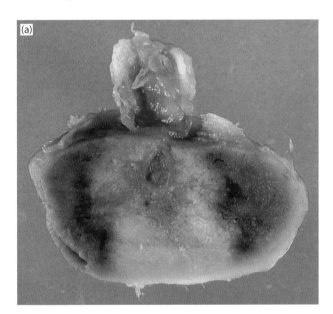

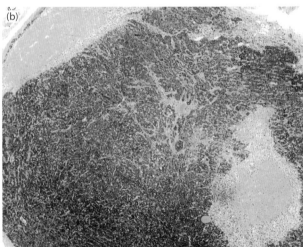

2.106 Pituitary gland in non-perfused brain. (a) Infarcted pituitary in a 55-year-old man who survived 2 weeks after a motor-vehicle accident and developed multisystem failure. The brain showed the typical dusky brown discolouration of non-perfused brain. The green staining of the pituitary tissue was caused by jaundice related to liver failure. **(b)** Pituitary infarction in a 75-year-old woman who had a three-vessel coronary bypass 19 days before death, with clamp time 79 minutes and pump time 98 minutes. The brain showed the dusky brown discolouration of a non-perfused brain, and the pituitary is seen here to include foci of infarction.

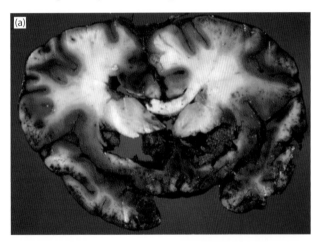

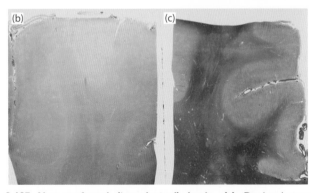

2.105 Non-perfused ('respirator') brain. (a) Dusky brown discolouration and friability are the gross features. **(b,c)** Microscopically, comparison with a normal brain **(c)** shows that the tissue of the non-perfused brain **(b)** stains poorly, with impaired demarcation between the grey and white matter. Luxol fast blue and PAS.

and cardiac arrest can occur after hours to days, but cases with survival of several months are reported.[775] In such cases, non-perfused brains show a superficial leukocytic tissue reaction, with white blood cells infiltrating the brain parenchyma from the perfused skull and meninges.[775,914]

Hypoglycaemic Brain Damage

Introduction

As far back as the early years following the discovery of insulin, poorly controlled administration of insulin for diabetes and the use of insulin shock 'therapy' for treating schizophrenia[894] provided evidence of the permanent brain damage in the form of neuronal loss and gliosis that resulted from hypoglycaemia with long-term survival.[587] The neuropathological findings contributed to the eventual abandonment of insulin therapy.[661] Hypoglycaemic brain damage is still seen today in several settings. One is accidental or intentional (i.e. homicidal or suicidal) insulin overdose.[507] Another is medication error, with insulin or another hypoglycaemic agent being inadvertently substituted for another drug or given in an erroneously high dose.[507] The commonest setting, however, is in the treatment of diabetes mellitus.[177] It has been estimated that 5–8 per cent of children with type 1 diabetes mellitus suffer bouts of severe hypoglycaemia (i.e. accompanied by stupor or coma).[107] Clinical trials have shown benefit of tight blood sugar control in reducing complications of diabetes mellitus but possibly an increase in the incidence of hypoglycaemic brain damage.[243,245,414]

Early work demonstrated the presence or absence of hypoglycaemic brain damage and its distribution. Over time, however, attention turned to the mechanism. Insulin itself was first suspected to be neurotoxic. Morphological studies of cellular lesions focused on early microvacuolation, shown ultrastructurally to be due to swollen mitochondria.[13,15,16,509] The inappropriate label 'ischaemic cell change' was used because of the identical eosinophilic appearance of the necrotic neurons to those in ischaemia.[135,136]

The cause of the hypoglycaemia is immaterial to the resulting brain damage. Oral hypoglycaemic agents such as sulphonylureas release endogenous insulin and produce hypoglycaemia, which has identical pathogenetic effects on the brain to those of hypoglycaemia due to exogenous insulin administration.[507,1053] The absolute level of blood sugar is also not important once it drops below about 1.5 mM. The presence or absence of cerebral EEG isoelectricity (flat EEG) predicts the presence or absence of brain damage. The flat EEG is preceded by the appearance of large, high-amplitude delta waves in the 1–4 Hz frequency range; the clinical counterpart of this is stupor.[674] Without EEG silence (equivalent to coma), brain damage in the form of neuronal necrosis does not occur.

The brain utilizes principally glucose as a fuel, with few exceptions. The infant brain is capable of oxidizing lactate,[412,417,1002] bestowing some protection against hypoglycaemia. This capability disappears with the progressive inability of lactate to enter the brain with maturation,[1002] although some lactate can probably be burned during hypoglycaemia, even by the adult brain.[330,725,917] Starvation is accompanied by brain utilization of ketone bodies, which

circulate at elevated levels.[404] The severely hypoglycaemic brain also burns ketones.[330,337] Lastly, hypoglycaemia, if profound enough to cause a flat EEG, can cause the breakdown of endogenous brain substrates,[14,337] which can be used for energy derived mainly from glycolysis.

Hypoglycaemia was long thought to injure the brain simply by starving the neurons of glucose. Such a theory suggests that depriving neurons of oxygen or glucose should produce identical patterns of damage. Yet in GCI, large neurons are, in general, most vulnerable; however, hypoglycaemia does not preferentially affect large neurons in the brain. As in ischaemic injury, in hypoglycaemia the pathogenesis of selective neuronal necrosis involves the release of excitatory and toxic (excitotoxic) compounds in the brain. It was shown that hypoglycaemic neuronal death, like ischaemic damage, could be blocked by NMDA receptor antagonists, which reduced neuronal necrosis despite exposure to identical levels of hypoglycaemia in untreated animals.[1089] These findings were accompanied by morphological demonstration of selective abnormalities in the dendritic tree of hippocampal neurons, where excitatory receptors are located, sparing axons of passage.[74] Thus, although internal cellular mechanisms may initiate hypoglycaemic neuronal necrosis, at least part of the pathogenesis of hypoglycaemic neuronal death involves ligand binding to external neuronal receptors. Indeed, the pathogenetic chain leading from a low blood sugar to neuronal death is longer and more indirect than might be predicted.

Biochemical Basis of Hypoglycaemic Brain Damage

Severe hypoglycaemia is accompanied by complex biochemical alterations.[67,940] Energy failure is reflected by a reduction in the energy charge to about 25–35 per cent of normal. Protein and lipid catabolism occurs. The flux through the glycolytic pathway is of course reduced, with a consequent reduction in tissue levels of lactate and pyruvate. Proton production in the Krebs cycle is also reduced. The shortage of protons and organic acids together leads to a tissue alkalosis,[782] in contrast to ischaemia in which these events do not take place.

Declining decarboxylation of pyruvate to yield CO_2 and acetate results in a shortage of acetyl-CoA, with which oxaloacetate condenses, and leads to a buildup of oxaloacetate. Oxaloacetate is the α-keto-acid in a transamination reaction with glutamate, yielding aspartic acid and α-ketoglutarate: oxaloacetate + glutamate = aspartate + α-ketoglutarate. The excess oxaloacetate resulting from reduced glycolytic flux drives this reaction to the right by the chemical law of mass action (Le Chatelier's principle). The result is a depletion of glutamate and a marked increase in aspartate.[12,13,99] Inhibiting glycolytic flux, even without peripheral hypoglycaemia, causes an identical increase in aspartate.[900] As a result of general failure of cells to maintain chemical compartmentation in severe hypoglycaemia, both aspartate and, to a lesser degree, glutamate increase in the extracellular space of the brain.[899]

Pharmacology of Hypoglycaemic Brain Damage

The aspartate and some glutamate released into the extracellular space in severe hypoglycaemia is present in the interstitial space of the brain and in the CSF. The ratio of aspartate to glutamate is the reverse of that in ischaemia, glutamate predominating in ischaemia, aspartate in hypoglycaemia. Because aspartate is more selective than glutamate for NMDA receptors, pharmacological blockade of NMDA receptors may be more effective in hypoglycaemia than in ischaemia. This has been the case experimentally.[773,1085,1089] Although AMPA antagonists have not been studied in hypoglycaemic brain damage, voltage-sensitive calcium channel blockade was ineffective[751] or even detrimental,[70] again in contrast with ischaemia, in which a protective effect of calcium antagonists is generally seen.[68] Another difference is that the selective destruction of GABAergic neurons seen in hypoxia or ischaemia (see Hypoxia and Brain Synapses) does not take place in hypoglycaemia.[859] Hypothermia has only a slight effect on hypoglycaemic brain damage,[18] compared with its powerful protective effect in ischaemia.

Morphology of Hypoglycaemic Brain Damage

Hypoglycaemic brain damage, like ischaemic brain damage, affects the hippocampus and cerebral cortex, causing selective neuronal necrosis. In human brain, it is often impossible to distinguish hypoglycaemic from ischaemic brain damage.[729,846] These two insults have long been considered pathologically identical.[136,218] However, controlled animal experiments delivering insults of hypoglycaemia and ischaemia to the brain under close physiological monitoring demonstrated morphological differences between hypoglycaemic and ischaemic brain damage.[71]

Lactate and pyruvate production is reduced in hypoglycaemia due to reduced glycolytic flux. In ischaemia anaerobic glycolysis is stimulated. Proton production by the Krebs cycle is reduced in hypoglycaemia but increased in ischaemia. This, in conjunction with ammonia production from deamination of amino acids in hypoglycaemia, contributes to a tissue alkalosis. The pH change is opposite

TABLE 2.9 Types and causes of cerebral amyloid angiopathy

Disease		Age of onset (year)	Gene / mutation	Precursor protein/ mutation	Function of precursor protein	Amyloid protein/ peptide
Sporadic Aβ-CAA						
	AD-associated	>60	*APP*	APP	Unknown	Aβ
	Non-AD associated	>60	'	'	'	'
	Down syndrome	>20	'	'	'	'
Hereditary Aβ-CAA						
	HCHWA-Dutch type	50	*APP*	p.Glu693Gln	Unknown	Aβ
	HCHWA-Flemish type	45	'	p.Ala692Gly	'	'
	HCHWA-Italian type	50	'	p.Glu693Lys	'	'
	HCHWA-Iowa type	50-66	'	p.Asp694Asn	'	'
	HCHWA-Arctic type	60	'	p.Glu693Gly; p.Glu693Δ	'	'
	HCHWA-Piedmont type	50-70	'	p.Leu705Val	'	'
HCHWA-Icelandic type (I)		20-30	*CYST C*	Cystatin C	Protease inhibitor	ACys
FAF			*GEL*	Gelsolin	Actin binding	AGel
FBD		45-50	*BRI2*	ABri precursor protein	Unknown	ABri
FDD		30	*BRI2*	ADan precursor protein	Unknown	ADan
FAP/MVA		35	*TTR*	Transthyretin	Transport protein	ATTR
PrP-CAA		38	*PRNP*	Prion protein	Infectious agent	APrP

AD, Alzheimer disease; APP, Amyloid precursor protein; CAA, cerebral amyloid angiopathy; FAF, familial amyloidosis, Finnish type; FAP/MVA, familial amyloidotic polyneuropathy/meningovascular amyloidosis; FBD, familial British dementia; FDD, familial Danish dementia; HCHWA, hereditary cerebral haemorrhage with amyloid angiopathy; PrP-CAA, prion protein cerebral amyloid angiopathy.

Data from Revesz *et al*.[839]

in ischaemia, where an acidosis develops. This lack of acidosis in hypoglycaemia likely accounts for the impossibility of producing hypoglycaemic pan-necrosis. Pan-necrosis is commonly seen in more severe tissue ischaemia, in which acidosis is probably a critical part of the pathogenesis of brain infarction. The lack of infarction in hypoglycaemic brain damage thus probably has a simple biochemical explanation. Even in the most severe form of pure hypoglycaemic brain damage, infarction does not occur; damage is limited to selective neuronal necrosis.

Neuronal necrosis in hypoglycaemia usually involves the cerebral cortex, hippocampus[75,507] and caudate nucleus.[507] The spinal cord[1007] is rarely involved but the cerebellum[635] is never affected. The granule cells of the dentate gyrus are usually conspicuously normal in ischaemic brain damage, being the last neuronal type within the hippocampus to be affected by ischaemia. In hypoglycaemia, possibly because of the extracellular overflow of large quantities of aspartate into the CSF,[899] necrosis of dentate granule cells, which contain excitatory receptors on the superficial molecular layer close to the ventricular fluid, occurs in experimental animals[74,1077] and man.[75,635] (Figure 2.107).

Another potentially distinguishing feature is the presence of Purkinje cell necrosis in ischaemia but not in hypoglycaemia. This may relate to the glucose transporter of the cerebellum, which is more efficient than elsewhere in the brain.[574] Reversible changes, such as mitochondrial swelling seen by electron microscopy in experimental material, do occur in the cerebellum in severe hypoglycaemia.[11]

The distribution and degree of neuronal necrosis in the cerebral cortex also tend to differ between hypoglycaemia and ischaemia. A superficial distribution of neuronal necrosis has been described in hypoglycaemic brain damage in humans[587] and animals.[294,362] This contrasts with an intracortical distribution of neuronal necrosis to the middle cortical laminae in global ischaemic insults. Hypoglycaemic brain damage also gives a paucity of intracortical necrosis, with a widespread even distribution of necrotic neurons over the hemisphere,[75,507] accounting for the high CBF in persistent coma.[17] The hemispheric distribution of neuronal necrosis does not show the predilection for arterial boundary zones seen in ischaemic brain damage (see Table 2.3).

Ultrastructurally, axon-sparing lesions imply that excitatory compounds are present in the extracellular space, binding selectively to dendrites and causing selective dendritic swelling due to receptor activation, followed by ion and water fluxes across the dendritic membrane. Dendritic swelling is a salient feature of hypoglycaemic neuronal death (Figure 2.15), unlike in ischaemia, in which glutamate[106] rather than aspartate[899] excitatory action predominates.

Neonatal Hypoglycaemia and Cerebral Palsy

Low glucose levels in the neonatal period are physiological and there is no good evidence for the concept of hypoglycaemic cerebral palsy. The few articles alleging neuronal death in neonatal hypoglycaemia have other causes discernible by history alone, most frequently ischaemia.[43] In a medicolegal case (personal communication with Ronald N Auer), a blood glucose level of almost zero (1 mg/dL)

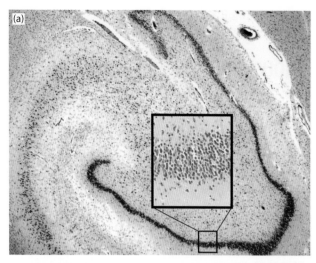

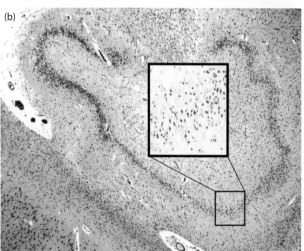

2.107 Hypoglycaemic brain damage. Asymmetrical damage is common in hypoglycaemic brain damage. This case shows a normal thick band of nuclei in the dentate gyrus of the right hippocampus **(a)** but near-total loss of the dentate granule cell band on the left side **(b)**. Such loss of dentate granule cells in ischaemia would be accompanied by loss of selective vulnerability and total necrosis of all neuronal elements, but here the dentate and CA1 are first affected.

was found in an awake and normally crying infant, alleged at that time to have been undergoing hypoglycaemic brain damage.

The likely cause of abnormal neurological function or structural brain abnormalities that are associated with neonatal hypoglycaemia is related to the overt or covert gestational diabetes in the mothers. The fetus produces insulin in response to maternal hyperglycaemia, as glucose crosses the placenta. The reactive hyperinsulinaemia often, if prolonged, leads to proliferation of the β-cells of the pancreas and other cells producing insulin in ectopic locations. The high insulin is secreted in a physiological attempt to counteract the fetal hyperglycaemia but causes macrosomia, as insulin is the growth hormone of the fetus (after birth, growth hormone, from the pituitary, takes over from insulin in determining body size and morphology). The fetal high insulin levels affect the growth and development of the fetal brain to result in abnormalities, often even macroscopic, with high morbidity.[240]

VENOUS THROMBOSIS AND INFARCTION

Clinical Picture and Incidence

Cerebral venous thrombosis (CVT) was long considered a rare, usually fatal disease, with a characteristic clinical picture of headache, seizures, focal signs, increased intracranial pressure, papilloedema and coma. The diagnosis was often established only at autopsy. Use of angiographic methods has changed our understanding of CVT.[563] Three-dimensional MR flow imaging[33] has shown that CVT is more common and often less serious than previously thought. The symptoms may be non-specific and mild. Headache and a clinical picture of benign intracranial hypertension should prompt appropriate imaging investigations (Figure 2.108).[799] The incidence of CVT is not known.[39] The risk is considered to be approximately equal for both genders. CVT can occur in all age groups, from neonates to the elderly. Among women there seems to be a peak in younger age groups, which probably reflects the association of CVT with pregnancy and the use of oral contraceptives.[39,168,259] The detection of milder forms of CVT and recognition that most patients, whose disease would not previously have been recognized, recover after recanalization of the thrombosed blood vessel, has contributed to the decrease in mortality to 5.5–13 per cent.[39,168,1001] Neuropathological experience is largely confined to infrequent fatal cases.

Aetiology of Venous Thromboses

CVT can be caused by a multitude of different conditions. In the past, infections were by far the most common cause[564] but in a 1992 series of 110 cases only 8.2 per cent had an infective aetiology.[39] The decrease is attributable

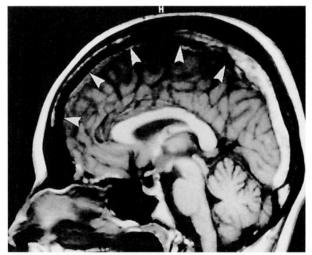

2.108 Superior sagittal sinus thrombosis. T1-weighted gadolinium-enhanced magnetic resonance image of a patient with thrombosis of the anterior part of the sagittal sinus (arrowheads).

Scan courtesy of O Salonen, Helsinki University Central Hospital, Helsinki, Finland.

to effective antimicrobial and anticoagulant therapy and improved diagnosis of non-infective CVT. The most common site of septic CVT is the cavernous sinus and the most common infective agent *Staphylococcus aureus*, spreading from an infection in the middle third of the face, sphenoid or ethmoid air sinus, or a dental abscess. Otitis media and mastoiditis may induce septic thrombosis in the lateral sinus, and infections of the scalp may extend via the diploic and emissary veins through the skull bone to the sagittal sinus. In immunosuppressed and chronically debilitated patients, various opportunistic microorganisms, including fungi and cytomegalovirus, can cause CVT.[254,675]

The altered hormonal status in young women during pregnancy, the puerperium or associated with the intake of oral contraceptives is an important risk factor for CVT.[39,168,259] Similarly, an altered state of coagulability associated with any surgery or condition affecting the patient's general health, such as malignancy and malnutrition, predisposes to CVT, as do systemic connective tissue and inflammatory diseases.[32] Behçet's disease was responsible for 10 in a series of 40 cases from Saudi Arabia.[229] Thrombosis may be promoted simply by stagnation of blood flow, e.g. in congenital or congestive heart disease or dehydration. Certain haematological disorders, including thrombocythaemia and sickle cell disease, increase the risk of CVT. Disorders leading to thrombophilia, e.g. Factor V Leiden; deficiencies of antithrombin III, protein C, protein S or plasminogen;[224,902,918,989] or the presence of acquired factors inducing hypercoagulability, e.g. antiphospholipid antibodies,[595,815] are rare causes but should be considered, particularly when no other cause is obvious.

Pathogenesis and Pathology

The cerebral venous network includes many well-developed collaterals (see earlier). Thrombosis has to be extensive, i.e. to involve a sinus or a major part of such, before venous return and capillary blood flow are sufficiently restricted to cause ischaemic damage. Vasogenic oedema develops and is aggravated by the engorgement of arterial blood vessels proximal to the occlusion. The combination of ischaemic injury and arterial engorgement results in haemorrhagic infarction. Thrombolytic therapy in CVT does not carry the same risk of expansion of haemorrhage as in arterial thrombosis. This is because recanalization leads to lowered intravascular pressure in the region of haemorrhagic infarction.

The most common sites of CVT are the superior sagittal sinus (72 per cent; Figures 2.108 and 2.109), lateral sinuses (70 per cent combined) and straight sinus (13 per cent), but thrombosis commonly extends to several sinuses or veins.[39] The anatomical variation of the left lateral sinus can cause diagnostic confusion; thrombosis of the sinus can cause infarction of the ipsilateral basal ganglia but CVT may be misdiagnosed if, as sometimes happens, the sinus lacks connection with the confluence of sinuses and its transverse portion is hypoplastic. Localized thrombosis of cerebral veins, especially cortical ones, rarely results in tissue damage. However, CVT of the deep internal veins and the great vein of Galen is likely to cause severe damage to the basal ganglia and brain stem.[824]

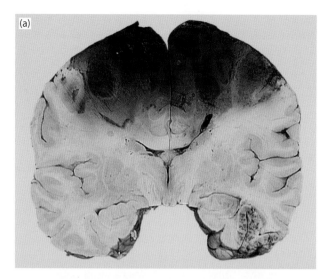

2.109 (a) Thrombosis of the superior sagittal sinus has caused haemorrhagic necrosis and intraparenchymal haemorrhage in the parasagittal brain parenchyma. **(b)** The occluding thrombus is seen within the sinus.

The most familiar pathological findings in CVT are those caused by superior sagittal sinus thrombosis: haemorrhagic infarction, parasagittal haemorrhages extending to the white matter, and marked oedema (Figure 2.109). The appearances on microscopy are largely the same as in any haemorrhagic infarct but there may be more profuse leukocytic invasion, because the patent arteries allow ready inflow of reactive inflammatory cells.

SMALL VESSEL DISEASES OF THE BRAIN

Sporadic and Familial Cerebral Amyloid Angiopathy

Cerebral amyloid angiopathy (CAA) comprises a group of protein-misfolding disorders characterized by the extracellular deposition of fibrillar proteins with amyloid properties in the walls of blood vessels of the brain and meninges.[837,838] It is an important cause of brain hemorrhage in the elderly.[1052] CAA may also lead to circulatory disturbances and cognitive impairment. To date, over 25 unrelated proteins are known to generate different types of amyloidosis.[876] Despite their biochemical heterogeneity, all amyloid fibrils share certain unique physicochemical properties. Amyloid proteins

acquire a β-sheet secondary structure and, through an intermediate stage of protofibrils, form highly insoluble cross-β-sheet quaternary fibrils. These bind dyes such as Congo red (in which case, they show apple-green birefringence under polarized light) and thioflavin S and T (which fluoresce under ultraviolet light) (Figure 2.110). Because of the characteristic binding of Congo red to amyloid-laden vessels, the condition was originally described as congophilic angiopathy. The term CAA is more appropriate and widely used, although the angiopathy can involve blood vessels in the cerebellum and brain stem as well as the cerebrum. Among the different amyloidogenic proteins, seven are associated with CAA (Table 2.9). In sporadic CAA, the amyloid fibrils are composed of Aβ, whereas in the rare hereditary CAAs all seven proteins are implicated.

Amyloid β Angiopathies

Sporadic Amyloid Angiopathies

General Aspects

The most common types of CAA are those associated with deposition of Aβ,[65,623] a cleavage product of amyloid-β precursor protein (APP), encoded on chromosome 21. Deposition of Aβ in the walls of cerebral blood vessels is seen in sporadic CAA, both associated with and independent of sporadic Alzheimer disease (AD), and in Down

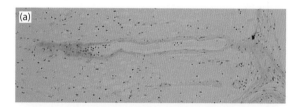

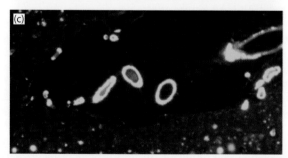

2.110 Cerebral amyloid angiopathy (CAA). A penetrating artery from a non-demented patient who died of an intracerebral haematoma. The arterial walls are thickened by amorphous substance, which is coloured red with alkaline Congo stain **(a)** and gives typical apple-green birefringence **(b)**. **(c)** Leptomeningeal blood vessels with abundant amyloid in their walls. Amyloid plaques are present in the adjacent cerebral cortex. Thioflavin S fluorescence.

TABLE 2.10 Molecular genetics of small vessel diseases of the brain

Disorder	Onset Age (y)†	Gene	Chromosome locus	Protein	Protein type/ function	Predicted dysfunction(s)
CADASIL (most common); autosomal dominant)	20–60	NOTCH3	19p13.2-p13.1	NOTCH3	Transmembrane cell signalling receptor	Aberrant cell-cell signalling, activates unfolded protein response and impaired gene transcription (NICD)
CARASIL (Maeda syndrome); autosomal recessive	20–30	HTRA1	10q25.3-q26.2	HTRA1	Serine protease	Promotes serine-protease-mediated cell death, suppresses TGFβ expression
RVCL disorders: HERNS (Chinese descent); CRV (cerebroretinal vasculopathy); HVR (hereditary vascular retinopathy)	30–50	TREX1	3p21.3-p21.2	TREX1	3′→5′-prime exonuclease DNase III	Disruption of cell death mechanisms, impaired DNA degradation and repair
COL4A1 and COL42-related disorders (stroke syndromes); autosomal dominant	14–49	COL4A1; COL4A2	13q34	COL4A1 and A2	Collagen IV, α1 and α2 chains, constituent of BM	Weakening of vascular basement membranes
Hereditary multi-infarct dementia of the Swedish type	28–38	Not known	No linkage to NOTCH3	-	-	-
PADMAL (pontine autosomal dominant microangiopathy and leukoencephalopathy)/subcorticalangiopathicencephalopathy	12–50	Not known	No linkage to NOTCH3	-	-	-
Hereditary small vessel disease of the brain (SVDB)*	36–52	Not known	Linkage to Chromosome 20	-	-	-

*Several other disorders prominently characterized by leukoencephalopathy and cognitive impairment have been described in isolated families. †Age of onset signifies when first cerebrovascular event or gait disturbance due to spasticity was recorded.
BM, basement membrane; CADASIL, cerebral autosomal dominant arteriopathy with subcortical infarcts and leukoencephalopathy; CARASIL, cerebral autosomal recessive arteriopathy with subcortical infarcts and leukoencephalopathy; RVCL, autosomal dominant retinal vasculopathy with cerebral leukodystrophy; NICD, NOTCH intracellular domain.

syndrome. Cerebrovascular Aβ accumulation has also been reported in patients with dementia pugilistica[1006] and in both cerebral and spinal vascular malformations.[399]

Epidemiology

Sporadic Aβ-CAA rarely affects people under 60 years of age. In those over 60, the prevalence is a little over 30 per cent,[279] increasing with age from just under 15 per cent in the seventh decade to 45 per cent after 80 years.[622,987] The high prevalence of CAA in AD indicates that these two disorders are interrelated: in AD, the proportion of patients with CAA increases with advancing age, reaching 80–90 per cent.[185,279,812] Furthermore, genetic polymorphisms or mutations that increase the risk of AD are also risk factors for CAA.[733] APOE ε4 allele in AD favours vascular over parenchymal deposition of Aβ in a dose-dependent manner.[185,827] The odds ratio for CAA increased approximately 3-fold for ε4 heterozygotes and 14–17 fold for ε4 homozygotes,[366,812] relative to individuals without the ε4 allele. There is also more severe CAA in ε4 homozygotes but not ε4 heterozygotes, than in people lacking an ε4 allele.[1063] The extent to which ε4 confers an increased risk of Aβ-CAA in non-AD patients is unclear; Love et al.[622] found no

association between ε4 and the prevalence of Aβ-CAA in the absence of AD or cerebral haemorrhage. Although protective against AD (see Chapter 16), APOE ε2 is a risk factor for CAA.[624,996] Sporadic Aβ-CAA is responsible for about 5–12 per cent of primary non-traumatic PBHs and is the most common cause of lobar haemorrhage.[345] However, in a large retrospective neuropathological study of brains from elderly individuals (mean age 78 years), a lower prevalence of CAA in those with than without PBH did not support the concept that CAA is the most important risk factor for PBH in the aged, and suggested that other risk factors including hypertension play a larger role even in this group.[480] The association of APOE genotype with Aβ-CAA-related haemorrhage (CAAH) has also been somewhat controversial. In an unselected population, APOE genotype did not associate with lobar PBH.[924] A meta-analysis found only marginally significantly increased likelihood of PBH among people with APOE ε2 and a non-significant trend toward increased PBH among those with ε4.[973]

Clinical Diagnosis

Sporadic CAA is predominantly a silent disease without overt clinical symptoms. Despite the successes in the development of positron emission tomography (PET) methods to detect CNS amyloid with tracers such as Pittsburgh compound B (PiB) there are no established methods to detect Aβ-CAA.[367,489] However, PiB binding was moderately increased in most patients with probable CAA-related intracerebral hemorrhage.[632] Non-AD associated CAA manifests most often when it becomes severe enough to cause rupture of a diseased small artery, with resulting PBH. In the *intra vitam* search for the cause of PBH, gradient echo (T2*-weighted) MRI has become an important non-invasive tool, because it detects micro-haemorrhages[1051] and the presence of these is used in the Boston criteria for estimating the clinical likelihood that PBH has resulted from CAA.[364,544] Micro-haemorrhages or microbleeds detected by gradient echo MR imaging occur preferentially in regions of high amyloid load as detected by PiB PET.[249] *Definite* diagnosis necessitates pathological verification. *Probable* CAAH can be diagnosed if two or more acute or chronic lobar macro-haemorrhages or micro-haemorrhages are identified without any other definite cause of intracerebral haemorrhage, such as excessive warfarin treatment, antecedent head injury, stroke, neoplasm, vascular malformation, vasculitis or blood dyscrasia/coagulopathy. CAAH is designated as *possible* when a single lobar haemorrhage is encountered in a patient older than 55 years with no other obvious cause of cerebral haemorrhage. Clinicopathological correlation in a series of 39 cases of PBH, confirmed CAAH in all 13 patients fulfilling 'probable' diagnostic criteria but in only 16 (62 per cent) of those who had been diagnosed as having 'possible' CAAH.[544] Rarely, Aβ-CAA may present as a mass lesion.[132]

Severe Aβ-CAA is an occasional cause of rapidly progressive dementia, associated with ischaemic damage to the white matter, and petechial cortical haemorrhages or infarcts (see also Aβ-related angiitis).[361,612]

Pathology

The appearance of brains with CAA depends largely on the consequences of the vascular disease. Major CAAHs tend to occur in the frontal or frontoparietal regions and are usually of lobar type. Cerebellar haemorrhages occur but less often.[1048] In addition to lobar haemorrhages, CAA may cause cortical petechial haemorrhages or small infarcts, and focal or diffuse white matter ischaemic changes. There may be accompanying AD pathology.[365,618]

The CAA preferentially affects small arteries and arterioles (Figure 2.110). Most often, CAA has a patchy distribution and, in contrast to the distribution of CAAHs, tends to be most severe in the occipital and parietal meninges and cortex.[1049] The hippocampus is usually spared and the white matter and basal ganglia are involved only rarely. The arterial walls are thickened by the deposition of amyloid. This appears amorphous, intensely eosinophilic, periodic acid-Schiff (PAS)-positive, Congo red-positive and birefringent, and fluorescent under ultraviolet light in sections stained with thioflavin S or T. In mildly affected blood vessels the amyloid is deposited in a reticular pattern in the basement membrane surrounding smooth muscle cells in the tunica media.[1082] More severe disease is associated with confluent accumulation of amyloid in the tunica media accompanied by destruction of the smooth muscle cells, and deposition of Aβ amyloid in the adventitia as well. Concentric

separation of the amyloid-laden media and adventitia often causes severely affected vessels to have a 'lumen within a lumen' appearance. Aβ can also accumulate in the walls of capillaries, albeit less often, predominantly in association with APOE ε4[624,996] and mainly in the deeper laminae in the parahippocampal and occipital cortex. In the meninges, larger blood vessels, particularly veins, sometimes show accumulation of amyloid in the outer adventitia only. The presence of abnormally round, thick-walled blood vessels in the meninges or cortical parenchyma should always signal possibility of CAA. The presence of Aβ can be demonstrated immunocytochemically. The predominant form of Aβ in arterioles and arteries is Aβ40, that in capillaries mainly Aβ42.[765] Electron microscopy reveals extracellular deposits of randomly oriented, straight, unbranched filaments of indefinite length with a diameter of approximately 6–9 nm (Figure 2.111). CAA tends to be associated with accentuated perivascular neurofibrillary pathology, particularly if the CAA is severe. Systematic morphometric analysis of sections of frontal, temporal and parietal cortex from 51 AD brains revealed that phospho-tau labelling of neurites around Aβ-laden arteries and arterioles significantly exceeded that around non-Aβ-laden blood vessels, which was, in turn, greater than the immunolabelling of cortex away from blood vessels.[1093]

Aβ-CAA may also be associated with an inflammatory reaction, of which two main patterns have been distinguished. The first consists of the accumulation around Aβ-laden vessels of lymphocytes, macrophages and microglia, and occasionally multinucleated giant cells, but without actual vasculitis. Eng *et al.*[278] described the clinical and pathological findings in 7 such patients (from a consecutive series of 42 cases of CAA). The patients all presented with subacute cognitive decline or seizures, most had white matter abnormalities on neuroimaging and 5 were homozygous for APOE ε4. Most improved clinically and radiologically after immunosuppressive treatment. Other patients may develop a granulomatous vasculitis similar to that in primary angiitis of the

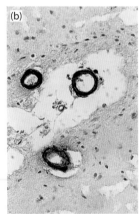

2.111 CAA. (a) Electron micrograph of approximately 10-nm-thick amyloid filaments in the wall of a cerebral artery from a patient with Aβ-cerebral amyloid angiopathy (Aβ-CAA). **(b)** The walls of intracerebral arteries from a patient with familial British dementia are strongly immunopositive with an antibody to amyloid *BRI*.

(b) Courtesy of Dr T Revesz, Institute of Neurology, London, UK.

CNS (PACNS).[41] The fairly stereotyped clinical, neuroradiological and neuropathological features in these patients, and other evidence of an immune response to Aβ, prompted Scolding *et al.* to define a clinical entity of Aβ-related angiitis (ABRA).[922] This is described in more detail earlier (see Inflammatory Diseases of the Cerebral Vasculature). ABRA manifests at a later age (60–70 years) than PACNS but may show a similar, if sometimes short-lived, beneficial response to immunosuppressive therapy.[922] Aβ is usually abundant in affected blood vessels but scant in the cerebral cortex, where activated microglia containing cytoplasmic Aβ may be present, occasionally in a plaque-like distribution. White matter oedema and rarefaction are common.

Pathogenesis

The pathogenesis of CAA is still an intensive topic of research. Three possible sources of Aβ accumulation in vessel walls may be considered. (i) *Systemic:* Aβ is derived from cells throughout the body and is carried in plasma and transported bidirectionally to and from the brain parenchyma by specific receptors in the vessel walls.[232] (ii) *Vascular:* Aβ is produced locally by vascular smooth muscle cells, endothelium and pericytes,[506] all of which express APP.[717] (iii) *Drainage:* Aβ accumulates because of impairment of its clearance from the CNS. Aβ formed by neurons within the CNS is cleared from the interstitial fluid by several processes, including enzymatic degradation within the brain parenchyma and the walls of blood vessels,[683,714] transcytosis across the BBB, with endothelial cell uptake mediated by specific receptors including lipoprotein receptor-related protein 1,[933] and drainage, together with other constituents of the interstitial fluid, along the perivascular extracellular matrix, to meningeal arteries and probably cervical lymph nodes.[1084] This drainage may be impaired in older people, as vascular disease reduces arterial pulsations (thought to supply the motive force for perivascular drainage), with resulting accumulation of Aβ in the arterial wall, further impeding vascular pulsatility.[1084] Precipitation within the perivascular extracellular matrix of amyloidogenic solutes such as Aβ in the course of their removal from the brain is probably the cause of most types of CAA.[1082] Reduced Aβ-degrading enzyme activity within the vessel wall may be a contributory factor: neprilysin activity was lowest in meningeal blood vessels from patients with most severe CAA, even after adjusting for smooth muscle content.[684] Raising or lowering neprilysin activity respectively decreased or increased the death of cultured human cerebrovascular smooth muscle cells on exposure to Aβ. Exclusively neuronal production of Aβ is sufficient to cause CAA.[419,420]

Affected blood vessels in Aβ-CAA may show segmental dilation or fibrinoid necrosis.[643] Serial sectioning and computer-assisted three-dimensional image analysis suggest that the following sequential steps lead to blood vessel rupture and haemorrhage:[639] (i) accumulation of amyloid in the arterial wall, (ii) death of smooth muscle cells, (iii) dilation (formation of micro-aneurysms) of the artery and (iv) breakdown of the BBB, (v) deposition of plasma proteins in the vessel wall (fibrinoid necrosis), and finally (vi) rupture and haemorrhage. Fibrinoid change was more marked than deposition of amyloid at sites of dilation and rupture.[639] Fibrinoid change was also significantly associated with possession of APOE

ε2,[663] which carries an increased risk of haemorrhage in people with Aβ-CAA. The death of vascular smooth muscle cells in CAA seems to depend at least partly on their uptake of Aβ and this, in turn, is influenced by *APOE* genotype.[886] In transgenic mice with the Swedish double mutation in APP, degradation of extracellular matrix proteins by MMP-9 is pivotal in the rupture of Aβ-laden arteries.[592]

Hereditary Amyloid β-Peptide Cerebral Amyloid Angiopathy

Aβ-CAA is familial in the syndromes of hereditary cerebral haemorrhage with amyloid angiopathy of the Dutch[631] (HCHWA-D) and Flemish[413] (HCHWA-F) types and in the different forms of familial Alzheimer disease (FAD). Mutations in APP that cause haemorrhages are mostly located within the Aβ domain and involve substitutions of amino acids 21–23 of Aβ, whereas those responsible for FAD are most often located in APP next to but outside of Aβ. In HCHWA-D the mutation is E693Q,[599] and in HCHWA-F it is A692G. Autosomal dominant HCHWA-D causes recurrent brain haemorrhage in middle-aged normotensive individuals.[126] The CAAH is fatal in about two-thirds of patients; survivors develop a vascular type of dementia that is independent of AD plaques and tangles.[718] CAA is also prominent in those with the Arctic E693G, Italian E693K and Iowa D694N APP mutations, considered to be primarily FADs. In the latter two, CAAH is a common feature, along with dementia. In those FADs associated with presenilin-1 mutations, CAA has been reported to be more common if the presenilin-1 mutation is located beyond codon 200.[645]

Other Cerebral Amyloid Angiopathies

Icelandic HCHWA

Hereditary cerebral haemorrhage with amyloid angiopathy of the Icelandic type (HCHWA-I, hereditary cystatin C amyloid angiopathy) is a rare autosomal dominant CAA associated with fatal brain haemorrhages in young and middle-aged normotensive adults.[5,484] In addition to cerebral and meningeal arteries, extracerebral tissues, including the skin, show deposition of ACys-amyloid.[102] The amyloidogenic protein is a 120-amino-acid cysteine proteinase inhibitor, cystatin C, with a glutamine-to-leucine substitution in codon 68 due to a CTG to CAG point mutation in exon 2.[340,598] The low extracellular concentration of mutated inhibitor of cysteine proteinases in the brain may contribute to the destruction of the amyloidotic blood vessels.[1000]

BRI2 Gene-Related Cerebral Amyloid Angiopathies

An autosomal dominant CAA with non-neuritic plaque and neurofibrillary tangle formation, clinically characterized by dementia, spastic tetraparesis and cerebellar ataxia with onset around the sixth decade was originally described by Worster–Drought *et al.*[665,803] The disease was called familial British dementia (FBD) when the defective gene *BRI2* was discovered.[1046] A mutation at another site on the same *BRI2* gene causes the disorder that was first described in a Danish family as heredopathia ophthalmo-oto-encephalica,[837] but later renamed familial Danish dementia (FDD). It is characterized by cataracts and ocular haemorrhages in the third decade, followed by hearing problems and, in the fourth to fifth decades, cerebellar ataxia and dementia.

Although CAA in both FBD (with deposition of ABri) and FDD (with deposition of ADan) is severe, with concentric splitting and occlusion of affected vessels (Figure 2.111b), haemorrhages are rare. CAA also affects small arteries and arterioles in white matter as well as in systemic organs. In FDD, Aβ is sometimes co-deposited with ADan in both vessels and parenchyma. In both disorders, amyloid plaques, hyperphosphorylated tau-positive neurofibrillary tangles and neurites are also present in the brain parenchyma.

FBD and FDD are both caused by a novel amyloidogenic mechanism.[876] The *BRI2* gene is located on chromosome 13 and encodes a putative type II single-spanning transmembrane precursor protein of 266 amino acids with a molecular weight of ~30 kDa. In both FBD and FDD, the precursor molecule BRI-PP is elongated by 11 amino acids. In FBD, a single base substitution in the stop codon of *BRI2* gene results in a larger, 277-residue precursor; release of its 34 C-terminal amino acids generates the highly insoluble ABri amyloid subunit with a relative molecular weight of 4 kDa. In FDD, a decamer duplication insertion before the normal stop codon 267 also results in a 277-residue precursor molecule with a 34-amino-acid amyloidogenic peptide, ADan.[1047]

Amyloid Transthyretin-Related Cerebral Amyloid Angiopathy

The *transthyretin* (*TTR*) gene is located on chromosome 18 and encodes a transporter protein for thyroid hormone and retinol-binding protein. Amyloid transthyretin (ATTR) derives from a partially folded monomer of mutated TTR that is prone to form amyloid fibrils. Currently more than 100 different mutations have been identified in TTR. In addition to systemic ATTR deposition, common consequences of which include peripheral neuropathy and cardiomyopathy, nine mutations are associated with oculoleptomeningeal amyloidosis with prominent meningeal CAA.[115,640] This is especially marked in patients with the Hungarian (D18G) and Ohio (V30G) mutations and produces a rusty brown colour to the pia mater.[334,791] These patients suffer from dementia, cerebellar ataxia, motor dysfunction, and decreased vision and hearing.[640]

Gelsolin-related Cerebral Amyloid Angiopathy

In gelsolin-related familial amyloidosis of the Finnish type,[390] the amyloid AGel is deposited systemically, particularly in the skin, peripheral nerves and cornea. In the CNS, deposition of AGel is widespread in spinal, cerebral, and meningeal blood vessels, and extensive extravascular deposits are present in the dura, spinal nerve roots and sensory ganglia.[542] Clinical manifestations include atrophic bulbar palsy, ataxia, minor cognitive impairment, facial palsy, peripheral neuropathy and corneal lattice dystrophy.

Gelsolin is an 80-kDa (cytoplasmic) to 83-kDa (plasma) protein involved in the gel–sol transformation of actin. It is encoded by the *GEL* gene on chromosome 9. The two mutations in AGel amyloidoses are both in codon 654, either G:A or G:T, resulting in substitutions D187N or D187Y. The cleaved product AGel is composed of amino acids 173–243.

Prion Protein-Related Cerebral Amyloid Angiopathy

CAA in patients with prion disease usually results from vascular deposition of Aβ. CAA due to prion protein (PrP) deposition is rare and associated with stop mutations in *PRNP*. In one family with Gerstmann–Sträussler–Scheinker syndrome, PrP amyloid (APrP) was extensively deposited in parenchymal and leptomeningeal vessels and in the surrounding neuropil.[339] A TAT to TAG point mutation in codon 145 (Y145STOP) resulted in an N- and C-terminally truncated amyloidogenic 7.5-kDA PrP fragment composed of 70 amino acids. Y163STOP, Y226STOP and Q227STOP *PRNP* mutations were also reported to cause APrP-CAA.[477,839]

Hereditary Small Vessel Diseases

Molecular genetic studies have identified several monogenic conditions that involve small vessels and predispose to ischaemic and haemorrhagic strokes and diffuse white matter disease (Table 2.10). Cerebral autosomal dominant arteriopathy with subcortical infarcts and leukoencephalopathy (CADASIL) remains the most common hereditary SVD. Other hereditary SVDs include cerebral autosomal recessive arteriopathy with subcortical infarcts and leukoencephalopathy (CARASIL), retinal vasculopathy with cerebral leukodystrophy (RVCL), and the collagen type IV, α1 (COL4A1)-related disorders.[246,1104] The hereditary SVDs, albeit with variations in phenotype, demonstrate convergent effects of microangiopathy on cerebral grey and white matter, leading to cognitive impairment.

CADASIL

CADASIL as we know it today was probably first described in 1955 in a family that was reported to have Binswanger's disease.[1035] Subsequently, several families with multiple

TABLE 2.11 Causes of non-traumatic parenchymal brain haemorrhage

Cause	Percent
Hypertension	50
Cerebral amyloid angiopathy	12
Anticoagulants	10
Tumours	8
Illicit and licit drugs	6
Arteriovenous malformations and aneurysms	5
Miscellaneous	9

Adapted from Feldman.[286]

strokes and dementia were reported under various designations. In 1993, the disease was linked to a locus in chromosome 19 in two French families and given the acronym CADASIL.[1013] The defective gene on chromosome 19 and its protein product were identified in 1996.[492] To date, there are likely more than 700 CADASIL families worldwide among all ethnicities, with an estimated prevalence of 4–5 per 100 000.[508,715,831]

Clinical Features

The cardinal clinical features of CADASIL are (i) migraine with aura, often exceptionally severe, sometimes to the extent of causing hemiparesis; (ii) recurrent ischaemic strokes; (iii) psychiatric symptoms; and (iv) cognitive decline, with eventual dementia.[181] CADASIL patients may have their first stroke before the age of 30 years but the peak is around 40–50 years. White matter changes are detectable on T2-weighted MRI well before clinical symptoms, temporopolar hyperintensities being the most characteristic finding (Figure 2.112). Decreased cerebral circulation can be demonstrated by PET[1027] and bolus-tracking MRI.[148,180] Multiple small infarcts, detectable on T1-weighted MRI, cause cognitive decline between 40 and 70 years, primarily in executive frontal lobe functions, followed by impairment of memory and other cognitive functions, leading to the development of subcortical dementia. About 80 per cent of CADASIL patients aged over 65 years are demented.[247] Only symptomatic therapy is available, and death usually ensues within 15–25 years of the first strokes.

Pathology

The characteristic neuropathological feature is non-arteriosclerotic, non-amyloid arteriopathy,[511,881] mainly affecting penetrating small and medium-sized arteries of the white matter but also present in leptomeningeal blood vessels. The symptoms are almost exclusively neurological although the arteriopathy is generalized and also affects, albeit less severely, blood vessels in systemic organs, including skin, skeletal muscle and peripheral nerve.[881] The vascular smooth muscle cells degenerate with accompanying accumulation of basophilic, PAS-positive granules in the tunica media. The vessel walls become markedly thickened and fibrotic (Figure 2.113), particularly in the white matter, leading to narrowing of the lumina, impaired circulation and multiple infarcts (Figure 2.114).[676] Electron microscopy shows destruction of vascular smooth muscle cells and deposition of pathognomonic clumps of granular osmiophilic material (GOM) between the degenerating vascular smooth muscle cells and in plasmalemmal indentations of these cells, particularly on the abluminal aspect (Figure 2.115). It is not exactly known how early GOM appears but it has been detected in skin biopsies of asymptomatic patients before the age of 20 years and was abundant in the cerebral arterioles of a 32-year-old patient.[676] Vessels in the cerebral white matter are most severely affected and where most infarcts occur. Cortical grey matter and the retina are relatively spared but infarcts in the basal ganglia are common (Figure 2.114) and infarcts also occur in the brain stem.

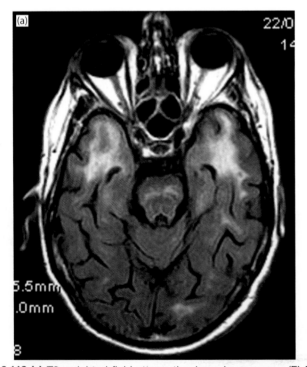 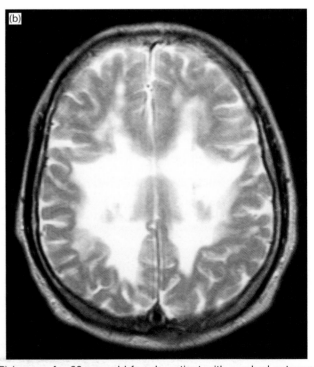

2.112 (a) T2-weighted fluid attenuation inversion recovery (FLAIR) image of a 62-year-old female patient with cerebral autosomal dominant arteriopathy with subcortical infarcts and leukoencephalopathy (CADASIL) who had several subcortical ischaemic events. The temporopolar hyperintensities are pathognomonic for CADASIL. **(b)** In a moderately demented homozygous CADASIL patient with a history of several strokes, the MRI shows extensive confluent hyperintensities.

Scan (b) courtesy of T Kurki, Turku University Hospital, Turku, Finland.

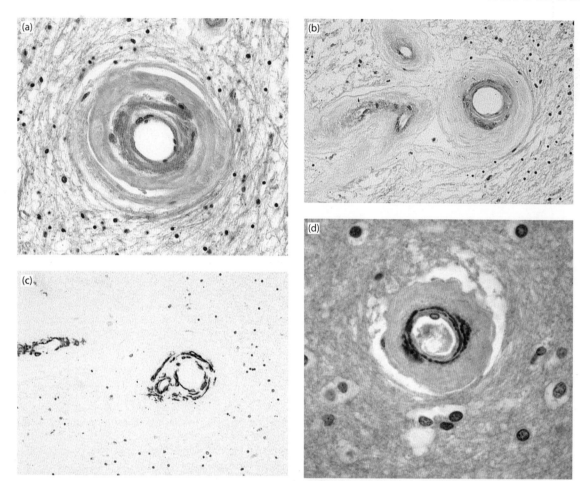

2.113 Cerebral autosomal dominant arteriopathy with subcortical infarcts and leukoencephalopathy (CADASIL). The media and the adventitia of a penetrating artery from the parietal white matter of the patient in Figure 2.112b are markedly thickened, with accumulation of basophilic **(a)**, periodic acid-Schiff (PAS)-positive **(b)**, granular non-amyloid material in the media, and fibrosis of the adventitia. Immunolabelling for smooth muscle actin **(c)** shows irregular degeneration of the smooth muscle cells. **(d)** Accumulation of material that is immunopositive for Notch-3 extracellular domain (N3ECD) in association with degenerating vascular smooth muscle cells.

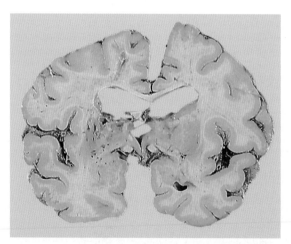

2.114 Cerebral autosomal dominant arteriopathy with subcortical infarcts and leukoencephalopathy (CADASIL). The brain of a 63-year-old female patient, who had presented with adult-onset migrainous headache with visual aura and who had her first stroke at the age of 52 years. Thereafter, strokes recurred several times per year and she became demented, with other stroke-related deficits. There are multiple subcortical infarcts, most obvious in the deep grey matter and left parasagittal frontal white matter.

In accord with other SVDs (see later), CADASIL is associated with alterations in retinal microvessel: arteriolar narrowing can be revealed by fundoscopy and fluorescence angiography.[181] Autopsy shows thickened vessel walls with fibrosis, accumulation of granular material, pericyte degeneration and loss of vascular smooth muscle cells in the central retinal artery and branches.[397]

Demonstration of GOM by electron microscopy in dermal biopsies offers a means of *intra vitam* diagnosis (Figure 2.115)[511,881] and was reported to be completely congruent with genetic screening.[1004] Immunohistochemical demonstration of Notch3 extracellular domains (N3ECD; Figure 2.115c) in skin arteries is also helpful in establishing diagnosis[495] provided clinical symptoms are consistent with the disease. These methods are practical when diagnosing new families with suspected disease, because genetic screening can be labour intensive and costly.

Genetics

The aberrant gene in CADASIL is *NOTCH3*, located at chromosome 19p13. In humans, *NOTCH3* spans 33 exons and encodes a single transmembrane receptor protein, NOTCH3, of 2321 amino acids (Figure 2.116). NOTCH3

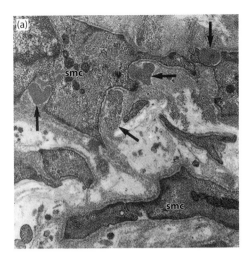
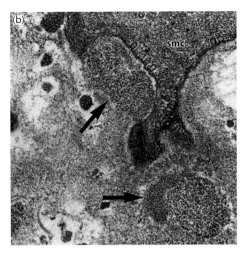
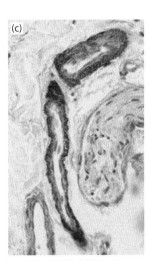

2.115 Cerebral autosomal dominant arteriopathy with subcortical infarcts and leukoencephalopathy (CADASIL). (a) Low-power electron micrograph of a small dermal artery in a skin biopsy. There are numerous deposits of granular osmiophilic material (GOM; arrows) in plasmalemmal indentations of smooth muscle cells (smc) some of which appear to be degenerating. The basal lamina of the cells is irregularly thickened. **(b)** Higher magnification of a biopsy from a 19-year-old patient demonstrates the typical granular appearance of GOM (arrows), and the pinocytotic vacuoles that are often seen in the adjacent smooth muscle cell cytoplasm (visible here beneath the upper GOM deposit). **(c)** Immunostaining with specific antibody demonstrates the accumulation of N3ECD in the walls of small arteries deep in the dermis.

(c) Reproduced with permission from Kalimo and Kalaria.[508] © 2005 Wiley-Blackwell.

has pivotal functions in stem cell renewal, proliferation, differentiation, and apoptosis during organogenesis, including vasculogenesis. *NOTCH* family genes are highly conserved within all species: orthologous genes have been identified in organisms from nematodes to humans.[57,583] In adult human tissues, NOTCH3 appears to be expressed predominantly in vascular smooth muscle cells.[494] Upon ligand (transmembrane Delta and Jagged molecules) binding, NOTCH33 is cleaved at sites S2 and S3, and the released intracellular domain enters the nucleus to regulate transcription (Figure 2.116). The ligands and the functional responses to NOTCH signalling in adult human tissues are not fully known.

More than 200 different mutations in *NOTCH3* segregate with CADASIL. The vast majority (~70 per cent) are missense point mutations. Almost all of the mutations result in either substitution of a wild-type cysteine by another amino acid or vice versa, in one of the 34 epidermal growth factor-like (EGF) repeats in N3ECD (Figure 2.116). In addition, small deletions have been described, which cause loss of one or three cysteine residues:[248,493] instead of the normal six cysteine residues, the mutant EGF repeats contain three, five or seven cysteine residues. Two *de novo* mutations as well as two non-cysteine mutations have been reported.[1104]

Pathogenesis

The pathogenesis of CADASIL remains elusive. The uneven number of cysteine residues affects the formation of disulphide bridges and changes the three-dimensional structure and NOTCH3 receptor.[493,494] The formation of abnormal disulphide bridges could affect receptor trafficking, processing, specificity for ligand binding and/or signal transduction. *In vitro* studies suggested that apart

from mutations in the ligand-binding site of the NOTCH3 receptor, mutations in *NOTCH3* do not appreciably impair signal transduction activity, NOTCH3 processing, or signalling to CBF1/RBP-Jκ activation (EGF repeats 10–11; Figure 2.116) Thus, loss of signalling by mutated NOTCH3 may not always be the direct cause of the disease. Over the years, the accumulation of misfolded, non-degradable N3ECDs that accumulate within GOM[1105] may affect the function of vascular smooth muscle cells and interfere with perivascular drainage.[173] Proteomic studies on cultured vascular smooth muscle cells from a CADASIL patient revealed differences in the expression of proteins involved in protein degradation and folding, indicating that mutant NOTCH3 causes endoplasmic reticulum stress and activates the unfolded protein response.[455,978] In HEK293 cell lines, mutant forms of NOTCH3 were prone to aggregation and retention in the endoplasmic reticulum; they reduced cell proliferation and increased sensitivity to proteasome inhibition.[978]

CARASIL (Maeda Syndrome)

Cerebral autosomal recessive arteriopathy with subcortical infarcts and leukoencephalopathy (CARASIL) was first described as Maeda syndrome.[55] The acronym CARASIL was applied later but neglects the skeletal pathology and could cause confusion with CADASIL. The predominantly normotensive patients, generally from families with consanguinity, have severe non-amyloid arteriopathy, leukoencephalopathy and lacunar infarcts together with characteristic spinal anomalies and alopecia. Thus far, few Japanese families with CARASIL have been described.[38,737] CARASIL has also been reported in China and exists in other countries.

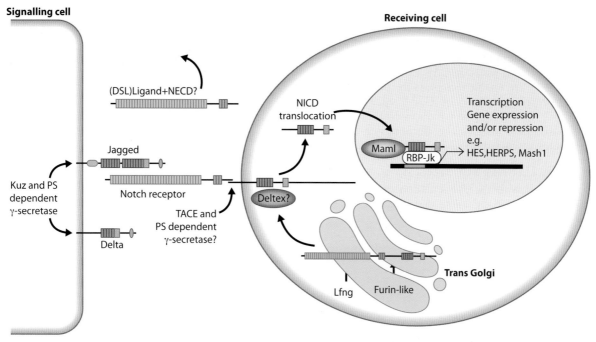

2.116 Schematic illustrates NOTCH receptor generation and signalling mechanism. The putative Notch signalling pathway involves a complex series of proteolytic cleavage events eventually mediating transcriptional activation of a target gene. NOTCH is glycosylated by O-fucosetransferase1 (O-FucT-1) and Fringe, in the endoplasmic reticulum (ER) and the trans-Golgi network, respectively, after translation. NOTCH first undergoes constitutive cleavage at a site designated S1, by a furin-like convertase in the trans-Golgi network, to form a functional heterodimeric receptor, which is presented on the cell surface. The leucine zipper domain in NOTCH extracellular domain (NECD) may regulate dimerization. NOTCH ligands such as Jagged and Delta are activated by endocytosis to interact with NOTCH, and NECD is trans-endocytosed by the ligand-presenting cell in a Mib1/Neur1-dependent process. The NECD dissociation enables second cleavage at S2 site by ADAM-10, a̲disintegrin a̲nd m̲etalloprotease/TNF-α converting enzyme (TACE). The S2-cleaved NOTCH receptor is ubiquitinated (Ub) and endocytosed in a clathrin/AP2/epsin1-dependent or -independent way to be either cleaved at S3 site by γ-secretase to generate an intracellular domain (ICD), or destined for lysosomal degradation. ICD translocates to the nucleus and binds to a DNA binding protein, CSL (for CBF1/RBP-Jκ, Suppressor of Hairless and Lag-1). The following recruitment of the coactivator, Maml (mastermind-like), promotes transcriptional activity of its target genes, such as the hairy/enhancer of split (Hes).

Diagram redrawn courtesy of Y Yamamoto, Yamaguchi University Graduate School of Medicine, Japan.

Clinical Features

Maeda syndrome begins at the age of 20–45 years and usually leads to death within about 7.5 years.[318,1108] It is more common in males and manifests with recurrent lacunar strokes, predominantly involving basal ganglia and brain stem structures. Most patients develop cognitive impairment. Patients also have orthopaedic manifestations, including intervertebral disc herniations, kyphosis, ossification of spinal ligaments and various osseous deformities. Alopecia is a characteristic feature. T2-weighted MRI reveals diffuse high-signal abnormalities in the white matter (most marked in the deep white matter), and multiple lacunar infarcts in the basal ganglia, thalamus brain stem and cerebellum.[318,1108]

Pathology

Small penetrating arteries and arterioles (100–400 μm in calibre) in the white matter and basal ganglia are most severely affected. The intima shows marked fibrous thickening and the internal elastica is fragmented. Vascular smooth muscle cells in the media are destroyed and adventitia is thinned. GOM deposits are not present. The lumen is usually narrowed but segmental dilation can occur. There is reduced myelin staining of cerebral white matter, with relative sparing of U-fibres. Multiple small foci of softening and cavitation are seen in the cerebral white matter, basal ganglia, thalamus and brain stem.[55,1108] Maeda syndrome appears to be a generalized disorder but the vascular pathology in other organs is less severe. Skin biopsy is not helpful for diagnosis.

Genetics

CARASIL co-segregates with the 2.4-Mb region on chromosome 10q, which contains mutations in the *HTRA1* gene.[393] Thus encodes the serine protease HTRA1, known to influence multiple processes including transforming growth factor β (TGF-β) signalling and perhaps the metabolism of APP, which includes several HTRA1 cleavage sites. Both nonsense and missense mutations of the *HTRA1* gene have been identified in CARASIL families. The mutations result in reduced HTRA1 protease activity (21–50 per cent) or loss of the protein.[393]

Pathogenesis

In CARASIL, the adventitial layers of vessels are profoundly thin, with reduced immunolabelling of type I, III and VI collagens. Degeneration of both the medial and adventitial layers is likely to impair autoregulation and functional hyperaemia. Immunohistochemical analysis of the cerebral vessels reveals increased extra domain-A region of fibronectin and versican in the thickened intima and TGF-β1 in the media of arterial walls. In CARASIL, the mutant variants of HTRA1 are unable to repress the phosphorylation of SMAD proteins, effectors of TGF-β signalling (Figure 2.117).[393,1104] TGF-β signalling induces synthesis of extracellular matrix proteins including fibronectin and versican and promotes vascular fibrosis.[991] TGF-β signalling is also involved in angiogenesis and vascular remodelling.

The findings in CARASIL extend the spectrum of diseases associated with the dysregulation of TGF-β signalling. Defective TGF-β signalling due to mutations in the TGF-β receptors leads to hereditary haemorrhagic telangiectasia, whereas activation of TGF-β signalling contributes to Marfan's syndrome and associated disorders.[991]

Dysregulation of inhibition of TGF-β signalling has been linked to alopecia and spondylosis, which are characteristic of CARASIL. A single-nucleotide polymorphism in the promoter region of *HTRA1* that results in elevated levels of HTRA1 is a genetic risk factor for the neovascular form of age-related macular degeneration.

Retinal Vasculopathies with Cerebral Leukodystrophy (RVCL)

Hereditary endotheliopathy with retinopathy, nephropathy and stroke (HERNS), cerebroretinal vasculopathy (CRV) and hereditary vascular retinopathy (HVR) were reported independently but linkage analysis demonstrated that they are allelic disorders or different phenotypes of same disease spectrum.[761] They map to the same locus on chromosome 3p21.1-p21.3 and are now described as autosomal

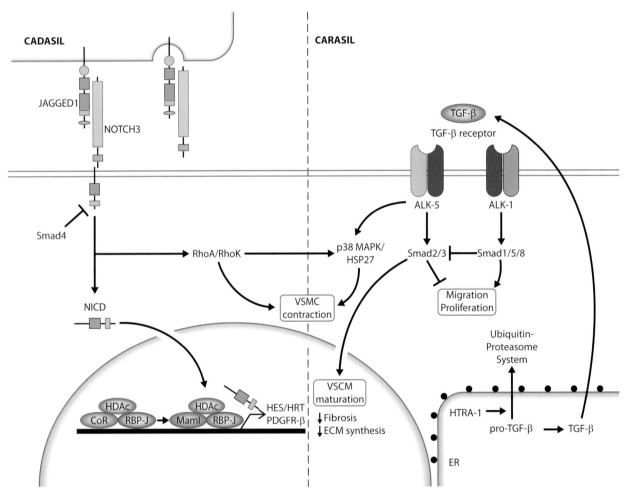

2.117 Signalling pathways in cerebral autosomal dominant arteriopathy with subcortical infarcts and leukoencephalopathy (CADASIL) and cerebral autosomal recessive arteriopathy with subcortical infarcts and leukoencephalopathy (CARASIL). Schematic shows putative crosstalk and mediators between NOTCH3 and TGF-β signalling pathways, which converge to cause changes in vascular smooth muscle cell (VSMC) maturation, fibrosis and extracellular matrix (ECM) synthesis. In NOTCH3 pathway, after final cleavage by γ-secretase from the receptor the NOTCH extracellular domain (NICD) translocates to the nucleus to form a complex with RBPJ to induce transcriptional activation of target genes. In the TGF-β pathway, HTRA1 protease activity inhibits TGF-β signalling via ALK5/Smad 2/3, leading to cell maturation and control of extracellular matrix production. ALK, activin receptor-like kinase or active TGF-β type I receptor; ECM, extracellular matrix; NECD, NOTCH extracellular domain; NICD, NOTCH intracellular domain; Rho, Rho protein kinases; Smad, mediators of transcriptional activation; VSMC, vascular smooth muscle cells.

Diagram courtesy of Y Yamamoto, Yamaguchi University Graduate School of Medicine, Japan.

dominant retinal vasculopathy with cerebral leukodystrophy (RVCL).[246,845]

Clinical Features

The RVCLs all cause progressive central visual impairment. In CRV, symptoms begin during the fourth or fifth decade. Subjects experience migraine-like episodes, transient ischemic attacks and strokes with motor and sensory loss, headaches, personality changes, depression and anxiety. In the late stages, subjects show cognitive impairment. Death occurs within 10 years of onset. Neuroimaging shows diffuse white matter changes consistent with SVD. Renal disease occurs in HERNS and CRV, whereas HVR is associated with Raynaud phenomenon.[1104] Neurological complications of CRV and HERNS lead to death before 55 years, whereas HVR patients live longer.[1104] Common ophthalmological findings in these disorders include capillary dropout and vascular tortuosity, aneurysms and telangiectasia. No specific therapy is available.

Pathology

Lesions occur in the pons, cerebellum and basal ganglia in addition to the frontal and parietal lobes, and consist of foci of necrosis with negligible inflammation. Blood vessels have multilayered endothelial basement membranes and thickened adventitia. Affected vessels are often occluded by fibrin thrombi, resulting in small infarcts in the surrounding white matter.[761] In the retina, there are micro-aneurysms of arterioles and foci of capillary telangiectasia; in advanced stages, the damaged arteries become occluded, giving rise to retinal infarcts. In CRV, cerebral pathology is similar to that of HERNS:[761] fibrosis of the walls of small and medium-sized arteries and veins within the white matter and basal ganglia lead to obliteration and foci of parenchymal necrosis. Abnormalities of the microvasculature in the liver and kidney are not as prominent as in the brain.

Genetics

SVDs in the RVCL group are caused by carboxyl terminal truncations causing frameshifts in the *TREX1* gene, which encodes an autonomous (non-processive) 3′–5′ DNA-specific exonuclease found in all mammalian cells. Whilst heterozygous mutations in *TREX1* cause RVCL, homozygous mutations in the same gene are linked to the typical autosomal recessive form of Aicardi–Goutières syndrome (AGS).[845] AGS manifests as a progressive encephalopathy of early onset, brain atrophy, demyelination, basal ganglia calcifications and chronic lymphocytic proliferation in the CSF.

Pathogenesis

Mutations in the carboxyl terminus of *TREX1* do not diminish the DNA-specific exonuclease activity of the protein but disrupt the predicted transmembrane domain and prevent TREX1 from localizing to the normal perinuclear site; the mutant proteins instead diffuse freely throughout the cytoplasm.[524] It is unclear how this leads to the phenotype of RVCL.

Rarer Vasculopathies and Angiopathies

Several individual families with hereditary SVDs have been described that are not explained by any of the known gene defects. Hereditary systemic angiopathy (HSA) manifests with cerebral calcifications, retinopathy, progressive nephropathy, hepatopathy and microangiopathy reminiscent of a RVCL variant.[1094] Patients present in their mid-forties with visual impairment, migraine-like headaches, skin rashes, seizures, motor paresis and cognitive decline. The retinal microvessels undergo progressive occlusion leading to ischaemic retinopathy with subsequent optic disc atrophy and formation of capillary aneurysms. Pathological changes include foci of necrosis in the white matter, with prominent perivascular inflammation, oedema and astrocytic gliosis. There is evidence of proliferation of microvessels, many with hyperplastic endothelium and severely thickened walls, and some showing fibrinoid necrosis or thrombosis.[1094]

SVD Associated with Collagen Type IV, A1 (COL4A1) and A2 (COL4A2) Mutations

The recently recognized spectrum of COL4A1and 2-related disorders encompasses a number of conditions with features of SVD of varying severity.[20,801,1034] The *COL4A1* gene is located on chromosome 13q34 and consists of 52 exons spanning approximately 158 kb. Mutations are associated with four major phenotypes:[579] (i) perinatal haemorrhage with porencephaly in survivors, (ii) hereditary infantile hemiparesis, retinal arteriolar tortuosity and leukoencephalopathy (HIHRATL),[1033] (iii) SVD with Axenfeld–Rieger anomaly (anterior segment dysgenesis of the eye) and (iv) hereditary angiopathy with nephropathy, aneurysms (typically of the internal carotid artery) and muscle cramps (HANAC). Other clinical manifestations can include Raynaud's phenomenon, kidney defects and cardiac arrhythmias. *COL4A1* mutations cause variable degrees of retinal arteriolar tortuosity, and abnormalities of endothelial basement membranes in the skin and cerebral vasculature have been documented.

Neurological manifestations of *COL4A1* and *COL4A2* mutations depend on the age of disease onset but can vary even within families. Affected individuals may present with infantile hemiparesis, seizures, visual loss, dystonia, strokes, migraine, mental retardation, or cognitive impairment and dementia. Single or recurrent intracranial haemorrhages may occur in non–hypertensive young adults: spontaneously, subsequent to trauma or as a result of anticoagulant use. Stroke is often the first presentation of the disease, with a mean age of onset of 36 years.[579] Neuroimaging shows typical features of SVD including diffuse leukoencephalopathy with most severe involvement of posterior periventricular regions, subcortical infarcts, cerebral microbleeds, and dilated perivascular spaces. Cases with porencephaly have large fluid-filled cavities that appear as paraventricular cysts.[579] Autosomal dominant *COL4A1*-related disease is described in at least 12 European caucasian families, with 100 per cent penetrance. Several other familial disorders with an SVD phenotype have been reported but the genetic basis not yet established.[1104]

HAEMORRHAGIC STROKE AND CONSEQUENCES

Parenchymal Brain Haemorrhage

The terms parenchymal brain haemorrhage (PBH) and intracerebral haemorrhage (ICH) are used interchangeably, although PBH is preferred, as the haemorrhage can involve the cerebellum and/or brain stem as well as the cerebrum.

Epidemiology and Causes

The relative frequency of PBH in first-time stroke varies widely between different populations. East–West differences were evident even prior to the reduction in stroke incidence over recent years (Table 2.8).[300] Annual incidence estimates in studies of white populations vary between 11 and 20 (crude) or 16 and 32 (age-adjusted) per 100000.[303,735] The incidence of PBH is dependent on the ethnic background and is, for example, markedly higher in African-Americans (in Kentucky, USA, 48.9 versus 26.6 per 100000)[300] and Far Eastern populations (47 (adjusted) in Japan[462] and over 60 in Taiwan).[443] Estimates of primary ICH and SAH in low- to middle-income countries during the period 2000–2008 were almost twice the rates in high-income countries (primary IAH 22 per 100000 versus 10 per 100000 and SAH 7 per 100000 versus 4 per 100000, respectively).[284] In several populations, the incidence of PBH has declined over the past three decades (Figure 2.118).[586,877,994] The mortality rate in PBH is higher than for ischaemic strokes: the 1-month case-fatality rate is estimated at 35–50 per cent.[283]

Modern imaging methods provide reliable *intra vitam* identification of PBH and have markedly improved the accuracy of diagnosis (Figure 2.119). T2*-weighted MRI allows the detection of not only the larger haematomas but also microbleeds: small, perivascular, clinically silent extravasations of blood.[368,546,863] Micro-bleeds can be detected in about 5 per cent of older adults without clinical cerebrovascular disease.[546] Micro-bleeds are demonstrable in association with 47–80 per cent of larger spontaneous PBHs and 18–78 per cent of people with ischaemic cerebrovascular disease.[546] They are thought to be harbingers of future major PBH.[27,546,1051]

Hypertension and CAA are numerically the most important causes of PBH (Table 2.11). Hypertension is thought to be responsible for about half of PBHs but the numbers have been decreasing with effective antihypertensive treatment. The cause of PBH affects its location in the CNS (Table 2.13).[286] The common, large parenchymal bleeds are often associated with micro-bleeds and subacute ischaemic lesions.[370]

Hypertensive Haemorrhage

The decline in the incidence of PBH is largely attributable to improvements in diagnosis and treatment of hypertension.[140,519,822,877] Hypertension was the cause of almost 90 per cent of all PBHs in the 1940s, 70–80 per cent in the 1970s[319] and only 45–60 per cent in the 1980s and 1990s (Table 2.13).[919,1097] PBH constitutes a relatively high proportion of strokes in Far Eastern populations, reflecting the high prevalence of hypertension.[527,554,931] PBH is often considered to be a one-time event, with recurrent bleeding being rare compared with that from saccular aneurysms, vascular malformations or CAA.[519] Rebleeding has, however, become more common with increased survival of high-risk patients. Recurrence rates from 1.8 to 6 per cent are reported, recurrence being more common in people with persistence of high diastolic blood pressure.[50]

Hypertensive PBH most commonly occurs in the deep cerebral grey matter (putamen and thalamus, about 60

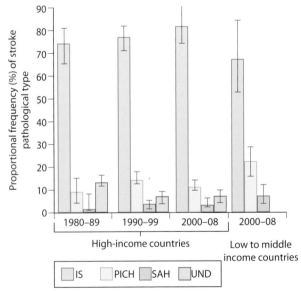

2.118 Worldwide trends in relative frequency of intracerebral and subarachnoid haemorrhage relative to ischaemic stroke. The graph compares the frequencies of stroke types in high-income countries and low- to middle-income countries across different study periods (pooled estimates). IS, ischaemic stroke; PICH, posterior intracerebral haemorrhage; SAH, subarachnoid haemorrhage; UND, undefined pathological type of stroke.

Graph courtesy of V Feigin, AUT University, Auckland, New Zealand.

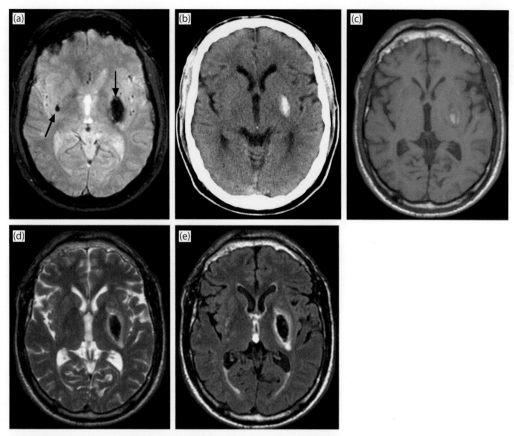

2.119 Images in a patient with an acute haematoma in the left putamen (down arrow) and clinically silent micro-bleeds (up arrow) in the basal ganglia. The micro-bleeds are detectable only in the transverse T2*-weighted gradient-echo image **(a)** as foci of low signal intensity. (b,c) show, in the same plane of section as (a), a computed tomography (CT) scan **(b)** and T1-weighted spin echo (SE) **(c)**, T2-weighted SE **(d)** and fluid-associated inversion recovery (FLAIR) **(e)** magnetic resonance (MR) images.

TABLE 2.12 Sites of non-traumatic parenchymal brain haemorrhage					
Haemorrhage		**n**	**per cent**	**per cent**	**per cent**
Lobar		65	31		
	One lobe			46	
	Frontal				17
	Parietal				11
	Temporal				9
	Occipital				9
	Two lobes			42	
	Three lobes			12	
Deep supratentorial		107	51		
	Putamen			48	
	Thalamus			43	
	Caudate			9	
Deep infratentorial		37	18		
	Cerebellum			70	
	Pons			30	
Total		209	100		

Adapted from Massaro AR *et al*.[652] With permission from Lippincott Williams & Wilkins/Wolters Kluwer Health.

TABLE 2.13 Proportional distribution of parenchymal brain haemorrhage (PBH) by location in different populations

Study	Total number of PBH	Lobar (per cent)	Deep cerebral (per cent)	Brain stem (per cent)	Cerebellum (per cent)
Cincinnati, USA	1038	35	49	6	10
Izumo, Japan	350	15	69	9	7
Southern Sweden	341	52	36	4	9
Jyväskylä, Finland	158	34	49	7	11
Dijon, France	87	18	67	6	9
Perth, Australia	60	30	52	7	10

Results obtained from several sources.[42,300,303,352,462,735]

per cent), the hemispheres (lobar, 20 per cent), cerebellum (13 per cent) and pons (7 per cent). The diameter of the arterial vessels at these sites ranges from 50 to 200 μm. Rupture of non-aneurysmal arterioles damaged by hypertension is regarded as the main cause of hypertensive haemorrhages; the role of Charcot–Bouchard microaneurysms as a source of PBH is uncertain (see Hypertensive Angiopathy, earlier).[183,1062] Generalized degenerative change in small arteries and arterioles, with perivascular accumulation of haemosiderin (microbleeds), is readily demonstrable pathologically, whereas microaneurysms are relatively rare. Identifying the source of a particular hypertensive bleed is generally not possible because the ruptured part of the affected small artery is likely to have been destroyed by the haemorrhage.

Cerebral Amyloid Angiopathy

Sporadic CAA is responsible for up to 12 per cent of primary non-traumatic PBH. Sporadic CAA is typically associated with lobar (i.e. peripherally located) ICH (it is responsible for about 30 per cent), particularly in elderly normotensive patients (Figures 2.110, 2.112 and 2.120). Severe CAA is associated with high risk of PBH.[1057] Microhaemorrhages are demonstrable in up to 80 per cent of CAA patients on T2*-weighted MRI.[863] The 'Boston criteria' estimate the clinical likelihood of PBH being CAAH.[364,544] A diagnosis of definite CAA requires pathological verification. Probable CAAH can be diagnosed if two or more acute or chronic lobar macrohaemorrhages or microhaemorrhages are identified, without any other definite cause of PBH (such as excessive warfarin treatment, antecedent head injury, stroke, neoplasm, vascular malformation, vasculitis or blood dyscrasia/coagulopathy). Possible CAAH is diagnosed when a single lobar haemorrhage occurs in a patient over 55 years with no other obvious cause of cerebral haemorrhage. In one prospective series with pathological follow-up, a diagnosis of probable CAA proved to be to 100 per cent accurate but CAA could be confirmed in only 62 per cent of cases diagnosed as possible.[544]

CAAH can occur at multiple sites and progress over time. CAAH tends to be superficial and is quite often associated with secondary SAH (Figure 2.120). Rupture of amyloid-laden arteries may sometimes occur directly into the subarachnoid space, causing primary SAH (see

Subarachnoid Haemorrhage). The rigid and fragile arterial walls may make haemostasis difficult to achieve after neurosurgical evacuation of the haematoma.[369,1062]

Antiplatelet and anticoagulant medication and head injury probably increase the risk of CAAH.[662] CAA increases the risk of PBH after fibrinolytic treatment of acute myocardial infarction.[786] It is unclear whether or not hypertension increases the risk of CAAH, although antihypertensive medication tended to lower the risk of probable CAAH (defined according to the Boston criteria, see earlier) in patients who had previously experienced a stroke or TIA.[54] The risk of CAAH is marginally increased in *APOE* ε2 carriers and shows a trend towards association with the ε4 allele.[973]

Haemorrhage Caused by Anticoagulants and Antithrombolytics

Nearly 1 per cent of patients treated with anticoagulants are reported to experience intracranial haemorrhage during therapy. The risk of PBH is increased about 8–11-fold.[401,519] During treatment of ischaemic stroke with anticoagulants, not only may the infarct not only undergo haemorrhagic transformation (see earlier), but bleeding from damaged blood vessels may also produce sizable haematomas. Anticoagulation prolongs the time for the breached vessels to be resealed, making haematomas larger and their prognosis worse. The mortality of PBH patients on anticoagulation is about twice the overall mortality of PBH.[301]

Thrombolytic therapy may similarly cause PBH. The incidence of PBH after thrombolysis for acute myocardial infarction is about 0.3–0.8 per cent.[468] The risk of PBH is considerably higher when the thrombolytic treatment is given for acute cerebral ischaemia. The incidence of PBH (often fatal) was reported to be 3.5 times higher among 5216 patients treated with thrombolysis than in placebo-treated patients. Despite this, there was a significant net reduction in the proportion of patients who died or became dependent on carers.[1068] The frequency of haemorrhagic events increases with delay of treatment beyond 3 hours.[1031]

Tumour Haemorrhage

Both primary and metastatic intracranial tumours can cause major PBH. In about half such cases, PBH is the first manifestation of the tumour.[608,642] The overall frequency of

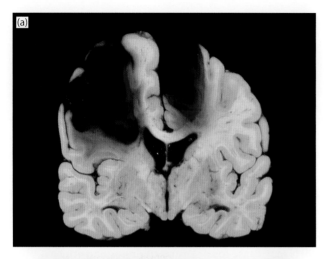

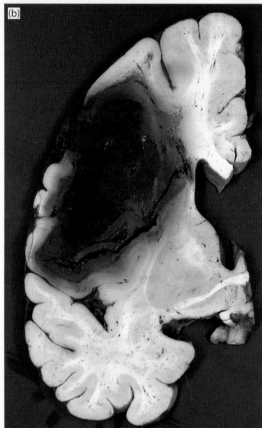

2.120 Haemorrhage associated with cerebral amyloid angiopathy (CAA). (a) Typical lobar and superficial distribution of haematomas caused by rupture of Aβ-laden blood vessels. Both haematomas are largely intraparenchymal but communicate with the subarachnoid space. **(b)** Large lobar haematoma complicating CAA.

Image (b) kindly provided by T Polvikoski, Newcastle University, UK.

spontaneous PBH in intracranial neoplasms ranges from 1.4 to 10 per cent, with a mean of 2–3 per cent. Conversely, the frequency of intracranial neoplasms in spontaneous PBH ranges from 0.8–7.4 per cent.[913] Abnormal blood vessels in the tumours are thought to be responsible for the haemorrhage. The rich vascularity helps in differential diagnosis, because ring-enhancement on CT scanning is common in

tumour haemorrhage but not acutely in non-neoplastic PBH. Metastases are the most common cause of tumour-related PBH. Among primary tumours associated with PBH, glioblastomas predominate, in approximate proportion to their relative frequency (it is perhaps surprising that PBH does not occur more often in these tumours in view of the profusion of abnormal blood vessels).[913] Bronchogenic carcinoma, choriocarcinoma, melanoma and renal cell carcinoma are the metastatic tumours most often responsible for PBH.[519]

Drugs and Intraparenchymal Haemorrhage

The risk of PBH among drug abusers is markedly increased, especially among users of cocaine, heroin and sympathomimetics such as phenylpropanolamine, phencyclidine and amphetamines.[596,949] Drug abuse was identified as a predisposing factor in 47 per cent of stroke patients under 35 years.[503] In drug abusers and even in alcoholics, the haematomas are commonly lobar, in the subcortical white matter. The bleeding usually develops within minutes to a few hours after drug use. Two main pathogenetic mechanisms have been proposed. The more likely is that the haemorrhage results from an acute rise in the blood pressure. Arteritis has also been reported but usually diagnosed on the angiographic appearance rather than histology, and in some reported cases with pathological confirmation the changes may reflect medial necrosis due to vasospasm rather than inflammatory vascular damage.[934] Ethanol also increases the risk, at the high end of the J-shaped relationship between haemorrhage and consumption. The relative risk for PBH was 2.4 if more than 400 g of ethanol was consumed per week.[347]

Haemorrhage from Arteriovenous Malformations, Angiomas and Aneurysms

The common location of arteriovenous malformations (AVMs) at the interface of the parenchyma and the subarachnoid or intraventricular space increases the risk of haemorrhage in any or all of these three compartments. SAH from AVMs is considerably less common than intraventricular and intracerebral bleeds.[47] Cavernous angiomas are less often a source of major PBH than are arteriovenous malformations, but minor, usually subclinical bleeding that results in local accumulation of haemosiderin is almost invariable. Clinically significant haemorrhage is reported in 9–25 per cent of sporadic[241,809] and 9 per cent of familial cavernomas.[573] The surgical specimen of an arteriovenous malformation rarely presents diagnostic difficulties. Some arteriovenous malformations may be difficult to detect within a PBH, but an otherwise unusual location of a small haematoma may indicate the diagnosis. For microscopic analysis sampling of all suspected areas within the wall of the haematoma cavity is recommended.[1062]

Rupture of a saccular aneurysm may cause a PBH when the fundus of the aneurysm is embedded in the parenchyma. Because the parent artery of these aneurysms lies in the subarachnoid space, the haematoma originates close to the basal surface of the brain and extends to a variable depth into the parenchyma, sometimes all the way to the ventricle (Figure 2.121). Bacterial emboli that cause an infective aneurysm commonly lodge within small intraparenchymal

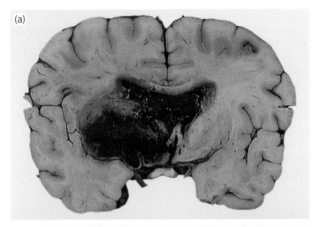

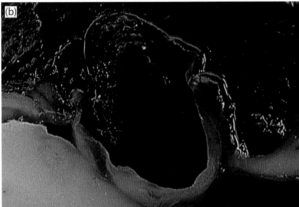

2.121 Intraparenchymal and intraventricular haemorrhage caused by ruptured saccular aneurysm. (a) The tip of the ruptured aneurysm in the intracranial part of the left internal carotid artery is embedded in the overlying parenchyma. The massive haemorrhage has extended through the basal ganglia into the lateral ventricle. **(b)** A close-up view shows the thinning of the wall of the aneurysm, which ruptured at its tip.

arteries. PBH caused by an infective aneurysm may look indistinguishable from a hypertensive haemorrhage (Figure 2.51b). The risk of rupture of infective aneurysms has been estimated at 3–7 per cent for patients with endocarditis. The rupture tends to occur within the first 5 weeks of endocarditis, and carries a mortality rate of the order of 80 per cent. Infective aneurysms are small in size and fragile in structure. They are often destroyed by the haemorrhage and difficult to identify post mortem. The clinical history and associated findings suggestive of infection, especially in the cardiac valves (Figure 2.51a), may help in determining the cause of the bleed.

SAH may be caused by aneurysms or arteriovenous malformations (see Subarachnoid Haemorrhage).

Clinical Features and Imaging of Parenchymal Brain Haemorrhage

In general, smaller PBHs manifest with focal neurological deficits. In larger supratentorial PBHs, the clinical picture is usually related to the mass effect of the haematoma. The bleeding may be monophasic and relatively brief, being stopped by quick clotting and tamponade. However,

enlargement of the haematoma can continue for up to 20 hours.[142] The expansion is ascribed to continued bleeding from the primary site and is significantly associated with elevated systolic blood pressure.[53] Thus, neurological deterioration after PBH may result not only from secondary oedema and ischaemia but also from continued enlargement of the haematoma.

Motor deficits occur in both deep and lobar PBHs but are more common in deep haemorrhages. Severe headaches and seizures are frequent in lobar haemorrhages,[179,519] whereas visual deficits are more often associated with deep bleeds.[652] Decreased levels of consciousness (55–60 per cent) and coma (20 per cent) are almost as frequent in lobar and deep PBH. Raised intracranial pressure and the risk of transtentorial herniation relate to the size of the haematoma and the surrounding oedema. Small pontine PBHs may cause focal sensory or motor impairments,[516,519] whereas larger pontine PBHs usually cause coma leading rapidly to death from compression of the vital centres. Cerebellar haemorrhages are characteristically associated with vertigo and nausea. They may rapidly obstruct the circulation of CSF, with consequent acute life-threatening hydrocephalus.[516,519]

In routine practice, CT scans are used for acute diagnosis of PBH (Figure 2.119a). CT shows a hyperdense region for 1–2 weeks, after which it is transformed into a hypodense lesion and occasionally becomes calcified.[532,565,1079] MRI, especially T2*-weighted, reveals acute haemorrhages as well as CT,[532] but in an emergency situation may be less accessible. In visualizing older haemorrhages, T2*-weighted MRI is superior (Figure 2.119b).[27,938]

The prognosis of PBH is dependent on the size and location of the haemorrhage, the patient's age and whether there is intraventricular extension. The 30-day case-fatality after supratentorial PBHs varies in population-based studies from 25 to 72 per cent, with a weighted mean of 48 per cent. In hospital-based series the numbers range from 27 per cent to 54 per cent, with a weighted mean of 35 per cent.[303] Deep PBHs have a 4–5 per cent higher fatality rate, probably reflecting the greater likelihood of intraventricular extension. Intraventricular extension increases the fatality rate to 78 per cent.[302,652] An increase in the size of the haematoma from below 20 mL to over 80 mL raises the fatality rate from 16 to 82 per cent.[303] The mortality of cerebellar haemorrhages is similar to or greater than that of supratentorial PBH. Over 80 per cent of patients with large pontine haemorrhages succumb, whereas only about 5 per cent of unilateral tegmental haemorrhages are fatal.[202]

Anatomical Aspects of Parenchymal Brain Haemorrhages

Supratentorial Haemorrhage

Large PBHs are usually divided into lobar haemorrhages, and deep haemorrhages in the basal ganglia and thalamus.[519,822] The ratio of lobar to deep haemorrhages varies considerably between different countries and ethnic groups (Table 2.14).[303,652] Hypertension is common among patients with deep PBHs, occurring in up to 80 per cent, compared with a prevalence of 31–55 per cent in patients with lobar haemorrhages. The relatively higher frequency of deep PBHs in black compared to white people at younger ages (35–54 years) has been ascribed mainly to higher rates of untreated hypertension.[300] As far as lobar PBHs are concerned, CAA, arteriovenous malformations (often of very small size) and leukaemia (Figure 2.122) are more common causes than hypertension.[407,516,652] On average,

TABLE 2.14   Cumulative mortality rates after aneurysmal subarachnoid haemorrhage

Time after bleed	Cumulative mortality (per cent)
Before medical care	15
Day 1	25–30
Week 1	40
Month 1	55
Month 6	60
Year 1	63
Year 5	65

Adapted from Broderick *et al.*[141] and Fogelholm R *et al.*[304]

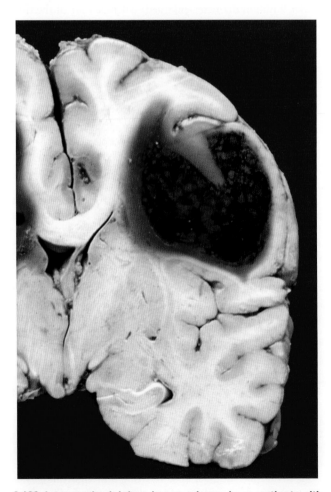

2.122 Intracerebral lobar haemorrhage in a patient with leukaemia. The presence of leukaemic cells gives the haematoma a variegated appearance.

patients with lobar haemorrhages are 4–9 years older than those with deep haemorrhages (65–68 years vs 59–61.5 years),[300,652] which probably reflects the increase in prevalence of CAA with age. This is consistent with a Swedish study in an aged population in which lobar PBHs were much more common than deep PBHs (Table 2.12). The association with CAA may also explain why lobar haemorrhages are more common in women, who live longer than men. Lobar PBHs are, on average, larger than deep PBHs, reflecting the greater volumes of the hemispheres than the deep nuclei. In the Stroke Data Bank the majority of the lobar PBHs exceeded 50 mL, whereas most deep PBHs were smaller than 15 mL.[652] The proximity of the ventricular system to deep haemorrhages accounts for their propensity to extend into the ventricular system.[652]

Infratentorial Haemorrhage

Cerebellar and brain stem haemorrhages constitute approximately 15–18 per cent of all primary PBHs in both Western and Far Eastern populations (Table 2.13). Most cerebellar haemorrhages are hemispheric. The great majority of brain stem haemorrhages are restricted to the pons. The medulla oblongata is a rare primary site.[90] Pontine haemorrhages usually destroy large parts of the basal or tegmental pons. CT scanning has also allowed *intra vitam* detection of small, non-fatal, infratentorial haemorrhages. In the cerebellum, these small PBHs are usually located in the vicinity of the fourth ventricle. In the pons they occur unilaterally in the tegmentum.[202,535] Primary pontine haemorrhages have been subclassified on the basis of their size and anatomical location.[202,516]

Pathology and Pathogenesis of Parenchymal Brain Haemorrhage

Pathologists encounter PBHs mostly at autopsy. Surgical removal of the haematoma is recommended only in selected patients.[407,523] If an operation is performed, however, careful sampling of the haematoma cavity and detailed histopathological analysis reveal the definite cause of the PBH in a surprisingly high proportion of patients (e.g. Figures 2.37 and 2.36).[1062] At post-mortem examination, PBHs are readily identified. Even small old lesions can be recognized by their orange tinge from haemosiderin. However, determination of a definite cause of the bleeding may be difficult, and even a thorough search of multiple sections may be inconclusive. The site, size and multiplicity of the PBHs should be considered, as well as the presence of any structural vascular abnormalities such as aneurysms, arteriovenous malformations or CAA.[1079]

Regardless of their cause, haematomas have a similar appearance. The time course of haematoma resorption and the response of the surrounding parenchyma have been analyzed systematically in experimental animals and the findings shown to correlate well with those in humans.[481,980] Fresh haematomas are sharply demarcated, with only limited spread of erythrocytes into the adjacent parenchyma. A rim of surrounding neurons and glia undergoes necrosis during the first day and oedema increases. The inflammatory reaction in the surrounding tissue is similar to that in infarcts. However, polymorphonuclear leukocytes may appear only a little later, by 48 hours, possibly because blood flow immediately surrounding the haematoma is impeded by compression. Macrophages assume the appearance of siderophages instead of foam cells, and blood-derived pigment may also be seen within astrocytes. The time course for the formation of haemosiderin in the CNS is the same as elsewhere in the body. It can be detected by Perls' method (Prussian blue stain) 1 day after commencement of phagocytosis. Haemosiderin may still be present several years later at the site of the haematoma within the astrocytes and macrophages in the walls of the cavity. It has been estimated that the clot is resorbed at about 0.7 mm/day. Astrocyte proliferation around the haematoma begins within 1 week, with simultaneous neovascularization, causing characteristic ring enhancement around the haematoma on CT.[980]

The pathogenesis of the tissue damage in PBH has been less studied than that of the damage in cerebral ischaemia. Mechanisms responsible for primary injury at or around the site of the haemorrhage include direct disruption of neurons and axons. Given the considerable size of the haematoma, the extent of this type of injury is often surprisingly limited, as evident from the remarkable abatement or even complete disappearance of PBH on CT scans.[565] This has been attributed to the splitting rather than destruction of fibre tracts. Brain ischaemia, both in the vicinity of the haematoma and global, plays a key role in the development of permanent damage in PBH.[481,536,667,711,722] By comparing the effects of injected autologous blood and inflated balloons of the same volume in the caudate nucleus, researchers demonstrated that the mass effect alone caused a reduction in local blood flow below the ischaemic threshold but that ischaemia was aggravated by substances derived from the blood clot even without a significant rise in the intracranial pressure. PBH with intraventricular extension raised the intracranial pressure still further and decreased cerebral perfusion pressure. After balloon inflation, the ischaemic damage progressed even when the balloon was deflated, indicating the existence of a surrounding penumbral zone, although if the deflation occurred within 10 minutes, the extent of the lesion at 24 hours was smaller than that after longer inflation times. Calcium-channel blockers and immunosuppression were reported to ameliorate the ischaemic injury, suggesting that there is a therapeutic window for PBH, although the period seems to be short.[667]

Subarachnoid Haemorrhage

General Aspects

In subarachnoid haemorrhage (SAH), the bleeding occurs in the subarachnoid space, alone or in conjunction with bleeding elsewhere in the CNS. In most populations, primary non-traumatic SAH represents about 5–9 per cent of all strokes (Table 2.14). The annual incidence of SAH from verified intracranial aneurysms is about 10–11 per 100 000 in most Western countries (range 6–17), with higher numbers reported from Japan, some parts of the USA and Finland but lower numbers from New Zealand, some parts of the USA and Scandinavia, excluding Finland.[123,461,668,901] Unlike the incidence of most other types of stroke, that of SAH has not decreased over the past 30 years. The incidence of aneurysmal SAH increases almost linearly with age, from below 1 per 100 000 before the age of 20 years to about 40 per 100 000 after 65 years. The median age of

onset of a first SAH is 50–60 years. Deaths caused by ruptured aneurysms constitute 16–24 per cent of all patients dying from cerebrovascular diseases.[304]

Modern neuroimaging techniques allow demonstration of the cause of SAH in the great majority of cases. Rupture of a saccular aneurysm is by far the most frequent cause of non-traumatic SAH and accounts for about 85 per cent of cases (range 57–94 per cent). AVMs are responsible for about 5–10 per cent. In approximately 10–15 per cent, the source of bleeding cannot be identified by neuroimaging,[304] perimesencephalic haemorrhage being the most common among these.[1037] Secondary SAH may occur in connection with PBH, blood being forced through the cortex into the subarachnoid space (Figure 2.120) or with intraventricular haemorrhage, when the blood follows CSF routes into the basal cisterns. In some patients with CAAH there is associated bleeding into the subarachnoid space and in a small proportion the haemorrhage is exclusively subarachnoid.[1101]

Rupture of Saccular Aneurysms and Aneurysmal Subarachnoid Haemorrhage

The rate of rupture of saccular aneurysms has been estimated at approximately 1–2 per cent per year. Larger aneurysms, in particular, tend to increase in size with time. In one series, over a median period of 47 months, none of 53 aneurysms under 9 mm in diameter enlarged, whereas four of the nine aneurysms with initial diameters of 9 mm or larger increased in size, as measured by MR angiography.[794] Size is a major independent risk factor for rupture, and a diameter of about 9–10 mm seems to represent a critical watershed.[502,970] In patients with multiple aneurysms, when ruptured ones are clipped, the size of other aneurysms that rupture later increases significantly during the follow-up period.[502] The growth of the aneurysm fundus seems to occur by passive yield to blood pressure with simultaneous reactive formation of granulation tissue.[970] It has been proposed that atherosclerosis contributes to the growth, and associated inflammation accelerates the rupture of aneurysms.[522,555]

About 10–43 per cent of major aneurysmal ruptures are preceded by clinical warning symptoms, usually within the preceding 1–3 weeks: headache, nausea, neck pain, cranial nerve palsies or visual defects.[235,499,808] These may be caused by local effects of aneurysmal expansion or possibly by 'warning leaks'. The rupture of a critically weakened aneurysm wall, most commonly at its fundus (Figures 2.48 and 2.121), frequently follows an acute rise in blood pressure during physical stress. It occurs significantly more often during waking hours, particularly in the morning because of diurnal variations in blood pressure. Much higher transient pressure peaks are evident during the waking hours.[1044] The variations in blood pressure are accentuated in elderly people and in hypertensive subjects, owing to diminished compliance of vessel walls. In a series of 250 patients with aneurysmal SAH, Vlak et al.[1055] identified 8 trigger factors for rupture of saccular aneurysms, all known to cause surges in blood pressure: coffee consumption, cola consumption, anger, being startled, straining for defaecation, sexual intercourse, nose blowing and vigorous physical exercise. The rupture most often causes SAH, which may be complicated by vasospasm and infarction.[5,423] Less frequently, the rupture results in PBH (Figure 2.121).[1045]

A ruptured aneurysm can usually be identified *in vivo* by angiography, CT or MRI (Figure 2.123). The distribution of subarachnoid blood may indicate the source of haemorrhage. Rupture of an aneurysm in the circle of Willis usually gives rise to blood in the basal cisterns. Bleeding from an aneurysm of the ACoA or an ACA often results in a haematoma between the frontal lobes, which may extend caudally above the corpus callosum. Blood from a ruptured MCA aneurysm tends to have an asymmetrical distribution, in and around the affected sylvian fissure. When the aneurysm (arising from the middle cerebral, anterior cerebral or anterior communicating artery) is embedded in the surrounding brain parenchyma, blood from the ruptured aneurysm may penetrate into brain tissue, resulting in PBH, which may even extend into the ventricular system (Figure 2.121). After SAH, blood often refluxes into the ventricular system; conversely, after PBH, intraventricular blood may spread throughout the ventricular system and into the subarachnoid space via the foramina of the fourth ventricle. Occasionally, SAH may penetrate spontaneously into the subdural space.

Histologically, the blood in the subarachnoid space is contained by the arachnoid membrane and the pia mater surrounding the blood vessels. The pia mater on the surface of the brain appears to be continuous with the leptomeningeal ensheathment of the blood vessels in the subarachnoid space. Such structural organisation seems usually to

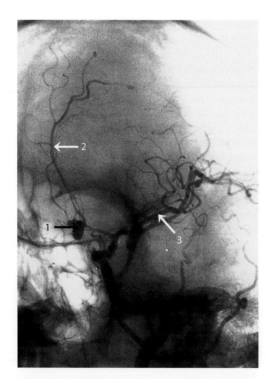

2.123 Saccular aneurysm. X-ray arteriography (an oblique projection) of a patient with subarachnoid haemorrhage discloses a ruptured saccular aneurysm of the anterior communicating artery (arrow 1). The secondary vasospasm in the pericallosal artery (arrow 2) has rendered it markedly narrower than the middle cerebral artery (arrow 3).

Scan courtesy of M Porras, Helsinki University Central Hospital, Helsinki, Finland.

prevent direct passage of red blood cells from the subarachnoid space to the perivascular spaces within the brain.[452]

Aneurysmal SAH has a high mortality. It accounts for about 16–24 per cent of all deaths from cerebrovascular disease.[901] About 40 per cent of patients die from the initial haemorrhage. Cumulative mortality rates after SAH[304] show that about 40 per cent of patients die from the initial haemorrhage (Table 2.15), 30 per cent of those within the first 24 hours. Without surgical or endovascular intervention, one-third of patients who survive the initial haemorrhage die from recurrent bleeding within 6 months of the initial episode.[498,871] Factors influencing survival include age, history of hypertension and, particularly, the amount of subarachnoid blood evident on early neuroimaging.[424,750] About 70 per cent of patients who are still alive 6 months after the haemorrhage may return to normal life, whereas 20 per cent are partially and 10 per cent severely disabled.[304]

Rebleeding

About 15–20 per cent of patients experience further bleeding, particularly during the first 24 hours.[460,498,1037] The cumulative risk of rebleeding after primary SAH is approximately 15 per cent by 7 days, 25 per cent by 14 days and 50 per cent by 6 months. After 6 months, the rate of rebleeding is about 3 per cent per year. After 10 years the risk approaches that of an unruptured aneurysm. Approximately two-thirds of the rebleeds are fatal, mortality increasing with successive bleeds.

Platelets are responsible for primary haemostasis after the rupture of an aneurysm. Coagulation factors and collagen subsequently secure the scar. A reduced ability of platelets to aggregate after primary SAH has been invoked as an explanation for the early rebleeds.[317,500] Later rebleeds have been attributed to lysis of the plugging clot by increased fibrinolytic activity in the CSF and plasma. Antifibrinolytic therapy decreases the risk of rebleeding by about 50 per cent, but the resulting reduction in mortality from rebleeding is offset by deaths from delayed cerebral ischaemia.[521,864]

Non-aneurysmal Subarachnoid Haemorrhages

In about 10–15 per cent of patients with definite SAH, no aneurysm or AVM is detected on digital subtraction

TABLE 2.15 Comparison of the pathological changes in global ischaemia and hypoglycaemia		
	Global ischaemia	**Hypoglycaemia**
Cerebral cortex: gross distribution	Watershed	Uniform
Cerebral cortex: layers involved	Middle laminae	Superficial laminae
Hippocampus	CA1, CA3 (dentate only if severe)	CA1, dentate
Cerebellum	Watershed	Absent
Brain stem	Can be involved	Absent

angiography.[849] In two-thirds of these patients, (approximately 7–10 per cent of all SAHs), the accumulation of blood is predominantly anterior to the midbrain and pons without spread to the sylvian or anterior interhemispheric fissure. This entity is described as perimesencephalic haemorrhage. It is reported to be the second most common type of SAH[1037] and usually affects young, male, non-hypertensive patients. The clinical course is much more benign than that of aneurysmal SAH. The headache begins more slowly, i.e. over minutes rather than in seconds. Seizures, focal neurological signs or unconsciousness are exceptional. Neither rebleeding nor delayed ischaemia occurs, and the prognosis is favourable with conservative treatment.[1037] The precise source of the haemorrhage is difficult to identify but detailed imaging may suggest leakage from veins or capillaries around the midbrain or bleeding from a small arterial aneurysm.[656,1036] In 2.5–5 per cent of patients with this perimesencephalic pattern of blood accumulation, the bleeding originates from a posterior fossa aneurysm.

Other suggested sources of SAH in the absence of a detectable aneurysm or other vascular abnormality include lenticulostriate or thalamoperforating vessels,[29] micro-aneurysms or micro-angiomas obliterated by the haemorrhage, saccular aneurysms undergoing spontaneous thrombosis after rupture, and segmental defects or necrosis of the tunica media of small arteries.[29,354,849] SAH patients without a detectable aneurysm or other vascular abnormality have a better prognosis than those with aneurysmal haemorrhages, even in the absence of any specific treatment. Rebleeding or delayed cerebral ischaemia is rare.[851,920,1037]

AVMs seldom cause only SAH and bleeding is often associated with intraventricular and PBH. These are more common complications of AVMs than ruptured saccular aneurysms.[47,860,1037] Saccular aneurysms sometimes develop on the feeding artery of an AVM (in approximately 10–20 per cent) and these may also rupture, giving rise more often to PBH than SAH.[1037] The annual risk of rupture of previously unruptured AVMs has been estimated at about 2–3 per cent.[357] Enlarging malformations and those with a single draining vein have a higher risk of bleeding.[689,1060]

Complications of Subarachnoid Haemorrhage

Arterial Vasospasm and Delayed Cerebral Ischaemia

Vasospasm of the cerebral arteries, and associated delayed cerebral ischaemia and infarction are important causes of morbidity and mortality in SAH, affecting about 20–30 per cent of patients.[4] The risk of ischaemia and infarction may be diminished by improved therapy, including nimodipine and maintenance of cerebral perfusion pressure.[4,293,1038] The clinical manifestations of delayed cerebral ischaemia do not usually begin before the fourth day after the initial SAH and reach their maximum around the seventh day. Delayed cerebral ischaemia is a serious complication. About 30 per cent of patients with delayed cerebral ischaemia die and another 30 per cent become permanently

disabled as a result of brain infarction, with only 40 per cent of the deficits being reversible.

Arterial vasospasm is usually demonstrable angiographically (Figure 2.123) but transcranial Doppler sonography is a useful method for non-invasive repeated examinations to monitor the spasm.[376] Vasospasm may persist even after death and can be demonstrated by post-mortem angiography.[515] Severe diffuse vasospasm with more than 50 per cent reduction in vessel calibre is usually associated with reduced global and regional CBF. Reduction in CBF correlates more closely with ischaemic symptoms than does the severity of angiographic vasospasm.[1092] In spite of severe vasospasm, CBF may be maintained by compensatory mechanisms, such as an increase in blood pressure and collateral circulation. CBF may be critically reduced, however, even in the absence of angiographic vasospasm, if SAH is associated with intracerebral haematoma, oedema or hydrocephalus. A minimum global CBF of 30–33 mL/100g/minute and regional CBF of 15–20 mL/100g/minute are needed to prevent the development of infarcts and permanent neurological deficits.[655]

The amount of subarachnoid blood in the basal cisterns and fissures within 3 days of the initial haemorrhage is highly predictive of the risk of delayed cerebral ischaemia and brain infarction. However, the site of these complicating lesions is not always related to the location of the maximum subarachnoid blood.[424,750] The volume of intraventricular blood is an independent predictor of ischaemia, as is the duration of the initial unconsciousness.[436]

Although the presence of blood in the CSF seems to be the crucial initiating event, the pathogenesis of arterial vasospasm is incompletely understood.[215,250,633] A host of compounds released from the subarachnoid blood clot has been proposed to be vasoconstrictive. Vasoconstrictive substances may also come from the circulation; morphological abnormalities suggesting increased BBB permeability have been described after SAH.[292,1092] These include increased endothelial pinocytosis and channel formation, opening of interendothelial tight junctions, endothelial detachment and destruction, and intraluminal platelet adhesion and aggregation on to the damaged endothelium within a few hours of SAH.[292] Contrast enhancement in the basal cisterns indicates increased arterial permeability in patients with SAH.[1092] Protracted contraction of the smooth muscle cells may result from the effect of vasoconstrictive agents (e.g. oxyhaemoglobin, endothelin-1, thromboxane A2, catecholamines, serotonin) and/or impairment of vasodilatation by mediators such as prostacyclin, endothelium-derived relaxing factor and NO.[215,228]

Increased Intracranial Pressure and Hydrocephalus

During the first few days after SAH, the majority of patients have increased intracranial pressure, which does not necessarily correlate with the amount of subarachnoid or intraventricular blood.[424] It seems to be caused largely by impaired reabsorption of CSF owing to the presence of blood in the subarachnoid space.[139] Increased intracranial pressure after SAH may also be caused by the mass effect of a space-occupying haematoma or brain oedema.

Ventricular dilatation occurs in about 20 per cent of patients during the acute phase and is related to the amount of intraventricular rather than subarachnoid blood.[418] About 10 per cent of patients develop delayed communicating hydrocephalus, often with symptoms of 'normal-pressure hydrocephalus', which can usually be alleviated by shunting.[418] This post-haemorrhagic hydrocephalus may result from meningeal fibrosis.[892]

Other Complications

Hypothalamic lesions (both perivascular haemorrhages and microinfarcts) have been described in patients dying soon after SAH. These may be associated with fluid and electrolyte imbalance. Electrocardiographic abnormalities and elevations of serum creatine kinase activity are common after SAH. These may be associated with a reduction in cardiac output, which increases the risk of cerebral ischaemia if vasospasm ensues.[660] Repeated SAH may cause leptomeningeal siderosis.

VASCULAR DISEASES OF THE SPINAL CORD

In general, the types of disease that affect blood vessels in the brain and brain stem also involve the spinal cord vasculature. In this section, the differences and special aspects of spinal cord circulatory diseases are discussed. The distribution of vascular diseases of the spinal cord reflects the anatomy of its vasculature (see earlier).

Atherosclerosis of Spinal Cord Arteries

Atherosclerosis does not usually affect spinal arteries severely. They may escape the brunt of this disease process because of their smaller size. However, spinal perfusion may be affected secondarily because the feeding arteries are often involved by atherosclerosis, which may reduce blood flow to the spinal cord. For example, the blood flow through the radicular arteries from a sclerotic aorta may be compromised at their aortic origin. Spinal cord infarction occasionally complicates surgical treatment of atherosclerotic aortic aneurysms (see later).

Vascular Malformations of the Spinal Cord

Similar types of malformation occur in the intracranial and intraspinal compartments, but their relative frequencies vary markedly. The variation is probably due to the smaller size of the blood vessels within the spinal canal and haemodynamic differences in different parts of the CNS.

Aneurysms

Isolated aneurysms not associated with AVMs are very rare in spinal arteries, although saccular and dissecting aneurysms and even giant aneurysms are reported.[651,954,967] They occur in the larger spinal arteries, for example the anterior spinal artery[954] or the artery of Adamkiewicz.[1050] Aneurysms associated with intraspinal arteriovenous malformations are more common, however (see Spinal Dural Arteriovenous Fistulae, below).[967] The high-pressure, high-volume turbulent flow in the feeding vessels to AVM is thought to be responsible.[868] Haemodynamic factors also explain the occurrence of aneurysms in association

with aortic coarctation or bilateral vertebral occlusion. In these cases spinal artery flow is increased because of their recruitment into the collateral circulation that bypasses the obstruction.[651]

Vascular Malformations

There is a lack of consensus on the definitions and classification of spinal vascular malformations.[581,967] The 1987 scheme of Rosenblum et al.[868] is often used. A new classification was proposed by Spetzler et al.[967] but has been criticized.

Spinal Dural Arteriovenous Fistulae

Arteriovenous fistulae are low-flow malformations.[868] The anatomy of these was characterized by selective angiography in the 1980s. The most common intraspinal vascular malformation (accounting for about one-third) is the spinal dural arteriovenous fistula. This has also been termed type I spinal AVM, angioma racemosum venosum, Foix–Alajouanine syndrome and angiodysgenetic necrotizing myelopathy.[238] Onset is usually after the third decade, with the peak during the fifth and sixth decades. The clinical picture is commonly one of progressive, often painless paraparesis. In a dural arteriovenous fistula, the shunting of blood occurs from a normal dural artery, most commonly at the thoracic or upper lumbar level (T4 to L3), through a fistula inside the dura sleeve of the emerging spinal nerve, into a single, structurally abnormal vein of the perimedullary venous plexus (Figure 2.124a).[103] On the basis of their age distribution and histopathology, arteriovenous fistulae are considered to be acquired malformations, in many cases probably related to trauma. Under increased intravascular pressure, the veins dilate and elongate, with variable thickening and fibrosis of their walls (arterialization), and form a meandering venous conglomerate, which is more prominent on the dorsal surface of the cord (Figure 2.124b). Dural arteriovenous fistulae almost never bleed. They cause ischaemic damage to the cord that can, if untreated, lead to extensive cord necrosis (Figure 2.124c). The ischaemia is ascribed mainly to spinal venous hypertension. Arteriovenous fistulae can usually be treated successfully by endovascular or surgical occlusion.

Spinal Intradural Arteriovenous Malformations

Intradural AVMs are a heterogeneous group of malformations that are often congenital. Many manifest during the first decade, with the peak age of presentation in the second and third decades.[868] They can affect any part of the spinal cord. The malformations have a nidus: a glomus of abnormal vessels either intramedullary or both extramedullary and intramedullary, and in the juvenile type occupying the whole spinal canal. They are generally high pressure, high flow malformations, supplied by single or multiple branches from spinal (medullary) arteries of the cord. This explains their tendency to bleed (in about one-third of symptomatic patients) and the occasional associated spinal bruit. These malformations have a tendency to

(a)

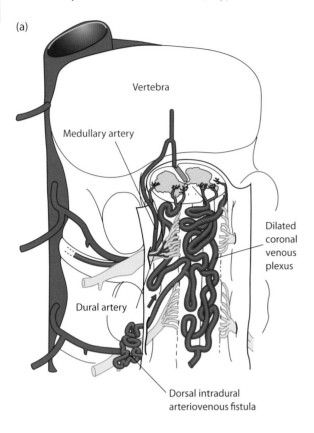

(b)

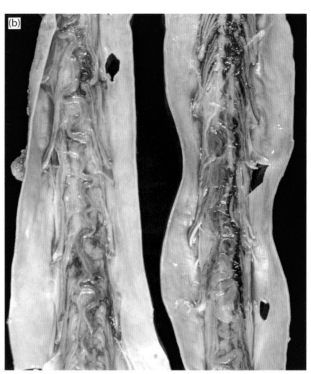

2.124 Spinal dural arteriovenous fistula. (a) Schematic drawing and **(b)** *ex vivo* appearance. The increased pressure at which blood is shunted to veins causes marked dilation and elongation of the veins, which become tortuous, with their walls thickened and fibrotic (arterialized). **(c)** In transverse section, the dilated veins are seen posterior to the extensively necrotic cord. Luxol-fast-blue–cresyl violet.

(a) Adapted with permission from Rosenblum et al.[868]

(c)

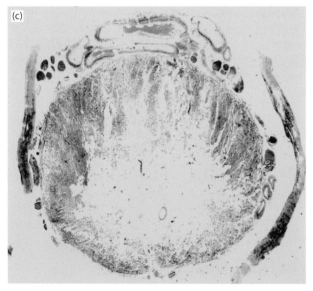

cause SAH of acute onset accompanied by back and root pain. They may present with progressive paraparesis and bowel and bladder dysfunction. Structurally, spinal AVMs are similar to their intracranial counterparts (Figure 2.52).

Spinal Cord Ischaemia

Infarcts in the spinal cord are less common than those in the brain. Their pathogenesis differs in several respects, although our present understanding is limited by the scarcity of epidemiological studies.[193] The smaller volume of the spinal cord is one explanation for the lower frequency of spinal infarcts. More importantly though, the smaller spinal arteries are relatively spared from atherosclerosis and thromboemboli. Most often, the infarct is caused by major vascular disease, the operative correction of a circulatory problem involving the aorta, or by vascular malformations of the spinal vasculature. Because of the variable and complicated anatomy of the vascular supply to the spinal cord and technical difficulties in collecting specimens at autopsy, the ultimate cause of spinal cord infarcts often remains undetermined.

Ischaemic Lesions Due to Aortic Diseases

Surgery-Associated Ischaemic Myelopathies

Vascular operations requiring cross-clamping of thoracic aorta carry considerable risk of spinal infarction and consequent paralysis.[359,798] Repair of thoracoabdominal aneurysms is one of the main reasons for such operations.[193] Until the early 1980s, these carried a risk of paralysis of up to 41 per cent in the most complex cases.[975] The main reason for paralysis was the loss of tributary blood flow to the lower spinal cord through intercostal or lumbar arteries below the level of T8, particularly if there was loss of flow through the important great radicular artery of Adamkiewicz (Figure 2.10). If these were oversewn during the operation, the risk of paralysis increased to 61 per cent, whereas their preservation and reimplantation reduced the risk considerably. The development of elaborate surgical, and electrophysiological and polarographic monitoring of cord function during the operation; hypothermia; distal aortic perfusion by atriofemoral shunting; CSF drainage and postoperative hypertension have reduced the frequency of paraparesis to 3–4 per cent.[975] In the late 1990s, the development of endovascular stent grafts provided a method for repair of thoracic aortic aneurysms and aortic dissections in which aortic clamping is not used. However, intercostal arteries are usually covered by the stent, which may cause spinal cord ischaemia. To prevent this stents that allow reimplantation of intercostal arteries are being developed.[194]

In correcting aortic coarctation associated with a patent ductus arteriosus (when the aorta is cross-clamped), the risk of paraparesis is much lower, because of the well-developed collaterals in these patients. In neonates and infants, the risk of ischaemic spinal damage is about 0.3 per cent. In older children and adults it is about 2.6 per cent.[238]

The ischaemic lesion commonly occurs below the midthoracic watershed region, because the flow through the upper spinal arteries is not sufficient to sustain this part of the cord without the supply through the great radicular artery (Figure 2.10). In less severe cases the ventral cord shows ischaemic damage but in severe cases most of the cord may undergo necrosis (Figure 2.125).

Aortic Dissection

Aortic dissection usually occurs as a complication of degenerative disease of the aortic wall. This may be atherosclerosis (often also associated with aneurysmal dilation) or a connective tissue disease such as Marfan's syndrome. It has also been reported in association with pregnancy, hypothyroidism and aortic stenosis. A tear in the intima allows blood to penetrate into the media and, often boosted by hypertension, to dissect the aortic wall. In about 2–8 per cent of

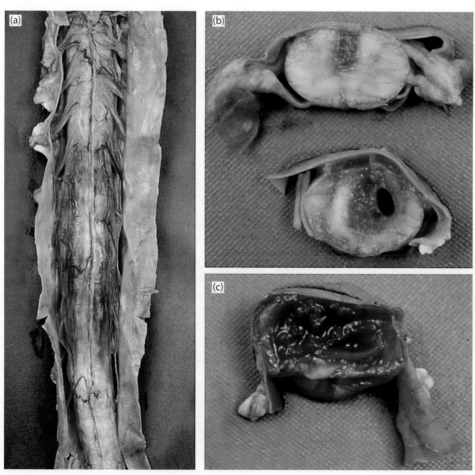

2.125 (a) Ischaemic lesion in the lower lumbar spinal cord. The lesion was caused by a dissecting aortic aneurysm. **(b)** Extensive necrosis of the cord above and below the **(c)** completely necrotic segment of the cord.

patients, aortic dissection presents with paraplegia or paraparesis and sensory loss associated with bowel and bladder dysfunction.[1120] As the dissection extends downwards, often from the aortic arch, it may shear off or occlude intercostal or lumbar arteries, cutting off blood flow to the spinal cord through the dorsal branches of these arteries.

The ischaemic lesions are variable in extension and topography. They tend to occur at the midthoracic T4–T6 level, the watershed region and a level at which the intercostal arteries are often involved (Figure 2.10). If the dissection extends downward to involve the artery of Adamkiewicz, the lesion is usually maximal at the T10–L1 level. The upper part of the cord is rarely damaged because it receives its blood supply from the vertebral arteries and costocervical or thyrocervical trunk. The latter is not affected by the dissection. The damage primarily involves the grey matter of the cord.

Embolic Occlusion

Emboli to the spinal vasculature are most commonly dislodged from an atherosclerotic aorta. The risk is increased when the aorta is subjected to operative procedures. The topography and extent of the ischaemic lesions depend on the anatomic pattern of spinal vasculature. As single events the symptoms may be trivial, but recurrent atheromatous embolism cause progressive myelopathy.[238]

A special type of embolic disease of the spinal cord is that caused by fibrocartilaginous emboli (FCE).[238,272] This is uncommon and only about 30 cases have been verified histopathologically (Figure 2.126). FCE has a bimodal age distribution, one peak in young adulthood and another in late middle age, with predominance in females.[272,1012] Its pre-mortem diagnosis is difficult because spinal biopsy is hardly ever performed and neither the clinical features nor neuroimaging is specific. Spinal cord infarction of unknown origin but occurring a few hours or days after minor spinal trauma should raise the suspicion of FCE. This occurs in the absence of a major vertebral bone lesion or of evidence of herniation of the nucleus pulposus into the vertebral body.

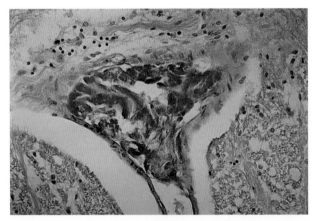

2.126 Fibrocartilaginous embolus in the anterior spinal artery. This embolus caused infarction of the medulla and anterior part of the cervical spinal cord.

Reproduced from Kase et al.[517] With permission from Lippincott Williams & Wilkins/Wolters Kluwer Health.

The pathogenesis of FCE has not been established. The spinal trauma may force disc material into vertebral body sinusoids.

A rare, neoplastic embolic process is angiotropic lymphoma (see Chapter 40). The intravascular growth and embolic spread are because of impaired inability of the malignant lymphoid cells that lack surface adhesion molecule CD18a to extravasate.[476] The neurological symptoms of IML are highly variable and non-specific but ischaemic myelopathy is relatively common.

Hypotensive Spinal Ischaemia

Hypotension is often an aggravating factor in ischaemic myelopathy complicating aortic surgery or dissection.[359,741,798] Other low-flow conditions such as cardiac arrest and severe hypotensive shock[238] and perinatal hypoxia–ischaemia[947] can also result in ischaemic spinal cord injury. The anatomical pattern of the spinal arteries results in an arterial border zone of ischaemic vulnerability in the lower thoracic region. Clinical studies have indicated that the mean level of the arterial watershed in global spinal ischaemia is T9 (in contrast with the classical view of a watershed at T4).[193] However, the pattern of the arterial supply is so variable that the vulnerable zone may even be lumbosacral (Figure 2.125).[78] In perinatal cases, lumbosacral segments were affected most severely.[947] Watershed infarction is thought to be less common in the spinal cord than the brain. This suggests that its development requires additional local circulatory factors besides the systemic arterial hypotension. However, patients with cerebral watershed infarcts are often obtunded and not examined thoroughly for spinal lesions; and in the event of death, post-mortem examination of the CNS is often limited to the brain.

Within the cord there is another watershed zone between the anterior and posterior spinal arteries. After severe ischaemia the cord may be almost completely necrotic (Figure 2.125). After less severe ischaemia or in less severely affected regions, the anterior grey matter bears the brunt of the damage, with better preservation of the surrounding white matter. In mildly affected cases and regions, there may be selective necrosis of the anterior grey matter only.

Venous Infarction

The most common type of venous infarction of the spinal cord is probably that associated with AVMs, usually in association with dural arteriovenous fistulae. The shunting of blood to the venous side through the fistula raises the venous pressure in the confined space of the spinal canal to such an extent that ischaemia ensues. Venous congestion rather than venous occlusion occurs. The infarction is thus non-haemorrhagic. Non-haemorrhagic venous infarction may also occur in the absence of a malformation but the pathogenesis in these cases is unclear.

Acute venous infarction can be caused by thrombosis of intramedullary or meningeal veins, e.g. due to cancer-related thrombophilia. In such cases, the infarct is usually haemorrhagic, the onset sudden and associated with pain.

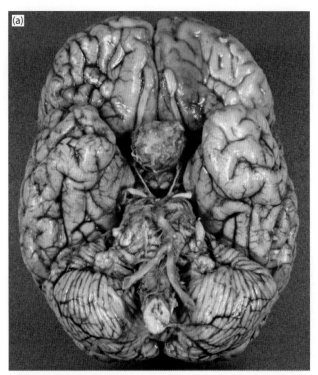

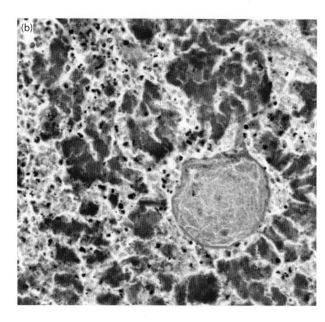

2.127 Pituitary apoplexy. (a) A large chromophobe macro-adenoma underwent necrosis, leading to the patient's death as a consequence of increased intracranial pressure and panhypopituitarism. **(b)** The adenoma cells are necrotic and polymorphonuclear leukocytes have invaded the tissue.

Images courtesy of A Paetau, Helsinki University Central Hospital, Helsinki, Finland.

Progression is rapid and the prognosis poor. The central parts of the cord are affected most severely. There may be an intramedullary haematoma.

A specific form of venous ischaemia occurs in association with brain death. The venous drainage in the upper cord is upwards into the cranial cavity. In brain death, the high intracranial pressure prevents this flow. If survival continues for some days the cervical and upper thoracic cord undergo venous infarction with circumferential necrosis in the cervical cord and radially oriented perivenous haemorrhages in the thoracic cord.[238]

Spinal Haemorrhage

Haemorrhages within the spinal canal can occur in the same four compartments with respect to the meninges as in the cranial cavity, i.e. intraparenchymal, subarachnoid, subdural and extradural.[238,816]

Intramedullary haemorrhage (haematomyelia) is often caused by trauma (see Chapter 10). Other causes are spinal AVMs, which are often located completely or partially within the cord and thus may bleed into the parenchyma. Bleeding diatheses (including due to anticoagulant therapy), intramedullary neoplasms or a syrinx may also be a cause or source of haemorrhage. On the other hand, intramedullary haemorrhages are seldom attributable to the common causes of brain haemorrhage, such as hypertension and amyloid angiopathy. A peculiarity of the spinal cord is the occurrence of small petechial haemorrhages of unknown pathogenesis. These are regularly found in asymptomatic patients.[238]

Spinal aneurysms may rupture, as do their intracranial counterparts, but are very uncommon on intraspinal arteries unless associated with AVMs. They are responsible for blood in the spinal subarachnoid space in under 1 per cent of all SAHs.[967] More commonly, ruptured AVMs cause SAH of primary intraspinal origin. Often the blood in the spinal subarachnoid space is a result of downward spread from an intracranial SAH.

Spinal subdural and epidural haemorrhages have the same predisposing factors as spinal haemorrhages.[816,883] In addition, lumbar puncture is a fairly common precipitating factor. Although spinal epidural haemorrhages are caused by trauma less often than their intracranial counterparts, minor trauma and straining can cause rupture of the thin-walled veins of the epidural plexus in the loose connective tissue surrounding the cord, especially posteriorly.

VASCULAR DISEASES OF THE PITUITARY GLAND

Ischaemic or haemorrhagic necrosis of the pituitary gland can result from a number of insults: local, such as trauma, neoplasms and compression by adjacent aneurysms (usually arising from the anterior part of the circle of Willis), and systemic or more generalized, such as hypotension, irradiation, ICH and increased intracranial pressure.

Pituitary apoplexy describes haemorrhagic necrosis of the pituitary. The term is reserved for the full-blown syndrome, evolving over hours to 1–2 days (see later). Most commonly, apoplexy occurs in the context of a pituitary adenoma, diagnosed or previously unknown, without predilection for a specific immunotype but often relatively large. In addition, apoplexy rarely complicates non-adenomatous pituitary disorders, such as abscess, metastatic tumour and lymphocytic hypophysitis.[266] It is uncommon, occurring as an acute serious complication in 1.6–2.8 per cent of pituitary adenomas. Lesser foci of haemorrhage or necrosis are sometimes detected by imaging or histology in adenomas from patients with minor or no related symptoms, and may explain the reported incidence of up to 13 per cent. An outlying incidence was reported of 21 per cent.[734] The clinical symptoms include severe headache, nausea and vomiting, motor ophthalmologic and visual disturbances resulting from compression of optic chiasm, and confusion or disturbed consciousness. Neutrophils infiltrating the necrotic tissue can spread into the CSF and, in previously undiagnosed cases of adenoma, may simulate purulent meningitis (Figure 2.127). Histopathological findings may be difficult to evaluate, because of extensive necrosis or haemorrhage, and infiltration by inflammatory cells. Upon immunohistochemical examination of the glandular remnants and taking into account the clinical history, it is often possible to determine the type of underlying adenoma.

Sheehan's syndrome is another disorder caused by necrosis of the pituitary gland. It is usually related to severe hypotension as a result of postpartum haemorrhage.[929]

With improvements in individual medical care, Sheehan's syndrome has become an extreme rarity under normal obstetric conditions. For example, Sheehan's syndrome was not detected among 55 patients with excessive obstetric haemorrhage.[285] The clinical symptoms reflect panhypopituitarism and include amenorrhoea, failure of lactation, weight loss, weakness, loss of pubic hair, breast atrophy and psychiatric disturbances.[285] Histopathological examination of the pituitary gland typically discloses extensive central necrosis with a rim of viable cells at the periphery. Sometimes lesser degrees of destruction may be evident. The neurohypophysis is usually spared.

ACKNOWLEDGEMENTS

This chapter updates and combines considerable information from Chapters 2 and 3 in the eighth edition. The previous authors (R Auer, G Sutherland, H Kalimo and M Kaste) are thanked for their substantial contributions, scholarship and generosity to allow us to re-use the material. We also thank A E Oakley, Newcastle University, for several illustrations. The contribution by Sylvia Asa to the section on vascular diseases of the pituitary gland is also gratefully acknowledged.

REFERENCES

1. Abbott NJ. Evidence for bulk flow of brain interstitial fluid: significance for physiology and pathology. *Neurochem Int* 2004;**45**:545–52.
2. Abbott NJ, Mendonca LL, Dolman DE. The blood–brain barrier in systemic lupus erythematosus. *Lupus* 2003;**12**:908–15.
3. Abdul–Rahman A, Dahlgren N, Ingvar M, *et al*. Local versus regional CBF in the rat at high (hypoxia) and low (phenobarbital anesthesia) flow rates. *Acta Physiol Scand* 1979;**106**:53–60.
4. Adams HP, Davis PH. Aneurysmal subarachnoid hemorrhage. In: Mohr JP, Choi DW, Grotta JC, *et al*. eds. *Stroke: pathophysiology, diagnosis, and management*. New York: Churchill Livingstone, 2004:377–96.
5. Adams HP Jr, Kassell NF, Torner JC, Haley EC Jr. Predicting cerebral ischaemia after aneurysmal subarachnoid hemorrhage: influences of clinical condition, CT results, and antifibrinolytic therapy. A report of the Cooperative Aneurysm Study. *Neurology* 1987;**37**:1586–91.
6. Adams RD, Jéquier M. The brain death syndrome: hypoxemic panencephalopathy. *Schweiz Med Wochenschr* 1969;**99**:65–73.
7. Adams RD, Cammermeyer J, Fitzgerald PJ, *et al*. Neuropathological aspects of thrombocytic acro-angiothrombosis;clinico-anatomical study of generalized platelet thrombosis. *J Neurol Neurosurg Psychiatry* 1948;**11**:27–43.
8. Adelson L, Sunshine I. Fatal hydrogen sulfide intoxication. *Arch Pathol* 1966;**81**:375–80.
9. Åden U, Bona E, Hagberg H, Fredholm BB. Changes in c-fos mRNA in the neonatal rat brain following hypoxic ischaemia. *Neurosci Lett* 1994;**180**:91–5.
10. Adhami F, Schloemer A, Kuan CY. The roles of autophagy in cerebral ischaemia. *Autophagy* 2007;**3**:42–4.
11. Agardh C–D, Siesjö BK. Hypoglycemic brain injury: phospholipids, free fatty acids, and cyclic nucleotides in the cerebellum of the rat after 30 and 60 minutes of severe insulin-induced hypoglycemia. *J Cereb Blood Flow Metab* 1981;**1**:267–75.
12. Agardh C–D, Folbergrová J, Siesjö BK. Cerebral metabolic changes in profound insulin-induced hypoglycemia, and in the recovery period following glucose administration. *J Neurochem* 1978;**31**:1135–42.
13. Agardh C–D, Kalimo H, Olsson Y, Siesjö BK. Hypoglycemic brain injury. I: metabolic and light microscopic findings in rat cerebral cortex during profound insulin-induced hypoglycemia and in the recovery period following glucose administration. *Acta Neuropathol (Berl)* 1980;**50**:31–41.
14. Agardh C–D, Chapman AG, Nilsson B, Siesjö BK. Endogenous substrates utilized by rat brain in severe insulin-induced hypoglycemia. *J Neurochem* 1981;**36**:490–500.
15. Agardh C–D, Kalimo H, Olsson Y, Siesjö BK. Hypoglycemic brain injury: metabolic and structural findings in rat cerebellar cortex during profound insulin-induced hypoglycemia and in the recovery period following glucose administration. *J Cereb Blood Flow Metab* 1981;**1**:71–84.
16. Agardh C–D, Chapman AG, Pelligrino D, Siesjö BK. Influence of severe hypoglycemia on mitochondrial and plasma membrane function in rat brain. *J Neurochem* 1982;**38**:662–8.
17. Agardh C–D, Rosén I, Ryding E. Persistent vegetative state with high cerebral blood flow following profound hypoglycemia. *Ann Neurol* 1983;**14**:482–6.
18. Agardh C–D, Smith M–L, Siesjö BK. The influence of hypothermia on hypoglycemia-induced brain damage in the rat. *Acta Neuropathol (Berl)* 1992;**83**:379–85.
19. Ahdab–Barmada M, Moossy J, Nemoto EM, Lin MR. Hyperoxia produces neuronal necrosis in the rat. *J Neuropathol Exp Neurol* 1986;**45**:233–46.
20. Alamowitch S, Plaisier E, Favrole P, *et al*. Cerebrovascular disease related to COL4A1 mutations in HANAC syndrome. *Neurology* 2009;**73**:1873–82.
21. Albaum HG, Noell WK, Chinn HI. Chemical changes in the rabbit brain during anoxia. *Am J Physiol* 1953;**174**:408–12.
22. Albers GW, Caplan LR, Easton JD, *et al*. Transient ischemic attack: proposal for a new definition. *N Engl J Med* 2002;**347**:1713–16.
23. Alberts M. Subarachnoid hemorrhage and intracranial aneurysms. In: Alberts M ed. *Genetics of cerebrovascular disease*. New York: Futura, 1999:237–59.
24. Albin RL, Greenamyre JT. Alternative excitotoxic hypotheses. *Neurology* 1992;**42**:733–8.
25. Albrecht C, Soumian S, Amey JS, *et al*. ABCA1 expression in carotid atherosclerotic plaques. *Stroke* 2004;**35**:2801–6.
26. Alderete JF, Jeri FR, Richardson EP Jr, *et al*. Irreversible coma: a clinical, electroencephalographic and

neuropathological study. *Trans Am Neurol Assoc* 1968;93:16–20.

27. Alemany M, Stenborg A, Terent A, Sonninen P, Raininko R. Coexistence of microhemorrhages and acute spontaneous brain hemorrhage: correlation with signs of microangiopathy and clinical data. *Radiology* 2006;238:240–47.

28. Alexander EL. Neurologic disease in Sjögren's syndrome: mononuclear inflammatory vasculopathy affecting central/peripheral nervous system and muscle. A clinical review and update of immunopathogenesis. *Rheum Dis Clin North Am* 1993;19:869–908.

29. Alexander MS, Dias PS, Uttley D. Spontaneous subarachnoid hemorrhage and negative cerebral pananogiography: review of 140 cases. *J Neurosurg* 1986;64:537–42.

30. Alg VS, Sofat R, Houlden H, Werring DJ. Genetic risk factors for intracranial aneurysms: a meta-analysis in more than 116,000 individuals. *Neurology* 2013;80:2154–65.

31. Allan LM, Rowan EN, Firbank MJ, *et al.* Long term incidence of dementia, predictors of mortality and pathological diagnosis in older stroke survivors. *Brain* 2012;134:3716–27.

32. Allroggen H, Abbott RJ. Cerebral venous sinus thrombosis. *Postgraduate medical journal* 2000;76:12–5.

33. Al-Shahi R, Bhattacharya JJ, Currie DG, *et al.* Prospective, population-based detection of intracranial vascular malformations in adults: the Scottish Intracranial Vascular Malformation Study (SIVMS). *Stroke* 2003;34:1163–9.

34. Al-Shahi R, Fang JS, Lewis SC, *et al.* Prevalence of adults with brain arteriovenous malformations: a community-based study in Scotland using capture-recapture analysis. *J Neurol Neurosurg Psychiatry* 2002;73:547–51.

35. Altamura C, Reinhard M, Vry MS, *et al.* The longitudinal changes of BOLD response and cerebral hemodynamics from acute to subacute stroke. A fMRI and TCD study. *BMC Neurosci* 2009;10:151.

36. Amara SG, Fontana AC. Excitatory amino acid transporters: keeping up with glutamate. *Neurochem Int* 2002;41:313–18.

37. Amarenco P, Bogousslavsky J, Caplan LR, Donnan GA, Hennerici MG. Classification of stroke subtypes. *Cerebrovasc Dis* 2009;27:493–501.

38. Amarenco P, Bogousslavsky J, Caplan LR, Donnan GA, Hennerici MG. New approach to stroke subtyping: the A–S–C–O (phenotypic) classification of stroke. *Cerebrovasc Dis* 2009;27:502–8.

39. Ameri A, Bousser MG. Cerebral venous thrombosis. *Neurol Clin* 1992;10:87–111.

40. Amin–Hanjani S, Robertson R, Arginteanu MS, Scott RM. Familial intracranial arteriovenous malformations: case report and review of the literature. *Pediatr Neurosurg* 1998;29:208–13.

40b. Amyry-Moghaddam M, Ottersen OP. The molecular basis of water transport in the brain. *Nat Rev Neurosci* 2003;4:991–1001.

41. Anders KH, Wang ZZ, Kornfeld M, *et al.* Giant cell arteritis in association with cerebral amyloid angiopathy: immunohistochemical and molecular studies. *Hum Pathol* 1997;28:1237–46.

42. Anderson CS, Chakera TMH, Stewart–Wynne EG, Jamrozik KD. Spectrum of primary intracerebral hemorrhage in Perth, Western Australia, 1989–90: incidence and outcome. *J Neurol Neurosurg Psychiatry* 1994;57:936–40.

43. Anderson JM, Milner RD, Strich SJ. Effects of neonatal hypoglycaemia on the nervous system: a pathological study. *J Neurol Neurosurg Psychiatry* 1967;30:295–310.

44. Anson J, Crowell RM. Cervicocranial arterial dissection. *Neurosurgery* 1991;29:89–96.

45. Antiphospholipid Antibodies in Stroke Study (APASS). The Antiphospholipid Antibodies in Stroke Study (APASS) group. Anticardiolipin antibodies are an independent risk factor for first ischemic stroke. *Neurology* 1993;43:2069–73.

46. Aoki K, Uchihara T, Tsuchiya K, *et al.* Enhanced expression of aquaporin 4 in human brain with infarction. *Acta Neuropathol* 2003;106:121–4.

47. Aoki N. Do intracranial arteriovenous malformations cause subarachnoid haemorrhage? Review of computed tomography features of ruptured arteriovenous malformations in the acute stage. *Acta Neurochir* 1991;112:92–5.

48. Aoki N, Mizutani H. Does moyamoya disease cause subarachnoid hemorrhage? Review of 54 cases with intracranial hemorrhage confirmed by computerized tomography. *J Neurosurg* 1984;60:348–53.

49. APASS. Anticardiolipin antibodies are an independent risk factor for first ischemic stroke. The Antiphospholipid Antibodies in Stroke Study (APASS) Group. *Neurology* 1993;43:2069–73.

50. Arakawa S, Saku Y, Ibayashi S, *et al.* Blood pressure control and recurrence of hypertensive brain hemorrhage. *Stroke* 1998;29:1806–9.

51. Arboix A, Besses C. Cerebrovascular disease as the initial clinical presentation of haematological disorders. *Eur Neurol* 1997;37:207–11.

52. Arboix A, Marti–Vilalta JL. Lacunar stroke. *Expert Rev Neurother* 2009;9:179–96.

53. Arima H, Anderson CS, Wang JG, *et al.* Lower treatment blood pressure is associated with greatest reduction in hematoma growth after acute intracerebral hemorrhage. *Hypertension* 2010;56:852–8.

54. Arima H, Tzourio C, Anderson C, *et al.* Effects of perindopril-based lowering of blood pressure on intracerebral hemorrhage related to amyloid angiopathy: the PROGRESS trial. *Stroke* 2010;41:394–6.

55. Arima K, Yanagawa S, Ito N, Ikeda S. Cerebral arterial pathology of CADASIL and CARASIL (Maeda syndrome). *Neuropathology* 2003;23:327–34.

56. Arnaud L, Haroche J, Mathian A, Gorochov G, Amoura Z. Pathogenesis of Takayasu's arteritis: a 2011 update. *Autoimmun Rev* 2011;11:61–7.

57. Artavanis-Tsakonas S, Rand MD, Lake RJ. Notch signalling: cell fate control and signal integration in development. *Science* 1999;284:770–76.

58. Arumugam TV, Magnus T, Woodruff TM, *et al.* Complement mediators in ischaemia-reperfusion injury. *Clin Chim Acta* 2006;374:33–45.

59. Asahara T, Kawamoto A, Masuda H. Concise review: Circulating endothelial progenitor cells for vascular medicine. *Stem Cells* 2012;29:1650–5.

60. Asahi M, Hoshimaru M, Uemura Y, *et al.* Expression of interleukin-1β converting enzyme gene family and bcl–2 gene family in the rat brain following permanent occlusion of the middle cerebral artery. *J Cereb Blood Flow Metab* 1997;17:11–18.

61. Ashwal S, Schneider S, Tomasi L, Thompson J. Prognostic implications of hyperglycemia and reduced cerebral blood flow in childhood near-drowning. *Neurology* 1990;40:820–3.

62. Asplund K, Tuomilehto J, Stegmayr B, Wester PO, Tunstall–Pedoe H. Diagnostic criteria and quality control of the registration of stroke events in the MONICA project. *Acta Med Scand Suppl* 1988;728:26–39.

63. Assaf SY, Chung SH. Release of endogenous Zn2+ from brain tissue during activity. *Nature* 1984;308:734–6.

64. Astrup J, Siesjö BK, Symon L. Thresholds in cerebral ischaemia: the ischemic penumbra. *Stroke* 1981;12:723–5.

65. Attems J, Jellinger K, Thal DR, Van Nostrand W. Review: sporadic cerebral amyloid angiopathy. *Neuropathol Appl Neurobiol* 2011;37:75–93.

66. Attia W, Tada T, Hongo K, *et al.* Microvascular pathological features of immediate perinidal parenchyma in cerebral arteriovenous malformations: giant bed capillaries. *J Neurosurg* 2003;98:823–7.

67. Auer RN. Progress review: hypoglycemic brain damage. *Stroke* 1986;17:699–708.

68. Auer RN. Calcium channel antagonists in cerebral ischaemia: a review. *Drug Dev* 1993;2:307–17.

69. Auer RN. Hypoglycemic brain damage. In: Kalimo H ed. *Cerebrovascular diseases*. Basel: ISN Neuropath Press, 2005:273.

70. Auer RN, Anderson LG. Hypoglycaemic brain damage: effect of a dihydropyridine calcium channel antagonist in rats. *Diabetologia* 1996;39:129–34.

71. Auer RN, Siesjö BK. Biological differences between ischaemia, hypoglycemia, and epilepsy. *Ann Neurol* 1988;24:699–707.

72. Auer R, Kalimo H, Olsson Y, Wieloch T. The dentate gyrus in hypoglycemia: pathology implicating excitotoxin-mediated neuronal necrosis. *Acta Neuropathol (Berl)* 1985;67:279–88.

73. Auer RN, Kalimo H, Olsson Y, Siesjö BK. The temporal evolution of hypoglycemic brain damage. I: light- and electron-microscopic findings in the rat cerebral cortex. *Acta Neuropathol (Berl)* 1985;67:13–24.

74. Auer RN, Kalimo H, Olsson Y, Siesjö BK. The temporal evolution of hypoglycemic brain damage. II: light- and electron-microscopic findings in the hippocampal gyrus and subiculum of the rat. *Acta Neuropathol (Berl)* 1985;67:25–36.

75. Auer RN, Hugh J, Cosgrove E, Curry B. Neuropathologic findings in three cases of profound hypoglycemia. *Clin Neuropathol* 1989;8:63–8.

76. Augustin HG, Koh GY, Thurston G, Alitalo K. Control of vascular morphogenesis and homeostasis through

the angiopoietin-TIE system. *Nat Rev Mol Cell Biol* 2009;**10**:165–77.

77. Awad IA, Robinson JR Jr, Mohanty S, Estes ML. Mixed vascular malformations of the brain: clinical and pathogenetic considerations. *Neurosurgery* 1993;**33**:179–88.

78. Azzarelli B, Roessmann U. Diffuse 'anoxic' myelopathy. *Neurology* 1977;**27**:1049–52.

79. Bacigaluppi M, Comi G, Hermann DM. Animal models of ischemic stroke. Part two: modelling cerebral ischaemia. *Open Neurol J* 2010;**4**:34–8.

80. Back T, Hoehn–Berlage M, Kohno K, Hossmann KA. Diffusion nuclear magnetic resonance imaging in experimental stroke: correlation with cerebral metabolites. *Stroke* 1994;**25**:494–500.

81. Badaut J, Hirt L, Granziera C, et al. Astrocyte-specific expression of aquaporin-9 in mouse brain is increased after transient focal cerebral ischaemia. *J Cereb Blood Flow Metab* 2001;**21**:477–82.

82. Bailey DM. Radical dioxygen: from gas to (unpaired!) electrons. *Adv Exp Med Biol* 2003;**543**:201–21.

83. Bailey EL, Smith C, Sudlow CL, Wardlaw JM. Pathology of lacunar ischaemic stroke in humans - a systematic review. *Brain Pathol* 2012;**22**(5):583–91.

84. Baldelli RJ, Green FHY, Auer RN. Sulfide toxicity: mechanical ventilation and hypotension determine survival rate and brain necrosis. *J Appl Physiol* 1993;**75**:1348–53.

85. Baldi S, Mounayer C, Piotin M, et al. Balloon-assisted coil placement in wide-necked bifurcation aneurysms by use of a new, compliant balloon microcatheter. *Am J Neuroradiol* 2003;**24**:1222–5.

86. Balentine JD. Pathogenesis of central nervous system lesions induced by exposure to hyperbaric oxygen. *Am J Pathol* 1968;**53**:1097–109.

87. Balentine JD. Ultrastructural pathology of hyperbaric oxygenation in the central nervous system, observations in the anterior horn gray matter. *Lab Invest* 1974;**31**:580–92.

88. Balentine JD, Greene WB. Myelopathy induced by lactic acid. *Acta Neuropathol (Berl)* 1987;**73**:233–9.

89. Barer GR, Fairlie J, Slade JY, et al. Effects of NOS inhibition on the cardiopulmonary system and brain microvascular markers after intermittent hypoxia in rats. *Brain Res* 2006;**1098**:196–203.

90. Barinagarrementeria F, Cantu C. Primary medullary hemorrhage. Report of four cases and review of the literature. *Stroke* 1994;**25**:1684–7.

91. Barker R, Wellington D, Esiri MM, Love S. Assessing white matter ischemic damage in dementia patients by measurement of myelin proteins. *J Cereb Blood Flow Metab* 2013;**33**:1050–7.

92. Barker R, Ashby EL, Wellington D, et al. Pathophysiology of white matter perfusion in Alzheimer's disease and vascular dementia. *Brain* 2014;**137**(Pt 5):1524–32.

93. Baron JC. Mapping the ischaemic penumbra with PET: implications for acute stroke treatment. *Cerebrovasc Dis* 1999;**9**:193–201.

94. Barry DI, Strandgaard S, Graham DI, et al. Cerebral blood flow in rats with renal and spontaneous hypertension: resetting of lower limit of autoregulation. *J Cereb Blood Flow Metab* 1982;**2**:347–53.

95. Basnyat B, Wu T, Gertsch JH. Neurological conditions at altitude that fall outside the usual definition of altitude sickness. *High Alt Med Biol* 2004;**5**:171–9.

95a. Batra S, Lin D, Recinos PF, et al. Cavernous malformations: natural history, diagnosis and treatment. *Nat Rev Neurol* 2009;**5**(12):659–70.

96. Bauer I, Pannen BH. Bench-to-bedside review: Carbon monoxide – from mitochondrial poisoning to therapeutic use. *Crit Care* 2009;**13**:220.

97. Bazille C, Keohane C, Gray F. Systemic diseases and drug-induced vasculitis. In: Kalimo H ed. *Pathology and genetics: cerebrovascular diseases*. Basel: ISN Neuropath Press, 2005:151–62.

98. Beck H, Plate KH. Angiogenesis after cerebral ischaemia. *Acta Neuropathol* 2009;**117**:481–96.

99. Behar KL, den Hollander JA, Petroff OAC, et al. Effect of hypoglycemic encephalopathy upon amino acids, high-energy phosphates, and pHi in the rat brain in vivo: detection by sequential 1H and 31P NMR spectroscopy. *J Neurochem* 1985;**44**:1045–55.

100. Behnke AR, Johnson FS, Poppen JR, Motley EP. The effects of oxygen on man at pressures from one to four atmospheres. *Am J Physiol* 1935;**110**:565–72.

101. Bellenguez C, Bevan S, Gschwendtner A, et al. Genome-wide association study identifies a variant in HDAC9 associated with large vessel ischemic stroke. *Nat Genet* 2012;**44**:328–33.

102. Benedikz E, Blöndal H, Gudmundsson G. Skin deposits in hereditary cystatin C amyloidosis. *Virchows Arch A Pathol Anat Histopathol* 1990;**417**:325–31.

103. Benhaiem N, Poirier J, Hurth M. Arteriovenous fistulae of the meninges draining into the spinal veins: a histological study of 28 cases. *Acta Neuropathol (Berl)* 1983;**62**:103–11.

104. Benseler SM, Schneider R. Central nervous system vasculitis in children. *Curr Opin Rheumatol* 2004;**16**:43–50.

105. Benseler SM, Silverman E, Aviv RI, et al. Primary central nervous system vasculitis in children. *Arthritis Rheum* 2006;**54**:1291–7.

106. Benveniste H, Drejer J, Schousboe A, Diemer NH. Elevation of the extracellular concentrations of glutamate and aspartate in rat hippocampus during transient cerebral ischaemia monitored by intracerebral microdialysis. *J Neurochem* 1984;**43**:1369–74.

107. Bergada I, Suissa S, Dufresne J, Schiffrin A. Severe hypoglycemia in IDDM children. *Diabetes Care* 1989;**12**:239–44.

108. Bergeron M, Ferriero DM, Vreman HJ, et al. Hypoxia-ischaemia, but not hypoxia alone, induces the expression of heme oxygenase-1 (HSP32) in newborn rat brain. *J Cereb Blood Flow Metab* 1997;**17**:647–58.

109. Berlet HH. Hypoxic survival of normoglycaemic young adult and adult mice in relation to cerebral metabolic rates. *J Neurochem* 1976;**26**:1267–74.

110. Berlit P. Diagnosis and treatment of cerebral vasculitis. *Ther Adv Neurol Disord* 2010;**3**:29–42.

111. Bernatsky S, Clarke A, Gladman DD, et al. Mortality related to cerebrovascular disease in systemic lupus erythematosus. *Lupus* 2006;**15**:835–9.

112. Bick RL, Arun B, Frenkel EP. Antiphospholipid–thrombosis syndromes. *Haemostasis* 1999;**29**:100–10.

113. Birnbaum J, Hellmann DB. Primary angiitis of the central nervous system. *Arch Neurol* 2009;**66**:704–9.

114. Black PM. Brain death. *N Engl J Med* 1978;**299**:338–44.

115. Blevins G, Macaulay R, Harder S, et al. Oculoleptomeningeal amyloidosis in a large kindred with a new transthyretin variant Tyr69His. *Neurology* 2003;**60**:1625–30.

116. Blin J, Ray CA, Chase TN, Piercey MF. Regional cerebral glucose metabolism compared in rodents and humans. *Brain Res* 1991;**568**:215–22.

117. Bliss TVP, Lømo T. Long-lasting potentiation of synaptic transmission in the dentate area of the anaesthetized rabbit following stimulation of the perforant path. *J Physiol* 1973;**232**:331–56.

118. Blumbergs PC, Byrne E. Hypotensive central infarction of the spinal cord. *J Neurol Neurosurg Psychiatry* 1980;**43**:751–3.

119. Boche D, Zotova E, Weller RO, et al. Consequence of Aβ immunization on the vasculature of human Alzheimer's disease brain. *Brain* 2008;**131**:3299–310.

120. Boero JA, Ascher J, Arregui A, et al. Increased brain capillaries in chronic hypoxia. *J Appl Physiol* 1999;**86**:1211–19.

121. Bogousslavsky J. Subcortical infarcts. In: Fisher M and Bogousslavsky J eds. *Current review of cerebrovascular disease*. Philadelphia, PA: Current Medicine, 1993:31–40.

122. Boiten J, Lodder J, Kessels F. Two clinically distinct lacunar infarct entities? A hypothesis. *Stroke* 1993;**24**:652–6.

123. Bonita R, Beaglehole R. Stroke mortality. In: Whisnant J ed. *Stroke: populations, cohorts, and clinical trials*. Oxford: Butterworth Heinemann, 1993:59–79.

124. Bonita R, Beaglehole R, North JD. Event incidence and case fatality rates of cerebrovascular disease in Auckland, New Zealand. *Am J Epidemiol* 1984;**120**:236–43.

125. Bonner H, Erslev A. The blood and the lymphoid organs. In: Rubin E, Farber J eds. *Pathology*, 2nd edn. Philadelphia, PA: JB Lippincott, 1994:994–1096.

126. Bornebroek M, Haan J, Maat-Schieman, et al. Hereditary cerebral hemorrhage with amyloidosis-Dutch type (HCHWA-D): I. A review of clinical, radiologic and genetic aspects. *Brain Pathol* 1996; **6**:111–14.

127. Boyajian RA, Otis SM. Acute effects of smoking on human cerebral blood flow: a transcranial Doppler ultrasonography study. *J Neuroimaging* 2000;**10**:204–8.

128. Boysen G. Primary intracerebral hemorrhage. In: Fisher M, Bogousslavsky J eds. *Current review of cerebrovascular disease*. Philadelphia, PA: Current Medicine, 1993:78–88.

129. Brandt T, Hausser I, Orberk E, et al. Ultrastructural connective tissue abnormalities in patients with

spontaneous cervicocerebral artery dissections. *Ann Neurol* 1998;**44**:281–5.

130. Brega KE, Seltzer WK, Munro LG, et al. Genotypic variations of type III collagen in patients with cerebral aneurysms. *Surg Neurol* 1996;**46**:253–6.

131. Brey RL, Escalante A. Neurological manifestations of antiphospholipid antibody syndrome. *Lupus* 1998;**7**(Suppl 2):67–74.

132. Briceno CE, Resch L, Bernstein M. Cerebral amyloid angiopathy presenting as a mass lesion. *Stroke* 1987;**18**:234–9.

133. Brierley JB, Cooper JE. Cerebral complications of hypotensive anaesthesia in a healthy adult. *J Neurol Neurosurg Psychiatry* 1962;**25**:24–30.

134. Brierley JB, Adams JH, Graham DI, Simpson JA. Neocortical death after cardiac arrest: a clinical, neurophysiological, and neuropathological report of two cases. *Lancet* 1971;**2**:560–5.

135. Brierley JB, Brown AW, Meldrum BS. The nature and time course of the neuronal alterations resulting from oligaemia and hypoglycemia in the brain of *Macaca mulatta*. *Brain* Res 1971;**25**:483–99.

136. Brierley JB, Meldrum BS, Brown AW. The threshold and neuropathology of cerebral 'anoxic-ischemic' cell change. *Arch Neurol* 1973;**29**:367–74.

137. Brierley JB, Prior PF, Calverley J, Brown AW. Cyanide intoxication in Macaca mulatta. *J Neurol Sci* 1977;**31**:133–57.

138. Brierley JB, Prior PF, Calverley J, Brown AW. Profound hypoxia in *Papio anubis* and *Macaca mulatta*: physiological and neuropathological effects. I. Abrupt exposure following normoxia. II: abrupt exposure following moderate hypoxia. *J Neurol Sci* 1978;**37**:1–29.

139. Brinker T, Seifert V, Stolke D. Acute changes in the dynamics of the cerebrospinal fluid system during experimental subarachnoid hemorrhage. *Neurosurgery* 1990;**27**:369–72.

140. Broderick J, Phillips S, Whisnant JP, et al. Incidence rates of stroke in the eighties: the end of the decline in stroke? *Stroke* 1989;**20**:577–82.

141. Broderick JP, Brott TG, Duldner JE, Tomsick T, Huster G. Volume of intracerebral hemorrhage. A powerful and easy-to-use predictor of 30-day mortality. *Stroke* 1993;**24**:987–93.

142. Brott T, Broderick J, Kothari R, et al. Early hemorrhage growth in patients with intracerebral hemorrhage. *Stroke* 1997;**28**:1–5.

143. Broughton BR, Reutens DC, Sobey CG. Apoptotic mechanisms after cerebral ischaemia. *Stroke* 2009;**40**:e331–9.

144. Brouns R, De Deyn PP. The complexity of neurobiological processes in acute ischemic stroke. *Clin Neurol Neurosurg* 2009;**111**:483–95.

145. Brown AW, Brierley JB. Evidence for early anoxic-ischaemic cell damage in the rat brain. *Experientia* 1966;**22**:546–7.

146. Brown AW, Brierley JB. The earliest alterations in rat neurones and astrocytes after anoxia-ischaemia. *Acta neuropathol* 1973;**23**:9–22.

147. Brown WR, Thore CR. Review: cerebral microvascular pathology in ageing and neurodegeneration. *Neuropathol Appl Neurobiol* 2011;**37**:56–74.

148. Bruening R, Dichgans M, Berchtenbreiter C, et al. Cerebral autosomal dominant arteriopathy with subcortical infarcts and leukoencephalopathy: decrease in regional cerebral blood volume in hyperintense subcortical lesions inversely correlates with disability and cognitive performance. *Am J Neuroradiol* 2001;**22**:1268–74.

149. Brugniaux JV, Hodges AN, Hanly PJ, Poulin MJ. Cerebrovascular responses to altitude. *Respir Physiol Neurobiol* 2007;**158**:212–23.

150. Bruick RK. Expression of the gene encoding the proapoptotic Nip3 protein is induced by hypoxia. *Proc Natl Acad Sci U S A* 2000;**97**:9082–7.

151. Bruno A, Adams JP Jr, Biller J, et al. Cerebral infarction due to moyamoya disease in young adults. *Stroke* 1988;**19**:826–33.

152. Buckley NA, Juurlink DN, Isbister G, Bennett MH, Lavonas EJ. Hyperbaric oxygen for carbon monoxide poisoning. *Cochrane Database Syst Rev* 2011:CD002041.

153. Budabin M. Neurologic complications of open heart surgery. *Mt Sinai J Med* 1982;**49**:311–13.

154. Budka H, Gray F. HIV induced central nervous system pathology. In: Gray F ed. *Atlas of the neuropathology of HIV infection*. Oxford: Oxford University Press, 1993:1–36.

155. Burger PC, Vogel FS. Hemorrhagic white matter infarction in three critically ill patients. *Hum Pathol* 1977;**8**:121–32.

156. Burnett WW, King EG, Grace M, Hall WF. Hydrogen sulfide poisoning: a review of 5 years' experience. *Can Med Assoc J* 1977;**117**:1277–81.

157. Busto R, Dietrich WD, Globus MY–T, Ginsberg MD. The importance of brain temperature in cerebral ischemic injury. *Stroke* 1989;**20**:1113–14.

158. Buttner A. Review: The neuropathology of drug abuse. *Neuropathol Appl Neurobiol* 2011;**37**:118–34.

159. Cacciapuoti F. Some considerations about the hypercoagulable states and their treatments. *Blood Coagul Fibrinolysis* 2011;**22**:155–9.

160. Calabrese LH. Vasculitis and infection with the human immunodeficiency virus. *Rheum Dis Clin North Am* 1991;**17**:131–47.

161. Calabrese LH, Duna GF. Evaluation and treatment of central nervous system vasculitis. *Curr Opin Rheumatol* 1995;**7**:37–44.

162. Calder IM, Palmer AC, Hughes JT, et al. Spinal cord degeneration associated with type II decompression sickness:case report. *Paraplegia* 1989;**27**:51–7.

163. Calhoun CL, Mottaz JH. Capillary bed of the rat cerebral cortex. The fine structure in experimental cerebral infarction. *Arch Neurol* 1966;**15**:320–8.

164. Camejo G, Hurt Camejo E, Wilkund O, et al. Association of Apo N lipoproteins with arterial proteoglycans: pathological significance and molecular basis. *Atherosclerosis* 1998;**139**:205–22.

165. Cammermeyer J. The importance of avoiding 'dark' neurons in experimental neuropathology. *Acta Neuropathol (Berl)* 1961;**1**:245–70.

166. Cammermeyer J. Is the solitary dark neuron a manifestation of postmortem trauma to the brain inadequately fixed by perfusion? *Histochemistry* 1978;**56**:97–115.

167. Campbell JA. Diet and resistance to oxygen want. *Quart J Exp Physiol* 1939;**29**:259–75.

168. Cantu C, Barinagarrementeria F. Cerebral venous thrombosis associated with pregnancy and puerperium. Review of 67 cases. *Stroke* 1993;**24**:1880–4.

169. Caplan LR. Dilatative arteriopathy (dolichoectasia): what is known and not known. *Ann Neurol* 2005;**57**:461–71.

170. Caplan RC. Brain embolism, revisited. *Neurology* 1993;**43**:1281–7.

171. Caranci F, Briganti F, Cirillo L, Leonardi M, Muto M. Epidemiology and genetics of intracranial aneurysms. *Eur J Radiol* 2013;**82**:1598–605.

172. Carare RO, Bernardes–Silva M, Newman TA, et al. Solutes, but not cells, drain from the brain parenchyma along basement membranes of capillaries and arteries: significance for cerebral amyloid angiopathy and neuroimmunology. *Neuropathol Appl Neurobiol* 2008;**34**:131–44.

173. Carare RO, Hawkes CA, Jeffrey M, Kalaria RN, Weller RO. Cerebral amyloid angiopathy, Prion angiopathy, CADASIL and the spectrum of Protein Elimination-Failure Angiopathies (PEFA) in neurodegenerative disease with a focus on therapy. *Neuropathol Appl Neurobiol* 2013;**39**(6):593–611.

173a. Carlsson LE, Santoso S, Spitzer C, et al. The alpha2 gene coding sequence T807/A873 of the platelet collagen receptor integrin alpha2beta1 might be a genetic risk factor for the development of stroke in younger patients. *Blood* 1999;**93**:3583–6

174. Carmeliet P, Ferreira V, Breier G, et al. Abnormal blood vessel development and lethality in embryos lacking a single VEGF allele. *Nature* 1996;**380**:435–9.

175. Carr SC, Farb A, Pearce WH, et al. Activated inflammatory cells are associated with plaque rupture in carotid artery stenosis. *Surgery* 1997;**122**:757–63.

176. Carson CW, Beall LD, Hunder GG, Johnson CM, Newman W. Serum ELAM-1 is increased in vasculitis, scleroderma, and systemic lupus erythematosus. *J Rheumatol* 1993;**20**:809–14.

177. Casparie AF, Elving LD. Severe hypoglycemia in diabetic patients: frequency, causes, prevention. *Diabetes Care* 1985;**8**:141–5.

178. Celesia GG, Grigg MM, Ross E. Generalized status myoclonicus in acute anoxic and toxic-metabolic encephalopathies. *Arch Neurol* 1988;**45**:781–4.

179. Cervoni L, Artico M, Salvati M, et al. Epileptic seizures in intracerebral hemorrhage: a clinical and prognostic study of 55 cases. *Neurosurg Rev* 1994;**17**:185–8.

180. Chabriat H, Pappata S, Ostergaard L, et al. Cerebral hemodynamics in CADASIL before and after acetazolamide challenge assessed with MRI bolus tracking. *Stroke* 2000;**31**:1904–12.

181. Chabriat H, Joutel A, Dichgans M, Tournier–Lasserve E, Bousser MG. CADASIL. *Lancet Neurol* 2009;**8**:643–53.

182. Challa VR, Bell MA, Moody DM. A combined hematoxylin–eosin, alkaline phosphatase and high-resolution

microradiographic study of lacunes. *Clin Neuropathol* 1990;9:196–204.

183. Challa VR, Moody DM, Bell MA. The Charcot–Bouchard aneurysm controversy: impact of a new histologic technique. *J Neuropathol Exp Neurol* 1992; 51:264–71.

184. Challa VR, Moody DM, Brown WR. Vascular malformations of the central nervous system. *J Neuropathol Exp Neurol* 1995;54:609–21.

185. Chalmers K, Wilcock GK, Love S. APOE ε4 influences the pathological phenotype of Alzheimer's disease by favouring cerebrovascular over parenchymal accumulation of A beta protein. *Neuropathol Appl Neurobiol* 2003;29:231–8.

186. Chan PH. Oxygen radicals in focal cerebral ischaemia. *Brain Pathol* 1994;4:59–65.

187. Chang CL, Donaghy M, Poulter M. Migraine and stroke in young women: case-control study. The World Health Organization Collaborative Study of Cardiovascular Disease and Steroid Hormone Contraception. *Br Med J* 1999;318:13–18.

188. Chappell JC, Bautch VL. Vascular development: genetic mechanisms and links to vascular disease. *Curr Top Dev Biol* 2010;90:43–72.

189. Chen H, Kim GS, Okami N, Narasimhan P, Chan PH. NADPH oxidase is involved in post-ischemic brain inflammation. *Neurobiol Dis* 2011;42:341–8.

190. Chen H, Yoshioka H, Kim GS, et al. Oxidative stress in ischemic brain damage: mechanisms of cell death and potential molecular targets for neuroprotection. *Antioxid Redox Signal* 2011;14:1505–17.

191. Chen XY, Wong KS, Lam WW, Zhao HL, Ng HK. Middle cerebral artery atherosclerosis: histological comparison between plaques associated with and not associated with infarct in a postmortem study. *Cerebrovasc Dis* 2008;25:74–80.

192. Cherici G, Alesiani M, Pellegrini–Giampietro DE, Moroni F. Ischaemia does not induce the release of excitotoxic amino acids from the hippocampus of newborn rats. *Dev Brain Res* 1991;60:235–40.

193. Cheshire WP, Santos CC, Massey EW, Howard JF Jr. Spinal cord infarction: etiology and outcome. *Neurology* 1996;47:321–30.

194. Chiesa R, Melissano G, Marrocco–Trischitta MM, et al. Spinal cord ischaemia after elective stent-graft repair of the thoracic aorta. *J Vasc Surg* 2005;42:11–17.

195. Chievitz E, Thiede T. Complications and causes of death in polycythaemia vera. *Acta Med Scand* 1962;172:513–23.

196. Chiu D, Shedden P, Bratina P, Grotta JC. Clinical features of moyamoya disease in the United States. *Stroke* 1998;29:1347–51.

197. Chobanian AV, Prescott MF, Haudenschild CC. Recent advances in molecular pathology: the effects of hypertension on the arterial wall. *Exp Mol Pathol* 1984;41:153–69.

198. Choi DW. Glutamate neurotoxicity and diseases of the nervous system. *Neuron* 1988;1:623–34.

199. Chopp M, Li Y, Zhang ZG, Freytag SO. p53 expression in brain after middle cerebral artery occlusion in the rat. *Biochem Biophys Res Comm* 1992;182:1201–7.

200. Chow FC, Marra CM, Cho TA. Cerebrovascular disease in central nervous system infections. *Semin Neurol* 2011;31:286–306.

201. Christofferson DE, Yuan J. Necroptosis as an alternative form of programmed cell death. *Curr Opin Cell Biol* 2010;22:263–8.

202. Chung CS, Park CH. Primary pontine hemorrhage: a new CT classification. *Neurology* 1992;42:830–34.

203. Clarke M. Systematic review of reviews of risk factors for intracranial aneurysms. *Neuroradiology* 2008;50:653–64.

204. Cloft HJ, Kallmes DF, Kallmes MH, et al. Prevalence of cerebral aneurysms in patients with fibromuscular dysplasia: a reassessment. *J Neurosurg* 1998;88:436–40.

205. Cohen EB, Pappas GD. Dark profiles in the apparently-normal central nervous system: a problem in the electron microscopic identification of early anterograde axonal degeneration. *J Comp Neurol* 1969;136:375–96.

206. Cohen NR, Tan TS, Barker CS. Intracerebral hemorrhage secondary to metastasis from presumed non-small cell lung carcinoma. *Neuropathol Appl Neurobiol* 2004;30:419–22.

207. Cohen PJ, Alexander SC, Smith TC, et al. Effects of hypoxia and normocarbia on cerebral blood flow and metabolism in conscious man. *J Appl Physiol* 1967;23:183–9.

208. Cohen Tervaert CJW, Kallenberg C. Neurologic manifestations of systemic vasculitides. *Rheum Dis Clin North Am* 1993;19:13–40.

209. Cole G, Cowie VA. Long survival after cardiac arrest: case report and neuropathological findings. *Clin Neuropathol* 1987;6:104–9.

210. Collaco–Moraes Y, Aspey BS, de Belleroche JS, Harrison MJ. Focal ischaemia causes an extensive induction of immediate early genes that are sensitive to MK-801. *Stroke* 1994;25:1855–60.

211. Collins R, Armitage J, Parish S, et al. Effects of cholesterol-lowering with simvastatin on stroke and other major vascular events in 20 536 people with cerebrovascular disease or other high-risk conditions. *Lancet* 2004;363:757–67.

212. Collins RC, Olney JW. Focal cortical seizures cause distant thalamic lesions. *Science* 1982;218:177–9.

213. Connett MC, Lausche JM. Fibromuscular hyperplasia of the internal carotid artery: report of a case. *Ann Surg* 1965;162:59–62.

214. Connor MD, Walker R, Modi G, Warlow CP. Burden of stroke in black populations in sub-Saharan Africa. *Lancet Neurol* 2007;6:269–78.

215. Cook DA. Mechanisms of cerebral vasospasm in subarachnoid haemorrhage. *Pharmacol Ther* 1995;66:259–84.

216. Cornog JL Jr, Gonatas NK, Feierman JR. Effects of intracerebral injection of ouabain on the fine structure of rat cerebral cortex. *Am J Pathol* 1967;51:573–90.

217. Correia SC, Moreira PI. Hypoxia-inducible factor 1: a new hope to counteract neurodegeneration? *J Neurochem* 2009;112:1–12.

218. Courville CB. Late cerebral changes incident to severe hypoglycemia (insulin shock): their relation to cerebral anoxia. *Arch Neurol Psychiat (Chic)* 1957;78:1–14.

219. Cragg B, Phillips S. A search for brain damage in a rat model of alcoholic sleep apnea. *Exp Neurol* 1984;84:219–24.

220. Craig HD, Gunel M, Cepeda O, et al. Multilocus linkage identifies two new loci for a mendelian form of stroke, cerebral cavernous malformation, at 7p15–13 and 3q25.2–27. *Hum Mol Genet* 1998;7:1851–8.

221. Craigie EH. On the relative vascularity of various parts of the central nervous system of the albino rat. *J Comp Neurol* 1920;31:429–64.

222. Craven C, Chinn H, MacVicar R. Effect of carrot diet and restricted feeding on the resistance of the rat to hypoxia. *J Aviat Med* 1950;21:256–8.

223. Cravioto H, Feigin I. Noninfectious granulomatous angiitis with a predilection for the nervous system. *Neurology* 1959;9:599–609.

224. Cros D, Comp PC, et al. Superior sagittal sinus thrombosis in a patient with protein S deficiency. *Stroke* 1990;21:633–6.

225. Croughan–Minihane MS, Petitti DB, Gordis L, Golditch I. Morbidity among breech infants according to method of delivery. *Obstet Gynecol* 1990;75:821–5.

226. Crumrine RC, Thomas AL, Morgan PF. Attenuation of p53 expression protects against focal ischemic damage in transgenic mice. *J Cereb Blood Flow Metab* 1994;14:887–91.

227. Czurko A, Nishino H. 'Collapsed' (argyrophilic, dark) neurons in rat model of transient focal cerebral ischaemia. *Neurosci Lett* 1993;162:71–4.

228. Dahlbäck B, Hildebrand B. Inherited resistance to activated protein C is corrected by anticoagulant factory activity found to be a property of factor V. *Proc Natl Acad Sci U S A* 1994;91:1396–400.

229. Daif A, Awada A, al–Rajeh S, Abduljabbar M, al Tahan AR, Obeid T, Malibary T. Cerebral venous thrombosis in adults. A study of 40 cases from Saudi Arabia. *Stroke: a journal of cerebral circulation* 1995;26:1193–5.

230. Dayes LA, Gardiner N. The neurological implications of fibromuscular dysplasia. *Mt Sinai J Med* 2005;72:418–20.

231. D'Cruz DP, Khamashta MA, Hughes GR. Systemic lupus erythematosus. *Lancet* 2007;369:587–96.

232. Deane R, Du Yan S, Submamaryan RK, et al. RAGE mediates amyloid-beta peptide transport across the blood–brain barrier and accumulation in brain. *Nat Med* 2003;9:907–13.

233. De Bilbao F, Guarin E, Nef P, et al. Cell death is prevented in thalamic fields but not in injured neocortical areas after permanent focal ischaemia in mice overexpressing the anti-apoptotic protein Bcl-2. *Eur J Neurosci* 2000;12:921–34.

234. De Courten–Myers GM, Fogelson HM, Kleinholz M, Myers RE. Hypoxic brain and heart injury thresholds in piglets. *Biomed Biochim Acta* 1989;48:S143–8.

235. De Falco FA. Sentinel headache. *Neurol Sci* 2004;25(Suppl 3):215–17.

236. De la Torre JC, Fortin T. Partial or global rat brain ischaemia: the SCOT model. *Brain Res Bull* 1991;26:365–72.

237. Dean JB, Mulkey DK, Garcia AJ 3rd, et al. Neuronal sensitivity to hyperoxia,

hypercapnia, and inert gases at hyperbaric pressures. *J Appl Physiol* 2003;**95**:883–909.

238. DeGirolami U, Kim RC. Spinal cord vascular disorders. In: Kalimo H ed. *Pathology and genetics: cerebrovascular diseases.* Basel: ISN Neuropath Press, 2005:336–44.

239. De Girolami U, Seilhean D, Hauw JJ. Neuropathology of central nervous system arterial syndromes. Part I: the supratentorial circulation. *J Neuropathol Exp Neurol* 2009;**68**:113–24.

240. Dekaban AS, Magee KR. Occurrence of neurologic abnormalities in infants of diabetic mothers. *Neurology* 1958;**8**: 193–200.

241. Del Curling O Jr, Kelly DL Jr, Elster AD, Craven TE. An analysis of the natural history of cavernous angiomas. *J Neurosurg* 1991;**75**:702–8.

242. Del Zoppo G, Ginis I, Hallenbeck JM, *et al.* Inflammation and stroke: putative role for cytokines, adhesion molecules and iNOS in brain response to ischaemia. *Brain Pathol* 2000;**10**:95–112.

243. Dempsey RJ, Combs DJ, Edwards Maley M, *et al.* Moderate hypothermia reduces postischemic edema development and leukotriene production. *Neurosurgery* 1987;**21**:177–81.

244. Deng Y, Tsao BP. Genetic susceptibility to systemic lupus erythematosus in the genomic era. *Nat Rev Rheumatol* 2010;**6**:683–92.

245. Diabetes Control and Complications Trial Research Group. The effect of intensive treatment of diabetes on the development and progression of long-term complications in insulin-dependent diabetes mellitus. Diabetes Control and Complications Trial Research Group. *N Engl J Med* 1993;**329**:977–86.

246. Dichgans M. Genetics of ischaemic stroke. *Lancet Neurol* 2007;**6**:149–61.

247. Dichgans M, Mayer M, Uttner I, *et al.* The phenotypic spectrum of CADASIL: clinical findings in 102 cases. *Ann Neurol* 1998;**44**:731–9.

248. Dichgans M, Herzog J, Gasser T. NOTCH3 mutation involving three cysteine residues in a family with typical CADASIL. *Neurology* 2001;**57**:1714–17.

249. Dierksen GA, Skehan ME, Khan MA, *et al.* Spatial relation between microbleeds and amyloid deposits in amyloid angiopathy. *Ann Neurol* 2010;**68**:545–8.

250. Dietrich HH, Dacey RG Jr. Molecular keys to the problems of cerebral vasospasm. *Neurosurgery* 2000;**46**:517–30.

251. DiFrancesco JC, Brioschi M, Brighina L, *et al.* Anti-Aβ autoantibodies in the CSF of a patient with CAA-related inflammation: a case report. *Neurology* 2011;**76**:842–4.

252. Dinsdale H. Hypertensive encephalopathy. In: Barnett H, Mohr J, Stein B, Yatsu F eds. *Stroke.* New York: Churchill Livingstone, 1993:787–92.

253. Dinsdale HB, Robertson DM, Haas RA. Cerebral blood flow in acute hypertension. *Arch Neurol* 1974;**31**:80–87.

254. Dinubile MJ. Septic thrombosis of the cavernous sinuses: neurological review. *Arch Neurol* 1988;**45**:567–74.

255. Dirnagl U, Becker K, Meisel A. Preconditioning and tolerance against cerebral ischaemia: from experimental strategies to clinical use. *Lancet Neurol* 2009;**8**:398–412.

256. Dirnagl U, Simon RP, Hallenbeck JM. Ischemic tolerance and endogenous neuroprotection. *Trends Neurosci* 2003;**26**:248–54.

257. Donald KW. Oxygen poisoning in man: part I. *Br Med J* 1947;**1**:667–72.

258. Donald KW. Oxygen poisoning in man: part II. *Br Med J* 1947;**1**:712–17.

259. Donaldson JO, Lee NS. Arterial and venous stroke associated with pregnancy. *Neurol Clin* 1994;**12**:583–99.

260. Donegan JH, Traystman RJ, Koehler RC, *et al.* Cerebrovascular hypoxic and autoregulatory responses during reduced brain metabolism. *Am J Physiol* 1985;**249**:H421–9.

261. Doppenberg EM, Zauner A, Bullock R, *et al.* Correlations between brain tissue oxygen tension, carbon dioxide tension, pH, and cerebral blood flow: a better way of monitoring the severely injured brain? *Surg Neurol* 1998;**49**:650–4.

262. Doppenberg EM, Zauner A, Watson JC, Bullock R. Determination of the ischemic threshold for brain oxygen tension. *Acta Neurochir Suppl (Wien)* 1998;**71**:166–9.

263. Doubal FN, MacLullich AM, Ferguson KJ, Dennis MS, Wardlaw JM. Enlarged perivascular spaces on MRI are a feature of cerebral small vessel disease. *Stroke* 2010;**41**:450–4.

264. Drake CG, Peerless SJ. Giant fusiform intracranial aneurysms: a review of 120 patients treated surgically from 1965 to 1992. *J Neurosurg* 1997;**87**:141–62.

265. Dreier JP. The role of spreading depression, spreading depolarization and spreading ischaemia in neurological disease. *Nat Med* 2011;**17**:439–47.

266. Dubuisson AS, Beckers A, Stevenaert A. Classical pituitary tumour apoplexy: clinical features, management and outcomes in a series of 24 patients. *Clin Neurol Neurosurg* 2007;**109**:63–70.

267. Duffy TE, Nelson SR, Lowry OH. Cerebral carbohydrate metabolism during acute hypoxia and recovery. *J Neurochem* 1972;**19**:959–77.

268. Duna GF, Calabrese LH. Limitations of invasive modalities in the diagnosis of primary angiitis of the central nervous system. *J Rheumatol* 1995;**22**: 662–7.

269. Dunn JF, Grinberg O, Roche M, *et al.* Noninvasive assessment of cerebral oxygenation during acclimation to hypobaric hypoxia. *J Cereb Blood Flow Metab* 2000;**20**:1632–5.

270. Duong TQ, Fisher M. Applications of diffusion/perfusion magnetic resonance imaging in experimental and clinical aspects of stroke. *Curr Atherosclerosis Rep* 2004;**6**:267–73.

271. Duprez T, Nzeusseu A, Peeters A, *et al.* Selective involvement of the choroid plexus on cerebral magnetic resonance images:a new radiological sign in patients with systemic lupus erythematosus with neurological symptoms. *J Rheumatol* 2001;**28**:387–91.

272. Duprez TP, Danvoye L, Hernalsteen D, *et al.* Fibrocartilaginous embolization to the spinal cord: serial MR imaging monitoring and pathologic study. *Am J Neuroradiol* 2005;**26**:496–501.

273. Edelman GJ, Hoffman WE, Charbel FT. Cerebral hypoxia after etomidate administration and temporary cerebral artery occlusion. *Anesth Analg* 1997;**85**:821–5.

274. Edgell RC, Abou–Chebl A, Yadav JS. Endovascular management of spontaneous carotid artery dissection. *J Vasc Surg* 2005;**42**:854–60.

275. Eftekhar B, Dadmehr M, Ansari S, *et al.* Are the distributions of variations of circle of Willis different in different populations? Results of an anatomical study and review of literature. *BMC Neurol* 2006;**6**:22.

276. Emerich DF, Skinner SJ, Borlongan CV, Vasconcellos AV, Thanos CG. The choroid plexus in the rise, fall and repair of the brain. *Bioessays* 2005;**27**:262–74.

277. Endres M, Namura S, Shimizu–Sasamata M, *et al.* Attenuation of delayed neuronal death after mild focal ischaemia in mice by inhibition of the caspase family. *J Cereb Blood Flow Metab* 1998;**18**:238–47.

278. Eng JA, Frosch MP, Choi K, Rebeck GW, Greenberg SM. Clinical manifestations of cerebral amyloid angiopathy-related inflammation. *Ann Neurol* 2004;**55**: 250–6.

279. Esiri MM, Wilcock GK Cerebral amyloid angiopathy in dementia and old age. *J Neurol Neurosurg Psychiatry* 1986;**49**:1221–6.

280. Eskenasy–Cottier AC, Leu HJ, Bassetti C, *et al.* A case of dissection of intracranial cerebral arteries with segmental mediolytic 'arteries'. *Clin Neuropathol* 1994;**13**:329–37.

281. Espinosa G, Cervera R. Antiphospholipid syndrome: frequency, main causes and risk factors of mortality. *Nat Rev Rheumatol* 2010;**6**:296–300.

282. Feekes JA, Cassell MD. The vascular supply of the functional compartments of the human striatum. *Brain* 2006;**129**:2189–201.

283. Feigin I, Prose P. Hypertensive fibrinoid arteritis of the brain and gross cerebral hemorrhage: a form of 'hyalinosis'. *Arch Neurol* 1959;**1**:98–110.

284. Feigin VL, Lawes CM, Bennett DA, Barker–Collo SL, Parag V. Worldwide stroke incidence and early case fatality reported in 56 population-based studies: a systematic review. *Lancet Neurol* 2009;**8**:355–69.

285. Feinberg EC, Molitch ME, Endres LK, Peaceman AM. The incidence of Sheehan's syndrome after obstetric hemorrhage. *Fertil Steril* 2005;**84**:975–9.

286. Feldman E. Intracerebral hemorrhage. In: Fisher M ed. *Clinical atlas of cerebriovascular disorders.* Chicago, IL: Wolfe, 1994:11–17.

287. Fellgiebel A, Muller MJ, Ginsberg L. CNS manifestations of Fabry's disease. *Lancet Neurol* 2006;**5**:791–5.

288. Ferrer I, Planas AM. Signalling of cell death and cell survival following focal cerebral ischaemia:life and death struggle in the penumbra. *J Neuropathol Exp Neurol* 2003;**62**:329–39.

289. Ferro JM. Vasculitis of the central nervous system. *J Neurol* 1998;**245**:766–76.

290. Fertrin KY, Costa FF. Genomic polymorphisms in sickle cell disease: implications for clinical diversity and treatment. *Expert Rev Hematol* 2011;**3**:443–58.

291. Fiermonte G, Aloe Spiriti MA, Latagliata R, *et al.* Polycythaemia vera and cerebral blood flow: a preliminary study with transcranial Doppler. *J Intern Med* 1993;**234**:599–602.

292. Findlay JM, Weir BKA, Kanamaru K, Espinosa F. Arterial wall changes in cerebral vasospasm. *Neurosurgery* 1989;**25**:736–46.

293. Findlay JM, Macdonald RL, Weir BK. Current concepts of pathophysiology and management of cerebral vasospasm following aneurysmal subarachnoid hemorrhage. *Cerebrovasc Brain Metab Rev* 1991;**3**:336–61.

294. Finley KH, Brenner C. Histologic evidence of damage to the brain in monkeys treated with Metrazol and insulin. *Arch Neurol Psychiatry* 1941;**45**:403–38.

295. Fischer EG. Impaired perfusion following cerebrovascular stasis: a review. *Arch Neurol* 1973;**29**:361–6.

296. Fisher CM. Pathological observations in hypertensive cerebral hemorrhage. *J Neuropathol Exp Neurol* 1971;**30**: 536–50.

297. Fisher CM. Cerebral miliary aneurysms in hypertension. *Am J Pathol* 1972;**66**:313–30.

298. Fisher CM. Lacunar infarcts: a review. *Cerebrovasc Dis* 1991;**1**:311–20.

299. Fisher M. *Clinical atlas of cerebrovascular disorders*. London:Wolfe, 1994.

300. Flaherty ML, Woo D, Haverbusch M, *et al.* Racial variations in location and risk of intracerebral hemorrhage. *Stroke* 2005;**36**:934–7.

301. Flibotte JJ, Hagan N, O'Donnell J, *et al.* Warfarin, hematoma expansion, and outcome of intracerebral hemorrhage. *Neurology* 2004;**63**:1059–64.

302. Fogelholm R, Murros K. Cigarette smoking and subarachnoid haemorrhage: a population-based case-control study. *J Neurol Neurosurg Psychiatry* 1987;**50**:78–80.

303. Fogelholm R, Nuutila M, Vuorela AL. Primary intracerebral haemorrhage in the Jyvaskyläregion, central Finland, 1985–89:incidence, case fatality rate, and functional outcome. *J Neurol Neurosurg Psychiatry* 1992;**55**:546–52.

304. Fogelholm R, Hernesniemi J, Vapalahti M, *et al.* Impact of early surgery on outcome after aneurysmal subarachnoid hemorrhage: a population-based study. *Stroke* 1993;**24**:1649–54.

305. Folberg R, Hendrix MJ, Maniotis AJ. Vasculogenic mimicry and tumor angiogenesis. *Am J Pathol* 2000;**156**:361–81.

306. Forster DM, Kunkler IH, Hartland P. Risk of cerebral bleeding from arteriovenous malformations in pregnancy: the Sheffield experience. *Stereotact Funct Neurosurg* 1993;**61**(Suppl):20–22.

307. Francis A, Pulsinelli W. The response of GABAergic and cholinergic neurons to transient cerebral ischaemia. *Brain Res* 1982;**243**:271–8.

308. Frank L, Bruhn T, Diemer NH. The effect of an AMPA antagonist (NBQX) on postischemic neuron loss and protein synthesis in the rat brain. *Exp Brain Res* 1993;**95**:70–6.

309. Frederickson CJ, Hernandez MD, McGinty JF. Translocation of zinc may contribute to seizure-induced death of neurons. *Brain Res* 1989;**480**:317–21.

310. Fredriksson K, Auer RN, Kalimo H, *et al.* Cerebrovascular lesions in stroke-prone spontaneously hypertensive rats. *Acta Neuropathol* 1985;**68**:284–94.

311. Fredriksson K, Nordborg C, Kalimo H, *et al.* Cerebral microangiopathy in stroke-prone spontaneously hypertensive rats: an immunohistochemical and ultrastructural study. *Acta Neuropathol* 1988;**75**: 241–52.

312. Freireich EJ, Thomas LB, Frei E III, *et al.* A distinctive type of intracerebral hemorrhage associated with 'blastic crisis' in patients with leukemia. *Cancer* 1960;**13**:146–54.

313. Frerichs KU, Kennedy C, Sokoloff L, Hallenbeck JM. Local cerebral blood flow during hibernation, a model of natural tolerance to 'cerebral ischaemia'. *J Cereb Blood Flow Metab* 1994;**14**:193–205.

314. Freund TF, Buzsáki G, Leon A, *et al.* Relationship of neuronal vulnerability and calcium binding protein immunoreactivity in ischaemia. *Exp Brain Res* 1990;**83**:55–66.

315. Friede RL. The histochemical architecture of Ammon's horn as related to selective vulnerability. *Acta Neuropathol (Berl)* 1966;**6**:1–13.

316. Frostegard J, Ulfgren AK, Nyberg P, *et al.* Cytokine expression in advanced human atherosclerotic plaques: dominance of pro-inflammatory (Th1) and macrophage stimulating cytokines. *Atherosclerosis* 1999;**145**:343.

317. Fujii Y, Takeuchi S, Sasaki O, *et al.* Ultra-early rebleeding in spontaneous subarachnoid hemorrhage. *J Neurosurg* 1996;**84**:35–42.

318. Fukutake T. Cerebral autosomal recessive arteriopathy with subcortical infarcts and leukoencephalopathy (CARASIL): from discovery to gene identification. *J Stroke Cerebrovasc Dis* 2011;**20**:85–93.

319. Furlan AJ, Whisnant JP, Elveback LR. The decreasing incidence of primary intracerebral hemorrhage:a population study. *Ann Neurol* 1979;**5**:367–73.

320. Gaal EI, Salo P, Kristiansson K, *et al.* Intracranial aneurysm risk locus 5q23.2 is associated with elevated systolic blood pressure. *PLoS Genet* 2012;**8**:e1002563.

321. Gaarskjaer FB. The hippocampal mossy fiber system of the rat studied with retrograde tracing techniques: correlation between topographic organization and neurogenetic gradients. *J Comp Neurol* 1982;**203**:717–35.

322. Gadoth N, Hirsch M. Primary and acquired forms of moyamoya syndrome. *Israel J Med Sci* 1980;**16**:370–77.

323. Gaengel K, Genove G, Armulik A, Betsholtz C. Endothelial-mural cell signalling in vascular development and angiogenesis. *Arterioscler Thromb Vasc Biol* 2009;**29**:630–8.

324. Gallyas F, Zoltay G, Dames W. Formation of 'dark' (argyrophilic) neurons of various origin proceeds with a common mechanism of biophysical nature (a novel hypothesis). *Acta Neuropathol (Berl)* 1992;**83**:504–9.

325. Gallyas F, Zoltay G, Horváth Z, *et al.* An immediate morphopathologic response of neurons to electroshock: a reliable model for producing 'dark' neurons in experimental neuropathology. *Neurobiology (Bp)* 1993;**1**:133–46.

326. Gao E, Young WL, Pile–Spellman J, *et al.* Mathematical considerations for modelling cerebral blood flow autoregulation to systemic arterial pressure. *Am J Physiol* 1998;**274**:H1023–31.

327. Garcia JH. Morphology of global cerebral ischaemia. *Crit Care Med* 1988;**16**:979–87.

328. Garcia JH, Lassen NA, Weiller C, *et al.* Ischemic stroke and incomplete infarction. *Stroke* 1996;**27**:761–5.

329. Gardiner M, Smith M–L, Kågström E, *et al.* Influence of blood glucose concentration on brain lactate accumulation during severe hypoxia and subsequent recovery of brain energy metabolism. *J Cereb Blood Flow Metab* 1982;**2**:429–38.

330. Gardiner RM. The effects of hypoglycaemia on cerebral blood flow and metabolism in the new-born calf. *J Physiol* 1980;**298**:37–51.

331. Garland H, Pearce J. Neurological complications of carbon monoxide poisoning. *Quart J Med* 1967;**36**:445–55.

332. Garland H, Greenberg J, Harriman DG. Infarction of the spinal cord. *Brain* 1966;**89**:645–62.

333. Garner A, Ashton N, Tripathi R, *et al.* Pathogenesis of hypertensive retinopathy: an experimental study in the monkey. *Br J Ophthalmol* 1975;**59**:3–44.

334. Garzuly F, Vidal R, Wisniewski T, *et al.* Familial meningocerebrovascular amyloidosis, Hungarian type, with mutant transthyretin (TTR Asp18Gly). *Neurology* 1996;**47**:1562–7.

335. Gegelashvili G, Schousboe A. High affinity glutamate transporters: regulation of expression and activity. *Mol Pharmacol* 1997;**52**:6–15.

336. Geisler BS, Brandhoff F, Fiehler J, *et al.* Blood–oxygen-level-dependent MRI allows metabolic description of tissue at risk in acute stroke patients. *Stroke: a journal of cerebral circulation* 2006;**37**:1778–84.

337. Ghajar JBG, Plum F, Duffy TE. Cerebral oxidative metabolism and blood flow during acute hypoglycemia and recovery in unanaesthetised rats. *J Neurochem* 1982;**38**:397–409.

338. Gherardi R, Belec L, Mhiri C, *et al.* The spectrum of vasculitis in human immunodeficiency virus-infected patients: a clinicopathologic evaluation. *Arthritis Rheum* 1993;**36**:1164–74.

339. Ghetti B, Piccardo P, Spillantini MG, *et al.* Vascular variant of prion protein cerebral amyloidosis with tau-positive neurofibrillary tangles: the phenotype of the stop codon 145 mutation in PRNP. *Proc Natl Acad Sci U S A* 1996;**93**:744–8.

340. Ghiso J, Jensson O, Frangione B. Amyloid fibrils in hereditary cerebral hemorrhage with amyloidosis of Icelandic type is a variant of γ-trace protein (cystatin C). *Proc Natl Acad Sci U S A* 1986;**83**:2974–8.

341. Giang DW. Central nervous system vasculitis secondary to infections, toxins, and neoplasms. *Semin Neurol* 1994;**14**:313–19.

342. Giannini C, Salvarani C, Hunder G, Brown RD. Primary central nervous system vasculitis:pathology and mechanisms. *Acta Neuropathol* 2012;**123**(6):759–72.

343. Gibbs FA, Gibbs EL, Lennox WG. Changes in human cerebral blood flow consequent to alterations in blood gases. *Am J Physiol* 1935;**111**:557–63.

344. Gidday JM. Cerebral preconditioning and ischaemic tolerance. *Nat Rev Neurosci* 2006;**7**:437–48.

345. Gilbert JJ, Vinters HV. Cerebral amyloid angiopathy: incidence and complications in the aging brain: I. Cerebral hemorrhage. *Stroke* 1983;**14**:915–23.

346. Gilden DH, Kleinschmidt–DeMasters BK, LaGuardia JJ, *et al*. Neurologic complications of the reactivation of varicella zoster virus. *N Engl J Med* 2000;**342**:635–45.

347. Gill JS, Shipley MJ, Tsementzis SA, *et al*. Alcohol consumption: a risk factor for hemorrhagic and non-hemorrhagic stroke. *Am J Med* 1991;**90**:489–97.

348. Gill R. The pharmacology of α-amino-3-hydroxy-5-methyl-4-isoxazole propionate (AMPA)/kainate antagonists and their role in cerebral ischaemia. *Cerebrovasc Brain Metab Rev* 1994;**6**:225–56.

349. Gilles FH, Nag D. Vulnerability of human spinal cord in transient cardiac arrest. *Neurology* 1971;**21**:833–9.

350. Gillilan LA. Veins of the spinal cord: anatomic details – suggested clinical applications. *Neurology* 1970;**20**:860–68.

351. Gilmer B, Kilkenny J, Tomaszewski C, Watts JA. Hyperbaric oxygen does not prevent neurologic sequelae after carbon monoxide poisoning. *Acad Emerg Med* 2002;**9**:1–8.

352. Giroud M, Gras P, Chadab N, *et al*. Cerebral hemorrhage in a French prospective population study. *J Neurol Neurosurg Psychiatry* 1991;**54**:595–8.

353. Giwa MO, Williams J, Elderfield K, *et al*. Neuropathologic evidence of endothelial changes in cerebral small vessel disease. *Neurology* 2012;**78**:167–74.

354. Gomez PA, Lobato RD, Rivas JJ, *et al*. Subarachnoid haemorrhage of unknown aetiology. *Acta Neurochir* 1989;**101**:35–41.

355. Gorelick PB. Distribution of atherosclerotic cerebrovascular lesions: effects of age, race, and sex. *Stroke* 1993;**24**(Suppl I):16–19.

356. Gossman MD, Berlin AJ, Weinstein MA, *et al*. Spontaneous direct carotid-cavernous fistula in childhood. *Ophthal Plast Reconstr Surg* 1993;**9**:62–5.

357. Graf CJ, Perret GE, Torner JC. Bleeding from cerebral anteriovenous malformations as part of their natural history. *J Neurosurg* 1983;**58**:331–7.

358. Graham DI, Mendelow AD, Tuor U, Fitch W. Neuropathologic consequences of internal carotid artery occlusion and hemorrhagic hypotension in baboons. *Stroke* 1990;**21**:428–34.

359. Gravereaux EC, Faries PL, Burks JA, *et al*. Risk of spinal cord ischaemia after endograft repair of thoracic aortic aneurysms. *J Vasc Surg* 2001;**34**(6):997–1003.

360. Gray FD Jr, Horner GJ. Survival following extreme hypoxemia. *J Am Med Assoc* 1970;**211**:1815–17.

361. Gray F, Dubas F, Roullet E, Escourolle R. Leukoencephalopathy in diffuse hemorrhagic cerebral amyloid angiopathy. *Ann Neurol* 1985;**18**:54–9.

362. Grayzel DM. Changes in the central nervous system due to convulsions due to hyperinsulinism. *Arch Intern Med* 1934;**54**:694–701.

363. Greenberg ME, Ziff EB, Greene LA. Stimulation of neuronal acetylcholine receptors induces rapid gene transcription. *Science* 1986;**234**:80–3.

364. Greenberg SM. Clinical aspects and diagnostic criteria of sporadic CAA-related haemorrhage. In: Verbeek MM, de Waal RMW, Vinters HV, eds. *Cerebral amyloid angiopathy in Alzheimer's disease and related disorders*. Dordrecht: Kluwer Academic Publishers 2000;3–19.

365. Greenberg SM, Vonsattel JP, Stakes JW, *et al*. The clinical spectrum of cerebral amyloid angiopathy:presentations without lobar hemorrhage. *Neurology* 1993;**43**:2073–9.

366. Greenberg SM, Rebeck GW, Vonsattel JP, *et al*. Apolipoprotein E ε4 and cerebral hemorrhage associated with amyloid angiopathy. *Ann Neurol* 1995;**38**:254–9.

367. Greenberg SM, Grabowski T, Gurol ME, *et al*. Detection of isolated cerebrovascular beta-amyloid with Pittsburgh compound B. *Ann Neurol* 2008;**64**:587–91.

368. Greenberg SM, Vernooij MW, Cordonnier C, *et al*. Cerebral microbleeds: a guide to detection and interpretation. *Lancet Neurol* 2009;**8**:165–74.

369. Greene GM, Godersky JC, Biller J, *et al*. Surgical experience with cerebral amyloid angiopathy. *Stroke* 1990;**21**:1545–9.

370. Gregoire SM, Charidimou A, Gadapa N, *et al*. Acute ischaemic brain lesions in intracerebral haemorrhage:multicentre cross-sectional magnetic resonance imaging study. *Brain* 2012;**134**:2376–86.

371. Grenwood DL, Gitlis VM, Alderuccio F, *et al*. Autoantibodies in neuropsychiatric lupus. *Autoimmunity* 2002;**35**:79–86.

372. Gretarsdottir S, Thorleifsson G, Manolescu A, *et al*. Risk variants for atrial fibrillation on chromosome 4q25 associate with ischemic stroke. *Ann Neurol* 2008;**64**:402–9.

373. Grinker RR. Über einen Fall von Leuchtgasvergiftung mit doppelseitiger Pallidumerweichung und schwerer Degeneration des tieferen Grosshirn Marklagers. *Z Ges Neurol Pschiatr* 1925;**98**:433–56.

374. Grohn OH, Kauppinen RA. Assessment of brain tissue viability in acute ischemic stroke by BOLD MRI. *NMR Biomed* 2001;**14**:432–40.

375. Gross PM, Sposito NM, Pettersen SE, *et al*. Topography of capillary density, glucose metabolism, and microvascular function within the rat inferior colliculus. *J Cereb Blood Flow Metab* 1987;**7**:154–60.

376. Grosset DG, Straiton J, McDonald I, Bullock R. Angiographic and Doppler diagnosis of cerebral artery vasospasm following subarachnoid haemorrhage. *Br J Neurosurg* 1993;**7**:291–8.

377. Grunnet ML, Paulson G. Pathological changes in irreversible brain death. *Dis Nerv Syst* 1971;**32**:690–4.

378. Gschwendtner A, Bevan S, Cole JW, *et al*. Sequence variants on chromosome 9p21.3 confer risk for atherosclerotic stroke. *Ann Neurol* 2009;**65**:531–9.

379. Gudbjartsson DF, Walters GB, Thorleifsson G, *et al*. Many sequence variants affecting diversity of adult human height. *Nat Genet* 2008;**40**:609–15.

380. Guo DC, Papke CL, Tran-Fadulu V, *et al*. Mutations in smooth muscle alpha-actin (ACTA2) cause coronary artery disease, stroke and moyamoya disease, along with thoracic aortic disease. *Am J Hum Genet* 2009;**84**:617–27.

381. Gurdjian ES, Stone WE, Webster JE. Cerebral metabolism in hypoxia. *Arch Neurol Psychiatry* 1944;**5**:472–7.

382. Hacke W, Donnan G, Fieschi C, *et al*. Association of outcome with early stroke treatment:pooled analysis of ATLANTIS, ECASS, and NINDS rt-PA stroke trials. *Lancet* 2004;**363**:768–74.

383. Hacke W, Kaste M, Fieschi C, *et al*. Intravenous thrombolysis with recombinant tissue plasminogen activator. European Cooperative Acute Stroke Study (ECASS). *J Am Med Assoc* 1995;**274**:1017–25.

384. Hackett PH, Roach RC. High altitude cerebral edema. *High Alt Med Biol* 2004;**5**:136–46.

385. Hadfield MG, Aydin F, Lippman HR, *et al*. Neuro-Behcet disease. *Clin Neuropathol* 1997;**16**:55–60.

386. Hager H, Hirschberger W, Scholz W. Electron microscopic changes in the brain tissue of Syrian hamsters following acute hypoxia. *Aerosp Med* 1960; **31**: 379–87.

387. Hajj-Ali RA, Singhal AB, Benseler S, Molloy E, Calabrese LH. Primary angiitis of the CNS. *Lancet Neurol* 2011;**10**:561–72.

388. Hakim AM. The induction and reversibility of cerebral acidosis in thiamine deficiency. *Ann Neurol* 1984;**16**:673–9.

389. Haltia M, Iivanainen M, Majuri H, Puranen M. Spontaneous occlusion of the circle of Willis (moyamoya syndrome). *Clin Neuropathol* 1982;**1**:11–22.

390. Haltia M, Ghiso J, Prelli F, *et al*. Amyloid in familial amyloidosis, Finnish type, is antigenically and structurally related to gelsolin. *Am J Pathol* 1990;**136**:1223–8.

391. Hamilton MG, Spetzler RF. The prospective application of a grading system for arteriovenous malformations. *Neurosurgery* 1994;**34**:2–6.

392. Hansson GK. Immune mechanisms in atherosclerosis. *Artheriosl Thromb Vasc Biol* 2001;**21**:1876–90.

393. Hara K, Shiga A, Fukutake T, *et al*. Association of HTRA1 mutations and familial ischemic cerebral small-vessel disease. *N Engl J Med* 2009;**360**:1729–39.

394. Haraldseth O, Nygård Ø, Grønås T, *et al*. Hyperglycemia in global cerebral ischaemia and reperfusion: a 31-phosphorus NMR spectroscopy study in rats. *Acta Anaesthesiol Scand* 1992;**36**:25–30.

395. Harb R, Whiteus C, Freitas C, Grutzendler J. *In vivo* imaging of cerebral microvascular plasticity from birth to death. *J Cereb Blood Flow Metab* 2012;**33**:146–56.

396. Harik SI, Hritz MA, LaManna JC. Hypoxia-induced brain angiogenesis in the adult rat. *J Physiol* 1995;**485**:525–30.

397. Haritoglou C, Hoops JP, Stefani FH, *et al*. Histopathological abnormalities in ocular blood vessels of CADASIL patients. *Am J Ophthalmol* 2004;**138**:302–5.

398. Harmon DL, Doyle RM, Meleady R, *et al*. Genetic analysis of the thermolabile variant of 5,10-methylenetetrahydrofolate reductase as a risk factor for ischemic stroke. *Arterioscler Thromb Vasc Biol* 1999;**19**:208–11.

399. Hart MN, Merz P, Bennett–Gray J, *et al*. β-Amyloid protein in Alzheimer's disease is found in cerebral and spinal cord vascular malformations. *Am J Pathol* 1988;**132**:167–72.

400. Hart R, Kanter M. Hematologic disorders and ischemic stroke. A selective review. *Stroke* 1990;**21**:1111–21.

401. Hart RG, Boop BS, Anderson DC. Oral anticoagulants and intracranial

hemorrhage: facts and hypotheses. *Stroke* 1995;**26**:1471–7.

402. Hartkamp MJ, van Der Grond J, van Everdingen KJ, et al. Circle of Willis collateral flow investigated by magnetic resonance angiography. *Stroke* 1999;**30**:2671–8

403. Hauw J. The history of lacunes. In: Donnan G, Norrving B, Bamford J, Bogousslavsky J eds. *Lacunar and other subcortical infarctions*. New York: Oxford University Press, 1995:3–15.

404. Hawkins RA, Williamson DH, Krebs HA. Ketone-body utilization by adult and suckling rat brain *in vivo*. *Biochem J* 1971;**122**:13–18.

405. Haymaker W, Ginzler AM, Ferguson RL. Residual neuropathological effects of cyanide poisoning: a study of the central nervous system of 23 dogs exposed to cyanide compounds. *Military Surgeon* 1952;**III**:231–46.

406. Heinzer S, Kuhn G, Krucker T, et al. Novel three-dimensional analysis tool for vascular trees indicates complete micro-networks, not single capillaries, as the angiogenic endpoint in mice overexpressing human VEGF(165) in the brain. *Neuroimage* 2008;**39**:1549–58.

407. Heiskanen O. Treatment of spontaneous intracerebral and intracerebellar hemorrhages. *Stroke* 1993;**24**(Suppl I):94–5.

408. Heiss WD. Experimental evidence of ischemic thresholds and functional recovery. *Stroke* 1992;**23**:1668–72.

409. Heiss WD, Podreka I. Imaging investigation in cerebral ischaemia: PET and SPECT. In: Kalimo H ed. *Pathology and genetics: cerebrovascular diseases*. Basel: ISN Neuropath Press, 2005:194–200.

410. Heiss WD, Graf R, Wienhard K, et al. Dynamic penumbra demonstrated by sequential multitracer PET after middle cerebral artery occlusion in cats. *J Cereb Blood Flow Metab* 1994;**14**:892–902.

411. Helgadottir A, Thorleifsson G, Magnusson KP, et al. The same sequence variant on 9p21 associates with myocardial infarction, abdominal aortic aneurysm and intracranial aneurysm. *Nat Genet* 2008;**40**:217–24.

412. Hellmann J, Vannucci RC, Nardis EE. Blood–brain barrier permeability to lactic acid in the newborn dog: lactate as a cerebral metabolic fuel. *Pediatr Res* 1982;**16**:40–4.

413. Hendriks L, van Duijn DC, Cras P, et al. Presenile dementia and cerebral haemorrhage linked to a mutation at codon 692 of the beta-amyloid precursor protein gene. *Nat Genet* 1992;**1**:218–21.

414. Henry RR, Gumbiner B, Ditzler T, et al. Intensive conventional insulin therapy for type II diabetes: metabolic effects during a 6–mo outpatient trial. *Diabetes Care* 1993;**16**:21–31.

415. Hermán P, Trübel HK, Hyder F. A multiparametric assessment of oxygen efflux from the brain. *J Cereb Blood Flow Metab* 2006;**26**:79–91.

416. Hermann DM, Keyvani K, van de Nes J, et al. Brain-reactive β-amyloid antibodies in primary CNS angiitis with cerebral amyloid angiopathy. *Neurology* 2011;**77**:503–5.

417. Hernández MJ, Vannucci RC, Salcedo A, Brennan RW. Cerebral blood flow and metabolism during hypoglycemia in newborn dogs. *J Neurochem* 1980;**35**:622–8.

418. Heros RC. Acute hydrocephalus after subarachnoid hemorrhage. *Stroke* 1989;**20**:715–17.

419. Herzig MC, Van Nostrand WE, Jucker M. Mechanism of cerebral β-amyloid angiopathy:murine and cellular models. *Brain Pathol* 2006;**16**:40–54.

420. Herzig MC, Winkler DT, Burgermeister P, et al. Aβ is targeted to the vasculature in a mouse model of hereditary cerebral hemorrhage with amyloidosis. *Nat Neurosci* 2004;**7**(9):954–60.

421. Heytens L, Verlooy J, Gheuens J, Bossaert L. Lazarus sign and extensor posturing in a brain-dead patient: case report. *J Neurosurg* 1989;**71**:449–51.

422. Higa M, Davanipour Z. Smoking and stroke. *Neuroepidemiology* 1991;**10**:211–22.

423. Hijdra A, Brouwers PJAM, Vermeulen M, van Gijn J. Grading the amount of blood on computed tomograms after subarachnoid hemorrhage. *Stroke* 1990;**21**:1156–61.

424. Hijdra A, van Gijn J, Nagelkerke NJ, et al. Prediction of delayed cerebral ischaemia, rebleeding and outcome after aneurysmal subarachnoid hemorrhage. *Stroke* 1988;**19**:1250–56.

425. Hills CP. Ultrastructural changes in the capillary bed of the rat cerebral cortex in anoxic-ischemic brain lesions. *Am J Pathol* 1964;**44**:531–43.

426. Hinchey J, Chaves C, Appignani B, et al. A reversible posterior leukoencephalopathy syndrome. *N Engl J Med* 1996;**334**:494–500.

427. Hochachka PW. Patterns of O2-dependence of metabolism. *Adv Exp Med Biol* 1988;**222**:143–51.

428. Hochachka PW, Clark CM, Brown WD, et al. The brain at high altitude: hypometabolism as a defense against chronic hypoxia? *J Cereb Blood Flow Metab* 1994;**14**:671–9.

429. Hoffman GS, Kerr GS, Leavitt RY, et al. Wegener granulomatosis: an analysis of 158 patients. *Ann Intern Med* 1992;**116**:488–98.

430. Hoffman WE, Charbel FT, Edelman G. Brain tissue oxygen, carbon dioxide, and pH in neurosurgical patients at risk for ischaemia. *Anesth Analg* 1996;**82**:582–6.

431. Hoffman WE, Charbel FT, Edelman G, Ausman JI. Brain tissue oxygen pressure, carbon dioxide pressure and pH during hypothermic circulatory arrest. *Surg Neurol* 1996;**46**:75–9.

432. Hoffman WE, Charbel FT, Edelman G, et al. Brain tissue oxygen pressure, carbon dioxide pressure and pH during ischaemia. *Neurol Res* 1996;**18**:54–6.

433. Hoffman WE, Charbel FT, Munoz L, Ausman JI. Comparison of brain tissue metabolic changes during ischaemia at 35° and 18°C. *Surg Neurol* 1998;**49**:85–8.

434. Holliday EG, Maguire JM, Evans TJ, et al. Common variants at 6p21.1 are associated with large artery atherosclerotic stroke. *Nat Genet* 2012;**44**:1147–51.

435. Holmes C, Boche D, Wilkinson D, et al. Long-term effects of Aβ2 immunisation in Alzheimer's disease: follow-up of a randomised, placebo-controlled phase I trial. *Lancet* 2008;**372**:216–23.

436. Hop JW, Rinkel GJ, Algra A, van Gijn J. Initial loss of consciousness and risk of delayed cerebral ischaemia after aneurysmal subarachnoid hemorrhage. *Stroke* 1999;**30**:2268–71.

437. Horie R. Studies on stroke in relation to cerebrovascular atherogenesis in stroke-prone spontaneously hypertensive rats (SHRSP). *Arch Jpn Chir* 1977;**46**:191–213.

438. Horn M, Schlote W. Delayed neuronal death and delayed neuronal recovery in the human brain following global ischaemia. *Acta Neuropathol (Berl)* 1992;**85**:79–87.

439. Hossmann KA. Glutamate-mediated injury in focal cerebral ischaemia: the excitotoxin hypothesis revised. *Brain Pathol* 1994;**4**:23–36.

440. Hossmann KA. Experimental models of focal brain ischaemia. In: Kalimo H ed. *Pathology and genetics: cerebrovascular diseases*. Basel: ISN Neuropath Press, 2005:227–35.

441. Howell GA, Welch MG, Frederickson CJ. Stimulation-induced uptake and release of zinc in hippocampal slices. *Nature* 1984;**308**:736–8.

442. Howells DW, Porritt MJ, Rewell SS, et al. Different strokes for different folks: the rich diversity of animal models of focal cerebral ischaemia. *J Cereb Blood Flow Metab* 2010;**30**:1412–31.

443. Hu HH, Sheng WY, Chu FL, et al. Incidence of stroke in Taiwan. *Stroke* 1992;**23**:1237–41.

444. Hugg JW, Duijn JH, Matson GB, et al. Elevated lactate and alkalosis in chronic human brain infarction observed by 1H and 31P MR spectroscopic imaging. *J Cereb Blood Flow Metab* 1992;**12**:734–44.

445. Hughson MD, McCarty GA, Sholer GM, et al. Thrombotic cerebral arteriopathy in patients with the antiphospholipid syndrome. *Mod Pathol* 1993;**6**:644–53.

446. Huisman MV, Rosendaal F. Thrombophilia. *Curr Opin Hematol* 1999;**6**:291–7.

447. Humphries SE, Morgan L. Genetic risk factors for stroke and carotid atherosclerosis: insights into pathophysiology from candidate gene approaches. *Lancet Neurol* 2004;**3**:227–35.

448. Hunt SP, Pini A, Evan G. Induction of c-fos-like protein in spinal cord neurons following sensory stimulation. *Nature* 1987;**328**:632–4.

449. Hunter S, Ballinger WE, Greer M. Nitrogen inhalation in the human. *Acta Neuropathol (Berl)* 1985;**68**:115–21.

450. Hurst RW, Haskal ZJ, Zager, et al. Endovascular stent treatment of cervical internal carotid artery aneurysms with parent vessel preservation. *Surg Neurol* 1998;**50**:313–17.

451. Hurst RW, Judkins A, Bolger W, et al. Mycotic aneurysm and cerebral infarction resulting from fungal sinusitis: imaging and pathologic correlation. *Am J Neuroradiol* 2001;**22**:858–63.

452. Hutchings M, Weller R. Anatomical relationships of the pia mater to cerebral blood vessels in man. *J Neurosurg* 1986;**65**:316–25.

453. Iadecola C, Anrather J. The immunology of stroke: from mechanisms to translation. *Nat Med* 2011;**17**:796–808.

454. Ichiyama T, Nishikawa M, Hayashi T, et al. Cerebral hypoperfusion during acute Kawasaki disease. *Stroke* 1998;**29**:1320–21.

455. Ihalainen S, Soliymani R, Iivanainen E, et al. Proteome analysis of cultivated

vascular smooth muscle cells from a CADASIL patient. *Mol Med* 2007;**13**:305–14.

456. Iliff JJ, Wang M, Liao Y, *et al.* A paravascular pathway facilitates CSF flow through the brain parenchyma and the clearance of interstitial solutes, including amyloid beta. *Sci Transl Med* 2012;**4**:147ra11.

457. Iliff JJ, Lee H, Yu M, *et al.* Brain-wide pathway for waste clearance captured by contrast-enhanced MRI. *J Clin Invest* 2013;**123**:1299–309.

458. Imaizumi H, Ujike Y, Asai Y, *et al.* Spinal cord ischaemia after cardiac arrest. *J Emerg Med* 1994;**12**:789–93.

459. IMS Study Investigators. Combined intravenous and intra-arterial recanalization for acute ischemic stroke: the interventional management of stroke study. *Stroke* 2004;**35**:904–12.

460. Inagawa T, Kamiya K, Ogasawara H, Yano T. Rebleeding of ruptured intracranial aneurysms in the acute stage. *Surg Neurol* 1987;**28**:93–9.

461. Inagawa T, Ishikawa S, Aoki H, *et al.* Aneurysmal subarachnoid hemorrhage in Izumo City and Shimane Prefecture of Japan: incidence. *Stroke* 1988;**19**:170–75.

462. Inagawa T, Ohbayashi N, Takechi A, *et al.* Primary intracerebral hemorrhage in Izumo City, Japan: incidence rates and outcome in relation to the site of hemorrhage. *Neurosurgery* 2003;**53**:1283–98.

463. Inamura K, Olsson Y, Siesjö BK. Substantia nigra damage induced by ischaemia in hyperglycemic rats:a light and electron microscopic study. *Acta Neuropathol (Berl)* 1987;**75**:131–9.

464. Inamura K, Smith M–L, Olsson Y, Siesjö BK. Pathogenesis of substantia nigra lesions following hyperglycemic ischaemia: changes in energy metabolites, cerebral blood flow, and morphology of pars reticulata in a rat model of ischaemia. *J Cereb Blood Flow Metab* 1988;**8**:375–84.

465. Ingvar M, Folbergrová J, Siesjö BK. Metabolic alterations underlying the development of hypermetabolic necrosis in the substantia nigra in status epilepticus. *J Cereb Blood Flow Metab* 1987;**7**:103–8.

466. Ingvar M, Morgan PF, Auer RN. The nature and timing of excitotoxic neuronal necrosis in the cerebral cortex, hippocampus and thalamus due to flurothyl-induced status epilepticus. *Acta Neuropathol (Berl)* 1988;**75**:362–9.

467. Ishikawa T, Kazumata K, Ni–iya Y, *et al.* Subarachnoid hemorrhage as a result of fungal aneurysm at the posterior communicating artery associated with occlusion of the internal carotid artery: case report. *Surg Neurol* 2002;**58**:261–5.

468. ISIS group. A randomised comparison of streptokinase *versus* tissue plasminogen activator vs anistreplase and of aspirin plus heparin vs aspirin alone among 41 299 cases of suspected acute myocardial infarction. *Lancet* 1992;**339**:753–70.

469. Ito U, Spatz M, Walker JT Jr, Klatzo I. Experimental cerebral ischaemia in Mongolian gerbils. I: light microscopic observations. *Acta Neuropathol (Berl)* 1975;**32**:209–23.

470. Ito Y, Niwa H, Iida T, *et al.* Post-transfusion reversible posterior leukoencephalopathy syndrome with cerebral vasoconstriction. *Neurology* 1997;**49**:1174–6.

471. Iwamoto H, Kiyohara Y, Fujishima M, *et al.* Prevalence of intracranial saccular aneurysms in a Japanese community based on a consecutive autopsy series during a 30-year observation period. The Hisayama study. *Stroke* 1999;**30**:1390–95.

472. Jacewicz M, Tanabe J, Pulsinelli WA. The CBF threshold and dynamics for focal cerebral infarction in spontaneously hypertensive rats. *J Cereb Blood Flow Metab* 1992;**12**:359–70.

473. Jackson C, Sudlow C. Comparing risks of death and recurrent vascular events between lacunar and non-lacunar infarction. *Brain* 2005;**128**:2507–17.

474. Jacob H. Über die diffuse Hemisphärenmarkerkrankung nach Kohlenoxydvergiftung bei Fallen mit Klinisch intervallere Verlaufsform. *Z Neurol Psychiat* 1939;**167**:161–79.

475. Jahan R, Murayama Y, Gobin YP, *et al.* Embolization of arteriovenous malformations with onyx: clinicopathological experience in 23 patients. *Neurosurgery* 2001;**48**:984–97.

476. Jalkanen S, Aho R, Kallajoki M, *et al.* Lymphocyte homing receptors and adhesion molecules in intravascular malignant lymphomatosis. *Int J Cancer* 1989;**44**:777–82.

477. Jansen C, Parchi P, Capellari S, Vermeij AJ, Corrado P, Baas F, Strammiello R, van Gool WA, van Swieten JC, Rozemuller AJ. Prion protein amyloidosis with divergent phenotype associated with two novel nonsense mutations in PRNP. *Acta neuropathol* 2010;**119**:189–97.

478. Janzer RC, Friede RL. Hypotensive brain stem necrosis of cardiac arrest encephalopathy. *Acta Neuropathol (Berl)* 1980;**50**:53–6.

479. Jason GW, Pajurkova EM, Lee RG. High-altitude mountaineering and brain function: neuropsychological testing of members of a Mount Everest expedition. *Aviat Space Environ Med* 1989;**60**:170–3.

480. Jellinger KA, Lauda F, Attems J. Sporadic cerebral amyloid angiopathy is not a frequent cause of spontaneous brain hemorrhage. *Eur J Neurol* 2007;**14**:923–8.

481. Jenkins A, Maxwell WL, Graham DI. Experimental intracerebral haematoma in the rat: sequential light microscopical changes. *Neuropathol Appl Neurobiol* 1989;**15**:477–86.

482. Jennette JC, Falk RJ, Milling DM. Pathogenesis of vasculitis. *Semin Neurol* 1994;**14**:291–306.

483. Jennings GH, Newton MA. Persistent paraplegia after repeated cardiac arrests. *Br Med J* 1969;**3**:572–3.

484. Jensson O, Gudmundsson G, Arnason A, *et al.* Hereditary cystatin C (γ-trace) amyloid angiopathy of the CNS causing cerebral hemorrhage. *Acta Neurol Scand* 1987;**76**:102–14.

485. Jiang BH, Semenza GL, Bauer C, Marti HH. Hypoxia-inducible factor 1 levels vary exponentially over a physiologically relevant range of O_2 tension. *Am J Physiol* 1996;**271**:C1172–80.

486. Jin R, Yang G, Li G. Inflammatory mechanisms in ischemic stroke: role of inflammatory cells. *J Leukoc Biol* 2010;**87**:779–89.

487. Johansen FF, Jørgensen MB, Diemer NH. Resistance of hippocampal CA-1 interneurons to 20 min of transient cerebral ischaemia in the rat. *Acta Neuropathol (Berl)* 1983;**61**:135–40.

488. Johansen FF, Zimmer J, Diemer NH. Early loss of somatostatin neurons in dentate hilus after cerebral ischaemia in the rat precedes CA-1 pyramidal cell loss. *Acta Neuropathol (Berl)* 1987;**73**:110–14.

489. Johnson KA, Gregas M, Becker JA, *et al.* Imaging of amyloid burden and distribution in cerebral amyloid angiopathy. *Ann Neurol* 2007;**62**:229–34.

490. Johnson MW, Hammond RR, Vinters HV. Fusiform, infectious and other aneurysmal lesions. In: Kalimo H ed. *Pathology and genetics: cerebrovascular diseases.* Basel: ISN Neuropath Press, 2005:112–18.

491. Johnston SC, Halbach VV, Smith WS, Gress DR. Rapid development of giant fusiform cerebral aneurysms in angiographically normal vessels. *Neurology* 1998;**50**:1163–6.

492. Joutel A, Corpechot C, Ducros A, *et al.* Notch3 mutations in CADASIL, a hereditary late-onset condition causing stroke and dementia. *Nature* 1996;**383**:707–10.

493. Joutel A, Vahedi K, Corpechot C, *et al.* Strong clustering and stereotyped nature of Notch3 mutations in CADASIL patients. *Lancet* 1997;**350**:1511–15.

494. Joutel A, Andreux F, Gaulis S, *et al.* The ectodomain of the Notch3 receptor accumulates within the cerebrovasculature of CADASIL patients. *J Clin Invest* 2000;**105**:597–605.

495. Joutel A, Favrole P, Labauge P, *et al.* Skin biopsy immunostaining with a Notch3 monoclonal antibody for CADASIL diagnosis. *Lancet* 2001;**358**:2049–51.

496. Jung JE, Kim GS, Chen H, *et al.* Reperfusion and neurovascular dysfunction in stroke: from basic mechanisms to potential strategies for neuroprotection. *Mol Neurobiol* 2010;**41**:172–9.

497. Juurlink DN, Stanbrook MB, McGuigan MA. Hyperbaric oxygen for carbon monoxide poisoning. *Cochrane Database Syst Rev* 2000;CD002041.

498. Juvela S. Rebleeding from ruptured intracranial aneurysms. *Surg Neurol* 1989;**32**:323–6.

499. Juvela S. Minor leak before rupture of an intracranial aneurysm and subarachnoid hemorrhage of unknown etiology. *Neurosurgery* 1992;**30**:7–11.

500. Juvela S, Kaste M. Reduced platelet aggregability and thromboxane release after rebleeding in patients with subarachnoid hemorrhage. *J Neurosurg* 1991;**74**:21–6.

501. Juvela S, Hillbom M, Numminen H, Koskinen P. Cigarette smoking and alcohol consumption as risk factors for aneurysmal subarachnoid hemorrhage. *Stroke* 1993;**24**:639–46.

502. Juvela S, Porras M, Heiskanen O. Natural history of unruptured intracranial aneurysms: a long-term follow-up study. *J Neurosurg* 1993;**79**:174–82.

503. Kaku DA, Lowenstein DH. Emergence of recreational drug abuse as a major risk factor for stroke in young adults. *Ann Intern Med* 1990;**113**:821–7.

504. Kalaria RN, Kalimo H. Non-atherosclerotic cerebrovascular disorders. *Brain Pathol* 2002;**12**:337–42.

505. Kalaria RN, Premkumar DR, Pax AB, Cohen DL, Lieberburg I. Production and increased detection of amyloid beta protein and amyloidogenic fragments in brain microvessels, meningeal vessels and choroid plexus in Alzheimer's disease. *Brain Res Mol Brain Res* 1996;**35**:58–68.

506. Kalaria RN, Perry RH, O'Brien J, Jaros E. Atheromatous disease in small intracerebral vessels, microinfarcts and dementia. *Neuropathol Appl Neurobiol* 2012;**38**(5):505–8.

507. Kalimo H, Olsson Y. Effect of severe hypoglycemia on the human brain. *Acta Neurol Scand* 1980;**62**:345–56.

508. Kalimo H, Kalaria R. Hereditary forms of vascular dementia. In: Kalimo H ed. *Pathology and genetics: cerebrovascular diseases.* Basel: ISN Neuropath Press, 2005:324–34.

509. Kalimo H, Agardh C–D, Olsson Y, Siesjö BK. Hypoglycemic brain injury. II: electron microscopic findings in rat cerebral neurons during profound insulin-induced hypoglycemia and in the recovery period following glucose administration. *Acta Neuropathol (Berl)* 1980;**50**:43–52.

510. Kalimo H, Fredriksson K, Nordborg C, *et al*. The spread of brain oedema in hypertensive brain injury. *Med Biol* 1986;**64**:133–7.

511. Kalimo H, Ruchoux MM, Viitanen M, Kalaria RN. CADASIL: a common form of hereditary arteriopathy causing brain infarcts and dementia. *Brain Pathol* 2002;**12**:371–84.

512. Kamenar E, Burger PC. Cerebral fat embolism: a neuropathological study of a microembolic state. *Stroke* 1980;**11**:477–84.

513. Kandel ER, Schwartz JH, Jessel TM, *et al*., eds. *Principles of Neural Science,* 5th ed. New York: McGraw-Hill, 2012.

514. Kannel WB, Gordon T, Wolf PA, McNamara P. Hemoglobin and the risk of cerebral infarction: the Framingham Study. *Stroke* 1972;**3**:409–20.

515. Karhunen PJ, Mannikko A, Penttila A, Liesto K. Diagnostic angiography in postoperative autopsies. *Am J Forensic Med Pathol* 1989;**10**:303–9.

516. Kase C. Cerebral amyloid angiopathy. In: Kase C, Caplan L eds. *Intracerebral hemorrhage.* Boston, MA: Butterworth Heinemann, 1994:179–200.

517. Kase CS, Varakis JN, Stafford JR, Mohr JP. Medial medullary infarction from fibrocartilaginous embolism to the anterior spinal artery. *Stroke* 1983;**14**:413–8.

518. Kase CS, Norrving B, *et al*. Cerebellar infarction: clinical and anatomic observations in 66 cases. *Stroke* 1993;**24**:76–83.

519. Kase CS, Mohr JP , Caplan LRl. Intracerebral hemorrhage. In: Mohr JP, Choi DW, Grotta JC, *et al*. eds. *Stroke: pathophysiology, diagnosis, and management.* New York: Churchill Livingstone, 2004:327–76.

520. Kassam A, Horowitz M, Chang YF, *et al*. Altered arterial homeostasis and cerebral aneurysms: a review of the literature and justification for a search of molecular biomarkers. *Neurosurgery* 2004;**54**:1199–212.

521. Kassell NF, Torner JC. The international cooperative study on timing of aneurysm surgery: an update. *Stroke* 1984;**15**:566–70.

522. Kataoka K, Taneda M, Asai T, *et al*. Structural fragility and inflammatory response of ruptured cerebral aneurysms: a comparative study between ruptured and unruptured cerebral aneurysms. *Stroke* 1999;**30**:1396–401.

523. Kaufman HH. Treatment of deep spontaneous intracerebral hematomas. *Stroke* 1993;**24**(Suppl I):101–6.

524. Kavanagh D, Spitzer D, Kothari PH, *et al*. New roles for the major human 3'-5' exonuclease TREX1 in human disease. *Cell Cycle* 2008;**7**:1718–25.

525. Kawahara N, Mishima K, Higashiyama S, *et al*. The gene for heparin-binding epidermal growth factor-like growth factor is stress-inducible: its role in cerebral ischaemia. *J Cereb Blood Flow Metab* 1999;**19**:307–20.

526. Kay AC, Solberg LA Jr, Nichols DA, Petitt RM. Prognostic significance of computed tomography of the brain in thrombotic thrombocytopenic purpura. *Mayo Clin Proc* 1991;**66**:602–7.

527. Kay R, Woo J, Kreel L, *et al*. Stroke subtypes among Chinese living in Hong Kong: the Shatin Stroke Registry. *Neurology* 1992;**42**:985–7.

528. Keogh JM, Badawi N. The origins of cerebral palsy. *Curr Opin Neurol* 2006;**19**:129–34.

529. Kepes JJ, Malone DG, Griffin W, *et al*. Surgical 'touch artifacts' of the cerebral cortex: an experimental study with light and electron microscopic analysis. *Clin Neuropathol* 1995;**14**:86–92.

530. Keramatipour M, McConnell RS, Kirkpatrick P, *et al*. The ACE I allele is associated with increased risk for ruptured intracranial aneurysms. *J Med Genet* 2000;**37**:498–500.

531. Khan TA, Shah T, Prieto D, *et al*. Apolipoprotein E genotype, cardiovascular biomarkers and risk of stroke: Systematic review and meta-analysis of 14 015 stroke cases and pooled analysis of primary biomarker data from up to 60 883 individuals. *Int J Epidemiol* 2013;**42**:475–92.

532. Kidwell CS, Saver JL, Mattiello J, *et al*. Thrombolytic reversal of acute human cerebral ischemic injury shown by diffusion/perfusion magnetic resonance imaging. *Ann Neurol* 2000;**47**:462–9.

533. Kiessling M, Hossmann KA. Focal cerebral ischaemia: molecular mechanisms and new therapeutic strategies. *Brain Pathol* 1994;**4**:21–2.

534. Kilduff TS, Miller JD, Radeke CM, *et al*. 14C-2-deoxyglucose uptake in the ground squirrel brain during entrance to and arousal from hibernation. *J Neurosci* 1990;**10**:2463–75.

535. Kim JS, Lee JH, Lee MC. Small primary intracerebral hemorrhage: clinical presentation of 28 cases. *Stroke* 1994;**25**:1500–506.

536. Kingman TA, Mendelow AD, Graham DI, *et al*. Experimental intracerebral mass: time-related effects on local cerebral blood flow. *J Neurosurg* 1987;**67**:732–8.

537. Kinney HC, Korein J, Panigrahy A, *et al*. Neuropathological findings in the brain of Karen Ann Quinlan: the role of the thalamus in the persistent vegetative state. *N Engl J Med* 1994;**330**:1469–75.

538. Kinouchi H, Sharp FR, Hill MP, *et al*. Induction of 70-kDa heat shock protein and hsp70m RNA following transient focal cerebral ischaemia in the rat. *J Cereb Blood Flow Metab* 1993;**13**:105–15.

539. Kirino T, Tamura A, Sano K. Delayed neuronal death in the rat hippocampus following transient forebrain ischaemia. *Acta Neuropathol (Berl)* 1984;**64**:139–47.

540. Kitanaka C, Tanaka J, Kuwahara M, *et al*. Nonsurgical treatment of unruptured intracranial vertebral artery dissection with serial follow-up angiography. *J Neurosurg* 1994;**80**:667–74.

541. Kittner SJ, Gorelick PB. Antiphospholipid antibodies and stroke: an epidemiological perspective. *Stroke* 1992;**23**(Suppl I):19–22.

542. Kiuru S, Salonen O, Haltia M. Gelsolin-related spinal and cerebral amyloid angiopathy. *Ann Neurol* 1999;**45**:305–11.

543. Kiyohara Y, Ueda K, Hasuo Y, *et al*. Hematocrit as a risk factor of cerebral infarction: long-term prospective population survey in a Japanese rural community. *Stroke* 1986;**17**:687–92.

544. Knudsen KA, Rosand J, Karluk D, Greenberg SM. Clinical diagnosis of cerebral amyloid angiopathy: validation of the Boston criteria. *Neurology* 2001;**56**:537–9.

545. Koehler RC, Traystman RJ, Rosenberg AA, *et al*. Role of O2–hemoglobin affinity on cerebrovascular response to carbon monoxide hypoxia. *Am J Physiol* 1983;**245**:H1019–23.

546. Koennecke HC. Cerebral microbleeds on MRI: prevalence, associations, and potential clinical implications. *Neurology* 2006;**66**:165–71.

547. Koike M, Shibata M, Tadakoshi M, *et al*. Inhibition of autophagy prevents hippocampal pyramidal neuron death after hypoxic-ischemic injury. *Am J Pathol* 2008;**172**:454–69.

548. Kokame K, Miyata T. Genetic defects leading to hereditary thrombotic thrombocytopenic purpura. *Sem Hematol* 2004;**41**:34–40.

549. Kokmen E, Whisnant JP, O'Fallon WM, Chu CP, Beard CM. Dementia after ischemic stroke: a population-based study in Rochester, Minnesota (1960–1984). *Neurology* 1996;**46**:154–9.

550. Kol S, Ammar R, Weisz G, Melamed Y. Hyperbaric oxygenation for arterial air embolism during cardiopulmonary bypass. *Ann Thorac Surg* 1993;**55**:401–3.

551. Kombian SB, Reiffenstein RJ, Colmers WF. The actions of hydrogen sulfide on dorsal raphe serotonergic neurons *in vitro. J Neurophysiol* 1993;**70**:81–96.

552. Komuro T, Borsody MK, Ono S, *et al*. The vasorelaxation of cerebral arteries by carbon monoxide. *Exp Biol Med (Maywood)* 2001;**226**:860–65.

553. Kondo S, Hashimoto N, Kikuchi H, *et al*. Apoptosis of medial smooth muscle cells in the development of saccular cerebral aneurysms in rats. *Stroke* 1998;**29**:181–8.

554. Korean Neurological Association. Epidemiology of cerebrovascular disease in Korea: a collaborative study, 1989–1990. *J Korean Med Sci* 1993;**8**:281–9.

555. Kosierkiewicz TA, Factor SM, Dickson DW. Immunocytochemical studies of atherosclerotic lesions of cerebral berry

aneurysms. *J Neuropathol Exp Neurol* 1994;**53**:399–406.

556. Kovanen P. Pathogenesis of carotid atherosclerosis: molecular and genetic aspects. In: Kalimo H ed. *Pathology and genetics: cerebrovascular diseases.* Basel: ISN Neuropath Press, 2005:74–84.

557. Kozniewska E, Weller L, Höper J, *et al.* Cerebrocortical microcirculation in different stages of hypoxic hypoxia. *J Cereb Blood Flow Metab* 1987; **7**:464–70.

558. Kraig RP, Chesler M. Astrocytic acidosis in hyperglycemic and complete ischaemia. *J Cereb Blood Flow Metab* 1990;**10**:104–14.

559. Kraig RP, Pulsinelli WA, Plum F. Carbonic acid buffer changes during complete brain ischaemia. *Am J Physiol* 1986;**250**:R348–57.

560. Kraig RP, Petito CK, Plum F, Pulsinelli WA. Hydrogen ions kill brain at concentrations reached in ischaemia. *J Cereb Blood Flow Metab* 1987;**7**:379–86.

561. Krainik A, Hund–Georgiadis M, Zysset S, von Cramon DY. Regional impairment of cerebrovascular reactivity and BOLD signal in adults after stroke. *Stroke* 2005;**36**:1146–52.

562. Krakauer J. *Into thin air: a personal account of the Mount Everest disaster.* New York: Macmillan, 1997.

563. Krayenbuhl H. Cerebral venous thrombosis: the diagnostic value of cerebral angiography. *Schweiz Arch Neurol Neurochir Psychiatry* 1954;**74**:261–87.

564. Krayenbuhl H. Cerebral venous and sinus thrombosis. *Clin Neurosurg* 1967; **14**:1–24.

565. Kreel L, Kay R, Woo J, *et al.* The radiological (CT) and clinical sequelae of primary intracerebral haemorrhage. *Br J Radiol* 1991;**64**:1096–100.

566. Kristensen B, Malm J, Nilsson TK, *et al.*≈Increased fibrinogen levels and acquired hypofibrinolysis in young adults with ischemic stroke. *Stroke* 1998;**29**:2261–7.

567. Krug T, Gabriel JP, Taipa R, *et al.* TTC7B emerges as a novel risk factor for ischemic stroke through the convergence of several genome-wide approaches. *J Cereb Blood Flow Metab* 2012;**32**:1061–72.

568. Kubes P, Ward PA. Leukocyte recruitment and the acute inflammatory response. *Brain Pathol* 2000;**10**:127–35.

569. Kudo T. Spontaneous occlusion of circle of Willis: a disease apparently confined to Japanese. *Neurology* 1968;**18**:485–96.

570. Kuivaniemi H, Prockop DJ, Wu Y, *et al.* Exclusion of mutations in the gene for type III collagen (COL3A1) as a common cause of intracranial aneurysms or cervical artery dissections. *Neurology* 1993;**43**:2652–8.

571. Kunz U, Mauer U, Waldbaur H, Oldenkott P. Früh- und Spätkomplikationen nach Schädel–Hirn Trauma: Chronisches Subduralhämatom/ Hygrom, Karotis-Sinus-cavernosus-Fistel, Abszedierung, Meningitis und Hydrozephalus. *Unfallchirurgie* 1993;**96**:595–603.

572. Labauge P, Enjolras O, Bonerandi JJ, *et al.* An association between autosomal dominant cerebral cavernomas and a distinctive hyperkeratotic cutaneous vascular malformation in 4 families. *Ann Neurol* 1999;**45**:250–4.

573. Labauge P, Brunereau L, Levy C, Laberge S, Houtteville JP. The natural history of familial cerebral cavernomas: a retrospective MRI study of 40 patients. *Neuroradiology* 2000;**42**:327–32.

574. LaManna JC, Harik SI. Regional comparisons of brain glucose influx. *Brain Res* 1985;**326**:299–305.

575. LaManna JC, Vendel LM, Farrell RM. Brain adaptation to chronic hypobaric hypoxia in rats. *J Appl Physiol* 1992; **72**:2238–43.

576. Lammie GA, Brannan F, Slattery J, Warlow C. Nonhypertensive cerebral small-vessel disease. An autopsy study. *Stroke* 1997;**28**:2222–9.

577. Lammie GA, Brannan F, Wardlaw JM. Incomplete lacunar infarction (type Ib lacunes). *Acta Neuropathol* 1998;**96**:163–71.

578. Lamy C, Oppenheim C, Meder JF, Mas JL. Neuroimaging in posterior reversible encephalopathy syndrome. *J Neuroimaging* 2004;**14**:89–96.

579. Lanfranconi S, Markus HS. COL4A1 mutations as a monogenic cause of cerebral small vessel disease: a systematic review. *Stroke* 2010;**41**:e513–8.

580. Lapresle J, Fardeau M. The central nervous system and carbon monoxide poisoning. II: anatomical study of brain lesions following intoxication with carbon monoxide (22 cases). *Prog Brain Res* 1967;**24**:31–74.

581. Lasjaunias P, Gracia–Monaco R, Rodesch G, *et al.* Vein of Galen malformation: endovascular management of 43 cases. *Childs Nerv Syst* 1991;**7**:360–7.

582. Lasjaunias P. Spinal cord vascular lesions. *J Neurosurg* 2003;**98** (1 suppl):119–20.

583. Lasky JL, Wu H. Notch signalling, brain development, and human disease. *Pediatr Res* 2005;**57**:104–9R.

584. Lassen NA, Agnoli A. The upper limit of autoregulation of cerebral blood flow: on the pathogenesis of hypertensive encephalopathy. *J Clin Lab Invest* 1973;**30**:113–16.

585. Lauritzen M, Dreier JP, Fabricius M, *et al.* Clinical relevance of cortical spreading depression in neurological disorders: migraine, malignant stroke, subarachnoid and intracranial hemorrhage, and traumatic brain injury. *J Cereb Blood Flow Metab* 2011;**31**:17–35.

586. Lawlor DA, Smith GD, Leon DA, *et al.* Secular trends in mortality by stroke subtype in the 20th century: a retrospective analysis. *Lancet* 2002;**360**:1818–23.

587. Lawrence RD, Meyer R, Nevin S. The pathological changes in the brain in fatal hypoglycemia. *Quart J Med* 1942;**11**: 181–201.

588. Lax NZ, Pienaar IS, Reeve AK, *et al.* Microangiopathy in the cerebellum of patients with mitochondrial DNA disease. *Brain* 2012;**135**:1736–50.

589. Leão AAP. Spreading depression of activity in the cerebral cortex. *J Neurophysiol* 1944;**7**:359–90.

590. Leblanc GG, Golanov E, Awad IA, Young WL. Biology of vascular malformations of the brain. *Stroke* 2009;**40**:e694–702.

591. Ledingham JGG, Rajagopalan B. Cerebral complications in the treatment of accelerated hypertension. *Q J Med* 1979;**48**:25–41.

592. Lee JM, Yin K, Hsin I, *et al.* Matrix metalloproteinase-9 in cerebral-amyloid-angiopathy-related hemorrhage. *J Neurol Sci* 2005;**229–230**:249–54.

593. Levine S, Stypulkowski W. Experimental cyanide encephalopathy. *Arch Pathol* 1959;**67**:306–23.

594. Levine S. Anoxic-ischemic encephalopathy in rats. *Am J Pathol* 1960;**36**:1–17.

595. Levine SR, Twyman RE, Gilman S. The role of anticoagulation in cavernous sinus thrombosis. *Neurology* 1988;**38**:517–22.

596. Levine SR, Brust JCM, Futrell N, *et al.* A comparative study of the cerebrovascular complications of cocaine: alkaloidal versus hydrochloride a review. *Neurology* 1991;**41**:1173–7.

597. Levine SR, Brey RL, Joseph CL, Havstad S. Risk of recurrent thromboembolic events in patients with focal cerebral ischaemia and antiphospholipid antibodies. *Stroke* 1992;**23**(Suppl I): 29–32.

598. Levy E, Lopez-Otin C, Ghiso J, *et al.* Stroke in Icelandic patients with hereditary amyloid angiopathy is related to a mutation in a cystatin C gene, an inhibitor of cysteine proteases. *J Exp Med* 1989;**169**:1771–8.

599. Levy E, Carman MD, Fernandez–Madrid IJ, *et al.* Mutation of the Alzheimer's disease amyloid gene in hereditary cerebral hemorrhage, Dutch type. *Science* 1990;**248**:1124–6.

600. Li L, Lundkvist A, Andersson D, *et al.* Protective role of reactive astrocytes in brain ischaemia. *J Cereb Blood Flow Metab* 2008;**28**:468–81.

601. Li M, Ransohoff RM. The roles of chemokine CXCL12 in embryonic and brain tumour angiogenesis. *Semin Cancer Biol* 2009;**19**:111–5.

602. Liang D, Bhatta S, Gerzanich V, Simard JM. Cytotoxic edema: mechanisms of pathological cell swelling. *Neurosurg Focus* 2007;**22**:E2.

603. Liang P, Hoffman GS. Advances in the medical and surgical treatment of Takayasu arteritis. *Curr Opin Rheumatol* 2005;**17**:16–24.

604. Libby P. Inflammation in atherosclerosis. *Nature* 2002;**420**:868–74.

605. Libby P, Ridker PM, Hansson GK. Progress and challenges in translating the biology of atherosclerosis. *Nature* 2011;**473**:317–25.

606. Lie JT. Illustrated histopathologic classification criteria for selected vasculitis syndromes. American College of Rheumatology Subcommittee on Classification of Vasculitis. *Arthritis Rheum* 1990;**33**:1074–87.

607. Lindstedt KA, Leskinen MJ, Kovanen PT. Proteolysis of the pericellular matrix:a novel element determining cell survival and death in the pathogenesis of plaque erosion and rupture. *Arterioscler Thromb Vasc Biol* 2004;**24**:1567–77.

608. Little JR, Dial B, Belanger G, Carpenter S. Brain hemorrhage from intracranial tumour. *Stroke* 1979;**10**:283–9.

609. Liu C, Gao Y, Barrett J, Hu B. Autophagy and protein aggregation after brain ischaemia. *J Neurochem* 2010;**115**:68–78.

610. Lo EH. A haemodynamic analysis of intracranial arteriovenous malformations. *Neurol Res* 1993;**15**:51–5.

611. Loeb C. Binswanger's disease is not a single entity. *Neurol Sci* 2000;**21**:343–8.

612. Loes DJ, Biller J, Yuh WT, *et al.* Leukoencephalopathy in cerebral amyloid angiopathy: MR imaging in four cases. *Am J Neuroradiol* 1990;**11**:485–8.

613. Longstreth WT Jr, Inui TS. High blood glucose level on hospital admission and poor neurological recovery after cardiac arrest. *Ann Neurol* 1984;**15**:59–63.

614. Lopez A, Prior MG, Reiffenstein RJ, Goodwin LR. Peracute toxic effects of inhaled hydrogen sulfide and injected sodium hydrosulfide on the lungs of rats. *Fundam Appl Toxicol* 1989;**12**:367–73.

615. Lorente de Nó R. Studies on the striation of the cerebral cortex. II: continuation of the study of the Ammonic system. *J Psychol Neurol* 1934;**46**:113–77.

616. Love S. Oxidative stress in brain ischaemia. *Brain Pathol* 1999;**9**:119–31.

617. Love S. Apoptosis and brain ischaemia. *Prog Neuropsychopharmacol Biol Psychiatry* 2003;**27**:267–82.

618. Love S. Contribution of cerebral amyloid angiopathy to Alzheimer's disease. *J Neurol Neurosurg Psychiatry* 2004;**75**:1–4.

619. Love S, Barber R. Expression of P-selectin and intercellular adhesion molecule-1 in human brain after focal infarction or cardiac arrest. *Neuropathology and applied neurobiology* 2001; 27: **465**-73.

620. Love S, Barber R, Srinivasan A, Wilcock GK. Activation of caspase-3 in permanent and transient brain ischaemia in man. *Neuroreport* 2000;**11**:2495–9.

621. Love S, Barber R, Wilcock GK. Neuronal death in brain infarcts in man. *Neuropathol Appl Neurobiol* 2000;**26**:55–66.

622. Love S, Nicoll JA, Hughes A, Wilcock GK. APOE and cerebral amyloid angiopathy in the elderly. *Neuroreport* 2003;**14**:1535–6.

623. Love S, Miners S, Palmer J, Chalmers K, Kehoe P. Insights into the pathogenesis and pathogenicity of cerebral amyloid angiopathy. *Front Biosci* 2009;**14**:4778–92.

624. Love S, Chalmers K, Ince P, *et al.* Development, appraisal, validation and implementation of a consensus protocol for the assessment of cerebral amyloid angiopathy in post-mortem brain tissue. *Am J Neurodegen Dis* 2014;3(1):19–32.

625. Lovelock CE, Molyneux AJ, Rothwell PM. Change in incidence and aetiology of intracerebral haemorrhage in Oxfordshire, UK, between 1981 and 2006:a population-based study. *Lancet Neurol* 2007;**6**:487–93.

626. Lownie SP. Intracranial dural arteriovenous fistulas: endovascular therapy. *Neurosurg Clin North Am* 1994;**5**:449–58.

627. Lu YM, Yin H–Z, Weiss JH. Ca21 permeable AMPA/kainate channels permit rapid injurious Ca21 entry. *Neuroreport* 1995;**6**:1089–92.

628. Lugaresi A, Montagna P, Morreale A, Gallassi R. 'Psychic akinesia' following carbon monoxide poisoning. *Eur Neurol* 1990;**30**:167–9.

629. Lundy EF, Klima LD, Huber TS, *et al.* Elevated blood ketone and glucagon levels cannot account for 1,3-butanediol induced cerebral protection in the Levine rat. *Stroke* 1987;**18**:217–22.

630. Luscher C, Nicoll RA, Malenka RC, Muller D. Synaptic plasticity and dynamic modulation of the postsynaptic membrane. *Nat Neurosci* 2000;**3**:545–50.

631. Luyendijk W, Bots GT, Vegeter–van der Vlis M, *et al.* Hereditary cerebral haemorrhage caused by cortical amyloid angiopathy. *J Neurol Sci* 1988;**85**:267–80.

632. Ly JV, Donnan GA, Villemagne VL, *et al.* 11C-PIB binding is increased in patients with cerebral amyloid angiopathy-related hemorrhage. *Neurology* 2010;**74**:487–93.

633. Macdonald RL, Weir BKA. A review of hemoglobin and the pathogenesis of cerebral vasospasm. *Stroke* 1991;**22**:971–82.

634. Mackay KB, Kusumoto K, Graham DI, McCulloch J. Effect of the kappa-1 opioid agonist CI-977 on ischemic brain damage and cerebral blood flow after middle cerebral artery occlusion in the rat. *Brain Res* 1993;**629**:10–18.

635. MacKeith SA, Meyer A. A death during insulin treatment of schizophrenia; with pathological report. *J Ment Sci* 1939;**85**:96–105.

636. Mackert BM, Staub F, Peters J, *et al.* Anoxia *in vitro* does not induce neuronal swelling or death. *J Neurol Sci* 1996;**139**:39–47.

637. MacMillan VH. Cerebral energy metabolism in cyanide encephalopathy. *J Cereb Blood Flow Metab* 1989;**9**:156–62.

638. Macrez R, Ali C, Toutirais O, Le Mauff B, *et al.* Stroke and the immune system: from pathophysiology to new therapeutic strategies. *Lancet Neurol* 2011;**10**:471–80.

639. Maeda A, Yamada M, Itoh Y, *et al.* Computer-assisted three-dimensional image analysis of cerebral amyloid angiopathy. *Stroke* 1993;**24**:1857–64.

640. Maia LF, Magalhaes R, Freitas J, *et al.* CNS involvement in V30M transthyretin amyloidosis: clinical, neuropathological and biochemical findings. *J Neurol Neurosurg Psychiatry* 2014:10.1136/jnnp-2014-308107.

641. Mancuso MR, Kuhnert F, Kuo CJ. Developmental angiogenesis of the central nervous system. *Lymphat Res Biol* 2008;**6**:173–80.

642. Mandybur TI. Intracranial hemorrhage caused by metastatic tumours. *Neurology* 1977;**27**:650–55.

643. Mandybur TI. Cerebral amyloid angiopathy:the vascular pathology and complications. *J Neuropathol Exp Neurol* 1986;**45**:79–90.

644. Manley GT, Binder DK, Papadopoulos MC, *et al.* New insights into water transport and edema in the central nervous system from phenotype analysis of aquaporin-4 null mice. *Neuroscience* 2004;**129**:983–91.

645. Mann DM, Pickering–Brown SM, Takeuchi A, Iwatsubo T. Amyloid angiopathy and variability in amyloid beta deposition is determined by mutation position in presenilin-1-linked Alzheimer's disease. *Am J Pathol* 2001;**158**:2165–75.

646. Marchal G, Serrati C, Rioux P, *et al.* PET imaging of cerebral perfusion and oxygen consumption in acute ischaemic stroke: relation to outcome. *Lancet* 1993;**341**:925–7.

647. Marie C, Bralet AM, Gueldry S, Bralet J. Fasting prior to transient cerebral ischaemia reduces delayed neuronal necrosis. *Metab Brain Dis* 1990;**5**:65–75.

648. Marie P. Des foyers lacunaires de disintegration et de differents etats cavitaires du cerveau. *Rev Med (Paris)* 1901;**21**:281–98.

649. Marion DW, Darby J, Yonas H. Acute regional cerebral blood flow changes caused by severe head injuries. *J Neurosurg* 1991;**74**:407–14.

650. Marti HJ, Bernaudin M, Bellail A, *et al.* Hypoxia-induced vascular endothelial growth factor expression precedes neovascularization after cerebral ischaemia. *Am J Pathol* 2000;**156**:965–76.

651. Massand MG, Wallace RC, Gonzalez LF, *et al.* Subarachnoid hemorrhage due to isolated spinal artery aneurysm in four patients. *Am J Neuroradiol* 2005;**26**:2415–9.

652. Massaro AR, Sacco LR, Mohr JP *et al.* Clinical discriminators of lobar and deep hemorrhage: the Stroke Data Bank. *Neurology* 1991;**41**:1881–5.

653. Matijevic N, Wu KK. Hypercoagulable states and strokes. *Curr Atheroscler Rep* 2006;**8**:324–9.

654. Matsell DG, Keene DL, Jimenez C, Humphreys P. Isolated angiitis of the central nervous system in childhood. *Can J Neurol Sci* 1990;**17**:151–4.

655. Matsuda M, Shiino A, Handa J. Sequential changes of cerebral blood flow after aneurysmal subarachnoid haemorrhage. *Acta Neurochir* 1990;**105**:98–106.

656. Matsumaru Y, Yanaka K, Muroi A, *et al.* Significance of a small bulge on the basilar artery in patients with perimesencephalic nonaneurysmal subarachnoid hemorrhage:report of two cases. *J Neurosurg* 2003;**98**:426–9.

657. Matsumoto H, Terada T, Tsura M, *et al.* Basic fibroblast growth factor released from a platinum coil with a polyvinyl alcohol core enhances cellular proliferation and vascular wall thickness: an in vitro and in vivo study. *Neurosurgery* 2003;**53**:402–7.

658. Mawad ME, Rivera V, Crawford S, *et al.* Spinal cord ischaemia after resection of thoracoabdominal aortic aneurysms: MR findings in 24 patients. *Am J Roentgenol* 1990;**155**:1303–7.

659. Maxwell PH, Pugh CW, Ratcliffe PJ. Inducible operation of the erythropoietin 39 enhancer in multiple cell lines: evidence for a widespread oxygen-sensing mechanism. *Proc Natl Acad Sci U S A* 1993;**90**:2423–7.

660. Mayer SA, Lin J, Homma S, *et al.* Myocardial injury and left ventricular performance after subarachnoid hemorrhage. *Stroke* 1999;**30**:780–86.

661. Mayer–Gross W. Insulin coma therapy of schizophrenia: some critical remarks on Dr. Sakel's report. *J Ment Sci* 1951;**97**:132–5.

662. McCarron MO, Nicoll JA, Ironside JW, *et al.* Cerebral amyloid angiopathy-related hemorrhage. Interaction of APOE ε2 with putative clinical risk factors. *Stroke* 1999;**30**:1643–6.

663. McCarron MO, Nicoll JA, Stewart J, *et al.* The apolipoprotein E ε2 allele and the pathological features in cerebral amyloid angiopathy-related hemorrhage. *J Neuropathol Exp Neurol* 1999;**58**:711–8.

664. McPherson RW, Zeger S, Traystman RJ. Relationship of somatosensory evoked potentials and cerebral oxygen

consumption during hypoxic hypoxia in dogs. *Stroke* 1986;**17**:30–36.

665. Mead S, James–Galton M, Revesz T, *et al*. Familial British dementia with amyloid angiopathy: early clinical, neuropsychological and imaging findings. *Brain* 2000;**123**:975–9.

666. Melzer N, Harder A, Gross CC, *et al*. CD4+ T cells predominate in cerebrospinal fluid and leptomeningeal and parenchymal infiltrates in cerebral amyloid beta-related angiitis. *Arch Neurol* 2012;**69**(6):773–7.

667. Mendelow AD. Mechanisms of ischemic brain damage with intracerebral hemorrhage. *Stroke* 1993;**24**(Suppl I):115–17.

668. Menghini VV, Brown RD Jr, Sicks JD, *et al*. Incidence and prevalence of intracranial aneurysms and hemorrhage in Olmstead County, Minnesota, 1965 to 1995. *Neurology* 1998;**51**:405–11.

669. Meschia JF, Worrall BB, Rich SS. Genetic susceptibility to ischemic stroke. *Nat Rev Neurol* 2011;**7**:369–78.

670. Mettler FA, Sax DS. Cerebellar cortical degeneration due to acute azide poisoning. *Brain* 1972;**95**:505–16.

671. Metz R, Bogousslavsky J. Lacunar stroke. In: Fisher M, Bogousslavsky J eds. *Current review of cerebrovascular disease*. Boston, MA: Butterworth Heinemann, 1999:93–105.

672. Meyer A. Über die Wirkung der Kohlenoxydvergiftung auf das Zentralnervensystem. *Z Neurol Psychiat* 1926;**100**:201–47.

673. Meyer A. Intoxications. In: Greenfield JG, Blackwood W eds. *Greenfield's neuropathology*, 2nd edn. London: Edward Arnold, 1963:235–87.

674. Meyer JS, Portnoy HD. Localized cerebral hypoglycemia simulating stroke. *Neurology* 1958;**8**:601–14.

675. Meyohas MC, Roullet E. Cerebral venous thrombosis and dual primary infection with human immunodeficiency virus and cytomegalovirus. *J Neurol Neurosurg Psychiatry* 1989;**52**:1010–16.

676. Miao Q, Paloneva T, Tuominen S, *et al*. Fibrosis and stenosis of the long penetrating cerebral arteries: the cause of the white matter pathology in CADASIL. *Brain Pathology* 2004;**14**:358–64.

677. Michaeli J, Mittelman M, Grisaru D, Rachmilewitz EA. Thromboembolic complications in beta thalassaemia. *Acta Haematol* 1992;**87**:71–4.

678. Mies G, Kohno K, Hossmann KA. Prevention of periinfarct direct current shifts with glutamate antagonist NBQX following occlusion of the middle cerebral artery in the rat. *J Cereb Blood Flow Metab* 1994;**14**:802–7.

679. Miller DV, Maleszewski JJ. The pathology of large-vessel vasculitides. *Clin Exp Rheumatol* 2011;**29**:S92–8.

680. Miller DV, Salvarani C, Hunder GG, *et al*. Biopsy findings in primary angiitis of the central nervous system. *Am J Surg Pathol* 2009;**33**:35–43.

681. Miller JR, Myers RE. Neuropathology of systemic circulatory arrest in adult monkeys. *Neurology* 1972;**22**:888–904.

682. Miller RF, Isaacson PG, Hall–Craggs M, *et al*. Cerebral CD1lymphocytosis in HIV-1 infected patients with immune restoration induced HAART. *Acta Neuropathol* 2004;**108**:17–23.

683. Miners JS, Barua N, Kehoe PG, Gill S, Love S. Aβ-degrading enzymes: potential for treatment of Alzheimer disease. *J Neuropathol Exp Neurol* 2011;**70**:944–59.

684. Miners JS, Kehoe P, Love S. Neprilysin protects against cerebral amyloid angiopathy and Aβ-induced degeneration of cerebrovascular smooth muscle cells. *Brain Pathol* 2011;**21**:594–605.

685. Miralles P, Berenguer J, Lacruz C, *et al*. Inflammatory reactions in progressive multifocal leucoencephalopathy after highly active retroviral therapy. *AIDS* 2001;**15**:1900–902.

686. Mitani A, Kataoka K. Critical levels of extracellular glutamate mediating gerbil hippocampal delayed neuronal death during hypothermia: brain microdialysis study. *Neuroscience* 1991;**42**:661–70.

687. Mitchell P, Kerr R, Mendelow AD, Molyneux A. Could late rebleeding overturn the superiority of cranial aneurysm coil embolization over clip ligation seen in the International Subarachnoid Aneurysm Trial? *J Neurosurg* 2008;**108**:437–42.

688. Miyamoto O, Auer RN. Hypoxia, hyperoxia, ischaemia, and brain necrosis. *Neurology* 2000;**54**:362–71.

689. Miyasaka Y, Yada K. *et al*. An analysis of the venous drainage system as a factor in hemorrhage from arteriovenous malformations. *J Neurosurg* 1992;**76**:239–43.

690. Miyata T. Takayasu arteritis. In: Kalimo H ed. *Pathology and genetics: cerebrovascular diseases*. Basel: ISN Neuropath Press, 2005:140–46.

691. Mohr J. Lacunes. In: Barnett H, Mohr J, Stein B, Yatsu F eds. *Stroke: pathophysiology, diagnosis, and management*. New York: Churchill Livingstone, 1992:539–60.

692. Molinari GF, Smith L, Goldstein MN, Satran R. Pathogenesis of cerebral mycotic aneurysms. *Neurology* 1973;**23**:325–32.

693. Monaghan DT, Cotman CW. Distribution of N-methyl-d-aspartate-sensitive l-[3H]glutamate-binding sites in rat brain. *J Neurosci* 1985;**5**:2909–19.

694. Moore P, Calabrese LH. Neurologic manifestations of systemic vasculitides. *Semin Neurol* 1994;**14**:300–306.

695. Moossy J. Pathology of cerebral atherosclerosis: influence of age, race, and gender. *Stroke* 1993;**24** (Suppl I):22–3.

696. Mori E, del Zoppo GJ, Chambers JD, *et al*. Inhibition of polymorphonuclear leukocyte adherence suppresses no-reflow after focal cerebral ischemia in baboons. *Stroke* 1992;**23**:712–18.

697. Morii S, Ngai AC, Ko KR, Winn HR. Role of adenosine in regulation of cerebral blood flow: effects of theophylline during normoxia and hypoxia. *Am J Physiol* 1987;**253**:H165–75.

698. Morris JG, Singh S, Fisher M. Testing for inherited thrombophilias in arterial stroke: can it cause more harm than good? *Stroke* 2011;**41**:2985–90.

699. Moskowitz MA, Lo EH, Iadecola C. The science of stroke: mechanisms in search of treatments. *Neuron* 2010;**67**:181–98.

700. Moulin T, Tatu L, Vuillier F, *et al*. Role of a stroke data bank in evaluating cerebral infarction subtypes: patterns and outcome of 1776 consecutive patients from the Besancon stroke registry. *Cerebrovasc Dis* 2000;**10**:261–71.

701. Moustafa RR, Baron JC. Pathophysiology of ischaemic stroke: insights from imaging, and implications for therapy and drug discovery. *Br J Pharmacol* 2008;**153**(Suppl 1):S44–54.

702. Moutsopoulos HM, Sarmas JH, Talan N. Is central nervous system involvement a systemic manifestation of primary Sjögren's syndrome? *Rheum Dis Clin North Am* 1993;**19**:909–12.

703. Muir KW, Lees KR. Excitatory amino acid antagonists for acute stroke. *Cochrane Database Syst Rev* 2003;CD001244.

704. Müller C, Rahn BA, Pfister U, Meinig RP. The incidence, pathogenesis, diagnosis, and treatment of fat embolism. *Orthopaed Rev* 1994;**23**:107–17.

705. Murayama S, Bouldin TW, Suzuki K. Selective sparing of Betz cells in primary motor area in hypoxic-ischemic encephalopathy. *Acta Neuropathol (Berl)* 1990;**80**:560–2.

706. Murayama Y,Tateshima Y, Tateshima S, *et al*. Matrix and bioabsorbable polymeric coils accelerate healing of intracranial aneurysms: long term experimental study. *Stroke* 2003;**34**:2031–7.

707. Murray V, Norrving B, Sandercock PA, *et al*. The molecular basis of thrombolysis and its clinical application in stroke. *J Intern Med* 2010;**267**:191–208.

708. Naff NJ, Wemmer J, Hoenig–Rigamonti K, Rigamonti DR. A longitudinal study of patients with venous malformations: documentation of a negligible hemorrhage risk and benign natural history. *Neurology* 1998;**50**:1709–14.

709. Nag S, Kapadia A, Stewart DJ. Review: molecular pathogenesis of blood–brain barrier breakdown in acute brain injury. *Neuropathol Appl Neurobiol* 2011;**37**:3–23.

710. Nagai Y, Naruse S, Weiner MW. Effect of hypoglycemia on changes of brain lactic acid and intracellular pH produced by ischaemia. *NMR Biomed* 1993;**6**:1–6.

711. Nahed BV, Seker A, Guclu B, *et al*. Mapping a mendelian form of intracranial aneurysm to 1p34.3–p36.13. *Am J Hum Genet* 2005;**76**:172–9.

712. Nakanishi S. Molecular diversity of glutamate receptors and implications for brain function. *Science* 1992;**258**:597–603.

713. Nakano S, Kogure K, Fujikura H. Ischaemia-induced slowly progressive neuronal damage in the rat brain. *Neuroscience* 1990;**38**:115–24.

714. Nalivaeva NN, Beckett C, Belyaev ND, Turner AJ. Are amyloid-degrading enzymes viable therapeutic targets in Alzheimer's disease? *J Neurochem* 2012;**120**(Suppl 1):167–85.

715. Narayan SK, Gorman G, Kalaria RN, Ford GA, Chinnery PF. The minimum prevalence of CADASIL in northeast England. *Neurology* 2012;**78**:1025–7.

716. Nastanski F, Gordon WI, Lekawa ME. Posttraumatic paradoxical fat embolism to the brain: a case report. *J Trauma* 2005;**58**:372–4.

717. Natte R, de Boer WI, Maat–Schieman ML, *et al*. Amyloid beta precursor protein-mRNA is expressed throughout cerebral vessel walls. *Brain Res* 1999;**828**:179–83.

718. Natte R, Maat–Schieman ML, Haan J, et al. Dementia in hereditary cerebral haemorrhage with amyloidosis-Dutch type is associated with cerebral amyloid angiopathy but is independent of plaques and neurofibrillary tangles. *Ann Neurol* 2001;**50**:765–72.

719. Nedergaard M, Hansen AJ. Spreading depression is not associated with neuronal injury in the normal brain. *Brain Res* 1988;**449**:395–8.

720. Nedergaard M, Goldman SA, Desai S, Pulsinelli WA. Acid-induced death in neurons and glia. *J Neurosci* 1991;**11**:2489–97.

721. Nedergaard M, Kraig RP, Tanabe J, Pulsinelli WA. Dynamics of interstitial and intracellular pH in evolving brain infarct. *Am J Physiol* 1991;**260**:R581–8.

722. Nehls DG, Mendelow AD, Graham DI, et al. Experimental intracerebral hemorrhage: progression of hemodynamic changes after production of a spontaneous mass lesion. *Neurosurgery* 1988;**23**:439–44.

723. Nelson D, Goetzl S, Robins S, Ivy AC. Carrot diet and susceptibility to acute 'anoxia'. *Proc Soc Exp Biol Med* 1943;**52**:1–2.

724. Nelson KB. The epidemiology of cerebral palsy in term infants. *Ment Retard Dev Disabil Res Rev* 2002;**8**:146–50.

725. Nemoto EM, Hoff JT. Lactate uptake and metabolism by brain during hyperlactatemia and hypoglycemia. *Stroke* 1974;**5**:48–53.

726. Nencini P, Baruffi MC, Abbate R, et al. Lupus anticoagulant and anticardiolipin antibodies in young adults with cerebral ischaemia. *Stroke* 1992;**23**:189–93.

727. Neuberger KT, Clarke ER. Subacute carbon monoxide poisoning with cerebral myelinopathy and multiple myocardial necroses. *Rocky Mountain Med J* 1945;**42**:29–34.

728. Nevander G, Ingvar M, Auer RN, Siesjö BK. Status epilepticus in well–oxygenated rats causes neuronal necrosis. *Ann Neurol* 1985;**18**:281–90.

729. Ng T, Graham DI, Adams JH, Ford I. Changes in the hippocampus and the cerebellum resulting from hypoxic insults:frequency and distribution. *Acta Neuropathol (Berl)* 1989;**78**:438–43.

730. Ni JW, Matsumoto K, Li HB, et al. Neuronal damage and decrease of central acetylcholine level following permanent occlusion of bilateral common carotid arteries in rat. *Brain Res* 1995;**673**: 290–96.

731. Niazi GA, Awada A, al Rajeh S, Larbi E. Hematological values and their assessment as risk factor in Saudi patients with stroke. *Acta Neurol Scand* 1994;**89**:439–45.

732. Nicoll JA, Wilkinson D, Holmes C, Steart P, Markham H, Weller RO. Neuropathology of human Alzheimer disease after immunization with amyloid-β peptide: a case report. *Nat Med* 2003;**9**:448–52.

733. Nicoll JA, Yamada M, Frackowiak J, et al. Cerebral amyloid angiopathy plays a direct role in the pathogenesis of Alzheimer's disease: pro-CAA position statement. *Neurobiol Aging* 2004;**25**:589–97.

734. Nielsen EH, Lindholm J, Bjerre P, et al. Frequent occurrence of pituitary apoplexy in patients with non-functioning pituitary adenoma. *Clin Endocrinol (Oxf)* 2006;**64**:319–22.

735. Nilsson OG, Lindgren A, Stahl N, et al. Incidence of intracerebral and subarachnoid haemorrhage in southern Sweden. *J Neurol Neurosurg Psychiatry* 2000;**69**:601–7.

736. Nishimoto A, Takeuchi S. Abnormal cerebrovascular network related to the internal carotid arteries. *J Neurosurg* 1968;**29**:255–60.

737. Nishimoto Y, Shibata M, Nihonmatsu M, et al. A novel mutation in the HTRA1 gene causes CARASIL without alopecia. *Neurology* 2011;**76**:1353–5.

738. Nobili F, Rodriguez G, Marenco S, et al. Regional cerebral blood flow in chronic hypertension: a correlative study. *Stroke* 1993;**24**:1148–53.

739. Nordborg E, Nordborg C. Giant cell arteritis: epidemiological clues to its pathogenesis and an update on its treatment. *Rheumatology* 2003;**42**:413–21.

740. Noshita N, Sugawara T, Fujimura M, et al. Manganese superoxide dismutase affects cytochrome c release and caspase-9 activation after transient focal cerebral ischaemia in mice. *J Cereb Blood Flow Metab* 2001;**21**:557–67.

741. Novy J, Carruzzo A, Maeder P, Bogousslavsky J. Spinal cord ischaemia: clinical and imaging patterns, pathogenesis, and outcomes in 27 patients. *Arch Neurol* 2006;**63**:1113–20.

742. Nowak TS Jr. Localization of 70 kDa stress protein mRNA induction in gerbil brain after ischaemia. *J Cereb Blood Flow Metab* 1991;**11**:432–9.

743. Nowak–Gottl U, Strater R, Heinecke A, et al. Lipoprotein (a) and genetic polymorphisms of clotting factor V, prothrombin, and methylenetetrahydrofolate reductase are risk factors of spontaneous ischemic stroke in childhood. *Blood* 1999;**94**:3678–82.

744. Nussmeier NA, Arlund C, Slogoff S. Neuropsychiatric complications after cardiopulmonary bypass: cerebral protection by a barbiturate. *Anesthesiology* 1986;**64**:165–70.

745. O'Donnell JA, Emery CL. Neurosyphilis: a current review. *Curr Infect Dis Rep* 2005;**7**:277–84.

746. Oehmichen M, Auer RN, König HG. Forensic types of ischaemia and asphyxia. In: *Forensic neuropathology and neurology*. Heidelberg: Springer–Verlag, 2006:293–317.

747. Ogata J, Yutani C, Imakita M, et al. Autolysis of the granular layer of the cerebellar cortex in brain death. *Acta Neuropathol (Berl)* 1986;**70**:75–8.

748. Ogata J, Yutani C, Otsubo R, et al. Heart and vessel pathology underlying brain infarction in 142 stroke patients. *Ann Neurol* 2008;**63**:770–81.

749. Ohene–Frempong K, Weiner SJ, et al. Cerebrovascular accidents in sickle cell disease: rates and risk factors. *Blood* 1998;**91**:288–94.

750. Öhman J, Servo A, Heiskanen O. Risk factors for cerebral infarction in good-grade patients after aneurysmal subarachnoid hemorrhage and surgery: a prospective study. *J Neurosurg* 1991;**74**:14–20.

751. Ohta S, Smith M–L, Siesjö BK. The effect of a dihydropyridine calcium antagonist (isradipine) on selective neuronal necrosis. *J Neurol Sci* 1991;**103**:109–15.

752. Okada Y, Copeland BR, Mori E, et al. P–selectin and intercellular adhesion molecule-1 expression after focal brain ischaemia and reperfusion. *Stroke* 1994;**25**:202–11.

753. Okeda R, Funata N, Song S–J, et al. Comparative study on pathogenesis of selective cerebral lesions in carbon monoxide poisoning and nitrogen hypoxia in cats. *Acta Neuropathol (Berl)* 1982;**56**:265–72.

754. Olin JW, Sealove BA. Diagnosis, management, and future developments of fibromuscular dysplasia. *J Vasc Surg* 2011;**53**:826–36 e1.

755. Olney JW. Glutamate-induced retinal degeneration in neonatal mice: electron microscopy of the acutely evolving lesions. *J Neuropathol Exp Neurol* 1969;**28**:455–74.

756. Olney JW. Brain lesions, obesity, and other disturbances in mice treated with monosodium glutamate. *Science* 1969;**164**:719–21.

757. Olney JW, Ho OL, Rhee V. Cytotoxic effects of acidic and sulphur containing amino acids on the infant mouse central nervous system. *Exp Brain Res* 1971;**14**:61–76.

758. Ongini E, Adami M, Ferri C, et al. Adenosine A2A receptors and neuroprotection. *Ann N Y Acad Sci* 1997;**825**:30–48.

759. Öörni K, Pentikäinen MO, Ala–Korpela M, et al. Aggregation, fusion and vesicle formation of modified LDL particles: molecular mechanisms and effects on matrix interactions. *J Lipid Res* 2000;**41**:1703–14.

760. Oorschot DE, Black MJ, Rangi F, Scarr E. Is Fos protein expressed by dying striatal neurons after immature hypoxic-ischemic brain injury? *Exp Neurol* 2000;**161**:227–33.

761. Ophoff RA, DeYoung J, Service SK, et al. Hereditary vascular retinopathy, cerebroretinal vasculopathy, and hereditary endotheliopathy with retinopathy, nephropathy, and stroke map to a single locus on chromosome 3p21.1–p21.3. *Am J Hum Genet* 2001;**69**:447–53.

762. Oppenheimer BS, Fishberg AM. Hypertensive encephalopathy. *Arch Intern Med* 1928;**41**:264–78.

763. Oppert M, Gleiter CH, MÜller C, et al. Kinetics and characteristics of an acute phase response following cardiac arrest. *Intensive Care Med* 1999;**25**:1386–94.

764. Osborne AG, Anderson RE. Angiographic spectrum of cervical and intracranial fibromuscular dysplasia. *Stroke* 1977;**8**:617–26.

765. Oshima K, Akiyama H, Tsuchiya K, et al. Relative paucity of tau accumulation in the small areas with abundant Aβ42-positive capillary amyloid angiopathy within a given cortical region in the brain of patients with Alzheimer pathology. *Acta neuropathol* 2006;**111**:510–8.

766. Ott E. Hyperviscosity syndromes. In: Toole J ed. *Handbook of clinical neurology*. Amsterdam: Elsevier, 1989:483–92.

767. Oztürk A, Demirci F, Yavuz T, et al. Antenatal and delivery risk factors and

prevalence of cerebral palsy in Duzce (Turkey). *Brain Dev* 2007;**29**:39–42.

768. Palmer AC. Target organs in decompression sickness. *Prog Underwater Sci* 1990;**15**:15–23.

769. Palmer AC, Calder IM, Hughes JT. Spinal cord degeneration in divers. Lancet 1987;**2**:1365–6.

770. Pajunen P, Pääkkönen R, Hämäläinen H, *et al.* Trends in fatal and nonfatal strokes among persons aged 35 to 85 years during 1991–2002 in Finland. *Stroke* 2005;**36**:244–8.

771. Pan J, Konstas AA, Bateman B, Ortolano GA, Pile–Spellman J. Reperfusion injury following cerebral ischaemia: pathophysiology, MR imaging and potential therapies. *Neuroradiology* 2007;**49**:93–102.

772. Papadopoulos DP, Mourouzis I, Thomopoulos C, Makris T, Papademetriou V. Hypertension crisis. *Blood Press* 2010;**19**:328–36.

773. Papagapiou MP, Auer RN. Regional neuroprotective effects of the NMDA receptor antagonist MK-801 (dizocilpine) in hypoglycemic brain damage. *J Cereb Blood Flow Metab* 1990;**10**:270–76.

774. Parisi JE. Fibromuscular dysplasia. In: Kalimo H ed. *Pathology and genetics: cerebrovascular diseases.* Basel: ISN Neuropath Press, 2005:164–8.

775. Parisi JE, Kim RC, Collins GH, Hilfinger MF. Brain death with prolonged somatic survival. *N Engl J Med* 1982;**306**:14–16.

776. Parlea L, Fahrig R, Holdsworth DW, Lownie SP. An analysis of the geometry of saccular intracranial aneurysms. *Am J Neuroradiol* 1999;**20**:1079–89.

777. Paschen W, Djuricic B, Mies G, *et al.* Lactate and pH in the brain: association and dissociation in different pathophysiological states. *J Neurochem* 1987;**48**:154–9.

778. Patel–Hett S, D'Amore PA. Signal transduction in vasculogenesis and developmental angiogenesis. *Int J Dev Biol* 2011;**55**:353–63.

779. Pavlakis SG, Frank Y, Chusid R. Hypertensive encephalopathy, reversible occipitoparietal encephalopathy, or reversible posterior leukoencephalopathy: three names for an old syndrome. *J Child Neurol* 1999;**14**:277–81.

780. Pearigen P, Gwinn R, Simon RP. The effects of in vivo hypoxia on brain injury. *Brain Res* 1996;**725**:184–91.

781. Pearson TC, Wetherley–Mein G. Vascular occlusive episodes and venous hematocrit in primary proliferative polycythemia. *Lancet* 1978;**2**:1219–22.

782. Pelligrino D, Almquist L–O, Siesjö BK. Effects of insulin-induced hypoglycemia on intracellular pH and impedance in the cerebral cortex of the rat. *Brain Res* 1981;**221**:129–47.

783. Pelz DM. Interventional neuroradiology. In: Kalimo H ed. *Pathology and genetics: cerebrovascular diseases.* Basel: ISN Neuropath Press, 2005:125–32.

784. Pendlebury ST. Stroke-related dementia: rates, risk factors and implications for future research. *Maturitas* 2009;**64**: 165–71.

785. Pendlebury ST, Rothwell PM. Prevalence, incidence, and factors associated with pre-stroke and post-stroke dementia: a systematic review

and meta-analysis. *Lancet Neurol* 2009;**8**:1006–18.

786. Pendlebury WW, Iole ED, Tracy RP, Dill BA. Intracerebral hemorrhage related to cerebral amyloid angiopathy and t-PA treatment. *Ann Neurol* 1991;**29**:210–13.

787. Penn DL, Komotar RJ, Sander Connolly E. Hemodynamic mechanisms underlying cerebral aneurysm pathogenesis. *J Clin Neurosci* 2012;**18**:1435–8.

788. Penney DG, Helfman CC, Dunbar JC Jr, McCoy LE. Acute severe carbon monoxide exposure in the rat: effects of hyperglycemia and hypoglycemia on mortality, recovery, and neurologic deficit. *J Physiol* 1991;**69**:1168–77.

789. Peral B, Gamble V, Strong C, *et al.* Identification of mutations in the duplicated region of the polycystic kidney disease 1 gene (PKD1) by a novel approach. *Am J Hum Genet* 1997;**60**:1399–410.

790. Peters DG, Kassam A, St Jean PL, *et al.* Functional polymorphism in the matrix metalloproteinase-9 promoter as a potential risk factor for intracranial aneurysm. *Stroke* 1999;**30**:2612–16.

791. Petersen RB, Goren H, Cohen M, *et al.* Transthyretin amyloidosis: a new mutation associated with dementia. *Ann Neurol* 1997;**41**:307–13.

791a. Petito CK, Feldmann E, Pulsinelli WA, Plum F. Delayed hippocampal damage in humans following cardiorespiratory arrest. *Neurology* 1987;**37**:1281–6.

792. Petroni A, Borghi A, Blasevich M, *et al.* Effects of hypoxia and recovery on brain eicosanoids and carbohydrate metabolites in rat brain cortex. *Brain Res* 1987;**415**:226–32.

793. Pettigrew HD, Teuber SS, Gershwin ME. Polyarteritis nodosa. *Compr Ther* 2007;**33**:144–9.

794. Phan TG, Huston J, 3rd, Brown RD, Jr., Wiebers DO, Piepgras DG. Intracranial saccular aneurysm enlargement determined using serial magnetic resonance angiography. *J Neurosurg* 2002;**97**:1023–8.

795. Pharoah PO, Cooke T, Rosenbloom I, Cooke RW. Trends in birth prevalence of cerebral palsy. *Arch Dis Child* 1987;**62**:379–84.

796. Phillis JW, DeLong RE, Towner JK. Adenosine deaminase inhibitors enhance cerebral anoxic hyperemia in the rat. *J Cereb Blood Flow Metab* 1985;**5**:295–9.

797. Pico F, Labreuche J, Touboul PJ, *et al.* Intracranial arterial dolichoectasia and small-vessel disease in stroke patients. *Ann Neurol* 2005;**57**:472–9.

798. Picone AL, Green RM, Ricotta JR, May AG, DeWeese JA. Spinal cord ischaemia following operations on the abdominal aorta. *J Vasc Surg* 1986;**3**:94–103.

799. Piette JC, Wechsler B, Vidailhet M. Idiopathic intracranial hypertension: don't forget cerebral venous thrombosis. *Am J Med* 1994;**97**:200.

800. Pignataro G, Scorziello A, Di Renzo G, Annunziato L. Post-ischemic brain damage: effect of ischemic preconditioning and postconditioning and identification of potential candidates for stroke therapy. *FEBS J* 2009;**276**:46–57.

801. Plaisier E, Gribouval O, Alamowitch S, *et al.* COL4A1 mutations and hereditary angiopathy, nephropathy, aneurysms,

and muscle cramps. *N Engl J Med* 2007;**357**:2687–95.

802. Planas AM, Soriano MA, Estrada A, *et al.* The heat shock stress response after brain lesions: induction of 72 kDa heat shock protein (cell types involved, axonal transport, transcriptional regulation) and protein synthesis inhibition. *Prog Neurobiol* 1997;**51**:607–36.

803. Plant GT, Revesz T, Barnard RO, *et al.* Familial cerebral amyloid angiography with nonneuritic amyloid plaque formation. *Brain* 1990;**113**:721–47.

804. Plate K. Neoformation and repair mechanisms of CNS blood vessels. In Kalimo H ed. *Pathology and genetics. Cerebrovascular diseases.* Basel: ISN Neuropath Press, 2005:9–13.

805. Plouin PF, Perdu J, La Batide–Alanore A, Boutouyrie P, Gimenez–Roqueplo AP, Jeunemaitre X. Fibromuscular dysplasia. *Orphanet J Rare Dis* 2007;**2**:28.

806. Plum F, Posner JB, Hain RF. Delayed neurological deterioration after anoxia. *Arch Intern Med* 1962;**110**:56–63.

807. Polkinghorne PJ, Sehmi K, Cross MR, *et al.* Ocular fundus lesions in divers. *Lancet* 1988;**2**:1381–3.

808. Polmear A. Sentinel headaches in aneurysmal subarachnoid haemorrhage: what is the true incidence? A systematic review. *Cephalalgia* 2003;**23**:935–41.

809. Porter PJ, Willinsky RA, Harper W, Wallace MC. Cerebral cavernous malformations: natural history and prognosis after clinical deterioration with or without hemorrhage. *J Neurosurg* 1997;**87**:190–97.

810. Postal M, Costallat LT, Appenzeller S. Neuropsychiatric manifestations in systemic lupus erythematosus: epidemiology, pathophysiology and management. *CNS Drugs* 2011;**25**: 721–36.

811. Poursines Y, Alliez J, Toga M. [Study of cortical lesions in a case of carbon monoxide poisoning.] [in French.] *Rev Neurol (Paris)* 1956;**94**:731–5.

812. Premkumar DR, Cohen DL, Hedera P, *et al.* Apolipoprotein E-epsilon4 alleles in cerebral amyloid angiopathy and cerebrovascular pathology associated with Alzheimer's disease. *Am J Pathol* 1996;**148**:2083–95.

813. Pritz MB. Cerebral aneurysm classification based on angioarchitecture. *J Stroke Cerebrovasc Dis* 2010;**20**:162–7.

814. Prockop LD, Chichkova RI. Carbon monoxide intoxication: an updated review. *J Neurol Sci* 2007;**262**:122–30.

815. Provenzale JM, Heinz ER, Ortel TL, *et al.* Antiphospholipid antibodies in patients without systemic lupus erythematosus: neuroradiologic findings. *Radiology* 1994;**192**:531–7.

816. Pullarkat VA, Kalapura T, Pincus M, Baskharoun R. Intraspinal hemorrhage complicating oral anticoagulant therapy: an unusual case of cervical hematomyelia and a review of the literature. *Arch Intern Med* 2000;**160**:237–40.

817. Pulsinelli WA, Waldman S, Rawlinson D, Plum F. Moderate hyperglycemia augments ischemic brain damage: a neuropathologic study in the rat. *Neurology* 1982;**32**:1239–46.

818. Pulsinelli WA, Levy DE, Sigsbee B, *et al.* Increased damage after ischemic stroke in patients with hyperglycemia with or

without diabetes mellitus. *Am J Med* 1983;**74**:540–44.

819. Puyal J, Ginet V, Grishchuk Y, Truttmann AC, Clarke PG. Neuronal autophagy as a mediator of life and death: contrasting roles in chronic neurodegenerative and acute neural disorders. *Neuroscientist* 2012;**18**(3):224–36.

820. Qi Y, Xue QM. Ganglioside levels in hypoxic brains from neonatal and premature infants. *Mol Chem Neuropathol* 1991;**14**:87–97.

821. Queiroz L de Sousa, Eduardo RMP. Occurrence of dark neurons in living mechanically injured rat neocortex. *Acta Neuropathol (Berl)* 1977;**38**:45–8.

822. Qureshi AI, Suri MF, Yahia AM, *et al*. Risk factors for subarachnoid hemorrhage. *Neurosurgery* 2001;**49**: 607–12.

823. Radwan W, Sawaya R. Intracranial haemorrhage associated with cerebral infections: a review. *Scand J Infect Dis* 2011;**43**:675–82.

824. Rahman NU, al Tahan AR. Computed tomographic evidence of an extensive thrombosis and infarction of the deep venous system. *Stroke* 1993;**24**:744–6.

825. Rami A. Review: autophagy in neurodegeneration: firefighter and/ or incendiarist? *Neuropathol Appl Neurobiol* 2009;**35**:449–61.

826. Rami A, Kogel D. Apoptosis meets autophagy-like cell death in the ischemic penumbra: two sides of the same coin? *Autophagy* 2008;**4**:422–6.

827. Rannikmae K, Samarasekera N, Martinez–Gonzalez NA, Al–Shahi Salman R, Sudlow CL. Genetics of cerebral amyloid angiopathy: systematic review and meta-analysis. *J Neurol Neurosurg Psychiatry* 2013;**84**(8):901–8.

828. Raskin N, Mullaney OC. The mental and neurological sequelae of carbon monoxide asphyxia in a case observed for 15 years. *J Nerv Ment Dis* 1940;**92**:640–59.

829. Ratcliffe PJ, O'Rourke JF, Maxwell PH, Pugh CW. Oxygen sensing, hypoxia-inducible factor-1 and the regulation of mammalian gene expression. *J Exp Biol* 1998;**201**:1153–62.

830. Raymond J, Guilbert F, Weill A, *et al*. Long-term angiographic recurrences after selective endovascular treatment of aneurysms with detachable coils. *Stroke* 2003;**34**;421–7.

831. Razvi SS, Davidson R, Bone I, Muir KW. The prevalence of cerebral autosomal dominant arteriopathy with subcortical infarcts and leucoencephalopathy (CADASIL) in the west of Scotland. *J Neurol Neurosurg Psychiatry* 2005;**76**:739–41.

832. Redzic ZB, Preston JE, Duncan JA, Chodobski A, Szmydynger–Chodobska J. The choroid plexus-cerebrospinal fluid system: from development to aging. *Curr Top Dev Biol* 2005;**71**:1–52.

833. Reiffenstein RJ, Hulbert WC, Roth SH. Toxicology of hydrogen sulfide. *Annu Rev Pharmacol Toxicol* 1992;**32**:109–34.

834. Reiss Y, Machein MR, Plate KH. The role of angiopoietins during angiogenesis in gliomas. *Brain Pathol* 2005;**15**:311–7.

835. Reuner KH, Ruf A, Grau A, *et al*. Prothrombin gene G20210A transition is a risk factor for cerebral venous thrombosis. *Stroke* 1998;**29**:1765–9.

836. Révész T, Geddes JF. Symmetrical columnar necrosis of the basal ganglia and brain stem in an adult following cardiac arrest. *Clin Neuropathol* 1988;**7**:294–8.

837. Revesz T, Holton JL, Lashley T, *et al*. Sporadic and familial cerebral amyloid angiopathies. *Brain Pathol* 2002;**12**:343–57.

838. Revesz T, Ghiso J, Lashley T, *et al*. Cerebral amyloid angiopathies: a pathologic, biochemical, and genetic view. *J Neuropathol Exp Neurol* 2003;**62**:885–98.

839. Revesz T, Holton JL, Lashley T, *et al*. Genetics and molecular pathogenesis of sporadic and hereditary cerebral amyloid angiopathies. *Acta Neuropathol* 2009;**118**:115–30.

840. Reynolds K, Lewis B, Nolen JD, *et al*. Alcohol consumption and risk of stroke: a meta-analysis. *J Am Med Assoc* 2003;**289**:579–88.

841. Rhiannon JJ. Systemic lupus erythematosus involving the nervous system: presentation, pathogenesis, and management. *Clin Rev Allergy Immunol* 2008;**34**:356–60.

842. Ribe EM, Serrano–Saiz E, Akpan N, Troy CM. Mechanisms of neuronal death in disease: defining the models and the players. *Biochem J* 2008;**415**:165–82.

843. Ribourtout E, Raymond J. Gene therapy and endovascular treatment of intracranial aneurysms. *Stroke* 2004;**35**:786–93.

844. Richards A, Graham D, Bullock R. Clinicopathological study of neurological complications due to hypertensive disorders of pregnancy. *J Neurol Neurosurg Psychiatry* 1988;**51**: 416–21.

845. Richards A, van den Maagdenberg AM, Jen JC, *et al*. C-terminal truncations in human 3′–5′ DNA exonuclease TREX1 cause autosomal dominant retinal vasculopathy with cerebral leukodystrophy. *Nat Genet* 2007;**39**:1068–70.

846. Richardson JC, Chambers RA, Heywood PM. Encephalopathies of anoxia and hypoglycemia. *Arch Neurol* 1959;**1**: 178–90.

847. Rie MA, Bernad PG. Prolonged hypoxia in man without circulatory compromise fails to demonstrate cerebral pathology. *Neurology* 1980;**30**:443.

848. Rigamonti D, Hadley MN, Drayer BP, *et al*. Cerebral cavernous malformations: incidence and familial occurrence. *N Engl J Med* 1988;**319**:343–7.

849. Rinkel GJ, Djibuti M, Algra A, van Gijn J. Prevalence and risk of rupture of intracranial aneurysms: a systematic review. *Stroke* 1998;**29**:251–6.

850. Rinkel GJ, Wijdicks EF, Vermeulen M, *et al*. Nonaneurysmal perimesencephalic subarachnoid hemorrhage: CT and MR patterns that differ from aneurysmal rupture. *Am J Neuroradiol* 1991;**12**: 829–34.

851. Rinkel GJ, Wijdicks EF, Vermeulen M, *et al*. The clinical course of perimesencephalic nonaneurysmal subarachnoid hemorrhage. *Ann Neurol* 1991;**29**:463–8.

852. Rinne J, Hernesniemi J, Puranen M, Saari T. Multiple intracranial aneurysms in a defined population: prospective

angiographic and clinical study. *Neurosurgery* 1994;**35**:803–8.

853. Risau W. Mechanisms of angiogenesis. *Nature* 1997;**386**:671–4.

854. Robertson GS, Pfaus JG, Atkinson LJ, *et al*. Sexual behavior increases c–fos expression in the forebrain of the male rat. *Brain Res* 1991;**564**:352–7.

855. Robertson KM, Hramiak IM, Gelb AW. Endocrine changes and haemodynamic stability after brain death. *Transplant Proc* 1989;**21**:1197–8.

856. Roher AE, Tyas SL, Maarouf CL, *et al*. Intracranial atherosclerosis as a contributing factor to Alzheimer's disease dementia. *Alzheimers Dement* 2011;**7**:436–44.

857. Rolett EL, Azzawi A, Liu KJ, *et al*. Critical oxygen tension in rat brain: a combined 31P-NMR and EPR oximetry study. *Am J Physiol Regul Integr Comp Physiol* 2000;**279**:R9–16.

858. Romanul F. Examination of the brain and spinal cord. In: Tedeschi C ed. *Neuropathology: methods and diagnosis*. Boston, MA: Little, Brown & Co, 1970:131–214.

859. Romijn HJ, de Jong BM. Unlike hypoxia, hypoglycemia does not preferentially destroy GABAergic neurons in developing rat neocortex explants in culture. *Brain Res* 1989;**480**:58–64.

860. Ronkainen A, Hernesniemi J. Subarachnoid haemorrhage of unknown aetiology. *Acta Neurochir* 1992;**119**: 29–34.

861. Ronkainen A, Hernesniemi J, Tromp G. Special features of familial intracranial aneurysms: report of 215 familial aneurysms. *Neurosurgery* 1995;**37**:43–7.

862. Ronkainen A, Hernesniemi J, Puranen M, *et al*. Familial intracranial aneurysms. *Lancet* 1997;**349**:380–84.

863. Roob G, Fazekas F. Magnetic resonance imaging of cerebral microbleeds. *Curr Opin Neurol* 2000;**13**:69–73.

864. Roos YB, de Haan RJ, Beenen LF, *et al*. Complications and outcome in patients with aneurysmal subarachnoid haemorrhage: a prospective hospital based cohort study in the Netherlands. *J Neurol Neurosurg Psychiatry* 2000;**68**:337–41.

865. Rose M. Die sogenannte Riechrinde beim Menschen und beim Affen. *J Psychol Neurol* 1926;**34**:261–401.

866. Rosen CL, DePalma L, Morita A. Primary angiitis of the central nervous system as the primary manifestation in Hodgkin's disease: a case report and review of the literature. *Neurosurgery* 2000;**46**:1504–8.

867. Rosenberg GA. Ischemic brain edema. *Progr Cardiovasc Dis* 1999;**42**:209–16.

868. Rosenblum B, Oldfield EH, Doppman JL, Di Chiro G. Spinal arteriovenous malformations: a comparison of dural arteriovenous fistulas and intradural AVM's in 81 patients. *J Neurosurg* 1987;**67**:795–802.

869. Rosenblum WI. Miliary aneurysms and 'fibrinoid' degeneration of cerebral blood vessels. *Hum Pathol* 1977;**8**:133–9.

870. Rosenblum WI. The importance of fibrinoid necrosis as the cause of cerebral hemorrhage in hypertension: commentary. *J Neuropathol Exp Neurol* 1993;**52**:11–13.

871. Rosenørn J, Eskesen V, Schmidt H, Ronde R. The risk of rebleeding from ruptured intracranial aneurysms. *J Neurosurg* 1987;**67**:329–32.

872. Ross DT, Duhaime AC. Degeneration of neurons in the thalamic reticular nucleus following transient ischaemia due to raised intracranial pressure: excitotoxic degeneration mediated via non-NMDA receptors? *Brain Res* 1989;**501**:129–43.

873. Ross DT, Graham DI. Selective loss and selective sparing of neurons in the thalamic reticular nucleus following human cardiac arrest. *J Cereb Blood Flow Metab* 1993;**13**:558–67.

874. Ross R. Atherosclerosis: an inflammatory disease. *N Engl J Med* 1999;**340**:115–26.

875. Ross R, Enevoldson T. Unusual types of ischemic stroke. In: Fisher M, Bogousslavsky J eds. *Current review of cerebrovascular disease*. Philadelphia, PA: Current Medicine, 1993:63–77.

876. Rostagno A, Holton JL, Lashley T, Revesz T, Ghiso J. Cerebral amyloidosis:amyloid subunits, mutants and phenotypes. *Cell Mol Life Sci* 2010;**67**:581–600.

877. Rothwell PM, Coull AJ, Giles MF, *et al.* Change in stroke incidence, mortality, case-fatality, severity, and risk factors in Oxfordshire, UK from 1981 to 2004 (Oxford Vascular Study). *Lancet* 2004;**363**:1925–33.

878. Rothwell PM, Algra A, Amarenco P. Medical treatment in acute and long-term secondary prevention after transient ischaemic attack and ischaemic stroke. *Lancet* 2011;**377**:1681–92.

879. Roubey RA, Hoffman M. From antiphospholipid syndrome to antibody-mediated thrombosis. *Lancet* 1997;**350**:1491–3.

880. Rouiller EM, Wan XS, Moret V, Liang F. Mapping of c-fos expression elicited by pure tones stimulation in the auditory pathways of the rat, with emphasis on the cochlear nucleus. *Neurosci Lett* 1992;**144**:19–24.

881. Ruchoux MM, Maurage CA. CADASIL: cerebral autosomal dominant arteriopathy with subcortical infarcts and leukoencephalopathy. *J Neuropathol Exp Neurol* 1997;**56**:947–64.

882. Ruigrok YM, Rinkel GJE, Wijmenga G. Genetics of intracranial aneurysms. *Lancet Neurol* 2005;**4**:179–89.

883. Russell NA, Benoit BG. Spinal subdural hematoma: a review. *Surg Neurol* 1983;**20**:133–7.

884. Russell R, Wade J. Haematological causes of cerebrovascular disease. In: Toole J ed. *Handbook of clinical neurology*, Vol. 11. Amsterdam: Elsevier, 1989:463–81.

885. Ruth–Sahd LA, Zulkosky K, Fetter ME. Carbon monoxide poisoning: case studies and review. *Dimens Crit Care Nurs* 2012;**30**:303–14.

886. Ruzali WAW, Kehoe PG, Love S. Influence of LRP-1 and apoE on Aβ uptake and toxicity to cerebrovascular smooth muscle cells. *J Alzheimers Dis* 2013;**33**:95–110.

887. Sacco R. Frequency and determinants of stroke. In: Fisher M ed. *Clinical atlas of cerebrovascular disorders*. London: Wolfe, 1994:1.2–16.

888. Sacco R. Classification of stroke. In: Fisher M ed. Clinical atlas of cerebrovascular disorders. London: Mosby-Wolfe, 1994:2.2–25.

889. Sadler JE, Moake JL, Miyata T, George JN. Recent advances in thrombotic thrombocytopenic purpura. *Hematology Am Soc Hematol Educ Program* 2004:407–23.

890. Sadove MS, Yon MK, Hollinger PH, *et al.* Severe prolonged cerebral hypoxic episode with complete recovery. *J Am Med Assoc* 1961;**175**:1102–4.

891. Sagoh M, Hirose Y, Murakami H, *et al.* Late hemorrhage from persistent pseudoaneurysm in vertebral artery dissection presenting with ischaemia: case report. *Surg Neurol* 1999;**52**:480–83.

892. Sajanti J, Björkstrand AS, Finnila S, *et al.* Increase of collagen synthesis and deposition in the arachnoid and the dura following subarachnoid hemorrhage in the rat. *Biochim Biophys Acta* 1999;**1454**:209–16.

893. Sakatani K, Murata Y, Fujiwara N, *et al.* Comparison of blood–oxygen-level-dependent functional magnetic resonance imaging and near-infrared spectroscopy recording during functional brain activation in patients with stroke and brain tumours. *J Biomed Opt* 2007;**12**:062110.

894. Sakel M. The methodical use of hypoglycemia in the treatment of psychoses. *Am J Psychiatry* 1937;**94**: 111–29.

895. Salford LG, Siesjö BK. The influence of arterial hypoxia and unilateral carotid artery occlusion upon regional blood flow and metabolism in the rat brain. *Acta Physiol Scand* 1974;**92**:130–41.

896. Salvarani C, Brown RD, Jr., Calamia KT, *et al.* Primary central nervous system vasculitis presenting with intracranial hemorrhage. *Arthritis Rheum* 2011;**63**:3598–606.

897. Sammons V, Davidson A, Tu J, Stoodley MA. Endothelial cells in the context of brain arteriovenous malformations. *J Clin Neurosci* 2011;**18**:165–70.

898. Samson D, Batjer HH, Bowman G, *et al.* A clinical study of the parameters and effects of temporary arterial occlusion in the management of intracranial aneurysms. *Neurosurgery* 1994;**34**:22–8.

899. Sandberg M, Butcher SP, Hagberg H. Extracellular overflow of neuroactive amino acids during severe insulin-induced hypoglycemia: in vivo dialysis of the rat hippocampus. *J Neurochem* 1986;**47**:178–84.

900. Sandberg M, Nyström B, Hamberger A. Metabolically derived aspartate: elevated extracellular levels *in vivo* in iodoacetate poisoning. *J Neurosci Res* 1985;**13**:489–95.

901. Sarti C, Tuomilehto J, Salomaa V, *et al.* Epidemiology of subarachnoid haemorrhage in Finland from 1983 to 1985. *Stroke* 1991;**22**:848–53.

902. Sauron B, Chiras J, Chain G, Castaigne P. Thrombophlébite cérébelleuse chez un homme porteur d'un déficit familial en antithrombine III. *Rev Neurol (Paris)* 1982;**138**:685.

903. Savoiado M. The vascular territories of the carotid and vertebrobasilar system. *Ital J Neurol Sci* 1986;**7**:405–9.

904. Schaller B, Graf R. Cerebral ischaemia and reperfusion: the pathophysiologic concept as a basis for clinical therapy. *J Cereb Blood Flow Metab* 2004;**24**:351–71.

905. Schievink WI, Puumala MR, Meyer FB, *et al.* Giant intracranial aneurysm and fibromuscular dysplasia in an adolescent with alpha 1-antitrypsin deficiency. *J Neurosurg* 1996;**85**:503–6.

906. Schievink WI, Parisi JE, Piepgras DG, Michels VV. Intracranial aneurysms in Marfan's syndrome: an autopsy study. *Neurosurgery* 1997;**41**:866–70.

907. Schievink WI, Torres VE, Wiebers DO, Huston J 3rd. Intracranial arterial dolichoectasia in autosomal dominant polycystic kidney disease. *J Am Soc Nephrol* 1997;**8**:1298–303.

908. Schievink WI, Katzmann JA, Piepgras DG. Alpha-1-antitrypsin deficiency in spontaneous intracranial artery dissections. *Cerebrovasc Dis* 1998;**8**: 42–4.

909. Schievink WI, Meyer FB, Parisi JE, Wijdicks EF. Fibromuscular dysplasia of the internal carotid artery associated with alpha1-antitrypsin deficiency. *Neurosurgery* 1998;**43**:229–33.

910. Schiff D, Lopes MB. Neuropathological correlates of reversible posterior leukoencephalopathy. *Neurocrit Care* 2005;**2**:303–5

911. Schiffmann R. Fabry disease. *Pharmacol Ther* 2009;**122**:65–77.

912. Schmahl FW, Betz E, Dettinger E, Hohorst HJ. Energiestoffwechsel der Grosshirnrinde und Elektroencephalogramm bei Sauerstoffmangel. *Pflögers Arch Physiol* 1966;**292**:46–59.

913. Schrader B, Barth H, Lang EW, *et al.* Spontaneous intracranial haematomas caused by neoplasms. *Acta Neurochir (Wien)* 2000;**142**:979–85.

914. Schröder R. Later changes in brain death: signs of partial recirculation. *Acta Neuropathol (Berl)* 1983;**62**:15–23.

915. Schulz UG, Rothwell PM. Differences in vascular risk factors between etiological subtypes of ischemic stroke: importance of population-based studies. *Stroke* 2003;**34**:2050–59.

916. Schunkert H, Konig IR, Kathiresan S, *et al.* Large-scale association analysis identifies 13 new susceptibility loci for coronary artery disease. *Nat Genet* 2011;**43**:333–8.

917. Schurr A, West CA, Rigor BM. Lactate-supported synaptic function in the rat hippocampal slice preparation. *Science* 1988;**240**:1326–8.

918. Schutta HS, Williams EC, Baranski BG, Sutula TP. Cerebral venous thrombosis with plasminogen deficiency. *Stroke* 1991;**22**:401–5.

919. Schütz H, Bödeker RH, Damian M, *et al.* Age-related spontaneous intracerebral hematoma in a German community. *Stroke* 1990;**21**:1412–18.

920. Schwartz TH, Solomon RA. Perimesencephalic nonaneurysmal subarachnoid hemorrhage: review of the literature. *Neurosurgery* 1996;**39**:433–40.

921. Scofield RH. Vasculitis in Sjogren's syndrome. *Curr Rheumatol Rep* 2011;**13**:482–8.

922. Scolding NJ, Joseph F, Kirby PA, al. Ab-related angiitis: primary angiitis of the central nervous system associated with cerebral amyloid angiopathy. *Brain* 2005;**128**:500–515.

923. Scott RM, Smith ER. Moyamoya disease and moyamoya syndrome. *N Engl J Med* 2009;**360**:1226–37.

924. Seifert T, Lechner A, Flooh E, *et al.* Lack of association of lobar intracerebral hemorrhage with apolipoprotein E genotype in an unselected population. *Cerebrovasc Dis* 2006;**21**:266–70.

925. Seko Y, Sugishita K, Sato O, *et al.* Expression of co-stimulatory

molecules (4-1BBL and Fas) and major histocompatibility class I chain-related A (MICA) in aortic tissue with Takayasu's arteritis. *J Vasc Res* 2004;**41**:84–90.

926. Semenza GL. Hypoxia-inducible factor 1: master regulator of O2 homeostasis. *Curr Opin Genet Dev* 1998;**8**:588–94.

927. Semenza GL, Roth PH, Fang HM, Wang GL. Transcriptional regulation of genes encoding glycolytic enzymes by hypoxia-inducible factor 1. *J Biol Chem* 1994;**269**:23757–63.

928. Severinghaus JW, Chiodi H, Eger EI 2nd, *et al*. Cerebral blood flow in man at high altitude: role of cerebrospinal fluid pH in normalization of flow in chronic hypocapnia. *Circ Res* 1966;**19**:274–82.

929. Sheehan HL. Postpartum necrosis of the anterior pituitary. *J Pathol* 1937;**45**:189.

930. Sheffield EA, Weller RO. Age changes at cerebral artery bifurcations and the pathogenesis of berry aneurysms. *J Neurol Sci* 1980;**46**:341–52.

931. Shi FL, Hart RG, Sherman DG, Tegeler CH. Stroke in the People's Republic of China. *Stroke* 1989;**20**:1581–5.

932. Shi ZS, Loh Y, Walker G, Duckwiler GR. Endovascular thrombectomy for acute ischemic stroke in failed intravenous tissue plasminogen activator versus non-intravenous tissue plasminogen activator patients: revascularization and outcomes stratified by the site of arterial occlusions. *Stroke* 2010;**41**:1185–92.

933. Shibata M, Yamada S, Kumar SR, *et al*. Clearance of Alzheimer's amyloid-ss(1-40) peptide from brain by LDL receptor-related protein-1 at the blood–brain barrier. *J Clin Invest* 2000;**106**:1489–99.

934. Shibata S, Mori K, Sekine I, Suyama H. Subarachnoid and intracerebral hemorrhage associated with necrotizing angiitis due to methamphetamine abuse – an autopsy case. *Neurol Med Chir (Tokyo)* 1991;**31**(1):49–52.

935. Shimotake J, Derugin N, Wendland M, Vexler ZS, Ferriero DM. Vascular endothelial growth factor receptor-2 inhibition promotes cell death and limits endothelial cell proliferation in a neonatal rodent model of stroke. *Stroke* 2010;**41**:343–9.

936. Shinton R, Beevers G. Meta-analysis of the relation between cigarette smoking and stroke. *Br Med J* 1989;**298**:789–94.

937. Shivalkar B, Van Loon J, Wieland W, *et al*. Variable effects of explosive or gradual increase of intracranial pressure on myocardial structure and function. *Circulation* 1993;**87**:230–9.

938. Shoamanesh A, Kwok CS, Benavente O. Cerebral microbleeds: histopathological correlation of neuroimaging. *Cerebrovasc Dis* 2012;**32**:528–34.

939. Siebke H, Breivik H, Rod T, Lind B. Survival after 40 minutes' submersion without cerebral sequelae. *Lancet* 1975;**i**:1275–7.

940. Siesjö BK. Hypoglycemia, brain metabolism, and brain damage. *Diabetes Metab Rev* 1988;**4**:113–44.

941. Siesjö BK. Pathophysiology and treatment of focal cerebral ischemia. Part I: Pathophysiology. *J Neurosurg* 1992;**77**:169–84.

942. Siesjö BK. Pathophysiology and treatment of focal cerebral ischaemia. Part II:

mechanisms of damage and treatment. *J Neurosurg* 1992;**77**:337–54.

943. Simon RP. Hypoxia versus ischemia. *Neurology* 1999;**52**:7–8.

944. Simon RP, Aminoff MJ. Electrographic status epilepticus in fatal anoxic coma. *Ann Neurol* 1986;**20**:351–5.

945. Simpson JE, Fernando MS, Clark L, *et al*. White matter lesions in an unselected cohort of the elderly: astrocytic, microglial and oligodendrocyte precursor cell responses. *Neuropathol Appl Neurobiol* 2007;**33**:410–9.

946. Singh H, Tahir TA, Alawo DO, Issa E, Brindle NP. Molecular control of angiopoietin signalling. *Biochem Soc Trans* 2011;**39**:1592–6.

947. Sladky JT, Rorke LB. Perinatal hypoxic/ischemic spinal cord injury. *Pediatr Pathol* 1986;**6**:87–101.

948. Sliwa JA, Maclean IC. Ischemic myelopathy: a review of spinal vasculature and related clinical syndromes. *Arch Phys Med Rehabil* 1992;**73**:365–72.

949. Sloan M. Cerebrovascular disorders associated with licit and illicit drugs. In: Fisher M, Bogousslavsky J eds. *Current review of cerebrovascular disease*. Philadelphia, PA: Current Medicine, 1993:48–62.

950. Sloper JJ, Johnson P, Powell TPS. Selective degeneration of interneurons in the motor cortex of infant monkeys following controlled hypoxia: a possible cause of epilepsy. *Brain Res* 1980;**198**:204–9.

951. Sloviter RS. A selective loss of hippocampal mossy fiber Timm stain accompanies granule cell seizure activity induced by perforant path stimulation. *Brain Res* 1985;**330**:150–53.

952. Sloviter RS, Dempster DW. 'Epileptic' brain damage is replicated qualitatively in the rat hippocampus by central injection of glutamate or aspartate but not by GABA or acetylcholine. *Brain Res Bull* 1985;**15**:39–60.

953. Smith SE, Meldrum BS. Cerebroprotective effect of lamotrigine after focal ischemia in rats. *Stroke* 1995;**26**:117–22.

954. Smith BS, Penka CF, Erickson LS, Matsuo F. Subarachnoid hemorrhage due to anterior spinal artery aneurysm. *Neurosurgery* 1986;**18**:217–19.

955. Smith CM, Chen Y, Sullivan ML, Kochanek PM, Clark RS. Autophagy in acute brain injury: feast, famine, or folly? *Neurobiol Dis* 2010;**43**:52–9.

956. Smith L, Kruszyna H, Smith RP. The effect of methemoglobin on the inhibition of cytochrome c oxidase by cyanide, sulfide or azide. *Biochem Pharmacol* 1977;**26**:2247–50.

957. Smith ML, Auer RN, Siesjö BK. The density and distribution of ischemic brain injury in the rat following 2–10 min of forebrain ischaemia. *Acta Neuropathol (Berl)* 1984;**64**:319–32.

958. Smith ML, Kalimo H, Warner DS, Siesjö BK. Morphological lesions in the brain preceding the development of postischemic seizures. *Acta Neuropathol (Berl)* 1988;**76**:253–64.

959. Smith WS, Sung G, Saver J, *et al*. Mechanical thrombectomy for acute ischemic stroke: final results of the Multi MERCI trial. *Stroke* 2008;**39**:1205–12.

960. Sobata E, Ohkuma H, Suzuki S. Cerebrovascular disorders

associated with von Recklinghausen's neurofibromatosis: a case report. *Neurosurgery* 1988; **22**:544–9.

960a. Söderfeldt B, Kalimo H, Olsson Y, Siesjö BK. Pathogenesis of brain lesions caused by experimental epilepsy: light and electron microscopic changes in the rat cerebral cortex following bicuculline-induced status epilepticus. *Acta Neuropathol (Berl)* 1981;**54**:219–31.

961. Söderfeldt B, Kalimo H, Olsson Y, Siesjö BK. Bicuculline-induced epileptic brain injury: transient and persistent cell changes in rat cerebral cortex in the early recovery period. *Acta Neuropathol (Berl)* 1983;**62**:87–95.

962. Sokrab TEO, Johansson BB, Kalimo H, Olsson Y. A transient hypertensive opening of the blood–brain barrier can lead to brain damage:extravasation of serum proteins and cellular changes in rats subjected to aortic compression. *Acta Neuropathol* 1988;**75**:557–65.

963. Solenov EI, Vetrivel L, Oshio K, *et al*. Optical measurement of swelling and water transport in spinal cord slices from aquaporin null mice. *J Neurosci Meth* 2002;**113**:85–90.

964. Somer T, Finegold SM. Vasculitides associated with infections, immunization, and antimicrobial drugs. *Clin Infect Dis* 1995;**20**:1010–36.

965. Sommer W. Erkrankung des Ammonshorns als aetiologisches Moment der Epilepsie. *Arch Psychiat* 1880;**10**:631–75.

966. Spangler KM, Challa VR, Moody DM, Bell MA. Arteriolar tortuosity of the white matter in aging and hypertension: a microradiographic study. *J Neuropathol Exp Neurol* 1994;**53**:22–6.

967. Spetzler RF, Detwiler PW, Riina HA, Porter RW. Modified classification of spinal cord vascular lesions. *J Neurosurg* 2002;**96**(2 Suppl):145–56.

968. Starkstein SE, Berthier ML, Leiguarda R. Psychic akinesia following bilateral pallidal lesions. *Int J Psychiatry Med* 1989;**19**:155–64.

969. Stary HC. Natural history and histological classification of atherosclerotic lesions: an update. *Arterioscler Thromb Vasc Biol* 2000;**20**:1177–8.

970. Steiger HJ. Pathophysiology of development and rupture of cerebral aneurysms. *Acta Neurochir Suppl (Wien)* 1990;**48**:1–57.

971. Stevens SL, Leung PY, Vartanian KB, *et al*. Multiple preconditioning paradigms converge on interferon regulatory factor-dependent signalling to promote tolerance to ischemic brain injury. *J Neurosci* 2011;**31**:8456–16.

972. Stoltz E, Kaps M. Gas and fat embolism. In: Kalimo H ed. Pathology and genetics: cerebrovascular diseases. Basel: ISN Neuropath Press, 2005:280–84.

973. Sudlow C, Martinez Gonzalez NA, Kim J, Clark C. Does apolipoprotein E genotype influence the risk of ischemic stroke, intracerebral hemorrhage, or subarachnoid hemorrhage? Systematic review and meta-analyses of 31 studies among 5961 cases and 17,965 controls. *Stroke* 2006;**37**:364–70.

974. Suzuki J, Takaku A. Cerebrovascular 'moyamoya' disease: disease showing abnormal net-like vessels in base of brain. *Arch Neurol* 1969;**20**:288–99.

975. Svensson LG. Paralysis after aortic surgery: in search of lost cord function. *Surgeon* 2005;**3**:396–405.

976. Switzer JA, Hess DC, Nichols FT, Adams RJ. Pathophysiology and treatment of stroke in sickle-cell disease: present and future. *Lancet Neurol* 2006;**5**:501–12.

977. Szer IS, Miller JH, Rawlings D, et al. Cerebral perfusion abnormalities in children with central nervous system manifestations of lupus detected by single photon emission computed tomography. *J Rheumatol* 1993;**20**:2143–8.

978. Takahashi K, Adachi K, Yoshizaki K, Kunimoto S, Kalaria RN, Watanabe A. Mutations in NOTCH3 cause the formation and retention of aggregates in the endoplasmic reticulum, leading to impaired cell proliferation. *Hum Mol Genet* 2010;**19**:79–89.

979. Takahashi K, Oharaseki T, Yokouchi Y. Pathogenesis of Kawasaki disease. *Clin Exp Immunol* 2011;**164**(Suppl 1):20–2.

980. Takasugi S, Ueda S, Matsumoto K. Chronological changes in spontaneous intracerebral hematoma: an experimental and clinical study. *Stroke* 1985;**16**:651–8.

981. Takebayashi S, Kaneko M. Electron microscopic studies of ruptured arteries in hypertensive intracerebral hemorrhage. *Stroke* 1983;**14**:28–36.

982. Takenaka K, Sakai H, Yamakawa H, et al. Angiotensin-1 converting enzyme gene polymorphism in intracranial saccular aneurysm individuals. *Neurol Res* 1998;**20**:607–11.

983. Takeya Y, Popper JS, Shimizu Y, et al. Epidemiologic studies of coronary heart disease and stroke in Japanese men living in Japan, Hawaii and California: incidence of stroke in Japan and Hawaii. *Stroke* 1984;**15**:15–23.

984. Tanaka H, Fujita N, Enoki T, et al. Relationship between variations in the circle of Willis and flow rates in internal carotid and basilar arteries determined by means of magnetic resonance imaging with semiautomated lumen segmentation: reference data from 125 healthy volunteers. *AJNR Am J Neuroradiol* 2006;**27**:1770–5.

985. Taniguchi M, Yamashita T, Kumura E, et al. Induction of aquaporin-4 water channel mRNA after focal cerebral ischaemia in rat. *Mol Brain Res* 2000;**78**:131–7.

986. Tanne D, Triplett DA, Levine SR. Antiphospholipid-protein antibodies and ischemic stroke: not just cardiolipin any more. *Stroke* 1998;**29**:1755–8.

987. Tanskanen M, Makela M, Myllykangas L, et al. Prevalence and severity of cerebral amyloid angiopathy: a population-based study on very elderly Finns (Vantaa 85+). *Neuropathol Appl Neurobiol* 2012;**38**:329–36.

988. Tardy B, Page Y, Convers P, et al. Thrombotic thrombocytopenic purpura: MR findings. *Am J Neuroradiol* 1993;**14**:489–90.

989. Tarras S, Gadia C, Meister L, et al. Homozygous protein C deficiency in a newborn. Clinicopathologic correlation. *Arch Neurol* 1988;**45**:214–20.

990. Tatlisumak T, Fisher M. Hematologic disorders associated with ischemic stroke. *J Neurol Sci* 1996;**140**:1–11.

991. ten Dijke P, Arthur HM. Extracellular control of TGF-beta signalling in vascular development and disease. *Nat Rev Mol Cell Biol* 2007;**8**:857–69.

992. Terasaki M, Sugita Y, Arakawa F, et al. CXCL12/CXCR4 signalling in malignant brain tumours: a potential pharmacological therapeutic target. *Brain Tumour Pathol* 2011;**28**:89–97.

993. Terent A. Stroke morbidity. In: Whisnant J ed. *Stroke: populations, cohorts, and clinical trials.* Oxford: Butterworth Heinemann, 1993:37–58.

994. Terent A. Trends in stroke incidence and 10-year survival in Soderhamn, Sweden, 1975–2001. *Stroke* 2003;**34**:1353–8.

995. Terent A, Marke LA, Asplund K, et al. Costs of stroke in Sweden: a national perspective. *Stroke* 1994;**25**:2363–9.

996. Thal DR, Ghebremedhin E, Rub U, et al. Two types of sporadic cerebral amyloid angiopathy. *J Neuropathol Exp Neurol* 2002;**61**:282–93.

997. Thom SR, Bhopale VM, Fisher D, et al. Delayed neuropathology after carbon monoxide poisoning is immune-mediated. *Proc Natl Acad Sci U S A* 2004;**101**:13660–5.

998. Thomas WS, Mori E, Copeland BR, et al. Tissue factor contributes to microvascular defects after focal cerebral ischemia. Stroke 1993;**24**:847–54.

999. Thomsen AH, Gregersen M. Suicide by carbon monoxide from car exhaust gas in Denmark 1995–1999. *Forensic Sci Int* 2006;**161**:41–6.

1000. Thorsteinsson L, Georgsson G, Asgeirsson B, et al. On the role of monocytes/macrophages in the pathogenesis of central nervous system lesions in hereditary cystatin C amyloid angiopathy. *J Neurol Sci* 1992;**108**:121–8.

1001. Thron A, Wessel K, Linden D, et al. Superior sagittal sinus thrombosis: neuroradiological evaluation and clinical findings. *J Neurol* 1986;**233**:283–8.

1002. Thurston JH, Hauhart RE, Schiro J. Lactate reverses insulin-induced hypoglycemic stupor in suckling-weanling mice: biochemical correlates in blood, liver, and brain. *J Cereb Blood Flow Metab* 1983;**3**:498–506.

1003. Tibbles PM, Edelsberg JS. Hyperbaric-oxygen therapy. N Engl J Med 1996;**334**:1642–8.

1004. Tikka S, Mykkanen K, Ruchoux MM, et al. Congruence between NOTCH3 mutations and GOM in 131 CADASIL patients. *Brain* 2009;**132**:933–9.

1005. Tohgi H, Yamanouchi H, Murakami M, Kameyama M. Importance of the hematocrit as a risk factor in cerebral infarction. *Stroke* 1978;**9**:369–74.

1006. Tokuda T, Ikeda S, Yanagisawa N, et al. Re-examination of ex-boxers' brains using immunohistochemistry with antibodies to amyloid protein and tau protein. *Acta Neuropathol* 1991;**82**:280–85.

1007. Tom MI, Richardson JC. Hypoglycemia from islet cell tumour of pancreas with amyotrophy and cerebrospinal nerve cell changes. *J Neuropathol Exp Neurol* 1951;**10**:57–66.

1008. Tombaugh GC, Sapolsky RM. Mild acidosis protects hippocampal neurons from injury induced by oxygen and glucose deprivation. *Brain Res* 1990;**506**:343–5.

1009. Toole J. *Cerebrovascular disorders.* New York: Raven Press, 1990.

1010. Toschi V, Motta A, Castelli C, et al. Prevalence and clinical significance of antiphospholipid antibodies to noncardiolipin antigens in systemic lupus erythematosus. *Haemostasis* 1993;**23**:275–83.

1011. Toschi V, Motta A, Castelli C, et al. High prevalence of antiphosphatidylinositol antibodies in young patients with cerebral ischaemia of undetermined cause. *Stroke* 1998;**29**:1759–64.

1012. Tosi L, Rigoli G, Beltramello A. Fibrocartilaginous embolism of the spinal cord: a clinical and pathogenetic reconsideration. *J Neurol Neurosurg Psychiatry* 1996;**60**:55–60.

1013. Tournier–Lasserve E, Joutel A, Melki J, et al. Cerebral autosomal dominant arteriopathy with subcortical infarcts and leukoencephalopathy maps to chromosome 19q12. *Nat Genet* 1993;**3**:256–9.

1014. Towbin A. The respirator brain death syndrome. *Hum Pathol* 1973;**4**:583–94.

1015. Tower DB, Young OM. Interspecies correlations of cerebral cortical oxygen consumption, acetylcholinesterase activity and chloride content: studies on the brains of the fin whale (*Balaenoptera physalus*) and the sperm whale (*Physeter catadon*). *J Neurochem* 1973;**20**:253–67.

1016. Tower DB, Young OM. The activities of butyrylcholinesterase and carbonic anhydrase, the rate of anaerobic glycolysis, and the question of a constant density of glial cells in cerebral cortices of various mammalian species from mouse to whale. *J Neurochem* 1973;**20**:269–78.

1017. Towfighi J, Gonatas NK. Effect of intracerebral injection of ouabain in adult and developing rats: an ultrastructural and autoradiographic study. *Lab Invest* 1973;**28**:170–80.

1018. Traylor M, Farrall M, Holliday EG, et al. Genetic risk factors for ischaemic stroke and its subtypes (the METASTROKE collaboration): a meta-analysis of genome-wide association studies. *Lancet Neurol* 2012;**11**:951–62.

1019. Traystman RJ, Fitzgerald RS. Cerebrovascular response to hypoxia in baroreceptor- and chemoreceptor-denervated dogs. *Am J Physiol* 1981;**241**:H724–31.

1020. Traystman RJ, Fitzgerald RS, Loscutoff SC. Cerebral circulatory responses to arterial hypoxia in normal and chemodenervated dogs. *Circ Res* 1978;**42**:649–57.

1021. Trollmann R, Gassmann M. The role of hypoxia-inducible transcription factors in the hypoxic neonatal brain. *Brain Dev* 2009;**31**:503–9.

1022. Trump BF, McDowell E, Collan Y. Studies on the pathogenesis of ischemic cell injury. VI. Mitochondrial flocculent densities in autolysis. *Virchows Arch B Cell Pathol Incl Mol Pathol* 1981;**35**:189–99.

1023. Tsai HM. Current concepts in thrombotic thrombocytopenic purpura. *Annu Rev Med* 2006;**57**:419–36.

1024. Tseng SC, Kimchi–Sarfaty C. SNPs in ADAMTS13. *Pharmacogenomics* 2011;**12**:1147–60.

1025. Tu J, Stoodley MA, Morgan MK, Storer KP. Ultrastructure of perinidal capillaries in cerebral arteriovenous malformations. *Neurosurgery* 2006;58:961–70;discussion 70.

1026. Tufo HM, Ostfeld AM, Shekelle R. Central nervous system dysfunction following open-heart surgery. J *Am Med Assoc* 1970;212:1333–40.

1027. Tuominen S, Juvonen V, Amberla K, *et al.* The phenotype of a homozygous CADASIL patient in comparison to nine age-matched heterozygous patients with the same R133C Notch3 mutation. *Stroke* 2001;32:1767–74.

1028. Turmaine M, Raza A, Mahal A, *et al.* Nonapoptotic neurodegeneration in a transgenic mouse model of Huntington's disease. *Proc Natl Acad Sci U S A* 2000;97:8093–7.

1029. Turnbull F. Blood Pressure Lowering Treatment Trialists' Collaboration. Effects of different blood-pressure-lowering regimens on major cardiovascular events: results of prospectively-designed overviews of randomised trials. *Lancet* 2003;362:1527–35.

1030. Tzourio C, Tehindrazanarivelo A, Iglesias S, *et al.* Case–control study of migraine and risk of ischaemic stroke in young women. *Br Med J* 1995;310:830–33.

1031. Ueda T, Hatakeyama T, Kumon Y, *et al.* Evaluation of risk of hemorrhagic transformation in local intra-arterial thrombolysis in acute ischemic stroke by initial SPECT. *Stroke* 1994;25: 298–303.

1032. Usui C, Inoue Y, Kimura M, *et al.* Irreversible subcortical dementia following high altitude illness. *High Alt Med Biol* 2004;5:77–81.

1033. Vahedi K, Massin P, Guichard JP, *et al.* Hereditary infantile hemiparesis, retinal arteriolar tortuosity, and leukoencephalopathy. *Neurology* 2003;60:57–63.

1034. Vahedi K, Kubis N, Boukobza M, *et al.* COL4A1 mutation in a patient with sporadic, recurrent intracerebral hemorrhage. *Stroke* 2007;38:1461–4.

1035. Van Bogaert L. Encéphalopathie sius-corticale progressive (Binswanger) à évolution rapide chez deux soeurs. *Med Hellen* 1955;24:961–72.

1036. Van der Schaaf IC, Velthuis BK, Gouw A, Rinkel GJ. Venous drainage in perimesencephalic hemorrhage. *Stroke* 2004;35:1614–8.

1037. Van Gijn J, Rinkel GJ. Subarachnoid haemorrhage: diagnosis, causes and management. *Brain* 2001;124:249–78.

1038. Van Gijn J, Kerr RS, Rinkel GJ. Subarachnoid haemorrhage. *Lancet* 2007;369:306–18.

1039. Van Harreveld A, Fifková E. Light- and electron-microscopic changes in central nervous tissue after electrophoretic injection of glutamate. *Exp Molec Pathol* 1971;15:61–81.

1040. Van't Hooft FM, von Bahr SJ, Silveira A, *et al.* Two common, functional polymorphisms in the promoter region of the beta-fibrinogen gene contribute to regulation of plasma fibrinogen concentration. *Arterioscler Thromb Vasc Biol* 1999;19:3063–70.

1041. Vandenabeele P, Galluzzi L, Vanden Berghe T, Kroemer G. Molecular mechanisms of necroptosis:an ordered cellular explosion. *Nat Rev Mol Cell Biol* 2010;11:700–14.

1042. Vasudevan A, Long JE, Crandall JE, Rubenstein JL, Bhide PG. Compartment-specific transcription factors orchestrate angiogenesis gradients in the embryonic brain. *Nat Neurosci* 2008;11:429–39.

1043. Veldhuisen B, Saris JJ, Haij S, *et al.* A spectrum of mutations in the second gene for autosomal dominant polycystic kidney disease (PKD2). *Am J Hum Genet* 1997;61:151–60.

1044. Vermeer SE, Rinkel GJ, Algra A. Circadian fluctuations in onset of subarachnoid hemorrhage:new data on aneurysmal and perimesencephalic hemorrhage and a systematic review. *Stroke* 1997;28:805–8.

1045. Vermeulen M, van Gijn J. The diagnosis of subarachnoid haemorrhage. *J Neurol Neurosurg Psychiatry* 1990;53:365–72.

1046. Vidal R, Frangione B, Rostagno A, *et al.* A stop-codon mutation in the BRI gene associated with familial British dementia. *Nature* 1999;399:776–81.

1047. Vidal R, Revesz T, Rostagno A, *et al.* A decamer duplication in the 39 region of the BRI gene originates an amyloid peptide that is associated with dementia in a Danish kindred. *Proc Natl Acad Sci U S A* 2000;97:4920–25.

1048. Vinters HV. Cerebral amyloid angiopathy:a critical review. *Stroke* 1987;18:311–24.

1049. Vinters HV, Gilbert JJ. Cerebral amyloid angiopathy: incidence and complications in the aging brain: II. The distribution of amyloid vascular changes. *Stroke* 1983;14:924–8.

1050. Vishteh AG, Brown AP, Spetzler RF. Aneurysm of the intradural artery of Adamkiewicz treated with muslin wrapping: technical case report. *Neurosurgery* 1997;40:207–9.

1051. Viswanathan A, Chabriat H. Cerebral microhemorrhage. *Stroke* 2006;37: 550–55.

1052. Viswanathan A, Greenberg SM. Cerebral amyloid angiopathy in the elderly. *Ann Neurol* 2011;70:871–80.

1053. Vital Cl, Picard J, Arné L, *et al.* Pathological study of three cases of hypoglycemic encephalopathy (one of which occurred after sulfamidotherapy). *Le Diabète* 1967;15:291–6.

1054. Vlak MH, Algra A, Brandenburg R, Rinkel GJ. Prevalence of unruptured intracranial aneurysms, with emphasis on sex, age, comorbidity, country, and time period: a systematic review and meta-analysis. *Lancet Neurol* 2011;10:626–36.

1055. Vlak MH, Rinkel GJ, Greebe P, van der Bom JG, Algra A. Trigger factors and their attributable risk for rupture of intracranial aneurysms: a case-crossover study. *Stroke* 2011;42:1878–82.

1056. Voll CL, Auer RN. Insulin attenuates ischemic brain damage independent of its hypoglycemic effect. *J Cereb Blood Flow Metab* 1991;11:1006–14.

1057. Vonsattel JPG, Myers RH, Hedley–Whyte ET, *et al.* Cerebral amyloid angiopathy without and with cerebral hemorrhages:a comparative histological study. *Ann Neurol* 1991;30:637–49.

1058. Voorhoeve RJ, Remeika JP, Freeland PE, Matthias BT. Rare-earth oxides of manganese and cobalt rival platinum for the treatment of carbon monoxide in auto exhaust. *Science* 1972;177:353–4.

1059. Wagner KR, Kleinholz M, Myers RE. Delayed neurologic deterioration following anoxia: brain mitochondrial and metabolic correlates. *J Neurochem* 1989;52:1407–17.

1060. Wakabayashi S, Ohno K, Shishido T, *et al.* Marked growth of a cerebral arteriovenous malformation: case report and review of the literature. *Neurosurgery* 1991;29:920–23.

1061. Wakai S, Nagai M. Histological verification of microaneurysm as a cause of cerebral haemorrhage in surgical specimens. *J Neurol Neurosurg Psychiatry* 1989;52:595–9.

1062. Wakai S, Kumakura N, Nagai M. Lobar intracerebral hemorrhage: a clinical, radiographic and pathological study of 29 consecutive operated cases with negative angiography. *J Neurosurg* 1992;76:231–8.

1063. Walker LC, Pahnke J, Madauss M, *et al.* Apolipoprotein E4 promotes the early deposition of Aβ 42 and then Aβ 40 in the elderly. *Acta Neuropathol* 2000;100: 36–42.

1064. Walker MT, Kilani RK, Toye LR, *et al.* Central and peripheral fusiform aneurysms six years after left atrial myxoma resection. *J Neurol Neurosurg Psychiatry* 2003;74:277–82.

1065. Wang QK. Update on the molecular genetics of vascular anomalies. *Lymphat Res Biol* 2005;3:226–33.

1066. Wang Q, Tang XN, Yenari MA. The inflammatory response in stroke. *J Neuroimmunol* 2007;184:53–68.

1067. Ward NL, Moore E, Noon K, *et al.* Cerebral angiogenic factors, angiogenesis, and physiological response to chronic hypoxia differ among four commonly used mouse strains. *J Appl Physiol* 2007;102:1927–35.

1068. Wardlaw JM, del Zoppo G, Yamaguchi T. Thrombolysis for acute ischaemic stroke. *Cochrane Database Syst Rev* 2000;CD000213.

1069. Wardlaw JM, Doubal F, Armitage P, *et al.* Lacunar stroke is associated with diffuse blood–brain barrier dysfunction. *Ann Neurol* 2009;65:194–202.

1070. Warlow CP. Epidemiology of stroke. *Lancet* 1998;352(Suppl III):1–4.

1071. Warlow C, Sudlow C, Dennis M, *et al.* Stroke. *Lancet* 2003;362:1211–24.

1072. Watanabe I, Tomita T, Hung K–S, Iwasaki Y. Edematous necrosis in thiamine-deficient encephalopathy of the mouse. *J Neuropathol Exp Neurol* 1981;40:454–71.

1073. Watanabe T, Tsuchida T, Kanda N, *et al.* Anti-fodrin antibodies in Sjögren syndrome and lupus erythematosus. *Arch Dermatol* 1999;135:535–9.

1074. Watkins JC, Evans RH. Excitatory amino acid transmitters. *Annu Rev Pharmacol Toxicol* 1981;21:165–204.

1075. Watnick T, Phakdeekitcharoen B, Johnson A, *et al.* Mutation detection of PKD1 identifies a novel mutation common to three families with aneurysms and/or very-early-onset disease. *Am J Hum Genet* 1999;65:1561–71.

1076. Weaver LK. Clinical practice. Carbon monoxide poisoning. *N Engl J Med* 2009;360:1217–25.

1077. Weil A, Liebert E, Heilbrunn G. Histopathologic changes in the brain in experimental hyperinsulinism. *Arch Neurol Psychiat* (Chic) 1938;39:467–81.

1078. Weinberg DG, Arnaout OM, Rahme RJ, *et al*. Moyamoya disease: a review of histopathology, biochemistry, and genetics. *Neurosurg Focus* 2011;30:E20.

1079. Weir B. The clinical problem of intracerebral hematoma. *Stroke* 1993;24(Suppl I):93.

1080. Welch GN, Loscalzo J. Homocysteine and atherothrombosis. *N Engl J Med* 1998;338:1042–50.

1081. Weller RO. Drainage pathways of CSF and interstitial fluid. In Kalimo H ed. *Pathology and genetics: cerebrovascular diseases.* Basel: ISN Neuropath Press, 2005:50–55.

1082. Weller RO, Djuanda E, Yow HY, Carare RO. Lymphatic drainage of the brain and the pathophysiology of neurological disease. *Acta Neuropathol* 2009;117: 1–14.

1083. Weller RO, Galea I, Carare RO, Minagar A. Pathophysiology of the lymphatic drainage of the central nervous system: Implications for pathogenesis and therapy of multiple sclerosis. *Pathophysiology* 2009;17: 295–306.

1084. Weller RO, Subash M, Preston SD, Mazanti I, Carare RO. Perivascular drainage of amyloid-peptides from the brain and its failure in cerebral amyloid angiopathy and Alzheimer's disease. *Brain Pathol* 2008;18:253–66.

1085. Westerberg E, Kehr J, Ungerstedt U, Wieloch T. The NMDA-antagonist MK-801 reduces extracellular amino acid levels during hypoglycemia and prevents striatal damage. *Neurosci Res Comm* 1988;3:151–8.

1086. Weyand CM, Goronzy JJ. Medium- and large-vessel vasculitis. *N Engl J Med* 2003;349:160–69.

1087. Weyand CM, Goronzy JJ. Giant cell arteritis and polymyalgia rheumatica. *Ann Intern Med* 2003;139:505–15.

1088. Weyand CM, Ma–Krupa W, Goronzy JJ. Immunopathways in giant cell arteritis and polymyalgia rheumatica. *Autoimm Revs* 2004;3:46–53.

1089. Wieloch T. Hypoglycemia-induced neuronal damage prevented by an N-methyl-d-aspartate antagonist. *Science* 1985;230:681–3.

1090. Wienhard K, Dahlbom M, Eriksson L, *et al*. The ECAT EXACT HR: performance of a new high resolution positron scanner. *J Comput Assist Tomogr* 1994;18:110–18.

1091. Wijdicks EF, Parisi JE, Sharbrough FW. Prognostic value of myoclonus status in comatose survivors of cardiac arrest. *Ann Neurol* 1994;35:239–43.

1092. Wilkins RH. Attempts at prevention or treatment of intracranial arterial spasm: an update. *Neurosurgery* 1986;18:808–25.

1093. Williams S, Chalmers K, Wilcock GK, Love S. Relationship of neurofibrillary pathology to cerebral amyloid angiopathy in Alzheimer's disease. *Neuropathol Appl Neurobiol* 2005;31:414–21.

1094. Winkler DT, Lyrer P, Probst A, *et al*. Hereditary systemic angiopathy (HSA) with cerebral calcifications, retinopathy, progressive nephropathy, and hepatopathy. *J Neurol* 2008;255:77–88.

1095. Wolf HK, Anthony DC, Fuller GN. Arterial border zone necrosis of the spinal cord. *Clin Neuropathol* 1990;9:60–65.

1096. Wong V, Yu YL, Liang RH, *et al*. Cerebral thrombosis in beta thalassemia/hemoglobin E disease. *Stroke* 1990;21:812–16.

1097. Woo D, Haverbusch M, Sekar P, *et al*. Effect of untreated hypertension on hemorrhagic stroke. *Stroke* 2004;35:1703–8.

1098. Woolfenden A, Albers G. Cardioembolic stroke. In: Fisher M, Bogousslavsky J eds. *Current review of cerebrovascular disease.* Boston, MA: Butterworth Heinemann, 1999:93–105.

1099. Wu HM, Huang SC, Hattori N, *et al*. Selective metabolic reduction in gray matter acutely following human traumatic brain injury. *J Neurotrauma* 2004;21:149–61.

1100. Wytrzes LM, Chatrian GE, Shaw CM, Wirch AL. Acute failure of forebrain with sparing of brain-stem function: electroencephalographic, multimodality evoked potential, and pathologic findings. *Arch Neurol* 1989;46:93–7.

1101. Yamada M, Itoh Y, Otomo E, Hayakawa M, Miyatake T. Subarachnoid haemorrhage in the elderly: a necropsy study of the association with cerebral amyloid angiopathy. *J Neurol Neurosurg Psychiatry* 1993;56:543–7.

1102. Yamamoto K, Hayakawa T, Mogami H, *et al*. Ultrastructural investigation of the CA1 region of the hippocampus after transient cerebral ischaemia in gerbils. *Acta Neuropathol (Berl)* 1990;80: 487–92.

1103. Yamamoto Y, Ihara M, Tham C, Low RW, Slade JY, Moss T, Oakley AE, Polvikoski T, Kalaria RN. Neuropathological correlates of temporal pole white matter hyperintensities in CADASIL. *Stroke* 2009;40:2004–11.

1104. Yamamoto Y, Craggs L, Baumann M, Kalimo H, Kalaria RN. Molecular genetics and pathology of hereditary small vessel diseases of the brain. *Neuropathol Appl Neurobiol* 2011; 37:94–113.

1105. Yamamoto Y, Craggs LJL, Watanabe A, *et al*. Brain Microvascular Accumulation and Distribution of the NOTCH3 ectodomain and GOM in CADASIL. *J Neuropathol Exp Neurol* 2013;72(5): 416–31.

1106. Yamashita M, Oka K, Tanaka K. Histopathology of the brain vascular network in moyamoya disease. *Stroke* 1983;14:50–58.

1107. Yan C, Chen J, Chen D, *et al*. Overexpression of the cell death suppressor Bcl-w in ischemic brain: implications for a neuroprotective role via the mitochondrial pathway. *J Cereb Blood Flow Metab* 2000;20:620–30.

1108. Yanagawa S, Ito N, Arima K, Ikeda S. Cerebral autosomal recessive arteriopathy with subcortical infarcts and leukoencephalopathy. *Neurology* 2002;58:817–20.

1109. Yang G, Chan PH, Chen J, *et al*. Human copper-zinc superoxide dismutase transgenic mice are highly resistant to reperfusion injury after focal cerebral ischaemia. *Stroke* 1994;25:165–70.

1110. Yao ZB, Li X, Xu ZC. GABAergic and asymmetrical synapses on somata of GABAergic neurons in CA1 and CA3 regions of rat hippocampus: a quantitative electron microscopic analysis. *Stroke* 1996;27:1411–16.

1111. Yoneyama T, Kasuya H, Onda H, *et al*. Collagen type I 2 (COL1A2) is the susceptible gene for intracranial aneurysms. *Stroke* 2004;35:443–8.

1112. Young GB, Gilbert JJ, Zochodne DW. The significance of myoclonic status epilepticus in postanoxic coma. *Neurology* 1990;40:1843–8.

1113. Younger DS. Vasculitis of the nervous system. *Curr Opin Neurol* 2004;17: 317–36.

1114. Yu MC, Bakay L, Lee JC. Ultrastructure of the central nervous system after prolonged hypoxia. I: neuronal alterations. *Acta Neuropathol (Berl)* 1972;22:222–34.

1115. Yu MC, Bakay L, Lee JC. Ultrastructure of the central nervous system after prolonged hypoxia. II: neuroglia and blood vessels. *Acta Neuropathol (Berl)* 1972;22:235–44.

1116. Yue WY, Chen ZP. Does vasculogenic mimicry exist in astrocytoma? *J Histochem Cytochem* 2005;53: 997–1002.

1117. Zanette EM, Roberti C, Mancini G, *et al*. Spontaneous middle cerebral artery reperfusion in ischemic stroke: a follow-up study with transcranial Doppler. *Stroke* 1995;26:430–33.

1118. Zannad F, Benetos A. Genetics of intima-media thickness. *Curr Opin Lipidol* 2003;14:191–200.

1119. Zhang F, Wu Y, Jia J. Exercise preconditioning and brain ischemic tolerance. *Neuroscience* 2011;177: 170–6.

1120. Zull DN, Cydulka R. Acute paraplegia: a presenting manifestation of aortic dissection. *Am J Med* 1988;84:765–70.

Disorders of the Perinatal Period

Rebecca D Folkerth and Marc R Del Bigio

INTRODUCTION

Consideration of perinatal brain disorders requires that the neuropathologist understands terminology used in obstetrics. A normal pregnancy lasts approximately 40 weeks from the last normal menstrual period (LNMP). In the 'ideal' 28 day menstrual cycle fertilization occurs on the 14th day of that cycle. Birth between 37 and 42 weeks after the LNMP is considered full term. Gestational age is the time from the LNMP to birth, postmenstrual age (typically used in the circumstance of premature birth) is the gestational age plus the chronological age (time elapsed from birth). Corrected age is the chronological age minus the number of weeks born before 40 weeks gestation.[153] (See Table 3.1.) Conceptional age (gestational age minus 2 weeks) is not a term recommended for use in obstetrics but is a necessary concept for understanding the terminology of comparative development (e.g. the Carnegie stages).[406] The embryonic period extends to the end of the 8th week after fertilization (Carnegie stage 23), and the fetal period extends thereafter until birth. The formal definition of the 'perinatal period' is from 20 gestational weeks to 28 postnatal days. In this chapter, however, we will consider pathological processes occurring through the first postnatal year (i.e. the end of infancy), because the neuropathology of this period is unique, given that this is a setting of rapid brain growth quite different from that in the mature brain. Very roughly, the human brain is 1/3 adult weight at full term birth, 2/3 adult weight at 1 year, and near adult weight by ~10 years, although the brain is not fully mature until the middle of the third decade of life (Figure 3.1; Box 3.1; Table 3.1).[587]

During the fetal and infantile period, there are rapid changes in neuronal differentiation, synaptogenesis, dendritic arborization, axonal elongation, gliogenesis and myelination. Quantitative magnetic resonance (MR) imaging studies underscore the dramatic changes occurring over the last half of human gestation in the volume, surface area and sulcation of the cerebral cortex, as well as in the microstructural organization of the white matter (Box 3.2). All of these organizational events occur after the ground plan for the central nervous system (CNS) is laid down in the embryonic period and after neuronal proliferation and migration in the cerebral hemispheres is largely complete, i.e. by the end of 20–22 gestational weeks.

In essence, the pathology from mid-gestation through infancy reflects cellular and tissue reactions brought about by a complex interplay of disease and rapidly changing developmental processes. Good evidence indicates that the most vulnerable time for cortical injury is toward the end of the gestational period, perhaps including the perinatal period, whereas the time associated with the greatest capacity for plasticity is 1 to 2 years of age.[297] The developmental changes occur so rapidly that responses to the same insult may vary considerably from mid-gestation through infancy, as demonstrated by hypoxic-ischaemic injury to cerebral white matter preferentially in the preterm infant and to grey matter preferentially after the neonatal period. Moreover, different types of brain lesion may be restricted to very specific time periods within this overall period (Figure 3.2). This latter phenomenon is due in part to the rapid changes in peak periods of maturation of different cellular and molecular components of the brain, e.g. the peak occurrence of neuronal migration and differentiation in early gestation compared with the peak period of myelination in infancy. The focus upon the period from mid-gestation to

3

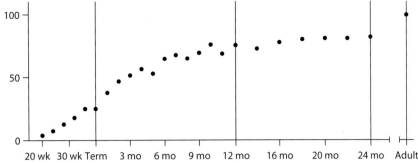

3.1 Graph showing brain growth expressed as a percentage of adult brain weight in early life. Mid-gestation through infancy (the first postnatal year of life) is a critical period in brain growth, at the end of which the brain attains 75 per cent of adult weight. Note that adult weight is reached by 10–15 years age.

TABLE 3.1 Definitions of preterm and post-term terminology		
Terminology	**Definition**	**Comment**
Post term	>42 weeks	213
Term	>37 to <42 completed weeks	Same
Near term	>37 to <39 weeks	381
Preterm	<37 completed weeks	World Health Organization ICD-10 Classification (1994)
Late preterm	>34 to <37 weeks	443
Early preterm	>23 to <34 weeks	381
Moderate preterm	32–33 weeks	Commonly used but not officially endorsed term[197]
Severe preterm	28–31 weeks	Same
Extreme preterm	<28 weeks	Same
Very early preterm	>18 to <23 weeks	381
Very low birth weight (VLBW)	1001–1500g	Approximately 28–32 weeks' gestation
Extreme low birth weight (ELBW)	501–1000g	Approximately 24–28 weeks' gestation
Extreme low birth weight (ELBW)	<500g	Approximately <24 weeks' gestation

are, to some extent, predictive of complications and outcomes. Recent opinions of international expert groups suggest 'splitting spontaneous preterm labor…may be difficult to justify'.[301] A comprehensive classification system for preterm birth requires expanded gestational boundaries that recognize the early origins of preterm delivery and emphasize fetal maturity over fetal age. Some evidence suggests that there may be greater morbidity among infants born at 37–38 weeks.[198] Five components recommended for description of preterm birth phenotypes are: (i) significant maternal conditions present before delivery, (ii) significant fetal conditions present before delivery, (iii) placental pathology, (iv) uterine signs of the initiation of parturition and (v) the mode of delivery (spontaneous labour, augmented labour, instrument assisted, Caesarian).[557] All of these factors can influence the newborn brain.

Birth and morbidity/mortality data are often reported according to birth weight, which roughly corresponds to gestational age but is influenced by many factors.[269] Brain maturation is more closely linked to gestational age[2,94,139] and morbidity data suggest that definition by age is preferable (see Table 3.1).[309]

BOX 3.2. Neuroimaging of the developing brain

Cranial ultrasonography still remains a mainstay of neonatal intensive care nursery, where the monitoring of premature, and often very sick, neonates is needed. In the last decade, however, many advances in more sophisticated neuroimaging techniques have vastly supplemented ultrasonography. These techniques have brought increased understanding of perinatal brain injury, although also raising new questions, and will be presented briefly here. The development of magnetic resonance (MR)-compatible neonatal incubators and of a neonatal head coil, for example, has enabled improved quality scanning of critically ill neonates compared to earlier eras.[421] This capability has allowed the refinement of imaging protocols, including 3-dimensional spoiled gradient echo, which reduces scan time and permits volumetric analysis.

Diffusion-weighted and diffusion tensor imaging (DTI) are based on the directional movement of water molecules, which varies depending upon the presumed 'microstructure' of the cells and tissues based on their water content. Thus, the qualitative and quantitative changes in the diffusivity and its directionality can be mapped and correlated with developmental sequences, such as myelination.[155] Mapping of selected axonal fibre bundles, such as the optic radiations, is possible by using computer-generated probabilistic algorithms, which are constantly being refined. Using high-resolution angular

BOX 3.1. Preterm delivery and newborn infant classifications

The term 'preterm' is preferred over 'premature' as a descriptor of birth at an early gestational age. The World Health Organization defines preterm birth as that occurring prior to 37 completed weeks (ICD-10 classification, 1994). In 2005, an estimated 9.6 per cent of all births worldwide were preterm, corresponding to about 12.9 million births, 85 per cent of which were concentrated in Africa and Asia. The rates of preterm birth were 11.9 per cent in Africa, 10.6 per cent in North America and 6.2 per cent in Europe.[45] For consistency it has been recommended that gestational age should be rounded off to the nearest completed week, not to the following week.[153]

There are no unanimously accepted definitions of preterm birth subclassification. These are nevertheless useful because they diffusion imaging, the tissue substrate for the radial coherence of signals from the hemispheres across intrauterine life through age 2 years has been correlated with the presence of radial glial fibres until the age of 30–31 postconceptional weeks. This pattern is gradually replaced by radial axons during the first postnatal year of life.[296,586] The eventual use of these methods is anticipated to include the detection of tract disruptions of even a few mm or less resulting from periventricular leukomalacia (PVL) or other pathologic processes in living infants.

Other methods based on movement of water molecules include perfusion MR with dynamic contrast, which provides information about relative cerebral blood volume, relative cerebral blood flow and mean transit time, and also about the integrity of the blood–brain barrier, by deriving an apparent diffusion coefficient (ADC) map of the imaged brain. Arterial spin labelling (ASL), in which a radiofrequency pulse is delivered that 'tags' a bolus of flowing blood by magnetization of its water molecules, enables quantitative cerebral perfusion studies without the use of contrast.[421] Blood oxygenation dependent (BOLD) techniques rely on the paramagnetic property of deoxyhemoglobin, and its ratio with oxyhemoglobin; combining this method with near-infrared spectroscopy allows estimation of cerebral oxygenation, usually used in an investigational setting.

Single voxel quantitative proton MR spectroscopy has the power to detect the presence of protons derived from certain molecules with differential distributions in normal versus injured brain. For example, N-acetyl-aspartate is synthesized by neurons and present in axons, as well as in immature oligodendrocytes, and increases dramatically around term; decreased amounts correlate with processes that cause neuronal or axonal loss.[421]

Volumetric and segmentation techniques using standard MR can parcellate different areas of the brain, e.g. myelinated from unmyelinated sites, by differing signal characteristics across development, in normal infants as well as in those with perinatal brain injury.[250,355] Surface-based techniques permit the analysis of cortical folding in normative and also pathologic conditions.[57,143]

Of significance is the changing profile of perinatal white matter injury and magnetic resonance imaging. Focal necrotic lesions evolving to cysts that are readily identified by cranial ultrasonography are no longer the principal feature of cerebral white matter damage in the premature infant.[565] Indeed, cystic periventricular leukomalacia (PVL) is now an uncommon finding by brain imaging in modern neonatal intensive care units.[565] Among infants born <28 weeks in the mid-2000s, magnetic resonance (MR) imaging studies show that 2–4 per cent had cystic PVL, whereas ~35 per cent had non-cystic white matter damage and mild atrophy.[242,258] This likely reflects improvement in patient management that has led to smaller and more subtle white matter injuries.

The white matter abnormalities are characterized by diffuse excessive high-signal intensities (DEHSI) on T_2-weighted images.[104] This diffuse change increases as a function of age from 21 per cent in preterm infants in the first postnatal week to 79 per cent at term equivalent.[337] However, DEHSI appears to resolve after a postmenstrual age of 50 weeks and is associated with normal neurologic outcome at a corrected age of 2 years; therefore, it seems to reflect a delayed developmental process rather than injury.[113] Premature infants at term equivalent often have decreased ADC values.[104] The ADC normally declines in cerebral white matter of premature infants, correlating with the period in which pre-myelinating oligodendrocytes begin to ensheath axons.[29,30] There has been considerable speculation about the correlate of white matter signal abnormalities, for example, that may indicate disrupted premyelinating oligodendroglia versus axonal damage.[92] An excellent autopsy study supports the idea that myelination failure is due to inhibited differentiation of preoligodendrocytes in diffuse astrogliotic lesions, and not axon loss.[74] It should be emphasized, however, that microscopic foci of periventricular necrosis may be found at autopsy when conventional MR imaging fails to demonstrate them. MR imaging may be used as a supplement to the autopsy, but is not a substitute for histologic examination of tissues.[98]

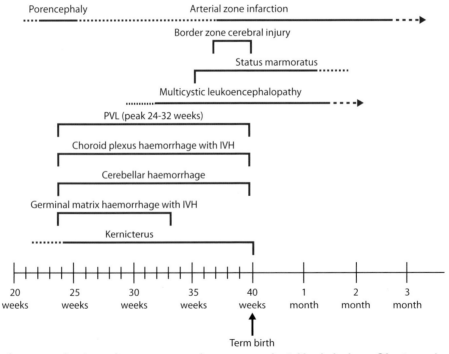

3.2 Timetable showing approximate peak occurrences of common perinatal brain lesions. Of note, periventricular leukomalacia (PVL) may originate in term infants and the first few postnatal months. Dashed lines represent uncertainty in the temporal limits. IVH, intraventricular haemorrhage.

the end of infancy as the purview of 'perinatal neuropathology' underscores the enormous impact that brain insults may play upon lifelong neurological morbidity.[302]

The principle of the interplay between disease processes, developmental programmes and plasticity is illustrated well in a study of perinatal injury in the developing human brain by Marin–Padilla.[349] He showed that undamaged cortical regions adjacent to injured sites retain their intrinsic vasculature, and undergo progressive post-injury transformations (termed 'acquired cortical dysplasia'), which adversely affect the structural and functional organization of developing neurons, axons, glial elements and microvasculature (Figure 3.3). He speculated that these transformations influence the neurological maturation of affected children and play a crucial role in the pathogenesis of ensuing neurological disorders, e.g. epilepsy, cerebral palsy, dyslexia, cognitive impairment and poor school performance.

In this chapter, the spectrum of perinatal lesions is highlighted, emphasizing the special features that distinguish perinatal lesions from those in the mature brain. The pathological features of malformations, perinatal degenerative disorders and inborn errors of metabolism are considered elsewhere (see Chapter 4, Malformations; Chapter 5, Metabolic and Neurodegenerative Diseases of Childhood; Chapter 6, Lysosomal Diseases; Chapter 7, Mitochondrial Diseases; and Chapter 8, Peroxisomal Diseases).

APPROACH TO EVALUATION OF PERINATAL SPECIMENS

Perinatal nervous system specimens originate from spontaneous or therapeutic abortions, stillbirths, term or premature infants and, according to the consideration explained earlier, infants throughout the first year of life. As in adult post-mortem examinations, the clinical details including family history, findings on prenatal ultrasound or magnetic resonance (MR) imaging, and maternal obstetric course, are critical to deciding the scope and focus of the examination. In particular, possible pathological involvement of skeletal muscles, peripheral nerves, cranial nerves, cerebrospinal fluid, spinal cord and the brain must be considered so that adequate samples are taken for analysis. Despite the ever-improving MR imaging capabilities, post-mortem imaging should be considered a supplement, not a substitute for, the conventional autopsy.[65]

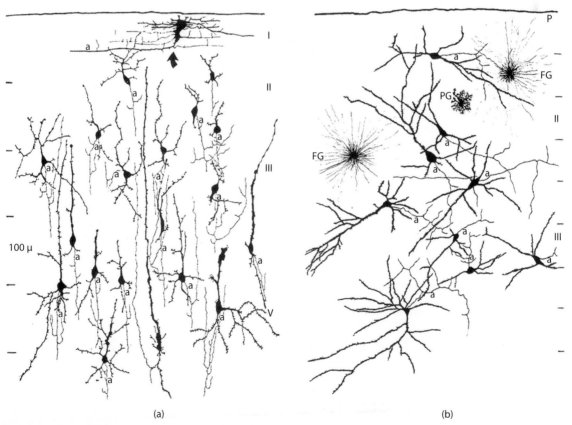

(a)   (b)

3.3 Perinatal damage affects subsequent neuronal organization and structure. Camera lucida drawings of rapid Golgi preparations showing neurons in superficial neocortex of brains from infants subjected to perinatal brain injury. **(a)** Detail from an atrophic gyrus of an 8-month child with developmental delay who was born at 24 weeks' gestation. Although the laminar and radial dendrite pattern is maintained, some neurons appear as small stellate cells and large pyramidal cells have lost their contact with layer I but retain their axons (a). The surviving Cajal–Retzius cells (arrow) have retained their typical morphologic features and long horizontal axons. **(b)** Detail from a 5-year child with cerebral palsy who was born at 32 weeks' gestation. The laminar pattern is obliterated and the neurons are markedly disorganized with disorientation of dendrites. Reactive astrocytes (FG) and non-reactive protoplasmic astrocytes (PG) are shown.

From Figure 4, Marin–Padilla.[349] With permission from Lippincott Williams & Wilkins/Wolters Kluwer Health.

The gross and histological examination of the immature brain demands a complete appreciation of potential artefactual changes. The immature brain is mechanically fragile owing to incompleteness of connective tissues associated with the leptomeninges and vasculature. Immature cells lack long processes and cell-to-cell connections that anchor them in place. Migrating cells, cells forming processes, and myelinating cells also make use of proteolytic enzymes (e.g. matrix metalloproteinases, calpains, and a variety of lysosomal enzymes).[84,350] When released post-mortem from their normal restraints, and particularly in stillbirths that remain at body temperature, these enzymes can rapidly digest brain tissue. These give rise to changes in the internal granular layer of the cerebellum,[399] irregularities of the neocortical surface that can mimic migrational abnormalities, and eventually outright tissue fragmentation and liquefaction. It should be noted though that even severely autolytic brain can offer insight into pathogenesis at the microscopic level (e.g. inflammatory/infectious processes, dying neurons). One must also take care not to mistake the 'dark neuron' artefact for dying cells.[279]

The perinatal post-mortem examination, especially when undertaken to explain a stillbirth or unexpected neonatal death, is incomplete without knowledge of the placental pathology, although the histopathological examination of the placenta falls outside of strict neuropathology expertise. Essential data include (i) the placental weight and size (expressed as a percentile of that expected for gestational age) as an indicator of possible placental insufficiency; (ii) the presence and extent of infarcts; (iii) the presence and extent of inflammation (chorioamnionitis, funisitis), which can be associated with fetal sepsis and white matter injury (see section on white matter injury, also Box 3.3); (iv) umbilical cord length, vascular abnormalities, and nuchal or extremity wrapping; and (v) evidence of placental abruption or praevia.

BRAIN GROWTH

Developmental neurobiology is an enormous field; the clinical neuropathologist cannot be expected to have a complete command of the knowledge base. However, there is a minimum appreciation of developmental cell biology and physiology necessary for diagnostic perinatal neuropathology. Among these is an understanding of the changes in the germinal layers of the growing neural tube, especially the idea that this is not a uniform layer either structurally or temporally in different parts of the central nervous system.

Growth of the fetal brain depends upon the availability of oxygen and nutrients and the appropriate trophic factors operating according to a genetic plan under the influence of the maternal environment. Of major relevance to the examination of the brains of fetuses and infants is the determination of the stage of brain growth and development. The stage of fetal development is classified by somatic weight and gestational age at birth, as per the definitions summarized in Table 3.1. Clearly, gestational estimations are not always exact and are subject to errors of 1–2 weeks or more, depending upon the time and method of assessment. Prematurity (<37 completed gestational weeks at birth) is subdivided by the gestational age at birth (Table 3.1). Three overall designations of morbidity and mortality risk (summarized in Table 3.1) are based on the corresponding age interval: infants at the limits of viability (≤25 weeks); very preterm infants, requiring intensive neonatal care to survive

---

BOX 3.3. Maternal factors, placental pathology and perinatal brain injury

Several studies have explored the epidemiological relationship between placental abnormalities and subsequent neurodevelopmental defects, including cerebral palsy. In part, these studies are driven by litigation by caretakers of neurologically impaired children against obstetrical providers and, thus, the need to demonstrate objective pathological evidence of an underlying gestational problem predisposing to brain injury. The placenta is not only a vital organ of the fetus but also a 'diary' of the pregnancy, reflecting antepartum conditions and potentially allowing distinction from intra- and postpartum events.

Intrauterine infection (premature rupture of membranes, maternal fever, positive amniotic fluid cultures) increases the risk of cerebral palsy[238] and perhaps more subtle brain damage such as that which leads to autism.[425] Various types of placental lesions, and the timing of them relative to delivery, are associated with neurological impairment in the infant.[450] In children born at term and later diagnosed with cerebral palsy or neurological impairment, several placental lesions were associated independently with cerebral palsy, including chorioamnionitis with fetal vasculitis, avascular villi and chorioamnionic haemosiderosis.[449]

In a study of neurologically impaired infants and controls, Redline further noted severe fetal placental vascular lesions (fetal thrombotic vasculopathy, chronic villitis with obliterative fetal vasculopathy, chorioamnionitis with severe fetal vasculitis, meconium-associated fetal vascular necrosis) in 51 per cent

versus 10 per cent.[449] In another study, recent non-occlusive thrombi of chorionic plate vessels and severe villous oedema were associated with cerebral palsy or other neurological disabilities.[451] In both of the latter two studies, the presence of more than one placental lesion increased the risk of neurological impairment. Amniotic sac inflammation (chorioamnionitis, umbilical vasculitis, amnion epithelial necrosis) was associated epidemiologically with intraventricular haemorrhage in a prospective cohort study of 1095 VLBW infants.[404] A unique study examined both neuropathology and placental pathology in order to ascertain the concurrence of antepartum (i.e. chronic) placental and brain lesions in 98 stillborn infants and infants surviving less than 1 hour.[205] An association was demonstrated between any type of central nervous system pathology (periventricular leukomalacia, neuronal death, germinal matrix haemorrhage) and placental chronic vascular problems, umbilical cord problems, funisitis and meconium. There was no correlation between chorioamnionitis and any form of brain injury in this cohort of non-survivors. New data also suggest that placental dysfunction can play a role in more subtle abnormalities of brain development.[245]

Examples of the timing of specific placental lesions associated with brain injury are (i) acute (defined as occurring within 0–6 hours of delivery): maternal hypotension, *abruptio placentae*, complete total umbilical cord obstruction and fetal vascular rupture; (ii) subacute (6 hours–7 days before delivery): cord entanglements, meconium-associated vascular necrosis, and fetomaternal haemorrhage; and (iii) chronic (greater than 1 week): maternal vascular underperfusion, villous infarcts villitis of unknown etiology, chronic abruption and fetal thrombotic vasculopathy.[450]

(26–33 weeks); and late preterm infants (34–36 weeks). The latter two categories of preterm births account for the majority and are of public health concern (the subject is reviewed in detail in reference 360). It is well recognized that survivors of prematurity are at risk of various health, behavioural and cognitive difficulties, beyond the scope of this chapter.

Infants may be of low birth weight for two reasons: they are appropriately developed, but preterm, or they have suffered intrauterine growth restriction (IUGR; formerly called intrauterine growth retardation). The definition of IUGR is a fetal birth weight and/or length below the 10th percentile for gestational age that is associated with a pathological restriction in growth.[37] The definition is based upon standard weight *versus* gestational age tables or, alternatively, individually 'customized' percentiles.[183] The causes of IUGR include chronic maternal illness (e.g. hypertension, diabetes), poor maternal nutrition, maternal cigarette smoking during pregnancy, immunological disorders, uteroplacental insufficiency (placental perfusion abnormalities) and fetal infections and malformations. Epigenetic factors also play a likely role in brain development as well as in response of the fetal brain to *in utero* insults (see Box 3.4). In 40–50 per cent of cases, the aetiology is unknown.[47] IUGR is associated with increased morbidity, including meconium aspiration, respiratory distress, hypothermia and hypoglycaemia; it is also associated with greater mortality.[184] Overall, perinatal mortality in IUGR without fetal anomalies is 100 per 1000 live births and, with fetal anomalies, 120 per 1000 live births.[47] The brains of the majority of growth-restricted infants have age-appropriate volume.[110]

However, although the 'brain sparing' effect often seen in IUGR might not be associated with microscopically detectable brain abnormalities, the subsequent presence of behavioural problems suggests that subtle abnormalities occur.[464] Charts of normal perinatal brain growth by age are provided in Tables 3.2 and 3.3.

BOX 3.4. Epigenetics and perinatal brain disorders

Epigenetic changes are heritable changes in gene expression (and cell phenotype) caused by mechanisms other than changes in the underlying DNA sequence. For example, enzymatic DNA methylation or histone deacetylation can suppress gene expression without altering the sequence of the silenced genes. Subtle brain changes, at the interface between acquired perinatal brain disorders and teratogenesis, result from epigenetic changes associated with maternal and perinatal factors.[33,538] Some may come to the attention of the neuropathologist, for example, sudden infant death or (rare) brain malformations associated with maternal alcohol ingestion.[405,446] Other functional changes include hormone exposure and sexual differentiation of the brain,[397] effect of leptin from maternal adipose tissue on the development of brain areas associated with feeding behaviour,[437] and effect of stress related glucocorticoids released during perinatal period on the later maturation of neurons and efficacy of progenitor cells.[226,229,400] Autopsy studies of premature infants showed that those receiving antenatal glucocorticoids had a lower density of hippocampal neurons and a lower density of blood vessels in the germinal matrix, although it was not clear if this is a causal relationship.[531,558] Many of these adverse brain effects seem to be mediated through epigenetic modification of progenitor cells.[156] Epigenetic factors can also negatively influence placental development, which may in turn have consequences to the fetal brain.[386]

TABLE 3.2 Fetal brain weights and head circumferences (5th, 50th and 95th percentiles) by gestational age

| Gestational age (wks) | Brain weights | | | Head circumferences | | |
	5th percentile (g)	50th percentile (g)	95th percentile (g)	5th percentile (mm)	50th percentile (mm)	95th percentile (mm)
12	0.35	0.51	0.68	48.8	63.0	77.1
13	3.35	3.68	4.011	63.0	78.0	93.1
14	5.44	7.72	10.01	76.7	92.7	108.7
15	8.36	12.65	16.93	90.1	107.0	123.9
16	12.11	18.45	24.80	103.1	120.9	138.6
17	16.67	25.14	33.60	115.7	134.3	153.0
18	22.07	32.70	43.33	127.9	147.4	166.9
19	28.29	41.14	54.00	139.7	160.0	180.4
20	35.33	50.47	65.61	151.1	172.3	193.5
21	43.19	60.67	78.15	162.2	184.1	206.1
22	51.88	71.75	91.62	172.8	195.6	218.4
23	61.40	83.72	106.03	183.1	206.6	230.2
24	71.74	96.56	121.38	192.9	217.3	241.6
25	82.90	110.28	137.66	202.4	227.5	252.6
26	94.89	124.88	154.87	211.5	237.3	263.2
27	107.70	140.36	173.02	220.2	246.8	273.3

Continued

TABLE 3.2 Fetal brain weights and head circumferences by gestational age (*Continued*)

Gestational age (wks)	Brain weights			Head circumferences		
	5th percentile (g)	50th percentile (g)	95th percentile (g)	5th percentile (mm)	50th percentile (mm)	95th percentile (mm)
28	121.34	156.72	192.10	228.6	255.8	283.0
29	135.80	173.96	212.12	236.5	264.4	292.3
30	151.09	192.08	233.07	272.6	301.2	251.2
31	167.20	211.08	254.96	251.2	280.5	309.7
32	184.13	230.96	277.79	258.0	287.9	317.7
33	201.89	251.72	301.55	264.4	294.9	325.4
34	220.47	273.36	326.24	270.4	301.5	332.6
35	239.88	295.87	351.87	276.0	307.7	339.4
36	260.11	319.27	378.43	281.2	313.5	345.7
37	281.17	343.55	405.93	286.0	318.9	351.7
38	303.05	368.71	434.36	290.5	323.8	357.2
39	325.75	394.74	463.73	294.5	328.4	362.3
40	349.28	421.66	494.03	298.2	332.6	367.0
41	373.64	449.45	525.27	301.5	336.4	371.3
42	398.81	478.13	557.44	304.4	339.8	375.2

Adapted from Phillips JB *et al*,[434] with permission from Lippincott Williams & Wilkins/Wolters Kluwer Health. Note that fixation can increase brain weight 5–15 per cent.[214,310] In addition, age correction of premature infants is usually associated with lower weight than term births; there may be racial differences; and it is important to use recent data because improved prenatal care and nutrition have been associated with increasing body size and organ weights.[239,529] Finally, head-circumference-to-brain-weight relationship is essentially linear over 24–44 gestational weeks.[321]

CELLULAR COMPONENTS OF THE DEVELOPING HUMAN CENTRAL NERVOUS SYSTEM

Neurons

The rate and timing of growth and differentiation of neurons in different central nervous system regions varies. Germinal layers in the spinal cord wax and wane earlier than in the forebrain.[273] Some evidence indicates that the earliest neurons appear in the neocortex even before the neural tube is completely closed.[75] Neuronal proliferation and migration occur in the first and early second trimester. The cortical plate begins to form from the arrival of migrating neurons as early as 8–9 weeks post-ovulation (10–11 weeks' gestation). Ventricular zone (VZ) cells give rise to radially migrating neurons that eventually have long distance connections. The VZ also generates cells that form the subventricular zone (SVZ), a secondary proliferative layer that generates GABAergic inhibitory interneuron precursors and glial precursors (Figure 3.4).[76,336] The peak period for neuronal proliferation is 2–4 months of gestation, and, for neuronal migration, 3–5 months of gestation.[44] Recent evidence indicates that neurogenesis continues up to 28 weeks' gestation in the ganglionic eminence. Most late born neurons are likely cortical and thalamic interneurons, although based on their molecular profile some might be glutamatergic neurons.[340,597,598] In the cerebellum,

Purkinje neurons are created by 20–23 weeks' gestation, but the external granular layer continues to generate internal granule layer neurons until 5–7 postnatal months.[3,313]

Perinatal neuropathology is concerned mainly with the organizational events that follow neuronal proliferation and migration. These organizational events occur from approximately the 20th gestational week until several years after birth. They include the establishment, differentiation and death of subplate neurons; the alignment, orientation and layering of cortical neurons; the determination of neuronal number, density, size, differentiation and final position; axonal outgrowth, collateralization and establishment of synaptic contacts; the development of the cytoskeleton; the formation and arborization of dendrites and the formation of spines; synaptogenesis; cell death (apoptosis) and selective elimination of neuronal processes and synapses; and the synthesis of neurotransmitters and the formation of the cellular apparatus for their function. The postmitotic neuron develops from an undifferentiated cell into a polarized cell with a complex shape characterized by a dendritic tree and an axon with collateralization. Neurons are likely committed to a particular phenotype and cortical address when they are born in the VZ.[444] In addition to innate genetic determinants of cell fate, there is a period in the cell cycle (late S-phase to mitosis) when neuroblasts are susceptible to environmental cues and can be induced to alter their programmed phenotype, laminar position and possibly even their subcortical connections.[359]

The cytoskeleton maintains the neuron's shape and polarity, which is reflected in differences between the morphology of dendrites and axons, the organization of the cytoskeleton and membrane specializations and the absence of protein synthesis in the axon. During development, the dendrites and axons grow out in a predetermined manner but the initiation of this is influenced by variable extrinsic conditions, including adhesion molecules, extracellular matrix molecules and growth factors produced by glial cells, blood vessels and axonal targets. Protein synthesis is likely controlled locally at growth cones and synapses.[77,352] It is beyond the scope of this chapter to discuss details of neuronal organization and, thus, only selected events will be discussed here.

Lamination

Lamination – the proper alignment, orientation and layering of cortical neurons – is one of the earliest cortical organizational events and occurs as neuronal migration ceases. For the neocortex, the initial cortical plate appears at 7–10 weeks of gestation and condensation into a six-layered cortex occurs at approximately 16 weeks of gestation.[444] This process is dependent on proper migration and function of the transient subplate neuron population, which directs migrating cortical plate neurons and targeting of their axonal projections (Figure 3.5).[188,363] During lamination, neurite outgrowth progresses, with the elaboration of dendritic and axonal ramifications. These major neuronal changes have been clearly demonstrated in the developing visual cortex (Figure 3.6). Lamina-specific antibodies are now available,[237] providing the means to study temporal

and spatial patterns of laminar development in the human cerebral cortex (Figure 3.7).

Synaptogenesis and Axonal Elongation

During development, synapses are formed over a protracted interval, beginning in the embryo and continuing into postnatal life.[570] Synaptogenesis is coupled closely to neuronal differentiation and the laying down of neuronal circuitry: following neuronal differentiation and the extension of axonal and dendritic processes (Figure 3.8), several of the genes encoding synaptic proteins are expressed). Correct neuronal connections are specified, *via* the mediation of cell-surface adhesion molecules, as initial, often transient, synapses are formed between growing neurites. Finally, some synapses are selectively eliminated as remodelling of cortical connections progresses. This elimination occurs *via* several mechanisms, including competition for trophic substances and increased electrical activity, and is mediated in part *via* ubiquitination pathways. The elimination of synapses in infancy has been reported in the human visual cortex.[252]

Historically, synapses have been examined with Golgi impregnation techniques that define spines and axonal boutons, and with electron microscopy, using sections stained with ethanolic phosphotungstic acid for visualizing synaptic junctions. Successful use of these techniques depends upon short post-mortem intervals. The development of antibodies that recognize synapse-associated molecules provides an alternative means for studying human synapses, allowing for immunohistochemical labelling in tissue obtained after

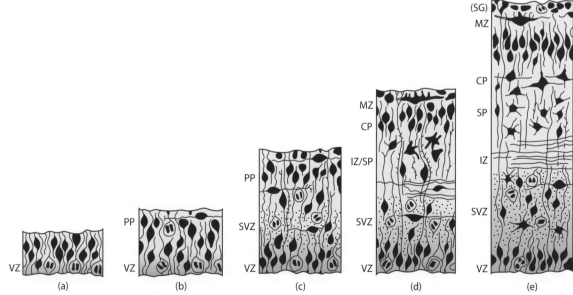

3.4 Neural tube layering during embryonic development. A proposed revision of the Boulder Committee's summary diagram of neocortical development shows the sequence of structural changes in the lateral part of the dorsal telencephalon at the approximate ages of embryonic day (E) 30 **(a)**; E31–E32 **(b)**, E45 **(c)**, E55 **(d)** and gestational week 14 **(e)**. The major layers are cortical plate (CP), intermediate zone (IZ), marginal zone (MZ), subventricular zone (SVZ), subpial granular layer (SG; part of the MZ), and the ventricular zone (VZ). The revised view incorporates transient compartments, including the preplate (PP) and the intermediate and subplate zones (IZ and SP).

TABLE 3.3 Brain weight in infants, up to age 24 months, in relation to postnatal age and body length

Age	Brain weight (g, mean ± SD)	Sample size
Newborn	325 ± 158	13
1 week	370 ± 78	2
2 weeks	456 ± 54	7
3 weeks	430 ± 120	6
1 month	492 ± 120	26
2 months	608 ± 248	22
3 months	672 ± 274	15
4 months	734 ± 240	13
5 months	687 ± 290	9
6 months	839 ± 290	8
7 months	880 ± 86	5
8 months	845 ± 280	4
9 months	905 ± 238	7
10 months	988 ± 280	10
11 months	893 ± 186	6
12 months	980 ± 156	3
14 months	944	
16 months	1010	
18 months	1042	
20 months	1050	
22 months	1059	
24 months	1064	

Data from 1 week to 12 months adapted from Thompson and Cohle,[529] and from 14 to 24 months adapted from Sunderman and Boerner.[508] Note that fixation can increase brain weight 5–15 per cent.[214,310] In addition, age-correction of premature infants is usually associated with lower weight than term births; there may be racial differences; and it is important to use recent data because improved prenatal care and nutrition have been associated with increasing body size and organ weights.[239,529]

post-mortem intervals of up to 48 hours. The molecules include growth-associated protein 43 (GAP-43), synapsin I and II, synaptotagmin, synaptobrevin, synaptophysin and synaptosome-associated protein of 25 000 daltons (SNAP-25).[471] They have varying roles, including molecular assembly of the synaptic junction and the delivery of pre- and postsynaptic components. For example, GAP-43 is a 'signal' protein that accumulates mainly in axon endings (growth cones and the presynaptic area of synapses) where it modulates axon growth, neurotransmitter release and synaptic plasticity.[379] As its expression occurs maximally during periods of growth and change, markers for GAP-43 protein or mRNA are highly informative regarding the temporospatial patterns of synaptogenesis *in vivo*. During development, high levels of GAP-43 appear along the entire length of the axon. After the establishment of stable synapses, most neurons cease expressing GAP-43 at high levels. In certain regions, however, high GAP-43 levels persist into adulthood, e.g. in the limbic and associative regions of the forebrain.

These presynaptic terminals in which GAP-43 levels remain high may represent sites that can undergo functional and possibly even structural changes (plasticity) in response to physiological activity throughout life.[49] GAP-43 immunostaining delineates the sequences of synaptogenesis and fibre tract elongation in the human telencephalon[229,371] and brain stem.[287] In the parietal white matter, GAP-43 expression peaks between 30 gestational weeks and term (37–40 gestational weeks) and corresponds to the onset of expression of phosphorylated neurofilament, a crucial cytoskeletal protein in the axon.[229] This critical period for axonal development in the human cerebral white matter coincides with the window of vulnerability to periventricular leukomalacia (PVL), the major underlying substrate of cognitive and motor abnormalities in survivors of prematurity (see later).

Radial Glia and Astrocytes

Radial glia represent a distinct cell population present in the early neural tube. They have a periventricular cell body and a long radial process that extends to the pial surface. They serve to guide neuron migration and later act as progenitors of astrocytes and as a source for neural stem cells.[12,378] Radial glial cells express glial fibrillary acidic protein (GFAP), brain lipid binding protein and the glutamate/aspartate transporter (GLAST), transforming into fibrillary or protoplasmic astrocytes.[95,107,561] The transcriptional regulation of the GFAP gene by ciliary neurotrophic factor, acting through the JAK/STAT (Janus kinases and their downstream signal transducer and activator of transcription) signalling pathway and cyclic adenine dinucleotide phosphate response element binding proteins (CREB),[473] is critical in the differentiation of precursor cells into astrocytes. Radial glia also demonstrate molecular features of neural precursor cells, such as intermediate filament proteins vimentin and nestin (Figure 3.9), and reaction centre type 2 (RC2).[85,226]

In the human fetus, radial glial processes are present at 12 weeks of gestation. Mature GFAP-positive astrocytes (Figure 3.9) are present in the brain stem at 15 weeks and increase until 20 weeks;[459] at 17 weeks they are detected in the hilus of the dentate gyrus.[459] At 20 weeks of gestation, GFAP-positive astrocytes are present in the neocortical plate and at 28–30 weeks the astrocytes of the outer molecular layer are prominent.[459] GFAP-positive glial cells are identifiable in the deep white matter of the cerebral hemispheres by 28 weeks[513] and increase in number until after birth.[203,228] The distal processes of radial glial fibres disappear from 20–28 weeks of gestation, possibly marking the time when migration on glial guides ceases.[345] In human brain sections treated with rapid Golgi silver impregnation, Marin–Padilla identified another glial precursor[346] on the outer surface of the neural tube; he postulated them to be involved in the maintenance of the external glial limiting membrane and he suggested that they gradually transform into protoplasmic astrocytes of the grey matter.

Mature astrocytes play major roles in potassium homeostasis, neurotransmitter uptake and synapse formation through an extensive network of processes that are placed near synapses and myelin internodes[189] *via* gap junctions that interconnect the astrocytes a metabolic syncytium is formed.[195,308] Astrocytes are also in communication with oligodendrocytes through gap junctions[409] and secreted molecules.[376] Astrocytes modulate development of the nervous system and regulate the blood–brain-barrier.[433] There is considerable regional variation in the morphology and function of astrocytes.[357,600]

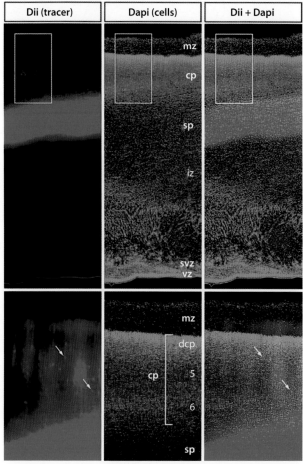

Dii (tracer)	Dapi (cells)	Dii + Dapi

3.5 Development of cortical connections at mid-gestation in human visual cortex (area 17). Dioctadecyl-tetramethylindocarbacyanine perchlorate (DiI-red) was injected into the optic radiations between the lateral geniculate nucleus and visual cortex in a fixed brain. Labelled thalamocortical axons project heavily into the subplate (sp). At this age, the cortical plate (cp) contains mainly neurons belonging to layers V and VI. The outer dense cortical plate (dcp) contains newly migrated neurons as a consequence of the fact that superficial cortical layers are 'born' later than those in the deep layers. The cerebral wall also consists of the ventricular zone (vz) and subventricular zone (svz), where precursors of neurons and glia proliferate; an intermediate zone (iz), where neurons migrate and axonal pathways develop; and a marginal zone (mz), which is the precursor of cortical layer I. At higher magnifications (bottom panels), the cortical plate is seen to contain some retrogradely labelled neuron somata (arrows), corresponding to corticobulbar and corticospinal projections from layer V and corticothalamic projections from layer VI. This experiment exemplifies the dynamic processes of cortical cell proliferation, migration and axon growth and pathfinding, which are susceptible to insults during fetal life. Fluorescent axons tracer DiI (red), counterstained with 4',6-diamidino-2-phenylindole (DAPI), a fluorescent DNA-binding molecule (blue nuclei).

From Hevner.[236] With permission from Lippincott Williams & Wilkins/Wolters Kluwer Health.

Astrocytes react to a broad range of pathological states and in doing so change their molecular and morphological features, becoming hypertrophic and re-entering proliferative phase of the cell cycle.[494] Although the changes represent an attempt to compensate for an altered nervous system environment, in the long-term, there is evidence that persisting changes can contribute to disease states such as epilepsy and the scarring that restricts potential regeneration following brain trauma.[219,307] In human fetal brain, astrocytes appear to multiply and differentiate independently of, and in many areas well before, myelinogenesis.[459,513] The capacity of astrocytes to 'react' with proliferation and hypertrophy develops after the first half of gestation. The immature astrocyte responds with less fibre formation than the astrocyte in the older brain. Reactive astrogliosis is first detectable with GFAP-positive immunostaining around 20 weeks and gemistocytic morphology is recognized around 23 weeks.[459]

Oligodendrocytes and Central Nervous System Myelination

Oligodendrocytes are the cells responsible for the formation of myelin in the central nervous system. The term 'myelination gliosis' was introduced in 1935 to describe the hypertrophic glial cells with eccentric nuclei characteristically present in the immature white matter.[457] Myelination glia are thought to represent a specific stage of immature oligodendrocytes prior to the deposition of myelin.[61,263] (Figure 3.9) They are strongly immunoreactive for GFAP and not to be mistaken for reactive astrocytes.

Oligodendrocytes derive from progenitors that migrate from the subventricular zone of the fetal brain (and potentially the adult brain) or from local oligodendrocyte precursor cells residing in the parenchyma.[159] Precursors might also arise directly from the ventricular zone in restricted regions of the spinal cord.[392] Following migration, a density-dependent feedback inhibition of proliferation reduces the responsiveness of the cells to growth factors; final matching of oligodendrocyte and axon number is accomplished through the local regulation of cell death.[407] Axons produce signals that regulate oligodendrocyte proliferation, survival, terminal differentiation and myelinogenesis.[487] Oligodendrocyte differentiation combines epigenetic repression of transcriptional inhibitors with direct transcriptional activation of myelin genes. The mechanisms include chromatin remodelling by histone deacetylases and gene silencing by non-coding RNAs.[151,325] The genetic basis underlying specification into the oligodendroglial cell lineage has been found to be a result of the Olig2 gene encoding a basic helix-loop-helix transcription factor.[507] Subsequent downstream induction of numerous transcription factors (Olig1, Ascl1, Nkx2.2, Sox10, YY1 and Tcf4) is required for generation of mature, postmitotic oligodendrocytes.[575] Using immunohistochemistry for Olig2 protein in the developing human brain, Jakovceski and Zecevic noted Olig2-positive cells appearing as early as 5 gestational weeks in the ventral neuroepithelium, and in the telencephalic proliferative zones and primitive white matter at 20 gestational weeks.[264] A note of caution: they reported that Olig2 also co-localizes with microtubule-associated protein 2, a neuronal marker, in a subset of neuronal precursor cells in the human fetal brain. In addition to coding for myelin proteins, the MBP and PLP genes encode for signals

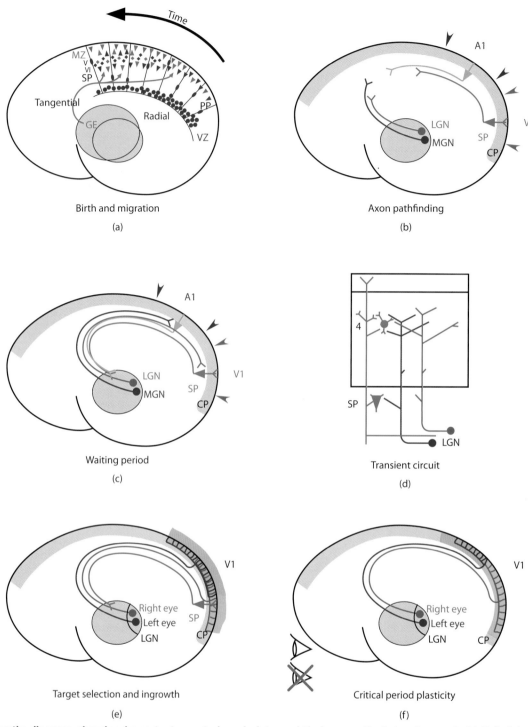

3.6 Schematic diagram showing important events in subplate and thalamocortical development. (a) Subplate (SP) neuron precursors are generated in the ventricular zone (VZ) or ganglionic eminence (GE; a concentrated region of the subventricular region), with subsequent radial (blue cells) and tangential (green cells) migration. The 'inside-out' laminar development of the cortical plate from the preplate (PP) proceeds from caudal to rostral (Time arrow). Cajal–Retzius cells (red) are present in the marginal zone (MZ). **(b,c)** Pioneer projections from the subplate (SP green) to the internal capsule occur during thalamocortical pathfinding. Axons from medial geniculate nucleus (MGN, blue) and lateral geniculate nucleus (LGN, red) establish primary connections in subplate during 'waiting period' before innervation of cortical plate. A1 = primary auditory cortex; V1 = primary visual cortex. **(d)** Details of the transient LGN-subplate-layer 4 circuit. **(e)** Thalamic axons proceed from the subplate neurons to cortical layer 4, with overlap of right eye (red) or left eye (blue) dominated inputs before formation of mature ocular dominance columns. **(f)** During the critical period for activity-dependent alteration in ocular dominance columns monocular deprivation (eye with 'X') results in atrophy of deprived eye columns (blue) and the expansion of active eye columns (red). Note that human subplate neurons appear prior to 15 weeks' gestational age and begin to regress at about 28 weeks, persisting in the postnatal state as subcortical interstitial neurons.[270,300]

Layer-Specific Markers in Human Neocortex

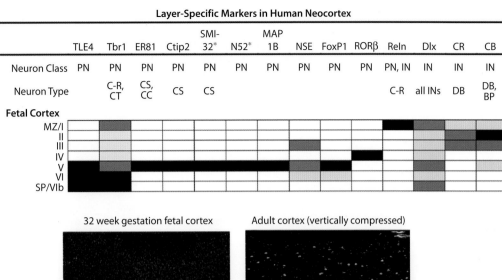

	TLE4	Tbr1	ER81	Ctip2	SMI-32*	N52*	MAP1B	NSE	FoxP1	RORβ	Reln	Dlx	CR	CB
Neuron Class	PN	PN	PN	PN	PN	PN	PN	PN	PN	PN	PN, IN	IN	IN	IN
Neuron Type		C-R, CT	CS, CC	CS	CS						C-R	all INs	DB	DB, BP

Fetal Cortex

Layers: MZ/I, II, III, IV, V, VI, SP/VIb

| 32 week gestation fetal cortex | Adult cortex (vertically compressed) |

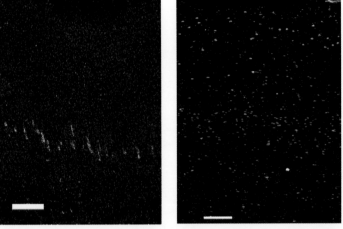

3.7 Layer-specific markers in developing human cerebral cortex. Neurons of each defined cortical layer have been shown to express different antigens that are detectable by antibodies under specified fixation conditions (darker represents stronger expression). The expression of various antigens is maturation dependent; MAP1B, NSE and FoxP1 have not been reported in adult cortex. MAP1B (yellow) is expressed in a subset of layer V projection neurons of the 32-week gestation fetal cortex. In adult cortex (compressed vertically for comparative purposes) calretinin (green) and ROR-A (red) are expressed mainly in layers II and IV, respectively.

Reproduced from Figures 2 and 3 of Hevner.[237] With permission from Lippincott Williams & Wilkins/Wolters Kluwer Health.

regulating oligodendrocyte number, amount of myelin and other aspects of oligodendrocyte behaviour.

The cellular sequences of myelination are defined by cell surface markers on oligodendrocyte precursors (O1 (galactocerebroside or GalC) precedes O4, which precedes myelin basic protein [MBP]), detected by immunohistochemistry,[182,441] and have been determined in the mid- and late gestational human fetus in the parietal white matter (Figure 3.10).[29,31] Growth factors are involved in the entire sequence of oligodendrocyte generation and maturation. Early on, platelet-derived growth factor (PDGF) and insulin-like growth factor 1 (IGF-1) control oligodendroglial precursor (O2A) proliferation; precursor cells compete for binding of these factors, which determine further progression through the lineage. Whereas IGF-1 increases the survival of both O2A progenitor cells and mature oligodendrocytes *in vitro*, PDGF increases survival of O2A cells and only those mature oligodendrocytes that retain PDGF receptors. Migration depends upon both secreted molecules and contact-mediated signalling.[64] Once migrated, discrete populations of immature oligodendrocytes (O4+O1+MBP–) act as 'initiators' of myelination, by generating juxta- and peri-axonal tubules along a restricted number of mature axons. These cells are detectable from 18 to 30 gestational weeks, when they clearly display a complex arbour of processes that contact multiple axons. The three sequential phases of myelinogenesis are: (i) the initial ensheathment of axons by pre-myelin tubules generated by immature oligodendrocytes (O4+O1+MBP–); (ii) the insertion of MBP into transitional early myelin; and (iii) the generation of mature, fully compacted myelin.[31] The precise nature of myelin lamination remains incompletely understood.[493] The appearance of O1 and CNPase correlates with terminal progression into the oligodendrocyte lineage.[182]

Mature myelin requires the developmentally regulated synthesis of myelin basic protein (MBP) and proteolipid protein (PLP), which are involved in membrane compaction; myelin-associated glycoprotein (MAG) and 2′,3′-cyclic-nucleotide-3′-phosphohydrolase (CNPase); and other proteins. The major lipid components of CNS myelin, including cholesterol, galactolipids (cerebrosides, sulfatides), phospholipids and sphingomyelin are also developmentally regulated and differ regionally (Figure 3.11).[288] The biochemical heterogeneity among myelinating sites is likely to contribute substantially to the regional variability and complexity of the histopathology of many inborn and acquired disorders of CNS white matter in the fetal, perinatal and infant periods.

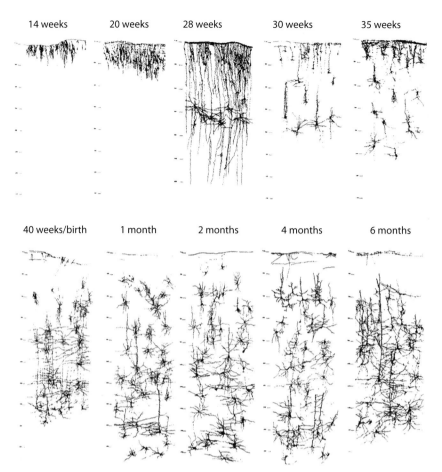

3.8 **Development of neocortex in human fetus and infant.** Camera lucida drawings of rapid Golgi preparations at 14, 20, 28, 30, 35 and 40 gestational weeks and at 1, 2, 4 and 6 postnatal months show progressively increasing complexity of neuron lamination and dendritic organization in visual cortex. Marks on the left are every 250 μm. In the 6-month brain, layer 3 neuron somata are located at ~1000 μm and layer 5 neuron somata are located at ~1750 μm. Quantitative analysis shows that apical dendrite length and complexity peak at ~24 months' postnatal age.[46]

Courtesy of F Chan, Hospital for Sick Children, Toronto, Canada.

The description of patterns of myelinogenesis has relied on histochemical staining of myelin in brains of different ages. Luxol fast blue is a very good dye for staining myelin.[294] However, the chemistry of staining is not specific; myelin contains a high proportion of non-polar amino acids and fatty acid residues that are not extracted by lipid solvents. When the water insoluble chemical is bound to the substrate in hydrophobic sites the diphenylguanidine dye is released.[97] Solochrome cyanin staining relies on a similar chemical reaction.[280] Human CNS myelination progresses in predictable sequences from caudal (spinal cord and brain stem) to rostral (telencephalon) (Figure 3.12). The rate of myelination in a particular pathway or region varies in a complex fashion, such that an early onset of myelination may not be necessarily associated with early completion of myelin maturation (Tables 3.4 and 3.5). Thus, white matter sites fall into eight subgroups based upon (i) the presence or absence of microscopic myelin at birth (group A or group B, respectively) and (ii) the age at which mature myelin is reached. Although beyond the scope of this chapter and perinatal brain disorders, it is important to note that the myelination in the thalamic connections and corpus callosum is not complete until 7–10 years, and the intracortical association bundles are not fully myelinated until the mid twenties; this has been shown using histologic techniques[68,288,587] as well as magnetic resonance imaging methods.[99]

Microglia

Microglia are immune effector cells and potential macrophages that play major roles in the defence and repair of CNS tissues in response to injury. Microglial precursors, which are monocyte-like cells, originate in the bone marrow and migrate into the brain during vasculogenesis, largely before formation of the blood–brain barrier.[375] They are detectable as early as 4.5 gestational weeks in the leptomeninges around the neural tube, and scattered throughout the parenchyma, by labelling with lectins (tomato lectin *Lycopersicon esculentum* or *Ricinus communis* agglutinin 1 [RCA-1]); a portion of these cells also label with antibody to the macrophage antigen CD68. Thereafter, these cells differentiate through a series of morphological transitions including the development of ramified processes. By 12 gestational weeks microglial cells are present in the thalamus and telencephalic wall,[456] and by about 20 gestational weeks in the germinal layer and subventricular and

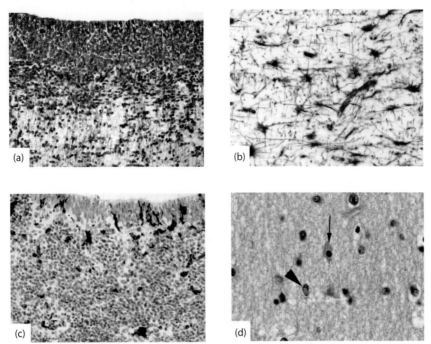

3.9 Glial cells in the developing brain. Cellular nuclei are readily apparent using routine staining methods such as haematoxylin and eosin. However, cell bodies and processes (and the identity of cells) may only be apparent using other staining approaches such as Golgi impregnation (see Figure 3.8) and immunohistochemistry. **(a)** Radial glia with cell bodies along the ventricle wall and long processes extending through the subventricular zone and intermediate zone are readily demonstrated in this 15-week frontal cerebrum using immunohistochemical detection of glial fibrillary acidic protein (GFAP) (brown; haematoxylin blue counterstain; 400× original magnification). **(b)** Microglia with elongated cell bodies and short processes can be found scattered in the normal immature brain and in increased quantities following any type of injury. Immunostain for HLA-DR (brown) in the ganglionic eminence of a 26-week gestation fetus (haematoxylin blue counterstain; 400× original magnification). **(c)** Astrocyte cell bodies and radiating processes are demonstrable in the corpus callosum of a 31-week fetus using immunostain for vimentin (brown). Note that some of the fibres might also be those of radial glia (haematoxylin blue counterstain; 600× original magnification). **(d)** Myelination glia are oligodendrocytes that hypertrophy prior to and during the time when they are actively producing myelin. In this example from the frontal white matter of a full term infant, these cells have prominent eosinophilic cytoplasm and small, round deeply basophilic nuclei (arrow). The nuclei of resting astrocytes have irregular, larger nuclei with granular chromatin (arrowhead) (haematoxylin and eosin; 1000× original magnification). Myelination glia may be immunoreactive for GFAP, but they should not be mistaken for reactive astrocytes, which tend to have longer processes and an eccentric large nucleus. As cells with this morphology are associated with the active phase of myelination, they appear in different parts of the central nervous system at different times (see Figures 3.10, 3.11 and 3.12).

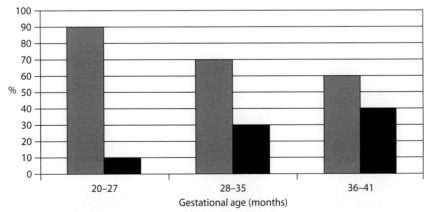

3.10 Bar graph demonstrating oligodendroglial maturational changes. The proportions of pre-oligodendrocytes (defined by immunoreactivity for O4 but not O1; grey bars) and immature oligodendrocytes (defined by immunoreactivity for O4 and O1; black bars) change relative to total number of oligodendrocyte lineage cells in the human fetal cerebrum. With increasing gestational age, the percentage of pre-oligodendrocytes decreases and the percentage of immature oligodendrocytes increases. At the peak age of periventricular leukomalacia, the pre-oligodendrocytes dominate the cerebral white matter.

Courtesy of Dr. Stephen Back.

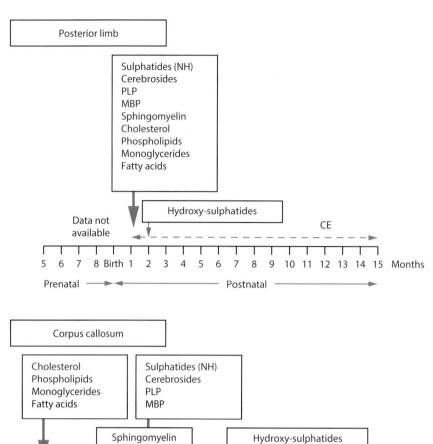

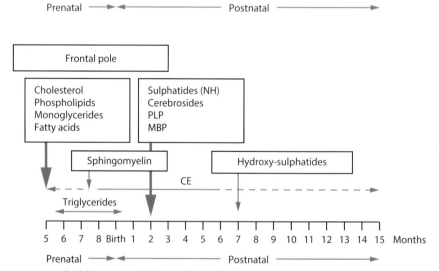

3.11 Sequence of appearance of white matter lipids and proteins in the frontal lobe white matter. This site myelinates relatively late. For cholesterol ester (CE), trace amounts are indicated by the dotted line. Biochemical sequences are identical in different sites, although they occur at different times. NH, nonhydroxy; MBP, myelin basic protein; PLP, proteolipid protein.

From Kinney et al.[288] With kind permission from Springer Science and Business Media.

ventricular zones (Figure 3.9).[51,202,204] The presence of acti-vated microglia in the germinal matrix may also reflect their phagocytic properties[161] in a region where apoptosis and programmed cell death occur.[322] As gestation progresses, microglia migrate sequentially through the germinal zone, subventricular zone, central white matter and, finally,

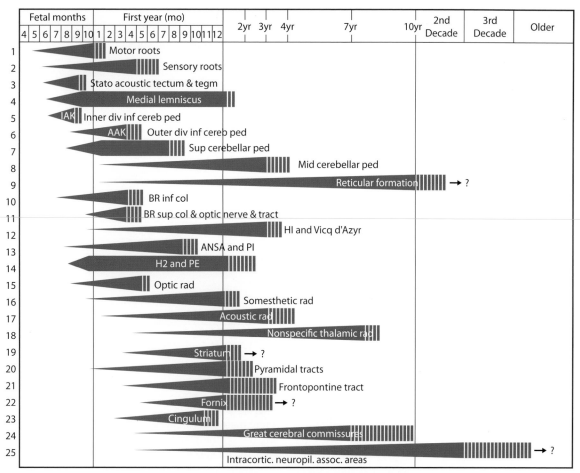

3.12 Pattern of myelin deposition in the developing human central nervous system. In this figure, the widening bars represent the period when myelin is being deposited. In general, myelin appears early in the spinal cord and brainstem (second half of gestation), and accumulates rapidly in the cerebrum during the first year of life. Many structures are not fully myelinated until later childhood years and the intracortical association bundles (especially those associated with frontal lobe connections) are not fully myelinated until the third decade of life. Note that the pattern of myelin deposition coincides with synaptic and functional maturation of related brain regions.

Reproduced from Yakovlev and Lecours (1967).[587] Greater detail appears in more recent publications.[68,284]

TABLE 3.4 White matter sites in Group A (onset of myelination before birth; see text for discussion)			
	A-1 (<68 weeks)	**A-2 (70–107 weeks)**	**A-3 (119–142 weeks)** **A-4 (>144 weeks)**
Sensory system			
Visual	Optic tract, optic chiasm		
Auditory			Brachium, inferior colliculus
Other			Tractus solitarius
Pyramidal system	Posterior limb, midbrain CST, pontine CST	Pyramid	Cervical CST, thoracic CST, lumbar CST
Extrapyramidal system	Hilum and amiculum of inferior olive, capsule of red nucleus, peridentate, middle cerebellar peduncle	Dentate hilum, ansa lenticularis, pontocerebellar fibres, cerebellum lateral hemisphere	Globus pallidus	Central tegmental tract
Central WM, commissures, capsules	Posterior limb, central corona radiata		
Limbic system		Stria medullaris thalami	Outer anterior commissure

CST, corticospinal tract; WM, white matter.

TABLE 3.5 White matter sites in Group B (onset of myelination after birth; see text for discussion)

	B-1 (<68 weeks)	B-2 (70–107 weeks)	B-3 (119–142 weeks)	B-4 (>144 weeks)
Sensory system				
Visual	Optic radiations, proximal and distal	SAF, calcarine cortex		Stripe of Gennari
Auditory	Auditory radiation, proximal	Heschl's gyrus		
Other			Lateral olfactory stria	
Pyramidal system				
Extrapyramidal system		Lateral crus penunculi	Medial crus penunculi	
Central WM		Distal radiation to precentral gyrus, posterior frontal, posterior parietal, occipital pole	Putamenal pencils, temporal lobe at LGN, temporal pole, frontal pole, SAF all sites	
Commissures, capsules	Corpus callosum body, splenium	Rostrum, external capsule, anterior limb	Anterior commissure, inner	Extreme capsules
Limbic system			Mammillothalamic tract, alveus, fimbria	Medial and lateral fornix

SAF, subcortical association fibres; WM, white matter; LGN, lateral geniculate nucleus.

cortex along radial glia pathways, white matter tracts and blood vessels.[453,455] The cerebral cortex of the human fetus and infant has negligible CD68-positive activated microglia compared with the white matter.[51,454] Although mature microglia are motile, even in the non-activated state, they do not leave the brain.[278,465] In contrast, a specific perivascular macrophage that resides between the endothelium and the glia limitans appears to move to and from the intravascular compartment.[413]

Under pathological conditions microglia become activated, migrate toward the offending site, release soluble factors (including cytotoxins, neurotrophins and immunomodulatory factors), and clear cellular debris by phagocytosis. Microglia are also phagocytic during developmental processes, especially at sites of synaptic elimination and cellular apoptosis. In these circumstances microglia appear to produce anti-inflammatory signals.[390] Macrophages in the parenchyma of the developing brain have a different morphology from those in the adult brain. In animal studies expression of surface antigens is age dependent.[112,431,552] A number of monocyte-macrophage antibodies (e.g. CD45, CD68, Iba1, HLA-DR) and lectins bind to microglia, but uniquely microglia-specific antibodies are not yet available.

A transient elevation in microglial cell density occurs in the cerebral white matter in the last few weeks of gestation as compared with postnatal ages.[51] This transient abundance raises the prospect that these cells participate in developmental processes in the organization of the white matter, including axonal development[229] and myelination.[220] The establishment and maintenance of CNS angiogenesis is thought to require, at least in part, the microglial expression of integrins.[149,370] Microglia participate in removal of degenerating axons and synapses that are 'pruned back' during the modelling of the developing brain.[58,260,584]

Microglia appear in developing fibre tracts throughout the white matter before myelinogenesis *in vivo*, and they stimulate the synthesis of sulfatide, a myelin-specific galactolipid, and myelin-specific proteins, including MBP, in oligodendrocytes.[220] Studies of axonal development and response to injury have suggested a role for activated microglia in the production of molecules critical for neurite outgrowth, including laminin and GAP-43.[354]

As the immunocompetent cells of the adult CNS, microglia express major histocompatibility complex (MHC) class II molecules.[51,256] Between 11 and 22 gestational weeks, MHC-II immunoreactivity is less prominent than CD68 expression in the cells in the choroid plexus, meninges and subependymal germinal zone.[154,454,580] The differences in expression between MHC-II and CD68 reflect the different cellular roles: microglial upregulation of CD68 during development prepares the cell to become phagocytic, whereas upregulation of MHC-II plays a role in immunity against foreign antigens in infection.[542]

Ependyma

In the developing nervous system, the first structure to differentiate is the floor plate, a specialized ependyma in the midline of the neural tube. The floor plate secretes retinoic acid and sonic hedgehog glycoprotein and has a ventralizing influence on the neural tube.[588] The immature ependyma is a pseudo-stratified secretory epithelium. It produces glycosaminoglycans and proteoglycans that repel growth cones, and netrin and S-100, which attract growing axons.[277] Immature ependymal cells lack cilia and have apical junctional complexes. They appear to provide structural, trophic and metabolic support for the underlying ventricular zone (VZ) and subventricular zone (SVZ).

As the ventricular zone (VZ) and radial glial cells regress, the central canal and ventricles are gradually covered by ciliated ependyma, the last site being the ganglionic eminence at 26–28 weeks' gestation.[122] Mature ependymal cells are capable of functioning as regulators of the transport of fluid, ions and small molecules between cerebrospinal fluid and brain parenchyma.[121] Some animal experiments suggest that focal populations of ependymal cells might be capable of acting as neural progenitors.[122]

Despite evidence for a barrier function, in human brain by 32 weeks the ependymal lining has areas of discontinuity, especially in the occipital horns. These changes must be distinguished from pathological responses to injury, as seen in hydrocephalus, hypoxic-ischaemic injury and ventriculitis due to infection. In the latter cases, the ependyma reacts to injury with stereotypical patterns of response, including denudation, subependymal gliosis, macrophage infiltrates and buried ependymal tubules.[470] Periventricular heterotopia, characterized by nodules of neurons ectopically located along the lateral ventricle walls, may have a genetic basis (e.g. as a result of mutations in the X-linked filamin A (*FLNA*) gene and the vesicle transport adenosine diphosphate (ADP)-ribosylation factor guanine exchange factor 2 (*ARFGEF2*) gene,[333] or they may arise secondary to disruption of the immature ependymal layer during fetal stage of life.[118]

Brain Vasculature

Given the importance of vascular lesions in the perinatal spectrum of neuropathology,[516] it is necessary to consider the development of the vascular system. At the macroscopic level, this was described long ago (Figure 3.13).[342,414,504] The formation of penetrating vascular trunks is terminated before global brain growth reaches a plateau; subsequent development of an appropriate capillary network is critical to normal brain function and development.[38] Experimental evidence suggests that the vascular compartment might influence neural progenitor cells during telencephalic neurogenesis, neuronal migration and neurite extension.[505] Interaction between glial progenitors and early blood vessels might also contribute to glial fate.[599] Emerging evidence suggests that microglia might play a role in development of blood vessels by regulating sprouting thorough a signalling mechanism distinct from VEGF dependent growth.[24] Interactions between astrocytes and endothelial cells are also critical in the development of the blood–brain barrier.[326] Vascular endothelial growth factor (VEGF) is undoubtedly important and the Wnt signaling pathway seems to play a role in vasculogenesis, however the signals involved in blood–brain barrier induction and maintenance remain unclear.[520] In conjunction with endothelia and astrocytes, pericytes (another cell of mesodermal origin that lies within the endothelial basement membrane) help to regulate the blood–brain barrier (BBB) and local blood flow.[138] Based upon the development of tight junction proteins and other observations, it is likely that the permeability of the BBB to macromolecules matures by 18–20 weeks' gestation.[1,36,559] Permeability to small molecules is likely greater than in adults, but it is not clear whether maturity is reached *in utero* or at full-term gestation.[152]

HYPOXIC-ISCHAEMIC INJURY TO THE DEVELOPING BRAIN

Hypoxic-ischaemic brain injury is the major neuropathological disorder of the perinatal period. Two mechanisms are operative, although they almost always occur in some combination that is not clinically separable: hypoxaemia, defined

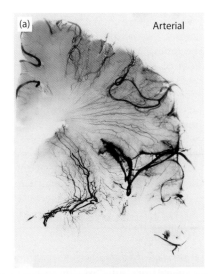

(a) Arterial (b) Venous

3.13 Demonstration of vasculature in the infant forebrain. A barium–gelatine mixture was injected into the major blood vessels in an unfixed autopsy brain. After fixation, the brain slices were X-rayed. **(a)** The arterial system is shown by injection into the intracranial internal carotid artery. Major arteries on the brain surface penetrate the brain and gradually taper as they advance toward the periventricular region. **(b)** The deep venous system is shown by injection into an internal cerebral vein. Deep veins converge on the periventricular region. Note the absence of filling near the external surface of the cerebrum, which is drained by a distinct venous system.

as a diminished amount of oxygen in the blood supply, and ischaemia, a diminished amount of blood perfusing the tissue. Hypoxaemia may arise *in utero* as a consequence of impaired placental function (Box 3.2), or intrapartum during prolonged or difficult labour. Whether pre-, intra- or post-partum, hypoxaemia and poor perfusion result in acidosis, and are referred to clinically as asphyxia. This condition is followed by a series of metabolic and biochemical reactions that can cause injury to brain and other organs, hypotension, hypoglycaemia and coagulopathy. Each of these insults has potential deleterious effects on the developing CNS and, for each of these, the gestational age of the infant and the duration of insult influence the damage inflicted upon the brain.

The neuropathology of hypoxic-ischaemic damage is complex, reflecting the many structural and functional variables that influence the reaction to injury of the developing human CNS. The transition at birth from a relatively hypoxic intrauterine environment to a normoxic or a relatively hyperoxic one in the circumstance of resuscitation may also contribute to a transient increase in free radical species in the brain in the immediate postnatal period (see section on antioxidants and free radical injury).[255]

Hypoxic-Ischaemic Brain Cell Injury

Most of the energy requirements of the brain are derived from the oxidation of glucose. Neither oxygen nor glucose is stored in the brain and astrocytic glycogen stores are minimal.[70] Deprivation, even for only a few minutes, will cause neuronal energy depletion and cell death. The pathophysiology of hypoxic-ischaemic damage is covered elsewhere in this book (Chapter 2, Vascular Disease, Hypoxia and Related Conditions). An abbreviated outline of the possible sequence of events leading to neuronal damage in the perinatal brain is provided here.

Briefly, hypoxia-ischaemia initially depresses adenosine triphosphate (ATP) synthesis, with failure of ion pumps and irreversible membrane failure, marked by rapid efflux of potassium ion and influx of sodium, calcium and chloride ions, along with water (cytotoxic or cellular oedema). Also affected are the energy-dependent glutamate uptake mechanisms in presynaptic nerve terminals and astrocytes, resulting in extracellular glutamate accumulation, and prolongation of the stimulation of the N-methyl-D-aspartate (NMDA), kainate and alpha-amino-3-hydroxy-5-methyl-4-isoxazole (AMPA) receptors (see later for details of excitotoxicity in hypoxic-ischaemic brain injury). Additional calcium and sodium ions then enter the neuron through open NMDA receptor channels, and sodium enters through open AMPA and kainate receptor channels. Further disruption of cellular organelles and function ensues when proteases, calpains, endonucleases, phospholipases and other catabolic enzymes are released (see Chapter 2, Vascular Disease, Hypoxia and Related Conditions). These changes in concert cause rapid cell death, which, when occurring in large fields, is visible as tissue infarction.

Vascular Zones and Failure of Autoregulation

Arteries penetrate the brain surface and have long, straight channels toward the white matter (ventriculopetal)

(Figure 3.13). Some anatomical studies reported the presence of ventriculofugal arteries (i.e. arteries that reflexively flow from the ventricle surface toward the white matter, arteries that pass from the choroid plexus into white matter and peristriatal arteries that pass dorsally. into white matter).[115,512,546] The presence of deep interarterial border zones was postulated to explain the distribution of PVL occurring in deep white matter.[128,461,512] However, other anatomical studies refute the existence of such arteries.[303,384,389,393] A careful review of the original studies by the present authors leads us to the conclusion that the existence of ventriculofugal arteries of significant length is unproven. The anastomotic capillary network differs between grey and white matter with respect to both density and pattern.[23] Venous channels arising in the deep white matter are directed both toward the brain surface and to the ventricle wall.

Cerebral blood flow is dependent on the balance between vasodilators and constrictors derived from the endothelium, neuronal innervations and perfusion pressure.[9] Autoregulation is the ability of brain blood vessels to maintain a constant cerebral blood flow in spite of fluctuations in the cerebral perfusion pressure, which depends upon the systemic blood pressure. Autoregulation is maintained by changes in the resistance of cerebral arterioles, the regulation of which depends upon constriction of smooth muscle in vascular walls, oxygen, carbon dioxide, pH, the sympathetic and cholinergic nervous system, prostaglandins, arginine vasopressin and vasointestinal polypeptide (VIP).[318] At the local level, glial neuronal interactions are also important.[25] A major contribution to perinatal brain injury is the propensity for the sick or premature neonate to exhibit a pressure-passive circulation, reflecting a disturbance of cerebral autoregulation.[193] In the preterm infant, vascular congestion and haemorrhage may occur with modest hypertension and there is increased vulnerability to ischaemic brain injury with modest hypotension.[318] Moreover, the mechanism of autoregulation is itself sensitive to hypoxaemia and hypercarbia in the preterm infant, and so persistent or severe asphyxia causes loss of autoregulation and the cerebral blood flow becomes passive to the changes in blood pressure.[318] Thus, loss of cerebrovascular autoregulation occurs in asphyxiated infants and is a major risk factor for perinatal hypoxic-ischaemic injury.[404,495] In an investigation of 88 infants born at <32 gestational weeks using near-infrared spectroscopic methods of cerebral intravascular oxygenation in tandem with mean arterial blood pressure determinations, O'Leary *et al* found a statistically significant association of impaired cerebrovascular autoregulation among the 37 per cent who developed severe germinal matrix haemorrhage with IVH.[404]

Antioxidant Systems and Free Radical Toxicity

Radicals (or free radicals) are atoms, molecules or ions with unpaired electrons, which cause radicals to be highly chemically reactive. Hypoxic-ischaemic injury generates free radicals in fetal and neonatal tissues during and after the insult, and these in turn perpetuate more cellular injury and opening of the blood–brain barrier.[439] Some investigators consider free radical injury to be the 'final common pathway' in brain injury, as it can result from hypoxia-ischaemia, excitotoxicity and cytokine toxicity.[568] Free radicals result in lipid membrane peroxidation and disruption, DNA oxidation or protein nitration and cell damage. Normally, more than 80 per cent of the oxygen consumed by the cell is reduced by cytochrome oxidase.[125,175] The remaining 10–20 per cent follows other oxidation-reduction reactions in the cytoplasm and mitochondria that produce the superoxide anion (O_2^-).[125,500] In the brain, the neurotransmitter and cell signalling molecule nitric oxide (NO, produced by nitric oxide synthase-containing microglia and some neurons) reacts avidly with superoxide to generate peroxynitrite ($ONOO^-$), itself a highly reactive species. Interaction of peroxide radical with free ferrous iron (Fe^{++}; present in sites of haemorrhage, as well as in oligodendrocytes about to undergo myelination[89,102] generates the potent hydroxyl radical (OH) *via* the Fenton reaction. Additional sources of free radicals include increased intracellular calcium and mitochondrial injury, activation of proteases leading to conversion of xanthine dehydrogenase to xanthine oxidase, and activation of phospholipase A2 leading to increased generation of oxygen free radicals from cyclooxygenase and lipoxygenase pathways. Free radical injury is the critical component of reperfusion injury: when oxygen is reintroduced into hypoxic-ischaemic tissues, a massive production of oxygen free radicals results, producing reperfusion injury over and above the damage already produced during the hypoxia. The primary antioxidant enzymes required by most cell types to inactivate free radicals are copper- and zinc-containing superoxide dismutase (SOD) (CuZnSOD) and manganese-containing SOD (MnSOD), catalase and glutathione peroxidase. CuZnSOD and MnSOD reduce superoxide anion radical to hydrogen peroxide (H_2O_2), which in turn is reduced to H_2O by catalase and peroxidases. Several clinical and experimental studies have examined the possibility of reducing neurological morbidity in perinatal hypoxic-ischaemic brain injury by two possible means: (i) preventing the formation of free radicals, by inhibition of xanthine oxidase, for example; and (ii) delivering antioxidants or free radical scavengers to sites of increased free radical production.[255,490]

The developing grey matter may be particularly susceptible to free radical injury in hypoxia-ischaemia because of a relative deficiency in the brain's antioxidant enzyme systems.[391] Based upon immunohistochemical detection of the proteins, CuZnSOD and manganese-containing MnSOD appear in the brain as early as 13 gestational weeks, becoming strongly positive after 23 gestational weeks through 2 years of postnatal age.[517] The premature infant also has low plasma levels of glutathione and a relative inability to sequester iron because of low transferrin levels.[491]

Immature oligodendroglia are particularly vulnerable to free radical toxicity. Among the mechanisms underlying this vulnerability is a rise in free radicals in excess of their antioxidant enzyme capacity.[27] The vulnerability of oligodendrocyte precursors to oxidative stress may be due in part to a lack of, or imbalance in the expression of, antioxidant enzymes early in development, before active myelin synthesis. In stage-specific rat oligodendrocyte cultures, O4+/O1+/MBP– negative cells die of free radical toxicity when raised in a medium depleted of cystine (leading to depletion of glutathione and accumulation of reactive oxygen species). In contrast, MBP-positive (mature) oligodendrocytes, as well as astrocytes, survive in such an environment.[28,30] Mature (MBP-positive) cells express higher levels of MnSOD, and introduction of this enzyme *via* an adenoviral vector into pre-myelinating oligodendrocytes confers protection of the mitochondrial membrane potential and thereby cell death from glutathione depletion.[43] Manganese-containing SOD further protects cells from damage by reactive nitrogen species, such as peroxynitrite,[200] which may also play a significant role in white matter damage in PVL.[228]

Catalase immunoreactive glial cells are not present in cerebral white matter from mid-gestation until 31–32 gestational weeks, at which time they are first visualized in the deep white matter. By term, all regions of the cerebral white matter (deep, intermediate and superficial) contain catalase-positive glia.[243] CuZn SOD immunoreactive glial cells appear in cerebellar white matter at 25–26 weeks of gestation and in temporal white matter at 31–32 weeks of gestation.[515] An analysis of antioxidant enzyme expression across human cerebral white matter development from 18 to 204 postmenstrual weeks, a developmental lag in the expression of both SODs compared with glutathione peroxidase and catalase was seen (Figure 3.14).[170] This lag suggests a dyssynchrony in tissue capability to sequentially break down superoxides. All antioxidant enzymes have higher-than-adult levels of expression during the peak period of postnatal myelin sheath synthesis in the cerebral white matter (i.e. 2–5 months of age), suggesting a need for maximal antioxidant capacity during high myelin lipid production.

Supporting a role for iron in triggering free-radical-mediated oligodendrocyte injury, cystine deprivation-induced death of oligodendrocyte precursors is prevented by pretreatment with the iron chelator desferrioxamine.[589]

Glutamate Receptors and Excitotoxicity

The excitatory transmitter glutamate may be released from injured neurons or axons, or by reversal of astrocyte glutamate transporters in the setting of hypoxia-ischaemia, to result in an excess of extracellular glutamate, which is toxic to many cell types, in a fashion that is developmentally regulated. Enhanced sensitivity of the immature brain to glutamate-induced toxicity reflects increased receptor density, altered receptor sensitivity (as a result of age-related differences in the molecular constitution of glutamate receptors) and/or differences in modulatory or compensatory mechanisms.[40] Regardless of the cell type, the mechanism is similar, involving calcium influx and subsequent generation of reactive oxygen and nitrogen species, participating in the 'final common pathway' of cellular damage discussed in the preceding section.

A developmental vulnerability to hypoxia-ischaemia in different grey matter regions of the brain is linked by animal data to age-related, transient elevations in glutamate receptors. The neurotoxicity of NMDA in the hippocampus, striatum and neocortex is maximal in the immature rat brain, peaking at postnatal days 6–7.[40] Autoradiographical receptor binding studies in rats indicate that in certain forebrain regions, the immature brain has a higher density of both NMDA and non-NMDA receptors compared with the adult brain.[261,369] Ligand affinities at NMDA receptors depend upon subunit composition, and there are marked developmental changes in the three major subunits – NMDAR1 (NR1), NMDAR2 (NR2A-B) and NMDAR3 – in experimental animals.[312] Non-NMDA glutamate receptors, e.g. AMPA receptor subunits, likewise demonstrate developmental regulation.[518]

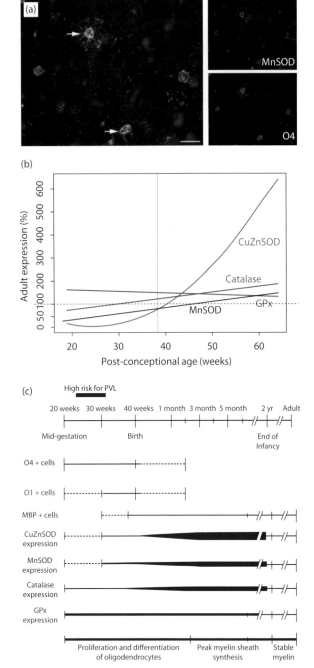

3.14 Dyssynchrony of antioxidant enzyme expression in the developing brain. (a) Expression of manganese superoxide dismutase (MnSOD, green immunofluorescence) by immature oligodendrocytes (O4, red immunofluorescence), demonstrated in the merged image. **(b)** Expression of the superoxide dismutases (MnSOD and CuZnSOD) reaches adult levels after birth. They lag behind the expression of catalase and glutathione peroxidase (GPx), which attain adult levels (or above) before term birth. **(c)** Developmental sequence of oligodendrocyte lineage cells relative to the expression of the major antioxidant enzymes in the developing cerebral (parietal) white matter of the human brain. Bar shows the peak period of vulnerability for periventricular leukomalacia (PVL). MBP, myelin basic protein.

From Folkerth et al.[170] With permission from Lippincott Williams & Wilkins/Wolters Kluwer Health.

In rat cerebral cortical neurons, the subunit GluR2 is relatively deficient, thereby conferring calcium permeability and greater vulnerability to excitotoxicity, in the early postnatal period. This period is a time of known susceptibility to hypoxia-ischaemia and subsequent development of seizures.

In cell culture, it has been shown that oligodendrocytes are vulnerable to micromolar concentrations of glutamate.[401] Moreover, the peak susceptibility to glutamate-induced cell death is in the developmental interval in which immature oligodendrocytes (O1+, MBP–) predominate.[401] The mechanism of glutamate-induced oligodendrocyte cell death involves both non-receptor- and receptor-mediated mechanisms. The receptor-independent mechanism involves glutamate transport into cells *via* glutamate/cystine exchange, resulting in depletion of intracellular cystine and, in turn, of glutathione, a key scavenger of oxygen free radicals, and finally death by intracellular oxygen free radicals.[401,589] Glutamate-induced toxicity to oligodendrocytes is also receptor-mediated.[592] In an oxygen-glucose deprivation system (an *in vitro* model for hypoxia-ischaemia), immature oligodendrocytes are more vulnerable than mature oligodendrocytes to excitotoxicity, an effect mediated by calcium-permeable AMPA receptors and blocked by AMPA antagonists.[126] The likely basis for this susceptibility is a preponderance of GluR2-deficient, and hence calcium-permeable, AMPA receptors in pre-myelinating oligodendrocytes.[518] The lack of GluR2 subunit expression on pre-myelinating oligodendrocytes has been seen in cerebral white matter of the developing human brain during the peak time frame of PVL.[518] In the setting of energy failure, glutamate transporters (Figure 3.15), which are physiologically responsible for clearing extracellular glutamate, operate in reverse and thereby release glutamate into the surrounding tissue, thus contributing to excessive levels of extracellular glutamate.

Immature oligodendroglia are also vulnerable to NMDA as well as non-NMDA receptor-mediated excitotoxicity.[274,368,468] In adult rats, NR1, NR2 and NR3 subunits are detected in myelin, indicating that all necessary subunits are present for the formation of functional NMDA receptors.[368] In perinatal mice, NMDA receptor subunits are expressed on the processes of oligodendrocytes, and NMDA receptor subunit mRNA is present in isolated white matter.[468] In a murine model of perinatal cerebral ischaemia, NMDA receptor activation results in rapid calcium-dependent detachment and disintegration of oligodendrocyte processes. In perinatal models, NMDA receptor subunits are expressed on oligodendrocyte processes, in contrast to the expression of AMPA receptors on somata.[468] Consistent with this observation, injury to oligodendrocyte somata is prevented by blocking AMPA/kainate receptors, whereas injury to processes is prevented by blocking NMDA receptors.[468] The functional significance of NMDA and non-NMDA receptor expression in pre-myelinating oligodendrocytes is unknown but may involve axon-glial signalling during myelination.[468] Calcium influx through activated NMDA receptors, for example, potentially affects cytoskeletal elements within oligodendrocyte processes and determines stabilization or retraction of the processes as they extend from the somata and make axonal contact.[468] Calcium influx has been shown to activate calpain, which can degrade cytoskeletal proteins and activate Bax, a mediator of apoptosis.[469] Details of this topic are reviewed by Volpe.[568]

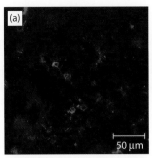 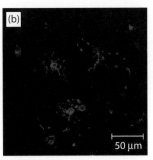 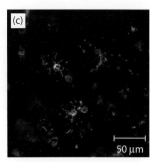

3.15 Glutamate transporter expression in the developing brain. Expression of the high affinity glial glutamate transporter GLT1 (also known as solute carrier family SLC1A2, or excitatory amino acid transporter 2; EAAT2) in immature oligodendrocytes with double label immunofluorescence and antibodies to GLT1 **(a,** green) and O4 **(b,** red) in cerebral white matter from a human fetus at 31 gestational weeks. **(c)** Merged image. SLC1A2 is closely related to glutamate/aspartate transporter high affinity sodium-dependent GLAST1 (SLC1A3), as well as EAAC1 (SLC1A1), which is typically thought of as a neuronal transporter. Glutamate transporters on immature oligodendrocytes may serve a critical role in maintaining glutamate homeostasis at a time when unmyelinated callosal axons are engaging in glutamatergic signalling with glial progenitors.

From DeSilva et al.[129] *With permission from the Society for Neuroscience.*

In human perinatal brain, studies of glutamate receptor subtypes have included a comprehensive cellular localization of AMPA receptor subtypes in the human cerebral cortex and telencephalic white matter.[518] As in the rat, a relative deficiency of GluR2 receptor subunit expression in the human cerebral cortex coincides with the interval of susceptibility (late gestation, term and early neonatal life) to hypoxia-ischaemia and seizures. The developmental profile of glutamate receptor subtypes has been mapped in the human brain stem using tissue autoradiography from midgestation to early infancy, and compared with the adult as the index of maturity.[418] There is almost no NMDA receptor binding in the human fetal brain stem at mid-gestation, suggesting that the vulnerability of the fetal brain stem to hypoxia-ischaemia is due to the high concentrations of kainate/AMPA receptors, and not NMDA receptors, at that age. NMDA receptors appear around birth and thereafter in infancy, whereas AMPA receptor binding declines sharply to significantly low levels thereafter. The finding that NMDA receptor/channel binding is almost undetectable in all regions of the human fetal brain stem at mid-gestation is unexpected, given the trophic role for NMDA in early CNS maturation in experimental animals. The brain stem data also suggest a differential development of components of the NMDA receptor/channel complex across early development.[419] Kainate binding is transiently elevated in the fetal and/or infant periods in the basis pontis, the inferior olive, the reticular core and the inferior colliculus, regions all thought to be particularly vulnerable to perinatal but not adult hypoxia-ischaemia.

Cytokine Toxicity

Cytokines are a heterogeneous group of polypeptide mediators that activate the immune response and inflammatory responses.[241] In the CNS, microglia induce reactive astrocytosis *via* release of TNF, IL-1, IL-6 and interferon-γ (IFN-γ).[394,479] Of relevance to perinatal white matter injury, these cytokines are soluble and diffusible, suggesting that they play a role, at least in part, in triggering the diffuse reactive gliosis adjacent to focal PVL, the immediate site of inflammation. Microglia are potently stimulated by endotoxin to produce

IL-1β which secondarily stimulates astrocytic expression of both TNF-α and IL-6.[315] *In vitro* studies indicate that TNF-α and IFN-γ have toxic effects upon mature oligodendrocytes.[332,366,458,478,553] TNF exposure results in a cascade leading to oligodendrocyte cell death by apoptosis,[362] but only IFN-γ is directly toxic to immature oligodendrocytes;[553] its expression correlates with free radical adduct formation in these cells.[171] The potentiation of IFN-γ-mediated injury by TNF-α in developing oligodendrocytes may indicate an important role for the latter in human PVL.[7] Furthermore, TNF-α has been identified in hypertrophic astrocytes and microglial cells in PVL.[591] Other cytokines, including IL-2,[272] and IL-6[591] have also been identified in human PVL. It should be emphasized, however, that cytokines may be recruited by ischaemia or other insults, as well as infection.

A causal role for infection and cytokine toxicity in perinatal brain injury (PVL being the most extensively analysed) is indicated by several classic studies in developmental neuropathology: (i) neonates with bacteraemia[323] and neonates born to mothers with chorioamnionitis or premature rupture of membranes[430] are at risk for PVL; (ii) PVL occurs in several animal models of endotoxin-induced injury, with and without systemic hypotension;[19,191,192,594] (iii) cerebral white matter lesions occur in fetal rabbits after the induction of maternal intrauterine infection;[591] (iv) elevation in umbilical cord blood of the pro-inflammatory cytokine interleukin 6 (IL-6) is associated with an increased risk of PVL;[590] and (v) activated microglia are increased in number in PVL[228] and appear to play a central role in the damage related to these mediators.[568] Both the maternal and the fetal inflammatory responses may contribute to the presence of inflammatory mediators in the fetal or infant brain. The transfer to brain of pathogen-associated molecular patterns (PAMPs) in circulating blood, or the transfer of pathogen-activated immune cells into the brain may occur, perhaps facilitated by disruption of the blood–brain barrier by systemic cytokines, or the stimulatory effects of circulating cytokines on brain endothelial cells.[109,195]

As to the specific role of microglia in PVL, upregulation of toll-like receptors (TLRs) on activated microglia indicates a recently recognized capability for innate immunity.[319] These receptors are activated by PAMPs, most significantly those present on Gram-positive bacteria (TLR2) and Gram-negatives (TLR4). Binding to these receptors initiates the downstream release of reactive oxygen and nitrogen species as well as cytokines and glutamate, directly damaging to developing oligodendrocytes and their process, along with axons and neuronal cells (see earlier sections).

Modes of Brain Cell Death

In a review of the literature concerning mechanisms of cell death one writer concluded, 'there is no field of basic cell biology and cell pathology that is more confusing and more unintelligible than the area of apoptosis versus necrosis'.[160] Contributing to the problem is a plethora of overlapping terminology related to the morphology and the molecular mechanisms and a desire to simplify the issue in pursuit of therapeutic interventions. Compounding the problem in the immature brain is, first, that there is, to some extent, normal physiological cell death and, second, that brain cells (perhaps most obviously neurons) manifest different modes of cell death depending on their stage of maturation. Majno and Joris summarized the issues clearly; they recommended the concept of accidental cell death, which can lead to tissue necrosis through a variety of cellular mechanisms including cellular apoptosis.[338] In his landmark review concerning ischaemic cell death in neurons, Lipton echoed the weakness of the term 'necrosis' and highlighted the fact that neuron death mechanisms could coexist.[330] Conceptually he identified two phases of initiators and activators (loss of energy, membrane depolarization, excitotoxicity, free radical formation, calcium ion influx), followed by structural changes caused by a variety of perpetrators (calpain-induced proteolysis, phospholipase-induced membrane changes, mitochondrial dysfunction, caspase activation, lysosomal enzyme release including cathepsins and granzymes) and ultimately cell death end stages. He identified five different morphologic appearances of dying and dead neurons including oedematous (swollen pale-staining cells), ischaemic (the classic hypereosinophilic neuron), homogenized ('ghost cells' with nuclear karyolysis appearing as a speckling of basophilic particles), apoptotic (shrunken cells with densely condensed or karyorrhectic nuclear material) and autophagocytic (cytoplasmic vacuoles and clumped chromatin). In recent years, the molecular pathways through which the various forms of brain cell death have become understood in greater detail and it is apparent that they are not mutually exclusive in individual cells.[255,320,541,593] From a practical diagnostic standpoint, it is critical to appreciate that these mechanisms, and the associated morphologic appearance, may coexist (Figure 3.16).[440] In the immature brain, the apoptotic appearance is likely to predominate, with a shift to the hypereosinophilic dying neuron predominately at around full term age. It should also be noted that dead neurons are more likely to be replaced by fossilized remains of calcium (and other substances) in the immature brain than in the mature brain.

GREY MATTER LESIONS

The distribution of dying neurons in the immature brain is dependent on the age and associated pathological features. Some of the patterns of injury, such as thalamic-brain stem injury, pontosubicular 'necrosis' and status marmoratus,

are seen only in the perinatal period and not in the mature brain. Frequently, grey matter lesions will accompany, and may even be overshadowed by, the lesions of periventricular leukomalacia (see later), in a pattern termed 'encephalopathy of prematurity'.[567] The reasons for selective and shifting vulnerability of grey matter are multifactorial. Circulatory factors play a role, as evidenced by selective neuronal death in vascular border zones. There are other factors: differences in regional metabolism that predispose to hypoxic cell death have been identified in experimental systems, e.g. differences in anaerobic glycolytic capacity, energy requirements, lactate accumulations, calcium influx and free radical formation and scavenging capacity, regional distribution of glutamate receptors and the regional accumulation of excitotoxic amino acids.[267,581]

Cerebral Cortex

Cortical neuron death may be focal, bilateral, symmetrical or diffuse. With severe injury, the entire cortex may be involved. Necrosis may be limited to interarterial border-zone regions; here, lesions are thought to result from hypotension and consequent decreased perfusion of tissues served by end branches of major blood vessels. In term newborn and infant brains with cortical lesions, the only gross abnormality may be cerebral swelling, manifested by diffuse widening and flattening of cerebral gyri and partial obliteration of sulci. As the calvarial sutures are not fused in the infant, cerebral oedema is rarely severe enough to cause mesial temporal lobe herniation, although the ventricles may be compressed. Prior to cerebral myelination, cortical pallor may be accentuated against the congested, red-brown white matter. Total cerebral cortical necrosis results in severe brain softening, with fragmentation of the tissue upon removal at autopsy.

The microscopic features of acute cortical necrosis vary with the severity of damage (Figure 3.17). In the early stages, in preparations stained with haematoxylin and eosin or cresyl violet, there are blotchy or sinuous zones of tissue pallor or spongiosis that have well-delineated edges. The appearance of the dying neurons depends on the degree of maturation of the infant. Prior to approximately full-term gestational age, apoptotic forms with karyorrhectic nuclei predominate with a transition to pyknotic/hypereosinophilic neurons thereafter. During the transition phase the two morphologies may coexist. Studies in the human have noted an apoptotic index of less than 1 per cent in the cortical plate at 27 weeks,[445] versus a higher proportion of apoptotic cells (up to 8.3 per cent) in the cingulate cortex in term neonatal deaths with clinical manifestations of hypoxic-ischaemic encephalopathy and near-term stillbirths,[147] suggesting a dynamic process across development or a relationship to perinatal injury. In relatively mild insults, selective neuronal death may be associated predominantly with clusters of activated microglia. Following more severe insults that result in cortical necrosis, macrophages phagocytose debris and endothelial cells become hypertrophic and hyperplastic. Neurons in the area of necrosis or at the edge may retain their shape, become encrusted with granular periodic acid-Schiff (PAS)-positive material and occasionally stain for calcium or iron salts (ferrugination). The capillary bed within or adjacent to the areas of necrosis may also be ferruginized.

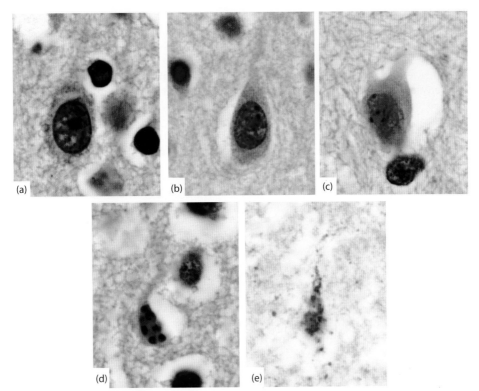

3.16 Neuron degeneration. Photomicrographs of neurons from immature brain exhibiting various types of degeneration and death features (all photographed 1000× oil immersion; haematoxylin and eosin stain). **(a)** Normal cortical layer 3 neuron in brain of an infant who died intrapartum at 40 weeks gestational age. The nucleolus is prominent and basophilic stippling in the cytoplasm (rough endoplasmic reticulum/Nissl substance) is apparent. **(b)** Layer 5 neuron, in same case, showing early eosinophilic change in the cytoplasm, loss of basophilia from the nucleolus and fragmentation of the chromatin. **(c)** Cortical neuron in the same case showing more advanced eosinophilia and homogenization of the cytoplasm, eosinophilia of the nucleolus and fine chromatin debris. The nuclear contour is retained. **(d)** Hippocampal neuron (CA1 sector) in the same case showing apoptotic features with clumping of the nuclear material and cytoplasmic eosinophilia. Note in this single brain that different modes of neuron death are seen in different regions. Eosinophilic change with pyknosis predominates in deep layers of the cerebral cortex, putamen, CA2 sector of the hippocampus, superior colliculi, dentate nucleus of the cerebellum, Purkinje neurons, tegmentum of the medulla oblongata and the inferior olivary nucleus. Apoptotic forms with nuclear karyorrhexis predominate in the superficial layers of the cerebral cortex, CA1 sector and dentate gyrus of the hippocampal formation, and the internal granular layer of the cerebellum. A mixed pattern is apparent in the basis pontis. The apoptotic mode of neuron death is more widespread in brains younger than ~35 weeks gestational age and is seldom seen following full term birth. **(e)** Dead cortical layer 5 neuron in brain of an infant who died *in utero* at 31 weeks gestational age and was delivered shortly thereafter with only mild autolytic features. This neuron exhibits features of necrosis with irregular fragments of the nucleus spread throughout the cell. Cell boundaries are not obvious. Note that a mix of necrotic and apoptotic forms can be seen in this circumstance.

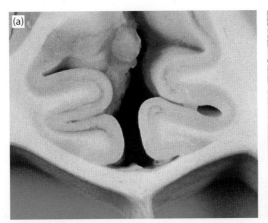
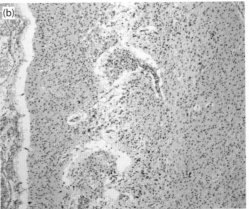

3.17 Laminar necrosis of the cerebral cortex. (a) Cingulate and superior cingulate gyri on the medial surface, from a 2-year-old child who had sustained severe hypoxic brain damage in the neonatal period of life. Yellow bands in the middle of the cortical layer represent a band of prior necrosis. **(b)** Photomicrograph of the frontal cortex from a 1-day-old infant who had placental insufficiency and was born at 38 weeks gestational age. The mid-cortex is hypocellular as a result of complete necrosis and the adjacent superficial and deeper layers are hypercellular as a result of accumulated reactive microglia and reactive astrocytes. There are very few neurons. 100× magnification; haematoxylin and eosin.

In the chronic phase, brain weight may be reduced. Cerebral atrophy results in a discrepancy between the relative sizes of the cerebrum and cerebellum. Severely damaged cerebral gyri are narrowed, sclerotic, cystic and paler than adjacent intact cortex. Associated white matter is also reduced in volume, sclerotic or cystic. Damaged gyri may have a mushroom-shaped appearance termed ulegyria if the deep portions are damaged and the crown is left intact as a consequence of regional blood flow differences. Sclerotic gyri may contain few or no neurons; the cortical ribbon is replaced by hypertrophic astrocytes and macrophages, which cluster in the neuropil or in the Virchow–Robin spaces. With survival, astrocytes become fibrillary, with small nuclei and a network of fibres with rare Rosenthal fibres. This chronic astroglial change may be demonstrated nicely with a modified phosphotungstic acid haematoxylin (PTAH) method.[344]

Hippocampus

The hippocampus of infants is susceptible to hypoxic-ischaemic injury, but this rarely occurs as an isolated lesion. Pathology of the infant hippocampus is difficult to interpret because it is immature at birth. Neurons of CA2, CA3 and the end plate have visually matured, but neurons of Sommer's sector (CA1), which is vulnerable to hypoxia-ischaemia, remain small and immature until 2 years of age.[176] Hippocampal neurons are relatively vulnerable from 22 weeks' gestation onward. Prior to term and up to 2 postnatal months the prevalent morphology of dying neurons is apoptotic with karyorrhectic nuclei. These changes, a component of so-called 'pontosubicular necrosis' (see later in the section on the brain stem and spinal cord), may be seen in the subiculum, the CA1 and CA3/4 sectors, and even the dentate gyrus. Later in infancy the interpretation of neuron death in Sommer's sector must be made cautiously, evaluating the distinction between immature and pyknotic neurons.[176] In mild insults, microglial aggregation along the inner margin of the dentate gyrus is a characteristic pathological finding in neonates and infants up to 8–9 months of age with known hypoxic or hypotensive episodes following perinatal asphyxia, congenital heart disease or chronic pulmonary dysfunction.[124] This change may also be present in stillborn fetuses. By standard microscopy, these infiltrating microglia appear as rod-shaped nuclei clustered or scattered diffusely in the polymorphous layer and adjacent granule layer of the dentate gyrus. Our experience is that they are strongly immunolabelled with anti-CD45 or anti-HAM56, but less robustly with anti-CD68 and anti-HLA-DR.

Basal Nuclei and Thalami

The neurons of the fetal and infant striatum (caudate and putamen), thalamus and the globus pallidus are frequently damaged by hypoxia-ischaemia. In association with PVL, neuron loss from the mediodorsal and reticular nuclei of the thalamus is common.[327] If there is considerable neuron loss, the striatum and thalamus (and less often the globus pallidus) may be atrophied and disorganized; mineralized neurons may be visible (Figure 3.18). Stained to highlight myelin, the white matter bundles are lost and replaced by a random marbled pattern of myelin tracts ('status

marmoratus'). Early authors suggested the disorganization was due to an insult prior to or during the phase of active myelination (i.e. perinatal to 6–9 months postnatal) that caused oligodendrocytes to inappropriately deposit myelin on structures that would not normally be myelinated. However, a careful analysis by electron microscopy revealed no inappropriate myelination and simply 'abnormal orientation of normally proportioned fibres scattering throughout the scars'.[177] Lesions in the basal ganglia occurring after the period of active myelination exhibit neuron loss along with reactive astroglial and microglial changes but not the profound structural disorganization.

Another lesion seen in the perinatal setting is mineralizing lenticulostriate vasculopathy, in which the walls of penetrating blood vessels in the putamen, globus pallidus, thalamus and caudate develop deposits of calcium and iron salts (Figure 3.19). The pathogenesis of this regional distribution is not entirely clear but may be related to the normal regional distribution of trace minerals. The striatum

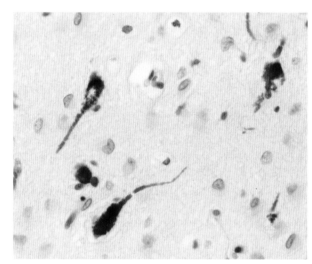

3.18 Mineralized neurons. Thalamus from a 9-week infant who sustained severe neonatal hypoxic-ischaemic brain damage. Many of the neurons have died and are replaced by mineralized 'fossils' that stain purple. 400× magnification; haematoxylin and eosin.

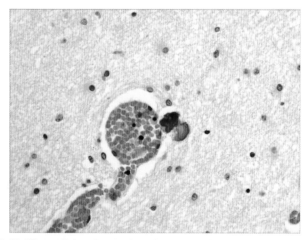

3.19 Mineralizing lenticulostriate vasculopathy. Focal mineral deposits along the wall of a small vein. Original magnification 400× haematoxylin and eosin.

and globus pallidus have the highest levels of the calcium binding proteins calbindin and calretinin parvalbumin,[377] along with average levels of free calcium and high levels of free iron and copper.[145,196] Such findings may be microscopic and focal, or widespread and visible as echogenic or 'bright' vessels on cranial ultrasound.[526] The typical settings in which this phenomenon is observed at autopsy are hypoxia-ischaemia with or without intraventricular haemorrhage (IVH), and infection (see later section, Infection, p. 250).[100,372] In isolation, the significance of this finding is unclear.

Cerebellum

Diffuse or isolated foci of cerebellar cortical necrosis most frequently coexist with hypoxic-ischaemic lesions elsewhere. Selective death of mature Purkinje cells does occur, usually in infants beyond 36 gestational weeks,[460] but it is uncommon. Selective granule cell death occurs infrequently in the postnatal period.[460] It also occurs *in utero*: the vulnerability of the cells may be due in part to their high metabolic rate.[460] The external granular layer exhibits a high proportion of proliferating cells until ~5 months postnatal,[3] Experimental studies indicate that hypoxia inhibits proliferative activity of the external granule cells and retards development of neuronal processes.[460,595] If destruction of the external granule cells is sufficiently severe, then cerebellar hypoplasia can result. Longstanding diffuse damage of the cerebellum consists grossly of atrophy; there is partial or complete loss of Purkinje cells, variable decrease in granule cells, tissue mineralization and proliferation of Bergmann glia. White matter may have a sclerotic appearance with gliosis and little or no myelin. The relationship to motor dysfunction is obvious and, given the putative role of the cerebellum in cognitive processing, damage may contribute to long-term intellectual impairment in the survivors of prematurity.

Brain Stem and Spinal Cord

Involvement of the brain stem is characteristic of hypoxic-ischaemic encephalopathy in the newborn and selective necrosis of brain stem nuclei is a recognized complication of asphyxia in experimental animals.[382,583] In the developing human brain, the structures injured most consistently include the inferior colliculus, nuclear populations in the tegmentum, basis pontis, inferior olivary nucleus and gracile and cuneate nuclei. Damage to the brain stem generally occurs in combination with lesions elsewhere. A particular pattern of symmetric neuron loss from the tegmental nuclei and 'reticular core' occurs in association with damage in the thalamus and basal ganglia of preterm and full-term infants subjected to hypoxia.[4,317,395,475] Vulnerability of the brain stem in the fetal period is attributed, at least in part, to the high density of kainate and AMPA receptors.[418,420]

Pontosubicular necrosis is characterized by dying neurons with nuclear karyorrhexis and cytoplasmic shrinking in the basis pontis and the subiculum (see earlier in Hippocampus, p. 234). Macrophages appear after 3–5 days and astrocytes proliferate without cavitation. Immunohistochemical demonstration of caspase-3 activation confirms that the mode of cell death is apoptosis.[463] This pattern of hypoxic-ischaemic injury may be apparent from 22 gestational weeks to 2 postnatal months and it typically coexists with tissue damage in other sites.

Spinal cord neurons may die following severe asphyxia or hypotension. Necrosis of the ventral horn is a typical component of the myeloencephalopathy that follows cardiorespiratory arrest and resuscitation.

WHITE MATTER LESIONS

Periventricular leukomalacia (PVL; 'white matter softening') is a developmental lesion of the cerebral white matter characterized by two key components: (i) focal or multifocal periventricular necrosis, which over time and if large enough may become cystic, and (ii) diffuse reactive astrogliosis and microglial activation of the surrounding white matter. The greatest period of risk for PVL is in the premature infant born at 24–32 weeks' gestational age (Figure 3.2); however, late preterm infants can develop an identical pattern of damage.[232] With improving intensive care of these at-risk infants, the incidence of cystic PVL is declining;[548] however, diffuse white matter injury remains a major finding at autopsy in premature infants.[435] PVL also occurs in full-term neonates, particularly those with congenital cardiac or pulmonary disease,[290] and may be found as an 'incidental' finding in previously healthy sudden infant death syndrome (SIDS) victims.[514] Two widely held (but not mutually exclusive) potential aetiologies for PVL are ischaemia-reperfusion and cytotoxic cytokines released during infection or ischaemia. In addition, a complex interplay of vascular factors predisposes to human periventricular white matter injury, including the presence of vascular end zones and a propensity for the sick premature infant to exhibit a pressure-passive circulation, reflecting a disturbance of cerebral autoregulation. As the window of vulnerability to perinatal cerebral white matter injury precedes active myelin synthesis in the cerebral hemispheres, specific characteristics of oligodendroglial precursors are postulated to play a role in white matter vulnerability.

The most commonly affected locations are anterior to the frontal horn, lateral corners of the lateral ventricles at the level of the foramen of Monro and lateral regions of the trigone and occipital horn, including the optic radiations. Macroscopically, the early changes may be quite subtle with the appearance of sliced brain normal or with only vague changes in the vascular pattern. The overlying cortical ribbon may appear pale and the white matter somewhat dusky. In more severe acute cases there are sharply circumscribed foci of softening, often with petechial haemorrhage or distended blood vessels, in the periventricular regions of the lateral ventricles (Figures 3.20–3.22).

Microscopically one may see reduced cellularity and a smudgy, hypereosinophilic appearance (as a result of pyknosis and loss of oligodendrocytes) in areas of very acute necrosis. Within the first 24 hours, one can identify tissue vacuolation (indicative of oedema), cellular eosinophilia with nuclear pyknosis and swollen axons. Necrosis affects all of the cellular elements including oligodendrocyte precursors, astrocytes, blood vessels and axons. Immunostaining may reveal clusters of damaged axons (e.g. with anti-amyloid precursor protein) and foci of activated microglia (e.g. with anti-HLA-DR). Within 48 hours, astrocytic proliferation and capillary hyperplasia develop, and within 3–7 days microglia proliferate and lipid-laden cells (macrophages) accumulate (Figure 3.21). In contrast to coagulative necrosis in systemic tissues, neutrophils do not enter the site of injury in PVL. Reactive astrocytes delineate PVL in its organizing

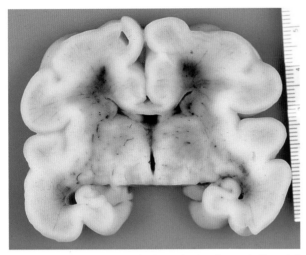

3.20 Acute periventricular leukomalacia. Coronal slice of the brain from a premature infant born at 29 weeks' gestational age and surviving 16 hours with severe hypoxia. The white matter superior to the frontal horns of the lateral ventricles has a dusky red discolouration due to hyperaemia and early tissue damage.

stages and the swollen axons may eventually mineralize, staining for calcium and/or iron. Cavitation occurs within a few weeks but small foci of necrosis may collapse into a solid glial scar. The healed lesions stain palely for myelin (e.g. with Luxol fast blue or immunocytologically with antibodies against MBP) because oligodendrocytes are absent.

In some instances, widespread necrosis of the cerebral white matter occurs.[316] The relationship of this widespread lesion to the more usual sharply circumscribed, small necrotic foci in the periventricular region is unclear but it likely represents the most severe end of a continuum of periventricular white matter damage. In these extreme cases, necrotic foci extend from the periventricular sites for a variable distance into the centrum semiovale and rarely as far as the subcortical white matter (Figure 3.21). In the chronic stage, the entire white matter becomes cavitated or atrophic with development of ventriculomegaly (hydrocephalus *ex vacuo*) and thinning of the corpus callosum.[128]

Diffuse white matter gliosis may occur on its own and invariably accompanies foci of necrosis. The typical appearance is of hypertrophic astrocytes (defined morphologically in H&E-stained sections by pale, vesicular nuclei and eosinophilic, irregular, hyaline cytoplasm forming processes) throughout the periventricular, central and intragyral white matter. Their cell bodies stain positively for GFAP; in practice, these may be difficult to distinguish from myelination glia but tend to be larger and have more star-like processes (Figure 3.9). Activated microglia, highlighted

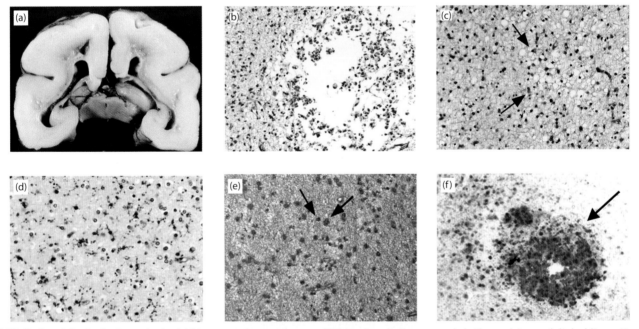

3.21 Periventricular leukomalacia. (a) Macroscopic appearance of bilateral cavitating necrosis in the parieto-occipital white matter in a 4-week-old infant delivered at 24 post-conceptual weeks. **(b)** Haematoxylin and eosin-stained section of organizing necrosis; note tissue loss, macrophage infiltrates and reactive astrocytes at the periphery. Original magnification 200×. **(c)** Diffuse white matter injury, with vacuolation and prominent astrocytes (arrows), surrounding focal necrosis. Haematoxylin and eosin stain; original magnification 200×. **(d)** CD68 immunostain, highlighting activated microglia in diffuse white matter injury. Original magnification 200×. **(e)** Immunostain for nitrotyrosine (marker of nitrative injury) in diffuse component, demonstrating nitration of larger glial cells with the morphology of astrocytes (arrowhead) and small glial cells most consistent with oligodendrocytes (arrow). Original magnification 200×. **(f)** Interferon-γ immunoreactivity in macrophages in a necrotic focus (arrow). Original magnification 400×.

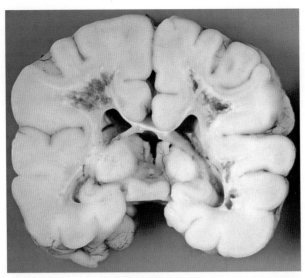

3.22 Subacute periventricular leukomalacia. Coronal slice of the brain from a premature infant who was born at 36 weeks' gestational age, suffered perinatal hypoxic insult and survived 2 weeks. There is bilateral, roughly symmetrical cavitation of the cerebral white matter superior and lateral to the bodies of the lateral ventricles.

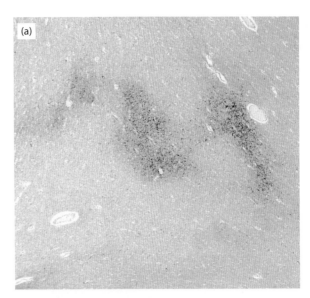

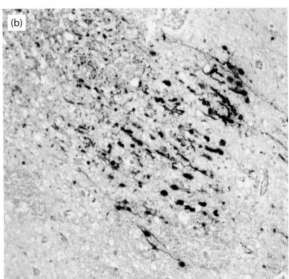

3.23 Axonal damage in white matter. Injured axons are detected by immunostaining for beta-amyloid precursor protein (beta-APP, brown), the anterograde transport of which is arrested at sites of axonal injury. **(a)** Sinuous clusters of damaged axons in parietal white matter of a 2-month infant who had been born at full term and died of pneumonia. Original magnification 40×. **(b)** Swollen axons in the parietal white matter of an infant who was born at 36 weeks' gestational age and survived 11 days after hypoxic insult. Original magnification 400×.

by CD68 or other markers, are prominent. Acutely damaged glia have pyknotic, dense nuclei; it can be difficult to determine if these are astrocytes or oligodendrocytes.[193] Immunohistochemistry with antibodies to pre-myelinating oligodendrocytes (O4) and markers of free radical damage (nitrotyrosine, malondialdehyde and hydroxynonenal adducts) demonstrates oligodendrocyte precursors that are modified by exposure to reactive nitrogen and oxygen species.[32,228] Some oligodendrocyte precursors are TUNEL-positive, suggesting that apoptosis occurs. PVL with both focal necrosis and diffuse gliosis, and diffuse white matter gliosis alone, may represent a continuum of ischaemic damage to cerebral white matter in the perinatal period. At the most severe end of the continuum is diffuse necrosis extending beyond the periventricular region into central and subcortical white matter, followed by combined focal necrosis and diffuse gliosis; at the least severe end is white matter gliosis alone. Nevertheless, the relationship of cerebral white matter gliosis to the diffuse component of PVL (i.e. whether they are the result of identical pathogenetic processes) is unknown.

Axonal Injury in Periventricular Leukomalacia

It is likely that axonal damage contributes to white matter atrophy or hypotrophy, along with the ensuing neurological handicaps. Axonal injury in PVL is identified histologically by swollen axons (spheroids) (Figure 3.23). This is highlighted by beta-amyloid precursor protein (beta-APP) immunoreactivity in spheroids around the edges of the necrotic foci of PVL.[22] In the peripheral and central nervous system, beta-APP undergoes fast axonal transport, which is ATP dependent. At sites of physical damage or metabolic injury (e.g. ischaemia), axonal transport stalls and beta-APP (along with other proteins) accumulates. In a study

of axonal injury in PVL involving 85 PVL patients and 30 non-PVL patients, ranging in age from 22 to 41 gestational weeks, white matter damage was divided into focal, widespread and diffuse categories. In this schema, 'diffuse' referred to necrosis extending from the periventricular to the subcortical areas. They found beta-APP immunopositivity in the axons around the foci of all types of necrosis in 76 per cent of PVL patients.[120] In extensive white matter injury with widespread cavitation, axons are obviously destroyed.[230] A recent autopsy study highlighted the axon damage in noncystic PVL and showed that Olig2 expressing cells were not depleted.[556]

Long-Term Consequences of Perinatal White Matter Damage

In a study utilizing diffusion tensor MR imaging, relative anisotropy was 25 per cent lower in the central white matter at term in the premature infants with white matter damage during gestation.[251] The investigators suggested that such disturbances potentially reflect degeneration of the axons or indicate impairment of axon growth or of myelin ensheathment.[251] By neuroimaging, myelinated white matter volume is reduced at term in premature infants with perinatal white matter injury.[257] In a study of seven very preterm infants with PVL at 44 gestational weeks, van de Bor et al. used an MR myelin grading score and found impaired myelination compared with age-matched controls.[545] In a study of premature infants with cystic PVL analysed in the neonatal period and between the ages of 9 and 16 months, De Vries et al. reported that MR imaging scans at 9 months or older showed dilated occipital horns and delayed myelination in contrast with age-matched controls, most marked in the occipito-thalamic radiation in five of six infants; on subsequent scanning in four infants, there was 'good progress in myelination'.[116] Dubowitz et al. reported the evolution of severe PVL by ultrasonography in the perinatal period and by MR imaging in the postnatal period in three surviving infants, in whom the development and subsequent resolution of cysts was followed by the appearance of ventricular enlargement and delayed myelination.[144] The delay was most noticeable in the optico-thalamic region, the site of the most extensive lesions observed on ultrasonography. Progress in myelination was observed in one infant in whom a repeat scan was performed, although the degree of myelination was still less than expected for age.

The aetiology of the failure of normal myelination is thought to be a combination of oligodendroglial cell loss, loss of oligodendrocyte processes and failure of differentiation of the oligodendrocyte progenitors (reviewed by Volpe).[568] In the modern era of neonatal intensive care, in which the picture of 'cystic' PVL has been replaced by diffuse high signal intensity changes on neuroimaging and microscopic versus macroscopic injury, recent work indicates that it is not the burden of tissue necrosis nor the accompanying axonopathy responsible for the observed myelination failure; rather, the presence of extensive gliosis and extracellular matrix remodelling, reflected in hyaluronic acid elevation, results in an expansion of the premyelinating oligodendrocyte pool and inhibition of progression through the normal stages of maturation.[74]

Pathologically, reports of survivors beyond the neonatal period with documented PVL have gross and/or histological evidence of deficient white matter volume and/or impaired myelination at autopsy. DeVries et al., for example, reported two subjects with PVL documented by ultrasound at 4 weeks and 10 days, respectively, who lived to 7 months and 3 years: at autopsy, the brains were remarkable for ventricular dilation, gliosis and a generalized lack of white matter; periventricular cystic cavities, seen on ultrasound in the perinatal period, had disappeared.[116] Rorke reported anatomical evidence of retarded myelination in telencephalic white matter in 19 autopsied high-risk infants, 15 (79 per cent) of whom had gliosis in

the white matter.[460] Iida et al. addressed the question of delayed myelination in PVL with an immunocytochemical study: antibodies to myelin basic protein (MBP, a marker of myelinated fibres) and to ferritin (present in oligodendrocytes) were used.[254] Semi-quantitative assessment did not reveal any difference in these markers between nine infants with only focal PVL (35 gestational weeks to 10 postnatal months at death) and 23 control infants without neuropathological lesions (23 gestational weeks to 12 postnatal months). In contrast, myelination was considered impaired in four of nine patients with widespread necrotic foci or marked gliosis (29 gestational weeks to 10 postnatal months at death): the degree of myelination was decreased relative to controls in the internal capsule and the deep, intermediate and subcortical white matter of the precentral gyrus in Luxol fast blue-stained sections more strongly than MBP-immunostained sections. Hypomyelination was not found by either method in the optic radiation or corpus callosum. In the focal PVL brains, the number and shape of ferritin-containing cells in non-necrotic areas were similar to those in controls; in contrast, in some of the brains with widespread foci of necrosis or diffuse gliosis, the number of ferritin-containing cells appeared decreased in number. This study suggests that myelination is not affected in focal PVL but is affected in at least some brains in which there is severe, diffuse damage.

In one study of 18 PVL cases and 18 perinatal brains without white matter damage, Billiards et al. found that the density of cells expressing pan-oligodendrocyte lineage marker Olig2 was not decreased in PVL, and there was no decrease in MBP immunostaining. However, there was a qualitative abnormality of MBP immunostaining in both the focal and diffuse components of PVL, such that MBP was concentrated around the nucleus, rather than being present in cell processes, suggesting damage to the processes or impaired axonal signalling for movement of MBP into the processes.[53]

With regard to damage to neuronal cells and their processes in PVL, Golgi preparations of the overlying cerebral cortex show that the subsequent differentiation of the cerebral cortex may be deprived of afferent terminals from corticopetal and association fibres.[348] Moreover, some of the neurons will fail to reach their targets because their axons (corticofugal fibres) have also been destroyed by the underlying white matter lesion. Consequently, the post-injury structural and functional maturation of the partially isolated grey matter will be altered. Some deep layer axotomized pyramidal neurons develop long horizontal collaterals that expand above the border of the necrotic zone (Figure 3.24). In this regard, it was proposed that the neurological sequelae (e.g. epilepsy, cerebral palsy; see Box 3.5) after perinatal white matter lesions are a direct consequence of post-injury neuronal process remodelling.[349] The potential for neuronal repair following PVL is suggested by the finding of increased densities of double-cortin-immunoreactive postmitotic migrating neurons in the subventricular zone, subcortical white matter and around the necrotic foci in PVL cases as compared to controls.[231]

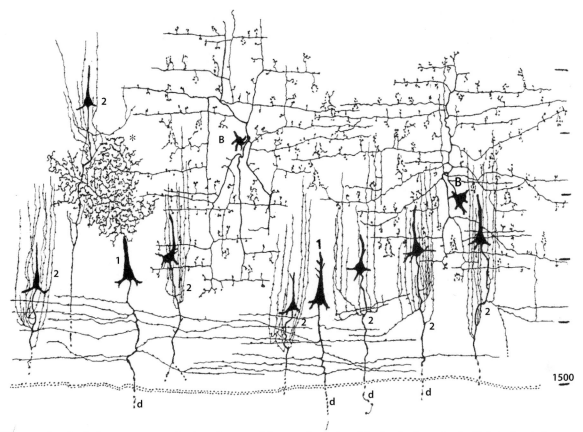

3.24 Microanatomical changes in cerebral cortex overlying early periventricular leukomalacia. This lesion is from an infant born at 38 weeks' gestational age who survived 12 days. This composite figure illustrates several post-injury neuronal alterations including axotomized pyramidal neurons (1) with long horizontal collaterals bordering the necrotic zone (at the bottom of the drawing); retrograde degeneration of axon profiles (d) of pyramidal neurons (2); and dispersed axon profiles of two hypertrophic basket cells (B) with long horizontal collaterals. Golgi preparations; camera lucida drawings.

From Marin–Padilla.[348] *With permission from Lippincott Williams & Wilkins/Woilters Kluwer Health.*

COMBINED GREY AND WHITE MATTER LESIONS

The preceding sections have focused on neuronal injury in predominantly grey matter regions and on oligodendrocyte/axon injury in predominantly white matter regions. Autopsy of infants who do not survive the perinatal period shows that grey and white matter injury frequently coexist, however, in a pattern sometimes referred to as 'encephalopathy of prematurity'.[567] Recent neuroimaging studies show a high incidence of reduced volumes in the cerebral cortex, thalamus, basal ganglia and hippocampus in premature infants at term-equivalent age or in childhood, whether or not there is associated white matter injury.[259,396,432] In the encephalopathy of prematurity, white matter damage is accompanied by oxidative damage to cortical and interstitial white matter neurons (which are remnants of the subplate).[172] In the cortex overlying PVL, the number of pyramidal neurons in layer V is diminished by 38 per cent in PVL *versus* controls.[18] Additionally, the granular subtype of white matter neuron is decreased by 54–80 per cent in PVL compared to controls, and this decrease is geographically distant from the site of necrosis.[292] Whether neuronal

loss results from axotomy and secondary 'dying back' of the neurons, direct free radical injury or other mechanisms remains unclear. Abnormalities in synaptogenesis, dendritic arborization and axonal elongation remain to be determined. Of note, this injury to grey matter sites in the setting of PVL can be detected by routine diagnostic methods and includes neuronal loss in the thalamus, globus pallidus and dentate nucleus in approximately a third of autopsied patients with white matter necrosis. In this population, gliosis is also seen in the basal nuclei in half, and in the basis pontis in 100 per cent, when compared with cases with only diffuse white matter gliosis.[435] Grey matter injury and PVL are also frequently detected in the brains of term infants with congenital heart disease dying after cardiac surgery (Figure 3.25).[291]

Advances in cardiac surgery since the 1980s have facilitated repair of congenital heart disease in increasingly younger patients.[268] As surgical and support techniques have improved, attention is increasingly focused upon the neurological outcome of survivors and the prevention of preoperative, intraoperative and postoperative hypoxic-ischaemic brain injury that may result in lifelong cognitive disturbances, seizures and motor disorders.[136,412,492] Newborns with complex congenital heart lesions are

BOX 3.5. Sequelae of perinatal brain injury

Preterm birth accounts for the majority of perinatal brain injury, the sequelae of which are often grouped under the rubric 'cerebral palsy'. Such an outcome is most prevalent among the very low birthweight (VLBW) population: in the United States alone, 1.5 per cent of the 4.3 million live births per year (or 65 000 individuals) are VLBW and, of these, 90 per cent survive as a result of advances in neonatal care.[234] Nevertheless, 5–10 per cent of these surviving infants will develop cerebral palsy, and 25–50 per cent will exhibit cognitive, attentional or behavioural deficits at school age.[360,566]

Thirty-four gestational weeks is considered a milestone beyond which intensive efforts to postpone premature births are not clinically performed.[108,127] Despite the perception that such infants are relatively healthy as they are 'near-term', the morbidity and mortality rates of infants born after 34 weeks but before 37 weeks (term) are substantially worse than those for infants born at term. Compared with full-term infants, late-preterm ('near-term') infants suffer higher frequencies of respiratory distress, feeding difficulties, apnoea, seizure disorders, temperature instability, hypoglycaemia, hyperbilirubinaemia, kernicterus and sudden infant death syndrome (SIDS).[14,52] Although a third of the increase in the prematurity rate in the United States is due to lower gestations, the balance is due to an increase in late preterm births.[52]

The common factor in prematurity is inadequate oxygenation from pulmonary immaturity, as well as blood pressure lability, leading to hypoxia-ischaemia; prolonged or difficult labours, chorioamnionitis and maternal conditions also predispose to hypoxia-ischaemia. Clinically, intrapartum hypoxia-ischaemia is defined clinically as a cord blood pH of <7.0 at delivery, occurring at an incidence of 2.5 per 1000 live births.[206] The incidence of cerebral palsy in clinically recognized settings of intrapartum hypoxia-ischaemia is 14.5 per cent.[206,221]

Clinically, the deficits of cerebral palsy encompass spastic or choreoathetotic motor disturbances, often diplegic (i.e. involving the lower extremities greater than the upper extremities, in turn greater than the face). Although defined as a 'non-progressive' motor deficit sustained in the perinatal period, the stability of the deficit in cerebral palsy is not absolute. Severely affected children exhibit functional decline in motor activity during adolescence.[222] This is generally attributed to musculoskeletal factors such as changing weight-to-strength ratio and prolonged spastic deformation.

The underlying neuropathology of cerebral palsy is complex, including not only hypoxic-ischaemic encephalopathy in the perinatal period but also malformations, traumatic lesions, infections and inborn metabolic disorders.[387] The clinicopathological associations of cerebral palsy were examined in the National Collaborative Perinatal Project in the USA, a study of 54 000 pregnancies. This study concluded that perinatal factors, including birth asphyxia, contributed little to the overall incidence of cerebral palsy; rather, factors operating before birth were responsible for most cases.[388] This observation has been supported by other studies.[532] Thus, neuropathologists must consider carefully the relationship of cerebral palsy to perinatal asphyxial lesions, as only 12–23 per cent cases can be related to intrapartum asphyxia.[388]

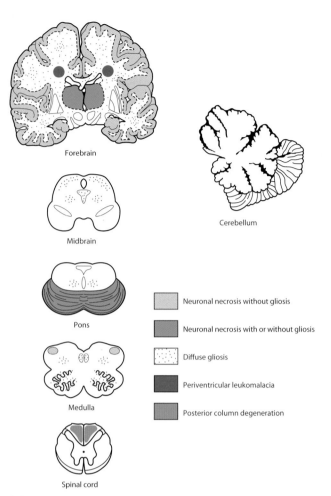

Forebrain

Midbrain

Pons

Medulla

Spinal cord

Cerebellum

	Neuronal necrosis without gliosis
	Neuronal necrosis with or without gliosis
	Diffuse gliosis
	Periventricular leukomalacia
	Posterior column degeneration

3.25 Composite pattern of central nervous system injury in infants with congenital heart disease dying after all types of cardiac surgery. The lesions shown are present in >50 per cent of the total cases. Diffuse gliosis (often accompanied by selective neuronal loss) in the brain stem is most common in the periaqueductal region including the reticular formation, the colliculi, the hypoglossal nucleus, the vestibular nucleus, the inferior olivary nucleus and the and basis pontis.

Redrawn from Kinney et al.[291] With kind permission from Springer Science and Business Media.

cortex (diffuse or focal) (67 per cent), (iii) acute hippocampal injury (CA1–4; 67 per cent), (iv) brain stem neuronal injury and gliosis (45–61 per cent), (v) diffuse neuronal injury and/or gliosis of the thalamus (58 per cent), and (vi) diffuse neuronal injury and/or gliosis in the basis pontis (Figure 3.25).[91] Thus, our thinking about PVL as strictly a 'white matter disease' must be revised.

Other clinical settings increase risk of both grey and white matter in newborns: congenital diaphragmatic hernia, meconium aspiration syndrome, severe pneumonia and septic shock with myocardial dysfunction may require treatment with extracorporeal membrane oxygenation (ECMO), a temporary life-support treatment for severe cardiorespiratory failure. The neuropathology of ECMO-treated patients includes hypoxic-ischaemic damage, such as multifocal cerebral cortical infarcts, neuronal necrosis of the cerebral cortex and other regions, thalamic damage with infarcts, parenchymal haemorrhages, PVL and

frequently born with brain immaturity.[511] In an analysis of infants dying after cardiac surgery, PVL occurred in 61 per cent of cases. In more than 50 per cent, hypoxic-ischaemic grey matter lesions in the surgically corrected cases were, in order of decreasing frequency, (i) inferior olivary gliosis (87 per cent), (ii) acute neuronal injury in the cerebral pontine and olivary necrosis and/or gliosis.[265,306,525] Infarcts may be associated with the ligation of the carotid artery for cannula placement, although infarcts or haemorrhages do not always lateralize to the right side.[265] Disseminated intravascular coagulation can occur, resulting in multiple, small cerebral infarcts and haemorrhages.[265] The systemic heparinization of ECMO-treated infants exacerbates bleeding, which frequently occurs during reperfusion in ischaemic areas. In these circumstances, larger zonal injury to adjacent grey and white matter structures can be ascribed to damage in particular vascular regions including major arterial zones, interarterial zones ('watershed') and deep venous zones. Such complex changes can be seen as well in very sick infants not subjected to ECMO therapy

(Figure 3.26). Some lesions are identified with potentially confusing (and in the opinion of some, antiquated) terminology such as schizencephaly (Greek schizein = to divide; abnormal clefts in the brain originally thought to be malformative) and porencephaly (Greek porus = small opening; originally described by Heschl as cysts in the brain communicating with the subarachnoid compartment, usually resulting from a destructive process). The current imaging literature has many examples of lesions described as porencephalic cysts arising from the ventricles but with no apparent communication with the brain surface. Furthermore, it should be noted that lesions predominantly affecting one compartment may in fact have multifocal effects.[383]

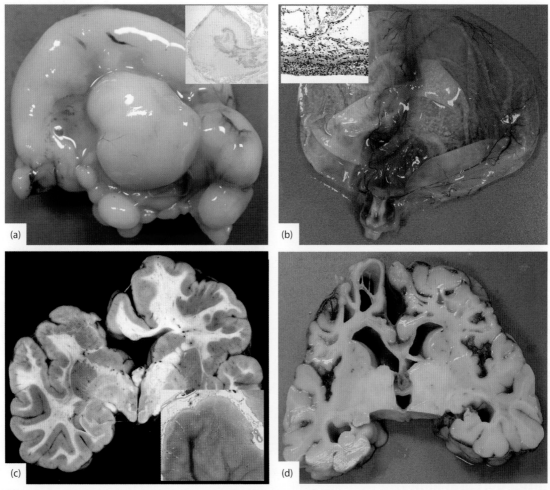

3.26 Ischaemic damage in mixed grey and white matter distributions. (a) *In utero* infarction involving almost the entire internal carotid artery distribution on one side. The timing is unknown but likely preceded birth at 20 weeks' gestational age by at least several weeks. The result is severe atrophy of the hemisphere with preservation of irregular nodules of tissue ('nodular porencephaly'). Inset shows disorganized structure of one of the small nodules. Haematoxylin and eosin, original magnification 12.5×. **(b)** *In utero* destruction of the entire brain with a small amount of residual brainstem tissue ('hydranencephaly'). Ultrasound imaging showed that this lesion occurred at least 2 weeks before birth at 24 weeks' gestational age. The precise pathogenesis is unclear; the absence of inflammation suggests a possible vascular insult. Inset shows a few layers of astroglial and microglial cells persisting along the leptomeninges. Haematoxylin and eosin; original magnification 400×. **(c)** Coronal slice through the brain of a 15-year-old child who sustained a cerebral infarct during the prenatal period in the distribution of the middle cerebral artery. The result is loss of tissue from the meninges to the ventricle resulting in a large cleft ('schizencephalic porencephaly'). Inset shows disorganized cortical tissue in the form of polymicrogyria at the edge of the infarct cavity. Solochrome cyanin and eosin; original magnification 40×. **(d)** Coronal slice through the brain of a 13-month infant who suffered a severe hypoxic-ischaemic insult following full term birth. The result is an asymmetric multicystic change with loss of grey and white matter and compensatory enlargement of the cerebral ventricles.

Arterial Region Ischaemic Damage

Regional ischaemic damage may clinically present as stroke in the perinatal period or may be identified some time later in association with seizures or cerebral palsy. Arterial infarcts, typically in the distribution of the middle cerebral artery,[117,295] are seen more commonly in full-term infants (Figure 3.2) and rarely in premature infants. The incidence of arterial stroke in full-term infants is rare, estimated at 1 in 10000 births. Arterial distribution lesions are relatively common, however, in children born at term with hemiplegic cerebral palsy (22 per cent) and among full-term infants with neonatal seizures (18 per cent). Case control studies show that the major risk factors are maternal fever (>38°C), Apgar score at 5 min <7, hypoglycaemia (<2.0 mmol/L) and early-onset sepsis/meningitis. Lesser risk factors include primiparity, fetal heart rate decelerations, meconium-stained amniotic fluid, emergency caesarean section and umbilical artery pH <7.10.[225,293,540] Coagulation defects have also been implicated in the pathogenesis of neonatal infarction,[218] including rare genetic abnormalities, e.g. factor V Leiden mutation,[530,552] prothrombin G20210 variant A, methyltetrahydrofolate C677T mutation, elevation of lipoprotein (a) and deficiencies of protein C, protein S, antithrombin III and plasmogen. Acquired prothrombotic conditions also increase the risk of stroke, for example, the presence of antiphospholipid (anticardiolipin) antibodies.[10]

The pathological appearance depends on the timing of the injury. Those that occur *in utero* generally appear as a region of focal atrophy corresponding to a major artery branch distribution with communication between the ventricle and the subarachnoid compartment (porencephalic cyst). The walls are smooth and the margins of these lesions are typically associated with an irregular cortical contour and simplified cortical structure (polymicrogyria). The distinction of so-called open and closed lip forms is likely of little value for understanding the pathogenesis; this is more a reflection of the size of the lesion. In some cases, the marginal surviving tissue may appear as clusters of small nodules (Figure 3.26). When these arterial distribution lesions occur near full term and in infants, the gross and histological appearance is similar to that seen in adults. Note that some lesions described as deep porencephalic cysts appear as focal ventricular expansions with normal overlying cerebral cortex; many of these are likely the end product of periventricular haemorrhagic lesions (see later).[207]

Parasagittal Cerebral Injury

Parasagittal cerebral injury refers to necrosis of the cerebral cortex and subjacent white matter in the parasagittal region with or without haemorrhage. It is often seen in the full-term infant with perinatal asphyxia (Figure 3.2). The necrotic areas are in border zones between the end fields of the major cerebral arteries. The most marked injury occurs in the posterior cerebrum in the border zone of the three major cerebral arteries. This lesion is due to reduced cerebral perfusion or oxygen delivery. Parasagittal cerebral injury typically does not occur in the premature infant.

Venous Sinus Thrombosis

Venous sinus thrombosis may occur in infants of all gestations. It is less common than arterial zone infarcts (0.67 cases per 100000 children per year). Infants may present with diffuse or focal neurologic signs. The critical diagnostic feature is demonstration of sinus occlusion by imaging. Risk factors include acute systemic illnesses (sepsis, dehydration), chronic systemic diseases (cardiac disease, indwelling catheters), prothrombotic states and head and neck disorders (local infection).[131] CT and MR venograms indicate that compression of the sinuses in the vicinity of the posterior fontanel may also be a risk factor.[521] Neonates tend to develop thrombosis in the superior sagittal sinus, lateral sinus, straight sinus, vein of Galen and internal cerebral veins, although in older children the most vulnerable regions are the superior sagittal sinus, lateral sinus and jugular veins.[211] The commonly associated pathological features are cerebral oedema, haemorrhagic venous infarcts and intraventricular haemorrhage. Thrombus, potentially with calcification, is identifiable in the affected sinuses (Figure 3.27).

Near-Total Asphyxia and Multicystic Encephalopathy

Near-total asphyxia is a pattern of brain injury that affects the putamen, thalamus and cerebral cortex, as well as the brain stem.[39] The event giving rise to this pattern of injury is often a relatively brief interval of intense asphyxia (such as umbilical cord compression) requiring resuscitation and is correlated with the later development of cerebral palsy.

'Multicystic encephalopathy' refers to lesions composed of large septated cavities throughout the cerebral cortex and white matter of both hemispheres and may, in some cases, represent the most severe end of the spectrum of near-total asphyxia (Figure 3.26). The cavities are variable in distribution, with a relative but not total sparing of the temporal lobe, basal ganglia and brain stem. Most lesions are thought to develop as the result of a hypoxic-ischaemic insult near the end of gestation or in the early postnatal period (Figure 3.2).[576] Multicystic encephalopathy is not, however, specific for hypoxic-ischaemic damage and can occur in other settings, e.g. herpes viral infection.[576]

CEREBRAL HAEMORRHAGES

Periventricular/Intraventricular Haemorrhage

Periventricular/intraventricular haemorrhage (PVH/IVH) is the most common haemorrhagic lesion in the premature infant (Figure 3.28).[547] Bleeding may arise from the germinal matrix (ganglionic eminence haemorrhage) or the white matter. Intraventricular haemorrhage (IVH) refers to blood collections within the ventricles; they may be extensions of ganglionic eminence haemorrhage or may arise from the choroid plexus. *In utero* bleeding related to fetal clotting disorders or trauma may rarely occur. Approximately 16 per cent of all births are premature (<37 weeks' gestation) and ~2 per cent are extremely premature (<31 weeks). The incidence of clinically significant brain haemorrhage in term

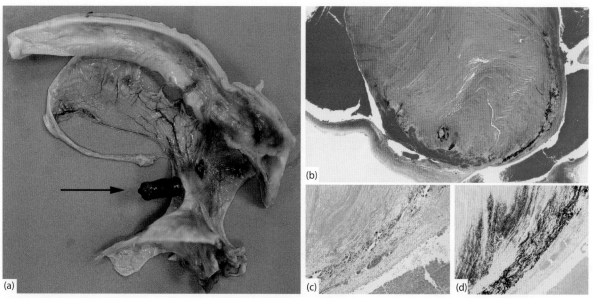

3.27 Sinus thrombosis in the perinatal brain. (a) Photograph of sagittal skull, falx cerebri and tentorium from an infant born at 31 weeks' gestational age who survived 3 weeks. Blood clot from within great vein of Galen (arrow) is attached to the thrombosed straight sinus; the transverse sinus was also thrombosed. **(b)** Histological section through the superior sagittal sinus of a 23-week gestational age fetus; thrombosis was detected *in utero* using magnetic resonance imaging. The lumen is occluded by clot with an obvious lamellar pattern and part of the sinus wall is calcified (dark regions). Haematoxylin and eosin; original magnification 12.5×. **(c)** and **(d)** show iron (blue; Perl's Prussian blue method) and calcium (black; von Kossa method) deposits in the sinus wall. Original magnification 100×.

(a) Courtesy of Dr S Krawitz, University of Manitoba, Canada.

infants is very low. Gestational age at the time of birth is the most important determinant of PVH/IVH. The likelihood rises steeply among births earlier than 32 gestational weeks with ~10 per cent at 30 weeks, ~40 per cent at 26 weeks and >50 per cent at 24 weeks.[91,96,194,235,486] Prospective studies utilizing repeated ultrasound imaging show that most haemorrhages occur in the first 3–24 hours after birth[133] and may enlarge thereafter.[417] Because of improved neonatal intensive care, the incidence of IVH has decreased for all gestational ages; however, the increasing survival of the smallest premature infants makes PVH a persistent clinical problem.[209]

IVH may present as a sudden clinical deterioration, with bulging fontanelle, hypotension, seizures, coma and apnoea, or it may be protracted over several days with subtle clinical signs.[423] IVH is usually diagnosed with cranial ultrasound and graded as follows: grade 1, haemorrhage limited to the germinal matrix; grade 2, blood filling the lateral ventricles without distension; grade 3, blood filling and distending the ventricular system; and grade 4, haemorrhage with parenchymal involvement. The definition of grade 4 IVH is controversial because haemorrhage in the periventricular white matter may occur without IVH.[416] About 15 per cent of cases of IVH have an intraparenchymal lesion characterized by haemorrhagic necrosis in the periventricular white matter, lateral and dorsal to the external angle of the lateral ventricles (Figure 3.28). The white matter lesion is typically asymmetric and may extend from the frontal to the parietal and occipital lobes. They are now considered to be venous infarcts of the deep white matter related to obstruction of the terminal veins by the large IVH, and not arising in the germinal matrix itself.

Although age at birth is the main determinant of PVH/IVH, there are secondary, often associated, risk factors.

Antepartum maternal risk factors include alloimmune and idiopathic thrombocytopenia, warfarin, cocaine, maternal seizures, severe abdominal trauma, amniocentesis, cholestasis of pregnancy and febrile disease. Fetal risk factors include congenital factor X and factor V deficiencies, congenital tumours, twin–twin transfusion and demise of co-twin.[79,484] Early-onset IVH in the premature infant is associated with lower gestational age, lower birth weight, fetal distress, fetal acidosis, premature rupture of amniotic membranes, amniotic infection, vaginal delivery, vertex presentation, resuscitation and mechanical ventilation.[466,555,578]

Most affected infants have been artificially ventilated because of lung immaturity; this is associated with hypoxaemia and blood pressure fluctuations, which are potentiating factors in the haemorrhagic process. Late-onset haemorrhages occurring after the first 24 hours are associated with respiratory distress syndrome, vigorous neonatal resuscitation, hypoxaemia, suctioning, acidosis, pneumothorax, seizures, heparin use, hypercarbia and blood-pressure fluctuation.[560,578] In mechanically ventilated preterm infants, Doppler studies have shown a marked continuous alteration in velocity of systolic and diastolic flow, related to the mechanics of ventilation.[404,426,428] Serial cranial ultrasound studies have demonstrated the relationship between fluctuating patterns of blood-flow velocity and the subsequent development of IVH.[562] Sick premature infants are at risk from harmful elevations in cerebral blood flow because there is a limited range of autoregulation and a pressure passive state regulates flow.[539,563] Increased flow, with resulting vascular congestion, may also occur with hypercapnia. Decreased flow, which can accompany asphyxia and hypotension, may injure the walls of blood vessels, causing then to rupture on reperfusion.

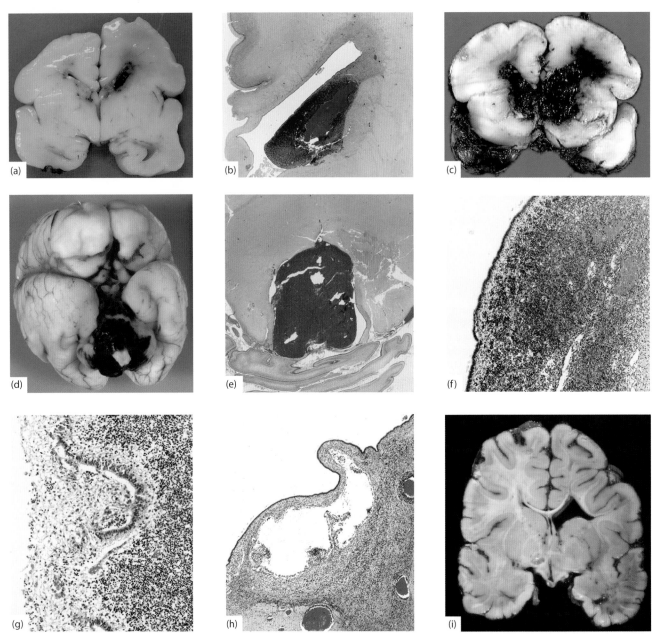

3.28 Germinal matrix and intraventricular haemorrhage. (a) Coronal slice through the brain of a 26-week gestational age premature infant who died within 2 hours of birth. A haematoma is apparent along the wall of the frontal horn of the lateral ventricle. **(b)** Histological section (coronal) through the region shows that the haematoma is confined to the ganglionic eminence; i.e. equivalent to small grade 1 haemorrhage by imaging criteria. Haematoxylin and eosin; original magnification 12.5×. **(c)** Coronal slice through the brain of a 26-week gestational age premature infant who died within 6 hours of birth. Blood fills and expands the lateral ventricles and a hematoma extends into the frontal periventricular region; i.e. equivalent to grade 4 haemorrhage by imaging criteria. **(d)** Photograph of the base of the brain of an infant who was born prematurely at 29 weeks' gestational age and survived 2 days with a grade 4 periventricular haemorrhage. Blood is apparent in the subarachnoid compartment surrounding the cerebellum (i.e. at the fourth ventricle outlets) and extending to the prepontine cistern and infundibulum. **(e)** Histological section (horizontal) through the pons and cerebellum of an infant who was born prematurely at 30 weeks' gestational age and survived 2 days with a grade 4 periventricular haemorrhage. Blood fills and expands the fourth ventricle. Haematoxylin and eosin; original magnification 12.5×. **(f)** Histological section of the ganglionic eminence from an infant who was born at 30 weeks' gestational age and survived 13 days following grade 3 intraventricular haemorrhage. Erythrocytes are intermixed with germinal matrix cells. Haematoxylin and eosin; original magnification 100×. **(g)** Histological section of the ganglionic eminence from an infant who was born at 25 weeks' gestational age and survived 8 weeks following grade 4 intraventricular haemorrhage. The ependymal layer is discontinuous and a mix of germinal cells, hemosiderin-containing macrophages (yellow) and new capillaries lie over the ependymal surface. Haematoxylin and eosin; original magnification 400×. **(h)** Histological section of the ganglionic eminence from an infant who was born at 27 weeks gestational age and survived 14 weeks following grade 3 intraventricular haemorrhage. The periventricular tissue shows cystic degeneration. Haematoxylin and eosin; original magnification 100×. **(i)** Coronal slice through the brain of a 2.5-year-old child who was born prematurely at 29 weeks' gestational age and suffered unilateral grade 4 intraventricular haemorrhage. The periventricular white matter on the affected side and the corpus callosum are severely hypotrophic.

The site of origin of PVH/IVH is the germinal matrix, a regional expansion of the subventricular zone that lines the immature ventricles. This is thickest along the lateral wall of the lateral ventricle, particularly at the ganglionic eminence between the head of the caudate and the thalamus (Figure 3.28). In the germinal matrix, progenitor cells give rise to immature neurons of the caudate, putamen and amygdala between 10 and 24 weeks of gestation and in the third trimester to immature glial cells. The matrix cells are densely packed with small amounts of cytoplasm and few processes. The periventricular germinal region has the highest density of blood vessels in the immature brain[35,86] and also has thin-walled veins that have relatively large lumina.[298] Factors contributing to vascular fragility include the high blood vessel density, a paucity of pericytes, and structural immaturity of the basal lamina.[34] Venules of the matrix drain into subependymal veins of the deep venous system, converging with the medullary, thalamostriate and choroidal veins as they form the terminal vein, the internal cerebral vein and the vein of Galen.[384] Blood vessels of the matrix may be particularly vulnerable to rupture because the surrounding tissue is inherently fragile, lacking in interwoven cell processes, because of the paucity of supportive extracellular matrix proteins and because of relatively high regional proteolytic activity related to cell migration.[190,563] Moreover, the subependymal veins, which are closest to the ventricle and the source of blood escaping into the ventricle, show progressive expansion of luminal diameter after 23 gestational weeks but do not undergo thickening of their walls until after 36 gestational weeks, suggesting a 'window of vulnerability' to rupture.[20] The architecture of blood vessels of the germinal matrix makes them vulnerable to mechanical distortion during birth and elevated venous pressure.

Studies of the blood vessels of the germinal matrix in preterm neonates have identified a venous origin for most haemorrhages.[186] In serial sections, germinal matrix haemorrhages are small, multifocal and caused by direct injury to the vascular endothelium.[347] The central, often large or dilated, blood vessel (venule) of almost every haemorrhagic focus shows focal endothelial cell death, rupture of the vascular wall and focal thrombosis. The injured blood vessel shows focal cytoplasmic and nuclear fragmentation of a few endothelial cells intermingled with fibrin products, although the remaining vessel wall is unaffected. These small haemorrhagic foci have a tendency to coalesce, thus evolving rapidly into larger destructive lesions. By the time most germinal matrix haemorrhages are studied histologically, it is generally too late to demonstrate these early vascular changes.[347]

Sequelae of Periventricular/Intraventricular Haemorrhage

PVH/IVH can have devastating permanent consequences on the brain.[334] Cerebral blood perfusion is decreased persistently for 2 weeks in affected infants, even in the presence of small haemorrhages.[554] Large haematomas may distort deep vascular structures leading to ischaemic lesions in the white matter and basal nuclei, the end result of which is focal ventricular enlargement with sparing of the cerebral cortex (Figure 3.28).[207] Recent analysis of autopsy material showed that perinatal PVH causes suppression of proliferation among germinal cells, which could result in impaired development of white matter and basal nuclei.[123] Proteolytic enzymes in blood plasma appear to be at least partly responsible for this

phenomenon.[271] This damage is associated with microglial activation.[509] Autopsy studies indicate that thyroid hormone receptors are altered in the vicinity of haemorrhagic lesions and, in a rabbit model, thyroid hormone supplementation reduces the associated white matter injury.[569] Most reports indicate that death is more likely in children with brain haemorrhage but recent clinical evidence suggests that PVH/IVH is associated with no more than a modest increase in the risk of adverse developmental outcome except when accompanied by white matter injury.[199,411]

Destruction of the germinal matrix by haemorrhage can lead to periventricular cyst formation with haemosiderin-laden macrophages, iron and reactive astrocytes.[106] In periventricular haemorrhagic injury, Golgi preparations indicate that the local destruction of radial glia stops cellular migration above the lesion (Figure 3.29),[349] and the precursor cells already travelling in damaged radial glia stop their migration, miss their target and form acquired heterotopias. The thickness of the neocortex overlying a repaired germinal matrix haemorrhage or a post-haemorrhagic hydrocephalus is reduced and the cytoarchitecture of the maturing grey matter may be secondarily altered (acquired neocortical dysplasia). As radial glial fibres are destroyed throughout the haemorrhagic site, all cellular migration will cease above the haemorrhage and many cells will miss their targets, resulting in focal heterotopias at any level of the cortex. Although any of these post-haemorrhagic changes could play a role in the pathogenesis of neurological sequelae,[347] it is critical to note that PVH/IVH seldom occurs in isolation; the premature infants are also at risk for grey and white matter lesions described earlier.

In addition to the local effects of haemorrhage, blood enters the ventricular system and subarachnoid space. At anatomically restricted sites, especially the cerebral aqueduct and fourth ventricle outlets, blood may initiate a reactive process that impairs flow of cerebrospinal fluid (CSF). This results in post-haemorrhagic hydrocephalus.[93] In rare cases, cryptic *in utero* PVH can cause obstruction of the cerebral aqueduct and lead to fetal hydrocephalus.[311]

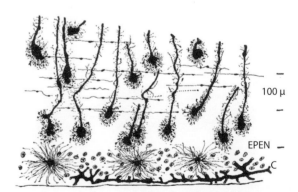

3.29 Cortical organization sequelae of periventricular haemorrhage. Camera lucida drawing of Golgi-impregnated tissue from the brain of an infant born at 28 weeks' gestational age who survived 10 days following periventricular haemorrhage. Near the ependymal surface (EPEN) are new capillaries (C) and stellate reactive astrocytes. Adjacent to, but not reaching, the ependymal surface are the bulbous ends of degenerating radial glial fibres that had been damaged by the haemorrhage.

From Marin–Padilla.[347] With permission from Lippincott Williams & Wilkins/Woolters Kluwer Health.

Rarely, periventricular haemorrhage may occur in term infants, some of whom have been shown to have mutations in endothelial structural proteins. (e.g. tight-junction protein JAM3).[373]

Choroid Plexus Haemorrhage

In premature infants, IVH may be caused by haemorrhage from the stroma of the choroid plexus. Careful inspection of the ventricle wall is required to exclude PVH, and the choroid plexus should be sampled for histologic analysis to demonstrate continuity between stromal blood and intraventricular blood collections (Figure 3.30). In term infants, IVH most frequently arises from bleeding in the choroid plexus, particularly at the level of the atrium of the lateral ventricle. One-third of patients with adequate perinatal histories have experienced difficult deliveries because of forceps rotations and breech extractions.[563]

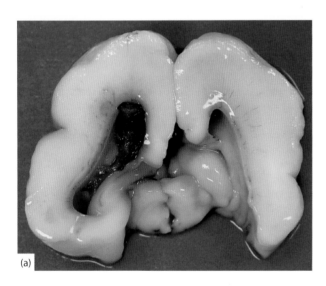

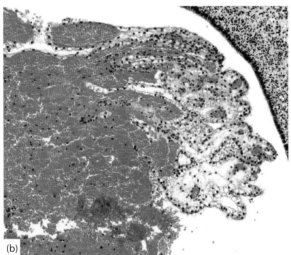

3.30 Choroid plexus haemorrhage. (a) Coronal slice through parietal region of brain from an infant born at 25 weeks' gestational age who lived a few hours. There is intraventricular blood, but careful inspection revealed no abnormalities of the ventricle wall. **(b)** Histological section from the same case showing blood erupting from the choroid plexus. Haematoxylin and eosin; original magnification 200×.

Cerebellar Haemorrhage

Cerebellar haemorrhages are more common in premature than term infants, with an incidence of 15–25 per cent in premature infants less than 32 weeks of gestation and/or with a birth weight of 1500 g.[178,210,353,422] Cerebellar haemorrhages are increasingly diagnosed as a result of improved neuroimaging techniques and survival of very small preterm infants.[329] Among 35 cases unilateral haemorrhage was present in 20 per cent, vermian haemorrhage in 20 per cent and combined cerebellar hemispheric and vermian haemorrhage in 9 per cent. There was associated supratentorial haemorrhage in 77 per cent of the cases and risk factors were essentially the same. Cerebellar haemorrhages are located in the cortex or subcortical white matter; they may be multiple and petechial or focal and large with contiguous blood in the subarachnoid space (Figure 3.31). Cerebellar haemorrhages have been documented by ultrasound before the onset of labour, and so antenatal factors must be considered in their pathogenesis.[216] A detailed autopsy study showed that neuron loss is an additional complication of cerebellar haemorrhage.[217]

Other Causes of Intracerebral Haemorrhage

Multiple small foci of grey and white matter haemorrhage may occur in term infants who have hypoxic brain damage and labile blood pressure.[415]

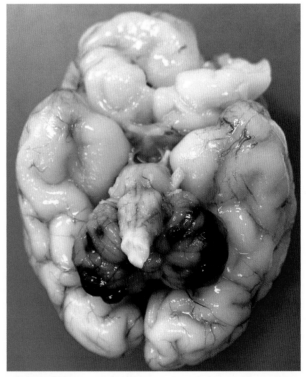

3.31 Cerebellar haemorrhage in a premature infant. Photograph shows haemorrhagic change and diffuse dark discolouration in the cerebellum of an infant born at 36 weeks' gestational age and who survived approximately 2 days with severe hypoxia.

Subdural/Subarachnoid Haemorrhage

Magnetic resonance imaging reveals a high incidence of asymptomatic small subdural / subarachnoid haemorrhage (~50 per cent) and choroid plexus haemorrhage (~30 per cent) following uncomplicated birth at term.[331,524] It is critical to recognize this when performing autopsies on neonates and infants; small intracranial blood collections at birth are a truly incidental finding and not necessarily related to a cause of death. The finding of a subdural haematoma in an infant raises the possibility of child abuse (see Chapter 10, Trauma); it is, however, important to keep perinatal trauma in the differential diagnosis.

BIRTH TRAUMA

Birth trauma refers to injury of the CNS or peripheral nervous system (PNS) in the premature or full-term newborn caused by mechanical factors during labour or delivery.[141,176,563] The occurrence of traumatic injuries has decreased significantly with improved obstetrical care, notably with the use of caesarean section,[11] however, use of instruments (vacuum extractors, forceps) continues to be associated with substantial risk for injuries to the head.[141] A comprehensive understanding of these birth-related injuries is critical because they must not be confused with injuries caused by child abuse. It is also important to remember that decomposition of stillborn infants can be associated with haemorrhage-like changes in the skull sutures.[452]

Scalp, Skull and Brain Injuries

Caput succedaneum is a common lesion consisting of localized haemorrhagic oedema of the subcutaneous tissue in the presenting part of the head; it resolves without problems. Subgaleal haematoma refers to a collection of blood between the periosteum and the aponeurosis of the scalp. The incidence is 0.2–3 per 1000 live births.[510] It is caused by trauma that tears the emissary veins connecting dural sinuses to scalp veins.[21] Subgaleal haemorrhage may dissect into the neck and be associated with considerable blood loss.[563] Subgaleal haematomas may rarely compress the underlying brain and require surgical evacuation.[13] Cephalhaematoma is a collection of blood between the periosteum and the outer surface of the skull, limited by the suture lines. It is associated with skull fracture in 25 per cent of cases.[276] Epidural haematomas, blood between the skull and the periosteum on its inner aspect, are rare in the neonatal period because the dura is tightly applied to the skull.

Skull fractures may be linear, depressed or related to occipital osteodiastasis.[146] Most resolve without difficulty. Growing skull fractures are caused when the dura is damaged allowing the subarachnoid compartment to communicate with the extraosseous tissue; the resulting leptomeningeal cyst expands and prevents the fracture from healing.[247]

Subdural haematoma is a collection of blood between the dura and the leptomeninges. Symptomatic subdural haematoma typically occurs in the full-term infants, although the relative proportion of affected premature infants has increased because of advances in obstetrical management of difficult full-term deliveries.[563] The source of haemorrhage may be (i) rupture of bridging superficial cerebral veins; (ii) cerebellar tentorial laceration, with rupture of the straight sinus, transverse sinus, vein of Galen or smaller infratentorial veins;[563] (iii) occipital osteodiastasis, with rupture of the occipital sinus; or (iv) laceration of the falx or tentorium, with rupture of the inferior sagittal sinus (Figure 3.32).[498] Subdural haemorrhage is more likely to occur when the infant head is relatively large compared to the birth canal; when the skull is compliant, as in premature infants; when the duration of the labour is unusually brief, not allowing enough time for dilation of the pelvic structures, or unusually long, subjecting the head to prolonged compression and moulding; or when instrumented extraction or rotational manoeuvres are required.[69,563] In these circumstances, there is excessive moulding of the head with stretching of dural and vascular structures. Subdural haemorrhage (and likely subarachnoid haemorrhage) occurs in small quantities in a large proportion of normal vaginal births. Magnetic resonance imaging studies of newborns show small blood collections acutely or small residual hemosiderin deposits in 6 to 29 per cent of infants following normal vaginal delivery and in up to 50 per cent of cases in the region of the tentorium.[331,524,579] These are frequently found during autopsies on newborns and some older infants (Figure 3.32).

Spinal Cord and Brain Stem Injury

Injury to the infant spinal cord during delivery results from excessive traction or rotation, unlike the usual compression injury characteristic of cord injuries in older patients. The newborn vertebral column is partially cartilaginous and more elastic with relatively hypotonic muscles. With longitudinal traction and hyperextension, the cord moves less than the spine and the cord 'ruptures' at its mobile sites, the lower cervical to upper thoracic levels. Contributing factors include fetal malposition, dystocia, prematurity, primiparity, precipitous delivery and fetal vertebral or foramen magnum malformations.[535]

The acute lesion of traumatic spinal cord injury is subdural and subarachnoid haemorrhage with intraspinal haemorrhagic necrosis, particularly involving the dorsal and central grey matter. The chronic lesion exhibits cystic cavitation, with disruption of cord architecture and fibrous adhesions between the dura, the leptomeninges and the cord. Vascular occlusions, perhaps developing as a post-traumatic event, may cause infarction of cord segments caudal to the level of the primary lesion. Lesions of the vertebral column are uncommon, consisting of fractures or dislocations and separation of the vertebral epiphysis. Fatal brain stem injury attributable to rotation or traction is sometimes seen in contiguity with upper cervical spinal cord injury.[535]

Nerve Injury

Peripheral and cranial nerve injury needs to be considered in neonates with stridor, respiratory distress, feeding difficulty, facial deformity and altered grasp reflexes. In one study of 19 370 consecutive deliveries the incidence of facial nerve injury was 0.6 per 1000 and the incidence of brachial plexus injury was 0.9 per 1000.[430] Facial nerve

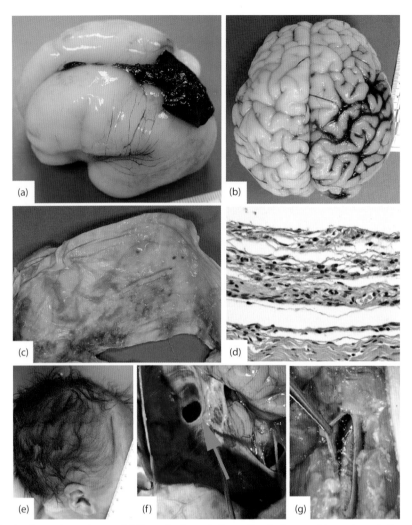

3.32 Birth trauma to the brain and meninges. (a) Superior/lateral view of brain from a 24-week gestational age infant who died at birth. A thick subdural hematoma lies on the brain surface and between the hemispheres. **(b)** Superior view showing subarachnoid blood collections on the surface of the brain from a full-term infant who died at birth during an emergency Caesarean delivery. **(c)** Internal surface of the dura mater from a 6-month infant who died in bed. There was no other evidence of recent or old trauma and no circumstantial reason to infer child abuse. Yellowish membranes that peel from the surface are indicative of resolving subdural blood collection. **(d)** Histological section from the same case shows neovascularization and haemosiderin-containing macrophages (yellow-brown). This is presumed to be a resolving hematoma sustained at birth (600×). **(e)** Photograph of the head of a full-term infant who died shortly following forceps-assisted vaginal birth. The side of the face has a reddened impression of the forceps blade. **(f)** Same case. There is a laceration in the tentorium cerebelli (arrow) that was associated with subdural and subarachnoid haemorrhage. **(g)** Same case. Opening of the dura mater of the cervical spinal cord showing subdural and subarachnoid haemorrhage.

injury results from compression by forceps or the maternal sacral promontory. Brachial plexus injury is associated with shoulder dystocia, other malpresentations and instrument-assisted vaginal delivery. Erb's palsy involves C5–6, Klumpke's paralysis involves C5–7 and Horner's syndrome (ptosis, miosis, anhidrosis) results from injury to the T1 root sympathetic fibres.[246] Phrenic nerve injury is usually associated with brachial plexus injury and may present with respiratory distress on the first day of life after some delay.[339]

These nerve injuries are mainly a functional problem, but may come to the attention of a neuropathologist as specimens obtained during nerve exploration and repair. Depending on the stage of injury, the trimmed nerve ends will exhibit traumatized nerves with haemorrhage and eventually neuroma formation.[474]

HYPOGLYCAEMIA

Glucose is essential for brain metabolism and its homoeostasis is complex. Experimental data suggest that the immature brain is more resistant than the mature brain to glucose reductions. Postulated compensatory mechanisms include increased blood flow and cerebral glucose extraction, enhanced ability to utilize alternative energy substrates (in particular, lactate) and low cerebral energy demands.[26,550,551] As the major source of brain glucose is *via* the blood supply, there is increased risk of brain injury if the blood glucose concentration is reduced pre- or postnatally. The mechanisms of cell injury in hypoglycaemia are complex and overlap with those due to hypoxia. Conditions associated with maternal hypoglycaemia and ketogenesis (diabetes, fasting, starvation) may have profound effects on fetal brain growth and other organs. Altered glucose states in the mother have been related to the health of the infant in later life.[544]

As 'normal' blood glucose levels are lower in premature infants, a precise laboratory definition of hypoglycaemia has been problematic; 40–45 mg/dL of serum has been proposed as a conservative definition.[476] The nonspecific clinical signs of hypoglycaemia (tremors, apnoea, hypotonia, lethargy, feeding difficulty and seizures) may not be apparent unless the glucose level is <20 mg/dL.[78] Hyperinsulinism occurs in infants of diabetic mothers and other rare conditions,[499] and in the setting of maternal drug therapy with chlorpropamide and benzothiazides.

There are few autopsy studies reporting brain damage associated with neonatal hypoglycaemia. Anderson *et al.* reported the effects of neonatal hypoglycaemia (<10 mg/dL) on the nervous system in six infants.[16] In three untreated cases, they reported a diffuse and widespread degeneration of large and small neurons. In the small neurons, there were indistinct or absent nuclear membranes and chromatin clumping. In large neurons, the nuclei were shrunken, opaque and stippled, with chromatolysis. Motor neurons of cranial nerve nuclei and spinal cord contained vacuoles. Nuclei of glial cells were pyknotic in two of three cases. The classic ischaemic neuronal changes (eosinophilic cytoplasm with an irregular pyknotic nucleus) rarely occurred. The occipital region of the cortex and the insula were the most extensively involved. It must be noted that several of the infants were reported to be cyanotic and there was no control group. Griffiths and Laurence studied brains from 17 infants with hypoxia plus hypoglycaemia (<20 mg/dL) and 17 with hypoxia alone.[208] Following a detailed examination of 10 brain regions, they concluded 'exposure to a limited degree of hypoglycaemia produces no detectable excess central nervous system damage at a histological level, over and above that produced by severe coincidental clinical hypoxia.' Fluge studied 19 autopsies and reported four cases with cerebral oedema and one case with intracranial haemorrhage;[167] details of the histology were not reported. In summary, based upon published autopsy material there are no conclusive data showing a pattern of brain injury specifically attributable to neonatal hypoglycaemia.

The imaging literature is more comprehensive. In an imaging and clinical study of 35 term infants with early brain MR imaging following symptomatic neonatal hypoglycaemia and without evidence of hypoxic-ischaemic encephalopathy, a wide range of injury patterns was observed. This included white matter damage (94 per cent; posterior predominant in 29 per cent, posterior limb of internal capsule in 11 per cent), cortical abnormalities (51 per cent), white matter haemorrhage (30 per cent), basal nuclei and thalamic lesions (40 per cent), and middle cerebral artery region infarctions (9 per cent).[72] This large study does not support prior reports of occipital predominant damage. The authors postulated that aberrant blood flow response might explain the white matter damage and (perhaps) secondary haemorrhage. Other clinical studies show that a combination of perinatal hypoglycaemia and hypoxia-ischaemia results in more severe damage to the sensorimotor cortex and corticospinal tracts than hypoxia-ischaemia alone.[519]

KERNICTERUS

Kernicterus (meaning yellow nuclei) was described by Schmorl, who observed yellow staining in specific nuclei in the brains of infants who died with severe neonatal jaundice.[481] Neonatal jaundice occurs in the transitional period after birth, a consequence of the breakdown of fetal haemoglobin (which is replaced with adult haemoglobin) and the relatively immature hepatic metabolism of bilirubin. Prematurity and sepsis, as well as other less common factors, can aggravate the situation.[501] Acute bilirubin encephalopathy is associated with poor feeding, lethargy, abnormal tone and posturing and seizures. The long-term sequelae include extrapyramidal abnormalities (choreoathetosis), spasticity, deafness, gaze abnormalities and enamel hypoplasia of deciduous teeth. Although in-hospital phototherapy is very effective at oxidizing bilirubin in cutaneous blood vessels, severe hyperbilirubinemia remains a rare but important cause of neurological damage, in part as a result of inadequate surveillance systems for safe newborn healthcare.[50]

The USA Kernicterus Registry (1992–2004) showed that among infants born >35 weeks' gestation, those who suffered acute bilirubin encephalopathy and/or post-icteric sequelae had total serum bilirubin levels of 20.7 to 59.9 mg per 100 ml.[266] The cellular and molecular basis of bilirubin-induced neurotoxicity remains incompletely understood. Most bilirubin is bound to albumin; however, free unconjugated bilirubin readily crosses the blood–brain barrier.[187] Experimental evidence indicates that low levels of unconjugated bilirubin can affect long-term potentiation in hippocampal neurons, with the possible long-term consequence of learning disability. Higher doses can kill neurons and possibly oligodendrocytes. The precise mechanisms are unclear, with excitotoxicity, inflammation and cytokines potentially contributing to apoptosis.[66,571]

Most of the pathologic features were defined in large series of perinatal autopsies from the 1960s to the 1980s. One of these reported kernicterus in 15 per cent of the autopsied neonates;[8] it is much less commonly encountered now. The macroscopic findings of kernicterus consist of symmetrical bright-yellow discolouration (due to unconjugated bilirubin) of selective neuronal groups (Figure 3.33). In term infants these include the globus pallidus, subthalamus and Ammon's horn, as well as less commonly the thalamus, striatum, cranial nerve nuclei, dentate nucleus, reticular formation, substantia nigra and spinal cord. In premature infants, the yellow colour is prominent in the thalamus, locus coeruleus and Purkinje cells.[8,485,601] The early microscopic lesions consist of vacuolization of neuronal cytoplasm with eosinophilic change, chromatolysis and spongy neuropil. Yellow granules of bilirubin within neurons may be observed in frozen sections but these are subtler in formalin-fixed paraffin-embedded tissue. In subacute lesions there is a loss of neurons and astrocytosis with macrophages. The dentate nucleus in kernicterus has dendritic swellings and eosinophilic bodies positive for synaptophysin and neurofilament antibodies.[305] Autopsy studies have also demonstrated damage in multiple brain stem auditory-related structures, such as the dorsal and ventral cochlear nuclei, superior olive and inferior colliculus, without significant abnormalities of the eighth nerve or inner ear structures.[8,142,185] Study of a single perinatal case suggests that the cerebellum may exhibit axon loss and increased capillary density.[67] Autopsies with

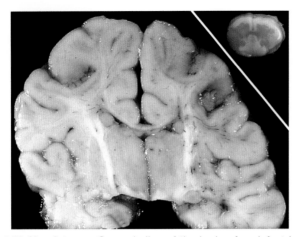

3.33 Kernicterus. Coronal slice of the brain of an infant born at full term who survived 3 weeks. There is symmetric bright-yellow discolouration of the thalami, putamen and tails of the caudate nuclei. Inset: horizontal slice from the cervical spinal cord of a different case showing severe yellow discoloration of the grey matter.

immunohistochemical characterization of neurons of acute neonatal cases and individuals who survived up to 29 years after bilirubin encephalopathy show loss of interneurons from the putamen and external pallidum. The presence of immunoreactivity for 8-hydroxy-2′-deoxyguanosine and 8-hydroxyguanosine suggests that oxidative damage to DNA occurs in lesions of the basal ganglia.[215]

INFECTION

In the perinatal period, viral, bacterial, fungal and protozoal organisms can infect the CNS. Infections may originate during gestation (transplacental), delivery (direct exposure to maternal blood or perineal organisms) or the first postnatal weeks (environmental in the context of an immature immune system). The pattern of brain injury depends upon the age of the infant. It may be difficult to distinguish with certainty the effects of an early infection from a primary malformative process (e.g. one with a genetic basis) or a vascular territory lesion, which can be secondary to infection. Teratogenic effects arise in the early fetal period when the basic structure of the CNS is under formation. Inflammatory destructive processes tend to occur from the second half of gestation onwards. Damage may be the consequence of a direct cytopathic effect or due to the inflammatory response. The microbiological aspects of infections have been reviewed elsewhere in this book (see Volume 2, Chapter 19, Viral Infections; Chapter 20, Bacterial Infections; Chapter 21, Parasitic Infections; and Chapter 22, Fungal Infections). The acronym TORCH was coined decades ago to cover the common congenital infectious agents TOxoplasmosis, Rubella, Cytomegalovirus and Herpes; some writers used STORCH to include Syphilis. In recent years the 'O' refers to 'other', an ever-expanding category including HIV, tuberculosis, *Listeria monocytogenes*, group B *Streptococcus*, leptospirosis, hepatitis B, enteroviruses, adenoviruses, varicella zoster virus and Epstein–Barr virus (see Table 3.6).[314] It should be noted that combined infections can occur.

From a practical diagnostic standpoint, the pathologist starts with an observed abnormality and then tries to determine the underlying cause. Fetuses or newborns with abnormally formed brains or ventriculomegaly plus calcifications are often the victims of transplacental infections in the TORCH category. If they are in the resolution phase there may be no inflammation aside from activated microglia. Diagnostic viral inclusions or infectious organisms are seldom found; sensitive techniques such as polymerase chain reaction (PCR) detection of organism DNA might identify a residual infectious agent. When there is inflammation in the meninges and/or the brain parenchyma, the task may be slightly easier. Evidence of organisms should be sought using histochemical and immunohistochemical techniques and possibly electron microscopy. If infection is suspected at the time of autopsy, samples should be obtained for bacterial and viral culture and tissue should be frozen for possible molecular probing. Although the inflammatory process is generally subtler than in the mature nervous system, the basic patterns are similar. A predominantly lymphocytic infiltrate is suggestive of a viral process, a neutrophilic infiltrate (including overt purulent meningitis) suggests bacterial infection and granulomatous inflammation suggests fungal or *Toxoplasma* infection. It is critical to remember that the brain is seldom the only target organ of congenital infections. Infection is among the most common causes of sudden unexpected deaths among neonates.[572]

Viral Infections

Many viruses can infect pregnant women with only minor adverse effects; however, some may cause increased rate of fetal demise or fetal damage.[5,408] Some viral infections are indiscriminately destructive, whereas others show tropism for specific neuronal, germinal or ependymal cell populations.[111,122,442] The more important ones, with respect to the developing CNS, are discussed here.

Rubella

Rubella virus, which is responsible for German measles, was identified as a cause of fetal cerebral damage in the early twentieth century.[503] The cell receptor for rubella has not been identified, but recent data suggest that myelin oligodendrocyte glycoprotein (MOG) can bind to the E1 envelope glycoprotein of rubella.[101] In cell culture, rubella infects astrocytes more effectively than neurons or oligodendrocytes.[87] This does not, however, explain the pathological findings, which include widespread damage to blood vessel walls.[573] Autopsies on infants with congenital rubella syndrome documented degenerative changes in leptomeningeal and intrinsic arteries and veins associated with foci of necrosis localized mainly in the deep white matter and basal nuclei.[462] Depending on the stage, there may be chronic inflammation in the leptomeninges and eyes.[488] Ocular cataracts, retinopathy and inner ear damage are well documented,[364] as well as retarded myelination.[275] Persistent calcification is a well-known consequence in survivors of *in utero* infection. The cerebral aqueduct may be occluded with resultant hydrocephalus (Figure 3.34).[472] Imaging studies show reduced total volume and focal white matter abnormalities of the brains in adults who had congenital rubella.[328] Rarely reported is a progressive encephalitis that appears as a late consequence of congenital rubella.[537]

TABLE 3.6 Critical pathological features of congenital/perinatal infections

Organism	Major route of infection	Typical timing of infection	CNS histologic features in active phase	CNS changes in chronic/resolution phase (imaging and gross findings)	Eye effects[367]
Coxsackie A and B virus	Transplacental (rare), intrapartum	Anytime	Depends on timing; mid gestation – ventriculitis with necrosis; late gestation – mild meningoencephalitis	Ventriculomegaly	
Cytomegalovirus (CMV)	Transplacental	First trimester has worse outcome	Meningoencephalitis, periventricular necrosis, cytomegalic cells with nuclear inclusions	Microencephaly, ventriculomegaly, polymicrogyria, periventricular calcification and heterotopias	Chorioretinitis and scar; anophthalmia
Herpes simplex virus (HSV type 1 and 2)	Transplacental; intrapartum	Late gestation has worse outcome	Meningoencephalitis; rare haemorrhagic necrosis; intranuclear inclusions	Calcification, infarction, atrophy	Chorioretinitis and scar; conjunctivitis; iridocyclitis; microphthalmia
Human immunodeficiency virus (HIV)	Intrapartum	Perinatal / intrapartum	Meningoencephalitis years after infection, multinucleate cells, myelin loss, calcific vasculopathy	Atrophy, neuron and myelin loss	Secondary infection CMV and toxoplasma
Rubella virus	Transplacental	First trimester	Meningoencephalitis, vasculitis with focal necrosis	Microencephaly, focal calcification, focal cavitation, hydrocephalus due to aqueduct scarring	Retinopathy; iris hypoplasia; microphthalmia; cataracts
Varicella zoster virus (VZV)	Transplacental	Anytime	Meningoencephalitis, ventriculitis, dorsal root ganglioneuritis, spinal motor neuron death	Ventriculomegaly	Chorioretinitis; microphthalmia; cataracts
Treponema pallidum (syphilis)	Transplacental	Late gestation	Lymphocytic meningitis, rare vasculitis	Focal infarction, hydrocephalus, juvenile paresis	Chorioretinitis and scar; conjunctivitis; iridocyclitis; cataracts
Toxoplasma gondii (protozoan)	Transplacental	Anytime	Meningoencephalitis, perivascular lymphocytes, eosinophils, granulomata, necrosis, organisms in tissue	Porencephaly, hydrocephaly, hydranencephaly, periventricular and basal nuclei calcification	Chorioretinitis and macular scars; microphthalmia

Cytomegalovirus

Cytomegalovirus (CMV) has a range of pathological effects, some of which lead to frank malformation of the brain. Glycoproteins on the viral surface bind with a putative gH/gL receptor on cell surface. This allows fusion or entry by pinocytosis, depending on the viral subtype and the cell type.[549] CMV can infect many cell types; neural progenitor cells including radial glia and endothelial cells are most vulnerable although differentiated neurons and microglia are relatively resistant.[90,489,528] Infected cells, identifiable by the characteristic inclusions, eventually undergo apoptosis. Destruction of the ventricular and subventricular zones of the developing brain results in malformations because progenitor cells fail to reach their destination. The timing of the infection determines the type of malformation. Autopsies on mid-gestation fetuses with CMV infection show that the brain is infected in approximately half of the cases. CMV can be found throughout the brain and is associated with inflammation and focal necrosis.[180] Similar destructive

lesions can be found in the inner ear.[527] Periventricular calcification and haemorrhage may occur along with cerebellar damage.[82,410,427,482] Children who survive *in utero* CMV infection may have severe neurological defects and brain malformations including microencephaly, lissencephaly, polymicrogyria and cerebellar hypoplasia.[62,135] More subtle abnormalities commonly include chorioretinitis and optic atrophy[17] and, more rarely, a diffuse leukoencephalopathy.[523]

Herpes Simplex

Herpes simplex viruses (HSVs) can bind to cells through several different receptors, but the most important for infection of neurons and radial glia is likely nectin-1.[438,467] Infected cells upregulate Toll-like receptors which, in turn, appear to mediate cell death through inflammatory processes.[262] Transplacental transmission of HSV to the fetus is rare. During the active process a necrotizing meningoencephalitis and/or retinitis may be seen. The pathologic lesions caused by HSV include inclusion-bearing cells (Cowdry type A),

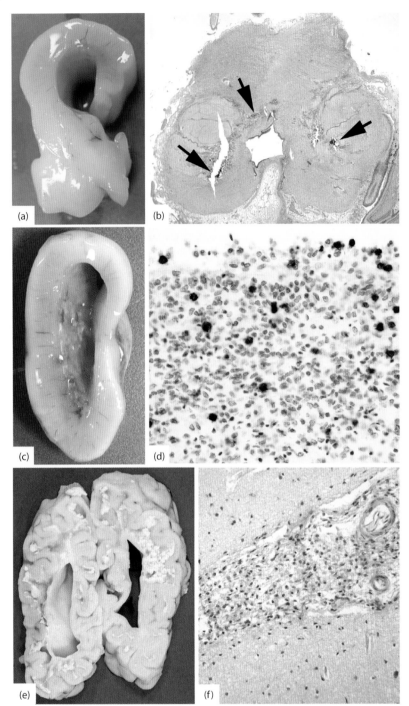

3.34 Viral infection of the fetal and perinatal brain. Enlarged cerebral ventricles (possibly along with other somatic abnormalities) detected on screening ultrasound may lead to termination of the pregnancy or be associated with ultimate fetal demise. In some of these cases, microscopic findings may be suggestive of a viral infection that has caused focal scarring and calcification (e.g. in the vicinity of the cerebral aqueduct or fourth ventricle) or severe widespread destruction associated with inflammation. An infectious cause cannot always be proven. **(a)** Half brain (frontal) from a 21-week fetus with enlarged, smooth-walled lateral ventricle. **(b)** Micrograph of the pons showing extensive calcification (arrows) surrounding upper fourth ventricle. Haematoxylin and eosin; original magnification 12.5×. There were activated microglia but no lymphocytes in the tissue; a viral cause was suspected but not proven. **(c)** Half brain (parietal) from a 24-week fetus with hydropic change and calcified liver, stillborn to a mother with Coxsackie B4 infection in the amniotic fluid (proven by polymerase chain reaction assay). The wall of the ventricle shows white patches due to calcification. **(d)** Microscopic examination showed loss of the ependymal layer with haemorrhage and inflammation in the periventricular region. Immunostaining for CD3, a marker of T lymphocytes; original magnification 400×. **(e)** Photograph of coronal slice through brain of a child born at 37 weeks' gestation who developed seizures and was found to have extensive brain calcification on the third day of life; he died at 15 weeks. The white matter is extensively calcified. **(f)** The leptomeninges exhibited fibrosis and scattered lymphocytes. Haematoxylin and eosin; original magnification 100×. Microscopic, ultrastructural and immunohistochemical studies were unrevealing; however, polymerase chain reaction assay of the brain tissue demonstrated the presence of herpes simplex type 1.

necrosis and inflammation. If the fetus does not die, healing is accompanied by microencephaly, cavitation (even hydran-encephaly) and extensive calcification (Figure 3.34).[253] Neonatal HSV infection usually is acquired during the birth process, as the neonate comes in contact with the virus during passage through an infected birth canal. Subsequent brain infection may occur 2–4 weeks later.[281]

Enteroviruses

Enteroviruses are members of the picornavirus family, a large and diverse group of small RNA viruses. Within the ten species are multiple serotypes including Coxsackieviruses A serotypes (human enteroviruses A and C), Coxsackievirus B and echovirus (human enterovirus B) and poliovirus (human enterovirus C). These are common causes of infection in neonates and infants. Coxsackieviruses enter cells using the coxsackievirus and adenovirus receptor (CAR) that is expressed during fetal development, particularly in the CNS and muscle, with a rapid downregulation after birth as cells mature.[173] In mice, type B Coxsackie viruses infect neuronal progenitor cells in the subventricular zone.[162] *In utero* infection with Coxsackie B4 at midgestation is associated with severe ventriculitis (Figure 3.34), although infection late in gestation causes more subtle leptomeningitis and mild encephalitis.[249] Another unspecified enterovirus was reported to cause subependymal haemorrhage in a 35-week fetus.[314] Echovirus may cause neonatal meningoencephalitis.[374] Rare autopsy cases of late congenital infections show nonspecific reactive glial changes, probably as a result of hypoxia, but no inflammation.[63,181]

Poliovirus is the human enterovirus that causes poliomyelitis. In the 1940s–50s, polio infection during pregnancy was a major problem, with increased fetal mortality.[42] Development of the vaccine has largely eradicated the infection, although it periodically surfaces in the developing world. Although feared in the early era, administration of the vaccine during pregnancy has no adverse effect on the fetus.[408] Transplacental transmission is extremely rare and intrapartum or neonatal transmission from an infected mother is more likely.[585] Rare cases of congenital poliomyelitis show that paralysis can develop within 2–3 days of birth.[81] Autopsies on neonates show lymphocytic infiltrate and dying motor neurons.[150]

HIV

Neurological manifestations of HIV are rare in the newborn period and do not present until later in infancy.[477] In approximately 85 per cent of paediatric cases of the acquired immunodeficiency syndrome (AIDS), HIV infection appears to be vertically transmitted from mother to infant during gestation or the perinatal period, with blood transfusion accounting for most of the remaining cases.[83] Abuse of intravenous drugs by the mother accounts for most cases of maternal infection.[83] HIV has been isolated from fetal tissues, including the CNS, as early as 15 weeks of gestation.[335,496] Longitudinal studies in children have documented neurological abnormalities in approximately 90 per cent of patients, mostly as a result of a direct HIV infection of the CNS, whereas a spectrum of opportunistic disorders accounts for a high percentage of neurological disease in adults.[73] Neurological manifestations of direct HIV infection in infants and young children

include primary HIV encephalopathy, with microcephaly in the congenital cases, pyramidal tract abnormalities and cognitive impairment characterized by loss of language and social skills. Morphologically, there is cerebral parenchymal atrophy, symmetrical ventricular dilation, calcification in the walls of blood vessels of the basal ganglia and white matter of frontal lobe, myelin alterations (broad, ill-defined areas of pallor in cerebral white matter to more discrete, well-defined areas of demyelination), white matter gliosis and scattered macrophages and inflammatory infiltrates with perivascular microglial nodules and multinucleated giant cells. Cerebral cortical alterations include neuronal depopulation, astrogliosis and scattered multinucleated giant cells. Spinal cord abnormalities include symmetrical pallor of the corticospinal tract.[132] Vacuolar myelopathy, reported in adults, is rare in children. In congenital cases, HIV encephalopathy is associated with microcephaly, but small brain size may be due to secondary destructive effects of the virus as well as a primary teratogenic effect upon growth. The incidence of encephalopathy after HIV infection is higher in infants than adults (9.9 per cent versus 0.3 per cent, respectively).[522] HIV encephalopathy in infancy is more likely an isolated symptom of AIDS than in adults and is associated with a reduction in intrauterine brain growth and very low levels of HIV-1 RNA in the cerebrospinal fluid. It also occurs at a higher level of immunocompetence than in adults and is not prevented by zidovudine treatment during gestation.[522] In short, HIV encephalopathy in infants is thought to have a different pathogenetic mechanism than that occurring in adults, attributed to intrauterine brain infection and the subsequent pathological effects of the virus on the developing brain.[522] The pathogenesis of neurological manifestations of HIV infection is discussed in detail in Volume 2, Chapter 19, Viral Infections.

Lymphocytic Choriomeningitis Virus (LCMV)

LCMV is a rodent-borne RNA virus that has been shown to be associated with a high incidence of fetal demise and may cause brain haemorrhage, microencephaly, pachygyria, porencephalic cysts, periventricular calcifications, ventriculomegaly, cerebellar hypoplasia and retinopathy.[15,41,60] Rare autopsy studies show lymphocytic infiltration, focal softening, reactive glial changes and perivascular oedema.[483]

Bacterial Infections

Treponema pallidum

T. pallidum is the spirochete bacterium that causes syphilis. Transplacental transmission is most likely in untreated pregnant women with primary syphilis and can occur as early as 9–10 weeks' gestation.[240] Congenital syphilis is less common than it once was, but remains a significant problem especially in the developing world.[227] The fetal effects are potentially severe, with the risk of stillbirth approximately 30 per cent.[358] Newborns may have spirochetes in the meninges and eyes associated with a lymphocyte- or macrophage-predominant inflammation.[212,223] Widespread ischaemic/haemorrhagic lesions have also been described.[166] Late presentation of congenital syphilis can be loss of hearing in infancy or childhood in association with abnormal tooth structure.[88]

Bacterial Meningitis in the Neonate

This term applies to the onset of meningitis in the first 30 days of life (Figure 3.35). Bacterial meningitis is the most common and serious type of neonatal intracranial bacterial infection.[566] It is more common in premature than full-term infants, and more common in the first months of life than in any succeeding month. The global incidence of bacterial meningitis has been estimated to be 0.7–1 per 1000 live births, with variable rates in different geographic regions.[502] Host factors that underlie the vulnerability include developmental deficiencies in immunity, obstetric complications, prematurity, prolonged rupture of membranes, maternal infection and chorioamnionitis.[480] The major organisms associated with neonatal meningitis, in order of frequency, are group B *Streptococcus*, *Escherichia coli* and *Listeria monocytogenes*. Remaining cases are secondary to other streptococcal and staphylococcal species, other Gram-negative enteric bacilli, *Neisseria meningitidis* and various unusual organisms. In the developing world, other organisms including Gram-negative bacilli (excluding *E. coli*), *Streptococcus pneumoniae*, *S. aureus* and *Haemophilus influenza* appear to be more important pathogens.[179] The prognosis for neonatal meningitis depends upon the microorganism, the age, the presence of seizures or coma, circulatory failure due to sepsis and the rapidity of diagnosis and therapy.[114] For group B streptococcal meningitis, severe neurological deficits have been reported in 21 per cent of cases.[148] Careful neuropathological investigation for bacterial meningitis is essential in sick premature infants that come to autopsy.

Two patterns of neonatal meningitis are differentiated by their times of onset.[244,429,543,577] Early-onset meningitis begins during the first week postpartum, usually in newborns with a history of complications during labour and delivery. The infection is acquired just before or during delivery and the responsible organisms are almost always group B *Streptococcus* or *E. coli*. Late-onset meningitis is due to postnatal contamination: it may begin as early as the fourth postpartum day, but it usually occurs after the first week. A variety of organisms, including *Staphylococcus*, *Pseudomonas* and *Klebsiella*, have been implicated, in addition to *E. coli* and group B *Streptococcus*. Because bacteraemia almost always precedes meningitis, a close association exists between the causes and rates of neonatal sepsis and meningitis. Septic dissemination may result from pneumonia, gastroenteritis, umbilical and skin infections and other systemic infections.

The study of cases of neonatal meningitis with variable lengths of survival has led to a basic understanding of its pathogenesis.[48,56] Bacteria invade the CNS *via* the blood. They appear to localize first in the choroid plexus and cause choroid plexitis, with entrance of bacteria into the ventricular system and subsequent movement to the leptomeninges *via* cerebrospinal fluid flow. Ventriculitis is a particularly common feature of neonatal meningitis. In the acute phase, i.e. the first week, the predominating cells in the subarachnoid space and ventricles are neutrophils, with scattered monocytes. Inflammatory infiltrates extend into the Virchow–Robin spaces. Bacteria are detectable free and within neutrophils and macrophages. Vasculitis occurs from the first week, with involvement of arteries and veins developing as an extension of the inflammatory reaction in the leptomeninges. Thrombosis with infarction may occur as early as the first week; the lesions may be venous or arterial distribution. The site of infarction is most often the cerebral cortex and subjacent white matter, although subependymal and deep white matter infarction may occur.[233] Cerebral oedema in the acute phase is common because cerebrovascular autoregulation is impaired, permeability of the blood–brain barrier is increased and an inflammatory response, with the synthesis and secretion of cytokines, is induced. The cerebrospinal fluid culture may be positive, with little or no evidence of leptomeningeal inflammation at autopsy, particularly with early-onset group B streptococcal infections; this discrepancy may reflect the rapidity of death or the immaturity of host responses. Experimental studies indicate that group B *Streptococcus* enters the brain by penetrating the blood–brain barrier *via* mechanisms that include cell membrane-anchored lipoteichoic acid expression on the group B streptococcal surface.[137]

In the second and third weeks, the proportion of neutrophils decreases to approximately 25 per cent of the total inflammatory population. Lymphocytes and plasma cells are present in relatively small numbers. Cranial nerves are infiltrated by exudate. Ventriculitis is characterized by an inflammatory exudate on the ependymal surface and the development of glial adhesions. After the third week, the exudate decreases and consists mainly of lymphocytes and macrophages. The exudate becomes organized, with proliferation of arachnoidal fibroblasts and formation of collagen strands. Dense fibrosis in the leptomeninges and ventricular system can obstruct cerebrospinal fluid flow and cause hydrocephalus

Bacterial Meningitis in the Infant

Haemophilus influenzae type b, *Streptococcus pneumoniae* and *Neisseria meningitidis* cause the majority of cases of meningitis in infants and young children up to two

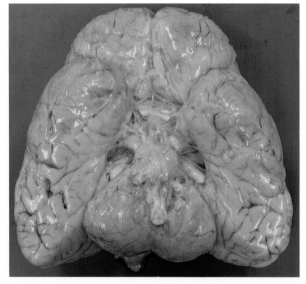

3.35 Neonatal bacterial (*Haemophilus influenzae*) meningitis. Yellow-green purulent exudates are present over the entire brain of a 5-month-old infant who presented with otitis media and pneumonia and progressed to meningitis and death within 1 day.

postnatal years of age.[6] The changes in susceptibility are explained in part by changes in antibody formation in the neonate. Antibodies to *Haemophilus*, for example, readily cross the placenta and grant passive immunity to the newborn; immunity subsides with the expected turnover and decay of antibodies. Vaccines are available for paediatric immunization against these organisms.

Fungal Infections

Fungal infection of the infant brain is much less common than viral or bacterial infection. The most common fungal organisms that infect the neonate, particularly the premature infant, are *Candida albicans, Mucor, Cryptococcus, Coccidioides* and *Aspergillus*.[140,351,448] The sick, premature newborn in the intensive care unit is particularly vulnerable to disseminated *Candida* and other fungal infections for the following reasons: developmental immune deficiencies, presence of prolonged indwelling vascular catheters and endotracheal tubes (breakdown of anatomical barriers to infection) and the use of multiple courses of broad-spectrum antibiotics (disruption of microbial flora). Approximately 5 per cent of premature infants develop candidiasis and one-third of these cases exhibit CNS involvement.[157,158] The CNS manifestations of candidiasis are purulent leptomeningitis with ventriculitis. Microscopically, fungal hyphae, neutrophilic/lymphocytic inflammation and necrotizing encephalitis are seen. Aspergillosis most commonly presents as abscess in infants and has a better prognosis than in adults.[140] Transplacental infection of the fetus by fungal organisms is extremely rare and confined to case reports.[361,436]

Toxoplasma and Other Parasites

Toxoplasmosis is a parasitic disease caused by the protozoan *Toxoplasma gondii* whose usual hosts are birds and small mammals, including domestic cats. Transmission to the fetus is most likely in the circumstance of newly infected pregnant women who seroconvert before 25–30 weeks' gestation.[103] Approximately 1:1000–1:8000 pregnancies are affected, with considerable regional variation.[130,201] Congenital toxoplasmosis can have severe adverse effects on the developing nervous system, including eye abnormalities (choroidoretinitis) (92 per cent), brain calcifications (79 per cent), and hydrocephalus (67 per cent).[403] The brain changes can be detected *in utero* at 24–34 weeks' gestation by ultrasound imaging.[341] Once it has entered the fetal circulation, *T. gondii* infects circulating monocytes and gains access to the brain (and other organs, including eyes) where it infects glial cells and neurons. A cytokine response is initiated with ensuing cell death.[80] Microscopically, the inflammatory reaction consists of lymphocytes and macrophages, with fewer neutrophils and plasma cells, along with microglial reaction. Most or all of the bradyzoites liberated by cell rupture are destroyed by the immune process. Infarctive necrosis may occur if blood vessel walls are affected. Calcification occurs in the resolution phase. Even small foci of infection can have profound effects if they cause scarring near the cerebral aqueduct or fourth ventricle.[174]

The protozoan parasite *Trypanosoma cruzi*, which causes Chagas disease prevalent in Latin America, can be transmitted to the fetus in approximately 5 per cent of infected pregnancies.[533] Brain examination of congenital infection in fetuses reveals meningoencephalitis, free parasites and reactive gliosis.[402]

Systemic Effects of Maternal and Neonatal Infection

Of increasing importance is the recognition that maternal/placental infection in the fetal period or systemic infection in the neonatal period may compromise subsequent neurological development, irrespective of direct CNS infection. There are strong epidemiologic associations between maternal infection and the development of cerebral palsy and intrauterine growth retardation.[398] At the histopathological level, many of the changes associated with premature birth (discussed earlier) can be seen. There is also weak evidence, but a vocal proponent base, that maternal infections are risk factors for autism, schizophrenia and other neurological disorders. The broad hypothesis is that cytokines enter the fetal circulation and then the brain with subsequent adverse effects on neuronal and glial development. Many possible routes of potentiation, including microglial activation and epigenetic changes, are postulated.[59,71] There are no specific correlates to this in the conventional neuropathological examination, although activated microglia can be identified in many cases.

Systemic infection in the neonate may have similar adverse effects. In a study involving 6093 ELBW survivors, the majority (65 per cent) had at least one infection and 5 per cent had a positive cerebrospinal fluid culture over their hospital course.[502] The adverse outcomes included cerebral palsy, low mental and psychomotor development indices, and visual impairment.[502] Microbes and/or their products in systemic sites may stimulate the production of cytokines, which have been shown experimentally to be toxic to developing oligodendrocytes.[564]

SUDDEN INFANT DEATH SYNDROME

Sudden infant death syndrome (SIDS) is defined as 'the sudden death of an infant under one year of age that remains unexplained after a thorough case investigation, including performance of a complete autopsy, examination of the death scene and review of the clinical history'.[582] As a syndrome, it may have more than one underlying cause.[285] As a 'diagnosis' of exclusion, the term is inconsistently used and some medicolegal jurisdictions are now avoiding it.[55] SIDS is temporally associated with sleep, leading to the premise that it occurs during sleep or transitions between sleep and waking. SIDS peaks at 2–4 months of age and 90 per cent of cases occur under 6 months of age, suggesting that developmental mechanisms play a major role in its pathogenesis. The time frame of SIDS coincides with the period in which dramatic and rapid changes occur as the newborn adapts to extrauterine life.[286] Despite an almost 50 per cent fall in the incidence of SIDS in the USA following the 1994 national recommendation for the supine sleeping position, SIDS remains a leading cause of postneonatal infant mortality, with an overall incidence of 0.57 per 1000 live births. Yet, the disparity in SIDS rates among racial and ethnic groups persists, likely reflecting differences in genetic and environmental factors as well as in infant sleep practices. An understanding of the underlying cause(s) and pathogenesis of SIDS is essential to provide biological plausibility to risk-reduction recommendations, and to promote their adoption by health providers and caretakers.

In order to conceptualize SIDS, a triple-risk model has been proposed.[164] According to this model, SIDS occurs when three factors impinge simultaneously upon the infant: (i) an underlying vulnerability in the infant's homoeostatic control (an intrinsic factor), (ii) a critical developmental period in autonomic and respiratory control and state maturation (another intrinsic factor) and (iii) an exogenous stressor, e.g. hypoxia, hypercapnia, asphyxia or reflex apnoea from the prone sleep position or hyperthermia from over-bundling (an extrinsic factor). Only when these three factors come together will SIDS occur. The intrinsic factors explain why not all babies die when placed prone: healthy babies do not die from SIDS; rather, only infants of a certain age with an underlying pathophysiological process ('vulnerability') die from SIDS.

With regard to the extrinsic risk factors, the mechanism that may operate during the prone sleeping position involves rebreathing of expired gases with hypoxia, hypercapnia or asphyxia, upper airway obstruction, impaired arousal thresholds in the prone position that hamper effects to turn the head, compromised upper airway reflexes, hyperthermia due to heat trapping in the face-down position or altered sensory/vestibular influences on blood pressure. All of these mechanisms potentially involve the brain stem. An additional risk factor for SIDS includes co-sleeping, which may increase ambient temperature (thermal stress) or contrybute to rebreathing in proximity to others in the bed.[54] Pacifiers have been shown to reduce the risk for SIDS, by potentially changing the passage of air surrounding the nose and mouth or by enhancing the development of neural pathways that control upper airway patency.[324]

Intrinsic factors include genetic polymorphisms in the serotonin transporter (5-HTT)[385,574] and cytokine interleukin-10 (IL-10)[506] For the 5-HTT polymorphism, the long (l) allele is more frequent than the short (s) allele in SIDS cases compared with controls. This polymorphism is associated with an increased 5-hydroxytryptamine (5-HT) reuptake, which in turn presumably leads to decreased 5-HT at the synapse. The potential relevance of 5-HTT polymorphisms to SIDS is strengthened by reports of 5-HT abnormalities in SIDS brain stems. Other intrinsic risk factors are suggested by three lines of evidence: (i) maternal and pregnancy-related factors associated with increased SIDS risk; (ii) neonatal abnormalities in neurological or autonomic function in at least some infants who subsequently die from SIDS; and (iii) subtle CNS and/or systemic abnormalities in autopsied cases of SIDS.[286] Epidemiological risk factors include prematurity, low birth weight, young maternal age, maternal anaemia, maternal cocaine and heroin abuse during pregnancy and maternal cigarette smoking during pregnancy. Late preterm infants have a two-fold greater risk for SIDS than term infants (1.4 per 1000 cases at 33–36 gestational weeks compared with 0.7 per 1000 at more than 37 gestational weeks). Relevant to the brain stem hypotheses in SIDS, brain stem function is less mature and breathing abnormalities are more frequent in preterm infants compared with term infants. The maternal risk factors further point to a suboptimal

intrauterine environment, i.e. contributions to SIDS risk from factors in fetal life.

A variety of autonomic, respiratory and sleep/wake state irregularities have been reported in epidemiological studies.[286] Other epidemiological risks for SIDS comprise diminished ventilatory responsiveness and/or impaired arousal to hypercarbia or hypoxia.[248] Analysis of cardiorespiratory recordings of the time of death in home-monitored SIDS and other infants show a profound bradycardia, as well as terminal gasping, suggesting that failure to autoresuscitate may be an important contributing factor in these deaths.[365]

Neuropathological abnormalities in cases of SIDS (Figure 3.36) include increased brain weight, subtle brain stem scarring, altered dendritic and spine density development in the brain stem, increased synaptic density in the central reticular nucleus of the medulla, altered brain stem neurotransmitter parameters, subtle hypomyelination within and rostral to the brain stem, cerebral white matter gliosis, increased neuropil in the hypoglossal nucleus, delayed development of the cervical vagus nerve, inferior olivary abnormalities, delay in the maturation of the external granular layer of the cerebellum and increased substance P in trigeminal fibres, reviewed by Kinney and others.[282,283,424] These findings are non-specific, and it is not clear which, if any, are related directly to the pathogenesis of sudden death. Several of these findings are considered secondary to pre- or perinatal hypoxia-ischaemia, which may play a role in the pathogenesis of SIDS. Other findings (e.g. delayed myelination, delayed cerebellar granule cell maturation, altered dendrite spine development) suggest abnormalities in developmental programmes.

One neural hypothesis is that SIDS, or a subset, results from a failure of the medullary serotonergic (5-HT) system in mediating homoeostatic responses to life-threatening challenges (e.g. asphyxia, hypoxia, hypercapnia) during sleep, as the vulnerable infant passes through the critical developmental period (triple-risk model).[282] The medullary 5-HT system is composed of 5-HT neurons intermingled with non-5-HT cells in the midline raphe, lateral extra-raphe regions and ventral surface (arcuate nucleus). The 5-HT neurons in this system project widely to other brain stem and spinal cord sites and are involved in the integration of homeostatic responses, including in respiration, chemosensitivity, blood pressure and temperature changes, upper airway reflexes and arousal.[282] The human arcuate nucleus at the ventral medullary surface is postulated to be homologous to chemosensitive respiratory zones in experimental animals at this same site.[165] The presence of 5-HT and glutamatergic neurons embedded within it strengthens this idea, as these neuronal phenotypes at the ventral surface have been shown to be chemosensitive to carbon dioxide in different experimental paradigms.[424] Severe hypoplasia or absence of the arcuate nucleus has been reported in a subset of SIDS,[163,304,343,356] correlated in one case with clinically recognized insensitivity to carbon dioxide.[168] Anatomic connections between the arcuate nucleus and the caudal raphe suggest a potential functional link between these two regions.[596]

Abnormalities in cell number as well as neurotransmitter binding levels in the medullary 5-HT system in SIDS cases have been reported.[282,289,593] The data further suggest

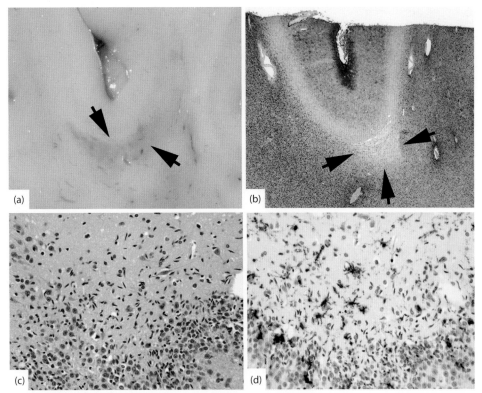

3.36 Neuropathological findings in sudden infant death syndrome (SIDS). By definition, major abnormalities are not found, however 'incidental' findings that suggest prior hypoxic damage are not uncommon. Small foci of white matter damage may be found with careful inspection, as shown in (a) and (b). Foci of microglial activation, particularly along the inner margin of the dentate gyrus of the hippocampal formation (as shown in [c]) and [d]) suggest mild hypoxic insults in the days prior to death.[124] **(a)** Photograph showing small focus of subcortical discolouration and softening (arrows) in the parietal white matter of a 5-month-old SIDS case. **(b)** Micrograph showing S100 immunoreactivity in the parietal white matter of a 4-month-old who was born at 36 weeks' gestational age. A small subcortical area (arrows) has reduced glial cell density. Original magnification 40×; the same region had reduced myelin staining. **(c)** Micrograph showing an abundance of rod-shaped nuclei along the dentate gyrus of a 4-month-old SIDS case. Haematoxylin and eosin; original magnification 400×. **(d)** Many, but not all, of the rod-shaped cells are immunoreactive for HLA-DR (a marker of activated microglia). Original magnification 400×. In some cases, immunostains for CD45 or CD68 may label the cells more uniformly; the significance of this is not known.

that prenatal exposures to alcohol and cigarette smoking are associated with reduced 5-HT binding in this system, providing a potential biologic basis for 'intrinsic' 5-HT deficits increasing the risk for SIDS in at least of subset of cases (Figure 3.37).[290] However, recent data have failed to show differences between some of these brainstem neurochemical abnormalities when comparing infant deaths with and without asphyxia-generating circumstances.[447] Detailed assessment of blood vessels in the cerebellum showed no differences between SIDS and control cases; this raises questions about the concept that SIDS infants are subjected to chronic hypoxia.[380]

EPILOGUE

This chapter is by no means comprehensive. It is meant to highlight the important neuropathological aspects of the more commonly encountered perinatal brain abnormalities. The reader should remember that there are entire texts devoted to the subject, some of them old but still relevant. Among these are books by Abraham Towbin,[534,536] Cyril Courville,[105] Lucy Rorke,[460] Floyd Gilles[193] and Waney Squier.[497]

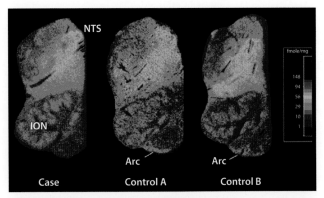

3.37 5-HT receptor binding in SIDS. Autoradiograms showing 5-hydroxytryptamine (5-HT) receptor binding in the hemisected medulla oblongata at the level of the inferior olivary nucleus (ION). In a case of sudden infant death syndrome (SIDS) dying at 2 weeks of age (Case) the arcuate nucleus (Arc) appears to have no binding, compared with 44 and 46 post-menstrual week controls (A and B) who died suddenly of known cause. In all samples, the nucleus of the solitary tract (NTS) shows strong binding. Scale in femtomoles/milligram of tissue.

From Kinney et al.[290] With permission from Lippincott Williams & Wilkins/Woilters Kluwer Health.

REFERENCES

1. Abbott NJ, Patabendige AA, Dolman DE, Yusof SR, Begley DJ. Structure and function of the blood–brain barrier. *Neurobiol Dis* 2010;**37**(1):13–25.

2. Abe S, Takagi K, Yamamoto T, Okuhata Y, Kato T. Assessment of cortical gyrus and sulcus formation using MR images in normal fetuses. *Prenat Diagn* 2003;**23**:225–231.

3. Abraham H, Tornoczky T, Kosztolanyi G, Seress L. Cell formation in the cortical layers of the developing human cerebellum. *Int J Dev Neurosci* 2001;**19**(1):53–62.

4. Adams RD, Prod'hom LS, Rabinowicz T. Intrauterine brain death. Neuraxial reticular core necrosis. *Acta Neuropathol* 1977;**40**(1):41–49.

5. Adams Waldorf KM, McAdams RM. Influence of infection during pregnancy on fetal development. *Reproduction* 2013;**146**(5):R151–62.

6. Agrawal S, Nadel S. Acute bacterial meningitis in infants and children: epidemiology and management. *Paediatr Drugs* 2011;**13**:385–400.

7. Agresti C, Bernardo A, Del Russo N, *et al.* Synergistic stimulation of MHC class I and IRF-1 gene expression by IFN-gamma and TNF-alpha in oligodendrocytes. *Eur J Neurosci* 1998;**10**(9):2975–83.

8. Ahdab–Barmada M, Moossy J. The neuropathology of kernicterus in the premature neonate: diagnostic problems. *J Neuropathol Exp Neurol* 1984;**43**(1):45–56.

9. Ainslie PN, Ogoh S. Regulation of cerebral blood flow in mammals during chronic hypoxia: a matter of balance. *Exp Physiol* 2010;**95**(2):251–262.

10. Akanli LF, Trasi SS, Thuraisamy K, *et al.* Neonatal middle cerebral artery infarction: association with elevated maternal anticardiolipin antibodies. *Am J Perinatol* 1998;**15**(6):399–402.

11. Alexander JM. Fetal injury associated with cesarean delivery. *Obstet Gynecol* 2006;**108**:885–890.

12. Alexander JM, Gilstrap LC, Cox SM, McIntire DM, Leveno KJ. Clinical chorioamnionitis and the prognosis for very low birth weight infants. *Obstet Gynecol* 1998;**91**(5 Pt 1):725–9.

13. Amar AP, Aryan HE, Meltzer HS, Levy ML. Neonatal subgaleal hematoma causing brain compression: report of two cases and review of the literature. *Neurosurgery* 2003;**52**(6):1470–4; discussion 1474.

14. Amiel–Tison C, Allen MC, Lebrun F, Rogowski J. Macropremies: underprivileged newborns. *Ment Retard Dev Disabil Res Rev* 2002;**8**(4):281–292.

15. Anderson JL, Levy PT, Leonard KB, *et al.* Congenital lymphocytic choriomeningitis virus: when to consider the diagnosis. *J Child Neurol* 2013; [Epub ahead of print].

16. Anderson JM, Milner RD, Strich SJ. Effects of neonatal hypoglycaemia on the nervous system: a pathological study. *J Neurol Neurosurg Psychiatry* 1967;**30**(4):295–310.

17. Anderson KS, Amos CS, Boppana S, Pass R. Ocular abnormalities in congenital cytomegalovirus infection. *J Am Optom Assoc* 1996;**67**(5):273–278.

18. Andiman SE, Haynes RL, Trachtenberg FL, *et al.* The cerebral cortex overlying periventricular leukomalacia: analysis of pyramidal neurons. *Brain Pathol* 2010;**20**(4):803–814.

19. Ando M, Takashima S, Mito T. Endotoxin, cerebral blood flow, amino acids and brain damage in young rabbits. *Brain Dev* 1988;**10**(6):365–370.

20. Anstrom JA, Brown WR, Moody DM, *et al.* Subependymal veins in premature neonates: implications for haemorrhage. *Pediatr Neurol* 2004;**30**(1):46–53.

21. Anton J, Pineda V, Martin C, Artigas J, Rivera J. Posttraumatic subgaleal hematoma: a case report and review of the literature. *Pediatr Emerg Care* 1999;**15**(5):347–349.

22. Arai Y, Deguchi K, Mizuguchi M, Takashima S. Expression of beta-amyloid precursor protein in axons of periventricular leukomalacia brains. *Pediatr Neurol* 1995;**13**(2):161–163.

23. Arisio R, Bonissone M, Piccoli E, Panzica G. Central nervous system microangioarchitecture in the human fetus. *Adv Clin Path* 2002;**6**(3–4):125–129.

24. Arnold T, Betsholtz C. Correction: The importance of microglia in the development of the vasculature in the central nervous system. *Vasc Cell* 2013;**5**(1):12–824X–5–12.

25. Attwell D, Buchan AM, Charpak S, *et al.* Glial and neuronal control of brain blood flow. *Nature* 2010;**468**(7321):232–243.

26. Auer RN. Progress review: hypoglycemic brain damage. *Stroke* 1986;**17**(4):699–708.

27. Back SA, Volpe JJ. Cellular and molecular pathogenesis of periventricular white matter injury. *Ment Retard Dev Dis* 1997;**3**:96–207.

28. Back SA, Gan X, Li Y, Rosenberg PA, Volpe JJ. Maturation-dependent vulnerability of oligodendrocytes to oxidative stress-induced death caused by glutathione depletion. *J Neurosci* 1998;**18**(16):6241–53.

29. Back SA, Luo NL, Borenstein NS, *et al.* Late oligodendrocyte progenitors coincide with the developmental window of vulnerability for human perinatal white matter injury. *J Neurosci* 2001;**21**(4):1302–1312.

30. Back SA, Han BH, Luo NL, *et al.* Selective vulnerability of late oligodendrocyte progenitors to hypoxia-ischaemia. *J Neurosci* 2002;**22**(2):455–63.

31. Back SA, Luo NL, Borenstein NS, *et al.* Arrested oligodendrocyte lineage progression during human cerebral white matter development: dissociation between the timing of progenitor differentiation and myelinogenesis. *J Neuropathol Exp Neurol* 2002;**61**:197–211.

32. Back SA, Luo NL, Mallinson RA, *et al.* Selective vulnerability of preterm white matter to oxidative damage defined by F2-isoprostanes. *Ann Neurol* 2005;**58**(1):108–20.

33. Bale TL, Baram TZ, Brown AS, *et al.* Early life programming and neurodevelopmental disorders. *Biol Psychiatry* 2010;**68**(4):314–319.

34. Ballabh P. Pathogenesis and prevention of intraventricular haemorrhage. *Clin Perinatol* 2014;**41**(1):47–67.

35. Ballabh P, Braun A, Nedergaard M. Anatomic analysis of blood vessels in germinal matrix, cerebral cortex and white matter in developing infants. *Pediatr Res* 2004;**56**(1):117–124.

36. Ballabh P, Hu F, Kumarasiri M, Braun A, Nedergaard M. Development of tight junction molecules in blood vessels of germinal matrix, cerebral cortex, and white matter. *Pediatr Res* 2005;**58**(4):791–798.

37. Bamberg C, Kalache KD. Prenatal diagnosis of fetal growth restriction. *Semin Fetal Neonatal Med* 2004;**9**(5):387–94.

38. Bar T. Patterns of vascularization in the developing cerebral cortex. *Ciba Found Symp* 1983;**100**:20–36.

39. Barkovich AJ, Westmark K, Partridge C, Sola A, Ferriero DM. Perinatal asphyxia: MR findings in the first 10 days. *AJNR Am J Neuroradiol* 1995;**16**(3):427–38.

40. Barks JD, Silverstein FS. Excitatory amino acids contribute to the pathogenesis of perinatal hypoxic-ischaemic brain injury. *Brain Pathol* 1992;**2**(3):235–243.

41. Barton LL, Mets MB. Congenital lymphocytic choriomeningitis virus infection: decade of rediscovery. *Clin Infect Dis* 2001;**33**:370–374.

42. Bates T. Poliomyelitis in pregnancy, fetus and newborn. *AMA Am J Dis Child* 1955;**90**:189–195.

43. Baud O, Haynes RF, Wang H, *et al.* Developmental up-regulation of MnSOD in rat oligodendrocytes confers protection against oxidative injury. *Eur J Neurosci* 2004;**20**(1):29–40.

44. Bayer SA, Altman J, Russo RJ, Zhang X. Timetables of neurogenesis in the human brain based on experimentally determined patterns in the rat. *Neurotoxicology* 1993;**14**(1):83–144.

45. Beck S, Wojdyla D, Say L, *et al.* The worldwide incidence of preterm birth: a systematic review of maternal mortality and morbidity. *Bull World Health Organ* 2010;**88**:31–38.

46. Becker LE, Armstrong DL, Chan F, Wood MM. Dendritic development in human occipital cortical neurons. *Brain Res* 1984;**315**(1):117–124.

47. Beckman CRB, Ling FW, Laube DW, *et al. Obstetrics and gynecology.* Baltimore: Lippincott Williams and Wilkins, 2004.

48. Bellinger DC, Jonas RA, Rappaport LA, *et al.* Developmental and neurologic status of children after heart surgery with hypothermic circulatory arrest or low-flow cardiopulmonary bypass. *N Engl J Med* 1995;**332**(9):549–555.

49. Benowitz LI, Routtenberg A. GAP-43: an intrinsic determinant of neuronal development and plasticity. *Trends Neurosci* 1997;**20**(2):84–91.

50. Bhutani VK, Johnson L. Synopsis report from the pilot USA Kernicterus Registry. *J Perinatol* 2009;**1**:S4–S7.

51. Billiards SS, Haynes RL, Folkerth RD, *et al.* Development of microglia in the cerebral white matter of the human fetus and infant. *J Comp Neurol* 2006;**497**(2):199–208.

52. Billiards SS, Pierson CR, Haynes RL, Folkerth RD, Kinney HC. Is the late preterm infant more vulnerable to gray matter injury than the term infant? *Clin Perinatol* 2006;**33**(4):915–33; abstract x–xi.

53. Billiards SS, Haynes RL, Folkerth RD, *et al.* Myelin abnormalities without oligodendrocyte loss in periventricular leukomalacia. *Brain Pathol* 2008;**18**(2):153–163.

3

54. Blair PS, Sidebotham P, Berry PJ, Evans M, Fleming PJ. Major epidemiological changes in sudden infant death syndrome: a 20-year population-based study in the UK. *Lancet* 2006;**367**(9507):314–9.

55. Blair PS, Byard RW, Fleming PJ. Sudden unexpected death in infancy (SUDI): suggested classification and applications to facilitate research activity. *Forensic Sci Med Pathol* 2012;**8**(3):312–315.

56. Blake D, Lombes A, Minetti C. MINGIE syndrome: report of 2 new patients (Abstract). *Neurology* 1990;**40**(Suppl):294.

57. Boardman JP, Counsell SJ, Rueckert D, *et al. Neuroimage* 2006;**32**:70–78.

58. Bodeutsch N, Thanos S. Migration of phagocytotic cells and development of the murine intraretinal microglial network: an *in vivo* study using fluorescent dyes. *Glia* 2000;**32**(1):91–101.

59. Boksa P. Effects of prenatal infection on brain development and behavior: a review of findings from animal models. *Brain Behav Immun* 2010;**24**(6):881–897.

60. Bonthius DJ, Wright R, Tseng B, *et al.* Congenital lymphocytic choriomeningitis virus infection: spectrum of disease. *Ann Neurol* 2007;**62**(4):347–355.

61. Borit A, McIntosh GC. Myelin basic protein and glial fibrillary acidic protein in human fetal brain. *Neuropathol Appl Neurobiol* 1981;**7**(4):279–287.

62. Bosnjak VM, Dakovic I, Duranovic V, *et al.* Malformations of cortical development in children with congenital cytomegalovirus infection – a study of nine children with proven congenital cytomegalovirus infection. *Coll Antropol* 2011;**35**(Suppl 1):229–234.

63. Boyd MT, Jordan SW, Davis LE. Fatal pneumonitis from congenital echovirus type 6 infection. *Pediatr Infect Dis J* 1987;**6**:1138–1139.

64. Bradl M, Lassmann H. Oligodendrocytes: biology and pathology. *Acta Neuropathol* 2010;**119**(1):37–53.

65. Breeze AC, Jessop FA, Set PA, *et al.* Minimally-invasive fetal autopsy using magnetic resonance imaging and percutaneous organ biopsies: clinical value and comparison to conventional autopsy. *Ultrasound Obstet Gynecol* 2011;**37**(3):317–323.

66. Brites D. Bilirubin injury to neurons and glial cells: new players, novel targets and newer insights. *Semin Perinatol* 2011;**35**(3):114–120.

67. Brito MA, Zurolo E, Pereira P, *et al.* Cerebellar axon/myelin loss, angiogenic sprouting, and neuronal increase of vascular endothelial growth factor in a preterm infant with kernicterus. *J Child Neurol* 2012;**27**(5):615–624.

68. Brody BA, Kinney HC, Kloman AS, Gilles FH. Sequence of central nervous system myelination in human infancy. I. An autopsy study of myelination. *J Neuropathol Exp Neurol* 1987;**46**(3):283–301.

69. Brouwer AJ, Groenendaal F, Koopman C, *et al.* Intracranial haemorrhage in full-term newborns: a hospital-based cohort study. *Neuroradiology* 2010;**52**(6):567–576.

70. Brown AM, Ransom BR. Astrocyte glycogen and brain energy metabolism. *Glia* 2007;**55**:1263–1271.

71. Burd I, Balakrishnan B, Kannan S. Models of fetal brain injury, intrauterine inflammation and preterm birth. *Am J Reprod Immunol* 2012;**67**(4):287–294.

72. Burns CM, Rutherford MA, Boardman JP, Cowan FM. Patterns of cerebral injury and neurodevelopmental outcomes after symptomatic neonatal hypoglycemia. *Pediatrics* 2008;**122**(1):65–74.

73. Burns DK. The neuropathology of pediatric acquired immunodeficiency syndrome. *J Child Neurol* 1992;**7**(4):332–346.

74. Buser JR, Maire J, Riddle A, *et al.* Arrested preoligodendrocyte maturation contributes to myelination failure in premature infants. *Ann Neurol* 2012;**71**(1):93–109.

75. Bystron I, Rakic P, Molnar Z, Blakemore C. The first neurons of the human cerebral cortex. *Nat Neurosci* 2006;**9**(7):880–886.

76. Bystron I, Blakemore C, Rakic P. Development of the human cerebral cortex: Boulder Committee revisited. *Nat Rev Neurosci* 2008;**9**(2):110–122.

77. Cajigas IJ, Tushev G, Will TJ, *et al.* The local transcriptome in the synaptic neuropil revealed by deep sequencing and high-resolution imaging. *Neuron* 2012;**74**(3):453–466.

78. Caksen H, Guven AS, Yilmaz C, *et al.* Clinical outcome and magnetic resonance imaging findings in infants with hypoglycemia. *J Child Neurol* 2011;**26**(1):25–30.

79. Canapicchi R, Cioni G, Strigini FA, *et al.* Prenatal diagnosis of periventricular haemorrhage by fetal brain magnetic resonance imaging. *Childs Nerv Syst* 1998;**14**(12):689–692.

80. Carruthers VB, Suzuki Y. Effects of *Toxoplasma gondii* infection on the brain. *Schizophr Bull* 2007;**33**:745–751.

81. Carter HM. Congenital poliomyelitis; report of a case. *Obstet Gynecol* 1956;**8**:373–374.

82. Ceballos R, Ch'ien LT, Whitley RJ, Brans YW. Cerebellar hypoplasia in an infant with congenital cytomegalovirus infection. *Pediatrics* 1976;**57**(1):155–157.

83. Centers for Disease Control and Prevention. *HIV/AIDS Surveillance Report.* Atlanta, GA: Centers for Disease Control, 1992.

84. Chakrabarti AK, Banik NL, Lobo DC, Terry EC, Hogan EL. Calcium-activated neutral proteinase (calpain) in rat brain during development: compartmentation and role in myelination. *Brain Res Dev Brain Res* 1993;**71**(1):107–113.

85. Chanas–Sacre G, Thiry M, Pirard S, *et al.* A 295-kDA intermediate filament-associated protein in radial glia and developing muscle cells *in vivo* and *in vitro*. *Dev Dyn* 2000;**219**(4):514–25.

86. Chang H, Cho KH, Hayashi S, *et al.* Site- and stage-dependent differences in vascular density of the human fetal brain. *Childs Nerv Syst* 2014;**30**(3):399–409.

87. Chantler JK, Smyrnis L, Tai G. Selective infection of astrocytes in human glial cell cultures by rubella virus. *Lab Invest* 1995;**72**:334–340.

88. Chau J. A systematic review of pediatric sensorineural hearing loss in congenital syphilis. *Int J Pediatr Otorhinolaryngol* 2009;**73**:787–792.

89. Cheepsunthorn P, Palmer C, Menzies S, Roberts RL, Connor JR. Hypoxic/ischaemic insult alters ferritin expression and myelination in neonatal rat brains. *J Comp Neurol* 2001;**431**(4):382–396.

90. Cheeran MC, Lokensgard JR, Schleiss MR. Neuropathogenesis of congenital cytomegalovirus infection: disease mechanisms and prospects for intervention. *Clin Microbiol Rev* 2009;**22**(1):99–126, Table of Contents.

91. Chen HJ, Wei KL, Zhou CL, *et al.* Incidence of brain injuries in premature infants with gestational age ≤ 34 weeks in ten urban hospitals in China. *World J Pediatr* 2013;**9**(1):17–24.

92. Cheong JL, Thompson DK, Wang HX, *et al.* Abnormal white matter signal on MR imaging is related to abnormal tissue microstructure. *AJNR Am J Neuroradiol* 2009;**30**(3):623–628.

93. Cherian S, Whitelaw A, Thoresen M, Love S. The pathogenesis of neonatal post-hemorrhagic hydrocephalus. *Brain Pathol* 2004;**14**(3):305–311.

94. Chi JG, Dooling EC, Gilles FH. Gyral development of the human brain. *Ann Neurol* 1977;**1**:86–93.

95. Choi BH, Kim RC, Lapman LW. Do radial glia give rise to both astroglial and oligodendroglial cells? *Dev Brain Res* 1983;**8**:119–30.

96. Claris O, Besnier S, Lapillonne A, Picaud JC, Salle BL. Incidence of ischaemic-hemorrhagic cerebral lesions in premature infants of gestational age ≤ 28 weeks: a prospective ultrasound study. *Biol Neonate* 1996;**70**(1):29–34.

97. Clasen RA, Simon RG, Scott R, *et al.* The staining of the myelin sheath by luxol dye techniques. *J Neuropathol Exp Neurol* 1973;**32**(2):271–283.

98. Cohen MC, Paley MN, Griffiths PD, Whitby EH. Less invasive autopsy: benefits and limitations of the use of magnetic resonance imaging in the perinatal postmortem. *Pediatr Dev Pathol* 2008;**11**(1):1–9.

99. Colby JB, Van Horn JD, Sowell ER. Quantitative in vivo evidence for broad regional gradients in the timing of white matter maturation during adolescence. *Neuroimage* 2011;**54**(1):25–31.

100. Coley BD. Importance of hypoxic/ischaemic conditions in the development of cerebral lenticulostriate vasculopathy. *Pediatr Radiol* 2000;**30**:846–855.

101. Cong H, Jiang Y, Tien P. Identification of the myelin oligodendrocyte glycoprotein as a cellular receptor for rubella virus. *J Virol* 2011;**85**(21):11038–11047.

102. Connor JR, Menzies SL. Relationship of iron to oligodendrocytes and myelination. *Glia* 1996;**17**(2):83–93.

103. Cortina-Borja M, Tan HK, Wallon M, *et al.* Prenatal treatment for serious neurological sequelae of congenital toxoplasmosis: an observational prospective cohort study. *PLoS Med* 2010;**7**(10):pii:e1000351.

104. Counsell SJ, Allsop JM, Harrison MC, *et al.* Diffusion-weighted imaging of the brain in preterm infants with focal and diffuse white matter abnormality. *Pediatrics* 2003;**112**(1 Pt 1):1–7.

105. Courville CB. *Birth and brain damage. An investigation into the problems of antenatal and paranatal anoxia and allied disorders and their relation to the many lesion-complexes residual thereto.* Pasadena, CA: MF Courville, 1971.

106. Craver RD. The cytology of cerebrospinal fluid associated with neonatal intraventricular haemorrhage. *Pediatr Pathol Lab Med* 1996;**16**(5):713–719.

107. Culican SM, Baumrind NL, Yamamoto M, Pearlman AL. Cortical radial glia: identification in tissue culture and evidence

for their transformation to astrocytes. *J Neurosci* 1990;**10**(2):684–692.

108. Cunningham FG, Leveno KJ, Bloom SL, *et al. Williams obstetrics*. 22nd edn. New York: McGraw–Hill, 2005.

109. D'Mello C. Cerebral microglia recruit monocytes into the brain in response to tumor necrosis factor signalling. *J Neurosci* 2009;**29**:2089–2102.

110. Damodaram MS, Story L, Eixarch E, *et al*. Fetal volumetry using magnetic resonance imaging in intrauterine growth restriction. *Early Hum Dev* 2012;**88**(Suppl 1):S35–40.

111. Das S, Basu A. Viral infection and neural stem/progenitor cell's fate: implications in brain development and neurological disorders. *Neurochem Int* 2011;**59**:357–366.

112. Davis EJ, Foster TD, Thomas WE. Cellular forms and functions of brain microglia. *Brain Res Bull* 1994;**34**(1):73–78.

113. de Bruine FT, van den Berg–Huysmans AA, Leijser LM, *et al*. Clinical implications of MR imaging findings in the white matter in very preterm infants: a 2-year follow-up study. *Radiology* 2011;**261**(3):899–906.

114. de Jonge RC, van Furth AM, Wassenaar M, Gemke RJ, Terwee CB. Predicting sequelae and death after bacterial meningitis in childhood: a systematic review of prognostic studies. *BMC Infect Dis* 2010;**10**:232.

115. De Reuck JL. Cerebral angioarchitecture and perinatal brain lesions in premature and full-term infants. *Acta Neurol Scand* 1984;**70**(6):391–395.

116. De Vries LS, Connell JA, Dubowitz LM, *et al*. Neurological, electrophysiological and MRI abnormalities in infants with extensive cystic leukomalacia. *Neuropediatrics* 1987;**18**(2):61–66.

117. de Vries LS, Groenendaal F, Eken P, *et al*. Infarcts in the vascular distribution of the middle cerebral artery in preterm and fullterm infants. *Neuropediatrics* 1997;**28**(2):88–96.

118. de Wit OA, den Dunnen WF, Sollie KM, *et al*. Pathogenesis of cerebral malformations in human fetuses with meningomyelocele. *Cerebrospinal Fluid Res* 2008;**5**:4.

119. Degos V, Favrais G, Kaindl AM, et al. Inflammation processes in perinatal brain damage. *J Neural Transm* 2010;**117**(8):1009–1017.

120. Deguchi K, Oguchi K, Matsuura N, Armstrong DD, Takashima S. Periventricular leukomalacia: relation to gestational age and axonal injury. *Pediatr Neurol* 1999;**20**(5):370–4.

121. Del Bigio MR. The ependyma: a protective barrier between brain and cerebrospinal fluid. *Glia* 1995;**14**(1):1–13.

122. Del Bigio MR. Ependymal cells: biology and pathology. *Acta Neuropathol* 2010;**119**(1):55–73.

123. Del Bigio MR. Cell proliferation in human ganglionic eminence and suppression after prematurity-associated haemorrhage. *Brain* 2011;**134**(Pt 5):1344–1361.

124. Del Bigio MR, Becker LE. Microglial aggregation in the dentate gyrus: a marker of mild hypoxic-ischaemic brain insult in human infants. *Neuropathol Appl Neurobiol* 1994;**20**(2):144–51.

125. Deli* vria–Papadopoulos M, Mishra OP. Mechanisms of cerebral injury in perinatal asphyxia and strategies for prevention. *J Pediatr* 1998;**132**(3 Pt 2):S30–4.

126. Deng W, Rosenberg PA, Volpe JJ, Jensen FE. Calcium-permeable AMPA/kainate receptors mediate toxicity

and preconditioning by oxygen-glucose deprivation in oligodendrocyte precursors. *Proc Natl Acad Sci U S A* 2003;**100**(11):6801–6.

127. DePalma RT, Leveno KJ, Kelly MA, Sherman ML, Carmody TJ. Birth weight threshold for postponing preterm birth. *Am J Obstet Gynecol* 1992;**167**(4 Pt 1):1145–9.

128. DeReuck J, Chattha AS, Richardson EP, Jr. Pathogenesis and evolution of periventricular leukomalacia in infancy. *Arch Neurol* 1972;**27**(3):229–236.

129. DeSilva TM, Kabakov AY, Goldhoff PE, Volpe JJ, Rosenberg PA. Regulation of glutamate transport in developing rat oligodendrocytes. *J Neurosci* 2009;**29**(24):7898–7908.

130. Desmonts G, Couvreur J. Congenital toxoplasmosis. A prospective study of 378 pregnancies. *N Engl J Med* 1974;**290**:1110–1116.

131. deVeber G, Andrew M, Adams C, *et al*. Cerebral sinovenous thrombosis in children. *N Engl J Med* 2001;**345**(6):417–423.

132. Dickson DW, Belman AL, Kim TS, Horoupian DS, Rubinstein A. Spinal cord pathology in pediatric acquired immunodeficiency syndrome. *Neurology* 1989;**39**(2 Pt 1):227–235.

133. Dolfin T, Skidmore MB, Fong KW, Hoskins EM, Shennan AT. Incidence, severity and timing of subependymal and intraventricular haemorrhages in preterm infants born in a perinatal unit as detected by serial real-time ultrasound. *Pediatrics* 1983;**71**(4):541–546.

134. Dommergues M, Petitjean J, Aubry MC, *et al*. Fetal enteroviral infection with cerebral ventriculomegaly and cardiomyopathy. *Fetal Diagn Ther* 1994;**9**(2):77–78.

135. Doneda C, Parazzini C, Righini A, *et al*. Early cerebral lesions in cytomegalovirus infection: prenatal MR imaging. *Radiology* 2010;**255**(2):613–621.

136. Donofrio MT, Duplessis AJ, Limperopoulos C. Impact of congenital heart disease on fetal brain development and injury. *Curr Opin Pediatr* 2011;**23**(5):502–511.

137. Doran KS, Engelson EJ, Khosravi A, *et al*. Blood–brain barrier invasion by group B Streptococcus depends upon proper cell-surface anchoring of lipoteichoic acid. *J Clin Invest* 2005;**115**(9):2499–507.

138. Dore–Duffy P. Pericytes: pluripotent cells of the blood brain barrier. *Curr Pharm Des* 2008;**14**(16):1581–1593.

139. Dorovini–Zis K, Dolman CL. Gestational development of brain. *Arch Pathol Lab Med* 1977;**101**:192–195.

140. Dotis J, Iosifidis E, Roilides E. Central nervous system aspergillosis in children: a systematic review of reported cases. *Int J Infect Dis* 2007;**11**:381–393.

141. Doumouchtsis SK, Arulkumaran S. Head trauma after instrumental births. *Clin Perinatol* 2008;**35**(1):69–83,viii.

142. Dublin WB. Neurologic lesions of erythroblastosis fetalis in relation to nuclear deafness. *Am J Clin Pathol* 1951;**21**(10):935–9.

143. Dubois J, Benders M, Borradori–Tolsa C, *et al*. Primary cortical folding in the human newborn: an early marker of later functional development. *Brain* 2008;**131**(Pt 8):2028–2041.

144. Dubowitz LM, Bydder GM, Mushin J. Developmental sequence of periventricular

leukomalacia. Correlation of ultrasound, clinical, and nuclear magnetic resonance functions. *Arch Dis Child* 1985;**60**(4):349–355.

145. Duflou H, Maenhaut W, De Reuck J. Regional distribution of potassium, calcium and six trace elements in normal human brain. *Neurochem Res* 1989;**14**(11):1099–1112.

146. Dupuis O, Silveira R, Dupont C, *et al*. Comparison of 'instrument-associated' and 'spontaneous' obstetric depressed skull fractures in a cohort of 68 neonates. *Am J Obstet Gynecol* 2005;**192**:165–170.

147. Edwards AD, Yue X, Cox P, *et al*. Apoptosis in the brains of infants suffering intrauterine cerebral injury. *Pediatr Res* 1997;**42**(5):684–9.

148. Edwards MS, Rench MA, Haffar AA, *et al*. Long-term sequelae of group B streptococcal meningitis in infants. *J Pediatr* 1985;**106**(5):717–722.

149. Eliceiri BP, Cheresh DA. The role of alpha v integrins during angiogenesis: insights into potential mechanisms of action and clinical development. *J Clin Invest* 1999;**103**(9):1227–30.

150. Elliot GB, Mcallister JE. Fetal poliomyelitis. *Am J Obstet Gynecol* 1956;**72**:896–902.

151. Emery B. Regulation of oligodendrocyte differentiation and myelination. *Science* 2010;**330**(6005):779–782.

152. Engelhardt B. Development of the blood–brain barrier. *Cell Tissue Res* 2003;**314**(1):119–129.

153. Engle WA. Age terminology during the perinatal period. American Academy of Pediatrics Committee on Fetus and Newborn. *Pediatrics* 2004;**114**: 1362–1364.

154. Esiri MM, al Izzi MS, Reading MC. Macrophages, microglial cells and HLA-DR antigens in fetal and infant brain. *J Clin Pathol* 1991;**44**(2):102–6.

155. Evans AC, Brain Development Cooperative Group. The NIH MRI study of normal brain development. *Neuroimage* 2006;**30**(1):184–202.

156. Evrard P, Marret S, Gressens P. Environmental and genetic determinants of neural migration and postmigratory survival. *Acta Paediatr Suppl* 1997;**422**: 20–26.

157. Faix RG. Systemic *Candida* infections in infants in intensive care nurseries: high incidence of central nervous system involvement. *J Pediatr* 1984;**105**:616–22.

158. Faix RG, Kovarik SM, Shaw TR, Johnson RV. Mucocutaneous and invasive candidiasis among very low birth weight (less than 1500 grams) infants in intensive care nurseries: a prospective study. *Pediatrics* 1989;**83**(1):101–107.

159. Fancy SP, Zhao C, Franklin RJ. Increased expression of Nkx2.2 and Olig2 identifies reactive oligodendrocyte progenitor cells responding to demyelination in the adult CNS. *Mol Cell Neurosci* 2004;**27**(3): 247–254.

160. Farber E. Programmed cell death: necrosis versus apoptosis. *Mod Pathol* 1994;**7**(5):605–609.

161. Ferrer I, Bernet E, Soriano E, del Rio T, Fonseca M. Naturally occurring cell death in the cerebral cortex of the rat and removal of dead cells by transitory phagocytes. *Neuroscience* 1990;**39**(2):451–8.

162. Feuer R, Pagarigan RR, Harkins S, *et al*. Coxsackievirus targets proliferating

neuronal progenitor cells in the neonatal CNS. *J Neurosci* 2005;**25**(9):2434–44.

163. Filiano JJ, Kinney HC. Arcuate nucleus hypoplasia in the sudden infant death syndrome. *J Neuropathol Exp Neurol* 1992;**51**(4):394–403.

164. Filiano JJ, Kinney HC. A perspective on neuropathologic findings in victims of the sudden infant death syndrome: the triple-risk model. *Biol Neonate* 1994;**65**(3–4):194–197.

165. Filiano JJ, Choi JC, Kinney HC. Candidate cell populations for respiratory chemosensitive fields in the human infant medulla. *J Comp Neurol* 1990;**293**(3):448–465.

166. Filippi L, Serafini L, Dani C, *et al*. Congenital syphilis: unique clinical presentation in three preterm newborns. *J Perinat Med* 2004;**32**(1):90–94.

167. Fluge G. Clinical aspects of neonatal hypoglycaemia. *Acta Paediatr Scand* 1974;**63**(6):826–832.

168. Folgering H, Kuyper F, Kille JF. Primary alveolar hypoventilation (Ondine's curse syndrome) in an infant without external arcuate nucleus. Case report. *Bull Eur Physiopathol Respir* 1979;**15**(4):659–665.

169. Folkerth RD. Neuropathologic substrate of cerebral palsy. *J Child Neurol* 2005;**20**(12):940–9.

170. Folkerth RD, Haynes RL, Borenstein NS, *et al*. Developmental lag in superoxide dismutases relative to other antioxidant enzymes in premyelinated human telencephalic white matter. *J Neuropathol Exp Neurol* 2004;**63**(9):990–9.

171. Folkerth RD, Keefe RJ, Haynes RL, *et al*. Interferon-gamma expression in periventricular leukomalacia in the human brain. *Brain Pathol* 2004;**14**(3):265–74.

172. Folkerth RD, Trachtenberg FL, Haynes RL. Oxidative injury in the cerebral cortex and subplate neurons in periventricular leukomalacia. *J Neuropathol Exp Neurol* 2008;**67**(7):677–686.

173. Freimuth P, Philipson L, Carson SD. The coxsackievirus and adenovirus receptor. *Curr Top Microbiol Immunol* 2008;**323**:67–87.

174. Frenkel JK. Pathology and pathogenesis of congenital toxoplasmosis. *Bull N Y Acad Med* 1974;**50**:182–191.

175. Fridovich I. The biology of oxygen radicals. *Science* 1978;**201**(4359): 875–880.

176. Friede R. Haemorrhages in asphyxiated premature infants. In: Friede R ed. *Developmental neuropathology*. 2nd edn. Berlin: Springer, 1989:44–58.

177. Friede RL, Schachenmayr W. Early stages of status marmoratus. *Acta Neuropathol* 1977;**38**(2):123–127.

178. Fumagalli M, Bassi L, Sirgiovanni I, *et al*. From germinal matrix to cerebellar haemorrhage. *J Matern Fetal Neonatal Med* 2013 [Epub ahead of print].

179. Furky JS, Swann O, Molyneux E. Systematic review: neonatal meningitis in the developing world. *Trop Med Int Health* 2011;**16**:672–679.

180. Gabrielli L, Bonasoni MP, Lazzarotto T, *et al*. Histological findings in fetuses congenitally infected by cytomegalovirus. *J Clin Virol* 2009;**46**(Suppl 4):S16–21.

181. Garcia AG, Basso NG, Fonseca ME, Outani HN. Congenital echo virus infection– morphological and virological study of fetal and placental tissue. *J Pathol* 1990;**160**(2):123–127.

182. Gard AL, Pfeiffer SE. Two proliferative stages of the oligodendrocyte lineage (A2B5+O4– and O4+GalC–) under different mitogenic control. *Neuron* 1990;**5**(5):615–625.

183. Gardosi J. Customized fetal growth standards: rationale and clinical application. *Semin Perinatol* 2004;**28**(1):33–40.

184. Garite TJ, Clark R, Thorp JA. Intrauterine growth restriction increases morbidity and mortality among premature neonates. *Am J Obstet Gynecol* 2004;**191**(2):481–7.

185. Gerrard J. Nuclear jaundice and deafness. *J Laryngol Otol* 1952;**66**(1):39–46.

186. Ghazi–Birry HS, Brown WR, Moody DM, *et al*. Human germinal matrix: venous origin of haemorrhage and vascular characteristics. *AJNR Am J Neuroradiol* 1997;**18**(2):219–229.

187. Ghersi–Egea JF, Gazzin S, Strazielle N. Blood–brain interfaces and bilirubin-induced neurological diseases. *Curr Pharm Des* 2009;**15**:2893–2907.

188. Ghosh A, Antonini A, McConnell SK, Shatz CJ. Requirement for subplate neurons in the formation of thalamocortical connections. *Nature* 1990;**347**(6289):179–81.

189. Giaume C, Koulakoff A, Roux L, Holcman D, Rouach N. Astroglial networks: a step further in neuroglial and gliovascular interactions. *Nat Rev Neurosci* 2010;**11**(2):87–99.

190. Gilles FH, Price RA, Kevy SV, Berenberg W. Fibrinolytic activity in the ganglionic eminence of the premature human brain. *Biol Neonate* 1971;**18**(5):426–432.

191. Gilles FH, Leviton A, Kerr CS. Endotoxin leucoencephalopathy in the telencephalon of the newborn kitten. *J Neurol Sci* 1976;**27**(2):183–191.

192. Gilles FH, Averill DR, Jr, Kerr CS. Neonatal endotoxin encephalopathy. *Ann Neurol* 1977;**2**(1):49–56.

193. Gilles F, Leviton A, Dooling E, eds. The developing human brain: growth and epidemiologic neuropathology. Boston, MA: John Wright–PSG; 1983.

194. Gleissner M, Jorch G, Avenarius S. Risk factors for intraventricular haemorrhage in a birth cohort of 3721 premature infants. *J Perinat Med* 2000;**28**(2):104–110.

195. Goldberg M, De Pitta M, Volman V, Berry H, Ben–Jacob E. Nonlinear gap junctions enable long-distance propagation of pulsating calcium waves in astrocyte networks. *PLoS Comput Biol* 2010;**6**(8):pii: e1000909.

196. Goldberg WJ, Allen N. Determination of Cu, Mn, Fe and Ca in six regions of normal human brain, by atomic absorption spectroscopy. *Clin Chem* 1981;**27**(4):562–564.

197. Goldenberg RL, Culhane JF, Iams JD, Romero R. Epidemiology and causes of preterm birth. *Lancet* 2008;**371**(9606): 75–84.

198. Goldenberg RL, Gravett MG, Iams J, *et al*. The preterm birth syndrome: issues to consider in creating a classification system. *Am J Obstet Gynecol* 2012;**206**:113–118.

199. Goldstein RF, Cotten CM, Shankaran S, *et al*. Influence of gestational age on death and neurodevelopmental outcome in premature infants with severe intracranial haemorrhage. *J Perinatol* 2013;**33**(1): 25–32.

200. Gonzalez–Zulueta M, Ensz LM, Mukhina G, *et al*. Manganese superoxide dismutase protects nNOS neurons from NMDA and nitric oxide-mediated neurotoxicity. *J Neurosci* 1998;**18**(6):2040–55.

201. Gordon N. Toxoplasmosis: a preventable cause of brain damage. *Dev Med Child Neurol* 1993;**35**:567–573.

202. Gould SJ, Howard S. An Immunohistochemical study of the germinal layer in the late gestation human fetal brain. *Neuropathol Appl Neurobiol* 1987;**13**(6):421–437.

203. Gould SJ, Howard S. Glial differentiation in the germinal layer of fetal and preterm infant brain: an immunocytochemical study. *Pediatr Pathol* 1988;**8**(1):25–36.

204. Gould SJ, Howard S. An immunohistochemical study of macrophages in the human fetal brain. *Neuropathol Appl Neurobiol* 1991;**17**(5):383–390.

205. Grafe M. The correlation of prenatal brain damage with placental pathology. *J Neuropathol Exp Neurol* 1994;**53**: 407–15.

206. Graham EM, Ruis KA, Hartman AL, Northington FJ, Fox HE. A systematic review of the role of intrapartum hypoxia-ischaemia in the causation of neonatal encephalopathy. *Am J Obstet Gynecol* 2008;**199**(6):587–595.

207. Grant EG, Kerner M, Schellinger D, *et al*. Evolution of porencephalic cysts from intraparenchymal haemorrhage in neonates: sonographic evidence. *AJR Am J Roentgenol* 1982;**138**(3):467–470.

208. Griffiths AD, Laurence KM. The effect of hypoxia and hypoglycaemia on the brain of the newborn human infant. *Dev Med Child Neurol* 1974;**16**(3):308–319.

209. Groenendaal F, Termote JU, van der Heide–Jalving M, van Haastert IC, de Vries LS. Complications affecting preterm neonates from 1991 to 2006: what have we gained? *Acta Paediatr* 2010;**99**(3):354–358.

210. Grunnet ML, Shields WD. Cerebellar haemorrhage in the premature infant. *J Pediatr* 1976;**88**(4 Pt 1):605–608.

211. Grunt S, Wingeier K, Wehrli E, *et al*. Cerebral sinus venous thrombosis in Swiss children. *Dev Med Child Neurol* 2010;**52**(12):1145–1150.

212. Guarner J, Greer PW, Bartlett J, *et al*. Congenital syphilis in a newborn: an immunopathologic study. *Mod Pathol* 1999;**12**(1):82–87.

213. *Guidelines for perinatal care*. 5th edn. American Academy of Pediatrics and the American College of Obstetricians and Gynecologists, 2005.

214. Guihard–Costa AM, Ménez F, Delezoide AL. Organ weights in human fetuses after formalin fixation: standards by gestational age and body weight. *Pediatr Dev Pathol* 2002;**5**:559–578.

215. Hachiya Y, Hayashi M. Bilirubin encephalopathy: a study of neuronal subpopulations and neurodegenerative mechanisms in 12 autopsy cases. *Brain Dev* 2008;**30**:269–278.

216. Hadi HA, Finley J, Mallette JQ, Strickland D. Prenatal diagnosis of cerebellar haemorrhage: medicolegal implications. *Am J Obstet Gynecol* 1994;**170**(5 Pt 1):1392–1395.

217. Haines KM, Wang W, Pierson CR. Cerebellar hemorrhagic injury in premature infants occurs during a vulnerable developmental period and is associated with wider neuropathology. *Acta Neuropathol Commun* 2013;**1**(1):69.

218. Hajnal BL, Sahebkar–Moghaddam F, Barnwell AJ, Barkovich AJ, Ferriero DM. Early prediction of neurologic outcome after perinatal depression. *Pediatr Neurol* 1999;**21**(5):788–793.

219. Hamby ME, Sofroniew MV. Reactive astrocytes as therapeutic targets for CNS disorders. *Neurotherapeutics* 2010;**7**(4):494–506.

220. Hamilton SP, Rome LH. Stimulation of *in vitro* myelin synthesis by microglia. *Glia* 1994;**11**(4):326–35.

221. Hankins GD, Speer M. Defining the pathogenesis and pathophysiology of neonatal encephalopathy and cerebral palsy. *Obstet Gynecol* 2003;**102**(3):628–636.

222. Hanna SE, Rosenbaum PL, Bartlett DJ, *et al*. Stability and decline in gross motor function among children and youth with cerebral palsy aged 2 to 21 years. *Dev Med Child Neurol* 2009;**51**(4):295–302.

223. Hardy JB, Hardy PH, Oppenheimer EH, Ryan SJ, Jr, Sheff RN. Failure of penicillin in a newborn with congenital syphilis. *JAMA* 1970;**212**(8):1345–1349.

224. Harris A, Seckl J. Glucocorticoids, prenatal stress and the programming of disease. *Horm Behav* 2011;**59**(3):279–289.

225. Harteman JC, Groenendaal F, Kwee A, *et al*. Risk factors for perinatal arterial ischaemic stroke in full-term infants: a case-control study. *Arch Dis Child Fetal Neonatal Ed* 2012;**97**(6):F411–6.

226. Hartfuss E, Galli R, Heins N, Gotz M. Characterization of CNS precursor subtypes and radial glia. *Dev Biol* 2001;**229**(1):15–30.

227. Hawkes S. Effectiveness of interventions to improve screening for syphilis in pregnancy: a systematic review and meta-analysis. *Lancet Infect Dis* 2011;**11**:684–691.

228. Haynes RL, Folkerth RD, Keefe RJ, *et al*. Nitrosative and oxidative injury to premyelinating oligodendrocytes in periventricular leukomalacia. *J Neuropathol Exp Neurol* 2003;**62**(5):441–50.

229. Haynes RL, Borenstein NS, Desilva TM, *et al*. Axonal development in the cerebral white matter of the human fetus and infant. *J Comp Neurol* 2005;**484**(2):156–67.

230. Haynes RL, Billiards SS, Borenstein NS, Volpe JJ, Kinney HC. Diffuse axonal injury in periventricular leukomalacia as determined by apoptotic marker fractin. *Pediatr Res* 2008;**63**(6):656–661.

231. Haynes RL, Xu G, Folkerth RD, Trachtenberg FL, Volpe JJ, Kinney HC. Potential neuronal repair in cerebral white matter injury in the human neonate. *Pediatr Res* 2011;**69**(1):62–67.

232. Haynes RL, Sleeper LA, Volpe JJ, Kinney HC. Neuropathologic studies of the encephalopathy of prematurity in the late preterm infant. *Clin Perinatol* 2013;**40**(4):707–722.

233. Hernandez MI, Sandoval CC, Tapia JL, *et al*. Stroke patterns in neonatal group B streptococcal meningitis. *Pediatr Neurol* 2011;**44**(4):282–288.

234. Heron M, Sutton PD, Xu J, *et al*. Annual summary of vital statistics: 2007. *Pediatrics* 2010;**125**(1):4–15.

235. Hesser U, Katz–Salamon M, Mortensson W, Flodmark O, Forssberg H. Diagnosis of intracranial lesions in very-low-birthweight infants by ultrasound: incidence and association with potential risk factors. *Acta Paediatr Suppl* 1997;**419**:16–26.

236. Hevner RF. Development of connections in the human visual system during fetal mid-gestation: a DiI-tracing study. *J Neuropathol Exp Neurol* 2000;**59**(5):385–392.

237. Hevner RF. Layer-specific markers as probes for neuron type identity in human neocortex and malformations of cortical development. *J Neuropathol Exp Neurol* 2007;**66**(2):101–109.

238. Himmelmann K, Ahlin K, Jacobsson B, Cans C, Thorsen P. Risk factors for cerebral palsy in children born at term. *Acta Obstet Gynecol Scand* 2011;**90**(10):1070–1081.

239. Ho KC, Roessmann U, Hause L, Monroe G. Correlation of perinatal brain growth with age, body size, sex and race. *J Neuropathol Exp Neurol* 1986;**45**:179–188.

240. Hollier LM, Cox SM. Syphilis. *Semin Perinatol* 1998;**22**:323–331.

241. Hopkins SJ, Rothwell NJ. Cytokines and the nervous system. I: Expression and recognition. *Trends Neurosci* 1995;**18**(2):83–88.

242. Horsch S, Hallberg B, Leifsdottir K, Skiold B, Nagy Z, Mosskin M, *et al*. Brain abnormalities in extremely low gestational age infants: a Swedish population based MRI study. *Acta Paediatr* 2007;**96**(7):979–984.

243. Houdou S, Kuruta H, Hasegawa M, *et al*. Developmental immunohistochemistry of catalase in the human brain. *Brain Res* 1991;**556**(2):267–270.

244. Hristeva L, Booy R, Bowler I, Wilkinson AR. Prospective surveillance of neonatal meningitis. *Arch Dis Child* 1993;**69**(1 Spec No):14–18.

245. Hsiao EY, Patterson PH. Placental regulation of maternal-fetal interactions and brain development. *Dev Neurobiol* 2012;**72**(10):1317–1326.

246. Hughes CA, Harley EH, Milmoe G, Bala R, Martorella A. Birth trauma in the head and neck. *Arch Otolaryngol Head Neck Surg* 1999;**125**(2):193–199.

247. Huisman TA, Fischer J, Willi UV, Eich GF, Martin E. 'Growing fontanelle': a serious complication of difficult vacuum extraction. *Neuroradiology* 1999;**41**(5):381–383.

248. Hunt CE. Impaired arousal from sleep: relationship to sudden infant death syndrome. *J Perinatol* 1989;**9**(2):184–187.

249. Hunt JC, Schneider C, Menticoglou S, Herath J, Del Bigio MR. Antenatal and postnatal diagnosis of coxsackie b4 infection: case series. *AJP Rep* 2012;**2**(1):1–6.

250. Hüppi PS. MR imaging and spectroscopy of brain development. *Magn Res Imaging Clin N Am* 2001;**9**:1–17.

251. Hüppi PS, Murphy B, Maier SE, *et al*. Microstructural brain development after perinatal cerebral white matter injury assessed by diffusion tensor magnetic resonance imaging. *Pediatrics* 2001;**107**(3):455–60.

252. Huttenlocher PR, de Courten C, Garey LJ, Van der Loos H. Synaptogenesis in human visual cortex – evidence for synapse elimination during normal development. *Neurosci Lett* 1982;**33**(3):247–252.

253. Hutto C. Intrauterine herpes simplex virus infections. *J Pediatr* 1987;**110**:97–101.

254. Iida K, Takashima S, Ueda K. Immunohistochemical study of myelination and oligodendrocyte in infants with periventricular leukomalacia. *Pediatr Neurol* 1995;**13**(4):296–304.

255. Ikonomidou C, Kaindl AM. Neuronal death and oxidative stress in the developing brain. *Antioxid Redox Signal* 2011;**14**(8):1535–1550.

256. Imamura K, Hishikawa N, Sawada M, *et al*. Distribution of major histocompatibility complex class II-positive microglia and cytokine profile of Parkinson's disease brains. *Acta Neuropathol (Berl)* 2003;**106**(6):518–26.

257. Inder TE, Hüppi PS, Warfield S, *et al*. Periventricular white matter injury in the premature infant is followed by reduced cerebral cortical gray matter volume at term. *Ann Neurol* 1999;**46**(5):755–760.

258. Inder TE, Anderson NJ, Spencer C, Wells S, Volpe JJ. White matter injury in the premature infant: a comparison between serial cranial sonographic and MR findings at term. *AJNR Am J Neuroradiol* 2003;**24**(5):805–9.

259. Inder TE, Warfield SK, Wang H, Hüppi PS, Volpe JJ. Abnormal cerebral structure is present at term in premature infants. *Pediatrics* 2005;**115**(2):286–94.

260. Innocenti GM, Clarke S, Koppel H. Transitory macrophages in the white matter of the developing visual cortex. II. Development and relations with axonal pathways. *Brain Res* 1983;**313**(1):55–66.

261. Insel TR, Miller LP, Gelhard RE. The ontogeny of excitatory amino acid receptors in rat forebrain – I. N-methyl-D-aspartate and quisqualate receptors. *Neuroscience* 1990;**35**(1):31–43.

262. Jackson AC, Rossiter JP, Lafon M. Expression of Toll-like receptor 3 in the human cerebellar cortex in rabies, herpes simplex encephalitis and other neurological diseases. *J Neurovirol* 2006;**12**(3):229–234.

263. Jagadha V, Halliday WC, Becker LE. Glial fibrillary acidic protein (GFAP) in oligodendrogliomas: a reflection of transient GFAP expression by immature oligodendroglia. *Can J Neurol Sci* 1986;**13**(4):307–311.

264. Jakovcevski I, Zecevic N. Olig transcription factors are expressed in oligodendrocyte and neuronal cells in human fetal CNS. *J Neurosci* 2005;**25**(44):10064–73.

265. Jarjour IT, Ahdab–Barmada M. Cerebrovascular lesions in infants and children dying after extracorporeal membrane oxygenation. *Pediatr Neurol* 1994;**10**(1):13–19.

266. Johnson L, Bhutani VK, Karp K, Sivieri EM, Shapiro SM. Clinical report from the pilot USA Kernicterus Registry (1992–2004). *J Perinatol* 2009;**1**:S25–S45.

267. Johnston MV. Excitotoxicity in perinatal brain injury. *Brain Pathol* 2005;**15**(3):234–40.

268. Jonas R, Newburger J, Volpe J. *Brain injury and pediatric cardiac surgery*. Boston, MA: Butterworth–Heinemann, 1996.

269. Joseph KS, Liu S, Rouleau J, *et al*. Influence of definition based versus pragmatic birth registration on international comparisons of perinatal and infant mortality: population based retrospective study. *BMJ* 2012;**344**:e746.

270. Judas M, Sedmak G, Pletikos M, Jovanov–Milosevic N. Populations of

subplate and interstitial neurons in fetal and adult human telencephalon. *J Anat* 2010;**217**(4):381–399.

271. Juliet PA, Frost EE, Balasubramaniam J, Del Bigio MR. Toxic effect of blood components on perinatal rat subventricular zone cells and oligodendrocyte precursor cell proliferation, differentiation and migration in culture. *J Neurochem* 2009;**109**:1285–1299.

272. Kadhim H, Tabarki B, De Prez C, Rona AM, Sebire G. Interleukin-2 in the pathogenesis of perinatal white matter damage. *Neurology* 2002;**58**(7): 1125–8.

273. Kahle W. Studien uber die Matrixphasen und die ortlichen Reifungsunterschiede im emryonalen menschlichen Gehirn. Die matrixphasen im allgemeinen. *Dtsch Z Nervenheilkd* 1951;**166**:272–302.

274. Karadottir R, Cavelier P, Bergersen LH, Attwell D. NMDA receptors are expressed in oligodendrocytes and activated in ischaemia. *Nature* 2005;**438**(22): 1162–1166.

275. Kemper TL. Retardation of the myelo- and cytoarchitectonic maturation of the brain in the congenital rubella syndrome. *Res Publ Assoc Res Nerv Ment Dis* 1973;**51**:23–62.

276. Kendall N, Woloshin H. Cephalhematoma associated with fracture of the skull. *J Pediatr* 1952;**41**(2):125–132.

277. Kennedy TE, Serafini T, de la Torre JR, Tessier–Lavigne M. Netrins are diffusible chemotropic factors for commissural axons in the embryonic spinal cord. *Cell* 1994;**78**(3):425–435.

278. Kettenmann H, Hanisch UK, Noda M, Verkhratsky A. Physiology of microglia. *Physiol Rev* 2011;**91**(2):461–553.

279. Kherani ZS, Auer RN. Pharmacologic analysis of the mechanism of dark neuron production in cerebral cortex. *Acta Neuropathol* 2008;**116**(4):447–452.

280. Kiernan JA. Chromoxane cyanine R. I. Physical and chemical properties of the dye and of some of its iron complexes. *J Microsc* 1984;**134**(Pt 1):13–23.

281. Kimberlin DW, Whitley RJ. Neonatal herpes: what have we learned? *Semin Pediatr Infect Dis* 2005;**16**(1):7–16.

282. Kinney HC. Abnormalities of the brainstem serotonergic system in the sudden infant death syndrome: a review. *Pediatr Dev Pathol* 2005;**8**(5):507–24.

283. Kinney HC. Brainstem mechanisms underlying the sudden infant death syndrome: evidence from human pathologic studies. *Dev Psychobiol* 2009;**51**(3):223–233.

284. Kinney JS, Kumar ML. Should we expand the TORCH complex? A description of clinical and diagnostic aspects of selected old and new agents. *Clin Perinatol* 1988;**15**:727–44.

285. Kinney HC, Thach BT. The sudden infant death syndrome. *N Engl J Med* 2009;**361**(8):795–805.

286. Kinney HC, Filiano JJ, Harper RM. The neuropathology of the sudden infant death syndrome. A review. *J Neuropathol Exp Neurol* 1992;**51**(2):115–126.

287. Kinney HC, O'Donnell TJ, Kriger P, White WF. Early developmental changes in [3H] nicotine binding in the human brainstem. *Neuroscience* 1993;**55**(4):1127–1138.

288. Kinney HC, Karthigasan J, Borenshteyn NI, Flax JD, Kirschner DA. Myelination in the developing human brain:

biochemical correlates. *Neurochem Res* 1994;**19**(8):983–996.

289. Kinney HC, Randall LL, Sleeper LA, *et al.* Serotonergic brainstem abnormalities in Northern Plains Indians with the sudden infant death syndrome. *J Neuropathol Exp Neurol* 2003;**62**(11):1178–91.

290. Kinney HC, Myers MM, Belliveau RA, *et al.* Subtle autonomic and respiratory dysfunction in sudden infant death syndrome associated with serotonergic brainstem abnormalities: a case report. *J Neuropathol Exp Neurol* 2005;**64**(8): 689–94.

291. Kinney HC, Panigrahy A, Newburger JW, Jonas RA, Sleeper LA. Hypoxic-ischaemic brain injury in infants with congenital heart disease dying after cardiac surgery. *Acta Neuropathol (Berl)* 2005;**110**(6): 563–578.

292. Kinney HC, Haynes RL, Xu G, *et al.* Neuron deficit in the white matter and subplate in periventricular leukomalacia. *Ann Neurol* 2012;**71**(3):397–406.

293. Kirton A, Shroff M, Pontigon AM, deVeber G. Risk factors and presentations of periventricular venous infarction *vs* arterial presumed perinatal ischaemic stroke. *Arch Neurol* 2010;**67**(7):842–848.

294. Kluver H, Barrera E. A method for the combined staining of cells and fibres in the nervous system. *J Neuropathol Exp Neurol* 1953;**12**(4):400–403.

295. Koelfen W, Freund M, Varnholt V. Neonatal stroke involving the middle cerebral artery in term infants: clinical presentation, EEG and imaging studies and outcome. *Dev Med Child Neurol* 1995;**37**(3):204–212.

296. Kolasinski J, Takahashi E, Stevens AA, *et al.* Radial and tangential neuronal migration pathways in the human fetal brain: anatomically distinct patterns of diffusion MRI coherence. *Neuroimage* 2013;**79**:412–422.

297. Kolb B, Gibb R, Gorny G. Cortical plasticity and the development of behavior after early frontal cortical injury. *Dev Neuropsychol* 2000;**18**(3):423–444.

298. Korinthenberg R, Sauer M, Ketelsen UP, *et al.* Congenital axonal neuropathy caused by deletions in the spinal muscular atrophy region. *Ann Neurol* 1997;**42**(3): 364–368.

299. Korosi A, Naninck EF, Oomen CA, et al. Early-life stress mediated modulation of adult neurogenesis and behavior. *Behav Brain Res* 2012;**227**(2):400–409.

300. Kostovic I, Judas M, Sedmak G. Developmental history of the subplate zone, subplate neurons and interstitial white matter neurons: relevance for schizophrenia. *Int J Dev Neurosci* 2011;**29**(3):193–205.

301. Kramer MS, Papageorghiou A, Culhane J, *et al.* Challenges in defining and classifying the preterm birth syndrome. *Am J Obstet Gynecol* 2012;**206**:108–112.

302. Kruse M, Michelsen SI, Flachs EM, *et al.* Lifetime costs of cerebral palsy. *Dev Med Child Neurol* 2009;**51**(8):622–628.

303. Kuban KC, Gilles FH. Human telencephalic angiogenesis. *Ann Neurol* 1985;**17**(6):539–548.

304. Kubo S, Orihara Y, Gotohda T, *et al.* Immunohistochemical studies on neuronal changes in brain stem nucleus of forensic autopsied cases. 1. Sudden infant death syndrome. *Nihon Hoigaku Zasshi* 1998;**52**(6):350–354.

305. Kumada S, Hayashi M, Umitsu R, *et al.* Neuropathology of the dentate nucleus in developmental disorders. *Acta Neuropathol* 1997;**94**(1):36–41.

306. Kupsky WJ, Kinney HC, Lidov HGW. Neuropathology of infants dying after extracorporal membrane oxygenation (ECMO). *J Neuropathol Exp Neurol* 1989;**48**:307.

307. Laird MD, Vender JR, Dhandapani KM. Opposing roles for reactive astrocytes following traumatic brain injury. *Neurosignals* 2008;**16**(2–3):154–164.

308. Langer J, Stephan J, Theis M, Rose CR. Gap junctions mediate intercellular spread of sodium between hippocampal astrocytes *in situ*. *Glia* 2012;**60**(2):239–252.

309. Lapeyre D, Klosowski S, Liska A, *et al.* Very preterm infant (<32 weeks) vs very low birth weight newborns (1500 grammes): comparison of two cohorts. *Arch Pediatr* 2004;**11**(5):412–416.

310. Larroche JC. *Developmental pathology of the neonate.* Amsterdam: Excerpta Medica, 1977.

311. Lategan B, Chodirker BN, Del Bigio MR. Fetal hydrocephalus caused by cryptic intraventricular haemorrhage. *Brain Pathol* 2010;**20**(2):391–398.

312. Laurie DJ, Bartke I, Schoepfer R, Naujoks K, Seeburg PH. Regional, developmental and interspecies expression of the four NMDAR2 subunits, examined using monoclonal antibodies. *Brain Res Mol Brain Res* 1997;**51**(1–2):23–32.

313. Lavezzi AM, Ottaviani G, Terni L, Matturri L. Histological and biological developmental characterization of the human cerebellar cortex. *Int J Dev Neurosci* 2006;**24**(6):365–371.

314. Ledger WJ. Perinatal infections and fetal/ neonatal brain injury. *Curr Opin Obstet Gynecol* 2008;**20**:120–124.

315. Lee SC, Liu W, Dickson DW, Brosnan CF, Berman JW. Cytokine production by human fetal microglia and astrocytes. Differential induction by lipopolysaccharide and IL-1 beta. *J Immunol* 1993;**150**(7):2659–2667.

316. Leech RW, Alvord EC, Jr. Morphologic variations in periventricular leukomalacia. *Am J Pathol* 1974;**74**(3):591–602.

317. Leech RW, Alvord EC, Jr. Anoxic-ischaemic encephalopathy in the human neonatal period. The significance of brain stem involvement. *Arch Neurol* 1977;**34**(2):109–113.

318. Legido A. Pathophysiology of perinatal hypoxic-ischaemic encephalopathy. *Acta Neuropediatr* 1994;**1**:97–110.

319. Lehnardt S. Innate immunity and neuroinflammation in the CNS: the role of microglia in Toll-like receptor-mediated neuronal injury. *Glia* 2010;**58**(3): 253–263.

320. Leist M, Jäättelä M. Four deaths and a funeral: from caspases to alternative mechanisms. *Nat Rev Mol Cell Biol* 2001;**2**:589–598.

321. Lemons JA, Schreiner RL, Gresham EL. Relationship of brain weight to head circumference in early infancy. *Hum Biol* 1981;**53**:351–354.

322. Levison SW, Rothstein RP, Brazel CY, Young GM, Albrecht PJ. Selective apoptosis within the rat subependymal zone: a plausible mechanism for determining which lineages develop from neural stem cells. *Dev Neurosci* 2000;**22**(1–2):106–15.

323. Leviton A, Gilles FH. Acquired perinatal leukoencephalopathy. *Ann Neurol* 1984;**16**(1):1–8.

324. Li DK, Willinger M, Petitti DB, *et al*. Use of a dummy (pacifier) during sleep and risk of sudden infant death syndrome (SIDS): population based case-control study. *BMJ* 2006;**332**(7532):18–22.

325. Li H, He Y, Richardson WD, Casaccia P. Two-tier transcriptional control of oligodendrocyte differentiation. *Curr Opin Neurobiol* 2009;**19**(5):479–485.

326. Liebner S, Czupalla CJ, Wolburg H. Current concepts of blood–brain barrier development. *Int J Dev Biol* 2011;**55**(4–5):467–476.

327. Ligam P, Haynes RL, Folkerth RD, *et al*. Thalamic damage in periventricular leukomalacia: novel pathologic observations relevant to cognitive deficits in survivors of prematurity. *Pediatr Res* 2009;**65**(5):524–529.

328. Lim KO, Beal DM, Harvey RL, Jr, *et al*. Brain dysmorphology in adults with congenital rubella plus schizophrenialike symptoms. *Biol Psychiatry* 1995;**37**(11):764–776.

329. Limperopoulos C, Benson CB, Bassan H, *et al*. Cerebellar haemorrhage in the preterm infant: ultrasonographic findings and risk factors. *Pediatrics* 2005;**116**(3):717–24.

330. Lipton P. Ischaemic cell death in brain neurons. *Physiol Rev* 1999;**79**(4):1431–1568.

331. Looney CB, Smith JK, Merck LH, *et al*. Intracranial haemorrhage in asymptomatic neonates: prevalence on MR images and relationship to obstetric and neonatal risk factors. *Radiology* 2007;**242**(2):535–541.

332. Louis JC, Magal E, Takayama S, Varon S. CNTF protection of oligodendrocytes against natural and tumor necrosis factor-induced death. *Science* 1993;**259**(5095):689–692.

333. Lu J, Tiao G, Folkerth R, *et al*. Overlapping expression of ARFGEF2 and Filamin A in the neuroependymal lining of the lateral ventricles: insights into the cause of periventricular heterotopia. *J Comp Neurol* 2006;**494**(3):476–84.

334. Luu TM, Ment LR, Schneider KC, *et al*. Lasting effects of preterm birth and neonatal brain haemorrhage at 12 years of age. *Pediatrics* 2009;**123**(3):1037–1044.

335. Lyman WD, Kress Y, Kure K, *et al*. Detection of HIV in fetal central nervous system tissue. *AIDS* 1990;**4**(9):917–920.

336. Ma T, Wang C, Wang L, *et al*. Subcortical origins of human and monkey neocortical interneurons. *Nat Neurosci* 2013;**16**(11):1588–1597.

337. Maalouf EF, Duggan PJ, Counsell SJ, *et al*. Comparison of findings on cranial ultrasound and magnetic resonance imaging in preterm infants. *Pediatrics* 2001;**107**(4):719–27.

338. Majno G, Joris I. Apoptosis, oncosis and necrosis. An overview of cell death. *Am J Pathol* 1995;**146**(1):3–15.

339. Malessy MJ, Pondaag W. Obstetric brachial plexus injuries. *Neurosurg Clin N Am* 2009;**20**(1):1–14, v.

340. Malik S, Vinukonda G, Vose LR, *et al*. Neurogenesis continues in the third trimester of pregnancy and is suppressed by premature birth. *J Neurosci* 2013;**33**(2):411–423.

341. Malinger G, Werner H, Rodriguez Leonel JC, *et al*. Prenatal brain imaging in congenital toxoplasmosis. *Prenat Diagn* 2011;**31**(9):881–886.

342. Mall FP. On the development of the blood-vessels of the brain in the human embryo. *Am J Anat* 1905;**4**:1–18.

343. Mallard C, Tolcos M, Leditschke J, Campbell P, Rees S. Reduction in choline acetyltransferase immunoreactivity but not muscarinic-m2 receptor immunoreactivity in the brainstem of SIDS infants. *J Neuropathol Exp Neurol* 1999;**58**(3):255–264.

344. Manlow A, Munoz DG. A non-toxic method for the demonstration of gliosis. *J Neuropathol Exp Neurol* 1992;**51**(3):298–302.

345. Marin–Padilla M. Prenatal and early postnatal ontogenesis of the human motor cortex: a golgi study. I. The sequential development of the cortical layers. *Brain Res* 1970;**23**(2):167–183.

346. Marin–Padilla M. Prenatal development of fibrous (white matter), protoplasmic (gray matter), and layer I astrocytes in the human cerebral cortex: a Golgi study. *J Comp Neurol* 1995;**357**(4):554–572.

347. Marin–Padilla M. Developmental neuropathology and impact of perinatal brain damage. I: hemorrhagic lesions of neocortex. *J Neuropathol Exp Neurol* 1996;**55**(7):758–773.

348. Marin–Padilla M. Developmental neuropathology and impact of perinatal brain damage. II: white matter lesions of the neocortex. *J Neuropathol Exp Neurol* 1997;**56**(3):219–235.

349. Marin–Padilla M. Developmental neuropathology and impact of perinatal brain damage. III: gray matter lesions of the neocortex. *J Neuropathol Exp Neurol* 1999;**58**(5):407–429.

350. Marks N, Stern F, Lajtha A. Changes in proteolytic enzymes and proteins during maturation of the brain. *Brain Res* 1975;**86**(2):307–322.

351. Martin FP, Lukeman JM, Ranson RF, Geppert LJ. Mucormycosis of the central nervous system associated with thrombosis of the internal carotid artery. *J Pediatr* 1954;**44**(4):437–442.

352. Martin KC. Martin Local protein synthesis during axon guidance and synaptic plasticity. *Curr Opin Neurobiol* 2004;**14**:305–310.

353. Martin R, Roessmann U, Fanaroff A. Massive intracerebellar haemorrhage in low-birth-weight infants. *J Pediatr* 1976;**89**(2):290–293.

354. Masuda–Nakagawa LM, Muller KJ, Nicholls JG. Axonal sprouting and laminin appearance after destruction of glial sheaths. *Proc Natl Acad Sci U S A* 1993;**90**(11):4966–70.

355. Mathur AM, Neil JJ, Inder TE. Understanding brain injury and neurodevelopmental disabilities in the preterm infant: the evolving role of advanced magnetic resonance imaging. *Semin Perinatol* 2010;**34**:57–66.

356. Matturri L, Biondo B, Mercurio P, Rossi L. Severe hypoplasia of medullary arcuate nucleus: quantitative analysis in sudden infant death syndrome. *Acta Neuropathol* 2000;**99**(4):371–375.

357. Matyash V, Kettenmann H. Heterogeneity in astrocyte morphology and physiology. *Brain Res Rev* 2010;**63**(1–2):2–10.

358. Mavrov GI, Goubenko TV. Clinical and epidemiological features of syphilis in pregnant women: the course and outcome of pregnancy. *Gynecol Obstet Invest* 2001;**52**:114–118.

359. McConnell SK, Kaznowski CE. Cell cycle dependence of laminar determination in developing neocortex. *Science* 1991;**254**(5029):282–285.

360. McCormick MC, Litt JS, Smith VC, Zupancic JA. Prematurity: an overview and public health implications. *Annu Rev Public Health* 2011;**32**:367–379.

361. McGregor JA, Kleinschmidt–DeMasters BK, Ogle J. Meningoencephalitis caused by *Histoplasma capsulatum* complicating pregnancy. *Am J Obstet Gynecol* 1986;**154**:925–931.

362. McLaurin J, D'Souza S, Stewart J, *et al*. Effect of tumor necrosis factor alpha and beta on human oligodendrocytes and neurons in culture. *Int J Dev Neurosci* 1995;**13**(3–4):369–381.

363. McQuillen PS, Ferriero DM. Perinatal subplate neuron injury: implications for cortical development and plasticity. *Brain Pathol* 2005;**15**(3):250–60.

364. Menser MA, Reye RD. The pathology of congenital rubella: a review written by request. *Pathology* 1974;**6**:215–222.

365. Meny RG, Carroll JL, Carbone MT, Kelly DH. Cardiorespiratory recordings from infants dying suddenly and unexpectedly a t home. *Pediatrics* 1994;**93**(1):44–49.

366. Merrill JE. Effects of interleukin-1 and tumor necrosis factor-alpha on astrocytes, microglia, oligodendrocytes and glial precursors *in vitro*. *Dev Neurosci* 1991;**13**:130–7.

367. Mets MB, Chhabra MS. Eye manifestations of intrauterine infections and their impact on childhood blindness. *Surv Ophthalmol* 2008;**53**:94–111.

368. Micu I, Jiang Q, Coderre E, *et al*. NMDA receptors mediate calcium accumulation in myelin during chemical ischaemia. *Nature* 2006;**439**(7079):988–992.

369. Miller LP, Johnson AE, Gelhard RE, Insel TR. The ontogeny of excitatory amino acid receptors in the rat forebrain – II. Kainic acid receptors. *Neuroscience* 1990;**35**(1):45–51.

370. Milner R, Campbell IL. Developmental regulation of beta1 integrins during angiogenesis in the central nervous system. *Mol Cell Neurosci* 2002;**20**(4):616–26.

371. Milosevic A, Kanazir S, Zecevic N. Immunocytochemical localization of growth-associated protein GAP-43 in early human development. *Brain Res Dev Brain Res* 1995;**84**(2):282–6.

372. Mittendorf R, Covert R, Pryde PG, *et al*. Association between lenticulostriate vasculopathy (LSV) and neonatal intraventricular haemorrhage (IVH). *J Perinatol* 2004;**24**(11):700–705.

373. Mochida GH, Ganesh VS, Felie JM, *et al*. A homozygous mutation in the tight-junction protein JAM3 causes hemorrhagic destruction of the brain, subependymal calcification and congenital cataracts. *Am J Hum Genet* 2010;**87**(6):882–889.

374. Modlin JF. Perinatal echovirus and group B coxsackievirus infections. *Clin Perinatol* 1988;**15**:233–246.

375. Monier A, Adle–Biassette H, Delezoide AL, *et al*. Entry and distribution of microglial cells in human embryonic and fetal cerebral cortex. *J Neuropathol Exp Neurol* 2007;**66**(5):372–382.

376. Moore CS, Abdullah SL, Brown A, Arulpragasam A, Crocker SJ. How factors secreted from astrocytes impact myelin repair. *J Neurosci Res* 2011;**89**(1):13–21.

377. Morel A, Loup F, Magnin M, Jeanmonod D. Neurochemical organization of the human basal ganglia: anatomofunctional territories defined by the distributions of calcium-binding proteins and SMI-32. *J Comp Neurol* 2002;**443**(1):86–103.

378. Mori T, Buffo A, Gotz M. The novel roles of glial cells revisited: the contribution of radial glia and astrocytes to neurogenesis. *Curr Top Dev Biol* 2005;**69**:67–99.

379. Mosevitsky MI. Nerve ending 'signal' proteins GAP-43, MARCKS and BASP1. *Int Rev Cytol* 2005;**245**:245–325.

380. Muller–Starck J, Buttner A, Kiessling MC, *et al.* No changes in cerebellar microvessel length density in sudden infant death syndrome: implications for pathogenetic mechanisms. *J Neuropathol Exp Neurol* 2014;**73**(4):312–323.

381. Myatt L, Eschenbach DA, Lye SJ, *et al.* International Preterm Birth Collaborative (PREBIC) Pathways and Systems Biology Working Groups, A standardized template for clinical studies in preterm birth. *Reprod Sci* 2012;**19**:474–482.

382. Myers RE. Four patterns of perinatal brain damage and their conditions of occurrence in primates. *Adv Neurol* 1975;**10**:223–234.

383. Nagasunder AC, Kinney HC, Bluml S, *et al.* Abnormal microstructure of the atrophic thalamus in preterm survivors with periventricular leukomalacia. *AJNR Am J Neuroradiol* 2011;**32**(1):185–191.

384. Nakamura Y, Okudera T, Hashimoto T. Vascular architecture in white matter of neonates: its relationship to periventricular leukomalacia. *J Neuropathol Exp Neurol* 1994;**53**(6):582–589.

385. Narita N, Narita M, Takashima S, *et al.* Serotonin transporter gene variation is a risk factor for sudden infant death syndrome in the Japanese population. *Pediatrics* 2001;**107**(4):690–2.

386. Nelissen EC, van Montfoort AP, Dumoulin JC, Evers JL. Epigenetics and the placenta. *Hum Reprod Update* 2011;**17**(3):397–417.

387. Nelson K, Ellenberg J. *The pathology of cerebral palsy.* Springfield, IL: Charles C Thomas, 1960.

388. Nelson KB, Ellenberg JH. Antecedents of cerebral palsy. Multivariate analysis of risk. *N Engl J Med* 1986;**315**(2):81–86.

389. Nelson MD, Jr, Gonzalez–Gomez I, Gilles FH. Dyke Award. The search for human telencephalic ventriculofugal arteries. *AJNR Am J Neuroradiol* 1991;**12**(2):215–222.

390. Neumann H, Kotter MR, Franklin RJ. Debris clearance by microglia: an essential link between degeneration and regeneration. *Brain* 2009;**132**(Pt 2):288–295.

391. Nishida A, Misaki Y, Kuruta H, Takashima S. Developmental expression of copper, zinc-superoxide dismutase in human brain by chemiluminescence. *Brain Dev* 1994;**16**(1):40–43.

392. Noll E, Miller RH. Oligodendrocyte precursors originate at the ventral ventricular zone dorsal to the ventral midline region in the embryonic rat spinal cord. *Development* 1993;**118**(2):563–573.

393. Nonaka H, Akima M, Hatori T, *et al.* Microvasculature of the human cerebral white matter: arteries of the deep white matter. *Neuropathology* 2003;**23**(2):111–118.

394. Norenberg MD. Astrocyte responses to CNS injury. *J Neuropathol Exp Neurol* 1994;**53**(3):213–220.

395. Norman MG. Antenatal neuronal loss and gliosis of the reticular formation, thalamus and hypothalamus. A report of three cases. *Neurology* 1972;**22**(9):910–6.

396. Nosarti C, Al–Asady MH, Frangou S, *et al.* Adolescents who were born very preterm have decreased brain volumes. *Brain* 2002;**125**(Pt 7):1616–23.

397. Nugent BM, McCarthy MM. Epigenetic underpinnings of developmental sex differences in the brain. *Neuroendocrinology* 2011;**93**(3):150–158.

398. O'Callaghan ME, MacLennan AH, Gibson CS, *et al.* Epidemiologic associations with cerebral palsy. *Obstet Gynecol* 2011;**118**(3):576–582.

399. Ogata J, Yutani C, Imakita M, *et al.* Autolysis of the granular layer of the cerebellar cortex in brain death. *Acta Neuropathol* 1986;**70**(1):75–78.

400. Oitzl MS, Champagne DL, van der Veen R, de Kloet ER. Brain development under stress: hypotheses of glucocorticoid actions revisited. *Neurosci Biobehav Rev* 2010;**34**(6):853–866.

401. Oka A, Belliveau MJ, Rosenberg PA, Volpe JJ. Vulnerability of oligodendroglia to glutamate: pharmacology, mechanisms, and prevention. *J Neurosci* 1993;**13**(4):1441–1453.

402. Okumura M, Aparecida dos Santos V, Camargo ME, Schultz R, Zugaib M. Prenatal diagnosis of congenital Chagas' disease (American trypanosomiasis). *Prenat Diagn* 2004;**24**(3):179–181.

403. Olariu TR, Remington JS, McLeod R, Alam A, Montoya JG. Severe congenital toxoplasmosis in the United States: clinical and serologic findings in untreated infants. *Pediatr Infect Dis J* 2011;**30**(12):1056–1061.

404. O'Leary H, Gregas MC, Limperopoulos C, *et al.* Elevated cerebral pressure passivity is associated with prematurity-related intracranial haemorrhage. *Pediatrics* 2009;**124**(1):302–309.

405. O'Leary CM, Watson L, D'Antoine H, Stanley F, Bower C. Heavy maternal alcohol consumption and cerebral palsy in the offspring. *Dev Med Child Neurol* 2012;**54**(3):224–230.

406. O'Rahilly R, Muller F. *The embryonic human brain: an atlas of developmental stages.* Wilmington, DE: Wiley–Liss 2006:358–358.

407. Orentas DM, Miller RH. Regulation of oligodendrocyte development. *Mol Neurobiol* 1998;**18**(3):247–259.

408. Ornoy A, Tenenbaum A. Pregnancy outcome following infections by coxsackie, echo, measles, mumps, hepatitis, polio and encephalitis viruses. *Reprod Toxicol* 2006;**21**:446–457.

409. Orthmann–Murphy JL, Abrams CK, Scherer SS. Gap junctions couple astrocytes and oligodendrocytes. *J Mol Neurosci* 2008;**35**(1):101–116.

410. Ortiz JU, Ostermayer E, Fischer T, *et al.* Severe fetal cytomegalovirus infection associated with cerebellar haemorrhage. *Ultrasound Obstet Gynecol* 2004;**23**(4):402–406.

411. O'Shea TM, Allred EN, Kuban KC, *et al.* Intraventricular haemorrhage and developmental outcomes at 24 months of age in extremely preterm infants. *J Child Neurol* 2012;**27**(1):22–29.

412. Owen M, Shevell M, Majnemer A, Limperopoulos C. Abnormal brain structure and function in newborns with complex congenital heart defects before open heart surgery: a review of the

evidence. *J Child Neurol* 2011;**26**:743–755.

413. Owens T, Bechmann I, Engelhardt B. Perivascular spaces and the two steps to neuroinflammation. *J Neuropathol Exp Neurol* 2008;**67**(12):1113–1121.

414. Padget DH. The cranial venous system in man in reference to development, adult configuration and relation to the arteries. *Am J Anat* 1956;**98**(3):307–355.

415. Pahlavan PS, Sutton W, Buist RJ, Del Bigio MR. Multifocal haemorrhagic brain damage following hypoxia and blood pressure lability: case report and rat model. *Neuropathol Appl Neurobiol* 2012;**38**(7):723–733.

416. Paneth N. Classifying brain damage in preterm infants. *J Pediatr* 1999;**134**(5):527–529.

417. Paneth N, Pinto–Martin J, Gardiner J, *et al.* Incidence and timing of germinal matrix/intraventricular haemorrhage in low birth weight infants. *Am J Epidemiol* 1993;**137**(11):1167–1176.

418. Panigrahy A, White WF, Rava LA, Kinney HC. Developmental changes in [3H] kainate binding in human brainstem sites vulnerable to perinatal hypoxia-ischaemia. *Neuroscience* 1995;**67**(2):441–54.

419. Panigrahy A, Filiano J, Sleeper LA, *et al.* Decreased serotonergic receptor binding in rhombic lip-derived regions of the medulla oblongata in the sudden infant death syndrome. *J Neuropathol Exp Neurol* 2000;**59**(5):377–384.

420. Panigrahy A, Rosenberg PA, Assmann S, Foley EC, Kinney HC. Differential expression of glutamate receptor subtypes in human brainstem sites involved in perinatal hypoxia-ischaemia. *J Comp Neurol* 2000;**427**(2):196–208.

421. Panigrahy A, Borzage M, Blüml S. *Semin perinatol* 2010;**34**(1):3–19.

422. Pape KE, Armstrong DL, Fitzhardinge PM. Central nervous system patholgoy associated with mask ventilation in the very low birthweight infant: a new etiology for intracerebral haemorrhages. *Pediatrics* 1976;**58**(4):473–483.

423. Papile LA, Burstein J, Burstein R, Koffler H. Incidence and evolution of subependymal and intraventricular haemorrhage: a study of infants with birth weights less than 1500 gm. *J Pediatr* 1978;**92**(4):529–534.

424. Paterson DS, Trachtenberg FL, Thompson EG, *et al.* Multiple serotonergic brainstem abnormalities in sudden infant death syndrome. *JAMA* 2006;**296**(17):2124–2132.

425. Patterson PH. Maternal infection and immune involvement in autism. *Trends Mol Med* 2011;**17**(7):389–394.

426. Perlman JM. The relationship between systemic hemodynamic perturbations and periventricular-intraventricular haemorrhage – a historical perspective. *Semin Pediatr Neurol* 2009;**16**(4):191–199.

427. Perlman JM, Argyle C. Lethal cytomegalovirus infection in preterm infants: clinical, radiological and neuropathological findings. *Ann Neurol* 1992;**31**:64–68.

428. Perlman JM, McMenamin JB, Volpe JJ. Fluctuating cerebral blood-flow velocity in respiratory-distress syndrome. Relation to the development of intraventricular haemorrhage. *N Engl J Med* 1983;**309**(4):204–209.

429. Perlman JM, Rollins N, Sanchez PJ. Late-onset meningitis in sick, very-low-birth-weight infants. Clinical and sonographic observations. *Am J Dis Child* 1992;**146**(11):1297–1301.

430. Perlow JH, Wigton T, Hart J, *et al*. Birth trauma. A five-year review of incidence and associated perinatal factors. *J Reprod Med* 1996;**41**(10):754–60.

431. Perry VH, Gordon S. Macrophages and microglia in the nervous system. *Trends Neurosci* 1988;**11**(6):273–277.

432. Peterson BS, Anderson AW, Ehrenkranz R, *et al*. Regional brain volumes and their later neurodevelopmental correlates in term and preterm infants. *Pediatrics* 2003;**111**(5 Pt 1):939–48.

433. Petzold GC, Murthy VN. Role of astrocytes in neurovascular coupling. *Neuron* 2011;**71**(5):782–797.

434. Phillips JB, Billson VR, Forbes AB. Autopsy standards for fetal lengths and organ weights of an Australian perinatal population. *Pathology* 2009;**41**(6): 515–526.

435. Pierson CR, Folkerth RD, Billiards SS, *et al*. Gray matter injury associated with periventricular leukomalacia in the premature infant. *Acta Neuropathol* 2007;**114**(6):619–631.

436. Popova NI. [Pathomorphology of the intrauterine fungus infections of the brain]. [in Russian]. *Vopr Okhr Materin Det* 1974;**19**:29–37.

437. Power ML. The human obesity epidemic, the mismatch paradigm and our modern 'captive' environment. *Am J Hum Biol* 2012;**24**(2):116–122.

438. Prandovszky E, Horvath S, Gellert L, *et al*. Nectin-1 (HveC) is expressed at high levels in neural subtypes that regulate radial migration of cortical and cerebellar neurons of the developing human and murine brain. *J Neurovirol* 2008;**14**(2):164–172.

439. Pun PB, Lu J, Moochhala S. Involvement of ROS in BBB dysfunction. *Free Radic Res* 2009;**43**(4):348–364.

440. Puyal J, Ginet V, Clarke PG. Multiple interacting cell death mechanisms in the mediation of excitotoxicity and ischaemic brain damage: a challenge for neuroprotection. *Prog Neurobiol* 2013;**105**:24–48.

441. Raff MC, Mirsky R, Fields KL, Lisak RP, Dorfman SH, Silberberg DH, *et al*. Galactocerebroside is a specific cell-surface antigenic marker for oligodendrocytes in culture. *Nature* 1978;**274**(5673):813–816.

442. Raine CS, Fields BN. Neurotropic viruses and the developing brain. *N Y State J Med* 1973;**73**:1169–1179.

443. Raju TN, Higgins RD, Stark AR, Leveno KJ. Optimizing care and outcome for late-preterm (near-term) infants: a summary of the workshop sponsored by the National Institute of Child Health and Human Development. *Pediatrics* 2006;**118**(3):1207–14.

444. Rakic P. Specification of cerebral cortical areas. *Science* 1988;**241**(4862):170–176.

445. Rakic S, Zecevic N. Programmed cell death in the developing human telencephalon. *Eur J Neurosci* 2000;**12**(8):2721–34.

446. Ramsay M. Genetic and epigenetic insights into fetal alcohol spectrum disorders. *Genome Med* 2010;**2**(4):27.

447. Randall BB, Paterson DS, Haas EA, *et al*. Potential asphyxia and brainstem abnormalities in sudden and unexpected death in infants. *Pediatrics* 2013;**132**(6):e1616–25.

448. Rao S, Ali U. Systemic fungal infections in neonates. *J Postgrad Med* 2005;**51**(Suppl 1):S27–S29.

449. Redline RW. Severe fetal placental vascular lesions in term infants with neurologic impairment. *Am J Obstet Gynecol* 2005;**192**(2):452–7.

450. Redline RW. Disorders of placental circulation and the fetal brain. *Clin Perinatol* 2009;**36**(3):549–559.

451. Redline RW, Wilson–Costello D, Borawski E, Fanaroff AA, Hack M. Placental lesions associated with neurologic impairment and cerebral palsy in very-low-birth-weight infants. *Arch Pathol Lab Med* 1998;**122**(12):1091–8.

452. Reichard R. Birth injury of the cranium and central nervous system. *Brain Pathol* 2008;**18**:565–570.

453. Rezaie P, Male D. Mesoglia and microglia – a historical review of the concept of mononuclear phagocytes within the central nervous system. *J Hist Neurosci* 2002;**11**(4):325–74.

454. Rezaie P, Cairns NJ, Male DK. Expression of adhesion molecules on human fetal cerebral vessels: relationship to microglial colonisation during development. *Brain Res Dev Brain Res* 1997;**104**(1–2):175–89.

455. Rezaie P, Patel K, Male DK. Microglia in the human fetal spinal cord – patterns of distribution, morphology and phenotype. *Brain Res Dev Brain Res* 1999;**115**(1):71–81.

456. Rezaie P, Bohl J, Ulfig N. Anomalous alterations affecting microglia in the central nervous system of a fetus at 12 weeks of gestation: case report. *Acta Neuropathol (Berl)* 2004;**107**(2):176–80.

457. Roback HH, Scherer H. Uber die feinere Morphologie des fruhkindlichen Gehirns unter besonderer Berucksichtigung der Glaientwicklung. *Virchows Archiv* 1935;**294**:365–413.

458. Robbins DS, Shirazi Y, Drysdale BE, *et al*. Production of cytotoxic factor for oligodendrocytes by stimulated astrocytes. *J Immunol* 1987;**139**(8):2593–2597.

459. Roessmann U, Gambetti P. Astrocytes in the developing human brain. An immunohistochemical study. *Acta Neuropathol* 1986;**70**(3–4):308–313.

460. Rorke LB. *Pathology of the perinatal brain injury*. New York: Raven Press, 1982.

461. Rorke LB. Anatomical features of the developing brain implicated in pathogenesis of hypoxic-ischaemic injury. *Brain Pathol* 1992;**2**(3):211–221.

462. Rorke LB, Spiro AJ. Cerebral lesions associated with congenital rubella syndrome. *J Neuropathol Exp Neurol* 1967;**1**:115–117.

463. Rossiter JP, Anderson LL, Yang F, Cole GM. Caspase-3 activation and caspase-like proteolytic activity in human perinatal hypoxic-ischaemic brain injury. *Acta Neuropathol (Berl)* 2002;**103**(1):66–73.

464. Roza SJ, Steegers EA, Verburg BO, *et al*. What is spared by fetal brain-sparing? Fetal circulatory redistribution and behavioral problems in the general population. *Am J Epidemiol* 2008;**168**(10):1145–1152.

465. Saijo K, Glass CK. Microglial cell origin and phenotypes in health and disease. *Nat Rev Immunol* 2011;**11**(11):775–787.

466. Salafia CM, Minior VK, Rosenkrantz TS, *et al*. Maternal, placental, and neonatal associations with early germinal matrix/ intraventricular haemorrhage in infants born before 32 weeks' gestation. *Am J Perinatol* 1995;**12**(6):429–436.

467. Salameh S, Sheth U, Shukla D. Early events in herpes simplex virus lifecycle with implications for an infection of lifetime. *Open Virol J* 2012;**6**:1–6.

468. Salter MG, Fern R. NMDA receptors are expressed in developing oligodendrocyte processes and mediate injury. *Nature* 2005;**438**(7071):1167–71.

469. Sanchez–Gomez MV, Alberdi E, Perez–Navarro E, Alberch J, Matute C. Bax and calpain mediate excitotoxic oligodendrocyte death induced by activation of both AMPA and kainate receptors. *J Neurosci* 2011;**31**(8):2996–3006.

470. Sarnat HB. Ependymal reactions to injury. A review. *J Neuropathol Exp Neurol* 1995;**54**(1):1–15.

471. Sarnat HB, Flores–Sarnat L, Trevenen CL. Synaptophysin immunoreactivity in the human hippocampus and neocortex from 6 to 41 weeks of gestation. *J Neuropathol Exp Neurol* 2010;**69**(3):234–245.

472. Sarwar M. Aqueductal occlusion in the congenital rubella syndrome. *Neurology* 1974;**24**:198–201.

473. Sauvageot CM, Stiles CD. Molecular mechanisms controlling cortical gliogenesis. *Curr Opin Neurobiol* 2002;**12**(3):244–9.

474. Schaakxs D, Bahm J, Sellhaus B, Weis J. Clinical and neuropathological study about the neurotization of the suprascapular nerve in obstetric brachial plexus lesions. *J Brachial Plex Peripher Nerve Inj* 2009;**4**:15.

475. Schneider H, Ballowitz L, Schachinger H, Hanefeld F, Droszus JU. Anoxic encephalopathy with predominant involvement of basal ganglia, brain stem and spinal cord in the perinatal period. Report on seven newborns. *Acta Neuropathol (Berl)* 1975;**32**(4):287–98.

476. Schwartz RP. Neonatal hypoglycemia: how low is too low? *J Pediatr* 1997;**131**(2):171–173.

477. Scott G. *Natural History of HIV Infection in Children*. HRS-D-MC-87, 1987.

478. Selmaj KW, Raine CS. Tumor necrosis factor mediates myelin and oligodendrocyte damage *in vitro*. *Ann Neurol* 1988;**23**:339–46.

479. Selmaj KW, Farooq M, Norton WT. Proliferation of astrocytes in vitro in response to cytokines. A primary role for tumor necrosis factor. *J Immunol* 1990;**144**:129–35.

480. Shah DK, Daley AJ, Hunt RW, Volpe JJ, Inder TE. Cerebral white matter injury in the newborn following *Escherichia coli* meningitis. *Eur J Paediatr Neurol* 2005;**9**(1):13–7.

481. Shapiro SM. Bilirubin toxicity in the developing nervous system. *Pediatr Neurol* 2003;**29**(5):410–21.

482. Shaw CM, Alvord EC, Jr. Subependymal germinolysis. *Arch Neurol* 1974;**31**: 374–381.

483. Sheinbergas MM, Ptashekas RS, Pikelite RL, Tuliavichene I, Sverdlov I. Clinical and pathomorphologic findings in hydrocephalus caused by prenatal infection with lymphocytic choriomeningitis virus. *Zh Nevropatol Psikhiatr Im S S Korsakova* 1977;**77**(7):1004–1007.

484. Sherer DM, Anyaegbunam A, Onyeije C. Antepartum fetal intracranial haemorrhage, predisposing factors and

prenatal sonography: a review. *Am J Perinatol* 1998;**15**(7):431–441.

485. Sherwood AJ, Smith JF. Bilirubin encephalopathy. *Neuropathol Appl Neurobiol* 1983;**9**(4):271–285.

486. Sheth RD. Trends in incidence and severity of intraventricular haemorrhage. *J Child Neurol* 1998;**13**(6):261–264.

487. Simons M, Trajkovic K. Neuron-glia communication in the control of oligodendrocyte function and myelin biogenesis. *J Cell Sci* 2006;**119**(Pt 21):4381–4389.

488. Singer DB, Rudolph AJ, Rosenberg HS, Rawls WE, Boniuk M. Pathology of the congenital rubella syndrome. *J Pediatr* 1967;**71**(5):665–675.

489. Sinzger C, Digel M, Jahn G. Cytomegalovirus cell tropism. *Curr Top Microbiol Immunol* 2008;**325**:63–83.

490. Slemmer JE, Shacka JJ, Sweeney MI, Weber JT. Antioxidants and free radical scavengers for the treatment of stroke, traumatic brain injury and aging. *Curr Med Chem* 2008;**15**(4):404–414.

491. Smith CV, Hansen TN, Martin NE, McMicken HW, Elliott SJ. Oxidant stress responses in premature infants during exposure to hyperoxia. *Pediatr Res* 1993;**34**(3):360–365.

492. Snookes SH, Gunn JK, Eldridge BJ, et al. A systematic review of motor and cognitive outcomes after early surgery for congenital heart disease. *Pediatrics* 2010;**125**: e818–827.

493. Sobottka B, Ziegler U, Kaech A, Becher B, Goebels N. CNS live imaging reveals a new mechanism of myelination: the liquid croissant model. *Glia* 2011;**59**(12): 1841–1849.

494. Sofroniew MV, Vinters HV. Astrocytes: biology and pathology. *Acta Neuropathol* 2010;**119**(1):7–35.

495. Soul JS, Hammer PE, Tsuji M, et al. Fluctuating pressure-passivity is common in the cerebral circulation of sick premature infants. *Pediatr Res* 2007;**61**(4):467–473.

496. Sprecher S, Soumenkoff G, Puissant F, Degueldre M. Vertical transmission of HIV in 15-week fetus. *Lancet* 1986;**2**(8501):288–289.

497. Squier W. *Acquired damage to the developing brain: timing and causation*. London: Arnold, 2002.

498. Squier W, Mack J. The neuropathology of infant subdural haemorrhage. *Forensic Sci Int* 2009;**187**:6–13.

499. Stanley CA, DeLeeuw S, Coates PM, et al. Chronic cardiomyopathy and weakness or acute coma in children with a defect in carnitine uptake. *Ann Neurol* 1991;**30**(5):709–716.

500. Stevens MK, Yaksh TL. Systematic studies on the effects of the NMDA receptor antagonist MK-801 on cerebral blood flow and responsivity, EEG and blood–brain barrier following complete reversible cerebral ischaemia. *J Cereb Blood Flow Metab* 1990;**10**(1):77–88.

501. Stevenson DK, Vreman HJ, Wong RJ. Bilirubin production and the risk of bilirubin neurotoxicity. *Semin Perinatol* 2011;**35**:121–126.

502. Stoll BJ, Hansen NI, Adams–Chapman I, et al. Neurodevelopmental and growth impairment among extremely low-birth-weight infants with neonatal infection. *JAMA* 2004;**292**(19):2357–2365.

503. Strauss H. Recently discovered causes of congenital cerebral defects; maternal rubella during pregnancy. *Proc Rudolf Virchow Med Soc City N Y* 1945;**4**:81.

504. Streeter GL. The development of the venous sinuses of the dura mater in the human embryo. Am J Anat 1915;**18**: 145–178.

505. Stubbs D, DeProto J, Nie K, et al. Neurovascular congruence during cerebral cortical development. *Cereb Cortex* 2009;**19**(Suppl 1):i32–41.

506. Summers AM, Summers CW, Drucker DB, et al. Association of IL-10 genotype with sudden infant death syndrome. *Hum Immunol* 2000;**61**(12):1270–3.

507. Sun T, Echelard Y, Lu R, et al. Olig bHLH proteins interact with homeodomain proteins to regulate cell fate acquisition in progenitors of the ventral neural tube. *Curr Biol* 2001;**11**(18):1413–20.

508. Sunderson FW, Boerner F. *Normal values in clinical medicine*. Philadelphia: W.B. Saunders Company, 1949.

509. Supramaniam V, Vontell R, Srinivasan L, et al. Microglia activation in the extremely preterm human brain. *Pediatr Res* 2013;**73**(3):301–309.

510. Swanson AE. Subgaleal haemorrhage: risk factors and outcomes. *Acta Obstet Gynecol Scand* 2012;**91**:260–263.

511. Tabbutt S, Gaynor JW, Newburger JW. Neurodevelopmental outcomes after congenital heart surgery and strategies for improvement. *Curr Opin Cardiol* 2012;**27**(2):82–91.

512. Takashima S, Tanaka K. Development of cerebrovascular architecture and its relationship to periventricular leukomalacia. *Arch Neurol* 1978;**35**(1): 11–16.

513. Takashima S, Becker LE. Developmental neuropathology in bronchopulmonary dysplasia: alteration of glial fibrillary acidic protein and myelination. *Brain Dev* 1984;**6**(5):451–457.

514. Takashima S, Armstrong D, Becker L, Bryan C. Cerebral hypoperfusion in the sudden infant death syndrome? Brainstem gliosis and vasculature. *Ann Neurol* 1978;**4**(3):257–262.

515. Takashima S, Kuruta H, Mito T, et al. Immunohistochemistry of superoxide dismutase-1 in developing human brain. *Brain Dev* 1990;**12**(2):211–213.

516. Takashima S, Itoh M, Oka A. A history of our understanding of cerebral vascular development and pathogenesis of perinatal brain damage over the past 30 years. *Semin Pediatr Neurol* 2009;**16**(4): 226–236.

517. Takikawa M, Kato S, Esumi H, et al. Temporospatial relationship between the expressions of superoxide dismutase and nitric oxide synthase in the developing human brain: immunohistochemical and immunoblotting analyses. *Acta Neuropathol (Berl)* 2001;**102**:572–580.

518. Talos DM, Fishman RE, Park H, et al. Developmental regulation of alpha-amino-3-hydroxy-5-methyl-4-isoxazole-propionic acid receptor subunit expression in forebrain and relationship to regional susceptibility to hypoxic/ischaemic injury. I. Rodent cerebral white matter and cortex. *J Comp Neurol* 2006;**497**(1):42–60.

519. Tam EW, Haeusslein LA, Bonifacio SL, et al. Hypoglycemia is associated with increased risk for brain injury and adverse neurodevelopmental outcome in neonates

at risk for encephalopathy. *J Pediatr* 2012;**161**(1):88–93.

520. Tam SJ, Watts RJ. Connecting vascular and nervous system development: angiogenesis and the blood–brain barrier. *Annu Rev Neurosci* 2010;**33**:379–408.

521. Tan M, Deveber G, Shroff M, et al. Sagittal sinus compression is associated with neonatal cerebral sinovenous thrombosis. *Pediatrics* 2011;**128**(2): e429–35.

522. Tardieu M, Le Chenadec J, Persoz A, et al. HIV-1-related encephalopathy in infants compared with children and adults. French Pediatric HIV Infection Study and the SEROCO Group. *Neurology* 2000;**54**(5):1089–95.

523. Tatli B, Ozmen M, Aydinli N, Caliskan M. Not a new leukodystrophy but congenital cytomegalovirus infection. *J Child Neurol* 2005;**20**(6):525–527.

524. Tavani F, Zimmerman RA, Clancy RR, Licht DJ, Mahle WT. Incidental intracranial haemorrhage after uncomplicated birth: MRI before and after neonatal heart surgery. *Neuroradiology* 2003;**45**(4):253–258.

525. Taylor GA, Fitz CR, Kapur S, Short BL. Cerebrovascular accidents in neonates treated with extracorporeal membrane oxygenation: sonographic-pathologic correlation. *AJR Am J Roentgenol* 1989;**153**(2):355–361.

526. Teele RL, Hernanz–Schulman M, Sotrel A. Echogenic vasculature in the basal ganglia of neonates: a sonographic sign of vasculopathy. *Radiology* 1988;**169**(2): 423–427.

527. Teissier N. Inner ear lesions in congenital cytomegalovirus infection of human fetuses. *Acta Neuropathol* 2011;**122**: 763–774.

528. Teissier N, Fallet–Bianco C, Delezoide AL, Laquerriere A, Marcorelles P, Khung–Savatovsky S, et al. Cytomegalovirus-induced brain malformations in fetuses. *J Neuropathol Exp Neurol* 2014;**73**(2): 143–158.

529. Thompson WS, Cohle SD. Fifteen-year retrospective study of infant organ weights and revision of standard weight tables. *J Forensic Sci* 2004;**49**:575–585.

530. Thorarensen O, Ryan S, Hunter J, Younkin DP. Factor V Leiden mutation: an unrecognized cause of hemiplegic cerebral palsy, neonatal stroke and placental thrombosis. *Ann Neurol* 1997;**42**(3): 372–375.

531. Tijsseling D, Wijnberger LD, Derks JB, et al. Effects of antenatal glucocorticoid therapy on hippocampal histology of preterm infants. *PLoS One* 2012;**7**(3):e33369.

532. Torfs CP, van den Berg B, Oechsli FW, Cummins S. Prenatal and perinatal factors in the etiology of cerebral palsy. *J Pediatr* 1990;**116**(4):615–619.

533. Torrico F, Alonso–Vega C, Suarez E, et al. Maternal *Trypanosoma cruzi* infection, pregnancy outcome, morbidity and mortality of congenitally infected and non-infected newborns in Bolivia. *Am J Trop Med Hyg* 2004;**70**(2):201–209.

534. Towbin A. *The pathology of cerebral palsy: the causes and underlying nature of the disorder*. Springfield, IL: Charles C Thomas, 1960.

535. Towbin A. Spinal cord and brain stem injury at birth. *Arch Pathol* 1964;**77**: 620–32.

536. Towbin A. *Brain damage in the newborn and its neurologic sequels: pathologic and clinical correlation.* Danvers, MA: PRM Publishing Co., 1998.

537. Townsend JJ, Baringer JR, Wolinsky JS, et al. Progressive rubella panencephalitis. Late onset after congenital rubella. *N Engl J Med* 1975;292(19):990–993.

538. Tozuka Y, Wada E, Wada K. 'Biocommunication' between mother and offspring: lessons from animals and new perspectives for brain science. *J Pharmacol Sci* 2009;110(2):127–132.

539. Tsuji M, Saul JP, du Plessis A, et al. Cerebral intravascular oxygenation correlates with mean arterial pressure in critically ill premature infants. *Pediatrics* 2000;106(4):625–32.

540. Tuckuviene R, Christensen AL, Helgested J, et al. Infant, obstetrical and maternal characteristics associated with thromboembolism in infancy: a nationwide population-based case-control study. *Arch Dis Child Fetal Neonatal Ed* 2012;97(6):F417–22.

541. Uchiyama Y, Koike M, Shibata M. Autophagic neuron death in neonatal brain ischaemia/hypoxia. *Autophagy* 2008;4(4):404–408.

542. Ulfig N, Bohl J, Neudorfer F, Rezaie P. Brain macrophages and microglia in human fetal hydrocephalus. *Brain Dev* 2004;26(5):307–15.

543. Unhanand M, Mustafa MM, McCracken GH, Jr, Nelson JD. Gram-negative enteric bacillary meningitis: a twenty-one-year experience. *J Pediatr* 1993;122(1):15–21.

544. Van Assche FA, Holemans K, Aerts L. Fetal growth and consequences for later life. *J Perinat Med* 1998;26(5):337–346.

545. van de Bor M, Guit GL, Schreuder AM, Wondergem J, Vielvoye GJ. Early detection of delayed myelination in preterm infants. *Pediatrics* 1989;84(3):407–411.

546. Van den Bergh R. Centrifugal elements in the vascular pattern of the deep intracerebral blood supply. *Angiology* 1969;20(2):88–94.

547. Van den Broeck C, Himpens E, Vanhaesebrouck P, Calders P, Oostra A. Influence of gestational age on the type of brain injury and neuromotor outcome in high-risk neonates. *Eur J Pediatr* 2008;167(9):1005–1009.

548. van Haastert IC, Groenendaal F, Uiterwaal CS, et al. Decreasing incidence and severity of cerebral palsy in prematurely born children. *J Pediatr* 2011;159(1):86–91.e1.

549. Vanarsdall AL, Johnson DC. Human cytomegalovirus entry into cells. *Curr Opin Virol* 2012;2:37–42.

550. Vannucci RC, Nardis EE, Vannucci SJ. Cerebral metabolism during hypoglycemia and asphyxia in newborn dogs. *Biol Neonate* 1980;38(5–6): 276–86.

551. Vannucci RC, Nardis EE, Vannucci SJ, Campbell PA. Cerebral carbohydrate and energy metabolism during hypoglycemia in newborn dogs. *Am J Physiol* 1981;240(3):R192–9.

552. Varelas PN, Sleight BJ, Rinder HM, Sze G, Ment LR. Stroke in a neonate heterozygous for factor V Leiden. *Pediatr Neurol* 1998;18(3):262–264.

553. Vartanian T, Li Y, Zhao M, Stefansson K. Interferon-gamma-induced oligodendrocyte cell death: implications for the pathogenesis of multiple sclerosis. *Mol Med* 1995;1(7):732–743.

554. Verhagen EA, Ter Horst HJ, Keating P, et al. Cerebral oxygenation in preterm infants with germinal matrix-intraventricular haemorrhages. *Stroke* 2010;41(12):2901–2907.

555. Verma U, Tejani N, Klein S, et al. Obstetric antecedents of intraventricular haemorrhage and periventricular leukomalacia in the low-birth-weight neonate. *Am J Obstet Gynecol* 1997;176(2):275–281.

556. Verney C, Pogledic I, Biran V, et al. Microglial reaction in axonal crossroads is a hallmark of noncystic periventricular white matter injury in very preterm infants. *J Neuropathol Exp Neurol* 2012;71(3):251–264.

557. Villar J, Papageorghiou AT, Knight HE, et al. The preterm birth syndrome: a prototype phenotypic classification. *Am J Obstet Gynecol* 2012;206:119–123.

558. Vinukonda G, Dummula K, Malik S, et al. Effect of prenatal glucocorticoids on cerebral vasculature of the developing brain. *Stroke* 2010;41(8):1766–1773.

559. Virgintino D, Errede M, Robertson D, et al. Immunolocalization of tight junction proteins in the adult and developing human brain. *Histochem Cell Biol* 2004;122(1):51–59.

560. Vohr B, Ment LR. Intraventricular haemorrhage in the preterm infant. *Early Hum Dev* 1996;44:1–16.

561. Voigt T. Development of glial cells in the cerebral wall of ferrets: direct tracing of their transformation from radial glia into astrocytes. *J Comp Neurol* 1989;289(1):74–88.

562. Volpe JJ. Intraventricular haemorrhage in the premature infant – current concepts. Part I. *Ann Neurol* 1989;25(1):3–11.

563. Volpe J. Hypoglycemia and brain injury. In: Volpe J ed. *Neurology of the newborn.* 3rd edn. Philadelphia: WB Saunders, 1995:467–98.

564. Volpe JJ. Neurobiology of periventricular leukomalacia in the premature infant. *Pediatr Res* 2001;50(5):553–62.

565. Volpe JJ. Cerebral white matter injury of the premature infant – more common than you think. *Pediatrics* 2003;112(1 Pt 1):176–80.

566. Volpe JJ. *Neurology of the newborn.* 5th ed. Philadelphia: Elsevier, 2008.

567. Volpe JJ. Brain injury in premature infants: a complex amalgam of destructive and developmental disturbances. *Lancet Neurol* 2009;8(1):110–124.

568. Volpe JJ, Kinney HC, Jensen FE, Rosenberg PA. The developing oligodendrocyte: key cellular target in brain injury in the premature infant. *Int J Dev Neurosci* 2011;29(4):423–440.

569. Vose LR, Vinukonda G, Jo S, et al. Treatment with thyroxine restores myelination and clinical recovery after intraventricular haemorrhage. *J Neurosci* 2013;33(44):17232–17246.

570. Waites CL, Craig AM, Garner CC. Mechanisms of vertebrate synaptogenesis. *Annu Rev Neurosci* 2005;28:251–74.

571. Watchko JF. Kernicterus and the molecular mechanisms of bilirubin-induced CNS injury in newborns. *Neuromolecular Med* 2006;8:513–529.

572. Weber MA, Ashworth MT, Risdon RA, et al. Sudden unexpected neonatal death in the first week of life: autopsy findings from a specialist centre. *J Matern Fetal Neonatal Med* 2009;22(5):398–404.

573. Webster WS. Teratogen update: congenital rubella. *Teratology* 1998;58:13–23.

574. Weese–Mayer DE, Zhou L, Berry–Kravis EM, et al. Association of the serotonin transporter gene with sudden infant death syndrome: a haplotype analysis. *Am J Med Genet A* 2003;122(3):238–45.

575. Wegner M. A matter of identity: transcriptional control in oligodendrocytes. *J Mol Neurosci* 2008;35(1):3–12.

576. Weidenheim KM, Kress Y, Epshteyn I, Rashbaum WK, Lyman WD. Early myelination in the human fetal lumbosacral spinal cord: characterization by light and electron microscopy. *J Neuropathol Exp Neurol* 1992;51(2):142–14.

577. Weisman LE, Stoll BJ, Cruess DF, et al. Early-onset group B streptococcal sepsis: a current assessment. *J Pediatr* 1992;121(3):428–433.

578. Wells JT, Ment LR. Prevention of intraventricular haemorrhage in preterm infants. *Early Hum Dev* 1995;42(3):209–233.

579. Whitby EH, Griffiths PD, Rutter S, et al. Frequency and natural history of subdural haemorrhages in babies and relation to obstetric factors. *Lancet* 2004;363(9412):846–851.

580. Wierzba–Bobrowicz T, Kosno–Kruszewska E, Gwiazda E, Lechowicz W. Major histocompatibility complex class II (MHC II) expression during the development of human fetal cerebral occipital lobe, cerebellum and hematopoietic organs. *Folia Neuropathol* 2000;38(3):111–8.

581. Williams RS, Lott IT, Ferrante RJ, Caviness VS, Jr. The cellular pathology of neuronal ceroid-lipofuscinosis. A golgi-electronmicroscopic study. *Arch Neurol* 1977;34(5):298–305.

582. Willinger M, James LS, Catz C. Defining the sudden infant death syndrome (SIDS): deliberations of an expert panel convened by the National Institute of Child Health and Human Development. *Pediatr Pathol* 1991;11(5):677–684.

583. Windle WF, Jacobson HN, Robert DE, et al. Structural and functional sequelae of asphyxia neonatorum in monkeys (*Macaca mulatta*). *Res Publ Assoc Res Nerv Ment Dis* 1962;39:169–182.

584. Witting A, Muller P, Herrmann A, Kettenmann H, Nolte C. Phagocytic clearance of apoptotic neurons by microglia/brain macrophages in vitro: involvement of lectin-, integrin-, and phosphatidylserine-mediated recognition. *J Neurochem* 2000;75(3):1060–70.

585. Wyatt HV. Poliomyelitis in the fetus and the newborn. A comment on the new understanding of the pathogenesis. *Clin Pediatr* 1979;18:33–38.

586. Xu G, Takahashi E, Folkerth RD, et al. Radial coherence of diffusion tractography in the cerebral white matter of the human fetus: neuroanatomic insights. *Cereb Cortex* 2014;24(3): 579–92.

587. Yakovlev P, Lecours A. The myelogenetic cycles of regional maturation of the brain. In: Minkowski A ed. *Regional development of the brain in early life.* Oxford: Blackwell, 1967:3–70.

588. Yamada T, Placzek M, Tanaka H, Dodd J, Jessell TM. Control of cell pattern in the developing nervous system: polarizing activity of the floor plate and notochord. *Cell* 1991;64(3):635–647.

589. Yonezawa M, Back SA, Gan X, Rosenberg PA, Volpe JJ. Cystine deprivation induces oligodendroglial death: rescue by free radical scavengers and by a diffusible glial factor. *J Neurochem* 1996;**67**(2):566–573.

590. Yoon BH, Romero R, Yang SH, *et al.* Interleukin-6 concentrations in umbilical cord plasma are elevated in neonates with white matter lesions associated with periventricular leukomalacia. *Am J Obstet Gynecol* 1996;**174**(5): 1433–1440.

591. Yoon BH, Kim CJ, Romero R, *et al.* Experimentally induced intrauterine infection causes fetal brain white matter lesions in rabbits. *Am J Obstet Gynecol* 1997;**177**(4):797–802.

592. Yoshioka A, Hardy M, Younkin DP, *et al.* Alpha-amino-3-hydroxy-5-methyl-4-isoxazolepropionate (AMPA) receptors mediate excitotoxicity in the oligodendroglial lineage. *J Neurochem* 1995;**64**(6):2442–2448.

593. Young C, Tenkova T, Dikranian K, Olney JW. Excitotoxic versus apoptotic mechanisms of neuronal cell death in perinatal hypoxia/ischaemia. *Curr Mol Med* 2004;**4**(2):77–85.

594. Young RS, Yagel SK, Towfighi J. Systemic and neuropathologic effects of *E. coli* endotoxin in neonatal dogs. *Pediatr Res* 1983;**17**:349–53.

595. Yu MC, Yu WH. Effect of hypoxia on cerebellar development: morphologic and radioautographic studies. *Exp Neurol* 1980;**70**(3):652–664.

596. Zec N, Filiano JJ, Kinney HC. Anatomic relationships of the human arcuate nucleus of the medulla: a DiI-labelling study. *J Neuropathol Exp Neurol* 1997;**56**: 509–22.

597. Zecevic N, Chen Y, Filipovic R. Contributions of cortical subventricular zone to the development of the human cerebral cortex. *J Comp Neurol* 2005;**491**(2):109–122.

598. Zecevic N, Hu F, Jakovcevski I. Interneurons in the developing human neocortex. *Dev Neurobiol* 2011;**71**(1):18–33.

599. Zerlin M, Goldman JE. Interactions between glial progenitors and blood vessels during early postnatal corticogenesis: blood vessel contact represents an early stage of astrocyte differentiation. *J Comp Neurol* 1997;**387**(4):537–546.

600. Zhang Y, Barres BA. Astrocyte heterogeneity: an underappreciated topic in neurobiology. *Curr Opin Neurobiol* 2010;**20**(5):588–594.

601. Zuelzer WW, Mudgett RT. Kernicterus; etiologic study based on an analysis of 55 cases. *Pediatrics* 1950;**6**:452–474.

3

Malformations

Brian N Harding and Jeffrey A Golden

INTRODUCTION

Malformations of the central nervous system (CNS) are of major clinical importance, leading to considerable mortality and morbidity, both prenatally and postnatally. The birth prevalence of CNS malformations is between 5 and 10 per 1000 births and appears to have remained fairly stable over the past 50 years.[535,732] Data collected from Europe and the USA between 1940 and 1990 show that 8–10 per cent of stillbirths and 5–6 per cent of early neonatal deaths are caused primarily by malformation of the CNS.[535] Moreover, CNS malformations are present in around 15 per cent of infants dying from causes associated with birth defects.[1113]

These figures probably underestimate the prevalence of the most and least severe malformations. Defects such as anencephaly, which have a higher prevalence among spontaneous abortions than in term pregnancies,[207] are probably more common than is suggested by the epidemiological data, which are based predominantly on liveborn infants and stillbirths. By contrast, subtle malformations such as neuronal migration defects are often not recognized at birth or in the first year of life and so may not be included in the epidemiological surveys. According to the epidemiological studies, the most common CNS malformations, in declining order of prevalence, are microcephaly, hydrocephaly, macrocephaly, myelomeningocele, anencephaly and encephalocele.

The disease spectrum includes gross structural malformations such as anencephaly and myelomeningocele (spina bifida) that threaten life directly, more subtle structural defects such as lissencephaly and microencephaly, in which epilepsy and mental retardation are common consequences, and functional brain deficits that cause learning difficulties and behavioural disturbance. In this chapter on CNS malformations, principles that are emerging from contemporary studies of the genetic, molecular, cellular and developmental biology of nervous system development in both humans and animal models will be reviewed. This will be followed by detailed consideration of the neuropathology of the main classes of malformation in the light of the main embryonic and fetal events of CNS development.

PRINCIPLES OF NERVOUS SYSTEM DEVELOPMENT

The causal factors (aetiology) and the embryonic and fetal processes (pathogenesis) that underlie the development of malformations of the CNS are the subjects of an extremely rapidly growing area of research. To gain an appreciation of the improved understanding that has emerged over the past few years, it is necessary to consider some of the major advances, which have included the isolation of causative genes, the development of animal (usually mouse) disease models and the elucidation of cellular and molecular mechanisms of embryonic and fetal development. These advances provide a starting point for understanding how genes and environmental factors can disrupt embryonic development to yield CNS malformations. Defining the scientific basis of CNS malformations is important, not only as an adjunct to the pathological analysis of the defects (see the section Pathology of Malformations, p. 293) but also as a preliminary to developing improved diagnostic techniques and, ultimately, for primary prevention of the malformations.

Aetiology

Both genetic and environmental factors have been implicated in the aetiology of CNS malformations. Although, for convenience, these categories are considered separately, it is important to bear in mind that, in reality, most birth defects are likely multifactorial, representing a combination of genetic, epigenetic and environmental factors (see later).

Genes

Dramatic progress has been made over the past two decades in determining the genetic basis of single-gene disorders in humans. The next challenge is to determine the genetic basis of those conditions, quantitatively more important, in which polygenic control is implicated. Table 4.1 lists the main diseases and syndromes that involve CNS malformations, including those for which a gene has been mapped or cloned. The accelerating pace of disease gene discovery is illustrated by the more than doubling (from 24 to 67; see Table 4.1) of the entries

with a definite gene identification since the seventh edition of this book in 2002. Four complementary strategies have contributed to this progress in identifying disease genes: positional cloning, analysis of candidate genes, the use of animal models and, most recently, deep sequencing methodologies.

Positional Gene Identification

The discovery of large numbers of highly polymorphic DNA markers that can be readily analysed by many techniques, including the recent employment of high density

TABLE 4.1 Single-gene disorders and other syndromes involving central nervous system (CNS) malformations						
Disease or locus name	CNS malformations involved	Gene	Function of gene product	Chromosome location	OMIM number*	Mouse model or homologue
Acrocallosal syndrome	Macrocephaly, agenesis of corpus callosum; overlaps with Greig's cephalopolysyndactyly	GLI3[306]	Transcription factor in Sonic hedgehog pathway	7p13	200990	Gli3 mutation disrupts dorsoventral patterning in telencephalon[1013]
Aicardi syndrome	Agenesis of corpus callosum, cerebral heterotopias	ND	ND	Xp22	304050	ND
Aicardi–Goutieres syndrome 1	Cerebral atrophy, leukoencephalopathy, intracranial calcifications, CSF lymphohocytosis	TREX1	Three prime repair exonuclease 1	3p21.3–p21.2		
Aicardi–Goutieres syndrome 2	Cerebral atrophy, leukoencephalopathy, intracranial calcifications, CSF lymphohocytosis	RNASEH2B	Heterotrimeric type II ribonuclease H enzyme	13q14.1	610181	
Aicardi–Goutieres syndrome 3	Cerebral atrophy, leukoencephalopathy, intracranial calcifications, CSF lymphohocytosis	RNASEH2C	Heterotrimeric type II ribonuclease H enzyme	11q13.2	610329	
Aicardi–Goutieres syndrome 4	Cerebral atrophy, leukoencephalopathy, intracranial calcifications, CSF lymphohocytosis	RNASEH2A	Heterotrimeric type II ribonuclease H enzyme	19p13.13	610333	
Aicardi–Goutieres syndrome 5	Cerebral atrophy, leukoencephalopathy, intracranial calcifications, CSF lymphohocytosis	SAMHD1	Role in innate immune response	20q11.2	612952	
Angelman syndrome	Microcephaly and cerebellar atrophy	UBE3A[558,675]	Human papilloma virus E6-associated protein: a ubiquitin protein ligase, expressed in brain only after maternal transmission	15q11	105830	Maternal inheritance of targeted UBE3a mutation causes Angelman-like phenotype[523]
Aniridia	Severe cerebral neuronal migration disorders in homozygotes	PAX6[429]	Paired-box-containing transcription factor	11p13	106210	Mutation of Pax6 in Sey mutant causes microphthalmia[464]

Continued

TABLE 4.1 Single-gene disorders and other syndromes involving central nervous system (CNS) malformations (*Continued*)

Disease or locus name	CNS malformations involved	Gene	Function of gene product	Chromosome location	OMIM number*	Mouse model or homologue
Anophthalmia or microphthalmia	Small or absent eyes	SOX2[319]	SRY-box-containing transcription factor	3q27	206900	Targeted mutation of *Sox2* is early embryonic lethal, with normal hetero-zygotes[33]
Anophthalmia or microphthalmia	Small or absent eyes with defects of optic nerve, optic chiasm and hippocampal anomalies	OTX2[839]	Homeobox-containing transcription factor	14q21–22	600037	Targeted mutation of *Otx2* causes agnathia and holoprosencephaly in heterozygotes[460]
Anophthalmia or microphthalmia	Small or absent eyes	CHX10[808]	Transcription factor	14q32	142993	Spontane-ous mutant of *Chx10* has microphthalmia[133]
Apert syndrome	Cerebellar anomalies, agenesis of corpus cal-losum, limbic defects	FGFR2[1076]	Fibroblast growth factor receptor 2: cell surface receptor for fibroblast growth factors	10q25–26	101200	Gain of func-tion mutation in *Fgfr2-IIIc* models aspects of Apert syndrome[418]
Arthrogryposis	Congenital contrac-tures syndrome, distal, type 1 and 2b	TPM2	Tropomysin 2 (beta)	9p13.2–p13.1		
Arthrogryposis, lethal	Congenital contrac-tures syndrome, distal, type 1 (LCCS1) with anterior horn deficits	GLE1	Require for RNA export from nucleus	9q34	611890	
Arthrogryposis, lethal	Congenital contrac-tures syndrome, type 2 (LCCS2)	ERBB3	EGFR family of rece-potr tyrosine kinases, B3	12q13	607598	
Arthrogryposis, lethal	Congenital contrac-tures syndrome, type 3 (LCCS3) with anterior horn deficits	PI51C	Type I phosphatidylino-sitol 4-phosphate 5-kinase	19p13.3	611369	
Cerebral cavern-ous malformation (CCM1)	Cerebral cavernous angioma	KRIT1[1051]	Krev interaction trapped-1: involved in integrin-mediated angiogenesis	7q11.2–q21	116860	ND
Cerebral cavern-ous malformation (CCM2)	Cerebral cavernous angioma	MGC4607[623]	Malcavernin: involved in integrin-mediated angiogenesis	7p13	603284	ND
Cerebral cavern-ous malformation (CCM3)	Cerebral cavernous angioma	PDCD10[76]	Cell-death-associated protein	3q26.1	603285	ND
Cerebrooculofacio-skeletal syndrome, type 1	Microcephaly and corpus callosum atrophy; also Cock-anye syndrome	ERCC6	DNA-binding protein that is important in transcription-coupled excision repair	10q11	214150; 133540	
Cerebrooculofacio-skeletal syndrome, type 2	Microcephaly and corpus callosum atrophy	ERCC2	DNA-binding protein that is important in transcription-coupled excision repair	19q13.2–13.3	610756	
Cerebrooculofacio-skeletal syndrome, type 3	Microcephaly and corpus callosum atrophy; also Cockanye-like syndrome	ERCC5	DNA-binding protein that is important in transcription-coupled excision repair	13q33		

TABLE 4.1 Single-gene disorders and other syndromes involving central nervous system (CNS) malformations (*Continued*)

Disease or locus name	CNS malformations involved	Gene	Function of gene product	Chromosome location	OMIM number*	Mouse model or homologue
Cerebrooculofacio-skeletal syndrome, type 4	Microcephaly and corpus callosum atrophy	ERCC1	DNA-binding protein that is important in transcription-coupled excision repair	19q13.2–13.3	610758	
Cockayne syndrome, type A	Microcephaly	ERCC8	DNA-binding protein that is important in transcription-coupled excision repair	5q12	216400	
Congenital central hypoventilation syndrome (CCHS)	Brain stem autonomic reflex disorder	PHOX2B[23]	Homeobox-containing transcription factor	4p12	209880	Targeted mutation of *Phox2b* gene exhibits noradrenergic defects in hindbrain[798]
Dandy–Walker syndrome	Hypoplasia and upward rotation of cerebellar vermis with dilatation of fourth ventricle	ZIC1 and ZIC4 contiguous gene deletion[404]	Zinc finger transcription factors	3q24	220200	Compound heterozygosity for targeted mutations of Zic1 and Zic4 causes Dandy–Walker-like appearance[404]
Feingold Syndrome, type 1	Microcephaly, duodenal atresia, mental retardation and limb anomalies	MYNC	MYC family basic helix-loop-helix domain transcription factor	2p24.3	164280	
Holoprosencephaly (HPE1)	Alobar holoprosencephaly	ND	ND	21q22.3	236100	ND
Holoprosencephaly (HPE2)	Alobar or semi-lobar holoprosencephaly	SIX3[1064]	Homologue of sine oculis gene of *Drosophila*: homeobox-containing transcription factor	2p21	157170	Targeted mutation of Six3 gene has truncation of forebrain[579]
Holoprosencephaly (HPE3)	Holoprosencephaly	SHH (Sonic hedgehog)[875]	Secreted signalling molecule; neural inducer	7q36	142945	Targeted mutation of *Shh* gene has holoprosencephaly in addition to many other defects[151]
Holoprosencephaly (HPE4)	Holoprosencephaly	TGIF[405]	Homeodomain protein functioning as repressor of TGF-β	18p11.3	142946	Targeted mutation of *Tgif* gene produces no visible phenotype[524]
Holoprosencephaly (HPE5; 13q32 deletion syndrome)	Holoprosencephaly, exencephaly	ZIC2[124]	Transcription factor encoded by homologue of odd paired gene of *Drosophila*	13q32	609637	Targeted mutation of *Zic2* gene has holoprosencehaly[734]
Holoprosencephaly (HPE6)	Holoprosencephaly	ND	ND	2q37.1	605934	ND
Holoprosencephaly (HPE7)	Holoprosencephaly	PTCH1[703]	Patched: membrane receptor for Sonic hedgehog protein	9q22.3	601309	Targeted mutation of *Ptch1* causes medulloblastoma in heterozygotes and neural tube defects in homozygotes[390]
Holoprosencephaly (HPE8)	Holoprosencephaly	ND	ND	14q13	609408	ND

Continued

TABLE 4.1 Single-gene disorders and other syndromes involving central nervous system (CNS) malformations (*Continued*)

Disease or locus name	CNS malformations involved	Gene	Function of gene product	Chromosome location	OMIM number*	Mouse model or homologue
Holoprosencephaly	Holoprosencephaly, dysplastic forebrain, midline anomalies	*TDGF1*[248]	TDGF/Cripto, an EGF-CFC protein enhancing activity of nodal signalling	3p23–p21	187395	Targeted mutation of *Tdgf/Cripto* causes early embryonic lethality[275]
Holoprosencephaly with pituitary anomalies	Holoprosencephaly, pituitary anomalies	*GLI2*[876]	Transcription factor in Sonic hedgehog pathway	2q14	165230	Targeted mutation of *Gli2* produces pituitary and forebrain defects in conjunction with loss of *Gli1* function[793]
Horizontal gaze palsy with progressive scoliosis (HGPPS)	Pontine and medulla structural anomalies, uncrossed corticospinal tracts	*ROBO3*[508]	Transmembrane receptor mediating axon guidance; homologue of roundabout gene of *Drosophila*	11q23–q25	607313	Targeted mutation of *Rig1/Robo3* gene causes failure of precerebellar axons to cross midline[649]
Hydrancephaly, Fowler type	Proliferative vasculopathy with hydrancephaly	*FLVCR2*	Transmembrane protein/calcium transporter	14q24.3	225790	
Hydrolethalus syndrome	Hydrocephalus and loss of dorsal midline structures of the CNS	*HYLS1*	Poorly characterized cytoplasmic protein	11q24.2	610693	
Joubert syndrome 1	Hypoplasia of cerebellar vermis, 'molar tooth sign', mental retardation	*INPP5E*	Inositol polyphosphate-5-phosphatase, 72 kDa	9q34.3	213300	ND
Joubert syndrome 2	Hypoplasia of cerebellar vermis, with renal and ocular anomalies (also Meckel syndrome type 2)	*TMEM216*	Transmembrane domain-containing protein	11p12–q13.3	608091	ND
Joubert syndrome 3	Hypoplasia of cerebellar vermis	*AHI1*[323]	Abelson helper integration site 1, transcription factor	6p23.3	608629	ND
Joubert syndrome 4	Hypoplasia of cerebellar vermis, with renal anomalies	*NPHP1*[792]	Nephrocystin: cytoskeleton-related protein localizing to primary cilia	2q13	609583	ND
Joubert syndrome 5	Hypoplasia of cerebellar vermis with retinal and renal anomalies, also Meckel syndrome type 4, Leber congenital amaurosis, type 10, Senior–Loken syndrome type 6, and Bardet–Biedl syndrome type 14	*CEP290*	Centrosomal protein	12q21.3	610188	
Joubert syndrome 6	Hypoplasia of cerebellar vermis, with renal and ocular anomalies (also Meckel syndrome type 3 and COACH syndrome)	*TMEM67*	Transmembrane domain-containing protein localized to the primary cilium	8q21.11		

TABLE 4.1 Single-gene disorders and other syndromes involving central nervous system (CNS) malformations (*Continued*)

Disease or locus name	CNS malformations involved	Gene	Function of gene product	Chromosome location	OMIM number*	Mouse model or homologue
Joubert syndrome 7	Hypoplasia of cerebellar vermis, with renal and ocular anomalies (also Meckel syndrome type 5 and COACH syndrome)	*RPGRIP1L*	Localized on basal-body of primary cilia and centrosomes	16q12.2	611560; 611561; 216360	
Joubert syndrome 8	Hypoplasia of cerebellar vermis, with renal and ocular anomalies	*ARL13B*	ADP-ribosylation factor	3q11.1	612291	
Joubert syndrome 9	Hypoplasia of cerebellar vermis, with renal and ocular anomalies (also Meckel syndrome type 6 and COACH syndrome)	*CC2D2A*	Coiled-coil and C2 domain-containing protein 2A, critical for cilia formation	4p15.32	612285	
Joubert syndrome 10	Hypoplasia of cerebellar vermis, with renal and ocular anomalies	*OFD1*	Ciliary protein with unknown function	Xp22.3–22.2	300804	
Kallmann syndrome 1	Agenesis of olfactory lobes, absence of GnRH-secreting neurons	*KAL1*[249,962]	Anosmin-1: secreted protein interacting with heparan sulphate proteoglycans and fibroblast growth factor signalling	Xp22.3	308700	ND
Kallmann syndrome 2	Kallmann syndrome with cleft lip or palate	*FGFR1* (loss of function)[280]	Cell-surface receptor for fibroblast growth factors	8p11.1–p11.2	147950	Targeted mutation of *Fgfr1* gene is early embryonic lethal[261,1109]
Kallmann syndrome 3	Agenesis of olfactory lobes, absence of GnRH-secreting neurons	*PROKR2*	G protein-coupled receptor for prokineticins	20p13	244200	
Kallmann syndrome 4	Agenesis of olfactory lobes, absence of GnRH-secreting neurons	*PROK2*	Expressed in the suprachiasmatic nucleus and may function in circadian clock	3p13	610628	
Kallmann syndrome 5	Agenesis of olfactory lobes, absence of GnRH-secreting neurons; also CHARGE syndrome and hypogonadotropic hypogonadism	*CHD7*	Chromodomain helicase DNA binding protein 7	8q12.1	612370; 214800; 146110	
Kallmann syndrome 6	Agenesis of olfactory lobes, absence of GnRH-secreting neurons	*FGF8* (androgen-induced)	Member fibroblast growth factor family	10q24	612702	
Lissencephaly (type I): autosomal recessive (Norman–Roberts type)	Lissencephaly with low sloping forehead and prominent nasal bridge	*RELN*[482]	Reelin: extracellular matrix protein produced by Cajal-Retzius cells required for neuronal migration	7q22	257320	Mutation of *Rln* gene in *reeler* mutant mouse causes cerebellar and cerebral cortical lamination anomalies[231]

Continued

TABLE 4.1 Single-gene disorders and other syndromes involving central nervous system (CNS) malformations (*Continued*)

Disease or locus name	CNS malformations involved	Gene	Function of gene product	Chromosome location	OMIM number*	Mouse model or homologue
Lissencephaly (type I): Miller–Dieker syndrome	Lissencephaly, cerebral heterotopias, facial dysmorphism	*LIS1* and *14-3-3ε;* (contiguous gene deletion)[1003]	LIS1: Non-catalytic subunit of brain platelet-activating factor acetyl hydrolase (PAFAH)	17p13.3	247200	Targeted loss of function alleles of *Pafah1b1* gene[464] and *14-3-3ε*[1003] causes neuronal migration disorders, with increasing severity in double heterozygote14-3-3ε: phosphoserine and phosphothreonine binding protein
Lissencephaly (type I): isolated lissencephaly sequence (ILS)	Lissencephaly	*LIS1* deletion alone	LIS1: as above	17p13.3	607432	Targeted loss of function alleles of *Pafah1b1* gene causes neuronal migration disorders[464]
Lissencephaly (type I): X-linked isolated	Lissencephaly with agenesis of corpus callosum in males; subcortical band heterotopia in females	*DCX*[270,379]	Doublecortin: microtubule-associated protein that interacts with non-receptor tyrosine kinases, including Abl	Xq22.3–q23	300067	Targeted mutation of doublecortin in mice disturbs hippocampal lamination;[196] suppression of doublecortin expression by RNAi inhibits neuronal migration in rat neocortex[35]
Lissencephaly (type I): X-linked (XLAG)	Lissencephaly with ambiguous genitalia	*ARX*[559]	Aristaless-related homeodomain transcription factor	Xp22.13	300215	Targeted mutation of *Arx* causes small brain, with disturbance of neuronal migration[559]
Lissencephaly (type II): Fukuyama congenital muscular dystrophy	Cobblestone lissencephaly, polymicrogyria	*FCMD*[563]	Fukutin: gene interrupted by retrotransposon insertion. A secreted protein, which may function as a glycosyl transferase in the Golgi	9q31	253800	Targeted mutation of *FCMD* gene causes muscular dystrophy and cortical dysplasia[997]
Lissencephaly (type II): muscle-eye-brain disease, type A, 5; type B, 5; type C, 5	Cobblestone lissencephaly, congenital myopia, glaucoma, retinal hypoplasia, mental retardation, hydrocephalus	*FKRP*	Protein targeted to the medial Golgi apparatus and necessary for posttranslational modification of dystroglycan	19q13.3	613153; 606612; 607155	
Lissencephaly (type II):Walker–Warburg syndrome	Agyria, cobblestone lissencephaly, cerebellar dysplasia and vermal agenesis, hydrocephaly, occipital encephalocele	*POMT1*[69]	O-mannosyl transferase 1: first enzyme in synthetic pathway of O-mannosyl glycans	9q31–q33	236670	Large(myd) mutant and targeted mutation of α-*dystroglycan* gene provide models of Walker-Warburg syndrome[481,717]
Lissencephaly (type II):Walker–Warburg syndrome	As for *POMT1*-related disease	*POMT2*[1043]	O-mannosyl transferase 1: may function with *POMT1* enzyme	14q24.3	607439	As for *POMT1*

TABLE 4.1 Single-gene disorders and other syndromes involving central nervous system (CNS) malformations (*Continued*)

Disease or locus name	CNS malformations involved	Gene	Function of gene product	Chromosome location	OMIM number*	Mouse model or homologue
Lissencephaly (type II): muscle-eye-brain disease	Cobblestone lissencephaly, congenital myopia, glaucoma, retinal hypoplasia, mental retardation, hydrocephalus	POMGnT1[1116]	O-mannose β-1,2-N-acetyl glucosaminyl transferase: second enzyme in synthetic pathway of O-mannosyl glycans	1p34–p33	253280	Targeted mutation of POMGnT1 gene causes phenotype resembling muscle-eye-brain disease[626]
Lissencephaly (type II): muscle-eye-brain disease	Cobblestone lissencephaly, congenital myopia, glaucoma, retinal hypoplasia, mental retardation, hydrocephalus	LARGE				
Lissencephaly (type III)	Agyria or pachygyria or laminar heterotopia, severe mental retardation, motor delay, variable presence of seizures, and abnormalities of corpus callosum, hippocampus, cerebellar vermis and brain stem	TUBA1A	Tubulin, alpha 1a	12q12–q14	611603	
Meckel–Gruber syndrome (Meckel syndrome type 1)	Occipital encephalocele, microcephaly, cerebral and cerebellar hypoplasia, Dandy–Walker syndrome	MKS1[575]	MKS1 protein implicated in ciliary function	17q21–q24	249000	ND
Meckel–Gruber syndrome (Meckel syndrome type 1)	As for MKS1-related disease	MKS3[955]	Meckelin: novel transmembrane protein of unknown function	17q21–q24	249000	Mutation of MKS3 orthologue in rat wpk mutant produces phenotype with some features of Meckel–Gruber syndrome[955]
Meckel syndrome type 2	Occipital meningoencephalocele	ND	ND	11q13	603194	ND
Megalencephalic leukoencephalopathy with subcortical cysts 1	Swollen white matter with subcortical cysts	MLC1	Unknown	22q13.33	604004	
Mental retardation with agensis of the corpus callosum (MRXS28)	X-linked cognitive disability with agenesis of the corpus callosum	IGBP1	B-cell signal transduction molecule	Xq13.1	300472	
Mental retardation with stereotypic movements, epilepsy and cerebral malformations		MEF2C	Myocyte enhancer factor 2C	5q14	613443	
Megalencephaly-capillary malformation syndrome (MCAP)	Primary megalencephaly with prenatal overgrowth and cutaneous vascular malformations	PIK3CA	PI3K/AKT signaling pathway	3q26.32	602501	
Megalencephaly-polymicrogyria-polydactyly-hydrocephalus (MPPH)	Complex and variable CNS and extra-CNS malformation syndrome	AKT3, PIK3CA and PIK3R2	PI3K/AKT signaling pathway	1q43–q44; 3q26.32; 19p13.11	611223; 171834; 603157	

Continued

TABLE 4.1 Single-gene disorders and other syndromes involving central nervous system (CNS) malformations (*Continued*)

Disease or locus name	CNS malformations involved	Gene	Function of gene product	Chromosome location	OMIM number*	Mouse model or homologue
Microcephalic osteodysplastic primordial dwarfism, type II	Severe microcephaly	PCNT	Centrosomal protein that binds calmodulin	21q22.3	210720	
Microcephaly	With cortical malformation including microgyria and hypoplasia of the corpus callosum with mental retardation	WDR62	WD repeat protein 62	19q13.12	600176	
Microcephaly (MCPH1)	Autosomal recessive primary microcephaly, mental retardation	Microcephalin[504]	BRCA-domain-containing protein involved in cell cycle control and DNA repair	8p2–pter	251200	ND
Microcephaly (MCPH2)	Primary microcephaly	ND	ND	19q13.1–13.2	604317	ND
Microcephaly (MCPH3)	Primary microcephaly	CDK5RAP2[98]	Centrosomal protein involved in mitosis	9q33.3	604804	ND
Microcephaly (MCPH4, MCPH9)	Primary microcephaly, also Seckel syndrome, type 5	CEP152	Involved in centrosomal function	15q11	604321; 614852; 613823	ND
Microcephaly (MCPH5)	Primary microcephaly	ASPM[97]	Microtubule binding protein involved in mitotic spindle	1q31	608716	ND
Microcephaly (MCPH6)	Primary microcephaly	CENPJ[98]	Centrosomal protein involved in mitosis	13q12.2	608393	ND
MICPCH	Mental retardation, microcephaly with pontine and cerebellar hypoplasia	CASK	Calcium/calmodulin-dependent serine protein kinase; a MAGUK (membrane-associated guanylate kinase) protein family member	Xp11.4	300749	Perinatal lethal with inhibitory neuron synaptic dysfunction
Neu–Laxova syndrome	Microcephaly, lissencephaly, agenesis of corpus callosum, atrophy of cerebrum, cerebellum and pons, absence of olfactory bulbs	ND	ND	ND	256520	ND
	Microcephaly, secondary (MCPHSBA)	Progressive microcephaly with seizures and brain atrophy	MED17	Mediator of RNA polymerase II	613668	
Pallister–Hall syndrome	Hypothalamic hamartoblastoma	GLI3 (frameshift mutations)	Transcription factor in Sonic hedgehog signalling pathway	7p13	146510	Dorsoventral patterning disrupted in *extra toes* (*Xt*) mouse, which has a *Gli3* mutation[1013]
Pelizaeus–Merzbacher syndrome	Dysmyelination of corticospinal tracts	PLP[846]	Proteolipid protein, a myelination component	Xq21.3–22	312080	*jp (jimpy) mutant* mouse exhibits phenotypic similarities to Pelizaeus-Merzbacher syndrome[745]

TABLE 4.1 Single-gene disorders and other syndromes involving central nervous system (CNS) malformations (*Continued*)

Disease or locus name	CNS malformations involved	Gene	Function of gene product	Chromosome location	OMIM number*	Mouse model or homologue
Periventricular heterotopia, autosomal recessive	Periventricular nodular heterotopia with microcephaly	ARFGEF2[933]	Brefeldin A-inhibited GEF2 protein (BIG2), involved in vesicular trafficking from Golgi	20q13.13	608097	ND
Periventricular heterotopia, X-linked	Periventricular nodular heterotopia with epilepsy	FLNA[345]	Filamin-1: actin-binding protein associated with cytoskeleton	Xq28	300049	ND
Pettigrew syndrome	Dandy–Walker malformation, basal ganglia anomalies	ND	ND	Xq25–27	304340	ND
Polymicrogyria	Seizures with optic nerve hypoplasia	TUBA8	Tubulin, alpha 8	22q11	613180	
Polymicrogyria, asymmetric	Seizures	TUBB2A	Tubulin, beta 2A class lia	6p25.2	610031	
Polymicrogyria, asymmetric	Seizures	TUBB2B	Tubulin, beta 2B class lia	6p25.2	610031	
Polymicrogyria, autosomal recessive	Bilateral parieto-frontal polymicrogyria	GPR56[817]	G-protein-coupled receptor	16q13	606854	ND
Polymicrogyria, bilateral occipital	Bilateral occipital polymicrogyria	ND	ND	6q16.3–q22.1	612691	ND
Pontocerebellar hypoplasia, type 1	Small cerebellum and brain stem, central and peripheral motor dysfunction from birth, gliosis and anterior horn cell degeneration resembling infantile spinal muscular atrophy	VRK1	Vaccinia-related serine/threonine kinase 1	14q32	607596	
Pontocerebellar hypoplasia, types 2A and 4	Small cerebellum and brain stem, progressive microcephaly from birth combined with extrapyramidal dyskinesia and chorea, epilepsy, and normal spinal cord findings	TSEN54	tRNA splicing endonuclease 54	17q25.1	277470; 225753	
Pontocerebellar hypoplasia, type 2B	Small cerebellum and brain stem, progressive microcephaly from birth combined with extrapyramidal dyskinesia and chorea, epilepsy, and normal spinal cord findings	TSEN2	tRNA splicing endonuclease 2	3p25.1	612389	
Pontocerebellar hypoplasia, type 2C	Small cerebellum and brain stem, progressive microcephaly from birth combined with extrapyramidal dyskinesia and chorea, epilepsy, and normal spinal cord findings	TSEN34	tRNA splicing endonuclease 34	19q13.4	612390	
Porencephaly, type 1		COL4A1	Collagen, type IV, alpha 1	13q34	175780	

Continued

TABLE 4.1 Single-gene disorders and other syndromes involving central nervous system (CNS) malformations (*Continued*)

Disease or locus name	CNS malformations involved	Gene	Function of gene product	Chromosome location	OMIM number*	Mouse model or homologue
Prader–Willi syndrome	Microcephaly	Contiguous gene deletion syndrome of *SNRPN* and *necdin*, when paternally transmitted[378]	SNRPN: small nuclear ribonucleo-protein polypeptide, with role in pre-mRNA splicing; *necdin:* interacts with cell-cycle proteins	15q11–q13	176270	Targeted mutation of *necdin* gene produces hypothalamic and behavioural effects similar to Prader-Willi syndrome[730]
Sacral agenesis (Currarino triad)	Meningocele	*HLXB9*[885]	Homeobox-containing transcription factor, coding for HB9 protein	7q36	176450	Targeted mutation of *Hlxb9* gene does not cause sacral defects in mice[443,615]
Seckel syndrome, type 1	Microcephaly	*ATR*	Serine-threonine protein kinase activated by DNA damage	3q23	601215	
Septo-optic dysplasia	Hypoplastic optic nerves, absent septum pellucidum, hypoplasia of hypothalamic nuclei, agenesis of corpus callosum	*HESX1*[230]	Paired-like homeodomain protein expressed in embryonic forebrain	3p21.2–p21.1	182230	Targeted mutation of *Hesx1* gene produces phenotype closely similar to septo-optic dysplasia[234]
Septo-optic dysplasia	Hypoplasia of infundibulum with hypopituitarism	*SOX3*[1097]	SRY-box-containing transcription factor	Xq27.1	313430	Targeted mutation of *Sox3* gene produces defects of hypothalamus-pituitary axis[866]
Smith–Lemli–Opitz syndrome	Microcephaly, holoprosencephaly	*DHCR7* (sterol delta-7-reductase)[1067]	Enzyme that converts 7-dihydro-cholesterol to cholesterol	11q12–q13	270400	Targeted mutation of *Dhcr7* gene produces phenotype with many similarities to human syndrome[1068]
Thanatophoric dysplasia, types I and II	Temporal lobe dysplasia, polymicrogyria	*FGFR3* (fibroblast growth factor receptor 3)[1023]	Cell-surface receptor for fibroblast growth factors	4p16.3	187600	Targeted mutations of *Fgfr3* gene produce phenotypes resembling human syndrome[503]
Tuberous sclerosis (TSC1)	Hamartomas of CNS and other tissues	*TSC1*[1044]	Hamartin, protein that interacts physically with tuberin	9q34	605284	Targeting of *Tsc1* gene produces renal and extra-renal tumour development in heterozygotes[787]
Tuberous sclerosis (TSC2)	Hamartomas of CNS and other tissues	*TSC2*[315]	Tuberin, GTPase activating protein for RAP1	16p13.3	191092	Targeting of *Tsc2* gene produces renal and extra-renal tumour development in heterozygotes[787]
Waardenburg syndrome type I	Neural crest and inner ear disorders, with myelomeningocele in homozygous individuals	*PAX3*[1003]	Paired-box-containing transcription factor	2q35	193500	*Sp* (*splotch*) mutants encoding disruptions of *Pax3* gene cause a phenotype resembling human syndrome[310]

TABLE 4.1 Single-gene disorders and other syndromes involving central nervous system (CNS) malformations (*Continued*)

Disease or locus name	CNS malformations involved	Gene	Function of gene product	Chromosome location	OMIM number*	Mouse model or homologue
X-linked congenital cerebellar hypoplasia (MRX60)	Cerebellar hypoplasia, mental retardation	*OPHN1*[78,815]	Oligophrenin-1	Xq12	300486	ND
X-linked hydrocephalus	Stenosis of aqueduct of Sylvius, agenesis of corpus callosum and septum pellucidum, fusion of thalami, hypoplasia of corticospinal tracts (CRASH syndrome)	*L1-CAM*[884]	Cell adhesion molecule, expressed especially on migrating neurons	Xq28	308840	Targeted mutation of *L1* gene causes CNS malformations resembling human syndrome[223]
X-linked mental retardation (MRXSO)	Mental retardation associated with cerebellar hypoplasia and facial dysmorphism	*OPHN1*	Rho-GTPase-activating protein that promotes GTP hydrolysis of Rho subfamily members	Xq12	300486	
Zellweger syndrome	Microgyria, polymicrogyria	*PEX* gene family members[1065]	Peroxin proteins participating in peroxisome function	Various loci	214100	Targeted mutation of *Pex2* and *Pex5* genes provide mouse models[34,322]

*Catalogue number in Online Mendelian Inheritance in Man (www.ncbi.nlm.nih.gov/entrez/query.fcgi?db=OMIM), from which further details on the data in this table can be found.

EGF-CFC, epidermal growth factor-Cripto/FRL-1/Cryptic; GnRH, gonadotrophin-releasing hormone; ND, not determined/available; RNAi, RNA interference; TGF-β, transforming growth factor β.

microarray technology, has made it possible to construct a detailed physical map of the human genome.[1073] The 2001 publication of the nearly complete six billion nucleotides of the human genome has permitted the accelerated rate of discovery, including the use of deep sequencing methology, also known as next generation sequencing.[582,1048] This enormous technological achievement has enabled the identification of the majority of transcribed genes in the genome, making disease gene identification a much more rapid task than was possible previously (Fig. 4.1). Despite this achievement, however, the sequencing of the human and mouse genomes stops short of defining the functions of the genes.[113,475]

Candidate Genes

The catalogue of mapped and cloned genes is growing rapidly, and it is now usually possible to draw up a list of candidate genes for a given genetic disease, based on its map position together with the putative function of the disease gene. The candidacy of genes can then be assessed by genetic mapping (finding genetic recombinations between the gene and the disease locus usually rules out its candidacy) or by searching for mutations in the coding region of the gene in patients with the disease. A lack of coding region mutations does not disprove candidacy, because regulatory mutations may be located at a considerable distance from the coding region. A full evaluation of a candidate gene may also require, therefore, evaluation of gene expression at the level of messenger RNA (mRNA), by northern blotting, reverse-transcription polymerase chain reaction (PCR) and *in situ* hybridization. Such candidate gene approachs were used to identify *fibroblast growth factor*

receptor 2 as the gene mutated in Apert syndrome[1076] and to implicate the *Sonic hedgehog* gene in holoprosencephaly (HPE3).[68,875]

Animal Models

An alternative approach to the isolation of human disease genes is to progress from an animal model and clone the human gene by homology. This is especially useful when informative human families are not readily available. In the case of diseases controlled by multiple genes or poorly penetrant genes, the analysis of animal models may have special advantages.[189] A large number of homologous loci have been identified in humans and mice,[242,925] some of which cause diseases involving CNS malformations when they are mutated (Tables 4.1 and 4.2). The strategy for cloning mouse genes is essentially that outlined for human genes, but with the added advantage that the number of informative individuals that can be scored in the linkage analysis is almost unlimited, thereby permitting more rapid and precise genetic location of genes. An important further advantage of using animal models is the possibility of performing detailed descriptive and experimental analysis of the pathogenesis of the disease phenotype, particularly at prenatal stages when human embryonic and fetal material is not generally available for study (see later, Pathogenesis, p. 286). Animal models that have aided our understanding of the genetic basis of human CNS disease include the *jimpy* mouse, which led to the identification of the *Proteolipid Protein* (*PLP*) gene as mutated in Pelizaeus–Merzbacher disease,[745,1098] and the *splotch* mouse, which implicated the *Pax3* gene in Waardenburg syndrome types I and III.[39,1003]

TABLE 4.2 Mouse mutations producing central nervous system (CNS) malformations*

Malformation	Mutant genes	Gene knockouts
Anencephaly	Extra toes (*Xt*; mutation of *Gli3*), open brain (*opb*; mutation of *Rab23*), splotch (*Sp*; mutation of *Pax3*)	*Apolipoprotein B, twist, MARCKS, noggin* and many others
Arhinencephaly	Polydactyly Nagoya (*pdn*; mutation of S100β protein)	*NCAM*
Basal ganglia	ND	*Nkx2.1, Dlx1/2*
Cerebellum: agenesis, generalized defects	Lurcher (*Lc*; mutation of δ2 glutamate receptor), swaying (*sw*; mutation of *Wnt-1* gene), Snell dwarf (*Pit1dw*; mutation of *Pit1* gene) and others	*Wnt-1, En-1, ErbB3, BDNF, Gbx2, Zic1*
Cerebellum: vermis defects	ND	*L1*
Cerebellum: granule cell anomalies	Staggerer (*sg*; mutation in RORα gene), weaver (*wv*; mutation of *Girk2* gene)	*Math1, cyclin D, NeuroD*
Cerebellum: Purkinje cell anomalies	Purkinje cell degeneration (*pcd*; mutation of *Nna1* gene)	Ca^{2+}/*calmodulin kinase IV, engrailed-2*
Corpus callosum defects	ND	*Mrp, Emx-1, Nfia*
Cranial nerves, ganglia and hindbrain defects	Kreisler (*kr*; mutation in novel leucine zipper protein)	*Krox20, Hoxa1, Hoxb1*
Cranio-rachischisis	Loop-tail (*Lp*; mutation of *Vangl2*), circletail (*Crc*; mutation of *Scrb1*), crash (*Crsh*; mutation of *Celsr1*)	*dishevelled-1/2* double mutant, *Ptk7*
Encephalocele	Forebrain overgrowth (*Fog*; mutation of *Apaf1*)	ND
Generalized or multiple defects	ND	*Brain derived neurotrophic factor receptor (trkB); Retinoblastoma*
Hydrocephalus	Congenital hydrocephalus (*ch*; mutation of *Mf1*), hydrocephalus 3 (*hy3*; mutation of *Hydin*)	*Apolipoprotein B, Otx2, NCAM, E2F-5, L1, Nfia*
Megalencephaly	ND	*Kv1.1*[810]
Myelination defects	Jimpy (*jp*; mutation of *Plp* gene encoding myelin proteolipid protein), shiverer (*shi*; mutation of myelin basic protein gene), ducky (*du*; mutation of *Cacna2d2*) and others	*GFAP*
Myelomeningocele (open spina bifida)	Bent tail (*Bn*; mutation of *Zic3* gene), curly tail (*ct*; mutation of *Grhl3*), splotch (*Sp*; mutation of *Pax3* gene)	*Mrp, noggin, Zic2*
Neuronal migration disorders	Small eye (*sey*; mutation of *Pax6* gene); reeler (*rl*; mutation of *reelin* gene), scrambler and yotari (*dbl*; mutation of *disabled* gene), dreher (*dr*; mutation of *Lmx1a* gene)	*Pafah1b, cdk5, p35, Dlx1/2, MARCKS, Pex2, Presenilin-1, Emx2*

*Further details on the data in this table can be found at Mouse Genome Informatics (www.informatics.jax.org) and from the following review articles: Hatten *et al.*,[449] Copp and Harding[191] and Juriloff and Harris.[532] See also Table 4.1.

ND, not determined.

Inheritance of Mutant Genes

Although many genetic diseases conform to the rules of mendelian inheritance, several variations on the theme have been highlighted by recent studies. Certain syndromes appear to result from simultaneous loss of two or more genes that are located adjacent to each other in the genome. These contiguous gene deletions may represent a combination of the effects of loss of function of each gene individually or could result from a defect in an interaction between the gene products that is required for normal development (see later, Environmental Factors, p. 284). A dominant form of hydrocephalus has been suggested to form part of a contiguous gene deletion syndrome on chromosome 8q926,

whereas co-deletion of the *LIS1* and *14-3-3* genes on the short arm of chromosome 17 has been shown to cause Miller–Dieker lissencephaly.[1020]

A second principle governing the inheritance of some genetic diseases is the role of genomic imprinting. This refers to the observation that the parent of origin of a mutation can determine the nature of the disease phenotype observed. Imprinting involves the inactivation by chemical modification, most commonly through DNA methylation, of gene sequences as a result of passage through either the male or female germline.[849] Angelman syndrome results from loss of function of *UBE3A*, a gene on chromosome 15q11 that is active in the brain only after maternal transmission. The disease appears either when both copies

(1) Analyse family pedigrees for recombination events

(2) Map gene on chromosome

(3) Create genomic DNA 'contig' between flanking genetic markers

(4) Refine map using new markers derived from the DNA contig

(5) Screen libraries of cDNA fragments to isolate genes present in the mapped region

(6) Evaluate candidate genes by mutation screening, database sequence analysis, expression pattern, transgenic rescue, etc.

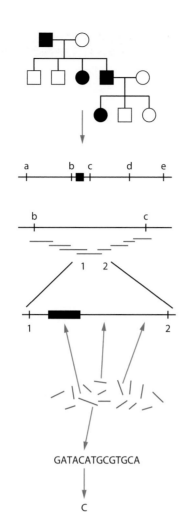

4.1 Diagram illustrating the principles of positional cloning to identify disease genes. (1) Linkage analysis is performed in one or more family pedigrees to identify recombinant genotypes that can be used to identify a chromosomal location for the disease gene. (2) Polymorphic DNA markers are positioned to identify DNA markers that map closely on either side of the gene. The aim is to define genetically a fragment of DNA that contains the gene of interest and that is sufficiently small to be able to search realistically for candidate coding sequences within it. In practice, this amounts to identifying flanking markers that are no more than 1–2 cM apart (1 cM is defined as 1 per cent genetic recombination, roughly amounting to one million nucleotide base pairs of human DNA). (3) Genomic DNA fragments are isolated from libraries in yeast artificial chromosomes (YACs) or bacterial artificial chromosomes (BACs), which span the region between the flanking markers. Such an overlapping set of YAC or BAC inserts is known as a genomic 'contig'. (4) The genetic map is refined by generating new DNA markers from the genomic fragments. (5) Nucleotide sequences are sought that encode proteins within the genomic contig. (6) Each identified sequence is evaluated as a candidate for the gene of interest. Features that characterize protein coding sequences include evolutionary conservation and association with DNA motifs called CpG islands.[203] Moreover, gene coding sequences can be isolated by selecting for DNA fragments containing splice donor/acceptor sites that are normally found in association with protein coding sequences, a technique called 'exon trapping'.[130] The most conclusive demonstration that a gene has been isolated comes from the finding of a pattern of tissue- and stage-specific expression appropriate for the disease phenotype, and the demonstration of mutations, predicted to cause a change in gene function, in patients with the disease. Rescue of a mutant mouse model, by transgenic insertion of the wild-type gene sequence, is a further demonstration that the gene has been cloned.

of *UBE3A* are inherited from the father (i.e. uniparental disomy) or when a deletion encompassing *UBE3A* is transmitted through the maternal line. Prader–Willi syndrome was originally linked to the same region of chromosome 15q11–q13 as Angelman syndrome, although the two conditions have quite different phenotypes. In Prader–Willi syndrome, a contiguous gene syndrome has been described, involving the *SNRPN* and *necdin* genes, which have proven to be active only after paternal transmission. Maternal uniparental disomy or paternal transmission of a 15q11 deletion leads to this disease.[754]

Loss and Gain of Function Mutations

It has been traditional to distinguish between dominant and recessive mutations. As individual genes are studied in greater detail, however, it is often found that recessive mutations have minor effects in the heterozygous state, whereas dominant genes may produce additional phenotypes when homozygous. A more useful distinction may be between mutations that act through loss or gain of function. Loss of function mutations abolish normal function of the gene product, as is seen in gene deletions and in cases where single base changes truncate, or otherwise inactivate, the gene product. If halving the gene dosage is

not limiting, then the mutation appears recessive, whereas if the gene product is needed in full dose, then the gene is said to exhibit haploinsufficiency and the mutation acts as a dominant, for instance as in Miller–Dieker lissencephaly. Gain of function mutations impart a novel function to the mutant gene product, as is seen in Apert syndrome, where mutations confer constitutive ('always on') function to fibroblast growth factor receptor 2 (FGFR-2). Dominant negative mutations involve alterations of the gene product, causing an inhibitory effect that overcomes function of the product of the normal allele, as is seen in the *jimpy* mouse model of Pelizaeus–Merzbacher disease, where a wild-type transgene is unable to overcome the mutant phenotype.[919] Yet another mutation type is represented by those mutations that confer a new function on the protein, permitting it to function in pathways not modulated by the normal gene product.

Chromosomal Disorders

In addition to the submicroscopic chromosomal alterations discussed earlier, several large-scale chromosomal anomalies cause CNS malformations as part of their phenotype (Table 4.3). In these cases, it is usually not possible to attribute specific features of the phenotype to particular

TABLE 4.3 Chromosomal disorders involving central nervous system (CNS) malformations

Chromosome	Excess	Deficiency
4	4p+ **Microcephaly**, agenesis of corpus callosum	4p− **Microcephaly**
5		5p− (Cri du chat syndrome) **Microcephaly**
8	8+ **Microcephaly**, agenesis of corpus callosum	
9	9+ **Microcephaly**, meningocele 9p+ **Macrocephaly**, hydrocephaly	Ms9p **Microcephaly**
13	13+ (Patau's syndrome) **Holoprosencephaly**, agenesis of corpus callosum, hydrocephaly, fusion of basal ganglia, cerebellar hypoplasia, myelomeningocele	13q− **Microcephaly, holoprosencephaly**
18	18+ (Edwards syndrome) **Microcephaly**, microgyria, cerebellar hypoplasia, agenesis of corpus callosum, arhinencephaly	18p− **Microcephaly**, holoprosencephaly, hydrocephaly, myelomeningocele 18q− **Microcephaly**
20	20p+ **Hydrocephaly**	
21	21+ (Down syndrome) **Microcephaly**	
x	XXXY, XXXXY, XXXX, XXXXX **Microcephaly**	XO (Turner syndrome) Defects of cerebellum, basal ganglia
Triploidy	**Holoprosencephaly, hydrocephaly, Arnold–Chiari** (often mosaic) **malformation, microcephaly,** myelomeningocele	

Malformations in **bold type** are present in the majority of cases. Malformations in normal type are present in a minority of cases. For further details, see Smith.[950]

genes because large parts of the genome are present in either increased or decreased copy number. However, progress is being made towards delineating the features of Down syndrome that result from trisomy of particular regions of chromosome 21, permitting a more concentrated search for candidate genes.[455] The development of techniques for constructing and maintaining artificial mammalian chromosomes[934,1119] has enabled a 'transchromosomic' mouse line to be prepared, carrying almost the entire human chromosome 21, as a model of Down syndrome.[770] In addition to these well defined chromosomal alternations, a number of structural changes including translocations, inversions, isocentric inversions, deletions, duplication, among others are increasingly being recognized as the basis for human developmental anomalies.

Environmental Factors

The action of exogenous influences in perturbing development *in utero*, leading to congenital malformations, has been studied for many decades under the broad umbrella of developmental toxicology and teratology. Research has been aimed at identifying teratogenic factors, evaluating their risk for human pregnancies and determining the mechanisms of teratogen action. In view of the public health implications, particularly in the aftermath of the thalidomide episode, the identification of teratogens has taken highest priority. This is perpetuated today by the increasing confinement of developmental toxicological research to the industrial pharmaceutical sector, where commercial pressures demand efforts aimed mainly at the production of safe new products rather than the mechanistic evaluation of teratogenic agents. Thus, our understanding of the mechanisms of action of teratogenic drugs and other environmental factors has not increased at the same rate as progress in understanding the genetic aetiology of malformations. Nevertheless, significant progress has been made in understanding the molecular basis of action of a small number of teratogenic agents (see later, Molecular Pathogenic Events, p. 286). Table 4.4 lists some of the teratogenic agents that are known, or suspected, to cause malformations of the CNS in humans and experimental animals.

Gene–Gene and Gene–Environment Interactions

The concentration of research studies on the identification of single genes, or the testing of single environmental factors in the aetiology of malformations, tends to obscure the important principle that CNS defects probably always result from complex interactions between factors. As an example, Figure 4.2 illustrates how such interactions determine the propensity of mouse embryos to develop neural tube defects (NTDs). Three types of interaction may be considered:

- *major genes*, which cause NTDs when homozygous, can interact with each other in doubly heterozygous individuals. These are called epistatic interactions (see later, Cascades of Developmental Events, p. 286). Interactions of this type between non-allelic mutations in humans could yield a relatively high frequency of malformations, even in the presence of low carrier gene frequencies;
- expression of major genes is modulated in incidence and severity by *modifier genes*, which, by themselves, do not produce NTDs. Studies of inbred mouse

TABLE 4.4 Teratogens known or suspected to produce central nervous system (CNS) malformations in humans*

Teratogenic agent	Malformations
Alcohol	Microcephaly, occasional meningomyelocele and hydrocephaly
Carbamazepine	Myelomeningocele
Cytomegalovirus	Hydrocephalus, microencephaly with cortical microgyria, occasional cerebellar microgyria
Diabetes mellitus	Neural tube defects: elevated (maternal) incidence
Herpes simplex	Microcephaly, hydranencephaly
Hyperthermia	Neuronal heterotopias, microcephaly, possible neural tube defects
Methyl mercury	Microcephaly
Phenylketonuria	Microcephaly (maternal)
Phenytoin	Microcephaly, holoprosencephaly
Retinoids	Hydrocephaly, microcephaly, neuronal migration defects, cerebellar agenesis/hypoplasia
Rubella	Microcephaly, occasional hydrocephalus and agenesis of corpus callosum
Toxoplasmosis	Necrotizing meningoencephalitis leading to hydrocephalus and calcification; polymicrogyria and hydranencephaly in some cases
Valproic acid	Myelomeningocele
Varicella	Necrotizing encephalitis with polymicrogyria
Warfarin	Microcephaly, hydrocephalus, Dandy–Walker cyst, agenesis of corpus callosum
X-irradiation	Microcephaly, pachygyria, cerebellar microgyria

*Further details on the data in this table are available.[115,534,580,581,949,1052]

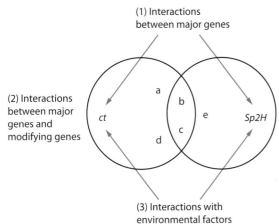

4.2 Aetiological interactions during the development of neural tube defects in the mouse. Three types of interaction have been demonstrated. (1) Interactions between major genes: each of the mutant genes *curly tail* (*ct*) and *splotch* (*Sp2H*) causes spina bifida when homozygous. Embryos that carry both mutations are affected more severely than embryos that are mutant at only one of the genetic loci,[310] demonstrating a summation of the effects of these non-allelic mutations. (2) Interactions between major and modifying genes: each of the major genes is expressed differently, with varying incidence and severity of defects, on different inbred genetic backgrounds. This demonstrates the existence of a varying complement of polymorphic modifying genes in the different inbred strains. Some of these modifying genes have been mapped.[739] (3) Interactions between major genes and environmental factors: exogenous agents such as folic acid and myo-inositol interact with the mutant genes to modify the incidence and severity of defects. For instance, myo-inositol reduces the frequency of spinal neural tube defects in *curly tail* mouse embryos,[393] whereas folic acid is effective in preventing neural tube defects in *splotch* mutant mice.[332]

strains with differing susceptibility to a genetically determined cranial NTD (anencephaly) identified a strain-to-strain variation in the precise rostrocaudal position at which the neural tube closes in the mouse brain (Figure 4.3). Susceptibility to anencephaly is determined by where along the body axis this closure event occurs. Strains that initiate closure at a caudal position within the midbrain are resistant to anencephaly, whereas those that begin brain closure rostrally within the forebrain are relatively susceptible.[337] The genes that control this variation in brain development in the mouse are predicted to have homologues that determine the risk of anencephaly in humans;

- as well as interactions between genes, *environmental factors* interact with both major and modifier genes, a reflection of the principle elaborated by Wilson that teratogenesis depends on both maternal and embryonic genotype.[1081] Gene–environment interactions can be either exacerbating or ameliorating with respect to the incidence and severity of the malformations induced. Examples of ameliorating effects include folic acid, which can prevent NTDs in humans and in the *splotch* mouse,[336,1062] and *myo*-inositol, which reduces the frequency of NTDs in the folate-resistant *curly tail* mouse.[400]

This interacting model of CNS malformation aetiology fits well with the observed multifactorial causation of conditions such as human NTDs.[145] Diseases that show a more closely mendelian transmission, or a more strict dependence on the action of a teratogen, may also be usefully

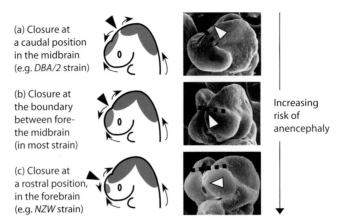

(a) Closure at a caudal position in the midbrain (e.g. *DBA/2* strain)

(b) Closure at the boundary between fore- the midbrain (in most strain)

(c) Closure at a rostral position, in the forebrain (e.g. *NZW* strain)

Increasing risk of anencephaly

4.3 Neural tube closure. Variations in the pattern of neural tube closure in the brain of different mouse strains, as shown diagrammatically (left) and in scanning electron micrographs (right). This variation determines susceptibility to anencephaly, as shown on the right-hand side. Arrowheads mark the site where brain closure begins (closure 2). In the diagrams, green and red shading indicate open regions of neural tube situated rostral and caudal, respectively, to the closure point. Arrows show the directions in which closure spreads throughout the brain. Dotted lines mark the boundary between the forebrain and midbrain. DBA/2 and NZW are inbred mouse strains that exhibit resistance and susceptibility, respectively, to anencephaly.

Reproduced with modification from Fleming and Copp.[333]

viewed within this framework, because interactions with subsidiary aetiological factors can account for variations in the penetrance and severity of the defects observed.

Pathogenesis

The completion of the Human Genome Project, with the decoding of the genetic material, heralds the start of the post-genomic era. Attention now turns from merely identifying genes to the more difficult task of determining their function. When applied to the study of CNS malformations, the challenge becomes one of understanding the molecular and cellular mechanisms that underlie CNS development, which, when disturbed by a genetic or an environmental perturbation, yield a malformation phenotype. Hence, the elucidation of pathogenic mechanisms becomes the focus of research. This is not solely to satisfy scientific curiosity. A thorough knowledge of the molecular pathogenesis of a disease process provides opportunities to develop interventional approaches to primary prevention, as exemplified by the use of folic acid to prevent NTDs.

Cascades of Developmental Events

The pathogenesis of a CNS malformation can be viewed as a cascade of events (Figure 4.4a) in which aetiological factors, either genetic or environmental, initiate a sequence of molecular, cellular and tissue alterations, culminating in the development of a pathological phenotype. The resulting pathogenic cascade is a variant of the normal developmental programme: a CNS component is generally formed, but with anomalous structure or function. The grouping of molecular events into developmental cascades has been placed on a firm foundation in recent years by the emergence of a new type of genetic analysis, called epistasis. In this experimental procedure, two distinct but related genetic defects are combined in the same individual to determine whether the genetic lesions summate (i.e. the resulting phenotype is more severe than with either lesion alone) or whether one lesion dominates (i.e. is epistatic) over the other. Summation is evidence that the two lesions operate in distinct pathways, whereas epistasis suggests a single pathway involving both genes. Originally developed in lower organisms such as *Drosophila* and the nematode *Caenorhabditis elegans*,[32] epistatic analysis is now being applied increasingly to mammalian development. For instance, a mutation in the gene *bcl-x* causes increased death of post-mitotic neurons by apoptosis in the mouse brain,[720] whereas inactivation of the *caspase-3* gene has the opposite effect, reducing neuronal death.[571] Despite the contrasting phenotypes of these knockout mice, the brain phenotype of mice lacking both *bcl-x* and *caspase-3* is similar to the *caspase-3* mutant alone, indicating an epistatic relationship between the genes. *Caspase-3* appears to act downstream of *bcl-x* in the pathway regulating neuronal cell death.[892]

Pathogenic cascades have so far been defined for very few CNS malformations. Usually, only fragmentary information is available on one or more aspects of the molecular and morphogenetic processes. For this reason, the following discussion considers pathogenic events at the molecular and cellular/tissue levels separately, citing instances in which particular malformations are known to involve a disturbance of a particular developmental process.

Molecular Pathogenic Events

Figure 4.4b is a schematic summary of the main molecular components of intercellular and intracellular signalling systems. Pathways of this basic type are being found to play a central role in a wide variety of normal and pathological cellular processes, including development of the nervous system.

4

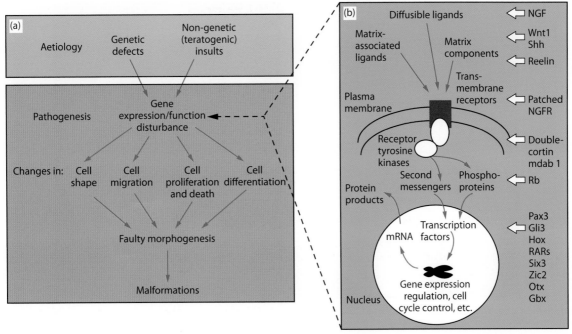

4.4 Cascades of pathogenic events leading to central nervous system (CNS) malformations. (a) Relationship between aetiology and pathogenesis, with a schematic representation of the molecular, cell/tissue and morphogenetic levels of organization. **(b)** Stylized representation of the main categories of molecular signalling event that mediate development of the nervous system. The sites of action of some genetic and environmental insults that cause CNS malformations, as discussed elsewhere in this chapter, are indicated on the right. Cells signal to each other *via* molecules that may be freely diffusible, associated with the extracellular matrix or part of the extracellular matrix. Receptors for these signalling ligands are connected, directly or indirectly, to intracellular signalling systems that transduce the extracellular signal and alter cellular metabolism and/or gene expression. Protein kinases are enzymes located on the cytoplasmic surface of the plasma membrane, which typically operate as dimers. They have been found to mediate a variety of extracellular signals *via* protein phosphorylation. A series of phosphorylation events ensues, yielding second messenger molecules (e.g. inositol phosphates) that have the ability to alter gene expression patterns within the cell. A number of genes that are regulated by activation of the signal transduction systems encode protein products, termed transcription factors, that serve to regulate the expression of other genes, so yielding cascades of gene regulation events.

Genetic Mechanisms

Once a novel gene has been identified as the likely cause of a CNS birth defect, the putative function of the protein product can often be predicted from its amino acid structure. For instance, transmembrane receptors have multiple hydrophobic amino acid stretches that span the lipid bilayer, whereas transcription factors have characteristic DNA binding motifs. This information enables a schematic depiction of the position in the cell signalling scheme occupied by the various genes that cause CNS malformations (Figure 4.4b, right-hand side).

Extracellular signalling molecules can be freely diffusible (e.g. nerve growth factor), diffusible but requiring association with the extracellular matrix for their function (e.g. FGF, Wnt-1, Sonic hedgehog) or part of the extracellular matrix (e.g. reelin). Mutations in the mouse *Wnt-1* gene lead to faulty specification of the developing brain, so that the posterior midbrain and anterior hindbrain fail to develop (Figure 4.5). *Sonic hedgehog* function is necessary for correct specification of the telencephalic vesicles, leading to holoprosencephaly when deficient,[151] whereas reelin, the protein product of the gene mutated in the *reeler* mouse, is required for the normal migration of neuroblasts during formation of the laminated structure of the cerebral cortex.[231,471]

Cell-surface receptors act as conduits, transmitting extracellular signals into the cell. Mutations in the *trkB* gene, which acts as a receptor for brain-derived neurotrophic factor (BDNF), produce abnormalities of cranial motor neurons, ganglia and cerebral cortex.[956] Fibroblast growth factor (FGF) receptors transduce signals from the large number of FGF ligands and are required for normal skull and brain development. FGFR-2 mutations have been identified in Apert syndrome whereas FGFR-3 mutations are found in thanatophoric dysplasia.[459a,1076]

Inside the cell, a large number of proteins participate in the transduction of extracellular signals. The *mdab1* gene encodes an adaptor protein that binds non-receptor tyrosine kinases such as Abl and Fyn. It appears to participate in the *reelin* pathway.[219]

Extracellular signals are generally passed to nuclear transcription factors that control the expression of other genes. A transition between extracellular signalling and modulation of gene expression are proteins like the *GLI* family of transcription factors. These molecules live primarily in the cytoplasm, but upon activation, in the case of GLIs through Sonic hedgehog (shh) signalling, they translocate to the nucleus where they activate or repress gene expression directly. Disturbance of *GLI3* function has been implicated in both Greig's cephalopolysyndactyly and

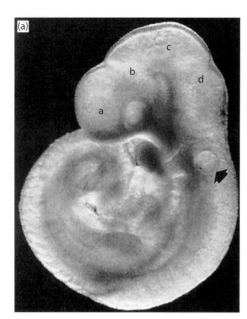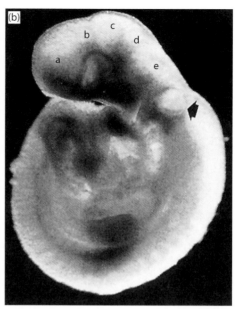

4.5 Hypoplasia of midbrain and rostral hindbrain. Midbrain and rostral hindbrain hypoplasia in the mouse resulting from a targeted mutation in the *Wnt-1* gene. Mouse embryos at 10.5 days of gestation (27-somite stage) from a mating between *Wnt-1* +/− heterozygotes. The regions of the developing brain are indicated: a, telencephalon; b, diencephalon; c, mesencephalon; d, metencephalon; e, myelencephalon. **(a)** *Wnt-1* +/− embryo with normal appearance; **(b)** *Wnt-1* −/− mutant embryo showing hypoplasia of midbrain and metencephalic region of the hindbrain. Magnification approximately ×20.

Reproduced by permission from McMahon and Bradley.[684] With permission from Elsevier.

Pallister–Hall syndrome.[550] Interestingly, another member of the GLI transcription factor family, GLI2, has been implicated in a different CNS malformation, specifically holoprosencephaly.[876] Pure transcriptions factors, proteins restricted to the nucleus, are also known to affect brain development. Heterozygotes for mutations in the transcription factor *PAX3* develop Waardenburg syndrome types I and III, in which disturbance of neural crest development is a prominent feature.[39,1003] In the mouse, loss of *Pax3* function yields the *splotch* phenotype, comprising anencephaly, myelomeningocele and neural crest defects.[708] Similarly, mutations in *ARX* result in a spectra of neurodevelopmental anomalies including lissencephaly.

A key intracellular event of importance in neuronal development is regulation of the cytoskeleton, the filamentous cytoplasmic system that underlies the ability of cells to adopt specific shapes and to move in a directed manner. Studies of the genes responsible for neuronal migration disorders, such as lissencephaly and heterotopias (Table 4.1), have focused attention on the molecular regulation of microtubule dynamics in migrating neuroblasts. For example, when the closely neighboring genes *LIS1* and *14-3-3?*are co-deleted in individuals they exhibit the Miller–Dieker lissencephaly syndrome.[1020] These two proteins interact with other proteins, including Ndel1, to regulate the microtubule motor protein dynein. This regulates mitotic spindle orientation during cell division and motility of the nucleus during neuroblast migration. Underproduction of neurons and defective neuronal migration result when the interacting system is disrupted.[82]

Teratogenic Mechanisms

As we learn more about the molecular basis of teratogen action, it is becoming apparent that exogenous agents, like genetic lesions, can disrupt key signalling events to produce CNS malformations. For example, retinoic acid was previously considered to exert its teratogenic effect by direct disruption of plasma membrane function, owing to its lipophilic nature. However, retinoic acid is now known to be an endogenous molecule that initiates a complex signalling pathway of key importance in embryonic development.[649,678] Retinol, the principal dietary retinoid, interacts during normal development with the cytoplasmic retinol binding protein (CRBP) during its metabolism by retinaldehyde dehydrogenase (RALDH) enzymes to the metabolically active all-trans-retinoic acid. The latter is then bound by cellular retinoic acid binding protein (CRABP), which presents it to nuclear receptors. Two classes of receptor exist: retinoic acid receptors (RARs), which bind all-trans-retinoic acid most avidly, and retinoid X receptors (RXRs), which have greater affinity for another retinoid, 9-cis retinoic acid. Occupancy of the receptors by their ligands stimulates them to act as transcription factors, regulating the expression of other genes that contain retinoic acid response elements (RAREs) in their promoter regions, thus initiating cascades of molecular events. A final crucial step in the pathway is the inactivation of retinoids by catabolizing CYP26 enzymes, which form the 'sink' in gradients of retinoic acid activity in the developing embryo.

Although the physiological roles of retinoids in development are not understood fully, it is becoming clear that excess retinoic acid produces malformations by a mechanism that involves upregulation of the expression of RARs, especially the β-isoform,[522] presumably leading to supranormal stimulation of RAR pathways. Conversely, retinol deficiency is teratogenic by depleting the developing embryo of activity within this important signalling pathway. Thus, the teratogenic action of an important class of CNS teratogens is becoming understood in terms of the intracellular molecular mechanisms.

Another class of teratogenic agent of which the function has come under scrutiny comprises the *Veratrum* alkaloids cyclopamine and jervine, which inhibit cholesterol biosynthesis. When applied to experimental animals these agents produce forebrain defects resembling human holoprosencephaly (HPE). The significance of this finding came into focus following the discovery of mutations in the *Sonic hedgehog* gene in patients with HPE and that cholesterol biosynthesis is critical for normal Shh signalling.[875] Cholesterol substitution of the N-terminal cleavage fragment of the Sonic hedgehog protein is essential for its function, and a proportion of patients with Smith–Lemli–Opitz syndrome, in which cholesterol biosynthesis is disturbed, also develop HPE.[545] Although the *Veratrum* alkaloids do not appear to interfere with the cholesterol modification of the Sonic hedgehog protein, they may inhibit the response of the Patched receptor through its 'sterol-sensing domain',[184] thereby inhibiting transmission of the Sonic hedgehog signal.

Pathogenesis at the Cellular and Tissue Levels

Most CNS malformations result from abnormalities of morphogenesis. In normal CNS development, molecular signalling events (see earlier, Molecular Pathogenic Events, p. 286) regulate and coordinate a series of cellular changes (Figure 4.4a), the net effect of which is embryonic shaping. Some key cellular events are: (i) cell shape change, for instance, as occurs during the bending of the neuroepithelium as the neural tube closes; (ii) cell migration, as in the migration of neuroblasts during the generation of the layered structure of the cerebral cortex; (iii) cell proliferation, as occurs in the ventricular zone of the developing brain and spinal cord; (iv) programmed cell death, which is responsible for removal of transient CNS structures, such as the subplate neurons that participate in early cortical neurogenesis; and (v) cell differentiation, the process responsible for the generation of the diverse types of neuron and glia in the CNS. This section provides examples of each of these cellular/tissue events, to illustrate how their disturbance may contribute to the pathogenesis of CNS malformations.

Cell Shape Change

The conversion of part of the primitive neuroepithelium (the neural plate) into active bending sites is a key feature of the closure of the neural folds during neurulation (Figure 4.6). Bending occurs in midline neuroepithelial cells (the median hinge point), under the influence of factors from the underlying notochord, including Sonic hedgehog protein, and at paired dorsolateral hinge points, under the influence of unidentified signals from the overlying surface (epidermal) ectoderm (Figure 4.6c,e). Bending at these sites is achieved by an interruption of the basal to apical to basal progression of nuclear translocation, which occurs as neuroepithelial cells progress through the cell cycle.[953] Cells with basally located nuclei accumulate, leading to bending *via* a local reduction in the apical surface area of the neuroepithelium (Figure 4.6d). The folded configuration of the bending neural plate is stabilized further by contraction of apically arranged actin microfilaments.[1114]

Failure of neural tube closure results in NTDs (anencephaly and myelomeningocele; see later, Neural Tube Defects: Dysraphic Disorders, p. 293). Although the underlying cellular pathogenesis of human NTDs has not been determined, NTDs in mice are known to result from disturbance of the bending sites in the folding neural plate. For instance, pathological enlargement of the median hinge point in the *looptail* mutant interrupts the initiation of closure in the upper spine,[401] leading to the severe NTD craniorachischisis (Figure 4.6a,b), whereas excessive ventral curvature of the caudal embryonic region inhibits formation of dorsolateral hinge points,[940] thereby disturbing low spinal closure in the *curly tail* mutant, leading to myelomeningocele (Figure 4.6f–h). The enhanced curvature of the body axis in *curly tail* embryos results from defective cell proliferation in non-neural tissues, the notochord and hindgut epithelium (Figure 4.7), emphasizing the principle that CNS malformations may derive from cellular abnormalities outside the neural tube.

Cell Migration

The stratified structures of the cerebral cortex arise through a process of tightly regulated migration of post-mitotic neuroblasts from the inner ventricular zone towards more superficial regions of the primitive neural tube (see later, Migration and Differentiation of Neuroblasts, p. 315). Radial migration of neuroblasts along radial glial fibres appears to be the mechanism mainly responsible for the generation of glutamatergic pyramidal neurons in the cerebral cortex, whereas tangential migration from the medial and lateral ganglionic eminence provides many of the γ-aminobutyric acid (GABA)-ergic interneurons.[794] Defective migration, particularly along radial glial fibres, has been identified as a pathogenic mechanism underlying disturbed lamination of the cerebral cortex and cerebellum in mice homozygous for the *reeler* mutation. Strikingly, the mutant phenotype involves an apparent reversal of the polarity of the normal cortical layers. *Reeler* neuroblasts begin their centrifugal migration along the radial glial fibres, but they are unable to pass postmigratory neurons in the deeper cortical layers.[823] *Reeler* mice lack an extracellular matrix molecule, named reelin,[221] which exhibits similarities to molecules involved in cellmatrix adhesion, such as tenascin. Reelin is expressed by neuronal cells but not radial glial cells, supporting the idea that the *reeler* phenotype results from a defect in adhesion between early post-migratory neurons. Other proteins that participate in the reelin pathway have been identified. Mice with mutations in the *mdab1* gene, and those with mutations in both very low-density lipoprotein (VLDL) receptor and apolipoprotein E (ApoE) receptor 2, develop very similar phenotypes to *reeler*. Molecular studies have shown that VLDL receptor and ApoE receptor 2 bind the reelin protein,

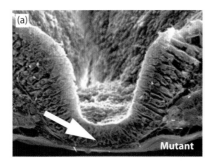
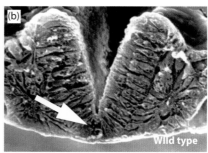

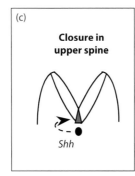
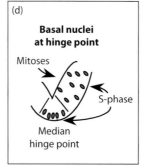
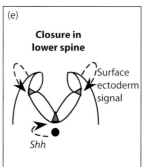

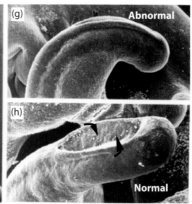

4.6 Pathogenesis of neural tube closure defects in two genetic mouse models. At upper spinal levels, bending at the midline of the neural plate **(b,c)** is achieved by concentration of cells with basally located nuclei **(d)**. In the *loop-tail* mutant mouse, an excessively broad midline region is formed (compare arrows in a and b), preventing apposition of the neural folds in the dorsal midline and leading to the development of craniorachischisis.[394] Lower spinal levels depend on bending at dorsolaterally positioned locations **(e)**. In the *curly tail* mutant, excessive ventral curvature of the body axis **(g)** (see also Figure 4.7) inhibits the formation of dorsolateral hinge points **(f)**, leading to the development of myelomeningocele.[1026] Normal embryos achieve apposition of the neural folds in the caudal region, as a result of dorsolateral bending **(h)**.

Reproduced from Van Straaten et al.[1045] and Greene et al.[401] with modification. With permission from Elsevier.

whereas the mdab1 protein acts intracellularly to transduce the reelin signal.[232,461] Hence, reelin appears to regulate an important genetic pathway controlling neuroblast migration during CNS development.

Cell Proliferation

CNS development occurs against a backdrop of a massive expansion in cell numbers. Cells pass through multiple rounds of proliferation within the ventricular and subventricular zones of the brain and spinal cord, exiting the cell cycle as they embark upon their migration to the cortical plate or mantle layer. In recent years, the molecular machinery that regulates the cell cycle has been elucidated in great detail.[888] Key roles are played by the cyclins, proteins that

vary in abundance at specific phases of the cell cycle, the cyclin-dependent kinases, which are activated by specific interactions with cyclins, and cell cycle phase-specific inhibitors, which can induce cells to exit the cell cycle to become quiescent. Null mutations in these cell-cycle-machinery genes may produce early embryonic lethality, as with the *cyclin A2* gene,[729] presumably indicating the essential general requirement of this gene for embryonic cell proliferation. Alternatively, there may be no discernible effect on development, as in the case of the G1 checkpoint inhibitor *p21CIP1/WAF1*, perhaps suggesting an overlapping function (redundancy) with another gene that plays a compensatory role.[262] A third type of outcome is the causation of specific CNS defects, as when the gene encoding cyclin D2 is inactivated. Null mutants develop specific abnormalities

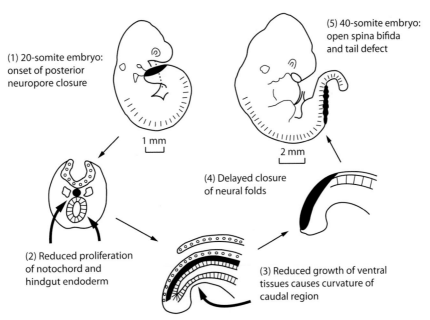

(1) 20-somite embryo: onset of posterior neuropore closure

1 mm

(5) 40-somite embryo: open spina bifida and tail defect

2 mm

(4) Delayed closure of neural folds

(2) Reduced proliferation of notochord and hindgut endoderm

(3) Reduced growth of ventral tissues causes curvature of caudal region

4.7 Spina bifida. Experimental analysis of the pathogenic sequence of events underlying spina bifida in the mutant *curly tail* mouse. This mutation causes lumbosacral spina bifida and/or tail flexion defects in around 50 per cent of homozygotes. The defects result from an imbalance of cell proliferation in the caudal embryonic region where growth of certain non-neural tissues, the notochord and hindgut endoderm, is reduced in affected embryos, whereas the neuroepithelium is unimpaired in its rate of proliferation. The notochord and hindgut are midline structures firmly attached to the ventral surface of the neuroepithelium. Their slow proliferation produces a mechanical distortion of the body axis, which curves ventrally, thereby opposing dorsolateral bending and inhibiting closure of the neuropore. Spina bifida can be prevented in this mutant either by correcting the cell proliferation imbalance, by inserting a splint into the caudal embryonic region to prevent the development of ventral curvature, or by treating embryos during neuropore closure with myo-inositol.[120,189,393]

of cerebellar development, with reduced numbers of granule cells and stellate interneurons.[492] Global defects of cell proliferation and/or cell death (see later, Programmed Cell Death, p. 291) have long been assumed to underlie such CNS defects as microcephaly and megalencephaly, but only recently has definitive evidence emerged to support this idea. Primary microcephaly (Table 4.1) has been found to result from mutation of several genes that encode proteins, for example microcephalin, that participate in mitotic spindle function during cell proliferation.[1096] At present, it appears that microcephaly is a disorder of deficient neurogenesis during brain development. Furthermore, studies using conditional gene targeting in mice have revealed a critical role for Wnt signalling, *via* β-catenin, in regulating the balance between cell cycle re-entry (to produce additional neuronal precursors) and exit of the cell cycle (to produce differentiated neurons and glia). Overexpression of stabilized β-catenin results in increased brain size, resembling megalencephaly.[150]

Programmed Cell Death

Programmed or physiological cell death is the phenomenon in which cells die during normal development. It is an expression of terminal differentiation, involves new gene expression, and is usually equated with the morphological process of apoptosis, in which cells die by nuclear condensation and fragmentation into membrane-bound bodies without release of cytoplasmic contents. This contrasts with necrosis, a pathological process, in which cells rupture and cytoplasmic contents are released. All cells are probably programmed to die by

apoptosis, being kept alive only by the constant presence of survival factors in the extracellular environment.[506] Moreover, teratogenic agents such as ethanol, which are capable of inducing CNS defects, appear to trigger the apoptotic pathway by inhibiting neurotransmitter receptors.[499]

The genes that specify which cells are destined to die or to survive during development were identified initially in the nematode *C. elegans*.[305] A number of corresponding mammalian genes have now been found to participate in the regulation of apoptosis. For instance, the *bcl-2* and *bcl-x* genes encode proteins that inhibit execution of the apoptotic pathway, by preventing the activation of downstream apoptotic enzymes called caspases.[906] The action of these genes is opposed by pro-apoptotic genes, including *Bad* and *Bax*. When overexpressed, *bcl-2* can prevent the degeneration of neurons in response to deprivation of neurotrophic factors *in vitro*[15] or cutting their axons *in vivo*,[287] although *bcl-x* seems likely to play a more important role than *bcl-2* in protecting neurons from programmed cell death *in vivo*.[388] Conversely, disruption of the *caspase-3* gene in mice leads to decreased apoptosis, producing heterotopic neuronal masses in the developing cerebral cortex.[571] Hence, details are emerging of the intracellular regulation of programmed cell death and the ways in which this pathway is integrated into normal CNS development.

Cell Differentiation and Regional Patterning

Coincident with the morphogenetic events of CNS development, the various regions of the presumptive brain and spinal cord become regionally specified. As gastrulation proceeds, the neuroepithelium is induced from naive ectoderm through interactions with underlying chordamesoderm and by the transmission of inductive signals in the plane of the neural plate.[1083] Bone morphogenetic proteins (BMPs) induce an epidermal default state, whereas molecules such as chordin and noggin antagonize BMP action and promote neural differentiation.[818,1082,1121] In birds and mammals, there is also a critical role for Wnt and FGF signalling in neural induction.[976] The newly induced neuroepithelium rapidly becomes regionally specified in both the anteroposterior (i.e. rostrocaudal) and dorsoventral axes, so that subsequent cellular differentiation occurs appropriate to the position within the CNS.

Anteroposterior Axis Neuroepithelial cells at different levels of the body axis express different combinations of homeobox genes, which encode transcription factors bearing the characteristic homeodomain DNA binding motif. In the hindbrain and spinal cord, a particular class of homeobox genes, the *Hox* genes, play a key role in specifying regional identity. The combination of *Hox* genes expressed in each hindbrain segment, or rhombomere, determines the developmental character of that segment, a mechanism referred to as the *Hox* code.[496] Treatment of gastrulation-stage mouse embryos with retinoic acid alters the pattern of *Hox* gene expression, causing cells of rhombomeres 2 and 3 to express *Hoxb1*, which is normally expressed only by cells of rhombomeres 4 and 5. The regional character of rhombomeres 2 and 3 is altered so that they now give rise to a nerve resembling cranial nerve VII (facial) rather than cranial nerve V (trigeminal) as would normally occur. Thus, the facial nerve is duplicated in the retinoic acid-treated embryos.[667] Different classes of homeobox genes impart regional identity to the midbrain and forebrain, a topic that is discussed further later in this chapter (see later, Disorders of Forebrain Induction, p. 304).

Dorsoventral Axis Specification of neuronal cell types along the dorsoventral axis of the brain and spinal cord depends on mutually antagonistic signals emanating from the midventral region (initially the notochord, subsequently the floor plate) and the dorsal midline (initially the apical ectoderm, subsequently the roof plate). Sonic hedgehog (Shh) is the ventral signal, and the manner in which this extracellular signalling molecule specifies cell identity in the dorsoventral axis is now understood in considerable detail.[312,500] A proteolytic N-terminal fragment of the Shh protein (Shh-N) is the active signal, requiring cholesterol addition for its activity. Shh-N interacts with a transmembrane receptor encoded by the *Patched* gene, which, in the absence of bound Shh ligand, serves to repress an intracellular pathway that is activated by the membrane-bound product of the *smoothened* gene. Binding of Shh-N to Patched de-represses the pathway, enabling the activation of downstream signalling, in which Gli proteins play an important role. The central significance of the Shh pathway is evident from the striking developmental abnormalities that result from loss of function of the mouse genes, and from the association of genes in the pathway with human developmental disease: *SHH* and *GLI2* mutants develop holoprosencephaly in both humans and mice, *PATCHED* mutants develop Gorlin syndrome in humans and severe CNS defects in mice, whereas

GLI3 mutations are implicated in Greig cephalopolysyndactyly and Pallister–Hall syndrome in humans and brain abnormalities in mice.

The protein Shh exerts a concentration-dependent influence over the differentiative fate of cells in the spinal cord, with high concentrations promoting the formation of ventral neuronal types and low concentrations promoting the differentiation of more dorsal cell types. Precisely how this concentration dependence is achieved appears to involve the regulation by Shh of the expression of a series of homeobox-containing genes that constitute a combinatorial code, analogous to the Hox code. Each neuronal type in the ventral spinal cord develops from cells that express a unique combination of these spinal cord homeobox genes.[116] Dorsal neuronal types are known to be specified by signals belonging to the *BMP* and *Wnt* gene families, with important roles for BMP antagonists, such as *noggin*,[657,958] although the details of this dorsal specification process are less well understood than in the ventral neural tube.

The Question of Cell Autonomy in Pathogenesis

It is often important to query through which cell type a defect of morphogenesis is mediated. A cell autonomous defect is one in which function of the mutated gene is required within the cell type(s) principally affected. In the converse situation of a non-cell autonomous defect, the mutated gene is required in cells other than those exhibiting overt pathology. Making the distinction between cell autonomy and cell non-autonomy is useful as it points the investigator towards the precise location of the pathogenic alteration. For instance, non-autonomous defects indicate a possible defect in an extracellular signal or a survival factor.

A functional approach to this question is provided by the analysis of chimaeras or mosaics, individuals in which cells of two different genotypes coexist. Chimaeras are individuals derived from two or more original zygotes, usually as a result of experimental interventions such as aggregating a pair of pre-implantation embryos or injecting early embryonic cells into another embryo. Mosaics are individuals in which genotypically different cells arise through somatic recombination or X-chromosome inactivation.

Chimaeric Studies in Animal Models

The chimaeric approach has been used, for instance, to investigate the development of defects of Purkinje cell development in the mouse mutants *purkinje cell degeneration* (*pcd*) and *reeler*, in which cerebellar Purkinje cells exhibit reduced survival and abnormal position, respectively.[722] Chimaeric individuals containing both mutant and normal cells show that *pcd* Purkinje cells die, irrespective of whether wild-type cells are also present in the developing cerebellum, indicating that the *pcd* defect is cell autonomous. In contrast, positioning of both wild-type and *reeler* Purkinje cells is abnormal in chimaeras,[1008] indicating that the *reeler* defect is expressed outside the Purkinje cell (i.e. a non-cell autonomous defect). This finding is consistent with the observation that reelin, the product of the *reeler* gene, is not expressed in Purkinje cells.[917] Abnormal migration of

Purkinje and granule cells occurs to regions outside the cerebellum in *Unc5h3* mutants, which lack a receptor for the netrin-1 ligand. Chimaera analysis has shown this effect to depend on the pioneering influence of the granule cells and not on the Purkinje neurons.[384]

Mosaic Studies in Humans

The neuronal migration defect present in individuals with X-linked *doublecortin* mutations has been suggested to be cell autonomous, based on the finding of both normal-migrating and abnormal-migrating neurons in the brains of *doublecortin* heterozygous females. In contrast, all neurons appear to be affected in hemizygous males. Female cells inactivate an X chromosome at random during development, and so females heterozygous for X-linked diseases are mosaic, with some cells expressing the normal allele and others expressing the mutant allele.[643] The normal cortical band in *doublecortin* females may arise solely from cells expressing the wild-type allele, whereas the heterotopic band may comprise cells expressing the mutant allele. Such a cell-autonomous defect would be consistent with the intracellular localization of the protein encoded by the *doublecortin* gene.[270,379]

A GUIDE TO INTERPRETING THE PATHOGENESIS OF MALFORMATIONS

For the neuropathologist faced with the variety and complexity of malformations of the CNS, the challenge is not only description and perhaps diagnosis, but also the urgent practical task of providing accurate and useful information for genetic counselling. Malformations resulting from primary disturbance of embryonic and fetal development need to be distinguished from disruptions and deformations, where there is secondary compromise of development owing to factors such as vascular interruption, necrosis caused by infectious agents and compression of the embryo or fetus by external mechanical influences. Experimental studies in laboratory animals have demonstrated the capacity of vascular disruptions to produce birth defects that mimic primary malformations. Interest in this aetiology for human defects has been raised by the association between chorionic villus sampling (CVS) performed earlier than 10 weeks after the most recent menstrual period and a variety of limb and craniofacial defects.[332] Moreover, mechanical trauma, for instance inflicted by amniotic bands, can produce a range of human birth defects.[462] It is of great importance for purposes of genetic counselling that secondary defects are distinguished from primary malformations, because evaluations and recurrence risk analyses differ significantly. To this end, detailed methods of fetal postmortem analysis can provide a rational basis for deciding whether a particular defect is a primary malformation or a secondary disruption or deformation.[544]

In practice, however, the distinction between primary malformations and secondary disruptions is far from simple. This applies particularly to the CNS, for a variety of interrelated reasons that depend ultimately on the extended organogenesis of the brain, which continues right through intrauterine life and, at least for myelination and for the late-migrating granule cells of the cerebellum and temporal lobe, well into postnatal existence. This extended period of brain development is marked not only by rapid structural change but also by changes in the individual selective vulnerabilities of cells and tissues and in the form of the repair mechanisms that can be mounted by microglia and neuroglia to any deleterious influence. Macrophage responses can antedate recognizable astrocytic responses, and so resorption of necrotic tissue may occur without a trace of glial repair. Kershman observed macrophages in human brain as early as 11 weeks,[549] and immunohistochemical methods have demonstrated them early in the second trimester.[395] By contrast, astrocytosis was first detected at 20–23 weeks of gestation.[878] A further issue is that, at least in mouse embryos, phagocytic fibroblasts can 'stand in' for macrophages when the latter are absent, as in the PU.1 null mutant.[1095] Consequently, the morphological end result of any given noxious influence can vary greatly, depending on the time of its operation, which may be brief, prolonged or repetitive. Moreover, the repertoire of responses is limited. Circumstantial clinical evidence and experimental manipulations indicate that the same anomaly can be produced by several different causes, both genetic and environmental. Viewed from another standpoint, there appear to be temporal (or perhaps temporospatial) windows when development appears particularly at risk and, consequently, particular types of structural anomaly tend to be grouped together, for instance, midline malformations or neuronal migration defects. These various lines of argument underscore the importance of the dynamic aspects of the immature nervous system for a proper understanding of developmental neuropathology.

PATHOLOGY OF MALFORMATIONS

Neural Tube Defects: Dysraphic Disorders

These disorders must be set in the context of the normal development of (i) the neural tube, the progenitor of the entire central nervous system, and (ii) the axial skeleton, which becomes modelled around the neural tube.

Neurulation can be divided broadly into primary and secondary phases. In primary neurulation, which occurs throughout the future brain and spinal cord down to the upper sacral level, the neural tube is formed by neural folding. Secondary neurulation, which produces the neural tube in the lower sacral and coccygeal regions, occurs by a quite different process involving canalization of a solid cord of cells rather than neural folding.[606]

Primary neurulation begins with neural induction, leading to the appearance of the neural plate, a thickened dorsal midline ectodermal structure. The lateral edges of the neural plate then elevate, beginning at about 17–18 days post-fertilization, defining a longitudinal neural groove that deepens with progressive elevation of the sides of the neural plate. The neural folds converge towards the midline and fuse, forming the neural tube, beginning at the future cervical/occipital boundary (designated closure site 1, Figure 4.8a) on day 22. Fusion proceeds in cranial and caudal directions from this level. Studies in the mouse have shown that fusion occurs separately, soon after this initial closure, at two other sites within the developing brain. Closure of the cranial neural tube then proceeds in a discontinuous bidirectional fashion (Figure 4.8a). Fusion spreads simultaneously along the future spinal region from closure site 1, being completed with closure of the posterior neuropore in the upper sacral region around days 26–28. This multisite closure process has been suggested also to occur in human embryos, because it can explain the variation in level of the body axis affected by neural tube defects in different individuals.[1038] By contrast, direct studies of neurulation-stage embryos are equivocal about the precise details of the sequence of closure events in humans.[776,794,989,1038] Studies of neural tube defects in the mouse show that each element of the neural tube closure sequence has its own distinct requirement for gene expression, so that mutations in different genes specifically disturb different events during neural tube closure.[188]

Axial skeletal development begins when the sclerotomal component of the mesodermal somites migrate to surround the neural tube, soon after its closure has been completed. In the cranial region, only the vault of the skull is formed by axial mesoderm, whereas the skull base and facial skeleton are derived from the neural crest.[1009] By contrast, in the occipital and spinal regions, sclerotomal cells undergo skeletal differentiation to form the entire vertebrae. At the lowest spinal levels, an apparently multipotential population of cells, the tail bud, is the sole source of all non-epidermal

tissues, including the neural tube and vertebrae. Therefore, anomalies of the sacral and coccygeal regions are often found to embrace several tissue types.

Studies in the mouse have shown that different genes are implicated in the development of open NTDs versus axial mesodermal defects.[193,566] In terms of pathogenic mechanisms, therefore, it is important to distinguish between axial malformations that result from (i) failure of neural tube closure, in which secondary bony defects occur owing to faulty skeletal modelling around the malformed neural tube (e.g. anencephaly, craniorachischisis, myelomeningocele), and (ii) lesions that result principally from bony defects that reflect primary abnormalities of axial mesodermal development, without a persistently open neural tube (Figure 4.8b). This latter group comprises two subgroups: those defects in which there is herniation of the neural tube through the bony defect (e.g. encephalocele, meningocele) and 'closed' defects that are invariably skin covered (e.g. spina bifida occulta, diastematomyelia).

Neural Tube Closure Defects

Craniorachischisis

Craniorachischisis is the most severe form of dysraphism. Brain and spinal cord are exposed to the surrounding amniotic fluid, resulting in necrosis, degeneration and angioma-like formations. It is noticeable that many cases of craniorachischisis exhibit a relatively well-developed optic system, and a similar finding has been reported for a mouse model of craniorachischisis, the *loop-tail* (*Lp*) mutant. In this mouse, neural tube fusion fails at closure site 1 (Figure 4.8a), resulting in

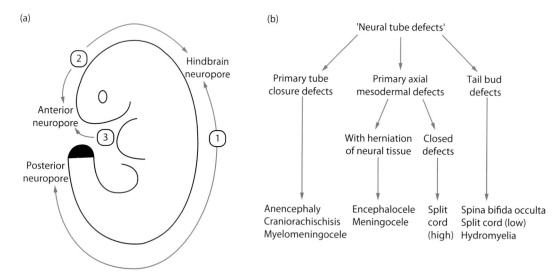

4.8 Stages of neurulation and classification of neural tube defects. (a) Stages of neurulation as demonstrated for the mouse[376] and inferred for humans.[1021] Primary neural tube closure is initiated at the cervical/hindbrain boundary (closure site 1) and separately, soon afterwards, at the boundary between the future mesencephalon and prosencephalon (closure site 2) and at the rostral extremity of the future prosencephalon (closure site 3). Fusion spreads bidirectionally between closures 1 and 2, and between closures 2 and 3, completing cranial neural tube formation at the hindbrain neuropore and anterior neuropore, respectively. Whereas cranial neural tube formation is occurring, neurulation progresses from the point of closure 1 in a caudal direction, through the spinal region, with completion of closure at the posterior neuropore, which is situated in the future upper sacral region. Below this level, secondary neurulation occurs, with formation of all non-epidermal tissues from a multipotential stem cell population in the tail bud (shaded). **(b)** A classification of neural tube defect malformations based on the embryonic tissue that appears primarily affected. See text for explanation.

Modified after Copp et al.[195] and Copp and Bernfield.[189] With permission from Lippincott Williams & Wilkins/Wolters Kluwer Health and Elsevier.

the severe dysraphic disorder, but fusion occurs normally in the cranial region, at closure sites 2 and 3,[195] yielding a relatively well-formed prosencephalon and optic vesicles.

Studies have demonstrated that craniorachischisis in mice results from disturbance of a specific signalling cascade, the planar cell polarity (PCP) pathway.[194] Wnt ligands interact with frizzled receptors to generate intracellular signals that are mediated *via* the cytoplasmic protein dishevelled. These signals follow the 'non-canonical' pathway, not involving β-catenin but rather being transduced *via* Rho GTPases to the cytoskeleton. Mutations in the PCP genes *Vangl2* (mutated in the *Lp* mouse), *Celsr1* (mutated in the *Crsh* mouse), *dishevelled-1* and *dishevelled-2* all give rise to a similar craniorachischisis phenotype. The PCP pathway is required for cells to participate in a directional planar array and, during late gastrulation, planar cell polarity underlies the process of convergent extension, by which the oval-shaped pre-neurulation embryo becomes transformed into the narrow, elongated neurulation-stage embryo.[1015] Failure of convergent extension yields a short, broad embryo, in which the neural folds are widely spaced, precluding initiation of closure. Although it seems very likely that a similar pathogenetic sequence underlies craniorachischisis in humans, it remains to be determined whether defective function of human PCP genes, including the *VANGLl2*, *CELSR1* and *DISHEVELLED* genes, are responsible.

Exencephaly

Exencephaly and anencephaly are different stages of the same developmental anomaly. Exencephaly has been described rarely in human fetal pathology,[662] probably because of the rapid necrosis of brain tissue exposed to amniotic fluid, leading to anencephaly. This phenomenon has been demonstrated directly in the retinoic-acid-treated rat, where an initial exencephalic appearance is converted to anencephaly by late gestation.[1094] The degenerative process is not rapid, however, as demonstrated by a study of surgically created neural tube lesions in the sheep, which can be corrected by a skin flap covering several weeks later in gestation, with only minimal functional deficit at birth.[693]

Anencephaly

Anencephaly was known in Egyptian antiquity, and in 1761 Morgagni compared the 'human monster' to a toad. It was also described by Geoffroy Saint-Hilaire early in the nineteenth century.[367,368] The calvaria is hypoplastic or absent, the base of the skull is thick and flattened, and there is a constant anomaly of the sphenoid bone resembling 'a bat with folded wings'.[661] The orbits are shallow, causing protrusion of the eyes. Attached to the skull base is a dark-reddish irregular mass of vascular tissue with multiple cavities containing cerebrospinal fluid (CSF), the area cerebrovasculosa. The mass is cystic, with a midline dorsal aperture opening to the exterior. No recognizable neural tissue can be found in the anterior and middle fossae, except for the trigeminal ganglia and limited lengths of the second to fifth cranial nerves. A hypoplastic anterior pituitary is present in a shallow sella, but the intermediate and posterior lobes are missing. The residual

amount of brain tissue varies. If the foramen magnum is intact, then a considerable proportion of the medulla is visible. Usually the foramen is deficient, with cervical spina bifida, short neck and deformed pinnae; then, only the caudal part of the medulla is present along with the distal parts of the lower cranial nerves. The pons, the cerebellum and the midbrain are grossly absent, although microscopic fragments of cerebellar folia may be seen.[65] Spinal involvement varies from non-fusion of the upper cervical arches without accompanying skull defect to complete rachischisis (Figure 4.9).[56]

There is controversy regarding the histological interpretation of the area cerebrovasculosa, which includes irregular masses of neural tissue, mainly glia with some neuroblasts or neurons, ependyma and tufts of choroid plexus and numerous thin-walled blood vessels. Although some consider that this tissue bears no resemblance to forebrain,[378] others studying well-oriented sections have observed vascular meninges surrounding a bilaterally symmetrical cavity, interpreted as forebrain ventricles, containing choroid plexus, and its walls composed of gliovascular tissue and lined partly by ependyma.[65] Covering the cystic area cerebrovasculosa is non-keratinizing squamous epithelium, which is laterally continuous with normal skin.

The medulla may be reasonably well preserved, but, as with the spinal cord, there is aplasia of descending tracts. The spinal leptomeninges are excessively vascular and contain islands of heterotopic neuroglial tissue.[67]

The absence of neurohypophysis and hypothalamus is associated with a hypoplastic adrenal cortex.[26,1027] Various other visceral abnormalities have been reported in cases of anencephaly,[737] a large thymus and hypoplastic lungs being the most frequent. Polyhydramnios is observed in about half of the cases, but its pathogenesis is unclear.

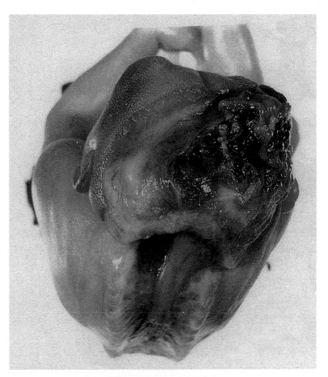

4.9 Anencephaly with complete rachischisis. Oblique view from above and behind.

Epidemiological studies have shown a high incidence of anencephaly in Ireland and Wales: 1–6 per 1000 live births compared with about 0.5–2 per 1000 live births in the USA and 0.5 per 1000 live births in France.[752] Females are affected more frequently than males. Anencephaly is the most common congenital malformation of the brain in human fetuses. Familial cases of anencephaly and/or spina bifida have been observed, but transmission is poorly understood. The role of diet and social class is still debated.

Early prenatal detection of NTDs was developed in the 1970s, based on the estimation of α-fetoprotein (AFP) in the amniotic fluid.[118] This protein is present in the fetal choroid plexus,[505] in addition to the fetal liver, and its elevated level in the amniotic fluid probably reflects direct leakage from the CSF. AFP is now measured routinely in maternal serum, forming the basis of prenatal screening for NTDs, in addition to ultrasound, which is becoming a widespread method for prenatal diagnosis for NTDs early in pregnancy.[218]

Myelomeningocele

This is a severe malformation in which both meninges and spinal cord herniate through a large vertebral defect. The fluctuant mass consists of a distended meningeal sac filled with CSF and covered by a thin membrane or by skin (Figure 4.10). The spinal cord may be closed, floating on the posterior surface of the arachnoid cyst. The central canal is often dilated and the posterior part herniates with the meninges. In other cases, often referred to as myelocele, the malformation appears as a flat, open lesion with CSF leaking on to the exposed area. The spinal cord at the site of the bony defect forms a flat, discoidal, highly vascular mass, the area medullovasculosa, which becomes epithelialized after birth. The posterior surface of the spinal cord is open and the central canal blends into the skin. Peripheral nerves end blindly within the vascular mass,

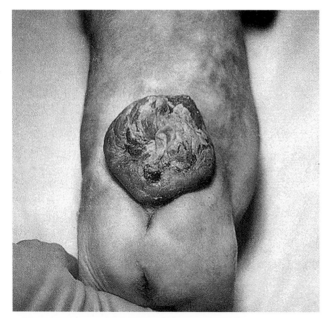

4.10 Lumbar myelomeningocele, showing ulceration of the surface, in an infant with hydrocephalus and paraplegia.

which comprises highly vascular connective tissue and islands of CNS, including neurons, glia and ependyma.

Hydrocephalus occurs in many infants with one of these defects, usually in association with the Chiari type II (Arnold–Chiari) malformation (see later, Chiari Malformations, p. 300). In addition, anomalies in the spinal cord above the myelomeningocele are common and include syringomyelia, hydromyelia, diastematomyelia, a double or multiple central canal and diplomyelia.[308]

The level of the defect has been the subject of particular attention for a number of reasons. Neurological disturbance depends largely on the level of defect, and the latter has been used as a prognostic indicator in terms of both the likely benefit of caesarean delivery before the onset of labour, which has been demonstrated to be beneficial in cases of relatively low, mild lesions,[639] and the outcome of surgery to close the defect in the neonatal period.[264,596,632] Recently, myelomeningocele closure *in utero* has been undertaken in the USA, with improved neurological function and a decreased requirement for postnatal hydrocephalic shunting.[3,130]

Epidemiologically, high and low defects show marked differences with, for instance, lesions above T12 exhibiting frequent association with malformations in other systems and a female preponderance, whereas low lesions are more often solitary malformations that have a more equal sex incidence or even a male preponderance. This led to the hypothesis that high and low defects arise by pathogenic mechanisms, involving neural folding (primary neurulation) and canalization (secondary neurulation), respectively.[189] However, this hypothesis appears unlikely in the light of findings that the transition from primary to secondary neurulation occurs in the upper sacral region of the embryo,[190,723] indicating that the vast majority of cases of myelomeningocele, both high and low, arise from disturbance of primary neurulation. The epidemiological differences between high and low lesions almost certainly result from heterogeneity in the mechanism of primary neurulation along the body axis (see Figure 4.8a).

Axial Mesodermal Defects with Herniation of the Neural Tube

Encephalocele and cranial meningocele consist of a protrusion of brain or meninges through a cranial defect. They occur most frequently in the occipital region, with 75–80 per cent of cases occurring there, whereas the frontal and lateral parts of the skull are affected much less often.

Occipital Encephalocele

The herniation occurs through the occipital bone, with or without involvement of the foramen magnum. The mass of tissue is often voluminous (Figure 4.11), attached to one of the cerebral hemispheres by a narrow pedicle and partially covered with normal skin and hair. Ulceration of the skin and secondary infection are frequent. Smaller encephaloceles may contain only fragments of disorganized CNS tissue, glia and ependyma. Larger herniations may include large portions of the cerebral hemispheres with ventricular cavities[147,541] as well as parts of the brain stem and cerebellum.

In the series of Karch and Urich, the herniation was always asymmetrical (Figure 4.12) and, although the gross convolutional pattern was normal and the cortex apparently not malformed, there was severe distortion, displacement and asymmetry of the basal ganglia.[541] Other features included anomalies of the hippocampi and commissural system, aberrant neural tissue in the cavity of the ventricles, distortion of the brain stem and agenesis of cranial nerve nuclei, absence of the vermis and near or complete absence of the cerebellar hemispheres. Karch and Urich drew particular attention to the presence of a persistent fetal vasculature in the form of an extensive plexus of thin-walled sinusoidal blood vessels in the leptomeninges. Similar abnormalities are present in the authors' material but, in addition, on several occasions the cerebral cortex has shown polymicrogyria (Figure 4.13).

Meckel–Gruber Syndrome

Occipital encephalocele is an important component of the Meckel–Gruber syndrome, a lethal autosomal recessive disorder with a characteristic phenotype including sloping forehead, occipital encephalocele, polydactyly, polycystic kidneys and hepatic fibrosis with bile duct proliferation.[7] The variability of neuropathological findings among the mass of case reports has prompted two extended series in an attempt to define a consistent pattern of abnormality. From observations in 59 cases, and detailed neuropathology in 10 cases, Paetau *et al.* noted encephalocele in 90 per cent, other prominent features being olfactory aplasia, midline defects and migration disorders such as polymicrogyria.[780] Another autopsy study of seven fetal or neonatal cases revealed a consistent pattern of malformations: (i) prosencephalic dysgenesis, arhinencephaly–holoprosencephaly and other midline anomalies; (ii) occipital exencephalocele taking the form of extrusion of parts of the rhombic roof through the posterior fontanelle; and (iii) rhombic dysgenesis, notably supracerebellar cyst, vermal agenesis, stenosis of the aqueduct and flattening and dysplasia of the brain stem.[7] The genetic basis of one type of Meckel–Gruber syndrome, linked to chromosome 17q21–q24, has been determined with the discovery of a novel gene, *MKS3*, which encodes meckelin.[955] Although very little is currently known of the function of meckelin, its identification can be expected to shed new light on the pathogenesis of Meckel–Gruber syndrome.

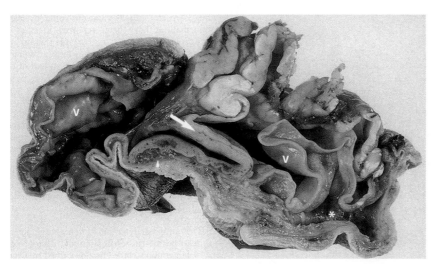

4.11 Encephalocele. Cut surface of a large surgical specimen. In this example, the two attenuated occipital lobes have herniated with their ventricular cavities (V). The cortex varies in thickness and is partly fused (arrow). Between brain and skin there is vascular meningeal tissue (arrowhead) and a cystic space (*).

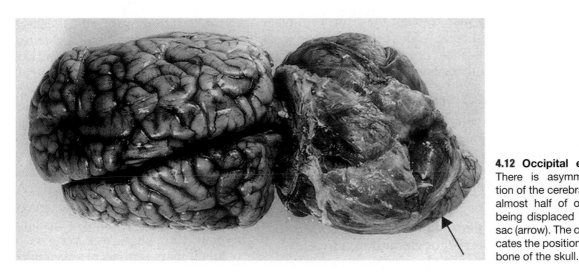

4.12 Occipital encephalocele. There is asymmetrical herniation of the cerebral hemispheres, almost half of one hemisphere being displaced into the hernial sac (arrow). The constriction indicates the position of the occipital bone of the skull.

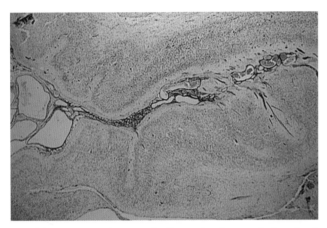

4.13 Microscopy of a surgically excised encephalocele. A malformed polymicrogyric cortex and persistent fetal vasculature in the leptomeninges are seen. Luxol fast blue–cresyl violet.

Parietal Encephalocele or Meningocele

Parietal encephalocele or meningocele occurs only occasionally.[682] Deformities of the brain are usually present and are not confined to the ipsilateral hemisphere. They include asymmetry, distortion of the ventricular walls, agenesis of the corpus callosum and hydrocephalus.

Anterior Encephalocele

Anterior encephalocele (Figure 4.14) is most commonly found at the fronto-ethmoidal junction. It is usually visible at the bridge of the nose (in 60 per cent of cases) as a bulging subcutaneous nodule and mild hypertelorism or as a mass of brain tissue. The encephalocele may expand into the nasal cavity (30 per cent of cases), ethmoidal or sphenoidal air sinuses, pharynx or orbit.[1122] The abnormal mass may contain disorganized brain tissue or gliotic cerebral cortex. As with occipital encephaloceles, the cerebral hemispheres within the intracranial cavity may be markedly skewed, with non-register of the basal ganglia and commissural anomalies. The clinical diagnosis may be difficult if only the meninges protrude through the cribriform plate of the ethmoid bone. Cerebrospinal fluid passing into the nasal cavity is indicative of a free communication between the subarachnoid space and the encephalocele.

As fronto-ethmoidal meningocele and encephalocele are rare in Western Europe but relatively common in South East Asia, genetic and/or teratogenic factors may be of importance in their aetiology.[335,991]

Meningocele

This is usually classified as a variant of spina bifida cystica, in which there is a vertebral defect combined with a cystic lesion of the back, most often in the lumbosacral region. Both the dura and arachnoid herniate through a vertebral defect, the spinal cord remaining in a normal position in the spinal canal, although it may show hydromyelia, diastematomyelia or tethering. The cyst is covered by skin, which has atrophic epidermis and lacks rete pegs and skin appendages. The wall of the cyst contains thin-walled blood vessels and islands of arachnoidal tissue, a narrow channel connecting the cyst with the vertebral canal.

4.14 Anterior (frontal) encephalocele (arrow).

Spina Bifida Occulta: Defects of Tail Bud Development

This is the least severe group of NTDs, which are sometimes grouped together under the general term 'occult spina bifida'. Although the spinal cord abnormality may be a prominent feature, there are often accompanying defects of skeletal (e.g. sacral agenesis), anorectal and urogenital systems. The spinal cord abnormalities may comprise overdistension of the central canal (hydromyelia), longitudinal duplication or splitting of the spinal cord (diplomyelia, diastematomyelia) and tethering of the lower end of the cord. The defects are most often located in the low lumbar and sacral regions, broadly corresponding to the region of secondary neurulation. Because neural folding is not involved at this level, the defects are always of the closed type. All of the abnormalities can be traced to a disturbance of development of the embryonic tail bud. Defective separation of neuroepithelial and mesodermal tissues during differentiation of the tail bud in animal models commonly yields a split cord.[172,286] Similarly, tethering of the cord within the vertebral canal probably represents the incomplete separation of neural from mesodermal components. The association of low spinal lesions with sacrococcygeal teratoma and lipoma is another manifestation of aberrant differentiation of the tail bud, which comprises a multipotential cell population.[999]

Occurrence of split cord at higher levels seems most likely to reflect secondary injury to the closed neural tube from malformed vertebral elements, as evidenced by the frequent association of diastematomyelia with a bony spur. These higher defects should probably be considered a malformation of axial mesodermal differentiation.

Hydromyelia

Overdistension of the central canal can result either from incomplete fusion of the posterior columns[969] or as a persistence of the primitive large canal of the embryo. The dilation may be focal and is often more pronounced in the lumbar region. In the neonate, isolated hydromyelia is usually asymptomatic and is an incidental finding at autopsy. It may be found only on serial slices of the spinal cord.[72] Hydromyelia may also be one anomaly in a more complex syndrome. It is, for instance, associated with the Chiari II malformation in approximately 40 per cent of cases.[648] The central canal may be lined by normal ependyma, which becomes replaced by glial tissue (Figure 4.15).

Split Cord

Although diplomyelia (duplication of the cord) and diastematomyelia (coexistence of two hemi-cords) are often distinguished (Figure 4.16), Pang and Dias proposed an alternative classification based on whether the hemi-cords have separate dural sacs, separated by a rigid osseocartilaginous septum (split cord malformation type I) or whether

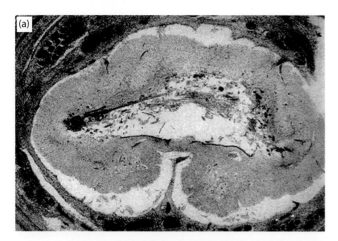

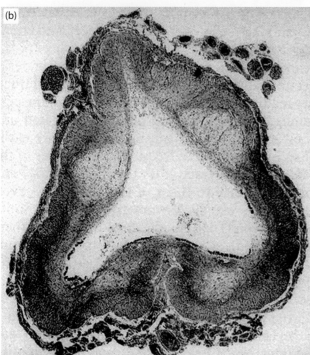

the hemi-cords coexist within a single dural sac, separated only by a non-rigid fibrous median septum (split cord malformation type II).[790] Both types of lesion can occur, in tandem, in the same patient. Split cord is more likely to be symptomatic in adults than in children,[788] and in neonates it may be an incidental finding.

Tethered Cord

Defects of the spinal cord can be associated with tethering within the vertebral canal, whatever the level of cord affected, as, for instance, in cervical myelomeningocele.[789] However, the tethered cord syndrome is usually reserved for lumbosacral defects in which there are variable combinations of thickening of the filum terminale, low or dilated conus medullaris, spinal lipoma, dermoid cyst, split cord, hydromyelia and sacral agenesis. Clinical signs associated with cord tethering include lower limb motor and sensory deficits and neuropathic bladder. The severity of symptoms increases with age, and patients are frequently treated surgically by untethering of the cord. Follow-up studies to determine the long-term effects of surgery have shown a good outcome in terms of maintained cord mobility and symptomatic improvement in some cases, in terms of resolution of upper motor neuron signs and enhanced bladder function.[199,341]

Prevention of Neural Tube Defects

The prospect for primary prevention of NTDs by folic acid was raised by a randomized controlled clinical trial conducted by the Medical Research Council (MRC) in the UK.[1062] This study demonstrated a 70 per cent reduction in the recurrence of NTDs after periconceptional supplementation with 4 mg folic acid per day. Evidence has also accumulated to suggest a preventive effect of folic acid on the first occurrence of NTDs.[79,221] Although 70 per cent of defects were prevented by folic acid in the MRC trial, up to 30 per cent of defects appeared resistant.

Mothers of affected fetuses either have normal red cell and serum folate levels or are mildly deficient, whereas mildly elevated levels of homocysteine are present in maternal blood and in the amniotic fluid of defective fetuses.[702,973] This finding suggests a key role for the enzyme methionine synthase, which catalyses the remethylation

4.15 Hydromyelia. (a) In a case with myelomeningocele and Arnold–Chiari malformation, the ependyma is intact anteriorly but posteriorly is partially destroyed by necrosis and haemorrhage. Fragments of gauze in the cavity are derived from dressings over the ulcerated meningomyelocele nearby. **(b)** In a newborn child with an Arnold–Chiari malformation, early gliosis replaces the ependymal lining in the posterior half of the cavity.

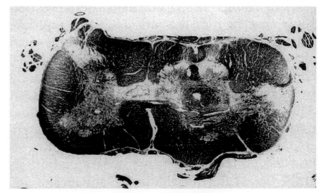

4.16 Partial duplication of the spinal cord.

Figure courtesy of Dr C Keohane, University of Cork, Cork, Ireland.

of homocysteine. Indeed, Kirke *et al.* showed that folate and vitamin B12 levels are independent risk factors for NTDs.[557] Alternatively, elevated homocysteine could reflect subnormal function of the enzyme 5,10-methylene-tetrahydrofolate reductase (MTHFR), which catalyses the production of 5-methyltetrahydrofolate, the methyl donor in the methionine synthase reaction. Patients with NTDs, and their parents, are more likely to possess a thermolabile variant of MTHFR than normal controls,[1042] suggesting that the C677T MTHRF polymorphism is a genetic risk factor for NTDs.

Mouse models of NTDs fall into two groups with respect to prevention by folic acid. NTDs in *splotch* mice are preventable by folic acid treatment, either *in vivo* or *in vitro*.[336] During neurulation, *splotch* embryos exhibit a decreased supply of folate-related metabolites for pyrimidine synthesis, which can be ameliorated by administration of folic acid or thymidine, but not by methionine. *Curly tail* mutant mice, in contrast, are resistant to folic acid, but low spinal defects in this system can be prevented by another vitamin-like molecule, inositol, administered either *in vivo* or *in vitro*.[400] Both *myo-* and D-*chiro*-inositol are effective in preventing spinal NTDs in the *curly tail* mouse,[173] with an essential role for the β1 and γ isoforms of protein kinase C, stimulation of which is part of the preventive mechanism.[174] The possibility arises that inositol may be a useful adjunct to folic acid therapy, perhaps allowing prevention of a larger proportion of NTDs than is possible with folic acid alone.

Chiari Malformations

In 1891, Chiari defined three types of cerebellar deformity associated with hydrocephalus.[152] In a subsequent paper, he added a fourth type, cerebellar hypoplasia,[153] which most authors now regard as a separate entity. The pathogenic relationship among Chiari's three types is controversial, but the morphological classification remains valid.

Chiari Type I

In Chiari's original description, conical elongations of the tonsils and neighbouring parts of the cerebellar hemispheres extend into the vertebral canal. These prolongations could be histologically normal, 'softened' or sclerosed, whereas the medulla was either unchanged or flattened by the cerebellar tongues. His index case was a girl aged 17 years, asymptomatic during life, in whom there was some widening of the lateral and third ventricles but without enlargement of the head. Some confusion has arisen by including chronic tonsillar herniation due to space occupying lesions or chronic hydrocephalus in this category,[353] or mistaking unilateral tonsillar protrusion for herniation of the vermis (Chiari II) at surgery. Nevertheless, the term 'Chiari type I' remains apposite for cerebellar herniation restricted solely to the tonsils (Figure 4.17), which may be atrophic, sclerotic and connected by fibrous adhesions to the sides and back of the slightly elongated medulla. The course of the upper cervical roots has been disputed, but in several cases examined at the National Hospital for Neurology and Neurosurgery, London, UK, upper cervical roots were angled in an upward direction (Figure 4.18). This has also been the conclusion in some surgical series.[42]

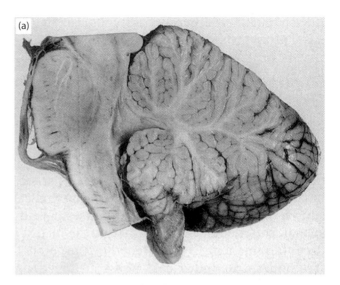

4.17 Chiari type I malformation. (a) Mid-sagittal section through the hindbrain. The peg-like protrusion of the cerebellar hemisphere is quite separate from the normal vermis. **(b)** Chiari type I malformation in a 10-month-old child presenting with polydactyly, hemi-hypertrophy and hemi-megalencephaly. A bifid tongue of tonsillar tissue extended to 2.5 cm below the olivary bulge.

Although often asymptomatic, Chiari type I is a not uncommon cause of late-onset hydrocephalus, with adult patients presenting with cerebellar ataxia, neck pain, pyramidal syndrome or dissociated sensory loss indicative of syringomyelia. It is increasingly recognized in the paediatric age group, owing to the increased availability of non-invasive neuroimaging, notably magnetic resonance (MR) imaging. In young infants of 2 years or under, presenting symptoms include headache and neck pain[236] or apnoeic episodes including near-miss sudden infant death syndrome.[294] By contrast, older children exhibit scoliosis and motor weakness associated with syringomyelia and syringobulbia.[232] An MR imaging study of 16 such children aged under 15 years showed an asymmetry of the syrinx towards the convex side of the scoliosis,[502] suggesting that asymmetric damage to the anterior horn may be the pathological substrate for the unbalanced strength of the paravertebral

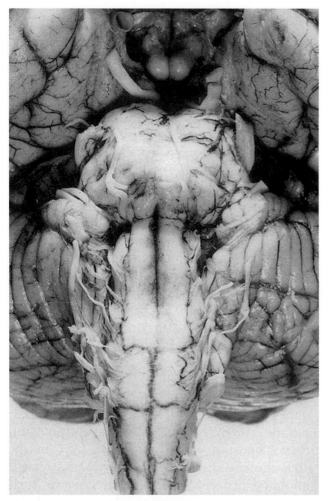

4.18 Chiari type I malformation showing upwardly directed cervical spinal roots.

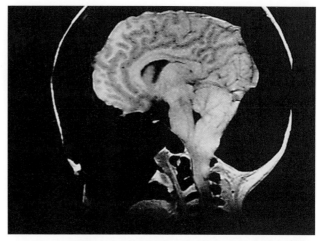

4.19 Chiari type I malformation. Magnetic resonance image showing tonsillar herniation to C2 in a patient with massive skull thickening due to craniometaphyseal dysplasia.

Figure courtesy of Dr K Chong, Great Ormond Street Hospital, London, UK.

patients, diagnosed by supine myelography, surgery and computed tomography (CT) scans.[927] The posterior fossa in these patients was significantly smaller than in controls, and there was an inverse relationship between the size of the posterior fossa and the degree of cerebellar herniation. The small size of the posterior fossa was also a notable feature in the family described by Coria *et al.*: autosomal dominant occipital dysplasia with or without Chiari type I anomaly affected three generations.[197] These studies suggest a primary role for occipital dysplasia in the pathogenesis of the Chiari type I malformation, a hypothesis that parallels that derived from the experimental model of Chiari type II malformation.[664]

muscles responsible for scoliosis. Lower cranial nerve palsies[297] may lead to sleep apnoea and vocal cord paralysis[899] or speech defects associated with velopharyngeal insufficiency.[370] For neuropathologists, a further important presentation is in connection with sudden unexpected death in childhood. Friede and Roessmann collected seven cases, aged from 7 months to 17 years, who before death were without neurological deficit.[353] Personal experience includes two such instances of sudden death associated with Chiari type I in the contexts of craniodiaphyseal dysplasia (Figure 4.19) and hemi-megalencephaly (Figure 4.17b).

The relationship of type I anomaly to syringomyelia is close: half the patients in clinical series of type I have syringomyelia,[42,288,711,927] whereas about 90 per cent of patients with idiopathic syringomyelia have a Chiari anomaly.[30,366,936] Familial occurrence of syringomyelia and Chiari type I has also been reported.[376] An association with craniocervical bony malformations has been increasingly recognized, notably the presence of platybasia, basilar impression, suboccipital dysplasia and Klippel–Feil anomaly,[31,42,239,611,711,799,927] but Chiari type I may also occur in conjunction with craniosynostosis (Figure 4.19).[353]

Using detailed measurements of lateral skull radiographs, Shady *et al.* demonstrated occipital dysplasia in over two-thirds of a large series of adult Chiari type I

Chiari Type II

Cleland first recorded herniation of the vermis combined with deformities of the medulla and tectal plate in an infant with a meningocele,[169] now generally known as the Chiari type II malformation.[923] It has been the subject of numerous pathological studies.[66,139–141,229,374,804,805] Most commonly found in infants in association with myelomeningocele and hydrocephalus, its essential components are elongation of the inferior vermis and brain stem and their displacement into the cervical spinal canal (Figure 4.20).

Bony and dural anomalies are characteristic and important for radiological diagnosis. Craniolacunia (Figure 4.21), irregular patches of thinning or complete erosion of the cranial vault, is frequent; in one radiological study it was present in 90 per cent of neonates with Chiari type II malformations and myelomeningocele.[993] The falx is short and fenestrated.[139] The posterior fossa is shallow, the torcula low and the clivus concave and thinned. The tentorial hiatus is widened, but the tentorial insertion is low, near the edge of an enlarged foramen magnum. The herniated cerebellar tissue varies from a short peg to a long tail and involves the nodulus, pyramis and uvula in that order.[229] It may extend down as far as the upper dorsal vertebral segments. Rarely, there is no cerebellar displacement, or the

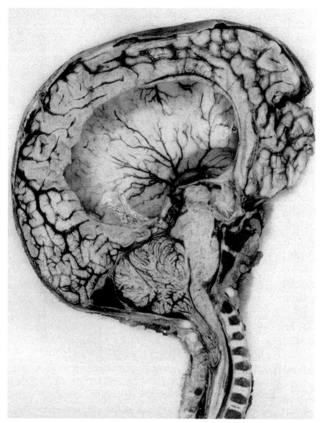

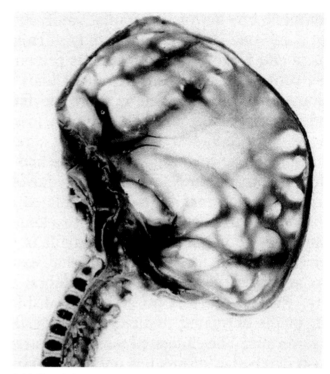

4.21 Craniolacunia of the skull from an infant with Arnold–Chiari malformation.

4.20 Arnold–Chiari malformation (Chiari type II) and hydrocephalus in a 7-week-old infant. Note the downward displacement of the cerebellar vermis and tonsils through the foramen magnum into the spinal canal and the beak-like deformity of the quadrigeminal plate.

herniated tissue includes tonsils as well as vermis. The elongated tongue of flattened whitish cerebellar vermis, often associated with choroid plexus, lies on the dorsal surface of the lower medulla and cord, bound to them firmly by fibrous meningeal adhesions. Its upper end is often grooved by the edge of the foramen magnum. The brain stem, particularly the medulla, the fourth ventricle and its choroid plexus, are elongated and displaced caudally. The cerebellar tail may cover the roof of the ventricle or may be intraventricular. In 50 per cent of cases, just caudal to the ventricle, the lower medulla below the gracile and cuneate nuclei forms an S-shaped curve or kink over the cervical cord (Figures 4.22 and 4.23).

Microscopically, the herniated cerebellar tissue shows Purkinje and granule cell depletion, with shrinkage and gliosis of the folia and absence of myelin. The presence of focal cortical dysplasia and grey heterotopias in the hemispheric white matter is well recognized, as well as distortion of brain stem tracts and nuclei. Gilbert *et al.* called attention to hypoplasia or agenesis of cranial nerve nuclei, olivary nuclei and pontine nuclei in their series of young infants.[374] Additional abnormalities are numerous. The cerebellar hemispheres are often asymmetrical and flattened dorsally; the vermis may be buried between the hemispheres, which can extend around the brain stem over its ventral surface, sometimes meeting in the

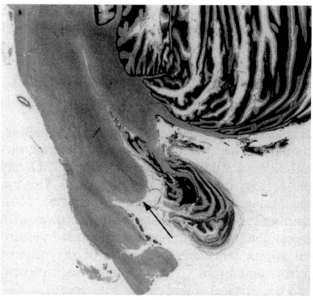

4.22 Sagittal section of brain stem and cerebellum in Arnold–Chiari malformation. The herniated cerebellar tissue lies posterior to the S-shaped kink (arrow) at the junction of the medulla and spinal cord.

midline.[229] The pontomedullary junction is ill defined, with an elongated rod-shaped pons. Also common is a beak-like deformity of the corpora quadrigemina, which is directed backwards and downwards to a point formed by the fusion of the inferior colliculi.[169] The upper four to six cervical spinal roots may be angled upwards towards their intervertebral foramina; lower roots are normally

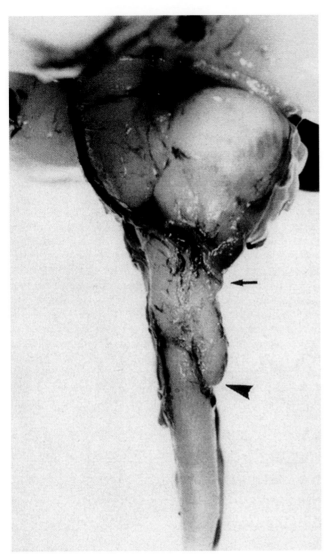

4.23 Arnold–Chiari malformation in a 20-week fetus. Lateral aspect of the hindbrain showing only slight cerebellar herniation (arrow) but marked elongation of the brain stem, with a prominent S-shaped curve over the upper cervical cord (arrowhead).

produce a pyocephalus, whereas the cerebral subarachnoid space is spared. The dilated cerebral hemispheres often show an abnormal convolutional pattern consisting of an excessive number of small gyri and shallow sulci, most appropriately termed polygyria because usually the normal cytoarchitecture is preserved, unlike the laminar abnormalities present in polymicrogyria, although several authors have also described true polymicrogyria.[307,805] McLendon *et al.* carried out a quantitative study on a series of 15 cases of Chiari type II malformation and hydrocephalus; all had polygyria macroscopically, a significant increase in sulcal length per unit area of cortex compared with controls and histologically normal lamination.[683] By contrast, more than half of the 25 cases discussed by Gilbert *et al.* showed disordered cortical lamination, with polymicrogyria in 40 per cent, a result that may reflect the particular population examined, comprising infants dying before the age of 2 years.[374]

Conflicting theories abound concerning the pathogenesis of Chiari type II, not surprisingly in view of the multiplicity of associated anomalies and the need for any satisfactory theory to explain the almost constant association of spina bifida. The earliest theories invoked pressure from above[153] in the form of hydrocephalus forcing the hindbrain downwards. Gardner suggested that a primary malformation of the fourth ventricle led to hydrocephalus and, subsequently, rupture of the already closed neural tube,[364] whereas Cameron proposed that the hydrocephalus was secondary to failure of neural tube closure, excessive drainage of CSF and apposition and then fusion of the ependyma, resulting in aqueduct stenosis.[139,140] Against such theories are the absence of hydrocephalus in some fetal cases[66] and in calves with Chiari type II malformations.[154] Traction from below was seen by Lichtenstein as a consequence of cord tethering by the meningocele.[618] Against this hypothesis are the normal alignment of lower cervical and thoracic roots,[55] the S-shape deformity of the lower brain stem and the rare examples of Chiari type II anomaly without dysraphism.[289,803] Developmental arrest of the pontine flexure was proposed by Daniel and Strich[229] and Peach,[804] but the expected disorganization of the topography of the brain stem nuclei does not occur.

A disproportion between the growth of the posterior fossa and its contents has been the focus of many recent theories. Barry *et al.* thought that the cerebellum had overgrown,[55] but others have noted reduced cerebellar weights[21,1046] and also a reduction in volume of the posterior fossa.[119] The smallness of the posterior fossa is the basis of the hypothesis proposed by Marin–Padilla and Marin–Padilla.[664] Using an experimental model induced by vitamin A administration to pregnant hamsters, they have shown that the basichondrocranium is shorter than normal, causing a reduction in size of the posterior fossa. All the neurological anomalies are considered to be secondary to the skeletal defects. Compression of the developing medulla within a small posterior fossa causes the abnormalities of the pontine and cervical flexures, whereas the relatively late growth spurt of the cerebellum means that it is neither pushed nor pulled but required to grow into the spinal canal. This explanation would be consistent with observations in early human fetuses. Evidence from the literature[55,66,289] and personal experience (see Figure 4.23) indicate that in the second trimester the cerebellar herniation is slight, even when the medullary abnormality is extensive. More recently a two-hit hypothesis has emerged:

placed. Other frequently described anomalies are subependymal nodular grey heterotopias in the lateral ventricles and thickening of the massa intermedia. Spina bifida is almost invariably present; exceptions are exceedingly rare.[803] Myelomeningocele is more common than meningocele and typically occurs at lumbar or lumbosacral level. Other associated spinal anomalies include hydromyelia, usually at C8,[648] syringomyelia just below the cervicomedullary junction, diastematomyelia and diplomyelia.[374]

Hydrocephalus is usually present and may be due to obstruction of the aqueduct or at the foramen magnum. Aqueduct atresia, forking and gliosis have been described.[139,645] In Emery's series obstruction was never complete, and the shortening and angulation of the aqueduct suggested a valve-like mechanism.[307] Another hypothesis is that the deformed hindbrain may plug the foramen magnum, and, because the exit foramina are situated within the spinal canal, CSF is prevented from reaching the intracranial subarachnoid space. The resultant hydrocephalus is communicating in type; ascending spinal meningitis may thus that primary congenital anatomic abnormalities allow a normal spinal cord to become secondarily damaged through one or a variety of factors such as amniotic fluid pressure, direct trauma or hydrodynamic pressure. Animal models supporting this concept were studied by Adzick and his colleagues and have led to pioneering attempts at *in utero* surgical repair of myelomeningocele defects in human fetuses. Early results suggest that fetal surgery can reverse hindbrain herniation and reduce shunt-dependent hydrocephalus.[2,109]

Chiari Type III

It comprises an occipitocervical or a high cervical bony defect, with herniation of cerebellum through the bony defect into the encephalocele. Other features may include beaking of the tectum, elongation and kinking of the brain stem and lumbar spina bifida (Figure 4.24). Overall, this is very rare.[153,805]

Disorders of Forebrain Induction

The various disorders of telencephalic evagination, bilateral outgrowth and commissure formation are associated with a number of distinctive clinical syndromes. Their complex terminology, overlapping morphological features and close association with craniofacial anomalies present a particular challenge to the neuropathologist. However, recent advances in genetics and developmental biology have provided important insights into the early embryonic events of forebrain formation and the pathogenesis of holoprosencephaly and associated conditions. Hence, we are beginning to appreciate how the spectrum of forebrain malformations may arise during development, an advance that should provide a more rational basis for the classification and understanding of these conditions in future.

Induction and Early Development of the Forebrain

The forebrain is induced from the anterior-most part of the embryo through a series of tissue interactions during

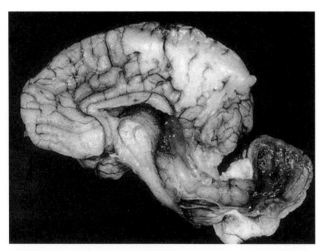

4.24 Chiari type III malformation. Cerebellar tissue is herniated into an occipitocervical encephalocele. Also present in this 5-day-old neonate was a lumbosacral meningomyelocele and a thoracic syrinx.

gastrulation in the third week post fertilization. First, the naive epiblast interacts with the anterior visceral endoderm, an underlying extra-embryonic tissue, which imparts anterior characteristics to the epiblast.[873] A second interaction then occurs with the prechordal plate, a region of underlying midline mesoderm and endoderm that develops as a rostral continuation of the notochord above the dorsal lip of the foregut. Sonic hedgehog (Shh) is the key signalling molecule in this second inductive interaction,[896] most likely through a complex and cooperative intersection with a second signalling molecule, BMP7.[224]

Shh induces the formation of ventral midline (floor plate) cells in the forebrain primordium, which subsequently become Shh-secreting themselves. Ventrally, the influence of Shh is to inhibit expression of the transcription factor Pax6 in the midline. Pax6 becomes restricted to two ventrolateral regions that subsequently develop as the optic primordia.[616] In the absence of Shh, or its downstream signalling function, Pax6 expression is not inhibited in the midline, and a single optic vesicle develops, giving rise to cyclopia.[1015]

Shh also exerts a critical influence on dorsal forebrain development, inhibiting expression of another transcription factor gene, *Emx1*, from the dorsal midline. Absence of *Emx1* is essential for the creation of a dorsal midline invagination, which is necessary for separation of the telencephalic vesicles and formation of the cerebral hemispheres. In the absence of Shh signalling, *Emx1* expression is continuous across the dorsal midline; midline structures such as the hippocampus and choroid plexus fail to form, and the telencephalic vesicles do not separate, yielding holoprosencephaly.[452]

Patterning the Forebrain Along the Anteroposterior Axis

Other signalling centres, apart from the prechordal plate, are essential for normal development and patterning of the forebrain. A diffusible signalling molecule, FGF8, is produced by the anterior tip of the neural plate (the anterior neural ridge) and also by the isthmus, a signalling centre at the boundary between the midbrain and hindbrain.[216] FGF8 appears to induce competence within the neural plate to respond to the Shh signal, thereby ensuring the orderly formation of forebrain and midbrain regions. For example, FGF8 primes the anterior-most neural plate to respond to Shh by expressing Foxg1 (also called brain factor 1, BF1), which is essential for formation of the cerebral hemispheres, as evinced by a lack of these forebrain structures in mice without Foxg1 function.[1102] As with the hindbrain, there is also an important role for homeobox-containing genes in specification of the forebrain and midbrain (see also earlier, Pathogenesis at the Cellular and Tissue Levels, p. 289). Genes belonging to the *Otx* family are required for formation of both forebrain and midbrain: mice lacking *Otx2* function exhibit a deficiency in both structures. The caudal boundary of Otx2 expression appears to define the position of the isthmus.[117] More caudally, the homeobox gene *Gbx2* appears essential for midbrain specification.[571] Hence, the subdivision of the primitive brain neural tube is achieved by signals emanating from the underlying prechordal plate and from within the neural plate itself, which serve to induce region-specific expression of homeobox genes that impose and maintain identity within the antero-posterior axis of the brain.

Later Development of the Forebrain

Concomitant with closure of the neural tube in the pros-encephalic region, the paired optic vesicles begin their outgrowth and, shortly afterwards, around 4–5 weeks, the lateral evagination of the paired cerebral or telen-cephalic vesicles from the single prosencephalic cavity becomes visible. The olfactory vesicles appear as paired outgrowths from the base of the brain at 6 weeks. They are induced to differentiate into olfactory bulbs and tracts by contact with nerve fibres from the olfactory placode, which have grown towards them through a perforant zone in the developing ethmoid bone. The anterior wall of the prosencephalon is relatively inert, so exerting a midline constraint on the rapid growth of the telence-phalic vesicles and resulting in their bilateral outgrowth. Around week 7, the choroid plexus invaginates each lat-eral vesicle, and gradually the broad connections into the central cavity narrow to become the foramina of Monro, whereas the central telencephalic cavity becomes the third ventricle. The commissures arise from the anterior wall of this central structure, the telencephalon medium. It stretches in the midline dorsally from the chiasmatic plate to the paraphysis. Its ventral portion, the lamina ter-minalis, remains thin, becoming the anterior wall of the third ventricle. The cellular and thickened dorsal portion, the lamina reuniens, has two parts.[844] Ventrally, the area precommissuralis will form the septal area and anterior commissure; dorsally, there is an area of complex growth and infolding, which is the anlage of the fornix, hippo-campus, psalterium, corpus callosum and septum pellu-cidum. As early as 9–10 weeks' gestation, the precursor fibres of the anterior commissure meet and decussate in the ventral part of the lamina reuniens, and the bilateral primordia of the fornices appear in the dorsal part of the lamina reuniens and grow dorsally towards the primordia of the hippocampi, which are by now clearly demarcated from the hemispheric isocortex in the medial walls of the hemispheres. By median infolding of the dorsal lamina reuniens a groove develops, the sulcus medianus telecephal-ali medii (SMTM), in the floor of the interhemispheric fissure. The banks of the groove fuse to form the massa commissuralis, into which the fibres of the hippocampal commissure grow at about 10–11 weeks, and dorsal to them the anlage of the corpus callosum at 11–12 weeks, possibly guided by a preformed glial sling.[945] The corpus callosum grows caudally with the growth of the hemi-sphere; beneath it, and between the banks of the SMTM, a pocket appears at around 13–14 weeks. Eventually, the anteroventral extension of the rostrum of the corpus callosum seals this pocket, forming the cavum, whereas its walls and the banks of the SMTM become elongated and thinned into the leaflets of the septum pellucidum by the rapid growth of hemispheres and callosum. The hip-pocampus also extends backwards and rotates with the growth of the hemispheres, then anteroventrally into the temporal lobe, but its rostral part regresses into a vestigial

remnant, the indusium griseum. From this synopsis it will be readily appreciated that a large variety of anomalies may result, depending on the timing and extent of the growth disturbance within the anterior wall of the telen-cephalon, and particularly in the midline.[603]

Holoprosencephaly

Anomalies of prosencephalic outgrowth form a wide spec-trum, from cyclopia to unilateral olfactory agenesis. The nomenclature of this malformation has evolved through holotelencephaly,[1103] telencephalosynapsis[256] to eventually received general acceptance as holoprosencephaly.[259] The last is favoured as it underscores the failure of cleavage of both telencephalon and diencephalon. The term 'arhin-encephaly' still persists as a generic term and in the more restricted sense of absence of olfactory bulbs, tracts and tubercles, for which the term 'olfactory aplasia' seems preferable.[350]

In holoprosencephaly, the degree of failure of hemi-spheric separation is variable, but pathological obser-vations are biased towards the most severe forms, and the more subtle variants are less frequently documented. Nevertheless, De Myer and Zeman subdivided holo-prosencephaly into three types; alobar, semi-lobar and lobar.[259] (For further, more recent, reviews see Cohen and Sulik,[179] Leech and Shuman,[603] Probst[836] and Siebert et al.[943])

Alobar Holoprosencephaly

In alobar holoprosencephaly, the most severe form, the brain is quite small and contains a monoventricular forebrain, a 'holosphere' undivided into hemispheres or lobes and with a bizarre convolutional pattern. The holo-sphere may be spherical or more flattened into a helmet or pancake shape. It can be tilted forwards within the cranial cavity or retroverted.[259,560] Seen from below, the orbital/ventral surface of the holosphere is smooth or shows abnormally wide, rudimentary gyri (Figure 4.25) with an anomalous vascular pattern (see later). The inter-hemispheric fissure and gyri recti are absent, and there is aplasia of the olfactory bulbs, tracts and tubercles and, depending on the state of the eyes, optic nerve hypopla-sia, unilateral aplasia or a midline fusion of the ante-rios visual structures resulting in cyclopia. Viewed from above and behind, the dorsal surface of the holosphere and its convolutions appear roughly horseshoe-shaped. Attached to its distal edge and extending backwards towards the tentorium is a membrane with associated choroid plexus, which partly roofs the single ventricular cavity. This membranous roof, thought to be evaginated tela choroidea,[620,629] often balloons out into a dorsally situated cyst (Figure 4.26), its thin and diaphanous wall easily torn by the prosector to reveal the single ventricu-lar cavity of the holosphere. The floor consists of fused basal ganglia and thalami, and the lateral edges, running in a complete arch beneath the attachment of the roof membrane to the holosphere, is the hippocampal forma-tion (Figure 4.27). Behind the thalami is the opening into

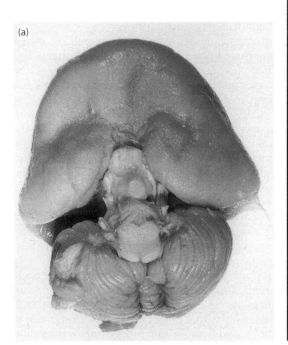

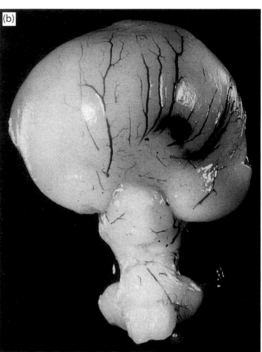

4.25 Alobar holoprosencephaly. (a) Inferior view of the brain of an 18-month-old child with normal chromosomes. Single fused holosphere without interhemispheric fissure and olfactory aplasia. Anterior and middle cerebral arteries are in shallow gutters over the surface. **(b)** 17-week fetus. The horseshoe-shaped holosphere seen from below shows olfactory aplasia and an aberrant vascular pattern.

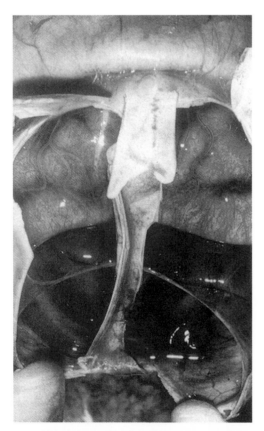

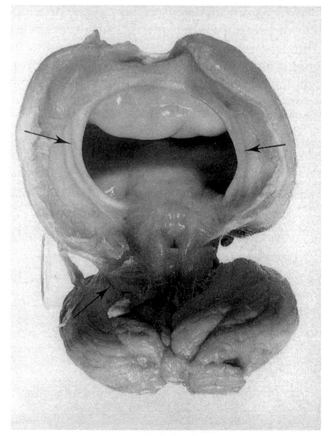

4.26 Holoprosencephaly. The delicate roofing membrane of the single ventricle has been photographed *in situ* within the skull to preserve its integrity. The membrane is continuous with the posterior part of the holosphere. The head is viewed from above.

4.27 Alobar holoprosencephaly (same case as in Figure 4.25) viewed from behind. Rupture of the cystic posterior roof of the holosphere (arrowhead) reveals the single ventricular cavity, midline fusion of basal ganglia and thalamus and the arch-like hippocampal formation running around the ventricular aperture (arrows).

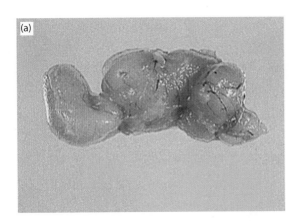

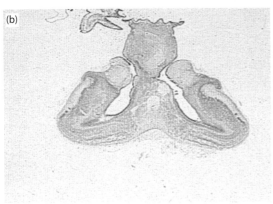

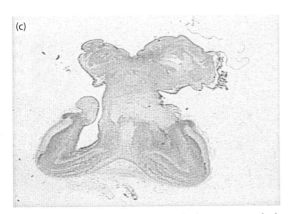

4.28 Extreme microcephaly with holoprosencephaly at 20 weeks of gestation (see text). (a) The 7 g brain viewed from its lateral aspect, anterior to the left. **(b)** Adjacent coronal sections showing bilateral hippocampal formations and temporal horns that communicate, and with a (disrupted) cyst lined by ependyma and choroid plexus **(c)**.

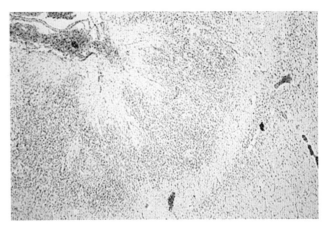

4.29 Four-layer polymicrogyric cortex in alobar holoprosencephaly. A case of trisomy 13. Cresyl violet.

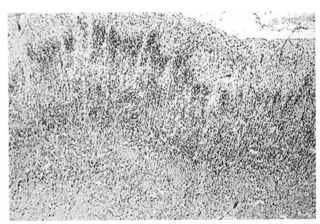

4.30 Holoprosencephaly. Abnormal cortex comprising segmentation of superficial layers and deeply placed paucicellular 'glomeruli'. Cresyl violet.

the aqueduct. There is no corpus callosum or septum, and on sectioning the pallium usually consists of cortex and only a thin layer of white matter. Additional abnormalities may include absent corpora striata, fused mamillary bodies or a single cerebral peduncle, and there are occasional reports of associated glioependymal cysts[892] and aqueductal atresia.[1057]

At first sight, it might appear that the parietal and occipital cortex are missing, but a cytoarchitectonic analysis suggested the opposite, a primary loss of frontal cortex.[1103] This extensive study of serial sections of holoprosencephalic

brains demonstrated a concentric arrangement of cortical zones radiating out from the ventricular cavity: first prepyriform and Ammon's horn, then entorhinal and presubicular cortex, surrounded again by limbic (cingulate) cortex and insula. The motor cortex occupied an anteromedian position in the holosphere, and granular type sensory cortex occupied the ventroposterior borders. Between them the parietotemporal association cortex was represented bilaterally by two small areas on the orbital surface, but the premotor frontal cortex was entirely missing. From this viewpoint, holoprosencephaly is not simply a failure of hemispheric cleavage (independent outgrowth of the two hemispheres) but also a severe hypoplasia of the neopallium. The small size of holoprosencephalic brains is especially notable in fetal cases. An extreme example of holoprosencephalic microcephaly is shown in Figure 4.28: at 20 weeks' gestation, the brain weight was only 7 g (normal: 50 g), the tiny globular forebrain exhibiting a rostral and central mass of heterotopic grey matter, basal polymicrogyria, bilateral hippocampal formations, and a horizontal slit-like ventricle with recognizable temporal horns communicating *via* two lateral apertures with a delicate cyst lined by ependyma and choroid plexus.

Growth failure in holoprosencephaly is further indicated by abnormalities of cortical structure and paralleled by aberrations of the circle of Willis. Although the cortical cytoarchitecture was generally well-preserved in the cases studied by Yakovlev,[1103] not infrequently holospheric cortex shows various forms of disorganized cortical lamination, including polymicrogyria (Figure 4.29), or periventricular and leptomeningeal heterotopia.[515,706,707,836] In six alobar cases, Mizuguchi and Morimatsu demonstrated abnormally thick cortex and disturbed neuronal migration, including segmentation of superficial neurons into irregular groups and deeply placed glomerular structures devoid of nerve cells and composed of thin dendrites and axons: similar findings were also observed by the authors (Figure 4.30).[706] In a detailed analysis of basal arterial patterns, normal circles of Willis were noted in arhinencephaly (olfactory aplasia), contrasting with eight cases of holoprosencephaly in which the anterior part of the circle was lacking whereas the posterior part appeared normal. The anterior and middle cerebral arteries were replaced by anteriorly directed branches from either one or both internal carotid arteries, and very large-calibre choroidal arteries supplied the wall of markedly vascular dorsal cysts.[777] These anomalous patterns appear to reflect the developmental stage reached before normal development ceased and the further functional demands imposed by the later modified growth pattern of the brain.[777]

Semi-lobar Holoprosencephaly

Semi-lobar holoprosencephaly is intermediate in severity, with less reduction in brain weight, rudimentary lobar structure and partial formation of a shallow interhemispheric fissure, usually most preserved occipitally. However, the cortex is broadly continuous across the midline (Figure 4.31). Olfactory structures are usually, but not invariably, absent.[620]

Lobar Holoprosencephaly

Despite almost normal brain size, in lobar holoprosencephaly there is separation of the hemispheres except at the most rostral midline where continuity is found most commonly along the orbital surface or over the corpus callosum (cingulosynapsis).[186,338,742] The state of the olfactory system is variable, the corpus callosum is absent or hypoplastic, and there is often heterotopic grey matter in the roof of the ventricle (Figure 4.32). De Myer and Zeman included in the lobar group patients with olfactory aplasia and callosal agenesis but with completely separated hemispheres.[259] This is terminologically confusing, and such cases are probably best categorized as commissural defects.

Interhemispheric Variant of Holoprosencephaly

This recently described variant most commonly exhibits separation of the cerebral hemispheres at the frontal and occipital poles but has a large grey matter mass replacing the body of the corpus callosum. The ventral forebrain structures are often spared in this condition whereas the pallium is partially or completely replaced by a heterogeneous and completely disorganized neural-glial mass.

Epidemiology and Associated Malformations

Holoprosencephaly is usually sporadic, but familial cases including affected twins have been documented.[13,259,465,550,868]

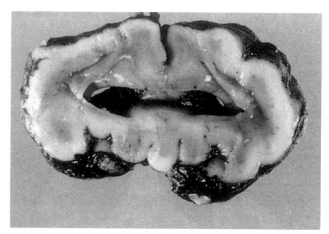

4.31 Semi-lobar holoprosencephaly in a 32-week fetus. There is a suggestion of lobar organization and partial separation of the hemispheres by a shallow interhemispheric fissure, although the orbital surface is completely fused.

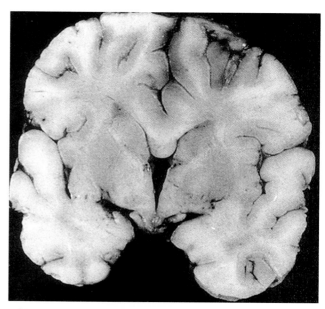

4.32 Lobar holoprosencephaly. The two hemispheres are well formed but the cingulate cortex is continuous across the interhemispheric fissure. Note the heterotopic grey matter in the roof of the ventricle.

Figure courtesy of Professor S Love, Frenchay Hospital, Bristol, UK.

Estimates for incidence vary from 1 in 31 000 births[354] to 1 in 16 000 live births and 1 in 250 abortions.[176] Increasingly, intrauterine diagnosis is made by ultrasound.[1030] Severe forms of holoprosencephaly are often stillborn, whereas surviving infants present with facial dysmorphism, varying degrees of psychomotor retardation, spasticity, apnoeic attacks and disturbance of temperature regulation.

So frequent is the association of certain craniofacial malformations with holoprosencephaly that De Myer *et al.* wrote 'The face predicts the brain',[260] but there are always exceptions.[797] The development of prechordal mesoderm and forebrain are related intimately, and consequently

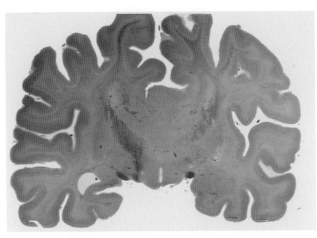

4.33 Interhemispheric variant of holoprosencephaly.

mid-facial hypoplasia is a regular concomitant of holoprosencephaly (Figure 4.34). The most severe anomaly is cyclopia, fusion of the orbits with a single eye or two eyes close together or fused. A small nasal protuberance or proboscis projects above the orbits; there may be absence of the jaw (agnathia) and fusion of the ears under the eye (synotia, otocephaly). Half the reported cases of agnathia show holoprosencephaly and there is also an association with situs inversus.[604,800] Lesser deformities include closely spaced orbits (hypotelorism), microphthalmia, a flattened nose with single nostril (cebocephaly) or a proboscis (ethmocephaly).

Other anomalies include cleft lip or palate, which is usually a true midline cleft and not the more common unilateral cleft, and sometimes trigonocephaly, in which premature closure of the metopic suture produces a narrow pointed forehead, triangular from above. Microfeatures include a single central incisor, a flat phyltrum and absence of the frenulum. Occasional patients even have hypertelorism,[893] although frontonasal dysplasia (median cleft face syndrome), which includes hypertelorism, a broad nasal ridge, median cleft lip, nose or palate and cranium bifidum occultum, is generally associated with normal mental development and not with holoprosencephaly.[258] Usually in alobar or semi-lobar holoprosencephaly the skull base is short and narrow, crista galli and lamina cribrosa are absent, the sella turcica is absent or shallow and there may be varying degrees of hypoplasia of nasal bones (see Figure 4.34). The falx and sagittal sinus are also usually missing. Anomalies of other organs have been described, but these are quite variable and seem to be more prominent in patients with chromosome anomalies: notable are pituitary aplasia and thyroid or adrenal hypoplasia.

Aetiology of Holoprosencephaly

This is heterogeneous. Well-known environmental causes include an increased incidence of holoprosencephaly in the offspring of women with maternal diabetes[54,177,244,536] and a notable association with maternal toxoplasmosis, syphilis and rubella. Fetal alcohol syndrome has also been implicated.[202] In animals, holoprosencephaly can be induced experimentally by a variety of mechanical and chemical

techniques;[179,568,880] perhaps the most well-known teratogenic incident concerned ewes grazing in Idaho, USA, which ingested *Veratrum californicum*[90] (see also earlier, Teratogenic Mechanisms, p. 288).

Chromosomal aberrations, although common, are not invariable; about 50 per cent of cases have a normal karyotype.[704] The most frequent aberration is trisomy 13;[796,797,868,957] holoprosencephaly is present in 70 per cent of individuals with trisomy 13, but it has also been associated with trisomy 18[135,299,868] and triploidy. Several structural anomalies are associated non-randomly with holoprosencephaly, including del(18p), del(7)(q36), dup(3)(p24-pter), del(2)(21) and del(21)(q22.3).[727]

The analysis of families with autosomal recessive, autosomal dominant and X-linked pedigrees[46,176,179,726] has culminated in the identification of several genetic (HPE) loci responsible for particular forms of holoprosencephaly (see Table 4.1). In the years since holoprosencephaly with cyclopia was described in the *Shh*-null mouse,[151] mutations have been identified in a series of *Shh*-related genes among humans with holoprosencephaly. First came the demonstration that the HPE3 locus, already known to map to chromosome 7q36, corresponded to a disruption of the human *SHH* gene.[875] In fact, heterozygosity for *SHH* mutations accounts for a significant proportion of cases of human holoprosencephaly.[741] This contrasts with the situation in mice, where heterozygous *Shh* mutant mice are normal and only homozygotes exhibit the holoprosencephalic phenotype. The reason for the haploinsufficiency of *SHH* mutation in humans is unknown. More recently, mutations in the gene encoding *PTCH*, the receptor for SHH, have been implicated as a rare cause of human holoprosencephaly (HPE7 locus),[703] and mutations in the *GLI2* gene, which is an element of the downstream Shh signalling pathway, have been associated with holoprosencephaly plus pituitary anomalies.[876] It appears that any genetic defect that decreases the strength of Shh signalling is likely to yield a holoprosencephalic phenotype.

Several other genes have been implicated in the aetiology of holoprosencephaly in humans (Table 4.1). The homeobox-containing gene *SIX3* is mutated in HPE2,[1064] a transforming growth factor (TGF)-β repressor is implicated in HPE4,[405] and the zinc-finger-containing transcription factor *ZIC2* is responsible for holoprosencephaly in patients with deletions of 13q32 (HPE5).[124] Mutations in *TDGF1* have also been identified in cases with dysplastic forebrain and midline anomalies.[248] Each of these genes is expressed in the early cranial neural plate and may function in relation to Shh signalling or the related *NODAL* pathway,[452] but precisely how these genes cause holoprosencephaly is not yet clear.

Olfactory Aplasia

Absence of olfactory bulbs, tracts, trigone and anterior perforated substance may occur in isolation and maybe an incidental post-mortem finding unsuspected in life. Anomalies of the orbital frontal convolutions, especially absence of the normal gyrus rectus, are a useful confirmation that the olfactory tract has not been inadvertently avulsed at autopsy. Olfactory aplasia may also

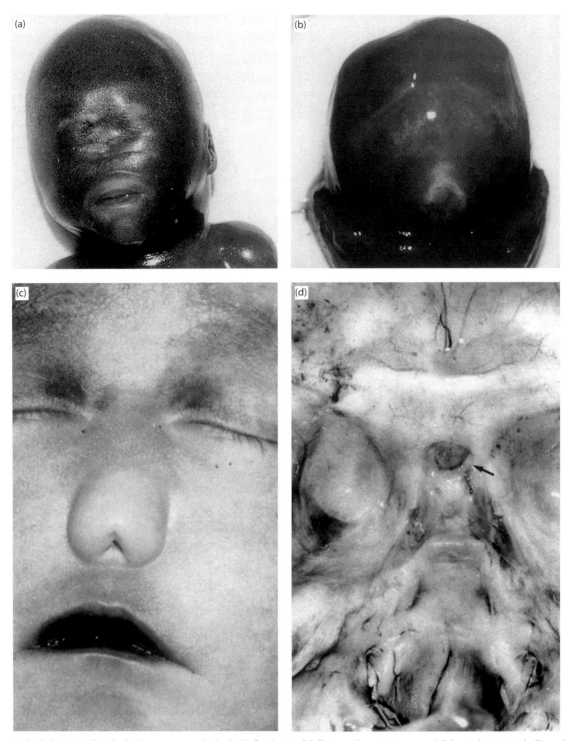

4.34 Craniofacial anomalies in holoprosencephaly. (a,b) Cyclopia. **(a)** Externally no eyes are visible and a central slit replaces the nose. **(b)** Removing the scalp reveals a single midline orbit. **(c)** Hypotelorism and cebocephaly. The base of the skull in this patient **(d)** is without an ethmoid plate. Optic nerves (arrow) are hypoplastic.

be associated with callosal agenesis or septo-optic dysplasia (see Figure 4.41), is present in Kallman syndrome (Table 4.1) and is frequent in Meckel–Gruber syndrome.[780] A neonate with Apert's syndrome showed absent olfactory bulbs and tracts, fusion of the olfactory tubercles and dysplasia of the hippocampus.[650] Bilateral olfactory aplasia is more common than the rare unilateral arhinencephaly (Figure 4.35).[13]

Atelencephaly/Aprosencephaly

First described by Iivanainen *et al.*[498] and Garcia and Duncan,[362] the exact nosological position of this rare and extreme form of microcephaly, which has features in common with both anencephaly and holoprosencephaly, is still in question. There are now about a dozen published clinical or morphological observations,[230,442,553,638,670,942,1018]

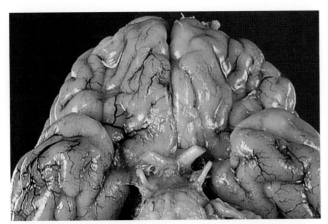

4.35 Unilateral olfactory aplasia. The right olfactory bulb, tract and sulcus are absent and the orbital surface has an anomalous gyral pattern in contrast to the normal left side.

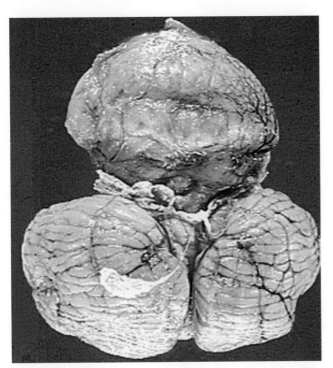

4.36 Atelencephaly. A 9-month-old infant with total brain weight of only 95 g, olfactory aplasia and a tiny, uncleaved, globular forebrain without ventricular cavity.

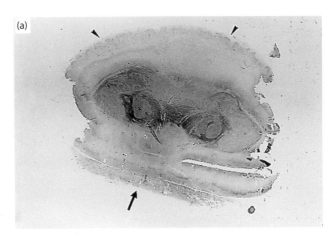

(a)

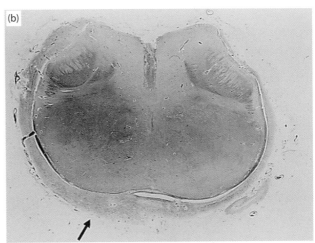

(b)

4.37 Atelencephaly. Low-power microscopy of coronal sections at **(a)** anterior and **(b)** posterior levels through the forebrain. Note the near-symmetry, the thick meningeal cuff of gliomesodermal tissue (arrow) and the thin undulating cortex (arrowhead). Luxol fast blue–cresyl violet.

together with three personally studied cases. Forebrain structures such as the cerebral cortex, basal ganglia and ventricles are virtually absent in atelencephaly, but in contradistinction to anencephaly the calvaria is intact and slopes sharply to a pointed vertex. Brain weights are some of the lowest on record, e.g. 8 g at 35 weeks' gestation and 105 g at 13 months postpartum. In some patients, diencephalic structures, including the eyes, optic nerves, mamillary bodies, hypothalamus and hypophysis, are also affected,[638,670] and in these cases the term 'aprosencephaly' has been deemed appropriate. Facial dysmorphism has similarities with holoprosencephaly, including hypertelorism or hypotelorism and cyclopia, and there may be extracranial anomalies of genitalia and limbs.[638] The tiny globular (Figure 4.36) or multinodular

forebrain remnant comprises disorganized grey and white matter and prominent leptomeningeal gliomesodermal proliferation, with calcifications and even calcific vasculopathy,[553] arguing for a destructive basis for the disorder. Some cases also show a residual bilaterality,[1018] ependymal-lined tubules or spaces,[498,1018] and disorganized or polymicrogyric cortex (Figure 4.37).[362] In one case there were also three intracranial cysts: an ependymal cyst, a pigmented epithelial cyst and a Rathke's cleft cyst.[553] The hindbrain is largely preserved, although cerebellar hypoplasia has been reported.[553] Given the severity of the malformation it is not surprising that lifespan is short (less than 2 years), but seizures have been recorded in several patients despite the lack of forebrain. The aetiology is unknown: all cases so far have been sporadic. In one infant, Towfighi *et al.* demonstrated a chromosomal deletion within 13q and numerous eosinophilic cytoplasmic inclusions in Purkinje cells.[1018]

Agenesis of the Corpus Callosum

Agenesis of the corpus callosum may form part of a more extensive malformation complex, such as holoprosencephaly, or the callosum may be totally or

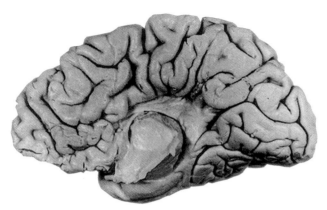

4.38 Agenesis of the corpus callosum. Note the radiating pattern of gyri and sulci.

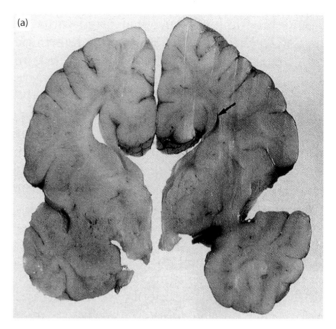

(a)

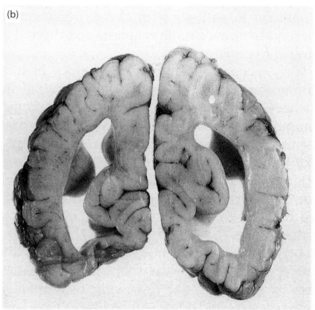

(b)

4.39 Agenesis of the corpus callosum in a 3-month-old child with maple syrup urine disease and a cardiomyopathy. (a) Coronal section showing absence of corpus callosum, midline septum and anterior commissure, the characteristic bat's wing-shaped lateral ventricles and Probst's bundles. **(b)** Markedly dilated occipital horns.

partially absent or hypoplastic in an otherwise normal brain. There are several extensive descriptions and reviews.[257,330,514,515,519,628,629,658,795] In partial agenesis, the posterior portion is usually missing, whereas the rostrum and genu remain. Wherever the callosum is absent, the medial interhemispheric surface has an abnormal gyral pattern. There is no cingulate gyrus and the gyri have a radiating pattern extending perpendicularly to the roof of the third ventricle (Figure 4.38). On coronal sections of the brain no structure seems to separate the lateral ventricles in the midline. The lateral ventricles have a membranous roof and upturned, pointed corners, often compared to a bat's wing. A prominent bundle of fibres, the Probst bundle,[774,837] is usually situated in the lateral part of this roof near to the apex of the ventricle (Figure 4.39a), but occasionally it is more medial. It runs in a longitudinal direction, is myelinated at the time when the corpus callosum should be myelinated, and is thought to include the misdirected callosal fibres, although its volume is considerably less than the normal callosum. Experimental studies in acallosal mice would support this contention.[944] The membranous roof of the third ventricle is often distended and bulges into the interhemispheric fissure, displacing the fornices laterally. The septum pellucidum may appear to be absent, but Loeser *et al.* demonstrated that its widely separated leaves do not run vertically but incline laterally from the fornices to the Probst bundles.[629] The occipital horns are often markedly dilated (Figure 4.39b). Regarding the other commissures, the posterior commissure is always present and the psalterium absent, but the anterior commissure is variable.

Rarely, a local callosal defect is associated with a midline mass: a meningioma, cyst,[748] hamartoma or, more often, lipoma.[624] Interhemispheric lipomata and callosal defects may be closely contiguous, a dorsal lipoma overlying a hypoplastic callosum, wrapping round it or associating with partial agenesis (Figure 4.40), and both are regularly associated with intraventricular choroid plexus lipoma.[1035] It has been suggested that rests of residual meningeal tissue differentiate into adipose tissue and cause mechanical obstruction to the growth of the corpus callosum.[1022,1115]

Patients with agenesis of the corpus callosum have a high incidence of associated anomalies, both cerebral and visceral. Of 11 cases reported by Parrish *et al.*, nine had cerebral malformations whereas eight had malformations

in the rest of the body.[795] These findings were confirmed and extended in a larger series in which hydrocephalus, migration defects and rhinencephalic defects were especially frequent.[515]

Agenesis of the corpus callosum may be sporadic or familial.[736] In addition to the associated anomalies described earlier, there are now several well-defined syndromes in which callosal agenesis is an important feature. In the X-linked dominant Aicardi's syndrome,[10,490,855] it is combined with infantile spasms, chorioretinopathy and depigmented lacunae, mental retardation and vertebral anomalies; polymicrogyria and cerebral heterotopias are also prominent features.[89,329] Menkes *et al.* described an autosomal recessive syndrome in which seizures were a prominent early feature.[692] Andermann *et al.* described an autosomal recessive disorder of callosal agenesis, sensorimotor neuropathy and dysmorphic features in a large kindred in Quebec, Canada.[24] Acrocallosal syndrome,[748,918] which includes polydactyly, macrocephaly and mental retardation, has been found to result from mutation of the *GLI3* gene.[306] Pineda *et al.* described two siblings with callosal

agenesis, hypothermia and apnoeic spells, and reviewed five further sporadic cases.[822] Finally, defects of midline commissures are common in Meckel–Gruber syndrome[780] and hydrolethalus syndrome.[904]

Clinical symptomatology varies and may depend largely on associated malformations. Agenesis of the corpus callosum can be entirely asymptomatic[108] (Figure 4.40), and subtle perceptual deficits may come to light only with sophisticated psychological testing.[393,513] The diagnosis is readily confirmed radiologically[556] or by fetal ultrasound.[412]

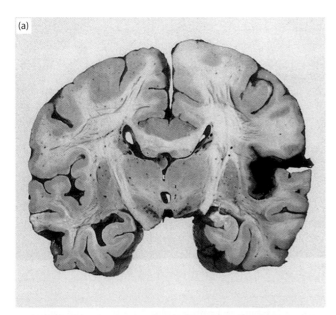

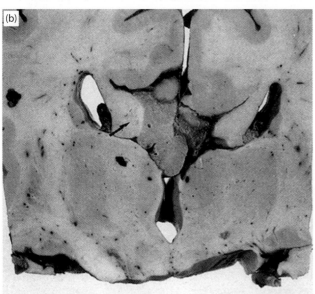

4.40 Partial callosal agenesis associated with a lipoma in an asymptomatic adult. (a) At the anterior thalamic level the thin callosum is closely adherent to the lipoma on both dorsal and ventral surfaces. **(b)** A little further posterior the callosum is completely interrupted by lipomatous tissue and small longitudinal Probst bundles can be discerned (arrow).

Figure courtesy of Dr C Torre, Rome, Italy.

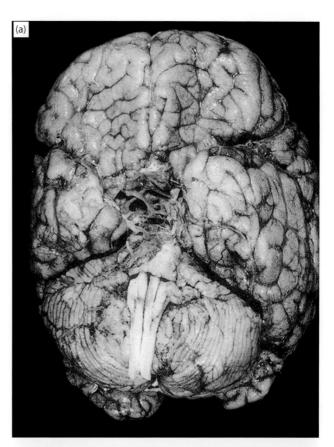

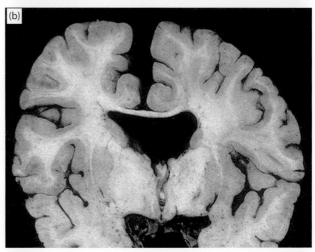

4.41 (a) Base of the brain showing olfactory aplasia and hypoplastic optic nerves and chiasm. **(b)** Coronal section showing absence of the septum pellucidum. The corpus callosum is thin and has a smooth ventricular surface.

Anomalies of the Septum Pellucidum

True primary agenesis of the septum pellucidum must be distinguished from secondary destruction, resulting from inflammation or hydrocephalus, when a fenestrated septum or small fragments of tissue remain attached beneath the corpus callosum. Absence of the septum may be the only abnormality in a brain,[281,350] or it may be associated with callosal agenesis, holoprosencephaly or other complex syndromes. One suggested association, with porencephaly, microgyria and heterotopias,[9,408] does not seem to be determined genetically.[350]

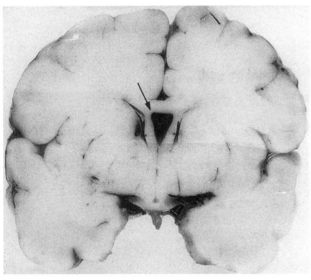

4.42 Cavum septi pellucidi (arrow) in a coronal section of a neonatal brain.

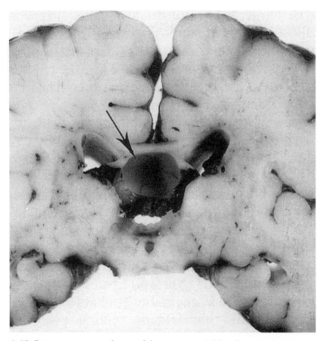

4.43 Cavum vergae (arrow) in a neonatal brain.

Septo–optic Dysplasia

This name was first used by De Morsier in a case report and review of patients with optic hypoplasia and septal aplasia.[255] It is now recognized that hypopituitarism is an equally important component of the syndrome.[491] Judging by the many clinical and radiological reports, the syndrome is not rare, but clinical manifestations are variable and often one part of the clinical triad is missing; in particular, the septum is present in many cases.[971] Neuropathological studies are very few and mostly incomplete. Roessmann *et al.* reported three cases: in all there was optic nerve and lateral geniculate hypoplasia and dysplasia of hypothalamic nuclei, but only one was without a septum.[879] Olfactory aplasia was noted in two of their cases and other findings included posterior pituitary hypoplasia, cerebral heterotopias and cerebellar dysplasias. Similarly, in a personally examined case, in addition to the full triad of septal, optic and hypothalamic involvement, there was widespread polymicrogyria and olfactory aplasia (Figure 4.41). Molecular analysis has identified mutations of the *HESX1* and *SOX2* mutations in patients with septo–optic dysplasia,[234] whereas mutations in *SOX3* have been identified in patients with infundibular hypoplasia and hypopituitarism.[1097] Environmental aetiologies are also possible in view of the report by Coulter *et al.* of septo–optic dysplasia in combination with fused frontal lobes (semi-lobar holoprosencephaly) in a 2.5-month-old girl exposed to maternal binge alcohol abuse during the first trimester.[202]

Cavum Septi Pellucidi and Cavum Vergae

Sometimes misnamed the fifth and sixth ventricles,[227] these midline cavities are bounded above by the corpus callosum and laterally by the two leaves of the septum pellucidum and the fornices. Anteriorly, the cavum septi pellucidi (Figure 4.42) lies between the transverse fibres of the genu of the corpus callosum and the anterior commissure. As the corpus callosum develops caudally, the cavity stretches backwards. Its posterior limit is ill defined and there is considerable individual variation. At times it is prolonged posteriorly by another cavity, the less common cavum vergae (Figure 4.43), which, in neonates, often bulges below the splenium of the corpus callosum. The cavum septi pellucidi and cavum vergae usually communicate freely but may be separated by a bridge of cerebral tissue. The cavum septi pellucidi is a constant feature in the human fetus but tends to become obliterated towards term.[588,930] In adults the cavity is found in about 20 per cent of brains studied at autopsy.[924] The cavum vergae has never been described alone.

The walls of the cavities are lined by neuroglial tissues without a definite epithelial structure. At times, macrophages are found in the walls or within the lumen. Under normal conditions, the cavity does not communicate with the lateral ventricles. In premature infants presenting with germinal layer and intraventricular haemorrhage, the veins of the septum are often overdistended and blood may be found in the cavum.

Cavum septi pellucidi and cavum vergae have been demonstrated *in utero* by ultrasound as early as 16 weeks.[1011] In premature and full-term infants the midline cavities are readily detectable on CT scan and ultrasound.[647]

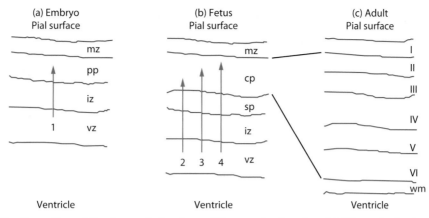

4.44 Main stages in the formation of the layered structure of the cerebral cortex. (a) The earliest migrating post-mitotic neuroblasts (1) leave the ventricular zone (vz), crossing the intermediate zone (iz) to form the preplate (pp). **(b)** The preplate is subsequently split by the settling of later migrating neuroblasts. Successive waves of centrifugally migrating neuroblasts (2–4) establish the 'inside–out' structure of the cortical plate (cp), situated between the subplate (sp), the remnant of the inner aspect of the preplate and the marginal zone. **(c)** Mature layered structure of the cerebral cortex, demonstrating the origin of layers II–VI from the cortical plate, whereas layer I (marginal zone; mz) Cajal–Retzius neurons derive largely from the outer portion of the preplate.

Cavum Veli Interpositi

Although this structure is not a commissural formation but a leptomeningeal cistern of minor importance, it seems appropriate to describe it here because of its relationship with the cavum vergae. This cistern, well known when pneumoencephalograms were commonly used, is normally patent in premature infants and becomes progressively sealed off. Sandwiched between the floor of the cavum vergae above and the roof of the third ventricle below, it communicates posteriorly with the cistern of the great vein of Galen. On X-ray it is difficult to distinguish it from a cavum vergae.[588] Occasionally, haemorrhage has been described in this area, collecting over the roof of the third ventricle and leading to obstructive hydrocephalus.[634]

Neuronal Migration Disorders

The process by which the mitotically active pseudostratified neuroepithelium of the 4-week-old neural tube is transformed into the mature cerebral cortex is a complex morphogenetic process that has been the subject of intense study for decades. The events of neuronal migration have been described in detail anatomically, but recent advances in molecular genetics and developmental biology have added a new dimension to our knowledge of neuronal migration and its pathology.

Development of the Cerebral Cortex

Migration and Differentiation of Neuroblasts

Between 6 and 8 weeks of gestation, post-mitotic neuroblasts begin to migrate across the pallium, so that from 6 weeks a three-layered structure is evident: outer marginal, intermediate and inner ventricular zones. Much of this early neuroblast migration represents a delamination from the pseudo-stratified epithelium lining the ventricle. As the pallial wall thickens, the migration from the pallial progenitor zone converse to being guided by radial glial processes

stretches from the ventricle to the pial surface. A second and important tangential migration of cells originating in the ganglionic eminence contributes most if not all GABAergic interneurons to the cerebral cortex (see earlier, Pathogenesis at the Cellular and Tissue Levels, p. 289).

Preplate and Subplate

The earliest migrating neuroblasts form a transient subpial structure called the primitive plexiform layer or preplate (Figure 4.44a).[663] With progression of neuroblast migration, the preplate splits into two components, an outer portion related closely to the marginal zone and an inner portion designated the subplate (Figure 4.44b). The neurons of both regions ultimately undergo programmed cell death (apoptosis) in large numbers, although some outer preplate cells survive as Cajal–Retzius neurons of the marginal zone. The subpial granular layer is a transient structure that appears in basal allocortex at around 12 weeks, migrates over the cortical surface and disappears by 24 weeks.[126] It may correspond to the outer portion of the preplate. During its brief existence, the subplate appears to subserve several important functions, for instance forming some of the earliest functioning neuronal connections, including the formation of pioneer corticothalamic axons. Moreover, subplate neurons serve as early targets for thalamocortical afferents, participating in the maturation of layer IV.[14]

Cortical Plate

The splitting of the preplate is achieved by the settling of later-migrating neuroblasts that form the definitive cortical plate, the precursor of the six-layered cerebral cortex. The process of cortical plate formation begins around 7 weeks of human embryonic development and continues until approximately 20 weeks. Successive waves of neuroblasts migrate past their predecessors, through an increasingly cellular environment, to take up progressively more superficial positions in the cortical plate (Figure 4.44c). Although most neuroblast proliferation is finished by 16 weeks, and migration is largely completed by 18 weeks (some cells continue to migrate until several months after birth. The inside–out pattern of neuronal migration, in which the earliest generated neurons come to occupy the deepest layer,[28,841] is central to the development of the cortical plate and is frequently disrupted in the diverse range of neuropathological conditions grouped under the general term 'neuronal migration disorders'.

Positioning and Differentiation of Neuronal and Glial Cell Types

Different cortical layers are characterized by neurons of different types, and the layer in which a migrating neuroblast settles appears to be determined even before the parent cell exits the ventricular zone, before the onset of its migration.[679] Deeper layers are primarily derived from progenitor cells in the ventricular zone that correspond to radial glia[510] whereas more superficial layer neurons are primarily derived from a second progenitor zone known as the subventricular zone (SVZ). The SVZ is populated by a second progenitor cell known as the transient amplifying cell. These cells are derived from the radial glial progenitors and undergo one to several divisions prior to generating neuroblasts. How neuroblasts 'know' intrinsically that they are destined to settle in specific cortical layers remains enigmatic but likely has to do with the spatial and temporal expression of a unique set of transcription factors. The fate of neuroblasts, with respect to their pathway of differentiation into neuronal or glial lineages, also appears to be determined while they are still resident in the ventricular zone.[409] Hence, defects involving abnormal cell types present in the cerebral cortex (dysplasia) may arise very early in the process of neuronal migration.

Genetics of Neuronal Migration Disorder

Recent years have witnessed dramatic advances in our knowledge of the genetic causation of neuronal migration disorders, in particular types I and II lissencephaly and neuronal heterotopias (Table 4.1).

Lissencephaly (Type I)

Neuronal migration to the cerebral cortex is disturbed on a large scale in type I lissencephaly, of which several genetic types are recognized. Isolated lissencephaly sequence (ILS) occurs in patients with deletions of the *LIS1* gene, which encodes the non-catalytic subunit of platelet activating factor acetyl hydrolase (PAFAH).[450,631] Miller–Dieker lissencephaly also results from deletion of *LIS1*, but with simultaneous deletion of several adjacent genes, of which the functionally most important is 14-3-3ε.[1020] Hence, Miller–Dieker lissencephaly is an example of a 'contiguous gene deletion syndrome' (see earlier, Inheritance of Mutant Genes). Lissencephaly also occurs in males with mutations in the X-linked gene *doublecortin*, whereas heterozygous females exhibit subcortical band heterotopia, with an apparently normal cortical ribbon and an additional abnormal grey matter zone in the subcortical white matter.[270,379] Because females heterozygous for *doublecortin* mutations are X-inactivation mosaics, it seems likely that the

normal cortical ribbon in such females is formed by neurons expressing the wild-type *doublecortin* allele, whereas cells expressing the mutant allele form the subcortical heterotopia. This strongly suggests that doublecortin functions cell-autonomously during neuronal migration. Recent work has identified the genetic basis of additional type I lissencephalies: an autosomal recessive form resulting from mutations in the *RLN* (reelin) gene,[482] and lissencephaly with ambiguous genitalia, caused by mutations in the *ARX* gene.[559]

Cobblestone Lissencephaly (Type II)

In type II lissencephaly, the brain surface has a cobblestone appearance, as a result of heterotopic neurons in the outermost (marginal) zone. Based on the function of all of the identified genes that have been found to be associated with type II lissencephalys, it appears the pathogenesis of this condition involves a defect in O-glycosylation and specifically of dystroglycan. Five genes are now known to be associated with the spectrum of disorders that all include type II lissencephaly. The first is the Walker–Warburg syndrome, which is associated with mutations in *POMT1* and *POMT2*. Patients with this syndrome also frequently exhibit cerebellar dysplasia, hydrocephaly and occipital encephalocele.[69,1043] Mutations in in *POMGnT1* have been identified in patients with muscle–eye–brain disorder,[1116] whereas the majority of Fukuyama muscular dystrophy (FMD, largely found in Japan), results from a retrotransposon integration into the *FCMD* gene, which encodes the fukutin protein.[563] Most recently mutations in *LARGE* have also been associated with a type II lissencephaly. The gene product from this gene is also required for O-glycosylation.[170]

Periventricular Heterotopia

Heterotopic neurons in a periventricular location are found in patients with mutations of the X-linked gene *filamin-1*.[345] As with *doublecortin*-related lissencephaly, females are X-inactivation mosaics, potentially explaining the coexistence in the brain of normally migrated as well as heterotopic neurons. On the other hand, patients with an autosomal recessive form of periventricular heterotopia plus microcephaly, resulting from mutation of the *ARFGEF2* gene,[933] also exhibit both normally migrating and heterotopic neurons, a phenomenon that is not yet understood.

Molecular and Cellular Mechanisms of Neuronal Migration Disorders

As the genetic basis of neuronal migration disorders has gradually unravelled, so experimental studies in mouse fetuses, brain slice cultures and cell cultures are beginning to shed light on the underlying molecular and cellular mechanisms. Although the evidence is as yet fragmented, it is already possible to identify critical events that are disturbed in the generation of the various types of neuronal migration disorder.[82,448]

Disturbance of Cell Proliferation and Exit from the Ventricular Zone

The ventricular and subventricular zones of the developing brain contain proliferative stem cells that ultimately give rise to the entire complement of neurons and glia. Originally, it was thought that neuronal stem cells represent a fundamentally distinct lineage from coexisting radial glia, cells whose fibrillary processes extend from the ventricular lumen to the pial surface and guide the radial migration of the neuroblasts. However, it is now clear that radial glia are progenitors for both neurons and glia.[142]

Whereas neuronal stem cells reside in the proliferative ventricular zone, their cell processes contact both apical and basal surfaces of the neuroepithelium and their nuclei move up and down the length of the cell, between apical and basal positions, with progression through the cell cycle. This process, called interkinetic nuclear migration, requires microtubular function and is dependent on a protein complex that includes the *LIS1* gene product together with binding protein partners Ndel1 and dynactin. This complex binds to the dynein heavy chain, enabling microtubular function.[939] Interkinetic nuclear migration is necessary not only for neuroblast proliferation but also to allow cells to exit from the ventricular zone and cross the subventricular zone. When *Lis1* function is experimentally diminished within the ventricular zone, interkinetic nuclear migration is disrupted and neuroblasts cross the subventricular zone more slowly or fail to do so.[1023] Hence, one mechanism leading to type I lissencephaly involving loss of LIS1 function disrupting microtubule-dependent nuclear motility.

Failure of Neuronal Migration on Radial Glia

Once neurons have left the ventricular zone and crossed the subventricular zone, a second phase begins in which they migrate along radial glial fibres to the cortical plate. This phase requires molecules, including astrotactin, that mediate the adhesion of neuroblasts to radial glial fibres.[1] Neuroblasts must also exert an intrinsic migratory activity along the radial glial fibre, a process that depends on remodelling of the actin cytoskeleton. Filamin-1, the product of the gene mutated in X-linked periventricular heterotopia, is an actin cross-linking phosphoprotein that interacts with Filamin-B and Filamin-A-interacting protein (FILIP) to regulate the actin reorganization necessary for neuronal migration.[735,932] Other proteins, for example Mig-2 and migfilin, also interact with filamin to link the actin cytoskeleton to the cell surface where interactions with the extracellular matrix occur.[1026] Hence, function of the actin cytoskeleton depends on the assembly of a protein complex that, when disturbed, leads to defective neuronal migration.

In addition to its role in regulating nuclear position during the proliferative cycle in the ventricular zone, the *LIS1* gene product has been implicated in ongoing neuronal migration, for example *via* its interaction with Rho GTPases in regulating cytoskeletal reorganization.[551] Microtubule dynamics also play an important role in radial glial guided migration, not only *via* LIS1 but also through a role for doublecortin, the product of the gene mutated in X-linked lissencephaly. Doublecortin is a microtubule-associated protein whose function depends on phosphorylation by kinases such as cyclin-dependent kinase 5 (Cdk5). Phosphorylation diminishes doublecortin's affinity for microtubule proteins, an effect that is necessary for its participation in neuronal migration.[1002] Interestingly,

gene targeting of *Cdk5* in mice yields disturbance of cortical lamination, similar to that seen in human males with *doublecortin* mutations.[771]

Defects of Neuronal Positioning in the Cerebral Cortex

Large-scale neuronal migration disorders often exhibit disturbed lamination of the cerebral cortex and cerebellum. In mice, this phenotype is exemplified by the brain of the *reeler* homozygote, in which the mutant phenotype involves an apparent reversal of the normal cortical layers. *Reeler* neuroblasts begin their centrifugal migration along radial glial fibres but are unable to pass post-migratory neurons in the deeper cortical layers.[823] The *reeler* mutation affects an extracellular matrix molecule, RELN, which exhibits similarities to molecules involved in cell-matrix adhesion, such as tenascin.[231] Reelin is expressed by Cajal–Retzius neurons of the preplate and marginal zone. In contrast, radial glia do not produce reelin, although they express reelin receptors.[637] It is thought that the *reeler* phenotype results from a defect in adhesion between early post-migratory neurons. Mice with mutations in the *mdab1* gene, and targeted mutations of the genes for VLDL receptor and ApoE receptor 2, all show very similar phenotypes to *reeler*. Molecular studies demonstrate that VLDL receptor and ApoE receptor 2 bind reelin protein, with mdab1 protein acting intracellularly to transduce the reelin signal following its phosphorylation by cytoplasmic kinases.[232,461] Autosomal recessive lissencephaly in humans is caused by mutations in the *RELN* gene, which encodes the human orthologue of reelin,[482] presumably with a similar underlying mechanism to that in the mouse. Important for the neuropathologic analysis, brains with a RELN mutation also exhibit extreme cerebellar hypoplasia.[482]

Overmigration of Neurons

Overmigration of neurons destined for layers II and III, to populate the marginal zone, appears to be a relatively common pathogenic mechanism underlying type II (cobblestone) lissencephaly and isolated leptomeningeal heterotopia. This can arise as a result of a disturbance of the interaction between the end feet of the radial glial fibres, which contact the pial surface of the brain, and the extracellular matrix of the glial limiting membrane (GLM). Marginal zone heterotopias are seen in primary generalized epilepsy,[687] developmental dyslexia[361] and type II lissencephaly. Indeed, abnormalities of the GLM have been described in FMD.[1110] It will be interesting to discover how the glycosyltransferases that are mutated in type II lissencephalies (see Table 4.1) are required for normal termination of neuronal migration at the pial-glial border. Animal models provide further evidence for a role of the GLM in the pathogenesis of marginal zone heterotopias. Breaches of the GLM are present in mice with autoimmune conditions that predispose to marginal zone heterotopias[935] and in mice with targeted mutations of the *Macs* gene, in which leptomeningeal heterotopias are observed.[93] Similar GLM defects occur in the *dreher* mouse, in which mutations of the *Lmx1a* gene lead to a phenotype that includes marginal zone heterotopias. In this case, birth-dating studies have demonstrated that the heterotopic neurons, which are mainly confined to the marginal zone, closely resemble layer II and III cells, pointing to overmigration as a pathogenic mechanism.[191]

Defective Tangential Migration

Although studies of neuronal migration disorders have concentrated largely on mechanisms relating to radial migration, there is accumulating evidence that disturbance of tangential migration from the lateral and medial ganglionic eminence can also be disrupted in cortical lamination defects. In particular, mutation of the X-linked gene *ARX*, a cause of lissencephaly with ambiguous genitalia, appears to be a candidate tangential neuronal migration disorder. The evidence comes from the distribution of GABAergic interneurons in the *Arx* knockout mouse. In affected males, they take up abnormal positions within the cortical plate, whereas radial migration from the subventricular zone appears normal.[559] As the positioning of GABAergic interneurons depends upon tangential migration, this is strong evidence for such a defect in *Arx* mutant mice. Other genes known to regulate tangential migration have been implicated in a variety of neurocognitive disorders such as autism.[896]

Disturbance of Programmed Cell Death

Aberrant attenuation of programmed cell death (apoptosis) has been implicated in the pathogenesis of certain types of neuronal migration disorder, particularly heterotopias.[881,1059] This proposal followed the demonstration that early-migrating preplate neuroblasts die by apoptosis during brain development.[14,833] Support for this idea comes from the finding of heterotopic neuronal masses in the cerebral cortex of mice with a targeted mutation in *caspase-3*, a critical gene in the molecular pathway leading to programmed cell death.[571] Thus, persistence of cells that normally die during brain development can yield neuronal heterotopias. Such apoptosis-related heterotopias may be difficult to distinguish from those arising as a result of disturbed neuronal migration, and it is unclear to what extent failure of programmed cell death should be considered a critical pathogenetic event in the origin of human neuronal migration disorders.

Abnormalities of Cytodifferentiation

The lesions of focal cortical dysplasia show clear morphological signs of disturbed cytodifferentiation, with aberrant expression of markers of neuronal and glial differentiation in dysplastic lesions. It is possible that dysplastic lesions are generated when neuroepithelial cells become arrested at various stages in the progression towards increasing cell specialization. The early stage at which determination of cell lineage appears to occur, for example between neuronal and glial differentiation among cells of the ventricular zone,[409] suggests that dysplastic lesions containing cells with intermediate neuronal/glial characteristics must have a very early origin during prenatal development. It is now accepted, however, that stem cells exist in the postnatal nervous system and retain the ability to differentiate along either neuronal or glial lineages,[681] raising the possibility that dysplastic lesions can arise at almost any stage of development, as a result of the aberrant differentiation of groups of previously multipotential stem cells.

A further possible mechanism of cortical dysplasia is the occurrence of somatic genetic change in otherwise normally differentiated cortical cells. This could account for the finding that cells within dysplastic lesions often re-express genes normally expressed only by cells at an earlier stage of differentiation or in a different cell lineage. Somatic mutation is commonly observed in the generation of various tumours[562] and has also been inferred from the finding of 'loss of heterozygosity' for polymorphic DNA markers in the lesions of tuberous sclerosis.[399] It remains to be determined whether somatic mutation or chromosomal damage, followed by clonal expansion, may play a significant role in the development of dysplastic cortical lesions.

Agyria (Lissencephaly) and Pachygyria

Agyria and pachygyria denote macroscopic abnormalities of the cortical surface associated microscopically with a thickened cortical ribbon. The term 'macrogyria' is used in the neuroradiological literature to denote reduced numbers of widened convolutions[50,412,413,573] but is of little value in neuropathology in view of its non-specificity. Various histological patterns, including polymicrogyria, pachygyria and cortical dysplasia, may contribute to such a macroscopic appearance; despite recent developments in neuroimaging techniques, differentiation between these patterns is still at the limit of resolution, especially in early life.[8] Friede considered 'agyria' preferable to 'lissencephaly',[350] a term introduced by Owen to denote the smooth brains of lower mammalian species.[778] Agyria implies absence of gyri, whereas pachygyria implies reduced numbers of broadened gyri; the difference is one of degree. However, complete absence of gyri is very rare in practice and the use of these terms is subjective. Several reviews and case reports are available.[58,209,225,271,278,516,570,759,801,802,982]

In agyria and pachygyria, the calvaria is usually small, misshapen and thickened, and the brain smaller and lighter than normal, but on rare occasions it may be excessively heavy.[25] The agyric brain (Figure 4.45) is almost completely smooth, with no primary sulci and only poorly defined rolandic and sylvian fissures, and there is a complete failure of opercularization or demarcation of the insula. In pachygyria, there are abnormally few widened convolutions and shallow sulci (Figure 4.46). Occasionally, pachygyria is combined with polymicrogyria (see Figure 4.65, p. 328). On coronal sectioning (Figure 4.47) the cortical ribbon is usually greatly thickened and the underlying white matter markedly reduced. The claustrum and extreme capsule are absent, whereas the lateral ventricles are enlarged (sometimes described as colpocephaly), often with nodules of heterotopic grey matter in the ventricular wall.

The most characteristic histological appearance is a four-layered cortex overlying a thin periventricular rim of white matter in which there may be numerous grey heterotopias. The four-layer cortex comprises a molecular layer, a relatively thin superficial neuron layer and a sparse cellular layer with a tangential myelin fibre plexus, beneath which is a thick neuronal layer, which may break up into bands or plumes of cells descending into the white matter (Figure 4.48).[209] At transitions with normally laminated cortex, only the two superficial layers coalesce with the normal cortical ribbon.[209] A radial alignment of neurons in columns is often evident, but the four-layer arrangement may not always be discernible. Neonates lack the tangential myelin fibres and show a single widened band of grey matter, and more complex horizontal lamination (Figure 4.49) has also

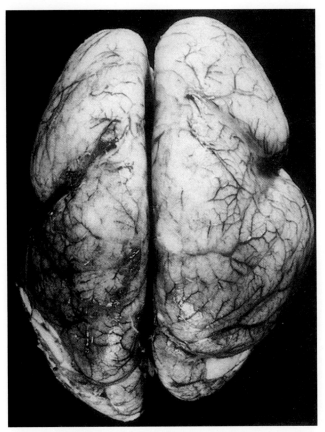

4.45 Agyria in a case of Miller–Dieker syndrome.

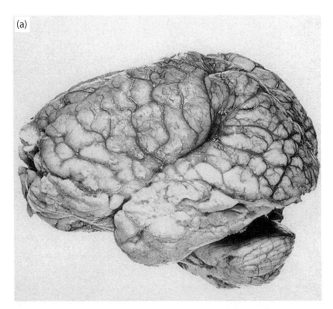

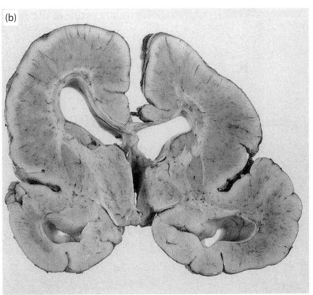

4.46 Agyria/pachygyria. The difference between agyria and pachygyria is one of degree. This brain of an 8-month-old boy weighed 500 g. Although most of the frontal and parietal lobes are smooth **(a)**, a coronal section **(b)** shows shallow cingulate and temporal sulci. Note the great thickness of the cortical ribbon, and the very thin subjacent white matter and corpus callosum.

been described.[516,725] The four-layer types of lissencephaly are found most commonly with mutations in *LIS1* and *DCX*.[343] Three-layer and two-layer forms have also been described, the former associated with *ARX* mutations and the latter without clear genetic correlate.[343] Inferior olivary heterotopia (see Figure 4.119, p. 356) and hypoplasia of the pyramidal tracts are frequent associated findings, whereas dentate dysplasia, cerebellar heterotopia and granule cell ectopia are less common.

Agyria may be sporadic or familial. Several familial syndromes have been described, the best delineated being Miller–Dieker syndrome.[235,271,277,699] Clinical features comprise microcephaly, bitemporal hollowing, small jaw, diminished spontaneous activity, profound mental and motor retardation, feeding problems, early hypotonia, later hypertonia and seizures. More than 90 per cent of patients with Miller–Dieker syndrome have a visible or submicroscopic deletion in a critical 350-kilobase region in chromosome 17p13.3;[601] this is a classic contiguous gene syndrome resulting from the loss of *LIS1*, the first lissencephaly gene identified and the loss of *14-3-3ε*.[852] Smaller deletions in this region effecting on *LIS1* with sporadic occurrence give rise to a more restricted phenotype known as the lissencephaly sequence.[277] A personal case of Miller–Dieker syndrome with the deletion resulting from a ring chromosome 17 is illustrated in Figures 4.45, 4.47 and 4.119.[928]

An X-linked form of lissencephaly occurring in male progeny of mothers exhibiting band or laminar heterotopia has now been mapped to *Xq22*; the gene has been cloned and named *doublecortin* (*DCX*). Morphological study in a

2-year-old boy revealed a smooth brain and an excessively thick cortex, which, microscopically, was partly four-layered and partly without laminations.[75] Simplified discontinuous olives and dysplastic dentate nuclei were also reported. In a comparative imaging study of children with *LIS-1* and *DCX*-lissencephaly, there were opposing gradients of severity, with *LIS-1* associated with more severe malformation posteriorly and *DCX*-lissencephaly malformations worse anteriorly.[279] Further genetically defined syndromes include autosomal recessive lissencephaly with cerebellar and hippocampal hypoplasia caused by *RELN* mutation[482] and X-linked lissencephaly with ambiguous genitalia associated with *ARX* mutation.[559]

Other families with agyria have been reported with slightly different dysmorphology, severe microcephaly[58,759,801] and

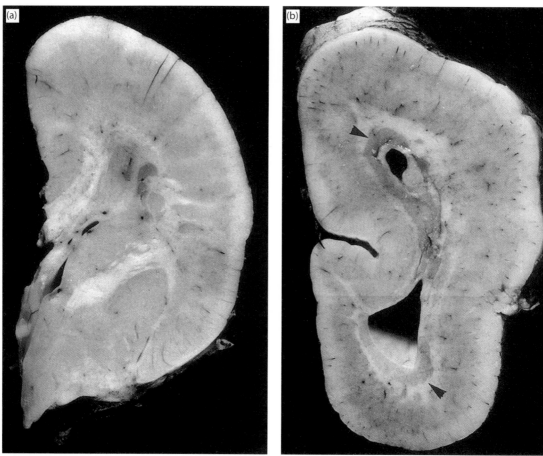

4.47 Agyria. (a) Frontal and **(b)** occipital coronal sections of the brain depicted in Figure 4.45. The cortical ribbon is three times the normal thickness, and in the small residuum of white matter there are numerous grey heterotopias (arrowheads), which are subependymal and almost continuous around the occipital horn.

probable autosomal recessive inheritance. A lethal syndrome combining pachygyria, joint contractures and facial abnormalities was described by Winter *et al.*,[1084] followed by two further similar clinical reports.[1025] Morphological findings are available only for the index case, a 430 g brain of a fetus of 38 weeks' gestation showing extreme brachycephaly and broad, simplified (macrogyric) convolutions, which histologically had an abnormally thick cortex expanded by one or two poorly cellular layers (Figure 4.49).[1084] The olives were not dysplastic. Another familial lissencephaly has been defined by a presentation with cleft palate and extreme cerebellar hypoplasia with brain stem disorganization.[548]

A comparative morphological study has begun to delineate more precisely the genotype–phenotype correlations in classic or type 1 lissencephaly. Forman *et al.* have distinguished at least four distinct subtypes and proposed a new subclassification.[343] The cortex associated with *LIS1* mutation was designated LIS-4LP, that is four-layered with a posterior predominance, in contrast to *DCX* cases with more anterior disturbance (LIS-4LA). A three-layered cortex (LIS-3L) has been linked to *ARX* mutations, whereas a two-layered cortex (LIS-2L) has been observed in other patients without definable genetic disorder.

Cerebro-ocular Dysplasias

Another group of agyric brains is characterized histologically by a quite distinctive form of cerebral cortical dysplasia, first described in detail by Walker[1063] and more recently termed 'lissencephaly type II'.[225] In Europe and the USA, these cases fall into several rare and overlapping familial syndromes, combining complex cerebral and ocular malformations and muscular dystrophy. A subtly different disorder, known as Fukuyama congenital muscular dystrophy,[359] is the next most common form of muscular dystrophy after Duchenne dystrophy in Japan.

Patients with Walker–Warburg syndrome, also known mnemonically as HARD+E syndrome (hydrocephalus, agyria, retinal dysplasia, encephalocele),[782,1066] show profound psychomotor retardation and hydrocephalus from birth and a variety of ocular anomalies and developmental defects; they die in infancy. Walker–Warburg syndrome has been shown to arise, in some cases, from mutation of the *POMT1* and *POMT2* genes, which encode an O-mannosyl transferase that catalyses the first step in the synthesis of O-mannosyl glycans.[69,1043] The cerebro-ocular dysplasia–muscular dystrophy syndrome (COD–MD) has very similar clinical features but adds muscle disease, as shown by a sharply rising creatinine phosphokinase and electromyographic evidence of a myopathy.[1017] Many cases have now been reported,[105,148,369,591,697,967,994,1063,1080] with increasing recognition that these two syndromes may be identical,[278] for muscle disease is always present when it is sought. In a review of over 60 patients, the most

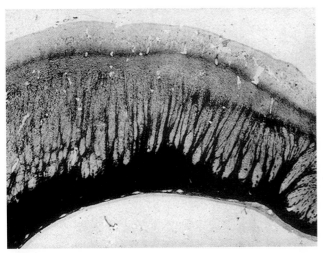

4.48 Agyria. Section through the hemispheric wall showing the plume-like radial fibres traversing the deeper layer of the grey matter. Kultschitsky–Pal method.

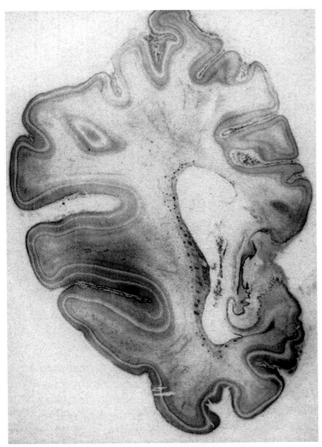

4.49 Pachygyria. Coronal section in a neonatal brain. The thick cortex has a varied laminar pattern with either one or two cell-poor layers. Cresyl violet.

consistent abnormalities were type II lissencephaly, cerebellar malformation, retinal malformation and congenital muscular dystrophy.[278] A further variant with possibly milder phenotype has also been reported as muscle–eye–brain disease.[612–4,907,1036]

In COD–MD, ocular abnormalities affect both anterior and posterior chambers and include central corneal opacity, abnormalities of the iris, cataract, retinal detachment and retinal dysplasia. Occipital meningocele or encephalocele is frequent. Skeletal muscle changes are those of a dystrophy with variability of fibre size, marked endomysial fibrosis, fibre degeneration and regeneration, as well as inflammatory infiltrates in some cases. Usually, the cerebral hemispheres are enlarged, but microcephaly has also been recorded. The brain is nearly always agyric, its smoothness accentuated by adherent thick white leptomeninges (Figure 4.50). There may be fusion of the medial parts of the frontal lobes and absent or hypoplastic olfactory bulbs and tracts (Figure 4.51). Optic nerves and chiasm are usually thin and grey. The cerebellum is small and flattened, its surface coarsely nodular without discernible folia; the vermis is usually partially or completely absent (Figure 4.52). On sectioning, hydrocephalus is marked, even massive, the aqueduct is patent or dilated and the fourth ventricle is enlarged (Figure 4.51). The corpus callosum is thin or not identified.

Histologically, the leptomeninges show a remarkable mesodermal proliferation with extensive glioneuronal heterotopia, obliterating the subarachnoid space (Figure 4.53) and fusing with the poorly cellular superficial layer of the cerebral cortex, the most likely cause of hydrocephalus.[1080] The cortical ribbon varies: in places it is thin and undulating, somewhat like polymicrogyria (Figure 4.54a), notably on the medial surface and where hemispheric fusion occurs, but for the most part it is markedly thickened and disorganized (Figure 4.54b). There appears to be an important direct association between the overlying abnormal leptomeningeal proliferation and the underlying thickened and dysplastic cortex. Squier demonstrated an abrupt transition between normal and thickened meninges in register with the subjacent normal and abnormal cortical plate.[967] From the surface, fibrovascular septa, which are well demonstrated on reticulin preparations,[967] extend inwards and separate the cortical neurons into irregular groups and clusters; areas of greater and lesser cell density can give a wave-like appearance at low magnification (Figure 4.54c). Normal lamination is absent: larger neurons have a tendency to be rather more superficial than expected, but radial alignment can be discerned in some places. Beneath these thicker areas of cortical plate there is usually a narrow, poorly cellular zone and then an archipelago-like arrangement of heterotopic grey matter running parallel with the cortical surface (Figure 4.54b,c; see also Figure 4.57a, p. 324),[369,967] which can give a striking 'double cortical layer' appearance on MR imaging.[1108] These ectopic grey islands are often roughly semicircular with a rounded deeper edge, and a linear upper surface, with tangentially aligned thin-walled blood vessels on its superficial aspect (see Figure 4.57b, p. 324).[369] Several observers have noted migrating neurons in fetal cases streaming through the gaps between these islands where the tangential blood vessels appear absent (for a similar personal observation, see Figure 4.57b, p. 324).[369,967]

In the cerebellum, cortical dysplasia is normally extensive (see Figure 4.122, p. 357), with numerous neuronal heterotopias in the white matter and simplification or fragmentation of the dentate nuclei. The optic nerves are hypoplastic and the lateral geniculate nuclei disorganized and not laminated. The brain stem is hypoplastic and invested

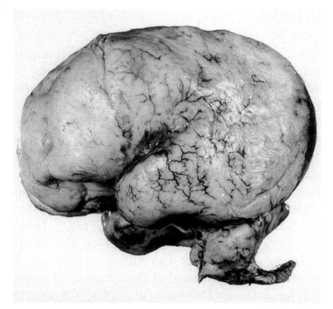

4.50 Cerebro-ocular dysplasia. The smoothness of the agyric hemispheres is accentuated by the thickened leptomeninges. Note the short, shallow sylvian fissure and the small flattened cerebellum.

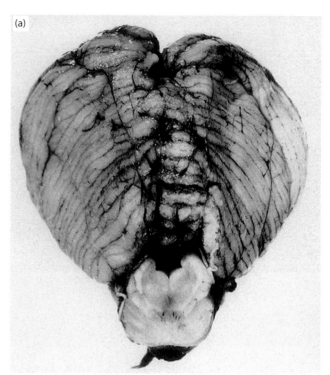

(a)

(b)

4.51 Cerebro-ocular dysplasia. Coronal section at the level of the striatum showing enlarged ventricles, smooth cortical surface and shallow interhemispheric fissure, beneath which the medial frontal lobes are interdigitated and fused. The cortical ribbon is abnormally thick but poorly demarcated from the white matter.

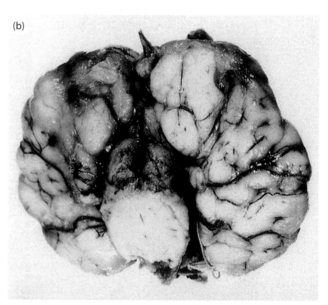

4.52 Cerebro-ocular dysplasia. Superior view of the hindbrain **(b)** compared with normal control **(a)**. The vermis is absent. The hemispheres are flattened and irregular, with a coarse knobbly surface devoid of folia.

by a thick mantle of fibrous and glial tissue, which is particularly prominent over the midbrain tectum (Figure 4.55). Pyramidal tracts are virtually absent: small aberrant bundles are sometimes present laterally in the tegmentum of the midbrain. The inferior olives are usually dysplastic and poorly convoluted.

Analysis of the multiplicity of malformations suggests a prolonged disruptive process active during the second and third trimesters. Early authors favoured a chronic fetal

meningoencephalitis, the infective agent being transmitted transplacentally through consecutive pregnancies,[148,1080] but the frequency of familial cases,[105,278] including affected cousins,[697] has become compelling evidence for autosomal recessive inheritance. Ultrasound evidence of hydrocephalus and encephalocele is specifically sought in siblings of known affected cases, leading to diagnosis in the second trimester; indeed, affected abortuses as young as 18 weeks' gestation show evidence of type II lissencephaly and cerebellar cortical dysplasia (Figure 4.56).[433,967] The availability of such fetal cases has led to several detailed reports

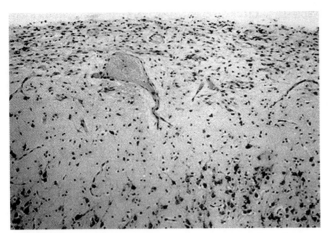

4.53 Cerebro-ocular dysplasia. Microscopic section of superficial cortex. The subarachnoid space is obliterated by diffuse glioneuronal heterotopia.

attempting to analyse the pathogenesis of the cortical disturbance. Miller *et al.*,[697] Squier[967] and Larroche and Nessmann[591] all considered the tangential blood vessels overlying the deeper neuronal heterotopic clusters to be of meningeal origin, and the disorganized cortex to be a consequence of massive overmigration of neuronoglial precursors through a disrupted pial–glial barrier. According to Squier, the blood vessels just above the deeply placed heterotopias marked the pial–glial boundary and these deep clusters became less numerous with increasing gestational age as the more superficial disorganized cortical ribbon became more prominent, a view contrary to personal observations (see Figure 4.54) and those of Gelot *et al.*[369] indicating persistence of prominent deeply placed heterotopias well into postnatal life. For Gelot and colleagues, the poorly cellular layer above the heterotopias that selectively stains with glial fibrillary acidic protein (GFAP) represents subcortical

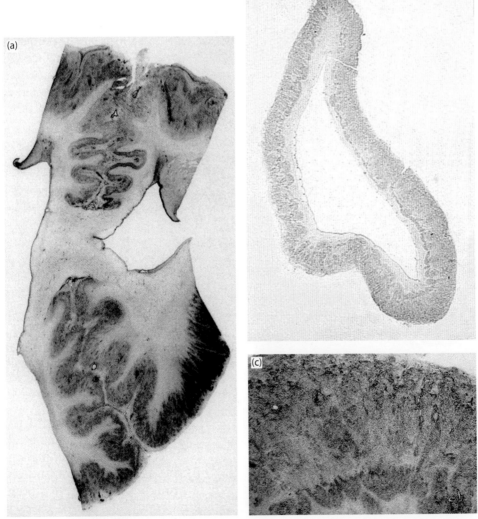

4.54 Cerebro-ocular dysplasia. Same case as in Figures 4.49 and 4.50. Low-power microscopy. **(a)** The cortex of the medial walls of the frontal lobes is thin, undulating and fused, whereas more laterally it is abnormally thick and disorganized. Luxol fast blue–cresyl violet. **(b)** In the occipital lobe, an extensive linear array of heterotopias lies beneath the thickened agyric cortex. Luxol fast blue–cresyl violet. **(c)** Higher power of (b). Varying cell density in lissencephaly II cortex gives a wave-like appearance. Deeply placed semicircular heterotopias are separated by a narrow paucicellular zone from the bulk of the cortex.

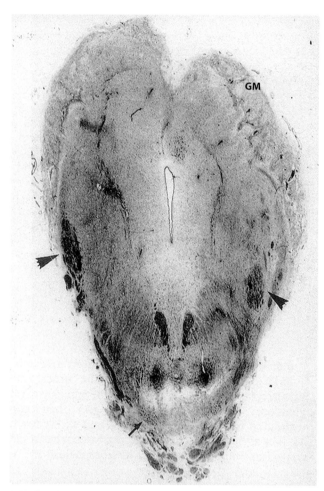

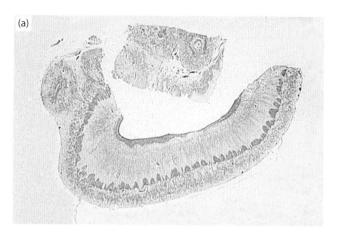

(a)

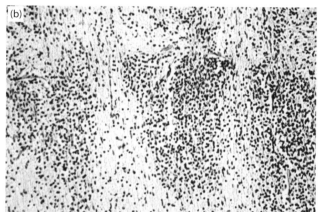

(b)

4.55 Cerebro-ocular dysplasia. Horizontal section of the midbrain, which is severely distorted as well as hypoplastic and surrounded by a thick cuff of gliomesodermal tissue (GM). Cerebral peduncles are absent from their normal position anterior to the substantia nigra (arrow). Ectopic fibre bundles are present dorsolaterally (arrowheads). Dorsal to the aqueduct is a focus of neuroblastic tissue. Luxol fast blue–cresyl violet.

4.57 Lissencephaly type II. (a) Low-power microscopy of the same case as in Figure 4.55. Lissencephalic cortex with prominent deeply placed heterotopias. **(b)** At higher magnification, neuroblasts can be seen streaming through a gap between the deep heterotopias; above these islands are prominent tangential blood vessels.

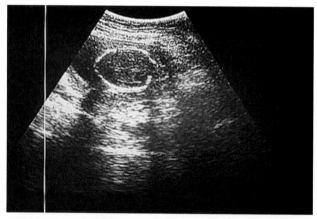

4.56 Lissencephaly type II in a fetal brain of 18 weeks' gestation. Ultrasound showing hydrocephaly and encephalocele.

white matter, and the tangential thin-walled blood vessels just superficial to the heterotopias, although reminiscent of the normal horizontal plexus seen in the deep white matter from mid-gestation, are abnormally large and appear

to be an obstacle to migration (Figure 4.57b). In Gelot and colleagues' view, the pathogenesis of lissencephaly type II is more complex and involves two distinct developmental events: an early disruption of radial migration and hypermigration into the meninges, due to both a disrupted pial–glial barrier and an abnormal deep tangential vascular plexus, is followed by a later disturbance in the growth of the cerebral surface, with meningeal proliferation and fibrovascular invasion of the cortical plate causing progressive fragmentation, burying and fusion of the surface. All these mechanisms in Gelot and colleagues' hypothesis can be ascribed to a primary meningeal abnormality.

In Fukuyama congenital muscular dystrophy,[359,591] cerebral malformations are similar but less severe: there is microcephaly with large areas of unlayered polymicrogyria and smaller areas of type II lissencephaly.[995] The survival is generally longer, and hydrocephalus and ocular abnormalities are less prominent.

Neu–Laxova Syndrome

Neu–Laxova syndrome is a rare lethal autosomal recessive syndrome comprising severe intrauterine growth retardation, microcephaly, characteristic grotesque facies, flexion deformities of the limbs, skin dysplasia and hypoplasia of various organs.[302,597,599,750] Several family studies have reported normal karyotype. The longest survivor was 7 weeks old. Neuropathological changes are prominent but thus far have received relatively limited study. Microcephaly is extreme; total brain weight as low as 10 g was recorded for a 37-week fetus.[724] Most reports mention lissencephaly, agenesis of the corpus callosum and cerebellar hypoplasia. There are two recent brief reports: one describes lissencephalic brains with evidence of abnormal migration and excessive neuronal loss suggested to result from a defect in lipid metabolism,[220] whereas the other reports polymicrogyria and growth failure of cortical neurons in neonates as well as excessive cell death in the ventricular matrix zone in a 14-week fetus.[760]

Polymicrogyria

This term, derived from Bielschowsky,[83] denotes multiple malformed convolutions. The term should not be confused with polygyria (see Figure 4.162, p. 377), i.e. excessive superficial sulcation with normal microscopic architecture associated with hydrocephalus. The surface configuration of polymicrogyric cortex may belie its convolutional complexity. The miniature gyri are fused together or piled one above the other, producing a smooth or irregularly bumpy surface, like cobblestones or morocco leather (Figure 4.58),

covering broad irregular convolutions, which may simulate pachygyria. Coronal sections reveal the heaped-up or submerged gyri, which contribute to a thickening of the cortical ribbon (Figure 4.59).

Polymicrogyria is not rare. Crome identified 27 examples out of 500 severely subnormal patients,[213] and 50 examples were found over a 15-year period in the Neuropathology Department at Great Ormond Street Hospital for Children, London, UK.

The extent of the lesion varies greatly and, with it, the degree of neurological disability. Widespread involvement of both hemispheres usually accompanies microcephaly and profound psychomotor retardation, whereas involvement of the centrosylvian cortex results in hypoplastic pyramidal tracts and spastic diplegia. A developmental form of Foix–Chavany–Marie syndrome, or facio-pharyngo-glossomasticatory diplegia, has been shown to result from anterior opercular or perisylvian polymicrogyria.[64,937] In recent years, several clinical and radiological studies of epileptic patients have documented a congenital bilateral perisylvian syndrome, in which pseudo-bulbar palsy and mental retardation are combined with various types of seizure disorder. Initial MR imaging studies demonstrated a central 'macrogyria', which most investigators now accept to be due to polymicrogyria,[574] a malformation also reported in some autistic individuals.[824]

Polymicrogyria may be bilateral and symmetrical and may correspond to a particular arterial territory, especially the middle cerebral. However, unilateral or markedly asymmetrical lesions are also known and can give rise to hemiplegia and epilepsy necessitating hemispherectomy (see later, Hemi-megalencephaly, p. 333). The cingulate and striate

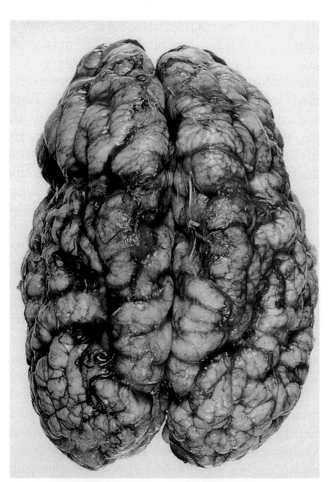

4.58 Polymicrogyria affecting most of the cortex but particularly obvious in the frontal lobe. The broad convolutions have a bumpy or cobblestone-like surface.

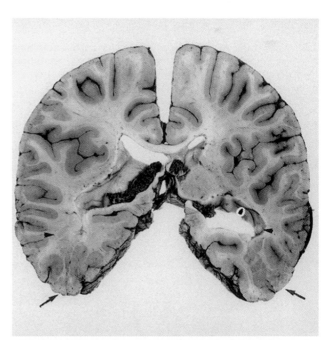

4.59 Bilateral symmetrical polymicrogyria (arrows) in the inferior parts of the temporal lobes. The cortical ribbon appears irregularly thickened because of the piling up of miniconvolutions. Nearby are nodular grey heterotopias (arrowheads), one of which is bulging into the ventricular cavity. The asymmetry of the ventricular system has resulted from surgical shunting for chronic hydrocephalus.

cortices and the hippocampus are spared. Polymicrogyric cortex also often surrounds porencephalic or hydranencephalic defects (Figure 4.60; see also Figure 4.86, p. 337)[245] Small foci of polymicrogyria may be incidental findings in neurologically normal individuals (Figure 4.61) and not infrequently are hidden in the depths of the insula (Figure 4.62). Associated anomalies include neuronal heterotopia in the cerebral and cerebellar white matter (Figure 4.59) and cerebellar cortical dysplasia. On occasion, polymicrogyria and pachygyria may occur together.

Polymicrogyria has a variety of histological patterns, but in essence the cortical ribbon is abnormally thin and laminated, is excessively folded and shows fusion of adjacent gyri. The most common arrangement, unlayered polymicrogyria, is a thin, undulating ribbon composed only of a molecular layer and a neuronal layer without laminar organization.[89,325,509] The molecular layers between adjacent mini-convolutions appear fused, and median blood vessels suggest an apparent line of fusion (Figure 4.63; see also Figure 4.13, p. 298). These structures form cell-sparse

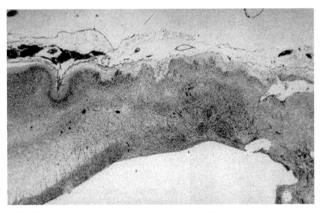

4.60 Early stage of polymicrogyria in the cortex adjoining a hydranencephalic cyst in a fetus of 22 weeks' gestation. Note the abrupt transition from the normal cortex to the thin and undulating polymicrogyric ribbon.

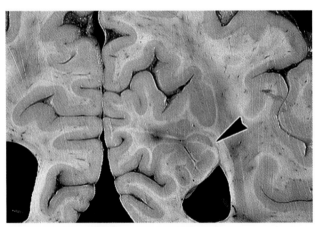

4.61 Incidental findings in a 57-year-old who died from acquired immunodeficiency syndrome (AIDS) were callosal agenesis and polymicrogyria (arrowhead) in the right occipital lobe adjacent to the striate cortex.

Figure courtesy of the Department of Neuropathology, National Hospital for Neurology and Neurosurgery, London, UK.

fingers pointing perpendicularly away from the surface into the grey matter, or several may radiate out like the fingers of a glove from a narrow surface zone. More complex pseudoglandular arrangements may arise as these molecular layer stalks branch and rejoin with adjacent stalks to produce completely submerged gyri or a map-like pattern of irregular neuronal clusters and cell-free areas. The undulating cell ribbon has no defined laminae and neurons are often immature with little cytoplasm. An abnormal tangential layer of myelin fibres may run superficially in the molecular layer and, as with many malformed brains, patches of ectopic glioneuronal tissue may thicken the overlying leptomeninges.

A different pattern, sometimes coexistent with that previously described, is the four-layer polymicrogyric cortex (Figure 4.64).[83,208,845] Many authors have pronounced it the most typical pattern,[317,350] and it has been the subject of much pathogenic discussion, but it is relatively uncommon, being present in only 1 in 10 examples in the collection of the Great Ormond Street Hospital for Children, London, UK. The four layers comprise a molecular layer and two layers of neurons separated by an intermediate layer of few cells and myelinated fibres. This cortex is thinner than normal, but at its usually abrupt transition with normal cortex the outer cell layer may be seen to be continuous with normal laminae II, III and IV, the cell-sparse zone with layer V and the inner layer with lamina VI.[80,250,509,755] This four-layer cortex is also abnormally folded but to a variable degree, in some cases only the superficial cell layer being markedly undulating and overlying a relatively flat deep cell layer; alternatively, the whole thickness of the ribbon shows marked perturbations.

The aetiology of polymicrogyria is heterogeneous: it includes examples of intrauterine ischaemia, a close relation with encephaloclastic lesions and twinning, and intrauterine infection with cytomegalovirus (CMV),[87,212,352,659,666] toxoplasmosis,[250,317] syphilis[317] and varicella zoster[435] (Figure 4.64). However, not only is polymicrogyria sporadic, but it also occurs in a number of familial syndromes and metabolic diseases. Unlayered polymicrogyria and subcortical heterotopias are prominent in Aicardi's syndrome, an X-linked dominant disorder.[10] Other familial

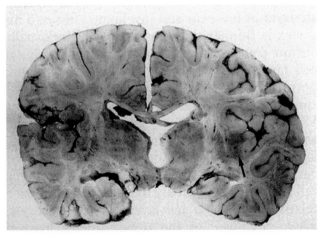

4.62 Polymicrogyria hidden beneath the insular cortex (arrow).

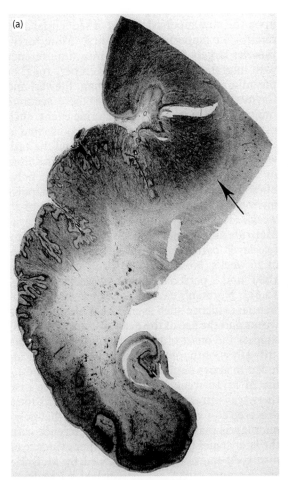

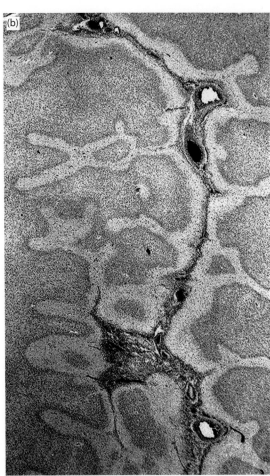

4.63 Unlayered polymicrogyria. (a) Irregularly branching fingers of molecular layer point perpendicularly away from the surface into the cortical grey matter, which can appear thickened by the piling-up of completely submerged convolutions (arrow). **(b)** Complex branching of the aberrant molecular layer, with median blood vessels suggesting fusion of adjacent convolutions.

occurrences of polymicrogyria, relatively few in number, are probably autosomal recessive. Indeed, it has been reported that frontoparietal polymicrogyria can result from recessive mutation of the *GPR56* gene, which encodes a G-coupled receptor; however, further studies suggest this is a unique malformation more closely related to bilateral lobar type II lissencephaly.[817] There is a single report of dermatomyositis and polymicrogyria in two sisters with mental retardation,[241] and personal experience includes three sibling pairs with polymicrogyria, two showing prominent intracranial calcification.[48,134,850]

The occurrence of polymicrogyria has been reported in Pelizaeus–Merzbacher disease,[768,786] glutaric acidaemia type II,[96,181] maple syrup urine disease[669] and histidinaemia,[198] in association with Leigh's syndrome and mitochondrial respiratory chain deficiency,[906] and in two peroxisomal diseases, Zellweger's cerebrohepatorenal syndrome and neonatal adrenoleukodystrophy. Whereas in neonatal adrenoleukodystrophy several authors have described polymicrogyria and subcortical heterotopias,[507,830,1033] Zellweger's syndrome regularly shows a more complex malformation involving pachygyria and polymicrogyria (Figure 4.65) and grey matter heterotopias,[5,316,1060] as well as olivary dysplasia (see Figure 4.129). Cytoarchitectonic studies suggest a partial disturbance of migration of multiple neuronal classes throughout the greater part of the migratory epoch.[316] Disturbed

migration was apparent in fetuses of 14 and 22 weeks' gestation,[832] the abnormality being limited to a wide region of the lateral parietal, frontal and temporal convexity.

The pathogenesis and time of occurrence of polymicrogyria have excited considerable debate and a diversity of causation has been proposed. Bielschowsky favoured an arrest of migration[83] but his hypothesis fails because it is based upon the now superseded idea of sequential migration, i.e. superficial layers first. More recent theories fall largely into two groups: interference with migration[676] or post-migrational necrosis.[861] Only rarely is there accurate timing of a catastrophic intrauterine event, such as maternal splenic rupture at 16 weeks[175] or two incidents of maternal coal gas poisoning in the fifth month[419] and at 24 weeks,[44] which resulted in large cerebral defects fringed by polymicrogyria. However, early stages of four-layer polymicrogyria have been described in fetuses of 27 weeks,[676,861] and unlayered polymicrogyria has been observed at 26 weeks' gestation,[756] both twin fetuses of 25 weeks[104] and 24 weeks,[350] and in a personal case of 22 weeks (see Figure 4.60). Norman's case was one of twins;[756] circumstantial evidence of maternal bleeding at 12 weeks and the age of the dead co-twin (17 weeks) suggest the onset of polymicrogyria to be in the third to fourth month, supported by a similar personal case. This timing suggests that polymicrogyria may result from an interference with the later stages of

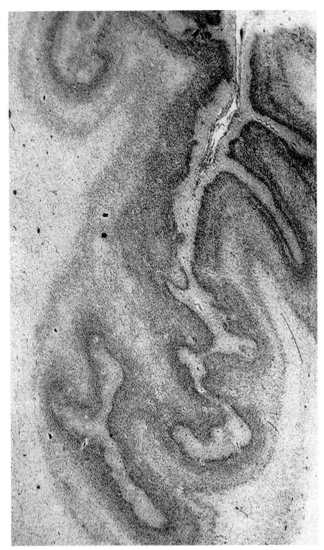

4.64 Four-layered polymicrogyria. Varicella zoster embryopathy. Luxol fast blue–cresyl violet.

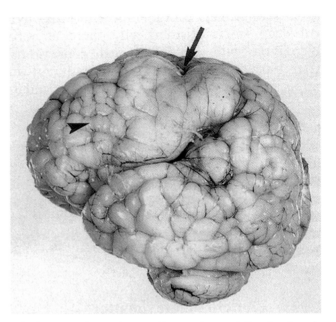

4.65 Zellweger's syndrome. Abnormal convolutional pattern in a 1-month-old child. Note areas of both polymicrogyria (arrowhead) and pachygyria (arrow).

migration, a view supported by an experimental model of four-layer polymicrogyria in rats produced by contact freezing of the cortical surface.[295,296] Golgi studies by Ferrer and co-workers of unlayered polymicrogyrias associated with porencephaly and in Aicardi's syndrome are consistent with this concept.[325,326,329] Although cell bodies and dendrites were abnormally oriented, the different neuronal types were situated at their appropriate cortical levels, suggesting that a partial necrosis of the cortical plate had occurred just before the end of the fifth month. However, another Golgi study in Aicardi's syndrome gave somewhat divergent findings. Billette de Villemeur *et al.* also observed a disorientated dendritic tree with apical dendrites bent tangential to the surface, towards and across the perpendicular fingers of fused molecular layers; in their three cases, however, each neuronal type, including the large pyramidal cells, was present through the whole thickness of the cortex, suggesting a severe disturbance of the migratory process.[89] The added presence of numerous nodular subcortical and periventricular heterotopias was further evidence for interference with migration.

An opposing and widely held hypothesis is that polymicrogyria is the result of a post-migrational destructive event.[245,250,509,642] Using cytoarchitectonic analysis and Golgi preparations, Caviness and colleagues demonstrated that, within the four-layer cortex, cells of layers II, III, IV and VI are present in their normal post-migrational positions, whereas only occasional layer V pyramidal cells can be found in the cell-sparse third layer.[609,861,1079] This organization implies that migration has been completed and that polymicrogyria is the result of post-migrational midcortical destruction. This hypothesis is also supported by studies using laminar makers indicating laminae in polymicrogyria are normally arranged from inside to outside but a loss neurons in most layers was observed. The authors of the later study suggested a primary defect in the glial-pial limitans with secondary neuronal defects in the postmigrational organization of the cortex.[530]

The laminar nature and topographical arrangement of the lesions, the frequency of bilateral symmetry, with middle cerebral artery distribution or juxtaposition to porencephaly, as well as historical data, suggest a hypoxic–ischaemic pathogenesis or transient intrauterine perfusion failure. As an adjunct to this hypothesis, Richman *et al.* developed a mechanical model of convolutional folding that predicts the excessive buckling of the polymicrogyric cortex.[862] Most recently mutations in a number of genes (*PAX6*, *TBR2*, *KIAA1279*, *RAB3GAP1* and *COL18A1*) have suggested a clear genetic pathogenesis to many cases, although the neurobiology underlying the development of polymicrogyria in these cases remains uncertain.[382,411]

Chondrodysplasias

Abnormal gyral patterns including polymicrogyria are a feature of certain forms of chondrodysplasia. Thanatophoric dwarfism,[665] a lethal congenital chondrodysplasia, is characterized by micromelia, narrow thorax and a large

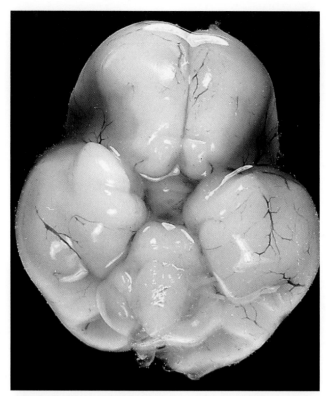

4.66 Thanatophoric dysplasia in a 20-week fetus. Note the abnormally large and hyperconvoluted temporal lobes.

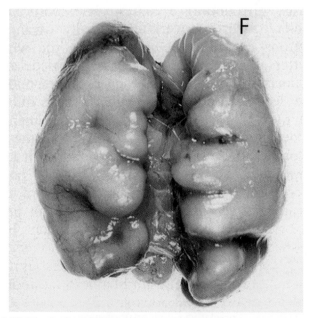

4.67 Short rib polydactyly syndrome in an 18-week fetus. The cerebral hemispheres, viewed from above, have lost their normal smooth contours and are deeply indented by irregular clefts. F, frontal.

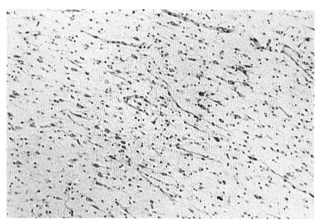

4.68 Diffuse neuronal heterotopia in a 2-month-old child with congenital nephrotic syndrome. Large numbers of mature neurons are scattered through the white matter. Luxol fast blue–cresyl violet.

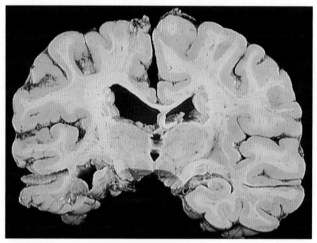

4.69 Neuronal heterotopia. Nodular subependymal heterotopias bulging into the frontal horns. A 51-year-old female with a 30-year history of epilepsy.

Figure courtesy of the Department of Neuropathology, National Hospital for Neurology and Neurosurgery, London, UK.

head with or without trilobed skull (clover-leaf) anomaly. Diagnostic neurological findings include flattened vertebral bodies, widened intervertebral disc spaces, thinning of the metaphyses of long bones and 'telephone receiver' femora. Neuropathological findings have been described by a number of authors.[201,396,474,485,1092] Principal features include megalencephaly, sometimes with mild hydrocephalus, posterior fossa hypoplasia and, in particular, convolutional anomalies and cortical dysplasia in the temporal lobes. The last are abnormally protuberant with broad gyri and deep sulci frequently oriented as radial spokes from the midbrain out along the ventral aspect of the temporal lobes (Figure 4.66). Histologically, there is polymicrogyria with leptomeningeal glioneuronal heterotopia and complete disorganization of Ammon's horns. Dysplastic thalamic and caudate nuclei and hyperconvoluted dentate and olivary nuclei are other findings. Mutations in *fibroblast growth factor receptor-3 (FGFR3)* have been identified in patients with thanatophoric dysplasia types I and II.[1006]

A still more bizarre convolutional pattern occurs in some cases of short rib polydactyly syndrome. In a personal case, the cerebral hemispheres of an 18-week fetus were distorted by numerous deep irregular clefts (Figure 4.67), the cerebral mantle was severely disorganized and there was hippocampal dysplasia and olivary heterotopia.

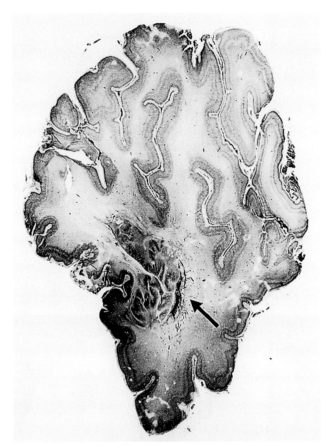

4.70 Within the fused frontal lobes of a 1-week-old child with a huge occipital encephalocele and arhinencephaly there is a large heterotopia forming a serpiginous grey band (arrow). Luxol fast blue–cresyl violet.

Neuronal Heterotopias within the Cerebral White Matter

These take three separate forms, classified descriptively into diffuse, nodular and laminar, which may occur separately or together, and both with and without other cerebral malformations. Published clinicopathological data have until recently been sparse, for these lesions are associated with a low mortality, but modern non-invasive investigative techniques for epilepsy are beginning to redress this situation.

Diffuse Neuronal Heterotopia

Ectopic neurons scattered haphazardly through the gyral and central white matter require cautious interpretation, particularly in early life when modest numbers are a normal occurrence, especially just beneath the cortex.[350] Slightly excessive numbers (compared with controls) have been reported in epileptic subjects and termed microdysgenesis (see later). Obviously, excessive numbers of neurons scattered diffusely through the cerebral white matter are occasionally associated with nodular heterotopias and other cerebral malformations. There are also rare reports of diffuse neuronal heterotopia as the principal finding in infants with early myoclonic epilepsy. In Spreafico's cases,[966] many of these ectopic cells were large fusiform spiny neurons, immunohistochemically negative to antiglutamic acid decarboxylase antibody, and were interpreted

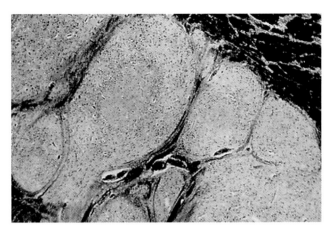

4.71 Neuronal heterotopia. Nodular grey heterotopias showing a simple radial organization of neurons around a central cell-poor zone. Luxol fast blue–cresyl violet.

as abnormally persistent interstitial or subplate neurons,[163] suggesting a failure of programmed cell death. Other pathological features were megalencephaly and olivary dysplasia. A similar situation is illustrated in Figure 4.68.

Nodular Heterotopia

Nodular masses of ectopic grey matter are found in a wide range of pathological and clinical situations, and their occurrence in epileptic subjects is increasingly recognized.[847] Heterotopic nodules can be sited deep within the centrum semiovale or within gyral cores, but more usually they are found close to the wall of the ventricle, often bulging into its cavity (Figures 4.59 and 4.69). They may occur singly or in large groups and vary in size from small discrete neuronal clusters to large multinodular conglomerates or extensive irregularly serpiginous bands (Figure 4.70). The histological composition of nodular heterotopias is quite variable. Some are simple collections of neurons with apparently random orientation and size distribution, whereas others exhibit a patterning suggestive of cortical lamination,[89,434] for example a bullseye appearance of a central, poorly cellular molecular zone surrounded by concentric rings of smaller and then larger nerve cells (Figure 4.71). Some functional aspects of nodular heterotopias in children were addressed in a study by Hannan et al.[425] Calretinin- and neuropeptide Y-positive interneurons were present in the nodules and were clustered abnormally in the overlying cortex, whereas fibre tracing using dioctadecyl-tetramethylindocarbacyanine perchlorate (DiI) demonstrated fibres exiting the nodules and one example of connectivity between nodules.

Directly correlative clinicopathological observations regarding the occurrence, detailed structure and topography of nodular heterotopia are rare. A retrospective review of autopsy material in the Department of Neuropathology, Great Ormond Street Hospital for Children, London, UK, covering a 15-year period (1975–90), retrieved 25 autopsy reports of nodular heterotopia in fetuses and children aged from 18 weeks' gestation to 14 years of age.[434] Surveying the available clinical data, there is a 2:1 excess of both females and epileptic patients. Clinicopathological diagnoses included (i) aetiologically specific peroxisomal, mitochondrial and chromosomal disorders, such as Zellweger,

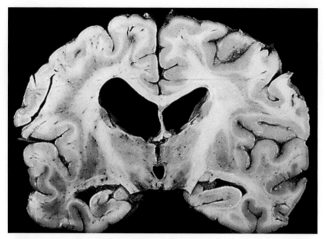

4.72 Laminar heterotopia. Coronal slice at midthalamic level in a 37-year-old female with a 27-year history of epilepsy. Bilateral, almost symmetrical bands of ectopic grey matter extend widely through the frontal and temporal lobes, sparing medial temporal areas. A well-defined zone of white matter separates the heterotopia from the overlying macroscopically normal cortex.

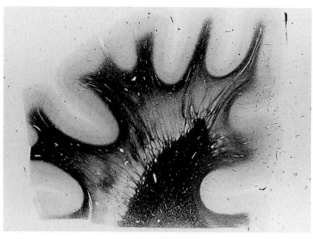

4.73 Laminar heterotopia. Low-power microscopic section from the frontal lobe. There is a tendency for the ectopia to fragment into islands in its deepest part. Luxol fast blue–cresyl violet.

Miller–Dieker and cerebral lactic acidosis due to pyruvate dehydrogenase (PDH) deficiency; (ii) well-characterized dysmorphic syndromes, for example Meckel–Gruber syndrome and septo-optic dysplasia; and (iii) known associations, such as with familial nephrotic syndrome.[783] Some patients had no significant preterminal neurological signs, their heterotopias being apparently incidental and unconnected with the terminal illness. The majority exhibited extensive CNS malformations, whereas in some the only additional abnormality was microcephaly and in two cases nodular heterotopia was the only malformation present. Microcephaly was significant in 14 cases and megalencephaly in 3 cases.

The female predominance is even more striking in clinical reports linking subependymal heterotopia with epilepsy (Figure 4.69). Raymond and colleagues reported 13 patients, 12 of them female with normal developmental milestones and normal intelligence and onset of epilepsy predominantly in the second decade.[846] The subependymal heterotopias were single nodules or coalesced into lumpy bands around the ventricle: trigones and occipital horns were involved most frequently. X-linked dominant inheritance with prenatal lethality in hemizygous males has been suggested in familial pedigrees of subependymal periventricular heterotopia.[452,539] The locus for this disorder has been mapped to Xq28 in females presenting only with epilepsy,[303] whereas a duplication of Xq28 has been demonstrated in a boy with a more severe phenotype including mental retardation, epilepsy, syndactyly, cerebellar hypoplasia and bilateral periventricular heterotopia.[331] The defective gene is *Filamin 1 (or FLNA)*, which links cell surface receptors with actin, and, thus, is important for cytoskeletal movement and thus cell migration.[345]

This is entirely consistent with earlier hypotheses that heterotopia probably results from a disruption of neuroblast migration. Because they contain neurons of various types (i.e. cells of different 'birthdays') when examined conventionally or with Golgi impregnation,[89,329] there must be a fundamental fault in the migratory process or an early focal insult to the germinal zone. Extrinsic insults associated with

neuronal ectopia include sustained maternal hyperthermia in the first trimester,[825] fetal exposure to methyl mercury poisoning,[157] and the atomic bomb at Nagasaki in a survivor who received an estimated dose of 1.57 Gy in the third week of gestation.[749] Experimental induction in rats has been achieved using X-irradiation.[328,864]

Laminar Heterotopia

Laminar or band heterotopia consists of extensive plates or bands of grey matter situated between cortex and ventricle but well separated from both. Only a very few neuropathological reports are available, in archival German and Italian literature,[508,673,1053,1074] but modern MR imaging investigations of epileptic patients have discovered a 'double cortex syndrome',[51,627,784,785,860] for which the genetic basis has been established in most cases.

Personal observations are of three autopsy cases, all female: epilepsy began in the second decade, intellectual deterioration varied from minimal to severe and, in two cases, there was a family history of siblings with epilepsy. The remarkably consistent findings were a normal surface convolutional pattern and, on coronal slices, bilateral, roughly symmetrical and extensive bands of heterotopic grey matter, involving most cortical regions (Figure 4.72) but sparing the striate (Figure 4.73), cingulate, fusiform and medial temporal gyri. The heterotopic bands were situated just beneath, running parallel with the cortex, but separated from it by a narrow but well-defined layer of white matter. They ranged in shape from a thin strip, through archipelago-like clusters to thick, wedge-like sheets. The overlying cortex and deep grey nuclei appeared normal, except for the claustrum, which was incorporated into the heterotopia.

Histologically, the cerebral cortex in these three cases appeared qualitatively normal, but stereological measurements in one indicated an excessively thick cortex of increased neuronal number.[432] This cortex overlay a variable but substantial zone of well-myelinated white matter including U-fibres, whereas the heterotopia emerged gradually from beneath the deepest part of this white matter. Neurons were arranged haphazardly in the outer,

4.74 Laminar heterotopia. There is a suggestion of columnar organization in the deeper parts of the ectopic grey matter. Bielschowsky's silver impregnation.

superficial zone of the heterotopia, but gave the impression of a columnar organization in the deeper part (Figure 4.74), emphasized by thin, radially arranged bundles of myelinated fibres running through it. Many of the nerve cells were pyramidal but generally smaller and more loosely scattered than those in the true cortex, and some appeared upside down. Closer to the ventricle, the heterotopia in some areas fragmented into islands separated by thicker bands of white matter (see Figure 4.73). Between the heterotopias and the lateral ventricle there was yet another broad zone of white matter. These observations are broadly in agreement with previous reports.[508,673,1053,1074] However, the cortex overlying the heterotopia is in some cases normal and in others pachygyric.[508,1074]

All autopsy reports and most MR imaging studies of laminar heterotopias are in female patients. Families are recorded where mothers and daughters present with laminar heterotopia whereas sons show lissencephaly.[821] This syndrome has been mapped to Xq22.3–Xq23,[270,379] and this mutation has also been demonstrated in a sporadic occurrence in a boy with laminar heterotopia.[820] The gene encodes doublecortin, the same as was described earlier in males with X-linked lissencephaly.[379] Cases have also been observed in patients with mutations in *LIS1*, these having an autosomal dominant inheritance pattern.

Microdysgenesis

Subtle structural abnormalities of cortical architecture have been highlighted by several authors studying autopsy material [688,689,1047] and temporal lobe specimens[432] from epileptic subjects. The suggestion that these abnormalities may have functional significance in seizure genesis[688] has proved controversial, because some of the abnormalities are present in normal, non-epileptic brains.[641] However, large-scale observations with numerous controls have provided some evidence for the pathological relevance of these morphological changes.[689] The precise definitions of microdysgenesis vary. Meencke and Veith's comprehensive description includes neuronal ectopias in the molecular layer and subcortical white matter, undulations of the second layer and architectural disturbances in deeper cortical layers: these

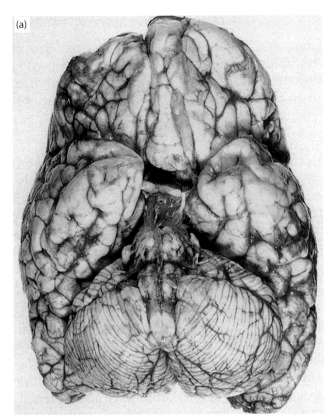

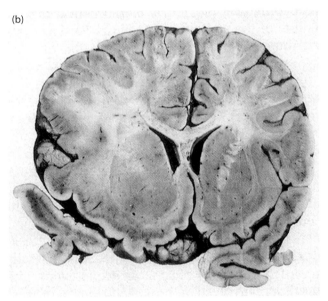

4.75 Hemi-megalencephaly with cortical dysplasia. (a) Inferior view of the brain, showing left-sided asymmetrical enlargement of the frontal lobe and olfactory tract. **(b)** Coronal section showing expanded convolutions with loss of demarcation between grey and white matter on the left side. The striatum is also greatly expanded and its white matter poorly defined.

were particularly prominent in subjects with primary generalized epilepsy and were also present in Lennox–Gastaut syndrome.[689] Hardimann *et al.*, in a study of 50 lobectomy specimens resected for intractable temporal lobe epilepsy and 33 carefully age- and sex-matched controls, defined microdysgenesis in terms of (i) severe neuronal ectopia in the subcortical white matter (more than eight neurons per 2 mm^2) or (ii) neuronal clusters (three or more cells) neighbouring bare areas without nerve cells.[431] Present in 42 per cent and 28 per cent of the epilepsy specimens, respectively, but not in controls, these morphological changes showed a statistically significant positive correlation with a favourable clinical outcome following surgery.

Cortical Dysplasia–Hemi-megalencephaly

'Cortical dysplasia', a somewhat unsatisfactory term presently used for a characteristic combination of disturbed neuronal migration and abnormal neuronoglial differentiation, occurs in three distinctive clinicopathological settings. The restricted lesions of localized cortical dysplasia are described in Chapter 12, Extrapyramidal Diseases of Movement; a diffuse and multifocal form associated with systemic hamartomata is seen in tuberous sclerosis (see section later, Tuberous Sclerosis [Bourneville's disease]). In this section, discussion is confined to a distinctive form of hemi-megalencephaly with the cytoarchitectural changes of cortical dysplasia originally described in isolated post-mortem case studies[88,226,240,282,595,656,1012,1019] but now, as a result of improved neurosurgical techniques and advances in neuro-imaging, also described in extensive resections, either complete or functional hemispherectomy or multilobar resection of epileptogenic tissue.[265,266,293,320,555,872,1054,1056]

Cortical dysplasia–hemi-megalencephaly in the absence of hemi-hypertrophy or viscerocutaneous stigmata of phakomatosis is sporadic, patients presenting with intractable seizures in early life. In the past, the long-term outlook was bleak, but early results with hemispherectomy are encouraging.

Recorded brain weights vary widely from well below to well above normal. One hemisphere (usually, although not always, the pathological hemisphere) is considerably larger than the other (Figures 4.75a and 4.76a), and this may involve all or some of the lobes (Figure 4.75a). Gyri are greatly expanded and very firm; after peeling off the meninges, the surface appears finely pitted, like orange skin. The convolutional pattern may be relatively preserved, or coarse and macrogyric. Unilateral enlargement of the olfactory tract has also been reported (Figure 4.75a).[240] On coronal sectioning the abnormal hemisphere shows irregular thickening of the cortical ribbon but with poor demarcation from the underlying white matter. The centrum semiovale is expanded, as well as the basal ganglia in some cases (Figure 4.75b).

Histologically, sulcation can vary from nearly normal to lissencephalic, and the transition from normal to abnormal cortex may be quite abrupt (Figure 4.77). The abnormal cortical ribbon is widened, although in many cases there is poor or non-existent demarcation (Figure 4.77) from the pathological subcortical 'white matter'. There is complete loss of horizontal cortical lamination, and on rare occasions an undulating pattern of the superficial part of the cortical plate indented by stubby fingers of paucicellular molecular layer, slightly reminiscent of polymicrogyria (Figure 4.78). Neuronal density appears qualitatively reduced,[871] although morphometric

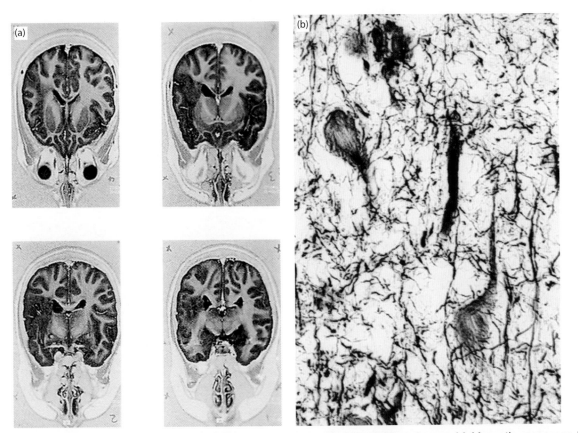

4.76 Unilateral cortical dysplasia in a 15-year-old girl presenting with intractable seizures. (a) Magnetic resonance image shows an extensive unilateral abnormality, but unusually this is in the smaller hemisphere. **(b)** Histology of the functional hemispherectomy specimen shows tangle formation within large dysplastic neurons in the cortex. Bielschowsky's silver impregnation.

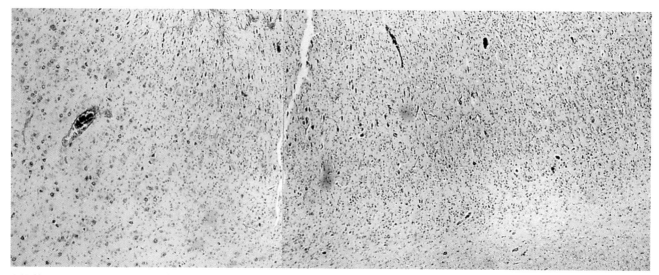

4.77 Hemi-megalencephaly with cortical dysplasia. The sharp transition from normal to dysplastic cortex is demonstrated in this low-power montage. Normal cortex is present on the right, the central part of the field contains scattered large dysplastic neurons, whereas on the left the cortex is completely replaced by abnormal cells. Luxol fast blue–cresyl violet.

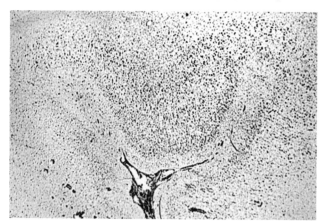

4.78 An undulating cortical ribbon with fused and buried gyri is a rare finding in cortical dysplasia. Luxol fast blue–cresyl violet.

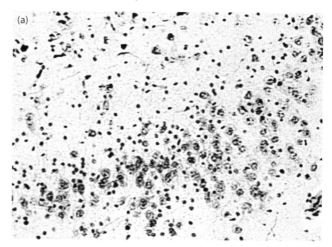

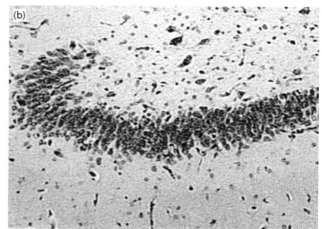

4.79 Hemi-egalencephaly with cortical dysplasia. Involvement of the granule cells from the dentate fascia. Compare their large size and dispersion **(a)** with a normal age-matched control **(b)**.

analyses are inconclusive, because both decreased[1012] and normal[266] cell density has been reported. Particularly striking are the cytological abnormalities of neurons and astrocytes, which may also be present in the basal ganglia, the brain stem and the olfactory tract. Many authors have noted the considerable numbers of abnormally large neurons, some larger than Betz cells, scattered through the cortex and present ectopically in the subcortical white matter (Figure 4.77). Hippocampal pyramidal cells and granule cells may also show neuronal cytomegaly (Figure 4.79).[871] Formal morphometry has confirmed these observations.[266,871] Nerve cells are misaligned, often bizarrely shaped, globose or multipolar, sometimes with multilobed or multiple nuclei, and occasionally even vacuolated. The nucleus is usually central, often partly outlined by a crescentic condensation of Nissl stain, contrasting with a central chromatolysis-like clearing of the Nissl substance. Using Golgi impregnation, Robain *et al.* demonstrated increased size and complex recurrent branching of the dendritic tree, suggesting that the hypertrophic neurons are both deafferented and polyploid,[872] in keeping with quantitative estimates of increased

nuclear DNA.[88,656] Vinters' group has emphasized the cytoskeletal abnormalities in the giant neurons,[266,320,1056] which react intensively with silver impregnations and immunomarkers for various neurofilament epitopes,

4

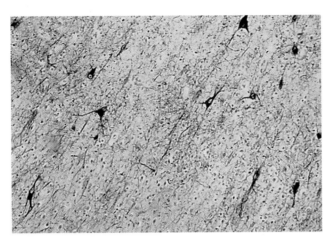

4.80 Cortical dysplasia. Upregulation of neurofilament antibody in large dysplastic neurons present diffusely in the cortex of a neonatal brain.

Case referred by Dr R Janzer, Lausanne, Switzerland.

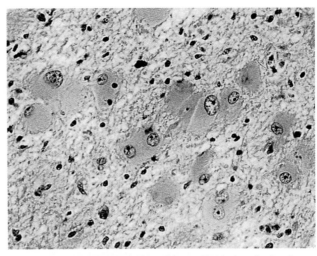

4.82 Hemi-megalencephaly with cortical dysplasia. Large globular balloon cells of indeterminate phenotype in gliotic demyelinated white matter.

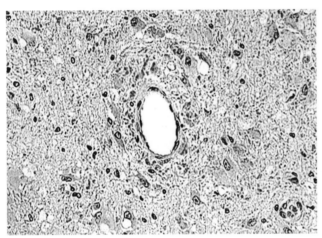

4.81 Hemi-megalencephaly with cortical dysplasia. Rosenthal fibres abundantly scattered in abnormal white matter and clustered around blood vessels are somewhat reminiscent of Alexander's disease.

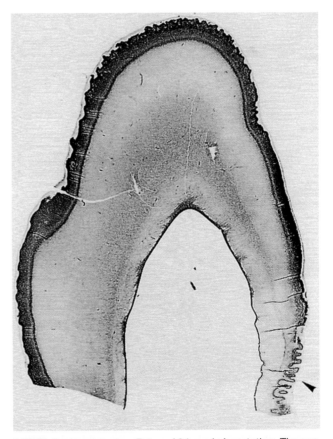

4.83 Status verrucosus. Fetus of 24 weeks' gestation. The cortex shows status verrucosus over the vertex, whereas polymicrogyria is evident on the medial surface (arrowhead). Luxol fast blue–cresyl violet.

and show tau- and ubiquitin-positive tangle-like formations reminiscent of Alzheimer tangles, except for the absence, both immunologically and ultrastructurally, of paired helical filaments (Figure 4.80).[293] Personal experience in both biopsy and autopsy specimens, including a neonate of 38 weeks' gestation (see Figures 4.76b and 4.78), who also showed olivary dysplasia, supports these observations.

Glial involvement varies from minimal to massive, even tumour-like on occasion,[1019] but usually swollen astrocytes are scattered individually or disposed in clusters in both cortex and white matter. They are large, with glassy cytoplasm, and round eccentrically placed, often nucleolated, nuclei, whereas multinucleated cells are common. Astrocytosis may be intense. Rosenthal fibres have been noted in the affected white matter;[555,856,872] rarely, their presence in very large numbers is associated with cystic rarefaction (Figure 4.81),[856] which could

be confused by the unwary with Alexander's disease. Perhaps more common are focal or perivascular calcifications. An important feature of the dysplasia is large globular or 'balloon' cells, of indeterminate phenotype (Figure 4.82), with round nucleolated nuclei, lacking

Nissl substance but having a variable reaction to GFAP. Immunohistochemical co-localization of both GFAP and vimentin[266] and GFAP and synaptophysin[1056] has been demonstrated in these cells. A morphometric study using silver impregnation for nuclear organizer regions and proliferating cell nuclear antigen (PCNA) immunohisto-chemistry suggested that balloon cells are unlikely to be undergoing proliferative activity.[266]

Thus, there are many cytological similarities between this form of hemi-megalencephaly and the more restricted lesions of focal cortical dysplasia,[511,1007] as well as tuberous sclerosis, the genes for which have now been mapped.[315,399] A further variant that appears to be a transition between the focal and hemi-megalencephalic forms has been delineated by MR imaging as focal transmantle dysplasia.[52]

Leptomeningeal Glioneuronal Heterotopia

Isolated islands of CNS tissue, including neurons and glia, within the leptomeninges, or focal protrusions of similar tissue from the surface into the meninges, are a common finding in brains harbouring a great variety of malformations, but especially in holoprosencephaly and migration defects, and are particularly extensive in cerebro-ocular dysplasias and atelencephaly. Indeed, the frequency of these lesions may have deterred investigators from establishing their incidence. However, Hirano *et al.* observed leptomeningeal glioneuronal heterotopia in 31 per cent of 129 autopsied infants, 7 months of age or younger.[469] Significantly, they were present in 65 per cent of individuals with CNS malformations including meningocele and holoprosencephaly, but especially polymicrogyria and neuronal heterotopia. They also occurred in 8 per cent of those with only extra-CNS malformations and in 4 per cent of infants without any other congenital anomalies.

Hypotheses of pathogenesis include excessive proliferation of the superficial granular layer,[350] glial proliferation resulting from external stimulation such as meningeal inflammation,[716] and implantation of germinal cells in the leptomeninges.[1000] Disruption of the pial–glial barrier has been postulated as the cause of the massive leptomeningeal heterotopia, which is such a prominent feature in cerebro-ocular dysplasias and type II lissencephaly.[369,967] The importance of the pial–glial barrier is emphasized further by Choi and Matthias in an elegant study of an unusual cortical dysplasia in which numerous focal extrusions of neural tissue into the meninges occurred through multiple glial bridges overlying narrow strips of dysplastic cortex.[156]

Brain Warts: Nodular Cortical Dysplasia

This is an uncommon anomaly with confusing terminology, including names such as 'status verrucosus',[845] also being used for polymicrogyria. The best accounts are those of Jakob[509] and Crome[211] (brain warts) and Morel and Wildi[718] (dysgénésie nodulaire disséminée de l'écorce). These superficial cortical nodules, 1–5 mm in diameter, are scattered over the surface of the hemispheres, with 2–16 per brain, with a predilection for the frontal lobe. The overlying pia mater is often adherent to the nodule, which is usually situated on the crown of a gyrus or the bank of a sulcus. Histological examination shows the nodule to be a herniation of laminae II and III through the molecular layer, which is thin or absent. Neurons of varying size are grouped around a radial bundle of myelinated fibres, sometimes with a central blood vessel. Nodules can be found in an otherwise normal cortex but occasionally are present in microcephalic brains with polymicrogyria.[80,211] Cortical nodules have been produced experimentally in rats by freezing the cortex.[296] The pathogenesis of the deformity is not clear, but the associated vascular pattern suggests a close relationship between vascular and tissue dysplasia.

Status Verrucosus Simplex or Status Pseudo-verrucosus

Not to be confused with macroscopic artefacts such as those caused by fixing fetal brains in a gauze bag[983] or by peeling off the leptomeninges before fixation, which appear as a fine shrinkage of the surface of the brain, status verrucosus simplex[857] is found only on microscopic examination in the brains of some fetuses of about 20 weeks' gestation. The second layer of the cortex forms irregular protrusions into the molecular layer (Figure 4.83), the external surface remaining smooth. The lesion is controversial: some believe it an artefact of autopsy handling, Larroche remarked that it was severe and diffuse in macerated brains but that it was not associated with other cortical anomalies.[589]

Hippocampal Anomalies

Malformations of the hippocampus occur in a variety of clinicopathological settings. Hypoplasia with a dysplastic dentate gyrus is common in trisomy 18.[696,990] Severe disorganization of the medial temporal lobe is well known in thanatophoric dysplasia and includes hypoplasia or aplasia of the dentate gyrus.[201,474,485,1092]

Houser and colleagues have described various degrees of dispersion of the granular layer of the dentate gyrus, including a bilaminar organization, in a series of

4.84 Duplication of the dentate fascia. This 2.5-year-old boy presented with global developmental delay and seizures beginning at 6.5 months. In addition to bilateral hippocampal sclerosis and duplication of the dentate fascia, there were nodular heterotopias in the temporal and insular white matter, cerebellar heterotopia, subtle tract anomalies in the brain stem and unilateral olfactory hypoplasia. Luxol fast blue–cresyl violet.

patients with severe temporal lobe epilepsy and a history of febrile convulsions early in life.[487,488] Migration of granule cells occurs late in development, continuing over a relatively long period and even postnatally,[27] and Houser and colleagues suggest that severe seizures early in life could conceivably disturb this late migration. Indeed, granule cell dispersion has been produced in a pilocarpine model of experimental epilepsy in rats.[691] However, dispersion or duplication of the dentate gyrus is not confined to classical temporal lobe epilepsy. It has been observed in association with dysembryoplastic neuroepithelial tumour in the temporal lobe[848] and by the author in microdysgenesis, Sturge–Weber syndrome, and migrational errors and infarcts responsible for extratemporal seizures, as well as on three occasions as a bilateral anomaly (Figure 4.84) at autopsy with and without epilepsy.[438]

Encephaloclastic Defects

The lesions discussed in this section exemplify the uncertain dividing line between true or 'primary' malformations–defects resulting from an intrinsically abnormal developmental process[965]–and disruptions or 'secondary' malformations. Encephaloclastic processes occurring in the first half of gestation may result in smooth-walled cavities masquerading at birth as primary malformations. This is because macrophage responses in the fetal brain mature long before reactive astrocytosis becomes detectable. Confusion is compounded further because there is often evidence of deranged development with disturbed neuronal migration near these lesions. However, in the interests of accurate genetic counselling, it is essential to recognize the true nature of these acquired lesions. Series of cases have been described.[107,210,214,245,352,420,584,642,721,1106]

Porencephaly

Under this heading, Heschl described defects in the wall of the cerebral hemisphere communicating between the ventricle and the surface.[457,458] He considered these defects to be congenital, the result of destructive processes occurring during fetal life, and differing only in extent from hydranencephaly. Unfortunately, the term came to be applied indiscriminately, particularly in the clinical and radiological literature, along with its corruption, polyporencephaly. There has now been a return, following Friede,[350] to a more precise definition close to Heschl's original concept. Thus, a

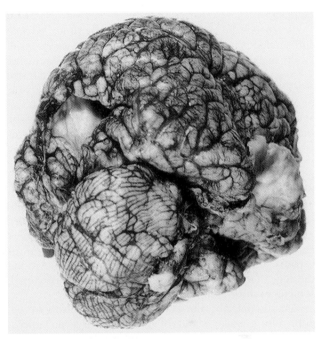

4.85 Porencephaly. Two defects are present: a shallow dimple on the orbital surface, and a porus in the occipital lobe allowing communication between the ventricle and subarachnoid space.

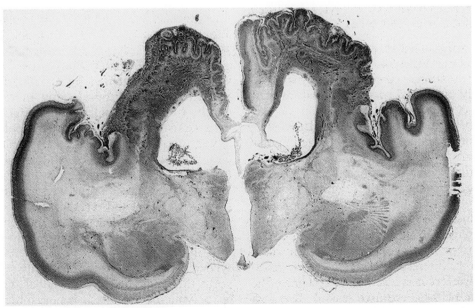

4.86 Porencephaly. Bilateral clefts with disorganized neuroblasts in their depths, surrounded by undulating polymicrogyric cortex. Twin of 32 weeks' gestation. Co-twin deceased at 19 weeks. Luxol fast blue–cresyl violet.

porus is best defined as a circumscribed hemispheric defect originating during fetal life and antedating the acquisition of a mature astroglial response or completion of convolutional development. Clinical manifestations are varied: some patients survive into adult life, but most exhibit severe mental retardation, blindness and sometimes epilepsy, tetraplegia or even decerebrate rigidity. The incidence is difficult to ascertain: one study, from the University of Colorado, USA, accumulated 11 examples out of 18 000 patients with seizure disorder or neurodevelopmental delay investigated with cranial CT over a 3-year period, an incidence of 1 per 1650 patients.[698] Whereas diagnosis used to rely on pneumoencephalography, it is now readily attainable with MR imaging.[49]

Porencephalic defects are always smooth-walled and surrounded by an abnormal gyral pattern but vary considerably in extent and depth, from small indentations (Figure 4.85) to deep clefts (Figure 4.86), which may be separated from the ventricular cavity only by a thin layer of tissue, or extensive defects through the full thickness of the hemispheric wall (Figure 4.85), which very rarely may approach the extent of hydranencephaly.[721] The most typical porus is a full-thickness defect with no inner membrane walling off the ventricle but its exterior closed over by a delicate membrane, often torn during autopsy. Pori are commonly bilateral, roughly symmetrical and centred around the sylvian fissures or central sulci. When the defect is unilateral (Figure 4.85) there may be a convolutional abnormality, particularly polymicrogyria, in a topographically congruent part of the contralateral hemisphere. Occasionally a porus is parasagittal,[245] orbital or occipital (Figure 4.85). Around the edge of a porus the convolutional pattern is abnormal; sometimes the gyri form a radiating pattern.

Microscopically, the cortical ribbon is disorganized, either into irregular islands of grey matter or as polymicrogyria (Figure 4.86). Mineralizations are frequently found identified within this abnormal cortex and can be individual nodules or linear arrays. This abnormal cortex extends over the edge of the porus and descends into the cleft, where it is co-terminous with the ventricular wall, which extends up into the cleft to meet it. This junction has been termed a pia-ependymal seam, but this is a misnomer because the ventricular surface is largely devoid of ependyma and instead is covered by a thick glial feltwork, which may extend for a short length over the abutting cortex before contributing to the inner part of the covering membrane, the outside being arachnoid.

Nodular grey heterotopias are often present beneath the ventricular wall, especially near the cortical junction. Elsewhere in the brain there may be large areas of polymicrogyria. The septum pellucidum is usually absent or at least partially deficient.[9] The basal ganglia may show focal scarring and mineralization but are usually normal, as are the brain stem and the cerebellum, but the thalamus is small owing to atrophy or hypoplasia of the cortical projection nuclei.[985]

Basket Brain

When bilateral porencephalic defects are extensive, only a thin central arch of tissue, residual cingulate and adjacent gyri may connect the occipital and frontal parts of the

brain, giving the appearance of a basket with a high handle.[769] Similar 'basket brains' have been described by many authors,[180,352,767,769,1106,1107] and they are best considered as intermediate between porencephaly and hydranencephaly.

Hydranencephaly

Cruveilhier first described fluid-filled bubble-like hemispheres,[217] or hydranencephaly.[584,1069] Most of the cerebral mantle is replaced by a thin, partly translucent membrane without a surface convolutional pattern. Although the inferior parts of the temporal and occipital lobes are often preserved, and sometimes also the orbital frontal area (Figure 4.87),

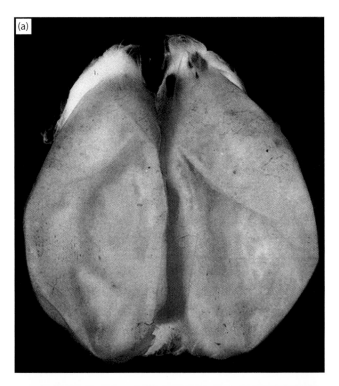

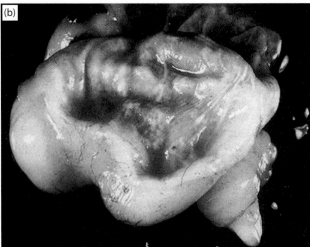

4.87 Hydranencephaly in a twin of 22 weeks' gestation. Estimated age of deceased co-twin was 17 weeks. **(a)** Bubble-like hemispheres photographed under water and viewed from above. **(b)** Lateral view. The cyst has collapsed out of water. Note the preservation of orbital frontal, and inferior temporal and occipital cortex.

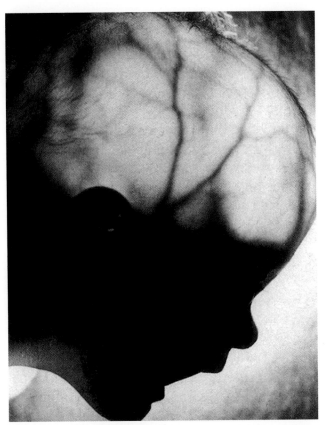

4.88 Transillumination of the head of a hydrancephalic infant.

on occasion only the hippocampi remain intact. The membrane consists of an outer connective tissue layer overlying an irregular patchy layer of glial fibres with occasional residual neurons, mineralized debris and collections of haemosiderin-laden macrophages. Rarely is there recognizable ependyma. At its insertion into the surviving cerebral mantle, the glial layer fuses with the molecular layer of the adjacent cortex, covering it for a variable distance. The cortex may show normal lamination or polymicrogyria. The deep grey masses are often rotated outwards: the thalamic relay nuclei are hypoplastic, whereas the basal ganglia may be normal or disorganized or show focal destruction. There is also aplasia of the corticospinal tracts.

Clinical manifestations vary: major involvement of the basal ganglia and hypothalamus causes severe impairment of thermoregulation, sleep pattern, sucking and swallowing, so survival is rarely for more than a few weeks. If the deep nuclei are preserved, lifespan may be prolonged for several years but there is spasticity, epilepsy and minimal psychomotor development. Head size is normal at birth but increases over the first few months. This increasing hydrocephalus has been variously ascribed to aqueduct atresia, gliosis and stenosis of the foramina of Monro.[214,584,1069] Transillumination may occasionally be dramatic (Figure 4.88), but MR imaging is now of particular value in diagnosis,[424] even *in utero*.[6]

The pathogenesis of these lesions has excited much argument. Heschl favoured a destructive rather than an aplastic process.[457,458] Yakovlev and Wadsworth, in an extensive study of bilateral clefts and full-thickness porencephaly,

coined the term 'schizencephaly' to embody their concept of a failure of growth and differentiation of circumscribed areas of the cerebral wall, occurring, they believed, during the first 2 months of intrauterine life.[1106,1107] However, their interpretation of the structure of the apposed walls of the cleft and the bridging membrane over the defect is no longer convincing, although for obscure reasons it lingers in the radiological literature.[49,698] There is now ample clinical, morphological and experimental evidence favouring a destructive origin for these lesions. Their timing is suggested by the coincident polymicrogyria to be between the fourth and sixth months; indeed, they have been observed as early as 24 and 26 weeks of gestation[350,756] and even at 22 weeks of gestation (see Figures 4.60 and 4.87). The topography and symmetry of porencephaly and hydranencephaly, and their preference for the territories of the carotid or the anterior and middle cerebral arteries,[245,584] suggest ischaemia and infarction as a likely pathogenic mechanism. Hydranencephaly has been produced in animal experiments by blocking or ligating the carotid arteries,[63,731] but in humans the major arteries are usually patent, suggesting that perfusion failure rather than arterial insufficiency is responsible. The porus can be perceived as the epicentre of the disturbance, with the surrounding polymicrogyria reflecting less intense ischaemia.[609,642] Norman's case is instructive in this regard: bilateral symmetrical fissures in a twin of 26 weeks' gestation had at their centres completely disorganized groups of neuroblasts mixed with macrophages and a focus of calcification, whereas the edges of the lesion showed polymicrogyria in the formative stage.[756] A personal case shows similar appearances at 32 weeks' gestation (see Figure 4.86).

Various aetiological factors have been proposed. In early series a high incidence of illegitimate births and attempted abortion was noted;[59,245] porencephaly has recently been documented following intramuscular injection of benzol as an attempted abortifacient in the first trimester.[106] Case reports of hydranencephaly include clinical histories of maternal poisoning or attempted suicide with domestic gas at 24 weeks and butane at 27 weeks and oestrogen ingestion at 18 weeks.[44,95,324] The evolution of porencephaly has been examined in sequential ultrasound scans following an episode of severe maternal hypoxia at 20 weeks.[389] Multicystic encephalopathy (see later, Multicystic Encephalopathy, p. 340) has also been reported after attempted suicide with butane gas at 30 weeks[391] and, following maternal bee sting, anaphylaxis at 35 weeks of gestation.[311]

There is an undoubted association between encephaloclastic lesions and twinning, particularly monozygotic, monochorionic twins,[11,531,592] often but not always in a twin to a macerated fetus (see Figures 4.86 and 4.87).[11,592,756] Intrauterine death of one twin, and the appearance of multicystic encephalopathy (see later), have been documented in serial CT scans.[493] Most of the suggested mechanisms are based on the twin-to-twin transfusion syndrome, because of the high incidence of interfetal vascular anastomoses in monochorionic twins.[592,646] Following the intrauterine death of a donor twin, in whom the stigmata of ischaemic damage may occasionally be found,[592] embolization of necrotic tissue or transfer of thromboplastin-rich blood from a dead to

(a)

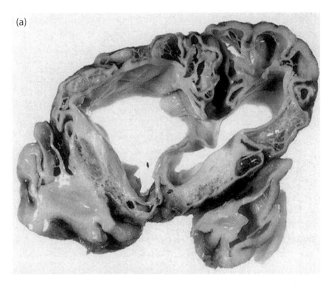

(b)

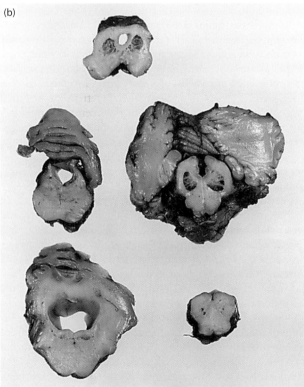

4.89 Multicystic encephalopathy. (a) White matter and deeper parts of the cortex as well as basal ganglia are transformed into a gliovascular meshwork. **(b)** Bilateral symmetrical cystic necrosis extending through the hindbrain.

4.90 Global hemispheric necrosis following severe birth asphyxia.

Figure courtesy of Dr T Moss, Frenchay Hospital, Bristol, UK.

a surviving recipient twin may cause disseminated intravascular coagulopathy[74,715] and bring in its wake multiple organ infarction.[333] Alternatively, massive blood loss from the surviving (recipient) twin into the tissues of the dead co-twin may have similar consequences. In other cases it is the smaller, presumed donor twin who is affected.[654,952] In a personally examined case, unilateral multicystic encephalopathy was present in one of a pair of conjoined twins who had shared a common pericardial cavity.

Another association is with fetal infection, especially toxoplasmosis and CMV, but other viruses and even intrauterine purulent meningitis are implicated occasionally.[352,484] Similar lesions in animals have been reported in association with intrauterine viral infection.[445]

Persistent or recurrent infection was also postulated by Bordarier and Robain to explain the occurrence of encephaloplastic lesions in two siblings, the first showing hydranencephaly, polymicrogyria and subcortical heterotopias, the second showing only polymicrogyria and heterotopias.[103] However, recent clinical and neuroimaging reports without pathological confirmation of familial porencephaly[463,486,874] and familial schizencephaly[398] have suggested a genetic abnormality.

Finally, it should be stressed that Fowler's familial hydranencephaly syndrome, although indistinguishable macroscopically and ultrasonographically from encephaloclastic hydranencephaly, is easily recognizable on histological examination (see later, Fowler's Syndrome: Proliferative Vasculopathy and Hydranencephaly–Hydrocephaly, p. 365).

Multicystic Encephalopathy

Although sharing the same clinical antecedents and probably the same pathogenesis as the smooth-walled porencephalic defects described above, these lesions originating during the third trimester are, by contrast, ragged, irregular and without accompanying cortical malformation (Figure 4.89).[210,327] The white matter and deep cortical layers of large parts of both hemispheres may be destroyed and replaced by a sponge-like arrangement of myriad glial-lined cysts intersected by thin gliovascular strands enmeshing collections of lipid-containing macrophages. Unilateral lesions can also occur.[283] The basal ganglia and the brain stem may be normal but often show extensive bilateral cystic necrosis (Figure 4.89).[952] Similar pathology may also occur at or after birth: the often very severe destruction (Figure 4.90) has been termed 'global hemispheric necrosis'.[350,621,636,1071] It is known to occur postnatally in the clinical context of hyperpyrexia and sudden collapse, can be followed in

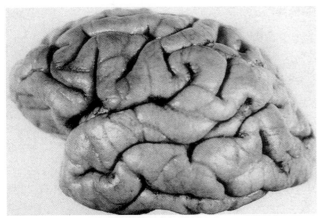

4.91 Familial microcephaly. Note the simple convolutional pattern formed by the relatively broad gyri.

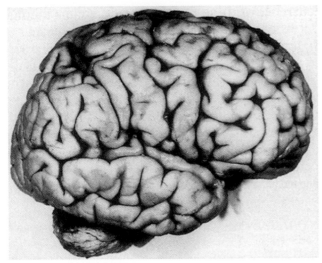

4.92 Right hemisphere of the brain in Down's syndrome. The cerebellum is small and the occipital contour flatter than normal. The frontal lobe is reduced in its anteroposterior diameter and the superior temporal gyrus is poorly developed.

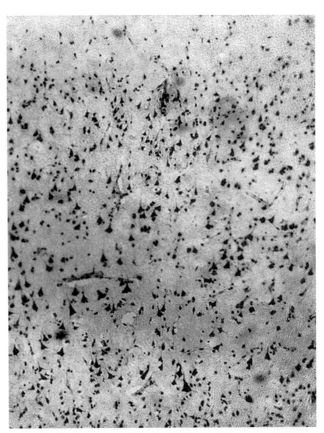

4.93 Section of the angular gyrus in Down's syndrome, showing diffuse loss of neurons in the third layer. Cresyl violet.

sequential neuroradiological studies and is an important consideration in the legal minefield surrounding claims for inadequate obstetric care.

Microcephaly

Etymologically, 'microcephaly' means a small head and, therefore, to describe a small brain the term 'micrencephaly' would be more appropriate. However, because brain and skull usually grow in parallel, the term 'microcephaly' is generally used to indicate smallness of the brain, occurring usually as a malformation. Normative data are available for head circumference from birth to adulthood[301,747] and for brain weight during fetal[410,590,829] and postnatal life.[246] Figures of two or more standard deviations below the mean are abnormal. Using this definition, microcephaly is a common, but not invariable, occurrence in malformed brains.

Microcephaly is a purely descriptive term and does not refer to a particular aetiology; classification is therefore difficult. Ross *et al.* proposed two large groups, with or without associated malformations.[887] Microcephaly with associated malformations includes genetic disorders (chromosomal or single gene defects), environmental causes and cases of unknown aetiology. Microcephaly without associated malformations includes primary microcephaly, several types of which have been linked to mutations in genes implicated in the function of the cellular mitotic apparatus (see Table 4.1). Other types of primary microcephaly result from genetic inborn errors of metabolism or environmental factors or are of unknown aetiology. Familial cases may show severe cortical anomalies (Figure 4.91),[869] but most cases of recessive microcephaly do not. In the neonatal period, sporadic microcephaly without associated malformations should arouse suspicion of intrauterine infection.[428,853]

Chromosomal Disorders

Trisomy 21: Down's Syndrome

The external appearance of the brain in a person with Down's syndrome is much more diagnostic than are the microscopic findings. The weight of the fully grown brain seldom exceeds 1200 g and is usually nearer 1000 g. The brain is abnormally rounded and short, with a steeply rising, almost vertical, occipital contour. The convolutional pattern usually shows no gross departure from normality, but there may be asymmetries on the two sides, a poverty of secondary sulci and exposure of the insula. Very characteristic in Down's syndrome, although not peculiar to it, is narrowness of the superior temporal gyrus (Figure 4.92).

The anomaly is often bilateral and is present in about half the cases. Conspicuous smallness of the cerebellum and brain stem in comparison with the cerebrum as a whole is also a common feature.[215] Benda proposed that most of these abnormalities of shape have been impressed on the developing brain by the retarded growth of the skull and pointed out that the crowns of gyri may be flattened, that the sagittally arranged sulci may show S-shaped distortion and that fibrous union may bind contiguous parts of the convolutions together.[70]

Many microscopic anomalies have been described in the syndrome, but they are not specific. Davidoff reported patchy cerebral calcifications and a poverty of nerve cells in the third cortical layer (Figure 4.93).[237] This finding is not very common, but irregularities of grouping certainly occur and so relatively acellular areas adjoin those that display an increased density due to the presence of clusters of small nerve cells. Systematic cell counts in random sections indicate that the nerve cells are, in fact, often increased on the average per unit area. This seeming paradox may be related to the small size of the brain as a whole, because, as a general rule, the more underdeveloped the state of the brain, the closer together are the cortical neurons. Active destruction of nerve cells associated with anoxia and oedema takes place in patients with Down's syndrome dying in infancy.[70] Several, more recent morphometric studies have shown a marked reduction in neuron numbers, up to 50 per cent compared with controls, from birth onwards, but not in fetal life,[184,1088] which seems particularly to involve granule layers.[889] There also appears to be developmental disorganization of the six-layer neocortex.[383] The Golgi method has shown changes in dendritic arborization and decreased spine numbers of neonatal and infantile patients with Down's syndrome, whereas fetal cases appeared normal.[996]

Benda also emphasized the poor myelination of the grey and white matter and demonstrated a striking poverty of myelin in the U-fibres.[70] Meyer and Jones were the first to demonstrate a conspicuous fibrillary gliosis of the central white matter in many, but by no means all, cases.[694] There was often no corresponding pallor in myelin preparations, although patchy areas of demyelination were sometimes seen, as they were in Davidoff's cases. Benda also reported frequent minor malformations in the spinal cord, including fusion of the columns of Clarke, so the nerve cells occupy a strip of grey matter lying behind the central canal. Except for undoubted signs of delayed or mildly disrupted development, the brain thus shows a few pathological changes that cannot be explained on the basis of intercurrent disease. Whether such fortuitous lesions are in some way related to abnormal constitutional factors, such as general circulatory embarrassment or endocrine anomalies, has not been determined precisely. Certainly the frequent association of congenital heart disease is an important factor. Arterial and venous thromboses are well-known complications of Fallot's teratology, as are embolism and cerebral abscess due to detachment of vegetations from congenitally malformed or infected valves. The important influence of congenital heart disease on myelination was also emphasized in the large series of Wisniewski and Schmidt–Sidor.[1087] Myelination was not delayed in fetuses, but of 129 cases from birth to age 6 years, 23 per cent showed myelination delay compared with controls. In Down's syndrome infants with congenital heart disease, this figure rose to 48 per cent. The well known neurodegenerative changes found in the brains of aging individuals with Down's syndrome are described in Chapter 15, Ageing of the Brain, and Chapter 16, Dementia.

Three types of chromosomal abnormality are responsible for Down's syndrome: trisomy 21 (in 95 per cent of cases), translocations (less than 5 per cent of cases) and mosaicism (less than 1 per cent of cases). The relative incidence of different chromosomal types has been remarkably constant in recorded series. The incidence of trisomy increases with advancing maternal age, whereas a higher proportion of translocations occurs in children of younger mothers. Most familial cases are associated with translocations. The pathogenesis of the condition, or the mechanism by which an additional chromosome, which in 95 per cent is maternal in origin, can cause a wide range of abnormalities with variable expressivity, has been the subject of abundant speculation and yet remains obscure.

Trisomy 13: Patau's Syndrome

Holoprosencephaly and arhinencephaly occur in about two-thirds of cases.[414] Cyclopia and microcephaly are other features. Cerebellar heterotopias and excessive numbers of neuroblasts in dentate and cochlear nuclei are also notable.

Trisomy 18: Edwards' Syndrome

Studies by Sumi[990] and Michaelson and Gilles[696] documented a variety of abnormalities in addition to microcephaly, particularly anomalies of the gyral pattern, dysplasia of the hippocampus, disorganization of the lateral geniculate nucleus and olivary dysplasia. Cerebellar heterotopias are less extensive than in trisomy 13.

Deletions

Deletions of part of one chromosome are often associated with intellectual disability and microcephaly; examples include 5p– in the cri du chat syndrome, 4p– in the Wolf–Hirschhorn syndrome associated with convolutional defects, heterotopias, dysplasia of the lateral geniculate, dentate and olivary nuclei and hyperplasia of the corpus callosum,[394] and 17p13– in Miller–Dieker syndrome[601] (see earlier, Agyria [Lissencephaly] and Pachygyria).

Fragile X Syndrome

Fragile X syndrome is now accepted as the second most frequent chromosomal disorder associated with developmental disability. An estimated incidence of 1 in 1000 liveborn males makes it the most common familial form of mental retardation. The fragile X site at position Xq27 can be induced in cells cultured at low folic acid and thymidine concentrations.[775] Distinguishing features are macroorchidism and a long face with prominent forehead and large ears. The scanty neuropathological information available includes microcephaly, neuronal heterotopias[292] and, in one autopsied case studied by light and electron microscopy, dendritic spine abnormalities associated with synaptic immaturity.[898]

Environmental Factors (see also Table 4.4)

Microcephaly may be associated with intrauterine growth retardation,[586,590] a heterogeneous condition with diverse associations such as malformation (visceral and/or cerebral) and infections, multiple pregnancy and placental insufficiency with maternal toxaemia or renal diseases. Hypotrophic infants usually have a large brain in relation to body weight and within normal limits for gestational age. However, in only a very few cases of a large series of infants showing retarded intrauterine growth was the brain weight two standard deviations below the mean, i.e. microcephalic by definition. The external configuration of the brain corresponds to gestational age and cytoarchitecture and myelination are not significantly modified in the neonatal period. Placental deficiency, which usually develops during the second half of pregnancy, probably does not affect the number of neurons but may possibly damage glial cells and impair subsequent myelination, which could play a part in the pathogenesis of so-called minimal brain damage. Dobbing and Sands emphasized the vulnerability of the fetal brain at certain periods of its development and it is possible that microcephaly may be induced by undernutrition.[276] In multiple pregnancies, fetal brain growth may be more retarded

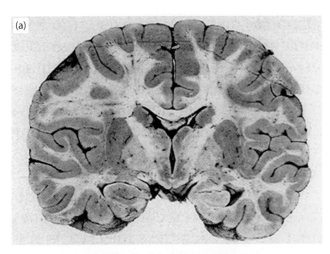

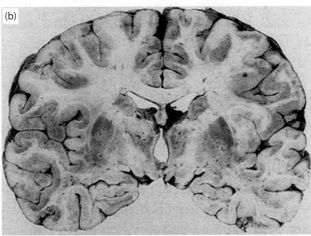

4.94 Microcephaly. Brain weight at 15 months was 700 g. **(a)** Coronal slice. Compare with age-matched control **(b)** of normal size. Maternal phenylketonuria: diet began at 8 weeks' gestation.

than body weight.[477] This may be very pronounced in cases of monochorionic placenta with the twin-to-twin transfusion syndrome.

Offspring of women with phenylketonuria often have microcephaly (Figure 4.94), mental retardation and a low birthweight.[346] This may occur even if a low phenylalanine diet is started during the first trimester.[951] In an international collaborative study, only those infants whose mothers were on a strict diet at the time of conception were reported to be normocephalic.[285] Histological examination of the microcephalic brain is usually unrewarding,[951] but abnormalities of pyramidal cells have been described in a Golgi study of a 4-month-old infant.[576]

Microcephaly is also an important feature of the fetal alcohol syndrome, along with a wide variety of other anomalies, including leptomeningeal glioneuronal heterotopias, agenesis of the corpus callosum and rare migration disorders of the cerebrum, cerebrum and hindbrain.[168,814,1086]

Irradiation

Deep X-ray therapy to the pelvis during the first 4 months of pregnancy has produced many cases of microcephaly. Anomalies of the eye (optic atrophy, choroidoretinitis, abnormal retinal pigmentation) are much more common in these cases than in microencephaly of genetic origin. Four histologically examined cases in which therapeutic irradiation had been carried out during pregnancy were quoted by Cowen and Geller.[205] Various anomalies of the eyes, cerebral hemispheres and thalamus were present in Johnson's cases.[525] Courville and Edmondson reported microcephaly with diffuse loss of cortical nerve cells.[203] Cerebellar microgyria with some structural disorientation was the main feature of the cases of Miskolczy[705] and Bogaert and Radermecker.[1039] At autopsy, three of these brains exhibited meningoencephalitis. To these cases may be added that of Uiberrack, which showed pachygyria of the cerebral cortex and defective development of the vermis.[1031] Studies on children exposed *in utero* to the atomic explosions at Hiroshima and Nagasaki revealed a relatively high incidence of microcephaly and mental retardation, but no specific malformations,[700,826,1093,1112] apart from one report of periventricular heterotopia in a microcephalic individual whose mother was located 1147 m from the hypocentre of the Nagasaki atomic bomb in the third gestational week.[749]

Maternal Infection

The deleterious effects of maternal viral infection have become well recognized since Gregg's original observations on the adverse effects of rubella.[403] Intrauterine viral infection can result in two different, but sometimes overlapping, types of pathological sequel: necrosis and inflammation on the one hand, and developmental interference, i.e. reduced brain growth or malformation, on the other.

Rubella

CNS involvement is common (as high as 80 per cent) in clinical series.[269] In addition to ocular defects, including cataracts, pigmentary retinopathy and microphthalmos,[403,895,1091] and sensorineural deafness,[806] neuropathological findings include chronic meningoencephalitis,[269] microcephaly[733] and retarded myelination and cytoarchitectonic development,[546,882] although the evidence for a causal relationship with specific malformations remains inconclusive.

Cytomegalovirus

The reported incidence of intrauterine CMV infection varies from 0.4 to 2.4 per cent,[585,970] but infection may not result in overt disease. Less than 5 per cent of infected neonates have a rapidly fatal systemic disorder, with involvement of the brain reported in 10–80 per cent.[427,428,680,977,1072] Common clinical findings are microcephaly, mental retardation, epilepsy, diplegia, chorioretinitis and intracerebral calcification. Neuropathologically, there is often microcephaly and sometimes hydrocephalus. Severe necrotizing lesions of the ependyma and periventricular tissue are notable[453,686] and may form cavitating lesions, such as porencephaly or hydranencephaly.[212,272,746] Intracytoplasmic viral inclusions may be very sparse, but their identification is assisted by immunocytochemistry or *in situ* hybridization. Perivascular calcifications are usually found in periventricular tissue but may be present in the cortex and basal ganglia. The association of intrauterine CMV infection with malformations, notably polymicrogyria and cerebellar cortical dysplasia, is now well accepted.[87,212,352,659,666]

Other Viruses

In one series, herpes simplex infection was associated with chorioretinitis, microcephaly, hydranencephaly and microphthalmia. Maternal varicella zoster infection in the first or second trimester may, on some occasions, produce a characteristic embryopathy involving skin, muscle, eye and brain.[290] The dermatomal distribution of cutaneous scarring and other various anomalies suggest a fetal zoster, and there is electrophysiological and neuropathological evidence of cord and dorsal root ganglion involvement.[435,968] A few autopsy studies have described severe destruction of the brain, and one personal case, in addition to a widespread necrotizing encephalitis, showed polymicrogyria bilaterally in the insular cortex (see Figure 4.64).[435]

Other Organisms

Fetal infection with toxoplasmosis causes a necrotizing meningoencephalitis that results in hydrocephalus and widespread calcification. As with CMV, polymicrogyria and cortical defects such as hydranencephaly have also been recorded. An early stage of hydranencephaly was the result of a rare case of intrauterine purulent encephalitis.[484]

Megalencephaly

The accepted definition of megalencephaly is a brain weight 2.5 standard deviations above the mean for the age and sex. Various descriptive terms have been used, such as 'cerebral gigantism' and 'macrocephaly'. Megalencephaly includes a variety of clinical and pathological conditions and no classification is entirely satisfactory.[263,342,762]

Dekaban and Sakuragawa subdivided the megalencephalies into three main categories and various subgroups, based on their own and previously reported cases: primary

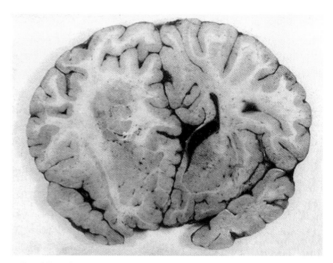

4.95 Megalencephaly with multiple cerebral anomalies, including absence of corpus callosum and an abnormal gyral pattern. A child aged 3 years with total brain weight of 1554 g. Coronal section showing a large collection of heterotopias in the white matter on the left and longitudinal callosal (Probst) bundle on the right.

and secondary megalencephaly and hemi-megalencephaly.[247] Primary megalencephaly may be an isolated finding or associated with achondroplasia and endocrine disorders, or it may be familial. Males are affected twice as often as females and, in the majority of patients, the large head is noticed within the first year of life. In about 25 per cent of cases there was no obvious abnormality of the brain other than its size; of the remainder, half showed some cytoarchitectonic or neuronal abnormalities, and the other half severe macroscopically visible malformations (Figure 4.95). Mental retardation and neurological disorders of some sort were present in the majority of patients. Personal observations in several autistic subjects have disclosed megalencephaly as well as olivary heterotopia.[36] Secondary megalencephaly is associated with genetic disorders, such as the sphingolipidoses and mucopolysaccharidoses, and with various leukodystrophies (including Alexander's and Canavan's diseases) and neurocutaneous syndromes. Megalencephaly may be present at birth or may become manifest during the early postnatal years.

A particular form of megalencephaly described as cerebral gigantism was reported by Sotos *et al*.[961] Evident at birth are large hands, feet and cranium with moderate prognathism, and advanced osseous maturation for the age. There have been sporadic as well as familial cases,[426] and autosomal dominant inheritance has been described.[1123] Moderate to severe mental retardation is present. In a detailed clinical study of 22 patients, CT scan abnormalities included ventricular dilation, midline cava and widening of the sylvian fissures.[1089]

Hemi-megalencephaly

This term denotes manifest asymmetrical enlargement, particularly, but not exclusively, of one cerebral hemisphere. The nosology is unclear, although there is considerable overlap with neurocutaneous disorders. Indeed, hemi-megalencephaly may be a feature of tuberous sclerosis[521] and linear sebaceous naevi of Jadassohn (Solomon's) syndrome.[57,1117] In an autopsy study of one such patient, Choi and Kudo observed asymmetry of the limbs, extensive skin lesions, and hemi-megalencephaly histologically characterized by disturbed neuronal migration, abnormal cortical architecture and glial proliferation.[155] Somewhat similar is a small group of patients in whom hemi-megalencephaly is associated with ipsilateral hemi-hypertrophy, multiple haemangiomas and vascular naevi, the Klippel–Trenaunay–Weber syndrome. In a 17-year-old girl, the right cerebral and cerebellar hemispheres and the right side of the pons were enlarged, possibly resulting from an absolute increase in neuronal numbers, but no malformations were found.[674]

With advances in MR imaging and surgical management of intractable epilepsy in the paediatric population, neuropathologists are increasingly faced with resections (partial or total) of asymmetrically large hemispheres. Morphological findings are varied and include polymicrogyria, pachygyria and cortical dysplasia.[266,871] Recent studies from several labs have identified somatic mutations of genes in the PIK3-AKT3-mTOR pathway in a variety of subjects with various phenotypes including megalencephaly[865] and hemimegalencephaly.[602,827] The implication from these various studies is that there is somatic gain of function in the mTOR pathway, which is known to affect cell growth, and is involved in the pathogenesis of tuberous sclerosis.

Anomalies of the Wall of the Lateral Ventricles

Adhesions between two facing ventricular walls have been described under various names such as coarctation[238] and coaptation[61] of the walls, goniosynapses[200] and transventricular adhesions.[765] The sites of adhesion are between the caudate nucleus and the corpus callosum, or between opposing walls of the anterior horn of the lateral ventricle or, less often, of the occipital and temporal horns. The affected ventricle may be reduced in size and the contralateral cavity enlarged. The pathogenesis is obscure, although a developmental impairment, taking place between weeks 9 and 15 of gestation, has been implicated.[238] However, in two children aged 9 months and 13 years and presenting with transventricular adhesions, Norman and McMenemey described haemosiderin-laden macrophages in the ependymal wall and the caudate nucleus, suggestive of birth injury with residual haemorrhage and scarring.[765] Similar conclusions can be inferred from the case of a 2-year-old girl, born prematurely at 34 weeks of gestation with a complex congenital heart defect, rapidly progressive postnatal hydrocephalus and evidence of intraventricular haemorrhage (Figure 4.96). In the line of obliteration of the ventricle astrocytosis, fibrillary gliosis, ependymal tubules and iron-laden macrophages were observed.

Third Ventricular Obliteration

Neonatal-onset hydrocephalus in association with congenital fusion of the thalamus is extremely rare: it has been reported in two siblings[162] and in a 6-month-old boy with a unique combination of anomalies including atresia of the aqueduct and upper fourth ventricle and rhomboencephalosynapsis.[547]

Malformations of the Cerebellum

The classification of cerebellar anomalies remains difficult and the literature is confused. The tendency is to follow phylogenetic principles, but this approach cannot always encompass the great variety of transitional forms that straddle the normally accepted categories. Moreover, the correlation between phylogenetic and morphological subdivision is not precise. Additional complications are the heterogeneity of associated malformations, the multiplicity of postulated aetiological factors and, most importantly, the very prolonged development of the cerebellum, which extends well into postnatal life. The dividing line between malformative and degenerative processes becomes increasingly blurred in the later stages of gestation. Noxious agents that would be purely destructive elsewhere in the brain may influence granule cell proliferation, differentiation and the late growth spurt of the cerebellum; conversely, secondary atrophy in the wake of perinatal hypoxic–ischaemic injury, epilepsy or a consequence of anticonvulsant therapy may contribute significantly to the final morphological result. A brief embryological summary is pertinent to the discussion. Detailed reviews and experimental studies are available.[16,18–20,348,593,607,695,941]

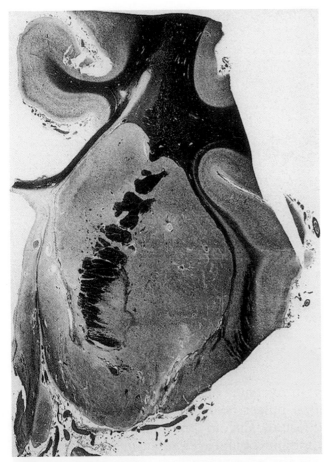

4.96 Anomaly of the lateral ventricle. Coronal section of the striatum and frontal lobe showing obliteration of the frontal horn and body of the lateral ventricle, with adhesion between the corpus callosum and head of the caudate nucleus. Luxol fast blue–cresyl violet.

During gestational week 4, with the closure and segmentation of the neural tube, the rhombencephalon becomes temporarily the largest part of the brain. Differential growth in the rhombencephalon during week 5 of gestation results in formation of the pontine flexure, widening the neural tube at this point with thinning of its roof, which becomes transversely creased. Within this crease, or plica choroidea, the choroid plexus will develop, whereas caudal to it the membranous roof of the fourth ventricle forms a pouch-like evagination in many species;[94] in humans this perforates, forming the foramen of Magendie by 12 weeks,[119] whereas the foramina of Luschka open later, probably around 16 weeks. The roof rostral to the plica, the anterior membranous area, becomes briefly permeable to CSF and is later incorporated into the developing vermis. Also anterior to the plica, the lateral parts of the alar plates undergo intense neuroblastic proliferation, enlarging to form the rhombic lips, the paired primordia of the cerebellum that gradually extend dorsomedially, meeting the roof of the fourth ventricle at around 6 weeks of gestation and then fusing together in the midline during month 3. Cerebellar growth, which has been entirely intraventricular, now becomes extraventricular and the various subdivisions begin to appear: first the posterolateral or flocculonodular fissure at 9 weeks, demarcating the vestibular or archicerebellum from the rest, and then at 12 weeks the primary fissure separating anterior from posterior lobes, the spinocerebellum or palaeocerebellum from the pontocerebellum or neocerebellum. It is this last, phylogenetically youngest, portion of the cerebellum that becomes predominant in mammals, and its various fissures form

4–8 weeks after those of the vermis and flocculonodular lobes. The neurons of the cerebellar cortex and deep nuclei as well as the pontine and inferior olivary nuclei all derive from the alar plates, ventral migrations into the pontine grey and olivary ribbons, and lateral migration into the rhombic lips from where there are two divergent pathways. The first is inwards through the cerebellar plate, the path followed by the Purkinje cells and neurons of the deep nuclei. The second pathway forms the external granular layer, *via* migration around the inferior border of the rhombic lip and then cephalad and lateral over the surface of the developing cerebellum.[451] This rapidly proliferating layer first appears in week 9, covers the whole surface by 14 weeks, reaches maximum thickness at 24 weeks, and is normally gone by the end of the first 9 months. The cells from this layer migrate inward along Bergmann glia, past the Purkinje cell layer to form the internal granular layer. It is this population of neurons that has a major influence on folial development.[840]

Cerebellar Agenesis

Total absence of the cerebellum is rare. Early reports, such as those of Combette and Priestley describing, respectively, an 11-year-old child with epilepsy and mental retardation[185] and an infant aged 4 months with hydrocephalus, spasticity

4.97 Cerebellar agenesis.

Case of Anton and Zingerle.[28]

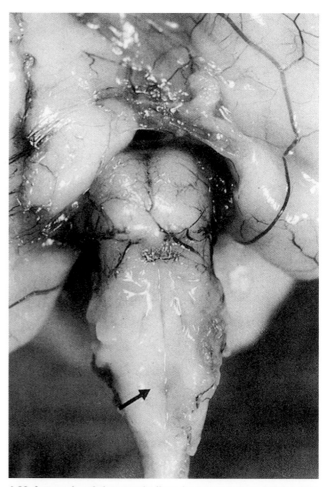

4.98 Agenesis of the cerebellum in a premature infant. The floor of the fourth ventricle (arrow) is exposed by the total absence of cerebellar tissue.

and spina bifida,[835] were without histological verification. Stewart[981] and later Macchi and Bentivoglio,[644] reviewing the literature, pointed out that, in most cases, some remaining cerebellar tissue, especially the flocculonodular lobe, could be demonstrated (Figure 4.97).[29] In addition, related structures such as the pontine nuclei and inferior olives are hypoplastic or dysplastic. Ricardi and Marcus[859] reported two brothers with congenital hydrocephalus who died early in infancy: autopsy of one child showed cerebellar agenesis. Larroche's case (Figure 4.98) was a premature stillborn male with hydrocephalus; a large arachnoid cyst filled the interhemispheric fissure, and the corpus callosum and septum pellucidum were absent. Patients with total or near-total absence of the cerebellum usually have developmental disability, both mental and physical, but the cerebellar defect may not be suspected during life.[897,979] Total agenesis of the cerebellum is also a feature of large occipital encephaloceles.[541] Absence of one hemisphere is much less rare[984] (Figure 4.99) and is associated with changes in the contralateral inferior olive and pons, which may take the form of secondary atrophy rather than maldevelopment.[619,644]

Aplasia of the Vermis

Developmental defects of the palaeocerebellum, which predominantly involve the vermis, occur in a variety of disorders. The Dandy–Walker and Joubert's syndromes are the best known, but several others will be mentioned briefly. The clinical and radiological diagnosis of these syndromes and the complexities of their differential diagnosis are reviewed by Bordarier and Aicardi.[102] In tectocerebellar dysraphia with occipital encephalocele,[349,608,779,954] the three principal features are partial or total vermal agenesis, a severe deformation of the midbrain tectum and a cerebelloencephalocele. Rhomboencephalosynapsis[407,501,547,913] is a rare form of cerebellar hypoplasia in which the hemispheres and underlying dentate nuclei are fused across the midline in the absence of a vermis (Figure 4.100). Other structures are absent or dysplastic, including the palaeocerebellar roof nuclei, dorsal accessory olive and inferior olive. The cerebellar anomaly may be combined with other midline anomalies, septo-optic dysplasia,[913] commissural abnormalities[369] and fusion of the thalami.[547] Partial or complete absence of the vermis is also a feature of the Walker–Warburg Syndrome (see earlier, Cerebro-ocular Dysplasias, p. 320; and see Figure 4.52, p. 322).

Dandy–Walker Syndrome

Case studies by Dandy and Blackfan[228] and Taggart and Walker[992] were amalgamated into the well-known eponym by Benda.[71] Large series and general reviews are available.[123,222,373,444,998] There are three essential elements for diagnosis: partial or complete agenesis of the vermis, cystic dilation of the fourth ventricle and enlargement of the posterior fossa. Hydrocephalus is a frequent but not constant accompaniment.

In a minority of patients the vermis is completely absent, but in most patients the superior part remains, usually anteriorly rotated, and becomes attenuated inferiorly, where it blends with the membranous roof of the cystic fourth ventricle (Figures 4.101 and 4.102). This is often hugely dilated, its cystic roof sometimes herniating upwards through the

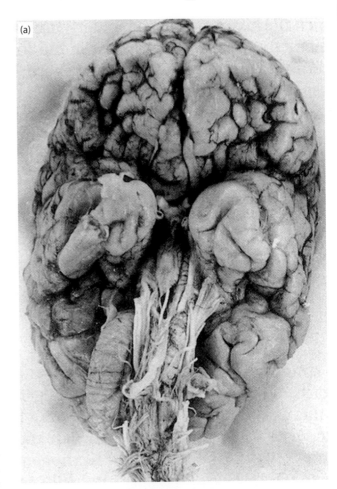

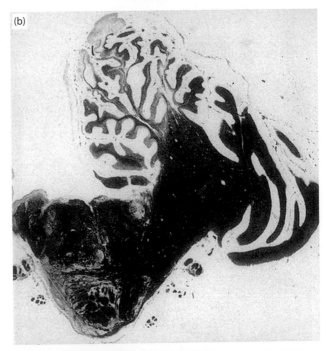

4.99 Malformation of the cerebellum. Absence of the left cerebellar hemisphere and middle cerebellar peduncle and marked reduction in the size of the contralateral pontine nuclei.

tentorial hiatus towards the splenium of the corpus callosum. The diaphanous roof membrane (Figure 4.103), too readily

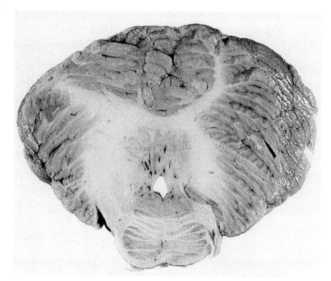

4.100 Rhomboencephalosynapsis. Horizontal section of the hindbrain showing absence of the vermis and fusion of the hemispheric cortex and dentate nuclei.

Figure courtesy of Dr PV Best, Aberdeen, UK.

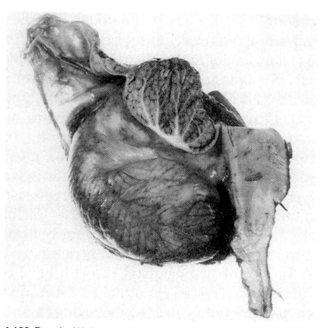

4.102 Dandy–Walker malformation. Sagittal section of brain stem and cerebellum, showing preservation of the superior vermis and incorporation of the rudimentary inferior vermis into the cyst wall.

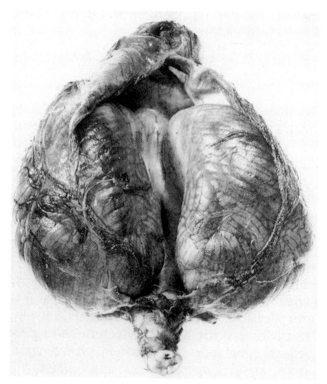

4.101 Dandy–Walker malformation. Inferior surface of the cerebellum, showing the defect in the vermis and the line of attachment of the tent-like cyst.

torn by the prosector, is attached laterally to the cerebellar hemispheres and caudally to the medulla. Histologically, it comprises an outer layer of leptomeningeal fibrous tissue and an inner layer of glia, including ependyma and occasionally cerebellar remnants (Figure 4.104). The inner aspects of the cerebellar hemispheres–deep white matter overlaid by attenuated ependyma–form the smooth white lateral walls of the

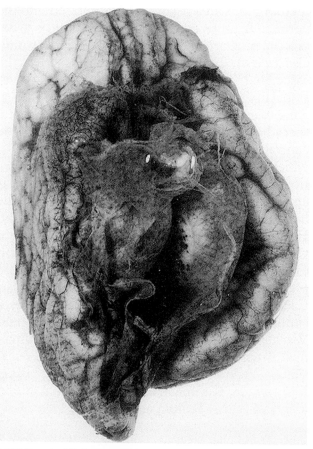

4.103 Dandy–Walker malformation. The delicate membrane of the posterior fossa cyst ruptures readily during the dissection. Here it is demonstrated by photography under water.

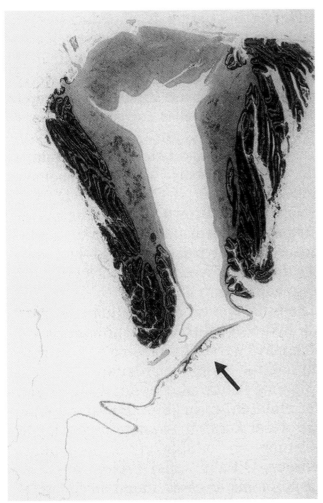

4.104 Dandy–Walker malformation. Horizontal section of brain stem and cerebellum. There is complete absence of the vermis, and remnants of cerebellar tissue (arrow) are incorporated into the cyst wall.

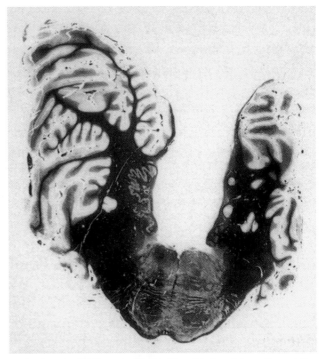

4.105 Dandy–Walker malformation. Horizontal section of the hindbrain showing absence of vermis, and hypoplastic right cerebellar hemisphere and dentate nucleus. Luxol fast blue–cresyl violet.

cyst, the floor of which is the dorsum of the brain stem. The choroid plexus is abnormally positioned in the lateral recesses and bordering the medullary insertion of the roof membrane. The patency of the fourth ventricle foramina has been much argued over, but in a majority of cases they are patent.[222,444] Enlargement of the posterior fossa is important diagnostically: the skull is characteristically dolichocephalic,[71] with an occipital bulge, the attachment of the tentorium is high and almost vertical, whereas the lateral sinuses and torcula are abnormally elevated.

Apart from hydrocephalus and polygyria (see Figure 4.162), a great variety of neural and non-neural malformations are associated with the Dandy–Walker syndrome. In the series of Hart *et al.*, two-thirds had other anomalies of the CNS and one-quarter had systemic defects.[444] Callosal agenesis is particularly common,[222,444,998] and other anomalies include cerebral cortical dysplasias, polymicrogyria and pachygyria, microcephaly, aqueduct stenosis, infundibular hamartomata, syringomyelia and occipital meningocele. Also important are other hindbrain abnormalities (Figure 4.105): cerebellar hypoplasia,[512] cerebellar heterotopias, cerebellar cortical dysplasia,[992] dentate dysplasia, olivary dysplasia and

heterotopia[71,422] and anomalies of pyramidal tract decussation.[222,577] Systemic malformations include polydactyly and syndactyly, cleft palate, Klippel–Feil and Cornelia di Lange syndromes, polycystic kidneys and spina bifida.

The usual clinical presentation is early in life with hydrocephalus and a prominent occiput, which may be transilluminable posteriorly. Poor head control, motor retardation, spasticity and respiratory failure also occur. Clinical signs in older children–nystagmus and ataxia–simulate those of a posterior fossa space-occupying lesion. Presentation in adulthood is also described,[363] as well as asymptomatic cases found incidentally post mortem.[622]

MRI is the modality of choice to image the posterior fossa and angiography often demonstrates a downward displacement, hypoplasia or absence of the cerebellar arteries.[594] Diagnosis *in utero* is possible with ultrasound.[447,753] Differentiation must be made from Arnold–Chiari malformation, in which the posterior fossa is small and the tentorial insertion low, and from retrocerebellar arachnoid cysts, which compress the brain stem but do not communicate with the fourth ventricle.

Various pathogenic theories have been advanced. Early workers attached great importance to foraminal atresia, believing that this caused hydrocephalus and subsequent bulging of the anterior membranous area. There are two insurmountable objections: the fourth ventricle foramina are more often patent and, embryologically, they become patent only after the paired cerebellar primordia fuse and the anterior membranous area becomes incorporated into the vermis. A more acceptable explanation is a developmental arrest of the hindbrain, which would also account for the atretic foramina as well as the associated brain stem anomalies and the occasional involvement of the cerebellar hemispheres. This would also be in keeping with more widespread evidence of arrested development in the CNS and elsewhere. All of this suggests that the Dandy–Walker malformation originates before the third month. There is also supporting experimental evidence from the hydrocephalic mouse,[99,120] in which the anterior membranous area persists and expands between the vermis and the choroid plexus, comparable to the Dandy–Walker cyst. Galactoflavin administration to mice induced a similar defect and callosal agenesis.[537] The aetiology of most, if not all, cases of human Dandy–Walker malformation appear to be genetic in origin. Mutations in *NID1*, *LAMC1* and *AP1S2* have been found in patients with the Dandy–Walker malformation and additional genes are expected to be found.[137,233]

Joubert's Syndrome

A familial syndrome of episodic hyperpnoea, abnormal eye movements, ataxia and mental retardation, associated with agenesis of the vermis, was first recognized by Joubert *et al.*[508] Recently, this syndrome has been definitively associated with a series of related complex malformation syndromes that include renal and retinal anomalies. This family has its basis in mutations in ciliary structural proteins and thus falls into ciliopathies.[910] Very few cases have been studies post mortem,[138,146,351,440,529,554] however, on this evidence, the cerebral hemispheres are largely unaffected. The ventricular system may be moderately dilated, including the fourth ventricle, although this is slight compared with the cystic dilation in the Dandy–Walker malformation. The vermis is either completely

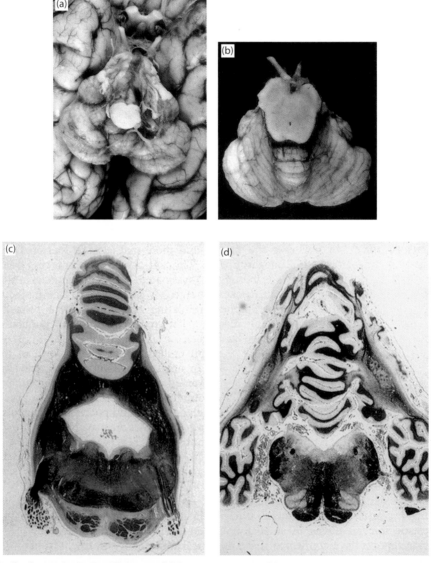

4.106 Pontoneocerebellar hypoplasia: familial case. (a) Base of the brain; **(b)** superior surface of the cerebellum. The tiny cerebellar hemispheres have few, rudimentary folia, whereas the vermis and flocculi are much better preserved. **(c,d)** Horizontal sections of the hindbrain. Note the disparity between neocerebellum and vermis and flocculi, the hypoplastic middle cerebellar peduncles and basis pontis, the dentate nuclei broken into islands and the simplified olives. Luxol fast blue–cresyl violet.

absent or represented by a few rudimentary folia but, histologically, the cerebellar cortex is normal. In the subcortical and deep cerebellar white matter there are numerous heterotopias of large nerve cells, the dentate nucleus is dysplastic and segmented and the roof nuclei cannot be found. In the medulla, olivary dysplasia, in the form of a C-shaped band, and anomalies of the pyramidal tracts and cranial nerve nuclei have been described.[351,440,554] In one case the midbrain tegmentum was unsegmented[138] (see Table 4.1).

Pontoneocerebellar Hypoplasia

Under this title, Brun, as part of a larger study of cerebellar anomalies, described microcephaly and severe mental retardation in two children of 11 months and 15 months who showed rudimentary cerebellar hemispheres with relatively well-preserved palaeocerebellum, a peculiar segmentation of the dentate nucleus, severely hypoplastic nuclei pontis, absent arcuate nuclei and hypoplastic inferior olives.[127–9] A small number of later case reports confirmed these findings,[59,86,122,360,543,567,668,813,870] and, with three personally studied cases,[59] form the basis of the following account.

Microcephaly is notable in most of these patients and may be profound,[870] although there are no corresponding histological abnormalities. Even so, the hindbrain is disproportionately small, often 3 per cent or less of the total weight, on account of the slender brain stem (particularly the pons) and extremely small cerebellar hemispheres. The surface of the lateral lobes may be reduced to a few coarse convolutions or may be almost smooth, whereas the vermis and flocculonodular lobes are more nearly normal in size and foliation (Figure 4.106).

Histologically, the cortex, associated white matter and roof nuclei of the vermis and archicerebellum are normal. In the hypoplastic cerebellar hemispheres, the cortex varies from virtual normality, through modest neuronal depletion, to complete absence of Purkinje and granule cells and associated fibre plexuses, with, in their stead, a loose gliotic tissue (Figure 4.107); both forms may coexist closely, with either a gradual transition or a sharp boundary between them, and there may also be foci of dysplastic cortex. The central white matter is very small and poorly myelinated. The dentate nuclei in all these cases are grossly disorganized, lacking their normal undulating ribbon, hilum or proper amiculum. The reduced neuronal population is gathered into small nests (Figure 4.108), islands of neuropil

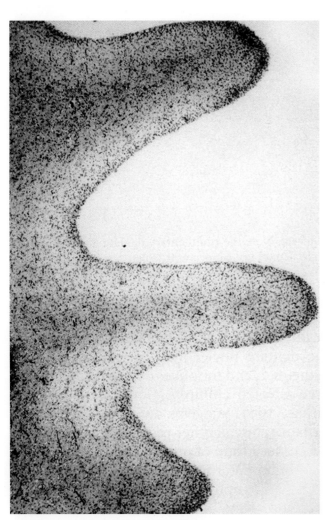

4.107 Pontoneocerebellar hypoplasia. Hypoplastic cerebellar folia are devoid of nerve cells. Luxol fast blue–cresyl violet.

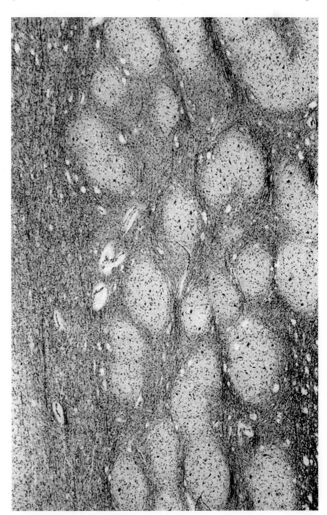

4.108 Pontoneocerebellar hypoplasia. The dentate nucleus is broken up into small islands of nerve cells. Luxol fast blue–cresyl violet.

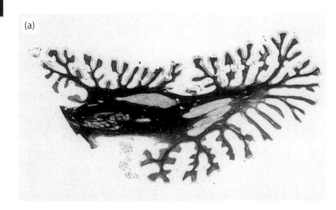

(a)

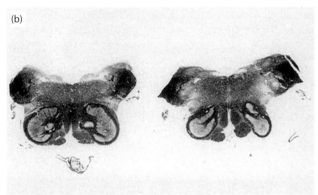

(b)

4.109 Cerebellar hypoplasia. (a) There is global cortical hypoplasia. Note the rudimentary dentate nucleus and larger grey matter heterotopias as well as **(b)** olivary dysplasia in the medulla. Luxol fast blue–cresyl violet.

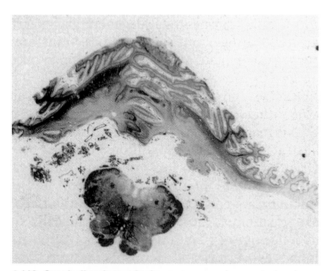

4.110 Cerebellar hypoplasia accompanying anterior horn cell degeneration. Horizontal section of hindbrain.

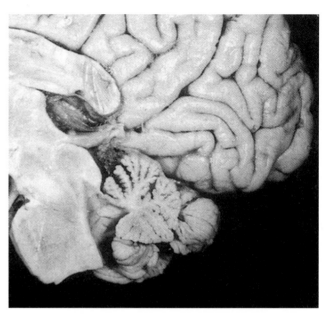

4.111 Granular layer aplasia. Midsagittal section showing small cerebellum and shrinkage of the superior vermis.

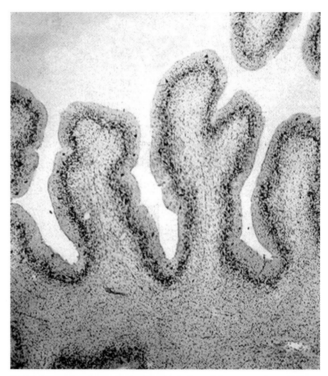

4.112 Granular layer aplasia. Atrophy of the granule cell layer and preservation of the Purkinje cells, many of which are present in the molecular layer. Cresyl violet.

embedded in a meshwork of myelin fibres, or small groups of neurons rimmed by glial cells. Sometimes the neurons appear rather large and rounded. The superior and middle cerebellar peduncles are thin and poorly myelinated, whereas the restiform bodies are usually better preserved. The pontine base is very shallow, with few transverse fibres and markedly hypoplastic nuclei pontis, often without significant gliosis (see Figure 4.106c). The inferior olives may be mildly hypoplastic or dysplastic (see Figure 4.106d), with a variable amount of cell loss and fibrillary gliosis, but the arcuate nuclei are usually absent.

Clinical abnormality is present from birth, with microcephaly, severe psychomotor retardation, feeding difficulties, choreiform and other abnormal movements, myoclonic jerks and seizures. Death usually occurs before the age of 2 years, although one of the authors' cases survived until 9 years (see Figure 4.106a). The aetiology is still a mystery;

most cases are sporadic, but familial examples (see Figure 4.106) are known.[59,437,668] Most authors agree that the disparity between neocerebellar and palaeocerebellar cortex suggests a developmental disturbance at the end of the third month, but the widespread involvement of other anatomically connected structures has excited considerable debate. Brun believed that the 'string-of-beads' arrangement of the dentate indicated an arrest at an early stage of development,[127-9] but this is not borne out by developmental studies or personal experience.[86,360,728] Biemond considered the dentate anomaly to be the primary malformation from which all else followed,[86] but more recent authors explain dentate and brain stem changes as secondary linked atrophy, both anterograde and retrograde, after initial interference with neocerebellar cortical development.[59,543,870] Whether anterograde transneuronal degeneration in early life could bring about such a radical transformation of the dentate nucleus is still an open question.

Other Examples of Cerebellar Hypoplasia

These are more difficult to classify. One example is the first case of Norman and Urich,[766] discussed and illustrated in previous editions of this book,[1034] in which the crenated outline of the dentate ribbon and its amiculum are preserved. In Rubinstein and Freeman's patient, remarkable clinically for his longevity of 72 years and lack of cerebellar signs, the vermis and hemispheres were both severely hypoplastic whereas dentate and olivary nuclei were rudimentary.[897] In case 1 of Vogt and Astwaraturow, the vermis and neocerebellar cortex were both hypoplastic, but cerebellar white matter was well preserved and small, although properly convoluted dentate ribbons were clearly recognizable, accompanied by very large grey matter heterotopias.[1058] The example in Figure 4.109 similarly shows palaeocerebellar and neocerebellar hypoplasia: rudimentary segmented dentate nuclei and large heterotopias are present in the cerebellar white matter and the olivary nuclei are irregularly thickened arcs.

Cerebellar hypoplasia can also present in combination with anterior horn cell degeneration resembling Werdnig–Hoffmann disease.[253,397,538,1058,1070] In these infants, hypoplasia affects the hemispheres and vermis to a much more equal extent than in pontoneocerebellar hypoplasia, and secondary cortical atrophy is particularly severe in the inferior parts of the hemispheres (Figure 4.110). The dentate nuclei may be simplified, but not segmented, and the olives dysplastic. A sibship of three affected children suggests autosomal recessive inheritance,[397] but all other cases have been sporadic.

Another group of infants with a clinical illness resembling Werdnig–Hoffmann disease showed cerebellar atrophy rather than hypoplasia.[253,397] The nosological position of this disorder remains to be clarified, but it seems both clinically and pathologically distinct from Fazio–Londe disease, which encompasses cerebellar atrophy, although this is overshadowed by bulbospinal motor neuron degeneration.[685,974]

Granular Layer Aplasia

The development of the external granular layer occurs relatively late in fetal life. Granular layer aplasia, a direct consequence of interference with external granule cell proliferation or migration, thus stands uniquely among malformations at the interface between primary maldevelopment and secondary atrophy. It is rare and the small literature is far outweighed by the numerous experimental animal models that simulate the human disorder.

The first description of granular layer aplasia was in a litter of cats,[456] followed by human cases.[81,468,761,908,1032] The brain is usually small, but the gyral pattern and histology of the cerebrum are unremarkable. The brain stem has a relatively normal appearance, whereas the cerebellum is greatly reduced in size: its overall convolutional pattern is retained but individual folia are shrunken and sclerotic which might infer a secondary degeneration (Figure 4.111). Parts of the vermis or flocculi may appear to be better preserved.

Histologically, the short, stubby folia are composed of a thin molecular layer overlying a rather crowded line of Purkinje cells and almost no internal granular layer (Figure 4.112). Many Purkinje cells are scattered ectopically through the molecular layer, 'tout à fait au hazard, mais le plus souvent

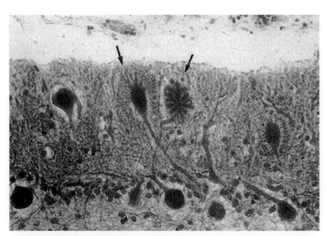

4.113 Granular layer aplasia. Dendritic expansion, with fine terminal brushwork of fibres. Bielschowsky's silver method.

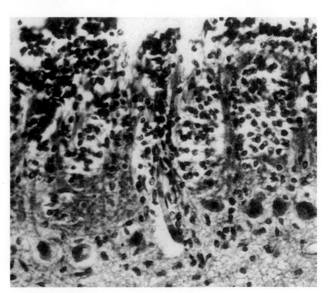

4.114 Granular layer aplasia. Ectopic granule cells filling the molecular layer.

jetées comme un paquet de grains'.[1039] These dislocated Purkinje cells are abnormally shaped, with 'weeping-willow' or more horizontal dendritic arborization, and striking cactus-like expansions of terminal dendrites covered in fine spikes (Figure 4.113). They lack distinct pericellular baskets and their axons, which may run horizontally rather than perpendicular to the surface, often have numerous torpedo expansions. In young infants, the external granular layer is severely deficient, whereas in older patients small foci of the external granular layer may persist. More striking, however, are groups of ectopic granule cell somata stranded at any level in the molecular layer, hanging below the thin external granular layer or just above the Purkinje layer or scattered at random. In one personal case, these

cells almost fill the molecular layer apart from a narrow acellular zone just above the Purkinje layer (Figure 4.114). A diffuse fibrillary gliosis extends through the cortex and white matter, although there is minimal myelin depletion. Dentate and olivary neurons are usually preserved within a gliotic neuropil.

The main clinical features–severe mental retardation and cerebellar ataxia–present early but remain stationary, with some patients surviving well into adult life. Norman's cases were familial,[761] but all others have been sporadic. The mother of van Bogaert and Radermecker's patient received radium treatment for cancer between months 5 and 6 of gestation,[1039] a remarkably close parallel to the irradiation experiments that are discussed next.

Our understanding of the pathogenesis of granular layer aplasia has been assisted greatly by numerous experimental models. In various animals, the superficial granular layer can be destroyed and granule cell ectopia produced by neonatal X-irradiation[17,18] or by exposure to various antimitotic agents.[470,744,938,1111] Intrauterine viral infection can also effect the same result.[552] The regenerative capacity of the fetal granular layer is such that a single destructive dose is insufficient and multiple doses must be given.[17] With prolonged dosing, Purkinje cell heterotopia may be produced; allowing regeneration after a few days causes granule cell ectopia. Another intensively studied model is the mutant mouse *weaver* (*wv, Kcnj6 mutation*), in which a very small cerebellum, granule layer aplasia and ectopic Purkinje and granule cells have been shown to result from both an abnormality of the Bergmann glia and a failure of granule cell migration.[842,843] Functional mossy fibre terminals abnormally situated on Purkinje dendrites have been demonstrated in *wv* cerebellum,[960] and in humans a similar phenomenon has been suggested by the abnormal distribution of synaptophysin immunostaining in the molecular layer.[392]

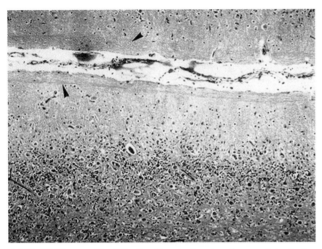

4.115 GM2 gangliosidosis. Cerebellar involvement includes granule cell hypoplasia and a dense layer of horizontal fibres superficial to the molecular layer (arrowhead).

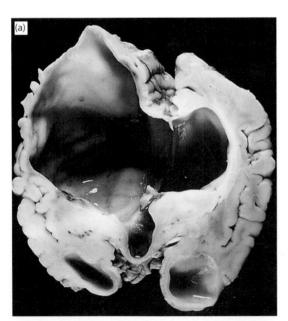
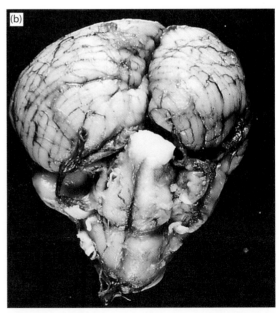

4.116 Crossed cerebellar atrophy in a 5-month-old infant, a twin noted to have ventriculomegaly at 32 weeks' gestation. There is massive old destruction of the left cerebral hemisphere and thalamus **(a)**, ipsilateral atrophy of the pontine base and contralateral cerebellar atrophy **(b)**.

Case referred by Dr PJ Luthert, Institute of Ophthalmology, London, UK.

In many respects the weaver model is closer to the human situation than is the irradiation model, which by contrast shows a layer of tightly packed horizontal fibres superficial to the molecular layer, into which Purkinje dendrites do not penetrate. This is not found in pure granular layer aplasia but is present in some cases of GM2 gangliosidosis (Figure 4.115).[84,347] Coincidental granule layer aplasia has also been described in Pelizaeus–Merzbacher disease.[273,916,1010]

Crossed Cerebellar Atrophy

Unilateral cerebellar atrophy (Figure 4.116), usually observed many years after an extensive lesion in the contralateral cerebral hemisphere, is a capricious, variable and contentious condition. It is capricious in its unpredictability and variable in histological appearance and, despite extensive case literature, it still excites considerable debate. The initial cerebral injury usually occurs in infancy or childhood, but there are occasional observations of initial cerebral damage occurring well into adult life.[62] In addition, the destruction may be intrauterine (Figure 4.116).

Urich and colleagues have provided a helpful critique of the pathogenic basis of crossed cerebellar atrophy[984,1001] and have defined three types of lesion. The first, reduced size of the cerebellar hemisphere without histological changes to the cortex, results from transneuronal atrophy of the nuclei pontis and thence the middle cerebellar peduncle following on from lesions to the frontopontine and temporopontine tracts. The second pattern, in which granular layer degeneration predominates, is considered to be the result of anterograde transneuronal degeneration, because it is associated with atrophy of the contralateral nuclei pontis.[984] Alternatively, in a few cases, cerebellar cortical degeneration has been linked with retrograde transneuronal degeneration (thalamus, red nucleus, superior cerebellar peduncle, dentate and Purkinje cells).[164,630] A last pattern is lobular sclerosis, for which pathological and circumstantial evidence strongly implicates epileptic seizures as the principal pathogenic factor.[1001]

Cerebellar Heterotopias

Foci of ectopic grey matter within the cerebellar white matter are relatively frequent, incidental findings in infants. More common in the hemispheres than in the vermis, they vary in size from just a few cells to large islands or sheets of grey matter (Figure 4.117; see also Figures 4.108, p. 351, and 4.126, p. 359). Two forms occur: clusters of large cells reminiscent of Purkinje or dentate neurons surrounded by a thin corona of neuropil, and islands of heterotopic cortex, the individual layers being more or less well organized. Heterotopias are

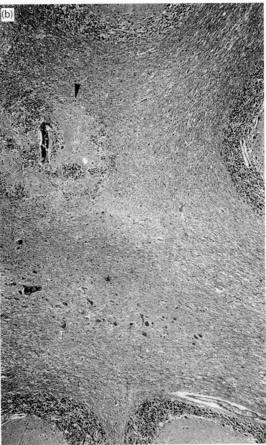

4.117 Cerebellar heterotopias. (a) Small circumscribed collection of large neurons in the folial white matter of a cerebellar hemisphere. Luxol fast blue–cresyl violet. **(b)** Two small heterotopias are incidental findings in an adult: one is a collection of large neurons, the other an island of heterotopic cortex.

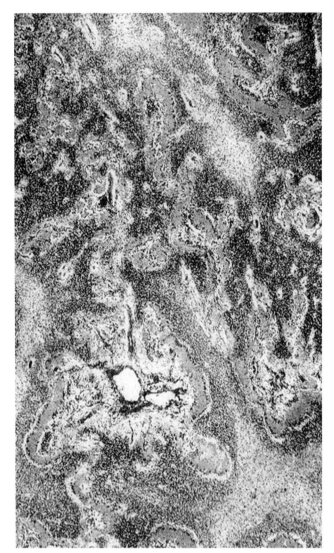

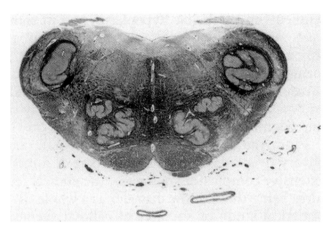

4.119 Miller–Dieker syndrome. Horizontal section of the medulla showing multiple olivary heterotopias. Their folded shape and myelin fibre sheaths are reminiscent of the normal nucleus. Here the inferior olives are small and dysplastic. Luxol fast blue–cresyl violet.

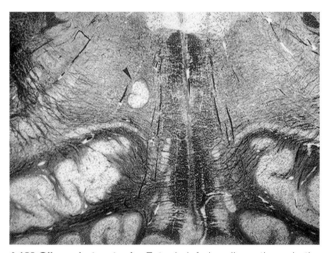

4.118 Cerebellar cortical dysplasia in an infant with cerebro-ocular dysplasia. Despite the chaotic appearance of the cortex and lack of foliation, the individual layers remain correctly positioned with one another and adjacent molecular layers seem to be fused. Luxol fast blue–cresyl violet.

4.120 Olivary heterotopia. Ectopic inferior olivary tissue in the medial part of the medullary tegmentum. Case of trisomy 13. Luxol fast blue–cresyl violet.

Figure courtesy of Drs RO Barnard and T Revesz, National Hospital for Neurology and Neurosurgery, London, UK.

notably common in trisomy 13,[763] cerebellar hypoplasias (see Figure 4.109) and brain stem dysplasias and in conjunction with other migration disorders. However, ectopic neuronal clusters are relatively frequent in normal infants: the incidence may be over 50 per cent[883] and roughly equivalent in infants with non-neural malformations. Ectopic neurons are rarely observed in adults (Figure 4.117b), implying either sampling error or involution.

Cerebellar Cortical Dysplasia

Brun's term 'heterotaxia' for disorganized cortical tissue within the cerebellar cortex has not been universally accepted,[127–9] but other proffered titles such as 'polymicrogyria' and 'pachygyria' are equally unsuitable. Small foci of dysplastic cortex are not rare, occurring, for example, in 14 per cent of 147 normal infants[883] in the flocculonodular lobes and tonsils and adjacent to the cerebellar

peduncles. Cerebellar cortical dysplasia is notable in postencephalitic porencephaly,[250,352] but by far the most extensive examples, replacing most of the normal cortex, are found in the cerebro-ocular dysplasias (Figures 4.52 and 4.118). The folial pattern is completely obliterated, with a smooth or irregularly fissured surface, which on sectioning is a very thick, grey layer. Microscopically, the cortical layers are in correct register but the folia are scrambled together with an apparent fusing of apposed molecular layers (Figure 4.118). Deeper parts of the cortex may be spared. In a quantitative study of three examples, Schalch and Friede found a relative excess of granule cells and a deficit of Purkinje cells, suggesting damage to the Purkinje cells before the formation of the external granular layer.[914] In a fourth lesion, however, the internal granular layer was deficient, an observation similar to that of De León *et al.*, who concluded that a relatively late interference with the

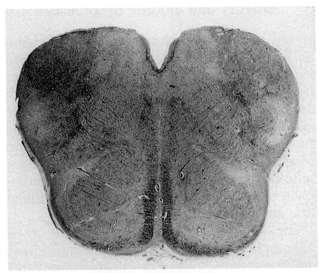

4.121 Olivary heterotopia. Olivary dysplasia in a patient with C trigonocephaly syndrome. The olivary nucleus is C shaped, and its dorsal part is greatly thickened. Luxol fast blue–cresyl violet.

external granular layer was responsible.[251] Superficial destruction might be expected in inflammatory cases, an idea that receives support from experiments in rats and hamsters, where selective destruction of cerebellar meningeal cells by 6-hydroxydopamine appears to destabilize the surface, disrupt the glial scaffold and give rise to cerebellar cortical dysplasia.[561]

Brain Stem Malformations

Olivary Heterotopia

Ectopic segments of inferior olivary nucleus may be found anywhere along the migration route taken by their neuroblastic precursors from the rhombic lip to the ventral medulla.[473] Single or multiple, they comprise small groups of typical olivary neurons and surrounding neuropil, which may also be folded and ensheathed by myelinated fibres in a manner reminiscent of the normal nucleus (Figure 4.119), although the main part of the nucleus may be dysplastic. Most heterotopias occur laterally, near the inferior cerebellar peduncle, but some are placed more medially near the rootlets of the hypoglossal nucleus (Figure 4.120). Autoradiographical studies in the rat demonstrate the majority of olivary neurons to originate laterally in the alar plate,[304] but some arise more medially in the basal plate, which accounts for the medially placed heterotopias. It is presumed that arrested migration leaves stranded precursors to differentiate in an ectopic site. Because migration takes place before the end of the third month,[313] it is not surprising that olivary heterotopia regularly associates with agyria or pachygyria[209,235,271,699,759] (Figure 4.119) rather than with polymicrogyria. Other associations are with the Dandy–Walker syndrome,[422] with cerebral lactic acidosis due to pyruvate dehydrogenase deficiency,[160] and with three cases of megalencephaly, two of trisomy 13 and one with chondrodysplasia and a bizarre gyral pattern (personal observations).

(a)

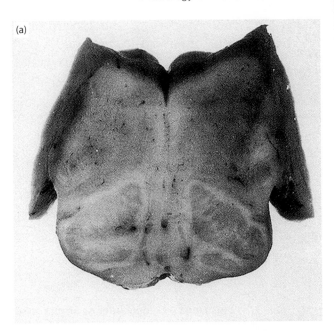

(b)

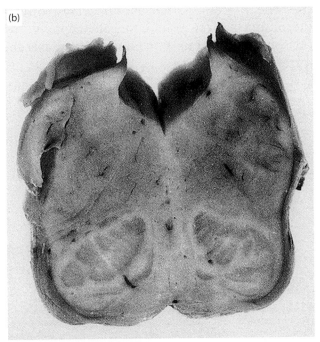

4.122 Zellweger's syndrome. Adjacent horizontal sections through the medulla **(a,b)**. The inferior olivary nuclei are poorly convoluted and fragmented, but their basic shape is a dorsally thickened C.

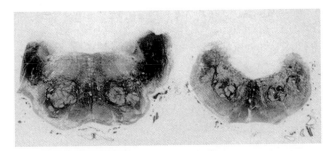

4.123 Olivary dysplasia. Horizontal sections of upper and lower medulla. Both inferior olives are broken into a series of convoluted fragments. Luxol fast blue–cresyl violet.

(a)

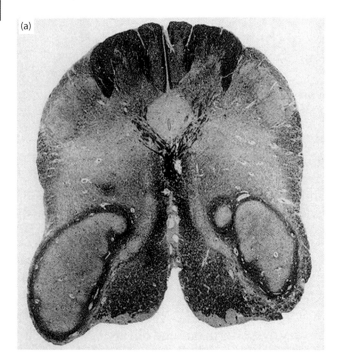

(b)

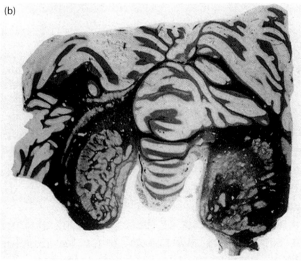

4.124 Dentato-olivary dysplasia. Identical malformations were present in identical twins suffering a rapidly fatal seizure disorder beginning at 5 months. Another sibling has since presented with a similar clinical disorder. Horizontal sections of medulla **(a)** and cerebellum **(b)**. The inferior olives are replaced by elliptical masses, whereas the dentate nucleus is completely disorganized. Luxol fast blue–cresyl violet.

Olivary and Dentate Dysplasias

In view of their common ancestry from the rhombic lip, it is predictable that malformations of the dentate and the inferior olivary nuclei tend to occur together to form dentato-olivary dysplasia, but each may also occur separately. The terminology offered is far from satisfactory; terms such as 'pachygyria' and 'polymicrogyria' are quite unsuited to this particular context, necessitating simple description.

The olive may be hyperconvoluted, as in thanatophoric dysplasia,[474,1092] or hypoconvoluted, as in cerebellar aplasia or hypoplasia (see earlier, Dandy–Walker Syndrome, p. 347;

(a)

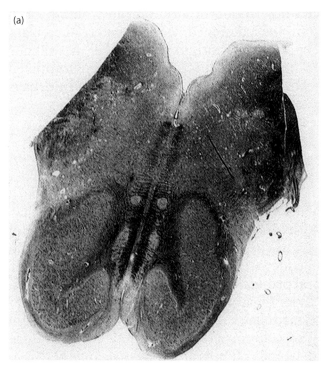

(b)

4.125 Dentato-olivary dysplasia in a neonate. Sections of medulla **(a)** and cerebellar hemisphere **(b)**. The conformation of both olivary and dentate nuclei is reminiscent of that found in a second trimester fetus. Luxol fast blue–cresyl violet.

Pontocerebellar Hypoplasia, p. 351; and Other Examples of Cerebellar Hypoplasia, p. 353). It may form a simple C-shaped band without folds as described in Joubert's syndrome.[351] Alternatively, the dorsal part of the C may be greatly thickened, sometimes with a peripheral capsule of neurons, a finding noted in trisomy 18[990] and in a personal observation of C-trigonocephaly syndrome with callosal agenesis and polymicrogyria (Figure 4.121). Olivary dysplasia is particularly prominent in Zellweger syndrome. Poverty of convolutions, peripheral margination of neurons and fragmentation of the nucleus have been described in neonates[252,316,1060] and in fetal brains.[831] All these features are combined with an overall C-shape and dorsal thickening in several personally studied Zellweger brains (Figure 4.122). Other varieties of olivary dysplasia include complete disorganization into a group of unconnected segments (Figure 4.123),[926,1063] a solid mass of cells (Figure 4.124a) and a horseshoe band resembling the initial stage of fetal development (Figure 4.125a).

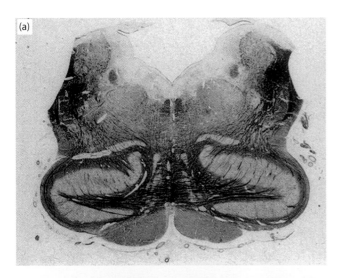

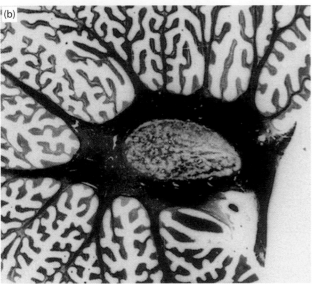

4.126 Dentato-olivary dysplasia. Sections of medulla **(a)** and cerebellar hemisphere **(b)**. This particular combination of coarse, hook-shaped olive and apparently solid dentate nucleus lacking a hilum is associated with intractable seizure disorder.[428]

As far as the dentate nucleus is concerned, it also can be hyperconvoluted in thanatophoric dysplasia, and simplified or broken into islands or segments in cerebellar hypoplasias and in some cases of trisomy 13[761] and 18.[696] Occasionally, it forms a thick plate without convolution[728] or with interconnected bands and masses of cells (Figure 4.126b), or it may be unrecognizable (Figure 4.124b).

Thus, dentate and olivary dysplasias are often only part of a more extensive complex of anomalies, the clinical features being those of the disorder as a whole. In a review of 50 cases of callosal agenesis collected over 40 years, Jellinger *et al.* observed four examples of dentato-olivary dysplasia, but detailed descriptions were not given.[515] Sometimes, however, dentato-olivary dysplasia is the sole significant morphological finding. A remarkably stereotyped dentato-olivary dysplasia has been observed in five children presenting with intractable seizures from early infancy.[112,436] There was severe developmental delay and a variety of seizures occurred, with tonic seizures predominating: the severely abnormal

electroencephalograms (EEGs) showed a burst-suppression pattern in the early months of life. Typically, the inferior olives are hook-shaped, coarse and lacking undulations, whereas the dentate nuclei form a compact or club-shaped mass of interconnected grey islands irregularly separated by myelin fibres (Figure 4.126). Another, similar case has since been reported by Robain and Dulac,[867] and three further autopsied examples of this association have been examined, one of whom has a younger sister presenting a strikingly similar clinical picture, suggesting autosomal inheritance.[671]

The pathogenesis of these lesions is obscure. For some types there is evidence for developmental arrest in the second trimester (Figure 4.125). In the only detailed study of normal development, Murofushi demonstrated the metamorphosis of these nuclei from a hook-shaped plate (olive) or diffuse mass of cells (dentate) at 3.5 months of gestation into their typically folded conformations by about 7 months.[728]

Moebius Syndrome

Congenital facial diplegia with bilateral abducens palsies produces a striking clinical picture in the neonate: a mask-like, expressionless face with internal strabismus.[709] Other oculomotor nerves and lower cranial nerves, including XII, may also be involved. Skeletal abnormalities, absent muscles and mental retardation are also described:[454] Poland's anomaly (absent pectoralis muscle and symbrachydactyly) is a well-known association. Towfighi *et al.* reported one autopsied case and reviewed 14 others, concluding that Moebius syndrome was pathologically heterogeneous.[1016] They classified the available morphological reports into four groups: aplasia or hypoplasia of cranial nerve nuclei; primary peripheral nerve involvement; focal necrosis and calcification in brain stem nuclei, possibly secondary to fetal infection or anoxia; and myopathy. In the first group, lack of necrosis or evidence of degenerative change and the presence of other brain stem anomalies, such as olivary dysplasia, are strong evidence for a primary malformation.[459,863] Two further autopsied cases have been described briefly by Sudarshan and Goldie; one showed brain stem necrosis and the other brain stem hypoplasia.[988] In two personally examined cases demonstrating aplasia of nuclei of cranial nerves VII and XII, there was olivary dysplasia in one and olivary heterotopia in the other; in a third patient a circumscribed area of old necrosis and calcification involved the brain stem tegmentum including VI, VII, X and XII cranial nerve nuclei and reticular formation (Figure 4.127), ultimately leading to sudden death through sleep apnoea. Harbord *et al.* give a clinical and MR imaging description of an infant in whom Moebius syndrome combined with unilateral cerebellar hypoplasia: the authors suggest that these abnormalities could result from a vascular disruption in the basilar artery between 33 and 40 days of gestation.[430]

Abnormalities of the Pyramidal Tracts

Complete or more often subtotal absence of the corticospinal tracts is commonly found in anencephaly, holoprosencephaly, porencephaly and hydranencephaly. An important

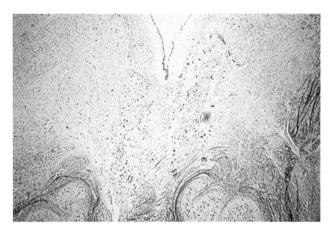

4.127 Moebius syndrome. There is atrophy, gliosis and calcification in the medullary tegmentum. Luxol fast blue–cresyl violet.

association has been reported with X-linked congenital aqueduct stenosis where the absence is predominately observed at the level of the pons and medulla.[144] The cerebral peduncles and basis pontis are small, and the inferior olivary nuclei form the ventral border of the medulla (Figure 4.128). In the spinal cord the dorsal columns are laterally rotated, the ventral and lateral columns are very small and there is an accessory dorsolateral sulcus.

Unilateral hypertrophy is rare.[40,125,791,911,1049,1050] Unilateral lesions of the sensorimotor cortex and/or internal capsule in early life may lead to unilateral hypotrophy of the ipsilateral corticospinal tract and hypertrophy of the contralateral fibre bundles. There is obvious asymmetry of the pyramidal tracts throughout the brain stem, which continues into the spinal cord, whereas in the medulla the abnormally large pyramid may displace the inferior olive dorsally (Figure 4.129). Fibre counts suggest that fibre number rather than size is increased in the hypertrophic tract.[350,909]

Fasciculation of the pyramids into discrete bundles is also occasionally present in malformed brains, as illustrated from a case of polymicrogyria, olivary dysplasia and asymmetry of the pyramidal tracts (Figure 4.130).

Malformations of the Spinal Cord

In addition to the severe gross malformations described earlier (see Neural Tube Defects: Dysraphic Disorders, p. 393) and usually obvious in the neonatal period, other, less severe abnormalities may not be suspected on gross examination. Histological examination at many levels of the cord may be required for their identification. These anomalies, usually described in the adult, also raise the question as to whether they are primary malformations or acquired lesions.

Syringomyelia

The term 'syringomyelia' denotes tubular cavitation of the spinal cord extending over many segments. It may be impossible even at autopsy to distinguish syringomyelia from hydromyelia, although the distinction may have important aetiological implications. Cavities situated in the medulla, syringobulbia, are often associated with syringomyelia. Rarely, cavitation extends into the pons and, exceptionally, it may reach the midbrain and even the internal capsule.[964]

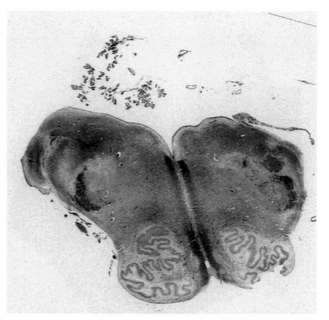

4.128 Absence of the pyramids in a case of X-linked hydrocephalus. Luxol fast blue–cresyl violet.

Clinically, the disease usually begins during the second and third decades and is slowly progressive, or the symptoms may increase at first rapidly and then more slowly. They may cease to progress at any time. Unless associated with bulbar symptoms, the disease rarely causes death directly, but considerable disability is produced by the weakness of limbs and trunk and by the almost invariable spinal deformity. Syringomyelia is usually encountered in adults, but it has occurred occasionally in infants, the youngest reported being 5 weeks old.[291] About 90 per cent of patients with idiopathic syringomyelia have Chiari type I malformation.[936]

When exposed at operation or autopsy, the spinal cord appears swollen and tense in the cervical region and may fill the spinal canal, an enlargement readily demonstrable radiologically. Externally, apart from the swelling, the spinal cord appears normal and there is no leptomeningeal thickening. The syrinx is filled with a clear fluid,[365] usually of a similar composition to CSF or yellow with a high protein content. When, at autopsy, the fluid within the syrinx is allowed to escape, the cord becomes flattened, most often in its anteroposterior diameter. A thorough examination of the syrinx may require many transverse sections. The cavity is usually found to be largest in the cervical region but is often absent from the first cervical segment. The syrinx commonly extends through the upper thoracic segments for a varying distance, but the lumbosacral enlargement is rarely involved. In typical cases, the cavity in the cervical enlargement extends transversely across the cord, involving the more posterior parts of the ventral horns and passing across the midline behind the central canal. It often extends into the posterior horns, also (Figure 4.131a). When very large, it occupies most of the cross-sectional area of the cord; the more anterior groups of motor nerve cells lie in front of it but, otherwise, little grey matter remains and the lateral and posterior white columns are reduced by compression to a narrow zone of fibres (Figure 4.131b). Extensions in the midline or, more laterally, into the

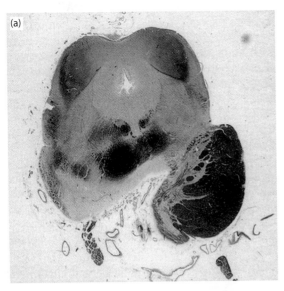

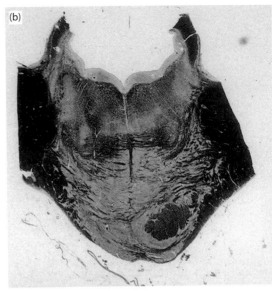

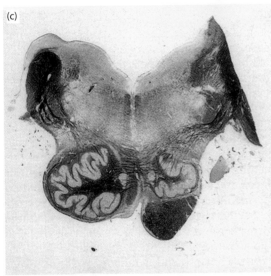

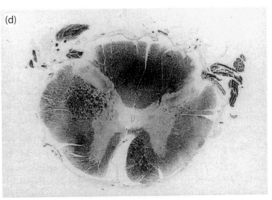

4.129 Unilateral hypertrophy of the corticospinal tracts. Horizontal sections of the brain stem and spinal cord. Luxol fast blue–cresyl violet.

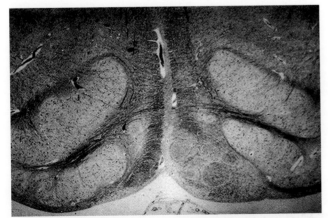

4.130 Abnormality of the corticospinal tracts. Section through the ventral part of the medulla, showing fasciculation and asymmetry of the pyramids. Luxol fast blue–cresyl violet.

posterior columns are common and the anterior white commissure is often destroyed either by pressure or by a midline anterior extension of the cavity. In the thoracic cord the cavity commonly lies in the posterior horns and is often unilateral. When bilateral, the cavities may be separate or may be joined in the region of the grey commissure and so form a single U-shaped cavity. Serial sections show that cavities that are double at one level usually join into a single cavity at some point above, and it is usual for the cavity on one side to end at a higher level than the other. More rarely, two cavities join in the lower thoracic region. Although the grey matter is the common site of cavitation, extensions into the posterior or lateral columns or across the white commissure are not unusual, and the cavity may reach the pial surface at the tips of the dorsal horns at any level.

The walls of the cavity vary greatly in character, especially from case to case but also in different parts of the same cavity. Greenfield believed that these variations depended on the age of the cavity and the degree of tension within it. Where there is recent extension, the wall is irregular and consists of degenerated neuroglial and neural elements. Myelinated nerve fibres enclosed by sheaths of Schwann cells are commonly found in the wall of the syrinx and have been thought

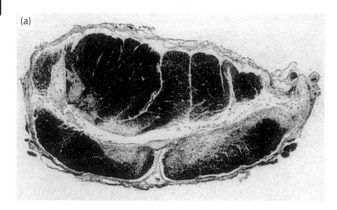

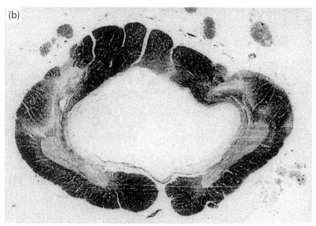

4.131 Syringomyelia. Transverse sections of the spinal cord stained to demonstrate myelin. **(a)** Cervical level: the slit-like cavity extends into both posterior horns. **(b)** Thoracic level: the cavity merges with the central canal. Loyez's method.

to arise by regeneration from damaged posterior nerve roots.[494] The myelin around the syrinx stains poorly, as it does in the oedematous white matter, and the appearances suggest tearing of the tissues and transudation of serous fluid into them. When the cavity has been established for a long time, there is surrounding astrocytic hyperplasia with large fibre-forming astrocytes lying chiefly in a tangential direction to form a dense concentric wall up to 1–2 mm in thickness. It is common to find a thin layer of collagen covering some part of the wall. Thicker strands of collagen, or blood vessels with hyalinized walls, may be seen passing across the cavity from one wall to another. Where the cavity communicates with the central canal, as it does not uncommonly, especially in the cervical region, part of the central wall of the cavity is lined with ependymal cells, but in most places these do not take part in the formation of the wall, a feature distinguishing syringomyelia from hydromyelia. When in syringomyelia, there is a lining of ependymal; the layer of glial tissue that is deep to the ependymal layer is usually thinner than that surrounding the rest of the cavity.

Syringobulbia

Slit-like cavities in the medulla usually lie in one of three positions:[518]

- the most common is a slit running out in an anterolateral direction from the floor of the fourth ventricle external

to the hypoglossal nucleus. It may communicate with the cavity of the ventricle but sometimes begins anterior to this. It passes outwards and forwards for a variable distance towards the descending tract of the trigeminal nerve, usually destroying the fasciculus solitarius and the fibres that pass dorsally from the nucleus ambiguus to join those arising in the dorsal vagal nucleus (Figure 4.132c). A cavity in this position is most commonly limited to the lower half or two-thirds of the medulla, where it interrupts the fibres passing from the gracile and cuneate nuclei to form the decussation of the medial lemniscus. It may also descend low enough to interrupt many of the decussating pyramidal fibres (Figure 4.132d). At this level, it extends transversely from the grey matter lateral to the central canal to a position anterior to the substantia gelatinosa and the descending tract of the trigeminal nerve. Cavities in this position usually have thin walls of neuroglial tissue. Occasionally, the slit is replaced by a neuroglial scar, which interrupts the fibre system as completely as a slit or cavity. Such appearances are not uncommon at the upper and lower ends of a cavity and may arise from secondary fusion of its walls. These cavities are usually unilateral, but they may be bilateral and roughly symmetrical (Figure 4.132b), although extending over differing levels of the cord;

- almost equally common is an extension of the fourth ventricle along the median raphe for a variable distance. This is usually lined by ependyma, but occasionally it is replaced by a neuroglial scar containing ependymal cells or small ependyma-lined tubules, like those seen in atresia of the aqueduct (Figure 4.132b). By interrupting the decussation of fibres passing from the descending vestibular nucleus to the median longitudinal fasciculus, these slits may cause nystagmus but are otherwise asymptomatic. When this median extension is small and is not associated with other cavitation in the medulla, it scarcely merits the name of 'syringobulbia';

- a rarer position for a cavity is between the pyramid and the inferior olive, where it interrupts the emerging fibres of the hypoglossal nerve (Figure 4.132a). These cavities are usually unilateral; in Spiller's case they were bilateral, although only one was large enough to cause atrophic palsy of the tongue.[964] A cavity in this situation may also damage the anteromedian part of the olive or the posterior part of the pyramid.

Cavities in the pons usually lie in the tegmentum, where they may destroy the fibres of cranial nerves VI or VII or the central tegmental tract. They may pass down posterior to the olive for a short distance. Extensions of the cavity to a higher level are extremely rare. In Spiller's case the cavity in one pyramid passed up among the corticospinal fibres, destroying also the substantia nigra, and ended in the internal capsule and the caudate and lentiform nuclei.

Secondary Degenerations

Secondary degenerations follow destruction of tracts and fibre systems both in syringomyelia and in syringobulbia. The pyramidal tracts may be pressed on by an anterior cavity between the pyramid and medullary olive, or a slit at a lower level may interrupt pyramidal fibres during their

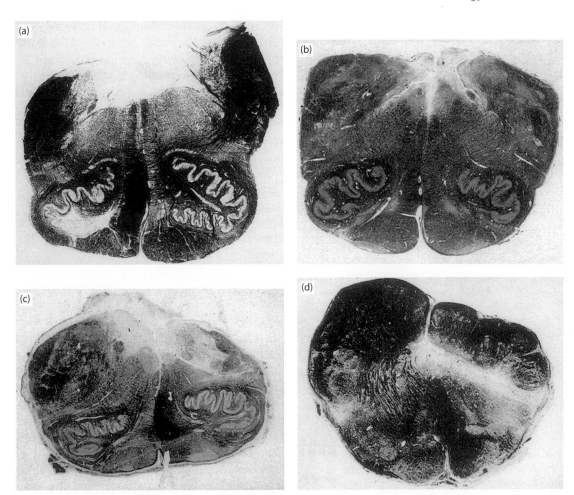

4.132 Syringobulbia. Transverse sections of medulla. **(a)** Case with ventral slit between olive and pyramid and dorsolateral slit at a lower level that has produced degeneration of the contralateral medial lemniscus. **(b)** Case with bilateral dorsolateral and dorso-medial slits. **(c)** Case with dorsolateral slit producing degeneration of the medial lemniscus. **(d)** Section at the level of the pyramidal decussation. The lateral extension of the cavity has destroyed the crossed pyramidal fibres. Loyez's method for myelin.

decussation. In the latter case, there is commonly some retrograde degeneration of the pyramidal fibres in the medulla. Alternatively, a large cavity in the cervical region may destroy much of the pyramidal tract by compression. The spinocerebellar tracts may degenerate owing to destruction of their cells of origin in the grey matter of the cord or of the fibres leaving these. The fibres passing in the posterior columns may be compressed or destroyed by extensions of the cavity in the cord. In cases of syringobulbia the fibres relayed to the thalamus from the gracile and cuneate nuclei are commonly destroyed as they pass towards the decussation of the medial lemniscus, so there is more or less complete absence of the medial lemniscus on the side opposite to the lesion (Figure 4.132a,c); retrograde degeneration occurs in the nerve cells of these nuclei. Fibres passing to the spinothalamic tracts are very commonly involved either at or near their cells of origin in the posterior horns or as they cross the midline in the central commissure. The dorsolateral cavities in the medulla commonly interrupt some of the fibres passing inwards from the descending trigeminal nucleus to the medial lemniscus.

A great variety of clinical signs are associated with these lesions. Wasting of the muscles of the hands and forearms, dissociated or complete anaesthesia, paralysis of the lower

cranial nerves and facial or corneal analgesia, oculomotor palsies and nystagmus occur, depending on the level and extent of the lesions.

Secondary Syringomyelia

This term should be reserved for longitudinal cysts secondary to clearly evident causes. Tumours, trauma, adhesive arachnoiditis, haematomyelia and vascular lesions account for most cases. Most of the cysts are small, but in tumours the associated cystic lesion may be extensive and progressive. Post-traumatic syringomyelia is dealt with in Chapter 10, Trauma.

Aetiology and Pathogenesis

The inclusion of syringomyelia in the large group of spinal cord malformations appears justified at first sight on anatomical grounds, but this may not be correct and brings us little nearer to an understanding of its cause. One difficulty is the distinction of syringomyelia from hydromyelia. Gardner summarized his conclusions in the statement that 'syringomyelia is symptomatic hydromyelia'.[365] This view was derived from the finding of a malformation of the hindbrain obstructing the foramen of Magendie in a large proportion of cases. Gardner proposed that this obstruction causes distension initially of the central canal, which may then rupture, allowing CSF to escape into the substance of the spinal cord. In hydromyelia, there is an undoubted association with neural tube malformations. It is also possible in some cases to trace the development of a syringomyelic cavity from hydromyelia–for example, in cases of Chiari's malformation in the adult–but this is true for only a minority of cases. Syringomyelia is only occasionally familial.

Several different factors may be combined in the pathogenesis of cavitation in syringomyelia: (i) Instability in the lines of junction of the alar and basal laminae with each other is suggested by the position of the cavities, particularly in the medulla, along lines of fusion that occur relatively late in fetal life. (ii) The constant movements of flexion and torsion to which the cervical cord and lower medulla are exposed when the neck is bent or the head is turned[915] must impose stresses that may well cause small tears of the tissue in the centre of the cord, explaining the tendency for cavities to begin and attain greatest size in the cervical region. (iii) Once cavitation in the spinal cord has begun, it may be enlarged by transudation of fluid into it under pressure. This expansion is more likely to occur in the grey matter than in the firmer columns of white matter and, owing to the restricting investment of pia mater, must take place chiefly upwards and downwards. It is favoured by anything that increases venous congestion in the body cavity, such as muscular effort, because there are no valves on the veins draining the spinal cord. The fact that a syringomyelic cavity usually ceases at the C2 level and has no communications with the fourth ventricle may also be of considerable importance.[964]

Asymmetry of Crossing of the Corticospinal Tracts

Failure to decussate is rare,[1050] but asymmetrical decussation is a not uncommon finding in neonates (Figure 4.133). The classic concept of the decussation of the pyramidal tract and of anterior (direct) and lateral (crossed) fibres was revised by Yakovlev, who studied fetal and neonatal brains in serial sections stained for myelin.[1104] The pyramidal tracts are readily identified in the spinal cord of the very young because they are poorly myelinated and their size and shape can be studied more easily than in adults. Most frequently, the fibres of the left pyramidal tract cross the midline first and pass anteriorly to those of the right. Moreover, the cord is usually asymmetric. Nathan *et al.* found three-quarters of these asymmetric cords to be larger on the right side owing to a greater number of corticospinal fibres crossing to the right.[743] Crossing from left to right occurs at a more cranial level than the reverse. Two other rare anomalies have been described: a superficially placed lateral tract[350,1050] and fibres crossing into the dorsal columns,[1050] where they are normally found in rodents.

Arthrogryposis Multiplex Congenita

The clinical presentation of multiple congenital contractures can result from a multiplicity of prenatal disorders that cause fetal hypokinesia.[710] Clinicopathological studies of large series of cases indicate most to be neurogenic and fewer myopathic.[43,417] Neurogenic causes include many

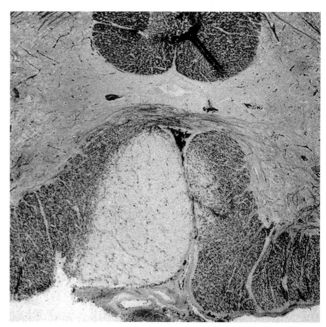

4.133 Malformation of the corticospinal tracts. Transverse section of cervical spinal cord. There is marked asymmetry of the direct pyramidal tracts (unmyelinated in the neonate and therefore pale in section stained for myelin).

examples of developmental brain disease, either acquired destructive lesions or malformation syndromes (trisomy 18, Arnold–Chiari syndrome, Meckel–Gruber syndrome, Marden–Walker syndrome, Moebius syndrome). Anterior horn cell degeneration is a frequent cause, but morphological studies suggest subtle differences from Werdnig–Hoffmann disease.[167] It is also important to distinguish the cerebral and spinal pathology of fetal dyskinesia syndromes, including those that comprise camptodactyly, multiple ankyloses, facial dysmorphism, polyhydramnios, fetal growth retardation and pulmonary hypoplasia.[41,416]

Vascular Malformations

Aneurysmal Malformation of the Vein of Galen

This is the most frequent form observed in the neonate. First recognized in 1923,[1090] more than 100 cases have been reviewed by Lagos.[578] The malformation is not a true aneurysm but an arteriovenous fistula, the vein of Galen undergoing aneurysmal dilation.[466] The dilated vein of Galen (Figure 4.134) is generally fed by blood from one or both posterior cerebral arteries or one of their branches (Figure 4.135) and, less frequently, from small posterior branches of the middle cerebral arteries. The aneurysm may also be fed by anomalous branches of the carotid and/or basilar circulation.[625] The blood vessels may have a normal architecture but, more often, a lace-like network of tortuous vessels empties into the saccular vein of Galen, which may be up to several centimetres in diameter. The entire venous system, including transverse and straight sinuses, is dilated. If the shunt is large, clinical signs may develop soon after birth or within a few weeks. Silverman *et al.*[946] and Pollock and Laslett[828] were the first to point out that cerebral arteriovenous fistula can lead to cardiomegaly and congestive cardiac failure in the newborn and many cases have since been recognized.[387,478,633] through the fistula.[610,633] In addition, twitching or generalized tonic convulsions[166] and hydrocephalus may occur as a result of compression of the aqueduct.[171,714,901]

Various cerebral lesions, either in the territory of the corresponding arteries or elsewhere, have been described in association with this vascular malformation, including periventricular infarction and intraventricular haemorrhage due to compression by the tumour-like vascular mass[757,921] and periventricular calcification.[809] In addition, the aneurysm itself may become calcified[948] or thrombosed.[598]

Other arteriovenous malformations may be located elsewhere. In the neonate, communication between the anterior or middle cerebral arteries and the sagittal sinus[978] or between a branch of the middle cerebral artery and the lateral sinus[144] has been described. An arteriovenous malformation involving cerebellar veins and arteries in a complicated network of blood vessels over the vermis was found in a premature infant.[37] These vascular malformations probably result from failure of the primitive capillary plexus to differentiate into mature arterial and venous channels.

Sturge–Weber Syndrome (Encephalofacial Angiomatosis)

This neurocutaneous syndrome, characterized by naevus formation in the skin in the territories of the sensory branches of cranial nerve V, and by ocular angioma, was first described clinically by Sturge in 1879.[987] In 1897, Kallischer demonstrated the meningeal angiomatosis,[533] and in 1923 Dimitri described skull calcification on X-ray, later shown to be intracerebral.[274] There is excessive vascularity of the meninges, the small veins being tortuous and increased in number, giving a dark purple colour to the cerebral surface. On sectioning the brain, calcification may be readily visible beneath the hypervascularized meninges in the outer cortical layers. On microscopic examination (Figure 4.136), the walls of the blood vessel are encrusted with deposits of iron and calcium, and calcific granular deposits of varying size lie freely in the parenchyma. Polymicrogyria and heterotopias may also be found in the cerebrum and cerebellum. The port-wine stain of the face is present at birth, as is buphthalmos in 70 per cent of cases.[1037] Intracranial calcification has exceptionally been reported in neonates[931] and diagnosed on CT scan.[772]

Focal neurological signs and cerebral atrophy may be evident at birth, but most of the symptoms, such as hemiparesis, hemiplegia, epilepsy and mental retardation, begin within the first year of life[870] or in early childhood.[159,406] The pathogenesis of the disease is not known, but incomplete involution of the embryonal vasculature has been proposed.[159] Although familial cases have been reported, there is no clear evidence for hereditary transmission.

Fowler's Syndrome: Proliferative Vasculopathy and Hydranencephaly–Hydrocephaly

It is most important to differentiate between this rare familial syndrome, first described in five siblings by Fowler, and the more typical, sporadically occurring, encephaloclastic form of hydranencephaly from which it is ultrasonographically and macroscopically indistinguishable. To date there have been four published reports.[344,439,441,758]

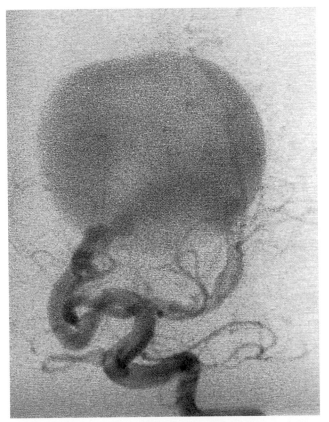

4.134 Aneurysm of the great vein of Galen. This angiogram demonstrates at least two fistulae from the markedly hypertrophic posterior choroidal arteries.

Figure courtesy of Dr W Taylor, National Hospital for Neurology and Neurosurgery, London, UK.

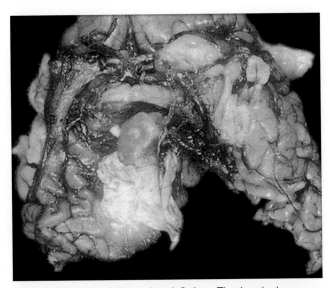

4.135 Aneurysm of the vein of Galen. The hemispheres are splayed apart to reveal the aneurysm and its feeding blood vessels from the posterior cerebral artery.

The differential diagnosis from congenital heart defect is important. In arteriovenous fistula, auscultation over the surface of the skull may reveal a continuous murmur in up to 80 per cent of cases,[478] indicating the flow of blood

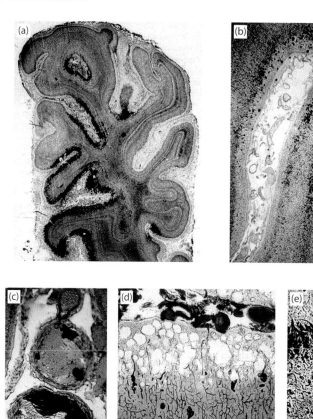

4.136 Sturge–Weber disease. (a) Low-power view of section through the occipital pole, showing the meningeal venous angioma and the darkly stained dense calcification mainly in the outer layers of the cortex. Carbol azure. **(b)** A higher-power view of the meninges and cortex of the occipital pole. The calcification affects deeper as well as superficial layers of the cortex. Carbol azure. **(c)** Calcified meningeal arteries. Carbol azure. **(d)** Abnormal convoluted blood vessels in the non-calcified layer of the cerebral cortex. Masson's trichrome. **(e)** Adventitial fibrosis occasionally found in areas of laminar calcification. The calcium deposits are peculiarly dense and powdery in these areas. Gomori's reticulin.

Clinical and pathological findings are remarkably consistent. Recurrent intrauterine death or therapeutic interruption of pregnancy following ultrasound demonstration of 'hydranencephaly' as early as 13 weeks' gestation reveals an abortus with severe arthrogryposis, pterygia and muscular hypoplasia and massive cystic dilation of the ventricles (Figure 4.137). Although extremely attenuated, the cerebral wall retains some degree of normal organization, with a thin but recognizable ventricular zone and persistence of radial glia.[439] However, the cortical plate is very thin, interrupted or folded; there is a greatly reduced population of neurons, scattered calcospherites and a most remarkable and pathognomonic glomeruloid vascular proliferation (Figure 4.138). These striking vascular structures, composed of inclusion-bearing endothelial cells (Figure 4.139a),[439] are found in all parts of the CNS, including the germinal matrix and the spinal cord (Figure 4.139b), and, in Fowler's case, also in the retina but not in the leptomeninges or other tissues. Calcifications are particularly massive near the ventricular surface of the basal ganglia, and only here is there minor evidence of necrosis.

The presence of pterygia[758] and ultrasound evidence of well-established hydranencephaly by 13 weeks of gestation[439] date the onset of this autosomal recessive disorder to the first trimester; Norman and McGillivray considered a primary failure of neuroectodermal proliferation to occur before the seventh week.[758] Alternatively, because vascular channels first appear in the pallium at about the time when neuronal migration begins, and may be critical to the survival of neuronoglial precursors, it could be argued that glomeruloid changes set in motion a destructive process.[439] Massive cystic ventricular dilation could then be a consequence of an altered hydrodynamic equilibrium between CSF pressure and the attenuated and weakened cerebral wall, because CSF pressure is known to exert a powerful effect on embryonic brain enlargement.[268]

Diffuse Meningocerebral Angiodysplasia and Renal Agenesis

This rare association has been documented in three isolated reports,[136,564,1037] and a fourth example is depicted in Figure 4.140. All were premature stillbirths with bilateral renal agenesis and other features typical of the oligohydramnios sequence, namely Potter's facies, contraction

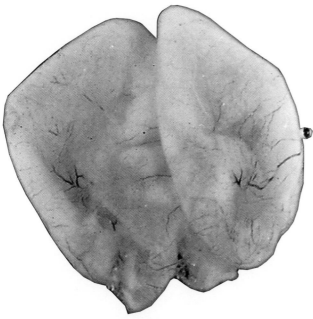

4.137 Fowler's syndrome, or proliferative vasculopathy and hydranencephaly–hydrocephaly. Cystic cerebral hemispheres in a 17-week fetus.

Reproduced with permission from Harding.[439] *Copyright © 2008, John Wiley and Sons.*

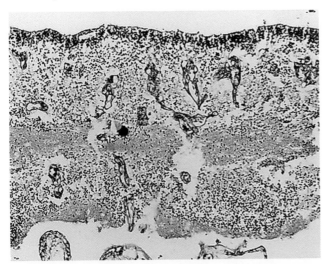

4.138 Same case as in Figure 4.140. Histologically the pallium is severely attenuated and contains numerous glomeruloid vascular structures. Gordon and Sweet silver impregnation for reticulin fibres.

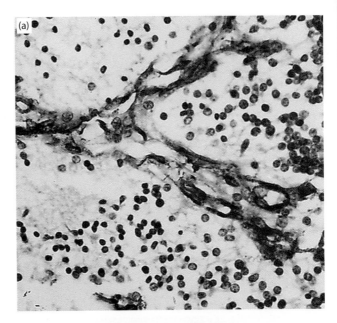

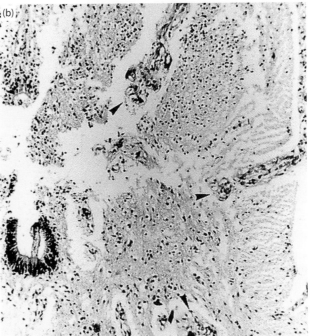

4.139 Same case as in Figure 4.140. (a) The constituent cells of the glomeruli are decorated by antibodies to the endothelial marker *Ulex europaeus* lectin. **(b)** Vascular glomeruloids are widespread in central nervous system tissue: several (arrowheads) are shown in the spinal cord.

deformities of the limbs and bilateral pulmonary hypoplasia. A diffuse angiodysplasia, of dilated and thrombosed capillaries and venules, involves the leptomeninges and parenchyma of the cerebral hemispheres and is associated with extensive infarction and calcification.

Arachnoid Cysts

Arachnoid cysts have been the subject of an abundant literature.[931] However, there seems to be some confusion concerning the definition itself, as well as the pathogenesis of the lesions. Anomalies as diverse as dilated cisterns,

ependymal cysts of the cavum vergae, diverticula of the ventricles and cysts of neural origin have been described under the term 'congenital arachnoid cyst'. Strictly speaking, the so-called primary arachnoid cyst should be regarded as a developmental abnormality of the arachnoid.[406,772,972] Ultrastructural studies have demonstrated that the cyst is formed by splitting of the arachnoid membrane, which is reinforced by a thick layer of collagen.[159,854,912] The wall of the cyst is totally independent of the inner layer of the dura mater. The fine structure of the cells is similar to that of normal trabecular arachnoid cells.

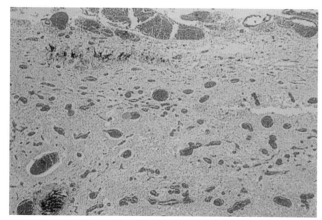

4.140 Diffuse meningocerebral angiodysplasia and renal agenesis. Leptomeninges and cerebral parenchyma are studded with numerous dilated and thrombosed capillaries. Infarction and calcification ensue.

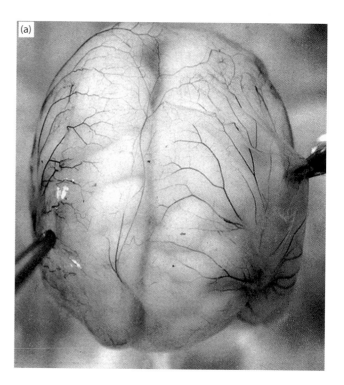

(a)

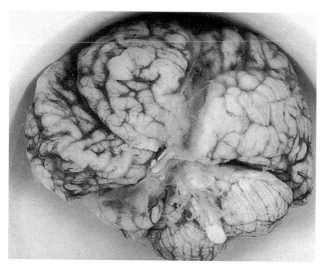

4.141 Bilateral sylvian fissure arachnoid cysts in a 9-month-old child. Photographed in water.

The distribution of the cysts in over 200 cases was: sylvian fissure, 49 per cent (Figure 4.141); cerebellopontine angle, 11 per cent; supracollicular area, 10 per cent; vermis, 9 per cent; interhemispheric fissure, 5 per cent; cerebral convexity, 4 per cent; and the clival and interpenduncular area, 3 per cent.[854]

Arachnoid cysts are rare in neonates. Larroche found five in about 3000 consecutive fetal and neonatal autopsies: two extended over the convexity of the hemispheres (Figure 4.142a) or filled the interhemispheric fissure (Figure 4.142b) (in one there was an associated absence of the septum pellucidum), one occurred over the temporal pole, and two occurred in the posterior fossa over a cerebellar hemisphere, clearly differing from a Dandy–Walker malformation.[587] Other types of cyst have been reported in the neonatal period. Loeser and Alvord described a cluster of interhemispheric cysts attached to the falx, with absence of the corpus callosum.[628] These were lined by ependyma and contained choroid plexus, features that might be found in diverticula of the third

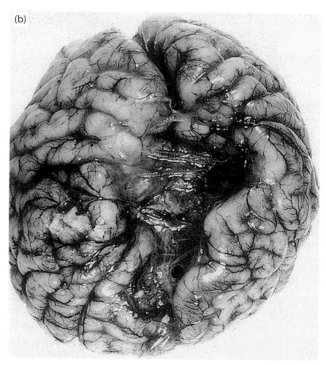

(b)

4.142 Arachnoid cysts. (a) Cystic dilation of subarachnoid space extending over the convexities of both hemispheres in a neonatal brain. **(b)** Arachnoid cyst in the interhemispheric fissure.

ventricle. A dilated cavum veli interpositi may be mistaken for a cyst or a posterior fossa tumour.[1005] Haemorrhage in this area with secondary hydrocephalus due to compression of the aqueduct of Sylvius was reported by Lourie and Berne,[634] and a paracollicular plate cyst described in a baby born after a difficult labour and containing stigmata of old haemorrhage[569] may represent a link between primary and secondary arachnoid cysts. The possibility exists that the anomalies may be overlooked at autopsy in fetuses and neonates, or certain cysts may develop and become symptomatic only after birth. Occasional cases have been diagnosed in neonates by CT scan[446] and by ultrasound.[647]

Phakomatosis

Tuberous Sclerosis (Bourneville's Disease)

This complex disorder of protean manifestations was first described by Recklinghausen in a neonate with cardiac rhabdomyomata,[851] but the term 'tuberous sclerosis' was invoked by Bourneville[110] in a 3-year-old girl with mental retardation, seizures and facial angiofibromas.[101] More recent clinical and morphological reviews include that of Fryer and Osborne[355] and the monograph edited by Gomez.[386]

The brain is the most frequently affected organ in tuberous sclerosis. Brain weights, although usually normal, can range widely from microcephalic to megalencephalic. The characteristic lesions are cortical tubers, subependymal nodules and heterotopias in the white matter, all of which have been recorded in fetuses as early as 28 weeks of gestation,[158,929] and in the author's experience as early as 19 weeks' gestation. In the unfixed brain, cortical tubers are readily recognizable as firm nodules projecting slightly above the surface of the cortex, but they are even more striking after the leptomeninges have been stripped (Figures 4.143 and 4.144). In fetal and neonatal brains, tubers are not so visually prominent but are obvious on palpation. Varying in size from millimetres to several centimetres, tubers are rounded or wart-like protrusions of single or adjacent gyri, very firm to the touch and pale in colour. They may be wide and flat or round and dimpled (Figures 4.143 and 4.144).[807] Scattered randomly over the cortical surface, up to 40 have been observed in a single brain. On coronal section, tubers greatly expand the gyri and blur the margin between grey and white matter; they may also be present in the depths of sulci. Histologically, the normal cortical architecture is effaced by collections of large, bizarre cells with stout processes, peripheral vacuolation, prominent nucleoli and sometimes multiple nuclei (Figures 4.145–4.148). With conventional stains, they may be characterized as atypical astrocytes or abnormal maloriented neurons but many are indeterminate. Neurofibrillary tangles, argentophilic globules and granulovacuolar degeneration have also been described.[467] Clusters of these abnormal cells may also be found scattered widely in the deep white matter and macroscopically normal cortex. The tuber may show a marked fibrillary gliosis, particularly in the subpial zone where fibres are condensed into sheaf-like structures (Figure 4.149). In older individuals, myelin staining stops abruptly like a flat plate beneath the abnormal cortex, whereas the gyral core is depleted of myelin and is gliotic (Figure 4.150). Tubers may calcify and multiple 'brain stones' can be seen radiologically.[1105] Tubers are occasionally present in the cerebellum;[1040] disorganized cortex, abnormal astrocytes and Purkinje cells and calcification are the main features (Figure 4.151).

Subependymal nodules may occur in the third and fourth ventricles, and even the aqueduct, but most are found in the lateral ventricles, particularly near the sulcus terminalis, with their deeper parts embedded in the caudate nucleus or thalamus. They are firm, or stony hard due to calcification, and form round or elongated protrusions into the ventricles, either singly or in rows, when they have been likened to candle gutterings (Figure 4.152). Nodules at the foramen of Monro are of particular clinical importance because they may obstruct the foramen to cause hydrocephalus. Histologically (Figure 4.153), beneath a covering layer of ependyma there is a mixture of elongated or markedly swollen glial cells and their processes, giant or multinucleated cells, and often marked calcium deposition, whereas in neonates there may be scattered clusters of neuroblasts. A sequential CT

4.143 Tuberous sclerosis. A cortical tuber (Pellizzi type 1) is shown as a widened and flattened region of a gyrus. The surface is slightly granular, and the tuber is pale and firm to the touch. The leptomeninges have been stripped off.

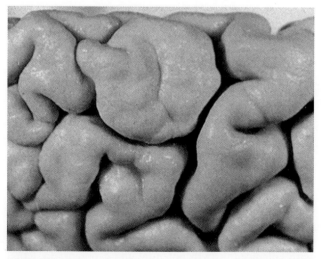

4.144 Tuberous sclerosis. A cortical tuber (Pellizzi type 2) appears as a rounded flattened nodule with a rough dimpled surface. The tuber is elevated above the surrounding brain, from which it is demarcated by a sulcus.

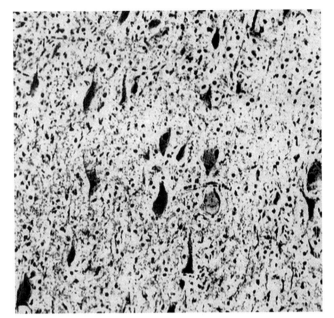

4.145 Tuberous sclerosis. Section of a cortical tuber, showing giant cells. Hortega's silver carbonate method for astrocytes.

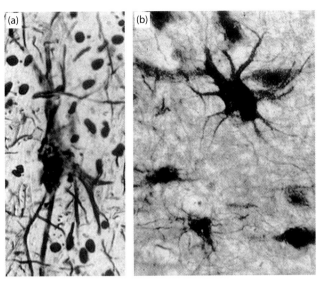

4.147 Tuberous sclerosis. Giant cells of astrocytic type in a cortical tuber. **(a)** Hortega's method; **(b)** Cajal's gold sublimate method.

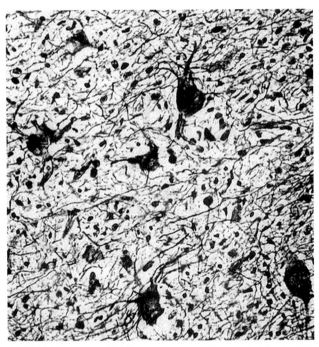

4.146 Tuberous sclerosis. A group of large abnormal cells in a cortical tuber. They are stained by Bielschowsky's silver method, suggesting a neuronal phenotype.

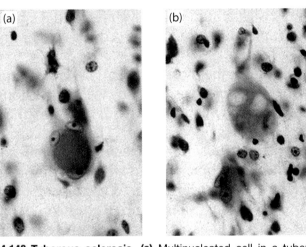

4.148 Tuberous sclerosis. (a) Multinucleated cell in a tuber. Carbol azure. **(b)** Large cells with vacuolated cytoplasm. Carbol azure.

study has documented the gradual progression of a subependymal nodule into a subependymal giant cell astrocytoma (Figure 4.154) (see also Chapter 27, Astrocytic Tumours).[719] Histologically, the distinction between the two is far from clear but, for the clinician, signs of raised intracranial pressure herald the presence of a tumour with enhanced growth potential and a potentially lethal location.

Much discussion has attended the ontogenesis of the bizarre giant cells in tuberous sclerosis. Ultrastructural studies have indicated both astrocytic and neuronal features.[73,243,858,947,1021] Several immunohistochemical studies, however, have shown a paucity of staining for GFAP in the subependymal giant cells (Figure 4.155),[100,158,739,929,975] and some authors have noted a high proportion of GFAP-positive cells in subcortical lesions and cortical tubers, suggesting that acquisition of GFAP is associated with migration.[158,929] Concomitant expression of neurofilament[100,158] and galactocerebroside[158] has also been reported. It seems that both migration and differentiation are disturbed in these apparently pluripotential cells.

There is a high incidence of seizures in patients with tuberous sclerosis, usually commencing within the first few months of life, when their significance may be underestimated in the absence of skin signs.[386,712] Mental retardation is the second most common neurological manifestation. Evidence of raised intracranial pressure is found in about

4

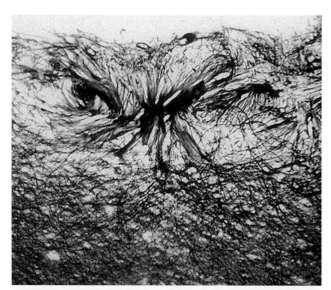

4.149 Tuberous sclerosis. Sheaf-like bundles of neuroglial fibres in the superficial part of a cortical tuber. Holzer's method.

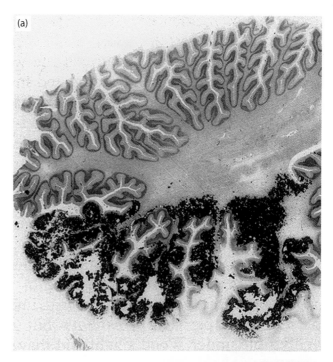

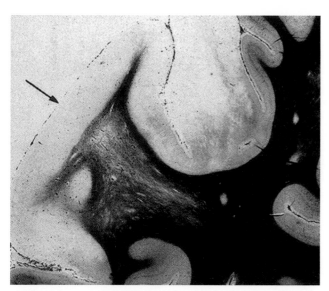

4.150 Tuberous sclerosis. A cortical tuber (arrow) overlies the flat margin of the central core of white matter, which shows a paucity of myelin and is also gliosed. Heidenhain's method for myelin.

5 per cent of cases. Behavioural problems are common and include hyperactivity, screaming, destructiveness, aggression, sleeplessness, self-mutilation and autism spectrum disorder.

Skin manifestations, the most common clinical findings in tuberous sclerosis, are of five types. Hypomelanotic macules, 1–3 cm in diameter, are present in 90 per cent of patients but may be visible only under Wood's light and may not be found in early life. Facial angiofibromas (adenoma sebaceum) form a butterfly rash over the cheeks, nose, lower lip and chin and appear between 2 and 5 years of age. Periungual or subungual fibromas, more often on the toes than the fingers, rarely occur before puberty and may be the sole manifestation of the disease. Shagreen patches, which are

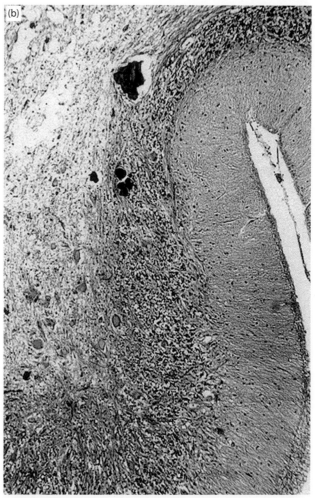

4.151 Tuberous sclerosis. (a) A cerebellar tuber with extensive calcification in the inferior part of the hemisphere. **(b)** Higher magnification shows disorganized cortex and destruction of the white matter. There are large astrocytes and calcospherites.

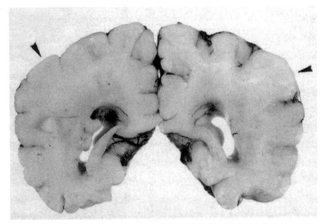

4.152 Tuberous sclerosis in a premature infant. There are multiple subependymal nodules and cortical tubers (arrowheads).

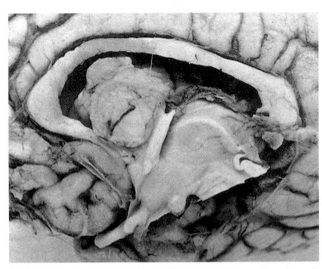

4.154 A large tumour arises from a subependymal nodule over the caudate nucleus in an adult case of tuberous sclerosis.

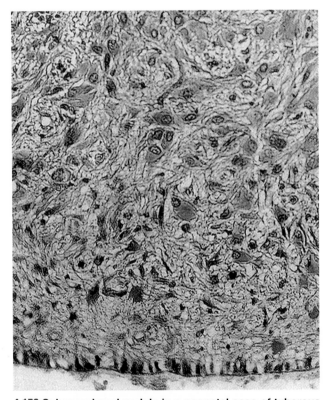

4.153 Subependymal nodule in a neonatal case of tuberous sclerosis. The cells are large with a single rounded nucleus and abundant eosinophilic cytoplasm, and are intermingled with leashes or whorls of glial fibres.

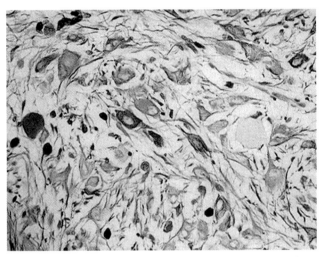

4.155 Tuberous sclerosis. Subependymal nodule. Immunohistochemistry for glial fibrillary acidic protein (GFAP) shows marked variation in reactivity. Only a minority of giant cells are strongly positive.

fibrous hamartomas of dorsal surfaces, are rarely seen before puberty, whereas fibrous plaques on the forehead and scalp can be the earliest skin sign. Poliosis and leucotrichia are also common.

In the Mayo Clinic series, 50 per cent of patients had ocular involvement.[386] The characteristic lesion is the retinal giant cell astrocytoma, but other diagnostically significant lesions include hypopigmented iris spots, white eyelashes and hamartomata of the eyelids and conjunctivae.

Cardiac rhabdomyoma, whether single or multiple, is common: *in vivo* studies using two-dimensional echocardiography found an incidence of 50 per cent in the population attending a paediatric neurology clinic[60] and of 64 per cent in a combined child and adult population.[372] Rhabdomyomata have been reported in a fetus of 6 months' gestation[600] and in several other fetal cases presenting with hydrops.[158] In general, these tumours are slow-growing and asymptomatic but may cause obstruction to flow, myocardial dysfunction, cardiac arrhythmias and sudden death.[149,1024,1041] The tumours grow within the ventricular walls and, although not encapsulated, are sharply demarcated from surrounding myocardium. Some form large, pedunculated masses that protrude into the chambers of the heart, obstructing flow (Figure 4.156). Histologically, the tumour cells are large, up to 20 μm in diameter, with

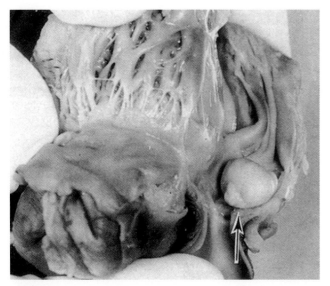

4.156 Dissection of the heart of a neonate with tuberous sclerosis. A large rhabdomyoma obstructs the left ventricular outflow tract (arrow).

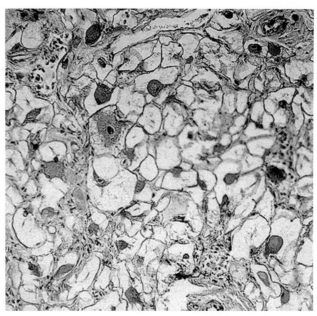

4.157 Typical microscopic appearance of one of the multiple rhabdomyomatous nodules found in the tuberous sclerosis complex.

eccentrically placed single or multiple nuclei, and vacuolated glycogen-containing cytoplasm with fine cytoplasmic strands (spider cells) (Figure 4.157).

In the Mayo Clinic series, 55 per cent of living patients had renal lesions.[386] Angiomyolipomas, usually multiple, are the most common, but they are rare under 10 years of age. Renal cysts are less frequent but may present in infancy. Angiomyolipomas can rarely be found in the lungs, liver, adrenal glands, thyroid, gonads, teeth, gums or bones.

In recent years, there has been increased recognition of milder forms of tuberous sclerosis, and older prevalence figures have been radically revised: the Oxford prevalence study suggested that the true prevalence for 0–5 years could be as high as 1 in 10 000.[495] Tuberous sclerosis is inherited as an autosomal dominant trait with very high penetrance.[47] It was thought that the large number of new cases was explicable in terms of a high mutation rate, but studies of asymptomatic family members with cranial CT scanning have identified a significant proportion with abnormalities.[339]

Linkage studies have shown locus heterogeneity with disease-determining genes mapped to chromosome 9q34 (designated *TSC1* gene) and 16p13.3 (*TSC2* gene).[315,356] Allele loss (i.e. loss of heterozygosity) for 16p13.3 has been demonstrated in hamartomas, a cortical tuber and a giant cell astrocytoma from patients with tuberous sclerosis, consistent with the hypothesis that *TSC2* acts as a tumour suppressor gene.[399]

Hypomelanosis of Ito

This rare neurocutaneous syndrome has a variable phenotype.[380] The few neuropathological reports indicate some similarities with tuberous sclerosis and cortical dysplasia, although the process extends more widely and shows a lesser degree of cytological dysplasia.[652,886] Megalencephaly or hemimegalencephaly, firm macrogyric convolutions, loss of cortical layering and widespread neuronal heterotopia,

TABLE 4.5 Scheme for hydrocephalus in childhood

A. Imbalance between production and drainage of CSF
  1. Overproduction of CSF: choroid plexus papilloma
  2. Interference with CSF movement
    a. Reduced propulsion: ciliary dysplasia
    b. Physical block to flow
      i. Ventricular system
        • Tumours
        • Intrinsic block
        • Extrinsic compression
        • Malformation
        • Membranous obstruction to foramen of Monro
        • Obliteration of third ventricle
        • Connatal obstruction of aqueduct of Sylvius (atresia, stenosis, vascular anomaly)
        • Obstruction to foramina of Luschka and Magendie (atresia, Dandy–Walker, arachnoid cyst)
        • Inflammation and haemorrhage and their consequences
        • Adhesions in lateral ventricle, obliteration of foramen of Monro
        • Gliosis or septum of aqueduct
        • Gliosis or fibrosis of fourth ventricle outlets
      ii. Subarachnoid space
        • Chiari malformation at foramen magnum
        • Diffuse obliteration: cerebro-ocular dysplasias
        • Fibrosis: after haemorrhage, meningitis, mucopolysaccharidosis
  3. Failure of absorption
    • Absence of arachnoid granulations and cilia
    • Functional changes: cranial dysplasias
B. Developmental anomalies: ventriculomegaly of uncertain pathogenesis
C. Following destruction or degeneration of brain tissue
    • Hypoxic–ischaemic (prenatal or postnatal)
    • Degenerative (grey or white matter)
D. Mixed due to A and C

CSF, cerebrospinal fluid.

with reactive (not giant) astrocytes and Rosenthal fibres, are the chief morphological features. A *de novo* somatic mutation in mTOR has been reported in one case.[602]

Von Recklinghausen's Disease

See see also Volume II, Chapters 26–45 on tumours of the nervous system.

Hydrocephalus

An overview of the causes of hydrocephalus in childhood is given in Table 4.5; the content of this section roughly follows this plan. However, because general aspects of the subject are dealt with in Chapter 3, Diseases of the Perinatal Period, only developmental lesions producing hydrocephalus are described here, concentrating on those lesions not covered earlier in the chapter.

Imbalance between Production and Drainage of Cerebrospinal Fluid

Most frequently, this results from obstruction to the flow of CSF:[900] extrinsic tumours may compress the ventricular system, whereas intraventricular tumours, malformations, haemorrhage and inflammation directly impede flow.

Obstruction to the Foramina of Monro

This is usually the result of neoplasia, neonatal meningitis and ventriculitis, or the aftermath of intraventricular haemorrhage, but there are rare reports of unilateral hydrocephalus secondary to membranous obstruction of one foramen,[1075] including intrauterine diagnosis by ultrasound at 29 weeks of gestation.[738] In one case, the obstructive glial–ependymal membrane and other subtle anomalies suggested an early developmental onset.[812]

Obliteration of the Third Ventricle

This is exceptional. Fusion of the thalami has been demonstrated post mortem in one of two siblings presenting with neonatal hydrocephalus[162] and in a unique association with aqueductal atresia and rhomboencephalosynapsis.[547]

Connatal Obstruction of the Aqueduct of Sylvius

Russell emphasized the importance of the aqueduct, the narrowest part of the ventricular system, as a site of maldevelopmental causes of hydrocephalus,[900] but her attempts to define clear-cut categories for the lesions, stenosis, forking (atresia) and septum formation, which she regarded as malformative, in contrast to gliotic lesions, which were considered a sequel to acquired inflammation, have excited much controversy. Subsequent investigators have varied in their use of this terminology, and individual cases may defy simple categorization.[284] Several considerations necessitate modification of Russell's schema. It has become increasingly realized that the distinction between congenital and acquired lesions is somewhat artificial. The aqueduct develops as a gradual narrowing of the neural tube,[1028] which has *ab initio* a lumen, so all obstructions are, strictly speaking, acquired. Furthermore, absence of gliosis does not

militate against an infectious aetiology, as demonstrated by experimental induction of aqueduct stenosis in hamsters inoculated with mumps virus.[526] Finally, aqueduct stenosis may be the result, rather than the cause, of hydrocephalus. From a study of resin casts of the aqueduct in children with spina bifida, Williams postulated that the tectal plate was compressed by the expanding hydrocephalic hemispheres.[1078] There is experimental support for this idea. Communicating hydrocephalus followed by stenosis of the aqueduct was described in the hydrocephalic mutant mouse *oh*.[107] Masters *et al.* demonstrated in mice inoculated with reovirus that hydrocephalus was related to the degree of inflammation and fibrosis within CSF pathways and aqueduct stenosis was a secondary consequence of midbrain compression.[672] Findings in X-linked hydrocephalus in humans would also favour this concept (see later, Stenosis).

Stenosis To substantiate a diagnosis of aqueduct stenosis, that is, a greatly reduced lumen in the absence of significant histological abnormality or gliosis (Figure 4.158), serial sections may be required. The aqueduct is a curved, irregular tube that varies in calibre along its length, having two constrictions either side of a central ampulla. In adults, the narrowest part ranges in area from 0.4 to 1.5 mm^2.[1099]

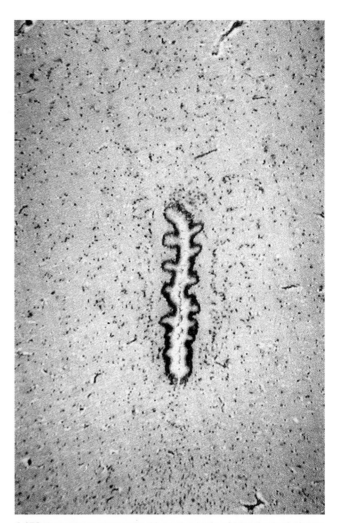

4.158 Aqueduct stenosis. Note the lack of histological abnormality in the adjacent neuropil.

Emery and Staschak found a wide range of calibre in normal children: the mean diameter at the narrowest point was 0.5 mm[2].[309] A rare form of X-linked recessive hydrocephalus and aqueduct stenosis was first described by Bickers and Adams.[79] Perhaps 2 per cent of all cases of congenital hydrocephalus are of this type.[46] Some 30 families have now been reported.[299,402,583] Hydrocephalus affects only boys, who are usually stillborn or show neonatal macrocephaly, but there is considerable variability of expression within families, with some male siblings showing more subtle changes in head circumference, mental retardation and survival into adult life.[298,1077] Adduction–flexion deformity of the thumbs is present in about 25 per cent of cases.[318] Neuropathological examination has revealed a very narrow stenotic aqueduct in some cases[299,480] but a normally patent aqueduct in others,[421,583,1077] leading to the suggestion that aqueduct stenosis is a secondary phenomenon. Congenital absence of the pyramids is an important association.[161,299,583] X-linked hydrocephalus is the result of mutations in the *L1CAM* gene, which is further implicated in X-linked spastic paraplegia and MASA syndrome[357,884] A very rare autosomal recessive form of aqueduct stenosis has also been reported.[489]

Atresia/Forking Both terms have their detractors: 'atresia' was unacceptable to Russell because a small lumen was always present;[900] 'forking' was abandoned by Friede because of confusion with normal anatomical variation and embryological inconsistency.[350] Baker and Vinters made the suggestion of 'aqueductal dysgenesis', which also carries pathogenetic implications that may not prove to be justified.[38] In view of our knowledge of experimental, virally induced aqueduct block, the term 'obliteration' seems more appropriate at present.

A variable portion of the aqueduct may be invisible to the naked eye, but histologically there are groups of ependymal cells forming small rosettes or tiny ependymal canals scattered irregularly across the midbrain tegmentum (Figure 4.159). The normal outline of the aqueduct cannot be discerned, but there is no gliosis surrounding the aqueductules. Aqueduct atresia may be associated with Arnold–Chiari malformation, hydranencephaly and cases of craniosynostosis[37] or may occur in isolation. Mumps infection is a possible aetiological factor in view of the histological similarity of experimentally induced aqueduct atresia[526] and reports of aqueduct stenosis in children following mumps meningoencephalitis.[91,963]

Gliosis By contrast with atresia, in aqueductal gliosis the contours of the aqueduct remain recognizable as an interrupted ring of ependymal cells, rosettes and tubules. Dense fibrillary subependymal gliosis surrounds this ring and largely fills the area within it. There may be one or two small central channels, but they are without ependymal lining. Widespread ependymitis, especially of the fourth ventricle, suggests that the lesion is either post-inflammatory or post-haemorrhagic; proliferation of the subependymal glia and organization of pus or haematoma blocking the aqueductal lumen are possible mechanisms.[350] Such cases are characterized clinically by the gradual onset of hydrocephalus in early childhood or occasionally in adult life.

Septum Occlusion of the aqueduct by a thin septum is rare. Turnbull and Drake described four cases and

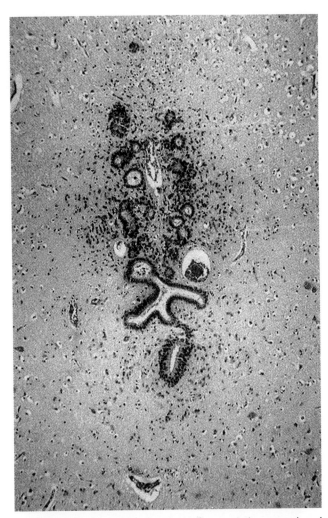

4.159 Aqueduct atresia. Small tubules and tiny ependymal canals are irregularly disposed in the expected position for the aqueduct in the midbrain tegmentum.

reviewed 12 others: one was 4 months old, and the others ranged in age from 2 to 46 years.[1029] A thin, translucent membrane interrupts the aqueduct at its lower end (Figure 4.160a), and sometimes there is a pinhole opening. Histologically, the membrane is composed of loose fibrillary glial tissue. Turnbull and Drake suggested that the membrane was derived from a glial plug at the caudal end of the aqueduct, which had become attenuated by prolonged pressure from above.[1029] The notion that septum formation is a variant of aqueductal gliosis is supported by several personal observations in which the thin glial membrane is surrounded almost completely by a ring of ependymal tissue (Figure 4.160b).

Vascular Malformation The aqueduct may be compressed by an aneurysm of the great vein of Galen situated over the quadrigeminal plate[901] or can be blocked directly by a vascular malformation.[204,894]

Fourth Ventricular Foramina

Vuia[1061] and Friede[350] have drawn an important distinction between the Dandy–Walker malformation (see earlier,

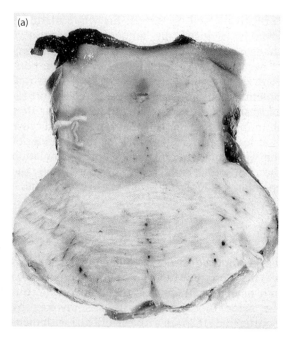

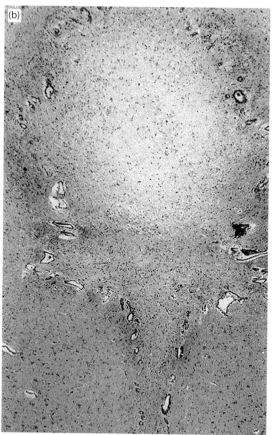

4.160 Occlusion of the aqueduct by a membranous septum.
(a) Section through the junction of the midbrain and pons.
(b) The membrane comprises loose glial tissue surrounded by a ring of ependymal canals.

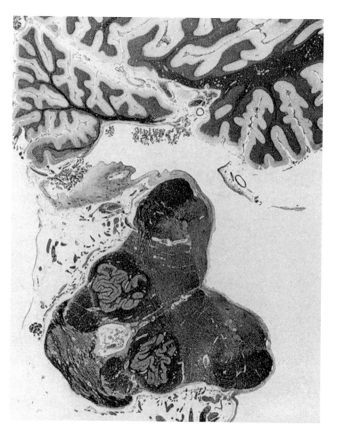

4.161 Atresia of the foramina of Luschka. Note the choroid plexus within the widened foramina, which are closed over with glial membranes.

Dandy–Walker Syndrome), in which the fourth ventricular foramina may or may not be closed, and rare reports of atresia of these foramina and hydrocephalus without vermal aplasia. Each outlet is more than sufficient for CSF drainage,[53] and so all three must be obstructed to produce hydrocephalus. Several authors have described membranous pouches closing over the foramina. Histologically, either the membrane is fibroblastic, suggesting a haemorrhagic or inflammatory causation,[22] or it consists of a delicate glial membrane lined on the inside by ependyma (Figure 4.161).[479]

Arachnoid Granulations

Congenital aplasia of arachnoid granulations has been rarely described in hydrocephalic children.[375,410] Hydrocephalus complicating certain forms of cranial dysplasia, craniosynostosis, achondroplasia, and Apert's, Crouzon's and Pfeiffer's syndromes may be related to functional changes in CSF absorption by arachnoid granulations.[178,183,324,476,1085] Physiological studies in such patients, simultaneously recording pressures in the lateral ventricle, superior sagittal sinus and jugular vein while manipulating the intracerebral pressure, suggest that the superior sagittal sinus venous pressure resulting from anatomical obstruction is increased independently of changes in intracranial pressure and is the cause of the hydrocephalus.[819,902]

Reduced Propulsion of Cerebrospinal Fluid

Bulk flow of CSF in humans is the result of both continuous secretion from the choroid plexus and the cervical pressure wave; a contributory role for the beating of ependymal cilia is less certain.[1100] There are, however, rare reports of hydrocephalus in children associated with primary ciliary dyskinesia related ultrastructurally to absence of the outer dynein arms of each cilium,[402] and with ciliary aplasia.[267] There is also a mutant mouse (*hpy/hpy*) that shows a generalized disorder of cilia and hydrocephalus.[4]

Developmental Abnormalities in Which Ventriculomegaly is an Essential Feature but of Uncertain Pathogenesis

Hydrocephalus is a cardinal feature of the hydrolethalus syndrome, an autosomal recessive lethal disorder confined largely to Finland, where it was first recognized in the course of a large-scale study of Meckel–Gruber syndrome.[903,904] The complex phenotype includes micrognathia, polydactyly, congenital heart defects and genitourinary anomalies. A preliminary neuropathological report on eight cases showed a keyhole-shaped foramen magnum, hypoplastic hindbrain, arhinencephaly, midline dysgenesis and fusion of the thalami.[781] The widely separated hemispheres sit in the base of the skull covered by a distended and torn arachnoid membrane, allowing communication of the large ventricles with the subarachnoid space, the result of a massive intracranial accumulation of CSF.

Hydrocephalus is also present in severe forms of Smith–Lemli–Opitz syndrome,[635] X-linked orofacial–digital syndrome,[45] occasionally in Meckel–Gruber syndrome (see earlier, Dandy–Walker Syndrome), and in a rare malformation syndrome associated with congenital cerebral lactic acidosis and pyruvate dehydrogenase deficiency, characterized morphologically by microcephaly, hypoplastic pyramids, olivary heterotopia and hydrocephalus.[160] For a more complete list of malformations associated with hydrocephalus, see Stevenson and Hall.[980]

General Aspects

The distended hemispheres are fragile and easily torn post mortem and collapse when CSF escapes. In many cases, however, even after complete draining of CSF, the brain is heavier than that of an infant of the same age. This may be due to infiltration of fluid into the brain tissue. In addition, the external surface of the brain may show increased numbers of small gyri and shallow sulci without any underlying cytoarchitectonic abnormality, best termed 'polygyria' (Figure 4.162). Postnatal growth of the cortical surface seems to proceed normally in spite of the expanding hydrocephalus and Friede suggested that with the abnormal distension of the hemisphere a great portion of intrasulcal cortex is exposed, resulting in redundant gyration.[350]

Sections of the brain show the enlarged cavities, a grossly distended infundibulum (Figure 4.163) reduced to a thin, transparent membrane, and a fenestrated or absent septum pellucidum. The corpus callosum becomes extremely thin. The anterior horns of the ventricles are usually less dilated than the temporal and occipital horns, where the cerebral mantle may be reduced in thickness to a few millimetres. Although the cortex is relatively well preserved, the white matter is severely reduced in amount. Degenerative changes

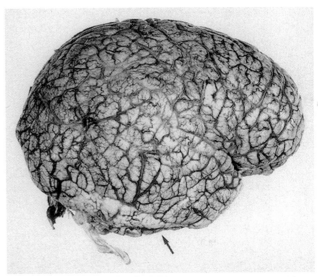

4.162 Hydrocephalus with polygyria, i.e. increased numbers of normal but small convolutions. Note the marked contrast between the polygyria, which involves most of the surface of the hemisphere and a small area of polymicrogyria affecting the inferior part of the temporal and occipital lobes (arrow).

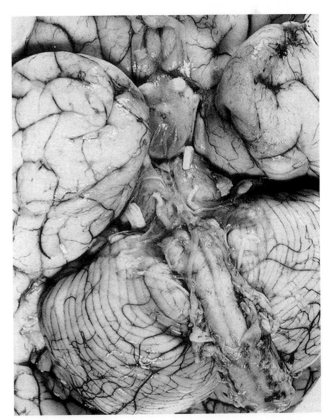

4.163 Hydrocephalus. Close-up of the base of the brain, showing cystic distension of the infundibulum.

may be due to the stretching and tearing of nerve fibres and diffusion of CSF into the periventricular white matter causing interstitial oedema or chronic compression. Remarkable re-expansion of the cerebral mantle may occur after shunting, suggesting that a mechanical cause is at least partly responsible for the thinning of the white matter. On microscopic examination, the ventricular walls are denuded and the ependyma is replaced by a layer of glial tissue.

CONGENITAL DISORDERS OF THE PITUITARY

Ectopic Adenohypophysis

Residual adenohypophysial tissue can be trapped anywhere along the path followed by Rathke's pouch during fetal development.[111,165,690] The sphenoid sinus is the most common site for ectopic pituitary, followed by the immediate suprasellar region;[182,483] these ectopic foci may be incidental findings or they can undergo hyperplastic or neoplastic change.

Pituitary Aplasia/Hypoplasia

This disorder is usually associated with severe congenital malformations such as those seen in the Cornelia de Lange syndrome[92] or Arnold–Chiari malformation.[358] The result is hypopituitarism with subsequent thyroid and adrenal hypoplasia or aplasia.[300,542,565,713,816] One form of this disorder associated with septo-optic dysplasia has been attributed to a mutation in *Rpx-1* (also known as *Hesx-1*), *Sox2* or *Sox3*.[234,677]

Inactivating mutations of *PROP-1* result in combined pituitary hormone deficiency.[340,1101] Pituitary dwarfism with hypothyroidism occurs in patients with mutations of the *pit-1* gene,[811,838,1004] likely due to hypoplasia of somatotrophs, lactotrophs and thyrotrophs.[617]

Pituitary Duplication or Dystopia

These lesions are also usually associated with other congenital malformations.[834,877]

Empty Sella Syndrome

A defective or absent diaphragma sella results in increased CSF pressure that causes sellar enlargement and flattens the pituitary against the floor of the sella turcica.[528] Pituitary function is usually unaffected.[77] However, some patients have hyperprolactinaemia, attributed to distortion of the pituitary stalk and interference with the tonic dopaminergic inhibition of prolactin.[371] This can be difficult to distinguish from a prolactinoma that develops in the setting of an empty sella.

ACKNOWLEDGEMENTS

The section Congenital Disorders of the Pituitary was kindly contributed by Sylvia Asa.

REFERENCES

1. Adams NC, Tomoda T, Cooper M, *et al*. Mice that lack astrotactin have slowed neuronal migration. *Development* 2002;**129**:965–72.
2. Adzick NS. Fetal myelomeningocele: natural history, pathophysiology, and *in utero* intervention. *Semin Fetal Neonatal Med* 2010;**15**(1):9–14.
3. Adzick NS, Sutton LN, Crombleholme TM, Flake AW. Successful fetal surgery for spina bifida. *Lancet* 1998;**352**:1675–6.
4. Afzelius BA. The immotile cilia syndrome and other ciliary diseases. *Int Rev Exp Pathol* 1979;**19**:1–43X.
5. Agamanolis DP, Patre S. Glycogen accumulation in the central nervous system in cerebro-hepatorenal syndrome: report of a case with ultrastructural studies. *J Neurol Sci* 1979;**41**:325–42.
6. Aguirre–Villa–Coro A, Dominguez R. Intrauterine diagnosis of hydranencephaly by magnetic resonance. *Magn Res Imag* 1989;**7**:105–7.
7. Ahdab–Barmada M, Claassen D. A distinctive triad of malformations of the central nervous system in the Meckel–Gruber syndrome. *J Neuropathol Exp Neurol* 1990;**49**:610–20.
8. Aicardi J. The agyria–pachygyria complex: a spectrum of cortical malformations. *Brain Dev* 1991;**13**:1–8.
9. Aicardi J, Goutières F. The syndrome of absence of the septum pellucidum with porencephalies and other developmental defects. *Neuropediatrics* 1981;**12**:319–29.
10. Aicardi J, Lefebvre J, Lerique–Koechlin A. A new syndrome: spasms in flexion,

callosal agenesis, ocular abnormalities. *Electroencephalogr Clin Neurophysiol* 1965;**19**:609–10.
11. Aicardi J, Goutières F, Verbois AH de. Multiple encephalomalacia of infants and its relationship to abnormal gestation and hydranencephaly. *J Neurol Sci* 1972;**15**:357–73.
12. Aicardi J, Castello–Branco M, Roy C. Le syndrome de Joubert: a propos de cinq observations. Arch Fr Pediatr 1983;**40**:625–9.
13. Aleksic S, Budzilovich G, Reuben R, *et al*. Unilateral arhinencephaly in Goldenhar–Gorlin syndrome. *Dev Med Child Neurol* 1975;**17**:498–504.
14. Allendoerfer KL, Shatz CJ. The subplate, a transient neocortical structure: its role in the development of connections between thalamus and cortex. *Annu Rev Neurosci* 1994;**17**:185–218.
15. Allsopp TE, Wyatt S, Paterson HF, Davies AM. The proto-oncogene *bcl-2* can selectively rescue neurotrophic factor-dependent neurons from apoptosis. *Cell* 1993;**73**:295–307.
16. Altman J. Autoradiographic and histological studies of postnatal neurogenesis: II. A longitudinal investigation of the kinetics, migration and transformation of cells incorporation tritiated thymidine in infant rats, with special reference to postnatal neurogenesis in some brain regions. *J Comp Neurol* 1966;**128**:431–74.
17. Altman J, Anderson WJ. Experimental reorganization of the cerebellar cortex: II. Effects of elimination of

most microneurons with prolonged X-irradiation started at four days. *J Comp Neurol* 1973;**149**:123–52.
18. Altman J, Anderson WJ, Wright KA. Reconstitution of the external granular layer of the cerebellar cortex in infant rats after low-level X-irradiation. *Anat Rec* 1969;**163**:453–72.
19. Altman J, Anderson WJ, Bayer SA. Prenatal development of the cerebellar system in the rat:I. Cytogenesis and histogenesis of the deep nuclei and the cortex of the cerebellum. *J Comp Neurol* 1978;**179**:23–48.
20. Altman J, Anderson WJ, Bayer SA. Embryonic development of the rat cerebellum: III. Regional differences in the time of origin, migration and settling of Purkinje cells. *J Comp Neurol* 1985;**231**:42–65.
21. Alvord EC Jr, Fields WS, Desmond MM eds. *The pathology of hydrocephalus*. Springfield, IL: CC Thomas, 1961.
22. Amacher AL, Page LK. Hydrocephalus due to membranous obstruction of the fourth ventricle. *J Neurosurg* 1971;**35**:672–6.
23. Amiel J, Laudier B, Attié–Bitach T, *et al*. Polyalanine expansion and frameshift mutations of the paired-like homeobox gene *PHOX2B* in congenital central hypoventilation syndrome. *Nat Genet* 2003;**33**:459–61.
24. Andermann E, Andermann F, Bergeron D, *et al*. Familial agenesis of the corpus callosum with sensori-motor neuronopathy: genetic and epidemiological studies of over 170 patients. *Can J Neurol Sci* 1979;**6**:400.

25. Andrews PI, Hulette CM. An infant with macrocephaly, abnormal neuronal migration and persistent olfactory ventricles. *Clin Neuropathol* 1993;**12**:13–18.

26. Angevine DM. Pathologic anatomy of hypophysis and adrenals in anencephaly. *Arch Pathol* 1938;**26**:507–18.

27. Angevine JB. Time of neuron origin in the hippocampal region:an autoradiographic study in the mouse. *Exp Neurol* 1965;**13** (Suppl 2):1–70.

28. Angevine JB, Sidman RL. Autoradiographic study of cell migration during histogenesis of cerebral cortex in the mouse. *Nature* 1961;**192**:766–8.

29. Anton G, Zingerle H. Gename Beschreibung eines Falles von beiderseitigem Kleinhirnmangel. *Arch Psychiat Nervenkrankheit* 1914;**54**:8–75.

30. Appleby A, Foster JB, Hankinson J, Hudson P. The diagnosis and management of the Chiari anomalies in adult life. *Brain* 1968;**91**:131–40.

31. Archer CR, Horenstein S, Sudaram M. The Arnold–Chiari malformation presenting in adult life: a report of 13 cases and review of the literature. *J Chronic Dis* 1977;**30**:369–82.

32. Avery L, Wasserman S. Ordering gene function: the interpretation of epistasis in regulatory hierarchies. *Trends Genet* 1992;**8**:312–16.

33. Avilion AA, Nicolis SK, Pevny LH, *et al.* Multipotent cell lineages in early mouse development depend on SOX2 function. *Genes Dev* 2003;**17**:126–40.

34. Baes M, Gressens P, Baumgart E, *et al.* A mouse model for Zellweger syndrome. *Nat Genet* 1997;**17**:49–57.

35. Bai J, Ramos RL, Ackman JB, *et al.* RNAi reveals doublecortin is required for radial migration in rat neocortex. *Nat Neurosci* 2003;**6**:1277–83.

36. Bailey A, Luthert P, Bolton P, *et al.* Autism and megalencephaly. *Lancet* 1993;**341**:1225–6.

37. Baird WF, Stitt D. Arteriovenous aneurysm of the cerebellum in a premature infant. *Pediatrics* 1959;**24**:455–7.

38. Baker DW, Vinters HV. Hydrocephalus with cerebral aqueductal dysgenesis and craniofacial anomalies. *Acta Neuropathol (Berl)* 1984;**63**:170–73.

39. Baldwin CT, Hoth CF, Amos JA, *et al.* An exonic mutation in the *HuP2* paired domain gene causes Waardenburg's syndrome. *Nature* 1992;**355**:637–8.

40. Balthasar K, Schlagenhauff R. Unilateral agenesis of pyramidal system. In: Lüthy F, Bischoff A eds. *Vth International Congress of Neuropathology.* Amsterdam: Excerpta Medical, 1966:881–7.

41. Bamshad M, Van Heest AE, Pleasure D. Arthrogryposis: a review and update. *J Bone Joint Surg Am* 2009;**91** Suppl 4:40–6.

42. Banerji NK, Millar JHD. Chiari malformation presenting in adult life: its relationship to syringomyelia. *Brain* 1974;**97**:157–68.

43. Banker BQ. Neuropathologic aspects of arthrogryposis multiplex congenita. *Clin Orthopaed* 1985;**144**:30–43.

44. Bankl H, Jellinger J. Zentralnervöse Schäden nach fätaler Kohlenoxydvergiftung. *Beitr Pathol Anat Pathol* 1967;**135**:350–76.

45. Baraitser M. Syndrome of the mouth: the orofaciodigital syndromes. *J Med Genet* 1986;**23**:116–19.

46. Baraitser M. *The genetics of neurological disorders*, 2nd edn. Oxford: Oxford University Press, 1990.

47. Baraitser M, Patton MA. Reduced prevalence in tuberous sclerosis. *J Med Genet* 1985;**22**:29–31.

48. Baraitser M, Brett EM, Piesowicz AT. Microcephaly and intracranial calcification in two brothers. *J Med Genet* 1983;**20**:210–12.

49. Barkovich AJ, Norman D. MR imaging of schizencephaly. *Am J Neuroradiol* 1988;**9**:297–302.

50. Barkovich AJ, Chuang SH, Norman D. MR of neuronal migration anomalies. *Am J Neuroradiol* 1987;**8**:1009–17.

51. Barkovich AJ, Jackson DE, Boyer RS. Band heterotopias:a newly recognised neuronal migration anomaly. *Radiology* 1989;**171**:455–8.

52. Barkovich AJ, Kuzniecky RI, Bollen AW, Grant PE. Focal transmantle dysplasia:a specific malformation of cortical development. *Neurology* 1997;**49**:1148–52.

53. Barr ML. Observations on the foramen of Magendie in a series of human brains. *Brain* 1948;**71**:281–9.

54. Barr M Jr, Hansen JW, Currey K, *et al.* Holoprosencephaly in infants of diabetic mothers. *J Pediatr* 1983;**102**:565–8.

55. Barry A, Patten BM, Stewart BH. Possible factors in the development of the Arnold–Chiari malformation. *J Neurosurg* 1957;**14**:285–301.

56. Barson AJ. Spina bifida: the significance of the level and extent of the defect to the morphogenesis. *Dev Med Child Neurol* 1970;**12**:129–44.

57. Barth PG, Valk J, Kalsbeek GL, Blom A. Organoid nevus syndrome (linear nevus sebaceous of Judassohn). Clinical and radiological study of a case. *Neuropediatrics* 1977;**8**:418–28.

58. Barth PG, Mullaart R, Stam FC, Sloof JL. Familial lissencephaly with extreme neopallial hypoplasia. *Brain Dev* 1982;**4**:145–51.

59. Barth PG, Vrensen GFJM, Uylings HBM, *et al.* Inherited syndrome of microcephaly, dyskinesia and pontocerebellar hypoplasia: a systemic atrophy with early onset. *J Neurol Sci* 1990;**97**:25–42.

60. Bass JL, Breningstall GN, Swaiman KF. Echocardiographic evidence of cardiac rhabdomyoma in tuberous sclerosis. *Am J Cardiol* 1985;**55**:1379–82.

61. Bates JI, Netsky MG. Developmental anomalies of the horns of the lateral ventricles. *J Neuropathol Exp Neurol* 1995;**14**:316–25.

62. Baudrimont M, Gray F, Meininger V, Escourolle R, Castaine P. Atrophie cérébelleuse croisée après lésion hémisphérique survenue àl'âge adulte. *Rev Neurol* 1983;**139**:485–95.

63. Becker H. Cüber Hirngefässausschaltungen II:Intrakranielle Gefässverschlüssw. Cüber experimentelle Hydranencephalie (Blasenhirn). *Dtsch Z Nervenheilkunde* 1949;**161**:446–505.

64. Becker PS, Dixon AM, Troncoso JC. Bilateral opercular polymicrogyria. *Ann Neurol* 1989;**25**:90–2.

65. Bell JE, Green RJL. Studies on the area cerebrovasculosa of anencephalic fetuses. *J Pathol* 1982;**137**:315–28.

66. Bell JE, Gordon A, Maloney AFJ. The association of hydrocephalus and Arnold–Chiari malformation with spina bifida in the fetus. *Neuropathol Appl Neurobiol* 1980;**6**:29–39.

67. Bell JE, Gordon A, Maloney AFJ. Abnormalities of the spinal meninges in anencephalic fetuses. *J Pathol* 1981;**133**:131–44.

68. Belloni E, Muenke M, Roessler E, *et al.* Identification of *Sonic hedgehog* as a candidate gene responsible for holoprosencephaly. *Nat Genet* 1996;**14**:353–6.

69. Beltran–Valero de Bernabe D, Currier S, Steinbrecher A, *et al.* Mutations in the O-mannosyltransferase gene POMT1 give rise to the severe neuronal migration disorder Walker–Warburg syndrome. *Am J Hum Genet* 2002;**71**:1033–43.

70. Benda CE. *Mongolism and cretinism*. New York: Grune and Stratton, 1947.

71. Benda CE. The Dandy–Walker syndrome or the so-called atresias of the foramen of Magendie. *J Neuropathol Exp Neurol* 1954;**13**:14–29.

72. Benda CE. Dysraphic states. *J Neuropathol Exp Neurol* 1959;**18**:56–74.

73. Bender BL, Yunis EJ. Central nervous system pathology of tuberous sclerosis in children. *Ultrastruct Pathol* 1980;**1**:287–99.

74. Benirschke K. Twin placenta in perinatal mortality. *N Y State J Med* 1961;**61**:1499–508.

75. Berg MJ, Schifitto G, Powers JM, *et al.* X-linked female band heterotopia–male lissencephaly syndrome. *Neurology* 1998;**50**:1143–6.

76. Bergametti F, Denier C, Labauge P, *et al.* Mutations within the programmed cell death 10 gene cause cerebral cavernous malformations. *Am J Hum Genet* 2005;**76**:42–51.

77. Bergeron C, Kovacs K, Bilbao JM. Primary empty sella: a histologic and immunocytologic study. *Arch Intern Med* 1979;**139**:248–9.

78. Bergmann C, Zerres K, Senderek J, *et al.* Oligophrenin 1 (OPHN1) gene mutation causes syndromic X-linked mental retardation with epilepsy, rostral ventricular enlargement and cerebellar hypoplasia. *Brain* 2003;**126**:1537–44.

79. Berry RJ, Li Z, Erickson JD, *et al.* China–US Collaborative Project Neu:prevention of neural-tube defects with folic acid in China. *N Engl J Med* 1999;**341**:1485–90.

80. Bertrand I, Gruner J. The status verrucosus of the cerebral cortex. *J Neuropathol Exp Neurol* 1955;**14**:331–47.

81. Bickers DS, Adams RD. Hereditary stenosis of the aqueduct of Sylvius as a cause of congenital hydrocephalus. *Brain* 1949;**72**:246–62.

82. Bielas S, Higginbotham H, Koizumi H, *et al.* Cortical neuronal migration mutants suggest separate but intersecting pathways. *Annu Rev Cell Dev Biol* 2004;**20**:593–618.

83. Bielschowsky M. über Mikrogyrie. *J Psychol Neurol* 1916;**22**:1–47.

84. Bielschowsky M. Zur Histopathologie und Pathogenese der amaurotischen Idiotie mit besonderer Berücksichtigung der zerebellären Veränderungen. *J Psychol Neurol* 1920;**26**:123–244.

85. Bielschowsky M. Cüber die Oberflächengestaltung des Grosshirnmantels bei Pachgyrie, Mikrogyrie und bei normaler Entwicklung. *J Psychol Neurol* 1923;**30**:29–76.

86. Biemond A. Hypoplasia ponto-neocerebellaris, with malformation of the dentate nucleus. *Folia Psychiat Neurol Neurochirurg Neerland* 1955;**58**:2–7.

87. Bignami A, Appicciutoli L. Micropolygyria and cerebral calcification in cytomegalic inclusion disease. *Acta Neuropathol (Berl)* 1964;**4**:127–37.

88. Bignami A, Palladini G, Zappella M. Unilateral megalencephaly with nerve cell hypertrophy. An anatomical and quantitative histochemical study. *Brain Res* 1968;**9**:103–14.

89. Billette de Villemeur T, Chiron C, Robain O. Unlayered polymicrogyria and agenesis of the corpus callosum: a relevant association? *Acta Neuropathol (Berl)* 1992;**83**:265–70.

90. Binns W, James LF, Shupe JL, Thacker EJ. Cyclopian-type malformation in lambs. *Arch Environ Health* 1962;**5**:106–8.

91. Bistrian B, Phillips CA, Kaye IS. Fatal mumps meningoencephalitis:isolation of virus premortem and postmortem. *J Am Med Assoc* 1972;**222**:478–9.

92. Björklöf K, Brundelet PJ. Typus degenerativus amstelodamensis (Cornelia de Lange first syndrome): congenital hypopituitarism due to a cyst of Rathke's cleft? *Acta Pediatr Scand* 1965;**54**:275–287.

93. Blackshear PJ, Silver J, Nairn AC, et al. Widespread neuronal ectopia associated with secondary defects in cerebrocortical chondroitin sulfate proteoglycans and basal lamina in MARCKS-deficient mice. *Exp Neurol* 1997;**145**:46–61.

94. Blake JA. The roof and lateral recesses of the fourth ventricle, considered morphologically and embryologically. *J Comp Neurol* 1900;**10**:79–108.

95. Blanc JF, Lapillonne A, Pouillaude JH, Badinand N. Hydranencephaly and ingestion of estrogens during pregnancy:fetal cerebral vascular complication? *Arch Fr Pediatr* 1988;**45**:483–5.

96. Böhm N, Uy J, Kiessling M, Lehnert W. Multiple acyl-CoA dehydrogenation deficiency (glutaric acidemia type II), congenital polycystic kidneys and symmetrical warty dysplasia of the cerebral cortex in two newborn brothers: II. Morphology and pathogenesis. *Eur J Pediatr* 1982;**139**:60–65.

97. Bond J, Roberts E, Mochida GH, et al. ASPM is a major determinant of cerebral cortical size. *Nat Genet* 2002;**32**:316–20.

98. Bond J, Roberts E, Springell K, et al. A centrosomal mechanism involving CDK5RAP2 and CENPJ controls brain size. *Nat Genet* 2005;**37**:353–5.

99. Bonnevie K, Brodal A. Hereditary hydrocephalus in the house mouse: IV. The development of the cerebellar anomalies during fetal life with notes on the normal development of the mouse cerebellum. *Lkr Norske Vidensk Akad Oslo 1 Matj-nat KI* 1946;**4**:4–60.

100. Bonnin JM, Rubinstein LJ, Papasozomenos SCh, Marangos PJ. Subependymal giant cell astrocytoma:significance and possible cytogenic implications of an immunohistochemical study. *Acta Neuropathol (Berl)* 1984;**62**:185–93.

101. Borberg A. Clinical and genetic investigations into tuberous sclerosis and Recklinghausen's neurofibromatosis. *Acta Psychol Neurol Scand* 1951;**71** (Suppl):11–239.

102. Bordarier C, Aicardi J. Dandy–Walker syndrome and agenesis of the cerebellar vermis: diagnostic problems and genetic counselling. *Dev Med Child Neurol* 1990;**32**:285–94.

103. Bordarier C, Robain O. Familial recurrence of prenatal encephaloclastic damage: anatomo-clinical report of two cases. *Neuropediatrics* 1989;**20**:103–6.

104. Bordarier C, Robain O. Microgyric and necrotic cortical lesions in twin fetuses: original cerebral damage consecutive to twinning? *Brain Dev* 1992;**14**:174–8.

105. Bordarier C, Aicardi J, Goutières F. Congenital hydrocephalus and eye abnormalities with severe developmental brain defects: Warburg's syndrome. *Ann Neurol* 1984;**16**:60–65.

106. Bordarier C, Robain O, Ponsot G. Bilateral porencephalic defect in a newborn after injection of benzol during pregnancy. *Brain Dev* 1991;**13**:126–9.

107. Borit A, Sidman RL. New mutant mouse with communicating hydrocephalus and secondary aqueductal stenosis. *Acta Neuropathol (Berl)* 1972;**21**:316–31.

108. Bossy JG. Morphological study of a case of complete, isolated and asymptomatic agenesis of the corpus callosum. *Arch Anat Histol Embryol* 1970;**53**:289–340.

109. Bouchard S, Davey MG, Rintoul NE, et al. Correction of hindbrain herniation and anatomy of the vermis after in utero repair of myelomeningocele in sheep. *J Pediatr Surg* 2003;**38**:451–458.

110. Bourneville DM. Sclérose tubéreuse des circonvolutions cérébrales:idiotie et épilepsie hémiplégique. *Arch Neurol* 1880;**1**:81–91.

111. Boyd JD. Observations of the human pharyngeal hypophysis. *J Endocrinol* 1956;**14**:66–77.

112. Boyd SG, Harding BN. Intractable seizures, intermittent EEG activity and dentato-olivary dysplasia. *Electroencephalogr Clin Neurophysiol* 1991;**78**:85P.

113. Bradley A. Mining the mouse genome: we have the draft sequence–but how do we unlock its secrets? *Nature* 2002;**420**:512–14.

114. Branda CS, Dymecki SM. Talking about a revolution: the impact of site-specific recombinases on genetic analyses in mice. *Dev Cell* 2004;**6**:7–28.

115. Brent RL. Radiation teratogenesis. *Teratology* 1980;**21**:281–98.

116. Briscoe J, Pierani A, Jessell TM, Ericson J. A homeodomain protein code specifies progenitor cell identity and neuronal fate in the ventral neural tube. *Cell* 2000;**101**:435–45.

117. Broccoli V, Boncinelli E, Wurst W. The caudal limit of Otx2 expression positions the isthmic organizer. *Nature* 1999;**401**:164–8.

118. Brock DJH, Sutcliffe RG. Alpha-fetoprotein in the antenatal diagnosis of anencephaly and spina bifida. *Lancet* 1972;**ii**:197–8.

119. Brocklehurst G. The development of the human cerebrospinal fluid pathway with particular reference to the roof of the fourth ventricle. *J Anat* 1969;**105**:467–75.

120. Brodal A, Bonnevie K, Harkmark W. Hereditary hydrocephalus in the house mouse: II. The anomalies of the cerebellum, and partial defective development of the vermis. *Skr Norske Vidensk-Acad I Mat-Nov K2* 1944.

121. Brook FA, Shum AS, Van Straaten HW, Copp AJ. Curvature of the caudal region is responsible for failure of neural tube closure in the curly tail (ct) mouse embryo. *Development* 1991;**113**:671–8.

122. Brouwer B. Hypoplasia ponto-neocerebellaris. *Psychiat Neurol Bladen* 1924;**6**:461–9.

123. Brown JR. The Dandy–Walker syndrome. In: Vinken PJ, Bruyn GW eds. *Handbook of clinical neurology*. Amsterdam: North-Holland, 1977:623–46.

124. Brown SA, Warburton D, Brown LY, et al. Holoprosencephaly due to mutations in ZIC2, a homologue of Drosophila odd-paired. *Nat Genet* 1998;**20**:180–83.

125. Bruggen J van der. über Ersatz der Pyramidenbahnfunktion. *Disch Z Nervenheilkunde* 1930;**113**:250–77.

126. Brun A. Marginal glioneural heterotopias of the central nervous system. *Acta Pathol Microbiol Scand* 1965;**65**:221–33.

127. Brun R. Zur Kenntnis der Bildungsfehler des Kleinhirns. *Schweizer Arch Neurol Psychiatrie* 1917;**1**:61–123.

128. Brun R. Zur Kenntnis der Bildungsfehler des Kleinhirns. *Schweizer Arch Neurol Psychiatrie* 1918;**2**:48–105.

129. Brun R. Zur Kenntnis der Bildungsfehler des Kleinhirns. *Schweizer Arch Neurol Psychiatrie* 1918;**3**:13–88.

130. Bruner JP, Tulipan N, Paschall RL, et al. Fetal surgery for myelomeningocele and the incidence of shunt-dependent hydrocephalus. *J Am Med Assoc* 1999;**282**:1819–25.

131. Buckler AJ, Chang DD, Graw SL, et al. Exon amplification: a strategy to isolate mammalian genes based on RNA splicing. *Proc Natl Acad Sci U S A* 1991;**88**:4005–9.

132. Burk U, Hayek HW, Zeider U. Holoprosencephaly in monozygotic twins: clinical and computed tomographic findings. *Am J Med Genet* 1981;**9**:13–17.

133. Burmeister M, Novak T, Liang MY, et al. Ocular retardation mouse caused by Chx10 homeobox null allele: impaired retinal progenitor proliferation and bipolar cell differentiation. *Nat Genet* 1996;**12**:376–84.

134. Burn J, Wickramasinghe T, Harding BN, Baraitser M. A syndrome of intracerebral calcification and microcephaly in 2 siblings, resembling intrauterine infection. *Clin Genet* 1986;**30**:112–16.

135. Butler LJ, Snodgrass G, France N, et al. Trisomy syndrome: analysis of 13 cases. *Arch Dis Child* 1965;**40**:600–611.

136. Byrnes RL, Boellaard JW. Renal agenesis and meningocerebral angiomatosis. *Arch Pathol* 1958;**66**:23–31.

137. Cacciagi P, Desvignes JP, Girard N, et al. AP1S2 is mutated in X-linked Dandy–Walker malformation with intellectual disability, basal ganglia disease and seizures (Pettigrew syndrome). *Eur J Hum Genet* 2014;**22**(3):363–8.

138. Calogero JA. Vermian agenesis and unsegmented midbrain tectum. *J Neurosurg* 1977;**47**:605–8.

139. Cameron AH. The Arnold–Chiari and the neuroanatomical malformations associated with spina bifida. *J Pathol Bacteriol* 1957;**73**:195–211.

140. Cameron AH. Malformations of the neuro-spinal axis, urogenital tract and foregut in spina bifida attributable to disturbances of the blastopore. *J Pathol Bacteriol* 1957;**73**:213–21.

141. Cameron AH, Jones EL, Smith WT, Wood BSB. Combined hepatic and cerebral degeneration in infancy. *J Pathol Bacteriol* 1968;96:227–30.

142. Campbell K, Gotz M. Radial glia: multi-purpose cells for vertebrate brain development. *Trends Neurosci* 2002;25:235–8.

143. Capecchi MR. Altering the genome by homologous recombination. *Science* 1989;244:1288–92.

144. Carrea R, Girado JM. Angiomatous and fistulous arteriovenous aneurysm in children. In:Luyendijk W ed. *Progress in brain research*. Amsterdam:Elsevier, 1968:433–9.

145. Carter CO. Clues to the aetiology of neural tube malformations. *Dev Med Child Neurol* 1974;16 (Suppl 32):3–15.

146. Casaer P, Ules JSH, Devlieger H, et al. Variability of outcome in Joubert syndrome. *Neuropediatrics* 1985:16:43–5.

147. Caviness VS Jr, Evrard P. Occipital encephalocele: a pathologic and anatomic analysis. *Acta Neuropathol (Berl)* 1975;32:245–55.

148. Chan CC, Egbert PR, Herrick MK, Urich H. Oculocerebral malformations:a reappraisal of Walker's 'lissencephaly'. *Arch Neurol* 1980;37:104–8.

149. Chan HSL, Sonley JJ, Moës CAF, et al. Primary and secondary tumours of childhood involving the heart, pericardium and great vessels. *Cancer* 1985;56:825–36.

150. Chenn A, Walsh CA. Regulation of cerebral cortical size by control of cell cycle exit in neural precursors. *Science* 2002;297:365–9.

151. Chiang C, Litingtung Y, Lee E, et al. Cyclopia and defective axial patterning in mice lacking *Sonic hedgehog* gene function. *Nature* 1996;383:407–13.

152. Chiari H. ber Veränderungen des Kleinhirns, des Pons und der Medulla oblongata infolge von congenitaler Hydrocephalie des Grosshirns. *Dtsch Med Wochenschr* 1891;27:1172–5.

153. Chiari H. ber die Veränderungen des Kleinhirns, des Pons, und der Medulla oblongata infolge von congenitaler Hydrocephalie des Grosshirns. *Denkschr Akad Wissenschaft Wien* 1896;63:71–116.

154. Cho DY, Leipold HW. Arnold–Chiari malformation and associated anomalies in calves. *Acta Neuropathol (Berl)* 1977;39:129–34.

155. Choi BH, Kudo M. Abnormal neuronal migration and gliomatosis cerebri in epidermal nevus syndrome. *Acta Neuropathol (Berl)* 1981;53:319–25.

156. Choi BH, Matthias SC. Cortical dysplasia associated with massive ectopia of neurons and glial cells within the subarachnoid space. *Acta Neuropathol (Berl)* 1987;73:105–9.

157. Choi BH, Lapham LW, Amin–Zaki L, Saleem T. Abnormal neuronal migration, deranged cerebral cortical organization, and diffuse white matter astrocytosis of human fetal brain; a major effect of methylmercury poisoning *in utero*. *J Neuropathol Exp Neurol* 1978;37:719–33.

158. Chou TM, Chou SM. Tuberous sclerosis in the premature infant: a report of a case with immunohistochemistry on the CNS. *Clin Neuropathol* 1989;8:45–52.

159. Choux M, Raybaud C, Puisard N, et al. Intracranial supratentorial cysts in children excluding tumour and parasitic cysts. *Childs Brain* 1978;4:15–32.

160. Chow CH, Anderson RM, Kelly GCT. Neuropathology in cerebral lactic acidosis. *Acta Neuropathol (Berl)* 1987;74:393–6.

161. Chow CH, Halliday JL, Anderson RM, et al. Congenital absence of pyramids and its significance in genetic diseases. *Acta Neuropathol (Berl)* 1985;65:313–17.

162. Chow CW, McKelvie PA, Anderson RM, et al. Autosomal recessive hydrocephalus with third ventricle obstruction. *Am J Med Genet* 1990;35:310–13.

163. Chun JJM, Shatz CJ. Interstitial cells of the adult neocortical white matter are the remnant of the early generated subplate neuron population. *J Comp Neurol* 1989;282:555–69.

164. Chung HD. Retrograde crossed cerebellar atrophy. *Brain* 1985;108:881–95.

165. Ciocca DR, Puy LA, Stati AO. Identification of seven hormone-producing cell types in the human pharyngeal hypophysis. *J Clin Endocrinol Metab* 1985;60:212–16.

166. Claireaux ER, Newman CGH. Arterio-venous fistula (aneurysm) of the great vein of Galen with heart failure in the neonatal period. *Arch Dis Child* 1960;35:605–12.

167. Clarren SK, Hall JG. Neuropathologic findings on the spinal cords of 10 infants with arthrogryposis. *J Neurol Sci* 1983;58:89–102.

168. Clarren SK, Alvord EC Jr, Sumi SM, et al. Brain malformations related to prenatal exposure to alcohol. *J Pediatr* 1978;92:64–7.

169. Cleland J. Contribution to the study of spina bifida, encephalocele and anencephalus. *J Anat Physiol* 1883;17:257–92.

170. Clement E, Mercuri E, Godfrey C, et al. Brain involvement in muscular dystrophies with defective dystroglycan glycosylation. *Ann Neurol* 2008;64:573–582.

171. Clément R, Gerbeaux J, Combes-Hamelle A, et al. Anévrysmes arterio-veineux de l'ampoule de Galien chez le nourrisson. Leur rôle dans l'hydrocéphalie communicante. *Presse Med* 1954;62:658–61.

172. Cogliatti SB. Diplomyelia: caudal duplication of the neural tube in mice. *Teratology* 1986;34:343–52.

173. Cogram P, Tesh S, Tesh J, et al. D-chiro-inositol is more effective than myo-inositol in preventing folate-resistant mouse neural tube defects. *Hum Reprod* 2002;17:2451–8.

174. Cogram P, Hynes A, Dunlevy LPE, et al. Specific isoforms of protein kinase C are essential for prevention of folate-resistant neural tube defects by inositol. *Hum Mol Genet* 2004;13:7–14.

175. Cohen M, Roessmann U. *In utero* brain damage: relationship of gestational age to pathological consequences. *Dev Med Child Neurol* 1994;36:263–8.

176. Cohen MM Jr. Perspectives on holoprosencephaly: part I. Epidemiology, genetics, and syndromology. *Teratology* 1989;40:211–35.

177. Cohen MM Jr. Perspectives on holoprosencephaly: part III. Spectra, distinctions, continuities and discontinuities. *Am J Med Genet* 1989;34:271–88.

178. Cohen MM Jr, Kreiborg S. The central nervous system in Apert syndrome. *Am J Med Genet* 1990;35:36–45.

179. Cohen MM Jr, Sulik KK. Perspectives on holoprosencephaly: part II. Central nervous system, craniofacial anatomy, syndrome commentary, diagnostic approach, and experimental studies. *J Craniofac Genet Dev Biol* 1992;12:196–244.

180. Cohn R, Neumann MA. Porencephaly: a clinicopathologic study. *J Neuropathol Exp Neurol* 1946;5:257–70.

181. Colevas AD, Edwards J, Hruban RH, et al. Glutaric acidemia type II: comparison of pathologic features in two infants. *Arch Pathol Lab Med* 1988;112:1133–9.

182. Colohan ART, Grady MS, Bonnin JM, et al. Ectopic pituitary gland simulating a suprasellar tumor. *Neurosurgery* 1987;20:43–8.

183. Collmann H, Sörensen N, Krauss J, Mühling J. Hydrocephalus in craniosynostosis. *Childs Nerv Syst* 1988;4:279–85.

184. Colon EY. The structure of the cerebral cortex in Down's syndrome: quantitative analysis. *Neuropediatrics* 1972;3:362–76.

185. Combette M. Absence complète du cervelet, des pédoncules postérieurs et de la protubérance cérébrale chez une jeune fille morte dans sa onzième année. *Bull Soc Anat Paris* 1831;5:148–53.

186. Constantinidis J. Cingulosynapsis (continuitéinterhémisphérique du cortex cingulaire). *Arch Suisse Neurol Neurochurg Psychiatry* 1969;104:137–49.

187. Cooper MK, Porter JA, Young KE, Beachy PA. Teratogen-mediated inhibition of target tissue response to *Shh* signaling. *Science* 1998;280:1603–7.

188. Copp AJ. Genetic models of mammalian neural tube defects. In: Bock G, Marsh J eds. *Neural tube defects*. Chichester: John Wiley & Sons, 1994:118–34.

189. Copp AJ, Bernfield M. Etiology and pathogenesis of human neural tube defects: insights from mouse models. *Curr Opin Pediatr* 1994;6:624–31.

190. Copp AJ, Brock FA. Does lumbosacral spina bifida arise by failure of neural folding or by defective canalisation? *J Med Genet* 1989;26:160–6.

191. Copp AJ, Harding BN. Neuronal migration disorders in humans and in mouse models—an overview. *Epilepsy Res* 1999;36:133–41.

192. Copp AJ, Crolla JA, Brook FA. Prevention of spinal neural tube defects in the mouse embryo by growth retardation during neurulation. *Development* 1988;104:297–303.

193. Copp AJ, Brook FA, Estibeiro P, et al. The embryonic development of mammalian neural tube defects. *Prog Neurobiol* 1990;35:363–403.

194. Copp AJ, Greene NDE, Murdoch JN. The genetic basis of mammalian neurulation. *Nat Rev Genet* 2003;4:784–93.

195. Copp AJ, Checiu I, Henson JN. Developmental basis of severe neural tube defects in the *loop-tail (Lp)* mutant mouse:use of microsatellite DNA markers to identify embryonic genotype. *Dev Biol* 1994;165:20–9.

196. Corbo JC, Deuel TA, Long JM, et al. Doublecortin is required in mice for lamination of the hippocampus but not the neocortex. *J Neurosci* 2002;22:7548–57.

197. Coria F, Quintana F, Rebollo M, *et al.* Occipital dysplasia and Chiari type I deformity in a family. *J Neurol Sci* 1983;**62**:147–58.

198. Corner BD, Holton JB, Norman RM, Williams PM. A case of histidinemia controlled with a low histidine diet. *Pediatrics* 1968;**41**:1074–81.

199. Cornette L, Verpoorten C, Lagae L, *et al.* Tethered cord syndrome in occult spinal dysraphism: timing and outcome of surgical release. *Neurology* 1998;**50**:1761–5.

200. Costoulas G. Un type peu connu d'anomalies des ventricules cérébraux:les goniosynapses (étude anatomo-clinique et statistique). *Ann Anat Pathol* 1958;**3**:268–83.

201. Coulter CL, Leech RW, Brumback RA, Bradley Schaefer G. Cerebral abnormalities in thanatophoric dysplasia. *Childs Nerv Syst* 1991;**7**:21–6.

202. Coulter CL, Leech RW, Schaefer GB, *et al.* Midline cerebral dysgenesis, dysfunction of the hypothalamic–pituitary axis, and fetal alcohol effects. *Arch Neurol* 1993;**50**:771–5.

203. Courville CB, Edmondson HA. Mental deficiency from intrauterine exposure to radiation. *Bull LA Neurol Soc* 1958;**23**:11–20.

204. Courville CB. Obstructive internal hydrocephalus incidental to small vascular anomaly of the midbrain. *Bull LA Neurol Soc* 1961;**26**:41–5.

205. Cowen D, Geller LM. Long-term pathological effects of prenatal X-irradiation on the central nervous system of the rat. *J Neuropathol Exp Neurol* 1960;**19**:488–527.

206. Craig JM, Bickmore WA. The distribution of CpG islands in mammalian chromosomes. *Nat Genet* 1994;**7**:376–82.

207. Creasy MR, Alberman ED. Congenital malformations of the central nervous system in spontaneous abortions. *J Med Genet* 1976;**13**:9–16.

208. Crome L. Microgyria. *J Pathol Bacteriol* 1952;**64**:479–95.

209. Crome L. Pachygyria. *J Pathol Bacteriol* 1956;**71**:335–52.

210. Crome L. Multilocular cystic encephalopathy of infants. *J Neurol Neurosurg Psychiatry* 1958;**21**:146–52.

211. Crome L. Brain warts. *J Ment Def Res* 1969;**13**:360–5.

212. Crome L, France N. Microgyria and cytomegalic inclusion disease in infancy. *J Clin Pathol* 1959;**12**:427–34.

213. Crome L, Stern J. *Pathology of mental retardation*, 2nd edn. Edinburgh: Churchill Livingstone, 1972.

214. Crome L, Sylvester PE. Hydranencephaly (hydrencephaly). *Arch Dis Child* 1958;**33**:235–45.

215. Crome L, Cowie V, Slater RE. A statistical note on cerebellar and brain-stem weight in mongolism. *J Ment Def Res* 1966;**10**:69–72.

216. Crossley PH, Martinez S, Martin GR. Midbrain development induced by FGF8 in the chick embryo. *Nature* 1996;**380**:66–8.

217. Cruveilhier J. *Anatomie pathologique du corps humain*. Paris:Baillière, 1829.

218. Cuckle HS. Screening for neural tube defects. In: Bock G, Marsh J eds. *Neural tube defects*. Chichester: John Wiley & Sons, 1994:253–66.

219. Curran T, D'Arcangelo G. Role of reelin in the control of brain development. *Brain Res Rev* 1998;**26**:285–94.

220. Curtis MT, Furth EE, Rorke LB. Cortical dysplasia and lipids in Neu–Laxova syndrome. *Brain Pathol* 1994;**4**:456.

221. Czeizel AE, Dudás I. Prevention of the first occurrence of neural-tube defects by periconceptional vitamin supplementation. *N Engl J Med* 1992;**327**:1832–5.

222. D'Agostino AN, Kernohan JN, Brown JR. The Dandy–Walker syndrome. *J Neuropathol Exp Neurol* 1963;**22**:450–70.

223. Dahme M, Bartsch U, Martini R, *et al.* Disruption of the mouse *L1* gene leads to malformations of the nervous system. *Nat Genet* 1997;**17**:346–9.

224. Dale JK, Vesque C, Lints TJ, *et al.* Cooperation of BMP7 and SHH in the induction of forebrain ventral midline cells by prechordal mesoderm. *Cell* 1997;**90**:257–69.

225. Dambska M, Wisniewski K, Sher JH. Lissencephaly:two distinct clinico-pathological types. *Brain Dev* 1983;**5**:302–10.

226. Dambska M, Wisniewski K, Sher JH. An autopsy case of hemimegalencephaly. *Brain Dev* 1984;**6**:60–4.

227. Dandy WE. Congenital cerebral cysts of the cavum septi pellucidi (fifth ventricle) and cavum vergae (sixth ventricle). *Arch Neurol Psychiatry* 1931;**25**:44–66.

228. Dandy WE, Blackfan KD. Internal hydrocephalus, an experimental clinical and pathological study. *Am J Dis Child* 1914;**8**:406–82.

229. Daniel PM, Strich SJ. Some observations on the congenital deformity of the central nervous system known as the Arnold–Chiari malformation. *J Neuropathol Exp Neurol* 1958;**17**:255–66.

230. Danner R, Shewmon DA, Sherman MP. Seizures in an atelencephalic infant: is the cortex essential for neonatal seizures? *Arch Neurol Psychiatry* 1985;**42**:1014–16.

231. D'Arcangelo G, Miao GG, Chen S-C, *et al.* A protein related to extracellular matrix proteins deleted in the mouse mutant reeler. *Nature* 1995;**374**:719–23.

232. D'Arcangelo G, Homayouni R, Keshvara L, *et al.* Reelin is a ligand for lipoprotein receptors. *Neuron* 1999;**24**:471–9.

233. Darbro BW, Mahajan VB, Gakhar L, *et al.* Mutations in extracellular matrix genes NID1 and LAMC1 cause autosomal dominant Dandy–Walker malformation and occipital cephaloceles. *Hum Mutat* 2013 Aug;**34**(8):1075–9.

234. Dattani MT, Martinez–Barbera JP, Thomas PQ, *et al.* Mutations in the homeobox gene *HESX1/Hesx1* associated with septo-optic dysplasia in human and mouse. *Nat Genet* 1998;**19**:125–33.

235. Daube JR, Chou SM. Lissencephaly: two cases. *Neurology* 1966;**16**:179–91.

236. Dauser RC, DiPietro MA, Venes JL. Symptomatic Chiari I malformation in childhood: a report of 7 cases. *Pediatr Neurosci* 1984;**14**:184–90.

237. Davidoff LM. Brain in mongolian idiocy: report of 10 cases. *Arch Neurol Psychiatry* 1928;**20**:1229–57.

238. Davidoff LM. Coarctation of the walls of the lateral angles of the lateral cerebral ventrices. *J Neurosurg* 1946;**3**:250–6.

239. Davies WW. Radiological changes associated with Arnold–Chiari malformation. *Br J Radiol* 1967;**40**:262–9.

240. Davis RL, Nelson E. Unilateral ganglioglioma in a tuberosclerotic brain. *J Neuropathol Exp Neurol* 1961;**21**:571–81.

241. De Bleecker J, De Reuck J, Martin JJ, *et al.* Autosomal recessive inheritance of polymicrogyria and dermatomyositis with paracrystalline inclusions. *Clin Neuropathol* 1990;**9**:299–304.

242. DeBry RW, Seldin MF. Human/mouse homology relationships. *Genomics* 1996;**33**:337–51.

243. De Chadarevian JP, Hollenberg RD. Subependymal giant cell tumor of tuberose sclerosis:a light and ultrastructural study. *J Neuropathol Exp Neurol* 1979;**38**:419–33.

244. Dekaban A. Arhinencephaly in an infant born to a diabetic mother. *J Neuropathol Exp Neurol* 1959;**18**:620–6.

245. Dekaban A. Large defects in cerebral hemispheres associated with cortical dysgenesis. *J Neuropathol Exp Neurol* 1965;**24**:512–30.

246. Dekaban AS, Sadowsky D. Changes in brain weights during the span of human life: relation of brain weights to body heights and body weights. *Ann Neurol* 1978;**4**:345–56.

247. Dekaban AS, Sakuragawa N. Megalencephaly. In: Vinken PJ, Bruyn GW eds. *Handbook of clinical neurology*. Amsterdam: North-Holland, 1977:647–60.

248. De la Cruz JM, Bamford RN, Burdine RD, *et al.* A loss-of-function mutation in the CFC domain of TDGF1 is associated with human forebrain defects. *Hum Genet* 2002;**110**:422–8.

249. Del Castillo I, Cohen-Salmon M, Blanchard S, *et al.* Structure of the X-linked Kallmann syndrome gene and its homologous pseudogene on the Y chromosome. *Nat Genet* 1992;**2**:305–10.

250. De Leon G. Observations on cerebral and cerebellar microgyria. *Acta Neuropathol (Berl)* 1972;**20**:278–87.

251. De Leon GA, Grover WD, Mestre GM. Cerebellar microgyria. *Acta Neuropathol (Berl)* 1976;**35**:81–5.

252. De Leon GA, Grover WD, Huff DS, *et al.* Globoid cells, glial nodules and peculiar fibrillary changes in the cerebro-hepato-renal syndrome of Zellweger. *Ann Neurol* 1977;**2**:473–84.

253. De Leon GA, Grover WD, D'Cruz CA. Amyotrophic cerebellar hypoplasia: a specific form of infantile spinal atrophy. *Acta Neuropathol (Berl)* 1984;**63**:282–6.

254. Deloukas P, Schuler GD, Gyapay G, *et al.* A physical map of 30 000 human genes. *Science* 1998;**282**:744–6.

255. De Morsier G. Agénésie du septum lucidum avec malformations du tractus optique –la dysplasie septo-optique. *Schweizer Arch Neurol Psychiatrie* 1956;**77**:267–92.

256. De Morsier G. études sur les dysraphies crânioencéphaliques:VI. Télencéphalosynapsis:hémisphères cérébraux incomplètement séparés. *Psychiat Neurol Basel* 1961;**141**: 239–74.

257. De Morsier G, Mozer JJ. Agénésie compleète de la commissure calleuse et troubles du développement de l'héisphère gauche avec hémiparesie droite et intégritémentale. (Le syndrome embryonnaire précoce de l'artère cérébrale antérieure.) *Schweizer Arch Neurol Psychiatrie* 1935;**35**:64–95.

258. De Myer W. The median cleft face syndrome. *Neurology* 1967;**17**:961–71.

259. De Myer W, Zeman W. Alobar holoprosencephaly (arhinencephaly) with median cleft lip and palate:clinical, electroencephalographic and nosologic considerations. *Confinia Neurol Basel* 1963;**23**:1–36.

260. De Myer W, Zeman W, Palmer C. The face predicts the brain; diagnostic significance of median facial anomalies for holoprosencephaly (arhinencephaly). *Pediatrics* 1964;**34**:256–63.

261. Deng CX, Wynshaw–Boris A, Shen MM, *et al.* Murine FGFR-1 is required for early postimplantation growth and axial organization. *Genes Dev* 1994;**8**:3045–57.

262. Deng CX, Zhang PM, Harper JW, *et al.* Mice lacking p21CIP1/WAF1 undergo normal development, but are defective in G1 checkpoint control. *Cell* 1995;**82**:675–84.

263. Dennis JP, Rosenberg HS, Alvord EC Jr. Megalencephaly, internal hydrocephalus and other neurological aspects of achondroplasia. *Brain* 1961;**84**:427–45.

264. Deonna T. Prognostic des myeloméningocèles. *J Gynecol Obstet Biol Reprod* 1981;**10**:181–4.

265. De Rosa MJ, Farrell MA, Burke MM, *et al.* An assessment of the proliferative potential of 'balloon cells' in focal cortical resections performed for childhood epilepsy. *Neuropathol Appl Neurobiol* 1992;**18**:566–74.

266. De Rosa MJ, Secor DL, Barsom M, *et al.* Neuropathologic findings in surgically treated hemimegalencephaly: immunohistochemical, morphometric, and ultrastructural study. *Acta Neuropathol (Berl)* 1992;**84**:250–60.

267. De Santi MM, Magni A, Valletta EA, *et al.* Hydrocephalus bronchiectasis and ciliary aplasia. *Arch Dis Child* 1990;**65**:543–4.

268. Desmond ME, Jacobson AG. Embryonic brain enlargement requires cerebrospinal fluid pressure. *Dev Biol* 1977;**57**:188–98.

269. Desmond MM, Wilson G, Melnick J, *et al.* Congenital rubella encephalitis. *J Pediatr* 1967;**71**:311–31.

270. Des Portes V, Pinard JM, Billuart P, *et al.* A novel CNS gene required for neuronal migration and involved in X-linked subcortical laminar heterotopia and lissencephaly syndrome. *Cell* 1998;**92**:51–61.

271. Dieker H, Edwards RH, zuRhein G, *et al.* The lissencephaly syndrome. *Birth Defects Orig Artic Ser* 1969;**5**:53–64.

272. Diezel P. Mikrogyrie infolge cerebraler Speicheldrüsenvirusinfektion im Rahmen einer generalisierten Cytomegalie bei einem Säugling zugleich ein Beitrag zur Theorie der Windungsbildung. *Virchows Arch Pathol Anat* 1954;**325**:109–30.

273. Diezel PB, Fritsch H, Jakob H. Leukodystrophie mit orthochromatischen Abbaustoffen:Ein Beitrag zur Pelizaeus-Merzbacherschen Krankheit. *Virchows Arch Pathol Anat Physiol Klin Med* 1965;**338**:371–94.

274. Dimitri V. Tumor cerebral congénito (angioma cavernoso). *Revista del Asociacion Medica Argentina* 1923;**36**:1029.

275. Ding J, Yang L, Yan YT, *et al.* Cripto is required for correct orientation of the anterior-posterior axis in the mouse embryo. *Nature* 1998;**395**:702–7.

276. Dobbing J, Sands J. Vulnerability of developing brain: IX. The effect of nutritional growth retardation on the timing of the brain growth spurt. *Biol Neonate* 1971;**19**:363–78.

277. Dobyns WB, Stratton RF, Greenberg F. Syndromes with lissencephaly: I. Miller–Dieker and Norman–Roberts syndromes. *Am J Med Genet* 1984;**18**:509–26.

278. Dobyns WB, Pagon RA, Armstrong D, *et al.* Diagnostic criteria for Walker–Warburg syndrome. *Am J Med Genet* 1989;**32**:145–210.

279. Dobyns WB, Truwit CL, Ross ME, *et al.* Differences in the gyral pattern distinguish chromosome 17-linked and X-linked lissencephaly. *Neurology* 1999;**53**:270–77.

280. Dodé C, Levilliers J, Dupont JM, *et al.* Loss-of-function mutations in FGFR1 cause autosomal dominant Kallmann syndrome. *Nat Genet* 2003;**33**:463–5.

281. Dolgopol VB. Absence of the septum pellucidum as the only anomaly in the brain. *Arch Neurol Psychiatry* 1938;**40**:1244–8.

282. Dom R, Brucher JM. Hamartoblastome (gangliocytome diffuse) unilateral de l'écorce cérébrale. *Rev Neurol* 1969;**120**:307–18.

283. Doornik MC van, Hennekam RCM. Hemihydranencephaly with favourable outcome. *Dev Med Child Neurol* 1992;**34**:454–8.

284. Drachman DA, Richardson EP Jr. Aqueductal narrowing congenital and acquired. *Arch Neurol* 1961;**5**:552–9.

285. Droghari E, Smith I, Beaseley M, Lloyd JK. Timing of strict diet in relation to fetal damage in maternal phenylketonuria. *Lancet* 1987;**ii**:927–30.

286. Dryden RJ. Duplication of the spinal cord: a discussion of the possible embryogenesis of diplomyelia. *Dev Med Child Neurol* 1980;**22**:234–43.

287. Dubois-Dauphin M, Frankowski H, Tsujimoto Y, *et al.* Neonatal motoneurons overexpressing the *bcl*-2 protooncogene in transgenic mice are protected from axotomy-induced cell death. *Proc Natl Acad Sci U S A* 1994;**91**:3309–13.

288. Du Boulay G, Shah JH, Currie JC, Logue V. The mechanism of hydromyelia in Chiari I malformations. *Br J Radiol* 1974;**47**:579–87.

289. Duckett S. Foetal Arnold–Chiari malformation. *Acta Neuropathol (Berl)* 1966;**7**:175–9.

290. Dudgeon JA. Varicella-zoster infections. In: Dudgeon JA, Marshall WC eds. *Viral diseases of the fetus and newborn*, 2nd edn. Philadelphia, PA: WB Saunders, 1985:161–74.

291. Duffy PE, Ziter FA. Infantile syringobulbia: a study of its pathology and a proposed relationship to neurogenic stridor in infancy. *Neurology* 1964;**14**:500–509.

292. Dunn HG, Renpenning H, Gerrard JW, *et al.* Mental retardation as a sex-linked defect. *Am J Ment Def* 1963;**64**:827–48.

293. Duong T, De Rosa MJ, Poukens V, *et al.* Neuronal cytoskeletal abnormalities in human cerebral cortical dysplasia. *Acta Neuropathol (Berl)* 1994;**87**:493–503.

294. Dure LS, Percy AK, Cheek WR, Laurent JP. Chiari type I malformation in children. *J Pediatr* 1989;**15**:573–6.

295. Dvorák K, Feit J. Migration of neuroblasts through partial necrosis of the cerebral cortex in newborn rats: contribution to the problems of morphological development and developmental period of cerebral microgyria. *Acta Neuropathol (Berl)* 1977;**38**:203–12.

296. Dvorák K, Feit J, Juránková Z. Experimentally induced focal microgyria and status verrucosus deformity in rats: pathogenesis and interrelations – histological and autoradiological study. *Acta Neuropathol (Berl)* 1978;**44**:121–9.

297. Dyste GN, Menezes AH. Presentation and management of pediatric Chiari malformations without myelodysplasia. *Neurosurgery* 1988;**23**:589–97.

298. Edwards JH. The syndrome of sex-linked hydrocephalus. *Arch Dis Child* 1961;**36**:486–93.

299. Edwards JH, Norman RM, Roberts JM. Sex-linked hydrocephalus: report of a family with 15 affected members. *Arch Dis Child* 1961;**36**:481–5.

300. Ehrlich RM. Ectopic and hypoplastic pituitary with adrenal hypoplasia. *J Pediatr* 1957;**51**:377–84.

301. Eichorn DH, Bayley N. Growth in head circumference from birth through young adulthood. *Child Dev* 1962;**33**:257–71.

302. Ejeckam GG, Wadhwa JK, Williams JP, Lacson AG. Neu–Laxova syndrome:report of two cases. *Pediatr Pathol* 1986;**5**:195–306.

303. Eksioglu YZ, Scheffer IE, Cardenas P, *et al.* Periventricular heterotopia: an X-linked dominant epilepsy locus causing aberrant cerebral cortical development. *Neuron* 1996;**16**:77–87.

304. Ellenberger C Jr, Hanaway J, Netsky MG. Embryogenesis of the inferior olivary nucleus in the rat:a radiographic study and a re-evaluation of the rhombic lip. *J Comp Neurol* 1969;**137**:71–87.

305. Ellis HM, Horvitz HR. Genetic control of programmed cell death in the nematode C. *elegans*. *Cell* 1986;**44**:817–29.

306. Elson E, Perveen R, Donnai D, *et al.* De novo GL13 mutation in acrocallosal syndrome: broadening the phenotypic spectrum of GL13 defects and overlap with murine models. *J Med Genet* 2002;**39**:804–6.

307. Emery JL. Deformity of the aqueduct of Sylvius in children with hydrocephalus and myelomeningocoele. *Dev Med Child Neurol* 1974;**16** (Suppl 32):40–48.

308. Emery JL, Lendon RG. The local cord lesion in neurospinal dysraphism (meningomyelocele). *J Pathol* 1973;**110**:83–96.

309. Emery JL, Staschak MC. The size and form of the cerebral aqueduct in children. *Brain* 1972;**95**:591–8.

310. Epstein DJ, Vekemans M, Gros P. Splotch (Sp^{2H}), a mutation affecting development of the mouse neural tube, shows a deletion within the paired homeodomain of Pax-3. *Cell* 1991;**67**:767–74.

311. Erasmus C, Blackwood W, Wilson J. Infantile multicystic encephalomalacia after maternal bee sting anaphylaxis during pregnancy. *Arch Dis Child* 1982;**57**:785–7.

312. Ericson J, Briscoe J, Rashbass P, *et al.* Graded sonic hedgehog signaling and the specification of cell fate in the ventral neural tube. *Cold Spring Harb Symp Quant Biol* 1997;**62**:451–66.

313. Essick CR. The development of the nuclei pontis and nucleus arcuatus in man. *Am J Anat* 1912;**13**:25–54.

314. Estibeiro JP, Brook FA, Copp AJ. Interaction between splotch (*Sp*) and curly

tail (*ct*) mouse mutants in the embryonic development of neural tube defects. *Development* 1993;**119**:113–21.

315. European Chromosome 16 Tuberous Sclerosis Consortium. Identification and characterization of the tuberous sclerosis gene on chromosome 16. *Cell* 1993;**75**:1305–15.

316. Evrard P, Caviness VS, Prats–Vinas J, Lyon G. The mechanism of arrest of neuronal migration in the Zellweger malformation: an hypothesis based upon cytoarchitectonic analysis. *Acta Neuropathol (Berl)* 1978;**41**:109–17.

317. Evrard P, Saint–Georges P de, Kadhim HJ, *et al.* Pathology of prenatal encephalopathies. In: French JH, Hard S, Casaer P eds. *Child neurology and developmental disabilities.* Baltimore, MD:Paul H Brookes, 1989;153–76.

318. Faivre J, Lemarec B, Bretagne J, Pecker J. X-linked hydrocephalus, aqueductal stenosis, mental retardation, and adduction–flexion deformity of the thumbs: report of a family. *Childs Brain* 1976;**2**:226–33.

319. Fantes J, Ragge NK, Lynch SA, *et al.* Mutations in SOX2 cause anophthalmia. *Nat Genet* 2003;**33**:461–3.

320. Farrell MA, DeRosa MJ, Curran JG, *et al.* Neuropathologic findings in cortical resections (including hemispherectomies) performed for the treatment of intractable childhood epilepsy. *Acta Neuropathol (Berl)* 1992;**83**:246–59.

321. FauréC, Lepintre J, Lyon G. étude radiologique du syndrome de Dandy–Walker. *Acta Radiol* 1963;**1**:843–56.

322. Faust PL, Hatten ME. Targeted deletion of the PEX2 peroxisome assembly gene in mice provides a model for Zellweger syndrome, a human neuronal migration disorder. *J Cell Biol* 1997;**139**:1293–305.

323. Ferland RJ, Eyaid W, Collura RV, *et al.* Abnormal cerebellar development and axonal decussation due to mutations in AHI1 in Joubert syndrome. *Nat Genet* 2004;**36**:1008–13.

324. Fernandez F, Perez–Higueras A, Hernandez R, *et al.* Hydranencephaly after maternal butane-gas intoxication. *Dev Med Child Neurol* 1986;**28**:361–3.

325. Ferrer I. A Golgi analysis of unlayered polymicrogyria. *Acta Neuropathol (Berl)* 1984;**65**:69–76.

326. Ferrer I, Catala I. Unlayered polymicrogyria:structural and developmental aspects. *Anat Embryol* 1991;**184**:517–28.

327. Ferrer I, Navarro C. Multicystic encephalomalacia of infancy. Clinico-pathological report of 7 cases. *J Neurol Sci* 1978;**38**:179–89.

328. Ferrer I, Xumeira A, Santamaria J. Cerebral malformation induced by prenatal X-irradiation: an autoradiographic and Golgi study. *J Anat* 1984;**138**:81–93.

329. Ferrer I, Cusi MV, Liarte A, Campistol J. A Golgi study of the polymicrogyric cortex in Aicardi syndrome. *Brain Dev* 1986;**8**:518–25.

330. Field M, Ashton R, White K. Agenesis of the corpus callosum: report of two preschool children and review of the literature. *Dev Med Child Neurol* 1978;**20**:47–61.

331. Fink JM, Dobyns WB, Guerrini R, Hirsch BA. Identification of a duplication of Xq28 associated with bilateral periventricular nodular heterotopia. *Am J Hum Genet* 1997;**61**:379–87.

332. Firth HV, Boyd PA, Chamberlain P, *et al.* Severe limb abnormalities after chorion villus sampling at 56–66 days' gestation. *Lancet* 1991;**337**:762–3.

333. Fisher JE, Siongco A. Complications from *in utero* death of a monozygous co-twin. *Pediatr Pathol* 1989;**9**:765–71.

334. Fishman MA, Hogan GR, Didge PR. The concurrence of hydrocephalus and craniosynostosis. *J Neurosurg* 1971;**34**:621–9.

335. Flatz G, Sukthomya C. Fronto-ethmoidal encephalomeningoceles in the population of northern Thailand. *Humangenetik* 1970;**11**:1–9.

336. Fleming A, Copp AJ. Embryonic folate metabolism and mouse neural tube defects. *Science* 1998;**280**:2107–9.

337. Fleming A, Copp AJ. A genetic risk factor for mouse neural tube defects: defining the embryonic basis. *Hum Mol Genet* 2000;**9**:575–81.

338. Fleming GWTH, Norman RM. Arhinencephaly with incomplete separation of the cerebral hemispheres. *J Ment Sci* 1942;**88**:341–3.

339. Fleury P, Groot WP de, Delleman JW, *et al.* Tuberous sclerosis:the incidence of sporadic cases versus familial cases. *Brain Dev* 1979;**2**:107–17.

340. Fofanova O, Takamura N, Kinoshita E, *et al.* Compound heterozygous deletion of the prop-1 gene in children with combined pituitary hormone deficiency. *J Clin Endocrinol Metab* 1998;**837**:2601–4.

341. Fone PD, Vapnek JM, Litwiller SE, *et al.* Urodynamic findings in the tethered spinal cord syndrome:does surgical release improve bladder function. *J Urol* 1997;**157**:604–9.

342. Fontan A, Battin JJ. La Mégalencéphalie primitive. *Arch Fr Pediatr* 1965;**22**:521–9.

343. Forman MS, Squier W, Dobyns WB, Golden JA. Genotypically defined lissencephalies show distinct pathologies. *J Neuropathol Exp Neurol* 2005;**64**:847–57.

344. Fowler M, Dow R, White TA, Greer CH. Congenital hydrocephalus-hydranencephaly in five siblings with autopsy studies:a new disease. *Dev Med Child Neurol* 1972;**14**:173–88.

345. Fox JW, Lamperti ED, Eksioglu YZ, *et al.* Mutations in filamin 1 prevent migration of cerebral cortical neurons in human periventricular heterotopia. *Neuron* 1998;**21**:1315–25.

346. Frankenberg WK, Duncan BR, Coffelt RW, *et al.* Maternal phenylketonuria: implications for growth and development. *J Pediatr* 1968;**75**:560–70.

347. Friede RL. Arrested cerebellar development: a type of cerebellar degeneration in amaurotic idiocy. *J Neurol Neurosurg Psychiatry* 1964;**27**:41–5.

348. Friede RL. Dating the development of the human cerebellum. *Acta Neuropathol (Berl)* 1973;**23**:48–58.

349. Friede RL. Uncommon syndromes of cerebellar vermis aplasia. II: Tecto-cerebellar dysplasia with occipital encephalocele. *Dev Med Child Neurol* 1978;**20**:764–72.

350. Friede RL. *Developmental neuropathology*, 2nd edn. Berlin: Springer, 1989.

351. Friede RL, Boltshauser E. Uncommon syndromes of cerebellar vermis aplasia. I. Jourbert syndrome. *Dev Med Child Neurol* 1978;**20**:758–63.

352. Friede RL, Mikolasek J. Postencephalitic porencephaly hydranencephaly or polymicrogyria. A review. *Acta Neuropathol (Berl)* 1978;**43**:161–8.

353. Friede RL, Roessmann S. Chronic tonsillar herniation: an attempt at clarifying chronic herniations at the foramen magnum. *Acta Neuropathol (Berl)* 1986;**34**:219–35.

354. Frutiger P. Zur Frage der Arhinencephalie. *Acta Anat* 1969;**73**:410–30.

355. Fryer AE, Osborne JP. Tuberous sclerosis: a clinical appraisal. *Pediatr Res Commun* 1987;**1**:239–55.

356. Fryer AE, Chalmers A, Connor JM, *et al.* Evidence that the gene for tuberous sclerosis is on chromosome 9. *Lancet* 1987;**i**:659–61.

357. Fryns JP, Spaepen A, Cassiman JJ, Van den Berghe H. X linked complicated spastic paraplegia, MASA syndrome, and X linked hydrocephalus owing to congenital stenosis of the aqueduct of Sylvius: variable expression of the same mutation at Xq28. *J Med Genet* 1991;**28**:429–431.

358. Fujita K, Matsuo N, Mori O, *et al.* The association of hypopituitarism with small pituitary, invisible stalk, type 1 Arnold–Chiari malformation, and syringomyelia in several patients born in breech position: a further proof of birth injury theory on the pathogenesis of 'idiopathic hypopituitarism'. *Eur J Pediatr* 1992;**151**:266–70.

359. Fukuyama Y, Osawa M, Suzuki H. Congenital progressive muscular dystrophy of the Fukuyama type: clinical, genetic and pathological considerations. *Brain Dev* 1981;**3**:1–29.

360. Gadisseux JF, Rodriguez J, Lyon G. Pontoneocerebellar hypoplasia: a probable consequence of prenatal destruction of the pontine nuclei and a possible role of phenytoin intoxication. *Clin Neuropathol* 1984;**3**:160–7.

361. Galaburda AM. Neuroanatomic basis of developmental dyslexia. *Neurol Clin* 1993;**11**:161–73.

362. Garcia CA, Duncan C. Atelencephalic microphaly. *Dev Med Child Neurol* 1977;**19**:227–32.

363. Gardner E, O'Rahilly R, Prolo D. The Dandy–Walker and Arnold–Chiari malformations: clinical developmental and teratological considerations. *Arch Neurol* 1975;**32**:393–407.

364. Gardner WJ. Rupture of the neural tube: the cause of myelomeningocele. *Arch Neurol* 1961;**4**:1–7.

365. Gardner WJ. *The dysraphic states.* Amsterdam: Excerpta Medica, 1973.

366. Gardner WJ, Karnosh LJ, Angel L. Syringomyelia: a result of embryonal atresia of foramen of Magendie. *Trans Am Neurol Assoc* 1957;**82**:144–5.

367. Geoffroy Saint–Hilaire I. Sur de nouveaux anencéphales humains confirmant par leur fait d'organisation la derniére théorie sur les monstres. *Mém Muséum d'Hist Nat (Paris)* 1825;**12**:233–56.

368. Geoffroy Saint–Hilaire I. *Histoire générale et particulière des anomalies de l'organisation chez l'homme et les anomaux.* Paris:JB Baillère, 1832.

369. Gelot A, Billette de Villemeur T, Bordarier C, *et al.* Developmental aspects of type II lissencephaly: comparative study of dysplastic lesions in fetal and

post-natal brains. *Acta Neuropathol (Berl)* 1995;**89**:72–84.

370. Gerard CL, Dugas M, Narcy P, Hertz-Pannier J. Chiari malformation type I in a child with velopharyngeal insufficiency. *Dev Med Child Neurol* 1992;**34**:174–6.

371. Gharib H, Frey HM, Laws ER Jr, Randall RV, Scheithauer BW. Coexistent primary empty sella syndrome and hyperprolactinemia:report of 11 cases. *Arch Intern Med* 1983;**143**:1383–6.

372. Gibbs JL. The heart and tuberous sclerosis. *Br Heart J* 1985;**54**:596–9.

373. Gibson JB. Congenital hydrocephalus due to atresia of the foramen of Magendie. *J Neuropathol Exp Neurol* 1955;**14**:244–62.

374. Gilbert JN, Jones KC, Rorke LB, *et al.* Central nervous system anomalies associated with meningomyelocele, hydrocephalus and the Arnold–Chiari malformations: reappraisal of thesis regarding the pathogenesis of posterior neural tube closure defects. *Neurosurgery* 1986;**18**:559–63.

375. Gilles FH, Davidson RI. Communicating hydrocephalus associated with deficient dysplastic parasagittal arachnoidal granulations. *J Neurosurg* 1971;**35**:421–6.

376. Gimenez–Roldani S, Benito C, Mateo D. Familial communicating syringomyelia. *J Neurol Sci* 1987;**36**:135–46.

377. Gingrich JR, Roder J. Inducible gene expression in the nervous system of transgenic mice. *Annu Rev Neurosci* 1998;**21**:377–405.

378. Giroud A. Causes and morphogenesis of anencephaly. In: Wolstenholime GEW, O'Connor CM eds. *Congenital malformations*. London: Churchill, 1960:199–212.

379. Gleeson JG, Allen KM, Fox JW, *et al.* Doublecortin, a brain-specific gene mutated in human X-linked lissencephaly and double cortex syndrome, encodes a putative signaling protein. *Cell* 1998;**92**:63–72.

380. Glover MT, Brett EM, Atherton DJ. Hypomelanosis of Ito: spectrum of the disease. *J Pediatr* 1989;**115**:75–80.

381. Golden JA, Chernoff GF. Intermittent pattern of neural tube closure in two strains of mice. *Teratology* 1993;**47**:73–80.

382. Golden JA, Harding BN. Cortical malformations: unfolding polymicrogyria. *Nat Rev Neurol* 2010;**6**(9):471–2.

383. Golden JA, Hyman BT. Development of the superior temporal neocortex is anomalous in trisomy 21. *J Neuropathol Exp Neurol* 1994;**53**(5):513–20.

384. Goldowitz D, Hamre KM, Przyborski SA, *et al.* Granule cells and cerebellar boundaries: analysis of Unc5h3 mutant chimeras. *J Neurosci* 2000;**20**:4129–37.

385. Goldstone AP. Prader–Willi syndrome:advances in genetics, pathophysiology and treatment. *Trends Endocrinol Metab* 2004;**15**:12–20.

386. Gomez MR. *Tuberous sclerosis*. New York: Raven Press, 1979.

387. Gomez MR, Whitten CF, Nolke A, *et al.* Aneurysmal malformation of the great vein of Galen causing heart failure in early infancy: report of 5 cases. *Pediatrics* 1963;**31**:400–11.

388. González-García M, García I, Ding L, *et al.* bcl-x is expressed in embryonic and postnatal neural tissues and functions to

389. prevent neuronal cell death. *Proc Natl Acad Sci U S A* 1995;**92**:4304–8.

389. Goodlin RC, Heidrick WP, Papenfuss HL, Kubitz RL. Fetal malformations associated with maternal hypoxia. *Am J Obstet Gynecol* 1984;**49**:228–9.

390. Goodrich LV, Milenkovic L, Higgins KM, *et al.* Altered neural cell fates and medulloblastoma in mouse patched mutants. *Science* 1997;**277**:1109–13.

391. Gosseye S, Golaire M, Larroche JC. Cerebral, renal and splenic lesions due to fetal anoxia and their relationship to malformations. *Dev Med Child Neurol* 1982;**24**:510–18.

392. Goto S, Hirano A, Rojas-Corona RR. A comparative immunocytochemical study of human cerebellar cortex in X-chromosome-linked copper malabsorption (Menkes–kinky hair disease) and granule cell type cerebellar degeneration. *Neuropathol Appl Neurobiol* 1989;**15**:419–31.

393. Gott PS, Saul RE. Agenesis of the corpus callosum: limits of functional compensation. *Neurology* 1978;**28**:1272–9.

394. Gottfried M, Lavine L, Roessmann U. Neuropathological findings in Wolf–Hirschhorn (4p-) syndrome. *Acta Neuropathol (Berl)* 1981;**55**:163–5.

395. Gould S, Howard S. An immunohistological study of macrophages in the human fetal brain. *Neuropathol Appl Neurobiol* 1990;**16**:261–2.

396. Goutières F, Aicardi J, Farkas–Bargeton E. Une malformation cérébrale particulière associée au nanisme thanatophore. *Rev Neurol* 1971;**125**:435–46.

397. Goutières F, Aicardi J, Farkas E. Anterior horn cell disease associated with pontocerebellar hypoplasia in infants. *J Neurol Neurosurg Psychiatry* 1977;**40**:370–78.

398. Granata T, Farina L, Faiella A, *et al.* Familial schizencephaly associated with EMX2 mutation. *Neurology* 1997;**48**:1403–6.

399. Green AJ, Smith M, Yates JRW. Loss of heterozygosity on chromosome 16p13.3 in hamartomas from tuberous sclerosis patients. *Nat Genet* 1994;**6**:193–6.

400. Greene NDE, Copp AJ. Inositol prevents folate-resistant neural tube defects in the mouse. *Nat Med* 1997;**3**:60–66.

401. Greene NDE, Gerrelli D, Van Straaten HWM, Copp AJ. Abnormalities of floor plate, notochord and somite differentiation in the *loop-tail (Lp)* mouse: a model of severe neural tube defects. *Mech Dev* 1998;**73**:59–72.

402. Greenstone MA, Jones RWA, Dewar A, *et al.* Hydrocephalus and primary ciliary dyskinesia. *Arch Dis Child* 1984;**59**:481–2.

403. Gregg NM. Congenital cataract following German measles in the mother. *Trans Ophthalmol Soc Austral* 1941;**3**:35–46.

404. Grinberg I, Northrup H, Ardinger H, *et al.* Heterozygous deletion of the linked genes ZIC1 and ZIC4 is involved in Dandy–Walker malformation. *Nat Genet* 2004;**36**:1053–5.

405. Gripp KW, Wotton D, Edwards MC, *et al.* Mutations in TGIF cause holoprosencephaly and link NODAL signalling to human neural axis determination. *Nat Genet* 2000;**25**:205–8.

406. Grollmus JM, Wilson CB, Newton TH. Paramesencephalic arachnoid cysts. *Neurology* 1976;**26**:128–34.

407. Gross H. Die Rhombencephalosynapsis, eine systemisierte Kleinhirnfehlbildung. *Arch Psychiat Nervenkrankheit* 1959;**199**:537–52.

408. Gross H, Hoff H. Sur les dysraphies cranioencéphaliques. In: Heuyer G, Feld M, Gjuner J. eds. *Malformations congénitales du cerveau.* Paris: Masson, 1959:287–96.

409. Grove EA, Williams BP, Li D–Q, *et al.* Multiple restricted lineages in the embryonic rat cerebral cortex. *Development* 1993;**117**:553–61.

410. Gruenwald P, Minh H. Evaluation of body and organ weights in perinatal pathology: I. Normal standard derived from autopsies. *Am J Clin Pathol* 1960;**39**:247–53.

411. Guerrini R, Parrini E. Neuronal migration disorders. *Neurobiol Dis* 2010;**38**(2):154–66.

412. Guerrini R, Dravet C, Raybaud C, *et al.* Epilepsy and focal gyral anomalies detected by MRI: electroclinico-morphological correlations and follow-up. *Dev Med Child Neurol* 1992;**34**:706–18.

413. Guerrini R, Dravet C, Raybaud C, *et al.* Neurological findings and seizure outcome in children with bilateral opercular macrogyric-like changes detected by MRI. *Dev Med Child Neurol* 1992;**34**:694–705.

414. Gullotta F, Rehder H, Gropp A. Descriptive neuropathology of chromosomal disorders in man. *Hum Genet* 1981;**57**:337–44.

415. Gutierrez Y, Friede RL, Kaliney WJ. Agenesis of arachnoid granulations and its relationship to communicating hydrocephalus. *J Neurosurg* 1975;**43**:553–8.

416. Hageman G, Willemse J, van Ketel BA, *et al.* The heterogeneity of the Pene–Shokeir syndrome. *Neuropaediatrics* 1987;**45**:18–50.

417. Hageman G, Willemse J, van Ketel BA, Verdonck AFMM. The pathogenesis of fetal hypokinesia: a neurological study of 72 cases of congenital contractures with emphasis on cerebral lesions. *Neuropediatrics* 1987;**18**:22–33.

418. Hajihosseini MK, Wilson S, De Moerlooze L, *et al.* A splicing switch and gain-of-function mutation in FgfR2-IIIc hemizygotes causes Apert/Pfeiffer-syndrome-like phenotypes. *Proc Natl Acad Sci U S A* 2001;**98**:3855–60.

419. Hallervorden J. über eine Kohlenoxydvergiftung im Fötalleben mit Eintwicklungstörung der Hirnrinde. *Allg Z Psychiatrie* 1949;**124**:289–98.

420. Halsey JH, Allen N, Chamberlin HR. The morphogenesis of hydranencephaly. *J Neurol Sci* 1971;**12**:187–217.

421. Hanau J, Franc B, Faivre J, Foncin JF. Hydrocéphalie genetique liée au sexe:étude anatomique. *Rev Neurol* 1978;**134**:437–42.

422. Hanaway J, Netsky M. Heterotopias of the inferior olive: relation to Dandy–Walker malformation and correlation with experimental data. *J Neuropathol Exp Neurol* 1968;**30**:380–9.

423. Hanaway J, Lee SI, Netsky MG. Pachygyria:relation of findings to modern embryological concepts. *Neurology* 1968;**18**:791–9.

424. Hanigan WC, Aldrich WM. MRI and evoked potentials in a child with hydranencephaly. *Pediatr Neurol* 1988;**4**:185–7.

425. Hannan AJ, Servotte S, Katsnelson A, *et al.* Characterization of nodular neuronal heterotopia in children. *Brain* 1999;**122**:219–38.

426. Hansen FJ, Friis B. Familial occurrence of cerebral gigantism, Sotos' syndrome. *Acta Paediatr Scand* 1976;**65**:387–9.

427. Hanshaw JB. Congenital and acquired cytomegalovirus infection. *Pediatr Clin North Am* 1966;**13**:279–93.

428. Hanshaw JB, Dudgeon JA, Marshall WC. *Viral diseases of the fetus and newborn.* 2nd edn. Philadelphia, PA: WB Saunders, 1985.

429. Hanson I, Van Heyningen V. Pax6: more than meets the eye. *Trends Genet* 1995;**11**:268–72.

430. Harbord MG, Finn JP, Hall–Craggs MA, *et al.* Moebius syndrome with unilateral cerebellar hypoplasia. *J Med Genet* 1989;**26**:579–82.

431. Hardimann O, Burke T, Phillips J, *et al.* Microdysgenesis in resected temporal neocortex: incidence and clinical significance in focal epilepsy. *Neurology* 1988;**38**:1041–7.

432. Harding B. Laminar heterotopia: a possible failure of programmed cell death? *J Neuropathol Exp Neurol* 1996;**55**:1.

433. Harding BN. Cerebro-ocular dysplasia with muscular dystrophy. *Neuropathol Appl Neurobiol* 1988;**14**:258.

434. Harding BN. Gray matter heterotopia. In: Guerrini R, Pfanner P, Roger J, *et al.* eds. *Dysplasias of cerebral cortex in childhood-onset epilepsy.* Philadelphia, PA: Lippincott–Raven, 1996:81–8.

435. Harding BN, Baumer JA. Congenital varicellazoster: a serologically proven case with necrotizing encephalitis and malformation. *Acta Neuropathol (Berl)* 1988;**76**:311–15.

436. Harding BN, Boyd S. Intractable seizures from infancy can be associated with dentato-olivary dysplasia. *J Neurol Sci* 1991;**104**:157–65.

437. Harding BN, Erdohazi M. Cerebellar disease in childhood: pontoneocerebellar hypoplasia. *Neuropathol Appl Neurobiol* 1989;**15**:294.

438. Harding B, Thom M. Bilateral hippocampal granule cell dispersion: autopsy study of 3 infants. *Neuropathol Appl Neurobiol* 2001;**27**:245–51.

439. Harding BN, Ramani P, Thurley P. The familial syndrome of proliferative vasculopathy and hydranencephaly–hydrocephaly: immunocytochemical and ultrastructural evidence for endothelial proliferation. *Neuropathol Appl Neurobiol* 1995;**21**:61–7.

440. Harmant–van Rijckevorsel G, Aubert–Tulkens G, Moulin D, Lyon G. Le Syndrome de Jourbert: étude clinique et anatome-pathologique. *Rev Neurol* 1983;**139**:715–24.

441. Harper C, Hockey A. Proliferative vasculopathy and an hydranencephalic–hydrocephalic syndrome: a neuropathological study of two siblings. *Dev Med Child Neurol* 1983;**25**:232–9.

442. Harris CP, Townsend JJ, Norman MG, *et al.* Atelencephalic aprosencephaly. *J Child Neurol* 1994;**9**:412–16.

443. Harrison KA, Thaler J, Pfaff SL, *et al.* Pancreas dorsal lobe agenesis and abnormal islets of Langerhans in Hlxb9-deficient mice. *Nat Genet* 1999;**23**:71–5.

444. Hart MN, Malamud N, Ellis WG. The Dandy–Walker syndrome: a clinicopathological study based on 28 cases. *Neurology* 1972;**22**:771–80.

445. Hartley WJ, Saram WG de, Della–Porta AJ, *et al.* Pathology of congenital bovine epizootic arthrogyposis and hydranencephaly and its relationship to Akabane virus. *Austral Vet J* 1977;**53**:319–25.

446. Harwood–Nash DC, Fitz CR. *Neuroradiology in infants and children.* 3rd edn. St Louis, MO: Mosby, 1976.

447. Hatjis CG, Horber JD, Anderson GG. The *in utero* diagnosis of a posterior fossa intracranial cyst (Dandy–Walker cyst). *Am J Obstet Gynecol* 1981;**140**:473–4.

448. Hatten ME. LIS-less neurons don't even make it to the starting gate. *J Cell Biol* 2005;**170**:867–71.

449. Hatten ME, Alder J, Zimmerman K, Heintz N. Genes involved in cerebellar cell specification and differentiation. *Curr Opin Neurobiol* 1997;**7**:40–7.

450. Hattori M, Adachi H, Tsujimoto M, *et al.* Miller–Dieker lissencephaly gene encodes a subunit of brain platelet-activating factor. *Nature* 1994;**370**:216–18.

451. Hausmann B, Sievers J. Cerebellar external granule cells are attached to the basal lamina from the onset of migration up to the end of their proliferative activity. *J Comp Neurol* 1985;**241**:50–62.

452. Hayhurst M, McConnell SK. Mouse models of holoprosencephaly. *Curr Opin Neurol* 2003;**16**:135–41.

453. Haymaker W, Girdany BR, Stephens G, *et al.* Cerebral involvement with advanced periventricular calcification in generalised cytomegalic inclusion disease in the newborn. *J Neuropathol Exp Neurol* 1954;**13**:562–86.

454. Henderson JL. The congenital facial diplegia syndrome: clinical features, pathology and etiology. *Brain* 1939;**62**:381–403.

455. Hernandez D, Fisher EMC. Down syndrome genetics: unravelling a multifactorial disorder. *Hum Mol Genet* 1996;**5**:1411–16.

456. Herringham WP, Andrewes FW. Two cases of cerebellar disease in cats, with staggering. *St Bartholomews Hosp Rep* 1888;**24**:241–8.

457. Heschl R. Gehirndefect und Hydrocephalus. *Vierteljahrschrift für praktikale Heilkunde Prague* 1859;**61**:59–74.

458. Heschl R. Ein neuer Fall von Porencephalie. *Vierteljahrschrift für praktikale Heilkunde Prague* 1861;**72**:102–4.

459. Heubner O. Ueber angeborene Kernmangel (infantiler Kernschwund, Moebius). *Charitéannalen* 1900;**25**:211–43.

459a. Hevner RF. The cerebral cortex malformation in thanatophoric dysplasia: neuropathology and pathogenesis. *Acta Neuropathol (Berl)* 2005;**110**, 208–21.

460. Hide T, Hatakeyama J, Kimura–Yoshida C, *et al.* Genetic modifiers of otocephalic phenotypes in Otx2 heterozygous mutant mice. *Development* 2002;**129**:4347–57.

461. Hiesberger T, Trommsdorff M, Howell BW, *et al.* Direct binding of Reelin to VLDL receptor and ApoE receptor 2 induces tyrosine phosphorylation of disabled-1 and modulates tau phosphorylation. *Neuron* 1999;**24**:481–9.

462. Higginbottom MC, Jones KL, Hall BD, Smith DW. The amniotic band disruption complex: timing of amniotic rupture and variable spectra of consequent defects. *J Pediatr* 1979;**95**:544–9.

463. Hilburger AC, Willis JK, Bouldin E, Henderson–Tilton A. Familial schizencephaly. *Brain Dev* 1993;**15**:234–6.

464. Hill RE, Favor J, Hogan BLM, *et al.* Mouse Small eye results from mutations in a paired-like homeobox-containing gene. *Nature* 1991;**354**:522–5.

465. Hintz RL, Menking M, Sotos JF. Familial holoprosencephaly with endocrine dysgenesis. *J Pediatr* 1968;**72**:81–7.

466. Hirano A, Solomon S. Arteriovenous aneurysm of the vein of Galen. *Arch Neurol* 1960;**3**:589–93.

467. Hirano A, Tuazon R, Zimmerman HM. Neurofibrillary changes, granulovacuolar bodies and argentophilic globules observed in tuberous sclerosis. *Acta Neuropathol (Berl)* 1968;**11**:257–61.

468. Hirano A, Dembitzer HM, Ghatak NKI, Zimmerman HM. On the relationship between human and experimental granule cell type cerebellar degeneration. *J Neuropathol Exp Neurol* 1973;**32**:493–502.

469. Hirano S, Houdou S, Hasegawa M, *et al.* Clinicopathologic studies on leptomeningeal glioneuronal heterotopia in congenital anomalies. *Pediatr Neurol* 1992;**8**:441–4.

470. Hirono I, Shibuya C, Hayashi K. Induction of a cerebellar disorder with cycasin in newborn mice and hamsters. *Proc Soc Exp Biol Med* 1969;**131**:593–600.

471. Hirotsune S, Takahara T, Sasaki N, *et al.* The reeler gene encodes a protein with an EGF-like motif expressed by pioneer neurons. *Nat Genet* 1995;**10**:77–83.

472. Hirotsune S, Fleck MW, Gambello MJ, *et al.* Graded reduction of Pafah1b1 (Lis1) activity results in neuronal migration defects and early embryonic lethality. *Nat Genet* 1998;**19**:333–9.

473. His W. Die Entwicklung des menschlichen Rautenhirns vom Ende des ersten bis zum Beginn des dritten Monats. I Verlängstes Mark. *Abhandlungen des Königlich Sächsischen Gesellschaften der Wissenschaften Leipzig* 1891;**29**:1–74.

474. Ho KL, Chang CH, Yang SS, Chason JL. Neuropathologic findings in thanatophoric dysplasia. *Acta Neuropathol (Berl)* 1984;**63**:218–28.

475. Hochgeschwender U, Brennan MB. The impact of genomics on mammalian neurobiology. *BioEssays* 1999;**21**:157–63.

476. Hoffman HJ, Hendrick EB. Early neurosurgical repair in craniofacial dysmorphism. *J Neurosurg* 1979;**51**:769–803.

477. Hofman MA. Energy metabolism and relative brain size in human neonates from single and multiple gestations: an allometric study. *Biol Neonate* 1984;**45**:157–64.

478. Holden AM Jr, Fyler DC, Shillito J, Nadas AS. Congestive heart failure from intracranial arteriovenous fistula in infancy. *Pediatrics* 1972;**49**:30–39.

479. Holland HC, Graham WL. Congenital atresia of the foramina of Luschka and Magendie with hydrocephalus: report of a case in an adult. *J Neurosurg* 1958;**15**:688–94.

480. Holmes L, Nash A, Zu Rhein G, et al. X-linked aqueductal stenosis: clinical and morphological findings in two families. *Pediatrics* 1973;**51**:697–704.

481. Holzfeind PJ, Grewal PK, Reitsamer HA, et al. Skeletal, cardiac and tongue muscle pathology, defective retinal transmission, and neuronal migration defects in the Largemyd mouse defines a natural model for glycosylation-deficient muscle-eye-brain disorders. *Hum Mol Genet* 2002;**11**:2673–87.

482. Hong SE, Shugart YY, Huang DT, et al. Autosomal recessive lissencephaly with cerebellar hypoplasia is associated with human RELN mutations. *Nat Genet* 2000;**26**:93–6.

483. Hori A. Suprasellar peri-infundibular ectopic adenohypophysis in fetal and adult brains. *J Neurosurg* 1985;**62**:113–15.

484. Hori A, Minwegen J. Intrauterine purulent encephalitis with early stage of hydranencephaly. Case report. *Acta Neuropathol (Berl)* 1984;**64**:72–4.

485. Hori A, Friede RL, Fischer G. Ventricular diverticles with localised dysgenesis of the temporal lobe in cloverleaf skull anomaly. *Acta Neuropathol (Berl)* 1983;**60**:132–6.

486. Hosley MA, Abroms IF, Ragland RL. Schizencephaly: case report of familial incidence. *Pediatr Neurol* 1992;**8**:148–50.

487. Houser CR. Granule cell dispersion in the dentate gyrus of humans with temporal lobe epilepsy. *Brain Res* 1990;**535**:195–204.

488. Houser CR, Swartz BE, Walsh GO, et al. Granule cell dispersion in the dentate gyrus:possible alterations of neuronal migration in human temporal lobe epilepsy. In: Engel JJ, Wasterlain C, Cavalheiro EA, et al., eds. *Molecular biology of epilepsy.* Amsterdam: Elsevier, 1992:41–9.

489. Howard FH, Till K, Carter CO. A family study of hydrocephalus resulting from aqueduct stenosis. *J Med Genet* 1981;**18**:252–5.

490. Hoyt CS, Billson F, Ouvrier R, Wise G. Ocular features of Aicardi's syndrome. *Arch Ophthalmol* 1978;**96**:291–5.

491. Hoyt WF, Kaplan SL, Grumbach MM, Glaser JS. Septo-optic dysplasia and pituitary dwarfism. *Lancet* 1970;**i**:893–4.

492. Huard JMT, Forster CC, Carter ML, et al. Cerebellar histogenesis is disturbed in mice lacking cyclin D2. *Development* 1999;**126**:1927–35.

493. Hughes HE, Miskin M. Congenital microcephaly due to vascular disruption: *in utero* documentation. *Pediatrics* 1986;**78**:85–7.

494. Hughes JT, Brownell B. Aberrant nerve fibres within the spinal cord. *J Neurol Neurosurg Psychiatry* 1963;**26**:528–34.

495. Hunt A, Lindenbaum RH. Tuberous sclerosis: a new estimate of prevalence within the Oxford region. *J Med Genet* 1984;**21**:272–7.

496. Hunt P, Krumlauf R. *Hox* codes and positional specification in vertebrate embryonic axes. *Annu Rev Cell Biol* 1992;**8**:227–56.

497. Huttenlocher RR, Taravath S, Mojtahedi S. Periventricular heterotopia and epilepsy. *Neurology* 1994;**44**:51–5.

498. Iivanainen M, Haltia M, Lydecken K. Atelencephaly. *Dev Med Child Neurol* 1977;**19**:663–8.

499. Ikonomidou C, Bittigau P, Ishimaru MJ, et al. Ethanol-induced apoptotic neurodegeneration and fetal alcohol syndrome. *Science* 2000;**287**:1056–60.

500. Ingham PW. Transducing hedgehog: the story so far. *EMBO J* 1998;**17**:3505–11.

501. Isaac M, Best P. Two cases of agenesis of the vermis of cerebellum, with fusion of the dentate nuclei and cerebellar hemispheres. *Acta Neuropathol (Berl)* 1987;**74**:278–80.

502. Isu T, Chono Y, Iwasaki Y, et al. Scoliosis associated with syringomyelia presenting in children. *Childs Nerv Syst* 1992;**8**:97–100.

503. Iwata T, Chen L, Li CI, et al. A neonatal lethal mutation in FGFR3 uncouples proliferation and differentiation of growth plate chondrocytes in embryos. *Hum Mol Genet* 2000;**9**:1603–13.

504. Jackson AP, Eastwood H, Bell SM, et al. Identification of microcephalin, a protein implicated in determining the size of the human brain. *Am J Hum Genet* 2002;**71**:136–42.

505. Jacobsen M, Jacobsen GK, Clausen PP, et al. Intracellular plasma proteins in human fetal choroid plexus during development: II. The distribution of prealbumin, albumin, α-fetoprotein, transferrin, IgG, IgA, IgM, and alpha1-antitrypsin. *Dev Brain Res* 1982;**3**:251–62.

506. Jacobson MD, Weil M, Raff MC. Programmed cell death in animal development. *Cell* 1997;**88**:347–54.

507. Jaffe R, Crumrine P, Hashida Y, Moser HW. Neonatal adenoleukodystrophy: clinical, pathological and biochemical delineation of a syndrome affecting both males and females. *Am J Pathol* 1982;**108**:100–111.

508. Jakob H. Faktoren bei der Entstehung der normalen und der entwicklungs-gestörten Hirnrinde. *Z Neurol Psychiatrie* 1936;**155**:1–39.

509. Jakob H. Die feinere Oberflächengestaltung der Hirnwindungen, die Hirnwarzenbildung und die Mikropolygrie. *Z Neurol Psychiatrie* 1940;**170**:64–84.

510. Lui JH, Hansen DV, Kriegstein AR. Development and the human neocortex. *Cell* 2011;**146**:18–36.

511. Janota I, Polkey CE. Cortical dysplasia in epilepsy: a study of material from surgical resections for intractable epilepsy. In: Pedley TA, Meldrum BS eds. *Recent advances in epilepsy.* 5th edn. London: Churchill Livingstone, 1992:37–49.

512. Janzer RC, Friede RL. Dandy–Walker syndrome with atresia of the fourth ventricle and multiple rhombencephalic malformations. *Acta Neuropathol (Berl)* 1982;**58**:81–6.

513. Jeeves MA. Agenesis of the corpus callosum. In: Boller F, Grafman J eds. *Handbook of neuropsychology.* 4th edn. New York: Elsevier, 1990:99–114.

514. Jellinger K, Gross H. Congenital telencephalic midline defects. *Neuropädiatrie* 1973;**4**:446–52.

515. Jellinger K, Rett A. Agyria–pachygyria (lissencephaly syndrome). *Neuropädiatrie* 1976;**7**:66–91.

516. Jellinger K, Gross H, Kaltenbäck E, Grisold W. Holoprosencephaly and agenesis of the corpus callosum: frequency of associated malformations. *Acta Neuropathol (Berl)* 1981;**55**:1–10.

517. Jen JC, Chan WM, Bosley TM, et al. Mutations in a human ROBO gene disrupt hindbrain axon pathway crossing and morphogenesis. *Science* 2004;**304**:1509–13.

518. Jenkyn LR, Roberts DW, Merlis AL, et al. Dandy–Walker malformation in identical twins. *Neurology* 1981;**31**:337–41.

519. Jeret JS, Serur D, Wisniewski KE, Lubin RA. Clinicopathological findings associated with agenesis of the corpus callosum. *Brain Dev* 1987;**9**:255–64.

520. Jervis GA. Early senile dementia in mongoloid idiocy. *Am J Psychiatry* 1948;**105**:102–6.

521. Jervis GA. Spongioneuroblastoma and tuberous sclerosis. *J Neuropathol Exp Neurol* 1954;**13**:105–16.

522. Jiang H, Gyda M III, Harnish DC, et al. Teratogenesis by retinoic acid analogs positively correlates with elevation of retinoic acid receptor-β2 mRNA levels in treated embryos. *Teratology* 1994;**50**:38–43.

523. Jiang YH, Armstrong D, Albrecht U, et al. Mutation of the Angelman ubiquitin ligase in mice causes increased cytoplasmic p53 and deficits of contextual learning and long-term potentiation. *Neuron* 1998;**21**:799–811.

524. Jin JZ, Gu S, McKinney P, et al. Expression and functional analysis of Tgif during mouse midline development. *Dev Dyn* 2006;**235**:547–53.

525. Johnson FE. Injury of the child by Roentgen ray during pregnancy: report of a case. *J Pediatr* 1938;**13**:894–901.

526. Johnson RT, Johnson K. Hydrocephalus following virus infection: the pathology of aqueductal stenosis developing after experimental mumps virus infection. *J Neuropathol Exp Neurol* 1968;**27**:591–606.

527. Jonesco–Sisesti N. *La syringobulbie.* Paris: Masson, 1932.

528. Jordan RM, Kendall JW, et al. The primary empty sella syndrome. Analysis of the clinical characteristics, radiographic features, pituitary function and cerebrospinal fluid adenohypophysial hormone concentrations. *Am J Med* 1977;**62**:569–80.

529. Joubert M, Eisenring JJ, Robb JP, Andermann F. Familial agenesis of the cerebellar vermis: a syndrome of episodic hyperpnea, abnormal eye movements, ataxia and retardation. *Neurology* 1969;**19**:813–25.

530. Judkins AR, Martinez D, Ferreira P, Dobyns WB, Golden JA. Polymicrogyria includes fusion of the molecular layer and decreased neuronal populations but normal cortical laminar organization. *J Neuropathol Exp Neurol* 201170(6):438–43.

531. Jung JH, Graham JM Jr, Schultz N, Smith DW. Congenital hydranencephaly/porencephaly due to vascular disruption in monozygotic twins. *Pediatrics* 1984;**73**:467–9.

532. Juriloff DM, Harris MJ. Mouse models for neural tube closure defects. *Hum Mol Genet* 2000;**9**:993–1000.

533. Kallischer S. Demonstration des Gehirns eines Kindes mit Telangectasie der links-seitigen Gesicht und Kopfhaut und der Hirnoberfläche. *Kindische Wochenschrift* 1897;**34**:1059.

534. Kalter H. *Teratology of the central nervous system.* Chicago, IL: University of Chicago Press, 1968.

535. Kalter H. Five-decade international trends in the relation of perinatal mortality and congenital malformations: stillbirth and neonatal death compared. *Int J Epidemiol* 1991;**20**:173–9.

536. Kalter H. Case reports of malformations associated with maternal diabetes: history and critique. *Clin Genet* 1993;**43**:174–9.

537. Kalter H, Warkany J. Congenital malformations: etiologic factors and their role in prevention. *N Engl J Med* 1983;**308**:424–31.

538. Kamoshita S, Takei Y, Miyao M, *et al.* Pontocerebellar hypoplasia associated with infantile motor neuron disease (Norman's disease). *Pediatr Pathol* 1990;**10**:133–42.

539. Kamuro K, Tenokuchi Y. Familial periventricular nodular heterotopia. *Brain Dev* 1993;**15**:237–41.

540. Kang S, Graham JM Jr, Olney AH, Biesecker LG. GLI3 frameshift mutations cause autosomal dominant Pallister–Hall syndrome. *Nat Genet* 1997;**15**:266–8.

541. Karch SB, Urich H. Occipital encephalocele: a morphological study. *J Neurol Sci* 1972;**15**:89–112.

542. Kauschansky A, Genel M, Smith GJ. Congenital hypopituitarism in female infants. Its association with hypoglycemia and hypothyroidism. *Am J Dis Child* 1979;**133**:165–9.

543. Kawagoe T, Jacob H. Neocerebellar hypoplasia with systemic combined olivo-ponto-dentatal degeneration in a 9-day-old baby: contribution to the problem of relations between malformation and systemic degeneration in early life. *Clin Neuropathol* 1986;**5**:203–8.

544. Keeling JW, Kjær I. Diagnostic distinction between anencephaly and amnion rupture sequence based on skeletal analysis. *J Med Genet* 1994;**31**:823–9.

545. Kelley RI, Roessler E, Hennekam RCM, *et al.* Holoprosencephaly in RSH/Smith–Lemli–Opitz syndrome: does abnormal cholesterol metabolism affect the function of *Sonic Hedgehog. Am J Med Genet* 1996;**66**:478–84.

546. Kemper TL, Lecours AR, Gates MG, Yakovlev PI. Retardation of the myelo- and cyto-architectonic maturation of the brain in the congenital rubella syndrome in early development. *Res Publ Pub Assoc Res Nerv Ment Dis* 1973;**51**:23–62.

547. Kepes JJ, Clough C, Villanueva A. Congenital fusion of the thalami (atresia of the third ventricle) and associated anomalies in a 6 month old infant. *Acta Neuropathol (Berl)* 1969;**13**:97–104.

548. Kerner B, Graham JM Jr, Golden JA, *et al.* Familial lissencephaly with cleft palate and severe cerebellar hypoplasia. *Am J Med Genet* 1999;**87**:440–45.

549. Kershman J. Genesis of microglia in the human brain. *Arch Neurol Psychiatry* 1939;**41**:24–50.

550. Khan M, Rozdilsky B, Gerrard JW. Familial holoprosencephaly. *Dev Med Child Neurol* 1970;**12**:71–6.

551. Kholmanskikh SS, Dobrin JS, Wynshaw-Boris A, *et al.* Disregulated RhoGTPases and actin cytoskeleton contribute to the migration defect in Lis1-deficient neurons. *J Neurosci* 2003;**23**:8673–81.

552. Kilham L, Margolis G. Cerebellar ataxia in hamsters inoculated with rat virus. *Science* 1964;**143**:1047–8.

553. Kim TS, Cho S, Dickson DW. Aprosencephaly: review of the literature and report of a case with cerebellar hypoplasia, pigmented epithelial cyst and Rathke's cleft cyst. *Acta Neuropathol (Berl)* 1990;**79**:424–31.

554. King MD, Dudgeon J, Stephenson JBP. Joubert's syndrome with retinal dysplasia: neonatal tachypnoea as the clue to a genetic brain–eye malformation. *Arch Dis Child* 1984;**59**:709–18.

555. King MD, Stephenson JBP, Ziervogel M, *et al.* Hemimegalencephaly: a case for hemispherectomy. *Neuropediatrics* 1985;**16**:46–55.

556. Kingsley DPE. Neuro-imaging. In: Brett EM ed. *Paediatric neurology.* London: Churchill Livingstone, 1991:836–7.

557. Kirke PN, Molloy AM, Daly LE, *et al.* Maternal plasma folate and vitamin B12 are independent risk factors for neural tube defects. *Q J Med* 1993;**86**:703–8.

558. Kishino T, Lalande M, Wagstaff J. UBE3A/E6-AP mutations cause Angelman syndrome. *Nat Genet* 1997;**15**:70–73.

559. Kitamura K, Yanazawa M, Sugiyama N, *et al.* Mutation of ARX causes abnormal development of forebrain and testes in mice and X-linked lissencephaly with abnormal genitalia in humans. *Nat Genet* 2002;**32**:359–69.

560. Kitanaka C, Iwasaki Y, Yamada H. Retroflexion of holoprosencephaly: report of two cases. *Childs Nerv Syst* 1992;**8**:317–21.

561. Knebel–Doeberitz C, Sievers J, Sadler M, *et al.* Destruction of meningeal cells in the newborn hamster cerebellum with 6 hydroxydopamine prevents foliation and lamination in the rostral cerebellum. *Neuroscience* 1986;**17**:409–26.

562. Knudson AJ. Genetics of human cancer. *Annu Rev Genet* 1986;**20**:231–51.

563. Kobayashi K, Nakahori Y, Miyake M, *et al.* An ancient retrotransposal insertion causes Fukuyama-type congenital muscular dystrophy. *Nature* 1998;**394**:388–92.

564. Köhler U. Sturge-Webersche Krankheit bei einer Frühgeburt. *Zentralblatt Allg Pathol Anat* 1940;**75**:81–5.

565. Kosaki K, Matsuo N, Tamai S, Miyama S, Momoshima S. Isolated aplasia of the anterior pituitary as a cause of congenital panhypopituitarism. *Horm Res* 1991;**35**:226–8.

566. Koseki H, Wallin J, Wilting J, *et al.* A role for *Pax-1* as a mediator of notochordal signals during the dorsoventral specification of vertebrae. *Development* 1993;**119**:629–60.

567. Koster S. Two cases of hypoplasia pontoneocere-bellaris. *Acta Psychiat Kobenhavn* 1926;**1**:47–76.

568. Kotzot D, Weigl J, Huk W, Rott HD. Hydantoin syndrome with holoprosencephaly:a possible rare teratogenic effect. *Teratology* 1993;**48**:15–19.

569. Kruyff E. Paracollicular plate cysts. *Am J Roentgenol* 1965;**95**:899–916.

570. Kuchelmeister K, Bergmann M, Gullotta F. Neuropathology of lissencephalies. *Childs Nerv Syst* 1993;**9**:394–9.

571. Kuida K, Zheng TS, Na SQ, *et al.* Decreased apoptosis in the brain and premature lethality in CPP32-deficient mice. *Nature* 1996;**384**:368–72.

572. Kundrat H. *Arhinencephalie als typische Art von Missbildung.* Graz: Luschner and Lubensky, 1882.

573. Kuzniecky R, Andermann F, Tampieri D, *et al.* Bilateral central macrogyria: epilepsy, pseudobulbar palsy, and mental retardation –a recognizable neuronal migration disorder. *Ann Neurol* 1989;**25**:547–54.

574. Kuzniecky R, Andermann F, Guerrini R. The epileptic spectrum in the congenital bilateral perisylvian syndrome. CBPS Multicenter Collaborative Study. *Neurology* 1994;**44**:379–85.

575. Kyttala M, Tallila J, Salonen R, *et al.* MKS1, encoding a component of the flagellar apparatus basal body proteome, is mutated in Meckel syndrome. *Nat Genet* 2006;**38**:155–7.

576. Lacey DJ, Terplan K. Abnormal cerebral cortical neurons in a child with maternal PKU syndrome. *J Child Neurol* 1987;**2**:201–4.

577. Lagger RL. Failure of pyramidal tract decussation in the Dandy–Walker syndrome: report of two cases. *J Neurosurg* 1979;**50**:382–7.

578. Lagos JC. Congenital aneurysms and arteriovenous malformations. In: Vinken PJ, Bruyn GW eds. *Handbook of clinical neurology.* Amsterdam: North-Holland, 1977:137–209.

579. Lagutin OV, Zhu CQC, Kobayashi D, *et al.* Six3 repression of Wnt signaling in the anterior neuroectoderm is essential for vertebrate forebrain development. *Genes Dev* 2003;**17**:368–79.

580. Lammer EJ, Chen DT, Hoar RM, *et al.* Retinoic acid embryopathy. *N Engl J Med* 1985;**313**:837–41.

581. Lammer EJ, Sever LE, Oakley GP. Teratogen update: valproic acid. *Teratology* 1987;**35**:465–73.

582. Lander ES, Linton LM, Birren B, *et al.* Initial sequencing and analysis of the human genome. *Nature* 2001;**409**:860–921.

583. Landrieu P, Ninane J, Ferrière G, Lyon G. Aqueductal stenosis in X-linked hydrocephalus: a secondary phenomenon? *Dev Med Child Neurol* 1979;**21**:637–52.

584. Lange–Cosack H. Die Hydranencephalie (Blasenhirn) als Sonderform der Grosshirnlosigkeit. *Arch Psychiat Nervenkrankheit* 1944;**117**:1–51.

585. Larke RPB, Wheatley E, Saroj S, Chernesky M. Congenital cytomegalovirus infection in an urban Canadian community. *J Infect Dis* 1980;**142**:647–53.

586. Larroche J. *Developmental pathology of the neonate.* Amsterdam: Excerpta Medica, 1977.

587. Larroche J. Malformations of the nervous system. In: Hume Adams J, Corsellis JAN, Duchen LW eds. *Greenfield's Neuropathology.* 4th edn. London: Edward Arnold, 1984:385–450.

588. Larroche JC, Baudey J. Cavum septi lucidi, cavum vergae, cavum veli interpositi: cavités de la ligne médiane (étude anatomique et pneumoencephalographique dans la période néonatale). *Biol Neonate* 1961;**3**:193–236.

589. Larroche JC, Encha–Razavi F. The central nervous system. In: Wigglesworth JS, Singer DB eds. *Textbook of fetal and perinatal pathology.* Oxford: Blackwell, 1991:778–842.

590. Larroche JC, Maunoury T. Analyse statistique de la croissance pondérale des foetus et des viscères pendant la vie intrautérine. *Arch Fr Pediatr* 1973;**30**:927–49.

591. Larroche JC, Nessmann C. Focal cerebral anomalies and retinal dysplasia in a 23–24-week-old fetus. *Brain Dev* 1993;**15**:51–6.

592. Larroche JC, Droulle P, Delezoide AL, *et al*. Brain damage in monozygous twins. *Biol Neonate* 1990;**57**: 261–78.

593. Larsell O. The development of the cerebellum in man in relation to its comparative anatomy. *J Comp Neurol* 1947;**87**:85–129.

594. La Torre E, Fortuna A, Occhipinti E. Angiographic differentiation between Dandy–Walker cyst and arachnoid cyst of the posterior fossa in newborn infants and children. *J Neurosurg* 1973;**38**:298–308.

595. Laurence KM. A case of unilateral megalencephaly. *Dev Med Child Neurol* 1964;**6**:585–90.

596. Laurence KM, Tew BJ. Follow-up of 65 survivors from the 425 cases of spina bifida born in South Wales between 1956 and 1962. *Dev Med Child Neurol* 1967;**13** (Suppl):1–13.

597. Laxova R, O'Hara PT, Timothy JAD. A further example of a lethal autosomal recessive condition in sibs. *J Ment Def Res* 1971;**16**:139–43.

598. Lazar ML. Vein of Galen aneurysm: successful excision of a completely thrombosed aneurysm in an infant. *Surg Neurol* 1974;**2**:22–4.

599. Lazjuk GI, Lurie IW, Ostrowskaja TI, *et al*. Brief clinical observations: the Neu–Laxova syndrome –a distinct entity. *Am J Med Genet* 1979;**3**:261–7.

600. Leach WB. Primary neoplasms of the heart. *Acta Pathol* 1947;**44**:198–204.

601. Ledbetter SA, Kuwano A, Dobyns WB, Ledbetter DH. Microdeletions of chromosome 17p13 as a cause of isolated lissencephaly. *Am J Hum Genet* 1992;**50**:182–9.

602. Lee JH, Huynh M, Silhavy JL, *et al*. De novo somatic mutations in components of the PI3K-AKT3-mTOR pathway cause hemimegalencephaly. *Nat Genet* 2012;**44**(8):941–5.

603. Leech RW, Shuman RM. Holoprosencephaly and related midline cerebral anomalies: a review. *J Child Neurol* 1986;**1**:3–18.

604. Leech RW, Bowlby LS, Brumback RA, Schaefer GB Jr. Agnathia, holoprosencephaly and situs inversus: report of a case. *Am J Med Genet* 1988;**29**:483–90.

605. Legouis R, Hardelin J–P, Levilliers J, *et al*. The candidate gene for the X-linked Kallman syndrome encodes a protein related to adhesion molecules. *Cell* 1991;**67**:423–35.

606. Lemire RJ. Variations in development of the caudal neural tube in human embryos (Horizons XIV–XXI). *Teratology* 1969;**2**:361–70.

607. Lemire RJ, Loeser JD, Leech RW, *et al*. *Normal and abnormal development of the human nervous system*. Hagerstown, MD: Harper & Row, 1975.

608. Leong ASY, Shaw CM. The pathology of occipital encephalocele and a discussion of the pathogenesis. *Pathology* 1979;**11**:223–34.

609. Levine DN, Fisher MA, Caviness VS. Porencephaly with microgyria: a pathological study. *Acta Neuropathol (Berl)* 1974;**29**:99–113.

610. Levine OR, Jameson A, Nellhaus G, Gold AP. Cardiac complications of cerebral arterio-venous fistula in infancy. *Pediatrics* 1962;**30**:563–75.

611. Levy WS, Mason L, Hahn JF. Chiari malformation presenting in adults: a surgical experience in 127 cases. *Neurosurgery* 1983;**12**:377–89.

612. Leyten QH, Renkawek K, Renier WO, *et al*. Neuropathological findings in muscle–eye–brain disease (MEB-D). Neuropathological delineation of MEB-D from congenital muscular dystrophy of the Fukuyama type. *Acta Neuropathol (Berl)* 1991;**83**:55–60.

613. Leyten QH, Gabreels FJ, Renier WO, *et al*. Congenital muscular dystrophy with eye and brain malformations in six Dutch patients. *Neuropediatrics* 1992;**23**:316–20.

614. Leyten QH, Renkawek K, Renier WO. Neuropathological findings in muscle–eye–brain disease (MEB-D). Neuropathological delineation of MEB-D from congenital muscular dystrophy of Fukuyama type. *Acta Neuropathol (Berl)* 1999;**83**:55–60.

615. Li H, Arber S, Jessell TM, Edlund H. Selective agenesis of the dorsal pancreas in mice lacking homeobox gene *Hlxb9*. *Nat Genet* 1999;**23**:67–70.

616. Li HS, Tierney C, Wen L, *et al*. A single morphogenetic field gives rise to two retina primordia under the influence of the prechordal plate. *Development* 1997;**124**:603–15.

617. Li S, Crenshaw EB III, Rawson EJ, *et al*. Dwarf locus mutants lacking three pituitary cell types result from mutations in the POU-domain gene *pit-1*. *Nature* 1990;**347**:528–33.

618. Lichtenstein BW. Distant neuroanatomic complications of spina bifida (spinal dysraphism). *Arch Neurol Psychiatry* 1942;**47**:195–214.

619. Lichtenstein BW. Maldevelopments of the cerebellum. *J Neuropathol Exp Neurol* 1943;**2**:164–77.

620. Lichtenstein BW, Maloney JE. Malformation of the forebrain with comments on the so-called dorsal cyst, the corpus callosum and the hippocampal structures. *J Neuropathol Exp Neurol* 1954;**13**:117–28.

621. Lindenberg R, Swanson PD. Infantile hydranencephaly:a report of five cases of infarction of both cerebral hemispheres in infancy. *Brain* 1967;**90**:839–50.

622. Lipton HL, Prezios TJ, Moses H. Adult onset of the Dandy–Walker syndrome. *Arch Neurol* 1978;**35**:672–4.

623. Liquori CL, Berg MJ, Siegel AM, *et al*. Mutations in a gene encoding a novel protein containing a phosphotyrosine-binding domain cause type 2 cerebral cavernous malformations. *Am J Hum Genet* 2003;**73**:1459–64.

624. List CF, Holt JE, Everett M. Lipoma of the corpus callosum: a clinicopathological study. *Am J Radiol* 1946;**55**:125–34.

625. Litvak J, Yahr MD, Ransohoff J. Aneurysms of great vein of Galen and midline cerebral arteriovenous anomalies. *J Neurosurg* 1995;**17**:945–54.

626. Liu J, Ball SL, Yang Y, *et al*. A genetic model for muscle-eye-brain disease in mice lacking protein O-mannose 1,2-N-acetylglucosaminyltransferase (POMGnT1). *Mech Dev* 2006;**123**:228–40.

627. Livingston JH, Aicardi J. Unusual MRI appearances of diffuse subcortical heterotopia or 'double cortex' in two children. *J Neurol Neurosurg Psychiatry* 1990;**53**:617–20.

628. Loeser JD, Alvord EC. Clinicopathological correlation in agenesis of the corpus callosum. *Neurology* 1968;**18**: 745–56.

629. Loeser JD, Ellsworth CA, Alvord EC. Agenesis of the corpus callosum. *Brain* 1968;**91**:533–70.

630. Loiseau P, Vital C, DeBoucard P, *et al*. Etude anatomo-clinique d'un cas d'hémiplégie–épilepsie avec mouvements anormaux. Atrophie cérébelleuse croisée. *Rev Neurol* 1968;**118**:77–82.

631. Lo Nigro C, Chong CS, Smith AC, *et al*. Point mutations and an intragenic deletion in LIS1, the lissencephaly causative gene in isolated lissencephaly sequence and Miller–Dieker syndrome. *Hum Mol Genet* 1997;**6**:157–64.

632. Lorber J. Results of treatment of myelomeningocele: an analysis of 524 unselected cases, with special reference to possible selection for treatment. *Dev Med Child Neurol* 1971;**13** (Suppl 25):279.

633. Loth P, Casasoprana A, Thibert M. Une cause rare de défaillance cardiaque néonatale:l'anévrysme arterio-veineux intracranien. *Arch Fr Pediatr* 1972;**24**:255–68.

634. Lourie H, Berne A. A contribution on the etiology and pathogenesis of congenital communicating hydrocephalus. *Biol Neonate* 1961;**15**:815–22.

635. Lowry RB. Variability in the Smith–Lemli–Opitz syndrome:overlap with the Meckel syndrome. *Am J Med Genet* 1983;**14**:429–33.

636. Lumsden CE. Multiple cystic softening of the brain in the newborn. *J Neuropathol Exp Neurol* 1950;**9**:119–37.

637. Luque JM, Morante-Oria J, Fairén A. Localization of ApoER2, VLDLR and Dab1 in radial glia: groundwork for a new model of reelin action during cortical development. *Dev Brain Res* 2003;**140**:195–203.

638. Lurie IW, Nedzved MK, Lazjuk GI, *et al*. Aprosencephaly–atelencephaly and the aprosencephaly (XK) syndrome. *Am J Med Genet* 1979;**3**:303–9.

639. Luthy DA, Wardinsky T, Shurtleff DB, *et al*. Cesarean section before the onset of labor and subsequent motor function in infants with meningomyelocele diagnosed antenatally. *N Engl J Med* 1991;**324**: 662–6.

640. Lynn RB, Buchanan DG, Fenichel GM, Freeman FR. Agenesis of the corpus callosum. *Arch Neurol* 1980;**37**:444–5.

641. Lyon G, Gastaut H. Considerations on the significance attributed to unusual cerebral histological findings recently described in eight patients with primary generalised epilepsy. *Epilepsia* 1985;**26**:365–7.

642. Lyon G, Robain O. étude comparative des encéphalopathies circulatoires prénatales et paranatales (hydranencéphalies, porencéphalies et encéphalomalacias kystiques de la substance blanche). *Acta Neuropathol (Berl)* 1967;**9**:79–98.

643. Lyon MF. Some milestones in the history of X-chromosome inactivation. *Annu Rev Genet* 1992;**26**:17–28.

644. Macchi G, Bentivoglio M. Agenesis or hypoplasia of cerebellar structures. In: Vinken PJ, Bruyn G eds. *Handbook of clinical neurology*. Amsterdam:Elsevier, 1977:367–93.

645. MacFarlane A, Maloney AFJ. The appearance of the aqueduct and its relationship to hydrocephalus in the Arnold–Chiari malformation. *Brain* 1957;**80**:479–91.

646. Machin GA. Twins and their disorders. In: Reed GB, Claireaux AE, Cockburn F eds. *Diseases of the fetus and newborn:pathology, imaging, genetics and management*. 2nd edn. London: Chapman & Hall, 1995:201–25.

647. Mack LD, Rumack CM, Johnson ML. Ultrasound evaluation of cystic intracranial lesions in the neonate. *Radiology* 1980;**37**:451–55.

648. MacKenzie NG, Emery JL. Deformities of the cervical cord in children with neurospinal dysraphism. *Dev Med Child Neurol* 1971;**13** (Suppl 25):58–61.

649. Maden M. Role and distribution of retinoic acid during CNS development. *Int Rev Cytol* 2001;**209**:1–77.

650. Maksem A, Roessmann U. Apert's syndrome with central nervous system anomalies. *Acta Neuropathol (Berl)* 1979;**48**:59–61.

651. Malamud N, Gaitz CM. Neuropathology of organic brain syndromes associated with aging. In: Malamud N, Gaitz CM eds. *Aging and the brain*. New York: Plenum Press, 1972:**63**–87.

652. Malherbe V, Pariente D, Tardieu M, *et al*. Central nervous system lesions in hypomelanosis of Ito: an MRI and pathological study. *J Neurol* 1993;**240**:302–4.

653. Mallamaci A, Mercurio A, Muzio L, *et al*. The lack of Emx2 causes impairment of Reelin signalling and defects of neuronal migration in the developing cerebral cortex. *J Neurosci* 2000;**20**:1109–19.

654. Manterola A, Towbin A, Yakovlev PI. Cerebral infarction in the human fetus near term. *J Neuropathol Exp Neurol* 1966;**25**:471–88.

655. Manya H, Chiba A, Yoshida A, *et al*. Demonstration of mammalian protein O-mannosyltransferase activity: coexpression of POMT1 and POMT2 required for enzymatic activity. *Proc Natl Acad Sci U S A* 2004;**101**:500–505.

656. Manz HJ, Phillips TM, McCullough DC, Rowden G. Unilateral megalencephaly, cerebral cortical dysplasia, neuronal hypertrophy, and heterotopia:cytomorphometric, fluorometric cytochemical, and biochemical analyses. *Acta Neuropathol (Berl)* 1979;**45**:97–103.

657. Manzanares M, Krumlauf R. Developmental biology: raising the roof. *Nature* 2000;**403**:720–21.

658. Marburg O. So-called agenesis of the corpus callosum (callosal defect). *Arch Neurol Psychiatry* 1949;**61**:297–312.

659. Marie J, See G, Gruner J, *et al*. Manifestations cérébrales de la maladie des inclusions cytomégaliques. *Ann Pediatr* 1957;**25**:248–56.

660. Marillat V, Sabatier C, Failli V, *et al*. The slit receptor Rig-1/Robo3 controls midline crossing by hindbrain precerebellar neurons and axons. *Neuron* 2004;**43**:69–79.

661. Marin–Padilla M. Study of the skull in human cranioschisis. *Acta Anat* 1965;**62**:1–20.

662. Marin–Padilla M. Morphogenesis of anencephaly and related malformations. *Curr Topics Pathol* 1970;**51**:145–74.

663. Marin–Padilla M. Cajal–Retzius cells and the development of the neocortex. *Trends Neurosci* 1998;**21**:64–71.

664. Marin–Padilla M, Marin–Padilla TM. Morphogenesis of experimentally induced Arnold–Chiari malformation. *J Neurol Sci* 1981;**50**:29–55.

665. Maroteaux P, Lamy M. Robert JM. La nanisme thanatophore. *Presse Med* 1967;**75**:2519–24.

666. Marques Dias MJ, Harmant-van Rijckevorsel G, Landrieu P, Lyon G. Prenatal cytomegalovirus disease and cerebral microgyria: evidence for perfusion failure, not disturbance of histogenesis, as the major cause of fetal cytomegalovirus encephalopathy. *Neuropediatrics* 1984;**15**:18–24.

667. Marshall H, Nonchev S, Sham MH, *et al*. Retinoic acid alters hindbrain *Hox* code and induces transformation of rhombomeres 2/3 into a 4/5 identity. *Nature* 1992;**360**:737–41.

668. Martin F. Ueber eine vestibulo-cerebelläre Entwicklungshemmung im Rahmen ausgedehnter osteo-neuraler Dysgenesien. *Acta Psychiat Neurol Scand* 1949;**24**:207–22.

669. Martin JK, Norman RM. Maple syrup urine disease in an infant with microgyria. *Dev Med Child Neurol* 1967;**9**:152–9.

670. Martin RA, Carey JG. A review and case report of aprosencephaly and the XK aprosencephaly syndrome. *Am J Med Genet* 1982;**11**:369–71.

671. Martland BN, Harding RE, Morton, Young I. Dentato-olivary dysplasia in sibs: an autosomal recessive disorders? *J Med Genetics* 1997;**34**:1021–3.

672. Masters C, Alpers M, Kakulas B. Pathogenesis of Reovirus type 1 hydrocephalus in mice: significance of aqueductal changes. *Arch Neurol* 1977;**34**:18–28.

673. Matell M. Ein Fall von Heterotopie der grauen Substanz in den beiden Hemisphären des Grosshirns. *Arch Psychiat Nervenkrankheit* 1893;**25**:124–36.

674. Matsubara O, Tanaka M, Ida T, Okeda R. Hemimegalencephaly with hemihypertrophy (Klippel–Trénaunay–Weber syndrome). *Virchows Arch Pathol Anat* 1983;**400**:155–62.

675. Matsuura T, Sutcliffe JS, Fang P, *et al*. De novo truncating mutations in the E6-AP ubiquitin-protein ligase gene (UBE3A) in Angelman syndrome. *Nat Genet* 1997;**15**:74–7.

676. McBride MC, Kemper TL. Pathogenesis of fourlayered microgyric cortex in man. *Acta Neuropathol (Berl)* 1982;**57**:93–8.

677. McCabe MJ, Alatzoglou KS, Dattani MT. Septo-optic dysplasia and other midline defects: the role of transcription factors: HESX1 and beyond. *Best Pract Res Clin Endocrinol Metab* 2011;**25**(1):115–24.

678. McCaffery PJ, Adams J, Maden M, *et al*. Too much of a good thing: retinoic acid as an endogenous regulator of neural differentiation and exogenous teratogen. *Eur J Neurosci* 2003;**18**:457–72.

679. McConnell SK, Kaznowski CE. Cell cycle dependence of laminar determination in developing neocortex. *Science* 1991;**254**:282–5.

680. McCracken G, Shinefield H, Cobb K, *et al*. Congenital cytomegalic inclusion disease. *Am J Dis Child* 1969;**117**:522–39.

681. McKay R. Stem cells in the central nervous system. *Science* 1997;**276**:66–71.

682. McLaurin RL. Parietal cephaloceles. *Neurology* 1964;**14**:764–74.

683. McLendon RE, Crain BJ, Oakes WJ, Burger PC. Cerebral polygyria in the Chiari type II (Arnold–Chiari) malformation. *Clin Neuropathol* 1985;**4**:200–205.

684. McMahon AP, Bradley A. The *Wint-l* (*int-l*) proto-oncogene is required for development of a large region of the mouse brain. *Cell* 1990;**62**:1073–85.

685. McShane MA, Boyd S, Harding B, *et al*. Progressive bulbar paralysis of childhood: a reappraisal of Fazio–Londe disease. *Brain* 1992;**115**:1889–900.

686. Mediaris DN. Cytomegalic inclusion disease: an analysis of the clinical features based on the literature and six additional cases. *Pediatrics* 1957;**19**:466–80.

687. Meencke HJ. Neuron density in the molecular layer of the frontal cortex in primary generalized epilepsy. *Epilepsia* 1985;**26**:450–54.

688. Meencke HJ, Janz D. Neuropathological findings in primary generalised epilepsy: a study of eight cases. *Epilepsia* 1984;**25**:8–21.

689. Meencke HJ, Veith G. Migration disturbances in epilepsy. In: Engel JJ, Wasterlain C, Cavalheiro EA, *et al*., eds. *Molecular biology of epilepsy*. Amsterdam: Elsevier, 1992:31–40.

690. Melchionna RH, Moore RA. The pharyngeal pituitary gland. *Am J Pathol* 1938;**14**:763–71.

691. Mello LEAM, Cavalheiro EA, Tan AI, *et al*. Granule cell dispersion in relation to mossy fiber sprouting, hippocampal loss, silent period and seizure frequency in the pilocarpine model of epilepsy. In: Engel JJ, Wasterlain C, Cavalheiro EA, *et al*. eds. *Molecular biology of epilepsy*. Amsterdam: Elsevier, 1992:51–60.

692. Menkes JH, Philippart M, Clark DE. Hereditary partial agenesis of the corpus callosum. *Arch Neurol* 1964;**11**:198–208.

693. Meuli M, Meuli-Simmen C, Hutchins GM, *et al*. *In utero* surgery rescues neurological function at birth in sheep with spina bifida. *Nat Med* 1995;**1**:342–7.

694. Meyer A, Jones TB. Histological changes in brain in mongolism. *J Ment Sci* 1939;**85**:206–21.

695. Miale IL, Sidman RL. An autoradiographic analysis of histogenesis in the mouse cerebellum. *Exp Neurol* 1961;**4**:277–96.

696. Michaelson PS, Gilles FH. Central nervous system abnormalities in trisomy E (17–18) syndrome. *J Neurol Sci* 1972;**15**:193–208.

697. Miller G, Ladda RL, Towfighi J. Cerebro-ocular dysplasia–muscular dystrophy (Walker–Warburg) syndrome: findings in a 20 week fetus. *Acta Neuropathol (Berl)* 1991;**82**:234–8.

698. Miller GM, Stears JC, Guggenheim MA, Wilkening GN. Schizencephaly: a clinical and CT study. *Neurology* 1984;**34**:997–1001.

699. Miller J. Lissencephaly in two siblings. *Neurology* 1963;**13**:841–50.

700. Miller RW. Effects of ionizing radiation from the atomic bomb on Japanese children. *Pediatrics* 1968;**41**:257–63.

701. Millett S, Campbell K, Epstein DJ, *et al*. A role for *Gbx2* in repression of *Otx2* and positioning the mid/hindbrain organizer. *Nature* 1999;**401**:161–4.

702. Mills JL, McPartlin JM, Kirke PN, *et al.* Homocysteine metabolism in pregnancies complicated by neural-tube defects. *Lancet* 1995;**345**:149–51.

703. Ming JE, Kaupas ME, Roessler E, *et al.* Mutations in PATCHED-1, the receptor for SONIC HEDGEHOG, are associated with holoprosencephaly. *Hum Genet* 2002;**110**:297–301.

704. Ming PL, Goodner DM, Park TS. Cytogenetic variants in holoprosencephaly: report of a case and review of the literature. *Am J Dis Child* 1976;**130**:864–7.

705. Miskolczy D. Ein Fall von Kleinhirnmissbildung. *Arch Psychiat Nervenkrankheit* 1931;**93**:596–615.

706. Mizuguchi M, Morimatsu Y. Histopathological study of alobar holoprosencephaly: 1. Abnormal laminar architecture of the telencephalic cortex. *Acta Neuropathol (Berl)* 1989;**78**:176–82.

707. Mizuguchi M, Maekawa S, Kamoshita S. Distribution of leptomeningeal glioneuronal heterotopia in alobar holoprosencephaly. *Arch Neurol* 1994;**51**:951–4.

708. Moase CE, Trasler DG. Splotch locus mouse mutants: models for neural tube defects and Waardenburg syndrome type I in humans. *J Med Genet* 1992;**29**:145–51.

709. Möbius PJ. Ueber angeborene doppelseitige Abducens-Facialis-Lähmung. *Münchner Med Wochenschr* 1888;**35**:108–11.

710. Moessinger AC. Fetal akinesia deformation sequence: an animal model. *Pediatrics* 1983;**72**:857–63.

711. Mohr PD, Strang FA, Sambrook MA, Bodie HG. The clinical and surgical features of 40 patients with primary cerebellar ectopia (adult Chiari malformation). *Q J Med* 1977;**181**:85–96.

712. Monaghan HP, Krafchik BR, Macgregor DL, Fitz CR. Tuberous sclerosis complex in children. *Am J Dis Child* 1981;**135**:912–17.

713. Moncrieff MW, Hill DS, Archer J, Arthur LJ. Congenital absence of pituitary gland and adrenal hypoplasia. *Arch Dis Child* 1972;**47**:136–7.

714. Montoya G, Dohn DF, Mercer RD. Arteriovenous malformation of the vein of Galen as a cause of heart failure and hydrocephalus in infants. *Neurology* 1971;**21**:1054–8.

715. Moore CM, McAdams AJ, Sutherland J. Intrauterine disseminated intravascular coagulation: a syndrome of multiple pregnancy with a dead twin fetus. *J Pediatr* 1969;**74**:523–8.

716. Moore GR, Raine CS. Leptomeningeal and adventitial gliosis as a consequence of chronic inflammation. *Neuropathol Appl Neurobiol* 1986;**12**:371–8.

717. Moore SA, Saito F, Chen J, *et al.* Deletion of brain dystroglycan recapitulates aspects of congenital muscular dystrophy. *Nature* 2002;**418**:422–5.

718. Morel F, Wildi E. Dysgénésie nodulaire disséminée de l'écorce frontale. *Rev Neurol* 1952;**87**:251–70.

719. Morimoto K, Mogami H. Sequential CT study of subependymal giant-cell astrocytoma associated with tuberous sclerosis. *J Neurosurg* 1986;**65**:874–7.

720. Motoyama N, Wang F, Roth KA, *et al.* Massive cell death of immature hematopoietic cells and neurons in Bcl-x-deficient mice. *Science* 1995;**267**:1506–10.

721. Muir CS. Hydranencephaly and allied disorders: a study of cerebral defect in Chinese children. *Arch Dis Child* 1959;**34**:231–46.

722. Mullen RJ, Herrup K. Chimeric analysis of mouse cerebellar mutants. In: Breakfield XO ed. *Neurogenetics: genetic approaches to the nervous system.* New York: Elsevier, 1979:173–96.

723. Müller F, O'Rahilly R. The development of the human brain, the closure of the caudal neuropore, and the beginning of secondary neurulation at stage 12. *Anat Embryol* 1987;**176**:413–30.

724. Muller LM, De Jong G, Mouton SCE, *et al.* A case of Neu–Laxova syndrome: prenatal ultrasound monitoring in the third trimester and the histopathological findings. *Am J Med Genet* 1987;**26**:421–9.

725. Münchoff C, Noetzel H. Ueber eine nahezu totale Agyrie bei einem 6 Jahre alt gewordenen Knaben. *Acta Neuropathol (Berl)* 1965;**4**:469–75.

726. Münke M. Clinical, cytogenetic, and molecular approaches to the genetic heterogeneity of holoprosencephaly. *Am J Med Genet* 1989;**34**:237–46.

727. Münke M, Emanuel BS, Zackai EH. Holoprosencephaly: association with interstitial deletion of 2p and review of the cytogenetic literature. *Am J Med Genet* 1988;**30**:929–38.

728. Murofushi K. Normalentwicklung und Dysgenesien von Dentatum und Oliva inferior. *Acta Neuropathol (Berl)* 1974;**27**:317–28.

729. Murphy M, Stinnakre MG, Senamaud-Beaufort C, *et al.* Delayed early embryonic lethality following disruption of the murine cyclin A2 gene. *Nat Genet* 1997;**15**:83–6.

730. Muscatelli F, Abrous DN, Massacrier A, *et al.* Disruption of the mouse Necdin gene results in hypothalamic and behavioral alterations reminiscent of the human Prader–Willi syndrome. *Hum Mol Genet* 2000;**9**:3101–10.

731. Myers RE. Brain pathology following fetal vascular occlusion: an experimental study. *Invest Ophthalmol* 1969;**8**:41–50.

732. Myrianthopoulos NC. Our load of central nervous system malformations. *Birth Defects* 1979;**15**:1–18.

733. Naeye RL. Brain stem and adrenal abnormalities in the sudden infant death syndrome. *Am J Clin Pathol* 1966;**66**:526–39.

734. Nagai T, Aruga J, Minowa O, *et al.* Zic2 regulates the kinetics of neurulation. *Proc Natl Acad Sci U S A* 2000;**97**:1618–23.

735. Nagano T, Morikubo S, Sato M. Filamin A and FILIP (Filamin A-Interacting Protein) regulate cell polarity and motility in neocortical subventricular and intermediate zones during radial migration. *J Neurosci* 2004;**24**:9648–57.

736. Naiman J, Frazer FC. Agenesis of the corpus callosum. *Arch Neurol Psychiatry* 1955;**74**:182–5.

737. Nakado KK. Anencephaly: a review. *Dev Med Child Neurol* 1973;**15**:383–400.

738. Nakamura S, Makiyama H, Miyagi A, *et al.* Congenital unilateral hydrocephalus. *Childs Nerv Syst* 1989;**5**:367–70.

739. Nakamura Y, Becker LE. Subependymal giant-cell tumour: astrocytic or neuronal? *Acta Neuropathol (Berl)* 1983;**60**:271–7.

740. Nakatsu T, Uwabe C, Shiota K. Neural tube closure in humans initiates at multiple sites: evidence from human embryos and implications for the pathogenesis of neural tube defects. *Anat Embryol* 2000;**201**:455–66.

741. Nanni L, Ming JE, Bocian M, *et al.* The mutational spectrum of the Sonic Hedgehog gene in holoprosencephaly: SHH mutations cause a significant proportion of autosomal dominant holoprosencephaly. *Hum Mol Genet* 1999;**8**:2479–88.

742. Nathan PW, Smith MC. Normal mentality associated with a maldeveloped 'rhinencephalon'. *Neurol Neurosurg Psychiatry* 1950;**13**:191–7.

743. Nathan PW, Smith MC, Deacon P. The corticospinal tracts in man: course and location of fibres at different segmental levels. *Brain* 1990;**113**:303–24.

744. Nathanson N, Cole G, Van der Loos H. Heterotopic cerebellar granule cells following administration of cyclophosphamide to suckling rats. *Brain Res* 1969;**15**:532–6.

745. Nave KA, Lai C, Bloom FE, Milner RJ. Jimpy mutant mouse: a 74-base deletion in the mRNA for myelin proteolipid protein and evidence for a primary defect in RNA splicing. *Proc Natl Acad Sci U S A* 1986;**83**:9264–8.

746. Navin JJ, Angevine JM. Congenital cytomegalic inclusion disease with porencephaly. *Neurology* 1968;**18**:470–72.

747. Nellhaus G. Head circumference from birth to eighteen years: practical composite international and interracial graphs. *Pediatrics* 1968;**41**:106–19.

748. Nelson MM, Thompson AJ. The acrocallosal syndrome. *Am J Med Genet* 1982;**12**:195–9.

749. Neriishi S, Matsumura H. Morphological observation of the central nervous system in an *in utero* exposed autopsy case. *J Radiat Res* 1983;**24**:18.

750. Neu RL, Kajii T, Gardner LI, *et al.* A lethal syndrome of microcephaly with multiple congenital anomalies in three siblings. *Pediatrics* 1971;**47**:610–12.

751. Neumann PE, Frankel WN, Letts VA, *et al.* Multifactorial inheritance of neural tube defects: localization of the major gene and recognition of modifiers in ct mutant mice. *Nat Genet* 1994;**6**:357–62.

752. Nevin NC, Johnston WP, Merrett JD. Influence of social class on the risk of recurrence of anencephalus and spina bifida. *Dev Med Child Neurol* 1981;**23**:155–9.

753. Newman GCI, Buschi A, Sugg NK, *et al.* Dandy–Walker syndrome diagnosed *in utero* by ultrasonography. *Neurology* 1982;**32**:180–84.

754. Nicholls RD, Saitoh S, Horsthemke B. Imprinting in Prader–Willi and Angelman syndromes. *Trends Genet* 1998;**14**:194–200.

755. Nieuwenhuijse P. Zur Kenntnis der Mikrogyrie. *Psychiat Neurog Bladen Amsterdam* 1913;**17**:9–53.

756. Norman MG. Bilateral encephaloclastic lesions in a 26 week gestation fetus: effect on neuroblast migration. *Can J Neurol Sci* 1989;**7**:191–4.

757. Norman MG, Becker LE. Cerebral damage in neonates resulting from arterio-venous malformations in the vein of

Galen. *J Neurol Neurosurg Psychiatry* 1974;**37**:252–8.

758. Norman MG, McGillivray B. Fetal neuropathology of proliferative vasculopathy and hydranencephaly–hydrocephaly with multiple limb pterygia. *Pediatr Neurosci* 1988;**14**:301–6.

759. Norman MG, Becker LE, Sirois J, Tremblay LJM. Lissencephaly. *Can J Neurol Sci* 1976;**3**:39–46.

760. Norman MG, White VA, Dimmick JE. Pathology of Neu–Laxova syndrome. *Brain Pathol* 1994;**4**:394.

761. Norman RM. Primary degeneration of the granular layer of the cerebellum; an unusual form of familial cerebellar atrophy occurring in early life. *Brain* 1940;**63**:365–79.

762. Norman RM. Cerebral birth injury. In: Greenfield JG, Blackwood W, McMenemey WH, Meyer A, Norman RM eds. *Neuropathology*. London: Edward Arnold, 1958:354–68.

763. Norman RM. Neuropathological findings in trisomies 13–15 and 17–18 with special reference to the cerebellum. *Dev Med Child Neurol* 1966;**8**:170–7.

764. Norman RM, Kay JM. Cerebello-thalamo-spinal degeneration in infancy: an unusual variant of Werdnig–Hoffmann disease. *Arch Dis Child* 1965;**40**:302–8.

765. Norman RM, McMenemey WH. Transventricular adhesion in association with birth injury of the caudate nucleus. *J Neuropathol Exp Neurol* 1955;**14**:85–91.

766. Norman RM, Urich H. Cerebellar hypoplasia associated with systemic degeneration in early life. *J Neurol Neurosurg Psychiatry* 1958;**21**:159–66.

767. Norman RM, Urich H, Woods GE. The relationship between prenatal porencephaly and the encephalomalacias of early life. *J Ment Sci* 1958;**104**:758–71.

768. Norman RM, Tingey AH, Harvey PW, Gregory AM. Pelizaeus–Merzbacher disease;a form of sudanophil leukodystrophy. *J Neurol Neurosurg Psychiatry* 1966;**29**:521–9.

769. Obersteiner H. Ein porencephalisches Gehirn. *Arbeit Neurol Inst Wien* 1902;**8**:1–66.

770. O'Doherty A, Ruf S, Mulligan C, *et al.* An aneuploid mouse strain carrying human chromosome 21 with Down syndrome phenotypes. *Science* 2005;**309**:2033–7.

771. Ohshima T, Ward JM, Huh CG, *et al.* Targeted disruption of the cyclin-dependent kinase 5 gene results in abnormal corticogenesis, neuronal pathology and perinatal death. *Proc Natl Acad Sci U S A* 1996;**93**:11173–8.

772. Oliver LC. Primary arachnoid cysts: report of two cases. *Br Med J* 1958;**2**:1147–9.

773. Olson MI, Shaw CM. Presenile dementia and Alzheimer's disease in mongolism. *Brain* 1969;**92**:147–56.

774. Onufrowicz W. Das balkenlose Microcephalengehirn Hofmann: Ein Beitrag zur pathologischen und normalen Anatomie des menschlichen Gehirnes. *Arch Psychiat Nervenkrankheit* 1887;**18**:305–28.

775. Opitz JM, Sutherland GR. Conference report: international workshop on the fragile X and X-linked mental retardation. *Am J Med Genet* 1984;**17**:5–94.

776. O'Rahilly R, Muller F. Bidirectional closure of the rostral neuropore

in the human embryo. *Am J Anat* 1989;**184**:259–68.

777. Overbeeke JJ van, Hillen B, Vermeij–Keers C. The arterial pattern at the base of arhinencephalic and prosencephalic brains. *J Anat* 1994;**185**:51–63.

778. Owen C. *On the anatomy of the vertebrates*. London: Longmans & Green, 1868.

779. Padget DH, Lindenberg R. Inverse cerebellum morphogenetically related to Dandy–Walker and Arnold–Chiari syndromes: bizarre malformed brain with occipital encephalocele. *Johns Hopkins Med J* 1972;**131**:228–46.

780. Paetau A, Salonen R, Haltia M. Brain pathology in the Meckel syndrome: a study of 59 cases. *Clin Neuropathol* 1985;**4**:56–62.

781. Paetau A, Salonen R, Herva R. Neuropathology of hydrolethalus syndrome (Abstract). *Brain Pathol* 1994;**4**:392.

782. Pagon RA, Chandler JW, Collie WR, *et al.* Hydrocephalus, agyria, retinal dysplasia, encephalocele (HARD E) syndrome: an autosomal recessive condition. *Birth Defects* 1995;**14**:233–41.

783. Palm L, Hägerstrand I, Kristoffersson U, *et al.* Nephrosis and disturbances of neuronal migration in male siblings:new hereditary disorder? *Arch Dis Child* 1986;**61**:545–8.

784. Palmini A, Andermann F, Aicardi J, *et al.* Diffuse cortical dysplasia, or the 'double cortex' syndrome: the clinical and epileptic spectrum in 10 patients. *Neurology* 1991;**41**:1656–62.

785. Palmini A, Andermann F, de Grissac H, *et al.* Stages and patterns of centrifugal arrest of diffuse neuronal migration disorders. *Dev Med Child Neurol* 1993;**35**:331–9.

786. Pamphlett R, Silberstein P. Pelizaeus–Merzbacher disease in a brother and sister. *Acta Neuropathol (Berl)* 1986;**69**:343–6.

787. Pan D, Dong J, Zhang Y, *et al.* Tuberous sclerosis complex: from Drosophila to human disease. *Trends Cell Biol* 2004;**14**:78–85.

788. Pang D. Split cord malformation: part II. Clinical syndrome. *Neurosurgery* 1992;**31**:481–500.

789. Pang D, Dias MS. Cervical myelomeningoceles. *Neurosurgery* 1993;**33**:363–73.

790. Pang D, Dias MS, Ahab-Barmada M. Split cord malformation: part I. A unified theory of embryogenesis for double spinal cord malformations. *Neurosurgery* 1992;**31**:451–80.

791. Papez JW, Vonderahe AR. Infantile cerebral palsy of hemiplegic type. *J Neuropathol Exp Neurol* 1947;**6**:244–52.

792. Parisi MA, Bennett CL, Eckert ML, *et al.* The NPHP1 gene deletion associated with juvenile nephronophthisis is present in a subset of individuals with Joubert syndrome. *Am J Hum Genet* 2004;**75**:82–91.

793. Park HL, Bai C, Platt KA, *et al.* Mouse Gli1 mutants are viable but have defects in SHH signaling in combination with a Gli2 mutation. *Development* 2000;**127**:1593–605.

794. Parnavelas JG. The origin and migration of cortical neurones: new vistas. *Trends Neurosci* 2000;**23**:126–31.

795. Parrish ML, Roessmann U, Levinsohn MW. Agenesis of the corpus callosum: a study of the frequency of associated malformations. *Ann Neurol* 1979;**6**:349–54.

796. Patau K, Smith DW, Therman E, *et al.* Multiple congenital anomaly caused by an extra autosome. *Lancet* 1960;i:790–93.

797. Patel H, Dolman CL. Byrne MA. Holoprosencephaly with median cleft lip. *Am J Dis Child* 1972;**124**:217–21.

798. Pattyn A, Morin X, Cremer H, *et al.* The homeobox gene Phox2b is essential for the development of autonomic neural crest derivatives. *Nature* 1999;**399**:366–70.

799. Paul K, Lye RH, Strang FA, Dutton J. Arnold–Chiari malformation: review of 71 cases. *J Neurosurg* 1983;**58**:183–7.

800. Pauli RM, Graham JM, Barr M. Agnathia, situs inversus and associated malformations. *Teratology* 1981;**23**:85–93.

801. Pavone L, Gullotta F, Incorpora G, *et al.* Isolated lissencephaly: report of four patients from two unrelated families. *J Child Neurol* 1990;**5**:52–9.

802. Pavone L, Rizzo R, Dobyns WB. Clinical manifestations and evaluation of isolated lissencephaly. *Childs Nerv Syst* 1993;**9**:387–90.

803. Peach B. Arnold–Chiari malformation with normal spine. *Arch Neutrol* 1964;**10**:497–501.

804. Peach B. Arnold–Chiari malformation: morphogenesis. *Arch Neurol* 1965;**12**:527–35.

805. Peach B. Arnold–Chiari malformation: anatomic features of 20 cases. *Arch Neurol* 1965;**12**:613–21.

806. Peckham CS, Martin JAM, Marshall WC, Dudgeon JA. Congenital rubella deafness: a preventable disease. *Lancet* 1978;i:258–61.

807. Pelizzi GB. Contributo allo studio dell'idiozia. *Rivista Spermentali di Freniatria e Medicina Legale delle Alienazioni Mentale* 1901;**27**:265–9.

808. Percin EF, Ploder LA, Yu JJ, *et al.* Human microphthalmia associated with mutations in the retinal homeobox gene CHX10. *Nat Genet* 2000;**25**:397–401.

809. Perez–Fontan JJ, Herrera M, Fina A, Peguero G. Periventricular calcifications in a newborn associated with aneurysm of the great vein of Galen. *Pediatr Radiol* 1982;**12**:249–51.

810. Petersson S, Persson AS, Johansen JE, *et al.* Truncation of the Shaker-like voltage-gated potassium channel, Kv1.1, causes megencephaly. *Eur J Neurosci* 2003;**18**:3231–40.

811. Pfäffle RW, DiMattia GE, Parks JS, *et al.* Mutation of the POU-specific domain of Pit-1 and hypopituitarism without pituitary hypoplasia. *Science* 1992;**257**:1118–21.

812. Pfeiffer G, Friede RL. Unilateral hydrocephalus from early developmental occlusion of one foramen of Monro. *Acta Neuropathol (Berl)* 1984;**64**:75–7.

813. Pfeiffer J, Pfeiffer RA. Hypoplasia ponto-neocerebellaris. *J Neurol* 1977;**215**:241–51.

814. Pfeiffer J, Majewski F, Fischbach H, *et al.* Alcohol embryopathy and fetopathy: neuropathology of 3 children and 3 fetuses. *J Neurol Sci* 1979;**41**:125–37.

815. Philip N, Chabrol B, Lossi AM, *et al.* Mutations in the oligophrenin-1 gene (OPHN1) cause X linked congenital

cerebellar hypoplasia. *J Med Genet* 2003;**40**:441–6.

816. Pholsena M, Young J, Couzinet B, Schaison G. Primary adrenal and thyroid insufficiencies associated with hypopituitarism: a diagnostic challenge. *Clin Endocrinol (Oxf)* 1994;**40**:693–5.

817. Piao X, Hill RS, Bodell A, *et al.* G protein-coupled receptor-dependent development of human frontal cortex. *Science* 2004;**303**:2033–6.

818. Piccolo S, Sasai Y, Lu B, De Robertis EM. Dorsoventral patterning in *Xenopus*: inhibition of ventral signals by direct binding of Chordin to BMP-4. *Cell* 1996;**86**:589–98.

819. Pierre-Kahn A, Hirsch J, Renier D, *et al.* Hydrocephalus and achondroplasia: a study of 25 observations. *Childs Brain* 1981;**7**:205–19.

820. Pilz DT, Kuc J, Matsumoto N, *et al.* Subcortical band heterotopia in rare affected males can be caused by missense mutations in DCX (XLIS) or LIS-1. *Hum Mol Genet* 1999;**8**:1757–60.

821. Pinard JM, Motte J, Chiron C, *et al.* Subcortical laminar heterotopia and lissencephaly in two families: a single X linked dominant gene. *J Neurol Neurosurg Psychiatry* 1994;**57**:914–20.

822. Pineda M, Gonzalez A, Fabreques I, *et al.* Familial agenesis of the corpus callosum with hyperthermia and apnoeic spells. *Neuropediatrics* 1984;**15**:63–7.

823. Pinto-Lord MC, Evrard P, Caviness VS Jr. Obstructed neuronal migration along radial glial fibers in the neocortex of the reeler mouse: a Golgi-EM analysis. *Dev Brain Res* 1982;**4**:379–93.

824. Piven J, Berthier ML, Starkstein SE, *et al.* Magnetic resonance imaging evidence for a defect of cerebral cortical development in autism. *Am J Psychiatry* 1990;**147**:734–9.

825. Pleet H, Graham JM Jr, Smith DW. Central nervous system and facial defects associated with maternal hyperthermia at four to 14 weeks' gestation. *Pediatrics* 1981;**67**:785–9.

826. Plummer C. Anomalies occurring in children exposed *in utero* to atomic bomb in Hiroshima. *Pediatrics* 1952;**10**:687–93.

827. Poduri A, Evrony GD, Cai X, *et al.* Somatic activation of AKT3 causes hemispheric developmental brain malformations. *Neuron* 2012;**74**:41–8.

828. Pollock AQ, Laslett PA. Cerebral arteriovenous fistula producing cardiac failure in the newborn infant. *J Pediatr* 1958;**53**:731–6.

829. Potter EL, Potter EL, Craig JM eds. *Pathology of the fetus and infant.* 3rd edn. Chicago, IL: Year Book, 1976.

830. Powers JM. Adreno-leukodystrophy (adreno-testiculo-leuko-myelo-neuropathic complex). *Clin Neuropathol* 1985;**4**:181–99.

831. Powers JM, Moser HW, Moser AB, *et al.* Fetal cerebrohepatorenal (Zellweger) syndrome: dysmorphic, radiologic, biochemical, and pathologic findings in four affected fetuses. *Hum Pathol* 1985;**16**:610–20.

832. Powers JM, Tummons RC, Caviness VS, *et al.* Structural and chemical alteration in the cerebrohepato-renal (Zellweger) syndrome. *J Neuropathol Exp Neurol* 1989;**48**:270–89.

833. Price DJ, Aslam S, Tasker L, Gillies K. Fates of the earliest generated cells in the developing murine neocortex. *J Comp Neurol* 1997;**377**:414–22.

834. Priesel A. Uber die dystopie der neurohyophyse. *Virchows Arch Pathol Anat Physiol Klin Med* 1927;**266**:407–15.

835. Priestley DB. Complete absence of the cerebellum. *Lancet* 1920;**ii**:1302–17.

836. Probst FP. *The prosencephalies.* Berlin: Springer, 1979.

837. Probst M. über den Bau des vollständig balkenlosen Grosshirns sowie über Mikrogyrie und Heterotopie der grauen Substanz. *Arch Psychiat Nervenkrankheit* 1901;**34**:709–86.

838. Radovick S, Nations M, Du Y, *et al.* A mutation in the POU-homeodomain of Pit-1 responsible for combined pituitary hormone deficiency. *Science* 1992;**257**:1115–18.

839. Ragge NK, Brown AG, Poloschek CM, *et al.* Heterozygous mutations of OTX2 cause severe ocular malformations. *Am J Hum Genet* 2005;**76**:1008–22.

840. Rakic P. Neuron–glia relationship during granule cell migration in developing cerebellar cortex: a Golgi and electron microscopic study in Macacus rhesus. *J Comp Neurol* 1971;**141**:283–312.

841. Rakic P. Mode of cell migration to the superficial layers of fetal monkey neocortex. *J Comp Neurol* 1972;**145**:61–84.

842. Rakic P, Sidman RL. Sequence of developmental abnormalities leading to granule cell deficit in cerebellar cortex of weaver mutant mice. *J Comp Neurol* 1973;**152**:103–32.

843. Rakic P, Sidman RL. Organization of cerebellar cortex secondary to deficit of granule cells in weaver mutant mice. *J Comp Neurol* 1973;**152**:133–62.

844. Rakic P, Yakovlev PI. Development of the corpus callosum and cavum septi in man. *J Comp Neurol* 1968;**132**:45–72.

845. Ranke O. Beiträge zur Kenntnis der normalen und pathologischen Hirnrindenbildung. *Beit Pathol Anat Pathol* 1910;**47**:51–125.

846. Raskind WH, Williams CA, Hudson LD, *et al.* Complete deletion of the proteolipid protein gene (PLP) in a family with X-linked Pelizaeus–Merzbacher disease. *Am J Hum Genet* 1991;**49**:1355–60.

847. Raymond AA, Fish DR, Stevens JM, *et al.* Subependymal heterotopia: a distinct neuronal migration disorder associated with epilepsy. *J Neurol Neurosurg Psychiatry* 1994;**57**:1195–202.

848. Raymond AA, Halpin SFS, Alsanjari N, *et al.* Dysembryoplastic neuroepithelial tumour: features in 16 patients. *Brain* 1994;**117**:461–75.

849. Razin A, Cedar H. DNA methylation and genomic imprinting. *Cell* 1994;**77**: 473–6.

850. Reardon W, Hockey A, Silberstein P, *et al.* Autosomal recessive congenital intrauterine infection-like syndrome of microcephaly, intracranial calcification and CNS disease. *Am J Med Genet* 1994;**52**:58–65.

851. Recklinghausen F von. *über die multiplen Fibrome der Haut und ihre Beziehung zu den multiplen Neuromen.* 20th edn. Berlin: Hirschwald, 1882.

852. Reiner O, Carrozzo R, Shen Y, *et al.* Isolation of a Miller–Dieker lissencephaly gene containing G protein beta-subunit-like repeats. *Nature* 1993;**364**:717–21.

853. Remington JS, Klein JO. *Infectious disease of the fetus and newborn infant.* Philadelphia, PA: WB Saunders, 1976.

854. Rengachary SS, Watanabe I. Ultrastructure and pathogenesis of intracranial arachnoid cysts. *J Neuropathol Exp Neurol* 1982;**40**:61–83.

855. Renier W, Gabreels FJM, Mol L, Korten J. Agenesis of the corpus callosum, chorioretinopathy and infantile spasms (Aicardi syndrome). *Psychiat Neurol Neurochirurg* 1973;**76**:39–45.

856. Renowden SA, Squier MV. Unusual magnetic resonance and neuropathological findings in hemimegalencephaly: report of a case following hemispherectomy. *Dev Med Child Neurol* 1994;**36**:357–69.

857. Retzius G. *Das Menschenhird.* Stockholm, 1895.

858. Ribadeau–Dumas JL, Poirier J, Escourolle R. Etude ultrastructurale des lésions cérébrales de la sclérose tubéreuse de Bourneville. *Acta Neuropathol (Berl)* 1973;**25**:259–70.

859. Riccardi VM, Marcus ES. Congenital hydrocephalus and cerebellar agenesis. *Clin Genet* 1978;**13**:443–7.

860. Ricci S, Cusmai R, Fariello G, *et al.* Double cortex: a neuronal migration anomaly as a possible cause of Lennox Gastaut syndrome. *Arch Neurol* 1992;**49**:61–4.

861. Richman DP, Stewart RM, Caviness VS. Cerebral microgyria in a 27 week fetus: an architectonic and topographic analysis. *J Neuropathol Exp Neurol* 1974;**33**:374–84.

862. Richman DP, Stewart RM, Hutchinson JW, Caviness VS. Mechanical model of brain convolutional development. *Science* 1975;**189**:18–21.

863. Richter RB. Unilateral congenital hypoplasia of the facial nucleus. *J Neuropathol Exp Neurol* 1960;**19**:33–41.

864. Riggs HE, McGrath JJ, Schwarz HP. Malformation of the adult brain (albino rat) resulting from prenatal irradiation. *J Neuropathol Exp Neurol* 1956;**15**: 432–47.

865. Rivière JB, Mirzaa GM, O'Roak BJ, Beddaoui M, *et al.* De novo germline and postzygotic mutations in AKT3, PIK3R2 and PIK3CA cause a spectrum of related megalencephaly syndromes. *Nat Genet* 2012;**44**:934–940.

866. Rizzoti K, Brunelli S, Carmignac D, *et al.* SOX3 is required during the formation of the hypothalamo-pituitary axis. *Nat Genet* 2004;**36**:247–55.

867. Robain O, Dulac O. Early epileptic encephalopathy with suppression bursts and olivary-dentate dysplasia. *Neuropediatrics* 1992;**23**:162–4.

868. Robain O, Gorce F. Arhinencéphalie. *Arch Fr Pediatr* 1972;**29**:861–79.

869. Robain O, Lyon G. Les microcéphalies familiales par malformations cérébrales. *Acta Neuropathol (Berl)* 1972;**20**:96–109.

870. Robain O, Dulac O, Lejeune J. Cerebellar hemispheric agenesis. *Acta Neuropathol (Berl)* 1987;**74**:202–6.

871. Robain O, Floquet C, Heldt N, Rozenberg F. Hemimegalencephaly: a clinicopathological study of four cases. *Neuropathol Appl Neurobiol* 1988;**14**:125–35.

872. Robain O, Chiron C, Dulac O. Electron microscopic and Golgi study in a case of hemimegalencephaly. *Acta Neuropathol (Berl)* 1989;**77**:664–6.

873. Robb L, Tam PP. Gastrula organiser and embryonic patterning in the mouse. *Semin Cell Dev Biol* 2004;**15**:543–54.

874. Robinson RO. Familial schizencephaly. *Dev Med Child Neurol* 1991;**33**:1010–12.

875. Roessler E, Belloni E, Gaudenz K, *et al.* Mutations in the human *Sonic Hedgehog* gene cause holoprosencephaly. *Nat Genet* 1996;**14**:357–60.

876. Roessler E, Du YZ, Mullor JL, *et al.* Loss-of-function mutations in the human GLI2 gene are associated with pituitary anomalies and holoprosencephaly-like features. *Proc Natl Acad Sci U S A* 2003;**100**:13424–9.

877. Roessmann U. Duplication of the pituitary gland and spinal cord. *Arch Pathol Lab Med* 1985;**109**:518–20.

878. Roessmann U, Gambetti P. Astrocytes in the developing human brain. *Acta Neuropathol (Berl)* 1986;**70**:309–13.

879. Roessmann U, Velasco ME, Small EJ, Hori A. Neuropathology of 'septo-optic dysplasia' (de Morsier syndrome) with immunohistochemical studies of the hypothalamus and pituitary gland. *J Neuropathol Exp Neurol* 1987;**46**:597–608.

880. Rogers KT. Experimental production of perfect cyclopia in chick by means of LiCl with a survey of the literature on cyclopia produced experimentally by various means. *Dev Biol* 1963;**8**:129–50.

881. Rorke LB. The role of disordered genetic control of neurogenesis in the pathogenesis of migration disorders. *J Neuropath Exp Neurol* 1994;**53**:105–17.

882. Rorke LB, Spiro AJ. Cerebral lesions in congenital rubella syndrome. *J Pediatr* 1967;**70**:243–55.

883. Rorke LB, Fogelson MH, Riggs H. Cerebellar heterotopia in infancy. *Dev Med Child Neurol* 1968;**10**:644–50.

884. Rosenthal A, Jouet M, Kenwrick S. Aberrant splicing of neural cell adhesion molecule L1 mRNA in a family with X-linked hydrocephalus. *Nat Genet* 1992;**2**:107–12.

885. Ross AJ, Ruiz–Perez V, Wang Y, *et al.* A homeobox gene, HLXB9, is the major locus for dominantly inherited sacral agenesis. *Nat Genet* 1998;**20**:358–61.

886. Ross DL, Liwnicz BH, Chun RWM, Gilbert E. Hypomelanosis of Ito (incontinentia pigmenti achromians): a clinicopathologic study: macrocephaly and gery matter heterotopias. *Neurology* 1982;**32**:1013–16.

887. Ross JJ, Frias JL, Vinken PJ, Bruyn GW eds. Microcephaly. In: *Handbook of clinical neurology.* Amsterdam: North-Holland, 1977:507–24.

888. Ross ME. Cell division and the nervous system: regulating the cycle from neural differentiation to death. *Trends Neurosci* 1996;**19**:62–8.

889. Ross MH, Galaburda AM, Kemper TL. Down's syndrome: is there a decreased population of neurons? *Neurology* 1984;**34**:909–16.

890. Rossant J. Manipulating the mouse genome: implications for neurobiology. *Neuron* 1990;**4**:323–34.

891. Rossing R, Friede RL. Holoprosencephaly with retroprosencephalic extracerebral cyst. *Dev Med Child Neurol* 1992;**34**:177–81.

892. Roth KA, Kuan CY, Haydar TF, *et al.* Epistatic and independent functions of Caspase-3 and Bcl-XL in developmental programmed cell death. *Proc Natl Acad Sci U S A* 2000;**97**:466–71.

893. Roubicek M, Spranger J, Wende S. Frontonasal dysplasia as an expression of holoprosencephaly. *Eur J Pediatr* 1981;**137**:229–31.

894. Rowbotham GF. Small aneurysm completely obstructing lower end of aqueduct of Sylvius. *Arch Neurol Psychiatry* 1938;**40**:1241–3.

895. Roy FH, Hiatt RL, Korones SB, Roane J. Ocular manifestations of congenital rubella syndrome. *Arch Ophthalmol* 1966;**75**:601–7.

896. Rubenstein JL. Annual Research Review: Development of the cerebral cortex: implications for neurodevelopmental disorders. *J Child Psychol Psychiatry* 2011;**52**(4):339–55.

897. Rubenstein JLR, Beachy PA. Patterning of the embryonic forebrain. *Curr Opin Neurobiol* 1998;**8**:18–26.

898. Rubinstein HS, Freeman W. Cerebellar agenesis. *J Nerv Ment Dis* 1940;**92**:489–502.

899. Rudelli RD, Brown WT, Wisniewski K, *et al.* Adult fragile X syndrome: clinico-neuropathologic findings. *Acta Neuropathol* 1985;**67**:289–95.

900. Ruff ME, Oakes WJ, Fisher SR, Spock A. Sleep apnea and vocal cord paralysis secondary to type I Chiari malformation. *Pediatrics* 1987;**80**:231–4.

901. Russell D. *Observations on the pathology of hydrocephalus.* MRC Special Report Series No. 265. London: HMSO, 1949.

902. Russell DS, Nevin S. Aneurysm of the great vein of Galen causing internal hydrocephalus: report of two cases. *J Pathol Bacteriol* 1940;**51**:375–83.

903. Sainte–Rose LH, LaCombe J, Pierre–Khan A, *et al.* Intracranial venous sinus hypertension: cause or consequence of hydrocephalus in infants? *J Neurosurg* 1984;**60**:727–36.

904. Salonen R, Herva R. Hydrolethalus syndrome. *J Med Genet* 1990;**27**:756–9.

905. Salonen R, Herva R, Norio R. The hydrolethalus syndrome: delineation of a 'new' lethal malformation syndrome based on 28 patients. *Clin Genet* 1981;**19**:321–30.

906. Salvesen GS, Dixit VM. Caspases: intracellular signaling by proteolysis. *Cell* 1997;**91**:443–6.

907. Samsom JF, Barth PG, Vries JIP de, *et al.* Familial mitochondrial encephalopathy with fetal ultrasonographic ventriculomegaly and intracerebral calcifications. *Eur J Pediatr* 1994;**153**:510–16.

908. Santavuori P, Somer H, Saino K. Muscle-eye–brain disease (MEB). *Brain Dev* 1989;**11**:147–53.

909. Sarnat HB, Alcala H. Human cerebellar hypoplasia: a syndrome of diverse causes. *Arch Neurol* 1980;**37**:300–305.

910. Sattar S, Gleeson JG. The ciliopathies in neuronal development: a clinical approach to investigation of Joubert syndrome and Joubert syndrome-related disorders. *Dev Med Child Neurol* 2011;**53**:793–798.

911. Scales DA, Collins GH. Cerebral degeneration with hypertrophy of the contralateral pyramid. *Arch Neurol* 1972;**26**:186–90.

912. Schachenmayr W, Friede RL. Fine structure of arachnoid cysts. *J Neuropathol Exp Neurol* 1979;**38**:434–46.

913. Schachenmayr W, Friede RL. Rhombencephalosynapsis: a Viennese malformation? *Dev Med Child Neurol* 1982;**24**:178–82.

914. Schalch E, Friede RL. A quantitative study of the composition of cerebellar cortical dysplasias. *Acta Neuropathol (Berl)* 1979;**47**:67–70.

915. Schaltenbrandt G. *Die Nervenkrankheiten.* Stuttgart: Thieme, 1951.

916. Scheffer IE, Baraitser M, Wilson J, *et al.* Pelizaeus–Merzbacher disease: classical or connatal? *Neuropediatrics* 1991;**22**:71–8.

917. Schiffmann SN, Bernier B, Goffinet AM. Reelin mRNA expression during mouse brain development. *Eur J Neurosci* 1997;**9**:1055–71.

918. Schinzel A, Schipke R, Riege D, Scoville WB. Postaxial polydactyly, hallux duplication, absence of the corpus callosum, macrencephaly and severe mental retardation: a new syndrome? Acute subdural haemorrhage at birth. *Helv Paediat Acta* 1954;**14**:468–73.

919. Schneider A, Griffiths IR, Readhead C, Nave K–A. Dominant-negative action of the jimpy mutation in mice complemented with an autosomal transgene for myelin proteolipid protein. *Proc Natl Acad Sci U S A* 1995;**92**:4447–51.

920. Schochet SS, Lampert PW, McCormick WF. Neurofibrillary tangles in patients with Down's syndrome: a light and electron-microscopic study. *Acta Neuropathol (Berl)* 1973;**23**:342–6.

921. Schum TR, Meyer GA, Gransz JP, Glaspey JC. Neonatal intraventricular hemorrhage due to an intracranial arteriovenous malformation: a case report. *Pediatrics* 1979;**64**:242–4.

922. Schwaiger H, Weirich HG, Brunner P, *et al.* Radiation sensitivity of Down's syndrome fibroblasts might be due to overexpressed Cu/Zn-superoxide dismutase. *Eur J Cell Biol* 1989;**48**:79–87.

923. Schwalbe E, Gredig M. über Entwicklungströrungen des Kleinhirns, Hirnstamms und Halsmarks bei Spina Bifida (Arnold'sche und Chiari'sche Missbildung). *Beit Pathol Anat* 1907;**40**:132–94.

924. Schwidde JT. Incidence of cavum septi pellucidi and cavum vergae in 1032 human brains. *Arch Neurol Psychiatry* 1952;**67**:625–32.

925. Searle AG, Edwards JH, Hall JG. Mouse homologues of human hereditary disease. *J Med Genet* 1994;**31**:1–19.

926. Sees JN, Towfighi J, Robins DB, Ladda RL. 'Marden–Walker syndrome': neuropathologic findings in two siblings. *Pediatr Pathol* 1990;**10**:807–18.

927. Shady W, Metcalfe RA, Butler P. The incidence of craniocervical bony anomalies in the adult Chiari malformation. *J Neurol Sci* 1987;**82**:193–203.

928. Sharief N, Craze J, Summers D, *et al.* Miller–Dieker syndrome with ring chromosome 17. *Arch Dis Child* 1991;**667**:10–12.

929. Sharp D, Robertson DM. Tuberous sclerosis in an infant of 28 weeks gestational age. *Can J Neurol Sci* 1983;**10**:59–62.

930. Shaw CM, Alvord EC. Cava septi pellucidi et vergae: their normal and pathological states. *Brain* 1969;**92**:213–24.

931. Shaw CM, Alvord EC. 'Congenital arachnoid' cysts and their differential

diagnosis. In: Vinken PJa, Bruyn GW eds. *Handbook of clinical neurology.* Amsterdam: North-Holland, 1977:75–135.

932. Sheen VL, Feng YY, Graham D, *et al.* Filamin A and Filamin B are co-expressed within neurons during periods of neuronal migration and can physically interact. *Hum Mol Genet* 2002;**11**:2845–54.

933. Sheen VL, Ganesh VS, Topcu M, *et al.* Mutations in ARFGEF2 implicate vesicle trafficking in neural progenitor proliferation and migration in the human cerebral cortex. *Nat Genet* 2004;**36**:69–76.

934. Shen MH, Yang J, Loupart ML, *et al.* Human mini-chromosomes in mouse embryonal stem cells. *Hum Mol Genet* 1997;**6**:1375–82.

935. Sherman GF, Morrison L, Rosen GD, *et al.* Brain abnormalities in immune defective mice. *Brain Res* 1990;**532**:25–33.

936. Sherman JL, Barkovich AJ, Citrin CM. The MR appearances of syringomyelia: new observations. *Am J Neuroradiol* 1986;**7**:985–95.

937. Shevell MI, Carmant L, Meagher Villemure K. Developmental bilateral perisylvian dysplasia. *Pediatr Neurol* 1992;**8**:299–302.

938. Shimada M, Langman J. Repair of the external granular layer of the hamster cerebellum after prenatal and postnatal administration of methylazoxymethanol. *Teratology* 1970;**3**:119–39.

939. Shu TZ, Ayala R, Nguyen MD, *et al.* Ndel1 operates in a common pathway with LIS1 and cytoplasmic dynein to regulate cortical neuronal positioning. *Neuron* 2004;**44**:263–77.

940. Shum ASW, Copp AJ. Regional differences in morphogenesis of the neuroepithelium suggest multiple mechanisms of spinal neurulation in the mouse. *Anat Embryol* 1996;**194**:65–73.

941. Sidman RL, Rakic P. Neuronal migration, with special reference to the developing human brain: a review. *Brain Res* 1973;**62**:1–35.

942. Siebert JR, Warkany J, Lemire RJ. Atelencephalic microcephaly in a 21 week human fetus. *Teratology* 1986;**34**:9–19.

943. Siebert JR, Cohen MM Jr, Sulik KK, *et al. Holoprosencephaly: an overview and atlas of cases.* New York: Wiley–Liss, 1990:48–93.

944. Silver J, Ogawa MY. Postnatally induced formation of the corpus callosum in acallosal mice on glia-coated cellulose bridges. *Science* 1983;**220**:1067–9.

945. Silver J, Lorenz SE, Wahlstein D, Coughlin J. Axonal guidance during development of the great cerebral commissures: descriptive and experimental studies, *in vivo*, on the role of preformed glial pathways. *J Comp Neurol* 1982;**210**:10–29.

946. Silverman BK, Brechx T, Craig J, Nadas A. Congestive failure in the newborn, caused by cerebral arterio-venous fistula. *Am J Dis Child* 1955;**89**:539–43.

947. Sima AAF, Robertson DM. Subependymal giant-cell astrocytoma: case report with ultrastructural study. *J Neurosurg* 1979;**50**:240–45.

948. Siqueira EB, Murray KJ. Calcified aneurysms of the vein of Galen: report of a presumed case and review of the literature. *Neurochirurgia* 1972;**15**:106–12.

949. Smith DW. The fetal alcohol syndrome. *Hosp Pract* 1979;**14**:121–8.

950. Smith DW. *Recognizable patterns of human malformations: genetic, embryologic and clinical aspects.* Philadelphia, PA: WB Saunders, 1988.

951. Smith I, Erdohazi M, Macartney FJ, *et al.* Fetal damage despite low-phenylalanine diet after conception in a phenylketonuric woman. *Lancet* 1979;**i**:17–19.

952. Smith JF, Rodeck C. Multiple cystic and focal encephalomalacia in infancy and childhood with brain stem damage. *J Neurol Sci* 1975;**25**:377–88.

953. Smith JL, Schoenwolf GC. Role of cell-cycle in regulating neuroepithelial cell shape during bending of the chick neural plate. *Cell Tissue Res* 1988;**252**:491–500.

954. Smith MT, Huntington HW. Inverse cerebellum and occipital encephalocele: a dorsal fusion defect uniting the Arnold–Chiari and Dandy–Walker spectrum. *Neurology* 1977;**27**:246–51.

955. Smith UM, Consugar M, Tee LJ, *et al.* The transmembrane protein meckelin (MKS3) is mutated in Meckel–Gruber syndrome and the wpk rat. *Nat Genet* 2006;**38**:191–6.

956. Snider WD. Functions of the neurotrophins during nervous system development: What the knockouts are teaching us. *Cell* 1994;**77**:627–38.

957. Snodgrass G, Butler J, France N, *et al.* The 'D' (13–15) trisomy syndrome: an analysis of 7 examples. *Arch Dis Child* 1966;**42**:250–61.

958. Sokol SY. Wnt signaling and dorso-ventral axis specification in vertebrates. *Curr Opin Genet Dev* 1999;**9**:405–10.

959. Solitare GB, Lamarche JB, Solt LC, *et al.* Alzheimer's disease and senile dementia as seen in mongoloids: neuropathological observations. Interhemispheric cyst of neuroepithelial origin in association with partial agenesis of the corpus callosum: case report and review of the literature. *Am J Ment Def* 1980;**52**:399–403.

960. Sotelo C. Anatomical physiological and biochemical studies of the cerebellum from mutant mice: II. Morphological study of cerebellar cortical neurons and circuits in the Weaver mouse. *Brain Res* 1975;**94**:19–44.

961. Sotos JF, Dodge PR, Muirhead D, *et al.* Cerebral gigantism in childhood. *N Engl J Med* 1964;**271**:109–16.

962. Soussi–Yanicostas N, De Castro F, Julliard AK, *et al.* Anosmin-1, defective in the X-linked form of Kallmann syndrome, promotes axonal branch formation from olfactory bulb output neurons. *Cell* 2002;**109**:217–28.

963. Spataro RF, Lin SR, Horner FA, *et al.* Aqueductal stenosis and hydrocephalus: rate sequelae of mumps virus infection. *Neuroradiology* 1976;**12**:11–13.

964. Spiller WG. Syringomyelia, extending from the sacral region of the spinal cord through the medulla oblongata, right side of the pons and right cerebral peduncle to the upper part of the right internal capsule (syringobulbia). *Br Med J* 1906;**2**:1017–21.

965. Spranger JJ, Benirschke K, Hall JG, *et al.* Errors of morphogenesis concepts and terms: recommendations of an international working group. *J Pediatr* 1982;**100**:160–65.

966. Spreafico R, Angelini L, Binelli S, *et al.* Burst suppression and impairment of neocortical ontogenesis: electroclinical and neuropathological findings in two infants with early myoclonic encephalopathy. *Epilepsia* 1993;**34**: 800–8.

967. Squier MV. Fetal type II lissencephaly: a case report. *Childs Nerv Syst* 1993;**9**:400–402.

968. Srabstein JC, Morris N, Larke RPB, *et al.* Is there a congenital varicella syndrome? *J Pediatr* 1974;**84**:239–43.

969. Staemmler M. *Hydromyelie, Syringomyelie und Gliose.* Berlin: Springer, 1942.

970. Stagno S, Reynolds DW, Huang E, *et al.* Congenital cytomegalovirus infection: occurrence in an immune population. *N Engl J Med* 1977;**296**:1254–8.

971. Stanhope R, Preece MA, Brook CGD. Hypoplastic optic nerves and pituitary dysfunction: a spectrum of anatomical and endocrine abnormalities. *Arch Dis Child* 1984;**59**:111–14.

972. Starkman SP, Brown TC, Linell EA. Cerebral arachnoid cysts. *J Neuropathol Exp Neurol* 1958;**17**:484–500.

973. Steegers–Theunissen RPM, Boers GHJ, Trijbels FJM, Eskes TKAB. Neural-tube defects and derangement of homocysteine metabolism. *N Engl J Med* 1991;**324**:199–200.

974. Steimann GS, Rorke LB, Brown MJ. Infantile neuronal degeneration masquerading as Werdnig–Hoffmann disease. *Ann Neurol* 1980;**8**:317–24.

975. Stephansson K, Wollmann R. Distribution of glial fibrillary acidic protein in central nervous system lesions of tuberous sclerosis. *Acta Neuropathol (Berl)* 1980;**52**:135–40.

976. Stern CD. Induction and initial patterning of the nervous system: the chick embryo enters the scene. *Curr Opin Genet Dev* 2002;**12**:447–51.

977. Stern H, Elek SD, Booth JC, Fleck DG. Microbial causes of mental retardation: the role of prenatal infection with cytomegalovirus, rubella virus and toxoplasma. *Lancet* 1969;**ii**:443–8.

978. Stern L, Ramos AD, Wigglesworth FW. Congenital heart failure secondary to cerebral arteriovenous aneurysm in the newborn infant. *Am J Dis Child* 1968;**115**:581–7.

979. Sternberg C. Ueber vollständigen Defekt des Kleinhirnes. *Verhandlungen Dtsch Pathol Ges* 1912;**15**:353–9.

980. Stevenson RE, Hall JG. *Human malformations and related anomalies.* 2nd edn. Cary, NC: Cary Press, 2006, Chapter 15.

981. Stewart RM. Cerebellar agenesis. *J Ment Sci* 1956;**102**:67–77.

982. Stewart RM, Richman DP, Caviness VS Jr. Lissencephaly and pachygyria, an architectonic and topographical analysis. *Acta Neuropathol (Berl)* 1975;**31**:1–12.

983. Streeter G. The cortex of the brain in the human embryo during the fourth month with special reference to the so-called 'papillae of Retzius'. *Am J Anat* 1912;**7**:337–44.

984. Strefling AM, Urich H. Crossed cerebellar atrophy: an old problem revisited. *Acta Neuropathol (Berl)* 1982;**57**:197–202.

985. Strefling AM, Urich H. Prenatal porencephaly: the pattern of secondary lesions. *Acta Neuropathol (Berl)* 1986;**71**:171–5.

986. Strong OS. A case of unilateral cerebellar agenesia. *J Comp Neurol* 1915;**25**:361–74.

987. Sturge WA. A case of partial epilepsy, apparently due to a lesion of one of the vasomotor centres of the brain. *Trans Clin Soc Lond* 1879;**12**:162–7.

988. Sudarshan A, Goldie WD. The spectrum of congenital facial diplegia (Moebius syndrome). *Pediatr Neurol* 1985;**1**:180–84.

989. Sulik KK, Zuker RM, Dehart DB, *et al.* Normal patterns of neural tube closure differ in the human and the mouse. *Proc Greenwood Genet Cent* 1998;**18**:129–30.

990. Sumi SM. Brain malformations in the trisomy 18 syndrome. *Brain* 1970;**93**:821.

991. Suwanwela C, Suwanwela N. A morphological classification of sincipital encephalomeningoceles. *J Neurosurg* 1972;**36**:201–11.

992. Taggart JK, Walker AE. Congenital atresia of the foramina of Luschka and Magendie. *Arch Neurol Psychiatry* 1942;**48**:583–612.

993. Tajima M, Yamada H, Kageyama N. Craniolacunia in newborn with myelomeningocele. *Childs Brain* 1977;**3**:297–303.

994. Takada K, Becker LE, Takashima S. Walker–Warburg syndrome with skeletal muscle involvement: a report of three patients. *Pediatr Neurosci* 1987;**13**:202–9.

995. Takada K, Nakamura H, Takashima S. Cortical dysplasia in Fukuyama congenital muscular dystrophy (FCMD): a Golgi and angioarchitectonic analysis. *Acta Neuropathol (Berl)* 1988;**76**:170–78.

996. Takashima S, Becker LE, Armstrong DL, Chan F. Abnormal neuronal development in the visual cortex of the human fetus and infant with Down's syndrome: a quantitative and qualitative Golgi study. *Brain Res* 1981;**225**:1–21.

997. Takeda S, Kondo M, Sasaki J, *et al.* Fukutin is required for maintenance of muscle integrity, cortical histiogenesis and normal eye development. *Hum Mol Genet* 2003;**12**:1449–59.

998. Tal Y, Freigung B, Dunn HG, *et al.* Dandy–Walker syndrome analysis of 2 cases. *Dev Med Child Neurol* 1980;**22**:184–201.

999. Tam PPL. The histogenetic capacity of tissues in the caudal end of the embryonic axis of the mouse. *J Embryol Exp Morphol* 1984;**82**:253–66.

1000. Tamagawa K, Scheidt P, Friede RL. Experimental production of leptomeningeal heterotopias from dissociated fetal tissue. *Acta Neuropathol (Berl)* 1989;**78**:153–8.

1001. Tan N, Urich H. Postictal cerebral hemiatrophy: with a contribution to the problem of crossed cerebellar atrophy. *Acta Neuropathol (Berl)* 1984;**62**:332–9.

1002. Tanaka T, Serneo FF, Tseng HC, *et al.* Cdk5 phosphorylation of doublecortin ser297 regulates its effect on neuronal migration. *Neuron* 2004;**41**:215–27.

1003. Tassabehji M, Read AP, Newton VE, *et al.* Waardenburg's syndrome patients have mutations in the human homologue of the *Pax-3* paired box gene. *Nature* 1992;**355**:635–6.

1004. Tatsumi K, Miyai K, Notomi T, *et al.* Cretinism with combined hormone deficiency caused by a mutation in the Pit-1 gene. *Nat Genet* 1992;**1**:56–58.

1005. Taveras JM, Poser CM. Roentgenologic aspects of cerebral angiography in children. *Am J Roentgenol* 1959;**82**:371–91.

1006. Tavormina PL, Shiang R, Thompson LM, *et al.* Thanatophoric dysplasia (types I and II) caused by distinct mutations in fibroblast growth factor receptor 3. *Nat Genet* 1995;**9**:321–8.

1007. Taylor DC, Falconer MA, Bruton CJ, Corsellis JAN. Focal dysplasia of the cerebral cortex in epilepsy. *J Neurol Neurosurg Psychiatry* 1971;**34**:369–87.

1008. Terashima T, Inoue K, Inoue Y, *et al.* Observations on the cerebellum of normal–reeler mutant mouse chimera. *J Comp Neurol* 1986;**252**:264–78.

1009. Thorogood P. Differentiation and morphogenesis of cranial skeletal tissues. In: Hanken J, Hall BK eds. *The Skull.* Chicago, IL: University of Chicago Press, 1994:113–52.

1010. Thulin B, McTaggart D, Neuberger KT. Demyelinating leukodystrophy with total cortical cerebellar atrophy. *Arch Neurol* 1968;**18**:113–22.

1011. Timor–Tritsch IE, Monteagudo A, Warren WB. Transvaginal ultrasonographic definition in CNS in first and early second trimester. *Am J Obstet Gynecol* 1991;**164**:497–503.

1012. Tjiam AT, Stefanko S, Schenk VMD, Vlieger M. Infantile spasms associated with hemihypsarrhthmia and hemimegalencephaly. *Dev Med Child Neurol* 1978;**20**:779–98.

1013. Tole S, Ragsdale CW, Grove EA. Dorsoventral patterning of the telencephalon is disrupted in the mouse mutant *extra-toes*. *Dev Biol* 2000;**217**:254–65.

1014. Ton CCT, Hirvonen H, Miwa H, *et al.* Positional cloning and characterization of a paired box- and homeobox-containing gene from the aniridia region. *Cell* 1991;**67**:1059–74.

1015. Torban E, Kor C, Gros P. Van Gogh-like2 (Strabismus) and its role in planar cell polarity and convergent extension in vertebrates. *Trends Genet* 2004;**20**:570–77.

1016. Towfighi J, Marks K, Palmer E, Vannucci R. Moebius syndrome. Neuropathologic observations. *Acta Neuropathol (Berl)* 1979;**48**:11–17.

1017. Towfighi J, Sassani JW, Suzuki K, Ladda RL. Cerebro-ocular dysplasia–muscular dystrophy (COD-MD) syndrome. *Acta Neuropathol (Berl)* 1984;**65**:110–23.

1018. Towfighi J, Ladda RL, Sharkey FE. Purkinje cell inclusions and 'atelencephaly' in 13q-chromosomal syndrome. *Arch Pathol Lab Med* 1987;**111**:146–50.

1019. Townsend JJ, Nielsen SL, Malamud N. Unilateral megalencephaly: hamartoma or neoplasm? *Neurology* 1975;**25**:448–53.

1020. Toyo-oka K, Shionoya A, Gambello MJ, *et al.* 14-3-3e is important for neuronal migration by binding to NUDEL: a molecular explanation for Miller–Dieker syndrome. *Nat Genet* 2003;**34**:274–85.

1021. Trombley IK, Mirra SS. Ultrastructure of tuberous sclerosis: cortical tuber and subependymal tumor. *Ann Neurol* 1981;**9**:174–81.

1022. Truwit CL, Barkovich AJ. Pathogenesis of intracranial lipoma: an MR study

in 42 patients. *Am J Neuroradiol* 1987;**11**:665–74.

1023. Tsai JW, Chen Y, Kriegstein AR, *et al.* LIS1 RNA interference blocks neural stem cell division, morphogenesis, and motility at multiple stages. *J Cell Biol* 2005;**170**:935–45.

1024. Tsarkraklides V, Burke B, Mastri A, *et al.* Rhabdomyomas of the heart. *Am J Dis Child* 1974;**128**:639–46.

1025. Tsukahara M, Sugio Y, Kajii T, *et al.* Pachygyria, joint contractures, and facial abnormalities: a new lethal syndrome. *J Med Genet* 1990;**27**:532.

1026. Tu Y, Wu S, Shi X, *et al.* Migfilin and Mig-2 link focal adhesions to filamin and the actin cytoskeleton and function in cell shape modulation. *Cell* 2003;**113**:37–47.

1027. Tuchmann–Duplessis H, Larroche JC. Anencéphalie et atrophie cortico-surrénale. *C R Soc Biol* 1958;**152**:300–302.

1028. Turkewitsch N. La constitution anatomique de l'Aquaeductus cerebri de l'homme. *Arch Anat Histol Embryol* 1936;**21**:323–57.

1029. Turnbull IM, Drake CG. Membranous occlusion of the aqueduct of Sylvius. *J Neurol* 1966:**24**:24–33.

1030. Twining P, Zuccollo J. The ultrasound markers of chromosomal disease: a retrospective study. *Br J Radiol* 1993;**66**:408–14.

1031. Uiberrack F. Demonstration eines Falles von Ganglienzellerkrankung bei einem Fötus nach Röntgenbestrahlung der Mutter. *Zentralblatt Allg Pathol Pathol Anat* 1942;**80**:187.

1032. Ule G. Kleinhirnrindenatrophie vom Körnertyp. *Dtsch Z Nervenheilkunde* 1952;**168**:195–226.

1033. Ulrich J, Herschkowitz N, Heitz P, *et al.* Adenoleukodystrophy: preliminary report of a connatal case: light and electron microscopical, immunohistochemical and biochemical findings. *Acta Neuropathol (Berl)* 1978;**43**:77–83.

1034. Urich H. Malformations of the nervous system. In: Blackwood W, Corsellis JAN eds. *Greenfield's Neuropathology.* 3rd edn. London: Edward Arnold, 1977:361–469.

1035. Vade A, Horowitz SW. Agenesis of the corpus callosum and intraventricular lipomas. *Pediatr Neurol* 1992;**8**:307–9.

1036. Valanne L, Pihko H, Katevuo K, *et al.* MRI of the brain in muscle–eye–brain (MEB) disease. *Neuroradiology* 1994;**36**:473–6.

1037. Valdivieso EMB, Scholtz CL. Diffuse meningocerebral angiodysplasia. *Pediatr Pathol* 1986;**6**:119–26.

1038. Van Allen MI, Kalousek DK, Chernoff GF, *et al.* Evidence for multi-site closure of the neural tube in humans. *Am J Med Genet* 1993;**47**:723–43.

1039. Van Bogaert L, Radermecker MA. Une dysgénésie cérébelleuse chez un enfant du radium. *Rev Neurol* 1955;**93**:65–82.

1040. Van Bogaert L, Paillas JE, Berard-Badier M, Payan H. Etude sur la sclérose tubéreuse de Bourneville a forme cérébelleuse. *Rev Neurol* 1958;**98**:673–89.

1041. Van der Hauwaert LG. Cardiac tumours in infancy and childhood. *Br Heart J* 1971;**33**:125–32.

1042. Van der Put NMJ, Eskes TKAB, Blom HJ. Is the common 677C T mutation in the methylenetetrahydrofolate reductase gene a risk factor for neural tube defects? A meta-analysis. *Q J Med* 1997;**90**:111–15.

1043. Van Reeuwijk J, Janssen M, van den EC, *et al.* POMT2 mutations cause alpha-dystroglycan hypoglycosylation and Walker–Warburg syndrome. *J Med Genet* 2005;**42**:907–12.

1044. Van Slegtenhorst M, De Hoogt R, Hermans C, *et al.* Identification of the tuberous sclerosis gene TSC1 on chromosome 9q34. *Science* 1997;**277**:805–8.

1045. Van Straaten HWM, Hekking JWM, Consten C, Copp AJ. Intrinsic and extrinsic factors in the mechanism of neurulation: effect of curvature of the body axis on closure of the posterior neuropore. *Development* 1993;**117**:1163–72.

1046. Variend S, Emery JL. The weight of the cerebellum in children with myelomeningocele. *Dev Med Child Neurol* 1973;**15**(Suppl 29):77–83.

1047. Veith G, Wicke R. *Cerebrale Differenzierungsstörungen bei Epilepsie Jahrbuch 1968.* Köln-Opladen: Westdeutscher, 1968.

1048. Venter JC, Adams MD, Myers EW, *et al.* The sequence of the human genome. *Science* 2001;**291**:1304–51.

1049. Verhaart W. Hypertrophy of peduncle and pyramid as a result of degeneration of contralateral corticofugal fiber tracts. *J Comp Neurol* 1950;**19**:1–15.

1050. Verhaart W, Kramer W. The uncrossed pyramidal tract. *Acta Psychiat Neurol Scand* 1952;**27**:181–200.

1051. Verlaan DJ, Davenport WJ, Stefan H, *et al.* Cerebral cavernous malformations: mutations in Krit1. *Neurology* 2002;**58**:853–7.

1052. Vestermark V. Teratogenicity of carbamazepine:a review of the literature. *Dev Brain Dysfunct* 1993;**6**:266–78.

1053. Viani F, Strada GP, Riboldi A, *et al.* Aspetti neuropatologici della sindrome di Lennox-Gastaut: considerazioni su tue casi. *Riv Neurol* 1977;**4**:413–52.

1054. Vigevano F, Bertini E, Boldrini R, *et al.* Hemimegalencephaly and intractable epilepsy: benefits of hemispherectomy. *Epilepsia* 1989;**30**:833–43.

1055. Vincent C, Kalatzis V, Compain S, *et al.* A proposed new contiguous gene syndrome on 8q consists of branchio-oto-renal (BOR) syndrome, Duane syndrome, a dominant form of hydrocephalus and trapeze aplasia: implications for the mapping of the BOR gene. *Hum Mol Genet* 1994;**3**:1859–66.

1056. Vinters HV, Fisher RS, Cornford ME, *et al.* Morphological substrates of infantile spasms: studies based on surgically resected cerebral tissue. *Childs Nerv Syst* 1992;**8**:8–17.

1057. Vogel H, Gessaga EC, Horoupian DS, Urich H. Aqueductal atresia as a feature of arhinencephalic syndromes. *Clin Neuropathol* 1990;**9**:191–5.

1058. Vogt H, Astwararuturow M. Ueber angeborene Kleinhirnerkrankungen mit Beiträgen zur Entwicklungsgeschichte des Kleinhirns. *Arch Psychiat Nervenkrankheit* 1912;**49**:175–203.

1059. Volpe JJ. Subplate neurons –missing link in brain injury of the premature infant. *Pediatrics* 1996;**97**:112–13.

1060. Volpe JJ, Adams RD. Cerebro-hepatorenal syndrome of Zellweger: an inherited disorder of neuronal migration. *Acta Neuropathol (Berl)* 1972;**20**:175–98.

1061. Vuia O. Malformation of the paraflocculus and atresia of the foramina Magendie and Luschka in a child. *Psychiat Neurol Neurochirurg Amsterdam* 1973;**76**:261–6.

1062. Wald N, Sneddon J, Densem J, *et al.* Prevention of neural tube defects: results of the Medical Research Council Vitamin Study. *Lancet* 1991;**338**:131–7.

1063. Walker AE. Lissencephaly. *Arch Neurol Psychiatry* 1942;**48**:13–29.

1064. Wallis DE, Roessler E, Hehr U, *et al.* Mutations in the homeodomain of the human SIX3 gene cause holoprosencephaly. *Nat Genet* 1999;**22**:196–8.

1065. Wanders RJ, Waterham HR. Peroxisomal disorders I: biochemistry and genetics of peroxisome biogenesis disorders. *Clin Genet* 2005;**67**:107–33.

1066. Warburg M. Hydrocephaly, congenital retinal nonattachment and congenital falciform fold. *Am J Ophthalmol* 1978;**85**:88–94.

1067. Wassif CA, Maslen C, Kachilele Linjewile S, *et al.* Mutations in the human sterol delta7-reductase gene at 11q12-13 cause Smith-Lemli-Opitz syndrome. *Am J Hum Genet* 1998;**63**:55–62.

1068. Wassif CA, Zhu P, Kratz L, *et al.* Biochemical, phenotypic and neurophysiological characterization of a genetic mouse model of RSH/Smith–Lemli–Opitz syndrome. *Hum Mol Genet* 2001;**10**:555–64.

1069. Watson EH. Hydranencephaly: report of two cases which combine features of hydrocephalus and anencephaly. *Am J Dis Child* 1944;**67**:282–7.

1070. Weinberg AG, Kirkpatrick JB. Cerebellar hypoplasia in Werdnig–Hoffmann disease. *Dev Med Child Neurol* 1975;**17**:511–16.

1071. Weiss MH, Young HF, McFarland DE. Hydranencephaly of postnatal origin. *J Neurosurg* 1970;**32**:715–20.

1072. Weller TH, Macauley JC, Craig JH, Wirth P. Isolation of intranuclear inclusion producing agents from infants with illnesses resembling cytomegalic inclusion disease. *Proc Sec Exp Biol Med* 1957;**94**:4–12.

1073. Wicking C, Williamson B. From linked marker to gene. *Trends Genet* 1991;**7**:288–93.

1074. Wiest WD, Hallerworden J. Migrationshemmung in Gross- und Kleinhirn. *Dtsch Z Nervenheilkunde* 1958;**178**:244–38.

1075. Wilberger JE Jr, Vertosick FT Jr, Vries JK. Unilateral hydrocephalus secondary to congenital atresia of the foramen of Monro: case report. *J Neurosurg* 1983;**59**:899–901.

1076. Wilkie AOM, Slaney SF, Oldridge M, *et al.* Apert syndrome results from localized mutations of *FGFR2* and is allelic with Crouzon syndrome. *Nat Genet* 1995;**9**:165–72.

1077. Willems PJ, Brouwer OF, Dijkstra I, Wilminik J. X-linked hydrocephalus. *Am J Med Genet* 1987;**27**:921–8.

1078. Williams B. Is aqueduct stenosis a result of hydrocephalus? *Brain* 1973;**93**:399–412.

1079. Williams RS, Ferrante RJ, Caviness VS. The cellular pathology of microgyria. *Acta Neuropathol (Berl)* 1976;**36**:269–83.

1080. Williams RS, Swisher CN, Jennings M, *et al.* Cerebro-ocular dysgenesis (Walker–Warburg syndrome):neuropathologic and etiologic analysis. *Neurology* 1984;**34**:1531–41.

1081. Wilson JG. *Environment and birth defects.* New York: Academic Press, 1973:11–34.

1082. Wilson PA, Hemmati-Brivanlou A. Induction of epidermis and inhibition of neural fate by Bmp-4. *Nature* 1995;**376**:331–3.

1083. Wilson SI, Edlund T. Neural induction: toward a unifying mechanism. *Nat Neurosci* 2001;**4** (Suppl):1161–8.

1084. Winter RM, Harding BN, Hyde J. Unknown syndrome: pachygyria, joint contractures and facial abnormalities. *J Med Genet* 1989;**26**:788–9.

1085. Wise BL, Sondheimer F, Kaufman S. Achondroplasia and hydrocephalus. *Neuropadiatrie* 1971;**3**:106–13.

1086. Wisniewski K, Dambska M, Sher JH, Qazi Q. A clinical neuropathological study of the fetal alcohol syndrome. *Neuropaediatrics* 1995;**14**:197–201.

1087. Wisniewski KE, Schmidt–Sidor B. Postnatal delay of myelin formation in brains from Down syndrome infants and children. *Clin Neuropathol* 1989;**8**:55–62.

1088. Wisniewski KE, Laure–Kamionowska M, Wisniewski HM. Evidence of arrest of neurogenesis and synaptogenesis in Down syndrome brains. *N Engl J Med* 1984;**311**:1187–8.

1089. Wit JM, Beemes FE, Barth FC, *et al.* Cerebral gigantism (Sotos syndrome): compiled data of 22 cases. Analysis of clinical features, growth and palsma somatomedin. *Eur J Pediatr* 1985;**144**:131–40.

1090. Wohak H. Ein Fall von Varix der Vena magna Galeni bei einem Neugeborenen. *Virchows Arch Pathol Anat* 1923;**242**:58–68.

1091. Wolf SM. The ocular manifestations of congenital rubella: a prospective study of 328 cases of congenital rubella. *Pediatr Ophthalmol* 1973;**10**:101–41.

1092. Wongmongkolrit T, Bush M, Roessmann U. Neuropathological findings in thanatophoric dysplasia. *Arch Pathol Lab Med* 1983;**107**:132–5.

1093. Wood JW, Johnson KG, Omori Y. *In utero* exposure to the Hiroshima atomic bomb: an evaluation of head size and mental retardation twenty years later. *Pediatrics* 1967;**39**:385–92.

1094. Wood LR, Smith MT. Generation of anencephaly: 1. Aberrant neurulation and 2. Conversion of exencephaly to anencephaly. *J Neuropathol Exp Neurol* 1984;**43**:620–33.

1095. Wood W, Turmaine M, Weber R, *et al.* Mesenchymal cells engulf and clear apoptotic footplate cells in macrophageless PU.1 null mouse embryos. *Development* 2000;**127**:5245–52.

1096. Woods CG, Bond J, Enard W. Autosomal recessive primary microcephaly (MCPH): a review of clinical, molecular and evolutionary findings. *Am J Hum Genet* 2005;**76**:717–28.

1097. Woods KS, Cundall M, Turton J, *et al.* Over- and underdosage of SOX3 is associated with infundibular hypoplasia and hypopituitarism. *Am J Hum Genet* 2005;**76**:833–49.

1098. Woodward K, Malcolm S. Proteolipid protein gene: Pelizaeus–Merzbacher disease in humans and neurodegeneration in mice. *Trends Genet* 1999;**15**:125–8.

1099. Woollam DHM, Millen JW. Anatomical considerations in the pathology of stenosis of the cerebral aqueduct. *Brain* 1953;**76**:104–12.

1100. Worthington WC, Cathcart RS. Ciliary currents on ependymal surfaces. *Ann N Y Acad Sci* 1965;**130**:944–50.

1101. Wu W, Cogan JD, Pfaffle RW, *et al.* Mutations in PROP1 cause familial combined pituitary hormone deficiency. *Nat Genet* 1998;**18**:147–9.

1102. Xuan S, Baptista CA, Balas G, *et al.* Winged helix transcription factor BF-1 is essential for the development of the cerebral hemispheres. *Neuron* 1995;**14**:1141–52.

1103. Yakovlev PI. Pathoarchitectonic studies of the cerebral malformations: III. Arhinencephalies (holotelencephalies). *J Neuropathol Exp Neurol* 1959;**18**:22–55.

1104. Yakovlev PI. A proposed definition of the limbic system. In: Hockman CH ed. *Limbic system mechanisms and autonomic function.* Springfield, IL: CC Thomas, 1972:241–83.

1105. Yakovlev PI, Corwin W. Roentgenographic sign in cases of tuberous sclerosis of the brain (multiple 'brain stones'). *Arch Neurol Psychiatry* 1939;**40**:1030–37.

1106. Yakovlev PI, Wadsworth RC. Schizencephalies: a study of the congenital clefts in the cerebral mantle. I. Clefts with fused lips. *J Neuropathol Exp Neurol* 1946;**5**:116–30.

1107. Yakovlev PI, Wadsworth RC. Schizencephalies: a study of the congenital clefts in the cerebral mantle. II. Clefts with hydrocephalus and lips separated. *J Neuropathol Exp Neurol* 1946;**5**:169–206.

1108. Yamaguchi E, Hayashi T, Kondoh H, *et al.* A case of Walker–Warburg syndrome with uncommon findings: double cortical layer, temporal cyst and increased serum IgM. *Brain Dev* 1993;**15**:61–5.

1109. Yamaguchi TP, Harpal K, Henkemeyer M, *et al.* fgfr-1 is required for embryonic growth and mesodermal patterning during mouse gastrulation. *Genes Dev* 1994;**8**:3032–44.

1110. Yamamoto T, Toyoda C, Kobayashi M, *et al.* Pial–glial barrier abnormalities in fetuses with Fukuyama congenital muscular dystrophy. *Brain Dev* 1997;**19**:35–42.

1111. Yamano T, Shimada M, Abe Y, *et al.* Destruction of external granular layer and subsequent cerebellar abnormalities. *Acta Neuropathol (Berl)* 1983;**59**:41–7.

1112. Yamazaki JN, Wright SW, Wright PM. Outcome of pregnancy in women exposed to atomic bomb in Nagasaki. *Am J Dis Child* 1954;**87**:448–63.

1113. Yang QH, Khoury MJ, Mannino D. Trends and patterns of mortality associated with birth defects and genetic diseases in the United States, 1979–1992: an analysis of multiple-cause mortality data. *Genet Epidemiol* 1997;**14**:493–505.

1114. Ybot–Gonzalez P, Copp AJ. Bending of the neural plate during mouse spinal neurulation is independent of actin microfilaments. *Dev Dyn* 1999;**215**:273–83.

1115. Yock DH Jr. Choroid plexus lipomas associated with lipoma of the corpus callosum. *J Comput Assist Tomogr* 1980;**4**:678–82.

1116. Yoshida A, Kobayashi K, Manya H, *et al.* Muscular dystrophy and neuronal migration disorder caused by mutations in a glycosyltransferase, POMGnT1. *Dev Cell* 2001;**1**:717–24.

1117. Zaremba J, Wislawski J, Bidzinsky J, *et al.* Jadassohn's naevus phakomatosis: a report of two cases. *J Ment Def* 1978;**22**:91–102.

1118. Zemlan FP, Thienhaus OJ, Bosmann HB. Superoxide dismutase activity in Alzheimer's disease: possible mechanism for paired helical filament formation. *Brain Res* 1989;**476**:160–62.

1119. Zheng BH, Sage M, Cai WW, *et al.* Engineering a mouse balancer chromosome. *Nat Genet* 1999;**22**:375–8.

1120. Ziegler AL, Deonna T, Calame A. Hidden intelligence of a multiple handicapped child with Joubert syndrome. *Dev Med Child Neurol* 1990;**32**:261–6.

1121. Zimmerman LB, De Jesús–Escobar JM, Harland RM. The Spemann organizer signal noggin binds and inactivates bone morphogenetic protein 4. *Cell* 1996;**86**:599–606.

1122. Ziter F, Bramwit D. Nasal encephaloceles and gliomas. *Br J Radiol* 1970;**43**:136.

1123. Zonana J, Sotos JF, Romshe CA, *et al.* Dominant inheritance of cerebral gigantism. *J Pediatr* 1977;**91**:251–6.

Metabolic and Neurodegenerative Diseases of Childhood

Thomas S Jacques and Brian N Harding

INTRODUCTION

In this chapter, childhood disorders that do not fit neatly into other chapters are reviewed. They are a mixture of single gene, metabolic and neurodegenerative diseases. The chapter has four parts: grey matter/neuronal disorders, white matter disorders, amino acid and related disorders and miscellaneous metabolic disorders.

NEURONAL DISORDERS

Alpers–Huttenlocher Syndrome

Definition

Alpers syndrome (progressive neuronal degeneration of childhood with liver disease [PNDC]) is an autosomal recessive disorder defined by combined cerebral cortical and hepatic degeneration.

Genetics

Although the cause has been subject to historical uncertainty, PNDC is now regarded as the most severe manifestation of mitochondrial DNA depletion.[343]

In genetically confirmed cases, it is typically caused by homozygous or compound heterozygous mutations in the nuclear gene that encodes for mitochondrial DNA polymerase gamma (*POLG*).[84,111,239,240]

Related hepatocerebral syndromes can be caused by mutations in other genes, such as the nuclear gene *C10orf2* that encodes for twinkle and twinky proteins.[309] Similar neuropathology has been reported in patients with combined oxidative phosphorylation deficiency-14 (COXPD14; OMIM: 614946; http://www.ncbi.nlm.nih.gov/omim) due to mutation in the nuclear gene *FARS2* (OMIM: 611592) that encodes for phenylalanyl-tRNA synthetase-2.[103]

Clinical Features

One Swedish study found an overall incidence of mitochondrial encephalomyopathies in preschool children (<6 years) of 1 in 11 000, with the incidence of Alpers syndrome in particular to be 1 in 51 000.[82] There are many instances of sibling involvement (12 of 26 families in Harding's series,[134] including one pair of twins and two sets of three siblings) in keeping with autosomal recessive inheritance.

Patients typically present in the first 2 years of life.[307,341] The onset is usually insidious, with seizures and developmental stagnation, although presentation with generalized or focal status epilepticus can occur. With progression, epilepsy becomes prominent, including myoclonus and other seizure types, especially partial continuous epilepsy. Dementia, blindness and spasticity develop and, usually late in the course, there is progressive liver failure. The electroencephalogram (EEG) is often helpful, with large slow waves and asymmetrical runs of polyspikes. Visual evoked potentials (VEPs) are often asymmetrical early on. Neuroimaging shows progressive cerebral atrophy, especially of the occipital lobes, without signal change.

Pathology

Systemic Pathology

Hepatic disease (Figure 5.1) is constant. In a small minority of patients, there is only fatty change, occasionally mid-zonal necrosis,[136] but most have a characteristic pathological appearance: severe and diffuse micro-vesicular fatty change, portal inflammation, hepatocyte necrosis and collapse of liver cell plates, massive haphazard bile duct proliferation or transformation, and bridging fibrosis and nodular cirrhosis. Patchy oncocytic change is sometimes observed and its sharp contrast with adjacent groups of pale fat-laden cells produces a characteristic 'moth-eaten' appearance. In some instances, however, there is merely destruction and fibrosis of the liver. Examination of sequential biopsies

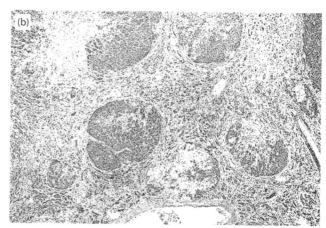

5.1 Hepatic pathology in Alpers disease. (a) Nodular cirrhosis in a post-mortem specimen. **(b)** Histological features include architectural disarray, bile duct proliferation, nodular regeneration and fat deposition, yielding a 'moth-eaten' appearance within cirrhotic nodules.

followed by autopsy in the same patient has demonstrated progression from focal necrosis, inflammation and fatty change through nodule formation to the full-blown cirrhosis, bile duct proliferation and fat accumulation.[134]

It has been suggested that valproate may exacerbate the disease, hastening its progression.[150,283] Indeed mutations in *POLG* are present in some patients with valproate-induced hepatotoxicity but no known PNDC.[236,337] Valproate is therefore contraindicated in PNDC.[151] However, similar pathological changes occur irrespective of treatment with sodium valproate. Indeed, clinical, histologic and biochemical evidence of liver disease occur before anticonvulsant therapy.[38,102]

Neuropathology

Macroscopic appearances may be normal even on close examination, but characteristically there are patches of thinned, granular discoloured cortex, and sometimes laminar disruption. Selective involvement of the medial occipital lobe, especially the calcarine cortex, is typical (Figure 5.2). Therefore, extensive sampling of this region is recommended in suspected cases. Cerebral white matter and deep grey nuclei are usually unremarkable, but circumscribed softening of the occipital white matter has been documented in one adult patient.[139] In one extreme example, global cerebral atrophy was present.

Microscopic changes (Figure 5.3) are usually more widespread than the naked-eye appearances suggest. The changes are patchy and do not respect vascular territories or sulcal topography. The distribution is particularly helpful in distinguishing the changes from ischaemia. The pathological process, devoid of specific cytological markers or inflammatory changes, shows graded intensification through the depths of the cortical ribbon. The mildest lesions demonstrate superficial astrocytosis, but more severely affected areas show increasing vacuolation, neuronal loss and astrocytosis spreading down through the ribbon until the whole cortex is thin, replaced by hypertrophic astrocytes, glial fibres and prominent capillaries. In the end, there is a glial scar, often demonstrating considerable neutral fat in frozen sections. If liver failure is present, Alzheimer type II astrocytes may be prominent.

The striate cortex bears the brunt of the damage: in one series, the visual cortex showed complete devastation in 17 of 22 cases, and in only three was it affected less severely than other cortical areas.[134] The two hemispheres need not be affected equally. White matter changes, by contrast, are usually slight. Neuronal loss, vacuolation and astrocytosis can involve the thalamus, lateral geniculate body, amygdala, substantia nigra and dentate nuclei. Hippocampal sclerosis and focal or diffuse cerebellar cortical lesions are common, whereas neuronal loss in the brain stem and tract degeneration in the spinal cord are occasional findings.

Epileptic Encephalopathies

Definition

Epileptic encephalopathies are disorders with epilepsy and progressive neurological deficits where it is hypothesized that the seizures themselves contribute to the progressive disturbance in cerebral function.[104] Over the last few years, many genes have been identified that cause early onset epileptic encephalopathies (early infantile epileptic encephalopathy [EIEE]) (Table 5.1). The neuropathology has only been described in a relatively small number of cases.[248,273] EIEE can also be occur in a range of other genetic disorders, including GLUT1 deficiency syndrome, glycine encephalopathy, Aicardi–Goutières syndrome and in males with *MECP2* mutations; it is frequently associated with malformations.[273]

Pathology

Dravet Syndrome

Dravet syndrome, or severe myoclonic epilepsy of infancy, is characterized by onset in early life (usually in the first year) with febrile hemiclonic or generalized seizures in a previously well child.[61,96,289] These are followed by the development of multiple seizure types and developmental regression. Dravet syndrome can be caused by mutations in the *SCN1A* gene that encodes for a subunit of a voltage-gated sodium channel.[66,96] *SCN1A* mutations are associated with a spectrum of epileptic disorders that includes Dravet

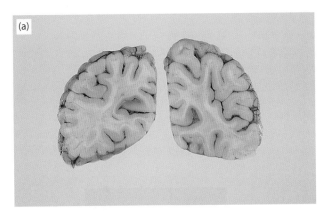

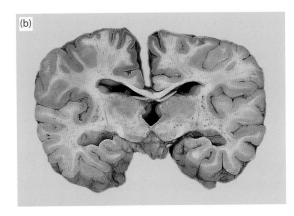

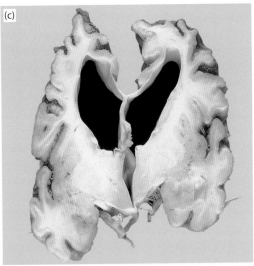

5.2 Alpers disease: macroscopic features. (a) Patches of granular and brown shrunken cortex mostly involve medial (calcarine) occipital cortex. **(b)** Patchy lesions involving the frontal cortex. **(c)** A rare case showing global cortical destruction.

syndrome, generalized epilepsy with febrile seizures plus type 2 (GEFS+2)[106] and some cases of migrating partial seizure of infancy.[56,117]

Patients are at significant risk of early death including sudden unexplained death in infancy.[120,166,190] In a series of paediatric and adult post-mortem cases, there were frequent cerebellar atrophy, infrequent hippocampal sclerosis, normal cortical development and occasional myelopathy affecting the dorsal columns.[61] In contrast, in a separate report of a case of sudden death in Dravet syndrome, there was a 'multifocal micronodular dysplasia' of the temporal lobe on one side and bilateral end folial sclerosis.[190] In addition, there are rare reports, predating genetic investigation, one of which describes a spinal cord malformation.[289]

Migrating Partial Seizures of Infancy (Epilepsy of Infancy with Migrating Focal Seizures)

This syndrome is characterized by the onset before 6 months of age of partial seizures that on EEG show changing ('migrating') ictal foci. The disease becomes complicated by developmental delay and regression and has a poor clinical outcome. It can be caused by gain of function mutations in the *KCNT1* gene, which encodes for a potassium channel,[26] or with *SCN1A* mutations (see earlier),[56,117] *PLCB1* (beta-1 phospholipase C) deletions[276] or mutations in *SLC25A22* (which encodes a mitochondrial transporter).[277]

The reported neuropathology shows much variation. In two cases, we found hippocampal sclerosis, one associated with granule cell dispersion. Both patients had some generalized cerebral volume loss, but one had particularly severe damage to the putamina with neuronal loss, atrophy and gliosis.[214]

Two of the cases from the first description of the disease came to autopsy and in both cases, the main neuropathological findings was hippocampal sclerosis.[72] There is a report of a patient with no significant neuropathological findings.[384] There is one report of a patient who in addition to hippocampal sclerosis was described as having focal cortical dysplasia, type IIIa and polymicrogyria.[108]

CDKL5-related Disorders

The *CDKL5* gene is an X-linked gene, mutations in which cause a diverse range of epileptic syndromes that include atypical Rett syndrome, early onset epileptic encephalopathies and milder forms.[22,60,313,376] We have described a child who suffered from SUDEP (sudden unexpected death in epilepsy) with a proven *CDKL5* mutation and a syndrome of infantile spasms and Rett-like syndrome. The neuropathology shows subtle malformations (prominent brachycephaly, small superior temporal gyri and heterotopias in the vermis) and prominent Purkinje cell degeneration as demonstrated by frequent axonal torpedoes.[263]

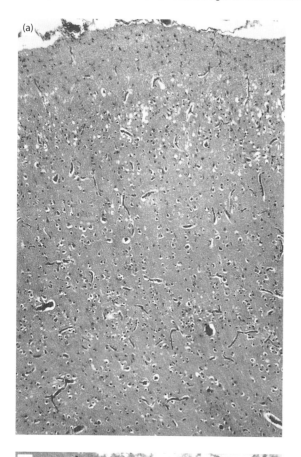

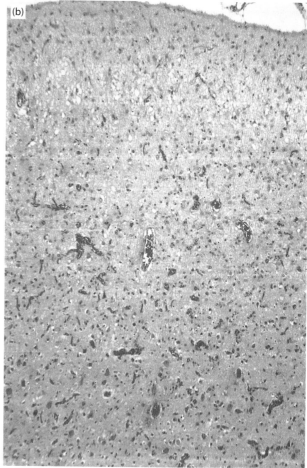

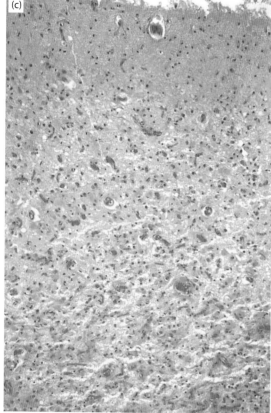

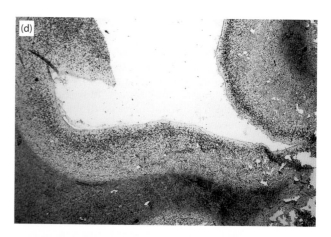

5.3 Alpers disease: CNS histology. (a) At its mildest, there are superficial spongy changes and astrocytosis in layers I and II. **(b)** Neuronal loss spreads gradually more deeply, until **(c)** the whole cortical ribbon is affected and **(d)** sudanophilic lipid is readily demonstrated in frozen sections.

TABLE 5.1 Major genetic loci associated with EIEE (see Nordli[248] and Pavone *et al.*[273] and OMIM)

Disorder	Synonyms	Gene	Protein
EIEE1	X-linked infantile spasm syndrome 1 (ISSX1)	*ARX*	Aristaless-related homeobox protein
EIEE2	X-linked infantile spasm syndrome 2 (ISSX2)	*CDKL5*	Cyclin-dependent kinase like-5 (serine/threonine protein kinase (STK9)
EIEE3		*SLC25A22*	Solute carrier family 25 (mitochondrial carrier, glutamate), member 22
EIEE4		*STXBP1*	Syntaxin binding protein 1
EIEE5		*SPTAN1*	Spectrin, alpha, non-erythrocytic 1
EIEE6	Dravet syndrome, severe myoclonic epilepsy of infancy, mutations in the same gene cause generalized epilepsy with febrile seizures +2 (GEFS+2)	*SCN1A*	Sodium channel, neuronal type I, alpha subunit
EIEE7		*KCNQ2*	Potassium channel, voltage-gated, KQT-subfamily, member 2
EIEE8	Hyperekplexia and epilepsy	*ARHGEF9*	Rho guanine nucleotide exchange factor 9
EIEE9	Epilepsy, female-restricted, with mental retardation, Juberg–Hellman syndrome	*PDCH19*	Protocadherin 19
EIEE10	Microcephaly, seizures and developmental delay	*PNKP*	Polynucleotide kinase 3-prime phosphatase
EIEE11		*SCN2A*	Sodium channel, voltage-gated type II, alpha subunit
EIEE12		*PLCB1*	Phospholipase C, beta-1
EIEE13		*SCN8A*	Sodium channel, voltage-gated type VIII, alpha subunit
EIEE14		*KCNT1*	Potassium channel, subfamily T, Member 1
EIEE15		*ST3GAL3*	ST3 beta-galactosidase alpha-2, 3-sialyltransferase 3
EIEE16		*TBC1D24*	TBC1 domain family, member 24

STXBP1

The *STXBP1* gene encodes for a syntaxin-binding protein that is involved in synaptic vesicle release. Mutations cause an early onset epileptic encephalopathy.[305] The pathology of one patient who underwent surgery for epilepsy has been reported to show prominent micro-columnar architecture considered to be focal cortical dysplasia, type Ia.[377]

Neuronal Brain Iron Accumulation Disorders (Axonal Dystrophies)

Definitions

Neurodegeneration with brain iron accumulation (NBIA) represents a series of genetic disorders characterized by iron accumulation in the brain and frequently axonal dystrophy. The major forms have previously been described under the names Hallevorden–Spatz Syndrome (HSS) and infantile neuroaxonal dystrophy (iNAD). However, the identification of the responsible genes, pantothenate kinase 2 (*PANK2*)[395] and phospholipase A2, Group VI (*PLA2G6*)[126,171,227] has redefined the field. The major forms are now regarded as pantothenate kinase-associated neurodegeneration (PKAN) and PLA2G6-associated neurodegeneration (PLAN). A differential diagnosis of NBIA is given in Table 5.2. The discussion is restricted to the two common forms that present in childhood.

Pathogenesis

The pathogenesis of these disorders is poorly understood. Pantothenate kinases catalyze the conversion of pantothenate (vitamin B₅) to phosphopantothenate, the first step in the synthesis of coenzyme A (CoA), a crucial acyl carrier in a large number of enzyme reactions. PANK2 localizes to mitochondria of neurons, particularly within the cortex, globus pallidus and the basal nucleus of Meynert,[180] raising the possibility of a defect of mitochondrial lipid synthesis. There are a number of metabolic defects including abnormalities of mitochondrial function in PKAN.[196] Fly models suggest that a defect in tubulin and histone acetylation may be important.[323]

PL2AG6 encodes for a group VI A calcium-independent phospholipase A2 known as calcium-independent phospholipase A2 beta (iPLA2β). The enzyme hydrolyses the sn-2 acyl chain of phospholipids to fatty acid and phospholipids and may have a role in remodelling membrane phospholipids and related signalling pathways.[316]

TABLE 5.2 Major neurodegenerative syndromes associated with brain iron accumulation[141,316]

Disease	Clinical syndromes	Gene	Gene product
NBIA type 1	Hallevorden–Spatz syndrome/ PANK-associated neurodegeneration (PKAN)	*PANK2*	Pantothenate kinase 2
NBIA type 2	PLA2G6 associated neurodegeneration (PLAN). Infantile neuroaxonal dystrophy (iNAD). PARK14	*PLA2G6*	Phospholipase A2, group VI
FAHN	FA2H-associated neurodegeneration (FAHN)/SPG35	*FA2H*	Fatty acid 2 hydroxylase
Mitochondrial membrane associated neurodegeneration	MPAN	*C19orf12*	C19orf12
Kufor–Rakeb disease	PARK9	*ATP13A2*	Lysosomal ATPase
Aceruloplasmin		*Ceruloplasmin*	Ceruloplasmin
Neuroferritonopathy		*FTL*	Ferritin light chain
Static encephalopathy (of childhood) with neurodegeneration in adulthood (SENDA syndrome)		Not known	
Idiopathic cases			

Interestingly, recent data suggest that mutations associated with iNAD cause a reduction in enzyme activity; in contrast, mutations in the same gene that cause dystonia-parkinsonism (PARK14), preserve enzyme activity.[105]

Clinical Features

PKAN

Classic PKAN is a progressive disorder with onset before 6 years of age, often with gait difficulties.[13,141,142,316] The clinical picture is then complicated by both pyramidal and extrapyramidal symptoms, such as parkinsonism. Orofacial dyskinesia may be prominent. Eye movement abnormalities are common. Cognitive and behavioural abnormalities may occur. A pigmentary retinopathy can also be encountered. T2-weighted MR images show central hyperintensity within a surrounding area of hypointensity centred on the anterior globus pallidus (the 'eye of the tiger' sign). Adult onset cases may have atypical clinical features, e.g. with less prominent motor symptoms.

PLAN

Early onset cases usually present early in the second year of life.[227,264,316] Development stagnates, severe truncal hypotonia develops and a squint or visual failure appears. The disorder then progresses with dementia, visual failure, spastic tetraparesis and areflexia, and amyotrophy developing before 4 years of age. Seizures are not a prominent feature. Later onset may be atypical. A syndrome of later onset parkinsonism with dystonia may occur with some *PLA2G6* mutations (PARK14).

Pre-genetic Neuropathology

Most pathological descriptions predate the identification of responsible genes.

Both PKAN and PLAN are associated with iron accumulation and dystrophic axonal spheroids. Dystrophic axonal swellings may occur in the central nervous system (CNS) and peripheral nervous system (PNS) under many conditions, both physiological as part of the normal ageing process in the gracile and cuneate nuclei, the substantia nigra and pallidum, and pathological, either as the core feature of a disease or as a secondary reactive phenomenon.

The axonal swellings (Figure 5.4) on which diagnosis depends range in diameter from 20 to 120 μm, are round, oval or elongated in outline, and stain quite variably with eosin from a subtle or granular pallor to a dense eosinophilia. Sometimes, a dense core and peripheral pallor or a narrow cleft can be seen. Silver methods readily demonstrate the spheroids, but with irregular intensity, vacuolation or whorling. In brain biopsies (now performed rarely), small cortical spheroids are elusive on light microscopy. Spheroids are immunoreactive with neurofilament and ubiquitin antibodies, although larger swellings (>30 μm) may be unreactive.

With electron microscopy (Figure 5.5), the swellings, developing preferentially in terminal axons and presynaptic endings, are filled with various organelles including synaptic vesicles, but most characteristically granulovesicular and tubulomembranous material, sometimes with a paracrystalline appearance, often with a central cleft.[88] This is the method of choice for definitive histological diagnosis and can be undertaken in PLAN from various tissues, including the brain, peripheral nerve, muscle, conjunctiva, rectum and skin (for preference near sweat glands, where nerve terminals are plentiful).[21,172,260,385]

Pre-genetic descriptions of what were termed iNAD, report that the macroscopic appearances (Figure 5.6) vary from normal to severe cerebral atrophy, but cerebellar atrophy is common. Axonal spheroids are present widely in the brain, spinal cord and PNS, but are particularly abundant

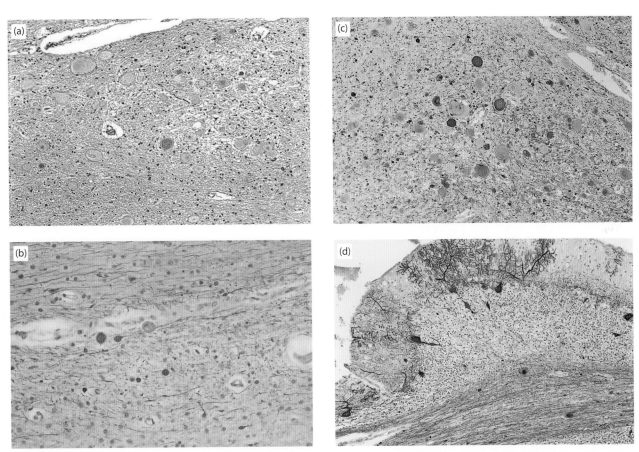

5.4 Infantile neuroaxonal dystrophy. Microscopic appearances of dystrophic axonal swellings in the posterior columns identified on **(a)** haematoxylin and eosin, **(b)** Glees silver method, and **(c)** anti-neurofilament immunohistochemistry (note that large swellings >35μm in diameter stain only irregularly). **(d)** In the cerebellum, anti-neurofilament immunohistochemistry demonstrates Purkinje torpedo swellings in the cortex and dystrophic axons in the deep white matter.

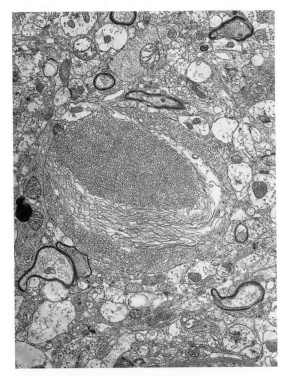

5.5 Infantile neuroaxonal dystrophy. Electron micrograph of cerebral biopsy. The axonal spheroid is bound by a single membrane and packed with tubules cut in transverse or longitudinal planes.

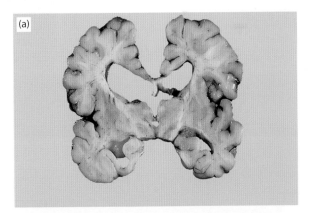

5.6 Infantile neuroaxonal dystrophy. (a) Marked cerebral cortical atrophy and ventriculomegaly, with **(b)** striking shrinkage of cerebellar folia.

in the brain stem tegmentum, posterior columns and spinal grey matter, cerebellar grey and white matter, pallidum, thalamus, substantia nigra, hypothalamus and the cerebral cortex. Cerebellar cortical atrophy and myelin pallor are other features. Patients with neonatal onset may show prominent cerebellar or brain stem involvement.[161,169,346]

In the pre-genetic age, instances of neuroaxonal dystrophy with late infantile or juvenile onset caused nosological difficulties, particularly as in many patients the preferential involvement of the pallidum and substantia nigra and the presence of iron-containing pigment in addition to spheroids suggested a close similarity to more typical cases of Hallervorden–Spatz syndrome (Figure 5.7).

Neuropathology of Genetically Confirmed Cases

PKAN

Kruer *et al.*[181] described six cases of genetically confirmed PKAN, where the pathology was restricted to the globus pallidus with some extension into adjacent structures, but not the more extensive pathology seen in genetically confirmed *PL2AG6* cases. Macroscopically, the major abnormality was a rust-coloured atrophic globus pallidus. The pathology consisted of axonal spheroids, degenerating neurons, iron deposition and gliosis. Notably, in contrast to some of the pre-genetic descriptions, synuclein pathology was not seen. The degenerating neurons stained for ubiquitin and some showed faint staining for tau.

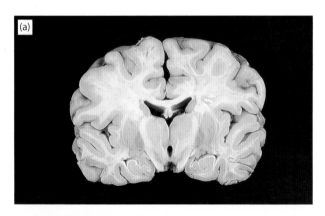

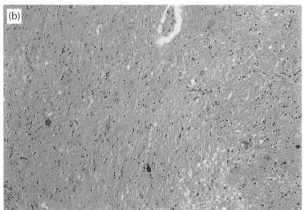

5.7 NBIA of the PKAN type. (a) Characteristic yellow-brown discolouration and atrophy of globus pallidus. **(b)** Axon spheroids, calcospherites and brown pigment deposits in the globus pallidus.

Li *et al.*[199] described three additional cases, confirming the relative restriction to the globus pallidus and medial putamen, with synucleinopathy not a feature. They also suggested that a population of spheroids in the pallidum were similar to ovoid bodies seen in areas of chronic haemorrhage.

PLAN

In a series of genetically confirmed cases, peripheral pathology was widespread with 87 per cent of cases having axonal dystrophy in peripheral nerve samples.[126] One of the cases in this series came to post-mortem analysis and showed widespread atrophy of the cerebral cortex, white matter, basal ganglia and cerebellum. There was brown discolouration of the substantia nigra and globus pallidus. In contrast to the PKAN cases, there was widespread neuronal loss and gliosis and axonal spheroids at multiple sites. There was cerebellar degeneration. There was iron deposition in the substantia nigra and globus pallidus. Also in contrast to PKAN cases, there was extensive synuclein pathology and tau-related pathology.

Paisan-Ruiz *et al.*[265] confirmed these findings in a further five cases.

Recent data indicate that there are clinical abnormalities of olfaction in NBIA and that there are pathological abnormalities of the olfactory bulb with ferritin deposition.[98]

Rett Syndrome

Definition

Rett syndrome is a neurodevelopmental disorder presenting between 6 and 18 months, almost always affecting girls and characterized by developmental arrest and regression, loss of speech and stereotypical movements, particularly of the hands.[241]

Genetics

Approximately three-quarters of patients with Rett syndrome, typical or atypical (retained speech or without microcephaly), have *de novo* mutations in the *MECP2* gene,[11] making this an X-linked dominant disorder. Mutations in *CDKL5* can produce a Rett-like syndrome[64,107,208,303,313] and mutations in *FOXG1* cause a congenital form.[15]

Pathogenesis

MECP2 encodes a methyl-CpG-binding protein that recruits histone deacetylase to repress gene transcription.[81,170] The pathogenesis of Rett syndrome and the complex relationship between genotype and phenotype are poorly understood, but have been significantly enhanced by a number of transgenic models.[62] MeCP2 is required for adult neural function as shown by development of a Rett syndrome-like phenotype with selective knockout of the gene in mature mice.[213] MeCP2 has a major role in neuronal regulation and at least in part is activated in relation to neuronal activity.[100] Indeed, in a model using ES cell-derived neurons, there is a global reduction in transcription.[202] Furthermore, *MECP2* mutations increase expression of retrotransposons in developing neural progenitors, raising the intriguing possibility that mutations in *MECP2* may cause secondary somatic mutations during neural development.[233] However, MeCP2 may also function in glial cells as shown by the rescue of the knockout phenotype by wild-type astrocytes or microglia.[93,203]

Clinical Features

This disorder almost exclusively affects girls.[152,241] Although early development is not entirely normal, first symptoms are usually noticed between 6 and 18 months of age and consist of developmental stagnation, hypotonia, retarded head growth and disinterest. This phase is followed after a few months by rapid regression, with irritability, loss of purposeful hand movements, and development of hand stereotypies (e.g. 'knitting' movements, fingers to mouth) and loss of communication (language and social). After a few weeks of regression, girls enter a more stable stage that lasts several years, with severe learning difficulties, retarded head growth (or frank microcephaly), hand stereotypies, ataxia or apraxia and an abnormal respiratory pattern. Patients often develop epilepsy and scoliosis. The final stage develops after the age of 10 years. It consists of upper and lower motor neuron signs, decreased motility, progressive scoliosis and, often, reduced seizures.

Pathology

Morphological abnormalities are subtle. They have been reviewed extensively by Armstrong.[16] The most obvious changes are in somatic and brain growth. Decreased head growth is observed in the first few months of life, followed by slowing of body growth in terms of both height and weight. Particularly small feet in girls with Rett syndrome have been attributed to autonomic dysfunction. All organs except for the adrenal glands weigh less than normal for age, but in relation to body length only the brain is small – around 900 g and static – without a significant increase or progressive decline after 4 years of age.[19]

In the first neuropathological examination, Jellinger and Seitelberger[162] noted reduced melanin in the pars compacta of the substantia nigra, which was confirmed in later studies and was associated with a mild reduction in tyrosine hydroxylase.[163]

Studies of the cerebral cortex have indicated a range of cytoarchitectural changes including reduced neuronal size and increased packing density,[31] reductions in pyramidal neurons, dendritic spines, the numbers of afferent axons and in synaptophysin expression.[33] There is a significant reduction in dendritic branching in frontal motor and temporal cortex, even greater than that seen in the brain in Down syndrome.[17,18] Abnormalities of columnar architecture have also been reported.[59]

Other findings include variable pathology in the basal ganglia and hippocampus,[97,163,197] progressive loss of cerebellar Purkinje cells[253] and abnormalities of the olive.[31] Degenerative changes are described in the white and grey matter of the spinal cord.[252]

A distal axonal neuropathy and abnormalities of spinal ganglia have been described.[164,252] A range of changes has been described in muscle biopsies (reviewed in Duncan Armstrong[97]).

There is clinical evidence of autonomic dysfunction. Abnormalities of substance P and monoamines in the brain stem have been implicated in the pathogenesis of this clinical problem.[89,271,299]

LEUKODYSTROPHIES AND LEUKOENCEPHALOPATHIES

Approach to Leukodystrophies

A distinction is sometimes made between leukoencephalopathies, defined as diseases of the brain predominantly involving white matter, and leukodystrophies, diseases characterized by a failure to form or maintain myelin. The pathological literature often predates genetic confirmation and there has been a significant expansion of the number of genes and disorders, many of which do not yet have well-described pathology. However, some white matter diseases have distinctive pathological features, which are summarized in Table 5.3.

Alexander Disease

Definition

This disorder is characterized by diffuse or focal demyelination in the presence of astrocytic inclusions, Rosenthal fibres. These are formed by intermediate filaments composed of glial fibrillary acidic protein (GFAP) conjugated with abnormally phosphorylated and partly ubiquitinated heat-shock protein, α-B-crystallin.

Genetics

Alexander disease is caused by heterozygous mutations in the *GFAP* gene.[46,234,311] The mutations usually cause amino acid substitutions affecting the conserved rod domains. Most mutations are *de novo*, but there are some families with autosomal dominant inheritance.[234]

Pathogenesis

Alexander disease appears to be the first example of a primary genetic disease in astrocytes.[46,218] Following the observation that overexpression of human GFAP in a transgenic mouse is lethal and produces inclusion bodies indistinguishable from Rosenthal fibres, direct sequence analysis of DNA from patients with Alexander disease identified *de novo* dominant gain-of-function mutations in the gene encoding GFAP.[297] Subsequently, mouse models mimicking the human mutations have been described (reviewed in Messing et al.[219]), but notably loss-of-function mice do not develop features of the disease. Indeed, most pathologically proven cases of Alexander disease are associated with *de novo* dominant gain-of-function mutations.[219] It is likely that overexpression of GFAP is a common consequence of the mutations. The phenotype appears to be modified by heat shock proteins, often co-localized with Rosenthal fibres.[219]

TABLE 5.3 The differential diagnosis of developmental white matter diseases and their most distinctive pathological findings

Disease	Distinctive pathological features	Presumed pathogenesis
Adrenoleukodystrophy	Patchy perivascular demyelination with chronic inflammation, swollen striated cells in adrenal cortex	Peroxisomal disorder
Adult onset leukoencephalopathy with axonal spheroids and pigmented glia	Axonal spheroids, pigmented glia	Disorder of microglia
Aicardi–Goutières syndrome	Calcification, vasculopathy	Overproduction of alpha-interferon
Alexander disease	Frequent Rosenthal fibres	Disorder of astrocytes
Canavan disease	Spongiform leukodystrophy	Amino acid disorder
Cockayne syndrome	Dysmorphic features, cerebral atrophy, calcification, discontinuous myelin loss	DNA repair disorder
Krabbe disease	Globoid cells	Lysosomal storage disorder
Megalencephalic leukoencephalopathy with subcortical cysts	Megalencephaly, cyst formation	Uncertain pathogenesis possibly due to regulation of osmotic balance
Metachromatic leukodystrophy	Metachromatic storage material	Lysosomal storage disorder
Pelizaeus–Merzbacher disease	Selective involvement of CNS myelin, discontinuous myelin loss	Disorder of the major protein component of myelin
Vanishing white matter disease	Cystic loss of white matter without gliosis	Uncertain but possibly due to over-activation of the unfolded protein response

Clinical Features

Most clinical descriptions recognize three forms (infantile, juvenile, adult).[8,92,200,231,235,243,284,302,311,312,348,365] However, alternative clinical classifications have been proposed.[284,391] Infantile is the typical form, presenting before 2 years of age with megalencephaly and seizures and ultimately progressing to quadriparesis. Hydrocephalus may be a presenting feature and may be associated with stenosis of the cerebral aqueduct.[8] The juvenile form shows a more bulbar and cerebellar presentation with later cognitive impairment. The adult form shows slow progression with ataxia, quadriparesis, an eye movement disorder, palatal myoclonus and late cognitive decline. Finally, the later onset forms may present with a posterior fossa mass and the differential diagnosis with pilocytic astrocytoma may be problematic even after biopsy.[365,371]

Pathology

Because of its rarity, descriptions rely upon individual case reports following Alexander's original observation,[8] a few reviews[45,282,302] and personal experience. Most of these descriptions predate genetic confirmation. Megalencephaly, a characteristic presenting sign in early infancy, may give way to atrophy by the time of autopsy. Macroscopic changes may be minimal, particularly juvenile-onset cases dominated by bulbar signs, but infantile examples usually present with varying degrees of ventricular dilation and white matter atrophy or destruction (Figure 5.8a). Hydrocephalus may be striking in younger patients and related to aqueduct compression.[8,118,353] The cerebral white matter appears granular, yellow, sunken or cavitated (Figure 5.8a); this may be extensive or more confined to anterior parts of the frontal and temporal lobes. Grey matter and hindbrain structures are usually unremarkable.

The principal microscopic feature is extensive accumulation of Rosenthal fibres (Figure 5.8b) throughout the brain. Demyelination is variable, being most extensive in infantile cases, where it largely parallels in intensity the degree of Rosenthal fibre formation in the white matter. Fat-laden macrophages, cyst formation, occasional perivascular lymphocytic cuffs, and calcospherites may also occur. Demyelination is less prominent in juvenile cases. With conventional stains, the rod- and club-shaped hyaline eosinophilic structures isolated from cell bodies and identical to those first described by Rosenthal in 1898, differ only in degree and extent from those of other lesions, including low-grade tumours and cortical dysplasia. Ultrastructurally, they comprise amorphous osmophilic granular material surrounded by 10-nm filaments; immunocytochemistry demonstrates peripheral staining with antibodies to GFAP, α-B crystallin, 27 kDa heat-shock protein and ubiquitin, although the centre is left unstained. Rosenthal fibres are most numerous in subependymal, subpial and perivascular locations (Figure 5.8b) with marked involvement of cerebral white matter, thalamus and basal ganglia; the cerebral cortex is largely spared. Rosenthal fibre formation may be intense in brain stem nuclei and tracts, as well as the cerebellar white matter and dentate nucleus, but usually the cerebellar cortex is spared. Illustrated in Figure 5.8e is a rare exception where Rosenthal fibres developed in the cerebellar cortex and penetrated the leptomeninges. Rosenthal fibre formation is also notable in the proximal parts of the optic nerves and both grey and white matter of the spinal cord.

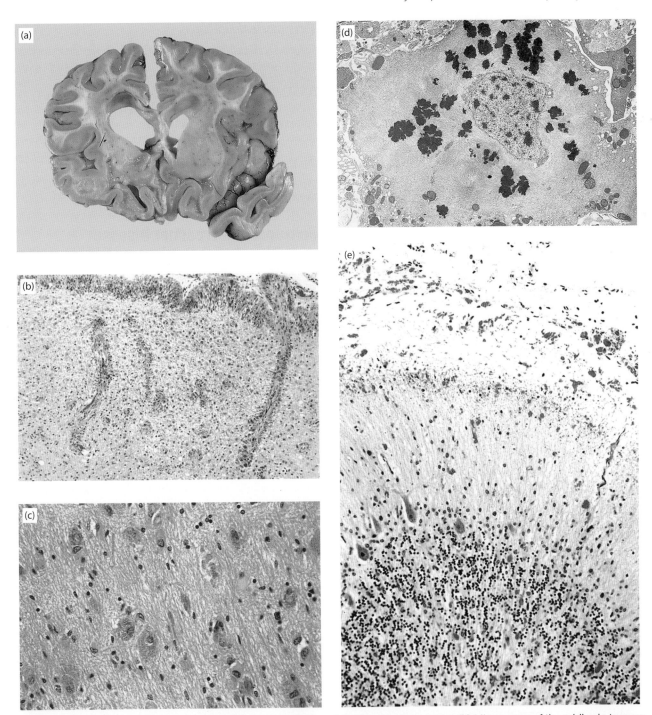

5.8 Alexander disease. (a) Gelatinous dissolution and cavitation of the frontal white matter. **(b)** Microscopy of the midbrain tegmentum, showing myriad Rosenthal fibres beneath the pia and surrounding blood vessels. **(c)** In younger patients, typical Rosenthal fibres may be lacking, but eosinophilic granules fill the juxtanuclear cytoplasm. **(d)** Ultrastructurally, these intracytoplasmic structures are osmiophilic densities coated with intermediate filaments, i.e. miniature Rosenthal bodies. **(e)** Very unusual involvement of the cerebellum, with massive Rosenthal fibre deposition in the cortex and the leptomeninges.

In general, neurons are preserved, even in severely affected areas, but cerebellar degeneration has been described. An important feature in infantile cases is the presence of swollen astrocytes containing small hyaline granules within their cell bodies, sometimes forming a peripheral ring in the cytoplasm (Figure 5.8c); and identical ultrastructurally (Figure 5.8d) to Rosenthal fibres. These cells can show nuclear atypia and mitotic activity suggesting a neoplasm, particularly in a biopsy.

Megalencephalic Leukoencephalopathy with Subcortical Cysts

Definition

Megalencephalic leukoencephalopathy (MLC) is a rare genetic disorder characterized by a large head and a leukoencephalopathy with progressive cyst formation.

Genetics

The disorder is inherited in an autosomal recessive manner. Two genes have been described in three forms of the disease: MLC1 caused by mutation in *MLC1* and MLC2A and 2b caused by mutations in *HEPCAM* (*GliaCAM*)(reviewed in van der Knaap *et al.*[366]).

Pathogenesis

The pathogenesis has been reviewed[206,366] and is hypothesized to represent a defect in the regulation of osmotic pressure by astrocytes causing white matter oedema.

Clinical Features

In the classic form, asymptomatic megalencephaly develops during the first year of life.[366] Neuroimaging at this stage reveals a diffuse abnormality of cerebral hemispheric white matter, and there may already be cysts in temporal and parietal lobes. Onset of a spastic–ataxic motor disorder usually occurs by the fifth year of life, but occurs later exceptionally. The motor disorder is progressive, and seizures and dementia develop later, usually in the teenage years. Atypical cases associated with *HEPCAM* mutations (MLC type 2B) may be static or improve clinically.

Pathology

Knowledge of the neuropathology of this condition is limited to a handful of cases.

Biopsy Findings

A spongiform leukoencephalopathy with gliosis, intact myelinated fibres and many oil red O-positive macrophages was described by van der Knaap *et al.*[357] Ultrastructurally, there were many vacuoles containing debris and surrounded by five-layered membranes equivalent to myelin lamellae; splitting of myelin at the intraperiod line and vacuolation in the outer part of the myelin sheath was also observed. The large vacuoles were positive for myelin basic protein (MBP).

By contrast, in a biopsy from one of two affected siblings, the white matter was rarefied and pale staining, but not spongy.[133] On electron microscopy, there was a hugely increased extracellular space and abnormally thin myelin surrounding some axons. The myelin was normally compacted.

Miles *et al.*[222] reported the findings in a biopsy from a patient with proven *MLC1* mutations. They found frequent uniform white matter vacuoles (2–4 μm), which were mostly empty and lined by a single-layer membrane. However, some vacuoles contained organelles (mitochondria) or were lined by two- or three-layered membranes. They noted blebs on the outer layer (and sometimes inner layer) of the myelin and suggested these might be the source of the vacuoles.

Finally, Pascual-Castroviejo *et al.*[269] described the biopsy findings in an additional patient with proven *MLC1* gene mutations. This showed fine vacuolation of the white matter. On electron microscopy, there were increased extracellular spaces, decreased numbers of axons, and thin myelin sheaths lacking intramyelinic oedema.

Post-mortem Findings

Lopez-Hernandez and colleagues[206] have described the post-mortem neuropathology of a patient with homozygous *MLC1* mutations. They found reduced white matter, particularly in frontal and temporal lobes. There was relative sparing of the U-fibres, internal capsule, corpus callosum, anterior commissure and optic nerves and tracts. The cerebellar white matter was normal. Affected areas showed myelin loss and vacuolation, decreased astrocyte and oligodendrocyte numbers, prominent corpora amylacea, and small foci of perivascular lymphocytes and macrophages.

Pelizaeus–Merzbacher Disease

Definition

Pelizaeus-Merzbacher disease (PMD) is a CNS myelin disorder caused by abnormalities in the major structural protein of CNS myelin, proteolipid protein (PLP).

Genetics

In PMD, there are mutations in the *proteolipid protein-1* gene at Xq21–22 encoding PLP. In one-third of patients, there are point mutations and deletions, the rest harbouring gene duplications.[387]

Pathogenesis

The *PLP* gene encodes for two major splice variants, PLP1 and DM20, thought to have a role in the compaction in the extracellular surface of the myelin.[387] Gene expression is mostly limited to the CNS, explaining the sparing of the PNS in PMD patients. The most common mutation (gene duplication) leads to increased expression.

Clinical Features

In the past, PMD was subtyped on a morphological and clinical basis into connatal, classical variants and a mild form that is allelic with X-linked spastic paraplegic type 2 (SPG2).[387] The essential pathological process is similar in all cases. Although the clinical onset is almost always in the first year of life, the rate of progression is variable, and not surprisingly the degree of myelin loss and subsequent astrocytosis reflects this.

Boys present with nystagmus, hypotonia, progressive spastic paraparesis and a movement disorder without evidence of peripheral demyelination. Thereafter, the disorder is variable, with a progressive motor disorder, mental retardation and often dementia. MRI of the brain shows no myelin formation in the cerebral hemispheres, although some may be present infratentorially. Lifespan is shortened and appears to be determined by the severity of the disorder.[145,238]

Pathology

The brain is usually small and below normal weight by one-third to one-half. On sectioning, the grey matter appears intact, and ventricular dilation is proportionate to white matter loss. The central and digitate white matter, corpus callosum, capsules, commissures and fornix are grey-brown, gelatinous or firm and sunken (Figure 5.9a). There may be patchy U-fibre sparing and sometimes white streaks in the centrum semi-ovale. The optic nerves and chiasm are grey and thin, although other cranial nerves are normal. The white matter tracts of the cerebellum, brain stem and cord are shrunken, grey and gelatinous, in contrast to the plump and white cranial and spinal nerve roots (Figure 5.9b).

Histology confirms the preserved grey matter apart from cerebellar cortical atrophy, with either predominant granule or Purkinje cell degeneration. Polymicrogyria has been reported occasionally.[268] In the hemispheres, white matter varies from almost complete absence of myelin, particularly when death occurs in infancy, to the classical tigroid or discontinuous pattern with preserved perivascular islets, when the clinical course is more protracted (Figure 5.9d). Oligodendrocytes are markedly reduced or absent. There is astrocytosis, fibrillary gliosis and usually only sparse sudanophilic lipid in perivascular macrophages. Axons are preserved, and although most are naked, myelin discontinuities can be observed ultrastructurally (Figure 5.9e). Spinal and cranial nerve roots, which have a different myelin structural protein (PMP-22), are normally myelinated (Figure 5.9c). In well-oriented sections, individual axons can be followed in continuity from an unmyelinated state centrally through the root transition zone into the normally myelinated root.

PMD-like Diseases

There are many other dysmyelinating disorders. Some, such as 18q-syndrome, 3-phosphoglycerate dehydrogenase deficiency and merosin-deficient congenital muscular dystrophy, are part of recognized diseases. An increasing range of genes has been associated with PMD-like disorder/hypomyelinating leukodystrophies (HLD) (see Table 5.4).

Spongy Leukodystrophy (Canavan or Van Bogaert–Bertrand Disease)

Definition

This disease is characterized by a spongy white matter vacuolation.

Genetics

The disorder is caused by deficiency of aspartoacylase (ASPA), an enzyme necessary for the catabolism of N-acetylaspartate (NAA) and N-acetylaspartylglutamate (NAAG). Two mutations account for most (98 per cent) of cases in Ashkenazi Jewish patients and one mutation accounts for up to 60 per cent of non-Ashkenazi patients (reviewed in Matalon and Michals-Matalon[211]).

Pathogenesis

ASPA is present in oligodendroglial cell bodies.[29] It acts to hydrolyse NAA to acetate and aspartate. NAA is an amino acid synthesized in neurons, which also synthesize the neurotransmitter derivative of NAA, NAAG. However, NAA and NAAG are not catabolized by neurons. NAAG is hydrolysed to NAA by astrocytes and NAA is hydrolysed by ASPA in oligodendrocytes.[29,182] The mechanism that leads to abnormalities of myelin is not well understood but proposed hypotheses include that there is a requirement for NAA-derived acetate by myelinating oligodendrocytes (the acetate transport hypothesis) or that the disease arises from the osmotic consequence of accumulated NAA (the osmotic-hydrostatic hypothesis).[29]

Clinical Features

Onset of disease is normally in the first 6 months of life, although exceptionally it may be later. First symptoms include poor visual fixation, irritability and poor sucking. Developmental stagnation (or regression), nystagmus and macrocephaly develop during the first year of life in the majority of patients. Epilepsy usually manifests later. A clinical hallmark is raised NAA in the CSF.[211]

Pathology

In the typical infantile form, brain weight is usually 50 per cent heavier in the first 2 years of life, but later decreases to normal levels as cerebral atrophy progresses. Little myelin is evident in coronal sections, and the loss of U-fibres blurs the cortical grey-white junction. The white matter appears grey, gelatinous and sunken in, occasionally with cystic breakdown (Figure 5.10a). The capsules and corpus callosum are also affected. The cortex, basal ganglia and brain stem appear normal, but cerebellar and spinal cord white matter is soft, retracted and grey.

Histologically, a fine vacuolation with myriad empty, histochemically negative spaces up to 200 μm in diameter extensively involves the white matter of the cerebral hemispheres (Figure 5.10b), including the corpus callosum, capsules and fornix, the optic nerves, cerebellum and long tracts of the brain stem and spinal cord. Myelin staining is not seen, but immunohistochemistry may demonstrate myelin basic protein lining the vacuoles.

Sudanophilia is not usually present, but there is astrocytosis including both Alzheimer type I and type II glia. Oligodendroglia are present until quite late in the process. Axons are largely preserved, but spheroids have been demonstrated.[168] Rarely, in the authors' experience, the central cerebral and cerebellar white matter is replaced by a loose mass of astrocytes, macrophages and capillaries. Vacuole formation is particularly intense at the cortical grey–white junction and spreads into the deeper layers of the cortex, where reactive astrocytes and Alzheimer II glia are present but neurons are spared. Alzheimer II glia are also notable in the basal ganglia. Another zone of prominent spongy change is the boundary between the cerebellar molecular and granule cell layers, which may result in a line of cleavage, with Purkinje cells and Bergmann glia clinging to the molecular layer.

Ultrastructural studies show that the vacuoles in the white matter are large electron-lucent spaces surrounded by myelin leaflets, which appear to be formed by splitting of myelin lamellae at the intraperiod line. In the cortex, cell bodies and processes of astrocytes are markedly swollen, and their watery cytoplasm contains extremely elongated mitochondria, 12–15 μm in length, with a diagnostically characteristic structure of central crystalline cores surrounded by a ladder-like array of abnormal cristae (Figure 5.10c).

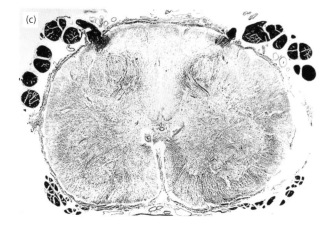

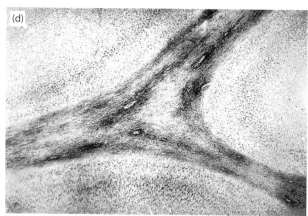

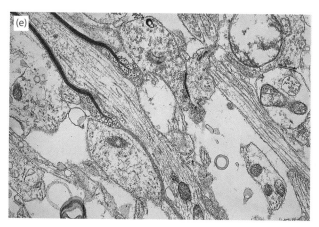

5.9 Pelizaeus–Merzbacher disease. (a) Coronal section of the frontal lobes shows no normal white matter, but central and digitate white matter, corpus callosum and capsules are grey-brown and sunken in. **(b)** Anterior spinal cord. There is a striking disparity between the normal bulk and whiteness of the spinal roots and the grey unmyelinated spinal cord. **(c)** The central spinal cord white matter is completely unstained by Luxol fast blue, although peripheral myelin in nerve roots stains normally. **(d)** Preserved perivascular myelin islets in cerebral white matter. Luxol fast blue. **(e)** Electron micrograph of a biopsy in which there is discontinuous myelination as the myelin sheath stops abruptly at a hemi-node, leaving the naked axon to continue through the neuropil.

Vanishing White Matter Disease

Definition

This leukodystrophy is characterized by loss of central myelin leaving large pseudocystic spaces. It has also been reported under the names, childhood ataxia with central hypomyelination and myelinopathia centralis diffusa.

Genetics

Mutations in five genes cause vanishing white matter disease (vWMD), each of which encodes a subunit of eukaryocytic initiation factor 2B: *EIF2B1*, *EIF2B2*, *EIF2B3*, *EIF2B4* and *EIF2B5*.[193,363]

TABLE 5.4 Genes associated with hypomyelinating leukodystrophies (HLD)

Disorder	Synonyms	Gene	Protein
HLD1	Pelizaeus–Merzbacher disease/spastic paraplegia 2	*PLP1*	PLP-1
HLD2	PMD-like disease 1	*GJC2/GJA12*	Gamma-2 GAP junction protein
HLD3		*AIMP1*	Aminoacyl-tRNA synthetase complex interacting multifunctional protein 1
HLD4	Mitochondrial HPS60 chaperonopathy	*HSPD1*	Heat shock protein 60
HLD5		*FAM126A*	Family with sequence similarity 126, member A
HLD6		*TUBBA4*	Beta-tubulin, class IVA
HLD7	4H syndrome	*POLR3A*	RNA polymerase III, subunit A
HLD8		*POLR3B*	RNA polymerase III, subunit A
Allan–Herndon–Dudley syndrome		*SLC16A2*	Monocarboxylate acid transporter 8

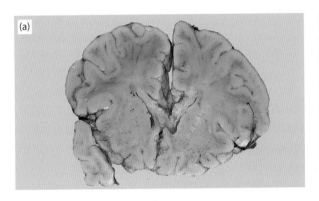
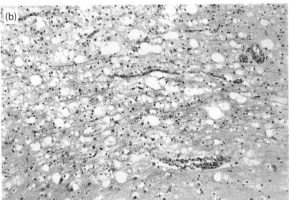
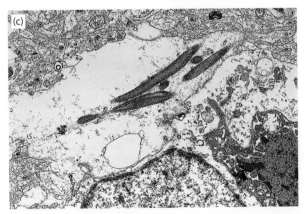

5.10 Canavan disease. (a) Coronal section through the frontal lobes showing extensive lack of myelin, brown discolouration and 'softening' involving white matter and U-fibres, as well as the corpus callosum and internal capsule. **(b)** The cerebral white matter and especially the U-fibres are demyelinated and finely vacuolated. Luxol fast blue. **(c)** Electron microscopy of a swollen astrocytic process in the cortex containing extremely elongated mitochondria with abnormally structured cristae.

Pathogenesis

The pathogenesis of vWMD has been reviewed.[51] All the genes involved encode for components of eIF2B. eIF2B regulates the rate-limiting step for the initiation of translation by catalyzing the conversion of GDP associated with eIF2, a factor required to initiate translation, to the active form, GTP. Indeed vWMD-associated mutations reduce eIF2B activity.[201,293] However, cell culture data have suggested that disease is not caused by an overall effect on protein translation.[51,369] An alternative mechanism is that eIF2 and eIF2B have roles in the regulation of the unfolded protein response (UPR), a cellular stress mechanism that responds to misfolded or denatured proteins,[51] one aspect of which is to regulate cell survival. The UPR system is activated in the glia of patients with vWMD.[367,370] It is possible that further UPR activation in times of cellular stress may lead to an exaggerated stress response,[51] ultimately leading to cell death.

Clinical Features

This autosomal recessive disorder is characterized by the onset of neurological symptoms following minor head injury or incidental infection, and a diagnostic appearance on MR imaging. Early development is normal, and the first symptoms usually develop between 18 months and 5 years of age. Presentation with cerebellar ataxia and less prominent spasticity is typical. The initial neurological feature is that of a progressive ataxic–spastic disorder, and later bulbar involvement, optic atrophy and occasionally epilepsy. A severe early infantile form (Cree encephalopathy) affects the Chippewayan and Cree indigenous populations in Canada.

Adult onset and adolescent cases are recognized with epileptic seizures, complicated migraine, cognitive impairment and psychiatric symptoms. Women may develop ovarian failure; ovarioleukodystrophy.[51,359,361,362]

MRI shows diffuse cerebral white matter lesions with variable brain stem involvement in pre-symptomatic individuals. Once symptoms develop, there is progression, with areas having the signal intensity of CSF and involvement of the dentate nucleus. Clinical diagnosis depends principally on characteristic white matter signal changes.

Pathology

The pathology has been summarized in the literature.[43,115,116,279,296,315,342,359,361] Significant atrophy or reduction in brain weight is not typical, but an undue softness on palpation results from the remarkable cavitation of the hemispheric white matter (Figure 5.11a). The cortex and basal ganglia are unaffected, but the cerebral white matter is mostly cavitated and gelatinous or has the appearance of a greyish, lacy cobweb punctuated by yellow-white spots. Subcortical U-fibres are only partially spared, whereas the corpus callosum, capsules and commissures are variably affected. The cerebellar cortex is normal or slightly atrophied, but its white matter is diffusely grey. The brain stem is also shrunken and its white matter tracts are diffusely grey, but in some cases focal chalky white spots stand out in the pontine tegmentum.

Microscopic examination confirms the general preservation of grey structures, with the exception of the cerebellar cortex, which is subject to Purkinje cell degeneration and depletion. The large cavities in the centrum semi-ovale are surrounded by cellular tissue in which there are many naked axons, few myelin fragments, reactive astrocytes, but only modest gliosis, macrophages containing sudanophilic material and in particular oligodendrocytes, which appear increased in density (Figure 5.11b–d). The vacuoles stain for myelin proteins.[315,359] Demyelination extends well beyond the area of cavitation to include digitate white matter,

capsules, corpus callosum, commissures, fornix and cerebral peduncles, as well as the cerebellum. In these areas, oligodendrocyte numbers are increased. However, myelin loss in the brain stem tracts and spinal cord is apparently not accompanied by similar oligodendrocyte increases. Sheets of lipidized phagocytes in the pontine tegmentum explain the discrete lesions observed macroscopically.[296] These are the presumed morphological correlate of symmetrical foci of high signal intensity on MR imaging, characteristic of this disorder.[359]

Glial abnormalities characterize the disease. There are foamy oligodendrocytes with unusually abundant cytoplasm.[115,386] They can be demonstrated to express proliferative and pro- and anti-apoptotic markers.[49,368] There are increased oligodendrocytes and oligodendrocyte progenitor numbers with relatively little astrocytic reaction.[52,116,368]

Abnormalities of astrocyte morphology include blunted processes, increased proliferation (*in vivo* and *in vitro*) and an abnormal and immature phenotype.[52] Glial cells cultured from a patient with vWMD (*eIF2B5*) showed normal oligodendrocytes, but impaired and abnormal astrocytic differentiation.[94]

Hippocampal damage has been described in Cree leukoencephalopathy.[115] A peripheral neuropathy has been described in a case of vWMD.[109] Pathological studies of organs outside the nervous system are sparse, but occasional descriptions of dysplastic ovaries are reported.[43,364]

Adult Onset Leukoencephalopathy with Axonal Spheroids and Pigmented Glia

Definition

Adult onset leukoencephalopathy with axonal spheroids and pigmented glia (ALSP) is a proposed term incorporating diseases previously described as pigmentary orthochromatic leukodystrophy (POLD) and hereditary diffuse leukoencephalopathy with axonal spheroids (HLDS). Both syndromes have similar clinicopathological features and are caused by mutations in the same gene, leading to suggestions that they be regarded as a single entity.[9,244,383] The disease is exceptionally rare in children, but is included here for completeness.

Genetics

Mutations in the kinase domain of the *CSF1R* gene (colony-stimulating factor receptor type 1) have been shown to cause HDLS and more recently POLD.[129,224,244,285]

Pathogenesis

CSF1R encodes for a receptor tyrosine kinase whose ligand is colony-stimulating factor, which regulates mononuclear cell lineages including microglia.[332] The mutations in ALSP affect the tyrosine kinase domain with evidence that they affect autophosphorylation in response to ligand.[244,285] Rademakers *et al.*[285] have suggested that ALSP may be in a spectrum of microglial disorders along with Nasu–Hakola disease. The latter presents with dementia and bone cysts thought to be due to abnormalities of microglia and osteoclasts, respectively.[34] Nasu–Hakola disease is

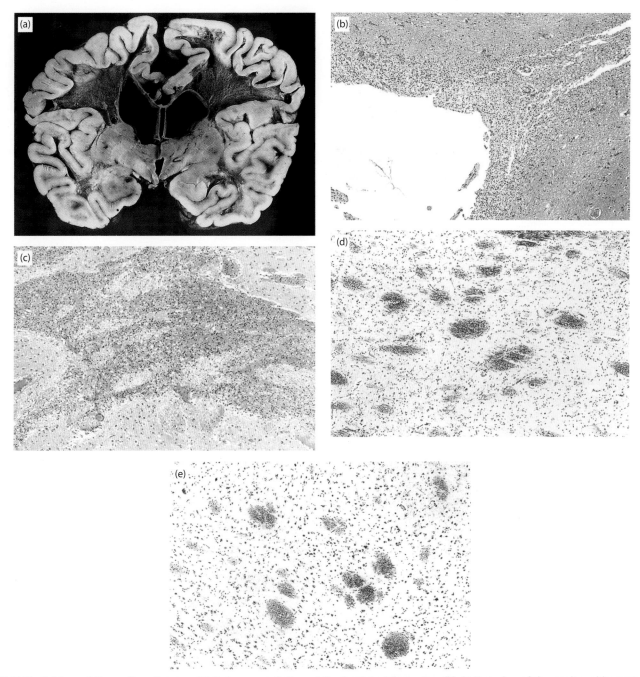

5.11 Vanishing white matter disease. (a) Extreme cavitation of the frontal white matter. **(b)** At the edge of the cavity, white matter is hypercellular, with increased numbers of oligodendroglia. **(c)** Increased density of oligodendrocytes in the internal capsule. **(d)** Compare the oligodendroglial hypercellularity in the pencil fibres of the basal ganglia with **(e)** the age-matched control. Luxol fast blue–cresyl violet.

caused by mutations in the genes, *TREM2* or *TRYOBP* (*DAP12*)[34] and the latter is implicated in CSF1R signalling.[256] This suggests that both diseases act through the same signalling pathway in microglia and may be variants of microglial disorders.

Clinical Features

ALSP is predominantly a disease of adult onset. In a large cohort of genetically confirmed cases, age of onset varied between 40 and 59 years.[129] *CSF1R* mutations were found in 11 per cent of adult patients with a leukodystrophy. Childhood cases are exceptional.[137]

Wider *et al.*[383] reviewed the clinical features of patients diagnosed with HDLS and POLD. Presentation is typically with psychiatric problems and early emotional lability and abnormal behaviour, followed by progressive intellectual deterioration, frontal lobe syndrome, motor signs, rigidity, seizures, tetraparesis, incontinence, mutism and decerebration.

Pathology

Most historical cases were labelled as HDLS or POLD.[125,251] In a direct comparison, no distinctive pathological features were recognized between the two 'diseases'.[9] Both showed loss of white matter with relative axonal preservation, pigment in microglia (and occasionally other glia) and axonal spheroids. Similar features are described in genetically confirmed cases.[129,244,285]

Brain weights vary from normal to profoundly reduced. The surface of the fresh brain may have a greenish tinge. On sectioning, the lateral ventricles are dilated and hemispheric white matter is shrunken and grey-brown. Subcortical U-fibres, optic nerves, cortex, deep grey matter, cerebellum and brain stem all appear normal.

Microscopic demyelination is extensive, sparing only the U-fibres, optic nerves and hindbrain. Areas of complete myelin loss show severe gliosis and few macrophages; axons are relatively spared, but there are frequent axonal swellings. In other areas, demyelination is not complete and there are large numbers of globular macrophages containing dark yellow-brown pigment, which is autofluorescent, birefringent, non-metachromatic, sudanophilic and stained by Luxol fast blue, periodic acid–Schiff (PAS), Masson Fontana for melanin, Schmorl and, in some cases, Perls' stain. Similar pigment is present in astrocytes and in the authors' paediatric case was observed in microglial cells. Other electron-microscopic studies have shown similar findings. Cytoplasmic inclusions of two types, some membrane-bound, are present in macrophages, astrocytes and oligodendroglia. There are mixed structures invoking lipofuscin, and fingerprint or multilamellar bodies interpreted as ceroid pigment. Okeda et al.[251] observed similar inclusions in the peripheral nerve of one patient but without evidence of hypomyelination.

Aicardi–Goutières Leukoencephalopathy

Definition

This disorder is a familial syndrome of calcification, diffuse demyelination, brain atrophy and a mild CSF pleocytosis.

Genetics

Aicardi–Goutières syndrome is genetically heterogeneous with at least six genes identified (see Table 5.5). It is usually inherited in an autosomal recessive manner, but *de novo* dominant mutations are recognized in *TREX1*.[6]

Pathogenesis

The disorder is thought to arise from the overproduction of α-interferon (α-IFN) within the CNS. All the genes have functions in clearing nucleic acid (see Table 5.5) and one hypothesis is that there is a failure to clear nucleic acids that triggers a type of immune response, including α-IFN, that normally clears viral nucleic acids. α-IFN then mediates the downstream brain injury.[7,77,80]

Several features reflect this immune dysregulation. First, there are many features that overlap with intracranial congenital infection. Second, inflammation is a pathological feature. Finally, there is overlap in the genes, immunology and clinical phenotypes with forms of systemic lupus erythematosus (SLE).

Clinical Features

Clinical heterogeneity is common.[6,290,336] In classical cases, onset is often in the first 4 months of life, with microcephaly developing in the first year. Development stagnates, hypotonia is punctuated by opisthotonus, and the child becomes decerebrate over a few years. Calcification develops within the basal ganglia and white matter. In addition, imaging shows cerebral atrophy and white matter changes. There is CSF pleocytosis and raised α-IFN. Even within the same family, there is considerable variation in the symptoms and rate of evolution of the brain calcification and atrophy. There may be chilblains and sterile pyurias. The expansion of genes associated with AGS is reflected in the wide clinical phenotype of the disease including overlap with SLE, which some children develop in an early onset form.

Pathology

The original case reports of Aicardi and Goutières[5] did not include morphology, although the pathology is inferred from other publications.[288,355] The authors have examined several similar cases. Microcephaly is striking, brain weights being two-thirds to one-half expected, but the hindbrain is disproportionately small. Coronal sections demonstrate atrophic friable gyri, pronounced ventricular dilation, shrunken central white matter, thin corpus callosum, and atrophic basal ganglia and thalamus (Figure 5.12a,b). The brain stem is slender and firm and the cerebellum small, with marked cortical atrophy.

Microscopic calcification is widespread (Figure 5.12c), being slight in the cortex, and more severe in white matter, basal ganglia, thalamus and cerebellum. Spherical lamellated PAS-, iron- and calcium-positive concretions are scattered freely in the parenchyma or clustered around

TABLE 5.5 The genetic subtypes of Aicardi–Goutières syndrome

Disease	Gene	Protein	Reference
AGS1	*TREX1*	3′–5′ exonuclease	78
AGS2	*RNASEH2B*	Subunit B of RNAse H2	79
AGS3	*RNASEH2C*	Subunit C of RNAse H2	79
AGS4	*RNASEH2A*	Subunit A of RNAse H2	79
AGS5	*SAMHD1*	SAM domain- and HD domain-containing protein 1	291
AGS6	*ADAR1*	RNA specific adenosine deaminase	292

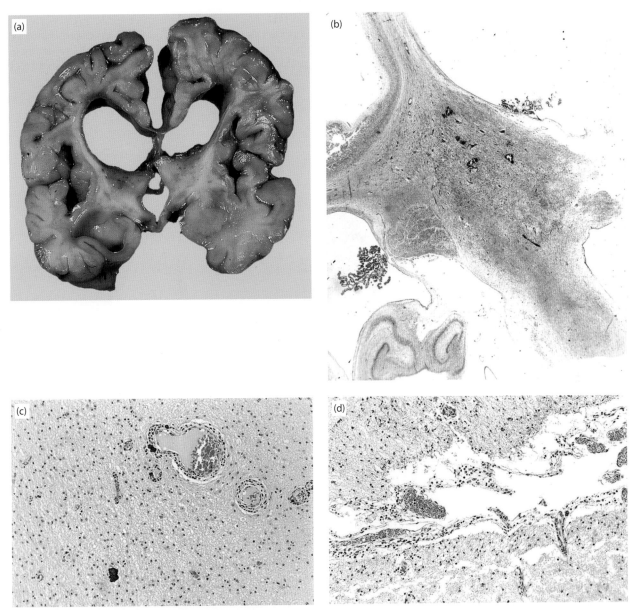

5.12 Aicardi–Goutières leukoencephalopathy. (a) Profound grey- and white-matter atrophy. **(b)** Low-power view of the thalamus and temporal lobe. The shrunken white matter lacks myelin, the cortex and thalamus are severely atrophic, and there are central thalamic mineralizations. Luxol fast blue–cresyl violet. **(c)** Calcospherites and thin perivascular lymphocytic cuffs in the white matter. **(d)** Slight lymphocytic infiltrate in the leptomeninges.

blood vessels. Mineralization affects the dendritic tree of cerebellar Purkinje cells. Vascular calcification and cerebral infarction is recognized.[28] Hemispheric, cerebellar and brain stem white matter tracts are greatly reduced in volume (Figure 5.12b) and lack myelin staining or show fragmentation of myelinated fibres. Oligodendroglial populations are preserved, but there is marked fibrillary gliosis; many hypertrophic astrocytes show intracytoplasmic neutral lipid. In the cerebral grey matter, neuronal loss parallels the degree of calcification. Other features include hippocampal sclerosis, cerebellar cortical degeneration involving Purkinje cells with axonal torpedoes and dendritic asteroid deformities, and destruction and calcification of the olivary and dentate nuclei. Scanty lymphocytic infiltrates can be found in the leptomeninges and sometimes the parenchyma (Figure 5.12c,d).

Pathological descriptions of genetically confirmed cases are rare. Ramesh *et al.*[286] described the autopsy findings in a patient with *SMADH1* mutation who died following subarachnoid haemorrhage. They describe predominantly vascular pathology with a vasculitis of the cerebral and leptomeningeal vessels, calcification of the walls of small blood vessels, a ruptured aneurysm at the origin of the left middle and anterior cerebral arteries, and an intact aneurysm of the right middle cerebral artery.

Band-like Calcification with Simplified Gyration and Polymicrogyria

Band-like calcification with simplified gyration and polymicrogyria (BLCPMG) is an autosomal recessive disease that resembles congenital infection, hence the alternative term 'pseudo-TORCH'. O'Driscoll *et al.*[249] have found mutations in the gene encoding the tight junction protein, occludin. The disease is included here given historical debate regarding overlap between Aicardi–Goutières syndrome and BLCPMG, but the distinctive clinicopathological and genetic features suggests that they are best regarded as distinct disorders.

Patients present with early onset seizures, developmental arrest and progressive microcephaly, characterized by intracranial calcification and malformations of cortical development (principally polymicrogyria). Patients also have thrombocytopenia, hepatosplenomegaly and hepatic dysfunction.

Post-mortem descriptions that predate genetic confirmation describe the combination of intracranial calcification with malformations of cortical development that include polymicrogyria and pachygyria.[53] In a genetically confirmed post-mortem case, calcification was most marked around blood vessels in the deep cerebral and cerebellar cortex; there was also bilateral peri-Sylvian polymicrogyria and widespread gliosis.[249]

Cockayne Syndrome

Definition

This disorder is characterized by poor growth, dysmorphism, sensitivity of the skin to the sun and neurological abnormalities.

Genetics

This autosomal recessive disorder results from mutations of the *CSA/ERCC8* gene on 5q, the *CSB/ERCC6* gene on 10q, or the xeroderma pigmentosum-related genes, *XPB*, *XPD* and *XPG*.[189]

Pathogenesis

The underlying defect is believed to be failure of transcription-coupled repair of oxidized DNA bases. Because neurons are rapidly metabolizing cells that produce high levels of reactive oxygen species, this eventually causes a general failure of DNA transcription, with reduced messenger RNA (mRNA) formation and a strong pro-apoptotic signal.[23,70,86,189]

Clinical Features

Several reviews are published.[194,237,259,287] Neurological abnormalities are normally apparent after the first year, although some are present from birth and consist of microcephaly, structural ocular abnormalities, cataracts, pigmentary retinopathy, deafness, a peripheral neuropathy, and progressive pyramidal tract and cerebellar signs secondary to a leukodystrophy. Neuroimaging shows brain atrophy and calcification of the basal ganglia. Unlike other sun-sensitive conditions, tumour formation is not a feature.

Pathology

Many case reports and small series have been reviewed by Weidenheim *et al.*[380] Microcephaly and cerebral atrophy are marked. The brain weight is often half that expected for age, but the hindbrain is even smaller. The cortex looks

normal but the lateral ventricles are greatly enlarged, whereas the central white matter is considerably reduced in volume and has a striking appearance of alternating grey and white patches (Figure 5.13a),[272] which also involves the

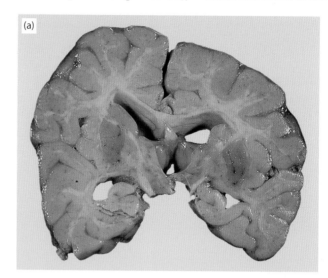

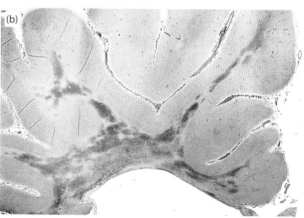

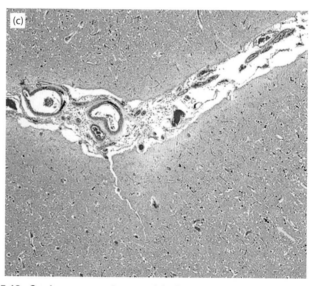

5.13 Cockayne syndrome. (a) Central white matter and corpus callosum are reduced and have a mottled appearance. **(b)** Tigroid demyelination in the frontal lobe. Luxol fast blue. **(c)** There are small calcospherites in the atrophied cortex and linear calcification in leptomeningeal vessel walls.

corpus callosum, capsules and optic nerves. There may be thalamic atrophy and visible calcifications in the deep grey matter. The brain stem is shrunken and the cerebellar folia are often atrophied, although their white matter is narrow and grey. The principal histological features are discontinuous demyelination (Figure 5.13b) with gliosis throughout the supratentorial and infratentorial white matter. There is a debate as to what extent the myelin defect represents loss of existing myelin versus abnormalities in myelin development.[380] Numerous mulberry-like calcific concretions affect both the grey and white matter (Figure 5.13c) and sometimes also vascular walls and Purkinje cell bodies and dendritic trees. Cerebellar cortical degeneration may be prominent, producing a plethora of Purkinje axon torpedoes and dendritic asteroid expansions.

Segmental demyelination also affects the peripheral nervous system.[225,226,310,380] Involvement of the auditory and visual system is common with hair cell and neuronal loss in the inner ear and ocular abnormalities (including pigmentary retinopathy and optic atrophy).[287,380]

A range of vascular abnormalities are described including mineralization, 'string vessels', premature arteriosclerosis and subdural haemorrhages.[140,287,380]

Occasional findings include multinucleated neurons,[192] abnormal astrocytes[380] and neurofibrillary tangles.[347,380] Cerebral malformations are only present in exceptional cases.[287,380]

DISORDERS OF AMINO ACID METABOLISM AND RELATED DISORDERS

Introduction

This section covers a range of inborn errors of metabolism due to defects in the breakdown of amino acids or related organic acids. Most are rare, recessively inherited disorders. Most are recognized biochemically, genetically and clinically, but relatively few have detailed neuropathology. The discussion here is limited to those with well-defined neuropathology.

The catabolism of amino acids begins with the transfer of the amino group to an alpha keto-acid (e.g. alpha ketoglutarate). This product enters the urea cycle. The carbon skeleton is broken down by a series of different enzymatic reactions that return the carbon components to the Krebs cycle. Therefore, in broad terms, the disorders of amino acid metabolism may be considered as defects either of the urea cycle (discussed at the end of this section) or of breakdown of the carbon skeleton (discussed later).

A related terminology covers the organic acidurias. These are defined as disorders of metabolism that lead to urinary accumulation of non-amino organic acids. The main group are those due to defects in the catabolism of the branched chain amino acids including propionic acidaemia, methylmalonic aciduria and maple syrup urine disorder.

General Neuropathology

Many of these disorders present as a catastrophic severe neonatal illness and need to be considered in the differential diagnosis of a child with a severe neonatal-onset neurological disorder. However, some are more chronic (e.g. phenylketonuria or homocystinuria) or show a milder phenotype, which may be punctuated by episodes of decompensation. Biochemical screening makes completely undiagnosed cases relatively rare.

Although there are specific features characterizing some amino acid metabolic disorders, the common defining neuropathologic feature is a spongy myelinopathy. This is characterized by vacuoles developing in central myelin, often at an early stage in the brain stem, cerebellum and spinal cord. The split in the myelin develops along the intraperiod line (i.e. the extracellular space). The differential diagnosis of spongy myelinopathy includes Canavan disease, mitochondrial disease (e.g. Kearns–Sayre), galactosaemia and toxins (e.g. hexachlorophene, isoniazid, cuprizone, triethyltin). In addition, non-specific vacuolation (e.g. in the context of cerebral oedema) must be excluded. The major exceptions are the urea cycle disorders and homocystinuria, where spongy myelinopathy is not typical.

The molecular mechanisms that lead to neurological disease are poorly understood, but possible contributory mechanisms include accumulation of toxic intermediate metabolites (e.g. accumulation of branched amino acids or their ketoacids), secondary disorders of carnitine (e.g. due to complex formation with coenzyme A derivatives), secondary disorders of mitochondrial respiratory chain function, energy deficiency during metabolic crises, excitotoxicity of gluataric acid metabolites (acting at NMDA receptors) or hyperammonaemia (secondary).

More common disorders are included in the following sections. For rarer disorders or those with less well-characterized neuropathology (see Table 5.6).

The Hyperphenylalaninaemic Syndromes

The hyperphenylalaninaemic syndromes comprise L-phenylalanine 4-mono-oxygenase (phenylalanine hydroxylase) deficiency (classic phenylketonuria) and deficiency of its essential cofactor tetrahydrobiopterin (Figure 5.14).

Phenylketonuria

Biochemical Defect

Phenylketonuria (PKU) is caused by deficiency of the hepatic enzyme phenylalanine mono-oxygenase (PHA, phenylalanine hydroxylase). The biochemical consequences are the accumulation of phenylalanine and its metabolites in the blood and a relative deficiency of tyrosine, which becomes an essential amino acid (Figure 5.14). The main impact of the disease is upon the brain, which normally does not contain phenylalanine mono-oxygenase. Untreated, the early manifestations are microcephaly, severe mental retardation and epilepsy; in the second or third decade, there is the emergence or progression of a motor disorder.

There are several potential mechanisms of the neurological consequences, but central is the accumulation of phenylalanine in the blood.[177] This causes an increase in brain phenylalanine concentration and deficiencies of the other large neutral amino acids (by competition for transport into the brain). As a consequence, brain protein synthesis, myelin turnover and biogenic amine neurotransmission are disturbed.

TABLE 5.6 Rarer disorders of amino acid metabolism and related disorders

Disorder	Enzyme defect	Neuropathology
Organic acid disorders		
Isovaleric acidaemia	Isovaleryl CoA dehydrogenase	Limited descriptions
		Cerebellar haemorrhage and glial cell degeneration have been described.[113] The brain may be normal[242]
3-Methylcrotonyl-CoA carboxylase deficiency	Methylcrotonyl-CoA carboxylase deficiency: • Alpha-subunit-type 1 • Beta-subunit-type 2	No neuropathology described. Cases are clinically variable[127] but can include strokes,[333] leukodystrophy[87] and a necrotizing encephalopathy[32]
Ketothiolase deficiency (mitochondrial acetoacetyl-CoA thiolase deficiency)	Acetyl-CoA acetyltransferase-1	No neuropathology described. MRI data describe involvement of the globus pallidus and posterio-lateral putamen[254,258]
3-Hydroxy-3-methylglutaryl-CoA (HMG-CoA) lyase deficiency	HMG-CoA lyase	Limited descriptions. A single brain biopsy revealed reactive gliosis, spongiosis, and increased intracellular astrocytic glycogen concentration in the white matter.[396] The clinical literature describes abnormalities of the cerebral white matter, the corticospinal tracts, the pons, the caudate nucleus, thalamus and dentate nucleus[360,390]
Miscellaneous disorders with distinctive patterns of organic aciduria and neurological presentations		
Biotinidase deficiency	Biotinidase	Clinical descriptions show a spectrum of MRI changes including myelopathy and leukoencephalopathy. One autopsy study showed extensive necrotizing lesions (similar to Leigh's) but with hippocampal and parahippocampal involvement[147]
Glutaric acidaemia type II (multiple Acyl-CoA dehydrogenase deficiency)	Alpha (ETFA) and beta (ETFB) subunits of electron transfer flavoprotein, and electron transfer flavoprotein dehydrogenase (ETFDH)	Acquired changes:[65,122,123,345] 1. Cerebral oedema 2. White matter damage with gliosis 3. Striatal degeneration Malformations: 4. Malformations of cortical development,[42] including pachygyria, 'warty' dysplasia, white matter heterotopias 5. Hypoplasia of the temporal lobes[338]
MCAD deficiency	Medium-chain acyl-coenzyme A dehydrogenase	Cerebral oedema[324]
Mevalonic aciduria	Mevalonate kinase gene	Perinatal cases may show absent septum pellucidum, periventricular cavitation with ventricular strands.[334] Cerebellar and cerebral atrophy is recognized clinically[300]
Krebs cycle disorders presenting with neurological symptoms and acidosis		
Fumarase deficiency	Fumarate hydratase gene	No neuropathology described. Radiological appearances have suggested:[10,204,304] periventricular cysts, ventricular enlargement, agenesis of the corpus callosum, polymicrogyria, lissencephaly, reduced white matter and a small brain stem
2-Ketoglutarate dehydrogenase deficiency	May occur as part of dihydrolipoamide dehydrogenase deficiency	Not described

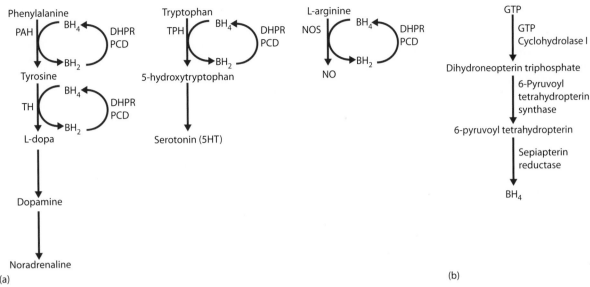

(a)   (b)

5.14 This simplified schematic highlights the major pathways involving tetrahydrobiopterin (BH₄) metabolism. Panel **(a)** shows the three clinically relevant pathways dependent on BH₄ (PAH-phenylalanine hydroxylase/mono-oxygenase; TH-tyrosine hydroxylase; TPH-tryptophan hydroxylase; NOS-nitric oxide synthetase). In addition, BH₄ is required for the action of alkylglycerol mono-oxygenase (not shown). Panel **(b)** shows the enzymes involved in BH₄ synthesis. Mutations in PAH cause classic phenylketonuria. Deficiencies of BH₄ are caused by mutations in either the enzymes that recycle BH₄ (pterin-4-α-carbinolamine dehydratase [PCD] and dihydropteridine reductase [DHPR], panel [a]) or the enzymes responsible for its synthesis (guanosine triphosphate [GTP]-cyclohydrolase, 6-pyruvoyltetrahydropterin synthase, sepiapterin reductase [panel b]).

Intriguing data show that high concentrations of phenylalanine can form amyloid-like fibrils that are neurotoxic, implying a novel model of amino-acid aggregation in the pathogenesis.[2]

Clinical Features

Severe neurological disability can be largely prevented by a strict reduced phenylalanine diet started in early infancy. Although treated patients avoid the severe neurological complications, they suffer lower intelligence, possible neuropsychological or neurological deficits and cerebral white matter abnormalities.[326,354,378] There is also a risk of neurological decline in adults who have relaxed their diet.[212,352] Furthermore, several studies have raised concerns over the long-term metabolic consequences of PKU and its treatment, e.g. oxidative stress.[295] There is growing interest in additional supplementation (for example, by the use of sapropterin, the orally active form of the cofactor for PHA, tetrahydrobiopterin [BH4]) in the treatment of PKU with the aim of improving the diet-unresponsive complications.[329]

Untreated maternal PKU leads to severe teratogenic effects.[175,191,281,325] Many of these complications are related to phenylalanine levels and the risk can be reduced by dietary control in pregnancy.

Pathology

Untreated PKU

The neuropathology of untreated PKU is documented in the historical literature.[30,75,209] The most consistent feature is microcephaly with brains typically around 80 per cent of the normal weight.[153] In addition, patients may have myelin defects and abnormalities of neuronal maturation.

In many but not all patients, there are variable degrees of white matter disturbance, ranging from spongiosis and delayed myelination in younger children to focal myelin pallor or even breakdown with neutral fat accumulation in adults.[30,75,153,209] In addition, there may be fibrillary gliosis.[75,209] The white matter changes may involve the centrum semi-ovale, optic nerves and long tracts of the brain stem. Furthermore, there is evidence of a biochemical disturbance of myelin.[215]

Abnormalities of neuronal maturation have been reported but are less frequent.[153] For example, in three adults with profound mental retardation studied by Bauman and Kemper,[30] in addition to myelin pallor in hemispheric and brain stem white matter, there were increased density of cortical neurons, reduced neuronal size and Nissl content, and poorly developed dendritic trees and spines. Similar abnormalities have also been described in animal models of PKU.[73,153]

Treated PKU

Data on the structural consequences of treated PKU come mostly from the radiological literature.[35,36,68,69,212] In many, there are white matter changes (reviewed in Bick et al.[36]) with inconsistent correlation to dietary history. In addition, long-term radiological studies of patients treated early show reduced brain volumes with larger putamina than patients without PKU.[41] The clinical significance of these neuroimaging changes is uncertain.

The Maternal PKU Syndrome

The offspring of mothers with PKU are at risk of the maternal PKU syndrome, which may include microcephaly and cardiac anomalies.[175,191,281,325] Interestingly, in some reports, the white matter changes of PKU are not a feature.[198] For example, Lacey and Terplan[186] described a child with maternal PKU syndrome in whom there was microcephaly, neuronal loss in layer III of the cerebral cortex and immature cortical neurons. In contrast, Koch *et al.*[176] described white matter changes in a child born to a mother with poorly controlled PKU. The child was one of twins and had congenital heart disease. The brain was small with enlarged ventricles and marked reduction in white matter bulk with delayed myelination.

Disorders of Tetrahydrobiopterin

Biochemical Defect

Tetrahydrobiopterin (BH4) is a cofactor required for phenylalanine mono-oxygenase (the gene mutated in classic phenylketonuria) activity (Figure 5.14)[381] and therefore mutations affecting BH4 metabolism can cause an atypical form of PKU (accounting for up to 2 per cent).[39,205]

BH4 is also a cofactor for other enzymes, including tyrosine and tryptophan mono-oxygenase and nitric oxidase.[381] The former are required for synthesis of monoamine neurotransmitters (including dopamine and serotonin) and therefore, movement disorders (parkinsonism and dystonia) are also a feature of BH4 metabolic errors.

Recessively inherited BH4 deficiencies can be caused by mutations in the enzymes involved in recycling BH4, for example pterin-4-α-carbinolamine dehydratase (5 per cent of cases) and dihydropteridine reductase (30 per cent of cases) or mutations in enzymes responsible for its synthesis (guanosine triphosphate [GTP]-cyclohydrolase, 5 per cent) and 6-pyruvoyltetrahydropterin synthase (60 per cent).[39,205,381] Sepiapterin reductase deficiencies produce a selective BH4 deficiency in the brain, as they are thought to be compensatory mechanisms in other organs.[205]

A dominantly inherited, less severe form of GTP-cyclohydrolase I deficiency may present later in life with symptoms of dopamine deficiency alone (dopa-responsive dystonia).[24,25,155]

Pathology

Patients with dihydropteridine reductase deficiency have been reported to be at risk of sudden death,[205] and there are case reports describing the neuropathology.[221,349] Common features are widespread demyelination with spongy vacuolation. In addition, there are areas of neuronal loss with abnormal vasculature and perivascular calcification, which may affect basal ganglia, cerebral cortex or thalamus. These changes correlate with those in the radiological literature (reviewed in Longo[205]).

In one case, there was abnormal neuronal orientation with abnormalities of dendrites and dendritic spines[349] and in another, prominent white matter neurons.[221]

Non-ketotic Hyperglycinaemia

Biochemical Defect

Non-ketotic hyperglycinaemia (also known as glycine encephalopathy) is a recessively inherited disorder caused by a defect in the breakdown of glycine, namely a deficiency of the intramitochondrial glycine cleavage enzyme (GCE)

system (Figure 5.15). The term 'non-ketotic' is used to distinguish these from disorders of organic acid metabolism (e.g. propionic acidaemia) that can also produce hyperglycinaemia associated with ketosis. The disorder is rare overall, but is more common in particular groups (e.g. in Finland[183]).

The GCE system consists of four subunits, three of which are associated with mutations in patients: the P-protein (encoded by the *GLDC* gene, mutations in which account for 70–75 per cent of cases), the T-protein (encoded by the *AMT* genes, mutations in which account for approximately 20 per cent of cases) and the H-protein (encoded for by *GCSH*, mutations in which account for <1 per cent of disease).[132] Rare cases are not explained by mutations in these subunits.

In addition to its role in glycine metabolism, one of the products of the GCE system is 5,10-methylene folate, which is a donor for one carbon metabolism. Therefore several mechanisms may contribute to the pathogenesis of this disorder; the early vegetative symptoms can be understood as arising from excessive stimulation of brain stem inhibitory glycine receptors. The later symptoms are thought to arise from excessive stimulation of excitatory glutamate N-methyl-D-aspartate (NMDA) receptors, which require glycine as a co-agonist. Later still, the reduced supply of single carbon groups to brain metabolism might result in myelin abnormalities.

Clinical Features

Most patients present with a severe neonatal-onset form, although milder cases present later in infancy or even in childhood.[44,131,132,184,207] The phenotype often remains true in individual families. In the neonatal-onset form, most develop symptoms in the first 2 days of life, becoming profoundly hypotonic (with preserved or brisk tendon reflexes) and lethargic, with abnormalities of eye movement. The encephalopathy progresses to coma, with the development of segmental myoclonic jerks, hiccups and apnoea. The EEG shows bursts of spike-wave complexes without normal background

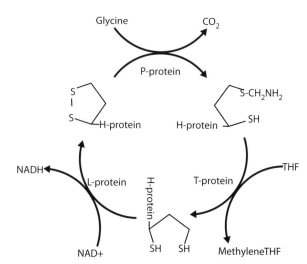

5.15 The glycine cleavage enzyme system consists of four proteins (H, L, P and T-protein) of which mutations in the genes encoding for H, P and T-protein cause non-ketotic hyperglycinaemia.

activity. Many die at this stage. Survivors regain respiration at around 3 weeks of age; intractable epilepsy develops after about 3 months and infants and children have profound impairment, with no adaptive or social behaviour. Brain MRI shows a severe leukoencephalopathy. Glycine accumulates in all body fluids, but is preferentially elevated in the CSF. A CSF/plasma glycine ratio of greater than 0.08 is diagnostic. The normal physiological hyperglycinuria of the newborn renders urinary glycine difficult to interpret. Despite multiple therapeutic endeavours, outcome remains disappointingly poor.

Pathology

Abnormalities of myelination are the most frequently reported pathological feature. For example, Shuman *et al.*[321] described a 3-year-old whose cerebral white matter and brain stem tracts stained poorly for myelin and were gliotic, the centrum semi-ovale being severely vacuolated. Agamanolis *et al.*[3] described spongy myelinopathy in a series of neonates; although the amount of myelin did not differ from normal controls, all areas had a spongy appearance. The cerebellar white matter, corticospinal and optic tracts were particularly affected. Ultrastructurally, the vacuoles were lined by myelin and appeared to form by intraperiod splitting. Anderson *et al.*[12] described spongy change in the white matter of a neonate with hyperglycinaemia. Long tract MRI changes have been demonstrated in the spinal cord of a late-onset case.[379]

Malformations are also recognized. Two-thirds of Dobyns' cases showed anomalies of gyration, callosal agenesis and cerebellar hypoplasia.[95] Neuronal damage, as demonstrated by ferrugination, is also described in one series.[3]

The Homocystinurias

Biochemical Defect

The homocystinurias are defined by increased homocysteine excretion in the urine as a result of defects in methionine metabolism (Figure 5.16). Methionine is converted to homocysteine in a series of reactions that donate a methyl group to an acceptor molecule. The homocysteine can then either be converted to cystathionine by cystathionine β-synthase or be remethylated to regenerate methionine in a vitamin B_{12}-dependent reaction.[40] Therefore, homocystinuria can be caused by deficiency of cystathionine β-synthase (classical homocystinuria) or by defects in remethylation. In cystathionine β-synthase deficiency, plasma methionine is raised, whereas in the remethylation defects, it is reduced.

Clinical Features

Cystathionine β-Synthase Deficiency

Cystathionine β-synthase deficiency is a multisystem disorder with ocular, skeletal, vascular and CNS involvement.[40,314] It often presents with high myopia or lens dislocation. The skeletal abnormalities become more obvious around puberty, with arachnodactyly, dolichostenomelia and enlargement of both metaphyses and epiphyses. There may be marfanoid features. Most have CNS manifestations, including mental retardation, epilepsy or psychiatric disturbance. The risk of a thromboembolic event increases with age. These are more commonly venous than arterial, although both may occur.

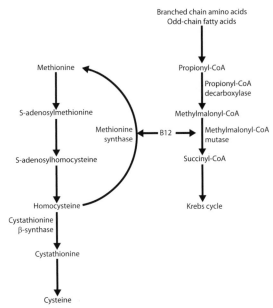

5.16 Two major pathways depend on vitamin B_{12}. On the left, the methionine metabolic pathway leads to homocysteine generation. Homocysteine can then be recycled to methionine in a vitamin B_{12}-dependent reaction (catalyzed by methionine synthase) or converted to cysteine through a pathway that includes the enzyme cystathionine β-synthase. Mutations affecting the latter cause classic homocystinuria. Mutations in methionine synthase or a range of enzymes involved in vitamin B_{12} metabolism can also cause homocystinuria. On the right, the pathway shows that propionyl-CoA is generated from odd-chain fatty acids or branched chain amino acids. It is converted to methylmalonyl CoA by propionyl-CoA decarboxylase, mutations in which cause propionic acidaemia. Methylmalonyl CoA is converted by methylmalonic CoA mutase to succinyl-CoA, which enters the Krebs' cycle. Methylmalonic CoA mutase is vitamin B_{12}-dependent and mutations affecting this enzyme or vitamin B_{12} metabolism can cause methylmalonic acidaemia.

Thromboembolism in childhood is commonly precipitated by an intercurrent illness that causes dehydration.

Approximately half of the patients with cystathionine β-synthase deficiency respond to pharmacological doses of pyridoxine (the coenzyme for cystathionine β-synthase) and subsequently do well. The other patients are more difficult to treat and often require dietary methionine restriction and an alternative methyl group donor, such as betaine; their outcome is less favourable. Patients diagnosed at birth and treated prospectively do not develop complications.

The ocular, skeletal and vascular complications of cystathionine β-synthase deficiency appear to be secondary to the accumulation of homocysteine. The pathogenesis of the neural complications is less clear, but involves metabolic mechanisms as well as cerebrovascular disease.

In contrast to the remethylation defects (see later), cystathionine β-synthase deficiency is not usually associated with white matter disease, although this has occasionally been described.[47]

Remethylation Defects

The remethylation defects consist of several groups of disorders that interfere with metabolism of vitamin B_{12}

(cobalamin) and are associated with homocystinuria, in some cases combined with methylmalonic acidaemia. The main groups are shown in Table 5.7. Several of these cobalamin (cbl) disorders have been classified according to their complementation in cell culture. Each group has similar clinical features that vary according to age of onset (and, presumably, defect severity).[57,314] The most common forms are the cblC defect and MTHFR deficiency,[314] which dominate the clinical descriptions.

Some forms can have prenatal presentation. For example, cblC may present with intrauterine growth retardation (IUGR), congenital heart disease/cardiomyopathy, microcephaly and dysmorphism (reviewed in Carrillo-Carrasco et al.[57]). Infants with inborn errors of cobalamin metabolism can also present with a micro-angiopathy with widespread organ involvement.

In the early infantile-onset remethylation defects, the child often suffers an acute neurological deterioration, which may be fatal. In late infantile/early childhood-onset, there is usually a neurological presentation with slowing of brain growth and development, followed by progressive neurological disorder. During the progressive phase, the child dements, may develop an ataxic–spastic or extrapyramidal movement disorder and often has signs of a peripheral neuropathy. There may be bouts of unexplained lethargy and coma.[57]

In late childhood/adult-onset, remethylation defects may present in a similar way to those in younger patients. However, there may be asymptomatic siblings, stroke, unexplained psychosis and symptoms of subacute combined degeneration.[314]

The inborn errors of cobalamin metabolism may be accompanied by a macrocytic anaemia. A high index of suspicion and a raised plasma total homocysteine with a reduced (or low normal) plasma methionine concentration are the initial clues to a remethylation defect.[314]

Treatment outcome for the remethylation defects (with vitamin supplements including cobalamin) depends on the underlying defect and how early treatment is instigated (reviewed in Schiff and Blom[314]).

Pathology

Cystathionine β-Synthase Deficiency

In classic homocystinuria, the principal lesions are ischaemic: thromboembolic occlusive vascular disease in arteries, veins and dural sinuses leads to cerebral, cerebellar, thalamic and midbrain infarcts.[54,58,118,121,382] Degenerative changes in vascular muscle and elastic coats with intimal thickening are prominent.

Remethylation Defects

Appearances suggestive of subacute combined degeneration have been reported in a patient with an inborn error of cobalamin metabolism,[85] although the spinal cord was not examined. A similar diffuse leukoencephalopathy of focal perivascular demyelination coalescing into a large area of myelin loss in the centrum semi-ovale combined with typical changes of subacute combined degeneration throughout the cord has been observed in two patients with 5,10-methylenetetrahydrofolate reductase deficiency (Figure 5.17).[67] A further case presented by Nishimura et al.[245] revealed perivascular white matter disease, axonal swellings and a peripheral neuropathy.

The neuropathology of a patient with the cblE form has been reported. The disease was complicated by haemolytic uraemic syndrome, which undoubtedly influenced the neuropathological findings.[266] The neuropathology showed haemorrhage in the putamen, thalamus and cerebellum. There was necrosis in the cerebellum. There was diffuse white matter involvement with spongiosis and necrosis, and severe astrocytosis. Microangiopathy was also observed

TABLE 5.7 Classification of defects of cobalamin metabolism by complementation groups[57,71,314,375]

Disease group	Complementation group	Gene	Functional enzyme defect
Methylmalonic acidaemia and homocystinuria, cblD type	cblD	*MMADHC*	Decreased activity of methylmalonyl-CoA mutase and methionine synthase
Methylmalonic acidaemia and homocystinuria, cblC type	cblC	*MMACHC*	Decreased activity of methylmalonyl-CoA mutase and methionine synthase
Homocystinuria-megaloblastic anaemia due to defect in cobalamin metabolism, cblE type	cblE	Methionine synthase reductase (*MTRR*)	Functional defect in methionine synthase
Homocystinuria due to deficiency of n(5,10)-methylenetetrahydrofolate reductase activity	Not applicable	Methylenetetrahydrofolate reductase deficiency (*MTHFR*)	Defect in methylenetetrahydrofolate reductase
Methylmalonic acidaemia and homocystinuria, cblJ type	cblJ	*ABCD4* (ATP-binding cassette, subfamily D, member 4)	Decreased activity of methylmalonyl-CoA mutase and methionine synthase
Methylcobalamin deficiency, cblG type	cblG	Methionine synthase	Methionine synthase
Methylmalonic aciduria and homocystinuria, cblF type	cblF	*LMBRD1* (encoding for a putative lysosomal cobalamin exporter)	Decreased activity of methylmalonyl-CoA mutase and methionine synthase

with increased vessel wall thickness, reduced arterial inner diameter and capillary oedema.

Powers *et al.*[278] reported the neuropathology of two patients with the cblC form. One patient had morphological evidence of an axonal neuropathy. Both patients showed perivascular demyelination of the cerebrum and one had features of subacute combined degeneration of the spinal cord. Smith *et al.*[327] described a case of spongy myelopathy (subacute degeneration of the cord) in a case of cblC.

(a)

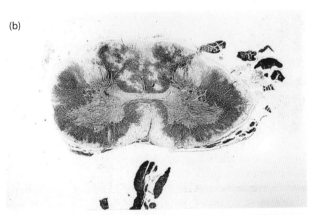

(b)

5.17 5,10-Methylenetetrahydrofolate reductase deficiency. **(a)** Multiple foci of perivascular demyelination in the cerebral white matter. **(b)** Subacute combined degeneration of the spinal cord. Luxol fast blue–cresyl violet.

Propionic Acidaemia

Biochemical Defect

Propionic acidaemia is an accumulation of propionic acid and its metabolites caused by a deficiency of propionyl-CoA carboxylase (Figure 5.16). Propionic acid is produced from metabolism of branched chain amino acids (valine, methionine, isoleucine, threonine) and from odd chain fatty acids. Propionyl-CoA carboxylase converts propionyl-CoA to methylmalonyl-CoA.[91,317] The latter is then converted to succinyl-CoA via methylmalonyl-CoA mutase, mutations in which cause methylmalonic acidaemia (see later). Succinyl-CoA then enters the Krebs' cycle.

Mutations occur in the genes encoding for both subunits, PCCA and PCCB.[91,317] Accumulation of propionic acid and its metabolites in body fluids can also be caused by a functional deficiency of propionyl-CoA carboxylase due to inborn errors of biotin metabolism or, very rarely, biotin deficiency.[261,393]

Clinical Features

Most children with propionic acidaemia present in the early neonatal period with a relentlessly progressive encephalopathy resulting in coma.[91,128,274,275,319] Later-onset forms are usually characterized by episodes of acute encephalopathy on a background of anorexia, failure to thrive and developmental delay. Cardiomyopathy is common. Investigation reveals neutropenia and thrombocytopenia with acidosis, ketosis, hyperammonaemia, mild or moderate hyperglycinaemia, reduced free carnitine and raised propionylcarnitine, and the excretion of propionyl-glycine and methylcitrate in the urine (especially during an acute episode).

Despite treatment (e.g. with protein restriction and carnitine[91]), the outlook is poor with a significant risk of neurological impairment and other complications.[128,318,319]

Pathology

Spongy degeneration of the white matter and especially of the globus pallidus is the chief finding with neonatal onset and death during infancy.[146,321,335] However, spongiosis may not present until later in the disease[335] or with later-onset group, where pathology centres on the basal ganglia with neuronal loss and gliosis, or even bilateral marbling (Figure 5.18).[138] These changes correspond to computed tomography (CT) lucencies seen in the basal ganglia during episodic decompensation and underlie the severe movement disorder commonly present.[344] In some cases, there is prominent cerebral cortical vacuolation.[110] In addition, occasional cases present with cerebral haemorrhage.[83] Finally, occasional cases of optic neuropathy are described.[154]

Methylmalonic Acidaemia

Biochemical Defect

Accumulation of methylmalonic acid in body fluids is caused by actual (mut[0] or mut[−]) or functional deficiency of methylmalonyl-CoA mutase. Methylmalonyl-CoA mutase catalyzes the conversion of methylmalonyl-CoA (product of propionic acid metabolism) to succinyl-CoA, which enters the Krebs' cycle (Figure 5.16).[91]

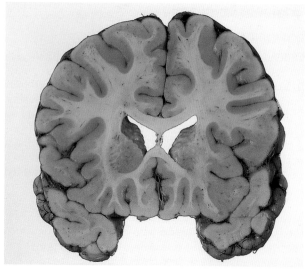

5.18 Propionic acidaemia. Atrophy and marbling of the striatum are apparent.

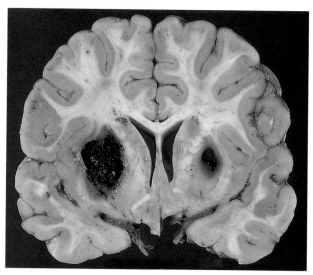

5.19 Methylmalonic acidaemia. There are bilateral putaminal haemorrhages.

Methylmalonyl-CoA mutase requires vitamin B_{12}-derived cofactors. An isolated functional defect in methylmalonyl-CoA mutase can be caused by inborn errors of cobalamin metabolism affecting its coenzyme adenosylcobalamin (e.g. cblA [MMA gene], cblB [MMAB encodes cob(I)alamin transferase]).[91,375] In addition, other disorders of cobalamin may produce methylmalonic acidaemia often associated with homocystinuria (see earlier).

Clinical Features

As might be anticipated by the biochemistry, the clinical presentation of methylmalonic acidaemia is very similar to that of propionic acidaemia, with the exception that large amounts of methylmalonic acid are excreted in the urine.[91,375] Once the patient is started on a protein-restricted diet and receiving supplemental carnitine, responsiveness to hydroxocobalamin treatment is determined by measuring methylmalonic acid excretion. The neurological outcome is determined by the age at onset and cobalamin responsiveness.

Pathology

Neuropathological changes may be non-specific. Dave *et al.* describe multiple small cerebellar haemorrhages.[83] Larnaout *et al.* described necrotizing lesions of the globus pallidus with spongiosis in the subthalamic nuclei, mammillary bodies, internal capsule, superior cerebellar peduncle and tegmentum of the brain stem.[187] An example of putaminal haemorrhages is shown in Figure 5.19.

Sarnat and Flores-Sarnat[308] have described a case showing immature cortical organization with prominent microcolumns.

Maple Syrup Urine Disease (Branched-Chain Oxo-Acid Dehydrogenase Complex Deficiency)

Biochemical Defect

Maple syrup urine disease (MSUD) is an organic aciduria due to a defect in the metabolism of branched-chain amino acids (leucine, isoleucine and valine). It is caused by mutations in the genes encoding for the four subunits of the branched-chain alpha-keto acid dehydrogenase complex (BCKDC): E1-alpha subunit (*BCKDHA* gene, MSUD type IA),[394] E1-beta subunit (*BCKDHB* gene, MSUD type IB[246]), E2 subunit (*DBT* gene, MSUD type II)[114,144] and E3 (*DLD* gene, dihydrolipoamide dehydrogenase deficiency, or sometime MSUD type III).[148] The latter enzyme is also part of the pyruvate dehydrogenase complex and the alpha-ketoglutarate complex and therefore mutations in this gene cause an overlapping more severe clinical phenotype. There is also a milder variant of MSUD (type V) caused by mutations in the regulatory gene *PPM1K*.[257]

Clinical Features

The clinical presentation of MSUD is similar to that of methylmalonic and propionic acidaemias, with severe neonatal-onset and less severe later-onset forms.[339] There is a controversial thiamine-responsive form. With the development of neurological symptoms, the infant's urine emits a sweet, caramel-like smell (similar to maple syrup or fenugreek), caused by the accumulation of sotolone. Long-term treatment depends on dietary modifications with some patients coming to liver transplantation.

Pathology

Neuropathological observations from fatalities in early infancy include widespread white matter sponginess with loss of oligodendroglia, astrocytosis and reduced myelination,[76] but usually no evidence of myelin destruction. Lipid patterns in the white matter at this early stage are normal,[216] but in untreated older patients, cerebrosides and proteolipid protein are markedly reduced compared with treated patients.[280] Experimental studies of myelinating cerebellar cultures have suggested that the metabolite α-keto-isocaproic acid is a myelination inhibitor that is toxic to glial cells.[322]

There is a single report of abnormal dendritic development in a 6-year-old MSUD patient with mild mental retardation.[167] A case series described by Revillo *et al.*[294] showed that cerebral oedema with tonsillar herniation may be an acute cause of death.

Glutaric Aciduria Type 1

Biochemical Defect

Glutaric aciduria type 1 (GAD1) is a recessively inherited disorder caused by deficiency of the mitochondrial enzyme, glutaryl-CoA dehydrogenase. The enzyme catalyzes the conversion of glutaryl-CoA to crotonyl-CoA and is involved in the breakdown of lysine, hydroxylysine and tryptophan.[159]

Clinical Features

Clinical variability is recognized.[143,185] During the first year of life, mild developmental delay, irritability and macrocephaly may develop. At this stage, neuroimaging shows characteristic frontotemporal atrophy often in the presence of a large head (microencephalic macrocephaly). Symptoms develop towards the end of the first year with approximately two-thirds of patients developing an encephalitic crisis (often mistaken for encephalitis), subsequently recovering with a complex movement disorder consisting of dystonia and chorea. The remaining one-third develop an insidiously progressive dystonia without an encephalitic crisis. Neuroimaging at this stage shows basal ganglia lesions (striatal necrosis) in addition to frontotemporal atrophy. Thereafter, there may be further encephalitic crises resulting in an increasingly severe movement disorder, spasticity, mental retardation and epilepsy, or these may develop insidiously. Rarely, infants present with acute subdural and retinal haemorrhage, which may be misdiagnosed as non-accidental head injury.[37,55,119,230]

The child excretes excessive amounts of urinary glutaric and 3-hydroxyglutaric acids, whereas plasma free carnitine is reduced and glutarylcarnitine is raised.[178] Fibroblast glutaryl-CoA dehydrogenase activity is reduced. Published treatment guidelines principally consist of low lysine diet and carnitine supplementation.[178] If treatment is started early, the outcome is good.[74,178]

Pathology

The few neuropathological reports all describe basal ganglia degeneration, predominantly involving the striatum,[173,195,328] but additional features include frontotemporal atrophy and status spongiosus of the white matter (Figure 5.20). The possibility of GAD1 needs to be considered in the context of a child with subdural haemorrhage. One possible contributory mechanism is the prominent cerebral atrophy in the context of an enlarged head, causing an enlarged space to be traversed by the emissary veins,[124] which become vulnerable to rupturing.

Urea Cycle Disorders

Biochemical Defects

These comprise disorders affecting the five reactions of the urea cycle: carbamoyl phosphate synthase 1 deficiency, ornithine carbamoyl transferase deficiency, argininosuccinate synthase deficiency (causing classic citrullinaemia), argininosuccinate lyase deficiency (arginosuccinic aciduria) and arginase deficiency (Figure 5.21). This group also

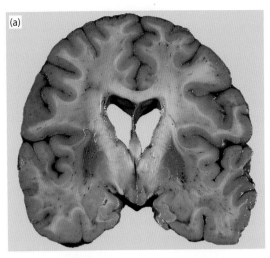

(a)

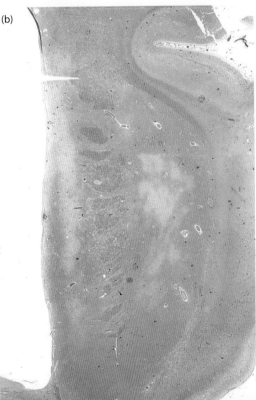

(b)

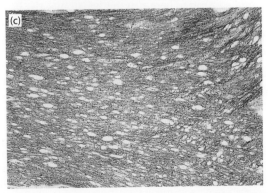

(c)

5.20 Glutaric acidaemia type I. (a) The caudate and putamen are shrunken and grey. **(b)** Linear gliotic scars are seen in the atrophied striatal nuclei. **(c)** Status spongiosus in the optic nerve. Luxol fast blue.

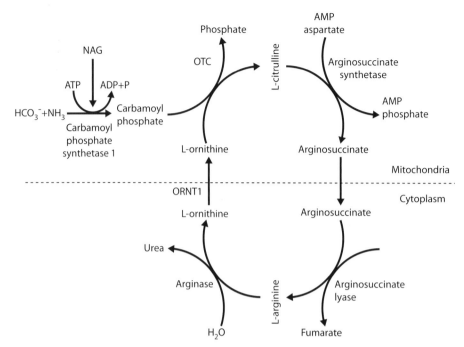

5.21 The urea cycle takes excess ammonia and generates urea. It occurs partly within the mitochondria and partly within the cytoplasm. Mutations involve each of the major enzymes: carbamoyl phosphate synthase 1 (CPS1), ornithine carbamoyltransferase/ornithine transcarbamoylase (OTC), argininosuccinate synthase (causing classic citrullinaemia), argininosuccinate lyase deficiency (arginosuccinic aciduria) and arginase. Mutations can also cause N-acetylglutamate synthase loss leading to a deficiency of N-acetylglutamate (NAG), an essential activator of CPS1. Finally, mitochondrial ornithine transporter (ORNT1, *SLC25A15*) mutations cause the hyperornithinaemia-hyperammonaemia-homocitrullinuria (HHH) syndrome.

encompasses deficiencies of N-acetylglutamate synthase (NAGS) leading to a deficiency of N-acetylglutamate (NAG), an essential activator of CPS1, the disorder mimicking CPS1 deficiency. Finally, it includes mutations in the mitochondrial ornithine transporter (ORNT1, *SLC25A15*), causing the hyperornithinaemia-hyperammonaemia-homocitrullinuria (HHH) syndrome.[1,130] Ornithine carbamoyltransferase deficiency is the most common and is X-linked with manifesting heterozygotes. The others are all recessively inherited.

Clinical Features

The clinical features are variable.[130] Many patients present in the neonatal period with a hyperammonaemic encephalopathy, with loss of appetite, vomiting, lethargy, ataxia, coma and seizures. There may be hepatomegaly. In later-onset forms, there is often a preceding history of protein avoidance, vomiting, failure to thrive and mental retardation. Children with arginosuccinic aciduria develop brittle hair with trichorrhexis nodosa. Although children with arginase deficiency can have episodes of acute hyperammonaemic encephalopathy, most present with a progressive diplegia and dementia. Dystonia, ataxia and seizures also commonly develop. The key investigations are plasma ammonia and amino acids, and urinary orotic acid. These determine the severity of hyperammonaemia (and, therefore, the acute management) and point towards the site of the defect. Depending upon the disease, the diagnosis is confirmed enzymatically or by mutation analysis.

Management can be divided into emergency versus long-term care. The principles of the latter are the restriction of dietary protein intake, promoting nitrogen excretion via alternative pathways utilizing sodium benzoate and sodium phenylbutyrate, and supplementation, e.g. with arginine, which, excepting arginase deficiency, becomes an essential amino acid. Liver transplantation is considered curative in some cases.[130]

Pathology

Autopsy studies in newborns dying rapidly from fulminant hyperammonaemia show relatively non-specific findings: brain swelling, spongy changes in white matter and basal ganglia, and cortical neuronal loss.[99,179] A more specific and frequent, although by no means consistent, observation is the presence of Alzheimer type II astrocytes, which appear to correlate with the level of hyperammonaemia. Massive cystic cerebral hemispheric destruction with severe atrophy and ulegyria (Figure 5.22) and mineralization of deep grey matter neurons have been described in severely affected females with X-linked ornithine carbamoyltransferase deficiency surviving beyond the neonatal period,[50,135,179] and in a girl with carbamoyl phosphate synthetase deficiency.[99] The relative contributions of hypotension and cardiac arrest to the pathogenesis are disputed, but it is notable that the Ammon's horn and cerebellum may be spared.[99,179]

The occurrence in some cases of hypomyelination and cerebellar heterotopias, as well as very early evidence of cystic necrosis suggests that damage may begin in fetal life.[112,135] We recently described a case of OTC deficiency with evidence of acute brain injury, that additionally showed malformations in the form of dentate nucleus dysplasia, granule cell dispersion and large cerebellar heterotopias. This raises the possibility of *in utero* effects on the fetus, despite heterozygous compensation from the mother.[262]

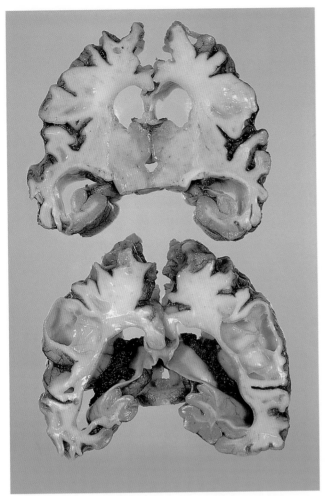

5.22 Ornithine carbamoyltransferase deficiency. Global cystic destruction has resulted in a 'walnut brain' with gliotic knife-edge gyri.

MISCELLANEOUS METABOLIC DISORDERS

Congenital Disorders of Glycosylation

Definition

These are a heterogeneous group of disorders characterized by defects affecting post-translational glycosylation of proteins and other macromolecules. Protein glycosylation may be N-linked or O-linked and disorders affect one or both. In addition, there are disorders of lipid glycosylation. Given the rapidly expanding list of complex glycosylation disorders, a new classification was proposed in 2009.[158] Each disorder is named with the official gene name followed by '-CDG'. This supersedes the previous classification based on transferrin isoelectric focussing pattern (CDG-1 and CDG-II).

Among the disorders of O-linked glycosylation are the dystroglycanopathies presenting with muscular dystrophy, type 2 lissencephaly and ocular anomalies.

Disorders of N-linked glycosylation have similarly expanded (see Theodore and Morava[351]) with disorders being due to either sugar chain assembly (type 1) or processing (type 2). The clinical features of these disorders are diverse, but neurological complications (developmental delay, ataxia and visual disorders), dysmorphic facial features, failure to thrive and coagulopathies are frequent.[351] Abnormalities of isoelectric focussing of transferrin are common and are used in the clinical investigation.

Descriptions of the neuropathological findings are limited and restricted to the more common neurological disorders, particularly PIMM2-CDG (phosphomannose mutase 2 deficiency). A number of case reports also predate genetic classification. Some but not all of these showed a severe pattern of olivopontocerebellar atrophy.[14,149,340,358,388]

Pathology

PIMM2-CDG

This recessively inherited multisystem disorder due to a deficiency of phosphomannose mutase (PMM) 2 is the most common CDG. It usually has severe CNS, but variable multisystem involvement.[20,351] In infancy, developmental delay, rolling eye movements and an alternating squint, hypotonia and depressed reflexes develop. Later, ataxia and moderate to severe mental retardation become evident. In childhood, retinitis pigmentosa, stroke-like episodes and epilepsy may occur, but dementia is not evident. Mild facial dysmorphism, abnormal subcutaneous fat distribution on the buttocks and inverted nipples may be apparent early. There is great variability in the involvement of other organs, even between siblings. Neuroradiology shows marked early-onset cerebellar atrophy. The diagnosis is suspected from abnormalities of serum transferrin isoelectric focusing and confirmed on leukocyte phosphomannomutase activity. There is no effective treatment.

Aronica et al. described in detail the neuropathology of a genetically confirmed case.[20] This patient had a small brain, but with particularly severe cerebellar involvement, and relative preservation of the pons and inferior olivary nucleus. There was marked cerebellar atrophy, particularly affecting the anterior lobe and sparing the flocculonodular lobe. Loss of Purkinje cells accompanied occasional axonal torpedoes and some dendritic swellings. The autopsy showed multisystem disease with hepatosplenomegaly and hypertrophic cardiomyopathy. Muscle samples showed mild myopathic changes with focal myofibrillary disarray on electron microscopy.

Wurm et al.[389] described two cases that came to autopsy, one with a small but structurally normal cerebrum, the other with agenesis of the corpus callosum.

There are cases reported that have come to autopsy following presentation as hydrops fetalis with extra-CNS involvement. There was sparing of the cerebellum in one[356] and partial atrophy in another.[247]

Miscellaneous Forms

Vesla et al. described a patient with ALG8-CDG (CDG Ih).[372] ALG8-CDG is caused by deficiency of N-glucosyltransferase. The child was born preterm and presented in the neonatal period with oedema, hypotonia, seizures and coagulopathy. At post-mortem evaluation, there was mild cerebellar atrophy but no cortical malformations.

Sun et al. described the post-mortem findings in a patient with ALG3-CDG (CDG Id) (deficiency of dolichyl-P

(Dol-P)-mannose (Man):Man$_5$-N-acetyl glucosamine (GlcNAc)$_2$-PP-Dol-alpha-1,3-mannosyltransferase). The child had optic atrophy, severe hypotonia, dysmorphism, hypoglycaemia and skeletal abnormalities. The post-mortem brain showed Dandy–Walker malformation with vermal agenesis and agenesis of the corpus callosum.

Haemorrhagic lesions of the cerebral cortex and subcortical white matter were noted in a genetically undefined case

(CDG-x) as reported by Agarwal *et al.*[4] In addition, there were prominent Alzheimer type II cells.

Miscellaneous Paediatric Disorders

There has been a large and on-going expansion in the number of genetic and metabolic disorders in children, many with relatively limited neuropathological descriptions. A partial summary is given in Table 5.8.

TABLE 5.8 Miscellaneous paediatric disorders

Disorder	Enzyme defect	Neuropathology
Epilepsy, pyridoxine-dependent (EPD)	*ALDH7A1* gene that encodes for an alpha-aminoadipic semialdehyde (AASA) dehydrogenase (antiquitin)	Not described, but predicted to show malformations of cortical development, ventriculomegaly and cerebral haemorrhage based on MRI descriptions[217,223]
Pyridox(am)ine 5′-phosphate oxidase deficiency	*PNPO* gene	Not described
Molybendum cofactor deficiency	*MOCS1, MOCS2, GEPH* (gephyrin)	Severe destructive lesions of the cerebrum focussed on the cortex or white matter with cystic change in several reports. Enlarged ventricles. Variable posterior fossa involvement[48,156,298,301,306,350]
Multiple carboxylase deficiency	Holocarboxylase synthetase	Not described. Subependymal cysts described on neuroradiology[330]
GLUT1 deficiency syndrome	GLUT1 gene (*SCL2A1*)	Not described
Serine biosynthesis defects	3-Phosphoglycerate dehydrogenase, phosphoserine aminotransferase, phosphoserine phosphatase	Not described
Menkes kinky hair disease. Allelic with occipital horn syndrome[267] and trichopoliodystrophy	*ATP7A* (alpha-polypeptide Cu-transporting ATPase)	Cerebral and cerebellar atrophy, Purkinje cell loss, axonal torpedoes, somatic spikes, vasculopathy, mitochondrial abnormalities, spinal cord demyelination, abnormalities of hair morphology[27,160,210,228,250,392]
Adenylosuccinate lyase deficiency	*ADSL*	Microcystic encephalopathy with destruction of all cellular elements[220,232]
GABA transaminase deficiency	*GABAT*	Leukodystrophy[157]
L-2-hydroxy glutaric aciduria	*L2HGDH*	Spongiosis affecting pallidum, dentate and white matter.[63,165,188,320] Possible predisposition to brain tumours[229,270]
D-2-hydroxy glutaric aciduria	*D2HGDH, IDH2*	Saccular aneurysm, vasculopathy, cerebral white matter loss[101]
Succinic semialdehyde dehydrogenase deficiency (SSADHD)	*ALDH5A1*	Discolouration of globus pallidus with congestion and mineralization[174]
Lesch–Nyhan syndrome	Hypoxanthine-guanine phosphoribosyltransferase (*HPRT*)	Posterior column degeneration, cerebellar granule cell degeneration, peripheral neuropathy, destructive lesions of the cerebrum, small brain. Historical literature suggests variable histology with remarkable basal ganglia preservation; Del Bigio's review raised the possibility of cerebellar damage as being the most consistent feature[90,255,331,374,375]

REFERENCES

1. Adeva MM, Souto G, Blanco N, Donapetry C. Ammonium metabolism in humans. *Metabolism* 2012;**61**:1495–511.

2. Adler-Abramovich L, Vaks L, Carny O, *et al*. Phenylalanine assembly into toxic fibrils suggests amyloid etiology in phenylketonuria. *Nat Chem Biol* 2012;**8**:701–6.

3. Agamanolis DP, Potter JL, Herrick MK, Sternberger NH. The neuropathology of glycine encephalopathy: a report of five cases with immunohistochemical and ultrastructural observations. *Neurology* 1982;**32**:975–85.

4. Agarwal B, Ahmed A, Rushing EJ, *et al*. Congenital disorder of glycosylation-X: clinicopathologic study of an autopsy case with distinct neuropathologic features. *Hum Pathol* 2007;**38**:1714–19.

5. Aicardi J, Goutières F. A progressive familial encephalopathy in infancy with calcifications of the basal ganglia and chronic cerebrospinal fluid lymphocytosis. *Ann Neurol* 1984;**15**:49–54.

6. Aicardi J, Crow YJ, Stephenson JB. Aicardi–Goutières syndrome. In: Pagon RA, Adam MP, Bird TD, *et al*. editors. *GeneReviews*. University of Washington, Seattle 2012.

7. Akwa Y, Hassett DE, Eloranta ML, *et al*. Transgenic expression of IFN-alpha in the central nervous system of mice protects against lethal neurotropic viral infection but induces inflammation and neurodegeneration. *J Immunol* 1998;**161**:5016–26.

8. Alexander WS. Progressive fibrinoid degeneration of fibrillary astrocytes associated with mental retardation in a hydrocephalic infant. *Brain* 1949;**72**:373–81.

9. Ali ZS, Van Der Voorn JP, Powers JM. A comparative morphologic analysis of adult onset leukodystrophy with neuroaxonal spheroids and pigmented glia – a role for oxidative damage. *J Neuropathol Exp Neurol* 2007;**66**:660–72.

10. Allegri G, Fernandes MJ, Scalco FB, *et al*. Fumaric aciduria: an overview and the first Brazilian case report. *J Inherit Metab Dis* 2010;**33**:411–19.

11. Amir RE, Van den Veyver IB, Wan M, *et al*. Rett syndrome is caused by mutations in X-linked MECP2, encoding methyl-CpG-binding protein 2. *Nat Genet* 1999;**23**:185–8.

12. Anderson JM. Spongy degeneration in the white matter of the central nervous system in the newborn: pathological findings in three infants, one with hyperglycinaemia. *J Neurol Neurosurg Psychiatr* 1969;**32**:328–37.

13. Antonini A, Goldwurm S, Benti R, *et al*. Genetic, clinical, and imaging characterization of one patient with late-onset, slowly progressive, pantothenate kinase-associated neurodegeneration. *Mov Disord* 2006;**21**:417–18.

14. Antoun H, Villeneuve N, Gelot A, Panisset S, Adamsbaum C. Cerebellar atrophy: an important feature of carbohydrate deficient glycoprotein syndrome type 1. *Pediatr Radiol* 1999;**29**:194–8.

15. Ariani F, Hayek G, Rondinella D, *et al*. FOXG1 is responsible for the congenital variant of Rett syndrome. *Am J Human Genet* 2008;**83**:89–93.

16. Armstrong DD. Neuropathology of Rett syndrome. *Ment Retard Dev Disabil Res Rev* 2002;**8**:72–6.

17. Armstrong DD, Dunn K, Antalffy B. Decreased dendritic branching in frontal, motor and limbic cortex in Rett syndrome compared with trisomy 21. *J Neuropathol Exp Neurol* 1998;**57**:1013–17.

18. Armstrong D, Dunn JK, Antalffy B, Trivedi R. Selective dendritic alterations in the cortex of Rett syndrome. *J Neuropathol Exp Neurol* 1995;**54**:195–201.

19. Armstrong DD, Dunn JK, Schultz RJ, *et al*. Organ growth in Rett syndrome: a postmortem examination analysis. *Pediatr Neurol* 1999;**20**:125–9.

20. Aronica E, Kempen AAMW, Heide M, *et al*. Congenital disorder of glycosylation type Ia: a clinicopathological report of a newborn infant with cerebellar pathology. *Acta Neuropathol* 2005;**109**:433–42.

21. Arsénio-Nunes ML, Goutières F. Diagnosis of infantile neuroaxonal dystrophy by conjunctival biopsy. *J Neurol Neurosurg Psychiatr* 1978;**41**:511–15.

22. Bahi-Buisson N, Villeneuve N, Caietta E, *et al*. Recurrent mutations in the CDKL5 gene: Genotype-phenotype relationships. *Am J Med Genet* 2012;**158A**:1612–9.

23. Balajee AS, Bohr VA. Genomic heterogeneity of nucleotide excision repair. *Gene* 2000;**250**:15–30.

24. Bandmann O, Nygaard TG, Surtees R, *et al*. Dopa-responsive dystonia in British patients: new mutations of the GTP-cyclohydrolase I gene and evidence for genetic heterogeneity. *Hum Mol Genet* 1996;**5**:403–6.

25. Bandmann O, Valente EM, Holmans P, *et al*. Dopa-responsive dystonia: a clinical and molecular genetic study. *Ann Neurol* 1998;**44**:649–56.

26. Barcia G, Fleming MR, Deligniere A, *et al*. De novo gain-of-function KCNT1 channel mutations cause malignant migrating partial seizures of infancy. *Nat Genet*;2012;**44**:1255–9.

27. Barnard RO, Best PV, Erdohazi M. Neuropathology of Menkes' disease. *Dev Med Child Neurol* 1978;**20**:586–97.

28. Barth PG, Walter A, van Gelderen I. Aicardi–Goutières syndrome: a genetic microangiopathy? *Acta Neuropathol* 1999;**98**:212–16.

29. Baslow MH, Guilfoyle DN. Are astrocytes the missing link between lack of brain aspartoacylase activity and the spongiform leukodystrophy in Canavan disease? *Neurochem Res* 2009;**34**:1523–34.

30. Bauman ML, Kemper TL. Morphologic and histoanatomic observations of the brain in untreated human phenylketonuria. *Acta Neuropathol* 1982;**58**:55–63.

31. Bauman ML, Kemper TL, Arin DM. Pervasive neuroanatomic abnormalities of the brain in three cases of Rett's syndrome. *Neurology* 1995;**45**:1581–6.

32. Baykal T, Gokcay GH, Ince Z, *et al*. Consanguineous 3-methylcrotonyl-CoA carboxylase deficiency: early-onset necrotizing encephalopathy with lethal outcome. *J Inherit Metab Dis* 2005;**28**:229–33.

33. Belichenko PV, Hagberg B, Dahlström A. Morphological study of neocortical areas in Rett syndrome. *Acta Neuropathol* 1997;**93**:50–61.

34. Bianchin MM, Martin KC, de Souza AC, de Oliveira MA, de Mello Rieder CR. Comment on microglia in neurodegenerative disease. Correspondence. *Nat Rev Neurol* 2010;**6**:1–2.

35. Bick U, Fahrendorf G, Ludolph AC, *et al*. Disturbed myelination in patients with treated hyperphenylalaninaemia: evaluation with magnetic resonance imaging. *Eur J Pediatr* 1991;**150**:185–9.

36. Bick U, Ullrich K, Stöber U, *et al*. White matter abnormalities in patients with treated hyperphenylalaninaemia: magnetic resonance relaxometry and proton spectroscopy findings. *Eur J Pediatr* 1993;**152**:1012–20.

37. Bishop FS, Liu JK, McCall TD, Brockmeyer DL. Glutaric aciduria type 1 presenting as bilateral subdural hematomas mimicking nonaccidental trauma. *J Neurosurg* 2007;**106**(3 Suppl):222–6.

38. Blackwood W, Buxton PH, Cumings JN, Robertson DJ, Tucker SM. Diffuse cerebral degeneration in infancy (Alpers' disease). *Arch Dis Child* 1963;**38**:193–204.

39. Blau N, Hennermann JB, Langenbeck U, Lichter-Konecki U. Molecular genetics and metabolism. *Mol Genet Metab* 2011;**104**:S2–S9.

40. Blom HJ, Smulders Y. Overview of homocysteine and folate metabolism. With special references to cardiovascular disease and neural tube defects. *J Inherit Metab Dis* 2010;**34**:75–81.

41. Bodner KE, Aldridge K, Moffitt AJ, *et al*. A volumetric study of basal ganglia structures in individuals with early-treated phenylketonuria. *Mol Genet Metab* 2012;**107**:302–7.

42. Böhm N, Uy J, Kiessling M, Lehnert W. Multiple acyl-CoA dehydrogenation deficiency (glutaric aciduria type II), congenital polycystic kidneys, and symmetric warty dysplasia of the cerebral cortex in two newborn brothers. II. Morphology and pathogenesis. *Eur J Pediatr* 1982;**139**:60–1.

43. Boltshauser E, Barth PG, Troost D, Martin E, Stallmach T. 'Vanishing white matter' and ovarian dysgenesis in an infant with cerebro-oculo-facio-skeletal phenotype. *Neuropediatrics* 2002;**33**:57–62.

44. Boneh A, Degani Y, Harari M. Prognostic clues and outcome of early treatment of nonketotic hyperglycinemia. *Pediatr Neurol* 1996;**15**:137–41.

45. Borrett D, Becker LE. Alexander's disease. A disease of astrocytes. *Brain* 1985;**108**(Pt 2):367–85.

46. Brenner M, Johnson AB, Boespflug-Tanguy O, *et al*. Mutations in GFAP, encoding glial fibrillary acidic protein, are associated with Alexander disease. *Nat Genet* 2001;**27**:117–20.

47. Brenton JN, Matsumoto JA, Rust RS, Wilson WG. White matter changes in an untreated, newly diagnosed case of classical homocystinuria. *J Child Neurol* 2014;**29**:88–92.

48. Brown GK, Scholem RD, Croll HB, Wraith JE, McGill JJ. Sulfite oxidase deficiency: clinical, neuroradiologic, and biochemical features in two new patients. *Neurology* 1989;**39**(2 Pt 1):252–7.

49. Brück W, Herms J, Brockmann K, Schulz-Schaeffer W, Hanefeld F. Myelin-

opathia centralis diffusa (vanishing white matter disease): evidence of apoptotic oligodendrocyte degeneration in early lesion development. *Ann Neurol* 2001;**50**:532–6.

50. Bruton CJ, Corsellis JA, Russell A. Hereditary hyperammonaemia. *Brain* 1970;**93**:423–34.

51. Bugiani M, Boor I, Powers JM, Scheper GC, van der Knaap MS. Leukoencephalopathy with vanishing white matter: a review. *J Neuropathol Exp Neurol* 2010;**69**:987–96.

52. Bugiani M, Boor I, van Kollenburg B, *et al.* Defective glial maturation in vanishing white matter disease. *J Neuropathol Exp Neurol* 2011;**70**:69–82.

53. Burn J, Wickramasinghe HT, Harding B, Baraitser M. A syndrome with intracranial calcification and microcephaly in two sibs, resembling intrauterine infection. *Clin Genet* 1986;**30**:112–16.

54. Cardo E, Campistol J, Caritg J, *et al.* Fatal haemorrhagic infarct in an infant with homocystinuria. *Dev Med Child Neurol* 1999;**41**:132–5.

55. Carman KB, Aydogdu SD, Yakut A, Yarar C. Glutaric aciduria type 1 presenting as subdural haematoma. *J Paediatr Child Health* 2012;**48**:712.

56. Carranza Rojo D, Hamiwka L, McMahon JM, *et al. De novo* SCN1A mutations in migrating partial seizures of infancy. *Neurology* 2011;**77**:380–3.

57. Carrillo-Carrasco N, Chandler RJ, Venditti CP. Combined methylmalonic acidemia and homocystinuria, cblC type. I. Clinical presentations, diagnosis and management. *J Inherit Metab Dis* 2011;**35**:91–102.

58. Carson NA, Dent CE, Field C, Gaull GE. Homocystinuria: clinical and pathological review of ten cases. *J Pediatr* 1965;**66**:565–83.

59. Casanova MF, Buxhoeveden D, Switala A, Roy E. Rett syndrome as a minicolumnopathy. *Clin Neuropathol* 2003;**22**:163–8.

60. Castrén M, Gaily E, Tengström C, *et al.* Epilepsy caused by CDKL5 mutations. *Eur J Paediatr Neurol* 2011;**15**:65–9.

61. Catarino CB, Liu JYW, Liagkouras I, *et al.* Dravet syndrome as epileptic encephalopathy: evidence from long-term course and neuropathology. *Brain* 2011;**134**(Pt 10):2982–3010.

62. Chao H-T, Zoghbi HY. MeCP2: only 100% will do. *Nat Neurosci* 2012;**15**:176–7.

63. Chen E, Nyhan WL, Jakobs C, *et al.* L-2-Hydroxyglutaric aciduria: neuropathological correlations and first report of severe neurodegenerative disease and neonatal death. *J Inherit Metab Dis* 1996;**19**:335–43.

64. Chen Q, Zhu Y-C, Yu J, *et al.* CDKL5, a protein associated with rett syndrome, regulates neuronal morphogenesis via Rac1 signaling. *J Neurosci* 2010;**30**:12777–86.

65. Chow CW, Frerman FE, Goodman SI, *et al.* Striatal degeneration in glutaric acidaemia type II. *Acta Neuropathol* 1989;**77**:554–6.

66. Claes L, Del-Favero J, Ceulemans B, *et al. De novo* mutations in the sodium-channel gene SCN1A cause severe myoclonic epilepsy of infancy. *Am J Hum Genet* 2001;**68**:1327–32.

67. Clayton PT, Smith I, Harding B, *et al.* Subacute combined degeneration of the cord, dementia and parkinsonism due to an inborn error of folate metabolism. *J Neurol Neurosurg Psychiatr* 1986;**49**:920–7.

68. Cleary MA, Walter JH, Wraith JE, *et al.* Magnetic resonance imaging of the brain in phenylketonuria. *Lancet* 1994;**344**:87–90.

69. Cleary MA, Walter JH, Wraith JE, *et al.* Magnetic resonance imaging in phenylketonuria: reversal of cerebral white matter change. *J Pediatr* 1995;**127**:251–5.

70. Cleaver JE, Thompson LH, Richardson AS, States JC. A summary of mutations in the UV-sensitive disorders: xeroderma pigmentosum, Cockayne syndrome, and trichothiodystrophy. *Hum Mutat* 1999;**14**:9–22.

71. Coelho D, Kim JC, Miousse IR, *et al.* Mutations in ABCD4 cause a new inborn error of vitamin B12 metabolism. *Nat Genet* 2012;**44**:1152–5.

72. Coppola G, Plouin P, Chiron C, Robain O, Dulac O. Migrating partial seizures in infancy: a malignant disorder with developmental arrest. *Epilepsia* 1995;**36**:1017–24.

73. Cordero ME, Trejo M, Colombo M, Aranda V. Histological maturation of the neocortex in phenylketonuric rats. *Early Hum Dev* 1983;**8**:157–73.

74. Couce ML, López-Suárez O, Báveda MD, *et al.* Glutaric aciduria type I: Outcome of patients with early- versus late-diagnosis. *Eur J Paediatr Neurol* 2013;**17**:383–9.

75. Crome L. The association of phenylketonuria with leucodystrophy. *J Neurol Neurosurg Psychiatr* 1962;**25**:149–53.

76. Crome L, Dutton G, Ross CF. Maple syrup urine disease. *J Pathol Bacteriol* 1961;**81**:379–84.

77. Crow YJ, Rehwinkel J. Aicardi–Goutières syndrome and related phenotypes: linking nucleic acid metabolism with autoimmunity. *Hum Mol Genet* 2009;**18**:R130–6.

78. Crow YJ, Hayward BE, Parmar R, *et al.* Mutations in the gene encoding the 3′–5′ DNA exonuclease TREX1 cause Aicardi–Goutières syndrome at the AGS1 locus. *Nat Genet* 2006;**38**:917–20.

79. Crow YJ, Leitch A, Hayward BE, *et al.* Mutations in genes encoding ribonuclease H2 subunits cause Aicardi-Goutières syndrome and mimic congenital viral brain infection. *Nat Genet* 2006;**38**:910–16.

80. Cuadrado E, Jansen MH, Anink J, *et al.* Chronic exposure of astrocytes to interferon-α reveals molecular changes related to Aicardi–Goutières syndrome. *Brain* 2013;**136**(Pt 1):245–58.

81. Damen D, Heumann R. MeCP2 phosphorylation in the brain: from transcription to behavior. *Biol Chem* 2013;**394**:1595–605.

82. Darin N, Oldfors A, Moslemi AR, Holme E, Tulinius M. The incidence of mitochondrial encephalomyopathies in childhood: clinical features and morphological, biochemical, and DNA abnormalities. *Ann Neurol* 2001;**49**:377–83.

83. Dave P, Curless RG, Steinman L. Cerebellar hemorrhage complicating methylmalonic and propionic acidemia. *Arch Neurol* 1984;**41**:1293–6.

84. Davidzon G, Mancuso M, Ferraris S, *et al.* POLG mutations and Alpers syndrome. *Ann Neurol* 2005;**57**:921–3.

85. Dayan AD, Ramsey RB. An inborn error of vitamin B12 metabolism associated with cellular deficiency of coenzyme forms of the vitamin. Pathological and neurochemical findings in one case. *J Neurol Sci* 1974;**23**:117–28.

86. de Boer J, Hoeijmakers JH. Nucleotide excision repair and human syndromes. *Carcinogenesis* 2000;**21**:453–60.

87. de Kremer RD, Latini A, Suormala T, *et al.* Leukodystrophy and CSF purine abnormalities associated with isolated 3-methylcrotonyl-CoA carboxylase deficiency. *Metab Brain Dis* 2002;**17**:13–18.

88. De Leon GA, Mitchell MH. Histological and ultrastructural features of dystrophic isocortical axons in infantile neuroaxonal dystrophy (Seitelberger's disease). *Acta Neuropathol* 1985;**66**:89–97.

89. Deguchi K, Antalffy BA, Twohill LJ, *et al.* Substance P immunoreactivity in Rett syndrome. *Pediatr Neurol* 2000;**22**:259–66.

90. Del Bigio MR, Halliday WC. Multifocal atrophy of cerebellar internal granular neurons in Lesch-Nyhan disease: case reports and review. *J Neuropathol Exp Neurol* 2007;**66**:346–53.

91. Deodato F, Boenzi S, Santorelli FM, Dionisi-Vici C. Methylmalonic and propionic aciduria. *Am J Med Genet* 2006;**142C**:104–12.

92. Deprez M, D'Hooghe M, Misson JP, *et al.* Infantile and juvenile presentations of Alexander's disease: a report of two cases. *Acta Neurol Scand* 1999;**99**:158–65.

93. Derecki NC, Cronk JC, Lu Z, *et al.* Wild-type microglia arrest pathology in a mouse model of Rett syndrome. *Nature* 2012;**484**:105–9.

94. Dietrich J, Lacagnina M, Gass D, *et al.* EIF2B5 mutations compromise GFAP+ astrocyte generation in vanishing white matter leukodystrophy. *Nat Med* 2005;**11**:277–83.

95. Dobyns WB. Agenesis of the corpus callosum and gyral malformations are frequent manifestations of nonketotic hyperglycinemia. *Neurology* 1989;**39**:817–20

96. Dravet C. How Dravet syndrome became a model for studying childhood genetic epilepsies. *Brain* 2012;**135**:2309–11

97. Duncan Armstrong D. Neuropathology of Rett syndrome. *J Child Neurol* 2005;**20**:747–53.

98. Dziewulska D, Doi H, Fasano A, *et al.* Olfactory impairment and pathology in neurodegenerative disorders with brain iron accumulation. *Acta Neuropathol* 2013;**126**:151–3.

99. Ebels EJ. Neuropathological observations in a patient with carbamylphosphate-synthetase deficiency and in two sibs. *Arch Dis Child* 1972;**47**:47–51.

100. Ebert DH, Gabel HW, Robinson ND, *et al.* Activity-dependent phosphorylation of MECP2 threonine 308 regulates interaction with NcoR. *Nature* 2013;**499**:341–5.

101. Eeg-Olofsson O, Wei Wei Zhang, Olsson Y, Jagell S, Hagenfeldt L. D-2-Hydroxyglutaric aciduria with cerebral, vascular, and muscular abnormalities in a 14-year-old boy. *J Child Neurol* 2000;**15**:488–92.

102. Egger J, Harding BN, Boyd SG, Wilson J, Erdohazi M. Progressive neuronal degeneration of childhood (PNDC) with liver disease. *Clin Pediatr (Phila)* 1987;**26**:167–73.

103. Elo JM, Yadavalli SS, Euro L, *et al.* Mitochondrial phenylalanyl-tRNA synthetase mutations underlie fatal infantile Alpers encephalopathy. *Hum Mol Genet* 2012;**21**:4521–9.

104. Engel J, International League Against Epilepsy (ILAE). A proposed diagnostic scheme for people with epileptic seizures and with epilepsy: report of the ILAE Task Force on Classification and Terminology. *Epilepsia [Internet]* 2001;**42**:796–803.

105. Engel LA, Jing Z, O'Brien DE, Sun M, Kotzbauer PT. Catalytic function of

PLA2G6 Is impaired by mutations associated with infantile neuroaxonal dystrophy but not dystonia-parkinsonism. *PLoS ONE* 2010;**5**:e12897.

106. Escayg A, MacDonald BT, Meisler MH, *et al*. Mutations of SCN1A, encoding a neuronal sodium channel, in two families with GEFS+2. *Nat Genet* 2000;**24**:343–5.

107. Evans JC, Archer HL, Colley JP, *et al*. Early onset seizures and Rett-like features associated with mutations in CDKL5. *Eur J Hum Genet* 2005;**13**:1113–20.

108. Fasulo L, Saucedo S, Caceres L, Solis S, Caraballo R. Migrating focal seizures during infancy: a case report and pathologic study. *Pediatr Neurol* 2012;**46**:182–4.

109. Federico A, Scali O, Stromillo ML, *et al*. Peripheral neuropathy in vanishing white matter disease with a novel EIF2B5 mutation. *Neurology* 2006;**67**:353–5.

110. Feliz B, Witt DR, Harris BT. Propionic acidemia: a neuropathology case report and review of prior cases. *Arch Pathol Lab Med* 2003;**127**:e325–8.

111. Ferrari G, Lamantea E, Donati A, *et al*. Infantile hepatocerebral syndromes associated with mutations in the mitochondrial DNA polymerase-gammaA. *Brain* 2005;**128**(Pt 4):723–31.

112. Filloux F, Townsend JJ, Leonard C. Ornithine transcarbamylase deficiency: neuropathologic changes acquired *in utero*. *J Pediatr* 1986;**108**:942–5.

113. Fischer AQ, Challa VR, Burton BK, McLean WT. Cerebellar hemorrhage complicating isovaleric acidemia: a case report. *Neurology* 1981;**31**:746–8.

114. Fisher CW, Lau KS, Fisher CR, *et al*. A 17-bp insertion and a Phe215- Cys missense mutation in the dihydrolipoyl transacylase (E2) mRNA from a thiamine-responsive maple syrup urine disease patient WG-34. *Biochem Biophys Res Commun* 1991;**174**:804–9.

115. Fogli A, Wong K, Eymard-Pierre E, *et al*. Cree leukoencephalopathy and CACH/VWM disease are allelic at theEIF2B5 locus. *Ann Neurol* 2002;**52**:506–10.

116. Francalanci P, Eymard-Pierre E, Dionisi-Vici C, *et al*. Fatal infantile leukodystrophy: a severe variant of CACH/VWM syndrome, allelic to chromosome 3q27. *Neurology* 2001;**57**:265–70.

117. Freilich ER. Novel SCN1A Mutation in a proband with malignant migrating partial seizures of infancy SCN1A in migrating partial seizures. *Arch Neurol* 2011;**68**:665–71.

118. Friede RL. *Developmental neuropathology*, 2nd edn. Berlin: Springer-Verlag, 1989.

119. Gago LC, Wegner RK, Capone A, Williams GA. Intraretinal hemorrhages and chronic subdural effusions: glutaric aciduria type 1 can be mistaken for shaken baby syndrome. *Retina* 2003;**23**:724–6.

120. Genton P, Velizarova R, Dravet C. Dravet syndrome: The long-term outcome. *Epilepsia* 2011;**52**:44–9.

121. Gibson JB, Carson NA, Neill DW. Pathological findings in homocystinuria. *J Clin Pathol* 1964;**17**:427–37.

122. Goodman SI, Norenberg MD, Shikes RH, Breslich DJ, Moe PG. Glutaric aciduria: biochemical and morphologic considerations. *J Pediatr* 1977;**90**:746–50.

123. Goodman SI, Reale M, Berlow S. Glutaric acidemia type II: a form with deleterious intrauterine effects. *J Pediatr* 1983;**102**:411–13.

124. Gordon N. Glutaric aciduria types I and II. *Brain Dev* 2006;**28**:136–40.

125. Gray F, Destee A, Bourre JM, *et al*. Pigmentary type of orthochromatic leukodystrophy (OLD): a new case with ultrastructural and biochemical study. *J Neuropathol Exp Neurol* 1987;**46**:585–96.

126. Gregory A, Westaway SK, Holm IE, *et al*. Neurodegeneration associated with genetic defects in phospholipase A2. *Neurology* 2008;**71**:1402–9.

127. Grünert SC, Stucki M, Morscher RJ, *et al*. 3-methylcrotonyl-CoA carboxylase deficiency: clinical, biochemical, enzymatic and molecular studies in 88 individuals. *Orphanet J Rare Dis* 2012;**7**:31.

128. Grünert SC, Müllerleile S, De Silva L, *et al*. Propionic acidemia: clinical course and outcome in 55 pediatric and adolescent patients. *Orphanet J Rare Dis* 2013;**8**:6.

129. Guerreiro R, Kara E, Le Ber I, *et al*. Genetic analysis of inherited leukodystrophies: genotype-phenotype correlations in the CSF1R gene. *JAMA Neurol* 2013;**70**:875–82.

130. Häberle J, Boddaert N, Burlina A, *et al*. Suggested guidelines for the diagnosis and management of urea cycle disorders. *Orphanet J Rare Dis* 2012;**7**:32.

131. Hamosh A, Maher JF, Bellus GA, Rasmussen SA, Johnston MV. Long-term use of high-dose benzoate and dextromethorphan for the treatment of nonketotic hyperglycinemia. *J Pediatr* 1998;**132**:709–13.

132. Hamosh A, Scharer G, Van Hove J. Glycine encephalopathy. In: Pagon RA, Adam MP, Bird TD, *et al*. (eds). *GeneReviews*. University of Washington, Seattle, WA, 1993.

133. Harbord MG, Harden A, Harding B, Brett EM, Baraitser M. Megalencephaly with dysmyelination, spasticity, ataxia, seizures and distinctive neurophysiological findings in two siblings. *Neuropediatrics* 1990;**21**:164–8.

134. Harding BN. Progressive neuronal degeneration of childhood with liver disease (Alpers–Huttenlocher syndrome): a personal review. *J Child Neurol* 1990;**5**:273–87.

135. Harding BN, Leonard JV, Erdohazi M. Ornithine carbamoyl transferase deficiency: A neuropathological study. *Eur J Pediatr* 1984;**141**:215–20.

136. Harding BN, Egger J, Portmann B, Erdohazi M. Progressive neuronal degeneration of childhood with liver disease. A pathological study. *Brain* 1986;**109**(Pt 1):181–206.

137. Harding BN, Donley D, Wilson ER. Pigmentary orthochromatic leucodystrophy – a new familial case with onset in early infancy. *Neuropathol Appl Neurobiol* 1990;**16**:270–1.

138. Harding BN, Leonard JV, Erdohazi M. Propionic acidaemia: a neuropathological study of two patients presenting in infancy. *Neuropathol Appl Neurobiol* 1991;**17**:133–8.

139. Harding BN, Alsanjari N, Smith SJ, *et al*. Progressive neuronal degeneration of childhood with liver disease (Alpers' disease) presenting in young adults. *J Neurol Neurosurg Psychiatr* 1995;**58**:320–5.

140. Hayashi M, Miwa-Saito N, Tanuma N, Kubota M. Brain vascular changes in Cockayne syndrome. *Neuropathology* 2011;**32**:113–17.

141. Hayflick SJ, Hogarth P. As iron goes, so goes disease? *Haematologica* 2011;**96**:1571–2.

142. Hayflick SJ, Westaway SK, Levinson B, *et al* Genetic, clinical, and radiographic delineation of Hallervorden–Spatz syndrome. *N Engl J Med* 2003;**348**:33–40.

143. Hedlund GL, Longo N, Pasquali M. Glutaric acidemia type 1. *Am J Med Genet* 2006;**142C**:86–94.

144. Herring WJ, Litwer S, Weber JL, Danner DJ. Molecular genetic basis of maple syrup urine disease in a family with two defective alleles for branched chain acyl-transferase and localization of the gene to human chromosome 1. *Am J Hum Genet* 1991;**48**:342–50.

145. Hodes ME, Zimmerman AW, Aydanian A, *et al*. Different mutations in the same codon of the proteolipid protein gene, PLP, may help in correlating genotype with phenotype in Pelizaeus-Merzbacher disease/X-linked spastic paraplegia (PMD/SPG2). *Am J Med Genet* 1999;**82**:132–9.

146. Hommes FA, Kuipers JR, Elema JD, Jansen JF, Jonxis JH. Propionicacidemia, a new inborn error of metabolism. *Pediatr Res* 1968;**2**:519–24.

147. Honavar M, Janota I, Neville BG, Chalmers RA. Neuropathology of biotinidase deficiency. *Acta Neuropathol* 1992;**84**:461–4.

148. Hong YS, Kerr DS, Craigen WJ, *et al*. Identification of two mutations in a compound heterozygous child with dihydrolipoamide dehydrogenase deficiency. *Hum Mol Genet* 1996;**5**:1925–30.

149. Horslen SP, Clayton PT, Harding BN, *et al*. Olivopontocerebellar atrophy of neonatal onset and disialotransferrin developmental deficiency syndrome. *Arch Dis Child* 1991;**66**:1027–32.

150. Horvath R. Phenotypic spectrum associated with mutations of the mitochondrial polymerase gene. *Brain* 2006;**129**:1674–84.

151. Hunter MF, Peters H, Salemi R, Thorburn D, Mackay MT. Alpers syndrome with mutations in POLG: clinical and investigative features. *Pediatr Neurol* 2011;**45**:311–8.

152. Huppke P, Laccone F, Krämer N, Engel W, Hanefeld F. Rett syndrome: analysis of MECP2 and clinical characterization of 31 patients. *Hum Mol Genet* 2000;**9**:1369–75.

153. Huttenlocher PR. The neuropathology of phenylketonuria: human and animal studies. *Eur J Pediatr* 2000;**159**:S102–6.

154. Ianchulev T, Kolin T, Moseley K, Sadun A. Optic nerve atrophy in propionic acidemia. *Ophthalmology* 2003;**110**:1850–4.

155. Ichinose H, Ohye T, Takahashi E, *et al*. Hereditary progressive dystonia with marked diurnal fluctuation caused by mutations in the GTP cyclohydrolase I gene. *Nat Genet* 1994;**8**:236–42.

156. Irreverre F, Mudd SH, Heizer WD, Laster L. Sulfite oxidase deficiency: Studies of a patient with mental retardation, dislocated ocular lenses, and abnormal urinary excretion of S-sulfo-l-cysteine, sulfite, and thiosulfate. *Biochem Med* 1967;**1**:187–217.

157. Jaeken J, Casaer P, De Cock P, *et al*. Gamma-aminobutyric acid-transaminase deficiency: a newly recognized inborn error of neurotransmitter metabolism. *Neuropediatrics* 1984;**15**:165–9.

158. Jaeken J, Hennet T, Matthijs G, Freeze HH. CDG nomenclature: Time for a change! *Biochim Biophys Acta* 2009;**1792**:825–6.

159. Jafari P, Braissant O, Bonafé L, Ballhausen D. Molecular genetics and metabolism. *Mol Genet Metab* 2011;**104**:425–37.

160. Jankov RP, Boerkoel CF, Hellmann J, *et al.* Lethal neonatal Menkes' disease with severe vasculopathy and fractures. *Acta Paediatr* 1998;**87**:1297–300.

161. Janota I. Neuroaxonal dystrophy in the neonate. A case report. *Acta Neuropathol* 1979;**46**:151–4.

162. Jellinger K, Seitelberger F. Neuropathology of Rett syndrome. *Am J Med Genet Suppl* 1986;**1**:259–88.

163. Jellinger K, Armstrong D, Zoghbi HY, Percy AK. Neuropathology of Rett syndrome. *Acta Neuropathol* 1988;**76**:142–58.

164. Jellinger K, Grisold W, Armstrong D, Rett A. Peripheral nerve involvement in the Rett syndrome. *Brain Dev* 1990;**12**:109–14.

165. Jequier Gygax M, Roulet-Perez E, Meagher-Villemure K, *et al.* Sudden unexpected death in an infant with L-2-hydroxyglutaric aciduria. *Eur J Pediatr* 2008;**168**:957–62.

166. Kalume F. Sudden unexpected death in Dravet syndrome: Respiratory and other physiological dysfunctions. *Respir Physiol Neurobiol* 2013;**189**:324–8.

167. Kamei A, Takashima S, Chan F, Becker LE. Abnormal dendritic development in maple syrup urine disease. *Pediatr Neurol* 1992;**8**:145–7.

168. Kamoshita S, Reed GB, Aguilar MJ. Axonal dystrophy in a case of Canavan's spongy degeneration. *Neurology* 1967;**17**:895–8.

169. Kamoshita S, Neustein HB, Landing BH. Infantile neuroaxonal dystrophy with neonatal onset. Neuropathologic and electron microscopic observations. *J Neuropathol Exp Neurol* 1968;**27**:300–23.

170. Katz DM, Berger-Sweeney JE, Eubanks JH, *et al.* Preclinical research in Rett syndrome: setting the foundation for translational success. *Dis Model Mech* 2012;**5**:733–45.

171. Khateeb S, Flusser H, Ofir R, *et al.* PLA2G6 mutation underlies infantile neuroaxonal dystrophy. *Am J Hum Genet* 2006;**79**:942–8.

172. Kimura S, Sasaki Y, Warlo I, Goebel H-H. Axonal pathology of the skin in infantile neuroaxonal dystrophy. *Acta Neuropathol* 1987;**75**:212–15.

173. Kimura S, Hara M, Nezu A, *et al.* Two cases of glutaric aciduria type 1: clinical and neuropathological findings. *J Neurol Sci* 1994;**123**:38–43.

174. Knerr I, Gibson KM, Murdoch G, *et al.* Neuropathology in succinic semialdehyde dehydrogenase deficiency. *Pediatr Neurol* 2010;**42**:255–8.

175. Koch R, Hanley W, Levy H, *et al.* The maternal phenylketonuria international study: 1984–2. *Pediatrics* 2003;**112**:1523–9.

176. Koch R, Verma S, Gilles FH. Neuropathology of a 4-month-old infant born to a woman with phenylketonuria. *Dev Med Child Neurol* 2008;**50**:230–3.

177. Kölker S, Sauer SW, Hoffmann GF, *et al.* Pathogenesis of CNS involvement in disorders of amino and organic acid metabolism. *J Inherit Metab Dis* 2008;**31**:194–204.

178. Kölker S, Christensen E, Leonard JV, *et al.* Diagnosis and management of glutaric aciduria type I – revised recommendations. *J Inherit Metab Dis* 2011;**34**:677–94.

179. Kornfeld M, Woodfin BM, Papile L, Davis LE, Bernard LR. Neuropathology of ornithine carbamyl transferase deficiency. *Acta Neuropathol* 1985;**65**:261–4.

180. Kotzbauer PT. Altered neuronal mitochondrial coenzyme a synthesis in neurodegeneration with brain iron accumulation caused by abnormal processing, stability, and catalytic activity of mutant pantothenate kinase 2. *J Neurosci* 2005;**25**:689–98.

181. Kruer MC, Hiken M, Gregory A, *et al.* Novel histopathologic findings in molecularly-confirmed pantothenate kinase-associated neurodegeneration. *Brain* 2011;**134**(Pt 4):947–58.

182. Kumar S, Mattan NS, de Vellis J. Canavan disease: a white matter disorder. *Ment Retard Dev Disabil Res Rev* 2006;**12**:157–65.

183. Kure S, Takayanagi M, Narisawa K, Tada K, Leisti J. Identification of a common mutation in Finnish patients with non-ketotic hyperglycinemia. *J Clin Invest* 1992;**90**:160–4.

184. Kure S, Rolland MO, Leisti J, *et al.* Prenatal diagnosis of non-ketotic hyperglycinaemia: enzymatic diagnosis in 28 families and DNA diagnosis detecting prevalent Finnish and Israeli-Arab mutations. *Prenat Diagn* 1999;**19**:717–20.

185. Kyllerman M, Skjeldal OH, Lundberg M, *et al.* Dystonia and dyskinesia in glutaric aciduria type I: clinical heterogeneity and therapeutic considerations. *Mov Disord* 1994;**9**:22–30.

186. Lacey DJ, Terplan K. Abnormal cerebral cortical neurons in a child with maternal PKU syndrome. *J Child Neurol* 1987;**2**:201–4.

187. Larnaout A, Mongalgi MA, Kaabachi N, *et al.* Methylmalonic acidaemia with bilateral globus pallidus involvement: a neuropathological study. *J Inherit Metab Dis* 1998;**21**:639–44.

188. Larnaout A, Hentati F, Belal S, *et al.* Clinical and pathological study of three Tunisian siblings with L-2-hydroxyglutaric aciduria. *Acta Neuropathol* 2013;**88**:367–70.

189. Laugel V. Cockayne syndrome: The expanding clinical and mutational spectrum. *Mech Ageing Dev* 2013;**134**:161–70.

190. Le Gal F, Korff CM, Monso-Hinard C, *et al.* A case of SUDEP in a patient with Dravet syndrome with SCN1A mutation. *Epilepsia* 2010;**51**(9):1915–8.

191. Lee PJ, Ridout D, Walter JH, Cockburn F. Maternal phenylketonuria: report from the United Kingdom Registry 1978–97. *Arch Dis Child* 2005;**90**:143–6.

192. Leech RW, Brumback RA, Miller RH, *et al.* Cockayne syndrome: clinicopathologic and tissue culture studies of affected siblings. *J Neuropathol Exp Neurol* 1985;**44**:507–19.

193. Leegwater PA, Vermeulen G, Könst AA, *et al.* Subunits of the translation initiation factor eIF2B are mutant in leukoencephalopathy with vanishing white matter. *Nat Genet* 2001;**29**:383–8.

194. Lehmann AR, Thompson AF, Harcourt SA, Stefanini M, Norris PG. Cockayne's syndrome: correlation of clinical features with cellular sensitivity of RNA synthesis to UV irradiation. *J Med Genet* 1993;**30**:679–82.

195. Leibel RL, Shih VE, Goodman SI, *et al.* Glutaric acidemia: a metabolic disorder causing progressive choreoathetosis. *Neurology* 1980;**30**:1163–8.

196. Leoni V, Strittmatter L, Zorzi G, *et al.* Molecular genetics and metabolism. *Mol Genet Metab* 2012;**105**:463–71.

197. Leontovich TA, Mukhina JK, Fedorov AA, Belichenko PV. Morphological study of the entorhinal cortex, hippocampal formation, and basal ganglia in Rett syndrome patients. *Neurobiol Dis* 1999;**6**:77–91.

198. Levy HL, Lobbregt D, Barnes PD, Poussaint TY. Maternal phenylketonuria: magnetic resonance imaging of the brain in offspring. *J Pediatr* 1996;**128**:770–5.

199. Li A, Paudel R, Johnson R, *et al.* Pantothenate kinase-associated neurodegeneration is not a synucleinopathy. *Neuropathol Appl Neurobiol* 2013;**39**:121–31.

200. Li R, Johnson AB, Salomons G, *et al.* Glial fibrillary acidic protein mutations in infantile, juvenile, and adult forms of Alexander disease. *Ann Neurol* 2005;**57**:310–26.

201. Li W, Wang X, van der Knaap MS, Proud CG. Mutations linked to leukoencephalopathy with vanishing white matter impair the function of the eukaryotic initiation factor 2B complex in diverse ways. *Mol Cell Biol* 2004;**24**:3295–306.

202. Li Y, Wang H, Muffat J, *et al.* Global transcriptional and translationalrepression in human embryonic stem cell-derived Rett syndrome neurons. *Stem Cell* 2013;**13**:446–58.

203. Lioy DT, Garg SK, Monaghan CE, *et al.* A role for glia in the progression of Rett's syndrome. *Nature* 2011;**475**:497–500.

204. Loeffen J, Smeets R, Voit T, Hoffmann G, Smeitink J. Fumarase deficiency presenting with periventricular cysts. *J Inherit Metab Dis* 2005;**28**:799–800.

205. Longo N. Disorders of biopterin metabolism. *J Inherit Metab Dis* 2009;**32**:333–42.

206. Lopez-Hernandez T, Sirisi S, Capdevila-Nortes X, *et al.* Molecular mechanisms of MLC1 and GLIALCAM mutations in megalencephalic leukoencephalopathy with subcortical cysts. *Hum Mol Genet* 2011;**20**:3266–77.

207. Lu FL, Wang P-J, Hwu W-L, Tsou Yau K-I, Wang T-R. Neonatal type of nonketotic hyperglycinemia. *Pediatr Neurol* 1999;**20**:295–300.

208. Mari F, Azimonti S, Bertani I, *et al.* CDKL5 belongs to the same molecular pathway of MeCP2 and it is responsible for the early-onset seizure variant of Rett syndrome. *Hum Mol Genet* 2005;**14**:1935–46.

209. Martin JJ, Schlote W. Central nervous system lesions in disorders of amino-acid metabolism. A neuropathological study. *J Neurol Sci* 1972;**15**:49–76.

210. Martin JJ, Flament-Durand J, Farriaux JP, *et al.* Menkes kinky-hair disease. A report on its pathology. *Acta Neuropathol* 1978;**42**:25–32.

211. Matalon R, Michals-Matalon K. Canavan disease. In: Pagon RA, Bird TD, Dolan CR, Fong CT, Stephens K (eds). *GeneReviews.* University of Washington, Seattle, WA, 2011.

212. McCombe PA, McLaughlin DB, Chalk JB, *et al.* Spasticity and white matter abnormalities in adult phenylketonuria. *J Neurol Neurosurg Psychiatr* 1992;**55**:359–61.

213. McGraw CM, Samaco RC, Zoghbi HY. Adult neural function requires MeCP2. *Science* 2011;**333**:186–6.

214. McTague A, Appleton R, Avula S, *et al.* Migrating partial seizures of infancy: expansion of the electroclinical, radiological and pathological disease spectrum. *Brain* 2013;**136**:1578–91.

215. Menkes JH. Cerebral proteolipids in phenylketonuria. *Neurology* 1968;**18**:1003–3.

216. Menkes JH, Philippart M, Fiol RE. Cerebral lipids in maple syrup disease. *J Pediatr* 1965;**66**:584–94.

217. Mercimek-Mahmutoglu S, Horvath GA, Coulter-Mackie M, *et al.* Profound neonatal hypoglycemia and lactic acidosis caused by pyridoxine-dependent epilepsy. *Pediatrics* 2012;**129**:e1368–72.

218. Messing A, Head MW, Galles K, *et al.* Fatal encephalopathy with astrocyte inclusions in GFAP transgenic mice. *Am J Pathol* 1998;**152**:391–8.

219. Messing A, Brenner M, Feany MB, Nedergaard M, Goldman JE. Alexander disease. *J Neurosci* 2012;**32**:5017–23.

220. Mierzewska H, Schmidt-Sidor B, Jurkiewicz E, *et al.* Severe encephalopathy with brain atrophy and hypomyelination due to adenylosuccinate lyase deficiency – MRI, clinical, biochemical and neuropathological findings of Polish patients. *Folia Neuropathol* 2009;**47**:314–20.

221. Miladi N, Larnaout A, Dhondt JL, *et al.* Dihydropteridine reductase deficiency in a large consanguineous Tunisian family: clinical, biochemical, and neuropathologic findings. *J Child Neurol* 1998;**13**:475–80.

222. Miles L, DeGrauw TJ, Dinopoulos A, *et al.* Megalencephalic leukoencephalopathy with subcortical cysts: a third confirmed case with literature review. *Pediatr Dev Pathol* 2009;**12**:180–6.

223. Mills PB, Footitt EJ, Mills KA, *et al.* Genotypic and phenotypic spectrum of pyridoxine-dependent epilepsy (ALDH7A1 deficiency). *Brain* 2010;**133**:2148–59.

224. Mitsui J, Matsukawa T, Ishiura H, *et al.* CSF1R mutations identified in three families with autosomal dominantly inherited leukoencephalopathy. *Am J Med Genet* 2012;**159B**:951–7.

225. Mizuguchi M, Itoh M. A 35-year-old female with growth and developmental retardation, progressive ataxia, dementia and visual loss. *Neuropathology* 2005;**25**:103–6.

226. Moosa A, Dubowitz V. Peripheral neuropathy in Cockayne's syndrome. *Arch Dis Child* 1970;**45**:674–7.

227. Morgan NV, Westaway SK, Morton JEV, *et al.* PLA2G6, encoding a phospholipase A2, is mutated in neurodegenerative disorders with high brain iron. *Nat Genet* 2006;**38**:752–4.

228. Morgello S, Peterson H de C, Kahn LJ, Laufer H. Menkes kinky hair disease with 'ragged red' fibers. *Dev Med Child Neurol* 1988;**30**:812–16.

229. Moroni I, Bugiani M, D'Incerti L, *et al.* L-2-hydroxyglutaric aciduria and brain malignant tumors: a predisposing condition? *Neurology* 2004;**62**:1882–4.

230. Morris AA, Hoffmann GF, Naughten ER, *et al.* Glutaric aciduria and suspected child abuse. *Arch Dis Child* 1999;**80**:404–5.

231. Moser HW. Alexander disease: combined gene analysis and MRI clarify pathogenesis and extend phenotype. *Ann Neurol* 2005;**57**:307–8.

232. Mouchegh K, Zikánová M, Hoffmann GF, *et al.* Lethal fetal and early neonatal presentation of adenylosuccinate lyase deficiency: observation of 6 patients in 4 families. *J Pediatr* 2007;**150**:57–61.e2.

233. Muotri AR, Marchetto MCN, Coufal NG, *et al.* L1 retrotransposition in neurons is modulated by MeCP2. *Nature* 2010;**468**:443–6.

234. Namekawa M, Takiyama Y, Aoki Y, *et al.* Identification of GFAP gene mutation in hereditary adult-onset Alexander's disease. *Ann Neurol* 2002;**52**:779–85.

235. Namekawa M, Takiyama Y, Honda J, *et al.* A novel adult case of juvenile-onset Alexander disease: complete remission of neurological symptoms for over 12 years, despite insidiously progressive cervicomedullary atrophy. *Neurol Sci* 2012;**33**:1389–92.

236. Nanau RM, Neuman MG. Adverse drug reactions induced by valproic acid. *Clin Biochem* 2013;**46**:1323–38.

237. Nance MA, Berry SA. Cockayne syndrome: review of 140 cases. *Am J Med Genet* 1992;**42**:68–84.

238. Nance MA, Boyadjiev S, Pratt VM, *et al.* Adult-onset neurodegenerative disorder due to proteolipid protein gene mutation in the mother of a man with Pelizaeus–Merzbacher disease. *Neurology* 1996;**47**:1333–5.

239. Naviaux RK, Nguyen KV. POLG mutations associated with Alpers' syndrome and mitochondrial DNA depletion. *Ann Neurol* 2004;**55**:706–12.

240. Naviaux RK, Nguyen KV. POLG mutations associated with Alpers syndrome and mitochondrial DNA depletion. *Ann Neurol* 2005;**58**:491.

241. Neul JL, Kaufmann WE, Glaze DG, *et al.* Rett syndrome: Revised diagnostic criteria and nomenclature. *Ann Neurol* 2010;**68**:944–50.

242. Newman CG, Wilson BD, Callaghan P, Young L. Neonatal death associated with isovalericacidaemia. *Lancet* 1967;**2**:439–42.

243. Ni Q, Johns GS, Manepalli A, Martin DS, Geller TJ. Infantile Alexander's disease: serial neuroradiologic findings. *J Child Neurol* 2002;**17**:463–6.

244. Nicholson AM, Baker MC, Finch NA, *et al.* CSF1R mutations link POLD and HDLS as a single disease entity. *Neurology* 2013;**80**:1033–40.

245. Nishimura M, Yoshino K, Tomita Y, *et al.* Central and peripheral nervous system pathology of homocystinuria due to 5,10-methylenetetrahydrofolate reductase deficiency. *Pediatr Neurol* 1985;**1**:375–8.

246. Nobukuni Y, Mitsubuchi H, Akaboshi I, *et al.* Maple syrup urine disease. Complete defect of the E1 beta subunit of the branched chain alpha-ketoacid dehydrogenase complex due to a deletion of an 11-bp repeat sequence which encodes a mitochondrial targeting leader peptide in a family with the disease. *J Clin Invest* 1991;**87**:1862–6.

247. Noelle V, Knuepfer M, Pulzer F, *et al.* Unusual presentation of congenital disorder of glycosylation type 1a: congenital persistent thrombocytopenia, hypertrophic cardiomyopathy and hydrops-like aspect due to marked peripheral oedema. *Eur J Pediatr* 2005;**164**:223–6.

248. Nordli DR. Epileptic encephalopathies in infants and children. *J Clin Neurophysiol* 2012;**29**:420–4.

249. O'Driscoll MC, Daly SB, Urquhart JE, *et al.* Recessive mutations in the gene encoding the tight junction protein occludin cause band-like calcification with simplified gyration and polymicrogyria. *Am J Hum Genet* 2010;**87**:354–64.

250. Okeda R, Gei S, Chen I, *et al.* Menkes' kinky hair disease: morphological and immunohistochemical comparison of two autopsied patients. *Acta Neuropathol* 1991;**81**:450–7.

251. Okeda R, Matsuo T, Kawahara Y, *et al.* Adult pigment type (Peiffer) of sudano-philic leukodystrophy. Pathological and morphometrical studies on two autopsy cases of siblings. *Acta Neuropathol* 1989;**78**:533–42.

252. Oldfors A, Hagberg B, Nordgren H, Sourander P, Witt-Engerström I. Rett syndrome: spinal cord neuropathology. *Pediatr Neurol* 1988;**4**:172–4.

253. Oldfors A, Sourander P, Armstrong DL, *et al.* Rett syndrome: cerebellar pathology. *Pediatr Neurol* 1990;**6**:310–14.

254. O'Neill ML, Kuo F, Saigal G. MRI of pallidal involvement in Lesch-Nyhan disease. *J Neuroimag* 2012; doi: 10.1111/j.1552-6569.2012.00772.x.

255. Origuchi Y, Miyoshino S, Mishima K, Mine K. Quantitative histologic study of the sural nerve in Lesch-Nyhan syndrome. *Pediatr Neurol* 1990;**6**:353–5.

256. Otero K, Turnbull IR, Poliani PL, *et al.* Macrophage colony-stimulating factor induces the proliferation and survival of macrophages via a pathway involving DAP12 and beta-catenin. *Nat Immunol* 2009;**10**:735–44.

257. Oyarzabal A, Martínez-Pardo M, Merinero B, *et al.* A novel regulatory defect in the branched-chain α-keto acid dehydrogenase complex due to a mutation in the PPM1K gene causes a mild variant phenotype of maple syrup urine disease. *Hum Mutat* 2012;**34**:355–62.

258. Ozand PT, Rashed M, Gascon GG, *et al.* 3-Ketothiolase deficiency: a review and four new patients with neurologic symptoms. *Brain Dev* 1994;**16**:38–45.

259. Ozdirim E, Topçu M, Ozön A, Cila A. Cockayne syndrome: review of 25 cases. *Pediatr Neurol* 1996;**15**:312–16.

260. Ozmen M, Caliçkan M, Goebel H-H, Apak S. Infantile neuroaxonal dystrophy: diagnosis by skin biopsy. *Brain Dev* 1991;**13**:256–9.

261. Pacheco-Alvarez D, Solárzano-Vargas RS, Del Río AL. Biotin in metabolism and its relationship to human disease. *Arch Med Res* 2002;**33**:439–47.

262. Paine SML, Grünewald S, Jacques TS. Antenatal neurodevelopmental defects in ornithine transcarbamylase deficiency. *Neuropathol Appl Neurobiol* 2012;**38**:509–12.

263. Paine SML, Munot P, Carmichael J, *et al.* The neuropathological consequences of CDKL5 mutation. *Neuropathol Appl Neurobiol* 2012;**38**:744–7.

264. Paisan-Ruiz C, Bhatia KP, Li A, *et al.* Characterization of PLA2G6 as a locus for dystonia-parkinsonism. *Ann Neurol* 2009;**65**:19–23.

265. Paisan-Ruiz C, Li A, Schneider SA, *et al.* Widespread Lewy body and tau accumulation in childhood and adult onset dystonia-parkinsonism cases with PLA2G6 mutations. *Neurobiol Aging* 2010;**33**:814–23.

266. Palanca D, Garcia-Cazorla A, Ortiz J, *et al. cblE*-Type homocystinuria presenting with features of haemolytic-uremic syndrome in the newborn period. JIMD Reports. Berlin, Heidelberg: Springer, 2012: 57–62.

267. Palmer CA, Percy AK. Neuropathology of occipital horn syndrome. *J Child Neurol* 2001;**16**:764–6.

268. Pamphlett R, Silberstein P. Pelizaeus–Merzbacher disease in a brother and sister. *Acta Neuropathol* 1986;**69**:343–6

269. Pascual-Castroviejo I, van der Knaap MS, Pronk JC, *et al.* Vacuolating megalen-

cephalic leukoencephalopathy: 24 year follow-up of two siblings. *Neurologia* 2005;**20**:33–40.

270. Patay Z, Mills JC, Lobel U, *et al.* Cerebral neoplasms in L-2 hydroxyglutaric aciduria: 3 new cases and meta-analysis of literature data. *Am J Neuroradiol* 2012;**33**:940–3.

271. Paterson DS, Thompson EG, Belliveau RA, *et al.* Serotonin transporter abnormality in the dorsal motor nucleus of the vagus in Rett syndrome: potential implications for clinical autonomic dysfunction. *J Neuropathol Exp Neurol* 2005;**64**:1018–27.

272. Patton MA, Giannelli F, Francis AJ, *et al.* Early onset Cockayne's syndrome: case reports with neuropathological and fibroblast studies. *J Med Genet* 1989;**26**:154–9.

273. Pavone P, Spalice A, Polizzi A, Parisi P, Ruggieri M. Ohtahara syndrome with emphasis on recent genetic discovery. *Brain Dev* 2012;**34**:459–68.

274. Peña L, Burton BK. Survey of health status and complications among propionic acidemia patients. *Am J Med Genet* 2012;**158A**:1641–6.

275. Peña L, Franks J, Chapman KA, *et al.* Molecular genetics and metabolism. *Mol Genet Metab* 2012;**105**:5–9.

276. Poduri A, Chopra SS, Neilan EG, *et al.* Homozygous PLCB1 deletion associated with malignant migrating partial seizures in infancy. *Epilepsia* 2012;**53**:e146–50.

277. Poduri A, Heinzen EL, Chitsazzadeh V, *et al.* SLC25A22 is a novel gene for migrating partial seizures in infancy. *Ann Neurol* 2013;**74**:873–82.

278. Powers JM, Rosenblatt DS, Schmidt RE, *et al.* Neurological and neuropathologic heterogeneity in two brothers with cobalamin C deficiency. *Ann Neurol* 2001;**49**:396–400.

279. Prass K, Brück W, Schröder NWJ, *et al.* Adult-onset leukoencephalopathy with vanishing white matter presenting with dementia. *Ann Neurol* 2001;**50**:665–8.

280. Prensky AL, Carr S, Moser HW. Development of myelin in inherited disorders of amino acid metabolism. *Arch Neurol* 1968;**19**:552–8.

281. Prick BW, Hop WC, Duvekot JJ. Maternal phenylketonuria and hyperphenylalaninemia in pregnancy: pregnancy complications and neonatal sequelae in untreated and treated pregnancies. *Am J Clin Nutr* 2012;**95**:374–82.

282. Pridmore CL, Baraitser M, Harding B, *et al.* Alexander's disease: clues to diagnosis. *J Child Neurol* 1993;**8**:134–44.

283. Pronicka E, Weglewska-Jurkiewicz A, Pronicki M, *et al.* Drug-resistant epilepsia and fulminant valproate liver toxicity. Alpers–Huttenlocher syndrome in two children confirmed post mortem by identification of p.W748S mutation in POLG gene. *Med Sci Monit* 2011;**17**:CR203–9.

284. Prust M, Wang J, Morizono H, *et al.* GFAP mutations, age at onset, and clinical subtypes in Alexander disease. *Neurology* 2011;**77**:1287–94.

285. Rademakers R, Baker M, Nicholson AM, *et al.* Mutations in the colony stimulating factor 1 receptor (CSF1R) gene cause hereditary diffuse leukoencephalopathy with spheroids. *Nat Genet* 2011;**44**:200–5.

286. Ramesh V, Bernardi B, Stafa A, *et al.* Intracerebral large artery disease in Aicardi–Goutières syndrome implicates SAMHD1 in vascular homeostasis. *Dev Med Child Neurol* 2010;**52**:725–32.

287. Rapin I, Weidenheim K, Lindenbaum Y, *et al.* Cockayne syndrome in adults: review with clinical and pathologic study of a new case. *J Child Neurol* 2006;**21**:991–1006.

288. Razavi-Encha F, Larroche JC, Gaillard D. Infantile familial encephalopathy with cerebral calcifications and leukodystrophy. *Neuropediatrics* 1988;**19**:72–9.

289. Renier WO, Renkawek K. Clinical and neuropathologic findings in a case of severe myoclonic epilepsy of infancy. *Epilepsia* 1990;**31**:287–91.

290. Rice G, Patrick T, Parmar R, *et al.* Clinical and molecular phenotype of Aicardi–Goutières syndrome. *Am J Hum Genet* 2007;**81**:713–25.

291. Rice GI, Bond J, Asipu A, *et al.* Mutations involved in Aicardi–Goutières syndrome implicate SAMHD1 as regulator of the innate immune response. *Nat Genet* 2009;**41**:829–32.

292. Rice GI, Kasher PR, Forte GMA, *et al.* Mutations in ADAR1 cause Aicardi–Goutières syndrome associated with a type I interferon signature. *Nat Genet* 2012;**44**:1243–8.

293. Richardson JP, Mohammad SS, Pavitt GD. Mutations causing childhood ataxia with central nervous system hypomyelination reduce eukaryotic initiation factor 2B complex formation and activity. *Mol Cell Biol* 2004;**24**:2352–63.

294. Riviello JJ, Rezvani I, DiGeorge AM, Foley CM. Cerebral edema causing death in children with maple syrup urine disease. *J Pediatr* 1991;**119**(1 Pt 1):42–5.

295. Rocha JC, Martins MJ. Oxidative stress in phenylketonuria: future directions. *J Inherit Metab Dis* 2011;**35**:381–98.

296. Rodriguez D, Gelot A, Gaspera della B, *et al.* Increased density of oligodendrocytes in childhood ataxia with diffuse central hypomyelination (CACH) syndrome: neuropathological and biochemical study of two cases. *Acta Neuropathol* 1999;**97**:469–80.

297. Rodriguez D, Gauthier F, Bertini E, *et al.* Infantile Alexander disease: spectrum of GFAP mutations and genotype-phenotype correlation. *Am J Hum Genet* 2001;**69**:1134–40.

298. Rosenblum WI. Neuropathologic changes in a case of sulfite oxidase deficiency. *Neurology* 1968;**18**:1187–96.

299. Roux J-C, Villard L. Biogenic amines in Rett syndrome: the usual suspects. *Behav Genet* 2009;**40**:59–75.

300. Ruiz Gomez A, Couce ML, Garcia-Villoria J, *et al.* Clinical, genetic, and therapeutic diversity in 2 patients with severe mevalonate kinase deficiency. *Pediatrics* 2012;**129**:e535–9.

301. Rupar CA, Gillett J, Gordon BA, Ramsay DA. Neuropathologic changes in a case of sulfite oxidase deficiency. *Neurology* 1968;**18**:1187–96.

302. Russo LS, Aron A, Anderson PJ. Alexander's disease: a report and reappraisal. *Neurology* 1976;**26**:607–14.

303. Russo S, Marchi M, Cogliati F, *et al.* Novel mutations in the CDKL5 gene, predicted effects and associated phenotypes. *Neurogenetics* 2009;**10**:241–50.

304. Saini AG, Singhi P. Infantile metabolic encephalopathy due to fumarase deficiency. *J Child Neurol* 2013;**28**:535–7.

305. Saitsu H, Kato M, Mizuguchi T, *et al.* De novo mutations in the gene encoding STXBP1 (MUNC18-1) cause early infantile epileptic encephalopathy. *Nat Genet* 2008;**40**:782–8.

306. Salman MS, Ackerley C, Senger C, Becker L. New insights into the neuropathogenesis of molybdenum cofactor deficiency. *Can J Neurol Sci* 2002;**29**:91–6.

307. Saneto RP, Cohen BH, Copeland WC, Naviaux RK. Alpers–Huttenlocher syndrome. *Pediatr Neurol* 2013;**48**:167–78.

308. Sarnat HB, Flores-Sarnat L. Radial microcolumnar cortical architecture: maturational arrest or cortical dysplasia? *Pediatr Neurol* 2013;**48**:259–70.

309. Sarzi E, Goffart S, Serre V, *et al.* Twinkle helicase (PEO1) gene mutation causes mitochondrial DNA depletion. *Ann Neurol* 2007;**62**:579–87.

310. Sasaki K, Tachi N, Shinoda M, *et al.* Demyelinating peripheral neuropathy in Cockayne syndrome: a histopathologic and morphometric study. *Brain Dev* 1992;**14**:114–17.

311. Sawaishi Y, Yano T, Takaku I, Takada G. Juvenile Alexander disease with a novel mutation in glial fibrillary acidic protein gene. *Neurology* 2002;**58**:1541–3.

312. Sawaishi Y. Review of Alexander disease: Beyond the classical concept of leukodystrophy. *Brain Dev* 2009;**31**:493–8.

313. Scala E, Ariani F, Mari F, *et al.* CDKL5/STK9 is mutated in Rett syndrome variant with infantile spasms. *J Med Genet* 2005;**42**:103–7.

314. Schiff M, Blom H. Treatment of inherited homocystinurias. *Neuropediatrics* 2012;**43**:295–304.

315. Schiffmann R, Moller JR, Trapp BD, *et al.* Childhood ataxia with diffuse central nervous system hypomyelination. *Ann Neurol* 1994;**35**:331–40.

316. Schneider SA, Dusek P, Hardy J, *et al.* Genetics and pathophysiology of neurodegeneration with brain iron accumulation (NBIA). *Curr Neuropharmacol* 2013;**11**:59–79.

317. Scholl-Bürgi S, Sass JO, Zschocke J, Karall D. Amino acid metabolism in patients with propionic acidaemia. *J Inherit Metab Dis* 2010;**35**:65–70.

318. Schreiber H, Rowley DA. Cancer. Quo vadis, specificity? *Science* 2008;**319**:164–5.

319. Schreiber J, Chapman KA, Summar ML, *et al.* Neurologic considerations in propionic acidemia. *Mol Genet Metab* 2012;**105**:10–15.

320. Seijo-Martínez M, Navarro C, Castro del Río M, *et al.* L-2-hydroxyglutaric aciduria: clinical, neuroimaging, and neuropathological findings. *Arch Neurol* 2005;**62**:666–70.

321. Shuman RM, Leech RW, Scott CR. The neuropathology of the nonketotic and ketotic hyperglycinemias: three cases. *Neurology* 1978;**28**:139–46.

322. Silberberg DH. Maple syrup urine disease metabolites studied in cerebellum cultures. *J Neurochem* 1969;**16**:1141–6.

323. Siudeja K, Srinivasan B, Xu L, *et al.* Impaired Coenzyme A metabolism affects histone and tubulin acetylation in Drosophila and human cell models of pantothenate kinase associated neurodegeneration. *EMBO Mol Med* 2011;**3**:755–66.

324. Smith ET Jr, Davis GJ. Medium-chain acyl-coenzyme-a dehydrogenase deficiency: not just another Reye syndrome. *Am J Forensic Med Pathol* 1993;**14**:313–18.

325. Smith I, Erdohazi M, Macartney FJ, *et al.* Fetal damage despite low-phenylalanine diet after conception in a phenylketonuric woman. *Lancet* 1979;**1**:17–19.

326. Smith I, Beasley MG, Ades AE. Intelligence and quality of dietary treatment

in phenylketonuria. *Arch Dis Child* 1990;**65**:472–8.

327. Smith SE, Kinney HC, Swoboda KJ, Levy HL. Subacute combined degeneration of the spinal cord in cblC disorder despite treatment with B12. *Mol Genet Metab* 2006;**88**:138–45.

328. Soffer D, Amir N, Elpeleg ON, et al. Striatal degeneration and spongy myelinopathy in glutaric acidemia. *J Neurol Sci* 2002;**107**:199–.

329. Somaraju UR, Merrin M. Sapropterin dihydrochloride for phenylketonuria. *Cochrane Database Syst Rev* 2012;**12**:CD008005.

330. Squires L, Betz B, Umfleet J, Kelley R. Resolution of subependymal cysts in neonatal holocarboxylase synthetase deficiency. *Dev Med Child Neurol* 1997;**39**:267–9.

331. Stall JN, Hawley DA, Hagen MC, Bull MJ, Hattab EM. Posterior column degeneration in the cervical/thoracic spinal cord in Lesch–Nyhan syndrome (LNS): a case report. *Neuropathol Appl Neurobiol* 2010;**36**:680–4.

332. Stanley ER, Berg KL, Einstein DB, et al. Biology and action of colony-stimulating factor-1. *Mol Reprod Dev* 1997;**46**:4–10.

333. Steen C, Baumgartner ER, Duran M, et al. Metabolic stroke in isolated 3-methylcrotonyl-CoA carboxylase deficiency. *Eur J Pediatr* 1999;**158**:730–3.

334. Steiner LA, Ehrenkranz RA, Peterec SM, et al. Perinatal onset mevalonate kinase deficiency. *Pediatr Dev Pathol* 2011;**14**:301–6

335. Steinman L, Clancy RR, Cann H, Urich H. The neuropathology of propionic acidemia. *Dev Med Child Neurol* 1983;**25**:87–94.

336. Stephenson JBP. Aicardi–Goutières syndrome (AGS). *Eur J Paediatr Neurol* 2008;**12**:355–8.

337. Stewart JD, Horvath R, Baruffini E, et al. Polymerase γ gene POLG determines the risk of sodium valproate-induced liver toxicity. *Hepatology* 2010;**52**:1791–6.

338. Stöckler S, Radner H, Karpf EF, Hauer A, Ebner F. Symmetric hypoplasia of the temporal cerebral lobes in an infant with glutaric aciduria type II (multiple acyl-coenzyme A dehydrogenase deficiency). *J Pediatr* 1994;**124**:601–4.

339. Strauss KA, Puffenberger EG, Morton DH. Maple syrup urine disease. In: Pagon RA, Adam MP, Bird TD, et al. (eds). *GeneReviews*. University of Washington, Seattle, WA, 1993.

340. Strømme P, Maehlen J, Strøm EH, Torvik A. Postmortem findings in two patients with the carbohydrate deficient glycoprotein syndrome. *Acta Paediatr* 1991;**80**:55–62.

341. Stumpf JD, Saneto RP, Copeland WC. Clinical and molecular features of POLG-related mitochondrial disease. *Cold Spring Harb Perspect Biol* 2013;**5**:a011395–5.

342. Sugiura C, Miyata H, Oka A, et al. A Japanese girl with leukoencephalopathy with vanishing white matter. *Brain Dev* 2001;**23**:58–61.

343. Suomalainen A, Isohanni P. Mitochondrial DNA depletion syndromes – many genes, common mechanisms. *Neuromuscul Disord* 2010;**20**:429–37.

344. Surtees RA, Matthews EE, Leonard JV. Neurologic outcome of propionic acidemia. *Pediatr Neurol* 1992;**8**:333–7.

345. Sweetman L, Nyhan WL, Trauner DA, Merritt TA, Singh M. Glutaric aciduria Type II. *J Pediatr* 1980;**96**:1020–6.

346. Tachibana H, Hayashi T, Kajii T, Takashima S, Sasaki K. Neuroaxonal dystrophy of neonatal onset with unusual clinicopathological findings. *Brain Dev* 1986;**8**:605–9.

347. Takada K, Becker LE. Cockayne's syndrome: report of two autopsy cases associated with neurofibrillary tangles. *Clin Neuropathol* 1986;**5**:64–8.

348. Takanashi J, Sugita K, Tanabe Y, Niimi H. Adolescent case of Alexander disease: MR imaging and MR spectroscopy. *Pediatr Neurol* 1998;**18**:67–70.

349. Takashima S, Chan F, Becker LE. Cortical dysgenesis in a variant of phenylketonuria (dihydropteridine reductase deficiency). *Pediatr Pathol* 1991;**11**:771–9.

350. Tan W-H, Eichler FS, Hoda S, et al. Isolated sulfite oxidase deficiency: a case report with a novel mutation and review of the literature. *Pediatrics* 2005;**116**:757–66.

351. Theodore M, Morava E. Congenital disorders of glycosylation. *Curr Opin Pediatr* 2011;**23**:581–7.

352. Thompson AJ, Youl BD, Kendall B, et al. Neurological deterioration in young adults with phenylketonuria. *Lancet* 1990;**336**:602–5.

353. Townsend JJ, Wilson JF, Harris T, Coulter D, Fife R. Alexander's disease. *Acta Neuropathol* 1985;**67**:163–6.

354. Trefz F, Maillot F, Motzfeldt K, Schwarz M. Molecular genetics and metabolism. *Mol Genet Metab* 2011;**104**:S26–S30.

355. Troost D, van Rossum A, Veiga Pires J, Willemse J. Cerebral calcifications and cerebellar hypoplasia in two children: clinical, radiologic and neuropathological studies – a separate neurodevelopmental entity. *Neuropediatrics* 1984;**15**:102–9.

356. van de Kamp JM, Lefeber DJ, Ruijter GJG, et al. Congenital disorder cf glycosylation type Ia presenting with hydrops fetalis. *J Med Genet* 2007;**44**:277–80.

357. van der Knaap MS, Barth PG, Vrensen GF, Valk J. Histopathology of an infantile-onset spongiform leukoencephalopathy with a discrepantly mild clinical course. *Acta Neuropathol* 1996;**92**:206–12.

358. van der Knaap MS, Wevers RA, Monnens L, et al. Congenital nephrotic syndrome: a novel phenotype of type I carbohydrate-deficient glycoprotein syndrome. *J Inherit Metab Dis* 1996;**19**:787–91.

359. van der Knaap MS, Barth PG, Gabreëls FJ, et al. A new leukoencephalopathy with vanishing white matter. *Neurology* 1997;**48**:845–55.

360. van der Knaap MS, Bakker HD, Valk J. MR imaging and proton spectroscopy in 3-hydroxy-3-methylglutaryl coenzyme A lyase deficiency. *AJNR Am J Neuroradiol* 1998;**19**:378–82.

361. van der Knaap MS, Kamphorst W, Barth PG, et al. Phenotypic variation in leukoencephalopathy with vanishing white matter. *Neurology* 1998;**51**:540–7.

362. van der Knaap MS, Wevers RA, Kure S, et al. Increased cerebrospinal fluid glycine: a biochemical marker for a leukoencephalopathy with vanishing white matter. *J Child Neurol* 1999;**14**:728–31.

363. van der Knaap MS, Leegwater PAJ, Könst AAM, et al. Mutations in each of the five subunits of translation initiation factor eIF2B can cause leukoencephalopathy with vanishing white matter. *Ann Neurol* 2002;**51**:264–70.

364. van der Knaap MS, van Berkel CGM, Herms J, et al. eIF2B-related disorders: antenatal onset and involvement of multiple organs. *Am J Hum Genet* 2003;**73**:1199–207.

365. van der Knaap MS, Salomons GS, Li R, et al. Unusual variants of Alexander's disease. *Ann Neurol* 2005;**57**:327–38.

366. van der Knaap MS, Boor I, Estevez R. Megalencephalic leukoencephalopathy with subcortical cysts: chronic white matter oedema due to a defect in brain ion and water homoeostasis. *Lancet Neurol* 2012;**11**:973–85.

367. Van Der Voorn JP, van Kollenburg B, Bertrand G, et al. The unfolded protein response in vanishing white matter disease. *J Neuropathol Exp Neurol* 2005;**64**:770–5.

368. Van Haren K, Van Der Voorn JP, Peterson DR, van der Knaap MS, Powers JM. The life and death of oligodendrocytes in vanishing white matter disease. *J Neuropathol Exp Neurol* 2004;**63**:618–30.

369. van Kollenburg B, Thomas AAM, Vermeulen G, et al. Regulation of protein synthesis in lymphoblasts from vanishing white matter patients. *Neurobiol Dis* 2006;**21**:496–504.

370. van Kollenburg B, van Dijk J, Garbern J, et al. Glia-specific activation of all pathways of the unfolded protein response in vanishing white matter disease. *J Neuropathol Exp Neurol* 2006;**65**:707–15.

371. Van Poppel K, Broniscer A, Patay Z, Morris EB. Alexander disease: an important mimicker of focal brainstem glioma. *Pediatr Blood Cancer* 2009;**53**:1355–6.

372. Vesela K, Honzik T, Hansikova H, et al. A new case of ALG8 deficiency (CDG Ih). *J Inherit Metab Dis* 2009;**32**:259–64.

373. Visser JE, Bär PR, Jinnah HA. Lesch–Nyhan disease and the basal ganglia. *Brain Res Brain Res Rev* 2000;**32**:449–75.

374. Wada Y, Arakawa T, Koizumi K. Lesch–Nyhan syndrome: autopsy findings and *in vitro* study of incorporation of 14C-8-inosine into uric acid, guanosine-monophosphate and adenosine-monophosphate in the liver. *Tohoku J Exp Med* 1968;**95**:253.

375. Watkins D, Rosenblatt DS. Inborn errors of cobalamin absorption and metabolism. *Am J Med Genet* 2011;**157**:33–44.

376. Weaving LS, Christodoulou J, Williamson SL, et al. Mutations of CDKL5 cause a severe neurodevelopmental disorder with infantile spasms and mental retardation. *Am J Hum Genet* 2004;**75**:1079–93.

377. Weckhuysen S, Holmgren P, Hendrickx R, et al. Reduction of seizure frequency after epilepsy surgery in a patient with STXBP1 encephalopathy and clinical description of six novel mutation carriers. *Epilepsia* 2013;**54**:e74–80.

378. Weglage J, Pietsch M, Fünders B, Koch HG, Ullrich K. Neurological findings in early treated phenylketonuria. *Acta Paediatr* 1995;**84**:411–15.

379. Wei SH, Weng WC, Lee NC, Hwu WL, Lee WT. Unusual spinal cord lesions in late-onset non-ketotic hyperglycinemia. *J Child Neurol* 2011;**26**:900–3.

380. Weidenheim KM, Dickson DW, Rapin I. Neuropathology of Cockayne syndrome: Evidence for impaired development, premature aging, and neurodegeneration. *Mech Ageing Dev* 2009;**130**:619–36.

381. Werner ER, Blau N, Thöny B. Tetrahydrobiopterin: biochemistry and pathophysiology. *Biochem J* 2011;**438**:397–414.

382. White HH, Rowland LP, Araki S, Thompson HL, Cowen D. Homocystinuria. *Arch Neurol* 1965;**13**:455–70.

383. Wider C, Van Gerpen JA, Dearmond S, *et al*. Leukoencephalopathy with spheroids (HDLS) and pigmentary leukodystrophy (POLD): a single entity? *Neurology* 2009;**72**:1953–9.

384. Wilmshurst JM, Appleton DB, Grattan-Smith PJ. Migrating partial seizures in infancy: two new cases. *J Child Neurol* 2000;**15**:717–22.

385. Wisniewski K, Wisniewski HM. Diagnosis of infantile neuroaxonal dystrophy by skin biopsy. *Ann Neurol* 1980;**7**:377–9.

386. Wong K, Armstrong RC, Gyure KA, *et al*. Foamy cells with oligodendroglial phenotype in childhood ataxia with diffuse central nervous system hypomyelination syndrome. *Acta Neuropathol* 2000;**100**:635–46.

387. Woodward KJ. The molecular and cellular defects underlying Pelizaeus–Merzbacher disease. *Expert Rev Mol Med* 2008;**10**:e14.

388. Worthington S, Arbuckle S, Nelson P, *et al*. Carbohydrate deficient glycoprotein syndrome type I: a cause of cerebellar vermis hypoplasia. *J Paediatr Child Health* 1997;**33**:531–4.

389. Wurm D, Hänsgen A, Kim Y-J, *et al*. Early fatal course in siblings with CDG-Ia (caused by two novel mutations in the PMM2 gene): clinical, molecular and autopsy findings. *Eur J Pediatr* 2007;**166**:377–8.

390. Yalçinkaya C, Dinçer A, Gündüz E, *et al*. MRI and MRS in HMG-CoA lyase deficiency. *Pediatr Neurol* 1999;**20**:375–80.

391. Yoshida T, Nakagawa M. Clinical aspects and pathology of Alexander disease, and morphological and functional alteration of astrocytes induced by GFAP mutation. *Neuropathology* 2012;**32**:440–6.

392. Yoshimura N, Kudo H. Mitochondrial abnormalities in Menkes' kinky hair disease (MKHD). Electron-microscopic study of the brain from an autopsy case. *Acta Neuropathol* 1983;**59**:295–303.

393. Zempleni J, Hassan YI, Wijeratne SS. Biotin and biotinidase deficiency. *Expert Rev Endocrinol Metab* 2008;**3**:715–24.

394. Zhang B, Edenberg HJ, Crabb DW, Harris RA. Evidence for both a regulatory mutation and a structural mutation in a family with maple syrup urine disease. *J Clin Invest* 1989;**83**:1425–9.

395. Zhou B, Westaway SK, Levinson B, *et al*. A novel pantothenate kinase gene (PANK2) is defective in Hallervorden–Spatz syndrome. *Nat Genet* 2001;**28**:345–9.

396. Zoghbi HY, Spence JE, Beaudet AL, *et al*. Atypical presentation and neuropathological studies in 3-hydroxy-3-methylglutaryl-CoA lyase deficiency. *Ann Neurol* 1986;**20**:367–9.

Lysosomal Diseases

Steven U Walkley, Kinuko Suzuki and Kunihiko Suzuki

INTRODUCTION

Lysosomes are subcellular membrane-bound organelles containing primary acid hydrolases and other proteins responsible for the catabolism of naturally occurring and phagocytized intracellular macromolecules. Lysosomes are in direct continuity with both endosomal and autophagosomal streams in cells, thus forming what has been referred to as the greater lysosomal system and comprising the predominant recycling system of cells.[545] As a result of lysosomal malfunction, various molecules accumulate within lysosomes and late endosomes, forming abnormal intracellular inclusions or storage bodies.[302] The concept of lysosomal storage disease was established in 1963 when H.G. Hers discovered that in type II glycogenosis (Pompe disease), glycogen accumulated within vacuoles secondary to a defect of an α-1,4-glucosidase normally present in lysosomes. Based on this finding, Hers hypothesized that other so-called storage disorders (e.g. Tay–Sachs and Niemann–Pick diseases) could have a similar causation.[209] Hers' concept of inborn lysosomal disorders included two easily detectable characteristics: (i) a single lysosomal hydrolytic enzyme must be genetically deficient; and (ii) the substrate(s) of that enzyme would, as a consequence, accumulate within pathologically altered secondary lysosomes. In the ensuing years, however, the concept of genetic lysosomal disease evolved to include disorders caused by genetic abnormalities of proteins that are not lysosomal enzymes. These include glycoprotein cofactors for lysosomal enzymes and related soluble proteins essential for lysososmal function, non-lysosomal enzymes involved in the post-translational modification or transport of lysosomal enzymes, and transmembrane proteins responsible for egress of metabolites out of lysosomes and other functions. It is anticipated that this list will grow as a more detailed understanding of lysosomal diseases emerge.[388] Today more than 50 distinct

genetic lysosomal diseases have been identified. Although most types of individual lysosomal diseases are rare, when taken as a group their prevalence is 1 per 7700 live births, making them one of the major families of genetic disease.[330] Most involve the central nervous system (CNS). They are autosomal recessive disorders, with the exceptions of Fabry, Danon and Hunter diseases, which are X-linked recessive disorders. Lysosomal storage diseases are traditionally subclassified according to the nature of the primary storage material accumulating within cells (i.e. sphingolipidoses, mucopolysaccharidoses, glycoproteinoses). They are a clinically heterogeneous group of diseases with complex pathogenic cascades and with multiple clinical phenotypes known in each group (Box 6.1).

Many lysosomal storage diseases affect the nervous system but can be diagnosed by biopsy studies of non-neural tissues (Box 6.2).[73,166] Prenatal diagnosis can be carried out by morphological examination of chorionic villi or cultured amniotic cells.[283] In recent years, in addition to naturally occurring animal models, many mouse models of human lysosomal diseases have been generated by targeted disruption of specific genes.[481] Readily available animal models enable investigators to conduct experiments aimed at understanding pathogenesis, discovering biomarkers and developing therapeutic strategies.[67,231,323,388,483,485,568]

In this chapter, the focus is on the morphological and pathological aspects of lysosomal storage diseases. Information on the biochemistry and molecular genetics of the individual diseases is provided to a limited extent. Readers who wish to learn more details on the molecular genetics, biochemistry and pathogenesis of lysosomal diseases are referred to Valle and colleagues,[522] many chapters in which are cited here, as well as Platt and Walkley.[388] Readers interested in the historical aspects of lysosomes and storage diseases should consult Hers and van Hoof.[208]

BOX 6.1. Lysosomal diseases: why so complex?

Representing more than 50 individual disorders, lysosomal diseases inevitably present a complex picture to the uninitiated. The complexity begins with their nomenclature, which by historical precedent is based in part on the type of major storage material identified as accumulating. This is true both for group names of these disorders (e.g. the sphingolipidoses, mucopolysaccharidoses, etc.) and for some individual diseases (e.g. aspartylglucosaminuria and cholesterol ester storage disease). Some of the original group naming schemes are now largely misleading, such as the neuronal ceroid lipofuscinoses or even the mucolipidoses, yet they are unlikely to be dropped given their use for decades. Because many of the lysosomal diseases were described more than a century ago, some are named according to their discoverers; hence Tay–Sachs, Pompe, Gaucher and others. Although attempts to organize these diseases by molecular defect have been useful,[388] the classical system of nomenclature is clearly here to stay. Perhaps an even more important reason for lysosomal disease complexity is the nature of the disorders themselves. Many exhibit widespread systemic effects involving bone, cartilage and viscera (e.g. many of the mucopolysaccharidoses), and most affect the brain, where effects can be serious, leading to a plethora of clinical consequences from intellectual disability and dementia to motor and sensory disturbances, seizures, and so on. Many lysosomal diseases affecting the brain share features with more common neurodegenerative diseases, including Alzheimer's, Parkinson's and Huntington's. Included here is the presence of tauopathy, TDP43 aggregation, α-synuclein accumulation and Lewy body formation, cholesterol dysregulation, selective neuronal vulnerability, neuroinflammation and so on. Recently, similarities between lysosomal diseases affecting brain and other genetic conditions of intellectual disability (Christianson and Rett syndromes) have also begun to emerge.

An important feature of most lysosomal diseases is that they are progressive in nature and ultimately fatal, although early on in life individuals may appear completely normal. As a consequence a major question that has plagued the lysosomal disease field since its inception has been why disturbances in the lysosome and resulting storage are so deleterious to cell and organ function. In recent years the concept of cytoxicity secondary to storage has been in most instances replaced with a more modern view focused on analysis of specific disease cascades secondary to the initiating molecular defect. Such impact on lysosomal function is now recognized as a disturbance in what has been referred to as the greater lysosomal system or network.[545] The latter includes a host of endosomal and retrosomal signal transduction events, as well as autophagosomal processing with possible ties to the other major degradative system of cells, the ubiquitin-proteasomal system. The concept of storage-induced cytotoxicity has also been re-examined with the view that accumulating substrates may be more deleterious though their compromise in salvage, depriving cells of vital precursors to synthetic pathways, rather than to 'storage' *per se*. Clearly, the lysosomal system, whether healthy or diseased, is central to the normal function of all cells. Consequently, understanding lysosomal disease pathogenesis has the potential not only to reveal the contributions of this vital recycling centre to normal cells, but also to provide novel insights into the pathogenesis of more common neurodegenerative and intellectual disability disorders.

BOX 6.2. Lysosomal diseases: diagnostic overview

The wide variety of clinical, radiological, biochemical and genetic changes found in lysosomal diseases poses diagnostic challenges. Originally defined by Hers in 1963 as diseases caused by a genetic defect of lysosomal enzymes, and resulting in intralysosomal accumulation of their substrate materials, our understanding of the biological underpinnings of these diseases has changed over the past four decades.[388] Most of these diseases are autosomal recessive and affect infants or young children, manifesting as growth retardation and progressive neurological deterioration with seizures. Gangliosidoses, Niemann–Pick disease types A and B, Krabbe disease and metachromatic leukodystrophy all belong to this classical group. The morphology and biochemistry of lysosomal diseases have been investigated intensely. Previously, diagnosis was primarily dependent on histopathological and electron-microscopic studies of biopsy specimens from brain, myenteric plexus and, in some cases, skin. Currently, biochemical identification of the enzyme deficiency is a standard approach to diagnosis. Enzyme activity can be determined on any fresh or frozen tissue, but cultured fibroblasts from skin biopsies are most commonly used. Many of the defective genes have been identified, and mutation analysis of these genes is now common practice. Importantly, genetic analysis can also be used for prenatal diagnosis. Diagnostic challenges still exist, however, in some of the chronic/adult-onset lysosomal disorders. Although relatively rare, late-onset forms often manifest as chronic progressive neurological diseases. For example, late-onset chronic GM2 gangliosidosis may clinically mimic amyotrophic lateral sclerosis or spinocerebellar degeneration, and choreoathetosis has been the major clinical manifestation of patients with chronic GM1 gangliosidosis. Dementia may be the presenting sign of adult metachromatic leukodystrophy, and progressive peripheral neuropathy can be the predominant problem in adult globoid cell leukodystrophy (Krabbe disease). A close association of parkinsonism with type I Gaucher disease has been documented.

Importantly, a significant portion of lysosomal diseases is now known not to be caused by enzyme defects. For some of these diseases, electron-microscopic investigation remains an effective tool to suggest a diagnosis. In this setting, ultrastructural analysis usually involves examination of skin biopsy or cultured skin fibroblasts. In addition, in some of the lysosomal diseases, such as Niemann–Pick disease type C and sphingolipid activator deficiencies in which genetic defects affect a structural protein or transport mechanism, enzyme assays do not contribute to diagnosis. For Niemann–Pick type C disease detection of unesterified cholesterol by filipin stain of fibroblasts cultured with LDL has been the most reliable diagnostic procedure. The recent discovery, however, of oxidized cholesterol compounds in the serum of Niemann–Pick C patients and animal models has raised the possibility of a far more definitive diagnostic tool for this disorder and possibly one also useful as a biomarker for clinical drug trials.[391] Although neuroimaging may be helpful to monitor disease progression and possibly to evaluate the effects of therapy, most lysosomal diseases are in serious need of identifiable biomarkers that could be used in this way.

SPHINGOLIPIDOSES AND RELATED DISORDERS

Sphingolipidoses are a group of lysosomal storage diseases caused by deficient activity of lysosomal enzymes or defects in other proteins crucial for the degradation and recycling of sphingolipids (Figure 6.1). Such protein defects cause blockage in the degradation and salvage pathways and excessive storage of one or more sphingolipids. Intraneuronal accumulation of unesterified cholesterol is also typically an accompanying feature of these diseases. Table 6.1 lists the diseases included in this group.

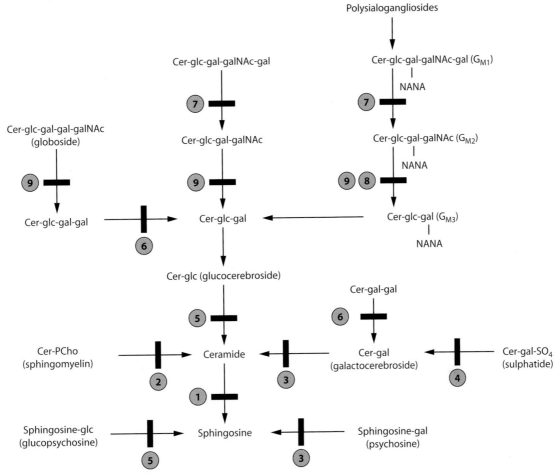

6.1 Chemical and metabolic relationship among the major sphingolipids. Arrows indicate normal pathways within the lysosome. Biosynthesis of these lipids occurs in the Golgi/*trans*-Golgi network in essentially the reverse direction. Locations of genetic metabolic blocks of known diseases are indicated with rectangles. The numbers correspond to the following diseases: (1) Farber disease; (2) Niemann–Pick disease (types A and B); (3) globoid cell leukodystrophy (Krabbe disease); (4) metachromatic leukodystrophy; (5) Gaucher disease; (6) Fabry disease; (7) GM1 gangliosidosis; (8) GM2 gangliosidosis (Tay–Sachs disease, B1 variant, AB variant); (9) GM2 gangliosidosis (Sandhoff disease).

Reproduced from Siegel G, et al (eds). Basic neurochemistry, 6th edn. Philadelphia: Lippincott Williams & Wilkins, 1999, by permission of the copyright holder, American Society for Neurochemistry.

Gangliosidoses

Gangliosidoses are a group of diseases with excessive accumulation of gangliosides in the CNS and to some extent in the viscera, caused by a defect in their degradation. GM1 and GM2 gangliosidoses are known.[179,476,487]

GM1 Gangliosidosis and Morquio Disease (Syndrome) Type B

These diseases are allelic in that they are both caused by deficient activity of acid β-galactosidase. GM1 gangliosidosis affects predominantly the CNS, but in severe infantile cases additional bony and visceral involvements are prominent. The clinical phenotype of Morquio disease type B is similar to that of Morquio disease type A (mucopolysaccharidosis; MPS IV), hence the name, and is characterized by bony abnormalities and corneal clouding but without CNS involvement.[487]

History and Clinical Features

The first detailed report of a probable case of GM1 gangliosidosis is that by Norman and colleagues in 1959.[361] Described therein is a 9-month-old male infant who presented with generalized flaccidity, an enlarged liver, a facies with depressed nasal bones, slight corneal haze and lumbar kyphosis, all of which are features suggestive of an MPS disease. Cherry-red spots were present at the macula. Histopathologically, there was massive neuronal storage and infiltration of foamy macrophages in the bone marrow and visceral organs, suggestive of Niemann–Pick disease. However, biochemical analysis revealed a large excess of a new type of sphingolipid, GM1 ganglioside, which was distinct from the material accumulating in the brain in Tay–Sachs disease. Because of clinicopathological complexity of the disease, the original patient by Norman and colleagues was reported as 'Tay–Sachs disease with visceral involvement'. Similar cases were also reported as 'pseudo-Hurler

TABLE 6.1 Major sphingolipidoses and related disorders

Clinical diagnosis	Affected compounds	Protein defect
GM1 gangliosidosis	GM1 ganglioside, galactose-rich fragments of glycoproteins	GM1 ganglioside β-galactosidase
GM2 gangliosidosis		
Tay–Sachs disease, B variant	GM2 ganglioside	β-Hexosaminidase A
B1 variant	GM2 ganglioside	β-Hexosaminidase, α-subunit (see text)
AB variant	GM2 ganglioside	GM2 activator protein
Sandhoff disease, O variant	GM2 ganglioside, asialo GM2 ganglioside, globoside	β-Hexosaminidases A and B, β-subunit (see text)
Niemann–Pick disease types A and B	Sphingomyelin	Acid sphingomyelinase
Niemann–Pick disease type C	Cholesterol GM2 ganglioside GM3 ganglioside Lactosylceramide Glucosylceramide	NPC1 or NPC2 proteins
Gaucher disease	Glucosylceramide, glucosylsphingosine	Glucosylceramidase
Farber disease	Ceramide	Acid ceramidase
Fabry disease	Trihexosylceramide	α-Galactosidase A
Metachromatic leukodystrophy	Sulfatide	Arylsulfatase A
Multiple sulphatase deficiency	Sulfatide and other compounds (see text)	Arylsulfatase A, B, C and others (see text)
Globoid cell leukodystrophy (Krabbe disease)	Galactosylceramide, galactosylsphingosine	Galactosylceramidase
Total Sap deficiency	Multiple sphingolipids[41]	Sap-A to Sap-D
Sap-B deficiency	Sulfatide and others	Sap-B (sulphatide activator)
Sap-C deficiency	Glucosylceramide	Sap-C

Sap, sphingolipid activator protein/saposin.

disease', familial neurovisceral lipidosis, generalized gangliosidosis etc., in earlier days[487] Today, the disease is termed GM1 gangliosidosis and three clinical types are recognized: infantile (type 1), late infantile/juvenile (type 2) and chronic/adult (type 3). The clinical features of the infant reported by Norman and colleagues[361] are typical of the infantile type. The late infantile/juvenile type has a more heterogeneous clinical course, and in some cases the cherry-red spots or visceromegaly may not be apparent. Progressive intellectual and motor retardation begin usually between the ages of 1 and 5 years. Dysmorphic features that are pronounced in the infantile type are not usually present in late infantile/juvenile type.[164,172] Chronic (or adult) type patients have a more slowly progressive pyramidal and extrapyramidal disease without dysmorphic features, visceromegaly or macular cherry-red spots. Intellectual deterioration is usually mild. Dystonia is the most pronounced clinical feature of the adult type of GM1 gangliosidosis.[348,584] Morquio disease type B is primarily a disease of the skeletal system and presents clinically as a mild phenotype of Morquio A disease. All patients with Morquio disease type B have normal intelligence without any evidence of mental deterioration, although neurological manifestations may occur secondary to bony deformities.[15]

Biochemistry

The cause of GM1 gangliosidosis and Morquio B disease is a genetic deficiency in activity of lysosomal β-galactosidase. In the infantile type, three to five times the normal amounts of total brain ganglioside could be accumulated at the terminal stage. GM1 ganglioside, which constitutes approximately 22–25 molar per cent of total ganglioside in the normal grey matter, is present at 90–95 molar per cent in the grey matter of affected infants. Thus, the increase of GM1 ganglioside itself could be up to 20 times more than normal in the grey matter.[476] The main storage materials in the systemic organs are heterogeneous and are derived from glycoprotein, keratan sulphate (sulfate) and other carbohydrate-containing tissue constituents.[476] An abnormal increase in GM1 ganglioside, although the absolute amount is relatively small, is also detected in visceral organs. In the adult type, an accumulation of GM1 ganglioside is limited to brain regions where morphological evidence of neuronal storage is present.[256] All patients with GM1 gangliosidosis excrete galactose-rich fragments of glycoproteins and keratan sulphate into the urine, similar to patients with Morquio A disease (MPS IVA). Little is known about the analytical chemistry of the brain in Morquio B disease.

However, biochemical analysis suggests that there is a considerable overlap for storage material and enzyme activity between GM1 gangliosidosis and Morquio B disease.

Molecular Genetics

The human β-galactosidase gene (*GLB1*) maps to chromosome 3 and produces two alternatively spliced transcripts encoding lysosomal enzyme β-galactosidase (GLB1) and the elastin-binding protein (EBP).[487] GLB1 is catalytically active in its precursor form, whereas EBP is not lysosomal in location and does not display β-galactosidase activity. Defects in EBP may, however, contribute to the pathogenesis of some forms of GM1 gangliosidosis and the reader is referred to a recent review for additional details.[487] Over 100 mutations responsible for GM1 gangliosidosis and Morquio B disease have been identified.[68,370,487] The mutations include missense/nonsense mutation, duplication/insertion and insertion causing splicing defects. The molecular basis for the distinct phenotypes of GM1 gangliosidosis and Morquio B disease has not been clarified. However, common mutations have been found in certain clinical phenotypes (e.g. R482H in Italian patients with infantile type, R208C in American patients with infantile type, R201C in Japanese patients with juvenile type, I51T in Japanese patients with adult type, and W273L in white patients with Morquio B disease).[373] Recent studies have described the detailed crystal structure of the human enzyme, analysed relative to mutations directly affecting ligand recognition as well as the protein core and its surface.[370]

Neuroimaging

Cranial computed tomography (CT)[96] and magnetic resonance (MR) imaging demonstrate increased density in the thalami and low density in the cerebral white matter; as the disease progresses, diffuse cerebral atrophy (generalized cortical atrophy and ventricular enlargement) and loss of myelin in the white matter are pronounced in early onset patients.[186,487] Magnetic resonance imaging shows that the thalami may be slightly bright in T1-weighted images,

whereas they are hypointense in T2-weighted images (Figure 6.2). Indeed, hypointensity of thalami in T2-weighted images appears to be a common finding across many lysosomal diseases.[26] In adult cases of GM1 gangliosidosis, CT reveals diffuse mild cerebral atrophy and/or localized atrophy of the head of the caudate nucleus.[69,170,521]

Pathology

The histopathology of infantile and late infantile types was well documented in early reports, and the detailed histology of cases has been reviewed.[487] Hepatosplenomegaly, bony abnormalities and facial dysmorphism similar to those found in MPS are usually present in the infantile type.[284] These changes are milder or absent in the late infantile/juvenile type. Visceromegaly and bony changes are not features of the adult (or chronic) type.[170,474,527] Infiltration of storage cells is present in the spleen, in both red pulp and the lymphoid follicles, in lymph nodes, in hepatic sinusoids and in bone marrow. The storage materials are water soluble and, as a consequence, the cytoplasm of storage cells often appears empty (foamy) in routine histological sections or electron micrographs. The cytoplasm of the renal glomerular epithelium is often vacuolated.[491] Vacuoles are also found in the sweat gland epithelium, skin fibroblasts and lymphocytes. Storage cells in the hepatic sinusoids are only weakly eosinophilic on routine haematoxylin and eosin (H&E) staining (Figure 6.3) and strongly positive with periodic acid–Schiff (PAS), and are Alcian blue-positive. Fine filamentous or tubular structures are found within these cytoplasmic vacuoles at the ultrastructural level (Figure 6.4).[164,474] Similar tubular structures are also noted in some adult type cases.[170] Various degrees of cardiomyopathy with infiltration of storage cells in the myocardium and mitral valve have been reported in some infantile type cases.[164]

The neuropathology of infantile and late infantile/juvenile types of GM1 gangliosidosis is that of a typical neuronal storage disease, similar or almost identical to that of GM2 gangliosidosis. The megalencephaly sometimes reported in Tay–Sachs disease (infantile GM2 gangliosidosis), however,

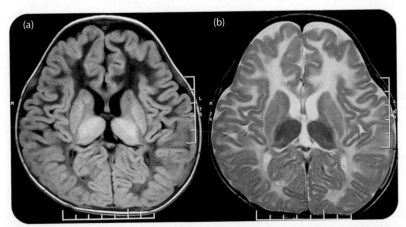

6.2 GM1 gangliosidosis, infantile type. (a) T1- and **(b)** T2-weighted images of a 13-month old boy reveal diffuse dysmyelination of white matter with bilateral thalamic changes such that on the T1-weighted image this area appears hyperintense whereas on the T2-weighted image appears hypointense. The latter is reported characteristic of many lysosomal diseases.

Reproduced from Sharma S, Sankhyan N, Kabra M, Gulati S. Teaching neuroImages: T2 hypointense thalami in infantile GM1 gangliosidosis. Neurology 2010;74:e47, with permission from Lippincott Williams & Wilkins/Wolters Kluwer Health.

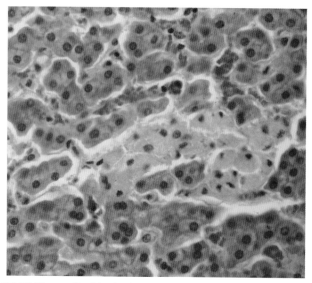

6.3 GM1 gangliosidosis, late infantile type. A cluster of macrophages is seen among normal-appearing hepatocytes in liver.

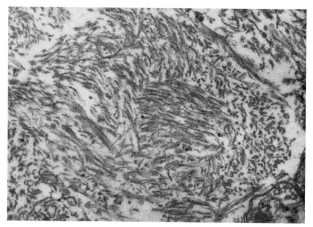

6.4 GM1 gangliosidosis, late infantile type. The storage material in the macrophage in the liver consists of fine tubular/filamentous material morphologically totally different from the storage material in neurons. Electron micrograph.

Reprinted from Suzuki, K (1979).[473] With permission from The McGraw-Hill companies.

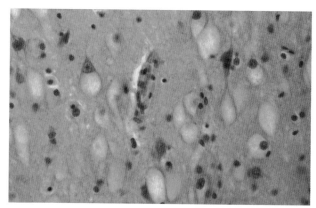

6.5 GM1 gangliosidosis, infantile type. Diffuse neuronal storage with swollen perikarya is evident in cerebral cortex.

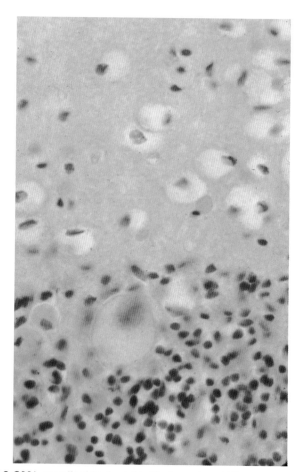

6.6 GM1 gangliosidosis, infantile type. In addition to storage in the perikarya of Purkinje cells, many macrophages with storage material are identified in the molecular layer.

has not been reported. At the terminal stage, the brain is diffusely atrophic, with dilated ventricles and markedly reduced weight. The brain may be firmer than normal. Histologically, enlarged neurons with cytoplasmic storage materials are seen throughout the cerebrum (Figure 6.5), cerebellum (Figure 6.6), brain stem and spinal cord, and even in the dorsal root ganglia, peripheral autonomic ganglia and myenteric plexuses in the intestine. In routine paraffin sections, the storage materials in neurons are variably and weakly stained with PAS and/or Luxol fast blue (LFB). In contrast, glial cells are strongly positive with the PAS stain. In frozen sections, however, neuronal as well as glial storage materials both stain strongly with PAS (Figure 6.7) and mildly positive with Sudan black. The storage materials (inclusions) are strongly positive for acid phosphatase activity, indicating their lysosomal nature. They also stain with filipin, consistent with

accumulation of unesterified cholesterol.[548] The inclusions show the ultrastructural features of membranous cytoplasmic bodies (MCBs) (Figures 6.8 and 6.9).[172] The MCBs in GM1 gangliosidosis are similar to those of GM2 gangliosidosis morphologically,[503] although the major constituent is GM1 ganglioside in the former and GM2 ganglioside in the latter.[476] With Golgi preparation, abnormally enlarged axon hillock regions, known as meganeurites, are readily recognized in cerebral cortical pyramidal neurons.[397,542] Swollen

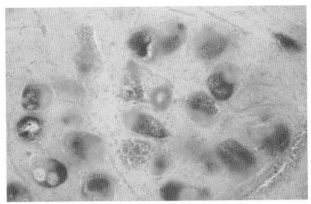

6.7 GM1 gangliosidosis, late infantile type. Frozen section histology of anterior horn neurons in the spinal cord. Neuronal storage material stains strongly with periodic acid–Schiff (PAS) stain on frozen section, whereas no or only faint PAS stain is detected on paraffin-embedded tissue.

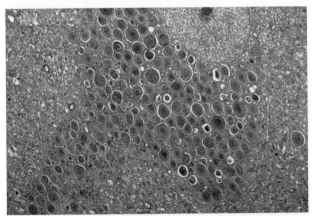

6.8 GM2 gangliosidosis, infantile type. Cytoplasm of affected neurons is packed with membranous cytoplasmic bodies (MCBs). Electron micrograph.

Reprinted from Brain Res 116; Purpura DP, Suzuki K. Distortion of neuronal geometry and formation of aberrant synapses in neuronal storage disease, 1–21. Copyright 1976 with permission from Elsevier.

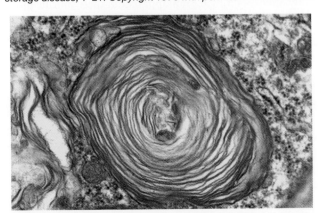

6.9 GM2 gangliosidosis. Higher magnification of a membranous cytoplasmic body (MCB), consisting of multilayered concentric lamellae. Electron micrograph.

(a) 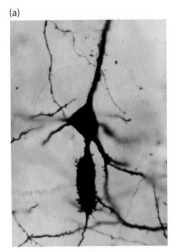 (b)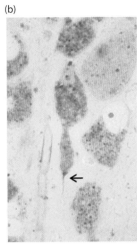

6.10 GM1 gangliosidosis (feline model) showing cellular changes in cerebral cortex. (a) Golgi-impregnated cortical pyramidal neuron in GM1 gangliosidosis showing an enlarged axon hillock (meganeurite) covered with short dendritic sprouts. **(b)** Cerebral cortical neuron stained with ferric ion-ferrocyanide and with safranin red, demonstrating dark-stained initial segment distal to meganeurite (arrow), indicating that meganeurites are expansions of the axonal hillock region proximal to the initial segment and that they are therefore distinct from axonal enlargements (spheroids).

Reproduced with permission from Walkley SU. Cellular pathology of lysosomal storage disorders. Brain Pathol 1998;8:175–93. © 2006, John Wiley and Sons.

(a) 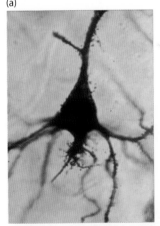 (b)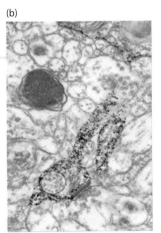

6.11 GM1 gangliosidosis (feline) showing cellular changes in cerebral cortex. (a) Golgi-impregnated pyramidal neuron with an enlarged axon hillock covered with ectopic dendritic sprouting. **(b)** Electron-microscopic view of an ectopic dendrite showing the presence of a large asymmetrical synapse.

(a) Reproduced with permission from Walkley SU.[542] © 2006, John Wiley and Sons. (b) Reproduced with permission from Walkley SU, Wurzelmann S. Alterations in synaptic connectivity in cerebral cortex in neuronal storage diseases. Mental Retard Develop Disabil 1995;1:183–192. Copyright © 1995 Wiley-Liss, Inc.

axon hillocks and meganeurites, the latter occurring proximal to axonal initial segments (Figure 6.10), are accompanied by the presence of ectopic dendritic spines and neurites, which form asymmetrical (excitatory) synapses with local axons (Figure 6.11). This same phenomenon is recognized in a variety of lysosomal diseases with primary or secondary ganglioside storage, but has not been found in other neurodegenerative or neurodevelopmental disorders.[542]

In agreement with CT and MR imaging indicating a myelin deficit, the cerebral white matter is abnormal in the infantile type. Paucity of myelin staining, gliosis, axonal degeneration and macrophage infiltration have been documented.[143] Proteolipid protein and myelin basic protein are profoundly deficient and the numbers of oligodendrocytes markedly decreased in the myelin-deficient white matter.[523] Notable ocular pathological features are corneal haze and cherry-red spots in the maculae. The cytoplasm of retinal ganglion cells is filled with MCBs. In the cornea, Alcian blue-positive and colloidal iron-positive materials accumulate, and corneal epithelial cells are vacuolated. This corneal pathology is similar to that in MPS.[127] The neuropathological investigation of the adult/chronic type of GM1 gangliosidosis is very limited.[170,474,591] However, the pathology of the brain in all reported cases is similar. The brain shows localized atrophy of the caudate nucleus, with mild to moderate dilation of the lateral ventricles. Histologically, enlarged neurons are identified mainly in the basal ganglia (Figure 6.12) (caudate nucleus, putamen and globus pallidus), accompanied by a marked neuronal loss. Storage materials label with anti-GM1 ganglioside antibody.[584] MCBs and other morphologically heterogeneous inclusions are demonstrated ultrastructurally. Meganeurites are noted in some neurons in the basal ganglia,[170] but storage neurons are rare in the cerebral cortex. Purkinje cells may be moderately reduced in number and some display focal swelling of dendrites with storage materials.[170] MCBs may be present in neurons in the myenteric plexuses. MCBs have been described in neurons in the spinal anterior horn, dorsal root ganglion, myenteric plexus and retina of an affected fetus as early as 22–24 weeks of gestational age.[46,580]

Pathogenesis

The pathogenetic mechanism of neuronal storage diseases is not well elucidated. GM1 ganglioside accumulates during disease progression and thus it is reasonable to assume that an accumulation of the ganglioside or related glycolipids is in some manner responsible for the clinical deterioration. In fact, reduction of gangliosides by treating with an inhibitor of the glycosphingolipid synthesis provided clinicopathological improvement in gangliosidosis model mice.[389] As mentioned earlier, neurons with ganglioside storage display abnormal dendritic sprouting and new synapse formation.[543] Certainly, it is likely that these abnormalities contribute to neuronal dysfunction in neuronal storage diseases such as the gangliosidoses. Interestingly, GM1 ganglioside has been considered to have a neuroprotective role and has been considered in the past as a therapy for treating neuro-injury and a variety of neurodegenerative disorders.[189] Its lyso-compounds, however, are known to be cytotoxic, and thus an accumulation of the lyso-form of ganglioside has been suggested to be responsible for the neuronal degeneration.[255] An important role of inflammation is also suggested in neurodegeneration in the gangliosidoses.[233] However, the exact pathogenic mechanism for how the accumulation of the specific metabolites triggers neurodegeneration remains to be clarified. More recently, an activation of an unfolded protein response (UPR) induced by accumulation of sialoglycolipid GM1 in the endoplasmic reticulum has been suggested as the cause of neuronal loss via apoptosis.[504] As described earlier (see Box 6.1), likely a host of complex disease cascade events involving endosomal, autophagosomal and salvage processes all contribute in some manner to neuronal and glial cell dysfunction.[545]

Animal Models

GM1 gangliosidosis is widespread in the animal kingdom and has been reported in cats,[542] cows and dogs.[347,388] The neuropathological features are similar to those of human infantile/late infantile GM1 gangliosidosis. The feline model has been used extensively in studies on aberrant dendritogenesis in the neuronal storage diseases.[542] The extent of the visceral involvement in these models is variable. More recently, with the targeted disruption of the β-galactosidase gene, mouse models of GM1 gangliosidosis were generated independently by two groups.[193,319] The neuropathology of these murine models is that of typical neuronal storage disease. Unlike human and other naturally occurring animal models, however, visceral and bony abnormalities are not apparent in these murine models. In addition, the murine models do not exhibit the ectopic dendritogenesis so evident in humans and higher mammals.[553]

GM2 Gangliosidoses

History and Clinical Features

The infantile type of hexosaminidase A deficiency, or Tay–Sachs disease, is the prototype of neuronal storage disease in humans. This disease was first reported by Tay in the 1880s as a case of intellectual disability with the macular cherry-red spots; later, widespread neurological manifestations were described by Sachs.[140,179] The classic infantile Tay–Sachs disease occurs among Ashkenazi Jews with an unusually high frequency.[76] Affected infants appear normal at birth but develop progressive psychomotor retardation several months later. Clinical symptoms are almost exclusively neurological, with poor head control, hypotonia and hyperacusis. Megalencephaly and tonic–clonic or minor motor seizures may occur in the later course of the disease. Death usually results within a few years.[179,476,537]

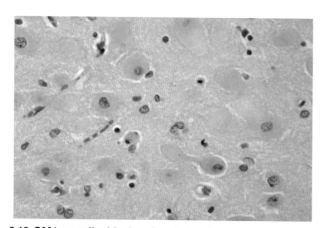

6.12 GM1 gangliosidosis, chronic/adult type. Affected neurons in the putamen.

The clinical phenotype of infantile Sandhoff disease (hexosaminidase B deficiency) is essentially indistinguishable from that of Tay–Sachs disease. Despite some visceral storage, as described later, there are usually no detectable organomegaly or skeletal abnormalities. The AB variant is exceedingly rare. The clinical phenotype of reported cases is similar to that of Tay–Sachs disease.[101] Unlike the high frequency of Tay–Sachs disease among Ashkenazi Jews (with ancestors from eastern Poland/western Russia),[76] no ethnic or geographic preponderance has been noted in Sandhoff disease or the AB variant. The clinical signs and symptoms of infantile GM2 gangliosidosis are usually stereotypical, but among late-onset cases clinical symptoms may vary considerably.[207,356,459] The onset can be at any time, from late infantile to adult age. The onset of the late infantile type is after 18 months of age, with progressive dementia and seizures. The juvenile form becomes symptomatic between ages 4 and 6 years, with dementia. The first patient described as having juvenile GM2 gangliosidosis[482] was later recognized as having the enzymologically unique B1 variant form, as described below.[493] Several additional B1 variant cases have since been reported; all are of the juvenile type.[311,512] The highest incidence of this type has been described in Portugal.[512] Patients with the chronic or adult forms usually develop neurological signs and symptoms in early childhood, with a protracted course. A majority of patients are Ashkenazi Jews.[356] Problems with balance and climbing stairs, signs of either cerebellar or anterior motor neuron involvement, are the most frequent presenting complaints. Unlike the infantile type, the visual functions and optic fundi of adult patients are normal but may show abnormalities in saccadic movement.[414] At least 35 per cent of patients have their first symptoms before age 10 years. Progression of clinical signs and symptoms is usually slow. In some cases, mentation may be well preserved, but psychosis and depression have been reported. Cherry-red spots are detected less frequently. Dystonia, choreoathetosis and other extrapyramidal signs, and ataxia–signs that are reminiscent of spinocerebellar degeneration–may be present. Some patients with courses resembling Friedreich ataxia or motor neuron disease have been well documented.[117,265] The clinical features may vary between and within families, and classic infantile Tay–Sachs disease has been reported within the families of late-onset adult type patients. The majority of these late-onset cases are B variant, but late-onset Sandhoff disease has also been reported.[429]

Biochemistry

GM2 ganglioside is a very minor constituent in the normal adult brain. In GM2 gangliosidoses, GM2 ganglioside accumulates in the brain and, to a lesser extent, in visceral organs, resulting from a defective degradation of GM2 ganglioside. In the infantile type the accumulation of GM2 ganglioside is massive, but in late-onset cases GM2 ganglioside accumulation is much less. Degradation of GM2 ganglioside and related compounds by the β-hexosaminidase system is a complex process. This system (Figure 6.13) consists of two major isozymes, β-hexosaminidase A and B, one minor isozyme S and the GM2 activator protein.

The isozymes are formed by different combinations of two subunits, α and β. Hexosaminidase A is a heterodimer consisting of the α- and β-subunits (αβ), whereas hexosaminidase B is a homodimer of two β-subunits (ββ). The minor form, hexosaminidase S, is a homodimer of the α-subunit (αα). Degradation of GM2 ganglioside *in vivo* requires both β-hexosaminidase A and GM2 activator protein. The α- and β-subunits and the GM2 activator protein are encoded by three distinct genes; abnormalities in any one of the three genes result in defective catabolism of GM2 ganglioside and related lipids and thus in genetically different forms of GM2 gangliosidosis. In Tay–Sachs disease and its variant cases, hexosaminidase A is genetically absent or defective but hexosaminidase B is intact (B variant).[371,421] In Sandhoff disease, resulting from a genetic defect of the β-subunit, activity of both hexosaminidase A and B is deficient (O variant).[421] As a result, all natural sphingolipid and glycolipid substrates cannot be degraded, and other sphingoglycolipids such as globoside are additionally accumulated in the systemic organs. GM2 activator is a polypeptide necessary for degradation of GM2 ganglioside. Thus, genetic defects in the GM2 activator protein cause conditions similar to the deficiency of β-hexosaminidase A activity, resulting in a clinical disease that closely mimics the hexosaminidase A deficiency despite the normal enzyme activity. This type is called AB variant.[421] Hexosaminidase A has a rare variant, the B1 variant, in which the structure of the α-subunit is defective.[322,493] As a consequence, hexosaminidase A in this variant shows normal activity against the artificial substrate 4-methylumbelliferyl-2-acetamido-2-deoxy-β-D-glucopyranoside (4MU-GlcNAc) but is inactive against its sulphated form (4MU-GlcNAc sulphate) and the natural substrate GM2 ganglioside.[179,476,512]

Molecular Genetics

The gene coding for the α-subunit of β-hexosaminidase (*HEXA*) is located on chromosome 15. The gene coding for the β-subunit (*HEXB*) and the gene coding for the GM2 activator protein (*GM2A*) are on chromosome 5. Currently, no fewer than 92 specific *HEXA*, 26 *HEXB* and 4 *GM2A* mutations have been characterized; for details, see Gravel and colleagues.[179]

Numerous disease-causing abnormalities are known in the β-hexosaminidase α-subunit gene.[179] Some mutations are associated with specific ethnic groups. For example, the specific point mutation found in more than 80 per cent of the abnormal alleles among Japanese patients with infantile Tay–Sachs disease has not been reported anywhere else.[494] The point mutation that causes the typical juvenile B1 variant form has been traced to northern Portugal.[115,512] On the other hand, mutations in the classic infantile Tay–Sachs disease among the Ashkenazi Jewish population are heterogeneous.[179,476]

Neuroimaging

Computed tomography and MR imaging of patients with infantile type, regardless of the genotype, are very similar to those of infantile GM1 (see Figure 6.2, p. 443) and may show bilaterally increased density in the thalami and basal ganglia. These regions may be hypointense on

The GM2 gangliosidoses

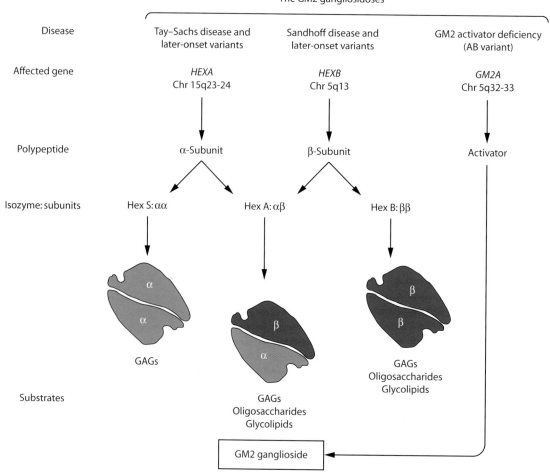

6.13 Diagram of GM2 gangliosidosis and the β-hexosaminidase system in humans.

From Suzuki K, Proia RL, Suzuki K. Mouse models of human lysosomal diseases. Brain Pathol 1998;8:195–215, with permission. Copyright © 2006, John Wiley and Sons.

T2-weighted MR imaging. The involvement of the basal ganglia and thalamus is earlier; cortical atrophy with ventricular enlargement becomes evident at later stages. The cerebral white matter shows patchy areas of increased T2 signal intensity. In a juvenile type (B1 variant), supratentorial structures showed progressive cortical and white matter atrophy; no basal ganglia or thalamic abnormalities were observed.[181] Magnetic resonance spectroscopy of Sandhoff disease revealed findings indicating widespread demyelination.[8] In late-onset cases, cerebellar atrophy, particularly of the vermis, with a normal-appearing cerebrum has been reported to be a prominent feature.[467]

Pathology

Pathology is limited to the nervous system in Tay–Sachs disease, B1 variant and AB variant. In Sandhoff disease (O variant), however, visceral storage is present in the liver, pancreas, spleen and kidney. Vacuolation of hepatocytes and pancreatic acinar cells, PAS-positive Kupffer cells in the hepatic sinusoids, and histiocytes in the germinal centres in the spleen have been reported. PAS-positive materials are also present in the renal tubular epithelium.[188,486]

Details of the neuropathology of Tay–Sachs disease have been described well by Volk and colleagues.[537] The gross changes of the brain may vary with the duration of the clinical course. During the first 12–14 months, the brain is atrophic, with moderately dilated ventricles. The brain weight increases gradually during the period between 15 and 24 months, and marked enlargement of the brain occurs in patients who survive beyond 24 months of age. The brain is rubbery and firm in consistency. With progression of the disease, the white matter becomes depressed and translucent, and the grey–white junction becomes blurred. The optic nerves, cerebellum and brain stem become atrophic. Histopathologically, neuronal storage is present throughout the CNS (Figures 6.14 and 6.15) and the normal cytoarchitecture is markedly distorted. Because of massive storage material in neurons, the Nissl substance is pushed to the periphery of the neuronal perikarya. The storage materials show strong PAS positivity in frozen sections (see Figure 6.7, p. 445). The axon hillock regions of cortical pyramidal neurons are dramatically enlarged by the accumulation of storage materials, forming meganeurites. When visualized with Golgi staining, meganeurites are seen to be accompanied by synapse-covered spines and ectopic dendrites identical to those described for GM1 gangliosidosis (see Figures 6.10 to 6.11 and Figure 6.16).[396,397,543,546] Normal

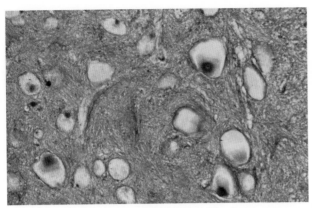

6.14 GM2 gangliosidosis, infantile type (Tay–Sachs disease). Affected neurons in the anterior horn of the spinal cord: paraffin section stained with periodic acid–Schiff (PAS). (Compare with frozen section with PAS stain in Figure 6.7.)

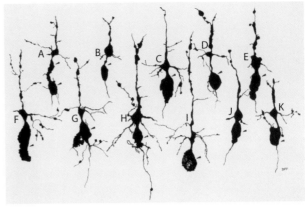

6.16 GM2 gangliosidosis, infantile type. Camera lucida drawing of Golgi-impregnated cortical pyramidal neurons with meganeurites.

Reprinted from Brain Res 116; Purpura DP, Suzuki K. Distortion of neuronal geometry and formation of aberrant synapses in neuronal storage disease, 1–21. Copyright 1976 with permission from Elsevier.

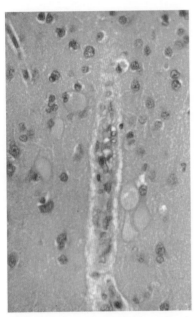

6.15 GM2 gangliosidosis, juvenile type. Unlike in the infantile type, the extent of neuronal storage in cerebral cortex varies among individual neurons. Meganeurites are evident in some neurons.

apical and basal dendrites become distorted and atrophic as the disease progresses. Ultrastructurally, the perikarya of affected neurons are packed with concentrically lamellated lipid inclusions of approximately 1 μm in diameter. These inclusions are MCBs and are again essentially identical to those in GM1 gangliosidosis (see Figures 6.8 and 6.9).[503] Biochemical analysis of MCBs indicates that they consist of gangliosides (mostly GM2 ganglioside), cholesterol and phospholipids, with minor amounts of protein. Loss of myelin in the white matter is notable, and in some patients demyelination may involve almost the entire white matter, resembling leukodystrophy. Astrocytic as well as microglial/macrophage responses are extensive. Degenerative changes are pronounced in the cerebellum, with loss of the Purkinje and granular cells. Purkinje cell axons, as in other lysosomal diseases in which these cells die,[551] exhibit numerous spheroids or swellings,

a phenomenon referred to as neuroaxonal dystrophy. Remaining Purkinje cells show distended perikarya and in some cases antler-like expansions of their dendrites by storage materials. Apoptotic neuronal death has been demonstrated by terminal transferase-mediated nick-end labelling (TUNEL) in Tay–Sachs disease and Sandhoff disease.[219] Similar cerebellar changes are seen across a wide spectrum of lysosomal diseases, for example, Niemann–Pick disease type C. Neuronal storage is also noted in the ganglion cells in the retina, in the autonomic ganglia, in the dorsal root ganglia and in the myenteric plexus.

The neuropathology of infantile Sandhoff disease is similar to that of Tay–Sachs disease. Neurons contain MCBs that are morphologically identical to those in Tay–Sachs disease. However, isolated MCBs from the brain of patients with Sandhoff disease contain much higher accumulations of ceramide trihexoside than the MCBs in Tay–Sachs disease.[476] The AB variant is extremely rare, and neuropathological examination has been carried out on a very limited number of cases. The clinical and biochemical phenotypes are very similar to those of Tay–Sachs disease, and the neuropathology is essentially indistinguishable from that of Tay–Sachs disease and Sandhoff disease by light-microscopic examination.[101,171] Progressive formation of meganeurites with aberrant neurites was originally discovered in this variant.[396] Electron-microscopic investigation has shown, in addition to typical MCBs, 'zebra bodies' within the perikarya of some cortical neurons (Figure 6.17). In addition to neurons, various heterogeneous inclusions have been noted in astrocytes, oligodendroglia and microglia.[101,171] Thus, these fine-structural findings of AB variant appear somewhat more complicated than those of Tay–Sachs disease and Sandhoff disease.

Neuropathological investigation of late-onset GM2 gangliosidosis has been very limited. In a juvenile B1 variant patient, Benninger and colleagues reported an atrophic brain with markedly dilated ventricles.[40] In that patient, the neurons were moderately to severely enlarged, and the

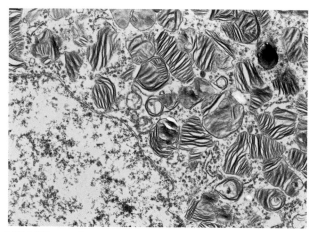

6.17 GM2 gangliosidosis, AB variant. Cerebral cortical neuron containing zebra bodies. Electron micrograph.

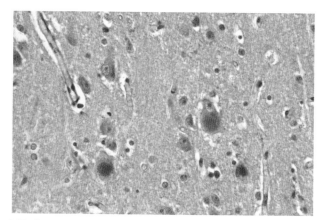

6.19 GM2 gangliosidosis, chronic/adult type. Cerebral cortex shows selective neuronal storage of periodic acid–Schiff-positive material in pyramidal neurons. Periodic acid–Schiff.

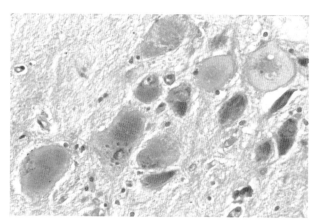

6.18 GM2 gangliosidosis, chronic/adult type. Neurons in the substantia nigra, demonstrating heterogeneous neuronal storage materials. Periodic acid–Schiff.

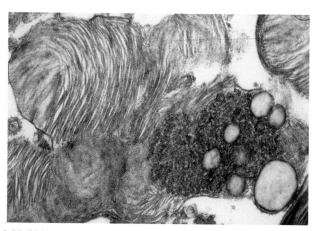

6.20 GM2 gangliosidosis, juvenile type. Electron micrograph of a neuron containing membranous cytoplasmic bodies (MCBs) and lipofuscin.

perikarya distended by a granular, strongly LFB-positive storage material and contained MCB and zebra bodies at the ultrastructural level. Marked cerebral and cerebellar atrophy was present in a 14-year-old patient reported by Suzuki and colleagues.[482] Microscopically, neuronal storage was observed throughout the brain but was less than that of the infantile type. Light- and electron-microscopic features of the chronic or adult type are very similar to those of the juvenile type reported by Suzuki and colleagues.[482] Neuronal storage material is more heterogeneous, in particular in the deep cerebral nuclei and brain stem (Figure 6.18). Neuronal storage material is coarsely granular, showing strong PAS positivity even on paraffin histology (Figure 6.19). The electron-microscopic features of the storage materials are heterogeneous and complex. In addition to MCBs (Figure 6.20), membranous vesicular bodies and electron-dense conglomerates of morphologically heterogeneous lipofuscin-like materials (Figure 6.21) are found in many neurons. The patient reported by Rapin and colleagues was a 16-year-old female, the youngest of three affected siblings with the spinocerebellar phenotype.[405] Jellinger and co-workers reported a 67-year-old whose clinical symptoms were of chronic neuromuscular disease beginning at age 19 years.[229] In both patients, heterogeneous inclusions were found throughout the cerebrum, cerebellum and brain stem. A 44-year-old female reported by Kornfeld had the motor neuron phenotype; later, she developed a peculiar pain syndrome with dysaesthesia, paraesthesia and hyperalgesia affecting most of her body, and psychiatric symptoms.[265] In this patient, neuronal storage was most extensive in the spinal cord neurons and dorsal root ganglia. There was minimal storage in the cerebrum and cerebellum. Enlarged neurons containing well-formed MCBs are identified in the myenteric plexus neurons in all types, including the adult or chronic type of GM2 gangliosidosis (Figure 6.22), despite heterogeneous clinical phenotypes.

Pathogenesis

Like other lysosomal diseases, the pathogenetic mechanisms that lead to the perturbed neuronal circuitry and eventual neurodegeneration in GM2 gangliosidosis are not well understood. Gangliosides are normal cell constituents, but excessive lysosomal accumulation of GM2 is likely to lead to a complex cascade of events involving many aspects of lysosome and cell function. GM2 ganglioside occurs only in minute amounts in normal mature brain, most likely as part of transitory synthetic (Golgi/*trans*-Golgi network [TGN]) and catabolic

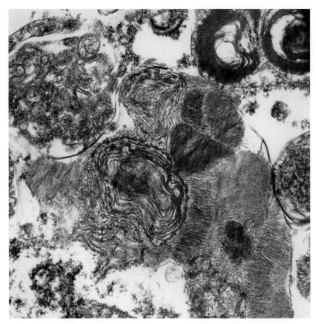

6.21 GM2 gangliosidosis, chronic/adult type. Neuronal inclusions consisting of heterogeneous electron-dense conglomerates of storage materials. Electron micrograph.

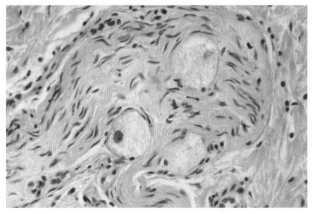

6.22 GM2 gangliosidosis, chronic/adult type. Affected neurons in the myenteric plexus show marked perikaryal storage, as noted in the infantile type.

(lysosome) processing. Whether lysosomal storage of GM2 leads to leakage and redistribution of GM2 to other sites in the neuron, or its lack of salvage to the Golgi/TGN leads to compensatory upregulation in complex ganglioside synthesis, is presently unknown.[595] However, either mechanism has the potential to alter signal transduction events at endosomal and plasmalemmal locations. Indeed, the formation of ectopic dendrites and abnormal synaptic connections may be a manifestation of such abnormal signalling.[543] The resulting alterations in neural connectivity could, in turn, underlie a variety of clinical neurological symptoms from intellectual disability to seizures. Neuroaxonal dystrophy affecting Purkinje cells and other neurons in lysosomal diseases has also recently emerged as a possible contributor to brain dysfunction as well as later neurodegeneration rather than simply being a result of cell death.[551] Formation of toxic compounds such as lysoganglioside GM2,[255] and participation of inflammation[233,349] are also thought to be important factors causing neuronal dysfunction and eventual degeneration. Finally, lysosomal storage in GM2 gangliosidosis, as well some other lipidoses, is also accompanied by accumulation of α-synuclein, a well-documented component of Lewy bodies and related inclusions in Parkinson's disease and certain dementias.[478]

Animal Models

GM2 gangliosidosis has been reported in the dog,[12,95] cat,[58,89,313] Jakob sheep,[390] pig,[268,386] deer,[145] and flamingo.[85] In all species, the typical feature of neuronal storage with swollen perikarya was noted throughout the brain. The feline models are equivalent to Sandhoff disease (hexosaminidase β-subunit deficiency) and to the AB variant (activator protein deficiency). The recently discovered Jakob sheep model is equivalent to Tay–Sachs disease. GM2 gangliosidosis in Muntjac deer is a deficiency in the activity of β-hexosaminidase A and thus is equivalent to B variant. Accumulation of GM2 ganglioside is well documented in canine and porcine models, but the data for further classification are not systematically available. Meganeurites and ectopic dendritogenesis as seen in human GM2 gangliosidosis have been reported in the canine and feline models (see Figures 6.10 to 6.11). Ectopic neurite growth is correlated positively with the extent of accumulation of GM2 ganglioside, both in GM2 gangliosidosis as well as other lysosomal diseases.[543] In addition to these naturally occurring models, murine models of Tay–Sachs disease, Sandhoff disease and AB variant have been generated by targeted disruption of the genes for the α-subunit,[86,385,495,578] the β-subunit[385,424] and the GM2 activator protein.[299] Unlike Tay–Sachs disease in humans, the mouse model of Tay–Sachs disease lacks a clinical phenotype by at least 12 months of age, and neuronal storage occurs in relatively localized regions. Most notably, there is no or very little evidence of neuronal storage in the cerebellum and spinal cord.[495] However, a disease phenotype may appear after 2 years, reminiscent of the late-onset form in human disease.[232,332] Diffuse neuronal storage is present in the murine model of Sandhoff disease. GM2 activator-deficient mice (AB variant mice) show localized neuronal storage in the cerebrum similar to the Tay–Sachs model, but unlike the Tay–Sachs model storage is also present in the cerebellar Purkinje cells.[481]

Niemann–Pick Disease Types A and B

Niemann–Pick disease types A and B are neurovisceral lysosomal lipid storage diseases caused by deficiency of acid sphingomyelinase. The diseases that were subclassified as types C or D in the past have now been shown clearly to be the result of a very different mechanism and are described separately following this section.

History and Clinical Course

The first patient with Niemann–Pick disease type A was described by a German paediatrician, Albert Niemann. The patient was an Ashkenazi Jewish infant with massive hepatosplenomegaly and rapidly progressive neurological symptoms and died at age 18 months. Several years later, the unique pathological features of a disease similar to the infant described by Niemann were delineated by Ludwick Pick; thus, this disease entity bears both their names.[431] The incidence of type A Niemann–Pick disease is high in the Ashkenazi Jewish population. Clinically, the two phenotypes are distinguished by the presence or absence of nervous system involvement. The clinical presentation of type A (neurovisceral form) is relatively stereotypical, and the patient described by Niemann is a classic example. About 50 per cent of patients with type A develop cherry-red spots as noted in patients with Tay–Sachs disease, but hyperacusis and macrocephaly apparently do not occur and seizures are rare. In addition to hepatosplenomegaly, diffuse reticular infiltration is found in the lungs. The majority of patients with type A are infantile, but late infantile and juvenile variants have also been documented.[431,526,527] In contrast, patients with type B (visceral form) present with massive hepatosplenomegaly and typically without any significant neurological abnormalities. The hepatosplenomegaly may be detected within the first year or later, usually later in childhood, but the clinical course is considered to be slowly progressive and patients usually survive to adulthood. Ocular manifestations (macular halos, cherry-red maculae) have been reported in some patients with type B.[327,381] Pulmonary infiltration is commonly detected radiologically. Unlike in Gaucher disease, bone involvement is not a serious complication. Compared with type A, the phenotypic presentation of type B may be more variable. Intermediate types between A and B with some atypical neurological or psychiatric symptoms have been reported, suggesting a phenotypic continuum within neurovisceral type A and visceral type B.[381]

Biochemistry

In Niemann–Pick disease types A and B, the lysosomal hydrolase acid sphingomyelinase is deficient.[431] Analysis of the lipid content shows up to 50-fold accumulation of sphingomyelin, associated with an increase in unesterified cholesterol and other phospholipids accumulating in the visceral organs such as the liver and spleen. In the brain of patients with type A, sphingomyelin is increased 5- to 10-fold in the grey matter. The ganglioside pattern in grey matter is altered, with a significant increase in GM2 and GM3 gangliosides.[238] In white matter, myelin lipids are reduced in general but sphingomyelin is relatively increased. The lipid profile in brains of patients with type B is normal, without an increase in sphingomyelin.[411,526] Increased levels of lysosphingomyelin (sphingosylphosphorylcholine) are reported in the liver and spleen in both types A and B. Massive accumulation of this lipid is documented in the brain of patients with type A.[411] Because lysosphingolipids have been implicated in the biochemical pathogenesis of genetic lysosomal sphingolipidoses,[199,198] this elevation of lysosphingomyelin in type A could play a role in the pathophysiology of brain dysfunction in type A.

Molecular Genetics

Type A and B Niemann–Pick diseases are caused by mutations of the acid sphingomyelinase gene (*SMPD1, NPD*) and are transmitted by a Mendelian autosomal recessive inheritance. The gene is assigned to the chromosome region 11p15.1–15.4. More than 100 mutations causing Niemann–Pick disease have been identified to date. Missense, nonsense and frame-shift mutations and in-frame deletions of specific codons have been reported. Three common mutations (R496L, L302P, fsP330) have been identified in Ashkenazi Jewish patients with type A. The 677delT is common in Israeli Arab patients with type A. Q292K seems associated with late-onset cases with mild manifesting neurological symptoms.[201] In type B, a single mutation, >R608, has been reported to occur commonly. Homozygosity for the >608 mutations seems to be associated with a milder disease phenotype. Four additional mutations have been reported in non-Jewish patients with type B.[327,431,526]

Neuroimaging

The imaging findings in infantile patients with type A are not specific and are similar to those of infantile GM1 and GM2 gangliosidoses. The most common finding is cortical atrophy with enlarged ventricles. In the later stages of the disease, the corpus callosum appears thin and the white matter may show patchy areas of increased T2 signal intensity. One cranial MR imaging study of Niemann–Pick disease documented in the literature reported two siblings with type A with a protracted course. Pronounced cerebellar and mild cerebral atrophy were seen in both patients, although neurological manifestations including cerebellar signs and symptoms were noted in only one patient.[366]

Pathology

Type A is a neurovisceral lipid storage disease, whereas in type B storage is primarily limited in the visceral organs. The liver and spleen are grossly enlarged. Histologically, numerous foamy macrophages (Figure 6.23) are found in many organs, including bone marrow on routinely processed paraffin sections. Cholesterol can be demonstrated in the cytoplasm of these cells by Schultz or perchloric acid naphthoquinone (PAN) stain on frozen sections, or by

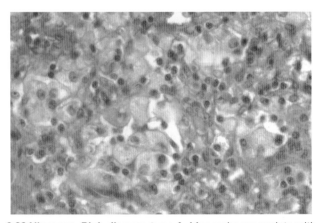

6.23 Niemann–Pick disease type A. Macrophages replete with stored material within the liver.

filipin labelling of frozen or vibratome sections. Vacuolation in lymphocytes has been reported. The foamy macrophages are usually mononuclear but rarely multinuclear, filling the red pulp of the spleen, hepatic sinusoids, pulmonary alveolar spaces, thymus and lymph nodes. In older patients with type B, macrophages may be less vacuolated and instead their cytoplasm may be filled with small granules intensely stained blue with Giemsa or Wright stain (sea-blue histiocytes). Ultrastructural features of the storage materials are multiple electron-lucent vacuoles and small concentric electron-dense lamellar structures. The cytoplasmic granules in older patients consist of electron-dense and more tightly packed lamellar or vesicular structures.

The brain of patients with type A is diffusely atrophic. The cerebellum is often severely affected and small. Histopathology is typical of a neuronal storage disease and indistinguishable from that of infantile GM1 and GM2 gangliosidosis (see Figures 6.5, 6.6, 6.17). Storage neurons with distended perikarya are seen throughout the nervous system, including the autonomic ganglia, dorsal root ganglia and myenteric plexus. The neuronal inclusions closely resemble the MCBs of gangliosidoses but are smaller and less compact. Lamellar membranes are often arranged concentrically around vacuoles or electron-dense small MCB-like structures.[526] In addition to neuronal storage of sphingomyelin, GM2 and GM3 gangliosides are also known to accumulate. Lipid storage is also conspicuous in macrophages/microglia. Vacuolated macrophages similar to those of visceral organs are also found in the leptomeninges and choroid plexus. Lipid storage is also recognized in the retinal ganglion cells, amacrine and Müller cells, and conjunctival and corneal epithelial cells.[409,526] Peripheral neuropathy with infiltration by foamy macrophages has also been reported.[285]

Pathogenesis

Like other sphingolipidoses, the pathogenesis of the cellular dysfunction is not well understood. However, lysosphingolipids that lack the fatty acid component of sphingolipids are considered to play some role in the cellular dysfunction, as lysosphingolipids are potent reversible inhibitors of protein kinase C activity.[199,198] An accumulation of a lysosphingolipid, sphingosylphosphocholine, has been shown in type A (see Biochemistry).[431] Importantly, sphingomyelin, like complex gangliosides and cholesterol, is known to be a constituent of specialized patches of membrane microdomains referred to as rafts, which are also known to function as signalling platforms for a variety of receptors. The common involvement of sphingomyelin and gangliosides in Niemann-Pick type A disease may signal defects in raft formation and in the endocytosis and processing of such membrane components in neurons, which could explain aspects of pathogenesis.[155]

Animal Models

Naturally occurring models of type A Niemann–Pick disease have been described in the poodle dog[65] and Siamese cat.[559] More recently, with targeted disruption of the acid sphingomyelinase gene, murine models of

Niemann–Pick disease type A have been generated by two groups.[217,274,275,334,374] The phenotypes of these two murine models appear somewhat different from each other, however. Light- and electron-microscopic features of the pathological processes in these mice closely resemble those of human Niemann–Pick disease type A.[217,274,275]

Niemann–Pick Disease Type C

History and Clinical Course

In 1961, Allen Crocker classified Niemann–Pick disease into four groups, A, B, C and D, on the basis of clinical phenotypes and biochemical analysis of stored lipids in the tissue. Types A and B are discussed in the previous section. Type C was thought to be a sphingomyelin storage disease with slowly progressive neurological illness, whereas type D closely resembled type C except that it is a genetic isolate in Nova Scotia. Now it has been established clearly that Niemann–Pick disease type C (NPC) is not a primary sphingomyelin storage disorder due to acid sphingomyelinase deficiency, as originally thought, but is an intracellular cholesterol-sphingolipid trafficking abnormality caused by mutation of the *NPC1* or *NPC2* gene. Niemann–Pick disease type D is allelic to type C.[297,380,526,527] Clinical manifestations are heterogeneous, in particular in late-onset slowly progressive phenotypes. Thus, NPC was described with a multitude of descriptive names in the past, as listed in Table 6.2. The age of presentation may vary considerably, and initial manifestations can be hepatic, neurological or psychiatric. Systemic (liver, spleen, lung) involvement and neurological symptoms are not always correlated. Clinically, the disease is classified by the age of onset and presenting neurological symptoms. The most common types (60–70 per cent of cases)

TABLE 6.2 Earlier terms associated with Niemann–Pick disease types C and D

Juvenile dystonic lipidosis
Juvenile dystonic idiocy without amaurosis
Atypical cerebral lipidosis
Atypical juvenile lipidosis
Subacute Niemann–Pick disease
Juvenile Niemann–Pick disease
Ophthalmoplegic lipidosis
Neurovisceral storage diseases with visceral supranuclear ophthalmoplegia
Neville–Lake syndrome
Neville's disease
Subacute neurovisceral lipidosis
Lactosylceramidosis
Sea-blue histiocyte disease
Chronic reticuloendothelial cell storage disease
Nova Scotian variant of Niemann–Pick disease

are those with late infantile and juvenile neurological onset forms, with splenomegaly or hepatosplenomegaly, although no clinically detectable organomegaly has been reported in at least 10 per cent of cases. Ataxic gait and poor school performance due to intellectual impairment and impaired fine movement are often the initial presenting symptoms, followed by onset of seizures, cataplexy and supranuclear vertical gaze palsy (downward or upward, or both). Prolonged neonatal cholestatic jaundice associated with progressive hepatosplenomegaly is present in nearly half of these patients, but usually this is self-limiting and resolves spontaneously by 2–4 months of age. NPC has been reported as the second most common genetic cause of liver disease in infancy in the UK, after α-1-antitrypsin deficiency. About 10 per cent of patients with neonatal jaundice, however, may develop a rapidly fatal liver failure and die before age 6 months without any neurological symptoms. There is a rare severe infantile form with hepatosplenomegaly and severe neurological symptoms. Hypotonia and delayed developmental milestones are the presenting neurological problems around the age of 1–1.5 years, and affected children die before the age of 5 years. This form is more frequent in southern Europe and the Middle East than in the USA. Psychosis and/or progressive dementia is a common presentation of adult-onset patients.[444] Visceromegaly is not clinically detectable in nearly half of the patients. Vertical gaze paresis may not be present.[380]

Biochemistry

Storage compounds in NPC are complex. In the liver and spleen, sphingomyelin, unesterified cholesterol, glucosylceramide, lactosylceramide, GM3 ganglioside and lyso(bis)-phosphatidic acid are moderately increased and, to a lesser extent, other phospholipids and glycolipids are also present. In the brain, however, neither cholesterol nor sphingomyelin is increased when analysed biochemically. In the cerebral cortex, concentrations of total cholesterol, sphingomyelin and gangliosides again are within the normal range. In the white matter, loss of galactosylceramide and other myelin lipids (including cholesterol) is extensive in infantile or late infantile forms with severe clinical phenotype, whereas only a slight decrease is noted in the late-onset or chronic form. Glycolipid compositions are markedly abnormal. Glucosylceramide and lactosylceramide are increased significantly[380,529] and the normally minor constituents, GM2 and GM3 gangliosides, are relatively increased, as is sphingosine. The elevation of GM2 and GM3 gangliosides follows a pattern described in a number of lysosomal disorders including Niemann–Pick type A,[380,529] several of the MPS diseases and even α-mannosiosis.[544]

These changes in glycolipid composition in the brain are apparently concomitant with onset of neurological symptoms, in contrast to the earlier accumulation of glycolipids in visceral organs.[529] The sum of sphingomyelin, galactosylceramide and sulphatide (sulfatide) constitutes more than 90 per cent of normal brain sphingolipids, whereas total gangliosides are less than 5 per cent. The relative 'increase' in GM2 and GM3 is much less than 1 per cent of total brain sphingolipids, and even the increase in glucosylceramide and lactosylceramide does not add more than a few per cent to the total sphingolipids. Although the quantitative aspects of the brain glycolipid abnormalities should be kept in perspective, it is also clear from confocal microscopy studies[603] that these increases in GM2 and GM3 occur extensively in neurons and may signal the presence of significant lysosomal processing defects in these cells. Additionally, because of significant myelin loss, combined galactosylceramide, sulphatide and sphingomyelin are actually reduced by a third or more such that total sphingolipids in the whole brain of NPC patients are reduced by a third or more.

Unlike Niemann–Pick disease types A and B, and in many other lysosomal storage diseases, NPC is not caused by the deficiency of a specific enzyme, and the storage of various lipid compounds is believed to result from a unique error in cellular trafficking. Endocytosed cholesterol is sequestered in lysosomes, and intracellular transport to the plasma membrane, the endoplasmic reticulum and elsewhere is retarded.[380] This observed defect in cholesterol egress has proven useful for diagnosis of NPC.[458] After an addition of low-density lipoprotein (LDL) in the media, cultured fibroblasts of patients with NPC accumulate unesterified cholesterol in perinuclear lysosomes that is easily detected by filipin staining. In the liver and spleen, sphingomyelin and cholesterol are moderately (twofold to fivefold) increased. There is an accumulation of bis-(monoacylglycero)-phosphate, glucosylceramide and lactosylceramide. The storage disease reported as lactosylceramidosis in the past because of marked accumulation of lactosylceramide in the brain has been shown to be a case of NPC biochemically and by mutation analysis.

Molecular Genetics

Niemann–Pick disease type C is panethnic. It consists of two distinct genetic groups, NPC1 and NPC2.[380] More than 95 per cent of patients with NPC belong to the NPC1 group. The estimated prevalence of NPC1 is about 1 in 150 000 but the real prevalence could be higher because of difficulty in clinical diagnosis due to the heterogeneous clinical phenotypes. Type C is more common than types A and B combined. A regionally high incidence is reported in Nova Scotia, Canada (formerly known as type D). The NPC1 gene is located on chromosome 18q11–12.[70,71] More than 300 mutations, including nonsense and missense mutations as well as insertions, deletions and duplications, have been described. NPC2 is a previously known gene that encodes a cholesterol binding protein, HE1. HE1/NPC2 has been mapped to chromosome 14q24.3.[355] A total of 20 different mutations have been identified with good genotype/phenotype correlation; for details, see Millat and colleagues[333] and Vanier and Millat.[525] The precise physiological functions of the NPC1 and NPC2 proteins remain elusive.[433,525] However, recent studies have offered a model whereby unesterified cholesterol in late endosomes is shuttled by the soluble NPC2 protein to the membrane-bound NPC1 protein, which subsequently transfers the cholesterol across the membrane to other membrane sites in the cell.[557]

Neuroimaging

Magnetic resonance imaging and related brain imaging findings are variable. Some brains are normal but others show diffuse atrophy with enlarged ventricles and widened sulci. In some patients, white matter abnormalities have been detected. Proton MR spectroscopic imaging of the brain has shown a significant decrease in the *N*-acetylaspartate and creatine ratio (NA/Cre) in the frontal and parietal cortices, centrum semiovale and caudate nucleus. The choline/creatine ratio was significantly increased in the frontal cortex and centrum semiovale.[500] More recent neuroimaging studies that included positron emission tomography (PET) have revealed bilateral hypometabolism in their prefrontal cortex and dorsomedial thalamus, and hypermetabolism in the parietal-occipital white matter, areas of the basal ganglia, cerebellum and pons.[220] These overall results are consistent with a diffuse degenerative process involving the brain.

Pathology

Niemann–Pick disease type C is a neurovisceral lipid storage disease. Reflecting the clinical heterogeneity, the extent of the pathology may differ considerably between patients. In general, however, younger patients with an acute clinical course tend to have more severe hepatosplenomegaly and marked neuronal storage than patients with a slowly progressive clinical course. Visceral organs are variously infiltrated with foamy storage macrophages; in older patients, macrophages containing basophilic granules (sea-blue histiocytes) are also present. These storage cells tend to be clustered in the red pulp in the spleen, and within the hepatic sinusoids. The histological features of visceral organs are closely similar to those of types A and B Niemann–Pick disease. In early onset cases, hepatic pathology is usually conspicuous, with cholestasis and giant cell transformation of hepatocytes, and thus these cases are not uncommonly diagnosed as 'giant cell hepatitis'.[415,436] In an ultrastructural study of a 20-week fetus, crystalline structures consistent with cholesterol were demonstrated in Kupffer cells and other macrophages. Also reported were large pleomorphic inclusions and hyperplasia of pericanalicular microfilaments in hepatocytes, suggestive of early cholestasis.[119] Severe pulmonary involvement has been reported in some infantile cases (Figure 6.24).[270,387] These early lethal pulmonary cases appear to belong to NPC2.[430] With ultrastructural investigation, storage materials (inclusions) can be detected in a variety of cell types other than macrophages (e.g. hepatocytes, conjunctival epithelium, endothelial cells, pericytes, keratinocytes, lens epithelial cells, Schwann cells, smooth muscle cells and fibroblasts). These inclusions are pleomorphic, consisting of various-sized electron-lucent vacuoles and electron-dense amorphous or curved short membranous structures (Figure 6.25).

The brain is usually atrophic, but the extent of atrophy is variable. Cerebellar atrophy may be particularly pronounced in some cases. Histopathological examination shows neuronal storage characterized by swollen perikarya with storage materials. The neuronal storage is more extensive and widespread in younger patients with

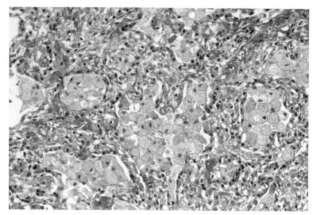

6.24 Niemann–Pick disease type C. Massive accumulation of storage macrophages within alveolar spaces in lung.

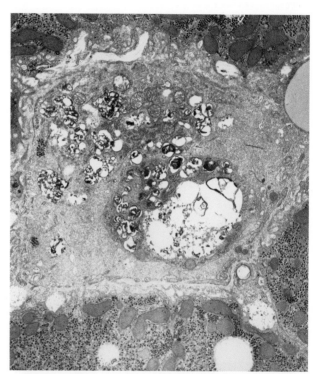

6.25 Niemann–Pick disease type C. Macrophage in the liver, containing multiple membrane-bound vacuoles and electron-dense membranous structure. Electron micrograph.

shorter clinical courses than chronic progressive adult patients. In the latter, neuronal storage may be more localized in the deep cerebral nuclei, brain stem and spinal cord (Figure 6.26).[479,480] Perikarya of affected neurons are packed with fine vacuoles and stain variably with PAS and/or LFB. In frozen sections, neuronal storage materials stain strongly positive with PAS.[113,547] Studies using antibodies and histochemical (filipin and BCθ labelling) techniques to determine the distribution of gangliosides and cholesterol, respectively, have revealed their extensive accumulation in neurons in all areas of the brain.[603] Remarkably, confocal analysis has revealed that although individual neurons accumulate both GM2 and GM3 gangliosides, they tend to be sequestered in vesicles that are independent of one another.[603] A similar pattern of staining is observed in

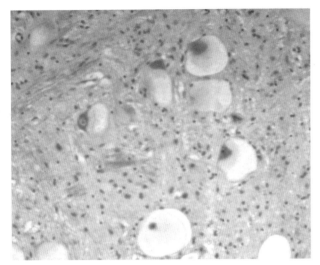

6.26 Niemann–Pick disease type C. Ballooned storage neurons in the anterior horn of the spinal cord.

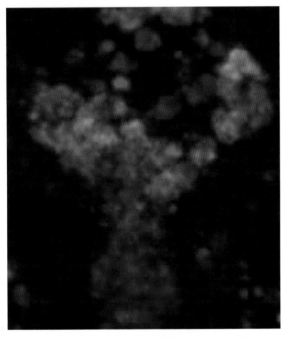

6.27 Niemann–Pick disease type C (murine model). Confocal microscopic image of a cortical pyramidal neuron bearing a meganeurite. Immunostaining for GM2 (red) and BCθ staining for unesterified cholesterol (green) reveal sequestration in separate vesicular compartments in the neuron.

Reproduced with permission from Zhou S, Davidson C, McGlynn R, et al. Endosomal/lysosomal processing of gangliosides affects neuronal cholesterol sequestration in Niemann–Pick disease type C. Am J Pathol 2011;179(2):890–902, with permission from Elsevier.

other lysosomal diseases with GM2 and GM3 storage (see Figure 6.65). Furthermore, cholesterol and GM2 ganglioside did not consistently co-localize to the same vesicles within individual neurons (Figure 6.27). This unexpected distribution is not limited to NPC but also occurs in MPS and other lysosomal diseases with GM2 and GM3 and cholesterol storage.[326,544] Ultrastructural features of the neuronal storage materials in NPC resemble abnormal

multivesicular bodies not unlike those described earlier in other cells in this disease (Figure 6.28). Swollen axon hillocks (meganeurites) and ectopic dendrites are conspicuous (Figure 6.29). Focal neuroaxonal dystrophies (Figure 6.30) (axonal spheroids) are also conspicuous in NPC disease, being most abundant in acute infantile forms and occurring less commonly in older patients. Such spheroids are particularly abundant in Purkinje neurons (Figure 6.31) where they may contribute to early Purkinje cell death.[551] Neurofibrillary tangles (NFTs) are detected consistently in NPC patients with slowly progressive chronic courses (Figures 6.32 and 6.33). These tangles consist of paired helical filaments (PHFs) that by immunocytochemistry and Western blot analysis are similar to the PHF tau in Alzheimer disease.[477,528,547] Patients homozygous for apolipoprotein E (ApoE) ε4 alleles appear to have more NFTs with some β-amyloid deposition.[420] Neurofibrillary tangles have also been reported in an NPC2 patient.[254] Diffuse immunoreactivity for α-synuclein can be demonstrated in swollen storage neurons as well as glial cells. The immunoreactivity is particularly enhanced in patients with the *APOE* ε4 allele. Well-defined Lewy bodies are identified in some cases.[419] Variable extent of neuronal loss, in particular loss of Purkinje cells in the cerebellum, is conspicuous in some cases. TUNEL shows neuronal as well as glial cell apoptosis.[575] As noted in other lysosomal storage diseases, a progressive increase in pro-inflammatory cytokines suggests a role of inflammation in the neurodegeneration.[36,575] Combined features of axonal and myelin degeneration with spheroid formation have been reported in peripheral nerves. Schwann cells contain membrane-bound multilobulated lysosomal inclusions in their distended cytoplasm.[190]

Pathogenesis

Determining the precise physiological functions of the NPC1 and NPC2 proteins has garnered intense effort by the medical community because of their apparent key role in cholesterol egress from late endosomes and lysosomes. Lack of either protein, or both proteins in double knockout mice, results in essentially identical disorders consistent with metabolic cooperativity of the two diverse proteins.[453] Recent studies provide evidence for a model in which NPC2 binds unesterified cholesterol in the late endosome/lysosome and passes it to the membrane-associated NPC1 protein for facilitated egress out of this organelle, thus the absence of either protein leads to the observed build-up of unesterified cholesterol in neurons.[557] Such increases in cholesterol in the cell body of neurons have suggested it may be deficient in cholesterol elsewhere (e.g. in axons), and evidence to support this view is available.[245] Failure of cholesterol egress due to NPC1 deficiency is known to cause compensatory increases in cholesterol synthesis in liver and other tissues, likely exacerbating the storage phenomenon.[576] Recent studies suggest that the accumulation of simple gangliosides (GM2, GM3) and neutral sphingolipids (lactosylceramide and glucosylceramide) may be occurring through a similar block in egress followed by possible compensatory increases in glycosphingolipid synthesis.[603] Interestingly, this and related studies clearly indicate that the restriction of complex ganglioside synthesis leads to

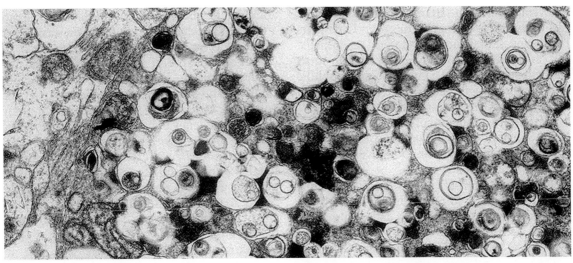

6.28 Niemann–Pick disease type C. Neuronal inclusions consist of polymorphous cytoplasmic bodies. Electron micrograph.

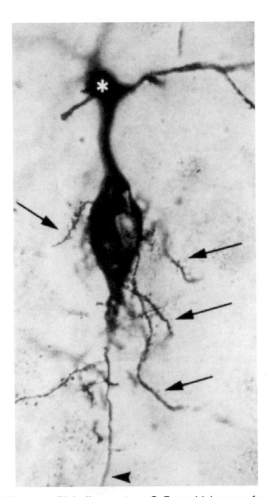

6.29 Niemann–Pick disease type C. Pyramidal neuron from the cerebral cortex of a 3-year-old child who succumbed to the disease. The cell has a meganeurite with numerous spine covered ectopic dendrites (arrows). Asterisk labels cell body; arrowhead, the axon. Golgi impregnation.

Reproduced from March PA, Thrall MA, Wurzelmann S, et al. Dendritic and axonal abnormalities in feline Niemann–Pick disease type C. Acta Neuropathologica 1997;94:164–72, with kind permission from Springer Science and Business Media.

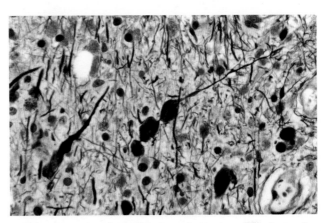

6.30 Niemann–Pick disease type C. Neuroaxonal dystrophy, with focal axonal swelling. Bielschowsky.

Reproduced with permission from Vanier MT, Suzuki K. Recent advances in elucidating Niemann–Pick C disease. Brain Pathol 1998;8:163–74. Copyright © 2006, John Wiley and Sons.

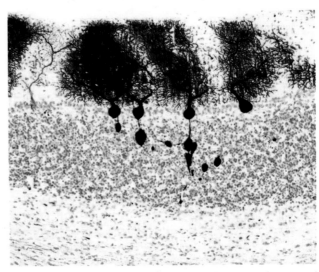

6.31 Niemann–Pick disease type C (feline model). Loss of Purkinje neurons in the cerebellum is accompanied by the presence of numerous axonal spheroids. Calbindin stain.

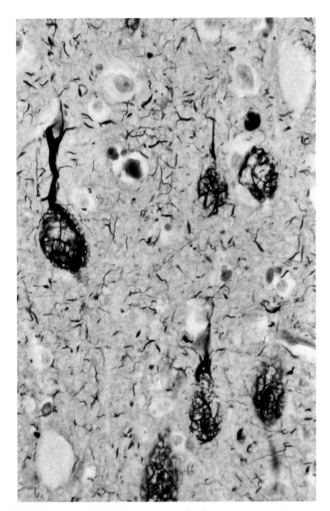

6.32 Niemann–Pick disease type C. Storage neurons in cerebral cortex (temporal lobe) showing neurofibrillary tangles in meganeurites. Numerous neuropil threads are present in neuropil. Bielschowsky.

Reproduced with permission from Vanier MT, Suzuki K. Recent advances in elucidating Niemann–Pick C disease. Brain Pathol 1998;8:163–74. Copyright © 2006, John Wiley and Sons.

substantially reduced cholesterol accumulation in neurons lacking NPC1 as well as NPC2 proteins.[174,301] This finding is consistent with there being close links between these two classes of storage compounds and the NPC1-2 egress mechanism involving late endosomes/lysosomes.[603]

Animal Models

Naturally occurring feline, canine and murine models are known. Mouse models of NPC1 and NPC2 have been generated, and the pattern of neurodegeneration is reported to resemble that observed in naturally occurring NPC1 model mice.[324,453] Light- and electron-microscopic features of these models are similar to those of type C patients with an acute or subacute clinical course. Infiltration by lipid-containing foamy macrophages is extensive in the lung, liver, spleen and lymph nodes. Hepatosplenomegaly is absent in the canine model. Neuroaxonal dystrophy is frequently observed in the feline and murine models but is relatively mild in the canine model. As in human patients, meganeurites with

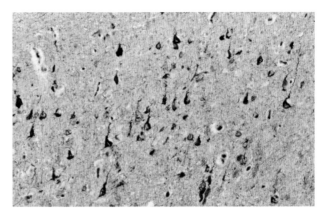

6.33 Niemann–Pick disease type C. Cerebral cortex (orbital gyrus), showing many flame-shaped Alzheimer-type neurofibrillary tangles without amyloid deposition. Bielschowsky.

ectopic dendrites occur in the feline model. In the murine models, cerebral and cerebellar atrophy occurs over time and orderly loss of Purkinje cells correlates with disease progression[211] and is extensive in older mice. Cholesterol and GM2 ganglioside accumulation is detected in the neurons of NPC1 mice before neurodegeneration.[406,509,603] Hyperphosphorylation of tau has been detected in an NPC1 mouse model[509] but, unlike in the human disease, no NFTs have been reported in any of these animal models.[547] Peripheral nerve degeneration, as noted in human patients, has been reported in a murine model.[212] Available animal models have facilitated therapy development for this disorder. A glycosphingolipid synthesis inhibitor has shown efficacy in the mouse[592] and cat[463,592] models, and more recently hydroxypropyl-β-cyclodextrin, an FDA-approved excipient has shown significant benefit in reducing intraneuronal storage of cholesterol and gangliosides.[100] NPC1 disease has also been prevented in the mouse by introduction of an Npc1 cDNA transgene.[304]

Gaucher Disease

History and Clinical Features

Gaucher disease is an autosomal recessive glycolipid storage disease characterized by an accumulation of glucosylceramide (glucocerebroside) within cells of the monocyte/macrophage system. It is caused by the deficient activity of a lysosomal hydrolase, acid β-glucosidase (glucosylceramidase). Although in reality encompassing a continuum of clinical features from perinatal lethality to asymptomatic adult forms, three separate clinical phenotypes – types I, II and III – have been long recognized.[60,178]

The first description of a type I non-neuronopathic patient was by Phillipe Charles Ernest Gaucher in 1882, who described a 32-year-old female with massive splenomegaly thought to be caused by a primary splenic neoplasm. Type II, or the acute neuronopathic type of Gaucher disease, was described more than two decades later. In 1959, slowly progressive subacute neuronopathic type III was described; this is known as the Norrbottnian type because patients were clustered in the Norrbotten province of northern Sweden. The most prominent clinical manifestation of type I is massive hepatosplenomegaly without neurological involvement. Although the majority of cases of this type occur in adults, children may also be affected. Children with type I Gaucher disease usually show massive hepatosplenomegaly with severe hepatic dysfunction and extensive skeletal abnormalities.[222] Neurological symptoms are usually absent. Clinical features of type II Gaucher disease are those of severe progressive neurological illness in addition to hepatosplenomegaly, usually affecting infants. Oculomotor abnormalities are often the initial clinical manifestation, and the majority of patients die within 2 years.[264] A severe neonatal form of Gaucher disease with a rapidly progressive course has been reported. Infants with this form may exhibit marked ichthyotic skin (collodion babies)[465,466] or hydrops fetalis.[446,496] The severity of the clinical manifestations in type III is intermediate between that of types II and I. Neurological symptoms occur later and are usually less severe than those of type II. Type III is divided further into three subgroups. Type IIIa is a progressive neurological disease with myoclonus and dementia.[378,533] In type IIIb, neurological manifestations are limited to horizontal supranuclear gaze palsy but there is severe skeletal and visceral involvement. Type IIIc has horizontal supranuclear gaze palsy and corneal opacities and cardiac valve calcification are present but there is little evidence of visceral involvement.[53] In recent years it has been shown that some patients with Gaucher disease, as well as some carriers, develop parkinsonism and that individuals with Parkinson's disease have an increased frequency of mutations in the gene encoding glucocerebrosidase, *GBA*.[168,497,530,561] The reason(s) for this association are unknown, but it is an area of intense investigation.[561] Patients with genetic deficiency in one of the sphingolipid activator proteins, saposin C (Sap-C), have a clinical phenotype similar to that of patients with neuronopathic Gaucher disease (see Pathology, this page).[377]

Biochemistry

The basic biochemical defect of Gaucher disease is a deficient activity of acid β-glucosidase (glucocerebrosidase), resulting in a massive accumulation of glucocerebroside in cells of the monocyte/macrophage system in all types, including Sap-C deficiency. In the brain, glucocerebroside is increased but remains in very small amounts. A metabolically related potentially toxic compound, glucosylsphingosine, also accumulates in the tissues. The level of splenic glucosylsphingosine bears no relation to the type of Gaucher disease or the age, genotype or clinical course of the patient. The hepatic level of glucosylsphingosine is lower than that in the spleen. Glucosylsphingosine is not present in normal brain, and its level is in the normal range in patients with type I disease. The levels are higher in type II than type III disease. The highest value detected is in fetuses with hydrops fetalis. Thus, an accumulation of glucosylsphingosine appears to be proportional to the extent of the involvement of the CNS[372] and may be responsible for extensive neuronal degeneration.

Molecular Genetics

All types of Gaucher disease are autosomal recessive and are caused by mutations in the gene coding for glucocerebrosidase (*GBA*). The gene is localized on chromosome 1q21. More than 200 disease-causing mutations have been identified, with the majority being missense mutations. Type I occurs more frequently in the Ashkenazi Jewish population; the most common mutation in this type is N370S.[497] No ethnic predominance is known for type II. A number of mutations have been identified in type II patients. Homozygosity for the L444P null mutation is usually associated with type II disease. The prevailing mutation in type III is the L444P point mutation. However, even patients with the same mutation may exhibit marked variability in clinical phenotypes. Currently there is no clear explanation why the same genotype for a disease-causing mutation may express as different clinical phenotypes; for more details, see Grabowski and colleagues.[178]

Pathology

In all types, massive infiltration by macrophages with glucocerebroside storage (Gaucher cells) is seen in visceral organs. Infiltration is particularly conspicuous in the liver (Figure 6.34) and spleen, resulting in hepatosplenomegaly. Gaucher cell infiltration in the bone marrow (Figure 6.35) often causes severe bone diseases such as avascular necrosis and pathological fracture. Severe bone disease appears to be much more frequent in splenectomized patients. Gaucher cells are characterized by the wrinkled appearance of their

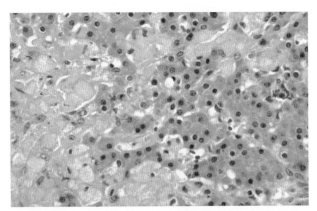

6.34 Gaucher disease, type II. Massive infiltration of Gaucher cells in liver.

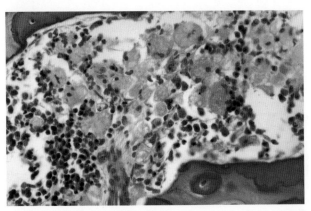

6.35 Gaucher disease, type II. Bone marrow is almost totally replaced by infiltrating Gaucher cells.

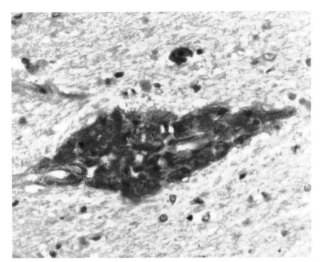

6.36 Gaucher disease, adult type I without neurological signs or symptoms. Perivascular Gaucher cells are stained strongly with periodic acid–Schiff (PAS) in the hypothalamic region. (PAS.)

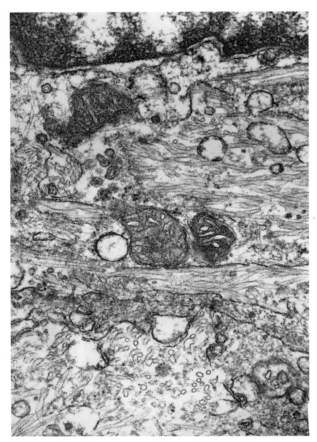

6.37 Gaucher disease, type II. Longitudinal tubular glucocerebroside inclusions within a Gaucher cell. Electron micrograph.

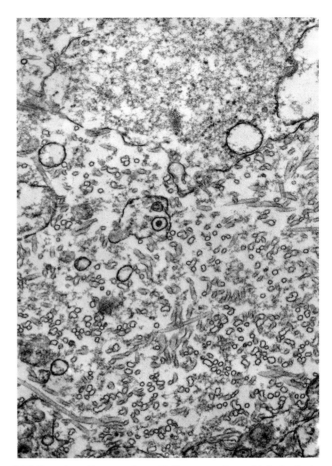

6.38 Gaucher disease, type II. Cross-section of tubular inclusions in a Gaucher cell. Electron micrograph.

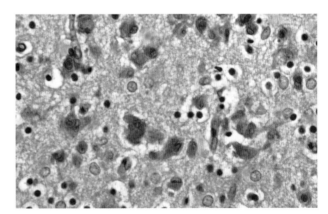

6.39 Gaucher disease, type II neuronopathic form. Individually scattered Gaucher cells with PAS-positive storage material within the occipital cortex. Luxol fast blue/periodic acid–Schiff.

cytoplasm. The cytoplasm is PAS-positive (Figure 6.36) in both paraffin and frozen sections. Ultrastructurally, the cytoplasm contains curved or twisted tubular structures characteristic of glucocerebroside (Figures 6.37 and 6.38). The brain of patients with type I is grossly normal. However, Gaucher cells may be detected in the leptomeninges as well as in the perivascular regions (Figure 6.36) within the brain, in association with an intense gliomesodermal reaction. In patients with type I associated with parkinsonism, numerous α-synuclein positive inclusions, similar to Lewy bodies, are found in hippocampal CA2–4 neurons.[574] In type III patients with severe neurological symptoms, such as oculomotor apraxia, myoclonus and progressive bulbar signs, marked infiltration by Gaucher cells (Figure 6.39) with neuronal degeneration and gliosis is well documented. Neuronal loss and degeneration are widespread. The deep cerebral nuclei, hypothalamus and brain stem are affected more severely than the cerebral cortex.[247] In the cerebellum, dentate nuclei are often severely involved. In the cerebral cortex, neuronal degeneration and Gaucher cell infiltration are most conspicuous in the occipital cortex.[247] Although rare, neurons containing typical tubular inclusions are detected. Some patients with type II may develop the disease prenatally, neonatally or later during the first year of life and may have associated hydrops fetalis and/or congenital ichthyosis.[446,474] In patients with type III disease, neuropathological findings are essentially similar to but less severe than those of patients with type II disease.[88,87,247] In some patients with myoclonus, severe selective neuronal degeneration in the cerebellar dentate nucleus and of the dentatorubrothalamic pathway has been reported;[87,533] in these cases, sparse Gaucher cells were detected in the occipital cortex only. Conradi and co-workers reported a child with late infantile Gaucher disease with oculomotor apraxia, progressive myoclonus and prominent bulbar signs; the dentate nuclei were severely involved, consistent with the clinical sign of myoclonus.[87] In addition, an extensive band-like intraparenchymal accumulation of Gaucher cells, as noted in infantile type II disease, was present. In Sap-C deficiency, Gaucher cells are present in the liver and spleen. Neuropathological features include neuronal lipid storage and neuronal loss, but no Gaucher cells are reported in the brain.[377]

Pathogenesis

Glucosylceramide accumulates massively in the visceral organs, but in the brain only relatively small amounts accumulate in all types. Glucosylsphingosine, a potentially toxic substrate of glucosylceramidase (glucocerebrosidase), accumulates in the visceral organs and brain. The levels of glucosylsphingosine in the spleen and liver do not appear to influence the pathological process. However, in the brain there appears to be a significant correlation of the glucosylsphingosine level and severity of neurodegeneration, suggesting that glucosylsphingosine may contribute to the nervous system involvement in Gaucher disease.[375] Furthermore, glucosylsphingosine has been shown to be toxic to cultured neuronal cells, and factors released by Gaucher cells, including pro-inflammatory cytokines, have been suggested as a mechanistic link between the pathology and diverse clinical manifestations.[91,432] Interestingly, genome-wide association studies have recently implicated CLN8 as a genetic factor that may be making major contributions to the phenotypic heterogeneity in Gaucher disease.[593]

Animal Models

Extensive efforts have been made to develop animal models of Gaucher disease, as recently reviewed.[134] With targeted disruption of the murine glucocerebrosidase gene, a murine model of Gaucher disease was generated. This null mouse died within 24 hours of birth, having less than 4 per cent of the normal glucosylceramidase activity. There was no hepatosplenomegaly. Accumulation of glucocerebroside was found in the liver, lung, brain and bone marrow biochemically, and glucocerebroside inclusions were identified in macrophages in the liver, spleen and bone marrow at the ultrastructural level.[567] The most significant pathological change in these mice was abnormally prominent rugation

with hyperkeratosis in the skin.[440] The phenotype of these mice was similar to that of the severe subtype of type II Gaucher disease ('collodion' babies).[445,496] Another group generated mice carrying mutations found among human patients–the RecNciI mutation, which can cause type II disease, and the L444P mutation, which is associated with type III. Mice homozygous for the RecNciI mutation had a little enzyme activity and accumulated glucocerebroside in the tissues. Mice homozygous for the L444P mutation had a higher level of enzyme activity, and there was no detectable accumulation of glucocerebroside in the tissues. These mice also died within 48 hours of birth, and had a clinical phenotype similar to the mice with the null mutation, despite their detectable residual glucocerebrosidase activities.[300] Point mutation mutants and conditional knockouts have also been made, the latter providing models of greater longevity.[293] In addition to mouse models, spontaneous disease has been identified in the Australian Sydney Silky breed of dogs but was lost to follow-up. More recently, missense mutations leading to reduced glucocerebrosidase activity, glucosylceramide accumulation and a neurological phenotype have been identified in Southdown sheep in Australia.[244]

Farber Disease

History and Clinical Features

Farber disease is a very rare, clinically heterogeneous disease, first described in 1952 by Sidney Farber and co-workers in three children with disseminated 'lipogranulomatosis'. The disease presents most commonly during the first few months after birth with a unique triad of clinical manifestations: (i) painfully and progressively deformed joints; (ii) subcutaneous nodules, particularly near joints and over pressure points; and (iii) progressive hoarseness due to laryngeal involvement. Patients develop swollen joints and a hoarse weak cry shortly after birth. Episodes of fever and dyspnoea, associated with pulmonary infiltration, occur frequently. Death usually occurs with intercurrent infection or inanition by 2–3 years of age. Outstanding clinical features are stiff, swollen and painful joints with amyotrophy and mucocutaneous nodules in the larynx, causing hoarse or faint voice, and in the scalp and abdominal or thoracic walls. Psychomotor retardation, myoclonus and tonic–clonic seizures and other signs of neurological manifestations are present in many cases. Cherry-red spots may be present in some neurologically progressive cases. Some patients may have a less aggressive course and may survive until adolescence or young adulthood. Late-onset patients may not show any obvious signs of systemic involvement. Levade and colleagues have described seven clinical subtypes: type 1 (classic form), types 2 and 3 (intermediate and mild forms), type 4 (neonatal visceral form), type 5 (neurological progressive), type 6 (combined Farber and Sandhoff disease) and type 7 (prosaposin deficiency).[293]

Biochemistry

Farber disease is caused by the deficient activity of an acid ceramidase, which results in an excess of ceramide in all tissues. High concentrations of ceramide are found in the subcutaneous nodules and the kidney. The severity of the disease appears to correlate with the extent of impaired ceramide turnover in cultured fibroblasts and lymphoid cells and cellular ceramide accumulation.[292,398] Diagnosis can be made by the assay of acid ceramidase in cultured fibroblasts and leukocytes.

Molecular Genetics

The mode of inheritance is autosomal recessive. Farber disease is very rare and its prevalence is unknown, but all ethnic groups appear to be involved. The human acid ceramidase (N-acylsphingosine amidohydrolase) gene (ASAH, AC) has been mapped to chromosomal region 8p21.3/22.[293] To date, 21 different sequence variations have been described in the acid ceramidase gene, including polymorphisms, point mutations, deletions and insertions.[293] Most or all appear to be private mutations and genotype/phenotype correlations have been difficult to establish.

Pathology

The cutaneous nodules are granulomatous lesions containing varying numbers of foam cells. These cells are distended with PAS-positive material. In later stages, fibrosis and infiltration with lymphocytes and plasma cells become apparent. Similar granulomata involve joints, subcutaneous tissue (usually at points subjected to pressure, such as knees and elbows), the larynx, the lungs, the kidney and, less commonly, the heart, liver and spleen. Hepatosplenomegaly, conjunctival nodules and macroglossia may be present. In the nervous system, the major pathological finding is neuronal storage with PAS-positive material. The neuronal storage is most pronounced in anterior horn cells and large neurons in the medulla, pons and cerebellum, and the least affected are in the cerebral cortex. In the patient reported by Moser and co-workers,[345] loss of neurons and gliosis were prominent in the cerebral cortex, and dying neurons and neuronophagia were noted in the anterior horns. The white matter shows focal myelin degeneration and gliosis. Neuronal storage is also noted in autonomic and posterior root ganglia. Macular cherry-red spots from retinal ganglion cell storage may be present.[345,590] Various types of ultrastructural feature have been reported in the storage materials. Short curvilinear tubular bodies are observed mainly in the fibroblasts and cells of the monocyte/macrophage system.[589] In the peripheral nerves, the cytoplasm of Schwann cells is often filled by electron-lucent lysosomal inclusions of various shapes with well-defined electron-dense rims; these are called 'banana bodies'.[382] Loss of large myelinated fibres has been reported.[382] Zebra bodies have been described in the neuronal perikarya.[589] A detailed review of the pathological lesions has been provided by Pellissier and colleagues.[382]

Animal Models

A knockout of the murine Asah1 gene resulted in a lethal homozygous phenotype with homozygous embryos living to the two-cell stage only.[294] Heterozygous mice exhibited no physical phenotype but at 6 months of age showed lipid inclusions in different organs including the liver, lung, skin and bone. More recently Medin and colleagues utilized gene-targeting technology to introduce a single nucleotide mutation into the murine Asah1 gene.[7a] This mutation had been previously reported in a patient who manifested a severe form of Farber disease. Mice homozygous for this mutation did not exhibit embryonic death but rather lived to 12 weeks of age. During this time they showed growth retardation, lethargy and weak forelimb strength. Lymphoid organs were enlarged and MRI scanning of the brain revealed hydrocephalus in several of the homozygous littermates. Ceramide analysis demonstrated high levels of ceramide in the spleen, liver, brain, lung, heart and kidney. Under the light microscope, organs were infiltrated by macrophages characterized by eosinophilic cytoplasm and foamy appearance. The characteristic Farber bodies (curvilinear tubules) were identified in a hepatic section under the electron microscope.

Fabry Disease

History and Clinical Features

Fabry disease is an X-linked recessive disorder caused by the deficiency of α-galactosidase A.[128] It was originally described in 1898 as a dermatological disease, angiokeratoma corporis diffusum, independently by two dermatologists, Johannes Fabry in Germany and William Anderson in England. Typical clinical features of hemizygous males with Fabry disease are the presence of angiokeratomata in the skin, painful peripheral neuropathy, transient ischaemic attacks and/or cerebral infarcts, myocardial infarcts and renal failure. Corneal and lenticular opacities are usually detected by slit-lamp examination. These clinical manifestations are due to progressive storage of the glycolipid in the vascular endothelial cells, resulting in ischaemia. The onset of these symptoms is usually during childhood or adolescence. Heterozygous females are usually asymptomatic, but some symptoms may manifest with increasing age.[107,150] A rare 'cardiac variant' of Fabry disease has been described in recent years. These patients are usually asymptomatic during most of their life and are diagnosed only after the onset of cardiac manifestations such as hypertrophic cardiomyopathy or myocardial infarction later in life.[121,351,539] Patients with these atypical milder variants have residual activity of α-galactosidase A. Recent studies have indicated that the incidence of Fabry disease may be much greater than previously believed.[353]

Biochemistry

The enzyme deficient in Fabry disease is the lysosomal enzyme α-galactosidase A. Consequently, glycosphingolipids with terminal α-galactosyl moieties such as globotriaosylceramide (ceramide trihexoside, gal-gal-glc-ceramide) and galabiosylceramide (gal-gal-ceramide) accumulate in the tissue. Classically affected hemizygotes have essentially no demonstrable α-galactosidase A activity. Female carriers have an intermediate level of enzyme activity. However, residual activity has been detected in an atypical variant form.[351,539]

Molecular Genetics

The α-galactosidase A gene (GLA) is localized to the q22.1 region of the X chromosome. Over 300 mutations have been identified. Gene mutations are heterogeneous, and most mutations have been private and found only in single pedigrees. Approximately 75 per cent of the mutations causing Fabry disease are missense or nonsense mutations. For more details, see Desnick *et al.*[107,128]

Neuroimaging

In patients older than 54 years, all had lesions indicating small vessel disease on MR imaging. Those with severe brain involvement had infarcts involving both grey and white matter. The periventricular white matter may be of abnormal signal intensity on T2-weighted MR imaging. With time, these abnormal areas of T2 signal become confluent and the brain becomes atrophic. In female heterozygous patients, the imaging findings tend to be subtle.[94,228] T1 hyperintensity in the pulvinar has been suggested as a key imaging finding in Fabry disease.[344,489] An MR spectroscopy study demonstrated widespread reduction of *N*-acetylaspartate, indicating a more widespread pattern of cerebral involvement than detected on MR imaging.[499]

Pathology

The pathological changes of Fabry disease are characterized by widespread tissue deposits of glycosphingolipids that show birefringence with 'Maltese crosses' under polarizing microscopy. These lipids are preferentially deposited in the endothelial and smooth muscle cells of blood vessels throughout the body, in eccrine sweat glands,[287] the pituitary gland, epithelial cells of the glomerulus (Figure 6.40) and the tubules of the kidney.[13] At the ultrastructural level, they are composed of electron-dense concentric tightly packed lamellar structures (Figure 6.41). The swollen vascular endothelial cells, often accompanied by endothelial proliferation, encroach upon the lumen, causing a focal increase in the intraluminal pressure and thrombus formation, resulting in ischaemia or frank infarction. Progressive aneurysmal dilation of the weakened vascular walls has been well documented in the retinal and conjunctival vessels and telangiectatic vessels in the skin. In the heart, lipids are deposited within the myocardial cells and many lipid-laden storage cells are found in the mitral and tricuspid valves.[239] Vascular involvement is also prominent in the

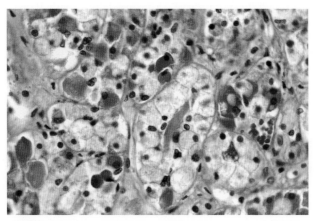

6.40 Fabry disease. Swollen epithelial cells with stored lipid in the anterior lobe of the pituitary gland.

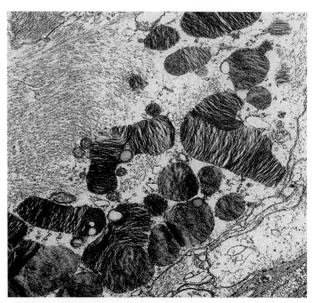

6.41 Fabry disease. Membrane-bound collection of lipid lamellae in an astrocyte. Electron micrograph.

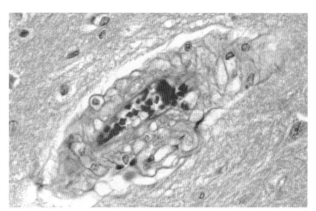

6.42 Fabry disease. Vacuolated vascular smooth muscle cells are very conspicuous in a cerebral vessel.

nervous system (Figure 6.42), often resulting in cerebral ischaemia/infarcts or peripheral nerve conduction abnormalities.[262] The lipid deposition is also noted in leptomeningeal cells, neurons of the peripheral and central autonomic nervous system, and primary neurons of somatic afferent pathways.[109,122,471] In the CNS, swollen storage neurons are detected in the amygdala, hypothalamic nuclei, hippocampus, entorhinal cortex and brain stem. Additionally, enlarged neurons are seen in some peripheral ganglia (Figure 6.43). Storage materials are also detected in astrocytes (Figure 6.41). The storage materials stain positively with LFB and are strongly immunoreactive with antibodies to ceramide trihexoside.[109] In the peripheral nerve, loss of myelinated and unmyelinated nerve fibres and lipid deposits in Schwann cells has been documented.[61]

Pathogenesis

As a result of the deficient activity of α-galactosidase A, α-galactosyl-containing glycosphingolipids accumulate in the vascular endothelial cells and smooth muscle and in some neurons. Mechanical compression of the vasculature

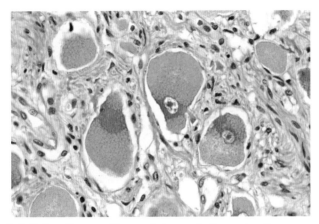

6.43 Fabry disease. Neuronal storage in the periaortic sympathetic ganglion.

and neuronal dysfunction by an accumulation of these lipids may be responsible for some clinical signs and symptoms. However, the reason for the widespread but selective involvement of neurons and other cells in brain is not understood and links between neuronal storage and clinical symptoms (e.g., pain manifestations[47]) have not been adequately addressed.

Animal Models

There is no known naturally occurring animal model of Fabry disease. By targeted disruption of the gene, mouse models have been generated independently by two groups.[368,556] These mice are phenotypically normal, although α-galactosidase A activity is undetectable. Lipid inclusions similar to those in human patients are demonstrated in the liver and kidney on electron microscopy as well as by lectin histochemistry with *Griffonia* (*Bandeiraea*) *simplicifolia*, which binds selectively to α-D-galactosyl residues. Accumulation of these lipids increases with age.[369] α-Galactosidase A transgenic mice carrying normal or mutant human α-galactosidase genes have also been generated.[443]

Metachromatic Leukodystrophy

History and Clinical Features

The first comprehensive clinicopathological report on metachromatic leukodystrophy (MLD) was by Juergen Pfeiffer in 1959, who demonstrated striking brown metachromasia, using the acetic acid–cresyl violet stain, in the brain of patients with juvenile-type leukodystrophy reported earlier by Scholz. Excessive accumulation of sulphatide in MLD tissues was discovered in 1958, and the deficiency of arylsulphatase A (ASA; arylsulfatase) as the underlying cause of the sulphatide accumulation was reported several years later. The heterogeneous clinical types have been customarily classified into three major subtypes: late infantile, juvenile and adult.[226] Rare congenital and infantile types have also been reported. The late infantile type reported by Greenfield in 1933 is the most frequent type. Patients have normal early milestones. The initial clinical presentation usually involves frequent falls with mild spasticity of the extremities during the first

and second years. Loss of motor function and language, spastic quadriplegia and cortical blindness follow rapidly to the point of a vegetative state, and patients survive only rarely beyond 5 years of age. Juvenile MLD has a more insidious onset between the ages of 4 and 12 years; clinical signs and symptoms tend to progress more slowly than in the late infantile type. Behavioural abnormalities and slurred speech precede extrapyramidal and motor dysfunction. Adult MLD may occur at any age after puberty. Progress is usually very slow, with changes in personality,[52] decreased mental status or other psychiatric symptoms.[37,276] In some patients with the adult type, peripheral neuropathy may be the presenting clinical symptom.[147] As described later and in a subsequent section, some patients with an MLD phenotype actually have a genetic deficiency of the sphingolipid activator protein known as saposin B (Sap-B). In addition, there is another rare lysosomal disorder known as multiple sulphatase (sulfatase) deficiency (MSD), which although genetically distinct from MLD shows late infantile MLD features combined with others resembling MPS disease.[216] MSD is discussed in a separate section later.

Biochemistry

The basic defect causing MLD is the deficiency of ASA, which results in an accumulation of sulphatide in the tissue. The sulphatide accumulation is highest in the central and peripheral nervous system and in the kidney, but is also found in other tissues such as the gall bladder and liver. Enzyme activity is extremely low in the late infantile type but some residual activity can be detected in late-onset cases.[290] Low ASA activity may be found in some healthy individuals (pseudo-deficiency). Because up to 2 per cent of Europeans may carry the gene for pseudo-deficiency, awareness of this entity is crucially important for the diagnosis of MLD. Reflecting the extensive loss of myelin in the white matter, concentrations of myelin lipids, cholesterol, phospholipids and cerebrosides are markedly decreased in the late infantile type, although the sulphatide content is increased. This very low cerebroside/sulphatide ratio in the cerebral white matter is typical of MLD. Increased sulphatide is also reported in isolated myelin.[363] Excess amounts of sulphatide are excreted into the urine.[261] Demonstration of abnormal sulphatide excretion in urine is the ultimate proof that the mutation of ASA caused the disease. Patients with adult types excrete significantly smaller amounts of sulphatides than those with late infantile or juvenile types. ASA pseudo-deficiency homozygotes, and compound heterozygotes with one disease-causing allele and one pseudo-deficiency allele, excrete only trace amounts of sulphatide and thus are differentiated clearly from MLD patients. ASA is a heat-labile component of the enzyme sulphatide sulphatase, which requires an additional heat-stable factor (Sab-B) to stimulate its activity *in vivo*. Therefore, ASA activity in patients with Sap-B deficiency shows normal levels.[427,441]

Molecular Genetics

MLD is a disease of autosomal recessive inheritance, and all types are caused by a primary defect in the ASA gene

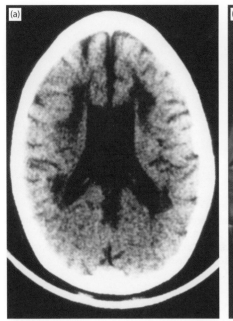
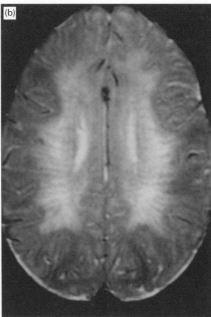
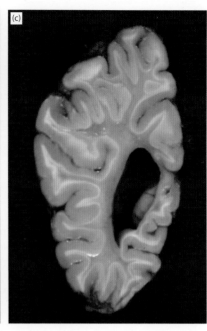

6.44 Metachromatic leukodystrophy (MLD). (a) MLD in an advanced stage. Axial computed tomography (CT) scan, showing low density in the periventricular white matter and diffuse cerebral cortical atrophy. **(b)** Axial T2-weighted image, showing increased and abnormal signal intensity in the centrum semiovale. **(c)** Coronal section of the occipital region of the cerebral hemisphere. The demyelinated white matter appears greyish, in contrast to preserved white matter in the subcortical region.

(*ARSA*) that maps to chromosome 22q13. More than 63 different disease-causing mutations have been identified; many are private. For details on the mutations, see Jaeken *et al.*[226] Reasonably good phenotype/genotype correlation exists. Homozygosity for the null mutation (O-alleles) causes the late infantile type. Alleles with residual enzyme activity (R-alleles) are found in juvenile and adult MLD patients. One copy of the R-allele causes a juvenile type and two copies result in a milder adult phenotype. One to two per cent of healthy individuals may have ASA activity sufficiently low for enzymatic diagnosis of MLD (pseudo-deficiency). It is important to know the existence of such a condition because individuals with the pseudo-deficient allele may be misdiagnosed as having MLD or being carriers.[162,243]

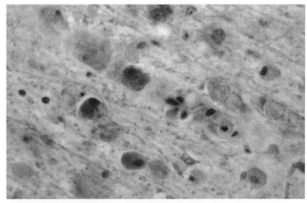

6.45 Metachromatic leukodystrophy. Demyelinating white matter, with scattered glial cells and macrophages.

Neuroimaging

Computed tomography of the brain shows symmetrical areas of low density, particularly in the periventricular white matter (Figure 6.44a). The periventricular lesion extends into the corpus callosum, the internal and external capsules and the brain stem tracts. Magnetic resonance imaging studies show diffuse symmetrical high signals on proton density and T2-weighted images throughout the white matter (Figure 6.44b). In addition, hypointense signal abnormalities showing radiating 'tigroid' and punctate 'leopard-skin' appearance can be demonstrated in the deep white matter on T2-weighted imaging in patients with late infantile and juvenile types.[132,251] The tigroid and leopard-skin appearance in the white matter is thought to represent islands of preserved myelin within the demyelinated areas and is similar to the appearance of the white matter in Pelizaeus–Merzbacher disease. The subcortical U-fibres

tend to be spared until late in the disease (Figure 6.44c). In the advanced and late stages of the disease, cerebral atrophy becomes prominent. Proton MR spectroscopy reveals a marked decrease in choline peaks.[437]

Pathology

The major pathology of MLD is demyelination in the central (Figures 6.44c and 6.45) and peripheral nervous systems and deposits of metachromatic materials in cells within the nervous system and visceral organs (Figures 6.45 to 6.48). The metachromatic deposits show pink metachromasia with toluidine blue (Figure 6.46) and characteristic brown metachromasia with acidic cresyl violet stain (Figures 6.47 and 6.48). The metachromatic granules from the brain of patients with MLD were isolated and their composition characterized. Chemical analysis shows that

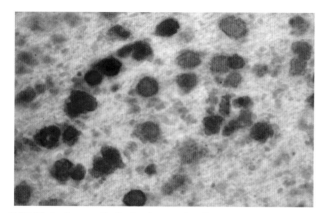

6.46 Metachromatic leukodystrophy. The storage material in the macrophages and glial cells in the demyelinating white matter, showing pink metachromasia on frozen section. Toluidine blue.

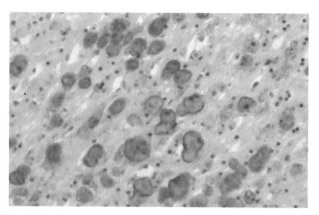

6.47 Metachromatic leukodystrophy. Macrophages in demyelinating cerebral white matter, demonstrating brown metachromasia on frozen section. Acidic cresyl violet.

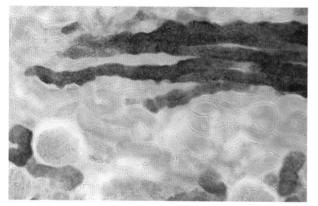

6.48 Metachromatic leukodystrophy. Brown metachromasia is also recognized in renal tubules on frozen section. Acidic cresyl violet.

the molar ratio of cholesterol to galactolipids to phosphatides is 1:1:1. Most of the galactolipids are sulphatides. Electron-microscopic studies have revealed various morphological features. Prismatic and tuff-stone inclusions are especially characteristic (Figure 6.49). Herringbone and honeycomb patterns are often well recognized as unique components of the prismatic inclusion (Figure 6.49). These

6.49 Metachromatic leukodystrophy. Characteristic storage material in macrophage, showing alternating fine electron-dense and lucent short prismatic lamellar structures arranged into herringbone pattern. Electron micrograph.

MLD inclusions can be reproduced *in vivo* or *in vitro* by overloading sulphatides onto the tissue or adding into the culture medium. In the adult type, MLD inclusions frequently form composites with lipofuscin. In the visceral organs, the metachromatic granules are found in macrophages in lymph nodes and spleen, in sweat gland epithelial cells, in Kupffer cells and the bile duct epithelium in the liver, in the epithelium and stroma in the gall bladder, in the adrenal medulla and in the islets of Langerhans in the pancreas. A rare occurrence of papillomatosis of the gall bladder has been reported.[365] In the kidney, the metachromatic material is present in the epithelial cells of the distal convoluted tubules (Figure 6.48), the thin limb of the loop of Henle and the collecting tubules. The brain of patients with the late infantile type shows extensive demyelination throughout the white matter (Figure 6.44), which is grey and firm. The cerebellum and brain stem are atrophic. Paucity of myelin, with presence of numerous reactive astrocytes and macrophages containing metachromatic granules (Figure 6.46), are the characteristic features of MLD. Axons are also decreased. Unlike adrenoleukodystrophy and orthochromic leukodystrophy, the granules in the macrophages are eosinophilic, PAS-positive and non-sudanophilic and show the characteristic brown metachromasia (Figure 6.47) in frozen sections. Some neurons in the motor cortex, thalamus, hypothalamus, basal ganglia, brain stem, dentate nucleus and anterior horn in the spinal cord contain inclusions with metachromasia.[383] At the ultrastructural level, these neuronal inclusions resemble the MCBs of gangliosidoses. In contrast to the relatively uniform neuronal inclusions, inclusions in the glial cells are more pleomorphic (Figure 6.50).[383] These neuropathological changes are similar but milder in juvenile and adult types. Segmental demyelination is found in the peripheral nerves (Figure 6.51). Large-diameter fibres tend to be affected more severely than small-diameter fibres. Loss of axons is relatively mild in the late infantile and juvenile types but is more pronounced in the adult type and often associated with endoneurial fibrosis. Schwann cells and endoneurial macrophages contain metachromatic granules. The different clinical variants do not show any striking differences in the type of inclusion.[506]

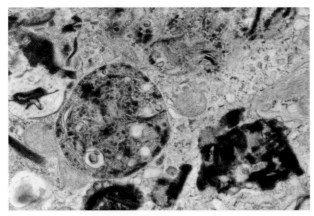

6.50 Metachromatic leukodystrophy. Pleomorphic inclusions in a glial cell. Electron micrograph.

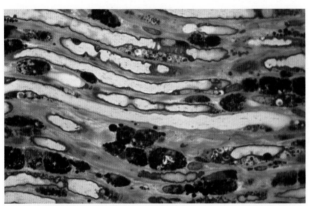

6.51 Metachromatic leukodystrophy. Demyelinating nerve fibres and macrophages, with storage material of cauda equina in a 1-μm-thick section. Toluidine blue.

Pathogenesis

Failure to degrade sulphatide is the central biochemical defect in all forms of MLD. Lysosomal sulphatide accumulation varies by cell type and is quite variable. Whereas some tissues (heart, lung, bones) show little sulphatide accumulation and lack evident pathology, others, particularly white matter of the CNS and peripheral nervous system, show severe sulphatide accumulation and degenerative changes. These changes, seen as demyelination, are reflective of severe oligodendrocyte involvement centrally and Schwann cells peripherally. Abnormal accumulation of sulphatide within these types of glial cell, and accompanying cellular dysfunction, precede and probably trigger the demyelination. Such loss of myelin would be expected to generate the characteristic clinical features. It is also believed that a cytotoxic compound, sulphogalactosylsphingosine (lysosulphatide), may accumulate and contribute to pathogenesis by damaging the myelingenerating glial cells.[226]

Animal Models

There is no known naturally occurring animal model of metachromatic leukodystrophy. ASA-deficient mice generated by homologous recombination exhibit significant impairment of neuromuscular coordination and hearing. The pattern of lipid storage in neuronal and non-neuronal tissues resembles that described in human patients with MLD.[339,571] Myelination is delayed.[577] However, the phenotype of the MLD mice is milder than that of human patients, and the mice have a normal lifespan without extensive demyelination.[161]

Multiple Sulphatase Deficiency

History, Clinical Features, Biochemistry and Molecular Genetics

There are eight different metabolic disorders that are known to be caused by deficiency of individual sulphatases. Six of these are lysosomal storage diseases, including metachromatic leukodystrophy and five different types of mucopolysaccharidosis (see pages 464 and 472). The other two disorders are due to deficiencies of non lysosomal sulphatases. In the disorder known as multiple sulfatase deficiency (MSD), the activity of all sulphatases is profoundly reduced. The first cases of MSD were described by J.H. Austin in the early 1960s,[216] and although rare, additional cases have since been described. Patients exhibit deficiencies of arylsulphatase A, B and C, steroid sulphatases, and various sulphatases related to degradation of mucopolysaccharides resulting in an accumulation of sulphatides and mucopolysaccharides in the brain, liver, kidney and urine. Similar to MLD, the proportion of sulphatide is higher in myelin and the pattern of gangliosides in the brain is abnormal in MSD patients, with increases in normally minor monosialogangliosides, GM2 and GM3. Generally, affected individuals exhibit a complex clinical phenotype that includes characteristics of single sulphatase deficiencies, such as rapid neurologic deterioration and developmental delay resembling, as mentioned earlier, late infantile MLD. Overall, the phenotype of MSD patients combines all of the clinical symptoms observed in each individual sulphatase deficiency.[598] MSD is caused by mutations in the *sumf1* gene, encoding sulphatase modifying factor 1 (also known as FGE, formylglycine-generating enzyme),[90,112] which is involved in the post-translational modification of sulphatases resulting in the conversion of a cysteine located in the catalytic site of sulphatases into formylglycine.

Pathology

The pathology of MSD, like the clinical features, is that of a combined MLD and mucopolysaccharidosis (MPS). Grossly, the leptomeninges are fibrotic, causing severe obstructive hydrocephalus in a case reported by Kepes and co-workers,[250] whereas the case reported by Guerra and colleagues[183] showed atrophy of both the cerebrum and the cerebellum. Perivascular dilation as noted in MPS is pronounced in the cerebral white matter. In the distended cortical neurons, zebra bodies similar to those in MPS and other types of granulo-membrano-vacuolar bodies are found at the ultrastructural level. The honeycomb structure with a hexagonal configuration as described in MLD is found in glial cells. In a rare patient who had only mild deficiencies of arylsulphatases but had severe deficiencies of iduronate sulphatase and heparansulphamidase, pathological changes were closer to those of MPS, with minimal deposition of metachromatic materials in the white matter.[310]

Animal Models

Ballabio and colleagues reported developing mice lacking the formylglycine-generating enzyme after knocking out the *SUMF1* gene. Homozygous affected mice displayed early mortality, congenital growth retardation, skeletal abnormalities, and neurological defects. Tissues revealed progressive cell vacuolation and significant lysosomal storage of glycosaminoglycans. Affected mice also exhibited a generalized inflammatory reaction manifesting as the presence of large numbers of highly vacuolated macrophages. These cells appeared to be the main site of lysosomal storage. Activated microglia were also found in the cerebral cortex and cerebellum along with astrogliosis and neuronal cell loss.[598] Subsequent studies revealed that cells from *SUMF1*[-/-] mice had increased numbers of autophagosomes, which was believed to suggest impaired autophagosome–lysosome fusion. Through such studies MSD as well as other lysosomal disorders are now viewed as disorders of autophagy.[296]

Krabbe Disease (Globoid Cell Leukodystrophy)

History and Clinical Features

In 1916, the Danish physician Knud Krabbe described clinical and pathological findings in five infants from two families who died of an 'acute infantile familial diffuse brain sclerosis'. These infants developed fits of violent crying and irritability beginning at age 4–6 months, followed by progressive muscular rigidity and violent tonic spasms evoked by stimuli such as noise, light and touching. Death occurred between the ages of 11 months and 1.5 years. The clinical course of these patients is typical of patients with infantile globoid cell leukodystrophy (GLD). Late-onset juvenile or adult cases with an insidious onset and slower course are also known.[260,263,309] The onset of juvenile type GLD is between the ages of 3 and 10 years. A gait disturbance, pes cavus or equinovarus deformity of the feet may be early symptoms and signs that lead to spastic paraparesis. Visual failure and an intellectual decline develop slowly but progressively.[309] The clinical course of the later onset disease, however, is less stereotypical than that of the infantile form, and some patients may attain close to the normal lifespan.[263] Adult-onset GLD is usually characterized by slowly developing asymmetric limb weakness, spastic gait, poor coordination and balance, and tremors, mimicking motor neuron diseases.[206,418] Dementia is present rarely. Targeted disruption of Sap-A in mice causes a late-onset chronic variant of GLD.[320] The first human case of GLD resulting from Sap-A mutation was reported in 2005.[460] Although the clinical course of infantile and late-onset GLD differs significantly, there are no apparent differences in the level of residual activity of galactosylceramidase activity among different types as measured *in vitro*.[560]

Biochemistry

The fundamental defect is a deficiency in the activity of galactosylceramidase (galactocerebrosidase), the lysosomal enzyme catalyzing the degradation of galactosylceramide to galactose and ceramide. The unique biochemical characteristic of infantile GLD is a lack of abnormal accumulation of galactosylceramide in the brain, despite the fact that galactosylceramide is the primary natural substrate of the genetically deficient degradative enzyme galactosylceramidase. This paradoxical phenomenon results from the unique localization of galactosylceramide in the myelin sheath and very rapid and early disappearance of the myelin-forming cells, oligodendrocytes, in the brain. Because the myelinating cells disappear at a very early stage of myelination, and because no further synthesis of galactosylceramide occurs, galactosylceramide does not accumulate beyond the level attained at the early stage of myelination. The accumulation of galactosylceramide is limited within the characteristic globoid cells. However, a consistent and perhaps the most important analytical finding is a large increase of a related toxic metabolite, psychosine (galactosylsphingosine), in the white matter. Psychosine is essentially non-detectable in the normal brain, and thus abnormal accumulation is considered the key mechanism in the pathogenesis of the disease (see Pathogenesis, p. 470).[560] The chemical composition of grey matter is relatively normal.

Molecular Genetics

GLD is an autosomal recessive disease caused by a deficiency of galactosylceramidase. The human galactosylceramidase (galactocerebrosidase) gene (*GALC*) maps to chromosome 14q24.3–32.1.[79] Over 60 disease-causing mutations, both nonsense and missense mutations, have been identified. One particular mutation, 502T/del, is quite common in patients from the US and from northern Europe. In Swedish infant patients, this mutation makes up 75 per cent of the mutant alleles. The G809A mutation has been found in patients with late-onset GLD. For more details on mutation analysis, see Wenger *et al.*[560]

Neuroimaging

The white matter of infantile GLD may be hypodense on CT and hyperintense on T2-weighted MR imaging. Deep grey matter, cerebellar white matter and pyramidal tract are constantly involved (Figure 6.52a,b).[303] The white matter abnormalities may have a 'linear' intrinsic pattern.[524] With time, the brain becomes atrophic. Occasionally, the optic nerves and chiasm may appear thick and enlarged (Figure 6.52c). The optic nerve MR imaging findings may be the initial imaging manifestation of the disease. The nerve roots of the cauda equina may also be thick and show enhancement after contrast. In late-onset GLD, imaging studies showed preferential involvement of the occipital white matter and the splenium of corpus callosum. In some cases, T2-weighted MR images showed selective increased signal intensity along the corticospinal tracts.[136] In proton MR spectroscopy studies, an infantile GLD showed

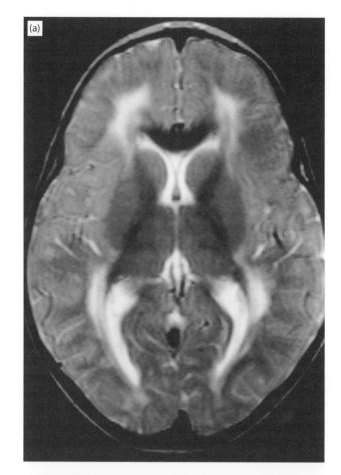

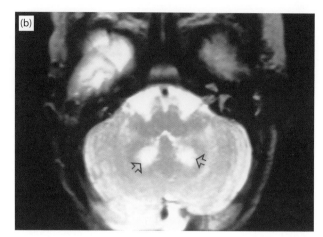

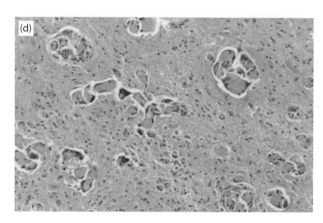

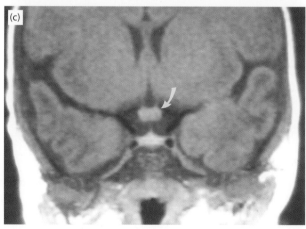

6.52 Globoid cell leukodystrophy. (a) Axial magnetic resonance (MR) T2-weighted image, showing increased and abnormal signal intensity in the white matter of the frontal and occipital lobes, particularly in a periventricular distribution. **(b)** The same patient as in **(a)**: there is abnormally increased signal intensity in the deep cerebellar white matter corresponding to the dentate nuclei (arrows). **(c)** Coronal T1-weighted image, showing enlarged optic chiasm (arrow), which was the first imaging abnormality in this patient. **(d)** Demyelinating cerebral white matter, with many clustered globoid macrophages.

(a–c) Courtesy of Dr Mauricio Castillo, University of North Carolina, Chapel Hill, NC, USA.

pronounced elevation of both myo-inositol and choline-containing compounds and a decrease of N-acetylaspartate in affected white matter. The changes in the grey matter were similar but much milder. In adult GLD, MR spectroscopy data of the white matter have been close to normal or have showed only a slight increase in choline and myo-inositol.[63,136]

Pathology

In GLD, pathology is limited to the nervous system. In the most common infantile type, the brain is atrophic, with firm, rubbery, gliotic white matter. Loss of myelin is nearly complete, with a possible exception of subcortical intergyral arcuate U-fibres. Microscopically, marked paucity of myelin with some axonal degeneration is seen throughout the brain. Extensive fibrillary gliosis and infiltration of numerous macrophages (Figures 6.52d and 6.53), some multinucleated (epithelioid and globoid cells), are unique characteristic features. The globoid cells are abundant in the region of active demyelination and often clustered around blood vessels (Figures 6.52d and 6.53). Oligodendrocytes are markedly reduced.[260,558] Correlative MR imaging and neuropathological studies have shown that the areas of marked hyperintensity on the T2-weighted MR images correspond to the areas of demyelination with globoid cell infiltration.[384] Globoid cells contain PAS-positive storage materials. At the

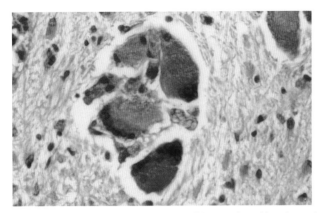

6.53 Globoid cell leukodystrophy. Clustered multinucleated globoid cells containing periodic acid–Schiff (PAS)-positive storage material. Periodic acid–Schiff.

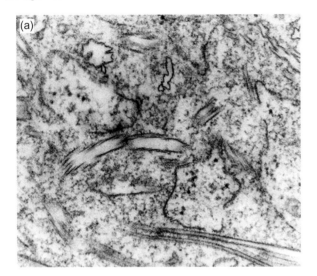

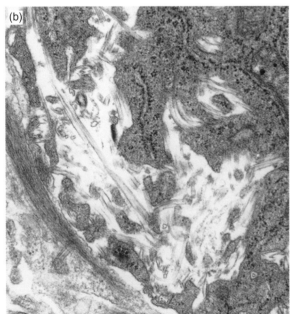

6.54 Globoid cell leukodystrophy. (a) Tubular inclusions in a globoid macrophage. The inclusions are polygonal in shape on cross-section. Electron micrograph. **(b)** Slender tubular inclusions within electron-lucent spaces in a globoid cell. Electron micrograph.

ultrastructural level, the globoid cells contain polygonal or slender tubular structures (Figure 6.54) that are structurally identical to chemically pure galactosylceramide (galactocerebroside). In long-surviving patients, the white matter may be totally gliotic and devoid of macrophages.[328] Neuronal loss is conspicuous in the thalamus, cerebellar Purkinje cell layer and dentate nucleus, pons and brain stem.[328,384] The optic nerves are usually atrophic, but in some cases they are markedly enlarged with extensive gliosis (Figure 6.55). The peripheral nerves may be grossly enlarged and firm. Marked endoneurial fibrosis and features of segmental demyelination and remyelination with onion-bulb formation are often present. Quantitative analysis demonstrates severe loss of large myelinated fibres, but the number of unmyelinated fibres is not diminished. Endoneurial macrophages and also Schwann cells contain tubular inclusions similar to those in the globoid cells in the cerebral white matter. Neuropathological reports of late-onset patients are limited. Choi and colleagues described the neuropathology of adult-onset GLD in 18-year-old twins;[81] both died of severe graft-versus-host disease 2 months after allogeneic bone marrow transplantation. The brains showed degeneration of the optic radiation and frontoparietal white matter, with corticospinal tract degeneration. Multiple necrotic foci with calcium deposits were found within the lesion. Globoid cell infiltration was noted in actively degenerating white matter. In the peripheral nerves of adult GLD, loss of myelinated fibres, disproportionately thin myelin sheaths and inclusions in Schwann cells have been described.[321,418]

Pathogenesis

An accumulation of psychosine, the toxic substrate of the missing enzyme galactosylceramidase, is a unique feature in the brain of patients with GLD. This toxic compound is considered to play a major role in degeneration of oligodendrocytes and myelin.[475] More recently, excessive production of nitric oxide (NO) by inducible nitric oxide synthase (iNOS) was suggested to play an important role in pathogenesis, because psychosine potentiates cytokine-mediated induction of iNOS.[165]

Animal Models

Genetic galactosylceramidase deficiency (Krabbe disease) occurs naturally in the mouse (twitcher), sheep, dogs (West Highland white terriers, Cairn terriers, blue-tick hounds, beagles) and Rhesus monkeys. Clinical and pathological

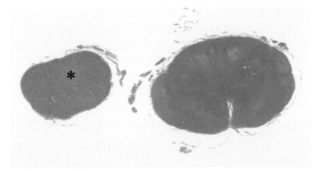

6.55 Globoid cell leukodystrophy. Cross-section of markedly enlarged optic nerve (*), compared with the spinal cord.

features of these models are closely similar to those of the human diseases.[475,558] The murine model of late infantile GLD, the twitcher mouse, has been used extensively for experimental therapeutic approaches, including bone marrow transplantation, gene therapy and substrate reduction therapy.[51,102,403,442] With gene targeting, chronic milder forms of GLD models have been generated.[308,320]

Sphingolipid Activator Protein Deficiencies

History, Clinical Features, Biochemistry and Molecular Genetics

Several lysosomal hydrolases need the assistance of small non-enzymatic glycoprotein cofactors to degrade their substrate sphingolipids. These glycoproteins, also localized within the lysosome, are referred to as activator proteins. If the activator protein for a certain enzyme is genetically defective, then the normal function of the enzyme is perturbed, resulting in a disease similar to the deficiency of the enzyme itself. Two genes code for all established and putative sphingolipid activator proteins. One encodes the GM2 activator (*GM2A*) and is located on human chromosome 5q31. The other is on chromosome 10q21; it codes for the Sap precursor protein (prosaposin, *PSAP*), which is post-translationally processed to four homologous activator proteins, saposin A (Sap-A), saposin B (Sap-B), saposin C (Sap-C) and saposin D (Sap-D).[422] The first sphingolipid activator protein, the sulphatide activator (now known as Sap-B) was discovered by Mehl and Jatzkewitz.[329] Deficiency of GM2 activator causes the AB variant of GM2 gangliosidosis (see GM2 gangliosidoses). Deficiencies of Sap-B and Sap-C result in MLD-like[191,558] and Gaucher-like[83] phenotypes, respectively. Sap-B had been called sulphatide/GM1 activator (Sap-1) and Sap-C had been termed Sap-2 in the past. Sap-A deficiency causes a phenotype of chronic late-onset Krabbe disease in the mouse.[320] The first Sap-A mutation causing Krabbe disease in an infant was reported in 2005.[460] Sap-D is suggested to be a lysosomal acid ceramidase activator but to date no human disease caused by deficiency of Sap-D has been reported. As described later (see Animal Models, this page), the mouse model of Sap-D deficiency causes a urinary system defect as well as cerebellar Purkinje cell degeneration.[318]

Of note, total deficiency of all four saposins due to a mutation in the initiation codon of the precursor has been reported.[59,200,428] The parents of these siblings were fourth cousins, both carrying the same mutant allele.[428] An adult patient with neurovisceral lipid storage disease originally reported as having Niemann–Pick disease type C was later found to have total saposin deficiency.[221] More recently, Elleder and co-workers reported an additional infant with total saposin deficiency (prosaposin deficiency). The infant had multifocal myoclonus, cyanotic hypoxaemia immediately after birth, massive ascites and grand-mal epilepsy, and died at 17 weeks of age.[123]

Pathology

The deficiency of GM2 activator protein causes the AB variant of GM2 gangliosidosis. The clinical phenotype and neuropathological findings are similar to those of GM2

gangliosidosis caused by deficiency of β-hexosaminidase A (see GM2 Gangliosidoses, p. 446).[101,171]

Pathological findings in Sap-B deficiency are similar to those of MLD caused by deficiency of ASA. In the sural nerve biopsy of a 21-year-old patient, Hahn and co-workers reported reduction of myelinated fibres, hypomyelination and onion-bulb formation, indicative of chronic recurring demyelination, and accumulation of metachromatic and orthochromic globules in Schwann cells and macrophages. These globules showed a hexagonal array or membrane-bound tuff-stone appearance, typical ultrastructural features of storage materials in MLD.[192] In addition to the storage of metachromatic materials in macrophages and glial cells in the white matter, scattered cortical neurons with swollen perikarya have been identified in the brain.[558] Such pathological changes would be expected, because Sap-B has activator functions for degradation of other sphingolipids, such as ceramide trihexoside. Brown metachromasia was demonstrated in the storage macrophages in a rectal biopsy of a 7-year-old child with Sap-B deficiency, but submucosal neurons demonstrated non-metachromatic but PAS-positive storage materials with ultrastructural features of MCBs.[427]

In a juvenile patient with Sap-C deficiency,[82] there was a striking increase of glucosylceramide and massive neuronal lipid storage in the brain.[377] In the neuropathological report of an 8-year-old boy with Sap-C deficiency, the most striking neuronal storage was found in the anterior horn neurons in the spinal cord. There was loss of neurons and neuronophagia similar to that seen in type II Gaucher disease. However, the neuronal storage materials were not the tubular inclusions of glucocerebroside but consisted of lipofuscin as well as concentrically arranged membranous structures at the ultrastructural level. Moderately enlarged neurons were found in the cerebral cortex, thalamus, reticular formation and pons. Perivascular Gaucher cells were absent.[377] Thus, neuropathological features of Sap-C deficiency are quite different from those of Gaucher disease, despite the clinical similarity.

One of the two infants with prosaposin deficiency (the other was aborted during pregnancy), who died at 16 weeks, had massive hepatosplenomegaly and the storage cells in the bone marrow resembled Gaucher cells. However, ultrastructural studies of the liver, nerve and skin revealed more complex inclusions and only a few structures reminiscent of Gaucher disease. Neuroimaging before death indicated an atrophic brain with hydrocephalus. Post-mortem examination was not conducted and thus no information on brain pathology is available.[200] An infant reported by Elleder and co-workers showed a pronounced generalized neuronal storage with severe depletion of cortical neurons and extreme paucity of myelin and oligodendroglia.[123] In an adult patient with prosaposin deficiency, neurovisceral storage appeared similar to that of Niemann–Pick disease type C.[126,221]

Animal Models

With targeted disruption of the mouse sphingolipid activator protein (prosaposin) gene, a murine model of total deficiency of Saps was generated.[150] In this model, widespread neurovisceral lipid storage and leukodystrophy with extensive axonal dystrophic changes are observed.[375] Targeted disruption of Sap-A generated the mouse model of chronic Krabbe disease.[320] No specific Sap-D deficiency is known in human patients, but Sap-D knockout mice showed degeneration of renal tubules and progressive loss of cerebellar Purkinje cells.[318] As indicated earlier (see GM2 Gangliosidoses), a naturally occurring cat model of GM2 gangliosidosis AB variant has been reported[313] and a mouse model has been generated by targeted disruption of the GM2 activator protein gene.[299]

MUCOPOLYSACCHARIDOSES

Mucopolysaccharidoses (MPSs) are a genetically heterogeneous group of inborn errors of lysosomal glycosaminoglycan (GAG) metabolism. The MPSs are autosomal recessive diseases, with the exception of Hunter disease (MPS type II), which is X-linked. Each disorder results from deficiency of an enzyme required at a step in the GAG degradation pathway, and partially degraded GAG accumulates excessively in almost all tissues. There are 11 known enzyme deficiencies that give rise to 7 distinct MPSs (Table 6.3). Clinical similarities occur among different enzyme deficiencies and, conversely, a wide spectrum of clinical severity exists within each enzyme deficiency. All the MPSs share many clinical features, such as a chronic and progressive course, multisystem involvement, organomegaly, dysostosis multiplex, abnormal facies, hearing loss, corneal clouding, and cardiovascular and joint problems. The pattern of the most prominent storage materials depends on the nature of the enzymatic defects. Accumulation of dermatan sulphate in Hurler, Hunter, Maroteaux–Lamy and Sly syndromes is associated with severe skeletal abnormalities, whereas accumulation of heparan sulphate in Hurler, Hunter and Sanfilippo syndromes correlates with intellectual disability (Figure 6.56). However, the major neuronal storage materials in MPS diseases are gangliosides rather than glycosaminoglycans.[236] The cellular pathology of affected tissues is generally similar in all cases of MPS.[357,563]

Mucopolysaccharidosis I (Hurler and Scheie Syndromes)

History and Clinical Features

MPS I is usually classified into three phenotypes: Hurler (MPS IH), Hurler–Scheie (MPS IH-S) and Scheie (MPS IS) syndromes. MPS IH, first reported by Gertrude Hurler in 1919, is often cited as the prototype of MPS disease. It is the most severe form of MPS I with patients characterized by coarse facies, hepatosplenomegaly, corneal clouding, severe skeletal abnormalities and intellectual disability. These features are recognized typically between the ages of 9 and 18 months, and affected children usually die before the age of 10 years. Acute cardiomyopathy may be a presenting symptom in some infants. MPS is at the mildest end of the spectrum in the MPS I phenotype, with normal intellect. The coarse facial features, corneal clouding and joint stiffness are usually recognized after 5 years of age; the patient may have a normal lifespan. MPS IS was previously classified as MPS V, but this term and designation are no longer used. Patients with MPS IH-S show varying phenotypes intermediate between MPS IH and IS. Survival to adulthood is common. In both MPS IH-S and MPS IS, compression of the cervical spinal cord by thickened dura mater may occur.

TABLE 6.3 Mucopolysaccharidoses

Clinical diagnosis	Affected glycosaminoglycan	Enzyme defect
Hurler (disease) (MPS I H)	Dermatan sulphate Heparan sulphate	α-ʟ-Iduronidase
Hurler/Scheie (MPS I H-S) Scheie (MPS I S)		
Hunter (MPS II)	Dermatan sulphate Heparan sulphate	α-Iduronate-2-sulphatase
Sanfilippo, type A (MPS IIIA)	Heparan sulphate	Heparan N-sulphatase
Sanfilippo, type B (MPS IIIB)	Heparan sulphate	α-N-acetyl-glucosaminidase
Sanfilippo, type C (MPS IIIC)	Heparan sulphate	Acetyl coenzyme A α-glucosaminide-N-acetyltransferase
Sanfilippo, type D (MPS IIID)	Heparan sulphate	N-acetylglucosamine 6-sulphatase
Morquio, type A (MPS IVA)	Keratan sulphate	Galactosamine-6-sulphatase
Morquio, type B (MPS IVB)	Keratan sulphate	β-Galactosidase
Maroteau–Lamy (MPS VI)	Dermatan sulphate	N-acetyl-galactosamine-4-sulphatase
Sly (MPS VII)	Dermatan sulphate Heparan sulphate Chondroitin sulphate	β-Glucuronidase
Hyaluronidase deficiency (MPS IX)	Hyaluronan	Hyaluronidase

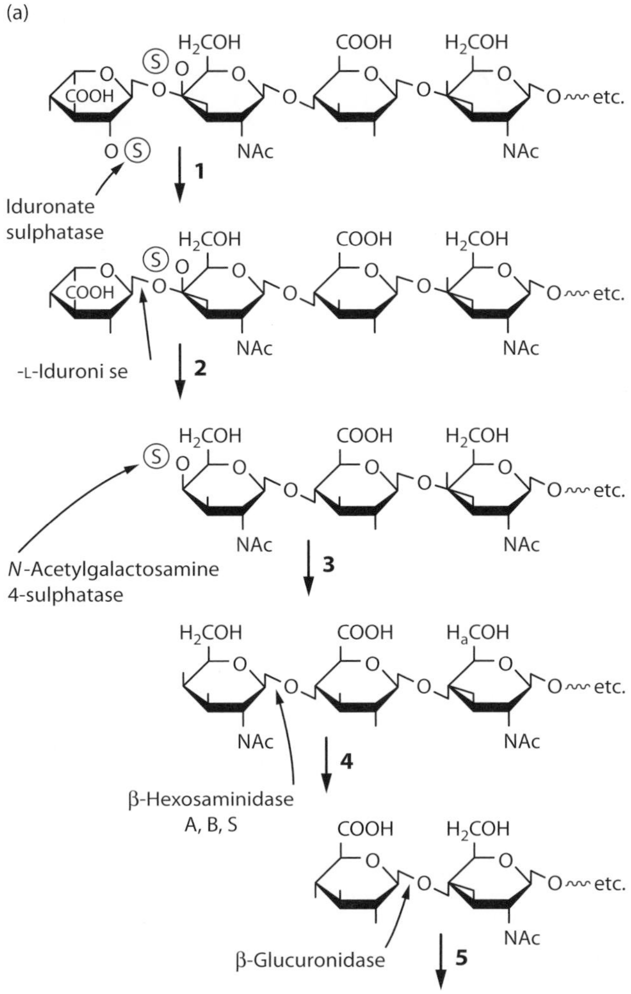

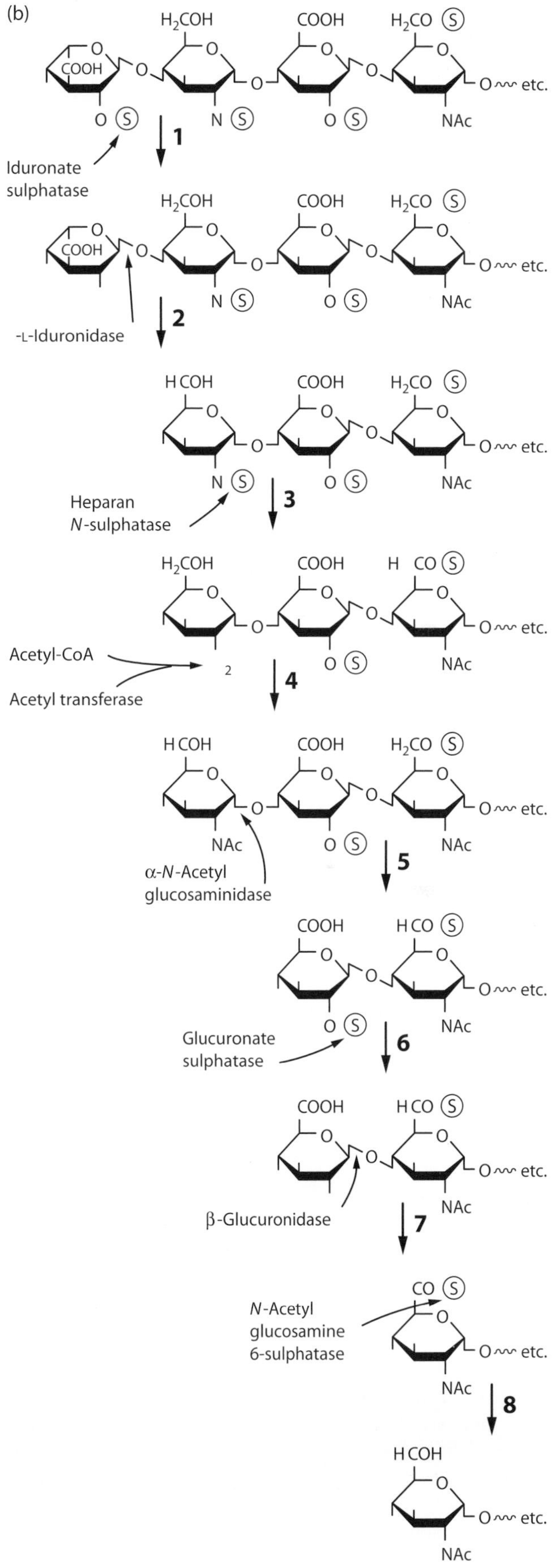

(a)

(b)

6.56 (a) Stepwise degradation of dermatan sulphate. The deficiency diseases corresponding to the numbered reactions are: (1) MPSII, Hunter syndrome; (2) MPSI, Hurler, Hurler–Scheie and Scheie syndromes; (3) MPSIV, Maroteaux–Lamy syndrome; (4) Sandhoff disease, for β-hexosaminidase A and B deficiency; (5) MPSVII, Sly syndrome. This schematic drawing depicts all structures known to occur within dermatan sulphate and does not imply that they occur in equal proportion. **(b)** Stepwise degradation of heparan sulphate. The deficiency diseases corresponding to the numbered reactions are: (1) MPSII, Hunter syndrome; (2) MPSI, Hurler, Hurler–Scheie and Scheie syndromes; (3) MPSIIIA, Sanfilippo syndrome type A; (4) MPSIIIC, Sanfilippo syndrome type C; (5) MPSIIIB, Sanfilippo syndrome type B; (6) no deficiency disease yet known; (7) MPSVII, Sly syndrome; (8) MPSIIID, Sanfilippo syndrome type D. The schematic drawing depicts all structures known to occur within heparan sulphate and does not imply that they occur stoichiometrically.

Continued

(c)

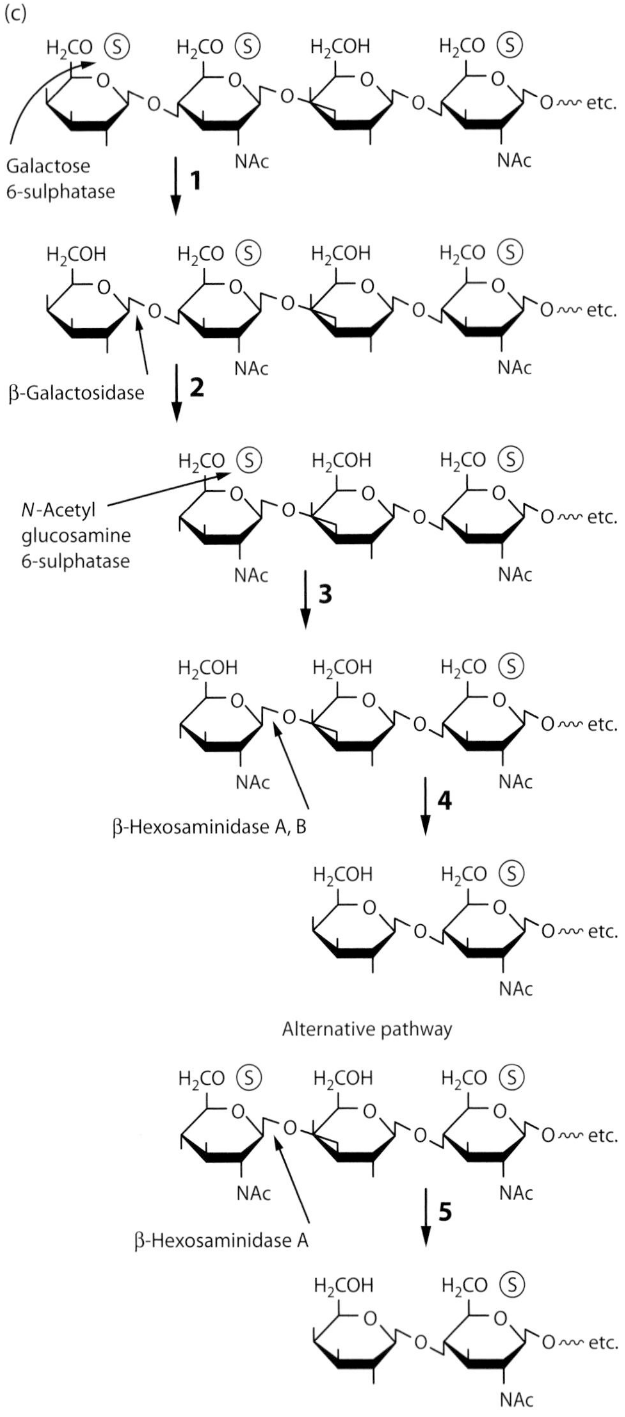

(d)

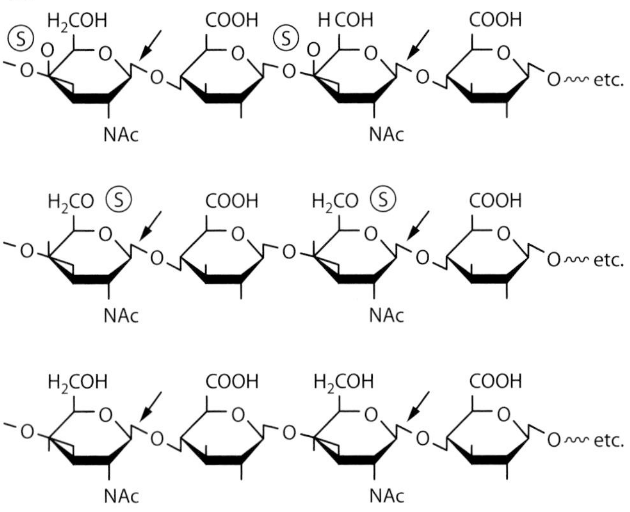

6.56 (*Continued*) (c) Stepwise degradation of keratan sulphate. The deficiency diseases corresponding to the numbered reactions are: (1) MPSIVA, Morquio syndrome type A; (2) MPS IVB, Morquio syndrome type B; (3) MPSIIID, Sanfilippo syndrome type D; (4) Sandhoff disease; (5) Tay–Sachs and Sandhoff diseases. The alternative pathway releases intact *N*-acetylglucosamine 6-sulphate, a departure from the usual stepwise cleavage of sulphate and sugar residues. **(d)** Endoglycosidase degradation of chondroitin sulphate and hyaluronan. Arrows show potential sites for hyaluronidase cleavage of chondroitin 4-sulphate (top), chondroitin 6-sulphate (middle) and hyaluronan (bottom) into oligosaccharide fragments. The oligosaccharides are hydrolyzed further by stepwise action of *N*-acetylgalactosamine 4-sulphatase or 6-sulphatase (for oligosaccharides derived from chondroitin 4- or 6-sulphate), β-hexosaminidase A or B, and β-glucuronidase.

Reproduced from Figures 136-2, 136-3, 136-4 and 136-5 in Scriver CR, Beaudet AI, Sly W, et al (eds). The metabolic and molecular basis of inherited disease, 9th edn. 2001, by permission of the copyright holder, McGraw-Hill.

Biochemistry

The cause of MPS I is a deficiency of α-L-iduronidase, resulting in tissue accumulation and excessive urinary excretion of dermatan and heparan sulphate (Table 6.3); thus, analysis of urinary glycosaminoglycan was the earliest biochemical procedure for the diagnosis of MPS. In the nervous system, there is a relative accumulation of the normally very minor gangliosides GM2 and GM3 within the neuronal perikarya.[236,326] The degree of accumulation is similar to that seen in Niemann–Pick type A, Niemann–Pick type C, MLD, several MPSs and other neurodegenerative diseases. The exact mechanism leading to the secondary ganglioside accumulation in MPS (as well as other lysosomal diseases) is not understood. A report suggested secondary inhibition of acid β-galactosidase activity by the accumulated acidic mucopolysaccharides,[252] but this would be expected to cause GM1 not GM2 and GM3 storage.

Molecular Genetics

The gene encoding α-L-iduronidase (*IDUA*, *IDA*) has been localized to chromosome 4p16.3. Over 46 disease-causing mutations are known.[357] MPS IH, MPS IS and MPS IH-S are allelic. As with a number of other lysosomal disorders (Niemann–Pick, Gaucher), why brain involvement is largely absent in some is unknown although it may be attributable to the presence of sufficient residual enzyme activity in brain to prevent or dramatically retard the onset of clinical disease.

Neuroimaging

Patients with MPS IH and MPS II may have a dolicocephalic skull. The frontal bones may also be slightly pointed in the midline. There are no distinguishing imaging findings among the different types of MPS. The white matter is abnormal and myelination may be delayed. Multiple cystic areas corresponding to enlarged perivascular spaces are noted particularly in the deep white matter and corpus callosum (Figure 6.57a);[439] they follow the course of the deep veins and may also be seen in the basal ganglia. The brain is atrophic, with dilated ventricles (Figure 6.57b).[151,317] The subarachnoid spaces are prominent (Figure 6.57b), and occasionally arachnoid cysts are seen. The skull is thick. In the spine, the vertebrae are malformed and show central beaking (Figure 6.57c). The most common site for compression of the spinal cord is at C1–C2 (Figure 6.57d). Deposition of mucopolysaccharides around the dens leads to laxity of the transverse ligament and subluxations. With time, this ligament thickens. In addition, the foramen magnum may be narrowed. The spinal cord may be compressed, and T2-weighted MR imaging may show increased signal intensity, which correlates with the presence of myelomalacia. The dura may also become thickened at other levels and lead to the spinal canal stenosis.

Pathology

Widespread accumulation of GAGs results in structural abnormalities in many organs. Hepatosplenomegaly, bony abnormalities (dysostosis multiplex) and hydrocephalus (Figure 6.57b) are the most pronounced gross pathological features in MPS IH patients. In routine histological examination of paraffin-embedded tissues, cells distended with large clear vacuoles ('clear cells') (Figure 6.58) are seen throughout mesodermal tissues such as cartilage, fascia, periosteum, blood vessels, heart valves (Figure 6.58), cornea, choroid plexus (Figure 6.59) and cerebral leptomeninges. These vacuoles are the sequelae of GAG extraction during the tissue processing and often appear empty or filled with only scanty amorphous or granular material at the ultrastructural level.[407,563] These 'clear cells' are often associated with extensive fibrosis in various visceral organs. The leptomeninges often show extensive fibrosis, and obstructive hydrocephalus is common. An association with arachnoid cysts has been reported.[563] Dilation of perivascular spaces with loosely packed fibrous tissues is conspicuous within the cerebral white matter (Figures 6.60 to 6.62).[104,105]

The basic cellular pathology of the CNS is that of the neuronal storage disease throughout the brain. However, a wide variation can be observed in the extent of the neuronal storage. A considerable degree of neuronal loss and gliosis is common, resulting in cytoarchitectural derangements. As in many lysosomal diseases, microglial activation is a prominent finding.[367] Structural analysis of storage neurons by use of Golgi preparations reveals meganeurites[397] with ectopic dendrites in human patients and in feline models.[542,543,552] The neuronal storage materials stain with PAS and are often metachromatic. Alcian blue stain is positive in neuronal perikarya as well as in the dilated perivascular regions. Antibody staining for GM2 and GM3 gangliosides and filipin staining for unesterified cholesterol reveal intraneuronal accumulation in widespread areas of the CNS, with the two gangliosides largely sequestered independently of one another.[582] This is similar to MPS IIIA disease discussed later (see Figure 6.65) as well as other lysosomal diseases with GM2 and GM3 storage such as NPC disease.[603] In the cerebellum, Purkinje cells are decreased in number. The remaining Purkinje cells show marked distension of the perikarya (Figure 6.63), with peculiar fusiform expansion of their dendrites.[141] The neuronal storage is also present in the peripheral ganglia. Vacuolation of the Schwann cell cytoplasm has been documented. The neuronal inclusions are morphologically distinct from MCBs of gangliosidoses[503] and are referred to as 'zebra bodies' (Figure 6.64). Zebra bodies appear to be the sites of storage of GM2 and GM3 gangliosides.[326] Reports of the neuropathology of milder forms are limited. Jellinger and co-workers reported a 42-year-old woman with a clinical presentation mimicking Friedreich ataxia. Neuronal storage was limited to the thalamus, hypothalamus, hippocampus, brain stem nuclei, spinal motor neurons and Purkinje cell dendrites. Neurons contained abundant 'ceroid' in their perikarya.[230] Conductive and sensorineural deafness is a common complication of MPS I. Degeneration of the organ of Corti and the neurons of the spiral ganglion with infiltration of vacuolated storage cells in the inner ear has been well documented.[148]

Animal Models

Naturally occurring feline and canine models are known.[388] Their clinical phenotype and pathology are closely similar to those of human MPS I. These models have been used for investigation of cellular pathogenesis[542] as well as therapeutic manipulation. Two murine MPS I models have been generated by targeted disruption of the α-L-iduronidase gene.[84,367]

Mucopolysaccharidosis II (Hunter Syndrome)

History and Clinical Features

Mucopolysaccharidosis II (MPS II) or Hunter syndrome is an X-linked recessive disorder, first described by Charles Hunter in 1917 in two brothers. Prominent clinical features are stiff joints, severe short stature, hepatosplenomegaly, mental retardation, progressive deafness and coarse face. Corneal clouding, however, is not a usual feature. In the majority of cases, clinical diagnosis can be made before age 3 years. Affected children usually die before reaching their early teens. A rare female case exists.[511] Patients with the mild form may develop

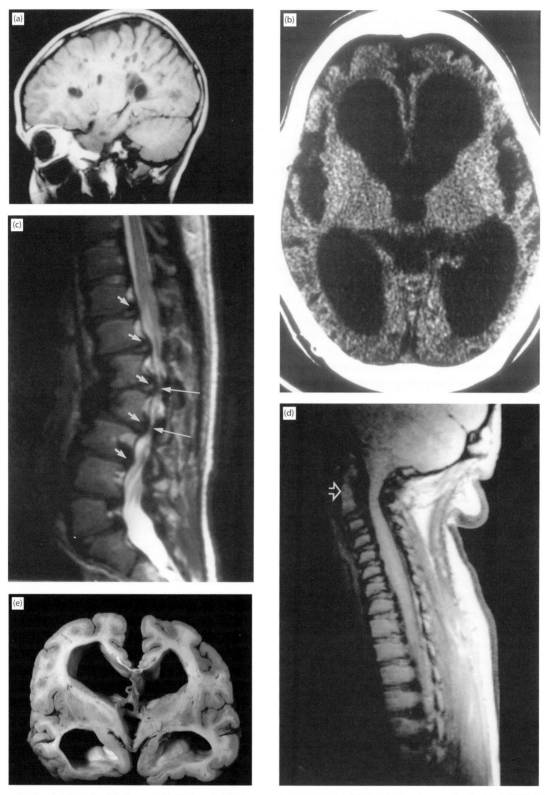

6.57 Mucopolysaccharidosis. (a) Parasagittal T1-weighted magnetic resonance (MR) imaging in a patient with MPSII, Hunter syndrome, showing cyst-like dilated perivascular spaces in the cerebral white matter. **(b)** Axial computed tomography (CT) image of a patient with MPSII, Hunter syndrome, showing cerebral atrophy, markedly enlarged ventricles and prominent subarachnoid spaces. **(c)** Midsagittal T2-weighted MR imaging in a patient with Hunter syndrome, showing deformed vertebrae, bulged discs (short arrows) and multilevel stenosis of the spinal canal (long arrows). **(d)** Midsagittal proton density image in a patient with MPSIH, Hurler syndrome, showing thick dura in the spinal canal and a deformed dens. At the C1–C2 level (arrow), the spinal cord is compressed. **(e)** Coronal section of the cerebrum in MPSIH, Hurler syndrome, showing marked hydrocephalus.

(a–d) Courtesy of Dr Mauricio Castillo, University of North Carolina, Chapel Hill, NC, USA.

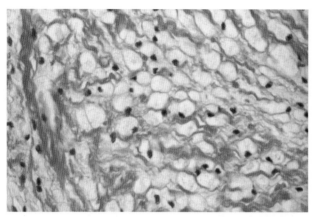

6.58 Mucopolysaccharidosis IH (Hurler). Heart valve, showing numerous clear cells.

6.60 Mucopolysaccharidosis IS (Scheie). Section of cerebrum, showing many holes representing dilated perivascular spaces.

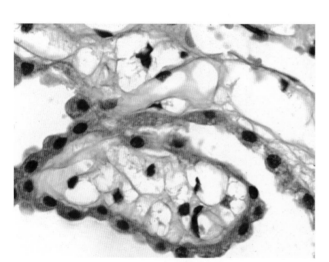

6.59 Mucopolysaccharidosis IH (Hurler). Many clear vacu-olated cells are seen in the stroma of choroid plexus.

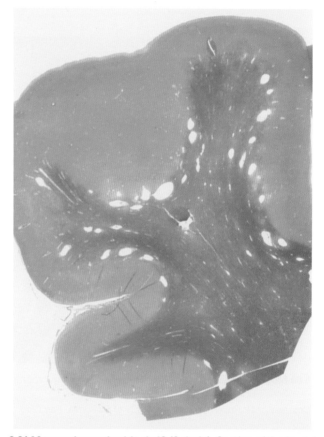

6.61 Mucopolysaccharidosis IS (Scheie). Section of the frontal lobe showing dilation of perivascular spaces, most pronounced at the corticomedullary junction. Solochrome and eosin.

symptoms later and survive longer into adulthood, with minimal or no neurological involvement.[586,587] As noted in the original cases reported by Hunter, nodules or pap-ules over the scapula are frequently observed in MPS II.[305]

Biochemistry

MPS II is caused by a deficiency of iduronate-2-sulphate sulphatase.[31] Dermatan sulphate and heparan sulphate accumulate in the tissues and are excreted in the urine (Table 6.3).

Molecular Genetics

MPS II is the only X-linked MPS. The gene (*IDS*, *MPS2*, *SIDS*) maps to Xq 28. The majority of disease-causing alterations are point mutations, with many missense and nonsense mutations having been identified. Rarely, female carriers of disease-causing mutations are symptomatic.[569]

Neuroimaging

Brain imaging findings are similar to MPS I disease. (See Mucopolysaccharidosis I.)

Pathology

Pathological findings in the visceral organs and brain in the severe form of MPS II are similar to those of MPS IH.[278] Asymptomatic papules on the surface of the skin of the thighs or arms are a unique feature in MPSII.[305] No infor-mation on the neuropathology of the mild form is available.

Animal Models

Naturally occurring MPS II has been reported in Labrador retrievers.[564] A mouse model of human MPS II has been generated by gene targeting.[346]

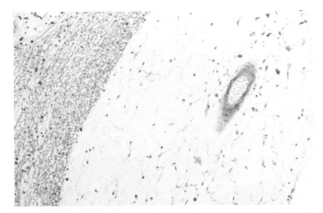

6.62 Mucopolysaccharidosis IS (Scheie). Dilated perivascular space containing loose mesh of fibrous tissue. Luxol fast blue and periodic acid–Schiff.

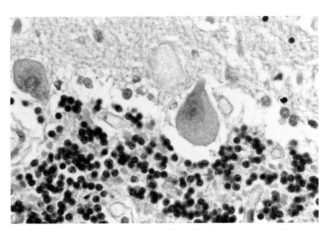

6.63 Mucopolysaccharidosis IH (Hurler). Swollen Purkinje cells with storage material in the cerebellum.

6.64 Mucopolysaccharidosis IH (Hurler). Electron micrograph of neuronal inclusion zebra bodies.

Courtesy of Dr Robert D Terry, University of California, San Diego, CA, USA.

Mucopolysaccharidosis III (Sanfilippo Disease)

History and Clinical Features

In 1963, Sylvester Sanfilippo and colleagues described a group of eight patients who had severe and progressive intellectual retardation with somatic features resembling, but less severe than, Hurler disease.[423] These patients excreted 'acid mucopolysaccharide' consisting primarily of heparan sulphate. Reports of additional patients with similar clinical manifestations followed. This group of the diseases is now called Sanfilippo disease or MPS III. Further biochemical investigations revealed that MPS III includes four distinct genetic diseases (A–D) caused by deficiencies of

four different enzymes (see Table 6.3, p. 472). Interestingly, a recent report indicates a possible fifth type of MPS III disease, MPS IIIE.[271] The various types of MPS III disease cannot be differentiated from each other clinically or pathologically and there is great variability in the clinical expression within this syndrome both among the different types and within each type. The onset of clinical signs, such as speech or behavioural problems, is usually between the ages of 2 and 6 years. Intractable behavioural problems are frequent clinical manifestations. Progressive cognitive and intellectual deterioration follows. Death usually occurs in the second decade of life.[85] Adult patients with mild intellectual impairment or late-onset dementia have been reported, some with associated retinitis pigmentosa.[42]

Biochemistry

Heparan sulphate is degraded by sequential action of several lysosomal enzymes. Deficiency of any one of the enzymes necessary for this degradation pathway results in the same clinicopathological phenotype. Therefore, the four established biochemical types, MPS III A, B, C and D, caused by defects in the four different enzymes (see Table 6.3, p. 472), cannot be distinguished clinically. Heparan sulphate accounts for most of the GAG accumulated in the tissue and excreted in the urine. In the brain, gangliosides GM2, GM3 and GD2 are elevated due in some manner to the secondary metabolic perturbation induced by the accumulation of heparan sulphate.[236] The mechanism causing secondary accumulation of ganglioside is not well understood (see Mucopolysaccharidosis I, p. 472).

Molecular Genetics

The genes for the deficient enzymes in type A (heparan N-sulphatase; SGSH), type B (α-N-acetyl-glucosaminidase; NAGLU), type C (acetyl coenzyme A α-glucosaminide-N-acetyltransferase; HGSNAT) and type D (N-acetylglucosamine-6-sulphatase;GNS) have been cloned, and many disease-causing mutations identified. The genes are located on chromosomes 17q25.3,[434] 17q21,[594] 8p11.1,[133,138] and 12q14,[410] respectively. Type C is the most recent of the genes identified and the missing enzyme, N-acetyltransferase, is unique in that it is not a hydrolase. Rather, it functions to acetylate the non-reducing terminal α-glucosamine residue of intralysosomal heparin or heparan sulphate, thus converting it into a substrate for lysosomal α-N-acetyl glucosaminidase.[133] As mentioned earlier, recent studies have identified a possible type E form of MPS III disease caused by defective activity of a lysosomal sulphatase known as arylsulphatase G.[271] It is believed that this enzyme is the glucosamine 3-O-sulphatase that is required for the complete degradation of heparan sulphate. No patients with this enzyme deficiency have yet been described; however, an arylsulphatase G mutation and deficiency have been reported in American Staffordshire terriers, but were believed to cause a form of neuronal ceroid lipofuscinosis disease based on histopathological criteria.[4]

Pathology

The pathological appearance of MPS III is similar regardless of the deficient enzymes.[236] Clear vacuolated cells are seen throughout the visceral organs but to a far lesser extent than in MPS I or MPS II. Neuronal storage is seen throughout the cerebrum and cerebellum. The extent and the distribution of storage materials vary among different neurons.[492] Storage materials are strongly positive with PAS and various stains for lipofuscin and possess intense autofluorescence under ultraviolet (UV) light.[116] Confocal microscopy studies of MPS IIIA neurons have revealed segregation of the accumulated GM2 and GM3 gangliosides to separate endolysosomal vesicles, with sequestered cholesterol being more closely associated with GM3 (Figure 6.65).[326] Ultrastructural features of neuronal storage materials include zebra bodies, MCBs and

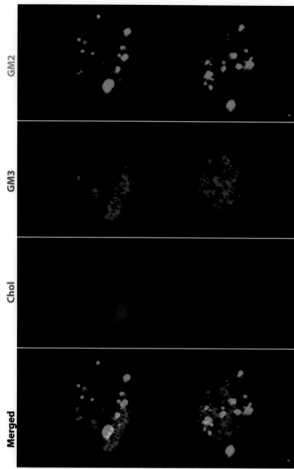

6.65 Mucopolysaccharidosis IIIA (murine model). Immunofluorescent staining for GM2 (green) and GM3 (red) gangliosides and filipin staining for unesterified cholesterol (blue) reveal sequestration in different vesicular compartments in adjacent cerebral cortical neurons.

Reproduced with permission from McGlynn R, Dobrenis K, Walkley SU. Differential subcellular localization of cholesterol, gangliosides, and glycosaminoglycans in murine models of mucopolysaccharide storage disorders. J Comp Neurol 2004;480:415–26. Copyright © 2004 Wiley-Liss Inc.

membrano-granulo-vacuolar inclusions.[205] In a long-surviving patient with type C, who became symptomatic at age 2 years and died at age 39 years, marked neuronal loss and gliosis were reported in the cerebral cortex.[279]

Animal Models

Naturally occurring caprine MPS IIID, canine MPS IIIA and avian MPS IIIB have been described. Similar to the human disease, each of these animal models shows predominantly neurological phenotypes.[142,163,234,235] A severe form of the caprine model, however, shows corneal clouding and bony and cartilaginous abnormalities. A spontaneous model of MPS IIIA has been discovered in mice.[44,45] With transgenic approaches, a mouse model of MPS IIIB has been generated.[295] The clinical phenotypes of these murine models are milder than those of their corresponding human diseases. Cytoplasmic vacuolation indicative of GAG storage is present in macrophages, hepatocytes, and renal tubular and glomerular epithelium. Electron-dense inclusions are detected in certain neurons. Similar to the human disease, there are relative increases in GM2 and GM3 gangliosides in the brain. As of this writing, a knockout of MPS IIIC disease is not yet available. A knockout mouse has been generated for arylsulphatase G, tentatively called MPS IIIE as described earlier, and has also been described in the dog. The knockout mouse was shown to accumulate heparan sulphate in viscera and brain similar to other MPS III diseases.

Mucopolysaccharidosis IVA (Morquio A Disease)

History and Clinical Features

The clinical phenotype of patients with MPS IVA consists of short trunk dwarfism, a unique form of skeletal dysplasia and fine corneal opacification. In addition, the connective tissue of the cornea, airways and heart valves may be affected, and patients may present with nocturnal dyspnoea and obstructive sleep apnoea, cardiac dysfunction and ophthalmological problems.[362] As in most patients with MPS, those with MPS IVA are normal at birth. Clinical signs of kyphosis, growth retardation and genu valgus become apparent by the age of 3.5 years. Odontoid hypoplasia is a universal bony abnormality, and cervical myelopathy develops as a grave consequence of the instability of the hypoplastic odontoid processes.[362,408] Intelligence is usually not affected. The phenotypically similar but genetically distinct disease, Morquio B disease, due to acid β-galactosidase deficiency and thus allelic to GM1 gangliosidosis, is discussed under GM1 Gangliosidosis and Morquio Disease (Syndrome) type B.

Biochemistry

MPS IVA is caused by defective degradation of keratan sulphate due to a genetic deficiency of N-acetylgalactosamine-6-sulphatase (galactose-6-sulphatase) (see Table 6.3, p. 472). A large quantity of keratan sulphate-derived oligosaccharide fragments is excreted into the urine.

Molecular Genetics

The gene encoding N-acetylgalactosamine-6-sulphatase (GALNS, MPS4A) maps to chromosome 16q24.3.[32] Mutation analysis has defined over 100 different mutations.

Pathology

Similar to the other types of MPS, clear vacuolated cells are seen in many organs in MPS IVA (Figure 6.66). However, MPS IVA is primarily a disease of the skeletal system. Cervical myelopathy resulting from the hypoplastic/dysplastic odontoid process is the most common and serious pathological change associated with MPS IVA.[362] Intelligence is usually normal or only slightly abnormal in patients with MPS IVA. However, localized neuronal storage in the cerebral cortex, thalamus, hypothalamus, basal ganglia and hippocampus (Figure 6.67) has

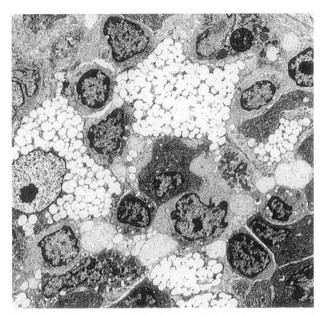

6.66 Mucopolysaccharidosis IV (Morquio). Electron micrograph of tonsil, showing histiocytes filled with empty membrane-bound vacuoles. Similar appearances are seen in fibroblasts and histiocytes in other mucopolysaccharidoses.

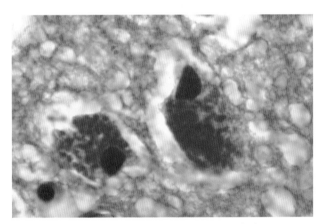

6.67 Mucopolysaccharidosis IVA (Morquio A). Storage neurons in hippocampus containing strongly periodic acid–Schiff-positive globular inclusions. Periodic acid–Schiff.

been reported.[269] In this case, the storage materials were PAS-positive (see Figure 6.56) and consisted of stacked, straight or tangled, loosely packed wavy membranes of various sizes (Figure 6.68) at the ultrastructural level.[269]

The activity of N-acetyl galactosamine-6-sulphate sulphatase was deficient in the liver and brain in the reported case, and thus the patient definitely had MPS IVA. However, caution is required in interpreting some descriptions because patients reported in the literature often do not have enzymatic or molecular information and thus it is unclear whether they have MPS IVA or MPS IVB.

Animal Models

Naturally occurring animal models of MPS IVA are not known. With a targeted disruption of the deficient gene, N-acetylgalactosamine-6-sulphate sulphatase, a murine model of Morquio A disease has been generated.[508]

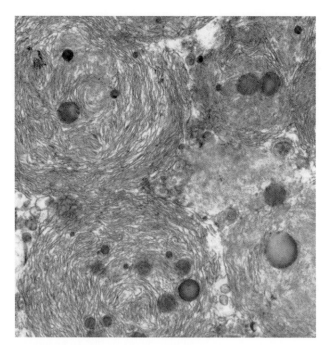

6.68 Mucopolysaccharidosis IVA (Morquio A). Neuronal inclusions consist of straight or wavy, thin, electron-dense membranes and spherical, moderately electron-dense lipid droplets. Electron micrograph.

Mucopolysaccharidosis VI (Maroteaux–Lamy Disease)

History and Clinical Features

MPS VI was first recognized in 1963 by Pierre Maroteaux and co-workers as a Hurler-like syndrome with preservation of intelligence and urinary excretion of predominantly dermatan sulphate. Clinically, MPS IV is characterized by extremely short stature, corneal opacities, dysostosis multiplex and a progressive cardiopulmonary disease. Growth retardation is usually recognized by 2–3 years of age. Hepatosplenomegaly is often present. Patients usually have normal intelligence without clinical evidence of neurological dysfunction. However, hydrocephalus and nerve entrapment may occur. Carpal tunnel syndrome and generalized restriction of articular movements are often associated with MPS VI. Spastic paraparesis from atlantoaxial subluxation consequent to hypoplasia of the odontoid processes and progressive thickening of the dura has been documented.[248,588] The majority of patients die of heart failure in the second or third decade of life.

Biochemistry

MPS VI is caused by a genetic deficiency in the activity of N-acetylgalactosamine-4-sulphatase (arylsulphatase B) (see Table 6.3).

Molecular Genetics

The gene (ARSB) is localized to 5q11–13.[298] More than 30 disease-causing mutations have been identified.

Pathology

Detailed descriptions of the pathology of MPS VI are limited. Keller and co-workers reported a 27-year-old patient with a severe form of MPS VI. In addition to compressive cervical myelopathy and defective enchondral ossification, the patient had thickened heart valves with endocardial fibrosis and 'increased amounts of GAGs in neurons in the CNS'.[248] Thickening of the sclera and the optic nerve sheaths and atrophy of the optic nerve have also been described.[468] Light- and electron-microscopic studies of the thickened dura showed an extensive accumulation of GAGs in ballooned cells.[588]

Animal Models

Feline and rat models of MPS VI are known.[110] The clinical phenotype and pathological lesions closely resemble those described in human patients.[203,277] In the feline model, lysosomal inclusions resembling zebra bodies or MCBs have been reported in occasional neurons and glial cells in the cerebral cortex.[552] A mouse model of MPS VI disease has also been made.[129]

Mucopolysaccharidosis VII (Sly Disease)

History and Clinical Features

MPS VII is a very rare form of MPS, first described by William Sly and colleagues in a child with physical features closely similar to those of Hurler disease (MPS IH) but with a deficiency of lysosomal β-glucuronidase.[455] Short stature, hepatosplenomegaly, dysostosis multiplex and intellectual impairment are the main clinical manifestations. The phenotype is variable, including isolated neonatal ascites and non-immune hydrops fetalis.[78,425] Granulocytes often show striking coarse metachromatic granules. Neonatal, infantile and juvenile forms are known. The juvenile form is the mildest, with only mild kyphosis or scoliosis.

Biochemistry

MPS VII is caused by deficiency of the lysosomal β-glucuronidase (see Table 6.3), resulting in an accumulation of GAGs containing terminal β-linked glucuronic acid residues: dermatan sulphate, heparan sulphate and chondroitin sulphate. The enzyme deficiency can be detected in leukocytes, in fibroblasts and in the brain.[536]

Molecular Genetics

The β-glucuronidase gene (GUSB) is localized to chromosome 7q21.11. More than three dozen disease-causing mutations have been identified. The original patient reported by Sly and co-workers was a compound heterozygote, and a small amount of β-glucuronidase activity was detected in that patient.[455]

Pathology

Reports of the pathology on MPS VII are sparse. Only one detailed post-mortem study of the patient originally described by Sly et al.[455] is reported in the literature. Grossly, dysostosis multiplex, arterial stenosis and fibrous thickening of the cardiac valves with calcification were found. Typical of MPS, lysosomal storage was found throughout visceral organs and in the CNS. Unlike MPS I, MPS II and MPS III, however, the extent of the neuronal storage in the CNS was variable; the most severe storage was in the hippocampus, subiculum, presubiculum, substantia nigra, inferior olive, and spinal anterior horn and intermediolateral neurons. The storage materials were autofluorescent, colloidal iron- and PAS-positive and weakly Alcian blue-positive. Storage was also noted occasionally in glial cells and foamy macrophages in the leptomeninges. Ultrastructurally, membrane-bound lysosomal inclusions consisting of whorls of membranous materials were admixed with a small amount of lipid droplets. The neurons in the Meissner and Auerbach plexuses were distended with cytoplasmic vacuoles containing GAGs. Vacuolation was also noted in the non-pigmented ciliary epithelium, corneal fibrocytes and lens epithelium in the eye. The neonatal form of MPS VII causes non-immune hydrops fetalis, and many foamy vacuolated cells are present in visceral organs and also in the Hofbauer cells of the placenta.[342]

Animal Models

Naturally occurring canine, feline and murine models of MPS VII are known.[204,448] These models show a severe clinical phenotype with marked skeletal abnormalities. The murine model has been used extensively for the investigation of disease pathogenesis and therapeutic strategies using enzyme or gene replacement.[50] In addition, murine models of specific human missense mutations have been generated by targeted mutagenesis.[507]

Mucopolysaccharidosis IX

The lysosomal hyaluronidase is required for the degradation of hyaluronan, a high-molecular-mass glycosaminoglycan that is abundant in the extracellular matrix of connective tissues. In 1996, a patient with a deficiency of serum hyaluronidase activity was discovered.[354] More recently, a single consanguineous family with three affected siblings has also been confirmed.[224] The original patient was a 14-year-old female who had a soft tissue mass over the lateral aspect of her ankle at age 7.5 years; additional periarticular masses

developed involving the finger, popliteal fossa, patella and lateral malleoli during the next year. Biopsy revealed histiocytes filled with numerous membrane-bound vacuoles containing a flocculent, medium-dense material and a small number of dense secondary lysosomes. The patient had no detectable plasma hyaluronidase activity and had mutations in the gene encoding lysosomal hyaluronidase (HYAL1).[510] She had mildly dysmorphic craniofacial features, with flattened nasal bridges, bifid uvula and a submucosal cleft palate. She is mildly short in stature. Her ophthalmological and neurological examination was normal and her academic performance was good. The more recent group of three patients in one family exhibited a phenotype limited to diffuse joint involvement with proliferative synovitis. This lysosomal storage disorder has been designated mucopolysaccharidosis IX.

GLYCOPROTEIN STORAGE DISORDERS

Glycoprotein storage disorders (Table 6.4) are a series of lysosomal disorders that affect primarily the degradation of carbohydrate chains of glycoproteins due to genetically defective lysosomal hydrolases.[505] The classification of this group of diseases is arbitrary to some extent. For example, the acid β-galactosidase deficiency discussed earlier (see GM1 Gangliosidosis and Morquio Disease [Syndrome] Type B, and Mucopolysaccharidosis IVA [Morquio A Disease]) can also be considered a glycoprotein disorder, because degradation of glycoprotein sugar chains with a terminal β-galactose residue is also impaired in this disorder.

Sialidase (Neuraminidase) Deficiency (Sialidosis)

History and Clinical Features

There are two distinct clinical phenotypes of sialidosis caused by sialidase (also known as neuraminidase) deficiency. Type 1 is characterized by macular cherry-red spots and myoclonus, whereas type 2 is more severe, exhibiting dysmorphic features, dysostosis multiplex, cherry-red spots, organomegaly and intellectual disability.[505] In the past, some type 2 patients were classified as having

TABLE 6.4 Glycoprotein storage disorders		
Clinical diagnosis	**Affected compounds**	**Enzyme defects**
Sialidosis (formerly MLI)	Sialyloligosaccharides	α-Neuraminidase (sialidase)
Galactosialidosis (combined sialidase-β-galactosidase deficiency)	Sialyloligosaccharides	Carboxypeptidase 'Protective protein' (see text)
α-Mannosidosis	Oligosaccharides with terminal α-mannose	α-Mannosidase
β-Mannosidosis	β-Mannosyl-glcNac, heparan sulphate	β-Mannosidase heparan-sulfamidase
α-Fucosidosis	Fucose-containing oligosaccharides and glycolipids	α-Fucosidase
Aspartylglucosaminuria	Aspartylglucosamine and glycoasparagines	Glycosylasparaginase
Schindler disease	α-N-acetyl-galactosamine	α-N-acetyl-galactosaminidase

a condition referred to as mucolipidosis I (MLI) because of the clinical similarity to MLII and III diseases (I-cell disease and pseudo-Hurler polydystrophy, respectively). Today the term MLI is no longer used. Also, many patients who were originally classified as having late-onset type 2 sialidosis (juvenile type 2) were later found to have galactosialidosis (see Galactosialidosis, this page). Thus, one should be aware of the confusing status of earlier reports on this group of diseases.[152] A 13-year-old girl reported by Gonatas and colleagues as having 'juvenile lipidosis' is the first clinicopathological report of a type I patient.[173] As noted in that patient, intractable myoclonus and decreased visual acuity with a cherry-red spot are the usual presenting symptoms in type 1 patients between 8 and 25 years of age.[404] Type 2 patients are recognized at or shortly after birth by their dysmorphic features and hepatosplenomegaly. Hydrops fetalis has also been reported as a congenital variant of type 2.[505]

Biochemistry

There are four mammalian neuraminidases (sialidases), designated as NEU1, NEU2, NEU3 and NEU4.[337] Based on differences in subcellular localization, pH optima and substrate preferences, these enzymes are designated cytosolic (NEU2), plasma membrane (NEU3) and lysosomal (NEU1), although variation in these locations has been reported. NEU4, for example, has been found in lysosomes, mitochondria and endoplasmic reticulum. The lysosomal α-neuraminidase (sialidase) activity known as NEU1 is deficient in sialidosis. The deficiency is more marked in sialidosis type 2 than in type 1 and emerging evidence suggesting both lysosomal and nonlysosomal functions for NEU1 may complicate understanding of disease pathogenesis.[394] The extent of the tissue accumulation and urinary excretion of sialyloligosaccharides and sialylglycopeptides appears to be correlated with clinical severity.

Molecular Genetics

Sialidosis is inherited in an autosomal recessive manner. There appears to be ethnic predilection in type 1, and the majority of such patients have been Italian. The gene for lysosomal α-neuraminidase (*NEU1*) is located on chromosome 6p21.3 in the region containing the major histocompatibility complex (HLA) locus. As of 2003, 34 disease-causing mutations had been identified.[440,505]

Pathology

The most conspicuous pathological feature is the presence of cytoplasmic vacuolation in neurons, oligodendrocytes, vascular endothelial cells and many cells in the visceral organs, in particular hepatocytes, Kupffer cells and histiocytes in the spleen and bone marrow.[10,579] These pathological changes are highly similar in both type 1 and 2 patients. At the ultrastructural level, the vacuoles are membrane-bound and usually contain floccular material. They stain variably with PAS and/or Alcian blue. Vacuolation of renal tubular epithelial cells is conspicuous in some cases. Vacuoles containing short stacked lamellae or small concentric myelin figures have been noted in the spinal neurons and in neurons in the myenteric plexus[579] and in

the Schwann cells. In a 22-year-old patient, Allegranza and co-workers described the fine cytoplasmic vacuolation of several neurons of the cortex, corpus callosum and thalamus, and of perineuronal as well as interfascicular oligodendrocytes.[10] They also observed unique large vacuoles that caused enlargement of the perikaryon of some motor neurons of the brain stem and of the anterior horns of the spinal cord. Similar vacuolar changes involve neurons in the dorsal root ganglion, autonomic ganglion and ganglion cells in the myenteric plexus. Additionally, they found neuronal intracytoplasmic storage of lipofuscin-like pigment throughout the brain, particularly conspicuous in the thalamus, the pulvinar and the periventricular and supraoptic nuclei. Lipofuscin-like bodies have been reported in the neurons and hepatocytes in other type 1 patients.[173] Neuronal loss has been reported in lateral geniculate nuclei of thalamus, substantia nigra and the gracile and cuneate nuclei. In the cerebellum, loss of Purkinje cells accompanied by Bergmann glia proliferation has been described.[10]

Animal Models

A spontaneous mouse mutant, SM/J, has a mutation in the *Neu-1* locus, causing a genetic defect in lysosomal sialidase.[413] A mouse model of lysosomal sialidase deficiency has also been generated by targeted disruption of the gene.[103] The *Neu1*$^{-/-}$ animals develop a severe sialidosis phenotype including profound hearing loss.

Galactosialidosis

History and Clinical Features

Galactosialidosis is a lysosomal storage disorder associated with combined deficiency of β-galactosidase and neuraminidase (NEU1) secondary to a defect of the lysosomal protective protein/cathepsin A (PPCA).[97] It is characterized clinically by dysmorphic features, psychomotor retardation, macular cherry-red spots and myoclonus, foam cells in the bone marrow and vacuolated lymphocytes. It is usually classified into three types: I (early [severe] infantile), II (late [mild] infantile) and III (juvenile/adult). Clinical signs and symptoms of the severe infantile type are similar to those of type II sialidosis. Coarse facies, hepatosplenomegaly, dysostosis multiplex and telangiectasia are common clinical features; affected infants usually do not survive beyond 1 year of age. Hydrops fetalis has been described. The late (mild) infantile type exhibits coarse facies, hepatosplenomegaly, dysostosis multiplex and corneal clouding, resembling a mucopolysaccharidosis. A macular cherry-red spot is present but seizures are rare. Patients may have mild intellectual disability but neurological manifestations appear only in the later years of life. Although all known patients with the late infantile type are Caucasian, a large majority of reported patients with the juvenile type are Japanese.[488,490] Patients usually exhibit a mild degree of coarse facies and bony abnormalities. Macular cherry-red spots and angiokeratomas are common, but hepatosplenomegaly is rare. Typical neurological features are myoclonus, cerebellar ataxia, seizures and slowly progressive intellectual disability.

Biochemistry and Molecular Genetics

Galactosialidosis is a rare autosomal recessive disorder caused by mutations in the gene *CTSA* (also known as *PPGB, GSL, NGBE, GLB2*), which maps to 20q13.1 and encodes a 32- or 20-kDa dimeric protective protein that has cathepsin A-like activity (PPCA). This protein forms a complex with the two lysosomal enzymes, acid β-galactosidase and sialidase, and stabilizes them. Thus, a deficiency of PPCA (CTSA, PPGB, GSL, NGBE, GLB2) secondarily causes a combined deficiency of both acid β-galactosidase and sialidase (α-neuraminidase). The deficiency of PPCA leads to intralysosomal proteolysis of these enzymes and an accumulation of sialyloligosaccharides in the lysosomes and excessive sialyloligosaccharides in urine.[97] Several disease-causing mutations have been identified.[97,180]

Pathology

The major pathological finding is cytoplasmic vacuolation in many types of cell, indicating storage of water-soluble materials such as oligosaccharides. In the severe infantile form, numerous clear cytoplasmic vacuoles are conspicuous features in hepatocytes, Kupffer cells, Schwann cells, fibroblasts, endothelial cells, lymphocytes and plasma cells. These vacuoles are membrane-bound and similar to those seen in sialidosis. The glomerular epithelial cells and renal tubular epithelial cells are also finely vacuolated. The neurons in the autonomic ganglion contain numerous MCBs and pleomorphic dense bodies.[581] Similar changes have also been reported in the late infantile/juvenile and adult types. Neuropathological investigations have been carried out in only a limited number of cases. In the CNS of a late infantile patient, gross atrophy of the optic nerve, thalamus, globus pallidus and lateral geniculate bodies, brain stem and cerebellum was reported. Histologically, marked neuronal loss and fibrillary gliosis were present in these grossly atrophic structures. Neuronal storage was found in only certain neurons: the Betz cells in the motor cortex, the neurons in the basal forebrain, the cranial nerve nuclei, the spinal anterior horn, and in the trigeminal and spinal ganglia.[376] Distribution of storage neurons was similar in other cases, although the extent of neuronal loss appears to differ in individual cases. In one of the early infantile patients (age 14 months), multiple cortical-subcortical infarcts were found in the brain, most likely caused by compromised circulation resulting from the narrowing of the lumen by storage vacuoles in endothelial cells. Numerous fine vacuoles were detected in neurons in semi-thin sections.[359] Electron-microscopic features of the neuronal storage materials are variable; they are described as MCBs, parallel, wavy-lamellar or tortuous tubular structures, lipofuscin-like irregular-shaped pleomorphic bodies, and cytoplasmic vacuoles with fine granules and lamellar materials.[359] Sural nerve biopsy in some cases revealed decreased nerve fibre density and vacuoles in Schwann cell cytoplasm.[338]

Animal Models

A sheep model reported as ovine GM1 gangliosidosis shows deficiency of both β-galactosidase and neuraminidase and is homologous to human galactosialidosis.[392] A transgenic mouse model of galactosialidosis has been generated.[596] Similar to the human disease, the mice show coarse facies, bony abnormalities and hepatosplenomegaly.

α-Mannosidosis and β-Mannosidosis

History and Clinical Features

α-Mannosidosis is classified as either a more severe type I or the milder type II. Type I is characterized by severe progressive psychomotor retardation, coarse facies, dysostosis multiplex, hepatosplenomegaly and deafness starting in early childhood. Patients usually die between 3 and 12 years of age. These clinical signs and symptoms can be confused with those of mucopolysaccharidoses. Patients with the milder type II typically present with speech and hearing disorders, tremor, muscle flaccidity and ataxia in early adolescence. Intellectual disability in type II patients is mild to moderate, and patients survive to adulthood. However, there is a broad clinical spectrum with considerable phenotypic variations, and thus some cases are difficult to classify.[187,519] There are many disease-causing mutations, but so far there is no apparent correlation between the type of mutation and clinical manifestation.[18] Characteristically, patients with α-mannosidosis are immunologically compromised and are susceptible to infection.[312] Gingival hyperplasia due to an accumulation of storage in histiocytes is a unique and conspicuous feature in some patients, and the diagnosis can be made by gingival biopsy.[111]

β-Mannosidosis is an extremely rare disease in humans. The clinical manifestations are heterogeneous. Hearing loss, frequent infections, intellectual disability and behavioural problems are relatively common features,[38,484,517] but other diverse clinical manifestations have been reported including spinocerebellar ataxia and atypical Gilles de la Tourette Syndrome.[281,435] Angiokeratoma corporis diffusum has also been documented in some cases.[484,517] β-Mannosidosis is a rare example of a human lysosomal disease that was discovered several years after an equivalent disease was identified in a non-human mammalian species (see Animal Models, this page).

Biochemistry

In α-mannosidosis, activity of the lysosomal enzyme acid α-mannosidase is deficient and a range of undegraded mannose-containing oligosaccharides accumulate in lysosomes. In β-mannosidosis, β-mannosidase activity is markedly deficient in various tissues, serum, leukocytes and cultured fibroblasts. An excess of mannosyl (1:4)-*N*-acetylglucosamine disaccharide accumulates in the tissue and is excreted in the urine.[505]

Molecular Genetics

Both α- and β-mannosidosis are autosomal recessive diseases. The gene for lysosomal α-mannosidosis (*MAN2B1, MANB*) maps to chromosome 19cen–q13.1. Many disease-causing mutations have been identified.[41,505] The gene for β-mannosidosis (*MANBA, MANB1*) is localized to chromosome 4q21 25.[9]

Pathology

The pathology of both α- and β-mannosidosis has been well investigated in animal models.[306,307,379] However, pathological studies of human patients are limited. The cardinal pathological feature of human α-mannosidosis is the presence of vacuoles, some of which contain coarse granules, within the cytoplasm of many types of cells including neutrophils and lymphocytes in the peripheral blood and in the bone marrow (Figure 6.69). These vacuolated cells, however, are not specific and have been recognized in mucopolysaccharidoses, mucolipidoses, sialidosis, galactosialidosis and other lysosomal glycoprotein disorders. Metachromasia, as noted in mucopolysaccharidoses, is not present. These storage vacuoles show various degrees of diastase-resistant PAS positivity and a positive periodic acid–silver methanamine reaction. They are membrane-bound and contain various amounts of reticular or floccular, slightly to moderately electron-dense material at the ultrastructural level. Some vacuoles may appear totally electron-lucent and empty.[111] Examination of the brain of patients with α-mannosidosis has been reported in only two cases.[253,472] In a 4.5-year-old child, Kjellman and colleagues observed ballooning of nerve cells with empty cytoplasm throughout the cerebral cortex, brain stem and spinal cord. The neuronal storage was inconspicuous or absent in the basal ganglia. In addition, they observed a widespread neuronal loss in the cerebral cortex, and loss of myelin and gliosis in the white matter. In the cerebellum, loss of Purkinje and granular cells was widespread.[253] Histopathological changes of neuronal storage in a 3.5-year-old girl reported by Sung and colleagues were similar to those in the previous report. However, the extent of neuronal storage was not uniform. The neuronal storage was seen throughout the nervous system, including neurons in the brain stem, spinal cord, and paravertebral sympathetic, trigeminal and dorsal spinal root ganglia. Cytoplasmic ballooning appeared more striking in larger neurons in the cerebral cortex and striatum. Neurons in the hypothalamus and subthalamic nuclei were severely affected, but neuronal storage was variable in different thalamic nuclei. The Purkinje cells in the cerebellum showed focal ballooning of their dendrites but perikaryal storage

was not extensive. Scattered macrophages containing PAS-positive materials and astrogliosis were also noted. Affected neurons were ballooned to varying degrees and appeared finely vacuolated in paraffin-embedded tissues.[472] Unlike the case reported by Kjellman et al.,[253] neither neuronal loss in the cerebral cortex nor demyelination in the cerebral white matter was observed in the case reported by Sung et al.[472] At the ultrastructural level, the perikarya of the affected neurons were filled with storage vacuoles of varying sizes between 1 μm and 2 μm in diameter, which were bound by a single membrane and contained loosely dispersed, fine reticulogranular material (Figure 6.70). Vacuoles were also observed in astrocytes, endothelial cells and pericytes but not in oligodendrocytes. On rare occasions, storage vacuoles having loose stacks of membranes were encountered.[472] The ultrastructural features of these storage vacuoles are similar to those described in bovine and feline α-mannosidosis. In the feline model, accumulation of GM2 and GM3 gangliosides has been reported in scattered neurons of the cerebral cortex and elsewhere, similar to that of MPS diseases.[176,544] Cortical pyramidal neurons with ganglioside accumulation also exhibited ectopic dendrites that appeared essentially identical to those in primary gangliosidoses and MPS diseases.[176,550] Neuroaxonal dystrophy, particularly affecting cerebellar Purkinje neurons, was also reported as a significant event in the animal models.[549]

A skin biopsy in β-mannosidosis has shown cytoplasmic vacuolation in various cell types, including endothelial cells, fibroblasts, the secretory portion of eccrine sweat glands, and keratinocytes.[517] No other reports on the pathology of human β-mannosidosis are available. However, detailed pathological reports on the caprine model of β-mannosidosis are available in the literature.[237,306,307,379] Basic pathological features in these animal models are the presence of storage

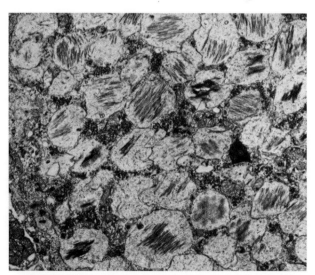

6.70 Mannosidosis. Electron micrograph of ventral horn cell of lumbar spinal cord. Storage vacuoles contain varying amounts of fine fibrils in stacks. Storage vacuoles interdigitate with one another.

Reproduced with permission from Sung JH, Hayano M, Desnick RJ. Mannosidosis: pathology of the nervous system. J Neuropathol Exp Neurol 1977;36:807–20, with permission from Lippincott Williams & Wilkins/Wolters Kluwer Health.

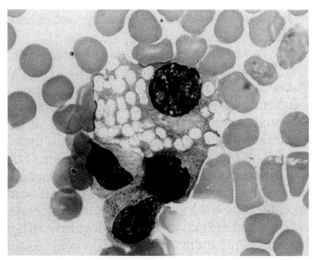

6.69 Mannosidosis. Bone marrow film. Many of the plasma cells contain large bold vacuoles. May–Grunwald–Giemsa.

vacuoles in various types of cells, including neurons, and extensive hypomyelination of the white matter.

Animal Models

Naturally occurring bovine, feline and guinea-pig α-mannosidosis and caprine and bovine β-mannosidosis are known. These models have been used extensively for the investigation of the basic pathological processes and for therapeutic experiments.[93,237,306,307,379,534,544] With transgenic approaches, mouse models of both α- and β-mannosidosis has been generated.[464,597] The biochemical and morphological features of the murine models appear similar to the mild forms of the human diseases.

Fucosidosis

History and Clinical Features

Fucosidosis is a rare autosomal recessive lysosomal storage disease first described in Italian siblings as cases of 'new mucopolysaccharide-lipid-storage disease' with progressive intellectual disability and rapid neurological deterioration and death before reaching 5 years of age. Because of its resemblance to mucopolysaccharidoses, α-fucosidosis was called mucopolysaccharidosis F in some earlier publications. The clinical course is quite variable. Some patients, including the original patients, had a rapid clinical course with neurological deterioration[154] but in the majority of patients the clinical course is slower, with survival well into the second or third decade of life.[223,225,565] The clinical manifestations include progressive neurological deterioration with intellectual disability, seizures, dysostosis multiplex, visceromegaly, angiokeratoma, telangiectasia, ocular abnormalities, hearing loss and recurrent infections. Because of the considerable clinical variation between patients, the disease is often classified into two major groups: type I (the severe form, as noted in the originally reported patients) and type II (the more common, less severe form). However, as more patients have been described, a clinically intermediate type has been recognized. These types most likely represent a continuous clinical spectrum of the disease. A comprehensive review of the clinical features of 77 patients was published in 1991.[565]

Biochemistry

The basic biochemical defect is a deficiency in the activity of the lysosomal α-L-fucosidase. L-Fucose is a common constituent of a great variety of biological molecules, and thus a large number of glycolipids and glycoproteins (fucoglycolipids, fucoglycoproteins) accumulate in the tissues and are also excreted in urine. The large accumulation of glycoproteins compared with that of oligosaccharides is considered by some to be unique to fucosidoses, as no significant accumulation of glycoproteins is noted in mannosidosis.[565]

Molecular Genetics

α-Fucosidosis is an autosomal recessive disease. The *FUCA1* gene maps to chromosome 1p34.1–36.1. The original patients were described in Italy and apparently the incidence is higher in Italy and among people of Italian descent. However, the disease has been reported in many different countries.[565] Over 23 different mutations have been reported.[505] The large number of mutations suggests genetic heterogeneity;[92] however, no genotype/phenotype correlation can be made, suggesting that phenotypic differences may be due to secondary, unknown factors.[566]

Neuroimaging

Reports of neuroimaging studies of fucosidosis are few.[154,225,393] High T2-signal intensity of white matter, as noted in fucosidosis, has been reported in other lysosomal storage diseases, but the combination of white matter hyperintensity in T2-weighted images and hypointense signals in the basal ganglia is characteristic of fucosidosis.[225,393] The brain may be enlarged or atrophic, depending on the stage of the disease.[120,565] Magnetic resonance imaging studies of a 13-year-old patient with chronic infantile fucosidosis (intermediate form) with unique extrapyramidal symptoms showed severe cerebral and cerebellar atrophy, especially in the frontal lobes. T1-weighted images demonstrated high intensity in the thalamus. T2-weighted images demonstrated increased signal in the white matter of the frontal and occipital lobes and low intensities in the bilateral thalami, striatum, substantia nigra, red nucleus and mammillary bodies.[225]

Pathology

Pathological studies of fucosidosis in humans are limited, with only sketchy information from a few autopsy reports. The cardinal pathological feature is a varying degree of cytoplasmic vacuolation in many cell types in the liver, spleen, lymph nodes, endocrine glands, peripheral nerve, brain, conjunctiva and cultured fibroblasts.[223,511] These vacuoles are similar to those seen in α-mannosidosis; however, ultrastructural features of storage vacuoles in α-fucosidosis are more complex. These vacuoles are single, membrane-bound and often contain two different components, moderately electron-dense reticular materials and lamellar inclusions with alternating electron-dense and electron-lucent lamellae. These lamellar structures are arranged concentrically or seen as bundles of flat lamellae. These types of inclusion are well described in the vacuolated hepatocytes, vascular endothelial cells, macrophages and Schwann cells. In rectal biopsy specimens from type II patients, α-L-fucose residue was clearly demonstrated histochemically in macrophages and vascular endothelial cells with lectin *Ulex europeus* agglutinin-1 (UEA-1).[223] Grossly, the brain may be atrophic[98] or megalencephalic.[120] Storage vacuoles are conspicuous in both neurons and glial cells throughout the brain. Loss of neurons from the thalamus, dentate nucleus and Purkinje cell layer in the cerebellum is well documented. Remaining neurons show marked enlargement of their perikarya, with membrane-bound vacuoles containing moderately electron-dense reticular materials. Exceptionally, membrane-bound vacuoles with bundles of parallel lamellae are observed. Electron-lucent vacuoles containing reticular materials and electron-dense inclusions, displaying either an irregular reticulum or lamellar inclusions, are found in astrocytes. The white matter shows marked demyelination and gliosis.[64] Notably, numerous Rosenthal fibres (Figure 6.71) may be conspicuous.[159,288] Only clear inclusions are found in oligodendrocytes. Macrophages and

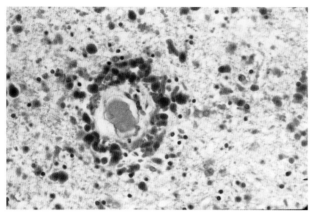

6.71 Fucosidosis. Perivascular Rosenthal fibres (from Alexander disease).

vascular endothelial cells contain clear and/or dense inclusions similar to those in astrocytes.[511] Vacuolar and lamellar inclusions are also described in biopsy specimens of angiokeratoma from patients with fucosidosis. Endothelial cells, pericytes, fibroblasts and dark cells of sweat glands contain almost exclusively vacuolar inclusions, whereas lamellar inclusions are in Schwann cells and myoepithelial cells in eccrine sweat glands.[266] Secondary storage of gangliosides, such as GM2 and GM3 described in many other lysosomal disorders from the MPSs to α-mannosidosis, has not been reported in fucosidosis.

Animal Models

α-Fucosidosis has been reported in springer spaniels from Australia, the UK and the USA, and has been used to explore pathogenesis issues as well as demonstrate partial effectiveness of bone marrow transplantation for this disorder.[449,498] As of this writing, no mouse models of fucosidosis have been developed.

Aspartylglucosaminuria (Aspartylglycosaminuria)

History and Clinical Features

Aspartylglucosaminuria (AGU) is a hereditary metabolic disorder with slowly progressive psychomotor deterioration from infancy, skeletal abnormalities, coarse facies and susceptibility to infection.[20,23] Seborrhoeic skin with erythema, facial angiofibroma and oedematous buccal mucosa and gingival overgrowths have been described as frequent features.[21] The disorder is considered to be the most common lysosomal storage disorder of glycoprotein degradation, but its occurrence is largely restricted to Finland.[22,343] It is characterized by urinary excretion and a massive accumulation of aspartylglucosamine (GlucNAc-Asn) and other glycoasparagines in body fluids and tissues.

Biochemistry

The basic biochemical defect in AGU is a profound deficiency of aspartylglucosaminidase (AGA) activity. This enzyme hydrolyzes the *N*-glycosidic linkage between *N*-acetylglucosamine and L-asparagine during the lysosomal degradation of asparagine-linked sugar chains of glycoprotein.[16] As a consequence of lacking glycosylasparaginase activity, aspartylglucosamine and other glycoasparagines accumulate in tissues, leukocytes and plasma.[343]

Molecular Genetics

Aspartylglucosaminuria is an autosomal recessive glycoprotein degradation disorder with worldwide distribution. The gene encoding aspartylglucosaminidase (*AGA*) is assigned to chromosome 4q 34–35. Many disease-causing mutations have been identified.[17,417] It is the most common lysosomal storage disorder in the Finnish population, with an estimated incidence of 1 in 18 000 in Finnish newborns. A single nucleotide change in the gene encoding AGA is responsible for the majority of Finnish cases. Finnish and non-Finnish patients have different mutations.[22] At least 15 different disease-causing mutations have been identified.

Neuroimaging and Pathology

Only limited descriptions of the pathology are available in the literature.[19,27,28,194] Overall, the basic pathology is one of neurovisceral storage and macroscopically the brain is atrophic. With brain imaging studies it has been shown that the thalamus is affected, particularly the pulvinar nuclei. Histopathologically, storage cells are characterized by the presence of numerous vacuoles in the cytoplasm. In the liver, hepatocytes contain large empty vacuoles in which finely granular or reticular materials dispersed with scattered small opaque 'lipid' droplets were identified at the ultrastructural level (Figure 6.72a,b). In the brain, neuronal storage, regional loss of neurons and gliosis have been described. The perikarya of storage neurons in the cerebral cortex are distended, with poorly defined vacuoles (Figure 6.72c). These vacuoles are smaller and more numerous than those in the hepatocytes. Some neurons contain lipofuscin-like granules in the perikaryal region. The white matter shows diffuse pallor of myelin and gliosis in some cases. Ultrastructurally, membrane-bound electron-lucent vacuoles and electron-dense residual bodies consisting of membrane-bound lumps of granular materials are noted (Figure 6.72d).[19,28,194] Electron-dense and electron-lucent inclusions are also found in macrophages. Pericytes contain many vacuoles, but vacuolation is less conspicuous in neuroglial cells and endothelial cells.

Animal Models

With targeted disruption of the glycosylasparaginase gene, murine models of aspartylglucosaminuria have been generated. These mice show many biochemical as well as pathological features found in human patients with AGU[175,502] and have been used for experimental studies, including enzyme replacement therapy.[249]

Schindler Disease (Kanzaki Disease)

History and Clinical Features

This is an extremely rare disease caused by lysosomal α-*N*-acetylgalactosaminidase deficiency. Three phenotypically distinct types are known. The phenotype of type I is that of

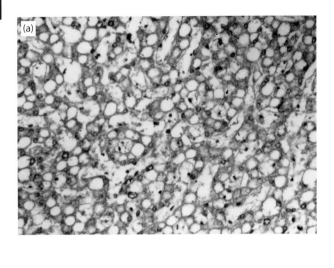

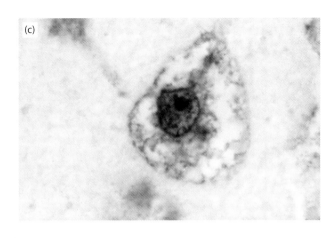

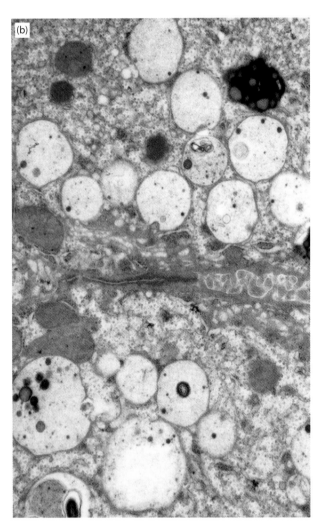

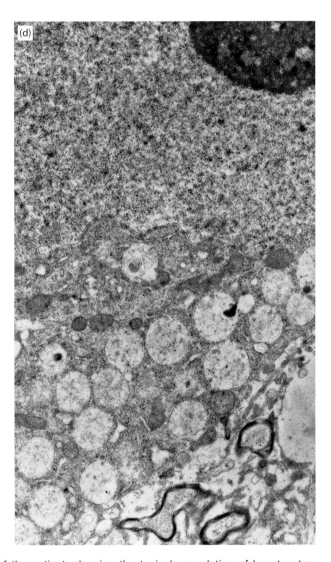

6.72 Aspartylglycosaminuria. (a) Light micrograph of the liver of the patient, showing the typical vacuolation of hepatocytes, resembling an extreme degree of fatty metamorphosis. **(b)** Two hepatocytes with numerous membrane-bound electron-lucent storage vesicles and some electron-dense bodies both within and outside the vesicles. Electron micrograph. **(c)** Light micrograph of a swollen cerebral cortical neuron with indistinct vacuolation, a characteristic feature even though it looks like an artefactual 'watery change'. Luxol fast blue–Cresyl violet. **(d)** Electron-lucent storage vesicles in the cytoplasm of a cerebral cortical neuron from the brain biopsy. Electron micrograph.

Courtesy of Dr Matti Haltia, University of Helsinki, Helsinki, Finland.

infantile neuroaxonal dystrophy (INAD). Type II is a milder adult-onset disorder characterized by angiokeratoma corporis diffusum, mild intellectual impairment and peripheral neuroaxonal degeneration. Type III is an intermediate form; its clinical manifestations range from epilepsy and mental retardation in infancy to autistic behaviour in early childhood. The first type I patients were two German brothers who presented with progressive psychomotor deterioration of an infantile onset, bilateral pyramidal tract signs with marked muscular hypotonia, nystagmus, myoclonus and seizures.[426] Light- and electron-microscopic examination of the brain biopsy from one of the patients showed typical axonal pathology as seen in INAD. However, normal activity of the enzyme in other patients with INAD indicates that the disease in these two brothers, characterized by the deficiency of α-N-acetylgalactosaminidase, differed from the usual type of INAD, despite the pathological similarities. The first case of a milder type II form was found in a 46-year-old Japanese woman with angiokeratoma corporis diffusum and glycoproteinuria.[241,242] This patient had slight dysmorphic features and a low IQ. Association of sensory motor polyneuropathy and hearing impairment,[518] and tortuous retinal as well as conjunctival vessels,[74] have been reported.

Biochemistry

The basic defect is the deficiency of α-N-acetylgalactosaminidase. As a consequence, urinary excretion of glycopeptides with terminal and internal α-N-acetylgalactosamininyl residues has been identified.[108] Excessive intralysosomal storage of α-N-acetylgalactosamine-containing material was demonstrated in cultured fibroblasts using a lectin from *Helix pomata*, which is specific for terminal α-N-acetylgalactosamine residues.

Molecular Genetics

This is an autosomal recessive disease. The gene for α-N-acetylgalactosaminidase (*NAGA*) has been assigned to chromosomal region 22q11. The gene has remarkable homology of its predicted amino acid sequences and positions of its intron–exon junctions with α-galactosidase A, the enzyme deficient in Fabry disease, suggesting a common ancestral gene.[108]

Pathology

The light- and electron-microscope features of the biopsy from the frontal lobe from one of the original type I patients described by Schindler and colleagues were characteristic of INAD.[426] At the light-microscopic level, abundant spheroids with characteristic straight or curved clear clefts were seen in the cerebral cortex (Figure 6.73). The spheroids appear to be preferentially formed in γ-aminobutyric acid (GABA)-ergic neurons.[108] Tubulovesicular and lamelliform membranous arrays, electron-lucent clefts and electron-dense axoplasmic matrices were the contents of these spheroids at the ultrastructural level.[573] Similar axonal changes were also present in the peripheral autonomic nerve fibres. However, there was no evidence of storage in neurons or other cell types. Remarkably, visceral

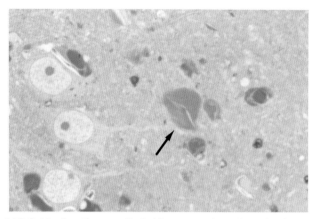

6.73 Schindler disease. Brain biopsy specimen from the frontal lobe, showing abundant discrete deposits or axonal spheroids (arrow) throughout the cortical neuropil. Toluidine-blue-stained 1-μm-thick plastic section.

Courtesy of Dr David Wolf, Mount Sinai School of Medicine, New York, NY, USA.

pathology was absent. Light- and electron-microscopic studies of peripheral blood cells, aspirated and biopsied bone marrow, conjunctiva, jejunal mucosa, muscle, liver and the sural nerve from the type I probands were all essentially normal.[108] In type II patients, whose clinical course is much milder, small clear cytoplasmic vacuoles are noted in granulocytes, monocytes and lymphocytes. Similar cytoplasmic vacuoles have been reported in several cell types, including endothelial cells, pericytes, fibroblasts, adipose tissues, Schwann cells, axons of the peripheral nerve, arrector muscles and eccrine sweat gland cells. The localized angiokeratomata in the skin described in the type II patient consisted of hyperkeratosis and dilated thin walled vessels.[241,242] In a type II patient with polyneuropathy, decreased density of myelinated fibres with axonal degeneration has been reported.[517] The CNS pathology of the type II or III patient has not been reported.

Animal Models

A transgenic murine model of the disease has been generated. Widespread lysosomal storage has been observed in various cell types in the CNS and various organs. Axonal spheroids are rarely found in the brain. Thus, the histology of the mouse model appears somewhat different from that of human patients.[108]

MUCOLIPIDOSES

The term 'mucolipidosis' (ML) was introduced decades ago to describe a group of lysosomal storage diseases with clinical features of both the mucopolysaccharidoses and the sphingolipidoses. They were classified into four clinical phenotypes, I, II, III and IV. Subsequently, however, it became apparent that MLI or sialidosis was more appropriately classified as a glycoproteinosis (as discussed earlier), and as a consequence the ML I name has been dropped. This has left MLII and III, which bear similarities to one another, and ML IV that is mechanistically completely unrelated but still carries the ML designation. Today, ML II/III disorders are often considered among the glycoproteinoses and the cellular lesions in ML IV have drawn parallels to sphingolipid storage diseases. Thus the grouping here of ML II/ML III and ML IV disorders is simply a reflection of historical events and in no way represents common disease mechanisms (Table 6.5).

I-Cell Disease and Pseudo-Hurler Polydystrophy (Mucolipidosis II and III)

History and Clinical Features

Originally believed to be separate disorders, studies have shown that mucolipidosis type II (I-cell disease, ICD) and mucolipidosis type III (pseudo-Hurler polydystrophy, PHP; also sometimes referred to as MLIIIA) are allelic variants of the same disorder. A variant form of ML III, referred to as IIIC, is caused by a different gene defect but clinically resembles ML IIIA (PHP) disease. (The reader is referred to Cathey and colleagues[72] for a more detailed explanation of ML II/III nomenclature.) Infants with ICD are usually smaller than normal at birth, indicating the disorder has already manifested *in utero*. The major clinical manifestations include thick swollen skin, particularly around the ears, congenital hernia and orthopaedic abnormalities, such as club foot, dislocation of the hip, kyphosis and other spine abnormalities, growth retardation of the skeletal system, progressive impairment of motion in all joints, hypotonia and mild to severe psychomotor retardation. Facial abnormalities similar to those of MPS IH are present, but hepatosplenomegaly and corneal clouding are usually not apparent. The majority of patients with ICD express clinical symptoms at or shortly after birth, with fatal outcome most often in early childhood. Clinical manifestations of PHP (as well as ML IIIC) are similar but milder, with a protracted course. The onset of clinical symptoms is usually around 3 years of age, but the clinical course can be quite variable.[400,513] Some patients with PHP survive well into adulthood.[462] In both patients with ICD and patients with PHP, excessive sialyloligosaccharides are excreted in the urine.[267] Interestingly, as described later, mutations in the genes coding for the proteins associated with ML II/III diseases have also been implicated in cases of persistent stuttering.[240]

Biochemistry

Lysosomal enzymes are post-translationally modified. For proper intracellular transport of newly synthesized lysosomal enzymes to the lysosome, the mannose 6-phosphate recognition marker is required. Defective synthesis of the mannose 6-phosphate recognition marker in these disorders results in impairment of proper targeting of lysosomal enzymes to the lysosome (their physiological site of function) and in leakage of newly synthesized enzymes out of the cells. This recognition marker is generated in the Golgi complex by a two-step reaction. The first step is the addition of N-acetylglucosamine phosphate to the mannose residue of the sugar chain of the enzymes by the action of uridine diphosphate (UDP)-N-acetylglucosamine-1-phosphotransferase. The second step is removal of the N-acetylglucosamine, which then exposes the terminal mannose 6-phosphate. The first phosphorylation step is defective in all known patients with ICD.[540] The consequence of this defect is that newly synthesized lysosomal enzymes do not reach the lysosome and are instead excreted out of the cell. Thus, activities of many lysosomal enzymes in solid tissues are defective, but they are abnormally elevated in body fluids, including serum. Cultured fibroblasts from these patients show defective activities of a large number of lysosomal acid hydrolases. However, in the culture media, the activity of these enzymes is increased greatly, and the enzymes are not phosphorylated.[202]

Molecular Genetics

ICD, PHP and ML IIIC are each inherited as autosomal recessive traits and involve functional defects in the same enzyme, UDP-N-acetylglucosamine-1-phosphotransferase. This enzyme consists of three subunits, α, β and γ. Subunits α and β are generated from a single polypeptide encoded by a single gene (GNPTAB) on chromosome 12. Defects in the αβ subunit cause ICD and PHP disease. Subunit γ is encoded by its own gene (GNPTG) on chromosome 16, and when defective causes ML IIIC disease.[72] There is also a second enzyme encoded for by a third gene that is closely linked to the phosphotransferase function. This enzyme removes the glucosamine residue and exposes the terminal mannose 6-phosphate and is known as N-acetylglucosamine-1-phosphodiester α-N-acetylglucosaminidase, or the 'uncovering enzyme' (UCE).

TABLE 6.5 Mucolipidoses[a]		
Clinical diagnosis	**Gene**	**Protein defects**
Mucolipidosis II (inclusion cell disease; I-cell disease; ICD; MLII alpha/beta)	GNPTAB	α–β subunit of UDP-N-acetylglucosamine-1-phosphotransferase
Mucolipidosis IIIA (pseudo-Hurler polydystrophy; PHP; MLIII alpha/beta)	GNPTAB	α–β subunit of UDP-N-acetylglucosamine-1-phosphotransferase
Mucolipidosis IIIC (MLIII gamma)	GNPTG	γ-subunit of UDP-N-acetylglucosamine-1-phosphotransferase
Mucolipidosis IV (MLIV)	MCOLN1	Mucolipin-1

[a]Classification is historical only; no mechanistic similarity between MLII/III and IV (see text).

It is encoded by the gene *NAGPA*. Mutations in *GNPTAB* and *GNPTG* cause all the presently known cases of MLII/III disease.[267,399,400]

As mentioned earlier, however, mutations in *NAGPA* as well as in *GNPTAB* and *GNPTG*, have been linked to persistent stuttering. Further, recent experimental studies have shown that the mutations in *NAGPA* linked to stuttering cause lower UCE activity.[289] Exactly how such reductions in the function of UCE, and presumably GNPT, lead to neuronal functional changes contributing to stuttering is unknown. ML II/III disorders have not been described as having persistent stuttering *per se* as a clinical feature, although speech deficits have been reported.

Pathology

The name 'I-cell disease' originated from the observation in phase-contrast microscopy of large numbers of granular inclusions in cultured skin fibroblasts from a patient with the disease. These fibroblasts were called 'inclusion cells' or 'I-cells', and thus the disorder was so-named. Later, a similar I-cell phenomenon was observed in PHP fibroblasts in culture. The pathological hallmark of ICD and PHP is intracytoplasmic membrane-bound vacuoles in lymphocytes and fibroblasts in various tissues, such as skin, conjunctiva, lymph nodes, spleen, gingiva, heart valves and bone in the regions of enchondral and membranous bone formation. Vacuoles were also reported in Schwann cells, perineural cells, endothelial cells and pericytes in the hepatocytes, myocardial fibres, epithelial cells of renal glomeruli and tubules. These vacuoles stain for colloidal iron and with PAS, Alcian blue and Sudan III and IV.[257] Neuronal and glial changes are minimal, and there is no massive neuronal storage. Only a mild accumulation of haematoxyphilic granules is present in the pontomedullary reticular formations and in neurons in the spinal anterior horn. These granules in neurons represent zebra bodies and MCBs at the ultrastructural level. A few membrane-bound vacuoles with fibrillogranular contents and some lamellar profiles or small lipofuscin granules have also been described in neurons. Electron-lucent vacuoles are found in astrocytes, oligodendrocytes and mesenchymal cells around blood vessels as well as in endothelial cells.[316] Sympathetic and parasympathetic neurons contain a few lipofuscin granules. Lamellar inclusions similar to MCBs have been described in the spinal ganglions.[352] To our knowledge, no autopsy studies of patients with PHP have been reported.

Animal Models

A domestic short-haired cat with a deficiency of *N*-acetylglucosamine-1-phosphotransferase has been reported.[325] Clinically and pathologically, both at the light- and electron-microscopic levels, this feline model is similar to human ICD. Mice deficient in GNPTAB were generated using gene trap methodology. Homozygous animals exhibited small size, cartilage defects, cytoplasmic inclusions in secretory organs, elevated plasma lysosomal enzymes and retinal degeneration.[160] Mice lacking UCE have also been generated by gene targeting.[54] In this case affected mice grew normally and lacked detectable tissue abnormalities. They did, however, exhibit elevated levels of several

lysosomal enzymes. Mice deficient in mannose 6-phosphate receptors have also been generated by crossing three mutant mice carrying null alleles for Igf2 and the 300-kDa and 46-kDa mannose 6-phosphate receptors Mpr 300 and Mpr 46, respectively. The triple-deficient mice phenotypically resemble human ICD.[114]

Mucolipidosis IV

History and Clinical Features

In 1974, Berman and colleagues reported a male infant with congenital corneal opacity but without any obvious skeletal or neurological abnormalities. Bone marrow aspirates, however, showed numerous histiocytes with their cytoplasm distended by sudanophilic materials. Hepatocytes and conjunctival epithelial cells contained MCBs similar to those of gangliosidoses. Cytoplasmic vacuoles containing finely granular material or lamellated structures similar to MPS disease were also noted in Kupffer cells and some conjunctival epithelial cells.[43] Accordingly, this new disorder was termed mucolipidosis type IV (ML IV) and the name has persisted. Subsequent reports indicated marked heterogeneity in the clinical symptoms, but major characteristics of this disorder are profound psychomotor retardation and ophthalmological abnormalities, including corneal opacity, retinal degeneration and strabismus. No organomegaly or skeletal abnormalities are present. These symptoms are pronounced at around age 2–3 years, but some patients show little if any deterioration for decades thereafter. Magnetic resonance imaging of the brain demonstrates a hypoplastic corpus callosum. Iron deficiency anaemia and increased blood gastrin levels caused by constitutive achlorhydria are common laboratory findings; this phenomenon is thought to result from an accumulation of lysosomal inclusions in gastric parietal cells.[14,30]

Biochemistry

A broad spectrum of lipid and acid mucopolysaccharides, including gangliosides, has been identified as the storage materials. However, the stored materials are typically normal constituents of cellular membranes, and the lysosomal hydrolases participating in the catabolism of these stored substances are normal. Thus, abnormal storage is likely to be due to an abnormal endocytic process, or disregulated membrane fusion event, resulting from a functional absence of the MLIV protein.[30,450] In normal cells, endocytosed membrane components, or degradation products, eventually reach the Golgi apparatus. This final route is apparently blocked in MLIV disease and endocytosed membrane components thus remain in the storage vesicles.[77]

Molecular Genetics

Mucolipidosis IV is an autosomal recessive disorder and the disease-causing gene (*MCOLN1*) maps to chromosome 19p13.2–13.3. Over 90 per cent of patients are Ashkenazi Jews, and the estimated heterozygote frequency in this population is 1 in 100. Fourteen mutations were known as of 2002, with two mutations identified among 95 per cent of the Ashkenazi ML IV alleles.[450] *MCOLN1* encodes muco-lipin 1, which is a membrane protein with homology to a group of calcium channels of the transient receptor poten-tial (TRP/TRPL) family.[30,33,470] It is presently unknown how mucolipin 1 is involved in endocytic and endosomal/lysosomal recycling process. A variety of functions from pH regulation to membrane fusion regulation have been suggested.[169]

Pathology

The tissue most often examined for diagnosis of MLIV is the epithelium of the conjunctiva. As reported in the original paper by Berman and colleagues, conjuncti-val biopsies reveal two types of characteristic inclusion: (i) single membrane-bound cytoplasmic vacuoles contain-ing fibrinogranular and small membranous structures; and (ii) concentric lamellar structures similar to MCBs of gangliosidoses.[43] The presence of these two types of inclusion in the conjunctival epithelium in a patient with intellectual disability is almost diagnostic of ML IV (Figures 6.74 and 6.75). In addition to vacuoles and MCBs, several other types of inclusion resembling cur-vilinear and fingerprint profiles have been noted in skin biopsy specimens.[34] There is a single autopsy report from a 23-year-old ML IV patient who had mild corneal cloud-ing, severe kyphoscoliosis, multiple joint contractures and bilateral short fourth metatarsals. Vacuolation was noted in the hepatic bile ductal and pancreatic ductal epithe-lium, Kupffer cells, macrophages in the lung, spleen and

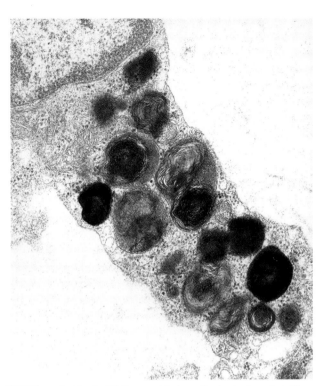

6.75 Mucolipidosis IV. Concentric lamellar inclusions in a fibro-blast. Electron micrograph.

lymph nodes, ganglion cells of the myenteric plexus, para-sympathetic and sympathetic ganglia, and in the vascular endothelial cells. The brain was atrophic and the corpus callosum was thin. Histologically, cortical cytoarchitec-ture was retained, but the parietal and inferior frontal gyri revealed a more columnar (early fetal) organization. Widespread neuronal loss and gliosis was noted in the thalamus, hippocampus (CA3), substantia nigra, basis pontis, inferior olivary nucleus, spinal anterior horn and the cerebellar Purkinje cell layer. Neurons and microglial cells and possibly astrocytes contained pigmented cyto-plasmic granules. These granules were brown on H&E staining, turquoise green with LFB, and intensely posi-tive with PAS and Sudan black. White matter and brain stem fibre tracts were stained weakly with LFB. The stor-age granules in cortical neurons were osmiophilic, amor-phous, granular material with a few lamellar membranes at the ultrastructural level. They differed from the lamellar and vacuolar inclusions noted in other cell types.[144] The neuronal inclusions with additional features, such as a compound body with lamellar and granular components, a dense strongly osmiophilic inclusion with tightly packed lamellar structures, bodies resembling MCBs and curvilin-ear bodies, have been reported in the brain biopsy study of a 24-month-old child with ML IV (Figure 6.76).[501] An MR imaging study detected hypoplastic and malforma-tive abnormalities in the corpus callosum and white mat-ter as well as cerebral and cerebellar atrophy, particularly in older patients.[146]

Animal Models

ML IV disease has been studied in *Caenorhabditis elegans* and *Drosophila*, and a mouse knockout has

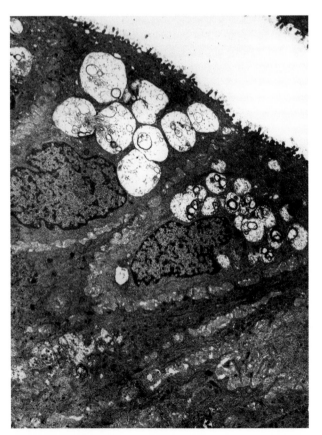

6.74 Mucolipidosis IV. Vacuolation and myelin figures in the conjunctival epithelial cells. Electron micrograph.

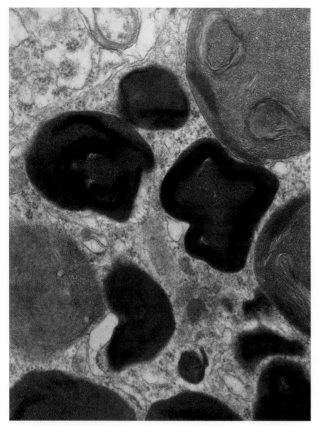

6.76 Mucolipidosis IV. Neuronal inclusions consisting of densely packed concentrically arranged membranes or more complexed crystalloid pattern. Electron micrograph.

recently been generated by Venugopal and colleagues.[532] Homozygous null animals developed gait abnormalities and hind limb paralysis by 6 months of age and died a month later. Widespread storage was seen in brain and the affected mice exhibited high gastrin levels, all consistent with the disease in humans. Brain disease included the presence of GM2 storage in neurons as well as neuroaxonal dystrophy and Purkinje cell loss in the cerebellum.[331]

GLYCOGEN STORAGE DISORDERS

Glycogen represents the key reservoir of glucose in most cells and not surprisingly, glycogen synthesis and degradation are tightly regulated. Abnormalities in this regulation are collectively referred to as 'glycogenoses', of which 15 separate diseases are presently known.[7] These diseases involve enzyme and protein defects in a number of metabolic pathways controlling both glycogen synthesis and degradation. Only 2 of the 15 diseases represent errors in lysosomal function. The most of common of these is glycogen storage disease (GSD) type II caused by the absence of a lysosomal glycogen degrading enzyme; the other is caused by a defect in a structural protein of the lysosome, the absence of which appears to lead to a defect in autophagy (Table 6.6).

Glycogen Storage Disease Type II: Acid α-Glucosidase (Acid Maltase) Deficiency (Pompe Disease)

History and Clinical Features

Joannes Cassianus Pompe in 1932 described the case of a 7-month-old female who died suddenly with cardiac hypertrophy of unknown origin and who also had massive accumulation of glycogen within vacuoles in the heart and in other tissues. Descriptions of additional cases followed quickly with many displaying cardiomegaly as well as hypotonia, hepatomegaly and macroglossia, and dying within the first year of life. In 1963 H.G. Hers discovered that acid α-glucosidase was deficient in this disorder and this discovery became the basis for establishing the very concept of lysosomal disease.[209] Pompe disease is the most severe form of GSD. Two major clinical types are known. The first is a generalized rapidly fatal disorder affecting young infants. The clinical presentations are massive cardiomegaly, macroglossia, hepatomegaly, progressive muscle weakness and hypotonia ('floppy baby'), with death before 1 year of age. In some infants, predominantly muscular symptoms without cardiomegaly have been noted. The second type is characterized by a slowly progressive proximal muscle weakness. The onset is usually in the childhood, but some adult-onset patients have been described. Between these two extremes are heterogeneous intermediate types, variously termed childhood, juvenile or muscular variants. These intermediate types usually affect skeletal muscle without cardiac involvement, and their clinical course is slowly progressive.[170]

Biochemistry

Acid α-glucosidase (acid maltase) activity in cultured fibroblasts and many other tissues is almost absent in the infantile type and is markedly reduced in late-onset cases. The enzyme deficiency results in intralysosomal accumulation of glycogen in many tissues. In general, the amounts of residual enzyme activity correlate inversely with the severity of the disease. Serum creatinine kinase is elevated in both types, but the infantile-onset type shows much higher increases. Response to epinephrine and glucagon is normal, indicating normal glucose metabolism. Glycogen content in the muscle is increased as much as ten-fold in the infantile-onset type.[214]

Molecular Genetics

Pompe disease is an autosomal recessive disorder, caused by mutations in the gene for acid α-glucosidase (*GAA*), located on chromosome 17q25. Over 40 mutations are known to date.[214] In western nations, Pompe disease accounts for approximately 15 per cent of GSDs. The incidence may vary in different ethnic groups and for different clinical groups. The highest incidence of the infantile type appears to be among African Americans (1 in 14 000) and Chinese (1 in 40 000–50 000).[214]

Pathology and Pathogenesis

The most conspicuous abnormality in infantile-onset patients is massive accumulation of glycogen in the liver,

TABLE 6.6 Glycogen storage disorders

Clinical diagnosis	Affected compounds	Protein defects
Pompe disease	Glycogen	Acid α-glucosidase
Danon disease	Glycogen; other compounds processed by autophagy	Lysosome-associated membrane protein-2

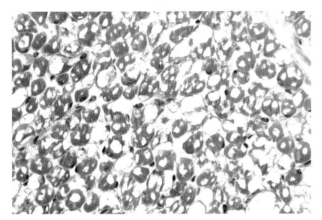

6.77 Pompe disease (infantile glycogenosis type II). Skeletal muscle, showing marked vacuolation within the myofibres due to loss of accumulated glycogen during processing.

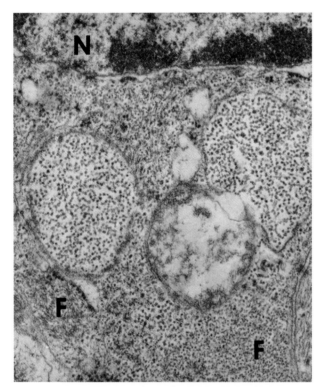

6.78 Pompe disease (infantile glycogenosis type II). Two vacuoles with glycogen accumulation are evident within this astrocyte. F, glial filaments; N, nucleus. Electron micrograph.

Reproduced from Gambetti P, DiMauro S, Baker L. Nervous system in Pompe's disease: ultrastructure and biochemistry. J Neuropathol Exp Neurol 1971;30:412–30, with permission from Lippincott Williams & Wilkins/Walters Kluwer Health.

heart and skeletal muscles (Figure 6.77). The entire heart is greatly enlarged, with thickened ventricular walls as a result of the glycogen accumulation. The glycogen accumulation is predominantly intralysosomal, largely as β-particles, but is also found in the cytoplasm. Glycogen accumulation is also present in neurons and glia (Figure 6.78) in the CNS. The accumulation is more marked in the brain stem and spinal cord neurons; however, neuronal dysfunction as a result has not been clearly established. Accumulation is also detected in Schwann cells, neurons in the dorsal spinal ganglia and myenteric plexus.[156,315] In late-onset cases, skeletal muscle is usually the only tissue affected by increased glycogen. Aneurysms and vacuolar degeneration of cerebral arteries have been reported in late-onset acid maltase deficiency.[272] Like other lysosomal diseases, the link between failure to degrade glycogen and cell dysfunction remains poorly understood in Pompe disease. Given that the transit of cytosolic glycogen into lysosomes would likely require an autophagic mechanism, the role played by macroautophagy in the ensuring disease cascade has been studied. Interestingly, crossing *GAA*-deficient mice with mice in which *Atg7*, a critical component in the macroautophagy cascade, is knocked out specifically in skeletal muscle leads to an amelioration of overall glycogen accumulation.[402] This finding is consistent with suppression of autophagy in muscle by dietary therapy as providing at least some level of therapeutic benefit in Pompe disease.[454]

Animal Models

There are several naturally occurring animal models of Pompe disease, including Brahman and Shorthorn cattle, Lapland dogs, Japanese quail, cats and sheep.[213,554,555] The Lapland dog model appears to be most analogous to the infantile-onset disease in humans.[554] As mentioned earlier, knockout mouse models of Pompe disease have also been generated.[48,49,401]

Lysosomal Glycogen Storage Disease without α-Glucosidase Deficiency (Danon Disease)

In 1981, M.J. Danon and colleagues described two unrelated 16-year-old boys with cardiomyopathy, skeletal myopathy and intellectual disability. The disorder had features of a primary GSD similar to Pompe, yet lacked acid maltase deficiency. Subsequent cases, in addition to heart, skeletal muscle and brain involvement,[99] also showed retinal, hepatic and pulmonary disease.[55] Early death is not uncommon in Danon disease patients and typically results from arrhythmia and severe cardiomyopathy. Phenotypic differences between men and women are attributable to the disorder having an X-linked inheritance. The causative gene is lysosome-associated membrane protein-2 (*LAMP2*), which encodes LAMP2, a lysosome-associated membrane glycoprotein[358] and is mapped to chromosome Xq24–q25.[218] The pathological hallmark is intracytoplasmic vacuoles containing acid phosphatase-positive autophagic material and glycogen in skeletal and cardiac muscle cells.[469] In a recent report summarizing 82 patients, all of the males were found to exhibit mild to moderate cognitive and learning deficits whereas this occurred in women 50% of the time.[55] Studies of CNS pathology have not been reported.

NEURONAL CEROID LIPOFUSCINOSIS (BATTEN DISEASE)

The term 'neuronal ceroid lipofuscinosis' (NCL) was coined in 1969 by Zeman and Dycken[591] to describe a clinically heterogeneous group of diseases that had previously been classified with Tay–Sachs disease and other disorders known as the 'amaurotic family idiocies'. The basis of this new categorization was the presence of specific neuropathological features, most notably the accumulation of autofluorescent ceroid lipofuscin-like materials (inclusions) in the cytoplasm of neurons and other cell types. Although the name was a misnomer due to storage not being limited to neurons and the terms 'ceroid' and 'lipofuscin' being poorly defined biochemically, it has persisted nonetheless. In the 1980s, largely out of the convenience of a simpler name, the term 'Batten disease' began to be used to refer to the entire family of NCL disorders. This term, however, is also used specifically for the juvenile form of NCL recognized by Frederick Batten in 1914; thus caution is needed when interpreting references to 'Batten disease' in the literature. Over the decades, the NCL family has expanded to include a host of disorders characterized by lysosomal storage bodies that resemble one another and vary from those of most other lysosomal diseases. Although efforts persist to identify a single thread linking all of these conditions to a common pathological mechanism, to date no such definitive link has emerged. Indeed, it is possible that these disorders are not as similar as first thought.

For many years, the NCLs were subclassified into four major groups – infantile neuronal ceroid lipofuscinosis (INCL), late infantile neuronal ceroid lipofuscinosis (LINCL), juvenile neuronal ceroid lipofuscinosis (JNCL) and adult neuronal ceroid lipofuscinosis (ANCL) types – by their clinical presentation and the ultrastructural morphology of the abnormal inclusions.[167,197,340] More recently, as many as 14 genetically distinct loci responsible for various types of NCL have been designated and are referred to as CLN1–14 (v).[340] This number is highly likely to continue to grow. CLN1, CLN2, CLN3 and CLN4 correspond to the traditional four major clinical subtypes, but in addition many variant forms occur. These include three variants of LINCL (defective genes *CLN5*, *CLN6* and *CLN7* – Finnish, Costa Rican and Turkish variants, respectively). However, it is now evident that these geographical labels are inaccurate because these forms of NCL disease are found worldwide. Mutations in CLN8 cause two very different clinical disorders, a late infantile variant and progressive epilepsy with

mental retardation (EPMR/northern epilepsy). Multiple adult-onset forms of NCL disease are also now recognized, with mutations in *CLN6* linked to some autosomal recessive cases and mutations in *DNAJC5* to some dominant forms, with the latter now also referred to as CLN4 disease. It is also apparent that milder mutations in some of the earlier onset forms of NCL can cause a delayed adult onset.

Although most of these defective genes are associated with lysosomal accumulation of the subunit c of mitochondrial ATP synthase (SCMAS), the diseases caused by CLN1 and CLN10 defects exhibit storage characterized by Saps (saposins) A and D.[130] The prominence of such protein accumulation led to the NCLs being considered 'lysosomal proteinoses'[137] but, as described later, the reasons for such substrate accumulation remain for the most part ill defined. CLN1 is a lysosomal palmitoyl protein thioesterase (PPT) and CLN2 is a lysosomal tripeptidyl peptidase (TPP1). CLN3 and CLN7 are proteins of unknown physiological functions with multiple membrane-spanning regions and both are physically associated with the lysosomal membrane, although CLN3 may also be expressed in the Golgi and/or endoplasmic reticulum (ER).[215,341] CLN5 is a soluble lysosomal glycoprotein, but it is unclear whether it has any enzyme activity.

The established and proposed human forms of NCL disease are shown in Table 6.7. Clinicopathological and ultrastructural details of neuronal storage bodies in the various types of NCL have been well documented.[167,197,340] In addition, an extremely rare example of congenital NCL exists in humans and in sheep, caused by mutation of cathepsin D.[149,447,515] Four new NCL diseases have also been recently designated (CLN11–CLN14) that involve defects in another cathepsin (F) as well as other proteins (Table 6.7).

Many generic animal models of NCL have been documented in mammalian species, including cattle, sheep, goats, cats, dogs and mice. Accumulation of SCMAS and Saps A and D has been documented in most of these models. More recently, transgenic mouse models have been generated.[336] Basic pathological processes noted in all types of NCL in humans as well as in animal models are neuronal storage and degeneration with subsequent loss of neurons, in association with glial cell response. The extent of neuronal loss differs between different types, but in general the loss is most extensive in earlier onset forms. The neuron loss displays remarkable selectivity, with GABAergic interneurons, thalamocortical relay neurons and cerebellar Purkinje neurons all targeted to different extents. Neuron loss is preceded by localized glial cell activation, and region-specific dendritic spine and presynaptic abnormalities leading to synapse loss have been reported. The pathogenetic mechanisms underlying such pathological processes, however, have not been elucidated. In animal models, apoptosis and/or autophagy have been suggested as possible mechanisms of neuronal cell death.[259,336]

CLN1 Disease

History and Clinical Features

CLN1 disease, also referred to as INCL, is an autosomal recessive disease that occurs predominantly in Finland.[25] There have been CLN1 cases reported from outside of

TABLE 6.7 Neuronal ceroid lipofuscinoses

Gene name	Alternative gene name	Clinical classification and eponym	Chromosome location	Defective protein	Inclusion body morphology (EM)	Stored Protein
CLN1	PPT1	Infantile NCL (Haltia–Santavuori)	1p32	PPT1	GROD	Saps
CLN2	TPP1	Late infantile NCL (Jansky–Bielschowsky)	11p15.5	TPP1	CL	SCMAS Saps
CLN3	–	Juvenile NCL (Batten; Spielmeyer–Sjogren)	16p12.1	CLN3	FP (CL,RL)	SCMAS Saps
CLN4	–	Adult NCL (Kufs type A; recessive inheritance)	ND	ND	CL, FP, GR, GROD, RL, ZB	SCMAS Saps
CLN4B	DNAJC5	Adult NCL (Parry; dominant inheritance)	20p13.33	CSPα	GROD, GR	SCMAS Saps
CLN5	–	CLN5 variant late infantile	13q22	CLN5	CL, FP, RL	SCMAS Saps
CLN6	–	CLN6 variant late infantile NCL (Lake–Cavanagh) and Adult NCL (Kufs type B)	15q21–23	CLN6	CL, FP, RL	SCMAS
CLN7	MFSD8	CLN7 variant late infantile NCL	4q28.1-28.2	MFSD8	CL, FP, RL	SCMAS
CLN8	–	CLN8 variant late infantile NCL Northern Epilepsy/EPMR	8p23	CLN8	CL, FP, RL	SCMAS Saps
(CLN9)	–	Early juvenile variant NCL	ND	ND	GROD, CL, FP	SCMAS
CLN10	CTSD	Congenital NCL	11p15.5	Cathepsin D	GROD	Saps
CLN11	GRN	Variant adult NCL	17q21.32	Progranulin	FP, RL	Unk.
CLN12	ATP13A2	Variant juvenile NCL	1p36	ATPase type 13A2	ND	ND
CLN13	CTSF	Variant Adult NCL (Kufs type B)	11q13	Cathepsin F	ND	ND
CLN14	KCTD7	Variant Infantile NCL (Progressive myoclonic epilepsy)	7q11.2	Potassium channel tetramerization domain containing protein 7	ND	ND

CL, curvilinear body; CSPα, cysteine-string protein α; FP, fingerprint body; EM, electron microscopy; fingerprint body; EPMR, progressive epilepsy with mental retardation; GR, granular inclusions; GROD, granular osmiophilic deposit; ND, not determined; PPT1, palmitoyl protein thioesterase 1; RL, rectilinear profile; Saps, sphingolipid activator proteins/saposin; SCMAS, subunit c, mitochondrial ATP synthase; TPP1, tripeptidyl peptidase 1; ZB, zebra body.

Adapted from Mole S, Williams RL, Goebel HH, eds. *The neuronal ceroid lipofuscinoses (Batten disease)*, 2nd edn. New York: Oxford University Press, 2011.

Finland, but the incidence is lower elsewhere. The disease is most commonly characterized by psychomotor retardation starting around 10–18 months of age. However, muscular hypotonia, clumsiness in fine motor control and retarded head growth may be detected earlier. Magnetic resonance imaging can detect hypointensity in T2-weighted images as an early sign. By the age of 2 years, the majority of patients show severe visual deterioration, with sluggish papillary reaction, progressive optic atrophy and retinal hypopigmentation. Hyperexcitability with poor sleep resulting from thalamic degeneration and seizures with frequent myoclonic jerks become apparent in the later stage. Most characteristically, the electroencephalogram (EEG) becomes essentially flat in later stages of the disease because of nearly complete loss of cortical neurons. Extremely rare neonatal/congenital cases, in which clinical manifestation is present at birth, have been reported.[35,158] For example, the term infant reported by Barohn and co-workers was microcephalic at birth and developed status epilepticus and died 36 hours later.[35] These cases are now considered to be a separate entity from CLN1 disease (see CLN10 disease).

Biochemistry

Patients with CLN1 disease have deficient activity of the lysosomal enzyme palmitoyl protein thioesterase I (PPT1). PPT1 is a housekeeping enzyme present in many tissues; it is particularly abundant in the spleen, brain and testis. Its expression is regulated developmentally and is increased with maturation of the brain. The predominant storage material in this disease, unlike most of the NCLs, includes Saps A and D. Prenatal and postnatal enzyme analysis for identification of CLN1 disease has been established.[215,341]

Molecular Genetics

CLN1 disease is caused by a genetic defect in the *CLN1* gene, which encodes PPT1 and is located on chromosome 1p32. Nearly 50 different gene mutations have been reported in patients with the infantile clinical phenotype and an accumulation of so-called granular osmiophilic deposits (GRODs).[341] All cases of CLN1 disease in Finland carry the common missense mutation, c.364A.T. In addition, *CLN1* mutations have also been identified in late infantile, juvenile and adult phenotypes.[129]

Pathology

The most significant gross abnormalities are limited to the brain and its coverings. The head circumference is slightly reduced and the calvarium is abnormally thick. A thick layer of gelatinous tissue covers the inner aspect of the cerebral dura mater. The brain is extremely atrophic, weighing only 250–450 g at the terminal stage. There is marked narrowing of the gyri and widening of the sulci throughout. Optic nerves are atrophic. Atrophy is also noted in the cerebellum and brain stem, whereas the spinal cord is well preserved. On sectioning, the cerebral cortex and cerebellar folia are markedly thinned and the white matter is shrunken and firm in consistency. The histopathology of the brain has been described in three stages.[195-197] Stage I (up to about 2.5 years of age) is characterized by neuronal storage of colourless or slightly yellowish granules and pronounced astrocytic hyperplasia. At stage II (2.5–4 years of age), the cerebral cortex shows an extensive loss of neurons in association with astrocytic hyperplasia and macrophage infiltration. In stage III (after age 4 years), cerebral cortical neurons are almost completely lost (Figure 6.79) and marked secondary white matter degeneration is present. Interestingly, however, giant neurons of Betz in the precentral gyrus and pyramidal neurons of the CA1 and CA4 sectors of the hippocampus and spinal anterior horn neurons are partially preserved. In the retina, the photoreceptor cells, the bipolar cells and the ganglion cells have completely disappeared and

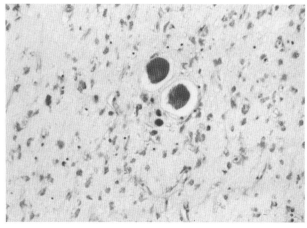

6.79 Infantile neuronal ceroid lipofuscinosis. Almost complete loss of cerebral cortical neurons associated with intense glial reaction. Periodic acid–Schiff.

Courtesy of Dr Matti Haltia, University of Helsinki, Helsinki, Finland.

are replaced by reactive gliosis. Accumulation of the storage materials is noted in neurons of the spinal ganglia and autonomic ganglia, many epithelial cells including those of eccrine sweat glands, thyroid follicles and testes, skeletal, cardiac and smooth muscle cells, endothelial cells and macrophages.[195-197] The storage granules show strong yellowish autofluorescence under UV light and are strongly positive for acid phosphatase activity, as well as for PAS, Sudan black B and LFB stains. They are immunostained with antibodies against Saps A and D and amyloid β peptide but not with the antibody against SCMAS.[514] The ultrastructural features of cytosomes of CLN1 disease are membrane-bound cytoplasmic inclusions of 1–3 μm in diameter (GRODs; Figures 6.80 and 6.81).[195,196] Detection of GRODs in neurons of the myenteric and submucosal plexuses and also in some non-neural cells, including muscle cells, endothelial cells and chorionic villi, was essential for diagnosis before the identification of PPT1 deficiency. Unlike classical late infantile and juvenile forms of NCL disease, no vacuolated lymphocytes are present in the peripheral blood, but GRODs can be detected in the cytoplasm by electron-microscopic investigation.

Pathogenesis

In spite of the identification of the gene and protein defects responsible for CLN1 disease, and the development of a mouse model exhibiting the characteristic disease phenotype (see later), little is understood as about the pathogenesis of this brain disease. Although the PPT1 enzyme has been shown to remove long-chain fatty acids from cysteine residues, the naturally occurring substrate(s) of this enzyme in brain and other tissues is not known. It is also not understood why Saps accumulate in neurons or why there is such striking, widespread and early death of neurons. In addition to infantile forms that can vary in terms of onset and duration, some cases of juvenile and adult forms have also been recognized. Current evidence does suggest that the severity of the disease course follows the amount of residual enzyme activity present in brain.[340]

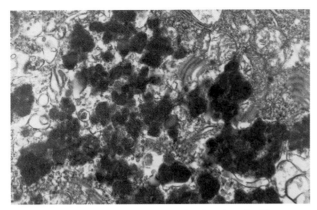

6.80 Late infantile neuronal ceroid lipofuscinosis. The neuron contains granular osmiophilic deposits, similar to those found in the infantile type. Electron micrograph.

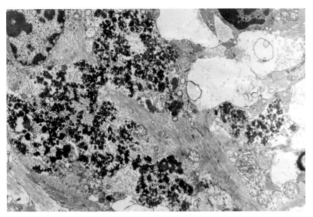

6.81 Late infantile neuronal ceroid lipofuscinosis. Granular osmiophilic deposits in an astrocyte. Electron micrograph.

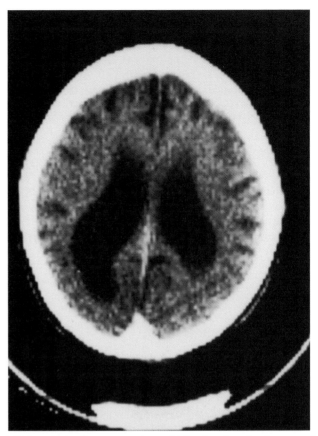

6.82 Late infantile neuronal ceroid lipofuscinosis. Computed tomography scan shows marked cerebral atrophy, with dilated ventricles.

Animal Models

Naturally occurring models of CLN1 disease have not been reported. *Ppt1* knockout mice have been made and show NCL-like pathology with an accumulation of autofluorescent lipopigment throughout the brain, associated with pronounced cerebral atrophy. They are fertile but die before 10 months of age.[184] Mice with a knockout of *Ppt2*, the homologue of *Ppt1*, show a milder clinical phenotype than *Ppt1* knockout mice.[184,185] As in other forms of NCL, there is evidence of glial activation and synaptic changes preceding neuron loss, and a low level of lymphocyte infiltration. The roles of these innate and immune responses are under investigation.

CLN2 Disease

History and Clinical Features

Classic late infantile NCL or CLN2 disease occurs worldwide in many different ethnic groups. However, clusters of cases have been reported in Newfoundland, Costa Rica and western Finland. The early symptoms, such as speech delay, hypotonia, ataxia and grand-mal seizures, usually appear between the ages of 2 and 4 years, with their onset preceded by a period of normal development. The seizures (partial or generalized), ataxia, myoclonus and developmental

problems progress gradually. Motor and cognitive functions deteriorate rapidly, and by age 5 years affected children are completely bedridden, mute and cachectic. Visual acuity declines gradually to blindness.[197,570] The EEG shows an occipital photosensitive response and persists until an advanced state of the illness. The electroretinogram (ERG) is diminished at presentation and soon becomes extinguished, whereas visual evoked potentials (VEPs) are grossly enhanced in keeping with the photosensitive EEG response. The pattern of these combined results is very characteristic of CLN2. VEPs persist until very late into the illness but eventually diminish when the child reaches a preterminal stage. Computed tomography (Figure 6.82) and MR imaging (Figure 6.83a,b) show varying degrees of cerebral and cerebellar atrophy, depending on the stages of the disease process, and may show hyperintensity of bilateral periventricular white matter on proton density and T2-weighted images. tripeptidyl peptidase I (TTP1) deficiency has also been implicated in some juvenile variant NCL cases.[75]

Biochemistry

Late infantile neuronal ceroid lipofuscinosis is caused by deficiency in the activity of lysosomal TPP1, a member of a family of serine-carboxyl proteinases.[75] *In vitro* studies have suggested that several peptide hormones including cholecystokinin and glucagon are substrates for TPP1. Importantly, TPP1 has also been shown to initiate

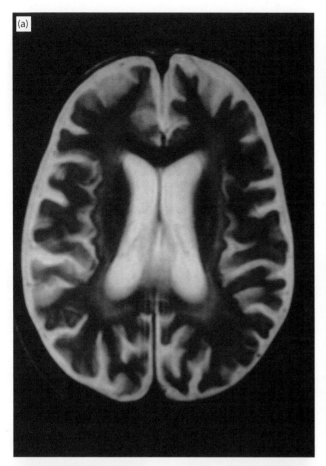

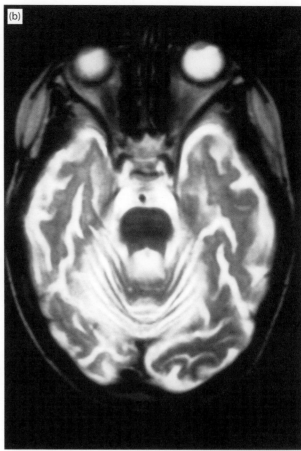

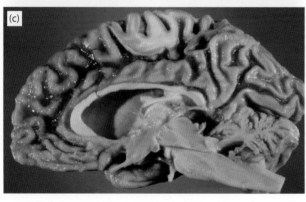

6.83 Late infantile neuronal ceroid lipofuscinosis. (a) Axial T2-weighted magnetic resonance (MR) imaging, showing diffuse cerebral cortical atrophy and greatly enlarged lateral ventricles. **(b)** T2-weighted MR imaging of the same patient as in (a), showing atrophy of the temporal and occipital lobes and significant atrophy of the cerebellum and brain stem. **(c)** The brain of the patient with marked atrophy of cerebrum and cerebellum.

(a,b) Courtesy of Dr Mauricio Castillo, University of North Carolina, Chapel Hill, NC, USA.

the degradation of SCMAS, the highly hydrophobic protein recognized as a major storage component of most of the NCL disorders. Cathepsin D is believed to complete SCMAS degradation after the initial TPP1 step and deficiency of cathepsin D is believed the cause of another form of NCL disease (CLN10). Small amounts of Saps A and D and dolichol are also present as components of the storage materials in CLN2 disease, although why this occurs is not understood.[167,197,516] TPP1 activity can be measured by a simple enzymatic assay.[585]

Molecular Genetics

The gene responsible for almost all classic late infantile NCL patients (*CLN2 or TPP1*) maps to chromosome 11p15.5 and encodes a lysosomal enzyme TPP1. Nearly

70 disease-causing mutations have been identified.[75,341] The gene is expressed in most cell types with TPP1 assuming a lysosomal localization in cells.

Pathology

The brain shows severe atrophy (Figure 6.83c). Brain weights of 500–700 g are not uncommon at the terminal stage. The cerebral cortex and the cerebellar folia are thinned. Ventricular dilation (hydrocephalus *ex vacuo*) with thinning of the corpus callosum and firm atrophic white matter reflect the severity of the degenerative changes. Similar to infantile NCL described earlier, the most pronounced changes noted in post-mortem brains are severe depletion of neurons and associated astrocytic gliosis and activation of microglia. Increased immunocytochemical staining of BCL-2 and TUNEL suggests that many neurons disappear by apoptosis.[286,395] In the cerebellum, both the granular and Purkinje cells are lost (Figures 6.84 and 6.85). The perikarya of remaining neurons are distended, with granular storage materials (Figures 6.86 and 6.87). Meganeurites are conspicuous in some cortical neurons; however, unlike those seen in gangliosidoses and Niemann–Pick type C disease, ectopic dendrite growth has not been

detected on meganeurites in NCL disease.[542] The storage material in neurons is similar to that in infantile NCL, emitting autofluorescence under UV light (Figure 6.88) and staining strongly with PAS (Figure 6.89), Sudan black and LFB. Immunoreactivity for SCMAS has also been demonstrated in the neuronal storage material.[167,197] At the ultrastructural level, the storage material consists of curvilinear inclusion bodies (Figures 6.90 and 6.91) enclosed by a single unit

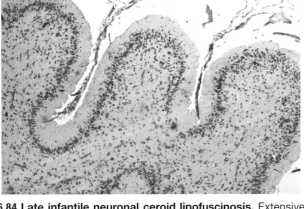

6.84 Late infantile neuronal ceroid lipofuscinosis. Extensive loss of neurons is evident in the cerebellum.

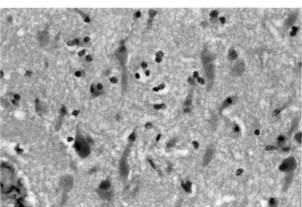

6.87 Late infantile neuronal ceroid lipofuscinosis. Affected neurons in cerebral cortex demonstrate many slender meganeurites.

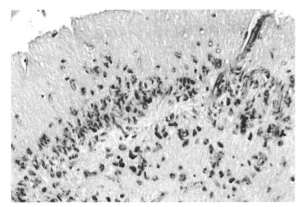

6.85 Late infantile neuronal ceroid lipofuscinosis. Extensive gliosis without any identifiable Purkinje or granular cells in the folium of the cerebellum. Periodic acid–Schiff.

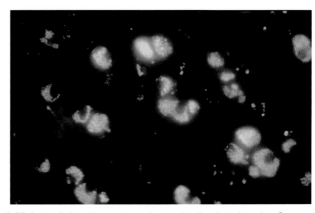

6.88 Late infantile neuronal ceroid lipofuscinosis. Storage material in neurons emits strong autofluorescence.

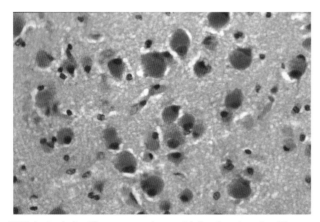

6.86 Late infantile neuronal ceroid lipofuscinosis. Storage neurons in the cerebral cortex show distension of the perikarya.

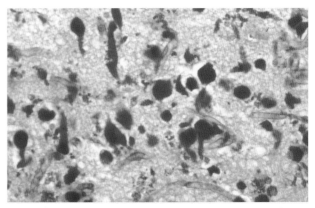

6.89 Late infantile neuronal ceroid lipofuscinosis. Storage material in neurons as well as glial cells stains strongly positive with periodic acid–Schiff.

6.90 Late infantile neuronal ceroid lipofuscinosis. Low magnification of curvilinear neuronal inclusions. Electron micrograph.

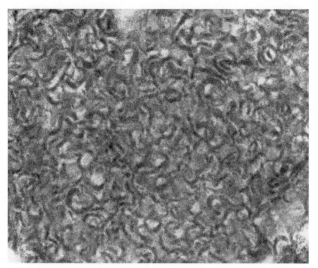

6.91 Late infantile neuronal ceroid lipofuscinosis. Higher magnification of the inclusions shown in Figure 6.90. Curvilinear profiles of the inclusion are identified clearly. Electron micrograph.

membrane and associated with SCMAS and acid phosphatase activity. A variable degree of retinal degeneration, with loss of neurons and gliosis, has been reported. Accumulation of sudanophilic, autofluorescent material with a curvilinear profile has been well documented in the extracerebral tissues. Thus, rectal and skin biopsies are useful for the diagnosis of LINCL.[166] In an electron-microscopic study of a skin biopsy, curvilinear inclusion bodies were found in the sweat gland, peripheral nerve, smooth muscle and endothelial cells. Peripheral lymphocytes may show clear cytoplasmic vacuoles that contain curvilinear bodies. Curvilinear inclusions may be detected in the patient's urine samples.

Pathogenesis

As in CLN1 and other forms of NCL disease, an understanding of the individual steps in the pathogenic cascade leading to widespread neuron death in CLN2 disease is lacking. Here, death of neurons is widespread, although not as severe as in CLN1, with astrocytic and microglial reactions predominating as the disease ensues. For this and other lysosomal diseases,[296] an alteration in macroautophagy has recently

emerged as a possible factor of importance. In view of the fact that the NCL disorders involve protein storage, including a key subunit of a mitochondrial enzyme (SCMAS), autophagy involvement would seem logical because mitochondria are degraded by this mechanism. Recent studies focused on neuron death in CLN2 disease have also implicated lysosomal membrane permeabilization as a potential player in the pathogenic cascade, with lysosomal rupture and release of SCMAS into the cytoplasm being followed by p62-associated protein aggregate formation.[334a]

Animal Models

Spontaneous animal models do not exist but a mouse model was generated by targeted disruption of *Cln2/Tpp1* gene.[452] These mice exhibit early clinical disease, including seizures, and typically die by 15 weeks of age. Loss of neurons is accompanied by gliosis as observed in the human disorders.

CLN3 Disease

History and Clinical Features

Juvenile neuronal ceroid lipofuscinosis (CLN3 disease) has been reported worldwide but appears to occur at a much higher incidence in populations with Scandinavian or northern European ancestry than in African American and Jewish populations. The eponym, Batten disease, was originally coined specifically for this disease, although it is now commonly used to designate all types of NCL (see earlier comments, p. 495). Visual difficulty is usually an early symptom starting at age 4–9 years because of progressive retinitis pigmentosa. A few years later, cognitive decline becomes noticeable gradually and progressively. Speech difficulty and seizures follow. By the mid-teens to age 20 years, many patients have almost completely lost their light perception and speech. The severity of the epilepsy is variable. By the mid-teens, many patients develop signs of parkinsonism. Angry outbursts, violent behaviour and hallucinations are other common clinical symptoms. Brain atrophy is evident on MR imaging in the later stages of the disease.[167,197]

Biochemistry

The CLN3 protein is a lysosomal glycoprotein with multitransmembrane topology of still unknown function. The major component of the storage material in JNCL is the same highly hydrophilic lipoprotein (SCMAS) that accumulates in most of the NCL disorders. Interestingly, in CLN3 disease, many lysosomal enzymes, including TPP1 and cathepsin D, which are believed involved in the degradation of SCMAS are elevated. Increased levels of Sap-A and D and phosphorylated dolichols are also found in the storage material.[215] However, the significance of these storage materials in the pathogenesis of CLN3 disease is not understood clearly, as they can be found in other types of NCL and in chronic forms of other lysosomal diseases.[167,215]

Molecular Genetics

The *CLN3* gene, encoding a lysosomal membrane protein,[292a] is assigned to chromosome 16p12.1. As of 2012, more than 40 mutations have been reported.[1,306] *CLN3* is highly conserved and mammalian homologues have been described in the mouse, rabbit, dog and sheep. However, the physiological functions of these homologous proteins are not known.

Pathology

The most notable laboratory finding is the presence of vacuolated lymphocytes in routine blood smears. Inclusions with a characteristic fingerprint profile are observed in these lymphocytes as well as in neurons in the brain on electron-microscopic examination.

Grossly, the brain shows moderate to severe atrophy, in proportion to the length of the disease processes. Microscopically, a variable degree of neuronal loss is present in the cerebrum and cerebellum. With pigment architectonic analysis, selective loss of neurons in layers II and V of the cerebral cortex and in the striatum and amygdaloid nucleus has been demonstrated.[57] The perikarya of remaining neurons are distended, with granular storage material that appears pale yellow in H&E sections and stains strongly with PAS, Sudan black and LFB. The neuronal storage material in this and other NCL diseases is strongly autofluorescent under UV light (Figure 6.88, p. 500) and immunoreactive for SCMAS (Figure 6.92). At the ultrastructural level, the storage material consists of inclusions with fingerprint profiles (Figures 6.93 to 6.95). Extensive neuronal loss occurs in the retina,

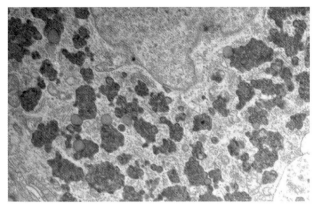

6.93 Juvenile neuronal ceroid lipofuscinosis. Low magnification of neuronal inclusions with fingerprint-like profiles. Electron micrograph.

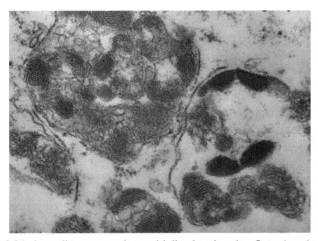

6.94 Juvenile neuronal ceroid lipofuscinosis. Cytoplasmic inclusions with fingerprint-like profiles in cerebral cortical neurons. Electron micrograph.

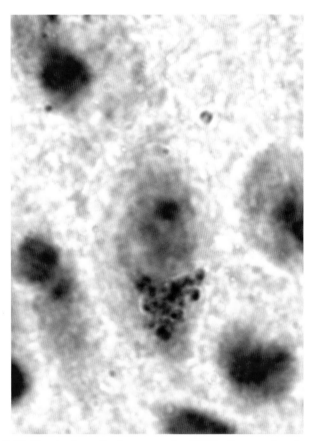

6.92 Cerebral cortical neuron in CLN3 disease (murine model) showing accumulation of subunit c of mitochondrial ATP synthase (SCMAS) revealed by antibody staining and peroxidase histochemistry. Nissl counterstain.

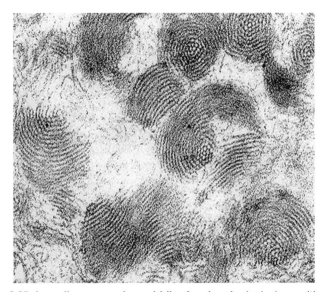

6.95 Juvenile neuronal ceroid lipofuscinosis. Inclusions with fingerprint-like profiles are present in a neuron from the gastrointestinal tract. Electron micrograph.

and apoptosis of photoreceptor cells has been demonstrated in a mouse model. Granular storage material is also found in many cell types throughout the visceral organs. The staining characteristics and ultrastructural features of storage material in visceral tissue may differ slightly from those in neurons; for more details see Mole *et al.*[1] In contrast to classic CLN2 disease, there is no immunohistochemical evidence of SCMAS in the liver, adrenal gland or endocrine pancreas.[125]

Pathogenesis

Remarkably, although the CLN3 gene was the first human NCL to be cloned, with this occurring nearly 20 years ago, the function of the CLN3 protein remains as enigmatic as ever.[1] implicated in lysosomal acidification, lysosomal arginine transport, membrane fusion and vesicular transport, and a variety of other cell operations, little consensus has emerged as to what it is actually doing in cells and why its absence is so devastating to neurons. Even its location in cells is controversial, though here most data support its location in membranes of late endosomes/lysosomes. Without a clearer understanding of the role of the protein in normal brain cells, understanding what happens in its absence has proved difficult. In many respects CLN3 disease shares features with the other NCLs including SCMAS accumulation and widespread neurodegeneration, albeit slower than that occurring in CLN1 and CLN2 diseases. One feature of CLN3 disease that appears unique is the development of an autoimmune response, yet even here whether this is impacting neurological disease progression is controversial.[1]

Animal Models

There are no naturally occurring animal models of CLN3 disease. However, murine models have been generated by targeted disruption of the mouse *Cln3* gene, with one recreating the major human disease-causing mutation.[246,273,335] Autofluorescent storage material is found in neurons and many other types of cell. The inclusions show rectilinear/fingerprint appearance resembling inclusions of CLN3 disease in humans, and contain SCMAS. These models, in spite of a milder disease course than seen in humans, have been invaluable in teasing apart features of pathogenesis. Neurodegeneration is widespread but appears early and most severe in thalamic regions of the brain.[1] As noted earlier for CLN2 disease, autophagy compromise has been cited for a number of the NCLs, with CLN3 among them.

CLN4 Disease

History and Clinical Features

Although one of the original major four groups of recognized NCLs, adult NCL is today a heterogeneous and poorly defined subgroup. Indeed, some cases in the literature reported as adult NCL may have been adult-onset neuronal lipid storage disease, such as Niemann–Pick type C disease or late-onset GM2 gangliosidosis.[340] Other adult variants may have been caused by defects in CTSD, PPT1 or TPP1, in which case they would be appropriately termed CLN10, CLN1 or CLN2, respectively. The patient originally described by Kufs had a slowly progressive dementing illness beginning at age 30 years. However, onset during adolescence has also been reported. Presenting clinical features include behavioural changes, seizures, ataxia, cerebellar signs and pyramidal/extrapyramidal movement disorders, followed by dementia. Unlike other NCLs, patients with adult-onset forms show no ophthalmological symptoms. Although most cases of CLN4 have been considered to be autosomal recessive like most other NCLs, a dominant form has also been recognized and is referred to as Parry disease. Recessive forms of CLN4 have been subclassified into two subtypes: phenotype A with progressive myoclonus epilepsy with dementia and motor signs and phenotype B with pronounced behavioural changes, dementia and progressive motor disturbances.[167,197,340] There is overlap, however, and it may be better to view the clinical subtypes as a continuum. The dominant form, Parry disease, exhibits features of both phenotypes A and B.

Biochemistry and Molecular Genetics

As with most other NCLs, SCMAS is a major storage material in neurons and brain; however, it is absent in other tissues. As mentioned earlier, adult-onset NCLs are genetically heterogeneous with some documented as autosomal recessive (most commonly referred to as Kufs disease) and others as having dominant inheritance (Parry disease) (see Table 6.7).[56,167,197,340] Some adult-onset cases have been linked to other known NCL genes, namely *CLN1* (*PPT1*),*CLN5* and *CLN10*. In addition, one patient showed two mutations in the *N*-sulphoglucosamine sulphohydrolase gene[56,451] – an enzyme whose deficiency is typically associated with MPS IIIA (see earlier). Such findings have suggested to some that the adult-onset NCLs are not distinct genetic entities but rather reflect extremes of a clinical spectrum involving low penetrance and variable expression of NCL mutations. However, Benitez and colleagues, using whole exome sequencing in individuals with autosomal dominant NCL (Parry disease) discovered a novel single nucleotide variation in the *DNAJC5* gene.[39] This gene codes for a synaptic vesicle protein known as cysteine-string protein-α (CSPα). Additional studies support this finding.[364,531] The involvement of CSPα is intriguing and may implicate specific aspects of synaptic dysfunction in the pathogenesis of adult-onset NCL disease. This gene defect is presently given as *CLN4B*. The gene responsible for typical autosomal recessive adult NCL disease (Kufs type A) has not yet been determined but has been designated *CLN4*. Recently it has been reported that mutations in *CLN6*, a gene associated with variant late infantile NCL, also cause a subset of these recessively inherited adult-onset NCL cases, and it is most likely that future studies will reveal a number of unique genes linked to adult-onset NCLs.

Pathology

The brain weight is reduced, with atrophy of the cerebrum, particularly prominent in the frontal and frontoparietal region of the cerebral cortex. The cerebellum may be atrophic as well. Microscopically, neuronal storage and variable degrees of neuronal loss are present.[340] On pigmentoarchitectonic analysis there appears to be a selective neuronal storage in the neocortex and also in the subcortical nuclei similar to that noted in CLN3 disease.[57] Secondary degeneration of white matter associated with astrogliosis and microglial activation has been observed. Staining characteristics of the neuronal storage material are similar to those of other types of NCL. Immunocytochemically, SCMAS and Saps A and D have been demonstrated in the storage material.[125] At the ultrastructural level, the storage material consists of heterogeneous membrane-bound inclusions containing various combinations of GRODs, curvilinear, rectilinear and fingerprint profiles and lipid droplets.[167,197]

In the majority of cases, no storage materials are detected in biopsies from conjunctiva, skin or peripheral nerves. However, in some cases, fingerprint profiles have been found in eccrine sweat gland epithelium, and curvilinear and rectilinear profiles have been noted in skeletal muscles.[314] Osmiophilic inclusions with or without fingerprint profiles have been noted in skin, muscle and rectal biopsies in some cases. Thus, unlike the childhood forms of NCL, storage in extracerebral tissues is inconsistent and ultrastructural features of storage material are quite heterogeneous in the adult-onset diseases.

Animal Models

A murine knockout for the *Dnajc5* gene leading to a loss the CSPα protein is available and exhibits progressive neurodegeneration and death by 2–4 months of age.[139] No other CLN4 models are currently available.

CLN5 Disease

CLN5 disease, as well as CLN6, CLN7 and CLN8 discussed subsequently, are variant types of NCL including so-called Finnish variant late infantile NCL, variant late infantile/early juvenile NCL, Turkish variant late infantile NCL, and northern epilepsy. These have been separated from the classic forms of NCL by genetic complementation studies, with the gene symbols being assigned accordingly (see Table 6.7). Thus, it is important to note that these are not clinical or pathological variants but are distinct genetically from each other and from the classic forms of NCL (CLN1-CLN4).

History, Clinical Features and Molecular Genetics

The signature case for CLN5 disease was originally identified in Finland and classified as a variant of late infantile NCL. The gene was cloned in 1998 and additional cases both in and outside of Finland were recognized that had a similar clinical presentation. To date, 27 different mutations have been identified in both Finnish and non-Finnish families.[2] CLN5 disease is rare and variable in terms of clinical features. Most affected individuals appear as normal initially and develop delayed psychomotor development at ages 5–7 years. In some cases onset is later and consists of visual deficits and tremors. Earlier and later (including adult) onset cases have also been observed. Most affected individuals die between 12 and 23 years of age. The CLN5 protein, like the CLN3 protein, has proved enigmatic and controversial. Current understanding is that it is a soluble glycoprotein that undergoes cleavage in the ER, and is transported to lysosomes via the mannose 6-phosphate mechanism. As recently reviewed,[2] there is some evidence that CLN5 interacts with other NCL proteins, including CLN1/PPT1, CLN2/TPP1, CLN3, CLN6 and CLN8, and that overexpression of CLN1 can rescue a CLN5 deficiency.

Pathology and Pathogenesis

Like most other NCLs, brain pathology in CLN5 disease is characterized by severe generalized cerebral and cerebellar atrophy. Intraneuronal storage of SCMAS as well as Sap-A and D has been reported. Storage in cortical pyramidal neurons can be severe and typically leads to the formation of meganeurites, though the presence of ectopic dendrites accompanying this change has not been reported. Storage material ultrastructurally is similar to that in other NCLs and consists of fingerprint bodies, curvilinear and rectilinear bodies. Pathogenic mechanisms in CLN5 disease are not understood and the function for the CLN5 protein has not been defined. Its interaction with other CLN proteins, as mentioned earlier and recently reviewed,[2] may hold the key to this understanding.

Animal Models

A mouse knockout model of *CLN5* disease has been generated that displays many of the same relatively slowly progressing features of NCL disease in other murine models, including a prominent early glial activation. A unique feature of the *CLN5* knockout is an early predominance of cortical neuron death over thalamic neuron involvement.[538]

CLN6 Disease

History, Clinical Features And Molecular Genetics

Originally known as early juvenile NCL and classed as a variant of the late infantile disease, the CLN6 gene was identified in 2002.[11,18] Clinically, this disease resembles classic CLN2 disease but tends to have slightly later onset and a milder course. Seizures and motor system dysfunction are typical early signs. Death generally occurs late in the second decade of life. To date 41 mutations in the *CLN6* gene have been described. The CLN6 protein is a highly conserved polytopic membrane protein possessing seven transmembrane protein domains and is believed localized to the ER. Its function has not been established.[11] Mutations in the *CLN6* gene were also recently reported to cause some forms of adult-onset NCL (see earlier).

Pathology and Pathogenesis

Brain atrophy with ventricular dilation resembling CLN2 AND other NCL diseases is characteristic. Neuron loss, particularly in cerebral cortex, is widespread and surviving neurons exhibit storage of typical autofluorescent material and SCMAS. Gliosis is widely evident. Ultrastructural appearance of storage material closely resembles that in CLN5 and CLN7 diseases, appearing as rectilinear, curvilinear and fingerprint bodies. The pathogenesis of CLN6 disease is not understood, and how defects in an ER protein cause lysosomal dysfunction is particularly problematic. The availability of several spontaneously occurring animal models has provided an important resource for addressing these questions.

Animal Models

CLN6 disease has been discovered in both South Hampshire and Merino sheep, as well as in mice known formerly as *Nclf*. The *Nclf* mice and ovine NCL in South Hampshire that are considered to be models for CLN6[157,562] are available for research, and display clinical and pathological features parallel to those in human cases.[11]

CLN7 Disease

History, Clinical Features and Molecular Genetics

Originally classified as a variant of late infantile NCL of diverse ethnic origin, the *CLN7* gene was cloned in 2007 and identified as *MFSD8*, which codes for a major facilitator superfamily protein. Affected individuals exhibit seizures and developmental regression before the age of 6 years, followed by a wide spectrum of motor, visual and behavioural abnormalities that may progress rapidly. *MFSD8*, also known as *CLN7*, codes for a 518 amino acid polypeptide. This polytopic integral membrane protein is believed to possess 12 membrane-spanning domains and to function as a transporter. However, its substrate specificity is unknown. The protein appears to be expressed at low levels in most tissues, and current evidence localizes it to lysosomes. To date 22 mutations have been reported, including missense, nonsense, insertions and deletions. Details of CLN7 disease can be found in a recent review.[124]

Pathology and Pathogenesis

Like other NCL disorders, brain atrophy is characteristic of CLN7 disease and includes both the cerebrum and cerebellum. Neurons contain autofluorescent material and SCMAS, and stain with PAS, Sudan black and LFB in a manner typical of other NCL diseases. Neuronal storage is particularly pronounced in the basal ganglia and thalamus. Ultrastructurally, inclusions are seen as fingerprint, rectilinear and curvilinear bodies. There is currently no understanding as to why defects in the CLN7 protein causes disease or what the relationship might be to other forms of NCL disease.

Animal Models

No naturally occurring animal models of CLN7 disease have been documented, but a knockout mouse model has been developed.[98a]

CLN8 Disease

History, Clinical Features and Molecular Genetics

Two strikingly different conditions reside within the CLN8 disease category. The first is northern epilepsy, so identified because of its occurrence in Finland[210] and also referred to as progressive EPMR. A second clinical condition subsequently identified in Turkey exhibited features of a variant late infantile form of NCL and these cases as well as others have been found to be caused by mutations in *CLN8*. This gene codes for a 33-kDa polytopic non-glycosylated membrane protein that is believed to reside in ER membranes as well as ER-Golgi membranes (ERGIC). Its function is unknown though a possible role in sphingolipid synthesis has been suggested. Readers are referred to a recent review for additional details.[6]

Pathology and Pathogenesis

Unlike other NCL disorders, inspection of gross brain pathology has not revealed evidence of striking cerebral or cerebellar atrophy. Storage material in neurons stained with PAS, Sudan black and LFB, a finding typical of NCL disease. Immunocytochemical studies also revealed storage of SCMAS and Saps. Ultrastructural analysis of storage granules showed the presence of curvilinear, rectilinear and fingerprint bodies. How defects in the CLN8 protein lead to an NCL-like disease phenotype is presently unknown.

Animal Models

A spontaneous neurological mutant in mice, known as motor neuron degeneration or *MND*, was found to harbour a defect in *CLN8*.[336] These mice display prominent abnormalities in the spinal cord, hence their original classification, but also exhibit neuron storage and cell loss in cortex, hippocampus and thalamus, with a prominent loss of GABAergic neurons.

CLN9 Disease

History, Clinical Features and Pathology

The CLN9 designation has been assigned to cases of juvenile NCL disease when patients do not harbour a defect in *CLN3*. Diagnoses have been made in German and Serbian families, and affected individuals did not exhibit defects in *CLN6*, *CLN7* or *CLN8*, in addition to *CLN3*. Pathologically, CLN9 disease resembles CLN3 disease. The gene and protein defects responsible for this disease are unknown and thus no identified animal model exists.

CLN10 Disease

History, Clinical Features, Biochemistry and Molecular Genetics

Congenital NCL is an extremely rare type and was first reported by Norman and Wood as a congenital form of amaurotic familial idiocy.[360] Since then, about ten sporadic cases have been reported.[35,149,158] Clinically, affected infants are usually born following an uneventful pregnancy but present with severe respiratory insufficiency and seizures; death usually occurs shortly after birth. Affected infants show extreme microcephaly: the brain weight may be 100–150 g or less, but body weight is usually within the normal range. Magnetic resonance imaging of an infant reported by Fritchie and colleagues showed generalized hypoplasia of the cerebral and cerebellar hemispheres, with abundant extra-axial cerebrospinal fluid spaces, minimal hydrocephalus and craniobasal dysplasia. Electroencephalography (EEG) revealed a pattern consistent with severe bihemispheric dysfunction without evidence of seizure activity.[149]

Biochemistry and molecular analyses of congenital NCL are limited to two reports by Siintola and colleagues[447] and by Fritchie and her colleagues.[149] The former reported identification of a novel homozygous nucleotide change, c.764dupA, in the cathepsin D gene in an affected infant and identified the same molecular change in a heterozygous form in the healthy father.[447] A c.299C>T (p.Ser100Phe) mutation in exon 3 of the cathepsin D gene in the homozygous form was found in the infant and in heterozygous form in the parents of the latter.[149]

Pathology and Pathogenesis

The brain is extremely small and firm on palpation. The gyral pattern suggests developmental delay. There is marked loss of neurons and replacement by numerous hypertrophic astrocytes and macrophages/microglia. The cerebellar cortex is notable for severe loss of Purkinje and granular cells. These cells contain PAS-positive and autofluorescent storage materials that are strongly immunoreactive with antibody to Sap-D but not for SCMAS. These storage materials accumulate in the perikarya as well as in the axonal hillock region, giving the appearance of meganeurites,[540] and show ultrastructural features of GRODs.[149]

Animal Models

Animal models of cathepsin D deficiency are known in Swedish Landrace lambs,[515] American bulldogs,[29] mice[258] and *Drosophila*.[350] The mouse model dies at approximately 28 days of age and displays a rapidly progressing NCL-like phenotype, with progressive neuron loss that is preceded by glial activation, synaptic pathology and defects in neurotransmission.

Other Variant Types of Neuronal Ceroid Lipofuscinosis

In addition to the 10 NCL diseases described earlier, studies in recent years have suggested four new members to this family of disorders. These are designated CLN11 to CLN14 (see Table 6.7).

CLN11 Disease

Smith and colleagues carried out a linkage analysis on two siblings with an NCL-like condition consisting of visual failure, seizures, mild cerebellar atrophy and cognitive decline.[457] A skin biopsy had revealed fingerprint profiles in membrane-bound vesicles. They found that both individuals harboured homozygous mutations in the progranulin gene (*GRN*). Interestingly, individuals heterozygous for mutations in this gene are known to develop a form of frontotemporal lobar degeneration (FTLD-TDP) in late life. Parents of the affected siblings were in their fifties and, although healthy, did have family history of early onset dementia. FTLD-TDP consists of late-onset (seventh to eighth decade) behavioural and cognitive decline; however, cerebellar ataxia, seizures and retinal degeneration are absent. Histopathologically, the condition consists of TDP-43-immunoreactive inclusion bodies in neurons and glia rather than lipopigment storage. Interestingly, studies carried out on mice deficient in *GRN* showed the presence of ubiquitin-positive autofluorescent lipofuscin which by electron microscopy showed abundant rectilinear bodies.[5]

Thus it appears that homozygosity for mutations in *GRN* are linked to a rare disease with features of NCL disease, whereas heterozygosity leads to a more common neurodegenerative condition.[457]

CLN12 Disease

Two recent reports examining a late-onset NCL disease in Tibetan terriers found that mutations in *ATP13A2* (also known as *CLN12, KRPPD, PARK9, HSA9947, RP1-37C10.4*) appeared to be the cause of disease in that breed of dogs.[135,572] The ATP13A2 protein is a member of the P-type superfamily of ATPases that are known to transport inorganic cations and other substrates across cell membranes, in this case lysosomal membranes. Interestingly, defects in this gene in humans had earlier been shown to cause Kufor–Rakeb syndrome (KRS), also referred to as Parkinson's disease 9.[520] KRS is a rare form of autosomal recessive hereditary parkinsonism with dementia and juvenile onset. The disease in dogs, however, did not resemble that of KRS. Subsequent studies by Bras and colleagues carrying out exome sequencing in a large family in Belgium with documented juvenile-onset NCL disease but no evidence of involvement of any previously identified CLN genes, found a single homozygous mutation in the *ATP13A2* gene.[62] These findings are believed to draw strong links between lysosomal function and parkinsonism as well as to reveal a new form of NCL disease (CLN12).

CLN13 Disease

As described earlier, type A Kufs disease presents with progressive myoclonic epilepsy and is caused by mutations in the *CLN4* gene. There has also been recognized a type B form of Kufs that along with the characteristic adult onset manifests with clinically evident dementia and motor system deterioration. Using a linkage analysis and exome sequencing to investigate two families with clinically diagnosed Kufs type B disease, Smith and colleagues found missense mutations in the gene for the lysosomal protease known as cathepsin F (CTSF).[456] These studies established mutations in *CTSF* as a cause of some cases of Kufs type B disease and this disease has been designated CLN13.

CLN14 Disease

Genetic studies were carried out on a large Mexican family with NCL disease documented by conservative clinical and histopathological studies. Family members exhibited infantile-onset progressive myoclonus epilepsy, visual loss, cognitive regression and motor system dysfunction, with fatality during their teenage years. Whole exome sequencing revealed a mutation in exon 4 of *KCD7* (potassium channel tetramerization domain-containing protein 7).[461] Other studies have linked mutations in this gene to progressive myoclonus epilepsy but not to NCL disease.

Other NCL diseases

Defects in a number of other genes have been reported to cause NCL-like phenotypes, but CLN gene designations have not been made. Most notable are defects in chloride channels.[227] These channels, which act as Cl⁻/H⁺ exchangers, are a highly conserved gene family. Mutations in *CLCN7*,

TABLE 6.8 Other lysosomal disorders

Clinical diagnosis	Gene	Affected compounds	Protein defects
Acid lipase deficiency (Wolman disease; cholesteryl ester storage disease)	LIPA	Cholesterol ester and triglycerides	Acid lipase
Free sialic acid storage disease (Salla disease; infantile sialic acid storage disease)	SLC17A5	Free sialic acid	SLC9A17 transporter
Cystinosis	CTNS	Cystine	Cystinosin

whose similarly named protein resides in late endosomes and lysosomes and appears to be critical for acidification, cause bone disease in children but also a variety of CNS effects including cortical atrophy. CLCN6 is similarly localized to late endosomes and mice lacking this protein show well-documented SCMAS and lipofuscin accumulation in neurons. CLCN3 is found in endosomes as well as synaptic vesicles and its functional absence leads to SCMAS accumulation and neuron loss in the hippocampus.

As described earlier, a defect in arylsulphatase G has been suggested to be a possible candidate cause of adult-onset NCL disease based on studies in American Staffordshire terriers,[4] but more recent work links this enzyme deficiency to MPS III disease (see page 479 for discussion). Another gene defect reported to cause adult-onset NCL disease involves SGSH, which codes for N-sulphoglucosamine sulphohydrolase, also known as sulphamidase. In this case, however, the patient is likely to have had Sanfilippo type A (MPS III type A) disease (see earlier discussion on page 478) with secondary lipopigment accumulation. It is perhaps worth noting that a number of lysosomal diseases, most notably MPS diseases, have in some cases been reported to accumulate SCMAS in some neurons in a manner similar to NCL disorders.[416]

OTHER LYSOSOMAL DISORDERS (TABLE 6.8)

Acid Lipase Deficiency (Wolman Disease, Cholesteryl Ester Storage Disease)

History and Clinical Features

Wolman disease is an infantile disorder characterized by hepatosplenomegaly, abdominal distension, steatorrhoea and other gastrointestinal signs and symptoms. Affected infants almost always die before reaching 1 year of age. Specific symptoms related to the CNS are uncommon, but affected infants may develop progressive mental deterioration. Radiological demonstration of calcified adrenal glands together with the abovementioned symptoms is almost pathognomonic. The clinical phenotype of Wolman disease is remarkably uniform. A milder form, known as cholesteryl ester storage disease (CESD), is also known. Clinical presentation of this milder form is relatively heterogeneous. Patients have hepatosplenomegaly and widespread lipid deposits in many organs. Hyperlipoproteinaemia is common and patients develop premature atherosclerosis.[177]

Biochemistry and Molecular Genetics

Cholesteryl esters and triglycerides accumulate in many organs, including the liver, adrenal gland and spleen. The activity of a lysosomal acid hydrolase catalyzing hydrolysis of cholesteryl esters and triglycerides (acid lipase) is severely deficient in all tissues, including the liver, spleen, lymph nodes, aorta, peripheral blood leukocytes and cultured skin fibroblasts. Wolman disease and CESD are allelic and inherited as autosomal recessive traits, involving the gene for lysosomal acid lipase (LIPA), located on chromosome 10q23.[177]

Pathology

The pathological manifestations of Wolman disease and CESD are similar. The liver, spleen and adrenal glands are always grossly enlarged, with diffuse or focal yellow or orange discolouration. Lipid droplets are found in hepatic parenchymal cells, adrenal cortical cells, Kupffer cells and macrophages (Figure 6.96). The adrenal glands often contain flecks of gritty calcified tissue (Figure 6.97). Sudanophilic lipid droplets are prominent in the muscular layer as well as in the vascular endothelium on frozen sections. Reported pathological examinations of the nervous system are limited. Swollen neurons in the medulla and retina, foamy cells in the leptomeninges and choroid plexus, accumulation of sudanophilic materials in microglia and perivascular histiocytes, and diffuse fibrillary gliosis have been described in some patients.[182] Storage of sudanophilic granules has also been reported in the ganglion cells of the myenteric plexus. In an ultrastructural study, lipid droplets were reported in Schwann cells, oligodendrocytes, astrocytes and endothelial cells, but not in CNS neurons.[66]

Animal Models

Naturally occurring autosomal recessive lysosomal acid lipase deficiency is known in Donryu rats. Similar to Wolman disease in humans, this model shows hepatosplenomegaly with many foamy lipid-containing macrophages. Biochemical analysis shows massive accumulation of cholesteryl esters and triglycerides and deficiency of acid lipase activity.[280,583] More recently, with targeted disruption of the mouse lysosomal acid lipase gene, a mouse model of acid lipase deficiency has been generated. This model is clinically similar to the milder form (CESD) but mimics Wolman disease biochemically and histopathologically.[118]

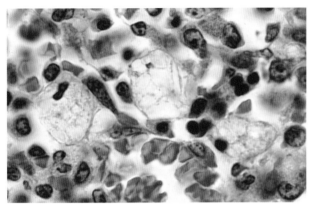

6.96 Wolman disease. Lipid-containing macrophages are present in the bone marrow smear. Giemsa.

Courtesy of Dr James Powers, University of Rochester, Rochester, NY, USA.

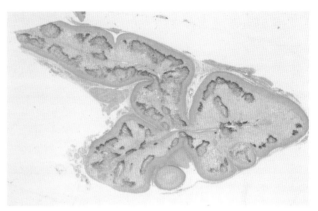

6.97 Wolman disease. Calcium deposition is conspicuous in the adrenal cortex.

Courtesy of Dr James Powers, University of Rochester, Rochester, NY, USA.

Free Sialic Acid Storage Disorder (Salla Disease and Infantile Sialic Acid Storage Disease)

History and Clinical Features

Genetic defects in the transporter protein that moves sialic acid across the lysosomal membrane result in lysosomal accumulation of free sialic acid. Thus although often classified as a membrane transport disorder, such defects also lead to lysosomal disease. Two distinct phenotypes are associated with this condition: the adult type or Salla disease, and the infantile sialic acid storage disease (ISSD).[22] Salla disease is known almost exclusively among the Finnish population. The eponym 'Salla' refers to the name of a small area in the north-eastern part of Finnish Lapland, where the first patients were discovered and where the prevalence of the disease is very high. The clinical signs of hypotonia and ocular nystagmus develop at around age 3–6 months. Afterwards, developmental milestones are delayed. Ataxia, dysarthria and difficulty in walking become gradually apparent. Patients with Salla disease have severe intellectual disability, and IQ levels below 20 are not uncommon. Physical growth is also retarded but without any significant dysmorphic features. There is no hepatosplenomegaly. The lifespan is not seriously affected. Unlike Salla disease, the allelic infantile form of the disease (ISSD) occurs in many ethnic groups. Infants with ISSD show a more fulminant course, with severe psychomotor retardation, dysmorphic features, hepatosplenomegaly, nephrotic syndrome and ascites; these patients usually die within the first year of life. Cardiomegaly may be present but corneal opacity and dysostosis multiplex are not present.[291] Significant numbers of patients present with hydrops fetalis. Several intermediate types have also been reported.[22]

It is worth noting the existence of another condition, sialuria, which although not a lysosomal disorder, bears similarity to sialic acid storage disease. This is a very rare disease caused by a defective feedback mechanism of intracellular synthesis of sialic acid, resulting in overproduction of sialic acid. The primary genetic defect has been identified as a mutation in the *GNE* (previously *GLCNE*) gene, encoding UDP-*N*-acetylglucosamine 2-epimerase, a key enzyme in sialic acid synthesis.[438] The pathogenetic mechanism is unusual in that the mutation occurs in such a way that it does not affect the catalytic activity of the enzyme but abolishes the ability of the gene to be regulated normally. The unregulated expression of the gene results in overproduction of sialic acid. Excessive sialic acid is, however, stored in the cytoplasm rather than in the lysosome. Because neither the enzyme nor the abnormal accumulation of the substrate is localized in the lysosome, this is not a lysosomal disorder as originally defined by Hers. It is mentioned here because it causes occasional confusion with the free sialic acid storage disease caused by deficiency of the sialic acid transporter. Clinically, excessive urinary excretion of sialic acid, variable developmental delay and hepatosplenomegaly are present.

Biochemistry

The basic metabolic defect is an impairment of an active proton-driven and substrate-specific transport system of free sialic acid across the lysosomal membrane, resulting in an accumulation of free sialic acid within the lysosome. The transporter is in the solute carrier family 17 and is referred to as SLC9A17. Thus, tissues and cultured fibroblasts from patients with Salla disease and ISSD contain a large amount of free sialic acid. Patients with Salla disease and ISSD store, respectively, 10 times and 100 times the normal amount of free (unbound) *N*-acetylneuraminic acid in their tissues.[22] Urinary excretion of free sialic acid is also increased. The diagnosis is made by detecting an increased amount of free sialic acid in urine and tissues by skin or conjunctival biopsies. Detection of increased intracellular free sialic acid in amniotic fluid cells and chorionic villi has been used for prenatal diagnosis.[282,283]

Molecular Genetics

Salla disease and ISSD are autosomal recessive diseases. ISSD is allelic to Salla disease. The Salla disease locus maps to 6q14–q15, and the gene (*SLC17A5, 51ASD, SLD*) has been cloned. Five Finnish patients with Salla disease were homoallelic for a mutation (R39C), and six different mutations were found in six ISSD patients of different ethnic origin.[22]

Neuroimaging

Early CT scans of patients with Salla disease showed cortical and basal atrophy, more pronounced in older patients. Magnetic resonance imaging demonstrated abnormalities in myelination in the cerebrum, but myelination in the cerebellum was normal. The corpus callosum was thin and hypoplastic. There appears to be a good correlation of severity in phenotype with the age of the patient and with MR imaging abnormalities; severely affected, older patients showed cortical and cerebellar atrophy and increased periventricular intensity as a sign of a destructive white matter process. Magnetic resonance spectroscopy showed a consistent increase of N-acetylaspartate and reduced choline signals.[22]

Pathology

The cardinal pathology is cytoplasmic storage of colloidal iron- and Alcian blue-positive materials in various types of cells. On routine paraffin histology, the storage materials are dissolved during processing and thus the cytoplasm of the storage cells appears vacuolated. The vacuoles are membrane-bound and often contain reticulogranular and occasional laminated material at the ultrastructural level. Notable vacuolated cells include lymphocytes, the proximal renal tubular epithelium, hepatocytes, Kupffer cells, macrophages, eccrine glandular and tubular epithelium, Schwann cells and endothelial cells. In the brain, neurons and glial cells show similar cytoplasmic storage. White matter degeneration of varying degrees is present, with loss of axons and extensive gliosis. Axonal spheroids and microcalcification may be detected.[24,291] In two patients with Salla disease who died at age 41 years, loss of Purkinje cells and abnormal amounts of lipofuscin and NFTs were found in the neocortex, presubiculum, nucleus basalis of Meynert and locus ceruleus.[24]

Animal Models

There are no known animal models of Salla disease or ISSD.

Cystinosis

History and Clinical Features

Cystinosis is a rare autosomal recessive lysosomal disorder resulting from defective transport of an amino acid, cystine, across the lysosomal membrane. The transport protein is termed cystinosin. There are two basic phenotypes: nephropathic and non-nephropathic. The majority of the nephropathic form occurs in infancy, although some late-onset cases are known. The first documented case was a 21.5-month-old boy who died of inanition. He had curious white crystal depositions in the liver on post-mortem examination. Biochemical analysis of the visceral organs revealed an abnormal accumulation of cystine.[3] Affected individuals are normal at birth. Gradually, at around

the age of 6–18 months, various clinical manifestations become apparent, including failure to thrive, the renal Fanconi syndrome characterized by polyuria and dehydration, glucosuria, hypokalaemia and hypophosphataemic rickets. Growth retardation is one of the most conspicuous clinical features,[153] and an average patient has a height around the third percentile at 1 year of age. By age 10 years, affected children lose almost all renal function and require dialysis or transplantation. Those who survive longer develop damage to other organs, including the thyroid, eye, CNS, pancreas and muscle. Computed tomography and MR imaging studies often reveal cerebral cortical atrophy. Impaired cognitive function and behavioural problems have been reported in these patients.[106] Patients with the non-nephropathic or benign form of cystinosis have crystalline deposits in their bone marrow and leukocytes but never develop renal disease; life expectancy is normal. These patients are usually discovered serendipitously when an ophthalmological examination reveals crystalline opacity in the cornea and conjunctiva (non-nephropathic ocular form).

Biochemistry

Cystinosis is an autosomal recessive disorder of lysosomal storage caused by a defective lysosomal integral membrane protein, cystinosin, which transports cystine across the lysosomal membrane. In affected individuals, free cystine accumulates to 10–1000 times the normal level and forms crystals within lysosomes. The rate of cystine accumulation is variable between different tissues. Increased cystine levels in the brain have been noted only in patients who survived longer. Pure cystine crystals can be rectangular or hexagonal. Because the cystine accumulation is intracellular, plasma and urine cystine concentration is usually not elevated, an important biochemical distinction from cystinuria, in which urinary cystine level is abnormally high. Elevated cystine content in polymorphonuclear leukocytes and fibroblasts has been used as the diagnostic marker.[153]

Molecular Genetics

Cystinosin is encoded by the *CTNS* gene on chromosome 17p13. As of 2011, more than 30 disease-causing mutations had been identified. The more severe mutations lead to the nephropathic infantile form that occurs with an incidence of 1 in 100 000–200 000 live births; the less severe mutations lead to the non-nephropathic late-onset form.[153]

Pathology

The kidney is the most severely affected organ (Figure 6.98). Pathologically, the cystinotic kidney shows different stages of destruction, with giant cell transformation of the glomerular epithelium, hyperplasia and hypertrophy of the juxtaglomerular apparatus, and occasional dark cells and cytoplasmic inclusions. Ultrastructurally, cystine crystals can be demonstrated as well-defined angular/rectangular electron-lucent spaces (Figure 6.99). Ultimately, the affected kidney shows classic pathology of end-stage renal disease, with scarring and fibrosis, chronic interstitial nephritis

6.98 Cystinosis. Reflectile cystine deposits are present in the kidney. Dark field microscopy.

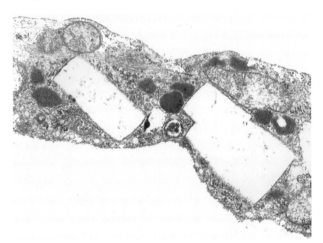

6.99 Cystinosis. The rectangular or polygonal profiles of cystine crystals are seen in a macrophage. Electron micrograph.

and tubular degeneration.[153] Degeneration of the retinal pigment epithelium and accumulation of cystine crystals in the retina, cornea and conjunctiva are the well-documented ophthalmic pathology of cystinosis. Cystine storage in the CNS is very limited in young patients. However, deposition of cystine crystals has been demonstrated in long-surviving patients.[535] Vogel

and co-workers examined the brain of a 28-year-old male who received kidney transplantation 18 years before his death. The patient had gait difficulty, intention tremor and progressive dementia. Diffuse ventriculomegaly with cerebral cortical atrophy was detected on CT and MR imaging. Calcification of the basal ganglia was also found on CT at age 20 years. At post-mortem examination, cystine crystals were present in moderate quantities in various organs. However, no cystine crystals were seen in the donor kidney. The cerebral cortex was atrophic and the shrunken white matter contained 'yellowish white flecks of gritty material'. Cystic necrosis was present in the globus pallidus. Cystine crystals were found in the frontal cortex and substantia nigra, mostly around capillaries when frozen section preparations were examined under polarized light. Numerous angular, needle-shaped and hexagonal crystalline spaces were identified in the cytoplasm of pericytes and possible oligodendrocytes at the ultrastructural level.[535] Necrotic foci and calcium deposits were also described in other patients. Cystine crystals were also reported in a brain biopsy sample of the arachnoid and cerebral cortex from a child who developed non-absorptive hydrocephalus.[412]

Animal Models

There are no known naturally occurring models of cystinosis, however, a mouse model of cystinosis has been made using a promoter trap approach. The presence of a mislocalized and non-functional cystinosin protein resulted in mice with cystine accumulation in most organs. Ocular, bone and behavioural problems were noted though the mice failed to develop renal failure.[80]

ACKNOWLEDGEMENTS

The authors wish to thank Drs Matthew Micsenyi, Brian Lake, Matti Haltia, Robert D Terry, David Wolf, James Powers and Mauricio Castillo for allowing us to use their light and electron micrographs and neuroimaging for illustration in this chapter. We also thank Drs Jonathan Cooper and Stuart Kornfeld for their comments on the sections on NCL disorders and mucolipidoses, respectively.

REFERENCES

1. Åberg L, Autti T, Braulke T, *et al.* CLN3. In: Mole S, Williams RL, Goebel HH, eds. *The neuronal ceroid lipofuscinoses (Batten disease)*, 2nd edn. Oxford: Oxford University Press, 2011:110–39.

2. Åberg L, Autti T, Cooper JD, *et al.* CLN5. In: Mole S, Williams RL, Goebel HH, eds. *The neuronal ceroid lipofuscinoses (Batten disease)*, 2nd edn. Oxford: Oxford University Press, 2011:140–58.

3. Abderhalden E. Familiare cystindiathese. *Z Physiol Chem* 1903;**38**:557–61.

4. Abitol M, Thibaud JL, Olby NJ, *et al.* A canine arylsulfatase G (ARSG) mutation leading to a sulfatase deficiency is associated with neuronal ceroid lipofuscinoses. *Proc Natl Acad Sci U S A* 2010;**107**:14775–80.

5. Ahmed Z, Sheng H, Xu YF, *et al.* Accelerated lipofuscinosis and ubiquitination in granulin knockout mice suggest a role for progranulin in successful aging. *Am J Pathol* 2010;**177**:311–24.

6. Aiello C, Cannelli N, Cooper JD. CLN8 In: Mole S, Williams RL, Goebel HH, eds. Mole S, Williams RL, Goebel HH, eds. *The neuronal ceroid lipofuscinoses (Batten disease)*, second edn. Oxford: Oxford University Press, 2011:140–58.

7. Akman HO, Raghavan A, Craigen WJ. Animal models of glycogen storage disorders. *Prog Mol Biol Transl Sci* 2011;**100**:369–88.

7a. Alayoubi A, Wang J, Au B, *et al.* Systemic ceramide accumulation leads to severe and varied pathological consequences. *EMBO Mol Med* 2013;**5**:827-42.

8. Alkan A, Kutlu R, Yakinci C, *et al.* Infantile Sandhoff's disease: multivoxel magnetic resonance spectroscopy findings. *J Child Neurol* 2003;**18**:425–8.

9. Alkhayat AH, Kraemer SA, Leipprandt JR, *et al.* Human beta-mannosidase cDNA characterization and first identification of a mutation associated with human beta-mannosidosis. *Hum Mol Genet* 1998;**7**:75–83.

10. Allegranza A, Tredici G, Marmiroli P, *et al.* Sialidosis type I: pathological study in an adult. *Clin Neuropathol* 1989;**8**:266–71.

11. Alroy J, Braulke T, Cismondi IA, CLN6. In: Mole S, Williams RL, Goebel HH, eds. Mole S, Williams RL, Goebel HH, eds. *The neuronal ceroid lipofuscinoses (Batten disease)*, second edn. Oxford: Oxford University Press, 2011:189–202.

12. Alroy J, Orgad U, DeGasperi R, *et al.* Canine GM1-gangliosidosis. A clinical, morphologic, histochemical, and biochemical comparison of two different models. *Am J Pathol* 1992;**140**:675–89.

13. Alroy J, Sabnis S, Kopp JB. Renal pathology in Fabry disease. *J Am Soc Nephrol* 2002;**13**: S134–8.

14. Altarescu G, Sun M, Moore DF, *et al.* The neurogenetics of mucolipidosis type IV. *Neurology* 2002;**59**:306–13.

15. Arbisser AL, Donnelly KA, Scott CI Jr, *et al.* Morquio-like syndrome with beta-galactosidase deficiency and normal hexosamine sulfatase activity: mucopolysaccharidosis IVB. *Am J Med Genet* 1977;**12**:195–205.

16. Aronson NN Jr. Asparagine-linked glycoproteins and their degradation. In: Mononen I, Aronson NN Jr eds. *Lysosomal storage disease: aspartylglycosaminuria.* Austin, TX: R.G. Landes/Heidelberg, Springer-Verlag, 1997:55–75.

17. Aronson NN Jr. Aspartylglycosaminuria: biochemistry and molecular biology. *Biochim Biophys Acta* 1999;**1455**:139–54.

18. Arsov T, Smith KR, Damiano J, *et al.* Kufs disease, the major form of neuronal ceroid lipofuscinoses, caused by mutations in CLN6. *Am J Hum Genet* 2011;**88**:566–73.

19. Arstila AU, Palo J, Haltia M, *et al.* Aspartylglucosaminuria: I. Fine structural studies of liver, kidney and brain. *Acta Neuropathol* 1972;**20**:207–16.

20. Arvio M, Autio S, Mononen T. Clinical manifestations of aspartylglycosaminuria. In: Mononen I, Aronson NN Jr eds. *Lysosomal storage disease: aspartylglycosaminuria.* Austin, TX: R.G. Landes/Heidelberg, Springer-Verlag, 1997:19–31.

21. Arvio P, Arvio M, Kero M, *et al.* Overgrowth of oral mucosa and facial skin, a novel feature of aspartylglucosaminuria. *J Med Genet* 1999;**36**:398–404.

22. Aula P, Gahl WA. Disorders of free sialic acid storage. In: Valle D, Beaudet AL, Vogelstein B, *et al.* eds. *The online metabolic and molecular bases of inherited disease.* New York: McGraw-Hill, 2011;Part 21, Chapter 200.

23. Aula P, Jalanko A, Peltonen L. Aspartylglucosaminuria. In: Valle D, Beaudet AL, Vogelstein B, *et al.* eds. *The online metabolic and molecular bases of inherited disease,* New York: McGraw-Hill, 2011. Part 16, Chapter 141.

24. Autio-Harmainen H, Oldfors A, Sourander P, *et al.* Neuropathology of Salla disease. *Acta Neuropathol* 1988;**75**:481–90.

25. Autti T, Cooper J, van DIggelen OP, *et al.* CLN1. In: Mole S, Williams RL, Goebel HH, eds. Mole S, Williams RL, Goebel HH, eds. *The neuronal ceroid lipofuscinoses (Batten disease),* second edn. Oxford: Oxford University Press, 2011:55–79.

26. Autti T, Joensuu R, Aberg L. Decreased T2 signal in the thalami may be a sign of lysosomal storage disease. *Neuroradiology* 2007;**49**:571–8.

27. Autti T, Lönnqvist T, Joensuu R. Bilateral pulvinar signal intensity decrease on T2-weighted images in patients with aspartylglucosaminuria. *Acta Radiol* 2008;**28**:687–92.

28. Autti T, Railinko R, Haltia M, *et al.* Aspartylglucosaminuria: radiologic course of the disease with histopathological correlation. *J Child Neurol* 1997;**12**:369–75.

29. Awano T, Katz ML, O'Brien DP, *et al.* A mutation in the cathepsin D gene (CTSD) in American bulldogs with neuronal ceroid lipofuscinosis. *Mol Genet Metab* 2006;**87**:341–8.

30. Bach G. Mucolipidosis type IV. Minireview. *Mol Genet Metab* 2001;**73**:197–203.

31. Bach G, Friedman R, Weissmann B, Neufeld EF. The defect in the Hurler and Sheie syndromes: deficiency of α-L-iduronidase. *Proc Natl Acad Sci U S A* 1972;**69**:2048–51.

32. Baker E, Guo XH, Orsborn AM, *et al.* The Morquio A syndrome (mucopolysaccharidosis IVA) gene maps to 16q24.3. *Am J Hum Genet* 1993;**52**:96–8.

33. Bargal R, Avidan N, Ben-Asher E, *et al.* Identification of the gene causing mucolipidosis type IV. *Nat Genet* 2000;**26**:120–23.

34. Bargal R, Goebel HH, Latta, Bach G. Mucolipidosis IV: novel mutation and diverse ultrastructural spectrum in the skin. *Neuropediatrics* 2002;**33**:199–202.

35. Barohn RJ, Dowd DC, Kagan-Hallet KS. Congenital ceroid-lipofuscinosis. *Pediatr Neurol* 1992;**8**:54–9.

36. Baudry M, Yao Y, Simmons D, *et al.* Postnatal development of inflammation in a murine model of Niemann–Pick type C disease: immunohistochemical observations of microglia and astroglia. *Exp Neurol* 2003;**184**:887–903.

37. Baumann N, Turpin J-C, Lefevre M, Colsch B. Motor and psycho-cognitive types in adult metachromatic leukodystrophy: genotype/phenotype relationship? *J Physiol Paris* 2002;**96**:301–6.

38. Bedilu R, Nummy KA, Cooper A, *et al.* Variable clinical presentation of lysosomal β-mannosidosis in patients with null mutations. *Mol Genet Metab* 2002;**77**:282–90.

39. Benitez BA, Alvarado D, Cai Y, *et al.* Exome-sequencing confirms DNAJC5 mutations as cause of adult neuronal ceroid-lipofuscinosis. *PLoS One* 2011;**6**:e26741.

40. Benninger C, Ullrich-Butt B, Zhan S-S, Schmitt HP. GM2 gangliosidosis B1 variant in a boy of German/Hungarian descent. *Clin Neuropathol* 1993;**12**:196–200.

41. Berg T, Riise HM, Hansen GM, *et al.* Spectrum of mutations in alpha-mannosidosis. *Am J Hum Genet* 1999;**64**:77–88.

42. Berger-Plantinga EG, Vanneste JAL, Groener JEM, van Schooneveld MJ. Adult onset dementia and retinitis pigmentosa due to mucopolysaccharidosis III-C in two sisters. *J Neurol* 2004;**251**:479–81.

43. Berman ER, Livni N, Shapira E, *et al.* Congenital corneal clouding with abnormal systemic storage bodies: a new variant of mucolipidosis. *J Pediatr* 1974;**84**:519–26.

44. Bhattacharyya R, Gliddon B, Beccari T, *et al.* A novel missence mutation in lysosomal sulfamidase is the basis of MPS IIIA in a spontaneous mouse mutant. *Glycobiology* 2001;**11**:99–103.

45. Bhaumik M, Muller VJ, Rozakis T, *et al.* A mouse model for mucopolysaccharidosis type IIIA (Sanfilippo syndrome). *Glycobiology* 1999;**9**:1389–96.

46. Bieber FR, Mortimer G, Kolodny E, *et al.* Pathologic findings in fetal GM1 gangliosidosis. *Arch Neurol* 1986;**43**:736–8.

47. Biegstraaten M, Hollak CEM, Bakkers M, *et al.,* Small fiber neuropathy in Fabry disease. *Mol Genet Metab* 2012;**106**:135–41.

48. Bijvoet AG, van de Kamp EH, Kroos M, *et al.* Generalized glycogen storage and cardiomegaly in a knockout mouse model of Pompe disease. *Hum Mol Genet* 1998;**7**:53–62.

49. Bijvoet AG, Van Hirtum H, Vermey M, *et al.* Pathologic features of glycogen storage disease type II highlighted in the knockout mouse model. *J Pathol* 1999;**89**:416–24.

50. Birkenmeier EH, Barker JE, Vogler CA, *et al.* Increased life span and correction of metabolic defects in murine mucopolysaccharidosis type VII after syngeneic bone marrow transplantation. *Blood* 1991;**78**:3081–92.

51. Biswas S, Biesiada H, Williams TD, LeVine SM. Substrate reduction intervention by l-cycloserine in twitcher mice (globoid cell leukodystrophy) on a B6;CAST/Ei background. *Neurosci Lett* 2003;**347**:33–6.

52. Black DN, Taber KH, Hurley RA. Metachromatic leukodystrophy: a model for the study of psychosis. *J Neuropsychiatry Clin Neurosci* 2003;**15**:289–93.

53. Bohlega S, Kambouris M, Shahid M, *et al.* Gaucher disease with oculomotor apraxia and cardiovascular calcification (Gaucher type IIIC). *Neurology* 2000;**54**:261–3.

54. Boonen M, Vogel P, Platt KA, *et al.* Mice lacking mannose 6-phosphate uncovering enzyme activity have a milder phenotype than mice deficient for *N*-acetylglucosamine-1-phosphotransferase activity. *Mol Biol Cell.* 2009;**20**:4381–9.

55. Boucek D, Jirikowic J, Taylor M. Natural history of Danon disease. *Genet Med* 2011;**13**:563–68.

56. Boustany R, Ceuterick-de Groote C, Goebel HH, *et al.* Genetically unassigned or unusual NCLs. In: Mole S, Williams RL, Goebel HH, eds. Mole S, Williams RL, Goebel HH, eds. *The neuronal ceroid lipofuscinoses (Batten disease),* Second edn. Oxford: Oxford University Press, 2011:213–36.

57. Braak H, Braak E. Pathoarchitectonic pattern of iso- and allocortical lesions in juvenile and adult neuronal ceroid-lipofuscinosis. *J Inherit Metab Dis* 1993;**16**:259–62.

58. Bradbury AM, Morrison NE, Hwang M, *et al.* Neurodegenerative lysosomal storage disease in European Burmese cats with hexosaminidase beta-subunit deficiency. *Mol Genet Metab* 2009;**97**:53–9.

59. Bradová V, Smid F, Ulrich-Bott B, *et al.* Prosaposin deficiency: further characterization of the sphingolipid activator protein-deficient sibs. Multiple glycolipid elevations (including lactosylceramidosis), partial enzyme deficiencies and ultrastructure of the skin in this generalized sphingolipid storage disease. *Hum Genet* 1993;**92**:143–52.

60. Brady RO. Gaucher disease. In: Moser HW ed. *Handbook of clinical neurology,* vol. 22. Amsterdam: Elsevier Science, 1996:123–32.

61. Brady RO, Schiffmann R. Fabry's disease. In: Dyck PJ, Thomas PK eds. *Peripheral neuropathy.* Philadelphia, PA: Elsevier Saunders, 2005:1893–904.

62. Bras J, Verloes A, Schneider SA, *et al.* Mutation of the parkinsonism gene

ATP13A2 causes neuronal ceroid-lipofuscinosis. *Hum Mol Genet.* 2012;21:2646–50.

63. Brockmann K, Dechent P, Wilken B, *et al.* Proton MRS profile of cerebral metabolic abnormalities in Krabbe disease. *Neurology* 2003;60:819–25.

64. Bugiani O, Borrone C. Fucosidosis: a neuropathological study. *Riv Patol Nerv Ment* 1976;97:133–41.

65. Bundza A, Lowden JA, Charlton KM. Niemann–Pick disease in a poodle dog. *Vet Pathol* 1979;16:530–38.

66. Byrd JC, Powers JM. Wolman's disease: ultrastructural evidence of lipid accumulation in central and peripheral nervous system. *Acta Neuropathol* 1979;45:37–42.

67. Cachon-Gonzalez MB, Wang SZ, Lynch A, *et al.* Effective gene therapy in an authentic model of Tay–Sachs related diseases. *Proc Natl Acad Sci U S A* 2006;105:10373–8.

68. Caciotti A, Garman SC, Rivera-Colón Y, *et al.* GM1 gangliosidosis and Morquio B disease: an update on genetic alterations and clinical findings. *Biochim Biophys Acta* 2011;1812:782–90.

69. Campdelacreu J, Munoz E, Gomez B, *et al.* Generalized dystonia with abnormal magnetic resonance imaging signal in the basal ganglia: a case of adult onset GM1 gangliosidosis. *Mov Disord* 2002;17:1095–7.

70. Carstea ED, Morris JA, Coleman KG, *et al.* Niemann–Pick C1 disease gene homology to mediators of cholesterol homeostasis. *Science* 1997;277:228–31.

71. Carstea ED, Polymeropoulos MH, Parker CC, *et al.* Linkage of Niemann–Pick disease type C to human chromosome 18. *Proc Natl Acad Sci U S A* 1993;90: 2002–4.

72. Cathey SS, Kudo M, Tiede S, *et al.* Molecular order in mucolipidosis II and III nomenclature. *Am J Med Genet Am* 2008;146A:512–3.

73. Ceuterick-de Groote C, Martin J-J. Extracerebral biopsy in lysosomal and peroxisomal disorders: ultrastructural findings. *Brain Pathol* 1998;8:121–32.

74. Chabas A, Coll MJ, Aparicio M, Rodriguez Diaz E. Mild phenotypic expression of α-N-acetylgalactosaminidase deficiency in two adult siblings. *J Inherit Metab Dis* 1994;17:724–31.

75. Chang M, Cooper J, Davidson BL, *et al.* CLN2. In: Mole S, Williams RL, Goebel HH, eds. *The neuronal ceroid lipofuscinoses (Batten disease),* 2nd edn. Oxford: Oxford University Press, 2011;80–109.

76. Charrow J. Ashkenazi Jewish genetic disorders. *Fam Cancer* 2004;3:201–6.

77. Chen CS, Bach G, Pagano RE. Abnormal transport along the lysosomal pathway in mucolipidosis type IV disease. *Proc Natl Acad Sci U S A* 1998;95:6373–8.

78. Chen Y, Verp MS, Knutel T, Hibbard JU. Mucopolysaccharidosis type VII as a cause of recurrent non-immune hydrops fetalis. *J Perinat Med* 2003;31:535–7.

79. Chen YQ, Rafi G, Degala G, Wenger DA. Cloning and expression of DNA encoding human galactocerebrosidase, the enzyme deficient in globoid cell leukodystrophy. *Hum Mol Genet* 1993;2:1841–5.

80. Cherqui S, Sevin C, Hamard G, *et al.* Intralysosomal cystine accumulation in mice lacking cystinosin, the protein defective in cystinosis. *Molec Cellul Biol* 2002;22:7622–32.

81. Choi KG, Sung JH, Clark HB, Krivit W. Pathology of adult-onset globoid cell leukodystrophy (GLD). *J Neuropathol Exp Neurol* 1991;50:336.

82. Christomanou H, Aignesberger A, Linke RP. Immunochemical characterization of two activator proteins stimulating enzymatic sphingomyelin degradation in vitro: absence of one of them in a human Gaucher disease variant. *Biol Chem Hoppe Seyler* 1986;367:879–90.

83. Christomanou H, Chabas A, Pampols T, Guardiola A. Activator protein deficient Gaucher's disease: a second patient with the newly identified lipid storage disorder. *Klin Wochenschr* 1989;67:999–1003.

84. Clark LA, Russel CS, Pownall S, *et al.* Murine mucopolysaccharidosis type I: targeted disruption of the murine α-L-iduronidase gene. *Hum Mol Genet* 1997;6:503–11.

85. Cleary MA, Wraith JE. Management of mucopolysaccharidosis type III. *Arch Dis Child* 1993;69:403–6.

86. Cohen-Tannoudji M, Marchand P, Akli S, *et al.* Disruption of murine HEXA gene leads to enzymatic deficiency and to neuronal lysosomal storage, similar to that observed in Tay–Sachs disease. *Mamm Genome* 1995;6:844–9.

87. Conradi N, Kyllerman M, Mansson J-E, *et al.* Late-infantile Gaucher disease in a child with myoclonus and bulbar signs: neuropathological and neurochemical findings. *Acta Neuropathol* 1991;82:152–7.

88. Conradi N, Sourander P, Nilsson O, *et al.* Neuropathology of the Norbottnian type of Gaucher disease: morphological and biochemical studies. *Acta Neuropathol* 1984;55:99–109.

89. Cork LC, Munnell JF, Lorenz MD, *et al.* GM2 ganglioside lysosomal storage disease in cats with beta-hexosaminidase deficiency. *Science* 1977;196:1014–7.

90. Cosma MP, Pepe S, Annunziata I, *et al.* The multiple sulfatase deficiency gene encodes an essential and limiting factor for the activity of sulfatases. *Cell* 2003;113:445–56.

91. Cox TM. Gaucher disease: understanding the molecular pathogeneses of sphingolipidoses. *J Inherit Metab Dis* 2001;24 (Suppl 2):106–21.

92. Cragg H, Williamson M, Young E, *et al.* Fucosidosis: genetic and biochemical analysis of eight cases. *J Med Genet* 1997;34:105–10.

93. Crawley AC, Walkley SU, Developmental analysis of CNS pathology in the lysosomal storage disease α-Mannosidosis. *J Neuropath Exp Neurol* 2007;66:287–96.

94. Crutchfield KE, Patronas NJ, Dambrosia JM, *et al.* Quantitative analysis of cerebral vasculopathy in patients with Fabry disease. *Neurology* 1998;50:1746–9.

95. Cummings JF, Wood PA, Walkley SU, *et al.* GM2 gangliosidosis in a Japanese spaniel. *Acta Neuropathol* 1985;67: 247–53.

96. Curless RG. Computed tomograph of GM1 gangliosidosis. *J Pediatr* 1984;105:964–6.

97. D'Azzo A, Andrea G, Strisciuglio P, Galjaard H. Galactosialidosis. In: Scriver CR, Beaudet AL, Sly WS, *et al.* eds. *The metabolic basis of inherited diseases,* 7th edn. New York: McGraw-Hill, 2001:3811–26.

98. Dahms BB, Davis R, Neustein HB, Landing BH. Fucosidosis: light, histochemical and ultrastructural findings. *Lab Invest* 1979;40:302–3.

98a. Damme M, Brandenstein L, Fehr S, *et al.* Gene disruption of *Mfsd8* in mice provides the first animal model for CLN7 disease. *Neurobiol Dis* 2014;65:12-24.

99. Danon MJ, Oh SJ, DiMauro S, *et al.* Lysosomal glycogen storage disease with normal acid maltase. *Neurology* 1981;31:51–7.

100. Davidson CD, Ali NF, Micsenyi MC, *et al.* Chronic cyclodextrin administration in Niemann–Pick C disease ameliorates intraneuronal cholesterol and glycosphingolipid storage and disease progression. *PLoS One* 2009;4:e6951.

101. De Baecque CM, Suzuki K, Rapin I, *et al.* GM2-gangliosidosis AB variant: clinicopathological study of a case. *Acta Neuropathol* 1975;33:207–26.

102. DeGasperi R, Friedrich VL, Perez GM, *et al.* Transgenic rescue of Krabbe disease in the twitcher mouse. *Gene Ther* 2004;11:1188–94.

103. de Geest N, Bonten E, Mann L, *et al.* Systemic and neurologic abnormalities distinguish the lysosomal disorders sialidosis and galactosialidosis in mice. *Hum Mol Genet.* 2002;11:1455–64.

104. Dekaban AS, Constantopoulos G. Mucopolysaccharidosis type I, II, IIIA and V. Pathological and biochemical abnormalities in the neural and mesenchymal elements of the brain. *Acta Neuropathol* 1977;39:1–7.

105. Dekaban AS, Constantopoulos G, Herman MM, Steusing JK. Mucopolysaccharidosis type V (Scheie syndrome). *Arch Pathol* 1976;100:237–45.

106. Delgado G, Schatz A, Nichols S, *et al.* Behavioral profiles of children with infantile nephropathic cystinosis. *Dev Med Child Neurol* 2005;47:403–7.

107. Desnick RJ, Brady R, Barranger J, *et al.* Fabry disease, an under-recognized multisystemic disorder: expert recommendations for diagnosis, management, and enzyme replacement therapy. *Ann Intern Med* 2003;138: 338–46.

108. Desnick RJ, Schindler D. α-N-Acetylgalactosaminidase deficiency: Schindler disease. Valle D, Beaudet AL, Vogelstein B, *et al.* eds. *The online metabolic and molecular bases of inherited disease,* New York: McGraw-Hill, 2011. Part 16, Chapter 139.

109. De Veber GA, Schwarting GA, Kolodny EH, Kowall NW. Fabry disease: immunochemical characterization of neuronal involvement. *Ann Neurol* 1992;31:409–15.

110. Di Natale P, Annella T, Daniele A, *et al.* Animal models for lysosomal storage diseases: a new case of feline mucopolysaccharidosis VI. *J Inherit Metab Dis* 1992;15:17–24.

111. Dickersin GR, Lott IT, Kolodny EH, Dvorak AM. A light and electron microscopic study of mannosidosis. *Hum Pathol* 1980;11:246–56.

112. Dierks T, Schmidt B, Borissenko LV, *et al.* Multiple sulfatase deficiency is caused by mutations in the gene encoding the human C-formylglycine generating enzyme. *Cell* 2003;113:435–44.

113. Deleted in proof.
114. Dittmer F, Hafner A, Ulbrich EJ, *et al.* I-cell disease-like phenotype in mice deficient in mannose 6-phosphate receptors. *Transgenic Res* 1998;7: 473–83.
115. Dos Santos MR, Tanaka A, Sa Miranda MC, *et al.* GM2-gangliosidosis B1 variant: analysis of β-hexosaminidase α gene mutations in 11 patients from a defined region in Portugal. *Am J Hum Genet* 1991;49:886–90.
116. Dowspon JH, Wilton-Cox H, Oldfors A, Sourander P. Autofluorescence emission spectra of neuronal lipopigment in mucopolysaccharidosis (Sanfilippo's syndrome). *Acta Neuropathol* 1989;77:426–9.
117. Drory VE, Birnbaum M, Peleg L, *et al.* Hexosaminidase A deficiency is an uncommon cause of a syndrome mimicking amyotrophic lateral sclerosis. *Muscle Nerve* 2003;28:109–12.
118. Du H, Duanmu M, Witte D, Grabowski GA. Targeted disruption of the mouse lysosomal acid lipase gene: long-term survival with massive cholesteryl ester and triglyceride storage. *Hum Mol Genet* 1998;7:1347–54.
119. Dumontel C, Girod C, Dijoud F, *et al.* Fetal Niemann–Pick disease type C: ultrastructural and lipid findings in liver and spleen. *Virchows Arch A Pathol Anat* 1993;422:253–9.
120. Durand P, Borrone C, Della Cella G. Fucosidosis. *J Pediatr* 1969;75:665–74.
121. Elleder M, Bradova V, Smid F, *et al.* Cardiocyte storage and hypertrophy as a sole manifestation of Fabry's disease. *Virchows Arch A Pathol Anat Histopathol* 1990;417:449–55.
122. Elleder M, Christomanou H, Kustermann-Kuhn B, Harzer K. Leptomeningeal lipid storage patterns in Fabry disease. *Acta Neuropathol* 1994;88:579–82.
123. Elleder M, Jerabkova M, Befekadu A, *et al.* Prosaposin deficiency: a rarely diagnosed, rapidly progressing, neonatal neurovisceral lipid storage disease. Report of a further patient. *Neuropediatrics* 2005;36:171–80.
124. Elleder M, Kousi M, Lehesjoki A-E, *et al.* CLN7. In: Mole S, Williams RL, Goebel HH, eds. Mole S, Williams RL, Goebel HH, eds. *The neuronal ceroid lipofuscinoses (Batten disease)*, Second edn. Oxford: Oxford University Press, 2011:176–89.
125. Elleder M, Sokola J, Hebiek M. Follow-up study of subunit c of mitochondrial ATP synthase (SCMAS) in Batten disease and in unrelated lysosomal disorders. *Acta Neuropathol* 1997;93:379–90.
126. Elleder M, Vanier MT, Harzer K, Suzuki K. Sphingolipid activator protein deficiency. In: Golden JA, Harding BN eds. *Pathology and genetics: developmental neuropathology*. Basel: ISN Neuropath Press, 2004:266–9.
127. Emery JM, Green WR, Wyllie RG, Howell RR. GM1-gangliosidosis: ocular and pathological manifestations. *Arch Ophthalmol* 1971;85:177–87.
128. Eng C, Ioannou Y, Desnick RJ. α-Galactosidase A deficiency: Fabry disease. In: Valle D, Beaudet AL, Vogelstein B, *et al.* eds. *The online metabolic and molecular basis of inherited disease*, New York: McGraw-Hill, 2011;Part 16, Chapter 150.

129. Evers M, Saftig P, Schmidt P, *et al.* Targeted disruption of the arylsulfatase B gene results in mice resembling the phenotype of mucopolysaccharidosis VI. *Proc Natl Acad Sci U S A* 1996;93: 8214–9.
130. Ezaki J, Kominami E. The intracellular location and function of proteins of neuronal ceroid lipofuscinosis. *Brain Pathol* 2004;14:77–85.
131. Deleted in proof.
132. Faerber EN, Melvin J, Smergel E. MRI appearances of metachromatic leukodystrophy. *Pediatr Radiol* 1999;29:669–72.
133. Fan X, Zhang H, Zhang S, *et al.*, Identification of the gene encoding the enzyme deficient in mucopolysaccharidosis IIIC (Sanfilippo disease type C). *Am J Hum Genet* 2006;79:738–44.
134. Farfel-Becker T, Vitner EB, Futerman AH. Animal models for Gaucher disease research. *Dis Model Mech* 2011;4:746–52.
135. Farias FH, Zeng R, Johnson GS, *et al.* A truncating mutation in ATP13A2 is responsible for adult-onset neuronal ceroid lipofuscinosis in Tibetan terriers. *Neurobiol Dis* 2011;42:468–474.
136. Farina L, Bizzi A, Finocchiaro G, *et al.* MR imaging and proton MR spectroscopy in adult Krabbe disease. *AJNR Am J Neuroradiol* 2000;21:1478–82.
137. Fearnley IM, Walker JE, Martinus RD, *et al.* The sequence of the major protein stored in ovine ceroid lipofuscinosis is identical with that of the dicyclohexylcarbodiimide-reactive proteolipid of mitochondrial ATP synthase. *Biochem J* 1990;268:751–8.
138. Feldhammer M, Durand S, Mr?zov? L, *et al.* Sanfilippo syndrome type C: Mutation spectrum in the heparan sulphate acetyl-CoA: α-glucosaminide N-acetyltransferase (*HGSNAT*) gene. *Hum Mutat* 2009;30:918–25.
139. Fernández-Chacón R, Wölfel M, Nishimune H, *et al.* The synaptic vesicle protein CSP alpha prevents presynaptic degeneration. *Neuron.* 2004;42:237–51.
140. Fernandes Filho JA, Shapiro BE. Tay–Sachs disease. *Arch Neurol* 2004;61:1466–8.
141. Ferrer I, Cusi V, Pineda M, *et al.* Focal dendritic swelling in Purkinje cells in mucopolysaccharidoses types I, II and III: a Golgi and ultrastructural study. *Neuropathol Appl Neurobiol* 1988;14:315–23.
142. Fischer A, Carmichael KP, Munnell JF, *et al.* Sulfamidase deficiency in a family of Dachshunds: a canine model of mucopolysaccharidosis IIIA (Sanfilippo A). *Pediatr Res* 1998;44:74–82.
143. Folkerth RD, Alroy J, Bhan I, Kaye EM. Infantile Gm1 gangliosidosis: complete morphology and histochemistry of two autopsy cases, with particular reference to delayed central nervous system myelination. *Pediatr Dev Pathol* 2000;13:73–86.
144. Folkerth RD, Alroy J, Lomakina I, *et al.* Mucolipidosis IV: morphology and histochemistry of an autopsy case. *J Neuropathol Exp Neurol* 1995;54:154–64.
145. Fox J, Li Y-T, Dawson G, *et al.* Naturally occurring GM2 gangliosidosis in two muntjak deer with pathological and biochemical features of human classical Tay–Sachs disease (type B GM2 gangliosidosis). *Acta Neuropathol* 1999;97:57–62.

146. Frei KP, Patronas NJ, Crutchfield KE, *et al.* Mucolipidosis type IV: characteristic MRI findings. *Neurology* 1998;51:565–9.
147. Fressinaud C, Vallat J, Mason M, *et al.* Adult-onset metachromatic leukodystrophy presenting as isolated peripheral neuropathy. *Neurology* 1992;42:1396–8.
148. Friedmann I, Spellacy E, Crow J, Watts RW. Histopathological studies of the temporal bones in Hurler's disease [mucopolysaccharidosis (MPS) IH]. *J Laryngol Otol* 1985;99:29–41.
149. Fritchie K, Thorne LB, Armao DM, *et al.* Novel mutation and the first prenatal screening of cathepsin D deficiency (CLN10). *Acta Neuropathol* 2009;117(2);201–8.
150. Fujita N, Suzuki K, Vanier MT, *et al.* Targeted disruption of the mouse sphingolipid activator protein gene: a complex phenotype, including severe leukodystrophy and widespread storage of multiple sphingolipids. *Hum Mol Genet* 1996;5:711–23.
151. Gabrielli O, Polonara G, Regnicolo L, *et al.* Correlation between cerebral MRI abnormalities and mental retardation in patients with mucopolysaccharidoses. *Am J Med Genet* 2004;125A:224–31.
152. Gahl WA, Krasnewich D, Williams JC. Sialidoses. In: Moser HW ed. *Handbook of clinical neurology*, Vol. 22. Amsterdam: Elsevier, 1996:359–75.
153. Gahl WA, Thoene JG, Schneider JA. Cystinosis: a disorder of lysosomal membrane transport. In: Valle, D., Beaudet AL, Vogelstein, B., *et al.* eds. *The online metabolic and molecular bases of inherited disease*, New York: McGraw-Hill, 2011. Part 21, Chapter 200.
154. Galluzzi P, Rufa A, Balestri P, *et al.* MR brain imaging of fucosidosis type I. *AJNR Am J Neuroradiol* 2001;22:777–80.
155. Galvan C, Camoletto PG, Cristofani F, *et al.* Anomalous surface distribution of glycosyl phosphatidyl inositol-anchored proteins in neurons lacking acid sphingomyelinase. *Mol Biol Cell* 2008;19:509–22.
156. Gambetti P, DiMauro S, Baker L. Nervous system in Pompe's disease: Ultrastructure and biochemistry. *J Neuropathol Exp Neurol* 1971;30:412–30.
157. Gao H, Boustany R-M, Espinola JA, *et al.* Mutations in a novel *CLN6*-encoded transmembrane protein cause variant neuronal ceroid lipofuscinosis in man and mouse. *Am J Hum Genet* 2002;70:324–35.
158. Garborg I, Torvik A, Hals J, *et al.* Congenital neuronal ceroid lipofuscinosis: a case report. *Acta Pathol Microbiol Immunol Scand* 1987;95:119–25.
159. Garcia CP, McGarry PA, Duncan CM. Fucosidosis and Alexander's leukodystrophy. *J Neuropathol Exp Neurol* 1980;39:353.
160. Gelfman CM, Vogel P, Issa TM, *et al.* Mice lacking alpha/beta subunits of GlcNAc-1-phosphotransferase exhibit growth retardation, retinal degeneration, and secretory cell lesions. *Invest Ophthalmol Vis Sci* 2007;48:5221–8.
161. Gieselmann V, Matzner U, Hess B, *et al.* Metachromatic leukodystrophy: molecular genetics and an animal model. *J Inherit Metab Dis* 1998;21:564–74.
162. Gieselmann V, Polten A, Kreysing J, *et al.* Molecular genetics of metachromatic

leukodystrophy. *Dev Neurosci* 1991;**13**:222–7.

163. Giger U, Shivaprasad H, Wang P, *et al.* Mucopolysaccharidosis type IIIB (Sanfilippo B syndrome) in emus *Veterinary Pathol* 1997;**34**:5.

164. Gilbert EF, Varakis J, Opitz JM, *et al.* Generalized gangliosidosis type II (juvenile GM1 gangliosidosis). *Eur J Pediatr* 1975;**120**:151–80.

165. Giri S, Jatana M, Rattan R, *et al.* Galactosylsphingosine (psychosine)-induced expression of cytokine-mediated inducible nitric oxide synthases via AP-1 and C/EBP: implications for Krabbe disease. *FASEB J* 2002;**16**:661–72.

166. Goebel HH. Neurodegenerative diseases: biopsy diagnosis in children. In: Galcia JH ed. *Neuropathology: the diagnostic approach*. St Louis, MO: Mosby, 1997:581–635.

167. Goebel HH, Wisniewski KE. Current state of clinical and morphological features in human NCL. *Brain Pathol* 2004;**14**:61–9.

168. Goker-Alpan O, Schiffmann R, LaMarca ME, *et al.* Parkinsonism among Gaucher disease carriers. *J Med Genet* 2004;**41**:937–40.

169. Goldin E, Slaugenhaupt SA, Smith JA, Schiffmann R. Mucolipidosis type IV. In: Valle D, Beaudet AL, Vogelstein B, *et al.* eds. *The online metabolic and molecular bases of inherited disease*, New York: McGraw-Hill, 2011; Part 16, Chapter 138.1.

170. Goldman JE, Katz D, Rapin I, *et al.* Chronic GM1 gangliosidosis presenting as dystonia: I. Clinical and pathological features. *Ann Neurol* 1981;**9**:465–75.

171. Goldman JE, Yamanaka T, Rapin I, *et al.* The AB variant of GM2-gangliosidosis. *Acta Neuropathol* 1980;**52**:189–202.

172. Gonatas K, Gonatas J. Ultrastructural and biochemical observations on a case of systemic late infantile lipidosis and its relationship to Tay–Sachs disease and gargoylism. *J Neuropathol Exp Neurol* 1965;**24**:318–40.

173. Gonatas NK, Terry RD, Winkler R, *et al.* A case of juvenile lipidosis: the significance of electron microscopic and biochemical observations of a cerebral biopsy. *J Neuropathol Exp Neurol* 1963;**22**:557–80.

174. Gondré-Lewis M, Dobrenis K, Walkley SU. Cholesterol accumulation in NPC1-deficient neurons is ganglioside dependent. *Curr Biol* 2003;**13**:1324–29.

175. Gonzalez-Gomez I, Mononen I, Heisterkamp N, *et al.* Progressive neurodegeneration in aspartylglucosaminuria mice. *Am J Pathol* 1998;**155**:1293–300.

176. Goodman LA, Livingston PO, Walkley SU. Ectopic dendrites occur only on cortical pyramidal neurons containing elevated GM2 ganglioside in α-mannosidosis. *Proc Natl Acad Sci U S A* 1991;**88**:11330–4.

177. Grabowski GA, Charnas L, Du H. Lysosomal acid lipase deficiency: The Wolman disease/cholesteryl ester storage disease spectrum. In: Valle D, Beaudet AL, Vogelstein B, *et al.* eds. *The online metabolic and molecular basis of inherited disease,* New York: McGraw-Hill, 2011; Part 16, Chapter 142.

178. Grabowski G, Petsko GA, Kolodny EH. Gaucher disease. In: Valle D, Beaudet AL, Vogelstein B, *et al.* eds. *The online metabolic and molecular bases of inherited*

disease, New York: McGraw-Hill, 2011; Part 16, Chapter 146.

179. Gravel RA, Kaback M, Proia, R.L., *et al.* The GM2 gangliosidoses. In: Valle, D., Beaudet AL, Vogelstein, B., *et al.* eds. *The online metabolic and molecular bases of inherited disease*, New York: McGraw-Hill, 2011. Part 16, Chapter 153.

180. Groener J, Maaswinkel-Mooy P, Smit V, *et al.* New mutations in two Dutch patients with early infantile galactosialidosis. *Mol Genet Metab* 2003;**78**:222–8.

181. Grosso S, Farnetani MA, Berardi R, *et al.* GM2 gangliosidosis variant B1: neuroradiological findings. *J Neurol* 2003;**250**:17–21.

182. Guazzi GC, Martin JJ, Philippart M, *et al.* Wolman's disease. *Eur Neurol* 1968;**1**:334–62.

183. Guerra WF, Verity MA, Fluharty AL, *et al.* Multiple sulfatase deficiency: clinical, neuropathological, ultrastructural and biochemical studies. *J Neuropathol Exp Neurol* 1990;**49**:406–23.

184. Gupta P, Soyombo AA, Atashband A, *et al.* Disruption of PPT-I or PPT-2 causes neuronal ceroid lipofuscinosis in knockout mice. *Proc Natl Acad Sci U S A* 2001;**98**:13566–71.

185. Gupta P, Soyombo AA, Shelton JM, *et al.* Disruption of PPT-2 in mice causes an unusual lysosomal storage disorder with neurovisceral features. *Proc Natl Acad Sci U S A* 2003;**100**:12325–30.

186. Gururaj A, Sztrihe L, Hertecant J, *et al.* Magnetic resonance imaging findings and novel mutations in GM1 gangliosidosis. *J Child Neurol* 2005;**20**:57–60.

187. Gutschalk A, Harting I, Cantz M, *et al.* Adult alpha-mannosidosis: clinical progression in the absence of demyelination. *Neurology* 2004;**63**:1744–6.

188. Hadfield MG, Mamunes P, David RB. The pathology of Sandhoff's disease. *J Pathol* 1977;**23**:137–44.

189. Hadjiconstantinou M, Neff NH. GM1 ganglioside: In vivo and in vitro actions on central neurotransmitter trophic system. *J Neurochem* 1998;**70**:1335–45.

190. Hahn AF, Gilbert JJ, Kwarciak C, *et al.* Nerve biopsy findings in Niemann–Pick type II (NPC). *Acta Neuropathol* 1994;**87**:149–54.

191. Hahn AF, Gordon BA, Feleki V, *et al.* A variant form of metachromatic leukodystrophy without arylsulfatase deficiency. *Ann Neurol* 1982;**12**:33–6.

192. Hahn AF, Gordon BA, Gilbert JJ, Hinton GG. The AB-variant of metachromatic leukodystrophy (postulated activator protein deficiency): light and electron microscopic findings in sural nerve biopsy. *Acta Neuropathol* 1981;**55**:281–7.

193. Hahn CN, del Pilar Martin M, Schroder M, *et al.* Generalized CNS disease and massive GM1 ganglioside accumulation in mice defective in lysosomal acid β-galactosidase. *Hum Mol Genet* 1997;**6**:205–11.

194. Haltia M, Palo J, Autio S. Aspartylglucosaminuria: a generalized storage disease. *Acta Neuropathol* 1975;**31**:243–55.

195. Haltia M, Rapola J, Santavuori P, Kernen A. Infantile type of so-called neuronal ceroid-lipofuscinosis. Part 2: morphological and biochemical studies. *J Neurol Sci* 1973;**18**:269–85.

196. Haltia M, Rapola J, Santavuori P. Infantile type of so-called neuronal

ceroidlipofuscinosis: histological and electron microscopical studies. *Acta Neuropathol* 1973;**26**:157–70.

197. Haltia M. The neuronal ceroid-lipofuscinoses. *J Neuropathol Exp Neurol* 2003;**62**:1–13.

198. Hannun YA, Bell RM. Functions of sphingolipid breakdown products in cellular regulation. *Science* 1989;**243**:500–507.

199. Hannun YA, Bell RM. Lysosphingolipids inhibit protein kinase C: implications for sphingolipidoses. *Science* 1987;**235**:670–74.

200. Harzer K, Paton BC, Poulos A, *et al.* Sphingolipid activator protein (SAP) deficiency in a 16 week-old atypical Gaucher disease patient and his fetal siblings: biochemical signs of combined sphingolipidoses. *Eur J Pediatr* 1989;**149**:31–9.

201. Harzer K, Rolfs A, Bauer P, *et al.* Niemann–Pick disease type A and B are clinically but also enzymatically heterogeneous: pitfall in the laboratory diagnosis of sphingomyelinase deficiency associated with the mutation Q292 K. *Neuropediatrics* 2003;**34**:301–6.

202. Hasilik A, Neufeld EF. Biosynthesis of lysosomal enzymes in fibroblasts. *J Biol Chem* 1980;**255**:4937–45.

203. Haskins ME, Aguirre GD, Jezyk PF, Patterson DF. The pathology of the feline model of mucopolysaccharidosis VI. *Am J Pathol* 1980;**101**:657–74.

204. Haskins ME, Giger U, Patterson DF. Animal models of lysosomal storage diseases: their development and clinical relevance. In: Mehta A, Beck M, Sunder-Plassmann G eds. *Fabry Disease: Perspectives from 5 years of FOS.* Oxford PharmaGenesis Ltd, 2006.

205. Haust MD, Gordon BA. Ultrastructural and biochemical aspects of the Sanfilippo syndrome: type III genetic mucopolysaccharidosis. *Connect Tissue Res* 1986;**15**:57–64.

206. Henderson RD, MacMillan JC, Bradfield JM. Adult onset Krabbe disease may mimic motor neuron disease. *J Clin Neurosci* 2003;**10**:638–9.

207. Hendriksz CJ, Corry PC, Wraith JE, *et al.* Juvenile Sandhoff disease: nine new cases and a review of the literature. *J Inherit Metab Dis* 2004;**27**:241–9.

208. Hers HG, van Hoof F eds. *Lysosomes and storage diseases.* New York and London: Academic Press, 1973.

209. Hers HG. Inborn lysosomal diseases. *Gastroenterology* 1965;**48**:625–33.

210. Herva R, Tyynelä J, Hirvasniemi A, *et al.* Northern epilepsy: a novel form of neuronal ceroid-lipofuscinosis. *Brain Pathol* 2000;**10**:215–22.

211. Higashi Y, Murayama S, Pentchev PG, Suzuki K. Cerebellar degeneration in the Niemann–Pick type C mouse. *Acta Neuropathol* 1993;**85**:175–84.

212. Higashi Y, Murayama S, Pentchev PG, Suzuki K. Peripheral nerve pathology in Niemann–Pick type C mouse. *Acta Neuropathol* 1995;**90**:158–63.

213. Higuchi I, Nonaka I, Usuki F, *et al.* Acid maltase deficiency in the Japanese quail: early morphological event in skeletal muscle. *Acta Neuropathol* 1987;**73**:32–7.

214. Hirschhorn R and Reuser AJJ. Glycogen storage disease type II: acid α-glucosidase (acid maltase) deficiency. In: Valle, D., Beaudet AL, Vogelstein, B., *et al.* eds. *The online metabolic and molecular bases of*

inherited disease, New York: McGraw-Hill, 2011. Part 16, Chapter 135.

215. Hoffmann S, Peltonen L. The neuronal ceroid lipofuscinosis. In Valle, D., Beaudet AL, Vogelstein, B., *et al.* eds. *The online metabolic and molecular bases of inherited disease*, New York: McGraw-Hill, 2011. Part 16, Chapter 154.

216. Hopwood JJ, Ballabio A. Multiple sulfatase deficiency and the nature of the sulfatase family. In: Valle, D., Beaudet AL, Vogelstein, B., *et al.* eds. *The online metabolic and molecular bases of inherited disease*, New York: McGraw-Hill, 2011. Part 16, Chapter 149.

217. Horinouchi K, Erlich S, Perl DP, *et al.* Acid sphingomyelinase deficient mice: a model of types A and B Niemann–Pick disease. *Nat Genet* 1995;**10**:288–93.

218. Horvath J, Ketelsen UP, Geibel-Zehender A, *et al.* Identification of a novel LAMP2 mutation responsible for X-chromosomal dominant Denon disease. *Neuropediatrics* 2003;**34**:270–73.

219. Huang JQ, Trasler JM, Igdoura S, *et al.* Apoptotic cell death in mouse models of GM2 gangliosidosis and observations on human Tay–Sachs and Sandhoff diseases. *Hum Mol Genet* 1997;**6**:1879–85.

220. Huang J-Y, Peng S-F, Yang C-C, *et al.* Neuroimaging findings in a brain with Niemann–Pick type C disease. *J Formos Med Assoc* 2011;**110**:537–42.

221. Hulkova H, Cervenkova M, Ledvinova J, *et al.* A novel mutation in the coding region of the prosaposin gene leads to a complete deficiency of prosaposin and saposins, and is associated with a complex sphingolipidoses dominated by lactosylceramide accumulation. *Hum Mol Genet* 2001;**10**:927–40.

222. Ida H, Rennert OM, Kato S, *et al.* Severe skeletal complications in Japanese patients with type 1 Gaucher disease. *J Inherit Metab Dis* 1999;**22**:63–73.

223. Ikeda S, Kondo K, Oguchi K, *et al.* Adult fucosidosis: histochemical and ultrastructural studies of rectal mucosa biopsy. *Neurology* 1984;**34**:451–6.

224. Imundo L, Leduc CA, Guha S *et al.* A complete deficiency of hyaluronoglucosaminidase 1 (HYAL1) presenting as familial juvenile idiopathic arthritis. *J Inhert Metab Dis* 2011;**34**:1013–22.

225. Inui K, Akagi M, Nishigaki T, *et al.* A case of chronic infantile type of fucosidosis: clinical and magnetic resonance image findings. *Brain Dev* 2000;**22**:47–9.

226. Jaeken J, Gieselmann V, von Figura K. Metachromatic leukodystrophy. In: Valle D, Beaudet AL, Vogelstein B, *et al.* eds. *The online metabolic and molecular bases of inherited disease*, New York: McGraw-Hill, 2011; Part 16, Chapter 148.

227. Jalanko A, Braulke T. Neuronal ceroid lipofuscinoses. *Biochim Biophys Acta* 2009;**1793**:697–709.

228. Jardim L, Vedolin L, Schwartz IV, *et al.* CNS involvement in Fabry disease: clinical and imaging studies before and after 12 months of enzyme replacement therapy. *J Inherit Metab Dis* 2004;**27**:229–40.

229. Jellinger K, Anzil AP, Seemann D, Bernheimer H. Adult GM2 gangliosidosis masquerading as slowly progressive muscular atrophy: motor neuron disease phenotype. *Clin Neuropathol* 1982;**1**: 31–44.

230. Jellinger K, Paulus W, Grisold W, Paschke E. New phenotype of adult alpha-l-iduronidase deficiency (mucopolysaccharidosis I) masquerading as Friedreich's ataxia with cardiopathy. *Clin Neuropathol* 1990;**9**:163–9.

231. Jeyakumar M, Butters TD, Dwek RA, Platt FM. Glycosphingolipid lysosomal storage diseases: therapy and pathogenesis. *Neuropathol Appl Neurobiol* 2002;**28**:343–57.

232. Jeyakumar M, Smith D, Eliott-Smith E, *et al.* An inducible mouse model of late onset Tay–Sachs disease. *Neurobiol Dis* 2002;**10**:201–10.

233. Jeyakumar M, Thomas R, Elliot-Smith ZE, *et al.* Central nervous system inflammation is a hallmark of pathogenesis in mouse models of GM1 and GM2 gangliosiudosis. *Brain* 2003;**126**:974–87.

234. Jolly RD, Johnston AC, Norman EJ, *et al.* Pathology of mucopolysaccharidosis IIIA in Huntaway dogs. *Vet Pathol* 2007;**44**:569–78.

235. Jones MZ, Alroy J, Boyer PJ, *et al.* Caprine mucopolysaccharidosis-IIID: clinical, biochemical, morphological and immunohistochemical characteristics. *J Neuropathol Exp Neurol* 1998;**57**:148–57.

236. Jones MZ, Alroy J, Ruledge JC, *et al.* Human mucopolysaccharidosis III D; clinical, biochemical, morphological and immunohistochemical characteristics. *J Neuropathol Exper Neurol* 1997;**56**:1158–67.

237. Jones MZ, Cunningham JG, Dade AW. Caprine β-mannosidosis: clinical and pathological features. *J Neuropathol Exp Neurol* 1983;**42**:268–85.

238. Kamoshita S, Aaron AM, Suzuki K, Suzuki K. Infantile Niemann–Pick disease: a chemical study with isolation and characterization of membranous cytoplasmic bodies and myelin. *Am J Dis Child* 1969;**117**:379–94.

239. Kampmann C, Wiethoff CM, Perrot A, *et al.* The heart in Anderson Fabry disease. *Z Kardiol* 2002;**91**:786–95.

240. Kang C, Riazuddin S, Mundorff J, *et al.* Mutations in the lysosomal enzyme–targeting pathway and persistent stuttering. *N Engl J Med* 2010;**362**:677–85.

241. Kanzaki T, Wang AM, Desnick RJ. Lysosomal α-N-acetylgalactosaminidase deficiency, the enzymatic defect in angiokeratoma corporis diffusum with glycopeptiduria. *J Clin Invest* 1991;**88**:707–11.

242. Kanzaki T, Yokota M, Mizuno N, *et al.* Novel lysosomal glycoaminoacid storage disease with angiokeratoma corporis diffusum. *Lancet* 1989;**1**:875–77.

243. Kappler J, Watts RWE, Conzelmann E, *et al.* Low arylsulfatase A activity and choreoathetotic syndrome in three siblings: differentiation of pseudodeficiency from metachromatic leukodystrophy. *Eur J Pediatr* 1991;**150**:287–90.

244. Karageorgos L, Lancaster MJ, Nimmo JS, Hopwood JJ. Gaucher disease in sheep. *J Inherit Metab Dis* 2011;**34**:209–15.

245. Karten B, Vance DE, Campenot RB, Vance JE. Cholesterol accumulates in cell bodies, but is decreased in axons, of Niemann–Pick C1-deficient neurons. *J Neurochem* 2002;**83**:1154–63.

246. Katz ML, Shibuya H, Lin P-C, *et al.* A mouse gene knockout model for juvenile ceroid-lipofuscinosis (Batten disease). *J Neurosci Res* 1999;**57**:551–6.

247. Kaye EM, Ullman MD, Wilson ER, Barranger JA. Type 2 and type 3 Gaucher disease: a morphological and biochemical study. *Ann Neurol* 1986;**20**:223–30.

248. Keller C, Briner J, Schneider J, *et al.* Mukopolysaccharidose type VI-A (Morbus Maroteaux–Lamy): Korrelation der klinischen und pathologisch-anatomischen Befunde bei einem 27 jahringen Patienten. *Helv Paediatr Acta* 1987;**42**:317–33.

249. Kelo E, Dunder U, Mononen I. Massive accumulation of Man2GlcNAc2-Asn in nonneuronal tissues of glycosylasparaginase-deficient mice and its removal by enzyme replacement therapy. *Glycobiology* 2005;**15**:79–85.

250. Kepes JJ, Berry A 3rd, Zacharias DL. Multiple sulfatase deficiency: bridge between neuronal storage diseases and leukodystrophies. *Pathology* 1988;**20**:285–91.

251. Kim TS, Kim I-O, Kim WS, *et al.* MR of childhood metachromatic leukodystrophy. *AJNR Am J Neuroradiol* 1997;**18**:733–8.

252. Kint JA, Dacremont G, Carton D, *et al.* Mucopolysaccharidosis: secondarily induced abnormal distribution of lysosomal isoenzymes. *Science* 1973;**181**:352–4.

253. Kjellman B, Gamstrop I, Brun A, *et al.* Mannosidosis: a clinical and histopathologic study. *J Pediatr* 1969;**75**:366–73.

254. Klünemann HH, Elleder M, Kaminski WE, *et al.* Frontal lobe atrophy due to a mutation in the cholesterol binding protein HE1/NPC2. *Ann Neurol* 2002;**52**:743–9.

255. Kobayashi T, Goto I, Okada S, *et al.* Accumulation of lysosphingolipids in tissues from patients with GM1 and GM2 gangliosidoses. *J Neurochem* 1992;**59**:1452–58.

256. Kobayashi T, Suzuki K. Chronic GM1 gangliosidosis presenting as dystonia: II. Biochemical studies. *Ann Neurol* 1981;**9**:476–83.

257. Koga M, Ishihara T, Hoshii Y, *et al.* Histochemical and ultrastructural studies of inclusion bodies found in tissues from three siblings with I-cell disease. *Pathol Int* 1994;**44**:223–9.

258. Koike M, Nakanishi H, Saftig P, *et al.* Cathepsin D deficiency induces lysosomal storage with ceroid lipofuscin in mouse CNS neurons. *J Neurosci* 2000;**15**:6898–906.

259. Koike M, Shibata M, Waguri S, *et al.* Participation of autophagy in storage of lysosomes in neurons from mouse models of neuronal ceroid-lipofuscinoses (Batten disease). *Am J Pathol* 2005;**167**: 1713–28.

260. Kolodny EH. Globoid cell leukodystrophy. In: Moser HW ed. *Handbook of clinical neurology: neurodystrophies and neurolipidosis*. Amsterdam: Elsevier Science, 1996:187–210.

261. Kolodny EH, Fluharty AL. Metachromatic leukodystrophy and multiple sulfatase deficiency: sulfatide lipidosis. In: Scriver CR, Beaudet AL, Sly WS, *et al.* eds. *The metabolic and molecular basis of inherited disease*, 7th edn. New York: McGraw-Hill, 1995:2693–739.

262. Kolodny EH, Pastores GM. Anderson Fabry disease: extra renal, neurologic manifestations. *J Am Soc Nephrol* 2002;**13**: S150–53.

263. Kolodny EH, Raghavan S, Krivit W. Late onset Krabbe disease (globoid cell leukodystrophy): clinical and biochemical

features of 15 cases. *Dev Neurosci* 1991;**13**:232–9.

264. Kolodny EH, Ulman MD, Mankin HJ, *et al*. Phenotypic manifestations of Gaucher disease: clinical features in 48 biochemically verified type I patients and comment on type II patients. In: Desnick RJ, Gatt S, Grabowski GA eds. *Gaucher disease: a century of delineation and research*. New York: Alan R. Liss, 1982:33–65.

265. Kornfeld M. Neuropathology of chronic GM2 gangliosidosis due to hexosaminidase deficiency. *Clin Neuropathol* 2008;**27**(5):302–8.

266. Kornfeld M, Snider RD, Wenger DA. Fucosidosis with angiokeratoma: electron microscopic changes in the skin. *Arch Pathol Lab Med* 1977;**101**:478–85.

267. Kornfeld S, Sly WS. I-cell disease and pseudo-Hurler polydystrophy: disorders of lysosomal enzyme phosphorylation and localization. In: Valle, D., Beaudet AL, Vogelstein, B., *et al*. eds. *The online metabolic and molecular bases of inherited disease*, New York: McGraw-Hill, 2011. Part 16, Chapter 138.

268. Kosanke SD, Pierce KR, Read WK. Morphogenesis of light and electron microscopic lesions in porcine GM2-gangliosidosis. *Vet Pathol* 1979;**16**:6–17.

269. Koto A, Horwitz AL, Suzuki K, *et al*. The Morquio syndrome: neuropathology and biochemistry. *Ann Neurol* 1978;**4**:26–36.

270. Kovesi TA, Lee J, Shuckett B, *et al*. Pulmonary infiltration in Niemann–Pick disease type C. *J Inherit Metab Dis* 1996;**19**:792–3.

271. Kowalewski B, Lamanna WC, Lawrence R, *et al*., Arylsulfatase G inactivation causes loss of heparan sulfate 3-O-sulfatase activity and mucopolysaccharidosis in mice. *Proc Natl Acad Sci U S A* 2012;**109**:10310–5.

272. Kretzschmar HA, Wagner H, Hubner G, *et al*. Aneurysms and vacuolar degeneration of cerebral arteries in late-onset acid maltase deficiency. *J Neurol Sci* 1990;**98**:169–83.

273. Kriscenski-Perry E, Applegate CD, Serour A, *et al*. Altered flurothyl seizure induction latency, phenotype and subsequent mortality in a mouse model of juvenile neuronal ceroid lipofuscinosis/Batten disease. *Epilepsia* 2002;**43**:1137–40.

274. Kuemmel TA, Schroeder R, Stoffel W. Light and electron microscopic analysis of the central and peripheral nervous system of acid sphingomyelinase-deficient mice resulting from gene targeting. *J Neuropathol Exp Neurol* 1997;**56**:171–9.

275. Kuemmel TA, Thiele J, Schroeder R, Stoffel W. Pathology of visceral organs and bone marrow in an acid sphingomyelinase deficient knockout mouse line, mimicking human Niemann–Pick disease type A: a light and electron microscopic study. *Pathol Res Pract* 1997;**193**:663–71.

276. Kumperscak HG, Paschke E, Gradisnik P, *et al*. Adult metachromatic leukodystrophy: disorganized schizophrenia-like symptoms and postpartum depression in 2 sisters. *J Psychiatry Neurosci* 2005;**30**:33–6.

277. Kunieda T, Simonaro CM, Yoshida M, *et al*. Mucopolysaccharidosis type VI in rats: isolation of cDNAs encoding arylsulfatase B, chromosomal localization of the gene, and identification of the mutation. *Genomics* 1995;**29**:582–7.

278. Kurihara M, Kumagai K, Goto K, *et al*. Severe type Hunter's syndrome: polysomnographic and neuropathological study. *Neuropediatrics* 1992;**23**:248–56.

279. Kurihara M, Kumagai K, Yagishita S. Sanfilippo syndrome type C: a clinicopathological autopsy study of a long-term survivor. *Pediatr Neurol* 1996;**14**:317–21.

280. Kuriwaki K, Yoshida H. Morphological characteristics of lipid accumulation in liver-constituting cells of acid lipase deficiency rats (Wolman's disease model rats). *Pathol Int* 1999;**49**:291–7.

281. Labauge P, Renard D, Castelnovo G, *et al*. β-Mannosidosis: a new cause of spinocerebellar ataxia. *Clin Neurol Neurosurg* 2009;**111**:109–10.

282. Lake BD, Young EP, Nicolaides K. Prenatal diagnosis of infantile sialic acid storage disease in a twin pregnancy. *J Inherit Metab Dis* 1989;**12**:152–6.

283. Lake BD, Young EP, Winchester BG. Prenatal diagnosis of lysosomal storage diseases. *Brain Pathol* 1998;**8**:132–49.

284. Landing BH, Silverman FN, Craig JM, *et al*. Familial neurovisceral lipidosis. *Am J Dis Child* 1964;**108**:503–22.

285. Landrieu P, Said G. Peripheral neuropathy in type A Niemann–Pick disease: a morphological study. *Acta Neuropathol* 1984;**63**:66–71.

286. Lane SC, Jolly RD, Schmechel DE, *et al*. Apoptosis as the mechanism of neurodegeneration in Batten's disease. *J Neurochem* 1996;**67**:677–83.

287. Lao LM, Kumakiri M, Mima H, *et al*. The ultrastructural characteristics of eccrine sweat glands in a Fabry disease patient with hypohidrosis. *J Dermatol Sci* 1998;**18**:109–17.

288. Larbrisseau A, Brochu P, Jasmine G. Fucosidose de type I: étude anatomique. *Arch Fr Pediatr* 1982;**39**:1013–25.

289. Lee WS, Kang C, Drayna D, Kornfeld S. Analysis of mannose 6-phosphate uncovering enzyme mutations associated with persistent stuttering. *J Biol Chem.* 2011;**286**:39786–93.

290. Leinekugel P, Michel S, Conzelmann E, Sandhoff K. Quantitative correlation between the residual activity of β-hexosaminidase A and arylsulfatase A and the severity of the resulting lysosomal storage disease. *Hum Genet* 1992;**88**:513–23.

291. Lemyre E, Russo P, Melancon SB, *et al*. Clinical spectrum of infantile free sialic acid storage disease. *Am J Med Genet* 1999;**82**:385–91.

292. Levade T, Moser HW, Fensom AH, *et al*. Neurodegenerative course in ceramidase deficiency (Farber disease) correlates with the residual lysosomal ceramide turnover in cultured living patient cells. *J Neurol Sci* 1995;**134**:108–14.

292a. Lerner T, Boustany R, Anderson J, *et al*. Isolation of a novel gene underlying Batten disease, *CLN3. Cell* 1995;82: 949-57.

293. Levade T, Sandhoff K, Schulze H, Medin JA. Acid ceramidase deficiency: Farber lipogranulomatosis. In: Valle D, Beaudet AL, Vogelstein B, *et al*. eds. *The online metabolic and molecular bases of inherited disease*, New York: McGraw-Hill, 2011; Part 16, Chapter 143.

294. Li CM, Park JH, Simonaro CM, *et al*. Insertional mutagenesis of the mouse acid ceramidase gene leads to early embryonic lethality in homozygotes and progressive lipid storage disease in heterozygotes. *Genomics* 2002;**79**:218–24. Erratum in: *Genomics* 2002;**79**:890.

295. Li H-H, Yu W-H, Rozengurt N, *et al*. Mouse model of Sanfilippo syndrome type B produced by targeted disruption of the gene encoding alpha-N-acetylglucosaminidase. *Proc Natl Acad Sci U S A* 1999;**96**:14505–10.

296. Lieberman AP, Puertollano R, Raben N, *et al*. Autophagy in lysosomal storage disorders. *Autophagy* 2012;**8**:719–30.

297. Liscum L, Sturley SL eds. Niemann–Pick type C disease. *Biochim Biophys Acta* 2004;**1685**:1–90.

298. Littjens T, Baker EG, Beckmann KR, *et al*. Chromosomal localization of *ARSB*, the gene for human N-acetylgalactosamine–4–sulphatase. *Hum Genet* 1989;**82**:67–8.

299. Liu Y, Hoffmann A, Grinberg A, *et al*. Mouse model of GM2 activator deficiency manifests cerebellar pathology and motor impairment. *Proc Natl Acad Sci U S A* 1997;**94**:8138–43.

300. Liu Y, Suzuki K, Reed JD, *et al*. Mice with type 2 and 3 Gaucher disease point mutations generated by a single insertion mutagenesis procedure (SIMP). *Proc Natl Acad Sci U S A* 1998;**95**:2503–8.

301. Liu Y, Wu YP, Wada R, *et al*., Alleviation of neuronal ganglioside storage does not improve the clinical course of the Niemann–Pick C disease mouse. *Hum Mol Genet* 2000;**9**:1087–92.

302. Lloyd JB, Mason RW. *Biology of the lysosome: subcellular biochemistry*, Vol. 27. New York: Plenum Press, 1996.

303. Loes DJ, Peters C, Krivit W. Globoid cell leukodystrophy: distinguishing early-onset from late-onset disease using a brain MR imaging score method. *AJNR Am J Neuroradiol* 1999;**20**:316–23.

304. Loftus S, Erickson RP, Walkley SU, *et al*. Rescue of neurodegeneration in Niemann–Pick C mice by a prion-promoter-driven *Npc1* cDNA transgene. *Hum Mol Genet* 2002;**11**:3107–14.

305. Lonergan CL, Payne AR, Wilson WG, *et al*. What syndrome is this? *Pediatr Dermatol* 2004;**21**:679–81.

306. Lovell KL, Jones MZ. Axonal and myelin lesions in β-mannosidosis: ultrastructural characteristics. *Acta Neuropathol* 1985;**65**:293–9.

307. Lovell KL, Jones MZ. Distribution of central nervous system lesions in beta-mannosidosis. *Acta Neuropathol* 1983;**62**:121–4.

308. Luzi P, Rafi MA, Zaka M, *et al*. Generation of a mouse with low galactocerebrosidase activity by gene targeting: a new model of globoid cell leukodystrophy (Krabbe disease). *Mol Genet Metab* 2001;**73**:211–23.

309. Lyon G, Hagberg B, Evrard PH, *et al*. Symptomatology of late onset Krabbe's leukodystrophy: the European experience. *Dev Neurosci* 1991;**13**:240–44.

310. Macaulay RJB, Lowry NJ, Casey RE. Pathologic findings of multiple sulfatase deficiency reflect the pattern of enzyme deficiencies. *Pediatr Neurol* 1998;**19**:372–6.

311. Maia M, Alves D, Ribeiro G, *et al*. Juvenile GM2 gangliosidosis variant B: clinical and biochemical study in seven patients. *Neuropediatrics* 1990;**21**:18–23.

312. Malm D, Halvorsen DS, Tranebjaerg L, Sjursen H. Immunodeficiency in alpha-mannosidosis: a matched case–control study on immunoglobulins, complement

factors, receptor density, phagocytosis and intracellular killing in leucocytes. *Eur J Pediatr* 2000;**159**:699–703.

313. Martin DR, Cox NR, Morrison NE, *et al.* Mutation of the GM2 activator protein in a feline model of GM2 gangliosidosis. *Acta Neuropathol* 2005;**110**:443–50.

314. Martin JJ, Ceuterick C. Adult neuronal ceroid-lipofuscinosis: personal observations. *Acta Neurol Belg* 1997;**97**:85–92.

315. Martin JJ, de Barsy TH, de Schrijver F, *et al.* Acid maltase deficiency (type II glycogenosis). *J Neurol Sci* 1976;**30**: 155–66.

316. Martin JJ, Leroy JG, Vaneygen M. Ceuterick C. I-cell disease: a further report on its pathology. *Acta Neuropathol* 1984;**64**:234–42.

317. Matheus MG, Castillo M, Smith JK, *et al.* Brain MRI findings in patients with mucopolysaccharidosis types I and II and mild clinical presentation. *Neuroradiology* 2004;**46**:666–72.

318. Matsuda J, Kido M, Tadano-Aritomi K, *et al.* Mutation in saposin D domain of sphingolipid activator protein gene causes urinary system defects and cerebellar Purkinje cell degeneration with accumulation of hydroxy fatty acid containing ceramide in mouse. *Hum Mol Genet* 2004;**13**:2709–23.

319. Matsuda J, Suzuki O, Oshima A, *et al.* β-Galactosidase-deficient mouse as an animal model for GM1-gangliosidosis. *Glycoconj J* 1997;**14**:729–36.

320. Matsuda J, Vanier MT, Saito Y, *et al.* A mutation in the saposin A domain of the sphingolipid activator protein (prosaposin) gene results in late-onset, chronic form of globoid cell leukodystrophy in the mouse. *Human Mol Genet* 2001;**10**:1191–9.

321. Matsumoto R, Oka N, Nagashima Y, *et al.* Peripheral neuropathy in late-onset Krabbe's disease: histochemical and ultrastructural findings. *Acta Neuropathol* 1996;**92**:635–9.

322. Matsuzawa F, Aikawa S, Sakuraba H, *et al.* Structural basis of the GM2 gangliosidosis B variant. *J Hum Genet* 2003;**48**:582–9.

323. Mattocks M, Bagovich M, De Rosa M, *et al.* Treatment of neutral glycosphingolipid lysosomal storage diseases via inhibition of the ABC drug transporter, MDR1: cyclosporin A can lower serum and liver globotriaosyl ceramide levels in the Fabry mouse model. *FEBS J* 2006;**273**:2064–75.

324. Maue RA, Burgess RW, Wang B, *et al.* A novel mouse model of Niemann–Pick type C disease carrying a D1005G-Npc1 mutation comparable to commonly observed human mutations. *Hum Mol Genet* 2012;**21**:730–50.

325. Mazrier H, van Hoeven M, Wang P, *et al.* Inheritance, biochemical abnormalities, and clinical features of feline, mucolipidosis II: the first animal model of human I-cell disease. *J Hered* 2003;**94**:363–73.

326. McGlynn R, Dobrenis K, Walkley SU. Differential subcellular localization of cholesterol, gangliosides, and glycosaminoglycans in murine models of mucopolysaccharide storage disorders. *J Comp Neurol* 2004;**480**:415–26.

327. McGovern MM, Wasserstein MP, Aron A, *et al.* Ocular manifestations of Niemann–Pick disease type B. *Ophthalmology* 2004;**111**:1424–7.

328. McKelvie P, Vine P, Hopkins I, Poulos A. A case of Krabbe's leukodystrophy without globoid cells. *Pathology* 1990;**22**:235–8.

329. Mehl E, Jatzkewitz H. Eine cerebrosidsulfatase aus Schweineniere. *Hoppe Seylers Z Physiol Chem* 1964;**339**:260–75.

330. Meikle PS, Hopwood JJ, Clague AE, Carey WF. Prevalence of lysosomal storage disorders. *JAMA* 1999;**281**:249–54.

331. Micsenyi MC, Dobrenis K, Stephney G, *et al.* Neuropathology of the Mcoln1$^{-/-}$ knockout mouse model of mucolipidosis Type IV. *J Neuropathol Exp Neurol* 2009;**68**:125–35.

332. Miklyaeva EI, Dong W, Bureau A, Fattahie R, *et al.* Late onset Tay–Sachs disease in mice with targeted disruption of the Hexa gene: behavioral changes and pathology of the central nervous system. *Brain Res* 2004;**1001**:37–50.

333. Millat G, Chikh K, Naureckiene S, *et al.* Niemann–Pick disease type C: spectrum of HE1 mutations and genotype/phenotype correlations in the NPC2 group. *Am J Hum Genet* 2001;**69**:1013–21.

334. Miranda SR, Erlich S, Friedrich VL, *et al.* Biochemical, pathological, and clinical response to transplantation of normal bone marrow cells into acid sphingomyelinase-deficient mice. *Transplantation* 1998;**65**:884–92.

335. Mitchison HM, Bernard DJ, Green ND, *et al.* Targeted disruption of the Cln3 gene provides a mouse model for Batten disease. *Neurobiol Dis* 1999;**6**:321–34.

334a. Micsenyi M, Sikora J, Stephney G, *et al.* Lysosomal membrane permeability stimulates protein aggregate formation in neurons of a lysosomal disease. *J Neurosci* 2013;33(26):10815-27.

336. Mitchison HM, Lim MJ, Cooper JD. Selectivity and types of cell death in the neuronal ceroid lipofuscinoses (NCLs). *Brain Pathol* 2004;**14**:86–96.

337. Miyagi T, Yamaguchi K. Mammalian sialidases: physiological and pathological roles in cellular functions. *Glycobiology* 2012;**22**:880–96.

338. Mochizuki A, Motoyoshi Y, Takeuchi M, *et al.* A case of adult type galactosialidosis with involvement of peripheral nerves. *J Neurol* 2000;**247**:708–10.

339. Molander-Melin M, Pernber Z, Franken S, *et al.* Accumulation of sulfatide in neuronal and glial cells of arylsulfatase A deficient mice. *J Neurocytol* 2004;**33**:417–27.

340. Mole S, Williams RL, Goebel HH, eds. *The neuronal ceroid lipofuscinoses (Batten disease)*, Second edn. Oxford: Oxford University Press, 2011.

341. Mole S. The genetic spectrum of human neuronal ceroid-lipofuscinoses. *Brain Pathol* 2004;**14**:70–76.

342. Molyneux AJ, Blair E, Coleman N, Daish P. Mucopolysaccharidosis type VII associated with hydrops fetalis: histopathological and ultrastructural features with genetic implications. *J Clin Pathol* 1997;**50**:252–4.

343. Mononen T, Mononen I. Biochemistry and biochemical diagrams of aspartylglycosaminuria. In: Mononen I, Aronson NN Jr eds. *Lysosomal storage disease: aspartylglycosaminuria*. Austin, TX: R.G. Landes, 1997:41–9.

344. Moore DF, Ye F, Schiffmann R, Butman JA. Increased signal intensity in the pulvinar on T1-weighted images: a pathognomonic MR imaging sign of Fabry disease. *AJNR Am J Neuroradiol* 2003;**24**:1096–101.

345. Moser HW, Linke T, Fenson AH, *et al.* Acid ceramidase deficiency: Farber lipogranulomatosis. In: Scriver CR, Beaudet AL, Sly WS, *et al.* eds. *The metabolic and molecular basis of inherited disease*, 8th edn. New York: McGraw-Hill, 2001:3573–88.

346. Muenzer J, Lamsa JC, Garcia A, *et al.* Enzyme replacement therapy in mucopolysaccharidosis type II (Hunter syndrome): a preliminary report. *Acta Paediatr Suppl* 2002;**91**:98–9.

347. Muller G, Alldinger S, Moritz A, *et al.* GM1 gangliosidosis in Alaskan huskies: clinical and pathologic findings. *Vet Pathol* 2001;**38**:281–90.

348. Muthane U, Chickabasaviah Y, Kaneski C, *et al.* Clinical features of adult GM1 gangliosidosis: report of three Indian patients and review of 40 cases. *Mov Disord* 2004;**19**:1334–41.

349. Myerowitz R, Lawson D, Mizukami H, *et al.* Molecular pathophysiology in Tay–Sachs and Sandhoff diseases as revealed by gene expression profiling. *Hum Mol Genet* 2002;**11**:1343–51.

350. Myllykangas L, Tyynelä J, Page-McCaw A, *et al.* Cathepsin D-deficient *Drosophila* recapitulate the key features of neuronal ceroid lipofuscinosis. *Neurobiol Dis* 2005;**19**:194–9.

351. Nagao Y, Nakashima H, Fukuhara Y, *et al.* Hypertrophic cardiomyopathy in late-onset variant of Fabry disease with high residual activity of α-galactosidase A. *Clin Genet* 1991;**39**:233–7.

352. Nagashima K, Sakakibara K, Endo H, *et al.* I-cell disease (mucolipidosis II): pathological and biochemical studies of an autopsy case. *Acta Pathol Jpn* 1977;**27**:251–64.

353. Nakao S, Takenaka T, Maeda M, *et al.* An atypical variant of Fabry's disease in men with left ventricular hypertrophy. *N Engl J Med* 1995;**333**:288–93.

354. Natowicz MR, Short MP, Wang Y, *et al.* Clinical and biochemical manifestations of hyaluronidase deficiency. *N Engl J Med* 1996;**335**:1029–33.

355. Naureckiene S, Sleat DE, Lackland H, *et al.* Identification of HE1 as the second gene of Niemann–Pick C disease. *Science* 2000;**290**:2298–301.

356. Neudorfer O, Pastores GM, Zeng BJ, *et al.* Late-onset Tay–Sachs disease: phenotypic characterization and genotypic correlations in 21 affected patients. *Genet Med* 2005;**7**:119–23.

357. Neufeld EF, Muenzer J. The mucopolysaccharidoses. In: Valle, D., Beaudet AL, Vogelstein, B., *et al.* eds. *The online metabolic and molecular bases of inherited disease*, New York: McGraw-Hill, 2011. Part 16, Chapter 136.

358. Nishino I, Fu J, Tanji K, *et al.* Primary LAMP-2 deficiency causes X-linked vacuolar cardiomyopathy and myopathy (Danon disease). *Nature* 2000;**406**:906–10.

359. Nordborg C, Kyllerman M, Conradi N, Månsson J-E. Early-infantile galactosialidosis with multiple brain infarctions: morphological, neuropathological and neurochemical findings. *Acta Neuropathol* 1997;**93**: 24–33.

360. Norman RM, Wood NA. A congenital form of amaurotic family idiocy. *J Neurol Neurosurg Psychiatry* 1941;**4**:175–90.

361. Norman RM, Urich H, Tingey AM, Goodbody RA. Tay–Sachs' disease with visceral involvement and its relationship to Niemann–Pick disease. *J Pathol Bacteriol* 1959;**78**:409–21.

362. Northover H, Cowie RA, Wrait JE. Mucopolysaccharidosis type IVA: a clinical review. *J Inherit Metab Dis* 1996;**19**:357–65.

363. Norton WR, Poduslo SE. Biochemical studies of metachromatic leukodystrophy in three siblings. *Acta Neuropathol* 1982;**57**:188–96.

364. Nosková L, Stránecký V, Hartmannová H,*et al.*, Mutations in DNAJC5, encoding cysteine-string protein alpha, cause autosomal-dominant adult-onset neuronal ceroid lipofuscinosis. *Am J Hum Genet* 2011;**89**:241–52.

365. Oak S, Rao S, Karmarkar S, *et al.* Papillomatosis of the gallbladder in metachromatic leukodystrophy. *Pathol Int* 1997;**12**:424–5.

366. Obenberger J, Seidl Z, Pavlù H, Elleder M. MRI in an unusually protracted neuronopathic variant of acid sphingomyelinase deficiency. *Neuroradiology* 1999;**41**:182–4.

367. Ohmi K, Greenberg DS, Rajavel KS, *et al.* Activated microglia in cortex of mouse models of mucopolysaccharidoses I and IIIB. *Proc Natl Acad Sci U S A* 2003;**100**:1902–07.

368. Ohshima T, Murray GJ, Swaim WD, *et al.* α-Galactosidase A deficient mice: a model of Fabry disease. *Proc Natl Acad Sci U S A* 1997;**94**:2540–44.

369. Ohshima T, Schiffmann R, Murray GJ, *et al.* Aging accentuates and bone marrow transplantation ameliorates metabolic defects in Fabry disease mice. *Proc Natl Acad Sci U S A* 1999;**96**:6423–7.

370. Ohto U, Usui K, Ochi T, *et al.* Crystal structure of human β-galactosidase: structural basis of GM1 gangliosidosis and morquio B diseases. *J Biol Chem* 2012;**287**:1801–12.

371. Okada S, O'Brien JS. Tay–Sachs disease: generalized absence of a β-d-N-acetylhexosaminidase component. *Science* 1969;**165**:698–700.

372. Orvisky E, Park JK, LaMarca ME, *et al.* Glucosylsphingosine accumulation in tissues from patients with Gaucher disease: correlation with phenotype and genotype. *Mol Genet Metab* 2002;**76**:262–70.

373. Oshima A, Yoshida K, Shimmoto M, *et al.* Human β-galactosidase gene mutation in Morquio B disease. *Am J Hum Genet* 1991;**49**:1091–3.

374. Otterbach B, Stoffel W. Acid sphingomyelinase-deficient mice mimic the neurovisceral form of human lysosomal storage disease (Niemann–Pick disease). *Cell* 1995;**81**:1053–61.

375. Oya Y, Nakayasu H, Fujita N, *et al.* Pathological study of mice with total deficiency of sphingolipid activator proteins (SAP knockout mice). *Acta Neuropathol* 1998;**96**:29–40.

376. Oyanagi K, Ohama E, Miyashita K, *et al.* Galactosialidosis: neuropathological findings in a case of the late-infantile type. *Acta Neuropathol* 1991;**82**:331–9.

377. Pampols T, Pineda M, Giros ML, *et al.* Neuronopathic juvenile glucosylceramidosis due to sap-C deficiency: clinical course, neuropathology and brain lipid composition in this Gaucher disease variant. *Acta Neuropathol* 1999;**97**:91–7.

378. Park JK, Orvisky E, Tayebi N, *et al.* Myoclonic epilepsy in Gaucher disease: genotype-phenotype insight from a rare patient subgroup. *Pediatr Res* 2003;**53**:387–95.

379. Patterson JS, Jones MZ, Lovell KL, Abbitt B. Neuropathology of bovine beta-mannosidosis. *J Neuropathol Exp Neurol* 1991;**50**:538–46.

380. Patterson MC, Vanier MT, Suzuki K, *et al.* Niemann–Pick disease type C: a lipid trafficking disorder. In: Valle, D., Beaudet AL, Vogelstein, B., *et al.* eds. *The online metabolic and molecular bases of inherited disease*, New York: McGraw-Hill, 2011. Part 16, Chapter 145.

381. Pavlù-Pereira H, Asfaw B, Poupètov\aacl; H, *et al.* Acid sphingomyelinase deficiency: phenotype variability with prevalence of intermediate phenotype in a series of twenty five Czech and Slovak patients. A multi-approach study. *J Inherit Metab Dis* 2005;**28**:203–27.

382. Pellissier JF, Berard-Badier M, Pinsard N. Farber's disease in two siblings, sural nerve and subcutaneous biopsies by light and electron microscopy. *Acta Neuropathol* 1986;**72**:178–88.

383. Peng L, Suzuki K. Ultrastructural study of neurons in metachromatic leukodystrophy. *Clin Neuropathol* 1987;**6**:224–30.

384. Percy AK, Odrezin GT, Knowles PD, *et al.* Globoid cell leukodystrophy: comparison of neuropathology with magnetic resonance imaging. *Acta Neuropathol* 1994;**88**:26–32.

385. Phaneuf D, Wakamatsu N, Huang J-Q, *et al.* Dramatically different phenotypes in mouse models of human Tay–Sachs and Sandhoff diseases. *Hum Mol Genet* 1996;**5**:1–14.

386. Pierce KR, Kosanke SD, Bay WW, Brides CH. Animal model: porcine cerebrospinal lipodystrophy (GM2 gangliosidosis). *Am J Pathol* 1976;**83**:419–22.

387. Pin I, Pradines S, Pincemaille O, *et al.* [A fatal respiratory form of type C Niemann–Pick disease.] *Arch Fr Pediatr* 1990;**47**:373–5.

388. Platt FM, Walkley SU eds. *Lysosomal disorders of the brain*. Oxford: Oxford University Press, 2004.

389. Platt FM, Jeyakumar M, Andersson U, *et al.* Substrate reduction therapy in mouse models of the glycosphingolipidoses. *Philos Trans R Soc Lond B Biol Sci* 2003;**358**:947–54.

390. Porter BF, Lewis BC, Edwards JF, *et al.* Pathology of GM2 gangliosidosis in Jacob sheep. *Vet Pathol* 2011;**48**:807–13.

391. Porter FD, Scherrer DE, Lanier MH, *et al.*, Cholesterol oxidation products are sensitive and specific blood-based biomarkers for Niemann–Pick C1 disease. *Sci Transl Med* 2010;**2**:56–81.

392. Prieur DJ, Ahern-Rindell AJ, Murnane RD. Animal model of human disease. *Am J Pathol* 1991;**139**:1511–13.

393. Provenzale JM, Barboriak DP, Sims K. Neuroradiologic findings in fucosidosis, a lysosomal storage disease. *AJNR Am J Neuroradiol* 1995;**16**:809–13.

394. Pshezhetsky AV, Hinek A. Where catabolism meets signalling: neuraminidase 1 as a modulator of cell receptors. *Glycoconj J* 2011;**28**:441–52.

395. Puranam KL, Guo WX, Qian WH, *et al.* CLN3 defines a novel antiapoptotic pathway in neurodegeneration that is mediated by ceramide. *Mol Genet Met* 1999;**66**:294–308.

396. Purpura DP. Ectopic dendritic growth in mature pyramidal neurons in human ganglioside storage disease. *Nature* 1978;**276**:520–21.

397. Purpura DP, Suzuki K. Distortion of neuronal geometry and formation of aberrant synapses in neuronal storage disease. *Brain Res* 1976;**116**:1–21.

398. Qualman SJ, Moser HW, Valle D, *et al.* Farber disease: pathologic diagnosis in sibs with phenotypic variability. *Am J Med Genet Suppl* 1987;**3**:233–41.

399. Raas-Rothschild A, Bargal R, Goldman O, *et al.* Genomic organization of the UDP-N-acetylglucosamine-1-phosphotransferase gamma subunit (GNPTAG) and its mutations in mucolipidosis III. *J Med Genet* 2004;**41**:e52.

400. Raas-Rothschild A, Cormier-Daire V, Bao M, *et al.* Molecular basis of variant pseudo-hurler polydystrophy (mucolipidosis IIIC). *J Clin Invest* 2000;**105**:673–81.

401. Raben N, Nagaraju K, Lee E, *et al.* Targeted disruption of the acid alpha-glucosidase gene in mice causes an illness with critical features of both infantile and adult human glycogen storage disease type II. *J Biol Chem* 1998;**273**:19086–92.

402. Raben N, Wong A, Ralston E, Myerowitz R. Autophagy and mitochondria in Pompe disease. *Am J Med Genet C Semin Med Genet* 2012;**160C**:13–21.

403. Rafi MA, Zhi Rao H, Passini, *et al.* AAV-mediated expression of galactocerebrosidase in brain results in attenuated symptoms and extended life span in murine models of globoid cell leukodystrophy. *Mol Ther* 2005;**11**: 734–44.

404. Rapin I, Goldfisher S, Katzman R, *et al.* The cherry-red spot-myoclonus syndrome. *Ann Neurol* 1978;**3**:234–42.

405. Rapin I, Suzuki K, Suzuki K, Valsamis MP. Adult (chronic) GM2 gangliosidosis. *Arch Neurol* 1976;**33**:120–30.

406. Reid PC, Sakashita N, Sugii S, *et al.* A novel cholesterol stain reveals early neuronal cholesterol accumulation in the Niemann–Pick type C1 mouse brain. *J Lipid Res* 2004;**45**:582–91.

407. Resnick JM, Whitley CB, Leonard S, *et al.* Light and electron microscopic features of the liver in mucopolysaccharidosis. *Hum Pathol* 1994;**25**:276–86.

408. Rigante D, Antuzzi D, Ricci R, Segni G. Cervical myelopathy in mucopolysaccharidosis type IV. *Clin Neuropathol* 1999;**18**:84–6.

409. Robb RM, Kuwabara T. The ocular pathology of type A Niemann–Pick disease: a light and electron microscopic study. *Invest Ophthalmol Vis Sci* 1973;**12**:366–77.

410. Robertson DA, Callen DF, Baker EG, *et al.* Chromosomal localization of the gene for human glucosamine-6-sulphatase to 12q14. *Hum Genet* 1988;**79**:175–8.

411. Rodriguez-Lafrasse C, Vanier MT. Sphingosylphosphorylcholine in Niemann–Pick disease brain: accumulation in type A but not type B. *Neurochem Res* 1999;**24**:199–205.

412. Ross DL, Strife CF, Towbin R, Bove KE. Nonabsorptive hydrocephalus associated with nephropathic cystinosis. *Neurology* 1982;**32**:1330–34.

413. Rottier RJ, Bonten E, d'Azzo A. A point mutation in the neu-1 locus causes the neuraminidase defect in the SM/J mouse. *Hum Mol Genet* 1998;**7**:313–21.

414. Rucker JC, Shapiro BE, Han YH, *et al.* Neuro-ophthalmology of late-onset Tay–Sachs disease (LOTS). *Neurology* 2004;**63**:1918–26.

415. Rutledge JC. Progressive neonatal liver failure due to type C Niemann–Pick disease. *Pediatr Pathol* 1989;**9**:779–84.

416. Ryazantsev S, Yu WH, Zhao HZ, *et al.* Lysosomal accumulation of SCMAS (subunit c of mitochondrial ATP synthase) in neurons of the mouse model of mucopolysaccharidosis III B. *Mol Genet Metab.* 2007;**90**:393–401.

417. Saarela J, Laine M, Oinonen C, *et al.* Molecular pathogenesis of a disease: structural consequences of aspartylglucosaminuria mutations. *Hum Mol Genet* 2001;**10**:983–95.

418. Sabatelli M, Quaranta L, Madia F, *et al.* Peripheral neuropathy with hypomyelinating features in adult-onset Krabbe's disease. *Neuromuscul Disord* 2002;**12**:386–91.

419. Saito Y, Suzuki K, Hulette CM, Murayama S. Aberrant phosphorylation of α-synuclein in human Niemann–Pick type C1 disease. *J Neuropathol Exp Neurol* 2004;**63**:323–8.

420. Saito Y, Suzuki K, Namba E, *et al.* Niemann–Pick type C disease: accelerated neurofibrillary tangle formation and amyloid β deposition associated with apolipoprotein E e4 homozygosity. *Ann Neurol* 2002;**52**:351–5.

421. Sandhoff K. Variation of β-acetylhexo-saminidase pattern in Tay–Sachs disease. *FEBS Lett* 1969;**4**:351–4.

422. Sandhoff K, Kolter T, Harzer K, *et al.* Sphingolipid activator proteins. In: Valle, D., Beaudet AL, Vogelstein, B., *et al.* eds. *The online metabolic and molecular bases of inherited disease*, New York: McGraw-Hill, 2011. Part 16, Chapter 134.

423. Sanfilippo SJ, Podosin R, Langer L, Good RA. Mental retardation associated with acid mucopolysacchariduria (heparitin sulfate type). *J Pediatr* 1963;**63**:837–8.

424. Sango K, Yamanaka S, Hoffmann A, *et al.* Mouse models of Tay–Sachs and Sandhoff diseases differ in neurologic phenotype and ganglioside metabolism. *Nat Genet* 1995;**11**:170–76.

425. Saxonhouse MA, Behnke M, Williams JL, *et al.* Type VII presenting with isolated neonatal ascites. *J Perinatol* 2003;**23**:73–5.

426. Schindler D, Bishop DF, Wolf DE, *et al.* Neuroaxonal dystrophy due to lysosomal α-N-acetylgalactosaminidase deficiency. *N Engl J Med* 1989;**320**:1735–40.

427. Schlote W, Harzer K, Christomanou H, *et al.* Sphingolipid activator protein 1 deficiency in metachromatic leucodystrophy with normal arylsulphatase A activity: a clinical, morphological, biochemical and immunological study. *Eur J Pediatr* 1991;**150**:584–91.

428. Schnabel D, Schroder M, Frust W, *et al.* Simultaneous deficiency of sphingolipid activator proteins 1 and 2 is caused by a mutation in the initiation cordon of their common gene. *J Biol Chem* 1992;**287**:3312–15.

429. Schnorf H, Bosshard NU, Gitzelmann R, *et al.* [Adult form of GM2-gangliosidosis: a man and 2 sisters with hexosaminidase-A and -B deficiency (Sandhoff disease) and literature review.] *Schweiz Med Wochenschr* 1996;**126**:757–64.

430. Schofer O, Mischo B, Puschel W, *et al.* Early-lethal pulmonary form of Niemann–Pick type C disease belonging to

a second, rare genetic complementation group. *Eur J Pediatr* 1998;**157**:45–9.

431. Schuchman EH, Desnick RJ. Niemann–Pick disease types A and B: acid sphingomyelinase deficiencies. In: Valle, D., Beaudet AL, Vogelstein, B., *et al.* eds. *The online metabolic and molecular bases of inherited disease*, New York: McGraw-Hill, 2011. Part 16, Chapter 144.

432. Schueler UH, Kolter T, Kaneski CR, *et al.* Toxicity of glucosylsphingosine (glucopsychosine) to cultured neuronal cells: a model system for assessing neuronal damage in Gaucher disease type 2 and 3. *Neurobiol Dis* 2003;**14**:595–601.

433. Scott C, Ioannou YA. The NPC1 protein: structure implies function. Review. *Biochim Biophys Acta* 2004;**1685**:8–13.

434. Scott HS, Blanch L, Guo XH, *et al.* Cloning of the sulphamidase gene and identification of mutations in Sanfilippo A syndrome. *Nat Genet* 1995;**11**:465–7.

435. Sedel F, Friderici K, Nummy K, *et al.* Atypical Gilles de la Tourette Syndrome with beta-mannosidase deficiency. *Arch Neurol* 2006;**63**:129–31.

436. Semeraro LA, Riely CA, Kolodny EH, *et al.* Niemann–Pick variant lipidosis presenting as 'neonatal hepatitis'. *J Pediatr Gastroenterol Nutr* 1986;**5**:492–500.

437. Sener RN. Metachromatic leukodystrophy: diffusion MR imaging and proton MR spectroscopy. *Acta Radiol* 2003;**44**:440–43.

438. Seppala R, Lehto VP, Gahl WA. Mutations in the human UDP-N-acetylglucamine 2-epimerase gene define the disease sialuria and the allosteric site of the enzyme. *Am J Hum Genet* 1999;**64**:1563–9.

439. Seto T, Kono K, Morimoto K, *et al.* Brain magnetic resonance imaging in 23 patients with mucopolysaccharidoses and the effect of bone marrow transplantation. *Ann Neurol* 2001;**50**:79–92.

440. Seyrantepe V, Poupetova H, Froissart R, *et al.* Molecular pathology of NEU1 gene in sialidosis. *Hum Mut* 2003;**22**:343–52.

441. Shapiro LJ, Aleck KA, Kaback MM, *et al.* Metachromatic leukodystrophy without arylsulfatase A deficiency. *Pediatr Res* 1979;**13**:1179–83.

442. Shen JS, Watabe K, Ohashi T, Eto Y. Intraventricular administration of recombinant adenovirus to neonatal twitcher mouse leads to clinicopathological improvements. *Gene Ther* 2001;**8**:1081–7.

443. Shimmoto M, Kase R, Itoh K, *et al.* Generation and characterization of transgenic mice expressing a human mutant α-galactosidase with R301Q substitution causing a variant form of Fabry disease. *FEBS Lett* 1997;**417**:89–91.

444. Shulman LM, David NJ, Weiner WJ. Psychosis as the initial manifestation of adult onset Niemann–Pick disease type C. *Neurology* 1995;**45**:1739–43.

445. Sidransky E, Sherer DM, Ginns EI. Gaucher disease in the neonate: a distinct Gaucher phenotype is analogous to a mouse model created by targeted disruption of the glucocerebrosidase gene. *Pediatr Res* 1992;**32**:494–8.

446. Sidransky E, Tayebi N, Stubblefield BK, *et al.* The clinical, molecular and pathological characterization of a family with two cases of lethal perinatal type 2 Gaucher disease. *J Med Genet* 1996;**33**:132–6.

447. Siintola E, Partanen S, Strömme P, *et al.* Cathepsin D deficiency underlies congenital human neuronal ceroid-lipofuscinosis. *Brain* 2006;**129**:1438–45.

448. Silverstein Dombrowski DC, Carmichael KP, Wang P, *et al.* Mucopolysaccharidosis type VII in a German shepherd dog. *J Am Vet Med Assoc* 2004;**224**:553–7.

449. Skelly BJ, Sargan DR, Winchester BG, *et al.* Genomic screening for fucosidosis in English springer spaniels. *Am J Vet Res* 1999;**60**:726–9.

450. Slaugenhaupt SA. The molecular basis of mucolipidosis type IV. *Curr Mol Med* 2002;**2**:445–50.

451. Sleat DE, Ding L, Wang S, *et al.* Mass spectrometry-based protein profiling to determine the cause of lysosomal storage diseases of unknown etiology. *Mol Cell Proteomics* 2009;**8**:1708–18.

452. Sleat DE, Wiseman JA, El-Banna M, *et al.* A mouse model of classical late-infantile neuronal ceroid lipofuscinosis based on targeted disruption of the CLN2 gene results in a loss of tripeptidyl-peptidase I activity and progressive neurodegeneration. *J Neurosci* 2004;**24**:9117–26.

453. Sleat DE, Wiseman JA, El-Banna M, *et al.* Genetic evidence for non-redundant functional co-operativity between NPC1 and NPC2 in lipid transport. *Proc Natl Acad Sci U S A* 2004;**101**:5886–91.

454. Slonim AE, Bulone L, Minikes J. Benign course of glycogen storage disease type IIb in two brothers: Nature or nurture? *Muscle Nerve* 2006;**33**:571–4.

455. Sly WS, Quinton BA, McAlister WH, Rimoin DL. Beta glucuronidase deficiency: report of clinical, radiologic and biochemical features of a new mucopolysaccharidosis. *J Pediatr* 1973;**82**:249–57.

456. Smith KR, Dahl H-H, Canafoglia L, *et al.* Mutations in the gene encoding cathepsin F are a cause of type B Kufs disease. 13th International Conference on Neuronal Ceroid Lipofuscinosis (Batten Disease). 2012;O28.

457. Smith KR, Damiano J, Franceschetti S, *et al.* Strikingly different clinicopatho-logical phenotypes determined by progranulin-mutation dosage. *Am J Hum Genet* 2012;**90**:1102–7.

458. Sokol J, Blanchette-Mackie J, Kruth GS, *et al.* Type C Niemann–Pick disease: lysosomal accumulation and defective intracellular mobilization of low density lipoprotein cholesterol. *J Biol Chem* 1988;**263**:3411–17.

459. Specola N, Vanier MT, Goutieres F, *et al.* The juvenile and chronic forms of GM2 gangliosidosis: clinical and enzymatic heterogeneity. *Neurology* 1990;**40**:145–50.

460. Spiegel R, Bach G, Sury V, *et al.* A mutation in the saposin A coding region of the prosaposin gene in an infant presenting as Krabbe disease: first report of saposin A deficiency in humans. *Mol Genet Metab* 2005;**84**:160–66.

461. Staropoli JF, Karaa A, Lim ET, *et al.* A homozygous mutation in KCTD7 links neuronal ceroid lipofuscinosis to the ubiquitin-proteasome system. *Am J Hum Genet* 2012;**91**:202–8.

462. Steet RA, Hullin R, Kudo M, *et al.* A splicing mutation in the α/ β GlucNAc-1-phosphotransferase gene results in an adult onset form of mucolipidosis III associated with sensory neuropathy and cardiomyopathy. *Am J Med Genet* 2005;**132A**:369–75.

463. Stein VM, Crooks A, Ding W, *et al.* Miglustat improves Purkinje cell survival

and alters microglial phenotype in feline Niemann–Pick disease type C. *J Neuropathol Exp Neurol* 2012;**71**:434–48.

464. Stinchi S, Lullmann-Rauch R, Hartmann D, *et al.* Targeted disruption of the lysosomal alpha-mannosidase gene results in mice resembling a mild form of human alpha-mannosidosis. *Hum Mol Genet* 1999;**8**:1365–72.

465. Stone DL, Carey WF, Christodoulou J, *et al.* Type 2 Gaucher disease: the collodion baby phenotype revisited. *Arch Dis Child Fetal Neonatal Ed* 2000;**82**:163–6.

466. Stone DL, van Diggelen OP, de Klerk J, *et al.* Is the perinatal lethal form of Gaucher disease more common than classic type 2 Gaucher disease? *Eur J Hum Genet* 1999;**7**:505–9.

467. Streifler JY, Gornish M, Hadar H, Gadoth N. Brain imaging in late-onset GM2 gangliosidosis. *Neurology* 1993;**43**:2055–8.

468. Sturmer J. Mukopolysaccharidose type VI-A (Morbus Maroteaux–Lamy). Klinisch-Pathologischer Fallbericht. *Klin Monatsbl Augenheilkd* 1989;**194**: 273–81.

469. Sugie K, Yamamoto A, Murayama K, *et al.* Clinicopathological features of genetically confirmed Danon disease. *Neurology* 2002;**58**:1773–8.

470. Sun M, Goldin E, Stahl S, *et al.* Mucolipidosis type IV is caused by mutations in a gene encoding a novel transient receptor potential channel. *Hum Mol Genet* 2000;**9**:2471–8.

471. Sung JH. Autonomic neurons affected by lipid storage in the spinal cord of Fabry's disease: distribution of autonomic neurons in the sacral cord. *J Neuropathol Exp Neurol* 1979;**38**:87–98.

472. Sung JH, Hayano M, Desnick RJ. Mannosidosis: pathology of the nervous system. *J Neuropathol Exp Neurol* 1977;**36**:807–20.

473. Suzuki, K. Metabolic disease. In: Johannessen JV ed. *Electron microscopy in human medicine*, Vol. 6. London: McGraw-Hill, 1979:3–53.

474. Suzuki K. Neuropathology of late onset gangliosidoses. *Dev Neurosci* 1991;**13**:205–10.

475. Suzuki K. Twenty-five years of the 'psychosine hypothesis': a personal perspective and its history and present status. *Neurochem Res* 1998;**23**:251–9.

476. Suzuki K, Suzuki K. The gangliosidoses. In: Moser HW ed. *Handbook of clinical neurology: neurodystropies and neurolipidosis.* Amsterdam: Elsevier, 1996:247–80.

477. Suzuki K, Suzuki K. Lysosomal diseases. In: Graham DL, Lantos PL eds. *Greenfield's neuropathology*, 7th edn. London: Arnold, 2002:666–735.

478. Suzuki K, Iseld E, Togo T, *et al.* Neuronal and Glial accumulation of alpha- and beta-synucleins in human lipidoses. *Acta Neuropathol* 2007;**114**:481–9.

479. Suzuki K, Parker CC, Pentchev PG, *et al.* Neurofibrillary tangles in Niemann–Pick disease type C. *Acta Neuropathol* 1995;**89**:227–38.

480. Suzuki K, Parker CC, Pentchev PG. Niemann–Pick disease type C: neuropathology revisited. *Dev Brain Dysfunct* 1997;**10**:306–20.

481. Suzuki K, Proia RL, Suzuki K. Mouse models of human lysosomal diseases. *Brain Pathol* 1998;**8**:195–215.

482. Suzuki K, Suzuki K, Rapin I, *et al.* Juvenile GM2-gangliosidosis. *Neurology* 1970;**20**:190–204.

483. Suzuki K, Vanier MT, Suzuki K. Lysosomal Disorders. In: Popko B ed. *Mouse models of human genetic neurological disease.* New York: Kluwer Academic/Plenum Publishers, 1999:245–83.

484. Suzuki N, Konohana I, Fukushige T, Kanzaki T. β-Mannosidosis with angiokeratoma corporis diffusum. *J Dermatol* 2004;**31**:931–5.

485. Suzuki Y. β-Galactosidase deficiency: an approach to chaperone therapy. *J Inherit Metab Dis* 2006;**29**:471–6.

486. Suzuki Y, Jacob JC, Suzuki K, *et al.* GM2 gangliosidosis with total hexosaminidase deficiency. *Neurology* 1971;**21**:313–28.

487. Suzuki Y, Namba E, Matsuda J. *et al.* β-Galactosidase deficiency (β-galactosidosis): GM1 gangliosidosis and Morquio B disease. In: Valle D, Beaudet AL, Vogelstein B, *et al.* eds. *The online metabolic and molecular bases of inherited disease*, New York: McGraw-Hill, 2011; Part 16, Chapter 147.

488. Suzuki Y, Namba E, Tsuji A, *et al.* Clinical heterogeneity in galactosialidosis. *Dev Brain Dysfunct* 1988;**1**:285–93.

489. Takahashi J, Barkovich AJ, Dillon WP, *et al.* T1 hyperintensity in the Purvinar: key imaging for diagnosis of Fabry disease. *AJNR Am J Neuroradiol* 2003;**24**:916–21.

490. Takano T, Shimmoto M, Fukuhara Y, *et al.* Galactosialidosis: clinical and molecular analysis of 19 Japanese patients. *Dev Brain Dysfunct* 1991;**4**:271–80.

491. Takebayashi S, Bassewitz H, Themann H. Feinstructurelle Veranderungen der Niere bei generalisierter Gangliosidose GM1. *Virchows Arch B Cell Pathol* 1970;**5**:301–13.

492. Tamagawa K, Morimatsu Y, Fujisawa K, *et al.* Neuropathological study and chemico-pathological correlation in sibling cases of Sanfilippo syndrome type B. *Brain Dev* 1985;**7**:599–609.

493. Tanaka A, Ohno K, Suzuki K. GM2 ganglioside B1 variant: a wide geographic and ethnic distribution of the specific β-hexosaminidase α-chain mutation originally identified in a Puerto Rican patient. *Biochem Biophys Res Commun* 1988;**156**:1015–19.

494. Tanaka A, Sakazaki H, Murakami H, *et al.* Molecular genetics of Tay–Sachs disease in Japan. *J Inherit Metab Dis* 1994;**17**:593–600.

495. Taniike M, Yamanaka S, Proia RL, *et al.* Neuropathology of mice with targeted disruption of Hexa gene, a model of Tay–Sachs disease. *Acta Neuropathol* 1995;**89**:296–304.

496. Tayebi N, Cushner SR, Kleijer W, *et al.* Prenatal lethality of a homozygous null mutation in the human glucocerebrosidase gene. *Am J Med Genet* 1997;**73**:41–7.

497. Tayebi N, Walker J, Stubblefield B, *et al.* Gaucher disease with parkinsonian manifestations: does glucocerebrosidase deficiency contribute to a vulnerability to parkinsonism? *Mol Genet Metab* 2003;**79**:104–9.

498. Taylor RM, Farrow BR, Stewart GJ, Healy PJ. Enzyme replacement in nervous tissue after allogeneic bone marrow transplantation for fucosidosid dogs. *Lancet* 1986 4;**2(8510)**:772–4.

499. Tedeschi G, Bonavita S, Banerjee TK, *et al.* Diffuse central neuronal involvement in Fabry disease: a proton MRS imaging study. *Neurology* 1999;**52**:1663–7.

500. Tedeschi G, Bonavita S, Barton NW, *et al.* Proton magnetic resonance spectroscopic imaging in the clinical evaluation of patients with Niemann–Pick type C disease. *J Neurol Neurosurg Psychiatry* 1998;**65**:72–9.

501. Tellez-Nagel I, Rapin I, Iwamoto T, *et al.* Mucolipidosis IV: clinical, ultrastructural, histochemical, and chemical studies of a case, including a brain biopsy. *Arch Neurol* 1976;**33**:828–35.

502. Tenhunen K, Uusitalo A, Autti T, *et al.* Monitoring the CNS pathology in aspartylglucosaminuria mice. *J Neuropathol Exp Neurol* 1998;**57**:1154–63.

503. Terry RD, Weiss M. Studies in Tay–Sachs disease II: ultrastructure of the cerebrum. *J Neuropathol Exp Neurol* 1963;**22**:18–55.

504. Tessitore A, del P Martin M, Sano R, *et al.* GM1-ganglioside-mediated activation of the unfolded protein response causes neuronal death in a neurodegenerative gangliosidosis. *Mol Cell* 2004;**15**:753–66.

505. Thomas GH. Disorders of glycoprotein degradation : α-mannosidosis, β-mannosidosis, fucosidosis and sialidosis. In: Valle, D., Beaudet AL, Vogelstein, B., *et al.*, eds. *The online metabolic and molecular bases of inherited disease*, New York: McGraw-Hill, 2011. Part 16, Chapter 140.

506. Thomas PK, Goebel HH. Lysosomal and peroxisomal disorders. In: Dyck PJ, Thomas PK eds. *Peripheral neuropathy*, 4th ed. Philadelphia, PA: Elsevier Saunders, 2005:1845–54.

507. Tomatsu S, Orii KO, Vogler C, *et al.* Missense models [Gustm (E536A) Sly, Gustm (E536Q) Sly, and Gustm (L175F) Sly] of murine mucopolysaccharidosis type VII produced by targeted mutagenesis. *Proc Natl Acad Sci U S A* 2002;**99**:14982–7.

508. Tomatsu S, Orii KO, Vogler C, *et al.* Mouse model of N-acetylgalactosamine-6-sulfate sulfatase deficiency (Galns2/2) produced by targeted disruption of the gene defective in Morquio A disease. *Hum Mol Genet* 2003;**12**:3349–58.

509. Treiber-Held S, Distl R, Meske V, *et al.* Spatial and temporal distribution of intracellular free cholesterol in brains of a Niemann–Pick type C mouse model showing hyperphosphorylated tau protein: implication for Alzheimer's disease. *J Pathol* 2003;**200**:95–103.

510. Triggs-Raine B, Salo TJ, Zhang H, *et al.* Mutations in *HYAL1*, a member of a tandemly distributed multigene family encoding disparate hyaluronidase activities, cause a newly described lysosomal disorder, mucopolysaccharidosis IX. *Proc Natl Acad Sci U S A* 1999;**96**:6296–300.

511. Troost J, Straks W, Willemse J. Fucosidosis II: ultrastructure. *Neuropadiatrie* 1977;**8**: 163–71.

512. Tutor JC. Biochemical characterization of the GM2 gangliosidosis B1 variant. *Braz J Med Biol Res* 2004;**37**:777–83.

513. Tylki-Szymanska A, Czartoryska B, Groener JEM, Lugowska A. Clinical variability in mucolipidosis III (pseudo-hurler polydystrophy). *Am J Med Genet* 2002;**108**:214–18.

514. Tyynelä J, Baumann M, Henseler M, *et al.* Sphingolipid activator proteins in the neuronal ceroid-lipofuscinosis: an immunological study. *Acta* 1995;**89**: 391–8.

515. Tyynelä J, Sohar I, Sleat DE, *et al.* A mutation in the ovine cathepsin D gene causes a congenital lysosomal storage disease with profound neurodegeneration. *EMBO J* 2000;**15**:2786–92.

516. Tyynelä J, Suopanki J, Baumann M, Haltia M. Sphingolipid activator proteins (SAPs) in neuronal ceroid lipofuscinosis (NCL). *Neuropediatrics* 1997;**28**:49–52.

517. Uchino Y, Fukushige T, Yotsumoto S, *et al.* Morphological and biochemical studies of human β-mannosidosis: identification of a novel β-mannosidase gene mutation. *Br J Dermatol* 2003;**149**:23–9.

518. Umehara F, Matsumuro K, Kurono Y, *et al.* Neurologic manifestations of Kanzaki disease. *Neurology* 2004;**62**:1604–6.

519. Urushihara M, Kagami S, Yasutomo K, *et al.* Sisters with α-mannosidosis and systemic lupus erythematosus. *Eur J Pediatr* 2004;**163**:192–5.

520. Usenovic M, Krainc D. Lysosomal dysfunction in neurodegeneration: the role of ATP13A2/PARK9. *Autophagy* 2012;**8**:987–88.

521. Uyama E, Terasaki T, Watanabe S, *et al.* Type 3 GM1 gangliosidosis: characteristic MRI findings correlated with dystonia. *Acta Neurol Scand* 1992;**86**:609–15.

522. Valle D, Beaudet AL, Vogelstein B, *et al.* eds. *The online metabolic and molecular bases of inherited disease,* New York: McGraw-Hill, 2011.

523. Van der Voorn JP, Kamphorst W, van der Knaap MS, Powers J. The leukoencephalopathy of infantile GM1 gangliosidosis: oligodendrocytic loss and axonal dysfunction. *Acta Neuropathol* 2004;**107**:539–45.

524. Van der Voorn JP, Pouwels PJ, Kamphorst W, *et al.* Histopathologic correlates of radial stripes on MR images in lysosomal storage disorders. *AJNR Am J Neuroradiol* 2005;**26**:442–6.

525. Vanier MT, Millat G. Structure and function of the NPC2 protein: review. *Biochim Biophys Acta* 2004;**1685**:14–21.

526. Vanier MT, Suzuki K. Niemann–Pick diseases. In: Moser HW ed. *Handbook of clinical neurology,* Vol. 22. Amsterdam: Elsevier Science, 1996:133–62.

527. Vanier MT, Suzuki K. Recent advances in elucidating Niemann–Pick C disease. *Brain Pathol* 1998;**8**:163–74.

528. Vanier MT, Saito Y, Murayama S, Suzuki K. Niemann–Pick type C disease. In: Golden JA, Harding BN eds. *Developmental neuropathology: pathology and genetics.* Basel: ISN Neuropath Press, 2004:283–6.

529. Vanier MT. Lipid changes in Niemann–Pick disease type C brain: personal experience and review of the literature. *Neurochem Res* 1999;**24**:481–9.

530. Varkonyi J, Rosenbaum H, Baumann N, *et al.* Gaucher disease associated with parkinsonism: four further case reports. *Am J Med Genet* 2003;**116A**:348–51.

531. Velinov M, Dolzhanskaya N, Gonzalez M, *et al.* Mutations in the gene DNAJC5 cause autosomal dominant Kufs disease in a proportion of cases: study of the Parry family and 8 other families. *PLoS One* 2012;**7**:e29729.

532. Venugopal B, Browning BF, Curcio-Morelli C, *et al.* Neurologic, gastric, and opthalmologic pathologies in a murine model of mucolipidosis type IV, *Am. J. Hum. Genet.* 2007;**81**:1070–83.

533. Verghese J, Goldberg RF, Desnick RJ, *et al.* Myoclonus from selective dentate nucleus degeneration type 3 Gaucher disease. *Arch Neurol* 2000;**57**:389–95.

534. Vite CH, McGowan JC, Braund KG, *et al.* Histopathology, electrodiagnostic testing, and magnetic resonance imaging show significant peripheral and central nervous system myelin abnormalities in the cat model of alpha mannosidosis. *J Neuropathol Exp Neurol* 2001;**60**:817–28.

535. Vogel DG, Malekzadeh MH, Cornford ME, *et al.* Central nervous system involvement in nephropathic cystinosis. *J Neuropathol Exp Neurol* 1990;**49**:591–9.

536. Vogler C, Levy B, Kyle JW, *et al.* Mucopolysaccharidosis VII: postmortem biochemical and pathological findings in a young adult with beta-glucuronidase deficiency. *Mod Pathol* 1994;**7**:132–7.

537. Volk BW, Schneck L, Adachi M. The clinic, pathology and biochemistry of Tay–Sachs disease. In: Vinken PJ, Bruyn GW eds. *Handbook of clinical neurology.* Amsterdam: North-Holland Publishing, 1970:385–426.

538. von Schantz C, Kielar C, Hansen SN, *et al.* Progressive thalamocortical neuron loss in Cln5 deficient mice: Distinct effects in Finish late infantile NCL. *Neurobiol Dis* 2009;**34**:308–319.

539. Von Scheidt W, Eng CM, Fitzmaurice TF, *et al.* An atypical variant of Fabry's disease confined to the heart. *N Engl J Med* 1991;**324**:395–9.

540. Waheed AB, Pohlman R, Hasilik A, *et al.* Deficiency of UDP-*N*-acetylglucosamine: lysosomal enzyme *N*-acetylglucosamine-1-phosphotransferase in organs of I-cell patients. *Biochem Biophys Res Commun* 1982;**105**:1052–8.

541. Wakabayashi K, Gustafson AM, Sidransky E, Goldin E. Mucolipidosis IV: an update. *Molec Genet Metab* 2011;**104**:206–13.

542. Walkley SU. Cellular pathology of lysosomal storage disorders. *Brain Pathol* 1998;**8**:175–93.

543. Walkley SU. Neurobiology and cellular pathogenesis of ganglioside storage diseases. *Philos Trans R Soc Lond B Biol Sci* 2003;**358**:893–904.

544. Walkley SU. Secondary accumulation of gangliosides in lysosomal storage disorders. *Semin Cell Dev Biol* 2004;**15**:433–44.

545. Walkley SU. Lysosomal diseases: why so complex? *J Inherit Met Dis* 2009;**32**:181–9.

546. Walkley SU, Pierok AL. Ferric ion-ferrocyanide staining in ganglioside storage disease establishes that meganeurites are of axon hillock origin and distinct from axonal spheroids. *Brain Res* 1986;**382**:379–86.

547. Walkley SU, Suzuki K. Consequences of NPC1 and NPC2 loss of function in mammalian neurons: molecular and cell biology of lipids. *Biochim Biophys Acta* 2004;**1685**:48–62.

548. Walkley SU, Vanier MT. Secondary lipid accumulation in lysosomal disease. *Biochim Biophys Acta* 2009;**1793**:726–36.

549. Walkley SU, Baker HJ, Rattazzi MC, Haskins ME, Wu J-Y. Neuroaxonal dystrophy in neuronal storage disorders: Evidence for major GABAergic neuron involvement. *J Neurol Sci* 1981;**104**:1–8.

550. Walkley SU, Blakemore WF, Purpura DP. Alterations in neuron morphology in feline mannosidosis: a Golgi study. *Acta Neuropathol* 1981;**53**:75–79.

551. Walkley SU, Sikora J, Micsenyi M., *et al.* Lysosomal compromise and brain dysfunction: examining the role of neuroaxonal dystrophy. *Biochem Soc Trans* 2010;**38**:1436–1441.

552. Walkley SU, Thrall MA, Haskins ME, *et al.* Abnormal neuronal metabolism and storage in mucopolysaccharidosis type VI (Maroteaux–Lamy) disease. *Neuropathol Appl Neurobiol* 2005;**31**:536–44.

553. Walkley SU, Zervas M, Wiseman S. Gangliosides as modulators of dendritogenesis in storage disease-affected and normal pyramidal neurons. *Cerebral Cortex* 2000;**10**:1028–1037.

554. Walvoort HC, Dormans JAMA, van den Ingh TSGAM. Comparative pathology of the canine model of glycogen storage type II (Pompe's disease). *J Inherit Metab Dis* 1985;**8**:38–46.

555. Walvoort HC. Glycogen storage diseases in animals and their potential value as models of human disease. *J Inherit Metab Dis* 1983;**6**:3–16.

556. Wang AM, Ioannou YA, Zeidner KM, *et al.* Fabry disease: generation of a mouse model with α-galactosidase A deficiency. *Am J Hum Genet* 1996;**59** (Suppl): A208.

557. Wang ML, Motamed M, Infante RE, *et al.,* Identification of surface residues on Niemann–Pick C2 essential for hydrophobic handoff of cholesterol to NPC1 in lysosomes. *Cell Metabol* 2010;**12**:166–73.

558. Wenger DA, DeGala G, Williams C, *et al.* Clinical, pathological, and biochemical studies on an infantile case of sulfatide/GM1 activator protein deficiency. *Am J Med Genet* 1989;**33**:255–65.

559. Wenger DA, Sattler M, Kudoh, T, *et al.* Niemann–Pick disease: a genetic model in Siamese cats. *Science* 1980;**208**:1471–3.

560. Wenger DA, Suzuki K, Suzuki Y, Suzuki K. Galactosylceramide lipidosis: globoid cell leukodystrophy (Krabbe disease). In: Valle, D., Beaudet AL, Vogelstein, B., *et al.* eds. The online metabolic and molecular bases of inherited disease, New York: McGraw-Hill, 2011. Part 16, Chapter 147.

561. Westbroek W, Gustafson AM, Sidransky E. Exploring the link between glucocerebrosidase mutations and parkinsonism. *Trends Mol Med* 2011;**17**:485–93.

562. Wheeler RB, Sharp JD, Schltz RA, *et al.* The gene mutated in variant late-infantile neuronal ceroid lipofuscinosis (*CLN6*) and in *nclf* mutant mice encodes a novel predicted transmembrane protein. *Am J Hum Genet* 2002;**70**:537–42.

563. Whitley CB. The mucopolysaccharidoses. In: Moser HW ed. *Handbook of clinical neurology,* Vol. 22. Amsterdam: Elsevier Science, 1996:281–327.

564. Wilkerson MJ, Lewis DC, Marks SL, Prieur DJ. Clinical and morphological features of mucopolysaccharidosis type II in a dog: naturally occurring model of Hunter syndrome. *Vet Pathol* 1998;**35**:230–33.

565. Willems PJ, Gatti R, Darby JK, *et al.* Fucosidosis revisited: a review of 77 patients. *Am J Med Genet* 1991;**38**:111–31.

566. Willems PJ, Seo HC, Coucke P, *et al.* Spectrum of mutations in fucosidosis. *Eur J Hum Genet* 1999;**7**:60–67.

567. Willemsen R, Tybulewicz E, Sidransky E, *et al.* A biochemical and ultrastructural evaluation of the type 2 Gaucher mouse. *Mol Chem Neuropath* 1995;**24**:179–92.

568. Winchester B, Vellodi A, Young E. The molecular basis of lysosomal storage

diseases and their treatment. *Biochem Soci Trans* 2000;**28**:150–54.

569. Winchester B, Young E, Geddes S, *et al.* Female twin with Hunter disease due to non-random inactivation of the X-chromosome: a consequence of twinning. *Am J Med Genet* 1992;**44**:834–8.

570. Wisniewski KE, Kida E, Golabek AA, *et al.* Neuronal ceroid lipofuscinosis: classification and diagnosis. In: Wisniewski KE, Zhong N eds. *Batten disease: diagnosis, treatment and research.* San Diego, CA: Academic Press, 2001:1–34.

571. Wittke D, Hartmann D, Gieselmann V, Lullmann-Rauch R. Lysosomal sulfatide storage in the brain of arylsulfatase A-deficient mice: cellular alterations and topographic distribution. *Acta Neuropathol* 2004;**108**:261–71.

572. Wohlke A, Philipp U, Bock P, *et al.* A one base pair deletion in the canine ATP13A2 gene causes exon skipping and late-onset neuronal ceroid lipofuscinosis in the Tibetan terrier. *PLoS Genet* 2011;**7**:e1002304.

573. Wolf D. Neuroaxonal dystrophy in infantile alpha-*N*-acetylgalactosaminidase deficiency. *J Neurol Sci* 1995;**132**:44–56.

574. Wong K, Sidransky E, Verma A, *et al.* Neuropathology provides clues to the pathophysiology of Gaucher disease. *Mol Genet Metab* 2004;**82**:192–207.

575. Wu Y-P, Mizukami H, Matsuda J, *et al.* Apoptosis accompanied by up-regulation of TNF-α death pathway genes in the brain of Niemann–Pick type C disease. *Mol Genet Metab* 2005;**84**:9–17.

576. Xie C, Turley SD, Pentchev PG, Dietschy JM. Cholesterol balance and metabolism in mice with loss of function of Niemann–Pick C protein. *Am J Physiol.* 1999;**276**:E336–44.

577. Yaghootfam A, Gieselmann V, Eckhardt M. Delay of myelin formation in arylsulfatase A-deficient mice. *Eur J Neurosci* 2005;**21**:711–20.

578. Yamanaka S, Johnson MD, Grinberg A, *et al.* Targeted disruption of the Hexa gene results in mice with biochemical and

pathologic features of Tay–Sachs disease. *Proc Natl Acad Sci U S A* 1994;**91**:9975–9.

579. Yamano T, Shimada M, Matsuzaki K, *et al.* Pathological study on a severe sialidosis (α-neuraminidase deficiency). *Acta Neuropathol* 1986;**71**:278–84.

580. Yamano T, Shimada M, Okada S, *et al.* Ultrastructural study on nervous system of fetus with GM1-gangliosidosis type 1. *Acta Neuropathol* 1983;**64**:15–20.

581. Yamano T, Shimada M, Sugino H, *et al.* Ultrastructural study on a severe infantile sialidosis (beta-galactosidase-alpha-neuraminidase deficiency). *Neuropediatrics* 1985;**16**:109–12.

582. Yanjanin NM, Velez JI, Gropman A, *et al.* Linear clinical progression, independent of age of onset, in Niemann–Pick disease, type C. *Am J Med Genetics B Neuropsychiatr Genet* 2010;**153** (Part B): 132–40.

583. Yoshida H Kuriyama M. 1990. Genetic lipid storage disease with lysosomal acid lipase deficiency in rats. *Lab Anim Sci* 40:486–9.

584. Yoshida K, Ikeda S, Kawaguchi K, Yanagisawa N. Adult GM1 gangliosidosis. *Neurology* 1994;**44**:2376–82.

585. Young EP, Worthington VC, Jackson M, Winchester BG. Pre- and postnatal diagnosis of patients with CLN1 and CLN2 by assay of palmitoyl-protein thioesterase and tripeptidyl peptidase I activities. *Eur J Pediatr Neurol* 2001;**5** (Suppl A):193–6.

586. Young ID, Harper PS, Archer IM, Newcombe RG. A clinical and genetic study of Hunter's syndrome. 1. Heterogeneity. *J Med Genet* 1982;**19**:401–7.

587. Young ID, Harper PS, Newcombe RG, Archer IM. A clinical and genetic study of Hunter's syndrome. 2. Differences between the mild and severe forms. *J Med Genet* 1982;**19**:408–11.

588. Young R, Kleinman G, Ojemann RG, *et al.* Compressive myelopathy in Maroteaux–Lamy syndrome: clinical and pathological findings. *Ann Neurol* 1980;**8**:336–40.

589. Zappatini-Tommasi L, Dumontel C, Guibaud P, Girod C. Farber disease: an ultrastructural study. *Virchows Arch A Pathol Anat Histopathol* 1992;**420**: 281–90.

590. Zarbin MA, Green WR, Moser HW, Morton SJ. Farber's disease: light and electron microscopic study of the eye. *Arch Ophthalmol* 1985;**103**:73–80.

591. Zeman W, Dyken P. Neuronal ceroid lipofuscinosis (Batten's disease): relationship to amaurotic family idiocy? *Pediatrics* 1969;**44**:570–83.

592. Zervas M, Somers K, Thrall MA, Walkley, SU. Critical role for glycosphingolipids in Niemann–Pick disease type C. *Curr Biol* 2001;**11**: 1283–87.

593. Zhang CK, Stein PB, Liu J, *et al.* Genome-wide association study of N370S homozygous Gaucher disease reveals the candidacy of CLN8 gene as a genetic modifier contributing to extreme phenotypic variation. *Am J Hematol* 2012;**87**:377–83.

594. Zhao HG, Li HH, Bach G, *et al.* The molecular basis of Sanfilippo syndrome type B. *Proc Natl Acad Sci U S A* 1996;**93**:6101–5.

595. Zhou S, Davidson C, McGlynn R, *et al.* Endosomal/lysosomal processing of gangliosides affects neuronal cholesterol sequestration in Niemann–Pick Disease Type C. *Am J Pathol* 2011;**179**:890–902.

596. Zhou XY, Morreau H, Rottier R, *et al.* Mouse model for the lysosomal disorder galactosialidosis and correction of the phenotype with overexpressing erythroid precursor cells. *Genes Dev* 1995;**9**: 2623–34.

597. Zhu M, Lovell KL., Patterson JS, *et al.* β-Mannosidosis mice: a model for the human lysosomal storage disease. *Hum Molecr Genet* 2006;**15**:493–500.

598. Zito E, Tacchetti C, Cosma M, *et al.*, Systemic inflammation and neurodegeneration in a mouse model of multiple sulfatase deficiency. *Proc Natl Acad Sci U S A* 2007;**104**:4506–11.

Mitochondrial Disorders

Patrick F Chinnery, Nichola Z Lax, Evelyn Jaros, Robert W Taylor,
Douglas M Turnbull and Salvatore DiMauro

INTRODUCTION

The first human mitochondrial disease was described in a patient with non-thyroidal hypermetabolism (Luft disease) nearly 50 years ago.[139] Although this disorder is exceptionally rare, the clinical description and biochemical studies paved the way for three decades of clinical and pathological research on cases of suspected mitochondrial disease. Patients were classified into groups based upon the pattern of clinical involvement, histological and ultrastructural abnormalities of mitochondria, and biochemical assays of mitochondrial function. It was clear that there were clinical similarities among some patients, allowing for the definition of syndromes such as the Kearns-Sayre syndrome (KSS) or chronic progressive external ophthalmoplegia (CPEO), but it was recognized that there was considerable phenotypic diversity and that many patients did not fit neatly into a specific diagnostic group. The inheritance pattern also varied. Some patients appeared to be sporadic cases, whereas others were clearly familial. It was known for some time that mitochondrial DNA (mtDNA) was maternally inherited, and whereas some families displayed a clear maternal inheritance pattern, others were sporadic cases and some families showed autosomal dominant and recessive transmission of the same trait. Attempts to classify the mitochondrial diseases were based upon the number and size of mitochondria in skeletal muscle, leading to terms such as pleoconial or megaconial myopathies,[221] and also on the pattern of respiratory chain involvement. Different groups attempted to subdivide suspected mitochondrial disease into discrete categories (the 'splitters')[199] and those who thought of all mitochondrial disease as a single, if wide, spectrum of disorders (the 'lumpers').[191] At this early stage it was apparent that mitochondrial disorders were a heterogeneous group: clinically, histologically, biochemically and, probably, genetically.

Following the discovery in the early 1960s that mitochondria contain their own DNA (mtDNA),[170] there were two major advances, both in the 1980s: the human mtDNA sequence was published in 1981,[4] and in 1988 the first pathogenic mtDNA mutations were identified.[80,264] The so-called 'Cambridge' reference sequence became the benchmark for all future sequence comparisons,[4] with minor revisions at a later stage[7] (referred to as the revised Cambridge reference sequence, or rCRS). The 1990s became the decade of the mitochondrial genome. Over 150 different pathogenic point mutations and a larger number of different rearrangements (i.e. partial deletions and duplications) of mtDNA were associated with disease,[215] with an expanding clinical phenotype accompanied by major advances in our understanding of the molecular pathophysiology.[213,263] Technological advances fuelled a growing interest in the role of the cell nucleus in human mitochondrial disease,[45] with the description of new nuclear genetic defects causing secondary defects of mtDNA affecting mtDNA quality (deletions) or quantity (depletion), mutations in genes coding for respiratory chain subunits, or mutations in genes encoding respiratory chain assembly components. More recent new categories include disorders of the lipid mitochondrial membrane (e.g. Barth syndrome), disorders of protein import (e.g. Mohr-Tranebjaerg syndrome) and disorders of mitochondrial fusion and fission (e.g. autosomal dominant optic atrophy and Charcot-Marie-Tooth disease type 2A).[269] Thus, the emphasis on disease classification has moved towards the molecular level, identifying discrete categories of genetic disease, with overlapping pathological mechanisms and clinical phenotypes. In many ways, the 'lumpers' and 'splitters' were both right and wrong.

Mitochondrial dysfunction plays an important role in the pathophysiology of a number of well-established nuclear genetic disorders, such as X-linked sideroblastic anaemia, Friedreich's ataxia (FRDA), hereditary spastic paraplegia (SPG7)[30] and Wilson's disease (ATP7B).[141] Mitochondrial dysfunction also appears to be an important factor in ageing and the pathophysiology of a number of more common neurodegenerative diseases, including Alzheimer's disease,

Parkinson's disease, Huntington's disease and amyotrophic lateral sclerosis (motor neuron disease) (for reviews see references 10, 38 and 212). These 'secondary' mitochondrial defects are not the subject of this chapter.

Mitochondria are intimately involved in cellular homeostasis. They play a part in intracellular signalling and apoptosis, intermediate metabolism and in the metabolism of amino acids, lipids, cholesterol, steroids and nucleotides, among other functions including one major apoptotic pathway. Mitochondria have a fundamental role in cellular energy metabolism, including fatty acid β oxidation and the urea cycle. However, the terms 'mitochondrial disorder' and 'mitochondrial disease', which are the subject of this chapter, refer to pathological defects of the final common pathway of energy production – the respiratory chain and oxidative phosphorylation.

MITOCHONDRIAL BIOLOGY: BIOCHEMISTRY AND GENETICS

Mitochondria are double-membrane subcellular organelles present in all nucleated mammalian cells. Their structure varies in different cell types and also within the same cell over time. Although they can appear as discrete cigar-shaped structures, it is more accurate to think of mitochondria as a budding and fusing network similar to the endoplasmic reticulum.

The mitochondrial respiratory chain consists of a group of five enzyme complexes situated on the inner mitochondrial membrane (Figure 7.1). Each complex is composed of multiple subunits, the largest being complex I with over 40 polypeptide components. Reduced cofactors (NADH and FADH$_2$) generated from the intermediary metabolism of carbohydrates, proteins and fats donate electrons to complex I and complex II. These electrons flow along the complexes down an electrochemical gradient, shuttled by complexes III and IV and by two mobile electron carriers, ubiquinone (ubiquinol, coenzyme Q$_{10}$) and cytochrome c. The electron-transfer function of complexes I-IV is accomplished via subunits harbouring prosthetic groups (iron-sulphur [sulfur] groups in complexes I, II and III, and heme iron in cytochrome c and in complex IV). The liberated energy is used by complexes I, III and IV to pump protons (H$^+$) out of the mitochondrial matrix into the intermembrane space. This proton gradient, which generates the bulk of the mitochondrial membrane potential (the asymmetric distribution of ions, such as Na$^+$, K$^+$ and Ca^{2+}, across the inner membrane makes up the 'chemical'

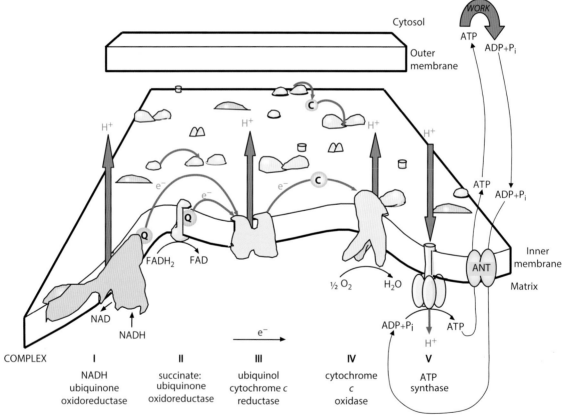

7.1 The respiratory chain. Schematic diagram of the respiratory chain. Reduced cofactors (NADH and FADH$_2$) are produced from the intermediary metabolism of carbohydrates, proteins and fats. These cofactors donate electrons (e$^-$) to complex I (NADH-ubiquinone oxidoreductase) and complex II (succinate-ubiquinone oxidoreductase). These electrons flow between the complexes down an electrochemical gradient (black arrow), shuttled by ubiquinone (Q) and cytochrome c (C), involving complex III (ubiquinol-cytochrome c oxidase reductase) and complex IV (cytochrome c oxidase, or COX). Complex IV donates an electron to oxygen, which results in the formation of water. Protons (H$^+$) are pumped from the mitochondrial matrix into the intermembrane space (red arrows). This proton gradient generates the mitochondrial membrane potential, which is harnessed by complex V to synthesise adenosine triphosphate (ATP) from adenosine diphosphate (ADP) and inorganic phosphate (P$_i$). ANT, adenine nucleotide translocator that exchanges ADP for ATP across the mitochondrial membrane.

portion of the gradient), is harnessed by complex V to synthesize adenosine triphosphate (ATP) from adenosine diphosphate (ADP) and inorganic phosphate. The overall process is called oxidative phosphorylation. ATP is the high-energy source used for essentially all active metabolic processes within the cell, and it must be released from the mitochondrion in exchange for cytosolic ADP. This is carried out by the adenine nucleotide translocator, which has a number of tissue specific isoforms.

The respiratory chain proteins are synthesized from two distinct genomes: mtDNA and nuclear DNA (nDNA). MtDNA is a small 16.6-kb circle of double-stranded DNA that specifies 13 respiratory chain polypeptides and 24 nucleic acids (two ribosomal RNAs [rRNAs] and 22 transfer RNAs [tRNAs]) that are required for intramitochondrial protein synthesis (Figure 7.2).[4] Nuclear genes code for the majority of mitochondrial respiratory chain polypeptides.[220] These polypeptides are synthesized in the cytoplasm with a mitochondrial targeting sequence that directs them through the translocation machinery spanning the outer and inner membranes. The targeting sequence is then cleaved before the subunit is assembled with its counterparts on the inner mitochondrial membrane. The components of the import machinery ('TOM' proteins in the outer membrane and 'TIM' in the inner membrane), the importation processing enzymes and the respiratory chain assembly proteins are all the products of nuclear genes (e.g. *SURF1, SCO1, SCO2, COX10, COX15, BCS1L, ATP12*).

Nuclear genes are also important for maintaining the mitochondrial genome, including those encoding the mitochondrial DNA polymerase γ (*POLG*) and factors that maintain an appropriate balance of free nucleotides within the mitochondrion (*TP, TK, DGK* and *ANT1*). *C10orf2(PEO1)* encodes the mitochondrial helicase Twinkle, which also appears to be important for mtDNA maintenance. Nuclear DNA also codes for essential factors needed for intramitochondrial transcription and translation, including *TFAM, TFBM1* and *TFBM2*, and mitochondrial RNA metabolism (e.g. *LRPPRC, EFG1, EFGu*). A disruption of either nuclear or mitochondrial genes can therefore cause mitochondrial dysfunction and human disease (see Table 7.2).[215]

CLINICAL PRESENTATION OF MITOCHONDRIAL DISORDERS

Neurological disease is one of the most consistently reported features of mitochondrial dysfunction, and often one of the most disabling aspects.[154] We describe some of the best recognized neurological syndromes later (Table 7.1), many of which are due to a specific mtDNA or nuclear defect (Table 7.2). However, it is important to recognize that a very large proportion of patients with mitochondrial disease do not conform to the clinical criteria for a particular syndrome or manifest only some of the features. These patients present

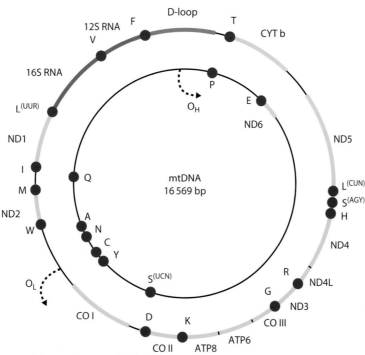

7.2 The human mitochondrial genome. The human mitochondrial genome (mtDNA) is a small 16,569-kb molecule of double-stranded DNA. MtDNA encodes for 13 essential components of the respiratory chain. ND1-ND6 and ND4L encode seven subunits of complex I (NADH-ubiquinone oxidoreductase). Cyt *b* is the only mtDNA encoded complex III subunit (ubiquinol-cytochrome *c* oxidase reductase). COX I to III encode for three of the complex IV (cytochrome *c* oxidase, or COX) subunits, and the ATP 6 and ATP 8 genes encode for two subunits of complex V (ATP synthase). Two ribosomal RNA genes (12S and 16S rRNA) and 22 transfer RNA genes are interspaced between the protein-encoding genes. These provide the necessary RNA components for intramitochondrial protein synthesis. D-loop, the 1.1-kb non-coding region that is involved in the regulation of transcription and replication of the molecule, is the only region not directly involved in the synthesis of respiratory chain polypeptides. O_H and O_L are the origins of heavy- and light-strand mtDNA replication.

TABLE 7.1 Classical mitochondrial clinical syndromes

	Primary features	Additional features
Alpers—Huttenlocher syndrome	Childhood onset refractory seizures, developmental delay and liver failure	Hypotonia, ataxia, sensitivity to valproate
Chronic progressive external ophthalmoplegia (CPEO)	External ophthalmoplegia and bilateral ptosis	Mild proximal myopathy
Infantile myopathy and lactic acidosis (fatal and non-fatal forms)	Hypotonia in the first year of life Feeding and respiratory difficulties	Fatal form may be associated with a cardiomyopathy and/or the Toni-Fanconi-Debre syndrome
Kearns-Sayre syndrome (KSS)	PEO onset before age 20 with pigmentary retinopathy Plus one of the following: CSF protein greater than 1 g/l, cerebellar ataxia, heart block	Bilateral deafness Myopathy Dysphagia Diabetes mellitus and hypoparathyroidism Dementia
Leber hereditary optic neuropathy (LHON)	Subacute painless bilateral visual failure Males:females approx. 4:1 Median age of onset 24 years	Dystonia Cardiac pre-excitation syndromes
Leigh syndrome (LS)	Subacute relapsing encephalopathy with cerebellar and brain-stem signs presenting during infancy	Basal ganglia lucencies
Mitochondrial neurogastrointestinal encephalomyopathy (MNGIE)	Chronic progressive external ophthalmoplegia, ptosis, gastrointestinal dysmotility (pseudo-obstruction), peripheral neuropathy and myopathy	Diffuse leukoencephalopathy
Mitochondrial encephalomyopathy with lactic acidosis and stroke-like episodes (MELAS)	Stroke-like episodes before age 40 years Seizures and/or dementia Ragged-red fibres and/or lactic acidosis	Diabetes mellitus Cardiomyopathy (hypertrophic leading to dilated) Bilateral deafness Pigmentary retinopathy Cerebellar ataxia
Myoclonic epilepsy with ragged-red fibres (MERRF)	Myoclonus Seizures Cerebellar ataxia Myopathy	Dementia, optic atrophy Bilateral deafness Peripheral neuropathy Spasticity Multiple lipomata
Neurogenic weakness with ataxia and retinitis pigmentosa (NARP)	Late childhood or adult onset peripheral neuropathy with associated ataxia and pigmentary retinopathy	Basal ganglia lucencies Abnormal electroretinogram Sensori-motor neuropathy
Pearson syndrome	Sideroblastic anemia of childhood Pancytopenia Exocrine pancreatic failure	Renal tubular defects

a considerable clinical and diagnostic challenge and many remain without a diagnosis for many years. Nonetheless, a mitochondrial disorder can be suspected in patients presenting with: (1) an unexplained combination of neuromuscular and/ or non-neuromuscular symptoms; (2) a progressive course; (3) involvement of an increasing number of seemingly unrelated organs or tissues as the disease progresses; and (4) occasional improvement or even disappearance of the initial symptoms as other organs become involved.

Chronic Progressive External Ophthalmoplegia

CPEO is defined by a slowly progressive ophthalmoplegia and ptosis.[123,164] Many patients first notice ptosis in their third or fourth decade, but the age of onset is variable.

Contrary to early textbook descriptions, marked asymmetry is not uncommon, although most patients go on to develop bilateral disease. Diplopia is commonly reported during the course of the disease, but tends to be mild. As the disease progresses, there are often associated proximal muscle weakness and fatigue, though rarely to a debilitating extent. Cardiac conduction defects and respiratory insufficiency are uncommon but can occur in severe cases (c.f. KSS).

The genetic defect in patients with CPEO is usually a single mtDNA deletion or multiple mtDNA deletions secondary to mutations in nuclear genes coding for factors with a role in mtDNA maintenance and replication. Single deletions of mtDNA are associated with sporadic PEO and with KSS.[217] In some rare instances, PEO is due to a point mutation of the mitochondrial genome.[26,223]

Families with either autosomal recessive (arPEO) or autosomal dominant (adPEO) CPEO are well recognized and

TABLE 7.2 Genetic classification of human mitochondrial disorders

Disorder	Inheritance pattern
Primary mitochondrial DNA disorders	
Rearrangements (large-scale partial deletions and duplications)	
CPEO	S or M
KSS	S or M
Diabetes and deafness	S
Pearson marrow-pancreas syndrome	S or M
Sporadic tubulopathy	S
Point mutations	
Protein-encoding genes	
LHON	M
NARP/LS	M
tRNA genes	M
MELAS	M
MERRF	M
CPEO	M
Myopathy	M
Cardiomyopathy	M
Diabetes and deafness	M
Encephalomyopathy	M
rRNA genes	
Non-syndromic sensorineural deafness	M
Aminoglycoside-induced non-syndromic deafness	M
Nuclear genetic disorders	
Disorders of mtDNA maintenance	
Autosomal progressive external ophthalmoplegia (with 2o multiple mtDNA deletions)	AD or AR
Mitochondrial neurogastrointestinal encephalomyopathy (with 2o multiple mtDNA deletions)	AR
Myopathy with mtDNA depletion	AR
Pure myopathy	AR
Encephalomyopathy	AR
Encephalopathy with liver failure	AR
Primary disorders of the respiratory chain	
Leigh syndrome	AR
Leukodystrophy and myoclonic epilepsy	AR
Cardioencephalomyopathy	AR
Optic atrophy and ataxia	AD or AR
Disorders of mitochondrial protein import	
Dystonia-deafness	XLR
Disorders of assembly of the respiratory chain	
Leigh syndrome	AR
Cardioencephalomyopathy	AR
Hepatic failure and encephalopathy	AR

Continued

TABLE 7.2 Genetic classification of human mitochondrial disorders (*Continued*)

Disorder	Inheritance pattern
Tubulopathy, encephalopathy and liver failure	AR
Encephalopathy	AR
Disorders of RNA metabolism	
Leigh syndrome	AR
Disorders of the lipid membrane	
Coenzyme Q10 deficiency	AR
Pure myopathy	
Renal tubulopathy	
Ataxia and/or seizures	
Barth syndrome	XLR

AD, autosomal dominant; AR, autosomal recessive; CPEO, chromic progressive external ophthalmoplegia; LHON, Leber hereditary optic neuropathy; KSS, Kearns-Sayre syndrome; M, maternal; MELAS, mitochondrial encephalomyopathy with lactic acidosis and stroke-like episodes; MERRF, myoclonic epilepsy with ragged-red fibres; NARP, neurogenetic weakness with ataxia and retinitis pigmentosa; S, sporadic; XLR, X-linked recessive.

are associated with multiple species of deleted mtDNA.[16,270] Dominant CPEO has been associated with mutations in *ANT1*,[109] *C10orf2* (Twinkle),[232] *POLG*,[259] *POLG2*,[135] *OPA1*[85] and *RRM2B*.[254] However, mutations in *POLG*,[1,124] *TK2*[252] and *RRM2B*[54] can also cause arPEO or sporadic cases. In patients with *POLG* mutations the clinical features are more variable than those seen in patients with mutations in *C10orf2* or *ANT1*. In addition to PEO, the clinical features in patients with *POLG* mutations include parkinsonism, psychiatric disorders, dysphagia, neuropathy, deafness, hypogonadism and ataxia.[52,143] In contrast, mutations in *ANT1* have been associated with a relatively mild, slowly progressive myopathy and few or no extramuscular symptoms.[109] Mutations in *C10orf2* (*PEO1*) have been associated with clinical presentations ranging from late-onset 'pure' PEO to PEO with proximal limb and facial weakness, dysphagia and dysphonia, mild ataxia and peripheral neuropathy.[233]

In addition to the typical clinical CPEO syndromes, external ophthalmoplegia can also be part of other clinical syndromes, including mitochondrial encephalomyopathy with lactic acidosis and stroke-like episodes (MELAS), myoclonic epilepsy with ragged-red fibres (MERRF) and mitochondrial neurogastrointestinal encephalomyopathy (MNGIE) (see later). PEO can also be associated with optic atrophy as part of a complex multisystem neurological disorder.[85] SANDO (sensory ataxic neuropathy with dysphagia and ophthalmoplegia)[49] can be caused by mutation in *POLG* and *C10Orf2*.[86,262] MIRAS (mitochondrial recessive ataxia syndrome) is also due to mutations in *POLG* and PEO may be a late feature of this disorder.[68,260]

Kearns-Sayre Syndrome

The onset of ophthalmoplegia and pigmentary retinopathy before the age of 20 years is characteristic of KSS. This predominantly sporadic condition is usually the result of either a large-scale single deletion or complex rearrangements of mtDNA. Other clinical features include cerebellar ataxia, proximal myopathy, complete heart block, cardiomyopathy, endocrinopathies, short stature, deafness, dysphagia,[119] deficits in visuospatial perception and executive function[17] and an elevated cerebrospinal fluid (CSF) protein. As might be predicted from the early onset of this multisystem disorder, life expectancy is considerably reduced.

Pearson (Bone Marrow Pancreas) Syndrome

Pearson (bone marrow pancreas) syndrome (PS) is a rare multisystem disorder resulting from large-scale rearrangements of mtDNA. The clinical features in these patients are dominated by sideroblastic anaemia with pancytopenia, vacuolation of marrow precursors and exocrine pancreatic dysfunction. The disease presents in infancy and frequently results in early death. Although survival through childhood leads to an improvement in anaemia, patients then develop the characteristic features of KSS.[197]

For PS, KSS and CPEO, the clinical severity appears to correlate with the tissue localization of mutated mtDNA. In PS (and to a lesser extent in KSS), mutated mtDNA can be demonstrated in a wide variety of tissues, whereas in CPEO the defective mtDNA is confined to skeletal muscle.

Mitochondrial Encephalopathy Lactic Acidosis and Stroke-like Episodes

This clinical syndrome is characterized by parieto-occipital stroke-like episodes, often associated (as the name implies) with concurrent encephalopathy and elevated plasma and CSF lactate.[77] These events may herald the onset of clinical disease in an otherwise apparently healthy young adult, or represent a new symptom in someone who already has a lengthy past medical history. Termed 'stroke-like' episodes, these events are not a result of thromboembolic disease but of the respiratory chain dysfunction within cerebral tissues. Consequently, the focal deficit is rarely of sudden onset and often develops over hours or even days. Radiological lesions frequently fail to conform to a single recognized vascular territory and may exhibit surprisingly rapid resolution associated with clinical improvement (Figure 7.3). Such events are commonly associated with a variable prodrome of headache, nausea and encephalopathic features, further distinguishing them from most thromboembolic causes of stroke. Seizures commonly accompany the acute episodes, sometimes preceding any focal deficit. These patients may have had migraine for many years. The presence of deafness, diabetes or myopathy may contribute to the diagnosis. It is not uncommon for patients to have been investigated for failure to thrive during childhood, and to have short stature and longstanding gastrointestinal complaints (reflecting bowel dysmotility) as adults.

The m.3243A>G mutation in the mitochondrial tRNA^Leu(UUR) gene was the first and is the most frequently described mtDNA mutation associated with this clinical phenotype.[63] However, other mutations often affecting MT-ND subunit genes[205] have been described in association with this presentation. The m.3243A>G mutation also causes other distinct clinical phenotypes such as maternally inherited diabetes and deafness (MIDD)[258] or CPEO.[163] Extraordinarily, such variations in phenotype can occur between individuals in the same family harbouring identical mutations.[34] A maternal mode of transmission is therefore not always apparent, and the mother of the patient may remain asymptomatic as a result of low levels of heteroplasmy in her tissues.

Myoclonic Epilepsy with Ragged-Red Fibres

MERRF was initially reported in association with an A>G mutation in the gene encoding mitochondrial tRNA^Lys at position 8344 and – although other mutations in the same tRNA have now been reported – this remains the most common cause of this disease.[219,222] In patients with MERRF there is progressive myoclonus, focal and generalized epilepsy, cerebellar ataxia and myopathy. Proximal muscle weakness can be pronounced, with gross wasting in a limb girdle distribution. Sensory ataxia is frequent, and the loss of proprioception is due to a progressive sensorimotor neuropathy. Occasionally, proprioception is lost without evidence of a neuropathy: in these cases, the pathology is thought to be more proximal and due to degeneration of the dorsal root ganglia or posterior columns. Pyramidal signs, such as hyperreflexia, clonus and up-going plantar responses can also occur. In the later stages of the disease, dementia is common and often exhibits prominent frontal features. It is perhaps not surprising that patients later confirmed to have MERRF had previously been diagnosed with Ramsay Hunt syndrome, spinal muscular atrophy, motor neurone disease or even limb girdle muscular dystrophy. Indeed, the m.8344A>G mutation has been identified as the cause of Ekbom's syndrome, characterized by cerebellar ataxia, photomyoclonus, skeletal deformities and lipomata.[125,249] The reported aversion to 'flickering' lights is almost uniform in these patients, whilst generalised seizures are surprisingly infrequent.

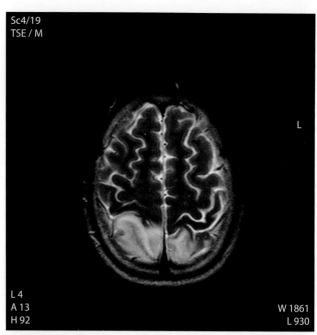

Sc4/19
TSE / M

L

L 4
A 13
H 92
W 1861
L 930

7.3 Neuroimaging in mitochondrial encephalomyopathy with lactic acidosis and stroke-like episodes. MRI showing occipital high signal during a stroke-like episode in MELAS.

With thanks to Dr Andrew Schaefer, University of Newcastle upon Tyne.

Loss of mobility, usually due to a combination of myopathy and cerebellar and sensory ataxia, rarely occurs earlier than two decades from the disease onset. Hypertrophic cardiomyopathy can also occur but re-entrant atrioventricular tachycardias, such as Wolff-Parkinson-White syndrome, are more commonly reported. Life expectancy is reduced significantly and pedigrees frequently include individuals who have suffered sudden and unexplained deaths before their sixth decade.

Leigh Syndrome

Leigh syndrome (LS) is a progressive neurodegenerative condition of infancy and childhood. The characteristic symmetric necrotic lesions distributed along the brain stem, diencephalon and basal ganglia were first described on post-mortem tissue,[131] but are readily visible on magnetic resonance imaging (MRI) or computerized tomography (CT) scan (Figure 7.4). Both clinical presentation and course vary considerably, but common features include signs of brain stem or basal ganglia dysfunction such as respiratory abnormalities, nystagmus, ataxia, dystonia, hypotonia and optic atrophy. Developmental delay or, more commonly, regression are prominent clinical features of this disorder and the latter may only be evident after a period of slow developmental progress. The clinical course can follow a stepwise deterioration with moderate recovery of developmental skills between episodes of regression or show a slowly progressive decline.

LS is due to severe failure of oxidative metabolism within the mitochondria of the developing brain owing to a variety of biochemical and molecular defects, including nuclear or mtDNA mutations.[195,272] Inheritance can

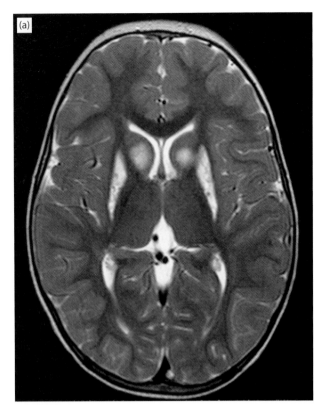
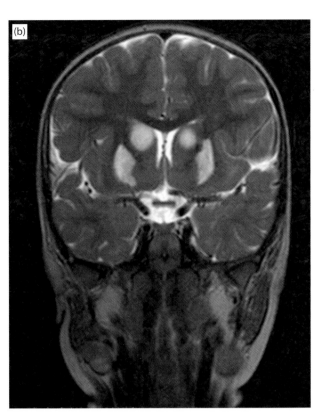

7.4 Neuroimaging in Leigh syndrome. (a) Axial T2 and **(b)** coronal FLAIR MR imaging from a child with Leigh syndrome showing hyperintensity of caudate and putamen.

With thanks to Dr Robert McFarland, University of Newcastle upon Tyne.

therefore be X-linked recessive, autosomal recessive, or maternal depending on the molecular defect.

Neuropathy, Ataxia and Retinitis Pigmentosa

First described as a variable combination of developmental delay, retinitis pigmentosa, dementia, seizures, ataxia, proximal neurogenic muscle weakness and sensory neuropathy in four members of a single family,[81] the phenotype has now been expanded to include cardiomyopathy and Leigh syndrome. The original family had a heteroplasmic T>G transversion at nucleotide pair 8993 in subunit 6 of mitochondrial ATPase (ATP6), probably impairing ATP synthesis. A T>C mutation at 8993 has been associated with a generally milder clinical phenotype, but higher frequency of ataxia. Mutations in the mitochondrial ATPase subunits do not affect cytochrome *c* oxidase (COX) activity and therefore neuropathy, ataxia and retinitis pigmentosa (NARP) patients will have no evidence of mitochondrial myopathy on routine histochemical analysis.

Leber Hereditary Optic Neuropathy

Leber hereditary optic neuropathy (LHON, also called Leber's optic atrophy) typically presents in young adults with sequential bilateral visual failure predominantly in males.[84,177] Over 95 per cent of LHON cases worldwide are due to one of three point mutations of mtDNA in genes that code for complex I of the respiratory chain: m.3460G>A in

MTND1, m.11778G>A in *MTND4* and m.14484T>C in *MTND6*.[142] Other rare mtDNA point mutations account for the remaining cases.[28] Biochemical findings in LHON are variable[21] and depend upon the tissues studied and the nuclear genetic background.[35] Biochemical studies in cell lines implicate apoptosis in the focal neuronal loss that characterizes the disorder.[40,268]

Only ~50 per cent of male and ~10 per cent of female mutation carriers develop symptoms, indicating that additional genetic or environmental factors are required for the phenotypic expression of LHON.[174] Family segregation analyses are consistent with a nuclear-encoded X-linked susceptibility allele in some pedigrees,[22,169] and are supported by recent linkage data.[87] However, the underlying causal variants in nuclear DNA have not been identified and environmental factors, such as oestrogens, may play a major role. Cigarette smoking is a major factor precipitating visual failure in carriers of LHON mtDNA mutations.[116]

Mitochondrial Neurogastrointestinal Encephalopathy Syndrome

This multisystem disorder is characterized by early adult onset of chronic progressive external ophthalmoplegia, ptosis, gastrointestinal dysmotility (pseudo-obstruction), diffuse leukoencephalopathy, peripheral neuropathy and myopathy. Histological and biochemical studies of MNGIE patients have confirmed the involvement of mitochondria in this disorder. The inheritance is autosomal recessive and it is due to mutations in the thymidine phosphorylase (*TYMP*) gene.[179,234] There is emerging evidence that bone marrow transplantation is effective in this disorder.

Polymerase Gamma Encephalopathies

The 16,569 base-pair human mitochondrial genome is replicated by the mitochondrial polymerase gamma (POLG), which consists of two subunits, a large 140 kDa catalytic subunit and a small 55 kDa accessory subunit. The catalytic subunit exhibits DNA polymerase and 3'-5' exonuclease activities, whereas the accessory subunit functions to increase DNA binding affinity, stimulate catalytic activity and enhance DNA processive activities.[103,134] The gene encoding for polymerase gamma (*POLG*) is located on chromosome 15q25 and over 160 mutations are associated with a broad clinical spectrum of disease.[83] The diverse clinical phenotypes include autosomal dominant PEO,[259] autosomal recessive PEO,[261] ataxia neuropathy spectrum (ANS),[68,260,266] Alpers–Huttenlocher spectrum (AHS)[173] and parkinsonism.[41]

POLG mutations causing ANS typically have an adult onset and are clinically characterized by ataxia and sensory neuropathy. ANS encompasses those disorders previously categorized as mitochondrial spinocerebellar ataxia and epilepsy, sensory ataxic neuropathy with dysarthria and ophthalmoplegia (SANDO) and mitochondrial inherited recessive ataxia syndrome.[37] Patients with ANS manifest with central symptoms of sensory neuropathy and ataxia. The neuropathy can have the features of a sensorimotor neuropathy, a sensory neuronopathy or a sensory ganglionopathy. *POLG* mutations also cause cerebellar ataxia associated with PEO or sensory neuropathy.[211] Seizures are also frequently described and manifest as complex partial, clonic or myoclonic, epilepsia partialis continua or convulsive status epilepticus. Seizures can be documented on EEG and may trigger stroke-like lesions with a posterior predilection, similar to those seen in patients with MELAS.[255]

The pathogenesis of *POLG* mutations is thought to occur through secondary defects in the maintenance of mtDNA (tissue-specific multiple mtDNA deletions) or in the replication of mtDNA (mtDNA depletion).

Mitochondrial DNA Depletion Syndromes

Mitochondrial DNA depletion syndromes (MDS) fall into three principal clinical groups: pure myopathy, pure encephalopathy and the hepatocerebral form. The pure myopathic form is caused by mutations in the gene encoding thymidine kinase, *TK2*[200] and *RRM2B*.[18] In some cases, a lower motor neuron phenotype has been reported in association with the myopathic form of MDS.[145,194] Patients with the cerebral form harbour mutations in genes coding for key Krebs cycle enzymes (*SUCLA2, SUCLG1*), which indirectly affect the respiratory chain.[188] Patients with the hepatocerebral form present with hepatic failure, severe failure to thrive, neurological abnormalities and hypoglycaemia[53,147] as a result of mutations in *DGUOK, MPV17, POLG, PEO1* or *SUCLG1*.[146,208,235] A distinct disorder characterized by hepatopathy and demyelinating neuropathy predominates in the Navajo population of the southwestern United States (Navajo neurohepatopathy, NNH) and is due to a homozygous MPV17 mutation. Most of these disorders are fatal in early neonatal or infantile life, with death from respiratory failure and associated infections.

Alpers–Huttenlocher Syndrome

AHS is an autosomal recessive hepatocerebral syndrome characterized clinically by severe developmental delay/psychomotor regression, intractable seizures, liver failure and death occurring usually in childhood.[3,88] Unlike patients with the hepatocerebral form of MDS, patients with AHS suffer from stepwise neurological deterioration along with myoclonic epilepsy unresponsive to treatment. Other typical features include cortical blindness with loss of visual-evoked potentials on the EEG; occipital atrophy on CT scan;[70] generalized severe cortical atrophy and multiple lesions on MRI; and progressive liver dysfunction and acute liver failure after exposure to sodium valproate.[15] Presentation in young adults has also been reported and is characterized by prominent intractable seizures associated with subacute encephalopathy, visual and sensory symptoms and signs, MRI evidence of lesions in the occipital lobes and thalamus, and chronic hepatitis.[72] AHS is due to mutations in the *POLG* gene.[42,50,118,173] Screening more than 30 AHS patients from different European and US diagnostic research centres showed that the majority of patients were compound heterozygotes, carrying two pathogenic *POLG* mutations, most commonly the Ala467Thr mutation associated with a variety of mutations in the other *POLG* allele.[50,83,175]

Infantile Onset Spinocerebellar Taxia

Autosomal recessive mutations, including the common founder mutation T457L in the gene encoding the mtDNA helicase, *Twinkle (C10ORF2)*, lead to a group of early onset disorders defined as infantile onset spinocerebellar ataxia (IOSCA). These are clinically characterized by muscle hypotonia, athetosis, ataxia, ophthalmoplegia, hearing deficits, sensory axonal neuropathy and epileptic encephalopathy.[176] IOSCA expands the spectrum of MDS with evidence of reduced mtDNA content in liver and brain tissues.[67,69,209]

Mitochondrial Myopathy

Myopathy is an important finding in patients with mitochondrial disease and a prominent clinical feature of the syndromes described later. A significant number of patients with mitochondrial disease present with non-specific symptoms, but proximal weakness and exercise intolerance stand out. Some patients with prominent myopathy show involvement of additional organ systems and become more obviously suggestive of a mitochondrial disorder, whereas others keep showing an isolated myopathy with gradual deterioration in power and involvement of muscle groups outside the shoulder and hip girdles, including the diaphragm. The finding of a proximal myopathy, in conjunction with other clinical features such as diabetes, sensory organ impairment or progressive multisystem disease, should prompt further investigation for mitochondrial disorders. Symptoms compatible with mtDNA disease in maternal relatives should also raise suspicion, even if the pedigree appears to exhibit variable penetrance.

INVESTIGATION OF SUSPECTED MITOCHONDRIAL DISEASE

Clinical Studies

The differential diagnosis for patients with mitochondrial disease is extensive as a result of the varied clinical presentation. In addition, the investigation of presumed mitochondrial disease is made more difficult not only because the same genetic or biochemical defect may present in a variety of different ways, but also because the same clinical syndrome may be due to a variety of different biochemical or molecular defects. The detailed history and examination of a patient with suspected mitochondrial disease are therefore crucial to both initiation and interpretation of the relevant investigations.

In some, the initial presentation alone may be sufficient to suggest a diagnosis of mtDNA disease. For example, stroke-like episodes (as described earlier in the text) are usually accompanied by a number of features, which help distinguish them from thromboembolic events. CPEO (often presenting as ptosis), is relatively specific to mtDNA disease and can usually be differentiated clinically from other forms of ophthalmoplegia.[196] In the majority of cases, however, presentations are sufficiently non-specific as to require the connection of several aspects of the history and examination to suggest the diagnosis. For example strokes, seizures, diabetes, deafness, migraine and gastrointestinal complaints are common complaints. Together, whether in an individual or a pedigree, they may represent mitochondrial disease. Family history is of great importance, but may also be misleading. Limited pedigrees may appear dominant if the absence of male transmission is not appreciated. Detailed pedigree analysis is therefore vital. Variable levels of mutation (heteroplasmy) throughout the family may mean that the mother and siblings of an adult with m.3243 A>G mutation (commonly associated with MELAS) remain asymptomatic, yet there is a history of deafness and failure to thrive in a maternal cousin. Mitochondrial pedigrees are therefore rarely as transparent as those due to nuclear defects. Reports of other neurological conditions within the family should also raise alert, particularly if their clinical course appears atypical.

Few individual features on examination are specific to mitochondrial disease and it is more frequently the combination of several clinical signs that point toward the diagnosis. CPEO is relatively specific for mitochondrial disease, though similarities do exist with other disorders of muscle, neuromuscular junction and supranuclear structures.[196] High-frequency sensorineural hearing loss is typical of mtDNA disease, but may occur in a number of other genetic disorders. Mitochondrial myopathies can be particularly difficult to identify, especially in their early stages. Often, proximal power appears normal but there may be a degree of fatigability. Ataxia, dementia and neuropathy rarely develop in isolation, more often forming part of a degenerative multisystem involvement.

Simple blood tests play only a minor role in the diagnosis of mtDNA disorders. Instead, their purpose is to provide supporting evidence for the clinical diagnosis and to detect potential complications associated with the disease. Initial investigations should therefore include creatine kinase, resting blood lactate, electrolytes, full blood count, thyroid and liver function, bone chemistry, fasting blood glucose and glycosylated haemoglobin (HbA1c). Creatine kinase levels can vary greatly, but are typically normal or only modestly elevated (below 500 U/L). Levels exceeding 1000 U/L are rare but can occur, particularly in the presence of renal disease or seizures or in patients with the myopathic presentation of the mtDNA depletion syndrome (MDS). It is vital that all patients have an ECG at baseline as cardiac conduction defects or cardiac hypertrophy may remain asymptomatic, even at an advanced stage. Chest radiography and echocardiography become appropriate investigations in those individuals with clinical or ECG evidence of cardiorespiratory involvement. Electromyography may be normal even in the presence of clinical myopathy, and nerve conduction studies can demonstrate either axonal or a mixed axonal-demyelinating peripheral sensorimotor neuropathy. The electroencephalogram (EEG) may reveal a pattern of generalized slow waves, indicative of a subacute encephalopathy, or subclinical seizure activity. Cognitive impairment, central neurological signs, movement disorder, or abnormal EEG all warrant some form of cerebral imaging. A variety of changes can be seen on computed tomography (CT) and magnetic resonance imaging (MRI) (see Figures 7.3 and 7.4), but abnormalities are frequently non-specific (e.g. cerebral or cerebellar atrophy). Certain findings are characteristic of specific mitochondrial disorders, for example the symmetrical hypodensities of brain stem, thalamus and basal ganglia seen in Leigh syndrome (Figure 7.4). More often, however, changes strongly suggestive of mitochondrial dysfunction (e.g. basal ganglia calcification) are not unique to any one phenotype (see Figures 7.3 to 7.5).

Even where these studies fail to reveal specific abnormalities, a strong clinical suspicion of mtDNA disease should prompt further investigation. Where a specific phenotype exists, molecular studies may be possible from blood or epithelial samples (see later). Where such studies are negative, or the phenotype is less clear, muscle biopsy is often warranted.

Systemic Metabolic Consequences of Mitochondrial Disease Dysfunction

Increased concentrations of organic acids (particularly lactic acid and TCA cycle intermediates) in blood, urine, cerebrospinal fluid, as well as in tissues, are frequently encountered in children with mitochondrial disease, but are of much less value in adults.[45,97,226] Increased lactic acid production is thought to result from pyruvate accumulation and the altered cellular redox state, under conditions where mitochondrial oxidation is impaired. This shifts the equilibrium of the pyruvate-lactate reaction catalysed by the NADH-dependent lactate dehydrogenase. Overproduction of lactic acid may cause lactic acidosis, resulting in a global disturbance of cellular pH.[167] Amino acid imbalances are also common in mitochondrial disorders. Alanine, like lactic acid, is converted from pyruvate under conditions of metabolic disturbance, because the equilibrium of the alanine aminotransferase reaction is dependent on pyruvate levels. However, great care must be taken in interpreting the results for these biochemical studies. Lactate levels in

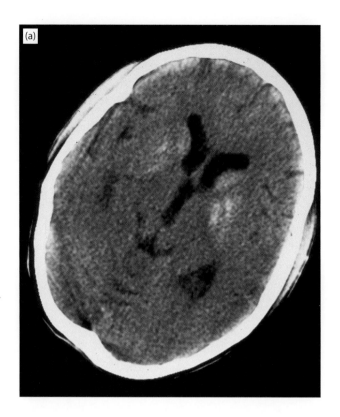

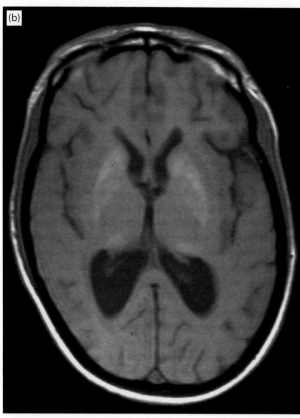

7.5 Basal ganglia calcification in mitochondrial disorders. (a) Brain CT showing basal ganglia calcification in a patient harbouring the m.3243A>G MELAS mutation. **(b)** Brain MRI from the same subject showing generalised atrophy and increased signal in the basal ganglia.

particular are perturbed by a number of factors (such as sepsis, strokes and seizures) and thus are not specific for mitochondrial defects.

Muscle Histopathology

Muscle is a commonly affected tissue in patients with mitochondrial disease (Table 7.3) and often shows specific pathological features even if there is relatively little clinical involvement. Some very well-recognized pathological hallmarks are both helpful in the diagnosis of mitochondrial disease and help explain some clinical features (see Chapter 25, Diseases of Skeletal Muscle). One such hallmark is fibres showing a subsarcolemmal collection of abnormal mitochondria – 'ragged-red fibres' – so named because of their reddish appearance with the Gomori trichrome stain (Figure 7.6d). Although this technique is still used in many pathology laboratories, more appropriate techniques to evaluate mitochondrial involvement are specific histochemical reactions for the mitochondrial enzymes succinate dehydrogenase (SDH) and cytochrome *c* oxidase (COX). SDH is part of complex II of the respiratory chain and contains subunits only encoded by the nuclear genome and is a better marker of subsarcolemmal mitochondrial accumulation in the presence of a mitochondrial defect. These 'ragged-blue' fibres (as they are often called) are the histochemical equivalent of ragged-red fibres (Figure 7.6c). The COX reaction is particularly useful in evaluating mitochondrial myopathies because COX contains subunits encoded by both the mitochondrial and nuclear genomes (Figure 7.6e), and thus no

activity is detected in the presence of many mtDNA defects. A mosaic pattern of SDH hyperactivity and COX deficiency is highly suggestive of a heteroplasmic mtDNA disorder and most ragged-red fibres are COX-deficient (Figure 7.6g). However, some patients with MELAS or point mutations in either *MTND* genes[6] or cytochrome *b*[5] have muscle fibres showing mitochondrial proliferation (ragged-blue) but normal COX activity. Although COX-deficient fibres may be relatively easy to detect especially if they are abundant, they may be difficult to detect when they are scarce. In this case, sequential COX-SDH histochemistry[214] is extremely valuable (Figure 7.6g). It is much easier to detect the blue-staining COX-deficient fibres, which might otherwise go unnoticed against a background of normal COX activity. In addition, this sequential COX/SDH histochemistry is extremely helpful in evaluating respiratory deficiency in other tissues. A global decrease in the activity of COX is usually suggestive of a nuclear mutation in one of the ancillary proteins required for COX assembly and function such as *SURF1*,[247,271] although a similar pattern is observed in some patients presenting with pathogenic, homoplasmic mitochondrial tRNA gene mutations[151] (Figure 7.6f).

The mitochondrial enzyme histochemistry can be very informative but it is not an infallible diagnostic technique. Histochemical and histopathological abnormalities are much more common in adults with mitochondrial disease than in children. This reflects the more common involvement of the mitochondrial genome in adults than in children. The SDH assay will also detect patients deficient in complex II,[246] but many patients with defects involving complex I,

TABLE 7.3 Summary of neurohistopathological diagnostic features of major mitochondrial disorders

	CPEO CPEO+	KSS	PS	MNGIE	MDS: MDS-Hc MDS-M	MELAS	MERRF	LS	NARP	LHON LHON+	AHS
Age (at onset)	Childhood (arPEO) and adulthood: 20–40 y (adPEO)	Childhood, juveniles, young adults	Infancy; survivors develop KSS	Childhood, juveniles, young adults	HC and M: Infancy	Childhood, juveniles, adults	Childhood, juveniles, young adults	Infancy, childhood; occasionally in juveniles/ adults	Infancy, juvenile, adults	Young adult-sor juvenile/ adults	Childhood, young adults
Muscle localization	Extraocular	Extraocular, cardiac and skeletal	As KSS, if it develops -	Smooth/ viscera, skeletal & extraocular	HC: None? M: Skeletal	Skeletal and extra-ocular	Skeletal and extra-ocular	Rare	Rare	LHON+ dystonia: Skeletal[b]	None[c]
RRF	+	+	As KSS, if it develops	+	M: +	+	+	– to –/+	–	–?	–
CNS localization *Major*	Rare, retina in some	WM: cerebrum, cerebellum and cord GM: brain stem	As KSS, if it develops	WM: cerebrum	HC: WM: cerebrum and cerebellum	GM: cerebrum especially at gyral crests in occipital, parietal and temporal lobes	GM: dentate and olivary nuclei and associated tracts	GM: brain stem and subcortical plus surrounding WM (mammillary bodies spared unlike Wernicke's)	Retina; widespread in cerebrum (unspecified)	Retinal ganglion cells and optic nerves	GM: striate cx
Minor	CPEO+: retina; variable - may have KSS-type lesions	BG			M: NA M: Lower motor neurons	WM: cerebellum			Some show overlap with LS	LHON+: WM in spinal cord and frontal lobe; GM in BG	GM: Subcortical, and olivary nuclei and cerebellar cx
CNS imaging[a]		Early bilateral lesions in WM of cerebrum, cerebellum; GM in brain stem, BG and thalamus; cortical, cerebellar cortical and brain stem atrophy develop		Diffuse leukoencephalopathy sparing CC	HC: NA M: either normal or cerebellar vermis atrophy and mild diffuse cx atrophy	Stroke-like lesions in cx, often in occipital lobes/ non-vascular distribution (may resolve over time)/ calcification in BG	Non-specific: cerebral or cerebellar atrophy or BG haemorrhage or lesions in subcortical WM	Symmetric or asymmetric lesions in brain stem, BG and dentate nucleus; acute lesions can fade	Lesion distribution depends on clinical presentation: 1. ponto-cerebellar atrophy, 2 Leigh-like, 3 MELAS-like	Optic nerve lesions; LHON +: depending on symptoms: 1 multiple WM lesions; 2 a-/ symmetrical BG lesions; 3 symmetrical lesions in frontal WM	Lesions in occipital lobes and thalamus

	CPEO CPEO+	KSS	PS	MNGIE	MDS: MDS-Hc MDS-M	MELAS	MERRF	LS	NARP	LHON LHON+	AHS
CNS histopathology pattern	Retinal pigmentary degeneration in some; CPEO+ retinal pigmentary degeneration and KSS-type lesions	Spongy degeneration of cerebral and cerebellar WM and spongy brain stem GM		Poorly defined leukoencephalopathy and BBB dysfunction	HC: Spongy degeneration of cerebral and cerebellar WM M: NA	Cortical laminar necrosis or cortical ablation: cell death, rapid infarct-like lesions	Neuron loss with tract degeneration	Cystic neuropil rarefaction and capillary prominence: *"vasculo-necrotic lesions"*	NA	Retinal ganglion cell loss and axonal loss in optic nerve LHON+: myelin loss in spinal cord and peripheral nerves, cystic degeneration of cerebral WM	Spongy degeneration in striate cortex; neuron loss in sensory ganglia and central-peripheral axonopathy
Spongy[d] cortical GM	-	-/+	-	-?	-	+	-/+	-		-	+++
Spongy subcortical GM	-	+	-	-?	-	-/+	-	+++		-	+
Spongy brain stem GM	-	++	-	-?	-	-/+	-/+	+++		-	-
Spongy WM	-	+++	-	-/+, may be endothelial loss	+	-/+	+	-/+		LHON+: +++	-
Neuron loss	-	as spongy GM: -/+ to ++	-	-?	HC: + Purkinje cells only	+++	+++	+++		LHON & LHON+: +++ Retinal ganglion cells	+ in cerebral & cerebellar cx
Astrogliosis	-	-/+	-	may be microgliosis	HC: +	++	++	++		LHON+: ++	++
Oligodendroglial loss	-	+	-	-?	-	++	+	-/+		LHON+: ++	-
Macrophages/microgliosis	-	-/+	-	-?	-	+++	+	++		LHON+: ++	-
Capillary prominence/proliferation	-	-/+	-	-?	-	++	+/-	+++		-	+ in cerebral cx
Mineralization of vessel walls	-	++ in GP and thalamus	-	-?	-	+++ BG	+/-	-		-	-

Continued

TABLE 7.3 Summary of neurohistopathological diagnostic features of major mitochondrial disorders *(Continued)*

	CPEO CPEO+	KSS	PS	MNGIE	MDS: MDS-Hc MDS-M	MELAS	MERRF	LS	NARP	LHON LHON+	AHS
PNS/ANS localization		Peripheral nerves		Visceral and peripheral nerves, autonomic ganglia, meningeal and peripheral nerve vessels	M: Lower motor neurons and projections	Peripheral nerves	Sensory ganglia and projections peripheral and central	Peripheral nerves	Peripheral nerves	LHON+: peripheral nerves	Sensory ganglia and peripheral and central projections

Severity of pathology: −, negative; ±, rare; +, mild; ++, moderate; +++, severe.

[a]Haas R, Dietrich R. Neuroimaging of mitochondrial disorders. *Mitochondrion* 2004 Sep;4(5–6):471–90.

[b]De Vries DD, Went LN, Bruyn GW, *et al.* Genetic and biochemical impairment of mitochondrial complex I activity in a family with Leber hereditary optic neuropathy and hereditary spastic dystonia. *Am J Hum Genet* 1996 Apr; 58(4):703–11.

[c]Gauthier-Villars M, Landrieu P, Cormier-Daire V, *et al.* Respiratory chain deficiency in Alpers syndrome. *Neuropediatrics* 2001 Jun;32(3):150–2; Tesarova M, Mayr JA, Wenchich L, *et al.* Mitochondrial DNA depletion in Alpers syndrome. *Neuropediatrics* 2004 Aug;35(4):217–23; Nguyen KV, Ostergard E, Ravin SH, *et al.* POLG mutations in Alpers syndrome. *Neurology* 2005;65:1493–5.

[d]Spongy is a neurohistopathological term used to describe the microcystic degeneration in grey matter and white matter that occurs in mitochondrial disorders.

adPEO, Autosomal dominant progressive external ophthalmoplegia; AHS, Alpers–Huttenlocher syndrome; ANS, autonomic nervous system; arPEO, autosomal recessive progressive external ophthalmoplegia; BBB, blood–brain barrier; BG, basal ganglia; CC, corpus callosum; CNS, central nervous system; CPEO, chronic progressive external ophthalmoplegia; CPEO+, CPEO with additional neurological features; CX, cortex; GM, grey matter; GP, globus pallidus; KSS, Kearns–Sayre syndrome; LHON, Leber hereditary optic neuropathy; LHON+, LHON with additional neurological features; LS, Leigh's syndrome; MELAS, mitochondrial encephalomyopathy, lactic acidosis, stroke-like episodes; MERRF, myoclonic epilepsy with ragged-red fibres; MDS, mitochondrial DNA depletion syndrome MDS-HC, hepatocerebral form of MDS, MDS-M, myopathic form of MDS; MNGIE, mitochondrial neurogastrointestinal encephalopathy syndrome; NA, information not available; NARP, neuropathy, ataxia and retinitis pigmentosa; PNS, peripheral nervous system; PS, Pearson's (bone marrow pancreas) syndrome; RRF, ragged-red fibres; WM, white matter.

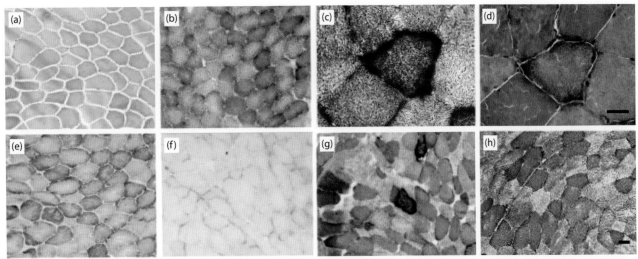

7.6 Muscle pathology in mitochondrial disorders. Cryostat sections (20 µm) of quadriceps skeletal muscle. **(a)** Haematoxylin and eosin showing increased angulation of the muscle fibres and an increase in the proportion of internal nuclei. **(b)** Succinate dehydrogenase (SDH) histochemistry in a patient with an heteroplasmic mtDNA defect, showing subsarcolemmal proliferation of mitochondria. **(c)** Higher power view of SDH histochemistry showing a classical ragged-red fibre. **(d)** Ragged-red fibre shown by Gomori trichrome staining. **(e)** Cytochrome *c* oxidase (COX) histochemistry in a normal subject. **(f)** Global reduction in COX activity seen in a patient with a nuclear gene defect. **(g)** Mosaic COX defect demonstrated by sequential COX-SDH histochemistry in a patient with heteroplasmic pathogenic mtDNA mutation. **(h)** COX-SDH histochemistry from an aged subject showing a single COX-deficient muscle fibre. (a,b,e,f,g,h: scale bar 50 µm as shown in h. c,d: scale bar 25 µm as shown in d.)

complex III or complex V may have normal biopsy findings because there are no reliable histochemical methods for these complexes. Immunocytochemistry may be valuable, but is not part of routine diagnostic screening. Even patients with well-recognized phenotypes such as MELAS can present with an apparently normal muscle biopsy. Patients with multiple mtDNA deletions due to *POLG* mutations may also present with normal muscle histochemistry[261] as can patients with classical chronic progressive external ophthalmoplegia (CPEO) and a single mtDNA deletion,[210] or patients in whom the clinically affected organ is not muscle (e.g. *MPV17* mutations causing mtDNA depletion in liver). With standard histology stains, increased number of lipid droplets within myofibres is a regular feature in patients with PEO and KSS but not in MELAS and MERRF.[207] At the electron microscope level, the extraocular muscles in CPEO reveal a 'mosaic-like' pattern caused by a selective damage of muscle fibres, whereas in LHON cases there is a diffuse increase in mitochondria with preservation of myofibrils.[29]

Finding low levels of COX-deficient muscle fibres in elderly subjects does not indicate mitochondrial disease. In ageing, acquired mtDNA mutations are present and these clonally expand to high levels in individual muscle fibres. This is manifest as a small number of COX-deficient fibres (usually 1–2 per cent, Figure 7.6h). However, because in some mtDNA diseases, the number of COX-deficient fibres can be quite low, there is often a diagnostic problem in determining whether the changes seen are due to ageing or to an mtDNA defect.[19]

Biochemical Assessment of Mitochondrial Function

Mitochondrial oxidative phosphorylation is an extremely complicated biochemical process and it is not surprising that

biochemical assessments are challenging. Measurements of oxidative phosphorylation in different tissues are also important in cases with multisystem involvement. The preparation of intact muscle mitochondria offers a wide range of diagnostic testing for mitochondrial biochemical abnormalities.[32] Rates of flux, substrate oxidation and ATP production are measured by polarography or using ^{14}C-labelled substrates.[251] However, as these tests require that biopsies be sent to specialist centres, most laboratories use frozen muscle samples. In these samples, it is possible to measure the activities of all respiratory chain complexes independently.

Biochemical assays are more important in the investigation of paediatric cases because many children have recessive mutations in nuclear-encoded structural or ancillary proteins that severely compromise enzyme activity. In adults, the biochemical defect may be more subtle and even undetectable in some patients with proven mtDNA defects. Isolated defects involving one complex may be due to mutations of specific subunits. Multiple enzyme defects involving complexes I, III and IV are sometimes seen in patients harbouring single, large-scale mtDNA deletions, mutations in mtDNA tRNA or nuclear genes involved in mtDNA translation.[98] High-throughput oxygraphy has recently been shown to be effective in a diagnostic context.[94]

Genetic Studies in Mitochondrial Disease

Molecular genetic testing of nuclear and mitochondrial DNA provides a fast and accurate means of diagnosing mitochondrial disorders. This can be straightforward in many cases, but can be difficult to interpret as a result of the inherent complexity of mtDNA, and to the huge spectrum of nuclear DNA defects described in patients with mitochondrial diseases.

If the history and examination are suggestive of a classic mitochondrial syndrome such as MERRF or LHON, then investigation for the common mtDNA point mutations known to cause these syndromes should be undertaken in blood. For patients with symptoms suggestive of MELAS, investigation of the m.3243A>G mutation in cells spun down from a urine sample is a much more reliable test, as levels of this mutation may be undetectable in blood.[150,216] Results of screening for specific mutations should, however, be interpreted with caution as negative results do not exclude other forms of mitochondrial disease in a patient in whom there is a high index of clinical suspicion. In such cases, analysis of DNA extracted from a muscle biopsy is important.

A clear autosomal inheritance pattern (usually recessive) supports an underlying nuclear genetic aetiology, but a clear family history is absent in most cases. Those with isolated complex IV deficiency may harbour mutations in genes identified thus far that encode accessory proteins necessary for assembly of the COX holoenzyme complex, for example: *SURF1*,[247,271] *SCO1*,[256] *SCO2*,[100] *COX10*,[257] *COX15*,[8] or *LRPPRC*.[162] The protein product of *LRPPRC* is required for the translation of mtDNA subunits. Children with isolated complex I deficiency and myopathy or Leigh syndrome are more likely to harbour mutations in one of the many nuclear-encoded structural subunits of this enzyme,[250] although accumulating data indicate that pathogenic mtDNA mutations are much more frequent in this paediatric population than previously predicted.[31,113–115,153] The number of nuclear genes associated with mitochondrial disease has been growing exponentially since the introduction of next generation sequencing technologies. The genes listed here are far from comprehensive.

Rearrangements of the mitochondrial genome, including single deletions, duplications and multiple mtDNA deletions, have classically been detected by Southern blot (Figure 7.7a). Southern blot analysis will also identify cases of mtDNA depletion if, in addition to a mitochondrial probe, the blot is hybridized simultaneously with a probe to detect a nuclear gene (commonly 18S rRNA[239]), although real-time PCR methods are also used routinely and are cheaper and faster. Though the technique of Southern blotting remains the 'gold standard' test and will certainly detect all cases of single mtDNA deletions, it may miss low levels of multiple mtDNA deletions in patients with only mild weakness.[44]

Numerous PCR-based assays now exist for the study of mtDNA deletions, and long-range PCR is routinely used by many laboratories as the initial screen for mtDNA rearrangements.[55] Being PCR-based, these assays are very sensitive (Figure 7.7b) and as such require care in their interpretation. First, unlike Southern blotting, which is quantitative down to its detection threshold of about 5 per cent mutated mtDNA, many of the commercially available enzymes for long-range PCR amplify preferentially smaller templates, making quantification impossible. This means that in patients with single mtDNA deletions, only the rearranged mtDNA molecules are amplified, even in the presence of residual full-length, 16.6-kb (wild-type) mtDNA. Second, the sensitive nature of the amplification process means that the PCR of skeletal muscle DNA from normal, elderly controls often reveals low levels of smaller amplicons, indistinguishable from patients with mitochondrial disease. Similar to the finding of COX-deficient fibres on histochemistry, this is consistent with the presence of age-related, somatic mtDNA deletions.[157]

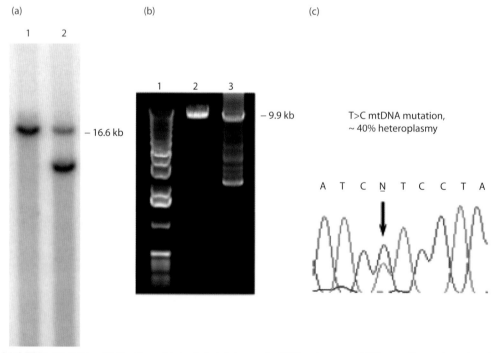

7.7 Mitochondrial DNA analysis. (a) Southern blot of skeletal muscle DNA using a mtDNA specific probe, showing (1) full-length 16.5 kb wild-type mtDNA from a control subject, and (2) an heteroplasmic mtDNA deletion in a patient with CPEO. **(b)** Long-range polymerase chain reaction (PCR) of skeletal muscle mtDNA showing (1) size marker, (2) full-length wild-type mtDNA, and (3) multiple deletions in a patient with autosomal dominant PEO. **(c)** Fluorescent sequence chromatogram showing an heteroplasmic mtDNA point mutation in skeletal muscle mtDNA.

Long-range PCR techniques using shorter extension times may be more valuable in differentiating the deletions seen in ageing from those observed in patients with multiple mtDNA deletions syndromes.[140] Finally, in some cases where the clinical and histochemical findings are suggestive of a multiple mtDNA deletion disorder, the relative amount of deleted mtDNA can be determined in individual COX-deficient and COX-positive muscle fibres by real-time PCR.[74]

The investigation of mtDNA point mutations uses standard molecular genetic techniques, although heteroplasmy complicates the issue because the level of a specific mtDNA mutation may be very low or undetectable in blood but present at high levels in muscle. The choice of tissue for investigation under these circumstances is crucial.

If no common mtDNA mutations are detected in patients with phenotypes typical of mitochondrial disease, then analysis of the entire mitochondrial genome is now routine (Figure 7.7c). However, both diagnostic and evolutionary studies have highlighted the extensive mtDNA sequence variation within human populations, with distinct clusters of sequence changes forming well-recognised haplogroups.[75] Because the majority of mtDNA sequence variants are neutral polymorphisms with no pathogenic significance, careful assessment of newly identified mutations must be made to establish a link with human disease. DiMauro and Schon[47] put forward canonical criteria that they suggest should be met in order to support a pathogenic role for a novel mtDNA mutation. More recently, there has been development of scoring schemes, which use available evolutionary, structural and clinical data to evaluate the likely pathogenicity of mutations in the mitochondrial genome.[152,159]

As mentioned earlier, next-generation sequencing technologies, including custom-designed captures of large number of mitochondrial genes[24,25] or whole exome sequencing,[66,149] have revolutionized the molecular diagnosis of Mendelian mitochondrial disease, leading to the identification of many new mitochondrial disease genes.

MITOCHONDRIAL ENCEPHALOPATHIES— NEUROPATHOLOGY

Kearns-Sayre Syndrome

Externally, the brains and optic nerves of KSS patients are often unremarkable, although atrophy of the cerebrum, cerebellum and optic nerves may be noted. This is in keeping with neuroimaging studies that consistently report diffuse cerebral and cerebellar atrophy. In addition, bilateral, symmetrical, T2 hyperintense MRI lesions are often observed in the cerebral and cerebellar white matter, globus pallidus substantia nigra, dorsal midbrain and thalami (Figure 7.8). Confluent grey-tan discolorations and translucent softenings in the white matter, and discolorations of the basal ganglia are seen in coronal slices. However, the white matter abnormality is more diffuse than generally appreciated neuropathologically.

The neuropathological hallmark of KSS is widespread spongy vacuolation of the white matter (spongiform leukoencephalopathy) in the cerebrum, brain stem, spinal cord, cerebellum, thalamus and basal ganglia.[228] The severity of the lesions can vary from slight myelin pallor, to mild

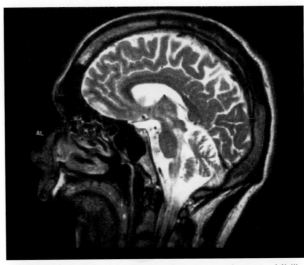

7.8 Neuroimaging in Kearns-Sayre syndrome. Midline T2-weighted image. Abnormal high signal is noted extending bilaterally from the medulla through the midbrain in a 30-year-old male with the Kearns Sayre syndrome due to a single 7-kb deletion of mtDNA.

Adapted from Chinnery PF, Elliot C, Green GR, et al. The spectrum of hearing loss due to mitochondrial DNA defects. Brain 2000;123: 82–92. By permission of Oxford University Press on behalf of The Guarantors of Brain.

spongy change, to coarse vacuolation resulting in a sieve-like appearance (Figure 7.9). Electron microscopic examination has revealed that the spongy alteration in the myelin is due to splits at the intraperiod line.[186] This alteration has been attributed to a preferential mitochondrial dysfunction in oligodendrocytes.[127,155,186,241] Recent investigations have shown reduced density of oligodendrocytes, preferential loss of proteins expressed by oligodendrocytes, including 2', 3'-cyclic nucleotide 3'-phosphodiesterase and myelin-associated glycoprotein, and evidence of respiratory chain deficiency.[127] Higher levels of deleted mtDNA could also be detected in white matter neurons compared to grey matter neurons with 76.9 per cent and 43.9 per cent of mtDNA molecules deleted, respectively. Morphologically, axons have been reported to be relatively preserved, although severe axonal loss tends to occur in the subcortical white matter at late pathological stages (Figure 7.10). Peripheral myelin loss has also been reported in cranial nerves and in spinal motor and sensory nerve roots of some KSS patients.

The cerebral cortex is relatively spared from spongiform degeneration and neuronal loss (Figure 7.10). In some KSS patients, neuronal depletion and spongiform change have been reported in brain stem nuclei, including the substantia nigra, cranial nerve nuclei, the red nucleus and – less markedly – in the thalamus, basal ganglia and spinal cord grey matter.[228] In the cerebellum, Purkinje cell loss and thicker dendritic trees ('dendritic cacti') of the remaining Purkinje cells have been observed. The neurons of the dentate nucleus are usually preserved, though spongiform change and degeneration of presynaptic terminals of Purkinje cells on dentate neurons have been reported in one KSS patient.[127,241] In contrast to the white matter, the spongiform changes in the grey matter are thought to result from mitochondrial dysfunction in astrocytes, resulting in a failure of ATP-dependent

(a)

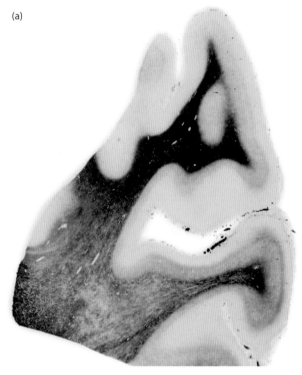

(b)

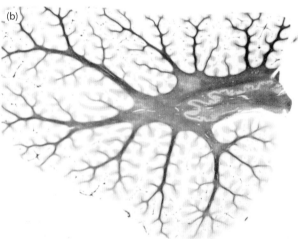

7.9 Kearns-Sayre syndrome. A 40-year-old female with a single mtDNA deletion. **(a)** Demyelination and spongiform change in central white matter adjacent to motor and sensory cortex. Peripheral gyri are less affected. **(b)** Cerebellar white matter showing similar features. Loyez stain for myelin.

ion and water transport across the astrocytic membranes.[241] Furthermore, an immunohistochemical study, which looked at the expression of respiratory chain subunits in the cerebellum from two KSS patients, noted decreased expression of mtDNA-encoded subunits in the dentate neurons without significant alterations of these subunits in the cerebellar cortex or in the inferior olivary neurons that innervate Purkinje cells.[245] Recently, immunohistochemical analysis of respiratory chain components in the olivo-cerebellum of another patient with KSS showed extensive loss of complex I and IV subunit expression in inferior olivary neurons, moderate loss in dentate nucleus neurons and mild loss in Purkinje cells.[127]Quantitative analysis of neuronal cell density has shown that Purkinje cells are most vulnerable with a 60 per

cent reduction in cell density, whereas the inferior olivary nucleus showed a 52 per cent reduction and the dentate nucleus only a 23 per cent reduction. Molecular analysis to investigate the distribution of deleted (δ) mtDNA at the single cell level revealed similar levels of δ-mtDNA in Purkinje cells and olivary neurons at 68 per cent and 68 per cent, respectively, whereas the dentate nucleus neurons showed 44 per cent heteroplasmy, which appears to correlate with the pattern of reduced cell density in these nuclei.

Astrogliosis, macrophage response and capillary proliferation within the areas of spongiform change are usually inconspicuous in KSS patients, though mineral (calcium) deposits are commonly observed in the microvasculature of the globus pallidus and thalamus. In addition, there may be iron encrustations (haemosiderosis) of vascular walls in the globus pallidus, and haemosiderin within astrocytes and microglia of the globus pallidus and caudate nucleus.

Oncocytic transformation of the choroid plexus is also common in KSS. Moreover, a study designed to investigate mtDNA and respiratory chain components in the epithelial cells of the choroid plexus in KSS patients revealed lack or marked reduction of the mtDNA-encoded subunit II of COX in these cells, contrasting with normal expression of the nDNA-encoded FeS subunit of complex II.[244] Furthermore, using *in situ* hybridization, the authors were able to show that δ-mtDNAs were the predominant mtDNA species within these respiratory chain deficient choroid plexus cells. The authors suggested that this mitochondrial dysfunction may lead to failure of the CSF-blood barrier, which, in turn, may account for the increased CSF protein and lactate levels, and the reduced CSF folate levels observed in KSS patients.[192,243] In particular, folates must enter the CSF from the blood, because they are not synthesized in the CNS. They are transferred from the blood into the CSF via the choroid plexus by active transport.[132] Therefore, it is entirely conceivable that mitochondrial dysfunction in the choroid plexus plays an important role in the reduced CSF folate levels in patients with KSS. The main CSF folate, 5-methyl tetrahydrofolate (5-CH$_3$THF), is essential for a number of brain metabolic pathways, including DNA and RNA synthesis, serotonin metabolism, and synthesis and methylation of membrane phospholipids. Indeed, its concentration in the CSF and the brain of mammals is fourfold higher than in plasma.[230,231] Therefore, reduced levels of 5-CH$_3$THF in the CSF may play an important role in the genesis of the characteristic spongiform leukoencephalopathy in this mitochondrial disorder. Despite this elegant hypothesis, the selective distribution of neuropathological changes remains unexplained.

Mitochondrial Encephalomyopathy with Lactic Acidosis and Stroke-like Episodes

Externally, the brain often shows generalized, non-specific atrophy and dilated ventricular spaces. However the hallmark neuropathological features of MELAS are infarct or 'infarct-like' lesions of the cerebral cortex and subcortical white matter.[89,122,160,166,183,218,240,251] The topographical localization involving gyral crests and subjacent white matter in the occipital/temporal lobes can be considered pathognomonic of MELAS.

These multifocal, often asymmetrical, lesions are visible on external examination of the brain where the meninges

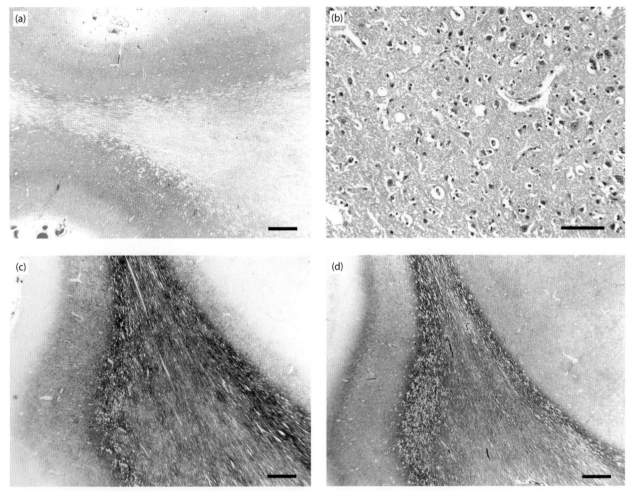

7.10 Kearns-Sayre syndrome. A 40-year-old female with a single mtDNA deletion. **(a)** Severe spongy change in the white matter and relatively intact cortex in posterior frontal lobe. Haematoxylin and eosin (H&E), bar = 850 μm. **(b)** Enlarged portion of cortex from (a), showing mild microvacuolation only. H&E, bar = 100 μm. **(c)** Severe myelin loss from white matter in posterior frontal lobe, serial section to (a). Loyez myelin stain, bar = 850 μm. **(d)** Severe axon loss from white matter in posterior frontal lobe, serial section to (a). Bielschowsky silver stain for axons, bar = 850 μm.

and denuded gyral crests produce a multiply indented surface appearance (Figure 7.11). In brain slices, identification of MELAS lesions may require careful gyral examination because actual loss of grey matter is not always very visible (Figure 7.12). These lesions are thought to relate to stroke-like episodes in patients. The lesions do not follow the vascular territories of specific cerebral arteries or border zone territories.[89,92] Microscopically, in many respects they resemble true infarcts in both the acute and chronic stages, and are characterized by extensive neuronal loss, neuronal eosinophilia, microvacuolation of neuropil and astrogliosis (Figure 7.13). As in Leigh syndrome, dilated small blood vessels with normal or swollen endothelial cells are conspicuous within and at the edge of lesions, but increased vascularity may also occur in viable cortex next to the lesions.

In cortical and subcortical regions not affected by infarct or infarct-like lesions, there may be extensive neuronal loss and neuropil microvacuolation. Specific neuronal depletion is observed in the cerebellum, the gracile and cuneate nuclei (Figure 7.14d), the inferior medullary olives, the pons and the dorsal nucleus of vagus.[160,183,228]

The second most frequent lesion observed in patients with MELAS is dystrophic mineralization (calcification) of the basal ganglia.[228] Mineralization is visible on neuroimaging and closer inspection has revealed that these deposits are actually within the walls of the basal ganglia vasculature.[76,183,228,237] Usually, the calcification is not associated with basal ganglia dysfunction, as the neuronal population is spared and ischaemic-like lesions are rarely observed at this site (Figure 7.14c). The mechanism leading to dystrophic calcification is uncertain, as is its relative specifically to MELAS compared with other mtDNA encephalopathies. Mineral deposits have also been reported in the globus pallidus and thalamus in several KSS patients,[186,228] but not in any MERRF patients. The mineralization in MELAS has a similar histological and radiological distribution to that found in normal aged individuals and basal ganglia calcification is known to follow a wide range of insults to the brain, such as hypoxia and radiation. In normal individuals,[237] it has been suggested that ageing and cellular stress lead to mitochondrial dysfunction, which results in calcium deposition in the basal ganglia vasculature It is possible that this process is enhanced or accelerated in patients with the 3243A>G mutation, but the precise pathogenetic mechanism for vessel calcification remains to be determined.

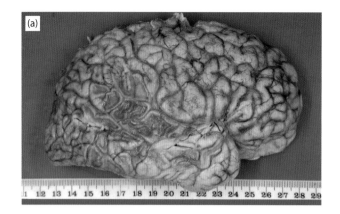

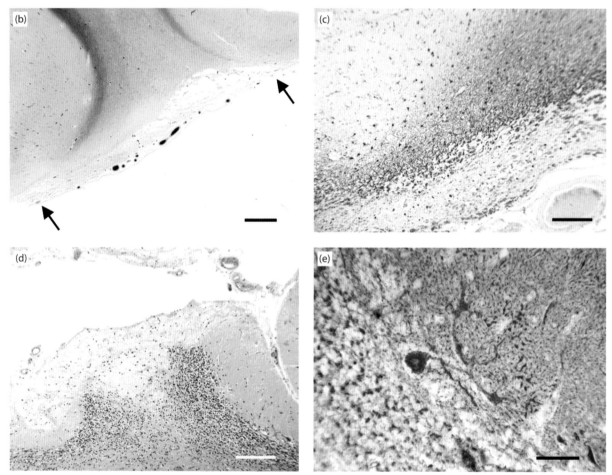

7.11 Mitochondrial encephalomyopathy with lactic acidosis and stroke-like episodes. (a) Extensive cortical ablation/laminar necrosis over the occipital, lateral parietal and lateral inferior temporal lobes in a fixed right cerebral hemisphere. Note the depressions of the pia-arachnoid-cortex at gyral crests and the more preserved cortex next to sulci. **(b)** Cortical ablation at the gyral crest (between arrows) and severe loss of myelinated fibres in the underlying white matter; the arcuate myelinated fibres underlying the unaffected cortex in the sulci are relatively spared; posterior temporal lobe. Loyez myelin stain, bar = 1000 μm. **(c)** Prominent astrogliosis at the interface between the ablated cortex (lower right corner) and white matter, where milder astrogliosis is present (upper left corner), glial fibrillary acidic protein immunohistochemistry (IHC). Haematoxylin counterstain, bar = 200 μm. **(d)** Infarct-like lesion involving the molecular, Purkinje and granular cell layers of the cerebellar cortex and the adjoining white matter. Haematoxylin and eosin, bar = 200 μm. **(e)** Purkinje cell in the vicinity of the infarct-like lesion showing abnormally disorganized dendrites with thickening of primary, secondary and tertiary dendritic branches, COX-I IHC. Haematoxylin counterstain, bar = 50 μm.

Adapted with permission from Betts et al.[13]

The cerebellum is often severely affected in MELAS, with features ranging from devastating ischaemic-like lesions, to microvacuolation of the cortex and white matter, atrophy of the folia and focal Purkinje cell loss.[128] Abnormal Purkinje cell dendrites, including cactus formations, have also been reported in several patients with the 3243A>G mutation.[240] These formations were first described as a characteristic finding in Menkes' kinky hair disease.[158] They are thought to be

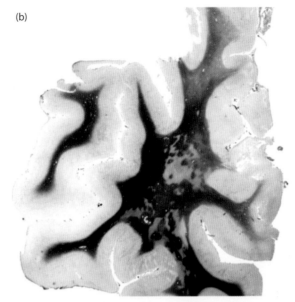

7.12 Mitochondrial encephalomyopathy with lactic acidosis and stroke-like episodes. (a) Temporal cortex with loss of gyral crests over the lateral and inferior surfaces of temporal gyri (arrows). Enlarged insert (double shaft arrows) illustrates surface grey matter ablation in the superior temporal gyrus with a rim of surviving grey matter at the edge. The latter is partially highlighted by photography. Cystic structures may form at the edge of the crest crater abutting the grey matter margin – ball and arrow heads. The floor of the crater-like formations consists of white matter covered with pia-arachnoid membrane. Smaller areas of surface gyral ablation are present at the lateral surface of the middle and inferior temporal gyri. These features are virtually pathognomonic of MELAS. The section is from a 52-year-old woman. **(b)** Frontal cortex and white matter showing central infarcts and demyelination in a 60-year-old woman. Loyez stain.

Adapted with permission from Betts et al.[13]

caused by abnormal accumulation of mitochondria within the dendrites, similar to the abnormalities in ragged-red skeletal muscle fibres.[165,267] It is unclear why cactus formations seem to occur only in the cerebellum. They may reflect a selective vulnerability of Purkinje cells to metabolic disturbances.

A number of different hypotheses have been proposed to account for the pathogenesis of the ischaemic-lesions and neuronal cell death in MELAS. One of the most commonly cited hypotheses suggests that abnormalities within the vascular smooth muscle and endothelial cells of the MELAS brain constitute the pathogenic basis of the brain lesions (the primary vascular hypothesis). This hypothesis is based on the observations of aggregated, enlarged mitochondria in smooth muscle and endothelial cells of the cerebral blood vessel in MELAS patients.[58,62,160,183,248] These changes are most marked in pial arterioles and small arteries, whose vasculature plays an important role in the autoregulation of cerebral blood flow. Further support to this hypothesis is provided by the observation of arterioles strongly reactive to succinate dehydrogenase (strongly succinate dehydrogenase-reactive blood vessels, SSVs) in MELAS patients.[73] In addition, an immunohistochemical study of MELAS brains demonstrated that, although the expression of the nDNA-encoded FeS subunit of complex III was normal, the expression of the mtDNA-encoded COX II subunit was markedly reduced in small arteries and arterioles.[242] It has been suggested that these changes in the cerebral blood vessels could lead to aberrant vascular tone, resulting in local ischaemia and the stroke-like episodes.[62]

Recent work has confirmed widespread respiratory chain deficiency in the cerebral blood vessels of the brains from MELAS patients.[13,129] COX-deficient blood vessels were observed in various brain regions from MELAS and MERRF patients harbouring the m.3243A>G and the m.8344A>G point mutation, respectively. In contrast, cerebral vessels were found to be COX-positive when similar regions from control brains were examined. Importantly, high levels of the m.3243A>G mutation were also found in COX-deficient vessels; a mutant load of 99 per cent was observed in large leptomeningeal arteries, compared to 95 per cent ±3 per cent in smaller cortical arterioles. The mutation load in surrounding normal cortex was lower (76 per cent).[13] These findings strengthen the suggestion that vascular changes are important in the pathogenesis of the ischaemic-like infarcts in MELAS. However, because respiratory chain deficient vessels were found in all regions of the brain examined, the deficits alone cannot explain cortical selectivity of the lesions.

An extension of the vascular hypothesis is the suggestion that changes in the blood–brain barrier (BBB) are important factors in the development of neuropathological changes in MELAS, and indeed other mitochondrial diseases.[108,182,183,243] Again, this hypothesis is based on the observations of abnormal mitochondria and respiratory chain deficiency in the cerebral vasculature in addition to the findings of similar changes in the epithelial cells of the choroid plexus in MELAS patients.[182,243] Furthermore, Tanji and colleagues showed by immunohistochemistry deficiency of the mtDNA-encoded subunit II of COX and normal expression of the nDNA-encoded FeS subunit of complex III in MELAS choroid epithelial cells. The authors suggested this implied that the m.3243A>G mutation was abundant in the epithelial cells and the mutation had reached the threshold level required to impair

7.13 Mitochondrial encephalomyopathy with lactic acidosis and stroke-like episodes. (a) Partial laminar necrosis at the gyral crest of occipital cortex proceeds from the upper to deeper cortical layers. Upper cortical layers show paler staining, prominent microvacuolation and severe neuron loss, whereas the deeper cortical layers show less prominent microvacuolation. Cresyl fast violet. **(b)** Enlarged portion of the deeper cortical layers boxed in (a), showing neurons at various stages of necrosis (empty arrows) and capillary hypertrophy (filled-in arrow). Cresyl fast violet. **(c)** Low concentration of the mtDNA-encoded cytochrome c oxidase (COX-I) mitochondrial enzyme in the microvacuolated upper cortical layers and a high concentration of COX-I in the deeper cortical layers, serial section to (a). COX-I immunohistochemistry. **(d)** Enlarged area portion of cortex boxed in (c) at an interface between the deeper cortical layers affected by the neuronal necrosis (high concentration of COX-I in right third of image) and a relatively intact cortex (moderate concentration of COX-I in the neuropil in left two-thirds of image); note a relatively intact pyramidal neuron (filled-in arrow) containing a high concentration of COX-I. COX-I immunohistochemistry. **(e)** Activated microglia (filled-in arrow) and necrotic neurons (empty arrows) in deeper cortical layers, serial section to (a) and (c). CD68 immunohistochemistry. **(f)** Macrophages (filled-in arrows) within neuropil and in the vicinity of a cortical blood vessel (chevron) in upper cortical layers, serial section to (a) and (c). CD68 immunohistochemistry. **(g)** Activated astroglia (filled-in arrows) in the vicinity of the partial laminar necrosis. Glial fibrillary acidic protein immunohistochemistry. (c–g Haematoxylin counterstain. a,c: bar = 300 μm; b: bar = 20 μm; d–g: bar = 30 μm.)

Adapted with permission from Betts et al.[13]

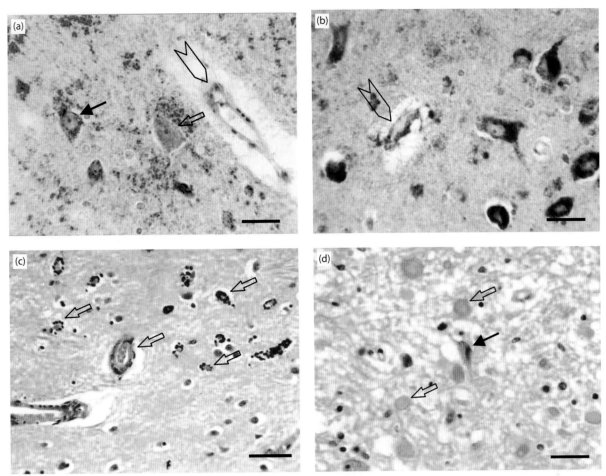

7.14 Mitochondrial encephalomyopathy with lactic acidosis and stroke-like episodes. (a) mtDNA-encoded COX-I mitochondrial enzyme is either absent (empty arrow) or reduced in concentration (filled-in arrow) in morphologically preserved polymorphic neurons of hippocampal CA4 segment; the fine granules in the neuropil surrounding the neurons represent synaptic labelling; the blood vessel wall (chevron) shows absence of COX-I, COX-I immunohistochemistry (IHC) with haematoxylin counterstain. **(b)** nDNA-encoded succinate dehydrogenase (SDH) mitochondrial enzyme is present at high concentration in all the hippocampal polymorphic neurons and in the blood vessel wall (chevron); the fine granules in the neuropil surrounding the neurons represent synaptic labelling; serial section to (a). SDH IHC and haematoxylin counterstain levels. **(c)** Dystrophic microcalcification in vessel walls of putamen H&E. **(d).** Severe microvacuolation, neuron loss, apoptotic neuron (filled-in arrow) and axonal spheroids (empty arrows) in the gracile nucleus. H&E. (a,b,d: bar = 30 µm; c: bar = 50 µm.)

Adapted with permission from Betts et al.[13]

mitochondrial protein synthesis. Based on these observations and on the presence of COX II deficiency in the microvasculature, the authors went on to ascertain if a breakdown in the BBB permeability had occurred in the cerebral cortex of MELAS patients. They found evidence of serum protein (fibrinogen) in the superficial and deep layers of both infarcted and non-infarcted cortex, indicating that a breakdown of the BBB does occur in MELAS patients.[243] Indeed, a recent study has shown that high levels of the m.3243A>G and m.8344A>G point mutations are associated with COX-deficiency within the endothelium. There is also evidence in MELAS brains of a loss of endothelial cell integrity with reduced immunoreactivity for tight junctional proteins, including occludin and zona-occludins 1 and extravasation of fibrinogen into the grey matter and white matter neuropil.[129]

An alternative hypothesis to explain the distribution of the infarct-like lesions and necrosis within specific CNS regions blames mitochondrial energy failure not only in the brain vasculature but also within the neurons (the metabolic hypothesis).[58] To support this hypothesis, Gilchrist and colleagues provided electron microscopic evidence of abnormal mitochondria within neuronal, smooth muscle and endothelial cells. According to the metabolic hypothesis, the distribution of the infarct-like lesions in the posterior temporal and occipital cortex may indicate a greater metabolic demand on energy-challenged cortical cells in these than in other brain regions. This is supported by MR spectroscopic data, which shows increased concentrations of lactate and impaired oxidative metabolism within the focal cortical lesions during an acute episode, with return to normal after clinical resolution.[104,148] Furthermore, positron emission tomography (PET) scanning revealed that impaired glucose uptake in MELAS patients, with or without CNS symptoms, was most prominent in the occipital and temporal regions. However, it was not established whether this decrease was due to impaired cerebral perfusion or to higher metabolic rate of neurons in these regions.[161]

Another explanation for the unusual topographical distribution of cortical neuropathology relates to different threshold levels for the m.3243A>G mutation of neurons in different brain regions. For example, neurons in the posterior cortex of MELAS patients may have a lower threshold for the m.3243A>G mutation and a higher vulnerability to the respiratory chain defect, thus explaining the selective distribution of the pathology. However, in one MELAS patient there was little correlation between the threshold levels in different neuronal populations and the distribution of neuropathological changes. The threshold for the m.3243A>G mutation was surprisingly low in the hippocampus (between 31 per cent and 43 per cent) whereas m.3243A>G levels greater than 70 per cent were observed in COX-positive neurons within the occipital cortex and cerebellum, indicating that the threshold level in these neurons was >70 per cent. These observations are inconsistent with the hypothesis that cell loss is determined by the threshold level for the 3243A>G mutation.

Yet another hypothesis posits a neurovascular cellular mechanism and was proposed by Iizuka and colleagues, who suggest that neuronal hyperexcitability or seizure activity is the predominant trigger of ischaemic-like lesions in mtDNA disease.[90,93] Evidence in support of this theory stems from clinical studies in patients with the m.3243A>G mutation, of whom 72 per cent suffer from headaches and 50 per cent are affected by epileptic seizures and hemianopia. Because EEG-recorded seizure activity often correlates with the lesion foci in the acute stages, this suggests that epileptic activity is responsible for driving the structural changes in the brain. It has been suggested that the cause of neuronal hyperexcitability is the presence of the mtDNA defect within neurons, astrocytes and microglia, which alters extracellular ion homeostasis and induces membrane instability, eventually resulting in impaired neuronal networking and manifesting as epilepsy. In conjunction with these changes, impaired mitochondrial function in the microvasculature is likely to cause impaired cerebrovascular reactivity and loss of BBB integrity, which could lead to mismatched neuronal activity and blood supply and may even perpetuate the seizure activity. The prolonged epileptic activities are considered likely to drive the progressive spread of the lesions. Neuroradiological imaging studies have provided some insight into the cellular mechanisms of the stroke-like lesions both in the acute and in the chronic stages, but many also show contradictory results. In the acute stages, measures of cerebrovascular perfusion using SPECT showed evidence of hypoperfusion hours after seizure activity,[117,181] whereas others have shown hyperperfusion in the days, and hypoperfusion in the months, after the event.[91] Measures using diffusion-weighted imaging (DWI) and apparent diffusion coefficient (ADC) to tease out the cellular and vascular contributions showed high DWI signals and either normal or decreased ADC,[96,184,265] suggestive of vasogenic and cytotoxic oedema. In the chronic stages, in the areas of MRI T2-hyperintensities, magnetic resonance spectroscopy (MRS) demonstrates decreased N-acetyl-aspartate and presence of lactate peaks, indicative of irreversible changes in cortical regions.[121,255] The conflicting imaging results could be due to a number of factors, including the definition of temporal events and interpretation bias.

The formation of infarct-like lesions in patients with mtDNA disease is increasingly being recognized as part of the disease process in patients harbouring genetic defects other than the m.3243A>G, including m.8344A>G and *POLG* mutations, impacting on disease morbidity and mortality. For this reason, understanding the physiopathology of stroke-like lesion is an important area under intense investigation.

Myoclonic Epilepsy with Ragged-Red Fibres

On external examination, the brain often appears unremarkable, although there may be brown discoloration and shrinkage of the dentate nucleus. Microscopically, severe neuronal loss and astrocytosis are observed in the dentate nucleus and may be accompanied by grumous degeneration (Figure 7.15). Neuronal loss is less severe in the red nuclei, pons and basal ganglia. Moderate loss of pallidal neurons, particularly of the outer segment, is common but is rarely associated with mineralization. The gracile and cuneate nuclei and Clarke's column in the spinal cord may also be affected. Within the cerebellum, loss of Purkinje cells is generally moderate. Quantitative assessment of neuronal cell density has revealed 85 per cent, 67 per cent and 75 per cent reduction in neuron density in the inferior olives, Purkinje cells and dentate nucleus, respectively.[127] In fact, ischaemic-like lesions have been documented in the cerebellar cortex in association with biochemical defects and pathological changes in the microvasculature.[129] Studies have also demonstrated the accumulation of abnormally large mitochondria in the cerebellar cortex and dentate nuclei of patients.[56,242,243] Despite PET studies showing evidence of decreased cortical metabolism with normal cortical blood flow and cerebral pH, loss of cerebral cortical neurons is rare.[12]

Biochemical defects in respiratory chain complexes I, III and IV are common in MERRF.[229] Immunohistochemical studies revealed selective and severe deficiency of mtDNA-encoded subunits of complex I and of subunits I and II of complex IV (COX-I and COX-II) in the surviving neurons of the dentate nucleus and medullary olive (Figure 7.15). However, this deficiency was also present in Purkinje cells and in scattered neurons within the medullary nuclei and cerebral cortex. A micro-dissection study of mutated mtDNA in a MERRF patient has documented that the percentage of m.8344A>G mutant DNAs was similar in neuronal somata and in adjacent neuropil and glia.[271] In this study, mutant mtDNA was approximately 97 per cent in Purkinje cells compared to 80 per cent in anterior horn cells. However, the distribution of mutant mtDNA did not correlate well with the degree of cell loss, which was 45 per cent in the dentate neurons and only 7 per cent in Purkinje cells. A similar study has shown high levels of mutated mtDNA in individual neurons from the inferior olives, Purkinje cells and dentate nucleus at 86.1, 91.9 and 87.4 per cent, respectively.[127] The authors concluded that additional factors probably contribute to cell death in MERRF. Indeed, a recent study has shown an increased mtDNA copy number in pathologically affected regions of the brain, including the putamen, hippocampus and caudate nucleus.[20] Whether this reflects a compensatory mechanism remains to be elucidated.

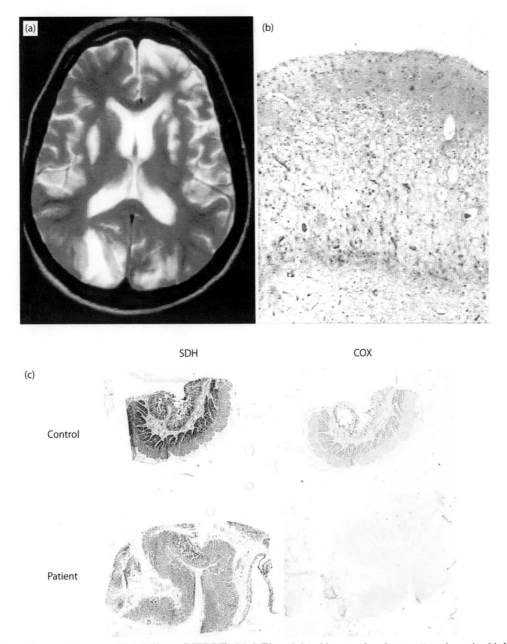

SDH   COX

7.15 **Myoclonic epilepsy with ragged-red fibres (MERRF). (a)** A T2-weighted image showing scattered cerebral infarcts, particularly in the occipital lobes. **(b)** Haematoxylin and eosin stain showing an infarcted region in the occipital cortex (×55). **(c)** Succinate dehydrogenase and cytochrome *c* oxidase histochemistry on sections of frozen ileum from a control and from a patient with MERRF (×2.5). Note the reduced stain for COX in all layers in the patient's sample.

Adapted from Tanji K, Gamez J, Cervera C, et al. The A8344G mutation in mitochondrial DNA associated with stroke-like episodes and gastrointestinal dysfunction. Acta Neuropathol [Burl], 2003;105:69–75. With kind permission of Springer Science and Business Media.

Leigh Syndrome

In LS cases coming to autopsy, neuropathological lesions may be varied and widespread, and involve grey and white matter from the optic nerves to the spinal cord. Cystic degeneration or areas of softening are characteristically visible as symmetrical lesions occurring in a distinct topographical distribution affecting particularly substantia nigra and periaqueductal grey matter in the midbrain, and the putamen in basal ganglia. In some cases, lesions may also be visible in the thalamus or subthalamic nuclei (Figure 7.16a). Microscopic lesions are more widespread and commonly involve the dorsal pons, inferior olives in the medulla, roof nuclei in the cerebellum, and posterior columns and anterior horns in cervical spinal cord (Figure 7.16b). In the subcortical nuclei, the putamen may be more affected than the caudate nucleus and the subthalamic nucleus may be selectively involved. The cerebral cortex and white matter are generally spared, as are the mammillary bodies. The lack of involvement of the mammillary bodies is an important feature in juvenile or adult cases because it helps differentiate Leigh syndrome from Wernicke's encephalopathy.

The microscopic features are distinctive and in established lesions consist of 'vasculonecrotic' lesions characterized by

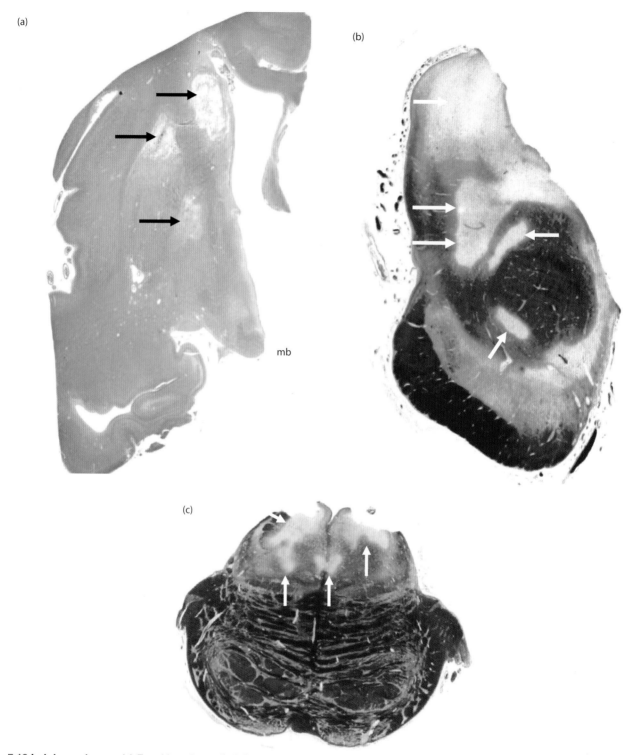

(a)

(b)

mb

(c)

7.16 Leigh syndrome. (a) Focal basal ganglia infarcts in caudate nucleus, dorsal putamen and globus pallidus (arrows) with sparing of mammillary body (mb). Similar infarcts in **(b)** midbrain and **(c)** pons involving periaqueductal tissue, red nucleus and colliculi (arrows). (Nissl and Loyez stains.)

a combination of tissue rarefaction with varying degrees of spongiform neuropil change, and capillary prominence suggesting either local capillary aggregation or capillary proliferation. This unusual capillary response is within or adjacent to affected grey or white matter. Recent lesions may show neuropil, probably cytotoxic, oedema and swollen cell processes. As the lesion evolves, this is followed by reactive neurone and astroglial cell changes, microglial proliferation and eventual tissue necrosis. In grey matter lesions, neuronal ischaemic-type change and neuron loss predominate, and in white matter demyelination, axonal loss, axonal spheroid formation and spongy change may be apparent.

The type of lesion found may be dependent on the brain area affected, and cytotoxic oedematous lesions in the putamen may differ from necrosis occurring in periaqueductal grey matter. Although neuronal loss and neuronal ischaemic change may be prominent in the affected areas, it is often possible to identify apparently intact individual neurones in areas showing severe rarefaction or neuropil degeneration. Surviving neurones are an important clue to the differential diagnosis of such infarct-like lesions, because their presence helps differentiate LS lesions from the more common infarcts associated with hypoxic-ischaemic insults.

Preferentially or selectively involved brain areas in LS include substantia nigra, inferior colliculi and periaqueductal grey, dorsal medulla, spinal cord (especially dorsal columns and anterior horns), cerebellar roof nuclei and adjacent white matter, pontine tegmentum, corpus striatum (especially putamen), inferior olives in medulla, subthalamic nuclei and thalami. In cerebellar folia, Purkinje cell abnormalities and loss may be associated with neurone loss in the inferior olives (Figure 7.16a-c).

Beyond childhood, documented LS cases are uncommon and their aetiology is more controversial. Nagashima and colleagues report a predilection for midbrain, brain stem and thalamic lesions in this age group and a more rapid clinical course.[168]

Leber Hereditary Optic Neuropathy

Histopathological data is only available on a few cases of LHON and none during the acute phase of visual loss.[27] The most striking finding in the molecularly confirmed cases[112,201–203] was the dramatic loss of the retinal ganglion cell layer and retinal nerve fibre layer, predominantly affecting the central fibres, with variable sparing of the periphery. Evidence of axoplasmic abnormalities was also observed with focal accumulation of mitochondria and cytoplasmic debris. No inflammatory cells were seen. Focal demyelination and occasional regions of re-myelination were associated with numerous glial cells and occasional macrophages filled with lipofuscin, suggestive of ongoing neurodegeneration long after the subacute visual loss. Additional features were seen in a patient with two mtDNA mutations,[111] in whom residual retinal ganglion cells contained swollen mitochondria and double-membrane bodies containing calcium. This family was unusual because some members exhibited additional clinical features similar to MELAS.

Additional clinical features in patients with LHON have also been documented, including peripheral neuropathy (see Mitochondrial Peripheral Neuropathies, p. 552) and a Leigh-like encephalopathy with dystonia.[102] Anita Harding first described a multiple sclerosis (MS)-like illness in eight women from families with LHON due to the m.11778G>A mutation of mitochondrial DNA (mtDNA).[71] Subsequent reports have described clinically definite MS in males[23,187] and females[71] with other LHON mtDNA mutations,[82,110] usually in patients presenting with prominent optic nerve dysfunction. It is currently not clear whether these rare cases arise purely by chance, or whether mtDNA mutations are aetiologically linked to the pathogenesis of the MS-like disorder. Recent pathological examination of one case with the m.14484T>C mutation (complicated by Hashimoto's thyroiditis) revealed a spectrum of neuropathological changes, including actively and inactively demyelinating plaques in the white matter and optic nerve, vacuolation and cystic necrosis with CD8-positive T-cells in the frontal lobe, axonal damage and vacuolation of white matter,[120] implicating immune mediated mechanisms. In addition, spinal cord degeneration has also been described in one LHON patient.[101]

Multiple mtDNA Deletions Disorders

The neuropathological changes in a single patient with multiple mtDNA deletions were originally described by Cottrell and colleagues[39] (Figure 7.17). At the time, the molecular defect was not known, but subsequent sequencing of nuclear DNA from postmortem samples revealed two autosomal recessive *POLG* mutations, p.Ala467Thr and p.X1240Gln. Macroscopically, the brain showed moderate atrophy, and the brain stem and the cerebellum were reduced in size. Neurohistopathological changes included massive neuronal depletion in the inferior olivary nuclei in the medulla and moderate to severe Purkinje cell loss with less neuronal depletion in the dentate nucleus. Moderate microgliosis was present in the red nuclei, while the pons remained unaffected. This distribution of pathology is indicative of cerebello-olivary atrophy and is likely to be responsible for the cerebellar ataxia observed in this patient and in other individuals with multiple mtDNA deletions. Massive myelin loss and associated axonal loss and astrogliosis were observed in the dorsal columns of the cervical spinal cord, and in dorsal spinal roots. Severe neuronal degeneration was seen in dorsal root ganglia and paraspinal sympathetic ganglia with evidence of respiratory chain deficiencies for complexes I and IV in remaining cells. Laser microcapture dissection of individual sensory neurons confirmed the presence of multiple mtDNA deletions, and quantification of mtDNA copy number revealed a significant reduction in mtDNA content.[128] Sural nerve showed features of chronic axonal degeneration and regeneration. In the spinal cord, the ventral and lateral myelin tracts and motor neurons were intact. These changes are likely to account for the sensory neuropathy afflicting this patient. Sections of the midbrain revealed severe neuronal depletion from the substantia nigra, without specific neuronal cytopathology or Lewy body formation. This pathology is likely to explain the parkinsonism symptoms observed in patients.

Cottrell and colleagues[39] observed that most COX-deficient neurons were found in the reticular formation, nucleus, ambiguous, caudate nucleus, putamen, globus pallidus and pontine nuclei. The lowest levels were found in the cerebellum, hippocampus, motor cortex and spinal cord. Thus, the proportional distribution of COX-deficient neurons did not always correlate directly with the degree of neuropathological damage, as regions of high neuronal loss had relatively low proportions of COX-deficient cells. Further, other clinically affected CNS regions had higher levels of COX-deficient neurons without significant cell loss. These findings indicate that additional factors must be involved in determining neuronal susceptibility to multiple mtDNA deletions in the pathogenesis of this disease. These factors may include neuronal dependence on oxidative phosphorylation or thresholds for apoptosis, in which mitochondria play a pivotal role.

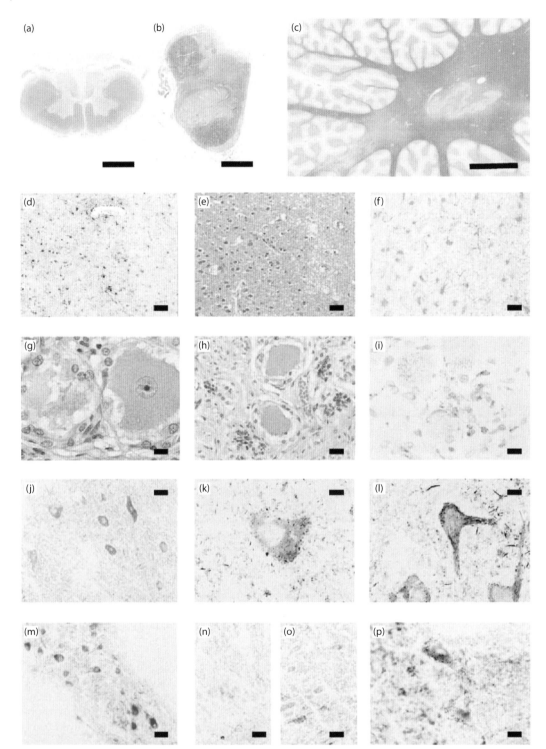

7.17 Multiple mtDNA deletion disorder. (a) Severe myelin pallor in the dorsal columns of the mid-cervical spinal cord. **(b)** Medulla oblongata showing loss of myelin from the inferior olivary nucleus. **(c)** Irregular myelin pallor of the central and foliar cerebellar white matter. **(d)** Diffuse astrocytic gliosis of the occipital cortex with sparing of the lower laminae. **(e)** Microvacuolation of superficial layers was prominent in all neocortical regions. **(f)** Reactive astrocytes in the depleted pyramidal layer of hippocampal sector CA1. **(g)** Dorsal root ganglion cells. **(h)** Satellite cell clusters (nodules of Nageotte) throughout the dorsal root ganglia indicating neuronal loss and focal macrophage immunoreactivity **(i)**. **(j)** Enzyme histochemistry of the spinal ventral horn showing cytochrome *c* oxidase (COX)-deficient/SDH positive anterior horn cells (blue) contrasting with the normal brown stain of the other large motor neurons. **(k)** A single COX-deficient lower motor neuron. **(l)** Normal COX staining in these neurons. **(m–p)** Varying populations of neurons (m, inferior olive; n, hippocampal dentate cells; o,p: hippocampal pyramidal cells). Stains: a,b,c: cresyl violet/Luxol fast blue; d,f: GFAP; e,g,h: haematoxylin and eosin; I: CD68; m,n,o,p: COX-SDH histochemistry. Scale bars: a,b: 3 μm; c: 5 μm; d,e: 100 μm; f,j,m,n,o: 50 μm; h,i,k,l,p: 25 μm; g: 12.5 μm.

Adapted from Cottrell DA, Ince PG, Blakely EL, et al. Neuropathological and histochemical changes in a multiple mitochondrial deletion disorder. J Neuropathol Exp Neurol, 2000; 59: 621–7. Reproduced with permission from Lippincott Williams & Wilkins/Wolters Kluwers Health.

The neuropathology of two siblings from a Swedish adPEO family has been investigated.[156] Both patients harboured the mtDNA *POLG* p.Tyr955Cys mutation and died at similar age, 60 and 61 years. Microscopic examination revealed severe neuronal loss of pigmented neurons in the substantia nigra of both patients, without the presence of Lewy bodies, in agreement with a recent study.[14] The cerebellum and white matter were without obvious changes. Interestingly, one patient also displayed pathology typical of Alzheimer's disease (AD), including numerous diffuse neuritic plaques throughout the cerebral cortex and neurofibrillary tangles in the hippocampus, entorhinal cortex, amygdala and nucleus of Meynert. It may be purely coincidental that this patient developed AD pathology in addition to changes associated with multiple mtDNA deletions. However, the accumulation of mtDNA point mutations and deletions is thought to play an important role in ageing and degenerative disorders, such as AD. Therefore, it is possible that the higher levels of mtDNA mutations present in this patient may have contributed to the early development of Alzheimer pathology in the brain of this patient.

Mitochondrial neurogastrointestinal encephalomyopathy (MNGIE) is an autosomal recessive disease associated with multiple deletions and depletion of mtDNA in skeletal muscle.[78,190] In 1976, Okamura and associates reported the first case as 'congenital oculoskeletal myopathy with abnormal muscle and liver mitochondria'.[185] Since then, more than 35 additional individual with MNGIE have been described, with several acronyms: myo-, neuro-, gastrointestinal encephalopathy (MNGIE),[9] polyneuropathy, ophthalmoplegia, leukoencephalopathy and intestinal pseudo-obstruction (POLIP);[224] oculogastrointestinal muscular dystrophy (OGIMD);[95] and mitochondrial encephalomyopathy with sensorimotor polyneuropathy, ophthalmoplegia and pseudo-obstruction (MEPOP).[198]

In 1999, Hirano and colleagues identified mutations in the gene encoding thymidine phosphorylase (*TYMP*), located on chromosome 13.32qter, as the cause of MNGIE. This enzyme usually catabolizes thymidine to thymine and 2-deoxy-d-ribose 1-phosphate. In MNGIE patients, mutations in *TYMP* severely reduce the enzyme activity in leukocytes and presumably in other tissues.[179] As a consequence of the *TYMP* defect, average plasma levels of thymidine are elevated nearly 50-fold in patients. The accumulation of thymidine is likely to alter deoxynucleoside and nucleotide pools and consequently impair mtDNA replication, repair, or both, leading to mtDNA abnormalities (depletion, multiple deletions and point mutations).

The most prominent and debilitating symptom of MNGIE is gastrointestinal dysmotility, which is due to neuromuscular dysfunction. Any portion of the enteric system, from the oropharynx through the small intestine, may be affected.[133] Histological abnormalities, including increased numbers of abnormal appearing mitochondria, are present in both intestinal smooth muscles and the enteric nervous system.[9,76,95,138] Recent studies document mtDNA depletion combined with mitochondrial proliferation and smooth muscle cell atrophy in the external layer of the muscularis propria in the stomach and small intestine with loss of interstitial cells of Cajal, pacemaker cells responsible for stimulating gut contraction.[59,60]

In contrast, the clinical neurological features of MNGIE are relatively mild. One of the most consistent features of MNGIE is a leukoencephalopathy observed as T2 hyperintensity on MRI. This striking leukoencephalopathy is not manifested neurologically or neuropsychiatrically as dementia or mental retardation. Furthermore, the MRI findings lack a pathological correlation, such as demyelination or gliosis at autopsy.[238] Szigeti and colleagues hypothesised that *TYMP* may play a role in BBB function, and that the loss of function of this enzyme results in BBB breakdown causing subtle oedema, which correlates with the white matter changes observed on MRI in patients.[238] Using albumin immunohistochemistry, they investigated the intracellular albumin immunoreactivity in astrocytes and neurons in two MNGIE cases and age-matched normal controls. They found a statistically significant difference in the number of albumin-positive cells between the MNGIE cases (26.08 ± 6.29) and the healthy control subjects (7.59 ± 3.17) (Figure 7.18c,d). Furthermore, the white matter capillaries and astroglial cells also indicated a striking loss of TP expression in MNGIE brains (Figure 7.18a,b). However, further investigation is required to elucidate the importance of TP in the maintenance of BBB, and it remains unknown whether BBB dysfunction is a primary or secondary phenomenon of MNGIE. The most severe neurological manifestation of MNGIE is the neuropathy and this is discussed later.

Mitochondrial DNA Depletion Syndromes

Neuropathological investigations of MDS have increased in recent years. A post-mortem study of an infant with a documented mutation in the *DGUOK* gene (4-bp GATT duplication: nucleotides 763–766 in exon 6) revealed hepatocerebral pathology. The liver showed cirrhosis with small and middle-sized nodules and peripheral fibrosis. In the central nervous system, foci of spongy degeneration in the white matter of cerebral and cerebellar hemispheres were associated with mild astrogliosis. The cerebral cortex and subcortical grey nuclei were normal without any evidence of axonal degeneration. The cerebellar cortex showed focal Purkinje cell loss with Bergmann gliosis, while the dentate nucleus was unaffected.[53] Muscle biopsies from members of two families with mutations in the TK2 gene (family one: homozygous Ala181Val substitution; family two: substitutions Cys108Trp and Leu257Pro) showed severe depletion of mtDNA and features of progressive dystrophic process, including ragged-red fibres, COX deficiency, great variability in fibre size and shape and increased content of connective and fat tissue; the central nervous system was not examined histologically.[57]

Alpers–Huttenlocher Syndrome

Macroscopically, the cerebral cortex is variably involved in AHS, showing patchy thinning and discoloration, with a striking predilection for the striate cortex. Microscopic changes affect all areas but affect the calcarine cortex most severely and include spongiosis, neuronal loss and astrogliosis progressing down the cortical layers frequently accompanied by signs of capillary proliferation.[227] In the liver,

MNGIE   Control

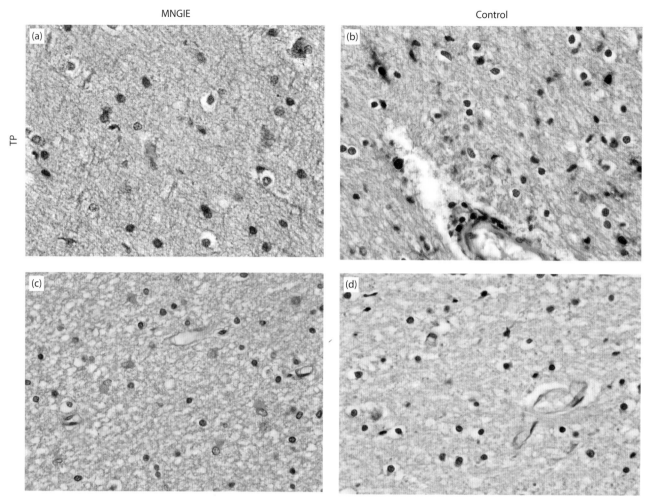

7.18 Mitochondrial neurogastrointestinal encephalomyopathy. Thymidine phosphorylase (TP) immunoreactivity is markedly decreased in the white matter of MNGIE patients **(a)** compared to controls **(b)**. Albumin-positive reactive astrocytes are rare in healthy control subjects **(d)**, but are prominent in MNGIE **(c)**, reflecting increased blood–brain barrier permeability.

Adapted from Szigeti K, Sule N, Adesina AM, et al. Increased blood–brain barrier permeability with thymidine phosphorylase deficiency. Ann Neurol 2004;56:881–6. With permission of John Wiley and Sons. Copyright © 2004 American Neurological Society.

the pathological changes include fatty change, hepatocyte loss, bile duct proliferation, fibrosis and often cirrhosis.[70] In addition to the cortex, lesions have been found in sub-cortical gray nuclei and cerebellar cortex of an infant with AHS.[225] Central-peripheral axonopathy affecting the deep sensation carried by the peripheral nerve fibres and the posterior tracts of the spinal cord, due to neuronal loss in the sensory ganglia, has also been reported in a juvenile patient with AHS.[225]

Infantile Onset Spinocerebellar Ataxia

Neuroradiological imaging typically reveals cerebellar cortical, olivopontocerebellar and spinocerebellar atrophy. Measures of the acute diffusion coefficient (ADC) following onset of epilepsy reveal multiple small hypointensities involving the whole hemisphere, thalamus and caudate nucleus. These hypointensities are suggestive of cortical oedema localized to non-vascular territories. DWI shows hyperintensities within the oedematous regions similar to early ischaemic changes. Neuropathological features

include patchy laminar cortical necrosis in the occipital cortex, basal ganglia, thalami and subthalamic nucleus.[137] Immunohistochemical studies have shown a selective loss of complex I subunits in the cerebellum and frontal cortex, whereas subunits of complexes II and IV remain unchanged. Molecular genetic investigation reveals mtDNA depletion in the cerebrum and cerebellum, with residual amounts of mtDNA at levels 5–20 per cent of control tissues.[67]

MITOCHONDRIAL PERIPHERAL NEUROPATHIES

Diseases of the central nervous system and muscle have taken central stage among disorders due to mitochondrial dysfunction ('mitochondrial encephalomyopathies') and have often overshadowed peripheral nerve involvement, which, however, is extremely frequent. This should come as no surprise because considerable energy is required for axonal transport and for the synthesis and deposition of the myelin sheath.

Clinically, mitochondrial peripheral neuropathies (PNs) are predominantly distal and sensory, with blunted or absent deep tendon reflexes. Electrophysiology is usually consistent with axonal PN and shows decreased sensory and motor amplitudes and relatively preserved conduction velocities, although some patients have mainly demyelinating PN, with prolonged distal latencies, slowed conduction velocities and absent F waves. Sural nerve biopsy often shows loss of large and small myelinated fibres, thinly remyelinated fibres and demyelinated axons. Abnormal mitochondrial, some containing paracrystalline inclusions, can be seen in Schwann cells, axons and in endothelial or smooth muscle cells of endoneurial and perineurial arterioles.

Peripheral nerves are rarely—if ever—affected in isolation by mitochondrial dysfunction, and PNs are usually part of syndromic disorders. A rational classification of the mitochondrial PNs can be based on a combination of genetic and clinical criteria: whether they are due to mutations in mitochondrial DNA (mtDNA) or in nuclear DNA (nDNA); and whether neuropathy is a major or minor clinical component.

PN Due to mtDNA Mutations

This group of disorders can, in turn, be subdivided according to the type of the mutation: some mutations, such as large-scale mtDNA rearrangements (single deletions, duplications, or both together) and point mutations in tRNA or rRNA genes, impair mitochondrial protein synthesis *in toto*, whereas mutations in protein-coding genes affect specifically the respiratory chain complex to which the mutated protein belongs. Rearrangements of mtDNA are usually *de novo* events and the related disorders are sporadic; point mutations in tRNA genes are usually maternally inherited; point mutations in protein-coding genes can be maternally inherited or *de novo* – and often somatic – events resulting in sporadic and tissue-specific disorders.

Neuropathy as a Major Clinical Component

Neuropathy dominates the clinical picture in three conditions, two (MERRF and MELAS) due to mutations in tRNA genes, the other (NARP) due to mutations in the gene encoding the ATPase 6 component of complex V.

Myoclonic Epilepsy with Ragged-Red Fibres

Neuropathy is present in about 30 per cent of patients with typical MERRF and the m.8344A>G mutation in tRNA[Lys46] but is even more frequent (80 per cent) in the MERRF variant associated with multiple symmetric lipomatosis (MSL), also known as Madelung disease.[171,172] These patients have multiple non-encapsulated, often disfiguring lipomas in the neck and shoulder-girdle region and a predominantly axonal sensorimotor neuropathy. Other symptoms include cerebellar ataxia, neurosensory hearing loss, optic atrophy and mitochondrial myopathy with ragged-red fibres, which are COX-negative histochemically. Sural nerve biopsies show a predominantly axonal neuropathy with loss of large myelinated axons.[33,193] Ultrastructural studies have shown abnormal mitochondria with amorphous matrix both in axons and in Schwann cells. A recent study found combined

central and peripheral demyelination with reduced motor conduction and absent sensory action potentials in a child with the m.8344A>G mutation.

Mitochondrial Encephalomyopathy, Lactic Acidosis, Stroke-like Episodes

In this most common of mtDNA-related and maternally inherited disorders, the frequency of PN was estimated at 22 per cent,[105] but a recent study of 30 patients with typical MELAS and the common m.3243A>G mutation in tRNA[Leu(UUR)] has revealed that 77 per cent had abnormal nerve conduction measures.[106] Of these, 43 per cent had sensory abnormalities only, 35 per cent had both sensory and motor abnormalities, and 22 per cent had motor abnormalities only. Electrophysiological changes indicated axonal or mixed neuropathy in 83 per cent of patients and demyelinating neuropathy in the remaining 17 per cent. Symptoms of PN were present in only half of the patients, but almost all had decreased reflexes or distal sensory abnormalities on exam, especially in the legs. Male gender and older age seemed to contribute to the genetic disposition to develop PN. The common occurrence of a subclinical PN in MELAS/m.3243A>G patients makes them vulnerable to dichloroacetate, a lactate-lowering agent often used anecdotally in mitochondrial patients. This risk became evident during a randomized, placebo controlled therapeutic trial of dichloroacetate, which had to be interrupted because of peripheral nerve toxicity.[107]

Neuropathy, Ataxia and Retinitis Pigmentosa

Neuropathy is a defining feature in this maternally inherited disorder, also characterized by developmental delay, ataxia, retinitis pigmentosa, seizures and dementia, and typically associated with the m.8993T>G mutation in *ATPase* 6 gene of mtDNA.[80] Patients have both proximal neurogenic and distal limb weakness, absent ankle jerks and loss of vibratory sensation. Muscle biopsies often show features of denervation but no ragged-red fibres. When the mutation load surpasses 90 per cent, the clinical presentation is in infancy and the CNS is predominantly affected (maternally inherited Leigh syndrome, MILS). Nerve conduction studies typically reveal a sensorimotor neuropathy with length-dependent axonal deficits.

Neuropathy as a Minor Clinical Component
Kearns-Sayre Syndrome

Despite the multisystem nature of this disorder due to single deletions of mtDNA, PN is usually subclinical and often revealed only by electrophysiological studies, possibly because post-mortem studies are rare. One early clinico-pathological study demonstrated peripheral myelin loss in spinal motor and sensory nerve roots and in cranial nerves of a KSS patient.[64]

Leber's Hereditary Optic Neuropathy

LHON is a maternally inherited subacute, painless loss of vision inexplicably more common in men than women and almost invariably associated with three mutations in genes encoding subunits of complex I, m.11778G>A, m.3460G>A and m.14484T>C. Although the optic nerve is the target organ in this disease, PN is rare.[178] Sural biopsy in one patient with the clinical diagnosis of LHON (not confirmed genetically), spastic paraplegia and PN, showed signs of axonal degeneration and demyelination.[189]

Peripheral Neuropathies Due to nDNA Mutations

From the genetic point of view, these disorders fall into multiple subgroups, including: (i) mutations in genes encoding respiratory chain subunits; (ii) mutations in genes encoding assembly proteins; (iii) defects of intergenomic signaling (mtDNA multiple deletions; mtDNA depletion; defective translation of mtDNA); and (iv) mutations in genes controlling mitochondrial motility, fusion and fission.

Neuropathy as a Major Clinical Component

Neuropathy characterizes the clinical picture in some defects of intergenomic signaling with multiple mtDNA deletions and mtDNA depletion (MNGIE, SANDO), and in two forms of Charcot-Marie-Tooth type 2A due to defects of mitochondrial fusion or motility.

Mitochondrial Neurogastrointestinal Encephalomyopathy

The defining clinical features of this autosomal recessive disorder due to mutations in the gene (*TYMP*) encoding the enzyme thymidine phosphorylase (TP) are gastrointestinal dysmotility, cachexia, ophthalmoplegia, PN and leukoencephalopathy.[179,180] PN is present to a more or less severe degree in all patients with MNGIE and, in a few, it is the predominant or the presenting symptom. The neuropathy involves the legs more than the arms and is manifested by stocking-glove sensory loss and areflexia. Nerve conduction studies show features of demyelination in about 75 per cent of patients and features of mixed axonal and demyelinating

neuropathy in 25 per cent. Muscle biopsy usually shows signs of denervation (fibre type grouping, group atrophy and target fibres), together with mitochondrial proliferation (ragged-red fibres, COX-negative fibres). Molecular studies of muscle show both multiple mtDNA deletions and some degree of mtDNA depletion. Absence of TP activity in the buffy coat and greatly increased blood levels of thymidine are useful diagnostic clues.[234] Nerve biopsies show loss of myelinated fibres, segmental demyelination and remyelination and occasional onion bulb formation (Figure 7.19). Ultrastructural studies have shown abnormal mitochondria in Schwann cells.[204] In a few patients with MNGIE, demyelination predominates and the clinical presentation may mimic chronic inflammatory demyelinating polyneuropathy (CIDP) or Charcot–Marie–Tooth disease.[11,204]

Sensory Ataxic Neuropathy with Dysarthria and Ophthalmoplegia

This autosomal recessive syndrome is one of the protean clinical expressions of mutations in the gene (*POLG*) encoding polymerase γ, an enzyme involved in the synthesis, replication and repair of mtDNA.[136] The cardinal features of SANDO are sensory ataxic neuropathy, dysarthria, and progressive external ophthalmoplegia (PEO). Depression is often part of this syndrome. The muscle biopsy shows COX-negative ragged-red fibres and multiple mtDNA deletions are detectable in muscle and other tissues. One patient with SANDO harboured a heterozygous mutation in the *C10orf2* gene (previously known as *Twinkle* and now renamed *PEO1*), suggesting that autosomal dominant inheritance is sometimes possible. Patients have loss of vibration and position sense in the legs, mildly decreased pinprick and temperature sensation, sensory ataxic gait and areflexia. Nerve conduction studies show absent sensory responses in all limbs, although motor responses and conduction velocities are relatively preserved. Sural nerve biopsies show loss of large and small myelinated axons, with regenerative clusters and endoneurial fibrosis, but without

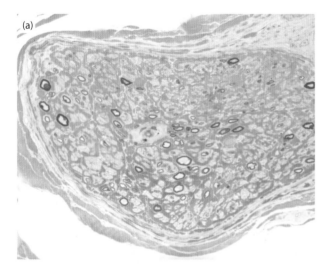
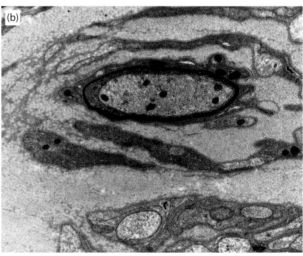

7.19 Mitochondrial neurogastrointestinal encephalomyopathy, sural nerve. (a) The nerve shows loss of myelinated fibres, scattered isolated thinly-myelinated fibres and a few small regenerative clusters. Semithin plastic section, toluidine blue. **(b)** This electron micrograph shows a small onion bulb. No ultrastructural abnormalities of mitochondria were encountered.

With thanks to Dr Hays, Columbia University Medical Center, New York.

abnormal mitochondria.[260] A recent study provides evidence that SANDO may be a sensory ganglionopathy with degeneration of sensory axons.[130]

CMT 2A

Mitochondria move on microtubular rails propelled by motor proteins, usually GTPases called kinesins. Nowhere is mitochondrial motility more important than in the peripheral nervous system, as axonal flow is essential to carry ATP-generating mitochondria all the way from the soma of an anterior horn motor neuron to the neuromuscular junction (to use but one example). Mitochondrial motility is strictly related to mitochondrial fusion and fission. The machinery for mitochondrial fusion requires several proteins, including two outer membrane GTPases, mitofusin 1 (MFN1) and mitofusin 2 (MFN2), and a third dynamin-related GTPase, OPA-1, which is located in the inner mitochondrial membrane. Similarly, the machinery for mitochondrial fission involves several proteins acting in concert, especially a GTPase called dynamin-related protein 1 (DRP-1). Interestingly, mutations in two genes, one (KIF1B) encoding a kinesin motor protein, the other (MFN2) encoding mitofusin 2, have been associated with CMT 2A, an autosomal dominant axonal neuropathy characterized by symmetric, distal, motor and sensory neuropathy, with normal or minimally slowed nerve conduction velocities.[126,273] Mutations in OPA1, encoding a fusion protein, cause autosomal dominant optic atrophy (AOA), the mendelian counterpart of LHON,[2,43] whereas mutations in KIF5A, encoding a kinesin motor protein, cause a variant of autosomal dominant spastic paraplegia.[51]

Neuropathy as a Minor Clinical Component
POLG Mutations

Although PN is a defining clinical component of SANDO, it is also very common and often a dominant feature in patients with autosomal dominant or recessive PEO and other mutations in POLG. 'Mild axonal neuropathy', 'stocking-glove numbness', 'impaired vibration' and 'impaired vibration, sensory ataxia, areflexia' have been described in patients harbouring heterozygous mutations,[52,143] and in two compound heterozygous patients.[52,143] Sural nerve biopsies in two patients showed loss of large and small myelinated fibres, axonal degeneration and fibrosis,[144] and demyelination and axonal loss.[79] Electrophysiological assessment of PN in 11 patients harbouring POLG mutations reveals a predominantly sensory neuronopathy with evidence of motor fibre involvement in later life. In agreement with this, post-mortem investigation of spinal cord and dorsal root ganglion tissues from a patients harbouring compound heterozygous autosomal recessive POLG mutations revealed striking neuronal cell loss from the dorsal root ganglia with evidence on-going degenerative changes.[129,155] In addition, severe respiratory chain deficiencies of complexes I and IV were found in remaining cells. Molecular analysis showed a striking reduction of mtDNA content in remaining neurons, which is likely responsible for the high levels of mitochondrial respiratory deficiency.[130]

Leigh Syndrome

Leigh syndrome is a neurodegenerative disorder of infancy or early childhood defined neuropathologically by bilateral symmetrical foci of spongiform degeneration and vascular proliferation in the basal ganglia, thalamus, and brain stem. Clinically, these children show developmental regression, brain stem dysfunction (recurrent vomiting, nystagmus, abnormal respiration) and seizures. Retinitis pigmentosa characterizes the maternally inherited form (MILS), which is often seen in families with NARP (see p. 530). LS has been associated with a great variety of biochemical mitochondrial defects. Aside from NARP/MILS, most forms of LS are due to mutations in nuclear genes encoding subunits of the pyruvate dehydrogenase complex (PDHC), subunits of respiratory chain complexes (complex I and complex II) or assembly proteins (complexes I, III, IV and V). Although PN is not a major clinical feature of LS (or is subclinical and overshadowed by the encephalopathy), 'acute polyneuropathy',[236] 'Guillain-Barré syndrome'[36] and polyneuropathy[65] were reported in three children with biochemically undefined LS. The biochemical or molecular aetiologies were also unknown in four children with neuropathologically defined LS, whose sural nerve biopsies showed primary demyelination and remyelination and loss of both myelinated and unmyelinated axons.[61] Demyelinating neuropathy was described in three children with LS and cardiomyopathy, who were homozygous for a mutation (E140K) in SCO2, but documentation of the PN was limited to neurogenic atrophy in muscle biopsies and decreased nerve conduction velocities in one case.[99] A sural nerve biopsy from a 5-year-old child with COX deficiency and a homozygous nonsense mutation in SURF1 showed loss of large diameter myelinated fibres and thin myelin sheaths in remaining fibres.[206]

REFERENCES

1. Agostino A, Valletta L, Chinnery PF, et al. Mutations of ANT1, Twinkle, and POLG1 in sporadic progressive external ophthalmoplegia (PEO). *Neurology*, 2003; 60: 1354–6.

2. Alexander C, Votruba M, Pesch UE, et al. OPA1, encoding a dynamin-related GTPase, is mutated in autosomal dominant optic atrophy linked to chromosome 3q28. *Nat Genet*, 2000; 26: 211–5.

3. Alpers B. Diffuse progressive degeneration of the gray matter of the cerebrum. *Arch Neurol Psychiatry*, 1931; 25: 469–505.

4. Anderson S, Bankier AT, Barrell BG, et al. Sequence and organization of the human mitochondrial genome. *Nature*, 1981; 290: 457–65.

5. Andreu AL, Hanna MG, Reichmann H, et al. Exercise intolerance due to mutations in the cytochrome *b* gene of mitochondrial DNA. *N Engl J Med*, 1999a; 341: 1037–44.

6. Andreu AL, Tanji K, Bruno C, et al. Exercise intolerance due to a nonsense mutation in the mtDNA ND4 gene. *Ann Neurol*, 1999b; 45: 820–3.

7. Andrews RM, Griffiths PG, Johnson MA, Turnbull DM. Histochemical localisation of mitochondrial enzyme activity in human optic nerve and retina. *Brit J Ophthalmol*, 1999; 83: 231–5.

8. Antonicka H, Mattman A, Carlson CG, et al. Mutations in COX15 produce

a defect in the mitochondrial heme biosynthetic pathway, causing early-onset fatal hypertrophic cardiomyopathy. *Am J Hum Genet*, 2003; **72**: 101–14.

9. Bardosi A, Creutzfeldt W, Dimauro S, et al. Myo-, neuro-, gastrointestinal encephalopathy (MNGIE syndrome) due to partial deficiency of cytochrome-c-oxidase. A new mitochondrial multisystem disorder. *Acta Neuropathol (Berl)*, 1987; **74** 248–58.

10. Beal MF. Mitochondria take center stage in ageing and neurodegeneration. *Ann Neurol*, 2005; **58** 495–505.

11. Bedlack RS, Vu T, Hammans S, et al. MNGIE neuropathy: five cases mimicking chronic inflammatory demyelinating polyneuropathy. *Muscle Nerve*, 2004; **29**: 364–8.

12. Berkovic SF, Shoubridge EA, Andermann F, et al. Clinical spectrum of mitochondrial DNA mutation at base pair 8344 [letter; comment]. *Lancet*, 1991; **338**: 457.

13. Betts J, Jaros E, Perry RH, et al. Molecular neuropathology of MELAS: level of heteroplasmy in individual neurones and evidence of extensive vascular involvement. *Neuropathol Appl Neurobiol*, 2006; **32**: 359–73.

14. Betts-Henderson J, Jaros E, Krishnan KJ, et al. Alpha-synuclein pathology and parkinsonism associated with POLG1 mutations and multiple mitochondrial DNA deletions. *Neuropathol Appl Neurobiol*, 2009; **35**: 120–4.

15. Bicknese AR, May W, Hickey WF, Dodson WE. Early childhood hepatocerebral degeneration misdiagnosed as valproate hepatotoxicity. *Ann Neurol*, 1992; **32**: 767–75.

16. Bohlega S, Tanji K, Santorelli FM, et al. Multiple mitochondrial DNA deletions associated with autosomal recessive ophthalmoplegia and severe cardiomyopathy. *Neurology*, 1996; **46**: 1329–34.

17. Bosbach S, Kornblum C, Schroder R, Wagner M. Executive and visuospatial deficits in patients with chronic progressive external ophthalmoplegia and Kearns-Sayre syndrome. *Brain*, 2003; **126**: 1231–40.

18. Bourdon A, Minai L, Serre V, et al. (2007) Mutation of RRM2B, encoding p53-controlled ribonucleotide reductase (p53R2), causes severe mitochondrial DNA depletion. *Nat Genet*, 2007; **39**: 776–80.

19. Brierley EJ, Johnson MA, Lightowlers RN, James OF, Turnbull DM. Role of mitochondrial DNA mutations in human aging: implications for the central nervous system and muscle. *Ann Neurol*, 1998; **43**: 217–23.

20. Brinckmann A, Weiss C, Wilbert F, et al. Regionalized pathology correlates with augmentation of mtDNA copy numbers in a patient with myoclonic epilepsy with ragged-red fibers (MERRF-syndrome). *PLoS One*, 2010: **5**: e13513.

21. Brown DT, Samuels DC, Michael EM, Turnbull DM, Chinnery PF. Random genetic drift determines the level of mutant mtDNA in human primary oocytes. *Am J Hum Genet*, 2000; **68**: 533–6.

22. Bu X, Rotter JI. X chromosomal-linked and mitochondrial gene control of Leber hereditary optic neuropathy: Evidence from segregation analysis for dependence on X-chromosome inactivation. *Proc Nat Acad Sci USA*, 1991; **88**: 8198–202.

23. Buhmann C, Gbadamosi J, Heesen C. Visual recovery in a man with the rare combination of mtDNA 11778 LHON

mutation and a MS-like disease after mitoxantrone therapy. *Acta Neurol Scand*, 2002; **106**: 236–9.

24. Calvo SE, Compton AG, Hershman SG, et al. Molecular diagnosis of infantile mitochondrial disease with targeted next-generation sequencing. *Sci Transl Med*, 2012; **4**: 118–10.

25. Calvo SE, Tucker EJ, Compton AG, et al. High-throughput, pooled sequencing identifies mutations in NUBPL and FOXRED1 in human complex I deficiency. *Nat Genet*, 2010; **42**: 851–8.

26. Campos Y, Garcia A, Lopez A, et al. Cosegregation of the mitochondrial DNA A1555G and G4309A mutations results in deafness and mitochondrial myopathy. *Muscle Nerve*, 2002; **25**: 185–8.

27. Carelli V, Ross-Cisneros FN, Sadun AA. Optic nerve degeneration and mitochondrial dysfunction: genetic and acquired optic neuropathies. *Neurochem Int*, 2002; **40**: 573–84.

28. Carelli V, Ross-Cisneros FN, Sadun AA. Mitochondrial dysfunction as a cause of optic neuropathies. *Prog Retin Eye Res*, 2004; **23**: 53–89.

29. Carta A, Carelli V, D'Adda T, Ross-Cisneros FN, Sadun AA. Human extraocular muscles in mitochondrial diseases: comparing chronic progressive external ophthalmoplegia with Leber's hereditary optic neuropathy. *Br J Ophthalmol*, 2005; **89**: 825–7.

30. Casari G, De Fusco M, Ciarmatori S, et al. Spastic paraplegia and OXPHOS impairment caused by mutations in paraplegin, a nuclear-encoded mitochondrial metalloprotease. *Cell*, 1998; **93**: 973–83.

31. Chol M, Lebon S, Benit P, et al. The mitochondrial DNA G13513A MELAS mutation in the NADH dehydrogenase 5 gene is a frequent cause of Leigh-like syndrome with isolated complex I deficiency. *J Med Genet*, 2003; **40**: 188–91.

32. Chretien D, Rustin P, Bourgeron T, et al. Reference charts for respiratory chain activities in human tissues. *Clin Chim Acta*, 1994; **228**: 53–70.

33. Chu CC, Huang CC, Fang W, et al. Peripheral neuropathy in mitochondrial encephalomyopathies. *Eur Neurol*, 1997; **37**: 110–5.

34. Ciafaloni E, Ricci E, Shanske S, et al. MELAS: clinical features, biochemistry, and molecular genetics. *Ann Neurol*, 1992; **31**: 391–8.

35. Cock HR, Tabrizi SJ, Cooper JM, Schapira AH. (1998) The influence of nuclear background on the biochemical expression of 3460 Leber's hereditary optic neuropathy. *Ann Neurol*, 1998; **44**: 187–93.

36. Coker SB. Leigh disease presenting as Guillain-Barre syndrome. *Pediatr Neurol*, 1993; **9**: 61–3.

37. Copeland WC. Defects in mitochondrial DNA replication and human disease. *Crit Rev Biochem Mol Biol*, 2012; **47**: 64–74.

38. Cottrell DA, Turnbull DM. Mitochondria and ageing. *Curr Opin Clin Nutr Metab Care*, 2000; **3**: 473–8.

39. Cottrell DA, Ince PG, Blakely EL, et al. Neuropathological and histochemical changes in a multiple mitochondrial deletion disorder. *J Neuropathol Exp Neurol*, 2000; **59**: 621–7.

40. Danielson S R, Wong A, Carelli V, et al. Cells bearing mutations causing Leber's hereditary optic neuropathy are sensitized

to Fas-induced apoptosis. *J Biol Chem*, 2002; **277**: 5810–5.

41. Davidzon G, Greene P, Mancuso M, et al. Early-onset familial parkinsonism due to POLG mutations. *Ann Neurol*, 2006; **59**: 859–62.

42. Davidzon, G., Mancuso, M., Ferraris, S., et al. POLG mutations and Alpers syndrome. *Ann Neurol*, 2005; **57**: 921–3.

43. Delettre C, Lenaers G, Pelloquin L, Belenguer P, Hamel CP. OPA1 (Kjer type) dominant optic atrophy: a novel mitochondrial disease. *Mol Genet Metab*, 2002; **75**: 97–107.

44. Deschauer M, Kiefer R, Blakely EL, et al. A novel Twinkle gene mutation in autosomal dominant progressive external ophthalmoplegia. *Neuromuscul Disord*, 2003; **13**: 568–72.

45. Di Mauro S. Mitochondrial encephalomyopathies: back to Mendelian genetics [editorial; comment]. *Ann Neurol*, 1999; **45**: 693–4.

46. Di Mauro S, Hirano M, Kaufmann P, Al E. Clinical features and genetics of myoclonic epilepsy with ragged red fibers. In Fahn S, Frucht SJ eds. *Myoclonus and paroxysmal dyskinesia*. Philadelphia, Lippincott Williams & Wilkins, 2002.

47. Di Mauro S, Schon EA. Mitochondrial DNA mutations in human disease. *Am J Med Genet*, 2001; **106**: 18–26.

48. Esteitie N, Hinttala R, Wibom R et al. Secondary metabolic effects in complex I deficiency. *Ann Neurol*, 2005; **58**: 544–52.

49. Fadic R, Russell JA, Vedanarayanan VV, et al. Sensory ataxic neuropathy as the presenting feature of a novel mitochondrial disease. *Neurology*, 1997; **49**: 239–245.

50. Ferrari G, Lamantea E, Donati A, et al. Infantile hepatocerebral syndromes associated with mutations in the mitochondrial DNA polymerase-gamma A. *Brain*, 2005; **128**: 723–31.

51. Fichera M, Lo Giudice M, Falco M, et al. Evidence of kinesin heavy chain (KIF5A) involvement in pure hereditary spastic paraplegia. *Neurology*, 2004; **63**: 1108–10.

52. Filosto M, Mancuso M, Nishigaki Y, et al. Clinical and genetic heterogeneity in progressive external ophthalmoplegia due to mutations in polymerase gamma. *Arch Neurol*, 2003; **60**: 1279–84.

53. Filosto M, Mancuso M, Tomelleri G, et al. Hepato-cerebral syndrome: genetic and pathological studies in an infant with a dGK mutation. *Acta Neuropathol (Berl)*, 2004; **108**: 168–71.

54. Fratter C, Raman P, Alston CL, et al. RRM2B mutations are frequent in familial PEO with multiple mtDNA deletions. *Neurology*, 2011; **76**: 2032–4.

55. Fromenty B, Manfredi G, Sadlock J, et al. Efficient and specific amplification of identified partial duplications of human mitochondrial DNA by long PCR. *Biochim Biophys Acta*, 1996; **1308**: 222–30.

56. Fukuhara N. MERRF: a clinicopathological study. Relationships between myoclonus epilepsies and mitochondrial myopathies. *Rev Neurol (Paris)*, 1991; **147**: 476–9.

57. Galbiati S, Bordoni A, Papadimitriou D, et al. New mutations in TK2 gene associated with mitochondrial DNA depletion. *Pediatr Neurol*, 2006; **34**: 177–85.

58. Gilchrist JM, Sikirica M, Stopa E, Shanske S. Adult-onset MELAS. Evidence for involvement of neurons as well as cerebral

vasculature in strokelike episodes. *Stroke*, 1996; **27**: 1420–3.

59. Giordano C, Sebastiani M, De Giorgio R, et al. Gastrointestinal dysmotility in mitochondrial neurogastrointestinal encephalomyopathy is caused by mitochondrial DNA depletion. *Am J Pathol*, 2008; **173**: 1120–8.

60. Giordano C, Sebastiani M, Plazzi G, et al. Mitochondrial neurogastrointestinal encephalomyopathy: evidence of mitochondrial DNA depletion in the small intestine. *Gastroenterology*, 2006; **130**: 893–901.

61. Goebel HH, Bardosi A, Friede RL, et al. Sural nerve biopsy studies in Leigh's subacute necrotizing encephalomyelopathy. *Muscle Nerve*, 1986; **9**: 165–73.

62. Goto Y. Clinical features of MELAS and mitochondrial DNA mutations. *Muscle Nerve*, 1995; **3**: S107–12.

63. Goto Y, Nonaka I, Horai S. A mutation in the tRNA(Leu)(UUR) gene associated with the MELAS subgroup of mitochondrial encephalomyopathies. *Nature*, 1990; **348**: 651–3.

64. Groothuis DR, Schulman S, Wollman R, Frey J, Vick NA. Demyelinating radiculopathy in the Kearns-Sayre syndrome: a clinicopathological study. *Ann Neurol*, 1980; **8**: 373–80.

65. Grunnet ML, Zalneraitis EL, Russman BS, Barwick MC. Juvenile Leigh's encephalomyelopathy with peripheral neuropathy, myopathy, and cardiomyopathy. *J Child Neurol*, 1991; **6**: 159–63.

66. Haack TB, Danhauser K, Haberberger B, et al. Exome sequencing identifies ACAD9 mutations as a cause of complex I deficiency. *Nat Genet*, 2010; **42**: 1131–4.

67. Hakonen AH, Goffart S, Marjavaara S, et al. Infantile-onset spinocerebellar ataxia and mitochondrial recessive ataxia syndrome are associated with neuronal complex I defect and mtDNA depletion. *Hum Mol Genet*, 2008; **17**: 3822–35.

68. Hakonen AH, Heiskanen S, Juvonen V, et al. Mitochondrial DNA polymerase W748S mutation: a common cause of autosomal recessive ataxia with ancient European origin. *Am J Hum Genet*, 2005; **77**: 430–41.

69. Hakonen AH, Isohanni P, Paetau A, et al. Recessive Twinkle mutations in early onset encephalopathy with mtDNA depletion. *Brain*, 2007; **130**: 3032–40.

70. Harding AE, Holt IJ, Cooper JM, et al. Mitochondrial myopathies: genetic defects. *Biochemical Society Transactions*, 1990; **18**: 519–22.

71. Harding AE, Sweeney MG., Miller DH, et al. Occurrence of a multiple sclerosis-like illness in women who have a Leber's hereditary optic neuropathy mitochondrial DNA mutation. *Brain*, 1992; **115**: 979–89.

72. Harding BN, Alsanjari N, Smith SJ, et al. Progressive neuronal degeneration of childhood with liver disease (Alpers' disease) presenting in young adults. *J Neurol, Neurosurg Psychiatry*, 1995; **58**: 320–5.

73. Hasegawa H, Matsuoka T, Goto Y, Nonaka I. Strongly succinate dehydrogenase-reactive blood vessels in muscles from patients with mitochondrial myopathy, encephalopathy, lactic acidosis, and stroke-like episodes. *Ann Neurol*, 1991; **29**: 601–5.

74. He L, Chinnery PF, Durham SE, et al. Detection and quantification of mitochondrial DNA deletions in individual cells by real-time PCR. *Nucleic Acids Res*, 2002; **30**: e68.

75. Herrnstadt C, Elson JL, Fahy E, et al. Reduced-median-network analysis of complete mitochondrial DNA coding-region sequences for the major African, Asian, and European haplogroups. *Am J Hum Genet*, 2002; **70**: 1152–71.

76. Hirano M, Pavlakis SG. Mitochondrial myopathy, encephalopathy, lactic acidosis, and strokelike episodes (MELAS): current concepts. *J Child Neurol*, 1994; **9**: 4–13.

77. Hirano M, Ricci E, Koenigsberger MR, et al. MELAS: an original case and clinical criteria for diagnosis. *Neuromuscul Disord*, 1992; **2**: 125–35.

78. Hirano M, Silvestri G, Blake DM, et al. Mitochondrial neurogastrointestinal encephalomyopathy (MNGIE): clinical, biochemical, and genetic features of an autosomal recessive mitochondrial disorder. *Neurology*, 1994; **44**: 721–7.

79. Hisama FM, Mancuso M, Filosto M, Di Mauro S. Progressive external ophthalmoplegia: a new family with tremor and peripheral neuropathy. *Am J Med Genet A*, 2005; **135**: 217–9.

80. Holt I, Harding AE, Morgan-Hughes JA. Deletion of muscle mitochondrial DNA in patients with mitochondrial myopathies. *Nature*, 1988; **331**: 717–9.

81. Holt IJ, Harding AE, Petty RK, Morgan-Hughes JA. A new mitochondrial disease associated with mitochondrial DNA heteroplasmy. *Am J Hum Genet*, 1990; **46**: 428–33.

82. Horvath R, Abricht A, Shoubridge EA, et al. Leber's hereditary optic neuropathy presenting as multiple-sclerosis like illness. *J Neurol*, 2000; **247**: 65–67.

83. Horvath R, Hudson G, Ferrari G, et al. Phenotypic spectrum associated with mutations of the mitochondrial polymerase gamma gene. *Brain*, 2006; **129**: 1674–84.

84. Howell N. Leber hereditary optic neuropathy: how do mitochondrial DNA mutations cause degeneration of the optic nerve? *J Bioenerget Biomemb*, 1997; **29**: 165–73.

85. Hudson G, Amati-Bonneau P, Blakely EL, et al. Mutation of OPA1 causes dominant optic atrophy with external ophthalmoplegia, ataxia, deafness and multiple mitochondrial DNA deletions: a novel disorder of mtDNA maintenance. *Brain*, 2008; **131**: 329–37.

86. Hudson G, Deschauer M, Busse K, Zierz S, Chinnery, P. F. Sensory ataxic neuropathy due to a novel C10Orf2 mutation with probable germline mosaicism. *Neurology*, 2005; **64**: 371–3.

87. Hudson G, Keers S, Man PY, et al. Identification of an x-chromosomal locus and haplotype modulating the phenotype of a mitochondrial DNA disorder. *Am J Hum Genet*, 2005; **77**: 1086–91.

88. Huttenlocher PR, Solitare GB, Adams G. Infantile diffuse cerebral degeneration with hepatic cirrhosis. *Arch Neurol*, 1976; **33**: 186–92.

89. Ihara Y, Namba R, Kuroda S, Sato T, Shirabe T. Mitochondrial encephalomyopathy (MELAS): pathological study and successful therapy with coenzyme Q10 and idebenone. *J Neurol Sci*, 1989; **90**: 263–71.

90. Iizuka T, Sakai F. Pathogenesis of stroke-like episodes in MELAS: analysis of neurovascular cellular mechanisms. *Curr Neurovasc Res*, 2005; **2**: 29–45.

91. Iizuka T, Sakai F, Ide T, et al. Regional cerebral blood flow and cerebrovascular reactivity during chronic stage of stroke-like episodes in MELAS -- implication of neurovascular cellular mechanism. *J Neurol Sci*, 2007; **257**: 126–38.

92. Iizuka T, Sakai F, Kan S, Suzuki N. Slowly progressive spread of the stroke-like lesions in MELAS. *Neurology*, 2003; **61**: 1238–44.

93. Iizuka T, Sakai F, Suzuki N, et al. Neuronal hyperexcitability in stroke-like episodes of MELAS syndrome. *Neurology*, 2002; **59**: 816–24.

94. Invernizzi F, D'Amato I, Jensen PB, et al. Microscale oxygraphy reveals OXPHOS impairment in MRC mutant cells. *Mitochondrion*, 2012; **12**: 328–35.

95. Ionasescu VV, Hart M, Di Mauro S, Moraes CT. Clinical and morphologic features of a myopathy associated with a point mutation in the mitochondrial *tRNA(Pro)* gene. *Neurology*, 1994; **44**: 975–7.

96. Ito S, Shirai W, Asahina M, Hattori T. Clinical and brain MR imaging features focusing on the brain stem and cerebellum in patients with myoclonic epilepsy with ragged-red fibers due to mitochondrial A8344G mutation. *Am J Neuroradiol*, 2008; **29**: 392–5.

97. Jackson MJ, Schaefer JA, Johnson MA, et al. Presentation and clinical investigation of mitochondrial respiratory chain disease. *Brain*, 1995; **118**: 339–357.

98. Jacobs HT, Turnbull DM. Nuclear genes and mitochondrial translation: a new class of genetic disease. *Trends Genet*, 2005; **21**: 312–4.

99. Jaksch M, Horvath R, Horn N, et al. Homozygosity (E140K) in SCO2 causes delayed infantile onset of cardiomyopathy and neuropathy. *Neurology*, 2001; **57**: 1440–6.

100. Jaksch M, Ogilvie I, Yao J, et al. Mutations in SCO2 are associated with a distinct form of hypertrophic cardiomyopathy and cytochrome c oxidase deficiency. *Hum Mol Genet*, 2000; **9**: 795–801.

101. Jaros E, Mahad DJ, Hudson G, et al. Primary spinal cord neurodegeneration in Leber hereditary optic neuropathy. *Neurology*, 2007; **69**: 214–6.

102. Jun AS, Brown MD, Wallace DC. A mitochondrial DNA mutation at nucleotide pair 14459 of the NADH dehydrogenase subunit 6 gene associated with maternally inherited Leber hereditary optic neuropathy and dystonia. *Proc Nat Acad Sci USA*, 1994; **91**: 6206–10.

103. Kaguni LS. DNA polymerase gamma, the mitochondrial replicase. *Annu Rev Biochem*, 2004; **73**: 293–320.

104. Kamada K, Takeuchi F, Houkin K, et al. Reversible brain dysfunction in MELAS: MEG, and (1)H MRS analysis. *J Neurol Neurosurg Psychiatry*, 2001; **70**: 675–8.

105. Karppa M, Syrjala P, Tolonen U, Majamaa K. Peripheral neuropathy in patients with the 3243A>G mutation in mitochondrial DNA. *J Neurol*, 2003; **250**: 216–21.

106. Kaufmann P, Anziska Y, Gooch CEA. Nerve conduction abnormalities in MELAS/3243 patients. *Arch Neurol*, 2006; **63**: 746–8.

107. Kaufmann P, Engelstad K, Wei Y-H, Al E. Dichloracetate causes toxic neuropathy in

MELAS: a randomized, controlled clinical trail. *Neurology*, 2006; **66**: 324–30.

108. Kaufmann P, Shungu DC, Sano MC, et al. Cerebral lactic acidosis correlates with neurological impairment in MELAS. *Neurology*, 2004; **62**: 1297–302.

109. Kaukonen J, Juselius JK, Tiranti V, et al. Role of adenine nucleotide translocator 1 in mtDNA maintenance. *Science*, 2000; **289**: 782–5.

110. Kellar-Wood H, Robertson N, Govan GG, Compston DA, Harding AE. Leber's hereditary optic neuropathy mitochondrial DNA mutations in multiple sclerosis. *Ann Neurol*, 1994; **36**: 109–12.

111. Kerrison JB, Howell N, Miller NR., Hirst L, Green WR. Leber hereditary optic neuropathy. Electron microscopy and molecular genetic analysis of a case [see comments]. *Ophthalmology*, 1995; **102**: 1509–16.

112. Kerrison JB, Miller NR, Hsu F, et al. A case-control study of tobacco and alcohol consumption in Leber hereditary optic neuropathy. *Am J Ophthalmol*, 2000; **130**: 803–12.

113. Kirby DM, Boneh A, Chow CW, et al. Low mutant load of mitochondrial DNA G13513A mutation can cause Leigh's disease. *Ann Neurol*, 2003; **54**: 473–8.

114. Kirby DM, Kahler SG, Freckmann ML, Reddihough D, Thorburn DR. Leigh disease caused by the mitochondrial DNA G14459A mutation in unrelated families. *Ann Neurol*, 2000; **48**: 102–4.

115. Kirby DM, McFarland R, Ohtake A, et al. Mutations of the mitochondrial ND1 gene as a cause of MELAS. *J Med Genet*, 2004; **41**: 784–9.

116. Kirkman MA, Yu-Wai-Man P, Korsten A, et al. Gene-environment interactions in Leber hereditary optic neuropathy. *Brain*, 2009; **132**: 2317–26.

117. Koga Y, Akita Y, Junko N, et al. Endothelial dysfunction in MELAS improved by l-arginine supplementation. *Neurology*, 2006; **66**: 1766–9.

118. Kollberg G, Moslemi AR, Darin N, et al. POLG1 mutations associated with progressive encephalopathy in childhood. *J Neuropathol Exp Neurol*, 2006; **65**: 758–68.

119. Kornblum C, Broicher R, Walther E, et al. Cricopharyngeal achalasia is a common cause of dysphagia in patients with mtDNA deletions. *Neurology*, 2001; **56**: 1409–12.

120. Kovacs GG, Hoftberger R, Majtenyi K, et al. Neuropathology of white matter disease in Leber's hereditary optic neuropathy. *Brain*, 2005; **128**: 35–41.

121. Kubota M, Sakakihara Y, Mori M, Yamagata T, Momoi-Yoshida M. Beneficial effect of L-arginine for stroke-like episode in MELAS. *Brain Dev*, 2004; **26**: 481–3; discussion 480.

122. Kuriyama M, Umezaki H, Fukuda Y, et al. Mitochondrial encephalomyopathy with lactate-pyruvate elevation and brain infarctions. *Neurology*, 1984; **34**: 72–7.

123. Laforet P, Lombes A, Eymard B, et al. Chronic progressive external ophthalmoplegia with ragged-red fibers: clinical, morphological and genetic investigations in 43 patients. *Neuromuscul Disord*, 1995; **5**: 399–413.

124. Lamantea E, Tiranti V, Bordoni A, et al. Mutations of mitochondrial DNA polymerase gammaA are a frequent cause of autosomal dominant or recessive

progressive external ophthalmoplegia. *Ann Neurol*, 2002; **52**: 211–9.

125. Larsson NG, Tulinius MH, Holme E, Oldfors A. Pathogenetic aspects of the A8344G mutation of mitochondrial DNA associated with MERRF syndrome and multiple symmetric lipomas. *Muscle & Nerve*, 1995; **3**: S102–6.

126. Lawson VH, Graham BV, Flanigan KM. Clinical and electrophysiologic features of CMT2A with mutations in the mitofusin 2 gene. *Neurology*, 2005; **65**: 197–204.

127. Lax NZ, Campbell GR, Reeve AK, et al. Loss of myelin associated glycoprotein in Kearns-Sayre syndrome. *Arch Neurol*, 2012a; **69**: 490–9.

128. Lax NZ, Hepplewhite PD, Reeve AK, et al. Cerebellar ataxia in patients with mitochondrial DNA disease: a molecular clinicopathological study. *J Neuropathol Exp Neurol*, 2012b; **71**: 148–161.

129. Lax NZ, Pienaar IS, Reeve AK., et al. Microangiopathy in the cerebellum of patients with mitochondrial DNA disease. *Brain*, 2012c; **135**: 1736–50.

130. Lax NZ, Whittaker RG, Hepplewhite PD, et al. Sensory neuronopathy in patients harbouring recessive polymerase gamma mutations. *Brain*, 2012d; **135**: 62–71.

131. Leigh D. Subacute necrotizing encephalomyelopathy in an infant. *J Neurol Neurosurg Psych*: 1951; **14**: 216–21.

132. Levitt M, Nixon PF, Pincus JH, Bertino JR. Transport characteristics of folates in cerebrospinal fluid: a study utilizing doubly labeled 5-methyltetrahydrofolate and 5-formyltetrahydrofolate. *J Clin Invest*, 1971; **50**: 1301–8.

133. Li V, Hostein J, Romero NB, et al. Chronic intestinal pseudoobstruction with myopathy and ophthalmoplegia. A muscular biochemical study of a mitochondrial disorder. *Dig Dis Sci*, 1992; **37**: 456–63.

134. Lim SE, Longley MJ, Copeland WC. The mitochondrial p55 accessory subunit of human DNA polymerase gamma enhances DNA binding, promotes processive DNA synthesis, and confers N-ethylmaleimide resistance. *J Biol Chem*, 1999; **274**: 38197–203.

135. Longley MJ, Clark S, Yu-Wai-Man C, et al. Mutant POLG2 disrupts DNA polymerase gamma subunits and causes progressive external ophthalmoplegia. *Am J Hum Genet*, 2006; **78**: 1026–34.

136. Longley MJ, Graziewicz MA, Bienstock RJ, Copeland WC. Consequences of mutations in human DNA polymerase gamma. *Gene*, 2005; **354**: 125–31.

137. Lonnqvist T, Paetau A, Nikali K, Von Boguslawski K, Pihko H. Infantile onset spinocerebellar ataxia with sensory neuropathy (IOSCA): neuropathological features. *J Neurol Sci*, 1998; **161**: 57–65.

138. Lowsky R, Davidson G, Wolman S, Jeejeebhoy KN, Hegele RA. Familial visceral myopathy associated with a mitochondrial myopathy. *Gut*, 1993; **34**: 279–83.

139. Luft R, Ikkos D, Palmieri G, Ernster L, Afzelius B. A case of severe hypermetabolism of nonthyroid origin with a defect in the maintenance of mitochondrial respiratory control: a correlated clinical, biochemical and morphological study. *J Clin Invest*, 1962; **41**: 1776–1804.

140. Luoma PT, Luo N, Loscher WN, et al. Functional defects due to spacer-region mutations of human mitochondrial DNA polymerase in a family with an ataxia-myopathy syndrome. *Hum Mol Genet*, 2005; **14**: 1907–20.

141. Lutsenko S, Cooper MJ. Localization of the Wilson's disease protein product to mitochondria. *Proc Nat Acad Sci USA*, 1998; **95**: 6004–9.

142. Mackey DA, Oostra R-J, Rosenberg T, et al. Primary pathogenic mtDNA mutations in multigeneration pedigrees with Leber hereditary optic neuropathy. *Am J Hum Genet*, 1996; **59**: 481–5.

143. Mancuso M, Filosto M, Bellan M, et al. POLG mutations causing ophthalmoplegia, sensorimotor polyneuropathy, ataxia, and deafness. *Neurology*, 2004a; **62**: 316–8.

144. Mancuso M, Filosto M, Oh SJ, Di Mauro S. A novel polymerase gamma mutation in a family with ophthalmoplegia, neuropathy, and parkinsonism. *Arch Neurol*, 2004b; **61**: 1777–9.

145. Mancuso M, Salviati L, Sacconi S, et al. Mitochondrial DNA depletion: mutations in thymidine kinase gene with myopathy and SMA. *Neurology*, 2002; **59**: 1197–202.

146. Mandel H, Szargel R, Labay V, et al. The deoxyguanosine kinase gene is mutated in individuals with depleted hepatocerebral mitochondrial DNA. *Nat Genet*, 2001a; **29**: 337–41.

147. Mandel H, Szargel R, Labay V, et al. The deoyguanosine kinase gene is mutated in individuals with depleted hepatocerebral mitochondrial DNA. *Nat Genet*, 2001b; **29**: 337.

148. Mathews PM, Andermann F, Silver K, Karpati G, Arnold DL. Proton MR spectroscopic characterization of differences in regional brain metabolic abnormalities in mitochondrial encephalomyopathies. *Neurology*, 1993; **43**: 2484–90.

149. Mayr JA, Haack TB, Graf E, et al Lack of the mitochondrial protein acylglycerol kinase causes Sengers syndrome. *Am J Hum Genet*, 2012; **90**: 314–20.

150. McDonnell MT, Schaefer AM, Blakely EL, et al. Noninvasive diagnosis of the 3243A>G mitochondrial mutation using urinary epithelial cells. *Eur J Hum Genet*, 2004; **12**: 778–81.

151. McFarland R, Clark KM, Morris AA, et al. Multiple neonatal deaths due to a homoplasmic mitochondrial DNA mutation. *Nat Genet*, 2002; **30**: 145–6.

152. McFarland R, Elson JL, Taylor RW, Howell N, Turnbull DM. Assigning pathogenicity to mitochondrial tRNA mutations: when "definitely maybe" is not good enough. *Trends Genet*, 2004a; **20**: 591–6.

153. McFarland R, Kirby DM, Fowler KJ, et al. De novo mutations in the mitochondrial ND3 gene as a cause of infantile mitochondrial encephalopathy and complex I deficiency. *Ann Neurol*, 2004; **55**:58–64.

154. McFarland R, Taylor RW, Turnbull DM. A neurological perspective on mitochondrial disease. *Lancet Neurol*, 2010; **9**: 829–40.

155. McKelvie PA, Morley JB, Byrne E, Marzuki S. Mitochondrial encephalomyopathies: a correlation between neuropathological findings and defects in mitochondrial DNA. *J Neurol Sci*, 1991; **102**: 51–60.

156. Melberg A, Nennesmo I, Moslemi AR, et al. Alzheimer pathology associated with POLG1 mutation, multiple mtDNA

deletions, and APOE4/4: premature ageing or just coincidence? *Acta Neuropathol (Berl)*, 2005; **110**: 315–6.

157. Melov S, Shoffner JM, Kaufman A, Wallace DC. Marked increase in the number and variety of mitochondrial DNA rearrangements in aging human skeletal muscle [published erratum appears in *Nucleic Acids Res* 1995 Dec 11;23(23):4938]. *Nucleic Acids Res*, 1995; **23**: 4122–6.

158. Menkes JH, Alter M, Steigleder GK, Weakley DR, Sung JH. A sex-linked recessive disorder with retardation of growth, peculiar hair, and focal cerebral and cerebellar degeneration. *Pediatrics*, 1962; **29**: 764–79.

159. Mitchell AL, Elson JL, Howell N, Taylor RW, Turnbull DM. Sequence variation in mitochondrial complex I Genes: Mutation or polymorphism? *J Med Genet* 2005; **43**: 175–9.

160. Mizukami K, Sasaki M, Suzuki T, et al. Central nervous system changes in mitochondrial encephalomyopathy: light and electron microscopic study. *Acta Neuropathol (Berl)*, 1992; **83**: 449–52.

161. Molnar MJ, Valikovics A, Molnar S, et al. Cerebral blood flow and glucose metabolism in mitochondrial disorders. *Neurology*, 2000; **55**: 544–8.

162. Mootha VK, LePage P, Miller K, et al. Identification of a gene causing human cytochrome c oxidase deficiency by integrative genomics. *Proc Natl Acad Sci U S A*, 2003; **100**: 605–10.

163. Moraes CT, Ciacci F, Silvestri G, et al. Atypical clinical presentations associated with the MELAS mutation at position 3243 of human mitochondrial DNA. *Neuromuscul Disord*, 1993; **3**: 43–50.

164. Moraes CT, Di Mauro S, Zeviani M., et al. Mitochondrial DNA deletions in progressive external ophthalmoplegia and Kearns-Sayre syndrome. *N Engl J Med*, 1989; **320**: 1293–9.

165. Morgello S, Peterson HD, Kahn LJ, Laufer H. Menkes kinky hair disease with 'ragged red' fibers. *Dev Med Child Neurol*, 1988; **30**: 812–6.

166. Mori O, Yamazaki M, Ohaki Y, et al. Mitochondrial encephalomyopathy with lactic acidosis and stroke like episodes (MELAS) with prominent degeneration of the intestinal wall and cactus-like cerebellar pathology. *Acta Neuropathol (Berl)*, 2000; **100**: 712–7.

167. Munnich A, Rustin P, Rotig A, et al. Clinical aspects of mitochondrial disorders. *J Inherit Metab Dis*, 1992; **15**: 448–55.

168. Nagashima T, Mori M, Katayama K, et al. Adult Leigh syndrome with mitochondrial DNA mutation at 8993. *Acta Neuropathol (Berl)*, 1999; **97**: 416–22.

169. Nakamura M, Fujiwara Y, Yamamoto M. The two locus control of Leber hereditary optic neuropathy and a high penetrance in Japanese pedigrees. *Hum Genet*, 1993; **91**: 339–41.

170. Nass S, Nass MMK. Intramitochondrial fibres with DNA characteristics. *J Cell Biol*, 1963; **19**: 593–629.

171. Naumann M, Kiefer R, Toyka KV, et al. Mitochondrial dysfunction with myoclonus epilepsy and ragged-red fibers point mutation in nerve, muscle, and adipose tissue of a patient with multiple symmetric lipomatosis. *Muscle Nerve*, 1997; **20**: 833–9.

172. Naumann M, Reiners K, Gold R, et al. Mitochondrial dysfunction in adult-onset myopathies with structural abnormalities. *Acta Neuropathol (Berl)*, 1995; **89**: 152–7.

173. Naviaux RK, Nguyen KV. POLG mutations associated with Alpers' syndrome and mitochondrial DNA depletion. *Ann Neurol*, 2004; **55**: 706–12.

174. Newman NJ. From genotype to phenotype in Leber hereditary optic neuropathy: still more questions than answers. *J Neuroophthalmol*, 2002; **22**: 257–61.

175. Nguyen KV, Ostergaard E, Ravn SH, et al. POLG mutations in Alpers syndrome. *Neurology*, 2005; **65**: 1493–5.

176. Nikali K, Suomalainen A, Saharinen J, et al. Infantile onset spinocerebellar ataxia is caused by recessive mutations in mitochondrial proteins Twinkle and Twinky. *Hum Mol Genet*, 2005; **14**: 2981–90.

177. Nikoskelainen EK, Huoponen K, Juvonen V, et al. Ophthalmologic findings in Leber hereditary optic neuropathy, with special reference to mtDNA mutations. *Ophthalmology*, 1996; **103**: 504–14.

178. Nikoskelainen EK, Marttila RJ, Huoponen K, et al. Leber's "plus": neurological abnormalities in patients with Leber's hereditary optic neuropathy. *J Neurol Neurosurg Psychiatry*, 1995; **59**: 160–4.

179. Nishino I, Spinazzola A, Hirano M. Thymidine phosphorylase gene mutations in MNGIE, a human mitochondrial disorder. *Science*, 1999; **283**: 689–92.

180. Nishino I, Spinazzola A, Papadimitriou A, et al. Mitochondrial neurogastrointestinal encephalomyopathy: an autosomal recessive disorder due to thymidine phosphorylase mutations. *Ann Neurol*, 2000; **47**: 792–800.

181. Nishioka J, Akita Y, Yatsuga S, et al. Inappropriate intracranial hemodynamics in the natural course of MELAS. *Brain Dev*, 2008; **30**: 100–5.

182. Ohama E, Ikuta F. Involvement of choroid plexus in mitochondrial encephalomyopathy (MELAS). *Acta Neuropathol (Berl)*, 1987; **75**: 1–7.

183. Ohama E, Ohara S, Ikuta F, et al. Mitochondrial angiopathy in cerebral blood vessels of mitochondrial encephalomyopathy. *Acta Neuropathol (Berl)*, 1987; **74**: 226–33.

184. Ohshita T, Oka M, Imon Y, et al. Serial diffusion-weighted imaging in MELAS. *Neuroradiology*, 2000; **42**: 651–6.

185. Okamura K, Santa T, Nagae K, Omae T. Congenital oculoskeletal myopathy with abnormal muscle and liver mitochondria. *J Neurol Sci*, 1976; **27**: 79–91.

186. Oldfors A, Fyhr IM, Holme E, Larsson NG, Tulinius M. Neuropathology in Kearns-Sayre syndrome. *Acta Neuropathol (Berl)*, 1990; **80**: 541–6.

187. Olsen NK, Hansen AW, Norby S, et al. Leber's hereditary optic neuropathy associated with a disorder indistinguishable from multiple sclerosis in a male harbouring the mitochondrial DNA 11778 mutation. *Acta Neurol Scand*, 1995; **91**: 326–9.

188. Ostergaard E, Christensen E, Kristensen E, et al. Deficiency of the alpha subunit of succinate-coenzyme A ligase causes fatal infantile lactic acidosis with mitochondrial DNA depletion. *Am J Hum Genet*, 2007; **81**: 383–7.

189. Pages M, Pages AM. Leber's disease with spastic paraplegia and peripheral neuropathy. Case report with nerve biopsy study. *Eur Neurol*, 1983; **22**: 181–5.

190. Papadimitriou A, Comi GP, Hadjigeorgiou GM, et al. Partial depletion and multiple deletions of muscle mtDNA in familial MNGIE syndrome. *Neurology*, 1998; **51**: 1086–92.

191. Petty RK, Harding AE, Morgan-Hughes JA. The clinical features of mitochondrial myopathy. *Brain*, 1986; **109**: 915–38.

192. Pineda M, Ormazabal A, Lopez-Gallardo E., et al. Cerebral folate deficiency and leukoencephalopathy caused by a mitochondrial DNA deletion. *Ann Neurol*, 2006; **59**: 394–8.

193. Pollock M, Nicholson GI, Nukada H, Cameron S, Frankish P. Neuropathy in multiple symmetric lipomatosis. Madelung's disease. *Brain*, 1988; **111**: 1157–71.

194. Pons R, Andreetta F, Wang CH, et al. Mitochondrial myopathy simulating spinal muscular atrophy. *Pediatr Neurol*, 1996; **15**: 153–8.

195. Rahman S, Blok RB, Dahl HH, et al. Leigh syndrome: clinical features and biochemical and DNA abnormalities. *Ann Neurol*, 1996; **39**: 343–51.

196. Richardson C, Smith T, Schaefer A, Turnbull D, Griffiths P. Ocular motility findings in chronic progressive external ophthalmoplegia. *Eye* 2004; **19**, 258–63.

197. Rotig A, Bourgeron T, Chretien D, Rustin P, Munnich A. Spectrum of mitochondrial DNA rearrangements in the Pearson marrow-pancreas syndrome. *Hum Mol Genet*, 1995; **4**: 1327–30.

198. Rowland LP, Blake DM, Hirano M, et al. Clinical syndromes associated with ragged red fibers. *Rev Neurol (Paris)*, 1991; **147**: 467–73.

199. Rowland LP, Hays AP, Di Mauro S, Al E. Diverse clinical disorders associated with abnomalities of mitochondria. In Scarlato G, Cerri C eds. *Mitochondrial pathology in muscle diseases*. Padova, Piccin, 1983.

200. Saada A, Shaag A, Mandel H, et al. Mutant mitochondrial thymidine kinase in mitochondrial DNA depletion myopathy. *Nat Genet*, 2001; **29**: 342–4.

201. Sadun AA, Carelli V, Bose S, et al. First application of extremely high-resolution magnetic resonance imaging to study microscopic features of normal and LHON human optic nerve. *Ophthalmology*, 2002; **109**: 1085–91.

202. Sadun AA, Kashima Y, Wuredeman AE, et al. Morphological findings in the visual system in a case of Leber's hereditary optic neuropathy. *Clin Neurosci*, 1994; **2**: 165–72.

203. Sadun AA, Win PH, Ross-Cisneros FN, Walker SO, Carelli V. Leber's hereditary optic neuropathy differentially affects smaller axons in the optic nerve. *Trans Am Ophthalmol Soc*, 2000; **98**: 223–32; discussion 232–5.

204. Said G, Lacroix C, Plante-Bordeneuve V, et al. Clinicopathological aspects of the neuropathy of neurogastrointestinal encephalomyopathy (MNGIE) in four patients including two with a Charcot-Marie-Tooth presentation. *J Neurol*, 2005; **252**: 655–62.

205. Santorelli FM, Tanji K, Kulikova R, et al. Identification of a novel mutation in the mtDNA ND5 gene associated with MELAS. *Biochem Biophys Res Comm*, 1997; **238**: 326–8.

206. Santoro L, Carrozzo R, Malandrini A, et al. A novel SURF1 mutation results in Leigh

syndrome with peripheral neuropathy caused by cytochrome c oxidase deficiency. *Neuromuscul Disord*, 2000; **10**: 450–3.

207. Sarnat HB, Marin-Garcia J. Pathology of mitochondrial encephalomyopathies. *Can J Neurol Sci*, 2005; **32**: 152–66.

208. Sarzi E, Bourdon A, Chretien D, et al. Mitochondrial DNA depletion is a prevalent cause of multiple respiratory chain deficiency in childhood. *J Pediatr*, 2007; **150**: 531–4, 534 e1–6.

209. Sarzi E, Goffart S, Serre V, et al. Twinkle helicase (PEO1) gene mutation causes mitochondrial DNA depletion. *Ann Neurol*, 2007; **62**: 579–87.

210. Schaefer AM, Blakely EL, Griffiths PG, Turnbull DM, Taylor RW. Ophthalmoplegia due to mitochondrial DNA disease: The need for genetic diagnosis. *Muscle Nerve* 2005; **32**: 104–7.

211. Schicks J, Synofzik M, Schulte C, Schols L. POLG, but not PEO1, is a frequent cause of cerebellar ataxia in Central Europe. *Mov Disord*, 2010; **25**: 2678–82.

212. Schon EA, Przedborski S. Mitochondria: the next (neurode) generation. *Neuron* 2011; **23**: 1033–53.

213. Schon EA, Bonilla E, Di Mauro S. Mitochondrial DNA mutations and pathogenesis. *J Bioenerg Biomemb*, 1997; **29**: 131–49.

214. Sciacco M, Bonilla E, Schon EA, Di Mauro S, Moraes CT. Distribution of wild-type and respiration-deficient mtDNA in normal and respiration-deficient muscle fibers from patients with mitochondrial myopathy. *HumMol Genet*, 1994; **3**: 13–9.

215. Servidei S. Mitochondrial encephalomyopathies: gene mutation. *Neuromuscul Disord*, 2004; **14**: 107–16.

216. Shanske S, Pancrudo J, Kaufmann P, et al. Varying loads of the mitochondrial DNA A3243G mutation in different tissues: implications for diagnosis. *Am J Med Genet*, 2004; **130A**: 134–7.

217. Shanske S, Tang Y, Hirano M, et al. Identical mitochondrial DNA deletion in a woman with ocular myopathy and in her son with pearson syndrome. *Am J Hum Genet*, 2002; **71**: 679–83.

218. Shapira Y, Cederbaum SD, Cancilla PA, Nielsen D, Lippe BM. Familial poliodystrophy, mitochondrial myopathy, and lactate acidemia. *Neurology*, 1975; **25**: 614–21.

219. Shoffner JM, Lott MT, Lezza AM, et al. Myoclonic epilepsy and ragged-red fiber disease (MERRF) is associated with a mitochondrial DNA tRNA(Lys) mutation. *Cell*, 1990; **61**: 931–7.

220. Shoubridge EA. Nuclear genetic defects of oxidative phosphorylation. *Hum Mol Genet*, 2001; **10**: 2277–84.

221. Shy GM, Gonatas NK, Perez M. Two childhood myopathies with abnormal mitochondria. I. Megaconial myopathy. II. Pleioconial myopathy. *Brain*, 1966; **89**: 133–158.

222. Silvestri G, Moraes CT, Shanske S, Oh SJ, Di Mauro S. A new mtDNA mutation in the tRNA(Lys) gene associated with myoclonic epilepsy and ragged-red fibers (MERRF). *Am J Hum Genet*, 1992; **51**: 1213–7.

223. Silvestri G, Servidei S, Rana M, et al. A novel mitochondrial DNA point mutation in the tRNA(Ile) gene is associated with progressive external ophthalmoplegia. *Biochem Biophys Res Comm*, 1996; **220**: 623–7.

224. Simon LT, Horoupian DS, Dorfman LJ, et al. Polyneuropathy, ophthalmoplegia, leukoecephalopathy, and intestinal pseudo-obstruction: POLIP syndrome. *Ann Neurol*, 1990; **28**: 349–60.

225. Simonati A, Filosto M, Tomelleri G, et al. Central-peripheral sensory axonopathy in a juvenile case of Alpers-Huttenlocher disease. *J Neurol*, 2003; **250**: 702–6.

226. Smeitink JA. Mitochondrial disorders: clinical presentation and diagnostic dilemmas. *J Inherit Metab Dis*, 2003; **26**: 199–207.

227. Sofou K, Moslemi AR, Kollberg G, et al. Phenotypic and genotypic variability in Alpers syndrome. *Eur J Paediatr Neurol*, 2012; **16**: 379–89.

228. Sparaco M, Bonilla E, Di Mauro S, Powers JM. Neuropathology of mitochondrial encephalomyopathies due to mitochondrial DNA defects. *J Neuropathol Exp Neurol*, 1993; **52**: 1–10.

229. Sparaco M, Schon EA, Di Mauro S, Bonilla E. Myoclonic epilepsy with ragged-red fibers (MERRF): an immunohistochemical study of the brain. *Brain Pathol*, 1995; **5**: 125–33.

230. Spector R. Micronutrient homeostasis in mammalian brain and cerebrospinal fluid. *J Neurochem*, 1989; **53**: 1667–74.

231. Spector R, Johanson CE. The mammalian choroid plexus. *Sci Am*, 1989; **261**: 68–74.

232. Spelbrink JN, Li FY, Tiranti V, et al. Human mitochondrial DNA deletions associated with mutations in the gene encoding Twinkle, a phage T7 gene 4-like protein localised in mitochondria. *Nat Genet*, 2001; **28**: 223–31.

233. Spinazzola A, Zeviani M. Disorders of nuclear-mitochondrial intergenomic signaling. *Gene*, 2005; **354**: 162–8.

234. Spinazzola A, Marti R, Nishino I, et al. Altered thymidine metabolism due to defects of thymidine phosphorylase. *J Biol Chem*, 2002; **277**: 4128–33.

235. Spinazzola A, Viscomi C, Fernandez-Vizarra E, et al. MPV17 encodes an inner mitochondrial membrane protein and is mutated in infantile hepatic mitochondrial DNA depletion. *Nat Genet*, 2006; **38**: 570–5.

236. Stickler DE, Carney PR, Valenstein ER. Juvenile-onset Leigh syndrome with an acute polyneuropathy at presentation. *J Child Neurol*, 2003; **18**: 574–6.

237. Sue CM, Crimmins DS, Soo YS, et al. Neuroradiological features of six kindreds with MELAS tRNA(Leu) A2343G point mutation: implications for pathogenesis. *J Neurol Neurosurg Psychiatry*, 1998; **65**: 233–40.

238. Szigeti K, Sule N, Adesina AM, et al. Increased blood–brain barrier permeability with thymidine phosphorylase deficiency. *Ann Neurol* 2004; **56**: 881–6.

239. Taanman JW, Bodnar AG, Cooper JM, et al. Molecular mechanisms in mitochondrial DNA depletion syndrome. *Hum Mol Genet*, 1997; **6**: 935–42.

240. Tanahashi C, Nakayama A, Yoshida M, et al. MELAS with the mitochondrial DNA 3243 point mutation: a neuropathological study. *Acta Neuropathol (Berl)*, 2000; **99**: 31–8.

241. Tanji K, Di Mauro S, Bonilla E. Disconnection of cerebellar Purkinje cells in Kearns-Sayre syndrome. *J Neurol Sci*, 1999; **166**: 64–70.

242. Tanji K, Gamez J, Cervera C, et al. The A8344G mutation in mitochondrial DNA associated with stroke-like episodes

and gastrointestinal dysfunction. *Acta Neuropathol (Berl)*, 2003; **105**: 69–75.

243. Tanji K, Kunimatsu T, Vu TH, Bonilla E. Neuropathological features of mitochondrial disorders. *Semin Cell Dev Biol*, 2001; **12**: 429–39.

244. Tanji K, Schon EA, Di Mauro S, Bonilla E. Kearns-sayre syndrome: oncocytic transformation of choroid plexus epithelium. *J Neurol Sci*, 2000; **178**: 29–36.

245. Tanji K, Vu TH, Schon EA, Di Mauro S, Bonilla E. Kearns-Sayre syndrome: unusual pattern of expression of subunits of the respiratory chain in the cerebellar system. *Ann Neurol*, 1999; **45**: 377–83.

246. Taylor RW, Birch-Machin MA, Schaefer J, et al. Deficiency of complex II of the mitochondrial respiratory chain in late-onset optic atrophy and ataxia. *Ann Neurol*, 1996; **39**: 224–32.

247. Tiranti V, Hoertnagel K, Carrozzo R, et al. Mutations of SURF-1 in Leigh disease associated with cytochrome c oxidase deficiency. *Am J Hum Genet*, 1998; **63**: 1609–21.

248. Tokunaga M, Mita S, Sakuta R, Nonaka I, Araki S. Increased mitochondrial DNA in blood vessels and ragged-red fibres in mitochondrial myopathy, encephalopathy, lactic acidosis, and stroke-like episodes (MELAS). *Ann Neurol*, 1993; **33**: 275–80.

249. Traff J, Holme E, Ekbom K, Nilsson BY. Ekbom's syndrome of photomyoclonus, cerebellar ataxia and cervical lipoma is associated with the tRNA(Lys) A8344G mutation in mitochondrial DNA. *Acta Neurol Scand*, 1995; **92**: 394–7.

250. Triepels RH, Van Den Heuvel L, Trijbels F, Smeitink JA. Respiratory chain complex I deficiency. *Am J Med Genet*, 2001; **106**: 37–45.

251. Trijbels FJ, Ruitenbeek W, Huizing M, et al. Defects in the mitochondrial energy metabolism outside the respiratory chain and the pyruvate dehydrogenase complex. *Mol Cell Biochem*, 1997; **174**: 243–7.

252. Tsuchiya K, Miyazaki H, Akabane H, et al. MELAS with prominent white matter gliosis and atrophy of the cerebellar granular layer: a clinical, genetic, and pathological study. *Acta Neuropathol (Berl)*, 1999; **97**: 520–4.

253. Tyynismaa H, Sun R, Ahola-Erkkila S, et al. Thymidine kinase 2 mutations in autosomal recessive progressive external ophthalmoplegia with multiple mitochondrial DNA deletions. *Hum Mol Genet*, 2012; **21**: 66–75.

254. Tyynismaa, H., Ylikallio, E., Patel, M., et al. A heterozygous truncating mutation in RRM2B causes autosomal-dominant progressive external ophthalmoplegia with multiple mtDNA deletions. *Am J Hum Genet*, 2009; **85**: 290–5.

255. Tzoulis C, Neckelmann G, Mork SJ, et al. Localized cerebral energy failure in DNA polymerase gamma-associated encephalopathy syndromes. *Brain*, 2010; **133**: 1428–37.

256. Valnot I, Osmond S, Gigarel N, et al. Mutations of the SCO1 gene in mitochondrial cytochrome c oxidase deficiency with neonatal-onset hepatic failure and encephalopathy. *Am J Hum Genet*, 2000; **67**: 1104–9.

257. Valnot I, Von Kleist-Retzow JC, Barrientos A, et al. A mutation in the human heme A:farnesyltransferase gene (COX10)

causes cytochrome c oxidase deficiency. *Hum Mol Genet*, 2000b; **9**: 1245–9.

258. van den Ouweland JWM, Lemkes HHPJ, Ruitenbeek K. Mutation in mitochondrial tRNA^Leu(UUR) gene in a large pedigree with maternally transmitted type II diabetes mellitus and deafness. *Nat Genet*, 1992; **1**: 368–71.

259. Van Goethem G, Dermaut B, Lofgren A, Martin JJ, Van Broeckhoven C. Mutation of POLG is associated with progressive external ophthalmoplegia characterized by mtDNA deletions. *Nat Genet*, 2001; **28**: 211–2.

260. Van Goethem G, Luoma P, Rantamaki M, et al. POLG mutations in neurodegenerative disorders with ataxia but no muscle involvement. *Neurology*, 2004; **63**: 1251–7.

261. Van Goethem G, Martin JJ, Dermaut B, et al. Recessive POLG mutations presenting with sensory and ataxic neuropathy in compound heterozygote patients with progressive external ophthalmoplegia. *Neuromuscul Disord*, 2003a; **13**: 133–42.

262. Van Goethem G, Schwartz M, Lofgren A, et al. Novel POLG mutations in progressive external ophthalmoplegia mimicking mitochondrial

neurogastrointestinal encephalomyopathy. *Eur J Hum Genet*, 2003; **11**: 547–9.

263. Wallace DC. Mitochondrial diseases in mouse and man. *Science*, 1999; **283**: 1482–8.

264. Wallace DC, Singh G, Lott MT, et al. Mitochondrial DNA mutation associated with Leber's hereditary optic neuropathy. *Science*, 1988; **242**: 1427–30.

265. Wang XY, Noguchi K, Takashima S, et al. Serial diffusion-weighted imaging in a patient with MELAS and presumed cytotoxic oedema. *Neuroradiology*, 2003; **45**: 640–3.

266. Winterthun S, Ferrari G, He L, et al. Autosomal recessive mitochondrial ataxic syndrome due to mitochondrial polymerase gamma mutations. *Neurology*, 2005; **64**: 1204–8.

267. Yoshimura N, Kudo H. Mitochondrial abnormalities in Menkes' kinky hair disease (MKHD). Electron-microscopic study of the brain from an autopsy case. *Acta Neuropathol (Berl)*, 1983; **59**: 295–303.

268. Zanna C, Ghelli A, Porcelli AM, et al. Caspase-independent death of Leber's hereditary optic neuropathy cybrids is driven by energetic failure and mediated by

AIF and Endonuclease G. *Apoptosis*, 2005; **10**: 997–1007.

269. Zeviani M, Di Donato S. Mitochondrial disorders. *Brain*, 2004; **127**: 2153–72.

270. Zeviani M, Sevidei S, Gallera C, Al E. An autosomal dominant disorder with multiple deletions of mitochondrial DNA starting in the D-loop region. *Nature*, 1989; **339**: 309–11.

271. Zhou L, Chomyn A, Attardi G, Miller CA. Myoclonic epilepsy and ragged red fibres (MERRF) syndrome: selective vulnerability of CNS neurons does not correlate with the level of mitochondrial tRNAlys mutation in individual neuronal isolates. *J Neurosci*, 1997; **17**: 7746–53.

272. Zhu, Z., Yao, J., Johns, T., et al. *SURF*1, encoding a factor involved in the biogenesis of cytochrome c oxidase, is mutated in Leigh syndrome. *Nat Genet*, 1998; **20**: 337–43.

273. Zuchner S, Mersiyanova IV, Muglia M, et al. Mutations in the mitochondrial GTPase mitofusin 2 cause Charcot-Marie-Tooth neuropathy type 2A. *Nat Genet*, 2004; **36**: 449–51.

Peroxisomal Disorders

Phyllis L Faust and James M Powers

INTRODUCTION

Peroxisomes are intracytoplasmic, single membrane-bound organelles that play major roles in diverse reactions of lipid metabolism as well as cellular scavenging of peroxides and reactive oxygen species. Many of their functions are highly conserved throughout evolution, although some are unique to particular species.[80] Peroxisomes and mitochondria are interrelated biochemically, and defects in both organelles have been reported in the Zellweger spectrum[62,158] and adreno-leukodystrophy (ALD). Both peroxisomes and mitochondria are involved in β-oxidation of fatty acids and the α-oxidation of phytanic acid.[191] Long-chain fatty acids are predominantly oxidized in mitochondria, whereas oxidation of very long-chain fatty acids (VLCFAs $\geq$ C22) and branched-chain fatty acids is initiated in peroxisomes and then completed in mitochondria.[194] In yeast and plants, fatty acid β-oxidation is confined to peroxisomes. Peroxisomes participate in both generation and scavenging of cellular reactive oxygen species (ROS).[162,192] Peroxisomes also have metabolic interactions with endoplasmic reticulum (ER), including breakdown of dicarboxylic acids generated by ω-oxidation of fatty acids in the ER, and activation of hepatic ER stress responses in peroxisomal disorders.[80,178] The number, size and function of peroxisomes change dramatically with cellular type and environmental stimuli and common transcriptional regulators, such as peroxisome proliferator-activated receptors, coordinately control enzyme levels in peroxisomes, mitochondria and ER.[80,163] Peroxisomes and mitochondria also share key components of their division machinery and increasing evidence supports a functional interaction of peroxisomes with ER. Peroxisome interactions with lipid droplets and the lysosomal-endosomal system are increasingly recognized. Novel peroxisome functions, such

as signalling platforms for antiviral innate immunity[39] and tuberous sclerosis complex regulation of mammalian target of rapamycin (mTORC1) and autophagy in response to ROS,[204] continue to be discovered.

Peroxisomal disorders have been estimated to be responsible for approximately 10 per cent of human heritable metabolic diseases. They produce systemic, multiorgan lesions, in which the central nervous system (CNS), eyes, skeleton, liver and adrenal glands are most severely involved.[135]

The diagnosis of peroxisomal disorders is based initially on clinical symptomatology, followed by confirmatory laboratory studies.[65,171,193] Most of the peroxisomal disorders are transmitted in an autosomal recessive pattern, except for ALD and adrenomyeloneuropathy (AMN), which are X linked. Dysmorphic facies or rhizomelia (shortening of forelimbs), particularly when accompanied by hypotonia and seizures in the neonatal period, or signs of central white matter disease (e.g. spasticity, ataxia, aphasia), myelopathy and neuropathy in male children or adults, should alert the clinician to the possibility of a peroxisomal disorder. Plasma VLCFA levels, particularly C26:0, are elevated in almost all peroxisomal disorders and are therefore an excellent, albeit technically challenging, initial laboratory-screening test. If VLCFA levels are abnormal and the patient is suspected of having a peroxisomal disorder other than ALD or AMN, then plasma levels of bile acid intermediates (dihydroxycholestanoic acid, trihydroxycholestanoic acid [THCA]), pristanic acid, phytanic acid, pipecolic acid and docosahexaenoic acid (DHA) should be measured. Diagnosis may require functional enzyme assays of *de novo* plasmalogen biosynthesis, fatty acid β-oxidation, phytanic acid α-oxidation and immunostaining for peroxisomal matrix/membrane proteins in fibroblast cultures.[96] Characteristic patterns of brain involvement seen in magnetic resonance imaging (MRI) may assist in diagnosis, particularly in biochemically atypical patients.[130,185] Prenatal diagnosis in at-risk families through analyses of VLCFA and molecular-genetic analyses is available for most peroxisomal disorders and is highly reliable. The identification of mutant peroxin (PEX) genes allows the identification of heterozygotes or carrier status and has become the definitive test of peroxisome biogenesis disorders (PBDs).[171,202] Newborn screening methods have been developed, although not yet widely implemented, for X-ALD and other peroxisomal β-oxidation disorders.[176]

If the diagnosis is not made antemortem and a peroxisomal disorder is suspected, then samples of plasma, brain, liver and adrenal gland should be frozen at autopsy for biochemical analysis. Sections of brain, liver and adrenal gland should be fixed for light and electron microscopy and immunohistochemistry,[36] small samples of skeletal muscle and spleen snap-frozen for molecular genetic studies, and a sample of skin removed for fibroblast culture. Even if such measures are not taken, such as in retrospective archival investigations, VLCFA and phytanic acid determinations can be made on formalin-fixed wet tissue samples of brain and adrenal gland[112] because of the postmortem stability and insolubility of these metabolites in aqueous solution.

PEROXISOMES

Peroxisomes, originally called 'microbodies' by electron microscopists, are found in all nucleated cells of plants (glyoxysomes) and mammals. Human peroxisomes are characterized ultrastructurally by a granular, moderately electron-dense matrix surrounded by a unit membrane, a shape that varies from primarily spherical, with a diameter of approximately 0.1–1.0 μm, to elongated tubular (reticulum) (Figure 8.1a). They were initially visualized by enhanced electron density after diaminobenzidine staining, as a result of peroxidatic activity of the matrix enzyme catalase.[62] Immunofluorescence microscopy with antibodies to peroxisomal matrix and membrane proteins now allows more specific detection.[88,157] The presence of membrane-bound particulate (i.e. peroxisomal) *versus* cytosolic (soluble) catalase activity has been the traditional biochemical marker. Much of our understanding of mammalian peroxisomes has been derived through studies of hepatic and renal tubular peroxisomes (reviewed by Depreter *et al.*[36]). The investigation of yeast and Chinese hamster ovary cell peroxisomes, highly homologous to those of mammals, yielded rapid and dramatic insights into peroxisomal biogenesis and diseases.[65,103,121] The peroxisomes in the CNS are smaller (micro-peroxisomes) than those in the liver and kidney[73] and are biochemically diverse, with functional differences between cell types and brain regions.[3,156] In the mature mammalian CNS peroxisomes are most abundant in oligodendrocytes, whereas in the developing CNS they are also abundant at the termini of developing neurons and have been implicated in the early determination of neural polarity.[18,73] In human fetuses, catalase-positive neurons are observed in the basal ganglia, thalamus and cerebellum at 27–28 weeks and in the frontal cortex at

35 weeks of gestation. As in other mammals, the prominence of peroxisomes in neurons decreases with postnatal age. Catalase-positive glia are identified in deep white matter at 31–32 weeks of gestation and, throughout the remainder of gestation, they appear to shift from deep to superficial white matter.[74]

The peroxisome is named for its peroxide-based respiration, in which a variety of oxidases generates hydrogen peroxide, which is decomposed by catalase or peroxidatically to yield O_2 and water (Figure 8.1b).[32,192] The relevance of peroxisomes in mammalian cells was only fully appreciated after morphological and biochemical abnormalities of this organelle were noted in a few rare human diseases. Peroxisomal (and mitochondrial) abnormalities were recognized first in the cerebro-hepato-renal syndrome of Zellweger (ZS),[62] which is the prototype PBD.[65] These diseases display biochemical defects in most or all of the known functions of peroxisomes in humans, which are executed by over 50 matrix enzymes, including: β-oxidation of long-chain fatty acids, VLCFA, pristanic acid, cholestanoic acids and eicosanoids; α-oxidation of phytanic acid; pipecolic acid degradation; glyoxylate detoxification; glutaryl-CoA metabolism; and the biosynthesis of etherphospholipids (plasmalogens), DHA, cholesterol and dolichol.[191,192,193] Additional and more current information is available on the European Peroxisome[DB] website (www.peroxisomedb.org).

PEROXISOMAL DISORDERS

Peroxisomal disorders have been classified into two main categories: the peroxisome biogenesis disorders (PBDs), in which the peroxisome lacks most or multiple content proteins and sometimes membrane, and single-enzyme disorders, in which there is loss of a specific component of a peroxisomal metabolic pathway (Table 8.1).

In PBDs, previously referred to as generalized peroxisomal diseases, morphologically identifiable or biochemically particulate peroxisomes are absent or severely deficient, owing to a fundamental defect in their assembly; this leads to loss of multiple peroxisomal enzymatic functions. Many peroxisomal matrix enzymes, such as those involved in β-oxidation, need to be imported if they are to be biologically active. Proteins that control peroxisome assembly and division are called peroxins (Pexp) and all are nuclear-encoded (by *PEX* genes). Distinct cellular machineries sort matrix and membrane proteins to peroxisomes. Matrix proteins are synthesized on free polyribosomes and are post-translationally imported into pre-existing peroxisomes.[80,103] Matrix proteins are sorted by two distinct sequences, known as peroxisomal targeting signal (PTS) type 1 and type 2, which are recognized by cytosolic import receptors, Pex5p and Pex7p, respectively, followed by translocation across the membrane through a complex import machinery. Peroxisomes import fully folded, cofactor-bound and even oligomeric matrix proteins. Most mammalian matrix proteins contain a carboxy-terminal serine–lysine–leucine (SKL) sequence characteristic of PTS1, whereas thiolase, alkyl-dihydroxyacetone phosphate (DHAP)

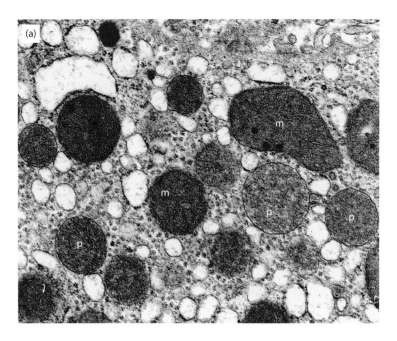

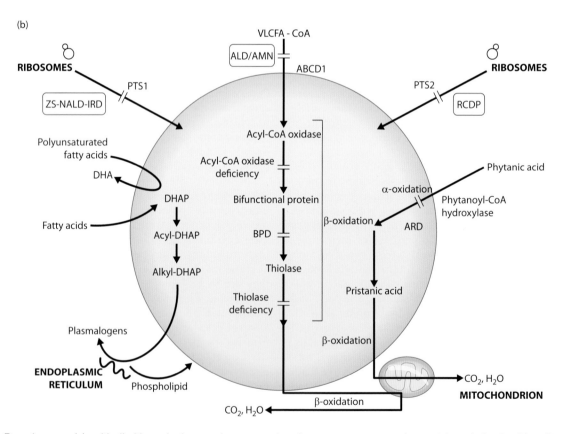

8.1 (a) Peroxisomes (p) with limiting single membranes and a homogeneous granular matrix admixed with mitochondria (m) displaying internal cristae, electron-dense granules and limiting double membranes in a hepatocyte. Electron microscopy of a biopsy. **(b)** Metabolic pathways of neuroperoxisomal disorders. ABCD1, adenosine triphosphate-binding cassette D1; ALD, adreno-leukodystrophy; AMN, adrenomyeloneuropathy; ARD, adult Refsum's disease; BPD, bifunctional protein deficiency; CoA, coenzyme A; DHA, docosahexaenoic acid; DHAP, dihydroxyacetone phosphate; IRD, infantile Refsum's disease; NALD, neonatal adreno-leukodystrophy; PTS1, peroxisomal targeting signal 1; PTS2, peroxisomal targeting signal 2; RCDP, rhizomelic chondrodysplasia punctata; VLCFA, very long-chain fatty acid; ZS, Zellweger syndrome.

TABLE 8.1 Classification of peroxisomal disorders

Group I: PBD and multiple peroxisomal functions

Cerebro-hepato-renal Zellweger syndrome (ZS)

Neonatal adreno-leukodystrophy (NALD)

Infantile Refsum disease (IRD)

Rhizomelic chondrodysplasia punctata (RCDP), type I, classical

Group II: Disorders with morphologically intact peroxisomes and a single protein deficiency

A. Pseudo-PBD

   1. Acyl-CoA oxidase deficiency (pseudo-NALD)

   2. Bifunctional protein deficiency (BPD)

   3. Di- and trihydroxycholestanoic deficiencies

   4. RCDP types II, III (DHAP acyltransferase, alkyl DHAP synthase deficiencies)

   5. Zellweger-like syndrome

B. ALD and AMN

C. Refsum's disease, classical, adult (ARD) and atypical Refsum's disease

D. Miscellaneous

   1. Mulibrey (muscle–liver–brain–eye) nanism

   2. Peroxisomal racemase deficiency

   3. Glutaric aciduria, type III (glutaryl-CoA oxidase deficiency)

   4. Hyperoxaluria type I (alanine glyoxylate aminotransferase deficiency)

   5. Acatalasaemia

Group III: Others

ALD, adreno-leukodystrophy; AMN, adrenomyeloneuropathy; CoA, coenzyme A; DHAP, dihydroxyacetone phosphate; PBD, peroxisome biogenesis disorder

synthase and phytanoyl-CoA hydroxylase display PTS2 amino-terminal sequences, which are proteolytically cleaved after import into the peroxisome. The import of peroxisomal membrane proteins (PMPs) remains less understood and controversial.[80,186] Some PMPs (class I) may be imported directly from cytosol to peroxisomes, with sorting via the Pex19 receptor/chaperone that recognizes internal membrane peroxisomal targeting sequences (mPTS). The Pex19p-PMP complex then interacts with Pex3p (and Pex16p) at peroxisomal membranes to mediate PMP insertion. However, recent studies demonstrate that many PMPs (class II) are initially trafficked through the ER into a specific pre-peroxisomal compartment, where interactions with Pex3p and Pex19p lead to budding of vesicles for subsequent delivery to peroxisomes. Peroxisomes can form both by growth and division (fission) of pre-existing organelles, with the ER providing membrane lipids, as well as de novo via the ER; this occurs in a multistep maturation pathway initiated by Pex11p.[163] Key components for both peroxisomal and mitochondrial fission are the dynamin-like GTPase DLP1/Drp1 and the DLP1 membrane adaptors Fis1 and Mff.

Early attempts to understand the pathogenesis of PBD were largely classical morphological studies,[45,190] followed shortly thereafter by biochemical approaches. At about the same time, complementation analysis, in which abnormal fibroblasts from two different patients are induced to fuse into a multinucleated heterokaryon, was applied to PBDs.[42,113] As with other diseases, correction of the abnormalities occurs only if one abnormal cell provides the gene product that is deficient in the other, thereby reflecting distinct genotypes. Complementation analysis has identified thirteen distinct genotypes in PBDs, with characteristic clinical phenotypes;[65,113,115] all display an impairment of matrix protein import and three have defects in peroxisome membrane biogenesis (Table 8.2).[103,130] At present, more than thirty different peroxins have been identified in various species; although many are conserved from lower to higher eukaryotes, there may be some functional redundancy. Complementation analysis may fail to identify other putative human PEX genes, such as a distinct peroxisomal and mitochondrial fission defect due to mutation in the DLP1 gene.[196] PBD patients with milder clinical phenotypes may be diagnostically challenging, with variably defective or even normal peroxisomal functions in plasma and/or fibroblasts. The majority of peroxins are involved in matrix protein import, forming distinct subcomplexes that mediate PTS receptor-cargo docking and translocation and receptor recycling and degradation machinery. PEX 3, 16 and 19 are critical for peroxisomal membrane protein import (Table 8.2).

Genotype–phenotype correlations have been poor because defects in the same gene can produce different phenotypes and defects in different genes can produce the same phenotype.[65,115,141] For example, the ZS phenotype is associated with all human PEX genes excepting PEX 7. The most common complementation group is group 1, because of a defect in PEX 1, in which the ZS, neonatal adreno-leukodystrophy (NALD) and infantile Refsum's disease (IRD) phenotypes (referred to as the Zellweger spectrum) are seen in decreasing order of prevalence and severity.[30] A defect in PEX 1 is responsible for about 60–70 per cent of ZS–NALD–IRD phenotypes; PEX6 and PEX12 (16 per cent and 9 per cent of PBD patients, respectively) comprise other more common genotypes.[42,202] The phenotypic variability with the same genotype largely correlates with the nature of the gene defect and the resultant deficiency of mRNA/protein, leading to a variable import defect of matrix proteins. For example, the G843D missense mutation in PEX1 has approximately 15 per cent residual matrix protein import activity and correlates with the mildest IRD phenotype, although the 2097inST frameshift mutation has no import activity and is associated with the most severe ZS phenotype.[30] A comparable situation has been found with PEX 5, 6, 7, 10 and 12.[65,115] The temperature sensitivity of peroxisomal permeability may also play a role.[79] Most patients with Zellweger spectrum, with an incidence of one per 25000–50000 births, have a defect in PTS1-matrix protein import. By contrast, rhizomelic chondrodysplasia punctata (RCDP) type 1, which is the only PBD associated with PEX 7, displays a selective defect in PTS2-matrix protein import. Over 90 per cent of RCDP patients have defects in PEX 7. A common mutant allele, PEX 7 L292ter, is found in almost 50 per cent of patients with RCDP; this causes reduced amounts of PEX 7 with no residual activity. On the other hand, some missense mutations result in PTS2

TABLE 8.2 Human *PEX*-genes and peroxins

PEX-gene	Peroxin function	Phenotype
PEX 1	AAA-ATPase; matrix protein import, PTS receptor recycling	ZS, NALD, IRD
PEX 2	RING-finger PMP; E3 ligase and PTS receptor ubiquitination	ZS, IRD
PEX 3	PMP biogenesis and Pex19 receptor; *de novo* formation	ZS
PEX 5	PTS1 receptor, predominantly cytoplasmic; matrix protein import	NALD, ZS, IRD
PEX 6	AAA-ATPase; matrix protein import, PTS receptor recycling	ZS, NALD
PEX 7	PTS2 receptor, predominantly cytoplasmic; matrix protein import	RCDP, ARD
PEX 10	RING-finger PMP; E3 ligase and PTS receptor ubiquitination	ZS, NALD
PEX 12	RING-finger PMP; E3 ligase and PTS receptor ubiquitination	ZS, NALD, IRD
PEX 13	PMP; matrix protein import, docking complex component	ZS, NALD
PEX 14	PMP; matrix protein import, docking complex component	ZS
PEX 16	PMP-targeting; proliferation; *de novo* formation	ZS
PEX 19	PMP receptor and chaperone; *de novo* formation	ZS
PEX 26	PMP; matrix protein import, membrane anchor for Pex6	ZS, NALD, IRD

AAA, ATPases associated with diverse cellular activities; ARD, adult Refsum's disease; ATP, adenosine triphosphate; IRD, infantile Refsum's disease; NALD, neonatal adreno-leukodystrophy; PMP, peroxisomal membrane protein; PTS1, peroxisomal targeting signal 1; PTS2, peroxisomal targeting signal 2; RCDP, rhizomelic chondrodysplasia punctata; ZS, Zellweger syndrome

receptors with some residual activity and milder RCDP phenotypes.[19,191]

Four cases of 'hyperpipecolic acidaemia' have been reported. They were initially assigned to the PBDs, but the legitimacy of this common biochemical abnormality as a separate disease entity among the PBDs has been rejected.[65] One, at least, has a PEX 1 mutation.[191] Hence, 'hyperpipecolic acidaemia' has been deleted from Table 8.1. The two reported autopsy cases, on patients who died at or before 27 months of age, are included here for historical continuity.[25,57] Minor dysmorphic features, hepatomegaly, developmental delay, hypotonia, pigmentary retinopathy and progressive neurological dysfunction are reported. Micro-nodular cirrhosis and hepatocytic glycolipid inclusions were seen, in addition to dilations of renal tubules. The neuropathological findings were discordant. In neither case were abnormal neuronal migrations or significant brain atrophy noted. In one case, the white matter of brain showed multiple areas of demyelination and astrocytosis, most prominently in the internal capsule, pons, medulla and cerebellum. Abnormal, often striated, material was observed in both macrophages and astrocytes. Macrophages also contained sudanophilic spherical droplets and the striated macrophages were periodic acid–Schiff (PAS)-positive and contained angulate lysosomes with spicules. Myelin was interpreted as degenerate ultrastructurally, but post-mortem autolysis may have complicated this interpretation.[57] In the other case, the cerebral white matter was hypoplastic (similar to RCDP type I case of Agamanolis and Novak[1] later in this chapter) and showed a moderate decrease in myelin staining without myelin breakdown. There was also a noteworthy accumulation of 1- to 1.5-mm, PAS-positive, diastase-resistant, alcian blue-negative, non-sudanophilic, non-fluorescent granules in satellite cells and astrocytes, including perivascular foot processes. This material seemed to correlate ultrastructurally with irregular membrano-vesicular profiles within astrocyte cytoplasm.[25] The adrenal glands in both cases were reported to be within normal limits, but rare PAS-positive striated macrophages were found within the adrenal cortex of the Gatfield case (personal observation, slide courtesy of Dr Daria Haust).

General Pathology and Neuropathological Overview of Peroxisomal Disorders

The major organ systems involved in peroxisomal disorders are the CNS, peripheral nervous system (PNS), skeleton, eyes, liver, adrenal gland and kidney.[135] In the PBDs and those single enzyme deficiencies that simulate them clinically (pseudo-PBDs), a range of extraneural lesions is usually seen: dysmorphic facies; stippled calcifications to shortened proximal long bones (rhizomelia); portal fibrosis to micro-nodular cirrhosis and steatosis; PAS-positive macrophages with angulate lysosomes containing trilaminar spicules (Figure 8.2); striated adrenocortical cells containing lamellae and lipid profiles to adrenal atrophy; and renal cortical micro-cysts to macrocysts.[37,135] The eyes typically display cataracts, atypical pigmentary retinopathy, degeneration of photoreceptor cells, ganglion cell loss and optic atrophy. Angulate lysosomes with spicular inclusions in retinal macrophages and electron-opaque membranous cytoplasmic bodies in ganglion cells are observed ultrastructurally.[29] Of the PBDs, ZS is the most severe clinically and biochemically and in general results in the most severe lesions. However, skeletal involvement is most impressive in RCDP and its phenocopies, hepatic macrophages in IRD and lymphoid/thymus macrophages in NALD.[135] The characteristic ultrastructural inclusion in PBD is the membrane-bound angulate lysosome. This contains rigid-appearing trilaminar spicules consisting of two electron-dense leaflets of 3–5 nm thickness, which are separated by a regular electron-lucent space of 6–12 nm. These structures are also seen in brain, adrenal, retina and rarely other macrophages of ALD and AMN.[29,58,143] In spite of their prominence in peroxisomal disorders, especially PBDs, they are non-specific.[38] For example, they have been reported in skin biopsies of other degenerative metabolic diseases, particularly in late infantile and juvenile neuronal ceroid lipofuscinoses.[41] Hence, their identification in macrophages of the CNS or an affected

visceral organ is of paramount diagnostic importance. They contrast morphologically with the most characteristic inclusions of ALD and AMN, which are linear to gently curved lamellae, and clear lipid profiles lying free in the cytoplasm of adrenocortical, Leydig and Schwann cells[138,139] and brain macrophages[133,159] (Figure 8.3). Lamellae are not membrane-bound and fundamentally consist of two 2.5-nm, electron-dense leaflets separated by a variable electron-lucent space of 1–7 nm. They often lie within or at the edge of large clear spaces (lamellar-lipid profiles). Cells that possess spicules or lamellae/lamellar-lipid profiles are birefringent and retain their birefringence after acetone extraction.[36,85] The lamellae/lamellar-lipid profiles, apparently in contrast to the spicules,[36] can be extracted with xylene or n-hexane. This non-polar lipid has been identified as abnormal cholesterol esters containing saturated VLCFA.[78] Additional details about morphology, composition and formation are available.[36,134,142,143] In view of the known facts that the only major biochemical defect in ALD is the accumulation of saturated (except for minimal monounsaturated) VLCFAs and that VLCFAs are increased in all PBDs and pseudo-PBDs, angulate lysosomes with spicules in peroxisomal disorders most probably contain biochemically modified VLCFA. Peroxisomes vary greatly in their morphology; they are almost undetectable, atrophic or hypertrophic in PBD, malformed to normal in pseudo-PBD, and normal in most other single enzymopathies. Mosaicism has also been documented.[36]

Neuropathological lesions in peroxisomal disorders are of three major types: (i) abnormalities in neuronal migration or positioning, which are characteristic of PBD and pseudo-PBD, particularly ZS; (ii) defects in the formation or maintenance of central white matter, the former typically seen in PBD and the latter in ALD and AMN and NALD; and (iii) post-developmental neuronal degenerations, which are most frequent in AMN, PBD and Refsum's disease.[141] Except for ZS, NALD, ALD and AMN, pathological data on most peroxisomal disorders are scant and should be considered provisional.[135] The prototypes of the PBD group, ZS and, of the single

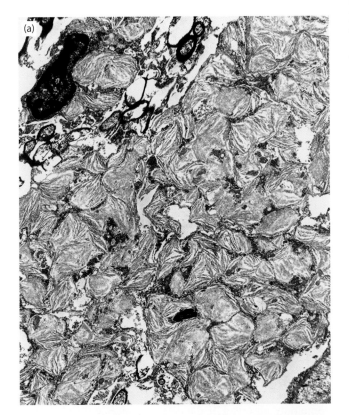

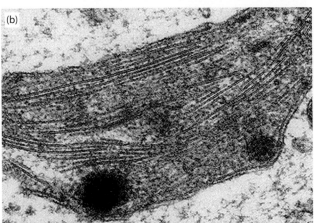

8.2 Inclusions in PBDs. (a) Angulate lysosomes containing trilaminar straight spicules in a central nervous system macrophage from a symptomatic adreno-leukodystrophy heterozygote. Electron microscopy of autopsy material. **(b)** Angulate lysosome containing trilaminar inclusions in hepatocyte of infantile Refsum's disease. Electron microscopy of a biopsy.

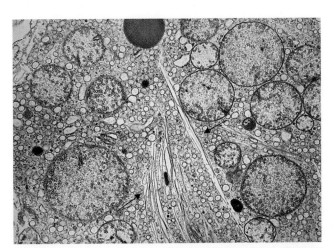

8.3 ALD. Electron micrograph of lamellae and lamellar-lipid profiles (arrows) among dilated smooth endoplasmic reticulum and variably sized mitochondria in a 22-week fetal zone adrenocortical cell of adreno-leukodystrophy. Autopsy.

protein deficient group, ALD and AMN, will be emphasized. Abnormal white matter is prominent in ZS and predominant in NALD and ALD; consequently, an overview of primary diseases of myelin[136] can provide a perspective for the white matter lesions described later.

LEUKODYSTROPHIES AND OTHER DEFECTS IN MYELINATION

The term 'leukodystrophy' is generally used to describe genetic (inherited) and progressive disorders that primarily and directly affect CNS myelin. The classic leukodystrophies have been historically classified as 'dysmyelinative' primary diseases of myelin. Other diseases displaying a comparable confluent loss of myelin but lacking the genetic, progressive or primary myelin involvement have been referred to as 'leukoencephalopathies'. The genetic defect in leukodystrophies may result in synthesis of biochemically abnormal myelin or in a molecular abnormality in myelin-forming cells, usually oligodendrocytes, which adversely impacts myelin in other ways. Irrespective of the biochemical/molecular abnormality, the end result is typically a confluent destruction, or failed development, of central white matter. In the latter and much less common 'hypomyelinative' leukodystrophies, such as Pelizaeus–Merzbacher disease (PMD), a molecular defect in oligodendrocytes impedes the formation of myelin. There also are a number of neonatal-infantile diseases, including PBD, in which it is difficult to decide whether there is hypomyelination and, if so, whether it represents a delay in CNS myelin formation or an arrest of myelination. Both of these situations are 'hypomyelinative' at autopsy and, at least theoretically, could be considered as 'hypomyelinative leukodystrophies'. However, a delay in CNS myelination is usually not progressive and, hence, it may be more appropriate to consider these as leukoencephalopathies. It takes considerable experience with numerous controls and a sound knowledge of the timing of regional myelinogenesis in the CNS[21,95] to make these determinations neuropathologically. MRI, particularly on T2-weighted images, has considerably improved our sensitivity of diagnosis of white matter disorders, and the same patterns of classic gross neuropathological lesions can be appreciated more easily.[130,161,184,185] Longitudinal studies using MRI, especially with modifications including magnetization transfer and diffusion anisotropy, provide powerful pathogenetic insights.

Most leukodystrophies fall into the dysmyelinative category, where myelin is initially formed to a variable extent but subsequently breaks down. The myelin may break down (i) because it is biochemically abnormal, such as in metachromatic leukodystrophy (MLD) and ALD; (ii) because an oligodendroglial toxin accumulates because of the molecular defect in the oligodendrocyte, such as has been proposed in globoid cell leukodystrophy (GLD, Krabbe disease); or (iii) for unknown reasons, such as in sudanophilic (orthochromatic) leukodystrophies (SLD). When myelin is catabolized, its protein and lipid constituents, particularly galactolipids (cerebroside, sulphatide [cerebroside sulfate]) and cholesterol, are liberated and can be appreciated with traditional carbohydrate (e.g. PAS) and

lipid (e.g. oil red O) stains. Galactolipids are PAS-positive; sulphatide, which contains anionic sulfate groups, is also metachromatic. When cells with normal catabolic enzymes degrade biochemically normal myelin, these staining reactions are ephemeral. However, if galactocerebrosidase or arylsulphatase (arylsulfatase) activity is absent or markedly diminished, these staining qualities persist. Consequently, in GLD lacking galactocerebrosidase the myelin debris that accumulates is PAS-positive, although in MLD it is PAS-positive and metachromatic. The liberated cholesterol is esterified primarily by macrophages and usually persists for much longer as cytoplasmic vacuoles (lipophages, gitter cells, compound granular corpuscles). Therefore, if the myelin is biochemically 'normal' and the host possesses normal catabolic enzymes, then the myelin debris consists primarily of cholesterol esters. Normal cholesterol esters, the 'floating fraction' of neurochemists, are stained convincingly with neutral lipid or sudanophilic dyes, hence the term 'sudanophilic'. As cholesterol esters are not metachromatic, they are also 'orthochromatic'.[136]

Histological examination reveals a dynamic process, with various cells participating at specific times. In general, the affected myelin sheaths display morphological changes of vacuolation, blebbing, fragmentation and loss of stainability with traditional myelin stains, which is accompanied by cellular reactions characteristic of each disease. Most commonly, in the dysmyelinative leukodystrophies, axons lacking myelin sheaths are admixed with numerous macrophages containing myelin debris, which is often diagnostically distinctive, and hypertrophic or reactive astrocytes. Later, the macrophages migrate to perivascular spaces around venules, either to die there by apoptosis or to exit the brain; the reactive astrocytes involute and produce a chronic astroglial scar, which may be anisomorphic or isomorphic. In the 'hypomyelinative' leukodystrophies, there is little need of a macrophage response but reactive astrocytosis is usually prominent.

The classic dysmyelinative leukodystrophies (ALD, MLD, GLD and probably some SLD) involve defects in myelin lipids that are qualitatively similar in the CNS and PNS; hence, involvement of both CNS and PNS myelin is commonly observed. If the defect involves myelin proteins that are largely restricted to one compartment (e.g. proteolipid protein, in CNS myelin), then the myelin lesions are likewise restricted (e.g. to the CNS in PMD). All leukodystrophies mentioned earlier display similar macroscopic CNS features: reduced brain weight, optic atrophy, bilaterally symmetrical diffuse loss or lack of deep cerebral and cerebellar white matter, which is replaced by firm tan to grey astrocytic gliosis, relative sparing of subcortical arcuate ('U') fibres, atrophy of the corpus callosum and compensatory (*ex vacuo*) hydrocephalus.

Light and electron microscopic examination shows that each leukodystrophy has its own characteristic and usually diagnostic lesions, in addition to the common features of reduced myelin staining, loss of oligodendrocytes, relative sparing of axons, macrophages containing myelin debris and reactive astrocytosis to fibrillary astrocytic gliosis. Axonal loss is greater and inflammatory cells, other than macrophages, are conspicuously absent in the leukodystrophies when compared with the prototypic demyelinative diseases, multiple sclerosis (MS) and acute disseminated (allergic) encephalomyelitis (ADEM). ALD is the exception, because it is markedly inflammatory and mimics MS.[133,136,159] ALD differs from MS, however, by the typical localization of lymphocytes just within (rather than at) the demyelinative edge, by the relative paucity of B-lymphocytes and plasma cells,[148] and by the T-helper cell-cytokine pattern.[109] In MS and ADEM, lymphocytic or lymphocytic/plasmacytic reactions, more axonal sparing, lesion asymmetry when bilateral, random involvement of subcortical fibres, sudanophilic myelin debris and restriction of lesions to the CNS reflect the immune destruction of biochemically normal central myelin, in which myelin proteins appear to be a major target.

There is an additional group of primary diseases of myelin, sometimes referred to as 'myelinolytic' or 'spongy myelinopathies', which share the common histopathological feature of spongy or vacuolated myelin due to intramyelinic oedema. The splits in myelin usually occur at the intraperiod (extracellular face) line. They typically elicit little to no macrophage or reactive astrocytic response. Despite these common histopathological features, their aetiologies are diverse, usually toxic-metabolic and include vitamin B12 deficiency, aminoacidurias and mitochondrial disorders.[136] Canavan's disease bridges this classification scheme by exhibiting a spongiform myelinopathy (i.e. myelinolytic) and a confluent, progressive and genetically determined myelin lesion (i.e. a leukodystrophy). Peroxisomal disorders are not characterized by spongy myelin, except in one family with atypical Refsum's disease, in which a mitochondrial defect was not completely excluded.[181] Table 8.3 summarizes the main histopathologic, biochemical and molecular features seen in peroxisomal disorders.

GROUP I: DISORDERS OF PEROXISOME BIOGENESIS

'Classic' Zellweger Syndrome

These infants usually present at birth with characteristic dysmorphic facies, seizures, severe hypotonia and profound psychomotor retardation. They usually die in the first year of life. Additional clinical and radiological findings include cataracts, pigmentary retinopathy, optic atrophy, sensorineural hearing deficits, equinovarus deformity, hepatomegaly and stippled calcifications of the patellae, femora and humeri. Systemic pathological findings may also include biliary dysgenesis, ventriculoseptal defects, islet cell hyperplasia and hypoplasia of the thymus and lung. Renal microcysts, predominantly cortical and varying from 0.1 to 0.5 cm, arise from both tubules and glomeruli.[17,124,127,203] The adrenal cortex displays scattered and infrequent adrenocortical striated cells with lamellar-lipid profiles and PAS-positive macrophages with spicules.[63] Many of these abnormalities have been documented in affected fetuses.[144] Hepatic peroxisomes originally had been reported as absent, and mitochondrial abnormalities were observed;[62] subsequently, immunofluorescent studies identified remnant peroxisomes present as 'membrane ghosts', representing enlarged vesicles with peroxisomal membrane proteins but lacking matrix proteins[157] (reviewed in Depreter et al.[36]). Deficiency of Pex3p, Pex16p or Pex19p impairs PMP biogenesis, and peroxisomal membranes do not form. As a result of multiple enzyme losses, elevations in VLCFA and phytanic, pristanic, pipecolic,

TABLE 8.3 Overview of peroxisomal disorders

	ZS	NALD	RCDP	BPD	ALD	AMN
Major localization	Cer	Cer	CNS	Cer	Cer	SC
	GM	WM		GM	WM	
Histopathology	Dysgenesis	De	Microencephaly	Dysgenesis	De	Axonopathy
Misguided neurons	+ + +	+	0 ?	+ +	0	0
Cer/Cbl myelination	Hypo	De	Hypo	Hypo/De	De	Dys/De
Oligodendroglial loss	±	+ +	±	±	+ + +	+ +
Astrocytosis	+ +	+ +	±	±	+ + +	+
Microgliosis	±	+ +	±	±	+ + +	+ + +
Inflammation	0	+ +	0	+	+ + +	±
Adrenal lamellae	+	+	0	+	+ + +	+ + +
Angulate lysosomes	+	+ +	0	+	+	+
Biochemical defect	Multiple	Multiple	↓ Plasmalogens	↑ VLCFA ↑ Cholestanoic acids	↑ VLCFA	↑ VLCFA
Molecular defect	PEX 1*	PEX 1*	PEX 7	D-BP	ABCD1	ABCD1
	Import	Import	Import	Activity	Import	Import

*PEX 1 most common gene defect. For other PEX genes, see Table 8.2. 0, negative; ±, rare; +, mild; + +, moderate; + + +, severe.

ABCD1, adenosine triphosphate-binding cassette D1; ALD, adreno-leukodystrophy; AMN, adrenomyeloneuropathy; BPD, bifunctional protein deficiency; Cbl, cerebellar; Cer, cerebral; CNS, central nervous system; D-BP, dextro isomer of bifunctional protein; De, demyelination; Dys, dysmyelination; GM, grey matter; Hypo, hypomyelination; NALD, neonatal adreno-leukodystrophy; PEX, peroxin; RCDP, rhizomelic chondrodysplasia punctata; SC, spinal cord; VLCFA, very long-chain fatty acid; WM, white matter; ZS, Zellweger syndrome.

di- and tri-hydroxycholestanoic acids with decreases in plasmalogens and DHA are evident. Phytanic acid is derived exclusively from dietary sources, whereas pristanic acid comes from phytanic acid breakdown and dietary sources; hence, the presence and prevalence of both may vary with age, diet and catabolic enzyme activity.[191,193]

Neuropathology of Zellweger Syndrome

The major abnormalities are in the CNS, where defective neuronal migrations dominate.[45,190] These infants classically show a unique combination of centrosylvian or parasylvian pachygyria (medial) and polymicrogyria (lateral and extending into the lateral frontal lobe and the lateral parieto-temporo-occipital region) (Figure 8.4); this localization for the major neocortical malformation is consistent but may be associated with other areas of polymicrogyria and pachygyria. There may also be an abnormal vertical tilt to the sylvian fissure. The limbic areas of the brain are typically spared. Coronal sections of the cerebrum exhibit a thickened cortex, with either excessive superficial plications or obvious subcortical heterotopias. Microscopy of the micro-polygyric cortex usually reveals fusion of the molecular layers and better preservation of the supragranular cortex (Figure 8.5). The outer cortex is typically occupied by medium to large pyramidal cells destined for deep cortex, whereas the usually superficial neuronal populations are detained in deep cortex and subcortical white matter. The pachygyric cortex has a similar but more severe alteration; the subcortical heterotopias are likewise more prominent (Figure 8.6). These cortical abnormalities differ from those of classic four-layered polymicrogyria or lissencephaly-pachygyria in that all neuronal classes seem affected, with those destined for the outer layers tending to be more impeded. The cerebellum is grossly unremarkable, but it is common to find heterotopic Purkinje cells (often polydendritic) or combinations of Purkinje cells and granule cells in the white matter, especially in the nodulus. There is dysplasia of claustra, medullary olives and often the dentate nuclei. These dysplasias are not true heterotopias but, rather, appear to reflect a problem with neuronal positioning or the terminal stages of migration.[135,141] For example, the olives and dentate nuclei may lack their normal serpiginous configuration or consist of discontinuous islands of neurons; the olives may display peripheral palisading of neurons (Figure 8.7). Striated and globose PAS-positive macrophages, containing abnormal lipid cytosomes, have been identified in grey and white matter of cerebrum and cerebellum (Figure 8.8).[2,33,200] These abnormal lipids are detectable with proton MR spectroscopy.[68] Morphological evidence of a restricted neuronal lipidosis is seen in the form of striated neuronal perikarya and dystrophic spheroids containing lamellar and lipid profiles in Clarke's and lateral cuneate nuclei (Figures 8.9–8.11).[146] Ventricular enlargement is common, as are periventricular cysts and ependymal abnormalities.[65]

Brains from fetuses at risk of developing ZS show neocortical migratory defects as early as postmenstrual estimated gestational age (EGA) 14 weeks, in the form of micro-polygyric ripples and subtle subcortical heterotopias. Thin abnormal cortical plates with more obvious subcortical heterotopias occur at EGA 22–24 weeks (Figure 8.12). Astrocytes, neuroblasts, immature neurons, radial glia and

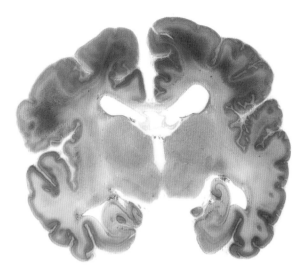

8.4 Zellweger syndrome. Coronal section of a 4-week-old PEX 1 male with Zellweger syndrome, demonstrating bilateral parasylvian medial pachygyria with prominent subcortical heterotopia and asymmetric lateral polymicrogyria, particularly of insular cortex. Nissl.

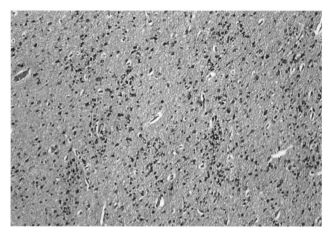

8.5 Zellweger syndrome. Micropolygyric heterotopia in Zellweger syndrome, with fewer pyramidal and more granular neurons than its pachygyric counterpart.

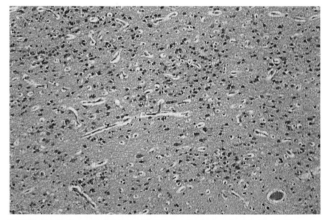

8.6 Zellweger syndrome. Pachygyric heterotopia in Zellweger syndrome, consisting of many pyramidal neurons.

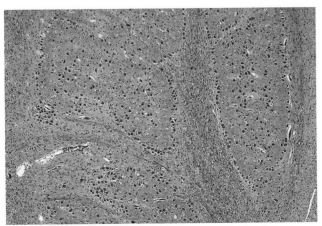

8.7 Zellweger syndrome. Discontinuous and simplified inferior olive with peripheral palisading of neurons in a 4-week-old *PEX 1* male with Zellweger syndrome.

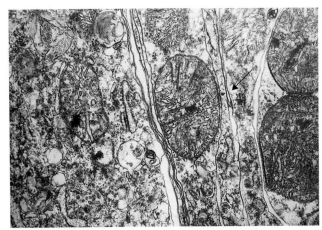

8.10 Zellweger syndrome. Lamellae and lamellar-lipid profiles (arrow) between mitochondria in dorsal nucleus of Clarke perikaryon of a 12-week-old male with Zellweger syndrome. Electron microscopy of autopsy material.

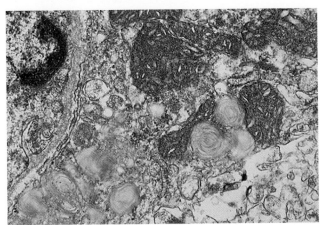

8.8 Zellweger syndrome. Electron-opaque membranous cytoplasmic bodies, typical of Zellweger syndrome, in an astrocyte of occipital cortex in a 13-week-old male with Zellweger syndrome. Autopsy.

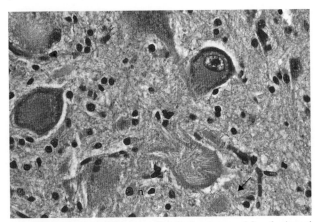

8.9 Zellweger syndrome. Striated neuron (n) of dorsal nucleus of Clarke, with spheroid (arrow) in the same patient as in Figure 8.8.

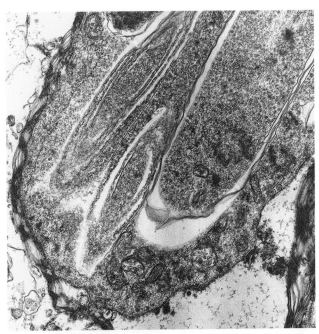

8.11 Zellweger syndrome. Electron micrograph of lamellae and lamellar-lipid profiles in axonal swelling within dorsal nucleus of Clarke of the same patient as in Figure 8.10. Autopsy.

PAS-positive macrophages contain abnormal pleomorphic cytosomes; these include electron-opaque membranous cytoplasmic bodies, perhaps representing gangliosides containing saturated VLCFA. Some neurites also exhibit lamellar and lipid profiles. Dysplastic alterations of the inferior olivary and dentate nuclei are present, as well as renal cortical microcysts, fetal zone adrenocortical striated cells and patellar mineralization.[147] These fetal lesions support the proposal that the insult (presumably metabolic) causing the neocortical migration defect is operating throughout the entire neocortical neuronal migratory period.

The CNS white matter in ZS is commonly abnormal (Figure 8.13) but does not contain inflammatory cells (lymphocytes or plasma cells). Although some authors have referred to this as a leukodystrophy,[127] most would not, and the nature of the defect is still unclear.[2,135] It appears to be dysmyelinative and primarily hypomyelinative, in that there is usually little sudanophilic lipid accumulation in macrophages. Reactive astrocytosis may be relatively inconspicuous in immature white matter or severe, particularly in

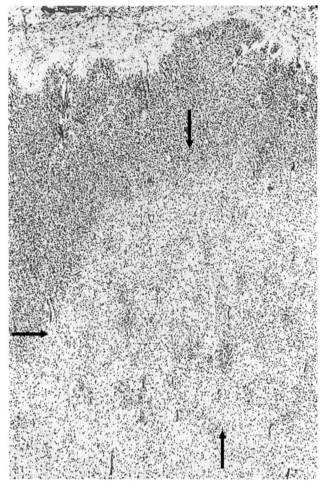

8.12 Zellweger syndrome. Abnormal cortical plate with subcortical heterotopia (arrows) in incipient pachygyric superior parietal gyrus of a 22-week-old male fetus with Zellweger syndrome.

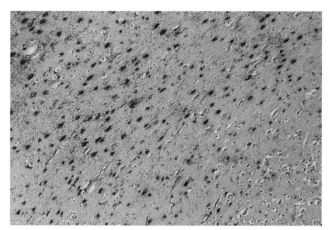

8.14 Zellweger syndrome. Subcortical cerebral white matter of another patient syndrome, with reactive astrocytosis. Glial fibrillary acidic protein.

areas with heterotopias (Figure 8.14); it can even extend throughout the white matter of the neuraxis. Many abnormal lipid cytosomes are present in astrocytes and oligodendrocytes,[200] some of which might be VLCFA-gangliosides, and plasmalogen deficiency could interfere with normal myelination.[135,141] Myelin pallor may reflect a decrease in the normal complement of myelinated axons, reflecting the severe pachygyria–polymicrogyria; this is supported by more severe involvement of the posterior limb of the internal capsule underlying areas of perisylvian polymicrogyria (Figure 8.13). Additionally, superimposed hypoxic–ischaemic–acidotic damage due to seizures, systemic metabolic abnormalities and chronic debilitation are probable complicating factors. It is important to emphasize that, irrespective of the precise pathogenesis of the white matter lesion in ZS, it is not inflammatory and differs morphologically from that of ALD. Many of the gross neuropathological features of ZS can be appreciated in living patients with MRI (Figure 8.15).[130,184,185]

Neonatal Adreno-Leukodystrophy

This highly variable PBD has features in common with both ZS and ALD but is less severe.[82,106,182] Clinically, infants with NALD resemble those with ZS, but they usually die around 36 months of age and survival can be highly variable, even extending into adolescence.[92] Despite its name, NALD is transmitted as an autosomal recessive trait and the coexistence of NALD and ALD in the same kindred has never been reported. Hepatic lesions are common. Peroxisomes in the liver have been reported as missing, decreased in size or enlarged, with associated biochemical abnormalities in all peroxisomal functions[75] (reviewed by Depreter et al.[36]). Mitochondrial abnormalities have also been documented.[75] Adrenocortical atrophy with striated adrenocortical cells mimics the adrenal pathology of ALD. The diffuse distribution of PAS-positive macrophages with angulate lysosomes containing spicules in the liver, thymus, spleen, lymph node, lung, gastrointestinal tract and adrenal gland contrasts with their hepatic predominance in IRD. The PNS in NALD has been unremarkable or has shown evidence of demyelination and thin myelin sheaths, and lamellae and lipid profiles (see ALD) have been visualized in

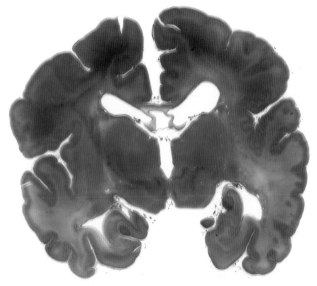

8.13 Zellweger syndrome. Coronal section of a 4-week-old *PEX 1* male, demonstrating an asymmetric paucity of myelin most notably in the posterior limbs. Luxol fast blue–periodic acid–Schiff.

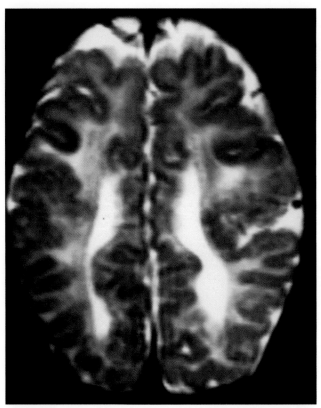

8.15 Zellweger syndrome. Axial T2-weighted image from a 3-month-old *PEX 1* male, showing sparse myelination, frontal pachygyria, parasylvian polymicrogyria and some ventriculomegaly.

Reprinted with kind permission of Springer Science and Business Media and Prof. MS van der Knaap from van der Knaap and Valk.[184]

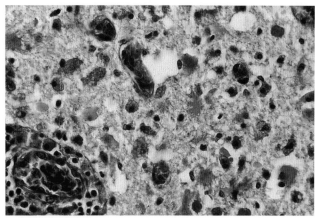

8.16 ALD. Inflammatory demyelinative lesion in occipital white matter of a 14-year-old male with neonatal adreno-leukodystrophy. Luxol fast blue–periodic acid–Schiff.

systemic PAS-positive macrophages, inflammatory demyelinative CNS lesions and increases in saturated C26:0; ZS exhibits chondrodysplasia, renal microcysts, pachygyria–polymicrogyria with dysplastic claustra, dentate nuclei and olives and the accumulation of both saturated and monounsaturated C26:0.[48,92]

Infantile Refsum's or Phytanic Acid Storage Disease

Patients with infantile Refsum's disease (IRD) have mental retardation, dysmorphic facies of a minor degree, retinitis pigmentosa, sensorineural hearing deficits, failure to thrive and hypocholesterolaemia.[131] All peroxisomal biochemical parameters are abnormal and peroxisomes are either deficient or reduced in number[75] (reviewed by Depreter *et al.*[36]). The clinical course is relatively mild compared to NALD, with presentation after the neonatal period and survival at least into the early teens. Some patients who are phenotypically IRD develop clinical and radiological evidence of a progressive leukoencephalopathy.[130] A distinct Zellweger spectrum phenotype is a late-onset, white matter disease with a central cerebellar predilection, clinically characterized by precipitous neurologic regression in late infancy, adolescence or adulthood.[8] The most conspicuous pathological alteration in IRD is the prominence of PAS-positive macrophages containing angulate lysosomes in liver and accumulation of spicular structures in hepatocytes.[166] Abnormal mitochondria have been noted.[75] Hepatomegaly with micronodular cirrhosis was reported at autopsy in a 12-year-old boy.[179] The adrenal glands were interpreted as hypoplastic but the appearance is more likely to have resulted from atrophy. CNS migration defects were not observed, but the cerebellar cortex was diffusely small; granule cells were preferentially reduced in number and Purkinje cells were abnormally situated in the molecular layer. MRI studies[184] and similar neuropathology in chronic RCDP indicate that this probably represents cerebellar atrophy.[149] The cerebral white matter demonstrated focal decreases in the number of myelinated axons, particularly in the periventricular region, corpus callosum, corticospinal tracts and optic nerves. MRI studies have shown

Schwann cells.[111] One case of NALD demonstrated severe atrophy of the auditory sensory epithelium and tectorial membrane.[77] Retinopathy with a 'leopard spot' appearance may be a distinctive clinical feature.[48]

The CNS in NALD, in contrast to ZS, shows greater involvement of white matter than disrupted neuronal migration. Both ZS and NALD are associated with a slight increase in brain weight and infants are often macrocephalic. Heterotopic Purkinje cells are observed. Neuronal loss has been reported in the olives, dentate nuclei and thalami of two patients with NALD. PAS-positive macrophages are present in the cortex of the cerebrum and, particularly, the cerebellum, similar to ZS, as are rare swollen neurons in the arcuate nucleus and perhaps in the pons.[182] Polymicrogyria may be diffuse, focal or multifocal and associated with subcortical heterotopias; pachygyria and dysplastic olives have not been reported. Diffuse dysmyelination/demyelination of cerebral and cerebellar white matter, often with a prominent perivascular lymphocytic reaction, is another distinguishing CNS feature and resembles ALD (Figure 8.16). PAS-positive macrophages containing both spicules and lamellae are present and variable degrees of reactive astrocytosis to chronic fibrillary astrogliosis are seen. Sparing of the arcuate fibres has been reported and sudanophilic macrophages are commonly observed within white matter lesions. In summary, NALD displays adrenal atrophy, more widespread cerebral involvement, including the posterior limb and pyramidal fibres in brain stem.[130,184] Inflammatory cells were not seen but there were numerous perivascular collections of non-sudanophilic and PAS-negative macrophages and dense gliosis. Occasional perivascular macrophages were also noted in brain stem and cerebellar white matter. Ultrastructurally, brain macrophages contained lamellae but lacked angulate lysosomes. Astrocytes also displayed lamellar inclusions. Two 'atypical' IRD siblings, who died at 3.5 years and 8 months, have also been reported. Marked adrenal atrophy with striated cells (not illustrated) and sparse PAS-positive macrophages was associated with displaced Purkinje cells in the molecular layer, peripheral palisading of neurons in the olive and cirrhosis with PAS-positive macrophages.[26]

With respect to the sensorineural hearing deficit, which is also seen in ZS, NALD, ARD, RCDP, acyl-CoA oxidase deficiency and bifunctional protein deficiency (BPD), one patient with IRD exhibited good preservation of ganglion cells and nerve fibres in the organ of Corti but severe atrophy of the sensory epithelium and stria vascularis.[179]

Rhizomelic Chondrodysplasia Punctata, Type I, Classic

Classic RCDP presents at birth with severe shortening and stippled calcifications of the humerus and femur, vertebral defects, joint contractures, dysmorphic facies, psychomotor and growth retardation, cataracts and optic atrophy, sensorineural hearing deficits and microcephaly.[169] Most also have ichthyosis. Survival in RCDP is better than once thought, with 90 per cent surviving to age 1 year and 50 per cent surviving to 6 years.[199] Some milder variants may have chondrodysplasia without rhizomelia, cataracts and less severe growth and mental deficiency.[130] Plasmalogen deficiency is even more severe than that seen in ZS. VLCFA are normal. Dihydroxyacetonephosphate acyltransferase (DAPAT) and alkylglycerone phosphate synthase (AGPS) enzymes, which initiate peroxisomal plasmalogen synthesis, are also deficient.[191] Pex7p imports AGPS into peroxisomes and DAPAT activity requires the presence of AGPS. The phytanic acid oxidation defect, because of deficient Pex7p import of phytanoyl-CoA hydroxylase (PhyH), approximates that of adult Refsum's disease and ZS. In some hepatocytes peroxisomes cannot be identified, whereas in others they are irregularly shaped and enlarged.[36,65] Few post-mortem examinations of infantile cases have been undertaken.[59,132] Although most RCDP infants are microencephalic, no satisfactory morphological correlation for their severe psychomotor retardation has been found. Microscopical examination of the CNS in patients who died at about 1 year of age was essentially unremarkable; however, in one these brains, the inferior olives showed focal discontinuities and were considered to be a mild form of the defect seen in ZS.[132] The white matter of a girl aged 3 years with RCDP was diffusely reduced in size but appeared to be normally myelinated, except in the occipital lobe; diffuse cerebellar degeneration and a corresponding neuronal loss in the olives were observed.[1] We confirmed the microencephaly (Figure 8.17) in two patients with chronic RCDP and focal dysplasia of the olive in one;

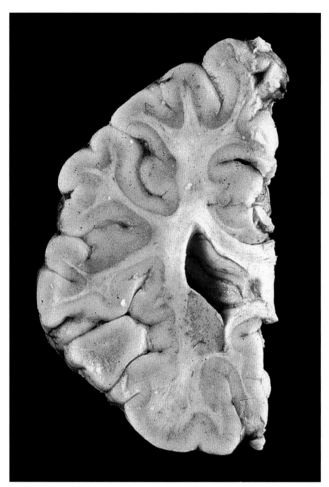

8.17 RCDP. Microencephalic brain of an 11-year-old boy with rhizomelic chondrodysplasia punctata, with reduced volume of frontal white matter.

they both displayed severe cerebellar atrophy because of losses of Purkinje and granule cells (Figure 8.18). Neuronal loss in the olives and a variable pallor of myelin with corresponding reactive astrocytosis were also noted. Thus, post-developmental cerebellar atrophy occurs in RCDP patients with prolonged survival. Phytanic acid accumulation (diet dependent), perhaps in concert with reduced tissue plasmalogen levels, has been proposed to cause the apoptotic death of Purkinje and granule cells by altering calcium homeostasis.[149] Neocortical migration defects are generally considered to be absent, supporting the hypothesis that the migration defects in PBD and pseudo-PBD are due, at least in part, to elevated VLCFA.[147] However, one patient with short humeri and femora, widespread stippled calcifications, collecting tubule and glomerular cysts and cataracts displayed pachygyria of the posterior frontal and pararolandic region and focal microgyria of the frontal pole; designation of this case as RCDP remains uncertain as it preceded the era of biochemical or genetic confirmation.[141] Another patient, clinically classified, although not genetically confirmed, as RCDP, had pachygyria and polymicrogyria of bilateral frontal lobes on neuroimaging.[61] MRI in severe RCDP phenotypes reveals delayed myelination, regressive white matter changes with a parietooccipital predominance, ventricular dilatation and cerebellar atrophy.[6,130]

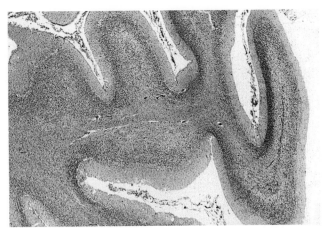

8.18 RCDP. Loss of granule neurons and Purkinje cells, worse distally, in atrophic cerebellum of a 9-year-old girl with rhizomelic chondrodysplasia punctata.

A PEX 7 knockout mouse provides some interesting insights into this disorder. Neocortical lamination abnormalities were detected, but so were elevations in VLCFA (deficient Pex7p-mediated import of thiolase [ACAA1]) that are not seen in human RCDP. The authors speculated that the neuronal migration defect was due to a combination of elevated VLCFA and decreased plasmalogens (reviewed in da Silva[31]).

GROUP II: DISORDERS WITH MORPHOLOGICALLY INTACT PEROXISOMES AND A SINGLE PROTEIN DEFICIENCY

Pseudo-Peroxisome Biogenesis Disorders

Acyl Coenzyme A Oxidase Deficiency (Pseudo-neonatal Adreno-Leukodystrophy)

Over thirty cases have been reported (reviewed in Wang et al.[195]).[100,193] Although dysmorphic features were not observed in some patients, mild anomalous facial features were noted in others. All paediatric patients displayed moderate to severe neonatal hypotonia, seizures and psychomotor retardation. Sensorineural hearing deficits and abnormal electroretinograms were reported. The only biochemical abnormality was elevated VLCFA; hepatic peroxisomes were heterogeneous in size, many enlarged and angulated, and they were increased in number (reviewed by Depreter et al.[36]). Macrophages with angulate lysosomes were not identified. Neuroradiological studies usually indicate an absence of migration defects, diffuse and progressive CT hypodensities in cerebral white matter, a thin corpus callosum and cerebellar 'hypoplasia' (probably atrophy). However, several patients have had cortication defects, such as perisylvian polymicrogyria or pachygyria. The white matter of one progressive case demonstrated abnormal contrast enhancement, suggesting an inflammatory component, such as in NALD and ALD. With MRI, the white matter abnormality was seen as high signal intensity in T2-weighted images but contrast material was not administered.[100] The distribution of the lesions was mainly cerebellum, cerebral

peduncles and brain stem in the early stages but also posterior cerebrum later. Neuropathologic data are rare. Two siblings, one of whom had haematopoietic cell transplantation, had inflammatory demyelination of cerebral white matter as severe as that in ALD with a comparable posterior predominance. Less inflammatory demyelination was observed in cerebellar white matter and basis pontis, again comparable to ALD. In keeping with the dual localization of acyl coenzyme A oxidase in glia and neurons,[47] an olivopontocerebellar-like degeneration was also noted. Severe neuronal losses in the cerebellar and cerebral (particularly motor) cortex, as well as dentate, olivary and basis pontine nuclei, resulted in severe cerebellar and pontine atrophy, moderate cerebrocortical atrophy and corresponding secondary tract degeneration of cerebral and cerebellar peduncles (see Wang et al.[195]). A knockout mouse model has been developed but neuropathological data were not provided.[5]

D-Bifunctional Protein Deficiency (2-Enoyl Coenzyme A Hydratase/D-3 Hydroxacyl Coenzyme A Dehydrogenase)

This deficiency is by far the most common of the pseudo-PBD single enzymopathies.[198] The deficient enzyme is d-bifunctional protein (D-BP; also referred to as multifunctional protein 2, multifunctional enzyme II or D-peroxisomal bifunctional enzyme), not L-bifunctional protein, as originally reported.[188,197] BPD is a more severe clinical phenotype than acyl-CoA oxidase deficiency.[198] Biochemically, these patients demonstrate elevations in VLCFA, bile acid intermediates and perhaps phytanic and pristanic acids but erythrocyte plasmalogens are normal.[191,193,198] Clinically, they often resemble the severe Zellweger phenotype, with neonatal hypotonia and seizures, facial dysmorphia, psychomotor retardation, neuronal migration defects (polymicrogyria) and hypomyelination (Figure 8.19). Most affected children die within the first two years of life. Milder BPD cases may have longer (≥ 7.5 years) survival and peroxisomal biochemical abnormalities may be lacking in plasma. The clinical, biochemical, imaging and (to some extent) neuropathological spectrum of a large cohort of these patients has been reported.[53] Peroxisomes may be reduced in number, enlarged or undetectable.[53]

The first reported patient with BPD had clinical features of NALD; he died at 5.5 months of age. Autopsy findings consisted of mild portal fibrosis, glomerular microcysts and adrenal atrophy with 'lipid-containing balloon cell'; striated cells were not reported. The CNS revealed polymicrogyria, focal heterotopias, 'demyelination' of cerebral white matter and periventricular cysts.[197] In another case, previously classified as atypical acyl-CoA oxidase deficiency, there was dysmyelination and inflammatory demyelination of the CNS white matter (occipital lobes and cerebellum), resembling that of NALD and ALD. There were also some mild cerebral and cerebellar heterotopias, microgyria (but not pachygyria or polymicrogyria) and focal dysplasia of olivary and Clarke's nuclei. PAS-positive macrophages with angulate lysosomes and adrenocortical atrophy with striated cortical cells were present.[119] A third patient displayed centrosylvian pachygyria-polymicrogyria, reminiscent of, but milder than, ZS, diffuse hemispheric hypomyelination with subcortical heterotopic neurons, Purkinje cell heterotopias and simplified

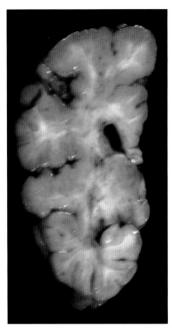

8.19 PBD. Polymicrogyric pattern in sylvian fissure and white matter abnormalities in an 8-month-old female with D-bifunctional protein deficiency.

convolutions of the dentate nucleus and inferior olive.[91] In summary, focal polymicrogyria and an inflammatory leukoencephalopathy, more consistent with NALD, as well as centrosylvian pachygyria-polymicrogyria, a non-inflammatory leukoencephalopathy and olivary dysplasia, more consistent with ZS, have been reported in BPD (Figure 8.19).

The single patient reported with thiolase deficiency (3-oxoacyl-CoA thiolase deficiency; pseudo-ZS) was later classified as BPD. This patient resembled classic ZS, with increased VLCFA and bile acid intermediates, dysmorphic facies, profound hypotonia and typical pathologies in liver, kidney and adrenal.[64,164] However, the inflammatory demyelination and astrocytosis of cerebellar white matter, heterotopic subcortical Purkinje cells, focal polymicrogyria,[135] and adrenal atrophy with striated cells were more reminiscent of NALD.

Additional clinical presentations are reported in BPD. A male aged 4 years with BPD demonstrated optico-cochleodentate degeneration as well as frontoparietal and insular microgyria (normal lamination), cerebellar atrophy, cerebral and particularly cerebellar white matter loss and a severe axonopathy of sural nerve.[165] Two siblings, the longest surviving females reported with a mild BPD mutation, developed ovarian dysgenesis, along with hearing loss and ataxia that were clinically defined as Perrault syndrome.[129] In view of the evidence of oxidative stress in ALD and the ZS knockout mice, it was somewhat surprising that it was also found in BPD but not in the PBD.[52] Phospholipid analysis of brain tissue from a BPD patient revealed declines in myelin lipids indicative of dysmyelination, as well as decreased plasmalogens in grey matter, despite intact plasmalogen synthesis enzymes in BPD (reviewed in da Silva[31]).

The remaining diseases in the pseudo-PBD category – Zellweger-like syndrome,[173] in which multiple peroxisomal enzymes were deficient, di- and tri-hydroxycholestanoic acidaemia[27] and RCDP type 3[34] – are rare and lack neuropathological data. Isolated deficiency of plasmalogen

biosynthesis enzymes in RCDP type 2 (DAPAT deficiency) or RCDP type 3 (AGPS deficiency) produces clinical phenotypes indistinguishable from classic RCDP. Neuroradiological abnormalities of CNS myelination have been found in RCDP type 2.[193] A neonatal patient with RCDP type 2 reportedly had some heterotopias, but the description 'multiple heterotopic foci of immature neurons in the vicinity of the third ventricle' suggests normal findings in an infant. The olive was also reported to be 'broadened'.[69]

Adreno-Leukodystrophy and Adrenomyeloneuropathy

X-linked ALD has two major phenotypes: juvenile (childhood cerebral) ALD and its adult variant AMN.[13,116] Much less frequent types (accounting for fewer than 15 per cent of cases) are adolescent and adult cerebral ALD, which differ from juvenile only in the age of onset; adrenal insufficiency-only (addisonian); asymptomatic; and rare olivopontocerebellar[99,123] and spinocerebellar[107] types. Approximately 65 per cent of heterozygote (female) carriers demonstrate the AMN phenotype, usually mild and later in life.[13,83]

The molecular genetics of ALD has been a field of intense activity in the past two decades, but genotype–phenotype correlations are lacking. The same genetic defect is associated with juvenile ALD, AMN and even addisonian-only ALD.[93,141] This has generated the hypothesis that a modifier gene and/or environmental factors are responsible for the phenotypic variation (reviewed by Moser).[114] CD1 gene polymorphisms (see later) do not seem to contribute to the phenotypic variation,[7] but head trauma, activity of the mitochondrial enzyme manganese superoxide dismutase (SOD2), and loss of AMP-activated protein kinase α-1 (AMPKα1), may.[22,155,204] It is also unclear how the genetic defect in ABCD1 (formerly ALDP) is related to the major biochemical abnormality identified in ALD – elevations of saturated VLCFA in blood and tissues. Previously, it was believed that the major determinant was decreased activity of VLCF acyl-CoA synthetase (lignoceroyl-CoA ligase) with a consequent elevation of VLCFA. Current evidence suggests that the major physiological function of ABCD1 is the transport of VLCFA-CoA across the peroxisomal membrane.[94] VLCFAs have been identified in many myelin components, but especially in cholesterol esters,[78] gangliosides,[76] phosphatidylcholine[172] and proteolipid protein.[16]

The *ALD* gene (ABCD1) is localized to the Xq28 region, occupies 21–26 kilobases (kb) of genomic DNA, contains 10 exons and encodes an mRNA of 3.7–4.3 kb to translate a protein of 745–750 amino acids. This ALD protein (ABCD1), instead of being the deficient synthetase, was found to be an integral membrane protein of peroxisomes with the properties of an ATP binding cassette (ABC) half-transporter.[35,94,118] ABCD1 (ALDP) mRNA is expressed in all tissues but is highest in the adrenal glands, intermediate in the brain and almost undetectable in the liver. ABCD1 is strongly expressed in microglia, astrocytes and endothelial cells; oligodendrocytes have little to no ABCD1, except those in the corpus callosum and internal capsule. ABCD1 is not detectable in the fibroblasts of about two-thirds of patients with ALD. A large number of mutations have been identified: approximately 54 per cent are missense, 25 per cent frameshift, 10 per cent nonsense and 7 per cent large deletions. A mutational hotspot is noted in exon 5. All mutations examined, except for about a third of the missense, do not express detectable ABCD1 in fibroblasts, although all have ABCD1 mRNA.[93]

The juvenile or childhood cerebral form usually presents between 6 and 9 years of age with behavioural, auditory, visual and gait abnormalities or adrenocortical insufficiency (Addison's disease). The disease is rapidly progressive, typically leading to death within 3 years.[159] MRI confirms the gross neuropathological findings described previously as well as the usual progression of the lesions from parieto-occipital to frontal lobes. Enhancement at the advancing edge of the demyelinative process is highly characteristic (reviewed by Moser[114] and van der Knaap and Valk[184]), and MRI patterns can predict the progression of the cerebral disease (Figure 8.20).[101] The adrenal cortex and testis are the only two non-neural organs that show significant lesions.[134,135,139]

Adrenocortical cells, particularly those of the inner fasciculata-reticularis, become ballooned and, of diagnostic import, striated as a result of accumulations of lamellae, lamellar-lipid profiles and fine lipid clefts (Figures 8.21–8.23).[138] The striated material, consisting of the same abnormal cholesterol esters as in CNS striated lipophages,[85] appears to lead to cell dysfunction, atrophy and apoptotic death.[137,142] Histoenzymatic decreases in mitochondrial α-glycerophosphate dehydrogenase, 3-β-hydroxysteroid dehydrogenase and reduced triphosphopyridine nucleotide (TPNH) diaphorase have been reported and they show excessive peripheral cytolysis under the electron microscope.

Moreover, the striated cells appear to adapt poorly to a tissue culture environment.[142] Ultimately, primary atrophy of the adrenal cortex ensues. Inflammatory cells are rarely observed and are probably an epiphenomenon; anti-adrenal antibodies are not detected. The adrenal cortex in AMN displays the same qualitative changes as most adrenal glands of patients with ALD but tends to be more atrophic; this is related to the longer duration of hypoadrenalism and consequent corticosteroid replacement therapy in AMN.[134] The same striated adrenocortical cells have been identified in ALD heterozygotes, both symptomatic and asymptomatic, but limited to small, multifocal clusters.[145] This resembles ZS but is in striking contrast to ALD or AMN and to the fetal adrenal zone in affected fetuses, where this lesion is diffuse.[143]

Testicular abnormalities in prepubertal males with ALD are usually present only at the ultrastructural level and consist of lamellae and lipid profiles in the interstitial cells of Leydig and their precursors. The testis in AMN and adult cerebral ALD demonstrates the same Leydig cell alterations as noted already in childhood ALD but there is also Leydig cell loss. No inflammatory cells have been seen, except in a single unpublished case of adult ALD. Degenerative changes in the seminiferous tubules in AMN appear indistinguishable from those of adult cerebral ALD and vary from a maturation arrest of spermatocytes to a Sertoli cell-only phenotype, in which all germ cells are depleted. Ultrastructurally, vacuolation of the Sertoli cell's endoplasmic reticulum followed by widened intercellular spaces appears to be the initial lesion of the seminiferous tubules.[140]

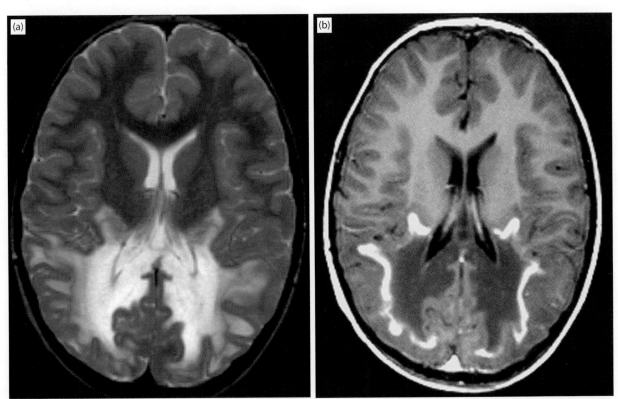

8.20 Magnetic resonance images from an 8-year-old male with juvenile (cerebral) adreno-leukodystrophy. **(a)** T2-weighted image, showing the typical pattern of symmetrical parieto-occipital lesions (bright white). **(b)** Contrast-enhanced T1-weighted image, showing intense enhancement of the rim of the lesions (bright white).

Reprinted with kind permission of Springer Science and Business Media and Prof. MS van der Knaap from van der Knaap and Valk.[184]

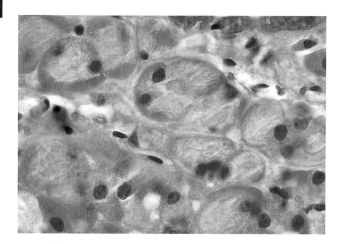

8.21 ALD. Ballooned adrenocortical cells of adreno-leukodystrophy and adrenomyeloneuropathy, with diagnostic striations.

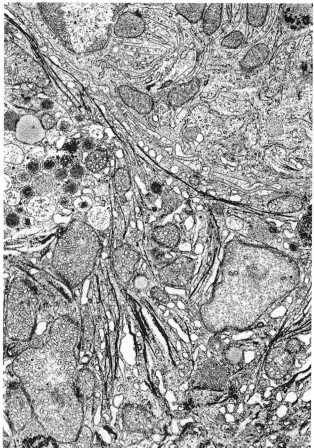

8.22 ALD. Lamellae and lamellar-lipid profiles in adrenocortical cells of juvenile adreno-leukodystrophy. Medullary cell granules on left. Biopsy.

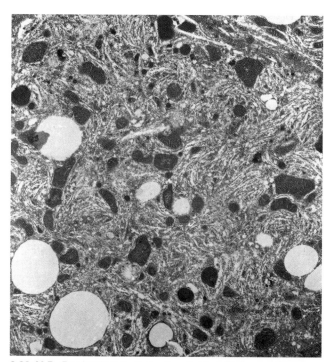

8.23 ALD. Predominantly fine, clear lipid clefts in an adrenocortical cell of another juvenile patient with adreno-leukodystrophy. Biopsy.

The Neuropathology of Adreno-Leukodystrophy

In addition to the gross and microscopic features common to most leukodystrophies, ALD has its own distinguishing characteristics.[134,159] Mild to moderate premature atherosclerosis can be seen in adult patients. The loss of myelin is almost always most prominent in the parieto-occipital region (Figure 8.24). The advancing edges, usually frontal, are more asymmetrical and may display a white to pink 'softening' that blends imperceptibly into normal white matter. Cavitation and calcification of white matter may be seen in severe cases. Arcuate fibres are relatively spared but the posterior cingulum, corpus callosum, fornix, hippocampal commissure, posterior limb of the internal capsule and optic tracts are typically involved (Figure 8.25). The cerebellar white matter usually exhibits a similar, but milder, confluent loss of myelin and sclerosis. Secondary corticospinal tract degeneration extending down through the peduncles, basis pontis, medullary pyramids and spinal cord is characteristic (Figure 8.26). The brain stem may also display primary demyelinative foci, especially in the basis pontis. The spinal cord is spared, except for the descending tract degeneration. Cerebral and cerebellar grey matter may

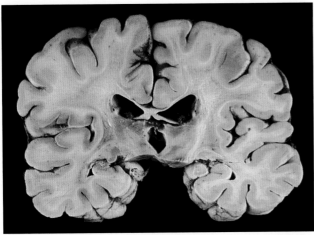

8.24 ALD. Bilaterally symmetrical, confluent demyelination of parietal white matter, particularly of the posterior limb of the internal capsule, with sparing of subcortical myelin in adrenomyeloneuropathy/adreno-leukodystrophy.

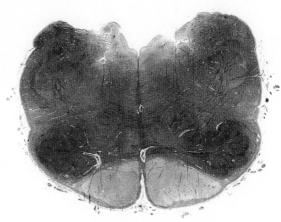

8.26 ALD. Loss of myelinated fibres in asymmetrical medullary pyramids of the same patient as in Figure 8.23. Luxol fast blue–haematoxylin and eosin.

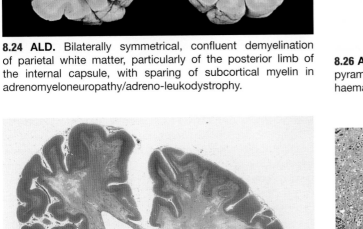

8.25 ALD. More severe loss of myelin, including most arcuate fibres, in juvenile adreno-leukodystrophy. Luxol fast blue–haematoxylin and eosin.

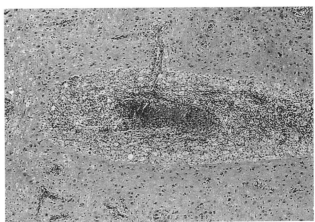

8.27 ALD. Subacute inflammatory demyelinative lesion of juvenile adreno-leukodystrophy.

be intact or may be atrophic if there is severe (e.g. cavitary) damage to the centrum semiovale or cerebellar white matter.

Histopathologically, there are marked losses of myelinated axons (myelin, axons) and oligodendrocytes. Apoptotic nuclear changes have not been seen in oligodendrocytes with traditional stains but some appear pyknotic. Random preservation of individual myelinated axons in foci of myelin loss is common but these myelin sheaths may be thin or irregular. The advancing edges of myelin loss are sites of intense perivascular inflammation by lymphocytes and lipophages and the demyelinated white matter shows depletion of oligodendrocytes and reactive astrocytosis (Figures 8.27 and 8.28). The predominant lipophage has granular to vacuolated cytoplasm, which is intensely sudanophilic and PAS-positive. The second type stains less intensely and usually demonstrates striated cytoplasm because of the presence of clear clefts. The number of striations generally correlates inversely with sudanophilia and PAS positivity. Striated macrophages retain birefringence after acetone extraction, although granular macrophages generally do not.[85] This non-polar lipid is cholesterol esterified with saturated

VLCFA,[78] and it does not accumulate until after the phase of myelin breakdown.[85,134,175] Small numbers of perivascular lymphocytes, hypertrophic astrocytes and lipophages are noted even in sclerotic lesions.

The lymphocytes display both T-cell, including CD4+ and B-cell phenotypes,[66] with the T-cell predominant in two studies.[81,148] Plasma cells are much less frequent. Immunoglobulins identified in pathological regions were elutable at neutral pH, implying that they are not bound to a tissue component and are probably due to a disruption of the blood–brain barrier.[15] Inflammatory foci are usually most intense immediately within the advancing edge, where myelin and oligodendrocytes have already been lost, axons are relatively spared and many interstitial lipophages are present. It was primarily this finding that prompted a two-stage pathogenetic theory:[134,135,139,159] first, a biochemical defect in the myelin membrane leads to its breakdown (dysmyelination); second, an inflammatory immune response directed at a CNS antigen exposed during this dysmyelination causes additional and more extensive destruction of myelin (demyelination). Dysmyelinative foci (myelin pallor and

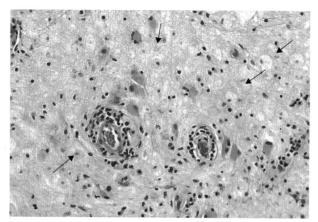

8.28 Inflammatory demyelinative lesion of ALD. Perivascular lymphocytes and lipophages, interstitial lipophages (arrows) and reactive astrocytes are prominent in demyelinated white matter displaying a reduction in oligodendrocytes.

a few PAS-positive macrophages), analogous to, but less common than, those in AMN, can be found in otherwise normal white matter far beyond the advancing demyelinative edge. Such abnormalities can be appreciated with proton MR spectroscopy, which also indicates axonal pathology.[44] It is plausible to speculate that these abnormalities correlate with those areas of biochemically affected, but grossly 'normal', white matter, particularly with VLCFA-phosphatidylcholine.[175] Perilesional white matter contains increased VLCFA in lyso-phosphatidyl-cholines, which induce microglia apoptosis and macrophage recruitment from the periphery.[13] CD8-positive cytotoxic lymphocytes (CTLs) are the most prevalent lymphocytes and appear to be associated intimately with interfascicular oligodendrocytes of 'normal' white matter, which suggests a prominent pathogenetic role for CTLs in the early myelin lesion of ALD.[81] Hypertrophy, and perhaps hyperplasia, of astrocytes and macrophage infiltration are also seen just outside the advancing edge, where some splitting of myelin sheaths and oedema is observed. These astrocytes and macrophages outside and at the edge show both tumour necrosis factor α (TNF-α) and interleukin 1 (IL-1) immunoreactivities. Intercellular adhesion molecule (ICAM) and class I and II upregulation are also noted. These data generated the additional hypotheses that cytokines, particularly TNF-α, initiate the secondary phase of inflammatory demyelination, in which macrophages and T-lymphocytes are the main effector cells.[148] Subsequent studies demonstrated an upregulation of TNF receptor II mRNA. The absence of the cell-mediated immunosuppressive cytokine interleukin 4 (IL-4) and the presence of γ-interferon aligned the T-helper cell response in ALD, closer to the TH1 subtype.[109] Others have provided evidence for TNF-α, interleukin 12 (IL-12) and γ-interferon.[102,125] Unstimulated ALD patient-derived lymphocytes have increased proinflammatory gene expression, including IL-6 and inducible nitric oxide synthase (iNOS), compared to control- or heterozygote carrier-patient derived lymphocytes.[168] Oxidative stress and damage in the form of an excess of iNOS and abnormal protein nitration have been identified in the inflammatory demyelination of ALD;[60,152] the

presence of the highly toxic peroxynitrite can be inferred from these latter findings. More recently, firm evidence for a TH1 response (γ-interferon and IL-12) and the participation of 4-hydroxynonenal (HNE), a toxic byproduct of lipid peroxidation that is a self-propagating element, has been provided.[152] The close association of CD8+ CTLs, macrophages, HNE and myelin debris in areas outside the active demyelinative edge also suggests an early role for oxidative damage.[152] The immunolabelling of astrocytes and macrophages for CD1b, CD1c and CD1d raised the possibility of lipid antigen presentation in the pathogenesis of the inflammatory demyelination and that CD1 may be a modifier gene[81] (reviewed by Heinzler *et al.*[70]). However, more recent evidence mitigates against CD1 as a modifier.[7] Abnormal myelin lipids and proteins, particularly gangliosides and proteolipid protein, containing saturated VLCFA, have been proposed as the cytokine trigger/immunologic target.[114,134,135,141] The death of oligodendrocytes in ALD appears to be mainly lytic (not apoptotic, in contrast to adrenocortical cells) and mediated by the granule exocytosis pathway of CD8 CTLs,[81] peroxynitrite and HNE.[152]

The wallerian-like degeneration of the corticospinal tracts exhibits equivalent losses of axons and myelin, mild hypertrophy of astrocytes and a few lipophages. In contrast to non-ALD secondary tract degeneration, ALD tract degeneration often contains foci of perivascular lymphocytes and striated lipophages.[134,159]

Ultrastructural examination of old, sclerotic lesions usually reveals little more than astrocytic processes filled with dense intermediate filaments. Rarely, crystals, consistent with hydroxyapatite, are seen within myelinated axons and in the extracellular space. Active lesions contain demyelinated intact axons, demyelinated degenerate axons, thinly myelinated axons, a variety of inflammatory cells and two populations of macrophages. The predominant macrophage contains myelin debris and opaque lipid droplets (cholesterol esters), whereas the other contains lipid droplets, lamellae, lipid profiles and angulate lysosomes containing spicules (Figure 8.29).[134] The latter macrophage is the type that retains birefringence after acetone extraction. Beyond this active edge of demyelination, the extracellular space is enlarged; splitting and fragmentation of myelin sheaths are observed.[159,172] Convincing evidence of lamellar inclusions in oligodendrocytes is rare.[174] The extent and intensity of CNS white matter lesions in ALD do not correlate with clinical or pathological involvement of the adrenal cortex. A possible role for androgens in triggering the onset of ALD and AMN has been suggested.[134]

Neuropathology of Adrenomyeloneuropathy

Typically, patients with AMN present with stiffness or clumsiness of the legs in their third or fourth decade, which progresses slowly but inexorably over the next few decades to severe spastic paraparesis.[67] Patients have been divided into 'pure' and 'cerebral', the latter term indicating the presence of brain lesions.[114]

In patients with AMN, as well as in symptomatic female heterozygotes, the spinal cord bears the brunt of the disease (Figure 8.30).[134,141,150,160] Loss of myelinated axons and a milder loss of oligodendrocytes are observed in the long

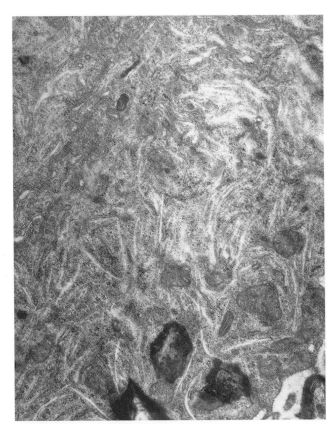

8.29 Electron microscopy in ALD. Central nervous system macrophage in juvenile adreno-leukodystrophy, containing predominantly fine, clear lipid clefts. Autopsy.

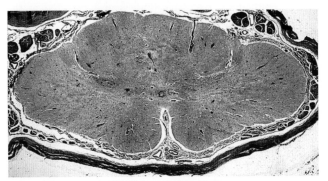

8.30 AMN. Severe loss of myelinated fibres in long tracts of adrenomyeloneuropathy spinal cord, with sparing of the propriospinal fibres. Luxol fast blue–periodic acid–Schiff.

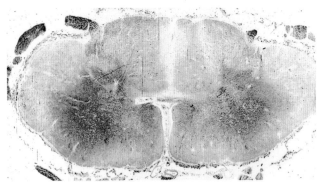

8.31 AMN. More localized loss of myelinated fibres in both the gracile tracts and the postero-lateral columns in another patient with adrenomyeloneuropathy. Modified Bielschowsky. Silver impregnation.

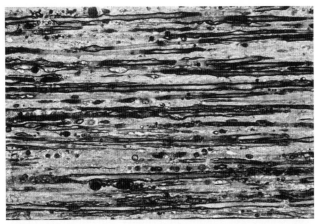

8.32 AMN. Loss of myelinated fibres and myelin ovoids in the gracile tract of adrenomyeloneuropathy. Semi-thin longitudinal section. Toluidine blue.

ascending and descending tracts of spinal cord, especially in the fasciculus gracilis and lateral corticospinal tracts (Figure 8.31). The pattern of fibre loss is consistent with a distal axonopathy, in that the greatest losses are observed in the lumbar corticospinal and cervical gracile and dorsal spinocerebellar tracts. Axonal loss is usually commensurate with, or greater than, myelin loss. Gangliosides containing VLCFA, present in AMN axonal membranes, are postulated to be the major pathogenic element,[134,135,141,150] perhaps by interfering with neurotrophic factor–receptor interactions. Sudanophilia and inflammation are minimal or absent. Astrocytic gliosis, usually isomorphic, is moderate, but the predominant reactive cells are activated microglia. Some sparing of individual myelinated fibres is noted, even in severely affected tracts. Perivascular accumulations of striated and granular PAS-positive lipophages are present, particularly in relatively preserved tracts. Spinal grey matter is unremarkable. In semi-thin and thin sections, the affected tracts may show segmental demyelination, myelin corrugation, axonal ovoids, axons with thin myelin sheaths and probably axonal atrophy (Figure 8.32).[150,153] Macrophages with pleomorphic cytoplasmic inclusions, mainly spicules, have been visualized. The severity of the myelopathy does not appear to correlate with the duration of neurological symptoms, presence or duration of endocrine abnormalities or extent of supraspinal neuropathological lesions. Atrophy of neurons in the dorsal root ganglia, predominantly involving the largest and without appreciable neuronal loss, and lipidic mitochondrial inclusions have been demonstrated.[151] Thus, this slowly progressive myelopathy with a late onset and perikaryal preservation should be amenable to therapeutic intervention.

The involvement of cerebral and cerebellar white matter in AMN is variable but it is usually minimal compared with ALD.[134,141,150,160] In some cases, there may be no lesions of white matter; most, if not all, probably contain at least microscopic dysmyelinative foci, measuring millimetres in diameter, of myelin pallor with relative to total axonal and oligodendrocytic sparing, activation of microglia, and the recruitment of PAS-positive, striated macrophages (Figure 8.33). Reactive astrocytosis and lymphocytes are

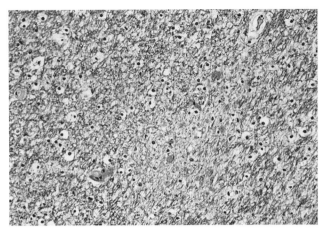

8.33 AMN. Dysmyelinative lesion of the cerebral white matter displaying myelin pallor and PAS-positive macrophages in adrenomyeloneuropathy. Luxol fast blue–periodic acid–Schiff.

absent or minimal. These abnormalities probably explain some of the neuropsychological deficiencies[43] and spectroscopic abnormalities[40] in 'pure' AMN. Other patients with AMN demonstrate confluent losses of myelin staining in the cerebrum and, to a lesser extent, the cerebellum, without significant inflammation and with sparing of arcuate fibres.[160] Still others display inflammatory demyelinative lesions with axonal loss, qualitatively similar to ALD but much more localized. Mixtures of these lesions may coexist in the same patient. Involvement of the posterior limbs can result in secondary corticospinal tract degeneration in the midbrain, pons, medulla and spinal cord, accompanied by mild reactive astrocytosis and a few lymphocytes. The concurrent finding of primary inflammatory demyelinative foci in the basis pontis can further complicate the interpretation of pyramidal tract degeneration. In AMN, however, the pyramidal tracts more commonly seem to undergo a dying-back axonopathy and pyramidal signs appear early in AMN, even 'pure' AMN. Thus, the corticospinal tracts in some patients with 'cerebral' AMN may undergo both anterograde and retrograde axonal degeneration, whereas patients with 'pure' AMN appear to develop only a retrograde, dying-back axonopathy. Finally, about 25 per cent of patients with AMN experience the myeloneuropathy for several decades before they develop confluent cerebral inflammatory demyelination qualitatively and usually quantitatively similar to childhood cerebral ALD (AMN/ALD); they usually die within a few years of this transition[148] (reviewed by Moser *et al.*[116]). In addition to these primary dysmyelinative and inflammatory demyelinative lesions, patients with AMN also demonstrate noninflammatory, bilateral, fairly symmetrical supraspinal lesions displaying comparable losses of axons and myelin (i.e. system degenerations). These involve the medial and lateral lemnisci, brachium conjunctivum, middle and inferior cerebellar peduncles, optic system, and particularly the geniculo-calcarine tracts.[134,160]

Refsum's Disease, Classic or Adult, and Atypical Refsum's Disease

Peroxisomal phytanoyl-coenzyme A hydroxylase (phytanic acid α-hydroxylase) is the enzyme deficient in most patients diagnosed clinically as having ARD.[84] In a small subset of patients, the defect is in *PEX 7*, not the hydroxylase.[19,183] In this subset, defects in plasmalogen synthesis and peroxisomal thiolase are also noted. In classical ARD, phytanic acid elevation is the only biochemical abnormality, as a result of deficiency in the initial α-oxidation of phytanic acid that occurs in peroxisomes; the subsequent β-oxidation of its degradative product, pristanic acid, in humans is essentially completed in the peroxisome.[191] Phytanic acid is a 20-carbon branched-chain fatty acid solely of dietary origin; hence, dietary restriction can be an effective treatment, particularly for the peripheral neuropathy. The typical patient with ARD presents before 20 years of age with decreased visual acuity because of pigmentary retinopathy, peripheral neuropathy, cerebellar ataxia, sensorineural hearing deficits, cardiac problems and dry skin to ichthyosis.[170] Many of these features are typical of PBD. Chronic RCDP has comparably elevated phytanic acid levels and ichthyosis but demonstrates low plasmalogens and cerebellar atrophy,[149] whereas ARD has normal plasmalogens and lacks cerebellar atrophy.

In view of the fact that dietary therapy is so successful, neuropathological studies are essentially restricted to those performed decades ago.[23] Neuropathological abnormalities are most prominent in peripheral nerve, where onion bulbs predominate. Osmiophilic granular, granulomembranous and crystalloid bodies are seen in Schwann cell cytoplasm.[46] The relationship of any of these structures to the increased phytanic acid found in peripheral nerve in this disease is unclear. Oil red O-positive lipid droplets, presumably phytanic acid,[23] in the CNS have been noted in the leptomeninges, subpial glia, ependymal cells and choroid plexus epithelium. Excessive lipid is also detected in the pallidum and around cerebral and retinal blood vessels.[180] Other CNS lesions include system degenerations, with marked loss of myelin (presumably myelinated fibres) in olivary and dentate nuclei and olivocerebellar fibres. Loss of myelin (presumably myelinated fibres) and gliosis also occur in the superior and middle cerebellar peduncles, pyramidal tracts and medial lemnisci. Neuronal loss is noted in the inferior olivary, dentate, cochlear, vestibular, gracile and cuneate nuclei. Thus, the CNS lesions of ARD seem to resemble those of AMN and myoclonus epilepsy with ragged-red fibres (MERRF; see Chapter 7, Mitochondrial Diseases), in that they are mainly tract degenerations but with superimposed lipid accumulations. Cerebral cortical neurons, Purkinje cells and spinal ganglion cells are also reduced in number. Degeneration and loss of anterior horn cells and ascending tract degeneration have been interpreted as secondary to the peripheral neuropathy. MRI studies are not available. The reason why myelin is the major site of disease in the PNS but not the CNS, is probably that phytanic acid concentrations in the PNS of ARD are much greater[104] and PNS myelin has a higher turnover rate. A knockout mouse model has been generated, which has exhibited signs of peripheral neuropathy and ataxia, Purkinje cell loss, astrocytosis and an upregulation of calcium-binding proteins in the CNS.[54]

Most of the few remaining conditions listed in Table 8.1 are rare. Patients have biochemical evidence of peroxisomal dysfunction but lack either neuropathological data or noteworthy neurological features. Mulibrey (muscle–liver–brain–eye) nanism (Perheentupa syndrome) is the most recently recognized peroxisomal disorder and manifests as severe perinatal growth retardation, dysmorphic features, pericardial constriction and hepatomegaly. A J-shaped sella, small yellow deposits in the ocular fundi, slight muscular weakness and ventriculomegaly are also seen. The disease is most common in Finnish people. Mutations have been identified in TRIM37, which encodes a RING-B-box-coiled-coil subfamily of the zinc-finger protein that is localized to peroxisomes.[86] TRIM37 functions as an ubiquitin E3 ligase;[87] markers of peroxisomal dysfunction in plasma and fibroblasts (e.g. excess VLCFA) are lacking. Three patients with a sensorimotor neuropathy and elevated levels of pristanic acid and C27 bile acid intermediates were found to have a deficiency of α-methylacyl-CoA racemase, which is responsible for the conversion of pristanoyl-CoA and the C27 bile acyl-CoAs to their (S)-stereoisomers.[51] A relapsing encephalopathy with seizures and cognitive decline has also been reported in three adults.[177]

GROUP III: OTHERS

An archival case of orthochromatic leukodystrophy with epithelioid cells of Norman–Gullotta (case 1) had elevated VLCFA and equivocally elevated phytanic acid as well as 'typical lamellar inclusions'.[112] One ataxic patient had increased concentrations of pristanic acid, phytanic acid and C27 bile acids;[28] two other patients had a peripheral neuropathy, one of whom was also ataxic, and they demonstrated panperoxisomal dysfunction.[105] Autosomal recessive cerebellar ataxia is rarely associated with mild PBDs, as recently identified in patients with mild *PEX2* gene mutations.[110,167] Another patient with a neuromuscular disorder resembling Werdnig–Hoffmann disease also had a panperoxisomal defect.[10] Finally, patients with the typical, but milder, biochemical abnormalities of ZS may present with pigmentary degeneration of the retina and sensorineural hearing loss alone.[65]

SPECIFIC TREATMENT OF PEROXISOMAL DISORDERS

Docosahexaenoic acid has been used in the treatment of PBD with reported success;[108,120] however, a double-blind trial did not show efficacy.[126] Small molecules with chaperone-like properties have enhanced residual protein activity in cells from PBD patients with intermediate and milder phenotypes;[12,205] this includes betaine, which is currently in clinical trials for patients with *Pex1-G843D* mutations (ClinicalTrials.gov Identifier: NCT01838941). Ether lipid precursors will rescue plasmalogen levels and organ pathology in Pex7 knockout mice and RCDP patient cells,[20,201] but efficacy in RCDP patients remains to be determined. Patients with ALD and AMN who are addisonian need glucocorticoid replacement therapy. Bone marrow transplantation in patients with ALD with mild neurological symptomatology has been its most effective therapy thus far.[128] Gene therapy is becoming a realistic possibility.[24] Immunomodulatory and immunosuppressive drugs have failed to prevent progression of cerebral neuroinflammation

in ALD. Dietary restriction and treatment with glyceryl trierucate/glyceryl trioleate (Lorenzo's oil), in spite of the publicity and its ability to reduce plasma VLCFA, has not had a statistically significant effect on the progression of AMN or ALD[187] but may have helped to delay the onset of cerebral disease in asymptomatic boys with normal MRI findings.[117] Modulation of microsomal elongases (ELOVL1, ELOVL6) have the potential to reduce VLCFA levels.[13,122] In view of the mitochondrial defects and presence of reactive species and oxidative damage in human ALD and mouse ABCD1 –/– tissues, antioxidant (e.g. N-acetylcysteine) and mitochondrial-enhancing therapies are under consideration.[14] Riboflavin appeared to have a favourable biochemical effect on the first patient reported with glutaric aciduria, type III.[11] Dietary restriction has had a profound impact on ARD. Phytanic acid restriction also had a positive clinical effect on an ataxic patient with di- and tri-hydroxycholestanaemia[27] and another ataxic patient with increased plasma levels of phytanic, pristanic and C27 bile acids.[28]

CONSIDERATIONS ON THE CELLULAR PATHOGENESIS OF PEROXISOMAL DISORDERS

Much has been learned about the morphological, biochemical, cellular and molecular intricacies of peroxisomes and their disorders. However, our insights on how these metabolic defects cause the various degenerative brain phenotypes remains limited. Based on biochemical/neuropathological correlative studies of human patients, primarily post-mortem, saturated VLCFA have been implicated as major pathogenic culprits in most peroxisomal disorders, particularly ALD and AMN. We now know the physiological role of ABCD1 and that ABCD1 deficiency leads to a failure in the import of VLCFA-CoA into the peroxisome and pathogenic elevations in saturated VLCFA. The relative insolubility of VLCFA at normal body temperature and their incorporation into membrane and myelin constituents, perhaps in particular phosphatidylcholine in 'normal' white matter,[175] have an adverse impact on the fluidity of these membranes by increasing their viscosity.[72,142] This is thought to lead to myelin instability with dysmyelination and, especially in ALD, to inflammatory (immune) demyelination.[81,148] Their incorporation into axonal membranes, particularly as gangliosides, may interfere with normal receptor–neurotrophic factor interactions, resulting in perikaryal and axonal atrophy with axonal loss in AMN.[150,151]

Mouse models have provided considerable insight into pathogenetic mechanisms in peroxisomal disorders. Several ZS knockouts (KOs) have been generated, including of *PEX2*, *PEX5* and *PEX13*; all display abnormal neocortical neuronal migrations and early neonatal lethality (reviewed in Baes[4,5]). ZS mice have mitochondrial abnormalities with evidence of oxidative stress.[9] In addition, defective hepatic and CNS peroxisomes both contribute to neuronal migration defects,[98] and systemic metabolic abnormalities, such as elevations of bile acid precursors, may impact on neuronal migration.[49,50] Deletion of *PEX5* in all neural cell types by Nestin-Cre causes cerebellar hypomyelination followed by diffuse demyelination, axon loss and neuroinflammation throughout the brain. Defective peroxisomal β-oxidation alone does not disturb neocortical migration in mice; however, D-BP KO mice do develop defects similar to *NES-PEX5–/–*,[189] but less severe and later in onset, suggesting that additional factors, possibly a lack of plasmalogens, contribute to the early phenotype of peroxisome deficiency in brain.[5] Plasmalogen deficiency in *DAPAT* KO leads to defects in myelination, paranode organization and cerebellar Purkinje cell innervation; mice with combined β-oxidation and plasmalogen deficiencies show more severe phenotypes, further suggesting a synergistic effect of these metabolic defects.[5,31,129] Pex1-G844D knock-in mouse recapitulates many classic features of mild PBD cases, providing a model for Pex1-G843D patients, with longer postnatal survival, retinopathy with cone photoreceptor degeneration, and normalization of peroxisomal β-oxidation in mutant fibroblasts with chaperone-like compounds.[71]

Three laboratories have generated KO mice lacking ABCD1; none of these has developed the cerebral inflammatory demyelination or the oxidative damage[152] typical of human ALD. Rather, spinal axonal degeneration, more reminiscent of AMN, occurs, particularly if ABCD2 is also knocked out.[154] We have demonstrated mitochondrial abnormalities in the *ABCD1* KO and found that the rate of peroxisomal VLCFA β-oxidation is related directly to the rate of mitochondrial long-chain fatty acid oxidation (reviewed by Heinzer *et al.*[70]). However, it has confirmed our belief that the most fundamental CNS abnormality is axonal and probably in cell membranes,[139] both of myelin sheaths and axons. Recent studies using the *ABCD1* KO have strongly supported the pathogenic role of saturated VLCFA and a prominent and early role for oxidative damage in the pathogenesis of ALD and AMN (reviewed in Galea[56]). Perhaps most importantly, it has provided another link to a concomitant mitochondrial abnormality in ALD and AMN,[55] which was first demonstrated in adrenocortical cells in the 1970s.[139,142] Our data also suggest that an alternative possibility – increased microsomal elongation[122] – is an additional explanation for the elevations in VLCFA.

Evidence for primary oligodendrocyte dysfunction in ALD pathogenesis is provided by mice with CNPase-Cre-directed, oligodendrocyte specific deletion of *PEX5*.[89] Similar to human X-ALD, these *CNP-PEX5–/–* mutants develop progressive symmetric subcortical demyelination, severe axonal loss, VLCFA accumulation and neuroinflammation, including infiltrating CD8+ T-cells and cytokine production. It has been postulated that ABCD1-deficient peroxisomes accumulate secondary peroxisomal changes over time that ignite the inflammatory response in brain; for instance, a concomitant depletion of plasmalogens is noted in human X-ALD demyelinated brain tissue.[31] Peroxisomes in oligodendrocytes are found concentrated in paranodal loops, which are sites of axon–glial interaction that may provide trophic support to axons independent of myelin. *CNP-PEX5–/–* mice also develop peripheral neuropathy, associated with vesicle filled swellings at the paranodes.[90]

A similar mechanism has been proposed for phytanic acid in the PNS myelin of ARD. This branched-chain fatty acid, with a 'thorny' configuration, may take the place of normal straight-chain fatty acids and also cause a membrane (myelin) instability.[104,170] In the cerebellar atrophy of chronic RCDP, and perhaps IRD, the tissue plasmalogen deficiency is viewed as a possible contributing factor.[149] Toxicity of phytanic acid and pristanic acid is mediated by multiple mitochondrial dysfunctions, generation of reactive oxygen species and dysregulation of intracellular $Ca2+$ signalling pathways in glial cells.[97] The importance of DHA to retinal photoreceptor cells suggests a pathogenetic role for its deficiency in the atypical pigmentary retinopathy of PBD.

Thus, there is considerable evidence that abnormal fatty acids accumulating in peroxisomal disorders, particularly saturated VLCFA and phytanic acid, are incorporated into cell membranes (including myelin and axons) and perturb their micro-environments. This would lead to dysmyelination and dysfunction, atrophy and death of vulnerable cells. Deficiencies in tissue plasmalogens and DHA may contribute to their vulnerability in PBD. However, whether these peroxisomal biochemical accumulations and/or deficiencies can fully explain the various tissue pathologies is still debated, and information on how peroxisome deficiencies may affect intracellular signalling pathways, including disturbance to membrane-raft domains and other protein-lipid modifications, awaits further research. Finally, there is increasing recognition of pathogenic roles for oxidative stress/damage and concomitant mitochondrial dysfunction in both groups of peroxisomal disorders.

DEDICATION

We dedicate this chapter to our friend and collaborator for three decades, Hugo W. Moser, MD, who passed away on 20 January 2007. His commitment to peroxisomal diseases, particularly ALD, and his establishing of the Peroxisomal Disease Center in the Kennedy Krieger Institute, enabled the study of these perplexing diseases and provided much of the pathological material displayed in this chapter.

REFERENCES

1. Agamanolis DP, Novak RW. Rhizomelic chondrodysplasia punctata: report of a case with review of the literature and correlation with other peroxisomal disorders. *Pediatr Pathol Lab Med* 1995;**15**:503–13.
2. Agamanolis DP, Robinson HB, Jr, Timmons GD. Cerebro-hepato-renal syndrome: report of a case with histochemical and ultrastructural

3. observations. *J Neuropathol Exp Neurol* 1976;**35**: 226–46.
4. Ahlemeyer B, Neubert I, Kovacs WJ, Baumgart–Vogt E. Differential expression of peroxisomal matrix and membrane proteins during postnatal development of mouse brain. *J Comp Neurol* 2007;**505**(1):1–17.
5. Baes M, Van Veldhoven PP. Generalised and conditional inactivation of *Pex*

genes in mice. *Biochim Biophys Acta* 2006;**1763**:1785–93.
6. Baes M, Van Veldhoven PP. Mouse models for peroxisome biogenesis defects and β-oxidation enzyme deficiencies. *Biochim Biophys Acta* 2012;**1822**:1489–500.
7. Bams–Mengerink AM, Majoie CB, Duran M, *et al.* MRI of the brain and cervical spinal cord in rhizomelic chondrodysplasia punctata. *Neurology* 2006;**66**:798–803.

7. Barbier M, Sabbagh A, Kasper E, *et al.* CD1 gene polymorphisms and phenotypic variability in X-linked adreno-leukodystrophy. *PLoS One* 2012;7:e29872.

8. Barth PG, Majoie CB, Gootjes J, *et al.* Neuroimaging of peroxisome biogenesis disorders (Zellweger spectrum) with prolonged survival. *Neurology* 2004;**10**:439–44.

9. Baumgart E, Vanhorebeek I, Grabenbauer M, *et al.* Mitochondrial alterations caused by defective peroxisomal biogenesis in a mouse model for Zellweger syndrome (PEX5 knockout mouse). *Am J Pathol* 2001;**159**:1477–94.

10. Baumgartner MR, Verhoeven NM, Jakobs C, *et al.* Defective peroxisome biogenesis with a neuromuscular disorder resembling Werdnig–Hoffmann disease. *Neurology* 1998;**51**:1427–32.

11. Bennett MJ, Pollitt RJ, Goodman SI, et al. Atypical riboflavin-responsive glutaric aciduria and deficient peroxisomal glutaryl-CoA oxidase activity: a new peroxisomal disorder. *J Inherit Metabol Dis* 1991;14:165–73.

12. Berendse K, Ebberink MS, Ijlst L *et al.* Arginine improves peroxisome functioning in cells from patients with a mild peroxisome biogenesis disorder. *Orphanet J Rare Dis* 2013;8:138.

13. Berger J, Forss–Petter S, Eichler FS. Pathophysiology of X-linked adreno-leukodystrophy. *Biochimie* 2014;98:135–42.

14. Berger J, Pujol A, Aubourg P, Forss–Petter S. Current and future pharmacological treatment strategies in X-linked adreno-leukodystrophy. *Brain Pathol* 2010;**20**:845–56.

15. Bernheimer H, Budka H, Müller P. Brain tissue immunoglobulins in adreno-leukodystrophy: a comparison with multiple sclerosis and systemic lupus erythematosus. *Acta Neuropathol* 1983;**59**:95–102.

16. Bizzozero OA, Zuniga G, Lees MB. Fatty acid composition of human proteolipid protein in peroxisomal disorders. *J Neurochem* 1991;**56**:872–8.

17. Bowen P, Lee CSN, Zellweger H, Lindenberg R. A familial syndrome of multiple congenital defects. *Bull Johns Hopkins Hosp* 1964;**114**:402–14.

18. Bradke F, Dotti CG. Neuronal polarity: vectorial cytoplasmic flow precedes axon formation. *Neuron* 1997;**19**:1175–86.

19. Braverman N, Chen L, Lin P, *et al.* Mutation analysis of PEX 7 in 60 probands with rhizomelic chondrodysplasia punctata and functional correlations of genotype with phenotype. *Hum Mutat* 2002;**20**:284–97.

20. Brites P, Ferreira AS, da Silva TF, *et al.* Alkyl-glycerol rescues plasmalogen levels and pathology of ether-phospholipid deficient mice. *PLoS One* 2011;6:e28539

21. Brody BA, Kinney HC, Kloman A, Gilles FH. Sequence of central nervous system myelination in human infancy: I. an autopsy study of myelination. *J Neuropathol Exp Neurol* 1987;**46**:283–30.

22. Brose RD, Avramopoulos D, Smith KD. SOD2 as a potential modifier of X-linked adreno-leukodystrophy clinical phenotypes. *J Neurol* 2012;**259**:1440–7.

23. Cammermeyer J. Refsum's disease. In: Vinken PJ, Bruyn GW eds. *Handbook of clinical neurology*, Vol. 21. Amsterdam: North Holland, 1975:231–61.

24. Cartier N, Hacein–Bey–Abina S, Bartholomae CC et al. Lentiviral hematopoietic cell gene therapy for X-linked adreno-leukodystrophy. *Methods Enzymol* 2012;**507**:187–98.

25. Challa VR, Geisinger KR, Burton BK. Pathologic alterations in the brain and liver in hyperpipecolic acidemia. *J Neuropathol Exp Neurol* 1983;**42**:627–38.

26. Chow CW, Poulos A, Fellenberg AJ, *et al.* Autopsy findings in two siblings with infantile Refsum disease. *Acta Neuropathol* 1992;**83**:190–95.

27. Christensen E, Van Eldere J, Brandt NJ, *et al.* A new peroxisomal disorder: di- and trihydroxycholestanoemia due to a presumed trihydroxycholestanoyl-CoA oxidase deficiency. *J Inherit Metabol Dis* 1990;**13**:363–6.

28. Clayton PT, Johnson AW, Mills KA, *et al.* Ataxia associated with increased plasma concentrations of pristanic acid, phytanic acid and C27 bile acids but normal fibroblast branched-chain fatty acid oxidation. *J Inherit Metabol Dis* 1996;**19**:761–8.

29. Cohen SMZ, Brown FR, Martyn L, *et al.* Ocular histopathologic and biochemical studies of the cerebrohepatorenal syndrome (Zellweger's syndrome) and its relationship to neonatal adreno-leukodystrophy. *Am J Ophthalmol* 1983;**96**:488–501.

30. Crane DI, Maxwell MA, Paton BC. PEX1 mutations in the Zellweger spectrum of the peroxisome biogenesis disorders. *Hum Mutat* 2005;**26**:167–75.

31. da Silva TF, Sousa VF, Malheiro1 AH *et al.* The importance of ether-phospholipids: a view from the perspective of mouse models. *Biochim Biophys Acta* 2012;**1822**:1501–8.

32. De Duve C, Baudhuin P. Peroxisomes (microbodies and related particles). *Physiol Rev* 1966;**46**:323–57.

33. De Leon GA, Grover WD, Huff DS, *et al.* Globoid cells, glial nodules, and peculiar fibrillary changes in the cerebro-hepato-renal syndrome of Zellweger. *Ann Neurol* 1977;2:473–84.

34. De Vet ECJM, Ijlst L, Oostheim W, et al. Alkyl-dihydroxyacetonephosphate synthase: fate in peroxisome biogenesis disorders and identification of the point mutation underlying a single enzyme deficiency. *J Biol Chem* 1998;**273**:10296–301.

35. Dean M, Annilo T. Evolution of the ATP-binding cassette (ABC) transporter superfamily in vertebrates. *Ann Rev Genom Hum Genet* 2005;6:123–42.

36. Depreter M, Espeel M, Roels F. Human peroxisomal disorders. *Microsc Res Tech* 2003;**61**:203–23.

37. Dimmick JE, Applegarth DA. Pathology of peroxisomal disorders. *Perspect Paediatr Pathol* 1993;**17**:45–98.

38. Dingemans KP, Mooi WJ, van den Bergh Weerman MA. Angulate lysosomes. *Ultrastruct Pathol* 1983;5:113–22.

39. Dixit E, Boulant S, Zhang Y, *et al.* Peroxisomes are signalling platforms for antiviral innate immunity. *Cell* 2010;**141**:668–81.

40. Dubey P, Fatemi A, Barker PB, *et al.* Spectroscopic evidence of cerebral axonopathy in patients with 'pure' adrenomyeloneuropathy. *Neurology* 2005;**64**:304–10.

41. Dumontel C, Rousselle C, Guigard M–P, Trouillas J. Angulate lysosomes in skin biopsies of patients with degenerative neurological disorders: high frequency in neuronal ceroid lipofuscinosis. *Acta Neuropathol* 1999;**98**:91–6.

42. Ebberink MS, Mooijer PA, Gootjes J, *et al.* Genetic classification and mutational spectrum of more than 600 patients with a Zellweger syndrome spectrum disorder. *Hum Mutat* 2011;**32**:59–69.

43. Edwin D, Speedie LJ, Kohler W, *et al.* Cognitive and brain magnetic resonance imaging findings in adrenomyeloneuropathy. *Ann Neurol* 1996;**40**:675–8.

44. Eichler FS, Itoh R, Barker PB, *et al.* Proton MR spectroscopic and diffusion tensor brain MR imaging in X-linked adreno-leukodystrophy: initial experience. *Radiology* 2002;**225**: 245–52.

45. Evrard P, Caviness VS, Prats–Vinas J, Lyon G. The mechanism of arrest of neuronal migration in the Zellweger malformation: an hypothesis based upon cytoarchitectonic analysis. *Acta Neuropathol* 1978;**41**:109–17.

46. Fardeau M, Engel WK. Ultrastructural study of a peripheral nerve biopsy in Refsum's disease. *J Neuropathol Exp Neurol* 1969;**28**:278–94.

47. Farioli–Vecchioli S, Moreno S, Ceru MP. Immunocytochemical localization of acyl-CoA oxidase in the rat central nervous system. *J Neurocytol* 2001;**30**:21–33.

48. Farrell DF. Neonatal adreno-leukodystrophy: a clinical, pathologic, and biochemical study. *Pediatr Neurol* 2012;**47**:330–6.

49. Faust PL, Su H–M, Moser A, Moser HW. The peroxisome deficient PEX2 Zellweger mouse. *J Molec Neurosci* 2001;**16**:289–97.

50. Faust PL, Banka D, Siriratsivawong R, *et al.* Peroxisome biogenesis disorders: the role of peroxisomes and metabolic dysfunction in developing brain. *J Inherit Metabol Dis* 2005;**28**:369–83.

51. Ferdinandusse S, Denis S, Clayton PT, *et al.* Mutations in the gene encoding peroxisomal a-methylacyl-CoA racemase cause adult-onset sensory motor neuropathy. *Nat Genet* 2000;**24**:188–91.

52. Ferdinandusse S, Finckh B, de Hingh YC, *et al.* Evidence for increased oxidative stress in peroxisomal d-bifunctional protein deficiency. *Molec Genet Metab* 2003;**79**:281–7.

53. Ferdinandusse S, Denis S, Mooyer PAW, *et al.* Clinical and biochemical spectrum of d-bifunctional protein deficiency. *Ann Neurol* 2006;**59**:92–104.

54. Ferdinandusse S, Zomer AW, Komen JC, *et al.* Ataxia with loss of Purkinje cells in a mouse model for Refsum disease. *Proc Natl Acad Sci U S A* 2008;**105**:17712–17.

55. Fourcade S, López–Erauskin J, Ruiz M, Ferrer I, Pujol A. Mitochondrial dysfunction and oxidative damage cooperatively fuel axonal degeneration in X-linked adreno-leukodystrophy. *Biochimie* 2014;98:143–9.

56. Galea E, Launay N, Portero–Otin M, *et al.* Oxidative stress underlying axonal degeneration in adreno-leukodystrophy: A paradigm for multifactorial neurodegenerative diseases? *Biochim Biophys Acta* 2012;**1822**:1475–88.

57. Gatfield PD, Taller E, Hinton GG, *et al.* Hyperpipecolatemia: a new metabolic disorder associated with neuropathy and hepatomegaly–a case study. *Can Med Assoc J* 1968;**99**:1215–33.

58. Ghatak NR, Nochlin D, Peris M, Myer EC. Morphology and distribution of cytoplasmic inclusions in adreno-leukodystrophy. *J Neurol Sci* 1981;**50**:391–8.

59. Gilbert EF, Opitz JM, Spranger JW, *et al.* Chondrodysplasia punctata: rhizomelic form. Pathologic and radiologic studies of three infants. *Eur J Paediatr* 1976;**123**:89–109.

60. Gilg AG, Pahan K, Singh K, Singh I. Inducible nitric oxide synthase in the central nervous system of patients with X-linked adreno-leukodystrophy. *J Neuropathol Exp Neurol* 2000;**59**:1063–9.

61. Goh S. Neuroimaging features in a neonate with rhizomelic chondrodysplasia punctata. *Pediatr Neurol* 2007;**37**:382–4.

62. Goldfischer S, Moore CL, Johnson AB, *et al.* Peroxisomal and mitochondrial defects in the cerebro-hepato-renal syndrome. *Science* 1973;**182**:62–4.

63. Goldfischer S, Powers JM, Johnson AB, *et al.* Striated adrenocortical cells in cerebro-hepato-renal (Zellweger) syndrome. *Virchows Arch A Pathol Anat Histopathol* 1983;**401**:355–61.

64. Goldfischer S, Collins J, Rapin I, *et al.* Pseudo-Zellweger syndrome: deficiencies in several peroxisomal oxidative activities. *J Paediatr* 1986;**108**:25–32.

65. Gould SJ, Raymond GV, Valle D. The peroxisome biogenesis disorders. In: Scriver CR, Beaudet AL, Sly WS, Valle D eds. *The metabolic and molecular bases of inherited disease*, 8th edn. New York: McGraw-Hill, 2001:3181–218.

66. Griffin DE, Moser HW, Mendoza Q, *et al.* Identification of the inflammatory cells in the central nervous system of patients with adreno-leukodystrophy. *Ann Neurol* 1985;**18**:660–64.

67. Griffin JW, Goren E, Schaumburg H, *et al.* Adrenomyeloneuropathy: a probable variant of adreno-leukodystrophy. I. Clinical and endocrinologic aspects. *Neurology* 1977;**27**:1107–13.

68. Groenendaal F, Bianchi MC, Battini R, *et al.* Proton magnetic resonance spectroscopy (1H-MRS) of the cerebrum in two young infants with Zellweger syndrome. *Neuropediatrics* 2001;**32**:23–7.

69. Hebestreit H, Wanders RJA, Schutgens RBH, *et al.* Isolated dihydroxyacetonephosphate-acyl-transferase deficiency in rhizomelic chondrodysplasia punctata: clinical presentation, metabolic and histological findings. *Eur J Paediatr* 1996;**155**:1035–9.

70. Heinzer AK, McGuinness MC, Lu J–F, *et al.* Mouse models and genetic modifiers in X-linked adreno-leukodystrophy. In: Roels F, Baes M, de Bie S eds. *Peroxisomal disorders and regulation of genes*. New York: Kluwer Plenum Press, 2003;1–18.

71. Hiebler S, Masuda T, Hacia JG, *et al.* The Pex1-G844D mouse: A model for mild human Zellweger spectrum disorder. *Mol Genet Metab* 2014;**111**(4):522–32.

72. Ho JK, Moser H, Kishimoto Y, Hamilton JA. Interactions of a very long chain fatty acid with model membranes and serum albumin: implications for the pathogenesis of adreno-leukodystrophy. *J Clin Invest* 1995;**96**:1455–63.

73. Holtzman E. Peroxisomes in nervous tissue. *Ann N Y Acad Sci* 1982;**386**:523–5.

74. Houdou S, Kuruta H, Hasegawa M. Developmental immunohistochemistry of catalase in the human brain. *Brain Res* 1991;**556**:267–70.

75. Hughes JL, Poulos A, Robertson E, *et al.* Pathology of hepatic peroxisomes and mitochondria in patients with peroxisomal disorders. *Virchows Arch A Pathol Anat Histopathol* 1990;**416**:255–64.

76. Igarashi M, Belchis D, Suzuki K. Brain gangliosides in adreno-leukodystrophy. *J Neurochem* 1976;**27**:327–8.

77. Igarashi M, Neely JG, Anthony PF, Alford BR. Cochlear nerve degeneration coincident with adrenocerebroleukodystrophy. *Arch Otolaryngol* 1976;**102**:722–6.

78. Igarashi M, Schaumburg HH, Powers J, *et al.* Fatty acid abnormality in adreno-leukodystrophy. *J Neurochem* 1976;**26**:851–60.

79. Imamura A, Tsukamoto T, Shimozawa N, *et al.* Temperature-sensitive phenotypes of peroxisome-assembly processes represent the milder forms of human peroxisome-biogenesis disorders. *Am J Hum Genet* 1998;**62**:1539–43.

80. Islinger M, Grille S, Fahimi HD, Schrader M. The peroxisome: an update on mysteries. *Histochem Cell Biol* 2012;**137**:547–74.

81. Ito M, Blumberg BM, Mock DJ, *et al.* Potential environmental and host participants in the early white matter lesion of adreno-leukodystrophy: morphologic evidence for CD8 cytotoxic T-cells, cytolysis of oligodendrocytes and CD1-mediated lipid antigen presentation. *J Neuropathol Exp Neurol* 2001;**60**:1004–19.

82. Jaffe R, Crumrine P, Hashida Y, Moser HW. Neonatal adreno-leukodystrophy: clinical, pathologic, and biochemical delineation of a syndrome affecting both males and females. *Am J Pathol* 1982;**108**:100–111.

83. Jangouk P, Zackowski KM, Naidu S, Raymond GV. Adreno-leukodystrophy in female heterozygotes: underrecognized and undertreated. *Mol Genet Metab* 2012;**105**:180–5.

84. Jansen GA, Wanders RJA, Watkins PA, Mihalik SJ. Phytanoyl-coenzyme A hydroxylase deficiency: the enzyme defect in Refsum's disease. *N Engl J Med* 1997;**337**:133–4.

85. Johnson AB, Schaumburg HH, Powers JM. Histochemical characteristics of the striated inclusions of adreno-leukodystrophy. *J Histochem Cytochem* 1976;**24**:725–30.

86. Kallijarvi J, Avela K, Lipsanen–Nyman M, *et al.* The *TRIM37* gene encodes a peroxisomal RING-B-box coiled-coil protein: classification of mulibrey nanism as a new peroxisomal disorder. *Am J Hum Genet* 2002;**70**:1215–28.

87. Kallijärvi J, Lahtinen U, Hämäläinen R, *et al.* TRIM37 defective in mulibrey nanism is a novel RING finger ubiquitin E3 ligase. *Exp Cell Res* 2005;**308**:146–55.

88. Kamei A, Houdou S, Takashima S, *et al.* Peroxisomal disorders in children: immunohistochemistry and neuropathology. *J Paediatr* 1993;**122**:573–9.

89. Kassmann CM, Lappe–Siefke C, Baes M, *et al.* Axonal loss and neuroinflammation caused by peroxisome-deficient oligodendrocytes. *Nat Genet* 2007;**39**:969–76.

90. Kassmann CM, Quintes S, Rietdorf J, *et al.* A role for myelin-associated peroxisomes in maintaining paranodal loops and axonal integrity. *FEBS Lett* 2011;**585**:2205–11.

91. Kaufmann WE, Theda C, Naidu TC, *et al.* Neuronal migration abnormality in peroxisomal bifunctional enzyme defect. *Ann Neurol* 1996;**39**:268–71.

92. Kelley RI, Datta NS, Dobyns WB, *et al.* Neonatal adreno-leukodystrophy: new cases, biochemical studies and differentiation from Zellweger and related peroxisomal polydystrophy syndromes. *Am J Med Genet* 1986;**23**:869–901.

93. Kemp S, Pujol A, Waterham HR, *et al.* ABCD1 mutations and the X-linked adreno-leukodystrophy mutation database: role in diagnosis and clinical correlations. *Hum Mutat* 2001;**18**:499–515.

94. Kemp S, Wanders R. Biochemical aspects of X-linked adreno-leukodystrophy. *Brain Pathol* 2010;**20**:831–7.

95. Kinney HC, Brody BA, Kloman A, Gilles FH. Sequence of central system myelination in human infancy: II. Patterns of myelination in autopsied infants. *J Neuropathol Exp Neurol* 1988;**47**:217–34.

96. Krause C, Rosewich H, Gärtner J. Rational diagnostic strategy for Zellweger syndrome spectrum patients. *Eur J Hum Genet* 2009;**17**:741–8.

97. Kruska N, Reiser G. Phytanic acid and pristanic acid, branched-chain fatty acids associated with Refsum disease and other inherited peroxisomal disorders, mediate intracellular Ca2+ signalling through activation of free fatty acid receptor GPR40. *Neurobiol Dis* 2011;**43**:465–72.

98. Krysko O, Hulshagen L, Janssen A, *et al.* Neocortical and cerebellar developmental abnormalities in conditions of selective elimination of peroxisomes from brain or from liver. *J Neurosci Res* 2007;**85**:58–72.

99. Kuroda S, Hirano A, Yuasa S. Adreno-leukodystrophy: cerebello-brainstem dominant case. *Acta Neuropathol* 1983;**60**:149–52.

100. Kyllerman M, Blomstrand S, Mansson JE, *et al.* Central nervous system malformations and white matter changes in pseudo-neonatal adreno-leukodystrophy. *Neuropediatrics* 1990;**21**:199–201.

101. Loes DJ, Fatemi A, Melhem ER, *et al.* Analysis of MRI patterns aids prediction of progression in X-linked adreno-leukodystrophy. *Neurology* 2003;**61**:369–74.

102. Lund TC, Stadem PS, Panoskaltsis–Mortari A, *et al.* Elevated cerebral spinal fluid cytokine levels in boys with cerebral adreno-leukodystrophy correlates with MRI severity. *PLoS One* 2012;**7**:e32218.

103. Ma C, Agrawal G, Subramani S. Peroxisome assembly: matrix and membrane protein biogenesis. *J Cell Biol* 2011;**193**:7–16.

104. MacBrinn MC, O'Brien JS. Lipid composition of the nervous system in Refsum's disease. *J Lipid Res* 1968;**9**:552–61.

105. MacCollin M, DeVivo DC, Moser AB, Beard M. Ataxia and peripheral neuropathy: a benign variant of peroxisome dysgenesis. *Ann Neurol* 1990;**28**:833–6.

106. Manz HJ, Schuelein M, McCullough DC, *et al.* New phenotypic variant of adreno-leukodystrophy. *J Neurol Sci* 1980;**45**:245–60.

107. Marsden CD, Obeso JA, Lang AE. Adrenoleukomyeloneuropathy presenting as spinocerebellar degeneration. *Neurology* 1982;**32**:1031–2.

108. Martinez M, Vazquez E. MRI evidence that docosahexaenoic acid ethyl ester improves myelination in generalized peroxisomal disorders. *Neurology* 1998;51:26–32.

109. McGuinness MC, Powers JM, Bias WB, *et al.* Human leukocyte antigens and cytokine expression in cerebral inflammatory demyelinative lesions of X-linked adreno-leukodystrophy and multiple sclerosis. *J Neuroimmunol* 1997;75:174–82.

110. Mignarri A, Vinciguerra C, Giorgio A, *et al.* Zellweger spectrum disorder with mild phenotype caused by *PEX2* gene mutations. *JIMD Rep* 2012;6:43–6.

111. Mito T, Takada K, Akaboshi S, *et al.* A pathological study of a peripheral nerve in a case of neonatal adreno-leukodystrophy. *Acta Neuropathol* 1989;77:437–40.

112. Molzer B, Gullotta F, Harzer K, *et al.* Unusual orthochromatic leukodystrophy with epitheloid cells (Norman–Gullotta): increase of very long chain fatty acids in brain discloses a peroxisomal disorder. *Acta Neuropathol* 1993;86:187–9.

113. Moser AB, Rasmussen M, Naidu S, *et al.* Phenotype of patients with peroxisomal disorders subdivided into sixteen complementation groups. *J Paediatr* 1995;127:13–22.

114. Moser HW. Adreno-leukodystrophy: phenotype, genetics, pathogenesis and therapy. *Brain* 1997;120:1485–508.

115. Moser HW. Minireview: genotype–phenotype correlations in disorders of peroxisome biogenesis. *Molec Genet Metab* 1999;68:316–27.

116. Moser HW, Smith KD, Watkins PA, *et al.* X-Linked adreno-leukodystrophy. In: Scriver CR, Beaudet AL, Sly WS, Valle D eds. *The metabolic and molecular bases of inherited disease*, 8th edn. New York: McGraw–Hill, 2001:3257–301.

117. Moser HW, Raymond GV, Lu S–E, *et al.* Follow-up of 89 asymptomatic patients with adreno-leukodystrophy treated with Lorenzo's oil. *Arch Neurol* 2005;62:1073–80.

118. Mosser J, Lutz Y, Stoeckel ME, *et al.* The gene responsible for adreno-leukodystrophy encodes a peroxisomal membrane protein. *Hum Molec Genet* 1994;3:265–71.

119. Naidu S, Hoefler G, Watkins PA, *et al.* Neonatal seizures and retardation in a girl with biochemical features of X-linked adreno-leukodystrophy: a possible new peroxisomal disease entity. *Neurology* 1988;38:1100–107.

120. Noguer MT, Martinez M. Visual follow-up in peroxisomal-disorder patients treated with docosahexaenoic acid ethyl ester. *Invest Ophthalmol Vis Sci* 2010;51:2277–85.

121. Nuttall JM, Motley A, Hettema EH. Peroxisome biogenesis: recent advances. *Curr Opin Cell Biol* 2011;23:421–6.

122. Ofman R, Dijkstra IM, van Roermund CW *et al.* The role of ELOVL1 in very long-chain fatty acid homeostasis and X-linked adreno-leukodystrophy. *EMBO Mol Med* 2010;2:90–7.

123. Ohno T, Tsuchida H, Fukuhara N, *et al.* Adreno-leukodystrophy: a clinical variant presenting as olivopontocerebellar atrophy. *J Neurol* 1984;231:167–9.

124. Opitz JM, ZuRhein GM, Vitale L, *et al.* The Zellweger syndrome (cerebro-hepato-renal syndrome). *Birth Defects* 1969;5:144–66.

125. Paintlia AS, Gilg AG, Khan M, *et al.* Correlation of very long chain fatty acid accumulation and inflammatory disease progression in childhood X-ALD: implications for potential therapies. *Neurobiol Dis* 2003;14:425–39.

126. Paker AM, Sunness JS, Brereton NH, *et al.* Docosahexaenoic acid therapy in peroxisomal diseases: results of a double-blind, randomized trial. *Neurology* 2010;75:826–30.

127. Passarge E, McAdams AJ. Cerebro-hepato-renal syndrome: a newly recognized hereditary disorder of multiple congenital defects, including sudanophilic leukodystrophy, cirrhosis of the liver and polycystic kidneys. *J Paediatr* 1967;71:691–702.

128. Peters C, Charnas LR, Tan Y, *et al.* Cerebral X-linked adreno-leukodystrophy: the international hematopoietic cell transplantation experience from 1982 to 1999. *Blood* 2004;104:881–8.

129. Pierce SB, Walsh T, Chisholm KM, *et al.* Mutations in the DBP-deficiency protein HSD17B4 cause ovarian dysgenesis, hearing loss and ataxia of Perrault Syndrome. *Am J Hum Genet* 2010;87:282–8.

130. Poll–The BT, Gärtner J. Clinical diagnosis, biochemical findings and MRI spectrum of peroxisomal disorders. *Biochim Biophys Acta* 2012;1822:1421–9.

131. Poll–Thé BT, Saudubray JM, Ogier HAM, *et al.* Infantile Refsum disease: an inherited peroxisomal disorder. Comparison with Zellweger syndrome and neonatal adreno-leukodystrophy. *Eur J Paediatr* 1987;146:477–83.

132. Poulos A, Sheffield L, Sharp P, *et al.* Rhizomelic chondrodysplasia punctata: clinical, pathologic and biochemical findings in two patients. *J Paediatr* 1988;113:685–90.

133. Powell H, Tindall R, Schultz P, *et al.* Adreno-leukodystrophy: electron microscopic findings. *Arch Neurol* 1975;32:250–60.

134. Powers JM. Review article: adreno-leukodystrophy (adreno-testiculo-leuko-myelo-neuropathic-complex). *Clin Neuropathol* 1985;4:181–99.

135. Powers JM. Presidential address: the pathology of peroxisomal disorders with pathogenetic considerations. *J Neuropathol Exp Neurol* 1995;54:710–19.

136. Powers JM. The leukodystrophies: overview and classification. In: Lazzarini R ed. *Myelin biology and disorders*, Vol. 2. San Diego, CA: Elsevier Academic Press, 2004:663–90.

137. Powers JM, Schaumburg HH. The adrenal cortex in adreno-leukodystrophy. *Arch Pathol* 1973;96:305–10.

138. Powers JM, Schaumburg HH. Adreno-leukodystrophy: similar ultrastructural changes in adrenal cortical and Schwann cells. *Arch Neurol* 1974;30:406–8.

139. Powers JM, Schaumburg HH. Adreno-leukodystrophy (sex-linked Schilder's disease). *Am J Pathol* 1974;76:481–500.

140. Powers JM, Schaumburg HH. The testis in adreno-leukodystrophy. *Am J Pathol* 1981;102:90–98.

141. Powers JM, Moser HW. Peroxisomal disorders: genotype, phenotype, major neuropathologic lesions, and pathogenesis. *Brain Pathol* 1998;8:101–20.

142. Powers JM, Schaumburg HH, Johnson AB, Raine CS. A correlative study of the adrenal cortex in adreno-leukodystrophy-evidence for a fatal intoxication with very long chain saturated fatty acids. *Invest Cell Pathol* 1980;3:353–76.

143. Powers JM, Moser HW, Moser AB, Schaumburg HH. Fetal adreno-leukodystrophy: the significance of pathologic lesions in adrenal gland and testis. *Hum Pathol* 1982;13:1013–19.

144. Powers JM, Moser HW, Moser AB, *et al.* Fetal cerebrohepatorenal (Zellweger) syndrome: dysmorphica, radiologic, biochemical and pathologic findings in four affected fetuses. *Hum Pathol* 1985;16:610–20.

145. Powers JM, Moser HW, Moser AB, *et al.* Pathologic findings in adreno-leukodystrophy heterozygotes. *Arch Pathol Lab Med* 1987;111:151–3.

146. Powers JM, Tummons RC, Moser AB, *et al.* Neuronal lipidosis and neuroaxonal dystrophy in cerebro-hepato-renal (Zellweger) syndrome. *Acta Neuropathol* 1987;73:333–43.

147. Powers JM, Tummons RC, Caviness VS, Jr, *et al.* Structural and chemical alterations in the cerebral maldevelopment of fetal cerebro-hepato-renal (Zellweger) syndrome. *J Neuropathol Exp Neurol* 1989;48:270–89.

148. Powers JM, Liu Y, Moser A, Moser H. The inflammatory myelinopathy of adreno-leukodystrophy. *J Neuropathol Exp Neurol* 1992;51:630–43.

149. Powers JM, Kenjarski TP, Moser AB, Moser HW. Cerebellar atrophy in chronic rhizomelic chondrodysplasia punctata: a potential role for phytanic acid and calcium in the death of its Purkinje cells. *Acta Neuropathol* 1999;98:129–34.

150. Powers JM, DeCiero DP, Ito M, *et al.* Adrenomyeloneuropathy: a neuropathologic review featuring its noninflammatory myelopathy. *J Neuropathol Exp Neurol* 2000; 59:89–102.

151. Powers JM, DeCiero D, Cox C, *et al.* The dorsal root ganglia in adrenomyeloneuropathy: neuronal atrophy and abnormal mitochondria. *J Neuropathol Exp Neurol* 2001;60:493– 501.

152. Powers JM, Pei Z, Heinzer AK, *et al.* Adreno-leukodystrophy: oxidative stress of mice and men. *J Neuropathol Exp Neurol* 2005;64:1067–79.

153. Probst A, Ulrich J, Heitz PU, Herschkowitz N. Adrenomyeloneuropathy: a protracted, pseudosystematic variant of adreno-leukodystrophy. *Acta Neuropathol* 1980;49:105–15.

154. Pujol A, Ferrer I, Camps C, *et al.* Functional overlap between ABCD1 (ALD) and ABCD2 (ALDR) transporters: a therapeutic target for X-adreno-leukodystrophy. *Hum Molec Genet* 2004;13:2997–3006.

155. Raymond GV, Seidman R, Monteith TS, *et al.* Head trauma can initiate the onset of adreno-leukodystrophy. *J Neurol Sci* 2010;290:70–74.

156. Roels F, Depreter M, Espeel M, *et al.* Peroxisomes during development and in distinct cell types. In: Roels F, Baes M, De Bie S eds. *Peroxisomal disorders and regulation of genes*. New York: Kluwer Academic/Plenum Publishers, 2003:39–54.

157. Santos MJ, Imanaka T, Shio H, *et al.* 1988. Peroxisomal membrane ghosts in Zellweger syndrome: aberrant organelle assembly. *Science* 1989;239:1536–8.

158. Sarnat H, Machin G, Darwish H, Rubin S. Mitochondrial myopathy of cerebro-hepato-renal (Zellweger) syndrome. *Can J Neurol Sci* 1983;**10**:170–77.

159. Schaumburg HH, Powers JM, Raine CS, *et al.* Adreno-leukodystrophy: a clinical and pathological study of 17 cases. *Arch Neurol* 1975;**33**:577–91.

160. Schaumburg HH, Powers JM, Raine CS, *et al.* Adrenomyeloneuropathy: a probable variant of adreno-leukodystrophy. II: general pathologic, neuropathologic and biochemical aspects. *Neurology* 1977;**27**:1114–19.

161. Schiffmann R, van der Knaap MS. Invited article: an MRI-based approach to the diagnosis of white matter disorders. *Neurology* 2009;**72**(8):750–9.

162. Schrader M, Fahimi HD. Peroxisomes and oxidative stress. *Biochim Biophys Acta* 2006;**1763**:1755–66.

163. Schrader M, Bonekamp NA, Islinger M. Fission and proliferation of peroxisomes. *Biochim Biophys Acta* 2012;**1822**:1343–57.

164. Schram AW, Goldfischer S, van Roermund CWT, *et al.* Human peroxisomal 3-oxoacyl-coenzyme A thiolase deficiency. *Proc Natl Acad Sci U S A* 1987;**84**: 2494–6.

165. Schruder JM, Hackel V, Wanders RJA, *et al.* Optico-cochleo-dentate degeneration associated with severe peripheral neuropathy and caused by peroxisomal d-bifunctional protein deficiency. *Acta Neuropathol* 2004;**108**:154–67.

166. Scotto JM, Hadchouel M, Odievre M, *et al.* Infantile phytanic acid storage disease, a possible variant of Refsum's disease: three cases, including ultrastructural studies of the liver. *J Inherit Metabol Dis* 1982;**5**:83–90.

167. Sevin C, Ferdinandusse S, Waterham HR, Wanders RJ, Aubourg P. Autosomal recessive cerebellar ataxia caused by mutations in the PEX2 gene. *Orphanet J Rare Dis* 2011;**6**:8.

168. Singh J, Giri S. Loss of AMP-activated protein kinase in X-linked adreno-leukodystrophy patient-derived fibroblasts and lymphocytes. *Biochem Biophys Res Commun* 2014;**445**:126–31.

169. Spranger JW, Opitz JM, Bidder U. Heterogeneity of chondrodysplasia punctata. *Humangenetik* 1971;**11**:190–212.

170. Steinberg D. Refsum disease. In: Scriver CR, Beaudet AL, Sly WS, Valle D eds. *The metabolic and molecular bases of inherited disease*, Vol. II. New York: McGraw–Hill, 1995:2351–69.

171. Steinberg S, Chen L, Wei L, *et al.* The PEX gene screen:molecular diagnosis of peroxisome biogenesis disorders in the Zellweger syndrome spectrum. *Molec Genet Metab* 2004;**83**:252–63.

172. Suzuki K, Grover WD. Ultrastructural and biochemical studies of Schilder's disease: I. Ultrastructure. *J Neuropathol Exp Neurol* 1970;**29**:392–404.

173. Suzuki Y, Shimozawa N, Orii T, *et al.* Zellweger-like syndrome with detectable hepatic peroxisomes: a variant form of peroxisomal disorder. *J Paediatr* 1988;**113**:841–5.

174. Takeda S, Ohama E, Ikuta F. Adreno-leukodystrophy: early ultrastructural changes in the brain. *Acta Neuropathol* 1989;**78**:124–30.

175. Theda C, Moser AB, Powers JM, Moser HW. Phospholipids in X-linked adreno-leukodystrophy white matter-fatty acid abnormalities before the onset of demyelination. *J Neurol Sci* 1992;**110**: 195–204.

176. Theda C, Gibbons K, Defor TE, *et al.* Newborn screening for X-linked adreno-leukodystrophy: further evidence high throughput screening is feasible. *Mol Genet Metab* 2014;**111**:55–7.

177. Thompson SA, Calvin J, Hogg S, *et al.* Relapsing encephalopathy in a patient with α-methylacyl-CoA racemase deficiency. *BMJ Case Rep* 2009;pii: bcr08.2008.0814.

178. Thoms S, Grostrokennborg S, Gärtner J. Organelle interplay in peroxisomal disorders. *Trends Mol Med* 2009;**15**: 293–302.

179. Torvik A, Torp S, Kase BF, *et al.* Infantile Refsum's disease: a generalized peroxisomal disorder–case report with postmortem examination. *J Neurol Sci* 1988;**85**:39–53.

180. Toussaint D, Danis P. An ocular pathologic study of Refsum's syndrome. *Am J Ophthalmol* 1971;**72**:342–7.

181. Tranchant C, Aubourg P, Mohr M, *et al.* A new peroxisomal disease with phytanic and pipecolic acid oxidation. *Neurology* 1993;**43**:2044–8.

182. Ulrich J, Herschkowitz N, Hertz P, *et al.* Adreno-leukodystrophy: preliminary report of a connatal case–light and electron microscopical, immunohistochemical and biochemical findings. *Acta Neuropathol* 1978;**43**:77–83.

183. Van den Brink DM, Brites P, Haasjes J, *et al.* Identification of PEX 7 as the second gene involved in Refsum disease. *Am J Hum Genet* 2003;**72**:471–7.

184. Van der Knaap MS, Valk J eds. *Magnetic resonance of myelination and myelin disorders*, 3rd edn. Berlin: Springer–Verlag, 2005.

185. van der Knaap MS, Wassmer E, Wolf NI, *et al.* MRI as diagnostic tool in early-onset peroxisomal disorders. *Neurology* 2012;**78**:1304–8.

186. van der Zand A, Braakman I, Tabak HF. Peroxisomal membrane proteins insert into the endoplasmic reticulum. *Mol Biol Cell* 2010;**21**(12):2057–65.

187. Van Geel BM, Assies J, Haverkort EB, *et al.* Progression of abnormalities in adrenomyeloneuropathy and neurologically asymptomatic X-linked adreno-leukodystrophy despite treatment with 'Lorenzo's oil'. *J Neurol Neurosurg Psychiatry* 1999;**67**:290–99.

188. Van Grunsven EG, van Berkel E, Mooijer PAW, *et al.* Peroxisomal bifunctional protein deficiency revisited: resolution of its true enzymatic and molecular basis. *Am J Hum Genet* 1999;**64**:99–107.

189. Verheijden S, Beckers L, De Munter S, Van Veldhoven PP, Baes M. Central nervous system pathology in MFP2 deficiency: Insights from general and conditional knockout mouse models. *Biochimie* 2014;**98**:119–26.

190. Volpe JJ, Adams RD. Cerebro-hepato-renal syndrome of Zellweger: an inherited disorder of neuronal migration. *Acta Neuropathol* 1972;**20**:175–98.

191. Wanders RJA. Metabolic and molecular basis of peroxisomal disorders: a review. *Am J Med Genet* 2004;**126A**:355–75.

192. Wanders RJ, Waterham HR. Biochemistry of mammalian peroxisomes revisited. *Annu Rev Biochem* 2006;**75**:295–332.

193. Wanders RJA, Barth PG, Heymans HSA. Single peroxisomal enzyme deficiencies. In: Scriver CR, Beaudet AL, Sly WS, Valle D eds. *The metabolic and molecular bases of inherited disease*, 8th edn. New York: McGraw–Hill, 2001:3219–48.

194. Wanders RJ, Ferdinandusse S, Brites P, Kemp S. Peroxisomes, lipid metabolism and lipotoxicity. *Biochim Biophys Acta* 2010;**1801**:272–80.

195. Wang R, Monuki E, Powers J, *et al.* Effects of hematopoietic stem cell transplantation on acyl-CoA oxidase deficiency: a sibling comparison study. *J Inherit Metab Dis* 2014;[Epub ahead of print].

196. Waterham HR, Koster J, van Roermund CW, *et al.* A lethal defect of mitochondrial and peroxisomal fission. *N Engl J Med* 2007;**356**:1736–41.

197. Watkins PA, Chen WW, Harris CJ, *et al.* Peroxisomal bifunctional enzyme deficiency. *J Clin Invest* 1989;**83**:771–7.

198. Watkins PA, McGuinness MC, Raymond GV, *et al.* Distinction between peroxisomal bifunctional enzyme and acyl-CoA oxidase deficiencies. *Ann Neurol* 1995;**38**:472–7.

199. White AL, Modaff P, Holland–Morris F, Pauli RM. Natural history of rhizomelic chondrodysplasia punctata. *Am J Med Genet* 2003;**118A**:332–42.

200. Wisniewski T, Powers J, Moser A, Moser H. Ultrastructural evidence for a gliopathy in cerebro-hepato-renal (Zellweger) syndrome. *J Neuropathol Exp Neurol* 1989;**48**:366.

201. Wood PL, Khan MA, Smith T, *et al. In vitro/in vivo* plasmalogen replacement evaluations in rhizomelic chrondrodysplasia punctata and Pelizaeus–Merzbacher disease using PPI-1011, an ether lipid plasmalogen precursor. *Lipids Health Dis* 2011;**10**:182.

202. Yik WY, Steinberg SJ, Moser AB, Moser HW, *et al.* Identification of novel mutations and sequence variation in the Zellweger syndrome spectrum of peroxisome biogenesis disorders. *Hum Mutat* 2009;**30**: E467–80.

203. Zellweger H. The cerebro-hepato-renal (Zellweger) syndrome and other peroxisomal disorders. *Dev Med Child Neurol* 1987;**29**: 821–9.

204. Zhang J, Kim J, Alexander A, *et al.* A tuberous sclerosis complex signalling node at the peroxisome regulates mTORC1 and autophagy in response to ROS. *Nat Cell Biol* 2013;**15**:1186–96.

205. Zhang R, Chen L, Jiralerspong S *et al.* Recovery of PEX1-Gly843Asp peroxisome dysfunction by small-molecule compounds. *Proc Natl Acad Sci U S A* 2010;**107**: 5569–74.

Nutritional and Toxic Diseases

Jillian Kril, Leila Chimelli, Christopher M Morris and John B Harris

INTRODUCTION

The clinical and pathological sequelae of nutrition deficiencies depend on the type and severity of the deficiency and the developmental stage at which the deficiency is experienced. The developing central nervous system is most vulnerable during periods of rapid growth, whereas, in many instances, the adult nervous system is less vulnerable because of the body's propensity to protect the physiological demands of the brain.

Nutritional deficiencies can result from either intrinsic or extrinsic factors. *Undernutrition* occurs when total protein-calorie intake is inadequate. Social and environmental reasons, as well as chronic disease, can result in inadequate dietary intake or impaired absorption. In developed countries, patients with malignancy or HIV-AIDS are commonly susceptible to undernutrition, whereas in developing countries external factors are often responsible for undernutrition. *Malnutrition* occurs where there is an imbalance of nutrients. This may be a highly specific deficiency of a trace element, amino acid or vitamin, or a more global deficiency such as protein malnutrition in the case of kwashiorkor (see later). Malnutrition can result from lifestyle factors such as vegetarianism or from a wide variety of medical and psychiatric conditions. Patients with gastroenterological disease or who have undergone gastroenterological surgery are particularly vulnerable to malnutrition because of impaired absorption of nutrients. Malnutrition can also occur in subjects with increased utilization of nutrients and in patients with chronic liver disease because the liver acts as a store for a number of nutrients, especially vitamins.

MALNUTRITION

Maternal nutritional status can affect the development of a number of organ systems in the infant, including the brain. There is also an expanding body of literature linking maternal undernutrition and later life risk of disease, especially cardio-vascular disease and diabetes. Severe maternal undernutrition, especially in the last trimester,[132] results in infants with lower birth weight and smaller circumference of head and abdomen whereas mild or moderate maternal undernutrition, or differences in maternal nutrient intake, appears to have little effect on brain size.[307] Cognitive performance measured at school age of low birth weight infants has been shown to be comparable with that of normal weight infants,[95,144] suggesting no permanent abnormality in these children.

Postnatal nutritional disorders are seen in the childhood diseases of *marasmus*, as a result of a severe reduction in protein and energy intake and *kwashiorkor*, due to insufficient protein intake accompanied by adequate energy intake. Although these disorders are considered ends of a spectrum of nutritional insufficiency, in many instances there is considerable overlap (marasmic kwashiorkor) and they are collectively referred to as protein-energy malnutrition (PEM). It is difficult to assess the nervous system consequences of PEM as such markedly abnormal nutritional states are almost invariably accompanied by deficiencies of essential amino acids, vitamins and minerals. Furthermore, the functional consequences of childhood nutritional deficiency are often not easily separated from other environmental factors such as socio-economic status, medical care and psychomotor stimulation.[190] The neurological features of PEM are alterations in mentation (irritability and apathy) and, during the recovery period, a myelopathy.[542] Reduced brain weight[59] and neuroimaging evidence of cerebral atrophy and ventricular dilatation[11,388] relative to age-matched controls have been reported in up to 75 per cent of PEM cases. The effects of PEM are most marked when they occur in the first 2 years of life because this is a period of maximum brain growth.[132] Complete reversal of the atrophy is seen following nutritional rehabilitation.[11]

Neuropathologically, the consensus from human and experimental animal studies is that postnatal undernutrition leads to impaired myelin development. Reduced myelination in infants with PEM has been shown using magnetic resonance imaging (MRI)[313] and lipid profiling.[337] In addition, the ratio of myelinated to unmyelinated axons is reduced in the corpus callosum and pyramidal tracts.[570] Other neuropathological changes in experimental animals following undernutrition are reviewed by Bedi.[36]

Undernutrition in the adult results from a variety of causes including famine, self-imposed starvation, cachexia accompanying chronic disease, anorexia nervosa and other eating disorders. Such conditions may result in one or more of the specific deficiencies described later or in a generalized brain atrophy that is largely reversible on nutritional rehabilitation.

Thiamine Deficiency

Thiamine (vitamin B_1) is integral to maintaining adequate cerebral energy supplies through its involvement in brain glucose metabolism. In its phosphorylated form thiamine is used as a cofactor by three major enzymes: α-ketoglutarate dehydrogenase, pyruvate dehydrogenase complex and transketolase.[67] In addition, thiamine has non-cofactor roles in nerve conduction and membrane transport.[230] The daily requirement for thiamine has been estimated at 1.0–1.5 mg[162] and as uptake and turnover rates are comparable,[67] thiamine deficiency can occur rapidly if body stores are depleted. In the human, thiamine deficiency results in beriberi or Wernicke's encephalopathy (WE).

Beriberi is most often seen in Asian populations where it has both cardiac and peripheral nervous system manifestations. Central nervous system involvement is rare.[293] Clinically, beriberi manifests as a sensorimotor polyneuropathy of weakness and dysaesthesia. Pathologically, there is axonal degeneration particularly of the large myelinated fibres and secondary segmental demyelination.[392]

WE is commonly associated with alcoholism because chronic alcohol ingestion results in increased thiamine utilization, reduced gastrointestinal uptake and impaired phosphorylation of thiamine to the phosphorylated form.[73] However, WE has also been reported in an array of conditions that affect nutrition including prolonged intravenous feeding,[159] malignancies, infections, AIDS,[72] starvation and hyperemesis gravidarum.[98] Although WE is usually reported in adults, nine cases of WE were reported in infants following feeding with a formula devoid of thiamine.[150] In recent years, the increasing frequency of bariatric procedures for the treatment of obesity has seen an increase in the number of cases of WE. Aasheim[1] reviewed 100 cases of WE identified following bariatric surgery and found that persistent vomiting, glucose loading and parenteral feeding are risk factors for developing WE in these patients. In another study of 318 patients having undergone gastric by-pass surgery, 18.6 per cent had thiamine deficiency 1 year after surgery,[107] indicating impaired thiamine status is common in this population. Impaired thiamine status is also reported in patients with heart failure[571] and diabetes[527] suggesting risk of WE is also high in non-alcoholic populations.

Clinically, WE is characterized by the 'classical triad' of ocular motor abnormalities, cerebellar dysfunction and altered mental state, which might manifest as disorientation, confusion or even coma.[211,554] However, it is well recognized that the pathology of WE may develop without recorded evidence of clinical signs and that the majority of patients with WE do not have all three triad signs.[211] Operational criteria with a high degree of specificity and sensitivity have been developed for the identification of WE.[78] These criteria require the presence of at least two of four signs: ocular motor abnormalities, cerebellar dysfunction, altered mental state or mild memory impairment, evidence of dietary deficiency defined as impaired thiamine status on laboratory measures, a BMI <2 SD below mean or grossly impaired dietary intake. Using these criteria, diagnostic accuracy improved to 85 per cent from 22 per cent using the classical triad.[78] Recently, these operationalized criteria were used to show that alcoholics, without the classical triad of WE, but who have one or more signs of WE have impaired cognitive function.[424]

A subset of WE patients develop a profound anterograde amnesia called Korsakoff's syndrome (KS). Although most commonly seen in alcoholics, KS has also been reported in non-alcoholic, nutritionally impaired individuals and in patients following infarction of the thalamus, head injury or tumours.[296] On routine neuropathological examination, it is not possible to distinguish WE patients with or without the amnesia of KS, so in most instances the disorder is referred to as the Wernicke–Korsakoff syndrome (WKS).

MRI has proved useful in the *in vivo* diagnosis of WE because both the acute and chronic pathology can be easily visualised.[402,512] In acute disease, symmetrical areas of hyperintensity are seen in the mammillary bodies, medial thalami, fornix and periaqueductal grey matter on T2-weighted late echo fast spin (FSE), fluid attenuated inversion recovery (FLAIR) and diffusion weighted (DWI) imaging. MRI of chronic WE shows atrophy of mammillary bodies and cerebellar vermis and enlargement of the third ventricle.[402,512]

Neuropathology: Wernicke–Korsakoff Syndrome (WKS)

The neuropathological appearance of WKS depends on the age and severity of the lesion.[123] In the acute phase, there are petechial haemorrhages that may be visible macroscopically in subcortical regions surrounding the third and fourth

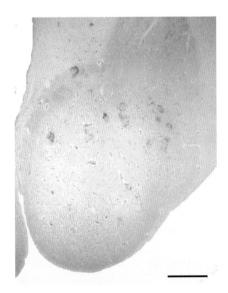

9.1 Acute Wernicke's encephalopathy. The mammillary bodies are of normal size, but appear oedematous and include multiple foci of recent haemorrhage. Scale bar, 1 mm.

ventricles. The medial hypothalamus, including the mammillary bodies (Figure 9.1), thalamus, periaqueductal grey matter and floor of the fourth ventricle are most commonly affected. Importantly, 25 per cent of cases of acute WKS have macroscopically normal mammillary bodies,[207] but in rare cases there may be necrosis of the periventricular regions.

Microscopically, acute WE is characterized by oedema, hypertrophy of endothelial cells and extravasation of erythrocytes (Figure 9.2). The earliest changes, visible after 1–2 days, are in the endothelial cells, which are followed by an astrocytic reaction. Vascular changes affect most vessel types whereas fibrinoid degeneration and haemorrhage affect arterioles and capillaries.[395] In the acute phase, morphological changes in the neurons are not apparent and there is an absence of an inflammatory reaction.

In the chronic state, there is shrinkage and brown discolouration of the mammillary bodies (Figure 9.3). Microscopic examination reveals loss of parenchyma within the mammillary bodies with astrocytosis (Figure 9.4a) and haemosiderin-laden macrophages (Figure 9.4b). There is relative preservation of neurons. There are also instances where the haemorrhages of acute WE are superimposed on

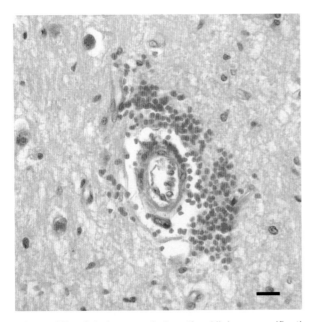

9.2 Acute Wernicke's encephalopathy. Higher magnification shows swelling of the endothelium and extravasation of erythrocytes. Scale bar, 20 μm.

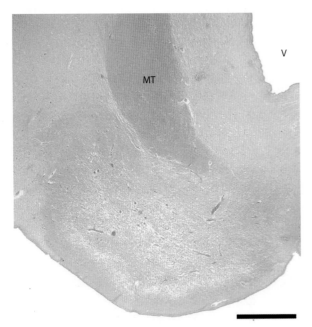

9.3 Chronic Wernicke's encephalopathy. The mammillary bodies are shrunken and centrally rarefied. Scale, 1 mm. MT, mammillothalamic tract; V, third ventricle.

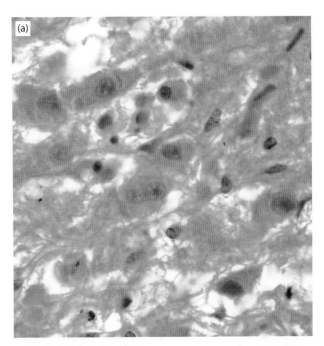

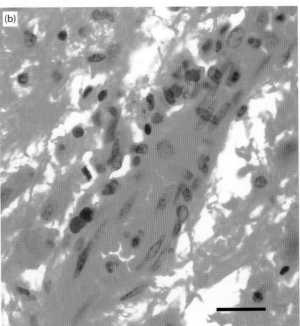

9.4 High-power photomicrographs of chronic Wernicke's encephalopathy, showing **(a)** rarefaction of neuropil and **(b)** gliosis, with perivascular accumulation of haemosiderin-laden macrophages. Scale bar, 20 μm.

the neuronal loss and cavitation of chronic WE, presumably as a result of recurrent bouts of thiamine deficiency.

Early work by Victor and colleagues[554] suggested that damage to the dorsomedial nucleus of the thalamus was the substrate of amnesia in KS. However, quantitative studies using unbiased stereological techniques in thiamine-deficient alcoholics with and without KS have shown this not to be the case. Alcoholics with KS, but not those with WE, have neuronal loss from the anterior principal nucleus of the thalamus, whereas neuronal loss from the mediodorsal nucleus of the thalamus and the medial mammillary nucleus of the hypothalamus is equivalent in WE and KS.[204] Other brain regions involved in memory circuits, such as the dorsolateral prefrontal cortex,[297] anterior cingulate cortex[295] and hippocampus,[205] show either no neuronal loss or equivalent loss to alcoholics without memory impairment. These findings suggest that damage to a number of sites involved in memory circuits is necessary for amnesia in KS.[296]

The role of thiamine deficiency in the aetiology of cerebellar degeneration remains unclear. Cerebellar degeneration is seen in approximately one third of alcoholics with WKS,[207,554] in non-alcoholic WE and also in alcoholics without WE. Consequently, the relative contribution of alcohol toxicity and thiamine deficiency is debated, although it is important to note that many alcoholics without overt WE have impaired or marginal thiamine status.[118] The involvement of thiamine in cerebellar disease is supported by a study that demonstrated a correlation between serum thiamine level and cerebellar volume on MRI that held true for thiamine levels within the normal range.[338] In addition, cerebellar degeneration is not related to daily or total alcohol consumption[146] arguing against a direct role for alcohol toxicity. Clinically, cerebellar degeneration manifests as instability of gait and incoordination of lower limbs and also contributes to cognitive impairment.[584] Cerebellar atrophy is readily identifiable on midsagittal MRI scans.

Macroscopically, there is atrophy of the anterior, superior aspect of the vermis involving predominantly lobes I–IV (Figure 9.5a).[422] Routine microscopic examination reveals patchy loss of Purkinje cells, thinning of the molecular layer and proliferation of Bergmann glia (Figure 9.5b). Quantitative studies show a decrease in the mean density of Purkinje cells of 20–30 per cent,[27,422,423] but no loss of granule cells.[27] Furthermore, features of the clinical triad of WE were found to correlate with Purkinje cell loss is specific cerebellar regions.[27]

B₁₂ Deficiency

B_{12} (cobalamin) is required by two central nervous system (CNS) enzymes (methylmalonyl coenzyme A mutase and folate-dependent methionine synthase), which are necessary for the production of cell membranes and myelin. Most commonly, deficiency results from malabsorption of B_{12} due to pernicious anaemia, an autoimmune disorder where the intrinsic factor necessary for B_{12} absorption is not produced. Other gastric causes, such as chronic gastritis, bariatric surgery, tumours and infections, have also been reported and the elderly are susceptible to B_{12} deficiency from gastric causes. Rarely, B_{12} deficiency results from dietary deficiency in vegans or vegetarians. Daily requirement is 1–3 μg and the liver stores 4–5 mg.

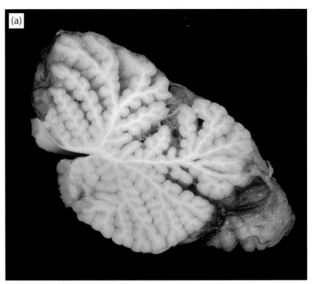

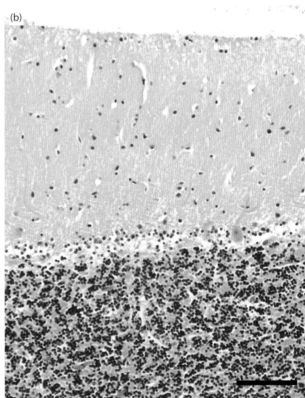

9.5 Midsagittal section of the cerebellar vermis from an alcoholic patient. (a) There is atrophy of the folia, which is most marked superiorly. **(b)** Microscopic section of the superior vermis, showing thinning of the molecular layer, patchy loss of Purkinje cells and mild loss of granule cells. Scale bar, 100 μm.

Clinically, B_{12} deficiency has haematological and neurological consequences, the most common being megaloblastic anaemia and subacute combined degeneration (SCD), respectively. SCD is a demyelinating disorder of the spinal cord that manifests initially as a progressive, symmetrical sensory disturbance of distal limbs. Severe cases progress to develop ataxia, spasticity and a loss of reflexes, proprioception and vibration senses. Cognitive abnormalities have also been reported.

The macroscopic appearance of SCD is atrophy and discolouration of the spinal cord, especially of the posterior and lateral columns. These changes can be seen as hyperintensities on T2-weighted MRI scans.[350] Earliest microscopic changes involve swelling of the myelin sheaths surrounding large diameter axons with preservation of the axon. In severe disease, there is demyelination and vacuolation of the white matter tracts (Figure 9.6), destruction of axons and the presence of foamy macrophages.

Together with folate and vitamin B_6 (pyridoxine), B_{12} is required for the methylation of homocysteine and a deficiency of either may result in elevated serum homocysteine levels that in turn have been linked to an increased risk of cardiovascular and cerebrovascular disease, depression and possibly dementia. Dietary supplementation reduces homocysteine levels.

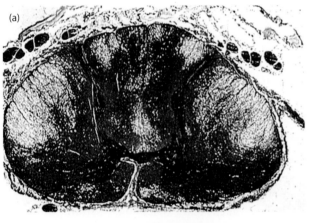

(a)

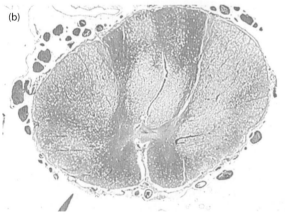

(b)

9.6 Vitamin B_{12} deficiency. (a) Thoracic spinal cord, showing spongy vacuolation in the posterior and lateral white matter columns, typical of combined subacute degeneration of the cord. Loyez myelin stain. **(b)** Acquired immunodeficiency syndrome (AIDS) myelopathy: note the confluent vacuolation in the posterior and lateral columns and milder vacuolation in the anterior columns. The appearances are similar to those in **(a)**.

Pellagra

Deficiency of vitamin B_3 (niacin) results in pellagra, a syndrome of light-sensitive dermatitis, diarrhoea and dementia, although neurological signs may occur in the absence of the skin and gastroenterological manifestations. Deficiency may result from insufficient dietary intake either of the vitamin itself or of its precursor tryptophan. Pellagra is rare, most commonly occurring in alcoholics but is also seen in vegetarians and vegans, in those with impaired intestinal uptake and following treatment with a number of drugs, most notably isoniazid used for the treatment of tuberculosis.[250] Hauw and colleagues[221] found 22 cases of pellagra in alcoholics among 8200 autopsies. This compared with 11 cases of WKS and 17 cases of Marchiafava–Bignami disease. The daily requirement of niacin is 13–20 mg, but high doses (>100 mg/day) can result in liver damage and skin irritation.

Macroscopically, the brain in pellagra is normal. Microscopically, there is chromatolysis of neurons characterized by eccentrically placed nuclei and loss of Nissl substance (Figure 9.7). Chromatolysis is most frequently found in the pontine nuclei and dentate of the cerebellum, although it is also seen in other brain stem nuclei and posterior horn cells.[221] In nutritionally derived pellagra, the cerebral cortex is rarely affected, whereas in isoniazid-induced pellagra the cerebral cortex, and in particular Betz cells, were affected in all cases in addition to the pontine nuclei.[250] The polyneuropathy found in patients with niacin deficiency has been shown to be due to demyelination, but in longstanding cases axon degeneration also occurs.[565]

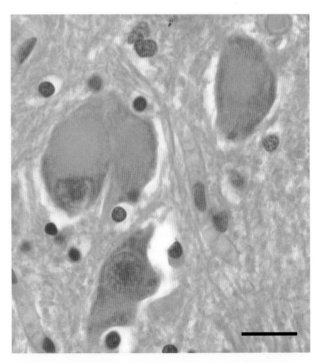

9.7 Chromatolytic neuron in the basis pontis in pellagra. The nucleus is eccentric and there is margination of Nissl substance. Scale bar, 20 μm.

Folic Acid

Folate (vitamin B_9) as its tetrahydrofolate derivatives is necessary for three important metabolic processes; the synthesis of purine nucleotides; the conversion of homocysteine to methionine in conjunction with vitamin B_{12}; and the synthesis of deoxythmidylate monophosphate which is required for DNA synthesis. Deficiency of folate during pregnancy has been associated with an increased incidence of neural tube defects[447] and supplementation has been shown to reduce the risk of birth defects.[432] Some medications with folate antagonist activity, such as antiepileptics, have also been associated with neural tube defects whereas others have not,[343] indicating specificity of action. In rare cases, congenitally impaired folate transport across the intestine and blood–brain barrier has been described in young children resulting in a progressive neurological disorder.[440]

The role of folate deficiency in CNS disease in the adult is controversial. Low serum folate levels are relatively common, especially in the elderly, alcoholics and those with gastroenterological disease. A wide variety of symptoms, including muscle weakness and depression, have been attributed to folate deficiency with little consensus. Nevertheless, Parry[405] did describe 20 patients with neurological abnormalities in which folate deficiency was demonstrated in the absence of B_{12} deficiency and in which symptoms improved with folate supplementation. Ten showed a peripheral neuropathy, eight had SCD of the cord and two had a myelopathy. Folate-responsive changes in mental function were also described in nine patients. The author suggested that each of these neurological deficits could be accounted for by disruption of folate-dependent methionine synthase. The link between folate deficiency, elevated serum homocysteine and vascular disease, depression and dementia has prompted renewed interest in the neurological sequelae of folate deficiency.

ALCOHOL INTOXICATION

Alcohol-Related Brain Damage

The neurological consequences of alcohol abuse are many and varied.[123] There are primary CNS effects of alcohol, such as intoxication and chronic toxicity, and secondary effects from medical and lifestyle factors. Many alcoholics are poorly nourished, have coexisting medical conditions, suffer from additional psychiatric conditions or are at greater risk of infections or trauma. Consequently, the neuropathological examination of the brain of an alcoholic patient may reveal a great deal of pathology, and distinguishing the direct effects of alcohol from these other factors may be difficult. In broad terms, the neuropathology can be divided into that which occurs in alcoholics with either nutritional deficiencies (e.g. WKS, pellagra) or liver disease (viz. hepatic encephalopathy) and that which occurs in alcoholics without other disease (alcohol-related brain damage, ARBD).[292]

Macroscopically, the brain of an ARBD patient shows atrophy. Harper and Blumbergs[208] showed a 70-g reduction in mean brain weight between alcoholics and controls. This has been confirmed in a number of other studies[209,498] and by a comparison of brain volume and intracranial cavity volume.[210] In non-alcoholic subjects, the mean pericerebral space is 8.3 per cent, whereas in ARBD, it is 11.3 per cent and 14.7 per cent in alcoholics with WKS. Quantitative morphometry has shown the loss of brain tissue is largely due to a decrease in white matter volume,[122,210,297] a finding that has been confirmed using volumetric MRI.[418] The mean white matter reduction is 14 per cent,[210] although this is most marked in alcoholics with WKS and less severe in ARBD where it is largely restricted to the frontal lobe.[297] This may, however, represent a dose phenomenon rather than an effect of thiamine deficiency as white matter volume is negatively correlated with maximum daily alcohol consumption.[297] MRI studies have also found a decrease in cortical volume in alcoholics,[258,418] especially in the frontal lobes.[420] The magnitude of atrophy is greater in older alcoholics than young alcoholics, despite similar drinking histories suggesting increased susceptibility of the ageing brain.[418,419] Abstinence from alcohol results in, at least partial, resolution of atrophic changes,[477,532] yet this capacity declines with advancing age.[294]

Neuronal loss in alcoholics occurs in discrete anatomical regions. In the cerebral cortex, the loss is restricted to the superior frontal cortex (BA8),[297] however the magnitude of loss (mean 23 per cent) is less than that which can be reliably identified using routine neuropathological examination. Neuronal loss does not occur from the primary motor, anterior cingulate or temporal cortices.[295,297] There is atrophy, but not neuronal loss in the hippocampus, which has been shown to relate to whether or not the patient was still drinking at the time of death.[205] In subcortical regions, there is neuronal loss from the supraoptic and paraventricular nuclei of the hypothalamus, but not from the mammillary bodies,[204] anterior principal or mediodorsal nuclei of the thalamus,[204] serotonergic dorsal raphe,[26] locus ceruleus,[197] basal forebrain[115] or cerebellum.[27] Neuronal loss from the hypothalamus is related to maximum daily alcohol consumption.[206] The reasons for this anatomical specificity of susceptibility to alcohol toxicity are unclear. However, regional variations in the subunit composition of amino acid neurotransmitter receptors has been hypothesized as one potential mechanism.[134]

The clinical manifestations of alcohol abuse are predominantly disorders of the frontal lobes reflecting the greater damage in this region.[585] Executive functions such as planning ability, self-regulation, goal setting and working memory have all been shown to be impaired in alcoholics.[31,172] Cerebellar ataxia and incoordination are common in cases with cerebellar atrophy.[554]

Central Pontine Myelinolysis and Marchiafava–Bignami Disease

In rare instances, chronic alcoholics can also develop other neurological conditions. Hepatic encephalopathy (see section later), central pontine myelinolysis (CPM) and Marchiafava–Bignami disease (MBD) are examples of such conditions. CPM is due to the rapid correction of electrolyte imbalances, especially chronic hyponatraemia,[336] but has been reported with other electrolyte imbalances such as hypophosphataemia[149] leading to the description of the osmotic demyelination syndromes.[336] CPM is usually seen in patients who are markedly malnourished or debilitated such as alcoholics, but has also been reported in patients with advanced liver disease, following liver transplantation, and in those with HIV-AIDS or severe burns. Limiting the rate of correction of chronic hyponatraemia to ≤8 mmol/L per day significantly reduces the occurrence of CPM.

Clinical manifestations depend on the severity of the lesion. MRI studies have identified cases with early disease and complete recovery. More severe cases result in a biphasic course where initially there are encephalopathy and seizures from hyponatraemia and then, following correction, deterioration occurs again several days later.[336] In the second phase, there are quadriparesis, dysarthria and dysphagia when lesions are restricted to the basis pontis, and additional oculomotor abnormalities if the tegmentum is also involved. When lesions are extensive a 'locked-in' syndrome may ensue.[336]

Pathologically, CPM is characterized by demyelinating lesions in the midline of the basis pontis. These are symmetrical, triangular or 'butterfly'-shaped lesions that appear as pallor on myelin-stained preparations (Figure 9.8a). Microscopically, there is loss of myelin with relative preservation of axons in early lesions (Figure 9.8b), but more long-standing lesions also show axonal degeneration and the presence of foamy macrophages. Extrapontine lesions were recognized many decades ago and have been described in up to 50 per cent of cases of CPM in the cerebellum, lateral geniculate body, external capsule, hippocampus, putamen and cerebral cortex.[175] Microscopically, these lesions are similar to those in the pons.

MBD is extremely rare and is mostly described in reports of isolated cases. In recent years, the greater availability of MRI and the utility of this technique in identifying discrete alterations in brain signal have meant a larger number of cases have been identified, but few of these have been examined pathologically. Lesions in the corpus callosum can be identified as hyperintensities on T2-weighted, FLAIR and diffusion-weighted sequences. In a study of the progression of radiological changes in MBD, areas of hyperintensity were seen to persist for 4 months, but to be largely resolved by 11 months.[166]

Classical descriptions of MBD describe three patterns of clinical symptoms: acute MBD, which is a severe neurological disorder of altered consciousness and seizures that rapidly progresses to death; subacute MBD, in which there are interhemispheric disconnection, abnormalities of higher cognitive function and gait disturbance; and a chronic form characterized by progressive dementia.[166] However, this classification has recently been challenged. In the largest review of MBD to date, Heinrich and colleagues[226] collated data from 50 cases arising from 41 publications. They found a predominance in males (M:F, 3.2:1) and a mean age of 46.6 years (range, 26–66). They also found two distinct subtypes using clinical and radiological parameters. Type A occurred in 19 cases (38 per cent) and was characterized by a major disturbance of consciousness (stupor or coma) and extensive lesions on MRI. These cases were more likely to die or to have a major disability.[226] Type B occurred in 29 (58 per cent) cases, had no or minor impairment of consciousness, partial callosal lesions on MRI and minimal impairment.[226]

Pathologically, the brain in MBD may be normal or may exhibit thinning and discolouration or cavitation of the corpus callosum. Microscopically, there is a spectrum of severity from demyelination with preservation of axons to necrosis and cystic cavitation. There is a loss of oligodendrocytes, upregulation of astrocytes and foamy macrophages. Rarely, there is involvement of other white matter tracts such as the centrum semiovale, cerebral peduncles or optic chiasm.

The aetiology of MBD is unknown. It was originally described in Italian red wine drinkers and attributed to an unidentified toxin in the wine, yet subsequent cases in

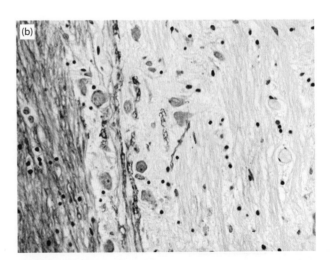

(a)

(b)

9.8 (a) Central pontine myelinolysis, showing the symmetrical, triangular region of myelin loss. **(b)** Higher magnification view, showing the loss of myelin within the region of myelin loss (towards the right), but preservation of neurons. Myelin in the adjacent part of the pons (towards the left) appears normal. Luxol fast blue and cresyl violet.

Slide courtesy of Dr Michael Rodriguez.

people who drink other types of alcoholic beverage have discounted this hypothesis. Clinical improvement following treatment with vitamin supplements[166] has led to the current theory that MBD is due to vitamin deficiency.

Hepatic Encephalopathy

Hepatic encephalopathy (HE) is a syndrome of neurological dysfunction that occurs in patients with acute or chronic liver disease. There are acute and chronic forms of HE depending on the time course and severity of the liver disease. The clinical pattern and pathology of HE is independent of the aetiology of the liver disease. HE associated with acute liver failure (HE type A[154]) is a disorder of rapid onset that occurs in the setting of fulminant liver failure, such as with paracetamol overdose. In the early stages, patients exhibit an increase in muscle tone and altered mental state that progresses to stupor and coma. Death is common secondary to raised intracranial pressure from cerebral oedema. Neuropathologically, there is brain swelling with a decrease in ventricular volume, flattening of gyri and herniation. Microscopically, oedema is evident and ultrastructural studies have revealed that this is due to swelling of perivascular astrocytes.[270]

Traditionally, chronic HE, often seen following chronic alcoholism, was often referred to a portal-systemic encephalopathy because it commonly occurs as a result of the shunting of portal blood into the systemic circulation.[69] This could happen either spontaneously as a result of portal hypertension in patients with cirrhosis or following surgery to alleviate variceal bleeding.[69] However, the recognition that shunting can occur in the absence of hepatic disease has led to the suggestion that HE be classified as type B if encephalopathy is associated with portal-systemic bypass in the absence of hepatic disease and type C if the encephalopathy is associated with cirrhosis and portal hypertension or portal-systemic shunting.[154]

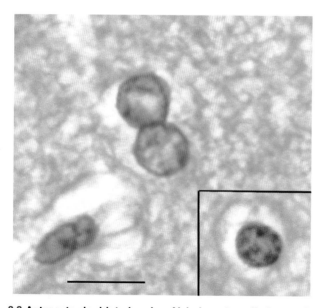

9.9 Astrocyte doublet showing Alzheimer type II change, in a case of hepatic encephalopathy. The astrocytes are larger than normal (inset) and show margination of nuclear chromatin. Scale bar, 10 μm.

The neuropathology of chronic HE is largely restricted to the astrocyte (Figure 9.9). Morphologically, astrocytes undergo Alzheimer type II change in which they show enlarged, pale nuclei with a rim of chromatin and prominent nucleoli. Astrocyte pairs and triplets are seen, but evidence of mitosis has not been found.[383] In severe cases, nuclei become lobulated and contain granules of glycogen. Astrocytes also lose their immunoreactivity for glial fibrillary acid protein and contain increased numbers of mitochondria.[383] Alzheimer type II astrocytes are most commonly seen in grey matter regions, such as the pons, cortical grey matter and putamen. In the cerebral cortex, they are often found in the deep layers. Alzheimer type II change in astrocytes is also seen other metabolic encephalopathies including uremia and hypercapnia and, in infants, hypoxia and hypoglycaemia.[383]

The pathogenesis of HE has been studied extensively in both humans and animal models following portacaval anastomosis and the underlying disorder is one of astrocyte-neuron trafficking.[71] Cirrhosis of the liver results in elevated blood and brain ammonia levels and it is ammonia toxicity that is the leading hypothesis for the causation of HE. Astrocytic changes similar to Alzheimer type II change can be induced by hyperammonaemia in experimental animals and tissue culture and is also seen in patients with congenital hyperammonaemia due to inherited disorders of enzymes involved in the urea cycle.[68] Millimolar concentrations of ammonia in the brain impairs postsynaptic inhibitory neurotransmission. Although there is substantial evidence for a central role of ammonia in HE, the mechanism by which ammonia causes the cerebral dysfunction is not entirely clear.[457]

Neuronal abnormalities in HE are not apparent on routine neuropathological examination and recovery of neurological function in HE patients following treatment of the underlying liver disease adds weight to the viewpoint that HE is a functional rather than a structural brain disorder. Nevertheless, there is evidence of persistent neurological abnormalities in some patients suggesting neuronal damage can occur.[70]

Uraemic Encephalopathy

Uraemic encephalopathy can manifest as a spectrum of neurological abnormalities from mild changes in cognitive function through to delirium and even coma. It can be associated with motor (e.g. tremor, asterixis, myoclonus), visual and sleep disturbances, alterations in consciousness and seizures.[58,303] Peripheral neuropathy is also frequently reported in patients with uraemia.[161,303] Neurological function is usually improved with successful treatment of the renal failure. Historically, cases with dialysis dementia, a progressive, often fatal syndrome of motor disturbance, personality change, psychosis and seizures, were described but these are now rare due to improvements in dialysis. Accumulation of aluminium was found to be the basis of dialysis dementia (see this chapter).

The pathogenesis of uraemic encephalopathy has not been fully elucidated. A wide variety of compounds has been implicated including the accumulation of urea and other metabolites, alterations in amino acid neurotransmitter systems and hormone imbalance, especially parathyroid hormone.[58,303] Macroscopically, the brain is often normal. In rare cases, CPM and extrapontine myelinolysis can occur. Microscopically, Alzheimer type II astrocytes are seen and in some cases areas of perivascular neuronal degeneration and demyelination.

DISORDERS OF THE HYPOTHALAMIC-PITUITARY AXIS

The hypothalamus and pituitary gland are considered together in view of their close structural and functional relationships. We have covered neither the neuropathology of vascular lesions nor the consequences of physical injury to the head nor the biological aspects of hypothalamic structure and function. For reference, we have tabulated the principal products of the hypothalamus and pituitary gland (Table 9.1). We have concentrated on the major clinical manifestations of disorders of the hypothalamic–pituitary axis and their neuropathological associations.

Precocious Puberty

Puberty is characterized by maturation of the hypothalamic–pituitary–gonadal (HPG) axis, the appearance of secondary sexual characteristics, acceleration of growth and, ultimately, the capacity for fertility. Disorders of pubertal development may occur at any of the steps in this maturational process, leading to either precocious or delayed puberty.[261]

Precocious puberty, defined as the premature development of secondary sexual characteristics and a postpubertal endocrine profile, is associated with elevated blood levels of gonadotrophin hormone releasing hormone (GnRH), follicle stimulating hormone (FSH) and luteinizing hormone (LH) and sex steroids. Early activation of the HPG axis gives rise to central precocious puberty, whereas lesions that secrete gonadotrophin-like substances and androgen- or oestrogen-producing neoplasms cause precocious pseudo-puberty. Misdiagnosis is a common problem. One survey of 104 cases referred for evaluation identified nine cases of true precocious puberty; the majority of children referred were benign normal variants of the general population.[268]

It is generally agreed that the age of onset of central precocious puberty (CPP) is under 8 years in females and 9 years in males, although it is suspected that it may be much earlier. The aetiology of organic CPP is multifactorial and is related directly to the location and type of lesion. Demonstrable hypothalamic pathology is seen in fewer than 10 per cent of cases of CPP. Despite the absence of a demonstrable lesion, idiopathic precocious puberty is presumably hypothalamic in origin. It is a prominent feature of the McCune–Albright syndrome, a disorder not associated with observable hypothalamic lesion.[288]

CPP can be divided into two major groups. The first includes lesions that elaborate GnRH–for example, hypothalamic hamartoma in which CPP is associated with a childhood epileptic syndrome characterized by gelastic seizures, agenesis of the corpus callosum, Dandy–Walker complex and grey matter heterotopias.[193,485] The second group includes endocrine-inactive lesions that non-specifically affect hypothalamic centres engaged in the control of sexual maturation. Reported lesions include germ-cell tumours,[352,381] pineal cysts,[130] gliomas,[454] infections, such as congenital human immunodeficiency virus (HIV) infection[239] and congenital toxoplasmosis,[482] suprasellar arachnoid cysts,[507] hydrocephalus, Angelman's syndrome and trauma.[48] Also implicated are craniopharyngioma (the tumour associated most often with pubertal delay), Rathke's cleft cyst,[354] duplication of the hypophysis with thickening of the hypothalamus,[127] sectioning of the pituitary stalk secondary to Langerhans cell histiocytosis,[367] neurofibromatosis type 1[556] and tuberous sclerosis complex.[7] Rutland and colleagues reported the third case of hypomelanosis of Ito associated with precocious puberty in a girl aged 5 years, with extensive cerebral involvement, suggesting a central mechanism of precocious puberty,[467] although two previously reported cases had abnormal gonads and responded to therapy, indicating a peripheral mechanism.

A prevalence of 27.5 per cent of familial cases among 147 patients with idiopathic CPP has been reported by

TABLE 9.1 Principal hormones of the hypothalamic-pituitary axis

Hypothalamic hormone	Pituitary hormone regulated	Location of hormone in pituitary
GHRH	GH	Anterior pituitary
Somatostatin	GH	
TRH	TH	Anterior pituitary
GnRH	FSH, LH	Anterior pituitary
CRH	ACTH	Anterior pituitary
ADH	Produced in hypothalamus, neurally transferred to pituitary	Posterior pituitary
OT	Produced in hypothalamus, neurally transferred to pituitary	Posterior pituitary

ACTH, adrenocorticotrophic hormone; ADH, antidiuretic hormone (vasopressin); CRH, corticotrophin releasing hormone; FSH, follicle stimulating hormone; GH, growth hormone; GHRH, growth hormone releasing hormone; GnRH, gonadotrophin hormone releasing hormone; LH, luteinizing hormone; OT, oxytocin; TH, thyrotrophic hormone; TRH, thyrotrophic hormone releasing hormone.

Phillip and Lazar.[421] Segregation analysis of this cohort suggested an autosomal dominant transmission with incomplete sex-dependent penetrance. Grosso reported three patients presenting with CPP in whom karyotype analysis demonstrated abnormal chromosomal patterns.[189] One showed XXX syndrome, which is commonly characterized by premature ovarian failure. The second patient had a chromosomal aberration involving an imprinted chromosomal region, deletion of which is commonly associated with Prader–Willi syndrome and hypogonadotrophic hypogonadism. The third patient was a boy carrying a rare duplication of chromosome 9. All had degrees of mental retardation, were treated with luteinizing hormone releasing hormone (LHRH) analogues and did not progress with precocious sexual development. Iliev and colleagues reported a girl aged 5 years with precocious puberty and high levels of testosterone as a result of a mixed gonadal dysgenesis with a testosterone-producing gonadoblastoma.[249]

Ectopic Production of Hypothalamic Hormones

Hormones of hypothalamic origin are referred to as releasing hormones (RH) or hypothalamic hormones.[192] A variety of tumours, most neuroendocrine in nature, are known to produce hormones indistinguishable from releasing hormones of hypothalamic origin in terms of their endocrine effects. Of these, the production and effects of ectopic GHRH and corticotrophin releasing hormone (CRH) have been best documented. GnRH and thyrotrophin releasing hormone (TRH) can also be produced ectopically. Vasopressin (antidiuretic hormone; ADH) is the principal posterior lobe-related hormone to be produced ectopically. Less obviously 'ectopic' in origin is tumour-associated hypersecretion of oxytocin, vasoactive intestinal peptide (VIP), substance P and somatostatin.

Growth Hormone Releasing Hormone

GHRH was first isolated and characterized as a 44-residue peptide from a rare tumour of the pancreas, which had induced acromegaly in the absence of a pituitary tumour. The physiology of GHRH and the diagnosis and treatment of GHRH-mediated acromegaly has been reviewed.[135] GHRH-producing neurons have been well characterized in the hypothalamus by immunostaining techniques. Hypothalamic tumours, including hamartomas, choristomas and gangliocytomas, may produce excessive GHRH, with subsequent GH hypersecretion and acromegaly. Immunoreactive GHRH is present in several neoplasms, including carcinoid (bronchial, thymic) tumours, pancreatic cell tumours, small cell lung carcinoma, medullary carcinoma of the thyroid, adrenal adenomas and phaeochromocytoma.[50,396] Clinical acromegaly due to tumour GHRH production in these patients is, however, very uncommon (Figure 9.10).[470]

Corticotrophin Releasing Hormone

CRH is expressed in the brain, pituitary and many peripheral tissues, including adrenal medulla (mostly in layers

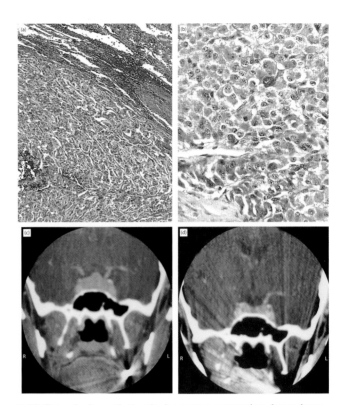

9.10 Ectopic hypothalamic hormone secretion by a bronchial carcinoid tumour. (a) Nests and sheets of tumour cells in a carcinoid tumour. **(b)** The cells are immunopositive for GHRH. **(c)** A contrast-enhanced computed tomography (CT) scan shows marked enlargement of the pituitary, as a result of growth hormone (GH) cell hyperplasia, which **(d)** remitted after resection of the tumour.

of cells adjacent to the cortex). In the brain, the presence of CRH in the cerebellum and the limbic system suggests a role distinct from that in the hypothalamic–pituitary axis.[192]

The evaluation of Cushing's syndrome is extremely complex and the ectopic production of CRH must be considered in its differential diagnosis. In one child, it was due to ectopic production of CRH by a ganglioneuroblastoma.[587] Because most CRH-producing tumours also secrete adrenocorticotrophic hormone (ACTH), the ectopic production may represent a paracrine phenomenon in addition to an endocrine phenomenon. Ectopic CRH may also indirectly provoke pituitary ACTH secretion. This dual mechanism of ACTH production may explain the resistance of the tumour to feedback inhibition and a CRH stimulation response indistinguishable from that observed in pituitary-dependent Cushing's syndrome.

Gonadotrophin Releasing Hormone

GnRH appears to be involved in synaptic transmission in sympathetic ganglia and may function as a neuromodulator in its multiple other locations in the nervous system. It has been shown that, during development, some GnRH neurons remain in the nasal cavity, including the olfactory and vomeronasal mucosa. GnRH receptors are also expressed by chemosensory neurons. The elucidation of GnRH neurons in this site, in addition to the hypothalamus, provides a link between the anosmia and isolated gonadotrophin deficiency in patients with some forms of Kallmann's syndrome (Figure 9.11) and related developmental abnormalities.[192]

Somatostatin

Somatostatin, an inhibitor of GH secretion, has multiple roles. There is extensive evidence that somatostatin originates from the immune system and has significant immunomodulatory activity. Thymic epithelial and dendritic cells synthesize somatostatin and express type 2 somatostatin receptors. Somatostatin is recognized to be part of an immunoregulatory circuit that inhibits production of interferon γ (IFN-γ), tumour necrosis factor α (TNF-α), CRH and substance P at sites of (chronic) inflammation.[566] Scintigraphy with labelled analogues of somatostatin has shown the presence of several types of receptor in pathological peripheral tissues, including the retro-orbital tissue in Graves' disease, breast cancer, malignant melanomas, small cell lung cancer and bronchial carcinoid tumours.[386] In fact, all so-called neural crest or neuroendocrine tumours (insulinoma, glucagonoma, gastrinoma) have been shown to express somatostatin receptors and to respond, sometimes dramatically, to long-acting analogues of somatostatin. In addition, prostatic tumours contain cellular elements that are immunoreactive for a variety of peptides, such as somatostatin, bombesin, TRH, chromogranin A and serotonin.[202]

Antidiuretic Hormone/Vasopressin

Ectopic ADH production is most often associated with pulmonary neoplasms and infection, particularly tuberculosis. In addition, adenocarcinoma of the prostate with ectopic ADH production has been reported.[252] Ectopic expression of vasopressin V1b and V2 receptors in the adrenal glands of patients with familial ACTH-independent macronodular adrenal hyperplasia has been reported.[315] Neither of these receptors is known to be normally expressed in the adrenal gland.

The syndrome of inappropriate antidiuretic hormone secretion (SIADH) (see Box 9.1) is not an uncommon

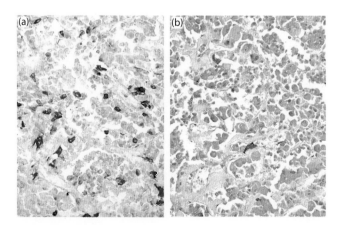

9.11 Kallmann's syndrome. (a) Normal density of gonadotrophs labelled with anti-FSH antibody. **(b)** Marked reduction in the density of FSH-positive gonadotrophs in Kallmann's syndrome – the effect of diminished hypothalamic stimulation.

disorder. It is characterized by symptomatic hyponatraemia, with high urine osmolarity, and can be caused by a variety of conditions (Table 9.2). Unlike diabetes insipidus, for which the anatomical basis is usually obvious, that of SIADH is poorly understood. Whereas stimulatory osmoreceptors lie close to the ADH-producing magnocellular nuclei, inhibitory pathways are dispersed, their centres lying as far afield as the brain stem.[547] As a result, lesions remote from the hypothalamus and posterior pituitary may result in ADH release. In addition, ectopic production of ADH, particularly by tumour cells (see earlier), may be responsible for SIADH. The treatment of ADH excess consists of water restriction and slow re-establishment of sodium balance; too rapid an elevation of sodium may result in central pontine myelinolysis.[384]

Diabetes Insipidus

Diabetes insipidus is a heterogeneous condition characterized by polyuria and polydipsia and caused by a lack of ADH secretion, its physiological suppression following excessive water intake, or the resistance of the kidney to its physiological actions.[171] Physiological water balance is dependent upon functioning hypothalamic osmoreceptors, the capacity of the hypothalamus to produce ADH, the structural and functional integrity of the pituitary stalk and posterior lobe, the presence of renal vasopressin receptors, and a normal response to thirst. Diabetes insipidus may, therefore, be central (i.e. pathology is present in the brain or pituitary gland, causing a deficiency of ADH synthesis in the hypothalamus and/or of secretion from the neurohypophysis) or

TABLE 9.2 Disorders associated with excessive antidiuretic hormone (ADH)

Neoplasia (extra-CNS)	Carcinoma of the lung, tonsil, duodenum, pancreas, ureter and prostate
	Lymphoma, leukaemia, Hodgkin's disease
	Mesothelioma
	Ewing's sarcoma
	Aesthesioneuroblastoma
CNS disease	Trauma, neurosurgery
	Infection: meningitis (tuberculous), encephalitis, brain abscess, malaria
	Tumours (glioma, etc.)
	Hydrocephalus
	Delirium tremens
	Guillain–Barré syndrome
	Multiple sclerosis
	Haemorrhage: intracerebral, subarachnoid, subdural
Pulmonary disease	Tuberculosis
	Cavitary aspergillosis
	Pneumonia (viral, bacterial or fungal)
	Positive-pressure ventilation[a]
Endocrine diseases	Addison's disease, myxoedema, hypopituitarism, lymphocytic panhypophysitis
Miscellaneous	Cirrhosis with ascites[a]
	Myocardial infarction[a]
	Congestive heart failure[a]
	Acute intermittent porphyria
	Postoperative state[a]
Drugs	Vasopressin, oxytocin, chlorpropamide, chlorothiazide, clofibrate, carbamazepine, nicotine, phenothiazines, cyclophosphamide, morphine, barbiturates, amiodarone

CNS, central nervous system.
[a]Disorders that may induce 'appropriate' ADH hypersecretion.

TABLE 9.3 Diabetes insipidus: variants and causes

Central diabetes insipidus	Primary	Idiopathic: autoimmune disease (?)
		Hereditary: autosomal dominant, sex-linked recessive
		Wolfram (DIDMOAD) syndrome
	Secondary	Hypothalamic disease
		Trauma (surgical trauma, 'stalk section', head injury)
		Neoplasia
		Primary (craniopharyngioma, glioma, germ cell tumour)
	Infection	Encephalitis, meningitis (bacteria, fungi)
	Vascular	Hypoxic tissue injury
		Postpartum pituitary necrosis
		Intraventricular haemorrhage
		CSF leak after transsphenoidal surgery
	Systemic disease	Sarcoidosis, Langerhans' cell histiocytosis
		Extramedullary haemopoiesis
Nephrogenic diabetes insipidus	Primary	Hereditary (X-linked): vasopressin-unresponsive
	Secondary	Electrolyte disturbance (hypokalaemia, hypercalcaemia)
		Chronic renal disease
		Drugs (lithium, colchicine, gentamicin)

CSF, cerebrospinal fluid; DIDMOAD, diabetes insipidus, diabetes mellitus, optic atrophy, neural deafness.

nephrogenic (when the kidney is unable to produce concentrated urine, because of the insensitivity of the distal nephron to ADH). Central diabetes insipidus may be idiopathic or caused by a variety of lesions (Table 9.3). A thorough investigation to rule out CNS lesions needs to be undertaken before considering a diagnosis of idiopathic central diabetes insipidus, because in many cases it is caused by a germinoma, craniopharyngioma, Langerhans' cell histiocytosis, sarcoidosis, local inflammatory, autoimmune or vascular diseases, or trauma.

Central diabetes insipidus is a common complication of transsphenoidal surgery, but it may also be caused by or associated with an intraoperative cerebrospinal fluid (CSF) leak, a craniopharyngioma, a Rathke cleft cyst or a microadenoma. The first three conditions are risk factors for persistent diabetes insipidus,[376] but diabetes insipidus is usually transient in nature.

Autoimmune central diabetes insipidus is likely in young patients with a clinical history of autoimmune disease and radiological evidence of pituitary stalk thickening.[425] It is probable that many cases of idiopathic central diabetes insipidus are actually autoimmune central diabetes insipidus.[425]

Nephrogenic diabetes insipidus, due to various causes (Table 9.3), may result in severe dehydration and electrolyte imbalances. Reversible nephrogenic diabetes insipidus results in resolution of the condition. Long-term treatment with lithium may result in irreversible nephrogenic diabetes insipidus.[168]

Hereditary diabetes insipidus with a variable age of onset is exceedingly rare, representing only 1 per cent of cases of central diabetes insipidus. Hereditary diabetes insipidus encompasses vasopressin-sensitive degenerative disorders of magnocellular neurons in the supraoptic and paraventricular nuclei[269] and may also be seen in Wolfram's syndrome,[137] a disorder characterized by atrophy of the hypothalamic nuclei, and degeneration of the optic pathway, pons and cerebellum.[85]

Given the variety of causes of central diabetes insipidus, the prognosis varies considerably. Resultant diabetes insipidus may be transient or permanent, depending upon the level of the lesion and its degree of tissue destruction. In most cases of hypothalamic involvement, both supraoptic and paraventricular nuclei are affected; a destructive lesion proximal in the neurosecretory system, such as the upper portion of the pituitary stalk, produces permanent diabetes insipidus due to retrograde axonal degeneration and atrophy of the magnocellular nuclei (Figure 9.12).[321] A low stalk lesion or destruction of the posterior pituitary produces only temporary diabetes insipidus because a small proportion of axons originating in the supraoptic nucleus terminates high in the median eminence and is consequently spared. Furthermore, destruction of the stalk at a low level permits axonal regeneration.[117] Biopsy of an enlarged pituitary stalk should be reserved for patients with a hypothalamic–pituitary mass and progressive thickening of the pituitary stalk, because spontaneous recovery may occur. Neuronal loss must be extensive and high, involving the infundibular region or median eminence to result in permanent deficits. Autopsy studies are few but demonstrate striking neuronal loss and gliosis.[54,185] Similar changes have been described in 'idiopathic diabetes insipidus'.[185] In occasional cases, magnocellular nuclei appear morphologically normal but lack vasopressin immunoreactivity.[370]

Hypothalamic Hypogonadism

This condition results from the deficient or dysrhythmic release of GnRH. The amplitude and rhythmic frequency of release regulates the secretion of LH and FSH. The term 'idiopathic hypothalamic hypogonadism' is used broadly to describe the condition in all patients with an unexplained abnormality of GnRH secretion; Kallman's syndrome describes an inherited condition of GnRH abnormality in association with anosmia.[77,528] The presenting features of patients with hypothalamic hypogonadism are delayed puberty and/or incomplete sexual development. In males, these include prepubertal testes, micropenis, cryptorchidism, and in females delayed breast bud development and abnormal pubic and axillary hair. The neuropathological consequences of hypothalamic hypogonadism are hypoplasia of the lateral tuberal nuclei, an increase in neurons in the subventricular nucleus and a marked increase in gonadotrophs in the anterior pituitary.[287] Pulsatile GnRH therapy is usually effective.

A variety of midline malformations may also be associated with GnRH deficiency. For example, hypogonadism is associated regularly with hypothalamic lesions due to sarcoidosis, Langerhans' histiocytosis and a variety of neoplasms. Hypogonadism is a common finding in idiopathic haemochromatosis, usually due to iron deposition in gonadotrophs (Figure 9.13).[39] One report however,

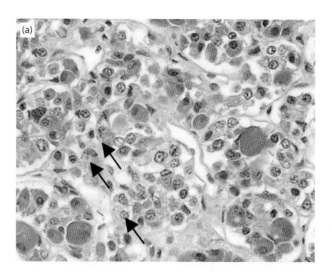

(a)

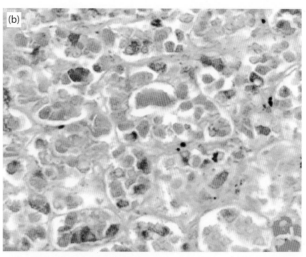

(b)

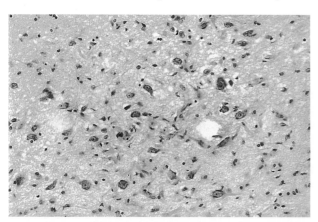

9.12 Familial central diabetes insipidus. Marked reduction in the size and number of secretory neurons. The patient, a 70-year-old male with lifelong diabetes insipidus and several affected family members, had a serum vasopressin level of 1.5 pg/mL, which did not rise with dehydration.

Courtesy of Dr C Bergeron, University of Toronto, Toronto, Ontario, Canada.

9.13 Pituitary haemosiderosis. (a) In the adenohypophysis of a patient with haemochromatosis, there is deposition of iron in scattered hormone-producing parenchymal cells. (b) Co-localization reveals that the majority of cells containing iron (blue) are gonadotrophs that contain β-follicle stimulating hormone (FSH) (brown).

described a clinically hypogonadal male with haemochromatosis and laboratory evidence for a combined defect in hypothalamic and pituitary function.[496]

Idiopathic hypothalamic hypogonadism has been associated with polyostotic fibrous dysplasia (Albright's syndrome).[241] Hypogonadotrophic hypogonadism may be associated with exercise-related and psychogenic amenorrhoea (also known as hypothalamic amenorrhoea) and anorexia, but the precise mechanisms involved are unclear. The evaluation of hypothalamic hypogonadism involves the exclusion of other systemic disorders and broader disorders of the hypothalamus, such as panhypopituitarism.[582]

Hypothalamic Dwarfism

In addition to pituitary disease and end-organ resistance to GH, dwarfism may result from hypothalamic dysfunction. In many instances, as a result of their proximity and shared circulation, the hypothalamus and pituitary are simultaneously involved by inflammatory, infectious, neoplastic, congenital or neonatal diseases associated with growth retardation. Underlying causes include bacterial meningitis, granulomatous disease, Langerhans' cell histiocytosis, hypothalamic neuronal hamartoma and neoplasms, particularly craniopharyngioma and germ cell tumours.

Aside from the more often idiopathic nature of hypothalamic growth retardation in early childhood, this may also be due to disorders in the pre-, peri-, and postnatal periods.[111] Midline developmental defects, such as septo-optic dysplasia, may also be responsible.[20,360,427] An autosomal recessive form of hypothalamic dwarfism is associated with failure to produce GHRH. In such cases, somatotrophs are structurally normal and produce GH.[452] Patients with mitochondrial encephalomyopathies, especially MELAS (mitochondrial encephalomyopathy, lactic acidosis, stroke-like episodes) and MERRF (myoclonus epilepsy with ragged red fibres), have also been reported to have hypothalamic–pituitary dysfunction (see also Chapter 7).[391]

Diencephalic Syndrome

The diencephalic syndrome is a rare but potentially lethal cause of failure to thrive in infants and young children. Clinical features are severe emaciation, normal linear growth, normal or precocious intellectual development, ophthalmological abnormalities, including late optic atrophy, and signs of hypothalamic dysfunction such as euphoria, hyperkinesis, hypertension and hypoglycaemia.[66,262,408,409,429] Although extremely rare, diencephalic syndrome can occur in adults.[351,483]

In almost all instances, the syndrome is associated with CNS tumours in the hypothalamic–optic pathway region, such as an optic pathway glioma (pilocytic astrocytoma),[157,429,557] craniopharyngioma[351,483] or germ cell tumours.[352] The syndrome usually improves with removal of the tumour.[121] Van der Wal and colleagues reported leptomeningeal spread of a chiasmatic pilocytic astrocytoma in a child presenting with diencephalic syndrome.[548] Despite treatment with chemotherapy and radiation, the tumour recurred and progressed to a high-grade astrocytoma. One rare case of pilocytic astrocytoma and diencephalic

syndrome occurring together in a 30-year-old woman with neurofibromatosis type 1 has been described.[281] MR imaging revealed a tumour in the chiasmatic–hypothalamic region, which invaded the thalamus, brain stem and cerebellum. Some reports[157,429] indicate these tumours are usually larger, occur at a younger age, and behave more aggressively than similarly located tumours not presenting with diencephalic syndrome. The suggestion has been made that in a child affected by diencephalic syndrome and hypothalamic juvenile pilocytic astrocytoma, the patient should be reviewed carefully for evidence of tumour dissemination.[412]

Developmental cysts and inflammatory lesions are less often the cause of diencephalic syndrome. Some cases are congenital. Involvement of the anterior hypothalamus is a common feature.[408] Accordingly, diencephalic syndrome should be considered in any child with emaciation despite adequate caloric intake and an inappropriately euphoric mood.[121]

Metabolic Disorders of the Pituitary

Amyloid deposition involving the pituitary gland occurs as part of a systemic disorder and is seen occasionally in pituitary adenomas, most commonly prolactinomas.[45,46,306,359,558] The amyloid is deposited in vessel walls or the interstitium, as an extracellular, amorphous, eosinophilic substance that shows apple green birefringence with Congo red under polarized light.

Haemosiderosis of the pituitary glands of patients with haemochromatosis (Figure 9.13) shows preferential deposition in gonadotrophs,[278,324] explaining the most frequent endocrine clinical manifestation of hypogonadism.

METABOLIC DISORDERS

Metabolic Encephalopathies

Metabolic encephalopathies are a diverse group of disorders. Primary metabolic encephalopathies are those resulting from inherited metabolic abnormalities; those resulting from hypoxic-ischaemic states, systemic diseases and toxins are secondary (Table 9.4). The brain has extensive demands for oxygen and energy substrates, requiring approximately 20 per cent of total body oxygen utilization and approximately 75 per cent of hepatic glucose production. An adequate blood supply is clearly essential, and limiting either oxygen or glucose can result in metabolic dysfunction. Furthermore, as aerobic oxidation of glucose is the sole source of energy in the brain under normal conditions, the consequences of hypoglycaemia are rapid and serious.

Primary metabolic disorders (inborn errors of metabolism) are normally rare, and unique abnormalities are reported on occasion.[5] The number of disorders described has increased steeply in recent decades as a result of improvements in the range, accuracy and availability of diagnostic tests and expansion of our knowledge of the molecular basis of brain function. In general, these conditions are characterized by the disruption of a metabolic pathway and the resulting accumulation of precursor substrates and the depletion of immediate or subsequent products. Such disorders can result from mutations or

TABLE 9.4 Causes of metabolic encephalopathy

Primary		Mitochondrial disorders
		Urea cycle disorders
		Amino acid metabolism disorders
		Organic acid metabolism disorders
		Fatty acid metabolism disorders
		Lysosomal disorders
		Peroxisomal disorders
		Porphyria
Secondary	Hypoxia	Anaemia
		Pulmonary disease
		Asphyxia
		Ischaemia
	Cardiac or cardiovascular disease	Hypotension
		Hypertension
		Microvascular disease
		Hypovolaemia
		Hyperviscosity
	Systemic disorders	Hepatic disease
		Renal disease
		Malnutrition
		Electrolyte imbalance
		Endocrine dysfunction
		Infection
		Toxins
		Alcohol
		Medications and other drugs
		Heavy metals
		Organic compounds
		Carbon monoxide

may have a later onset, more gradual course or little or no clinical deficit. The range of clinical signs is also broad, varying from focal deficits to global deficits involving all functional domains.

The clinical diagnosis of a primary metabolic encephalopathy can be difficult.[5] A combination of clinical history (including family history), clinical examination, biochemistry, histology and genetic analysis may be required in order to arrive at a diagnosis. In some situations, for example in storage disorders, electron microscopy of blood cells or skin may reveal morphological features that assist in diagnosis and alleviate the need for brain biopsy.

Secondary or acquired metabolic encephalopathies are encountered more frequently and, in many instances, are life-threatening complications of systemic disease (Table 9.4). Most result in global cerebral impairment, which may manifest as altered conscious level, epilepsy, or psychiatric and/or motor abnormalities.[299] Additional focal abnormalities and cerebral signs can also exist. Clinical signs in many of these encephalopathies may be reversed on correction of the underlying systemic disorder, and fluctuation in their clinical course may reflect variations in organ function. The course of secondary metabolic encephalopathies is also variable, with some presenting as coma or progressing rapidly to coma and others having a more indolent course.

Disorders of Amino Acid Metabolism

These disorders are due to inborn errors of amino acid metabolism, e.g. phenylketonuria, hyperglycinaemia, maple syrup urine disease, homocystinuria, disorders of the urea cycle and others. Almost all are autosomal recessive. Most manifest early in neonates, presenting with variable combinations of symptoms and signs, particularly vomiting, feeding difficulties, irritability, weight loss, fits, abnormal movements, respiratory distress, metabolic acidosis, lethargy and coma. In view of the non-specific nature of many of these signs, the process of investigation should be, first, the exclusion of conditions such as meningitis, intracranial haemorrhage, hypoxic-ischaemic disorders and cerebral malformations. The possibility of a disorder of amino acid metabolism should then be considered, particularly if there is a similar history in the family or parental consanguinity. Screening for metabolic diseases should target glycaemia, gasometry, ionogram, ammonaemia, ketonic acidaemia and the more specific assessment of enzymatic dosages in the urine, blood and other tissues. These diseases provoke great clinical interest, because early diagnosis often allows treatment that prevents future irreparable cerebral damage.

There have been relatively few recent advances in the neuropathology of disorders of amino acid metabolism since the reviews by Martin and Schlote[335] and Crome and Stern.[113] In most cases, the lesions are non-specific, the most common change being spongiosis and cavitation of the white matter and gliosis. Cavitation is relatively rare and its presence usually depends on the duration of the pathologic process. An exception is homocystinuria, in which the lesions are usually ischaemic in appearance. Disorders of urea cycle enzymes, which cause the accumulation of precursors of urea (ammonia and glutamine), induce widespread Alzheimer

other DNA abnormalities, and involve abnormalities of enzymes or other proteins involved in the regulation, activation or transport of metabolites. Disorders may be inherited in an autosomal, X-linked or mitochondrial pattern (see OMIM for current information, www.ncbi.nlm.nih.gov/omim).

The clinical presentation and age of onset of primary metabolic encephalopathies vary widely with the major determinant being the pathway involved. Abnormalities that affect essential pathways usually have an early age of onset and marked clinical manifestations, whereas others type II disease change in astrocytes. Important determinants of the neuropathological changes are the age of the patient (length of survival) and the efficacy of treatments, such as dietary restriction. A number of different mechanisms may play a role in causing the structural and functional deficits seen in these disorders. Some amino acids are neurotransmitters (e.g. glycine) or are required in the formation of neurotransmitters (e.g. tyrosine). Other amino acids, when present in excessive amounts, can interfere with mitochondrial function, and deficits of others may lead to abnormalities of lipid and protein metabolism. Furthermore, the ketoacids and other by-products of metabolic failure in amino acid metabolism (e.g. ammonia) may induce toxicity by interfering with various cellular functions, including neurotransmission.

Phenylketonuria

Phenylketonuria (PKU) is an autosomal recessive disorder, and one form of hyperphenylalaninaemia (defined as a plasma phenylalanine level above 0.12 mM/dL) which is associated with impaired cognitive development if not detected and treated during the first few weeks of life. This inborn error of metabolism results from a deficiency of the phenylalanine hydroxylase (PAH) gene, which synthesizes tyrosine from phenylalanine. Non-PKU hyperphenylalaninaemia is a benign condition as a result of a defect in the hydroxylation of phenylalanine to tyrosine. The different clinical manifestations in the two phenotypes suggest that the degree of phenylalanine excess influences pathogenesis.[560] There is great geographical and ethnic variation in the incidence of PKU, with 5–190 cases per 1 million births.[459] Numerous different mutations have been identified in a variety of populations with many being found at relatively high frequencies.[590]

A classical description of PKU involves a normal birth and an infant that seems to develop normally for several months, before development slows and then stops, when the child drifts into irreversible mental retardation.[92] Newborn screening for PKU began 35–40 years ago in most industrialized countries. Because of this initiative and the early institution of phenylalanine-restricted diets, there are now many young adults with this disease that have normal or near-normal intellectual function. Up to 10 per cent of adults with classic PKU, and possibly 50 per cent of those with milder variants, may not need treatment; after adolescence, intelligence does not appear to deteriorate, at least into early adulthood, even if diet therapy is discontinued or poorly controlled. Blood levels of phenylalanine are poor prognostic indicators, but remain the most important indicator that dietary control is needed; it has been suggested (but not rigorously tested) that brain levels may be more helpful. Despite dietary control, neuropsychological and psychosocial problems develop frequently, needing focused and intensive support by healthcare providers.[125]

MRI studies show white matter changes in a significant proportion of cases.[524] The severity of these changes correlates with blood phenylalanine concentrations, as well as with brain phenylalanine concentrations measured by MR spectroscopy.[353] Dezortova and colleagues used MRI to study the pathological changes observed in periventricular white matter. Known PKU lesions characterized by T2 enhancement in periventricular white matter were observed in all patients.[129] The MR spectra from the lesioned areas showed a significant decrease in choline concentration and support the hypothesis that the T2 increase in a PKU lesion reflects an increase in water mobility, which might be explained by changes in the volume of the extracellular space and possible abnormalities in myelin sheaths.

Neuropathological changes in untreated patients occur in regions that develop postnatally, particularly with respect to myelination of the subcortical white matter and spinal cord and the growth of axons, dendrites and synapses in the cerebral cortex.[32] Because of seizures, hippocampal pyramidal and cerebellar Purkinje cell loss is present.[284] In addition, a small minority of brains show evidence of progressive white matter leukodystrophy, with white matter spongiosis, gliosis and delayed myelination. Biochemical studies reveal diminished levels of cerebral lipids and proteolipids.[334] The pathological changes are thought to be due to toxic effects of phenylalanine and/or its metabolites, and it is assumed that they can be prevented by dietary therapy during infancy and childhood, but direct confirmation by neuropathological studies is lacking.[245] It is important to note that, despite dietary control, patients with PKU usually have moderate hyperphenylalaninaemia throughout life.

The fetus is sensitive to increased maternal levels of phenylalanine; an affected fetus shows facial dysmorphism, microcephaly, intrauterine growth retardation, developmental delay, congenital heart disease, mental retardation and other malformations.[319] In one UK study of the impact of a maternal phenylalanine-restricted diet in women with PKU,[314] of 228 live births, metabolic control for the first 12–16 weeks of gestation had most influence on outcome.[314,319] In a 15-year international study, women with PKU on a phenylalanine-restricted diet before conception showed microcephaly in 27 per cent of children, and 7 per cent with serious congenital heart disease.[280] Improvements in control of phenylalanine levels have reduced these problems.[426]

Hyperglycinaemia

Two distinct types of hyperglycinaemia are recognized, and both can present as a life-threatening illness in the newborn. In ketotic hyperglycinaemia (KHG), there is a primary metabolic block in the catabolism of some organic acids, leading to a severe acidosis and hyperglycinaemia. The second type, non-ketotic hyperglycinaemia (NKHG), also known as glycine encephalopathy, is an autosomal recessive disorder caused by a defect in the glycine cleavage system (GCS), a complex of four proteins encoded by separate genes.[19,236,301,539]

With hyperglycinaemia, large amounts of glycine accumulate in body fluids, including the CSF, and are classically associated with neonatal apnoea, lethargy, hypotonia, poor feeding, seizures and deep coma, followed by severe psychomotor retardation in those who survive. In a series of 65 patients, one-third of the subjects died, 8 during the neonatal period and 14 thereafter. Median age at death for boys was 2.6 years, which compares with less than 1 month for girls. Of 25 patients living more than 3 years, ten were able to walk and say or sign words; all were boys, revealing a striking gender difference in mortality and developmental progress.[236]

In NKHG, neuropathological changes are similar to those described in other amino acidopathies, with spongiosis and gliosis of the white matter. An abnormal corpus callosum and/or dilation of the ventricular system have been associated with especially poor motor and speech development. In a study of the brains of three infants, there was a dramatic reduction in the volume of the white matter of the cerebral hemispheres (Figure 9.14).[492] In a study of a 17-year-old patient, the myelinopathy appeared to be static when compared with neonatal cases, suggesting that the neurological deficits could be due to neurotransmitter abnormalities rather than damage to myelin.[9] Glycine is the major inhibitory neurotransmitter in the spinal cord and brain stem, but it may also play a role as a transmitter in the cerebral cortex, and the high levels that can be measured in CNS tissue in this disease may account for the clinical syndrome.[42,229]

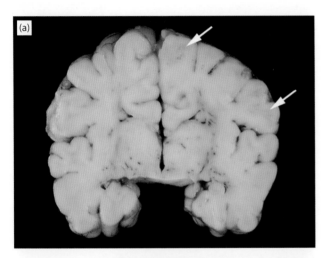

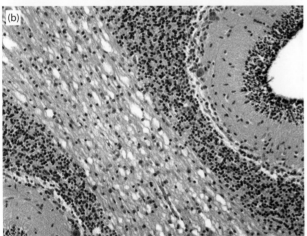

9.14 Non-ketotic hyperglycinaemia (glycine encephalopathy). Brain slice from a male infant who was normal at birth but fed poorly, became jittery and increasingly hypotonic and had recurrent seizures. Metabolic studies revealed non-ketotic hyperglycinaemia. He died at 1 month. **(a)** Macroscopic examination of the brain revealed partial agenesis of the corpus callosum (a common finding in this disease) and slight, ill-defined yellow discolouration of the white matter (arrows). **(b)** Histology showed marked spongiosus as a result of intramyelinic oedema (illustrated here in the cerebellar vermis), and reactive astrocytosis.

Courtesy of Seth Love, University of Bristol, UK.

Electron microscopy shows that the vacuoles in this condition lie within the myelin sheath.[8] The white matter lesions are bilateral and symmetrical and located in the dorsal brain stem, cerebral peduncles and posterior limbs of the internal capsule, the expected sites of abnormality in vacuolating myelinopathy. The lesions are more conspicuous and extensive on diffusion-weighted MR images than on T2-weighted images.[274,480] Using sequential MR imaging with T2-signal intensity and restricted diffusion, white matter changes have been documented.[364] At 3 weeks, the abnormalities were consistent with a vacuolating myelinopathy, and these increased in extent by 3 months. At 17 months, diffusion restriction had disappeared, probably because of the coalescing of myelin vacuoles, and there was evidence of axonal loss. Using MR spectroscopy in NKHG, the existence of glycine disposal pathways producing increases in CSF lactate, and increased creatine in CSF and brain from an early age.[114,555] The cerebral *N*-acetyl aspartate to *myo*-inositol-glycine ratio was identified as a prognostic indicator of the disease.

Three atypical variants of NKHG have been described: neonatal, infantile and late-onset.[131] Atypical variants have heterogeneous clinical presentations, in contrast to the uniform severe neurological symptoms in classical NKHG. In the neonatal variant, presentation is similar to the classical form, but the subsequent outcome is significantly better. Mental retardation and behavioural abnormalities are prevalent in both infantile and late-onset forms, although the phenotype in late-onset atypical NKHG is more heterogeneous. Hyperglycinaemia in atypical neonatal and infantile NKHG is caused by a deficient GCS. The cause of hyperglycinaemia in atypical late-onset NKHG is uncertain, although several mutations of the P-protein *GLDC* gene have been identified, along with the T-protein (*AMT*) and H-protein (*GCSH*). Some cases with GLDC mutation are associated with residual glycine decarboxylase activity, and early therapeutic intervention may be crucial in order to improve the outcome in patients harbouring such mutations. Identification of more mutations causing atypical NKHG and information about the mutations' effects on enzyme activity may help to predict the course of patients with a milder phenotype, as well as those who may respond to early therapeutic intervention.

Transient NKHG is rare.[16] It is characterized by clinical and biochemical findings similar to those seen in classic NKHG. Abnormalities in amino acids partially or completely remit in a period ranging from days to months. Because most patients exhibit normal development, distinguishing the transient form from the classic form is important.

Maple Syrup Urine Disease

Genetic disorders of branched-chain alpha-ketoacid (BCKA) dehydrogenase metabolism produce amino acidopathies and various forms of organic aciduria, with severe clinical consequences. A metabolic block in the oxidative decarboxylation of BCKA caused by mutations in the mitochondrial BCKA dehydrogenase complex (BCKDC) results in increased branched-chain amino acids (BCAA) and their ketoacids (BCKA), causing maple syrup urine disease (MSUD) and acute or chronic encephalopathy.[510]

Classically, this disease presents as a fulminating neurological disorder with vomiting, lethargy, hypertonicity, convulsions and a peculiar odour (maple syrup) of the urine. Rapid deterioration may lead to death, and mental retardation accompanies longer survival. The primary approach to therapy is to modify the diet early in life in order to reduce BCAA and BCKA, thus preventing or minimizing brain dysfunction. There are presently five known clinical phenotypes for MSUD: classic, intermediate, intermittent, thiamine-responsive and dihydrolipoamide dehydrogenase (E3)-deficient. This classification is based on severity of disease, response to thiamine therapy, and the affected gene.[103] Reduced glutamate, glutamine and γ-aminobutyric acid (GABA) concentrations in the cerebrum are considered to be the cause of MSUD encephalopathies. The long-term restriction of BCKA in the diet and orthotopic liver transplantation have been effective in controlling plasma BCKA levels and mitigating some of the neurological manifestations. To date, approximately 100 mutations have been identified in four of the six genes that encode the human BCKDC catalytic machine. Chuang and colleagues have documented a strong correlation between the presence of mutant E2 proteins and the thiamine-responsive MSUD phenotype.[103]

Changes in MRI in 14 juvenile and adult patients with MSUD consisted of an increased signal in white matter on T2-weighted images, which is compatible with a disturbance of water content and dysmyelination.

Areas affected most commonly were the mesencephalon, brain stem, thalamus and globus pallidus; supratentorial lesions were restricted to severe cases. The severity of dysmyelination does not correlate well with acute neurotoxicity but is correlated with median plasma BCAA concentrations over the 6–36 months prior to imaging.[475]

The principal neuropathological findings in untreated cases are spongiosis and gliosis of the white matter, aberrant orientation of neurons, and abnormalities of dendrites and dendritic spines.[263] Secondary effects are hypoxia and hypoglycaemia, which may be associated with classic crises and can complicate the pathological picture.

An acute axonal neuropathy was demonstrated in a woman aged 25 years with MSUD, who developed areflexia, generalized weakness and distal sensory loss over 1 week.[277] Electrodiagnostic studies indicated an acute axonal polyneuropathy, and sural nerve biopsy revealed acute wallerian degeneration without inflammation. Peripheral neuropathy associated with MSUD may become more common as chronic dietary restrictions and improved management of the disease allow survival into adulthood.[362]

MSUD has been identified in Poll-Hereford and Poll-Shorthorn cattle in Australia. The disease is first seen within 2 days of birth. Affected calves are weak and develop limb rigidity and opisthotonus by day 5. Intramyelinic vacuoles, myelin oedema and axonal swellings are reproducible findings (Figure 9.15).[212] Plasma and CSF contain high levels of BCAA.

Homocystinuria

Homocystinuria is an autosomal recessive inborn error of the methionine metabolic pathway. In its principal form, it is due to cystathionine-beta-synthase (CBS) deficiency

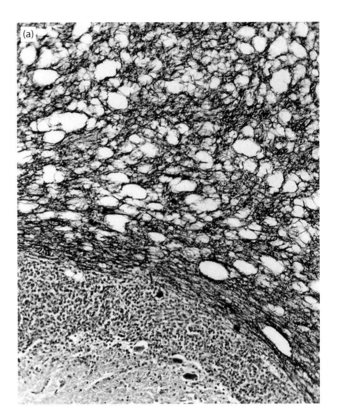

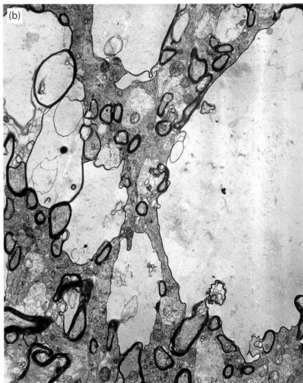

9.15 Section of cerebellum from a Poll–Hereford calf with maple syrup urine disease. (a) There is spongiosis of the white matter. The granule cell and molecular layers appear normal. Phosphotungstic acid-haematoxylin. **(b)** Electron microscopy from specimen in (a), showing that the vacuoles are intramyelinic.

Courtesy of Dr PAW Harper.[213] Copyright © 2008, John Wiley and Sons.

and is characterized by increased plasma homocysteine.[365] Homocystinuria may be more common than previously reported,[448] and screening of newborns for homocystinuria through mutation detection is recommended.

The clinical syndrome is more complex than with other aminoacidurias, and abnormalities can be present in the eye, skeleton, vasculature and CNS.[120] Mental retardation is common in infants who are not treated from birth, but a structural explanation for this is not evident. Although considered a disease of infancy or childhood, some individuals develop symptoms in adulthood.[436]

The predominant neuropathological lesions relate to thromboembolic disease. Involvement of cerebral vessels produces infarcts in the cerebrum, cerebellum, midbrain and thalamus.[365] Thrombi in the dural sinuses have also been reported.[108] Arterial walls often have fibrous intimal thickening, even in children.[38,318,567]

Combined methylmalonic aciduria and homocystinuria is the most common inborn error of cobalamin metabolism due to *MMACHC* gene mutation.[317,462] This complex disorder presents within 12 months of age with severe neurological, haematological and gastrointestinal abnormalities. Clinical manifestations include hypotonia, failure to thrive and poor feeding. Ophthalmological and dermatological changes are common.[116] Diffuse supratentorial white matter oedema and dysmyelination is the typical MR picture at presentation, whereas loss of white matter bulk characterizes later stages of the disease. White matter damage is probably caused by reduced methyl group availability and non-physiological fatty acid toxicity, whereas focal gliosis results from homocysteine-induced toxicity to the endothelium. Hydrocephalus, which can be a feature, may result from diffuse intracranial arterial stiffness, known as reduced arterial pulsation hydrocephalus. The disease is commonly believed to be a disease of infants, but adult-onset forms have been identified.[460]

Other disorders of amino acid metabolism, such as tyrosinaemia,[22,187,465] and disorders of histidine, proline and hydroxyproline metabolism have been reviewed extensively elsewhere.[113,335]

Urea Cycle Disorders

Various disorders cause hyperammonaemia during childhood, including those caused by inherited defects in urea synthesis and related metabolic pathways. These disorders can be grouped into two types: disorders of the enzymes that make up the urea cycle, and disorders of the transporters or metabolites of the amino acids related to the urea cycle.[145] Ornithine transcarbamylase (OTC) deficiency is probably the most common, but there are four other enzymes where mutation can result in elevated levels of blood ammonia as a result of the failure of conversion of ammonia to urea, giving rise, respectively, to arginaemia, arginosuccinic aciduria, citrullinaemia and hyperornithinaemia.

OTC shows an X-linked mode of inheritance, and most males with low residual OTC activity present with severe chronic hyperammonaemia in the neonatal period. Female carriers have a milder form of the disease. The syndrome results from a deficiency of the mitochondrial enzyme OTC, which catalyses the conversion of ornithine and carbamoyl phosphate to citrulline. OTC mutations can be divided into two groups: those with neonatal onset and completely abolished enzyme activity, and those with later onset and partial and variable enzyme deficiency.[181]

The clinical syndrome in urea cycle disorders (UCDs) is related to hyperammonaemia. Except for arginaemia, which can present as a progressive tetraplegia and mental retardation,[472] the UCDs have a similar clinical presentation. In the neonatal period and after an initial period when they appear well, infants develop poor feeding, drowsiness, lethargy, hypothermia, tachypnoea and apnoea. When OTC occurs among males in the neonatal period, it is likely to be lethal, and CNS injury in OTC deficiency may have occurred *in utero*. The diagnosis should be considered if coma with cerebral oedema and respiratory alkalosis occur for no obvious reason. The signs and symptoms of childhood UCDs are often vague but nevertheless recurrent; fulminant presentations associated with acute illness are common. A disorder of urea cycle metabolism should be considered in children who have recurrent symptoms, especially neurological abnormalities associated with periods of decompensation. In late infancy, affected individuals may also present with vomiting and mental changes, and in many cases the precipitant is a protein load, e.g. a change of diet or infection.[500] Additional disease-specific symptoms are related to the particular metabolic defect. These specific clinical manifestations are often due to an excess or lack of specific amino acids.[145] Nassogne and colleagues reviewed the clinical presentation of 217 patients with UCDs, including 121 patients with neonatal onset forms and 96 patients with late-onset forms.[373] The latter may present at any age and carry a 28 per cent mortality rate and a subsequent risk of disabilities.

Routine laboratory tests, including measurement of plasma ammonia concentration, can indicate a potential UCD; ammonia levels above 200 μmol/L are usually caused by inherited metabolic diseases. Specific metabolic testing and, ultimately, enzymatic or genetic confirmation are necessary, however, in order to establish a diagnosis. In the case of prenatal diagnosis, this is possible with a chorionic villus sample or amniotic fluid cells.[181,196]

Neuropathological findings in congenital OTC deficiency are variable, ranging from a microscopically normal brain with Alzheimer type II astrocytic change to severe damage in cerebral cortex and deep grey matter structures. For example, one report describes the cerebral hemispheres of a female aged 6 years as 'little more than bags of leptomeninges', whereas her cousin, aged 8 years and affected by the same disorder, had no macroscopic abnormality of her brain.[62] Epilepsy may well cause some of the more devastating neuropathological damage reported in these disorders. MRI imaging in urea cycle disorders shows changes in white matter connectivity.[188]

In Holstein–Friesian calves affected by citrullinaemia due to argininosuccinate synthetase deficiency, there is spongiosis of the grey matter in the cerebral cortex, with vacuolation of the cytoplasm of astroglial cells (Figure 9.16).[214] The acute clinical encephalopathy is associated with hyperammonaemia and a relative increase in glutamate-mediated excitatory activity.[133]

Treatment consists of a low protein diet and alternative pathway drugs, such as sodium benzoate and

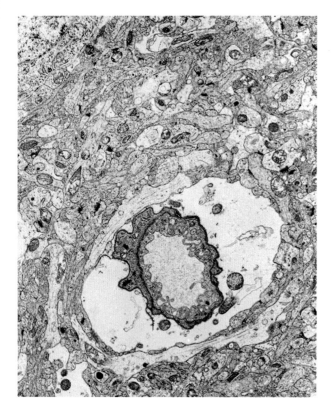

9.16 Section of cerebrum from a Holstein–Friesian calf with citrullinaemia. There is expansion of the cytoplasm of the astroglial processes in the neuropil and the astrocytic end feet surrounding blood vessels, with a reduced density of organelles.

Courtesy of Dr PAW Harper.[214] With permission from Elsevier.

phenylbutyrate.[61] Haemodialysis or haemofiltration should be instituted if ammonia concentrations exceed 500 μmol/L or if they do not fall promptly.[316] Administration of specific amino acids and use of alternative pathways for discarding excess nitrogen have been recommended, but although combinations of these treatments are employed extensively, the prognosis in severe cases remains unsatisfactory.[145]

Propionic Acidaemia

Propionic acidaemia is a disorder of branched-chain amino acid and odd-chain fatty acid metabolism. The clinical features typically begin shortly after birth, with rare cases presenting in young adulthood. Episodic decompensations are characterized by dehydration, lethargy, nausea and vomiting, as well as occasional neurological sequelae. Propionic acidaemia is one of the most frequent organic acidurias, but information is rather limited. Data on 49 patients with propionic acidaemia from Europe[471] identified by selective metabolic screening, showed 86 per cent were of early-onset propionic acidaemia with presentation within the first 90 days of life. The mortality rate was one-third.

The defect in the propionyl-coenzyme A carboxylase enzyme, causes accumulation of toxic organic acid metabolites. Neuropathological findings have not been characterized extensively but include white matter spongiosis in neonates. Widespread grey matter vacuolization may be seen at autopsy. Although a previously unreported finding in propionic acidaemia, diffuse grey matter vacuolization has been described in other fatty acid metabolic disorders.[153]

Porphyrias

The porphyrias are a heterogeneous group of disorders that result from inherited or acquired defects of metabolism characterized by overproduction and excretion of porphyrins or their precursors, with dysregulation of one of the eight enzymes in the haem biosynthetic pathway.[83,160] Each of the porphyrias is characterized by a unique pattern of overproduction, accumulation and excretion of haem biosynthesis intermediates.[382] They are classified as erythropoietic or hepatic, depending on the primary organ in which excess production takes place. Only the hepatic porphyries, including acute intermittent porphyria and porphyria cutanea tarda (PCT), produce neurological disease.

The porphyrias rarely manifest before puberty. Factors that may precipitate acute disease include certain drugs (e.g. barbiturates, sulphonamides, griseofulvin, meprobamate, phenytoin, succinimides, steroids), infections, starvation and menstruation (some women experience attacks just before menstruation). A comprehensive review of the history, pathophysiology, classification and treatment of porphyria should be consulted for further information.[385]

Acute intermittent porphyria (AIP) due to mutation in hydroxymethylbilane synthase (HMBS) is an inherited metabolic disease with an autosomal dominant pattern of inheritance with reduced penetrance where only 10–20 per cent of mutation carriers will develop symptoms.[416] Because the biochemical parameters of patients and their non-affected relatives overlap, the diagnosis may remain undetermined during the symptom-free phase, however, mutation detection in AIP provides high clinical benefit. The neurological manifestations of AIP commonly present as a peripheral autonomic and motor neuropathy.[35,345] Neuropsychiatric symptoms such as anxiety, depression, insomnia, disorientation, hallucinations and paranoia, as well as seizures and cranial nerve neuropathies, are also reported.[345] During an acute attack, which includes various neurovisceral symptoms, measurement of urinary porphobilinogen is the method of choice to confirm diagnosis. Rare cases with compound heterozygous or homozygous mutation of HMBS have been described with early severe motor and neurological manifestations, sensory nerve conduction deficits due to denervation, with delayed myelination seen on MRI and in one case porencephaly.[43,232,503]

Neuropathological studies of the porphyrias have revealed a range of inconsistent findings. Hierons reviewed the neuropathological literature on acute porphyria and presented five cases.[233,234] Lesions are either minimal or absent, except for those that can be explained on the basis of associated hypoxia during severe attacks. Suarez and colleagues presented the morphological findings of 35 patients with AIP and reported diffuse neuronal loss, occipital cerebral ischaemia, perivascular pigmentation, diffuse perivascular demyelination, cytolysis of Betz cells or no obvious abnormality.[511] Nuclear chromatolysis of cranial nerves has been reported, as has Purkinje cell loss. Two reports have documented neuronal loss in the supraoptic and paraventricular nuclei of the hypothalamus, which was proposed as a mechanism for the hyponatraemia observed in some porphyric patients. Central pontine myelinolysis has also been observed.[513]

MRI studies have shown multiple cerebral lesions during attacks of acute porphyria.[10,276,300] The lesions, which may be ischaemic, are multifocal and resolve after several weeks.[75] One hypothesis is that because nitric oxide synthase is a haem protein and nitric oxide is a vasodilator, unopposed cerebral vasoconstriction due to decreased nitric oxide from severe haem deficiency may occur during acute attacks.[300]

The peripheral autonomic, motor and sensory disturbances are due to an axonal neuropathy. Distal degeneration of posterior column fibres has been reported in some cases. Chromatolysis of neurons in peripheral ganglia and the spinal cord, particularly among the anterior horn cells, is reported frequently. Degeneration of various spinal cord tracts and dorsal root ganglia has also been reported.[234,478,578] The peripheral nerves themselves may show segmental demyelination and axonal swelling and fragmentation.[577]

NEUROTOXICOLOGY

In previous editions, the late David Ray provided a succinct summary of the subject: 'Neurotoxicology is the science dealing with adverse effects produced in the nervous system by synthetic and naturally occurring chemicals. It is not an exact science, in the sense that the adverse effects may be subtle, protracted and multifactorial in origin. The biological targets utilized by the toxins may result in no obvious structural pathology, despite causing significant behavioural change. In many cases, toxicity may be caused not by direct action on a neuron or its neurites, but on its metabolic state, the localized circulation, or the availability of glucose or oxygen.'

The general population is exposed to a very wide range of doses of potentially toxic agents, and little is known about the effects of long-term low-level exposure to neurotoxic agents or of the potential for reversibility of neurotoxicity, particularly in the ageing nervous system. Currently, there is little direct evidence that acute exposure to environmental toxins will lead to a decline in cognitive function or the development of psychiatric behaviour except in relatively specific cases (e.g. following exposure of the developing brain to lead or the memory loss that may follow severe domoic acid poisoning). It is however, difficult to be confident that prolonged exposure to low levels of neurotoxic agents is without subtle behavioural changes in a vulnerable subset of the population. Such possibilities present significant problems for the processes of diagnosis and prognosis, and also mean that neuropathological data are sparse.

For the majority of neurotoxins and neurotoxic exposures, descriptions of neuropathological changes are likely to be confined to animal models because the nature of human poisoning is rare and cases often do not come to autopsy or to specialist centres where appropriate neuropathological investigation can be undertaken. The lack of access to the brain, spinal cord and the peripheral nervous systems ensures that neuropathology tends only to describe and analyse tissues obtained at autopsy. The data, therefore, reflect end-stage disease and are often difficult to place in the context of aetiology and development. Although neuropathy may be a common finding, nerve conduction studies are the only likely investigation given the reduced frequency of (sural) nerve biopsy. Many of the studies on neurotoxic exposures are now almost of historical nature because we are more aware of the potential for specific chemicals to cause toxicity and therefore controls on their use have led to reduced exposure and consequently decreased incidence of clinical cases in developed countries. There are specific problems associated with certain neurotoxins in developing countries, where exposure is less regulated, or unique geography brings localized neurological conditions. That said, we are now entering an era where the effects of low level and chronic exposure to specific toxins are likely to become more apparent, and such exposures, although showing similar effects to acute exposures, may show subtle differences in clinical symptoms and neuropathological changes. Clinical indications are therefore likely to help to pinpoint the nature of the chemical exposure, and imaging-based techniques such as MRI, positron emission tomography (PET), single photon emission computed tomography (SPECT) will provide clues to the systems involved and thereby the possible toxic exposures. It cannot be overemphasized that very detailed occupational and specific lifestyle history becomes crucial for those investigations where industrial exposure (e.g. lead or manganese) or geography (e.g. arsenic) is suspected, and is key in identifying potential neurotoxic chemicals.

Neuropathological studies in animals need to be treated with some degree of caution because direct comparison can be confounded with differences in physiology and neural responses. Perhaps the exceptions here are neuropathological investigation in primates where the physiology is sufficiently similar although, as with human studies, these are understandably rare. Even when animals are used, note should be taken of the route of administration, its dose and formulation as well as the general suitability of the species itself.[141] These concerns are frequently ignored in academic studies but are rather more rigidly adhered to in the protocols laid down by US and OECD guidelines for the study of potentially neurotoxic agents. It is of interest that the guidelines identify the need to 'assess the effects of any substance on learning memory and performance'. As non-invasive structural and functional imaging becomes more available and more sensitive, it may become possible to directly visualize changes appearing as the result of exposure to neurotoxic agents, to make longitudinal studies of both patients and experimental animals, and therefore relate the neurotoxic assault to clinical tests of function and behaviour. To date, however, the application of imaging techniques to toxicology remains minimal. Notwithstanding these caveats, considerable detail on the potential mechanisms of neurotoxins and their pathological changes has been produced through small animal studies.

Neurotoxins Affecting Mitochondrial Energy Production

The high metabolic demands of the nervous system means that disturbances of energy supply either by altered oxygenation or by inhibition of cellular energy production can cause rapid changes in neural activity and over prolonged periods considerable neuropathological changes. Chemicals that can affect energy production can provide relatively selective damage to the nervous system as with certain mitochondrial toxins, or neuropathological changes in combination with peripheral effects for example with cyanide poisoning. In general, there is a relative selectivity for basal ganglia structures with necrosis of the globus pallidus and putamen being common, but cortical structures, and in particular the hippocampal formation can be equally affected. Given the high energy demands of the nervous system, high level exposures can often have rapid effects, though equally chronic lower level exposures can still cause neuropathological changes.

Carbon Monoxide

One of the most commonly encountered neurotoxins is carbon monoxide (CO) and in the UK alone there are between 200 to 500 deaths per year associated with CO intoxication, and perhaps 4000 or more associated with non-fatal exposure leading to hospitalization.[200] Given the numbers of individuals who may be inadvertently exposed to CO because of faulty heating systems or the use of CO as a means of self-harm, CO intoxication is often encountered in emergency situations and the consequences are well described. Individuals are often comatose or very drowsy following exposure, and breathing is shallow with tachycardia, and the individual may have seizures, ataxia, often with signs of cyanosis. Cardiac injury is common in moderate to severe intoxication (carboxyhaemoglobin levels in excess of 20 per cent) and is a considerable cause of mortality.[228] Prior symptoms, particularly in cases of accidental exposure, are persistent headache, nausea, abdominal pain, fatigue, confusion and cognitive impairment, and exacerbation of other medical conditions. These symptoms are however not specific and therefore the possibility of misdiagnosis exists, particularly given the problems of measuring CO levels in blood. Chronic low level exposure to CO not resulting in acute hospitalization may have a risk of long-term effects on the nervous system though this is lacking in any critical detail. Certainly, up to half of individuals who survive acutely debilitating CO exposure will have delayed or continued neurological or cognitive problems and the use of more fuel-efficient engines in cars means that it is more difficult to cause lethal CO poisoning in cases of self-harm resulting in a change in the spectrum of illness associated with CO exposure.[237,407] Hyperbaric oxygen therapy is suggested as the method of choice for recovery with improved outcome, although this is often unavailable and there may be disadvantages.[237]

The principal mode of action of CO is adduction to haemoglobin to form carboxyhaemoglobin (HbCO), which has approximately 200 times greater affinity for haemoglobin than oxygen. Consequently, there is decreased oxygen capacity of the blood and gradual anoxia. CO and HbCO

levels of over 20 per cent in blood are normally seen in those individuals with acute intoxication with levels in normal individuals of below 1 per cent, although over 10 per cent blood HbCO can be seen in smokers.[222] Blood levels of HbCO can rapidly drop following removal from exposure and within a few hours can return to normal making clinical investigation difficult where patients are not suspected as being exposed, because routine pulse oximetry is not able to detect HbCO.[228] Other haem-containing proteins are also adducted by CO, and respiratory chain enzymes in mitochondria are a prime site of inhibition by CO with inhibition of cytochrome c oxidase (Complex IV) persisting long after CO levels return to normal.[349] The neurological and neuropathological effects of CO are therefore due to a combination of acute hypoxia and more prolonged impairment of neuronal energy production. Similar biochemical changes also occur with cyanide exposure and the neuropathological changes also show parallels. CO also acts as an intracellular and local signalling molecule, being produced by cells in a manner analogous to nitric oxide and hydrogen sulphide and therefore acute local effects on cell signalling may be an additional factor in the mechanism of action of CO. One such finding is perhaps associated with the common frontal headache seen in CO intoxication, causing altered blood flow both cortically and in the basal ganglia, and which may be a contributory factor in the prominent necrosis seen in the globus pallidus.[363] The greater affinity of CO for fetal haemoglobin means that its actions on the fetus are greater than on the mother, and exposure to CO has been reported in children born to exposed mothers and also cigarette-smoking mothers.[285]

The outcome of severe CO exposure with recovery has most often been investigated by means of imaging using MRI, CT or SPECT/PET imaging. Typical changes include bilateral necrosis of the globus pallidus (Figure 9.17), although these changes are not universally present and may depend on exposure levels.[93] Haemorrhagic infarction of the globus pallidus can be seen on MRI,[44] along with ischaemic change in the putamen and medial parts of the thalamus,[474] as well as infarction of white matter and atrophy of the corpus callosum.[428] One study showed that just over half of cases showed globus pallidus necrosis and a quarter showed putaminal involvement.[389] Reduced basal ganglia volume appears to be associated with reduced verbal memory and is present in up to 88 per cent of some studies,[237] suggesting that the globus pallidus lesion is not necessarily a typical feature in all cases. Pallidoreticular damage and thalamic changes are seen in many severely affected cases.[93,167,272,537] This extended damage from the globus pallidus can be observed using susceptibility weighted imaging and is associated with more severe outcome including poorer performance on a wide range of neuropsychiatric tests, and parkinsonian features with concomitant reduction in dopaminergic imaging markers.[93] Extrapyramidal syndromes of varying presentation involving dystonia, parkinsonism, and chorea[100,101,402] are consequently common following high exposures to CO and show associated pathology. Delayed neurological impairment following recovery of consciousness after CO is frequent, involving cognitive impairment.[240,402]

With very severe cases of CO poisoning with recovery, there may be considerable cortical atrophy and prolonged

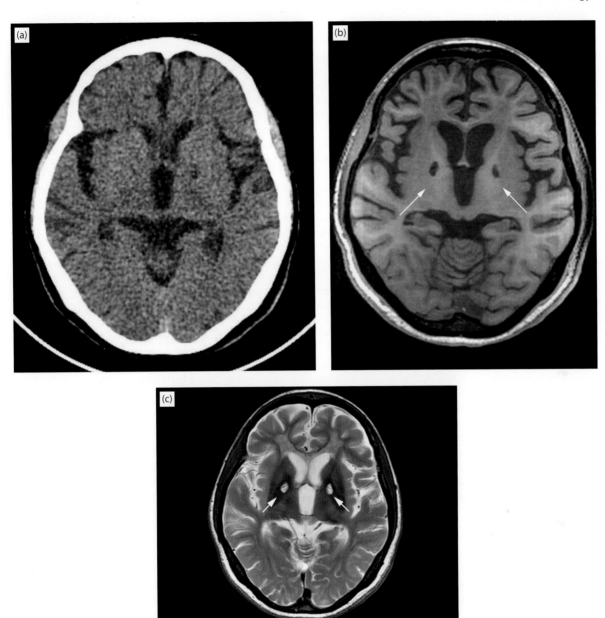

9.17 Pallidal lesions in a case of carbon monoxide intoxication. After recovery from acute CO exposure, only minimal abnormality is seen within the globus pallidus on CT imaging **(a)**, but there is a well-demarcated area of reduced signal (arrows) in the globus pallidus on T1-weighted imaging **(b)** corresponding with a region of necrosis and heterogeneous hyperintense signal on T2-weighted imaging (arrows in **c**).

Images courtesy of Prof Jiun-Jie Wang and Dr Chiung-Chih Chang. Adapted from Chang et al.[93] *By permission of Oxford University Press on behalf of The Guarantors of Brain.*

cognitive and neuropsychiatric impairment.[93,449] Similarly, decreased blood flow to frontal lobes can be observed using SPECT in CO exposed individuals even in the absence of major neuropsychiatric impairment[563] and appears to correlate with decreased neuropsychiatric and neurological performance.[102,481,540] Early study suggested up to 90 per cent of individuals will show signs of cognitive impairment in a variety of domains including attention and processing speed along with changes in mood, with up to 70 per cent showing signs of hippocampal atrophy.[227] Although MRI and also magnetic resonance spectroscopy (MRS) may indicate evidence of damage in key areas such as the basal ganglia (Figure 9.17),

pallidoreticular system and white matter, in some instances there is neurological and neuropsychiatric impairment in the absence of changes on imaging.[433] These imaging data indicate that pathological changes in cortical structures are relatively common following CO exposure and may reflect metabolic impairment due to CO in cortical neurons.

Damage to white matter and fibre tracts are common following CO intoxication and using prospective MRI data shows that white matter changes are present as a major finding following CO poisoning, even in the absence of any major neurological complications.[23,139] Bilateral hyperintense lesions in the white matter including centrum semiovale indicative of demyelination are frequently seen[275] although in one study only 12 per cent of CO-exposed individuals showed white matter hyperintensities with no differences in centrum semiovale hyperintensities.[404] This white matter damage may, however, be seen on FLAIR images even in the absence of changes on conventional T1 and T2 MRI images suggesting that simple structural MRI may not accurately reflect existing white matter change.[368] White matter damage has been followed in CO poisoning where increased choline and decreased N-acetylaspartate were associated with areas of demyelination.[369,469] The suggestion has been made that this white matter damage is an immune-mediated change involving production of myelin basic protein adducts similar to that seen in experimental autoimmune encephalomyelitis,[523] although in the absence of clinical support for this, this needs to be treated with caution. Early experimental studies in the cat suggest that rather than true primary demyelination, there is a central wallerian degeneration with die-back of axons.[238] Elevated S100B or myelin basic protein in the CSF or serum may be useful predictors of the extent of white matter damage in the CNS,[63] although the utility of serum measurements of these proteins has been questioned.[443]

Most often, individuals with CO exposure coming to autopsy will be acute exposure cases due to suicide with no or very limited survival period after exposure. Acute fatal exposure to CO shows chromatolysis of neurons in the hippocampus and in cerebellar Purkinje cells indicative of anoxic change. The seminal work of Lapresle and Fardeau describing neuropathological changes following CO exposure and subsequent periods of survival before death describes several groups of neuropathological changes.[311] Principal amongst these are the white matter changes involving the centrum semiovale, corpus callosum and within the cerebral peduncles where often confluent foci of demyelination and/or necrosis can be seen.[311] Four groups of cases showing white matter change were described with Group 1 cases of short post-exposure survival (4–5 days) showing small multifocal areas of necrosis in the white matter and corpus callosum, often associated with changes in the hippocampus, cerebellum, putamen and thalamus. Group 2 cases often of longer survival duration showed marked necrosis within the centrum semiovale. Group 3 cases of intermediate duration of 2–3 weeks, frequently showed recovery then relapse associated with confluent areas of spongy demyelination with sparing of axons and perivascular myelin ('Grinker's myelinopathy') and associated necrosis. A final Group 4 shows cases with limited changes often in the hippocampus and cortex of focal necrosis and spongy change perhaps indicating a more anoxic change. In all these groupings, however, there is a continuum of pathology with variable degrees of change. The white matter changes and axonal damage in CO intoxication show the presence of axonal bulbs and torpedoes, which can be seen using amyloid precursor protein immunostaining in the areas of demyelination.[136] Astrocytic hypertrophy and microglial activation typically accompanies these areas of necrosis and demyelination. Also within these areas, necrotic changes are observed involving lipid and haemosiderin-laden macrophages and in the globus pallidus and hippocampus calcium deposits within the neuropil surrounding vessels. Vascular changes involving the presence of enlarged

perivascular spaces, regions of capillary and arterial necrosis, endothelial loss and swelling, and accompanying red cell extravasation are common, particularly in areas of necrosis within the anterior globus pallidus. Experimental reproduction of some of the features of CO poisoning suggests that decreased blood flow to the globus pallidus may be a cause of the necrosis possibly due to CO action on the vasculature[504] and that the 'Grinker's myelinopathy' may be due to a combination of initial hypoxia due to CO followed by secondary hypotension at a later stage corresponding with the diphasic clinical presentation seen in the patients of Group 3 of Lapresle and Fardeau.[311,394] Corresponding with the areas of necrosis such as in the globus pallidus, putamen, substantia nigra and thalamus, there is associated neuronal loss and this can be seen in the hippocampus, particularly in CA1 and CA3, and also in patchy areas within the cerebral cortex and loss of cerebellar Purkinje cells.[4,136,311] Neuropathologic investigation shows oedema in the basal ganglia and hippocampus along with the presence of both apoptotic and necrotic neurones.[543]

N-Methyl-4-Phenyl-1,2,3,6-Tetrahydropyridine

Considerable attention has been given to the possibility that environmental exposure to a neurotoxin is associated with the development of Parkinson's disease. This hypothesis stems from the finding that human exposure to 1-methyl-4-phenyl-1,2,3,6-tetrahydropyridine (MPTP) can cause acute-onset parkinsonism[119,309] and from epidemiological studies of Parkinson's disease that indicate that disease risk may be increased by increased exposure to pesticides in particular. MPTP was initially identified as a contaminant in preparations of methyl-4-phenyl-4-propionoxypiperidine (desmethylprodine) a synthetic opioid used illegally, following identification of clinically acute onset parkinsonism in the synthetic chemist who had prepared desmethylprodine.[119] At autopsy 18 months later, the individual showed typical nigral degeneration with Lewy bodies, although the association with MPTP was not made at the time. Following development of acute onset L-dopa responsive parkinsonism in several drug users[309,572] and also in an industrial chemist exposed to MPTP,[308] MPTP was identified as the specific chemical, with reproduction of clinical symptoms in experimental animals with an estimated total dose of between 160 and 640 mg of MPTP over several weeks sufficient to cause parkinsonism in man.[309] Although humans and primates are sensitive to the effects of MPTP,[65,257,289,309] rodents are generally resistant with rats only showing reversible nigral pathology at near lethal systemic doses, and mice showing slightly higher sensitivity, though this is strain dependent.[199]

The selectivity of MPTP in causing parkinsonism is achieved by MPTP being able to cross the blood–brain barrier as a result of a relatively neutral charge, whereas in astrocytes it is converted by monoamine oxidase-B via the intermediate compound MPDP[86] to the active metabolite MPP+.[97,333] The generally accepted view is that MPP+ is selectively taken up by the dopamine transporter into dopaminergic neurones[256,333] and actively concentrated in mitochondria[441] where it acts as an inhibitor of the mitochondrial respiratory chain, selectively inhibiting complex I and causing depletion of cellular energy supply and cell death.[378,442] Although the pathological changes seen in MPTP-induced parkinsonism (nigral degeneration and Lewy bodies) are identical to typical Parkinson's disease,[119,308,309] it is unlikely that any further cases will ever be apparent to routine pathology. The active metabolite, MPP+, is, however, the chemical cyperquat, which is used as a herbicide along with similar compounds such as paraquat (1,1′-dimethyl-4,4′-bipyridinium), which are thought to act like MPTP by causing inhibition of complex I of the mitochondrial respiratory chain. In rodents, although paraquat shows nigral degeneration,[57,332,340,522] there are differences in mechanisms of action,[57,451] which suggest that paraquat produces its effects by redox cycling with indirect mitochondrial damage.[521] In man, paraquat toxicity, at least acutely, is not associated with direct neurological complications, although most reports show patchy intracerebral necrosis particularly in the white matter in paraquat fatalities.[184,242,366] This distribution of pathology may relate to the abnormal oxygen consumption caused by paraquat poisoning,[580] the severe lung damage and subsequent hypoxia, myocardial depression and depletion of cellular NADP. Of note, paraquat poisoning is difficult to treat when certain levels of paraquat have been ingested, and patients may survive for several days before death with supportive care. The neuropathology of poisoning may change with time. Similarly, diquat (1,1′-ethylene-2,2′-bipyridyldiylium) causes intracerebral haemorrhage, often of the brain stem[260,463,549] but also within the globus pallidus, similar to CO poisoning, with recoverable hemiparesis,[468] but may also cause persistent parkinsonism.[479] The long-term neurological consequences of paraquat or diquat following recovery from exposure are unknown.

Animal studies have provided considerable support for the concept of mitochondrial toxins and development of Parkinson's disease. In primates, MPTP exposure either acutely or chronically causes a relatively selective degeneration of dopaminergic neurones in the substantia nigra with accompanying gliosis,[65] but also of the paranigral (A10) group.[564] This cell loss also shows accompanying necrosis, possibly due to the rapid onset and appears either acutely[518,564] or chronically.[203,551] Mitochondrial inclusions are seen in macaques treated with MPTP[518] and in baboons this appears to involve the deposition of α-synuclein in the remaining neurons, although these do not have the appearance of typical Lewy bodies (Figure 9.18).[289] Depletion of neurons in other brain areas is also present with loss of dopaminergic neurons and accompanying gliosis in the hypothalamus of marmosets,[174] and in some instances locus caeruleus neurons are lost[518] and serotonergic neuron changes as evidenced by decreased serotonin markers in the striatum.[411]

The pesticide rotenone ((2R,6aS,12aS)-1,2,6,6a,12,12a-hexahydro-2-isopropenyl-8,9-dimethoxychromeno[3,4-b]furo[2,3-h]chromen-6-one) has received considerable attention recently as a toxin based model for Parkinson's disease. Rotenone is normally used as a general purpose non-specific insecticide and in some cases for the elimination of fish and is found naturally occurring in several different genera of leguminous (pea) plants such as *Derris eliptica*. As a highly specific and sensitive mitochondrial complex I inhibitor, rotenone is also used in biochemical assays of mitochondrial function[186] and chronic administration of rotenone has been used to model aspects of Parkinson's disease.[41,82] Pathology is reproduced in rats only near to the maximal

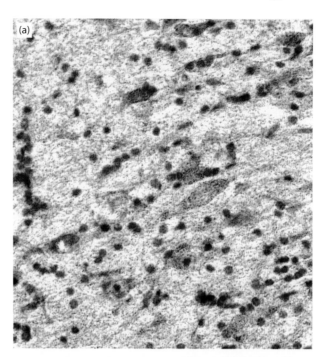

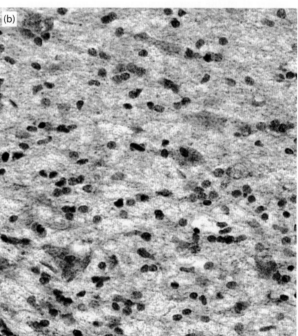

9.18 Sections through the substantia nigra of the common marmoset (*Callithrix jacchus*) after treatment with the neurotoxin MPTP. (a,b) The sections were labelled with anti-α-synuclein antibody and show diffuse and finely granular α-synuclein in the cytoplasm of remaining neurons along with proliferation of glial nuclei.

Images kindly supplied by Dr Sarah Salvage, Dr Atsuko Hikima and Professor Peter Jenner of Neurodegenerative Diseases Research Group, Institute of Pharmaceutical Science, School of Biomedical and Health Sciences, Guy's Campus, King's College, London UK.

tolerable dose and many animals die acutely showing cerebral perivascular haemorrhage and oedema in the CNS and similar changes in peripheral organs, comparable to other agents causing anoxia.[17] For some of the rats that survive older dosing regimens, there is selective degeneration of nigral neurons and the presence of amorphous α-synuclein positive inclusions[41,487] accompanied by early microglial activation.[486] A similar pathology can also be achieved by infusion of the rotenoid deguelin into rats.[76] Acute ingestion of rotenone and fatality in man is rare and does not appear to cause nigral degeneration, although multiple small haemorrhagic infarcts were described in one case[372] and in another.[406] In this respect, rotenone, at least acutely, shows some similarities to carbon monoxide and cyanide toxicity in causing acute anoxic change. In one case of poisoning in a child who died 8 hours after intake of a mixture containing 6 per cent rotenone, the brain showed signs of an 'acute hypoxic episode', but no specific details were given.[128]

3-Nitropropionic Acid

3-Nitropropionic acid is a fungal metabolite of the genus *Arthrinium* and was found to be responsible for human poisonings through consumption of contaminated sugarcane. 3-Nitropropionate is a structural analogue of succinic acid and irreversibly inhibits succinate dehydrogenase, complex II of the mitochondrial respiratory chain. Local injection of malonate, a reversible inhibitor of succinic dehydrogenase, produces a very similar lesion.[34] Cases, often children, develop convulsions followed by coma and death.[224,348] Dystonia often occurs following recovery from acute symptoms and CT scans show low-density lesions predominantly in the putamen.[224,348] No neuropathological findings on human cases have been reported in western literature. Pathology in animals has been studied owing to the similarity with Huntington's disease pathology, with mice,[182] rats[198] and primates showing lesions in a number of brain areas with a clear sequence of relative vulnerability (see Table 9.5). There is selective loss of medium-sized neurons in the putamen,[379] with vascular breakdown preceding neuronal loss, lesser involvement of the globus pallidus, and only occasional involvement of the caudate or claustrum,[164,328] although in some human cases a more diffuse lesion occurs.[224] Repeated lower-dose administration in rats[574] and primates leads to selective neuronal death, confined to GABAergic/substance P-expressing medium-sized spiny neurons in the rostrolateral caudate/putamen.[399,455] Lesions caused by 3-nitropropionic acid are characterized by mitochondrial swelling in neurons, first in processes and then in cell bodies. Axonal swelling and myelin splitting occur mainly in the internal capsule, resembling that seen after hypoxic/ischaemic damage. Astrocytic mitochondria were involved less prominently than those in neurons, with oligodendrocytes minimally affected. The vasculature was involved at all stages of lesion development, with a decrease in endothelial cell cytoplasmic density, thrombin accumulation in capillaries, aggregates of perivascular protein and petechial haemorrhages.[450] This vascular pathology described earlier[164,379,450] has not been noted by some,[33] possibly because they are examining end-stage lesions where there is vascular repair. There may be a useful analogy with the primarily vascular and astrocytic lesions produced by m-dinitrobenzene[456] or the primary astrocytic lesions produced by α-chlorohydrin,[89] which produce a secondary neuronal loss that becomes difficult to distinguish

TABLE 9.5 Topography of damage produced by metabolically acting neurotoxic agents or hypoglycaemia in the rat brain

Inferior colliculi[a,b,c]
Auditory cortex
Anterior cingulate cortex
Superior olives[a,b,c]
Sensorimotor cortex[f]
Frontal cortex[f]
Medial geniculate nuclei[a]
Parietal cortex
Lateral habenular nuclei
Cochlear nuclei[a,b,c]
Visual cortex
Caudate nuclei[b,d,e,f]
Vestibular nuclei[a,b,c,d]
Thalamic nuclei[a,d]
Superior colliculi
Cerebellar nuclei[b,c,d]
Hippocampus[b,d,f]
Red nuclei[a,b]
Inferior olives[a]
Septal nuclei
Pontine nuclei[a]
Globus pallidus[d]
Cerebellar cortex[f]
Substantia nigra[d,e]
Hypothalamus
Amygdala
Corpus callosum
Cerebellar white matter

Brain regions are listed in decreasing order of glucose phosphorylation rate in the normal conscious rat.[375]

Neurotoxic agents: [a]3-chloropropanediol; [b]6-aminonicotinamide; [c]1,3-dinitrobenzene; [d]3-nitropropionate; [e]rotenone; [f]hypoglycaemia.

from primary neuronal damage at survivals of greater than 2 days.

Cyanide Toxicity

Block of oxidative energy metabolism by hydrogen cyanide or cyanide salts can produce neuropathology by inhibition of mitochondrial cytochrome oxidase (Complex IV). Cyanide toxicity outside of self-harm is seen most commonly as a result of inhaling hydrogen cyanide from burning plastics,[12] although cyanide salts are used in electroplating, as nitroprusside in the treatment of hypertension, and cyanogen chloride has been used as a chemical warfare agent. In forensic cases, cyanide toxicity can be seen in victims of house fires where carbon monoxide will be an additional neurotoxicant.[12] The effects of cyanide poisoning closely resemble those of hypoxia when not acutely fatal and show similarities to carbon monoxide poisoning with parkinsonism or dystonia being a common finding.[151,183,410,458,546] MRI shows symmetrical striatal, cerebral and cerebellar lesions after cyanide poisoning, often with delayed onset,[437,461] with reduced dopaminergic markers in the striatum indicative of nigral pathology.[458,586] One report of pathology, however, showed no dopamine neuron loss in the substantia nigra despite parkinsonism, which may indicate that reduced striatal signal is due to striatal rather than nigral pathology.[544] Histopathology of cyanide survivors shows cerebral, striatal and cerebellar neuronal loss with one report showing pseudolaminar necrosis.[453] In individuals dying after house fires where cyanide or carbon monoxide may be contributors, the presence of intranuclear ubiquitinated Marinesco-like bodies and diffuse ubiquitinated staining in substantia nigra dopaminergic neurons has been reported.[435] Hydrogen sulphide also acts on cytochrome oxidase in a similar manner to cyanide, and acute poisoning with hydrogen sulphide (commonly encountered in the oil-refining industry) produces similar effects.

Methyl Bromide

Methyl bromide poisoning usually results from its use as a fumigant and antifungal agent, and despite aims to reduce its use in developed countries its use is nonetheless widespread. Methyl bromide and methyl chloride intoxication show acute headache, vomiting and nausea progressing to convulsions and coma.[244,488] The production of petechial haemorrhages, oedema and neuronal loss in the cerebral and cerebellar cortices, dentate and brain stem nuclei[220] suggests a metabolic basis for toxicity.[87,170] In some instances, mammillary body and inferior colliculus pathology is seen with a Wernicke's-like distribution,[505] although the mechanism is not understood. A distal sensorimotor polyneuropathy is also seen. Ataxia is noted at least acutely and MRI in several instances has shown signal changes in the splenium of the corpus callosum[28,266] and in the dentate nucleus.[515]

The bioactivation of methanol to the cytochrome oxidase inhibitor formate suggests that methanol toxicity may be due to impairment of energy metabolism.[377] Formate is normally present in the body, but elevated levels are associated with ocular lesions in primates[497,520] and folate deficiency enhances its neurotoxicity. Although methanol initially produces an acute intoxication similar to that of ethanol, this is succeeded by nausea, vomiting, headache, loss of visual fields, and extrapyramidal signs (rigidity, akinesia, dystonia). The extrapyramidal signs are associated with selective necrosis of the putamen, similar though not identical to the effects of hypoxia.[320,342,539] Methanol produces retinal oedema and axonal swelling in the anterior segment of the optic nerve along with considerable gliosis and loss of vascular markers.[484,539]

Organophosphorus Compounds

Some of the more widely used insecticides belong to the organophosphorus (OP) group, which are widely used in agriculture and sometimes in home use, but also as chemical warfare agents (e.g. sarin, soman). Given their widespread use, OP compounds are readily available and often used in cases of self-harm.[140] Although acute effects are well recognized and due to inhibition of acetylcholinesterase, long-term, low-grade use of some OP compounds results in organophosphate-induced delayed polyneuropathy (OPIDPN), characterized by a distal distribution and delayed onset,[326] though some evidence suggests low-dose exposure may have peripheral effects in the absence of overt neuropathy[506] and potentially neuropsychiatric problems.[29,40] A dying-back axonopathy affects sensory, motor and autonomic neurons, although in mild cases degeneration may be confined to the extreme distal (intramuscular) processes of motor fibres. Other than an accumulation of smooth endoplasmic reticulum, there are negligible morphological changes before the onset of axonal degeneration and functional loss 10–14 days after acute exposure.[52] Some large-diameter fibres are particularly sensitive, including the proprioceptive afferents from muscle spindles. Within the CNS, long tracts in the spinal cord, medulla and cerebellum are involved. Young people and animals are relatively resistant to OPIDPN. In a series of 12 high-dose intoxications, clinical improvement was observed in (peripheral) lower motor neuron function over 1–2 years, but not centrally.[552]

Mechanistic studies have shown that the nature of the neurotoxicity produced by an organophosphorus ester can be predicted by its relative ability to interact with either acetylcholinesterase (acute cholinergic toxicity) or neuropathy target esterase (OPIDPN).[358] It is possible to classify agents in terms of producing neuropathy at doses that are not acutely toxic (tri-o-cresyl phosphate), that are acutely toxic (mipafox) but survivable, that are survivable only with acute therapy (trichlorophon, dichlorvos), or that have little or no neuropathic potential at any dose (sarin, paraoxon). The first two classes are banned from use as pesticides, but some agents from the third class have produced neuropathy in people who have attempted suicide and have required intensive care therapy. No current pesticides produce OPIDPN at dose levels that fail to produce severe acute cholinergic signs, but there have been several hundred predominantly historical cases of OPIDPN due to the lubricant tri-o-cresyl phosphate.[562] The mechanism of OPIDPN is at present uncertain, because the physiological role of neuropathy target esterase (NTE) is unclear. Mutation in this gene (patatin-like phospholipase domain containing protein 6, PNPLA6) causes an autosomal recessive disorder with remarkable similarities to OPIDN with an upper and lower motor axonopathy and spinal cord lesions,[439] suggesting that impairment of NTE function causes OPIDN.

Metals

Exposure to different metals has been established as a cause of either generalized disease or specific symptoms that can often indicate the nature of toxic exposure. Given the nature of exposure to metals, toxicity often results from industrial use or, often with childhood exposures, where parental activities involving industrial exposure to metals leads to exposure of the infant. Regulation of exposure has largely reduced toxicity in developed countries, however, there are still circumstances, many accidental, which can lead to neurotoxicity.

Lead

Lead has long been established as having both acute and chronic effects on the nervous system. While in developed countries there are strict regulations on exposure, there are still instances, often because of poor occupational health monitoring, where lead neurotoxicity occurs. Lead toxicity is to a certain extent dependent on the chemical form of lead (organic versus inorganic lead salts), although the outcome is often the same.

High-level exposure to inorganic lead (greater than 100 μg/dL blood) can result in gastrointestinal problems ('lead colic'), encephalopathy and peripheral neuropathy. Exposure can occur from old paint, soil, water standing in lead pipes, glazes in non-commercial pottery, lead glass and leaded petrol. The encephalopathy is seen more commonly in children than in adults, who require much higher exposures to develop lead encephalopathy. Characteristic effects of acute lead poisoning are vomiting, ataxia, apathy, convulsions and coma. Morphological correlates are brain oedema,[430] particularly of the cerebellum, which can be observed on MRI[24,401]; focal neuronal necrosis with a disproportionate astrocytic reaction; swelling of endothelial cells associated with increased vascular permeability; and petechial haemorrhage with subsequent vascular proliferation observed experimentally.[431] These changes are prominent in the neocortex and cerebellum. Astrocytes appear to play an important role in limiting lead exposure, and astrocytic changes are a prominent feature of lead toxicity.[380] Acute lead toxicity in neonatal animals is also characterized by prominent oedema and petechial haemorrhage.[106] The strong binding of lead to plasma proteins means that its actions are restricted largely to the peripheral nervous system (PNS) and the brain vasculature, except in neonates. Organic lead compounds enter the CNS and produce prominent neuronal loss with mitochondrial swelling in the hippocampus, striatum and brain stem experimentally though human toxicity is rare.[79,177,559] Lead has a potent effect on the retina, retinal toxicity representing the most sensitive effect of lead in adult experimental animals with experimental evidence for apoptotic death of rods and bipolar cells.[158] The peripheral neuropathy produced by lead is a primary motor and segmental demyelination with secondary axonal loss. As with lead encephalopathy, vascular damage is a prominent feature. Lead neuropathy is now seen only rarely, although significant reductions in radial nerve conduction velocity have been seen in groups of workers with exposures close to the neuropathic threshold.[142]

Lead remains a common developmental neurotoxicant.[282] Although high-level neonatal lead exposures with marked effects on intellectual development are rare, lower exposure levels producing small but significant effects are more common and blood lead above 10 μg/dL requiring intervention.[282] Maternal blood lead levels in excess of 10 μg/dL are associated with demonstrable abnormalities in the electroretinogram at 7–10 years of age.[282] Damage due to lead is greatest in the prefrontal cortex, hippocampus and cerebellum.[156] Animal studies have shown sensitivity to be greatest in the later stages of brain development, with delayed synaptogenesis and impaired learning performance though gross pathology is not apparent.[374] The mechanisms of lead toxicity are not clear, but lead substitutes for calcium and interferes with NMDA receptor expression and function, which could perturb brain development.[372] Competition between lead and calcium for intestinal absorption leads to the greater bioavailability of lead, and hence neurotoxicity, if there is a lack of calcium in the diet.

Aluminium

Aluminium is the most abundant metal in the earth's crust and excessive exposure is associated with acute and chronic encephalopathy. Early use of tap water for renal haemodialysis resulted in a predominantly reversible encephalopathy in patients. This appeared as a slow onset progressive cognitive impairment with dysarthria and apraxia, myoclonus with specific EEG changes and eventual coma, which was subsequently found to be linked to aluminium in the tap water.[15] Patients dying during these acute episodes showed generalized cortical spongiform pathology, although few other changes.[64] Aluminium accumulates in cortical neurons via transferrin-mediated uptake with little further distribution providing the relative selectivity for cortical pathology.[361] Subsequent studies also showed cases on continuous ambulatory peritoneal dialysis,[499] but also the potential for further cases as a result of the use of aluminium containing phosphate binders used to control hyperphosphataemic osteomalacia in renal patients. These aluminium compounds have been replaced with calcium-, magnesium- and more recently lanthanum-containing compounds.[325] Contamination of a water supply in the UK with aluminium sulphate in 1988 was associated with some neuropsychological effects in the local population, but these could not be attributed directly to aluminium.[398] Chronic aluminium exposure has been proposed as a contributor to Alzheimer's disease as a result of the finding of raised aluminium in some studies and experimental evidence for a form of neurofibrillary degeneration following intracerebral administration of aluminium salts,[112] but no consistent epidemiological evidence has been found to support this.

Mercury

Mercury has long established its role as a major neurotoxin although, as with many metals, the picture of toxicity varies with the state of mercury as organic mercury, metallic mercury (Hg^0) or mercury salts (Hg^+). A particular place also exists within neuropathology owing to the use of mercuric chloride in Golgi impregnation methods and in neurosurgery in the past use of Zenker's fluid.

Metallic mercury, being poorly absorbed through the gut, has low toxicity when ingested, and shows low acute toxicity when injected during cases of self-harm or accident, but chronic exposure to mercury vapour or potentially metallic mercury is associated with a progressive neuropsychiatric disorder of insidious onset, characterized by nervousness, emotional instability, depression, anorexia, insomnia, cognitive problems, and development of intentional tremor with micrographia, myoclonus, and in terminal cases, coma and death. Gingival disease and 'erythism' is also seen in typical cases.[49] MRI has demonstrated generalized cerebral and cerebellar atrophy.[568] Metallic mercury is suggested to preferentially enter large cortical motor neurons, which may explain the symptoms of intentional tremor.[400] In one report of acutely fatal mercury vapour inhalation, perivascular inflammation and multiple areas of cortical and white matter necrosis with accompanying gliosis were seen.[267] Mercury injection has also been associated chronically with myoclonus.[438] Exposure to inorganic mercury salts, such as mercuric chloride (Hg^{2+}) or mercurous salts (Hg_2^{2+}), is regarded as more toxic than metallic mercury as a result of the enhanced absorption. Mercury salts have a long history involving a variety of uses as treatments for syphilis, skin whitening creams, and teething aids in children with consequent reports of toxicity. Mercuric chloride is acutely toxic causing major kidney damage and has a lowest lethal dose (LDLO) of about 30 mg/kg. One report showed white matter hyperintensities indicative of vascular damage following exposure in a child.[37] Peripheral polyneuropathy and accompanying central effects have been also been observed following chronic use with demyelination and axonal degeneration being found in nerve biopsy.[124]

Organic mercury is generally more toxic than mercury salts and has been associated with major toxicity. Ethyl mercury toxicity has been rarely reported, though in one report following ingestion of contaminated meat, progressive gait disturbance and ataxia developed with eventual deterioration, coma and death. In non-fatal cases, eventual recovery occurred with some constriction of visual fields suggesting damage to the primary visual cortex. Neuropathology on fatal cases showed demyelination of cranial nerves and extensive loss of motor neurons at multiple levels within the spinal cord along with chromatolysis of remaining neurons. Loss of granule cells in the cerebellum was a feature with axonal torpedoes in Purkinje cells. Cortical gliosis with loss of neurons was observed with the visual cortex being affected; additional small neuron loss was seen in the caudate and putamen with gliosis.[105] Dimethyl mercury is one of the most toxic compounds known with exceptionally rare fatalities and major nervous system pathology. The archetypal case, due to skin absorption of a few hundred microlitres of dimethyl mercury, resulted in progressive weight loss, nausea, diarrhoea, paraesthesia in the extremities, tinnitus, gait disturbance, progressive neuropsychiatric decline and eventual coma with death approximately 10 months later. The brain was grossly atrophic with cortical thinning, neuronal loss and accompanying gliosis, with marked cerebellar degeneration and loss of granule cells.[493]

Methyl mercury has been associated with several cases of mercury intoxication owing to its widespread use in industry as part of chlorine and sodium hydroxide production, plastics production, and use as a fungicide for seed storage. Large scale poisonings have been seen in Japan in 1956–60 after environmental contamination,[143] Iraq in 1972 after its use as a fungicide,[13] and in Canada following environmental contamination.[341] In adults, pathology is almost entirely confined to the nervous system with progressive symptoms including paraesthesia of the extremities and lips, decreased concentration, fatigue, speech impairment, 'erythism', dysphagia, constricted visual fields and eventual blindness, ataxia, tremors, incoordination, abnormal reflexes, severe hearing loss, permanent CNS damage, coma and death with an estimated 25 mg whole body exposure required to give paraesthesia.[143] The neuropathological damage can be very selective, with loss of small neurons of the anterior primary visual cortex, somatosensory cortex, and superior temporal cortex and deep cerebellar granule cell loss

with relative sparing of Purkinje cells with the presence of stellate bodies on staining with silver staining.[147,148,243] Peripheral neuropathy is seen, sensory nerves being more affected, with wallerian degeneration and degeneration of gracile tracts,[148] this being persistent but mild and possibly related to damage to the somatosensory cortex.[541] The exact mechanism of methyl mercury toxicity is unknown, but methyl mercury is converted slowly to thiol-bound inorganic mercury, affecting glutathione production and cysteine containing proteins.[195]

Methyl mercury is a potent developmental neurotoxicant, with multiple examples of fetal and infant exposure following outbreaks. Affected cases show cerebral atrophy with abnormal cortical layering as a consequence of disordered neuronal migration,[99] particularly in the primary visual cortex, but also in the precentral, post-central and lateral temporal cortices, cerebellar atrophy with Purkinje cell and focal granule cell loss, degeneration of gracile tract fibres, and a sensory peripheral neuropathy.[148] The pathology is least selective with earlier exposure and is always less selective than that seen in adults. Pathology is not seen in infants exposed to moderately elevated exposures to methyl mercury as a result of diet.[310]

Manganese

Although manganese is an essential element, overexposure via mining and the production of alloys and batteries, and potentially use of manganese as a flux in welding are occupational causes of exposure, with over 50 000 individuals in the UK exposed through the workplace. Manganese toxicity may also be a feature in intravenous drug abusers using 'ephedrone' (2-methylamino-1-phenylpropan-1-one, methcathinone) manufactured using potassium permanganate.[494] Hepatic disease, particularly hepatic encephalopathy (see earlier under Hepatic Encephalopathy, p. 596) when there is portal vein shunting also causes abnormal retention of manganese.[217,291] Several studies have shown neuropsychiatric disturbances in individuals exposed industrially to manganese without the presence of overt neurological symptoms with workers showing increased levels of anxiety, fatigue, decreased fine motor skills, with persistence of these effects even after exposure ends.[51,344] These symptoms are present in highly exposed individuals who also show variously dystonia, parkinsonism, or chorea with no or poor response to L-dopa. MRI in cases of hepatic disease or in welders exposed to high levels of manganese shows hyperintensity of the globus pallidus on T1-weighted imaging in many cases with cirrhosis[259,279,331] and a characteristic selectivity of manganese neurotoxicity (Figure 9.19). Damage is largely restricted to the globus pallidus, where there is neuron loss, demyelination and gliosis, and in the substantia nigra zona reticulata with relative sparing of the dopaminergic neurons in the zona compacta, unlike Parkinson's disease, although dopaminergic dysfunction is seen in non-human primates (suggesting that striatal dopamine release is compromised).[191] There is also some involvement of the subthalamic nucleus and mammillary bodies in post-mortem cases.[579] Once neuronal loss has developed, characteristic motor problems such as rigidity, ataxia and fine tremor persist, although chelation therapy has shown some reversal if brain manganese

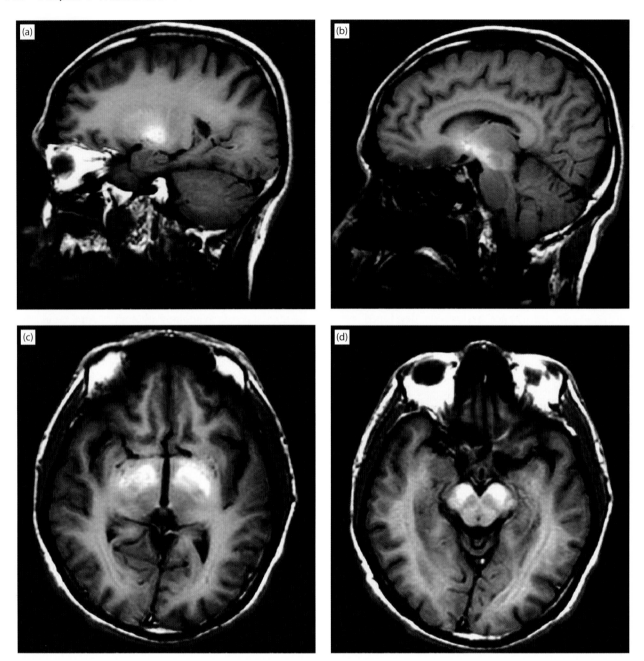

9.19 MR images from a patient with chronic manganese toxicity, showing strongly hyperintense signal in the globus pallidus and substantia nigra bilaterally. The images were obtained using a 1.5-T system in the sagittal (**a,b**) and axial (**c,d**) planes, with SE T1-weighted images 6.0 mm thickness/1.0 mm spacing.

Images kindly supplied by Dr MC Valentini, Chief of the Neuroradiological Department, University of Turin, Italy.

levels can be reduced.[231] Selective retention of manganese in the striatum and globus pallidus, interference with iron homeostasis and damage to mitochondria are likely mechanisms of pathogenesis.[21]

Arsenic

A distal polyneuropathy develops 10–20 days after a single exposure to inorganic arsenic, as occurs after malicious administration of arsenic salts, occupational exposure, burning high-arsenic-content coal[322] and drinking water from deep wells in arsenic-rich strata.[545] Sensory signs can sometimes be seen in advance of motor signs, but they are usually seen together. Cranial nerve involvement has also been described. Overt arsenic poisoning causes gastrointestinal effects followed by development of a distal motor and sensory neuropathy that is characterized by a dying back axonopathy of myelinated fibres with limited regeneration.[576] Functional recovery can take several years. Exposure to arsine gas (AsH_3) can also cause a similar neuropathy. The toxicity of arsenic is probably mediated by the trivalent form, which inhibits formation of acetyl co-enzyme A formation by pyruvate dehydrogenase by a mechanism involving redox reactions.[545]

Thallium

A predominantly peripheral, but also central, dying-back axonopathy with distal paraesthesia and hyperaesthesia followed by muscle weakness, cranial neuropathy, optic neuropathy with lens opacities, autonomic involvement, and alopecia with characteristic blackening of the hair roots, is seen after occupational or malicious poisoning by thallium.[88,466,516] Long, large-diameter sensory fibres are most sensitive, and central involvement is usually late and limited to long tracts. High doses produce poorly characterized neuropsychiatric features and encephalopathy,[525] plus choreoathetosis, tremor and dystonia, the morphological accompaniments of which include vascular damage and neuronal loss. The effects of thallium closely resemble those of arsenic, and they may share a common mechanism via inhibition of the pyruvate dehydrogenase complex.[88,525]

Tin

A selective excitotoxicity is produced by the fungicide trimethyl tin and also with triethyl lead. Trimethyl tin produces a selective and early increase in limbic system excitability in experimental animals.[444] The pattern of neuropathology in the hippocampus, pyriform cortex, amygdaloid nucleus, neocortex and various brain stem nuclei is typical of excitotoxicity,[14] with dendritic vacuolation and pyramidal neuronal necrosis. Not dissimilar changes have been seen in humans with white matter changes on MRI that resolved.[246,290] It has been proposed that trimethyl tin may be concentrated specifically in vulnerable neurons because of their greater expression of a mitochondrial protein, stannin (stanniocalcin), which binds trimethyl tin.[14] Disruption of the structure of central and peripheral myelin is a feature of poisoning with triethyl tin with myelin swelling found mainly in the CNS and optic nerve, with effects on peripheral nerve being restricted largely to the spinal nerve roots. Triethyltin is also directly ototoxic. Oligodendrocytes do not seem to be damaged directly, and the myelin swelling is completely reversible if secondary damage does not develop. In France in 1954, 290 people were poisoned by a preparation containing 5 per cent triethyl tin to treat boils. Of these, 110 died with central myelin swelling and raised intracranial pressure. Damage to myelin may be mediated by disruption of the charge interactions that hold myelin lamellae in tight apposition, but true demyelination is produced by chemicals damaging oligodendrocytes or Schwann cells. Both myelin disruption and demyelination impair nerve conduction and both are usually reversible, although the myelin sheath around remyelinated axons remains thin and intranodal lengths are often shorter than normal.[476]

Excitotoxins

The production of sustained neuronal hyperexcitation can generate neuronal death by compromising ionic homeostasis. Damage is usually focal and common features are postsynaptic swelling and selective neuronal necrosis, commonly caused by direct stimulation of neuronal glutamate receptors. The density of these largely determines vulnerability, but other factors such as inherent local excitability (e.g. high in limbic circuits) and local lack of a blood–brain barrier (e.g. in the area postrema) are also important. Exogenous glutamate, because of the high-activity reuptake systems in neurons and astrocytes that limit exposure, is neurotoxic only after exceptionally high doses. An excessively sustained release of endogenous glutamate as a consequence of brain injury, stroke or epilepsy, however, is a significant cause of secondary neurotoxic injury. The similarity of exogenous chemical-induced excitotoxicity and that of hypoxia–ischaemia (see Chapter 2) is a reflection of the importance of excitotoxic release of endogenous glutamate seen during reperfusion. Although in human material it is often difficult to dissociate the consequences of poisoning from incidental hypoxia, this is possible in experimental animal studies, enabling primary excitotoxicity to be recognized.

Domoic Acid

Domoic acid first came to light as a potential neurotoxin in man following an outbreak of amnesic shellfish poisoning on Prince Edward Island in Canada.[413] Affected individuals had eaten mussels contaminated with a marine diatom of *Pseudo-nitzschia* spp. that produce domoic acid and developed abdominal cramps, vomiting, diarrhoea, nausea, severe headache and, in some instances, short-term memory loss.[413] In a subgroup of exposed individuals, confusion, disorientation, seizures, myoclonus and coma were seen, with death in a small number, although comorbidities due to age may have been confounding factors.[413,519] Longer-term peripheral motor or sensorimotor neuronopathy and/or axonopathy are observed in severely intoxicated people.[519] Neuropathology in four individuals who showed neurological complications after exposure and died from 7 days to more than 3 years after exposure showed necrosis and loss of neurons in the hippocampus with some selectivity for the CA1, CA3 and CA4 neurons and neuron loss in the amygdala.[91,519] Additional sites of pathology, predominantly consisting of neuronal loss and accompanying gliosis, were the claustrum, olfactory cortex, septum, nucleus accumbens and, in some instances, the medial thalamus, insula cortex and orbitofrontal cortex, although the brain stem motor nuclei and spinal cord were spared.[519] Although CNS effects are major factors causing disability, peripheral effects on the heart and other organs by domoic acid may contribute to fatal outcomes.[434] Current regulatory control of shellfish quality and domoic acid levels makes the likelihood of outbreaks of domoic acid intoxication unlikely, although there are still instances of illegal harvesting of contaminated shellfish and mild poisoning that may cause long-term neuropsychiatric impairment. *Pseudo-nitzschia* spp. contamination of shellfish is almost global, with increasing reports of fish contamination.

Domoic acid is structurally similar to glutamate and kainic acid (found in red algae) and has its effects by acutely activating ionotropic glutamate AMPA/kainate receptors, which are most concentrated in limbic regions.[414] This causes influx of calcium into neurons and depolarization-induced glutamate release, which leads to excitotoxicity.[96] Toxicity is essentially limiting because domoic acid shows poor gut absorption and blood–brain barrier penetration. Rats and mice appear relatively insensitive to oral doses of domoic acid compared to human exposures with 1–5 mg/kg domoic acid thought to give symptoms in man[253,413,508] although in macaques, intraperitoneal doses that cause symptoms appear similar to the rat or mouse.[536] Pathology in primates and also rodents occurs principally in the limbic system with pyramidal neurons in the hippocampus, predominantly CA1 and CA3 neurons, affected following kindling seizure activity and shows striking similarity to human pathology.[18,473,501,509,514,535] Pathology is however seen mostly in animals that show seizures, which is also accompanied by microglial reactivity and astrogliosis.[18] In the macaque, mouse and rat, pathology is also seen in the hypothalamus and area postrema and other circumventricular organs.[60,534] In sea lions (*Zalophus californianus*), which are frequently affected off the coast of California as a result of consumption of contaminated anchovies, pathology is located within the limbic system with hippocampal atrophy and loss of CA1, CA3 and CA4 neurons and also granule cell loss, which may be unique to this species because it has not been observed in human cases. Chronic low-grade toxicity may also be a problem in sea lions in causing development of chronic epilepsy.[179]

Organophosphates

Excitotoxicity is not restricted to glutamatergic neurons of the limbic system, and there are several other examples of neuronal loss due to primary excitotoxicity. Cholinergic excitotoxicity has been described following severe poisoning with acetylcholinesterase inhibitors in animals. The lesions all involve primary neuronal necrosis. In rats, Soman produces brain lesions of a severity proportional to the signs of acute poisoning, and control of seizures markedly reduces damage.[490] The lesions in primates comprise diffuse cortical damage, and more focal neuronal loss and necrosis within the amygdala, piriform cortex, hippocampus, caudate nucleus and thalamus.[56] The nature of this damage, plus the association with seizures, suggests that the pathogenesis involves a combination of hypoxia secondary to breathing difficulties, direct cholinergic mediated excitotoxicity and secondary recruitment of glutamatergic excitotoxicity.[502] Direct toxicity is shown by the persistence of significant damage after effective seizure control and artificial ventilation.[490] Although there have been many clinical reports of poisoning, no morphological observations appear to have been made in survivors, though one study suggests subtle MRI changes with decreased grey matter and increased ventricular volume following possible low-dose sarin/cyclosarin exposure.[225]

In a large series of poisonings with organophosphorus pesticides with acute neurological deficits, persistent polyneuropathy, predominantly sensory, was present indicating long-term damage.[255] A report of acute sarin poisoning of sufficient severity to cause seizures and dyspnoea described memory impairment and learning problems 6 months after exposure suggestive of hippocampal damage.[219] Other organophosphorus poisonings have shown an asymmetrical blood flow deficit in the temporal lobe and basal ganglia, or in the primary visual cortex.[80,562] These effects follow high-dose exposure, but chronic low-level exposure to organophosphorus cholinesterase inhibitors has also been associated with impaired cognition.[264] A causal association has not been established and no morphological investigations have been made.

The origin of the hyperexcitation produced by the insect repellent N,N-diethyl-*m*-toluamide (DEET) is unclear, but it has been the cause of acute coma, seizures and other effects in humans after exceptionally large dermal exposures, mostly in young girls.[397] Pathology has not been characterized in humans, although in rats coma, EEG spiking and myelin vacuolation restricted entirely to the cerebellar motor nuclei have been reported.[553]

The abnormally high levels of nerve activity produced by pyrethroid insecticide poisoning in experimental animals can cause a peripheral axonal degeneration closely resembling wallerian degeneration.[81] This is seen only after exposures producing marked hyperexcitability, which is due to activity on sodium channels and is probably the axonal equivalent of central excitotoxicity.[53,445]

Very selective excitotoxicity is produced by some other agents in the cerebellum, an area not normally affected by excitotoxicity. The fungal tremogen penitrem-A found in mouldy foodstuffs, is associated with initial Purkinje cell dendritic changes, reactive vascular and astrocytic responses and eventual cellular necrosis,[90,327] related to its action on high-conductance calcium-dependent potassium channels. The industrial intermediate L-2-chloropropionic acid, damages only granule cells within the cerebellum, although it also produces neuronal loss in the medial habenular nucleus, pontine grey matter and inferior olivary nucleus, which is blocked by NMDA antagonists.[138] Some images illustrating selective pathology in experimental rats are shown in Figure 9.20. Reports of toxicity in man are unknown.

Neuropathies Associated with Solvents and Other Chemicals

The most important non-therapeutic chemicals causing peripheral neuropathy are the solvents *n*-hexane and methyl-*n*-butyl ketone. As with most other neurotoxins, the hazard is well recognized, and occupational poisoning is not seen in well-regulated industries, though abuse through 'glue sniffing' with coincident hypoxia due to rebreathing is seen. Additional hydrocarbon solvents, such as toluene, are often also involved. Prolonged exposure to *n*-hexane or methyl-*n*-butyl ketone results in a dying-back of distal sensorimotor axonopathy, with motor loss being more prominent. In man, this is characterized by axonal swellings and focal demyelination in large myelinated axons. Electrophysiological studies show that recovery is faster in sensory than motor nerves, although often delayed[94,138,302] and, as with remyelination, there is often thinning of new myelin sheets. There is also electrophysiological evidence for central auditory and visual fibre tract damage with little evidence of reversibility.[94] Experimental animal studies have shown that the first effect is a decrease in fast axoplasmic transport, followed by development of distal axonal swellings in central and peripheral nerves, with accumulation of neurofilaments. Axonal degeneration occurs distal to these swellings. With higher doses, the degeneration is more proximal. In myelinated fibres, the neurofilaments accumulate proximal to the nodes of Ranvier. Both *n*-hexane and methyl-*n*-butyl ketone produce their effects via a common metabolite, 2,5-hexanedione with the capacity to cyclize by reaction with lysine residues on proteins such as neurofilaments, which can then become cross-linked.[588]

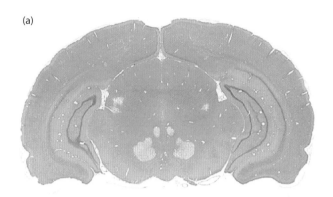

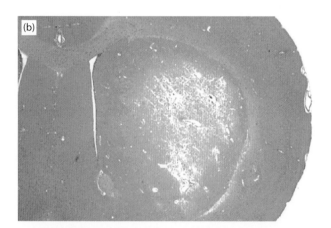

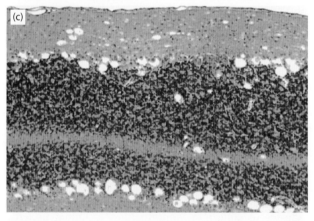

9.20 Illustrations of selective and vascular pathology in experimental rats. In each case, the distribution of the lesions is determined by inherent vulnerability of the susceptible cells or brain areas rather than by the local concentration of the chemical agent. **(a)** Regional selectivity of 3-chloropropanediol lesions within brain stem nuclei. **(b)** Lesions produced by a high intrathecal dose of gadolinium–diethylenetriamine penta-acetic acid (Gd-DTPA), a nuclear magnetic resonance imaging agent. Although the agent was given via the ventricular system, pathology was in deep, rather than periventricular, structures. **(c)** Selective loss of cerebellar Purkinje cells produced by penitrem-A.

Sections courtesy of Colin Willis and Christopher Nolan, MRC Applied Neuroscience Group, Nottingham, UK.

Because the cross-linking pyrrole–pyrrole bond requires an oxidized intermediate, the process is accelerated by oxidative stress and depletion of glutathione.[589] One report

suggests the occurrence of parkinsonian symptoms following chronic solvent exposure with nigral loss, although an absence of Lewy bodies.[417] A similar condition is seen with industrial exposure to high levels of carbon disulphide. This results in a distal dying-back of axonopathy and damage to cochlear hair cells leading to deafness.[110] Such exposures are now rare. The mechanism is thought to involve cross-linking of axonal proteins via dithiocarbamate metabolites of carbon disulphide.[495] This hypothesis is supported by the close similarity of carbon disulphide and n-hexane axonopathies in experimental animals.

Toluene is one of a very few organic solvents that have been shown to produce neuropathological damage following repeated high-level exposures,[155] but chronic lower level exposures also cause neuropathic changes.[489] These exposures have been associated with solvent abuse with hypoxia contributing, but because the lesion is a pure central myelinopathy, primary toxicity is likely. Magnetic resonance imaging has shown white matter loss, with cerebellar, cerebral and brain stem atrophy.[575] At the microscopic level, demyelination of major central tracts, especially of periventricular white matter, accompanied by astrocytosis and microglial activation, but without major axonal or neuronal loss, appears to be characteristic.[283] Evoked potential studies show corresponding delays in central conduction times. The mechanism of myelin loss appears to involve initial concentration of the lipophilic toluene in myelin, with oxidative metabolism of the toluene resulting in local redox stress.[339] Toluene and also xylene and styrene can cause ototoxicity. Toluene produces ototoxicity,[357] and in experimental animals these effects are seen at lower exposure levels when there is also exposure to noise.[329] Occupational measurements have strongly suggested that similar hearing loss may be seen in workers exposed to high, but not uncommon, levels of xylene or toluene if there is significant coexposure to noise.[355,356] Experimental animal studies indicate that repeated exposure to styrene at very high concentrations produces loss of outer hair cells in the mid-frequency range.[583] Mechanistic evidence suggests that cochlear hair cell toxicity may be related to local oxidative stress.[561] An analogous synergy in the central auditory pathway has been described between noise exposure and chemical toxicity in experimental rats given the redox disruptor *m*-dinitrobenzene.[446]

A slowly reversible trigeminal anaesthesia follows high-level acute and chronic exposure to trichloroethylene, a solvent and (formerly) an anaesthetic.[30,152,312] Conduction studies suggest a greater effect on small-diameter fibres. Axonopathy and some neuronal loss involving the trigeminal, facial, oculomotor and auditory nuclei have been described in one case.[74] In experimental animals, trigeminal nerve myelin is the primary target,[30] with lipid peroxidation mediated probably by oxidative metabolism to dichloroacetylene. The reason for the special susceptibility of the cranial nerves is unclear. Trichloroethylene has recently been linked to parkinsonism in an industrial setting[169] and potentially multiple system atrophy (Blain, personal communication), with an association being seen in epidemiological studies.[178] Experimental studies have shown reduced dopaminergic neurons in the substantia nigra[323] possibly through production of the mitochondria toxin 1-trichloromethyl-1,2,3,4-tetramthyl-beta-carbo-line (TaClo) via metabolism of trichloroethylene through chloral.[55,273]

Acrylamide (2-propenamide) is well characterized as being able to produce a central and peripheral axonopathy through occupational or accidental exposure. There have been multiple human poisoning incidents, but given the knowledge concerning acrylamide neurotoxicity very few since control of exposure was introduced.[25,176] Acute exposure to significant amounts of acrylamide produces seizures and cerebellar signs, which are followed by peripheral neuropathy. Acute poisoning shows a greater reversibility than chronic poisoning.[223] Despite a short half-life in the body, acrylamide shows cumulative toxic effects, with even weekly exposures producing a fully additive effect in experimental animals.[165] It induces a dying-back axonopathy of the deep touch and muscle spindle afferents before motor fibres become involved,[165] resulting in numbness, sweating, peeling of the skin on the hands and feet, some muscle weakness and an unsteady gait, with a marked sensory component. These effects are slowly reversible. One of the first effects seen in experimental animal studies is inhibition of fast axonal transport,[346] with subsequent distal paranodal accumulation of neurofilaments. Unmyelinated fibres are also susceptible, and a sympathetic and parasympathetic axonopathy has been produced. Cerebellar Purkinje cells show an accumulation of neurofilaments at the axon hillock. No human post-mortem studies have been reported, although sural nerve biopsy has shown loss of large-diameter fibres. Most clinical studies have been carried out by monitoring nerve conduction velocity or vibration sensitivity.[25]

Chemotherapy with cisplatin and related agents results in a dose-limiting peripheral distal large-fibre sensory neuropathy and ototoxicity, but its potential for producing a central axonopathy is limited by poor penetration of the blood–brain barrier. Selective dorsal root ganglion cell pathology has been shown in animal models,[529] although myelin breakdown and axonal loss is seen in sural nerve biopsy.[526] The mechanism of toxicity is unknown, but is associated with inhibition of axonal transport and selective uptake into hair cells via the organic cation transporter.[104,464] Similarly, chemotherapy with paclitaxel, vincristine or suramin can cause a mixed distal sensorimotor neuropathy, with loss of large myelinated fibres.[84] The early neuropathic pain produced by vincristine has been shown to be associated with hyperresponsiveness and microtubular disorganization (but not loss) of C-fibres in experimental animals.[531] Therapeutic use of thalidomide in multiple myeloma therapy causes neuropathy involving the dorsal root ganglia and is associated with degeneration in the dorsal columns detectable by MRI in some, but not all, patients.[173,251]

Therapeutic use of isoniazid causes a dose-dependent distal sensory and motor peripheral neuropathy affecting myelinated and unmyelinated nerves, and occasionally central fibres,[387] particularly in individuals less able to metabolize isoniazid via acetylation.[581] Acute neurotoxicity is associated with seizures. Although parenteral pyridoxine is an effective antidote, this is not always available.[517] In experimental animals, distal axonal degeneration is preceded by focal accumulation of smooth endoplasmic reticulum.[393] Intraneuronal oedema and loss of the blood–nerve barrier are seen along with regenerating fibres adjacent to degenerating fibres.[254]

Hexachlorophene is used widely and safely as a topical antimicrobial, but when first introduced and used at high concentrations (3–6 per cent) caused central and peripheral myelinopathy. Babies and young children were particularly susceptible. The central myelin damage was most marked in long tracts at the level of the brain stem.[491] Hexachlorophene also damages retinal photoreceptors.

Developmental Neurotoxicology

Several neurotoxins produce adverse effects on the developing brain after prenatal or postnatal exposure, often at lower doses than required in adults. Developmental neurotoxicity is usually irreversible with distinctive neuropathological features attributable either to the specific chemical or to the timing of the exposure relative to the stage of brain development. In the case of several agents, such as polychlorinated biphenyls, effects in animals and humans can only be seen functionally.

Valproate is an effective antiepileptic drug, but early toxicity studies in mice indicated its potential to produce developmental abnormalities, which have been confirmed in children born to mothers taking valproate during the first trimester of pregnancy. The children developed spina bifida, an effect subsequently shown to be dose related.[47] At lower doses, more subtle effects are seen with cognitive impairment and behavioural disturbances, an effect seen also with carbamazepine and lamotrigine.[109] Although valproate can interfere with folate metabolism, valproate embryo toxicity in mice is not preventable with folate supplementation,[201] and it is possible that inhibition of histone deacetylases, leading to altered gene transcription is key, an observation supported by the similar teratogenicity of the histone deacetylase inhibitor trichostatin A.[194] Another antiepileptic drug, phenytoin, also produces developmental neurotoxicity, with microcephaly and (in mice) damage to the cerebellum.[390]

Vitamin A (retinol) overdose during early pregnancy is teratogenic in both animals and humans, resulting in multiple defects, including microcephaly and disturbances of neuronal migration.[218] Similar effects are seen with isotretinoin for the treatment of acne.[305] Children and fetuses show a cluster of congenital abnormalities including underdeveloped jaw, heart defects, optic nerve and retinal degeneration, external ear and neural tube abnormalities. Lower doses in animals induce behavioural defects and oxidative stress.[126] By contrast, in older children and adults, headache and pseudo-tumour cerebri (increased intracranial pressure with no biochemical or morphological abnormalities) are the main dose-limiting effects when retinoic acid is used in tumour therapy.[6]

Bites and Stings

Envenoming bites and stings are inflicted by a very wide range of animals: snakes, spiders, scorpions, bees, wasps, cone snails and fishes are just a few of those implicated regularly in incidents of envenomation. The venom is a complex mixture of small polypeptides and other molecules elaborated in a specialized venom gland and inoculated via

9

a hollow or grooved fang or stinging part. Most venoms serve several functions: the capture of prey, defence and the initiation of a digestive process. Components targeting the PNS and the circulation are common constituents of venom, causing neuromuscular weakness, degeneration of peripheral nerves and skeletal muscle, and abnormalities of coagulation. Neurotoxic signs are particularly common following bites by snakes of the families *Elapidae*, a family of short-fanged snakes that includes kraits, mambas, coral snakes and cobras, and the *Hydrophidae*, the sea snakes. Envenoming bites by these snakes cause the classical signs of ptosis, dysphonia, and inability to smile and generalized neuromuscular weakness. Clotting times may be prolonged, but haemorrhage is rare. Necrosis may involve the skin and skeletal muscle. The venoms of vipers of the families *Viperidae* and *Crotalidae* are rarely neurotoxic, but coagulopathies and extensive haemorrhage are usual. The coagulopathy and systemic haemorrhages seen in association with many envenomings by vipers and crotalids (pit vipers) are usually of no special interest to the neuropathologist, unless the bleed is into the CNS. Central haemorrhages are relatively uncommon, but seem to occur most often following bites by the large vipers of South America and by Russell's viper (*Daboia* spp.) of South East Asia. With Russell's viper, a central bleed into the pituitary gland is not uncommon (Figure 9.21), giving rise to a delayed and irreversible pituitary failure.[286,538]

Cone snail stings can be fatal. The cone snails that are of particular interest in this context are the fish-eating cones, which produce venom that is powerfully neurotoxic; death results from neuromuscular paralysis. The venoms of spiders, scorpions and sea anemones contain toxins that cause neurotoxic signs of hyper- or hypoactivity in the PNS (summarized by Goonetilleke and Harris[180,215]). These toxins are described later according to their principle anatomical targets.

There are numerous ill-defined reports of long-lasting neurological problems resulting from envenomations, particularly following snake bites. It is probable that most problems arise when overtight ligatures have been applied to the bitten limb resulting in anoxic tissue damage. Formal studies of human neuropathology following envenoming have only rarely been reported in the clinical literature. All the data reported herein, unless explicitly stated otherwise, have been generated following work using experimental animals or isolated cells and tissues.

Post-Junctional Acetylcholine Receptor

Venoms of cone snails and the elapid and hydrophid snakes all contain toxins that target the α-subunit of the postsynaptic acetylcholine (ACh) receptor at the neuromuscular junction, thus preventing the binding of ACh and causing neuromuscular paralysis. Specific antivenoms accelerate the disassociation of toxin from receptor and reverse the paralysis. Anticholinesterases may accelerate recovery. In the absence of either of these therapies, assisted ventilation will keep the patient alive until disassociation occurs naturally at 12–24 hours post-envenoming. No pathological signs have ever been documented in association with the transient blockade of the neuromuscular ACh receptor.

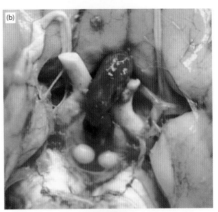

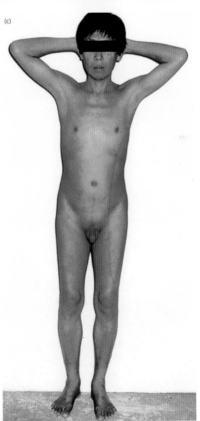

9.21 Panhypopituitarism following envenoming bite by Russell's viper in Burma. (a) *Daboia siamensis*, Tharawaddy, Burma. **(b)** Autopsy revealed haemorrhagic infarction of the anterior pituitary in a Burmese victim of an envenoming bite by *D. siamensis*. **(c)** Patient with signs of panhypopituitarism developing years after severe envenoming by *D. siamensis*.

MOTOR NERVE TERMINAL

Numerous toxins target the motor nerve terminal. Ω-Conotoxins in cone snail venoms block voltage-gated calcium channels, blocking transmitter release. The only pathological sign is the accumulation of synaptic vesicles in the nerve terminal. Neurotoxic phospholipases are common constituents of elapid venoms. These toxins cause the depletion of synaptic vesicles and the degeneration of both nerve terminal and axonal neurofilaments (Figures 9.22 and 9.23).

Recovery of function is slow and therapeutic intervention ineffective. Assisted ventilation may be required for more than a month. The recovery of function may be associated with abnormal patterns of skeletal muscle innervation and the delayed appearance of a poorly defined peripheral neuropathy.[216,298]

Toxins of scorpions, spiders and sea anemones target ion channels involved in action potential regulation, causing hyperactivity and enhanced transmitter release or reduced excitability, but there are no pathological features of note.

SKELETAL MUSCLE

Snake venoms from all major envenoming species may cause a severe rhabdomyolysis that results in acute and potentially fatal renal failure.[235]

Controversies and Uncertainties

Three areas of neurotoxicology are the subject of continuing uncertainty and so are grouped here. The first relates to the poorly defined encephalopathy that has been associated with long-term, low-level exposure to organic solvents.[533] Chronic solvent-related encephalopathy (also known as the psycho-organic syndrome) has been proposed as a lasting effect that is distinct from the compound-specific effects of high-level hexane exposure (axonopathy) or toluene abuse (encephalopathy). One useful feature is that there is little further progression of neuropsychiatric symptoms once occupational exposure to solvents has ceased.[163] White spirit (a petroleum distillate, also known as Stoddard fluid in the United States), styrene and xylene are the solvents most commonly cited. In some countries (notably Denmark, Finland, Norway and Sweden), impaired cognitive function associated with chronic solvent exposure has been considered diagnostic of an industrial injury if other CNS diseases can be excluded.[533] Although commendable from the social welfare viewpoint, this broad definition of causation has led to practical problems regarding differential diagnosis, and the causal relationship between organic solvents as a class and encephalopathy is not recognized in, for example, the USA or the UK, where regulatory limits are set solely by acute toxic potential. A number of epidemiological studies have, however, shown that neurobehavioural deficits are associated with estimates of past, but not current, solvent exposure level,[271,550] although additional factors, such as ethanol consumption and employment status, are also important.[163] Relatively few new cases are now being diagnosed, because exposure levels are falling.[533] There is no suggestion that medical use of transitory high doses of other organic solvents, such as trichloroethylene and halothane, as anaesthetics has led to a comparable condition. Neuropathological and imaging studies have failed to differentiate organic solvent syndrome from other conditions, such as dementia, seen in non-exposed people. A single case of chronic heptanone solvent exposure showing multiple small white matter foci by MRI has been reported.[568] Experimental animal studies of long-term exposure to white spirit have indicated that there are poorly reversible neurochemical effects and an increase

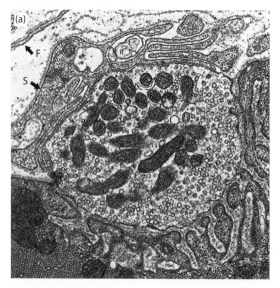
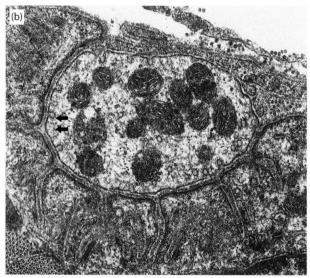

9.22 Electron micrographs of rat neuromuscular junction nerve terminals. (a) Normal terminal containing numerous synaptic vesicles. The nerve terminal is covered by Schwann cell (S) and fibroblast (F) processes. **(b)** Twenty four hours after exposure to the neurotoxic phospholipase A2. Note the loss of synaptic vesicles. A few vesicles remain fused to the nerve terminal membrane (arrows).

9

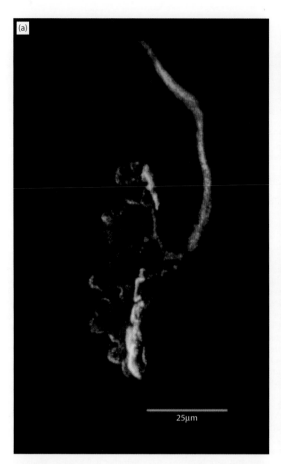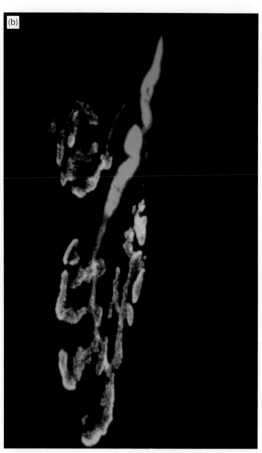

9.23 Neuromuscular junctions in soleus muscles of the rat. In the normal neuromuscular junction **(a)**, a single axon (red) inner-vates the end plate, within which it forms fine terminal branches that cover the acetylcholine receptors (green). In muscle **(b)**, 5 days after the axon had regenerated following exposure to phospholipase A2 (β-bungarotoxin), sprouts emerge from nodes of Ranvier to innervate three individual muscle fibres in close proximity to each other.

of GFAP expression (biochemical assay) in the cerebellum without overt cytopathology.[304]

Controversy is not restricted to observational clini-cal studies, where some ambiguity is to be expected, it is also a feature of some experimental studies. Thus, the pyrethroid insecticides are generally believed to be purely functional toxicants, producing hyperexcitability via reversible actions on ion channels.[445] Some published ani-mal studies, however, have indicated that repeated low-level exposure to permethrin has the potential to produce neuronal death indicated by scattered densely eosinophilic neurons and a reduction in neuronal counts, and also by a loss of neurofilament staining and reactive astrocyto-sis.[2] Quantification indicated hippocampal neuronal loss of 27–32 per cent,[3] and two other structurally and func-tionally unrelated chemicals, DEET and pyridostigmine bromide, also produced a similar pattern of neuropathol-ogy. Another pyrethroid, alpha-cypermethrin, produced a somewhat similar pattern of neuropathology in rats at a higher dose,[330] with neuron loss in CA3 of the hippo-campus, hypothalamus and cerebral cortex, with 'slight' loss of cerebellar Purkinje cells. A combination of severe neuronal loss in the hippocampus and neocortex and an increase in c-*Fos* and c-*Jun* immunoreactivity in the neo-cortex of rats given a single high dose of a third pyre-throid, deltamethrin, has been described,[573] though this was potentially related to seizure-associated excitotoxic-ity. In contrast to most of the earlier findings, regulatory studies using less sensitive measures, but higher pyrethroid exposure levels have yielded uniformly negative neuro-pathological findings after both single and repeated dose administration.[569] Hence, the consequences of repeated exposure to low levels of these agents remain somewhat unclear.

ACKNOWLEDGEMENTS

CMM is supported by the UK Health Protection Agency, Department of Health, and the UK Medical Research Council. JBH is funded through grants from the Wellcome Trust, the Royal College of Veterinary Surgeons of the United Kingdom and the Dubai Millennium Foundation.

We are indebted to the late David Ray who provided much of the section on neurotoxicology within this chapter. His wisdom and generosity will be missed.

REFERENCES

1. Aasheim ET. Wernicke encephalopathy after bariatric surgery: a systematic review. *Ann Surg* 2008;**248**:714–20.
2. Abdel-Rahman A, Shetty, AK, Abou-Donia MB. Subchronic dermal application of N,N-diethyl m-toluamide (DEET) and permethrin to adult rats, alone or in combination, causes diffuse neuronal cell death and cytoskeletal abnormalities in the cerebral cortex and the hippocampus, and Purkinje neuron loss in the cerebellum. *Exp Neurol* 2001;**172**:153–71.
3. Abdel-Rahman A, Dechkovskaia AM, Goldstein LB, et al. Neurological deficits induced by malathion, DEET, and permethrin, alone or in combination in adult rats. *J Toxicol Environ Health A* 2004;**67**:331–56.
4. Adam J, Baulac M, Hauw J-J, et al. Behavioral symptoms after pallido-nigral lesions: a clinico-pathological case. *Neurocase* 2008;**14**:125–30.
5. Adams RD, Victor M, Ropper AH. *Principles of neurology.* New York: McGraw-Hill, 1997.
6. Adamson PC, Widemann BC, Reaman GH, et al. A phase I trial and pharmacokinetic study of 9-cis-retinoic acid (ALRT1057) in pediatric patients with refractory cancer: a joint Pediatric Oncology Branch, National Cancer Institute, and Children's Cancer Group study. *Clin Cancer Res* 2001;**7**:3034–9.
7. Adhvaryu K, Shanbag P, Vaidya M. Tuberous sclerosis with hypothyroidism and precocious puberty. *Indian J Pediatr* 2004;**71**:273–5.
8. Agamanolis DP, Potter JL, Herrick MK, et al. The neuropathology of glycine encephalopathy: a report of five cases with immunohistochemical and ultrastructural observations. *Neurology* 1982;**32**:975–85.
9. Agamanolis DP, Potter JL, Lundgren DW. Neonatal glycine encephalopathy: biochemical and neuropathologic findings. *Pediatr Neurol* 1993;**9**:140–43.
10. Aggarwal A, Quint DJ, Lynch, JP, 3rd. MR imaging of porphyric encephalopathy. *AJR Am J Roentgenol* 1994;**162**:1218–20.
11. Akinyinka OO, Adekinka AO, Falade AG. The computed axial tomography of the brain in protein energy malnutrition. *Ann Trop Paediatr* 1995;**15**:329–33.
12. Alarie Y. Toxicity of fire smoke. *Crit Rev Toxicol* 2002;**32**:259–89.
13. Al-Damluji SF. Intoxication due to alkylmercury-treated seed-- 1971–72 outbreak in Iraq: clinical aspects. *Bull World Health Organ* 1976;**53**(Suppl):65–81.
14. Aldridge WN, Verschoyle RD, Thompson CA, Brown AW. The toxicity and neuropathology of dimethyltin and methyldiethyltin in rats. *Neuropathol Appl Neurobiol* 1987;**13**:55–69.
15. Alfrey AC, LeGendre GR, Kaehny WD. The dialysis encephalopathy syndrome. Possible aluminum intoxication. *N Engl J Med* 1976;**294**:184–8.
16. Aliefendioglu D, Tana Aslan Ay, Coskun T, et al. Transient nonketotic hyperglycinemia: two case reports and literature review. *Pediatr Neurol* 2003;**28**:151–5.
17. Allen AL, Luo C, Montgomery DL, et al. Vascular pathology in male Lewis rats following short-term, low-dose rotenone administration. *Vet Pathol* 2009;**46**:776–82.
18. Appel NM, Rapoport SI, O'Callaghan JP. Sequelae of parenteral domoic acid administration in rats: comparison of effects on different anatomical markers in brain. *Synapse* 1997;**25**:350–58.
19. Applegarth DA, Toone JR. Glycine encephalopathy (nonketotic hyperglycinemia): comments and speculations. *Am J Med Genet A* 2006;**140**:186–8.
20. Arslanian SA, Rothfus WE, Foley TP, Jr, Becker DJ. Hormonal, metabolic, and neuroradiologic abnormalities associated with septo-optic dysplasia. *Acta Endocrinol (Copenh)* 1984;**107**:282–8.
21. Aschner M, Erikson KM, Herrero Hernández E, Tjalkens R. Manganese and its role in Parkinson's disease: from transport to neuropathology. *Neuromolecular Med* 2009;**11**:252–66.
22. Ashorn M, Pitkänen S, Salo MK, Heinkinheimo M. Current strategies for the treatment of hereditary tyrosinemia type I. *Paediatr Drugs* 2006;**8**:47–54.
23. Aslan S, Karcioglu O, Bilge F, et al. Post-interval syndrome after carbon monoxide poisoning. *Vet Hum Toxicol* 2004;**46**:183–5.
24. Atre AL, Shinde PR, Shinde SN, et al. Pre- and posttreatment MR imaging findings in lead encephalopathy. *AJNR Am J Neuroradiol* 2006;**27**:902–3.
25. Bachmann M, Myers JE, Bezuidenhout BN. Acrylamide monomer and peripheral neuropathy in chemical workers. *Am J Ind Med* 1992;**21**:217–22.
26. Baker KG, Halliday GM, Kril JJ, Harper CG. Chronic alcoholics without Wernicke-Korsakoff syndrome or cirrhosis do not lose serotonergic neurons in the dorsal raphe nucleus. *Alcohol Clin Exp Res*1996;**20**61–6.
27. Baker KG, Harding AJ, Halliday GM, Kril JJ, Harper CG. Neuronal loss in functional zones of the cerebellum of chronic alcoholics with and without Wernicke's encephalopathy. *Neuroscience* 1999;**91**:429–38.
28. Balagopal K, Muthusamy K, Alexander M, Mani S. Methyl bromide poisoning presenting as acute ataxia. *Neurol India* 2011;**59**:768–9.
29. Baldi I, Filleul L, Mohammed-Brahim B, et al. Neuropsychologic effects of long-term exposure to pesticides: results from the French Phytoner study. *Environ Health Perspect* 2001;**109**:839–44.
30. Barret L, Torch S, Leray CL, Sarlière L, Saxod R. Morphometric and biochemical studies in trigeminal nerve of rat after trichloroethylene or dichloroacetylene oral administration. *Neurotoxicology* 1992;**13**:601–14.
31. Bates ME, Barry D, Bowden SC. Neurocognitive impairment associated with alcohol use disorders: implications for treatment. *Exp Clin Psychopharmacol* 2002;**10**:193–212.
32. Bauman ML, Kemper TL. Morphologic and histoanatomic observations of the brain in untreated human phenylketonuria. *Acta Neuropathol* 1982;**58**:55–63.
33. Beal MF, Brouillet E, Jenkins BG, Ferrante RJ, et al. Neurochemical and histologic characterization of striatal excitotoxic lesions produced by the mitochondrial toxin 3-nitropropionic acid. *J Neurosci* 1993;**13**:4181–92.
34. Beal MF, Henshaw DR, Jenkins BG, Rosen BR, Schulz JB. Coenzyme Q10 and nicotinamide block striatal lesions produced by the mitochondrial toxin malonate. *Ann Neurol* 1994;**36**:882–8.
35. Becker DM, Kramer S. The neurological manifestations of porphyria: a review. *Medicine (Baltimore)* 1977;**56**:411–23.
36. Bedi KS. *Lasting neuroanatomical changes following undernutrition during early life, in early nutrition and later achievement.* London: Academic Press, 1987.
37. Benz MR, Lee SH, Kellner L, Döhlemann C, Berweck S. Hyperintense lesions in brain MRI after exposure to a mercuric chloride-containing skin whitening cream. *Eur J Pediatr* 2010;**170**:747–50.
38. Berger JR, Dillon DA, Young BA, Goldstein SJ, Nelson P. Cystinosis of the brain and spinal cord with associated vasculopathy. *J Neurol Sci* 2009;**284**:182–5.
39. Bergeron C, Kovacs K. Pituitary siderosis. A histologic, immunocytologic, and ultrastructural study. *Am J Pathol* 1978;**93**:295–309.
40. Beseler C, Stallones L, Hoppin JA, et al. Depression and pesticide exposures in female spouses of licensed pesticide applicators in the agricultural health study cohort. *J Occup Environ Med* 2006;**48**:1005–13.
41. Betarbet R, Sherer T, MacKenzie G, et al. Chronic systemic pesticide exposure reproduces features of Parkinson's disease. *Nat Neurosci* 2000;**3**:1301–6.
42. Betz H. Glycine receptors: heterogeneous and widespread in the mammalian brain. *Trends Neurosci* 1991;**14**:458–61.
43. Beukeveld GJ, Wolthers BG, Nordmann Y, et al. A retrospective study of a patient with homozygous form of acute intermittent porphyria. *J Inherit Metab Dis* 1990;**13**:673–83.
44. Bianco F, Floris R. MRI appearances consistent with haemorrhagic infarction as an early manifestation of carbon monoxide poisoning. *Neuroradiology* 1996;**38**(Suppl 1):S70–72.
45. Bilbao JM, Horvath E, Hudson AR, Kovacs K, et al. Pituitary adenoma producing amyloid-like substance. *Arch Pathol* 1975;**99**:411–15.
46. Bilbao JM, Kovacs K, Horvath E, Higgins HP, Horsey WJ. Pituitary melanocorticotrophinoma with amyloid deposition. *Can J Neurol Sci* 1975;**2**:199–202.
47. Bjerkedal T, Czeizel A, Goujard J, et al. Valproic acid and spina bifida. *Lancet* 1982;**2**:1096.
48. Blendonohy PM, Philip PA. Precocious puberty in children after traumatic brain injury. *Brain Inj* 1991;**5**:63–8.
49. Bluhm RE, Bobbitt RG, Welch LW, et al. Elemental mercury vapour toxicity, treatment, and prognosis after acute, intensive exposure in chloralkali plant workers. Part I: History, neuropsychological findings and chelator effects. *Hum Exp Toxicol* 1992;**11**:201–10.
50. Boix, E, Picó A, Pinedo R, Aranda I, Kovacs K. Ectopic growth hormone-

releasing hormone secretion by thymic carcinoid tumour. *Clin Endocrinol (Oxf)* 2002;**57**:131–4.

51. Bouchard M, Mergler D, Baldwin M, *et al.* Neurobehavioral functioning after cessation of manganese exposure: a follow-up after 14 years. *Am J Ind Med* 2007;**50**:831–40.

52. Bouldin TW, Cavanagh JB. Organophosphorous neuropathy. II. A fine-structural study of the early stages of axonal degeneration. *Am J Pathol* 1979;**94**:253–70.

53. Bradberry SM, Cage SA, Proudfoot AT, Vale JA. Poisoning due to pyrethroids. *Toxicol Rev* 2005;**24**:93–106.

54. Braverman LE, Mancini JP, McGoldrick DM. Hereditary idiopathic diabetes insipidus. A case report with autopsy findings. *Ann Intern Med* 1965;**63**: 503–8.

55. Bringmann G, *et al.* Identification of the dopaminergic neurotoxin 1-trichloromethyl-1,2, 3,4-tetrahydro-beta-carboline in human blood after intake of the hypnotic chloral hydrate. *Anal Biochem* 1999;**270**:167–75.

56. Britt JO, Jr, *et al.* Histopathologic changes in the brain, heart, and skeletal muscle of rhesus macaques, ten days after exposure to soman (an organophosphorus nerve agent). *Comp Med* 2000;**50**:133–9.

57. Brooks AI, *et al.* Paraquat elicited neurobehavioral syndrome caused by dopaminergic neuron loss. *Brain Res* 1999;**823**:1–10.

58. Brouns R, De Deyn PP. Neurological complications in renal failure: a review. *Clin Neurol Neurosurg* 2004;**107**:1–16.

59. Brown RE. Organ weight in malnutrition with special reference to brain weight. *Dev Med Child Neurol* 1966;**8**:512–22.

60. Bruni JE, *et al.* Circumventricular organ origin of domoic acid-induced neuropathology and toxicology. *Brain Res Bull* 1991;**26**:419–24.

61. Brusilow SW, Horwich AL. Urea cycle enzymes. In: Scriver CR, *et al.* eds. *The metabolic and molecular basis of inherited disease*, 8th ed. New York: McGraw Hill, 2001:1909.

62. Bruton CJ, Corsellis JA, Russell A. Hereditary hyperammonaemia. *Brain* 1970;**93**:423–34.

63. Brvar M, *et al.* S100B protein in carbon monoxide poisoning: a pilot study. *Resuscitation* 2004;**61**:357–60.

64. Burks JS, *et al.* A fatal encephalopathy in chronic haemodialysis patients. *Lancet* 1976;**1**:764–8.

65. Burns RS, *et al.* A primate model of parkinsonism: selective destruction of dopaminergic neurons in the pars compacta of the substantia nigra by N-methyl-4-phenyl-1,2,3,6-tetrahydropyridine. *Proc Natl Acad Sci U S A* 1983;**80**:4546–50.

66. Burr IM, *et al.* Diencephalic syndrome revisited. *J Pediatr* 1976;**88**:439–44.

67. Butterworth RF. Effects of thiamine deficiency on brain metabolism: implications for the pathogenesis of the Wernicke-Korsakoff syndrome. *Alcohol Alcohol* 1989;**24**:271–9.

68. Butterworth RF. Portal-systemic encephalopathy: a disorder of neuron-astrocytic metabolic trafficking. *Dev Neurosci* 1993;**15**:313–19.

69. Butterworth RF. Hepatic encephalopathy. *Neurologist* 1995;**1**:95–104.

70. Butterworth RF. Neuronal cell death in hepatic encephalopathy. *Metab Brain Dis* 2007;**22**:309–20.

71. Butterworth RF. Altered glial-neuronal crosstalk: cornerstone in the pathogenesis of hepatic encephalopathy. *Neurochem Int* 2010;**57**:383–8.

72. Butterworth RF, *et al.* Thiamine deficiency and Wernicke's encephalopathy in AIDS. *Metab Brain Dis* 1991;**6**:207–12.

73. Butterworth RF, Kril JJ, Harper CG. Thiamine-dependent enzyme changes in the brains of alcoholics: relationship to the Wernicke-Korsakoff syndrome. *Alcohol Clin Exp Res* 1993;**17**:1084–8.

74. Buxton PH, Hayward M. Polyneuritis cranialis associated with industrial trichloroethylene poisoning. *J Neurol Neurosurg Psychiatry* 1967;**30**: 511–18.

75. Bylesjo I, *et al.* Brain magnetic resonance imaging white-matter lesions and cerebrospinal fluid findings in patients with acute intermittent porphyria. *Eur Neurol* 2004;**51**:1–5.

76. Caboni P, *et al.* Rotenone, deguelin, their metabolites, and the rat model of Parkinson's disease. *Chem Res Toxicol* 2004;**17**:1540–48.

77. Cadman SM, *et al.* Molecular pathogenesis of Kallmann's syndrome. *Horm Res* 2007;**67**:231–42.

78. Caine D, *et al.* Operational criteria for the classification of chronic alcoholics: identification of Wernicke's encephalopathy. *J Neurol Neurosurg Psychiatry* 1997;**62**:51–60.

79. Cairney S, *et al.* Saccade dysfunction associated with chronic petrol sniffing and lead encephalopathy. *J Neurol Neurosurg Psychiatry* 2004;**75**:472–6.

80. Callender TJ, Morrow L, Subramanian K. Evaluation of chronic neurological sequelae after acute pesticide exposure using SPECT brain scans. *J Toxicol Environ Health* 1994;**41**:275–84.

81. Calore EE, *et al.* Histologic peripheral nerve changes in rats induced by deltamethrin. *Ecotoxicol Environ Saf* 2000;**47**:82–6.

82. Cannon JR, *et al.* A highly reproducible rotenone model of Parkinson's disease. *Neurobiol Dis* 2009;**34**:279–90.

83. Cappellini MD, *et al.* Porphyrias at a glance: diagnosis and treatment. *Intern Emerg Med* 2010;**5**(Suppl 1):S73–80.

84. Carlson K, Ocean AJ. Peripheral neuropathy with microtubule-targeting agents: occurrence and management approach. *Clin Breast Cancer* 2011;**11**:73–81.

85. Carson MJ, Slager UT, Steinberg RM. Simultaneous occurrence of diabetes mellitus, diabetes insipidus, and optic atrophy in a brother and sister. *Am J Dis Child* 1977;**131**:1382–5.

86. Castagnoli N, Jr, Chiba K, Trevor AJ. Potential bioactivation pathways for the neurotoxin 1-methyl-4-phenyl-1,2,3,6-tetrahydropyridine (MPTP). *Life Sci* 1985;**36**:225–30.

87. Cavanagh JB. Methyl bromide intoxication and acute energy deprivation syndromes. *Neuropathol Appl Neurobiol* 1992;**18**:575–8.

88. Cavanagh JB, Nolan CC, Seville MP, Anderson VER, Leigh PN. Routes of excretion of neuronal lysosomal dense bodies after ventricular infusion of leupeptin in the rat: a study using ubiquitin and PGP 9.5 immunocytochemistry. *J Neurocytol* 1993;**22**:779–91.

89. Cavanagh JB, Nolan CC, Seville MP. The neurotoxicity of alpha-chlorohydrin in rats and mice: I. Evolution of the cellular changes. *Neuropathol Appl Neurobiol* 1993;**19**:240–52.

90. Cavanagh JB, *et al.* The effects of the tremorgenic mycotoxin penitrem A on the rat cerebellum. *Vet Pathol* 1998;**35**:53–63.

91. Cendes F, *et al.* Temporal lobe epilepsy caused by domoic acid intoxication: evidence for glutamate receptor-mediated excitotoxicity in humans. *Ann Neurol* 1995;**37**:123–6.

92. Centerwall SA, Centerwall WR. The discovery of phenylketonuria: the story of a young couple, two retarded children, and a scientist. *Pediatrics* 2000;**105**(1 Part 1): 89–103.

93. Chang CC, *et al.* Clinical significance of the pallidoreticular pathway in patients with carbon monoxide intoxication. *Brain* 2011;**134**(Part 12):3632–46.

94. Chang YC. Patients with n-hexane induced polyneuropathy: a clinical follow up. *Br J Ind Med* 1990;**47**:485–9.

95. Chaudhari S, *et al.* Pune low birth weight study - cognitive abilities and educational performance at twelve years. *Indian Pediatr* 2001;**41**:121–8.

96. Chavez AE, Singer JH, Diamond JS. Fast neurotransmitter release triggered by Ca influx through AMPA-type glutamate receptors. *Nature* 2006;**443**:705–8.

97. Chiba K, Trevor A, Castagnoli, N, Jr. Metabolism of the neurotoxic tertiary amine, MPTP, by brain monoamine oxidase. *Biochem Biophys Res Commun* 1984;**120**:574–8.

98. Chiossi G, *et al.* Hyperemesis gravidarum complicated by Wernicke encephalopathy: background, case report, and review of the literature. *Obstet Gynecol Surv* 2006;**61**:255–68.

99. Choi BH, *et al.* Abnormal neuronal migration, deranged cerebral cortical organization, and diffuse white matter astrocytosis of human fetal brain: a major effect of methylmercury poisoning in utero. *J Neuropathol Exp Neurol* 1978;**37**:719–33.

100. Choi IS. Parkinsonism after carbon monoxide poisoning. *Eur Neurol* 2002;**48**:30–33.

101. Choi IS, Cheon HY. Delayed movement disorders after carbon monoxide poisoning. *Eur Neurol* 1999;**42**:141–4.

102. Choi IS, *et al.* Evaluation of outcome of delayed neurologic sequelae after carbon monoxide poisoning by technetium-99m hexamethylpropylene amine oxime brain single photon emission computed tomography. *Eur Neurol* 1995;**35**:137–42.

103. Chuang DT, Chuang JL, Wynn RM. Lessons from genetic disorders of branched-chain amino acid metabolism. *J Nutr* 2006;**136**(1 Suppl):243S–9S.

104. Ciarimboli G, *et al.* Organic cation transporter 2 mediates cisplatin-induced oto- and nephrotoxicity and is a target for protective interventions. *Am J Pathol* 2010;**176**:1169–80.

105. Cinca I, *et al.* Accidental ethyl mercury poisoning with nervous system, skeletal muscle, and myocardium injury. *J Neurol Neurosurg Psychiatry* 1980;**43**:143–9.

106. Clasen RA, *et al.* Electron microscopic and chemical studies of the vascular changes and edema of lead encephalopathy.

A comparative study of the human and experimental disease. *Am J Pathol* 1974;**74**:215–40.

107. Clements RH, *et al.* Incidence of vitamin deficiency after laparoscopic Roux-en-Y gastric bypass in a university hospital setting. *Am Surg* 2006;**72**:1196–202; discussion 1203–204.

108. Cochran FB, Packman S. Homocystinuria presenting as sagittal sinus thrombosis. *Eur Neurol* 1992;**32**:1–3.

109. Cohen MJ, *et al.* Fetal antiepileptic drug exposure: motor, adaptive, and emotional/behavioral functioning at age 3 years. *Epilepsy Behav* 2011;**22**:240–46.

110. Colombi A, *et al.* Carbon disulfide neuropathy in rats. A morphological and ultrastructural study of degeneration and regeneration. *Clin Toxicol* 1981;**18**:1463–74.

111. Craft WH, Underwood LE, Van Wyk JJ. High incidence of perinatal insult in children with idiopathic hypopituitarism. *J Pediatr* 1980;**96**(3 Part 1):397–402.

112. Crapper DR, Krishnan SS, Dalton AJ. Brain aluminum distribution in Alzheimer's disease and experimental neurofibrillary degeneration. *Science* 1973;**180**:511–13.

113. Crome L, Stern J. *The pathology of mental retardation*, 2nd edn. Edinburgh: Churchill Livingstone, 1972.

114. Culjat M, *et al.* Magnetic resonance findings in a neonate with nonketotic hyperglycinemia: case report. *J Comput Assist Tomogr* 2010;**34**:762–5.

115. Cullen KM, *et al.* The nucleus basalis (Ch4) in the alcoholic Wernicke-Korsakoff syndrome: reduced cell number in both amnesic and non-amnesic patients. *J Neurol Neurosurg Psychiatry* 1997;**63**:315–20.

116. D'Alessandro G, Tagariello T, Piana G. Oral and craniofacial findings in a patient with methylmalonic aciduria and homocystinuria: review and a case report. *Minerva Stomatol* 2010;**59**:129–37.

117. Daniel PM, Prichard MM. Studies of the hypothalamus and the pituitary gland with special reference to the effects of transection of the pituitary stalk. *Acta Endocrinol Suppl (Copenh)* 1975;**201**:1–216.

118. Darnton-Hill I, Truswell AS. Thiamin status of a sample of homeless clinic attenders in Sydney. *Med J Aust* 1990;**152**:5–9.

119. Davis GC, *et al.* Chronic Parkinsonism secondary to intravenous injection of meperidine analogues. *Psychiatry Res* 1979;**1**:249–54.

120. Dawson PA, *et al.* Variable hyperhomocysteinaemia phenotype in heterozygotes for the Gly307Ser mutation in cystathionine beta-synthase. *Aust N Z J Med* 1996;**26**:180–85.

121. Dejkhamron P, Likasitwattankul S, Unachak K. Diencephalic syndrome: a rare and easily overlooked cause of failure to thrive. *J Med Assoc Thai* 2004;**87**:984–7.

122. de la Monte SM. Disproportionate atrophy of cerebral white matter in chronic alcoholics. *Arch Neurol* 1988;**45**:990–92.

123. de la Monte SM, Kril JJ. Human alcohol-related neuropathology. *Acta Neuropathol* 2014;**127**:71–90.

124. Deleu D, *et al.* Peripheral polyneuropathy due to chronic use of topical ammoniated mercury. *J Toxicol Clin Toxicol* 1998; **36**:233–7.

125. Demirkol M, *et al.* Follow up of phenylketonuria patients. *Mol Genet Metab* 2011;**104**(Suppl):S31–9.

126. de Oliveira MR, *et al.* Oxidative stress in the hippocampus, anxiety-like behavior and decreased locomotory and exploratory activity of adult rats: effects of sub acute vitamin A supplementation at therapeutic doses. *Neurotoxicology* 2007;**28**:1191–9.

127. de Penna GC, *et al.* Duplication of the hypophysis associated with precocious puberty: presentation of two cases and review of pituitary embryogenesis. *Arq Bras Endocrinol Metabol* 2005;**49**: 323–7.

128. De Wilde AR, Heyndrickx A, Carton D. A case of fatal rotenone poisoning in a child. *J Forensic Sci* 1986;**31**:1492–8.

129. Dezortova M, *et al.* MR in phenylketonuria-related brain lesions. *Acta Radiol* 2001;**42**:459–66.

130. Dickerman RD, *et al.* Precocious puberty associated with a pineal cyst: is it disinhibition of the hypothalamic–pituitary axis? *Neuro Endocrinol Lett* 2004;**25**:173–5.

131. Dinopoulos A, Matsubara Y, Kure S. Atypical variants of nonketotic hyperglycinemia. *Mol Genet Metab* 2005;**86**:61–9.

132. Dobbing J. Later development of brain and its vulnerability. In: Davis JA, Dobbing J eds. *Scientific foundations in pediatrics*. London: Heinemann Medical, 1981:565–76.

133. Dodd PR, *et al.* Glutamate and gamma-aminobutyric acid neurotransmitter systems in the acute phase of maple syrup urine disease and citrullinemia encephalopathies in newborn calves. *J Neurochem* 1992;**59**:582–90.

134. Dodd PR, *et al.* Glutamate-mediated transmission, alcohol, and alcoholism. *Neurochem Int* 2000;**37**:509–33.

135. Doga M, *et al.* Ectopic secretion of growth hormone-releasing hormone (GHRH) in neuroendocrine tumors: relevant clinical aspects. *Ann Oncol* 2001;**12**(Suppl 2): S89–94.

136. Dolinak D, Smith C, Graham DI. Global hypoxia per se is an unusual cause of axonal injury. *Acta Neuropathol* 2000; **100**:553–60.

137. Dreyer M, *et al.* The syndrome of diabetes insipidus, diabetes mellitus, optic atrophy, deafness, and other abnormalities (DIDMOAD-syndrome). Two affected sibs and a short review of the literature (98 cases). *Klin Wochenschr* 1982;**60**:471–5.

138. Duffell S, Lock EA. Re-evaluation of archival material for neuronal cell injury produced by L-2-chloropropionic acid in the rat brain. *Neurotoxicology* 2004;**25**:1031–40.

139. Durak AC, *et al.* Magnetic resonance imaging findings in chronic carbon monoxide intoxication. *Acta Radiol* 2005;**46**:322–7.

140. Eddleston M, *et al.* Predicting outcome using butyrylcholinesterase activity in organophosphorus pesticide self-poisoning. *QJM* 2008;**101**:467–74.

141. Eddleston M, *et al.* A role for solvents in the toxicity of agricultural organophosphorus pesticides. *Toxicology* 2012;**294**:94–103.

142. Ehle AL. Lead neuropathy and electrophysiological studies in low level lead exposure: a critical review. *Neurotoxicology* 1986;**7**:203–16.

143. Ekino S, *et al.* Minamata disease revisited: an update on the acute and chronic manifestations of methyl mercury poisoning. *J Neurol Sci* 2007;**262**:131–44.

144. Elgen I, Sommerfelt K, Ellertsen B. Cognitive performance in low birth weight cohort at 5 and 11 years of age. *Pediatr Neurol* 2003;**29**:111–16.

145. Endo F, *et al.* Clinical manifestations of inborn errors of the urea cycle and related metabolic disorders during childhood. *J Nutr* 2004;**134**(6 Suppl):1605S–609S; discussion 1630S–32S, 1667S–72S.

146. Estrin WJ. Alcoholic cerebellar degeneration is not a dose-dependent phenomenon. *Alcohol Clin Exp Res* 1987;**11**:372–5.

147. Eto K, *et al.* A fetal type of Minamata disease. An autopsy case report with special reference to the nervous system. *Mol Chem Neuropathol* 1992;**16**: 171–86.

148. Eto K, Marumoto M, Takeya M. The pathology of methyl mercury poisoning (Minamata disease). *Neuropathology* 2010;**30**:471–9.

149. Falcone N, *et al.* Central pontine myelinolysis induced by hypophosphatemia following Wernicke's encephalopathy. *Neurol Sci* 2003;**24**:407–10.

150. Fattal-Valevski A, *et al.* Outbreak of life-threatening thiamine deficiency in infants in Israel caused by a defective soy-based formula. *Pediatrics* 2005;**115**:e233–8.

151. Feldman JM, Feldman MD. Sequelae of attempted suicide by cyanide ingestion: a case report. *Int J Psychiatry Med* 1990;**20**:173–9.

152. Feldman RG, *et al.* Long-term follow-up after single toxic exposure to trichloroethylene. *Am J Ind Med* 1985;**8**:119–26.

153. Feliz B, Witt DR, Harris BT. Propionic acidemia: a neuropathology case report and review of prior cases. *Arch Pathol Lab Med* 2003;**127**:e325–8.

154. Ferenci P, *et al.* Hepatic encephalopathy-definition, nomenclature, diagnosis, and quantification: final report of the working party at the 11th World Congresses of Gastroenterology, Vienna, 1988. *Hepatology* 2002;**35**:716–21.

155. Filley CM, Halliday W, Kleinschmidt-DeMasters BK. The effects of toluene on the central nervous system. *J Neuropathol Exp Neurol* 2004;**63**:1–12.

156. Finkelstein Y, Markowitz ME, Rosen JF. Low-level lead-induced neurotoxicity in children: an update on central nervous system effects. *Brain Res Brain Res Rev* 1998;**27**:168–76.

157. Fleischman A, *et al.* Diencephalic syndrome: a cause of failure to thrive and a model of partial growth hormone resistance. *Pediatrics* 2005;**115**:e742–8.

158. Fox DA, Campbell ML, Blocker YS. Functional alterations and apoptotic cell death in the retina following developmental or adult lead exposure. *Neurotoxicology* 1997;**18**:645–64.

159. Francini-Pesenti F, *et al.* Wernicke's syndrome during parenteral feeding: not an unusual complication. *Nutrition* 2009; **25**:142–6.

160. Frank J, Christiano AM. The genetic bases of the porphyrias. *Skin Pharmacol Appl Skin Physiol* 1998;**11**:297–309.

161. Fraser CL, Arieff AI. Nervous system complications in uremia. *Ann Intern Med* 1988;**109**:143–53.

9

162. Freeman RM. *Rational use of vitamins in practice*. Toronto: JB Lippincott, 1979.

163. Friis L, Norback D, Edling C. Occurrence of neuropsychiatric symptoms at low levels of occupational exposure to organic solvents and relationships to health, lifestyle, and stress. *Int J Occup Environ Health* 1997;3:184–9.

164. Fu Y, et al. Consistent striatal damage in rats induced by 3-nitropropionic acid and cultures of arthrinium fungus. *Neurotoxicol Teratol* 1995;17:413–18.

165. Fullerton PM, Barnes JM. Peripheral neuropathy in rats produced by acrylamide. *Br J Ind Med* 1966;23:210–21.

166. Gambini A, et al. Marchiafava-Bignami disease: longitudinal MR imaging and MR spectroscopy study. *Am J Neuroradiol* 2003;24:249–53.

167. Gandini C, et al. Pallidoreticular-rubral brain damage on magnetic resonance imaging after carbon monoxide poisoning. *J Neuroimaging* 2002;12:102–3.

168. Garofeanu CG, et al. Causes of reversible nephrogenic diabetes insipidus: a systematic review. *Am J Kidney Dis* 2005;45:626–37.

169. Gash DM, et al. Trichloroethylene: Parkinsonism and complex 1 mitochondrial neurotoxicity. *Ann Neurol* 2008;63:184–92.

170. Geyer HL, Schaumburg HH, Herskovitz S. Methyl bromide intoxication causes reversible symmetric brainstem and cerebellar MRI lesions. *Neurology* 2005;64:1279–81.

171. Ghirardello S, et al. Current perspective on the pathogenesis of central diabetes insipidus. *J Pediatr Endocrinol Metab* 2005;18:631–45.

172. Giancola PR, Moss HB. Executive cognitive functioning in alcohol use disorders. *Recent Dev Alcohol* 1998;14:227–51.

173. Giannini F, et al. Thalidomide-induced neuropathy: a ganglionopathy? *Neurology* 2003;60:877–8.

174. Gibb WR, et al. Pathology of MPTP in the marmoset. *Adv Neurol* 1987;45:187–90.

175. Gocht A, Calmant HJ. Central pontine and extrapontine myelinolysis: a report of 58 cases. *Clin Neuropathol* 1987;6:262–70.

176. Goffeng LO, et al. Nerve conduction, visual evoked responses and electroretinography in tunnel workers previously exposed to acrylamide and N-methylolacrylamide containing grouting agents. *Neurotoxicol Teratol* 2008;30:186–94.

177. Goldings AS, Stewart RM. Organic lead encephalopathy: behavioral change and movement disorder following gasoline inhalation. *J Clin Psychiatry* 1982;43:70–72.

178. Goldman SM, et al. Solvent exposures and Parkinson disease risk in twins. *Ann Neurol* 2012;71:776–84.

179. Goldstein T, et al. Novel symptomatology and changing epidemiology of domoic acid toxicosis in California sea lions (*Zalophus californianus*): an increasing risk to marine mammal health. *Proc Biol Sci* 2008;275:267–76.

180. Goonetilleke A, Harris JB. Envenomation and consumption of poisonous seafood. *J Neurol Neurosurg Psychiatry* 2002;73:103–9.

181. Gordon N. Ornithine transcarbamylase deficiency: a urea cycle defect. *Eur J Paediatr Neurol* 2003;7:115–21.

182. Gould DH, Gustine DL. Basal ganglia degeneration, myelin alterations, and enzyme inhibition induced in mice by the plant toxin 3-nitropropanoic acid. *Neuropathol Appl Neurobiol* 1982;8:377–93.

183. Grandas F, Artieda J, Obeso JA. Clinical and CT scan findings in a case of cyanide intoxication. *Mov Disord* 1989;4:188–93.

184. Grant H, Lantos PL, Parkinson C. Cerebral damage in paraquat poisoning. *Histopathology* 1980;4:185–95.

185. Green JR, et al. Heredtary and idiopathic types of diabetes insipidus. *Brain* 1967;90:707–14.

186. Greenamyre JT, et al. Complex I and Parkinson's disease. *IUBMB Life* 2001;52:135–41.

187. Grompe M. The pathophysiology and treatment of hereditary tyrosinemia type 1. *Semin Liver Dis* 2001;21:563–71.

188. Gropman A. Brain imaging in urea cycle disorders. *Mol Genet Metab* 2010;100 (Suppl 1):S20–30.

189. Grosso S, et al. Central precocious puberty and abnormal chromosomal patterns. *Endocr Pathol* 2000;11:69–75.

190. Guesry P. The role of nutrition in brain development. *Prev Med* 1998;27:189–94.

191. Guilarte TR, et al. Nigrostriatal dopamine system dysfunction and subtle motor deficits in manganese-exposed non-human primates. *Exp Neurol* 2006;202:381–90.

192. Guillemin R. Hypothalamic hormones a.k.a. hypothalamic releasing factors. *J Endocrinol* 2005;184:11–28.

193. Gulati S, et al. Hypothalamic hamartoma, gelastic epilepsy, precocious puberty: a diffuse cerebral dysgenesis. *Brain Dev* 2002;24:784–6.

194. Gurvich N, et al. Association of valproate-induced teratogenesis with histone deacetylase inhibition in vivo. *FASEB J* 2005;19:1166–8.

195. Guzzi G, La Porta CA. Molecular mechanisms triggered by mercury. *Toxicology* 2008;244:1–12.

196. Haberle J, Koch HG. Genetic approach to prenatal diagnosis in urea cycle defects. *Prenat Diagn* 2004;24:378–83.

197. Halliday G, Ellis J, Harper C. The locus coeruleus and memory: a study of chronic alcoholics with and without the memory impairment of Korsakoff's psychosis. *Brain Res* 1992;598:33–7.

198. Hamilton BF, Gould DH. Nature and distribution of brain lesions in rats intoxicated with 3-nitropropionic acid: a type of hypoxic (energy deficient) brain damage. *Acta Neuropathol* 1987;72:286–97.

199. Hamre K, et al. Differential strain susceptibility following 1-methyl-4-phenyl-1,2,3,6-tetrahydropyridine (MPTP) administration acts in an autosomal dominant fashion: quantitative analysis in seven strains of *Mus musculus*. *Brain Res* 1999;828:91–103.

200. Hansard. *Carbon monoxide poisoning*. UK Parliament: London, 2009.

201. Hansen DK, et al. Effect of supplemental folic acid on valproic acid-induced embryotoxicity and tissue zinc levels in vivo. *Teratology* 1995;52:277–85.

202. Hansson J, Abrahamsson PA. Neuroendocrine pathogenesis in adenocarcinoma of the prostate. *Ann Oncol* 2001;12(Suppl 2):S145–52.

203. Hantraye P, et al. Stable parkinsonian syndrome and uneven loss of striatal dopamine fibres following chronic MPTP administration in baboons. *Neuroscience* 1993;53:169–78.

204. Harding A, et al. Degeneration of anterior thalamic nuclei differentiates alcoholics with amnesia. *Brain* 2000;123(Part 1): 141–54.

205. Harding AJ, et al. Chronic alcohol consumption does not cause hippocampal neuron loss in humans. *Hippocampus* 1997;7:78–87.

206. Harding AJ, et al. Loss of vasopressin-immunoreactive neurons in alcoholics is dose-related and time-dependent. *Neuroscience* 1996;72:699–708.

207. Harper CG. The incidence of Wernicke's encephalopathy in Australia – a neuropathological study of 131 cases. *J Neurol Neurosurg Psychiatry* 1983;46:593–8.

208. Harper CG, Blumbergs PC. Brain weights in alcoholics. *J Neurol Neurosurg Psychiatry* 1982;45:838–40.

209. Harper CG, Kril JJ. Neuropathological changes in alcoholics. In: Hunt WA, Nixon SJ eds. *Research monograph No. 22. Alcohol-induced brain damage*. Washington DC: National Institutes of Health, 1993:39–70.

210. Harper CG, Kril JJ, Holloway RL. Brain shrinkage in chronic alcoholics: a pathological study. *Br Med J (Clin Res Ed)* 1985;290:501–4.

211. Harper CG, Giles M, Finlay-Jones R. Clinical signs in the Wernicke–Korsakoff complex – a retrospective analysis of 131 cases diagnosed at autopsy. *J Neurol Neurosurg Psychiatry* 1986;49:341–5.

212. Harper PA, Healy PJ, Dennis JA. Maple syrup urine disease as a cause of spongiform encephalopathy in calves. *Vet Rec* 1986;119:62–5.

213. Harper PA, Dennis JA, Healy PJ, Brown GK. Maple syrup urine disease in calves: a clinical, pathological and biochemical study. *Aust Vet J* 1989;66:46–9.

214. Harper PA, Healy PJ, Dennis JA. Animal model of human disease. Citrullinemia (argininosuccinate synthetase deficiency). *Am J Pathol* 1989;135:1213–15.

215. Harris JB, Goonetilleke A. Animal poisons and the nervous system: what the neurologist needs to know. *J Neurol Neurosurg Psychiatry* 2004;75(Suppl 3):40–46.

216. Harris JB, Scott-Davey T. Secreted phospholipases A2 of snake venoms: effects on the peripheral neuromuscular system with comments on the role of phospholipases A2 in disorders of the CNS and their uses in industry. *Toxins (Basel)* 2013;5:2533–71.

217. Harris MK, et al. Neurologic presentations of hepatic disease. *Neurol Clin* 2010;28:89–105.

218. Hathcock JN, et al. Evaluation of vitamin A toxicity. *Am J Clin Nutr* 1990;52:183–202.

219. Hatta K, et al. Amnesia from sarin poisoning. *Lancet* 1996;347:1343.

220. Hauw JJ, et al. Postmortem studies on posthypoxic and post-methyl bromide intoxication: case reports. *Adv Neurol* 1986;43:201–14.

221. Hauw JJ, et al. Chromatolysis in alcoholic encephalopathies. *Brain* 1988;111:843–57.

222. Hawkins LH. Blood carbon monoxide levels as a function of daily cigarette consumption and physical activity. *Br J Ind Med* 1976;33:123–5.

223. He FS, et al. Neurological and electroneuromyographic assessment of the adverse effects of acrylamide on

occupationally exposed workers. *Scand J Work Environ Health* 1989;**15**:125–9.

224. He F, *et al*. Delayed dystonia with striatal CT lucencies induced by a mycotoxin (3-nitropropionic acid). *Neurology* 1995;**45**:2178–83.

225. Heaton KJ, *et al*. Quantitative magnetic resonance brain imaging in US army veterans of the 1991 Gulf War potentially exposed to sarin and cyclosarin. *Neurotoxicology* 2007;**28**:761–9.

226. Heinrich A, Runge A, Khaw AV. Clinicoradiologic subtypes of Marchiafava-Bignami disease. *J Neurol* 2004;**251**:1050–59.

227. Henke K, *et al*. Memory lost and regained following bilateral hippocampal damage. *J Cogn Neurosci* 1999;**11**:682–97.

228. Henry CR, *et al*. Myocardial injury and long-term mortality following moderate to severe carbon monoxide poisoning. *JAMA* 2006;**295**:398–402.

229. Hernandes MS, Troncone LR. Glycine as a neurotransmitter in the forebrain: a short review. *J Neural Transm* 2009;**116**:1551–60.

230. Héroux M, Butterworth RF. Animal models of the Wernicke-Korsakoff syndrome. In: Boulton A, Baker G, Butterworth R eds. *Animal models of neurological disease.* Clifton, NJ: Humana Press, 1992:95–131.

231. Herrero Hernandez E, *et al*. Follow-up of patients affected by manganese-induced parkinsonism after treatment with CaNa2EDTA. *Neurotoxicology* 2006;**27**:333–9.

232. Hessels J, *et al*. Homozygous acute intermittent porphyria in a 7-year-old boy with massive excretions of porphyrins and porphyrin precursors. *J Inherit Metab Dis* 2004;**27**:19–27.

233. Hierons R. Changes in the nervous system in acute porphyria. *Brain* 1957;**80**:176–92.

234. Hierons R. Acute intermittent porphyria. *Postgrad Med J* 1967;**43**:605–8.

235. Hood VL, Johnson JR. Acute renal failure with myoglobinuria after tiger snake bite. *Med J Aust* 1975;**2**:638–41.

236. Hoover-Fong JE, *et al*. Natural history of nonketotic hyperglycinemia in 65 patients. *Neurology* 2004;**63**:1847–53.

237. Hopkins RO, Woon FL. Neuroimaging, cognitive, and neurobehavioral outcomes following carbon monoxide poisoning. *Behav Cogn Neurosci Rev* 2006;**5**:141–55.

238. Horita N, *et al*. Experimental carbon monoxide leucoencephalopathy in the cat. *J Neuropathol Exp Neurol* 1980;**39**:197–211.

239. Horner JM, Bhumbra NA. Congenital HIV infection and precocious puberty. *J Pediatr Endocrinol Metab* 2003;**16**:791–3.

240. Hsiao CL, Kuo HC, Huang CC. Delayed encephalopathy after carbon monoxide intoxication: long-term prognosis and correlation of clinical manifestations and neuroimages. *Acta Neurol Taiwan* 2004;**13**:64–70.

241. Huang TS, *et al*. Idiopathic hypothalamic hypogonadism with polyostotic fibrous dysplasia: report of a case. *J Formos Med Assoc* 1990;**89**:310–13.

242. Hughes JT. Brain damage due to paraquat poisoning: a fatal case with neuropathological examination of the brain. *Neurotoxicology* 1988;**9**:243–8.

243. Hunter D, Russell DS. Focal cerebellar and cerebellar atrophy in a human subject due to organic mercury compounds. *J Neurol Neurosurg Psychiatry* 1954;**17**:235–41.

244. Hustinx WN, *et al*. Systemic effects of inhalational methyl bromide poisoning: a study of nine cases occupationally exposed due to inadvertent spread during fumigation. *Br J Ind Med* 1993;**50**:155–9.

245. Huttenlocher PR. The neuropathology of phenylketonuria: human and animal studies. *Eur J Pediatr* 2000;**159**(Suppl 2):S102–6.

246. Hwang CH. The sequential magnetic resonance images of tri-methyl tin leukoencephalopathy. *Neurol Sci* 2009;**30**:153–8.

247. Iida M, Takamoto S, Masuo M, *et al*. Transient lymphocytic panhypophysitis associated with SIADH leading to diabetes insipidus after glucocorticoid replacement. *Intern Med* 2003;**42**:991–5.

248. Ikegami H, Shiga T, Tsushima T, *et al*. Syndrome of inappropriate antidiuretic hormone secretion (SIADH) induced by amiodarone: a report on two cases. *J Cardiovasc Pharmacol Ther* 2002;**7**:25–8.

249. Iliev DI, Ranke MB, Wollmann HA. Mixed gonadal dysgenesis and precocious puberty. *Horm Res* 2002;**58**:30–33.

250. Ishii N, Nishihara Y. Pellagra encephalopathy among tuberculous patients: its relation to isoniazid therapy. *J Neurol Neurosurg Psychiatry* 1985;**48**:628–34.

251. Isoardo G, *et al*. Thalidomide neuropathy: clinical, electrophysiological and neuroradiological features. *Acta Neurol Scand* 2004;**109**:188–93.

252. Ito H, *et al*. Adenocarcinoma of the prostate with ectopic antidiuretic hormone production: a case report. *Hinyokika Kiyo* 2000;**46**:499–503.

253. Iverson F, *et al*. Domoic acid poisoning and mussel-associated intoxication: preliminary investigations into the response of mice and rats to toxic mussel extract. *Food Chem Toxicol* 1989;**27**:377–84.

254. Jacobs JM, *et al*. Studies on the early changes in acute isoniazid neuropathy in the rat. *Acta Neuropathol* 1979;**47**:85–92.

255. Jalali N, *et al*. Electrophysiological changes in patients with acute organophosphorous pesticide poisoning. *Basic Clin Pharmacol Toxicol* 2010;**108**:251–5.

256. Javitch JA, Snyder SH. Uptake of MPP(+) by dopamine neurons explains selectivity of parkinsonism-inducing neurotoxin, MPTP. *Eur J Pharmacol* 1984;**106**:455–6.

257. Jenner P, *et al*. 1-Methyl-4-phenyl-1,2,3,6-tetrahydropyridine-induced parkinsonism in the common marmoset. *Neurosci Lett* 1984;**50**:85–90.

258. Jernigan TL, *et al*. Reduced cerebral gray matter observed in alcoholics using magnetic resonance imaging. *Alcohol Clin Exp Res* 1991;**15**:418–27.

259. Josephs KA, *et al*. Neurologic manifestations in welders with pallidal MRI T1 hyperintensity. *Neurology* 2005;**64**:2033–9.

260. Jovic-Stosic J, Babic G, Todorovic V. Fatal diquat intoxication. *Vojnosanit Pregl* 2009;**66**:477–81.

261. Kakarla N, Bradshaw KD. Disorders of pubertal development: precocious puberty. *Semin Reprod Med* 2003;**21**:339–51.

262. Kalsbeck J. Diencephalic syndrome. In: Wilkins RH, Rengachary SS eds. *Neurosurgery.* Vol. 1. New York: McGraw-Hill, 1985:925–7.

263. Kamei A, *et al*. Abnormal dendritic development in maple syrup urine disease. *Pediatr Neurol* 1992;**8**:145–7.

264. Kamel F, *et al*. Neurologic symptoms in licensed pesticide applicators in the Agricultural Health Study. *Hum Exp Toxicol* 2007;**26**:243–50.

265. Kanda M, Omori Y, Shinoda S, *et al*. SIADH closely associated with non-functioning pituitary adenoma. *Endocr J* 2004;**51**:435–8.

266. Kang K, *et al*. Diffuse lesion in the splenium of the corpus callosum in patients with methyl bromide poisoning. *J Neurol Neurosurg Psychiatry* 2006;**77**:703–4.

267. Kanluen S, Gottlieb CA. A clinical pathologic study of four adult cases of acute mercury inhalation toxicity. *Arch Pathol Lab Med* 1991;**115**:56–60.

268. Kaplowitz P. Clinical characteristics of 104 children referred for evaluation of precocious puberty. *J Clin Endocrinol Metab* 2004;**89**:3644–50.

269. Kaplowitz PB, D'Ercole AJ, Robertson GL. Radioimmunoassay of vasopressin in familial central diabetes insipidus. *J Pediatr* 1982;**100**:76–81.

270. Kato M, *et al*. Electron microscopic study of brain capillaries in cerebral edema from fulminant hepatic failure. *Hepatology* 1992;**15**:1060–66.

271. Kaukiainen A, *et al*. Solvent-related health effects among construction painters with decreasing exposure. *Am J Ind Med* 2004;**46**:627–36.

272. Kawanami T, *et al*. The pallidoreticular pattern of brain damage on MRI in a patient with carbon monoxide poisoning. *J Neurol Neurosurg Psychiatry* 1998;**64**:282.

273. Keane PC, *et al*. Mitochondrial dysfunction in Parkinson's disease. *Parkinsons Dis* 2011;**2011**:716871.

274. Khong PL, *et al*. Diffusion-weighted MR imaging in neonatal nonketotic hyperglycinemia. *AJNR Am J Neuroradiol* 2003;**24**:1181–3.

275. Kim JH, *et al*. Delayed encephalopathy of acute carbon monoxide intoxication: diffusivity of cerebral white matter lesions. *AJNR Am J Neuroradiol* 2003;**24**:1592–7.

276. King PH, Bragdon AC. MRI reveals multiple reversible cerebral lesions in an attack of acute intermittent porphyria. *Neurology* 1991;**41**:1300–302.

277. Kleopa KA, *et al*. Acute axonal neuropathy in maple syrup urine disease. *Muscle Nerve* 2001;**24**:284–7.

278. Kletzky OA, *et al*. Gonadotropin insufficiency in patients with thalassemia major. *J Clin Endocrinol Metab* 1979;**48**:901–5.

279. Klos KJ, *et al*. Neurologic spectrum of chronic liver failure and basal ganglia T1 hyperintensity on magnetic resonance imaging: probable manganese neurotoxicity. *Arch Neurol* 2005;**62**:1385–90.

280. Koch R, *et al*. The Maternal Phenylketonuria International Study: 1984–2002. *Pediatrics* 2003;**112**(6 Part 2):1523–9.

281. Kohira I, *et al*. [Pilocytic astrocytoma and diencephalic syndrome in an adult with neurofibromatosis type 1]. *Rinsho Shinkeigaku* 2003;**43**:327–9.

282. Kordas K. Iron, lead, and children's behavior and cognition. *Annu Rev Nutr* 2010;**30**:123–48.

283. Kornfeld M, *et al*. Solvent vapor abuse leukoencephalopathy. Comparison to adrenoleukodystrophy. *J Neuropathol Exp Neurol* 1994;**53**:389–98.

284. Kornguth S, *et al*. Golgi-Kopsch silver study of the brain of a patient with

untreated phenylketonuria, seizures, and cortical blindness. *Am J Med Genet* 1992;**44**:443–8.

285. Kotimaa AJ, *et al*. Maternal smoking and hyperactivity in 8-year-old children. *J Am Acad Child Adolesc Psychiatry* 2003;**42**:826–33.

286. Kouyoumdjian JA, *et al*. Fatal extradural haematoma after snake bite (*Bothrops moojeni*). *Trans R Soc Trop Med Hyg* 1991;**85**:552.

287. Kovacs K, Sheehan HL. Pituitary changes in Kallmann's syndrome: a histologic, immunocytologic, ultrastructural, and immunoelectron microscopic study. *Fertil Steril* 1982;**37**:83–9.

288. Kovacs K, *et al*. Mammosomatotroph hyperplasia associated with acromegaly and hyperprolactinemia in a patient with the McCune-Albright syndrome. A histologic, immunocytologic and ultrastructural study of the surgically-removed adenohypophysis. *Virchows Arch A Pathol Anat Histopathol* 1984;**403**:77–86.

289. Kowall NW, *et al*. MPTP induces alpha-synuclein aggregation in the substantia nigra of baboons. *Neuroreport* 2000;**11**:211–13.

290. Kreyberg S, *et al*. Trimethyltin poisoning: report of a case with postmortem examination. *Clin Neuropathol* 1992;**11**:256–9.

291. Krieger D, *et al*. Manganese and chronic hepatic encephalopathy. *Lancet* 1995;**346**:270–74.

292. Kril JJ. The contribution of alcohol, thiamine deficiency and cirrhosis of the liver to cerebral cortical damage in alcoholics. *Metab Brain Dis* 1995;**10**:9–16.

293. Kril JJ. Neuropathology of thiamine deficiency disorders. *Metab Brain Dis* 1996;**11**:11–19.

294. Kril JJ, Halliday GM. Brain shrinkage in alcoholics: a decade on and what have we learned? *Prog Neurobiol* 1999;**58**:381–7.

295. Kril JJ, Harper CG. Neuronal counts from four regions of alcoholic brains. *Acta Neuropathol* 1989;**79**:200–204.

296. Kril JJ, Harper CG. Neuroanatomy and neuropathology associated with Korsakoff's syndrome. *Neuropsychol Rev* 2012;**22**:72–80.

297. Kril JJ, *et al*. The cerebral cortex is damaged in chronic alcoholics. *Neuroscience* 1997;**79**:983–8.

298. Kularatne SA. Common krait (*Bungarus caeruleus*) bite in Anuradhapura, Sri Lanka: a prospective clinical study, 1996–98. *Postgrad Med J* 2002;**78**:276–80.

299. Kunze K. Metabolic encephalopathies. *J Neurol* 2002;**249**:1150–59.

300. Kupferschmidt H, *et al*. Transient cortical blindness and biocipital brain lesions in two patients with acute intermittent porphyria. *Ann Intern Med* 1995;**123**: 598–600.

301. Kure S, *et al*. Heterozygous GLDC and GCSH gene mutations in transient neonatal hyperglycinemia. *Ann Neurol* 2002;**52**:643–6.

302. Kutlu G, *et al*. Peripheral neuropathy and visual evoked potential changes in workers exposed to n-hexane. *J Clin Neurosci* 2009;**16**:1296–9.

303. Lacerda G, Krummel T, Hirsch E. Neurologic presentations of renal diseases. *Neurol Clin* 2010;**28**:45–59.

304. Lam HR, *et al*. Inhalation exposure to white spirit causes region-dependent alterations in the levels of glial fibrillary acidic protein. *Neurotoxicol Teratol* 2000;**22**:725–31.

305. Lammer EJ, *et al*. Retinoic acid embryopathy. *N Engl J Med* 1985;**313**:837–41.

306. Landolt AM, Heitz PU. Differentiation of two types of amyloid occurring in pituitary adenomas. *Pathol Res Pract* 1988;**183**:552–4.

307. Langley-Evans AJ, Langley-Evans SC. Relationship between maternal nutrient intakes in early and late pregnancy and infants weight and proportions at birth: prospective cohort study. *J R Soc Health* 2003;**123**:210–16.

308. Langston JW, Ballard PA, Jr. Parkinson's disease in a chemist working with 1-methyl-4-phenyl-1,2,5,6-tetrahydropyridine. *N Engl J Med* 1983;**309**:310.

309. Langston JW, *et al*. Chronic parkinsonism in humans due to a product of meperidine-analog synthesis. *Science* 1983; **219**:979–80.

310. Lapham LW, *et al*. An analysis of autopsy brain tissue from infants prenatally exposed to methyl mercury. *Neurotoxicology* 1995;**16**:689–704.

311. Lapresle J, Fardeau M. The central nervous system and carbon monoxide poisoning. II. Anatomical study of brain lesions following intoxication with carbon monoxide (22 cases). *Prog Brain Res* 1967;**24**:31–74.

312. Leandri M, *et al*. Electrophysiological evidence of trigeminal root damage after trichloroethylene exposure. *Muscle Nerve* 1995;**18**:467–8.

313. Lecours AR, Mandujano M, Romero G. Ontogeny of brain and cognition: relevance to nutrition research. *Nutr Rev* 2001;**59**:S7–12.

314. Lee PJ, *et al*. Maternal phenylketonuria: report from the United Kingdom Registry 1978–97. *Arch Dis Child* 2005;**90**:143–6.

315. Lee S, *et al*. Ectopic expression of vasopressin V1b and V2 receptors in the adrenal glands of familial ACTH-independent macronodular adrenal hyperplasia. *Clin Endocrinol (Oxf)* 2005;**63**:625–30.

316. Leonard JV, Morris AA. Urea cycle disorders. *Semin Neonatol* 2002;**7**: 27–35.

317. Lerner-Ellis JP, *et al*. Identification of the gene responsible for methylmalonic aciduria and homocystinuria, cblC type. *Nat Genet* 2006;**38**:93–100.

318. Levine S, Paparo G. Brain lesions in a case of cystinosis. *Acta Neuropathol* 1982;**57**:217–20.

319. Levy HL, *et al*. Maternal mild hyperphenylalaninaemia: an international survey of offspring outcome. *Lancet* 1994;**344**:1589–94.

320. LeWitt PA, Martin SD. Dystonia and hypokinesis with putaminal necrosis after methanol intoxication. *Clin Neuropharmacol* 1988;**11**:161–7.

321. Lipsett MB, *et al*. An analysis of the polyuria induced by hypophysectomy in man. *J Clin Endocrinol Metab* 1956;**16**:183–95.

322. Liu J, *et al*. Chronic arsenic poisoning from burning high-arsenic-containing coal in Guizhou, China. *Environ Health Perspect* 2002;**110**:119–22.

323. Liu M, *et al*. Trichloroethylene induces dopaminergic neurodegeneration in Fisher 344 rats. *J Neurochem* 2011;**112**:773–83.

324. Livadas DP, *et al*. Pituitary and thyroid insufficiency in thalassaemic haemosiderosis. *Clin Endocrinol (Oxf)* 1984;**20**:435–43.

325. Loghman-Adham M. Safety of new phosphate binders for chronic renal failure. *Drug Saf* 2003;**26**:1093–115.

326. Lotti M, Moretto A. Organophosphate-induced delayed polyneuropathy. *Toxicol Rev* 2005;**24**:37–49.

327. Lu HX, *et al*. Toxin-produced Purkinje cell death: a model for neural stem cell transplantation studies. *Brain Res* 2008;**1207**:207–13.

328. Ludolph AC, *et al*. 3-Nitropropionic acid-exogenous animal neurotoxin and possible human striatal toxin. *Can J Neurol Sci* 1991;**18**:492–8.

329. Lund SP, Kristiansen GB. Hazards to hearing from combined exposure to toluene and noise in rats. *Int J Occup Med Environ Health* 2008;**21**:47–57.

330. Luty S, *et al*. Subacute toxicity of orally applied alpha-cypermethrin in Swiss mice. *Ann Agric Environ Med* 2000;**7**:33–41.

331. Maeda M, *et al*. Localization of manganese superoxide dismutase in the cerebral cortex and hippocampus of Alzheimer-type senile dementia. *Osaka City Med J* 1997;**43**:1–5.

332. Manning-Bog AB, *et al*. The herbicide paraquat causes up-regulation and aggregation of alpha-synuclein in mice: paraquat and alpha-synuclein. *J Biol Chem* 2002;**277**:1641–4.

333. Markstein R, Lahaye D. Neurochemical investigations in vitro with 1-methyl-4-phenyl-1,2,3,6-tetrahydropyridine (MPTP) in preparations of rat brain. *Eur J Pharmacol* 1984;**106**):301–11.

334. Martin JJ, Schlote W. Central nervous system lesions in disorders of amino-acid metabolism. A neuropathological study. *J Neurol Sci* 1972;**15**:49–76.

335. Martin JJ, Schlote W. Neuropathological study of aminoacidurias. *Monogr Hum Genet* 1972;**6**:64–78.

336. Martin RJ. Central pontine and extra-pontine myelinolysis: the osmotic demyelination syndromes. *J Neurol Neurosurg Psychiatry* 2004;**75**(Suppl III):22–8.

337. Martinez M. Myelin lipids in the developing cerebrum, cerebellum, and brain stem of normal and undernourished children. *J Neurochem* 1982;**39**:1684–92.

338. Maschke M, *et al*. Vermal atrophy of alcoholics correlate with serum thiamine levels but not with dentate iron concentrations as estimated by MRI. *J Neurol* 2005;**252**:704–11.

339. Mattia CJ, Adams JD, Jr, Bondy SC. Free radical induction in the brain and liver by products of toluene catabolism. *Biochem Pharmacol* 1993;**46**:103–10.

340. McCormack AL, *et al*. Environmental risk factors and Parkinson's disease: selective degeneration of nigral dopaminergic neurons caused by the herbicide paraquat. *Neurobiol Dis* 2002;**10**:119–27.

341. McKeown-Eyssen GE, Ruedy J. Methyl mercury exposure in northern Quebec. I. Neurologic findings in adults. *Am J Epidemiol* 1983;**118**:461–9.

342. McLean DR, Jacobs H, Mielke BW. Methanol poisoning: a clinical and pathological study. *Ann Neurol* 1980;**8**:161–7.

343. Meijer WM, *et al*. Folic acid sensitive birth defects in association with intrauterine

exposure to folic acid antagonists. *Reprod Toxicol* 2005;**20**:203–7.

344. Mergler D, *et al.* Nervous system dysfunction among workers with long-term exposure to manganese. *Environ Res* 1994;**64**:151–80.

345. Meyer UA, Schuurmans MM, Lindberg RL. Acute porphyrias: pathogenesis of neurological manifestations. *Semin Liver Dis* 1998;**18**:43–52.

346. Miller MS, Spencer PS. The mechanisms of acrylamide axonopathy. *Annu Rev Pharmacol Toxicol* 1985;**25**:643–66.

347. Mineta H, Miura K, Takebayashi S, *et al.* Immunohistochemical analysis of small cell carcinoma of the head and neck: a report of four patients and a review of sixteen patients in the literature with ectopic hormone production. *Ann Otol Rhinol Laryngol* 2001;**110**:76–82.

348. Ming L. Moldy sugarcane poisoning--a case report with a brief review. *J Toxicol Clin Toxicol* 1995;**33**:363–7.

349. Miro O, *et al.* Mitochondrial cytochrome c oxidase inhibition during acute carbon monoxide poisoning. *Pharmacol Toxicol* 1998;**82**:199–202.

350. Misra UK, Kalita J, Das A. Vitamin B12 deficiency neurological symdromes: a clinical, MRI and electrodiagnostic study. *Electromyogr Clin Neurophysiol* 2003;**43**:57–64.

351. Miyoshi Y, *et al.* Diencephalic syndrome of emaciation in an adult associated with a third ventricle intrinsic craniopharyngioma: case report. *Neurosurgery* 2003;**52**:224–7; discussion 227.

352. Mohan SM, *et al.* Suprasellar germ cell tumor presenting as diencephalic syndrome and precocious puberty. *J Pediatr Endocrinol Metab* 2003;**16**:443–6.

353. Moller HE, *et al.* Brain imaging and proton magnetic resonance spectroscopy in patients with phenylketonuria. *Pediatrics* 2003;**112**(6 Part 2):1580–83.

354. Monzavi R, Kelly DF, Geffner ME. Rathke's cleft cyst in two girls with precocious puberty. *J Pediatr Endocrinol Metab* 2004;**17**:781–5.

355. Morata TC, Campo P. Ototoxic effects of styrene alone or in concert with other agents: a review. *Noise Health* 2002;**4**:15–24.

356. Morata TC, Dunn DE, Sieber WK. Occupational exposure to noise and ototoxic organic solvents. *Arch Environ Health* 1994;**49**:359–65.

357. Morata TC, *et al.* Auditory and vestibular functions after single or combined exposure to toluene: a review. *Arch Toxicol* 1995;**69**:431–43.

358. Moretto A, Lotti M. Poisoning by organophosphorus insecticides and sensory neuropathy. *J Neurol Neurosurg Psychiatry* 1998;**64**:463–8.

359. Mori H, *et al.* Growth hormone-producing pituitary adenoma with crystal-like amyloid immunohistochemically positive for growth hormone. *Cancer* 1985;**55**:96–102.

360. Morishima A, Aranoff GS. Syndrome of septo-optic-pituitary dysplasia: the clinical spectrum. *Brain Dev* 1986;**8**:233–9.

361. Morris CM, *et al.* Comparison of the regional distribution of transferrin receptors and aluminium in the forebrain of chronic renal dialysis patients. *J Neurol Sci* 1989;**94**:295–306.

362. Morton DH, *et al.* Diagnosis and treatment of maple syrup disease: a study of 36 patients. *Pediatrics* 2002;**109**:999–1008.

363. Motterlini R, Otterbein LE. The therapeutic potential of carbon monoxide. *Nat Rev Drug Discov* 2010;**9**:728–43.

364. Mourmans J, *et al.* Sequential MR imaging changes in nonketotic hyperglycinemia. *AJNR Am J Neuroradiol* 2006;**27**:208–11.

365. Mudd SH, Levy HL, Kraus JP. Disorders of transsulfuration. In: Scriver CR *et al.* eds. *The metabolic and molecular basis of inherited disease*, 8 edn. New York: McGraw-Hill, 2001:2007–56.

366. Mukada T, Sasano N, Sato K. Autopsy findings in a case of acute paraquat poisoning with extensive cerebral purpura. *Tohoku J Exp Med* 1978;**125**:253–63.

367. Municchi G, *et al.* Central precocious puberty in multisystem Langerhans cell histiocytosis: a case report. *Pediatr Hematol Oncol* 2002;**19**:273–8.

368. Murata T, *et al.* Serial cerebral MRI with FLAIR sequences in acute carbon monoxide poisoning. *J Comput Assist Tomogr* 1995;**19**:631–4.

369. Murata T, *et al.* Neuronal damage in the interval form of CO poisoning determined by serial diffusion weighted magnetic resonance imaging plus 1H-magnetic resonance spectroscopy. *J Neurol Neurosurg Psychiatry* 2001;**71**:250–53.

370. Nagai I, *et al.* Two cases of hereditary diabetes insipidus, with an autopsy finding in one. *Acta Endocrinol (Copenh)* 1984;**105**:318–23.

371. Nakasaka Y, Atsumi M, Saigoh K, *et al.* Syndrome of inappropriate secretion of antidiuretic hormone (SIADH) associated with relapsing multiple sclerosis. *No To Shinkei* 2005;**57**:51–5.

372. Narongchai P, Narongchai S, Thampituk S. The first fatal case of yam bean and rotenone toxicity in Thailand. *J Med Assoc Thai* 2005;**88**:984–7.

373. Nassogne MC, *et al.* Urea cycle defects: management and outcome. *J Inherit Metab Dis* 2005;**28**:407–14.

374. Neal AP, Worley PF, Guilarte TR. Lead exposure during synaptogenesis alters NMDA receptor targeting via NMDA receptor inhibition. *Neurotoxicology* 2011;**32**:281–9.

375. Nehls DG, Park CK, MacCormack AG, McCulloch J. The effects of N-methyl-D-aspartate receptor blockade with MK-801 upon the relationship between cerebral blood flow and glucose utilisation. *Brain Res* 1990;**511**:271–9.

376. Nemergut EC, *et al.* Predictors of diabetes insipidus after transsphenoidal surgery: a review of 881 patients. *J Neurosurg* 2005;**103**:448–54.

377. Nicholls P. The effect of formate on cytochrome aa3 and on electron transport in the intact respiratory chain. *Biochim Biophys Acta* 1976;**430**:13–29.

378. Nicklas WJ, Vyas I, Heikkila RE. Inhibition of NADH-linked oxidation in brain mitochondria by 1-methyl-4-phenyl-pyridine, a metabolite of the neurotoxin, 1-methyl-4-phenyl-1,2,5,6-tetrahydropyridine. *Life Sci* 1985;**36**:2503–8.

379. Nishino H, *et al.* Acute 3-nitropropionic acid intoxication induces striatal astrocytic cell death and dysfunction of the blood–brain barrier: involvement of dopamine toxicity. *Neurosci Res* 1997;**27**:343–55.

380. Noack S, *et al.* Immunohistochemical localization of neuronal and glial calcium-binding proteins in hippocampus of chronically low level lead exposed rhesus monkeys. *Neurotoxicology* 1996;**17**:679–84.

381. Nogueira K, *et al.* hCG-secreting pineal teratoma causing precocious puberty: report of two patients and review of the literature. *J Pediatr Endocrinol Metab* 2002;**15**:1195–201.

382. Nordmann Y, Puy H. Human hereditary hepatic porphyrias. *Clin Chim Acta* 2002;**325**:17–37.

383. Norenberg MD. Astrocyte responses to CNS injury. *J Neuropathol Exp Neurol* 1994;**53**:213–20.

384. Norenberg MD, Leslie KO, Robertson AS. Association between rise in serum sodium and central pontine myelinolysis. *Ann Neurol* 1982;**11**:128–35.

385. Norman RA. Past and future: porphyria and porphyrins. *Skinmed* 2005;**4**:287–92.

386. O'Byrne KJ, *et al.* Somatostatin, its receptors and analogs, in lung cancer. *Chemotherapy* 2001;**47**(Suppl 2):78–108.

387. Ochoa J. Isoniazid neuropathy in man: quantitative electron microscope study. *Brain* 1970;**93**:831–50.

388. Odabas D, *et al.* Cranial MRI findings in children with protein energy malnutrition. *Int J Neurosci* 2005;**115**:829–37.

389. O'Donnell P, *et al.* The magnetic resonance imaging appearances of the brain in acute carbon monoxide poisoning. *Clin Radiol* 2000;**55**:273–80.

390. Ogura H, *et al.* Neurotoxic damage of granule cells in the dentate gyrus and the cerebellum and cognitive deficit following neonatal administration of phenytoin in mice. *J Neuropathol Exp Neurol* 2002;**61**:956–67.

391. Ohkoshi N, *et al.* Dysfunction of the hypothalamic-pituitary system in mitochondrial encephalomyopathies. *J Med* 1998;**29**:13–29.

392. Ohnishi A, *et al.* Beriberi neuropathy. Morphometric study of sural nerve. *J Neurol Sci* 1980;**45**:177–90.

393. Ohnishi A, Chua CL, Kuroiwa Y. Axonal degeneration distal to the site of accumulation of vesicular profiles in the myelinated fiber axon in experimental isoniazid neuropathy. *Acta Neuropathol* 1985;**67**:195–200.

394. Okeda R, *et al.* Comparative study on pathogenesis of selective cerebral lesions in carbon monoxide poisoning and nitrogen hypoxia in cats. *Acta Neuropathol* 1982;**56**:265–72.

395. Okeda R, *et al.* Vascular changes in acute Wernicke's encephalopathy. *Acta Neuropathol* 1995;**89**:420–24.

396. Osella G, *et al.* Acromegaly due to ectopic secretion of GHRH by bronchial carcinoid in a patient with empty sella. *J Endocrinol Invest* 2003;**26**:163–9.

397. Osimitz TG, *et al.* Adverse events associated with the use of insect repellents containing N,N-diethyl-m-toluamide (DEET). *Regul Toxicol Pharmacol* 2009;**56**:93–9.

398. Owen PJ, Miles DP. A review of hospital discharge rates in a population around Camelford in North Cornwall up to the fifth anniversary of an episode of aluminium sulphate absorption. *J Public Health Med* 1995;**17**:200–204.

399. Palfi S, *et al.* Delayed onset of progressive dystonia following subacute 3-nitropropionic acid treatment in Cebus apella monkeys. *Mov Disord* 2000;**15**:524–30.

400. Pamphlett R, Waley P. Uptake of inorganic mercury by the human brain. *Acta Neuropathol* 1996;**92**:525–7.

9

401. Pappas CL, *et al.* Lead encephalopathy: symptoms of a cerebellar mass lesion and obstructive hydrocephalus. *Surg Neurol* 1986;**26**:391–4.

402. Park S, Choi IS. Chorea following acute carbon monoxide poisoning. *Yonsei Med J* 2004;**45**:363–6.

403. Park SH, *et al.* Magnetic resonance reflects the pathological evolution of Wernicke encephalopathy. *J Neuroimaging* 2001;**11**:406–11.

404. Parkinson RB, *et al.* White matter hyperintensities and neuropsychological outcome following carbon monoxide poisoning. *Neurology* 2002;**58**:1525–32.

405. Parry TE. Folate responsive neuropathy. *Presse Medicale* 1994;**23**:131–7.

406. Patel F. Pesticidal suicide: adult fatal rotenone poisoning. *J Forensic Leg Med* 2011;**18**:340–42.

407. Pavese N, *et al.* Clinical outcome and magnetic resonance imaging of carbon monoxide intoxication. A long-term follow-up study. *Ital J Neurol Sci* 1999;**20**:171–8.

408. Pelc S. The diencephalic syndrome in infants. A review in relation to optic nerve glioma. *Eur Neurol* 1972;**7**:321–34.

409. Pelc S, Flament-Durand J. Histological evidence of optic chiasma glioma in the "diencephalic syndrome". *Arch Neurol* 1973;**28**:139–40.

410. Pentore R, Venneri A, Nichelli P. Accidental choke-cherry poisoning: early symptoms and neurological sequelae of an unusual case of cyanide intoxication. *Ital J Neurol Sci* 1996;**17**:233–5.

411. Perez-Otano I, *et al.* Extensive loss of brain dopamine and serotonin induced by chronic administration of MPTP in the marmoset. *Brain Res* 1991;**567**:127–32.

412. Perilongo G, *et al.* Diencephalic syndrome and disseminated juvenile pilocytic astrocytomas of the hypothalamic-optic chiasm region. *Cancer* 1997;**80**:142–6.

413. Perl TM, *et al.* An outbreak of toxic encephalopathy caused by eating mussels contaminated with domoic acid. *N Engl J Med* 1990;**322**:1775–80.

414. Perry EK, *et al.* Autoradiographic comparison of cholinergic and other transmitter receptors in the normal human hippocampus. *Hippocampus* 1993;**3**:307–15.

415. Petito CK, Navia BA, Cho ES, *et al.* Vacuolar myelopathy pathologically resembling subacute combined degeneration in patients with the acquired immunodeficiency syndrome. *N Engl J Med* 1985;**312**:874–9.

416. Petrides PE. Acute intermittent porphyria: mutation analysis and identification of gene carriers in a German kindred by PCR-DGGE analysis. *Skin Pharmacol Appl Skin Physiol* 1998;**11**:374–80.

417. Pezzoli G, *et al.* Clinical and pathological features in hydrocarbon-induced parkinsonism. *Ann Neurol* 1996;**40**:922–5.

418. Pfefferbaum A, *et al.* Brain gray and white matter volume loss accelerates with aging in chronic alcoholics: a quantitative MRI study. *Alcohol Clin Exp Res* 1992;**16**:1078–89.

419. Pfefferbaum A, *et al.* Increase in brain cerebrospinal fluid volume is greater in older than in younger alcoholic patients: a replication study and CT/MRI comparison. *Psychiatry Res* 1993;**50**:257–74.

420. Pfefferbaum A, *et al.* Frontal lobe volume loss observed with magnetic resonance imaging in older chronic alcoholics. *Alcohol Clin Exp Res* 1997;**21**:521–9.

421. Phillip M, Lazar L. Precocious puberty: growth and genetics. *Horm Res* 2005;**64**(Suppl 2):56–61.

422. Phillips SC, Harper CG, Kril JJ. A quantitative histological study of the cerebellar vermis in alcoholic patients. *Brain* 1987;**110**:301–14.

423. Phillips SC, Harper CG, Kril JJ. The contribution of Wernicke's encephalopathy to alcohol-related cerebellar damage. *Drug Alcohol Rev* 1990;**9**:53–60.

424. Pitel A-L, *et al.* Signs of preclinical Wernicke's encephalopathy and thiamine levels as predictors of neuropsychological deficits in alcoholism without Korsakoff's syndrome. *Neuropsychopharmacology* 2011;**36**:580–88.

425. Pivonello R, *et al.* Central diabetes insipidus and autoimmunity: relationship between the occurrence of antibodies to arginine vasopressin-secreting cells and clinical, immunological, and radiological features in a large cohort of patients with central diabetes insipidus of known and unknown etiology. *J Clin Endocrinol Metab* 2003;**88**:1629–36.

426. Platt LD, *et al.* The international study of pregnancy outcome in women with maternal phenylketonuria: report of a 12-year study. *Am J Obstet Gynecol* 2000;**182**:326–33.

427. Polizzi A, *et al.* Septo-optic dysplasia complex: a heterogeneous malformation syndrome. *Pediatr Neurol* 2006;**34**:66–71.

428. Porter SS, *et al.* Corpus callosum atrophy and neuropsychological outcome following carbon monoxide poisoning. *Arch Clin Neuropsychol* 2002;**17**:195–204.

429. Poussaint TY, *et al.* Diencephalic syndrome: clinical features and imaging findings. *AJNR Am J Neuroradiol* 1997;**18**:1499–505.

430. Powers JM, Rawe SE, Earlywine GR. Lead encephalopathy simulating a cerebral neoplasm in an adult. Case report. *J Neurosurg* 1977;**46**:816–19.

431. Press MF. Lead encephalopathy in neonatal Long-Evans rats: morphologic studies. *J Neuropathol Exp Neurol* 1977;**36**:169–93.

432. Prevention of neural tube defects: results of the Medical Research Council Vitamin Study. *Lancet* 1991;**338**:131–7.

433. Prockop LD, Chichkova RI. Carbon monoxide intoxication: an updated review. *J Neurol Sci* 2007;**262**:122–30.

434. Pulido OM. Domoic acid toxicologic pathology: a review. *Mar Drugs* 2008;**6**:180–219.

435. Quan L, *et al.* Intranuclear ubiquitin immunoreactivity in the pigmented neurons of the substantia nigra in fire fatalities. *Int J Legal Med* 2001;**114**:310–15.

436. Quere I, *et al.* [Homocystinuria in adulthood]. *Rev Med Interne* 2001;**22**(Suppl 3):347s–55s.

437. Rachinger J, *et al.* MR changes after acute cyanide intoxication. *AJNR Am J Neuroradiol* 2002;**23**:1398–401.

438. Ragothaman M, *et al.* Elemental mercury poisoning probably causes cortical myoclonus. *Mov Disord* 2007;**22**:1964–8.

439. Rainier S, *et al.* Motor neuron disease due to neuropathy target esterase gene mutation: clinical features of the index families. *Muscle Nerve* 2010;**43**:19–25.

440. Ramaekers VT. Cerebral folate deficiency. *Dev Med Child Neurol* 2004;**46**:843–851.

441. Ramsay RR, Dadgar J, Trevor A, Singer TP. Energy-driven uptake of N-methyl-4-phenylpyridine by brain mitochondria mediates the neurotoxicity of MPTP. *Life Sci* 1986;**39**:581–8.

442. Ramsay RR, Salach JI, Singer TP. Uptake of the neurotoxin 1-methyl-4-phenylpyridine (MPP+) by mitochondria and its relation to the inhibition of the mitochondrial oxidation of NAD+-linked substrates by MPP+. *Biochem Biophys Res Commun* 1986;**134**:743–8.

443. Rasmussen LS, *et al.* Biochemical markers for brain damage after carbon monoxide poisoning. *Acta Anaesthesiol Scand* 2004;**48**:469–73.

444. Ray DE. Electroencephalographic and evoked response correlates of trimethyltin induced neuronal damage in the rat hippocampus. *J Appl Toxicol* 1981;**1**:145–8.

445. Ray DE, Fry JR. A reassessment of the neurotoxicity of pyrethroid insecticides. *Pharmacol Ther* 2006;**111**:174–93.

446. Ray DE, *et al.* Functional/metabolic modulation of the brain stem lesions caused by 1,3-dinitrobenzene in the rat. *Neurotoxicology* 1992;**13**:379–88.

447. Refsum H. Folate, vitamin B12 and homocysteine in relation to birth defects and pregnancy outcome. *Br J Nutr* 2001;**85**(Suppl2):S109–13.

448. Refsum H, *et al.* Birth prevalence of homocystinuria. *J Pediatr* 2004;**144**:830–32.

449. Reynolds CR, Hopkins RO, Bigler ED. Continuing decline of memory skills with significant recovery of intellectual function following severe carbon monoxide exposure: clinical, psychometric, and neuroimaging findings. *Arch Clin Neuropsychol* 1999;**14**:235–49.

450. Reynolds DS, Morton AJ. Changes in blood–brain barrier permeability following neurotoxic lesions of rat brain can be visualised with trypan blue. *J Neurosci Meth* 1998;**79**:115–21.

451. Richardson JR, *et al.* Paraquat neurotoxicity is distinct from that of MPTP and rotenone. *Toxicol Sci* 2005;**88**:193–201.

452. Rimoin DL, Schechter JE. Histological and ultrastructural studies in isolated growth hormone deficiency. *J Clin Endocrinol Metab* 1973;**37**:725–35.

453. Riudavets MA, Aronica-Pollak P, Troncoso JC. Pseudolaminar necrosis in cyanide intoxication: a neuropathology case report. *Am J Forensic Med Pathol* 2005;**26**:189–91.

454. Rivarola, *et al.* Precocious puberty in children with tumours of the suprasellar and pineal areas: organic central precocious puberty. *Acta Paediatr* 2001;**90**:751–6.

455. Roitberg BZ, *et al.* Behavioral and morphological comparison of two nonhuman primate models of Huntington's disease. *Neurosurgery* 2002;**50**:137–45; discussion 145–6.

456. Romero I, *et al.* Vascular factors in the neurotoxic damage caused by 1,3-dinitrobenzene in the rat. *Neuropathol Appl Neurobiol* 1991;**17**:495–508.

457. Rose CF. Increase brain lactate in hepatic encephalopathy: cause or consequence? *Neurochem Int* 2010;**57**:389–94.

458. Rosenberg NL, Myers JA, Martin WR. Cyanide-induced parkinsonism: clinical, MRI, and 6-fluorodopa PET studies. *Neurology* 1989;**39**:142–4.

459. Rosenberg RN. *Neurogenetics: principles and practice*. New York: Raven Press, 1985.

460. Rosenblatt DS, *et al.* Clinical heterogeneity and prognosis in combined methylmalonic aciduria and homocystinuria (cblC). *J Inherit Metab Dis* 1997;**20**:528–38.

461. Rosenow F, *et al.* Neurological sequelae of cyanide intoxication--the patterns of clinical, magnetic resonance imaging, and positron emission tomography findings. *Ann Neurol* 1995;**38**:825–8.

462. Rossi A, *et al.* Early-onset combined methylmalonic aciduria and homocystinuria: neuroradiologic findings. *AJNR Am J Neuroradiol* 2001;**22**:554–63.

463. Rudez J, Sepcic K, Sepcic J. Vaginally applied diquat intoxication. *J Toxicol Clin Toxicol* 1999;**37**:877–9.

464. Russell JW, *et al.* Effect of cisplatin and ACTH4-9 on neural transport in cisplatin induced neurotoxicity. *Brain Res* 1995;**676**:258–67.

465. Russo PA, Mitchell GA, Tanguay RM. Tyrosinemia: a review. *Pediatr Dev Pathol* 2001;**4**:212–21.

466. Rusyniak DE, Furbee RB, Kirk MA. Thallium and arsenic poisoning in a small midwestern town. *Ann Emerg Med* 2002;**39**:307–11.

467. Rutland BM, Edgar MA, Horenstein MG. Hypomelanosis of Ito associated with precocious puberty. *Pediatr Neurol* 2006;**34**:51–4.

468. Saeed SA, Wilks MF, Coupe M. Acute diquat poisoning with intracerebral bleeding. *Postgrad Med J* 2001;**77**:329–32.

469. Sakamoto K, *et al.* Clinical studies on three cases of the interval form of carbon monoxide poisoning: serial proton magnetic resonance spectroscopy as a prognostic predictor. *Psychiatry Res* 1998;**83**:179–92.

470. Sano T, Asa SL, Kovacs K. Growth hormone-releasing hormone-producing tumors: clinical, biochemical, and morphological manifestations. *Endocr Rev* 1988;**9**:357–73.

471. Sass JO, *et al.* Propionic acidemia revisited: a workshop report. *Clin Pediatr (Phila)* 2004;**43**:837–43.

472. Scaglia F, Lee B. Clinical, biochemical, and molecular spectrum of hyperargininemia due to arginase I deficiency. *Am J Med Genet C Semin Med Genet* 2006;**142C**:113–20.

473. Scallet AC, *et al.* Domoic acid-treated cynomolgus monkeys (*M. fascicularis*): effects of dose on hippocampal neuronal and terminal degeneration. *Brain Res* 1993;**627**:307–13.

474. Schils F, *et al.* Unusual CT and MRI appearance of carbon monoxide poisoning. *JBR-BTR* 1999;**82**:13–15.

475. Schonberger S, *et al.* Dysmyelination in the brain of adolescents and young adults with maple syrup urine disease. *Mol Genet Metab* 2004;**82**:69–75.

476. Schroeder JM. *Pathology of peripheral nerves.* Heidelberg: Springer-Verlag, 2001.

477. Schroth G, *et al.* Reversible brain shrinkage in abstinent alcoholics, measured by MRI. *Neuroradiology* 1988;**30**:385–9.

478. Schwarz GA, Moulton JA. Porphyria; a clinical and neuropathologic report. *AMA Arch Intern Med* 1954;**94**:221–47.

479. Sechi GP, *et al.* Acute and persistent parkinsonism after use of diquat. *Neurology* 1992;**42**:261–3.

480. Sener RN. Nonketotic hyperglycinemia: diffusion magnetic resonance imaging findings. *J Comput Assist Tomogr* 2003;**27**:538–40.

481. Sesay M, *et al.* Regional cerebral blood flow measurements with xenon CT in the prediction of delayed encephalopathy after carbon monoxide intoxication. *Acta Neurol Scand Suppl* 1996;**166**:22–7.

482. Setian N, *et al.* Precocious puberty: an endocrine manifestation in congenital toxoplasmosis. *J Pediatr Endocrinol Metab* 2002;**15**:1487–90.

483. Sharma RR, Chandy MJ, Lad SD. Diencephalic syndrome of emaciation in an adult associated with a suprasellar craniopharyngioma--a case report. *Br J Neurosurg* 1990;**4**:77–80.

484. Sharpe JA, *et al.* Methanol optic neuropathy: a histopathological study. *Neurology* 1982;**32**:1093–100.

485. Shenoy SN, Raja A. Hypothalamic hamartoma with precocious puberty. *Pediatr Neurosurg* 2004;**40**:249–52.

486. Sherer TB, Betarbet R, Kim JH, Greenamyre JT. Selective microglial activation in the rat rotenone model of Parkinson's disease. *Neurosci Lett* 2003;**341**:87–90.

487. Sherer TB, Kim JH, Betarbet R, Greenamyre JT. Subcutaneous rotenone exposure causes highly selective dopaminergic degeneration and alpha-synuclein aggregation. *Exp Neurol* 2003;**179**:9–16.

488. Shield LK, Coleman TL, Markesbery WR. Methyl bromide intoxication: neurologic features, including simulation of Reye syndrome. *Neurology* 1977;**27**:959–62.

489. Shih HT, *et al.* Subclinical abnormalities in workers with continuous low-level toluene exposure. *Toxicol Ind Health* 2011;**7**:691–9.

490. Shih TM, Duniho SM, McDonough JH. Control of nerve agent-induced seizures is critical for neuroprotection and survival. *Toxicol Appl Pharmacol* 2003;**188**:69–80.

491. Shuman RM, Leech RW, Alvord, Jr, EC. Neurotoxicity of hexachlorophene in humans. II. A clinicopathological study of 46 premature infants. *Arch Neurol* 1975;**32**:320–5.

492. Shuman RM, Leech RW, Scott CR. The neuropathology of the nonketotic and ketotic hyperglycinemias: three cases. *Neurology* 1978;**28**:139–46.

493. Siegler RW, Nierenberg DW, Hickey WF. Fatal poisoning from liquid dimethylmercury: a neuropathologic study. *Hum Pathol* 1999;**30**:720–3.

494. Sikk K, *et al.* Irreversible motor impairment in young addicts: ephedrone, manganism or both? *Acta Neurol Scand* 2007;**115**:385–9.

495. Sills RC, *et al.* Characterization of carbon disulfide neurotoxicity in C57BL6 mice: behavioral, morphological, and molecular effects. *Toxicol Pathol* 2000;**28**:142–8.

496. Siminoski K, D'Costa M, Walfish PG. Hypogonadotropic hypogonadism in idiopathic hemochromatosis: evidence for combined hypothalamic and pituitary involvement. *J Endocrinol Invest* 1990;**13**:849–53.

497. Skrzydlewska E. Toxicological and metabolic consequences of methanol poisoning. *Toxicol Mech Methods* 2003;**13**:277–93.

498. Skullerud K. Variations in the size of the human brain. Influence of age, sex, body length, body mass index, alcoholism, Alzheimer changes, and cerebral atherosclerosis. *Acta Neurol Scand* 1985;**102**:1–94.

499. Smith DB, *et al.* Dialysis encephalopathy in peritoneal dialysis. *JAMA* 1980;**244**:365–6.

500. Smith W, *et al.* Urea cycle disorders: clinical presentation outside the newborn period. *Crit Care Clin* 2005;**21**(4 Suppl):S9–17.

501. Sobotka TJ, *et al.* Domoic acid: neurobehavioral and neurohistological effects of low-dose exposure in adult rats. *Neurotoxicol Teratol* 1996;**18**:659–70.

502. Solberg Y, Belkin M. The role of excitotoxicity in organophosphorous nerve agents central poisoning. *Trends Pharmacol Sci* 1997;**18**:183–5.

503. Solis C, *et al.* Acute intermittent porphyria: studies of the severe homozygous dominant disease provides insights into the neurologic attacks in acute porphyrias. *Arch Neurol* 2004;**61**:1764–70.

504. Song SY, *et al.* An experimental study of the pathogenesis of the selective lesion of the globus pallidus in acute carbon monoxide poisoning in cats. With special reference to the chronologic change in the cerebral local blood flow. *Acta Neuropathol* 1983;**61**:232–8.

505. Squier MV, Thompson J, Rajgopalan B. Case report: neuropathology of methyl bromide intoxication. *Neuropathol Appl Neurobiol* 1992;**18**:579–84.

506. Starks SE, *et al.* Peripheral nervous system function and organophosphate pesticide use among licensed pesticide applicators in the Agricultural Health Study. *Environ Health Perspect* 2012;**120**:515–20.

507. Starzyk J, *et al.* Suprasellar arachnoidal cyst as a cause of precocious puberty: report of three patients and literature overview. *J Pediatr Endocrinol Metab* 2003;**16**:447–55.

508. Stewart GR, *et al.* Domoic acid: a dementia-inducing excitotoxic food poison with kainic acid receptor specificity. *Exp Neurol* 1990;**110**:127–38.

509. Strain SM, Tasker RA. Hippocampal damage produced by systemic injections of domoic acid in mice. *Neuroscience* 1991;**44**:343–52.

510. Strauss KA, Puffenberger AG, Morton DH. Maple syrup urine disease. In: Pagon R, Bird T, Dolan C *et al.* eds. *GeneReviews.* Seattle, WA: University of Washington, 2006.

511. Suarez JI, *et al.* Acute intermittent porphyria: clinicopathologic correlation. Report of a case and review of the literature. *Neurology* 1997;**48**:1678–83.

512. Sullivan EV, Pfefferbaum A. Neuroimaging of the Wernicke-Korsakoff syndrome. *Alcohol Alcohol* 2009;**44**:155–65.

513. Susa S, *et al.* Acute intermittent porphyria with central pontine myelinolysis and cortical laminar necrosis. *Neuroradiology* 1999;**41**:835–9.

514. Sutherland RJ, Hoesing JM, Whishaw IQ. Domoic acid, an environmental toxin, produces hippocampal damage and severe memory impairment. *Neurosci Lett* 1990;**120**:221–3.

515. Suwanlaong K, Phanthumchinda K. Neurological manifestation of methyl bromide intoxication. *J Med Assoc Thai* 2008;**91**:421–6.

516. Tabandeh H, Crowston JG, Thompson GM. Ophthalmologic features of thallium poisoning. *Am J Ophthalmol* 1994;**117**:243–5.

517. Tai WP, Yue H, Hu PJ. Coma caused by isoniazid poisoning in a patient treated

with pyridoxine and hemodialysis. *Adv Ther* 2008;**25**:1085–8.

518. Tanaka J, *et al.* Neuropathological study on 1-methyl-4-phenyl-1,2,3,6-tetrahydropyridine of the crab-eating monkey. *Acta Neuropathol* 1988;**75**: 370–76.

519. Teitelbaum JS, *et al.* Neurologic sequelae of domoic acid intoxication due to the ingestion of contaminated mussels. *N Engl J Med* 1990;**322**:1781–7.

520. Tephly TR. The toxicity of methanol. *Life Sci* 1991;**48**:1031–41.

521. Thakar JH, Hassan MN. Effects of 1-methyl-4-phenyl-1,2,3,6-tetrahydropyridine (MPTP), cyperquat (MPP+) and paraquat on isolated mitochondria from rat striatum, cortex and liver. *Life Sci* 1988;**43**:143–9.

522. Thiruchelvam M, *et al.* Age-related irreversible progressive nigrostriatal dopaminergic neurotoxicity in the paraquat and maneb model of the Parkinson's disease phenotype. *Eur J Neurosci* 2003;**18**:589–600.

523. Thom SR, *et al.* Delayed neuropathology after carbon monoxide poisoning is immune-mediated. *Proc Natl Acad Sci U S A* 2004;**101**:13660–65.

524. Thompson AJ, *et al.* Brain MRI changes in phenylketonuria. Associations with dietary status. *Brain* 1993;**116**(Part 4):811–21.

525. Thompson C, Dent J, Saxby P. Effects of thallium poisoning on intellectual function. *Br J Psychiatry* 1988;**153**:396–9.

526. Thompson SW, *et al.* Cisplatin neuropathy. Clinical, electrophysiologic, morphologic, and toxicologic studies. *Cancer* 1984;**54**:1269–75.

527. Thornalley PJ, *et al.* High prevalence of low plasma thiamine concentration in diabetes linked to a marker of vascular disease. *Diabetologia* 2007;**50**:2164–70.

528. Toledo SP, Luthold W, Mattar E. Familial idiopathic gonadotropin deficiency: a hypothalamic form of hypogonadism. *Am J Med* 1983;**15**:405–16.

529. Tomiwa K, Nolan C, Cavanagh JB. The effects of cisplatin on rat spinal ganglia: a study by light and electron microscopy and by morphometry. *Acta Neuropathol* 1986;**69**:295–308.

530. Toone JR, *et al.* Novel mutations in the P-protein (glycine decarboxylase) gene in patients with glycine encephalopathy (non-ketotic hyperglycinemia). *Mol Genet Metab* 2002;**76**:243–9.

531. Topp KS, Tanner KD, Levine JD. Damage to the cytoskeleton of large diameter sensory neurons and myelinated axons in vincristine-induced painful peripheral neuropathy in the rat. *J Comp Neurol* 2000;**424**:563–76.

532. Trabert W, *et al.* Significant reversibility of alcoholic brain shrinkage within 3 weeks of abstinence. *Acta Psychiatr Scand* 1995;**92**:87–90.

533. Triebig G, Hallermann J. Survey of solvent related chronic encephalopathy as an occupational disease in European countries. *Occup Environ Med* 2001;**58**:575–81.

534. Tryphonas L, Iverson F. Neuropathology of excitatory neurotoxins: the domoic acid model. *Toxicol Pathol* 1990;**18**(1 Part 2):165–9.

535. Tryphonas L, Truelove J, Iverson F. Acute parenteral neurotoxicity of domoic acid in cynomolgus monkeys (*M. fascicularis*). *Toxicol Pathol* 1990;**18**:297–303.

536. Tryphonas L, Truelove J, Iverson F, Todd EC, Nera EA. Neuropathology of experimental domoic acid poisoning in non-human primates and rats. *Can Dis Wkly Rep* 1990;**16**(Suppl 1E):75–81.

537. Tuchman RF, Moser FG, Moshe SL. Carbon monoxide poisoning: bilateral lesions in the thalamus on MR imaging of the brain. *Pediatr Radiol* 1990;**20**: 478–9.

538. Tun P, *et al.* Acute and chronic pituitary failure resembling Sheehan's syndrome following bites by Russell's viper in Burma. *Lancet* 1987;**2**:763–7.

539. Turkmen N, *et al.* Glial fibrillary acidic protein (GFAP) and CD34 expression in the human optic nerve and brain in methanol toxicity. *Adv Ther* 2008;**25**:123–32.

540. Turner M, Kemp PM. Isotope brain scanning with Tc-HMPAO: a predictor of outcome in carbon monoxide poisoning? *J Accid Emerg Med* 1997;**14**:139–41.

541. Uchino M, *et al.* Clinical investigation of the lesions responsible for sensory disturbance in Minamata disease. *Tohoku J Exp Med* 2001;**195**:181–9.

542. Udani PM. Kwashiorkor myelopathy. *Indian J Child Health* 1962;**11**:498.

543. Uemura K, *et al.* Apoptotic and necrotic brain lesions in a fatal case of carbon monoxide poisoning. *Forensic Sci Int* 2001;**116**:213–19.

544. Uitti RJ, *et al.* Cyanide-induced parkinsonism: a clinicopathologic report. *Neurology* 1985;**35**:921–5.

545. Vahidnia A, van der Voet GB, de Wolff FA. Arsenic neurotoxicity: a review. *Hum Exp Toxicol* 2007;**26**:823–32.

546. Valenzuela R, Court J, Godoy J. Delayed cyanide induced dystonia. *J Neurol Neurosurg Psychiatry* 1992;**55**:198–9.

547. Valiquette G. The neurohypophysis. *Neurol Clin* 1987;**5**:291–331.

548. van der Wal EJ, Azzarelli B, Edwards-Brown M. Malignant transformation of a chiasmatic pilocytic astrocytoma in a patient with diencephalic syndrome. *Pediatr Radiol* 2003;**33**:207–10.

549. Vanholder R, *et al.* Diquat intoxication: report of two cases and review of the literature. *Am J Med* 1981;**70**:1267–71.

550. van Valen E, *et al.* The course of chronic solvent induced encephalopathy: a systematic review. *Neurotoxicology* 2009;**30**:1172–86.

551. Varastet M, *et al.* Chronic MPTP treatment reproduces in baboons the differential vulnerability of mesencephalic dopaminergic neurons observed in Parkinson's disease. *Neuroscience* 1994;**63**:47–56.

552. Vasilescu C, Florescu A. Clinical and electrophysiological study of neuropathy after organophosphorus compounds poisoning. *Arch Toxicol* 1980;**43**:305–15.

553. Verschoyle RD, *et al.* A comparison of the acute toxicity, neuropathology, and electrophysiology of N,N-diethyl-m-toluamide and N,N-dimethyl-2,2-diphenylacetamide in rats. *Fundam Appl Toxicol* 1992;**18**:79–88.

554. Victor M, Adams RD, Collins GH. *The Wernicke–Korsakoff syndrome and related neurological disorders of alcoholism and malnutrition*, 2nd edn. Contemporary Neurology Series. Philadelphia: FA Davis Co, 1989.

555. Viola A, *et al.* Magnetic resonance spectroscopy study of glycine pathways in nonketotic hyperglycinemia. *Pediatr Res* 2002;**52**: 292–300.

556. Virdis R, *et al.* Neurofibromatosis type 1 and precocious puberty. *J Pediatr Endocrinol Metab* 2000;**13**(Suppl 1):841–4.

557. Visrutaratna P, Oranratanachai K. Clinics in diagnostic imaging (81). Hypothalamic glioma with diencephalic syndrome. *Singapore Med J* 2003;**44**:45–50.

558. Voigt C, *et al.* Amyloid in pituitary adenomas. *Pathol Res Pract* 1988;**183**:555–7.

559. Walsh TJ, *et al.* Triethyl and trimethyl lead: effects on behavior, CNS morphology and concentrations of lead in blood and brain of rat. *Neurotoxicology* 1986;**7**:21–33.

560. Walter JH, *et al.* Biochemical control, genetic analysis and magnetic resonance imaging in patients with phenylketonuria. *Eur J Pediatr* 1993;**152**:822–7.

561. Wang J, *et al.* A peptide inhibitor of c-Jun N-terminal kinase protects against both aminoglycoside and acoustic trauma-induced auditory hair cell death and hearing loss. *J Neurosci* 2003;**23**:8596–607.

562. Wang L, *et al.* Thirteen-year follow-up of patients with tri-ortho-cresyl phosphate poisoning in northern suburbs of Xi'an in China. *Neurotoxicology* 2009;**30**:1084–7.

563. Watanabe N, *et al.* Statistical parametric mapping in brain single photon computed emission tomography after carbon monoxide intoxication. *Nucl Med Commun* 2002;**23**:355–66.

564. Waters CM, *et al.* An immunohistochemical study of the acute and long-term effects of 1-methyl-4-phenyl-1,2,3,6-tetrahydropyridine in the marmoset. *Neuroscience* 1987;**23**:1025–39.

565. Weber GA, Sloan P, Davies D. Nutritionally induced peripheral neuropathies. *Clin Podiatr Med Surg* 1990;**7**:107–28.

566. Weinstock JV, Elliott D. The somatostatin immunoregulatory circuit present at sites of chronic inflammation. *Eur J Endocrinol* 2000;**143**(Suppl 1):S15–19.

567. White HH, *et al.* Homocystinuria. *Arch Neurol* 1965;**13**:455–70.

568. White RF, *et al.* Magnetic resonance imaging (MRI), neurobehavioral testing, and toxic encephalopathy: two cases. *Environ Res* 1993;**61**:117–23.

569. WHO. Pesticide residues in food - 1999. Geneva: World Health Organization, 2001.

570. Wiggins R.C., Myelin development and nutritional insufficiency. *Brain Research* 1982;**257**:151–57.

571. Wooley JA. Characteristics of thiamin and its relevance to the management of heart failure. *Nutr Clin Pract* 2008;**23**:487–93.

572. Wright JM, *et al.* Chronic parkinsonism secondary to intranasal administration of a product of meperidine-analogue synthesis. *N Engl J Med* 1984;**310**:325.

573. Wu A, Liu Y. Prolonged expression of c-Fos and c-Jun in the cerebral cortex of rats after deltamethrin treatment. *Brain Res Mol Brain Res* 2003;**110**:147–51.

574. Wullner U, *et al.* 3-Nitropropionic acid toxicity in the striatum. *J Neurochem* 1994;**63**:1772–81.

575. Xiong J, *et al.* MR imaging of "spray heads": toluene abuse via aerosol paint inhalation. *AJNR Am J Neuroradiol* 1993;**14**:1195–9.

576. Xu Y, *et al.* Clinical manifestations and arsenic methylation after a rare subacute arsenic poisoning accident. *Toxicol Sci* 2008;**103**:278–84.

577. Yamada K, *et al*. A case of subacute combined degeneration: MRI findings. *Neuroradiology* 1998;**40**:398–400.

578. Yamada M, *et al*. An autopsy case of acute porphyria with a decrease of both uroporphyrinogen I synthetase and ferrochelatase activities. *Acta Neuropathol* 1984;**64**:6–11.

579. Yamada M, *et al*. Chronic manganese poisoning: a neuropathological study with determination of manganese distribution in the brain. *Acta Neuropathol* 1986;**70**:273–8.

580. Yamamoto I, *et al*. Correlating the severity of paraquat poisoning with specific hemodynamic and oxygen metabolism variables. *Crit Care Med* 2000;**28**:1877–83.

581. Yamamoto M, *et al*. Demonstration of slow acetylator genotype of N-acetyltransferase in isoniazid neuropathy using an archival hematoxylin and eosin section of a sural nerve biopsy specimen. *J Neurol Sci* 1996;**135**:51–4.

582. Yang CY, *et al*. Anterior pituitary failure (panhypopituitarism) with balanced chromosome translocation 46,XY,t(11;22)(q24;q13). *Zhonghua Yi Xue Za Zhi (Taipei)* 2001;**64**:247–52.

583. Yano BL, *et al*. Abnormal auditory brainstem responses and cochlear pathology in rats induced by an exaggerated styrene exposure regimen. *Toxicol Pathol* 1992;**20**:1–6.

584. Zahr NM, *et al*. Contributions of studies on alcohol use disorders to understanding cerebellar function. *Neuropsychol Rev* 2010;**20**:280–89.

585. Zahr NM, Kaufman KL, Harper CG. Clinical and pathological features of alcohol-related brain damage. *Nat Rev Neurol* 2011;**7**:284–94.

586. Zaknun JJ, *et al*. Cyanide-induced akinetic rigid syndrome: clinical, MRI, FDG-PET, beta-CIT and HMPAO SPECT findings. *Parkinsonism Relat Disord* 2005;**11**:125–9.

587. Zangeneh F, *et al*. Cushing's syndrome due to ectopic production of corticotropin-releasing hormone in an infant with ganglioneuroblastoma. *Endocr Pract* 2003;**9**:394–9.

588. Zhu M, *et al*. Formation and structure of cross-linking and monomeric pyrrole autoxidation products in 2,5-hexanedione-treated amino acids, peptides, and protein. *Chem Res Toxicol* 1994;**7**:551–8.

589. Zhu M, *et al*. Inhibition of 2,5-hexanedione-induced protein cross-linking by biological thiols: chemical mechanisms and toxicological implications. *Chem Res Toxicol* 1995;**8**:764–71.

590. Zschocke J. Phenylketonuria mutations in Europe. *Hum Mutat* 2003;**21**:345–56.

Trauma

Colin Smith, Susan Margulies and Ann-Christine Duhaime

INTRODUCTION

Epidemiology

Neurotrauma is a major cause of morbidity and mortality worldwide, and yet in many cases an avoidable one, being a reflection of the availability of fast, personal transport accounting for many road traffic accidents; the easy availability of firearms in many societies; casual violence in society, often fuelled by alcohol; and the lax approach to health and safety in some countries. In western societies, road traffic accident–related neurotrauma has been in decline over the past few decades. However, in developing countries, where medical care is often already overstretched, the incidence has been increasing.[364] The human brain is highly vulnerable to injury, and injury can compromise the quality of life through profound cognitive and neurobehavioral dysfunction.[86] Traumatic brain injury (TBI) is an overwhelming and major global public health problem and one of the most important causes of morbidity and mortality in both industrialized and developing countries.[415] Estimates by the World Health Organization (WHO) indicate 57 million people internationally have been hospitalized with one or more TBIs.[326]

In the United States, recent statistics from the National Center for Injury Prevention and Control (NCIPC) reveal that approximately 1.7 million cases of TBI were reported each year from 2002 to 2006 (www.cdc.gov/traumaticbraininjury/pdf/blue_book.pdf). Of those patients, about 1.36 million (80 per cent) were treated and discharged from emergency departments, 275 000 were hospitalized and 52 000 died from their injuries. The leading causes of TBI were falls (35.2 per cent), motor vehicle crashes (17.3 per cent), being struck by or against events (16.5 per cent), assaults (10 per cent) and unknown/other events (21 per cent). Children aged 0–4 years, adolescents aged 15–19 and adults over the age of 65 are the age groups most likely to sustain a TBI, and children aged 0–14 account for almost 500 000 emergency department visits per annum. Adults aged over 75 years have the highest rates of TBI-related hospitalization and death. At the beginning of 2005, 3.17 million people in the United States were living with permanent disability as a consequence of TBI,[478] highlighting the associated financial burden.

In the United Kingdom, more than 1 million patients attend hospital each year suffering from head injury.[218] Based on the Glasgow Coma Scale (GCS) (Table 10.1),[442] about 90 per cent of this group have a mild head injury, 5 per cent have moderate and 5 per cent severe head injury (GCS: mild 15–13, moderate 12–9, severe 8 or less). Approximately 20 per cent are admitted to hospital for observation and 5 per cent are transferred to specialist neurological care. Most serious injuries result from road traffic accidents (RTA), but most head injuries follow a fall (40 per cent) or an assault (20 per cent).

As the epidemiological data outlined earlier clearly demonstrate, traumatic brain injury remains a major cause of morbidity and mortality throughout the world, affecting young and old alike. Our understanding of the pathological consequences of head injury has been developed through observations on human autopsies and animal models developed to investigate pathophysiological mechanisms. One has to be aware of the limitations of both these approaches: autopsies mostly provide information relating to the most severe end of the clinical spectrum of TBI, fatal outcome; and animal models do not reproduce the polypathology of human brain injury, and there are likely to be significant differences in the anatomical basis of injury and cellular responses between species.

TABLE 10.1 Glasgow Coma Scale

Eye opening	Spontaneous	4
	To sound	3
	To pain	2
	None	1
Best verbal response	Oriented	5
	Confused conversation	4
	Inappropriate words	3
	Incomprehensible sounds	2
	None	1
Best motor response	Obeys commands	6
	Localises to pain	5
	Flexion-withdrawal	4
	Flexion abnormal	3
	Extension	2
	None	1

The injuries sustained and the outcome of TBI are determined and modified by many factors such as age and pre-existing illness, and genetic factors are also important.[214,251]

Classification of Traumatic Brain Injury

Classification systems can be pathological, clinical or mechanistic. Currently there is no single classification of traumatic brain injury that completely encompasses all the clinical, pathological and cellular/molecular features of this complex process. In 2007, a workshop convened by the National Institute of Neurological Disorders and Stroke (NINDS), supported by the Brain Injury Association of America, the Defense and Veterans Brain Injury Center and the National Institute of Disability and Rehabilitation Research, reviewed the current status of classification systems and arrived at recommendations for classifications to support translational and targeted therapies.[377] In addition, attempts have been made to define common data elements for TBI to help standardize clinical trial reporting.[111,268]

Severe TBI, as assessed by GCS, can be caused by a range of pathophysiological processes (Figure 10.1), and any therapy undergoing clinical trial is unlikely to be effective in managing all of these. Although some studies suggest TBI mortality has shown a steady decline over the past few decades,[262] because of increased use of a protocol-driven approach to the initial management of the head-injured patient, standardizing therapeutic management in the acute phase, more recent epidemiological studies indicate no clear decrease in TBI-related mortality or improvement in overall outcome.[372] Therapies derived from animal models have, to date, had little impact on outcome. The reasons for this are likely to be multifactorial, but the heterogeneity of human TBI is clearly a major factor.[267] A clear recommendation of the NINDS workshop was that a multidimensional classification system should be developed to support clinical trials in TBI, allowing targeted therapies for specific pathophysiological injuries to be tested.

Pathological classifications can be anatomical, describing injuries as focal or diffuse, or pathophysiological, in which injuries are subdivided into those that are primary (immediate direct consequences of the physical force of the trauma) and those that are secondary (delayed and/or

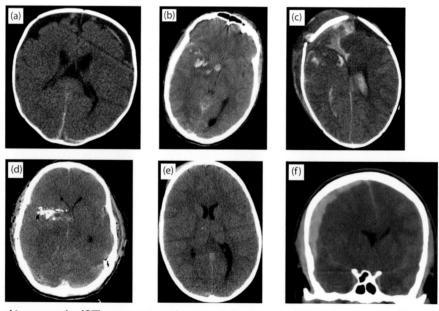

10.1 Varied computed tomography (CT) appearance of traumatic brain injury (TBI) patients presenting with unconsciousness after head injury, all graded as having severe head injury. A range of differing pathologies is seen. **(a)** 5-month-old with seizures and bilateral acute subdural haematoma (ASDH). **(b)** 16-year-old boy with extra-axial haemorrhage and multifocal contusions after a snowmobile crash. **(c)** 6-week-old struck by falling object, with depressed fracture, cortical laceration, and intracerebral and intraventricular hemorrhage. **(d)** 14-year-old boy with gunshot wound. **(e)** 6-year-old boy who was in a high-speed car crash and has radiological changes of diffuse axonal injury (DAI). **(f)** 16-year-old boy in car crash with ASDH (coronal view).

TABLE 10.2 Mechanisms of traumatic brain injury

Mechanism	Main pathology
Impact	Vascular (haemorrhages)
	Traumatic axonal injury
Inertial loading	Traumatic axonal injury
Penetrating	Local tissue necrosis
Blast	Brain swelling

TABLE 10.3 Classification of traumatic brain injury

Focal	Diffuse
Scalp lacerations	Global ischaemic injury
Skull fractures	Traumatic axonal injury/diffuse vascular injury
Contusions/lacerations	Brain swelling
Intracranial haemorrhage	
Focal lesions secondary to raised intracranial pressure	

indirect results of the trauma). A number of clinical classifications have been developed over the years, with the GCS being the most widely used,[442] although this scale is less useful in paediatric assessment and is a poor discriminator in mild head injury.

Mechanistic classifications describe impact, inertial loading, penetrating and blast injuries (Table 10.2). Impact injuries require the head to make contact with an object, with the forces potentially being transmitted to the brain. Injuries produced by inertial forces result from the differential movement of the brain relative to the cranial cavity, or of different parts of the brain relative to each other. It is worth noting that most clinical injuries have elements of both contact and inertial forces acting on the brain.

Penetrating injuries produce damage when an object passes through the protective covering of the skull resulting in direct parenchymal damage; in the case of firearm injuries, there is also a significant element of tissue damage caused by the pressure cavities produced by the projectile passing through brain tissue.[479]

Blast injuries are the least well described pathologically and are seen in an industrial, military or terrorist situation in which the shock waves from an explosive device can result in injuries to the brain parenchyma.[247]

This chapter will follow the standard outline for the neuropathological description of traumatic brain injury—focal and diffuse injuries (Table 10.3)—and will then detail the neuropathological features associated with penetrating and blast injuries, before considering spinal injuries and finally the specific paediatric injuries seen with obstetric TBI and abusive head trauma. We will then consider mild TBI, particularly in relation to sport, and finally the long-term consequences of TBI.

Experimental Models

Animal models of TBI have helped to advance the study of the cellular and molecular responses to TBI. Animal models allow an opportunity to control the physiological parameters of the model and to focus specifically on a single type of focal or diffuse injury. Human tissue typically shows polypathology, and many of the cellular and molecular responses are due to secondary injuries, such as ischaemia, rather than reflecting the primary tissue response, such as axonal injury. However, it is important not to overinterpret the data from animal models, and a number of clinical trials of promising pharmacological therapies have failed owing to a lack of recognition of the polypathology in human brain injury, and of differences in host factors.

Rodents remain the cornerstone of animal modelling of TBI, and a number of systems have been developed to model focal and diffuse pathologies.[471] However, large animal models, including dogs, sheep, swine and primates, have an important role to play in TBI research[103] and often show closer anatomical and physiological similarities to humans than do rodents. Indeed, it is surprising that so large a number of studies have addressed diffuse white matter injury in rodents, when rodents have so little white matter to study. Recent reports indicate that rodents show limited fidelity to human genomic and proteomic responses, injury timecourses, and grey and white brain matter distribution,[110,387] which implies that there are challenges in applying what is learned about injury in the rodent brain to the human. Regardless, animal models are a valuable tool for understanding how head impacts and sudden head movements translate to brain deformations and how brain deformations result in a spectrum of brain injuries, from mild to severe.

Significant advances have been made in detailing the functional anatomy of the brains of larger animals, allowing clearer association between the pathological assessment of injury and functional outcomes as assessed by a range of behavioural tests.[383] Studies using larger animals can also incorporate detailed post-injury neurocritical care monitoring,[133] providing valuable pathophysiological information post-injury.

Models of Focal Traumatic Brain Injury

The three main experimental models of focal TBI are weight drop, fluid percussion and controlled cortical impact.[316] The weight drop model[124] is most often a rodent model that uses a weight falling freely under gravity to produce a focal impact. The skull is usually not opened and fractures are common unless the skull is protected with a plate or disk. The controlled cortical impact method in rodents[95] uses a rigid impactor directly onto the exposed dura, causing an underlying contusion; in other variations, impact may occur directly onto the cortical surface. Midline fluid percussion injury[297] requires the skull to be opened by trephination over the sagittal suture. A fluid bolus is accelerated onto the dural surface, the force being modified by variation of height from which the pendulum used to accelerate the bolus is released. The fluid bolus also can be applied laterally for a different distribution of injury (see later).

These models are useful for modelling focal lesions such as contusions and haematomas. Rodents are most frequently used in these models, although they can be modified for larger animals.[22,272]

Diffuse Brain Injury Models

Models to study diffuse traumatic axonal injury have been developed for rodents and larger animals. Much of the initial work on diffuse axonal injury was done using non-human primates and the inertial acceleration brain injury model;[144] this method has also been used extensively in swine models.[131,373]

Some of the work relating to the basic science of axonal injury has been undertaken on the optic nerve stretch model; this model, developed in guinea pigs[147] and modified for rodents,[376] produces pure white matter injury that can be studied at different stages post-injury. *In vitro* models have been developed specifically to assess the pathophysiology of axonal stretch; the models allow a controlled one-directional stretch[466] and hold the potential for high-throughput analysis of potential therapies, more so than with established animal models.[412]

Other animal models to replicate inertial injury include automated and manual shaking of species from rodents to lambs,[125,414] models that aim to allow study of so-called 'shaken baby' syndrome and repetitive mild TBI.

Mixed Focal and Diffuse Brain Injury

The most widely used model for producing a mixed brain injury is the lateral fluid percussion model.[298] In this model, the craniotomy is moved from the sagittal suture (midline fluid percussion model) to a lateral position. Although the site of the impulse is unilateral, the pathology produced is bilateral.

Models of Penetrating Brain Injury

Historically, a number of animal models have been used to study ballistic brain injury, including non-human primates.[82] Most penetrating ballistic brain injury work currently undertaken in animals uses a rodent model.[462] This model does not use a fired projectile but rather an inflatable penetrating probe that can simulate the cavity produced by a missile. The probe extends into brain parenchyma in a controlled fashion from a stereotactic frame. Newer models use modified air rifle pellets,[350] and behavioural tests have been developed as part of the assessment in this model.[395]

Models of Blast Injury

Of all the animal models used to investigate traumatic brain injury, those modelling blast injury are the least well reported and characterized. Most use pressure generators to produce a blast wave within a shock tube and this blast wave can, to a degree, be controlled and scaled.[116,366] This approach has been used with smaller animals such as rats[63] and pigs,[91] and cognitive function has been shown to be affected in rodents even at low levels of blast injury.[380]

Primary Injuries

Primary injuries constitute various types of tissue disruption that occur as a direct consequence of the application of force to the nervous system and its coverings. These injuries in themselves cannot be avoided by therapeutic intervention, only prevented. They include scalp lacerations, skull fractures, several types of intracranial haemorrhage, cortical contusions and diffuse traumatic axonal injury. However, as will become apparent, the division between primary and secondary injury is to some extent artificial in that many of these processes are closely linked. In diffuse traumatic axonal injury, for example, although the initial damage is primary, the degenerative process evolves over a period of hours.

From the perspective of current and evolving therapies, it is useful to consider the concept of a delayed primary injury that represents tissue that although injured is not initially irreversibly injured, such that there is a window for therapeutic intervention: the primary forces associated with the TBI initiate molecular cascades that if untreated lead to irreversible damage. Most neuroprotective therapies target this delayed primary injury. The avoidance or limitation of secondary injuries, which are injuries not directly related to the initial trauma but arising as a consequence of downstream events such as an evolving mass lesion, are a major focus of neurosurgical intervention and aggressive critical care management, such as the prevention of seizures and the management of intracranial pressure.

In human neurotrauma, by far the most common type of force applied to the nervous system is dynamic loading, in which the forces are applied rapidly. These may be subdivided into impulsive and impact types of dynamic loads. Impulsive loading refers to the head being accelerated or decelerated without a direct impact, whereas in impact loading the head strikes an object. The consequences of such dynamic loading are dependent on many variables and form the basis of the study of the biomechanics of head injury.

Biomechanics

Biomechanical studies can provide insight into mechanisms of primary TBI, including the interrelationships among the forces experienced during impact, head and neck movements, tissue stiffness of the materials that compose the head/neck complex, deformation of structures at the macroscopic and microscopic level and biological responses to the various forces imposed on the head. The biological responses may be immediate or delayed, may be structural (torn vessels and axons) or functional (changes in blood flow or neurological status) and may differ with maturation. Biomechanical investigations typically include direct measurements of loading conditions and responses in humans, animals and anthropomorphic surrogates (i.e. crash test dummies); visualization of tissue responses to prescribed loads in order to characterize the responses of complex geometries or composite structures; mechanical property testing of individual components to identify changes with age; computational models to predict how tissues will deform during impact or rapid head rotations; and identification of the time-course of cell or tissue responses to specified deformations in order to define thresholds associated with various types of injuries.

Biomechanics investigators can use human data obtained prospectively, for example via sensors,[53,85,374] or retrospectively, by means of crash reconstructions, to help understand TBI scenarios. To obtain kinematic information in more controlled settings, human-like anthropomorphic surrogates (i.e. crash test dummies) and laboratory-based studies are used to re-enact film and witness accounts of sports-related events in order to estimate the forces of impact and head movements (kinematics), but surrogates cannot be used to predict or measure brain injuries or tissue distortions. Instead, results obtained using surrogates must be correlated with animal studies, autopsy reports, and patient records to infer biological responses to kinematic loading conditions or else with computational models to infer tissue deformations resulting from a head rotation or impact.

Computational models are used to estimate the tissue distortions and stresses that may result from a rapid head motion or head impact. Data on brain and skull tissue stiffness that can be incorporated into these models are available for young children (infants and toddlers) and adults,[74,355] but there are very limited data for older youths. Like surrogates, computational models cannot predict injury; rather, predicted tissue distortions are correlated with animal or human data. Early studies demonstrated that the brain tissue distortions and stresses in the skull that are associated, respectively, with axonal injury and skull fracture are smaller in young children than in adults.[74,358]

Biomechanical injury thresholds are most often used to re-enact actual or idealized scenarios to identify tissue distortions associated with acute injury, rather than long-term consequences, repeated exposures, or predisposing biological conditions. It is unknown if deformation injury thresholds for previously injured tissue, which may be hypoxic or metabolically compromised, are lower than for normally functioning tissue. The critical deformations associated with various types and severity of the specific brain injury of interest[377] are age and injury specific, such that the magnitude and rate of the distortion required to rupture a blood vessel are different from those required to injure an axon. It is widely accepted that smaller deformations may be associated with brief functional changes (deficits in synaptic transmission, signalling pathways and membrane permeability)[305] and that larger deformations may cause permanent structural changes.[60,117] Thus, tissue distortions and the rates of tissue deformation associated with concussion (with no lingering neural or vascular structural changes visible on radiological imaging or pathology) are probably lower than those for more severe brain injuries,[148] so it is inappropriate to rely on a single threshold for all head injuries.

Currently, because human data and computational models are limited, researchers use alternative idealized experimental preparations, such as animals, tissues and isolated cells, to create controllable settings with similar predisposing conditions among subjects and reproducible mechanical loads. Animal models are useful for measuring physiological responses, neuropathology and neurofunctional changes at prescribed time points after injury and have been discussed in more detail earlier in this chapter.

Using the tools described, researchers have determined that with or without a helmet, when the head contacts a stationary or moving object, there is a rapid change in velocity and a possible deformation of the skull. Skull deformation may produce a local contusion or haemorrhage if the deformations of the tissues exceed their injury thresholds. When the properties of the contact surfaces are softer or allow sliding or deformation, the rate of velocity change (acceleration or deceleration, depending on whether the velocity is increasing or decreasing) is lower. Similarly, if there is no head contact but only body contact, the deceleration of the moving head is usually lower than when the head is contacted directly.

After the initial rapid change in velocity caused by impact to the head or body, the subsequent motion of the head is influenced by the location of that initial point of contact and the interaction between the head, neck and body. There are three possible types of responses to head contact. First, if the contact is directed through the centre of the mass of the brain (centroid), there may be linear motion and no rotation of the head (e.g. a weight dropping down onto the top of the head or a blow to the back of the head that moves the ears and nose forward without neck flexion or extension). Animal studies have shown that these purely linear motions produce little brain motion or distortion and no concussion.[182,338] However, most often the contact force is not directed through the centroid of the brain, a situation that is referred to as a non-centroidal impact. After a non-centroidal contact, the head may rotate without a linear motion (e.g. as in shaking the head to indicate 'No'). This purely rotational motion produces a distortion of the brain's neural and vascular structures within the skull because the brain is softer than the skull and loosely coupled to it. More commonly, though, a head impact produces a change in head velocity that is associated with both linear acceleration and rotation of the head. This combined rotational and linear motion may occur because the contact is glancing (further away from the rotation centre), the body continues to move after the head is restrained by the contact surface, or the head bounces or rebounds after contact.

Internal structures of the head, such as the falx cerebri and tentorium cerebelli, influence how the brain moves within the skull and can cause local brain regions to be markedly deformed in certain directions of head rotation, so that sagittal and coronal rotations may produce more severe injuries in primates at lower accelerations and velocities.[144] In addition, animal and human studies have shown a general trend that higher rotational velocities and accelerations – rather than linear accelerations – tend to cause larger diffuse brain deformations and worse diffuse brain injuries[224] and that head injuries depend on the direction of head motion as well as on the magnitude of rotational kinematics.[119,145] Animal studies have indicated that it is important to limit the duration of exposure to acceleration, because research has shown that concussions occur when the duration of acceleration is increased.[339] Furthermore, animal studies have demonstrated that the location of brain deformation affects the resulting injury.[60,117,451]

Paediatric head injury, while sharing some similarities with adult injuries, differs with respect to the immaturity of many of the components of the developing nervous system. At birth, the bones of the cranial vault lack diploë and ossification is incomplete, with bony elements being joined by fibrous or cartilaginous tissue. Myelination of the human brain begins *in utero* and continues into early adult life. Myelination is particularly active in the first two years of life. As a result of these ongoing processes of maturation, the child's brain may respond differently to an adult brain for a given force.

The age at which the head injury is sustained was demonstrated to be important in determining the vulnerability to and recovery from a focal injury in a piglet model.[108,109] Piglets of different ages were injured using a scaled cortical impact model and then sacrificed 7 days after injury. Assessment of the brains demonstrated smaller lesions in the younger animals despite comparable injury inputs.[108] The authors concluded that vulnerability to focal mechanical trauma increases progressively during maturation. Magnetic resonance imaging (MRI) assessment of similar animals over a longer time period (24 hours, 1 week and 1 month post injury) showed that in the younger animals the lesions reached maximal volumes earlier and resolved more quickly.[108] Less damage was also found in younger animals with scaled subdural haematomas.[114] However, different results were obtained by Raghupathi *et al.* for inertial injuries, in which similar strains led to greater relative susceptibility in younger subjects.[358] Thus, relative vulnerability may vary with specific species, age and injury type.

In summary, further research is needed to define the direction-specific and brain region–specific thresholds for linear and rotational accelerations associated with TBI across the age spectrum.

Secondary Injury

It is apparent that the primary injury can underlie pathophysiological changes to the cerebral environment and initiate molecular and cellular changes that can have a significant impact on the outcome of the head-injured individual. TBI is heterogeneous; a patient who is in a coma with diffuse axonal injury may have very different pathophysiological perturbations from those of a patient with multiple cortical contusions, brain swelling and reduced cerebral perfusion. Two pathophysiological mechanisms that are key elements in secondary brain injury are energy depletion and disturbed calcium homeostasis.[67] In addition, neuroinflammation is a common response to TBI. The neuroinflammatory response may be beneficial initially, but damaging over time.[405]

Vascular and Metabolic Consequences

TBI tends to result in significant changes to cerebral blood flow (CBF). One study of 125 patients with severe TBI documented three different phases in CBF after TBI: an initial phase of hypoperfusion on the day of injury, followed by a hyperaemic phase lasting 1–3 days, and, finally, a phase of vasospasm lasting 4–15 days.[277] However, whether altered CBF in itself results in tissue injury is uncertain because TBI also results in a hypometabolic state,[450] such that the reduced cerebral perfusion does not in itself necessarily represent an ischaemic environment. Techniques to assess tissue oxygen extraction fractions are required to assess the impact of ischaemia properly in different brain regions after episodes of TBI.[76]

The forces associated with the primary injury lead to deformation of the neuronal cellular membrane resulting in alterations in membrane ion flows,[404] although this membrane disruption appears to be potentially reversible.[123] These changes at the cellular membrane can lead to potassium efflux and sodium and calcium influx, and the release of excitatory amino acids, particularly glutamate, that can activate N-methyl-D-aspartate (NMDA) receptors resulting in further calcium influx.[122] This alteration of the ionic homeostasis can lead to spreading membrane depolarization, a phenomenon reported in 50–60 per cent of severely head-injured patients,[183] exacerbating a non-ischaemic metabolic crisis.[239] Within minutes of brain trauma and ion movement, there are attempts to restore the ionic homeostasis. This is an energy-dependent process, and there is significant increase in a local glucose metabolism resulting in a localized increase in lactate production.[217] In cortical contusions, a pericontusional penumbra is described that also shows metabolic derangement, the derangement increasing closer to the contusion core.[469] Mitochondria undergo ultrastructural changes in different regions of human contusions, possibly in response to the changing metabolic activity.[25]

In the paediatric setting, particularly in neonates and infants, the metabolic responses are different.[24] The immature brain is more able to metabolize ketones, having a six-fold increase over the adult brain in the ability to metabolise β-hydroxybutyrate (β-OHB). It has been suggested that as the brain matures, the increase in local glucose metabolism reflects increased local synaptogenesis.[72] In addition, the immature brain has a greater vulnerability to excitotoxicity and ischaemia.[200] Studies in children have demonstrated hypoperfusion and low oxygen metabolic index in brain tissue after TBI, in keeping with mitochondrial dysfunction.[357]

Calcium influx associated with excessive NMDA receptor activation (excitotoxicity) after trauma can result in mitochondrial damage,[252] increased free radical production and activation of calcium-dependent proteases, such as caspases, calpains and phospholipases.[367] Damage to mitochondria can result in cellular necrosis, apoptosis or autophagy.[67] Apoptosis can be initiated by extrinsic and intrinsic pathways, the extrinsic pathway being activated by Fas or tumour necrosis factor-alpha (TNF-α) ligand binding, the intrinsic pathway by cytochrome-c release from the mitochondria-activating caspase-3. An early post-TBI increase was described in apoptosis-related proteins,[61,219] and in human TBI studies, morphological changes consistent with apoptosis were noted in contusions, peaking 24–48 hours after injury, although still detected at 10 days,[413] and in the white matter up to 12 months after an episode of TBI.[465]

Neuroinflammation

The inflammatory response develops rapidly after an episode of blunt force head injury. Neutrophils, lymphocytes and circulating monocytes infiltrate the damaged tissue, and there is local microglial and astrocytic activation. Blood–brain barrier dysfunction with cellular extravasation can be seen around contusions in the acute phase. The rodent weight-drop and cortical impact models of

TBI have shown that neutrophils accumulate in damaged tissue within 24 hours of trauma, as measured by myeloperoxidase (MPO) activity and immunohistochemistry.[73] In humans, vascular margination by neutrophils is seen in contusions by 24 hours, and by 3–5 days there is lymphocyte and monocyte infiltration, and both microglial and astrocytic activation.[191] Microglial activation can clearly be demonstrated by 72 hours post injury in human studies,[118] although lymphocyte numbers decline rapidly, suggesting they have a limited role in the overall inflammatory response.[150] However, microglial cells and macrophages can be seen associated with contusions beyond the acute phase, and the microglial response is not limited to the local environment of the contusion.[150] An imunohistochemical study of contusions from autopsy cases with survival times ranging from <11 to 334 hours demonstrated increasing pericontusional CD68 immuno-labelling, a marker of phagocytic activity, with survival time. A range of chemokines is produced by inflammatory cells at the site of tissue injury as part of a physiological response that maximizes tissue repair and limits further tissue injury.[480] Studies of human tissues and animal models showed that microglial activation can persist for a prolonged period after an episode of TBI.[84,150,212]

PATHOLOGY ASSOCIATED WITH FATAL HEAD INJURY

Blunt Force Head Injury; Focal Injury

Scalp and Skull Lesions

The scalp and skull may be injured by contact injury. The presence of scalp bruising is indicative of contact injury and, in some situations, may provide clues to the possible intracranial pathology; occipital bruising is typically associated with a backward fall and contrecoup contusions involving the frontal and temporal poles. Incised wounds are usually insignificant and easily managed in Accident and Emergency, but in some cases, especially in very young children, they may be associated with blood loss and hypotension, as well as possible associated brain injury.

The frequency of skull fractures is associated with the severity of the head injury. In one series, skull fractures were present in 80 per cent of cases with fatal head injury.[4] In clinical practice, one study recorded an incidence of 3 per cent in mild head injury presenting to Accident and Emergency, rising to 65 per cent in those requiring neurosurgical admission.[209] There is no direct correlation between the presence or absence of a skull fracture and underlying parenchymal brain injury in adults, unless the fracture is depressed and makes direct contact with the underlying tissue. However, as discussed later in this chapter, there is a correlation in adults between skull fracture and intracranial haemorrhage.[389]

Skull fractures are common in children after head injury, being seen in 12 per cent of one large cohort[213] and particularly within the youngest age groups. The majority are simple linear skull fractures not associated with underlying TBI.[151]

The development of the skull fracture is influenced by a combination of, among other parameters, impact velocity, impact surface characteristics, cortical bone thickness and cortical bone density.[181]

Different terms are used to describe skull fractures in clinical practice (Table 10.4). In general terms, impact against a flat surface, as may be seen in a fall, with distribution of the force over a large area, will produce linear fractures which can be extensive, whereas impact against smaller objects, such as a club or hammer, tends to cause localized, often depressed, fractures. Compound fractures are associated with scalp laceration. Linear fractures are the most commonly seen skull fractures in clinical practice. They most often occur over the convexity or across the base of the skull (basilar skull fracture). Basilar skull fractures

TABLE 10.4 Types of skull fracture	
Type of skull fracture	**Comment**
Linear	Inner and outer tables of skull broken
	Typically associated with a fall, the forces being distributed over a large area
	May be extensive
Basilar	Linear skull fractures that run across the base of the skull
	Require considerable force and usually associated with road traffic accidents or falls from height
	Typical clinical signs related to blood in sinuses, CSF leakage, orbital and perimastoid bruising
Depressed	Fractured bone moves downwards, such that the inner tables of the fracture and surrounding bone are no longer in continuity
	Usually related to force applied to small area
Comminuted	Fragmentation of skull bones
	Depressed fractures are often comminuted
Compound	Scalp laceration overlying skull fracture, such that the cranial cavity is exposed to the outside world
	Linear and depressed fractures can be compound
Ping-pong fracture	Concave deformation of skull bone, but no break of inner table of bone
	Seen in neonates and infants due to different biomechanical properties of developing skull bones
Diastatic	Fracture extends along suture lines
	Seen most commonly in neonates and infants
Growing	A growing skull fracture is seen in childhood and is due to dura and arachnoid extending between bone fragments, preventing union and healing
CSF, cerebrospinal fluid.	

may extend through the orbital roof of the frontal bone, or from side to side, often through the petrous temporal bones. In severe cases, a hinge-type fracture may develop such that there may be movement between the anterior/middle and posterior cranial parts of the skull at post-mortem examination. Basilar skull fractures often result in blood in the air sinuses and cerebrospinal fluid (CSF) leaks from the ears (otorrhea) and nose (rhinorrhea). Ecchymosis around the orbital tissues and around the mastoid process may also be seen (racoon's sign and Battle's sign).

In a depressed fracture, part of the skull is displaced inward such that the inner tables on each side of the fracture are no longer in continuity. In complex depressed fractures, the underlying dura is damaged. In some cases, linear radiating fracture lines extend from the central depressed fracture site. In severe cases, there may be a comminuted fracture, in which part of the skull has been fractured into multiple pieces. Comminuted fractures are frequently associated with parenchymal damage and are seen in high-velocity impacts, as might result from being struck by a moving vehicle or a fall from a height, or an assault with multiple blunt force impacts.

In neonates and infants, because of the less rigid nature of infant bone, 'ping-pong' fractures can be seen. These are depressed smooth fractures, their name reflecting the similarity in appearance to the defect produced when pushing into a ping-pong ball. They are not associated with a break in the inner table of the skull and are rarely associated with underlying parenchymal damage. Other childhood fractures include growing fractures and diastatic fractures. Growing fractures may occur when dural and arachnoid tissue is trapped between the edges of the fracture, preventing it from healing. In diastatic fractures, the fracture line extends along and separates one or more sutures in the skull. Diastatic fractures are more common in children but may be seen in adults.

Contusions and Lacerations

In simple terms, contusions represent bruising of the surface of the brain, the mechanical forces resulting in damage to small blood vessels and subsequent tissue haemorrhage. By definition, the pia mater is intact overlying contusions but torn in lacerations. Both types of injury typically involve the frontal poles, the inferior aspect of the frontal lobe including the gyrus rectus and the medial and lateral orbital gyri; the temporal poles and the lateral and inferior aspects of the temporal lobes; and the cortex above and below the Sylvian fissure. Fracture contusions may be seen at atypical sites in direct relationship to a skull fracture; the damaged bone ends become displaced and directly damage the underlying brain tissue. Contusions involving the occipital lobes and cerebellum are rare[7] because of the smooth inner surface of the posterior fossa of the skull (compare with the bony ridges of the anterior and middle fossae); when seen, they are usually associated with an adjacent skull fracture and result from direct contact to the head by an object.

Contusions may be non-haemorrhagic, although these are mostly described within the radiological literature.[188] However, in his detailed description of cortical contusions, Lindenberg[254] described the pathological appearances of non-haemorrhagic contusions. These are differentiated from ischaemic lesions on the basis of their anatomical distribution and the presence of subarachnoid haemorrhage overlying the lesions. However, it is likely that these lesions represent penumbral changes around typical haemorrhagic contusions.

Contusions typically involve the crests of gyri and are often superficial, involving the grey matter only. However, they may extend into underlying white matter and form a haematoma. In severe cases, extensive laceration injury with underlying parenchymal haemorrhage may be associated with subdural haemorrhage, a combination that constitutes a so-called burst lobe (Figure 10.2). This type of injury is most often seen involving the temporal and frontal lobes.

The contusions that occur as a result of impact with acceleration or deceleration, such as a fall, may occur underneath the point of impact (coup injuries) or distant from the point of impact (contrecoup injuries). In forward-fall coup contusions, scalp bruising is over the forehead, with the contusions involving the frontal and temporal lobes. It has been suggested that the pattern of contusional injury gives information as to the direction and magnitude of the force applied,[175] although more detailed studies[4,7] do not support these observations.

(a)

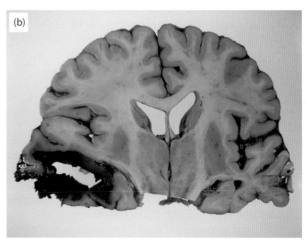

10.2 Burst lobe. This example is of a burst left temporal lobe in a chronic alcoholic who sustained a simple fall and associated skull fracture. **(a)** Lateral view, showing disruption of the cortical surface and blood clot extending out from the damaged brain parenchyma. This was continuous with an overlying acute subdural haematoma (ASDH). **(b)** Coronal section highlighting the haemorrhagic cavity within the left temporal lobe.

Contusions are dynamic lesions that evolve with time. They may become larger as a result of further haemorrhage some hours after the initial head injury, a phenomenon initially observed by computed tomography (CT) scanning,[173] but may cause rapid clinical deterioration in conscious patients in the absence of further haemorrhage,[345] particularly in relation to bifrontal contusions. Delayed traumatic intracerebral haemorrhage (DTICH) usually becomes apparent within 48 hours after the head injury.[417] The precise mechanism of this delayed haemorrhage is uncertain. One current theory is that post-traumatic coagulopathy results in continued or delayed microvascular haemorrhage; another, that the forces associated with the primary injury do not produce frank rupture of the microvessels at the time of injury but initiate molecular changes that result in subsequent structural failure.[230] Some treatments, such as mannitol, can transiently increase cerebral blood flow and have been suggested to exacerbate delayed contusional haemorrhage. Coagulopathy is common after TBI, although the actual incidence is difficult to ascertain owing to differences in severity of injury, time of testing and the tests used to assess coagulopathy in different studies.[425] The incidence of coagulopathy is directly correlated with the severity of TBI,[264] and in severe TBI, up to 45 per cent of patients become coagulopathic.[265] Although bleeding at the time of head injury is common, recent studies suggest that delayed contusional haemorrhage is a consequence of microvascular tissue failure at a time point after the head injury, i.e. that it is a secondary phenomenon. The sulphonylurea receptor 1 (SUR1)–regulated NC_{Ca-ATP} channel is upregulated after TBI and becomes active when intracellular adenosine triphosphate (ATP) is depleted.[64] In a rat model of focal cortical contusions, the regulatory unit of this ion channel (SUR1) was rapidly upregulated after injury, inducing endothelial apoptosis and capillary fragmentation.[401]

At autopsy, in the acute phase, contusions appear as punctate or confluent cortical haemorrhages (Figure 10.3) that can extend into underlying white matter. Overlying subarachnoid haemorrhage is commonly seen. Subsequently, the contusions shrink as the necrotic core of the lesion is absorbed, and the contusion takes on a golden brown colour secondary to haemosiderin deposition (Figure 10.4). Small haemorrhages are resorbed over 2–3 weeks, whereas larger haemorrhages may take significantly longer. Old contusions are a not-infrequent incidental autopsy finding, particularly in at-risk groups, such as chronic alcoholics. They can be differentiated from old ischaemic lesions in that contusions

are superficial and ischaemic lesions are typically found more deeply within the depths of sulci.

Histologically, acute contusions are haemorrhagic, the haemorrhage being predominantly perivascular. Central necrosis may develop after 24 hours. Vascular margination is evident by 24 hours, and by 3–5 days there is predominantly T-lymphocyte and monocyte infiltration and both microglial and astrocytic activation.[191] Neuronal cytoplasmic eosinophilia (red cell change) has been described within 1 hour, and neuronal incrustations within 3 hours.[19] Surrounding the central area of necrosis, there is a penumbral area in which reactive gliosis is seen but neuronal red cell change is absent. Macrophages phagocytose the degenerating red blood cells, the breakdown of haemoglobin giving rise to haemosiderin, which can be easily demonstrated with appropriate tinctorial stains. Haemosiderin-containing macrophages have been described within the haemorrhagic areas by 76 hours, and in surrounding cortex by 100 hours.[335] Macrophages are sparse within the first 48 hours after injury but increase in number by days 3 to 5 post injury.[19,260] Vascular changes have been described as part of the reparative response: factor VIII is upregulated within 3 hours of injury, tenascin within 1.6 days and thrombomodulin by 6.8 days.[185] Attempts have been made to provide detailed guidelines for the timing of histological changes in contusions, because this is an issue that may have importance in forensic practice.[331] An immunohistochemical study assessing subsets of macrophages described different temporal changes in haematoma and cortex: CD68 after 3 hours, LN-5 after 24 hours, HAM-56 after 36 hours, and 25F9 after 4 days in haematoma; CD68 after 12 hours, LN-5 and HAM-56 after 48 hours, and 25F9 after 10 days in adjacent cortex.[335]

In infants, contusions are uncommon. Subcortical tissue tears, also called slit-like lesions, have been described in this age group in a number of studies of childhood TBI,[52] although the actual mechanism of how these develop is uncertain. Indeed it may be inappropriate to use the term 'contusion' for these lesions, with alternative mechanisms including subcortical fluid collection and cyst formation.[420]

Gliding contusions were first described by Lindenberg and Freytag[255] and are predominantly white matter lesions

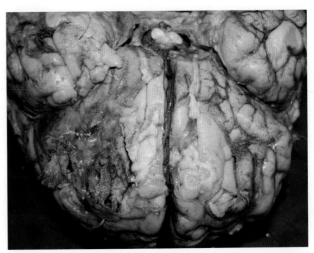

10.4 Old contusions involving the inferior aspect of the frontal lobes. There is degeneration of the cortical tissue and golden brown/orange discolouration of the surrounding parenchyma.

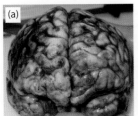
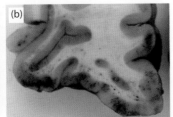

10.3 Acute contusions involving the frontal and temporal lobes. (a) There is discolouration of the cortical ribbon with focal subarachnoid haemorrhage. A coronal section through the parasagittal frontal lesion **(b)** shows small haemorrhages which, in this case, are mostly limited to the cortical ribbon.

10.5 Parasagittal white matter lesions. These form part of the spectrum of severe rotational injury and are most commonly seen in the frontal region. This example was from a road traffic accident, in a patient with a very short survival period.

seen in a parasagittal distribution (Figure 10.5). The term 'gliding contusion' is almost certainly incorrect as regards the pathogenesis; unlike typical cortical contusions that are a consequence of direct trauma, the mechanism is most likely rotational,[81] which is in keeping with clinical observations in a small case series.[391] The term 'gliding contusion' should be replaced by a more accurate descriptive term, such as parasagittal white matter haemorrhage, and should be considered part of the rotational injury spectrum, along with diffuse vascular injury and diffuse traumatic axonal injury. Parasagittal white matter haemorrhage is typically seen in road traffic accidents and is not associated with skull fractures or lucid intervals.[9]

Much of the literature relating to contusions is based on subjective interpretation of the extent of injury. A more rigorous method for assessing the extent of contusions in autopsy specimens was developed by Adams *et al.*[4] and subsequently modified.[7] This assesses the extent (scored 0–3) and depth (scored 0–4) of contusions in a variety of anatomical locators, producing a numerical score for each hemisphere that is then combined and interpreted as absent, mild, moderate or severe. The maximum score for an anatomical locator is 12 ($4 \times 3 = 12$). There are six anatomical locators on each side of the brain, and therefore the total contusion index (TCI) has a maximum value of 144 (each side $6 \times 12 = 72$, $2 \times 72 = 144$). Contusional injury is considered to be mild if the TCI is less than 20, moderate if between 20 and 37, and severe if greater than 37.[159] This approach confirmed that in fatal TBI, the distribution of contusions is most extensive in frontal and temporal regions, and in relation to the Sylvian fissure. A similar approach was used to develop a haemorrhagic lesion score.[375] This approach uses 116 separate brain sectors, and both macroscopic and microscopic haemorrhagic lesions in the brain are mapped in each sector using whole brain diagrams. An injury sector score between 0 and 116 can be derived for each case, providing a detailed overview of the anatomical distribution of all haemorrhagic injuries, including contusional injury. Both of these systems are research tools and rarely used in routine diagnostic practice.

INTRACRANIAL HAEMORRHAGES

Intracranial haemorrhages are classified by anatomical location as extradural–also known as epidural (EDH), subdural (SDH), subarachnoid (SAH) or intracerebral (ICH). They are the most common cause of clinical deterioration in patients who experience a lucid interval, 'talk and die' or 'talk and deteriorate after injury',[365] particularly in the case of EDH, which can occur with minimal primary brain injury. The clinical complications associated with a haematoma are related to the size/volume of the lesion, the anatomical location, and the rapidity with which the haematoma develops. The complications associated with a mass lesion are described later in this chapter.

Extradural (Epidural) Haematoma

Extradural haematomas (EDHs) are accumulations of blood within the extradural space. The incidence of EDH has been assessed in several large clinical and autopsy studies of head injury. Clinical studies report an incidence of between about 0.2 and 4 per cent of all head injuries.[136,205] Autopsy studies have reported an incidence of between 5 and 22 per cent, the incidence being highest in cases with a skull fracture,[130] although an EDH may occur in the absence of a fracture, especially in children.[136] A case series of 95 EDHs documented accidental falls as the leading cause (40 per cent of cases) followed by road traffic accidents (33 per cent) and assaults (21 per cent).[27]

A lucid interval is typically seen with EDHs but may also be seen with subdural haematomas and other expanding intracranial mass lesions.[40] In most case series of EDH, a lucid interval was described in no more than one third of all cases.[205,234,390] Arterial EDH generally becomes symptomatic before venous EDH, and EDH exerting pressure on the brain stem becomes symptomatic relatively rapidly. Earlier studies suggested that most EDHs reached a maximal size within minutes,[303] although more recent imaging studies and current clinical practice suggest that EDHs typically develop over hours,[36,99] and venous EDH over hours to days.

In Freytag's post-mortem study,[130] 98 per cent of EDH cases had an associated skull fracture. Posterior fossa EDHs were described as accounting for approximately 10 per cent of all EDHs in one childhood case series, but the overall numbers were small.[80]

EDHs most frequently (approximately 50 per cent) overlie the cerebral convexity in the frontotemporal region, associated with a fracture of the squamous temporal bone that results in damage to the underlying middle meningeal artery or vein. However, 20–30 per cent occur in other regions: 11 per cent are found in the anterior cranial fossa, related to anterior meningeal artery damage; 9 per cent occur in relation to the sagittal sinus; and 7 per cent occur in posterior cranial fossa, related to occipital meningeal artery or venous sinus damage.[205] Fatal EDHs in the posterior fossa are rare in adults, accounting for approximately 3 per cent of all EDHs seen at autopsy,[254] and are usually associated with occipital bone fractures. However, posterior fossa EDH accounts for a high proportion of non-arterial lesions, and in one study had a 13 per cent post-operative recurrence rate.[475]

Head impacts with associated skull fractures, particularly linear fractures, are common events in paediatric neurosurgery, being most frequent in infants. In infants and children, EDH is typically the result of venous bleeding from fractured skull bones or damaged dural sinuses. Because the dura is tightly adherent to the inner aspect of the skull, the haematoma accumulates slowly, and lucid intervals are more common. In one study, only 7 per cent of children with a confirmed EDH were unconscious at the time of presentation.[100] Because of the slow accumulation of most EDHs in children, prompt neurosurgical intervention often results in a good outcome.

Extradural bleeding strips the dura (periosteum) from the inner table of the skull, forming a circumscribed ovoid blood clot (Figure 10.6) that progressively flattens and indents the adjacent brain. Venous EDH, typically in children, usually originates from fractures and may be smaller and more difficult to distinguish by shape on CT scan from other extra-axial haemorrhages, such as subdural or subarachnoid blood. There may be little discernible damage to the underlying brain to the naked eye, although microscopic examination frequently demonstrates at least focal ischaemic injury in fatal cases. It has been suggested that blood entering the extradural space can leave via veins, forming an arteriovenous shunt, and that this shunt delays haemostasis and clinical symptoms, contributing to the lucid interval.[177] Over time, a pseudo-capsule of granulation tissue will extend from the dural surface around the haematoma, and chronic EDHs are described in approximately one quarter of surgical cases.[203]

Very rarely, an EDH may have a non-traumatic cause.[184] EDH has been described in association with paracranial infections, coagulopathies (both acquired and congenital), vascular malformations and neoplastic conditions, including metastatic disease. In addition, EDH may be seen in fire-related deaths; such lesions have a pink, foamy appearance and should not be considered to be a marker of ante-mortem trauma. They are typically associated with heat-related fissuring of the skull, although the actual mechanism of their formation is unknown.

Subdural Haematomas

A subdural haematoma (SDH) is a collection of blood within the potential space between dura and arachnoid and is most commonly venous in origin, but occasionally arterial. SDH can be acute (ASDH) or chronic (CSDH), although there is little evidence to suggest that a significant proportion of clinically important ASDHs become CSDHs; rather, CSDH seems to be a separate entity, possibly related to subclinical subdural bleeds. There has been considerable discussion around the anatomical basis of the subdural space. Haines and colleagues[179,180] demonstrated that there is no space or potential space, but rather that blood collects within a fissure that develops in the dural border cell layer. The dural border cell layer is the cell layer at the deep junction of the dura with the arachnoid. The cells that form this layer are only loosely adherent with no tight junctions. Bleeding does not separate the dural border cells from arachnoid cells, and SDHs have been shown to have dural border cells around their deep (cerebral) margin.[472]

Acute Subdural Haematoma

ASDH is most commonly seen after TBI but can also develop secondary to a number of non-traumatic causes (Table 10.5).[90,220] ASDH has been reported in 5 per cent of all hospital admissions for head injuries in clinical series.[115,206] The frequency increases with the severity of the head injury, and in autopsy studies ASDH is seen in 20–63 per cent of cases.[4,130] ASDH may be due to rupture of a bridging vein (a cortical vein passing from the cortical surface to the dural sinus), the so-called 'pure' subdural haematoma, or secondary to contusions with damage to cortical veins or arteries and overlying leptomeninges. ASDH due to cortical artery rupture after mild TBI in the absence of contusions has also been described and has a much better prognosis than other more common ASDH presentations.[284]

The distribution of bridging veins is considerably more extensive than the parasagittal convexity veins shown in many textbooks,[320] and many are not visible at standard post-mortem examination. That bridging veins can rupture to produce subdural bleeding has been disputed,[269] although anatomical,[180] biomechanical,[93] post-mortem[424] and clinical observations[233] offer strong support for this mechanism. Bridging veins have a consistent wall thickness as they pass across the subarachnoid space, but show marked variation in wall thickness as they enter the dural tissues along with increased circumferential, as opposed to longitudinal, collagen fibres.[473] These features make bridging veins more prone to damage within the dural border cell region.

ASDH resulting from cortical vein or artery damage usually is accompanied by underlying contusions or

10.6 An acute extradural haematoma (EDH) at autopsy. The haematoma is well circumscribed because of the tight adherence of the dura to the skull.

TABLE 10.5 Causes of subdural haemorrhage (SDH)

Cause of SDH	Comment
Trauma	SDH may be seen as a complication of blunt force head injury
Neurosurgical complications	SDH is a recognized complication of neurosurgical intervention in children. Neurosurgical management of hydrocephalus may be associated with SDH
Perinatal	SDH has rarely been described on fetal ultrasound scans and is well recognized as a complication of labour
Cerebrovascular disease, including vascular malformations	SDH can be associated with intracranial aneurysms, arteriovenous malformations and acquired vascular disease, such as cerebral amyloid angiopathy[127]
Coagulation and other haematological disorders	Both inherited and acquired coagulation disorders, and haematological malignancies can cause SDH
Metabolic disorders	Glutaric aciduria, Menkes disease and galactosaemia can cause SDH
Hypernatremia/severe dehydration	Elevated sodium levels may be secondary to a number of conditions including dehydration, intracranial trauma and intentional poisoning, and is often associated with intracerebral haemorrhages, including SDH
Cardiac malformation	SDH can be seen in cases of cardiac malformation[257]
Cerebrovenous sinus thrombosis (CVST)	SDH can rarely be associated with CVST, with or without associated venous infarction[15]

lacerations. In the most severe cases, there is considerable parenchymal damage, both to the cortex and to the underlying white matter, disruption of the pial surface, localized subarachnoid haemorrhage and ASDH. This combination is termed a 'burst lobe'. ASDH, with or without associated parenchymal damage, is most common in the temporal and frontal regions, and typically associated with falls and assaults. A study from the University of Pennsylvania with clinical data on all cases, and post-mortem data on all fatalities, reported ASDH to some degree in 65 of 434 cases (15 per cent), and reported that 39 of the 65 ASDH cases had bridging vein damage as the primary pathology.[144] Seventy-two per cent of ASDH cases were falls or assaults, while 24 per cent were road traffic accident cases. Posterior fossa ASDH is rare in adults, being reported in only ten cases out of 4315 TBI patients seen in one centre over a 14-year period;[437] these ASDHs were associated with occipital bone fractures and had a poor outcome. However, in children, although usually associated with occipital skull fracture, posterior fossa ASDH tends to present while the child is awake with slowly progressive symptoms, and it has a good prognosis.

ASDH has a high overall mortality. Most recent studies from neurosurgical centres quote a mortality rate of around 30–50 per cent for ASDH and a good outcome in about 30–50 per cent of cases.[245,384] A number of factors affect the outcome, including age, neurological status on admission and time to surgery,[388] but a key factor in determining outcome is the extent of associated parenchymal pathology, particularly contusions and brain swelling. The pathophysiology of brain swelling associated with ASDH is poorly defined. When present, swelling may be hemispheric or diffuse, involving both hemispheres.[384] Animal models suggest that the combination of ASDH and primary parenchymal injuries, such as diffuse traumatic axonal injury, exacerbates cerebral oedema.[447] Clinical studies have documented reduced blood flow in tissues underlying ASDH,[381] and post-mortem studies of ASDH have shown ischaemic injury underlying the haematoma.[156,161] However, although mass effect compromising blood flow to the underlying parenchyma is important in the genesis of brain swelling in large ASDH, relatively small, thin films of ASDH can also be associated with underlying brain swelling. Increased levels of extracellular glutamate were detected in a pig model of small ASDH,[308] and excitotoxicity can produce neuronal eosinophilia similar to that in ischaemia. The most extreme form of parenchymal injury associated with ASDH is the so-called 'big black brain' (Figure 10.7).[104] This devastating brain injury is seen at only a certain stage of human brain maturation, and the understanding of the pathophysiology remains incomplete. However, it has been hypothesized that subdural and subarachnoid blood with associated mass effect may result in decreased blood flow in the affected hemisphere. When this is accompanied by additional common secondary insults seen in TBI, such as apnoea and seizures (clinical or subclinical), which are noted in a significant proportion of infants and toddlers with this injury type, a mismatch of cerebral metabolic demand and substrate delivery may occur, leading to decompensation in one or both hemispheres and resulting in a pattern of widespread cortical and subcortical destruction.

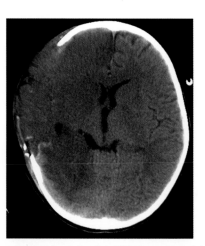

10.7 Unilateral 'big black brain'. This two-year-old presented with an acute subdural haematoma and underwent cranial decompression to decrease risk of subfalcine herniation. She recovered with a hemiparesis and mild to moderate cognitive impairment.

At post-mortem examination, the ASDH pours out as soon as the dura is reflected (Figure 10.8). With large lesions, there is deformation of the underlying cerebral hemisphere, often with accentuation of the gyral pattern on the same side as the haematoma, and flattening of the gyri on the contralateral side (Figure 10.9). ASDH consists of partially clotted blood and, in subacute cases, there may be early organization, with blood clot being at least partially adherent to the dura. Histological examination of the dura from cases of ASDH shows extravasated red blood cells in the dural tissues and lying within the subdural space (more accurately called the dural fissure). CD68-immunoreactive cells, presumed to be macrophages, are seen within 24–48 hours in the haematoma, and beyond 48 hours they are also seen in the dural tissues. MHC class II–expressing cells can be seen within dural tissues within 24 hours of injury.[16]

ASDH can resolve spontaneously, particularly in children, in whom the bleeding seen on CT and interpreted as subdural may actually be a variable mixture of subdural and subarachnoid blood.[108] Progression to a subacute SDH is associated with organization and resorption of the haematoma. In rare cases, it has been proposed that the blood may be diluted and removed by CSF through tears in the arachnoid that allow access to the subarachnoid space,[285] or that resorption is expedited by redistribution of blood clot across tears or fractures into the diploic bone of the skull or adjacent soft tissues, or as a result of its spreading out within the subdural space if there is brain atrophy.[241] The SDH becomes encapsulated with an inner and outer neomembrane, composed of fibroblasts and collagen. In animal models, the neomembrane could be seen within 48 hours of ASDH.[454] There is a fibroblastic reaction and growth of thin-walled sinusoids from the dural surface of the neomembrane. The classical histopathological descriptions of the evolution of SDH are of fibroblastic reaction and macrophage infiltration, followed by neovascularization and fibrosis of the haematoma (Figure 10.10).[325] By 3–6 months, the neomembrane is hyalinized, and by 12 months the haematoma has resolved and resembles normal dura. The sinusoids formed by neovascularization are fragile and are considered liable to rebleed, and occasional case reports support this contention.[198]

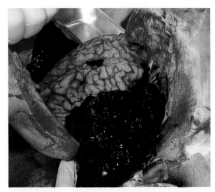

10.8 Acute subdural haematoma (ASDH) as seen at post-mortem examination. The blood has extended through the fissure created in the dural arachnoid border area, covering much of the underlying hemisphere. Compare this with acute extradural haematoma (EDH) (Figure 10.6).

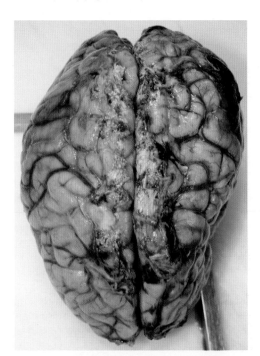

10.9 Deformation of the left cerebral hemisphere secondary to an acute subdural haematoma. The gyral pattern is accentuated on the side of the haematoma, whereas the contralateral hemisphere shows flattening of the gyri.

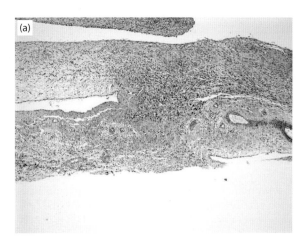

(a)

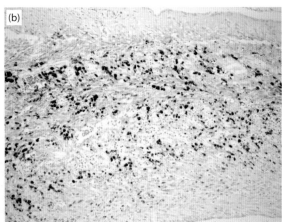

(b)

10.10 Bilateral chronic subdural haematomas. The haematomas are encapsulated and contain blood clot at various stages of organization. In addition there is frontotemporal atrophy.

Chronic Subdural Haematoma

An SDH is regarded as chronic (CSDH) when still present 3 weeks after an injury. However, the pathological entity CSDH should not be considered the end point of an organizing ASDH; rather CSDH appears to be a specific entity in itself, and its relationship with ASDH is not clear. CSDH appears to be an inflammatory angiogenic disorder and should not be considered to be a part of the typical evolution of an ASDH; most patients who survive an ASDH will show resorption of the haematoma with minimal residual membrane.[98] For the purposes of this discussion, subacute SDH refers to an ASDH that typically liquefies and resolves or, in some cases, can be drained through a burr hole, whereas CSDH refers to the formation of a thick-walled membrane encasing altered blood, prone to rebleeding and having a high incidence of recurrence. As such, CSDH should not be considered to be interchangeable with the thin subdural membrane seen as part of the resolution of an ASDH, but to be a specific reaction to subdural blood seen in a minority of patients.

The overall incidence of CSDH is 13 per 100 000, but there is a clear predominance in older age groups: 3.4 per 100 000 in those <65 years, rising to 58 per 100 000 in those >65 years.[229] CSDH often follows an episode of relatively trivial head injury, although in 25–50 per cent of cases there is no history of trauma.[276] CSDH may develop from an ASDH or from a subdural hygroma, or may be spontaneous with no history of a pre-existing lesion. Risk factors for a CSDH include an underlying coagulopathy, arachnoid cysts, neurosurgical shunts, metastatic carcinoma and long-term dialysis. Therapeutic anticoagulation is a major risk factor, accounting for almost 42 per cent of CSDH cases in one study.[400]

Pathologically, CSDH are typically found over the convexities and are bilateral in about 15 per cent of cases (Figure 10.11). They have a central core of altered blood with a liquid component, surrounded by an inner and an outer membrane. The membrane adjacent to the dura has a thickened fibroblastic layer, and vascular sinusoids extend from this region into the blood clot. Increased collagen is seen within the membranes. In some cases, prominent calcification can develop in the membranes of the CSDH.

The mechanism of enlargement of a CSDH is uncertain. Current evidence suggests that fragile new blood vessels within the evolving haematoma are susceptible to bleeding, resulting in repeated episodes of haemorrhage that enlarge the overall lesion.[474] There is an inflammatory response to repeated subclinical bleeds, and high levels of interleukin-6 (IL-6) and vascular growth factors are found in CSDH fluid.[192] The haematoma fluid and endothelial cells in the membranes contain high levels of fibrinolytic proteins that prevent clotting of the haematoma.[134]

Extra-axial Cerebrospinal Fluid and/or Proteinaceous Collections (Subdural Hygroma)

A subdural hygroma is a collection of CSF or serum, rather than blood, in the dural border cell layer, creating a fissure (space) similar to the anatomical location of an ASDH. The fluid may contain blood or blood-breakdown products, such that the fluid may be blood-stained or xanthochromic. The NINDS common data elements (CDE) project for TBI has attempted to standardize definitions and data elements, including standardized pathoanatomical terms,[310] particularly in relation to imaging. The recommendation is that the term 'subdural hygroma' should be replaced by 'extra-axial CSF' and/or 'proteinaceous collection'.

The fissure (space) may develop as a result of mild trauma or meningeal infection or after neurosurgical procedures. Subdural hygroma is seen in infants and children and in the elderly, particularly where there is brain atrophy. Trauma is the most common cause, and subdural hygroma accounts for 5–20 per cent of post-traumatic mass lesions,[242] although it may be more appropriate to consider them space-filling lesions because they typically do not cause an increase in pressure as assessed by lumbar puncture.[428]

The actual mechanism of formation of a subdural hygroma is uncertain; the favoured hypothesis is that a valve-like tear in the arachnoid allows CSF to access the dural border cells, creating the fissure. However, this has rarely been observed clinically. The hygromas are commonly bilateral[428] and usually within the frontal and temporal regions; posterior fossa examples are rare.

The vast majority of subdural hygromas are asymptomatic. A subdural hygroma will either resolve, owing to fluid absorption or expansion of the brain, or form a CSDH.

Subarachnoid Haemorrhage

Subarachnoid haemorrhage (SAH) is common in the setting of TBI. In a European study of moderate and severe TBI, 40 per cent of patients had SAH on imaging.[327] In the clinical setting, SAH is typically separated into traumatic and non-traumatic types, traumatic SAH (tSAH) covering all SAH associated with TBI. However, in pathological practice, the designation tSAH is often reserved for a specific entity: massive basal SAH with damage typically involving the vertebral artery, although other vessels can be involved. The differences in usage of terminology probably reflect the fact that SAH in the setting of fatal TBI is rarely, in itself, a major contributory factor and as such is not a major focus of the pathologist's investigations. However, basal SAH related to vertebral artery damage has a very high mortality, with patients often dying rapidly after sustaining the injury; it is often the only pathology found at post-mortem examination and of major significance in directing pathological investigations.

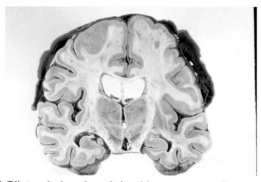

10.11 Bilateral chronic subdural haematomas. The haemtomas are encapsulated and contain blood clot at various stages of organization. In addition there is frontotemporal atrophy.

Isolated SAH, i.e. SAH with no other extracerebral haemorrhage or parenchymal injury, has been described in the setting of mild trauma. In one study, 478 patients were identified over a seven-year period with this diagnosis, and all had a good outcome, none requiring neurosurgical intervention.[356] In more severe TBI, SAH is associated with other parenchymal injuries, and clinical presentation and outcome are related to those other pathologies. SAH is typically associated with contusions and lacerations, the damaged cortical vessels, usually veins, bleeding into the subarachnoid space. tSAH may act as a space-occupying lesion with midline shift.[166] tSAH is also associated with higher intracranial pressure and a worse clinical outcome than is found in patients with similar other intracranial pathologies but no tSAH.[167] Other studies of TBI have shown that outcome is related to the GCS on admission, the amount of subarachnoid blood, and associated parenchymal lesions.[69] tSAH is associated with cerebral vasospasm, similar to that seen with non-traumatic ruptured aneurysms,[438] and there is a higher incidence of hypoxia and hypotension in this group.[167]

Massive basal tSAH is well recognized by the pathologist but much less well recognized in clinical trauma practice owing to the high incidence of early mortality. The haemorrhage is usually secondary to damage to the vertebral arteries but rarely may involve other vessels, such as the internal carotid[382] or basilar artery.[77] Damage usually follows a blow to the neck or, rarely, the head, and typically involves young, otherwise healthy males. Most cases are due to assault, but massive basal tSAH has been reported to result from a road traffic accident.[165] Collapse is usually very rapid with a short survival period.[77] The site of injury appears to be the intracranial segment of the vertebral artery as it passes through the dura,[165] and histological changes are seen at the junction between the extracranial and intracranial segments of the vertebral artery, particularly a decrease in the amount of elastic tissue in the intracranial portion,[461] which may account for this. A *COL3A1* gene mutation with associated segmental mediolytic arteriopathy was postulated as a possible risk factor for massive basal tSAH.[346] Although post-mortem dissection of the vertebral artery is technically difficult, a source of bleeding should be found in most cases and confirmed histologically. The main differential diagnosis is spontaneous rupture of an aneurysm. However, the temporal association between a blow to the head and sudden collapse, along with bruising in the neck muscles, is a strong indicator of massive basal tSAH.

The precise mechanism underlying this type of injury is uncertain. Hyperextension of the neck[315] and a rapid increase in intra-arterial pressure caused by a blow[441] have been suggested.

Intracerebral Haemorrhage

A traumatic intracerebral haematoma (ICH) is a parenchymal brain haematoma that does not extend through the cortical surface into the subarachnoid space. This definition excludes contusions and the haemorrhagic progression in contusions that was discussed previously. ICHs are most common in the frontal and temporal lobes and have been described in 15 per cent of fatal head injuries in one series.[130] In one study of 464 patients with TBI,

114 (25 per cent) had traumatic ICH: 85 had frontal haematomas, 51 had temporal haematomas and 25 had bilateral lesions.[467] Thirty-seven per cent of patients also had ASDH. A majority of patients with ICH have a poor outcome after severe TBI.[341]

Traumatic intracerebellar haematomas account for about 40 per cent of posterior fossa traumatic haematomas, which in turn account for about 1 per cent of all traumatic haemorrhagic lesions.[437] They are often associated with occipital bone fractures and have a poor clinical outcome, with 59 per cent resulting in fatality.[435] Outcome is related to GCS on admission and the size of the haematoma.[89]

The pathogenesis of ICH is unclear but is thought most likely to represent rupture of a parenchymal blood vessel at the time of injury, with immediate haemorrhage. ICH can be difficult to differentiate from a burst lobe or severe haemorrhagic contusion, and the delayed ICHs described in the literature are most likely within the spectrum of DTICH, discussed previously.

Cerebral amyloid angiopathy (CAA) is an important cause of non-traumatic lobar haemorrhage in the older population. It has been proposed that amyloid deposition in the blood vessel walls makes them stiffer and more prone to bleed after TBI.[342]

Traumatic basal ganglia haematomas, defined as larger than 2 cm in diameter, are seen in 2–3 per cent of patients with severe TBI,[37] with 94 per cent of cases resulting from road traffic accidents (Figure 10.12). The outcome for patients with traumatic basal ganglia haemorrhages is poor, with 59 per cent dying and only 16 per cent making a favourable recovery. A recent study showed 60 per cent to be due to road traffic accidents and 40 per cent due to falls, and confirmed the poor outcome, with 35 per cent of patients dying.[436] CT imaging showed 75 per cent of the haematomas to be in the putamen, 20 per cent in the thalamic region and 5 per cent in the globus pallidus.

A study of traumatic basal ganglia haematomas in children found that 52 per cent were due to high-velocity trauma and 38 per cent secondary to a fall from a height, with assault accounting for the remaining cases.[232] The authors found the same poor outcome in this population, with 52 per cent of patients dying. Small basal ganglia

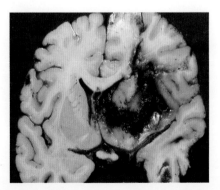

10.12 A traumatic intracerebral haematoma in the region of the basal ganglia. There is massive disruption of the surrounding tissues. Such lesions are typically seen in the setting of a high-velocity rotational head injury, in this case resulting from a road traffic accident.

ischaemic lesions have been described in children after mild trauma but appear to resolve in most cases with no significant long-term sequelae.[94]

Traumatic basal ganglia haemorrhages are thought to be due to angular acceleration forces damaging the deep penetrating arteries in the lenticulostriate and anterior choroidal distributions, causing haemorrhage.[322] A post-mortem study from the Glasgow database of TBI cases found traumatic basal ganglia haemorrhages in 10 per cent of the 635 cases studied.[8] In that study, most haematomas were in the thalamic nuclei. There was an increase in parasagittal white matter haemorrhages and diffuse traumatic axonal injury in these cases, all of these pathologies being associated with angular acceleration.

The haemorrhage begins immediately, but the haematoma may increase in size over 30–60 minutes post injury, the duration of bleeding being determined by blood pressure and any underlying coagulopathy.[37]

Traumatic Intraventricular Haemorrhage

Traumatic intraventricular haemorrhage (IVH) is found in 1.5–3 per cent of TBI cases and 10 per cent of fatal TBI cases (Figure 10.13).[283] Like traumatic basal ganglia haematomas, IVH is associated with angular acceleration forces, and haemorrhagic corpus callosum and brain stem lesions are often seen with IVH. In one autopsy series, IVH was seen in 17.6 per cent of cases, and in 10 per cent was massive.[286] IVH was most commonly seen in cases with high-energy impacts, such as road traffic accidents. In some cases, it represented the extension of a parenchymal haematoma into the ventricle or retrograde spread of subarachnoid blood from the infratentorial structures, although in a significant proportion haemorrhage was from structures in the periventricular region, including ruptured fornix/septum pellucidum, subependymal veins in the ventricular walls, choroid plexus or damaged corpus callosum.

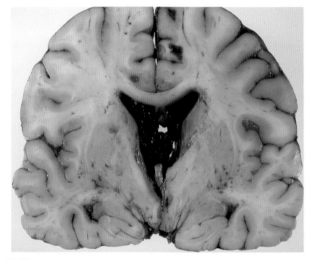

10.13 Massive intraventricular haemorrhage resulting from a road traffic accident. The forniceal structures are torn, and the haemorrhage may have arisen from small vessels within the septum pellucidum.

BRAIN INJURY SECONDARY TO RAISED INTRACRANIAL PRESSURE

As soon as the cranial sutures fuse, the skull is effectively a solid bony box. Although this arrangement is of great value in protecting the soft parenchyma of the brain from injury, the design allows little opportunity to accommodate enlarging mass lesions, such as expanding haematomas, within the cranial cavity. The increasing volume may be secondary to a diffuse process, such as brain swelling, or may be caused by a unilateral expanding mass lesion, such as a haematoma or contusion. Normal intracranial pressure (ICP) is in the range of 0–10 mmHg. Acute prolonged pressure greater than 20 mmHg is abnormal, greater than 40 mmHg is associated with neurological dysfunction and compromised cerebral circulation and above 60 mmHg is virtually always fatal. However, the ICP itself is not the entire explanation. Unlike conditions such as pseudotumour cerebri in which very high pressures can be tolerated because they are diffusely distributed and perfusion is maintained, in TBI most of the damage occurs from pressure gradients and resulting tissue shift, with compression of vessels and resultant infarction. Brain damage secondary to increased ICP, identified by 'notching' of the medial temporal lobe structures by the medial edge of the tentorium, was found in 75 per cent of fatal TBI cases in one post-mortem series.[158]

With a unilateral space-occupying lesion, displacement of fluid, such as cerebrospinal fluid and venous blood, serves initially to compensate for the mass effect of the lesion and to limit the increase in compartment pressure; however, ultimately a pressure differential develops that causes brain tissue to herniate into adjacent intracranial or intraspinal compartments. In diffuse brain swelling, systemic blood pressure (SBP) increases as a physiological response to maintain the cerebral perfusion pressure (CPP) in the face of increasing ICP. However, when these compensatory mechanisms have been exhausted, there is a sharp rise in ICP and corresponding fall in CPP, and finally vasomotor paralysis occurs: the resistance vessels of the cerebral circulation become flaccid, ICP equals CPP and there is cessation of cerebral blood flow.

Brain herniation may extend under the falx cerebri, damaging the cingulate gyrus (subfalcine or supracallosal hernia); under the tentorium cerebelli, damaging the parahippocampal gyrus/medial temporal lobe (tentorial or uncal hernia); or through the foramen magnum, damaging the tonsils of the cerebellum (tonsillar hernia) and the brain stem.

Subfalcine Herniation

A subfalcine hernia develops as a result of a supratentorial mass lesion. This type of hernia tends to be particularly prominent in relation to a convexity ASDH and is associated with midline shift. A subfalcine hernia may obstruct flow within the pericallosal artery (anterior cerebral circulation) resulting in infarction in the corpus callosum and cingulate gyrus. It can be difficult to differentiate between the haemorrhagic lesion associated with diffuse traumatic axonal

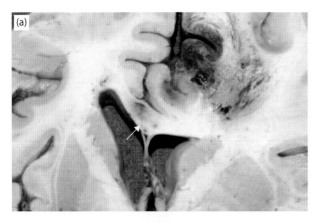

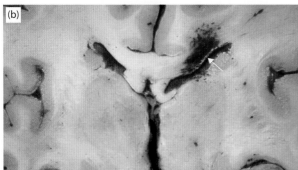

10.14 Corpus callosal haemorrhages in traumatic brain injury. (a) The haemorrhage is midline (arrow) and secondary to subfalcine herniation. **(b)** The haemorrhage is more laterally placed (arrow) and is a marker of a diffuse traumatic axonal injury of at least grade 2.

injury and that resulting from herniation (Figure 10.14); however, in subfalcine herniation, the lesion is usually midline, whereas in diffuse traumatic axonal injury the lesion is usually more laterally placed.

Tentorial Herniation

The medial part of the temporal lobe, including the parahippocampal gyrus, can herniate across the tentorium cerebelli, the tissue being pushed into the basal CSF cisterns. A wedge of necrosis usually develops in the inferior part of the temporal lobe at the point where it is in contact with the free edge of the tentorium cerebelli. In the acute stage, this can be seen as a wedge-shaped area of discolouration indicating early infarction, whereas with longer survival the lesion becomes a gliotic scar (Figure 10.15). The ipsilateral oculomotor nerve may be damaged and, when involved, appears kinked and discoloured. Branches of the ipsilateral posterior cerebral artery may be compromised by the tentorial hernia, resulting in infarction within the territory supplied by this artery, particularly involving the inferior part of the temporal lobe and medial occipital cortex.

Where there is massive tentorial herniation, the contralateral cerebral peduncle may be pushed against the tentorium cerebelli, resulting in a small area of haemorrhagic infarction, Kernohan's notch, which can result in a false localizing sign: weakness on the same side as the mass lesion.

Bilateral tentorial herniation is typically seen in cases of global brain swelling.

10.15 An old gliotic scar in the parahippocampal gyrus, indicative of previous tentorial herniation.

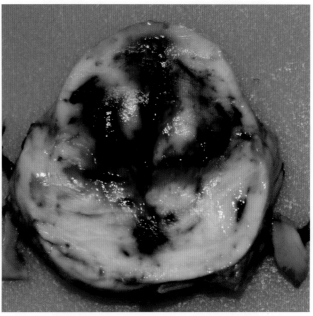

10.16 Extensive pontine haemorrhage secondary to axial displacement. This pathology is common in fatal traumatic brain injury cases with significant acute subdural or extradural haematomas.

Axial (Caudal) Displacement

Caudal displacement and elongation of the rostral brain stem may result in secondary brain stem (Duret) haemorrhage and infarction (Figure 10.16), a common terminal event in raised ICP, being seen in up to 51 per cent of brain-injured patients who die.[158] The haemorrhages are in the midline of the midbrain and pons, and very rarely in the medulla. The haemorrhages are most likely a consequence of vascular congestion within the brain stem parenchyma, and rupture of paramedian pontine branches of the basilar artery.[68] Although Duret haemorrhages are common terminal events in TBI patients with increased ICP, very occasionally they can be seen in patients who have a good outcome[427] and therefore should not be considered universally fatal. However, these cases have to be critically assessed. True Duret haemorrhages are due to axial displacement, with forces pushing down from above. However, brain stem haemorrhages can be seen

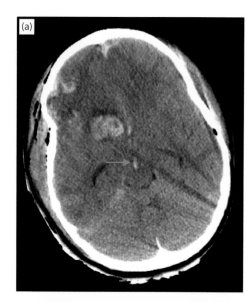

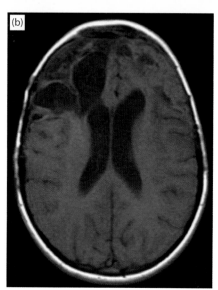

10.17 (a) Acute computed tomography (CT) brain scan of an eight-year-old girl who sustained bifrontal injuries and midbrain haemorrhage (arrow) in a road traffic accident. She had transient diabetes insipidus, related to forces through the anterior skull base and pituitary stalk, which resolved. **(b)** CT scan two years later, showing frontal encephalomalacia. The patient is an average student attending regular school classes, but has some attention deficits.

on imaging after TBI with predominantly frontal forces (often causing facial and anterior skull base fractures) resulting in frontal lobe contusions, pituitary stalk damage and midbrain haemorrhages (Figure 10.17). These are not true Duret haemorrhages and can be associated with a good outcome.

At autopsy Duret haemorrhages can be large, resulting in extensive damage to the midbrain and pons, such that in some cases the midbrain is replaced by a necrotic, haemorrhagic mass. The midline distribution should be differentiated from the dorsolateral distribution of haemorrhages in diffuse traumatic axonal injury, although often it can be difficult to differentiate between these two entities

macroscopically. Histological examination of cases with early brain stem displacement, particularly in large ASDH cases with a short post-injury survival time, often shows vascular engorgement, with expansion of the perivascular space by acellular fluid.

Diencephalic and Pituitary Injury

Pituitary and hypothalamic dysfunction are well-recognized complications of TBI.[477] Injury to these structures may be primary, as a result of direct shearing injuries in severe TBI with rotational forces, causing damage to the pituitary stalk and perforating choroidal branches, or secondary to ischaemia associated with increased ICP and axial displacement. There have been few pathological studies of these injuries,[83,398] but when looked for, these lesions are not uncommon in severe TBI, particularly road traffic accidents and falls from heights (Figure 10.18).

Tonsillar Herniation

The caudal displacement of the cerebellar tonsils occludes the foramen magnum, resulting in blockage of the CSF pathway and increasing ICP. There is necrosis of the tonsils, and often ischaemic damage is seen in the medulla. The combination of pressure and ischaemia on the medulla results in cardiorespiratory collapse. Assessment of tonsillar herniation can be difficult in the absence of necrosis, although usually at least microscopic haemorrhage can be identified.

Reverse Herniation

A mass lesion, such as ASDH or EDH, in the posterior fossa can result in rapid tonsillar herniation but will also

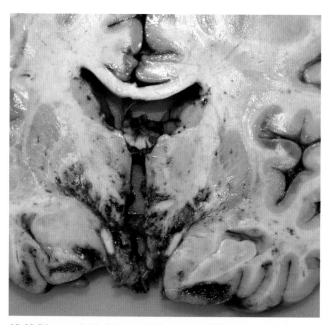

10.18 Diencephalic haemorrhages in a TBI case with associated temporal lobe contusions. The haemorrhages are due to damage to the perforating choroidal arteries. This type of injury is often associated with damage to the pituitary stalk.

often produce herniation of the superior part of the cerebellar hemispheres upward through the tentorium cerebelli. This may be associated with superior cerebellar artery territory infarction in the superior part of the cerebellar hemispheres.

BLUNT FORCE HEAD INJURY; DIFFUSE INJURY

Ischaemia

Ischaemic brain injury is a common finding in autopsy series of fatal traumatic brain injury, being seen in 91 per cent of cases in one study,[156] and in 88 per cent of a follow-up cohort.[161] There is a significant association between ischaemic brain injury and increased ICP. Early stabilization of patients reduces the incidence of hypotensive brain injury, but not of diffuse ischaemic injury. Diffuse ischaemic injury can develop as a consequence of reduced perfusion or metabolic mismatch. Reduced perfusion may be due to increasing cerebral swelling, secondary to cardiorespiratory arrest or a consequence of profound hypotension as a result of other injuries, particularly long bone fractures or solid organ injury. Metabolic mismatch occurs when metabolic demand increases significantly, although oxygen delivery may be compromised. This situation may be seen in seizures and other causes of excitotoxic stress.

The term 'ischaemia' means an absence of blood flow, but it is widely used to describe reduced blood flow that is insufficient to meet (often increased) metabolic demand. The end result of such mismatch is energy failure within a given population of cells. In the brain, neurons are particularly sensitive to reduced blood flow and are the first cell type to be injured. If the ischaemia is prolonged, then other cells are also damaged (glial cells, endothelial cells, smooth muscle cells) resulting in pan-necrosis, also known as infarction.

Histologically, ischaemia is recognized as cytoplasmic eosinophilia, and this change requires a survival of several hours from the time of injury. Attempts have been made to standardize an approach to grading ischaemic injury[161] as severe (diffuse, multifocal within different arterial territories, or a large lesion within a single arterial territory), moderate (lesions limited to boundary zone region, singly or in combination with subtotal infarction in the distribution of an arterial territory, or 5–10 subcortical lesions), or mild (5 or fewer small lesions).

Although ischaemic brain damage is a major component of TBI, it is discussed in more detail in Chapter 2 and will not be covered further here.

Diffuse Traumatic Axonal Injury

Axons can be damaged by many different aetiologies including trauma; metabolic encephalopathies such as hypoglycaemic encephalopathy; multiple sclerosis secondary to demyelination; ischaemia; or infection, e.g. by human immunodeficiency virus (HIV) or malaria parasites.[306]

Diffuse white matter injury was first described in the setting of TBI by Strich;[430] she described the pathology of five patients who were described as decerebrate for 5–15 months after severe TBI. The head injuries were described as uncomplicated, with no skull fracture or intracranial haemorrhage, no contusions, and no evidence of increased ICP. The pathology in these cases was described as diffuse white matter degeneration. White matter injury associated with TBI is widely accepted, although it has undergone a number of changes of terminology. The damage was initially thought to be an instantaneous shearing force at the time of the injury,[431] and was associated with what were termed 'axonal retraction balls'. The clinicopathological investigation of diffuse traumatic white matter injury was led by the Glasgow and Philadelphia groups, predominantly in the 1980s, and has continued to evolve. The group in Philadelphia used a non-impact non-human primate animal model to study the pathologies associated with TBI. They produced prolonged unconsciousness in the absence of an intracranial expanding mass lesion, particularly with angular or rotational acceleration, and with extended periods of acceleration.[146] This work was undertaken in conjunction with a review of the Glasgow human TBI database which led to the study of 45 cases of diffuse traumatic white matter injury.[5] Comparisons between the non-human primate and human studies showed many similar features: focal haemorrhagic lesions could be seen in the corpus callosum and dorsolateral quadrants of the rostral brain stem, and there was widespread axonal injury. In all of the human cases studied there was no lucid interval, i.e. they were unconscious from the time of injury until death. This same clinicopathological pattern was seen in the most severely injured non-human primates. Although the term diffuse axonal injury (DAI) had existed in the literature prior to this work, the terminology now acquired a precise meaning: diffuse axonal injury after an episode of trauma, with immediate, prolonged coma. The concept of DAI was expanded in further publications in the neuropathological literature.[8,11] However, confusion around the causes of axonal injury in TBI cases has existed in the literature, and as techniques for identifying axonal injury have evolved (discussed later), there was a need for terminology to reflect changing concepts. The term traumatic axonal injury (TAI) has been recommended for neuropathological practice[138] because it clearly defines the pathophysiological process underlying the axonal damage. In nontraumatic causes of axonal injury, the pathophysiological process should be clearly defined, e.g. as ischaemic axonal injury. Increasingly TAI is thought to underlie a spectrum of clinical states, extending from concussion with no loss of consciousness, through to immediate and prolonged coma. As such we strongly suggest that in neuropathological reports and studies, the term TAI be qualified as either focal or diffuse, with diffuse TAI equating to the original descriptions of DAI in the 1980s, in which TAI involves the central white matter, including the corpus callosum and parasagittal white matter, internal capsule and dorsolateral quadrant of the rostral brain stem. A recent publication suggested that the term TAI be reserved for localized axonal injury in animal models,[412] but we would encourage a uniform approach between diagnostic neuropathology and research studies. We propose that the designation DAI should be retained as a clinical diagnosis in a patient with immediate prolonged coma and supportive neuroradiological findings, described later; focal TAI should apply to the neuroradiological and histopathological assessment of localized trauma-induced axonal injury in humans and animal models; and diffuse TAI should be reserved for the histopathological description of diffuse trauma-induced axonal injury in humans and animal models. We would also strongly encourage the consideration of non-traumatic causes of axonal injury in animal models, because such consideration is often notably absent.

Clinicoradiological Diffuse Axonal Injury

In clinical usage, the term DAI typically refers to TBI that results in immediate loss of consciousness not due to a mass lesion or diffuse metabolic process, such as ischaemia, and that lasts more than a brief period of time. The exact time frame and clinical definitions have varied between authors and over time. Clinically, the severity of DAI has been assessed according to the duration of loss of consciousness, mild being associated with loss of consciousness for 6–24 hours, and moderate and severe for >24 hours, with severe also showing posturing and motor deficits.[246] DAI contributes to at least 35 per cent of cases of fatal TBI[142] and is the major cause of severe disability and persistent vegetative state after TBI.[157,225] Approximately 65 per cent of patients with clinically mild DAI have a good outcome, whereas only 15 per cent with severe DAI have a good outcome.

Because of the microscopic nature of the pathophysiological correlate of DAI, conventional imaging techniques lack the sensitivity to detect structural changes. CT has limited value in assessing DAI, and scans can appear normal in severe head injury.[258] However, very small haemorrhages, sometimes referred to as micro-bleeds, often co-localize with areas of axonal damage and can act as a surrogate for white matter injury. There are MRI sequences that are particularly suited to identifying blood products, such as T2* gradient echo, which can be useful in the clinical setting to support a diagnosis of DAI (Figure 10.19). Newer techniques do appear to offer greater sensitivity in assessment of DAI, although these techniques still require detailed validation, particularly with post-mortem correlation. Diffusion tensor imaging (DTI) is an MRI technique that measures restricted diffusion of water, building images of white matter tracts, and this technique has been shown to identify more lesions than other more routine MRI techniques, such as T2* gradient echo.[195] Susceptibility weighted imaging is used to demonstrate very small haemorrhages that are less well demonstrated by other imaging modalities (Figure 10.20). Because it is a relatively new technique, standardization of methodology is still evolving.[403] There are many papers in which DTI has been applied to animal models of TBI,[250] but to date there has been no correlation of DTI signal changes with actual microscopic tissue disruption in human brain tissue.

Histopathological Identification of Traumatic Axonal Injury

The macroscopic appearances of diffuse TAI are variable, ranging from an essentially macroscopically normal brain to extensive frontal petechial haemorrhages, parasagittal white matter lesions, focal corpus callosal haemorrhage, and haemorrhage in the dorsolateral midbrain and pons.

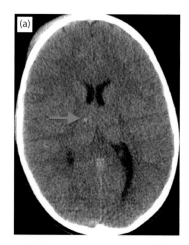

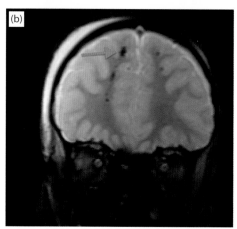

10.19 (a) Axial computed tomography (CT) scan in acute traumatic brain injury (TBI) showing areas of haemorrhage. The patient was unconscious at the time of admission, and both the clinical picture and CT appearances are consistent with diffuse axonal injury (DAI). **(b)** A T2* gradient echo magnetic resonance image (MRI) highlights the haemorrhages.

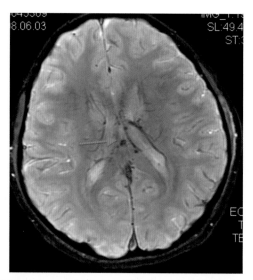

10.20 Magnetic resonance image (MRI) of a 14-year-old girl struck by a car, with immediate unconsciousness lasting several hours, followed by gradually improving lethargy. Susceptibility weighted imaging shows small haemorrhages that are seen as black dots (arrow) in the corpus callosum.

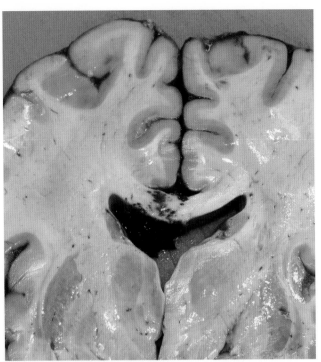

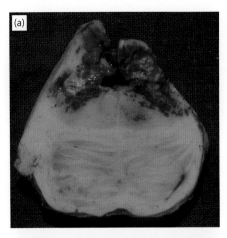

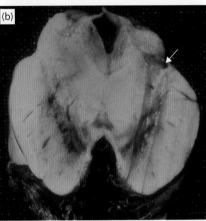

10.21 A more centrally placed haemorrhagic lesion in a case of diffuse traumatic axonal injury. The haemorrhagic lesion extends into the ventricular system, and there is damage in the region of the septum pellucidum and fornices. There is no evidence of subfalcine herniation, and there was no mass lesion.

10.22 (a) An acute bilateral dorsolateral haemorrhagic lesion involving the dorsolateral pons of a 35-year-old male who sustained a road traffic accident, was unconscious from the time of injury and died within a few hours of the accident, never recovering consciousness. **(b)** Midbrain of a 23-year-old male, victim of car traffic accident. Unconscious from the time of injury, the patient was maintained on a ventilator until death several weeks after the accident. An old haemorrhagic lesion (arrow) is seen in the dorsolateral midbrain.

Three degrees of diffuse TAI have been described: mild, moderate and severe.[10] In grade 1, there are microscopic changes in the white matter of the cerebral cortex, corpus callosum, dorsolateral midbrain and pons and the cerebellar peduncles. Grade 2 is distinguished by focal lesions, usually haemorrhagic, restricted to the corpus callosum; these lesions are typically laterally placed in the corpus callosum, but can extend medially to involve the interventricular septum and fornix (Figure 10.21), and are separate from the more midline corpus callosal haemorrhagic infarction seen with subfalcine herniation. In grade 3, additional focal haemorrhagic lesions are seen in the dorsolateral quadrants of the rostral brain stem. In the acute stage, the haemorrhages may be unilateral or bilateral (Figure 10.22a). If the patient has survived for several weeks, the lesions are granular and brown (Figure 10.22b), and with time become cystic and shrunken, being best seen using glial fibrillary acidic protein (GFAP) immunohistochemistry.

Of 122 cases of diffuse TAI identified from the Glasgow neurotrauma post-mortem archive, ten cases were of grade 1, 29 cases of grade 2 (in 11 of which the haemorrhagic lesions were only identified microscopically) and 83 cases of grade 3, with 34 of the cases having microscopic lesions only.[10] The more severe forms of diffuse TAI are associated with parasagittal white matter lesions[9] and deep basal ganglia haematomas.[8] An earlier study of diffuse TAI by the Glasgow group described the clinicopathological correlation of 45 cases, and found all to be unconscious from the time of injury and to remain in coma for a prolonged period.[5] However, 17 of the 122 cases reported in the subsequent study[10] had a

lucid interval, 15 having a partial lucid interval (confused but lucid) with grade 2 diffuse TAI, and two having a complete lucid interval with grade 1 diffuse TAI. These patients died from other pathology, mostly related to brain swelling. However, as discussed later, such data need to be interpreted with caution.

When assessing a brain post mortem for diffuse TAI, extensive sampling is required;[138] this should include the genu and splenium of corpus callosum, frontal parasagittal white matter, posterior limb of the internal capsule, cerebellar hemisphere, midbrain and pons, including the superior or middle cerebellar peduncle. TAI has been shown to be accentuated in the caudal part of the corpus callosum,[240] and it is recommended that both parts of this structure are sampled.

Several techniques have been used to identify damaged axons. As discussed later, trauma disrupts the normal axonal flow of proteins, causing proteins to accumulate at points where there is perturbation of normal axonal function, resulting in axonal swelling. These swellings occur along the length of an involved axon, giving an appearance called axonal

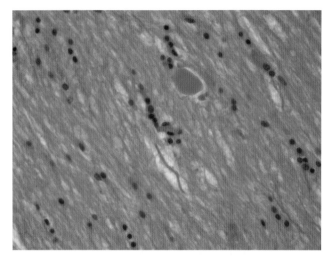

10.23 Eosinophilic axonal bulbs (axonal 'retraction balls') can be seen in areas of damaged white matter after about 15–18 hours' survival. Eosinophilic axonal bulbs may be seen several months after an episode of head injury, and in this setting they may not immunostain with antibody to β-APP. H&E ×40.

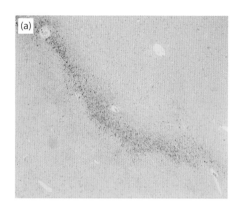

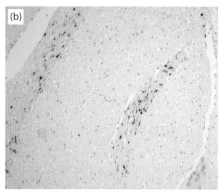

10.24 Different patterns of immunostaining can be identified to help differentiate between ischaemic and traumatic white matter injury. (a) β-APP immunostaining highlights a solid band of axonal injury, delineating an area of ischaemic damage (β-APP, ×4). **(b)** The susceptibility to axonal injury is determined by the orientation of the white matter bundles, typical of traumatic axonal injury. The example presented is from the pons, and the damaged axons are in one orientation, whereas the intervening undamaged axons are in another. β-APP, ×10.

varicosities. If the axonal pathology progresses to actual disconnection, a single swelling, called an axonal bulb or axonal 'retraction' bulb, is seen. These swellings are eosinophilic in haematoxylin- and eosin-stained sections (Figure 10.23), and can also be detected by silver stains, although a survival of 15–18 hours is required before axonal bulbs can be identified using these techniques. Immunohistochemistry is the most sensitive technique, and a range of proteins that move along the axon via the fast axonal transport system were assessed in human material.[396] This study found β-amyloid precursor protein (β-APP) to be the most specific and sensitive marker in human material, and immunohistochemistry for β-APP remains the most widely used method for detecting disruption to normal axonal flow. Older texts refer to a minimum survival of 2 hours before β-APP accumulation is seen, and 3 hours until axonal bulbs are identified in diffuse TAI.[301] However, owing to changes in the antibody clone used and advances in antigen retrieval techniques, the time frame for detection of axonal injury has changed significantly. More recent literature has described β-APP accumulation in paediatric cases 35–45 minutes after TBI[154] and 35 minutes after TBI in adults.[193] Indeed, one study of cases described as dead at the scene showed β-APP accumulating in the neuronal cytoplasm within minutes, and occasional beaded axons within the white matter.[318] The axonal pathology evolves over at least 24 hours and then plateaus, remaining easily identifiable for about 10–14 days after injury, after which the staining intensity lessens such that by 3–4 weeks after injury β-APP immunoreactivity is difficult to identify.[164]

As noted previously, β-APP accumulation is not specific to trauma and may be seen in axonal disruption of any aetiology, such as ischaemia[96] and hypoglycaemia.[97] When assessing a case of possible TBI, it is always important to be cognisant of other potential causes of axonal pathology, particularly ischaemia. Ischaemia in TBI cases is common, and can cause extensive β-APP accumulation throughout the white matter. However, ischaemic axonal injury is diffuse, and the accumulation of β-APP shows a geographic or

contiguous pattern, whereas diffuse TAI follows a specific anatomical distribution, often highlighting axonal damage in one white matter pathway, whereas an adjacent differently orientated pathway does not show axonal damage (Figure 10.24). Specific patterns of β-APP immunohistochemical staining have been described for TAI and ischaemic axonal injury, and it is possible to distinguish them from each other,[164,363] although in some cases the ischaemic axonal injury is so extensive it is not possible to comment on the presence or absence of underlying TAI.

It is important when reviewing the historical TAI literature to consider that papers describing axonal injury detected with silver stains and haematoxylin and eosin will have underrepresented the axonal pathology, and in many of the papers no consideration was given to other causes of axonal injury, particularly ischaemia. Papers describing axonal bulbs detected using silver stains in TBI cases with brain swelling and/or intracranial haemorrhage may reflect ischaemic damage rather than TAI. A review of cases with a diagnosis of diffuse TAI over three decades in the Glasgow neurotrauma post-mortem archive compared a 1968–1972 cohort (151 cases),[4] a 1981–1982 cohort (112 cases)[161] and a cohort from 1987–1999 (226 cases, unpublished). In the earlier cohorts diffuse TAI was diagnosed using haematoxylin and eosin and silver stains, whereas in the later cohort all cases were assessed using β-APP immunohistochemistry.

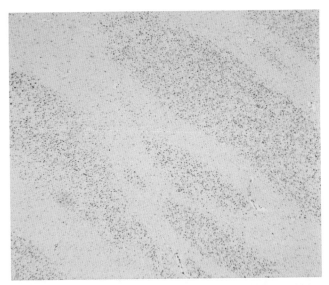

10.25 With longer survival (after several weeks) β-APP becomes an unreliable marker of axonal injury as the staining intensity fades. However, in these cases, CD68 immunoreactivity can be helpful in highlighting degenerating pathways (wallerian degeneration), again showing specific susceptibility related to axonal orientation. In this example from the pons, in a patient with seven months' survival in coma after an episode of traumatic brain injury (TBI), the corticospinal pathways are highlighted. CD68 immunostaining, ×10.

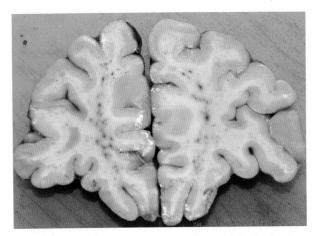

10.26 A coronal section through the frontal lobes, short survival (hours) after a road traffic accident. Petechial haemorrhages are widespread, scattered throughout the white matter.

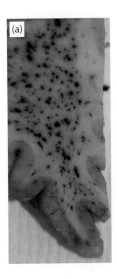

Diffuse TAI was reported in 41 per cent of the later cohort, compared to 18 per cent (1968–1972) and 33 per cent (1981–1982). Grading was compared between the 1987–1999 cohort and the published data from the 1968–1982 cohort:[10] grade 1, 8 versus 17 per cent; grade 2, 24 versus 7 per cent; grade 3, 68 versus 17 per cent. Although it is possible that the epidemiology of diffuse TAI changed between these cohorts, it is likely that the extensive axonal injury seen with silver stains in some of the older cohorts was at least partly attributable to ischaemic damage.

In long-term survival, the wallerian degeneration of the white matter tracts can be highlighted by immunohistochemistry for phagocytic markers, e.g. with antibody to CD68 (Figure 10.25). CD68 immunoreactivity can be seen in white matter pathways for many years after diffuse TAI.[407]

Haemorrhagic Lesions Associated with Rotational Injury

As described earlier, focal haemorrhagic tissue tears are seen in the corpus callosum and dorsolateral brain stem in severe rotational injury. Other focal haemorrhagic lesions are seen in severe TBI, as part of the spectrum of rotational injury. Diffuse vascular injury refers to extensive petechial haemorrhages extending through the white matter, particularly the frontal white matter, associated with immediate unconsciousness and a poor prognosis, often being seen in cases described as 'dead at the scene' (Figure 10.26). The pathology is thought to occur at the time of injury and to represent the shearing of many small parenchymal blood vessels.[4] Histological examination demonstrates perivascular haemorrhages (Figure 10.27). It is important to undertake histological examination in TBI cases with

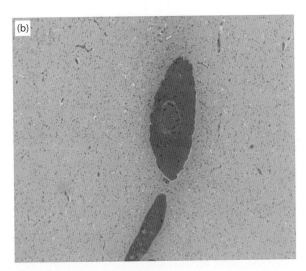

10.27 (a) A coronal section through the parasagittal frontal white matter showing extensive petechial haemorrhages. **(b)** A microscopic section from (a) showing prominent perivascular haemorrhage. The vessel can be seen within the centre of the haemorrhage. H&E, ×20.

a short survival period and no obvious mass lesion even if diffuse vascular injury cannot be identified macroscopically, because microscopic lesions are often present in the absence of obvious macroscopic lesions.[349] Parasagittal white matter lesions are often seen in cases with macroscopic or microscopic diffuse vascular injury. In addition, subcortical haemorrhages are a frequently neglected but important additional marker of a severe rotational head injury: these lesions lie at the grey–white matter interface, particularly in the inferior frontal regions, and are not associated with overlying cortical contusions. The final haemorrhagic lesion seen in severe rotational injuries is the traumatic basal ganglia haematoma described previously.

These haemorrhagic lesions are seen in severe rotational TBI cases, usually road traffic accidents or falls from a height, and predominantly occur in association with grade 3 dTAI, but also sometimes with grade 2 dTAI. As noted earlier, patients with these haemorrhagic lesions have a poor prognosis.

Clinicopathological Correlation

That there exists a spectrum of clinical and pathological changes associated with rotational forces has been clearly demonstrated. Clinically, this extends from mild concussion through to DAI, with immediate prolonged coma and poor prognosis. The pathological correlates of these clinical states are still being evaluated. It is clear that the clinical and radiological entity know as DAI is associated with severe diffuse TAI, with corpus callosum and dorsolateral brain stem lesions, and with a higher incidence of diffuse vascular injury and parasagittal white matter injury. A study of 14 cases of diffuse vascular injury found that all cases were road traffic accidents and had associated diffuse TAI grade 2 or 3, and all patients died within 24 hours of the injury.[349]

Studies using a porcine non-impact model suggest that it is the brain stem pathology that is the anatomical correlate of immediate unconsciousness and coma[42] and that the duration of coma is related to the extent of brain stem axonal pathology.[410] However, at the milder end of the clinical spectrum the association is less secure. The structural basis of concussion is unknown and discussed in more detail later. Focal axonal injury, particularly involving the corpus callosum, is increasingly recognized, but the clinical consequences are unknown. It is seen not infrequently in cases with other obvious causes of death, such as assaults with fatal stab wounds and road traffic accidents with massive thoracic and abdominal injuries, and the clinical significance of such a focal lesion is currently unknown. Likewise diffuse TAI in the absence of any lesions in the corpus callosum or dorsolateral brain stem has an uncertain clinical correlation. It has been reported in mild TBI with no loss of consciousness[33,34] and described in association with a lucid interval.[10] A neuropathologist can be confident when faced with a TBI case of immediate coma ultimately resulting in death, with no intracranial mass lesions but diffuse TAI and lesions in both the corpus callosum and the rostral brain stem, that the cause of unconsciousness is diffuse TAI. Presented with a similar case but with no focal haemorrhagic lesions, if the post-mortem examination is essentially negative but there is diffuse TAI, it is reasonable to suggest

diffuse TAI as the cause of unconsciousness. However, this grade 1 pathology does not necessitate a period of unconsciousness and is, in itself, a survivable injury. As such, the existing grading system while conceptually useful does not offer clear clinicopathological correlation between grades; grade 1 can be both a fatal and a survivable pathology.

Causes of Diffuse Traumatic Axonal Injury

Diffuse TAI is typically associated with high-velocity rotational forces, such as road traffic accidents and falls from a height.[10,33] Simple falls, i.e. a fall from one's own height or less, typically cause a rapid deceleration and ASDH and/or contusions. Severe diffuse TAI has not been demonstrated in simple falls, but is seen in falls from a height.[6,87] Diffuse TAI has been described in the setting of assaults, although it must be remembered that reliable descriptions of the incidents are often lacking in this setting. Graham *et al.*[162] reported 15 fatal assault cases and described grade 3 diffuse TAI in 10, grade 2 in 1, and grade 1 in 4. Other single case reports exist in the literature.[201] Since more sensitive clones of β-APP antibodies have been introduced, axonal injury is now being seen much more frequently in assault cases, particularly focal TAI;[138] however, the relevance of this pathology to the death of those concerned is uncertain.

The Pathophysiology of Axonal Damage in Traumatic Brain Injury

Initially, trauma-induced axonal injury described in human brains was thought to be the result of axons being disconnected by shearing forces at the time of the impact (primary axotomy) leading to axonal retraction and axoplasmic pooling.[431] We now know that only a minority of axons undergo primary axotomy, the majority being damaged as a consequence of focal axolemmal perturbations and degenerating over a period of time after the initial insult.[289] Primary axotomy, i.e. direct shearing through the axon at the time of injury, is uncommon, but has been identified in small diameter fibres by electron microscopy in the non-human primate.[288] However, most of the axonal damage is secondary and delayed. Morphological studies in a cat model showed an anterograde tracer, horseradish peroxidase (HRP), accumulating at sites of axonal dysfunction within 60 minutes of the injury, indicating altered axolemmal permeability, with axotomy developing between 6 and 12 hours post injury.[354] The time-course does appear to be species specific, being most rapid in rats and of longer duration in cats and pigs, the longest duration being described in man.[353]

How the rotational forces associated with TBI modify ionic regulation in focal axonal segments remains to be fully determined. The initial hypothesis developed from animal models suggested that pores were produced in the axonal membrane as a direct consequence of the forces associated with TBI, the process being referred to as mechanoporation. This was thought to occur within minutes and continue for several hours. *In vitro* studies have offered an alternative hypothesis. In an axonal stretch model, rapid influx of calcium occurred into the axon, but this appeared to be temporally related to a sodium influx through specific sodium channels.[466] This group suggested that stretch causes activation of mechanosensitive sodium channels, resulting in sodium influx and activation of voltage-gated calcium channels (VGCC), in turn causing calcium influx and an increased intra-axonal calcium concentration. Further work by this group suggests a negative feedback mechanism whereby calcium influx results in proteolysis of one unit of the sodium channel, causing persistently elevated intra-axonal levels of calcium.[204]

Sodium channels are found at the highest density at nodes of Ranvier, and the earliest structural changes occur at this site.[11] With failure of the sodium/calcium exchange system, additional calcium is released from intra-axonal stores, including the axoplasmic reticulum and mitochondria.[423] Fast axonal transport is a function of microtubules, and microtubule fragmentation occurs rapidly after stretching along multiple points of an axon. Undulations develop in the axons, and there is disruption to fast axonal transport, causing axonal swelling and a risk of fragmentation.[439] The varicosities associated with this process are described in human diffuse TAI,[440] and microtubule and neurofilament fragmentation have been demonstrated in other models.[287] Not all axons show the same level of vulnerability. Small-diameter unmyelinated axons are most susceptible to injury in both *in vitro* and *in vivo* systems.[422]

An increased calcium level causes the activation of calcium-dependent proteases (calpains) and caspases. This results in the modification of neurofilament subunits, leading to the accumulation of dephosphorylated neurofilaments and damage to other intra-axonal proteins, resulting in local impairment of axonal transport with resultant axonal swelling. Over time, this is followed by loss of continuity of the axon and wallerian degeneration. Calpains have been shown to be important contributors to ongoing cytoskeletal degeneration in wallerian degeneration.[266]

Immunohistochemistry for β-APP only identifies axons in which anterograde axoplasmic flow has been disrupted as a result of microtubule fragmentation, and there is likely to be a second population of axons in which structural changes are present but in which axonal transport continues. An important consequence of the activation of calcium-dependent proteases is the fragmentation of neurofilament proteins. Studies using animal models of TBI have shown that β-APP and RMO14, a marker of neurofilament compaction, highlight distinct populations of damaged axons.[275,429] This has been demonstrated in both piglets and humans in our laboratory (Figure 10.28). Neurofilament protein changes were shown to evolve over time in a pig TBI model: neurofilament-light (NF-L) accumulated in axons within 6 hours, but neurofilament-medium (NF-M) and neurofilament-heavy (NF-H) did not accumulate until 3 days after injury.[65] Along with alteration of the neurofilament structure, other components of the cytoskeleton, such as spectrin and ankyrin, become damaged, eventually resulting in axonal disconnection with closing of the damaged axolemmal membrane on each side of the disconnected axon.[412]

These observations highlight the need for a reappraisal of how traumatic axonal injury is diagnosed in humans, and how these different molecular changes correlate with clinical outcome. In particular β-APP-immunoreactive axons are not necessarily irreversibly damaged and may only be an indicator of transient dysfunction of axonal flow, whereas neurofilament compaction is more likely to represent an end-stage process.

10.28 Double-label immunofluorescence of traumatically injured axons in the human corpus callosum, showing intra-axonal β-APP accumulation (green) and phosphorylated neurofilament (SMI-34 antibody) immunoreactivity (red). Axons co-expressing β-APP and phosphorylated neurofilament protein appear yellow. Although there is some overlap, there are many axons showing changes consistent with neurofilament compaction that do not show β-APP accumulation.

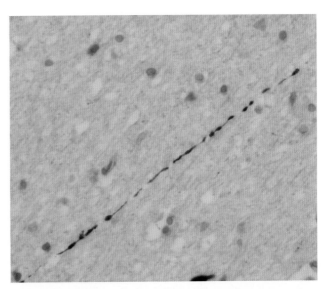

10.29 A damaged axon within the corpus callosum in a case of diffuse traumatic axonal injury (TAI). Varicosities have formed along the length of the axon, possibly related to microtubule fragmentation. β-APP immunostaining, ×40.

In the setting of human diffuse TAI, although the detection of β-APP is the most widely used marker of axonal injury, the neuropathologist needs to be careful in its interpretation. Accumulation of β-APP within axons in the absence of varicosities is common and is most likely a reflection of increased β-APP production within the cell body. It is particularly common in cases of global ischaemia with widespread neuronal cytoplasmic staining, and may not be a direct consequence of trauma. Varicosities along the length of an axon (Figure 10.29) appear to be related to microtubule fragmentation, an early step in axonal damage. Axonal bulbs (Figure 10.30) are likely to represent disconnection of the two ends of the damaged axon. Of all these

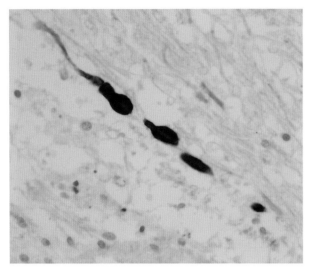

10.30 Swellings develop at the disconnected ends of damaged axons. β-APP immunostaining, ×40.

pathologies, it is possible that only the axonal bulb represents irreversible damage.

Brain Swelling

Brain swelling is a common finding in fatal TBI. Increased ICP associated with brain swelling remains the most common cause of death in severe TBI.[311] The swelling may be focal or diffuse and is mostly due to oedema, an increase in the water content of the brain tissue, and congestion, an increase in the cerebral blood volume, with oedema accounting for most brain swelling.[274] Oedema can be classified as cytotoxic, in which there is abnormal water retention by injured cells; vasogenic, in which blood–brain barrier (BBB) breakdown leads to the passage of plasma proteins and water into the extracellular compartment; and hydrocephalic (or interstitial), in which, as a result of increased intraventricular pressure, CSF is forced from the ventricle into the periventricular extracellular space.[328]

In the setting of TBI, the swelling may be focal, as a result of contusions or ICH; diffuse within one cerebral hemisphere, typically secondary to an overlying ASDH;[259] or diffuse in both cerebral hemispheres. Adjacent to contusions and ICH, there is physical disruption of the tissues, including BBB, and loss of the normal autoregulation within the local vasculature. The development of cerebral oedema appears to be principally due to severe disruption of the BBB involving endothelial cells, tight junctions, astrocytes or a combination of all components. MRI studies in rodents have shown that the BBB allows passage of fluid immediately after a closed blunt force head injury, but returns to normal function within 30 minutes.[26] After injury, there is altered expression of several proteins associated with BBB function including tight junction proteins and caveolin-1, a major component of the caveolae involved in transporting fluids.[329] Experimentally BBB proteins can be modulated to increase the movement of water from the brain parenchyma to the vascular system.[54] Aquaporin-4 (AQP-4) is an important molecule involved in water homeostasis in the brain, found predominantly in astrocyte cell membranes.[343] Interference with AQP-4 expression has been shown to reduce cerebral oedema in TBI models.[135]

Oedema can develop rapidly after TBI in animal models,[48,49] and rapid brain swelling within 20–30 minutes of injury has been described by CT scan.[226,476]

Diffuse swelling of one cerebral hemisphere is most typically associated with an adjacent ASDH. Even after this is removed surgically, the hemisphere may swell. The reasons for brain swelling in this context are not entirely understood but are likely to include a combination of a non-reactive vascular bed and local ischaemic injury. Hypermetabolism due to subclinical seizures or excitotoxic injury may also contribute.[44,104,202,231,312]

In the early CT era, the concept was introduced of 'diffuse brain swelling' as an entity seen primarily in paediatric head injury. Early studies suggested that this was due to cerebral hyperaemia with increased blood volume.[43] It was posited to reflect relative hyperaemia, although other studies suggested that the degree of hyperaemia was insufficient to cause cerebral swelling.[324] The increased prevalence in children is supported by post-mortem studies[160] where a specific type of 'malignant' brain swelling may be seen in the absence of significant ischaemic injury. However, the concept of 'malignant' oedema as an entity peculiar to childhood has been challenged. In a CT-based study, Lang et al.[235] found diffuse swelling of both cerebral hemispheres to be associated equally with paediatric and adult head injury, and to have a more aggressive course in adults. Bilateral brain swelling seems to be less common with modern TBI management, and this may relate to more care in the field to avoid hypoxia and ischaemia, more precise methods for fluid management with euvolemic dehydration, early ventriculostomy and decompression and improved imaging. Nonetheless, there are differences between normal cerebral blood flow and reactivity at different stages of maturation,[70,393,434,468] and these may influence the physiological propensity to brain swelling.

At autopsy, brain swelling is easily recognized by flattening of the gyri, sulcal compression, ventricular compression and midline shift if the swelling is unilateral. However, these features have not been found to be reliable for grading the severity of brain swelling,[186,271] and we suggest that brain swelling should simply be recorded as either present or absent, rather than as mild, moderate or severe.

Brain Stem Injuries

Brain stem lesions may be primary, as a direct consequence of the forces at the moment of injury, or secondary, as a result of the brain stem displacement associated with increased ICP. Brain stem lesions are common, being seen in 60 per cent of patients with severe TBI in one MRI study.[126] A good-prognosis group was defined on MRI with ventral lesions or superficial dorsal lesions, whereas a poor-prognosis group had deep dorsal brain stem lesions.[397]

Pathologically, primary brain stem injury has been considered to be a component of diffuse TAI, representing a severe form of the injury, discussed previously.[3] Al-Sarraj and colleagues[17] extended this concept and differentiated two separate forms of primary traumatic brain stem injury: that associated with diffuse TAI, and that which occurs in isolation, representing direct trauma to the brain stem, termed focal traumatic brain stem injury (FTBSI). The brain stem lesions of diffuse TAI were discussed earlier (pp. 656–657).

Twelve cases of FTBSI were identified in a cohort of 319 TBI cases.[17] The FTBSI was due to a complex fall from height or an accelerated fall (7/12), an assault (4/12) or severe impact on the top of the head (1/12). The longest documented survival was 2 days in this group, although most were found dead. Ten of the cases had skull fractures, eight involving the occipital bone extending into the posterior fossa, and two involving the base of the skull. The main pathological finding within the brain was haemorrhage in the brain stem.

Contusions and lacerations of the brain stem may be seen as a consequence of skull fractures around the foramen magnum, sometimes seen in the setting of extreme hyperextension of the neck.

In severe hyperextension of the neck, partial or complete pontomedullary or cervicomedullary avulsion can occur (Figure 10.31).[244] These lesions should not be dismissed as artefactual changes induced by poor post-mortem methodology, and most neuropathologists with experience in TBI examination will see cases. They are mostly related to road traffic accidents, and are seen particularly in pedestrians rather than in those in the vehicle, or in motorcyclists involved in accidents. Most individuals die immediately, but there are case reports of occasional examples where there is a survival of several days to weeks.[347] In one case series, 36 examples of pontomedullary avulsion were found in 988 TBI post-mortem cases, representing 3.6 per cent of the total.[402] Eight cases had pontomedullary avulsion in the absence of other brain pathology, 17 had avulsion with other brain stem lacerations and pathology elsewhere in the brain and 11 had brain stem lacerations outside the pontomedullary region, possibly fracture contusions/lacerations. Brain stem injuries were overrepresented in the motorcycle injuries group of this cohort, accounting for 41.7 per cent of the total.

Spinal Cord Injuries

Spinal cord injuries (SCIs) are common, the prevalence ranging from 11.5 to 57.8 cases per million people in the population.[1] The aetiology of SCI varies between countries, but road traffic accidents, falls, sports-related injuries and assaults are consistent causes between studies.[1] Age at the time of injury, neurological status at the time of injury and extent of injury were identified as predictors of survival.[449] In the paediatric population, SCI is uncommon, with cervical spine injuries accounting for 1.5 per cent of the National Pediatric Trauma Registry cases over a ten-year period.[21] Most cases involving younger children are due to road traffic accidents, whereas those involving adolescents are mostly sports related.[344] Younger children are more likely to have high spinal injuries, typically involving ligaments rather than bone, than will older children and adults.

Spinal traumatic EDH is exceptionally rare.[222] When spinal EDH is seen, it is usually spontaneous or related to therapeutic procedures. Cranial SDH can migrate to the spinal region,[249] but primary traumatic spinal SDH is also seen. It has been described in cervical[30] and thoracolumbar regions.[168] As with spinal EDH, there are a range of non-traumatic causes of spinal SDH, including coagulopathy, neoplasms, vascular malformations and therapeutic procedures.

Penetrating SCIs are rare outside military situations. Symptoms are related to the level of the injury, with high cervical lesions having a high mortality. Haemorrhage and tissue necrosis are seen in relation to the penetrating lesion.

Closed SCIs are the most common traumatic cord lesions in clinical practice, and are associated with fracture or dislocation of the spine (Figure 10.32). Cervical region SCI was seen in 5.4 per cent of patients with moderate or severe TBI.[190] SCI can be a consequence of hyperflexion/hyperextension movements, particularly in the

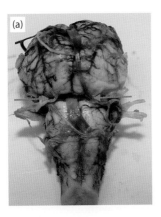

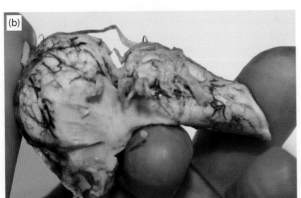

10.31 A severe brain stem injury caused by neck hyperextension during a motorcycle crash. There is partial separation of the brain stem at the pontomedullary junction (pontomedullary rent). Although the survival time was very short (minutes) haemorrhages could be seen in the tissue on either side of the tear.

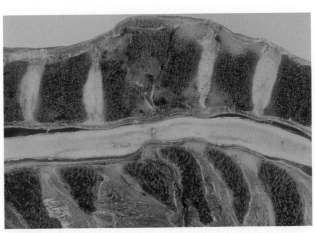

10.32 Fracture of the spinal column resulting in direct compression of the spinal cord.

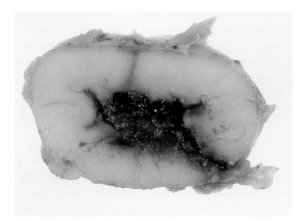

10.33 A cross-section of cervical spinal cord, showing central haemorrhagic infarction after a contusional spinal cord injury.

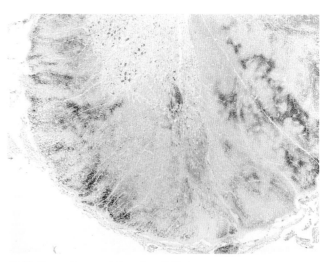

10.34 Extensive peripheral β-APP immunoreactivity seen in the high cervical spinal cord region. Although there was history of trauma, no associated soft tissue or bony injury was found at autopsy, and in this case β-APP accumulation has resulted from ischaemic rather than traumatic axonal injury. This part of the spinal cord is a recognized border-zone (watershed) region. It is important to be aware of this pattern of immunostaining to avoid potentially considering this a primary traumatic injury. β-APP immunostaining, ×2.

spinal region; compressive forces, such as a fall from height landing on the top of the skull; and rotational movements, causing fracture dislocations, most typically associated with thoracolumbar lesions.

Neuropathological examination in acute SCI may show compressive or contusional injuries. Compressive SCI may be due to fractures or displacement of intervertebral disc material, sometimes after mild trauma,[207] the pathology being a consequence of vascular compromise of the compressed tissue. In the early stages of compression, the cord shows venous congestion, progressing to necrosis and central cystic cavitation. Contusional SCI is more dramatic, typically associated with vertebral body fractures, in some cases with dislocation. The direct blunt force trauma to the spinal cord results in parenchymal haemorrhage and oedema that causes fusiform swelling of the cord. Bleeding is initially petechial, but these lesions coalesce to form more extensive haemorrhages, usually situated centrally within the spinal cord (Figure 10.33). Ischaemia develops, resulting in central infarction of the cord. The tissue damage extends for several levels above and below the site of direct injury. Axonal injury is prominent, and macrophage infiltration is seen after several days. In young children, distraction injuries can occur, typically involving the upper cervical spine or cervicomedullary junction. These are most often related to immobilization in restraints during high-speed road traffic accidents, or to crush injuries in which the head is run over by a vehicle.

Immunohistochemistry for axonal injury (β-APP) and macrophage infiltration (CD68) can be useful in defining SCI. However, extensive peripheral β-APP immunoreactivity in the upper cervical region may reflect a boundary-zone infarct, particularly in the setting of a 'respirator brain', rather than direct spinal cord trauma, and should be interpreted with caution (Figure 10.34).

Transection of the spinal cord is seen with extreme force, usually associated with dislocated fractures. In one study of 22 SCI cases, 38 per cent showed complete transection.[45]

Focal Vascular Injuries

We have discussed vertebral artery disruption as the cause of massive basal tSAH. However, a more common trauma-associated injury of the vertebral artery is dissection. Vertebral artery dissection develops when there is

disruption of the intima of the vessel, allowing thrombus to form, which can lead to infarction of part of the brain stem (particularly lateral medullary syndrome) or rarely the spinal cord.[194] The site of dissection is most commonly adjacent to the first and second cervical vertebrae.[256] Dissection can follow relatively minor trauma, and symptoms can develop over several days. Typical symptoms include ataxia, vertigo and nausea. An association has been suggested with neck manipulation.[187] Paediatric cases are also described.[62,319]

Internal carotid artery injuries secondary to trauma are well described, particularly after hyperextension of the neck.[23] In one study of 67 patients with blunt force carotid artery injury, 89 per cent were due to road traffic accidents, and 6 per cent to assaults.[120] One patient had a fatal transection of the internal carotid artery, while all other patients had dissection, with or without thrombus, cavernous sinus fistulae or pseudo-aneurysms. These injuries are often associated with skull fracture. Extracranial common carotid or external carotid artery injury is typically secondary to blunt force trauma to the neck and results in dissection with subsequent thrombosis.

Traumatic intracranial aneurysms can develop after blunt force or penetrating head injuries, and have a mortality of up to 50 per cent.[101] The aneurysms can be classified as follows: true aneurysms, in which incomplete vessel wall damage leads to subsequent dilatation of the damaged section of wall; false aneurysms (pseudo-aneurysms), which damage the full thickness of the vessel wall, with a false wall being formed by surrounding soft tissue structures; and mixed aneurysms, which histologically show a mixture of both, with a dilated section of vessel wall from which there has been bleeding, forming a haematoma that acts as a false wall.

A carotid cavernous fistula can develop after maxillofacial trauma, with direct communication between the internal carotid artery and the cavernous venous sinus. Presentation may be acute or up to several weeks following injury, and typically includes pulsatile proptosis, orbital and ocular erythema, headache and visual loss.

Cerebral venous sinus thrombosis (CVST) can be associated with TBI. However, one study concluded that CT venography should only be undertaken if a fracture extends across a dural venous sinus or jugular bulb, the risk being greater for occipital than parietal bone fractures.[92] Only 7 per cent of this high-risk study group developed venous infarction. At autopsy, it is important to remember that CVST and cortical vein thrombosis can be a consequence of reduced cerebral perfusion after TBI. In the most extreme situation, the so-called respirator brain,[330,460] secondary CVST and cortical vein thrombosis are very common. Radiologically, no association has been demonstrated between CVST and ASDH.[304]

PENETRATING INJURIES

Penetrating injuries are injuries in which an object enters the cranial cavity; in strict terms, a penetrating injury is one in which the missile enters the cranial cavity but does not exit, whereas a perforating injury is one where the missile also exits. The resulting pathology is very much determined by the nature of the missile. Sharp objects, such as knives, long nails or metal poles, may pierce the skull and extend into the underlying brain parenchyma causing local damage. In young children, objects may enter the cranial cavity through the orbital roof or nasopharynx, most often in association with a fall. They produce a haemorrhagic tract through the regions of parenchyma into which the object extends (Figure 10.35). High-velocity missiles, such as bullets, cause considerably more damage, the extent of the damage being related to the velocity of the missile; high-velocity military weapons produce greater tissue damage than small firearms.

Ballistic penetrating brain injuries are associated with a high mortality. A low GCS on admission, an associated intracranial haematoma, age >40 years, a trajectory passing through the ventricles and/or both hemispheres, and a unilaterally fixed dilated pupil are associated with a poor outcome.[18,278] A comparison of penetrating brain injuries between military and civilian groups found a significantly

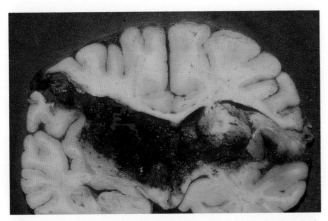

10.35 The haemorrhagic tract caused by a penetrating ballistic missile. There is cystic cavitation of the direct tract, but also extensive surrounding haemorrhagic infarction.

lower mortality in the military group (5.6 per cent military mortality, 47.9 per cent civilian mortality). There was a higher rate of intracranial pressure monitoring and neurosurgical intervention in the military group, although the difference in mortality between the two groups was probably due to a range of factors.[102] In civilian gunshot wounds, a good outcome can be seen in cases with rapid neurosurgical intervention, injury to the non-eloquent brain regions, and absence of injury to the brain stem and major vessels.[253]

As the missile passes through the brain parenchyma, it produces localized tissue damage, and in the wake of the missile a cavity forms, its size being determined by the energy of the projectile. The localized damage is a result of crushing of tissue by the missile passing through brain parenchyma. Studies in animal models have indicated that a penetrating ballistic injury causes local tissue damage including haemorrhage and necrosis, and initiates a biphasic inflammatory response: in the acute phase, there is cytokine expression and neutrophil infiltration; the delayed response involves white matter degeneration distant from the site of direct tissue injury and develops some days after the injury.[463,464] In addition, after a penetrating head injury, cerebral blood flow and cerebral metabolism are reduced.[227]

The tract of damaged tissue is roughly the same diameter as the projectile, unless there is a degree of yaw in the path of flight of the projectile, in which case the tract may be of greater diameter. A temporary cavity forms as the projectile passes through the brain and stretches surrounding tissue rather like the ripples spreading as a diver enters a swimming pool.[189] The final size of the cavity along the trajectory of the projectile is determined by the velocity and shape of the projectile, deformation of the projectile (such as mushrooming or flattening), fragmentation of the projectile and twisting or oscillation of the projectile about its flight axis (yaw).[121]

At post-mortem examination of ballistic penetrating brain injuries, fragments of the bullet may remain within brain parenchyma. If available, post-mortem radiology can be useful in identifying fragments prior to brain dissection. Three different zones have been described in penetrating missile injuries of the brain:[254] a central permanent cavity that contains necrotic brain tissue and blood, an intermediate zone with less tissue necrosis and parenchymal haemorrhages and a marginal zone with tissue discolouration. Both the intermediate and the marginal zones are related to the temporary cavity. Around the permanent cavity, there is axonal fragmentation and haemorrhagic extravasation. Axonal damage reduces radially, moving away from the permanent cavity.[332] By analysis of cases with a survival time beyond 2 hours, the tissue reaction has been further characterized.[333] CD68-immunoreactive macrophages demarcated a 1–2 mm necrotic zone around the permanent cavity, with β-APP immunoreactive damaged axons being seen in the surrounding tissues. In addition, β-APP immunoreactive axons could be seen remote from the permanent cavity.[334] A diffuse distribution of damaged axons was described in one series of 14 cases of gunshot wounds to the head, with extensive involvement of the brain stem in all cases.[228] Contusions may be seen at sites distant from the permanent cavity, particularly involving the lower cerebellum (owing to impact with the foramen magnum) and frontal and temporal lobes.

Blast Injuries

Traditionally, the study of blast injuries focused on the damage caused by blast waves to air-filled viscera, such as the lungs in the thoracic cavity. However, increasingly, and particularly in relation to the recent conflicts in Iraq and Afghanistan, attention has been focused on possible injuries to solid viscera, and the brain in particular. The abrupt pressure changes associated with a blast can lead to a mild head injury, and, in particular, symptoms suggestive of concussion. Long-term sequelae in the form of impaired concentration and memory problems have been described with a greater frequency after blast than non-blast traumatic brain injuries.[455]

The cellular basis of this injury is to date poorly defined, although there is considerable research activity in this field. The cellular responses produced by blast injuries, including microglial and astrocytic activation, were reviewed by Leung et al.[247] The presence or absence of white matter injury in the form of TAI is controversial. Diffusion tensor imaging has demonstrated diffuse disruption of white matter integrity in blast TBI.[88] In a non-human primate model exposed to blast TBI, distorted apical dendrites were described in the cortex, and there was loss of hippocampal CA1 pyramidal neurons, Purkinje cell dendritic degeneration, astrocytosis and upregulation of astrocytic aquaporin-4 and oligodendrocyte apoptosis, but no significant axonal injury.[263] Saljo et al.[379] described redistribution of phosphorylated neurofilaments from the axon to the neuronal cell body in an animal model of blast injury. A possible explanation as to why damage to axonal transport mechanisms has not been convincingly demonstrated in blast-injury models, as demonstrated in both blunt force and penetrating head injuries, may be either that a different white matter degenerative process occurs or that the current methods used to identify axonal damage are not sensitive enough to do so in thinner non-myelinated axons.[432]

Proteomics has been proposed as a powerful research tool that could be used to investigate the subcellular responses to blast injury,[14] although to date few data have been generated by this approach.

PERINATAL HEAD INJURY

Perinatal head injuries can develop as a result of excessive moulding of the cranial bones, or excessive force applied to the skull during delivery. Hyperextension of the neck during delivery may result in craniocervical injuries. This section will concentrate on perinatal skull fractures, traumatic intracranial haemorrhages and SCI. In his comprehensive monograph, Govaert et al.[155] urged caution in the assessment of incidence figures for cranial birth trauma, although he did note the clear decline in incidence from the 1950s through to the 1980s, from roughly 3 per 1000 to 0.5 per 1000 live births.

Linear skull fractures were described in up to 10 per cent of births in one series.[169] They are due to direct force on the bone, typically related to the use of forceps. Depressed (ping-pong) skull fractures are also associated with the use of forceps.[113] In themselves skull fractures, if appropriately managed, are not life-threatening. However, they can be associated with significant intracranial haemorrhage.

Perinatal EDH is rare, accounting for 2 per cent of perinatal traumatic haemorrhages in one post-mortem series.[433] Linear skull fracture is seen in most cases, and the bleeding is from meningeal arteries or veins. Other causes include parieto-temporal bone overlapping and excessive bending of the calvarial bone, rupturing underlying vessels.

Perinatal SDH is common, as discussed later in regard to abusive head trauma. However, clinically significant perinatal SDH is rare and is traumatic in aetiology. Tentorial laceration results in massive infratentorial ASDH. The tear may involve combinations of the vein of Galen, straight sinus or transverse sinus. Occipital osteodiastasis refers to the separation of the joints within the developing occipital bones, and this can result in damage to the occipital sinus and laceration of the cerebellum, causing infratentorial haemorrhage.[459] Laceration of the falx is less common than laceration of the tentorium cerebelli, and usually the tear is close to the junction of the falx and tentorium cerebelli. Bleeding is usually from the inferior sagittal sinus, with the haematoma forming above the corpus callosum. Finally, ASDH may be due to bridging vein rupture over the convexity of the brain. These lesions can be bilateral.[426]

Perinatal SCI results from excessive traction and/or rotation during delivery. SCI is uncommon with cephalic deliveries, but when it does occur it is a devastating injury involving the upper cervical spinal cord (Figure 10.36a) and in some cases causing complete transection of the cord.[399] In breech delivery, the site of injury is usually in the lower cervical/upper thoracic region.[270] In the acute phase, there is parenchymal haemorrhage, in some cases secondary to dislocation of the vertebral bodies, although the neonatal spinal column is particularly elastic.[452] The demonstration of a vital tissue reaction, such as macrophage infiltration or axonal swellings and axonal β-APP immunoreactivity, can be important in demonstrating that the injury was antemortem and does not represent a post-mortem artefact. If the neonate survives the acute injury, chronic degenerative changes develop, particularly cystic degeneration of the spinal cord parenchyma (Figure 10.36b).

ABUSIVE HEAD TRAUMA IN CHILDREN

Although head injury is relatively common in the paediatric population, the vast majority of cases are mild, causing few acute clinical concerns. As with adult head injury, the outcome in children is partly determined by the force of the injury and whether the injury involves primarily contact or inertial forces, and their magnitudes and distribution. Head injury in childhood may be due to a variety of causes including road traffic accidents, falls, injuries sustained in recreational and competitive activities, and assault. Of particular interest in this section is abusive head trauma (AHT).

The largest pathological study of fatal human paediatric traumatic brain injury looked at patients between the ages of 2 and 15 years; the results are referred to in the following discussion.[160] Skull fractures were recorded in 72 per cent of the cases, with the majority being linear. It is

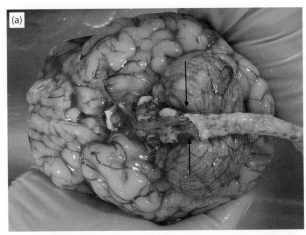

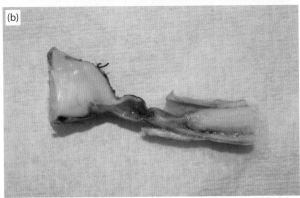

10.36 Perinatal spinal cord injury. This term infant had a complex delivery requiring forceps. Apgar scores were poor. The neonate was transferred to a neonatal intensive care unit and maintained on a ventilator for several days prior to death. At autopsy, there was obvious high-cervical spinal cord damage **(a)** which, on longitudinal section, showed cystic cavitation **(b)**.

Panel (a) courtesy of Dr Paul French, Glasgow, UK.

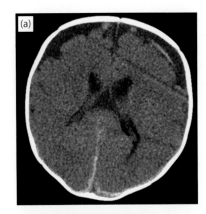

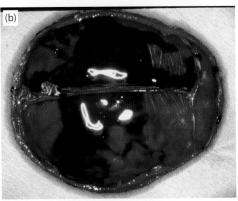

10.37 Bilateral thin film subdural haematomas (SDH) typical of that in abusive head trauma (AHT). Computed tomography (CT) imaging **(a)** and post-mortem **(b)** appearances. The skull cap has been removed with the dura intact, and thin acute subdural bleeds are seen bilaterally.

important not to confuse a linear skull fracture with sutures or Wormian bones, and the pathologist should be aware of this potential pitfall.[29] Contusions were seen in 92 per cent, large intracranial haematomas in 34 per cent and intracerebral haematomas in 16 per cent, of which 8 per cent were burst lobes, including two cases of posterior fossa haematoma. From this study, the incidence of major TBI pathological substrates is similar to that in adults. However, children under the age of one year were not represented in this study, and AHT has a peak incidence in infants (by definition aged 12 months or under), with an incidence of 28.9 per 100 000 per year in infants compared to an incidence of AHT of 4.1 per 100 000 per year in five-year-olds in one study.[386]

Child abuse can involve neglect, emotional abuse, physical abuse and sexual abuse. Scottish government statistics from 2002 recorded 1900 children as victims of child abuse, 33 per cent of whom suffered physical abuse. Of this group, 1.8 per cent had brain injury, with brain injury accounting for 0.6 per cent of all child abuse.[313]

The concept of abusive brain injuries in children has been known for many years,[50,176] and has had several names attached such as battered baby syndrome[51] and, more commonly, shaken baby syndrome (SBS). The latter name reflects the proposed mechanism resulting in the typical triad described in these cases: ASDH, usually thin film and bilateral (Figure 10.37); retinal haemorrhages; and ischaemic encephalopathy. Shaking was proposed to cause an acute whiplash-type injury. However, a biomechanical study highlighted the low forces generated by shaking compared to much higher forces generated by inflicted impact, the latter being more in line with the forces predicted to be required for ASDH and diffuse TAI.[105] The study suggested that shaking alone was unlikely to generate the forces required to produce the typical pathology. This was supported by post-mortem findings in this same study, which showed evidence of impact injury in all fatal cases. Biomechanical data have subsequently been conflicting as to whether shaking alone can cause a fatal injury.[79,340] Non-accidental trauma (NAT) and non-accidental head injury (NAHI) are two commonly used terms, although currently the term abusive head trauma (AHT) is recommended by the American Academy of Pediatrics;[71] all three terms avoid attributing injuries to any specific mechanism, recognizing that different mechanisms can cause the variety of injuries observed in this context.

There are numerous studies detailing the clinical and radiological characteristics associated with AHT. However, all such studies have problems with case selection, because often the diagnosis of AHT can be subjective and requires a critical assessment of a caregiver's history against the infant's injuries on presentation. In children <2 years of age, 24 per cent of all cases of TBI admitted to hospital were presumed to be AHT and 32 per cent were suspicious for AHT.[106] A systematic review of 24 studies assessing TBI in children up to six years old has been undertaken;[348] in 86 per cent of studies the mean age for AHT was <12 months, and in 43 per cent of studies the mean age for AHT was <6 months. By comparison the mean age for accidental TBI was <12 months in 50 per cent of studies, but <6 months in only 7 per cent. They found that ASDH, brain ischaemia, retinal haemorrhages, skull fracture with intracranial injury, metaphyseal fracture, long bone fracture, rib fractures, seizures, apnoea and no adequate history from caregiver were all significantly associated with AHT. A second systematic review found ASDH, cerebral ischaemia and brain swelling to be significantly associated with AHT, whereas EDH was significantly associated with accidental TBI.[221] Multiple convexity ASDH, interhemispheric haemorrhage and posterior fossa ASDH were particularly associated with AHT.

Pathological studies of AHT are limited, and are also complicated by case selection bias. Calder et al.[52] reported 12 cases aged 12 days to 2.5 years. Infants under 5 months had parasagittal white matter lesions (in this study called contusional tears). Infants and children older than 5 months had dorsolateral quadrant lesions, similar to those seen in adults. Vowles et al.[453] described diffuse axonal damage, described as similar to diffuse TAI and identified using silver stains, in ten infants <5 months old. However, it is likely that this study was in fact describing widespread axonal injury secondary to ischaemia rather than diffuse TAI. Hadley et al.[178] published a clinicopathological study of 13 infants, eight of whom died, with six post-mortem examinations. Cervicomedullary injury was seen in five of the infants who had post-mortem examination, in the form of EDH, ASDH and spinal cord contusion. Shannon et al.[392] reported 14 cases of AHT in the absence of skull fracture. They specifically assessed β-APP immunoreactivity and found extensive white matter deposits but concluded that much of this was due to ischaemia. However, they did note cervical nerve root β-APP immunoreactivity only in AHT cases and not in hypoxic-ischaemic encephalopathy control cases. Gleckman et al.[153] used β-APP immunohistochemistry to assess diffuse TAI in seven cases of AHT compared to three cases of accidental TBI. Diffuse TAI was reported in five of seven AHT cases, and two of three accidental TBI cases. However, this publication dismissed ischaemia as a potential cause of the axonal injury, although ischaemic axonal injury has since become widely recognized.

Geddes et al.[139,140] studied 37 infants (aged 20 days to 9 months) and 16 children (aged 13 months to 8 years), all suspected of being victims of AHT. They found that at age 2–3 months, the most common presenting symptom was apnoea or other breathing abnormalities, and most likely causes included skull fracture, thin film ASDH and axonal damage at the craniocervical junction. Extracranial injuries were typically absent in this group. Although ASDH was common in both groups, being seen in 72 and 71 per cent of cases, respectively, the volume and distribution were different. Seventy-eight per cent of the infant cases had global ischaemic injury identified microscopically. Twenty five of 37 cases in this group had β-APP immunoreactivity in the white matter, although in 35 per cent (13/25) the pattern was ischaemic. Only 5 per cent (2/37) of these cases had a pattern typical of diffuse TAI. In 22 per cent (8/37), focal TAI was seen in corticospinal bundles in the caudal pons and medulla,

and in seven of these cases, this focal TAI was the only β-APP immunoreactivity detected. Cervical nerve root β-APP immunoreactivity was seen in 8 per cent (3/37) of cases. The older age group had a higher incidence of extracranial injuries, particularly abdominal injuries, and had space-occupying ASDH. When present in the older age group, TAI showed the same diffuse pattern as in the adult. These studies demonstrated that the injuries sustained were influenced by the age of the child. Infants appeared to be susceptible to localized axonal injury at the cervicomedullary region, a feature not seen in older children. In both groups, global cerebral ischaemia was common, with associated cerebral swelling and increased intracranial pressure. The authors proposed that in infants, damage to the cervicomedullary region results in cardiorespiratory arrest and subsequent global cerebral ischaemia and swelling. Although this hypothesis might account for infants with bilateral damage, it does not provide an adequate explanation for those patients whose damage is predominantly or exclusively unilateral.

Reichard et al.[362] described β-APP immunoreactivity in the brain and spinal cord of 28 cases of AHT. Immunostaining was seen in 27, and in 22 cases the pattern was typical of ischaemia. Eight cases had a pattern typical of diffuse TAI, and 19 cases had a pattern typical of focal TAI, with combined ischaemic and traumatic patterns of axonal injury being seen in 16 cases.

One post-mortem study of AHT looked specifically at neck injuries.[39] Fifty-two children age two years and under, all homicide victims with clinical and post-mortem information, were identified over a nine-year period. Forty-one of 52 (79 per cent) were AHT cases. Twenty-nine of 41 had primary cervical cord injuries: parenchymal injuries (21 cases), meningeal haemorrhage (24 cases) and nerve root avulsion or dorsal ganglion haemorrhage (16 cases). Six cases had no evidence of a primary cranial impact site, and all six had SCI. No spinal fractures were seen, but six cases had soft tissue (ligamentous or muscular) neck injuries. The authors concluded that SCI was common in AHT in the absence of spinal fractures and often in the absence of significant soft tissue injury. They also found brain stem injury in 16/40 cases of AHT (one case was excluded from the overall group owing to poor condition of the tissue), and 14/16 brain stem injury cases had associated cervical SCI.

The presence of β-APP immunoreactivity in spinal cord nerve roots as a marker for AHT or other forms of TBI has been questioned.[421] Although this study highlights the need for research to address this issue, it was in itself flawed, including a case of TBI with skull fracture in the non-trauma study group.

ASDH is a common feature of AHT and is seen in virtually every case. The origin of ASDH in AHT has become an issue of some controversy.[418,419] The principal non-traumatic cause postulated to underlie infant ASDH involves a combination of increased intravascular pressure in a vascular system with hypoxic endothelial injury.[141] Intradural haemorrhage (IDH) and ASDH were described in a series of paediatric post-mortem examinations.[75,385] In the larger study, 636 paediatric autopsies were reviewed, and cases were categorised as non-macerated fetal, neonatal (0–4 weeks of age), infant (1–12 months) or child (1–3 years). ASDH was seen in 55/254 fetal cases and was associated with IDH in all cases. Seventy-two of 382 infants and children had ASDH, not always in association with IDH. The authors advocated a clear association between ASDH and hypoxic-ischaemic encephalopathy. However, in the fetal and neonatal groups, these data need to be interpreted with caution because ASDH is seen on MRI studies in asymptomatic neonates, resolving within one month in most cases,[261,458] although extending up to 7 weeks in one case.[371] The incidence of neonatal ASDH varies from 8 to 46 per cent in these MRI studies, suggesting that the post-mortem data reflect a common pathology in asymptomatic neonates that spontaneously resolves within the first few weeks of life. The cause of this subdural bleeding is mostly likely traumatic caused by the head being engaged during labour and, in vaginal delivery, passing through the birth canal and undergoing static loading and low-velocity strain to vessels. Hypoxia is not necessary to explain the presence of ASDH in this age group. Assessment of the older age group is difficult because no information is given in relation to other potential causes of ASDH such as coagulopathy in the setting of sepsis. As such, hypoxia remains an unproven hypothesis as a cause of ASDH in neonates, infants and children. Indeed, one pathologico-radiological study assessing infants and children <4 years after cardiorespiratory arrest found no evidence of ASDH in the non-traumatic group.[196] A single retrospective post-mortem review of ASDH in hypoxic-ischaemic encephalopathy cases found no causal association, but because this study was retrospective, small volumes of ASDH may have been missed.[47]

Retinal haemorrhages are a key element in the diagnosis of potential AHT. Detailed discussion of retinal haemorrhages is outside the scope of this chapter. The reader is referred to reviews of ocular pathology in AHT[470] and clinical studies that describe the importance of the extent and distribution of retinal haemorrhages in helping to differentiate between accidental TBI and AHT.[28,314]

Ultimately, it is not for the neuropathologist to make the diagnosis of AHT. Rather the neuropathologist should describe the post-mortem findings. If there is an absence of external injury or any other internal signs of trauma such as long bone or rib fractures, but the classical pathology associated with many inflicted injuries is seen (cases with only SDH, retinal haemorrhages and encephalopathy), it may be difficult to arrive at a diagnosis of AHT; the combination of SDH, retinal haemorrhage and encephalopathy is in itself not diagnostic of AHT. However, the presence of ASDH in an infant or child should always raise the possibility of trauma, and in the absence of any other natural disease, such as coagulopathy for whatever reason or metabolic disorder associated with ASDH, trauma must be the most likely diagnosis. In cases with an obvious impact site and/or skull fracture, the diagnosis of trauma is secure. However, the neuropathologist cannot diagnose intent, but rather his or her report is part of the overall investigation into the death of an infant or child with fatal TBI.

Fatal accidental TBI in childhood is usually the result of falls or road traffic accidents, often with massive crush fractures.[58] Low-level falls are common events in childhood and can be associated with significant head injury. Most household falls are benign,[106] and one study combining a clinical and biomechanics approach found no severe or life-threatening injuries after a household fall,[444] although it is widely acknowledged that EDH can occur from an accidental low-height fall in infants and young children. A large UK-based study analysed

data from 11466 infants <6 months old, and documented 3357 accidental falls in 2554 infants. Most falls resulted in no injury and serious injury was very rare. In particular, falls from sofas and beds did not result in serious injury in this series.[456] A review of childhood fall–related data from one paediatric trauma centre identified 729 cases[223] and found high-level falls (>15 feet [4.5 m]) had a mortality rate of 2.4 per cent, whereas low-level falls (<15 feet [4.5 m]) had a mortality rate of 1.0 per cent, although all fatal low falls had intracranial pathology. Fatal falls from 2–10 feet (0.5–3 m) have been described in the paediatric population, although the head to surface distance and initial velocities were likely higher than the stated height would suggest in many cases.[352] Most witnessed fatal short falls are associated with skull fracture, usually linear, and EDH or ASDH, the cause of death being the haemorrhage acting as a space-occupying lesion.[361] Age and fall height influence the types of head injuries seen; infants sustain skull fractures more commonly than toddlers, and low-level falls (<3 feet [1 m]) can cause intracranial injury in the absence of soft tissue injury or skull fracture, occurring in 6 per cent of infants and 16 per cent of toddlers.[199]

MILD TRAUMATIC BRAIN INJURY, CONCUSSION AND SPORTS-RELATED BRAIN INJURIES

Concussion refers to an immediate, usually reversible episode of brain dysfunction after TBI, typically, but not always, with sudden brief impairment of consciousness and loss of memory. It represents the mildest form of TBI, part of a clinical spectrum ranging from mild concussion, in which consciousness is often preserved, to severe diffuse TAI resulting in the vegetative state.[143] The terminology has evolved over time, and is not used by all authors or disciplines in the same way; for this reason, some authors advocate using the term 'concussion spectrum' to include many manifestations and constellations of the reported findings.[112] The main explanations that have been suggested for concussion can broadly be grouped as vascular, reticular, centripetal, pontine cholinergic and convulsive, the convulsive theory being best supported by the available scientific literature,[394] with neurophysiological pathology being considered the key event. Clinical guides for assessing and grading concussion have been developed,[56] but current recommendations regarding the evaluation and management of concussion in sport have moved away from grading systems.[152] In addition, this last report highlights the confusion around the terms concussion and mild TBI (mTBI), although the report was criticized and the value of existing tools, such as the Sports Concussion Assessment Tool 2, supported.[293]

The anatomical basis of concussion is uncertain. It may represent transient dysfunction of the neuron or may be associated with structural changes, such as focal TAI. Animal models have been developed to assess mTBI, including rodent models,[197] a piglet model[42] and rodent models of repetitive mTBI.[323] These models use both radiological and histopathological end points to assess structural changes in mTBI. Rodent models have demonstrated localized axonal damage after mTBI, with evolving anatomical distribution at different time points after injury,[416] but little other significant histopathological change has been described.

The literature on the neuropathology of human mTBI is sparse, owing to the limited material for study. Modern neuroimaging techniques cannot replace detailed histopathological examination, but do offer an *in vivo* assessment of pathology to the millimetre level[32] and have been extended to assess mTBI.[129] DTI in particular is well suited to assess *in vivo* white matter structural changes. Hylin *et al.*[197] correlated DTI changes with axonal injury, identified by silver stains, in a rodent model of mTBI. A meta-analysis of human mTBI DTI studies highlighted the posterior corpus callosum as being particularly susceptible to white matter changes.[20] MRI volumetric studies of mTBI patients showed loss of brain volume, which evolves over several months.[237]

The only neuropathological cohort used to study mTBI initially comprised five patients with mTBI (GCS 13–15) who had died of other causes, 2–99 days post-injury.[34] Axonal β-APP immunostaining, although sparse, was seen in the fornices and corpus callosum of all cases, and variably within the brain stem and cerebral white matter. This study was extended to six cases in a subsequent publication,[35] and the axonal injury was analysed in more detail using an axonal index sector score (AISS), a technique very similar to the haemorrhagic lesion score described earlier, in which 116 separate sectors within each brain are assessed to derive an overall axonal injury score. mTBI AISS ranged from 4 to 88, whereas the AISS in severe TBI cases ranged from 76 to 107, providing further evidence for a spectrum of axonal injury in human TBI.

Repetitive head injury is a particular concern within contact sports such as boxing, rugby, American football, Australian rules football and ice hockey. In addition, repetitive head injury is a high-profile research priority because of ongoing military conflicts with associated high incidence of repetitive head injury. The incidence of sport-related TBI in the United States is estimated at between 1.6 and 3.8 million episodes per year.[236]

Civilian mTBI and blast mTBI do appear to have similar effects on cognition.[248] Long-term cognitive dysfunction was reported in a rodent model of repetitive mTBI,[307] although an extended time between injuries limited the long-term decline, suggesting a period of vulnerability after each episode of mTBI.[273] Compared to single TBI, repeated TBI in a piglet model produced worsening performance on composite cognitive function tests, and increased mortality was seen in the repeated injury group, particularly with injuries 24 hours apart compared to injuries 1 week apart.[132] In addition, in piglets diffuse TAI was exacerbated by a second rotational injury 15 minutes after the first, with more areas of axonal damage being seen in the group with two injuries compared to the single-injury group.[360]

Second Impact Syndrome

Second impact syndrome (SIS) is a rare, often fatal, complication of mTBI in young athletes who have a second impact before the effects of an initial impact have fully resolved. A number of cases have been described, and the condition appears to involve adolescents, with American football being particularly overrepresented.[55,443] The reported

pathology includes thin-film ASDH[57] 'malignant' brain swelling possibly due to cerebral blood flow dysautoregulation and massive hyperaemia[457] and, in some cases, a combination of both.[317] However, the existence of the syndrome has been questioned and criteria were proposed in an attempt to standardize the reporting of cases.[295]

LONG-TERM SEQUELAE OF BRAIN INJURY

Vegetative State

The vegetative state refers to a loss of meaningful cognitive function and awareness in patients who retain spontaneous breathing and periods of wakefulness. The neuropathological basis of the vegetative state was explored in a post-mortem study of 49 patients in the vegetative state, 35 of whom had experienced TBI.[12] In the trauma-related cases, diffuse TAI of grade 2 or 3 was found in 71 per cent of cases, with thalamic pathology in 80 per cent of cases. In cases with minimal brain stem and cerebral cortical pathology, thalamic pathology was always present. Therefore, damage to the thalamic nuclei and/or the afferent and efferent white matter pathways of the thalamus appear to play a major role in the genesis of the vegetative state after head injury. The thalamic nuclei showed differing degrees of neuronal loss, with cognitive and executive function nuclei being most severely affected.[290,291] White matter (wallerian) degeneration is a consequence of severe diffuse TAI. The axonal loss results in gliosis and macrophage activation. A study assessing the neuropathology of the recently described unresponsive wakefulness syndrome highlighted the importance of brain stem pathology,[208] in keeping with the literature on DAI.

In contrast, the structural basis of moderate disability after TBI is more likely to be a focal lesion rather than diffuse brain pathology, usually an evacuated intracranial haematoma.[13] In a study of 30 severely disabled patients, 50 per cent had focal brain pathology only. Some severely disabled patients did show diffuse brain pathology similar to vegetative state patients, although it may be that there is a greater quantitative amount of damage in the vegetative cases.[210] In assessment of the pathology of moderate and severe disability, case selection may be important, and it must be remembered that post-mortem–based studies may not accurately reflect the clinical spectrum associated with moderate and severe disability.

Long-term Cognitive Problems

There is considerable retrospective epidemiological literature suggesting that TBI is associated with an increased risk of developing Alzheimer's disease in later life,[174,292] although not all studies have confirmed this association.[41] Prospective studies have also reported conflicting data, some studies showing an association,[351] whereas others show no association.[238] A meta-analysis of seven case-control studies[321] calculated a relative risk of developing Alzheimer's disease of 1.82 for head injury with loss of consciousness, only reaching statistical significance for males. Fleminger *et al.*[128] studied 15 case-control studies and calculated an odds ratio (OR) of 1.58. Again, however, this study showed that the association between head injury and Alzheimer's disease was statistically significant only for males (males, OR 2.26; females, OR 0.92), who form the majority of the head-injured population.

Follow-up of patients 15–25 years after admission to hospital with TBI provided further evidence of neurodegeneration at a late stage.[309] However, even mild head injury (acute GCS 13–15) is associated with a higher than expected incidence of disability (Glasgow Outcome Score: moderate or severe disability) one year post-injury.[445]

There is increasing evidence at a cellular and molecular level that there are similarities between the long-term responses to traumatic brain injury and Alzheimer's disease. Neuropathologically, Alzheimer's disease is characterized by the deposition of amyloid-β (Aβ) plaques, and the accumulation of tau-immunoreactive neurofibrillary tangles and neuritic threads (see Chapter 16). Studies have focused attention on neuroinflammation in the form of microglial activation, as a mechanism of potential relevance to neurodegeneration both in Alzheimer's disease and in the response to TBI.[405] Microglia are the principal cellular mediators of inflammatory processes in the central nervous system (CNS) and are a source of several of the proteins upregulated both in Alzheimer's disease and after TBI, including apolipoprotein E, APP and pro-inflammatory cytokines such as interleukin 1 (IL-1). This raises the possibility that some patients who sustain a head injury may have a microglial response that plays a role both in influencing their outcome following injury and in their increased susceptibility to Alzheimer's disease later in life, and that may be modified by susceptibility genes, such as *APOE*.

IL-1 is expressed in increased quantities in the cerebral cortex within hours of TBI,[171] and chronic overexpression of IL-1 is found in Alzheimer's disease.[170] Griffin *et al.*[172] proposed a 'cytokine cycle' in which TBI or other forms of brain injury, can, in susceptible individuals, initiate an overexuberant sustained inflammatory response that results in neurodegeneration. IL-1-positive microglial cells lie in close relation to APP-positive neurons and dystrophic neurites in head-injured patients and are also found in close apposition to neurofibrillary tangle-containing neurons in Alzheimer's disease.[171] In TBI, APP is upregulated in response to increased IL-1 and is known to be upregulated in Alzheimer's disease. Diffuse Aβ plaques have been identified in approximately 30 per cent of individuals who die shortly after a single episode of severe traumatic brain injury,[370] a higher proportion than in non–head injury controls. Most of the deposits consist of Aβ 42,[149] and the distribution does not correlate with other TBI-associated pathologies.[163] In addition to parenchymal Aβ plaques, intra-axonal co-localization of β-APP, Aβ and tau was described in a piglet model,[409] and Smith *et al.*[411] demonstrated Aβ accumulation within damaged axons in human brains after a single episode of fatal TBI. They postulated that damaged axons can act as a reservoir of Aβ that may then be involved in plaque formation.

Experimental studies suggest that the cellular pathology initiated by an episode of acute TBI may indeed be progressive, and the role of apoptosis after TBI was reviewed by Raghupathi *et al.*[359] Cell loss from neocortex, thalamus and hippocampus, with associated gliosis and ventriculomegaly, was demonstrated in rats after fluid-percussion injury, with tissue loss continuing up to 12 months after injury.[38,408] Studies of human TBI tissue demonstrated terminal deoxynucleotidyl transferase dUTP nick end labeling (TUNEL)–positive cells up to 12 months after TBI.[465] The majority of the cells were present in the white matter and were considered to be closely associated with wallerian degeneration. It is uncertain whether this is part of the normal reparative processes after axonal injury or represents a separate pathological process, with an episode of TBI initiating a delayed intrinsic cellular process resulting ultimately in cell death.

REPETITIVE HEAD INJURY AND CHRONIC TRAUMATIC ENCEPHALOPATHY

Clinically, two separate processes have been described in relation to the long-term effects of repetitive mTBI: chronic post-concussion syndrome (CPCS), with symptoms such as impaired attention, poor memory, and irritability persisting for years after mTBI; and chronic traumatic encephalopathy (CTE), a presumed neurodegenerative condition with behavioural, cognitive and/or motor symptoms.[215] CPCS is rare and is associated with premorbid anxiety and depression. The structural and neuropathological basis is unknown.

It has been known for many years that repetitive head injury is linked with neurodegeneration. The 'punch drunk' state was first described by Martland in 1928[279] and was later renamed 'dementia pugilistica'. However, neurodegeneration within the context of contact sports is widely reported, and the terminology has been updated to reflect this, the current descriptor being CTE. CTE as a term has been used for many decades but has become a subject of considerable interest since its detailed description in a National Football League (NFL) player.[336,337] The clinical descriptions of dementia pugilistica described a degree of intellectual deterioration, often with an associated movement disorder, usually parkinsonism, but in some cases predominantly ataxia. The largest study of this disorder clinically[368] examined 224 ex-boxers neurologically by electroencephalography and by simple psychometric testing. Seventeen per cent of the ex-boxers had varying degrees of movement disorder involving the cerebellar, pyramidal and extrapyramidal systems. Minor degrees of intellectual function were seen in several of the ex-boxers, although only two required long-term care as a result of their cognitive impairment, and Roberts concluded that the occurrence of encephalopathy increased significantly with the number of bouts and the length of the boxer's career. He also concluded, however, that the rate of cognitive decline was not greater than that associated with ageing alone. More recent studies[59,216,302] suggest that full-blown dementia pugilistica remains uncommon, although mild cognitive and movement disorders are still associated with boxing. Neuropsychological dysfunction was not seen in a study of 289 amateur boxers,[46] although a more recent study does report neurocognitive deficits.[282]

The risks associated with repetitive mTBI have been described in other sports such as football,[280,281] rugby union and Australian rules football,[296] American football[336] and ice hockey,[31] although in the absence of large prospective studies the absolute risk of cognitive impairment and of movement disorders secondary to repetitive mTBI sustained in these contact sports remains uncertain.

The largest pathological assessment of dementia pugilistica was the examination of the brains of 15 boxers, 11 of whom were diagnosed with dementia pugilistica in life.[78] This study reported three principal pathological features of the brain in dementia pugilistica: abnormalities of the septum pellucidum, with a fenestrated cavum septum being seen in 77 per cent of boxing cases and 3 per cent of controls; cerebellar damage, with gliosis of the inferior aspect of the cerebellar hemispheres adjacent to the tonsils in 10 of the 15 ex-boxers brains studied; and degeneration of the substantia nigra and locus coeruleus, where pigmented cell loss is often marked and neurofibrillary tangles can be seen in some of the remaining neurons. Lewy bodies are not a feature. Cortical contusions, a common finding in acute TBI, are not a significant feature of dementia pugilistica.[2,78]

In 2001, Schmidt *et al.* compared the molecular profiles of the neurofibrillary tangles in dementia pugilistica and Alzheimer's disease. They found that dementia pugilistica and Alzheimer's disease had a common tau isoform profile and phosphorylation state. They concluded that the mechanisms underlying both these conditions might be similar. Although not seen in the initial studies by Corsellis, extensive Aβ immunoreactive plaques were reported in cases of dementia pugilistica.[369] Neuritic plaques characteristic of Alzheimer's disease were absent in this series. However, neuritic plaques were reported in a later immunohistochemical study of dementia pugilistica.[446]

The neuropathology of repetitive head injury has been reported by Geddes *et al.*[137] They reported five cases with repetitive mTBI (two boxers, a footballer, a man with behavioural problems that involved repeated head banging, and an epileptic with recurrent mTBI during seizures) and described tau-immunoreactive neurofibrillary tangles and neuropil threads, particularly around blood vessels. The neuropathological features of TBI in boxing have been comprehensively reviewed.[448]

Recent studies have focused on the neuropathology of CTE.[299,300] The largest study assessed 85 cases of repetitive mTBI brains, mostly in athletes and military personnel, and described neuropathological changes considered by the authors to be pathognomonic of CTE in 68 cases. All cases were males with an age range of 17–98 years and a mean age of 59.5 years. The distribution of phosphorylated tau in CTE differed from that in Alzheimer's disease. CTE cases had subpial clusters of astrocytic tangles, neurofibrillary tangles predominantly within superficial cortical layers and within the depths of sulci, and perivascular distribution of both astrocytic and neuronal pathology. Four pathological stages were described: stage I–II, predominantly frontal perivascular pathology, and stage III–IV, widespread tau pathology in neocortical and subcortical regions, including brain stem. TAR-DNA binding protein 43 (TDP-43) immunoreactivity was found in 85 per cent of cases.[300] This was predominantly neuritic, but in CTE stage IV intraneuronal and intraglial inclusions were also described. Of the 68 cases with CTE, 43 (63 per cent) had a sole diagnosis of CTE, 8 (12 per cent) also had motor neuron disease, 7 (11 per cent) had associated Alzheimer's disease, 11 (16 per cent) had associated Lewy body disease and 4 (6 per cent) had frontotemporal lobar degeneration. The concurrence of CTE with other neurodegenerative disorders supports the concept that

repetitive mTBI results in widespread neurocellular protein dysfunction, at least in some examined patients. Interestingly, despite the observation that Aβ is deposited in acute TBI and in ex-boxers, Aβ does not appear to be a significant component of CTE. A study of human TBI cases with survival up to three years after TBI demonstrated ongoing axonal accumulation of APP with associated intra-axonal Aβ, but virtually no Aβ plaques, perhaps reflecting plaque dispersion after initial acute post-TBI Aβ plaque deposition.[66]

CTE remains an evolving concept, although current evidence suggests a novel tauopathy. However, there is a clear need to better define the association between neurodegenerative pathology and clinical phenotype, and to definitively demonstrate an association between repetitive mTBI and a specific pattern of neurodegenerative pathology; the post-mortem literature needs to be complemented by more detailed clinical data documenting the evolution of the clinical phenotype, supported by experimental evidence of a clear association between mTBI and abnormal tau accumulation. Causality between repetitive mTBI and CTE currently remains unproven, and any associated epidemiological risk factors have not been assessed.[294]

The distribution has been assessed of tau pathology after a single head injury with survival up to one month[406] and survival of 1–47 years.[211] In the acute phase, no significant tau pathology was seen. However, in the longer surviving group, neurofibrillary tangles highlighted by tau immunohistochemistry were seen with greater frequency in the TBI group than in age-matched controls, this difference being most apparent when cases aged 60 or under at the time of death were compared to age-matched controls: 34 per cent of the TBI cases had neurofibrillary tangles, compared to 9 per cent of the age-matched controls. The distribution of the neurofibrillary tangles was similar to that described for CTE: superficial frontal cortex, with clustering in the depths of sulci. This group also showed a trend towards a wider distribution and higher density of Aβ plaques than in age-matched controls.

The data relating to the neuroinflammatory response in long-term survivors of TBI are conflicting. One study looking at 52 cases of TBI survival ranging from 10 hours to 47 years described significant neuroinflammation related to ongoing white matter degradation, the neuroinflammation being most prominent with survival times between 1 and 17 years.[212] However, a similar study found that the neuroinflammatory response had returned to normal levels within a few years.[407] A single study of a dementia pugilistica case, a 55-year-old ex-boxer, described upregulation of HLA-DR microglia and neuronal expression of the complement factor C1q.[378] However, the possibilities remain that a single episode of TBI may lower the threshold for dementia in an individual and that neuroinflammation may play in role in ongoing neurodegeneration (Figure 10.38).

Whether these changes occur in all patients with a single episode of TBI or repetitive mTBI or predominantly in genetically or otherwise susceptible individuals, and whether there are effects of age or gender, are questions that remain incompletely understood. These issues have implications for recommendations for sports, military participation, and other activities undertaken by many individuals worldwide. Continued collaboration between neuropathologists and trauma experts in other fields is likely to play a key role in providing future insights into these important questions.

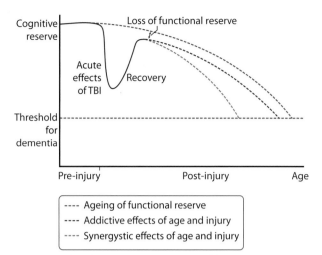

Cognitive reserve

Loss of functional reserve

Acute effects of TBI

Recovery

Threshold for dementia

Pre-injury   Post-injury   Age

---- Ageing of functional reserve
---- Addictive effects of age and injury
---- Synergystic effects of age and injury

10.38 A graphical representation of a postulated cognitive reserve and how head injury may increase the risk of cognitive decline. The broken green line [1] represents the 'normal' decline in cognitive function with age, until a threshold point is crossed (broken red line) resulting clinically in dementia. After an episode of traumatic brain injury, there is a significant decline in cognitive function that recovers, the degree of recovery being dependent on the severity of the head injury. Recovery is, however, not complete, resulting in a loss of functional reserve. After this point, the rate of cognitive decline may be as for normal ageing (broken blue line [2]) but with the dementia threshold being crossed earlier. Alternatively, pathogenic mechanisms initiated by the head injury may act synergistically with ageing to accelerate cognitive decline (broken purple line [3]).

Reproduced with permission from Smith C. Neuroinflammation after trauma. Neuropath Appl Neurobiol 2013:39:35–41. With permission of John Wiley and Sons. © 2012 British Neuropathological Society.

REFERENCES

1. Ackery A, Tator C, Krassioukov A. A global perspective on spinal cord injury epidemiology. *J Neurotrauma* 2004;**10**:1355–70.
2. Adams CW, Bruton CJ. The cerebral vasculature in dementia pugilistica. *J Neurol Neurosurg Psychiatry* 1989;**5**:600–4.
3. Adams H, Mitchell DE, Graham DI, Doyle D. Diffuse brain damage of immediate impact type. Its relationship to 'primary brain-stem damage' in head injury. *Brain* 1977;**3**:489–502.
4. Adams JH, Graham DI, Scott G, *et al.* Brain damage in fatal non-missile head injury. *J Clin Pathol* 1980;**12**:1132–45.
5. Adams JH, Graham DI, Murray LS, Scott G. Diffuse axonal injury due to nonmissile head injury in humans: an analysis of 45 cases. *Ann Neurol* 1982;**6**:557–63.
6. Adams JH, Doyle D, Graham DI, *et al.* Diffuse axonal injury in head injuries caused by a fall. *Lancet* 1984;**8417–8418**:1420–2.
7. Adams JH, Doyle D, Graham DI, *et al.* The contusion index: a reappraisal in human and experimental non-missile head injury. *Neuropathol Appl Neurobiol* 1985;**4**:299–308.
8. Adams JH, Doyle D, Graham DI, *et al.* Deep intracerebral (basal ganglia) haematomas in fatal non-missile head injury in man. *J Neurol Neurosurg Psychiatry* 1986;**9**:1039–43.
9. Adams JH, Doyle D, Graham DI, *et al.* Gliding contusions in nonmissile head injury in humans. *Arch Pathol Lab Med* 1986;**6**:485–8.
10. Adams JH, Doyle D, Ford I, *et al.* Diffuse axonal injury in head injury: definition, diagnosis and grading. *Histopathology* 1989;**1**:49–59.
11. Adams JH, Graham DI, Gennarelli TA, Maxwell WL. Diffuse axonal injury in non-missile head injury. *J Neurol Neurosurg Psychiatry* 1991;**6**:481–3.
12. Adams JH, Graham DI, Jennett B. The neuropathology of the vegetative state after an acute brain insult. *Brain* 2000;**123**:1327–38.
13. Adams JH, Graham DI, Jennett B. The structural basis of moderate disability after traumatic brain damage. *J Neurol Neurosurg Psychiatry* 2001;**4**:521–4.
14. Agoston DV, Gyorgy A, Eidelman O, Pollard HB. Proteomic biomarkers for blast neurotrauma: targeting cerebral edema, inflammation, and neuronal death cascades. *J Neurotrauma* 2009;**6**:901–11.
15. Akins PT, Axelrod YK, Ji C, *et al.* Cerebral venous sinus thrombosis complicated by subdural hematomas: case series and literature review. *Surg Neurol Int* 2013;**4**:85.
16. Al-Sarraj S, Mohamed S, Kibble M, Rezaie P. Subdural hematoma (SDH): assessment of macrophage reactivity within the dura mater and underlying hematoma. *Clin Neuropathol* 2004;**2**:62–75.
17. Al-Sarraj S, Fegan-Earl A, Ugbade A, *et al.* Focal traumatic brain stem injury is a rare type of head injury resulting from assault: a forensic neuropathological study. *J Forensic Leg Med* 2012;**3**:144–51.
18. Ambrosi PB, Valenca MM, Azevedo-Filho H. Prognostic factors in civilian gunshot wounds to the head: a series of 110 surgical patients and brief literature review. *Neurosurg Rev* 2012;**3**:429–35;discussion 435–6.
19. Anderson R, Opeskin K. Timing of early changes in brain trauma. *Am J Forensic Med Pathol* 1998;**1**:1–9.
20. Aoki Y, Inokuchi R, Gunshin M, *et al.* Diffusion tensor imaging studies of mild traumatic brain injury: a meta-analysis. *J Neurol Neurosurg Psychiatry* 2012;**9**:870–6.
21. Apple DF Jr, Anson CA, Hunter JD, Bell RB. Spinal cord injury in youth. *Clin Pediatr (Phil)* 1995;**2**:90–5.
22. Armstead WM, Kurth CD. Different cerebral hemodynamic responses following fluid percussion brain injury in the newborn and juvenile pig. *J Neurotrauma* 1994;**5**:487–97.
23. Arthurs ZM, Starnes BW. Blunt carotid and vertebral artery injuries. *Injury* 2008;**11**:1232–41.
24. Babikian T, Prins ML, Cai Y, *et al.* Molecular and physiological responses to juvenile traumatic brain injury: focus on growth and metabolism. *Dev Neurosci* 2010;**5–6**:431–41.
25. Balan IS, Saladino AJ, Aarabi B, *et al.* Cellular alterations in human traumatic brain injury: changes in mitochondrial morphology reflect regional levels of injury severity. *J Neurotrauma* 2013;**5**:367–81.
26. Barzo P, Marmarou A, Fatouros P, *et al.* Magnetic resonance imaging-monitored acute blood–brain barrier changes in experimental traumatic brain injury. *J Neurosurg* 1996;**6**:1113–21.
27. Baykaner K, Alp H, Ceviker N, *et al.* Observation of 95 patients with extradural hematoma and review of the literature. *Surg Neurol* 1988;**5**:339–41.
28. Bechtel K, Stoessel K, Leventhal JM, *et al.* Characteristics that distinguish accidental from abusive injury in hospitalized young children with head trauma. *Pediatrics* 2004;**1**:165–8.
29. Bellary SS, Steinberg A, Mirzayan N, *et al.* Wormian bones: a review. *Clin Anat* 2013;**26**:922–7.
30. Berhouma M, Al Dahak N, Messerer R, *et al.* A rare, high cervical traumatic spinal

subdural hematoma. *J Clin Neurosci* 2011;4:569–74.

31. Biasca N, Simmen HP, Trentz O. [Head injuries in ice hockey exemplified by the National Hockey League 'Hockey Canada' and European teams]. *Unfallchirurg* 1993;5:259–64.

32. Bigler ED. Neuroimaging biomarkers in mild traumatic brain injury (mTBI). *Neuropsychol Rev* 2013;3:169–209.

33. Blumbergs PC, Jones NR, North JB. Diffuse axonal injury in head trauma. *J Neurol Neurosurg Psychiatry* 1989;7:838–41.

34. Blumbergs PC, Scott G, Manavis J, *et al.* Staining of amyloid precursor protein to study axonal damage in mild head injury. *Lancet* 1994;**8929**:1055–6.

35. Blumbergs PC, Scott G, Manavis J, *et al.* Topography of axonal injury as defined by amyloid precursor protein and the sector scoring method in mild and severe closed head injury. *J Neurotrauma* 1995;4:565–72.

36. Borovich B, Braun J, Guilburd JN, *et al.* Delayed onset of traumatic extradural hematoma. *J Neurosurg* 1985;1:30–4.

37. Boto GR, Lobato RD, Rivas JJ, *et al.* Basal ganglia hematomas in severely head injured patients: clinicoradiological analysis of 37 cases. *J Neurosurg* 2001;2:224–32.

38. Bramlett HM, Dietrich WD, Green EJ, Busto R. Chronic histopathological consequences of fluid-percussion brain injury in rats: effects of post-traumatic hypothermia. *Acta Neuropathol* 1997;2:190–9.

39. Brennan LK, Rubin D, Christian CW, *et al.* Neck injuries in young pediatric homicide victims. *J Neurosurg Pediatr* 2009;3:232–9.

40. Bricolo AP, Pasut LM. Extradural hematoma: toward zero mortality. A prospective study. *Neurosurgery* 1984;1:8–12.

41. Broe GA, Henderson AS, Creasey H, *et al.* A case-control study of Alzheimer's disease in Australia. *Neurology* 1990;**11**:1698–707.

42. Browne KD, Chen XH, Meaney DF, Smith DH. Mild traumatic brain injury and diffuse axonal injury in swine. *J Neurotrauma* 2011;9:1747–55.

43. Bruce DA, Alavi A, Bilaniuk L, *et al.* Diffuse cerebral swelling following head injuries in children: the syndrome of 'malignant brain edema'. *J Neurosurg* 1981;2:170–8.

44. Bullock R, Butcher SP, Chen M, *et al.* Correlation of the extracellular glutamate concentration with extent of blood flow reduction after subdural hematoma in the rat. *J Neurosurgery* 1991;**74**:794–802.

45. Bunge RP, Puckett WR, Becerra JL, *et al.* Observations on the pathology of human spinal cord injury. A review and classification of 22 new cases with details from a case of chronic cord compression with extensive focal demyelination. *Adv Neurol* 1993;**59**:75–89.

46. Butler RJ. Neuropsychological investigation of amateur boxers. *Br J Sports Med* 1994;3:187–90.

47. Byard RW, Blumbergs P, Rutty G, *et al.* Lack of evidence for a causal relationship between hypoxic-ischemic encephalopathy and subdural hemorrhage in fetal life, infancy, and early childhood. *Pediatr Dev Pathol* 2007;5:348–50.

48. Byard RW, Bhatia KD, Reilly PL, Vink R. How rapidly does cerebral swelling follow trauma? Observations using an animal model and possible implications in infancy. *Leg Med (Tokyo)* 2009;S128–31.

49. Byard RW, Gabrielian L, Helps SC, *et al.* Further investigations into the speed of cerebral swelling following blunt cranial trauma. *J Forensic Sci* 2012;4:973–5.

50. Caffey J. On the theory and practice of shaking infants. Its potential residual effects of permanent brain damage and mental retardation. *Am J Dis Child* 1972;2:161–9.

51. Caffey J. The parent-infant traumatic stress syndrome (Caffey-Kempe syndrome), (battered babe syndrome). *Am J Roentgenol Radium Ther Nucl Med* 1972;2:218–29.

52. Calder IM, Hill I, Scholtz CL. Primary brain trauma in non-accidental injury. *J Clin Pathol* 1984;10:1095–100.

53. Camarillo DB, Shull PB, Mattson J, *et al.* An instrumented mouthguard for measuring linear and angular head impact kinematics in American football. *Ann Biomed Eng* 2013;9:1939–49.

54. Campbell M, Hanrahan F, Gobbo OL, *et al.* Targeted suppression of claudin-5 decreases cerebral oedema and improves cognitive outcome following traumatic brain injury. *Nat Commun* 2012;3:849.

55. Cantu RC. Second-impact syndrome. *Clin Sports Med* 1998;1:37–44.

56. Cantu RC. Posttraumatic retrograde and anterograde amnesia: pathophysiology and implications in grading and safe return to play. *J Athl Train* 2001;3:244–8.

57. Cantu RC, Gean AD. Second-impact syndrome and a small subdural hematoma: an uncommon catastrophic result of repetitive head injury with a characteristic imaging appearance. *J Neurotrauma* 2010;9:1557–64.

58. Case ME. Accidental traumatic head injury in infants and young children. *Brain Pathol* 2008;4:583–9.

59. Casson IR, Siegel O, Sham R, *et al.* Brain damage in modern boxers. *JAMA* 1984;2:2663–7.

60. Cater HL, Sundstrom LE, Morrison B 3rd. Temporal development of hippocampal cell death is dependent on tissue strain but not strain rate. *J Biomech* 2006;15:2810–8.

61. Cernak I, Chapman SM, Hamlin GP, Vink R. Temporal characterisation of pro- and anti-apoptotic mechanisms following diffuse traumatic brain injury in rats. *J Clin Neurosci* 2002;5:565–72.

62. Chamoun RB, Jea A. Traumatic intracranial and extracranial vascular injuries in children. *Neurosurg Clin N Am* 2010;3:529–42.

63. Chavko M, Koller WA, Prusaczyk WK, McCarron RM. Measurement of blast wave by a miniature fiber optic pressure transducer in the rat brain. *J Neurosci Methods* 2007;2:277–81.

64. Chen M, Dong Y, Simard JM. Functional coupling between sulfonylurea receptor type 1 and a nonselective cation channel in reactive astrocytes from adult rat brain. *J Neurosci* 2003;24:8568–77.

65. Chen XH, Meaney DF, Xu BN, *et al.* Evolution of neurofilament subtype accumulation in axons following diffuse brain injury in the pig. *J Neuropathol Exp Neurol* 1999;6:588–96.

66. Chen XH, Johnson VE, Uryu K, *et al.* A lack of amyloid beta plaques despite persistent accumulation of amyloid beta in axons of long-term survivors of traumatic brain injury. *Brain Pathol* 2009;2:214–23.

67. Cheng G, Kong RH, Zhang LM, Zhang JN. Mitochondria in traumatic brain injury and mitochondrial-targeted multipotential therapeutic strategies. *Br J Pharmacol* 2012;4:699–719

68. Chew KL, Baber Y, Iles L, O'Donnell C. Duret hemorrhage: demonstration of ruptured paramedian pontine branches of the basilar artery on minimally invasive, whole body postmortem CT angiography. *Forensic Sci Med Pathol* 2012;4:436–40

69. Chieregato A, Fainardi E, Morselli-Labate AM, *et al.* Factors associated with neurological outcome and lesion progression in traumatic subarachnoid hemorrhage patients. *Neurosurgery* 2005;4:671–80;discussion 671–80.

70. Chiron C, Raynaud C, Maziere B, *et al.* Changes in regional cerebral blood flow during brain maturation in children and adolescents. *J Nuclear Med* 1992;5:696–703.

71. Christian CW, Block R. Abusive head trauma in infants and children. *Pediatrics* 2009;5:1409–11.

72. Chugani HT. A critical period of brain development: studies of cerebral glucose utilization with PET. *Prev Med* 1998;2:184–8.

73. Clark RS, Schiding JK, Kaczorowski SL, *et al.* Neutrophil accumulation after traumatic brain injury in rats: comparison of weight drop and controlled cortical impact models. *J Neurotrauma* 1994;5:499–506.

74. Coats B, Margulies SS. Material properties of human infant skull and suture at high rates. *J Neurotrauma* 2006;8:1222–32.

75. Cohen MC, Scheimberg I. Evidence of occurrence of intradural and subdural hemorrhage in the perinatal and neonatal period in the context of hypoxic ischemic encephalopathy: an observational study from two referral institutions in the United Kingdom. *Pediatr Dev Pathol* 2009;3:169–76.

76. Coles JP, Fryer TD, Smielewski P, *et al.* Incidence and mechanisms of cerebral ischemia in early clinical head injury. *J Cereb Blood Flow Metab* 2004;2:202–11.

77. Contostavlos DL. Isolated basilar traumatic subarachnoid hemorrhage: an observer's 25 year re-evaluation of the pathogenetic possibilities. *Forensic Sci Int* 1995;1:61–74.

78. Corsellis JA, Bruton CJ, Freeman-Browne D. The aftermath of boxing. *Psychol Med* 1973;3:270–303.

79. Cory CZ, Jones BM. Can shaking alone cause fatal brain injury? A biomechanical assessment of the Duhaime shaken baby syndrome model. *Med Sci Law* 2003;4:317–33.

80. Costa Clara JM, Claramunt E, Ley L, Lafuente J. Traumatic extradural hematomas of the posterior fossa in children. *Childs Nerv Syst* 1996;3:145–8.

81. Courville CB. Traumatic intracerebral hemorrhages, with special reference to the mechanics of their production. *Bull Los Angel Neuro Soc* 1962;27:22–38.

82. Crockard HA, Brown FD, Johns LM, Mullan S. An experimental cerebral missile injury model in primates. *J Neurosurg* 1977;6:776–83.

83. Crompton MR. Hypothalamic lesions following closed head injury. *Brain* 1971;1:165–72.

84. Csuka E, Hans VH, Ammann E, *et al.* Cell activation and inflammatory

response following traumatic axonal injury in the rat. *Neuroreport* 2000;**11**:2587–90.

85. Daniel RW, Rowson S, Duma SM. Head impact exposure in youth football. *Ann Biomed Eng* 2012;**4**:976–81.

86. Dash PK, Zhao J, Hergenroeder G, Moore AN. Biomarkers for the diagnosis, prognosis, and evaluation of treatment efficacy for traumatic brain injury. *Neurotherapeutics* 2010;**1**:100–14.

87. Davceva N, Janevska V, Ilievski B, *et al.* Dilemmas concerning the diffuse axonal injury as a clinicopathological entity in forensic medical practice. *J Forensic Leg Med* 2012;**7**:413–18.

88. Davenport ND, Lim KO, Armstrong MT, Sponheim SR. Diffuse and spatially variable white matter disruptions are associated with blast-related mild traumatic brain injury. *Neuroimage* 2012;**3**:2017–24.

89. d'Avella D, Servadei F, Scerrati M, *et al.* Traumatic intracerebellar hemorrhage: clinicoradiological analysis of 81 patients. *Neurosurgery* 2002;**1**:16–25; discussion 25–7.

90. David TJ. Non-accidental head injury–the evidence. *Pediatr Radiol* 2008;**38**: S370–7.

91. de Lanerolle NC, Bandak F, Kang D, *et al.* Characteristics of an explosive blast-induced brain injury in an experimental model. *J Neuropathol Exp Neurol* 2011;**11**:1046–57.

92. Delgado Almandoz JE, Kelly HR, Schaefer PW, *et al.* Prevalence of traumatic dural venous sinus thrombosis in high-risk acute blunt head trauma patients evaluated with multidetector CT venography. *Radiology* 2010;**2**:570–7.

93. Depreitere B, Van Lierde C, Sloten JV, *et al.* Mechanics of acute subdural hematomas resulting from bridging vein rupture. *J Neurosurg* 2006;**6**:950–6.

94. Dharker SR, Mittal RS, Bhargava N. Ischemic lesions in basal ganglia in children after minor head injury. *Neurosurgery* 1993;**5**:863–5.

95. Dixon CE, Clifton GL, Lighthall JW, *et al.* A controlled cortical impact model of traumatic brain injury in the rat. *J Neurosci Methods* 1991;**3**:253–62.

96. Dolinak D, Smith C, Graham DI. Global hypoxia per se is an unusual cause of axonal injury. *Acta Neuropathol* 2000;**5**:553–60.

97. Dolinak D, Smith C, Graham DI. Hypoglycaemia is a cause of axonal injury. *Neuropathol Appl Neurobiol* 2000;**5**:448–53.

98. Dolinskas CA, Zimmerman RA, Bilaniuk LT, Gennarelli TA. Computed tomography of post-traumatic extracerebral hematomas: comparison to pathophysiology and responses to therapy. *J Trauma* 1979;**3**:163–9.

99. Domenicucci M, Signorini P, Strzelecki J, Delfini R. Delayed post-traumatic epidural hematoma. A review. *Neurosurg Rev* 1995;**2**:109–22.

100. dos Santos AL, Plese JP, Ciquini Junior O, *et al.* Extradural hematomas in children. *Pediatr Neurosurg* 1994;**1**:50–4.

101. Dubey A, Sung WS, Chen YY, *et al.* Traumatic intracranial aneurysm: a brief review. *J Clin Neurosci* 2008;**6**:609–12.

102. DuBose JJ, Barmparas G, Inaba K, *et al.* Isolated severe traumatic brain injuries sustained during combat operations: demographics, mortality outcomes, and lessons to be learned from contrasts to civilian counterparts. *J Trauma* 2011;**1**:1–6; discussion 16–18.

103. Duhaime AC. Large animal models of traumatic injury to the immature brain. *Dev Neurosci* 2006;**4–5**:380–7.

104. Duhaime AC, Durham S. Traumatic brain injury in infants: the phenomenon of subdural hemorrhage with hemispheric hypodensity ('Big Black Brain'). *Prog Brain Res* 2007;293–302.

105. Duhaime AC, Gennarelli TA, Thibault LE, *et al.* The shaken baby syndrome. A clinical, pathological, and biomechanical study. *J Neurosurg* 1987;**3**:409–15.

106. Duhaime AC, Alario AJ, Lewander WJ, *et al.* Head injury in very young children: mechanisms, injury types, and ophthalmologic findings in 100 hospitalized patients younger than 2 years of age. *Pediatrics* 1992;**2 Pt 1**:179–85.

107. Duhaime AC, Christian C, Armonda R, *et al.* Disappearing subdural hematomas in children. *Pediatr Neurosurg* 1996;**3**:116–22.

108. Duhaime AC, Margulies SS, Durham SR, *et al.* Maturation-dependent response of the piglet brain to scaled cortical impact. *J Neurosurg* 2000;**3**:455–62.

109. Duhaime AC, Hunter JV, Grate LL, *et al.* Magnetic resonance imaging studies of age-dependent responses to scaled focal brain injury in the piglet. *J Neurosurg* 2003;**3**:542–8.

110. Duhaime AC, Saykin AJ, McDonald BC, *et al.* Functional magnetic resonance imaging of the primary somatosensory cortex in piglets. *J Neurosurg* 2006;**4** (Suppl):259–64.

111. Duhaime AC, Gean AD, Haacke EM, *et al.* Common data elements in radiologic imaging of traumatic brain injury. *Arch Phys Med Rehabil* 2010;**11**:1661–6.

112. Duhaime AC, Beckwith JG, Maerlender AC, *et al.* Spectrum of acute clinical characteristics of diagnosed concussions in college athletes wearing instrumented helmets: clinical article. *J Neurosurg* 2012;**6**:1092–9.

113. Dupuis O, Silveira R, Dupont C, *et al.* Comparison of 'instrument-associated' and 'spontaneous' obstetric depressed skull fractures in a cohort of 68 neonates. *Am J Obstet Gynecol* 2005;**1**:165–70.

114. Durham SR, Duhaime AC. Maturation-dependent response of the immature brain to experimental subdural hematoma. *J Neurotrauma* 2007;**1**:5–14.

115. Echlin FA, Sordillo SV, Garvey TQ Jr. Acute, subacute, and chronic subdural hematoma. *J Am Med Assoc* 1956;**14**:1345–50.

116. Elder GA, Cristian A. Blast-related mild traumatic brain injury: mechanisms of injury and impact on clinical care. *Mt Sinai J Med* 2009;**2**:111–18.

117. Elkin BS, Morrison B 3rd. Region-specific tolerance criteria for the living brain. *Stapp Car Crash J* 2007;127–38.

118. Engel S, Schluesener H, Mittelbronn M, *et al.* Dynamics of microglial activation after human traumatic brain injury are revealed by delayed expression of macrophage-related proteins MRP8 and MRP14. *Acta Neuropathol* 2000;**3**:313–22.

119. Eucker SA, Smith C, Ralston J, *et al.* Physiological and histopathological responses following closed rotational head injury depend on direction of head motion. *Exp Neurol* 2011;**1**:79–88.

120. Fabian TC, Patton JH Jr, Croce MA, *et al.* Blunt carotid injury. Importance of early diagnosis and anticoagulant therapy. *Ann Surg* 1996;**5**:513–22;discussion 522–5.

121. Fackler ML, Bellamy RF, Malinowski JA. The wound profile: illustration of the missile-tissue interaction. *J Trauma* 1988;**1** (Suppl):S21–9.

122. Faden AI, Demediuk P, Panter SS, Vink R. The role of excitatory amino acids and NMDA receptors in traumatic brain injury. *Science* 1989;**4906**:798–800.

123. Farkas O, Lifshitz J, Povlishock JT. Mechanoporation induced by diffuse traumatic brain injury: an irreversible or reversible response to injury? *J Neurosci* 2006;**12**:3130–40.

124. Feeney DM, Boyeson MG, Linn RT, *et al.* Responses to cortical injury: I. Methodology and local effects of contusions in the rat. *Brain Res* 1981;**1**:67–77.

125. Finnie J, Lewis S, Manavis J, *et al.* Traumatic axonal injury in lambs: a model for paediatric axonal damage. *J Clin Neurosci* 1999;**1**:38–42.

126. Firsching R, Woischneck D, Diedrich M, *et al.* Early magnetic resonance imaging of brainstem lesions after severe head injury. *J Neurosurg* 1998;**5**:707–12.

127. Fischbein NJ, Wijman CA. Nontraumatic intracranial hemorrhage. *Neuroimaging Clin N Am* 2010;**4**:469–92..

128. Fleminger S, Oliver DL, Lovestone S, *et al.* Head injury as a risk factor for Alzheimer's disease: the evidence 10 years on; a partial replication. *J Neurol Neurosurg Psychiatry* 2003;**7**:857–62.

129. Fox WC, Park MS, Belverud S, *et al.* Contemporary imaging of mild TBI: the journey toward diffusion tensor imaging to assess neuronal damage. *Neurol Res* 2013;**3**:223–32.

130. Freytag E. Autopsy findings in head injuries from blunt forces. Statistical evaluation of 1,367 cases. *Arch Pathol* 1963;**75**:402–13.

131. Friess SH, Ichord RN, Owens K, *et al.* Neurobehavioral functional deficits following closed head injury in the neonatal pig. *Exp Neurol* 2007;**1**:234–43.

132. Friess SH, Ichord RN, Ralston J, *et al.* Repeated traumatic brain injury affects composite cognitive function in piglets. *J Neurotrauma* 2009;**7**:1111–21.

133. Friess SH, Ralston J, Eucker SA, *et al.* Neurocritical care monitoring correlates with neuropathology in a swine model of pediatric traumatic brain injury. *Neurosurgery* 2011;**5**:1139–47; discussion 1147.

134. Fujisawa H, Ito H, Saito K, *et al.* Immunohistochemical localization of tissue-type plasminogen activator in the lining wall of chronic subdural hematoma. *Surg Neurol* 1991;**6**:441–5.

135. Fukuda AM, Adami A, Pop V, *et al.* Posttraumatic reduction of edema with aquaporin-4 RNA interference improves acute and chronic functional recovery. *J Cereb Blood Flow Metab* 2013;**33**:1621–32.

136. Galbraith SL. Age-distribution of extradural haemorrhage without skull fracture. *Lancet* 1973;**7814**:1217–18.

137. Geddes JF, Vowles GH, Nicoll JA, Revesz T. Neuronal cytoskeletal changes are an early consequence of repetitive head injury. *Acta Neuropathol* 1999;**2**:171–8.

138. Geddes JF, Whitwell HL, Graham DI. Traumatic axonal injury: practical issues for diagnosis in medicolegal cases. *Neuropathol Appl Neurobiol* 2000;2:105–16.

139. Geddes JF, Hackshaw AK, Vowles GH, *et al.* Neuropathology of inflicted head injury in children. I. Patterns of brain damage. *Brain* 2001;Pt 7:1290–8.

140. Geddes JF, Vowles GH, Hackshaw AK, *et al.* Neuropathology of inflicted head injury in children. II. Microscopic brain injury in infants. *Brain* 2001;Pt 7:1299–306.

141. Geddes JF, Tasker RC, Hackshaw AK, *et al.* Dural haemorrhage in non-traumatic infant deaths: does it explain the bleeding in 'shaken baby syndrome'? *Neuropathol Appl Neurobiol* 2003;1:14–22.

142. Gennarelli TA. Head injury in man and experimental animals: clinical aspects. *Acta Neurochir Suppl (Wien)* 1983;32:1–13.

143. Gennarelli TA. Mechanisms of brain injury. *J Emerg Med* 1993;11 (Suppl 1):5–11.

144. Gennarelli TA, Thibault LE. Biomechanics of acute subdural hematoma. *J Trauma* 1982;8:680–6.

145. Gennarelli TA, Adams JH, Graham DI. Acceleration induced head injury in the monkey. I. The model, its mechanical and physiological correlates. *Acta Neuropathol Suppl* 1981;7:23–5.

146. Gennarelli TA, Thibault LE, Adams JH, *et al.* Diffuse axonal injury and traumatic coma in the primate. *Ann Neurol* 1982;6:564–74.

147. Gennarelli TA, Thibault LE, Tipperman R, *et al.* Axonal injury in the optic nerve: a model simulating diffuse axonal injury in the brain. *J Neurosurg* 1989;2:244–53.

148. Gennarelli TA, Pintar FA, Yoganandan N. Biomechanical tolerances for diffuse brain injury and a hypothesis for genotypic variability in response to trauma. Annual Proceedings from the Association of Advancement in Automotive Medicine 2003;624–8.

149. Gentleman SM, Greenberg BD, Savage MJ, *et al.* A beta 42 is the predominant form of amyloid beta-protein in the brains of short-term survivors of head injury. *Neuroreport* 1997;6:1519–22.

150. Gentleman SM, Leclercq PD, Moyes L, *et al.* Long-term intracerebral inflammatory response after traumatic brain injury. *Forensic Sci Int* 2004;2–3:97–104.

151. Giza CC. *Pediatric neurology*. Philadelphia, PA: Mosby/Elsevier, 2006.

152. Giza CC, Kutcher JS, Ashwal S, *et al.* Summary of evidence-based guideline update: evaluation and management of concussion in sports: report of the Guideline Development Subcommittee of the American Academy of Neurology. *Neurology* 2013;24:2250–7.

153. Gleckman AM, Bell MD, Evans RJ, Smith TW. Diffuse axonal injury in infants with nonaccidental craniocerebral trauma: enhanced detection by beta-amyloid precursor protein immunohistochemical staining. *Arch Pathol Lab Med* 1999;2:146–51.

154. Gorrie C, Oakes S, Duflou J, *et al.* Axonal injury in children after motor vehicle crashes: extent, distribution, and size of axonal swellings using beta-APP immunohistochemistry. *J Neurotrauma* 2002;10:1171–82.

155. Govaert P, Defoort P, Wigglesworth JS. *Cranial haemorrhage in the term newborn infant.* London: MacKeith, 1993.

156. Graham DI, Adams JH, Doyle D. Ischaemic brain damage in fatal non-missile head injuries. *J Neurol Sci* 1978;2–3:213–34.

157. Graham DI, McLellan D, Adams JH, *et al.* The neuropathology of the vegetative state and severe disability after non-missile head injury. *Acta Neurochir Suppl (Wien)* 1983;32:65–7.

158. Graham DI, Lawrence AE, Adams JH, *et al.* Brain damage in non-missile head injury secondary to high intracranial pressure. *Neuropathol Appl Neurobiol* 1987;3:209–17.

159. Graham DI, Lawrence AE, Adams JH, *et al.* Brain damage in fatal non-missile head injury without high intracranial pressure. *J Clin Pathol* 1988;1:34–7.

160. Graham DI, Ford I, Adams JH, *et al.* Fatal head injury in children. *J Clin Pathol* 1989;1:18–22.

161. Graham DI, Ford I, Adams JH, *et al.* Ischaemic brain damage is still common in fatal non-missile head injury. *J Neurol Neurosurg Psychiatry* 1989;3:346–50.

162. Graham DI, Clark JC, Adams JH, Gennarelli TA. Diffuse axonal injury caused by assault. *J Clin Pathol* 1992;9:840–1.

163. Graham DI, Gentleman SM, Lynch A, Roberts GW. Distribution of beta-amyloid protein in the brain following severe head injury. *Neuropathol Appl Neurobiol* 1995;1:27–34

164. Graham DI, Smith C, Reichard R, *et al.* Trials and tribulations of using beta-amyloid precursor protein immunohistochemistry to evaluate traumatic brain injury in adults. *Forensic Sci Int* 2004;2–3:89–96.

165. Gray JT, Puetz SM, Jackson SL, Green MA. Traumatic subarachnoid haemorrhage: a 10-year case study and review. *Forensic Sci Int* 1999;1:13–23.

166. Greene KA, Marciano FF, Johnson BA, *et al.* Impact of traumatic subarachnoid hemorrhage on outcome in nonpenetrating head injury. Part I: A proposed computerized tomography grading scale. *J Neurosurg* 1995;3:445–52.

167. Greene KA, Jacobowitz R, Marciano FF, *et al.* Impact of traumatic subarachnoid hemorrhage on outcome in nonpenetrating head injury. Part II: Relationship to clinical course and outcome variables during acute hospitalization. *J Trauma* 1996;6:964–71.

168. Greiner-Perth R, Mohsen Allam Y, Silbermann J, Gahr R. Traumatic subdural hematoma of the thoraco-lumbar junction of spinal cord. *J Spinal Disord Tech* 2007;3:239–41.

169. Gresham EL. Birth trauma. *Pediatr Clin North Am* 1975;2:317–28.

170. Griffin WS, Stanley LC, Ling C, *et al.* Brain interleukin 1 and S-100 immunoreactivity are elevated in Down syndrome and Alzheimer disease. *Proc Natl Acad Sci USA* 1989;19:7611–15.

171. Griffin WS, Sheng JG, Gentleman SM, *et al.* Microglial interleukin-1 alpha expression in human head injury: correlations with neuronal and neuritic beta-amyloid precursor protein expression. *Neurosci Lett* 1994;2:133–6.

172. Griffin WS, Sheng JG, Royston MC, *et al.* Glial-neuronal interactions in Alzheimer's disease: the potential role of a 'cytokine cycle' in disease progression. *Brain Pathol* 1998;1:65–72.

173. Gudeman SK, Kishore PR, Miller JD, *et al.* The genesis and significance of delayed traumatic intracerebral hematoma. *Neurosurgery* 1979;3:309–13.

174. Guo Z, Cupples LA, Kurz A, *et al.* Head injury and the risk of AD in the MIRAGE study. *Neurology* 2000;6:1316–23.

175. Gurdjian ES. Cerebral contusions: re-evaluation of the mechanism of their development. *J Trauma* 1976;1:35–51.

176. Guthkelch AN. Infantile subdural haematoma and its relationship to whiplash injuries. *Br Med J* 1971;5759:430–1.

177. Habash AH, Zwetnow NN, Ericson K, Lofgren J. Arterio-venous epidural shunting in epidural bleeding radiological and physiological characteristics. An experimental study in dogs. *Acta Neurochir (Wien)* 1983;3–4:291–313.

178. Hadley MN, Sonntag VK, Rekate HL, Murphy A. The infant whiplash-shake injury syndrome: a clinical and pathological study. *Neurosurgery* 1989;4:536–40.

179. Haines DE. On the question of a subdural space. *Anat Rec* 1991;1:3–21

180. Haines DE, Harkey HL, al-Mefty O. The 'subdural' space: a new look at an outdated concept. *Neurosurgery* 1993;1:111–20.

181. Hamel A, Llari M, Piercecchi-Marti MD, *et al.* Effects of fall conditions and biological variability on the mechanism of skull fractures caused by falls. *Int J Legal Med* 2013;1:111–18.

182. Hardy WN, Foster CD, Mason MJ, *et al.* Investigation of head injury mechanisms using neutral density technology and high-speed biplanar x-ray. *Stapp Car Crash J* 2001;45:337–68.

183. Hartings JA, Bullock MR, Okonkwo DO, *et al.* Spreading depolarisations and outcome after traumatic brain injury: a prospective observational study. *Lancet Neurol* 2011;12:1058–64.

184. Hassan MF, Dhamija B, Palmer JD, *et al.* Spontaneous cranial extradural hematoma: case report and review of literature. *Neuropathology* 2009;4:480–4.

185. Hausmann R, Betz P. The time course of the vascular response to human brain injury—an immunohistochemical study. *Int J Legal Med* 2000;5:288–92.

186. Hausmann R, Vogel C, Seidl S, Betz P. Value of morphological parameters for grading of brain swelling. *Int J Legal Med* 2006;4:219–25.

187. Haynes MJ, Vincent K, Fischhoff C, *et al.* Assessing the risk of stroke from neck manipulation: a systematic review. *Int J Clin Pract* 2012;10:940–7.

188. Hoelper BM, Reinert MM, Zauner A, *et al.* rCBF in hemorrhagic, non-hemorrhagic and mixed contusions after severe head injury and its effect on perilesional cerebral blood flow. *Acta Neurochir Suppl* 2000;76:21–5.

189. Hollerman JJ, Fackler ML, Coldwell DM, Ben-Menachem Y. Gunshot wounds: 1. Bullets, ballistics, and mechanisms of injury. *AJR Am J Roentgenol* 1990;4:685–90.

190. Holly LT, Kelly DF, Counelis GJ, *et al.* Cervical spine trauma associated with moderate and severe head injury: incidence, risk factors, and injury characteristics. *J Neurosurg* 2002;3 (Suppl):285–91.

191. Holmin S, Soderlund J, Biberfeld P, Mathiesen T. Intracerebral inflammation after human brain contusion. *Neurosurgery* 1998;**2**:291–8; discussion 298–9.

192. Hong HJ, Kim YJ, Yi HJ, et al. Role of angiogenic growth factors and inflammatory cytokine on recurrence of chronic subdural hematoma. *Surg Neurol* 2009;**2**:161–5;discussion 165–6.

193. Hortobagyi T, Wise S, Hunt N, et al. Traumatic axonal damage in the brain can be detected using beta-APP immunohistochemistry within 35 min after head injury to human adults. *Neuropathol Appl Neurobiol* 2007;**2**:226–37.

194. Hsu CY, Cheng CY, Lee JD, et al. Clinical features and outcomes of spinal cord infarction following vertebral artery dissection: a systematic review of the literature. *Neurol Res* 2013;**35**:676–83.

195. Huisman TA, Sorensen AG, Hergan K, et al. Diffusion-weighted imaging for the evaluation of diffuse axonal injury in closed head injury. *J Comput Assist Tomogr* 2003;**1**:5–11.

196. Hurley M, Dineen R, Padfield CJ, et al. Is there a causal relationship between the hypoxia-ischaemia associated with cardiorespiratory arrest and subdural haematomas? An observational study. *Br J Radiol* 2010;**993**:736–43.

197. Hylin MJ, Orsi SA, Zhao J, et al. Behavioral and histopathological alterations resulting from mild fluid percussion injury. *J Neurotrauma* 2013;**9**:702–15.

198. Hymel KP, Jenny C, Block RW. Intracranial hemorrhage and rebleeding in suspected victims of abusive head trauma: addressing the forensic controversies. *Child Maltreat* 2002;**4**:329–48.

199. Ibrahim NG, Wood J, Margulies SS, Christian CW. Influence of age and fall type on head injuries in infants and toddlers. *Int J Dev Neurosci* 2012;**3**:201–6.

200. Ikonomidou C, Mosinger JL, Salles KS, et al. Sensitivity of the developing rat brain to hypobaric/ischemic damage parallels sensitivity to N-methyl-aspartate neurotoxicity. *J Neurosci* 1989;**8**:2809–18.

201. Imajo T, Challener RC, Roessmann U. Diffuse axonal injury by assault. *Am J Forensic Med Pathol* 1987;**3**:217–19.

202. Inglis F, Kuroda Y, Bullock R. Glucose hypermetabolism after acute subdural hematoma is ameliorated by a competitive NMDA antagonist. *J Neurotrauma* 1992;**2**:75–84.

203. Iwakuma T, Brunngraber CV. Chronic extradural hematomas. A study of 21 cases. *J Neurosurg* 1973;**4**:488–93.

204. Iwata A, Stys PK, Wolf JA, et al. Traumatic axonal injury induces proteolytic cleavage of the voltage-gated sodium channels modulated by tetrodotoxin and protease inhibitors. *J Neurosci* 2004;**19**:4605–13.

205. Jamieson KG, Yelland JD. Extradural hematoma. Report of 167 cases. *J Neurosurg* 1968;**1**:13–23.

206. Jamieson KG, Yelland JD. Surgically treated traumatic subdural hematomas. *J Neurosurg* 1972;**2**:137–49.

207. Jang JW, Lee JK, Seo BR, Kim SH. Traumatic lumbar intradural disc rupture associated with an adjacent spinal compression fracture. *Spine* 2010;**15**:E726–9.

208. Jellinger KA. Neuropathology of prolonged unresponsive wakefulness syndrome after blunt head injury: review of 100 post-mortem cases. *Brain Inj* 2013;**7–8**:917–23.

209. Jennett B, Teasdale G. *Management of head injuries*. Philadelphia, PA: Davis, 1981.

210. Jennett B, Adams JH, Murray LS, Graham DI. Neuropathology in vegetative and severely disabled patients after head injury. *Neurology* 2001;**4**:486–90.

211. Johnson VE, Stewart W, Smith DH. Widespread tau and amyloid-beta pathology many years after a single traumatic brain injury in humans. *Brain Pathol* 2012;**2**:142–9.

212. Johnson VE, Stewart JE, Begbie FD, et al. Inflammation and white matter degeneration persist for years after a single traumatic brain injury. *Brain* 2013;**Pt 1**:28–42.

213. Johnstone AJ, Zuberi SH, Scobie WG. Skull fractures in children: a population study. *J Accid Emerg Med* 1996;**6**:386–9.

214. Jordan BD. Genetic influences on outcome following traumatic brain injury. *Neurochem Res* 2007;**4–5**:905–15.

215. Jordan BD. The clinical spectrum of sport-related traumatic brain injury. *Nat Rev Neurol* 2013;**4**:222–30.

216. Kaste M, Kuurne T, Vilkki J, et al. Is chronic brain damage in boxing a hazard of the past? *Lancet* 1982;**8309**:1186–8.

217. Kawamata T, Katayama Y, Hovda DA, et al. Lactate accumulation following concussive brain injury: the role of ionic fluxes induced by excitatory amino acids. *Brain Res* 1995;**2**:196–204.

218. Kay A, Teasdale G. Head injury in the United Kingdom. *World J Surg* 2001;**9**:1210–20.

219. Keane RW, Kraydieh S, Lotocki G, et al. Apoptotic and antiapoptotic mechanisms after traumatic brain injury. *J Cereb Blood Flow Metab* 2001;**10**:1189–98.

220. Kemp AM. Investigating subdural haemorrhage in infants. *Arch Dis Child* 2002;**2**:98–102.

221. Kemp AM, Jaspan T, Griffiths J, et al. Neuroimaging: what neuroradiological features distinguish abusive from non-abusive head trauma? A systematic review. *Arch Dis Child* 2011;**12**:1103–12.

222. Kessel G, Bocher-Schwarz HG, Ringel K, Perneczky A. The role of endoscopy in the treatment of acute traumatic anterior epidural hematoma of the cervical spine: case report. *Neurosurgery* 1997;**3**:688–90.

223. Kim KA, Wang MY, Griffith PM, et al. Analysis of pediatric head injury from falls. *Neurosurg Focus* 2000;**1**:e3.

224. Kimpara H, Iwamoto M. Mild traumatic brain injury predictors based on angular accelerations during impacts. *Ann Biomed Eng* 2012;**1**:114–26.

225. Kinney HC, Samuels MA. Neuropathology of the persistent vegetative state. A review. *J Neuropathol Exp Neurol* 1994;**6**:548–58.

226. Kobrine AI, Timmins E, Rajjoub RK, et al. Demonstration of massive traumatic brain swelling within 20 minutes after injury. Case report. *J Neurosurg* 1977;**2**:256–8.

227. Kordestani RK, Martin NA, McBride DQ. Cerebral hemodynamic disturbances following penetrating craniocerebral injury and their influence on outcome. *Neurosurg Clin N Am* 1995;**4**:657–67.

228. Koszyca B, Blumbergs PC, Manavis J, et al. Widespread axonal injury in gunshot wounds to the head using amyloid precursor protein as a marker. *J Neurotrauma* 1998;**9**:675–83.

229. Kudo H, Kuwamura K, Izawa I, et al. Chronic subdural hematoma in elderly people: present status on Awaji Island and epidemiological prospect. *Neurol Med Chir (Tokyo)* 1992;**4**:207–9.

230. Kurland D, Hong C, Aarabi B, et al. Hemorrhagic progression of a contusion after traumatic brain injury: a review. *J Neurotrauma* 2012;**1**:19–31.

231. Kuroda Y, Bullock R. Failure of cerebral blood flow-metabolism coupling after acute subdural hematoma in the rat. *Neurosurgery* 1992;**6**:1062–71.

232. Kurwale NS, Gupta DK, Mahapatra AK. Outcome of pediatric patients with traumatic basal ganglia hematoma: analysis of 21 cases. *Pediatr Neurosurg* 2010;**4**:267–71.

233. Kushi H, Saito T, Sakagami Y, et al. Acute subdural hematoma because of boxing. *J Trauma* 2009;**2**:298–303.

234. Kvarnes TL, Trumpy JH. Extradural haematoma. Report of 132 cases. *Acta Neurochir (Wien)* 1978;**1–3**:223–31.

235. Lang DA, Teasdale GM, Macpherson P, Lawrence A. Diffuse brain swelling after head injury: more often malignant in adults than children? *J Neurosurg* 1994;**4**:675–80.

236. Langlois JA, Rutland-Brown W, Wald MM. The epidemiology and impact of traumatic brain injury: a brief overview. *J Head Trauma Rehabil* 2006;**5**:375–8.

237. Lannsjo M, Raininko R, Bustamante M, et al. Brain pathology after mild traumatic brain injury: an exploratory study by repeated magnetic resonance examination. *J Rehabil Med* 2013;**8**:721–8.

238. Launer LJ, Andersen K, Dewey ME, et al. Rates and risk factors for dementia and Alzheimer's disease: results from EURODEM pooled analyses. EURODEM Incidence Research Group and Work Groups. European Studies of Dementia. *Neurology* 1999;**1**:78–84.

239. Lauritzen M, Dreier JP, Fabricius M, et al. Clinical relevance of cortical spreading depression in neurological disorders: migraine, malignant stroke, subarachnoid and intracranial hemorrhage, and traumatic brain injury. *J Cereb Blood Flow Metab* 2011;**1**:17–35.

240. Leclercq PD, McKenzie JE, Graham DI, Gentleman SM. Axonal injury is accentuated in the caudal corpus callosum of head-injured patients. *J Neurotrauma* 2001;**1**:1–9.

241. Lee CH, Kang DH, Hwang SH, et al. Spontaneous rapid reduction of a large acute subdural hematoma. *J Korean Med Sci* 2009;**6**:1224–6.

242. Lee KS. The pathogenesis and clinical significance of traumatic subdural hygroma. *Brain Inj* 1998;**7**:595–603.

243. Leestma JE. *Forensic neuropathology*. Boca Raton: CRC Press/Taylor & Francis, 2009.

244. Leestma JE, Kalelkar MB, Teas S. Ponto-medullary avulsion associated with cervical hyperextension. *Acta Neurochir Suppl (Wien)* 1983;**32**:69–73.

245. Leitgeb J, Mauritz W, Brazinova A, et al. Outcome after severe brain trauma due to acute subdural hematoma. *J Neurosurg* 2012;**2**:324–33.

246. Le Roux PCH, Andrews B. *Cerebral concussion and diffuse brain injury*. East Norwalk, CT: Appleton & Lange, 2000.

247. Leung LY, VandeVord PJ, Dal Cengio AL, *et al*. Blast related neurotrauma: a review of cellular injury. *Mol Cell Biomech* 2008;**3**:155–68.

248. Levin HS, Robertson CS. Mild traumatic brain injury in translation. *J Neurotrauma* 2013;**8**:610–17.

249. Li CH, Yew AY, Lu DC. Migration of traumatic intracranial subdural hematoma to lumbar spine causing radiculopathy. *Surg Neurol Int* 2013;**4**:81.

250. Li J, Li XY, Feng DF, Gu L. Quantitative evaluation of microscopic injury with diffusion tensor imaging in a rat model of diffuse axonal injury. *Eur J Neurosci* 2011;**5**:933–45.

251. Liaquat I, Dunn LT, Nicoll JA, *et al*. Effect of apolipoprotein E genotype on hematoma volume after trauma. *J Neurosurg* 2002;**1**:90–6.

252. Lifshitz J, Sullivan PG, Hovda DA, *et al*. Mitochondrial damage and dysfunction in traumatic brain injury. *Mitochondrion* 2004;**5–6**:705–13.

253. Lin DJ, Lam FC, Siracuse JJ, *et al*. 'Time is brain' the Gifford factor – or: why do some civilian gunshot wounds to the head do unexpectedly well? A case series with outcomes analysis and a management guide. *Surg Neurol Int* 2012;**3**:98.

254. Lindenberg R. *Trauma of meninges and brain*. New York: McGraw Hill, 1971.

255. Lindenberg R, Freytag E. The mechanism of cerebral contusions. A pathologic-anatomic study. *Arch Pathol* 1960;**69**:440–69.

256. Lleva P, Ahluwalia BS, Marks S, *et al*. Traumatic and spontaneous carotid and vertebral artery dissection in a level 1 trauma center. *J Clin Neurosci* 2012;**8**:1112–14.

257. Lo WD, Lee J, Rusin J, *et al*. Intracranial hemorrhage in children: an evolving spectrum. *Arch Neurol* 2008;**12**:1629–33.

258. Lobato RD, Sarabia R, Rivas JJ, *et al*. Normal computerized tomography scans in severe head injury. Prognostic and clinical management implications. *J Neurosurg* 1986;**6**:784–9.

259. Lobato RD, Sarabia R, Cordobes F, *et al*. Posttraumatic cerebral hemispheric swelling. Analysis of 55 cases studied with computerized tomography. *J Neurosurg* 1988;**3**:417–23.

260. Loberg EM, Torvik A. Brain contusions: the time sequence of the histological changes. *Med Sci Law* 1989;**2**:109–15.

261. Looney CB, Smith JK, Merck LH, *et al*. Intracranial hemorrhage in asymptomatic neonates: prevalence on MR images and relationship to obstetric and neonatal risk factors. *Radiology* 2007;**2**:535–41.

262. Lu J, Marmarou A, Choi S, *et al*. Mortality from traumatic brain injury. *Acta Neurochir Suppl* 2005;**95**:281–5.

263. Lu J, Ng KC, Ling G, *et al*. Effect of blast exposure on the brain structure and cognition in Macaca fascicularis. *J Neurotrauma* 2012;**7**:1434–54.

264. Lustenberger T, Talving P, Kobayashi L, *et al*. Early coagulopathy after isolated severe traumatic brain injury: relationship with hypoperfusion challenged. *J Trauma* 2010;**6**:1410–4.

265. Lustenberger T, Talving P, Kobayashi L, *et al*. Time course of coagulopathy in isolated severe traumatic brain injury. *Injury* 2010;**9**:924–8.

266. Ma M, Ferguson TA, Schoch KM, *et al*. Calpains mediate axonal cytoskeleton disintegration during Wallerian degeneration. *Neurobiol Dis* 2013;**56**:34–46.

267. Maas AI, Marmarou A, Murray GD, Steyerberg EW. Clinical trials in traumatic brain injury: current problems and future solutions. *Acta Neurochir Suppl* 2004;**89**:113–18.

268. Maas AI, Harrison-Felix CL, Menon D, *et al*. Common data elements for traumatic brain injury: recommendations from the interagency working group on demographics and clinical assessment. *Arch Phys Med Rehabil* 2010;**11**:1641–9.

269. Mack J, Squier W, Eastman JT. Anatomy and development of the meninges: implications for subdural collections and CSF circulation. *Pediatr Radiol* 2009;**3**:200–10.

270. MacKinnon JA, Perlman M, Kirpalani H, *et al*. Spinal cord injury at birth: diagnostic and prognostic data in twenty-two patients. *J Pediatr* 1993;**3**:431–7.

271. Madro R, Chagowski W. An attempt at objectivity of post mortem diagnostic of brain oedema. *Forensic Sci Int* 1987; **2–3**:125–9.

272. Madsen FF, Reske-Nielsen E. A simple mechanical model using a piston to produce localized cerebral contusions in pigs. *Acta Neurochir (Wien)* 1987;**1–2**:65–72.

273. Mannix R, Meehan WP, Mandeville J, *et al*. Clinical correlates in an experimental model of repetitive mild brain injury. *Ann Neurol* 2013;**74**:65–75.

274. Marmarou A, Fatouros PP, Barzo P, *et al*. Contribution of edema and cerebral blood volume to traumatic brain swelling in head-injured patients. *J Neurosurg* 2000;**2**:183–93.

275. Marmarou CR, Walker SA, Davis CL, Povlishock JT. Quantitative analysis of the relationship between intra-axonal neurofilament compaction and impaired axonal transport following diffuse traumatic brain injury. *J Neurotrauma* 2005;**10**:1066–80.

276. Marshall LF, Toole BM, Bowers SA. The National Traumatic Coma Data Bank. Part 2: Patients who talk and deteriorate: implications for treatment. *J Neurosurg* 1983;**2**:285–8.

277. Martin NA, Patwardhan RV, Alexander MJ, *et al*. Characterization of cerebral hemodynamic phases following severe head trauma: hypoperfusion, hyperemia, and vasospasm. *J Neurosurg* 1997;**1**:9–19.

278. Martins RS, Siqueira MG, Santos MT, *et al*. Prognostic factors and treatment of penetrating gunshot wounds to the head. *Surg Neurol* 2003;**2**:98–104;discussion 104.

279. Martland HS. Punch drunk. *J Am Med Assoc* 1928;**91**:1103–1107.

280. Matser JT, Kessels AG, Jordan BD, *et al*. Chronic traumatic brain injury in professional soccer players. *Neurology* 1998;**3**:791–6.

281. Matser EJ, Kessels AG, Lezak MD, *et al*. Neuropsychological impairment in amateur soccer players. *JAMA* 1999;**10**:971–3.

282. Matser EJ, Kessels AG, Lezak MD, *et al*. Acute traumatic brain injury in amateur boxing. *Phys Sportsmed* 2000;**1**:87–92.

283. Matsukawa H, Shinoda M, Fujii M, *et al*. Intraventricular hemorrhage on computed tomography and corpus callosum injury on magnetic resonance imaging in patients with isolated blunt traumatic brain injury. *J Neurosurg* 2012;**2**:334–9.

284. Matsuyama T, Shimomura T, Okumura Y, Sakaki T. Acute subdural hematomas due to rupture of cortical arteries: a study of the points of rupture in 19 cases. *Surg Neurol* 1997;**5**:423–7.

285. Matsuyama T, Shimomura T, Okumura Y, Sakaki T. Rapid resolution of symptomatic acute subdural hematoma: case report. *Surg Neurol* 1997;**2**:193–6.

286. Maxeiner H, Schirmer C. Frequency, types and causes of intraventricular haemorrhage in lethal blunt head injuries. *Leg Med (Tokyo)* 2009;**6**:278–84.

287. Maxwell WL, Graham DI. Loss of axonal microtubules and neurofilaments after stretch-injury to guinea pig optic nerve fibers. *J Neurotrauma* 1997;**9**:603–14.

288. Maxwell WL, Watt C, Graham DI, Gennarelli TA. Ultrastructural evidence of axonal shearing as a result of lateral acceleration of the head in non-human primates. *Acta Neuropathol* 1993;**2**:136–44.

289. Maxwell WL, Povlishock JT, Graham DL. A mechanistic analysis of nondisruptive axonal injury: a review. *J Neurotrauma* 1997;**7**:419–40.

290. Maxwell WL, Pennington K, MacKinnon MA, *et al*. Differential responses in three thalamic nuclei in moderately disabled, severely disabled and vegetative patients after blunt head injury. *Brain* 2004;**Pt 11**:2470–8.

291. Maxwell WL, MacKinnon MA, Smith DH, *et al*. Thalamic nuclei after human blunt head injury. *J Neuropathol Exp Neurol* 2006;**5**:478–88.

292. Mayeux R, Ottman R, Tang MX, *et al*. Genetic susceptibility and head injury as risk factors for Alzheimer's disease among community-dwelling elderly persons and their first-degree relatives. *Ann Neurol* 1993;**5**:494–501.

293. McCrory P. Traumatic brain injury: revisiting the AAN guidelines on sport-related concussion. *Nat Rev Neurol* 2013;**7**:361–2.

294. McCrory P, Meeuwisse WH, Kutcher JS, *et al*. What is the evidence for chronic concussion-related changes in retired athletes: behavioural, pathological and clinical outcomes? *Br J Sports Med* 2013;**5**:327–30.

295. McCrory PR, Berkovic SF. Second impact syndrome. *Neurology* 1998;**3**:677–83

296. McIntosh AS, McCrory P, Comerford J. The dynamics of concussive head impacts in rugby and Australian rules football. *Med Sci Sports Exerc* 2000;**12**:1980–4.

297. McIntosh TK, Noble L, Andrews B, Faden AI. Traumatic brain injury in the rat: characterization of a midline fluid-percussion model. *Cent Nerv Syst Trauma* 1987;**2**:119–34.

298. McIntosh TK, Vink R, Noble L, *et al*. Traumatic brain injury in the rat: characterization of a lateral fluid-percussion model. *Neuroscience* 1989;**1**:233–44.

299. McKee AC, Cantu RC, Nowinski CJ, *et al*. Chronic traumatic encephalopathy in athletes: progressive tauopathy after repetitive head injury. *J Neuropathol Exp Neurol* 2009;**7**:709–35.

300. McKee AC, Stein TD, Nowinski CJ, et al. The spectrum of disease in chronic traumatic encephalopathy. *Brain* 2013;**Pt 1**:43–64.

301. McKenzie KJ, McLellan DR, Gentleman SM, *et al*. Is beta-APP a marker of axonal

damage in short-surviving head injury? *Acta Neuropathol* 1996;6:608–13.

302. McLatchie G, Brooks N, Galbraith S, *et al.* Clinical neurological examination, neuropsychology, electroencephalography and computed tomographic head scanning in active amateur boxers. *J Neurol Neurosurg Psychiatry* 1987;1:96–9.

303. McLaurin RL, Ford LE. Extradural hematoma: statistical survey of forty-seven cases. *J Neurosurg* 1964;21:364–71

304. McLean LA, Frasier LD, Hedlund GL. Does intracranial venous thrombosis cause subdural hemorrhage in the pediatric population? *AJNR Am J Neuroradiol* 2012;7:1281–4.

305. Meaney DF, Smith DH. Biomechanics of concussion. *Clin Sports Med* 2011;1: 19–31, vii.

306. Medana IM, Esiri MM. Axonal damage: a key predictor of outcome in human CNS diseases. *Brain* 2003;Pt 3:515–30.

307. Meehan WP 3rd, Zhang J, Mannix R, Whalen MJ. Increasing recovery time between injuries improves cognitive outcome after repetitive mild concussive brain injuries in mice. *Neurosurgery* 2012;4:885–91.

308. Meissner A, Timaru-Kast R, Heimann A, *et al.* Effects of a small acute subdural hematoma following traumatic brain injury on neuromonitoring, brain swelling and histology in pigs. *Eur Surg Res* 2011;3:141–53.

309. Millar K, Nicoll JA, Thornhill S, *et al.* Long term neuropsychological outcome after head injury: relation to APOE genotype. *J Neurol Neurosurg Psychiatry* 2003;8:1047–52.

310. Miller AC, Odenkirchen J, Duhaime AC, Hicks R. Common data elements for research on traumatic brain injury: pediatric considerations. *J Neurotrauma* 2012;4:634–8.

311. Miller JD, Becker DP, Ward JD, *et al.* Significance of intracranial hypertension in severe head injury. *J Neurosurg* 1977;4:503–16.

312. Miller JD, Bullock R, Graham DI, *et al.* Ischemic brain damage in a model of acute subdural hematoma. *Neurosurgery* 1990;27:433–9.

313. Minns RA, Brown JK. *Shaking and other non-accidental head injuries in children.* London: MacKeith Press, 2005.

314. Minns RA, Jones PA, Tandon A, *et al.* Prediction of inflicted brain injury in infants and children using retinal imaging. *Pediatrics* 2012;5:e1227–34.

315. Miyazaki T, Kojima T, Chikasue F, *et al.* Traumatic rupture of intracranial vertebral artery due to hyperextension of the head: reports on three cases. *Forensic Sci Int* 1990;1:91–8.

316. Morales DM, Marklund N, Lebold D, *et al.* Experimental models of traumatic brain injury: do we really need to build a better mousetrap? *Neuroscience* 2005;4:971–89.

317. Mori T, Katayama Y, Kawamata T. Acute hemispheric swelling associated with thin subdural hematomas: pathophysiology of repetitive head injury in sports. *Acta Neurochir Suppl* 2006;96:40–3.

318. Morrison C, MacKenzie JM. Axonal injury in head injuries with very short survival times. *Neuropathol Appl Neurobiol* 2008;1:124–5.

319. Mortazavi MM, Verma K, Tubbs RS, Harrigan M. Pediatric traumatic carotid, vertebral and cerebral artery dissections: a review. *Childs Nerv Syst* 2011;12:2045–56.

320. Mortazavi MM, Denning M, Yalcin B, *et al.* The intracranial bridging veins: a comprehensive review of their history, anatomy, histology, pathology, and neurosurgical implications. *Childs Nerv Syst* 2013;Mar 2 [Epub ahead of print].

321. Mortimer JA, van Duijn CM, Chandra V, *et al.* Head trauma as a risk factor for Alzheimer's disease: a collaborative re-analysis of case-control studies. EURODEM Risk Factors Research Group. *Int J Epidemiol* 1991;20 (Suppl 2):S28–35.

322. Mosberg WH, Lindenberg R. Traumatic hemorrhage from the anterior choroidal artery. *J Neurosurg* 1959;2:209–21.

323. Mouzon B, Chaytow H, Crynen G, *et al.* Repetitive mild traumatic brain injury in a mouse model produces learning and memory deficits accompanied by histological changes. *J Neurotrauma* 2012;18:2761–73.

324. Muizelaar JP, Marmarou A, DeSalles AA, et al. Cerebral blood flow and metabolism in severely head-injured children. Part 1: Relationship with GCS score, outcome, ICP, and PVI. *J Neurosurg* 1989;1:63–71.

325. Munro D, Merritt HH. Surgical pathology of subdural hematoma–based on a study of one hundred and five cases. *Arch Neurol Psych* 1936;1:64–78.

326. Murray CJ, Lopez AD. Evidence-based health policy–lessons from the Global Burden of Disease Study. *Science* 1996;5288:740–3.

327. Murray GD, Teasdale GM, Braakman R, *et al.* The European Brain Injury Consortium survey of head injuries. *Acta Neurochir (Wien)* 1999;3:223–36.

328. Nag S, Manias JL, Stewart DJ. Pathology and new players in the pathogenesis of brain edema. *Acta Neuropathol* 2009;2:197–217.

329. Nag S, Venugopalan R, Stewart DJ. Increased caveolin-1 expression precedes decreased expression of occludin and claudin-5 during blood–brain barrier breakdown. *Acta Neuropathol* 2007;5:459–69.

330. Oehmichen M. Brain death: neuropathological findings and forensic implications. *Forensic Sci Int* 1994;3: 205–19.

331. Oehmichen M, Raff G. Timing of cortical contusion. Correlation between histomorphologic alterations and post-traumatic interval. *Z Rechtsmed* 1980;2:79–94.

332. Oehmichen M, Meissner C, Konig HG. Brain injury after gunshot wounding: morphometric analysis of cell destruction caused by temporary cavitation. *J Neurotrauma* 2000;2:155–62.

333. Oehmichen M, Meissner C, Konig HG. Brain injury after survived gunshot to the head: reactive alterations at sites remote from the missile track. *Forensic Sci Int* 2001;3:189–97.

334. Oehmichen M, Meissner C, Konig HG, Gehl HB. Gunshot injuries to the head and brain caused by low-velocity handguns and rifles. A review. *Forensic Sci Int* 2004; 2–3:111–20.

335. Oehmichen M, Jakob S, Mann S, *et al.* Macrophage subsets in mechanical brain injury (MBI)–a contribution to timing of MBI based on immunohistochemical methods: a pilot study. *Leg Med (Tokyo)* 2009;3:118–24.

336. Omalu BI, DeKosky ST, Minster RL, *et al.* Chronic traumatic encephalopathy in a National Football League player. *Neurosurgery* 2005;1:128–34;discussion 128–34.

337. Omalu BI, DeKosky ST, Hamilton RL, *et al.* Chronic traumatic encephalopathy in a national football league player: part II. *Neurosurgery* 2006;5:1086–92;discussion 1092–3.

338. Ommaya AK, Gennarelli TA. Cerebral concussion and traumatic unconsciousness. Correlation of experimental and clinical observations of blunt head injuries. *Brain* 1974;4:633–54.

339. Ommaya AK, Hirsch AE, Martinez JL. The role of whiplash in cerebral concussion. *Stapp Car Crash Conference* 1966;10:314–24.

340. Ommaya AK, Goldsmith W, Thibault L. Biomechanics and neuropathology of adult and paediatric head injury. *Br J Neurosurg* 2002;3:220–42.

341. Ono J, Yamaura A, Kubota M, *et al.* Outcome prediction in severe head injury: analyses of clinical prognostic factors. *J Clin Neurosci* 2001;2:120–3.

342. Opeskin K. Cerebral amyloid angiopathy. A review. *Am J Forensic Med Pathol* 1996;3:248–54.

343. Papadopoulos MC, Manley GT, Krishna S, Verkman AS. Aquaporin-4 facilitates reabsorption of excess fluid in vasogenic brain edema. *FASEB J* 2004;11:1291–3.

344. Parent S, Mac-Thiong JM, Roy-Beaudry M, *et al.* Spinal cord injury in the pediatric population: a systematic review of the literature. *J Neurotrauma* 2011;8:1515–24.

345. Peterson EC, Chesnut RM. Talk and die revisited: bifrontal contusions and late deterioration. *J Trauma* 2011;6:1588–92.

346. Pickup MJ, Pollanen MS. Traumatic subarachnoid hemorrhage and the COL3A1 gene: emergence of a potential causal link. *Forensic Sci Med Pathol* 2011;2:192–7.

347. Pilz P, Strohecker J, Grobovschek M. Survival after traumatic ponto-medullary tear. *J Neurol Neurosurg Psychiatry* 1982;5:422–7.

348. Piteau SJ, Ward MG, Barrowman NJ, Plint AC. Clinical and radiographic characteristics associated with abusive and nonabusive head trauma: a systematic review. *Pediatrics* 2012;2:315–23.

349. Pittella JE, Gusmao SN. Diffuse vascular injury in fatal road traffic accident victims: its relationship to diffuse axonal injury. *J Forensic Sci* 2003;3:626–30.

350. Plantman S, Ng KC, Lu J, *et al.* Characterization of a novel rat model of penetrating traumatic brain injury. *J Neurotrauma* 2012;6:1219–32.

351. Plassman BL, Havlik RJ, Steffens DC, *et al.* Documented head injury in early adulthood and risk of Alzheimer's disease and other dementias. *Neurology* 2000;8:1158–66.

352. Plunkett J. Fatal pediatric head injuries caused by short-distance falls. *Am J Forensic Med Pathol* 2001;1:1–12.

353. Povlishock JT, Christman CW. The pathobiology of traumatically induced axonal injury in animals and humans: a review of current thoughts. *J Neurotrauma* 1995;4:555–64.

354. Povlishock JT, Becker DP, Cheng CL, Vaughan GW. Axonal change in minor head injury. *J Neuropathol Exp Neurol* 1983;3:225–42.

355. Prange MT, Margulies SS. Regional, directional, and age-dependent properties of the brain undergoing large deformation. *J Biomech Eng* 2002;2:244–52.

356. Quigley MR, Chew BG, Swartz CE, Wilberger JE. The clinical significance of isolated traumatic subarachnoid hemorrhage. *J Trauma Acute Care Surg* 2013;2:581–4.

357. Ragan DK, McKinstry R, Benzinger T, et al. Alterations in cerebral oxygen metabolism after traumatic brain injury in children. *J Cereb Blood Flow Metab* 2013;1:48–52.

358. Raghupathi R, Margulies SS. Traumatic axonal injury after closed head injury in the neonatal pig. *J Neurotrauma* 2002;7:843–53.

359. Raghupathi R, Graham DI, McIntosh TK. Apoptosis after traumatic brain injury. *J Neurotrauma* 2000;10:927–38.

360. Raghupathi R, Mehr MF, Helfaer MA, Margulies SS. Traumatic axonal injury is exacerbated following repetitive closed head injury in the neonatal pig. *J Neurotrauma* 2004;3:307–16.

361. Reiber GD. Fatal falls in childhood. How far must children fall to sustain fatal head injury? Report of cases and review of the literature. *Am J Forensic Med Pathol* 1993;3:201–7.

362. Reichard RR, White CL 3rd, Hladik CL, Dolinak D. Beta-amyloid precursor protein staining of nonaccidental central nervous system injury in pediatric autopsies. *J Neurotrauma* 2003;4:347–55.

363. Reichard RR, Smith C, Graham DI. The significance of beta-APP immunoreactivity in forensic practice. *Neuropathol Appl Neurobiol* 2005;3:304–13.

364. Reilly P. The impact of neurotrauma on society: an international perspective. *Prog Brain Res* 2007;161:3–9.

365. Reilly PL, Graham DI, Adams JH, Jennett B. Patients with head injury who talk and die. *Lancet* 1975;7931:375–7.

366. Reneer DV, Hisel RD, Hoffman JM, et al. A multi-mode shock tube for investigation of blast-induced traumatic brain injury. *J Neurotrauma* 2011;1:95–104.

367. Ringger NC, Tolentino PJ, McKinsey DM, et al. Effects of injury severity on regional and temporal mRNA expression levels of calpains and caspases after traumatic brain injury in rats. *J Neurotrauma* 2004;7:829–41.

368. Roberts AH. *Brain damage in boxers;a study of the prevalence of traumatic encephalopathy among ex-professional boxers.* London: Pitman, 1969.

369. Roberts GW, Allsop D, Bruton C. The occult aftermath of boxing. *J Neurol Neurosurg Psychiatry* 1990;5:373–8.

370. Roberts GW, Gentleman SM, Lynch A, Graham DI. beta A4 amyloid protein deposition in brain after head trauma. *Lancet* 1991;8780:1422–3.

371. Rooks VJ, Eaton JP, Ruess L, et al. Prevalence and evolution of intracranial hemorrhage in asymptomatic term infants. *AJNR Am J Neuroradiol* 2008;6:1082–9.

372. Roozenbeek B, Maas AI, Menon DK. Changing patterns in the epidemiology of traumatic brain injury. *Nat Rev Neurol* 2013;4:231–6.

373. Ross DT, Meaney DF, Sabol MK, et al. Distribution of forebrain diffuse axonal injury following inertial closed head injury in miniature swine. *Exp Neurol* 1994;2:291–9.

374. Rowson S, Duma SM. Brain injury prediction: assessing the combined probability of concussion using linear and rotational head acceleration. *Ann Biomed Eng* 2013;5:873–82.

375. Ryan GA, McLean AJ, Vilenius AT, et al. Brain injury patterns in fatally injured pedestrians. *J Trauma* 1994;4:469–76.

376. Saatman KE, Abai B, Grosvenor A, et al. Traumatic axonal injury results in biphasic calpain activation and retrograde transport impairment in mice. *J Cereb Blood Flow Metab* 2003;1:34–42.

377. Saatman KE, Duhaime AC, Bullock R, et al. Classification of traumatic brain injury for targeted therapies. *J Neurotrauma* 2008;7:719–38.

378. Saing T, Dick M, Nelson PT, et al. Frontal cortex neuropathology in dementia pugilistica. *J Neurotrauma* 2012;6:1054–70.

379. Saljo A, Bao F, Haglid KG, Hansson HA. Blast exposure causes redistribution of phosphorylated neurofilament subunits in neurons of the adult rat brain. *J Neurotrauma* 2000;8:719–26.

380. Saljo A, Svensson B, Mayorga M, et al. Low-level blasts raise intracranial pressure and impair cognitive function in rats. *J Neurotrauma* 2009;8:1345–52.

381. Salvant JB Jr, Muizelaar JP. Changes in cerebral blood flow and metabolism related to the presence of subdural hematoma. *Neurosurgery* 1993;3:387–93; discussion 393.

382. Salvatori M, Kodikara S, Pollanen M. Fatal subarachnoid hemorrhage following traumatic rupture of the internal carotid artery. *Leg Med (Tokyo)* 2012;6:328–30.

383. Sauleau P, Lapouble E, Val-Laillet D, Malbert CH. The pig model in brain imaging and neurosurgery. *Animal* 2009;8:1138–51.

384. Sawauchi S, Abe T. The effect of haematoma, brain injury, and secondary insult on brain swelling in traumatic acute subdural haemorrhage. *Acta Neurochir (Wien)* 2008;6:531–6;discussion 536.

385. Scheimberg I, Cohen MC, Zapata Vazquez RE, et al. Nontraumatic intradural and subdural hemorrhage and hypoxic ischemic encephalopathy in fetuses, infants, and children up to three years of age: analysis of two audits of 636 cases from two referral centers in the United Kingdom. *Pediatr Dev Pathol* 2013;3:149–59.

386. Selassie AW, Borg K, Busch C, Russell WS. Abusive head trauma in young children: a population-based study. *Pediatr Emerg Care* 2013;3:283–91.

387. Seok J, Warren HS, Cuenca AG, et al. Genomic responses in mouse models poorly mimic human inflammatory diseases. *Proc Natl Acad Sci USA* 2013;9:3507–12.

388. Servadei F. Prognostic factors in severely head injured adult patients with acute subdural haematoma's. *Acta Neurochir (Wien)* 1997;4:279–85.

389. Servadei F, Ciucci G, Pagano F, et al. Skull fracture as a risk factor of intracranial complications in minor head injuries: a prospective CT study in a series of 98 adult patients. *J Neurol Neurosurg Psychiatry* 1988;4:526–8.

390. Servadei F, Piazza G, Seracchioli A, et al. Extradural haematomas: an analysis of the changing characteristics of patients admitted from 1980 to 1986. Diagnostic and therapeutic implications in 158 cases. *Brain Inj* 1988;2:87–100.

391. Sganzerla EP, Tomei G, Rampini P, et al. A peculiar intracerebral hemorrhage: the gliding contusion, its relationship to diffuse brain damage. *Neurosurg Rev* 1989;215–8.

392. Shannon P, Smith CR, Deck J, et al. Axonal injury and the neuropathology of shaken baby syndrome. *Acta Neuropathol* 1998;6:625–31.

393. Sharples PM, Stuart AG, Matthews DSF, et al. Cerebral blood flow and metabolism in children with severe head injury. Part 1: Relation to age, Glasgow coma score, outcome, intracranial pressure, and time after injury. *J Neurol Neurosurg Psychiatry* 1995;58:145–52.

394. Shaw NA. The neurophysiology of concussion. *Prog Neurobiol* 2002;4:281–344.

395. Shear DA, Lu XC, Bombard MC, et al. Longitudinal characterization of motor and cognitive deficits in a model of penetrating ballistic-like brain injury. *J Neurotrauma* 2010;10:1911–23.

396. Sherriff FE, Bridges LR, Gentleman SM, et al. Markers of axonal injury in post mortem human brain. *Acta Neuropathol* 1994;5:433–9.

397. Shibata Y, Matsumura A, Meguro K, Narushima K. Differentiation of mechanism and prognosis of traumatic brain stem lesions detected by magnetic resonance imaging in the acute stage. *Clin Neurol Neurosurg* 2000;3:124–8.

398. Shukla D, Mahadevan A, Sastry KV, Shankar SK. Pathology of post traumatic brainstem and hypothalamic injuries. *Clin Neuropathol* 2007;5:197–209.

399. Shulman ST, Madden JD, Esterly JR, Shanklin DR. Transection of spinal cord. A rare obstetrical complication of cephalic delivery. *Arch Dis Child* 1971;247:291–4.

400. Sim YW, Min KS, Lee MS, et al. Recent changes in risk factors of chronic subdural hematoma. *J Korean Neurosurg Soc* 2012;3:234–9.

401. Simard JM, Kilbourne M, Tsymbalyuk O, et al. Key role of sulfonylurea receptor 1 in progressive secondary hemorrhage after brain contusion. *J Neurotrauma* 2009;12:2257–67.

402. Simpson DA, Blumbergs PC, Cooter RD, et al. Pontomedullary tears and other gross brainstem injuries after vehicular accidents. *J Trauma* 1989;11:1519–25.

403. Singh M, Jeong J, Hwang D, et al. Novel diffusion tensor imaging methodology to detect and quantify injured regions and affected brain pathways in traumatic brain injury. *Magn Reson Imaging* 2010;1:22–40.

404. Singleton RH, Povlishock JT. Identification and characterization of heterogeneous neuronal injury and death in regions of diffuse brain injury: evidence for multiple independent injury phenotypes. *J Neurosci* 2004;14:3543–53.

405. Smith C. Review: the long-term consequences of microglial activation following acute traumatic brain injury. *Neuropathol Appl Neurobiol* 2013;1:35–44.

406. Smith C, Graham DI, Murray LS, Nicoll JA. Tau immunohistochemistry in acute brain injury. *Neuropathol Appl Neurobiol* 2003;5:496–502.

407. Smith C, Gentleman SM, Leclercq PD, et al. The neuroinflammatory response in humans after traumatic brain injury. *Neuropathol Appl Neurobiol* 2012;39:45–50.

408. Smith DH, Chen XH, Pierce JE, *et al.* Progressive atrophy and neuron death for one year following brain trauma in the rat. *J Neurotrauma* 1997;**10**:715–27.

409. Smith DH, Chen XH, Nonaka M, *et al.* Accumulation of amyloid beta and tau and the formation of neurofilament inclusions following diffuse brain injury in the pig. *J Neuropathol Exp Neurol* 1999;**9**:982–92.

410. Smith DH, Nonaka M, Miller R, *et al.* Immediate coma following inertial brain injury dependent on axonal damage in the brainstem. *J Neurosurg* 2000;**2**:315–22.

411. Smith DH, Chen XH, Iwata A, Graham DI. Amyloid beta accumulation in axons after traumatic brain injury in humans. *J Neurosurg* 2003;**5**:1072–7.

412. Smith DH, Hicks R, Povlishock JT. Therapy development for diffuse axonal injury. *J Neurotrauma* 2013;**5**:307–23.

413. Smith FM, Raghupathi R, MacKinnon MA, *et al.* TUNEL-positive staining of surface contusions after fatal head injury in man. *Acta Neuropathol* 2000;**5**:537–45.

414. Smith SL, Andrus PK, Gleason DD, Hall ED. Infant rat model of the shaken baby syndrome: preliminary characterization and evidence for the role of free radicals in cortical hemorrhaging and progressive neuronal degeneration. *J Neurotrauma* 1998;**9**:693–705.

415. Sosin DM, Sniezek JE, Thurman DJ. Incidence of mild and moderate brain injury in the United States, 1991. *Brain Inj* 1996;**1**:47–54.

416. Spain A, Daumas S, Lifshitz J, *et al.* Mild fluid percussion injury in mice produces evolving selective axonal pathology and cognitive deficits relevant to human brain injury. *J Neurotrauma* 2010;**8**:1429–38.

417. Sprick C, Bettag M, Bock WJ. Delayed traumatic intracranial hematomas–clinical study of seven years. *Neurosurg Rev* 1989;228–30.

418. Squier W. The 'Shaken Baby' syndrome: pathology and mechanisms. *Acta Neuropathol* 2011;**5**:519–42.

419. Squier W, Mack J. The neuropathology of infant subdural haemorrhage. *Forensic Sci Int* 2009;**1–3**:6–13.

420. Squier W, Austin T, Anslow P, Weller RO. Infant subcortical cystic leucomalacia: a distinct pathological entity resulting from impaired fluid handling. *Early Hum Dev* 2011;**6**:421–6.

421. Squier W, Scheimberg I, Smith C. Spinal nerve root beta-APP staining in infants is not a reliable indicator of trauma. *Forensic Sci Int* 2011;**1–3**:e31–5.

422. Staal JA, Vickers JC. Selective vulnerability of non-myelinated axons to stretch injury in an *in vitro* co-culture system. *J Neurotrauma* 2011;**5**:841–7.

423. Staal JA, Dickson TC, Gasperini R, *et al.* Initial calcium release from intracellular stores followed by calcium dysregulation is linked to secondary axotomy following transient axonal stretch injury. *J Neurochem* 2010;**5**:1147–55.

424. Stein KM, Ruf K, Ganten MK, Mattern R. Representation of cerebral bridging veins in infants by postmortem computed tomography. *Forensic Sci Int* 2006;**1–2**:93–101.

425. Stein SC, Smith DH. Coagulopathy in traumatic brain injury. *Neurocrit Care* 2004;**4**:479–88.

426. Stephens RP, Richardson AC, Lewin JS. Bilateral subdural hematomas in a newborn infant. *Pediatrics* 1997;**4**:619–21.

427. Stiver SI, Gean AD, Manley GT. Survival with good outcome after cerebral herniation and Duret hemorrhage caused by traumatic brain injury. *J Neurosurg* 2009;**6**:1242–6.

428. Stone JL, Lang RG, Sugar O, Moody RA. Traumatic subdural hygroma. *Neurosurgery* 1981;**5**:542–50.

429. Stone JR, Okonkwo DO, Dialo AO, *et al.* Impaired axonal transport and altered axolemmal permeability occur in distinct populations of damaged axons following traumatic brain injury. *Exp Neurol* 2004;**1**:59–69.

430. Strich SJ. Diffuse degeneration of the cerebral white matter in severe dementia following head injury. *J Neurol Neurosurg Psychiatry* 1956;**3**:163–85.

431. Strich SJ. Shearing of nerve fibres as a cause of brain damage due to head injury. *Lancet* 1961;**7200**:443–98.

432. Svetlov SI, Larner SF, Kirk DR, *et al.* Biomarkers of blast-induced neurotrauma: profiling molecular and cellular mechanisms of blast brain injury. *J Neurotrauma* 2009;**6**:913–21.

433. Takagi T, Nagai R, Wakabayashi S, *et al.* Extradural hemorrhage in the newborn as a result of birth trauma. *Childs Brain* 1978;**5**:306–18.

434. Takahashi T, Shirane R, Sato S, Yoshimoto T. Developmental changes of cerebral blood flow and oxygen metabolism in children. *Am J Neuroradiol* 1999;**5**:917–922.

435. Takeuchi S, Takasato Y, Masaoka H, Hayakawa T. Traumatic intra-cerebellar haematoma: study of 17 cases. *Br J Neurosurg* 2011;**1**:62–7.

436. Takeuchi S, Takasato Y, Masaoka H, *et al.* Traumatic basal ganglia hematomas: an analysis of 20 cases. *Acta Neurochir Suppl* 2013;**118**:139–42.

437. Takeuchi S, Wada K, Takasato Y, *et al.* Traumatic hematoma of the posterior fossa. *Acta Neurochir Suppl* 2013;**118**:135–8.

438. Taneda M, Kataoka K, Akai F, *et al.* Traumatic subarachnoid hemorrhage as a predictable indicator of delayed ischemic symptoms. *J Neurosurg* 1996;**5**:762–8.

439. Tang-Schomer MD, Patel AR, Baas PW, Smith DH. Mechanical breaking of microtubules in axons during dynamic stretch injury underlies delayed elasticity, microtubule disassembly, and axon degeneration. *FASEB J* 2010;**5**:1401–10.

440. Tang-Schomer MD, Johnson VE, Baas PW, *et al.* Partial interruption of axonal transport due to microtubule breakage accounts for the formation of periodic varicosities after traumatic axonal injury. *Exp Neurol* 2012;**1**:364–72.

441. Tatsuno Y, Lindenberg R. Basal subarachnoid hematomas as sole intracranial traumatic lesions. *Arch Pathol* 1974;**4**:211–15.

442. Teasdale G, Jennett B. Assessment of coma and impaired consciousness. A practical scale. *Lancet* 1974;**7872**:81–4.

443. Thomas M, Haas TS, Doerer JJ, *et al.* Epidemiology of sudden death in young, competitive athletes due to blunt trauma. *Pediatrics* 2011;**1**:e1–8.

444. Thompson AK, Bertocci G, Rice W, Pierce MC. Pediatric short-distance household falls: biomechanics and associated injury severity. *Accid Anal Prev* 2011;**1**:143–50.

445. Thornhill S, Teasdale GM, Murray GD, *et al.* Disability in young people and adults one year after head injury: prospective cohort study. *BMJ* 2000;**7250**:1631–5.

446. Tokuda T, Ikeda S, Yanagisawa N, *et al.* Re-examination of ex-boxers' brains using immunohistochemistry with antibodies to amyloid beta-protein and tau protein. *Acta Neuropathol* 1991;**4**:280–5.

447. Tomita Y, Sawauchi S, Beaumont A, Marmarou A. The synergistic effect of acute subdural hematoma combined with diffuse traumatic brain injury on brain edema. *Acta Neurochir Suppl* 2000;**76**:213–16.

448. Unterharnscheidt F, Taylor-Unterharnscheidt J. *Boxing: medical aspects.* Amsterdam: Academic Press, 2003.

449. van den Berg ME, Castellote JM, de Pedro-Cuesta J, Mahillo-Fernandez I. Survival after spinal cord injury: a systematic review. *J Neurotrauma* 2010;**8**:1517–28.

450. Verweij BH, Muizelaar JP, Vinas FC, *et al.* Impaired cerebral mitochondrial function after traumatic brain injury in humans. *J Neurosurg* 2000;**5**:815–20.

451. Vink R, Mullins PG, Temple MD, *et al.* Small shifts in craniotomy position in the lateral fluid percussion injury model are associated with differential lesion development. *J Neurotrauma* 2001;**8**:839–47.

452. Volpe JJ. *Neurology of the newborn.* Philadelphia, PA: Saunders, 2008.

453. Vowles GH, Scholtz CL, Cameron JM. Diffuse axonal injury in early infancy. *J Clin Pathol* 1987;**2**:185–9.

454. Wang D, Jiang R, Liu L, *et al.* Membrane neovascularization and drainage of subdural hematoma in a rat model. *J Neurotrauma* 2010;**8**:1489–98.

455. Warden D. Military TBI during the Iraq and Afghanistan wars. *J Head Trauma Rehabil* 2006;**5**:398–402.

456. Warrington SA, Wright CM, Team AS. Accidents and resulting injuries in premobile infants: data from the ALSPAC study. *Arch Dis Child* 2001;**2**:104–7.

457. Weinstein E, Turner M, Kuzma BB, Feuer H. Second impact syndrome in football: new imaging and insights into a rare and devastating condition. *J Neurosurg Pediatr* 2013;**3**:331–4.

458. Whitby EH, Griffiths PD, Rutter S, *et al.* Frequency and natural history of subdural haemorrhages in babies and relation to obstetric factors. *Lancet* 2004;**9412**:846–51.

459. Wigglesworth JS, Husemeyer RP. Intracranial birth trauma in vaginal breech delivery: the continued importance of injury to the occipital bone. *Br J Obstet Gynaecol* 1977;**9**:684–91.

460. Wijdicks EF, Pfeifer EA. Neuropathology of brain death in the modern transplant era. *Neurology* 2008;**15**:1234–7.

461. Wilkinson IM. The vertebral artery. Extracranial and intracranial structure. *Arch Neurol* 1972;**5**:392–6.

462. Williams AJ, Hartings JA, Lu XC, *et al.* Characterization of a new rat model of penetrating ballistic brain injury. *J Neurotrauma* 2005;**2**:313–31.

463. Williams AJ, Hartings JA, Lu XC, *et al.* Penetrating ballistic-like brain injury in the rat: differential time courses of hemorrhage, cell death, inflammation, and remote degeneration. *J Neurotrauma* 2006;**12**:1828–46.

464. Williams AJ, Wei HH, Dave JR, Tortella FC. Acute and delayed neuroinflammatory response following experimental penetrating ballistic brain injury in the rat. *J Neuroinflammation* 2007;**4**:17.

465. Williams S, Raghupathi R, MacKinnon MA, *et al. In situ* DNA fragmentation occurs in white matter up to 12 months after head injury in man. *Acta Neuropathol* 2001;**6**:581–90.

466. Wolf JA, Stys PK, Lusardi T, *et al.* Traumatic axonal injury induces calcium influx modulated by tetrodotoxin-sensitive sodium channels. *J Neurosci* 2001;**6**:1923–30.

467. Wong GK, Tang BY, Yeung JH, *et al.* Traumatic intracerebral haemorrhage: is the CT pattern related to outcome? *Br J Neurosurg* 2009;**6**:601–5.

468. Wootton R, Flecknell PA, John M. Accurate measurement of cerebral metabolism in the conscious, unrestrained neonatal piglet. I. Blood flow. *Biol Neonate* 1982;**5–6**:209–20.

469. Wu HM, Huang SC, Vespa P, *et al.* Redefining the pericontusional penumbra following traumatic brain injury: evidence of deteriorating metabolic derangements based on positron emission tomography. *J Neurotrauma* 2013;**5**:352–60.

470. Wygnanski-Jaffe T, Morad Y, Levin AV. Pathology of retinal hemorrhage in abusive head trauma. *Forensic Sci Med Pathol* 2009;**4**:291–7.

471. Xiong Y, Mahmood A, Chopp M. Animal models of traumatic brain injury. *Nat Rev Neurosci* 2013;**2**:128–42.

472. Yamashima T. The inner membrane of chronic subdural hematomas: pathology and pathophysiology. *Neurosurg Clin N Am* 2000;**3**:413–24.

473. Yamashima T, Friede RL. Why do bridging veins rupture into the virtual subdural space? *J Neurol Neurosurg Psychiatry* 1984;**2**:121–7.

474. Yamashima T, Yamamoto S. How do vessels proliferate in the capsule of a chronic subdural hematoma? *Neurosurgery* 1984;**5**:672–8.

475. Yilmazlar S, Kocaeli H, Dogan S, *et al.* Traumatic epidural haematomas of nonarterial origin: analysis of 30 consecutive cases. *Acta Neurochir* (*Wien*) 2005;**12**:1241–8;discussion 1248.

476. Yoshino E, Yamaki T, Higuchi T, *et al.* Acute brain edema in fatal head injury: analysis by dynamic CT scanning. *J Neurosurg* 1985;**6**:830–9.

477. Zaben M, El Ghoul W, Belli A. Post-traumatic head injury pituitary dysfunction. *Disabil Rehabil* 2013;**6**:522–5.

478. Zaloshnja E, Miller T, Langlois JA, Selassie AW. Prevalence of long-term disability from traumatic brain injury in the civilian population of the United States, 2005. *J Head Trauma Rehabil* 2008;**6**:394–400.

479. Zhang J, Yoganandan N, Pintar FA, Gennarelli TA. Temporal cavity and pressure distribution in a brain simulant following ballistic penetration. *J Neurotrauma* 2005;**11**:1335–47.

480. Ziebell JM, Morganti-Kossmann MC. Involvement of pro- and anti-inflammatory cytokines and chemokines in the pathophysiology of traumatic brain injury. *Neurotherapeutics* 2010;**1**:22–30.

Maria Thom and Sanjay Sisodiya

INTRODUCTION

Epilepsy is the second most common neurological disorder after stroke and affects 1–2 per cent of the population worldwide. According to the World Health Organization, approximately 50 million people worldwide have epilepsy. Although major advances in drug and surgical treatments have been made, in many patients, seizures remain uncontrolled, affecting the quality of life of patients and caregivers, with adverse psychological consequences, social stigma and the burden of excess morbidity and mortality associated with recurrent seizures. The healthcare costs of epilepsy are enormous. The ultimate goal for the future is the cure or prevention of epilepsy, or at least optimal control of seizures for the majority of the affected population.

CLASSIFICATION OF EPILEPSIES AND CLINICAL SYNDROMES

Epilepsy is a disorder of the brain characterized by an enduring predisposition to generate epileptic seizures and by the neurobiological, cognitive, psychological and social consequences of this condition. The definition of epilepsy requires the occurrence of at least one epileptic seizure.[154] The epilepsies comprise a heterogeneous group of conditions that share some common electro-clinical manifestations and responses to antiepileptic drugs. Overwhelmingly, at an individual level, the particular type of epilepsy determines outcome and guides management. Therefore, accurate evaluation of the features that characterize a patient's epilepsy is paramount. Aspects of the biology of epilepsies that might be considered important in their classification include causation (genetic and/or acquired), susceptibility factors, seizure types, natural history, response to treatment, consequences and eventual prognosis. The most

useful classification scheme would include only those factors that allow meaningful distinctions that might influence management and predict outcome.

Continued limitations in our understanding of the epilepsies are demonstrated by further efforts to organize the epilepsies in more meaningful ways. The scheme proposed in 2001 by the International League Against Epilepsy (ILEA),[131] was a five-axis classification as follows: axis 1, ictal phenomenology; axis 2, seizure type; axis 3, syndrome; axis 4, aetiology; and axis 5, impairment. This is an overarching and dynamic classification scheme: axes 1–4 each have their own dedicated and formalized subclassification schemes. Overall, the scheme was pragmatic and generated a framework for rational management of an individual patient's epilepsy. It is worth emphasizing that the scheme was intended to be flexible and to accommodate conceptual advances. It recognized that detailed descriptions of ictal phenomenology were not always necessary, and indeed that for some individuals with epilepsy, a syndromic diagnosis could not always be made. Of particular relevance, it suggested that classification schemes might well be appropriate in clinical trial and research environments, in which clinical context and neuropathological data could be combined to refine diagnostic schemes.

The scheme had not been adopted universally before the arrival in 2010 of a new organization[38] (Table 11.1). The organization brought revised terms and concepts, with a focus on stratification based strongly on genetics and experimental advances (e.g. functional magnetic resonance [MR] imaging data interpreted as showing the existence of epileptogenic networks in the human brain). Incorporation of emerging data is certainly a necessary development, rationalizing certain classes. However, this new organization has itself met with trenchant criticism.[20] Only time will tell if this scheme is adopted beyond specialist centres. For now, it is important that those in the epilepsy field appreciate that it is a work in progress, evolving with our understanding of

TABLE 11.1 Electro-clinical syndromes and other epilepsies

Neonatal period	Benign familial neonatal epilepsy (BFNE) Early myoclonic encephalopathy (BFNE) Ohtahara syndrome
Infancy	Epilepsy of infancy with migrating focal seizures West syndrome Myoclonic epilepsy in infancy (MEI) Benign infantile epilepsy Benign familial infantile epilepsy Dravet syndrome Myoclonic encephalopathy in non-progressive disorders
Childhood	Febrile seizures plus (FS+) (can start in infancy) Panayiotopoulos syndrome Epilepsy with myoclonic atonic (previously astatic) seizures Benign epilepsy with centrotemporal spikes (BECTS) Autosomal dominant nocturnal frontal lobe epilepsy (ADNFLE) Late onset childhood occipital epilepsy (Gastaut type) Epilepsy with myoclonic absences Lennox–Gastaut syndrome Epileptic encephalopathy with continuous spike-and-wave during sleep (CSWS) Landau–Kleffner syndrome (LKS) Childhood absence epilepsy (CAE)
Adolescence–Adult	Juvenile absence epilepsy (JAE) Juvenile myoclonic epilepsy (JME) Epilepsy with generalized tonic–clonic seizures alone Progressive myoclonus epilepsies (PME) Autosomal dominant epilepsy with auditory features (ADEAF) Other familial temporal lobe epilepsies
Less specific age relationships	Familial focal epilepsy with variable foci (childhood to adult) Reflex epilepsies
Distinctive constellations	Mesial temporal lobe epilepsy with hippocampal sclerosis (MTLE with HS) Rasmussen syndrome Gelastic seizures with hypothalamic hamartoma Hemiconvulsions–hemiplegia–epilepsy Epilepsies that do not fit these categories
Epilepsies attributed to structural-metabolic causes	Malformations of cortical development (hemi-megalencephaly, heterotopias, etc.) Neurocutaneous syndromes (tuberous sclerosis complex, Sturge–Weber, etc.) Tumour Infection Trauma Perinatal insults Stroke Vascular malformation
Epilepsies of unknown cause	
Conditions with epileptic seizures that are not forms of epilepsy	Benign neonatal seizures (BNS) Febrile seizures (FS)

Reproduced with permission from Berg et al.[38] Wiley Periodicals, Inc. © 2010 International League Against Epilepsy. The syndromes are arranged by age of onset and do not reflect aetiology.

underlying disease mechanisms. Neuropathology can still contribute to classification,[80] although the absence of tissue and systematic studies of many of the types of epilepsy listed later will always limit this contribution.

GENETICS

In the outbred human species, epilepsy is not a constitutive phenotype; however, there are certain genetic variants that either directly cause or contribute to such disease phenotypes. It is important to remember that 'epilepsy' is an umbrella diagnosis encompassing a wide range of different diseases and syndromes. An increasing number of rare, deleterious mutations that underlie specific epilepsies are being identified. To date, the majority of mutations leading to epilepsy as the main clinical consequence are found in genes encoding ion channels (Table 11.2).[294,304] Mutations in another group of genes generate epilepsies associated with malformations of cortical development, often with distinctive neuropathological appearances.[181] Although the progressive myoclonic epilepsies previously

TABLE 11.2 The genetics of monogenic idiopathic epilepsies and epileptic encephalopathies

Protein	Subunit	Gene	Gene locus	Phenotype
Neuronal nicotinic acetylcholinic receptor	α2-subunit α4-subunit β2-subunit	CHRNA2 CHRNA4 CHRNB2	8p21 20q13 1q21	ADNFLE ADNFLE ADNFLE
M-current protein channel	Kv7.2 Kv7.3	KCNQ2 KCNQ3	20q13 8q24	BFNS BFNS
Voltage gated sodium channel	α1-subunit	SCN1A	2q24	GEFS+ SMEI IGE-GTC MAE
	α2-subunit	SCN2A	2q23-q24.3	BFNIS BFIS SMEI
	β2-Subunit	SCN2B	19q13	GEFS+ EOAE+FS plus
GABA receptor	α1-subunit γ2-subunit	GABRA1 GABRG2	5q34-q35 5q34	JME GEFS+ CAE
Leucine-rich glioma inactivated 1		LG1	10q24	ADPEAF/ADLTE
Glucose transporter type1		SLC2A1	1p35-p31.3	EOAE
EF hand motif containing 1		EFHC1	6p12-p11	JME
Protocadherin		PCDH19	Xq22	EFMR
Cyclin-dependent kinase-like 5		CDKL5/STK9	Xq28	ISSX-RTT
Aristaless related homeobox		ARX	Xp22.13	OS
Sintaxin binding protein 1		STXBP1 (MUNC 18-1)	9q34.1	OS
Solute carrier family 25 member 22		SLC25A22	11p15.5	EME

Reproduced from Nicita F, De Liso P, Danti FR, et al. The genetics of monogenic idiopathic epilepsies and epileptic encephalopathies. Seizure 2012;21:3–11, with permission from Elsevier. Copyright © 2011 British Epilepsy Association. Genes and encoded proteins involved in the genetic epileptic syndromes and encephalopathies: ADLTE, autosomal dominant lateral temporal epilepsy; ADNFLE, autosomal dominant nocturnal frontal lobe epilepsy; ADPEAF, autosomal dominant partial epilepsy with auditory features; BFIS, benign familial infantile seizures; BFNIS, benign familial neonatal/infantile seizures; BFNS, benign familial seizures including benign familial neonatal seizures; CAE, childhood absence epilepsy; EFMR, epilepsy and mental retardation limited to females; EME, early myoclonic encephalopathy; EOAE, early onset absence epilepsy; FS plus, febrile seizures plus; GEFS+, genetic epilepsy with febrile seizures plus; IGE-GTC, idiopathic generalised epilepsy with generalised tonic clonic seizures; ISSX, X-linked West syndrome; JME, juvenile myoclonic epilepsy; MAE, myoclonic-astatic epilepsy; OS, Ohtahara syndrome; RTT, Rett syndrome; SMEI, severe myoclonic epilepsy of infancy (SMEI).

were distinguished by both clinical and neuropathological profiles, an increasing number of these epilepsies can now be defined genetically by mutations in a heterogeneous spectrum of genes.[125] Outside these clinicogenetic groupings, comparatively few genes have been identified in which mutations lead to disease with epilepsy as an important or sole manifestation; examples include LGI1[210] and STK9.[211] The neuropathology generated by mutations in these genes is unknown.

Most epilepsies are believed to be complex traits resulting from interactions between non-genetic and genetic factors, the latter thought to involve minor contributions from multiple genes (oligogenic or polygenic contributions). Patterns of inheritance in the majority of cases are thus complex and subtle. Although many association studies have been undertaken to examine the influence of common genetic variation on disease susceptibility in epilepsy, no genetic variants have been proven to underlie any common epilepsy.[387]

More recently, attention has focused on other types of genetic causation, especially structural variation. Larger chromosomal rearrangements, visible microscopically, have long been known in the epilepsies. Advances in technology have made possible the detection of submicroscopic rearrangements or copy number variation. Microdeletions and microduplications are now established risk factors, being found typically in 10–15 per cent of selected cases.[293] These contributors to epilepsy causation can highlight new genetic risk factors among the thousands of genes expressed in the brain. For most, however, the mechanism whereby copy number variation contributes to epilepsy is unknown, and the underlying pathological changes, if any, are also largely unknown, with some exceptions.[249]

The greatest excitement now lies with massively parallel sequencing such as exome and whole genome sequencing, which allows truly comprehensive evaluations of genetic susceptibility and causation. In the next few years, it is likely that there will be a much better understanding of the genetic landscape that will both inform, and benefit from, parallel neuropathological studies as has been illustrated in recent exome sequencing projects of epileptic encephalopathies identifying 329 de novo mutations.[2]

CLINICAL INVESTIGATIONS

In the assessment of epilepsy patients, the history takes primacy, ideally in conjunction with an eyewitness account. The diagnosis of epilepsy is clinical in the vast majority of cases, but a syndromic diagnosis is also possible on the basis of clinical symptoms and signs. Clinical evaluation also guides appropriate investigations and management. The clinical formulation, ideally with identification of an underlying cause or disease process, is the best guide to prognosis, as this is determined largely by aetiology.

Among the investigations that are likely to be indicated in an individual diagnosed with new-onset epilepsy are those directed at cause and those that further stratify the type of epilepsy or syndrome. Any of these investigations may also facilitate further assessment during the course of the disease. They include tests on venous blood (typically assays of haematopoietic, renal and hepatic function), a routine 12-lead electrocardiogram (ECG) and neuroimaging (ideally a high resolution MR imaging brain scan but rarely positron-emission tomography [PET] or single photon-emission computed tomography [SPECT]). Electroencephalography (EEG), optimally undertaken with video recording, may assist in confirming the diagnosis and type of epilepsy (focal versus generalized) and in localizing the area of seizure onset. Defining the epilepsy type is worthwhile, both for determining prognosis and making treatment decisions.

Typically, a routine EEG will be undertaken in the outpatient setting, but failure to acquire useful data may necessitate a further recording, possibly with sleep deprivation, or a 24-hour ambulatory EEG including periods of sleep. Abnormalities on EEG are often brought out during sleep, and such recordings are often more revealing than further routine recording.

The ability of high resolution MR imaging to identify most (approximately 80 per cent) causes of focal epilepsy has added significantly to our understanding of the disease. The most common abnormalities in patients with focal epilepsy are cerebrovascular disease, trauma, tumours, developmental and vascular abnormalities, and hippocampal sclerosis. Clearly, there are distinct implications for patient management, depending on the underlying pathology. Some disease processes require management in their own right.

Antiepileptic drugs produce satisfactory seizure control in most patients with epilepsy, but when seizures are intractable the possibility of surgical treatment should be considered. In this situation, the therapeutic aims should be to identify the underlying pathology (usually with a combination of clinical history, neuropsychometry and MR imaging), to prove that the pathology is related aetiologically to the epilepsy (usually with EEG, often prolonged scalp EEG video telemetry), to establish that resection of the focal pathology is unlikely to produce significant adverse effects (e.g. loss of motor, sensory, cognitive or mnemonic functions) when balanced against anticipated benefits, and then to offer focal resection. Most commonly, this takes the form of anterior temporal lobectomy for temporal lobe epilepsy due to hippocampal sclerosis. A range of other pathologies and surgical approaches is possible, sometimes informed by other investigations, including functional imaging (with functional magnetic resonance imaging [fMRI], PET or SPECT) and intracranial EEG recording. Seizure-free rates without significant morbidity of the order of 70 per cent (6–12 months after surgery) are reported following hippocampectomy for hippocampal sclerosis; only slightly lower seizure-free rates are reported for the second most commonly resected pathology, focal cortical dysplasia. A range of other surgical interventions exist for epilepsy, including biopsy, lesionectomy, multilobar resection, hemispherectomy, corpus callosotomy, multiple subpial transection, radiosurgery and the extracranial implantation of a vagus nerve stimulator.

EFFECTS OF SEIZURES

The effects of seizures on the brain are complex and have to be distinguished from the consequences of any primary neurological disease process that has led to seizure susceptibility. Although there is strong evidence to support the direct and detrimental effects of seizures on brain function and structure, brain impairment is not inevitable, for example in the 'benign' epilepsies.[190] Furthermore, injurious effects of seizures should be evaluated not only in terms of histologically evident structural changes, such as neuronal loss and gliosis, but also by assessing alterations at the cellular, synaptic and molecular levels, some of which may be reversible and others permanent. The potential clinical effects of seizure-related injury include worsening seizures or neurological, psychological and cognitive disability. Effects of seizures are influenced by the maturity of the brain (i.e. whether seizures are occurring in developing or adult brain) as well as their frequency, duration and cause. Epilepsy-induced neuronal injury is therefore an area of intense research with the ultimate goal of identifying neuroprotective treatments or other interventions to prevent or reverse functional deterioration.

Mechanisms of Neuronal Injury

The concept of excitotoxicity was first proposed in 1969 as a toxic effect of prolonged activation by excitatory amino acids, mediating neuronal injury in various neurological conditions. Following a prolonged seizure, there is excessive excitatory neurotransmitter release, with overstimulation of glutamate (N-methyl-D-aspartate [NMDA]) receptors and voltage-activated calcium channels, resulting in cellular influx of Ca^{2+} and mobilization of other ions (including K^+, Cl^-).[126] An increase in free calcium leads to mitochondrial dysfunction, release of mitochondrial Ca^{2+} and activation of various enzymes (lipases, endonucleases, proteases, catabolic enzymes) including MAPK/ERK (mitogen-activated protein kinase/extracellular signal-regulated kinase) activation.[430] Mitochondrial dysfunction contributes to seizure-mediated neuronal death.[227] Prolonged seizures may ultimately result in neuronal death by apoptotic or necrotic pathways. Based on classic morphological definitions, cell necrosis would appear to be the dominant mechanism.[248] Studies also support involvement of programmed cell death with activation of both the intrinsic and extrinsic pro-apoptotic pathways including altered expression of pro-apoptotic Bcl-2 family genes and increased expression of caspases.[66,194] Continued expression of apoptotic proteins, after the initial injury, may contribute to ongoing pathogenic mechanisms.[134] The early and late cellular effects of seizures on neurons, many of which are linked with the processes that promote further seizures or 'epileptogenesis' (see section on General Concepts of Epileptogenesis in Focal Epilepsy, p. 723), are summarized in Box 11.1.

Effects of Brief Seizures

One of the most pertinent clinical questions is whether a single brief seizure always causes neuronal damage and how this compares to the cumulative effects of repeated or prolonged seizures. Evidence that prolonged seizures, or status epilepticus, can result in neuronal loss is firmly established.[200] There are experimental data to show that even brief seizures may induce neuronal loss in the hippocampus.[78] In humans, longitudinal or serial MR imaging provide the most compelling evidence that, in some patients, repeated brief seizures can result in progressive hippocampal atrophy.[381,382] However, evidence from quantitative postmortem neuropathological studies demonstrates that hippocampal or cortical neuronal loss is not inevitable, even with a long history of frequent and prolonged seizures[411] or some epileptic encephalopathies, such as Dravet syndrome.[80]

Influence of Age and Brain Maturity

Infants and children have a higher risk of seizures, and although mostly benign, frequent seizures carry the risk of long-term developmental, cognitive and behavioural problems.[36,201] Although antiepileptic drugs or the underlying condition contribute to these problems, there is accumulating evidence that ongoing seizures play an important role. The immature or developing brain is affected differently by seizures[190] than adult brains, in terms of both the cellular/molecular responses and neuronal injury.[202] In experimental conditions, the developing brain has a lower threshold for seizures, but appears less vulnerable to neuronal loss, axonal sprouting and the toxic effects of glutamate. Seizure susceptibility peaks during the period of rapid brain growth and synaptogenesis. There are several explanations for this: (i) excitation predominates over inhibition in developing neuronal networks; (ii) glutamate receptor subunits, being maturationally regulated, promote excitability and the neurotransmitter γ-aminobutyric acid (GABA) also mediates excitatory rather than inhibitory hyperpolarizing effects; (iii) there are lower synaptic densities and reduced developmental regulation of voltage-gated ion channels, such as potassium (K_v) and sodium channels (Na_v) (dysfunction of the latter being implicated in many early life epilepsy syndromes); and (iv) rates of neurogenesis may be enhanced and astrocytic responses diminished in the developing brain in response to seizures.[338] Experiencing a seizure early in life may increase susceptibility to further seizures and the risk of epilepsy later in life, i.e. *'seizures beget seizures'*.[365]

Acquired (Secondary) Neuropathological Changes as a Consequence of Seizures and Post-Mortem Examination

In examining the brain of a patient with epilepsy, the neuropathologist addresses three main questions: (i) can a cause for the epilepsy be identified; (ii) are secondary changes of previous seizures present; and (iii) how has epilepsy contributed to the underlying cause for death? For the investigation of the first, tissue sampling is influenced by any macroscopic abnormality or by localizing clinical, electrophysiological and/or neuroimaging data. For secondary changes and investigations into cause of death, sampled regions of brain should include those most vulnerable to seizure-related injury, neuronal loss and gliosis: the hippocampus, neocortex, thalamus, amygdala, cerebellum and brain stem.

Secondary Cerebellar Pathology in Epilepsy

Macroscopic atrophy of the cerebellum has long been noted,[107] being present in 25 per cent of cases in one postmortem epilepsy series.[270] It is generally acknowledged that cerebellar atrophy is likely to be acquired during the course of the epilepsy, rather than a predisposing factor for seizures. MR imaging volumetric studies confirm a correlation between the severity of atrophy and duration of epilepsy.[314] Cerebellar atrophy has been observed in association with generalized and focal seizures syndromes, including temporal lobe epilepsy (TLE).

Neuropathological findings at autopsy may disclose symmetrical atrophy of the anterior lobes or, more commonly, the posterior lobes[111] (Figure 11.1). In mild cases, damage may be restricted to a folium, but in severe cases more generalized atrophy is observed. Crossed cerebellar atrophy (cerebellar diaschisis) is recognized in patients with contralateral destructive cerebral hemispheric lesions associated with seizures, including hemiatrophy. Patients with unilateral TLE, however, typically exhibit bilateral cerebellar atrophy.[314] Regardless of the lobar distribution, the histological findings are Purkinje cell loss and Bergmann gliosis, with relative preservation of basket cells (Figure 11.1). Occasional torpedo-like axonal swellings of Purkinje cells may be observed. The white matter typically appears normal.

BOX 11.1. Immediate and longer term cellular effects following a seizure

Immediate/early (minutes to hours)
- Increase [Ca^{2+}] (seconds/minutes)
- Altered kinase activity: phosphorylation or dephosphorylation of enzymes, receptors, ion channels
- MAPK/ERK2 pathway activation
- Immediate early gene expression (including c-fos, c-jun)

Intermediate (hours to days)
- Increased inflammatory mediators (cytokines, TNF, IL-1β, NF-B1)
- Protein synthesis (e.g. endogenous anti-convulsants, somatostatin, NPY)
- Growth factor expression (e.g. BDNF in hippocampal neurons)
- Alteration in subunit expression in glutamate and GABA receptors, with altered function
- Dendritic structural plasticity
- Altered synapse-associated gene expression
- Neuronal cell death

Late to chronic changes (days to months)
- Axonal sprouting e.g. mossy fibres (days, can be persistent)
- Enhanced neurogenesis (weeks)
- Reactive astrocytic gliosis (hypertrophy, proliferation and activation) (weeks/months)
- Altered or reduced capacity for neurogenesis (months)

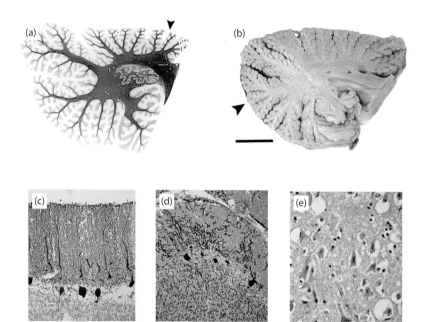

11.1 Acute and chronic effects of seizures. Cerebellar atrophy involving the anterior lobe folia mainly **(a)** and another case **(b)** with atrophy primarily involving the posterior lobe (arrowheads). Purkinje cell loss can be demonstrated with calbindin staining with normal preservation shown in some areas of the cortex **(c)** but elsewhere depletion of Purkinje cell bodies and reduction of their dendritic trees is noted **(d)**. The effects of status epilepticus seen in this case of a patient who died three days later showing extensive necrosis of cortical neurons with eosinophilic pyramidal cell **(e)**.

The cause of cerebellar atrophy has been attributed to seizure activity, in particular episodes of status epilepticus,[393] antiepileptic drug toxicity, particularly phenytoin, hypoxic-ischaemic injury, trauma as a result of seizures,[207] transneuronal degeneration and/or an underlying neurodevelopmental disorder. Of note, atrophy can be present in the absence of phenytoin exposure. The clinical consequences of cerebellar atrophy in the context of epilepsy are poorly understood.

Neuropathological Features Following Status Epilepticus

Status epilepticus (SE) is defined as an uninterrupted seizure activity lasting more than 30 minutes, and may be convulsive or non-convulsive. The incidence of SE is highest in the first year of life (including febrile SE) and in patients aged over 60 years.[190] SE can be seen with any seizure type and mortality increases to 19 per cent for seizures lasting more than 30 minutes.[207] Several MR imaging studies have described acute changes in the hippocampus, including swelling, oedema and T2 signal increase following SE.[331] In childhood, approximately 5–10 per cent of febrile convulsions may be prolonged and are regarded as a form of SE. Hippocampal swelling has also been shown in the acute period following febrile SE.[354]

In SE, normal inhibitory mechanisms fail and epileptic activity becomes self-sustaining. Identified pathomechanisms include endocytosis and downregulation of $GABA_A$ receptors,[207] along with recruitment of AMPA (α-amino-3-hydroxy-5-methyl-4-isoxazolepropionic acid) and NMDA receptors, with an overall proconvulsant effect. Over a longer time course, depletion of inhibitory peptides dynorphin, galanin, somatostatin and neuropeptide Y occurs, while proconvulsants including substance P increase, acting to self-sustain seizures.[98]

Neuropathological findings in fatal SE include acute, extensive necrosis of hippocampal pyramidal neurons,[382] which may be unilateral[284] or bilateral[331] (see Figure 11.1). Neuronal loss also occurs in CA1, CA3 and the hilus of the hippocampus (dentate granule cells and CA2 neurons may be spared),[331] amygdala (corticomedial and basolateral nuclei), neocortex (midcortical layers), entorhinal cortex, cerebellar Purkinje cells,[107] the dorsal medial thalamic nuclei,[163] mammillary bodies, basal ganglia,[331] and the brain stem.[417] Following recent SE, microgliosis and perivascular lymphocytic cuffs are noted. The neuronal damage may predominate unilaterally in some cases,[163] and with prolonged hemiconvulsions, cerebral hemiatrophy can ensue with striking unilateral laminar necrosis of the second to fourth cortical layers.[386] Extensive axonal swellings associated with neuronal injury have also been reported at 7 days following SE.[52] There is evidence from experimental models that hippocampal neuronal loss may continue after cessation of SE,[300,330] suggesting the continued need for neuroprotectants even after seizures have ceased. Hippocampal atrophy has also been observed following non-convulsive SE.[421] Importantly, MR imaging volumetric studies of patients over 1 year post-SE demonstrate that hippocampal or amygdala volume loss is not invariable.[346] Quantitative neuropathological studies in a subgroup of patients with long histories of generalized seizures, including frequent episodes of SE, also show that hippocampal and neocortical neuronal loss is not inevitable.[107,411]

Cerebral Trauma and Epilepsy

Post-Traumatic Epilepsy

Traumatic brain injury (TBI), accounts for 20 per cent of symptomatic epilepsy and 5 per cent of all epilepsy.[136]

Seizures are typically secondary and generalized, with or without focal symptoms. Early post-traumatic seizures occur within the first week, whereas late seizures mostly occur within 18 months[114] but can emerge up to 5 years after injury. The risk of developing epilepsy correlates with the severity of the injury. Acute early seizures can occur in up to 25 per cent of cases of moderate to severe TBI, with chronic epilepsy in 10–25 per cent of cases; incidences as high as 50 per cent have been recorded following missile injury.[255] Risk factors include alcoholism, age, penetrating dural injury, intracranial haemorrhage or contusion, depressed skull fracture, focal neurological deficit and early post-traumatic seizure. There is no clear risk association with ApoE genotype.[222] Disruption of the blood-brain barrier (BBB) following TBI has been implicated as an underlying etiologic mechanism.[364]

Animal models of TBI used to study post-traumatic epilepsy include partially isolated neocortex, fluid percussion, controlled cortical impact, iron injection, weight drop and penetrating ballistic-like injury models.[329] The problem with modelling epilepsy in TBI is that it is a heterogeneous disorder due to different types of injury.[329] In the partially isolated neocortex model, undercutting cortical fibres in the white matter induces acute and chronic seizures and demonstrates neuronal loss, gliosis with increased axonal sprouting and synaptic density and regressive changes to GABAergic cells;[334] this observed increase in cortical connectivity and excitability is considered to model some of the processes in cortical contusions and is similar to reorganizational changes described following early trauma in infancy.[272] In terms of studying disease-modifying strategies for prevention of post-traumatic epileptogenesis (from channel blockers to cell transplantation), the underlying mechanisms may be inextricably linked with cortical repair processes. As such, distinguishing adaptive repair from maladaptive repair leading to epilepsy is an area of intense research.[329]

Trauma as a Result of Epilepsy

Patients with established epilepsy have a higher incidence of minor and severe cerebral injuries, including cerebral haemorrhage and contusions. The risk is related to seizure frequency, type and control.[240,248] Head injuries have been reported in 24 per cent of epilepsy patients in one study.[71] In a post-mortem series of 138 patients with epilepsy, old TBI, most frequently frontotemporal contusions, was identified in 30 per cent of patients; only three cases reported a clinical diagnosis of post-traumatic epilepsy, supporting the notion that the trauma resulted from the seizures in most instances.[401] Furthermore, there was a significant association between Tau protein accumulation and the presence of TBI, reminiscent of chronic traumatic encephalopathy, as seen in athletes[280] (Figure 11.2).

Neurodegenerative Diseases: Epilepsy and Cognition

Epilepsy may form part of the clinical manifestations of various neurodegenerative diseases and other common neurological conditions such as strokes and inflammatory conditions including multiple sclerosis.[426] Patients with Alzheimer's disease are at increased risk for developing

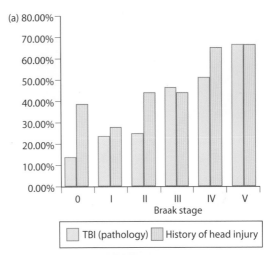

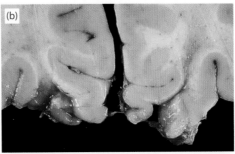

11.2 Traumatic brain injury. (a) The graph is adapted from a study[401] of 138 patients with chronic refractory epilepsy where the presence of a history of head injury or the identification of old head trauma at post mortem was correlated with the Braak stage. There was a correlation between the tangle load and the presence of old head trauma. **(b)** The figure shows an old frontobasal contusion from one of the patients in the study.

Data from Thom et al.[401]

epilepsy.[159] The underlying epileptogenic mechanism is unknown, but the toxic effects of amyloid-β on synaptic transmission have been considered.[159]

It is recognized that cognitive decline and neuropsychological comorbidities arise in some epilepsy patients.[195] There is a well-established risk of psychiatric comorbidity in patients with epilepsy.[150] In TLE, this may include a neurodevelopmental contribution, although cognitive deficits may progress further over time.[193,196] In a post-mortem study of ageing epilepsy patients with cognitive decline, common neurodegenerative diseases were identified in only half, suggesting that other, as yet undefined, processes lead to cognitive impairment.[401]

SURGICAL PATHOLOGY

Epilepsy surgery has been performed for over half a century, but has taken on a more important role in the management of refractory seizures in the last 20 years. Advances in structural and functional imaging and invasive electroencephalography have increased the precision of localizing the ictal onset zone and maximizing safe resections and seizure-free outcomes with minimal deficits. Overall 60–70 per cent of patients are seizure free 2 years after surgery, and long-term outcome studies suggest approximately half will remain seizure free 10 years after surgery.[124]

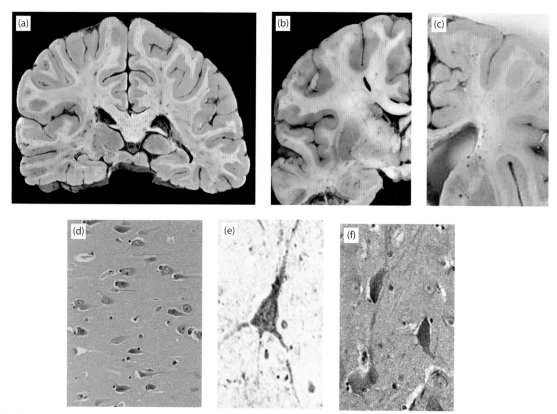

11.3 Sudden unexpected death in epilepsy (SUDEP). The top row shows three sections from SUDEP cases showing **(a)** left hippocampal sclerosis, **(b)** a DNT in the left middle temporal gyrus, and **(c)** focal coritical dysplasia type IIb in the right cingulate gyrus. The bottom row shows evidence of **(d)** acute eosinophilic change of neurons in CA1, **(e)** labelling of a pyramidal cell neuron in CA3 with HIF I-alpha, and **(f)** HSP-70 as evidence of acute neuronal injury in SUDEP. (See also Box 11.2.)

BOX 11.2. Sudden unexpected death in epilepsy (SUDEP)

Death in epilepsy may occur as a consequence of the disease process (e.g. tumour or stroke), or be seizure-related, including death from status epilepticus or accidental death during a seizure (e.g. drowning). SUDEP is defined as a sudden, unexpected, witnessed/unwitnessed, non-traumatic and non-drowning death in epilepsy with/without evidence of a seizure, excluding documented status epilepticus, and where autopsy does not reveal another cause of death. Based on a conservative annual SUDEP incidence figure of 1 in 200 with severe chronic epilepsy, 3000–5000 patients die of SUDEP each year in the USA alone. Deaths are likely to occur during or after seizures. Consistent risk factors are poor seizure control, frequent generalized tonic–clonic seizures, multiple anti-seizure drugs, and long-standing epilepsy. Deaths are often unwitnessed, often nocturnal, occurring during or just after a seizure, with the body found prone in or close to bed. Clinical evidence of a seizure is often present (bitten tongue or urinary incontinence). Witnessed deaths, for example in epilepsy monitoring units, occur within minutes after generalized tonic–clonic seizures, with central apnoea, bradycardia and EEG depression frequently seen.[344] Precise mechanisms of death are unknown, but may include: (i) autonomic dysregulation with cardiac arrhythmias, including ictal bradycardia and asystole (decreased heart rate variability has also been shown in SUDEP); (ii) central respiratory suppression/apnoea with additional airways compromise causing positional asphyxia; and (iii) cerebral and brain stem dysregulation or postictal 'cerebral electrical shutdown'.

In the UK, the majority of suspected SUDEP deaths are referred to the coroner and undergo a post-mortem examination. In SUDEP, by definition, no anatomical or toxicological cause of death is found. Neuropathological examination of the fixed whole brain may reveal mild degrees of cerebral oedema and congestion, but no significant swelling. In the few published pathology series of SUDEP, structural brain abnormalities, including malformations, contusions, dysplasia and hippocampal sclerosis were noted in 50–70 per cent of cases[43,362,398] (see Figure 11.3). Histological examination not infrequently reveals evidence of acute neuronal injury, particularly in the hippocampus (see Figure 11.3). Pathological changes in the heart, all presumed to be non-fatal, have been described in SUDEP, including increased heart weights and focal myocardial fibrosis, although pulmonary oedema has also been noted. As well as excluding drug overdose, toxicological investigations at autopsy can also give some indication of antiepileptic drug compliance. A diagnosis of definite SUDEP can only be made with a full post-mortem examination; the terms possible or probable SUDEP are reserved for cases with incomplete autopsy or the finding of a conflicting or contributing cause of death.[302]

In adults, the most common procedure is resection, with temporal lobectomies for the treatment of TLE accounting for 70–80 per cent of cases (including *en bloc* temporal lobectomies, anterior temporal lobectomy and selective amygdalohippocampectomy). In paediatric series, temporal lobectomies represent around 25 per cent of cases, with hemispherectomies, multilobar and frontal lobe resections more commonly performed.[188]

The three main pathology lesions are hippocampal sclerosis (HS), cortical dysplasias and low-grade tumours. In adult surgical series, HS is most common, whereas cortical dysplasias represent the main pathology in paediatric series (Table 11.3). Conventionally, the pathologies associated with TLE (or mesial temporal lobe epilepsy MTLE) have been divided into two main groups: HS in around 60–70 per cent of cases[54,123,133] and mass lesions in about 30–40 per cent of cases. HS in combination with another lesion (dysplasia, tumour), also known as 'dual pathology', accounts for a minority of cases. In a few cases, no specific lesion is identified other than gliosis; such cases have been termed 'pathology-negative', 'cryptogenic' or 'paradoxical' TLE.

Hippocampal Sclerosis

Historical Aspects

The earliest pathological descriptions of HS date as far back as 1825. Sommer's name is more generally linked with this pathology, following his review in 1880 of 90 post-mortem specimens from epilepsy patients,[397] describing the segmental hippocampal pyramidal cell loss in a region that was to become known as Sommer's sector. In 1889, the neurologist Hughlings Jackson linked the clinical symptoms of 'psychomotor epilepsy' with lesions in the hippocampus. Following the advent of EEG and the early epilepsy surgical programmes in the 1950s, 'Ammon's horn sclerosis' (as then termed) was recognized as the most common pathology. In 1956, Cavanagh and Meyer, based on their surgical series, described 'diffuse and disseminated' lesions associated with HS including neuronal loss and gliosis in the amygdala, parahippocampal gyrus and neocortex. From this, the term 'medial temporal sclerosis' was introduced to convey this more widespread sclerosis, with HS at the epicentre.[82]

Definition of Hippocampal Sclerosis

The pathological definition of HS is neuronal loss and gliosis mainly involving CA1 and CA4/3 subfields with sparing of the subiculum and relative resistance of CA2 pyramidal neurons and the granule cells (Figure 11.4). This pathology in patients with epilepsy, and typically TLE, is often accompanied by granule cell dispersion and axonal reorganization in the dentate gyrus, including mossy fibre sprouting. These cellular processes manifest as reduction in hippocampal volume (Box 11.3).

Neuropathological Features and Diagnosis of Hippocampal Sclerosis

The pathological diagnosis of HS on haematoxylin and eosin (H&E) sections is based on the identification of pyramidal neuron loss and gliosis involving mainly CA1, CA4 and CA3 subfields. This distinctive pattern of neuronal loss, apparent on qualitative histological examination and sometimes even macroscopically, is referred to as 'classical' hippocampal sclerosis (CHS). Where marked depletion of granule cell neurons is present, accompanied by more extensive pyramidal cell loss including CA2, this

is referred to as severe or total hippocampal sclerosis.[70] In HS associated with epilepsy, there is typically a sharp cut-off between CA1 and the spared neurons of the subiculum (Figure 11.5).

The extent and severity of neuronal loss can vary between cases. In a minority, the neuronal loss appears restricted to either the CA1 (termed CA1 HS) or the CA4 region (termed end folium HS or EF HS). These more limited, atypical patterns of sclerosis are less prevalent than classical HS (Table 11.4). Various semi-quantitative and quantitative scoring methods have been reported over the years in an aim to categorize the patterns of cell loss, as well as identify cases with more subtle neuronal loss (Table 11.4). A recent semi-quantitative scheme for the classification of HS, based on the severity of neuronal loss in hippocampal subfields, has been proposed and validated by the ILAE (International League Against Epilepsy).[52] This new classification groups all classical HS patterns into type 1, CA1 HS as type 2 and EF HS as type 3. Hippocampal gliosis alone, without neuronal loss, is not classified as a HS subtype. There is no evidence that type 2 or 3 progresses to type 1. However, the distribution of neuronal loss may also vary along the longitudinal hippocampal axis (from the pes to tail) on MR imaging, as well as neuropathology studies (Figure 11.4).[24]

In HS, glial fibrillary acidic protein (GFAP) stains confirm a dense fibrillary gliosis in the sclerosed subfields. Residual hilar neurons may show enlargement, coarse cytoplasmic staining with cresyl violet, argyrophylia on silver stains, accumulation of microtubules and neurofilaments and increased dendritic complexity.[55] Deposits of corpora amylacea may be pronounced in some cases in the collapsed stratum pyramidale with extensive gliosis and proliferation of the microvasculature may also be apparent.[123] Apoptotic cells or eosinophilic neurons are rarely seen in surgical specimens, and neuronophagia or focal infiltrates of microglia are occasional findings.

Granule Cell Dispersion (GCD)

The principal neurons of the dentate gyrus, the granule cells, form a densely packed layer approximately eight cells thick. Loss of this close apposition and dispersion of granule cells into the molecular layer, was first described in association with HS by Houser and colleagues.[205] This phenomenon, peculiar to seizure-induced hippocampal damage is encountered in 40–50 per cent of HS cases in surgical series.[48,435] In the presence of dispersion, granule cells often appear enlarged and more fusiform in shape, with increased cytoplasm and separation of cells. The border between granule cell and molecular layers becomes ill defined. Variations include clusters of granule cells in the molecular layer and a bilaminar pattern in 10–15 per cent of cases (Figure 11.6). The extent and pattern of GCD may vary both within and between cases and may alternate with regions of granule cell depletion. There is no agreed definition for GCD; a granule cell layer thicker than 10 cells[435] or 120 μm breadth[257] has been proposed but in many cases, the granule cell width may reach over 400 μm,[48] compared to approximately 100 μm in control subjects. Any assessment of GCD should be made on the straight sections of the granule cell layer (upper or lower blades), rather than at its inflections, where a false impression of broadening

TABLE 11.3 Relative incidence of focal pathologies in more recently reported epilepsy surgical series

Year	Reference	Study	Types of resection	Age of patients Mean/range	Number of cases	Time period of study	Malformations of cortical development (other than focal dysplasia)	Focal dysplasias	Vascular malformations	Tumours	Hippocampal sclerosis	Atrophic lesion (Old infarcts/traumatic lesions/scars)	Inflammatory pathology	Gliosis or no pathology	Dual pathology cases (HS+lesion)	Other
2002	Pasquier et al.[321]	Grenoble, France	All types of epilepsy resection	All ages	327	1990–2000	5%	18%		28.70%	25.90%	5%	1.2% (Rasmussen's encephalitis)	15.50%	5%	
2008	Harvey et al.[188]	Multicentre (20 centres)	All types of epilepsy resection	9 years (0–18)	413	2004 (1 year)		42.4% (5.1% TS)	1.5% cavernomas (2.9% Sturge-Weber)	19.1% (29% DNT, 28% GG, others 43%)	6.50%	9.9% (44% infarct, trauma 7%, unspecified 49%)	2.7% (Rasmussen's encephalitis)	6.30%	3.8%	3.6% hypothalamic hamartoma
2009	Blümcke[46]	German Epilepsy Reference Centre	All types of epilepsy resection	All ages	4512	1995–2007		12.70%	6%	27.30%	35.20%	5.20%	1.60%	6.80%	5%	
2010	Piao et al.[324]	Beijing, China	All types of epilepsy resection	All (0–57 years)	435	2005–2008		52.90%		11.70%	17%	22.80%				
2011	de Tisi et al.[124]	London, UK	All types of epilepsy resection	Adults	615	1990 to 2008		2.40%	5.3% (cavernoma)	12.8% DNT, other tumours 4%	66.10%			3.50%	Data not included	5.60%

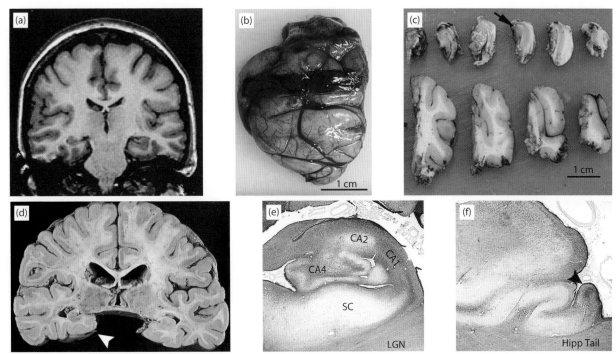

11.4 Magnetic resonance (MR) imaging and macroscopic appearances of hippocampal sclerosis. (a) MR imaging T1-weighted image demonstrated unilateral hippocampal volume loss. **(b)** A temporal lobe specimen received fresh from theatre and oriented with the middle temporal gyrus inked black. **(c)** Postfixation the temporal lobe specimen (lower row) and hippocampus are sliced at 3–5 mm intervals and inspected; reduction of CA1 volume may be apparent at this point (arrow). **(d)** A post-mortem specimen from a patient with long-standing epilepsy and unilateral hippocampal atrophy (arrowhead). **(e)** Post-mortem examination at the level of the lateral geniculate nucleus (LGN) confirms the classical pattern of hippocampal sclerosis (glial fibrillary acidic protein/cresyl violet stain), but in the hippocampal tail **(f)** gliosis is restricted to the end folium (arrowhead) confirming an anterior gradient for the hippocampal sclerosis.

BOX 11.3. Clinical history in temporal lobe epilepsy with hippocampal sclerosis

Temporal lobe epilepsy is divided into the common mesial temporal lobe epilepsy (MTLE) form and a rarer lateral temporal lobe subtype. MTLE is often associated with HS and drug-resistance, with characteristic natural history and seizure types.[435] In many cases, there is a report of an initial precipitating injury (IPI) occurring in the first five years of life, such as a complex febrile convulsion (a convulsion lasting more than 15 minutes, or recurring within 24 hours, or associated with lateralized signs), trauma, hypoxia or intracranial infection. This is followed by a latent interval (often several years) before the emergence of habitual seizures.[54,435]

Once habitual seizures have emerged, about 90% of patients with MTLE due to HS fail to respond to antiepileptic drugs.[355] Hippocampal sclerosis (HS) is often suspected on clinical grounds, especially for patients with long-standing epilepsy. High-resolution MR imaging (including thin contiguous slices appropriately oriented through the hippocampus) reliably identifies hippocampal sclerosis in most.[128] Scalp EEG patterns typically show ipsilateral anterior temporal interictal discharges and concordant ictal discharges, but a variety of EEG changes may be seen.

Following surgery for HS in appropriately selected patients, approximately two-thirds of patients remain seizure free in the first two to three years with roughly 57% seizure-free outcome at 5 years.[124] The cause for surgical failure in patients is unclear. Clinical risk factors include longer duration of epilepsy and preoperative history of secondary generalized seizures; the surgical approach and extent of resection may also contribute. Atypical patterns of HS, more extensive un-resected pathology involving adjacent mesial structures, neocortex or even subtle pathology in the contralateral hippocampus could form resurgent epileptogenic networks.[61,407]

may be present. In the presence of severe GCD it becomes difficult to assess any alteration in granule cell number. Golgi studies show that GCD is associated with a wider branching angle of the granule cell neuronal apical dendrites and recurrent basal dendrites.[156]

There is evidence linking the presence of GCD with early onset of epilepsy and febrile seizures (<4 years), suggesting it is an age-dependent phenomenon, as well as longer duration of epilepsy.[48] There is conflicting data regarding whether GCD signifies good outcome following surgery.[48,400] GCD has also been demonstrated in experimental models of TLE. For example, GCD is observed in the kainate model, first appearing at about 4 days following seizures, increasing over 8 weeks and persisting for at least 6 months.[383] Dispersion was prevented by application of glutamate antagonists and GABA$_A$ agonists.[383] In more recent studies where the dentate gyrus was experimentally transected in epilepsy models, GCD was blocked;[316] these observations support the hypothesis that GCD is directly linked to enhanced local excitability.

GCD is associated with radial gliosis (see Figure 11.6) in the granule cell layer; the length of the glial fibres has been shown to correlate with severity of GCD.[146] It has been proposed that the glial fibres act as a scaffold for granule cell migration, akin to their developmental role.[361] A few case reports have also identified GFAP and CD34 positive 'balloon cells' in GCD, reminiscent of those observed in focal cortical dysplasia (FCD) IIB, associated with striking rarefaction of the dentate hilus.[287,402]

Apart from rare reports of GCD occurring without seizures, but in the presence of widespread cortical malformations,[187] it is now generally accepted that in the context of

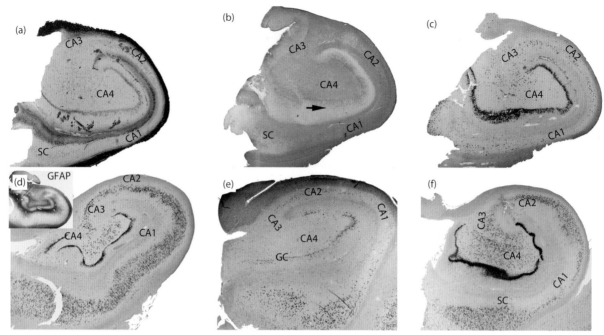

11.5 Classical and subtypes of hippocampal sclerosis based on the distribution of subfield neuronal loss. The pathological diagnosis of hippocampal sclerosis can be made using standard stains, such as haematoxylin & eosin or Luxol fast blue/cresyl Nissl (LFB/CV). Glial fibrillary acidic protein (GFAP) immunohistochemistry highlights extensive chronic fibrillary gliosis in CA1 subfields, in addition to the stellate reactive astrocytes in the granule cell layer (arrow in b). NeuN (or MAP2 and synaptophysin) immunohisto-chemistry helps to confirm the extent of pyramidal neuronal loss and is particularly useful in partial, fragmented and poorly oriented specimens, in which the definition of subfields is unclear. **(a–c)** Classical hippocampal sclerosis with LFB/CV, GFAP and NeuN confirming CA1 and CA4 as the most affected regions. **(d)** In end folium hippocampal sclerosis (EF HS) evidence of neuronal loss is restricted to CA4 neurons. Care must be taken in the diagnosis of EF HS; the diagnosis must not be based on the finding of gliosis alone in the hilus but definite evidence of CA4 pyramidal loss. **(e)** In severe or total HS, neuronal loss is noted in the granule cell (GC) layer as well as CA2. **(f)** CA1 HS is when the neuronal loss appears more restricted to CA1 sector. Note the sharp cut-off with the preserved subiculum (SC). (d–f, NeuN.)

TABLE 11.4 Classifications of patterns of hippocampal sclerosis with relative incidence in epilepsy surgical series and main findings

Series reference (n = number of cases)	Method of analysis	Classical HS patterns ILAE type 1		Atypical HS patterns ILAE type 2 and 3		No HS	Main observations from series
		Classical HS	Severe (total) HS	CA1 sclerosis	End folium (CA4) sclerosis		
Sagar and Oxbury (1987)[345] (n = 32)	Quantitative (CV)	6%	38%	10%	15%	31%	p<0.02 (significant difference reported in age onset between groups; younger in classical HS)
Bruton(1988)[70] (n = 122)	Qualitative assessment	50% (37% SF)	16%	No cases	3% (50% SF)	31% (20% SF)	
Davies et al. (1996)[116] (n = 98)	Semi-quantitative scheme (Wyler system)	60% (68% SF)		8% (66.7% SF)		32% (42.35 SF)	p<0.001 (age onset; younger in CHS) p<0.01 (outcome)
Van Paesschen et al. (1997)[419] (n = 59)	Stereological quantitation (CV and GFAP)	90% (81% SF)		No cases	10% (25% SF)	No cases	p<0.02 (age onset; older in EFS group) p<0.03 (FS: none in EFS group compared to 62% in CHS) p<0.04 (outcome)

TABLE 11.4 Classifications of patterns of hippocampal sclerosis with relative incidence in epilepsy surgical series and main findings (*Continued*)

| Series reference (*n* = number of cases) | Method of analysis | Classical HS patterns ILAE type 1 | | Atypical HS patterns ILAE type 2 and 3 | | | Main observations from series |
		Classical HS	Severe (total) HS	CA1 sclerosis	End folium (CA4) sclerosis	No HS	
Pasquier *et al.* (2002)[321] (*n* = 85)	Qualitative	72%		28%			
de Lanerolle *et al.* (2003)[122] (*n* = 99)	Quantitative (H&E)	73% (84.5%)		9% (77.8% SF)	No cases	18% (44% SF)	*p*<0.05 (age onset younger in CHS)
Blumcke *et al.* (2007)[49] (*n* = 178)	Automated quantitative analysis (NeuN) and cluster analysis	19% (72%)	53% (72.9%)	6% (66.7% SF)	4% (28%)	19% (58.6% SF)	*p*<0.05 (age onset; younger in CHS) *p*<0.05 (fewer cases in EFS and no HS with FS) *p*<0.04 (outcome)
Thom *et al.* (2010)[400] (*n* = 16)	Quantitative analysis (NeuN) and grouped using z scores compared to controls	36% (69%)	24% (71%)	6% (33%)	3% (100%)	10% (44%)	*p*<0.01 (age onset; older in EFS) *p*<0.05 (outcome) 21% the subtype was indeterminate

Figures represent the percentage of each subtype in the series, and the figures in brackets (parentheses) represent the percentage of patients seizure free following surgery. In most cases, the Engel system for outcome was used except for Thom *et al.*,[400] who used the ILAE system to classify outcome; the follow-up periods varied between series and, in some, not all patients were followed up. In general the atypical subtypes represent the minority. In some series, no cases of end folium sclerosis were reported and, in others, no CA1 sclerosis patterns. Differences in incidence between series may be accounted for by different definitions of subfield anatomical regions, different methodologies (quantitative versus qualitative evaluation in determination of neuronal loss) and different thresholds for cut-off points for significant neuronal loss compared to control values. These data (Blumcke *et al.*[49]) exclude dual pathology cases. There is some evidence that atypical HS cases are associated with worse seizure control following surgery and with older age of onset and less consistent history of febrile seizure. In most series, the cases designated as No HS tended to have the poorest outcomes. CHS, classic hippocampal sclerosis; CV, cresyl violet; EFS, end folium sclerosis; FS, febrile seizure; HS, hippocampal sclerosis; ILAE, International League Against Epilepsy; SF, seizure free following surgery.

HS, GCD is not a pre-existing or primary malformation. Two main pathogenic mechanisms underlying GCD have been debated in the light of complementary data from experimental models, where the temporal relationships of GCD and seizures can be better explored. These are that (i) GCD is a result of seizures enhancing local neurogenesis or (ii) it represents a migration of mature neurons as a result of local reelin deficiency following seizures.

The rate of adult neurogenesis in the subgranular zone dentate gyrus is altered under various pathological conditions, with normal physiological roles in learning and memory. Indeed, granule cell loss in the internal limb of the dentate gyrus in epilepsy[322] and loss of hippocampal regenerative capacity in the dentate gyrus, have been correlated with memory dysfunction.[106] There is ample evidence from experimental models that seizures increase neurogenesis, that new neurons can migrate to abnormal or ectopic positions, can be integrated into existing networks and acquire pro-epileptogenic physiology.[77,361,319,325] Experimentally, there is both evidence for and against a decline in the rate of dentate neurogenesis with chronicity of seizures.[189,319] In surgical tissues from patients with HS, higher numbers of cycling cells (MCM2 positive) were observed in HS patients with GCD, a proportion co-labelling for nestin but GFAP negative, supporting the notion of increased progenitor cell proliferation.[406] In another similar study, however, age-dependent decline in MCM2 and doublecortin expression was identified and independent of the extent of GCD, argued against a significant contribution of neurogenesis.[146] A more recent study confirms ongoing expression of neurogenesis genes in human HS cases, equivalent to control subjects, despite overall hippocampal neuronal loss.[135] Naturally, the limitations of human tissue samples in studying neurogenesis as the cause of GCD is that it only represents a single snapshot in time, mostly at the chronic stage of disease.

The extracellular matrix protein reelin, secreted by Cajal–Retzius cells in the hippocampus and developing cortex acts as a stop signal for migrating neurons. In the reeler mouse model, in addition to the widespread cortical abnormalities, GCD is present.[171] Decreased reelin protein and reelin-expressing cells have been shown in HS in MTLE patients,[162] implicating deficiency of this protein as a primary candidate mechanism in GCD. Supportive experimental studies have demonstrated diminished reelin following induction of seizures with the development of GCD.[184] Furthermore, dispersed neurons appear mature rather than newly generated. For example, in studies of calbindin expression, a marker of late granule cell maturation, the more dispersed cells in HS tend to retain expression, whereas more basal cells are calbindin negative.[1,11,261,275] Real-time imaging in mouse models has also confirmed movement of mature granule cells along their apical dendrites into the molecular layer, a process termed 'somatic translocation'.[297]

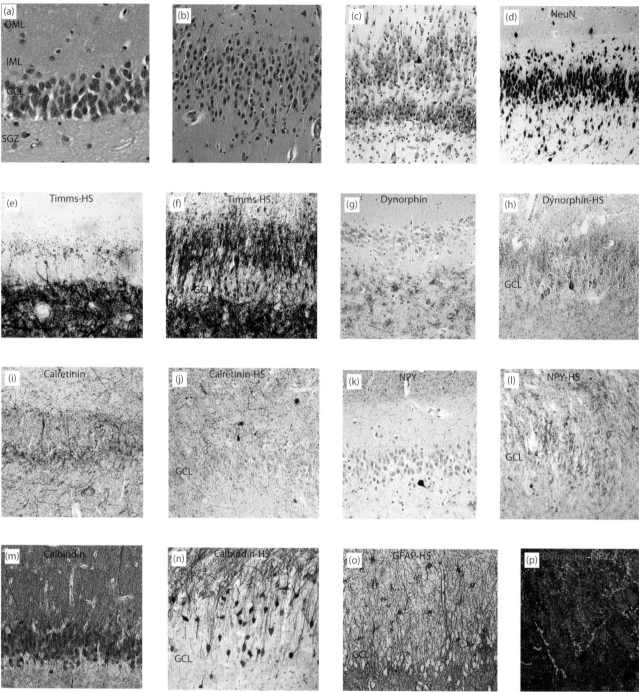

11.6 Dentate gyrus pathology in hippocampal sclerosis (HS). The orientations in all sections are shown as outer molecular layer (OML), inner molecular layer (IML), granule cell layer (GCL) and subgranular zone (SGZ). **(a)** Normal hippocampus (H&E). **(b)** Granule cell dispersion (GCD) in hippocampal sclerosis with migration and separation of the cells as they migrate out in the molecular layer. **(c)** Bilaminar pattern of GCD as seen on CV section and the granule cells appear more fusiform. **(d)** NeuN is useful to confirm GCD and in this field highlights GCD into the molecular layer as well as towards the hilus, which is a less common finding. **(e)** Timms staining where the majority of staining highlights the mossy fibre axons in the subgranular zone and hilus with infrequent silver staining in the inner molecular layer, suggesting early focal mossy fibre sprouting. **(f)** Timms staining, which demonstrates extensive labelling in the inner and outer molecular layer confirming mossy fibre sprouting. **(g)** Normal pattern of dynorphin staining identifies mossy fibre terminals in the CA4 region and in **(h)** in HS mossy fibre sprouting, dynorphin staining in the molecular layer is observed. **(i)** Normal pattern of calretinin staining with a dense fibre plexus around the GCL with **(j)** showing loss of this pattern in HS with radial sprouting of fibres through the GCL and IML. **(k)** Normal NPY pattern with labelling of scattered hilar neurons and a denser axon plexus in the outer molecular layer compared to the unlabelled inner molecular layer. In hippocampal sclerosis **(l)** extensive sprouting is observed through the GCL, IML and OML. **(m)** Normal pattern of calbindin expression in the granule cells and apical dendrites. **(n)** In HS, calbindin expression is often lost from the granule cells, or as here, shows expression in the most distal cells only with loss of expression in the basal cells. **(o)** GFAP staining typically highlights a radial fibre pattern and cellular staining in the dentate gyrus. **(p)** Confocal imaging with dynorphin and NPY confirms that there is no overlap in expression in these sprouted fibre networks.

Pathogenesis of Hippocampal Sclerosis

The earliest theories as to the cause of HS included vasomotor spasm during a seizure, birth trauma with compression of medial temporal lobe structures, and the result of falls during seizure.[397] Seizures are known to damage the hippocampus, particularly prolonged seizures and status epilepticus.[284,330,346,354] There is strong evidence that an initiating event, injury or insult, particularly prolonged febrile convulsions early in life (usually in the first 4 years), can 'prime' the immature hippocampus for the development of HS. This idea was first synthesized by Meyer in the 1950s (*Meyer's hypothesis*) and reconfirmed in modern epilepsy surgical series.[276] The timing of this first 'initial precipitating injury' (IPI) is likely to be critical, as well its nature, which may include trauma, infection or other event. Experimentally, electro-clinical TLE has been induced following prolonged febrile seizures,[127] and patterns of damage similar to human HS, with hippocampal seizures, can follow a single episode of hippocampal excitation,[309] providing similar models to the human condition. There is also evidence from some,[67,68] but not all,[203,250] longitudinal MR imaging studies in patients, that following the onset of chronic repetitive seizures, there is further progressive hippocampal atrophy.

As in status epilepticus, prolonged febrile seizures (as observed in prospective studies[245]) or repetitive seizures do not inevitably lead to HS, there are likely other factors influencing vulnerability, including genetic susceptibility. HS is considered a sporadic condition. Evidence from rare pedigrees suggests that there may be a common genetic basis for both febrile seizures and MTLE, and that one or more genes may contribute to its development[32,85,225] as shown in a recent study associating temporal lobe epilepsy with febrile seizures and SCN1A genotype.[215] As yet, no genetic susceptibility determinant to sporadic HS has been confirmed.[81,212,214,376] Genomic microdeletions, including 16p13.11, have been infrequently reported.[79] ApoE ε4 genotype has been associated with increased risk of bilateral HS.[75]

Human herpes virus 6 has been detected in astrocytes in HS[394] but not replicated in all studies.[305] Focal infiltration of T lymphocytes together with ICAM-1 and kallikrein expression in astrocytes in HS has been observed in support of an underlying neuro-inflammatory mechanism[301,366] and activation of both adaptive and innate immunity has been shown.[445] Pre-existing hippocampal developmental abnormalities have also been considered as a 'dysmaturational template' upon which HS occurs.[31,54,151,180,435]

Regarding patterns of sclerosis, there is no single explanation for the regional selectivity of pyramidal cell loss between hippocampal subfields; excitatory pathways, altered inhibitory input and the effectiveness of endogenous neuroprotective mechanisms are likely to be involved. More recently, subfield-specific regulation of microRNAs has been shown following seizures; these post-transcriptional regulators of gene expression may be critical for determining cell death pathways.[347] Indeed, there may be subtypes of HS with different causes, patterns of atrophy and outcomes.[61] Benign forms of TLE and HS exist with infrequent and well-controlled seizures, typically of adult onset.[235] Furthermore, HS is increasingly recognized on imaging and post mortem in the older and ageing population, without an epilepsy history but associated with memory impairment or cognitive deficit.[303] The pathological comparisons between HS associated with epilepsy and with ageing and dementia, including the dentate gyrus,[24] are summarized in Table 11.5.

TABLE 11.5 Comparison of the pathology of hippocampal sclerosis associated with epilepsy and HS associated with ageing and dementia

	HS in epilepsy (HS-e)	HS in dementia and ageing (HS-d)
Typical clinical presentation	Epilepsy, presenting in young adults; memory deficits and diffuse neuropsychological impairments may occur later	Memory loss/dementia, presenting in old age; may develop epilepsy in course of illness
Cytoarchitecture	Atrophy (neuronal loss and dense fibrillary gliosis) confined to CA1 with sharp demarcation from subiculum Granule cell dispersion in ~50% CA2 spared; neuronal loss usually present in CA4/3	Atrophy (neuronal loss) involves CA1 and subiculum with cellular gliosis Granule cell dispersion not reported CA2,3 and 4 spared
Axonal sprouting	Mossy fibre sprouting typically present	Mossy fibre sprouting infrequent and mild
Specific alterations to neurone/interneurons	CB: reduced expression in granule cells, particularly basal cells NPY: loss of cells + sprouting in DG CR: cell loss + sprouting DG PV: variable loss and complex chandelier terminals in DG	CB: loss in granule cells reported NPY/SS: loss in CA1 described and early loss in AD transgenic mice but sprouting not reported CR: reduction in hilar neurons PV: loss reported in dentate gyrus
Prevalence	35% in epilepsy surgical series 30–40% in post-mortem series	Prevalence in ageing population 0.4–26%

Continued

TABLE 11.5 Comparison of the pathology of hippocampal sclerosis associated with epilepsy and HS associated with ageing and dementia (*Continued*)

	HS in epilepsy (HS-e)	HS in dementia and ageing (HS-d)
Bilaterality	Bilateral in 48–56% in epilepsy PM series	Bilateral in 50–60%
Longitudinal extent	Can be localized to a region of the pes or body or extend throughout the rostrocaudal length	Can be localized or diffuse along the rostro-caudal length
Timing	Neuronal loss acquired early in life	Hippocampal neuronal loss occurs in old age
Causes of neuronal loss	Seizure mediated/excitotoxic neuronal injury	Heterogeneous causes: HS of ageing ('pure' HS) Cerebrovascular disease and vascular brain injury AD, FTLD with associated HS[a] Synucleinopathies with HS
TDP-43	TDP-43 not identified in surgical unilateral cases or bilateral HS-e at post mortem	TDP-43 inclusions in 89% of pure HS-d. Present bilaterally, even in unilateral HS

AD, Alzheimer's disease; CB, calbindin; CR, calretinin; DG, dentate gyrus; FTLD, frontotemporal lobe dementia; HS, hippocampal sclerosis; NPY, neuropeptide Y; PV, parvalbumin; TDP-43, transactive response (TAR) DNA binding protein-43; TLE, temporal lobe epilepsy. [a]The definition of HS associated with AD is pyramidal cell loss and gliosis in CA1 and subiculum of the hippocampal formation that is out of proportion to AD neuropathological changes in the same structures.

Epileptogenesis in Hippocampal Sclerosis

Challenges in addressing epileptogenesis in advanced-stage human pathology is to distinguish contributing processes that are pre-existing abnormalities from maladaptive reorganizational alterations, promoting hyperexcitable networks. Both mechanisms may be equally relevant in hippocampal epileptogenicity and have implications for preventive or treatment strategies. In HS, we are presented with the apparent paradox that an area of chronic scarring and neuronal loss has become epileptogenic. The main studies of epileptogenesis in HS have explored the possibilities of excitatory/inhibitory imbalance and neuronal network reorganization.

Altered Inhibitory Mechanisms in Hippocampal Sclerosis

The 'GABA hypothesis' of epilepsy proposes that a predisposition to seizures is maintained because of a relative deficit of inhibitory synaptic transmission. In support of this, both structural and functional changes to inhibitory networks have been demonstrated in HS and animal models of MTLE. Alterations to inhibitory pathways can be considered at three levels: (i) changes to specific populations of inhibitory interneurons; (ii) altered GABA receptor distribution or modulation of GABA receptor subunit composition; and (iii) functional changes to GABA neurotransmission (Figures 11.7 and 11.8).

Interneurons in Hippocampal Sclerosis

Loss of principal and excitatory pyramidal neurons in specific hippocampal subfields is the diagnostic hallmark of HS, but alteration of interneuronal populations may be more relevant to the events leading to epileptogenesis. Interneurons in the hippocampus are diverse in morphology and classified in tissue sections according to their location, dendritic and axonal projections, and protein content,[157] which closely correlates with their functional characteristics. They include neurons identified with calcium-binding proteins (calbindin, parvalbumin and calretinin), neuropeptide Y (NPY) and somatostatin. The majority of interneurons are inhibitory (GABAergic) and form complex local synaptic networks that influence responses in single neurons, controlling the time of pyramidal cell firing, or synchronously orchestrating responses in groups of neurons and interneurons. This is conducted through direct perisomatic (mainly parvalbumin-positive cells) or dendritic inhibition (mainly calbindin, NPY and somatostatin-positive cells) of principal cells as well as inhibition of inhibitory neurons that can synchronize groups of cells (mainly calretinin-positive cells). The functional heterogeneity of interneuronal cell types as well the remodelling that occurs in HS has important implications for pharmacotherapy directed at enhancing GABAergic stimulation. Animal models of epilepsy can inform the time course of interneuronal changes as well as functional alterations. However, there are important differences between hippocampal interneuronal populations in humans and animals and direct comparison is not always possible.

Stereotypical patterns of interneuronal cell loss, morphological changes and axonal reorganization have been demonstrated with immunohistochemistry in HS (Table 11.6). Many changes occur in parallel with the principal neuronal loss, with more subtle interneuronal changes noted in non-sclerotic TLE cases. Axonal sprouting of interneurons is a common alteration. Easily recognized in tissue sections, it can occur in the absence of cell loss, is significant to network changes[260] and is a useful tool in diagnostic practice. Sprouting of inhibitory networks, as NPY and calretinin in the dentate gyrus, tend to parallel each other as well as sprouting of excitatory networks (the mossy fibres; see Figure 11.6).[403] Fibre sprouting likely represents a compensatory or adaptive response to seizures. Transplantation of embryonic stem cell neural progenitors into the hippocampus in mice with TLE, show their maturation into several inhibitory cell types with axonal

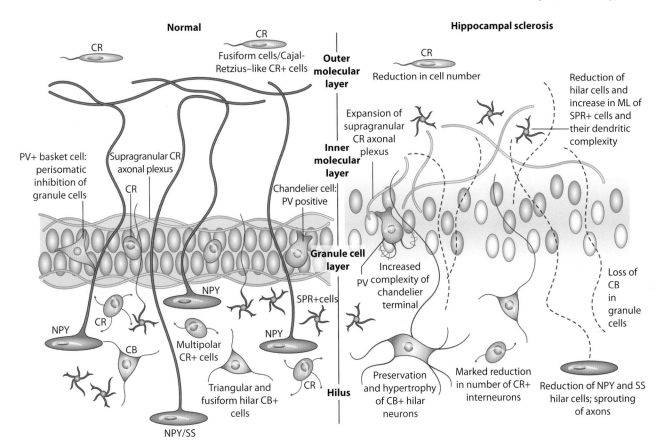

11.7 Diagram of the main morphological changes to interneuronal cell populations and fibre networks in the dentate gyrus in hippocampal sclerosis as demonstrated with immunohistochemistry and as described in Table 11.6. CB, calbindin; CR, calretinin; IN, interneuron; NPY, neuropeptide Y; PV, parvalbumin; SPR, substance P receptor positive cells; SS, somatostatin.

projections and physiological activity in the dentate gyrus, demonstrating a new potential therapeutic strategy.[264]

Reduced detection of calcium-binding proteins in histological tissues by immunohistochemistry may reflect reduced protein synthesis, altered protein conformation or true loss of cells. Furthermore, in surviving neurons, novel protein induction may occur. This may represent a neuroprotective strategy, with the buffering properties of calcium-binding proteins protecting cells from excitotoxic insults. In the calcium-binding protein knockout model, neuronal loss and susceptibility to seizures are present,[65] and experimental removal of hippocampal GABAergic cells results in hyperexcitability, CA1 neuronal loss and dispersion of dentate granule cells.[9] Neuropeptides (NPY, somatostatin, galanin, dynorphin, substance P and cholecystokinin) are known to influence neurotransmitter function, particularly glutamatergic transmission, although modulation of GABA-mediated inhibitory transmission may also occur. NPY is the most abundant neuropeptide in the CNS. Recent studies have highlighted the endogenous anticonvulsant properties of neuropeptides and their protective effects against epilepsy, mainly through inhibition of glutamate release, making these potential agents for the pharmacomodulation of seizure activity. Experimental studies confirm release, increased synthesis of NPY and somatostatin following sustained or brief seizures and increased transcription of their encoding genes.[424,425] Recent experimental studies confirm their anticonvulsant effects using viral vectors to induce

NPY overexpression in specific brain regions, demonstrating a promising therapeutic strategy.[308]

Mossy Cells

Loss of hilar mossy cells has been shown in human HS,[51] as well as in some animal models of MTLE. These multipolar cells, with their characteristic proximal thorny spine excrescences (as visualized on Golgi stain or biocytin injection) represent a significant proportion of all hilar neurons. Mossy cells are excitatory local neurons and receive afferent input from mossy fibre collaterals of granule cells and, in turn, form extensive axonal networks in the inner molecular layer of the dentate gyrus, synapsing with granule cell apical dendrites. In addition to this positive feedback loop, mossy cells also innervate inhibitory basket cells of the dentate gyrus. Thus they are unique cells in that they have both excitatory and inhibitory effects on granule cells.

Mossy cells are highly excitable neurons and considered to be particularly vulnerable to excitotoxic injury following a variety of cerebral insults including ischaemia, mild trauma and seizures.[339] These neurons have been identified in human epilepsy specimens with specific immunolabelling for mGluR7b, GluR1, GluR2/3 and CART (cocaine and amphetamine regulated transcription peptide).[51,239,427] There are relatively few quantitative studies concerning the loss of these cells in HS, but

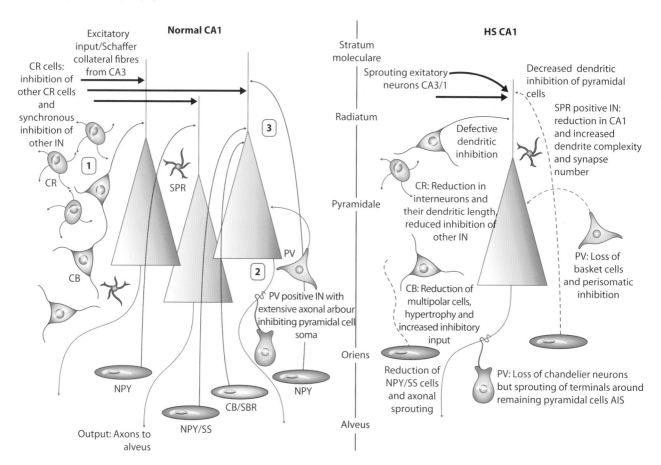

11.8 Diagram showing alteration of interneuronal number, morphology and networks reported in CA1 region in hippocampal sclerosis. This diagram illustrates some of the major changes reported that are detailed in Table 11.6. The three main types of CA1 inhibitory networks that are affected include: 1, interneuronal–interneuronal inhibition (mainly calretinin cell loss); 2, pyramidal cell perisomatic inhibition (altered parvalbumin networks); and 3, dendritic inhibition (loss of NPY and calbindin cells). This occurs in the setting of altered excitatory input from CA3 axonal collaterals and sprouted excitatory fibres from other subfields. AIS, axon initial segment; CB, calbindin; CR, calretinin; IN, interneuron; PV, parvalbumin; SPR, substance P receptor; SS, somatostatin.

recent evidence suggests that these cells are probably not more vulnerable than pyramidal neurons.[256,359] Loss of these cells is considered to be a major trigger of mossy fibre sprouting (see Enhanced Excitatory Mechanisms in Hippocampal Sclerosis, p. 703), and surviving mossy cells, because of their 'irritable' electrophysiological properties (high rates and potentiation of spontaneous excitatory postsynaptic potentials following depolarization), may in fact play a role in the amplification of seizure-like discharges in the damaged hippocampus.[339]

Changes to GABA Receptors in Hippocampal Sclerosis Classification of GABA receptors includes three major types: $GABA_A$ (ionotropic), $GABA_B$ (metabotropic) and $GABA_C$ (ionotropic). $GABA_A$ receptors are heteropentameric protein complexes that form ligand-gated, anion-selective channels, are widely distributed in the hippocampus, mainly in the postsynaptic element of inhibitory synapses, and mediate fast synaptic inhibition.[233] They are assembled from a combination of 16 possible subunits (α1–6, β1–3, γ1–3, δ, ε, θ and π) and there is coordinated expression of subunits in individual neurons, even in epilepsy. The β2, 3 and γ2 subunits are common to most $GABA_A$ receptors and the variation tends to occur in the α subunit, with different subunit composition conferring distinct pharmacological, kinetic and physiological properties. $GABA_B$ receptors are heterodimeric complexes (R1 and R2 subtypes) coupled to G-proteins and mediate slow and long-lasting inhibitory effects. Presynaptic receptors inhibit neurotransmitter release of either glutamate or GABA (which may therefore be anticonvulsant or proconvulsant), and postsynaptic receptors result in late inhibitory postsynaptic potentials (IPSPs), which may regulate interictal epileptiform discharges.[295] $GABA_{BR1}$ receptors are widely distributed in the normal hippocampus, including the granule cells and polymorphic layer of the dentate gyrus and the pyramidal cells in other subfields including CA1.[295]

Continued

TABLE 11.6 Morphological changes to hippocampal sclerosis interneuronal groups reported with immunohistochemistry in patients with TLE and HS

Cell type	Dentate gyrus and hilus[a]		CA1		Functional implications in HS
	Normal distribution	Reported findings in HS	Normal distribution	Reported findings in HS	
Calbindin	*Expression in granule cells*, their apical dendrites and mossy fibre pathway	Loss of expression in granule cells; either complete or restricted to basal cells[275] Conflicting reports regarding CB expression in sprouted mossy fibres[261,275]	*Horizontal interneurons* in stratum oriens: inhibition of proximal pyramidal cell dendrites	CB positive horizontal interneurons in stratum oriens probably better preserved than NPY/SS positive cells	Reduction of CB expression in granule cells; slower calcium entry and altered physiology
	Fusiform neurons and triangular/ multipolar neurons in hilar region and other layers	Preservation of hilar CB cells but increase in soma size of some interneurons, dendritic length and spines and sprouting[263]	*Multipolar cells with mixed morphology* in all layers with radial dendrites in stratum pyramidale	Reduction of cells in other layers	Altered CB expression in granule cells may have effects on cognition, memory, LTP and depression
		Novel CB expression in chandelier terminals noted in granule cell layer	Expression of CB reported in *pyramidal cells*	Hypertrophy and abnormal dendritic patterns in surviving CB positive cells[437]	Interneuron alterations to provide more effective inhibition in addition to enhanced synchronization of residual pyramidal cell neurons
				Increased inhibitory input to surviving CB cells (symmetric synapses) from other CB+ neurons	
				CB chandelier like terminals on AIS of surviving pyramidal cells reported[437] Loss of CB expression in pyramidal cells[260]	
Calretinin	*Multipolar cells* in polymorphic cell layer of hilus	Alteration in CR *Cajal–Retzius* neurons in molecular layer[47,415]	*Multipolar cells* in stratum moleculare and lacunosum and other layers	Reduction of 66% of cells compared to controls : no significant loss in non-HS epilepsy cases[415]	Loss of Cajal–Retzius cells in ML may result in local reelin deficiency.
	Fusiform neurons in molecular layer (some Cajal–Retzius-like cells with horizontal axis); many of these are reelin positive	Reduction of CR cells in the hilus by 50–68% ; GCL and ML by 65%[263]	*Bitufted cells* in stratum pyramidale, radiatum and oriens; many innervate other interneurons	Loss of dendrites in surviving cells	Defective network inhibition and synchronization of inhibitory cells acting on pyramidal cells ; potentiation of excitatory input
	Bitufted cells and small cells in dentate gyrus. Primarily CR cells inhibit other interneurons[415]	Expansion of CR axonal plexus into OML[263]		Reduced CR contacts on pyramidal cells, interneurons and zona adherentia[415]	
	Supragranular CR axonal plexus in IML (originating from supramammillary nucleus not local interneurons)				

11

TABLE 11.6 Morphological changes to hippocampal sclerosis interneuronal groups reported with immunohistochemistry in patients with TLE and HS (*Continued*)

Cell type	Dentate gyrus and hilus[a]		CA1		Functional implications in HS
	Normal distribution	Reported findings in HS	Normal distribution	Reported findings in HS	
Parvalbumin	*Chandelier cells*: Powerful inhibition of axon initial segment (AIS) of granule cells via chandelier terminals *Basket cells*: Perisomatic inhibition of granule cells	Reduction of chandelier cells and terminals in human and animal models in some studies but increase in complexity of terminals noted[11] Preferential survival of basket cells in some models ('Dormant Basket hypothesis' of hippocampal sclerosis)[370] Decreased PV-positive cells in hilus (CA4)/ reduction in labelling[437,448]	*Chandelier and basket cells* of pyramidal cell layer: direct Inhibition of pyramidal cell soma through extensive axonal arbors in pyramidal cell layer	Loss of PV-positive cells in region of neuronal loss and basket cell and chandelier terminals[11,436] Sprouting of chandelier terminals and basket formations around surviving neurons in CA1 borders with effective inhibition[11,436]	Hyper-innervation of remaining CA1 and GC neurons could be compensatory to balance increased excitation Reduction of paravalbumin staining may not always correlate with cell loss but loss of immunoreactivity
Neuro-peptide Y	*HIPP cells*: Located mainly in the hilus and granule cell layer; infrequent in ML. Form a dense plexus of axons in OML that synapse with granule cell dendrites Smaller proportion of NPY neurons in the hilus have a pyramidal morphology with molecular layer dendritic projections	Selective, early vulnerability of cells. Sprouting and beaded axons running perpendicular into the IML and OML, independent of hilar neuron loss[121,278,415] Increased mRNA for NPY is noted in residual hilar neurons[164]	*Horizontal interneurons* in stratum oriens and pyramidale of CA1: inhibition on pyramidal cell distal dendrites (mediate feedback inhibition)	Loss of CA1 NPY neurons but not as marked as hilar NPY neurons	Decreased dendritic inhibition of pyramidal cells. (Combined with more effective PV perisomatic inhibition and CA1 pyramidal cell sprouting – results in abnormal synchronization[436]
Somato-statin	*HIPP cells*: Deep hilus and the polymorphic cell layer; terminal axonal plexus in the outer molecular layer forming symmetrical synapses with dendrites of granule cells	Reduction in cell number Radial sprouting of fibres in the dentate gyrus[121,122,278]	Probable dendritic inhibition of pyramidal cells	Reduction in cell number but better preserved than hilar SS cells	Decreased dendritic inhibition of pyramidal cells
Substance P Receptor	Multipolar cells mainly in hilus with few in ML; hilar neurons send dendrites into ML	Reduction of interneurons in hilus and dendrites into ML; Increased multipolar cells in ML with increased complexity of dendritic branches[263]	Multipolar cells in stratum pyramidale mainly Horizontal cells in stratum oriens Co-localization with calbindin in approximately 20% Participate in dendritic inhibition in CA1[416]	Decrease in cell number and increased dendritic branches with increased synaptic input[416]	Dendritic sprouting and synaptic reorganization, independent of cell death[260]

GCL, granule cell layer; HIPP, hilar perforant pathway associated cells; HS, hippocampal sclerosis; IML, inner molecular layer; LTP, long-term potentiation; ML, molecular layer; OML, outer molecular layer.
[a]The term 'hilus' is used to denote the area between the blades of the dentate gyrus, a region that includes CA4 pyramidal cells in addition to interneurons including those in the polymorphic cell layer.

Alterations to GABA receptors have been shown in both experimental epilepsy models and human MTLE. In particular, the granule cells display remarkable plasticity including transcriptional dysregulation, changes in receptor density and subunit composition, resulting in altered kinetics, affinity and pharmacology. These changes can occur in the latent period in experimental systems and precede the onset of chronic seizures.[69,161] During status epilepticus, there is a reduction in the number of $GABA_A$ receptors. In human tissues, changes to GABA receptors have been reported in patients with TLE without HS.[165] Epilepsy itself, or possibly pharmacological agents, or altered connectivity resulting from neuronal loss, could modulate these receptors.

In addition, endocannabinoid presynaptic receptor system alterations have been found in epilepsy. This system modulates both glutamatergic and GABAergic synaptic transmission.[217] Decreases in cannabinoid type 1 receptor (CB1) and CB1 receptor binding protein mRNA and protein levels have been observed in human MTS,[256] but levels are increased on GABAergic fibres in the dentate gyrus,[262] suggesting primary modulation of GABAergic transmission.

Functional Changes in GABA Transmission The 'GABA hypothesis' of HS may extend further than defective inhibitory transmission, which is likely to be an oversimplification.[108] In addition to prevention of action potential firing, interneurons have important roles in network oscillations and neuronal synchronization, which may influence seizures more fundamentally.[21] There have also been studies demonstrating an inhibitory to excitatory shift of GABA in epilepsy,[35,101] which mirrors the functional role of GABA in the developing brain. Experimentally, the normally glutamatergic granule cells may synthesize and release GABA in epilepsy in addition to glutamate and zinc. GAD67 expression has been recently confirmed in the mossy fibres in human HS, the role of which may be to convert excessive glutamate released by mossy fibre terminals during epileptic seizures, to non-toxic GABA.[372] Furthermore, extracellular GABA concentration is also influenced by the rate of uptake of this neurotransmitter; altered levels of GABA transporters GAT-1 and GAT- 3 have been shown in TLE.[279]

Enhanced Excitatory Mechanisms in Hippocampal Sclerosis: Mossy Fibre Sprouting A converse hypothesis is that hippocampal hyperexcitability is not due to a primary loss of inhibition but, rather, excessive excitation.[123] Axonal sprouting is a common feature of the developing brain and also occurs in response to seizures, with remodelling of neuronal networks.[260] This capacity for plasticity, revived in adult tissue, is presumably a reparative response but may prove pro-epileptogenic. Abnormal connectivity and reorganization of the mossy fibre pathway was first observed in animal models and was subsequently demonstrated in human HS in a landmark paper by Sutula and colleagues.[380] The resulting enhanced excitability and synchronization of granule cells by abnormal sprouting has been argued to be a critical component in the development of recurrent seizures and HS.

Although their numbers are reduced in HS, granule cells appear relatively resistant to seizure-induced cell loss. With their high resting membrane potential, K^+ conductance and strong tonic and phasic GABA inhibition in the normal

state, they are considered a high-resistance 'gate' or filter to the propagation of seizures.[298] The mossy fibre pathway is part of the 'tri-synaptic' pathway: input from the entorhinal cortex via the perforant pathway innervates the apical dendrites of granule cells in the molecular layer, which extend their mossy fibre axons to CA4 and CA3 neurons, which in turn send (Schaffer) collateral axons to CA1 neurons. The mossy fibre axon originates from the hilar end of the soma, gives rise to collateral branches synapsing with hilar interneurons, with their main axons forming the characteristic giant (mossy) terminals on CA4 and CA3 pyramidal neurons. Normally fewer than 1 per cent of mossy fibres possess a recurrent axonal branch into the molecular layer.[348]

In animal models of MTLE (e.g. the kainate model) and in human HS, extensive recurrent projection of mossy fibre collaterals into the molecular layer of the dentate gyrus occurs, a process more commonly known as *mossy fibre spouting* (MFS). The majority (over 90 per cent) of these sprouted mossy fibres appear to make synaptic contact (excitatory asymmetric synapses)[84] with apical dendrites and spines of granule cells in the inner molecular layer. This therefore creates a recurrent excitatory circuit, potentially a pro-epileptogenic 'short-circuit'. MFS is best visualized (in both experimental and human tissue) with Timm silver method as the boutons contain high levels of zinc, sequestered in synaptic vesicles, released together with glutamate. In the normal hippocampus, dense Timm staining is seen in the hilus but not in the supragranular region. In the presence of extensive MFS, a dense confluent band of zinc-silver positive granules decorates the inner molecular layer of the dentate gyrus (see Figure 11.6). The Timm granules correspond to mossy fibre terminals on ultrastructural examination[380] and several granules may be present in a single mossy fibre synaptic terminal. MFS can also be demonstrated with immunohistochemical staining for dynorphin A (an opioid neuropeptide that is normally present in the granule cells and in the terminal fields of the mossy fibres[375]), chromogranin and synaptophysin.[326,336] In animal models of TLE, there is robust neo-expression of NPY in granule cells and sprouted mossy fibres,[299] which depresses dentate epileptiform activity. Similar expression in human MFS has not been confirmed.[403]

In animal models, MFS can be detected with Timm staining a week after status epilepticus, has been observed after a few partial seizures[379] and has been shown to precede the development of spontaneous seizures.[176] In the kindling model of seizures, in which repeated stimulation evokes progressive changes until spontaneous seizures occur, MFS occurs on day 4, increases with the development of seizures and is likely to be permanent.[83] Experimental models support the hypothesis that there is not a critical period for MFS and that the molecular cues that coordinate the axonal sprouting and synaptogenesis are sustained.[244] Indeed, in post-mortem studies, MFS has been confirmed in patients over 90 years of age.[403]

The mechanisms for MFS remain controversial. MFS has been demonstrated after a brief electrical stimulus, suggesting that abnormal electrical activity alone is sufficient. Loss of hilar neurons (mossy cells and neuropeptide inhibitory interneurons), which give rise to associational pathways in the supragranular zone, has been argued to be an important stimulus for MFS in epilepsy.[298] In this instance, MFS might be an attempt of the brain to restore function following neuronal damage. In support of this, in patients with refractory seizures due to temporal lobe lesions (tumours/old scars) but without significant CA4 cell loss, MFS is less pronounced.[278,326,336] Also, in animals developing spontaneous seizures, a correlation has been shown between the density of inner molecular layer sprouting and the severity of hilar neuronal loss.[307] However, sprouting can predate any apparent hippocampal neuronal loss in MTLE and of importance, MFS has not been reported in other (e.g. neurodegenerative) conditions with hippocampal neuronal loss. It has also been proposed that specific loss of hilar mossy cells acts as a trigger for MFS, through loss of cellular targets of mossy fibres leading to their auto-innervation. However, recent stereological studies in mouse models of TLE argue against this association.[427]

There is experimental evidence that newly generated cells form a significant contribution to mossy fibre sprouting and contribute to enhanced synchronization of granule cells.[77] New adult granule cells more often possess a basal dendrite, which provides a novel surface for innervation by recurrent mossy fibre axons.[299] Furthermore, experiments confirm mossy fibres as the main afferent input to newborn ectopic granule cells.[325] In human studies, MFS was not observed from dispersed, mature, calbindin-expressing neurons, but from more basal, phenotypically immature granule cells.[275] Expression of extracellular matrix proteins such as tenascin-C, laminin, fibronectin, phosphacan, neurocan[192] and GAP-43[335,336] may facilitate MFS in addition to N-cadherin expression, which has a function in neurite path finding and is upregulated prior to MFS.[247] There is also evidence for a role for mTORC1 in the process of MFS. Treatment with rapamycin reduces both seizures and MFS in animal models;[73,446] mTORC1 has roles in axonal growth, dendritic arborisation and synaptic plasticity.

The significance of MFS in hippocampal epileptogenicity is also controversial. There is experimental evidence to support the hypothesis that MFS acts both to synchronize GC firing and initiate or promote seizures.[298] Epileptic burst-like discharges originating from the entorhinal cortex are facilitated through mossy fibre sprouting, which may recruit and synchronize granule cell discharges.[379] Local application of glutamate to granule cells, in the presence of MFS in hippocampal slice preparations, can evoke excitatory postsynaptic potentials in remote granule cells, in support of recurrent excitatory circuits.[379] Normally, mossy fibres tend to be organized in a lamellar fashion, but in MFS aberrant fibres project longitudinally along the hippocampal axis, which facilitates longitudinal propagation of seizures. Recent experiments have shown that both MFS and epileptiform activity was blocked in a phosphatase and tensin homolog (PTEN) knockout model with rapamycin treatment.[378]

Against these arguments, however, suppression of mossy fibre growth using cyclohexamide did not affect the severity of spontaneous seizures in experimental animals,[252] and the induction of MFS in animal models is not necessarily associated with seizure development.[177,307] In patients, the diminished MFS observed in severe HS with granule cell depletion would indicate that it is not a prerequisite for temporal lobe seizures. Bilaterality of MFS, observed both experimentally[307] and in post-mortem studies in patients with long-term cessation of seizure activity,[403] support the notion that MFS is an epiphenomenon and a response to, or

facilitator of, seizures rather than a primary epileptogenic process. Sprouting of interneurons in HS has been shown in CA3 and CA1 pyramidal cell collateral axons,[243] indicating that it is a widespread phenomenon in response to seizures in surgical series. There is no evidence to support that the identification of MFS in surgical cases is an independent predictor of a more favourable seizure-free outcome.

Dual Pathology

It is well known that in a proportion of patients, HS can coexist with another lesion with more widespread epileptiform activity on EEG. This intriguing association raises the possibility that the HS is 'kindled' by epilepsy arising in the other lesion and often the observed hippocampal damage in these cases is less severe than in *isolated* HS. For strict definition purposes, the term *dual pathology* should be reserved for the confirmation of HS (either classical or atypical pattern) in association with a second pathology as a tumour, cavernoma or FCD type II. It should not include HS associated with FCD type IIIa or mild malformations of cortical development (MCD) in the temporal lobe (see Focal Cortical Dysplasia Type III). These entities were previously included as 'dual pathologies' and the quoted incidence was as high as 48–100 per cent (see Eriksson *et al.*[140] and Fauser *et al.*[148] for further details). Dual pathology also does not include the more extensive sclerosis of mesial temporal lobe structures (see Extrahippocampal Pathology with HS section, below) associated with HS or subtle anterior temporal lobe white matter changes on MR imaging.[199] Applying these criteria, true dual pathology is likely to represent 3–5 per cent of neuropathologically confirmed cases in large epilepsy surgical series.[46,188,321]

Extrahippocampal Pathology with HS ('HS-Plus')

Neuropathology studies have, from the outset, recognized that more extensive pathology may accompany HS,[82,270] the hippocampus being the epicentre of a wider process. In the recent years there has been increased interest in this extrahippocampal component and to move away from a 'hippocampocentric' view of TLE.[132,407] In the recent years this extended pathology has been studied utilizing quantitative neuroimaging methods.[61,218,246] The structures altered include those anatomically linked to the hippocampus, the amygdala, entorhinal cortex and thalamus. Proximity to the hippocampus may relate to severity of atrophy, whereas the degree of extrahippocampal atrophy may correlate with extent of HS, suggesting a common process.[291] The main rationale for wider investigations in TLE/HS are: (i) there is electrophysiological evidence to support origin of seizures from extrahippocampal structures;[209,238] (ii) the concept of wider disease may explain the surprisingly poor outcome following limited surgery; and (iii) the severity or progression of more widespread pathology may explain comorbidities, such as cognitive decline. The cellular and pathological processes in 'HS-plus' (although greatly underexplored compared to the hippocampus), typically manifest as local neuronal loss and gliosis. It remains to be established whether these more extensive changes arise as a result of the same initial 'insult' causing HS, if they are secondary to HS (i.e. a retrograde/anterograde degeneration), or if they arise independently.

Amygdala in Temporal lobe Epilepsy

Cavanagh and Meyer (1956) described amygdala sclerosis with loss of neurons in basal nuclei in association with HS, and Margerison and Corsellis reported amygdala damage in 15 out of 55 TLE patients, which was bilateral in 5 cases.[270] Gliosis and neuronal loss has been demonstrated in the lateral nucleus of the amygdala from patients with TLE, with the ventromedial aspects more severely affected.[443] In addition, basal nuclei, particularly the parvocellular division, may be involved[306] (Figure 11.9). In cases with severe neuronal loss and gliosis, the term 'amygdala sclerosis' may be applied but there is no strict histological definition for this diagnosis.[435] Amygdala sclerosis has also been reported in isolation without HS in MTLE.[435] Increased perineuronal satellitosis was noted in the lateral amygdala nuclei associated with HS, with 44 per cent representing NG-2 glia.[145] Enlargement of the amygdala has been touted as a common MR finding in otherwise 'image-negative' TLE. In one study however, pathological correlation disclosed frequent underlying pathologies, including low grade tumours and cortical dysplasias.[224]

The functional consequences of amygdala pathology are poorly understood, with pathophysiological studies largely based on experimental findings. In hippocampal–entorhinal–amygdala slices maintained *in vitro*, disconnection of Schaffer collaterals resulted in appearance of novel interictal discharges in amygdala and entorhinal cortex.[37] The basal may be more prone to generate epileptic activity than the lateral nucleus[306] and specific hippocampal–amygdala pathways may act as conduits for seizure activity.[219] In a recent study on human slices maintained from surgical resections, interictal activity recorded in the lateral nucleus was abolished with AMPA or GABA receptor blockade with altered receptor densities shown with autoradiography; this supports abnormal connectivities and a net decrease in GABA-mediated signalling, contributing to seizure discharges in the amygdala.[178] Interestingly, spontaneous discharges from the amygdala were much less likely when the amygdala was sclerosed. The amygdala is also a key structure for emotional modulation. Amygdala volume loss has been associated with dysphoric disorders in epilepsy including emotional instability[129] and high density of NPY neurons in the basolateral nuclei correlated with anxiety and depression in TLE.[160]

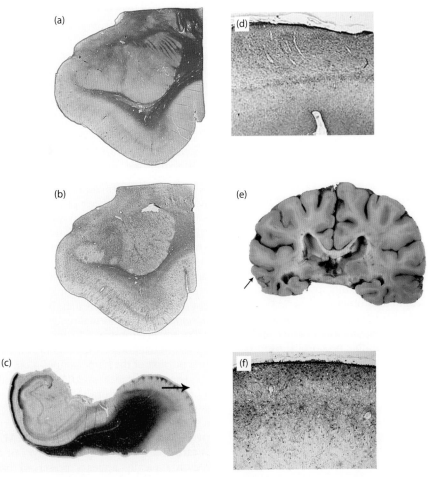

11.9 Hippocampal sclerosis (HS) 'plus'. Cases of HS with more widespread and disseminated neuropathological changes including **(a, b)** sclerosis of the lateral and basal amygdala nuclei, **(c, d)** laminar gliosis of the parahippocampal gyrus adjacent to the sclerosed hippocampus and **(e, f)** neocortical laminar gliosis in a post-mortem case with HS and unilateral hemispheric cortical atrophy. The glial fibrillary acidic protein (b,d,f) and Luxol fast blue (a,c) sections are shown.

Parahippocampal Gyrus Region and Entorhinal Cortex

The entorhinal cortex (EC) at the junction between hippocampus and neocortex acts as a conduit for incoming afferent information and reciprocal efferent signals from the hippocampus. Neurons from superficial layers (mainly layers II and III) send glutamatergic afferents, via the perforant pathway, to the dentate granule cells and CA1 neurons; subicular and CA1 pyramidal neurons have feedback connections to the deeper layers of the EC. Neuroimaging studies report volume reduction of parahippocampal gyral structures in TLE, either ipsilateral or bilateral to the seizure onset,[39,62] and abnormal epileptiform activity has been recorded in the EC,[120,209] which may sustain seizures. Studies support that the extent of parahippocampal resection (which includes the entorhinal cortex in its anterior part) dictates outcome following temporal lobe surgery.[61] Quantitative studies of the EC in HS surgical specimens report subtle and variable patterns of gliosis and neuronal loss, with destruction of an entire lamina being relatively uncommon[117,443] (see Figure 11.9). The subiculum, between the hippocampus proper and the parahippocampal cortex, appears generally well-preserved in surgical material, with no detectable neuronal loss[22,117] or alterations in synaptic density[4] in quantitative studies. Spontaneous rhythmic discharges have been reported in the subicular region from *in vitro* slices in both HS and non-sclerotic tissue.[101] As this is the sector for the major hippocampal efferent output, this might implicate this region in the spread of excitation.[123]

Neocortex

Quantitative imaging methods, including voxel-based morphometry[218,246] have confirmed cortical atrophy in association with HS and TLE in a series of patients compared to control groups. In general, temporal lobe volume reduction or atrophy is more often identified ipsilateral to the seizure focus, whereas extratemporal atrophy is more often bilateral. Extratemporal cortical changes can involve the cingulate, insular, occipitotemporal, orbitofrontal, parietal and dorsal frontal cortices,[59] with some variation in distribution dependent on the side of seizure onset. In addition, progressive atrophy has been shown in longitudinal MR imaging studies. In more extreme cases, evidence of hemiatrophy is observed in association with HS[312] (see Figure 11.9). White matter volume reduction and signal abnormalities, as well as loss of grey–white demarcation have also been reported in TLE, mainly ipsilaterally in the temporal lobe and pole, but also other regions.[352,442]

Many MR imaging studies have promoted the anatomical link between the cortical regions listed earlier and the hippocampus, favouring these limbic neocortical reciprocal network changes as a pathogenic mechanism for atrophy. Correlative neuropathological data are sparse. An early post-mortem study in patients with TLE and HS noted patchy cortical neuronal loss and gliosis involving frontal and occipital lobes in 22% of cases.[270] An investigation of widespread cortical pathology in HS with quantitative GFAP, NPY and CD68 staining, confirmed a lack a lateralization to the side of sclerosis and preferential frontotemporal polar atrophy. Subtle traumatic damage (as a consequence of head injury during seizure) rather than trans-synaptic hippocampal pathway degeneration, was proposed as the most likely mechanism.[45] The pathological correlate of white matter abnormalities on imaging remains poorly understood[139] and may relate to myelination abnormalities and reduction of white matter axons; in a recent series its presence did not influence postoperative outcome.[352]

Thalamus

The main hippocampal output pathway is through the fornix to frontal lobe, anterior nucleus of the thalamus and mammillary bodies; the dorsomedial nuclei of the thalamus also receive input from the amygdala and hippocampus with reciprocal connections with the prefrontal cortex. There are studies demonstrating associated ipsilateral[59] as well as bilateral thalamic atrophy[246] in quantitative MRI studies of patients with unilateral HS. Experimental data support a role for the mediodorsal thalamus in the induction and modulation of limbic seizures,[40] and synchronous discharges in the thalamus are regarded as a negative prognostic factor in TLE, indicative of wider network changes.[183] Margerison and Corsellis demonstrated thalamic atrophy post-mortem in 14 of 55 patients with TLE, more often in patients with HS and bilateral in 3 patients.[270] They did not notice any regional pattern of thalamic atrophy. A reduction in forniceal axon numbers has also been shown.[315]

Contralateral Hippocampus

There is evidence from imaging studies, that even in apparent unilateral MTLE there may be subtle atrophy in the contralateral hippocampus.[10,62,218] HS is reported to be bilateral PM in 48–56 per cent of epilepsy cases[283,401] and is often asymmetrical. In bilateral cases, the pattern and distribution of atrophy may determine which side is the 'generator' and which the 'receiver' of seizures.[60]

Hippocampal Malformations: Rotational Abnormalities

Rotational malformations, termed 'hippocampal malrotation' (HIMAL) or 'incomplete hippocampal inversion' (IHI), have been noted in epilepsy series studies. Usually studied by MR imaging, features include an abnormal round appearance, blurred internal structures, a vertical orientation of the collateral sulcus, mild dilation of the temporal horn and deviation of the fornix; the volume loss and signal changes typical of HS are absent. In many cases HIMAL is unilateral and isolated, but it may be bilateral or associated with developmental abnormalities, such as agenesis of the corpus callosum, heterotopia, polymicrogyria[28,357] or increased complexity of temporal cortex folding. HIMAL may be restricted to a portion of the hippocampus along its longitudinal axis.

The significance of HIMAL in epilepsy, in particular TLE, is still conjectural. Some imaging series suggest that it is more prevalent in epilepsy patients[28,167] but others show a similar frequency in healthy volunteers.[23] There are relatively few pathology studies of HIMAL. Superimposed HS is rare and the pyramidal cell layer of the CA1/subiculum region typically appears preserved but hyperconvoluted and 'heaped up' (Figure 11.10). Hippocampal malrotations do not necessarily evolve into HS,[357,369] and evidence of mossy fibre sprouting, GCD or the interneuronal alterations that typify HS are lacking.[31,357,409]

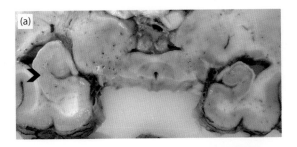

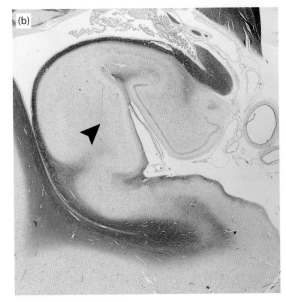

11.10 Hippocampal malrotational abnormality. (a) Macroscopic appearances with relatively normally shaped hippocampus on the right whereas the left side shows a more upright globular hippocampus with a vertical alignment of the subiculum (arrow). **(b)** A cresyl violet/Luxol fast blue stained section with complex, convolutional folding of the pyramidal cell layer in CA1 and subiculum (arrowhead) and incomplete inversion of the hippocampus proper into the dentate.

Malformations of the Cortex and Epilepsy

Malformations of cortical development (MCDs) occurring in patients with epilepsy have been increasingly recognized since the mid-1980s as a result of modern neuroimaging. Genetic studies have advanced our understanding of the aetiology of these lesions in the light of normal cortical development.[26] The range of MCD is enormous, with many malformations presenting early in life with developmental delay, encephalopathy or congenital motor problems; these are discussed in Chapter 4. In patients with more generalized MCDs in which epilepsy is the major clinical symptom, effective surgical treatment options are limited and the neuropathologist is more likely to encounter these lesions at autopsy. Localized or focal MCDs, by contrast, are more amenable to resection, can represent up to 50 per cent of epilepsy surgical specimens (see Table 11.3) and are therefore discussed in further detail here.

Focal Cortical Dysplasia

FCD was first introduced as a term by Taylor and colleagues over 40 years ago, for an epilepsy associated lesion that they considered distinct from the cortical lesions observed in tuberous sclerosis.[391] In the subsequent decades a wider variety of localized lesions became subsumed under the umbrella of 'cortical dysplasia' with the inclusion of varying abnormalities recognized with the increasing use of MR imaging in the assessment of patients with epilepsy, as well as more subtle abnormalities of cortical layering identified histologically. Various classification schemes have evolved over the years, which segregate dysplasia into subtypes based on their histological features with the aim to improve consistency in reporting among centres (Table 11.7). This has resulted in improved clinicopathological–radiological correlations with clear differences emerging among FCD subtypes on which to base informed, evidence-based preoperative decisions.[368] The Palmini system of 2004 divided cases into *FCD types I*

TABLE 11.7 Relative incidence of focal pathologies in more recently reported epilepsy surgical series

	NEUROPATHOLOGICAL FEATURE							
Classification	Balloon cells	Dysmorphic neurons	Hypertrophic neurons	Immature neurons	Cortical dyslamination	Excess of white matter neurons	Excess of layer I neurons	Cortical laminar abnormalities adjacent to a lesion
Taylor *et al.* (1971)[391] Meencke & Janz (1984)[282]	Focal cortical dysplasia					Microdysgenesis		
Tassi *et al.* (2002)[388]	'Taylor type' cortical dysplasia	Cytoarchitectural dysplasia				Architectural dysplasia		
Palmini *et al.* system (2004)[317]	FCD IIB	FCD IIA	FCD IB		FCD IA	Mild malformations of cortical development (MCD) type II	Mild MCD type I	FCDI (rarely FCD II)

Continued

TABLE 11.7 Relative incidence of focal pathologies in more recently reported epilepsy surgical series (*Continued*)

Classification	Balloon cells	Dysmorphic neurons	Hypertrophic neurons	Immature neurons	Cortical dyslamination	Excess of white matter neurons	Excess of layer I neurons	Cortical laminar abnormalities adjacent to a lesion
Blumcke *et al.* (2011)[53] ILAE classification 2011	FCD IIB	FCD IIA	FCD IA = abnormal radial lamination FCD IB = abnormal tangential lamination FCD IC = abnormal radial and tangential lamination			Mild MCD type II	Mild MCD type I	FCD III A to D (see bottom row)
FCD IIIA		FCD IIIB		FCD IIIC		FCD IIID		
Cortical lamination abnormalities in the temporal lobe associated with hippocampal sclerosis		Cortical lamination abnormalities adjacent to a glial or glioneuronal tumour		Cortical lamination abnormalities adjacent to vascular malformation		Cortical lamination abnormalities adjacent to any other lesion acquired during early life e.g. trauma, ischaemic injury, encephalitis		

FCD III (not otherwise specified, NOS): if a clinically/radiologically suspected principal lesion is not available for microscopic examination. Rare association of FCD type IIB with hippocampal sclerosis, tumours or vascular malformations should not be classified as FCD III.

and *II*.[317] The most recent ILAE classification of 2011 introduces a third subtype, *FCD III*, which separates dysplasias adjacent to lesions from isolated dysplasias.[53]

The best definition of FCD is a presumed developmental cortical malformation of the cortical plate, associated with clinical seizures, manifesting as abnormal cytoarchitecture, restricted in extent and with preservation of the gyral pattern. FCD represents the third most common structural abnormality in adult epilepsy surgical series after HS and tumours, comprising up to 13–20 per cent of cases, and is the most common abnormality in paediatric epilepsy series, representing 40–50 per cent of resected lesions[46,188,231] (see Table 11.3). The relative incidence of the FCD subtypes among series varies (Table 11.8); FCD I is the more commonly reported subtype but with remarkable variation in incidence relative to FCD II. Although there may be several reasons for this, it has been shown that interobserver concordance is greater for FCD II than FCD I or mild MCD.[91] Initial studies utilizing the new 2011 classification support improved consistency in the reporting of FCD lesions.[105]

Focal Cortical Dysplasia Type II

Clinical and Neuroradiological Features There is a strong association of FCD II with epilepsy; it has not been reported as an incidental, asymptomatic abnormality. Most patients with FCD type II have medically uncontrollable partial epilepsy, typically originating in childhood, and often with motor or secondary generalized seizures, episodes of status epilepticus or epilepsia partialis continua. There is a wide variability in associative cognitive impairment, with several series reporting it as a rare finding.[96] There are no clear clinical differences between FCD IIa and IIb.

MR imaging abnormalities are more frequent in type II than type I FCD. These include cortical thickening or grey–white matter blurring on T1-weighted images and abnormal signal intensity in both the cortex and white matter on T2 and FLAIR sequences.[265] The white matter signal changes frequently taper towards the ventricle (the transmantle sign; Figure 11.11). In Taylor's original series, the FCD lesions were preferentially located in the temporal lobe. Recent studies support that surgically resected FCD II is more often *extratemporal*, favouring the frontal lobe (55 per cent), around the central sulcus (16 per cent), insular cortex (5 per cent), parietal lobe (10 per cent), occipital lobe (7 per cent), temporal lobe (7 per cent) and multilobar (22 per cent)[96,102,232,241,429,433] and therefore often involving functional cortex.

The extent of the abnormality on imaging (and reflected in histological specimens) is variable; it can be limited to one gyrus, with a predilection for the bottom of the sulcus in some cases,[198] or show more extensive involvement of one lobe over several gyri. Multilobar or extensive hemispheric FCD is more common in patients less than 4 years old, but multilobar and discontinuous lesions of FCD II have also been rarely reported in adults and even involving opposite hemispheres in the absence of *TSC* gene mutations.[149] Of note, the MRI in pathology-proven FCD type II can be subtle or even negative. In MRI negative cases, invasive intracranial EEG recordings (SEEG), magnetoencephalography and functional imaging including FDG-PET may be used to better localize the lesion and the extent of the epileptogenic zone in relation to any undefined structural abnormality, prior to resective surgery. A characteristic EEG pattern reported in association with FCD II is continuous or subcontinuous rhythmic spikes or sharp waves.[317] Interictal activity in FCD II shows characteristic repetitive high amplitude fast spikes and slow waves.[53] fMRI in conjunction EEG may better identify and map the active seizure onset zones and the extent of resection required to remove all abnormal networks.

Surgical management is often indicated in FCD II and evidence supports an equally good outcome for surgery in adulthood as in childhood.[99] Completeness of excision (as assessed by postoperative MR imaging or histological

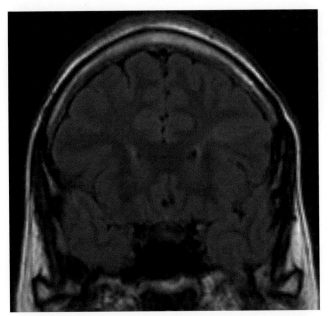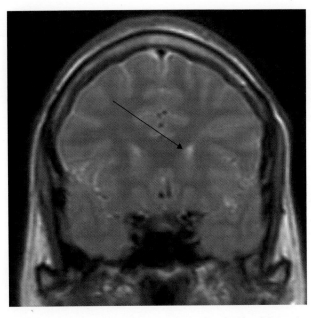

11.11 Magnetic resonance image of focal cortical dysplasia type IIB. The left frontal lesion is seen in the depth of the sulcus on T2 sequences (FLAIR on left) with a fine tail-like extension towards the lateral ventricle (arrow). This corresponds to the lesion resected and shown in Figure 11.12c–f.

margins) is associated with a better outcome.[99,223,230] It has been shown that excision of the cortical component, but not completeness of the white matter component, is necessary for seizure freedom.[428] Clinical improvements in seizures have also been reported for operated MR imaging-negative FCD cases.[95] Recent series report seizure-free incidences of up to 75[231] to 87.5 per cent.[223] Follow-up studies generally confirm more favourable outcome for FCD II than FCD I (see Table 11.8).

Pathology FCD type II may be visible macroscopically in surgical resected tissues as an area where the cortex appears thicker and less well defined from the white matter (Figure 11.12). The extent of cortex involved by FCD type II can be variable; it can be limited to the depth of a sulcus or extend for several centimetres along the cortical ribbon. Regions of FCD can be discontinuous with intervening regions of more normal cortical lamination.

The microscopic features that characterize cortical dysplasia in conventionally stained sections (H&E, cresyl violet/Luxol fast blue and silver stains) were first described by Bruton and Corsellis in a series of 10 patients undergoing surgery in 1971,[391] and remain the key diagnostic criteria. There is loss of the normal six-layered neocortical cytoarchitecture, with layer I remaining relatively cell free, although it may be broader than normal. The junction between the deep cortical layers and white matter is often ill defined. A lack of the normal radial, columnar organization of cortical neurons is also apparent. These cortical cytoarchitectural abnormalities are strikingly visualized with NeuN staining. Pallor of the white matter and myelin rarefaction is common, in some cases limited to the immediate subcortical white matter (see Figure 11.12) or trailing into the white matter towards the ventricle. Gliosis is variable but can be marked, and microgliosis may be evident. Rosenthal fibres are a rare finding in FCD II.

Profoundly abnormal cortical cell types are present in FCD II and define its subtypes (Figure 11.13; see Table 11.9). Dysmorphic neurons are present by definition and have increased soma size (16–43 μm compared to 12–25 μm diameter for normal pyramidal neurons),[53] abnormal shape and orientation. They mostly resemble pyramidal neurons but display abnormal dendritic patterns with tortuous, thick dendrites with decreased spine density. In cresyl violet stained sections, the Nissl substance appears abnormally clumped and eccentric thickening of nuclear membranes can be seen. Dysmorphic neurons are present throughout the cortex (but usually not layer I) and often trail into the underlying white matter. The cortical neuronal density may be altered and neurons with a normal appearance are interspersed among dysmorphic cells. Cytomegalic interneurons, as stained with calbindin, are also represented in the cortex and white matter in FCD II. Abnormal polarity of dysmorphic neurons ranges from slight rotation to complete inversion in relation to the pial surface.[90] In some cases dysmorphic neurons in the superficial cortex retain a more pyramidal morphology (hypertrophic pyramidal neurons) and in deeper layers display a more globose morphology with bipolar processes (see Figure 11.13). The dysplastic region fades out to normal cortex but scattered single dysmorphic or hypertrophic neurons may be at some distance away from the main lesion.

Balloon cells are the hallmark of type IIB FCD, which in most series is more common than type IIA (Table 11.9). Balloon cells have large round soma, with diameters of 20–90 μm and are generally larger and rounder than reactive astrocytes. The nucleus is often eccentric and the cytoplasm is pale pink and glassy in H&E preparations. Multinucleate giant cell forms and cells with polylobulated nuclei are frequent. Cells with an intermediate appearance, between balloon cells and neurons, can be identified. Following biocytin injection in slice preparations, typical balloon cells lack axons and dendritic spines. They tend to

TABLE 11.8 Relative incidence of focal cortical dysplasia (FCD) subtypes[a]

Series (n = number in entire epilepsy series where given)	FCD IA (% of FCD)	FCD IB (% of FCD)	FCD IIA (% of FCD)	FCD IIB (% of FCD)	Mild malformations of cortical development	Outcome postsurgery[b] (% seizure free)
Widdess-Walsh et al. (2005)[433] (n = 145)	n = 76 (52%)	n = 21 (14%)	n = 16 (12%)	n = 32 (22%)	Not stated	FCD IA=61%; FCD IB=38% FCD IIA=67%; FCD IIB=80%
Hudgins et al. (2005)[206] (n = 106)	n = 9 (60%)			n = 6 (40%)	Not stated	FCD IIB=100%; FCD I= 44%
Krsek et al. (2008)[231] (Miami)[c] (n = 567)	n = 55 (34%)	n = 39 (24%)	n = 35 (21%)	n = 35 (21%)	Type I n = 0 Type II n = 36	FCD IA=49%; FCD IB=45% FCD IIA=61%; FCD IIB=75%
Kim et al. (2009)[223]	n = 89 (61%)	n = 28 (19%)	n = 12 (8%)	n = 16 (12%)	n = 21	FCD IA=51.7%; FCD IB=60.7% FCD IIA=66.7%; FCD IIB=87.5%
Krsek et al. (2009)[232] (Erlangen) (n = 40)	n = 24 (60%)		n = 3 (7%)	n = 13 (33%)	Not stated	FCD II=75%; FCD I = 21%
Blumcke (2009)[46] (n = 4512)	n = 62 (22%)		n = 44 (16%)	n = 173 (62%)	Not stated	
Piao et al. (2010)[324] (n = 435)	n = 173 (81%)		n = 43 (19%)		n = 14	
Wagner et al. (2011)[428]	n = 0 (0%)		n = 17 (19%)	n = 74 (81%)		FCD IIA=59%; FCD IIB=74%
Average relative incidence of FCD subtype	**46%**	**19%**	**15%**	**39%**		**FCD IA=45%; FCD IB=49% FCD IIA=65%; FCD IIB= 84%**

[a] Reported in some larger epilepsy surgical series in which the Palmini system for classification has been applied or could be interpreted from the histological descriptions. These are mainly isolated dysplasias but in the Krsek[232,235] studies, hippocampal atrophy was noted in a proportion of cases on magnetic resonance imaging.

[b] Engel classification for postoperative outcome most often used (class 1 =Free of disabling seizures and 1A, completely seizure free after surgery). The follow up periods vary among series but was 2 years or more in the majority. Overall, FCD IA is the most commonly reported dysplasia type, although there was marked variation and in some centres no cases of FCD I were identified. In most series FCD IIA was less frequently reported than FCD IIB.

[c] Indicates paediatric only series.

aggregate in deeper cortical layers, imparting a 'moth eaten' appearance to the cortical–white matter junction, often in clusters in regions of poor myelination. Balloon cells can be the predominant cell type in dysplasia extending throughout the cortex, particularly layer I.

Maturation, Lineage and Developmental Origin of Abnormal Cell Types in FCD IIB Although these abnormal cell types are instantly recognizable in routine sections, immunohistochemistry allows further confirmation, particularly in smaller resections. Neurofilament gene expression has been shown to be augmented in FCD neurons.[392] Striking immunoreactivity of dysmorphic and hypertrophic neurons may be observed with non-phosphorylated, and less so with phosphorylated, neurofilament antibodies (see Table 11.9), which highlights their abnormal morphology, alignment and laminar position, equivalent to older silver preparations (see Figure 11.13). Neurofilament immunohistochemistry also demonstrates excessive and haphazard cortical networks of axons and coarse dendrites (see Figure 11.13).

Developmentally regulated cytoskeletal proteins are also strongly expressed in the abnormal cell types in FCD II (Table 11.9), including nestin, vimentin and Map1b. GFAP expression is variably expressed in balloon cells. The minor isoform GFAP δ, with roles in intermediate filament assembly and cell motility, usually restricted in expression to ventricular and marginal zone astrocytes, shows more prominent expression in balloon cells than conventional GFAP.[274] Studies of expression of stem cell and cortical layer-specific markers, to explore the maturity and lineage of the abnormal cell types in FCD II, have been carried out

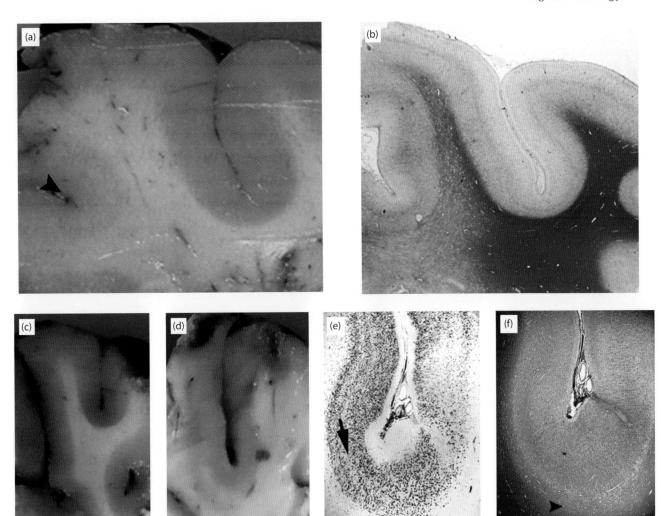

11.12 Macroscopic appearances of focal cortical dysplasia type IIB in surgical resection specimens. (a) The depth on one gyrus shows poor demarcation between the grey and white matter (arrowhead) compared to adjacent gyri. **(b)** On corresponding myelin-stained sections this correlates with increased myelin staining in the cortex and rarefaction of the underlying white matter myelin. **(c)** and **(d)** Slices from the same resection showing a normal cortical ribbon in (c), but in (d) the cortex appears poorly defined. This correlates to a region of dysplasia in the sulcal depth as seen on **(e)** NeuN staining (arrow) with loss of lamination, and **(f)** myelin stain in demonstrating subcortical myelin pallor (arrowhead).

(see Table 11.9). Coexpression of neuronal and glial markers by abnormal cell types has been demonstrated,[137,390] confirming aberrant neuroepithelial differentiation. There is evidence that balloon cells immunophenotypically most resemble radial glia, and dysmorphic neurons resemble intermediate progenitor cells/cortical pyramidal cells.[185,237]

Regarding growth regulation in FCD II, FGF-2 expression has been shown in dysmorphic cells in FCD IIB but not FCD IIA,[418] which could support developmental differences between these pathologies. In addition, vascular endothelial growth factor and its receptor have been shown in abnormal cells in FCD II indicating a potential autocrine effect.[58,311] Aberrant expression of apoptotic proteins Bcl-X$_L$, Bcl-2 and Bax has been shown in dysmorphic cells and balloon cells in FCD.[92] Studies of cell cycle proteins, including MCM2 and p53, confirm proliferative capacity of balloon cells, although they appear arrested in the early G1 phase, failing to either divide or die.[404,405] Reduced expression of CDK1 and cyclin A2 in FCD II also indicates a

decreased propensity for cells to enter the mitotic phase.[350] Recent *in vitro* studies of isolated balloon cells has confirmed that despite introduction of growth factors (such as epidermal growth factor and basic fibroblast growth factor), they do not pass from G1 to S-phase, supporting cell cycle arrest[441] and suggesting that they are a pathological or defective progenitor/stem cells. Neurofibrillary tangles and aberrant phosphorylation of Tau protein has been shown in the dysplastic neurons of older adults with FCD IIB, but not balloon cells (see Figure 11.13), suggesting vulnerability to premature ageing that might be mediated by dysregulation of CDK5.[358]

Aetiology FCD is a sporadic disorder without a known hereditary form. In terms of experimental models, there is no animal model that precisely resembles the human phenotype.[438] Current theories favour the concept that FCD II is a developmental abnormality arising in the later stages of gestation, or even in the postnatal period, with impaired

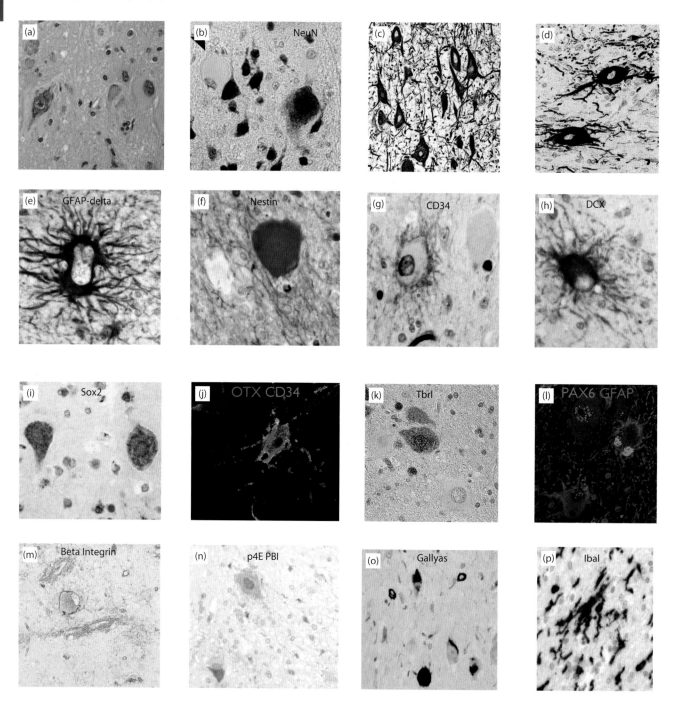

11.13 Cell types in focal cortical dysplasia type IIB (FCD IIB). **(a)** H&E section showing dysmorphic neurons and balloon cells and **(b)** NeuN staining highlighting the cytomegalic neurons with negative labelling of the balloon cells (arrow). **(c)** Neurofilament staining in layer II shows the mainly radial orientation of the dysmorphic cells, distended with neurofilament; also there is an increase in neurofilament positive processes (SMI32). **(d)** Neurofilament shows a more globoid, horizontal morphology of dysmorphic neurons in the deeper cortical layers. Balloon cells show positive labelling with developmental markers including **(e)** GFAP-delta, **(f)** nestin, **(g)** CD34, **(h)** doublecortin, and dysmorphic cells with **(i)** Sox2, **(j)** Otx and CD34, **(k)** Tbr1, **(l)** Pax6 and **(m)** beta-integrin. **(n)** In addition, activation of mTOR pathway proteins is seen in FCD II cells. **(o)** In older patients with FCD II neurofibrillary tangle pathology may be observed in the dysmorphic neurons. **(p)** In addition, in FCD II cases upregulation of microglial may be prominent. (See also Table 11.9 for details of proteins.)

Sections for figures (m) and (n) provided courtesy of Dr T Jacques, Great Ormond Street Hospital, London, UK.

TABLE 11.9 Immunohistochemistry profile highlighting the main cell types in focal cortical dysplasia (FCD) in relation to the proposed lineage of the cell type, maturation, abnormal function and proliferative status

Dysplasia type Cell type	FCDI	FCD II			FCD III
		Balloon cells	**Dysmorphic neurons**	**Other cell types**	
Mature, cell lineage markers		GFAP ↓ or → (small proportion co-express neuronal markers)	NP-NF (e.g. SMI31) ↑ P-NF (e.g. SMI32) ↑ NeuN →	Calbindin positive neurons → ↓ Parvalbumin-positive neurons → ↓ Calretinin-positive neurons→ ↓	FCD IIIa- NP-NF (SMI32) ↑ (layer II)
Cell immaturity; developmentally regulated and stem cell markers	DCX (layer I/II in younger patients) ↑ c-myc, Oct-4, FoxG1,KLF4, Nanog SOX2 and 3 →	Nestin, GFAP-delta ↑ DCX ↑, DCX-like protein, vimentin ↑ β1-integrin, Musashi-1 ↑,CD34 ↑ CD133 ↑, Pax6 ↑, SOX2 and 3, Oct-4↑ CRMP4, c-myc, FoxG1,KLF4, Nanog ↑[147,274,311,441]	Otx1↑, Pax6↑ DCX↑ DCX-like protein, nestin→ ↑ MASH1 ↑ SOX2&3, Oct-4↑ CRMP4 c-myc, FoxG1,KLF4, Nanog ↑		FCD IIIa-DCX multipolar cells layer II
Cortical layer markers	Otx1,Tbr1 ↑ (Layer II in younger patients)[185] RORBETA (layer V) ↑[342] ER81 (Layer IV-VI) ↑ NURR (layer V-VI) ↑	ER81 ↑ Otx1 ↑[185]	Tbr1 ↑, Map1b ↑		FCD IIIa-Tbr1, Map1↑ (layer II)[185] FCD IIIa-Parvalbumin ↓ (midcortical layers)[169] FCD IIIa- CUX2 (Layer II) ↑ RORBETA (layer V) ↑ ER81 (layer IV-VI) ↑ NURR (layer V-VI) ↑
Cellular transmission		SV2A ↓ GLT1 ↑ GS ↑ NR1. NR2A/B↑ GluR1/R3↑[175,310]	Synaptophysin ↑ VGlut ↑ NKCC1 (chloride channel) ↑,[14] AQP4↑ Cx43[168] NR2B↑ EEAT3 ↑		
Cell signalling cascades	mTOR pathway →	pS6, p4E-BP1 (mTOR pathway activation) ↑ pAKT (PiK3 pathway) ↑	Cdk5↑		
Inflammatory mediators/ signalling		AMOG ↑[57] Toll-like receptor2 ↑ IL-1β ↑	C1q and C3d (complement pathway) ↑, IL-1β ↑ Toll-like receptor 4 ↑	HLA-DR+ microglia ↑	
Growth factors		VEGF-A[58] FGF-2[418]	VEGF-A, B and VEGFR1 and 2 ↑		
Proliferation status		Mcm2 ↑ Cdk4, p53 ↑[404] bcl-X$_L$, bax, bcl2 ↑[92]	bcl-X$_L$, bax, bcl2 ↑		
Neurodegeneration			Phosphorylated Tau ↑ (in older patients)[358]		

↑, increased expression compared to normal adult cortex; ↓, reduced or no expression; →, expression not altered. AMOG, adhesion molecule on glia; AQPA, aquaporin; cdk5, cyclin-dependent kinase 5; CRMP4, collapsin response mediator protein 4; Cx43, connexin 43 gap junction protein; DCX, doublecortin; EEAT3, excitatory amino acid transporter; ER81, ETS transcription factor; FGF, fibroblast growth factor; FoxG1, winged helix transcription factor forkhead box G1; GLT1, glial glutamine transporter; GluR1/3, AMPA receptor subunit proteins; GS, glutamine synthetase; KLF4, Kruppel-like factor 4; Map, microtubule-associated protein; MASH1, mammalian achaete scute homolog-1; Mcm, mini chromosome maintenance protein; NKCC, sodium/potassium chloride co-transporter; NR1, NMDA receptor subunit protein 1; NR2A/B, NMDA receptor subunit protein 2A/B; NP-NF, non-phosphorylated neurofilament; Oct-4, octamer-4; Otx1, orthodenticle homolog 1; Pax6, paired box 6; P-NF, phosphorylated neurofilament; SM, Sternberger monoclonal antibody; SOX, sex-determining region Y-box; SV2A, synaptic vesicle protein 2A; Tbr, T-box-brain-gene; VEGF, vascular endothelial growth factor; vGlut, vesicular glutamate transporter.

cortical maturation and failure of normal cell differentiation. Abnormal retention of immature subplate cells in FCD has also been proposed through comparisons with the normal maturing brain.[8,88,277]

The histological similarities between FCD type II and cortical tubers in tuberous sclerosis (TSC) have led to investigations of common abnormalities of intracellular signalling pathways (see Table 11.10). TSC is caused by mutations in

TABLE 11.10 Distinction of focal cortical dyspasia (FCD) type IIB from tuberous sclerosus (TSC) lesions in focal epilepsy

	FCD IIB	Tuber in TSC
Localization	Tend to be solitary and localized; rarely multifocal	Typically multiple
Associated CNS pathology reported	Usually an isolated pathology Hypertrophy of the olive,[391] hippocampal sclerosis, LEATs occasionally reported as a dual or double pathology	SEGA Subependymal nodules Evidence for widespread microscopic cortical changes between tubers in magnetic resonance imaging[94] and post-mortem studies[268]
Other organ pathology	Absent[182]	Facial angiofibroma, periungual fibroma, hypopigmented macules, shagreen patch, multiple retinal nodular hamartomas, cardiac rhabdomyomas, lymphangioleiomyomatosis, renal angiomyolipoma
Clinical presentation/diagnosis	Virtually always associated with drug-resistant epilepsy Childhood to adult onset Cognitive deficits infrequent	Epilepsy in ~80%; drug-resistant in 50–80% Typically early onset seizures Epilepsy most common presenting symptom Infantile spasms in one-third of cases Semiology changes with age Association with autism, developmental delay, mental retardation
Genetics	TSC1 and TSC2 polymorphisms but no loss of function mutations	TSC1 or 2 mutations; no mutations in ~15%
Gross features	The cortical/WM boundaries are blurred and cortex may appear 'thicker'	Tuber often protrudes from the cortical surface as a hamartomatous growth
Cortical cellularity	Variable cellularity Impression of hypercellularity due to enlarged neurons	Paucicellular with bizarre neurons diffusely spaced[94]
Undifferentiated cells types	Balloon cells usually less predominant than dysmorphic neurons Balloon cells more typically located in white matter	Balloon cells (a.k.a. giant cells) predominate over dysmorphic neurons Balloon cells more often extend into cortex in addition to white matter
Astrocytic reaction	Variable astrogliosis	'Wheatsheafs' of superficial astrocytes; gliosis may be marked
Dystrophic tissue calcification	Rare/absent	May be present
mTOR pathway deregulation	Activation through PI3K/AKT pathway (unknown mechanism) ↑ p-S6 expression	Activation through loss of function of hamartin/tuberin complex ↑ p-S6, pS6K expression
Differences reported in immunohistochemistry patterns	αV integrin not expressed ; only β1 integrin ↑	αV integrin subunit expression↑[441]
Cellular electrophysiology within lesions	GABA: Glutamate synaptic activity higher; more robust responses to GABA[87] Electrophysiological properties of balloon cells similar in FCD and tubers	Spontaneous excitatory synaptic activity increased in TSC cortical neurons compared to FCD IIB[89] GABA: Glutamate synaptic activity lower; less robust responses to GABA[87]
Response to surgical treatment	Frequent improvement	More variable improvement but clear benefits from surgery (~50% seizure free)[143]

LEAT, Long-term epilepsy-associated tumour; SEGA, subependymal giant cell astrocytoma; WM, white matter.

one of two non-homologous tumour suppressor genes: *TSC1* (9q34) encoding hamartin and *TSC2* (16p13.3) encoding tuberin. There are approximately 700 allelic mutant gene variants and a clear genotype–phenotype correlation has not been established, although patients with *TSC2* mutations are often more severely affected.[110] As the clinical phenotype of TSC is highly variable, it has been called into question whether FCD IIB could represent a '*form fruste*' of TSC. Polymorphisms of *TSC1* have been identified in FCD type IIB lesions; in a cohort of 48 cases, two-thirds showed sequence alterations in *TSC1*, with two (in exons 5 and 17) resulting in amino acid exchange.[34] *TSC1* polymorphisms were not noted in FCD IIA samples,[265] although polymorphisms have not been replicated in all FCD IIB studies.[182] A similar polymorphism of exon 41 in *TCS2* has recently been demonstrated in FCD IIB, gangliogliomas and also 'non-lesional' epilepsy surgical cases, potentially supporting a more general role for this gene in epileptogenesis.[353]

Tuberin and hamartin are upstream regulators of mTOR (mammalian target of rapamycin), a highly conserved protein kinase that regulates protein synthesis, cellular metabolism, differentiation, growth and migration (Figure 11.14a).[142] In cortical tubers, mTOR activation has been demonstrated by excess phosphorylation of the p70-S6 kinase, ribosomal S6 protein and 4E-BP1 (eukaryotic initiation factor 4E binding protein-1) in balloon cells (also termed giant cells), and dysmorphic neurons compared to control cortex, using immunohistochemistry for phosphorylation specific epitopes. Aberrant mTOR signalling has also been demonstrated in FCD II with phosphorylation of S6 and eIF4E (eukaryotic translation initiation factor binding protein 1) in abnormal neurons and balloon cells (Figure 11.14b).[33,288] mTOR activation in FCD IIB likely occurs upstream in the PI3K-pathway[351] through activation of pAKT or PTEN,[349] although the exact mechanisms have not been determined and in one study pAKT activation in FCD was not confirmed.[288] A recent study has identified human papilloma virus (HPV) type 16 E6 protein in balloon cells in FCD IIB but not in other cell types or in TS lesions; as E6 protein is an activator of mTORC1 it is proposed that HPV infection during development could be the cause of FCD II.[97]

The functional importance of the mTOR pathway has been demonstrated through prevention of seizures in conditional knockout *TSC1* animal models that show some of the pathological features of TSC, by treatment with rapamycin, an mTOR inhibitor. Early clinical trials of mTOR inhibitors in the treatment of TSC are also encouraging with reduction in lesion size. However, the pathogenic mechanisms causing the development of the structural and cellular abnormalities of FCD and TSC are not necessarily the same as those promoting epileptogenesis; further study of the mTOR pathway influencing these processes is warranted.[440]

A further possibility is that FCD is a result of cortical reorganization and plasticity following a prior cerebro-cortical insult including early trauma.[113,228,251,271] Seizures themselves may also contribute to ongoing cytopathology. For example, in the cortical malformation model following prenatal exposure to methylazoxymethanol acetate, pilocarpine-induced status epilepticus results in further neuronal hypertrophy, cell clustering with excess neurofilament and increased NMDA NR2A/NR2B receptor subunit expression, replicating some of the changes observed in human

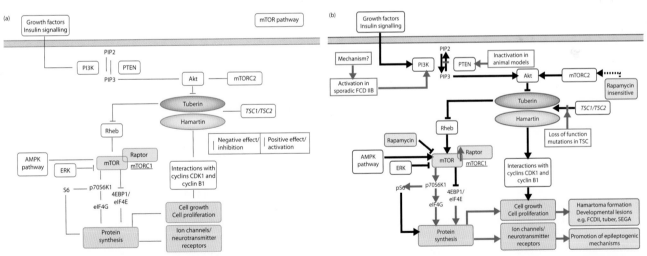

11.14 mTOR pathway in tuberous sclerosis and cortical dysplasia. (a) A complex series of interactions regulate hamartin and tuberin and activity of the mTOR pathway (mTORC1). Growth factors binding to transmembrane receptors activate PI3-kinase, and PIP3 (phosphatidylinositol triphosphate) activates the *AKT* proto-oncogene, which phosphorylates and inhibits tuberin. **(b)** Mutations of *TSC1* or *2* lead to loss, or reduction of function of tuberin/hamartin or activation of mTOR, with increased levels of activated/phosphorylated ribosomal S6-kinase and protein synthesis. In focal cortical dysplasia type IIB (FCD IIB), upstream activation in the mTOR pathway occurs through PI3K/AKT pathway activation due to unknown mechanisms. In mouse models of malformations, PTEN inactivation leads to activation of the mTOR pathway, and sequence alterations of PTEN have been shown in FCD IIB. Regardless of the initial upstream trigger, disinhibition or hyperactivation of the mTOR pathway leads to abnormally increased cell growth and proliferation. Rapamycin inhibits mTORC1 activation through dissociation of mTOR from Raptor. In (b) main changes are highlighted with red arrows. AMPK, adenosine monophosphate-activated protein kinase; 4E-BP1, eukaryotic initiation factor 4E binding protein-1; eIF4E, eukaryotic translation initiation factor binding protein 1; mTOR, mammalian target of rapamycin; mTORC1, mTOR complex 1; mTORC2, mTOR complex 2; p70S6K1, 70 kDa ribosomal protein S6 kinase; PTEN, phosphatase and tensin homolog; Raptor, regulatory unit of mTOR; S6, 40S ribosomal protein S6.

FCD IIB. This suggests that some of the neuronal alterations, including the synaptic remodelling, are an effect of seizures rather than a primary developmental abnormality.[103]

Mechanisms of Epileptogenesis in Focal Cortical Dysplasia Epilepsy is clearly associated with FCD lesions. Single cell recordings from dysplastic neurons have demonstrated abnormal intrinsic membrane properties and ion channel functions,[90] although no spontaneous epileptiform depolarizations have been shown, suggesting they are unlikely to operate as 'pacemaker' neurons.[86] Immature balloon cells do not display spontaneous synaptic current or action potentials[90] and lack synaptic contacts, suggesting that they are inert bystanders. Recent studies of glutamate transporters propose that balloon cells might exhibit a protective effect against local ictal activity, through increased glutamate clearance mechanisms.[175] Direct electrocorticographic recordings also support the theory that the balloon cell-rich centres of FCD are less active than other regions.[63]

Many microstructural and immunophenotypic abnormalities have been reported in FCD II that may underpin its epileptogenicity. Alterations of excitatory synaptic densities around dysmorphic neurons[3] and NMDA receptor assembly occurs in FCD II with NR2B subunit upregulation;[153,289,444] other studies show decreased receptor activity and NR2B expression.[6] AMPA receptor mRNA is also increased in dysplastic neurons.[220,317] These findings could correlate with reported electrophysiological excitability and abnormal neuronal bursting properties.[6,269,432]

Alteration of the GABAergic system is also evident in FCD II. Inhibitory interneurons, as demonstrated with glutamic acid decarboxylase (GAD), calcium-binding proteins or neuropeptides, are generally reduced in the region of FCD II.[5,7,25,317,373,399] Cytomegalic interneurons have also been identified in FCD IIB.[7,399] GABA$_A$ receptor subunit mRNAs (β1, β2, α1, α2) are altered in abnormal neurons[220] although increased GABA$_A$ receptor sensitivity has been recently shown.[87] Electrophysiological recordings, slice-culture and EM studies support overall preservation of GABA synaptic activity with hypertrophic inhibitory synaptic terminals seen surrounding dysplastic neurons.[3,7,170,373] Recordings from *in vitro* slice preparations of FCD have also implicated GABAergic mechanisms in initiation and synchronization of ictal discharges.[115] In summary, although there are fewer abnormal GABAergic cells represented, the remaining cells support GABA function and likely form abnormal, pro-epileptogenic networks.[7]

GABA is one of the earliest neurotransmitters in the CNS, and the developmentally regulated expression of cationic-chloride co-transporters that regulate intracellular chloride influence the capacity of GABA to generate hyperpolarizing inhibitory potentials. In the postnatal brain there is a switch from Na$^+$/K$^+$/2Cl$^-$ (NKCC1) to K$^+$/Cl$^-$ (KCC2) co-transporter with resultant reduction in intracellular Cl$^-$, pivotal for switching GABA from an excitatory to inhibitory neurotransmitter. Increased NKCC1 expression has been shown in dysplastic neurons[356] and balloon cells[14] in FCD IIB, with reduction of KCC2 mRNA in small neurons.[363] This could reflect either developmental dysmaturity or deregulation as response to seizures but may functionally contribute to local neuronal excitability[104] through enhanced depolarizing potentials and synchronizing local networks.

Reduced expression of synaptic vesicle protein SV2A has been shown in FCD II.[413] SV2A modulates synaptic transmission, knockout mice have spontaneous seizures and it is also the binding site of levetiracetam, an anti-epileptic drug. Studies of gap junction proteins, including Cx43, have also shown abnormal aggregates around balloon cells in FCD IIB, which could be of functional significance in the formation of abnormal local networks.[168] Furthermore, alteration in the extracellular space has been demonstrated in FCD II with abnormal diffusivity of molecules that could be permissive for non-synaptic transmission, enhancing local epileptogenicity.[420] Additionally, increased and altered distribution of aquaporin 4 in relation to the dysplastic neurons has been shown in FCD II that may influence local fluid homeostasis and modify neuronal function.[281] Finally, increasing interest has been directed at the contribution of pro-inflammatory mechanisms in FCD II in epileptogenesis (discussed in more detail in Inflammation and Epilepsy).

Differential Diagnosis The main differential diagnoses in FCD type II include gangliogliomas, tubers and hemimegalencephaly (HMEG). Neuroradiological correlation is paramount, particularly in small and poorly oriented specimens. Even in well-oriented specimens, cortical tubers in TSC can be indistinguishable from FCD. However, molecular, genetic and electrophysiology studies show differences (Table 11.10). HMEG is a sporadic disorder but can be syndromic, including the Proteus syndrome (*AKT1* mutation), Cowden syndrome (*PTEN* mutation), Klippel–Tranaunay–Weber syndrome, hypomelanosis of Ito, linear sebaceous nevus syndromes and, rarely, TSC. HMEG is characterized by unilateral hemispheric enlargement, which can be associated with abnormal gyri but cytoarchitectural abnormalities; diverse histopathological features have been described. Balloon cells and cytomegalic neurons, morphologically similar to those observed in FCD II and tubers, are recognized in many, but not all, cases. Abnormal expression of cyclin D1, Pp70S6K and P-S6 proteins has been demonstrated in sporadic HMEG, supporting activation of both mTOR and β-catenin/wnt cascades, which may be relevant to the hemispheric enlargement.[13]

Focal Cortical Dysplasia Type I
Clinical and Radiological Features There is some evidence for older mean age of onset of epilepsy and reduced seizure frequency in FCD I compared to FCD II, but this is not conclusive. An associated history of prenatal/perinatal risk factors (including prematurity, asphyxia, bleeding and brain injury) has been reported.[50,229] Low intelligence, behavioural problems and neurological symptoms other than epilepsy were also more common in FCD I than II in one series.[384] The EEG may show continuous irregular slowing. The MR imaging features of FCD I are less well defined and the lesions less often visible.[410] White matter signal changes, loss of white matter volume, lobar hypoplasia and atrophy have been reported.[285,384]

Multiple lobes, including the occipital and temporal lobes, may be involved[50] with a frontotemporal predominance in one surgical series.[389] Outcome data following surgery for FCD I suggest fewer patients with FCD I become seizure free compared to FCD II (see Table 11.8) and FCD IIIa/b.[389] The precise incidence of FCD I is unknown, ranging from 0 per cent to over 80 per cent of all FCD cases in epilepsy series (see Table 11.8); this may partly be due to interobserver differences in their recognition[91] and previous classification of FCD III cases as FCD I in the Palmini system (see Tables 11.7 and 11.8).

Pathological Features and Aetiology FCD type I is an isolated dysplasia in which abnormal cortical lamination is observed (dyslamination), unaccompanied by dysmorphic neurons or balloon cells. In the revised ILAE 2011 classification, three subtypes are recognized (see Table 11.7). In FCD IA, an exaggerated radial or microcolumnar pattern is observed with rows of small neurons arranged perpendicularly in the cortex (Figure 11.15).[197] FCD IB is defined by isocortex that lacks the normal six-layered tangential arrangement of neurons. The whole cortical thickness may be affected (a-laminar) or specific layers may be missing (e.g. cortical layers 2 or 4).[229] FCD IC has the combined components of types IA and B. FCD I may be accompanied by blurring of the grey–white matter boundary. Occasional immature neurons and hypertrophic neurons may be present in FCD I types. Immature neurons are defined as round to oval cells (diameter 10–12 μm) with a thin rim of cytoplasm, rudimentary dendrites and absent neurofilament expression. Hypertrophic neurons resemble large pyramidal neurons of layer V, but are abnormally located in the superficial cortical layers. Increased numbers of single, ectopic white matter neurons are also reported in the underlying white matter in FCD I.[232]

There are common pitfalls in the diagnosis of FCD I. The microcolumnar architecture of FCD IA has to be distinguished from regional cytoarchitectonic variations where a 'columnar organization' is a normal finding (e.g. in occipital and temporal lobes) and FCD IB, from normal a-granular cortex. Hypertrophic neurons may be over-interpreted as dysmorphic and identification of immature neurons is problematic in H&E sections alone. Laminar sclerosis (neuronal loss and gliosis) due to a previous insult (including status epilepticus) should not be misinterpreted as FCD I. The microcolumnar patterns in FCD IA may represent arrested maturation with persistence of developmental, vertically arranged pyramidal cells and their bundled projections of axons and dendrites. The identification of robust immunomarkers that could discriminate FCD IA from normal cortex would be of immense diagnostic value for routine practice but are limited at present. Immunophenotypic differences between FCD I and II have been noted, such as lack of mTOR pathway activation and an absence of stem cell markers expression (e.g. SOX2).[311] Regarding intrinsic epileptogenicity, electrophysiological studies support decreased GABA sensitivity in FCD I compared to normal cortex although there is no evidence for loss of interneurons.

Focal Cortical Dysplasia Type III

The type III dysplasia group comprises FCDs immediately associated with a principal pathological lesion, presumed to be the primary epileptogenic focus. FCD III was introduced

in the 2011 FCD consensus classification, to separate these dysplasias from the isolated type I FCDs, which they resemble. The clinical outcomes for the FCD III subtypes appear to be linked to the main lesion with which they are associated. The pathogenic mechanisms are likely to be different from FCD I, with FCD III representing postnatal maladaptive cortical reorganization rather than developmental abnormalities.[53] It remains to be established whether FCD III is independently epileptogenic or epiphenomenal.[368]

Focal Cortical Dysplasia Type IIIa FCD IIIa encompasses dysplasias associated with HS, arising in the adjacent temporal cortex. It includes laminar abnormalities not dissimilar to descriptions of FCD I (including isolated hypertrophic neurons), small 'lentiform' neuronal heterotopias in the subcortical white matter, and a distinct cortical abnormality also termed 'temporal lobe sclerosis' (next paragraph and Figure 11.15). An excess of single neurons in the white matter, which is a common finding in association with HS, is not interpreted as FCD IIIa because of its uncertain significance. With correct orientation of the surgical temporal lobe specimen (see Figure 11.4), assessment of dysplasia must be made in the knowledge of normal regional cytoarchitectonic variations between the temporal pole, posterior lobe and entorhinal cortex.

Temporal lobe sclerosis was first recognized by Meyer in the earliest epilepsy surgical series.[285] In a more recent series of 272 patients with HS, *temporal lobe sclerosis* was identified in 11 per cent and defined pathologically by a variable reduction of neurons from cortical layers II/III and superficial cortical gliosis; it is usually accompanied by additional architectural abnormalities of layer II, with abnormal neuronal orientation and aggregation[396] (see Figure 11.15). It may show a predilection for the temporopolar cortex or be more widespread; there is no definite predilection for involvement of sulcus versus gyral crown. Interneurons are normally distributed,[169,396] although reduction of parvalbumin-positive neurons has been reported.[169] This pathological change is always accompanied by HS and frequently associated with early onset febrile seizures. The neuronal clustering is highlighted with neurofilament immunohistochemistry and may be accompanied by abnormal horizontal bands of cortical myelinated fibres. It is currently regarded as an acquired, 'reorganizational dysplasia', possibly linked with the processes mediating HS. This subtle pathology is not visualized with current MR imaging[141] and its presence does not appear to influence postsurgical seizure outcome.

Focal Cortical Dysplasia Type IIIb FCD IIIb is characterized by cytoarchitectural abnormalities in the cortex adjacent to low grade glial and glioneuronal tumours associated with epilepsy (long-term epilepsy-associated tumours [LEATs]), particularly dysembryoplastic neuroepithelial tumours (DNTs) and gangliogliomas. One long-held interpretation of this association has been that this pre-existing cortical malformation forms the 'dysembryoplastic' zone from which the tumours arise. Arguing against this hypothesis, however, is that tumours developing from known cortical dysplasias are infrequent.[313] The main changes reported near LEATs include laminar architectural alterations (previously classified as FCD I[317]), layer I hypercellularity (which

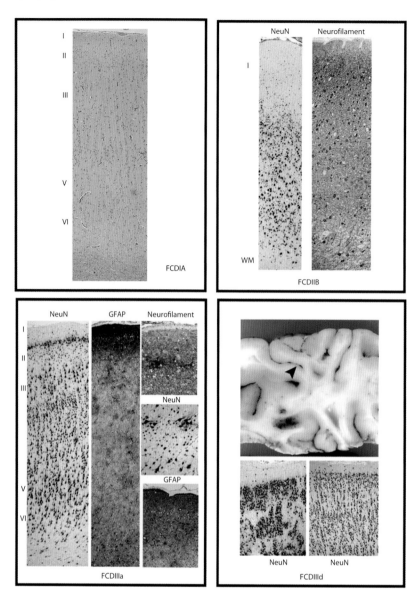

11.15 Dyslamination and alterations to cortical architecture in focal cortical dysplasia (FCD) subtypes. In FCD IA prominent microcolumns of cortical neurons are present (H&E). In FCD IIB, loss of radial and laminar alignment of neurons is present on neurofilament and NeuN stains with large, dysmorphic neurons through layers II to VI; layer I may appear wider (NeuN and neurofilament stain are from different cases). In FCD IIIa (FCD associated with hippocampal sclerosis), the architectural abnormalities are striking in the outer cortical layers with aggregation and clustering of neurons in the outermost part of layer II with associated gliosis and loss of larger pyramidal neurons in layers II and III (this constellation of features has also been termed temporal lobe sclerosis). The smaller images in the right of the panel illustrate the clustering and abnormal orientation of neurons in layer II. In FCD IIId (dysplasia adjacent to a perinatal infarct/ulegyria, arrow), there is abnormal layering and aggregation of neurons on the left compared to preserved lamination in normal cortex at the margins of surgical resection.

may extend a considerable distance beyond the main lesion) (see Figure 11.18) and small satellite clusters of immature cells in the cortex and white matter (also termed 'hamartias'). Some of these features are better visualized with CD34 rather than H&E (Figure 11.16). The interpretation of these histological features as tumour precursor (developmental) lesions or secondary tumour structures is still conjecture. White matter rarefaction present in the vicinity of DNTs and gangliogliomas is reminiscent of the hypomyelination observed in FCD IIB and can lead to preoperative misinterpretation on MR imaging (see Figure 11.18). As with any infiltrating tumour, the organization of the

bordering cortex may be disturbed, the layering may be indistinct and the cortex may appear paucicellular, with remaining neurons maloriented and even hypertrophic, simulating cortical dysplasia.

Reproducible diagnostic criteria for 'tumour-associated dysplasia' (FCD IIIb) are required, as exemplified by the marked variation in its current reporting. For example, the reported incidence of cortical dysplasia in DNT series varies from 0[74,410] to 83 per cent of cases.[384] The main pitfall is distinguishing (and excluding) what is part of the tumour versus real (either developmental or reorganizational) dysplasia. Incorporating NeuN and CD34 where possible, rather than evaluations based on H&E or cresyl violet staining alone, is likely to improve reporting accuracy (Figure 11.6). NeuN undoubtedly allows superior assessment of the six-layered architecture and CD34 as well as other tumour markers (IDH1, GFAP, Map2) and will allow identification of tumour components in the cortex and avoid overinterpretation of dysplasia. If this can be done reliably, the true incidence of FCD IIIb will emerge, allowing accurate evaluation of imaging characteristics, presurgical planning, and contributions to epileptogenesis.

Focal Cortical Dysplasias Types IIIc and d

FCD IIIc encompasses laminar architectural abnormalities occurring adjacent to vascular malformations, including cavernomas.[112,259] FCD IIId includes dysplasias (disorders of cortical layering or cytological composition, including hypertrophic neurons) in the context of a cortical lesion and acquired cortical lesions including perinatal ischaemic injury, trauma[271,272] or inflammatory process (e.g. Rasmussen's encephalitis).[385]

Mild Malformations of Cortical Development (mMCD)

The term mMCD was introduced in the Palmini system[317] to classify microscopic cortical architectural abnormalities reported in patients with epilepsy, previously subsumed under the broad term of 'microdysgenesis' (see Table 11.7). The mMCDs were subdivided into type I (excessive, heterotopic neurons in layer I) and type II (excess single neurons in the white matter) (Figure 11.17). These main categories are retained in the new ILAE classification.[53]

The adult cortical layer I normally contains sparse neurons, many being GABAergic. The potential aetiology of isolated layer I neuronal hypercellularity in type I mild MCD includes retention of residual preplate cells, (including reelin-secreting Cajal-Retzius cells), ongoing marginal zone proliferation from residual progenitor cells, radial migrating GABAergic neurons from the ganglionic eminence or

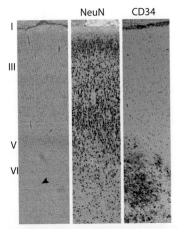

NeuN   CD34

11.16 Common pathological changes in the cortex adjacent to long-term epilepsy-associated tumours (including dys-embryoplastic neuroepithelial tumours and gangliogliomas) include apparent dyslamination on H&E, layer I hypercellularity at some distance away from the tumour, and ill-defined cellular nodules or hamartia-like clusters (arrow in H&E). NeuN is invaluable to confirm normal residual lamination of the cortex, and CD34 highlights hamartias as well as layer I hypercellularity.

heterotopic cortical plate neurons that have 'over migrated'. Precise (quantitative) anatomical definitions, functional implications, incidence, causes and associations of mMCD type I are needed. An excess of hypertrophic layer I neurons was associated with infantile spasms in one study.[8] Studies in NXSMD/EiJ mice with nodular layer I ectopia have demonstrated a composition of mixed neuronal types, including inhibitory neurons with connections to deeper cortical layers.[166]

Type II mMCD has most often been reported in the context of TLE, through several quantitative studies. Despite differences in methodologies regarding anatomical regions studied, types of neurons counted, cell counting method used (stereological versus two-dimensional counting methods), different underlying principal pathologies (e.g. HS, FCD I, FCD II), most studies have confirmed higher densities of single, morphologically normal appearing white matter neurons in epilepsy series when compared to control groups, albeit with varying numerical values[64,130,186,197,213,408] (Table 11.11). An increase in white matter neurons in the subcortical white matter can also be a striking component of FCD I and II, including dysmorphic neurons in FCD II.

As an isolated finding, there still remains uncertainty regarding the relationship of mMCD type II to epileptogenesis and any predictive value in terms of seizure outcomes following surgery. Although it has been proposed that white matter neurons might correlate with neuroimaging abnormalities in the temporal lobe white matter, including increased T2 signal and loss of grey–white matter definition, a recent study with precise coregistration between MR imaging and pathology specimens failed to show a correlation between stereologically acquired white matter neuronal densities and quantitative MR measures.[139]

An excess of white matter neurons in epilepsy may represent an epiphenomenon. Similar pathology has also been noted with other neurological conditions, such as schizophrenia. Normal interstitial or white matter neurons are diverse in cell type, including inhibitory interneurons (NPY, somatostatin, calbindin, calretinin, reelin, NADPH expressing neurons) (Figure 11.17); they extend axonal projections to the overlying cortex and are likely to be functionally integrated into cortical circuits.[414] A proportion represent remnants of subplate cells residing in the subcortical zone or gyral white matter. Critical for the establishment of thalamocortical pathways during development and vulnerable to injury in the perinatal period, their persistence and position in normal adult brain may indicate ongoing roles in 'gating' cortical inputs, coordinating inter-areal connectivity and regulating blood flow.[226,377] No study has yet addressed the relative diversity of white matter cell types in epilepsy compared to controls, although a recent study of $GABA_A$ receptors in TLE has shown selective changes in their subunit expression.[254] Possible explanations for excess white matter neurons in epilepsy therefore include a developmental abnormality (arrested migration of cortical plate cells), a maturational abnormality (e.g. surplus persistence of subplate neurons) or enhanced neurogenesis.

mMCD type III refers to the finding of excessive clustering of neurons in the cortex, although the diagnostic

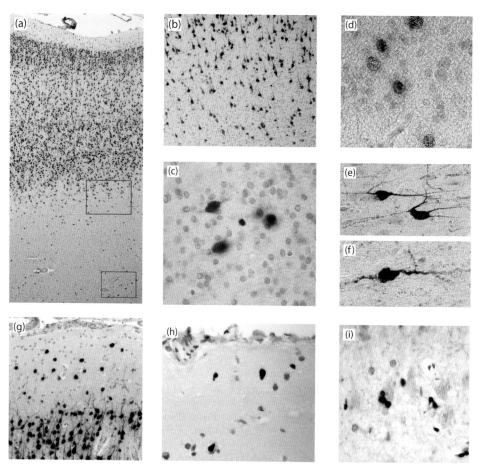

11.17 Mild malformations of cortical development. (a) NeuN-stained section in temporal lobe epilepsy with white matter neurons observed in the subcortical (subplate) region shown in **(b)** higher magnification and **(c)** single white matter neurons in the deeper white matter, shown in higher magnifications of varying size. Immunohistochemistry reveals varied types of white matter neurons including **(d)** Tbr I, **(e)** NPY and **(f)** calretinin-positive cells. **(g)** NeuN-positive cells are observed in layer I in temporal lobe epilepsy but without aggregation, as in control cortex; clustering of layer I neurons corresponds to type I mMCD. **(h)** Mcm2 labelling also highlights a proportion of layer I cells. **(i)** Doublecortin (DCX) immunohistochemistry highlights small cells in a satellite location around layer V pyramidal cells; increased perineuronal satellitosis corresponds to mMCD type IV.

criteria and significance remain poorly defined.[53] *mMCD IV* refers to exaggerated perineuronal or perivascular satellitosis by 'oligodendroglial cells', which is a frequent observation in the lower cortical layers, white matter and amygdala in TLE specimens.[145] Some satellite cells may represent immature doublecortin positive[374] or NG2 proliferative glia.[172]

Tumours

A wide variety of tumour types, particularly where there is cortical extension, can manifest clinically with focal seizures. The incidence of brain tumours in epilepsy patients is about 4 per cent and the frequency of epilepsy in brain tumour patients is 30 per cent or more, depending on the tumour type.[395] The term long-term epilepsy-associated tumour (LEAT) encompasses lesions in patients with long histories (often 2 years or more) of drug-resistant epilepsy. They are generally slowly growing, low grade, cortically based tumours, more often arising in younger patients and often exhibiting neuronal in addition to glial differentiation.

The main tumour types reported in epilepsy surgical series are summarized in Table 11.12. DNT and gangliogliomas (WHO grade I) together are most commonly reported, with mixed forms also described. The pathological features of these tumours are detailed in Chapter 32, and this section only addresses their biology in relation to epilepsy (Figure 11.18).

The prognosis for LEATs following surgical resection is, in the majority of cases, favourable from perspectives of improving both seizures and patient survival. However, in a proportion of cases, seizures continue postoperatively. Similarly, a minority of WHO grade I LEATs behave unpredictably and aggressively with local tumour regrowth or recurrence and, rarely, malignant transformation (e.g. ganglioglioma with anaplastic change). Combined immunohistochemical and molecular genetic studies (e.g. labelling patterns with CD34, nestin, MAP2 and mutant IDH1 protein, *BRAF* mutation/fusion and 1p/19q loss of heterozygosity studies) can help identify tumours less likely to behave favourably and distinguish more conventional, progressive gliomas.[395]

TABLE 11.11 Studies quantifying white matter neurons in epilepsy

Study	Pathology (n=number of studied cases)	Stain	Method of counting Anatomical region	Mean values in epilepsy (control values if tested)
Hardiman et al. (1988)[186]	Temporal lobe epilepsy (n = 49)	Nissl	Profile counting (2d) 'Deep' white matter	$4/mm^2$ ($<4/mm^2$)[a]
Emery et al. (1997)[130]	Temporal lobe epilepsy (n = 22)	Nissl	Profile counting (2D) All white matter	$4.11/mm^2$ ($2.35/mm^2$)[a]
Thom et al. (2001)[408]	Temporal lobe adjacent to HS (n = 31)	Nissl NeuN	Stereology (3D) All white matter	$1010/mm^3$ $2164/mm^{3\ a}$ ($1660/mm^3$)
Bothwell et al. (2001)[64]	Temporal lobe adjacent to HS (n = 10)	Nissl	Stereology (3D) All white matter	$1060/mm^{3\ a}$ ($750/mm^3$)
Hildebrandt et al. (2005)[197]	FCD I and II (n=25)	MAP2 (4 μm sections)	Cell profile (2D) $1mm^3$ of deep white matter	$21.4 \pm 6.8/mm^2$ ($10.5 \pm 2.9/mm^2$)[a]
Eriksson et al. (2006)[138]	Temporal lobe adjacent to HS (n = 10)	NeuN (7 μm sections)	Cell profile (2D) ROI in deep white matter	$23.4/mm^2$
Eriksson et al. (2007)[139]	Temporal lobe (mainly HS) (n = 9)	NeuN	Stereology (3d) ROI in deep white matter	$2480/mm^3$

[a] Indicates series where epilepsy cases were significantly different from control values.
HS, hippocampal sclerosis; ROI, region of interest.

Mechanisms of Epileptogenicity in Tumours

The propensity to develop seizures is higher in tumours located in the frontal, temporal or insular cortex, with tumour size and rate of growth potentially being relevant.[395] The prominent neuronal component (including immature neurons) of some LEATs is a possible explanation for their potent epileptogenicity, with a high density of hyperexcitable neurons associated with epileptiform discharges. Studies have revealed expression of ionotropic and metabotropic glutamate receptor subtypes in ganglioglioma and DNT as well as decreased glutamate transporters in glia, highlighting a role for enhanced glutamatergic transmission.[18,72] In addition, downregulation of $GABA_A$ receptor subunits, potassium channels and voltage-gated ion channels has been shown, which could promote excitability.[119] Activation of pro-inflammatory molecules and blood-brain barrier alterations may also play a role.

Peritumoural cortical changes may be equally important to epileptogenesis. This includes adjacent haemosiderin deposition, hypoxia, acidosis, alterations of the physiological and bursting properties of neighbouring neurons, high expression of GAP junction proteins (connexin-43 and -32), enzymatic changes (for example to adenosine kinase), and increased NKCC chloride transporters.[16,104,118,360] Furthermore, any adjacent tumour-associated cortical dysplasia, (FCD type IIIb), may be relevant to epileptogenesis and warrants further investigation. The slow growth rate of these tumours, compatible with long survival, is likely a critical factor in the induction of secondary cellular and network reorganization in adjacent cortex or even remote sites as the hippocampus.

Vascular Malformations and Epilepsy

Vascular malformations form up to 6 per cent of focal epileptogenic pathologies in recent surgical series (see Table 11.3). The main types are arteriovenous malformation (AVM) and cavernous angioma, with telangiectatic or angiodysgenetic malformations more rarely encountered. Epilepsy is a presenting feature in 17 per cent of AVMs and is the most common presenting symptom in cavernous angiomas (79 per cent)[290,447] These developmental lesions are typically surrounded by a rim of gliosis and haemosiderin. Early surgery is currently regarded as the optimal treatment, both for the epilepsy and to reduce risk of further haemorrhage; 84 per cent of patients with cavernous angiomas become seizure free.[290] Possible underlying mechanisms for the epilepsy include local ischaemia from arteriovenous shunting, peripheral gliosis, haemosiderin deposition or secondary epileptogenesis in the temporal lobe. In a recent study there was no association between neuronal discharges with intraoperative electrocorticography and the density of haemosiderin in the resected cavernoma.[152]

Hamartomas and Epilepsy

Hamartomas that cause epilepsy are poorly defined and form a small proportion of cases in surgical series. Glioneuronal hamartomas have been described involving the cortex, particularly temporal and frontal lobes, and represent circumscribed masses of mature but haphazardly arranged cell types, sometimes in association with adjacent cortical dysplasia. Oligodendroglial hamartomas have also been described.[2]

11

TABLE 11.12 The relative incidence of tumour types in reported epilepsy surgical series[a]

Series (n = number of cases) and era of collection	Low grade glioma or diffuse astrocytoma WHO grade II	Angiocentric glioma	Pilocytic astrocytoma	Pleomorphic xanthoastrocytoma	Oligodendroglioma WHO grade II	DNT (all types)	Ganglioglioma	Gangliocytoma	Mixed DNT/ganglioglioma	Glioneuronal tumour uncertain type	Other tumours (number in brackets/parentheses = % of series)
NHNN, London[b]; adults (n = 155) 1994–2010	16 (10.3%)		4 (2.5%)		7 (5.1%)	88 (56%)	12 (7.7%)	5 (3.2%)	10 (6.4%)	10[c] (6.4%)	Ependymoma (1), meningioma (1), anaplastic oligodendroglioma (1) (1.2%)
Kings, London; (n = 92) 1975–1999[204]	7 (7.6%)					74 (80%)	6 (6.5%)				Oligoastrocytoma (1), hamartoma (4) (5.4%)
Grenoble; all ages (n = 94) 1990–2000[321]				4 (4.2%)		61 (64.8%)	29 (30.8%)				
Cleveland; adults (n = 141) 1989–2009[332]	24 (17%)	1 (0.7%)	2 (1.4%)	2 (1.4%)	22 (15.6%)	10 (7.1%)	38 (27%)		3 (2.1%)	14 (9.9%)	Low-grade mixed glioma (13), hamartoma (5), meningioangiomatosis (2), gliofibroma (1), anaplastic ganglioglioma (1), anaplastic astrocytoma (1) (16%)
Cleveland; paediatric (n = 129) 1989–2009[333]	15 (11.6%)	3 (2.3%)	1 (0.8%)	2 (1.6%)	5 (3.9%)	17 (13.2%)	48 (37%)		3 (2.3%)	18 (14%)	Low-grade mixed glioma (8), hamartoma (2), meningioangiomatosis (4), other low-grade glioma (1) (11.6%)
Beijing; all ages (n = 51) 2004–2008[324]		1 (2%)		4 (7.8%)		10 (19.6%)	19 (37.7%)			13 (25%)	Meningioangiomatosis (1) (2%)
Illinois; (n = 39) 1981–2005[343 392]	4 (10.5%)			3 (7.9%)	3 (7.9%)	10 (26.3%)	14 (36%)				Mixed oligoastrocytoma (3), astroblastoma (1) (10.5%)
Erlangen, German epilepsy database[d] (n = 1354)	63 (4.7%)	3 (0.2%)	76 (5.6%)	35 (2.6%)	29 (2.1%)	246 (18.2%)	669 (49.4%)	5 (0.4%)		52 (3.8%)	Meningiomas (21), cysts (22), anaplastic gliomas (97), isomorphic astrocytomas (19), ependymomas (4), SEGA (13) (13.0%)

[a] Modified from Thom M, Blumcke I, Aronica E. Long-term epilepsy-associated tumors. *Brain Pathol* 2012;22(3):350–79, with permission from John Wiley and Sons. © 2012 The Authors; Brain Pathology © 2012 International Society of Neuropathology.

[b] These are unreported data taken from the epilepsy database kept at the National Hospital for Neurology and Neurosurgery London (NHNN) and include patients who have been investigated through the epilepsy surgical program during adulthood for drug-resistant focal epilepsy and where a surgical procedure was carried out.

[c] In these cases the glioneuronal tumour was difficult to unequivocally classify, mainly because of the small size of specimen.

[d] Unreported data from the German Epilepsy Brain Bank ,Erlangen. In the first three series, the category of diffuse DNT is included in the DNT numbers.

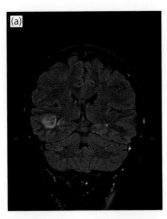
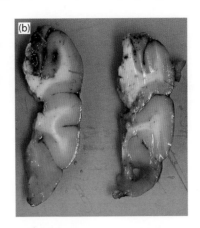
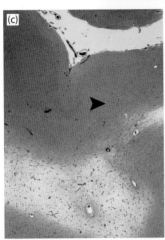
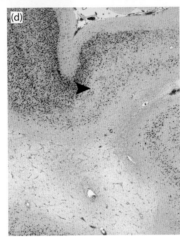
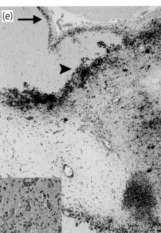

11.18 Long-term epilepsy-associated tumours. (a) Magnetic resonance imaging T2 sequence of lesion in a patient with an 8-year history of epilepsy; following resection, this was shown to be a complex dysembryoplastic neuroepithelial tumour (DNT). **(b)** Slices through a temporal lobe with a cystic lesion that correlated histologically to a diffuse type of DNT. **(c)** A further long-term epilepsy-associated tumour with a diffuse growth pattern through the cortex (edge of lesion shown at arrowhead); rarefaction of the underlying white matter was also striking on H&E. **(d)** In NeuN-stained sections, loss of neurons in the midcortical layers is striking at the margins of tumour infiltration but without a cortical dysplasia in adjacent cortex. **(e)** CD34 staining highlights the cortical infiltration, with satellite and layer I labelling (arrow) beyond the main lesion. The tumour cells have a rounded 'oligo-like' morphology throughout, without a dysplastic ganglion cell component and no true glioneuronal element (inset).

The hypothalamic hamartoma has a strong association with intrinsic subcortical epileptogenesis and gelastic seizures. It is composed of cytologically normal, small and large neurons[320] and may be causally linked to secondary cortical epileptogenesis.[221] The presumed hamartomatous cell proliferation of meningioangiomatosis does not clinically manifest with seizures when associated with NF2, but presents with epilepsy in over 80 percent of patients with sporadic meningioangiomatosis.[323] This more common sporadic form of meningioangiomatosis is typically solitary, and EEG data suggest that the epileptogenicity is confined to adjacent cortex, with seizures persisting in over half of patients postoperatively.[434]

GENERAL CONCEPTS OF EPILEPTOGENESIS IN FOCAL EPILEPSY

The term *epileptogenesis* encompasses the cascade of cellular events, following which a brain develops spontaneous seizures or epilepsy (Figure 11.19). Epileptogenesis is often divided into three stages: (i) the acute event (the triggering insult or initial seizure); (ii) a latent period (clinically silent); and (iii) spontaneous seizures. In humans, the latent period can last for months or years. These processes are most often applied to the study of 'acquired' or symptomatic epilepsies, estimated to represent up to 50 per cent of all epilepsies, but may also operate in genetic or idiopathic epilepsies.[173] The main challenge in studying the processes of epileptogenesis in advanced-stage human tissues is to distinguish underlying pre-existing abnormalities from secondary maladaptive reorganizational changes. Identifying the 'critical period' for potential intervention is better addressed through animal models. It is also likely that multiple epileptogenic mechanisms operate in concert. Understanding epileptogenesis is essential to identifying new therapeutic targets. At present, most available drugs are 'antiepilepsy' rather than 'antiepileptogenesis',[327] but promising new options, modifying cellular responses, could lead to prevention of epilepsy.[253,328]

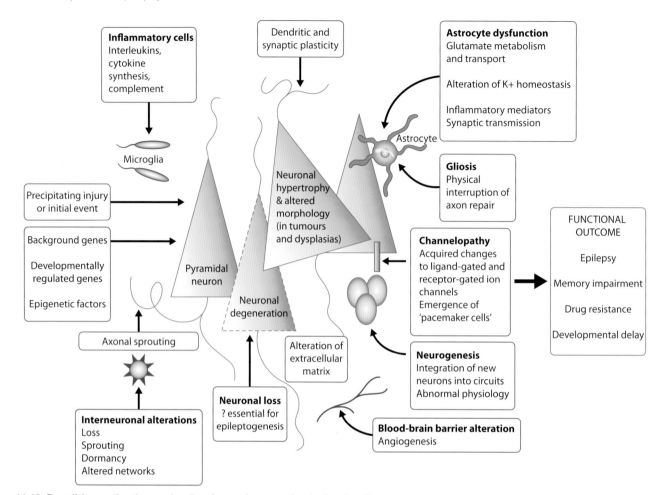

11.19 Possible mechanisms of epileptigenesis operating in focal epilepsy.

Lesional Versus Perilesional Tissue

The epileptogenic zone (which is the region of cortex or tissue mass that gives rise to seizures) may arise within the lesion (visualized with neuroimaging or macroscopically) and/or reside in the more normal appearing perilesional cortex.[439] This varies among pathologies and cases and is fundamental to the planning and individual tailoring of surgery aiming to cure seizures. For example, there is evidence that both tuberous and non-dysplastic cortex may generate seizures in TSC and similarly for localized FCDs.[439] This has been shown experimentally by electrophysiologic studies of slice preparations from resected focal lesions, showing differences in neuronal firing properties between the lesional and paralesional cortices in cortical dysplasia.[269]

Imbalance of Local Excitation to Inhibitory Networks

Excitatory/inhibitory imbalances have been extensively researched in the context of HS, FCD and tumours. Imbalances can occur at the network, cellular and synaptic levels. This includes reduction in inhibitory neurons and synapses, increased expression, assembly and function of excitatory neurotransmitter receptors or channels (e.g. glutamate receptors, calcium channels), and decreased expression of inhibitory mechanisms (e.g. GABA receptors, potassium channels) or of neurotransmitter transporter proteins (e.g. vesicular glutamate transporter, excitatory amino acid transporters [EAATs]). Experimental data confirm that synaptic remodelling can occur as a result of seizures rather than as a primary abnormality.[103] Pharmacological treatments administered prior to surgery may also affect neurotransmitter receptor status.

Abnormal synchronization of neurons may be a further mechanism relevant to the generation of seizures. Interneuronal populations, through connection with multiple cells, can enhance this process. Electrophysiological studies *in vivo* confirm that in focal lesions, the seizure onset zone may show abnormal neuronal synchronization, although functionally isolated from surrounding brain regions.[431] Furthermore, GABAergic inhibition can revert to excitation in epilepsy tissues, which may be mediated by changes in developmentally regulated, neuronal cation-chloride co-transporters (KCC2 and NKCC).[44] There is evidence for alteration of these co-transporters in human and experimental epileptic tissues.[27,296]

Inflammation and Epilepsy

In recent years, evidence has accumulated linking inflammatory processes with epilepsy and seizures. Febrile seizures usually occur with a rise in pro-inflammatory agents. Rasmussen's encephalitis and other autoimmune neurological disorders including paraneoplastic syndromes can cause seizures, with activation of the immune system that influences neuronal excitability. Investigators have also confirmed activation of inflammatory pathways in focal pathologies not classically linked with immunological dysfunction, such as FCD. Inflammation could be a consequence as well as a cause of seizures, but contribute to their perpetuation in a vicious loop.

Experimental data supports the hypothesis that seizures can induce brain inflammation and that recurrent seizures aggravate brain inflammation and may contribute to neuronal loss.[423] Although the CNS is an immune privileged site, activation of both adaptive (humoral and cell mediated immune responses) and innate immunity (microglial activation, cytokines, complement, and so on) have been demonstrated in epilepsy tissues. Several CNS cell types (including glia, neurons, endothelial cells) can express inflammatory mediators, including cytokines and their receptors, in particular IL-6, IL-1β and tumour necrosis factor (TNF). Altered permeability of the blood–brain barrier (BBB) can occur during seizures[144] and likely contributes to inflammatory responses. A genetic predisposition for sustained inflammatory reactions, such as gene polymorphisms, could also contribute.

In human TLE tissues, prominent and persistent activation of the interleukin system, IL-1β and IL-1 receptor, is observed in astrocytes, microglia and neurons, with an increase in hippocampal microglia, although B-cell and T-cell infiltrates are minimal and usually perivascular.[340,445] Inflammatory pathway activation in human TLE is supported by gene expression profiles.[12] Although it has been argued that inflammatory processes in TLE and HS might be driving forces in disease progression, experimental evidence also points to neuroprotective roles of infiltrating microglia and lymphocytes.

Activated microglia (HLA DR-positive), upregulation of TNF-α, NF-κB, IL-1β and complement signalling, and cytotoxic T (CD3+ and CD8+) cell infiltration have been found in TSC lesions.[56] Similarly, activation of microglia, complement and IL-Iβ signalling pathways, along with BBB impairment has been reported in FCD.[15] Recent studies also support the activation of Toll-like receptor (TLR) signalling in epilepsy; this inflammatory pathway is normally activated by pathogens, but overexpression of TLR4 and its ligand, HMGB1, has been shown in FCD and HS.[273,449]

The proconvulsant effects of inflammation on seizures may be mediated by alteration in receptor translation (e.g. NMDA receptor), inhibition of glutamate uptake, influences on cell death, neurogenesis, synaptic plasticity and BBB breakdown (Figure 11.20). There are currently no diagnostically useful CSF or serum inflammatory biomarkers detectable in patients with chronic refractory focal epilepsy.[422] Therapeutic targeting of the immune system to prevent epileptogenesis is an area of intensive research.

Autoimmune Encephalitides

Non-infectious autoimmune encephalitis associated with epilepsy can be considered as paraneoplastic encephalitis, non-paraneoplastic encephalitis and Rasmussen's encephalitis.[29] Paraneoplastic encephalitis is discussed elsewhere (see Chapter 45). In non-paraneoplastic encephalitides, serum antibodies against a variety of neuronal antigens (NMDA receptor subunits, voltage-gated potassium channels, GABA_B, AMPA and GAD) have been identified. Patients typically present with a subacute onset of seizures associated with neurological symptoms, cognitive and behavioural changes. There is relatively limited information regarding the composition and severity of the cellular infiltrates in these conditions, based on a few case reports. Inflammatory changes have been reported in locations in accordance with antigen distribution; for example, voltage-gated potassium channel-associated inflammation occurs in the limbic system, diffuse inflammation occurs with NMDAR antibodies, and the cortex, brain stem and spinal cord are involved in anti-GAD conditions. The immunological mechanisms are poorly understood, but autoantibodies against surface receptors may be pathogenic, supported by positive responses to plasmapheresis in some cases.

Rasmussen's Encephalitis

Rasmussen's encephalitis (RE) is a rare sporadic syndrome of unknown aetiology typically presenting in childhood with intractable seizures and associated with progressive unilateral hemispheric atrophy and neurological deficit (Figure 11.21). The mean age of onset is 6 years,[42] but RE may rarely arise in adulthood. Three clinical stages have been proposed: (i) an early 'prodromal' stage (average duration = 7 months) with low seizure frequency; (ii) an 'acute' stage (average duration = 8 months) characterized by increasing seizure frequency and neurological deficit; and (iii) a final 'residual' stage marked by permanent deficit and less frequent seizures.[42] Progressive neurological deficits include hemiparesis, hemianopia, cortical sensory loss, intellectual deterioration and speech problems, depending on which side of the cortex is affected. Seizures are typically refractory to drug treatment and include partial motor seizures, secondary generalized seizures and epilepsia partialis continua, the latter afflicting about half of patients. Adult-onset cases, which account for 10 per cent,[42] may show a more variable, focal, mild and protracted clinical course. Some younger patients may run a rapidly progressive course.

MR imaging shows progressive hemispheric atrophy, with altered signal intensity (on T2 and FLAIR) correlating histologically with regions of active inflammation. EEG changes over the affected hemisphere can be observed early and include epileptiform abnormalities and decreased background activity. CSF examination may be normal, or show raised cell counts and protein levels or oligoclonal bands.[179] Surgical procedures include diagnostic biopsies or functional neurosurgery to control seizures, including hemispherectomy.

The severity of the inflammatory process and the extent of the cortical scarring in RE vary with duration of disease. The early stages are characterized by more active chronic inflammation and less scarring than the later stages.[318] Inflammatory infiltrates in the cortex consist mainly of T

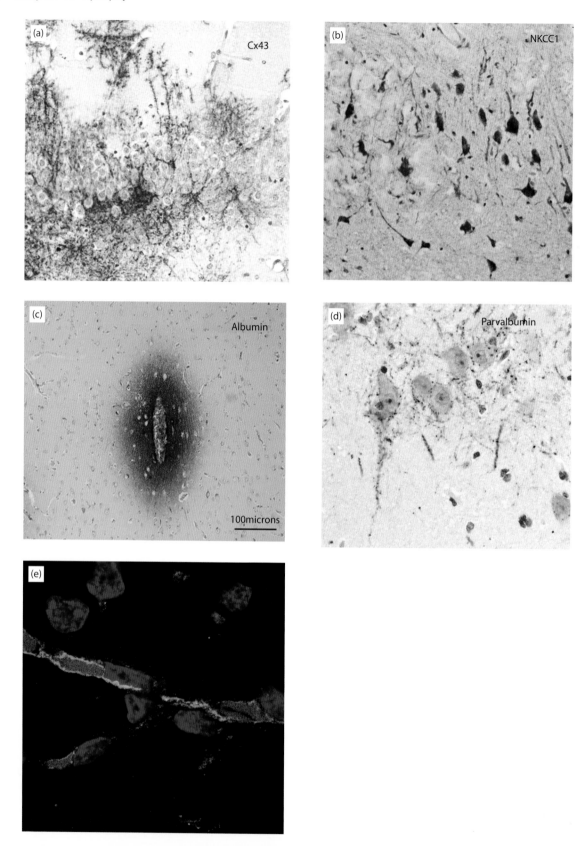

11.20 Investigations of cellular mechanisms in epileptigenesis in surgical tissues. (a) Increased connexion 43 positive glia in the dentate gyrus in temporal lobe epilepsy. **(b)** NKCC co-transporter expression in hippocampal pyramidal cells. **(c)** Increased parenchymal albumin deposition around capillary in TLE case, as evidence of loss of blood-brain barrier integrity. **(d)** Increased perosomatic parvalbumin-positive synapses around residual CA1 pyramidal cells in hippocampal sclerosis. **(e)** Coexpression of drug transporter proteins (p-glycoprotein; red, and CD34, green) in capillaries from a patient with epilepsy to investigate mechanisms of pharmacoresistance.

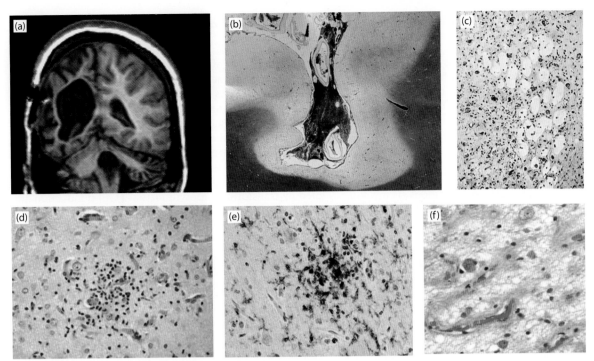

11.21 Rasmussen's encephalitis. (a) Magnetic resonance image showing hemispheric atrophy and cortical destruction. **(b)** Temporal lobe resection showing skip-like regions of cortical atrophy and collapse shown in **(c)** higher magnification with spongiosis (both LFB/CV stain). **(d)** In active lesions frequent inflammatory foci in the grey matter are seen with neurophagia as highlighted with **(e)** CD68 immunostaining. **(f)** Residual neurons may appear hypertrophic and dysmorphic and when these changes are pronounced in the perilesional cortex, this may be categorized as focal cortical dysplasia type IIId.

lymphocytes (CD8 > CD4), with perivascular and perineuronal clusters. B-lymphocytes are found less frequently in perivascular cuffs, although plasma cells are rare. Widespread microglial activation, as well as microglial clusters and nodules may be seen, but macrophage infiltrates are uncommon. Patchy neuronal degeneration, neuronophagia and neuronal dropout are present in early stages. With progressive damage, neuronal ballooning with distortion of cell shape, neurofilament accumulation and laminar disorganization may also be noted amounting to FCD type IIId (see Focal Cortical Dysplasia Type III, p. 717). Apoptotic neurons have also been identified.[41] In later stages, large areas of pan-laminar or patchy cortical necrosis are characterized by extensive neuronal loss, astrocytic gliosis and cortical spongiosis, with the inflammatory process being less prominent. Cortical scars may be extensive, involving a whole gyrus, or more 'punched out' wedge-like areas of destruction. The patchy and multifocal nature of RE explains why cortical biopsies may yield false-negative results. White matter inflammation and involvement of deep grey nuclei may also be present and inflammation may extend to the hippocampus, where HS may be seen. In cases where post-mortem tissue is available, true bilateral disease with associated inflammatory change is rare.[42,412]

There is no proven infectious cause for RE. An autoimmune disease process is considered most likely. Antibodies to a number of antigens have been described, including GluR3 and GAD. Autoantibodies could mediate injury by activation of excitatory receptors causing seizure discharges and excitotoxic neuronal death or complement activation. However, the GluR3 antibodies are not entirely specific, as not all RE patients have these antibodies and they have also been demonstrated in patients with focal seizures due to non-inflammatory causes.[266] Immunohistochemical studies do support a cytotoxic, T lymphocyte attack against neurons.[42] In addition, specific attack of astrocytes by cytotoxic T-cells has been demonstrated.[30] The epilepsy itself may also play some part in the initiation and progression processes and may explain the remarkable unilaterality of the inflammatory process, with breakdown of the BBB on the side of the recurrent partial seizures allowing local exposure to immunological attack.[42] Additionally, there are several reports of RE in association with a second pathology such as a cavernoma and low grade tumour[385]; it is possible that these lesions 'prime' the hemisphere for immunological attack possibly through disruption of the BBB.

Conventional antiepileptic drug treatments have limited effects in controlling seizures in RE, particularly epilepsia partialis continua. Functional hemispherectomy or extensive resection aimed at disconnecting the affected region is the current treatment of choice because it may halt disease progression with improvement in seizures and cessation of neurological deterioration. Early diagnosis (based on clinical, EEG, neuroimaging and biopsy features) and intervention are desirable in order to prevent permanent damage. Immunotherapies, including corticosteroids, plasma exchange and intravenous immunoglobulin infusion, are alternative treatments and are used when surgical intervention is less appropriate, for example with slower disease progression, minimal neurological deficits or bilateral disease.

Contribution of Astrocytes and Gliosis to Epileptogenesis

The study of epileptogenesis has always had a traditionally neurocentric focus but recent attention has been directed to astrocytes. Gliosis is a common finding in surgical resections in TLE, and may be the only finding. Over the last decade, there has been increasing awareness of the breadth of astrocytic function, beyond their acting as supporting cells, many of relevance to seizures and epilepsy[371]: they modulate neuronal excitability, glutamate clearance and release (via EAAT1 and EAAT and glutamine synthetase), have roles in ion homeostasis (in particular, extracellular K^+ concentrations via inwardly rectifying potassium channel Kir4.1), regulate blood flow, influence the BBB, have roles in memory and information processing, in addition to their established stem cell properties. Furthermore, there is some evidence astrocytes can act as antigen presenting cells, expressing inflammatory molecules and participating in the immune response in epilepsy.

Experimental data confirms that astrocytes may be functionally engaged in recurrent excitatory loops, sustaining seizure discharges as evidence of 'gliotransmission'.[174] Astroglial ATP, and its metabolite adenosine, regulate neuronal synaptic strength.[76] Adenosine is an inhibitory neuromodulator and endogenous anticonvulsant, largely regulated by astrocytes through its key metabolic enzyme adenosine kinase (ADK). Increased expression of astrocytic ADK has recently been shown in hippocampal astrocytes in TLE.[19] Astrocytes, rather than neurons, are also more extensively coupled via gap junctions, primarily connexin (Cx) 43, which mediate a local glial syncytium for Ca^{2+} signalling and K^+ homeostasis (Figure 11.20). Altered expression and distribution of Cx43 has been demonstrated in epilepsy tissues, including tumours, HS and FCD. This may support altered junctional communication, although some conflicting data are reported.[155,168] Nevertheless, gap junction blockers have been used as pharmacological targets in experimental models. Finally, filamin-positive astrocytic inclusions have been noted in a paediatric epilepsy series.[191]

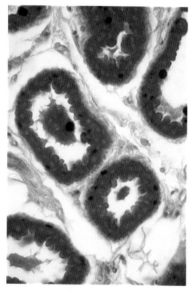

11.22 Lafora bodies. Lafora bodies seen within myoepithelial and epithelial cells in apocrine sweat glands in an axillary skin biopsy stained with Lugol's iodine.

Angiogenesis and the Blood–Brain Barrier Alteration

The BBB consists of endothelial cells with tight junctions, and its maintenance depends on normally functioning pericytes, astrocytes and basal lamina. Under normal conditions, the BBB protects the CNS by limiting the entry of plasma-borne substances and immune cells. Changes have been demonstrated in the physiology and structure of the BBB in status epilepticus and epilepsy,[100,267] with leakage of albumin (see Figure 11.20) or glutamate, potentially relevant to epileptogenesis. BBB dysfunction may be mediated by transforming growth factor (TGF)-β induction and inflammatory processes.

Angiogenesis, with increased microvessels, vascular endothelial growth factor receptor expression and impaired BBB with loss of tight junctions has been observed in TLE,[341] whereas other studies have demonstrated a reduction of microvasculature in HS.[216,292] Aquaporin 4 (AQP4) is primarily found in astrocytic end feet near capillaries, is undetectable in neurons and may play a role in regulating water movement. In FCD II, altered AQP4 distribution around capillaries and neurons was shown, which could alter the extracellular milieu and promote seizures.[281]

Pharmacoresistance or medically refractory epilepsy is epilepsy poorly controlled by antiepileptic drugs.[234] The target of the drug could be altered or, as the 'transporter hypothesis' proposes, multidrug transporters (that exist for the normal function of the BBB) are abnormally expressed, restricting antiepileptic drug access. This has been demonstrated in surgical and post-mortem samples through abnormal expression of p-glycoprotein and multidrug resistance-associated proteins[17] (see Figure 11.20).

EPILEPTIC ENCEPHALOPATHIES

The epileptic encephalopathies represent a group of devastating epileptic disorders that appear early in life.[109] They are characterized by pharmacoresistant generalized or focal seizures, persistent severe electroencephalography (EEG) abnormalities and cognitive dysfunction or decline. They include West syndrome, Lennon–Gastaut syndrome and Dravet syndrome among other childhood epilepsy syndromes. Ongoing seizures during critical periods of brain development in early life can have deleterious effects on brain maturation. Structural, congenital, genetic and metabolic causes underly some of these epileptic encephalopathies. Surgical intervention may be appropriate management in some cases.[158] There are limited neuropathological reports; one post-mortem series in patients with Dravet's syndrome with *SCN1a* mutations failed to identify specific neuropathological features.[80]

MYOCLONIC EPILEPSY

Progressive myoclonic epilepsies (PMEs) are a heterogeneous group of disorders characterized by myoclonus, epilepsy and progressive neurological dysfunction.[242] There are five major forms of PME: Unverricht–Lundborg disease,

Lafora's disease, neuronal ceroid lipofuscinosis, mitochondrial disorders and the sialidoses. In addition, there are many other rare PME disorders, typically presenting in childhood or adolescence,[150] including encephalopathy associated with neuroserpin inclusions.[344] The pathologies of Unverricht–Lundborg disease and Lafora's disease are described here, whereas neuronal ceroid lipofuscinosis, mitochondrial disorders and the sialidoses are presented in Chapters 6 and 7.

Unverricht–Lundborg Disease

Unverricht–Lundborg disease (EPM1) is an autosomal recessive disorder. The age of onset is 6–15 years,[52] presenting with slowly progressive, incapacitating, stimulus-sensitive myoclonus, in addition to generalized tonic–clonic seizures. Other neurological symptoms, such as dysarthria, ataxia and poor coordination, also develop. Pyramidal or extrapyramidal tract involvement is rare.[125] Dementia develops slowly and death usually follows after 40–50 years.

Post-mortem neuropathological examination of the brain reveals widespread non-specific degenerative changes. The most consistent neuropathological findings are losses of Purkinje and cerebellar granule cells, Bergmann gliosis and neuronal loss in the medial thalamus and spinal cord.[24] Recently, neuronal cytoplasmic inclusions containing lysosomal proteins, cathepsin-B and CD68 have been noted, in addition to labelling with TDP-43 and FUS and occasional FUS-positive ubiquitinylated neuronal intranuclear inclusions in support of neurodegeneration.[245] Loss of cortical GABA terminals has been recently shown, in support of cortical hyperexcitability.[215]

Mutations involve the *CSTB* gene on chromosome 21q22.3, encoding cystatin B, a ubiquitously expressed cystein protease inhibitor. There are various mutations reported, with the most common mutation, an unstable dodecamer expansion repeat unit in the promoter region, accounting for approximately 90 per cent of cases.[236] For this mutation, there is no genotype–phenotype correlation regarding age of onset or disease progression. Cystatin B is a small molecular weight protein that inhibits cysteine proteases and binds tightly to cathepsins B, H, L and S and is expressed in both the cytoplasm and nucleus of cells; its precise physiological function remains unknown,[242] although mutations likely result in loss of function.[208] Upregulation of cystatin B mRNA has been shown in the rat kindling model of epilepsy, which may support an endogenous neuroprotective role for this protein. Reduced mRNA levels for cystatin B occur in some cell lines from patients with EPM1 and are associated with a concomitant reduction in cystatin B activity and a corresponding increase in the activities of cathepsin S and L, which may be relevant to the pathogenesis of the observed neuronal degeneration. In animal models of the disease (knockout models or targeted mutations of the *CSTB* gene), apoptosis in cerebellar granule cells and hippocampal neurons has been shown, and these animals develop progressive ataxia and myoclonic seizures. Co-localization of cystatin B with a multiprotein complex in Purkinje cells in the adult rat cerebellum has been demonstrated, implying that this protein may have a specific function in the cerebellar region. Linkage analysis in a family with Unverricht–Lundborg disease has identified a further locus on chromosome 12 (designated EPM1B) associated with a younger age of disease onset.[158]

Lafora's Disease

The PME of Lafora's disease typically has its onset in late childhood or early adolescence, and is associated with rapid progression with death occurring within 10 years of diagnosis. Cases with a later onset and a more protracted course have also been described.[80] Seizures with visual symptoms are often present, in addition to generalized tonic–clonic seizures and myoclonic seizures.[125] Pyramidal tract involvement is common and cerebellar degeneration usually evolves later. Initially, seizures may respond well to medication, but then rapid progression differentiates this from benign myoclonus syndromes. Progressive dementia, apraxia and visual loss occur, leading ultimately to a vegetative state. Neuropathologically, inclusion bodies (Lafora bodies) distinguish Lafora's disease from other PME syndromes. These are periodic acid–Schiff (PAS)-positive, spherical to ovoid, intracytoplasmic inclusions (polyglucosan bodies) of 30–40 μm diameter, identified in the brain, spinal cord, striated muscle, heart, liver and skin. In the brain, the inclusions are found in the perikarya of neurons of cerebral cortex, thalamus, globus pallidus, substantia nigra, cerebellar cortex and dentate nucleus. Small dust-like PAS-positive granules may also be seen in the cytoplasm. These should be distinguished from corpora amylacea, a normal feature of the ageing brain, which are not found in neurons. In adult polyglucosan body disease, similar inclusions are present, but in axon hillocks rather than perikarya. Substantial neuronal loss is often observed post mortem. Prior to genetic testing, axillary skin biopsy was the diagnostic procedure of choice revealing Lafora bodies in peripheral sweat gland cells and apocrine myoepithelial cells (Figure 11.22), although the sensitivity was suboptimal with frequent false-negative results. Ultrastructural examination shows that Lafora bodies consist of granular and filamentous material, which is composed of insoluble glycogen, packed more densely in the core. The inclusions are closely associated with the endoplasmic reticulum.

Lafora's disease is caused by defects in one of three chromosomal loci. Up to 80 per cent are caused by mutations in the *EPM2A* gene, located on chromosome 6q24, which encodes two isoforms of the protein laforin, a protein phosphatase.[286] Mutations in a second gene, *EPM2B*, located on 6p22.3, encoding the ubiquitin ligase malin, also cause Lafora's disease, although there is evidence for a third locus.[93,367] Laforin and malin interact with each other as a complex and participate in diverse cellular pathways, including glycogen metabolism, in the ubiquitin–proteasome system and in the heat shock response. They interact with a diverse set of proteins, including GSK3.[367] The Lafora body may represent an end point structure that emerges as a neuronal response to major carbohydrate metabolism impairment.[258] In addition, impairment of autophagy–endosomal–lysosomal pathways likely contributes to the underlying progression.[337]

REFERENCES

1. Abraham H, Veszpremi B, Kravjak A, Kovacs K, Gomori E, Seress L. Ontogeny of calbindin immunoreactivity in the human hippocampal formation with a special emphasis on granule cells of the dentate gyrus. *Int J Dev Neurosci* 2009;27:115–27.
2. Allen AS, Berkovic SF, Cossette P, Delanty N, Dlugos D, Eichler EE, et al. De novo mutations in epileptic encephalopathies. *Nature* 2013;501:217–21.
3. Alonso-Nanclares L, Garbelli R, Sola RG, Pastor J, Tassi L, Spreafico R, et al. Microanatomy of the dysplastic neocortex from epileptic patients. *Brain* 2005;128:158–73.
4. Alonso-Nanclares L, Kastanauskaite A, Rodriguez JR, Gonzalez-Soriano J, Defelipe J. A stereological study of synapse number in the epileptic human hippocampus. *Front Neuroanat* 2011;5:8.
5. Andre VM, Cepeda C, Vinters HV, Huynh M, Mathern GW, Levine MS. Interneurons, GABAA currents, and subunit composition of the GABAA receptor in type I and type II cortical dysplasia. *Epilepsia* 2010;51(Suppl 3):166–70.
6. Andre VM, Flores-Hernandez J, Cepeda C, Starling AJ, Nguyen S, Lobo MK, et al. NMDA receptor alterations in neurons from pediatric cortical dysplasia tissue. *Cereb Cortex* 2004;14:634–46.
7. Andre VM, Wu N, Yamazaki I, Nguyen ST, Fisher RS, Vinters HV, et al. Cytomegalic interneurons: a new abnormal cell type in severe pediatric cortical dysplasia. *J Neuropathol Exp Neurol* 2007;66:491–504.
8. Andres M, Andre VM, Nguyen S, Salamon N, Cepeda C, Levine MS, et al. Human cortical dysplasia and epilepsy: an ontogenetic hypothesis based on volumetric MRI and NeuN neuronal density and size measurements. *Cereb Cortex* 2005;15:194–210.
9. Antonucci F, Alpar A, Kacza J, Caleo M, Verderio C, Giani A, et al. Cracking down on inhibition: selective removal of GABAergic interneurons from hippocampal networks. *J Neurosci* 2012;32:1989–2001.
10. Araujo D, Santos AC, Velasco TR, Wichert-Ana L, Terra-Bustamante VC, Alexandre V, Jr., et al. Volumetric evidence of bilateral damage in unilateral mesial temporal lobe epilepsy. *Epilepsia* 2006;47:1354–9.
11. Arellano JI, Munoz A, Ballesteros-Yanez I, Sola RG, DeFelipe J. Histopathology and reorganization of chandelier cells in the human epileptic sclerotic hippocampus. *Brain* 2004;127(Pt 1):45–64.
12. Aronica E, Gorter JA. Gene expression profile in temporal lobe epilepsy. *Neuroscientist* 2007;13:100–8.
13. Aronica E, Boer K, Baybis M, Yu J, Crino P. Co-expression of cyclin D1 and phosphorylated ribosomal S6 proteins in hemimegalencephaly. *Acta Neuropathol* 2007;114:287–93.
14. Aronica E, Boer K, Redeker S, Spliet WG, van Rijen PC, Troost D, et al. Differential expression patterns of chloride transporters, Na⁺-K⁺-2Cl⁻-cotransporter and K⁺-Cl⁻-cotransporter, in epilepsy-associated malformations of cortical development. *Neuroscience* 2007;145:185–96.
15. Aronica E, Crino PB. Inflammation in epilepsy: clinical observations. *Epilepsia* 2011;52(Suppl 3):26–32.
16. Aronica E, Gorter JA, Jansen GH, Leenstra S, Yankaya B, Troost D. Expression of connexin 43 and connexin 32 gap-junction proteins in epilepsy-associated brain tumors and in the perilesional epileptic cortex. *Acta Neuropathol* 2001;101:449–59.
17. Aronica E, Sisodiya SM, Gorter JA. Cerebral expression of drug transporters in epilepsy. *Adv Drug Deliv Rev* 2011.
18. Aronica E, Yankaya B, Jansen GH, Leenstra S, van Veelen CW, Gorter JA, et al. Ionotropic and metabotropic glutamate receptor protein expression in glioneuronal tumours from patients with intractable epilepsy. *Neuropathol Applied Neurobiol* 2001;27:223–37.
19. Aronica E, Zurolo E, Iyer A, de Groot M, Anink J, Carbonell C, et al. Upregulation of adenosine kinase in astrocytes in experimental and human temporal lobe epilepsy. *Epilepsia* 2011;52:1645–55.
20. Avanzini G. A sound conceptual framework for an epilepsy classification is still lacking. *Epilepsia* 2010;51:720–2.
21. Avoli M, de Curtis M. GABAergic synchronization in the limbic system and its role in the generation of epileptiform activity. *Prog Neurobiol* 2011;95:104–32.
22. Babb TL, Brown WJ, Pretorius J, Davenport C, Lieb JP, Crandall PH. Temporal lobe volumetric cell densities in temporal lobe epilepsy. *Epilepsia* 1984;25:729–40.
23. Bajic D, Wang C, Kumlien E, Mattsson P, Lundberg S, Eeg-Olofsson O, et al. Incomplete inversion of the hippocampus: a common developmental anomaly. *Eur Radiol* 2008;18:138–42.
24. Bandopadhyay R, Liu JY, Sisodiya SM, Thom M. A comparative study of the dentate gyrus in hippocampal sclerosis in epilepsy and dementia. *Neuropathol Applied Neurobiol* 2014;40:177–90.
25. Barinka F, Druga R, Marusic P, Krsek P, Zamecnik J. Calretinin immunoreactivity in focal cortical dysplasias and in non-malformed epileptic cortex. *Epilepsy Res* 2010;88:76–86.
26. Barkovich AJ, Guerrini R, Kuzniecky RI, Jackson GD, Dobyns WB. A developmental and genetic classification for malformations of cortical development: update 2012. *Brain* 2012;135:1348–69.
27. Barmashenko G, Hefft S, Aertsen A, Kirschstein T, Kohling R. Positive shifts of the GABAA receptor reversal potential due to altered chloride homeostasis is widespread after status epilepticus. *Epilepsia* 2011;52:1570–8.
28. Barsi P, Kenez J, Solymosi D, Kulin A, Halasz P, Rasonyi G, et al. Hippocampal malrotation with normal corpus callosum: a new entity? *Neuroradiology* 2000;42:339–45.
29. Bauer J, Bien CG. Encephalitis and epilepsy. *Semin Immunopathol* 2009;31:537–44.
30. Bauer J, Elger CE, Hans VH, Schramm J, Urbach H, Lassmann H, et al. Astrocytes are a specific immunological target in Rasmussen's encephalitis. *Ann Neurol* 2007;62:67–80.
31. Baulac M, De Grissac N, Hasboun D, Oppenheim C, Adam C, Arzimanoglou A, et al. Hippocampal developmental changes in patients with partial epilepsy: magnetic resonance imaging and clinical aspects. *Ann Neurol* 1998;44:223–33.
32. Baulac S, Gourfinkel-An I, Nabbout R, Huberfeld G, Serratosa J, Leguern E, et al. Fever, genes, and epilepsy. *Lancet Neurol* 2004;3:421–30.
33. Baybis M, Yu J, Lee A, Golden JA, Weiner H, McKhann G, 2nd, et al. mTOR cascade activation distinguishes tubers from focal cortical dysplasia. *Ann Neurol* 2004;56:478–87.
34. Becker AJ, Urbach H, Scheffler B, Baden T, Normann S, Lahl R, et al. Focal cortical dysplasia of Taylor's balloon cell type: mutational analysis of the TSC1 gene indicates a pathogenic relationship to tuberous sclerosis. *Ann Neurol* 2002;52:29–37.
35. Ben-Ari Y, Holmes GL. The multiple facets of gamma-aminobutyric acid dysfunction in epilepsy. *Curr Opin Neurol* 2005;18:141–5.
36. Ben-Ari Y, Holmes GL. Effects of seizures on developmental processes in the immature brain. *Lancet Neurol* 2006;5:1055–63.
37. Benini R, D'Antuono M, Pralong E, Avoli M. Involvement of amygdala networks in epileptiform synchronization in vitro. *Neuroscience* 2003;120:75–84.
38. Berg AT, Berkovic SF, Brodie MJ, Buchhalter J, Cross JH, van Emde Boas W, et al. Revised terminology and concepts for organization of seizures and epilepsies: report of the ILAE Commission on Classification and Terminology, 2005–2009. *Epilepsia* 2010;51:676–85.
39. Bernasconi N, Andermann F, Arnold DL, Bernasconi A. Entorhinal cortex MRI assessment in temporal, extratemporal, and idiopathic generalized epilepsy. *Epilepsia* 2003;44:1070–4.
40. Bertram EH, Zhang D, Williamson JM. Multiple roles of midline dorsal thalamic nuclei in induction and spread of limbic seizures. *Epilepsia* 2008;49:256–68.
41. Bien CG, Bauer J, Deckwerth TL, Wiendl H, Deckert M, Wiestler OD, et al. Destruction of neurons by cytotoxic T cells: a new pathogenic mechanism in Rasmussen's encephalitis. *Ann Neurol* 2002;51:311–8.
42. Bien CG, Granata T, Antozzi C, Cross JH, Dulac O, Kurthen M, et al. Pathogenesis, diagnosis and treatment of Rasmussen encephalitis: a European consensus statement. *Brain* 2005;128:454–71.
43. Black M, Graham DI. Sudden unexplained death in adults caused by intracranial pathology. *J Clin Pathol* 2002;55:44–50.
44. Blaesse P, Airaksinen MS, Rivera C, Kaila K. Cation-chloride cotransporters and neuronal function. *Neuron* 2009;61:820–38.
45. Blanc F, Martinian L, Liagkouras I, Catarino C, Sisodiya SM, Thom M. Investigation of widespread neocortical pathology associated with hippocampal sclerosis in epilepsy: a postmortem study. *Epilepsia* 2011;52:10–21.

46. Blumcke I. Neuropathology of focal epilepsies: a critical review. *Epilepsy Behav* 2009;**15**:34–9.

47. Blumcke I, Beck H, Suter B, Hoffmann D, Fodisch HJ, Wolf HK, et al. An increase of hippocampal calretinin-immunoreactive neurons correlates with early febrile seizures in temporal lobe epilepsy. *Acta Neuropathol* 1999;**97**:31–9.

48. Blumcke I, Kistner I, Clusmann H, Schramm J, Becker AJ, Elger CE, et al. Towards a clinico-pathological classification of granule cell dispersion in human mesial temporal lobe epilepsies. *Acta Neuropathol* 2009;**117**:535–44.

49. Blumcke I, Pauli E, Clusmann H, Schramm J, Becker A, Elger C, et al. A new clinico-pathological classification system for mesial temporal sclerosis. *Acta Neuropathol* 2007;**113**:235–44.

50. Blumcke I, Pieper T, Pauli E, Hildebrandt M, Kudernatsch M, Winkler P, et al. A distinct variant of focal cortical dysplasia type I characterised by magnetic resonance imaging and neuropathological examination in children with severe epilepsies. *Epileptic Disord* 2010;**12**:172–80.

51. Blumcke I, Suter B, Behle K, Kuhn R, Schramm J, Elger CE, et al. Loss of hilar mossy cells in Ammon's horn sclerosis. *Epilepsia* 2000;**41**(Suppl 6):S174–80.

52. Blumcke I, Thom M, Aronica E, Armstrong DD, Bartolomei F, Bernasconi A, et al. International consensus classification of hippocampal sclerosis in temporal lobe epilepsy: a Task Force report from the ILAE Commission on Diagnostic Methods. *Epilepsia* 2013;**54**:1315–29.

53. Blumcke I, Thom M, Aronica E, Armstrong DD, Vinters HV, Palmini A, et al. The clinicopathologic spectrum of focal cortical dysplasias: a consensus classification proposed by an ad hoc Task Force of the ILAE Diagnostic Methods Commission. *Epilepsia* 2011;**52**:158–74.

54. Blumcke I, Thom M, Wiestler OD. Ammon's horn sclerosis: a maldevelopmental disorder associated with temporal lobe epilepsy. *Brain Pathol* 2002;**12**:199–211.

55. Blumcke I, Zuschratter W, Schewe JC, Suter B, Lie AA, Riederer BM, et al. Cellular pathology of hilar neurons in Ammon's horn sclerosis. *J Comp Neurol* 1999;**414**:437–53.

56. Boer K, Jansen F, Nellist M, Redeker S, van den Ouweland AM, Spliet WG, et al. Inflammatory processes in cortical tubers and subependymal giant cell tumors of tuberous sclerosis complex. *Epilepsy Res* 2008;**78**:7–21.

57. Boer K, Spliet WG, van Rijen PC, Jansen FE, Aronica E. Expression patterns of AMOG in developing human cortex and malformations of cortical development. *Epilepsy Res* 2010;**91**:84–93.

58. Boer K, Troost D, Spliet WG, van Rijen PC, Gorter JA, Aronica E. Cellular distribution of vascular endothelial growth factor A (VEGFA) and B (VEGFB) and VEGF receptors 1 and 2 in focal cortical dysplasia type IIB. *Acta Neuropathol* 2008;**115**:683–96.

59. Bonilha L, Elm JJ, Edwards JC, Morgan PS, Hicks C, Lozar C, et al. How common is brain atrophy in patients with medial temporal lobe epilepsy? *Epilepsia* 2010;**51**:1774–9.

60. Bonilha L, Halford JJ, Morgan PS, Edwards JC. Hippocampal atrophy in temporal lobe epilepsy: the 'generator' and 'receiver'. *Acta Neurol Scand* 2012;**125**:105–10.

61. Bonilha L, Martz GU, Glazier SS, Edwards JC. Subtypes of medial temporal lobe epilepsy: influence on temporal lobectomy outcomes? *Epilepsia* 2012;**53**:1–6.

62. Bonilha L, Rorden C, Halford JJ, Eckert M, Appenzeller S, Cendes F, et al. Asymmetrical extra-hippocampal grey matter loss related to hippocampal atrophy in patients with medial temporal lobe epilepsy. *J Neurol Neurosurg Psychiatry* 2007;**78**:286–94.

63. Boonyapisit K, Najm I, Klem G, Ying Z, Burrier C, LaPresto E, et al. Epileptogenicity of focal malformations due to abnormal cortical development: direct electrocorticographic-histopathologic correlations. *Epilepsia* 2003;**44**:69–76.

64. Bothwell S, Meredith GE, Phillips J, Staunton H, Doherty C, Grigorenko E, et al. Neuronal hypertrophy in the neocortex of patients with temporal lobe epilepsy. *J Neurosci* 2001;**21**: 4789–800.

65. Bouilleret V, Schwaller B, Schurmans S, Celio MR, Fritschy JM. Neurodegenerative and morphogenic changes in a mouse model of temporal lobe epilepsy do not depend on the expression of the calcium-binding proteins parvalbumin, calbindin, or calretinin. *Neuroscience* 2000;**97**:47–58.

66. Bozzi Y, Dunleavy M, Henshall DC. Cell signaling underlying epileptic behavior. *Front Behav Neurosci* 2011;**5**: 45.

67. Briellmann RS, Berkovic SF, Syngeniotis A, King MA, Jackson GD. Seizure-associated hippocampal volume loss: a longitudinal magnetic resonance study of temporal lobe epilepsy. *Ann Neurol* 2002;**51**:641–4.

68. Briellmann RS, Newton MR, Wellard RM, Jackson GD. Hippocampal sclerosis following brief generalized seizures in adulthood. *Neurology* 2001;**57**:315–7.

69. Brooks-Kayal AR, Shumate MD, Jin H, Lin DD, Rikhter TY, Holloway KL, et al. Human neuronal gamma-aminobutyric acid(A) receptors: coordinated subunit mRNA expression and functional correlates in individual dentate granule cells. *J Neurosci* 1999;**19**:8312–8.

70. Bruton CJ. *The neuropathology of temporal lobe epilepsy.* Oxford: Oxford University Press; 1988.

71. Buck D, Baker GA, Jacoby A, Smith DF, Chadwick DW. Patients' experiences of injury as a result of epilepsy. *Epilepsia* 1997;**38**:439–44.

72. Buckingham SC, Campbell SL, Haas BR, Montana V, Robel S, Ogunrinu T, et al. Glutamate release by primary brain tumors induces epileptic activity. *Nat Med* 2011;**17**:1269–74.

73. Buckmaster PS, Ingram EA, Wen X. Inhibition of the mammalian target of rapamycin signaling pathway suppresses dentate granule cell axon sprouting in a rodent model of temporal lobe epilepsy. *J Neurosci* 2009;**29**:8259–69.

74. Burneo JG, Tellez-Zenteno J, Steven DA, Niaz N, Hader W, Pillay N, et al. Adult-onset epilepsy associated with dysembryoplastic neuroepithelial tumors. *Seizure* 2008;**17**:498–504.

75. Busch RM, Floden D, Lineweaver TT, Chapin JS, Unnwongse K, Wehner T, et al. Effect of apolipoprotein epsilon-4 allele on hippocampal and brain volume in intractable temporal lobe epilepsy. *Epilepsy Behav* 2011;**21**:88–90.

76. Butt AM. ATP: a ubiquitous gliotransmitter integrating neuron-glial networks. *Semin Cell Dev Biol* 2011;**22**:205–13.

77. Cameron MC, Zhan RZ, Nadler JV. Morphologic integration of hilar ectopic granule cells into dentate gyrus circuitry in the pilocarpine model of temporal lobe epilepsy. *J Comp Neurol* 2011;**519**:2175–92.

78. Cardoso A, Lukoyanova EA, Madeira MD, Lukoyanov NV. Seizure-induced structural and functional changes in the rat hippocampal formation: comparison between brief seizures and status epilepticus. *Behav Brain Res* 2011;**225**:538–46.

79. Catarino CB, Kasperaviciute D, Thom M, Cavalleri GL, Martinian L, Heinzen EL, et al. Genomic microdeletions associated with epilepsy: not a contraindication to resective surgery. *Epilepsia* 2011;**52**:1388–92.

80. Catarino CB, Liu JY, Liagkouras I, Gibbons VS, Labrum RW, Ellis R, et al. Dravet syndrome as epileptic encephalopathy: evidence from long-term course and neuropathology. *Brain* 2011;**134**:2982–3010.

81. Cavalleri GL, Lynch JM, Depondt C, Burley MW, Wood NW, Sisodiya SM, et al. Failure to replicate previously reported genetic associations with sporadic temporal lobe epilepsy: where to from here? *Brain* 2005;**128**:1832–40.

82. Cavanagh JB, Meyer A. Aetiological aspects of Ammon's horn sclerosis associated with temporal lobe epilepsy. *BMJ* 1956;**2**:1403–7.

83. Cavazos JE, Golarai G, Sutula TP. Mossy fiber synaptic reorganization induced by kindling: time course of development, progression, and permanence. *J Neurosci* 1991;**11**:2795–803.

84. Cavazos JE, Zhang P, Qazi R, Sutula TP. Ultrastructural features of sprouted mossy fiber synapses in kindled and kainic acid-treated rats. *J Comp Neurol* 2003;**458**:272–92.

85. Cendes F. Febrile seizures and mesial temporal sclerosis. *Curr Opin Neurol* 2004;**17**:161–4.

86. Cepeda C, Andre VM, Flores-Hernandez J, Nguyen OK, Wu N, Klapstein GJ, et al. Pediatric cortical dysplasia: correlations between neuroimaging, electrophysiology and location of cytomegalic neurons and balloon cells and glutamate/GABA synaptic circuits. *Dev Neurosci* 2005;**27**:59–76.

87. Cepeda C, Andre VM, Hauptman JS, Yamazaki I, Huynh MN, Chang JW, et al. Enhanced GABAergic network and receptor function in pediatric cortical dysplasia Type IIB compared with Tuberous Sclerosis Complex. *Neurobiol Dis* 2012;**45**:310–21.

88. Cepeda C, Andre VM, Levine MS, Salamon N, Miyata H, Vinters HV, et al. Epileptogenesis in pediatric cortical dysplasia: the dysmature cerebral developmental hypothesis. *Epilepsy Behav* 2006;**9**:219–35.

89. Cepeda C, Andre VM, Yamazaki I, Hauptman JS, Chen JY, Vinters HV, et al. Comparative study of cellular and synaptic abnormalities in brain tissue

samples from pediatric tuberous sclerosis complex and cortical dysplasia type II. *Epilepsia* 2010;**51(Suppl 3)**: 160–5.

90. Cepeda C, Hurst RS, Flores-Hernandez J, Hernandez-Echeagaray E, Klapstein GJ, Boylan MK, et al. Morphological and electrophysiological characterization of abnormal cell types in pediatric cortical dysplasia. *J Neurosci Res* 2003;**72**: 472–86.

91. Chamberlain WA, Cohen ML, Gyure KA, Kleinschmidt-DeMasters BK, Perry A, Powell SZ, et al. Interobserver and intraobserver reproducibility in focal cortical dysplasia (malformations of cortical development). *Epilepsia* 2009;**50**:2593–8.

92. Chamberlain WA, Prayson RA. Focal cortical dysplasia type II (malformations of cortical development) aberrantly expresses apoptotic proteins. *Appl Immunohistochem Mol Morphol* 2008;**16**:471–6.

93. Chan EM, Omer S, Ahmed M, Bridges LR, Bennett C, Scherer SW, et al. Progressive myoclonus epilepsy with polyglucosans (Lafora disease): evidence for a third locus. *Neurology* 2004;**63**: 565–7.

94. Chandra PS, Salamon N, Nguyen ST, Chang JW, Huynh MN, Cepeda C, et al. Infantile spasm-associated microencephaly in tuberous sclerosis complex and cortical dysplasia. *Neurology* 2007;**68**:438–45.

95. Chapman K, Wyllie E, Najm I, Ruggieri P, Bingaman W, Luders J, et al. Seizure outcome after epilepsy surgery in patients with normal preoperative MRI. *J Neurol Neurosurg Psychiatry* 2005;**76**:710–3.

96. Chassoux F, Landre E, Mellerio C, Turak B, Mann MW, Daumas-Duport C, et al. Type II focal cortical dysplasia: Electroclinical phenotype and surgical outcome related to imaging. *Epilepsia* 2012;**53**:349–58.

97. Chen J, Tsai V, Parker WE, Aronica E, Baybis M, Crino PB. Detection of human papillomavirus in human focal cortical dysplasia type IIB. *Ann Neurol* 2012;**72**:881–92.

98. Chen JW, Naylor DE, Wasterlain CG. Advances in the pathophysiology of status epilepticus. *Acta Neurol Scand* 2007;**115(4 Suppl)**:7–15.

99. Chern JJ, Patel AJ, Jea A, Curry DJ, Comair YG. Surgical outcome for focal cortical dysplasia: an analysis of recent surgical series. *J Neurosurg Pediatr* 2010;**6**:452–8.

100. Choi J, Koh S. Role of brain inflammation in epileptogenesis. *Yonsei Med J* 2008;**49**:1–18.

101. Cohen I, Navarro V, Clemenceau S, Baulac M, Miles R. On the origin of interictal activity in human temporal lobe epilepsy in vitro. *Science* 2002;**298**:1418–21.

102. Cohen-Gadol AA, Ozduman K, Bronen RA, Kim JH, Spencer DD. Long-term outcome after epilepsy surgery for focal cortical dysplasia. *J Neurosurg* 2004;**101**:55–65.

103. Colciaghi F, Finardi A, Frasca A, Balosso S, Nobili P, Carriero G, et al. Status epilepticus-induced pathologic plasticity in a rat model of focal cortical dysplasia. *Brain* 2011;**134**:2828–43.

104. Conti L, Palma E, Roseti C, Lauro C, Cipriani R, de Groot M, et al. Anomalous levels of Cl- transporters cause a decrease of GABAergic inhibition in human

peritumoral epileptic cortex. *Epilepsia* 2011;**52**:1635–44.

105. Coras R, de Boer OJ, Armstrong D, Becker A, Jacques TS, Miyata H, et al. Good interobserver and intraobserver agreement in the evaluation of the new ILAE classification of focal cortical dysplasias. *Epilepsia* 2012;**53**:1341–8.

106. Coras R, Siebzehnrubl FA, Pauli E, Huttner HB, Njunting M, Kobow K, et al. Low proliferation and differentiation capacities of adult hippocampal stem cells correlate with memory dysfunction in humans. *Brain* 2010;**133**:3359–72.

107. Corsellis JA, Bruton CJ. Neuropathology of status epilepticus in humans. *Adv Neurol* 1983;**34**:129–39.

108. Cossart R, Bernard C, Ben-Ari Y. Multiple facets of GABAergic neurons and synapses: multiple fates of GABA signalling in epilepsies. *Trends Neurosci* 2005;**28**:108–15.

109. Covanis A. Epileptic encephalopathies (including severe epilepsy syndromes). *Epilepsia* 2012;**53(Suppl 4)**: 114–26.

110. Crino PB. The pathophysiology of tuberous sclerosis complex. *Epilepsia* 2010;**51(Suppl 1)**: 27–9.

111. Crooks R, Mitchell T, Thom M. Patterns of cerebellar atrophy in patients with chronic epilepsy: a quantitative neuropathological study. *Epilepsy Res* 2000;**41**:63–73.

112. Cui Z, Luan G. A venous malformation accompanying focal cortical dysplasia resulting in a reorganization of language-eloquent areas. *J Clin Neurosci* 2011;**18**:404–6.

113. da Silva AV, Regondi MC, Cavalheiro EA, Spreafico R. Disruption of cortical development as a consequence of repetitive pilocarpine-induced status epilepticus in rats. *Epilepsia* 2005;**46 (Suppl 5)**: 22–30.

114. D'Ambrosio R, Perucca E. Epilepsy after head injury. *Curr Opin Neurol* 2004;**17**:731–5.

115. D'Antuono M, Louvel J, Kohling R, Mattia D, Bernasconi A, Olivier A, et al. GABAA receptor-dependent synchronization leads to ictogenesis in the human dysplastic cortex. *Brain* 2004;**127**:1626–40.

116. Davies KG, Hermann BP, Dohan FC, Jr., Foley KT, Bush AJ, Wyler AR. Relationship of hippocampal sclerosis to duration and age of onset of epilepsy, and childhood febrile seizures in temporal lobectomy patients. *Epilepsy Res* 1996;**24**:119–26.

117. Dawodu S, Thom M. Quantitative neuropathology of the entorhinal cortex region in patients with hippocampal sclerosis and temporal lobe epilepsy. *Epilepsia* 2005;**46**:23–30.

118. de Groot M, Iyer A, Zurolo E, Anink J, Heimans JJ, Boison D, et al. Overexpression of ADK in human astrocytic tumors and peritumoral tissue is related to tumor-associated epilepsy. *Epilepsia* 2012;**53**:58–66.

119. de Groot M, Reijneveld JC, Aronica E, Heimans JJ. Epilepsy in patients with a brain tumour: focal epilepsy requires focused treatment. *Brain* 2012;**135**:1002–16.

120. de Guzman P, D'Antuono M, Avoli M. Initiation of electrographic seizures by neuronal networks in entorhinal and perirhinal cortices in vitro. *Neuroscience* 2004;**123**:875–86.

121. de Lanerolle NC, Kim JH, Robbins RJ, Spencer DD. Hippocampal interneuron loss and plasticity in human temporal lobe epilepsy. *Brain Res* 1989;**495**:387–95.

122. de Lanerolle NC, Kim JH, Williamson A, Spencer SS, Zaveri HP, Eid T, et al. A retrospective analysis of hippocampal pathology in human temporal lobe epilepsy: evidence for distinctive patient subcategories. *Epilepsia* 2003;**44**:677–87.

123. de Lanerolle NC, Lee TS. New facets of the neuropathology and molecular profile of human temporal lobe epilepsy. *Epilepsy Behav* 2005;**7**:190–203.

124. de Tisi J, Bell GS, Peacock JL, McEvoy AW, Harkness WF, Sander JW, et al. The long-term outcome of adult epilepsy surgery, patterns of seizure remission, and relapse: a cohort study. *Lancet* 2011;**378**:1388–95.

125. Delgado-Escueta AV, Perez-Gosiengfiao KB, Bai D, Bailey J, Medina MT, Morita R, et al. Recent developments in the quest for myoclonic epilepsy genes. *Epilepsia* 2003;**44(Suppl 11)**:13–26.

126. Delorenzo RJ, Sun DA, Deshpande LS. Cellular mechanisms underlying acquired epilepsy: the calcium hypothesis of the induction and maintainance of epilepsy. *Pharmacol Ther* 2005;**105**:229–66.

127. Dube C, Richichi C, Bender RA, Chung G, Litt B, Baram TZ. Temporal lobe epilepsy after experimental prolonged febrile seizures: prospective analysis. *Brain* 2006;**129**:911–22.

128. Duncan JS. Imaging and epilepsy. *Brain* 1997;**120**:339–77.

129. Elst LT, Groffmann M, Ebert D, Schulze-Bonhage A. Amygdala volume loss in patients with dysphoric disorder of epilepsy. *Epilepsy Behav* 2009;**16**:105–12.

130. Emery JA, Roper SN, Rojiani AM. White matter neuronal heterotopia in temporal lobe epilepsy: a morphometric and immunohistochemical study. *J Neuropathol Exp Neurol* 1997;**56**:1276–82.

131. Engel J, Jr. A proposed diagnostic scheme for people with epileptic seizures and with epilepsy: report of the ILAE Task Force on Classification and Terminology. *Epilepsia* 2001;**42**:796–803.

132. Engel J, Jr., Thompson PM. Going beyond hippocampocentricity in the concept of mesial temporal lobe epilepsy. *Epilepsia* 2012;**53**:220–3.

133. Engel J, Jr., Wiebe S, French J, Sperling M, Williamson P, Spencer D, et al. Practice parameter: temporal lobe and localized neocortical resections for epilepsy: report of the Quality Standards Subcommittee of the American Academy of Neurology, in association with the American Epilepsy Society and the American Association of Neurological Surgeons. *Neurology* 2003;**60**:538–47.

134. Engel T, Henshall DC. Apoptosis, Bcl-2 family proteins and caspases: the ABCs of seizure-damage and epileptogenesis? *Int J Physiol Pathophysiol Pharmacol* 2009;**1**:97–115.

135. Engel T, Schindler CK, Sanz-Rodriguez A, Conroy RM, Meller R, Simon RP, et al. Expression of neurogenesis genes in human temporal lobe epilepsy with hippocampal sclerosis. *Int J Physiol Pathophysiol Pharmacol* 2011;**3**:38–47.

136. Englander J, Bushnik T, Duong TT, Cifu DX, Zafonte R, Wright J, et al. Analyzing risk factors for late posttraumatic seizures: a prospective, multicenter

investigation. *Arch Phys Med Rehabil* 2003;**84**:365–73.

137. Englund C, Folkerth RD, Born D, Lacy JM, Hevner RF. Aberrant neuronal-glial differentiation in Taylor-type focal cortical dysplasia (type IIA/B). *Acta Neuropathol* 2005;**109**:519–33.

138. Eriksson SH, Free SL, Thom M, Martinian L, Sisodiya SM. Methodological aspects of 3D and automated 2D analyses of white matter neuronal density in temporal lobe epilepsy. *Neuropathol Appl Neurobiol* 2006;**32**:260–70.

139. Eriksson SH, Free SL, Thom M, Martinian L, Symms MR, Salmenpera TM, *et al.* Correlation of quantitative MRI and neuropathology in epilepsy surgical resection specimens: T2 correlates with neuronal tissue in gray matter. *NeuroImage* 2007;**37**:48–55.

140. Eriksson SH, Nordborg C, Rydenhag B, Malmgren K. Parenchymal lesions in pharmacoresistant temporal lobe epilepsy: dual and multiple pathology. *Acta Neurol Scand* 2005;**112**:151–6.

141. Eriksson SH, Thom M, Symms MR, Focke NK, Martinian L, Sisodiya SM, *et al.* Cortical neuronal loss and hippocampal sclerosis are not detected by voxel-based morphometry in individual epilepsy surgery patients. *Hum Brain Map* 2009;**30**:3351–60.

142. Ess KC. Tuberous sclerosis complex: a brave new world? *Curr Opin Neurol* 2010;**23**:189–93.

143. Evans LT, Morse R, Roberts DW. Epilepsy surgery in tuberous sclerosis: a review. *Neurosurg Focus* 2012;**32**:E5.

144. Fabene PF, Bramanti P, Constantin G. The emerging role for chemokines in epilepsy. *J Neuroimmunol* 2010;**224**:22–7.

145. Faber-Zuschratter H, Huttmann K, Steinhauser C, Becker A, Schramm J, Okafo U, *et al.* Ultrastructural and functional characterization of satellitosis in the human lateral amygdala associated with Ammon's horn sclerosis. *Acta Neuropathol* 2009;**117**:545–55.

146. Fahrner A, Kann G, Flubacher A, Heinrich C, Freiman TM, Zentner J, *et al.* Granule cell dispersion is not accompanied by enhanced neurogenesis in temporal lobe epilepsy patients. *Exp Neurol* 2006.

147. Fauser S, Becker A, Schulze-Bonhage A, Hildebrandt M, Tuxhorn I, Pannek HW, *et al.* CD34-immunoreactive balloon cells in cortical malformations. *Acta Neuropathol* 2004;**108**:272–8.

148. Fauser S, Schulze-Bonhage A. Epileptogenicity of cortical dysplasia in temporal lobe dual pathology: an electrophysiological study with invasive recordings. *Brain* 2006;**129**:82–95.

149. Fauser S, Sisodiya SM, Martinian L, Thom M, Gumbinger C, Huppertz HJ, *et al.* Multi-focal occurrence of cortical dysplasia in epilepsy patients. *Brain* 2009;**132**:2079–90.

150. Fazel S, Wolf A, Langstrom N, Newton CR, Lichtenstein P. Premature mortality in epilepsy and the role of psychiatric comorbidity: a total population study. *Lancet* 2013;**382**:1646–54.

151. Fernandez G, Effenberger O, Vinz B, Steinlein O, Elger CE, Dohring W, *et al.* Hippocampal malformation as a cause of familial febrile convulsions and subsequent hippocampal sclerosis. *Neurology* 1998;**50**:909–17.

152. Ferrier CH, Aronica E, Leijten FS, Spliet WG, Boer K, van Rijen PC, *et al.* Electrocorticography discharge patterns in patients with a cavernous hemangioma and pharmacoresistent epilepsy. *J Neurosurg* 2007;**107**:495–503.

153. Finardi A, Gardoni F, Bassanini S, Lasio G, Cossu M, Tassi L, *et al.* NMDA receptor composition differs among anatomically diverse malformations of cortical development. *J Neuropathol Exp Neurol* 2006;**65**:883–93.

154. Fisher RS, van Emde Boas W, Blume W, Elger C, Genton P, Lee P, *et al.* Epileptic seizures and epilepsy: definitions proposed by the International League Against Epilepsy (ILAE) and the International Bureau for Epilepsy (IBE). *Epilepsia* 2005;**46**:470–2.

155. Fonseca CG, Green CR, Nicholson LF. Upregulation in astrocytic connexin 43 gap junction levels may exacerbate generalized seizures in mesial temporal lobe epilepsy. *Brain Res* 2002;**929**:105–16.

156. Freiman TM, Eismann-Schweimler J, Frotscher M. Granule cell dispersion in temporal lobe epilepsy is associated with changes in dendritic orientation and spine distribution. *Exp Neurol* 2011;**229**:332–8.

157. Freund TF, Buzsaki G. Interneurons of the hippocampus. *Hippocampus* 1996;**6**:347–470.

158. Fridley J, Reddy G, Curry D, Agadi S. Surgical treatment of pediatric epileptic encephalopathies. *Epilepsy Res Treat* 2013;**2013**: 720841.

159. Friedman D, Honig LS, Scarmeas N. Seizures and epilepsy in Alzheimer's disease. *CNS Neurosci Ther* 2012;**18**:285–94.

160. Frisch C, Hanke J, Kleineruschkamp S, Roske S, Kaaden S, Elger CE, *et al.* Positive correlation between the density of neuropeptide y positive neurons in the amygdala and parameters of self-reported anxiety and depression in mesiotemporal lobe epilepsy patients. *Biol Psychiatr* 2009;**66**:433–40.

161. Fritzschy JM, Kiener T, Bouilleret V, Loup F. GABAergic neurons and GABA(A)-receptors in temporal lobe epilepsy. *Neurochem Int* 1999;**34**:435–45.

162. Frotscher M, Haas CA, Forster E. Reelin controls granule cell migration in the dentate gyrus by acting on the radial glial scaffold. *Cereb Cortex* 2003;**13**:634–40.

163. Fujikawa DG, Itabashi HH, Wu A, Shinmei SS. Status epilepticus-induced neuronal loss in humans without systemic complications or epilepsy. *Epilepsia* 2000;**41**:981–91.

164. Furtinger S, Pirker S, Czech T, Baumgartner C, Ransmayr G, Sperk G. Plasticity of Y1 and Y2 receptors and neuropeptide Y fibers in patients with temporal lobe epilepsy. *J Neurosci* 2001;**21**:5804–12.

165. Furtinger S, Pirker S, Czech T, Baumgartner C, Sperk G. Increased expression of gamma-aminobutyric acid type B receptors in the hippocampus of patients with temporal lobe epilepsy. *Neurosci Lett* 2003;**352**:141–5.

166. Gabel LA. Layer I neocortical ectopia: cellular organization and local cortical circuitry. *Brain Res* 2011;**1381**: 148–58.

167. Gamss RP, Slasky SE, Bello JA, Miller TS, Shinnar S. Prevalence of hippocampal malrotation in a population without seizures. *AJNR Am J Neuroradiol* 2009;**30**:1571–3.

168. Garbelli R, Frassoni C, Condorelli DF, Trovato Salinaro A, Musso N, Medici V, *et al.* Expression of connexin 43 in the human epileptic and drug-resistant cerebral cortex. *Neurology* 2011;**76**:895–902.

169. Garbelli R, Meroni A, Magnaghi G, Beolchi MS, Ferrario A, Tassi L, *et al.* Architectural (Type IA) focal cortical dysplasia and parvalbumin immunostaining in temporal lobe epilepsy. *Epilepsia* 2006;**47**:1074–8.

170. Garbelli R, Munari C, De Biasi S, Vitellaro-Zuccarello L, Galli C, Bramerio M, *et al.* Taylor's cortical dysplasia: a confocal and ultrastructural immunohistochemical study. *Brain Pathol* 1999;**9**:445–61.

171. Gebhardt C, Del Turco D, Drakew A, Tielsch A, Herz J, Frotscher M, *et al.* Abnormal positioning of granule cells alters afferent fiber distribution in the mouse fascia dentata: morphologic evidence from reeler, apolipoprotein E receptor 2-, and very low density lipoprotein receptor knockout mice. *J Comp Neurol* 2002;**445**:278–92.

172. Geha S, Pallud J, Junier MP, Devaux B, Leonard N, Chassoux F, *et al.* NG2+/Olig2+ cells are the major cycle-related cell population of the adult human normal brain. *Brain Pathol* 20:399–411.

173. Giblin KA, Blumenfeld H. Is epilepsy a preventable disorder? New evidence from animal models. *Neuroscientist* 2010;**16**:253–75.

174. Gomez-Gonzalo M, Losi G, Chiavegato A, Zonta M, Cammarota M, Brondi M, *et al.* An excitatory loop with astrocytes contributes to drive neurons to seizure threshold. *PLoS Biol* 2010;**8**:e1000352.

175. Gonzalez-Martinez JA, Ying Z, Prayson R, Bingaman W, Najm I. Glutamate clearance mechanisms in resected cortical dysplasia. *J Neurosurg* 2011;**114**:1195–202.

176. Gorter JA, da Silva FH. Abnormal plastic changes in a rat model for mesial temporal lobe epilepsy: a short review. *Prog Brain Res* 2002;**138**:61–72.

177. Gorter JA, van Vliet EA, Aronica E, Lopes da Silva FH. Progression of spontaneous seizures after status epilepticus is associated with mossy fibre sprouting and extensive bilateral loss of hilar parvalbumin and somatostatin-immunoreactive neurons. *Eur J Neurosci* 2001;**13**:657–69.

178. Graebenitz S, Kedo O, Speckmann EJ, Gorji A, Panneck H, Hans V, *et al.* Interictal-like network activity and receptor expression in the epileptic human lateral amygdala. *Brain* 2011;**134**:2929–47.

179. Granata T, Gobbi G, Spreafico R, *et al.* Rasmussen's encephalitis: early characteristics allow diagnosis. *Neurology* 2003;**60**:422–5.

180. Grunewald RA, Farrow T, Vaughan P, Rittey CD, Mundy J. A magnetic resonance study of complicated early childhood convulsion. *J Neurol Neurosurg Psychiatry* 2001;**71**:638–42.

181. Guerrini R. Genetic malformations of the cerebral cortex and epilepsy. *Epilepsia* 2005;**46**(Suppl 1):32–7.

182. Gumbinger C, Rohsbach CB, Schulze-Bonhage A, Korinthenberg R, Zentner J, Haffner M, *et al.* Focal cortical dysplasia: a genotype-phenotype analysis of

polymorphisms and mutations in the TSC genes. *Epilepsia* 2009;**50**:1396–408.

183. Guye M, Regis J, Tamura M, Wendling F, McGonigal A, Chauvel P, et al. The role of corticothalamic coupling in human temporal lobe epilepsy. *Brain* 2006;**129**:1917–28.

184. Haas CA, Frotscher M. Reelin deficiency causes granule cell dispersion in epilepsy. *Exper Brain Res* 2010;**200**:141–9.

185. Hadjivassiliou G, Martinian L, Squier W, Blumcke I, Aronica E, Sisodiya SM, et al. The application of cortical layer markers in the evaluation of cortical dysplasias in epilepsy. *Acta Neuropathol* 2010;**120**:517–28.

186. Hardiman O, Burke T, Phillips J, Murphy S, O'Moore B, Staunton H, et al. Microdysgenesis in resected temporal neocortex: incidence and clinical significance in focal epilepsy. *Neurology* 1988;**38**:1041–7.

187. Harding B, Thom M. Bilateral hippocampal granule cell dispersion: autopsy study of 3 infants. *Neuropathol Appl Neurobiol* 2001;**27**:245–51.

188. Harvey AS, Cross JH, Shinnar S, Mathern BW. Defining the spectrum of international practice in pediatric epilepsy surgery patients. *Epilepsia* 2008;**49**:146–55.

189. Hattiangady B, Rao MS, Shetty AK. Chronic temporal lobe epilepsy is associated with severely declined dentate neurogenesis in the adult hippocampus. *Neurobiol Dis* 2004;**17**:473–90.

190. Haut SR, Veliskova J, Moshe SL. Susceptibility of immature and adult brains to seizure effects. *Lancet Neurol* 2004;**3**:608–17.

191. Hazrati LN, Kleinschmidt-DeMasters BK, Handler MH, Smith ML, Ochi A, Otsubo H, et al. Astrocytic inclusions in epilepsy: expanding the spectrum of filaminopathies. *J Neuropathol Exp Neurol* 2008;**67**:669–76.

192. Heck N, Garwood J, Loeffler JP, Larmet Y, Faissner A. Differential upregulation of extracellular matrix molecules associated with the appearance of granule cell dispersion and mossy fiber sprouting during epileptogenesis in a murine model of temporal lobe epilepsy. *Neuroscience* 2004;**129**:309–24.

193. Helmstaedter C, Elger CE. Chronic temporal lobe epilepsy: a neurodevelopmental or progressively dementing disease? *Brain* 2009;**132**:2822–30.

194. Henshall DC. Apoptosis signalling pathways in seizure-induced neuronal death and epilepsy. *Biochem Soc Trans* 2007;**35**:421–3.

195. Hermann B, Seidenberg M. Epilepsy and cognition. *Epilepsy Curr* 2007;**7**:1–6.

196. Hermann BP, Seidenberg M, Dow C, Jones J, Rutecki P, Bhattacharya A, et al. Cognitive prognosis in chronic temporal lobe epilepsy. *Ann Neurol* 2006;**60**:80–7.

197. Hildebrandt M, Pieper T, Winkler P, Kolodziejczyk D, Holthausen H, Blumcke I. Neuropathological spectrum of cortical dysplasia in children with severe focal epilepsies. *Acta Neuropathol* 2005;**110**:1–11.

198. Hofman PA, Fitt GJ, Harvey AS, Kuzniecky RI, Jackson G. Bottom-of-sulcus dysplasia: imaging features. *AJR Am J Roentgenol* 2011;**196**:881–5.

199. Hofman PA, Fitt G, Mitchell LA, Jackson GD. Hippocampal sclerosis and a second focal lesion: how often is it ipsilateral? *Epilepsia* 2011;**52**:718–21.

200. Holmes GL. Seizure-induced neuronal injury: animal data. *Neurology* 2002;**59**(Suppl 5):S3–6.

201. Holmes GL. Effects of early seizures on later behavior and epileptogenicity. *Ment Retard Dev Disabil Res Rev* 2004;**10**:101–5.

202. Holopainen IE. Seizures in the developing brain: cellular and molecular mechanisms of neuronal damage, neurogenesis and cellular reorganization. *Neurochem Int* 2008;**52**:935–47.

203. Holtkamp M, Schuchmann S, Gottschalk S, Meierkord H. Recurrent seizures do not cause hippocampal damage. *J Neurol* 2004;**251**:458–63.

204. Honavar M, Janota I, Polkey CE. Histological heterogeneity of dysembryoplastic neuroepithelial tumour: identification and differential diagnosis in a series of 74 cases. *Histopathology* 1999;**34**:342–56.

205. Houser CR. Granule cell dispersion in the dentate gyrus of humans with temporal lobe epilepsy. *Brain Res* 1990;**535**:195–204.

206. Hudgins RJ, Flamini JR, Palasis S, Cheng R, Burns CL, Gilreath CL. Surgical treatment of epilepsy in children caused by focal cortical dysplasia. *Pediatr Neurosurg* 2005;**41**:70–6.

207. Hunter G, Young GB. Status epilepticus: a review, with emphasis on refractory cases. *Can J Neurol Sci* 2012;**39**:157–69.

208. Joensuu T, Lehesjoki AE, Kopra O. Molecular background of EPM1-Unverricht-Lundborg disease. *Epilepsia* 2008;**49**:557–63.

209. Kahane P, Bartolomei F. Temporal lobe epilepsy and hippocampal sclerosis: lessons from depth EEG recordings. *Epilepsia* 2010;**51**(Suppl 1): 59–62.

210. Kalachikov S, Evgrafov O, Ross B, Winawer M, Barker-Cummings C, Martinelli Boneschi F, et al. Mutations in LGI1 cause autosomal-dominant partial epilepsy with auditory features. *Nat Genet* 2002;**30**:335–41.

211. Kalscheuer VM, Tao J, Donnelly A, Hollway G, Schwinger E, Kubart S, et al. Disruption of the serine/threonine kinase 9 gene causes severe X-linked infantile spasms and mental retardation. *Am J Hum Genet* 2003;**72**:1401–11.

212. Kanemoto K, Kawasaki J, Miyamoto T, Obayashi H, Nishimura M. Interleukin (IL)1beta, IL-1alpha, and IL-1 receptor antagonist gene polymorphisms in patients with temporal lobe epilepsy. *Ann Neurol* 2000;**47**:571–4.

213. Kasper BS, Stefan H, Buchfelder M, Paulus W. Temporal lobe microdysgenesis in epilepsy versus control brains. *J Neuropathol Exp Neurol* 1999;**58**:22–8.

214. Kasperaviciute D, Catarino CB, Heinzen EL, Depondt C, Cavalleri GL, Caboclo LO, et al. Common genetic variation and susceptibility to partial epilepsies: a genome-wide association study. *Brain* 2010;**133**:2136–47.

215. Kasperaviciute D, Catarino CB, Matarin M, Leu C, Novy J, Tostevin A, et al. Epilepsy, hippocampal sclerosis and febrile seizures linked by common genetic variation around SCN1A. *Brain* 2013;**136**:3140–50.

216. Kastanauskaite A, Alonso-Nanclares L, Blazquez-Llorca L, Pastor J, Sola RG, DeFelipe J. Alterations of the microvascular network in sclerotic hippocampi from patients with epilepsy. *J Neuropathol Exp Neurol* 2009;**68**:939–50.

217. Katona I, Freund TF. Endocannabinoid signaling as a synaptic circuit breaker in neurological disease. *Nat Med* 2008;**14**:923–30.

218. Keller SS, Roberts N. Voxel-based morphometry of temporal lobe epilepsy: an introduction and review of the literature. *Epilepsia* 2008;**49**:741–57.

219. Kemppainen S, Pitkanen A. Damage to the amygdalo-hippocampal projection in temporal lobe epilepsy: a tract-tracing study in chronic epileptic rats. *Neuroscience* 2004;**126**:485–501.

220. Kerfoot C, Vinters HV, Mathern GW. Cerebral cortical dysplasia: giant neurons show potential for increased excitation and axonal plasticity. *Dev Neurosci* 1999;**21**:260–70.

221. Kerrigan JF, Ng YT, Chung S, Rekate HL. The hypothalamic hamartoma: a model of subcortical epileptogenesis and encephalopathy. *Semin Pediatr Neurol* 2005;**12**:119–31.

222. Kharatishvili I, Pitkanen A. Posttraumatic epilepsy. *Curr Opin Neurol* 2010;**23**:183–8.

223. Kim DW, Lee SK, Chu K, Park KI, Lee SY, Lee CH, et al. Predictors of surgical outcome and pathologic considerations in focal cortical dysplasia. *Neurology* 2009;**72**:211–6.

224. Kim DW, Lee SK, Chung CK, Koh YC, Choe G, Lim SD. Clinical features and pathological characteristics of amygdala enlargement in mesial temporal lobe epilepsy. *J Clin Neurosci* 2012;**19**:509–12.

225. Kobayashi E, D'Agostino MD, Lopes-Cendes I, Berkovic SF, Li ML, Andermann E, et al. Hippocampal atrophy and T2-weighted signal changes in familial mesial temporal lobe epilepsy. *Neurology* 2003;**60**:405–9.

226. Kostovic I, Judas M, Sedmak G. Developmental history of the subplate zone, subplate neurons and interstitial white matter neurons: relevance for schizophrenia. *Int J Dev Neurosci* 2011;**29**:193–205.

227. Kovac S, Domijan AM, Walker MC, Abramov AY. Prolonged seizure activity impairs mitochondrial bioenergetics and induces cell death. *J Cell Sci* 2012;**125**:1796–806.

228. Kremer S, De Saint Martin A, Minotti L, Grand S, Benabid AL, Pasquier B, et al. [Focal cortical dysplasia possibly related to a probable prenatal ischemic injury]. *J Neuroradiol* 2002;**29**:200–3.

229. Krsek P, Jahodova A, Maton B, Jayakar P, Dean P, Korman B, et al. Low-grade focal cortical dysplasia is associated with prenatal and perinatal brain injury. *Epilepsia* 2010;**51**:2440–8.

230. Krsek P, Maton B, Jayakar P, Dean P, Korman B, Rey G, et al. Incomplete resection of focal cortical dysplasia is the main predictor of poor postsurgical outcome. *Neurology* 2009;**72**:217–23.

231. Krsek P, Maton B, Korman B, Pacheco-Jacome E, Jayakar P, Dunoyer C, et al. Different features of histopathological subtypes of pediatric focal cortical dysplasia. *Ann Neurol* 2008;**63**:758–69.

232. Krsek P, Pieper T, Karlmeier A, Hildebrandt M, Kolodziejczyk D, Winkler P, et al. Different presurgical characteristics and seizure outcomes in children with focal cortical dysplasia type I or II. *Epilepsia* 2009;**50**:125–37.

233. Kullmann DM, Ruiz A, Rusakov DM, Scott R, Semyanov A, Walker MC. Presynaptic, extrasynaptic and axonal GABAA receptors in the CNS: where and why? *Prog Biophys Mol Biol* 2005;87:33–46.

234. Kwan P, Arzimanoglou A, Berg AT, Brodie MJ, Allen Hauser W, Mathern G, *et al.* Definition of drug resistant epilepsy: consensus proposal by the ad hoc Task Force of the ILAE Commission on Therapeutic Strategies. *Epilepsia* 2010;51: 1069–77.

235. Labate A, Gambardella A, Andermann E, Aguglia U, Cendes F, Berkovic SF, *et al.* Benign mesial temporal lobe epilepsy. *Nature Rev Neurol* 2011;7:237–40.

236. Lalioti MD, Antonarakis SE, Scott HS. The epilepsy, the protease inhibitor and the dodecamer: progressive myoclonus epilepsy, cystatin b and a 12-mer repeat expansion. *Cytogenet Genome Res* 2003;100: 213–23.

237. Lamparello P, Baybis M, Pollard J, Hol EM, Eisenstat DD, Aronica E, *et al.* Developmental lineage of cell types in cortical dysplasia with balloon cells. *Brain* 2007;130:2267–76.

238. Laufs H, Richardson MP, Salek-Haddadi A, Vollmar C, Duncan JS, Gale K, *et al.* Converging PET and fMRI evidence for a common area involved in human focal epilepsies. *Neurology* 2011;77: 904–10.

239. Lavoie N, Jeyaraju DV, Peralta MR, 3rd, Seress L, Pellegrini L, Toth K. Vesicular zinc regulates the Ca2+ sensitivity of a subpopulation of presynaptic vesicles at hippocampal mossy fiber terminals. *J Neurosci* 2011;31:18251–65.

240. Lawn ND, Bamlet WR, Radhakrishnan K, O'Brien PC, So EL. Injuries due to seizures in persons with epilepsy: a population-based study. *Neurology* 2004;63:1565–70.

241. Lawson JA, Birchansky S, Pacheco E, Jayakar P, Resnick TJ, Dean P, *et al.* Distinct clinicopathologic subtypes of cortical dysplasia of Taylor. *Neurology* 2005;64:55–61.

242. Lehesjoki AE. Molecular background of progressive myoclonus epilepsy. *EMBO J* 2003;22:3473–8.

243. Lehmann TN, Gabriel S, Kovacs R, Eilers A, Kivi A, Schulze K, *et al.* Alterations of neuronal connectivity in area CA1 of hippocampal slices from temporal lobe epilepsy patients and from pilocarpine-treated epileptic rats. *Epilepsia* 2000;41 (Suppl 6): S190–4.

244. Lew FH, Buckmaster PS. Is there a critical period for mossy fiber sprouting in a mouse model of temporal lobe epilepsy? *Epilepsia* 2011;52:2326–32.

245. Lewis DV, Shinnar S, Hesdorffer DC, Bagiella E, Bello JA, Chan S, *et al.* Hippocampal sclerosis after febrile status epilepticus: The FEBSTAT study. *Ann Neurol* 2013.

246. Li J, Zhang Z, Shang H. A meta-analysis of voxel-based morphometry studies on unilateral refractory temporal lobe epilepsy. *Epilepsy Res* 2012;98:97–103.

247. Lin H, Huang Y, Wang Y, Jia J. Spatiotemporal profile of N-cadherin expression in the mossy fiber sprouting and synaptic plasticity following seizures. *Mol Cell Biochem* 2011;358:201–5.

248. Liou AK, Clark RS, Henshall DC, Yin XM, Chen J. To die or not to die for neurons in ischemia, traumatic brain injury and epilepsy: a review on the stress-activated signaling pathways and apoptotic pathways. *Prog Neurobiol* 2003;69:103–42.

249. Liu JY, Kasperaviciute D, Martinian L, Thom M, Sisodiya SM. Neuropathology of 16p13.11 deletion in epilepsy. *PLoS One* 2012;7:e34813.

250. Liu RS, Lemieux L, Bell GS, Sisodiya SM, Bartlett PA, Shorvon SD, *et al.* The structural consequences of newly diagnosed seizures. *Ann Neurol* 2002;52:573–80.

251. Lombroso CT. Can early postnatal closed head injury induce cortical dysplasia. *Epilepsia* 2000;41:245–53.

252. Longo BM, Mello LE. Blockade of pilocarpine- or kainate-induced mossy fiber sprouting by cycloheximide does not prevent subsequent epileptogenesis in rats. *Neurosci Let* 1997;226: 163–6.

253. Loscher W, Brandt C. Prevention or modification of epileptogenesis after brain insults: experimental approaches and translational research. *Pharmacol Rev* 2010;62:668–700.

254. Loup F, Picard F, Yonekawa Y, Wieser HG, Fritschy JM. Selective changes in GABAA receptor subtypes in white matter neurons of patients with focal epilepsy. *Brain* 2009;132:2449–63.

255. Lowenstein DH. Epilepsy after head injury: an overview. *Epilepsia* 2009;50 (Suppl 2): 4–9.

256. Ludanyi A, Eross L, Czirjak S, Vajda J, Halasz P, Watanabe M, *et al.* Downregulation of the CB1 cannabinoid receptor and related molecular elements of the endocannabinoid system in epileptic human hippocampus. *J Neurosci* 2008;28:2976–90.

257. Lurton D, El Bahh B, Sundstrom L, Rougier A. Granule cell dispersion is correlated with early epileptic events in human temporal lobe epilepsy. *J Neurol Sci* 1998;154:133–6.

258. Machado-Salas J, Avila-Costa MR, Guevara P, Guevara J, Duron RM, Bai D, *et al.* Ontogeny of Lafora bodies and neurocytoskeleton changes in Laforin-deficient mice. *Exp Neurol* 2012;236:131–40.

259. Maciunas JA, Syed TU, Cohen ML, Werz MA, Maciunas RJ, Koubeissi MZ. Triple pathology in epilepsy: coexistence of cavernous angiomas and cortical dysplasias with other lesions. *Epileps Res* 2010;91:106–10.

260. Magloczky Z. Sprouting in human temporal lobe epilepsy: excitatory pathways and axons of interneurons. *Epileps Res* 2010;89:52–9.

261. Magloczky Z, Halasz P, Vajda J, Czirjak S, Freund TF. Loss of Calbindin-D28K immunoreactivity from dentate granule cells in human temporal lobe epilepsy. *Neuroscience* 1997;76:377–85.

262. Magloczky Z, Toth K, Karlocai R, Nagy S, Eross L, Czirjak S, *et al.* Dynamic changes of CB1-receptor expression in hippocampi of epileptic mice and humans. *Epilepsia* 2010;51 (Suppl 3): 115–20.

263. Magloczky Z, Wittner L, Borhegyi Z, Halasz P, Vajda J, Czirjak S, *et al.* Changes in the distribution and connectivity of interneurons in the epileptic human dentate gyrus. *Neuroscience* 2000;96:7–25.

264. Maisano X, Litvina E, Tagliatela S, Aaron GB, Grabel LB, Naegele JR. Differentiation and functional incorporation of embryonic stem cell-derived GABAergic interneurons in the dentate gyrus of mice with temporal lobe epilepsy. *J Neurosci* 2012;32:46–61.

265. Majores M, Blumcke I, Urbach H, Meroni A, Hans V, Holthausen H, *et al.* Distinct allelic variants of TSC1 and TSC2 in epilepsy-associated cortical malformations without balloon cells. *J Neuropathol Exp Neurol* 2005;64:629–37.

266. Mantegazza R, Bernasconi P, Baggi F, Spreafico R, Ragona F, Antozzi C, *et al.* Antibodies against GluR3 peptides are not specific for Rasmussen's encephalitis but are also present in epilepsy patients with severe, early onset disease and intractable seizures. *J Neuroimmunol* 2002;131:179–85.

267. Marchi N, Tierney W, Alexopoulos AV, Puvenna V, Granata T, Janigro D. The etiological role of blood-brain barrier dysfunction in seizure disorders. *Cardiovasc Psychiatr Neurol* 2011;2011: 482415.

268. Marcotte L, Aronica E, Baybis M, Crino PB. Cytoarchitectural alterations are widespread in cerebral cortex in tuberous sclerosis complex. *Acta Neuropathol* 2012;123:685–93.

269. Marcuccilli CJ, Tryba AK, van Drongelen W, Koch H, Viemari JC, Pena-Ortega F, *et al.* Neuronal bursting properties in focal and parafocal regions in pediatric neocortical epilepsy stratified by histology. *J Clin Neurophysiol* 2010;27:387–97.

270. Margerison JH, Corsellis JA. Epilepsy and the temporal lobes. A clinical, electroencephalographic and neuropathological study of the brain in epilepsy, with particular reference to the temporal lobes. *Brain* 1966;89:499–530.

271. Marin-Padilla M, Parisi JE, Armstrong DL, Sargent SK, Kaplan JA. Shaken infant syndrome: developmental neuropathology, progressive cortical dysplasia, and epilepsy. *Acta Neuropathol* 2002;103:321–32.

272. Marin-Padilla M. Developmental neuropathology and impact of perinatal brain damage. III: gray matter lesions of the neocortex. *J Neuropathol Exp Neurol* 1999;58:407–29.

273. Maroso M, Balosso S, Ravizza T, *et al.* Toll-like receptor 4 and high-mobility group box-1 are involved in ictogenesis and can be targeted to reduce seizures. *Nat Med* 2010;16:413–9.

274. Martinian L, Boer K, Middeldorp J, Hol EM, Sisodiya SM, Squier W, *et al.* Expression patterns of glial fibrillary acidic protein (GFAP)-delta in epilepsy-associated lesional pathologies. *Neuropathol Applied Neurobiol* 2009;35:394–405.

275. Martinian L, Catarino CB, Thompson P, Sisodiya SM, Thom M. Calbindin D28K expression in relation to granule cell dispersion, mossy fibre sprouting and memory impairment in hippocampal sclerosis: A surgical and post mortem series. *Epileps Res* 2011.

276. Mathern GW, Adelson PD, Cahan LD, Leite JP. Hippocampal neuron damage in human epilepsy: Meyer's hypothesis revisited. *Prog Brain Res* 2002;135: 237–51.

277. Mathern GW, Andres M, Salamon N, Chandra PS, Andre VM, Cepeda C, et al. A hypothesis regarding the pathogenesis and epileptogenesis of pediatric cortical dysplasia and hemimegalencephaly based on MRI cerebral volumes and NeuN cortical cell densities. *Epilepsia* 2007;**48** (Suppl 5) : 74–8.

278. Mathern GW, Babb TL, Pretorius JK, Leite JP. Reactive synaptogenesis and neuron densities for neuropeptide Y, somatostatin, and glutamate decarboxylase immunoreactivity in the epileptogenic human fascia dentata. *J Neurosci* 1995;**15**:3990–4004.

279. Mathern GW, Mendoza D, Lozada A, Pretorius JK, Dehnes Y, Danbolt NC, et al. Hippocampal GABA and glutamate transporter immunoreactivity in patients with temporal lobe epilepsy. *Neurology* 1999;**52**:453–72.

280. McKee AC, Cantu RC, Nowinski CJ, Hedley-Whyte ET, Gavett BE, Budson AE, et al. Chronic traumatic encephalopathy in athletes: progressive tauopathy after repetitive head injury. *J Neuropathol Exp Neurol* 2009;**68**:709–35.

281. Medici V, Frassoni C, Tassi L, Spreafico R, Garbelli R. Aquaporin 4 expression in control and epileptic human cerebral cortex. *Brain Res* 2011;**1367**: 330–9.

282. Meencke HJ, Janz D. Neuropathological findings in primary generalized epilepsy: a study of eight cases. *Epilepsia* 1984;**25**:8–21.

283. Meencke HJ, Veith G, Lund S. Bilateral hippocampal sclerosis and secondary epileptogenesis. Epilepsy Res Suppl 1996;**12**: 335–42.

284. Men S, Lee DH, Barron JR, Munoz DG. Selective neuronal necrosis associated with status epilepticus: MR findings. *AJNR Am J Neuroradiol* 2000;**21**: 1837–40.

285. Meyer A, Falconer MA, Beck E. Pathological findings in temporal lobe epilepsy. *J Neurol Neurosurg Psychiatry* 1954;**17**:276–85.

286. Minassian BA, Ianzano L, Delgado-Escueta AV, Scherer SW. Identification of new and common mutations in the EPM2A gene in Lafora disease. *Neurology* 2000;**54**:488–90.

287. Miyahara H, Ryufuku M, Fu YJ, Kitaura H, Murakami H, Masuda H, et al. Balloon cells in the dentate gyrus in hippocampal sclerosis associated with non-herpetic acute limbic encephalitis. *Seizure* 2011;**20**:87–9.

288. Miyata H, Chiang AC, Vinters HV. Insulin signaling pathways in cortical dysplasia and TSC-tubers: tissue microarray analysis. *Ann Neurol* 2004;**56**: 510–9.

289. Moddel G, Jacobson B, Ying Z, Janigro D, Bingaman W, Gonzalez-Martinez J, et al. The NMDA receptor NR2B subunit contributes to epileptogenesis in human cortical dysplasia. *Brain Res* 2005;**1046**:10–23.

290. Moran NF, Fish DR, Kitchen N, Shorvon S, Kendall BE, Stevens JM. Supratentorial cavernous haemangiomas and epilepsy: a review of the literature and case series. *J Neurol Neurosurg Psychiatry* 1999;**66**:561–8.

291. Moran NF, Lemieux L, Kitchen ND, Fish DR, Shorvon SD. Extrahippocampal temporal lobe atrophy in temporal lobe epilepsy and mesial temporal sclerosis. *Brain* 2001;**124**:167–75.

292. Mott RT, Thore CR, Moody DM, Glazier SS, Ellis TL, Brown WR. Reduced ratio of afferent to total vascular density in mesial temporal sclerosis. *J Neuropathol Exp Neurol* 2009;**68**:1147–54.

293. Mulley JC, Mefford HC. Epilepsy and the new cytogenetics. *Epilepsia* 2011;**52**:423–32.

294. Mulley JC, Scheffer IE, Petrou S, Berkovic SF. Channelopathies as a genetic cause of epilepsy. *Curr Opin Neurol* 2003;**16**:171–6.

295. Munoz A, Arellano JI, DeFelipe J. GABABR1 receptor protein expression in human mesial temporal cortex: changes in temporal lobe epilepsy. *J Comp Neurol* 2002;**449**:166–79.

296. Munoz A, Mendez P, DeFelipe J, Alvarez-Leefmans FJ. Cation-chloride cotransporters and GABA-ergic innervation in the human epileptic hippocampus. *Epilepsia* 2007;**48**: 663–73.

297. Murphy BL, Danzer SC. Somatic translocation: a novel mechanism of granule cell dendritic dysmorphogenesis and dispersion. *J Neurosci* 2011;**31**: 2959–64.

298. Nadler JV. The recurrent mossy fiber pathway of the epileptic brain. *Neurochem Res* 2003;**28**:1649–58.

299. Nadler JV, Tu B, Timofeeva O, Jiao Y, Herzog H. Neuropeptide Y in the recurrent mossy fiber pathway. *Peptides* 2007;**28**:357–64.

300. Nairismagi J, Grohn OH, Kettunen MI, Nissinen J, Kauppinen RA, Pitkanen A. Progression of brain damage after status epilepticus and its association with epileptogenesis: a quantitative MRI study in a rat model of temporal lobe epilepsy. *Epilepsia* 2004;**45**:1024–34.

301. Nakahara H, Konishi Y, Beach TG, Yamada N, Makino S, Tooyama I. Infiltration of T lymphocytes and expression of icam-1 in the hippocampus of patients with hippocampal sclerosis. *Acta Histochem Cytochem* 2010;**43**:157–62.

302. Nashef L, So EL, Ryvlin P, Tomson T. Unifying the definitions of sudden unexpected death in epilepsy. *Epilepsia* 2012;**53**:227–33.

303. Nelson PT, Schmitt FA, Lin Y, Abner EL, Jicha GA, Patel E, et al. Hippocampal sclerosis in advanced age: clinical and pathological features. *Brain* 2011;**134**:1506–18.

304. Nicita F, De Liso P, Danti FR, Papetti L, Ursitti F, Castronovo A, et al. The genetics of monogenic idiopathic epilepsies and epileptic encephalopathies. *Seizure* 2012;**21**:3–11.

305. Niehusmann P, Mittelstaedt T, Bien CG, Drexler JF, Grote A, Schoch S, et al. Presence of human herpes virus 6 DNA exclusively in temporal lobe epilepsy brain tissue of patients with history of encephalitis. *Epilepsia* 2010;**51**: 2478–83.

306. Niittykoski M, Nissinen J, Penttonen M, Pitkanen A. Electrophysiological changes in the lateral and basal amygdaloid nuclei in temporal lobe epilepsy: an in vitro study in epileptic rats. *Neuroscience* 2004;**124**:269–81.

307. Nissinen J, Lukasiuk K, Pitkanen A. Is mossy fiber sprouting present at the time of the first spontaneous seizures in rat experimental temporal lobe epilepsy? *Hippocampus* 2001;**11**:299–310.

308. Noe F, Pool AH, Nissinen J, Gobbi M, Bland R, Rizzi M, et al. Neuropeptide Y gene therapy decreases chronic spontaneous seizures in a rat model of temporal lobe epilepsy. *Brain* 2008;**131**:1506–15.

309. Norwood BA, Bumanglag AV, Osculati F, Sbarbati A, Marzola P, Nicolato E, et al. Classic hippocampal sclerosis and hippocampal-onset epilepsy produced by a single "cryptic" episode of focal hippocampal excitation in awake rats. *J Comp Neurol* 2010;**518**:3381–407.

310. Oh HS, Lee MC, Kim HS, Lee JS, Lee JH, Kim MK, et al. Pathophysiologic characteristics of balloon cells in cortical dysplasia. *Childs Nerv Syst* 2008;**24**:175–83.

311. Orlova KA, Tsai V, Baybis M, Heuer GG, Sisodiya S, Thom M, et al. Early progenitor cell marker expression distinguishes type II from type I focal cortical dysplasias. *J Neuropathol Exp Neurol* 2010;**69**:850–63.

312. Orosz I, Hartel C, Gottschalk S, von Hof K, Bien CG, Sperner J. Cerebral hemiatrophy associated with hippocampal sclerosis following a single prolonged febrile seizure. *Eur J Pediatr* 2011;**170**: 789–94.

313. Ortiz-Gonzalez XR, Venneti S, Biegel JA, Rorke-Adams LB, Porter BE. Ganglioglioma arising from dysplastic cortex. *Epilepsia* 2011;**52**:e106–8.

314. Oyegbile TO, Bayless K, Dabbs K, Jones J, Rutecki P, Pierson R, et al. The nature and extent of cerebellar atrophy in chronic temporal lobe epilepsy. *Epilepsia* 2011;**52**:698–706.

315. Ozdogmus O, Cavdar S, Ersoy Y, Ercan F, Uzun I. A preliminary study, using electron and light-microscopic methods, of axon numbers in the fornix in autopsies of patients with temporal lobe epilepsy. *Anat Sci Int* 2009;**84**:2–6.

316. Pallud J, Haussler U, Langlois M, Hamelin S, Devaux B, Deransart C, et al. Dentate gyrus and hilus transection blocks seizure propagation and granule cell dispersion in a mouse model for mesial temporal lobe epilepsy. *Hippocampus* 2011;**21**:334–43.

317. Palmini A, Najm I, Avanzini G, Babb T, Guerrini R, Foldvary-Schaefer N, et al. Terminology and classification of the cortical dysplasias. *Neurology* 2004;**62**: S2–8.

318. Pardo CA, Vining EP, Guo L, Skolasky RL, Carson BS, Freeman JM. The pathology of Rasmussen syndrome: stages of cortical involvement and neuropathological studies in 45 hemispherectomies. *Epilepsia* 2004;**45**:516–26.

319. Parent JM, Yu TW, Leibowitz RT, Geschwind DH, Sloviter RS, Lowenstein DH. Dentate granule cell neurogenesis is increased by seizures and contributes to aberrant network reorganization in the adult rat hippocampus. *J Neurosci* 1997;**17**:3727–38.

320. Parvizi J, Le S, Foster BL, Bourgeois B, Riviello JJ, Prenger E, et al. Gelastic epilepsy and hypothalamic hamartomas: neuroanatomical analysis of brain lesions in 100 patients. *Brain* 2011;**134**:2960–8.

321. Pasquier B, Peoc HM, Fabre-Bocquentin B, Bensaadi L, Pasquier D, Hoffmann D, et al. Surgical pathology of drug-resistant partial epilepsy. A 10-year-experience

with a series of 327 consecutive resections. *Epileptic Disord* 2002;**4**:99–119.

322. Pauli E, Hildebrandt M, Romstock J, Stefan H, Blumcke I. Deficient memory acquisition in temporal lobe epilepsy is predicted by hippocampal granule cell loss. *Neurology* 2006;**67**:1383–9.

323. Perry A, Kurtkaya-Yapicier O, Scheithauer BW, Robinson S, Prayson RA, Kleinschmidt-DeMasters BK, *et al.* Insights into meningioangiomatosis with and without meningioma: a clinicopathologic and genetic series of 24 cases with review of the literature. *Brain Pathol* 2005;**15**:55–65.

324. Piao YS, Lu DH, Chen L, Liu J, Wang W, Liu L, *et al.* Neuropathological findings in intractable epilepsy: 435 Chinese cases. *Brain Pathol* 2010;**20**:902–8.

325. Pierce JP, Melton J, Punsoni M, McCloskey DP, Scharfman HE. Mossy fibers are the primary source of afferent input to ectopic granule cells that are born after pilocarpine-induced seizures. *Exp Neurol* 2005;**196**:316–31.

326. Pirker S, Czech T, Baumgartner C, Maier H, Novak K, Furtinger S, *et al.* Chromogranins as markers of altered hippocampal circuitry in temporal lobe epilepsy. *Ann Neurol* 2001;**50**:216–26.

327. Pitkanen A. Therapeutic approaches to epileptogenesis--hope on the horizon. *Epilepsia* 2010;**51**(Suppl 3): 2–17.

328. Pitkanen A, Lukasiuk K. Mechanisms of epileptogenesis and potential treatment targets. *Lancet Neurol* 2011;**10**:173–86.

329. Pitkanen A, Immonen RJ, Grohn OH, Kharatishvili I. From traumatic brain injury to posttraumatic epilepsy: what animal models tell us about the process and treatment options. *Epilepsia* 2009;**50** (Suppl 2): 21–9.

330. Pitkanen A, Nissinen J, Nairismagi J, Lukasiuk K, Grohn OH, Miettinen R, *et al.* Progression of neuronal damage after status epilepticus and during spontaneous seizures in a rat model of temporal lobe epilepsy. *Prog Brain Res* 2002;**135**: 67–83.

331. Pohlmann-Eden B, Gass A, Peters CN, Wennberg R, Blumcke I. Evolution of MRI changes and development of bilateral hippocampal sclerosis during long lasting generalised status epilepticus. *J Neurol Neurosurg Psychiatry* 2004;**75**:898–900.

332. Prayson RA. Brain tumors in adults with medically intractable epilepsy. *Am J Clin Pathol* 2011;**136**:557–63.

333. Prayson RA. Tumours arising in the setting of paediatric chronic epilepsy. *Pathology* 2010;**42**:426–31.

334. Prince DA, Parada I, Scalise K, Graber K, Jin X, Shen F. Epilepsy following cortical injury: cellular and molecular mechanisms as targets for potential prophylaxis. *Epilepsia* 2009;**50**(Suppl 2): 30–40.

335. Proper EA, Jansen GH, van Veelen CW, van Rijen PC, Gispen WH, de Graan PN. A grading system for hippocampal sclerosis based on the degree of hippocampal mossy fiber sprouting. *Acta Neuropathol* 2001;**101**:405–9.

336. Proper EA, Oestreicher AB, Jansen GH, Veelen CW, van Rijen PC, Gispen WH, *et al.* Immunohistochemical characterization of mossy fibre sprouting in the hippocampus of patients with pharmaco-resistant temporal lobe epilepsy. *Brain* 2000;**123**:19–30.

337. Puri R, Suzuki T, Yamakawa K, Ganesh S. Dysfunctions in endosomal-lysosomal and autophagy pathways underlie neuropathology in a mouse model for Lafora disease. *Hum Mol Genet* 2012;**21**: 175–84.

338. Rakhade SN, Jensen FE. Epileptogenesis in the immature brain: emerging mechanisms. *Nature Rev Neurol* 2009;**5**:380–91.

339. Ratzliff AH, Santhakumar V, Howard A, Soltesz I. Mossy cells in epilepsy: rigor mortis or vigor mortis? *Trends Neurosci* 2002;**25**:140–4.

340. Ravizza T, Gagliardi B, Noe F, Boer K, Aronica E, Vezzani A. Innate and adaptive immunity during epileptogenesis and spontaneous seizures: evidence from experimental models and human temporal lobe epilepsy. *Neurobiol Dis* 2008;**29**:142–60.

341. Rigau V, Morin M, Rousset MC, de Bock F, Lebrun A, Coubes P, *et al.* Angiogenesis is associated with blood-brain barrier permeability in temporal lobe epilepsy. *Brain* 2007;**130**:1942–56.

342. Rossini L, Moroni RF, Tassi L, Watakabe A, Yamamori T, Spreafico R, *et al.* Altered layer-specific gene expression in cortical samples from patients with temporal lobe epilepsy. *Epilepsia* 2011;**52**: 1928–37.

343. Ruban D, Byrne RW, Kanner A, Smith M, Cochran EJ, Roh D, *et al.* Chronic epilepsy associated with temporal tumors: long-term surgical outcome. *Neurosurg Focus* 2009;**27**:E6.

344. Ryvlin P, Nashef L, Lhatoo SD, Bateman LM, Bird J, Bleasel A, *et al.* Incidence and mechanisms of cardiorespiratory arrests in epilepsy monitoring units (MORTEMUS): a retrospective study. *Lancet Neurol* 20 13;**12**:966–77.

345. Sagar HJ, Oxbury JM. Hippocampal neuron loss in temporal lobe epilepsy: correlation with early childhood convulsions. *Ann Neurol* 1987;**22**:334–40.

346. Salmenpera T, Kalviainen R, Partanen K, Mervaala E, Pitkanen A. MRI volumetry of the hippocampus, amygdala, entorhinal cortex, and perirhinal cortex after status epilepticus. *Epileps Res* 2000;**40**:155–70.

347. Sano O, Reynolds JP, Jimenez-Mateos EM, Matsushima S, Taki W, Henshall DC. MicroRNA-34a upregulation during seizure-induced neuronal death. *Cell Death Dis* 2012;**3**:e287.

348. Scheibel ME, Crandall PH, Scheibel AB. The hippocampal-dentate complex in temporal lobe epilepsy. A Golgi study. *Epilepsia* 1974;**15**:55–80.

349. Schick V, Majores M, Engels G, Spitoni S, Koch A, Elger CE, *et al.* Activation of Akt independent of PTEN and CTMP tumor-suppressor gene mutations in epilepsy-associated Taylor-type focal cortical dysplasias. *Acta Neuropathol* 2006;**112**:715–25.

350. Schick V, Majores M, Fassunke J, Engels G, Simon M, Elger CE, *et al.* Mutational and expression analysis of CDK1, cyclinA2 and cyclinB1 in epilepsy-associated glioneuronal lesions. *Neuropathol Applied Neurobiol* 2007;**33**:152–62.

351. Schick V, Majores M, Koch A, Elger CE, Schramm J, Urbach H, *et al.* Alterations of phosphatidylinositol 3-kinase pathway components in epilepsy-associated glioneuronal lesions. *Epilepsia* 2007;**48**(Suppl 5):65–73.

352. Schijns OE, Bien CG, Majores M, von Lehe M, Urbach H, Becker A, *et al.* Presence of temporal gray-white matter abnormalities does not influence epilepsy surgery outcome in temporal lobe epilepsy with hippocampal sclerosis. *Neurosurgery* 2011;**68**:98–106;discussion 7.

353. Schonberger A, Gembe E, Grote A, Witt JA, Elger CE, Bien CG, *et al.* Genetic analysis of tuberous-sclerosis genes 1 and 2 in nonlesional focal epilepsy. *Epilepsy Behav* 2011;**21**:233–7.

354. Scott RC, Gadian DG, King MD, Chong WK, Cox TC, Neville BG, *et al.* Magnetic resonance imaging findings within 5 days of status epilepticus in childhood. *Brain* 2002;**125**:1951–9.

355. Semah F, Picot MC, Adam C, Broglin D, Arzimanoglou A, Bazin B, *et al.* Is the underlying cause of epilepsy a major prognostic factor for recurrence? *Neurology* 1998;**51**:1256–62.

356. Sen A, Martinian L, Nikolic M, Walker MC, Thom M, Sisodiya SM. Increased NKCC1 expression in refractory human epilepsy. *Epileps Res* 2007;**74**:220–7.

357. Sen A, Thom M, Martinian L, Dawodu S, Sisodiya SM. Hippocampal malformations do not necessarily evolve into hippocampal sclerosis. *Epilepsia* 2005;**46**:939–43.

358. Sen A, Thom M, Martinian L, Harding B, Cross JH, Nikolic M, *et al.* Pathological tau tangles localize to focal cortical dysplasia in older patients. *Epilepsia* 2007;**48**:1447–54.

359. Seress L, Abraham H, Czeh B, Fuchs E, Leranth C. Calretinin expression in hilar mossy cells of the hippocampal dentate gyrus of nonhuman primates and humans. *Hippocampus* 2008;**18**:425–34.

360. Shamji MF, Fric-Shamji EC, Benoit BG. Brain tumors and epilepsy: pathophysiology of peritumoral changes. *Neurosurg Rev* 2009;**32**:275–84;discussion 84–6.

361. Shapiro LA, Ribak CE. Integration of newly born dentate granule cells into adult brains: hypotheses based on normal and epileptic rodents. *Brain Res Brain Res Rev* 2005;**48**:43–56.

362. Shields LB, Hunsaker DM, Hunsaker JC, 3rd, Parker JC, Jr. Sudden unexpected death in epilepsy: neuropathologic findings. *Am J Forensic Med Pathol* 2002;**23**:307–14.

363. Shimizu-Okabe C, Tanaka M, Matsuda K, Mihara T, Okabe A, Sato K, *et al.* KCC2 was downregulated in small neurons localized in epileptogenic human focal cortical dysplasia. *Epileps Res* 2011;**93**:177–84.

364. Shlosberg D, Benifla M, Kaufer D, Friedman A. Blood-brain barrier breakdown as a therapeutic target in traumatic brain injury. *Nature Rev Neurol* 2010;**6**:393–403.

365. Silverstein FS, Jensen FE. Neonatal seizures. *Ann Neurol* 2007;**62**:112–20.

366. Simoes PS, Perosa SR, Arganaraz GA, Yacubian EM, Carrete H, Jr., Centeno RS, *et al.* Kallikrein 1 is overexpressed by astrocytes in the hippocampus of patients with refractory temporal lobe epilepsy, associated with hippocampal sclerosis. *Neurochem Int* 2011;**58**: 477–82.

367. Singh S, Ganesh S. Phenotype variations in Lafora progressive myoclonus epilepsy: possible involvement of genetic modifiers? *J Hum Genet* 2012;**57**:283–5.

368. Sisodiya S. Epilepsy: the new order-classifying focal cortical dysplasias. *Nature Rev Neurol* 2011;**7**:129–30.

369. Sloviter RS, Kudrimoti HS, Laxer KD, Barbaro NM, Chan S, Hirsch LJ, et al. "Tectonic" hippocampal malformations in patients with temporal lobe epilepsy. *Epilepsy Res* 2004;**59**:123–53.

370. Sloviter RS, Zappone CA, Harvey BD, Bumanglag AV, Bender RA, Frotscher M. "Dormant basket cell" hypothesis revisited: relative vulnerabilities of dentate gyrus mossy cells and inhibitory interneurons after hippocampal status epilepticus in the rat. *J Comp Neurol* 2003;**459**:44–76.

371. Sofroniew MV, Vinters HV. Astrocytes: biology and pathology. *Acta Neuropathol* 2010;**119**:7–35.

372. Sperk G, Wieselthaler-Holzl A, Pirker S, Tasan R, Strasser SS, Drexel M, et al. Glutamate decarboxylase 67 is expressed in hippocampal mossy fibers of temporal lobe epilepsy patients. *Hippocampus* 2012;**22**:590–603.

373. Spreafico R, Tassi L, Colombo N, Bramerio M, Galli C, Garbelli R, et al. Inhibitory circuits in human dysplastic tissue. *Epilepsia* 2000;**41**(**Suppl 6**):S168–73.

374. Srikandarajah N, Martinian L, Sisodiya SM, Squier W, Blumcke I, Aronica E, et al. Doublecortin expression in focal cortical dysplasia in epilepsy. *Epilepsia* 2009;**50**:2619–28.

375. Stengaard-Pedersen K, Fredens K, Larsson LI. Enkephalin and zinc in the hippocampal mossy fiber system. *Brain Res* 1981;**212**:230–3.

376. Stogmann E, Zimprich A, Baumgartner C, Aull-Watschinger S, Hollt V, Zimprich F. A functional polymorphism in the prodynorphin gene promotor is associated with temporal lobe epilepsy. *Ann Neurol* 2002;**51**:260–3.

377. Suarez-Sola ML, Gonzalez-Delgado FJ, Pueyo-Morlans M, Medina-Bolivar OC, Hernandez-Acosta NC, Gonzalez-Gomez M, et al. Neurons in the white matter of the adult human neocortex. *Front Neuroanat* 2009;**3**:7.

378. Sunnen CN, Brewster AL, Lugo JN, Vanegas F, Turcios E, Mukhi S, et al. Inhibition of the mammalian target of rapamycin blocks epilepsy progression in NS-Pten conditional knockout mice. *Epilepsia* 2011;**52**:2065–75.

379. Sutula T. Seizure-induced axonal sprouting: assessing connections between injury, local circuits, and epileptogenesis. *Epilepsy Curr* 2002;**2**:86–91.

380. Sutula T, Cascino G, Cavazos J, Parada I, Ramirez L. Mossy fiber synaptic reorganization in the epileptic human temporal lobe. *Ann Neurol* 1989;**26**:321–30.

381. Sutula TP, Hagen J, Pitkanen A. Do epileptic seizures damage the brain? *Curr Opin Neurol* 2003;**16**:189–95.

382. Sutula TP, Pitkanen A. More evidence for seizure-induced neuron loss: is hippocampal sclerosis both cause and effect of epilepsy? *Neurology* 2001;**57**:169–70.

383. Suzuki F, Heinrich C, Boehrer A, Mitsuya K, Kurokawa K, Matsuda M, et al. Glutamate receptor antagonists and benzodiazepine inhibit the progression of granule cell dispersion in a mouse model of mesial temporal lobe epilepsy. *Epilepsia* 2005;**46**:193–202.

384. Takahashi A, Hong SC, Seo DW, Hong SB, Lee M, Suh YL. Frequent association of cortical dysplasia in dysembryoplastic neuroepithelial tumor treated by epilepsy surgery. *Surg Neurol* 2005;**64**:419–27.

385. Takei H, Wilfong A, Malphrus A, Yoshor D, Hunter JV, Armstrong DL, et al. Dual pathology in Rasmussen's encephalitis: a study of seven cases and review of the literature. *Neuropathology* 2010;**30**:381–91.

386. Tan N, Urich H. Postictal cerebral hemiatrophy: with a contribution to the problem of crossed cerebellar atrophy. *Acta Neuropathol* 1984;**62**:332–9.

387. Tan NC, Mulley JC, Berkovic SF. Genetic association studies in epilepsy: "the truth is out there". *Epilepsia* 2004;**45**:1429–42.

388. Tassi L, Colombo N, Garbelli R, Francione S, Lo Russo G, Mai R, et al. Focal cortical dysplasia: neuropathological subtypes, EEG, neuroimaging and surgical outcome. *Brain* 2002;**125**:1719–32.

389. Tassi L, Garbelli R, Colombo N, Bramerio M, Lo Russo G, Deleo F, et al. Type I focal cortical dysplasia: surgical outcome is related to histopathology. *Epileptic Disord* 2010;**12**:181–91.

390. Tassi L, Pasquier B, Minotti L, Garbelli R, Kahane P, Benabid AL, et al. Cortical dysplasia: electroclinical, imaging, and neuropathologic study of 13 patients. *Epilepsia* 2001;**42**:1112–23.

391. Taylor DC, Falconer MA, Bruton CJ, Corsellis JA. Focal dysplasia of the cerebral cortex in epilepsy. *J Neurol Neurosurg Psychiatry* 1971;**34**:369–87.

392. Taylor JP, Sater R, French J, Baltuch G, Crino PB. Transcription of intermediate filament genes is enhanced in focal cortical dysplasia. *Acta Neuropathol* 2001;**102**:141–8.

393. Teixeira RA, Li LM, Santos SL, Zanardi VA, Guerreiro CA, Cendes F. Crossed cerebellar atrophy in patients with precocious destructive brain insults. *Arch Neurol* 2002;**59**:843–7.

394. Theodore WH, Epstein L, Gaillard WD, Shinnar S, Wainwright MS, Jacobson S. Human herpes virus 6B: a possible role in epilepsy? *Epilepsia* 2008;**49**:1828–37.

395. Thom M, Blumcke I, Aronica E. Long-term epilepsy-associated tumors. *Brain Pathol* 2012;**22**:350–79.

396. Thom M, Eriksson S, Martinian L, Caboclo LO, McEvoy AW, Duncan JS, et al. Temporal lobe sclerosis associated with hippocampal sclerosis in temporal lobe epilepsy: neuropathological features. *J Neuropathol Exp Neurol* 2009;**68**:928–38.

397. Thom M. Hippocampal sclerosis: progress since Sommer. *Brain Pathol* 2009;**19**:565–72.

398. Thorn M. Neuropathologic findings in postmortem studies of sudden death in epilepsy. *Epilepsia* 1997;**38**:S32–4.

399. Thom M, Harding BN, Lin WR, Martinian L, Cross H, Sisodiya SM. Cajal-Retzius cells, inhibitory interneuronal populations and neuropeptide Y expression in focal cortical dysplasia and microdysgenesis. *Acta Neuropathol* 2003;**105**:561–9.

400. Thom M, Liagkouras I, Elliot KJ, Martinian L, Harkness W, McEvoy A, et al. Reliability of patterns of hippocampal sclerosis as predictors of postsurgical outcome. *Epilepsia* 2010;**51**:1801–8.

401. Thom M, Liu JY, Thompson P, Phadke R, Narkiewicz M, Martinian L, et al.

402. Neurofibrillary tangle pathology and Braak staging in chronic epilepsy in relation to traumatic brain injury and hippocampal sclerosis: a post-mortem study. *Brain* 2011;**134**:2969–81.

402. Thom M, Martinian L, Caboclo LO, McEvoy AW, Sisodiya SM. Balloon cells associated with granule cell dispersion in the dentate gyrus in hippocampal sclerosis. *Acta Neuropathol* 2008;**115**:697–700.

403. Thom M, Martinian L, Catarino C, Yogarajah M, Koepp MJ, Caboclo L, et al. Bilateral reorganization of the dentate gyrus in hippocampal sclerosis: a postmortem study. *Neurology* 2009;**73**:1033–40.

404. Thom M, Martinian L, Sen A, Squier W, Harding BN, Cross JH, et al. An investigation of the expression of G1-phase cell cycle proteins in focal cortical dysplasia type IIB. *J Neuropathol Exp Neurol* 2007;**66**:1045–55.

405. Thom M, Martinian L, Sisodiya SM, Cross JH, Williams G, Stoeber K, et al. Mcm2 labelling of balloon cells in focal cortical dysplasia. *Neuropathol Appl Neurobiol* 2005;**31**:580–8.

406. Thom M, Martinian L, Williams G, Stoeber K, Sisodiya SM. Cell proliferation and granule cell dispersion in human hippocampal sclerosis. *J Neuropathol Exp Neurol* 2005;**64**:194–201.

407. Thom M, Mathern GW, Cross JH, Bertram EH. Mesial temporal lobe epilepsy: How do we improve surgical outcome? *Ann Neurol* 2010;**68**:424–34.

408. Thom M, Sisodiya S, Harkness W, Scaravilli F. Microdysgenesis in temporal lobe epilepsy. A quantitative and immunohistochemical study of white matter neurones. *Brain* 2001;**124**:2299–309.

409. Thom M, Sisodiya SM, Lin WR, Mitchell T, Free SL, Stevens J, et al. Bilateral isolated hippocampal malformation in temporal lobe epilepsy. *Neurology* 2002;**58**:1683–6.

410. Thom M, Toma A, An S, Martinian L, Hadjivassiliou G, Ratilal B, et al. One hundred and one dysembryoplastic neuroepithelial tumors: an adult epilepsy series with immunohistochemical, molecular genetic, and clinical correlations and a review of the literature. *J Neuropathol Exp Neurol* 2011;**70**:859–78.

411. Thom M, Zhou J, Martinian L, Sisodiya S. Quantitative post-mortem study of the hippocampus in chronic epilepsy: seizures do not inevitably cause neuronal loss. *Brain* 2005;**128**:1344–57.

412. Tobias SM, Robitaille Y, Hickey WF, et al. Bilateral Rasmussen encephalitis: postmortem documentation in a five-year-old. *Epilepsia* 2003;**44**: 127–30.

413. Toering ST, Boer K, de Groot M, Troost D, Heimans JJ, Spliet WG, et al. Expression patterns of synaptic vesicle protein 2A in focal cortical dysplasia and TSC-cortical tubers. *Epilepsia* 2009;**50**:1409–18.

414. Torres-Reveron J, Friedlander MJ. Properties of persistent postnatal cortical subplate neurons. *J Neurosci* 2007;**27**:9962–74.

415. Toth K, Eross L, Vajda J, Halasz P, Freund TF, Magloczky Z. Loss and reorganization of calretinin-containing interneurons in the epileptic human hippocampus. *Brain* 2010;**133**:2763–77.

416. Toth K, Wittner L, Urban Z, Doyle WK, Buzsaki G, Shigemoto R, et al.

Morphology and synaptic input of substance P receptor-immunoreactive interneurons in control and epileptic human hippocampus. *Neuroscience* 2007;**144**:495–508.

417. Tsuchida TN, Barkovich AJ, Bollen AW, Hart AP, Ferriero DM. Childhood status epilepticus and excitotoxic neuronal injury. *Pediatr Neurol* 2007;**36**: 253–7.

418. Ueda M, Sugiura C, Ohno K, Kakita A, Hori A, Ohama E, et al. Immunohisto-chemical expression of fibroblast growth factor-2 in developing human cerebrum and epilepsy-associated malformations of cortical development. *Neuropathology* 2011;**31**:589–98.

419. Van Paesschen W, Revesz T, Duncan JS, King MD, Connelly A. Quantitative neuropathology and quantitative magnetic resonance imaging of the hippocampus in temporal lobe epilepsy. *Ann Neurol* 1997;**42**:756–66.

420. Vargova L, Homola A, Cicanic M, Kuncova K, Krsek P, Marusic P, et al. The diffusion parameters of the extracellular space are altered in focal cortical dysplasias. *Neurosci Lett* 2011;**499**:19–23.

421. Vespa PM, McArthur DL, Xu Y, Eliseo M, Etchepare M, Dinov I, et al. Nonconvulsive seizures after traumatic brain injury are associated with hippocampal atrophy. *Neurology* 2010;**75**:792–8.

422. Vezzani A, Aronica E, Mazarati A, Pittman QJ. Epilepsy and brain inflammation. *Exp Neurol* 2011.

423. Vezzani A, French J, Bartfai T, Baram TZ. The role of inflammation in epilepsy. *Nature Rev Neurol* 2011;**7**:31–40.

424. Vezzani A, Hoyer D. Brain somatostatin: a candidate inhibitory role in seizures and epileptogenesis. *Eur J Neurosci* 1999;**11**:3767–76.

425. Vezzani A, Sperk G. Overexpression of NPY and Y2 receptors in epileptic brain tissue: an endogenous neuroprotective mechanism in temporal lobe epilepsy? *Neuropeptides* 2004;**38**:245–52.

426. Vincent A, Crino PB. Systemic and neurologic autoimmune disorders associated with seizures or epilepsy. *Epilepsia* 2011;**52**(Suppl 3): 12–7.

427. Volz F, Bock HH, Gierthmuehlen M, Zentner J, Haas CA, Freiman TM. Stereologic estimation of hippocampal GluR2/3- and calretinin-immunoreactive hilar neurons (presumptive mossy cells) in two mouse models of temporal lobe epilepsy. *Epilepsia* 2011;**52**: 1579–89.

428. Wagner J, Urbach H, Niehusmann P, von Lehe M, Elger CE, Wellmer J. Focal cortical dysplasia type IIb: completeness of cortical, not subcortical, resection is necessary for seizure freedom. *Epilepsia* 2011;**52**:1418–24.

429. Wagner J, Weber B, Urbach H, Elger CE, Huppertz HJ. Morphometric MRI analysis improves detection of focal cortical dysplasia type II. *Brain* 2011;**134**:2844–54.

430. Wang Y, Qin ZH. Molecular and cellular mechanisms of excitotoxic neuronal death. *Apoptosis* 2010;**15**:1382–402.

431. Warren CP, Hu S, Stead M, Brinkmann BH, Bower MR, Worrell GA. Synchrony in normal and focal epileptic brain: the seizure onset zone is functionally disconnected. *J Neurophysiol* 2010;**104**: 3530–9.

432. White R, Hua Y, Scheithauer B, Lynch DR, Henske EP, Crino PB. Selective alterations in glutamate and GABA receptor subunit mRNA expression in dysplastic neurons and giant cells of cortical tubers. *Ann Neurol* 2001;**49**:67–78.

433. Widdess-Walsh P, Kellinghaus C, Jeha L, Kotagal P, Prayson R, Bingaman W, et al. Electro-clinical and imaging characteristics of focal cortical dysplasia: Correlation with pathological subtypes. *Epileps Res* 2005;**67**:25–33.

434. Wiebe S, Munoz DG, Smith S, Lee DH. Meningioangiomatosis. A comprehensive analysis of clinical and laboratory features. *Brain* 1999;**122**:709–26.

435. Wieser HG. ILAE Commission Report. Mesial temporal lobe epilepsy with hippocampal sclerosis. *Epilepsia* 2004;**45**:695–714.

436. Wittner L, Eross L, Czirjak S, Halasz P, Freund TF, Magloczky Z. Surviving CA1 pyramidal cells receive intact perisomatic inhibitory input in the human epileptic hippocampus. *Brain* 2005;**128**:138–52.

437. Wittner L, Eross L, Szabo Z, Toth S, Czirjak S, Halasz P, et al. Synaptic reorganization of calbindin-positive neurons in the human hippocampal CA1 region in temporal lobe epilepsy. *Neuroscience* 2002;**115**:961–78.

438. Wong M. Animal models of focal cortical dysplasia and tuberous sclerosis complex: recent progress toward clinical applications. *Epilepsia* 2009;**50**(Suppl 9): 34–44.

439. Wong M. Mechanisms of epileptogenesis in tuberous sclerosis complex and related malformations of cortical development

with abnormal glioneuronal proliferation. *Epilepsia* 2008;**49**:8–21.

440. Wong M. Mammalian target of rapamycin (mTOR) activation in focal cortical dysplasia and related focal cortical malformations. *Exp Neurol* 2011;**244**:22–6.

441. Yasin SA, Latak K, Becherini F, Ganapathi A, Miller K, Campos O, et al. Balloon cells in human cortical dysplasia and tuberous sclerosis: isolation of a pathological progenitor-like cell. *Acta Neuropathol* 2010;**120**:85–96.

442. Yasuda CL, Valise C, Saude AV, Pereira AR, Pereira FR, Ferreira Costa AL, et al. Dynamic changes in white and gray matter volume are associated with outcome of surgical treatment in temporal lobe epilepsy. *NeuroImage* 2010;**49**:71–9.

443. Yilmazer-Hanke DM, Wolf HK, Schramm J, Elger CE, Wiestler OD, Blumcke I. Subregional pathology of the amygdala complex and entorhinal region in surgical specimens from patients with pharmacoresistant temporal lobe epilepsy. *J Neuropathol Exp Neurol* 2000;**59**:907–20.

444. Ying Z, Babb TL, Comair YG, Bingaman W, Bushey M, Touhalisky K. Induced expression of NMDAR2 proteins and differential expression of NMDAR1 splice variants in dysplastic neurons of human epileptic neocortex. *J Neuropathol Exp Neurol* 1998;**57**:47–62.

445. Zattoni M, Mura ML, Deprez F, Schwendener RA, Engelhardt B, Frei K, et al. Brain infiltration of leukocytes contributes to the pathophysiology of temporal lobe epilepsy. *J Neurosci* 2011;**31**:4037–50.

446. Zeng LH, Rensing NR, Wong M. The mammalian target of rapamycin signaling pathway mediates epileptogenesis in a model of temporal lobe epilepsy. *J Neurosci* 2009;**29**:6964–72.

447. Zhao J, Wang S, Li J, Qi W, Sui D, Zhao Y. Clinical characteristics and surgical results of patients with cerebral arteriovenous malformations. *Surg Neurol* 2005;**63**:156–61;discussion 61.

448. Zhu ZQ, Armstrong DL, Hamilton WJ, Grossman RG. Disproportionate loss of CA4 parvalbumin-immunoreactive interneurons in patients with Ammon's horn sclerosis. *J Neuropathol Exp Neurol* 1997;**56**:988–98.

449. Zurolo E1, Iyer A, Maroso M, et al. Activation of Toll-like receptor, RAGE and HMGB1 signalling in malformations of cortical development. *Brain* 2011;**134**:1015–32.

Extrapyramidal Diseases of Movement

Tamas Revesz, H Brent Clark, Janice L Holton,
Henry H Houlden, Paul G Ince and Glenda M Halliday

INTRODUCTION

This chapter provides a description of a major group of neurodegenerative diseases, which include i) akinetic-rigid and hyperkinetic movement disorders principally due to diseases affecting the basal ganglia (Table 12.1), ii) miscellaneous disorders affecting the basal ganglia and iii) disorders of central autonomic systems. Some of the diseases discussed in this chapter are of significant epidemiological importance and relatively common in the general neuropathological practice and these are best exemplified by Parkinson's disease (PD), which is the second most common neurodegenerative disease after Alzheimer's disease. An up-to-date summary of the genetic basis and pathogenesis of these diseases will be provided and, in addition to single gene mutations causing inherited neurodegenerative diseases, information about genetic risk factors is presented. Much of the knowledge about the genetic basis of diseases has been translated into clinical practice, as through genetic testing, ante-mortem diagnosis has become available in several of the inherited movement disorders. This is exemplified by diseases such as Huntington's disease, the ever-increasing number of familial forms of parkinsonism and dystonia, as well as several other disorders that are discussed in this chapter (see Tables 12.3, 12.15 and 12.19). Although several of the mutations may only occur in a relatively small number of families, for example, with an inherited form of parkinsonism, the knowledge deriving from these discoveries can help to map out important molecular pathways such as those relevant for the survival of the substantia nigra (SN) dopaminergic neurons or Lewy body formation. Since the last edition, the Braak hypothesis describing a stereotypic spread of α-synuclein pathology, together with data from experimental work and pathological observations of a host-to-graft spread of Lewy body pathology in PD patients treated with striatal transplants of embryonic mesencephalic neurons, provided a practical conceptual framework for a better understanding of disease progression in PD.[16,46,51,297,332]

In the absence of reliable biomarkers, pathological diagnosis, utilizing the knowledge that many of the degenerative movement disorders are characterized by accumulation of an abnormal protein, has remained the 'gold standard' for definite diagnosis. Therefore, we have continued to follow a clinicopathological classification in this chapter, which is further justified by the fact that distinct clinical phenotypes can have several underlying pathological causes. For example, in addition to a number of rather rare conditions, parkinsonism may be caused by at least three common or relatively common disorders; PD, progressive supranuclear palsy (PSP) and multiple system atrophy (MSA). Similarly, several different pathologies can underlie the clinical presentation of corticobasal syndrome. Equally important is the knowledge that the same neurodegenerative processes can have a number of clinicopathological variants that are well exemplified by the different subtypes of PSP or the Lewy body disorders, whose clinical phenotypes range from a movement disorder to dementia. However, widely used terms, such as 'synucleinopathy' for diseases characterized by accumulation of α-synuclein,[162,517] 'tauopathy' for diseases with accumulation of the tau protein[164] and 'polyglutamine diseases' with expanded polyglutamine tracts, indicate a common molecular basis for clinically and pathologically diverse conditions and point to a likely future molecular classification of diseases. With advances in molecular understanding of the pathogenesis, preservation of tissues for future studies of molecular and pathophysiological aspects of neurodegenerative diseases remains part of routine practice in brain banks.

TABLE 12.1 Classification of extrapyramidal disorders of movement

Akinetic-rigid movement disorders
  Parkinsonism
  Parkinson's disease
  Hereditary forms of parkinsonism with or without Lewy bodies
  Progressive supranuclear palsy
  Multiple system atrophy
  Corticobasal degeneration
  Postencephalitic parkinsonism
  Vascular parkinsonism
  Guamanian and other forms of western Pacific parkinsonism
  Neuronal intranuclear inclusion disease
  Stiff man syndrome

Hyperkinetic movement disorders
  Chorea
  Hemiballismus and ballism
  Myoclonus
  Dystonia
  Tics
  Tremor

While increasing the genetic and molecular information relevant for the diseases discussed in this chapter, we omitted some traditional texts such as neuroanatomy from the current edition because of space constraints. Therefore, readers are referred to previous editions of this textbook and other specialist textbooks. Online Mendelian Inheritance in Man (OMIM) numbers are used in some of the tables in this chapter and consulting the OMIM database (www.ncbi.nlm.nih.gov/entrez/query.fcgi?db=OMIM&cmd=Limits) could also provide further relevant information for the reader.

TRAUMA AND MOVEMENT DISORDERS

There has been medicolegal interest in the possibility that some movement disorders might be caused by trauma, because patients sometimes attribute onset of symptoms to a traumatic event. It is more common, however, for a movement disorder to have precipitated trauma through poor coordination. Only very rare cases of movement disorder have been reported to be associated with trauma.[96,105,241] Epidemiological studies linked head injury to PD, but the results have not been consistent.[311] Recent data suggest head injury may initiate/accelerate neurodegeneration in individuals with expansion of REP1, a polymorphic mixed-dinucleotide repeat in the SNCA promoter region, which has been reported to result in higher levels of α-synuclein expression.[167] An akinetic/rigid syndrome, originally described in boxers and termed dementia pugilistica and more recently chronic traumatic encephalopathy, is associated with repetitive head injury and characterized by neurofibrillary tangle pathology. In this the parkinsonian signs are thought to be due to degeneration of the pars compacta of the SN (SNpc).[360] The association of chronic traumatic encephalopathy with a number of neurodegenerative diseases has been raised recently.[361]

AKINETIC-RIGID MOVEMENT DISORDERS

Parkinson's Disease

The diagnosis of Parkinson's disease (PD) is currently undergoing conceptual revision as a result of the more pervasive pathology now known to occur, the variety of genes now known to be involved, and the focus of research on more diverse clinical features that are thought to appear prior to any dominant movement disorder. The term PD was coined by Charcot,[64] which was in recognition of the first medical description of the main neurological symptoms by James Parkinson in 1817.[434] For the clinical diagnosis of PD, classified as a hypokinetic movement disorder,[121] current diagnostic criteria require bradykinesia and at least one of either rigidity, rest tremor, or loss of postural reflexes.[224] A number of supportive prospective features include disease progression, a unilateral onset with some persistence over time, an excellent response to L-dopa for 5 or more years, severe L-dopa-induced choreiform movements over time, and a clinical course of 10 years or more.[224] Other clinical features can include secondary motor symptoms and non-motor symptoms (e.g. autonomic dysfunction, cognitive/neurobehavioral abnormalities and sleep disorders).[242] The certainty of clinical diagnosis is stratified into possible, probable and definite diagnostic categories.[153] Criteria for the definite diagnosis include postmortem confirmation.[153]

The neuropathological criteria for PD have been recently revised[102] to frame the definite diagnosis of PD into two parts, i) a description of the neuropathological findings and ii) a probabilistic statement about the likelihood that the pathology would be associated with a particular clinical syndrome. Historically, clinical PD has been associated with a loss of SN dopaminergic neurons accompanied by the characteristic neuronal intracytoplasmic inclusions, Lewy bodies (Figure 12.1). In 1893, Blocq and Marinesco suggested that unilateral parkinsonism was due to a small tuberculoma affecting the midbrain in a patient with tuberculosis.[433] Friedrich Heinrich Lewy first described a variety of eosinophilic neuronal cytoplasmic inclusions (Figure 12.1a–c) in the substantia innominata and dorsal motor nucleus of the vagus in cases with paralysis agitans in 1912.[219] They were designated 'corps de Lewy' when they were recognized subsequently in the SN by Constantin Trétiakoff, who was also the first to state that SN lesions were important in both PD and postencephalitic parkinsonism (PEP).[325] In 1938, Hassler's post-mortem study revealed that the caudate, putamen and globus pallidus were largely unaltered in PD, whereas dense clusters of neurons in the ventral part of the SN were severely affected (Figure 12.1d,e) with surviving cells often containing Lewy bodies.[207] In the 1950s, the recognition of the dopaminergic nature of the nigrostriatal neurons led to the discovery of the role of dopamine, a finding at the origin of L-dopa therapy.[240] Current neuropathological diagnostic criteria for definite PD require these two features, i) neuronal loss in the SN and ii) brain stem Lewy bodies (Figure 12.1).[102] The distribution of these pathologies will be discussed in detail later.

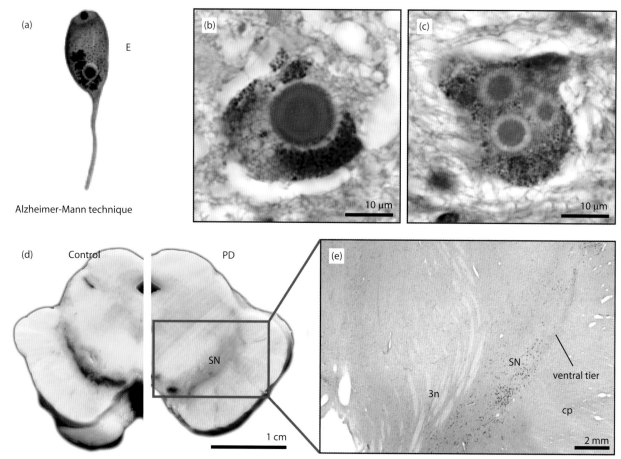

12.1 The two main diagnostic pathologies of Parkinson's disease (PD) are Lewy body inclusions (a–c) and depigmentation and cell loss in the ventrolateral substantia nigra (d,e). (a) Original drawing by Friedrich Lewy of the typical inclusions observed in PD. **(b,c)** Typical haematoxylin and eosin (H&E) stained Lewy bodies in neuromelanin pigmented neurons of the substantia nigra of patients with PD. **(d)** Photos of transverse sections through the midbrain showing the pigmented region of the substantia nigra (SN) in a control at left, and the same region depigmented in a patient with PD at right. **(e)** Transverse thin section through the midbrain at the level of the exiting third nerve fibres (3n) of a patient with PD stained with haematoxylin showing the reduction in neuromelanin pigmented neurons in the ventrolateral part of the SN. cp, cerebral peduncle. (a) Reproduced with permission from Lewy.[330a] With kind permission from Springer Science and Business Media.

In 1997, the recognition that mutation of the *SNCA* gene causes familial PD,[450] revolutionized the pathological construct of PD (see under Genetics). Subsequently the α-synuclein protein was found to be the main constituent of Lewy bodies,[518] as well as the glial cytoplasmic inclusions in MSA.[142,519] Because of the similar abnormal deposition of α-synuclein in PD, dementia with Lewy bodies (DLB) and MSA, these diseases are termed α-synucleinopathies,[163] which refer to a class of proteinopathies with the common properties of aggregated α-synuclein (Table 12.2). The two most common α-synucleinopathies are PD (discussed here) and DLB, with Lewy body pathology also commonly seen in association with other neurodegenerative diseases, such as Alzheimer's disease.

Epidemiology and Risk Factors

The mean age of onset of PD is around 60 years.[465] Patients with onset before the age of 40, who often have an identifiable genetic involvement, are divided into juvenile parkinsonism (onset before age 21 years) and young-onset parkinsonism (onset between the ages of 21 and 40 years).[488] The average disease duration is around 15 years, although this is variable and depends mainly on the age of onset (younger onset cases usually have longer disease duration).[133]

Age is the greatest risk factor for PD and both genetic and environmental factors appear to influence the risk of developing PD with increasing age. Data from longitudinal twin studies and nuclear families suggest that heritability accounts for around 35–40 per cent of cases (see Genetics of Parkinson's Disease),[195,587] implying that environmental factors dominate in approximately 60 per cent of cases. The role of environmental agents in the death of nigral neurons was reinforced by the finding of selective nigral neuronal death due to N-methyl-4-phenyl-1,2,3,6-tetrahydropyridine (MPTP).[313] The active metabolite of MPTP, 1-methyl-4-phenylpyridinium (MPP+), blocks complex I of the mitochondrial respiratory chain, resulting in dopamine-dependent oxidative stress.[522] Toxic pesticides and herbicides have been implicated in the development of human PD.[552] Paraquat, maneb and rotenone are currently used as agents to induce experimental PD with nigrostriatal dopaminergic degeneration with fibrillar α-synuclein inclusions.[44] Some

TABLE 12.2 Primary α-synucleinopathies

Syndrome	Clinical features (reference number)	Solubility and inclusion type	Regions of cell loss	Additional neuropathologies
Parkinson's disease	153,224	Highly insoluble, neuronal Lewy body	Ventrolateral substantia nigra	Rare
Dementia with Lewy bodies	363	Highly insoluble, neuronal Lewy body	Substantia nigra, nucleus basalis of Meynert, hippocampal presubiculum	Most have Aβ plaques, many have neurofibrillary tangles
Multiple system atrophy	161	More soluble, oligodendroglial cytoplasmic inclusions	Substantia nigra, putamen, pons, cerebellum	Rare
Neuroaxonal dystrophy due to mutations of the *PLA2G6* gene	176,425	Unknown, neuronal Lewy body	Variable	Basal ganglia iron accumulation

occupational exposures have been suggested to increase the risk of developing PD, including painters (organic solvents), wood workers (wood dust), teachers and medical workers (viral infections).[586] There is also evidence that increased intake of dairy products increases the risk of PD.[586]

The incidence of PD is higher in men than women,[538] which has been explained by the neuroprotective effect of oestrogen, also seen in MPTP animal models of PD.[10] Despite this, studies have shown an increased risk of PD with the long-term use of oral contraceptives,[392] implying that other environmental exposures may interact with any oestrogen effect.[586]

A number of pre-existing medical conditions are associated with an increased risk of PD including olfactory deficits, REM sleep behaviour disorder and depression, which are hypothesized to be early manifestations of PD. Others include infections/inflammation.[122] Patients with PD also have about a two-fold increase in CNS infections following diagnosis.[122] Gout has been shown to be protective for PD.[586] Uric acid/urate is a natural antioxidant and metal chelator that may alleviate oxidative stress in PD.[74] Also of interest is the decreased risk of cancer in patients with PD.[94] There are some lifestyle factors that also decrease the risk of PD including smoking, drinking coffee, taking vitamin E, non-steroidal anti-inflammatories and L-type calcium channel blockers, and being physically active.[79,438,586]

Genetics of Parkinson's Disease

The familial nature of PD is recognized in around 10 per cent of cases.[287,586,587] Initially mutations were identified in the *SNCA* gene[450] and subsequently in the *LRRK2* gene.[422,604] More recently rare autosomal dominant and autosomal recessive variants (Table 12.3) and a number of non-Mendelian genetic risk factors have been identified, such as the *glucocerebrosidase* (*GBA*) gene,[503] and polymorphic associations in the promoter region of the *SNCA* gene,[305] homozygosity for the H1 tau genotype[124] and the apolipoprotein E (*APOE*) ε2 or ε4 allele.[305] Recently genome-wide association (GWA) studies have identified significant loci at the *SNCA* and *MAPT* genes.[507]

Parkinson's Disease and α-Synuclein (PARK1)

As the first major discovery, the *A53T* missense mutation in *SNCA* was identified as a cause of autosomal dominant

PD in a large Greek-Italian (Contursi) kindred in 1997.[450] Subsequently an *A30P* mutation in a German family[304] and an *E46K* missense mutation in a Spanish family were reported.[598] *SNCA* gene triplication[510] and duplication,[65] resulting in 100 per cent and 50 per cent increases in genetic load, respectively, have also been identified.[509] Cases with different *SNCA* mutations show severe cell loss in pigmented brain stem nuclei with Lewy bodies. The *A53T* mutation shows conspicuous α-synuclein neuritic pathology, Lewy bodies, tau-positive neuritic and perikaryal inclusions and some with both tau and α-synuclein.[113] Widespread α-synuclein-positive Lewy bodies, Lewy neurites and glial aggregates have also been reported in the *A30P* mutation.[494] In α-synuclein gene triplication cases, there is severe neuronal loss in hippocampal CA2/3 subregions, pleomorphic Lewy bodies, α-synuclein-positive glial inclusions and widespread severe neuritic pathology.[185] The E46K pathology is similar to that seen in DLB,[598] although the neuropathology of the recently described *G51D* mutation resembles both the A53T and the *SNCA* gene multiplication cases, and it also includes MSA-like features, including α-synuclein-positive glial cytoplasmic inclusions.[280]

α-Synuclein is a 140-amino-acid protein, which is abundant in neural tissue and concentrates in synapses.[154] The function of α-synuclein is not understood, although it is likely to have a synaptic role.[477] Mutations of the *SNCA* gene were found to enhance the self-aggregation properties of α-synuclein.[78]

PARK2 (parkin)   A homozygous deletion of exons 3–7 of the *PARK2* (*parkin*) gene was initially identified in autosomal recessive juvenile PD.[286] Subsequently point mutations and deletions were found, which account for up to 10 per cent of early onset PD cases.[368] The association of heterozygous mutations with seemingly sporadic PD is unclear.[334]

Clinically, the phenotypes of *parkin* cases are usually indistinguishable from early onset L-dopa-responsive idiopathic PD. Atypical and later onset cases are described.[342]

There are a few pathologically examined cases with *PARK2* mutations. There is neuronal cell loss in pigmented nuclei of the brain stem. Lewy pathology has only been documented in a few cases.[123,455] Neurofibrillary tangles in the SN, locus coeruleus, red nucleus, posterior hypothalamus

TABLE 12.3 Monogenic genes identified in Parkinson's disease

Locus	Gene symbol	Gene product	OMIM number	Mode of inheritance	Age at onset and clinical features	Reference
PARK1/ PARK4	*SNCA*	α-synuclein	168601	Autosomal dominant	30–60 years missense. 30s triplication, 40/50s duplication. L-dopa responsive Parkinson's disease (PD) often with additional psychiatric features.	450,510
PARK2	*PARK2*	Parkin	600116	Autosomal recessive	Wide range from the 10 to 60 years old, usually in the 30s. L-dopa responsive parkinsonism often with prominent dystonia and late motor complications.	286
PARK6	*PINK1*	PTEN-induced kinase-1	605909	Autosomal recessive	Onset 30–50 years old. Typical L-dopa responsive parkinsonism.	551
PARK7	*PARK7*	DJ1	606324	Autosomal recessive	Onset 20s to 40 years, typical parkinsonism often with psychiatric features.	43
PARK8	*LRRK2*	Leucine-rich repeat kinase 2	607060	Autosomal dominant	Wide onset range from 30s to 50s. L-dopa response often with a slow less severe progression of disease.	422,604
PARK9 (Ku-for Rakeb)	*ATP13A2*	ATPase type 13A2 (ATP13A2)	606693	Autosomal recessive	Onset 10–22 years. Complicated parkinsonism, initially L-dopa responsive then additional motor features as well as prominent dystonia and pyramidal signs.	461
PARK14 or NBIA2	*PLA2G6*	Phospholipase A2, group VI	612953	Autosomal recessive	Classical type first decade. Parkinsonian type onset in the late teens or early 20s with initial L-dopa responsive parkinsonism then pyramidal signs, dystonia and ataxia.	423
PARK15	*FBXO7*	F-box protein 7	260300	Autosomal recessive	Onset is usually the late teens. L-dopa responsive but later progresses with prominent dystonia and pyramidal signs.	95
PARK17	*VPS35*	Vacuolar protein sorting 35 homolog	614203	Autosomal dominant	Onset 40s-50 years with typical L-dopa responsive parkinsonism.	556,602
NBIA1, no PARK number	*PANK2*	Pantothenate kinase 2	234200	Autosomal recessive	Classical type first decade. Parkinsonian type in the 20s. Initially L-dopa responsive then prominent dystonia, pyramidal signs and cognitive decline.	599

and cerebral cortex and tau-positive and argyrophilic thorn-shaped astrocytes have also been described.[377] A recent study examining five unrelated cases has found that, compared with PD, *PARK2* cases have more restricted morphological abnormality with predominant ventral nigral degeneration and absent or rare Lewy pathology.[106]

Parkin functions as a ubiquitin E3 ligase in the ubiquitin–proteosome pathway.[484] In drosophila models, parkin and PINK1 act in a common pathway in maintaining mitochondrial integrity and function of dopaminergic neurons.[75]

PARK6 (PINK1)   Homozygous mutations in the *PINK1* gene (PTEN-induced putative kinase 1) are the second most common cause of early onset autosomal recessive PD.[233,396,551]

Patient show slow disease progression and good response to L-dopa.

Recently neuropathological examination of the brain from an affected individual with *PINK1* mutation demonstrated neuronal loss in the SNpc with a few remaining neurons containing Lewy bodies. The Lewy pathology was sparse in the brain stem reticular formation and Meynert nucleus and was absent in the locus coeruleus, hippocampus, amygdala and neocortex.[479] As the Lewy pathology is atypical in this reported case, examination of further brains is required to establish the spectrum of neuropathology in *PINK1*-associated parkinsonism.

The *PINK1* gene encodes a putative serine/threonine kinase located in mitochondria. PINK1 protects cells against proteosome inhibition and apoptosis. It is ubiquitously expressed in human brain and is present in about 10 per cent of Lewy bodies in sporadic PD.[147] For the interaction between parkin and PINK1 see earlier.

PARK7 (DJ1)   Autosomal recessive mutations of the *PARK7* (DJ1) gene were identified as a cause of early-onset autosomal recessive parkinsonism.[43]*PARK7* mutations are rare and patients have L-dopa-responsive parkinsonism. No pathology data are available as yet for *DJ1*-related PD. DJ1 has been found in tau-positive inclusions,[471] but not in Lewy bodies.[31]

DJ1 may have several functions, including putative roles in neuronal oxidative stress response[572] and in protein folding and degradation.[463]

PARK8 (LRRK2, dardarin)   Autosomal dominant PD with linkage to a pericentromeric region on chromosome 12 was reported in a large Japanese pedigree.[141] Clinically, the disease resembled typical, L-dopa-responsive PD with an age of onset in the 50s. Subsequently, the underlying genetic cause in this and other families was shown to be mutation of the gene *LRRK2* (*leucine-rich repeat kinase 2*).[422,604]*LRRK2* mutations are the most common genetic cause of autosomal dominant, late-onset PD and have been observed in 0.5–3.0 per cent of 'sporadic' cases. The *p.G2019S* mutation[95,391] is relatively common across several populations and is responsible for PD in ~2 per cent of sporadic and ~5 per cent of familial PD cases in Northern European and North American populations.[160] This mutation was reported in ~20 per cent in Ashkenazi Jewish and in 40 per cent of North African Berber Arab PD patients.[327,419]

Most *LRRK2* cases described thus far exhibited typical Lewy body pathology, although cases with pure nigral degeneration, pathologies resembling PSP and MSA and cases with TDP43 pathology have also been reported.[143,203,336,353,582,604]

The presence of a kinase domain suggests a role for *LRRK2* in intracellular signalling pathways, and other structural features indicate possible interactions with other proteins.[330] Experimentally, *LRRK2* mutations increase kinase activity, supporting a gain-of-function mechanism.[544]

PARK9 (ATP13A2)   A homozygous mutation in the *ATP13A2* gene was identified in a large consanguineous family from Kufor Rakeb, Jordan with affected individuals presenting with atypical L-dopa-responsive juvenile PD.[461] Mutations were also identified in other cases with autosomal recessive PD.[424,461] Mutations in this gene are likely to interfere with the lysosomal degradation process.[424] No neuropathology of the central nervous system disease has as yet been described.

PARK14 (PLA2G6)   Genetic studies in families with early onset parkinsonism identified mutations in the *phospholipase A2, group VI (PLA2G6)* gene.[423] Mutations of *PLA2G6* are also associated with infantile neuroaxonal dystrophy or neurodegeneration with brain iron accumulation, type 2 (NBIA2) (see also later).[375]

Neuropathological data of patients with genetically proven disease indicate that, in addition to the extensive neuroaxonal dystrophy, there is cerebellar atrophy with or without evidence of brain iron accumulation and widespread

Lewy pathology. Neurofibrillary tangles and neuropil threads in medial temporal lobe structures and cerebral cortex have been documented.[176,425] The neuropathology is similar in the infantile onset and later onset parkinsonian cases.[425]

PARK15 (FBXO7)   Autosomal recessive, early onset PD was reported from a large Iranian family with a homozygous missense mutation in the *FBXO7* (*F-box only protein 7*) gene.[501] Since then, families with homozygous or heterozygous mutations have been reported from other countries.[424] No neuropathology has been reported in affected patients.

PARK17 (VPS35)   A *p.D620N* mutation in the *VPS35* gene have been recently identified in affected members of a Swiss kindred with late onset autosomal dominant PD.[556] This mutation has also been found in other families and in one sporadic PD case.[602]

Rarer Causes of Autosomal Dominant Parkinsonism

There are a number of disorders that often have L-dopa-responsive parkinsonism as part of the spectrum of their clinical phenotype. Examples include rare cases with *GCH1* mutations[213] or some kindreds with mutations in the *MAPT* gene.[23] Perry syndrome, another rare genetic cause of parkinsonism and characterized neuropathologically by pallidonigral TDP43 pathology, is due to mutation in the *dynactin 1* (*DCTN1*) gene.[125,581]

Recently, a founder mutation of the *EIF4G1* gene was identified in families. The phenotype is late onset autosomal dominant or sporadic L-dopa-responsive PD with Lewy body pathology.[66]

Parkinson's Disease Genetic Risk Factors and Loci

The REP1 polymorphism in the SNCA promoter region was shown to be associated with PD.[305,351]

Clinical observations indicate that patients and relatives of patients with Gaucher's disease are more often affected by PD than expected.[165,536] A strong association between single heterozygous *GBA* mutations and PD was subsequently confirmed and such mutations are present in ~15 per cent of Ashkenazi Jewish patients and ~3 per cent of non-Ashkenazi Jewish patients, compared to ~3 per cent and <1 per cent in matched controls, respectively.[2,503] Extensive Lewy pathology has been confirmed in PD cases with heterozygous *GBA* mutations[389] and classic Lewy pathology is also characteristic in Gaucher's patients who develop PD.[537,588]

Genome-wide Association (GWA) Studies

Genome-wide association (GWA) studies have provided a relatively efficient means to identify common genetic risk loci in PD. Not only have such studies demonstrated that there are risk loci at both the region of the *SNCA* gene and *MAPT* gene,[481,507] but they have also identified further loci containing genes such as *GAK*, *DGKQ* and the *HLA* locus (for a review see Houlden and Singleton[222]).

Pathogenesis

There are a variety of processes that have been considered in the pathogenesis of PD, including the initiation, sustaining and cell-damaging processes and processes involved in disease spread.

The detailed evaluation of aetiological and genetic features suggests that a variety of factors and cellular mechanisms may precipitate misfolding and aggregation of α-synuclein resulting in Lewy pathology of PD. Common themes environmentally and genetically include perturbation of mitochondria (*parkin* and *PINK1* associated genes and mitochondrial toxins), protein and lipid metabolism (*parkin, ATP13A2, NBIA, FBXO7* and *GBA* genes and the physical build-up of α-synuclein), and immune mechanisms (*LRRK2, EIF4G1* and *HLA* genes and association with infections and NSAIDs). Over time there is considerable cell damage in vulnerable neurons that leads to their death, although many neurons that succumb to disease do survive to be able to be observed with their Lewy pathology. The mechanisms eventually leading to the selective cell death observed in PD include the loss of sufficient mitochondria and the overwhelming build-up of intracellular debris, both of which appear to increase an inflammatory glial reaction finally resulting in cell dysfunction and death. The cellular mechanisms involved in the death-initiating pathways are those also involved in disease initiation (see earlier), with genetic and ongoing environmental factors potentially linked to the rapidity of the entire process.

Cellular attributes that sustain these initiating perturbations are also required, as neurons and neuronal systems have a variety of compensatory and repair mechanisms[18,49] with the disease sustained for a considerable time (estimated to be decades).[188] Increased oxidative stress and associated iron dysregulation is one such mechanism, which in nigral neurons manifests as increased by-products of oxidation of lipid, protein, DNA and RNA. Several factors may make these cells more vulnerable to ongoing oxidative stress, including their increased calcium levels through the use of L-type calcium channels for pacemaking[526] and their increased reliance on growth and survival factors.[468,560] There is some indication that the growth and survival factor oestrogen may be neuroprotective[10] and potentially explains the male predominance of PD. Factors affecting compensation and repair that are associated with older age must also be involved for the majority of cases and there is substantial evidence now to implicate growth factor signalling for insulin and genes important for mitochondrial electron transport chain function as responsible for detrimental changes precipitated with age.[466] A combination of age-related cellular changes with increased oxidative stress and iron dysregulation may overwhelm neuronal compensatory and repair mechanisms in vulnerable neurons to produce the ongoing cellular processes observed in PD.

Since the concept of a disease that spreads from the gut into the lower brain stem and up through the brain was proposed by Braak,[46] there has been considerable interest in the processes involved in disease spread. In 2008, the concept that the disease may spread from cell to cell took hold following the identification of Lewy bodies in foetal neurons grafted therapeutically into patients with PD.[297,332] Proposed mechanisms for this phenomenon include a prion-like process whereby α-synuclein aggregates released from one cell are taken-up by neighbouring cells where abnormal conformation of the host α-synuclein takes place *via* the process of permissive templating.[51,201]

Neuropathology of Parkinson's Disease

Pathological Definition

The external surface of the brain is generally similar to that seen in age-matched controls. The major macroscopic findings are the pale SN and locus coeruleus. The recent revision of the neuropathological diagnosis of PD[102] confirmed that the essential neuropathology requires: i) neuronal loss of the SNpc of a moderate to severe degree and ii) Lewy pathology (Figure 12.2). The optimal section of the midbrain for assessment is a transverse section perpendicular to the long axis of the brain stem at the level of the third cranial nerve, permitting evaluation of all of the pertinent nuclear groups at that level (Figure 12.2). Neuronal loss is assessed in haematoxylin and eosin (H&E) stained sections (or comparable histological stains) by comparing the neuronal density in the case to schematic representations showing examples of no, mild, moderate and severe neuronal loss in the pigmented SNpc (Figure 12.2). The presence of fibrillary astrocytosis and extraneuronal neuromelanin assists in the assessment of neuronal loss. Moderate to severe neuronal loss in the ventrolateral SN is related to the motor symptoms of PD. An important negative finding is the normal macroscopic and microscopic appearance of the globus pallidus, putamen, caudate and subthalamic nucleus.

Lewy pathology is the collective term for the constellation of α-synuclein immunoreactive inclusions, which are typified by neuronal Lewy bodies, a range of other neuronal perikaryal structures and α-synuclein immunoreactive inclusions in neuronal cell processes (see later) (Figure 12.3). Methodological differences among laboratories (e.g. section thickness, primary antibodies, antigen retrieval methods) can produce variable results, and although several studies have documented variability,[11,34] there is, as yet, no consensus on optimal staining procedures. Although the presence of Lewy pathology indicates PD, its density does not consistently relate to many clinical symptoms. For example, the density of Lewy pathology in the SN does not relate to motor deficits[174] and, in fact, remains stable over the course of PD.[175] Lewy pathology is not increased in relevant brain stem structures in patients with REM sleep behaviour disorder.[114] Olfactory bulb atrophy is not significant in PD despite considerable α-synuclein deposition.[226,381] Lewy pathology occurs in highest densities in the spinal cord and paraspinal ganglia with a rostro-caudal spread into end organs.[35,451] Notable exceptions to this lack of clinicopathological correlations are the strong correlations between Lewy pathology densities in temporal lobe regions and the presence of visual hallucinations[145,199] and the correlation between cortical Lewy pathology load and increasing cognitive impairment in the absence of other neurodegenerative pathologies.

Given that PD increases in frequency with age, it is important to assess other common age-related pathologies in every case. It is particularly vital to assess Alzheimer-type pathology because it is the most common co-occurring pathology that accompanies Lewy pathology. The new methods proposed by the National Institute on Aging-Alzheimer's Association guidelines for the neuropathological assessment of Alzheimer's disease are recommended for this purpose,[229,372] in association with the method recommended in the Third Consortium for Dementia with

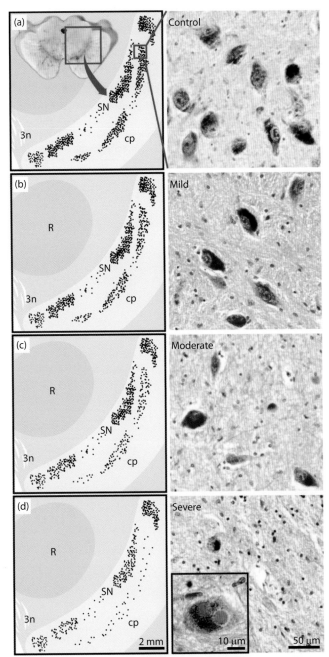

12.2 Schematic representation of the severity of cell loss in the pigmented substantia nigra (SN) and examples of high magnification haematoxylin and eosin (H&E) photos to assist with the standardization of the examination of cell loss. The midbrain is assessed at the level where there are visible exiting third nerve fibres (3n) ventromedial to the red nucleus (R) and dorsal to the cerebral peduncle (cp), as indicated in **(a)**. The clusters of pigmented neurons in the ventral and lateral regions of the SN are particularly affected with mild **(b)**, moderate **(c)** and severe **(d)** cell loss indicated compared with those unaffected **(a)**. Some remaining pigmented cells contain Lewy bodies (inset in d), although in cases with particularly severe cell loss, several sections may be needed to observe these.

Lewy Bodies.[363] In these protocols, α-synuclein immunohistochemistry is required, in addition to assessing the degree of Alzheimer-type pathology. In order to determine the likelihood that the Lewy pathology would contribute significantly to the cognitive disorder, a probability concept

has been developed.[102] There is a low probability of Lewy pathology relating to cognitive impairment if only brain stem Lewy pathology is present and a highest probability if widespread Lewy pathology is observed throughout the cortex (Table 12.4). The probability is mitigated in the presence of substantial Alzheimer-type pathology (Table 12.4).

Lewy Pathologies and Their Development

There is still quite a debate on how Lewy pathology develops and the relationship between the different types of Lewy pathology observed (Figure 12.3). Although there is no debate that α-synuclein has self-aggregation properties that are enhanced by genetic mutations associated with PD,[78] there is still debate around a number of initiating molecular aspects, particularly for sporadic PD. The finding that genetic multiplications cause familial PD with Lewy pathology could indicate that toxicity is caused by simply increasing the cellular monomeric form of α-synuclein. However, several recent studies have shown either no change or a progressive reduction in soluble α-synuclein in sporadic PD[459,600] with no change in α-Synuclein mRNA expression.[459,508] In PD there is an increase in membrane-associated α-synuclein,[600] which can accelerate fibril formation,[251] seed the aggregation of cytosolic α-synuclein[321] and have a role in inclusion formation.[26] Most α-synuclein in Lewy pathology is phosphorylated at serine 129 (S129),[14,139] and data suggest that this precedes Lewy pathology formation,[343,600] indicating an important pathogenic role. In addition, C-terminal truncation of α-synuclein has been highlighted as an important initiator of α-synuclein aggregation[456] and fibril formation.[329] α-Synuclein fibrils have been shown to pass between neurons and attract endogenous α-synuclein to form insoluble cellular aggregates.[344]

As discussed earlier, there is a variety of types of neuronal perikaryal Lewy pathology, including diffuse, granular, pleomorphic and Lewy body structures (Figure 12.3), as well as α-synuclein-immunoreactive inclusions in neuronal cell processes known as intraneuritic Lewy bodies, Lewy neurites, dot-like structures and axonal spheroids. These abnormal depositions of α-synuclein can be considered on a continuum of abnormal structures leading to Lewy body formation. Studies have revealed that α-synuclein aggregates are first evident microscopically as punctate material in cell bodies[309] and neurites.[266] In neurites, there is a coalescence of α-synuclein into loosely packed filaments considered to be premature or 'pale neurites' that frequently occur at the branch points of axon collaterals extending centripetally into proximal axonal segments.[267] Pale neurites coalesce and incorporate ubiquitin to form Lewy neurites.[267] Three-dimensional reconstructions show that 70 per cent of Lewy bodies show continuity with Lewy neurites, indicating that Lewy neurites aggregate and transform into Lewy bodies over time.[266,267] In cell bodies, the coalescence and incorporation of ubiquitin around punctate α-synuclein yields pale bodies, from which Lewy bodies also form.[309] Pale bodies are round areas of pale-staining eosinophilic material with disorganized fibrils interspersed with vacuoles and granular material ultrastructurally. Pale bodies appear to be the precursors of Lewy bodies via a 'compaction model',[309] whereas small aggregates of α-synuclein in neurites form larger Lewy neurites, which form Lewy

TABLE 12.4 Probability of Lewy pathology-related cognitive disorder

		NIA-Reagan Alzheimer-type pathology		
		NFT stage I–II/ CERAD possible	NFT stage III–IV/ CERAD probable	NFT stage V–VI/ CERAD definite
Lewy body type	Brain stem	Low	Low	Low
	Transitional	High	Intermediate	Low
	Diffuse	High	High	Intermediate

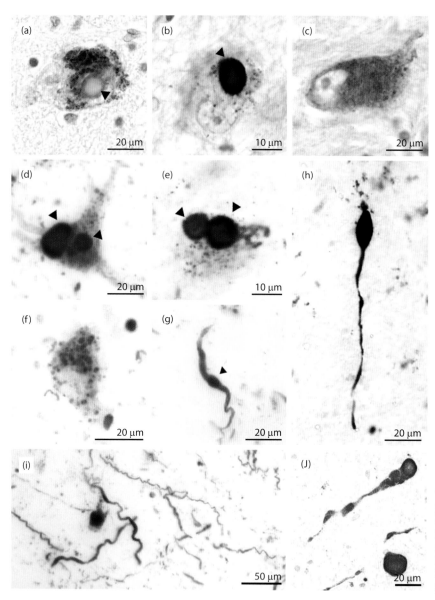

12.3 α-Synuclein-immunoreactive Lewy pathologies: intracellular neuronal Lewy bodies (arrowed in **a,b,d,e**), diffuse granular neuronal immunostaining **(b,c,d,e,f)**, pleomorphic neuronal perikaryal structures **(e)**, intraneuritic Lewy bodies (**h**, arrowed in **g**), Lewy neurites **(g,h,i)**, dot-like structures **(d,f,i)** and axonal spheroids **(j)**.

bodies via a 'growth model'.[266] A formal staging scheme for Lewy body formation has been proposed.[569] Stage 1 is observed as a diffuse, pale cytoplasmic α-synuclein staining, often seen in morphologically normal-looking neurons or their processes (Figure 12.4). Stage 2 is observed as an irregularly shaped, uneven α-synuclein staining of moderate intensity in neurons (Figure 12.4). Stage 3 is a discrete α-synuclein staining corresponding to pale bodies or neurites (Figure 12.4), which often display a peripheral condensation of α-synuclein. These 'early Lewy bodies' develop into typical Lewy bodies. Stage 4 is a ring-like staining of a typical Lewy body with a central core and a surrounding halo (Figure 12.4). Pale bodies outnumber Lewy bodies in the early stages of PD[569] with Lewy bodies suggested to survive on average around 6–16 months in degenerating neurons.[175] Ultrastructurally, the outer halo of classical Lewy bodies comprises ordered, radially arranged 7–20 nm fibrils associated with granular electron-dense coating material and vesicular structures. The core is composed of densely packed fibrils associated with dense granular material.

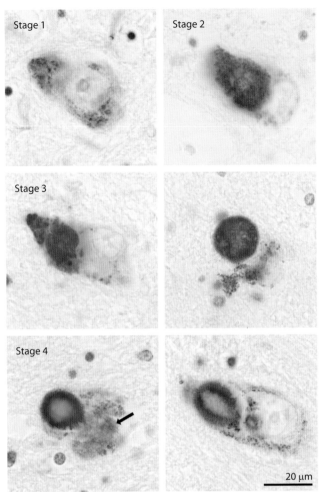

12.4 Examples of intraneuronal α-synuclein-immunoreactive aggregates representing different developmental stages of Lewy bodies, as proposed by Wakabayashi and colleagues. [569] Diffuse, pale cytoplasmic staining occurs in morphologically normal-looking neurons or their processes in Stage 1. Irregularly shaped, uneven staining of moderate intensity occurs in Stage 2. Discrete α-synuclein staining corresponding to pale bodies or neurites occurs in Stage 3 and by Stage 4 they have developed the ring-like staining of a typical Lewy body with a central core and a surrounding halo. With the continual accumulation of diffuse α-synuclein aggregations, additional pale bodies form (arrowed in Stage 4) to become mature Lewy bodies.

In addition to α-synuclein, a wide variety of proteins are demonstrable in Lewy pathology.[569] Many of these, which can be divided into several groups (Table 12.5), show modifications, including oxidization, nitration, ubiquitylation and phosphorylation.

Anatomical Distribution and Spread of Cell Loss and Lewy Pathologies

In PD both the cell loss and Lewy pathology occur in region-specific manners. The SNpc can be divided into ventral and dorsal tiers, which represent functionally distinct cell populations; each tier can be subdivided further into three regions.[189] In PD, the pattern of loss of nigral neuromelanin-containing neurons is the opposite of that found in ageing, being greatest in the lateral ventral tier (Figure 12.2), followed by the medial ventral tier, and being least in the dorsal tier.[128] Although a variety of vulnerability factors are proposed to differentiate these subdivisions of the SN,[109] there are limited data on the factors that explain the selective nature of the loss of only some dopamine neurons in this structure in PD.

The severity of dopamine cell loss in the SN correlates with the severity of symptoms identified using the motor sections of the Unified PD Rating Scale (UPDRS) and specifically to bradykinesia and rigidity.[174] Even in the elderly without overt PD, the degree of pigmented neuronal loss in the SN correlates with subtle changes in UPDRS scores.[52] From these studies it is clear that cell loss occurs in a preclinical phase with clinical symptoms sufficient for a diagnosis of PD occurring when 50–60 per cent of the ventrolateral nigral neurons have been lost (Figure 12.2).

Lewy pathology also occurs in a region specific manner. In the early 1980s, it was observed that Lewy pathology in the midbrain, basal forebrain and cortex in PD occurs in a stepwise fashion.[150] By the 1990s, many research groups had found evidence that Lewy pathology in PD was confined to particular brain regions and the concept that neurologically normal cases could have Lewy pathology (incidental Lewy body disease) was also proposed.[564] However, it was Braak's group in 2002 that showed that Lewy pathology in PD did not necessarily begin in the SN as previously thought, but in nuclei of the caudal brain stem, as well as in olfactory structures.[90] That Lewy pathology could occur in the absence of a defined clinical syndrome in aged brains led the way to the suggestion that such individuals have presymptomatic PD, although it is still debated whether all such individuals would inevitably develop PD.[134] The proportion of cross-sectional cases with incidental Lewy pathology varies widely (5–24 per cent) and may indicate the type of population recruited.[435] Overall, these studies paved the way for the pattern of abnormal deposition of α-synuclein in different autopsy populations to be used to define the progression of PD.[46]

Braak's theory for the progression proposes a characteristic pattern of deposition for α-synuclein over time in PD.[46] This theory is underpinned by three premises, that α-synuclein-positive Lewy pathology is essential for a diagnosis of PD, that multiple neuronal systems are involved as a result of pathological changes developing in a few susceptible types of nerve cells, and that pathology is detected presymptomatically and progresses in predictable stages to readily recognizable locations in the brain. The resulting six progressive stages of PD (Figure 12.5) have come to be widely utilized as a tool in neuropathological diagnosis and as a guiding principle in most research on PD, changing the clinical concept of the disease. Stages 1 and 2 involve Lewy pathology in the olfactory regions and lower brain stem (dorsal motor nucleus of the vagus nerve, the intermediate reticular zone and locus coeruleus) and are clinically presymptomatic. In stage 3, Lewy pathology occurs in the midbrain SNpc dopamine neurons without substantive neuronal loss and Lewy neurites also appear in the central subnucleus of the amygdala as a precursor for further limbic infiltration. The clinical symptoms of PD are first observed when midbrain dopamine cell loss is obvious in stage 4 when α-synuclein-positive Lewy neurites also appear in the basal forebrain and cortical areas, particularly in the CA2 region of the hippocampus and the parahippocampus.

By stages 5 and 6, there is substantial loss of dopaminergic neurons in the midbrain and α-synuclein deposition in higher order cortical association areas (such as the temporal, insular and anterior cingulate cortices). In the most severe cases, α-synuclein pathology affects the primary and secondary sensory cortices (Table 12.6). There is an assumption that α-synuclein pathology spreads progressively so that, for any grade, all the structures expected to be affected in lower stages will show pathology. The stages correspond roughly to the classification of Lewy body disorders proposed by Kosaka,[298] which was used in the consensus criteria for DLB.[363] From a purely neuropathological diagnostic perspective, it is difficult to differentiate end-stage PD from DLB.

There is a hot debate on how predictable the pathological staging of PD is. Braak reported divergence from the expected caudo-rostral sequence of pathology in ~6 per cent of the autopsy cases studied.[46] Several retrospective autopsy studies were performed to verify the anatomical progression, with 47 per cent of cases not fitting the staging scheme in a large study.[263] Others have reported no involvement of the lower brain stem, despite Lewy pathology in the higher brain stem or cortical regions.[247,436] In a study of cases aged 65 years and over, 36.5 per cent showed Lewy pathology but only 51 per cent of these conformed to the Braak staging scheme and 17 per cent had mainly cortical Lewy pathology.[596] Additional major criticisms have arisen from observations that many elderly individuals with widespread Lewy pathology lack the

TABLE 12.5 Groups of molecules found in Lewy pathology (LP)[569]

Group	Examples
Structural elements of the fibrils	α-Synuclein, neurofilaments
α-Synuclein binding proteins	Synphilin-1 and other chaperones and vesicle-binding proteins
Synphilin-1 binding proteins	Parkin and various ubiquitin associated proteins
Components of the ubiquitin-proteosome system	Ubiquitin, ubiquitin activating enzymes, ubiquitin conjugating enzymes, ubiquitin ligases, ubiquitin hydrolases, p62/sequestosome, proteasome subunits and activators
Proteins implicated in cellular responses	Hsc70-interacting proteins and molecular chaperones, heat shock, oxidative and cell stress proteins, clusterin, proteins involved in glycoxidation, interferon induced proteins
Proteins associated with phosphorylation and signal transduction	Kinases including LRRK2, proteins or enzymes associated with signal transduction
Cytoskeletal proteins	Microtubule associated proteins, tubulin, tubulin polymerization promoting protein
Cell cycle proteins	Cyclin, extracellular signal regulating kinases
Cytosolic proteins that passively diffuse into Lewy bodies	Tyrosine hydroxylase, etc.
Others	Lipids, mitochondria and mitochondrial proteins

TABLE 12.6 Practical guidelines for assessing stage/type of Lewy body pathology (adapted from Alafuzoff et al.[10a])

Anatomical region and structure	Brain stem regions					Basal forebrain/limbic structures					Neocortical regions		
	Medulla		Pons		Midbrain								
	dmv	irx	LC	R	SN	nbM	Hipp/CA2	Trent-c*	Amygdala	Cingulum	T-c	F-c	P-c
Braak stage[46]	1	1	2	2	3	3	4	4	4	5	5	5	6
DLB consensus stage[363]	Brain stem					Limbic					Neocortical		
Other									Amygdala predominant				

*Some strategies recommend the use of the temporo-occipital gyrus as it is involved to a similar extent.[10a]
dmv, dorsal motor nucleus of vagus; irx, intermediate reticular zone; LC, locus coeruleus; R, raphé nucleus; SN, substantia nigra; nbM, nucleus basalis of Meynert; Trent-c, transentorhinal cortex; T-c, temporal cortex; F-c, frontal cortex; P-c, parietal cortex.
Adapted from Alafuzoff et al.[11] With permission from Lippincott Williams & Wilkins/Wolters Kluwer Health.

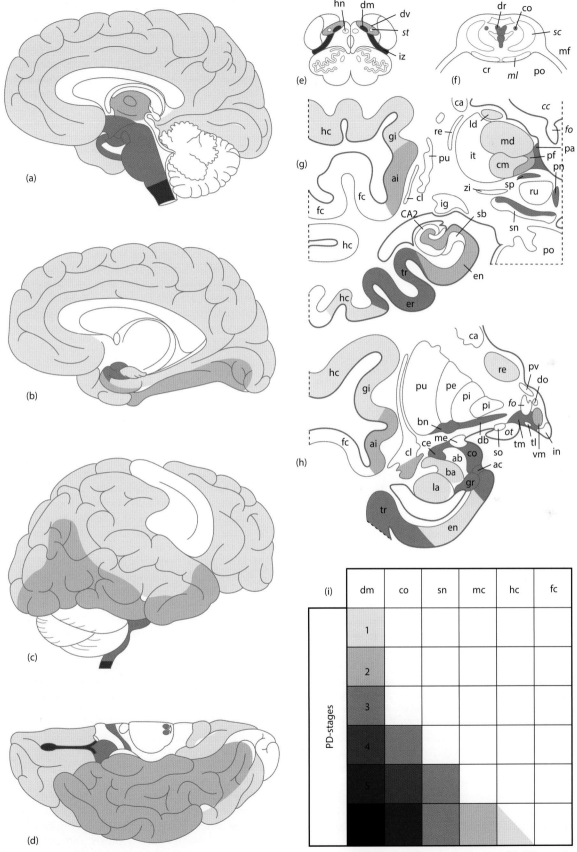

12.5 The characteristic progressive spread of α-synuclein Lewy pathology (LP) over time in patients with PD, as theorized by Braak and colleagues.[46]

Reproduced with permission from Braak et al.[46] With permission from Elsevier.

clinical features of PD.[56,436] However, the motor symptoms of PD relate best to nigral cell loss rather than Lewy pathology.[174,175] Although there will probably always be a number of outlying cases that do not conform to any staging scheme, the numbers with alternate distributions of Lewy pathology may be greater than expected for PD, but may also reflect the different types of cohorts examined or different technical issues. Prospective assessment of PD patients to autopsy, excluding those with early disease onset, shows that the progression of Lewy pathology in such typical cases is consistent with Braak PD staging, where brain stem Lewy pathology dominates in those surviving to 5 years; by 13 years, 50 per cent of cases have a transitional distribution; and by 18 years all have at least this pathological phenotype.[188,190]

Glial pathology is also present in PD, with evidence that glia are responsible for the progression of PD and play an important role in initiating the early tissue response.[193] Protoplasmic astrocytes are involved, they become non-reactive and accumulate α-synuclein (Figure 12.6).[515] Experimental evidence has shown that astrocytic α-synuclein deposition initiates the non-cell autonomous killing of neurons through microglial signalling.[322] There is evidence that microglia are activated early in PD (Figure 12.6) and possibly assist with the clearance of extracellular α-synuclein at this time.[406] Microglia transform to phagocytes and target neurons as the disease progresses but appear to become dysfunctional with increasing amounts of ingested debris.[191]

Variant Lewy Pathologies and Coexisting Diseases

Both retrospective[76,317,498] and prospective[188,190] studies have identified an older onset and more complex PD syndrome with shorter survival and more rapid cognitive decline due to an increased rate of progression. These older PD patients have higher amounts of cortical Lewy pathology and additional neuropathologies, particularly Alzheimer-type pathologies. Retrospectively assessing PD patients with dementia[17,76,198,317,498] reveals up to 10-fold higher amounts of cortical Lewy pathology compared to those without dementia and a correlation between the severity of Lewy pathology and Alzheimer-type pathology in such patients.[76,232,317] The more rapid course and higher amounts of pathological deposits in these cases with additional neuropathologies suggests an even faster rate of Lewy pathology deposition that appears to be linked to multiple pathologies in older onset PD patients. These cases have similarities to those of DLB.[363]

Lewy bodies occur as a secondary phenomenon in the amygdala and adjacent entorhinal cortex in both sporadic and familial Alzheimer's disease, Down's syndrome, Pick's disease, argyrophilic grain disease and Guam dementia complex.[193] Such secondary Lewy bodies may either be the direct consequence of aggregated tau or be due to cell stress resulting from the formation of primary tau-containing neuronal inclusions. Lewy bodies have been described in association with a number of other conditions where they may be considered an incidental finding. Lewy body pathology occurring in some of inherited forms of PD has been discussed earlier.

Treatments Including Cell Transplantation

A number of therapeutic strategies, in addition to dopamine replacement therapy, are currently recommended or undergoing active research. Silencing the cells that are responsible for inhibition of the thalamus are major neurosurgical targets and currently bilateral deep brain stimulation of the subthalamic nucleus is the preferred approach for the symptomatic treatment of PD with complications (especially dyskinesias) associated with prolonged L-dopa treatment.[48]

The concept that cells may be replaced to provide long-lasting relief of patients' symptoms has practically been implemented by the replacement of dopamine neurons. Clinical trials using transplantation of human foetal ventral mesencephalic tissue into the striatum of PD patients provided proof-of-principle that such grafts can restore striatal dopaminergic function. The transplants survive, reinnervate the striatum and generate adequate symptomatic relief in some patients for more than a decade following operation.[449] Current research is now focused on using stem cell derived neuroblasts for transplantation. However, a number of issues such as identifying the best source of cells, eliminating the risk of tumour formation and the development of graft-induced dyskinesias and standardizing dopaminergic cell production are still need to be solved.[148] The long-term survival of grafted cells may be compromised by a host-to-graft spread of α-synuclein pathology, documented in cases surviving for longer than a decade.[297,332]

Encephalitis Lethargica and Postencephalitic Parkinsonism

The designation 'encephalitis lethargica' (EL) was given by von Economo to a disease of the CNS that occurred in pandemic form in most parts of the world between 1915 and 1930. Outbreaks of EL coincided with an epidemic of influenza, but no consistent relationship has been demonstrated between these two clinically distinct illnesses.[358] Although antibodies to some strains of influenza virus bind to antigens in the SN and hypothalamus of patients with EL,[146] in other studies influenza virus RNA could not be detected in archival brain tissue.[357] The relationship between EL and postencephalitic parkinsonism (PEP) has been recently questioned and, according to one hypothesis, multiple causes may be behind PEP.[557]

Although the epidemic of EL declined rapidly after 1926 and ceased after 1930, sporadic cases of disease with similar clinical and pathological manifestations continued to occur.[152,282] In several of such cases, oligoclonal bands were demonstrated in the cerebrospinal fluid (CSF). Examination of recent EL-like cases found evidence of post-streptococcal autoimmune disease,[83] with anti-basal ganglia antibodies in the majority of these patients.[73] However, the possibility cannot be excluded that the antibodies are markers of a destructive encephalitic process, possibly infective, and not themselves responsible for the neurological disease.[558]

Several clinical variants of EL have been described. The most common presentation was somnolence, meningism and fever, often followed by the development of oculomotor nerve paralysis with ptosis and pupillary abnormalities. Some patients developed other cranial nerve and bulbar palsies or paralysis of one or more limbs. Some patients lapsed into stupor that lasted for periods ranging from days to years. Other neurological disturbances

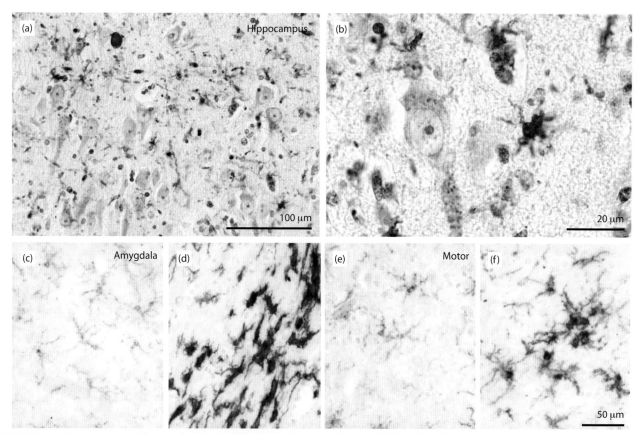

12.6 Photomicrographs of the involvement of glia in Parkinson's disease (PD). (a,b) α-Synuclein immunoreactive astrocytes and Lewy neurites in the hippocampus of a patient with PD. A number of immunoreactive astrocytes are visible, particularly at higher magnification in **(b)**. **(c–f)** HLA–DR immunoreactivity reveals quiescent microglia in unaffected amygdala **(c)** and pre-supplementary motor cortex **(e)** with significant upregulation of microglia in patients with PD **(d,f)**. These regions often have cell loss in end-stage patients with PD.[192,200,345]

included nystagmus, vertigo, tremor, pyramidal signs, dystonias, chorea, myoclonus, ataxia, behavioural disturbances, seizures, hallucinations, catatonia and mutism. About one-third of patients died during the acute illness, one-third made a complete recovery and one-third had residual neurological disease. The development of parkinsonism was usually gradual and often asymmetrical. In some patients, parkinsonism developed during recovery from the acute illness, whereas in other patients parkinsonism occurred only after an interval of several years. The parkinsonian features of rigidity, bradykinesia and tremor were occasionally complicated by episodes of oculogyric spasm, spasm of the neck muscles with deviation of the head and retraction of the eyelids or by other dystonias. Ophthalmoplegia and oculogyric crises have been less frequent manifestations in more recent cases of encephalitis lethargica.[38,83]

Neuropathology

In the acute illness, the brain appears normal or slightly swollen.[561] In the chronic phase of the disease, in patients with PEP, the brain may be mildly atrophic, but the most striking macroscopic finding is depigmentation of the SN. In contrast to the findings in idiopathic PD, the locus coeruleus is better pigmented.

In the acute phase, microscopy reveals perivascular infiltrates of lymphocytes, predominantly in the basal ganglia

and midbrain,[83] but, based on original descriptions, extensive cortical involvement is also likely to be a feature.[15] The infiltrates contain plasma cells and include morular cells containing multiple cytoplasmic immunoglobulin inclusion bodies.[282]

In PEP, the SN shows severe neuronal loss, neurofibrillary tangles (NFTs) and astrocytosis (Figure 12.7). NFTs predominate in the hippocampus and entorhinal cortex and are usually present in many other regions, including prefrontal and inferior temporal cortex where they preferentially localize to layers II and III (Table 12.7).[152,216] Evidence of inflammation including microglial activation is absent.[152] There are also tau-positive astrocytes in regions of neuronal degeneration.[231] Ultrastructurally, tangles appear as typical paired helical filaments and straight filaments are also present. The filaments in tau-positive, tufted-astrocyte-like structures are 15 nm wide.[197] Western blotting of insoluble tau demonstrates that both three-repeat tau (3R-tau) and four-repeat tau (4R-tau) isoforms are represented in the tau filaments.[54,93]

Drug- and Toxin-Related Parkinsonism

Drug-induced parkinsonism can resemble idiopathic PD clinically, causing bradykinesia, rigidity and, occasionally, tremor. Most cases are associated with neuroleptic drugs and caused by blockade of D2 dopamine receptors in the brain.[409]

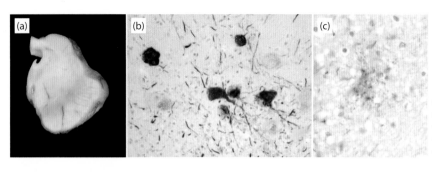

12.7 Post-encephalitic parkinsonism. (a) Severe pallor of the substantia nigra is a frequent feature. **(b)** Tau-immunopositive neurofibrillary tangles in the nigra are a prerequisite for the diagnosis. **(c)** Occasional tau-positive, tufted astrocyte-like structures are also found.

TABLE 12.7 Distribution of tangles in post-encephalitic parkinsonism

Tangles usually seen
    Substantia nigra
    Locus coeruleus
    Hippocampus
    Parahippocampal gyrus
    Nucleus basalis of Meynert

Tangles commonly seen
    Caudate nucleus and putamen
    Globus pallidus
    Subthalamic nucleus
    Dentate nucleus of cerebellum
    Central pontine nuclei
    Hypothalamus
    Thalamus
    Nuclei of reticular formation
    Anterior horn cells
    Insular, frontal, parietal and temporal neocortices

Tangles uncommonly seen or in low density
    Red nucleus
    Nuclei of basis pontis
    Inferior olivary nucleus
    Pedunculopontine nucleus
    Superior and inferior colliculi

Adapted from Geddes et al.[152] By permission of Oxford University Press on behalf of The Guarantors of Brain.

Drug-induced parkinsonism affects 15–60 per cent of patients treated with typical neuroleptics, depending on the type of neuroleptics, the dose used and the underlying susceptibility of the patient.[68,374] Severe and often fatal neuroleptic sensitivity occurs in elderly patients with confusion or dementia. Many such patients have Lewy body dementia and are predisposed to adverse drug effects by the relative dopamine deficiency associated with nigral cell loss.[363] Drug-induced parkinsonism can be associated with adverse reactions to other drugs, details of which are included in drug information resources and reviews.[42] Heroin addicts abusing meperidine developed a parkinsonian syndrome caused by MPTP, a contaminant of illicit chemical synthesis (see also earlier).[314]

Chronic use of 3,4-methylenedioxymethamphetamine (MDMA, ecstasy) and cocaine have also been associated with resting tremor and parkinsonism (see also later).[84,399] Severe parkinsonism was reported in poisoning with potassium cyanide and, at autopsy 19 months later, there was neuronal loss from the basal ganglia.[550] Other toxins thought to cause parkinsonism include manganese, paraquat, rotenone and other herbicides and pesticides (see also under Parkinson's Disease).[37,534,540]

Carbon monoxide intoxication may cause necrosis in the basal ganglia, which results in parkinsonism.[511]

Arteriosclerotic Disease and Parkinsonism

The term 'vascular parkinsonism' ('vascular pseudo-parkinsonism', 'arteriosclerotic pseudo-parkinsonism') implies an akinetic/rigid syndrome caused by lacunar infarction in the basal ganglia and associated with cerebral arteriolosclerosis. It is responsible for 2.5–5 per cent of the total cases of parkinsonism in various population based studies and clinical cohorts.[88] Clinically, the disease mainly affects the legs ('lower body parkinsonism'),[132] with sparing of the face and arms, and has a less satisfactory response to L-dopa therapy.[502] Bradykinesia and rigidity may be dominant, leading to an abnormal gait and postural instability, but resting tremor is unusual.[224] Patients usually have vascular risk factors, particularly hypertension. The chief neuropathological correlate of vascular parkinsonism is lacunar infarction affecting the basal ganglia, the deep white matter or the frontal lobes in the absence of alternative or coexisting pathological lesions linked with neurodegenerative disease.[601] Infarction of the SN is a rare cause of parkinsonism.[227]

Vascular disease in the basal ganglia is a common additional finding in idiopathic PD and should be evaluated as a possible supplementary pathogenic factor to the movement disorder.[225] The clinical syndromes of PSP and corticobasal degeneration (CBD) (see later) may also arise through vascular disease.[502]

Dementia and Parkinsonism

Parkinsonism is frequently associated with dementia. In some patients, the movement disorder precedes cognitive decline, whereas, in others, the reverse pertains. There are several candidate neuropathological correlates of a parkinsonism–dementia syndrome, the pathology of which is discussed in other sections (Table 12.8).

Guamanian and Other Forms of Western Pacific Parkinsonism

A very high incidence of amyotrophic lateral sclerosis (ALS) and parkinsonism/dementia complex (PDC) was recognized in three regions of the western Pacific: in the Marianas Islands of Guam and Rota, in the Kii peninsula, in Japan and in western New Guinea.[144,308,500] In the most extensively studied PDC of Guam, there is severe global cerebral atrophy, neuronal loss in the frontal and temporal lobes and insular cortex, CA1 hippocampal subregion, neostriatum, globus pallidus, SN and other brain stem nuclei. NFTs are most abundant in the hippocampus and entorhinal cortex, but also present in cerebral cortex, SN and brain stem and spinal cord.[370] In the

TABLE 12.8 Conditions with dementia and parkinsonism
Neurodegenerative diseases
Parkinson's disease with dementia
Familial Parkinson's disease (certain forms*)
Dementia with Lewy bodies
Progressive supranuclear palsy
Corticobasal degeneration
Multiple system atrophy
Frontotemporal dementia with parkinsonism linked to chromosome 17 due to mutations in the *MAPT* gene
Frontotemporal lobar degeneration with TDP-43 pathology
Frontotemporal lobar degeneration with FUS pathology (neuronal intermediate filament inclusion disease)
Pick's disease
Parkinsonism–dementia complex of Guam
Neurodegeneration with brain iron accumulation, type 2
Neuronal intranuclear inclusion disease
Other causes
Vascular parkinsonism
Chronic traumatic encephalopathy (dementia pugilistica)
Normal pressure hydrocephalus
Drug-induced parkinsonism/dementia

*See under Genetics of Parkinson's Disease.

cerebral cortex, NFTs show a predilection for layers II and III[215,217] and are present in laminae V and VI in more severe cases.[370] Their ultrastructural and biochemical features are similar to those in Alzheimer's disease.[53,239] Glial pathology is seen in PDC of Guam, including granular hazy astrocytes and crescent/coiled inclusions in oligodendroglia.[370,411,590] Lewy pathology is seen in the amygdala in up to ~50 per cent of the cases[370] and may also be found in other structures usually affected by α-synuclein pathology.[370] Aβ plaques are seen in a significant proportion of the cases and TDP43-positive neuronal and glial inclusions are common.[204,370]

Initial genetic investigation has failed to identify a single gene locus for the PDC of Guam,[380] although more recent genome-wide linkage and association analyses indicate the role of three loci, including the involvement of the MAPT region.[504]

Despite intense efforts, satisfactory explanation for these localized disorders has not been found. The cycad hypothesis has re-emerged postulating dietary consumption of cycad toxins, such as the excitotoxin β-N-methylamino-l-alanine (BMAA) or sterol glucosides, as causative.[30,516] The incidence of ALS/PDC of Guam has declined since the 1960s.[573]

Progressive Supranuclear Palsy

PSP was originally described as a neurodegenerative disorder characterized by postural instability, parkinsonism, ophthalmoplegia affecting chiefly vertical gaze, pseudobulbar palsy

and mild dementia with neuropathological changes consisting of neuronal loss, astrocytosis and NFTs primarily in basal ganglia and brain stem nuclei.[521] PSP is now defined as one of the atypical parkinsonian disorders and one of the primary tauopathies with deposition of predominantly 4R-tau into both neuronal and glial inclusions.[101] PSP is a sporadic disease, although it has genetic risk factors.[27,218] Epidemiological studies have not identified an environmental cause.

Characteristic PSP pathology may be seen in patients lacking the typical clinical features of PSP, who may present with one of a number of clinical syndromes (see under PSP variants).[103,583,585] In contrast, the classical clinical PSP syndrome may arise through a number of other pathological processes including CBD,[299,335] Lewy pathology,[407] Pick's disease,[407] cerebrovascular disease,[407] Creutzfeldt–Jakob disease,[257] frontotemporal dementia with parkinsonism linked to chromosome 17 (FTDP-17),[379] frontotemporal lobar degeneration with TDP43 inclusions (FTLD-TDP),[440] ALS[407] and MSA.[407]

Epidemiology, Clinical Features and Risk Factors

PSP is the most common form of atypical parkinsonism, comprising 2–6 per cent of patients with parkinsonism in specialist clinics.[273,338] The median age of onset is 64 years and males and females are affected equally.[57] The median disease duration is about 6 years.[583] The average annual incidence rate for ages 50 to 99 years is estimated at 5.3/100 000[45] and the prevalence rate ranges from 1.0 and 6.5/100 000 depending on the sample size.[388,489] A prevalence rate of 5.82/100 000 was reported from Japan,[275] which is higher in Japanese rural populations.[408] A high prevalence of atypical parkinsonism, reminiscent of PSP, was reported from Guadeloupe.[61]

The classical clinical presentation of PSP, designated as Richardson's syndrome (PSP-RS),[583] includes early postural instability, vertical supranuclear gaze palsy, a severe and L-dopa-unresponsive axial akinetic-rigid syndrome, early pseudobulbar dysarthria and dysphagia and fronto-subcortical cognitive impairment.[580,583] The National Institute of Neurological Disorders and the Society for Progressive Supranuclear Palsy (NINDS-PSPS) criteria developed for the clinical diagnosis are widely used. These are highly specific for the classical presentation, but have relatively low sensitivity.[338,339,580] On the basis of clinical diagnostic probability, the categories of possible, probable and definite PSP have been proposed. Definite PSP requires histological confirmation of PSP.[338,339]

For diagnostic purposes, identifying by MR imaging regional atrophy of the midbrain, brain stem tegmentum, superior cerebellar peduncles and frontal lobe along with T2 signal increase in structures such as the midbrain and inferior olives is useful.[401,520]

Neuropathology

Macroscopic Findings

In typical cases, there is atrophy of the subthalamic nucleus, midbrain and pontine tegmentum, marked pallor of the SN and, to a lesser extent, of the locus coeruleus. Atrophy of the superior cerebellar peduncle is common,[546] although it is preserved in some of the variants.[3,259] Dilatation of the third and fourth ventricles and aqueduct is a feature. Atrophy of the globus pallidus and the dentate nucleus is variable (Figure 12.8). Mild generalized or predominantly posterior frontal atrophy with involvement of the precentral and postcentral gyri can occur.

Microscopic Findings

Pathological criteria for the diagnosis of PSP have been proposed and validated, although an update to include findings of up-to-date tau immunohistochemistry and criteria for the diagnosis of PSP variants is awaited.[209,340] The NINDS neuropathological diagnostic criteria for PSP are shown in Table 12.9.[209] The diagnosis of PSP and other coexisting diseases should be made when there are Lewy bodies, α-synuclein-positive glial cytoplasmic inclusions, TDP43 or other inclusions or when infarcts are also present.[505,545,582,591]

The main pathology of PSP is accumulation of hyperphosphorylated tau in neurons and glia, neuronal loss and astrocytosis.[101,104,584] Therefore, routine histological examination should be supplemented by tau immunohistochemistry and, if necessary, silver stains such as the Gallyas impregnation (Figure 12.9). The 'globose' NFT is a characteristic finding in the brain stem, subthalamic nucleus and nucleus of Meynert (Figure 12.9). Cortical NFTs are more commonly flame-shaped, coiled or curvilinear. 'Pretangle' neurons show a diffuse or granular staining pattern for tau.[104]

In classical cases, neuronal loss, astrogliosis and neurofibrillary degeneration in PSP are most severe in the basal ganglia and brain stem nuclei (Tables 12.9–12.11), whereas cortical pathology predominates in the precentral gyrus and premotor cortex comprising NFTs, tau-immunoreactive glia and neuropil threads.[103,584]

Involvement of the striatum, globus pallidus, basal nucleus of Meynert and subthalamic nucleus by tau pathology is a constant feature (Figure 12.9). Together with the SN, the subthalamic nucleus is consistently affected by nerve cell loss and astrogliosis, whereas this is more variable in the globus pallidus. In the SN, neuronal loss is usually most severe in the ventrolateral tier of the SNpc, but it also affects non-pigmented neurons.[412] Degenerative changes are seen in the locus coeruleus, colliculi, midbrain and pontine tegmentum, periaqueductal grey matter, red nucleus, oculomotor complex, trochlear nucleus, nuclei implicated in the supranuclear organization of eye movements,[262,467] pontine nuclei, inferior olivary nucleus and cerebellar dentate nucleus. 'Grumose degeneration' implies eosinophilic granular material, representing degenerating axon terminals of Purkinje cells around neurons of the

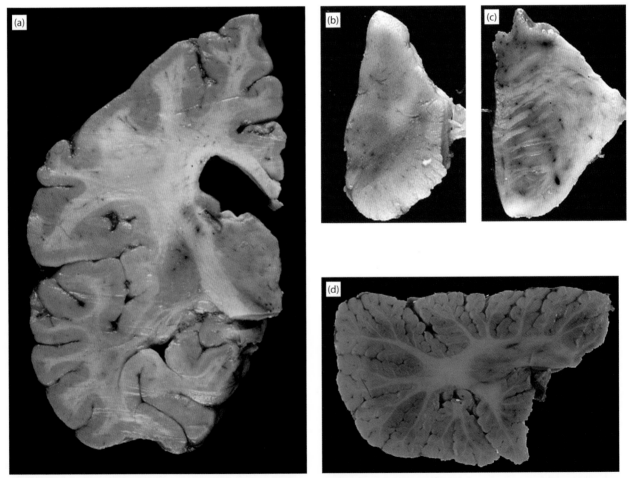

12.8 Characteristic macroscopic changes in a case of progressive supranuclear palsy. (a) Atrophy of the subthalamic nucleus is prominent. The midbrain tegmentum and pontine tegmentum are reduced in bulk **(b,c)**. There is marked pallor of the substantia nigra **(b)** and the outlines of the cerebellar dentate nucleus are blurred **(d)**.

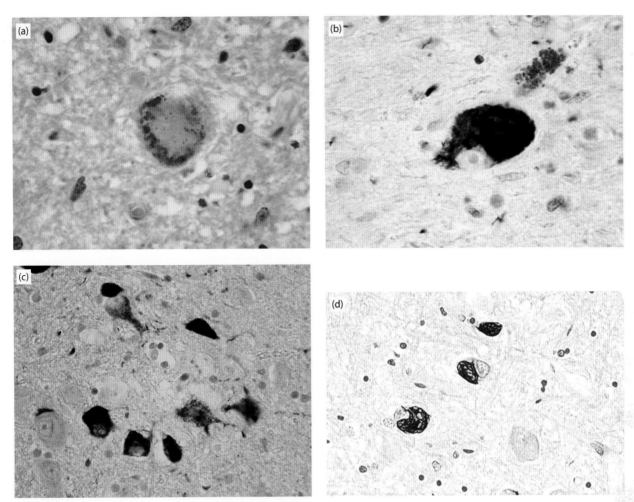

12.9 Progressive supranuclear palsy neurofibrillary tangles. (a) Typical globose tangle with faintly basophilic filamentous appearance in a pigmented neuron of the substantia nigra. **(b)** Antibody specific to four-repeat tau isoforms strongly labels a neurofibrillary tangle in the nigra. **(c)** Tau immunohistochemistry highlights numerous neurofibrillary tangles in the nigra. **(d)** Neurofibrillary tangles in neurons of the pontine nuclei are well demonstrated by Gallyas silver impregnation.

cerebellar dentate nucleus, some of which are swollen and achromatic.[237] It is thought to be related to pathology in the dentatorubrothalamic tract as well as the cerebellar white matter.[237] Hypertrophy of the inferior olivary nucleus is seen rarely in PSP.[272] Accumulation of non-filamentous tau in Purkinje cells and argyrophilic tau and ubiquitin-positive doughnut-shaped structures comprising tau filaments in radial processes of Bergmann glia may be seen.[445]

A distinctive form of astrocytic pathology in grey matter, designated as 'tufted astrocytes' can be demonstrated with tau immunohistochemistry or Gallyas silver impregnation (Figure 12.10).[101,104,294] Tufted astrocytes are stellate in shape with fine radiating processes and contrast in morphology with the 'astrocytic plaques' of CBD. Of the cortical regions, tufted astrocytes are most frequent in the primary motor and premotor cortices, less so in the superior and middle frontal gyri and parietal cortex and scanty in temporal and limbic cortices. They are also frequent in the caudate, putamen, superior colliculus and red nucleus, less so in the subthalamic nucleus, thalamus and globus pallidus and usually sparse, if any, in the SN, inferior olivary nucleus and dentate nucleus (Table 12.10).[208] The frequency of tufted astrocytes may vary substantially between cases.[260] Tufted astrocytes can be distinguished

morphologically from 'the thorn-shaped astrocytes' or 'spiny astrocytes', which possess several rather broad, cone-shaped tau-positive processes. These latter stain for glial fibrillary acidic protein (GFAP) and can be seen in elderly individuals in periventricular, subependymal and subpial areas.[492]

Coil or comma-shaped tau-immunoreactive and silver-positive oligodendroglial coiled bodies are a feature (Figure 12.10), which are also seen in other tauopathies such as argyrophilic grain disease, CBD and subtypes of FTDP-17.[101,104,584] In PSP they can be seen in the cerebral white matter, including internal capsule, cortex, subcortical grey nuclei, SN, cerebral peduncles, pontine tegmentum and base, red nucleus and also cerebellar white matter. Abnormal tau protein in neurons and glia contributes to widespread neuropil threads (Table 12.10).

Swollen, achromatic ballooned neurons may be seen in limbic areas and are indicative of coexistent argyrophilic grain disease (Table 12.11).[542,584] More widespread swollen neurons should raise the possibility of CBD (see under Corticobasal Degeneration).

Tau pathology is seen in the spinal cord.[271] Lesions in the Onufrowicz (Onuf's) nucleus may explain disorders of micturition.[483] Neuronal cell loss and NFTs were reported

TABLE 12.9 Diagnostic criteria for progressive supranuclear palsy (PSP)*

Inclusion criteria	Exclusion criteria
Typical PSP	
High density of neurofibrillary tangles and neuropil threads in the following areas: pallidum, subthalamic substantia nigra or pons; low to high density of neurofibrillary tangles or neuropil threads in at least three of the following areas: striatum, oculomotor complex, medulla or dentate nucleus†; and clinical history compatible with PSP	Large or numerous infarcts Marked diffuse or focal Lewy bodies Changes diagnostic of Alzheimer's disease Inclusions of multiple system atrophy Pick bodies Diffuse spongiosis Prion protein (PrP)-positive immunostaining of plaques
Combined PSP	
As for typical PSP, with the coexistence of infarcts in the brain stem and/or basal ganglia‡	

*Awaiting update to include clinicopathological variants.
†Tau-positive astrocyte processes or astrocyte cell bodies (tufted astrocytes) in the areas of neurofibrillary tangles and neuropil threads confirm the diagnosis.
‡The diagnosis should be typical PSP.
Reproduced from Hauw et al.[209] and Litvan et al.[339] With permission from Lippincott Williams & Wilkins/Wolters Kluwer Health.

TABLE 12.10 Distribution and severity of tau pathology in progressive supranuclear palsy*

Anatomical area	Neuronal	Threads	Coiled bodies	Tufted astrocytes
Temporal cortex	+	+	+	+
Frontoparietal cortex	+ +	+ +	+ +	+ + +
Motor cortex	+ +	+ +	+ +	+ + +
Striatum	+ +	++	+/+ +	+ + +
Pallidum	+ +	+ +	+ + +	+
Basal nucleus	+ + +	+ +	+	+
Thalamus, ventral	+ +	+ + +	+ + +	+/+ +
Thalamic fascicle	N/A	+ + +	+ + +	0
Subthalamic nucleus	+ +/+ + +	+ + +	+ +/+ + +	+/+ +
Hypothalamus	+ +/+ + +	+ +	+	+
Tectum	+ +	+ + +	+ +/+ + +	+ +/+ + +
Red nucleus	+ +	+ +/+ + +	+ +/+ + +	+/+ +
Substantia nigra	+ +/+ + +	+ +	+/+ +	+
Locus coeruleus	+ + +	+ +	+	0
Pontine tegmentum	+ +/+ + +	+ +/+ + +	+ +	+
Pontine nuclei (base)	+ +/+ + +	+ +	+	+
Medullary tegmentum	+ + +	+ + +	+/+ +	+
Inferior olive	+ +	+ +/+ + +	+	+
Dentate nucleus	+ +/+ + +	+ +	+	+
Cerebellar white matter	N/A	+/+ +	+ +	0

*In the clinicopathological variants of PSP,[3,255,259,337,547] the severity and distribution of the tau pathology will deviate from that shown in the table. Adapted from Dickson et al.[104] With permission from John Wiley and Sons. Copyright © 2011 International Society of Neuropathology.

to be most severe in the cervical cord[281] but were found throughout the cord by another study.[483]

There is microglial activation in PSP, which correlates with tau pathology and neuronal loss.[130,234]

In PSP argyrophilic grains have been reported to occur in ~19 per cent of the cases, whereas the incidence of TDP43 pathology varies between 0 and 26 per cent in the different reports.[542a,591]

Ultrastructurally, filaments from regions enriched in coiled bodies contain chiefly smooth 14 nm filaments, resembling abnormal thin tubules, whereas in regions rich in tufted astrocytes, irregular 22 nm filaments with jagged

TABLE 12.11 An approach to the differential diagnosis of sporadic parkinsonism with neurofibrillary tangle pathology

	Four-repeat tauopathies		Mixed four-and three-repeat tauopathies	
	PSP	**CBD**	**PEP**	**Guam PDC**
Genetic risk factors	*MAPT* (further risk factors have been identified by GWAS[218])	*MAPT*[223]	Not known	Two loci on chromosome 12 and a third in *MAPT*[504]
Tau western blot	Doublet pattern with a 33 kDa C-terminal fragment	Doublet pattern with a 37 kDa C-terminal fragment	Triplet pattern	Triplet pattern
Ultrastructure of tau filaments	SF and PHF	Twisted ribbons	PHF	PHF
Stereotypic macroscopic features (when present)	SN pallor, STN atrophy, midbrain and pontine tegmental and SCP atrophy, cerebellar dentate abnormality	SN pallor, frontal atrophy	SN pallor (near complete loss of neuromelanin pigment may occur)	SN pallor, severe cortical (frontotemporal) and brain stem atrophy
Distribution of NFT pathology	Cortex may be affected severely, but involvement of the basal ganglia and brain stem usually predominates	Usually severe in cortex, but also present in basal ganglia and brain stem nuclei; pretangles are common	The distribution resembles that seen in PSP. However, NFTs are fewer and are rare in IV, V and XII cranial nerve nuclei, pontine base, inferior olive, dentate nuclei, striatum and pallidum. Hippocampus and parahippocampus are commonly affected	The overall distribution is similar to that in PSP and PEP
Glial pathology	Tufted astrocytes, coiled bodies	Astrocytic plaques, coiled bodies	Not prominent, but tufted astrocyte-like structures are present	Not prominent, but granular hazy astrocytes, crescent/coiled oligodendroglial inclusions are seen
Threads	Present, often numerous	Extensive and characteristic	Present	Present
Ballooned (swollen) neurons	May be present in limbic regions when argyrophilic grains are usually seen[542]	Common in cerebral cortex	Not a diagnostic feature	Not a diagnostic feature

CBD, corticobasal degeneration; GWAS, genome-wide association study; PDC, parkinsonism/dementia complex; PEP, postencephalitic parkinsonism; PHF, paired helical filaments; PSP, progressive supranuclear palsy; SCP, superior cerebellar peduncle; SF, straight filament; SN, substantia nigra; STN, subthalamic nucleus.

contours predominate.[531] In NFTs, 15 nm straight filaments and Alzheimer's disease-type PHFs have been reported.[528]

PSP Variants (Atypical PSP)

Patients with PSP pathology may have a clinically atypical presentation. The variants include PSP with parkinsonism (PSP-P) with a PD-like disease onset.[583] Patients may present with pure akinesia with gait freezing with subsequent development of typical signs of PSP (PSP-PAGF).[585] In both the PSP-P and PSP-PAGF variants, the tau pathology is less severe than in PSP-RS.[584,585] Severe nerve cell loss and gliosis in the subthalamic nucleus, globus pallidus and SN were also documented in PSP-PAGF.[3] In large brain bank series, PSP may be the commonest cause of clinically diagnosed corticobasal syndrome (CBS) (PSP-CBS).[335] Patients with PSP may present with behavioural variant of frontotemporal dementia (bv-FTD)[107,205] or speech and language impairment (PSP-FTD).[255] The PSP-CBS and PSP-FTD variants are associated with an increased neocortical tau load.[103,337,547] In some patients, cerebellar ataxia is the initial and principal clinical presentation (PSP-CA). Tau-positive inclusions in Purkinje cells are more frequent and neuronal loss, gliosis and coiled body pathology are more severe in the cerebellar dentate nucleus than in other PSP cases.[265] In atypical PSP with corticospinal tract degeneration (PSP-CST), the clinical presentation is primary lateral sclerosis-like or CBS.[5,259] The tau pathology is severe in the motor cortex and present in frontal primary association areas, whereas the subcortical tau pathology and neuronal loss in the SN are mild. There is also microscopic evidence for involvement of the CST. A proportion of the oligodendroglial tau inclusions have a globular appearance. In this variant, the astrocytes show punctate or multiple small globular tau deposits and are Gallyas-negative.[5,259] Recently, PSP-CST has been classified as a variant of the novel, globular glial tauopathies.[5,7]

Tau Protein Involvement

In both PSP and CBD, tau filaments are composed mostly of 4R-tau isoforms[89,236,499] and, in both diseases, immunoblotting of detergent-insoluble tau shows two major bands at 68

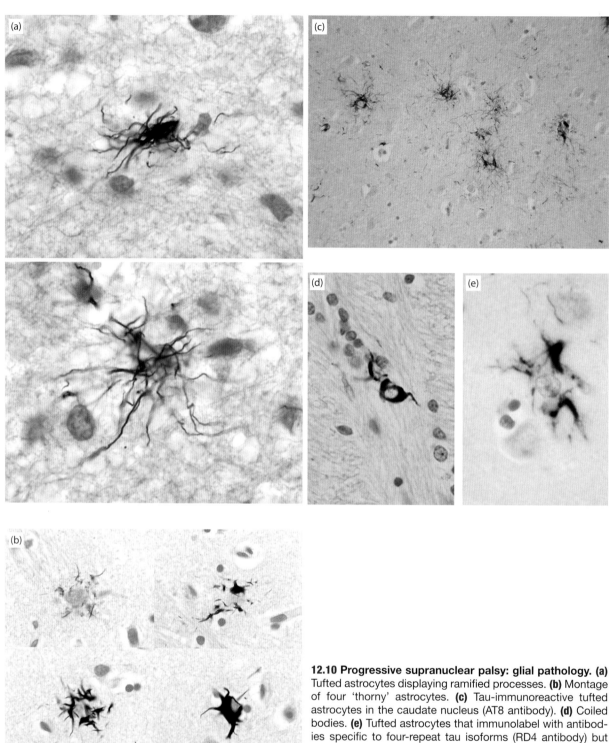

12.10 Progressive supranuclear palsy: glial pathology. (a) Tufted astrocytes displaying ramified processes. **(b)** Montage of four 'thorny' astrocytes. **(c)** Tau-immunoreactive tufted astrocytes in the caudate nucleus (AT8 antibody). **(d)** Coiled bodies. **(e)** Tufted astrocytes that immunolabel with antibodies specific to four-repeat tau isoforms (RD4 antibody) but not with antibodies to three-repeat tau isoforms. (a,b) Gallyas silver stain; (c) AT8; (d,e) RD4 immunohistochemistry.

kDa and 64 kDa. A 33 kDa C-terminal tau fragment is also present in PSP, in contrast to a 37 kDa fragment in CBD.[21]

Genetics

In Caucasian patients, a genetic link was shown initially between PSP and a dinucleotide repeat polymorphism in the intron between exons 9 and 10 of the *tau* (*MAPT*) gene.[77] Subsequently, of the two common MAPT haplotypes (H1, H2) an overrepresentation of H1 was shown.[27] The H1 subhaplotype C is strongly overrepresented in PSP, but the H1 subhaplotype B is not.[446] Familial PSP has been described, but these cases should be reclassified as FTDP-17.[228] A GWA study identified a number of signals associated with PSP risk at the *STX6*, *EIF2AK3* and *MOBP* genes.[218] Two independent variants in *MAPT* were confirmed affecting risk for PSP, one of which influences *MAPT* brain expression.[218]

Corticobasal Degeneration

The first comprehensive, clinicopathological description of CBD was by Rebeiz *et al.* in 1968,[464] but previous description is likely to exist.[29] The clinically diverse CBD is one of the tauopathies with widespread neuronal and glial 4R-tau inclusions.[40,300,335] The term CBS is used for the description of the classical clinical presentation, which includes progressive asymmetric rigidity and apraxia, dystonia, myoclonus, cortical sensory signs and alien limb phenomenon.[40,299,335] CBS may also be caused by a number of other conditions, including PSP,[299,335] Alzheimer's disease,[41,206] Pick's disease,[41] Creutzfeldt–Jakob disease,[41] neuronal intermediate filament inclusion disease (NIFID, FTLD with fused in sarcoma inclusions [FTLD-FUS])[256,318] and hereditary diffuse leukoencephalopathy with spheroids.[460]

CBD is a rare sporadic disorder of unknown aetiology sharing genetic risk factors with PSP.[223] There is no sex predilection. The average age of onset is ~60 years with disease duration of ~6–10 years.[470,541] Patients with 'classical' CBS develop asymmetrical clumsiness, stiffness or myoclonic jerks of a limb, more commonly an arm. Dystonic rigidity, akinesia and rhythmic myoclonus of the affected limb develop and some patients have the 'alien limb' phenomenon. Apraxia of leg movement together with pyramidal deficits causes difficulty with walking. Cortical sensory disturbance is common. Clinical features suggestive of PSP such as vertical supranuclear gaze palsy are increasingly recognized and can occur concurrently with CBS.[299,335] Although intellect initially was thought to be relatively well preserved, CBD is now considered a disorder of both movement and cognition.[39,40,300,335,473]

MRI studies of pathologically confirmed CBD cases indicate a characteristic pattern with posterior frontal and parietal cortical atrophy, significant involvement of the basal ganglia and atrophy of the middle corpus callosum.[261]

The characteristic pathological changes of CBD can be associated with a number of clinical presentations other than CBS,[299,335] including Richardson's syndrome,[299,335] behavioural variant of FTD,[278,335,473] primary progressive aphasia,[255,258,278,473] posterior cortical atrophy syndrome[250] and dementia with features similar to Alzheimer's disease.[300]

Neuropathology

Macroscopic Findings

Macroscopically, there is cortical atrophy of the posterior frontal and parietal regions with involvement of the precentral and postcentral gyri. The atrophy may be asymmetrical in cases clinically presenting with CBS with the more severe changes being seen contralateral to the most severely affected limbs. The cortical atrophy may show a parasagittal distribution and the temporal and occipital lobes are macroscopically unremarkable.[104,159,376] Cases with frontal dementia or primary progressive aphasia show a more generalized cortical atrophy.[100] There is variable dilatation of the ventricular system, the cerebral white matter may be reduced and the corpus callosum thinned. Atrophy of the subthalamic nucleus, prominent in PSP, is minimal. There is pallor of the SN and locus coeruleus. The spinal cord is macroscopically normal.

Microscopic Findings

Affected cortical areas are thinned because of neuronal loss with astrocytosis, often most severe in the superficial cortical laminae associated with superficial spongiosis. Severely affected cortex shows loss of laminar architecture, transcortical microvacuolation and marked astrocytosis. Ballooned ('achromatic') cortical neurons are a feature and are most frequent in cortical layers III, V and VI (Figure 12.11).[100] The SN shows cell loss and astrocytosis and, as in PSP, both pigmented and non-pigmented neurons are affected.[412] In residual nigral and locus coeruleus neurons, inclusions are visible in H&E-stained sections, often as variably basophilic globose tangles (Figure 12.12).[100] Neuronal loss, astrocytosis and occasional ballooned neurons are variable in the globus pallidus, striatum, subthalamic nucleus, thalamus, red nucleus and other brain stem nuclei. In contrast to PSP, the involvement of the subthalamic nucleus is mild.[100] The cerebellar dentate nucleus may be abnormal, including the presence of grumose degeneration, but this is more typical of PSP.[523] Degeneration of the corticospinal tracts occurs because of degeneration of the motor cortex[549] and there

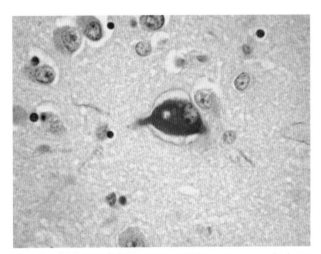

12.11 Corticobasal degeneration (CBD): swollen cortical neurons, which are not specific for CBD, are immunoreactive for αβ-crystallin.

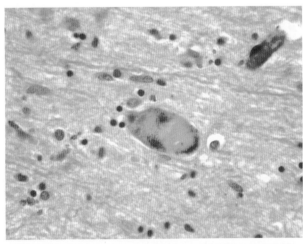

12.12 Corticobasal degeneration: basophilic neurofibrillary tangle with features of a globose tangle in the substantia nigra.

is myelin pallor in the cerebral white matter below areas of affected cortex.

Based on findings of tau immunohistochemistry, criteria for the neuropathological diagnosis of CBD have been established and validated, which, among others, emphasize the importance of tau-positive neuropil threads in grey and white matter of the cortex, basal ganglia, diencephalon and rostral brain stem.[100] Tau immunocytochemistry shows widespread neuronal and glial filamentous inclusions,[127] which specifically stain with antibodies recognising 4R-tau isoforms, but not with those to 3R-tau.[20] Small neurons in upper cortical layers are affected by tau-positive, granular or diffuse cytoplasmic inclusions designated as 'pretangles'. More compact inclusions resemble Pick bodies or small NFTs although others have a skein-like appearance (Figure 12.13).[100] Tau inclusions are generally not labelled with anti-ubiquitin,[565] variably stained with anti-p62 antibodies,[310] but can be well demonstrated with the Gallyas method.[19,532] Cortical neuronal inclusions are most prominent in areas of superficial microvacuolation in moderately affected cortex. Tau-immunoreactive neuronal inclusions are seen in subcortical grey nuclei, SN, other brain stem nuclei and cerebellar dentate nucleus (Figure 12.14).[100,104,565] Similar to PSP, accumulation of non-filamentous tau in Purkinje cells and the radial processes of Bergmann glia rarely occurs.[445]

The 'astrocytic plaque' is typical of the grey matter in CBD comprising a distinct annular array of tau and Gallyas-positive short stubby processes, which cluster around unstained neuropil (Figure 12.15a). The tau filaments are located in distal processes of astrocytes.[126,294] Astrocytic plaques are frequent in premotor, prefrontal and orbital cortices and can be found throughout the striatum (more frequently in the caudate than the putamen), but are uncommon in other subcortical, brain stem or cerebellar nuclei.[208] Astrocytic plaques are of significant value in the differential diagnosis.[100,300,335] Tau and Gallyas-positive oligodendroglial coiled bodies are widespread in cortices and white matter (Figure 12.15b).[126,565] Numerous

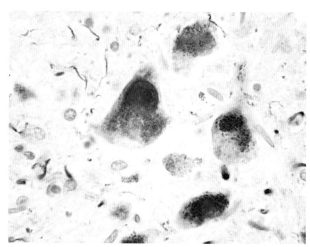

12.14 Corticobasal degeneration: tau-positive inclusions in pigmented neurons of the substantia nigra. AT8 immunohistochemistry.

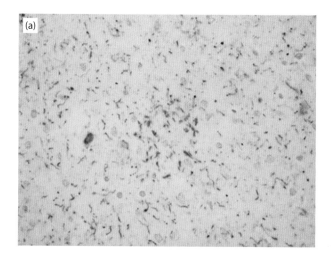

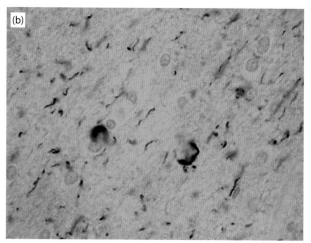

12.15 Glial pathology in corticobasal degeneration (CBD): (a) tau-immunopositive threads and astrocytic plaques (centre of field) are characteristic of CBD. **(b)** Tau-immunoreactive threads and oligodendroglial coiled bodies are also numerous in subcortical white matter. AT8 immunohistochemistry.

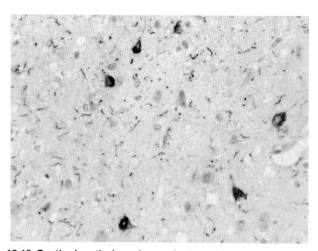

12.13 Cortical pathology in corticobasal degeneration: tau immunohistochemistry demonstrates numerous intraneuronal inclusions in small cortical neurons with frequent neuritic lesions. AT8 immunohistochemistry.

tau-positive and argyrophilic threads in grey and white matter are of significant differential diagnostic value in CBD (Figure 12.15b).[100,300,335]

Ballooned cortical neurons, strongly immunoreactive with antibodies to phosphorylated neurofilament proteins[98] and αB-crystallin,[341] (Table 12.11) are non-specific and of less diagnostic significance.[100,300]

The clinical heterogeneity of CBD is thought to be dependent on differences in the anatomical distribution of brain atrophy determined by neuronal loss and tau pathology.[299] In CBD cases, clinically presenting as CBS, tau pathology is severe in the primary motor and somatosensory areas and putamen, whereas in those with a PSP-like clinical presentation, limbic structures such as hippocampus, dentate gyrus and anterior nucleus of the thalamus, as well as the tegmentum of the medulla and cerebellar white matter, have more tau pathology than cases with CBS.[299] Depending on the nature of the speech disorder, marked atrophy of appropriate frontal and temporal areas[284] has been documented in cases with primary progressive aphasia.

NFTs in the SN and locus coeruleus and inclusions in oligodendroglia were described to contain bundles of 15 nm-wide straight tubules with NFTs also containing rare 'twisted tubules' ~20 nm in diameter with periodicity.[565] The twisted tubules, or twisted ribbons, in CBD are different from paired helical filaments in Alzheimer's disease as they are wider and have a longer periodicity.[306,307]

Argyrophilic grain disease pathology has been documented in ~40 per cent, and TDP43, predominantly glial pathology with staining of astrocytic plaque-like structures, may be seen in ~15 per cent of CBD cases.[542a,550a]

Genetics

Patients with PSP and CBD share a common risk factor: the extended H1 tau haplotype and, as in PSP, the H1 derived haplotype C is overrepresented in CBD. Both of these genetic similarities reinforce commonality in the molecular pathology of the two conditions.[223] Cases described in the literature as familial CBD are now regarded as FTDP-17 associated with *tau* gene mutation.[228] Because of small numbers, CBD cases were not included in the PSP GWA study.[218]

Differential Diagnosis

Table 12.11 summarizes an approach to the differential diagnosis of tauopathies with extrapyramidal movement disorders.

Multiple System Atrophy

MSA is a sporadic disorder formerly defined as three separate conditions: olivopontocerebellar atrophy (OPCA), with prominent cerebellar ataxia; Shy–Drager syndrome (SDS), characterized by autonomic dysfunction; and striatonigral degeneration (SND), with predominant parkinsonism. The pathology of MSA includes characteristic argyrophilic inclusion bodies in oligodendroglia, known as glial cytoplasmic inclusions (GCIs) or Papp–Lantos bodies,[432] which define MSA as a distinct pathological entity with a wide spectrum of clinical manifestations.[431] GCIs contain α-synuclein and, therefore, MSA is regarded as one of the synucleinopathies.[142,519,567] Clinical diagnostic consensus criteria for

MSA[161] advise that the term Shy–Drager syndrome is no longer used and recommend clinical classification into probable and possible MSA grouped according to the predominant motor disorder:

- MSA-P: patients with predominant parkinsonism;
- MSA-C: patients with predominant cerebellar features.

Previously, MSA encompassed several hereditary and sporadic system degenerations, but current use of the term is limited to disease associated with GCIs, whereas 'multisystem atrophy' is the preferred term for conditions in which there is degeneration of several systems in the absence of specific glial inclusions.[458]

MSA is a sporadic adult onset disease with a mean age at onset of ~58 years and equal distribution between males and females.[293] Disease duration is usually in the range 7–9 years, but rare cases may survive for up to 17 years.[442,490] The prevalence is estimated at 1.9 to 4.4/100 000 with an incidence of 3/100 000/year.[45,72,489] The clinical signs and motor symptoms correlate with the distribution of neuropathological changes.[193,578] In addition to a movement disorder, patients may experience non-motor symptoms and signs, which can precede the motor disorder and include rapid eye movement (REM) sleep behaviour disorder and autonomic failure with orthostatic hypotension, respiratory disturbance and urogenital dysfunction. Early autonomic failure has been associated with shorter survival in MSA.[400] It is uncommon for MSA to remain as a 'pure' syndrome and, usually, the clinical syndromes overlap as the disease progresses.[161,244,416] Death may be due to bulbar dysfunction and dysphagia, predisposing to aspiration pneumonia, impaired upper airway function associated with laryngeal stridor or may occur suddenly as a result of involvement of brain stem cardiovascular and respiratory centres.[493,527]

Neuropathology

Macroscopic Findings

The changes are prominent in the striatonigral (StrN) and olivopontocerebellar (OPC) regions often reflecting the clinical disease pattern. The putamen shows variable atrophy and dark discolouration, most severe in its dorsolateral aspect (Figure 12.16). The posterior putamen tends to be more affected, which is particularly apparent in early disease stages. The middle cerebellar peduncle, basis pontis and inferior olivary nucleus may be reduced in size and there is pallor of the SN and locus coeruleus (Figure 12.17a).[6] There is cerebellar cortical atrophy and the hemispheric white matter may also be atrophic and darkly discoloured although the superior cerebellar peduncle is well preserved (Figure 12.17b). Rarely, in cases designated minimal change MSA, macroscopic changes may be limited to pallor of the SN and locus coeruleus.[577] Cortical atrophy is sometimes apparent and may involve motor and premotor regions but ventricular dilatation is unusual.[566,593] The spinal cord usually appears normal, even in patients with severe autonomic dysfunction.

Histological Findings

Microscopic changes include neuronal loss, gliosis, myelin pallor and axonal degeneration. These are accompanied by α-synuclein-positive GCIs in oligodendrocytes, neuronal cytoplasmic inclusions (NCIs), neuronal nuclear inclusions, glial nuclear inclusions and neuropil threads (see later). Consensus criteria for the neuropathological diagnosis of

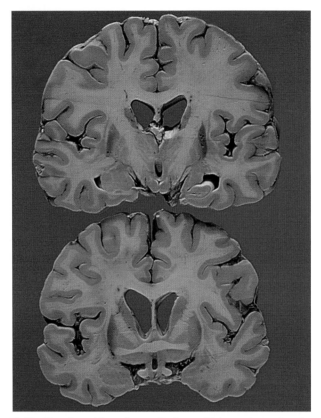

12.16 Atrophy of basal ganglia in multiple system atrophy. There is atrophy of the putamen, which also has a grey–green colour.

MSA require the presence of widespread GCIs together with neurodegeneration involving OPC and SN structures.[543]

Neuronal loss is most severe in the dorsolateral putamen, SNpc, locus coeruleus, Purkinje cell layer, basis pontis, inferior olivary nucleus, dorsal motor nucleus of vagus and intermediolateral spinal column.[431] Neuronal loss in olivopontocerebellar regions increases with disease duration, but this trend is not established as clearly as it is in the striatonigral system.[416] Although cortical neuronal loss is not readily apparent, there may be astrocytosis and this is particularly marked in the deeper laminae of the motor cortex.[138] The atrophic putamen shows an increase in iron deposition.[270]

Pallor or loss of myelin staining in white matter is most frequent in the external capsule, striatopallidal fibres, transverse pontine tracts, middle cerebellar peduncle and cerebellar white matter. Widespread myelin damage in MSA can be identified in otherwise apparently normal areas.[356] Increasing myelin loss is associated with decreased reactive microgliosis and, in contrast with grey matter, a decrease in number of GCIs.[238]

None of the cellular inclusions of MSA can be easily appreciated in H&E stained sections, although GCIs may be visualized as slightly basophilic cytoplasmic inclusions in oligodendrocytes. All inclusion types are readily demonstrated using silver stains or immunohistochemical staining for ubiquitin, p62 and, most specifically, α-synuclein (Figure 12.18). GCIs appear as flame- and sickle-shaped cytoplasmic structures superficially resembling small NFTs (Figures 12.18a and 12.18c). Ultrastructurally, GCIs are composed of randomly arranged 20–40 nm tubules, or filaments, associated with granular material. Two classes of filament are described in sarcosyl-extracted material: 5–18 nm-diameter twisted filaments with a periodicity of 70 nm and 10 nm straight filaments.[430,432,519] GCIs are widely distributed in the brain and spinal cord. In grey matter, the

(a)

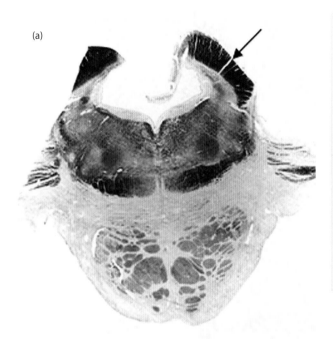

(b)

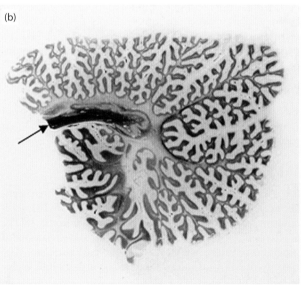

12.17 Multiple system atrophy. (a) There is loss of the transverse myelinated fibres forming the middle cerebellar peduncle in the pontine base. The superior cerebellar peduncle (arrow) is well preserved, as is the height of the pontine tegmentum. **(b)** There is loss of fibres in the cerebellar white matter, although the superior cerebellar peduncle (arrow) is preserved. Luxol fast blue (LFB)/cresyl violet.

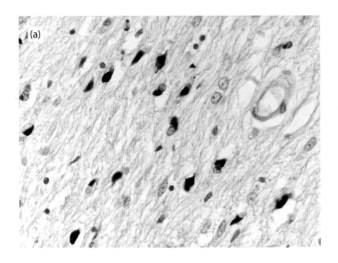

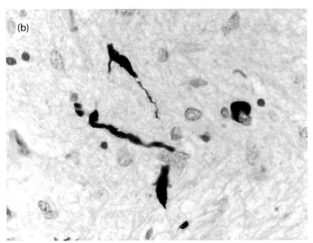

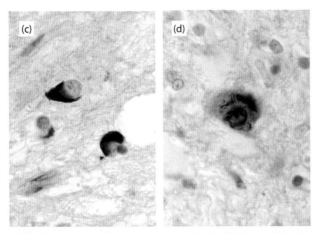

12.18 Numerous glial cytoplasmic inclusions are present in this case of multiple system atrophy (a,c). Neuronal processes containing α-synuclein are also found **(b)**. Neurons may contain both cytoplasmic and intranuclear inclusions, the latter having the appearance of a web of fine fibrils **(d)**. Pontine base, α-synuclein immunohistochemistry.

highest density is found in the primary motor and premotor cortices, putamen, globus pallidus, subthalamic nucleus, pretectal area, pontine nuclei, vestibular nuclei, motor nuclei of the Vth, VIIth and XIIth cranial nerves, pontomedullary reticular centres and the intermediolateral

column of the spinal cord. In white matter, they are most numerous in the internal and external capsules, corpus callosum, corticospinal tracts, middle cerebellar peduncle, cerebellar hemisphere and beneath the motor cortex.[315,431,432]

NCIs are argyrophilic, round, filamentous structures in neuronal cell bodies (Figure 12.18d) Ultrastructural examination of NCIs shows a meshwork of granule-associated filaments similar to GCIs.[22,269] NCIs are most numerous in the basis pontis and putamen, although they are also present in many other regions including subthalamic nucleus, arcuate nucleus, subiculum, amygdala, hippocampus, dentate fascia, SN, inferior olivary nucleus and brain stem reticular formation.[269,315,394,430] Neuronal and glial intranuclear inclusions are most frequent in the basis pontis and putamen. In neurons, they appear as a web of fine fibrils, often concentrated beneath the nuclear membrane (Figure 12.18d). Glial intranuclear inclusions are rod-shaped and sparsely distributed. Threads represent neuronal processes containing aggregated α-synuclein and can often be easily visualized in the basis pontis (Figure 12.18c).[430,519]

Neuropathological changes of MSA have occasionally been reported to occur in conjunction with histological features of other neurodegenerative diseases such as PSP.[505] Brain stem Lewy bodies affect 10–22 per cent of cases of MSA.[249,416] Alzheimer pathology is uncommon in MSA and appears to be less prominent than in age-matched controls.[249] TDP43 pathology is unusual in MSA cases.[157]

Neuropathological Subtypes of MSA

The pattern and severity of neuronal loss in OPC and StrN regions have been used as the basis for defining neuropathological subtypes of MSA. A study of 100 cases from the Queen Square Brain Bank, UK showed that, in 34 per cent of cases, StrN neuronal loss predominates (SND subtype); OPC predominant neuronal loss occurred in 17 per cent of cases (OPCA subtype); and that pathology was mixed in the remaining 49 per cent of cases (SND = OPCA subtype).[417] Consistent with the clinical literature, Japanese patients have been shown to have an increased frequency of the OPCA subtype.[417] A further approach to disease subtyping combined neuronal loss, GCIs, atrophy and gliosis in the grading of regional pathology.[248] In the rare minimal change MSA subtype, neuronal loss is largely restricted to the SN and locus coeruleus, although GCIs are more widespread.[577]

Clinicopathological Correlations

The distribution of neuropathological features with marked involvement of StrN and OPC structures together with the autonomic system has long been associated with the clinical phenotype of variable combinations of parkinsonism that is often poorly responsive to L-Dopa, cerebellar dysfunction and autonomic failure. Clinicopathological studies show a correlation between cases with StrN predominant pathology and parkinsonian features in life and between OPC predominant pathological changes and clinical cerebellar dysfunction.[193,248,416,578] L-Dopa response appears to relate to the degree of putaminal neuronal loss.[416]

Autonomic disturbance in MSA may precede the motor signs and is related to pathology in the autonomic system, including elements in the medulla, midbrain, pons, hypothalamus and spinal cord.[244,413] Respiratory and cardiac function may be compromised by involvement of autonomic nuclei in the medulla.[493,527] Degeneration of sympathetic neurons in the intermediolateral column of the spinal cord contributes to orthostatic hypotension.[578] Bladder, rectal and sexual dysfunction has been associated with cell loss from parasympathetic preganglionic nuclei, Onuf's nucleus and the inferior intermediolateral nucleus in the sacral cord.[296] The peripheral autonomic nervous system shows atrophy of the glossopharyngeal and vagus nerves.[32] Neuronal inclusions containing α-synuclein occur in the sympathetic ganglia, but not in Schwann cells, and pathology is not described in the visceral enteric plexus or in the innervation of glands and blood vessels.[393,513]

Pyramidal signs occur in around 54 per cent of MSA patients and are considered to be due to involvement of the motor cortex and pyramidal tracts.[156,161,431,548] Cognitive impairment has not, until recently, been regarded as compatible with a clinical diagnosis of MSA.[161,576] However, it is now recognized that impaired cognition occurs in up to 30 per cent of MSA patients and appears to be more commonly associated with MSA-P than MSA-C.[50,274,400] The neuropathological correlate of these findings is still uncertain.[24]

Biochemistry and Genetics

GCIs contain full-length α-synuclein showing post-translational modifications including phosphorylation at serine 129 and nitration.[111,112,139,158] α-Synuclein in MSA shows increased insolubility.[60,99] Numerous other proteins have also been described in GCIs with uncertain significance.[579]

No mutations have been identified in the α-synuclein gene in MSA patients.[414] In families with *SNCA* gene multiplications and the *G51D* mutation, both Lewy bodies and lesions with features resembling GCIs have been described.[185,280,402] An association with the H1 haplotype of MAPT has been described in MSA.[555] Although, according to several studies, the *SNCA* gene was associated with the disease,[9,475,487] a more recent worldwide GWA study did not confirm an association at the *SNCA* locus (Houlden and colleagues, unpublished data).

Disease Pathogenesis

The pathogenesis of MSA remains uncertain, but evidence from neuropathological studies, genetics and experimental models allows some hypotheses to be put forward. GCIs are regarded as important in the mechanism as they have been shown to increase in number with disease duration and there is a positive correlation between the GCI load and neuronal loss in SND and OPC regions.[416] The source of oligodendroglial α-synuclein in the human disease is uncertain,[415] although recent data raise the possibility of SNCA mRNA expression in oligodendrocytes.[25] Oligodendroglial dysfunction may precede α-synuclein accumulation also indicated by relocation of P25α from the myelin sheath to the oligodendroglial cell body.[514] The subsequent formation of a GCI may leave the oligodendrocyte vulnerable to further degeneration, possibly exacerbated by microglial activation, with loss of support for axons and consequent axonal and neuronal degeneration.[6] The emerging central role of the oligodendrocyte in the pathogenesis of MSA has led to the concept that this disease might be considered as an oligodendrogliopathy.[579]

Neuronal Intranuclear Inclusion Disease

Neuronal intranuclear inclusion disease (NIID), also known as neuronal intranuclear hyaline inclusion disease, is a rare, slowly progressive neurodegenerative condition characterized by eosinophilic neuronal intranuclear inclusions in central, peripheral and autonomic nervous systems.[194] NIID is usually diagnosed post mortem, although biopsy of peripheral tissues may provide diagnosis *in vivo*.[172,512]

The majority of reported cases are sporadic, but familial cases occur.[597] The clinical presentation is diverse. Patients with an infantile or juvenile onset usually show evidence of multisystem degeneration, including pyramidal signs, choreoathetosis, tremor, L-dopa-responsive dystonia (DRD), rigidity, oculogyric crisis, lower motor neuron abnormalities, cerebellar signs, personality change and learning difficulties.[254] Cardiomyopathy may be present.[359] In adult-onset disease, dementia may be prominent and combined with parkinsonism, cerebellar signs and autonomic dysfunction.[533,597]

In NIID, the brain may show varying degrees of generalized atrophy with ventricular enlargement and reduced thickness of the cerebral cortex. The brain stem may also show atrophy and the SN is often pale.[194]

Histologically, the most striking feature is the presence of neuronal intranuclear inclusions (Figure 12.19). These are sharply demarcated single or multiple rounded eosinophilic structures measuring 2–6 μm, although they may be larger in neurons of the basal ganglia and brain stem.[140,172,529] Variable numbers of small intranuclear inclusions may be present in astrocytes and are characteristic of late-onset disease.[533] In addition to intranuclear inclusions, there is neuronal loss, with modest astrocytic or microglial reaction, most noticeable in the SN, red nucleus, inferior olives, Purkinje cell layer, spinal anterior horns and Clarke's dorsal nuclei.[194,359] There is correlation between regional neuronal loss and clinical phenotype.[359,533] Neuronal intranuclear inclusions are found in peripheral and visceral neurons, including those in the dorsal root ganglia and myenteric plexus, and in parenchymal cells of the adrenal medulla.[359] Non-neuronal cells such as cardiomyocytes, adipocytes, fibroblasts and sweat gland cells may contain intranuclear inclusions.[359,512]

The neuronal intranuclear inclusions ultrastructurally are non-membrane-bound structures composed of 8–15 nm, tightly packed filaments and granular material,[530] and stained with antibodies to ubiquitin and other components of the ubiquitin proteosome system.[194,359,441] Occasional inclusions are immunostained with the 1C2 antibody for expanded polyglutamine tracts, which is, however, different from the strong ubiquitous staining observed in CAG repeat disorders.[529,530] Non-expanded ataxin-1, ataxin-2 and ataxin-3 are recruited into neuronal intranuclear inclusions,[530] which are also immunoreactive for SUMO-1[359,452] and contain NSF, dynamin-1 and Unc-18-1, involved in membrane trafficking of proteins.[453] Neuronal intranuclear inclusions are not labelled with Congo red, but show autofluorescence.[529]

Genetic studies and the absence of extended polyglutamine tracts in NIID inclusions help in differentiating it from

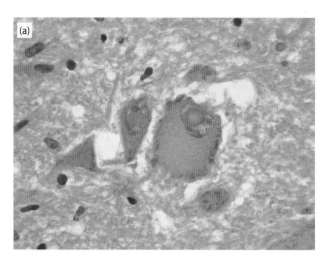 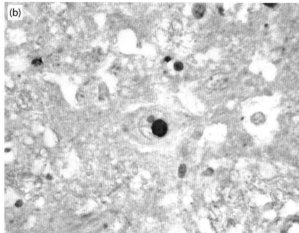

12.19 Neuronal intranuclear inclusion disease. (a) Intranuclear inclusion bodies in two neurons of the midbrain tegmentum. **(b)** An intranuclear inclusion body stained strongly with an anti-ubiquitin antibody.

trinucleotide repeat disorders. Intranuclear inclusions are also present in fragile X-associated tremor/ataxia syndrome caused by 'premutation' expansion (55–200 CGG repeats) in the *FMR1* gene.[149] The frequent ubiquitin-positive neuronal intranuclear inclusions in FTLD-TDP (type D) due to mutations in the *VCP* gene have different morphological characteristics and contain TDP43.

Conditions Characterized by Abnormal Stiffness

Several rare conditions are characterized by abnormal muscle stiffness and spasms. In the stiff-man or stiff-person syndrome (SPS) there is continuous muscle stiffness (mainly of the neck/axial muscles) with relative sparing of distal muscles. SPS mostly affects women in their 30s and 40s, but childhood onset is documented. In SPS, there is a relative lack of GABAergic inhibition in the spinal cord and brain stem.[367]

In the common cryptogenic form of SPS autoantibodies, directed against glutamic acid decarboxylase 65-kDA isoform (GAD65), occur in the majority of patients.[364,429] In paraneoplastic SPS associated with breast cancer and small cell lung cancer, there are autoantibodies to the neuronal protein amphiphysin.[364] Autoantibodies to gephyrin, another neuronal protein, have also been reported.[58]

In cases described as SPS, spinal cord histology shows loss and degeneration of nerve cells, with associated astrocytosis in motor nuclei of the anterior horns.[235,384] Other cases have shown lymphocytic perivascular infiltration in the spinal cord, brain stem or basal ganglia.[367] Muscle biopsy is non-diagnostic.

Progressive encephalomyelitis with rigidity and myoclonus (PERM) is part of the SPS spectrum. Patients with PERM show severe muscle stiffness, axial rigidity and excessive startle in response to various stimuli. PERM patients may have respiratory arrest. Some PERM patients have antibodies to GAD, but others have anti-glycine receptor alpha-1 antibodies.[559]

In PERM, the histological changes include neuronal loss, lymphoid and microglial infiltration in the brain stem and the spinal cord.[33]

CHOREA

Chorea is characterized by non-rhythmic rapid involuntary movements. It may be divided into two main groups, hereditary and sporadic, with a wide range of causes (Table 12.12).[421]

Huntington's Disease

Clinical Features

Huntington's disease (HD) is an autosomal dominant disorder that manifests classically as chorea, psychiatric symptoms progressing to dementia and cachexia. The mean age of onset is 40 years, but some patients have onset in infancy or old age. HD is associated with abnormal expansion in a CAG triplet repeat sequence. The responsible gene, which codes for a protein called huntingtin, is located on chromosome 4p16.3.

The prevalence of HD ranges from 5 to 10 per 100000 population. Certain populations, such as those of African and of Japanese ancestry, have a significantly lower rate of disease. The mean disease duration is 17 years.

TABLE 12.12 Causes of chorea	
Hereditary	**Sporadic**
Huntington's disease	Sydenham's chorea
Neuroacanthocytosis	Chorea associated with
Huntington's disease-like	pregnancy
syndromes	Focal lesions (vascular)
Benign hereditary chorea	Autoimmune diseases
Dentatorubropallidoluysial atrophy	Metabolic abnormalities
Paroxysmal choreoathetosis	Drug-related
Wilson's disease	
Neurodegeneration with brain iron	
accumulation	
(Pantothenate kinase-associated	
neurodegeneration)	
Lesch–Nyhan syndrome	
Ataxia telangiectasia	
Hereditary ferritinopathy	

Although HD is considered a hyperkinetic movement disorder, it can manifest with hypokinesia and rigidity. Patients with onset before age 20 years are more likely to have hypokinesia and rigidity at an early stage. The first clinical manifestation of the hyperkinetic form is chorea. Neuropsychological problems can antedate the onset of the movement disorder, especially in patients with late-onset disease. Generally, the disease progresses more quickly with an early onset than with a late onset. In the late stages of disease patients have cognitive decline, are increasingly rigid, dystonic and frequently have dysphagia.

Genetics, Clinical Correlations and Cell Biology

HD was initially mapped to chromosome 4p and subsequently the *huntingtin* gene, coding for the 348-kDa huntingtin protein, was identified. The first exon of the gene contains the CAG tandem repeat relevant to disease pathogenesis. Normal alleles contain approximately 20 repeats. Repeat lengths between 27 and 35 may be unstable without resulting in disease. There is variable penetrance of the disease between 36 and 39 repeats, most patients with HD have repeat expansions above 40, with an average in the mid-40s and a range of 36–180.[301] Repeat length in the affected allele is unstable, so that variable length is passed to affected offspring. This instability is particularly notable with paternal transmission, where the expansion is presumed to be responsible for anticipation with successive generations having an earlier disease onset. This reflects a negative correlation between repeat length and age of onset, and a positive correlation with severity of disease.[80,539] Later-onset disease is usually associated with maternal transmission. New sporadic cases can arise from a *de novo* expansion of the unstable, near-threshold CAG repeat domain in an unaffected parent, usually the father. The siblings of such *de novo* HD patients are at risk of disease through similar repeat expansion.[383]

In normal brain, huntingtin is located primarily in the somatodendritic compartment of neurons without nuclear staining.[184] The normal function of huntingtin is unknown, but it is important during embryogenesis. In HD, there is nuclear huntingtin immunoreactivity and there are also huntingtin-positive intranuclear inclusions.[480] Both murine models and human disease show mutational instability within the brain; CAG expansion is greater in striatum, which has implications for the selective vulnerability of neuronal populations.[276]

The toxicity of mutant huntingtin is thought to be due to a toxic gain of function, but there is evidence that loss of normal function may also contribute to pathogenesis.[110] Post-translational modification of mutant huntingtin has been shown to alter its toxicity.[181]

Neuropathology

Macroscopic Findings

Macroscopic changes in HD brains vary with the duration and stage of the disease. Early-stage cases may have no discernable gross changes, whereas late-stage cases often have severe atrophy, with brain weights ranging from 800 g to 1000 g. There is striking atrophy in the neostriatum with enlargement of the frontal horns of the lateral ventricles (Figure 12.20). In later-stage disease gross atrophy may be apparent in the pallidum, thalamus and even brain stem and cerebellum. The cerebral cortex and underlying white matter show reduced volume.

Microscopic Findings

A scheme for grading the severity of striatal pathology has been proposed and correlates with clinical severity (Table 12.13).[169,562] Histologically, there is loss of neurons, with associated astrocytosis and microgliosis in affected areas of the striatum (Figure 12.21). The pattern of striatal degeneration is stereotypical with neuronal loss progressing in a caudal to rostral direction in both caudate and putamen, dorsomedially to ventrolaterally in the caudate nucleus and dorsally to ventrally in the putamen. The nucleus accumbens is relatively spared.

The selective pattern of striatal degeneration in HD illuminates the pathogenesis of this movement disorder. The striosomes are affected early in the course of the disease, with particular loss of the GABA-/encephalin-containing medium spiny neurons projecting to the lateral pallidum.[13]

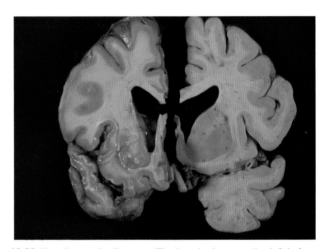

12.20 Huntington's disease. The hemisphere on the left is from a patient with Huntington's disease. Note the severe atrophy of the caudate and putamen as well as the discolouration of the grey matter. The hemisphere on the right is from a person of a similar age without neurological disease.

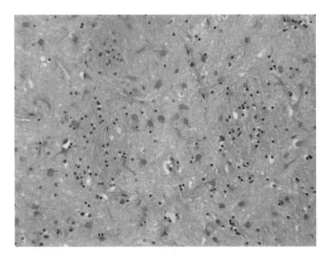

12.21 Huntington's disease. Photomicrograph of the putamen with severe neuronal loss and astrocytosis.

TABLE 12.13 Vonsattel grading of the striatum in Huntington's disease*

Grade		Description
0	Gross	No changes
	Microscopic	No qualitative changes, but 30–40 per cent neuronal loss by cell counting; ubiquitinated nuclear inclusions present
1	Gross	Minimal atrophy at level of caudate/putamen/accumbens and tail of caudate may be present
	Microscopic	Neuronal loss and astrocytosis in medial caudate and putamen; neostriatal neuronal loss. 50 per cent; nuclear inclusions present
2	Gross	Atrophy of putamen and head of caudate with preservation of convex lateral ventricular border; globus pallidus normal
	Microscopic	Neuronal loss and astrocytosis extends to ventrolateral caudate, with sparing of most of ventral putamen and nucleus accumbens; pencil fibres atrophic, with increased density of oligodendrocytes
3	Gross	Severe atrophy of putamen and head of caudate, with straight ventricular border; mild atrophy in pallidum
	Microscopic	Obvious neuronal loss and astrocytosis throughout caudate and putamen except for region near nucleus accumbens; astrocytosis and neuronal loss in pallidum; nucleus accumbens largely spared
4	Gross	Profound atrophy of putamen and head of caudate, concave ventricular border; pallidum obviously atrophic
	Microscopic	Severe neuronal loss and astrocytosis in all areas of caudate and putamen; milder involvement of nucleus accumbens

*Based on three samples, taken at the level of the caudate, nucleus accumbens and putamen; the globus pallidus; and the tail of the caudate.

In the hypokinetic–rigid form of HD, and also in later stages of the classic form, there is extensive loss of striatal neurons projecting to the internal pallidum and SN, contrasting with the preservation of these neurons in hyperkinetic HD. These findings suggest that chorea results from preferential loss of striatal neurons projecting to the lateral globus pallidus and that the hypokinetic–rigid form or late stage of HD results from the additional loss of striatal neurons projecting to the internal pallidum.[12] These models are insufficient to explain the non-motor features often seen in early HD.[252]

Morphometric studies show loss of large pyramidal neurons from the neocortex, entorhinal cortex and hippocampus without significant astrocytosis.[87] The cortical atrophy appears to progress from the motor and sensory cortices to the occipital, parietal and limbic cortices.[474] The globus pallidus is reduced in volume through neuronal loss

and loss of afferent fibres from the neostriatum. Reduced neuronal numbers are also described in the thalamus and subthalamus,[312] the lateral tuberal nucleus of the hypothalamus[302] and the SN.[410] The brain stem shows loss of large neurons in the nucleus pontis centralis caudalis and changes in the rostral interstitial nucleus of the medial longitudinal fasciculus.[292,326] Cerebellar atrophy with loss of Purkinje cells is described, particularly in patients with early-onset disease.

The presence of neuronal intranuclear inclusions composed of aggregates of abnormal huntingtin protein is diagnostically useful in HD. This can be demonstrated by immunohistochemistry using antibodies against huntingtin (especially against the amino-terminal region), ubiquitin or expanded polyglutamine tracts. Not all inclusions are positive for ubiquitin. These inclusions tend to be present in the cerebral cortex, the hippocampus and, to a lesser extent, the neostriatum, amygdala, dentate and red nuclei.[171,480] Abnormal neurites in the cortex, medial temporal lobe and striatum, with similar immunohistochemical properties, are more common than nuclear inclusions (Figure 12.22). Nuclear inclusions are more prevalent in patients with large repeat region expansion, whereas neuritic aggregates appear to be an age-related phenomenon.

Role of Autopsy in Huntington's Disease and Differential Diagnosis

The diagnosis of HD can be made readily in most patients by genetic testing for a CAG repeat expansion. However, there are several settings in which histopathological diagnosis remains important. One situation is when patients clinically appear to have HD but either have not been studied genetically or are found to have normal repeat lengths. There are several close mimics of HD, including Huntington's disease-like syndromes 1 and 2 (see later). Other diseases manifesting with chorea (Table 12.12) may be mistaken for hyperkinetic HD clinically. The differential diagnosis for the hypokinetic–rigid form of HD includes Wilson's disease, juvenile parkinsonism, neuroaxonal dystrophy and some neurometabolic storage disorders. Where possible, post-mortem studies should include the brain, spinal cord, peripheral nerve, skeletal muscle, liver, adrenal gland and bone marrow. With the exception of Huntington's disease-like 2 (which differs from HD primarily in the absence of huntingtin in the nuclear inclusions), the HD-mimicking conditions do not have the typical pathological findings of HD, particularly the characteristic topographical progression of striatal atrophy, and often have distinctive histopathological features that define them further. The other situation where post-mortem examination is particularly important occurs in patients who have not been diagnosed with HD before death but who demonstrate the findings of HD at autopsy. This scenario is more likely in patients in whom psychiatric symptoms or dementia overshadow any movement disorder.

Pathological mimics of HD may show significant atrophy of the striatum, but in most cases there are distinctive pathological features to separate them from HD. Some cases of frontotemporal lobar degenerations may have significant atrophy of the caudate, but cortical atrophy is more pronounced and there are characteristic inclusions. In MSA,

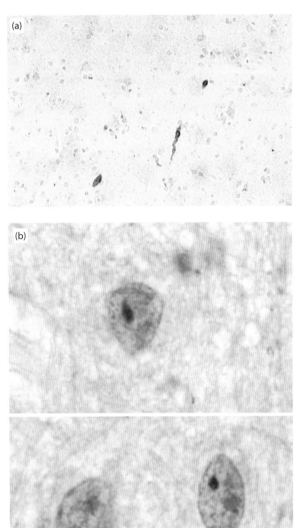

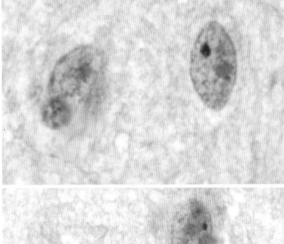

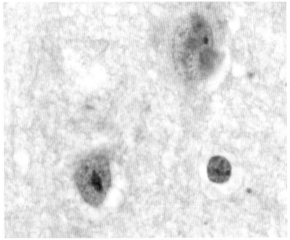

12.22　Huntington's disease: cortical pathology. (a) Anti-ubiquitin staining reveals numerous abnormal cortical neurites. **(b)** Ubiquitinated neuronal intranuclear inclusions. Anti-ubiquitin immunohistochemistry.

atrophy is accentuated in the putamen and there are GCIs and greater involvement of the SN, in addition to changes in the OPC system.

OTHER HEREDITARY CHOREAS

Neuroacanthocytosis

This condition also is known as chorea with acanthocytosis, although dystonia and even parkinsonism may be present in some patients. The severity of the neurological problems and the degree of acanthocytosis are not correlated. Genetic studies show this disorder to be heterogeneous. The most prevalent form is autosomal recessive chorea-acanthocytosis, with mutations in the *CHAC* gene (*VPS13A*; chromosome 9q21), which codes for chorein.[462] Other genetic conditions associated with the syndrome type neuroacanthocytosis include the X-linked McLeod syndrome, which is linked with the Kell blood group and mutations in the *XK* gene.[214] Huntington's disease-like syndrome type 2 may present with acanthocytosis, as can pantothenate kinase-associated neurodegeneration (PKAN; juvenile onset neuroaxonal dystrophy) and abetalipoproteinaemia. All of these subtypes have variable manifestations of dyskinesia, cognitive decline and progressive neurodegeneration primarily in the basal ganglia, but they are differentiated by genetic testing.

Neuropathological data are sparse for neuroacanthocytosis and not well correlated with genetics. Typically, gross atrophy of the neostriatum, with significant loss of small- and medium-sized neurons and accompanying astrocytosis, is present. Pallidal involvement can be severe. Milder neuronal loss and astrocytosis are reported in the thalamus, SN and spinal anterior horns. The absence of changes in the cerebral cortex, subthalamic nucleus and cerebellum may help in the differential diagnosis from HD.[469]

Huntington's Disease-like Syndromes

Approximately 1 per cent of patients with a clinical or pathological picture of HD have no mutation of the *huntingtin* gene. Three separate mutations or linkages have been identified within this population. Autosomal dominant Huntington disease-like type 1 (HDL1) is due to an expanded octapeptide repeat in the prion gene with valine at codon 129. Pathological examination shows atrophy of the basal ganglia, variable cortical atrophy and prion-specific changes, including typical prion plaques.[316]

Autosomal dominant Huntington disease-like type 2 (HDL2) is seen primarily in patients of African ancestry and is due to a CAG/CTG repeat expansion in an alternatively spliced exon in the *junctophilin-3* gene on chromosome 16q24.3. The disease resembles classical HD, with marked striatal atrophy and moderate cortical atrophy. Striatal neurodegeneration shows a dorsal-to-ventral gradient, and there are neuronal intranuclear inclusions that stain for both ubiquitin and expanded polyglutamine tracts,[352] but not for huntingtin.[476] It has been suggested that HDL2 is due to a repeat expansion in the CTG orientation and forms mRNA- CUG-repeat expansions, which in turn form separate intranuclear aggregates,[476] but also may result in a loss of function of junctophilin-3.[497]

Autosomal recessive Huntington disease-like type 3 (HDL3) may be linked close to the HD locus on chromosome 4p. Onset of symptoms is in the first decade of life. Pathological studies have not been reported.

Benign Hereditary Chorea

Benign hereditary chorea (BHC) is clinically and genetically heterogeneous. There are a number of kindreds with autosomal dominant inheritance related to mutations in the *TTF-1* (*thyroid transcription factor 1*) gene on chromosome 14q.[47] Most individuals have onset of chorea in childhood, with little progression and variable persistence into adult life. There is minimal or no cognitive deficit. The neuropathology of benign hereditary chorea is not well documented. A patient with a *TTF-1* mutation showed no significant gross or microscopic degeneration. Immunohistochemical studies of the striatum suggested a loss of those striatal interneurons whose tangential migration is mediated by TTF-1.[288] A second form, BHC2, recently has been described in two Japanese families with adult-onset minimally progressive chorea and genetic linkage to chromosome 8q21.3–q23.3. An autopsy of one patient found mild-to-moderate neuronal loss and gliosis in the striatum with gliosis and decreased volume of the cerebral hemispheric white matter. There were glial and neuronal 4R-tau-positive inclusions similar to those seen in PSP.[594]

Sydenham's Chorea

Sydenham's chorea typically arises following a group A Streptococcal infection, sometimes as a manifestation of rheumatic fever. It is characterized by bilateral or unilateral involuntary movements, dysarthria, affective changes, decreased tone and, less commonly, headache, seizures, weakness and sensory abnormalities. It is usually an acute monophasic condition, but recurrent episodes can occur. The neuropathology is not well described. Imaging studies suggest signal abnormalities in the basal ganglia, which sometimes persist.[119] Cross-reacting autoantibodies raised against streptococcal antigens are implicated.

MYOCLONUS

Myoclonus, a non-specific sign, defines a rapid, shock-like, involuntary contraction of a single muscle or muscle group. Myoclonus and dementia raise the possibility of Creutzfeldt–Jakob disease, but myoclonus may be present in a range of degenerative disorders, including Alzheimer's disease and CBD, infections such as acute encephalitides and subacute sclerosing panencephalitis and may be due to acquired metabolic causes, such as hepatic failure, renal failure and carbon dioxide retention.[63] Myoclonus and epilepsy characterize some epilepsy syndromes and also occur in mitochondrial disease, Batten's disease, sialidosis, Lafora body disease, dentatorubropallidoluysian atrophy (DRPLA) and Baltic myoclonus.

Arrhythmic myoclonus (stimulus-sensitive or action myoclonus) is precipitated by sensory stimulation in a wide range of CNS conditions, including hypoxia, trauma and drugs. Focal myoclonus (rhythmic myoclonus) occurs in lesions of the brain stem or spinal cord. Palatal myoclonus (or tremor) occurs in lesions of the central tegmental tract or dentate nucleus and may be associated with hypertrophy of the inferior olive. Such lesions may be degenerative or due to a range of pathologies, including infarction, neoplasia and demyelination. Segmental myoclonus is associated with inflammatory, traumatic or neoplastic diseases of the spinal cord. Brain stem myoclonus has been described in adults with infective disorders and cerebral lymphoma.[595] In children with neuroblastoma, it occurs in association with ocular myoclonus (opsoclonus).

HEMIBALLISMUS AND BALLISM

Ballism is characterized by violent involuntary flinging movements of the limbs. When unilateral, it is termed 'hemiballismus'. Most cases are caused by damage to the subthalamic nucleus or its outflow tracts, most commonly through infarcts or small haemorrhages, but rarely, infection, metastasis, demyelination or head injury may be responsible.[173] It may also occur as a complication of stereotactic operation for PD[333] or other neurosurgical procedure.

DYSTONIA

Definition and Classification

In 1911, Oppenheim proposed the term 'dystonia musculorum deformans', to describe a disorder in children with twisted postures, muscle spasms, gait abnormalities and clonic, rhythmic jerking movements with progression of symptoms leading to sustained fixed postural deformity.[405] Dystonia is now classified by age at onset, distribution and aetiology (Table 12.14). Elucidation of the disease gene underlying many dystonias has facilitated accurate molecular classification (Table 12.15).[135] In childhood onset cases dystonia tends to become more generalized, whereas in adult-onset cases the disease remains focal or segmental.[444] The prevalence of dystonia ranges from 15 to 30 per 100 000 population.[180]

Primary Dystonias

Primary dystonias include dystonias that are predominantly generalized and those with a tendency to remain focal.[444] The genetic background of these conditions is summarized in Table 12.15. Inherited dystonia can be separated into three categories: i) primary (idiopathic) torsion dystonia (PTD) where the dystonia is the only clinical sign except for tremor and this group includes DYT1, 2, 4, 6, 7, 13, 17 and 21, ii) primary plus dystonias; DYT3, 5/14, 11, 12, 15 and 16 and iii) primary paroxysmal dystonias; DYT8, 9/18, 10/19 and 20.

Primary Torsion Dystonias

This group includes autosomal dominant dystonias DYT1 and DYT4 and the autosomal recessive dystonia, DYT2. DYT6, 7, 13, 17 and 21. (For reviews see[135,439])

DYT1 is the most frequent member of the PTD group. Patients develop focal dystonia in childhood, which progresses

TABLE 12.14 Classification of dystonias

By aetiology	
1.Primary dystonias	
1a. Primary (idiopathic) torsion dystonias (also known as primary pure dystonia)	Torsion dystonia is the only clinical sign (apart from tremor) and there is no identifiable exogenous cause or other inherited or degenerative disease. Examples include DYT1, 2, 4, 6, 7, 13, 17 and 21
1b. Primary plus dystonias	Torsion dystonia is a prominent sign but is associated with another movement disorder, for example myoclonus or parkinsonism. There is no evidence of neurodegeneration. Includes: DYT3, 5/14, 11, 12, 15 and 16
1c. Primary paroxysmal dystonias	Torsion dystonia occurs in brief episodes. Includes: DYT8, 9/18, 10/19 and 20. These often have varying triggering factors: Paroxysmal non-kinesigenic dystonia (PNKD; DYT8) induced by alcohol, coffee, tea, etc. Paroxysmal kinesigenic choreoathetosis or dyskinesia (PKC or PKD; DYT10/19) induced by sudden movement
2. Heredodegenerative dystonias	Dystonia occurs in the context of a heredodegenerative disorder (see Table 12.16).
3. Secondary dystonia	Dystonia is a symptom of an identified neurological condition, such as a focal brain lesion, exposure to drugs or chemicals. Includes dystonia secondary to a brain tumour and off-period dystonia in Parkinson's disease.
By age at onset	
1. Early-onset (variably defined as ≤20–30 years)	Usually starts in a leg or arm and frequently progresses to involve other limbs and the trunk.
2. Late onset	Usually starts in the neck (including the larynx), the cranial muscles or one arm. Tends to remain localized with restricted progression to adjacent muscles.
By distribution	
1. Focal	Single body region, includes writer's cramp and blepharospasm.
2. Segmental	Contiguous body regions such as cranial and cervical or cervical and upper limb.
3. Multifocal	Non-contiguous body regions such as upper and lower limb or cranial and upper limb.
4. Generalized	Both legs and at least one other body region (usually one or both arms).
5. Hemidystonia	Half of the body, this is usually secondary to a structural lesion in the contralateral basal ganglia.

Adapted from Albanese *et al.*[11a] With permission from John Wiley and Sons. © 2010 The Author(s). European Journal of Neurology © 2010 EFNS.

to become generalized. The majority of DYT1 cases show an in-frame GAG trinucleotide deletion in the *TOR1A* gene encoding torsinA,[418] which is a member of the multifunctional AAA+ super-family of ATPases. TorsinA accumulates in neuronal cytoplasmic inclusions immunoreactive for ubiquitin and the nuclear envelope protein laminA/C in the pedunculopontine nucleus, cuneiform nucleus and periaqueductal grey matter in cases of DYT1 dystonia.[365,574]

Mutations in *THAP1*, encoding the protein thanatos-associated domain-containing apoptosis-associated protein 1 (THAP1), have been associated with DYT6, a form of PTD presenting in adolescence or adulthood.[136] THAP1 represses the expression of torsinA and mutations of *THAP1* result in increased expression of torsinA,[151] providing a link

between the pathogenesis of DYT1 and DYT6 dystonias. There have been no studies describing neuropathological features of DYT6 dystonia.

Primary Plus Dystonias

The group of primary plus dystonias includes dopa-responsive dystonia (DRD; DYT5/DYT14), inherited myoclonus-dystonia (IMD; DYT11 and DYT15), rapid onset dystonia–parkinsonism (RDP; DYT12), DYT3 and DYT16 (Tables 12.14 and 12.15).(For a review see Paudel *et al.*[439])

DRD (DYT5) is an autosomal dominant disorder, in which the dystonia may be focal or generalized and associated with parkinsonism. Patients show a marked therapeutic

12

TABLE 12.15 Monogenic genes identified in dystonia[(for a review see 439)]

Locus	Designation	Gene	OMIM	Chromosome	Inheritance	Age at onset and clinical features
DYT1	Early onset AD focal dystonia	Torsin A	128100	9q32-q34	Autosomal dominant	Early onset dystonia, begins in a limb and later becomes generalized
DYT2	Early onset AR dystonia	Unknown	224500	Unknown	Autosomal recessive	Early onset dystonia, segmental or generalized
DYT3	X-linked dystonia (Lubag)	TAF1	314250	Xq13.1	X-linked	Segmental or generalized dystonia with parkinsonism in half the cases
DYT4	Whispering dysphonia	Unknown	128101	Unknown	Autosomal dominant	Whispering dysphonia isolated or with torticollis
DYT5 and DYT14	Dopa responsive dystonia (DRD) (Segawa)	GTP-cyclo-hydrolase I (GTPCH I)	128230	14q22.1-q22.2	Autosomal dominant	Dystonia with parkinsonism, diurnal variation; very good response to L-dopa treatment
DYT6	Early onset AD focal dystonia	THAP1	602629	8p21-q22	Autosomal dominant	Adult onset, usually segmental, frequently laryngeal or oromandibular, rarely generalizes
DYT7	Early onset AD focal dystonia	Unknown	602124	8p	Autosomal dominant	Adult onset focal dystonia
DYT8	Paroxysmal non-kinesigenic dyskinesia (PKND)	MR-1	118800	2q33-2q35	Autosomal dominant	Attacks of dystonia or choreoathetosis, precipitated by stress, alcohol, coffee, fatigue, tobacco; treatment is to avoid precipitants, anticonvulsants often reduces attack frequency
DYT9 and DYT18	Paroxysmal choreoathetosis episodic ataxia (CSE)	Glucose transporter type 1 (GLUT1)	601042 612126	1p21-p13.3	Autosomal dominant	Similar to PKND with ataxia and/or spastic paraplegia between attacks; occasionally epilepsy, helped by anticonvulsants and the ketogenic diet
DYT10 and DYT19	Paroxysmal kinesigenic dyskinesia (PKC)	Proline-rich transmembrane protein 2 (PRRT2)	128200 611031	16p12-q12	Autosomal dominant	Attacks of dystonia or choreoathetosis, precipitated by sudden movement but also by stress; associated with migraine and epilepsy; carbamazepine is usually very effective
DTY11	Myoclonic dystonia	Epsilon-sarcoglycan (SGCE)	159900	7p21	Autosomal dominant	Responsive to alcohol with variable degrees of dystonia; benzodiazepines, antiepileptic and anticholinergic therapy can help some cases
DYT12	Rapid onset dystonia parkinsonism	ATP1A3	128235	19q12-q13.2	Autosomal dominant	Acute onset of generalized dystonia often with associated parkinsonism; often patients are treatment resistant
DYT13	Focal cranio-cervical dystonia	Unknown	607671	1p36.32-p36.13	Autosomal dominant	Focal or segmental dystonia with onset either in the craniocervical region or in the upper limbs
DYT15	Myoclonic dystonia	Unknown	607488	18p11	Autosomal dominant	Myoclonus and variable dystonia

Continued

TABLE 12.15 Monogenic genes identified in dystonia *(Continued)*

Locus	Designation	Gene	OMIM	Chromosome	Inheritance	Age at onset and clinical features
DYT16	DRD and parkinsonism	PRKRA	612067	2q31	Autosomal recessive	Dystonia-parkinsonism with bulbar problems and bradykinesia, some response to L-dopa
DYT17	Early onset AR dystonia	Unknown	612406	20p11.2-q13.12	Autosomal recessive	Focal onset dystonia with progression to segmental and then generalized dystonia
DYT20	Canadian PKND	Unknown	611147	2q31	Autosomal dominant	Canadian family with a phenotype identical to PKND, the MR-1 gene was negative
DYT21	Later onset AD focal dystonia	Unknown	614588	2q14.3-q21.3	Autosomal dominant	Swedish family with primary torsion dystonia, which later became generalized
Not given	DRD and parkinsonism	Tyrosine hydroxylase	605407	11p15.5	Autosomal recessive	Dystonia with parkinsonism, diurnal variation; very good response to L-dopa treatment
Not given	DRD and parkinsonism	Sepiapterin reductase deficiency	612716	2p14-p12	Autosomal recessive	Dystonia with parkinsonism, diurnal variation; very good response to L-dopa treatment

AD, autosomal dominant; AR, autosomal recessive; THAP1, thanatos-associated protein domain-containing apoptosis-associated protein; TAF1, TATA-binding protein-associated factor; MR-1, myofibrillogenesis regulator 1; ATP1A3, sodium/potassium-transporting ATPase subunit alpha-3; PRKRA, protein kinase, interferon-inducible double stranded RNA-dependent activator.

response to L-dopa. DRD is associated with mutations in the *GCH1* gene encoding guanosine triphosphate cyclohydrolase I (GTPCH I). DYT14 dystonia is also due to a deletion in the *GCH1* gene. Autosomal recessive DRD occurs with mutations of the tyrosine hydroxylase and sepiapterin reductase deficiency genes.[439] Neuropathological studies in DYT5 are very limited; neurons of the SN were described as hypomelanized, Lewy bodies were absent.[179]

IMD is autosomal dominant with a disease onset in childhood. The clinical phenotype includes myoclonic jerks with dystonia, usually torticollis or writer's cramp. It is associated with the DYT11 locus and is due to mutations in the *E-sarcoglycan* (*SGCE*) gene.[395,603] A further IMD locus (DYT15, chromosome 18p11) has been reported.[178]

The autosomal dominant RDP (DYT12) presents between late adolescence and the sixth decade with dystonic spasms, bradykinesia, postural instability, dysarthria and dysphagia. RDP is associated with mutations in the gene encoding the alpha-3 subunit of Na/K-ATPase (ATP1A3). Neuropathological examination in a single case showed no changes.[447]

DYT3 or X-linked dystonia–parkinsonism (XDP) occurs in the Philippines. There is focal dystonia becoming segmental or generalized and approximately half of the patients also develop parkinsonism. A variant was identified in the TATA-binding protein associated factor (*TAF1*) gene, which is known to interact with the dopamine receptor *D2* gene.[347] Neuropathology in XDP shows loss of the striosome projections to the globus pallidus and SN.[170]

Young onset dystonia-parkinsonism (DYT16) is caused by autosomal recessive mutations in *PRKRA*, which encodes protein kinase, interferon-inducible double-stranded RNA-dependent activator.[59]

Primary Paroxysmal Dystonias

The paroxysmal dyskinesias, classified according to precipitating events, include DYT8, 9, 10, 18, 19 and 20. In these conditions patients have an episodic movement disorder and are normal between episodes. DYT8 presents in childhood or early adulthood with episodes of dystonia, chorea and athetosis. It is associated with mutation of the myofibrillogenesis regulator 1 gene (*MR-1*).[323] Mutations in the *GLUT1* gene have been identified in DYT9 and DYT18.[524,575] DYT10/DYT19 may be sporadic or autosomal dominant with incomplete penetrance. They present in childhood with short attacks of involuntary movements. Mutations in the proline-rich transmembrane protein 2 (*PRRT2*) gene have recently been identified.[92,571] There is little neuropathological data on these conditions except for one report, which showed no abnormality.[277]

Idiopathic Sporadic Focal or Segmental Primary Dystonias

Sporadic cases of adult onset focal or segmental dystonia are frequent. A family history is present in ~25 per cent of patients. Cervical dystonia is most common, but this group includes blepharospasm, writer's cramp and other focal dystonias such as musician or dart players dystonia. Patients with blepharospasm or blepharospasm associated with oromandibular dystonia (Meige's syndrome) may develop parkinsonism and Lewy body pathology has been reported. Neuropathological studies, with the exception of Meige's syndrome and DYT1, have mostly shown no specific findings.[220]

Heredodegenerative Dystonias

This category includes neurodegenerative diseases in which dystonia may be a prominent feature, although patients usually present with other neurological symptoms. Some of these conditions are inherited in an autosomal dominant, autosomal recessive or X-linked manner, others are part of a mitochondrial disease and a further group do not have a known genetic cause. Examples are provided in (Table 12.16).[86]

The Autopsy in Dystonia

Neuropathological assessment may have important medicolegal implications when secondary dystonia is attributed to previous trauma or exposure to toxins. In all cases of dystonia other neurodegenerative diseases should be systematically excluded. Since there is little published autopsy data on primary dystonia, there is a pressing need to document pathological changes in such cases and for this systematic sampling and archiving of tissue is required. Fresh frozen tissue should also be stored to facilitate molecular/genetic investigations.

TIC DISORDERS

Tic disorders usually have their onset in childhood, often improving by adulthood. They are characterized by involuntary, sudden multiple, rapid and recurrent movements, classified as motor tics, and the uttering of brief sounds in the case of vocal tics. Tourette's syndrome (TS) is the best studied, which is characterized by multiple motor tics with one or more vocal tics occurring over a period of more than a year. TS is considered as part of a neurodevelopmental neuropsychiatric disorder.[320] It has been proposed that in a subset of cases with sudden onset of a tic disorder this may be due to a post-infectious autoimmune process secondary to group A streptococcal infection.[382] Genetic factors are regarded as important in the pathogenesis of TS,[457] but causative genes identified in small numbers of cases.[437]

The mechanisms underlying the development of tics are uncertain but the basal ganglia circuitry is implicated. Postmortem studies in TS are limited, but a stereological study has shown overall alterations in the neuronal number in components of the basal ganglia and altered proportions of parvalbumin-positive and cholinergic interneurons.[264,268] A further recent post-mortem study has provided evidence of dopaminergic dysfunction in the frontal cortex.[592]

MISCELLANEOUS DISORDERS AFFECTING THE DEEP GREY NUCLEI

A number of distinctive clinical and pathological conditions are described, primarily affecting basal ganglia or other subcortical grey matter regions that do not easily map to the conceptual framework around which conditions described elsewhere in this textbook are grouped. For several of these disorders, clinical, genetic and pathological constellations of data indicate that these can be securely regarded as specific conditions (e.g. aceruloplasminaemia, Hallervorden–Spatz disease, HARP syndrome, hereditary ferritinopathy or neuroferritinopathy, Menkes disease, Nasu–Hakola disease, Wilson's disease). Earlier editions of this text also described predominantly clinicopathological disorders with no specific pathology or relevant heredity. Given the increasing interval between the initial descriptions of such disorders and the modern era of molecular pathology, it is quite unclear what they represent. Regrettably, there is little data in the literature to indicate that such cases have been retrieved from the archive and subjected to modern molecular investigation and genetic analysis. The recent expansion in genetics, together with recognition of the anatomical and clinicopathological manifestations of cerebral small vessel

---

TABLE 12.16 Heredodegenerative dystonias*

Heredodegenerative dystonias with defined inheritance

Autosomal dominant	Autosomal recessive	X-linked	Mitochondrial
Huntington's disease	Wilson's disease	Lesch–Nyhan syndrome	MERRF
Parkinson's disease	Juvenile parkinsonism with	Rett's syndrome	MELAS
Dentatorubropallidoluysian	*parkin* mutations or the other	Pelizaeus–Merzbacher	Leber's hereditary optic
atrophy (DRPLA)	complex recessive forms such	disease	neuropathy
Other spinocerebellar	as those associated with muta-		
syndromes such as some of	tions in *PLA2G6* and *PANK2*		
the SCAs	Neurodegeneration with brain		
	iron accumulation		
	Ataxia-telangiectasia		

Heredodegenerative dystonias with undefined inheritance or sporadic cases

Parkinson's disease			
Progressive supranuclear palsy			
Multiple system atrophy			
Corticobasal degeneration			
Multiple sclerosis			

*Adapted from de Carvalho Aguiar and Ozelius.[86] © 2002, with permission from Elsevier. MERRF, myoclonic epilepsy with ragged red fibres; MELAS, mitochondrial encephalomyoathy with lactic acidosis and stroke-like episodes.

disease, tauopathies, prion diseases and other degenerative disorders now readily assignable to an underlying proteinopathy, raises considerable doubt as to whether cases reported before the era of routine immunocytochemistry would not now be assigned to a more specific diagnosis. For the purposes of the current edition of this text, this section focuses on the well-defined disorders. In the absence of any literature developments, relating to less specific miscellaneous basal ganglia disease, readers in pursuit of information about them are referred to older editions of the text.

Thalamus

Focal Lesions of the Thalamus

Associations of well-defined clinical disturbances with focal lesions in one or other of the basal ganglia are uncommon. Exceptions are the association of hemiballismus with lesions of the contralateral subthalamic nucleus (see earlier) and the thalamic syndrome.

The main feature of the thalamic syndrome is sensory disturbance affecting one side of the body. Sensory thresholds are raised, but stimulation of the affected side evokes a delayed, slowly subsiding, 'hyperpathic' response referred to a large area of the body. The lesion is commonly an infarct or haemorrhage involving the posterior ventral nucleus of the contralateral thalamus. Small, vascular lesions in other parts of the thalamus are a common finding in the brain of elderly subjects and may not be associated with a recognized clinical disturbance. The participation of individual thalamic nuclei in a wide range of degenerative and other diseases is well recognized.

Thalamic Degenerations

Degeneration of thalamic neurons with astrocytosis may be found in several conditions (listed later). Patients may have presented with behavioural disturbance, dementia or movement disorder or a combination of these. Care must be taken when considering autopsy strategy because some cases are prion diseases (see Chapter 18, Prion Diseases):

- Friedreich's ataxia;
- spinocerebellar degenerations;
- membranous lipodystrophy;
- Fukuyama muscular dystrophy;
- prion disorders: Creutzfeldt–Jakob disease (including vCJD), fatal familial insomnia.

Cases of isolated thalamic degenerations (thalamic atrophies) are rare, show features of neuronal loss and astrocytosis, and may also be related to prion disease.[243] Recent work has highlighted thalamic atrophy as a potential feature of 'pre-multiple sclerosis' patients presenting with a clinically isolated syndrome.[212]

Globus Pallidus

Pallidal Degenerations

Rare familial and sporadic diseases are described characterized by preferential degeneration in the globus pallidus, either alone or in combination with degeneration of other structures (Figure 12.23). Clinically, such conditions are manifest as a variety of movement disorders with or without dementia.[4,246] Some of the existing pathological accounts describe neuronal loss and astrocytosis without specific features. Before including a disease in this group, it is vital to exclude features associated with classifiable degenerative pathologies of the CNS, e.g. neuroaxonal dystrophy, inclusions of MSA, extramotor TDP43 inclusions, tauopathy and prion disease. Thus, older accounts of 'pallidal degenerations' are defined purely morphologically and it is not clear whether they truly represent distinct diseases. Pallidal degeneration occurs as part of the HARP syndrome together with hypobeta-lipoproteinaemia, acanthocytosis and retinitis pigmentosa.[348]

Striatum

Striatal Necrosis

Bilateral striatal necrosis has been associated with a wide range of aetiological factors (Table 12.17). Energy deprivation, supported by experimental studies,[166,354] emerges as a common mechanism. Striatal necrosis may be seen in infancy, as either a sporadic or a familial disorder with uncertain aetiology. Clinical features vary from acute encephalopathy to slowly progressive gait disturbance, dystonia, rigidity, ataxia and blindness. Neuropathological studies report neuronal loss, astrocytosis and microvacuolation.[328] Such cases have been classified into three subgroups: early acute onset, early gradual onset and late onset, although the aetiopathological validity of this is, as yet, uncertain. A variety of infectious, vascular and metabolic disorders may potentially underlie this heterogeneity. In view of the overlap with Leigh's disease and Leber's optic neuropathy, detailed study of mitochondrial metabolism is appropriate in such cases.

Neuroleptic Malignant Syndrome

The neuroleptic malignant syndrome occurs as an adverse reaction to certain drugs. It is characterized by hyperthermia, muscular rigidity with jerking, dysphagia, rhabdomyolysis and autonomic dysfunction. Consciousness is variably depressed. Investigations show a leukocytosis and elevated serum creatine kinase. Although originally described as an effect of neuroleptic drugs (most commonly haloperidol, fluphenazine and chlorpromazine), it can be caused

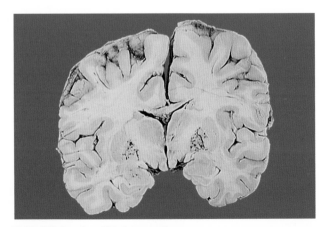

12.23 Bilateral pallidal necrosis.

TABLE 12.17 Causes of striatal necrosis

Causes
Hypoxia
Hypoglycaemia
Carbon monoxide poisoning
Methanol intoxication
Neonatal asphyxia
Mitochondrial cytopathy
Sodium nitroprusside
Haemolytic uraemic syndrome
Postinfective factors

TABLE 12.18 Disorders associated with mineralization of the basal ganglia*

Normal Ageing
Abnormal calcium metabolism
Hypoparathyroidism
Pseudohypoparathyroidism
Pseudopseudohypoparathyroidism
Hyperparathyroidism
Carbonic anhydrase II deficiency
Neuropsychiatric and neurodegenerative diseases
Alzheimer's disease
Down's syndrome
Parkinsonism
Non-parkinsonian movement disorder
Psychosis including schizophrenia
Nasu–Hakola disease
Möbius syndrome
Neurodegeneration with brain iron accumulation, type 1
Alexander's disease
Familial syndromes
Familial psychosis
Familial parkinsonism
Mitochondrial cytopathy
Infective and inflammatory disorders
HIV infection and AIDS
Meningoencephalitis
Lupus cerebritis
Toxic and hypoxic disorders
Chemotherapy and radiotherapy
Folate deficiency
Carbon monoxide poisoning
Hypoxia

*For a review see reference 61a.

by antidepressants, benzodiazepines, metaclopramide, amphetamines, cocaine or abrupt discontinuation of anti-parkinsonian drugs.[28,115,279] It has been suggested that this syndrome is most likely to develop in patients with pre-existing organic brain disease, in particular, dementia with Lewy bodies when they receive major tranquilizer therapy for behavioural problems.[362]

The pathogenesis of neuroleptic malignant syndrome is believed to be due to blockade of brain dopaminergic receptors. The syndrome is fatal in about 10 per cent of cases associated with multiorgan failure, autonomic dysfunction, renal failure, hyperthermia and rhabdomyolysis. At autopsy, there are changes consistent with general rhabdomyolysis leading to renal acute tubular necrosis caused by myoglobinuria and hypotension.[253] Changes in the nervous system, most often non-specific neuronal damage, are consistent with the mode of death in multiorgan failure. However, foci of recent necrosis in the anterior and lateral hypothalamic nuclei,[221] and cerebellar neuronal degeneration, as seen in heat-induced CNS injury, have also been described.[324]

A similar syndrome of hyperthermia and multiorgan failure may be seen after recreational ingestion of the drug 3,4-methylenedioxymethamphetamine (MDMA; 'ecstasy').[570] The psychotropic effects are through brain monoamines, particularly serotonin and dopamine, and it has been found to be toxic to serotonergic neurons.[506] MDMA-related death is also associated with water intoxication consequent upon massive drinking and exercise, characteristically during recreational use.[397]

Calcification of the Basal Ganglia

Calcification of subcortical nuclei, particularly the basal ganglia, is referred to by a variety of terms including bilateral striatopallidodentate calcification, brain calcinosis and Fahr's disease. Basal ganglia calcification (BGC) occurs increasingly with age and CT scanning has found it in 1 to 2 per cent of older patients.[129] Most cases are idiopathic, but it is associated with a wide variety of conditions (Table 12.18).[350] In one series a cause was found in just over 25 per cent of cases. The most common association is with disorders of calcium and phosphate metabolism, particularly hypoparathyroidism. BGC was found in 17 per cent of a series of mitochondrial cytopathy patients.[131]

The clinical significance of BGC is uncertain. In most cases, it appears to be incidental and it should not, in general, be used as an explanation for neurological disturbances. However, in a few cases, BGC has been linked to clinical manifestations including parkinsonism, hyperkinetic extrapyramidal disorders, cognitive impairment and psychiatric disturbances.[350] Familial cases of idiopathic basal ganglia calcification that are not secondary to known abnormalities, such as calcium metabolism defects, are also described. Most pedigrees appear to manifest an autosomal dominant pattern of inheritance and may be associated with parkinsonism, cognitive impairment and mild ataxia.[290] Study of one pedigree has established a link to a locus on chromosome 14q (designated IBGC-1).[155] Analysis of the locus is incomplete but has drawn attention to a single nucleotide polymorphisms in the gene *MGEA6/c-TAGE*.[404] However, this locus appears not to be involved in other families with autosomal dominant inheritance.[403]

Neuropathology

The distribution of mineralization is similar in all cases, irrespective of the aetiology. Calcification is symmetrical, affecting basal ganglia, thalamus, central white matter and dentate nucleus, with the globus pallidus being most affected (Figure 12.24). Small deposits may be found at the junction of cerebral cortex and white matter, the depths of cerebellar folia and cerebral cortical sulci (Figure 12.25).

Microscopy reveals diffuse mineralization in the tunica media of small blood vessels (Figure 12.26). Arteriolar walls may be thickened and vessels' lumina narrowed, but not occluded. Rows of small calcospherites may be seen along capillaries and there may be free parenchymal concretions. Large concretions are believed to develop from smaller deposits. The deposits appear basophilic in sections stained with H&E and they may appear lamellar. There is

deposition of calcium but iron may also be detected and traces of various other metals, including aluminium, manganese and zinc. Histochemical studies have revealed the presence of mucopolysaccharides. Immunohistochemical studies have also detected non-collagenous bone matrix proteins such as osteopontin in cases of BGC associated with neurodegenerative disorders.[129,137,350] Transmission electron microscopy has shown deposits in the cytoplasm of pericytes and in glial processes.[291]

Idiopathic cerebellar calcification may occur, generally accompanied by basal ganglia calcification. However, a case of pontocerebellar calcification without BGC has been described. Calcification was present in the neuropil, but in contrast to Fahr-type calcification, only rarely in association with vessels, and in association with positive immunoreactivity for calbindin D28K and parvalbumin.[478] The nosological and pathogenetic relationship of such variants to BGC is unclear.

Neuroaxonal Dystrophy, Disorders of Metal Biometabolism and Related Disorders

Several disorders are characterized by pathological changes of neuroaxonal dystrophy, accumulation or dysmetabolism of brain iron and copper and the syndrome of hepatocerebral degeneration (Table 12.19).[346,486] Most of these are genetically determined and include shared aspects of pathology and anatomical involvement. They remain difficult to classify and the interrelationships between them are the subject of ongoing research. For the purposes of this text they are grouped according to morphology (neuroaxonal dystrophies), abnormalities of cerebral iron metabolism (NBIA, neuroferritinopathy) and copper metabolism-associated hepatocerebral degeneration.

Neuroaxonal Dystrophy

Neuroaxonal dystrophy is a morphological abnormality of the axons in the central and peripheral nervous systems, defined as the occurrence of axon swellings, the larger of

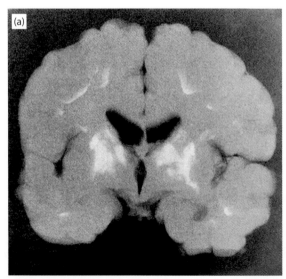
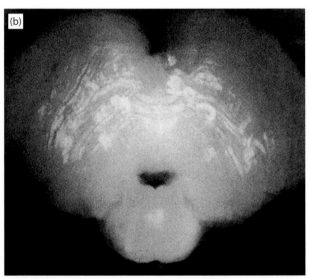

12.24 Mineralization in the brain. (a) X-ray of a brain slice at the level of the mammillary bodies. **(b)** X-ray slice of the pons and superior cerebellum.

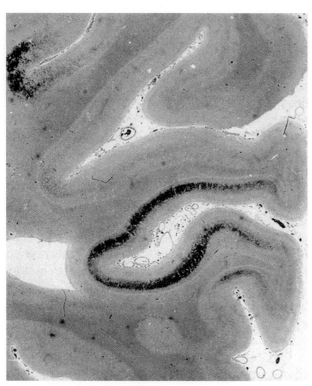

12.25 Cerebral mineralization. Calcification of the cortex in the striate area.

which are termed spheroids, composed of structured material (Figure 12.27). The pathogenesis of axonal swelling is uncertain. Neuroaxonal dystrophy has been subdivided into three types:

- physiological neuroaxonal dystrophy: potentially normal part of brain ageing;
- secondary neuroaxonal dystrophy: neuroaxonal dystrophy occurring as a 'reactive' process in another condition;
- primary neuroaxonal dystrophy: diseases in which neuroaxonal dystrophy is a significant pathology.

Dystrophic neuroaxonal swellings vary from 20 to 120 μm in diameter and have a pale, granular eosinophilic appearance, occasionally developing a more intensely stained eosinophilic core. Spheroids are generally well stained by silver techniques such as Bodian and may show unstained clefts and vacuoles, especially in the larger examples. Ultrastructural examination shows that the distended axon contains mitochondria, lysosome-related dense bodies, membrane-bound vesicles (tubulomembranous structures), amorphous matrix material and, generally, few neurofilaments.[196] Immunocytochemistry shows reactivity with antisera to neurofilament proteins and ubiquitin only in spheroids smaller than 30 μm.[373] Larger spheroids are unreactive in contrast to large spheroids in

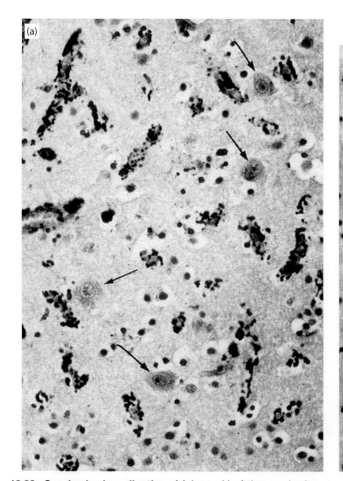

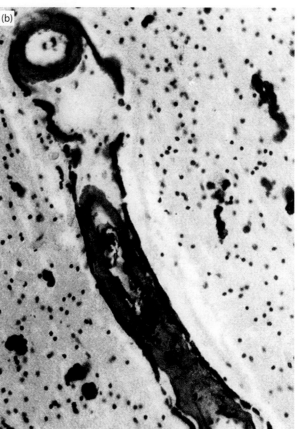

12.26 Cerebral mineralization. (a) Layer V of the cerebral cortex, showing pericapillary calcospherites (arrows indicate normal neurons). **(b)** Blood vessel in the cerebral white matter, showing medial calcification.

TABLE 12.19 Neuroaxonal dystrophies, accumulation or dysmetabolism of brain iron and copper, and copper-related hepatocerebral degeneration

Disorder	Synonyms	Gene symbol	Gene product	OMIM	Neuroax-onal dys-trophy	Iron dys-metabolism	Copper dysmetab-olism	Hepato-cerebral degenera-tion
Physiological neuroaxonal dystrophy					✓			
Neurodegen-eration with Brain iron accumulation, type 1 (NBIA1)	Pantothe-nate kinase-2-associated neurode-generation (Hallevor-den–Spatz syndrome)	PANK2	Panto-thenate kinase-2	234200	✓	✓		
Neurodegen-eration with brain iron accumulation, type 2 (NBIA2)	Seitelberg-er's disease or infantile neuroaxonal dystrophy; late infantile, juvenile, and adult neuroaxonal dystrophies; PARK14	PLA2G6	Phospho-lipase A2, group VI	612953	✓	✓		
Nasu–Hakola disease	Polycystic lipomem-branous osteodys-plasia with sclerosing leukoen-cephalopa-thy	DAP12/ TREM2	Triggering receptor expressed on myeloid cells 2	221770	✓			
Kufor–Rakeb disease (NBIA3)	PARK9	ATP13A2	ATPase type 13A2	606693	no data	✓		
FA2H-associ-ated neurode-generation	FAHN; SPG35	FA2H	Fatty acid 2-hydroxy-lase	612319	no data			
Neuroferriti-nopathy	Hereditary ferritinopa-thy	FTL	Ferritin light chain	606159	✓	✓		
Wilson's disease		ATP7B	ATP7B	277900	✓		✓	✓
Menkes dis-ease		ATP7A	ATP7A	309400				
Aceruloplas-minaemia		CP	Ceruloplas-min	604290		✓		

other conditions that generally react strongly with anti-sera to neurofilament proteins. Axonal spheroids also show immunoreactivity for α-synuclein.[568] Areas affected by neuroaxonal dystrophy show astrocytosis and accumulation of iron-containing pigment in spheroids and associated glia. This has led to the alternative descriptive term of pigment-spheroidal dystrophy for this type of pathological process.

Physiological Neuroaxonal Dystrophy

The development of neuroaxonal dystrophy is a common normal finding in certain parts of the CNS. Physiological neuroaxonal dystrophy is seen in the gracile and cuneate nuclei, the pars reticulata of the SN (SNpr) and the medial globus pallidus from the age of about 10 years.[496] The age-related incidence of axonal dystrophy in the first four decades in the gracile nucleus

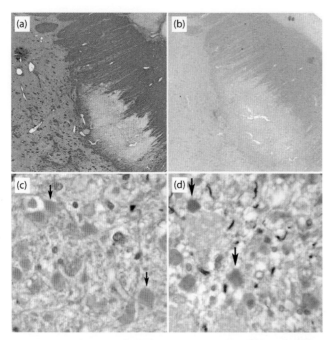

12.27 Neuroaxonal dystrophy in a case with neurodegeneration with brain iron accumulation, type 1 (pantothenate kinase-associated neurodegeneration, PKAN). (a) Rarefaction of the neuropil of the globus pallidus (b) with iron deposition. (c) Numerous axonal swellings in the globus pallidus, which (d) contain phosphorylated neurofilaments. (a,c) SMI31 immunohistochemistry; (b) Perls stain for iron; (c) H&E stain.

is 13, 53, 76 and 97 per cent, respectively.[525] Similar changes are seen in sympathetic ganglia with ageing.[485] The physiological development of neuroaxonal dystrophy may account for some of the neurological decline seen with ageing.[496]

Secondary Neuroaxonal Dystrophy

Neuroaxonal dystrophy may occur as a symptomatic phenomenon as a result of another disorder. This may be seen as accentuation of the severity of physiological neuroaxonal dystrophy or as the development of more widespread lesions.

Increased intensity of physiological neuroaxonal dystrophy is seen in idiopathic PD, Wilson's disease and chronic alcoholic liver disease.[245,495] In the medial pallidum and SNpr, the presence of accentuated axonal dystrophy is associated with an accumulation of brown pigment in dystrophic axons and glia, some being iron, some lipofuscin. This may impart a rusty discoloration to these areas macroscopically. This has also been described in association with small blood-vessel disease.[283]

Widespread neuroaxonal dystrophy may develop in several conditions as a symptomatic manifestation, including cystic fibrosis, congenital biliary atresia, Niemann–Pick disease type C, anticancer chemotherapy, Zellweger's syndrome and HTLV-1 infection.[62,118,454,589] In view of the animal models of neuroaxonal dystrophy caused by vitamin E deficiency, it has been proposed that malabsorption of vitamin E may have a pathogenic role in some cases of symptomatic neuraxonal dystrophy.[62]

Neurodegeneration with Brain Iron Accumulation, Type 1 (NBIA1, Formerly Hallervorden–Spatz Disease) or Pantothenate Kinase-Associated Neurodegeneration (PKAN)

PKAN is a progressive neurodegenerative condition with prominent neuroaxonal dystrophy associated with movement disorders. The disorder is associated with mutations in the pantothenate kinase 2 gene (*PANK2*).[202,599] Two mutations (*1231G>A*; *1253C>T*) are found in a third of the cases but mutations are reported in all 7 exons.[177] Missense mutations predominate, but deletions, duplications and splice-site mutations are also reported. Some mutations may produce milder phenotypes.[1] PANK2 protein regulates coenzymeA synthesis by catalyzing pantothenate (vitamin B5) phosphorylation. Interaction with mitochondria and the cellular iron exporter ferroportin iron are reported.[448] Older reports of Hallervorden–Spatz disease predate the genetic dissection of NBIA syndromes. Care should be taken when attributing clinicopathological characteristics on the basis of this older literature because some reported cases might actually represent NBIA2 or 3. A diagnosis of an NBIA subtype in new clinical and autopsy cases should now always include genotyping.[486]

There are clinical differences between cases with early and late onset. Classical early onset disease usually presents around 3–4 years of age, with 90 per cent occurring prior to 6 years.[210,211] Clinical features include pyramidal and extrapyramidal signs (chorea, parkinsonism) and symptoms. Dystonia is frequent and neuropsychiatric symptoms are described. Oculomotor and neuro-ophthalmic abnormalities are common.[117] In older onset cases (typically 2nd and 4th decade) extrapyramidal movement disorder may be less prominent. Early dystonic, behavioural and cognitive presentations are described. MRI appearances corresponding to 'eye-of-the-tiger' are reported in genetically verified cases.[486]

In genetically proven cases, affected areas macroscopically appear shrunken with rusty-brown discolouration. There is accumulation of pigment, which contains iron, lipofuscin and neuromelanin, and destruction of the globus pallidus, with variable involvement of adjacent structures (e.g. medial putamen and internal capsule) with axonal spheroids (Figure 12.27). The pigment is present within astrocytes and microglial cells, whereas some pigment appears free in the neuropil and often around blood vessels. Involvement of the SNpr is reported in the older literature, but recent reports emphasize the sparing of cortical, brain stem and the other deep grey nuclei.[303] Similarly, older reports of associated Lewy pathology have not been confirmed in genetically proven cases, in contrast to NBIA2 (see later).[303,331] Tauopathy is also more likely to be present in NBIA2.

Neurodegeneration with Brain Iron Accumulation, Type 2 (NBIA2, Formerly Infantile Neuroaxonal Dystrophy, Seitelberger's Syndrome) or PLA2G6-Associated Neurodegeneration (PLAN)

PLAN is due to mutations in the *PLA2G6* gene on chromosome 22q. This gene encodes a Ca2+-independent phospholipase A2 enzyme called iPLA2. Like NBIA1 the phenotype varies with age of onset. In addition, genotype-phenotype correlations suggest that mutations associated with loss of phospholipase activity produce infantile neuroaxonal dystrophy (INAD) syndromes whereas mutations with normal enzyme activity produce dystonia-parkinsonism syndromes (PARK14, see under Genetics of Parkinson's Disease).[120,423] Early onset cases characterized by INAD have weakness, hypotonia and areflexia. With disease progression rigidity, spasticity, cerebellar signs, deafness, optic atrophy and mental deterioration occur. Patients typically die before the age of 6 years.[386] As NBIA2 involves more prominent peripheral manifestations a diagnosis may be made, though largely superseded by genetic testing, from brain, peripheral nerve, conjunctival, skin or rectal biopsy if the characteristic axonal swellings are sought by immunohistochemical and ultrastructural techniques.[285,420] Macroscopically, the brain of affected infants shows cerebral and cerebellar atrophy with ventricular dilatation. Histologically, axonal spheroids are seen in the grey matter of the nervous system with degeneration of corticospinal and spinobulbar tracts associated with astrocytosis.[496] Severe diffuse Lewy pathology and neurofibrillary tangle pathology have been reported in the PLAN phenotypes (Figure 12.28).[425]

Kufor–Rakeb Disease (NBIA3; PARK9)

For genetic characteristics and clinical presentation see PARK9 (Genetics of Parkinson's Disease). In this disease, variable increase in basal ganglia iron content on brain imaging is seen and the designation NBIA3 proposed.[486] CNS pathology data are not yet published. Peripheral nerve changes include Schwann cell inclusions of possible lysosomal origin and this gene can cause a neuronal ceroid lipofuscinosis-type picture in animal models.

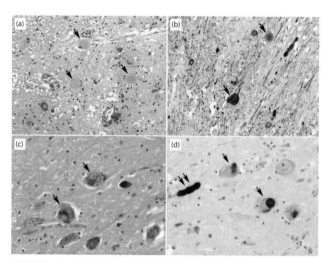

12.28 Neuropathological features of neurodegeneration with brain iron accumulation, type 2 (phospholipase A2, group VI, PLAN). (a) Numerous axonal swellings and spheroids in the posterior horn of the lumbar spinal cord, **(b)** many of which are also stained with an antibody to phosphorylated neurofilament. **(c,d)** Lewy pathology is present in the substantia nigra. (a,c) haematoxylin and eosin (H&E) stain; (b) RT97 immunohistochemistry; (d) α-synuclein immunohistochemistry.

FA2H-Associated Neurodegeneration (FAHN; SPG35)

Mutations in *FA2H* have been linked to leukodystrophy and a form of HSP (the SPG35 locus).[97,116] This fatty acid hydroxylase is part of ceramide/lipid metabolism. Recently another FA2H-linked phenotype has emerged in Italy and Albania associated with childhood onset of gait abnormality, spasticity, ataxia and dystonia resembling a neuraxonal dystrophy. MRI demonstrated globus pallidus iron accumulation and an NBIA-like picture. No pathological data are published.

Nasu–Hakola Disease

Nasu–Hakola disease (polycystic lipomembranous osteodysplasia with sclerosing leukoencephalopathy) is an autosomal recessive disease characterized by the development of repeated bone fractures caused by bone cysts, with onset in the second decade. Later there is degeneration of the cerebral white matter associated with mineralization of the basal ganglia, leading to progressive dementia of frontal lobe type, and extrapyramidal rigidity in the third decade.[186,187] Initially cases were reported from Japan and Scandinavia but cases are now recognized from elsewhere. The disorder is associated with mutations of *DAP12* (*TYROBP*) and *TREM2*, which constitute a receptor/adapter signalling complex expressed on osteoclasts, dendritic cells, macrophages and microglia.[295,426,428] *TREM2* mutations has been described in families with early-onset dementia and no bone cyst involvement.[71,182] The mutations cause fundamental disturbances of monocyte cell lineage biology.[385,387] It is of note that heterozygous rare variants in *TREM2* are associated with a significant increase in the risk of Alzheimer's disease.[183]

Histologically, the bone and fat show membranocystic lesions (Figure 12.29). There is demyelination of cerebral white matter with axonal spheroid formation. Severe neuronal loss with mineralization is seen in the basal ganglia and also in the thalamus in some cases.[427,535] Peripheral nerves may also be affected.[230]

Hereditary Ferritinopathy (Neuroferritinopathy)

Mutations in the gene encoding ferritin light (FTL) protein give rise to a multisystem CNS disorder. The clinical features include a wide range of signs and symptoms associated with basal ganglia and/or cerebellar degeneration. There is some evidence that genotype influences phenotype in the families so far characterized.

The genetic changes in the *FTL* gene associated with hereditary ferritinopathy are shown in Table 12.20. The first family (*FTL* mutation 1) described originated in the North West of England and shows two main patterns of disease.[70,82] The first is an HD-like choreiform disorder, but with no evidence of significant intellectual deterioration and much less rapidly progressive. The second pattern is that of a parkinsonian disorder with akinetic-rigid features. Other individuals may develop a dystonic illness.[81] Genetic studies have demonstrated a founder effect within all cases so far identified with this genetic mutation. The second family (*FTL* mutation 2), of French origin, includes a diverse phenotype of tremor, cerebellar ataxia, parkinsonism, pyramidal signs, behavioural disturbances and cognitive decline.[553,554] The third family (*FTL* mutation 3) is also of French origin and shows similar clinical features to the UK family.[349] All three genetic changes in *FTL* are point mutations leading to elongation of the C terminus of the protein by a novel peptide sequence unrelated to the normal C-terminus of FTL.

Hereditary ferritinopathy is characterized by a hallmark cytoplasmic lesion – the iron/ferritin body – which is eosinophilic, immunoreactive to ferritin and contains abundant stainable iron (Figure 12.30). The distribution of these lesions is dependent on the underlying genetic subtype and reflects the clinical features. In the UK mutation and the related French mutation, there is predominant involvement of the basal ganglia, especially the globus pallidus. Iron/ferritin bodies are probably initially formed in oligodendrocytes but more severely affected areas of the brain show them in all major cell classes and as, apparently, cell-free structures within the neuropil and white matter. Cystic degeneration of the basal ganglia, centred on the globus pallidus, is a characteristic feature apparent on radiology (Figure 12.31). In the family with a larger C-terminal missense peptide these lesions are especially prominent in the cerebellum. In all of the *FTL* gene mutations, they can be found in most regions of the central nervous system in established cases, the intensity of pathology increasing with duration of disease. For *FTL* mutation 1 the pattern of disease is that of a dentatorubropallidoluysian disorder but all other regions examined in advanced cases shows at least occasional iron/ferritin bodies.

The pathogenesis of the disease is not fully characterized. A transgenic mouse model expressing a pathogenic FTL mutation shows very similar pathology to the human disease. The brain of affected individuals shows iron accumulation but it is less clear that neurotoxicity is mediated by iron overload. The mutations in *FTL* are predicted to alter the C terminus conformation affecting the ability of the molecule to pack into the dodecahedron comprising each functional ferritin shell. It is likely that *FTL* mutations impair the stability of iron storage by ferritin. Excess free iron may drive over-expression of ferritin. The predilection for cerebral involvement may reflect the normal balance between FTL and FTH across the organs of the body because brain ferritin is rich in FTL. Ferritin and iron accumulate in nuclei and cytoplasm of other organs (e.g. liver) but are not associated with clinical features or significant change in extracerebral parameters of iron metabolism (e.g. serum transferrin). The possibility that the disorder may be diagnosed by muscle or skin biopsy has been proposed but has not been prospectively validated.[491] Recent work suggests a role for oxidative damage to mitochondrial DNA.[91]

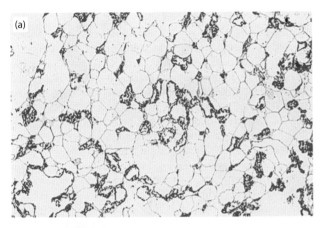

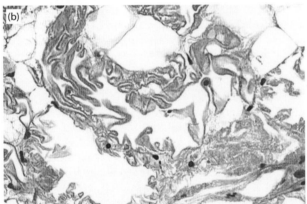

12.29 Nasu–Hakola disease. (a,b) Membranocystic lesion, with eosinophilic wrinkled thickening of the membrane of adipocytes in fat, is a key feature of the disease.

TABLE 12.20 Genetics of neuroferritinopathy (hereditary ferritinopathy)				
Origin	**DNA change**	**Wild type Peptide (175AA)**	**Mutant peptide**	**Main clinical features**
FTL1 'Cumbrian'	Insertion A @ 460–461	Codon 154 R – 21AAs– stop codon	Codon 154 K – 25 AA nonsense sequence – stop codon (179 AA)	Chorea, parkinsonism, focal dystonia, ataxia
FTL2 'French'	Insertion TC @ 498–499	Codon 168 F – 8AAs – stop codon	Codon 168 S – 24 AA nonsense sequence – stop codon (191AA)	Tremor, cerebellar signs, cognitive impairment, dyskinesia
FTL3 'New York'	Insertion C @ 646–647	Codon 148 H – 28AAs – stop codon	Codon 148 P – 32 AA nonsense sequence – stop codon (180AA)	Chorea, choreoathetosis, ataxia, pseudobulbar affect
FTL4 'Portuguese'	Substitution G→A 474	Codon 158 P	Codon 158 P→T	Psychosis, bradykinetic-rigidity

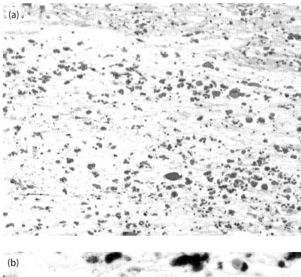

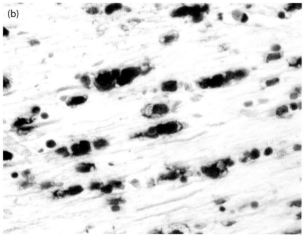

12.30   Iron/ferritin bodies in hereditary ferritinopathy.
(a) Eosinophilic bodies of variable size in the attenuated peri-cystic region of the globus pallidus. **(b)** Cords of iron/ferritin bodies in white matter suggest origin within oligodendrocytes. **(c)** Ferritin light chain immunoreactivity within iron ferritin bod-ies. (b) Diaminobenzidine (DAB)-enhanced Perls reaction; (c) ferritin immunohistochemistry.

Hepatocerebral Syndromes and Disorders of Copper Metabolism

Neurological symptoms are a conspicuous feature of a number of disorders that accompany liver disease. There is a

pathogenetic overlap, exemplified by Wilson's disease, with genetically determined syndromes in which there is a disturbance of copper metabolism. Extrapyramidal movement disorder, dystonia and cerebellar dysfunction are prominent among this diverse group of disorders but the neurological and neuropsychological morbidity that can occur is broad and protean. Pathogenetic explanation of the disease processes, and of the neurological dysfunction, is incompletely characterized even in those diseases associated with known genetic changes.

Chronic Acquired Hepatocerebral Degeneration

Movement disorders may arise in some patients after repeated episodes of hepatic encephalopathy but have also been observed in patients with chronic liver disease in the absence of known episodes of hepatic coma.[319] The pathophysiology is unknown, treatment response may be poor and permanent neurological dysfunction may occur. The disorder has been called 'pseudo-Wilson' syndrome. The clinical features of chronic acquired hepatocerebral syndrome (CAHS) include twitching (e.g. face, fingers, arms), athetosis (e.g. mouth, tongue, face), slow, slurred speech, cognitive dysfunction, tremor, ataxia, hyperreflexia and extrapyramidal rigidity.[398] Pathological data is largely confined to older literature and includes reports of astrocytic (reactive Alzheimer type 2 astrocytes) and neuronal (variable cell loss, microcavitation and laminar necrosis) changes in various brain regions, especially the cerebral cortex and adjacent white matter, cerebellar cortex, basal ganglia, dentate nucleus and brain stem.

Wilson's Disease (Hepatolenticular Degeneration)

This autosomal disorder is characterized by liver disease (cirrhosis) and cerebral degeneration associated with abnormal copper excretion. The underlying genetic abnormality is mutation of the *ATP7B* gene on chromosome 13 encoding a copper transporting ATPase (P-type ATPase).[55,443] Clinical manifestations are protean and vary with childhood or adult onset. A diagnostically important sign in the eye is the Kayser–Fleischer ring due to copper deposition in the limbus of the cornea. Neurological presentation includes parkinsonian features, tremor and dystonia, whereas neuropsychological features may include abnormal behaviours, personality change and schizophrenia-like symptoms.[8] Over 300 mutations are described but each population tends to have a predominant type. The *H1069Q* mutation is present in 37–63 per cent of western cases and may confer greater risk of neurologic manifestations.[85] The *R778L* mutation predominates in China. *ATP7B*, primarily expressed in the liver, is an essential part of the secretory mechanism of copper, involving its incorporation into ceruloplasmin and biliary elimination. Copper is an essential micronutrient and is used in a variety of enzymes in various tissues. Failure of copper elimination results in leakage of copper into plasma and its deposition in extrahepatic tissues.

Pathology shows copper deposition especially in the basal ganglia with cavitation, gliosis and neuronal loss. A small subgroup develop severe cortico-subcortical changes

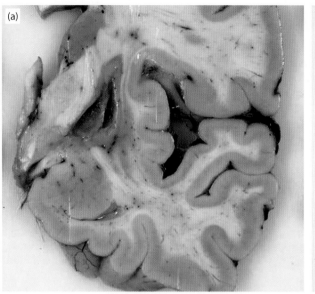
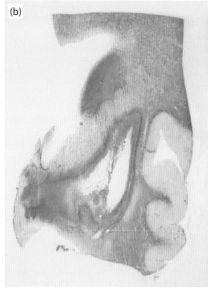

12.31 **(a)** Cavitation within the lentiform nucleus in hereditary ferritinopathy. **(b)** Iron deposition is widespread in the cerebral hemisphere, most prominently around the cavitated lesion. Diaminobenzidine (DAB)-enhanced Perls reaction.

of myelin degeneration, gliosis and profound neuronal loss of the deep cortical layers.[371] These myelin changes result in cavitating lesions of the subcortical gyral white matter, particularly in the gyri of the frontal convexity.

Menkes Disease

Menkes disease is an X-linked recessive disorder due to mutation of a gene encoding a cation transporting P-type ATPase with more than 75 per cent homology with the *ATP7B* gene.[69,563] The clinical picture is congenital and the most prominent features are abnormal hair, failure to thrive and cerebellar degeneration.[369] Neurological symptoms include mental retardation, seizures, hypothermia and hypotonia. Autopsy studies have shown diffuse cerebral and cerebellar atrophy with multifocal neuronal degeneration. A prominent feature is that of abnormal Purkinje cell dendritic arborisation with axonal swellings.

The differences in the clinical features of Wilson's disease and Menkes disease probably arise from the tissue specific functions of the gene products. The Menkes disease gene product is predominantly expressed in the placenta, GI tract and blood-brain barrier. Failure of copper transport across the placenta results in a deficiency of the activity of essential cuproenzymes, including lysyl oxidase, which results in the peripheral manifestations. It is speculated that the CNS features relate to deficiency of an as yet uncharacterized brain-specific cuproenzyme(s).

Aceruloplasminaemia

This autosomal recessive disorder is associated with mutations of the ceruloplasmin gene.[366] This abundant serum protein carries 95 per cent of plasma copper and functions as a ferroxidase to convert ferrous to ferric iron. Ceruloplasmin synthesis requires direct incorporation of 6 copper atoms in each molecule so that a lack of bioavailable copper (as in Wilson's disease) results in an unstable, copper-deficient form of ceruloplasmin that is rapidly degraded in the plasma.[482]

Neurological dysfunction in aceruloplasminaemia includes craniofacial dystonia (28 per cent of cases), cognitive impairment (42 per cent), cerebellar ataxia (46 per cent) and retinal degeneration (75 per cent), together with diabetes.[366] The clinical picture is associated with parenchymal accumulation of iron. This disorder, therefore, relates both to the pathophysiology of Wilson's disease and also hereditary ferritinopathy. T2-weighted MRI imaging shows increased iron content in the basal ganglia, and the limited pathology literature shows excess iron and selective neuronal loss.[378] Ceruloplasmin in plasma does not cross the BBB but there is astrocyte-specific gene expression in the CNS around blood vessels.[289] This astrocyte-derived ceruloplasmin oxidizes ferrous iron prior to incorporation into oligodendrocyte-derived transferrin.[346] The downstream pathogenesis of cerebral degeneration is not characterized but includes candidate mechanisms mediated by the toxic effects of excess iron such as loss of glial derived growth factors, oxidative injury or relative neuronal iron deficiency.

Abnormalities of Central Autonomic Systems

The many causes of autonomic failure may be divided into primary and secondary types (Table 12.21).[390]

Central causes of autonomic failure are clinically divided into those associated with pure progressive autonomic failure and others where autonomic failure is associated with parkinsonism (Shy–Drager syndrome).[36]

In the clinical investigation of autonomic dysfunction, tests of autonomic physiology are supplemented by neuroendocrine stimulation tests and measurements of catecholamines, allowing distinction between central and peripheral causes of autonomic dysfunction.[168]

Central Autonomic Failure

Patients who have autonomic failure with parkinsonism usually have a synucleinopathy or glial cytoplasmic

TABLE 12.21 Classification of autonomic failure

Primary autonomic failure
   Progressive autonomic failure with Lewy bodies
   Progressive autonomic failure with multiple system atrophy
   Progressive autonomic failure due to postganglionic pathology
   Dopamine α-hydroxylase deficiency

Secondary autonomic failure
   Structural lesions of corticolimbic or hypothalamic regions
   Structural lesions of spinal cord
   Structural lesions of brain stem
   Wernicke's encephalopathy
   Cerebrovascular disease
   Baroreceptor failure
   Botulism
   Acute autonomic neuropathy
   Peripheral neuropathy
      Diabetic
      Amyloid
      Inflammatory
      Alcoholic
      Toxic and drug-related
      Chronic renal failure
      Paraneoplastic
      Connective tissue disease
      Acute intermittent porphyria
      Familial neuropathy

inclusions characteristic of MSA. In fully developed MSA, autonomic failure may be accompanied by signs of SND and/or OPCA (see under Multiple System Atrophy). Autonomic failure caused by Lewy body pathology is usually associated with features of parkinsonism and may also be associated with cognitive dysfunction caused by cortical Lewy body disease. In both cases, autonomic dysfunction is related to loss of cells from the intermediolateral column of the spinal cord. In Lewy body-associated disease there is additional degeneration in peripheral sympathetic ganglia including cardiac innervation. In MSA, supraspinal autonomic nuclei become affected in addition to spinal pathology (see earlier, Multiple System Atrophy).

Primary progressive autonomic failure has several pathological causes:[67,355]

- Lewy body pathology in the absence of nigral involvement;
- MSA pathology in the absence of clinical involvement of other systems;
- a peripheral cause for autonomic failure of uncertain cause based on biochemical evidence of low plasma noradrenaline levels.

Bradbury–Eggleston syndrome implies 'pure' autonomic failure without an ascertained cause in life. It is, thus, a diagnosis of exclusion. Some patients with pure primary autonomic failure have reduced sympathetic efferent nerve impulses and reduced plasma noradrenaline levels in keeping with a loss of postganglionic sympathetic efferent neurons.[108] In non-diabetic patients with symptomatic autonomic failure both primary and secondary causes are frequently encountered. In a study of 100 patients with orthostatic hypotension, 38 per cent did not have autonomic failure, 35 per cent had secondary autonomic failure and 27 per cent had primary autonomic failure.[472]

Autopsy

Clinical information usually offers important clues as to the aetiology of autonomic failure, particularly peripheral causes associated with diabetes mellitus, alcoholism, renal failure, connective tissue disease and familial disorders. Even in the presence of a known clinical association, thorough investigation of central and peripheral autonomic pathways is always indicated.

Initial histological investigation of a central cause for autonomic failure should concentrate on synucleinopathy and the inclusions of MSA in spinal cord and brain. Assessment of a peripheral cause should initially concentrate on excluding the presence of a peripheral neuropathy, particularly diabetic neuropathy, inflammatory disease and amyloid.

ACKNOWLEDGEMENT

We thank authors of previous editions whose contributions to this chapter are gratefully acknowledged.

REFERENCES

1. Aggarwal A, Schneider S, Houlden H, *et al.* Indian-subcontinent NBIA: unusual phenotypes, novel *PANK2* mutations and undetermined genetic forms. *Mov Disord* 2010;**25**:1424–1431.
2. Aharon–Peretz J, Rosenbaum H, Gershoni–Baruch R. Mutations in the glucocerebrosidase gene and Parkinson's disease in Ashkenazi Jews. *N Engl J Med* 2004;**351**:1972–1977.
3. Ahmed Z, Josephs KA, Gonzalez J, *et al.* Clinical and neuropathologic features of progressive supranuclear palsy with severe pallido–nigro–luysial degeneration and axonal dystrophy. *Brain* 2008;**131**:460–472.
4. Ahmed Z, Tabrizi SJ, Li A, *et al.* A Huntington's disease phenocopy characterized by pallido–nigro–luysian degeneration with brain iron accumulation and p62-positive glial inclusions. *Neuropathol Appl Neurobiol* 2010;**36**:551–557.
5. Ahmed Z, Doherty KM, Silveira–Moriyama L, *et al.* Globular glial tauopathies (GGT)

presenting with motor neuron disease or frontotemporal dementia: an emerging group of 4-repeat tauopathies. *Acta Neuropathol* 2011;**122**:415–428.
6. Ahmed Z, Asi YT, Sailer A, *et al.* The neuropathology, pathophysiology and genetics of multiple system atrophy. *Neuropathol Appl Neurobiol* 2012;**38**:4–24.
7. Ahmed Z, Bigio EH, Budka H, *et al.* Globular glial tauopathies (GGT): consensus recommendations. *Acta Neuropathol* 2013;**126**:537–544.
8. Akil M, Schwartz J, Dutchak D, *et al.* The psychiatric presentations of Wilson's disease. *J Neuropsychiatry Clin Neurosci* 1991;**3**:377– 382.
9. Al–Chalabi A, Durr A, Wood NW, *et al.* Genetic variants of the alpha-synuclein gene SNCA are associated with multiple system atrophy. *PLoS One* 2009;**4**:e7114.
10. Al Sweidi S, Sanchez MG, Bourque M, *et al.* Oestrogen receptors and signalling

pathways: implications for neuroprotective effects of sex steroids in Parkinson's disease. *J Neuroendocrinol* 2012;**24**:48–61.
10a. Alafuzoff I, Ince PG, Arzberger T, *et al.* Staging/typing of Lewy body related alpha-synuclein pathology: a study of the BrainNet Europe Consortium. *Acta Neuropathol* 2009;**117**:635–652.
11. Alafuzoff I, Parkkinen L, Al–Sarraj S, *et al.* Assessment of alpha-synuclein pathology: a study of the BrainNet Europe Consortium. *J Neuropathol Exp Neurol* 2008;**67**:125–143.
11a. Albanese A, Asmus F, Bhatia KP, *et al.* EFNS guidelines on diagnosis and treatment of primary dystonias. *Eur J Neurol* 2011;**18**:5–18.
12. Albin RL, Reiner A, Anderson KD, *et al.* Striatal and nigral neuron subpopulations in rigid Huntington's disease: implications for the functional anatomy of chorea and rigidity-akinesia. *Ann Neurol* 1990;**27**:357–365.

13. Albin RL, Young AB, Penney JB, *et al.* Abnormalities of striatal projection neurons and N-methyl-D-aspartate receptors in presymptomatic Huntington's disease. *N Engl J Med* 1990;**322**:1293–1298.

14. Anderson JP, Walker DE, Goldstein JM, *et al.* Phosphorylation of Ser-129 is the dominant pathological modification of alpha-synuclein in familial and sporadic Lewy body disease. *J Biol Chem* 2006;**281**:29739–29752.

15. Anderson LL, Vilensky JA, Duvoisin RC. Review: neuropathology of acute phase encephalitis lethargica: a review of cases from the epidemic period. *Neuropathol Appl Neurobiol* 2009;**35**:462–472.

16. Angot E, Steiner JA, Hansen C, *et al.* Are synucleinopathies prion-like disorders? *Lancet Neurol* 2010;**9**:1128–1138.

17. Apaydin H, Ahlskog JE, Parisi JE, *et al.* Parkinson disease neuropathology: later-developing dementia and loss of the levodopa response. *Arch Neurol* 2002;**59**: 102–112.

18. Appel–Cresswell S, de la Fuente–Fernandez R, Galley S, McKeown MJ. Imaging of compensatory mechanisms in Parkinson's disease. *Curr Opin Neurol* 2010;**23**:407–412.

19. Arai N, Oda M. A variety of glial pathological structures by the modified Gallyas–Braak method. *Neuropathology* 1996;**16**:133–138.

20. Arai T, Ikeda K, Akiyama H, *et al.* Distinct isoforms of tau aggregated in neurons and glial cells in brains of patients with Pick's disease, corticobasal degeneration and progressive supranuclear palsy. *Acta Neuropathol* 2001;**101**:167–173.

21. Arai T, Ikeda K, Akiyama H, *et al.* Identification of amino-terminally cleaved tau fragments that distinguish progressive supranuclear palsy from corticobasal degeneration. *Ann Neurol* 2004;**55**:72–79.

22. Arima K, Murayama S, Mukoyama M, Inose T. Immunocytochemical and ultrastructural studies of neuronal and oligodendroglial cytoplasmic inclusions in multiple system atrophy. 1. Neuronal cytoplasmic inclusions. *Acta Neuropathol* 1992;**83**:453–460.

23. Arvanitakis Z, Witte RJ, Dickson DW, *et al.* Clinical-pathologic study of biomarkers in FTDP-17 (PPND family with N279K tau mutation). *Parkinsonism Relat Disord* 2007;**13**:230–239.

24. Asi YT, Ling H, Ahmed Z, *et al.* Neuropathological features of multiple system atrophy with cognitive impairment. *Mov Disord* 2004;**29**:884–888.

25. Asi YT, Simpson JE, Heath PR, *et al.* Alpha-synuclein mRNA expression in oligodendrocytes in MSA. *Glia* 2014;**62**:964–970.

26. Auluck PK, Caraveo G, Lindquist S. Alpha-synuclein: membrane interactions and toxicity in Parkinson's disease. *Annu Rev Cell Dev Biol* 2010;**26**:211–233.

27. Baker M, Litvan I, Houlden H, *et al.* Association of an extended haplotype in the tau gene with progressive supranuclear palsy. *Hum Mol Genet* 1999;**8**:711–715.

28. Bakheit A, Behan W. Pathology of neuroleptic malignant syndrome. *J Neurol Neurosurg Psychiatry* 1990;**53**:271.

29. Ballan G, Tison F. A historical case of probable corticobasal degeneration? *Mov Disord* 1997;**12**:1073–1074.

30. Banack SA, Cox PA. Biomagnification of cycad neurotoxins in flying foxes: implications for ALS–PDC in Guam 1. *Neurology* 2003;**61**:387–389.

31. Bandopadhyay R, Kingsbury AE, Cookson MR, *et al.* The expression of DJ-1 (PARK7) in normal human CNS and idiopathic Parkinson's disease. *Brain* 2004;**127**:420–430.

32. Bannister R, Oppenheimer DR. Degenerative diseases of the nervous system associated with autonomic failure. *Brain* 1972;**95**:457–474.

33. Barker RA, Revesz T, Thom M, *et al.* Review of 23 patients affected by the stiff man syndrome: clinical subdivision into stiff trunk (man) syndrome, stiff limb syndrome and progressive encephalomyelitis with rigidity. *J Neurol Neurosurg Psychiatry* 1998;**65**:633–640.

34. Beach TG, White CL, Hamilton RL, *et al.* Evaluation of alpha-synuclein immunohistochemical methods used by invited experts. *Acta Neuropathol* 2008;**116**:277–288.

35. Beach TG, Adler CH, Sue LI, *et al.* Multi-organ distribution of phosphorylated alpha-synuclein histopathology in subjects with Lewy body disorders. *Acta Neuropathol* 2010;**119**:689–702.

36. Benarroch EE. The central autonomic network: functional organization, dysfunction and perspective. *Mayo Clin Proc* 1993;**68**:988–1001.

37. Betarbet R, Sherer TB, MacKenzie G, *et al.* Chronic systemic pesticide exposure reproduces features of Parkinson's disease. *Nat Neurosci* 2000;**3**:1301–1306.

38. Blunt SB, Lane RJ, Turjanski N, Perkin GD. Clinical features and management of two cases of encephalitis lethargica. *Mov Disord* 1997;**12**:354–359.

39. Boeve BF. Links between frontotemporal lobar degeneration, corticobasal degeneration, progressive supranuclear palsy and amyotrophic lateral sclerosis. *Alzheimer Dis Assoc Disord* 2007;**21**:S31–38.

40. Boeve BF. The multiple phenotypes of corticobasal syndrome and corticobasal degeneration: implications for further study. *J Mol Neurosci* 2011;**45**:350–353.

41. Boeve BF, Maraganore DM, Parisi JE, *et al.* Pathologic heterogeneity in clinically diagnosed corticobasal degeneration. *Neurology* 1999;**53**:795–800.

42. Bondon–Guitton E, Perez–Lloret S, Bagheri H, *et al.* Drug-induced parkinsonism: a review of 17 years' experience in a regional pharmacovigilance center in France. *Mov Disord* 2011;**26**:2226–2231.

43. Bonifati V, Rizzu P, van Baren MJ, *et al.* Mutations in the *DJ-1* gene associated with autosomal recessive early-onset parkinsonism. *Science* 2003;**299**:256–259.

44. Bove J, Prou D, Perier C, Przedborski S. Toxin-induced models of Parkinson's disease. *NeuroRx* 2005;**2**:484–494.

45. Bower JH, Maraganore DM, McDonnell K, Rocca WA. Incidence of progressive supranuclear palsy and multiple system atrophy in Olmsted County, Minnesota, 1976 to 1990. *Neurology* 1997;**49**: 1284–1288.

46. Braak H, Del Tredici K, Rub U, *et al.* Staging of brain pathology related to sporadic Parkinson's disease. *Neurobiol Aging* 2003;**24**:197–211.

47. Breedveld GJ, Percy AK, MacDonald ME, *et al.* Clinical and genetic heterogeneity in benign hereditary chorea. *Neurology* 2002;**59**:579–584.

48. Bronstein JM, Tagliati M, Alterman RL, *et al.* Deep brain stimulation for Parkinson disease: an expert consensus and review of key issues. *Arch Neurol* 2011;**68**:165.

49. Brotchie J, Fitzer–Attas C. Mechanisms compensating for dopamine loss in early Parkinson disease. *Neurology* 2009;**72**:S32–38.

50. Brown RG, Lacomblez L, Landwehrmeyer BG, *et al.* Cognitive impairment in patients with multiple system atrophy and progressive supranuclear palsy. *Brain* 2010;**133**:2382–2393.

51. Brundin P, Li JY, Holton JL, *et al.* Research in motion: the enigma of Parkinson's disease pathology spread. *Nat Rev Neurosci* 2008;**9**:741–745.

52. Buchman AS, Shulman JM, Nag S, *et al.* Nigral pathology and parkinsonian signs in elders without Parkinson disease. *Ann Neurol* 2012;**71**:258–266.

53. Buee–Scherrer V, Buee L, Hof PR, *et al.* Neurofibrillary degeneration in amyotrophic lateral sclerosis/parkinsonism-dementia complex of Guam. Immunochemical characterization of tau proteins. *Am J Pathol* 1995;**146**:924–932.

54. Buee–Scherrer V, Buee L, Leveugle B, *et al.* Pathological tau proteins in postencephalitic parkinsonism: comparison with Alzheimer's disease and other neurodegenerative disorders. *Ann Neurol* 1997;**42**:356–359.

55. Bull P, Thomas G, Rommens J, *et al.* The Wilson disease gene is a putative copper transporting ATPase similar to the Menkes gene. *Nat Genet* 1993;**5**:327–337.

56. Burke RE, Dauer WT, Vonsattel JP. A critical evaluation of the Braak staging scheme for Parkinson's disease. *Ann Neurol* 2008;**64**:485–491.

57. Burn DJ, Lees AJ. Progressive supranuclear palsy: where are we now? *Lancet Neurol* 2002;**1**:359–369.

58. Butler MH, Hayashi A, Ohkoshi N, *et al.* Autoimmunity to gephyrin in stiff-man syndrome. *Neuron* 2000;**26**:307–312.

59. Camargos S, Scholz S, Simon–Sanchez J, *et al.* DYT16, a novel young-onset dystonia-parkinsonism disorder: identification of a segregating mutation in the stress-response protein PRKRA. *Lancet Neurol* 2008;**7**:207–215.

60. Campbell BC, McLean CA, Culvenor JG, *et al.* The solubility of alpha-synuclein in multiple system atrophy differs from that of dementia with Lewy bodies and Parkinson's disease. *J Neurochem* 2001;**76**:87–96.

61. Caparros–Lefebvre D, Sergeant N, Lees A, *et al.* Guadeloupean parkinsonism: a cluster of progressive supranuclear palsy-like tauopathy. *Brain* 2002;**125**:801–811.

61a. Casanova MF, Araque JM. Mineralization of the basal ganglia: implications for neuropsychiatry, pathology and neuroimaging. *Psychiatry Res* 2003;**121**:59–87.

62. Cavalier S, Gambetti P. Dystrophic axons and spinal cord demyelination in cystic fibrosis. *Neurology* 1981;**31**:714–718.

63. Caviness JN, Truong DD. Myoclonus. *Handb Clin Neurol* 2011;**100**:399–420.

64. Charcot JM. On Parkinson's disease. In: *Lectures on diseases of the nervous system delivered at the Salpêtrière*. London: New Sydenham Society, 1877:129–156.

65. Chartier–Harlin MC, Kachergus J, Roumier C, *et al.* Alpha-synuclein locus duplication as a cause of familial Parkinson's disease. *Lancet* 2004;**364**:1167–1169.

66. Chartier–Harlin MC, Dachsel JC, Vilarino–Guell C, et al. Translation initiator *EIF4G1* mutations in familial Parkinson disease. *Am J Hum Genet* 2011;**89**:398–406.

67. Chaudhuri KR. Autonomic dysfunction in movement disorders. *Curr Opin Neurol* 2001;**14**:505–511.

68. Chaudhuri KR. Drug-induced parkinsonism. In: Sethi KD ed. *Drug-induced movement disorders*. New York: Marcel Dekker, 2004:61–75.

69. Chelly J, Turner Z, Tonnesen T, et al. Isolation of a candidate gene for Menkes disease that encodes a potential heavy metal binding protein. *Nat Genet* 1993;**3**:14–19.

70. Chinnery P, Crompton D, Birchall D, et al. Clinical features and natural history of neuroferritinopathy caused by the *FTL1 460InsA* mutation. *Brain* 2007;**130**:110–119.

71. Chouery E, Delague V, Bergougnoux A, et al. Mutations in *TREM2* lead to pure early-onset dementia without bone cysts. *Hum Mutation* 2008;**29**:E194–204.

72. Chrysostome V, Tison F, Yekhlef F, et al. Epidemiology of multiple system atrophy: a prevalence and pilot risk factor study in Aquitaine, France. *Neuroepidemiology* 2004;**23**:201–208.

73. Church AJ, Cardoso F, Dale RC, et al. Anti-basal ganglia antibodies in acute and persistent Sydenham's chorea. *Neurology* 2002;**59**:227–231.

74. Cipriani S, Chen X, Schwarzschild MA. Urate: a novel biomarker of Parkinson's disease risk, diagnosis and prognosis. *Biomark Med* 2010;**4**:701–712.

75. Clark IE, Dodson MW, Jiang C, et al. Drosophila PINK1 is required for mitochondrial function and interacts genetically with parkin. *Nature* 2006;**441**:1162–1166.

76. Compta Y, Parkkinen L, O'Sullivan SS, et al. Lewy- and Alzheimer-type pathologies in Parkinson's disease dementia: which is more important? *Brain* 2011;**134**:1493–1505.

77. Conrad C, Andreadis A, Trojanowski JQ, et al. Genetic evidence for the involvement of tau in progressive supranuclear palsy. *Ann Neurol* 1997;**41**:277–281.

78. Conway KA, Lee SJ, Rochet JC, et al. Acceleration of oligomerization, not fibrillization, is a shared property of both alpha-synuclein mutations linked to early-onset Parkinson's disease: implications for pathogenesis and therapy. *Proc Natl Acad Sci U S A* 2000;**97**:571–576.

79. Costa J, Lunet N, Santos C, et al. Caffeine exposure and the risk of Parkinson's disease: a systematic review and meta-analysis of observational studies. *J Alzheimers Dis* 2010;**20** Suppl 1:S221–238.

80. Craufurd D, Dodge A. Mutation size and age at onset in Huntington's disease. *J Med Genet* 1993;**30**(12):1008–11.

81. Crompton DE, Chinnery PF, Fey C, et al. Neuroferritinopathy: a window on the role of iron in neurodegeneration. *Blood Cells Mol Dis* 2002;**29**:522–531.

82. Curtis AR, Fey C, Morris CM, et al. Mutation in the gene encoding ferritin light polypeptide causes dominant adult-onset basal ganglia disease. *Nat Genet* 2001;**28**:350–354.

83. Dale RC, Church AJ, Surtees RA, et al. Encephalitis lethargica syndrome: 20 new cases and evidence of basal ganglia autoimmunity. *Brain* 2004;**127**:21–33.

84. Daras M, Koppel BS, Atos–Radzion E. Cocaine-induced choreoathetoid movements ('crack dancing'). *Neurology* 1994;**44**:751–752.

85. de Bie P, Muller P, Wijmenga C, Klomp L. Molecular pathogenesis of Wilson and Menkes disease: correlation of mutations with molecular defects and disease phenotypes. *J Med Genet* 2007;**44**:673–688.

86. de Carvalho Aguiar PM, Ozelius LJ. Classification and genetics of dystonia. *Lancet Neurol* 2002;**1**:316–325.

87. De la Monte SM, Vonsattel JP, Richardson EP. Morphometric demonstration of atrophic changes in the cerebral cortex, white matter and neostriatum in Huntington's disease. *J Neuropathol Exp Neurol* 1988;**47**:516–525.

88. de Lau LM, Giesbergen PC, de Rijk MC, et al. Incidence of parkinsonism and Parkinson disease in a general population: the Rotterdam Study. *Neurology* 2004;**63**:1240–1244.

89. de Silva R, Lashley T, Gibb G, et al. Pathological inclusion bodies in tauopathies contain distinct complements of tau with three or four microtubule-binding repeat domains as demonstrated by new specific monoclonal antibodies. *Neuropathol Appl Neurobiol* 2003;**29**: 288–302.

90. Del Tredici K, Rub U, de Vos RA, et al. Where does parkinson disease pathology begin in the brain? *J Neuropathol Exp Neurol* 2002;**61**:413–426.

91. Deng X, Vidal R, Englander E. Accumulation of oxidative DNA damage in brain mitochondria in mouse model of hereditary ferritinopathy. *Neurosci Lett* 2010;**479**:44–48.

92. Depienne C, Brice A. Unlocking the genetics of paroxysmal kinesigenic dyskinesia. *Brain* 2011;**134**:3431–3434.

93. Deramecourt V, Maurage CA, Sergeant N, et al. An 88-year-old woman with long-lasting parkinsonism. *Brain Pathol* 2011;**21**:465–468.

94. Devine MJ, Plun–Favreau H, Wood NW. Parkinson's disease and cancer: two wars, one front. *Nat Rev Cancer* 2011;**11**:812–823.

95. Di Fonzo A, Dekker MC, Montagna P, et al. FBXO7 mutations cause autosomal recessive, early-onset parkinsonian-pyramidal syndrome. *Neurology* 2009;**72**:240–245.

96. Dick FD, De Palma G, Ahmadi A, et al. Environmental risk factors for Parkinson's disease and parkinsonism: the Geoparkinson study. *Occup Environ Med* 2007;**64**:666–672.

97. Dick K, Eckhardt M, Paisan–Ruiz C, et al. Mutation of *FA2H* underlies a complicated form of hereditary spastic paraplegia (SPG35). *Hum Mutat* 2010;**31**:E1250–1261.

98. Dickson DW, Yen SH, Suzuki KI, et al. Ballooned neurons in select neurodegenerative diseases contain phosphorylated neurofilament epitopes. *Acta Neuropathol* 1986;**71**:216–223.

99. Dickson DW, Liu W, Hardy J, et al. Widespread alterations of alpha-synuclein in multiple system atrophy. *Am J Pathol* 1999;**155**:1241–1251.

100. Dickson DW, Bergeron C, Chin SS, et al. Office of Rare Diseases neuropathologic criteria for corticobasal degeneration. *J Neuropathol Exp Neurol* 2002;**61**: 935–946.

101. Dickson DW, Rademakers R, Hutton ML. Progressive supranuclear palsy: pathology and genetics. *Brain Pathol* 2007;**17**:74–82.

102. Dickson DW, Braak H, Duda JE, et al. Neuropathological assessment of Parkinson's disease: refining the diagnostic criteria. *Lancet Neurol* 2009;**8**:1150–1157.

103. Dickson DW, Ahmed Z, Algom AA, et al. Neuropathology of variants of progressive supranuclear palsy. *Curr Opin Neurol* 2010;**23**:394–400.

104. Dickson DW, Hauw JJ, Agid Y, Litvan I. Progressive supranuclear palsy and corticobasal degeneration. In: Dickson DW, Weller RO eds. *Neurodegeneration: the molecular pathology of dementia and movement disorders*. Oxford: Wiley–Blackwell, 2011:135–155.

105. Doder M, Jahanshahi M, Turjanski N, et al. Parkinson's syndrome after closed head injury: a single case report. *J Neurol Neurosurg Psychiatry* 1999;**66**:380–385.

106. Doherty KM, Silveira–Moriyama L, Parkkinen L, et al. Parkin disease: a clinicopathologic entity? *JAMA Neurol* 2013;**70**:571–579.

107. Donker Kaat L, Boon AJ, Kamphorst W, et al. Frontal presentation in progressive supranuclear palsy. *Neurology* 2007;**69**:723–729.

108. Dotson R, Ochoa J, Marchettini P, et al. Sympathetic neural outflow directly recorded in patients with primary autonomic failure: clinical observations, microneurography and histopathology. *Neurology* 1990;**40**:1079–1085.

109. Double KL, Reyes S, Werry EL, Halliday GM. Selective cell death in neurodegeneration: why are some neurons spared in vulnerable regions? *Prog Neurobiol* 2010;**92**:316–329.

110. Dragatsis I, Levine MS, Zeitlin S. Inactivation of Hdh in the brain and testis results in progressive neurodegeneration and sterility in mice. *Nat Genet* 2000;**26**:300–306.

111. Duda JE, Giasson BI, Chen Q, et al. Widespread nitration of pathological inclusions in neurodegenerative synucleinopathies. *Am J Pathol* 2000;**157**:1439–1445.

112. Duda JE, Giasson BI, Gur TL, et al. Immunohistochemical and biochemical studies demonstrate a distinct profile of alpha-synuclein permutations in multiple system atrophy. *J Neuropathol Exp Neurol* 2000;**59**:830–841.

113. Duda JE, Giasson BI, Mabon ME, et al. Concurrence of alpha-synuclein and tau brain pathology in the Contursi kindred. *Acta Neuropathol* 2002;**104**:7–11.

114. Dugger BN, Murray ME, Boeve BF, et al. Neuropathological analysis of brainstem cholinergic and catecholaminergic nuclei in relation to rapid eye movement (REM) sleep behaviour disorder. *Neuropathol Appl Neurobiol* 2012;**38**:142–152.

115. Ebadi M, Pfeiffer RF, Murrin LC. Pathogenesis and treatment of neuroleptic malignant syndrome. *Gen Pharmacol* 1990;**21**:367–386.

116. Edvardson S, Hama S, Shaag A, et al. Mutations in the fatty acid 2-hydroxylase gene are associated with leukodystrophy with spastic paraparesis and dystonia. *Am J Hum Genet* 2008;**83**:643–648.

117. Egan R, Weleber R, Hogarth P, et al. Neuro-ophthalmologic and electroretinographic findings in pantothenate kinase-associated neurodegeneration (formerly Hallervorden–Spatz syndrome). *Am J Ophthalmol* 2005;**140**:267–274.

118. Elleder M, Jirasek A, Smid F, et al. Niemann Pick disease type C. Study on the nature of the cerebral storage process. *Acta Neuropathol* 1985;66:325–336.

119. Emery ES, Vieco PT. Sydenham Chorea: magnetic resonance imaging reveals permanent basal ganglia injury. *Neurology* 1997;48:531–533.

120. Engel L, Jing Z, O'Brien D, et al. Catalytic function of *PLA2G6* is impaired by mutations associated with infantile neuroaxonal dystrophy but not dystonia–parkinsonism. *PLoS ONE* 2011;5:e12897.

121. Fahn S. Classification of movement disorders. *Mov Disord* 2011;26:947–957.

122. Fang F, Wirdefeldt K, Jacks A, et al. CNS infections, sepsis and risk of Parkinson's disease. *Int J Epidemiol* 2012;41(4):1042–9.

123. Farrer M, Chan P, Chen R, et al. Lewy bodies and parkinsonism in families with parkin mutations. *Ann Neurol* 2001;50:293–300.

124. Farrer M, Skipper L, Berg M, et al. The tau H1 haplotype is associated with Parkinson's disease in the Norwegian population. *Neurosci Lett* 2002;322:83–86.

125. Farrer MJ, Hulihan MM, Kachergus JM, et al. *DCTN1* mutations in Perry syndrome. *Nat Genet* 2009;41:163–165.

126. Feany MB, Dickson DW. Widespread cytoskeletal pathology characterizes corticobasal degeneration. *Am J Pathol* 1995;146:1388–1396.

127. Feany MB, Mattiace LA, Dickson DW. Neuropathologic overlap of progressive supranuclear palsy, Pick's disease and corticobasal degeneration. *J Neuropathol Exp Neurol* 1996;55:53–67.

128. Fearnley JM, Lees AJ. Ageing and Parkinson's disease: substantia nigra regional selectivity. *Brain* 1991;114:2283–2301.

129. Fenelon G, Gray F, Paillard F, et al. A prospective study of patients with CT detected pallidal calcifications. *J Neurol Neurosurg Psychiatry* 1993;56:622–625.

130. Fernandez–Botran R, Ahmed Z, Crespo FA, et al. Cytokine expression and microglial activation in progressive supranuclear palsy. *Parkinsonism Relat Disord* 2011;17:683–688.

131. Finsterer J, Kopsa W. Basal ganglia calcification in mitochondrial disorders. *Metab Brain Dis* 2005;20:219–226.

132. FitzGerald PM, Jankovic J. Lower body parkinsonism: evidence for vascular etiology. *Mov Disord* 1989;4:249–260.

133. Forsaa EB, Larsen JP, Wentzel–Larsen T, Alves G. What predicts mortality in Parkinson disease?: a prospective population-based long-term study. *Neurology* 2010;75:1270–1276.

134. Frigerio R, Fujishiro H, Ahn TB, et al. Incidental Lewy body disease: do some cases represent a preclinical stage of dementia with Lewy bodies? *Neurobiol Aging* 2011;32:857–863.

135. Fuchs T, Ozelius LJ. Genetics of dystonia. *Semin Neurol* 2011;31:441–448.

136. Fuchs T, Gavarini S, Saunders–Pullman R, et al. Mutations in the *THAP1* gene are responsible for DYT6 primary torsion dystonia. *Nat Genet* 2009;41: 286–288.

137. Fujita D, Terada S, Ishizu H, et al. Immunohistochemical examination on intracranial calcification in neurodegenerative diseases. *Acta Neuropathol* 2003; 105:259–264.

138. Fujita T, Doi M, Ogata T, et al. Cerebral cortical pathology of sporadic olivopontocerebellar atrophy. *J Neurol Sci* 1993;116:41–46.

139. Fujiwara H, Hasegawa M, Dohmae N, et al. Alpha-synuclein is phosphorylated in synucleinopathy lesions. *Nat Cell Biol* 2002;4:160–164.

140. Funata N, Maeda Y, Koike M, et al. Neuronal intranuclear hyaline inclusion disease: report of a case and review of the literature. *Clin Neuropathol* 1990;9:89–96.

141. Funayama M, Hasegawa K, Kowa H, et al. A new locus for Parkinson's disease (PARK8) maps to chromosome 12p11.2–q13.1. *Ann Neurol* 2002;51:296–301.

142. Gai WP, Power JH, Blumbergs PC, Blessing WW. Multiple-system atrophy: a new alpha-synuclein disease? *Lancet* 1998;352:547–548.

143. Gaig C, Marti MJ, Ezquerra M, et al. G2019S *LRRK2* mutation causing Parkinson's disease without Lewy bodies? *J Neurol Neurosurg Psychiatry* 2007;78:626–628.

144. Gajdusek DC. Motor-neuron disease in natives of New Guinea. *N Engl J Med* 1963;268:474–476.

145. Gallagher DA, Parkkinen L, O'Sullivan SS, et al. Testing an aetiological model of visual hallucinations in Parkinson's disease. *Brain* 2011;134:3299–3309.

146. Gamboa ET, Wolf A, Yahr MD, et al. Influenza virus antigen in postencephalitic parkinsonism brain. Detection by immunofluorescence. *Arch Neurol* 1974;31:228–232.

147. Gandhi S, Muqit MM, Stanyer L, et al. PINK1 protein in normal human brain and Parkinson's disease. *Brain* 2006;129:1720–1731.

148. Ganz J, Lev N, Melamed E, Offen D. Cell replacement therapy for Parkinson's disease: how close are we to the clinic? *Expert Rev Neurother* 2011;11:1325–1339.

149. Garcia–Arocena D, Hagerman PJ. Advances in understanding the molecular basis of FXTAS. *Hum Mol Genet* 2010;19:R83–89.

150. Gaspar P, Gray F. Dementia in idiopathic parkinsons-disease – a neuropathological study of 32 cases. *Acta Neuropathol* 1984;64:43–52.

151. Gavarini S, Cayrol C, Fuchs T, et al. Direct interaction between causative genes of DYT1 and DYT6 primary dystonia. *Ann Neurol* 2010;68:549–553.

152. Geddes JF, Hughes AJ, Lees AJ, Daniel SE. Pathological overlap in cases of parkinsonism associated with neurofibrillary tangles. A study of recent cases of postencephalitic parkinsonism and comparison with progressive supranuclear palsy and Guamanian parkinsonism-dementia complex. *Brain* 1993;116:281–302.

153. Gelb DJ, Oliver E, Gilman S. Diagnostic criteria for Parkinson disease. *Arch Neurol* 1999;56:33–39.

154. George JM. The synucleins. *Genome Biol* 2002;3:REVIEWS3002.

155. Geschwind D, Loginov M, Stern J. Identification of a locus on chromosome 14q for idiopathic basal ganglia calcification (Fahr disease). *Am J Hum Genet* 1999;65: 764–772.

156. Geser F, Wenning GK, Seppi K, et al. Progression of multiple system atrophy (MSA): a prospective natural history study by the European MSA Study Group (EMSA SG). *Mov Disord* 2006;21:179–186.

157. Geser F, Malunda JA, Hurtig HI, et al. Subcortical TDP-43 pathology occurs infrequently in multiple system atrophy. *Neuropathol Appl Neurobiol* 2010;37:358–365.

158. Giasson BI, Duda JE, Murray IV, et al. Oxidative damage linked to neurodegeneration by selective alpha-synuclein nitration in synucleinopathy lesions. *Science* 2000;290:985–989.

159. Gibb WR, Luthert PJ, Marsden CD. Corticobasal degeneration. *Brain* 1989;112:1171–1192.

160. Gilks WP, Abou–Sleiman PM, Gandhi S, et al. A common *LRRK2* mutation in idiopathic Parkinson's disease. *Lancet* 2005;365:415–416.

161. Gilman S, Wenning GK, Low PA, et al. Second consensus statement on the diagnosis of multiple system atrophy. *Neurology* 2008;71:670–676.

162. Goedert M. Parkinson's disease and other alpha-synucleinopathies. *Clin Chem Lab Med* 2001;39:308–312.

163. Goedert M, Spillantini MG. Lewy body diseases and multiple system atrophy as alpha-synucleinopathies. *Mol Psychiatry* 1998;3:462–465.

164. Goedert M, Spillantini MG, Serpell LC, et al. From genetics to pathology: tau and alpha-synuclein assemblies in neurodegenerative diseases. *Philos Trans R Soc Lond B Biol Sci* 2001;356:213–227.

165. Goker–Alpan O, Lopez G, Vithayathil J, et al. The spectrum of parkinsonian manifestations associated with glucocerebrosidase mutations. *Arch Neurol* 2008;65:1353–1357.

166. Golden WC, Brambrink AM, Traystman RJ, et al. Failure to sustain recovery of Na,K-ATPase function is a possible mechanism for striatal neurodegeneration in hypoxic–ischemic newborn piglets. *Mol Brain Res* 2001;88:94–102.

167. Goldman SM, Kamel F, Ross GW, et al. Head injury, alpha-synuclein Rep1 and Parkinson's disease. *Ann Neurol* 2012;71:40–48.

168. Goldstein DS. Chemical mediators of autonomic nervous system activity. *Curr Opin Neurol Neurosurg* 1993;6:524–526.

169. Gomez–Tortosa E, MacDonald ME, Friend JC, et al. Quantitative neuropathological changes in presymptomatic Huntington's disease. *Ann Neurol* 2001;49:29–34.

170. Goto S, Lee LV, Munoz EL, et al. Functional anatomy of the basal ganglia in X-linked recessive dystonia-parkinsonism. *Ann Neurol* 2005;58:7–17.

171. Gourfinkel–An I, Cancel G, Trottier Y, et al. Differential distribution of the normal and mutated forms of huntingtin in the human brain. *Ann Neurol* 1997;42: 712–719.

172. Goutieres F, Mikol J, Aicardi J. Neuronal intranuclear inclusion disease in a child: diagnosis by rectal biopsy. *Ann Neurol* 1990;27:103–106.

173. Grandas F. Hemiballismus. *Handb Clin Neurol* 2011;100:249–260.

174. Greffard S, Verny M, Bonnet AM, et al. Motor score of the Unified Parkinson Disease Rating Scale as a good predictor of Lewy body-associated neuronal loss in the substantia nigra. *Arch Neurol* 2006;63:584–588.

175. Greffard S, Verny M, Bonnet AM, et al. A stable proportion of Lewy body bearing neurons in the substantia nigra suggests a model in which the Lewy body causes neuronal death. *Neurobiol Aging* 2010;31:99–103.

176. Gregory A, Westaway SK, Holm IE, *et al.* Neurodegeneration associated with genetic defects in phospholipase A(2). *Neurology* 2008;**71**:1402–1409.

177. Gregory A, Polster BJ, Hayflick SJ. Clinical and genetic delineation of neurodegeneration with brain iron accumulation. *J Med Genet* 2009;**46**:73–80.

178. Grimes DA, Han F, Lang AE, *et al.* A novel locus for inherited myoclonus-dystonia on 18p11. *Neurology* 2002;**59**:1183–1186.

179. Grotzsch H, Pizzolato GP, Ghika J, *et al.* Neuropathology of a case of dopa responsive dystonia associated with a new genetic locus, DYT14. *Neurology* 2002;**58**:1839–1842.

180. Group ESoDiEEC. A prevalence study of primary dystonia in eight European countries. *J Neurol* 2000;**247**:787–792.

181. Gu X, Greiner ER, Mishra R, *et al.* Serines 13 and 16 are critical determinants of full-length human mutant huntingtin induced disease pathogenesis in HD mice. *Neuron* 2009;**64**:828–840.

182. Guerreiro R, Bilgic B, Guven G, *et al.* A novel compound heterozygous mutation in TREM2 found in a Turkish frontotemporal dementia-like family. *Neurobiol Aging* 2013;**34**:2890 e2891–2895.

183. Guerreiro R, Wojtas A, Bras J, *et al.* TREM2 variants in Alzheimer's disease. *N Engl J Med* 2013;**368**:117–127.

184. Gutekunst CA, Li SH, Yi H, *et al.* Nuclear and neuropil aggregates in Huntington's disease: Relationship to neuropathology. *J Neurosci* 1999;**19**:2522–2534.

185. Gwinn–Hardy K, Mehta ND, Farrer M, *et al.* Distinctive neuropathology revealed by alpha-synuclein antibodies in hereditary parkinsonism and dementia linked to chromosome 4p. *Acta Neuropathol* 2000;**99**:663–672.

186. Hakola HP, Puranen M. Neuropsychiatric and brain CT findings in polycystic lipomembranous osteodysplasia with sclerosing leukoencephalopathy. *Acta Neurol Scand* 1993;**88**:370–375.

187. Hakola HPA, Jarvi OH, Sourander P. Osteodysplasia polycystica hereditaria combined with sclerosing leucoencephalopathy, a new entity of the dementia praesenilis group. *Acta Neurol Scand Suppl* 1970;**43**:79–80.

188. Halliday G, Hely M, Reid W, Morris J. The progression of pathology in longitudinally followed patients with Parkinson's disease. *Acta Neuropathol* 2008;**115**:409–415.

189. Halliday G, Reyes S, Double K. Substantia nigra, ventral tegmental area, and retrorubral fields. In: Mai JK, Paxinos G eds. *The Human Nervous System*. London: Elsevier, 2012:441–457.

190. Halliday GM, McCann H. The progression of pathology in Parkinson's disease. *Ann N Y Acad Sci* 2010;**1184**:188–195.

191. Halliday GM, Stevens CH. Glia: initiators and progressors of pathology in Parkinson's disease. *Mov Disord* 2011;**26**:6–17.

192. Halliday GM, Macdonald V, Henderson JM. A comparison of degeneration in motor thalamus and cortex between progressive supranuclear palsy and Parkinson's disease. *Brain* 2005;**128**:2272–2280.

193. Halliday GM, Holton JL, Revesz T, Dickson DW. Neuropathology underlying clinical variability in patients with synucleinopathies. *Acta Neuropathol* 2011;**122**:187–204.

194. Haltia M. Neuronal intranuclear inclusion disease. In: Dickson D eds. *Neurodegeneration: the molecular pathology of dementia and movement disorders*. Basel: ISN Neuropath Press, 2003:404–406.

195. Hamza TH, Payami H. The heritability of risk and age at onset of Parkinson's disease after accounting for known genetic risk factors. *J Hum Genet* 2010;**55**:241–243.

196. Hara M, Sasaki Y, Yagishita S. Ultrastructural figures of longitudinal sections of swollen axon, containing so-called spheroid body structure – a case of infantile neuroaxonal dystrophy. *J Clin Electron Microsc* 1989;**22**:395–399.

197. Haraguchi T, Ishizu H, Terada S, *et al.* An autopsy case of postencephalitic parkinsonism of von Economo type: some new observations concerning neurofibrillary tangles and astrocytic tangles. *Neuropathology* 2000;**20**:143–148.

198. Harding AJ, Halliday GM. Cortical Lewy body pathology in the diagnosis of dementia. *Acta Neuropathol* 2001;**102**:355–363.

199. Harding AJ, Broe GA, Halliday GM. Visual hallucinations in Lewy body disease relate to Lewy bodies in the temporal lobe. *Brain* 2002;**125**:391–403.

200. Harding AJ, Stimson E, Henderson JM, Halliday GM. Clinical correlates of selective pathology in the amygdala of patients with Parkinson's disease. *Brain* 2002;**125**:2431–2445.

201. Hardy J. Expression of normal sequence pathogenic proteins for neurodegenerative disease contributes to disease risk: 'permissive templating' as a general mechanism underlying neurodegeneration. *Biochem Soc Trans* 2005;**33**:578–581.

202. Hartig M, Hortnagel K, Garavaglia B, *et al.* Genotypic and phenotypic spectrum of PANK2 mutations in patients with neurodegeneration with brain iron accumulation. *Ann Neurol* 2006;**59**:248–256.

203. Hasegawa K, Stoessl AJ, Yokoyama T, *et al.* Familial parkinsonism: study of original Sagamihara PARK8 (I2020T) kindred with variable clinicopathologic outcomes. *Parkinsonism Relat Disord* 2009;**15**:300–306.

204. Hasegawa M, Arai T, Akiyama H, *et al.* TDP-43 is deposited in the Guam parkinsonism-dementia complex brains. *Brain* 2007;**130**:1386–1394.

205. Hassan A, Parisi JE, Josephs KA. Autopsy-proven progressive supranuclear palsy presenting as behavioral variant frontotemporal dementia. *Neurocase* 2011;**18**(6):478–88.

206. Hassan A, Whitwell JL, Josephs KA. The corticobasal syndrome-Alzheimer's disease conundrum. *Expert Rev Neurother* 2011;**11**:1569–1578.

207. Hassler R. Zur Pathologie der Paralysis agitans und des postencephalitschen Parkinsonismus. *J Psychol Neurol* 1938;**48**:387–476.

208. Hattori M, Hashizume Y, Yoshida M, *et al.* Distribution of astrocytic plaques in the corticobasal degeneration brain and comparison with tuft-shaped astrocytes in the progressive supranuclear palsy brain. *Acta Neuropathol* 2003;**106**:143–149.

209. Hauw JJ, Daniel SE, Dickson D, *et al.* Preliminary NINDS neuropathologic criteria for Steele–Richardson–Olszewski syndrome (progressive supranuclear palsy). *Neurology* 1994;**44**:2015–2019.

210. Hayflick S. Neurodegeneration with brain iron accumulation: from genes to pathogenesis. *Sem Pediatr Neurol* 2006;**13**:182–185.

211. Hayflick S, Westaway S, Levinson B, *et al.* Genetic clinical and and radiographic delineation of Hallervorden–Spatz syndrome. *N Engl J Med* 2003;**348**:33–40.

212. Henry R, Shieh M, Amirbekian B, *et al.* Connecting white matter injury and thalamic atrophy in clinically isolated syndromes. *J Neurol Sci* 2009;**282**:61–66.

213. Hjermind LE, Johannsen LG, Blau N, *et al.* Dopa-responsive dystonia and early-onset Parkinson's disease in a patient with GTP cyclohydrolase I deficiency? *Mov Disord* 2006;**21**(5):679–82.

214. Ho MF, Chalmers RM, Davis MB, *et al.* A novel point mutation in the McLeod syndrome gene in neuroacanthocytosis. *Ann Neurol* 1996;**39**:672–675.

215. Hof PR, Perl DP. Neurofibrillary tangles in the primary motor cortex in Guamanian amyotrophic lateral sclerosis/parkinsonism-dementia complex. *Neurosci Lett* 2002;**328**:294–298.

216. Hof PR, Charpiot A, Delacourte A, *et al.* Distribution of neurofibrillary tangles and senile plaques in the cerebral cortex in postencephalitic parkinsonism. *Neurosci Lett* 1992;**139**:10–14.

217. Hof PR, Perl DP, Loerzel AJ, *et al.* Amyotrophic lateral sclerosis and parkinsonism-dementia from Guam: differences in neurofibrillary tangle distribution and density in the hippocampal formation and neocortex. *Brain Res* 1994;**650**:107–116.

218. Hoglinger GU, Melhem NM, Dickson DW, *et al.* Identification of common variants influencing risk of the tauopathy progressive supranuclear palsy. *Nat Genet* 2011;**43**:699–705.

219. Holdorff B. Friedrich Heinrich Lewy (1885–1950) and his work. *J Hist Neurosci* 2002;**11**:19–28.

220. Holton JL, Schneider SA, Ganesharajah T, *et al.* Neuropathology of primary adult-onset dystonia. *Neurology* 2008;**70**:695–699.

221. Horn E, Lach B, Lapierre Y, *et al.* Hypothalamic pathology in the neuroleptic malignant syndrome. *Am J Psychiatry* 1988;**145**:617–620.

222. Houlden H, Singleton AB. The genetics and neuropathology of Parkinson's disease. *Acta Neuropathol* 2012;**124**:325–338.

223. Houlden H, Baker M, Morris HR, *et al.* Corticobasal degeneration and progressive supranuclear palsy share a common tau haplotype. *Neurology* 2001;**56**:1702–1706.

224. Hughes AJ, Daniel SE, Kilford L, Lees AJ. Accuracy of clinical diagnosis of idiopathic parkinsons-disease – a clinicopathological study of 100 cases. *J Neurol Neurosurg Psychiatry* 1992;**55**:181–184.

225. Hughes AJ, Daniel SE, Blankson S, Lees AJ. A clinicopathologic study of 100 cases of Parkinson's disease. *Arch Neurol* 1993;**50**:140–148.

226. Hummel T, Witt M, Reichmann H, *et al.* Immunohistochemical, volumetric and functional neuroimaging studies in patients with idiopathic Parkinson's disease. *J Neurol Sci* 2010;**289**:119–122.

227. Hunter R, Smith J, Thomson T, Dayan AD. Hemiparkinsonism with infarction of the ipsilateral substantia nigra. *Neuropathol Appl Neurobiol* 1978;**4**:297–301.

228. Hutton M, Lendon CL, Rizzu P, *et al.* Association of missense and 5'-splice-site mutations in tau with the inherited dementia FTDP-17. *Nature* 1998;**393**:702–705.

229. Hyman BT, Phelps CH, Beach TG, *et al.* National Institute on Aging-Alzheimer's Association guidelines for the neuropathologic assessment of Alzheimer's disease. *Alzheimers Dement* 2012;**8**:1–13.

230. Iannaccone S, Ferini SL, Nemni R, *et al.* Peripheral motor-sensory neuropathy in membranous lipodystrophy (Nasu's disease): a case report. *Clin Neuropathol* 1992;**11**:49–53.

231. Ikeda K, Akiyama H, Kondo H. Anti-tau-positive glial fibrillary tangles in the brain of postencephalitic parkinsonism of Economo type. *Neurosci Lett* 1993;**162**:176–178.

232. Irwin DJ, White MT, Toledo JB, *et al.* Neuropathologic substrates of Parkinson disease dementia. *Ann Neurol* 2012;**72**:587–598.

233. Ishihara–Paul L, Hulihan MM, Kachergus J, *et al.* PINK1 mutations and parkinsonism. *Neurology* 2008;**71**:896–902.

234. Ishizawa K, Dickson DW. Microglial activation parallels system degeneration in progressive supranuclear palsy and corticobasal degeneration. *J Neuropathol Exp Neurol* 2001;**60**:647–657.

235. Ishizawa K, Komori T, Okayama K, *et al.* Large motor neuron involvement in stiff-man syndrome: a qualitative and quantitative study. *Acta Neuropathol* 1999;**97**:63–70.

236. Ishizawa K, Ksiezak–Reding H, Davies P, *et al.* A double-labelling immunohistochemical study of tau exon 10 in Alzheimer's disease, progressive supranuclear palsy and Pick's disease. *Acta Neuropathol* 2000;**100**:235–244.

237. Ishizawa K, Lin WL, Tiseo P, *et al.* A qualitative and quantitative study of grumose degeneration in progressive supranuclear palsy. *J Neuropathol Exp Neurol* 2000;**59**:513–524.

238. Ishizawa K, Komori T, Arai N, *et al.* Glial cytoplasmic inclusions and tissue injury in multiple system atrophy: a quantitative study in white matter (olivopontocerebellar system) and gray matter (nigrostriatal system). *Neuropathology* 2008;**28**:249–257.

239. Itoh N, Ishiguro K, Arai H, *et al.* Biochemical and ultrastructural study of neurofibrillary tangles in amyotrophic lateral sclerosis/parkinsonism-dementia complex in the Kii peninsula of Japan. *J Neuropathol Exp Neurol* 2003;**62**:791–798.

240. Iversen SD, Iversen LL. Dopamine: 50 years in perspective. *Trends Neurosci* 2007;**30**:188–193.

241. Jankovic J. Can peripheral trauma induce dystonia and other movement disorders? Yes! *Mov Disord* 2001;**16**:7–12.

242. Jankovic J. Parkinson's disease: clinical features and diagnosis. *J Neurol Neurosurg Psychiatry* 2008;**79**:368–376.

243. Janssen JC, Lantos PL, Al Sarraj S, *et al.* Thalamic degeneration with negative prion protein immunostaining. *J Neurol* 2000;**247**:48–51.

244. Jecmenica–Lukic M, Poewe W, Tolosa E, Wenning GK. Premotor signs and symptoms of multiple system atrophy. *Lancet Neurol* 2012;**11**:361–368.

245. Jellinger K. Neuroaxonal dystrophy: its natural history and related diseases. *Prog Neuropathol* 1973;**2**:129–180.

246. Jellinger K. Pallidal, pallidonigral and pallidoluysionigral degenerations including association with thalamic and dentate degenerations. In: Vinken P, Bruyn G, Klawanseds H eds. *Handbook of clinical neurology: extrapyramidal disorders*. New York: Elsevier, 1986:445–464.

247. Jellinger KA. Lewy body-related alpha-synucleinopathy in the aged human brain. *J Neural Transm* 2004;**111**:1219–1235.

248. Jellinger KA, Seppi K, Wenning GK. Grading of neuropathology in multiple system atrophy: proposal for a novel scale. *Mov Disord* 2005;**20** Suppl 12:S29–36.

249. Jellinger KA. More frequent Lewy bodies but less frequent Alzheimer-type lesions in multiple system atrophy as compared to age-matched control brains. *Acta Neuropathol* 2007;**114**:299–303.

250. Jellinger KA, Grazer A, Petrovic K, *et al.* Four-repeat tauopathy clinically presenting as posterior cortical atrophy: atypical corticobasal degeneration? *Acta Neuropathol* 2011;**121**:267–277.

251. Jo E, McLaurin J, Yip CM, *et al.* alpha-Synuclein membrane interactions and lipid specificity. *J Biol Chem* 2000;**275**:34328–34334.

252. Joel D. Open interconnected model of basal ganglia-thalamocortical circuitry and its relevance to the clinical syndrome of Huntington's disease. *Mov Disord* 2001;**16**:407–423.

253. Jones EM, Dawson A. Neuroleptic malignant syndrome: a case report with post-mortem brain and muscle pathology. *J Neurol Neurosurg Psychiatry* 1989;**52**:1006–1009.

254. Josephs KA. Neuronal intranuclear inclusion disease: no longer a pain in the butt. *Neurology* 2011;**76**:1368–1369.

255. Josephs KA, Duffy JR. Apraxia of speech and nonfluent aphasia: a new clinical marker for corticobasal degeneration and progressive supranuclear palsy. *Curr Opin Neurol* 2008;**21**:688–692.

256. Josephs KA, Holton JL, Rossor MN, *et al.* Neurofilament inclusion body disease: a new proteinopathy? *Brain* 2003;**126**: 2291–2303.

257. Josephs KA, Tsuboi Y, Dickson DW. Creutzfeldt–Jakob disease presenting as progressive supranuclear palsy. *Eur J Neurol* 2004;**11**:343–346.

258. Josephs KA, Duffy JR, Strand EA, *et al.* Clinicopathological and imaging correlates of progressive aphasia and apraxia of speech. *Brain* 2006;**129**:1385–1398.

259. Josephs KA, Katsuse O, Beccano–Kelly DA, *et al.* Atypical progressive supranuclear palsy with corticospinal tract degeneration. *J Neuropathol Exp Neurol* 2006;**65**:396–405.

260. Josephs KA, Mandrekar JN, Dickson DW. The relationship between histopathological features of progressive supranuclear palsy and disease duration. *Parkinsonism Relat Disord* 2006;**12**:109–112.

261. Josephs KA, Whitwell JL, Dickson DW, *et al.* Voxel-based morphometry in autopsy proven PSP and CBD. *Neurobiol Aging* 2008;**29**:280–289.

262. Juncos JL, Hirsch EC, Malessa S, *et al.* Mesencephalic cholinergic nuclei in progressive supranuclear palsy. *Neurology* 1991;**41**:25–30.

263. Kalaitzakis ME, Graeber MB, Gentleman SM, Pearce RK. The dorsal motor nucleus of the vagus is not an obligatory trigger site of Parkinson's disease: a critical analysis of alpha-synuclein staging. *Neuropathol Appl Neurobiol* 2008;**34**:284–295.

264. Kalanithi PS, Zheng W, Kataoka Y, *et al.* Altered parvalbumin-positive neuron distribution in basal ganglia of individuals with Tourette syndrome. *Proc Natl Acad Sci U S A* 2005;**102**:13307–13312.

265. Kanazawa M, Shimohata T, Toyoshima Y, *et al.* Cerebellar involvement in progressive supranuclear palsy: a clinicopathological study. *Mov Disord* 2009;**24**:1312–1318.

266. Kanazawa T, Uchihara T, Takahashi A, *et al.* Three-layered structure shared between Lewy bodies and lewy neurites-three-dimensional reconstruction of triple-labeled sections. *Brain Pathol* 2008;**18**:415–422.

267. Kanazawa T, Adachi E, Orimo S, *et al.* Pale neurites, premature alpha-synuclein aggregates with centripetal extension from axon collaterals. *Brain Pathol* 2012;**22**:67–78.

268. Kataoka Y, Kalanithi PS, Grantz H, *et al.* Decreased number of parvalbumin and cholinergic interneurons in the striatum of individuals with Tourette syndrome. *J Comp Neurol* 2010;**518**:277–291.

269. Kato S, Nakamura H. Cytoplasmic argyrophilic inclusions in neurons of pontine nuclei in patients with olivopontocerebellar atrophy: immunohistochemical and ultrastructural studies. *Acta Neuropathol* 1990;**79**:584–594.

270. Kato S, Meshitsuka S, Ohama E, *et al.* Increased iron content in the putamen of patients with striatonigral degeneration. *Acta Neuropathol (Berl)* 1992;**84**:328–330.

271. Kato T, Hirano A, Weinberg MN, Jacobs AK. Spinal cord lesions in progressive supranuclear palsy: some new observations. *Acta Neuropathol* 1986;**71**:11–14.

272. Katsuse O, Dickson DW. Inferior olivary hypertrophy is uncommon in progressive supranuclear palsy. *Acta Neuropathol* 2004;**108**:143–146.

273. Katzenschlager R, Cardozo A, Avila Cobo MR, *et al.* Unclassifiable parkinsonism in two European tertiary referral centres for movement disorders. *Mov Disord* 2003;**18**:1123–1131.

274. Kawai Y, Suenaga M, Takeda A, *et al.* Cognitive impairments in multiple system atrophy: MSA–C vs MSA–P. *Neurology* 2008;**70**:1390–1396.

275. Kawashima M, Miyake M, Kusumi M, *et al.* Prevalence of progressive supranuclear palsy in Yonago, Japan. *Mov Disord* 2004;**19**:1239–1240.

276. Kennedy L, Evans E, Chen CM, *et al.* Dramatic tissue-specific mutation length increases are an early molecular event in Huntington disease pathogenesis. *Hum Mol Genet* 2003;**12**:3359–3367.

277. Kertesz A. Paroxysmal kinesigenic choreoathetosis. An entity within the paroxysmal choreoathetosis syndrome. Description of 10 cases, including 1 autopsied. *Neurology* 1967;**17**:680–690.

278. Kertesz A, McMonagle P, Blair M, *et al.* The evolution and pathology of frontotemporal dementia. *Brain* 2005;**128**:1996–2005.

279. Keyser DL, Rodnitzky RL. Neuroleptic malignant syndrome in Parkinson's disease after withdrawal or alteration of dopaminergic therapy. *Arch Intern Med* 1991;**151**:794–796.

280. Kiely AP, Asi YT, Kara E, *et al.* alpha-Synucleinopathy associated with *G51D SNCA* mutation: a link between Parkinson's disease and multiple system atrophy? *Acta Neuropathol* 2013;**125**:753–769.

281. Kikuchi H, Doh–Ura K, Kira J, Iwaki T. Preferential neurodegeneration in the cervical spinal cord of progressive supranuclear palsy. *Acta Neuropathol* 1999;**97**:577–584.

282. Kiley M, Esiri MM. A contemporary case of encephalitis lethargica. *Clin Neuropathol* 2001;**20**:2–7.

283. Kim RC, Ramachandran T, Parisi JE, *et al.* Pallidonigral pigmentation and spheroid formation with multiple striatal lacunar infarcts. *Neurology* 1981;**31**:774–777.

284. Kimura N, Kumamoto T, Hanaoka T, *et al.* Corticobasal degeneration presenting with progressive conduction aphasia. *J Neurol Sci* 2008;**269**:163–168.

285. Kimura S. Terminal axon pathology in infantile neuroaxonal dystrophy. *Pediatr Neurol* 1991;**7**:116–120.

286. Kitada T, Asakawa S, Hattori N, *et al.* Mutations in the *parkin* gene cause autosomal recessive juvenile parkinsonism. *Nature* 1998;**392**:605–608.

287. Klein C, Westenberger A. Genetics of Parkinson's disease. *Cold Spring Harb Perspect Med* 2012;**2**:a008888.

288. Kleiner–Fisman G, Calingasan NY, Putt M, *et al.* Alterations of striatal neurons in benign hereditary chorea. *Mov Disord* 2005;**20**:1353–1357.

289. Klomp L, Gitlin J. Expression of the ceruloplasmin gene in the human retina and brain: implications for a pathogenic model in aceruloplasminemia. *Hum Mol Genet* 1996;**5**:1989–1996.

290. Kobari M, Nogawa S, Sugimoto Y, Fukuuchi Y. Familial idiopathic brain calcification with autosomal dominant inheritance. *Neurology* 1997;**48**:645–649.

291. Kobayashi S, Yamadori I, Miki H, Ohmori M. Idiopathic nonarteriosclerotic cerebral calcification (Fahr's disease): an electron microscopic study. *Acta Neuropathol* 1987;**73**:62–66.

292. Koeppen AH. The nucleus pontis centralis caudalis in Huntington's disease. *J Neurol Sci* 1989;**91**:129–141.

293. Kollensperger M, Geser F, Ndayisaba JP, *et al.* Presentation, diagnosis and management of multiple system atrophy in Europe: final analysis of the European multiple system atrophy registry. *Mov Disord* 2010;**25**:2604–2612.

294. Komori T. Tau-positive glial inclusions in progressive supranuclear palsy, corticobasal degeneration and Pick's disease. *Brain Pathol* 1999;**9**:663–679.

295. Kondo T, Takahashi K, Kohara N, *et al.* Heterogeneity of presenile dementia with bone cysts (Nasu–Hakola disease): three genetic forms. *Neurology* 2002;**59**:1105–1107.

296. Konno H, Yamamoto T, Iwasaki Y, Iizuka H. Shy–Drager syndrome and amyotrophic lateral sclerosis. Cytoarchitectonic and morphometric studies of sacral autonomic neurons. *J Neurol Sci* 1986;**73**:193–204.

297. Kordower JH, Chu Y, Hauser RA, *et al.* Lewy body-like pathology in long-term embryonic nigral transplants in Parkinson's disease. *Nat Med* 2008;**14**:504–506.

298. Kosaka K, Tsuchiya K, Yoshimura M. Lewy body disease with and without dementia: a clinicopathological study of 35 cases. *Clin Neuropathol* 1988;**7**:299–305.

299. Kouri N, Murray ME, Hassan A, *et al.* Neuropathological features of corticobasal degeneration presenting as corticobasal syndrome or Richardson syndrome. *Brain* 2011;**134**:3264–3275.

300. Kouri N, Whitwell JL, Josephs KA, *et al.* Corticobasal degeneration: a pathologically distinct 4R tauopathy. *Nat Rev Neurol* 2011;**7**:263–272.

301. Kremer B, Goldberg P, Andrew SE, *et al.* A worldwide study of the Huntington's disease mutation: the sensitivity and specificity of measuring CAG repeats. *N Engl J Med* 1994;**330**(20):1401–1406.

302. Kremer HP, Roos RA, Dingjan G, *et al.* Atrophy of the hypothalamic lateral tuberal nucleus in Huntington's disease. *J Neuropathol Exp Neurol* 1990; **49**(4):371–382.

303. Kruer MC, Hiken M, Gregory A, *et al.* Novel histopathologic findings in molecularly-confirmed pantothenate kinase-associated neurodegeneration. *Brain* 2011;**134**:947–958.

304. Kruger R, Kuhn W, Muller T, *et al.* *Ala30Pro* mutation in the gene encoding alpha-synuclein in Parkinson's disease. *Nat Genet* 1998;**18**:106–108.

305. Kruger R, Vieira–Saecker AM, Kuhn W, *et al.* Increased susceptibility to sporadic Parkinson's disease by a certain combined alpha-synuclein/apolipoprotein E genotype. *Ann Neurol* 1999;**45**:611–617.

306. Ksiezak–Reding H, Morgan K, Mattiace LA, *et al.* Ultrastructure and biochemical composition of paired helical filaments in corticobasal degeneration. *Am J Pathol* 1994;**145**:1496–1508.

307. Ksiezak–Reding H, Tracz E, Yang LS, *et al.* Ultrastructural instability of paired helical filaments from corticobasal degeneration as examined by scanning transmission electron microscopy. *Am J Pathol* 1996;**149**:639–651.

308. Kurland LT, Mulder DW. Preliminary report on geographic distribution, with special reference to Mariana Islands, including clinical and pathologic observations. *Neurology* 1954;**4**:355–378.

309. Kuusisto E, Parkkinen L, Alafuzoff I. Morphogenesis of Lewy bodies: dissimilar incorporation of alpha-synuclein, ubiquitin and p62. *J Neuropathol Exp Neurol* 2003;**62**:1241–1253.

310. Kuusisto E, Kauppinen T, Alafuzoff I. Use of p62/SQSTM1 antibodies for neuropathological diagnosis. *Neuropathol Appl Neurobiol* 2008;**34**:169–180.

311. Lai BC, Marion SA, Teschke K, Tsui JK. Occupational and environmental risk factors for Parkinson's disease. *Parkinsonism Relat Disord* 2002;**8**:297–309.

312. Lange H, Thorner G, Hopf A, Schroder KF. Morphometric studies of the neuropathological changes in choreatic diseases. *J Neurol Sci* 1976;**28**:401–425.

313. Langston JW, Ballard P, Tetrud JW, Irwin I. Chronic Parkinsonism in humans due to a product of meperidine-analog synthesis. *Science* 1983;**219**:979–980.

314. Langston JW, Forno LS, Tetrud J, *et al.* Evidence of active nerve cell degeneration in the substantia nigra of humans years after 1–methyl-4-phenyl-1,2,3,6-tetrahydropyridine exposure. *Ann Neurol* 1999;**46**:598–605.

315. Lantos PL, Papp MI. Cellular pathology of multiple system atrophy: a review. *J Neurol Neurosurg Psychiatry* 1994;**57**:129–133.

316. Laplanche JL, Hachimi KH, Durieux I, *et al.* Prominent psychiatric features and early onset in an inherited prion disease with a new insertional mutation in the prion protein gene. *Brain* 1999;**122**:2375–2386.

317. Lashley T, Holton JL, Gray E. Cortical alpha-synuclein load is associated with amyloid-beta plaque burden in a subset of Parkinson's disease patients. *Acta Neuropathol* 2008;**115**:417–425.

318. Lashley T, Rohrer JD, Bandopadhyay R, *et al.* A comparative clinical, pathological, biochemical and genetic study of fused in sarcoma proteinopathies. *Brain* 2011;**134**:2548–2564.

319. Layrargues G. Movement dysfunction and hepatic encephalopathy. *Metabol Brain Dis* 2001;**16**:27–35.

320. Leckman JF. Tic disorders. *BMJ* 2012;**344**:d7659.

321. Lee HJ, Choi C, Lee SJ. Membrane-bound alpha-synuclein has a high aggregation propensity and the ability to seed the aggregation of the cytosolic form. *J Biol Chem* 2002;**277**:671–678.

322. Lee HJ, Suk JE, Patrick C, *et al.* Direct transfer of alpha-synuclein from neuron to astroglia causes inflammatory responses in synucleinopathies. *J Biol Chem* 2010;**285**(12):9262–72.

323. Lee HY, Xu Y, Huang Y, *et al.* The gene for paroxysmal non-kinesigenic dyskinesia encodes an enzyme in a stress response pathway. *Hum Mol Genet* 2004;**13**: 3161–3170.

324. Lee S, Merriam A, Kim TS, *et al.* Cerebellar degeneration in neuroleptic malignant syndrome: neuropathologic findings and review of the literature concerning heat-related nervous system injury. *J Neurol Neurosurg Psychiatry* 1989;**52**:387–391.

325. Lees AJ, Selikhova M, Andrade LA, Duyckaerts C. The black stuff and Konstantin Nikolaevich Tretiakoff. *Mov Disord* 2008;**23**:777–783.

326. Leigh RJ, Parhad IM, Clark AW, *et al.* Brainstem findings in Huntington's disease. Possible mechanisms for slow vertical saccades. *J Neurol Sci* 1985;**71**:247–256.

327. Lesage S, Durr A, Tazir M, *et al.* LRRK2 G2019S as a cause of Parkinson's disease in North African Arabs. *N Engl J Med* 2006;**354**:422–423.

328. Leuzzi V, Bertini E, De NA, *et al.* Bilateral striatal necrosis, dystonia and optic atrophy in two siblings. *J Neurol Neurosurg Psychiatry* 1992;**55**:16–19.

329. Levitan K, Chereau D, Cohen SI, *et al.* Conserved C-terminal charge exerts a profound influence on the aggregation rate of alpha-synuclein. *J Mol Biol* 2011;**411**:329–333.

330. Lewis PA. Assaying the kinase activity of LRRK2 *in vitro*. *J Vis Exp* 2012;**59**:pii: 3495.

330a. Lewy FH. Paralysis agitans. In: Lewandowski M (editor). *Pathologische Anatomie. Handbuch der Neurologie.* Berlin: Springer Verlag;1912;920–933.

331. Li A, Paudel R, Johnson R, *et al.* Pantothenate kinase-associated neurodegeneration is not a synucleinopathy. *Neuropathol Appl Neurobiol* 2012 Mar 15;doi:10.1111/j.1365–2990.2012.01269.x. [Epub ahead of print].

332. Li JY, Englund E, Holton JL, *et al.* Lewy bodies in grafted neurons in subjects with Parkinson's disease suggest host-to-graft disease propagation. *Nat Med* 2008;**14**:501–503.

333. Limousin P, Pollak P, Hoffmann D, *et al.* Abnormal involuntary movements induced by subthalamic nucleus stimulation in parkinsonian patients. *Mov Disord* 1996;**11**:231–235.

334. Lincoln SJ, Maraganore DM, Lesnick TG, *et al.* Parkin variants in North American Parkinson's disease: cases and controls. *Mov Disord* 2003;**18**:1306–1311.

335. Ling H, O'Sullivan SS, Holton JL, *et al.* Does corticobasal degeneration exist? A clinicopathological re-evaluation. *Brain* 2010;**133**:2045–2057.

336. Ling H, Kara E, Bandopadhyay R, *et al.* TDP-43 pathology in a patient carrying *G2019S LRRK2* mutation and a novel p.Q124E MAPT. *Neurobiol Aging* 2013;**34**(12):2889.e5–9;doi: 10.1016/j.neurobiolaging.2013.04.011.

337. Ling H, de Silva R, Massey LA, *et al.* Characteristics of progressive supranuclear palsy presenting with corticobasal syndrome: a cortical variant. *Neuropathol Appl Neurobiol* 2014;**40**:149–163.

338. Litvan I, Agid Y, Calne D, *et al.* Clinical research criteria for the diagnosis of progressive supranuclear palsy (Steele–Richardson–Olszewski syndrome): report of the NINDS–SPSP international workshop. *Neurology* 1996;**47**:1–9.

339. Litvan I, Agid Y, Jankovic J, *et al.* Accuracy of clinical criteria for the diagnosis of progressive supranuclear palsy (Steele–Richardson–Olszewski syndrome). *Neurology* 1996;**46**:922–930.

340. Litvan I, Hauw JJ, Bartko JJ, *et al.* Validity and reliability of the preliminary NINDS neuropathologic criteria for progressive supranuclear palsy and related disorders. *J Neuropathol Exp Neurol* 1996;**55**:97–105.

341. Lowe J, Errington DR, Lennox G, *et al.* Ballooned neurons in several neurodegenerative diseases and stroke contain alpha B crystallin. *Neuropathol Appl Neurobiol* 1992;**18**:341–350.

342. Lucking CB, Durr A, Bonifati V, *et al.* Association between early-onset Parkinson's disease and mutations in the *parkin* gene. *N Engl J Med* 2000;**342**:1560–1567.

343. Lue LF, Walker DG, Adler CH, *et al.* Biochemical increase in phosphorylated alpha-synuclein precedes histopathology of Lewy-type synucleinopathies. *Brain Pathol* 2012;**22**(6):745–56.

344. Luk KC, Kehm VM, Zhang B, *et al.* Intracerebral inoculation of pathological alpha-synuclein initiates a rapidly progressive neurodegenerative alpha-synucleinopathy in mice. *J Exp Med* 2012;**209**:975–986.

345. MacDonald V, Halliday GM. Selective loss of pyramidal neurons in the pre-supplementary motor cortex in Parkinson's disease. *Mov Disord* 2002;**17**:1166–1173.

346. Madsen E, Gitlin J. Copper and iron disorders of the brain. *Annu Rev Neurosci* 2007;**30**:317–337.

347. Makino S, Kaji R, Ando S, *et al.* Reduced neuron-specific expression of the *TAF1* gene is associated with X-linked dystonia-parkinsonism. *Am J Hum Genet* 2007;**80**:393–406.

348. Malandrini A, Cesaretti S, Mulinari M, *et al.* Acanthocytosis, retinitis pigmentosa, pallidal degeneration. Report of two cases without serum lipid abnormalities. *J Neurol Sci* 1996;**140**:129–131.

349. Mancuso M, Davidzon G, Kurlan R, *et al.* Hereditary ferritinopathy: a novel mutation, its cellular pathology, and pathogenetic insights. *J Neuropathol Exp Neurol* 2005;**64**:280–294.

350. Manyam B. What is and what is not 'Fahr's disease'. *Parkinsonism Relat Disord* 2005;**11**:73–80.

351. Maraganore DM, de Andrade M, Elbaz A, *et al.* Collaborative analysis of alpha-synuclein gene promoter variability and Parkinson disease. *JAMA* 2006;**296**:661–670.

352. Margolis RL, O'Hearn E, Rosenblatt A, *et al.* A disorder similar to Huntington's disease is associated with a novel CAG repeat expansion. *Ann Neurol* 2001;**50**:373–380.

353. Marti–Masso JF, Ruiz–Martinez J, Bolano MJ, *et al.* Neuropathology of Parkinson's disease with the *R1441G* mutation in *LRRK2 Mov Disord* 2009;**24**:1998–2001.

354. Martin LJ, Brambrink AM, Price AC, *et al.* Neuronal death in newborn striatum after hypoxia–ischemia is necrosis and evolves with oxidative stress. *Neurobiol Dis* 2000;**7**:169–191.

355. Mathias CJ, Polinsky RJ. Separating the primary autonomic failure syndromes, multiple system atrophy and pure autonomic failure from Parkinson's disease. *Adv Neurol* 1999;**80**:353–361.

356. Matsuo A, Akiguchi I, Lee GC, *et al.* Myelin degeneration in multiple system atrophy detected by unique antibodies. *Am J Pathol* 1998;**153**:735–744.

357. McCall S, Henry JM, Reid AH, Taubenberger JK. Influenza RNA not detected in archival brain tissues from acute encephalitis lethargica cases or in postencephalitic Parkinson cases. *J Neuropathol Exp Neurol* 2001;**60**:696–704.

358. McCall S, Vilensky JA, Gilman S, Taubenberger JK. The relationship between encephalitis lethargica and influenza: a critical analysis. *J Neurovirol* 2008;**14**:177–185.

359. McFadden K, Hamilton RL, Insalaco SJ, *et al.* Neuronal intranuclear inclusion disease without polyglutamine inclusions in a child. *J Neuropathol Exp Neurol* 2005;**64**:545–552.

360. McKee AC, Cantu RC, Nowinski CJ, *et al.* Chronic traumatic encephalopathy in athletes: progressive tauopathy after repetitive head injury. *J Neuropathol Exp Neurol* 2009;**68**:709–735.

361. McKee AC, Stein TD, Nowinski CJ, *et al.* The spectrum of disease in chronic traumatic encephalopathy. *Brain* 2013;**136**:43–64.

362. McKeith I, Fairbairn A, Perry R, *et al.* Neuroleptic sensitivity in patients with senile dementia of Lewy body type. *BMJ* 1992;**305**:673–678.

363. McKeith IG, Dickson DW, Lowe J, *et al.* Diagnosis and management of dementia with Lewy bodies: third report of the DLB Consortium. *Neurology* 2005;**65**:1863–1872.

364. McKeon A, Robinson MT, McEvoy KM, *et al.* Stiff-man syndrome and variants: clinical course, treatments, and outcomes. *Arch Neurol* 2012;**69**:230–238.

365. McNaught KS, Kapustin A, Jackson T, *et al.* Brainstem pathology in DYT1 primary torsion dystonia. *Ann Neurol* 2004;**56**:540–547.

366. McNeill A, Pandolfo M, Kuhn J, *et al.* The neurological presentation of ceruloplasmin gene mutations. *Eur Neurol* 2008;**60**:200–205.

367. Meinck HM, Ricker K, Hulser PJ, *et al.* Stiff man syndrome: clinical and laboratory findings in eight patients. *J Neurol* 1994;**241**:157–166.

368. Mellick GD, Siebert GA, Funayama M, *et al.* Screening PARK genes for mutations in early-onset Parkinson's disease patients from Queensland, Australia. *Parkinsonism Relat Disord* 2009;**15**:105–109.

369. Menkes JH. Kinky hair disease: twenty five years later. *Brain Dev* 1988;**10**:77–79.

370. Miklossy J, Steele JC, Yu S, *et al.* Enduring involvement of tau, beta-amyloid, alpha-synuclein, ubiquitin and TDP-43 pathology in the amyotrophic lateral sclerosis/parkinsonism-dementia complex of Guam (ALS/PDC). *Acta Neuropathol* 2008;**116**:625–637.

371. Mikol J, Vital C, Wassef M, *et al.* Extensive cortico-subcortical lesions in Wilson's disease: clinico-pathological study of two cases. *Acta Neuropathol (Berl)* 2005;**110**:451–458.

372. Montine TJ, Phelps CH, Beach TG, *et al.* National Institute on Aging–Alzheimer's Association guidelines for the neuropathologic assessment of Alzheimer's disease: a practical approach. *Acta Neuropathol* 2012;**123**:1–11.

373. Moretto G, Sparaco M, Monaco S, *et al.* Cytoskeletal changes and ubiquitin expression in dystrophic axons of Seitelberger's disease. *Clin Neuropathol* 1993;**12**:34–37.

374. Morgan JC, Sethi KD. Drug-induced tremors. *Lancet Neurol* 2005;**4**:866–876.

375. Morgan NV, Westaway SK, Morton JE, *et al.* PLA2G6, encoding a phospholipase A2, is mutated in neurodegenerative disorders with high brain iron. *Nat Genet* 2006;**38**:752–754.

376. Mori H, Nishimura M, Namba Y, Oda M. Corticobasal degeneration: a disease with widespread appearance of abnormal tau and neurofibrillary tangles and its relation to progressive supranuclear palsy. *Acta Neuropathol* 1994;**88**:113–121.

377. Mori H, Kondo T, Yokochi M, *et al.* Pathologic and biochemical studies of juvenile parkinsonism linked to chromosome 6q. *Neurology* 1998;**51**:890–892.

378. Morita H, Ikeda S, Yamamoto K, *et al.* Hereditary ceruloplasmin deficiency with hemosiderosis: a clinicopathological study of a Japanese family. *Ann Neurol* 1995;**37**:646–656.

379. Morris HR, Osaki Y, Holton J, *et al.* Tau exon 10 +16 mutation *FTDP-17* presenting clinically as sporadic young onset PSP. *Neurology* 2003;**61**:102–104.

380. Morris HR, Steele JC, Crook R, *et al.* Genome-wide analysis of the parkinsonism-dementia complex of Guam. *Arch Neurol* 2004;**61**:1889–1897.

381. Mundinano IC, Caballero MC, Ordonez C, *et al.* Increased dopaminergic cells and protein aggregates in the olfactory bulb of patients with neurodegenerative disorders. *Acta Neuropathol* 2011;**122**:61–74.

382. Murphy TK, Kurlan R, Leckman J. The immunobiology of Tourette's disorder, pediatric autoimmune neuropsychiatric disorders associated with Streptococcus and related disorders: a way forward. *J Child Adolesc Psychopharmacol* 2010;**20**:317–331.

383. Myers RH, MacDonald ME, Koroshetz WJ, *et al. De novo* expansion of a (CAG)n repeat in sporadic Huntington's disease. *Nat Genet* 1993;**5**:168–173.

384. Nakamura N, Fujiya S, Yahara O, *et al.* Stiff-man syndrome with spinal cord lesion. *Clin Neuropathol* 1986;**5**:40–46.

385. Nakano–Yokomizo T, Tahara–Hanaoka S, Nakahashi–Oda C, *et al.* The immunoreceptor adapter protein *DAP12* suppresses B lymphocyte-driven adaptive immune responses. *J Exp Med* 2011;**208**:1661–1671.

386. Nardocci N, Zorzi G, Farina L, *et al.* Infantile neuroaxonal dystrophy: clinical spectrum and diagnostic criteria. *Neurology* 1999;**52**:1472–1478.

387. Nataf S, Anginot A, Vuaillat C, *et al.* Brain and bone damage in KARAP/DAP12 loss-of-function mice correlate with alterations in microglia and osteoclast lineages. *Am J Pathol* 2005;**166**:275–286.

388. Nath U, Ben Shlomo Y, Thomson RG, *et al.* The prevalence of progressive supranuclear palsy (Steele–Richardson–Olszewski syndrome) in the UK. *Brain* 2001;**124**:1438–1449.

389. Neumann J, Bras J, Deas E, *et al.* Glucocerebrosidase mutations in clinical and pathologically proven Parkinson's disease. *Brain* 2009;**132**(Pt 7):1783–94.

390. Neurology TCCotAASatAAo. Consensus statement on the definition of orthostatic hypotension, pure autonomic failure, and multiple system atrophy. *Neurology* 1996;**46**:1470.

391. Nichols WC, Pankratz N, Hernandez D, *et al.* Genetic screening for a single common *LRRK2* mutation in familial Parkinson's disease. *Lancet* 2005;**365**:410–412.

392. Nicoletti A, Nicoletti G, Arabia G, *et al.* Reproductive factors and Parkinson's disease: a multicenter case-control study. *Mov Disord* 2011;**26**:2563–2566.

393. Nishie M, Mori F, Fujiwara H, *et al.* Accumulation of phosphorylated alpha-synuclein in the brain and peripheral ganglia of patients with multiple system atrophy. *Acta Neuropathol* 2004;**107**:292–298.

394. Nishie M, Mori F, Yoshimoto M, *et al.* A quantitative investigation of neuronal cytoplasmic and intranuclear inclusions in the pontine and inferior olivary nuclei in multiple system atrophy. *Neuropathol Appl Neurobiol* 2004;**30**:546–554.

395. Nishiyama A, Endo T, Takeda S, Imamura M. Identification and characterization of epsilon-sarcoglycans in the central nervous system. *Brain Res Mol Brain Res* 2004;**125**:1–12.

396. Nuytemans K, Theuns J, Cruts M, Van Broeckhoven C. Genetic etiology of Parkinson disease associated with mutations in the *SNCA, PARK2, PINK1, PARK7 LRRK2* genes: a mutation update. *Hum Mutat* 2010;**31**:763–780.

397. O'Connor A, Cluroe A, Couch R, *et al.* Death from hyponatraemia-induced cerebral oedema associated with MDMA ('Ecstasy') use. *N Z Med J* 1999;**112**(1091):255–6.

398. O'Reilly S. Neurologic disorders and liver disease. *Postgrad Med* 1971;**50**:126–133.

399. O'Suilleabhain P, Giller C. Rapidly progressive parkinsonism in a self-reported user of ecstasy and other drugs. *Mov Disord* 2003;**18**:1378–1381.

400. O'Sullivan SS, Massey LA, Williams DR, *et al.* Clinical outcomes of progressive supranuclear palsy and multiple system atrophy. *Brain* 2008;**131**:1362–1372.

401. Oba H, Yagishita A, Terada H, *et al.* New and reliable MRI diagnosis for progressive supranuclear palsy. *Neurology* 2005;**64**:2050–2055.

402. Obi T, Nishioka K, Ross OA, *et al.* Clinicopathologic study of a *SNCA* gene duplication patient with Parkinson disease and dementia. *Neurology* 2008;**70**:238–241.

403. Oliveira J, Spiteri E, Sobrido M, *et al.* Genetic heterogeneity in familial idiopathic basal ganglia calcification (Fahr disease). *Neurology* 2004;**63**:2165–2167.

404. Oliveira J, Sobrido M, Spiteri E, *et al.* Analysis of candidate genes at the IBGC1 locus associated with idiopathic basal ganglia calcification ('Fahr's disease'). *J Mol Neurosci* 2007;**33**:151–154.

405. Oppenheim H. Ueber eigenenartige Krampfkrankheit des kindlichen und jugendlichen Alters (Dysbasia lordotica progressiva, Dystonia Musculorum Deformans). *Neurol Centralbl* 1911;**30**:1090.

406. Orr CF, Rowe DB, Halliday GM. An inflammatory review of Parkinson's disease. *Prog Neurobiol* 2002;**68**:325–340.

407. Osaki Y, Ben–Shlomo Y, Lees AJ, *et al.* Accuracy of clinical diagnosis of progressive supranuclear palsy. *Mov Disord* 2004;**19**:181–189.

408. Osaki Y, Morita Y, Kuwahara T, *et al.* Prevalence of Parkinson's disease and atypical parkinsonian syndromes in a rural Japanese district. *Acta Neurol Scand* 2011;**124**:182–187.

409. Osser D. Neuroleptic induced pseudoparkinsonism. In: Joseph A, Young R eds. *Movement disorders in neurology and neuropsychiatry.* London: Blackwell Scientific Publications, 1992:70–80.

410. Oyanagi K, Takeda S, Takahashi H, *et al.* A quantitative investigation of the substantia nigra in Huntington's disease. *Ann Neurol* 1989;**26**:13–19.

411. Oyanagi K, Makifuchi T, Ohtoh T, *et al.* Distinct pathological features of the gallyas- and tau-positive glia in the parkinsonism-dementia complex and amyotrophic lateral sclerosis of Guam. *J Neuropathol Exp Neurol* 1997;**56**:308–316.

412. Oyanagi K, Tsuchiya K, Yamazaki M, Ikeda K. Substantia nigra in progressive supranuclear palsy, corticobasal degeneration and parkinsonism-dementia complex of Guam: specific pathological features. *J Neuropathol Exp Neurol* 2001;**60**:393–402.

413. Ozawa T. Morphological substrate of autonomic failure and neurohormonal dysfunction in multiple system atrophy: impact on determining phenotype spectrum. *Acta Neuropathol* 2007;**114**:201–211.

414. Ozawa T, Takano H, Onodera O, *et al.* No mutation in the entire coding region of the alpha-synuclein gene in pathologically confirmed cases of multiple system atrophy. *Neurosci Lett* 1999;**270**:110–112.

415. Ozawa T, Okuizumi K, Ikeuchi T, *et al.* Analysis of the expression level of alpha-synuclein mRNA using postmortem brain samples from pathologically confirmed cases of multiple system atrophy. *Acta Neuropathol* 2001;**102**:188–190.

416. Ozawa T, Paviour D, Quinn NP, *et al.* The spectrum of pathological involvement of the striatonigral and olivopontocerebellar systems in multiple system atrophy: clinicopathological correlations. *Brain* 2004;**127**:2657–2671.

417. Ozawa T, Tada M, Kakita A, *et al.* The phenotype spectrum of Japanese multiple system atrophy. *J Neurol Neurosurg Psychiatry* 2010;**81**:1253–1255.

418. Ozelius LJ, Hewett JW, Page CE, *et al.* The early-onset torsion dystonia gene (DYT1) encodes an ATP-binding protein. *Nat Genet* 1997;**17**:40–48.

419. Ozelius LJ, Senthil G, Saunders–Pullman R, *et al. LRRK2 G2019S* as a cause of Parkinson's disease in Ashkenazi Jews. *N Engl J Med* 2006;**354**:424–425.

420. Ozmen M, Caliskan M, Goebel HH, *et al.* Infantile neuroaxonal dystrophy: diagnosis by skin biopsy. *Brain Dev* 1991;**13**:256–259.

421. Padberg G, Bruyn G. Chorea: differential diagnosis. In: Vinken PJ, Bruyn GW, Klawans HL eds. *Extrapyramidal disorders.* Amsterdam: Elsevier, 1986: 549–64.

422. Paisan–Ruiz C, Jain S, Evans EW, *et al.* Cloning of the gene containing mutations that cause PARK8-linked Parkinson's disease. *Neuron* 2004;**44**:595–600.

423. Paisan–Ruiz C, Bhatia KP, Li A, *et al.* Characterization of PLA2G6 as a locus for dystonia-parkinsonism. *Ann Neurol* 2009;**65**:19–23.

424. Paisan–Ruiz C, Guevara R, Federoff M, *et al.* Early-onset L-dopa-responsive parkinsonism with pyramidal signs due to *ATP13A2, PLA2G6, FBXO7 spatacsin* mutations. *Mov Disord* 2010;**25**:1791–1800.

425. Paisan–Ruiz C, Li A, Schneider SA, *et al.* Widespread Lewy body and tau accumulation in childhood and adult onset dystonia-parkinsonism cases with *PLA2G6* mutations. *Neurobiol Aging* 2012;**33**:814–823.

426. Paloneva J, Kestila M, Wu J, *et al.* Loss-of-function mutations in *TYROBP (DAP12)* result in a presenile dementia with bone cysts. *Nat Genet* 2000;**25**:357–361.

427. Paloneva J, Autti T, Raininko R, *et al.* CNS manifestations of Nasu–Hakola disease: a frontal dementia with bone cysts. *Neurology* 2001;**56**:1552–1558.

428. Paloneva J, Manninen T, Christman G, *et al.* Mutations in two genes encoding different subunits of a receptor signalling complex result in an identical disease phenotype. *Am J Hum Genet* 2003;**71**:656–662.

429. Panzer J, Dalmau J. Movement disorders in paraneoplastic and autoimmune disease. *Curr Opin Neurol* 2011;**24**:346–353.

430. Papp MI, Lantos PL. Accumulation of tubular structures in oligodendroglial and neuronal cells as the basic alteration in multiple system atrophy. *J Neurol Sci* 1992;**107**:172–182.

431. Papp MI, Lantos PL. The distribution of oligodendroglial inclusions in multiple system atrophy and its relevance to clinical symptomatology. *Brain* 1994;**117**:235–243.

432. Papp MI, Kahn JE, Lantos PL. Glial cytoplasmic inclusions in the CNS of patients with multiple system atrophy (striatonigral degeneration, olivopontocerebellar atrophy and Shy–Drager syndrome). *J Neurol Sci* 1989;**94**:79–100.

433. Parent M, Parent A. Substantia nigra and Parkinson's disease: a brief history of their long and intimate relationship. *Can J Neurol Sci* 2010;**37**:313–319.

434. Parkinson J. *An essay on the shaking palsy.* Edited by London, Whittinham and Rowland for Sherwood, Needly and Jones, 1817.

435. Parkkinen L, Soininen H, Laakso M, Alafuzoff I. Alpha-synuclein pathology is highly dependent on the case selection. *Neuropathol Appl Neurobiol* 2001;**27**:314–325.

436. Parkkinen L, Pirttila T, Alafuzoff I. Applicability of current staging/categorization of alpha-synuclein pathology and their clinical relevance. *Acta Neuropathol* 2008;**115**:399–407.

437. Paschou P. The genetic basis of Gilles de la Tourette Syndrome. *Neurosci Biobehav Rev* 2013;**37**:1026–1039.

438. Pasternak B, Svanstrom H, Nielsen NM, *et al.* Use of calcium channel blockers and Parkinson's disease. *Am J Epidemiol* 2012;**175**:627–635.

439. Paudel R, Hardy J, Revesz T, *et al.* Review: genetics and neuropathology of primary pure dystonia. *Neuropathol Appl Neurobiol* 2012;**38**:520–534.

440. Paviour DC, Lees AJ, Josephs KA, *et al.* Frontotemporal lobar degeneration with ubiquitin-only-immunoreactive neuronal changes: broadening the clinical picture to include progressive supranuclear palsy. *Brain* 2004;**127**:2441–2451.

441. Paviour DC, Revesz T, Holton JL, *et al.* Neuronal intranuclear inclusion disease: report on a case originally diagnosed as dopa-responsive dystonia with Lewy bodies. *Mov Disord* 2005;**20**:1345–1349.

442. Petrovic IN, Ling H, Asi Y, *et al.* Multiple system atrophy-parkinsonism with slow progression and prolonged survival: a diagnostic catch. *Mov Disord* 2012;**27**:1186–1190.

443. Petrukhin K, Fischer S, Piratsu M, *et al.* Mapping, cloning and genetic characterisation of the region containing the Wilson disease gene. *Nature Genetics* 1993;**5**:338–343.

444. Phukan J, Albanese A, Gasser T, Warner T. Primary dystonia and dystonia-plus syndromes: clinical characteristics, diagnosis and pathogenesis. *Lancet Neurol* 2011;**10**:1074–1085.

445. Piao YS, Hayashi S, Wakabayashi K, *et al.* Cerebellar cortical tau pathology in progressive supranuclear palsy and corticobasal degeneration. *Acta Neuropathol* 2002;**103**:469–474.

446. Pittman AM, Myers AJ, bou–Sleiman P, *et al.* Linkage disequilibrium fine mapping and haplotype association analysis of the *tau* gene in progressive supranuclear palsy and corticobasal degeneration. *J Med Genet* 2005;**42**:837–846.

447. Pittock SJ, Joyce C, O'Keane V, *et al.* Rapid-onset dystonia-parkinsonism: a clinical and genetic analysis of a new kindred. *Neurology* 2000;**55**:991–995.

448. Poli M, Derosas M, Luscieti S, *et al.* Pantothenate kinase-2 (PANK2) silencing causes cell growth reduction, cell-specific ferroportin upregulation and iron deregulation. *Neurobiol Dis* 2010;**39**:204–210.

449. Politis M, Lindvall O. Clinical application of stem cell therapy in Parkinson's disease. *BMC Medicine* 2012;**10**:1.

450. Polymeropoulos MH, Lavedan C, Leroy E, *et al.* Mutation in the alpha-synuclein gene identified in families with Parkinson's disease. *Science* 1997;**276**:2045–2047.

451. Pouclet H, Lebouvier T, Coron E, *et al.* A comparison between rectal and colonic biopsies to detect Lewy pathology in Parkinson's disease. *Neurobiol Dis* 2012;**45**:305–309.

452. Pountney DL, Huang Y, Burns RJ, *et al.* SUMO-1 marks the nuclear inclusions in familial neuronal intranuclear inclusion disease. *Exp Neurol* 2003;**184**:436–446.

453. Pountney DL, Raftery MJ, Chegini F, *et al.* NSF, Unc-18-1, dynamin-1 and HSP90 are inclusion body components in neuronal intranuclear inclusion disease identified by anti-SUMO-1-immunocapture. *Acta Neuropathol* 2008;**116**:603–614.

454. Powers JM, Tummons RC, Moser AB, *et al.* Neuronal lipidosis and neuroaxonal dystrophy in cerebro-hepato-renal (Zellweger) syndrome. *Acta Neuropathol (Berl)* 1987;**73**:333–343.

455. Pramstaller PP, Schlossmacher MG, Jacques TS, *et al.* Lewy body Parkinson's disease in a large pedigree with 77 *Parkin* mutation carriers. *Ann Neurol* 2005;**58**:411–422.

456. Prasad K, Beach TG, Hedreen J, Richfield EK. Critical role of truncated alpha-synuclein and aggregates in Parkinson's disease and incidental Lewy body disease. *Brain Pathol* 2012;**22**(6):811–25.

457. Price RA, Kidd KK, Cohen DJ, *et al.* A twin study of Tourette syndrome. *Arch Gen Psychiatry* 1985;**42**:815–820.

458. Quinn N. Multiple system atrophy – the nature of the beast. *J Neurol Neurosurg Psychiatry* 1989;Suppl:78–89.

459. Quinn JG, Coulson DT, Brockbank S, *et al.* alpha-Synuclein mRNA and soluble alpha-synuclein protein levels in post-mortem brain from patients with Parkinson's disease, dementia with Lewy bodies and Alzheimer's disease. *Brain Res* 2012;**1459**:71–80.

460. Rademakers R, Baker M, Nicholson AM, *et al.* Mutations in the colony stimulating factor 1 receptor (*CSF1R*) gene cause hereditary diffuse leukoencephalopathy with spheroids. *Nat Genet* 2012;**44**:200–205.

461. Ramirez A, Heimbach A, Grundemann J, *et al.* Hereditary parkinsonism with dementia is caused by mutations in *ATP13A2*, encoding a lysosomal type 5 P-type ATPase. *Nat Genet* 2006;**38**:1184–1191.

462. Rampoldi L, Dobson–Stone C, Rubio JP, *et al.* A conserved sorting-associated protein is mutant in chorea-acanthocytosis. *Nat Genet* 2001;**28**:119–120.

463. Ramsey CP, Giasson BI. *L10p P158DEL DJ-1* mutations cause protein instability, aggregation and dimerization impairments. *J Neurosci Res* 2010;**88**:3111–3124.

464. Rebeiz JJ, Kolodny EH, Richardson EP, Jr. Corticodentatonigral degeneration with neuronal achromasia. *Arch Neurol* 1968;**18**:20–33.

465. Reider CR, Halter CA, Castelluccio PF, *et al.* Reliability of reported age at onset for Parkinson's disease. *Mov Disord* 2003;**18**:275–279.

466. Rera M, Azizi MJ, Walker DW. Organ-specific mediation of lifespan extension: More than a gut feeling? *Ageing Res Rev* 2013;**2**(1):436–44.

467. Revesz T, Sangha H, Daniel SE. The nucleus raphe interpositus in the Steele–Richardson–Olszewski syndrome (progressive supranuclear palsy). *Brain* 1996;**119**:1137–1143.

468. Reyes S, Fu Y, Double KL, *et al.* Trophic factors differentiate dopamine neurons vulnerable to Parkinson's disease. *Neurobiol Aging* 2013;**34**(3):873–86.

469. Rinne JO, Daniel SE, Scaravilli F, *et al.* The neuropathological features of neuroacanthocytosis. *Mov Disord* 1994;**9**:297–304.

470. Rinne JO, Lee MS, Thompson PD, Marsden CD. Corticobasal degeneration. A clinical study of 36 cases. *Brain* 1994;**117**:1183–1196.

471. Rizzu P, Hinkle DA, Zhukareva V, *et al.* DJ-1 co-localizes with tau inclusions: a link between parkinsonism and dementia. *Ann Neurol* 2004;**55**:113–118.

472. Robertson D, Robertson RM. Causes of chronic orthostatic hypotension. *Arch Intern Med* 1994;**154**:1620–1624.

473. Rohrer JD, Lashley T, Schott JM, *et al.* Clinical and neuroanatomical signatures of tissue pathology in frontotemporal lobar degeneration. *Brain* 2011;**134**: 2565–2581.

474. Rosas HD, Salat DH, Lee SY, *et al.* Cerebral cortex and the clinical expression of Huntington's disease: complexity and heterogeneity. *Brain* 2008;**131**:1057–1068.

475. Ross OA, Vilarino–Guell C, Wszolek ZK, *et al.* Reply to: SNCA variants are associated with increased risk of multiple system atrophy. *Ann Neurol* 2010;**67**:414–415.

476. Rudnicki DD, Pletnikova O, Vonsattel JP, *et al.* A comparison of Huntington disease and Huntington disease-like 2 neuropathology. *J Neuropathol Exp Neurol* 2008;**67**:366–374.

477. Saiki S, Sato S, Hattori N. Molecular pathogenesis of Parkinson's disease: update. *J Neurol Neurosurg Psychiatry* 2012;**83**:430–436.

478. Saito Y, Shibuya M, Hayashi M, *et al.* Cerebellopontine calcification: a new entity of idiopathic intracranial calcification? *Acta Neuropathol* 2005;**110**:77–83.

479. Samaranch L, Lorenzo–Betancor O, Arbelo JM, *et al.* PINK1-linked parkinsonism is associated with Lewy body pathology. *Brain* 2010;**133**:1128–1142.

480. Sapp E, Schwarz C, Chase K, *et al.* Huntingtin localization in brains of normal and Huntington's disease patients. *Ann Neurol* 1997;**42**:604–612.

481. Satake W, Nakabayashi Y, Mizuta I, *et al.* Genome-wide association study identifies common variants at four loci as genetic risk factors for Parkinson's disease. *Nat Genet* 2009;**41**:1303–1307.

482. Sato M, Gitlin J. Mechanisms of copper incorporation during the biosynthesis of human ceruloplasmin. *J Biol Chem* 1991;**266**:5128–5134.

483. Scaravilli T, Pramstaller PP, Salerno A, *et al.* Neuronal loss in Onuf's nucleus in three patients with progressive supranuclear palsy. *Ann Neurol* 2000;**48**:97–101.

484. Schlossmacher MG, Frosch MP, Gai WP, *et al.* Parkin localizes to the Lewy bodies of Parkinson disease and dementia with Lewy bodies. *Am J Pathol* 2002;**160**:1655–1667.

485. Schmidt RE, Dorsey D, Parvin CA, *et al.* Dystrophic axonal swellings develop as a function of age and diabetes in human dorsal root ganglia. *J Neuropathol Exp Neurol* 1997;**56**:1028–1043.

486. Schneider SA, Hardy J, Bhatia KP. Syndromes of neurodegeneration with brain iron accumulation (NBIA): an update on clinical presentations, histological and genetic underpinnings and treatment considerations. *Mov Disord* 2012;**27**:42–53.

487. Scholz SW, Houlden H, Schulte C, *et al.* SNCA variants are associated with increased risk for multiple system atrophy. *Ann Neurol* 2009;**65**:610–614.

488. Schrag A, Schott JM. Epidemiological, clinical, and genetic characteristics of early-onset parkinsonism. *Lancet Neurol* 2006;**5**:355–363.

489. Schrag A, Ben–Shlomo Y, Quinn NP. Prevalence of progressive supranuclear palsy and multiple system atrophy: a cross-sectional study. *Lancet* 1999;**354**:1771–1775.

490. Schrag A, Wenning GK, Quinn N, Ben–Shlomo Y. Survival in multiple system atrophy. *Mov Disord* 2008;**23**:294–296.

491. Schroder JM. Ferritinopathy: diagnosis by muscle or nerve biopsy, with a note on

other nuclear inclusion body diseases. *Acta Neuropathol* 2005;**109**:109–114.

492. Schultz C, Ghebremedhin E, Del Tredici K, *et al*. High prevalence of thorn-shaped astrocytes in the aged human medial temporal lobe. *Neurobiol Aging* 2004;**25**:397–405.

493. Schwarzacher SW, Rub U, Deller T. Neuroanatomical characteristics of the human pre-Botzinger complex and its involvement in neurodegenerative brainstem diseases. *Brain* 2011;**134**:24–35.

494. Seidel K, Schols L, Nuber S, *et al*. First appraisal of brain pathology owing to A30P mutant alpha-synuclein. *Ann Neurol* 2010;**67**:684–689.

495. Seitelberger F. Neuropathological conditions related to neuroaxonal dystrophy. *Acta Neuropathol* 1971;**5**:17–29.

496. Seitelberger F. Neuroaxonal dystrophy: its relation to aging and neurological diseases. In: Vinken PJ, Bruyn GW, Klawans HL eds. *Extrapyramidal disorders in handbook of clinical neurology* Vol 5 (49); Amsterdam: Elsevier Science Publishers BV, 1986:391–415.

497. Seixas AI, Holmes SE, Takeshima H, *et al*. Loss of junctophilin-3 contributes to Huntington disease-like 2 pathogenesis. *Ann Neurol* 2012;**71**:245–257.

498. Selikhova M, Williams DR, Kempster PA, *et al*. A clinico-pathological study of subtypes in Parkinson's disease. *Brain* 2009;**132**:2947–2957.

499. Sergeant N, Wattez A, Delacourte A. Neurofibrillary degeneration in progressive supranuclear palsy and corticobasal degeneration: Tau pathologies with exclusively 'exon 10' isoforms. *J Neurochem* 1999;**72**:1243–1249.

500. Shiraki H, Yase Y. Amyotrophic lateral sclerosis in Japan. In: Vinken PJ, Bruyn GW, Klawans HL eds. *Handbook of clinical neurology*, Vol 22; Amsterdam: North Holland, 1975:353–419.

501. Shojaee S, Sina F, Banihosseini SS, *et al*. Genome-wide linkage analysis of a Parkinsonian–pyramidal syndrome pedigree by 500 K SNP arrays. *Am J Hum Genet* 2008;**82**:1375–1384.

502. Sibon I, Fenelon G, Quinn NP, Tison F. Vascular parkinsonism. *J Neurol* 2004;**251**:513–524.

503. Sidransky E, Nalls MA, Aasly JO, *et al*. Multicenter analysis of glucocerebrosidase mutations in Parkinson's disease. *N Engl J Med* 2009;**361**:1651–1661.

504. Sieh W, Choi Y, Chapman NH, *et al*. Identification of novel susceptibility loci for Guam neurodegenerative disease: challenges of genome scans in genetic isolates. *Hum Mol Genet* 2009;**18**:3725–3738.

505. Silveira–Moriyama L, Gonzalez AM, O'Sullivan S, *et al*. Concomitant progressive supranuclear palsy and multiple system atrophy: More than a simple twist of fate? *Neurosci Lett* 2009;**467**:208–211.

506. Simantov R, Tauber M. The abused drug MDMA (Ecstasy) induces programmed death of human serotonergic cells. *FASEB J* 1997;**11**:141–146.

507. Simon–Sanchez J, Schulte C, Bras JM, *et al*. Genome-wide association study reveals genetic risk underlying Parkinson's disease. *Nat Genet* 2009;**41**:1308–1312.

508. Simunovic F, Yi M, Wang Y, *et al*. Gene expression profiling of substantia nigra dopamine neurons: further insights into Parkinson's disease pathology. *Brain* 2009;**132**:1795–1809.

509. Singleton A, Gwinn–Hardy K. Parkinson's disease and dementia with Lewy bodies: a difference in dose? *Lancet* 2004;**364**:1105–1107.

510. Singleton AB, Farrer M, Johnson J, *et al*. alpha-Synuclein locus triplication causes Parkinson's disease. *Science* 2003;**302**:841.

511. Sohn YH, Jeong Y, Kim HS, *et al*. The brain lesion responsible for parkinsonism after carbon monoxide poisoning. *Arch Neurol* 2000;**57**:1214–1218.

512. Sone J, Tanaka F, Koike H, *et al*. Skin biopsy is useful for the antemortem diagnosis of neuronal intranuclear inclusion disease. *Neurology* 2011;**76**:1372–1376.

513. Sone M, Yoshida M, Hashizume Y, *et al*. alpha-Synuclein-immunoreactive structure formation is enhanced in sympathetic ganglia of patients with multiple system atrophy. *Acta Neuropathol* 2005;**110**:19–26.

514. Song YJ, Lundvig DM, Huang Y, *et al*. p25alpha relocalizes in oligodendroglia from myelin to cytoplasmic inclusions in multiple system atrophy. *Am J Pathol* 2007;**171**:1291–1303.

515. Song YJ, Halliday GM, Holton JL, *et al*. Degeneration in different parkinsonian syndromes relates to astrocyte type and astrocyte protein expression. *J Neuropathol Exp Neurol* 2009;**68**:1073–1083.

516. Spencer PS, Palmer VS, Ludolph AC. On the decline and etiology of high-incidence motor system disease in West Papua (southwest New Guinea). *Mov Disord* 2005;**20** Suppl 12:S119–S126.

517. Spillantini MG, Goedert M. The alpha-synucleinopathies: Parkinson's disease, dementia with Lewy bodies, and multiple system atrophy. *Ann N Y Acad Sci* 2000;**920**:16–27.

518. Spillantini MG, Schmidt ML, Lee VMY, *et al*. Alpha-synuclein in Lewy bodies. *Nature* 1997;**388**:839–840.

519. Spillantini MG, Crowther RA, Jakes R, *et al*. Filamentous alpha-synuclein inclusions link multiple system atrophy with Parkinson's disease and dementia with Lewy bodies. *Neurosci Lett* 1998;**251**:205–208.

520. Stamelou M, Knake S, Oertel WH, Hoglinger GU. Magnetic resonance imaging in progressive supranuclear palsy. *J Neurol* 2011;**258**:549–558.

521. Steele JC, Richardson JC, Olszewski J. Progressive supranuclear palsy: a heterogeneous degeneration involving the brain stem, basal ganglia and cerebellum with vertical gaze and pseudobulbar palsy, nuchal dystonia and dementia. *Arch Neurol* 1964;**10**:333–359.

522. Storch A, Ludolph AC, Schwarz J. Dopamine transporter: involvement in selective dopaminergic neurotoxicity and degeneration. *J Neural Transm* 2004;**111**:1267–1286.

523. Su M, Yoshida Y, Hirata Y, *et al*. Degeneration of the cerebellar dentate nucleus in corticobasal degeneration: neuropathological and morphometric investigations. *Acta Neuropathol* 2000;**99**:365–370.

524. Suls A, Dedeken P, Goffin K, *et al*. Paroxysmal exercise-induced dyskinesia and epilepsy is due to mutations in *SLC2A1*, encoding the glucose transporter GLUT1. *Brain* 2008;**131**:1831–1844.

525. Sung JH, Mastri AR, Park SH. Axonal dystrophy in the gracile nucleus in children and young adults. Reappraisal of the

incidence and associated diseases. *J Neuropathol Exp Neurol* 1981;**40**:37–45.

526. Surmeier DJ, Guzman JN, Sanchez–Padilla J, Schumacker PT. The role of calcium and mitochondrial oxidant stress in the loss of substantia nigra pars compacta dopaminergic neurons in Parkinson's disease. *Neuroscience* 2011;**198**:221–231.

527. Tada M, Kakita A, Toyoshima Y, *et al*. Depletion of medullary serotonergic neurons in patients with multiple system atrophy who succumbed to sudden death. *Brain* 2009;**132**:1810–1819.

528. Takahashi H, Oyanagi K, Takeda S, *et al*. Occurrence of 15-nm-wide straight tubules in neocortical neurons in progressive supranuclear palsy. *Acta Neuropathol* 1989;**79**:233–239.

529. Takahashi J, Fukuda T, Tanaka J, *et al*. Neuronal intranuclear hyaline inclusion disease with polyglutamine-immunoreactive inclusions. *Acta Neuropathol* 2000;**99**:589–594.

530. Takahashi J, Tanaka J, Arai K, *et al*. Recruitment of nonexpanded polyglutamine proteins to intranuclear aggregates in neuronal intranuclear hyaline inclusion disease. *J Neuropathol Exp Neurol* 2001;**60**:369–376.

531. Takahashi M, Weidenheim KM, Dickson DW, Ksiezak–Reding H. Morphological and biochemical correlates of abnormal tau filaments in progressive supranuclear palsy. *J Neuropathol Exp Neurol* 2002;**61**:33–45.

532. Takahashi T, Amano N, Hanihara T, *et al*. Corticobasal degeneration: widespread argentophilic threads and glia in addition to neurofibrillary tangles. Similarities of cytoskeletal abnormalities in corticobasal degeneration and progressive supranuclear palsy. *J Neurol Sci* 1996;**138**:66–77.

533. Takahashi–Fujigasaki J. Neuronal intranuclear hyaline inclusion disease. *Neuropathology* 2003;**23**:351–359.

534. Takeda A. Manganese action in brain function. *Brain Res Rev* 2003;**41**:79–87.

535. Tanaka J. Nasu–Hakola disease: a review of its leukoencephalopathic and membranolipodystrophic features. *Neuropathology* 2000;**20** (Suppl):S25–29.

536. Tayebi N, Callahan M, Madike V, *et al*. Gaucher disease and parkinsonism: a phenotypic and genotypic characterization. *Mol Genet Metab* 2001;**73**:313–321.

537. Tayebi N, Walker J, Stubblefield B, *et al*. Gaucher disease with parkinsonian manifestations: does glucocerebrosidase deficiency contribute to a vulnerability to parkinsonism? *Mol Genet Metab* 2003;**79**:104–109.

538. Taylor KS, Cook JA, Counsell CE. Heterogeneity in male to female risk for Parkinson's disease. *J Neurol Neurosurg Psychiatry* 2007;**78**:905–906.

539. Telenius H, Kremer HP, Theilmann J, *et al*. Molecular analysis of juvenile Huntington disease: the major influence on (CAG)n repeat length is the sex of the affected parent. *Hum Mol Genet* 1993;**2**(10):1535–1540.

540. Thiruchelvam M, Richfield EK, Goodman BM, *et al*. Developmental exposure to the pesticides paraquat and maneb and the Parkinson's disease phenotype. *Neurotoxicology* 2002;**23**:621–633.

541. Thompson PD, Marsden CD. Corticobasal degeneration. In: Rossor MN eds. *Balliere's clinical neurology: unusual*

dementias. London: Balliere Tindall, Vol 1 (no 3), 1992:677–686.

542. Togo T, Dickson DW. Ballooned neurons in progressive supranuclear palsy are usually due to concurrent argyrophilic grain disease. *Acta Neuropathol* 2002;**104**: 53–56.

542a. Togo T, Sahara N, Yen SH, et al. Argyrophilic grain disease is a sporadic 4-repeat tauopathy. *J Neuropathol Exp Neurol* 2002;**61**:547–556.

543. Trojanowski JQ, Revesz T. Proposed neuropathological criteria for the post mortem diagnosis of multiple system atrophy. *Neuropathol Appl Neurobiol* 2007;**33**:615–620.

544. Tsika E, Moore DJ. Mechanisms of LRRK2-mediated neurodegeneration. *Curr Neurol Neurosci Rep* 2012;**12**(3):251–60.

545. Tsuboi Y, Ahlskog JE, Apaydin H, et al. Lewy bodies are not increased in progressive supranuclear palsy compared with normal controls. *Neurology* 2001;**57**:1675–1678.

546. Tsuboi Y, Slowinski J, Josephs KA, et al. Atrophy of superior cerebellar peduncle in progressive supranuclear palsy. *Neurology* 2003;**60**:1766–1769.

547. Tsuboi Y, Josephs KA, Boeve BF, et al. Increased tau burden in the cortices of progressive supranuclear palsy presenting with corticobasal syndrome. *Mov Disord* 2005;**20**:982–988.

548. Tsuchiya K, Ozawa E, Haga C, et al. Constant involvement of the Betz cells and pyramidal tract in multiple system atrophy: a clinicopathological study of seven autopsy cases. *Acta Neuropathol* 2000;**99**:628–636.

549. Tsuchiya K, Murayama S, Mitani K, et al. Constant and severe involvement of Betz cells in corticobasal degeneration is not consistent with pyramidal signs: a clinicopathological study of ten autopsy cases. *Acta Neuropathol* 2005;**109**: 353–366.

550. Uitti RJ, Rajput AH, Ashenhurst EM, Rozdilsky B. Cyanide-induced parkinsonism: a clinicopathologic report. *Neurology* 1985;**35**:921–925.

550a. Uryu K, Nakashima-Yasuda H, Forman MS, et al. Concomitant TAR-DNA-binding protein 43 pathology is present in Alzheimer disease and corticobasal degeneration but not in other tauopathies. *J Neuropathol Exp Neurol* 2008;**67**: 555–564.

551. Valente EM, Abou-Sleiman PM, Caputo V, et al. Hereditary early-onset Parkinson's disease caused by mutations in *PINK1* *Science* 2004;**304**:1158–1160.

552. van der Mark M, Brouwer M, Kromhout H, et al. Is pesticide use related to Parkinson disease? Some clues to heterogeneity in study results. *Environ Health Perspect* 2012;**120**:340–347.

553. Vidal R, Delisle MB, Rascol O, Ghetti B. Hereditary ferritinopathy. *J Neurol Sci* 2003;**207**:110–111.

554. Vidal R, Ghetti B, Takao M, et al. Intracellular ferritin accumulation in neural and extraneural tissue characterizes a neurodegenerative disease associated with a mutation in the ferritin light polypeptide gene. *J Neuropathol Exp Neurol* 2004;**63**:363–380.

555. Vilarino-Guell C, Soto-Ortolaza AI, Rajput A, et al. MAPT H1 haplotype is a risk factor for essential tremor and multiple system atrophy. *Neurology* 2011;**76**:670–672.

556. Vilarino-Guell C, Wider C, Ross OA, et al. VPS35 mutations in Parkinson disease. *Am J Hum Genet* 2011;**89**: 162–167.

557. Vilensky JA, Gilman S, McCall S. A historical analysis of the relationship between encephalitis lethargica and postencephalitic parkinsonism: a complex rather than a direct relationship. *Mov Disord* 2010;**25**:1116–1123.

558. Vincent A. Encephalitis lethargica: part of a spectrum of post-streptococcal autoimmune diseases? *Brain* 2004;**127**:2–3.

559. Vincent A, Bien CG, Irani SR, Waters P. Autoantibodies associated with diseases of the CNS: new developments and future challenges. *Lancet Neurol* 2011;**10**: 759–772.

560. von Bohlen und Halbach O, Unsicker K. Neurotrophic support of midbrain dopaminergic neurons. *Adv Exp Med Biol* 2009;**651**:73–80.

561. Von Economo C. Encephalitis lethargica. *Wien Klin Wochenschr* 1917;**30**:581–585.

562. Vonsattel JP, Myers RH, Stevens TJ, et al. Neuropathological classification of Huntington's disease. *J Neuropathol Exp Neurol* 1985;**44**(6):559–577.

563. Vulpe C, Levinson B, Whitney S, et al. Isolation of a candidate gene for Menkes disease and evidence that it encodes a copper-transporting ATPase. *Nat Genet* 1993;**3**:7–13.

564. Wakabayashi K, Takahashi H, Oyanagi K, Ikuta F. [Incidental occurrence of Lewy bodies in the brains of elderly patients – the relevance to aging and Parkinson's disease]. *No To Shinkei* 1993;**45**: 1033–1038.

565. Wakabayashi K, Oyanagi K, Makifuchi T, et al. Corticobasal degeneration: etiopathological significance of the cytoskeletal alterations. *Acta Neuropathol* 1994;**87**:545–553.

566. Wakabayashi K, Ikeuchi T, Ishikawa A, Takahashi H. Multiple system atrophy with severe involvement of the motor cortical areas and cerebral white matter. *J Neurol Sci* 1998;**156**:114–117.

567. Wakabayashi K, Yoshimoto M, Tsuji S, Takahashi H. Alpha-synuclein immunoreactivity in glial cytoplasmic inclusions in multiple system atrophy. *Neurosci Lett* 1998;**249**:180–182.

568. Wakabayashi K, Yoshimoto M, Fukushima T, et al. Widespread occurrence of alpha-synuclein/NACP-immunoreactive neuronal inclusions in juvenile and adult-onset Hallervorden–Spatz disease with Lewy bodies. *Neuropathol Appl Neurobiol* 1999;**25**:363–368.

569. Wakabayashi K, Tanji K, Mori F, Takahashi H. The Lewy body in Parkinson's disease: molecules implicated in the formation and degradation of alpha-synuclein aggregates. *Neuropathology* 2007;**27**:494–506.

570. Walsh T, Carmichael R, Chestnut J. A hyperthermic reaction to 'ecstasy'. *Br J Hosp Med* 1994;**51**:476.

571. Wang JL, Cao L, Li XH, et al. Identification of PRRT2 as the causative gene of paroxysmal kinesigenic dyskinesias. *Brain* 2011;**134**:3493–3501.

572. Wang Z, Zhang Y, Zhang S, et al. DJ-1 can inhibit microtubule associated protein 1 B formed aggregates. *Mol Neurodegener* 2011;**6**:38.

573. Waring SC, Esteban–Santillan C, Reed DM, et al. Incidence of amyotrophic lateral sclerosis and of the parkinsonism-dementia complex of Guam, 1950–1989. *Neuroepidemiology* 2004;**23**:192–200.

574. Warner TT, Granata A, Schiavo G. Torsin A and DYT1 dystonia: a synaptopathy? *Biochem Soc Trans* 2010;**38**:452–456.

575. Weber YG, Kamm C, Suls A, et al. Paroxysmal choreoathetosis/spasticity (DYT9) is caused by a GLUT1 defect. *Neurology* 2011;**77**:959–964.

576. Wenning GK, Brown R. Dementia in multiple system atrophy: does it exist? *Eur J Neurol* 2009;**16**:551–552.

577. Wenning GK, Quinn N, Magalhaes M, et al. 'Minimal change' multiple system atrophy. *Mov Disord* 1994;**9**:161–166.

578. Wenning GK, Tison F, BenShlomo Y, et al. Multiple system atrophy: A review of 203 pathologically proven cases. *Mov Disord* 1997;**12**:133–147.

579. Wenning GK, Stefanova N, Jellinger KA, et al. Multiple system atrophy: a primary oligodendrogliopathy. *Ann Neurol* 2008;**64**:239–246.

580. Wenning GK, Litvan I, Tolosa E. Milestones in atypical and secondary Parkinsonisms. *Mov Disord* 2011;**26**: 1083–1095.

581. Wider C, Dickson DW, Stoessl AJ, et al. Pallidonigral TDP-43 pathology in Perry syndrome. *Parkinsonism Relat Disord* 2009;**15**:281–286.

582. Wider C, Dickson DW, Wszolek ZK. Leucine-rich repeat kinase 2 gene-associated disease: redefining genotype-phenotype correlation. *Neurodegener Dis* 2010;**7**:175–179.

583. Williams DR, de Silva R, Paviour DC, et al. Characteristics of two distinct clinical phenotypes in pathologically proven progressive supranuclear palsy: Richardson's syndrome and PSP-parkinsonism. *Brain* 2005;**128**:1247–1258.

584. Williams DR, Holton JL, Strand C, et al. Pathological tau burden and distribution distinguishes progressive supranuclear palsy-parkinsonism from Richardson's syndrome. *Brain* 2007;**130**:1566–1576.

585. Williams DR, Holton JL, Strand K, et al. Pure akinesia with gait freezing: a third clinical phenotype of progressive supranuclear palsy. *Mov Disord* 2007;**22**: 2235–2241.

586. Wirdefeldt K, Adami HO, Cole P, et al. Epidemiology and etiology of Parkinson's disease: a review of the evidence. *Eur J Epidemiol* 2011;**26** Suppl 1:S1–58.

587. Wirdefeldt K, Gatz M, Reynolds CA, et al. Heritability of Parkinson disease in Swedish twins: a longitudinal study. *Neurobiol Aging* 2011;**32**:1923 e1921–1928.

588. Wong K, Sidransky E, Verma A, et al. Neuropathology provides clues to the pathophysiology of Gaucher disease. *Mol GenetMetab* 2004;**82**:192–207.

589. Wu E, Dickson DW, Jacobson S, et al. Neuroaxonal dystrophy in HTLV-1-associated myelopathy/tropical spastic paraparesis: neuropathologic and neuroimmunologic correlations. *Acta Neuropathol (Berl)* 1993;**86**:224–235.

590. Yamazaki M, Hasegawa M, Mori O, et al. Tau-positive fine granules in the cerebral white matter: a novel finding among the tauopathies exclusive to parkinsonism-dementia complex of Guam. *J Neuropathol Exp Neurol* 2005;**64**: 839–846.

591. Yokota O, Davidson Y, Bigio EH, *et al.* Phosphorylated TDP-43 pathology and hippocampal sclerosis in progressive supranuclear palsy. *Acta Neuropathol* 2010;**120**:55–66.

592. Yoon DY, Gause CD, Leckman JF, Singer HS. Frontal dopaminergic abnormality in Tourette syndrome: a postmortem analysis. *J Neurol Sci* 2007;**255**:50–56.

593. Yoshida M. Multiple system atrophy: alpha-synuclein and neuronal degeneration. *Neuropathology* 2007;**27**: 484–493.

594. Yoshida Y, Nunomura J, Shimohata T, *et al.* Benign hereditary chorea 2: Pathological findings in an autopsy case. *Neuropathology* 2012;**32**(5):557–65.

595. Young CA, MacKenzie JM, Chadwick DW, Williams IR. Opsoclonus-myoclonus syndrome: an autopsy study of three cases. *Eur J Med* 1993;**2**:239–241.

596. Zaccai J, Brayne C, McKeith I, *et al.* Patterns and stages of alpha-synucleinopathy: relevance in a population-based cohort. *Neurology* 2008;**70**:1042–1048.

597. Zannolli R, Gilman S, Rossi S, *et al.* Hereditary neuronal intranuclear inclusion disease with autonomic failure and cerebellar degeneration. *Arch Neurol* 2002;**59**:1319–1326.

598. Zarranz JJ, Alegre J, Gomez–Esteban JC, *et al.* The new mutation, *E46K*, of alpha-synuclein causes Parkinson and Lewy body dementia. *Ann Neurol* 2004;**55**:164–173.

599. Zhou B, Westaway SK, Levinson B, *et al.* A novel pantothenate kinase gene (PANK2) is defective in Hallervorden–Spatz syndrome. *Nat Genet* 2001;**28**:345–349.

600. Zhou J, Broe M, Huang Y, *et al.* Changes in the solubility and phosphorylation of alpha-synuclein over the course of Parkinson's disease. *Acta Neuropathol* 2011;**121**:695–704.

601. Zijlmans JC, Daniel SE, Hughes AJ, *et al.* Clinicopathological investigation of vascular parkinsonism, including clinical criteria for diagnosis. *Mov Disord* 2004;**19**: 630–640.

602. Zimprich A. Genetics of Parkinson's disease and essential tremor. *Curr Opin Neurol* 2011;**24**:318–323.

603. Zimprich A, Grabowski M, Asmus F, *et al.* Mutations in the gene encoding epsilon-sarcoglycan cause myoclonus-dystonia syndrome. *Nat Genet* 2001;**29**:66–69.

604. Zimprich A, Biskup S, Leitner P, *et al.* Mutations in *LRRK2* cause autosomal-dominant parkinsonism with pleomorphic pathology. *Neuron* 2004;**44**:601–607.

Degenerative Ataxic Disorders

H Brent Clark

INTRODUCTION

Ataxia refers to instability of posture or lack of coordination of movement occurring independently of motor weakness. Characteristic manifestations of ataxia include postural and gait abnormalities, incoordination of limb movements, dysarthria and oculomotor disturbances. Postural changes are typified by swaying instability of the trunk while standing or even sitting. A wide-based stance is assumed, but patients may still have difficulty maintaining balance. Gait is irregular, wide-based and staggering, with loss of the ability to tandem walk. Limb abnormalities include dysmetria, an inability to perform movements with proper timing and trajectory. Intention tremor, loss of fluidity of movement, inability to control force of movement and dysdiadochokinesia are features of dysmetria. The last of these refers to a reduced ability to perform rapid, alternating movements. Dysarthria is defined by abnormalities of rhythm, fluency and clarity of speech. Oculomotor changes may consist of disorders of ocular fixation, ocular alignment, visual pursuit, saccadic movements and nystagmus. Several clinical assessment scales have been developed, ICARS[108] and SARAS,[94] to evaluate the degree of clinical involvement in ataxic patients. A guideline for the clinical diagnosis and management of adult ataxias has been published by the European Federation of Neurological Societies.[111]

Although ataxic conditions can be separated into two groups, primary and secondary (Table 13.1), the primary group is the focus of this chapter, with particular emphasis on inherited forms (Table 13.2). The secondary ataxias and congenital ataxias are summarized briefly in Table 13.3.

CLASSIFICATION OF DEGENERATIVE ATAXIAS

Degenerative disorders resulting in ataxia typically involve the cerebellum, its afferent inputs and its efferent targets. Involvement of these neuroanatomical sites may be the principal manifestation of a neurodegenerative disease process or may be accompanied by neurodegeneration in other systems. Historically, classification of these conditions has been difficult because neither pathological nor clinical characterizations could reliably describe definitive disorders, particularly genetic forms of ataxia. Advances in the identification of the genetic bases of many of these conditions have resulted in a more rational basis of classification, but many of their clinicopathological features remain complicated by phenotypic heterogeneity.

Previous pathological classifications have focused on three major patterns of neurodegeneration:

- Cerebellar cortical degeneration is characterized by loss of Purkinje cells, with less or negligible involvement of cerebellar granule cells or cortical interneurons. Loss of neurons in the inferior olivary nuclei occurs frequently and is believed to result from retrograde trans-synaptic degeneration.
- Olivopontocerebellar degeneration (Fig. 13.1) is characterized by degeneration of the basal pontine nuclei and cerebellar cortex (principally Purkinje cells), with either primary or secondary degeneration of the inferior olivary nuclei. The cranial nerve nuclei, deep cerebellar nuclei, spinal cord, basal ganglia, substantia nigra and red nuclei also may be affected. The associated cerebellar white matter pathways show commensurate axonal loss (Fig. 13.1a).
- Spinocerebellar degeneration is characterized by loss of cerebellar afferent projections, with less or negligible involvement of the cerebellar cortex. The deep cerebellar nuclei may also be affected in some instances, such as in Friedreich's ataxia (FA).

Unfortunately, marked pathological heterogeneity results in overlap among these three, seemingly distinct, pathological subtypes. Furthermore, many of the pathological descriptions of disease entities were made before discovery of their genetic aetiologies, so that correlation of clinical, pathological and genetic information may be difficult.

TABLE 13.1 Classification of ataxias

Primary	Sporadic	Secondary
Inherited	Multiple system atrophy	Toxins, e.g. ethanol, drugs
Autosomal recessive	Idiopathic cerebellar cortical degeneration	Cerebrovascular disease
Autosomal dominant	Idiopathic olivopontocerebellar degeneration	Paraneoplastic/autoimmune disorders
X-linked		Neurometabolic disorders
		Prion diseases (CJD, GSS, kuru)
		Infectious disorders

CJD, Creutzfeldt–Jakob disease; GSS, Gerstmann–Sträussler–Scheinker disease.

TABLE 13.2 Classification of inherited ataxias

Autosomal recessive ataxia	Autosomal dominant ataxia	X-linked ataxias
Friedreich's ataxia	Spinocerebellar ataxias (SCAs) 1–36	Fragile X tremor/ataxia syndrome
Friedreich's ataxia, type 2	Dentatorubropallidoluysial atrophy (DRPLA)	Rare forms (see Table 13.6)
Cerebellar ataxia with early onset and retained tendon reflexes	Episodic ataxias	
Cerebellar ataxia with vitamin E deficiency		
Ataxia telangiectasia		
Ataxia telangiectasia-like disorder		
Early onset ataxia with oculomotor apraxia		
Other rare recessive ataxias (see Table 13.4)		

TABLE 13.3 Secondary and congenital causes of ataxia

Vascular disease or mass lesions of the cerebellum or its afferent/efferent pathways
Infective
　Abscesses
　Viral encephalitis
　Whipple's disease
　Ataxia associated with AIDS
Toxins and drugs
　Ethanol
　Phenytoin
　Cytosine arabinoside
Paraneoplastic syndromes
Langerhans cell histiocytosis
Multiple sclerosis
Other autoimmune disorders
　Gluten enteropathy
Glutamic acid decarboxylase deficiency
Miller–Fisher syndrome
Prion disorders
　GSS
　CJD (especially iatrogenic because of contaminated growth hormone)
Errors of metabolism
　Leukodystrophies
　Neuronal ceroid lipofuscinosis
Congenital malformations
　Dandy–Walker syndrome

AIDS, acquired immunodeficiency syndrome; CJD, Creutzfeldt–Jakob disease; GSS, Gerstmann–Sträussler–Scheinker disease.

Despite numerous patterns of regional involvement, the gross and histopathological features in individual areas are stereotypical. Cerebellar cortical atrophy is reflected in shrinkage of folia, predominantly due to loss of Purkinje cells (Figure 13.2). The early stages of Purkinje cell damage are associated with formation of eosinophilic, silver-positive, proximal axonal dilations, or 'torpedo bodies', usually within the granular layer. The axonal terminals from basket cells that envelop Purkinje cell somata often remain for some time after Purkinje cell loss, resulting in 'empty baskets'. Loss of Purkinje cells is typically accompanied by hypertrophy and hyperplasia of Bergmann glia, with shrinkage of the molecular layer, resulting in at least a transiently increased density of stellate and basket neurons. Loss of granular neurons is usually less severe than loss of Purkinje cells, but in longstanding disease there may be a significantly decreased density, often with an inconspicuous glial reaction. The interneurons of the molecular layer may be depleted in later stages of cortical degeneration. The reduction of Purkinje cell axon terminals in the deep cerebellar nuclei may induce gliosis, with little or no obvious neuronal loss, although in some conditions the dentate neurons are a primary target of the disease process. In the latter instance, there is often associated atrophy of the superior cerebellar peduncles. Inferior olivary atrophy usually accompanies chronic

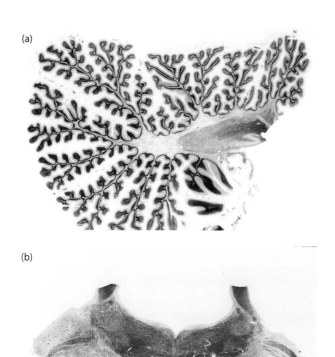

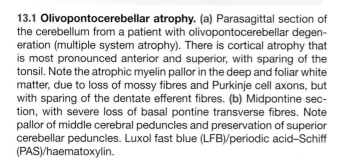

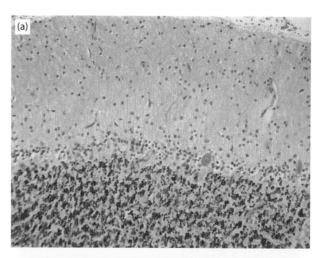

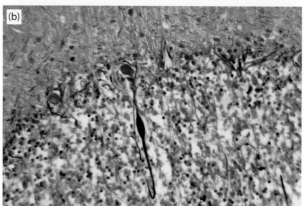

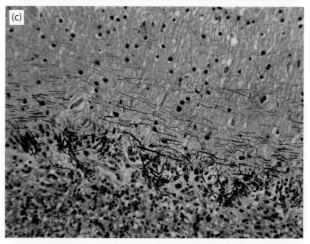

13.1 Olivopontocerebellar atrophy. (a) Parasagittal section of the cerebellum from a patient with olivopontocerebellar degeneration (multiple system atrophy). There is cortical atrophy that is most pronounced anterior and superior, with sparing of the tonsil. Note the atrophic myelin pallor in the deep and foliar white matter, due to loss of mossy fibres and Purkinje cell axons, but with sparing of the dentate efferent fibres. **(b)** Midpontine section, with severe loss of basal pontine transverse fibres. Note pallor of middle cerebral peduncles and preservation of superior cerebellar peduncles. Luxol fast blue (LFB)/periodic acid–Schiff (PAS)/haematoxylin.

13.2 Cerebellar cortical degeneration. (a) Severe loss of Purkinje cells with shrinkage of the molecular layer and Bergmann gliosis (SCA1). **(b)** Purkinje cell with a dilated proximal axon (torpedo). Bielschowsky. **(c)** 'Empty' basket fibres persisting in sites of degenerated Purkinje cells (SCA6). Bielschowsky.

Purkinje cell loss but can be surprisingly mild in some disorders that have severe cerebellar cortical atrophy. In some forms of ataxia, it is likely that olivary atrophy is a primary manifestation of the disease.

In more complex degenerations, with brain stem and spinal cord involvement, there are other characteristic features. Degeneration of the basal pontine nuclei is reflected by gross atrophy of the basis pontis, loss of transverse fibres, shrinkage of the middle cerebellar peduncles and loss of volume of the cerebellar white matter. Microscopically, these changes are not unique but show loss of neurons and fibre tracts, with accompanying astrocytosis. Similar histopathological changes can be seen in other brain stem and spinal cord nuclei and tracts.

New histopathological diagnostic tools have emerged from increased understanding of many of the degenerative ataxias. A number of the autosomal dominant spinocerebellar ataxias (SCAs) that are related to expanded polyglutamine residues within the mutant protein feature intraneuronal inclusions, which are immunopositive for the mutated protein or ubiquitin, or with antibodies directed against expanded polyglutamine tracts. Likewise, the glial cytoplasmic inclusions characteristic of multiple system atrophy, a common sporadic form of ataxia, are immunoreactive for α synuclein and ubiquitin. Calbindin is a useful marker for Purkinje neurons and transmitter-specific markers have utility in elucidating specific types of neuronal loss.

AUTOSOMAL RECESSIVE CEREBELLAR ATAXIAS

Most of the early-onset inherited ataxias are autosomal recessive, with the most common forms being FA and ataxia telangiectasia (AT). A few other rarer forms of recessive ataxia are discussed later and listed in Table 13.4. Guidelines for diagnosis and management of these disorders have been published.[1] Further information for these diseases may be found online at the Online Mendelian Inheritance in Man website (www.ncbi.nlm.nih.gov) and at the website of Washington University Department of Neurology (www.neuro.wustl.edu).

Friedreich's Ataxia

Clinical Features

Friedreich's ataxia is the most common hereditary ataxia in white populations, accounting for nearly half of all hereditary ataxias and up to 75 per cent of ataxias presenting before age 25 years.[41] The disease occurs almost exclusively in populations of European, Middle Eastern, Indian and North African ancestry, with a prevalence of 1–2 per 100 000 population. Onset typically occurs early in the second decade of life, but the range extends from 18 months of age to well into the third decade. The first manifestation of disease usually is instability of gait or a sense of clumsiness. Romberg's sign is an early feature, and choreiform motor restlessness sometimes occurs in the early stages. Truncal ataxia progresses to limb ataxia, dysarthria, areflexia and sensory neuropathy with distal loss of joint position and vibration sense. As the course progresses, there is lower extremity weakness with extensor plantar reflexes. Distal amyotrophy in the legs and hands and autonomic abnormalities, such as cold cyanotic feet, are common. Optic atrophy, often without significant visual loss, and sensorineural hearing loss are seen in 10–25 per cent of patients. Oculomotor changes are nearly always present, but nystagmus is relatively infrequent. Scoliosis and deformities of the feet are common. Most patients have a cardiomyopathy, which is frequently asymptomatic but is usually more severe in patients with early onset. Abnormal glucose tolerance is present in about a quarter of patients, 10 per cent overall having insulin-dependent diabetes mellitus. Discovery of the gene for FA has allowed the inclusion of some atypical cases within a broader spectrum of FA. There are variant forms of FA with a later onset and absence of such features as areflexia, reduced vibration sense or plantar extensor reflexes.

The course of FA is variable, but most patients require a wheelchair within 15 years of presentation. Death occurs most frequently in the fourth decade of life, but prolonged survival is common, particularly when diabetes and cardiac disease are absent or treated effectively.

Genetics and Cell Biology

The gene for FA, *FRDA*, is on chromosome 9q. It encodes an 18-kDa protein, frataxin,[13] which is expressed widely, with greatest levels in human heart and spinal cord; however, frataxin is also is present in brain, liver, skeletal muscle and pancreas. The mutation in all but a few patients is a GAA triplet repeat expansion in the first intron of the gene, affecting both copies. Normal alleles have fewer than 40 repeats, most having between 6 and 10 repeats. The range of repeat lengths associated with FA is from 70 to more than 1000, the most frequent number being 600–1000. The GAA repeat expansion interferes with frataxin expression, probably by repeat-induced conformational changes that inhibit transcription. The amount of expressed protein correlates inversely with repeat length and is influenced most by the size of the shorter of the two affected alleles. Protein expression also correlates with disease severity and the presence of cardiomyopathy, but the variability in age of onset is not

TABLE 13.4 Other rare autosomal recessive ataxias

Syndrome	OMIM number	Locus/gene
Ataxia with hypogonadotrophic hypogonadism (Gordon–Holmes syndrome)	212840	Unknown/unknown
Cayman ataxia	601238	19p13/caytaxin
Charlevoix–Saguenay spastic ataxia	270550	13q12/sacsin
Coenzyme Q10 deficiency	607426	9p13/unknown
Infantile-onset spinocerebellar ataxia	271245	10q24/unknown
Marinesco–Sjögren syndrome	248800	5q31/unknown
Refsum's disease	266500	10p/PAHX, 6q/PEX7
Salla disease (sialuria)	604369	6q14–q15/sialin
Spinocerebellar ataxia with axonal neuropathy	607250	14q31–q32/TDP1
Spinocerebellar ataxia with blindness and deafness	271250	6p23–p21/unknown
Spinocerebellar ataxia with childhood onset	609270	11p15/unknown

OMIM, Online Mendelian Inheritance in Man.

accounted for entirely by the expansion size. As with other nucleotide repeat expansions, the GAA repeat is unstable and may expand or contract during meiosis or even mitosis. In approximately 2–5 per cent of cases, a point mutation in the FRDA gene on one chromosome is coupled with a GAA expansion on the other. Most point mutations result in complete loss of function of frataxin, but several are associated with a milder phenotype, suggesting some preserved function.[17] Patients with FA with homozygous point mutations have not been identified. Some of the variant forms of FA can now be explained by the nature of the mutation, with milder and late-onset conditions associated with relatively short expansion of the GAA repeat or with point mutations that result in milder loss of frataxin function.

Although the normal function of frataxin is not completely understood, it has structural homologues in mammals, invertebrates, plants and yeast and is believed to be involved in iron transport in mitochondria. Gene knockout of the yeast homologue results in a shift of cytosolic iron into the mitochondria, resulting in a compensatory upregulation of iron transporters in the plasmalemma, which may result in an amplification of iron deposition. This deficit can be corrected by transfection with normal human frataxin.[119] Frataxin knockout in mice results in embryonic lethality, which can be rescued by transgenic human frataxin.[82] Iron accumulation has been noted in tissues of patients with FA, particularly in the heart and central nervous system (CNS). Increased mitochondrial iron may lead to oxidative injury or interference with oxidative phosphorylation. Clinical trials using antioxidants and enhancers of respiratory chain function have shown some cardiac protection.[42]

A recessive ataxia called Friedreich's ataxia type 2 (FRDA2; OMIM 601992) has been mapped to chromosome 9p23–p11. These patients are often clinically indistinguishable from those with classic FA but do not have the frataxin mutation. Other recessive ataxias in the differential diagnosis of FA include early-onset ataxia with retained reflexes and ataxia with isolated vitamin E deficiency (see Cerebellar Ataxia with Isolated Vitamin E Deficiency, this page).

Neuropathology

Friedreich's ataxia characteristically involves cerebellar afferents and deep cerebellar nuclear efferents, with minimal involvement of the cerebellar cortex. Grossly, the spinal cord appears to be abnormally small, often with obvious loss of myelinated fibres in the dorsal and lateral columns. The dorsal roots and their ganglia may be reduced in size. The cerebellum rarely shows gross atrophy, but the dentate nuclei may appear reduced in volume. The severity and distribution of microscopic pathology correlate with expansion of the GAA repeat. Microscopic changes in the spinal cord include myelinated fibre loss in the dorsal columns, with greater involvement of the gracile fasciculi; the corticospinal tracts, with greater involvement of the lateral columns; and the spinocerebellar tracts, with greater involvement of the dorsal pathways (Figure 13.3). There is severe loss of neurons in Clarke's columns, but other grey matter areas of the cord are spared. Dorsal root ganglia show neuronal loss, with frequent nodules of Nageotte, and

there is commensurate loss of myelinated axons in the dorsal roots.[62] The anterior roots are not affected in most cases. There is corticospinal tract degeneration at medullary and sometimes pontine levels (Figure 13.3d). The gracile and cuneate nuclei typically demonstrate obvious neuronal loss, as may the vestibular nuclei. Other cranial nerve, basal pontine and olivary nuclei are usually spared, although hypertrophic degeneration of the inferior olivary nuclei is occasionally present. The accessory olivary nuclei may be affected. Cell loss and astrocytosis are usually seen in the vestibular and cochlear nuclei and in the superior olives. Primary cerebellar involvement is in the dentate nuclei, with neuronal loss and degeneration of efferent fibres in the hilar zone and atrophy of the superior cerebellar peduncles (Figure 13.3e). Dentatoolivary projections by GABAergic neurons in the dentate nucleus are spared.[63] The cerebellar cortex may show mild to moderate loss of Purkinje cells and reduction in the thickness of the molecular layer. Purkinje cells may demonstrate dendritic abnormalities, axonal torpedoes and terminal axonal alterations that contribute to the grumose degeneration seen around neurons in the dentate nuclei. The red nuclei are rarely abnormal. There may be neuronal loss in the ventral-posterior thalamus, external globus pallidus and subthalamic nuclei. Most cases show mild degenerative changes in the optic nerves and tracts. The cerebral cortex is spared, although there may be loss of Betz cells in the primary motor cortex. Peripheral sensory axonopathy is very common, but motor neuropathy is infrequent. The cardiomyopathy is characterized by hypertrophy of myocytes and pleomorphic nuclei. There is granular deposition of iron, demonstrable with Prussian blue. The heart is often enlarged and has multiple foci of myofibre loss, with fibrosis and variable amounts of chronic inflammation. The endocrine pancreas may have selective β-cell atrophy.

Cerebellar Ataxia with Early Onset and Retained Tendon Reflexes

An early-onset cerebellar ataxia was described by Harding, in which many of the features of FA were absent and the prognosis was better.[40] These patients had preservation of deep tendon reflexes and a milder peripheral neuropathy, with no optic atrophy, skeletal changes, cardiomyopathy or diabetes. Pathological studies are not documented, but neuroimaging has shown atrophy, primarily in the cerebellum.[58] This condition has been linked to chromosome 13q12 in a Tunisian kindred;[76] however, it is likely to be heterogeneous.

Cerebellar Ataxia with Isolated Vitamin E Deficiency

A clinical mimic of FA, although typically without cardiomyopathy, has been described in patients with mutations of α transfer protein gene (TTPA), located on chromosome 8q13.[5] In some Mediterranean countries, this condition is as common as FA, but it lacks the GAA expansion in the frataxin gene. The TTPA mutation prevents proper utilization of dietary vitamin E, leading to deficiency. Most cases

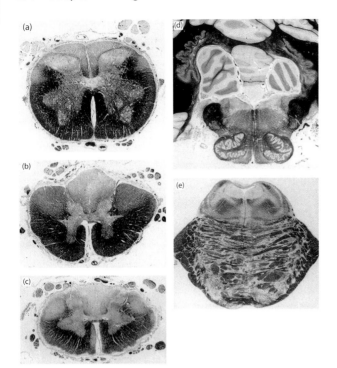

13.3 Friedreich's ataxia. (a) Sacral spinal cord from patient with Friedreich's ataxia who died at age 32 years. There is myelin pallor due to axonal loss in the posterior roots and columns and pyramidal tracts. **(b)** Thoracic cord from patient with Friedreich's ataxia who died at age 23 years. There is atrophic pallor of the posterior roots and columns, pyramidal tracts and spinocerebellar tracts. **(c)** Cervical cord from patient with Friedreich's ataxia who died at age 31 years. Atrophic pallor is present in the posterior roots and columns (gracile > cuneate), anterior and lateral pyramidal tracts and both posterior and anterior spinocerebellar tracts. **(d)** Medulla in patient with Friedreich's ataxia, with atrophic pallor of the pyramids. **(e)** Atrophic pallor of the superior cerebellar peduncles in the rostral pons. Stains for myelin.

are secondary to frameshift TTPA mutations, but milder cases have been associated with specific point mutations. Neuropathological studies have shown changes, such as posterior column degeneration with axonal spheroids, which can be seen in vitamin E deficiency in other clinical contexts. There is increased lipofuscin accumulation in a number of neuronal populations, including the dorsal root ganglia. Mild Purkinje cell loss has been reported. As this form of ataxia is treatable with dietary vitamin E supplementation, diagnostic studies for early-onset ataxias should include vitamin E levels. Kindreds with ataxia and vitamin E deficiency have been reported without TTPA mutations, suggesting that other gene mutations remain to be discovered.

Ataxia Telangiectasia

Clinical Features

Ataxia telangiectasia is the most common cause of progressive ataxia in the first five years of life. It has a prevalence of 1–2 per 100 000 population. Truncal and gait ataxia are usually present at the time the affected child is learning to walk, although late-onset variants are increasingly being recognized. Subsequently, there is limb ataxia, dysarthria

and a complex movement disorder usually dominated by chorea or dystonia.[10,120] Facial hypotonia, drooling and oculomotor apraxia are seen frequently, whereas areflexia, distal amyotrophy, weakness and sensory loss develop with disease progression. The characteristic conjunctival and cutaneous telangiectasias usually develop within a few years of the onset of ataxia. Abnormalities of the immune system, chiefly athymia or thymic hypoplasia and immunoglobulin deficiency, result in recurrent respiratory tract infections. Nearly half of patients develop a neoplasm, most commonly leukaemia or lymphoma, sometimes preceding the onset of ataxia. With increasing age, there is an increased risk of various carcinomas, malignant melanoma and primary brain neoplasms. There is an increased sensitivity to detrimental side effects of radiation exposure. Patients often have growth retardation and hypogonadal symptoms. There may be progeric changes of hair and skin. Laboratory investigations show elevated alpha-fetoprotein levels, cytogenetic evidence of chromosome fragility and increased radiosensitivity in cultured lymphocytes or fibroblasts. Most patients survive into the third decade of life.

Genetics and Cell Biology

Ataxia telangiectasia is due to mutations in the ataxia-telangiectasia mutated (ATM) gene, located on chromosome 11q.[91] The 350-kDa ATM protein is one of a family of phosphatidylinositol-3 kinases that respond to DNA damage by phosphorylating substrates involved in DNA repair or cell cycle control. The gene is large, with 150 kb and 66 exons; many different mutations have been identified throughout the gene, without clustering in mutational hot spots, so molecular diagnosis is difficult. Most of the known mutations result in either truncation of the protein or splicing abnormalities.[16] In addition to its role in double-stranded DNA repair, ATM appears to regulate signalling pathways related to tumour suppression, cell growth and differentiation.[98]

Other rare recessive disorders associated with DNA repair defects, such as xeroderma pigmentosum and Cockayne's syndrome, may have ataxia as part of the clinical syndrome, often accompanied by some of the systemic features of AT.

Neuropathology

The cerebellar cortex is grossly atrophic, with involvement of the vermis and the hemispheres. Microscopically, there is severe loss of Purkinje cells, with partial depletion of granular neurons. The stellate and basket cells are relatively preserved, and empty baskets are often prominent. Surviving Purkinje cells may have a heterotopic location in the molecular layer, aberrant dendritic morphology, eosinophilic dendritic or somatic inclusions and axonal torpedoes. There is retrograde degeneration in the inferior olivary nuclei and gliosis in the dentate nuclei, the latter probably due to loss of Purkinje cell terminals. Basal pontine nuclei are spared and the only other brain stem pathology may be an increased number of axonal spheroids in the posterior column nuclei in patients with longer survival. The spinal cord may have posterior column atrophy with the gracile tracts more affected. The dorsal root ganglia have nucleomegaly and dystrophic changes in satellite Schwann cells. Atrophy or loss of anterior horn neurons with neurogenic amyotrophy has been reported, usually in longstanding disease. The striatum is largely spared, but the substantia nigra may have loss of pigmented neurons and sometimes Lewy body-like inclusions. Progeric changes are present in older patients and consist of increased lipofuscin deposition in the forebrain and NFTs in cerebral cortex, hippocampus, basal ganglia and spinal cord. Older patients may also have small gliovascular nodules, consisting of dilated capillary loops with perivascular siderosis and fibrillary gliosis.[6]

Enlarged, bizarre, hyperchromatic nuclei occur in many systemic tissues. The anterior lobe of the pituitary is severely affected, but additional sites are increasingly affected with age. There is a widespread reduction of lymphoid tissue, and any residual thymus is dysplastic and without Hassall's corpuscles.

Ataxia Telangiectasia–like Conditions

Two separate entities have been described as 'ataxia telangiectasia-like'. One, reported in only a few kindreds, is called 'ataxia telangiectasia-like disorder' and presents in childhood with a later onset than AT. The major manifestations are ataxia, dysarthria and oculomotor apraxia. Clinical progression is slower and probably milder, with ambulation still present in the third decade of life. Radiosensitivity is present, but telangiectasias, as well as the immunological and neoplastic manifestations seen in AT, are typically absent.[29,106] The mutation has been mapped near the *ATM* gene on 11q, in the *MRE11* gene,[103] a member of the Mre11/Rad50/Nbs1 protein complex, which is involved in responses to cellular damage induced by ionizing radiation.

The second ataxia telangiectasia-like condition has been termed 'early-onset ataxia with oculomotor apraxia, type 1' (AOA1). Affected patients have onset in the middle of the first decade of life, with ataxia, oculomotor apraxia, choreoathetosis and peripheral neuropathy. Telangiectasias and the other systemic manifestations of AT are absent, but there is hypoalbuminaemia and hypercholesterolaemia. The mutation is on chromosome 9p13 in a gene called *APTX*, which codes for aprataxin, a member of the histidine triad super-family.[20] An additional condition, AOA2, results from a mutation in the *SETX* gene encoding senataxin on chromosome 9q34.[74] As oculomotor apraxia is not always present, this condition also has been designated as spinocerebellar ataxia recessive type 1 (SCAR1). These patients have onset in the second decade of life and often have amyotrophy as well as ataxia, oculomotor apraxia and neuropathy. None of these conditions has been well-characterized neuropathologically.[18]

Other Recessive Ataxias

There are various other rare forms of recessive ataxia, which are listed in Table 13.4. In addition, ataxia is a variably important manifestation of other recessively inherited disorders of the nervous system. Ataxia is associated with several types of inherited myoclonus. Unverricht–Lundborg disease, also known as Baltic myoclonus, presents with epilepsy and myoclonus, but ataxia develops later in its course. It is caused by a mutation in the gene for cystatin B.[80]

Atrophic changes in the dentate nuclei and spinal tracts may be present. Ramsay Hunt syndrome, or dyssynergia cerebellaris myoclonica, consists of myoclonic ataxia attributed to neuronal degeneration in the dentate nuclei. Two of the patients described by Hunt were twins with an apparent recessive inheritance pattern.[49] Ataxia may be a feature of mitochondrially transmitted disorders, such as myoclonic epilepsy with ragged red fibres (MERRF).

AUTOSOMAL DOMINANT CEREBELLAR ATAXIA

Even more difficult to classify than the recessive ataxias are the autosomal dominantly inherited ataxias, for which there is an ever-increasing number of forms, none of which is common. The combined incidence of all forms of autosomal dominant ataxia is 1–5 per 100 000 population. Because of extensive phenotypic variability within kindreds, both clinical and neuropathological classification schemes have been unsatisfactory. Pathological classifications tended to reflect neuroanatomical patterns, i.e. cerebellar cortical atrophy, multisystem atrophy, olivopontocerebellar atrophy (OPCA) or spinocerebellar atrophy, but these patterns may overlap within affected families, so that classification solely by pathological pattern is inadequate.[66] A clinical classification of autosomal dominant cerebellar ataxia (ADCA) was introduced by Harding.[41] Autosomal dominant cerebellar ataxia type I consists of cerebellar symptoms accompanied by numerous other neurological problems, including oculomotor, lower motor neuron, pyramidal, extrapyramidal, sensory, visual, auditory and cognitive problems. Pathologically, these patients tend to correspond to the multisystem or OPCA-like patterns. Autosomal dominant cerebellar ataxia type II resembles ADCA I, but patients have pigmentary retinal degeneration and are now known to correspond to SCA7. Patients with ADCA III have a more pure cerebellar syndrome, unaccompanied by other neurological problems.[69]

Classification, Pathology and Genetics

Genes have been mapped or identified for more than 30 forms of dominant ataxia since 1993, revolutionizing classification and leading to a greater understanding of molecular mechanisms. The new classification system lists the disorders as SCA types 1–37 (Table 13.5), with a few gaps in the numbering, as SCA9, SCA24 and SCA33 are unassigned and several separate designations may indicate the same condition (SCA15/16, and SCA19/22). Unfortunately, even this nomenclature is flawed, because a number of these disorders do not have both spinal and cerebellar involvement. Included in the autosomal dominant ataxias, but without the SCA designation, are DRPLA and the episodic ataxias. A common theme among a number of these diseases, including SCA1, SCA2, SCA3, SCA6, SCA7, SCA17 and DRPLA, is the presence of an expanded CAG repeat (coding for polyglutamine) within the affected gene. Descriptions of the following entities are necessarily brief, emphasizing the more typical presentations, but a number of these conditions may show a broad range of clinical and pathological manifestations. Some of this clinicopathological heterogeneity is

TABLE 13.5 Autosomal dominant spinocerebellar ataxias (SCAs)

Type	OMIM number	Locus/gene	Details
SCA1	164400	6q23/ataxin 1	CAG repeat, ADCA I, cerebellar cortex and dentate nuclei affected
SCA2	183090	12q24/ataxin 2	CAG repeat, ADCA I, dentate sparing
SCA3/MJD	109150	14q24–q32/ataxin 3	CAG repeat, ADCA I, cerebellar cortical sparing
SCA4	600223	16q22/unknown	Ataxia with sensory neuropathy
SCA5	600224	11p11–q11/beta-II-spectrin	ADCA III
SCA6	183086	19p13/CACNA1A	Small CAG repeat, ADCA III
SCA7	164500	3p21–p12/ataxin 7	CAG repeat, ADCA II
SCA8	608768	13q21/SCA8	Untranslated CTG repeat, ADCA III
SCA10	603516	22q13/ataxin 10	Intronic ATTCT repeat, seizures are common
SCA11	604432	15q14–q21/TTBK2	ADCA III, mild, late onset
SCA12	604326	5q21–q33/PPP2R2B	5′ untranslated CAG repeat, ADCA I, tremor
SCA13	605259	19q13.3–q13.4/KCNC3	Early onset, developmental delay
SCA14	605361	19q3.4/PKC-gamma	ADCA I or ACDA III depending on age of onset
SCA15/SCA16	606658	3p26.2-p25.3/ITPB1	ADCA III, slow progression
SCA17	607136	6q27/TBP	CAG repeat in TATA box binding protein, ADCA I with cognitive or psychiatric features
SCA18	607458	7q22–q32/IFRD1 is a candidate gene	Sensorimotor neuropathy
SCA19/SCA22	607346	1p21–q21/unknown	Cognitive impairment, SCA22 locus is the same but no cognitive features
SCA20	608687	11p13–q11/unknown	Maps near SCA5, dentate calcification and palatal myoclonus
SCA21	607454	7p21.3–p15.1/unknown	Extrapyramidal involvement is common
SCA23	610245	20p13–p12.3/ prodynorphin	Hyperreflexia, reduced vibration
SCA25	608703	2p21–p13/unknown	Early onset, sensory neuropathy
SCA26	609306	19p13.3/eEF2 is candidate gene	ADCA III
SCA27	609307	13q34/FGF-14	Fibroblast growth factor 14, early onset with tremor, later ataxia and cognitive changes
SCA28	610246	18p11.22–q11.2/AFG3L2	Early onset, slow progression, pyramidal signs
SCA29	117360	3p26/see SCA15/16	Dystonia, cognitive changes
SCA30	613371	4q34.3-q35.1/unknown	Late onset, mild ataxia, mild spasticity
SCA31	117210	16q21/BEAN	TGAAA-repeat expansion, late onset, hearing loss
SCA32	613909	7q32-q33/unknown	Variable cognitive impairment, male infertility
SCA34	133190	6p12.3-q16.2/unknown	Infantile onset, erythrokeratodermia, hyporeflexia
SCA35	613908	20p13/TGM6	Tremor, hyperreflexia, decreased proprioception
SCA36	614153	20p13/NOP56	Intronic GGCCTG-repeat expansion, late onset, motor neuron involvement
SCA37	Number not assigned	1p32/unknown	Late onset, slow progression, vertical gaze problems

ADCA, autosomal dominant cerebellar ataxia; MJD, Machado–Joseph disease; OMIM, Online Mendelian Inheritance in Man.

explained by genetic mechanisms, including variability in repeat expansion lengths in the polyglutamine disorders.

SCA1

SCA1 was the first autosomal dominant ataxia to be associated with an expanded CAG repeat. The gene for SCA1 is located on chromosome 6p23 and codes for ataxin 1, a protein of unknown function that is present in the nuclei of most neurons.[4,78] The clinical features are variable, depending largely on the size of the CAG repeat expansion. Onset ranges from childhood to late adult life, but the mean age is in the fourth decade. Truncal and limb ataxia and dysarthria are early features that are often followed by slowing or absence of saccadic eye movements, gaze paresis, bulbar motor symptoms, spasticity, peripheral neuropathy and mild cognitive impairment. Most patients are disabled within 5–10 years of onset and die 10–20 years after onset. The most frequent pathological pattern is olivopontocerebellar atrophy.[85] The cerebellum shows loss of Purkinje cells and neuronal loss in the dentate nuclei. Olivary neuronal loss is extensive and may be disproportionate to loss of Purkinje cells. Basal pontine neuronal atrophy is variable but may be severe. Other areas of brain stem involvement include the red nuclei and motor cranial nerve nuclei III, X and XII. The anterior horns, posterior columns and spinocerebellar tracts also have atrophy. The substantia nigra is usually normal, but the pallidum and even the neostriatum may be affected in some cases. Cerebral cortex and hippocampus may have mild neuronal loss. Intranuclear inclusions immunoreactive for ataxin 1 and ubiquitin are present in many areas of affected brains, but not typically in Purkinje cells. A recent study using thick-section techniques has indicated more wide-spread involvement in the brain stem, deep grey nuclei and cerebral cortex.[87]

SCA2

More common than SCA1, this condition is characterized by ataxia, tremor, slow saccadic eye movements, myoclonus, hyporeflexia and sensory neuropathy in nearly all cases. Parkinsonism, amyotrophy, oculomotor weakness, chorea and cognitive impairment may also be present. The age of onset is highly variable and ranges from infancy to later adult life. Most cases present before age 30 years. SCA2 results from a CAG repeat expansion in the ataxin-2 gene on 12q24.1.[83,107] Pathologically, there is an olivopontocerebellar atrophy in most patients. Purkinje cells are significantly depleted, but the dentate nuclei are relatively spared of neuronal loss. The substantia nigra has loss of neurons in the pars compacta, and there is usually neuronal loss in the neostriatum and pallidum. Atrophy is present in Clarke's nuclei, spinocerebellar tracts and dorsal columns, with less or no involvement of the corticospinal tracts. Thick-section techniques have found more widespread involvement of the brainstem and cerebral deep grey structures.[33,88] Cytoplasmic and nuclear inclusions, positive for ubiquitin and expanded ataxin 2, have been described, but are not as reliably present as in some other polyglutamine diseases. Significant anticipation can be seen in SCA2, with very long repeats sometimes resulting in infantile onset

in which the clinical-pathological appearance can be significantly different from the adult-onset disease.

SCA3

This condition also is known as Machado–Joseph disease (MJD), once thought to be a separate entity but sharing the same mutation, a CAG expansion in ataxin-3, which maps to chromosome 14q32.1.[100] It is a common form of hereditary ataxia in many European populations and in Japan. Gait ataxia, dysarthria, spasticity and nystagmus are common at presentation. Supranuclear ophthalmoplegia, bulbar amyotrophy, milder limb amyotrophy and loss of reflexes frequently ensue. Death is often related to dysphagia and a weakened cough. The age of onset is variable and often correlated with the clinical features. Classically, MJD was seen in patients of Portuguese–Azorean ancestry and was subdivided into different clinical types. Type 1 is the least common, having onset from 5 to 30 years of age and presenting with little ataxia but with spasticity, rigidity and bradykinesia. Ptosis or bulging of the eyes may be a feature in some patients. Type 2 is the most common and typically has onset in the fourth decade with progressive ataxia and spasticity. Type 3 has a mean onset in the fifth decade and is characterized by ataxia and sensorimotor neuropathy with amyotrophy and areflexia. Parkinsonism may be a predominant finding in some patients and is sometimes categorized as type 4. Neuropathological studies demonstrate involvement of cerebellar afferent and efferent pathways, extrapyramidal structures and lower motor neurons.[97] There is neuronal loss in Clarke's, vestibular, basal pontine, dentate and red nuclei and in the dorsal root ganglia, with concordant degeneration of their axonal tracts. The cerebellar cortex and olivary nuclei are largely spared. Extrapyramidal involvement includes the substantia nigra, the subthalamic nuclei and, to a lesser extent, the pallidum. The neostriatum and cerebral cortex are not significantly affected. There is also loss of motor neurons in the cranial nerve nuclei and anterior horns. Despite spasticity, there is no corticospinal tract atrophy. Cell-counting studies have shown loss of neurons in the lateral reticular, raphe interpositus and external cuneate nuclei.[97] Thick-section studies have found more extensive involvement of brain stem nuclei.[86] The degenerative changes often are accompanied by intranuclear neuronal inclusions that stain with antibodies against ataxin-3 or ubiquitin.[79] As for other polyglutamine diseases, the clinical severity and age of onset tend to correlate with the length of the expanded CAG repeat. The threshold for disease is higher than in some other CAG repeat diseases, with most affected patients having more than 60 CAG repeats. Anticipation is noted and is more associated with paternal transmission. Homozygous cases have been reported to have a more severe phenotype.

SCA4

Also known as 'hereditary ataxia with sensory neuropathy', SCA4 is a rare, progressive ataxia that maps to 16q22 but is not allelic to SCA31.[26] Presentation is in the third and fourth decades of life with ataxia and subsequent development of sensory neuropathy, which may not be noted by the patient at onset. There can also be auditory impairment. An autopsy study of one patient using thick-section analysis[44] found widespread degenerative changes in the cerebellum and brainstem. Neuronal loss was present in the substantia nigra and ventral tegmental area, central raphe and pontine nuclei, all auditory brainstem nuclei, in the abducens, principal trigeminal, spinal trigeminal, facial, superior vestibular, medial vestibular, interstitial vestibular, dorsal motor vagal, hypoglossal and prepositus hypoglossal nuclei, as well as in the nucleus raphe interpositus, all dorsal column nuclei and in inferior olives. There was marked loss of Purkinje cells as well as neuronal loss in the cerebellar fastigial nucleus, in the red, trochlear, lateral vestibular and lateral reticular nuclei, the reticulotegmental nucleus of the pons and the nucleus of Roller.

SCA5

Sometimes called the 'Lincoln ataxia' because it is present in kindred descended from the paternal grandparents of former United States president Abraham Lincoln, this condition has been mapped to the centromeric region of chromosome 11. SCA5 is a relatively mild form of ataxia, without an effect on lifespan. Truncal, gait and upper limb ataxia are typically present, along with dysarthria and abnormal eye movements. Mild sensory neuropathy and hyperreflexia are sometimes present. Pathology on a single case has been reported, changes being confined to cerebellar cortical degeneration with severe loss of Purkinje cells and mild to moderate loss of olivary neurons.[68] Mutations in the gene for beta-III-spectrin have been found in the Lincoln kindred and in two separate European families.[52] In a beta-III-spectrin knock-out model there were alterations demonstrated in Purkinje cell sodium currents and glutamate signalling.[81]

SCA6

This unusual type of polyglutamine disease is a common form of ataxia in people of northern European and Japanese ancestry. The affected gene, *CACNA1A* on 19p13,[122] encodes α subunit of a P/Q-type voltage-dependent calcium channel that is highly expressed in Purkinje cells. SCA6 presents most commonly after the age of 50 years and typically has a slow progression. As onset is so late, parents with the mutation may not have manifested the disease before death and an affected child could be thought to have a sporadic form of ataxia. Episodic ataxia may also be a feature early in the disease. Clinical features include dysarthria, truncal and limb ataxia, and abnormalities of eye movement, including nystagmus and an abnormal vestibulo-ocular reflex. Extrapyramidal symptoms have been described in some patients. Pathological studies have shown severe Purkinje cell loss that is more accentuated in the vermis and the superior portions of the cerebellar hemispheres. Surviving Purkinje cells have been described as having heterotopic, irregularly shaped nuclei and swollen dendrites with spiny protrusions. The cerebellar granular layer and the inferior olives have relatively mild neuronal loss but the severity appears to correlate with duration of disease. There is no obvious involvement of the pons, and although there is gliosis in the dentate nuclei, neuronal loss is not conspicuous. Despite reports of extrapyramidal symptoms in some

patients, pathological studies to date have shown minimal involvement of the substantia nigra and basal ganglia.[37,109] Analysis by thick sections has indicated mild neuronal loss in the deep cerebellar nuclei, brainstem and motor cortex that are not readily appreciated by classical neuropathological methods.[34] Ubiquitinylated inclusions are not present, but there have been reports of cytoplasmic inclusions in Purkinje cells that immunolabel for the CACNA1A protein.[54] The size of the normal allele ranges up to 18 CAG repeats, with affected individuals having expansion ranging from 19 to 31 repeats, compared with other CAG repeat diseases where the pathological expansion typically exceeds 40 repeats. Although there is inverse correlation between repeat length and age of onset, within families the repeat length tends to be stable and instances of apparent anticipation do not correlate well with further expansion of the CAG repeat. The polyglutamine tract within the CACNA1A protein is in the cytoplasmic tail at the C-terminal end. Of interest, truncation or missense mutations in *CACNA1A* are found in episodic ataxia type 2 (EA2) and missense mutations are found in familial hemiplegic migraine. Those conditions are categorized as channelopathies and there is evidence to suggest that the expanded polyglutamine seen in SCA6 may also result in channel dysfunction,[84] but it also has been shown that CACNA1A coordinates gene expression using a bicistronic mRNA bearing a cryptic internal ribosomal entry site (IRES). One cistron encodes the *alpha*1A subunit, whereas the second produces a transcription factor, *alpha*1ACT, which contains the polyglutamine tract and coordinates expression of genes involved in neural and Purkinje cell development. *Alpha*1ACT that contains an expanded polyQ tract, when expressed by itself, loses its function as a transcription factor and stimulator of neurite outgrowth, induces cell death in cultured neurons, and ataxia and cerebellar atrophy in transgenic mice.[25]

SCA7

Originally classified as ADCA II because of the association of ataxia with retinal macular degeneration, SCA7 is linked to chromosome 3p21.1–p14.1. The mutation is an expanded CAG repeat in the coding region of the *ataxin-7* gene.[22] Ataxin 7, a protein of unknown function, is expressed widely in neurons throughout the CNS. The normal allele has a CAG repeat length of 4–27 and the clearly pathological range is from 37 to greater than 300. Individuals with intermediate-range lengths of 28–36 repeats are asymptomatic but are at risk of transmitting expanded alleles to their offspring, resulting in apparently 'spontaneous' mutations. The age of onset ranges from infancy to over 70 years, but the mean is around 30 years. SCA7 shows anticipation and variability of phenotype that correlates with the length of the repeat expansion. Classical cases have onset with truncal and limb ataxia and dysarthria. Hyperreflexia and supranuclear ophthalmoplegia with slow saccades are also frequent findings. Less commonly, there may be extrapyramidal features, peripheral neuropathy and cognitive changes. Most patients eventually develop visual loss. Individuals with infantile onset have a rapid, severe course with early blindness. When onset is late, the visual symptoms may not develop until several decades after ataxic symptoms begin. Neuropathological changes in SCA7 consist of cerebellar atrophy, with marked loss of Purkinje cells, milder loss of granule cells and variable loss of neurons in the dentate nuclei. The inferior olives have severe gliosis and neuronal loss, but the basal pontine nuclei are usually less affected. The spinocerebellar and corticospinal tracts have axonal loss, but the posterior columns are relatively spared. There may be degeneration of motor neurons in the brain stem and anterior horns, and the subthalamic nuclei, the globus pallidus and substantia nigra are sometimes affected. Studies employing thick-section techniques have found neuronal loss in a number of sites in the brain stem and basal ganglia that are more difficult to appreciate by routine histopathology.[89] Retinal changes include severe loss of photoreceptors and ganglion cells, with attenuation of the nuclear and plexiform layers. Mutant ataxin-7-containing neuronal intranuclear inclusions are present in areas of neuronal loss and elsewhere. The inclusions are more likely to be ubiquitinylated in areas where degeneration is more pronounced, such as the inferior olives.

SCA8

Linked to chromosome 13q21, SCA8 appears to be caused by the expansion of a CTG trinucleotide repeat that is transcribed as part of an untranslated (17)RNA.[51,67] The age of onset ranges from congenital to over 70 years, but most patients present in the fourth or fifth decade. Gait and limb ataxia and dysarthria are invariably present, but oculomotor incoordination, spasticity, sensory loss and cognitive impairment may also be seen. The disease is usually slowly progressive, with ambulatory assistance required after two or more decades. Because of what appears to be incomplete penetrance, establishing the normal range of CTG-repeat lengths is difficult. Although most alleles have fewer than 40 repeats, individuals with alleles within the pathological range, generally 90–250 repeats, do not always manifest the disease. In addition, patients with other neurological or psychiatric conditions and patients without neurological abnormalities may have expansion within the *SCA8* gene. These observations have led some to question the pathogenic relationship of the gene to ataxia and to discourage the use of clinical testing for the gene in patients with apparent sporadic ataxia.[99] The neuropathology is not well-reported but there is Purkinje cell loss and olivary atrophy and one case reported to have nigral degeneration. Intranuclear inclusions labelling with 1C2, an antibody recognizing expanded polyglutamine residues, have been seen in human and transgenic mice.[75] There is evidence of bidirectional expression of the mutant gene, in which there is formation of untranslated CUG repeats that form intranuclear aggregates that may sequester splicing factors,[21] as well as translation in the reverse direction forming polyglutamine, polyalanine and polyserine residues through a unique repeat-associated non-ATG (RAN) translation mechanism.[123] These residues may have toxic effects in addition to the RNA-gain-of-function effects.

SCA10

A common form of ataxia in Mexico, SCA10 presents with gait and limb ataxia, nystagmus and dysarthria. Seizures occur in a number of patients. Onset ranges from 10 to 40 years, with anticipation often associated with paternal transmission. The neuropathology has not been characterized, but MR imaging studies reveal cerebellar atrophy without brain stem involvement. The mutation is a massive pentanucleotide (ATTCT) repeat expansion in intron 9 of the *ataxin-10* gene.[72] Ataxin 10 is highly expressed in olivocerebellar regions and experimental reduction of its function by small interfering RNAs results in increased apoptosis of cerebellar neurons.[71] Recent evidence indicates a toxic gain-of-function RNA mechanism with an expanded AUUCU repeat similar to the myotonic dystrophies, which are caused by non-coding CTG or CCTG repeat expansions.[118]

SCA11

Clinically, SCA11 is a mild, slowly progressive gait and limb ataxia with nystagmus and dysarthria. Onset is in the third decade of life. There may be hyperreflexia, but other neurological findings are absent. It has been mapped to chromosome 15q14–21.31585 and the mutation is in tau tubulin kinase 2 (TTK2). Neuropathological studies showed loss of Purkinje cells and granular neurons in the cerebellar cortex and neuronal loss in the dentate nucleus. There also was widespread cytoskeletal tau pathology with a different distribution than seen in Alzheimer's disease.[47]

SCA12

Tremor of the upper extremities and head may precede the onset of ataxia in this disorder. Increased deep tendon reflexes, hypokinesia, mild neuropathy and sometimes dementia ensue. Most affected individuals present in the fourth decade. Although it has not been characterized pathologically, neuroimaging reveals atrophy of the cerebellum and often the cerebral cortex. The gene, *PPP2R2B*, has been mapped to chromosome 5q31–q33, the mutation being an expanded CAG repeat in the 5'-untranslated region. Therefore, unlike the other CAG repeat diseases, there is no expanded polyglutamine tract in the affected protein, a brain-specific regulatory subunit of protein phosphatase PP2.[46] Correlation between age of onset and CAG repeat length is not as clear as what is seen with most other CAG repeat abnormalities.

SCA13

Seen in a single large French kindred, SCA13 is an early-onset, slowly progressive gait ataxia with dysarthria, mental retardation and delay in attaining developmental motor milestones. Nystagmus and pyramidal signs were features in some patients. The *SCA13* gene is linked to chromosome 19q13.3–q13.4.[45] The disease arises from mutations within the voltage-gated potassium channel, KCNC3. Magnetic resonance imaging in two patients showed moderate cerebellar and pontine atrophy.

SCA14

Families with this disorder have been described in Japan and Europe. The clinical presentation is variable. Most patients have gait and limb ataxia with dysarthria. Early-onset patients have more severe disease, often with myoclonus and tremor, although individuals with later onset are primarily ataxic. Imaging studies have shown severe cerebellar atrophy in several cases. A number of different mutations have been found in the affected gene, which is linked to chromosome 19q13.4 and codes for protein kinase C-γ (PKC-γ). This kinase is highly expressed in Purkinje cells, and there was a reduction of immunohistochemical staining for PKC-γ and ataxin-1 in the depleted Purkinje cells of an affected patient, with preservation of staining for calbindin.[14]

SCA15/SCA16

Once thought to be separate entities, these diseases are both caused by mutations, (SCA15: deletion, and SCA16: missense mutation), in the *ITPR-1* gene mapped to 3p24.2-pter, which encodes for an inositol 1,4,5-triphosphate receptor involved in the regulation of calcium release from the endoplasmic reticulum.[110] It has been proposed that the designation SCA16 be vacated.[31] Families have been identified in Australia and Japan. Patients usually present as adults with slowly progressive ataxia, dysarthria, oculomotor problems and sometimes action tremor of the head and upper extremities. Imaging has shown midline cerebellar atrophy without brain stem involvement.[32] A clinically distinct form of congenital non-progressive ataxia with cognitive impairment, currently designated SCA29, also has been mapped to this same chromosomal region 25a and shown to be due to a missense mutation in the same *ITPR-1* gene.[48]

SCA17

SCA17 is the result of a CAG repeat expansion in the *TATA box-binding protein (TBP)* gene on 6q27.[65] The age of onset ranges from the first to the seventh decade. Classic cases have a relatively pure cerebellar ataxia, which may be preceded by psychiatric problems or presenile dementia. Other patients may have rigidity, dystonia, dysphagia or even a Huntington disease-like clinical presentation, which has resulted in an alternative designation, Huntington disease-like 4. The distribution of neuropathological lesions has been variable, but areas of described involvement include neuronal loss in the cerebral cortex, caudate nucleus and medial thalamic nuclei, in addition to loss of neurons in the dentate nuclei, loss of Purkinje cells and hypertrophic degeneration of inferior olivary neurons.[9] Large intranuclear inclusions that label with antibodies against TBP, ubiquitin and polyglutamine are present in neurons in many areas. The CAG repeat length in normal alleles ranges from 25 to 42, and the pathological range is from 45 to 66. There is a relatively poor correlation between repeat length and age of onset. There appears to be greater stability of repeat lengths in SCA17 than in many of the other polyglutamine diseases, although intergenerational fluctuations do occur.

SCA18

An autosomal dominant ataxia linked to chromosome 7q22–q32 has been described in an American family of Irish ancestry. The gene has not been identified. Onset

of gait ataxia occurred in the teens and twenties, followed by dysmetria, reduced deep tendon reflexes and sensorimotor axonal neuropathy. Progression is slow but can result in loss of ambulation over time. Brkanac *et al.* termed this condition 'sensorimotor neuropathy with ataxia' (SMNA).[7] Imaging studies reveal mild cerebellar atrophy, but pathological findings have not been reported.

SCA19/SCA22

A four-generation Dutch kindred was described with linkage to chromosome 1p21.[93,113] Affected members had a relatively mild ataxia, with cognitive impairment, myoclonus and a low-frequency irregular postural tremor. Age of onset ranged from 20 to 45 years, with suggestion of anticipation. Magnetic resonance imaging studies revealed atrophy of the cerebellum with sparing of the brain stem. Mild cerebral atrophy was present in a few patients. A Chinese family was described with linkage to a locus on chromosome 1p21–q23. The age at onset ranged from 10 to 46 years. Clinical features included a slowly progressive gait and limb ataxia, dysarthria, nystagmus and hyporeflexia. Imaging studies were similar to those in the Dutch kindred. Earlier onset in succeeding generations suggests anticipation.[15] Although originally designated SCA22, the Chinese kindred has a linkage site overlapping that of SCA19, and it has been suggested that the two conditions are the same entity, with the differences in clinical phenotype between the two kindreds relating to unidentified factors.[92]

SCA20

Also known as dominantly inherited ataxia with dysphonia and dentate calcification,[59] SCA20 has been mapped to a 260 kilobase duplication in the pericentromeric region of chromosome 11.[60] Described in a single family of Anglo-Celtic descent, the age of onset was highly variable with anticipation of approximately one decade in subsequent generations. Dysarthria, gait and upper limb ataxia with dysphonia were characteristic features, and progression was slow and mild. There was no peripheral neuropathy, either clinically or electrophysiologically. Palatal myoclonus or tremor was present in the majority of cases, and CT scans revealed calcification of the dentate nuclei with sparing of the basal ganglia. Magnetic resonance imaging (MRI) studies have shown mild to moderate pan-cerebellar atrophy, but no pontine or supratentorial changes. Hypertrophic olivary changes have been imaged in several patients.[104]

SCA21

Mapped to chromosome 7p21.3–p15.1 in a French kindred, this condition is characterized by ataxia combined with extrapyramidal features. Onset ranged from the first to the third decade, with variable features of gait and limb ataxia, dysarthria, bradykinesia, tremor, rigidity and decreased tendon reflexes. Cognitive changes were present in some patients. Pathological findings have not been reported, but MR imaging studies revealed cerebellar atrophy without changes in the forebrain or brainstem.[23,115]

SCA23

A late-onset, slowly progressive ataxia that was mapped to chromosome 20p13–p12.3 in a Dutch family has been assigned SCA23. Different mutations have been identified in the prodynorphin gene, which encodes a precursor protein for the opioid neuropeptides and may cause toxic gain-of-function effects.[3] Gait and limb ataxia, with dysarthria, slowing of saccades, hyperreflexia and loss of vibratory sense are clinical features. Neuropathological findings in one patient included loss of Purkinje cells and neurons in the dentate nuclei and inferior olives, in addition to axonal degeneration in the posterior and lateral columns.[114]

SCA25

A form of ataxia with sensory involvement has been linked to chromosome 2p21–p13 in a French family.[101,102] Onset is usually in childhood or early adulthood and the clinical course is variable in severity and progression. Gait and limb ataxia with sensorimotor and autonomic neuropathy are typical findings. Hyporeflexia is common, but extensor plantar reflexes may be present. Axonal neuropathy has been shown pathologically, but brain pathology has not been reported. Imaging studies reveal cerebellar atrophy.

SCA26

This condition, linked to 19p13.3, was identified in a six-generation American family of Norwegian ancestry. This locus is near the genes for Cayman ataxia and SCA6, but there are no mutations in the *CACNA1A* gene. There is a single variant that co-segregates with the disease phenotype, which produces a single amino acid substitution in eukaryotic elongation factor 2, resulting in a P596H substitution at site that is critical for maintenance of reading-frame during translation.[43] Onset is from age 26 to 60 years, with slow progression of truncal and limb ataxia, dysarthria and abnormal visual pursuit. Imaging reveals cerebellar atrophy with sparing of the pons.[121] Personal observation of two patients at autopsy revealed severe loss of Purkinje neurons with milder loss of granular neurons. The inferior olivary nuclei had mild-to-moderate neuronal loss, but the basal pontine nuclei were uninvolved. There was gliosis but no neuronal loss in the deep cerebellar nuclei and mild gliosis was present in the periaqueductal grey matter without obvious neuronal loss. The posterior columns, spinocerebellar and corticospinal tracts were spared, as were bulbar and spinal motor neurons.

SCA27

A three-generation Dutch family has a form of slowly progressive ataxia linked to chromosome 13q34 and a missense mutation in the gene for fibroblast growth factor 14 (FGF14). FGF14 is expressed primarily in the CNS, and *FGF*-null mice have ataxia and dystonia.[117] There is onset of tremor in childhood, followed later by orofacial dyskinesia, slowly progressive limb and gait ataxia, dysarthria and nystagmus. Psychiatric and cognitive changes may be seen even later in the disease. Imaging studies have revealed mild cerebellar atrophy.[112] A patient from a different family with a truncating mutation in the same gene presented in adolescence with cognitive deficits, tremor, ataxia and dyskinesia.[19]

SCA28

An early-onset, slowly progressive form of ataxia was described in 11 members of a four-generation Italian family.[12] The affected gene is AFC3LE, which encodes a mitochondrial protease involved in the maintenance of the mitochondrial proteome and highly expressed in Purkinje cells.[24] Onset ranges from 12 to 36 years of age. Imbalance of stance and gait are the first symptoms. Other prominent features are gaze-evoked nystagmus, often followed by slow saccades, ophthalmoparesis and ptosis. Pyramidal signs are present in most patients, but the sensory system is intact. Most patients remain ambulatory for decades after onset. Imaging studies have shown cerebellar but not pontine atrophy.

SCA29

SCA29 has been shown to be due to a missense mutation in the *ITPR-1 (inositol triphosphate receptor type 1)* gene on 3p24, the same locus as SCA15 (see earlier).[48] It is an early-onset, non-progressive ataxia that is phenotypically distinct from SCA15. In addition to ataxia, the patients have variable cognitive deficits and dystonia. MRI studies have shown atrophy of the cerebellar vermis.

SCA30

SCA30 was originally observed in a single Anglo-Celtic family in Australia and has been linked to chromosomal region 4q34.3–q35.1. The clinical onset is later in life and presents with relatively pure slowly progressive ataxia with mild pyramidal signs. MRI reveals marked cerebellar atrophy with sparing of the flocculonodular regions and the brain stem.[105]

SCA31

This condition is common in Japan and presents with late-onset cerebellar ataxia, oculomotor problems and variable loss of hearing. Although it has a linkage similar to SCA4, the causative mutation of SCA31 is an insertion of an extensive pentanucleotide repeat expansion in an intron of the *BEAN (brain expressed associated with NEDD4)* gene. *In situ* hybridization reveals RNA foci in the nuclei of Purkinje cells suggesting an RNA toxic-gain-of-function disease mechanism.[90] There is cerebellar cortical atrophy with loss of Purkinje cells. The Purkinje cell degeneration is distinctive with formation of somatic sprouts as well as synaptophysin-positive halo-like structures surrounding the perikaryon.[73]

SCA32

Originally identified in a Chinese kindred, SCA32 has been linked to chromosomal region 7q32–q33. It is a variable-onset ataxia with cognitive impairment if onset is before age 40. Other non-cerebellar findings may be present in some patients. Onset in females generally is later than in males. Affected males are sterile with azoospermia. MRI shows cerebellar atrophy.[57] One unpublished neuropathological examination was performed personally. There was moderate-to-focally-severe cerebellar cortical atrophy with Purkinje cell loss and milder loss of granular neurons. The basal pons was spared and the deep cerebellar nuclei and inferior olives had gliosis with minimal neuronal loss. Only the rostral cervical cord was available for examination, but there was severe axonal loss in the gracile tracts with posterior root atrophy at the level sampled. The subthalamic nucleus, ventral and medial thalamus and periaqueductal grey matter had gliosis with mild neuronal loss.

SCA34

Originally described in a French-Canadian family as erythrokeratodermia with ataxia,[36] SCA34 has been linked to chromosomal region 16p12.3–q16.2. Soon after birth affected patients develop erythematous ichthyosiform plaques often on the extremities. The skin lesions were less apparent in the summer and often disappeared by age 25. By age 40 the rash reappears and patients develop slowly progressive but severe ataxic gait accompanied by hyporeflexia, nystagmus and dysarthria. There are no published studies of neuroimaging or neuropathology.

SCA35

Reported in a Chinese family with slowly progressive spinocerebellar ataxia arising in the fifth decade of life, SCA35 was found to be due to a mutation in the *transglutaminase 6* gene (*TGM6*). Early features included gait ataxia and dysarthria, with later upper limb dysmetria. Additional features included tremor, hyperreflexia, torticollis, ocular dysmetria and proprioceptive defects. None of the patients had nystagmus, ophthalmoplegia, peripheral neuropathy or cognitive decline. Two sisters from a second, unrelated Chinese family had a similar disorder and a different mutation on the same gene.[116] There is cerebellar atrophy on MRI studies.

SCA36

Found in five unrelated Japanese families, SCA36 is a late-onset disease that presents with gait and truncal ataxia, dysarthria and limb dysmetria. Patients with longer duration of disease often develop lower motor neuron degeneration particularly affecting the tongue, with swallowing relatively preserved. Amyotrophy and fasciculation in the limbs and trunk also may be present. Hyperreflexia is often present without spasticity. MRI shows mild cerebellar atrophy. The mutation is a hexanucleotide-repeat-expansion in intron 1 of the *NOP56* gene. *In situ* hybridization of lymphoblastoid cells from patients had nuclear RNA foci corresponding to the expanded repeat suggesting an RNA-mediated toxic gain-of-function mechanism of disease.[61] Neuropathologic examination of one patient found moderate degeneration of cerebellar Purkinje neurons with loss of neurons in the dentate nucleus and in brain stem and spinal lower motor neurons. There also was neurogenic amyotrophy. There was no abnormal immunohistochemical localization of the mutant gene product or TDP-43.[53]

SCA37

A Spanish kindred has been linked to an 11-megabase interval on chromosome 1p32.[96] The causative mutation could not be identified by exome-sequencing within this region. Clinically, the affected family members had late-onset, slowly progressing ataxia with selective alterations in vertical eye movements. The onset is characterized by falls, dysarthria and clumsiness and progresses to an uncomplicated cerebellar syndrome. Imaging studies reveal cerebellar atrophy without evidence of pontine involvement. No pathological studies have been performed.

Dentatorubropallidoluysian Atrophy (DRPLA)

Although not listed with the autosomal dominant spinocerebellar ataxias, DRPLA has many features in common with them. Dentatorubropallidoluysian atrophy has been linked to chromosome 12p13.31 and is caused by an expanded CAG repeat in the gene for atrophin-1, a 190-kDa protein of unknown function.[64] The disease is largely confined to populations of Japanese ancestry, but occasional cases have been seen in other ethnic groups, including an African American kindred, where it has been termed the Haw River syndrome.[11,28] The severity of the clinical syndrome and the age of onset are related to the length of the CAG repeat expansion. Patients with larger expansions and childhood onset tend to have myoclonic epilepsy as a prominent feature along with ataxia and cognitive decline. When onset is after 20 years of age, ataxia, chorea and dementia are the major features and seizures are less significant or absent. Adult-onset cases may be mistaken for Huntington's disease and other types of SCA.

At autopsy, the brain is often smaller than normal, with variable ventricular dilation but little cortical atrophy. There is gross atrophy with brown discolouration of the pallidum and subthalamic nucleus. The brain stem tegmentum is reduced in size, particularly in the pons. The cerebellar dentate nucleus and efferent fibres are discoloured and shrunken. Histologically, the pars externa of the globus pallidus has severe neuronal loss and gliosis. The subthalamic nucleus has severe gliosis, with better preservation of neurons, suggesting that the gliosis is secondary to loss of projections from the pallidum. The cerebellar cortex is largely spared, but there is severe neuronal loss in the dentate nucleus, with surviving neurons having grumose degeneration characterized by flocculent material around the perikaryon and proximal dendrites.[2] There is marked gliosis and fibre loss in the hilar projections and superior cerebellar peduncle and the red nucleus exhibits gliosis with relative preservation of neuronal somata. Less involvement has been described in the neostriatum, thalamus, substantia nigra and inferior olives. Axonal loss may occur in the spinocerebellar tracts and posterior columns of the spinal cord. Patients with the Haw River syndrome, who have a smaller expansion of the *atrophin-1* gene, have micro-calcification of the pallidum, neuroaxonal dystrophy of the nucleus gracilis and demyelination of the cerebral white matter, in addition to atrophy of the dentate nucleus and pallidum. In DRPLA, intranuclear inclusions that are immunoreactive for ubiquitin, atrophin-1 and expanded polyglutamine tracts are present in neurons and sometimes glia. Cytoplasmic inclusions with similar immunohistochemical properties are seen in neurons of the dentate nucleus. Diffuse staining of nuclei with antibodies against expanded polyglutamine tracts is more widespread than intranuclear inclusions, which are more concentrated in areas of degenerative change.

The pathogenic range of CAG repeats in DRPLA is 49–88, with normal alleles typically having fewer than 26 repeats. More unstable, intermediate-size alleles range from 37 to 48 repeats. Myelopathy with spastic paraparesis and truncal ataxia can be seen with homozygosity for intermediate size alleles.

The Episodic Ataxias

Seven forms of autosomal dominant episodic ataxia (EA) have been described, four of which have known genetic mutations. Three are due to mutations in ion channels, whereas another results from a mutation in an excitatory amino acid transporter protein.

Episodic ataxia type 1 (EA1) is due to missense point mutations in the gene on chromosome 12q13 for the KCNA1 voltage-gated potassium channel.[8] Onset is in the first two decades of life. Affected individuals have brief periods of ataxia lasting a few minutes, which are precipitated by exercise, emotional stimuli or postural change. There is no ataxia between episodes, but persistent myokymia of the face and extremities may be present. Some patients with mutations in the same gene may have partial epilepsy with myokymia in the absence of ataxia. Anticonvulsant therapy may be of benefit.

Episodic ataxia type 2 (EA2) is due to mutations in the *CACNA1A* gene on chromosome 19p13, which codes for a subunit of the P/Q voltage-gated calcium channel, the same gene implicated in SCA6 and familial hemiplegic migraine.[77] Most families have truncation or missense point mutations, but expanded repeats have also been described. Onset is in the first two decades of life. Ataxic episodes, persisting for 30 minutes to several hours, are triggered by emotional stress or physical activity. Gait ataxia, dysarthria and nystagmus are the chief clinical features, although dystonia or seizures are sometimes present. Between episodes, there may be persistent mild nystagmus and, with increasing age, ataxia. The neuropathology is not well studied, but imaging reveals cerebellar atrophy. Episodic ataxia type 2 appears to be a channelopathy with loss of function of the P/Q calcium channel. Spontaneous mutations of the same gene in mice result in the 'tottering' and 'leaner' phenotypes, characterized by ataxia and seizures.[30]

Episodic ataxias types 3, 4 and 7 have not been characterized genetically, but EA3 appears to respond to acetazolamide therapy while the others do not. Episodic ataxia type 5 (EA5) is caused by a mutation in the *CACNB4* gene, which codes for the beta-4 subunit of the voltage-gated L-type calcium channel.[27] Described in a French-Canadian kindred, it is clinically similar to EA2 and responds to acetazolamide. The same mutation in German kindred has a clinical phenotype of generalized epilepsy and praxis-induced seizures.

Episodic ataxia type 6 (EA6) is caused by a mutation in the gene for the excitatory amino acid transporter type 1 (EAAT1) and is also known as episodic ataxia, with seizures, migraine and alternating hemiplegia. Onset of ataxia occurs in the first year of life and may be induced by febrile episodes. There is usually mild persistent ataxia. Other features include paroxysmal seizures, migraine with hemianopsia and hemiplegia that alternately involves both cerebral hemispheres. This mutation appears to result in decreased uptake of synaptic glutamate.[56]

MITOCHONDRIAL ENCEPHALOPATHY

Maternally inherited mutations of mitochondrial DNA sometimes result in encephalopathic conditions in which cerebellar ataxia is one of a number of neurological findings. These findings may include peripheral neuropathy, myopathy, deafness, optic atrophy, myoclonus and dementia.[50] Neuropathologically, there is marked atrophy of the dentate and inferior olivary nuclei and milder changes in the red nuclei, cerebellar cortex and posterior columns of the spinal cord. Rarefactive changes, similar to those seen in Leigh's syndrome, may also be present in the brain stem. The neuropathological findings are not necessarily accompanied by the myopathic changes typical of mitochondrial cytopathies in muscle cells.

X-LINKED ATAXIAS

Fragile X Tremor/Ataxia Syndrome (FXTAS)

The fragile X syndrome, the most common form of inherited mental retardation, is caused by an expanded (>200) CGG trinucleotide repeat in the *FMR1* gene at Xq27.3. Males carrying repeat expansions that are considered premutation lengths (75–150) for the fragile X syndrome may develop a late-onset neurological syndrome of ataxia with a disproportionately great intention tremor. Other abnormalities may include memory and cognitive problems, parkinsonism, peripheral neuropathy, proximal lower limb weakness and autonomic dysfunction. Magnetic resonance imaging has shown increased T2-signal intensity in the middle cerebellar peduncles and deep cerebellar white matter and, sometimes, cerebellar atrophy. Neuropathological studies have shown major loss of Purkinje cells, with axonal torpedoes and Bergmann gliosis. White matter spongiosis as well as spongiosis of the middle cerebellar peduncles and atrophy of the basal pontine nuclei also has been described. Eosinophilic (ubiquitin-positive) intranuclear inclusions are present in neurons and astrocytes throughout the forebrain and brain stem, with the greatest density in the hippocampus and the frontal cortex. Intranuclear inclusions are rare to absent in Purkinje cells and basal pontine neurons, although they are present in a few neurons in the dentate nucleus and diffusely in cerebellar astrocytes. Antibodies against expanded polyglutamine tracts or the mutated protein fail to label the inclusions, which distinguishes them from the inclusions seen in ataxias due to CAG repeat expansion.[39] Sequestered in these inclusions is mRNA that contains expanded CGG repeats which bind to splicing factors such as muscleblind-like 1, suggesting an RNA gain-of-function mechanism.[95] Immunostaining

TABLE 13.6 X-linked ataxias

Disorder	OMIM number	Details
Fragile X tremor/ataxia syndrome	309550	See text
Sideroblastic anaemia and spinocerebellar ataxia	301310	ATP-binding cassette 7 transporter
X-linked cerebellar ataxia (CLA2)	302500	Xp11.21–q21.3
Cerebellar ataxia with extrapyramidal involvement	302600	No linkage identified
Arts syndrome, fatal X-linked ataxia with deafness and loss of vision	301835	Phosphoribosyl-pyrophosphate synthetase; early onset with death in childhood
Hoyerall–Hreidarsson syndrome	300240	DKC1; congenital cerebellar atrophy

ATP, adenosine triphosphate; OMIM, Online Mendelian Inheritance in Man.

for tau and α-synuclein is also negative.[38,55] Rare cases of female carriers with premutation-length expansion may develop a milder form of this condition.

Other forms of X-linked ataxia are listed in Table 13.6. Clinically, it is important to differentiate these conditions from atypical presentations of X-linked leukodystrophies.

SPORADIC DEGENERATIVE ATAXIA

Although genetic mutations are increasingly recognized as a major aetiological factor in ataxic syndromes, most adult patients who present with ataxia appear to have sporadic disease. Some are subsequently shown to have hereditary ataxias, particularly those forms due to trinucleotide repeat expansions, but most have disorders that are designated as idiopathic late-onset cerebellar ataxia (ILOCA). The most common of these is the cerebellar form of multiple system atrophy (MSA-C; see Chapter 12, Extrapyramidal Diseases of Movement).

Non-MSA forms of sporadic ataxia also exist. Other patients with ILOCA have cerebellar and pontine atrophy on imaging studies and may have other neurological symptoms besides ataxia, but do not typically progress to develop other features of MSA and do not have glial or neuronal inclusions at autopsy. Occasionally, patients with MSA-C present in a similar way to this form of sporadic OPCA, but usually develop other features of MSA within a few years. The clinical distinction is important, because non-MSA patients have a better prognosis.[35] There is little information on the specific neuropathology of these non-MSA forms, because most pathological descriptions of OPCA in all but the most recent literature have also included a mixture of inherited ataxias and MSA. Included in the ILOCA category are more pure cerebellar degenerations, with onset typically in the sixth decade or later. Cases such as these were reported by Marie *et al.* in 1922.[70] Pathology is usually restricted to the cerebellar cortex and inferior olives (probably as trans-synaptic degeneration), with or without degeneration of afferent spinal pathways. Such cases need to be distinguished from acquired ataxias secondary to toxic, paraneoplastic and other aetiologies.

REFERENCES

1. Anheim M, Tranchant C, Koenig M. The autosomal recessive cerebellar ataxias. *N Engl J Med* 2012;**366**(7):636–46.
2. Arai N, Amano N, Iwabuchi K, *et al.* Three categories of the degenerative appearance of the human cerebellar dentate nucleus: a morphometric and morphological study. *J Neurol Sci* 1988;**83**:129–43.
3. Bakalkin G, Watanabe H, Jezierska J, *et al.* Prodynorphin mutations cause the neurodegenerative disorder spinocerebellar ataxia type 23. *Am J Hum Genet* 2010;**87**(5):593–603.
4. Banfi S, Servadio A, Chung MY, *et al.* Identification and characterization of the gene causing type 1 spinocerebellar ataxia. *Nat Genet* 1994;**7**:513–20.
5. Ben Hamida M, Belal S, Sirugo G, *et al.* Friedreich's ataxia phenotype not linked to chromosome 9 and associated with selective autosomal recessive vitamin E deficiency in two inbred Tunisian families. *Neurology* 1993;**43**:2179–83.
6. Boder E. Ataxia-telangiectasia: an overview. *Kroc Found Ser* 1985;**19**:1–63
7. Brkanac Z, Fernandez M, Matsushita M, *et al.* Autosomal dominant sensory/motor neuropathy with ataxia (SMNA): linkage

to chromosome 7q22–q32. *Am J Med Genet* 2002;**114**:450–57.
8. Browne DL, Gancher ST, Nutt JG, *et al.* Episodic ataxia/myokymia syndrome is associated with point mutations in the human potassium channel gene, KCNA1. *Nat Genet* 1994;**8**:136–40.
9. Bruni AC, Takahashi–Fujigasaki J, Maltecca F, *et al.* Behavioural disorder, dementia, ataxia, and rigidity in a large family with TATA box-binding protein mutation. *Arch Neurol* 2004;**61**:1314–20.
10. Bundey S. Clinical and genetic features of ataxia–telangiectasia. *Int J Radiat Biol* 1994;**66** (Suppl 6):S23–9.
11. Burke JR, Ikeuchi T, Koide R, *et al.* Dentatorubral–pallidoluysian atrophy and Haw River syndrome. *Lancet* 1994;**344**:1711–12.
12. Cagnoli C, Mariotti C, Taroni F, *et al.* SCA28, a novel form of autosomal dominant cerebellar ataxia on chromosome 18p11.22–q11.2. *Brain* 2005;**129**:235–42.
13. Campuzano V, Montermini L, Molto MD, *et al.* Friedreich's ataxia: autosomal recessive disease caused by an intronic GAA triplet repeat expansion. *Science* 1996;**271**:1423–7.

14. Chen DH, Brkanac Z, Verlinde CL, *et al.* Missense mutations in the regulatory domain of PKC gamma: a new mechanism for dominant nonepisodic cerebellar ataxia. *Am J Hum Genet* 2003;**72**:839–49.
15. Chung MY, Lu YC, Cheng NC, Soong BW. A novel autosomal dominant spinocerebellar ataxia (SCA22) linked to chromosome 1p21–q23. *Brain* 2003;**126**:1293–9.
16. Concannon P, Gatti RA. Diversity of ATM gene mutations detected in patients with ataxia–telangiectasia. *Hum Mutat* 1997;**10**:100–107.
17. Cossee M, Durr A, Schmitt M, *et al.* Friedreich's ataxia: point mutations and clinical presentation of compound heterozygotes. *Ann Neurol* 1999;**45**:200–206.
18. Criscuolo C, Chessa L, Di Giandomenico S, *et al.* Ataxia with oculomotor apraxia type 2: a clinical, pathologic, and genetic study. *Neurology* 2006;**66**(8):1207–10.
19. Dalski A, Atici J, Kreuz FR, *et al.* Mutation analysis in the fibroblast growth factor 14 gene: frameshift mutation and polymorphisms in patients with inherited ataxias. *Eur J Hum Genet* 2005;**13**:118–20.
20. Date H, Onodera O, Tanaka H, *et al.* Early-onset ataxia with ocular motor apraxia and hypoalbuminemia is caused by

mutations in a new HIT superfamily gene. *Nat Genet* 2001;**29**:184–8.

21. Daughters RS, Tuttle DL, Gao W, *et al.* RNA gain-of-function in spinocerebellar ataxia type 8. *PLoS Genet* 2009;**5**(8):e1000600.

22. David G, Abbas N, Stevanin G, *et al.* Cloning of the SCA7 gene reveals a highly unstable CAG repeat expansion. *Nat Genet* 1997;**17**:65–70.

23. Devos D, Schraen–Maschke S, Vuillaume I, *et al.* Clinical features and genetic analysis of a new form of spinocerebellar ataxia. *Neurology* 2001;**56**:234–8.

24. Di Bella D, Lazzaro F, Brusco A, *et al.* Mutations in the mitochondrial protease gene AFG3L2 cause dominant hereditary ataxia SCA28. *Nat Genet* 2010;**42**(4):313–21.

25. Du X, Wang J, Zhu H, *et al.* Second cistron in CACNA1A gene encodes a transcription factor mediating cerebellar development and SCA6. *Cell* 2013;**154**(1):118–33.

25a. Dudding TE, Friend K, Schofield PW, *et al.* Autosomal dominant congenital non-progressive ataxia overlaps with the SCA15 locus. *Neurology* 2004;**63**:2288–92.

26. Edener U, Bernard V, Hellenbroich Y, Gillessen–Kaesbach G, Zuhlke C. Two dominantly inherited ataxias linked to chromosome 16q22.1:SCA4 and SCA31 are not allelic. *J Neurol* 2011;**258**(7):1223–7.

27. Escayg A, De Waard M, Lee DD, *et al.* Coding and noncoding variation of the human calcium-channel beta4-subunit gene CACNB4 in patients with idiopathic generalized epilepsy and episodic ataxia. *Am J Hum Genet* 2000;**66**:1531–9.

28. Farmer TW, Wingfield MS, Lynch SA, *et al.* Ataxia, chorea, seizures and dementia: pathologic features of a newly defined familial disorder. *Arch Neurol* 1989;**46**:774–9.

29. Fernet M, Gribaa M, Salih MA, *et al.* Identification and functional consequences of a novel MRE11 mutation affecting 10 Saudi Arabian patients with the ataxia telangiectasia-like disorder. *Hum Mol Genet* 2005;**14**:307–18.

30. Fletcher CF, Lutz CM, O'Sullivan TN, *et al.* Absence epilepsy in tottering mutant mice is associated with calcium channel defects. *Cell* 1996;**87**:607–17.

31. Gardner RJ. 'SCA16' is really SCA15. *J Med Genet* 2008;**45**(3):192.

32. Gardner RJ, Knight MA, Hara K, *et al.* Spinocerebellar ataxia type 15. *Cerebellum* 2005;**4**:47–50.

33. Gierga K, Burk K, Bauer M, *et al.* Involvement of the cranial nerves and their nuclei in spinocerebellar ataxia type 2 (SCA2). *Acta Neuropathol* 2005;**109**(6):617–31.

34. Gierga K, Schelhaas HJ, Brunt ER, *et al.* Spinocerebellar ataxia type 6 (SCA6): neurodegeneration goes beyond the known brain predilection sites. *Neuropathol Appl Neurobiol* 2009;**35**(5):515–27.

35. Gilman S, Little R, Johanns J, *et al.* Evolution of sporadic olivopontocerebellar atrophy into multiple system atrophy. *Neurology* 2000;**55**:527–32.

36. Giroux JM, Barbeau A. Erythrokeratodermia with ataxia. *Arch Dermatol* 1972;**106**(2):183–8.

37. Gomez CM, Thompson RM, Gammack JT, *et al.* Spinocerebellar ataxia type 6: gaze-evoked and vertical nystagmus, Purkinje cell degeneration and variable age of onset. *Ann Neurol* 1997;**42**:933–50.

38. Greco CM, Hagerman RJ, Tassone F, *et al.* Neuronal intranuclear inclusions in a new cerebellar tremor/ataxia syndrome among fragile X carriers. *Brain* 2002;**125**:1760–71.

39. Hagerman P. Fragile X-associated tremor/ataxia syndrome (FXTAS): pathology and mechanisms. *Acta Neuropathol* 2013;**126**(1):1–19.

40. Harding A. Early onset cerebellar ataxia with retained tendon reflexes: a clinical and genetic study of a disorder distinct from Friedreich's ataxia. *J Neurol Neurosurg Psychiatry* 1981;**44**:503–8.

41. Harding A. Clinical features and classification of inherited ataxias. *Adv Neurol* 1993;**61**:1–14.

42. Hart P, Lodi R, Rajagopalan B, *et al.* Antioxidant treatment of patients with Friedreich's ataxia: four-year follow-up. *Arch Neurol* 2005;**62**:621–6.

43. Hekman KE, Yu GY, Brown CD, *et al.* A conserved eEF2 coding variant in SCA26 leads to loss of translational fidelity and increased susceptibility to proteostatic insult. *Hum Mol Genet* 2012;**21**(26):5472–83.

44. Hellenbroich Y, Gierga K, Reusche E, *et al.* Spinocerebellar ataxia type 4 (SCA4): Initial pathoanatomical study reveals widespread cerebellar and brainstem degeneration. *J Neural Transm* 2006;**113**(7):829–43.

45. Herman Bert A, Stevanin G, Netter JC, *et al.* Mapping of spinocerebellar ataxia 13 to chromosome 19q13.3–q13.4 in a family with autosomal dominant cerebellar ataxia and mental retardation. *Am J Hum Genet* 2000;**67**:229–35.

46. Holmes SE, O'Hearn EE, McInnis MG, *et al.* Expansion of a novel CAG trinucleotide repeat in the 5 region of PPP2R2B is associated with SCA12. *Nat Genet* 1999;**23**:391–2.

47. Houlden H, Johnson J, Gardner–Thorpe C, *et al.* Mutations in TTBK2, encoding a kinase implicated in tau phosphorylation, segregate with spinocerebellar ataxia type 11. *Nat Genet* 2007;**39**(12):1434–6.

48. Huang L, Chardon JW, Carter MT, *et al.* Missense mutations in ITPR1 cause autosomal dominant congenital nonprogressive spinocerebellar ataxia. *Orphanet J Rare Dis* 2012;**7**:67.

49. Hunt JR. Dyssynergia cerebellaris myoclonica: primary atrophy of the dentate system. A contribution to the pathology and symptomatology of the cerebellum. *Brain* 1921;**44**:490–538.

50. Ihara Y, Namba R, Demiya M. A case of mitochondrial encephalomyopathy: mitochondrial myopathy, encephalopathy, lactic acidosis and strokelike episodes (MELAS). *Clin Neurol* 1987;**27**:969–75.

51. Ikeda Y, Shizuka Ikeda M, Watanabe M, *et al.* Asymptomatic CTG expansion at the SCA8 locus is associated with cerebellar atrophy on MRI. *J Neurol Sci* 2000;**182**:76–9.

52. Ikeda Y, Dick KA, Weatherspoon MR, *et al.* Spectrin mutations cause spinocerebellar ataxia type 5. *Nat Genet* 2006;**38**:184–90.

53. Ikeda Y, Ohta Y, Kobayashi H, *et al.* Clinical features of SCA36: a novel spinocerebellar ataxia with motor neuron involvement (Asidan). *Neurology* 2012;**79**(4):333–41.

54. Ishikawa K, Watanabe M, Yoshizawa K, *et al.* Clinical, neuropathological and molecular study in two families with spinocerebellar ataxia type 6 (SCA6). *J Neurol Neurosurg Psychiatry* 1999;**67**:86–9.

55. Jacquemont S, Hagerman RJ, Leehey M, *et al.* Fragile X premutation tremor/ataxia syndrome: molecular, clinical, and neuroimaging correlates. *Am J Hum Genet* 2003;**72**:869–78.

56. Jen JC, Wan J, Palos TP, *et al.* Mutation in the glutamate transporter EAAT1 causes episodic ataxia, hemiplegia, and seizures. *Neurology* 2005;**65**:529–34.

57. Jiang H, Zu H–P, Gomez CM. SCA32: an autosomal dominant cerebellar ataxia with azoospermia maps to chromosome 7q32-q33. *Mov Disord* 2010;**25**:S192.

58. Klockgether T, Petersen D, Grodd W, Dichgans J. Early onset cerebellar ataxia with retained tendon reflexes: clinical, electrophysiological and MRI observations in comparison with Friedreich's ataxia. *Brain* 1991;**114**:1559–73.

59. Knight MA, Gardner RJ, Bahlo M, *et al.* Dominantly inherited ataxia and dysphonia with dentate calcification: spinocerebellar ataxia type 20. *Brain* 2004;**127**:1172–81.

60. Knight MA, Hernandez D, Diede SJ, *et al.* A duplication at chromosome 11q12.2–11q12.3 is associated with spinocerebellar ataxia type 20. *Hum Mol Genet* 2008;**17**(24):3847–53.

61. Kobayashi H, Abe K, Matsuura T, *et al.* Expansion of intronic GGCCTG hexanucleotide repeat in NOP56 causes SCA36, a type of spinocerebellar ataxia accompanied by motor neuron involvement. *Am J Hum Genet* 2011;**89**(1):121–30.

62. Koeppen AH, Mazurkiewicz JE. Friedreich ataxia: neuropathology revised. *J Neuropathol Exp Neurol* 2013;**72**(2):78–90.

63. Koeppen AH, Davis AN, Morral JA. The cerebellar component of Friedreich's ataxia. *Acta Neuropathol* 2011;**122**(3):323–30.

64. Koide R, Ikeuchi T, Onodera O, *et al.* Unstable expansion of CAG repeat in hereditary dentatorubral–pallidoluysian atrophy (DRPLA). *Nat Genet* 1994;**6**:9–13.

65. Koide R, Kobayashi S, Shimohata T, *et al.* A neurological disease caused by an expanded CAG trinucleotide repeat in the TATA-binding protein gene: a new polyglutamine disease? *Hum Mol Genet* 1999;**8**:2047–53.

66. Konigsmark B, Weiner L. The olivopontocerebellar atrophies: a review. *Medicine* 1970;**49**:227–41.

67. Koob MD, Moseley ML, Schut LJ, *et al.* An untranslated CTG expansion causes a novel form of spinocerebellar ataxia (SCA8). *Nat Genet* 1999;**21**:379–84.

68. Liquori CL, Schut LJ, Clark HB, *et al.* *Spinocerebellar ataxia type 5.* Cambridge: Cambridge University Press, 2002.

69. Louis ED, Lynch T, Ford B, *et al.* Delayed-onset cerebellar syndrome. *Arch Neurol* 1996;**53**:450–54.

70. Marie P, Foix C, Alajouanine T. De l'atrophie cerebelleuse tardive a predominance corticale. *Revue Neurol* 1922;**38**:849–85, 1082–111.

71. Marz P, Probst A, Lang S, *et al.* Ataxin-10, the spinocerebellar ataxia type 10 neurodegenerative disorder protein, is essential for survival of cerebellar neurons. *J Biol Chem* 2004;**279**:35542–50.

72. Matsuura T, Yamagata T, Burgess DL, *et al.* Large expansion of the ATTCT pentanucleotide repeat in spinocerebellar ataxia type 10. *Nat Genet* 2000;**26**:191–4.

73. Mizusawa H. [Autosomal dominant spinocerebellar degeneration – new forms and

pathomechanisms]. *Rinsho Shinkeigaku* 2004;**44**(11):782–4.

74. Moreira MC, Klur S, Watanabe M, *et al.* Senataxin, the ortholog of a yeast RNA helicase, is mutant in ataxia-ocular apraxia 2. *Nat Genet* 2004;**36**:225–7.

75. Moseley ML, Zu T, Ikeda Y, *et al.* Bidirectional expression of CUG and CAG expansion transcripts and intranuclear polyglutamine inclusions in spinocerebellar ataxia type 8. *Nat Genet* 2006;**38**(7):758–69.

76. Mrissa N, Belal S, Hamida CB, *et al.* Linkage to chromosome 13q11–12 of an autosomal recessive cerebellar ataxia in a Tunisian family. *Neurology* 2000;**54**:1408–14.

77. Ophoff RA, Terwindt GM, Vergouwe MN, *et al.* Familial hemiplegic migraine and episodic ataxia type-2 are caused by mutations in the Ca2 channel gene CACNL1A4. *Cell* 1996;**87**:543–52.

78. Orr HT, Chung MY, Banfi S, *et al.* Expansion of an unstable trinucleotide CAG repeat in spinocerebellar ataxia type 1. *Nat Genet* 1993;**4**:221–6.

79. Paulson HL, Perez MK, Trottier Y, *et al.* Intranuclear inclusions of expanded polyglutamine protein in spinocerebellar ataxia type 3. *Neuron* 1997;**19**:333–44.

80. Pennacchio LA, Lehesjoki AE, Stone NE, *et al.* Mutations in the gene encoding cystatin B in progressive myoclonus epilepsy (EPM1). *Science* 1996;**271**:1731–4.

81. Perkins EM, Clarkson YL, Sabatier N, *et al.* Loss of beta-III spectrin leads to Purkinje cell dysfunction recapitulating the behaviour and neuropathology of spinocerebellar ataxia type 5 in humans. *J Neurosci* 2010;**30**(14):4857–67.

82. Pook M, Al-Mahdawi S, Carroll C, *et al.* Rescue of the Friedreich's ataxia knockout mouse by human YAC transgenesis. *Neurogenetics* 2001;**3**:185–93.

83. Pulst SM, Nechiporuk A, Nechiporuk T, *et al.* Moderate expansion of a normally biallelic trinucleotide repeat in spinocerebellar ataxia type 2. *Nat Genet* 1996;**14**:269–76.

84. Restituito S, Thompson RM, Eliet J, *et al.* The polyglutamine expansion in spinocerebellar ataxia type 6 causes a beta subunit-specific enhanced activation of P/Q-type calcium channels in Xenopus oocytes. *J Neurosci* 2000;**20**:6394–403.

85. Robitaille Y, Schut L, Kish SJ. Structural and immunocytochemical features of olivopontocerebellar atrophy caused by the spinocerebellar ataxia type 1 (SCA-1) mutation define a unique phenotype. *Acta Neuropathol (Berl)* 1995;**90**:572–81.

86. Rub U, Brunt ER, Petrasch–Parwez E, *et al.* Degeneration of ingestion-related brainstem nuclei in spinocerebellar ataxia type 2, 3, 6 and 7. *Neuropathol Appl Neurobiol* 2006;**32**(6):635–49.

87. Rub U, Burk K, Timmann D, *et al.* Spinocerebellar ataxia type 1 (SCA1): New pathoanatomical and clinico-pathological insights. *Neuropathol Appl Neurobiol* 2012;**38**:665–680.

88. Rub U, Del Turco D, Burk K, *et al.* Extended pathoanatomical studies point to a consistent affection of the thalamus in spinocerebellar ataxia type 2. *Neuropathol Appl Neurobiol* 2005;**31**(2):127–40.

89. Rub U, Schols L, Paulson H, *et al.* Clinical features, neurogenetics and neuropathology of the polyglutamine spinocerebellar ataxias type 1, 2, 3, 6 and 7. *Prog Neurobiol* 2013;**104**:38–66.

90. Sato N, Amino T, Kobayashi K, *et al.* Spinocerebellar ataxia type 31 is associated with 'inserted' penta-nucleotide repeats containing (TGGAA)n. *Am J Hum Genet* 2009;**85**(5):544–57.

91. Savitsky K, Bar–Shiva A, Gilad S, *et al.* A single ataxia telangiectasia gene with a product similar to PI-3 kinase. *Science* 1995;**268**:1749–53.

92. Schelhaas HJ, van de Warrenburg BP. Clinical, psychological, and genetic characteristics of spinocerebellar ataxia type 19 (SCA19). *Cerebellum* 2005;**4**:51–4.

93. Schelhaas HJ, Ippel PF, Hageman G, *et al.* Clinical and genetic analysis of a four-generation family with a distinct autosomal dominant cerebellar ataxia. *J Neurol* 2001;**248**:113–20.

94. Schmitz–Hubsch T, du Montcel ST, Baliko L, *et al.* Scale for the assessment and rating of ataxia: development of a new clinical scale. *Neurology* 2006;**66**(11):1717–20.

95. Sellier C, Rau F, Liu Y, *et al.* Sam68 sequestration and partial loss of function are associated with splicing alterations in FXTAS patients. *Embo J* 2010;**29**(7):1248–61.

96. Serrano–Munuera C, Corral–Juan M, Stevanin G, *et al.* New subtype of spinocerebellar ataxia with altered vertical eye movements mapping to chromosome 1p32. *JAMA Neurol* 2013;**70**(6):764–71.

97. Sesqueiros J, Coutinho P. *Epidemiology and clinical aspects of Machado–Joseph disease.* New York: Raven Press, 1993.

98. Shiloh Y. ATM (ataxia telangiectasia mutated): expanding roles in the DNA damage response and cellular homeostasis. *Biochem Soc Trans* 2001;**29**:661–6.

99. Sobrido MJ, Cholfin JA, Perlman S, *et al.* SCA8 repeat expansions in ataxia: a controversial association. *Neurology* 2001;**57**:1310–12.

100. Stevanin G, Le GE, Ravise N, *et al.* A third locus for autosomal dominant cerebellar ataxia type I maps to chromosome 14q24.3–qter: evidence for the existence of a fourth locus. *Am J Hum Genet* 1994;**54**:11–20.

101. Stevanin G, Bouslam N, Thobois S, *et al.* Spinocerebellar ataxia with sensory neuropathy (SCA25) maps to chromosome 2p. *Ann Neurol* 2004;**55**:97–104.

102. Stevanin G, Broussolle E, Streichenberger N, *et al.* Spinocerebellar ataxia with sensory neuropathy (SCA25). *Cerebellum* 2005;**4**:58–61.

103. Stewart GS, Maser RS, Stankovic T, *et al.* The DNA double-strand break repair gene *hMRE11* is mutated in individuals with an ataxia-telangiectasia-like disorder. *Cell* 1999;**99**:577–87.

104. Storey E, Knight MA, Forrest SM, Gardner RJ. Spinocerebellar ataxia type 20. *Cerebellum* 2005;**4**:55–7.

105. Storey E, Bahlo M, Fahey M, *et al.* A new dominantly inherited pure cerebellar ataxia, SCA 30. *J Neurol Neurosurg Psychiatry* 2009;**80**(4):408–11.

106. Taylor AM, Groom A, Byrd PJ. Ataxia-telangiectasia-like disorder (ATLD)–its clinical presentation and molecular basis. *DNA Repair (Amst)* 2004;**3**:1219–25.

107. Timmerman V, De Jonghe P, Simokovic S, *et al.* Distal hereditary motor neuropathy type II (distal HMN II): mapping of a locus to chromosome 12q24. *Hum Mol Genet* 1996;**5**:1065–9.

108. Trouillas P, Takayanagi T, Hallett M, *et al.* International Cooperative Ataxia Rating Scale for pharmacological assessment of the cerebellar syndrome. The Ataxia Neuropharmacology Committee of the World Federation of Neurology. *J Neurol Sci* 1997;**145**(2):205–11.

109. Tsuchiya K, Matsunaga T, Aoki M, *et al.* Familial amyotrophic lateral sclerosis with posterior column degeneration and basophilic inclusion bodies: a clinical, genetic and pathological study. *Clin Neuropathol* 2000;**20**:53–9.

110. van de Leemput J, Chandran J, Knight MA, *et al.* Deletion at ITPR1 underlies ataxia in mice and spinocerebellar ataxia 15 in humans. *PLoS Genet* 2007;**3**(6):e108.

111. van de Warrenburg BP, van Gaalen J, Boesch S, *et al.* EFNS/ENS Consensus on the diagnosis and management of chronic ataxias in adulthood. *Eur J Neurol* 2014;**21**(4):552–62.

112. Van Swieten JC, Brusse E, de Graaf BM, *et al.* A mutation in the fibroblast growth factor 14 gene is associated with autosomal dominant cerebellar ataxia. *Am J Hum Genet* 2003;**72**:191–9.

113. Verbeek DS, Schelhaas JH, Ippel EF, *et al.* Identification of a novel SCA locus (SCA19) in a Dutch autosomal dominant cerebellar ataxia family on chromosome region 1p21–q21. *Hum Genet* 2002;**111**:388–93.

114. Verbeek DS, van de Warrenburg BP, Wesseling P, *et al.* Mapping of the SCA23 locus involved in autosomal dominant cerebellar ataxia to chromosome region 20p13–12.3. *Brain* 2004;**127**:2551–7.

115. Vuillaume I, Devos D, Schraen–Maschke S, *et al.* A new locus for spinocerebellar ataxia (SCA21) maps to chromosome 7p21.3–p15.1. *Ann Neurol* 2002;**52**:666–70.

116. Wang JL, Yang X, Xia K, *et al.* TGM6 identified as a novel causative gene of spinocerebellar ataxias using exome sequencing. *Brain* 2010;**133** (Pt 12):3510–8.

117. Wang Q, Bardgett ME, Wong M, *et al.* Ataxia and paroxysmal dyskinesia in mice lacking axonally transported FGF14. *Neuron* 2002;**35**:25–38.

118. White M, Xia G, Gao R, *et al.* Transgenic mice with SCA10 pentanucleotide repeats show motor phenotype and susceptibility to seizure: a toxic RNA gain-of-function model. *J Neurosci Res* 2012;**90**(3):706–14.

119. Wilson R, Roof D. Respiratory deficiency due to loss of mitochondrial DNA in yeast lacking the frataxin homologue. *Nat Genet* 1997;**16**:352–7.

120. Woods C, Taylor A. Ataxia telangiectasia in the British Isles: the clinical and laboratory features of 70 affected individuals. *Q J Med* 1992;**82**:169–79.

121. Yu GY, Howell MJ, Roller MJ, *et al.* Spinocerebellar ataxia type 26 maps to chromosome 19p13.3 adjacent to SCA6. *Ann Neurol* 2005;**57**:349–54.

122. Zhuchenko O, Bailey J, Bonnen P, *et al.* Autosomal dominant cerebellar ataxia (SCA6) associated with small polyglutamine expansion in the alpha 1A-voltage-dependent calcium channel. *Nat Genet* 1997;**15**:62–9.

123. Zu T, Gibbens B, Doty NS, *et al.* Non-ATG-initiated translation directed by microsatellite expansions. *Proc Natl Acad Sci U S A* 2011;**108**(1):260–5.

Motor Neuron Disorders

Paul G Ince, J Robin Highley and Stephen B Wharton

INTRODUCTION

Disorders predominantly characterized by degeneration of motor neurons, either in the spinal cord anterior grey matter or brain stem/midbrain motor nuclei (lower motor neurons) or in the motor cortex (upper motor neurons), or a combination of both, are termed 'motor neuron diseases' and 'motor neuron disorders'. Use of the term motor neuron disease (MND) in the singular is generally accepted as a synonym for amyotrophic lateral sclerosis (ALS) and related diseases. Diseases that primarily affect motor axons are referred to as 'motor neuropathies' (see Chapter 24, Diseases of Peripheral Nerve). Distinguishing between disorders of the motor neuron and the motor axon presents a clinical diagnostic challenge such that it may be difficult to discern whether the neuronal perikaryon or the axon is the prime compartmental target of injury. Motor neuron disorders can be divided into three main groups:

- Primary motor neuron disorders
  - Idiopathic or heritable
- Secondary motor neuron disorders
  - Infective, metabolic, toxic and autoimmune
- Motor neuron degeneration as part of another multisystem neurodegenerative disease

The main diseases that predominantly affect motor neurons are shown in Table 14.1.

MOTOR NEURON DISEASE (AMYOTROPHIC LATERAL SCLEROSIS)

Clinical Features and Nomenclature

Motor neuron disease encompasses ALS and its variants. The condition is increasingly referred to as ALS/MND to resolve difficulties arising from differences in nomenclature in different parts of the world; in the USA, ALS is commonly referred to as 'Lou Gehrig's disease'. In ALS/MND, a primary degeneration of upper and lower motor neurons manifests as weakness and wasting of affected muscles, and evidence of corticospinal tract degeneration. Typically, there is preservation of eye movements and pelvic continence. The diagnosis of ALS/MND is made on the basis of clinical evidence of widespread combined upper and lower motor neuron damage that cannot be explained by other disorders, supported by electrophysiological evidence of anterior horn cell damage. Clinical criteria for the diagnosis of ALS have been developed (Table 14.2). The categories of clinically definite and probable ALS correlate well with pathological confirmation of the disease. Molecular pathology has demonstrated that the syndrome of progressive muscular atrophy, which is classified as 'suspected ALS' in the El Escorial criteria, is part of the same disease spectrum.[127] The prevailing hypothesis that ALS, progressive bulbar palsy (PBP), primary lateral sclerosis (PLS) and progressive muscular atrophy (PMA) are variants of a single clinicopathological syndrome also extends to include variants of frontotemporal dementia (Chapter 16, Dementia).[128,180]

Because several diseases can be associated with motor neuron loss, secondary causes of MND must be excluded. The finding of a characteristic inclusion body in anterior horn cells in this disease has facilitated precise pathological diagnosis.[173,211] Recent studies indicate that sporadic and most familial forms of ALS/MND are similar in terms of molecular pathology.

Early descriptions of MND from the nineteenth century[17,44,45] remain valid. The nomenclature and classification of ALS/MND derive partly from clinical features of the disease and partly from histopathology.

ALS, as defined by Charcot, is progressive, characterized by degeneration of both upper and lower motor neurons,

TABLE 14.1 Motor neuron disorders

Primary motor neuron disorders	Secondary motor neuron disorders
Idiopathic motor neuron diseases Sporadic ALS (sporadic motor neuron disease) ALS with frontal lobe dementia Western Pacific ALS–parkinsonism–dementia complex Guadeloupe form of parkinsonism–dementia with amyotrophy Monomelic motor neuron disease 'Madras'-type motor neuron disease	*Infective* Acute poliomyelitis/post-polio syndrome HIV infection HTLV-1 infection Syphilis
Heritable motor neuron diseases	*Radiation-induced motor radiculopathy associated with neurotoxins* Neurolathyrism 'Mantakassa' and related acute upper motor neuron syndromes Lead-induced motor neuropathy
Autosomal recessive Spinal muscular atrophy (Type 1 acute infantile) (Type 2 intermediate) (Type 3 juvenile, chronic) (Type 4 adult-onset, chronic) Neuroaxonal dystrophy Fazio–Londe disease Juvenile-onset ALS ('Tunisian' and related forms) Brown–Vialetto–van Laere syndrome (juvenile pseudo-bulbar palsy with sensorineural deafness)	*Metabolic disease* Hyperparathyroidism Hexosaminidase deficiency Hypoglycaemia-related motor neuronopathy *Autoimmune disease* Motor neuropathy with antiganglioside antibodies (multifocal motor neuropathy) Motor neuropathy with paraproteinaemia Motor neuron disease with lymphoma
X-linked Kennedy syndrome (bulbospinal neuronopathy; bulbar and spinal muscular atrophy)	*Motor neuron degeneration as part of multisystem neurodegenerative disease* Neuroacanthocytosis Machado–Joseph disease Huntington's disease Prion diseases
Autosomal dominant Familial ALS Spinocerebellar degenerations Distal spinal muscular atrophy Juvenile-onset ALS Familial ALS with frontal lobe dementia	

ALS, amyotrophic lateral sclerosis; HIV, human immunodeficiency virus; HTLV, human T-lymphotropic virus.

TABLE 14.2 Summary of the revised (1999) World Federation of Neurology criteria for the diagnosis of amyotrophic lateral sclerosis (ALS)

Diagnosis of ALS requires:

1. The presence of:
   (a) evidence of LMN degeneration by clinical, electrophysiological or neuropathological examination;
   (b) evidence of UMN degeneration by clinical examination;
   (c) progression of the motor syndrome within a region or to other regions, as determined by history or examination
and
2. The absence of:
   (a) electrophysiological and pathological evidence of other disease processes that might explain the signs of LMN or UMN degeneration and
   (b) neuroimaging evidence of other disease processes that might explain the observed clinical and electrophysiological signs.

Four diagnostic categories are recognized:*

1. *Clinically definite ALS*: on clinical grounds alone, evidence of UMN plus LMN signs in the bulbar region and in at least two spinal regions, or the presence of UMN signs in two spinal regions and LMN signs in three spinal regions;
2. *Clinically probable ALS*: on clinical grounds alone, UMN plus LMN signs in at least two regions with some UMN signs rostral to LMN signs;
3. *Probable, laboratory-supported ALS*: defined, after proper application of neuroimaging and clinical laboratory protocols has excluded other causes, as:
   (a) clinical evidence of UMN and LMN signs in only one region, or
   (b) UMN signs alone in one region and LMN signs defined by EMG criteria in at least two limbs;
4. *Possible ALS*: defined, once other diagnoses have been excluded, as:
   (a) UMN plus LMN signs in only one region
   (b) UMN signs alone in two or more regions
   (c) LMN signs found rostral to UMN signs.

*The category of suspected ALS has been discarded. It should be noted that these criteria were designed for use in clinical trials and not as rules for diagnosis in clinical practice. The revised criteria no longer exclude syndromes (e.g. ALS with parkinsonism or dementia) that are somewhat atypical of the core syndrome of ALS, but they do exclude primarily multisystem disorders that may have some involvement of upper and lower motor neurons.

EMG, electromyelography; LMN, lower motor neuron; UMN, upper motor neuron.

Reproduced from Brooks BR, Miller RG, Swash M, Munsat TL. El Escorial revisited: revised criteria for the diagnosis of amyotrophic lateral sclerosis. *Amyotroph Lateral Scler/ Other Motor Neuron Disord* 2000;1:293–9.

and is the most common pattern of ALS/MND syndromes. In about 20 per cent of cases, ALS begins with bulbar symptoms (dysarthria and dysphagia) progressing to neurogenic wasting of the tongue, and palatal paralysis, often termed progressive bulbar palsy (PBP). This motor syndrome can occur in demyelinating and cerebrovascular diseases, but the combination of inexorably progressive dysarthria with dysphagia, emotional lability and fasciculation and atrophy of the tongue almost always proves to be ALS/MND. The disease process soon spreads to involve the limbs and EMG studies will usually detect evidence of limb muscle denervation before clinical signs are evident.

Imaging is helpful in excluding other conditions, and new MRI modalities (e.g. diffusion weighted imaging, tractography) may detect involvement of motor and extramotor systems in the brain (Figures 14.1 and 14.2).

Progressive muscular atrophy (PMA) is characterized by lower motor neuron signs, loss of anterior horn cells and preservation of upper motor neurons and corticospinal tracts. This syndrome accounts for 10 per cent of presentations though many patients eventually develop upper motor neuron involvement.[127]

Primary lateral sclerosis (PLS) is uncommon and clinically characterized by upper motor neuron signs in the absence of lower motor neuron signs. The clinical course is frequently protracted and additional frontal lobe behaviours and non-cognitive features may be prominent over time. Atrophy and neuronal loss is restricted to the motor cortex and adjacent frontal cortex although autopsy data are sparse.[119,233,322]

The usual cause of death in ALS/MND is respiratory failure due to weakness of the diaphragm and accessory muscles of ventilation within 5 years of disease onset.[216] Patients with ALS/MND have been kept alive using invasive ventilation resulting eventually in a 'locked in' state of advanced multisystem neurodegeneration.[297] Most recently, advances in clinical care including non-invasive ventilatory support and percutaneous gastrostomy feeding have improved survival and terminal general physical condition of ALS/MND patients.[36,47] However, 15–20 per cent of patients with ALS survive for over 5 years and 5 per cent survive for longer than 10 years without assisted ventilation. Pathological findings in these atypical cases are variable. Dementia, characteristically frontal lobe-type, is a feature in a minority of patients (see Chapter 16, Dementia) and may precede, co-present or develop later than the motor features.

Epidemiology

ALS/MND has a worldwide distribution. Incidence rates are relatively constant throughout the world, with an average of about 2 per 100 000 and a prevalence of about

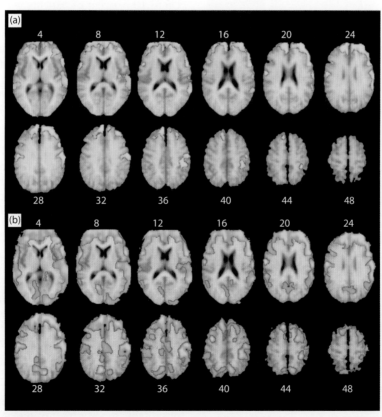

14.1 Motor neuron disease (MND). Positron-emission tomography (PET) using the γ-aminobutyric acid (GABA) A ligand [11C] flumazenil. Regions of decreased binding of the ligand in patients with MND relative to controls are superimposed on a normalized magnetic resonance image. Decreases in flumazenil binding involve both motor and extramotor areas, particularly the prefrontal cortex. The changes are shown at two levels of statistical significance: **(a)** P < 0.01; **(b)** P < 0.05. As expected, the decrease in binding appears more widespread at a lower level of statistical significance **(b)**.

From Lloyd et al.[169] Reproduced with permission of Oxford University Press on behalf of The Guarantors of Brain.

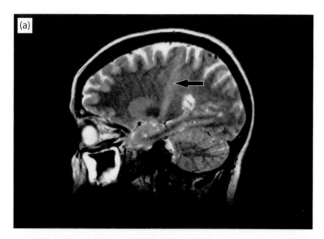

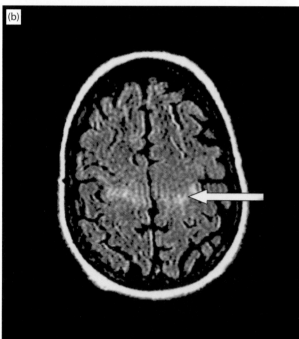

14.2 Motor neuron disease (MND). (a) T2-weighted image of the brain, showing altered signal (arrow) in the corticospinal tract in a patient with rapidly progressive MND and prominent upper motor neuron signs. **(b)** Axial fluid level-attenuated inversion recovery (FLAIR) image of the same patient, showing altered subcortical white matter signal in the primary motor area.

Reproduced from Ellis et al.[76] with permission from Lippincott Williams & Wilkins/Wolters Kluwer Health.

5 per 100 000.[170] The incidence may be rising, probably explained by better diagnosis and case ascertainment, and by increased life expectancy of the population. The prevalence of MND is much higher in certain areas of the Western Pacific (Guam, Irian Jaya, Kii peninsula).

The incidence of sporadic ALS/MND is higher in men than women (male:female ratio 1.5:1). The mean age of onset of sporadic ALS is around 60 years, but about 25 per cent of cases present under the age of 50 years, and a few in the third and fourth decades. Rare juvenile cases are described. The median survival of MND is 3–4 years, the prognosis being worse with bulbar onset disease. Younger patients (onset below the age of 50) and patients with PMA

often survive significantly longer than patients with typical late-onset ALS/MND.[106] There is no evidence that race significantly alters the risk of MND[18,19] although it may influence phenotype.[291]

Aetiology of ALS/MND

Around 10 per cent of ALS/MND is associated with clear mendelian inheritance. An increasing proportion (~50 per cent) of familial ALS/MND is attributable to defined genetic causes. Pathogenic mutations in these multiple genes arise at low frequency in apparently sporadic cases. The influence of environmental factors remains relevant in sporadic disease but is poorly characterized.

Environmental Factors

Some studies report higher rates of ALS in rural populations but there is no convincing evidence of significant clustering of cases.[19,50] A history of mechanical trauma is reported to be more common in ALS than in controls, including electrical trauma; however, this is controversial.[156] Although lead can cause a toxic motor neuropathy and an ALS-like syndrome, there is no firm evidence that lead or other trace metals cause ALS.[19]

Genetics

Up to 10 per cent of patients with ALS/MND have disease with a clear mendelian pattern of inheritance.[41] Most families show autosomal dominant transmission, but X-linked and recessive inheritance also occur (Table 14.3). The average age of onset of familial ALS (FALS) is about 10 years earlier than the sporadic disease, with a median survival time of about 24 months.[9] An increasing number of defined genetic abnormalities are described related to previously known ALS/MND genetic loci. Sixteen causative genes have so far been identified. The first description was of mutations of the copper zinc (Cu/Zn) superoxide dismutase (*SOD1*) gene (ALS1), shown to cause ALS/MND in 10–20 per cent of all familial cases.[240] Some of the most important causative genes have been fairly recently described, namely those encoding TAR DNA binding protein (*TARDBP*), fused in sarcoma (*FUS*) and *C9ORF72*. An uncommon juvenile variant of ALS/MND is linked to the gene *Alsin* (ALS2). Other genetic changes, such as those affecting the tail region of the neurofilament heavy chain (NF-H) subunit gene have been found in about 1 per cent of sporadic cases and are regarded as probable risk factor genes rather than directly causative.[80] However, such interpretation is problematic because one family is reported with ALS/MND related to an apparent mutation in NF-H.[6] The gene involved in autosomal recessive spinal muscular atrophy (*SMN1*) does not appear to be involved in ALS/MND.[223,226] However, the *SMN2* gene may be abnormal in some patients with ALS/MND so that the SMN locus may also confer phenotypic modification of disease.[306] The most important conceptual groupings to emerge from new genetic data in ALS/MND relate to the constellation of genes giving rise to intraneuronal TDP43 inclusions (*TARDBP*, *C9ORF72*, *VABP*, *VCP*, and *CHMP2B*), and to the constellation that can be mapped to potential mechanisms involving RNA metabolism (*TARDBP*, *FUS*, *C9ORF72*).

TABLE 14.3 Genetic loci causing familial ALS/MND or possibly associated with risk of sporadic disease in genetic association studies

		Autopsy data?	Chromosomal Locus	Gene Product			Reference
Familial							
	ALS1	✓	21q22	Superoxide dismutase 1	SOD1	AD,AR	240
	ALS2		2q33	Alsin	ALS2	AR	101, 320
	ALS3		18q21	?		AD	104
	ALS4		9q34	Senataxin	SETX	AD	49
	ALS5		15q15-22	Spastic paraplegia 11	SPG11	AR	113
	ALS6	✓	16q12.1-q12.2	Fused in sarcoma	FUS	AD	158, 305
	ALS7		20p13	?		AD	251
	ALS8	✓	20q13	Vesicle-associated membrane protein-associated protein B	VAPB	AD	215
	ALS9	✓	14q11	Angiogenin	ANG	AD	97
	ALS10	✓	1p36	TAR DNA binding protein 43	TDP43	AD	277
	ALS11		6q21	FIG4 homolog, SAC1 lipid phosphatase domain containing	FIG4	AD	52
	ALS12	✓	10p13	Optineurin	OPTN	AR and AD	186
	ALS13		12q23-24	Ataxin 2	ATXN2	Risk	75
	ALS14	✓	9p13	Valosin-containing protein	VCP	Unclear	132
	ALS15		Xp11-q12	Ubiquilin 2	UBQLN2	XD juvenile	67
	FTLD-ALS	✓	9q21-22	Chromosome 9 open reading frame 72	C9ORF72	AD/sporadic	65, 238
	FTLD- ALS	✓	3p12	Chromatin modifying protein 2B	CHMP2B	Unclear	227
			12q24	D-amino-acid oxidase	DAO	AD	196
Sporadic							
			2, 19q13	Neuron-specific kinesin heavy chain			292
			?5q13 (del)	NAIP			131
			?5q13	SMN			203
			6q25	Mn-SOD (SOD2)			303
			9p13	CNTF receptor α			121
			11p13	EAAT2			12
			11q12	CNTF			87
			14q11.2-q12	APEX nuclease			108
			19q13	ApoE			202
			22q12.1-q12.2	LIF			88
			6p21.3	VEGF			221
			22q12	NF-H			6
			Mitochondrial	Cytochrome c oxidase subunit1			55

AD, autosomal dominant; AR, autosomal recessive, XD, X-linked dominant; XR, X-linked recessive; FTLD-ALS, ALS with frontotemporal dementia; BSMA, spinobulbar muscular atrophy (Kennedy's disease); SMN, spinal muscular atrophy; Mn-SOD, manganese SOD; CNTF, ciliary neurotrophic factor; EAAT, excitatory amino-acid transporter; APEX, apurinic/apyrimidinic endonuclease; LIF, leukaemia inhibitory factor; VEGF, vascular endothelial growth factor; NF-h, neurofilament heavy subunit; ✓, see text for key references.

Pathobiology of *SOD1*-FALS

Mutations of the *copper zinc (Cu/Zn) superoxide dismutase (SOD1)* gene (ALS1) cause 10–20 per cent of all familial ALS/MND cases.[240] Reverse genetics established linkage of familial MND to chromosome 21q and sequencing of the *SOD1 Cu/ZnSOD* gene has revealed over 120 mutations in all of the five exons.[9] Most of the mutations identified are missense point mutations within exons. Several deletions and intronic mutations have also been identified. The mutations cause disease through a toxic gain-of-function mechanism often associated with evidence of abnormal protein folding and altered dimer interactions.

The clinical phenotype of *SOD1*-FALS varies considerably between kindreds with different mutations and even within a single kindred.[60,222,224,235] Some mutations are associated with short survival and little evidence of corticospinal tract damage (e.g. the A4V mutation, others with longer survival). The autosomal recessive D90A mutation, most commonly encountered in Scandinavia,[11] is typically associated with a slowly progressive ascending spastic paraparesis with later lower motor neuron and bulbar involvement. However, the D90A mutation can also behave as an autosomal dominant trait in some families more typical of *SOD1*-related disease. This suggests that the autosomal recessive form of D90A ALS/MND may be associated with a 'protective factor' that modifies the deleterious effect of the mutant protein.[5] The only other true recessive *SOD1* mutation (N86S) is associated with juvenile onset and rapid progression.[107] *SOD1* mutations are identified in apparently sporadic cases of ALS.[133]

The pathobiology of neurodegeneration associated with mutant *SOD1* remains unresolved. SOD1 is normally concerned with the removal of superoxide radicals and it was originally proposed that defective enzyme function might lead to free radical-mediated damage to motor neurons.[37] However, evidence suggests that other factors are involved:[4,33,39,299] the activity of mutant *SOD1*, in some cases, differs little from its normal (wild-type) counterpart; there is no clear relationship between the site of mutations in relation to the active (catalytic) site of the SOD1 molecule or with severity of the phenotype; transgenic *SOD1*-knockout mice do not develop MND. In contrast, mice expressing mutant *SOD1* show loss of spinal motor neurons and die from motor neuron degeneration.[39,100] Therefore, current evidence suggests that mutant *SOD1* gains a toxic property, although the nature of this toxic gain of function is unclear. Excitotoxicity, oxidative stress, increased peroxidase activity, nitrotyrosine formation, altered copper binding, cytoskeletal damage, protein aggregation and a pro-apoptotic effect of *SOD1* mutations have all been implicated.[10] Such mechanisms are not mutually exclusive and all may contribute to selective neuronal death.

The development of transgenic mouse models of SOD1-FALS (using non-cell specific promoters) has clarified some aspects of the disease, but has also raised new questions. Virtually all these models rely on significant genetic overexpression and their relevance to mechanisms in human disease is disputed. The concept of non-cell autonomous toxicity of mutant *SOD1* has arisen from experiments involving chimeric transgenes.[54]

RNA Metabolism and the Genetics of ALS/MND

Three recent genetic loci for ALS/MND have provided major advances in understanding the pathology and biology of ALS/MND. These are *TARDBP*, *FUS* and *C9ORF72*.[65,158,238,277,302,305] The *TARDBP* gene encodes for the protein TDP-43, which is concerned with RNA metabolism including regulation of RNA splicing and other mechanisms. TDP-43 binds a vast number of mRNA targets, many of which are themselves RNA processing genes.[232,263] TDP-43 has been shown to be the underlying proteinopathy associated with ubiquitylated inclusions in neurons and glia in sporadic ALS/MND and FTD. *FUS* encodes a protein that is also involved in RNA metabolism.[281] Mutations in both of these genes give rise to ALS/MND syndromes as a result of exonic mutations resulting in expression of mutant protein. Finally, the *C9ORF72* gene is implicated at high frequency in familial ALS/MND (~40 per cent) and a significant number of apparently sporadic cases (~7 per cent). The genetic change in this gene is an expansion of a GGGGCC repeat sequence in intron 1 between the two, non-coding exons 1 and 2. Most normal individuals have up to 3 repeat sequences and those presenting with ALS have more than 30. An intermediate range from 4 to 27 repeats is encountered and is of less certain pathogenetic significance. This expansion appears to sequester RNA processing molecules (see later). These three genetic changes triangulate around abnormal RNA metabolism as a potent cause of motor system degeneration and may induce axonal dysfunction through changes in intracellular RNA trafficking. They are also associated with frontotemporal lobar degeneration (FTLD) variants.

Neurotoxicity and Excitotoxicity

The evidence linking excitotoxicity (see Chapter 1, General Pathology of the Central Nervous System) to MND is indirect.[183,265] Raised plasma and cerebrospinal fluid levels of glutamate, impaired high-affinity glutamate uptake and selective loss of the glial glutamate transporter EAAT2 in the motor cortex and spinal cord have been reported in ALS/MND[231,242,264] but could be secondary to neuronal loss. Excitotoxic neuronal damage involves the excessive action of excitatory amino acids or excitotoxins on receptors expressed by susceptible neurons. The possibility that exon-skipping and exon-retaining mRNA variants of EAAT2 occur in ALS/MND tissue[167] has not been confirmed by other workers and these are considered to be normal splice variants that are also present in controls.[84,194] Editing of AMPA receptor RNA transcripts associated with expression of Ca2+-permeable excitatory receptors does map to vulnerable cell populations and may be relevant.[315] Vulnerable spinal motor neurons contain fewer calcium binding proteins such as parvalbumin and calbindin D_{28K} than do motor neurons that are spared.[14,123] Riluzole, a glutamate release inhibitor and sodium channel blocker, has been shown to prolong survival in ALS/MND[25,160] but lamotrigine, another glutamate release inhibitor, had no effect at low dosage.[74] The suggestion that the plant-derived excitotoxin β-methyl-amino-L-alanine (BMAA) may be the cause of the high incidence of MND seen in Western Pacific areas has been revived by recent claims regarding possible bioaccumulation in mammalian tissues.[59] However, alternative Cycad-derived toxins, including sterol glucosides, are plausible candidates for neurotoxicity.[146]

Neurofilaments and Cytoskeletal Abnormalities

Intra-axonal accumulations of neurofilaments (spheroids) are frequent in ageing and may be more prominent in ALS/MND. Less commonly, intraneuronal neurofilament accumulations are present, especially in SOD1-FALS. A few patients with MND have mutations of the gene coding for the neurofilament heavy chain (NF-H).[6] Neurofilaments (see Chapter 1) are important in determining axonal calibre and implicated in axonal transport.[135] Decreased expression of the neurofilament light chain (NF-L), necessary for neurofilament triplet formation and abnormal neurofilament phosphorylation in motor neuron perikarya, are reported in ALS/MND. Phosphorylation of side-arms is thought to affect neurofilament function. Glutamate has been shown to increase neurofilament phosphorylation and slow axonal transport.[1] Neurodegeneration may result from altered neurofilament structure (e.g. NF-H gene mutations), altered stoichiometry (e.g. decreased expression of NF-L), or excitotoxicity, nitrotyrosine formation or other post-translational modifications.[279] Knockout of mouse neurofilament genes and overexpression of mouse or human neurofilament genes has minimal effect on motor neuron function, even when overexpression results in massive accumulation of neurofilaments in motor neurons.[136,137] However, manipulation of neurofilament function can result in ALS/MND in mice (e.g. a point mutation in the conserved rod domain of NF-L). Knockout of the mouse NF-L gene in transgenic mutant *SOD1* mice prolongs survival, as does overexpression of human NF-H. The overexpression of peripherin, another intermediate filament, in mice leads to MND, and survival is further decreased if NF-L is also deficient.[135] Peripherin expression in motor neurons is upregulated by inflammatory cytokines and antisera to peripherin label compact but not skein-like neuronal cytoplasmic inclusions.[110] Further, there is evidence that TDP-43 can stabilise neurofilament mRNA.[308] In summary, there is good evidence that the neuronal cytoskeleton is altered in ALS/MND. Experimental studies suggest that the balance between neurofilament subunits and peripherin is important in motor neuron survival, and factors such as excitotoxicity and inflammatory responses also influence cytoskeletal function and may contribute to cell death.

Immune Mechanisms in ALS

Involvement of immune processes in ALS was raised by several findings. Autoantibodies to L-type calcium channels have been described in ALS[13,150] and infiltrates of lymphoid cells are seen in the spinal cord.[295] Motor neuron degeneration has also been associated with paraproteinaemia and lymphoma.[250] Trials of immunosuppressive therapy have been unsuccessful in MND.[244] Paraneoplastic motor neuron syndromes with an autoimmune basis are well recognized and can be clinically confused with ALS.[85]

In transgenic mice, overexpression of interleukin-3 causes a form of ALS/MND in which an immune response against spinal motor neurons is associated with their loss.[48] Microglial activation is readily demonstrable in ALS/MND tissue and forms the basis of experimental neuroimaging.[280] However, the complexities of microglial biology, especially the possibility that microglial activation may confer either a cytoprotective or a cytotoxic cytokine response, has hindered understanding of the significance of this finding in all neurodegenerative diseases.

Growth Factors

Motor neuron survival has been shown to depend on growth factors.[168] A primary defect in the availability of growth factors has not been shown in MND. Many trophic factors including nerve growth factor (NGF), glial-derived neurotrophic factor (GDNF), ciliary neurotrophic factor (CNTF) and brain-derived neurotrophic factor (BDNF) support neuronal survival and regrowth of cell processes. A change in a trophic factor expression or availability might render motor neurons more susceptible to programmed cell death and apoptosis. In transgenic mice, deletion of the hypoxia-response element in the vascular endothelial growth factor (VEGF) promoter causes adult-onset motor neuron degeneration, raising the possibility that growth factors responsible for microvascular maintenance may also be involved in motor neuron viability.[221] To date, all clinical trials of neurotrophic factors including insulin-like growth factor-1 (IGF-1), CNTF and BDNF have been negative, despite promising results in SOD1-FALS mice.[300]

Viral Infection

The suggestion of viral infection as a cause of MND stems from close similarities with poliomyelitis. A positive correlation between the incidence of paralytic poliomyelitis and deaths from ALS/MND 30–40 years later has been found but there is no evidence for a fall in ALS incidence or mortality in populations that have been immunized against poliomyelitis.[282] Studies have failed to reveal an infectious agent in ALS tissue[309] but others have claimed that viruses (specifically enterovirus) can be detected more commonly than in control tissue.[26]

Autopsy in Cases of ALS/MND

At autopsy, ALS/MND patients were traditionally very thin with little subcutaneous fat. However, modern clinical management of significant bulbar weakness includes percutaneous endoscopic gastrostomy nutrition. Such patients are likely to have a normal body weight or weight reduction commensurate with loss of muscle bulk through atrophy. Tracheostomy to protect the airway in cases with severe bulbar dysfunction has been replaced by non-invasive ventilation. The previous frequent association with changes of bronchopneumonia (often aspiration pneumonia) is less commonly encountered in the era of more intensive nutritional and respiratory care. Occasionally, ALS/MND may be confirmed at autopsy in patients with head trauma due to a fall.

Typically, widespread muscle atrophy affects proximal and distal muscles in the limbs, tongue, intercostal muscles and diaphragm. Symmetrical involvement of the intrinsic hand muscles is a useful indication of this at autopsy. Macroscopically, muscle bulk is reduced and the muscle is pale.

Where the diagnosis of ALS/MND is less certain clinically (e.g. a potential diagnosis of multifocal motor neuropathy with nerve conduction block), samples of peripheral nerve and muscle should be preserved for examination. Post-mortem blood samples may be used to look for the presence of autoantibodies in appropriate cases. In the light of emerging insights into the molecular pathogenesis of these disorders, and the possible need for genetic counselling, archiving of material for DNA or RNA analysis is becoming important.

Pathological Changes in the Brain and Spinal Cord

Macroscopically, the brain is generally unremarkable. In some cases, visible atrophy of the precentral gyrus is seen, best detected by removing the leptomeninges after fixation. This is particularly striking in primary lateral sclerosis.[147] In patients who have associated dementia, frontal or temporal lobe atrophy may be seen. There may be some atrophy of the medullary pyramids most obvious in horizontal slices through the medulla rather than externally. The spinal cord may be atrophic and the anterior nerve roots are generally shrunken and grey in comparison with the posterior sensory roots (Figure 14.3). Marked reduction in the number of large myelinated fibres is seen in ventral nerve roots, whereas peripheral nerves show axonal degeneration and

14.3 Motor neuron disease. (a) Normal subject. **(b)** Atrophy of nerve roots in the lumbosacral region of the spinal cord.

reduced numbers of large myelinated fibres.[228] Radicular glial bundles, which are pathological extensions of glia into nerve roots, may sometimes be observed in ventral and dorsal roots in ALS/MND. Although not specific to ALS/MND, they are a further indicator of root pathology.[151] In patients who have died early as a result of rapidly progressive bulbar dysfunction, such pathological changes may be anatomically restricted.

The key histological change is loss of motor neurons with associated astrocytosis in anterior horns of the spinal cord (Figure 14.4), the brain stem motor nuclei (cranial nerves: XII, motor VII, motor V) (Figure 14.5) and the motor cortex. A variety of inclusion bodies may be seen on haematoxylin and eosin (H&E) staining in surviving motor neurons and are discussed later.[126]

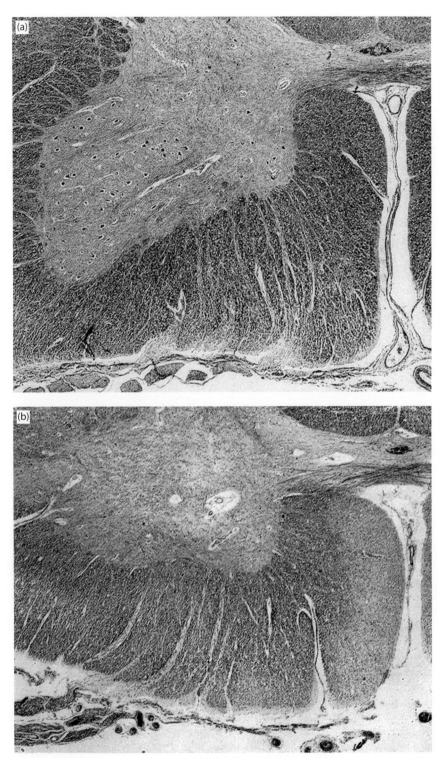

14.4 Motor neuron disease. Anterior horn of the spinal cord. **(a)** Normal subject. **(b)** Loss of motor cells from the sixth cervical segment. There is also loss of myelin from the uncrossed corticospinal tract. Klüver–Barrera.

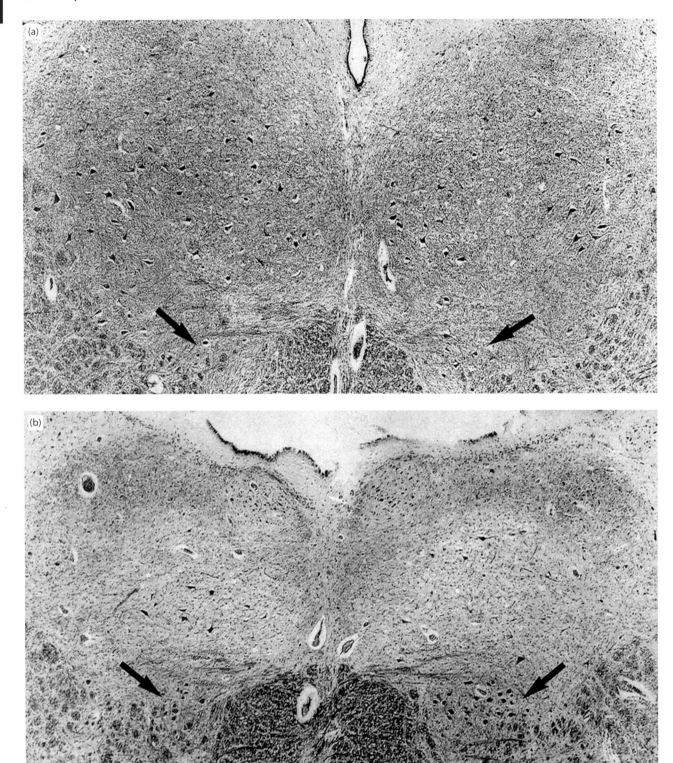

14.5 Motor neuron disease. Hypoglossal nuclei. **(a)** Normal subject. **(b)** Loss of motor cells from the hypoglossal nuclei. The nucleus of Roller (arrows) is unaffected. Klüver–Barrera.

Bunina bodies (Figure 14.6) are small eosinophilic inclusions of 2–5 µm, often arranged in small beaded chains. Despite ultrastructural characterization showing that they are likely to be of lysosomal derivation,[254] their origin and significance remain uncertain. They are cystatin-C immunoreactive.[219] Bunina bodies occur in at least 85 per cent of cases of ALS/MND but the true frequency is uncertain because there are no consensus guidelines for the scale and anatomical range of histological sampling.[230]

Currently considered the most important finding in ALS/MND is that of inclusion bodies in motor neurons detected by immunohistochemical staining for ubiquitin, ubiquilin-2 and p62.[172] The principal protein present in these inclusions is TDP-43.[15,212] These inclusions appear specific for sporadic ALS and some familial types, and can be found in both upper and lower motor neurons.[174] The most frequently encountered inclusion appears as an aggregate of thread-like structures (Figure 14.7) termed skeins.[125] Less common are more compact, spherical inclusions with or without a frayed filamentous margin (Figure 14.8) corresponding to previous designations such as hyaline or Lewy-like inclusions (Figure 14.9).[115] A minority of MND inclusions appear as very small scattered granules or single threads. Quantitative studies have shown a smaller number of inclusions in cases of long duration and highest numbers in cases with short duration, suggesting that the ALS inclusions are formed at an early stage of the disease.[214,258] Furthermore, cases with greater numbers of inclusions appear to show greater degrees of neuronal loss.[304] In addition to these forms, TDP-43 immunohistochemistry has revealed neurones with more diffuse, sometimes granular labelling of the cytoplasm, a loss of nuclear TDP-43 labelling, but no well-defined solid

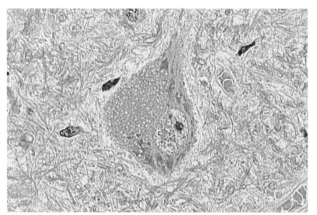

14.6 Motor neuron disease: Bunina body in anterior horn cell.

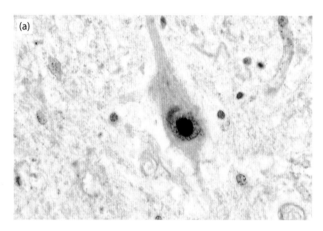

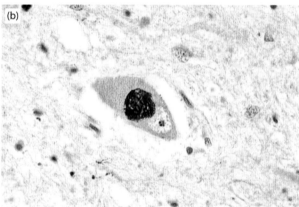

14.8 Motor neuron disease. (a,b) Spherical inclusions. Anti-ubiquitin immunoperoxidase, diaminobenzidine (DAB)/ haematoxylin.

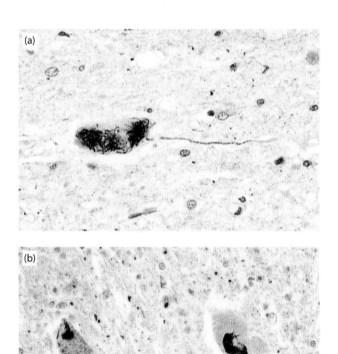

14.7 Motor neuron disease. (a,b) Thread-like structures form skein inclusions. Anti-ubiquitin immunoperoxidase, diaminobenzidine (DAB)/haematoxylin.

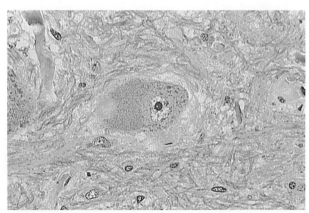

14.9 Motor neuron disease: pale eosinophilic hyaline or Lewy-like inclusion.

inclusion. These have been referred to as pre-inclusions and do not appear to be ubiquitylated (Figure 14.10).[68]

Ultrastructurally, skein inclusions are composed of arrays of 10–15 nm linear filaments associated with granules (Figure 14.11) and, frequently, Bunina bodies.[174] Post-translational phosphorylation and cleavage of TDP-43 are present. In healthy tissue, antibodies for normal TDP-43 show a widespread, predominantly nuclear staining pattern in many cell types. In ALS/MND the presence of cytoplasmic inclusions reactive for TDP-43 is associated with loss of the protein from the nucleus of affected cells. Antibodies for phospho-TDP-43 offer advantages because they seem to largely stain only pathological protein deposits.[212] However, the use of p62 is recommended as a primary screening method and to map the anatomical extent of neurodegeneration, with downstream use of a TDP-43 antibody to confirm the underlying molecular pathology in more limited anatomical sites.

A variety of other lesions have been identified in a few cases using conventional staining, especially 'basophil inclusions'.[298] This pathology is associated with FUS-pathology and linked to various dementia syndromes (Chapter 16, Dementia).[118]

Accumulation of neurofilament protein as large aggregates termed neurofilament conglomerates has been reported in association with certain *SOD1* mutations

(Figure 14.12).[128] Motor neuron swelling can be seen in sporadic cases associated with diffuse accumulation of phosphorylated neurofilaments and must be distinguished from neurofilament conglomerate inclusions. Such neuronal swelling appears to be non-specific and do not represent true central chromatolysis.[271] Golgi studies have shown abnormally thin axons emerging from spinal motor neurons with poor extension of the dendritic tree and thin dendrites.[143] Although there is a subjective impression that lipofuscin is increased in surviving neurons, this has not been confirmed in a quantitative study.[191] Axonal spheroids are frequently seen in the anterior horn of spinal cord in ALS.[163] Immunocytochemical and ultrastructural studies show that they are identical to spheroids seen in other diseases and in normal ageing and are associated with the intermediate filament peripherin.[58,195] Astrocytic gliosis is also seen in spinal cord grey matter with a distinctive distribution in relation to neuronal groups.[259]

In the brain stem, several motor neuron groups are affected, particularly the hypoglossal nucleus (Figure 14.5), nucleus ambiguous, motor nuclei of trigeminal nerve and facial nerves. In the majority of patients the nuclei of the IIIrd, IVth and VIth cranial nerves and Onufrowicz's nuclei appear normal and correlate with clinical features of preserved eye movements and sphincter control.[260] However,

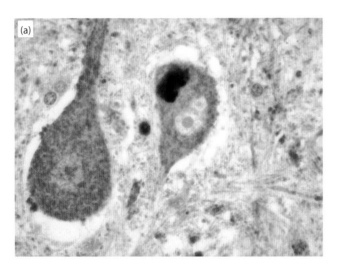

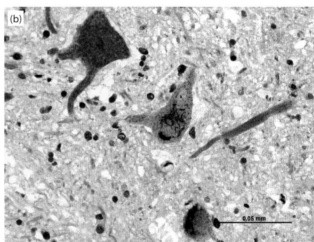

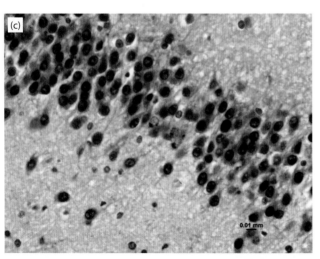

14.10 Pan-TDP43 immunocytochemistry shows a compact inclusion body together with diffuse cytoplasmic staining of this affected neuron and an adjacent neuron without inclusion formation. This 'pre-inclusion' staining represents un-ubiquitylated TDP43. It is associated with nuclear loss of TDP43 staining only in the presence of inclusion body formation **(a)**. Similar appearances are seen in **(b)** in relation to skein-like inclusion body material. In the hippocampal dentate granule cells inclusion bearing neurons demonstrate complete loss of nuclear TDP43 staining **(c)**.

neuronal cytoplasmic inclusions have been observed in these nuclei, indicating subclinical involvement.[149,220]

In the motor cortex, there may be a loss of Betz cells and a distinct pattern of astrocytic gliosis characterized by clusters of astrocytes, some forming a laminar band in the middle of the pyramidal cell layer.[205,207] Microvacuolation is seen in cases with marked neuronal loss.[213] Phagocytic cells

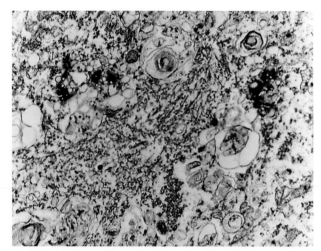

14.11 Motor neuron disease: inclusion (ultrastructural detail).

involved in neuronophagia may be identified by immunocytochemical techniques.[296] Interestingly, stereological studies that have focused on the total population of neurones (not limited to Betz cells) have revealed no changes in the number or size of neurones in the motor cortex.[95,289] Stereological analysis of Betz cells in isolation is yet to be done. Neuronal cytoplasmic inclusions may be seen but are infrequent compared with the spinal cord.[174,255] In primary lateral sclerosis the motor cortex bears the brunt of the disease, with other motor neuron groups being unaffected, conversely in progressive muscular atrophy the motor cortex (and descending corticospinal tracts) may be entirely normal.[127]

In ALS and PLS the white matter of the spinal cord shows myelin loss from the corticospinal tracts associated with astrocytic gliosis and the accumulation of microglia/macrophages.[127] The severity of myelin pallor is not directly related to loss of neurons from the motor cortex but, when severe, myelin loss can be traced up through the brain stem into the cerebral hemispheres. Myelin loss from corticospinal tracts, using conventional myelin staining, may be subtle in patients with early or mild upper motor neuron disease and it is possible that an axonopathy is the predominant feature of corticospinal tract degeneration in some patients. Thus, myelin loss is most evident in lower cord segments, supporting the hypothesis of a dying back degeneration of axons. The spinal cord may also show generalized pallor of myelin

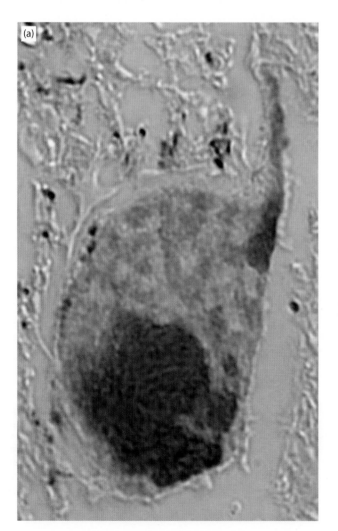

(a)

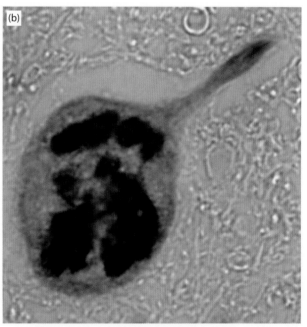

(b)

14.12 Neurofilament conglomerate inclusions in amyotrophic lateral sclerosis. These irregular masses of neurofilaments occupy much of the soma of lower motor neurons, but they are also present in many non-motor neurons across the brain and spinal cord. They are immunoreactive against both phosphorylated (a) and non-phosphorylated (b) neurofilament epitopes. Note the absence of generalized somatic staining for the non-phosphorylated epitope, suggesting sequestration of neurofilaments within the lesion. **(a)** SMI31 and **(b)** SMI32 immunohistochemistry.

staining in the anterior and lateral columns (Figure 14.13). Loss of myelin from the spinocerebellar tracts and posterior columns may also be seen in up to 50 per cent of sporadic cases and in familial cases. Myelin loss and degradation may be demonstrated by tinctorial methods, such as Luxol fast blue and the Marchi method (for degraded myelin), but, when most subtle, is best demonstrated using immunohistochemistry for the microglial lysosomal marker CD68 (Figure 14.14).[127] The concept of 'ALS with posterior column involvement' described in the Japanese literature does not appear to represent a distinct entity related to any particular genetic or clinicopathological subgroup of patients.

Neuropathology of SOD1-Familial ALS (FALS)

There is only limited data on the pathology of *SOD1*-FALS because of the lack of autopsies in a great many of the >130 different disease-related mutations in the *SOD1* gene thus far described. Published autopsy data on FALS patients that predate the era of molecular pathology and genetic screening of *SOD1* are of limited value because *SOD1*-FALS only constitutes up to 10 per cent of familial disease. To date, the cases predominantly fall into two groups: those with neurofilament conglomerate inclusions (Figure 14.12, e.g. A4V, and I113T) and those with apparently typical 'ubiquitin-only' inclusions (e.g. E110G, D101N, del125–126).[43,124,126,142,243] Comparison with sporadic disease indicates that TDP-43/ubiquitin inclusions are not usually encountered in *SOD1*-FALS.[179] In general, where anatomically comprehensive pathology is reported, the evidence suggests that *SOD1*-FALS is a multisystem disorder with extensive pathology throughout the central nervous system (CNS). However, there is undoubtedly a preponderance of pathology in the motor system associated with the prominent early clinical manifestations of motor system disease.

Neuropathology of *TARDBP* Familial ALS

As noted earlier, mutations of *TARDBP*, which encodes TDP-43 protein, may themselves cause ALS. The clinical and pathological phenotype of individuals with TDP-43 proteinopathy ranges from pure FTLD, FTLD with ALS to pure ALS. The pathological picture is of a classical ALS with extramotor pathology similar to that seen in sporadic disease.[91,154,285,302,321] Distinguishing features of disease due to *TARDBP* mutations have not been described.

Neuropathology of *FUS/TLS* Familial ALS

Mutations of the *FUS* gene account for approximately 3–5 per cent of familial ALS cases.[30,98,158,286,305] Immunohistochemistry reveals glial and neuronal cytoplasmic inclusions that are positive for antisera to FUS but not TDP-43 (Figure 14.15). Ubiquitylated cytoplasmic inclusions are usually absent or very few in number in most but not all cases.

Neuropathology of *C9ORF72* Familial ALS

A major development in ALS research was the discovery that intronic hexanucleotide repeat sequences in the *C9ORF72* gene may cause motor neurone disease and FTLD.[65,238] These cases show both classical ALS neuropathology with superimposed features that are classical of cases with the repeat expansion. Thus, there is lower motor neurone loss, ubiquitylated,

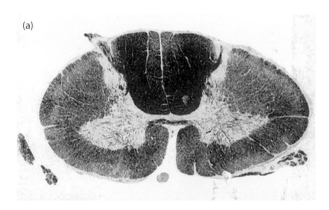

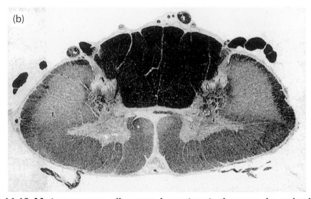

14.13 Motor neuron disease: long tract changes in spinal cord. Sections from cervical enlargement stained for myelin. **(a)** Patient with a 9-month history of bulbar palsy and no abnormal physical signs in the arms or legs. **(b)** Patient with wasting of the small muscles of the hand and spastic paraplegia.

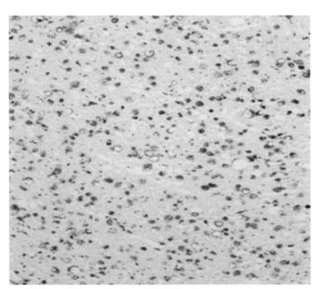

14.14 The corticospinal tract in the cord contains a diffuse population of phagocytic cells immunoreactive for the lysosomal marker CD68. This is the most sensitive method of demonstrating white matter degeneration in amyotrophic lateral sclerosis and other neurodegenerative disorders associated with relatively rapid clinical progression.

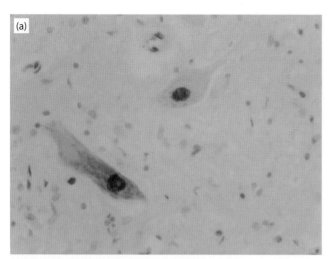
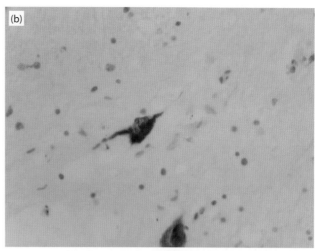

14.15 Immunohistochemistry for FUS protein in lower motor neurones normally shows predominantly nuclear, with some lesser cytoplasmic labelling **(a).** In cases with mutations of FUS, a number of motor neurones show the formation of FUS-positive neuronal cytoplasmic inclusions **(b).**

Reproduced with permission from Ince.[122] With kind permission from Springer Science and Business Media. Copyright © 2011, Springer-Verlag.

TDP-43-positive neuronal cytoplasmic inclusions and Bunina bodies. However, in addition many extra motor regions have p62- and ubiquilin-positive neuronal cytoplasmic inclusions that are largely negative for TDP-43 antisera. These are seen principally in the cerebellar granule cell layer, the neocortex, the hippocampal dentate fascia and CA4 and CA3 subregions. The latter region allows reliable differentiation between cases with and without the hexanucleotide repeat expansion (Figure 14.16).[7,27,32,38,56] More recent studies have found these TDP-43-negative inclusions to contain dipeptide repeats (DPR) translated from both sense and antisense RNA transcripts. Putative toxicity of these DPR aggregates is the basis of one theory of disease pathogenesis in these conditions.[20,89,199,200] Against this theory, it appears that neuropathological features of degeneration across the CNS correlate more closely with TDP-43 inclusion pathology as opposed to DPR pathology.[181] A second neuropathological feature seen using fluorescence *in situ* hybridization in *C9ORF72*-ALS is the presence of foci of RNA transcribed from the repeat sequence. These foci appear to be capable of sequestering RNA processing molecules, thereby depleting the cell of these agents and providing a second putative mechanism of toxicity.[57,71,252] A final potential pathogenic mechanism involves haploinsufficiency *via* reduced expression of the C9ORF72 protein, for which evidence is, thus far, inconsistent.[252]

Dementia and Non-motor Involvement

Involvement of non-motor systems is increasingly recognized in ALS/MND. It is particularly evident in long-term survivors with assisted ventilation, and anti-p62 immunostaining is recommended to detect inclusions. Such cases appear to map closely to those with a mutation in the *C9ORF72* gene.[65] Neuronal loss and inclusion formation may be seen in Clarke's column, the dorsal root ganglia, Onufrowicz's nucleus, the intermediolateral nucleus in the thoracic cord and the reticular formation.[141,148,208,218,253,283] Neuronal loss may be seen in subcortical structures including the basal ganglia, locus coeruleus, substantia nigra, thalamus,

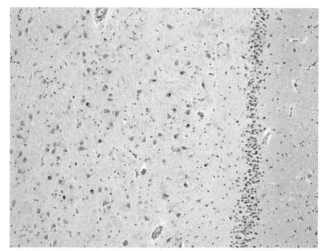

14.16 In C9ORF72-related motor neuron disease, both inclusions in the polymorphic neurons of hippocampal subfield CA4, and dentate granule cell inclusions, are conspicuous using immunocytochemistry to p62. This pathology in CA4 is not based on TDP43 aggregation but is immunoreactive for dinucleotide repeat proteins synthesized from the intronic hexanucleotide repeat region in the *C9ORF72* gene.

subthalamic nucleus, red nucleus, cerebellar dentate nucleus and the nuclei of pontine tegmentum. There may also be degeneration of non-motor tracts in the spinal cord and loss of myelinated fibres with segmental demyelination in the sensory nerves correlating with electrophysiological abnormalities.[111,198] TDP-43, p62 and ubiquitin immunohistochemistry has been used to show a continuum of pathology from pure ALS to ALS with FTLD to pure FTLD.[90,178]

The non-motor cortex, particularly in the anterior temporal, frontal and insular regions, may show microvacuolation in the superficial layers associated with neuronal loss and sometimes subcortical gliosis.[209] TDP-43/ubiquitin-reactive inclusions may be seen in small cortical neurons in affected areas as well as in dentate granule cells of the hippocampus (Figure 14.17).[314] This change is being increasingly

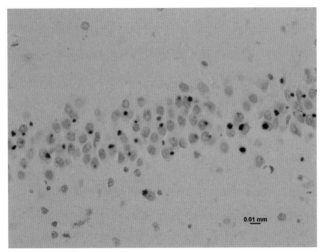

14.17 p62 immunoreactive dentate granule cell inclusions in a motor neuron disease case with significant extramotor pathology.

linked to the development of cognitive impairment in ALS/MND. In some patients, this change may be seen in the absence of clinical motor neuron disease, presenting as a primary dementia of frontal type. In other patients, dementia may be followed by the development of MND. ALS/MND and dementia are not uncommonly encountered with autosomal dominant transmission in a single family.[197,210] It is increasingly recognized that frontotemporal dementia with MND inclusions in non-motor cortex is a common form of non-Alzheimer primary degenerative dementia. Typically, such patients do not develop clinical features of ALS/MND in life.[145,155,241,290] A potential second genetic cause of frontotemporal dementia with ubiquitin inclusions has recently been identified in a large Danish family in whom the causative mutation affects the gene *CHMP2B*, which is involved in endosomal processing.[270] Recently, a few cases of ALS/MND (with probable familial disease) have also been linked to mutation of this gene. One of these cases, studied by the authors at autopsy, shows classical TDP-43/ubiquitin inclusion pathology. It remains unclear if this is a truly pathogenic genetic variation.

Differential Diagnosis of MND and Related Disorders

Several conditions can be clinically or pathologically confused with ALS/MND. Progressive syndromes of LMN degeneration such as late-onset spinal muscular atrophy (SMA type III) and X-linked bulbospinal neuronopathy (Kennedy's disease) and multifocal motor neuropathies with conduction block (sometimes associated with high titres of anti-ganglioside antibodies) may be clinically difficult to distinguish from PMA and PBP. In one study, just over 7 per cent of cases of ALS were rediagnosed as another condition, the most common mimics of ALS being multifocal motor neuropathy (22 per cent of misdiagnoses), and Kennedy's disease (13 per cent of misdiagnoses).[293] Another study with a high proportion of familial ALS cases suggested a 2 per cent misdiagnosis rate in sporadic and familial cases.[225] Juvenile cases of ALS have been described with a mean age of onset around 12 years, slow clinical progression and autosomal recessive inheritance.[24]

The familial syndrome of frontotemporal dementia and motor neuron loss linked to chromosome 17 has been subsequently linked to mutations in the *MAPT* gene, encoding tau protein, and is probably best regarded as a rare motor variant within the spectrum of familial tauopathies.[177]

Toxic and metabolic disorders affecting motor neurons and long tracts are discussed in Chapter 12, Extrapyramidal Diseases of Movement. MND in Guam is associated with both neurofibrillary tangles in the spinal cord and TDP-43/ubiquitin inclusions and has been recorded in sporadic non-Guam cases.[114] Paraneoplastic ALS/MND has been described in association with systemic malignancy[85] or lymphoproliferative disease.[94] Some patients with pathological features of MND overlap with pathological features seen in cases of multisystem degeneration.[247] When lower motor neuron involvement occurs in Creutzfeldt–Jakob disease, it is usually late and accompanied by signs of more fulminant cerebral and cerebellar involvement.[248]

Pathologically, the clinical syndrome of PLS has been linked with Lewy body pathology.[102] Upper motor neuron signs are also a feature of some cases of familial Alzheimer's disease due to presenilin gene mutations. The morphology and immunocytochemistry of lower motor neuron pathology in motor system disorders are summarised in Table 14.4.

POST-POLIO SYNDROME

Clinical Features

Late effects of polio still affect many individuals.[103] Post-polio syndrome (PPS) is a late-onset motor deterioration developing in approximately 20–85 per cent of individuals with previous acute poliomyelitis, often after many years of neurological stability.[18] Clinical features include fatigue, new onset of slowly progressive muscular weakness, pain and muscle atrophy. Respiratory muscle weakness may also occur.[93,117,294] As PPS is a diagnosis of exclusion, other causes of neurological deterioration should be excluded, including entrapment neuropathies, spondylotic myeloradiculopathies and incidental polymyositis or inclusion body myositis. PPS is more a slowly progressive disorder than ALS/MND; a mean yearly loss of neurological function of 1 per cent has been measured.[64] Risk factors for the development of the syndrome include greater severity of the initial acute poliomyelitis, permanent impairment after recovery, older age at the time of the acute illness, female sex, recent weight gain, the presence of other diseases and there is an inverse association with higher education.[236,294]

Pathogenesis

The development of PPS is thought to relate to new onset denervation resulting from degeneration of axonal terminals. Muscle biopsies in the stable post-polio state and in PPS show changes that reflect chronic denervation. In addition, muscle biopsies from PPS patients show angulated, esterase-positive fibres and muscle fibres expressing neural cell adhesion molecule, suggesting that newly denervated individual muscle fibres are present in PPS but not in the stable post-polio state.[63] Based on these findings and supported by electrophysiological data, it has been proposed that re-innervation following acute

TABLE 14.4 Cytopathology of the lower motor neuron in its common disorders

Disorder	Morphology	Inclusions	Neurofilament pattern
Sporadic ALS	Some atrophic motor neurons	Ubiquitin, p62 and TDP-43 positive cytoplasmic skein and compact inclusions Bunina bodies	
Familial ALS			
mt*C9ORF72*	Some atrophic motor neurons	As for sporadic ALS, but with ubiquitylated, TDP-43 negative inclusions in extra-motor regions especially hippocampus CA4 and cerebellar granule cells	
mtTARDBP	Some atrophic motor neurons	As for sporadic ALS	
mt*FUS*	Some atrophic motor neurons	FUS-positive neuronal cytoplasmic inclusions, no TDP43 inclusions	
mt*SOD1*	Some atrophic motor neurons	No TDP-43 positive inclusions but either a) ubiquitylated neuronal cytoplasmic inclusions (none-TDP43) or b) hyaline conglomerate inclusions	Hyaline conglomerate inclusions (b) stain with both phosphorylated and non-phosphorylated neurofilament antibodies
X-SBMA	Atrophy No chromatolysis	Intranuclear inclusions positive for androgen receptor, ubiquitin and poly-glutamine expansion	No alteration
SMA	Pale, ballooned neurons EM – accumulation of filaments	Diffuse abnormal staining with ubiquitin without inclusions	Abnormally phosphorylated neurofilaments

ALS, amyotrophic lateral sclerosis; EM, electron microscopy; TDP43, protein product of *TARDBP* gene; *C9ORF72*, chromosome 9 open reading frame 72 gene; CA4, Cornu Ammonis sector 4; *FUS*, fused in sarcoma gene; *SOD1*, *superoxide dismutase 1* gene, X-SBMA, X-linked spinobulbar muscular atrophy; SMA, spinal muscular atrophy.

poliomyelitis leads to the formation of enlarged motor units that are unstable, resulting in an impaired ability of these stressed motor neurons to maintain their distal axons leading to individual fibre denervation.[61] Inflammatory mechanisms may also operate in PPS and there is some evidence for poliovirus persistence, but the significance of these factors for pathogenesis remains unclear.[31,62,92]

Neuropathology

The spinal cord at autopsy shows similar features in both stable post-polio and PPS, with atrophy of the anterior horns, neuronal loss and gliosis. Inflammation, in the form of perivascular, parenchymal and meningeal lymphocytic infiltrates of mainly T-cell type, persists for many years after the acute infection.[139,229] In contrast, the corticospinal tracts are preserved.[83] The presence of occasional chromatolytic neurons in the spinal cord in PPS is an additional feature that implies on-going lower motor neuron pathology.[230] Lower motor neurons in PPS do not show evidence of ubiquitin immunoreactive inclusions, a useful feature in the pathological differential diagnosis from motor neuron disease.[129]

SPINAL AND BULBAR MUSCULAR ATROPHY (KENNEDY'S DISEASE)

Clinical Features

X-linked spinal and bulbar muscular atrophy (SBMA) is a predominantly lower motor neuron degeneration of males, affecting bulbar and spinal groups.[144] The disease usually shows clinical onset in the third to fifth decades and is slowly progressive. There may be an associated mild sensory neuronopathy, indicating that the nervous system pathology is not confined entirely to the motor system. In addition, features such as gynaecomastia reflect androgen insensitivity whereas abnormalities of serum biochemistry suggest that there is also visceral pathology (Table 14.5).[81]

Genetics and Pathogenesis

SBMA is a sex-linked disorder due to mutation of the androgen receptor (AR) gene located at Xq11-q12157.[159] This results in an expansion of a CAG trinucleotide repeat sequence in exon 1 of the gene from a maximum of approximately 35 repeats in the healthy population to a range of

TABLE 14.5 Clinical manifestations of X-linked spinobulbar atrophy (X-SBMA)

Inheritance	X-linked
Clinical onset	Third to fifth decades
Clinical features related to nervous system pathology	Muscle cramps Fasciculations Weakness (LMN type) of limb and atrophy of bulbar muscles
Features of androgen insensitivity	Gynaecomastia Reduced fertility Testicular atrophy
Abnormalities of serum biochemistry	Raised creatine kinase Impaired glucose tolerance Hepatic dysfunction Raised plasma lipids

LMN, lower motor neuron.

approximately 40 to 62 in association with the disease.[22] The phenotype resulting from this trinucleotide repeat expansion contrasts with that resulting from point mutations in the AR gene, which result in the androgen insensitivity syndrome. The size of the repeat expansion correlates with disease severity and inversely with age of onset of the disease but, in contrast to some other trinucleotide repeat disorders, the expansions appear to be relatively stable so that kindreds do not show anticipation as a feature.[8,72,120]

The normal androgen receptor is widely expressed in neurons in the nervous system and is not confined to the motor system.[164,182] It is also expressed in tissues outside the nervous system including muscle, testis and scrotal skin. The CAG repeat expansion in the mutated gene encodes an expanded polyglutamine tract that is expressed within the AR protein. Expanded polyglutamine tracts have a tendency to self-associate to form aggregates and neuropathological studies have demonstrated the formation of intranuclear inclusions in SBMA.[22,164]

Although androgens have trophic effects on motor neurons, expression of the mutant AR protein appears to cause disease through a gain-of-function mechanism. The pathogenic role of inclusion formation is unclear; diffuse nuclear accumulation of mutant protein correlates with CAG repeat length whereas nuclear inclusions do not.[2] Similarly, the most important biochemical forms of mutant AR are unclear. Protein aggregation and misfolding may be important, but there is also evidence that soluble oligomeric forms of the mutant AR protein are important toxic species, in common with other protein aggregation neurodegenerative disorders.[166]

A number of mechanisms have been implicated in neuronal toxicity. The inclusions sequester transcription factors and chaperone proteins, which may result in transcriptional dysregulation.[189] One CREB-binding protein transcription target that is affected may be vascular endothelial growth factor, which has been implicated in motor neuron survival.[276] Other putative mechanisms include impaired DNA binding, mitochondrial impairment, oxidative stress and impairment of axonal transport.[22] Transcriptional dysregulation of dynactin is implicated in the latter. A point mutation in dynactin has itself been associated with a form of (non-X-linked) spinal and bulbar muscular atrophy with laryngeal involvement.[234]

Neuropathology

Autopsy examination shows severe loss of lower motor neurons, accompanied by mild gliosis, from all levels of the spinal cord and from motor nuclei supplying cranial nerves V, VII, XI and XII; neurons of nuclei supplying extraocular muscles are spared[272] and there is no degeneration of the pyramidal tract. Loss of small neurons from the intermediate zone of spinal cord grey matter has also been demonstrated by quantitative studies. Remaining lower motor neurons show atrophy but not chromatolysis (Figure 14.18).

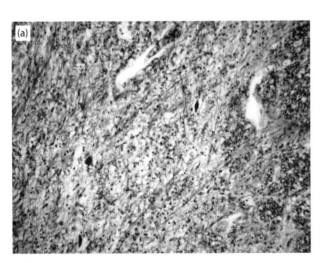

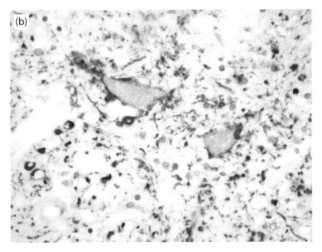

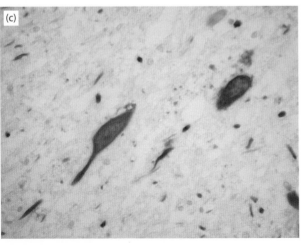

14.18 Spinal cord in X-linked spinobulbar muscular atrophy. (a) Anterior horn at C6 showing loss of motor neurons. Residual motor neurons are atrophic. Anterior white matter is seen to the right of the field. **(b)** Residual motor neurons show a normal pattern of staining for neurofilament proteins, with no expression of phosphorylated epitopes within the cell bodies. **(c)** There is normal cytoplasmic expression of non-phosphorylated neurofilament protein within motor neurons. (a) Luxol fast blue (LFB)/cresyl violet; (b) SMI31; (c) SMI32 immunohistochemistry.

Cytoplasmic ubiquitinated inclusions are not seen, in contrast to ALS, and neurofilament immunohistochemistry does not demonstrate an abnormal staining pattern.[273] Although sensory involvement is not prominent clinically, post-mortem examination reveals a central and peripheral axonopathy with myelinated fibre loss from fasciculus gracilis that is worse more rostrally (i.e. distal from the cell soma). Skeletal muscles, especially the tongue, show neurogenic atrophy.[8] Development of pre-senile dementia in a case of SBMA associated with subcortical frontal gliosis, neuronal depletion from the hippocampus, tauopathy and gliosis and microglial reaction in hippocampus and deep grey nuclei has been reported.[266] Testicular atrophy is present with hyalinized seminiferous tubules and arrested spermatogenesis.

Intraneuronal inclusions are an important neuropathological feature.[3] They range from 1 to 5 μm in diameter and ultrastructural examination shows non-membrane bound granular and filamentous material. The intranuclear inclusions are immunoreactive for the N-terminal portion of the AR protein but not the carboxy-terminal portion, suggesting abnormal processing of the protein, or epitope masking. Proteins are also ubiquitinated, so they may be demonstrated by ubiquitin immunohistochemistry. They may also be detected by immunohistochemistry with antibodies that recognize the expanded polyglutamine tract. This latter method appears to be the more sensitive and is of diagnostic use in autopsy material. Inclusions are present in neurons in a widespread distribution within the nervous system in SBMA and, as discussed earlier, there may also be diffuse nuclear and some cytoplasmic accumulation of mutant AR.[2] Inclusions are also found in non-CNS tissues.[165] Inclusion formation in scrotal skin correlates with disease, so anti-polyglutamine immunohistochemistry of scrotal skin biopsy may be a useful pathologic marker.[21]

SPINAL MUSCULAR ATROPHY

Spinal muscular atrophy (SMA) encompasses a large group of genetically-determined disorders characterised by slowly progressive degeneration of lower motor neurons (Table 14.6). SMA is a generally symmetrical, pure lower motor neuron disorder, without evidence of sensory or pyramidal tract involvement. Despite the name, there may also be involvement of bulbar neurons and, in certain rare forms (see later), this is the predominant feature. The most common forms are autosomal recessive, linked to chromosome 5q, but there are atypical, non-5q-linked forms of SMA with varying patterns of inheritance.

Autosomal Recessive Proximal SMA

The most common forms of SMA are autosomal recessive and linked to mutation of the *SMN1* gene on chromosome 5q13.[161] They are classified, according to age of onset and the highest degree of motor function achieved, into SMA type 1 (Werdnig–Hoffman's disease), SMA type 2 and SMA type 3 (Kugelberg–Welander disease).[184] SMA type 4 has been used for some late onset cases whilst SMA 0 has been introduced for infants with prenatal onset (Table 14.6). SMA affects approximately 1 in 10 000 live births, with

a carrier frequency of about 1 in 50.[217] For chronic SMA (types 2 and 3) estimated incidence is 1 in 24 000 live births with a working estimate 1 in 90 carriers.

SMA1 is an important cause of neonatal hypotonia (floppy baby syndrome), with predominantly proximal weakness that progresses to paralysis and with sparing of facial muscles. Prenatal onset occurs in about a third of cases, with reduced fetal movements. Whilst intercostal muscles are affected, there is relative early sparing of diaphragmatic function. The ability to test for *SMN* mutation is leading to recognition of broader phenotypes associated with 5q-linked mutation. More severe phenotypes of SMA, now designated SMA type 0, have been recognized with prenatal onset, severe weakness and asphyxia at birth and short survival.[73] A severe form has been described in three siblings with asphyxia, facial weakness and external ophthalmoplegia; sensory nerves showed a loss of myelinated fibres indicating involvement of afferent systems.[153] Some cases may be associated with arthrogryposis. This occurs in response to *in utero* muscle weakness and so may complicate a variety of disorders, including SMA.[28] *SMN* mutation is also found in association with a congenital cytoplasmic body myopathy, characterized by the presence of cytoplasmic bodies on muscle biopsy,[301] and a variant of SMA with early respiratory insufficiency, multiple joint contractures and bone fractures has been described.[86]

SMA2 is characterised by an onset in infancy and there is commonly survival into adolescence or adulthood. SMA3 starts in early childhood, but these patients generally become ambulant and have normal life expectancy. Late onset cases, even in adulthood, may be referred to as SMA4. Late-onset SMA should be included in the differential diagnosis of patients with a limb girdle syndrome, which includes limb girdle dystrophies, dystrophinopathies and metabolic myopathies. Late onset cases may also be confused with ALS/MND.

Non-5q SMA

A heterogeneous group of spinal muscular atrophies, with autosomal recessive, dominant and X-linked forms of inheritance, are not linked to mutation of *SMN1* on 5q (Table 14.6). These often have particular clinical features, allowing their recognition as atypical (SMA-plus) forms of SMA.[99,245,323] These include SMA with pontocerebellar hypoplasia, SMA with myoclonus epilepsy, scapuloperoneal SMA and a heterogeneous group of distal SMA, including a form with vocal cord paralysis. SMARD (SMA with respiratory distress) is a form of distal SMA, usually due to mutation of the *immunoglobulin μ-binding protein-2 gene (IGHMBP2)* on chromosome 11q13, in which, in contrast to SMA1, there is early involvement of the diaphragm.[138]

Genetics and Pathogenesis

Autosomal recessive 5q-linked SMA arises from mutation in the *SMN1* gene.[162] Most cases result from a homozygous *SMN1* deletion, but homozygous gene disruption may also result from mutations or rearrangements. The 5q chromosomal region contains an inverted duplication involving several genes (Figure 14.19). These include *GTF2H2 (p44)*, encoding a subunit of transcription factor TFIIH, *NAIP* (neuronal apoptosis inhibitory protein) and *SERF1(H4F5)*,

TABLE 14.6 Simplified classification of spinal muscular atrophy

5q-Linked SMA (AR)				
Type	Age at onset	Motor milestones	Prognosis	References
0	Prenatal	Respiratory support from birth	<6 months	
I Werdnig-Hoffmann	0–6 months	Never sits unaided	<2 years	
II Intermediate form	6–18 months	Sits independently	10–40 years	
III Kugelberg–Welander	>18 months	Stands and walks	Normal	
IV Adult SMA	2nd/3rd decade	Walks normally	Normal	
Atypical (non-5q) SMA				
Disease		Inheritance		
SMARD1		AR mutation of *Ighmb2*; chromosome 11q		138
SMA and pontocerebellar hypoplasia				99, 245
Distal SMA		AD or AR		99, 245
Scapuloperoneal SMA		AD; chromosome 12q24		99, 245
SMA and progressive myoclonus epilepsy		AD or AR		99, 245

SMA, spinal muscular atrophy; SMARD, SMA with respiratory distress; AR, autosomal recessive; AD, autosomal dominant.

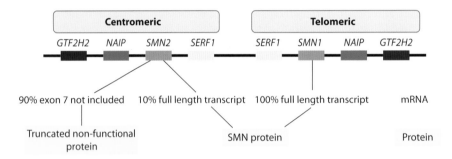

14.19 Arrangement of genes within the *SMN* gene locus on chromosome 5q. The telomeric copy (*SMN1*) is responsible for most SMN protein in normal individuals. SMA arises when this gene is defective so that gene expression is reliant on expression of the centromeric *SMN2* gene copy.

in addition to *SMN*.[176] The centromeric copy of *SMN*, designated *SMN2*, differs from *SMN1* by five nucleotides, including a C-to-T transition in the exonic splice enhancer of exon 7.[171] This causes exon 7 to be spliced out in the majority of transcripts from *SMN2*, resulting in production of mostly an unstable, non-functional protein from this centromeric copy of the gene. *SMN2* gene copy number varies in the population and this acts as a disease modifier. The transcripts derived from *SMN2*, although only having limited functionality, mean that variation in *SMN2* copy number affects clinical severity, with higher copy numbers associated with milder phenotypes. The type of *SMN1* mutation also affects phenotype, with mutations that disrupt exon 7 splicing (effectively converting *SMN1* to 2) also being associated with milder phenotypes. Mutations in other genes in this region may also act as modifiers.[184]

The *SMN* gene encodes a 294 amino acid, 38kDa protein that is expressed in both nucleus and cytoplasm. Although highly expressed in spinal motor neurons it is present in a widespread distribution[162] and is expressed from early fetal life, consistent with the *in utero* manifestation of the disease

in some *SMA1* cases.[288] Disease severity is proportional to the reduction in SMN protein. SMN protein is localized to nuclear structures called gems (gemini of coiled bodies) that associate closely with Cajal or coiled bodies. SMN, through this interaction, is important for assembly of the spliceosome, which catalyzes removal of introns from mRNA precursors. SMN protein is, therefore, important for pre-mRNA splicing.[317] The involvement of SMN protein in RNA metabolism suggests a common theme of abnormalities of RNA processing in motor neuron degeneration.[23]

Other mechanisms however may also contribute to pathogenesis. In a zebra-fish model, reduction of SMN proteins levels using anti-sense technology resulted in impaired motor axon outgrowth.[193] The ability of SMN protein to abrogate Bax and Fas induced apoptosis through interaction with Bcl-2 suggests that disturbance of apoptosis regulation may also be relevant.[130] SMN may also have functions in axonal RNA trafficking and at the neuromuscular junction, which may contribute toward the vulnerability of lower motor neurons with their long axons, in SMA.[184] Studies in animal models, showing early loss of presynaptic motor terminals and abnormal neurofilament accumulations, have suggested that the neuromuscular synapse is a vulnerable target in SMA.[206]

Neuropathology

The neuropathology of SMA1 has been the best studied. Macroscopically, there is atrophy of the anterior roots of the spinal cord. There is extensive loss of lower motor neurons from spinal cord anterior horns and brain stem motor nuclei, with sparing of autonomic nuclei in the thoracolumbar cord and Onufrowicz's nucleus in the sacral cord (Figure 14.20). Betz cells and the pyramidal tracts are preserved. Residual lower motor neurons have a swollen, pale appearance with loss of Nissl substance (Figure 14.21). Atrophic, shrunken neurons may be seen and occasional neurons may show neuronophagia. Anterior horns show gliosis and glial fibres around spaces from which neurons have been lost may form so-called empty beds[79,205,268] and ghost cells have been noted in Clarke's column.[205] Although superficially similar to chromatolytic neurons, the pale, ballooned neurons of SMA show ultrastructural and immunohistochemical differences from the neuronal swelling of true chromatolysis. Neurons in SMA1 show accumulation of intermediate filaments around the periphery of the cell corresponding to accumulation of abnormally phosphorylated neurofilaments. Ubiquitin immunostaining reveals an abnormal pattern of diffuse reactivity, distinct from the inclusions seen in motor neuron disease.[51,188,204] As well as in spinal motor neurons, ballooned neurons may also be observed in bulbar nuclei, including those of the extraocular muscles that are clinically spared.[140] They may be seen in Onufrowicz's nucleus, Clarke's nucleus of the spinal cord and the thalamus (Figure 14.22).[140] An additional feature is the presence of heterotopic (migratory) motor neurons, implying abnormal migration. These are mostly seen in the anterior horn white matter, in the ventral outflow region. Morphological and immunohistochemical features suggest that they are undifferentiated and their numbers decline exponentially with the age of the subject at death.[269]

Neuropathology reports of SMA2 also describe severe loss of motor neurons from spinal cord and nuclei of the hypoglossal and facial nerves.[16] Phrenic nerve motor neurons are relatively preserved, consistent with the relative preservation of diaphragmatic function characteristic of 5q13-linked SMA. In contrast to SMA1, ballooned neurons are not conspicuous and immunohistochemistry does not show the neurofilament and ubiquitin changes typical of SMA,[16] although few cases have been reported to date. SMA3 cases show anterior horn motor neuron loss and

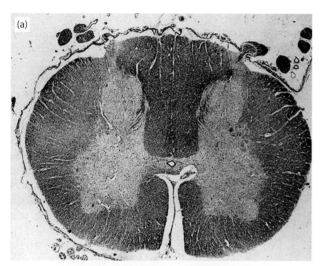

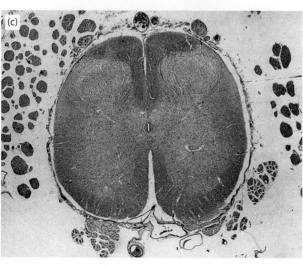

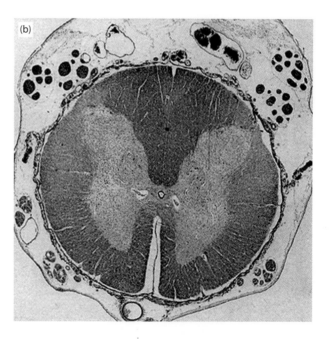

14.20 Spinal muscular atrophy: histology of spinal cord. Infant aged 3 months: **(a)** cervical, **(b)** thoracic (note preserved cells in dorsal nuclei) and **(c)** sacral (note relative atrophy of anterior roots) spinal cord. There is a severe loss of motor cells in all levels. Klüver–Barrera.

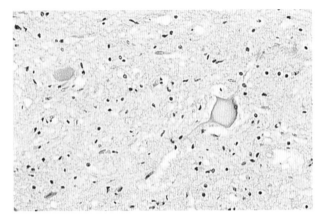

14.21 Spinal muscular atrophy (SMA): neuronal swelling.
Severe infantile SMA, showing swelling and achromasia of one
of the remaining motor neurons in the cord.

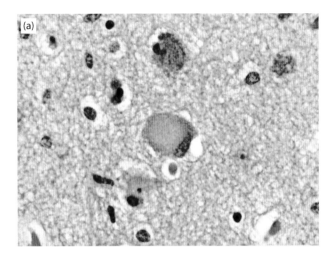

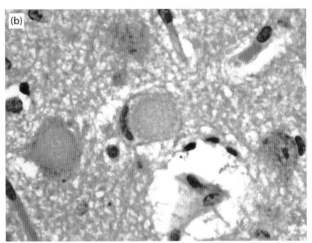

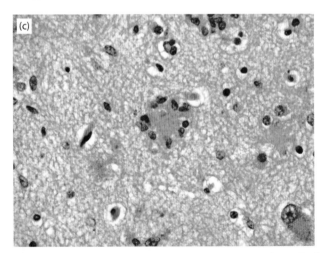

**14.22 Neuronal cytopathology in the thalamus in spinal
muscular atrophy type 1. (a)** Ballooned neuron. **(b)** Ballooned
neuron with pale cytoplasmic periphery. **(c)** Neuron undergoing
neuronophagia.

gliosis without upper motor or corticospinal tract changes.
A more recent, genetically confirmed, case demonstrated
peripheral perikaryal staining for neurofilament but no
ubiquitinated inclusions in anterior horn motor neurons.[157]

Neuropathological studies have attempted to address
the question of whether neuronal death in SMA occurs by
apoptosis. Studies using the TUNEL method, which labels
the DNA strand breaks that occur as a feature of apoptosis, have reported varying results for anterior horn cells in
SMA1 and there are questions over the specificity of this
technique. TUNEL positive cells have, however, also been
found in the ventrolateral thalamus, a site of cytopathology in SMA.[268] Ultrastructural studies in SMA1 have also
found apparent apoptotic changes and there is evidence for
altered expression of the apoptosis regulatory proteins Bcl-2
and p53. A high proportion of TUNEL-positive neurons
and alterations of Bcl-family proteins have been observed
in fetuses with SMA1 at around 10 to 20 weeks, a time
when there is increased physiological motor neuron loss
and development of neuromuscular synapses, suggesting
that there may be an excess of developmental programmed
cell death.[274,275] Changes of apoptosis have also been seen in
skeletal muscle.[79,287]

BULBOSPINAL MUSCULAR ATROPHY
OF CHILDHOOD

There are two very rare conditions of childhood in which there
is loss of neurons mainly from nuclei of lower cranial nerves.
Brown–Vialetto–van Laere syndrome presents with sensorineural deafness in late childhood or adolescence followed,
after a latent period, by palsies of the VIIth, IXth and XIIth
cranial nerves.[256] Other neurological features, including limb
weakness and respiratory compromise, develop and upper
motor neuron signs have been described in some cases. There
is believed to be clinical heterogeneity. Families with variability in age at onset are described and cases have been described
with onset in infancy. Inheritance is mostly autosomal recessive showing a female to male ratio of 3:1. There is cell loss
from the VIIth to XIIth cranial nerve nuclei with dense astrocytosis in the cochlear nuclei and loss of axons in cochlear
nerves. Degeneration of the spinocerebellar tract, Purkinje

cells, posterior columns and Clarke's column are described.
Motor neuron loss and anterior horn gliosis with depletion
of axons from motor roots have been observed in the spinal
cord.[69]

Brown–Vialetto–van Laere syndrome has been shown
to be associated with mutation of the *C20ORF54* gene

on chromosome 20p13.[96] This highly conserved gene is homologous to the rat gene encoding a riboflavin transport protein, and flavin deficiency in Brown–Vialletto–van Laere syndrome suggests that there is a riboflavin transporter defect.[35]

Fazio–Londe disease is distinguished from Brown–Vialletto–van Laere syndrome by the absence of deafness, but is otherwise similar. There appears to be genetic heterogeneity, with autosomal recessive forms with two patterns and a very rare autosomal dominant form. Neuropathological studies demonstrate severe cell loss from the hypoglossal, motor vagal and facial nuclei with occasional chromatolytic neurons. Anterior horn cell loss from the spinal cord, trigeminal motor and oculomotor nuclei is less marked. Neuronal loss has also been seen in the thalamus, striatum, dentate nucleus and cerebellar cortex.[192]

Rare cases of bulbar palsy associated with other clinical manifestations have been described, including retinitis pigmentosa, OPCA, optic atrophy, ophthalmoplegia and mental retardation. Progressive bulbar paralysis with a phenotype similar to Fazio–Londe disease has also be described in association with a mitochondrial respiratory chain defect,[239] as has a clinical phenocopy of primary muscular atrophy.[34]

HEREDITARY SPASTIC PARAPLEGIA

Clinical Features

Hereditary spastic paraplegia (HSP), or Strümpell–Lorrain syndrome, is a group of genetically determined disorders characterized by slowly progressive spastic paraparesis. HSP has a prevalence of approximately 3–10/100 000 and may be classified phenotypically into pure or complicated forms. In complicated HSP, spastic paraparesis may be accompanied by a variety of other neurological manifestations, including epilepsy, dementia, ataxia, extrapyramidal involvement, deafness, optic atrophy, retinopathy or cataracts. Even in cases of pure HSP, sensory and urinary manifestations may develop with age and there is a high incidence of sub-clinical lesions affecting various central nervous system functions.[249]

Classification and Pathogenesis

HSP is a genetically heterogeneous disorder and new gene loci continue to be linked to HSP. To date, 19 HSPs have been associated with autosomal dominant patterns of inheritance, 27 autosomal recessive and 5 X-linked and there are several unclassified forms, including one with a maternal pattern of inheritance.[82] Gene loci are designated SPG (SPastic parapleGia) though some loci are re-designated by the specific gene name where known (HUGO Gene Nomenclature Committee, www. genenames.org). Most pure HSP is autosomal dominant. Causative genes have been identified in eleven autosomal dominant forms, the most commonly associated mutations being in *SPAST, ATL1* and *REEP1*. Autosomal recessive forms are mostly complicated HSP and 16 causative genes have been identified. Three genes have

been identified amongst the X-linked loci (Table 14.7). A number of candidate cellular mechanisms are emerging for the corticospinal tract degeneration from the study of the function of these genes. These include defects in membrane trafficking and axonal transport, mitochondrial dysfunction and developmental abnormalities. The comparatively long length of corticospinal tract axons may render them particularly susceptible to mechanisms impairing axonal physiology.[249,278]

Mutation in the *SPAST* gene at the SPG4 locus is the most common cause of autosomal dominant HSP. It encodes spastin, a member of the AAA (ATPases associated with diverse cellular activities) group of proteins.[190] Spastin is a neuronal protein that appears to be a microtubule-severing protein with roles in endoplasmic reticulum morphogenesis and microtubule dynamics.[46,312] Mutation in spastin appear to result in loss of function through haploinsufficiency or dominant negative effects and might therefore perturb microtubular and cellular internal membrane function, thus affecting transport of organelles and macromolecules.[175] This would be expected to have a particular impact in the long axons of the corticospinal tracts. Alteration of transport function may also result from other mutations that result in HSP. Mutations at the *SPG10* locus involve the *kinesin heavy chain (KIF5A)* gene that encodes a molecular motor important for anterograde and retrograde axonal transport.[237] Atlastin, the product of the *ATL1* gene is a membrane GTPase localized to the endoplasmic reticulum. It is involved in membrane fusion and has a role in the structure and maintenance of the endoplasmic reticulum.[201] Spartin, encoded by the *SPG20* locus,[53] shows sequence homology to spastin and may be involved in endosomal trafficking. These and other gene mutations implicate impairment of axonal transport, membrane dynamics and maintenance of organelle form as common mechanisms leading to corticospinal tract degeneration.[29]

In another group of disorders, impairment of mitochondrial function may be an underlying mechanism. The gene product paraplegin, mutated in *SPG7*-related HSP, is localized to mitochondria, and muscle biopsy has shown changes suggestive of mitochondrial cytopathy.[42] In a mouse model of paraplegin-related HSP, early mitochondrial abnormalities associated with axonopathy suggested that local failure of mitochondrial function may be important in axonal degeneration.[77] Mutations in the gene encoding mitochondrial chaperonin Hsp60 is also associated with some cases of HSP.[105]

X-linked forms of HSP are associated with impaired development of the corticospinal tract and tend to have complicated phenotypes.[249,278] Mutations of the gene for *L1CAM*, a cell adhesion molecule and a member of the immunoglobulin gene superfamily, may result in spastic paraparesis associated with MASA syndrome (mental retardation, aphasia, shuffling gait and adducted thumbs), X-linked hydrocephalus or with CRASH syndrome (corpus callosum hypoplasia, retardation, adducted thumbs, spastic paraplegia, hydrocephalus).[134,307] X-linked HSP also occurs as a result of mutation of the gene for the major myelin protein, proteolipid protein (PLP), which is allelic to the demyelinating disorder, Pelizaeus–Merzbacher disease.[257]

TABLE 14.7 Genes and proteins implicated in hereditary spastic paraparesis[82,249]

SPG type	Gene	Protein	Inheritance
SPG3	*ATL1*	Atlastin	AD
SPG4	*SPAST*	Spastin	AD
SPG6	*NIPA1*	Non-imprinted in Prader–Willi/Angelman syndrome	AD
SPG8	*KIAA0196*	Strumpellin (KIAA0196)	AD
SPG10	*KIF5A*	Kinesin HC5A	AD
SPG12	*RNT2*	Reticulon 2	AD
SPG13	*SPGD1*	Heat shock 60KDa protein	AD
SPG17	*BSCL2*	Seipin	AD
SPG31	*REEP1*	Receptor expression-enhancing protein 1	AD
SPG33	*ZFYVE27*	Spastin binding protein	AD
SPG42	*SLC33A1*	Solute carrier family 33 (acetyl coA transporter) member 1	AD
SPG5	*CYP7B1*		
SPG7	*SPG7*	Paraplegin	AR
SPG11	*KIAA1840*	Spatacsin	AR
SPG15	*ZFYVE26*	Spastizin	AR
SPG18	*SPG20*	Spartin	AR
SPG20	*ACP33*	Maspardin	AR
SPG30	*KIF1A*	Kinesin 3	AR
SPG35	*FA2H*	FA2H	AR
SPG39	*NTE*	Neuropathy target esterase	AR
SPG44	*GJA12/GJC2*	Connexion 47	AR
SPG47	4 genes: *AP4, B1, E1, M1*	–	AR
SPG48	*KIAA0415*	AP5 protein	AR
SPG1	*L1CAM*	L1 cell adhesion molecule	XL
SPG2	*PLP1*	Proteolipoprotein 1	XL
SPG22	*SLC6A2*	Monocarboxylate thyroid hormone transporter 8	XL

AD, autosomal dominant; AR, autosomal recessive; XL, X-linked. The table does not include additional SPG loci implicated in HSP without currently identified genes/proteins.

Mutation of the *BSCL2* gene, which encodes seipin, is associated with Silver syndrome. This disorder is allelic with a form of distal hereditary motor neuropathy so that the term seipinopathies has been coined for this group of disorders.[128] Mutation of seipin, at least in cell culture models, causes improperly folded protein to accumulate in the endoplasmic reticulum, activating the unfolded protein response and endoplasmic reticulum stress. This raises pathogenetic mechanisms, potentially also linked to autophagy (SPG15[109]), similar to other neurodegenerative disorders where abnormal protein conformations are implicated.

Neuropathology

Motor System Pathology

The neuropathology of HSP is characterized by myelin pallor and axonal loss from the lateral and, to some extent, anterior corticospinal tracts.[246,261] Corticospinal tract degeneration is most severe in the thoraco-lumbar region. The spinocerebellar tracts and dorsal columns, particularly fasciculus gracilis at cervical levels, may also be involved. More severe degeneration in distal parts of long tracts suggests that HSP is a dying-back axonopathy (Figure 14.23). Quantitative studies have confirmed relative distal loss of axons, supporting this hypothesis. Axon loss appears to affect both large and small diameter fibres[66] and, at least in spastin-related HSP, is accompanied by a microglial reaction,[312] a feature that it shares in common with corticospinal tract degeneration in motor neuron disease.[127]

Although neuropathological studies of X-linked cases have been limited, hypoplasia or aplasia of the corticospinal tract associated with hypoplasia of the corpus callosum has been described in cases of X-linked hydrocephalus.[319] In a further case, most fibres of the corticospinal tract were found to have decussated normally.[70]

There may also be variable pathology of primary motor neuron cell bodies. Loss of Betz cells from the motor cortex has been described in some, but not all, studies.[78,262,313] Older studies report an absence of lower motor neuron pathology.[246] Recently, however, hyaline inclusions have been reported in lower motor neurons in *SPAST*-mutation related cases. Their significance is currently unclear, but altered staining patterns for cytoskeletal proteins and mitochondria in these cases provide evidence that the lower motor neuron may be involved in some cases.[311] Motor neuron loss from the spinal cord has also been reported in HSP associated with a *NIPA1* mutation associated with degeneration of the lateral corticospinal tract and neuronal loss in motor and frontal cortex. TDP43-positive skein and rounded inclusions were also observed in motor neurons and oligodendroglia, suggesting some pathogenic features in common with ALS/MND and FTLD-TDP.[185] Fibrillar, p62-positive inclusions and novel crystalloid oligodendroglial inclusions have been described in a case of autosomal dominant, complicated HSP.[316]

Muscle biopsy findings have been reported in a few cases of HSP. In paraplegin-mutation related cases, skeletal muscle has shown changes of mitochondrial pathology, with ragged red fibres, cytochrome oxidase negative fibres, peripheral accumulation of mitochondria and elevated succinate dehydrogenase activity.[42,190] Muscle biopsy has been reported as normal in chromosome 8q-linked HSP.[112]

Extra-Motor Pathology

Cognitive impairment and other non-motor manifestations in HSP predict pathology in non-motor areas of the CNS. In an HSP case with dementia, atrophy of pre- and post-central, superior temporal and frontal gyri was associated with decreased white matter, atrophy of corpus callosum and deep grey nuclei and depigmentation of substantia nigra. A loss of cortical neurons and immunohistochemical evidence of a decrease in calbindin D28K+ cells and parvalbumin reactive dendrites have been described.[78] Cognitive impairment is a particular feature of *SPAST*-mutation related HSP.[40,284,310] A *SPAST*-mutation related case with dementia showed depletion of hippocampal pyramidal cells, tau-positive tangles but not plaques, and α-synuclein positive Lewy bodies.[313] Tau pathology has been seen in other cases, especially in brain stem and glia,[311] suggesting that HSP may be associated with non-motor neurodegenerative changes in some *SPAST*-mutation families.

Differential Diagnosis

The clinical differential diagnosis of HSP is broad.[82,249] Results from clinical findings, family history and investigations in life, including findings from MRI examination, gene analysis, biochemical investigations and viral serology need to be taken into account in the autopsy assessment of a case of spastic paraparesis. Spastic paraparesis may be a component of a number of diseases, including B12 disorders, adrenomyeloneuropathy, metachromatic leukodystrophy, globoid leukodystrophy, Lesch–Nyhan syndrome or Sjögren–Larson syndrome. A more common differential diagnostic problem, the primary lateral sclerosis variant of ALS/MND, may present with a predominant spastic paraparesis but even in cases with little clinical evidence of lower motor neuron involvement, p62 and TDP-43 immunohistochemistry will identify the characteristic inclusions of ALS/MND at autopsy. In certain other neurodegenerative diseases, dementia may be associated with, and be preceded by, spastic paraparesis.[187] Some late onset cases of familial spastic paraparesis have been associated with mutation at codon 105 in the prion protein gene and may be variants of Gerstmann–Straussler–Scheinker syndrome, with the ensuing implications for taking appropriate precautions at autopsy.[152] In certain familial Alzheimer's disease cases due to mutation in the presenilin 1 gene, spastic paraparesis may precede dementia; these cases have characteristic 'cotton-wool' plaques in the cerebral cortex.[116]

ACKNOWLEDGEMENTS

We thank Dr Graham Lennox, Professor James Lowe and Professor Nigel Leigh for their contributions to this chapter in the 6th and 7th editions. Permission to use illustrations from these chapters and by Dr D R Oppenheimer and Dr Margaret Esiri from the 5th edition of this book is gratefully acknowledged.

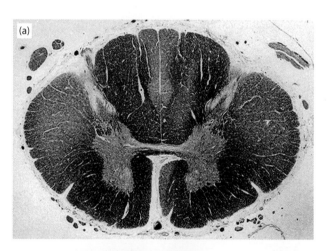

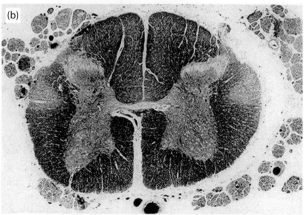

14.23 Familial spastic paraparesis. Dominant inheritance, early onset, death at age 74 years. **(a)** Fourth cervical level. **(b)** Lower lumbar level. The posterior columns are more affected at the higher level and the pyramidal tracts at the lower one. Myelin stain.

REFERENCES

1. Ackerly S, Grierson A, Brownlees J, et al. Glutamate slows axonal transport of neurofilaments in transfected neurons. *J Cell Biol* 2000;**150**:165–76.

2. Adachi H, Katsuno M, Minamiyama M, et al. Widespread nuclear and cytoplasmic accumulation of mutant androgen receptor in SBMA patients. *Brain* 2005;**128**:659–70.

3. Adachi H, Waza M, Katsuno M, et al. Pathogenesis and molecular targeted therapy of spinal and bulbar muscular atrophy. *Neuropathol Appl Neurobiol* 2007;**33**:135–51.

4. Al–Chalabi A,Leigh P. Recent advances in amyotrophic lateral sclerosis. *Curr Opin Neurol* 2000;**13**:397–405.

5. Al–Chalabi A, Andersen P, Chioza B, et al. Recessive amyotrophic lateral sclerosis families with the D90A *SOD1* mutation share a common founder: evidence for a linked protective factor. *Hum Mol Genet* 1998;**7**:2045–50.

6. Al–Chalabi A, Andersen P, Nilsson P, et al. Deletions of the heavy neurofilament subunit tail in amyotrophic lateral sclerosis. *Hum Mol Genet* 1999;**8**:157–64.

7. Al–Sarraj S, King A, Troakes C, et al. p62 positive, TDP-43 negative, neuronal cytoplasmic and intranuclear inclusions in the cerebellum and hippocampus define the pathology of C9orf72 linked FTLD and MND/ALS. *Acta Neuropathol* 2011;**122**:691–702.

8. Amato A, Prior T, Barohn R, et al. Kennedy's disease: a clinicopathologic correlation with mutations in the androgen receptor gene. *Neurology* 1993;**43**:791–4.

9. Andersen P. Genetics aspects of amyotrophic lateral sclerosis/motor neuron disease. In: Shaw P, Strong M eds, *Motor Neuron Disorders*. Philadelphia: Butterworth–Heinemann, 2005:207–36.

10. Andersen P. Amyotrophic lateral sclerosis associated with mutations in the CuZn superoxide dismutase gene. *Curr Neurol Neurosci Rep* 2006;**6**:37–46.

11. Andersen P, Forsgren L, Binzer M, et al. Autosomal recessive adult-onset amyotrophic lateral sclerosis associated with homozygosity for Asp90Ala CuZn-superoxide dismutase mutation. A clinical and genealogical study of 36 patients. *Brain* 1996;**119**:1153–72.

12. Aoki M, Lin C, Rothstein J, et al. Mutations in the glutamate transporter *EAAT2* gene do not cause abnormal EAAT2 transcripts in amyotrophic lateral sclerosis. *Ann Neurol* 1998;**43**:645–53.

13. Appel S, Smith RG, Engelhardt JI, et al. Evidence for autoimmunity in amyotrophic lateral sclerosis. *J Neurol Sci* 1993;**118**:169–74.

14. Appel S, Beers D, Siklos L, et al. Calcium: the Darth Vader of ALS. *Amyotroph Lateral Scler Other Motor Neuron Disord* 2001;**2**:S47–54.

15. Arai T, Hasegawa M, Akiyama H, et al. TDP-43 is a component of ubiquitin-positive tau-negative inclusions in frontotemporal lobar degeneration and amyotrophic lateral sclerosis. *Biochem Biophys Res Commun* 2006;**351**:602–11.

16. Araki S, Hayashi M, Tamagawa K, et al. Neuropathological analysis in spinal muscular atrophy type II. *Acta Neuropathol* 2003;**106**:441–8.

17. Aran F. Recherches sur une maladie non encore décrite du système musculaire (atrophie musculaire progressive). *Arch Gen Med* 1850;**24**:5–35, 172–214.

18. Armon C, Daube JR, Windebank AJ, et al. How frequently does classic amyotrophic lateral sclerosis develop in survivors of poliomyelitis? *Neurology* 1990;**40**:172–4.

19. Armon C, Kurland LT, Daube JR, et al. Epidemiologic correlates of sporadic amyotrophic lateral sclerosis. *Neurology* 1991;**41**:1077–84.

20. Ash PE, Bieniek KF, Gendron TF, et al. Unconventional translation of C9ORF72 GGGGCC expansion generates insoluble polypeptides specific to c9FTD/ALS. *Neuron* 2013;**77**:639–46.

21. Banno H, Adachi H. Mutant androgen receptor accumulation in spinal and bulbar muscular atrophy scrotal skin: a pathogenic marker. *Ann Neurol* 2006;**59**:520–6.

22. Banno H, Katsuno M. Pathogenesis and molecular targeted therapy of spinal and bulbar muscular atrophy (SBMA). *Cell Tissue Res* 2012;**349**:313–20.

23. Baumer D, Ansorge O. The role of RNA processing in the pathogenesis of motor neuron degeneration. *Exp Rev Mol Med* 2010;**12**:e21.

24. Ben Hamida M, Hentati F. Juvenile amyotrophic lateral sclerosis and related syndromes. *Adv Neurol* 1991;**56**:175–9.

25. Bensimon G, Lacomblez L, Meininger V, et al. A controlled trial of riluzole in amyotrophic lateral sclerosis. ALS/Riluzole Study Group. *N Engl J Med* 1994;**330**:585–91.

26. Berger M, Kopp N, Vital C, et al. Detection and cellular localisation of enterovirus RNA sequenes in spinal cord of patients with ALS. *Neurology* 2000;**54**:20–5.

27. Bigio E, Weintraub S, Rademakers R, et al. Frontotemporal lobar degeneration with TDP-43 proteinopathy and chromosome 9p repeat expansion in C9ORF72: clinicopathologic correlation. *Neuropathology* 2013;**33**:122–33.

28. Bingham P, Shen N, Rennert H, et al. Arthrogryposis due to infantile neuronal degeneration associated with deletion of the *SMNT* gene. *Neurology* 1997;**49**:848–51.

29. Blackstone C, O'Kane C. Hereditary spastic paraplegias: membrane traffic and the motor pathway. *Nat Rev Neurosci* 2011;**12**:31–42.

30. Blair I, Williams K, Warraich S, et al. FUS mutations in amyotrophic lateral sclerosis:-clinical, pathological, neurophysiological and genetic analysis. *J Neurol NeurosurgPsychiatry* 2010;**81**:639–45.

31. Blondel B, Duncan G, Couderc T, et al. Molecular aspects of poliovirus biology with a special focus on the interactions with nerve cells. *J Neurovirol* 1998;**4**:1–26.

32. Boeve B, Boylan K, Graff–Radford N, et al. Characterization of frontotemporal dementia and/or amyotrophic lateral sclerosis associated with the GGGGCC repeat expansion in C9ORF72. *Brain* 2012;**135**:765–83.

33. Borchelt DR, Lee MK, Slunt HS, et al. Superoxide dismutase 1 with mutations linked to familial amyotrophic lateral sclerosis possesses significant activity. *Proc Natl Acad Sci USA* 1994;**91**:8292–6.

34. Borthwick G, Johnson M, Ince P, Shaw P, Turnbull D. Mitochondrial enzyme activity in amyotrophic lateral sclerosis: implications for the role of mitochondria in neuronal cell death. *Ann Neurol* 1999;**46**:787–90.

35. Bosch A, Abeling N. Brown–Vialetto–Van Laere and Fazio Londe syndrome is associated with a riboflavin transporter defect mimicking mild MADD: a new inborn error of metabolism with potential treatment. *J Inherit Metab Dis* 2011;**34**:159–64.

36. Bourke S, Tomlinson M, Williams T, et al. Effects of non-invasive ventilation on survival and quality of life in patients with amyotrophic lateral sclerosis: a randomized controlled trial. *Lancet Neurol* 2006;**5**:140–7.

37. Bowling AC, Schulz JB, Brown RJ, et al. Superoxide dismutase activity, oxidative damage and mitochondrial energy metabolism in familial and sporadic amyotrophic lateral sclerosis. *J Neurochem* 1993;**61**:2322–5.

38. Brettschneider J, Van Deerlin V, Robinson J, et al. Pattern of ubiquilin pathology in ALS and FTLD indicates presence of C9ORF72 hexanucleotide expansion. *Acta Neuropathol* 2012;**123**:825–39.

39. Brown R. Amyotrophic lateral sclerosis: recent insights from genetics and transgenic mice. *Cell* 1995;**80**:687–92.

40. Byrne P, McMonagle P, Webb S, et al. Age-related cognitive decline in hereditary spastic paraparesis linked to chromosome 2P. *Neurology* 2000;**54**:1510–7.

41. Byrne S, Walsh C, Lynch C, et al. Rate of familial amyotrophic lateral sclerosis: a systematic review and meta-analysis. *J Neurol Neurosurg Psychiatry* 2011; **82**:623–7.

42. Casari G, De Fusco M, Ciarmatori S, et al. Spastic paraplegia and OXPHOS impairment caused by mutations in paraplegin, a nuclear-encoded mitochondrial metalloprotease. *Cell* 1998;**93**:973–83.

43. Cervenakova L, Protas I, Hirano A, et al. Progressive muscular atrophy variant of familial amyotrophic lateral sclerosis (PMA/ALS). *J Neurol Sci* 2000;**177**: 124–30.

44. Charcot JM. De la sclérose latérale amyotrophique. *Prog Med* 1874;**2**:325–32.7, 341–342, 453–5.

45. Charcot JM, Joffroy A. Deux cas d'atrophie musculaire progressive avec lésions de la substance grise et des fascicaux antérolateraux de la moelle épinière. *Arch Physiol (Paris)* 1869;**2**:354–367, 629–644, 744–60.

46. Charvin D, Cifuentes–Diaz C, Fonknechten N, et al. Mutations of SPG4 are responsible for a loss of function of spastin, an abundant neuronal proetin localized in the nucleus. *Hum Mol Genet* 2003;**12**:71–8.

47. Chavada G, El–Nayal A, Lee F, et al. Evaluation of two different methods for per-oral gastrostomy tube placement in patients with motor neuron disease (MND):PIG versus PEG procedures. *Amyotroph Lateral Scler Other Motor Neuron Disord* 2010;**11**:531–6.

48. Chavany C, Vicario Abejon C, et al. Transgenic mice for interleukin 3 develop motor neuron degeneration associated with

autoimmune reaction against spinal cord motor neurons. *Proc Natl Acad Sci USA* 1998;**95**:11354–9.

49. Chen Y–Z, Bennett C, Huynh H, *et al.* DNA/RNA helicase gene mutations in a form of juvenile amyotrophic lateral sclerosis (ALS4). *Am J Hum Genet* 2004;**74**:1128–35.

50. Chio A, Meineri P, Tribolo A, *et al.* Risk factors in motor neuron disease: a case–control study. *Neuroepidemiol* 1991;**10**:174–84.

51. Chou S, Wang H. Aberrant glycosylation/phosphorylation in chromatolytic motor-neurons of Werdnig–Hoffman disease. *J Neurol Sci* 1997;**152**:198–209.

52. Chow C, Landers J, Bergren S, *et al.* Deleterious variants of FIG4, a phospho-inositide phosphatase, in patients with ALS. *Am J Hum Genet* 2009;**84**:85–8.

53. Ciccarelli F, Patton M, McKusick V, Crosby A. SPG20 is mutated in Troyer syndrome, an hereditary spastic paraplegia. *Nat Genet* 2002;**31**:347–8.

54. Clement A, Nguyen M, Roberts E, *et al.* Wild-type non-neuronal cells extend survival of *SOD1* mutant motor neurons in ALS mice. *Science* 2003;**302**:113–7.

55. Comi G, Bordoni A, Salami S, *et al.* Cytochrome c oxidase subunit I microdeletion in a patient with motor neuron disease. *Ann Neurol* 1998;**43**:110–6.

56. Cooper–Knock J, Hewitt C, Highley J, *et al.* Clinico-pathological features in amyotrophic lateral sclerosis with expansions in C9ORF72. *Brain* 2012;**135**:694–715.

57. Cooper–Knock J, Walsh M, Higginbottom A, *et al.* Sequestration of multiple RNA recognition motif-containing proteins by C9ORF72 repeat expansions. *Brain* 2014;**137**:2040–51.

58. Corbo M, Hays AP. Peripherin and neuro-filament protein coexist in spinal spheroids of motor neuron disease. *J Neuropathol Exp Neurol* 1992;**51**:531–7.

59. Cox P, Banack S, Murch S. Biomagnification of cyanobacterial neurotoxins and neurodegenerative disease among the Chamorro people of Guam. *Proc Natl Acad Sci USA* 2003;**100**:13380–3.

60. Cudkowicz M, McKenna–Yasek D, Sapp P, *et al.* Epidemiology of mutations in superoxide dismutase in amyotrophic lateral sclerosis. *Ann Neurol* 1997;**42**:210–21.

61. Dalakas M. Pathogenetic mechanisms of post-polio syndrome: morphological, electrophysiological, virological and immunological correlations. *Ann NY Acad Sci* 1995;**753**:167–85.

62. Dalakas M. Pro-inflammatory cytokines and motor neuron dysfunction: is there a connection in post-polio syndrome? *J Neurol Sci* 2002;**205**:5–8.

63. Dalakas MC. Morphologic changes in the muscles of patients with postpoliomyelitis neuromuscular symptoms. *Neurology* 1988;**38**:99–104.

64. Dalakas MC. The post-polio syndrome as an evolved clinical entity. Definition and clinical description. *Ann N Y Acad Sci* 1995;**753**:68–80.

65. DeJesus–Hernandez M, Mackenzie I, Boeve B, *et al.* Expanded GGGGCC hexanucleotide repeat in noncoding region of C9ORF72 causes chromosome 9p-linked FTD and ALS. *Neuron* 2011;**72**:245–56.

66. DeLuca G, Ebers G, Esiri M. The extent of axonal loss in the long tracts in hereditary

67. spastic paraplegia. *Neuropathol Appl Neurobiol* 2004;**30**:576–84.

67. Deng H, Chen W, Hong S, Boycott K, *et al.* Mutations in *UBQLN2* cause dominant X-linked juvenile and adult-onset alS and ALS/dementia. *Nature* 2011;**477**:211–5.

68. Dickson D, Josephs K, Amador–Ortiz C. TDP-43 in differential diagnosis of motor neuron disorders. *Acta Neuropathol* 2007;**114**:71–9.

69. Dipti S, Childs A–M, Livingston J, *et al.* Brown–Vialetto–Van Laere syndrome;variability in age at onset and disease progression highlighting the phenotypic overlap with Fazio–Londe disease. *Brain Devel* 2005;**27**:443–6.

70. Dobson C, Villagra F, Clowry G, *et al.* Abnormal corticospinal function but normal axonal guidance in human *L1CAM* mutations. *Brain* 2001;**124**:2393–406.

71. Donnelly CJ, Zhang PW, Pham JT, *et al.* RNA toxicity from the ALS/FTD C9ORF72 expansion is mitigated by antisense intervention. *Neuron* 2013;**80**:415–28.

72. Doyu M, Sobue G, Mukai E, *et al.* Severity of X-linked recessive bulbospinal neuronopathy correlates with size of the tandem CAG repeat in the androgen receptor gene. *Ann Neurol* 1992;**32**:707–10.

73. Dubowitz V. Very severe spinal muscular atrophy (SMA type 0): an expanding clinical phenotype. *Eur J Paediat Neurol* 1999;**3**:49–51.

74. Eisen A, Stewart H, Schulzer M, *et al.* Anti-glutamate therapy in amyotrophic lateral sclerosis:a trial using lamotrigine. *Can J Neurol Sci* 1993;**20**:297–301.

75. Elden A, Kim H–J, Hart M, *et al.* Ataxin-2 intermediate-length polyglutamine expansions are associated with increased risk for ALS. *Nature* 2010;**466**:1069–75.

76. Ellis CM, Simmons A, Jones DK, *et al.* Diffusion tensor MRI assesses corticospinal tract damage in ALS. *Neurology* 1999;**53**(5):1051–8.

77. Ferreirinha F, Quattrini A, Pirozzi M, *et al.* Axonal degeneration in paraplegin-deficient mice is associated with abnormal mitochondria and impairment of axonal transport. *J Clin Invest* 2004;**113**:231–242.

78. Ferrer I, Olive M, Rivera R, *et al.* Hereditary spastic paraparesis with dementia, amyotrophy and peripheral neuropathy. A neuropathological study. *Neuropath Appl Neurobiol* 1995;**21**:255–61.

79. Fidzianska A, Goebel H, Warlo I. Acute infantile spinal muscular atrophy. Muscle apoptosis as a proposed pathogenetic mechanism. *Brain* 1990;**113**:433–45.

80. Figlewicz D, Krizus A, Martinoli M, *et al.* Variants of the heavy neurofilament subunit are associated with the development of amyotrophic lateral sclerosis. *Hum Mol Genet* 1994;**3**:1759–61.

81. Finsterer J. Bulbar and spinal muscular atrophy (Kennedy's disease): a review. *Eur J Neurol* 2009;**16**:556–61.

82. Finsterer J, Loscher W, Quasthoff S, *et al.* Hereditary spastic paraplegias with autosomal dominant, recessive, X–linked, or maternal trait of inheritance. *J Neurol Sci* 2012;**318**:1–18.

83. Fishman P. Late-convalescent poliomyelitis. *Arch Neurol* 1987;**44**:98–100.

84. Flowers J, Powell J, Leigh P, *et al.* Intron 7 retention and exon 9 skipping EAAT2 mRNA variants are not associated with

85. amyotrophic lateral sclerosis. *Ann Neurol* 2001;**49**:643–9.

85. Forsyth P, Dalmau J, Graus F, *et al.* Motor neuron syndromes in cancer patients. *Ann Neurol* 1997;**41**:722–30.

86. Garcia–Cabezas M, Garcia–Alix A, Martin Y, *et al.* Neonatal spinal muscular atrophy with multiple contractures, bone fractures, respiratory insufficiency and 5q13 deletion. *Acta Neuropathol* 2004;**107**:475–8.

87. Geiss R, Goetz R, Schrank B, *et al.* Potential implications of a ciliary neurotrophic factor gene mutation in a German population of patients with motor neuron disease. *Muscle Nerve* 1998;**21**:236–8.

88. Geiss R, Beck M, Goetz R, *et al.* Potential role of LIF as a modifier gene in the pathogenesis of amyotrophic lateral sclerosis. *Neurology* 2000;**54**:1003–5.

89. Gendron TF, Bieniek KF, Zhang YJ, *et al.* Antisense transcripts of the expanded C9ORF72 hexanucleotide repeat form nuclear RNA foci and undergo repeat-associated non-ATG translation in c9FTD/ALS. *Acta Neuropathol* 2013;**126**:829–44.

90. Geser F, Lee V–M, Trojanowski J. Amyotrophic lateral sclerosis and frontotemporal lobar degeneration: a spectrum of TDP-43 proteinopathies. *Neuropathology* 2010;**30**:103–12.

91. Gitcho MA, Bigio E, Mishra M, *et al.* TARDBP 3′-UTR variant in autopsy-confirmed frontotemporal lobar degeneration with TDP-43 proteinopathy. *Acta Neuropathol* 2009;**118**:633–45.

92. Gonzalez H, Khademi M, Andersson M, *et al.* Prior poliomyelitis: evidence of cytokine production in the central nervous system. *J Neurol Sci* 2002;**205**:9–13.

93. Gonzalez H, Olsson T, Borg K, *et al.* Management of post-polio syndrome. *Lancet Neurol* 2010;**9**:634–42.

94. Gordon P, Rowland L, Younger D, *et al.* Lymphoproliferative disorders and motor neuron disease: an update. *Neurology* 1997;**48**:1671–8.

95. Gredal O, Pakkenberg H, Karlsborg M, *et al.* Unchanged total number of neurons in motor cortex and neocortex in amyotrophic lateral sclerosis: a stereological study. *J Neurosci Methods* 2000;**95**:171–6.

96. Green P, Wiseman M, Crow YJ, *et al.* Brown–Vialetto–Van Laere syndrome, a ponto-bulbar palsy with deafness, is caused by mutations in C20orf54. *Am J Hum Genet* 2010;**86**:485–9.

97. Greenway M, Andersen P, Russ C, *et al.* ANG mutations segregate with familial and 'sporadic' amyotrophic lateral sclerosis. *Nat Genet* 2006;**38**:11–3.

98. Groen E, van Es M, van Vught P, *et al.* FUS mutations in familial amyotrophic lateral sclerosis in the Netherlands. *Arch Neurol* 2010;**67**:224–30.

99. Guillot N, Cuisset J–M, Cuvellier JC, *et al.* Unusual clinical features in infantile spinal muscular atrophies. *Brain Dev* 2008;**30**:169–78.

100. Gurney M, Pu H, Chiu A, *et al.* Motor neuron degeneration in mice that express a human Cu,Zn superoxide dismutase mutation. *Science* 1994;**264**:1772–5.

101. Hadano S, Hand C, Osuga H, *et al.* A gene encoding a putative GTPase regulator is mutated in familial amyotrophic lateral sclerosis 2. *Nature Genet* 2001;**29**:166–73.

102. Hainfellner JA, Pilz P, Lassmann H, *et al.* Diffuse Lewy body disease as substrate

of primary lateral sclerosis. *J Neurol* 1995;**242**:59–63.

103. Halstead L. A brief history of postpolio syndrome in the United States. *Arch Phys Med Rehab* 2011;**92**:1344–9.

104. Hand C, Khoris J, Salachas F, *et al.* A novel locus for familial amyotrophic lateral sclerosis on chromosome 18. *Am J Hum Genet* 2002;**70**:251–6.

105. Hansen J, Durr A, Cournu–Rebeix I, *et al.* Hereditary spastic paraplegia SPG13 is associated with a mutation in the gene encoding the mitochondrial chaperonin Hsp60. *Am J Hum Genet* 2002;**70**:1328–32.

106. Haverkamp L, Appel V, Appel S. Natural history of amyotrophic lateral sclerosis in a database population. Validation of a scoring system and a model for survival prediction. *Brain* 1995;**118**:707–19.

107. Hayward C, Brock D, Minns R, *et al.* Homozygosity for *Asn86Ser* mutation in the CuZn-superoxide dismutase gene produces a severe clinical phenotype in a juvenile onset case of familial amyotrophic lateral sclerosis. *J Med Genet* 1998;**35**:174.

108. Hayward C, Colville S, Swingler R, Brock D. Molecular genetic analysis of the APEX nuclease gene in amyotrophic lateral sclerosis. *Neurology* 1999;**52**:1899–901.

109. Hazan J, Lamy C, Melki J, *et al.* Autosomal dominant familial spastic paraplegia is genetically heterogeneous and one locus maps to chromosome 14q. *Nat Genet* 1993;**5**:163–7.

110. He CZ, Hays AP. Expression of peripherin in ubiquinated inclusions of amyotrophic lateral sclerosis. *J Neurol Sci* 2004;**217**:47–54.

111. Heads T, Pollock M, Robertson A, *et al.* Sensory nerve pathology in amyotrophic lateral sclerosis. *Acta Neuropathol* 1991;**82**:316–20.

112. Hedera P, DiMauro S, Bonilla E, *et al.* Phenotypic analysis of autosomal dominant hereditary spastic paraplegia linked to chromosome 8q. *Neurology* 1999;**53**:44–50.

113. Hentati A, Ouahchi K, Pericak–Vance M, *et al.* Linkage of a commoner form of recessive amyotrophic lateral sclerosis to chromosome 15q15–q22 markers. *Neurogenetics* 1998;**2**:55–60.

114. Hilton DA, Love S, Ferguson I, *et al.* Motor neuron disease with neurofibrillary tangles in a non-Guamanian patient. *Acta Neuropathol* 1995;**90**:101–6.

115. Hirano A. Cytopathology of amyotrophic lateral sclerosis. *Adv Neurol* 1991;**56**:91–101.

116. Houlden H, Baker M, McGowan E, *et al.* Variant Alzheimer's disease with spastic paraparisis and cotton wool plaques is caused by PS-1 mutations that lead to exceptionally high amyloid-β concentrations. *Ann Neurol* 2000;**48**:806–8.

117. Howard R. Poliomyelitis and the post-polio syndrome. *BMJ* 2005;**330**:1314–9.

118. Huang E, Zhang J, Geser F, *et al.* Extensive FUS-immunoreactive pathology in juvenile amyotrophic lateral sclerosis with basophilic inclusions. *Brain Pathol* 2010;**20**:1069–76.

119. Hudson A. Amyotrophic lateral sclerosis and its association with dementia, parkinsonism and other neurological disorders: a review. *Brain* 1981;**104**:217–47.

120. Igarashi S, Tanno Y, Onodera O, *et al.* Strong correlation between number of

CAG repeats in androgen receptor genes and the clinical onset of features of spinal and bulbar muscular atrophy. *Neurology* 1992;**42**:2300–2.

121. Imura T, Shimohama S, Kawamata J, Kimura J. Genetic variation in the ciliary neurotrophic factor receptor α gene and familial amyotrophic lateral sclerosis. *Ann Neurol* 1998;**43**:275.

122. Ince P, Highley JR, Kirby J, *et al.* Molecular pathology and genetic advances in amyotrophic lateral sclerosis: an emerging molecular pathway and the significance of glial pathology. *Acta Neuropathol* 2011;**122**:657–71.

123. Ince P, Stout N, Shaw P, *et al.* Parvalbumin and calbindin-28k in human motor system and in motor neuron disease. *Neuropathol Appl Neurobiol* 1993;**19**:291–9.

124. Ince P, Shaw P, Slade J, Jones C, Hudson P. Familial amyotrophic lateral sclerosis with a mutation in exon 4 of the Cu/Zn superoxide dismutase gene: pathological and immunocytochemical changes. *Acta Neuropathol* 1996;**92**:395–403.

125. Ince PG, Lowe J, Shaw PJ. Amyotrophic lateral sclerosis: current issues in classification, pathogenesis and molecular pathology. *Neuropathol Appl Neurobiol* 1998;**24**:104–17.

126. Ince P, Tomkins J, Slade J, Thatcher N, Shaw P. Amyotrophic lateral sclerosis associated with genetic abnormalities in the gene encoding Cu/Zn superoxide dismutase: molecular pathology of five new cases, and comparison with previous reports and 73 sporadic cases of ALS. *J Neuropathol Exp Neurol* 1998;**57**:895–904.

127. Ince P, Evans J, Knopp M, *et al.* Corticospinal tract degeneration in the progressive muscular atrophy variant of ALS. *Neurology* 2003;**60**:1252–8.

128. Ito D, Suzuki N. Seipinopathy: a novel endoplasmic reticulum stress-associated disease. *Brain Dev* 2009;**132**:8–15.

129. Ito H, Hirano A. Comparative study of spinal cord ubiquitin expression in post-poliomyelitis and sporadic amyotrophic lateral sclerosis. *Acta Neuropathol* 1994;**87**:425–9.

130. Iwahashi H, Eguchi Y, Yasuhara N, *et al.* Synergistic anti-apoptotic activity between Bcl-2 and SMN implicated in spinal muscular atrophy. *Nature* 1997;**390**:413–7.

131. Jackson M, Morrison K, Al–Chalabi A, *et al.* Analysis of chromosome 5q13 genes in amyotrophic lateral sclerosis: homozygous NAIP deletion in a sporadic case. *Ann Neurol* 1996;**39**:796–800.

132. Johnson J, Mandrioli J, Benatar M, *et al.* Exome sequencing reveals VCP mutations as a cause of familial ALS. *Neuron* 2010;**68**:857–64.

133. Jones CT, Swingler RJ, Brock DJ. Identification of a novel *SOD1* mutation in an apparently sporadic amyotrophic lateral sclerosis patient and the detection of Ile113Thr in three others. *Hum Mol Genet* 1994;**3**:649–50.

134. Jouet M, Rosenthal A, Armstrong G, *et al.* X-linked spastic paraplegia (SPG1), MASA syndrome and X-linked hydrocephalus result from mutations in the *L1* gene. *Nat Genet* 1994;**7**:402–7.

135. Julien JP. Neurofilament functions in health and disease. *Curr Opin Neurobiol* 1999;**9**:554–60.

136. Julien JP. Amyotrophic lateral sclerosis. Unfolding the toxicity of the misfolded. *Cell* 2001;**104**:581–91.

137. Julien JP, Beaulieu JM. Cytoskeletal abnormalities in amyotrophic lateral sclerosis: beneficial or detrimental effects? *J Neurol Sci* 2000;**180**:7–14.

138. Kaindl A, Guenther U, Rudnik–Schöneborn S, *et al.* Spinal muscular atrophy with respiratory distress type 1 (SMARD1). *J Child Neurol* 2008;**23**:199–204.

139. Kaminski H, Tresser N, Hogan R, Martin E. Spinal cord histopathology in long-term survivors of poliomyelitis. *Muscle Nerve* 1995;**18**:1208–9.

140. Kato S, Hirano A. Ubiquitin and phosphorylated neurofilament epitopes in ballooned neurons of the extraocular muscle nuclei in a case of Werdnig–Hoffman disease. *Acta Neuropathol* 1990;**80**:334–7.

141. Kato S, Hirano A. Involvement of the brain stem reticular formation in familial amyotrophic lateral sclerosis. *Clin Neuropathol* 1992;**11**:41–4.

142. Kato S, Shimoda M, Watanabe Y, *et al.* Familial amyotrophic lateral sclerosis with a two base pair deletion in superoxide dismutase 1 gene: multisystem degeneration with intracytoplasmic hyaline inclusions in astrocytes. *J Neuropathol Exp Neurol* 1996;**55**:1089–101.

143. Kato T, Hirano A, Donnenfeld H. A Golgi study of the large anterior horn cells of the lumbar cords in normal spinal cords and in amyotrophic lateral sclerosis. *Acta Neuropathol* 1987;**75**:34–40.

144. Kennedy WRA, Alter M, Sung JH. Progressive proximal spinal and bulbar muscular atrophy of late onset: a sex-linked recessive trait. *Neurology* 1968;**18**:671–80.

145. Kertesz A, Kawarai T, Rogaeva E, *et al.* Familial frontotemporal dementia with ubiquitin-positive, tau-negative inclusions. *Neurology* 2000;**54**:818–27.

146. Khabazian I, Bains J, Williams D, *et al.* Isolation of various forms of sterol b-D-glucoside from the seed of *Cycas circinalis*: neurotoxicity and implications for ALS-parkinsonism dementia complex. *J Neurochem* 2002;**82**:516–28.

147. Kiernan J, Hudson A. Frontal lobe atrophy in motor neuron diseases. *Brain* 1994;**117**:747–57.

148. Kihira T, Mizusawa H, Tada J, *et al.* Lewy body-like inclusions in Onuf's nucleus from two cases of sporadic amyotrophic lateral sclerosis. *J Neurol Sci* 1993;**115**:51–7.

149. Kihira T, Yoshida S, Yoshimasu F, *et al.* Involvement of Onuf's nucleus in amyotrophic lateral sclerosis. *J Neurol Sci* 1997;**147**:81–8.

150. Kimura F, Smith RG, Delbono O, *et al.* Amyotrophic lateral sclerosis patient antibodies label Ca2+ channel alpha 1 subunit. *Ann Neurol* 1994;**35**:164–71.

151. Kimura T, Budka H. Glial bundles in spinal nerve roots. An immunocytochemical study stressing their nonspecificity in various spinal cord and peripheral nerve diseases. *Acta Neuropathol* 1984;**65**:46–52.

152. Kitamoto T, Amano N, Terao Y, *et al.* A new inherited prion disease (PrP-P105L mutation) showing spastic paraparesis. *Ann Neurol* 1993;**34**:808–13.

153. Korinthenberg R, Sauer M, Ketelsen U–P, *et al.* Congenital axonal neuropathy caused by deletions in the spinal muscular atrophy region. *Ann Neurol* 1997;**42**:364–8.

154. Kovacs G, Murrell J, Horvath S, *et al.* TARDBP variation associated with frontotemporal dementia, supranuclear

gaze palsy, and chorea. *Mov Disord* 2009;**24**:1843–7.

155. Kovari E, Leuba G, Savioz A, *et al.* Familial frontotemporal dementia with ubiquitin inclusion bodies and without motor neuron disease. *Acta Neuropathol (Berl)* 2000;**100**:421–6.

156. Kurland LT, Radhakrishnan K, Smith GE, *et al.* Mechanical trauma as a risk factor in classic amyotrophic lateral sclerosis: lack of epidemiologic evidence. *J Neurol Sci* 1992;**113**:133–43.

157. Kuru S, Sakai M. An autopsy case of spinal muscular atrophy type III (Kugelberg–Welander disease). *Neuropathology* 2009;**29**:63–7.

158. Kwiatkowski T, Bosco D, Leclerc A, *et al.* Mutations in the FUS/TLS gene on chromosome 16 cause familial amyotrophic lateral sclerosis. *Science* 2009;**323**: 1205–8.

159. La Spada A, Wilson E, Lubahn D, Harding A, Fischbeck K. Androgen receptor gene mutations in X-linked spinal and muscular atrophy. *Nature* 1991;**352**:77–9.

160. Lacomblez L, Bensimon G, Leigh P, *et al.* Dose-ranging study of riluzole in amyotrophic lateral sclerosis. Amyotrophic lateral sclerosis study group II. *Lancet* 1996;**347**:1425–31.

161. Lefebvre S, Burglen L, Reboullet S, *et al.* Identification and characterization of a spinal muscular atrophy-determining gene. *Cell* 1995;**80**:1–5.

162. Lefebvre S, Burlet P, Liu Q, *et al.* Correlation between severity and SMN protein level in spinal muscular atrophy. *Nature Genet* 1997;**16**:265–9.

163. Leigh PN, Swash M. Cytoskeletal pathology in motor neuron diseases. *Adv Neurol* 1991;**56**:115–24.

164. Li M, Miwa S, Kobayashi Y, *et al.* Nuclear inclusions of the androgen receptor protein in spinal and bulbar muscular atrophy. *Ann Neurol* 1998;**44**:249–254.

165. Li M, Nakagomi Y, Kobayashi Y, *et al.* Non-neural nuclear inclusions of androgen receptor protein in spinal and bulbar muscular atrophy. *Am J Pathol* 1998;**153**:695–701.

166. Li M, Chevalier–Larsen E, *et al.* Soluble androgen receptor oligomers underlie pathology in a mouse model of spinobulbar muscular atrophy. *J Biol Chem* 2007;**282**:3157–64.

167. Lin C, Bristol L, Jin L, *et al.* Aberrant RNA processing in a neurodegenerative disease: the cause for absent EAAT2, a glutamate transporter, in amyotrophic lateral sclerosis. *Neuron* 1998;**20**:589–602.

168. Lindsay RM. Neuron saving schemes. *Neurobiology* 1995;**373**:289–344.

169. Lloyd CM, Richardson MP, Brooks DJ, *et al.* Extramotor involvement in ALS: PET studies with the GABAA ligand [11C] flumenazil. *Brain* 2000;**123**:2289–96.

170. Logroscino G, Traynor B, Hardiman O, *et al.* Descriptive epidemiology of amyotrophic lateral sclerosis: new evidence and unsolved issues. *J Neurol Neurosurg Psychiatry* 2008;**79**:6–11.

171. Lorson CL, Hahnen E, Androphy EJ, *et al.* A single nucleotide in the *SMN* gene regulates splicing and is responsible for spinal muscular atrophy. *Proc Natl Acad Sci USA* 1999;**96**:6307–11.

172. Lowe J. New pathological findings in amyotrophic lateral sclerosis. *J Neurol Sci* 1994;**124**:S38–51.

173. Lowe J, Lennox G, Jefferson D, *et al.* A filamentous inclusion body within anterior horn neurons in motor neuron disease defined by immunocytochemical localisation of ubiquitin. *Neurosci Lett* 1988;**93**:203–10.

174. Lowe J, Aldridge F, Lennox G, *et al.* Inclusion bodies in motor cortex and brainstem of patients with motor neuron disease are detected by immunocytochemical localization of ubiquitin. *Neurosci Lett* 1989;**105**:7–13.

175. Lumb J, Connell J, Allison R, Reid E. The AAA ATPase spastin links microtubule severing to membrane modelling. *Biochim Biophys Acta* 2012;**1823**:192–7.

176. Lunn M, Wang C. Spinal muscular atrophy. *Lancet* 2008;**371**:2120–33.

177. Lynch T, Sano M, Marder KS, *et al.* Clinical characteristics of a family with chromosome 17-linked disinhibition–dementia–parkinsonism–amyotrophy complex. *Neurology* 1994;**44**:1878–84.

178. Mackenzie I, Feldman H. Ubiquitin immunohistochemistry suggests classic motor neuron disease, motor neuron disease with dementia and frontotemporal dementia of the motor neuron disease type represent a clinicopathologic spectrum. *J Neuropathol Exp Neurol* 2005;**64**:730–739.

179. Mackenzie I, Bigio E, Ince P, *et al.* Pathological TDP-43 distinguishes sporadic amyotrophic lateral sclerosis from amyotrophic lateral sclerosis with *SOD1* mutaions. *Ann Neurol* 2007;**61**:427–34.

180. Mackenzie I, Neumann M, Bigio E, *et al.* Nomenclature and nosology for neuropathologic subtypes of frontotemporal lobar degeneration: an update. *Acta Neuropathol* 2010;**119**:1–4.

181. Mackenzie IR, Arzberger T, Kremmer E, *et al.* Dipeptide repeat protein pathology in C9ORF72 mutation cases: clinicopathological correlations. *Acta Neuropathol* 2013;**126**:859–79.

182. MacLean H, Warne G, Zajac J. Spinal and bulbar muscular atrophy: androgen receptor dysfunction caused by a trinucleotide repeat expansion. *J Neurol Sci* 1996;**135**:149–57.

183. Maragakis N, Rothstein J. Glutamate transporters in neurologic disease. *Arch Neurol* 2001;**58**:365–70.

184. Markowitz J, Singh P, Darras BT. Spinal muscular atrophy: a clinical and research update. *Pediatr Neurol* 2012;**46**:1–12.

185. Martinez–Lage M, Molina–Procel L, *et al.* TDP-43 pathology in a case of hereditary spastic paraplegia with a NIPA1/SPG6 mutation. *Acta Neuropathol* 2012;**124**:285–91.

186. Maruyama H, Morino H, Ito H, *et al.* Mutations of optineurin in amyotrophic lateral sclerosis. *Nature* 2010;**465**:223–6.

187. Masters C, Beyreuther K. The Worster–Drought syndrome and other syndromes of dementia with spastic paraparesis: the paradox of molecular pathology. *J Neuropathol Exp Neurol* 2001;**60**: 317–319.

188. Matsumoto S, Goto S, Kusaka H, *et al.* Ubiquitin-positive inclusion in anterior horn cells in subgroups of motor neuron diseases: a comparative study of adult-onset amyotrophic lateral sclerosis, juvenile amyotrophic lateral sclerosis and Werdnig–Hoffman disease. *J Neurol Sci* 1993;**115**:208–13.

189. McCampbell A, Taylor J, Taye A, *et al.* CREB-binding protein sequestration by expanded polyglutamine. *Hum Mol Genet* 2000;**9**:2197–202.

190. McDermott CJ, Dayaratne RK, Tomkins J, *et al.* Paraplegin gene analysis in hereditary spastic paraparesis (HSP) pedigrees in north east England. *Neurology* 2001;**56**:467–71.

191. McHolm GB, Aguilar MJ, Norris FH. Lipofuscin in amyotrophic lateral sclerosis. *Arch Neurol* 1984;**41**:1187–8.

192. McShane M, Boyd S, Harding B, Brett E, Wilson J. Progressive bulbar paralysis of childhood. A reappraisal of Fazio–Londe disease. *Brain* 1992;**115**:1889–900.

193. McWhorter M, Monani U, Burghes A, Beattie C. Knockdown of the survival motor neuron (SMN) protein in zebrafish causes defects in motor axon outgrowth and pathfinding. *J Cell Biol* 2003;**162**:919–31.

194. Meyer T, Fromm A, Munch C, *et al.* The RNA of the glutamate transporter EAAT2 is variably spliced in amyotrophic lateral sclerosis and normal individuals. *J Neurol Sci* 1999;**170**:45–50.

195. Migheli A, Pezzulo T, Attanasio A, *et al.* Peripherin immunoreactive structures in amyotrophic lateral sclerosis. *Lab Invest* 1993;**68**:185–91.

196. Mitchell J, Paula P, Chena H, *et al.* Familial amyotrophic lateral sclerosis is associated with a mutation in D-amino acid oxidase. *Proc Natl Acad Sci USA* 2010;**107**:7556–61.

197. Mitsuyama Y. Presenile dementia with motor neuron disease. *Dementia* 1993;**4**:137–142.

198. Mondelli M, Rossi A, Passero S, *et al.* Involvement of peripheral sensory fibres in amyotrophic lateral sclerosis: electrophysiological study of 64 cases. *Muscle Nerve* 1993;**16**:166–72.

199. Mori K, Arzberger T, Grasser FA, *et al.* Bidirectional transcripts of the expanded C9orf72 hexanucleotide repeat are translated into aggregating dipeptide repeat proteins. *Acta Neuropathol* 2013;**126**:881–93.

200. Mori K, Weng SM, Arzberger T, *et al.* The C9orf72 GGGGCC repeat is translated into aggregating dipeptide-repeat proteins in FTLD/ALS. *Science* 2013;**339**:1335–8.

201. Moss T, Daga A, McNew JA. Fusing a lasting relationship between ER tubules. *Trends Cell Biol* 2011;**21**:416–23.

202. Moulard B, Sefiani A, Laamri A, *et al.* Apolipoprotein E genotyping in amyotrophic lateral sclerosis: evidence for a major influence on clinical presentation and prognosis. *J Neurol Sci* 1996;**139** (Suppl):34–37.

203. Moulard B, Salachas F, Chassande B, *et al.* Association between centromeric deletions of the SMN gene and sporadic adult-onset lower motor neuron disease. *Ann Neurol* 1998;**43**:640–44.

204. Murayama S, Bouldin TW,Suzuki K. Immunocytochemical and ultrastructural studies of Werdnig–Hoffman disease. *Acta Neuropathol* 1991;**81**:408–17.

205. Murayama S, Inoue K, Kawakami H, *et al.* A unique pattern of astrocytosis in the primary motor area in amyotrophic lateral sclerosis. *Acta Neuropathol (Berl)* 1991;**82**:456–61.

206. Murray L, Talbot K, Gillingwater TH. Neuromuscular synaptic vulnerability in motor neurone disease: amyotrophic lateral sclerosis and spinal muscular atrophy. *Neuropathol Appl Neurobiol* 2010;**36**:133–56.

207. Nagy D, Kato T, Kushner PD. Reactive astrocytes are widespread in the cortical grey matter of amyotrophic lateral sclerosis. *J Neurosci Res* 1994;**38**:336–47.

208. Nakano I, Iwatsubo T, Hashizume Y, *et al.* Bunina bodies in neurons of the medullary reticular formation in amyotrophic lateral sclerosis. *Acta Neuropathol* 1993;**85**:471–4.

209. Neary D, Snowden JS, Mann DM. Familial progressive aphasia: its relationship to other forms of lobar atrophy. *J Neurol Neurosurg Psychiatry* 1993;**56**:1122–5.

210. Neary D, Snowden JS, Gustafson L, *et al.* Frontotemporal lobar degeneration: a consensus on clinical diagnostic criteria. *Neurology* 1998;**51**:1546–54.

211. Neumann M, Sampathu D, Kwong L, *et al.* Ubiquitinated TDP-43 in frontotemporal lobar degeneration and amyotrophic lateral sclerosis. *Science* 2006;**314**:130–3.

212. Neumann M, Kwong L, Lee E, *et al.* Phosphorylation of S409/410 of TDP–43 is a consistent feature in all sporadic and familial forms of TDP–43 proteinopathies. *Acta Neuropathol* 2009;**117**:137–49.

213. Nihei K, McKee AC, Kowall NW. Patterns of neuronal degeneration in the motor cortex of amyotrophic lateral sclerosis patients. *Acta Neuropathol* 1993;**86**:55–64.

214. Nishihira Y, Tan C–F, Hoshi Y, *et al.* Sporadic amyotrophic lateral sclerosis of long duration is associated with relatively mild TDP-43 pathology. *Acta Neuropathol* 2009;**117**:45–53.

215. Nishimura A, Mitne–Neto M, Silva HC, *et al.* A mutation in the vesicle-trafficking protein VAPB causes late-onset spinal muscular atrophy and amyotrophic lateral sclerosis. *Am J Hum Genet* 2004;**75**:822–31.

216. O'Brien T, Kelly M, Saunders C. Motor neurone disease: a hospice perspective. *BMJ* 1992;**304**:471–3.

217. Ogino S, Wilson R. Genetic testing and risk assessment for spinal muscular atrophy (SMA). *Hum Genet* 2002;**111**:477–500.

218. Okamoto K, Hirai S, Yamazaki Y, Sun X, Nakazoto Y. New ubiquitin-positive intraneuronal inclusions in the extra-motor cortices in patients with amyotrophic lateral sclerosis. *Neurosci Lett* 1991;**129**:233–6.

219. Okamoto K, Hirai S, Amari M, *et al.* Bunina bodies in amyotrophic lateral sclerosis immunostained with rabbit anti-cystatin C serum. *Neurosci Lett* 1993;**162**:125–8.

220. Okamoto K, Hirai S, Amari M, *et al.* Oculomotor nuclear pathology in amyotrophic lateral sclerosis. *Acta Neuropathol (Berl)* 1993;**85**:458–62.

221. Oosthuyse B, Moons L, Storkebaum E, *et al.* Deletion of the hypoxia-response element in the vascular endothelial growth factor promoter causes motor neuron degeneration. *Nat Genet* 2001;**28**:131–8.

222. Orrell R, Figlewicz D. Clinical implications of the genetics of ALS and other motor neuron disorders. *Neurology* 2001;**57**:9–17.

223. Orrell RW, Habgood JJ, Belleroche JS, *et al.* The relationship of spinal muscular atrophy to motor neuron disease: investigation of SMN and NAIP gene deletions in sporadic and familial ALS. *J Neurol Sci* 1997;**145**:55–61.

224. Orrell RW, Habgood JJ, Gardiner I, *et al.* Clinical and functional investigation of 10 missense mutations and a novel frameshift insertion mutation of the gene for copper–zinc superoxide dismutase in UK families with amyotrophic lateral sclerosis. *Neurology* 1997;**48**:746–51.

225. Parboosingh JS, Figlewicz DA, Krizus A, *et al.* Spinobulbar muscular atrophy can mimic ALS: the importance of genetic testing in male patients with atypical ALS. *Neurology* 1997;**49**:568–72.

226. Parboosingh JS, Meininger V, McKenna–Yasek D, Brown RH Jr, Rouleau GA. Deletions causing spinal muscular atrophy do not predispose to amyotrophic lateral sclerosis. *Arch Neurol* 1999;**56**:710–12.

227. Parkinson N, Ince P, Smith M, *et al.* ALS phenotypes with mutations in CHMP2B (charged multivesicular body protein 2B). *Neurology* 2006;**67**:1074–7.

228. Perrie WT, Lee GT, Curtis EM, *et al.* Changes in the myelinated axons of femoral nerve in amyotrophic lateral sclerosis. *J Neural Transm Suppl* 1993;**39**:223–33.

229. Pezeshkpour G, Dalakas M. Long-term changes in the spinal cords of patients with old poliomyelitis. *Arch Neurol* 1988;**45**:505–8.

230. Piao Y–S, Wakabayashi K, Kakita A, *et al.* Neuropathology with clinical correlations of sporadic amyotrophic lateral sclerosis: 102 autopsy cases examined between 1962 and 2000. *Brain Pathol* 2003;**12**:10–22.

231. Plaitakis A. Glutamate dysfunction and selective motor neuron degeneration in amyotrophic lateral sclerosis: a hypothesis. *Ann Neurol* 1990;**28**:3–8.

232. Polymenidou M, Lagier–Tourenne C, Hutt K, *et al.* Long pre-mRNA depletion and RNA missplicing contribute to neuronal vulnerability from loss of TDP-43. *Nat Neurosci* 2011;**14**:459–68.

233. Pringle CE, Hudson AJ, Munoz DG, *et al.* Primary lateral sclerosis: clinical features, neuropathology and diagnostic features. *Brain* 1992;**115**:495–520.

234. Puls I, Jonnakuty C, LaMonte B, *et al.* Mutant dynactin in motor neuron disease. *Nat Genet* 2003;**33**:455–6.

235. Radunovic A, Leigh P. Cu/Zn superoxide dismutase gene mutations in amyotrophic lateral sclerosis: correlation between genotype and clinical features. *J Neurol Neurosurg Psychiatry* 1995;**61**:565–72.

236. Ragonese P, Fierro B, Salemi G, *et al.* Prevalence and risk factors of post-polio syndrome in a cohort of polio survivors. *J Neurol Sci* 2005;**236**:31–5.

237. Reid E, Kloos M, Ashley–Kock A, *et al.* A kinesin heavy chain (KIF5A) mutation in hereditary spastic paraplegia (SPG10). *Am J Hum Genet* 2002;**71**:1189–94.

238. Renton A, Majounie E, Waite A, *et al.* A hexanucleotide repeat expansion in C9ORF72 is the cause of chromosome 9p21-Linked ALS-FTD. *Neuron* 2011;**72**:257–68.

239. Roeleveld–Versteerh A, Braun K, Smeitink J, Dorland L, De Koning T. Mitochondrial respiratory chain disease presenting as progressive bulbar paralysis of childhood. *J Inherit Metab Dis* 2004;**27**:281–3.

240. Rosen DR, Siddique T, Patterson D, *et al.* Mutations in Cu/Zn superoxide dismutase gene are associated with familial amyotrophic lateral sclerosis. *Nature* 1993;**362**:59–62.

241. Rossor MN, Revesz T, Lantos PL, *et al.* Semantic dementia with ubiquitin-positive tau-negative inclusion bodies. *Brain* 2000;**123**:267–76.

242. Rothstein JD, Tsai G, Kuncl RW, *et al.* Abnormal excitatory amino acid metabolism in amyotrophic lateral sclerosis. *Ann Neurol* 1990;**28**:18–25.

243. Rouleau GA, Clark AW, Rooke K, *et al.* SOD1 mutation is associated with accumulation of neurofilaments in amyotrophic lateral sclerosis. *Ann Neurol* 1996;**39**:128–31.

244. Rowland L. Amyotrophic lateral sclerosis: theories and therapies. *Ann Neurol* 1994;**35**:129–30.

245. Russman B. Spinal muscular atrophy: clinical classification and disease heterogeneity. *J Child Neurol* 2007;**22**:946–51.

246. Sack G, Huether C, Garg N. Familial spastic paraplegia – clinical and pathologic studies in a large kindred. *John Hopkins Med J* 1978;**143**:117–21.

247. Sakai M, Hashizume Y, Muroga T, *et al.* Neuropathological studies of familial amyotrophic lateral sclerosis with special reference to systemic degeneration of dentato–rubral system and neuronal loss of Onuf's nuclei. *Clin Neurol* 1988;**28**: 1197–204.

248. Salazar AM, Masters CL, Gajdusek C, *et al.* Syndromes of amyotrophic lateral sclerosis and dementia: relation to transmissible Creutzfeldt–Jakob disease. *Ann Neurol* 1983;**14**:17–26.

249. Salinas S, Proukakis C, Crosby A, Warner TT. Hereditary spastic paraplegia: clinical features and pathogenetic mechanisms. *Lancet Neurol* 2008;**7**:1127–38.

250. Sanders KA, Rowland LP, Murphy PL, *et al.* Motor neuron diseases and amyotrophic lateral sclerosis: GM1 antibodies and paraproteinemia. *Neurology* 1993;**43**:418–20.

251. Sapp P, Hosler B, McKenna–Yasek D, *et al.* Identification of two novel loci for dominantly inherited familial amyotrophic lateral sclerosis. *Am J Hum Genet* 2003;**73**:397–403.

252. Sareen D, O'Rourke JG, Meera P, *et al.* Targeting RNA foci in iPSC-derived motor neurons from ALS patients with a C9ORF72 repeat expansion. *Sci Transl Med* 2013;**5**:208ra149.

253. Sasaki S, Tsutsumi Y, Yamane K, *et al.* Sporadic amyotrophic lateral sclerosis with extensive neurological involvement. *Acta Neuropathol* 1992;**84**:211–15.

254. Sasaki S,Maruyama S. Ultrastructural study of Bunina bodies in the anterior horn neurons of patients with amyotrophic lateral sclerosis. *Neurosci Lett* 1993;**154**:117–20.

255. Sasaki S, Maruyama S. Immunocytochemical and ultrastructural studies of the motor cortex in amyotrophic lateral sclerosis. *Acta Neuropathol* 1994;**87**:578–85.

256. Sathasivam S. Brown–Vialetto-Van Laere syndrome. *Orphanet J Rare Dis* 2008;**3**:9.

257. Saugier–Veber P, Munnich A, Bonneau D, *et al.* X-linked spastic paraplegia and Pelizaeus–Mezbacher disease are allelic disorders at the proteolipid protein locus. *Nat Genet* 1994;**6**:257–62.

258. Schiffer D, Autilio–Gambetti L, Chio A, *et al.* Ubiquitin in motor neuron disease: a study at the light and electron microscopical level. *J Neuropathol Exp Neurol* 1991;**50**:463–73.

259. Schiffer D, Cordera S, Cavalla P, *et al.* Reactive astrogliosis of the spinal cord in amyotrophic lateral sclerosis. *J Neurol Sci* 1996;**139**(Suppl):27–33.

260. Schroder HD, Reske–Nielsen E. Preservation of the nucleus X-pelvic

floor motosystem in amyotrophic lateral sclerosis. *Clin Neuropathol* 1984;3:210–6.

261. Schwarz G. Hereditary (familial) spastic paraplegia. *Arch Neurol Psychiatry* 1952;68:655–82.

262. Schwartz G, Liu C. Hereditary (familial) spastic paraplegia: further clinical and pathologic observations. *Arch Neurol Psychiat* 1956;75:144–62.

263. Sephton C, Cenik C, Kucukural A, et al. Identification of neuronal RNA targets of TDP-43-containing ribonucleoprotein complexes. *J Biol Chem* 2010;286:1204–15.

264. Shaw P, Forrest V, Ince P, et al. CSF and plasma amino acid levels in motor neuron disease: elevation of CSF glutamate in a subset of patients. *Neurodegeneration* 1995;4:209–16.

265. Shaw P, Ince P. Glutamate, excitotoxicity and amyotrophic lateral sclerosis. *J Neurol* 1997;244(Suppl 2):S3–14.

266. Shaw P, Thagesen H, Tomkins J, et al. Kennedy's disease: unusual molecular pathologic and clinical features. *Neurology* 1998;51:252–5.

267. Shimura H, Hattori N, Kubo S, et al. Familial Parkinson disease gene product, parkin, is a ubiquitin–protein ligase. *Nat Genet* 2000;25:302–5.

268. Simic G, Seso–Simic D, Lucassen P, et al. Ultrastructural analysis and TUNEL demonstrate motor neuron apoptosis in Werdnig–Hoffmann disease. *J Neuropathol Exp Neurol* 2000;59:398–407.

269. Simic G, Mladinov M, Seso Simic D, et al. Abnormal motoneuron migration, differentiation, and axon outgrowth in spinal muscular atrophy. *Acta Neuropathol* 2008;115:313–26.

270. Skibinski G, Parkinson N, Brown J, et al. Mutations in the endosomal ESCRTIII-complex subunit CHMP2B in frontotemporal dementia. *Nat Genet* 2005;37:806–8.

271. Sobue G, Hashizume Y, Sahashi K, et al. Amyotrophic lateral sclerosis. Lack of central chromatolytic response of motor neurocytons corresponding to active axonal degeneration. *Arch Neurol* 1983;40:306–9.

272. Sobue G, Hashizume Y, Mukai E, et al. X-linked recessive bulbospinal neuronopathy. A clinicopathological study. *Brain* 1989;112:209–32.

273. Sobue G, Hashizume Y, Yasuda T, et al. Phosphorylated high molecular weight neurofilament protein in lower neurons in amyotrophic lateral sclerosis and other neurodegenerative diseases involving ventral horn cells. *Acta Neuropathol* 1990;79:402–8.

274. Soler–Botija C, Ferrer I, Gich I, Baiget M, Tizzano E. Neuronal death is enhanced and begins during foetal development in type I spinal muscular atrophy spinal cord. *Brain* 2002;125:1624–34.

275. Soler–Botija C, Ferrer I, Alvarez J, Baiget M, Tizzano E. Downregulation of Bcl-2 proteins in type 1 spinal muscular atrophy motor neurons during fetal development. *J Neuropathol Exp Neurol* 2003;62:420–6.

276. Sopher B, Thomas P, LaFevre–Bernt M, et al. Androgen receptor YAC transgenic mice recapitulate SBMA motor neuronopathy and implicate VEGF164 in the motor neuron degeneration. *Neuron* 2004;41:687–99.

277. Sreedharan J, Blair I, Tripathi V, et al. TDP-43 mutations in familial and sporadic amyotrophic lateral sclerosis. *Science* 2008;319:1668–72.

278. Stevanin G, Ruberg M, Brice A. Recent advances in the genetics of spastic paraplegias. *Curr Neurol Neurosci Rep* 2008;8:198–210.

279. Strong MJ. Neurofilament metabolism in sporadic amyotrophic lateral sclerosis. *J Neurol Sci* 1999;169:170–7.

280. Strong MJ, Wen W. Role of microglia in amyotrophic lateral sclerosis. In: Shaw P, Strong M eds. *Motor Neuron Disorders.* Philadelphia: Butterworth–Heinemann, 2005:341–55.

281. Strong MJ, Volkening K. TDP-43 and FUS/TLS: sending a complex message about messenger RNA in amyotrophic lateral sclerosis? *FEBS Journal* 2011;278:3569–77.

282. Swingler R, Fraser H, Warlow C. Motor neuron disease and polio in Scotland. *J Neurol Neurosurg Psychiatry* 1992;55:1116–20.

283. Takahashi H, Oyanagi K, Ikuta F, et al. Widespread multiple system degeneration in a patient with familial amyotrophic lateral sclerosis. *J Neurol Sci* 1993;120:15–21.

284. Tallaksen C, Guichart–Gomez E, Verpillat P, et al. Subtle cognitive impairment but no dementia in patients with spastin mutations. *Arch Neurol* 2003;60:1113–8.

285. Tamaoka A, Arai M, Itokawa M, et al. TDP-43 M337V mutation in familial amyotrophic lateral sclerosis in Japan. *Intern Med* 2010;49:331–4.

286. Tateishi T, Hokonohara T, Yamasaki R, et al. Multiple system degeneration with basophilic inclusions in Japanese ALS patients with *FUS* mutation. *Acta Neuropathol* 2010;119:355–64.

287. Tews D, Goebel H. Apoptosis-related proteins in skeletal muscle fibres of spinal muscular atrophy. *J Neuropathol Exp Neurol* 1997;56:150–6.

288. Tizzano E, Cabot C, Baiget M. Cell-specific survival motor neuron gene expression during human development of the central nervous system. *Am J Pathol* 1998;153:355–61.

289. Toft M, Gredal O, Pakkenberg B. The size distribution of neurons in the motor cortex in amyotrophic lateral sclerosis. *J Anat* 2005;207:399–407.

290. Tolnay M, Probst A. Frontotemporal lobar degeneration. An update on clinical, pathological and genetic findings. *Gerontology* 2001;47:1–8.

291. Tomik B, Nicotra A, Ellis C, et al. Phenotypic differences between African and white patients with motor neuron disease: a case control study. *J Neurol Neurosurg Psychiatry* 2000;69:251–3.

292. Toyoshima I, Sugawara M, Kato K, et al. Kinesin and cytoplasmic dynein in spinal spheroids with motor neuron disease. *J Neurol Sci* 1998;159:38–44.

293. Traynor B, Codd M, Corr B, et al. Amyotrophic lateral sclerosis mimic syndromes: a population-based study. *Arch Neurol* 2000;57:109–13.

294. Trojan D, Cashman N. Post-poliomyelitis syndrome. *Muscle Nerve* 2005;31:6–19.

295. Troost D, van den Oord J, Vianney de Jong JM. Immunohistochemical characterization of the inflammatory infiltrate in amyotrophic lateral sclerosis. *Neuropathol Appl Neurobiol* 1990;16:401–10.

296. Troost D, Claessen N, van den Oord J, et al. Neuronophagia in the motor cortex in amyotrophic lateral sclerosis. *Neuropathol Appl Neurobiol* 1993;19:390–7.

297. Tsuchiya K, Shintani S, Kikuchi M, et al. Sporadic amyotrophic lateral sclerosis of long duration mimicking spinal progressive muscular atrophy: a clinicopathological study. *J Neurol Sci* 1999;162:174–8.

298. Tsuchiya K, Matsunaga T, Aoki M, et al. Familial amyotrophic lateral sclerosis with posterior column degeneration and basophilic inclusion bodies: a clinical, genetic and pathological study. *Clin Neuropathol* 2000;20:53–9.

299. Tsuda T, Munthasser S, Fraser PE, et al. Analysis of the functional effects of a mutation in *SOD1* associated with familial amyotrophic lateral sclerosis. *Neuron* 1994;13:727–36.

300. Turner MR, Leigh PN. Disease modifying therapies in motor neuron disorders: the present position and potential future developments. In: Shaw P, Strong M eds. *Motor Neuron Disorders.* Philadelphia: Butterworth–Heinemann, 2005:497–544.

301. Vajsar J, Balslev T, Ray P, Siegel–Bartelt J, Jay V. Congenital cytoplasmic body myopathy with survival motor neuron gene deletion or Werdnig–Hoffmann disease. *Neurology* 1998;51:873–5.

302. Van Deerlin V, Leverenz J, Bekris L, et al. *TARDBP* mutations in amyotrophic lateral sclerosis with TDP-43 neuropathology: a genetic and histopathological analysis. *Lancet Neurol* 2008;7:409–16.

303. Van Lendeghem G, Tabatabaie P, Beckman L, et al. Mn–SOD signal sequence polymorphism associated with sporadic motor neuron disease. *Eur J Neurol* 1999;6:639–44.

304. van Welsem ME, Hogenhuis JA, Meininger V, et al. The relationship between Bunina bodies, skein-like inclusions and neuronal loss in amyotrophic lateral sclerosis. *Acta Neuropathol* 2002;103:583–9.

305. Vance C, Rogelj B, Hortobagyi T, et al. Mutations in *FUS*, an RNA processing protein, cause familial amyotrophic lateral sclerosis type 6. *Science* 2009;323:1208–11

306. Veldink JH, Berg LHvd, Cobben JM, et al. Homozygous deletion of the survival motor neuron 2 gene is a prognostic factor in sporadic ALS. *Neurology* 2001;56:749–52.

307. Vits L, Van Camp G, Coucke P, et al. MASA syndrome is due to mutations in the neural cell adhesion gene *L1CAMNat Genet* 1994;7:408–13.

308. Volkening K, Leystra–Lantz C, Yang W, et al. TAR DNA Binding Protein of 43 kDa (TDP-43), 14-3-3 proteins and copper/zinc superoxide dismutase (SOD1) interact to modulate NFL mRNA stability. Implications for altered RNA processing in amyotrophic lateral sclerosis (ALS). *Brain Res* 2009;1305:168–82.

309. Walter M, Schlaberg R, Hays A, et al. Absence of echovirus sequences in brain and spinal cord of amyotrophic lateral sclerosis patients. *Ann Neurol* 2001;49:249–53.

310. Webb S, Coleman D, Byrne P, et al. Autosomal dominant hereditary spastic paraparesis with cognitive loss linked to chromosome 2p. *Brain* 1998;121:601–9.

311. Wharton S, Ince P. Pathology of motor neuron disorders. In: Shaw P, Strong M eds. *Motor Neuron Disorders.* Philadelphia: Butterworth Heinemann, 2003:17–49.

312. Wharton S, McDermott C, Grierson A, et al. The cellular and molecular pathology of the motor system in hereditary spastic paraparesis due to mutation of the spastin gene. *J Neuropathol Exp Neurol* 2003;**62**:1166–77.

313. White K, Ince P, Lusher M, et al. Clinical and pathologic findings in hereditary spastic paraparesis with spastin mutation. *Neurology* 2000;**55**:89–94.

314. Wightman G, Anderson V, Martin J, et al. Hippocampal and neocortical ubiquitin-immunoreactive inclusions in amyotrophic lateral sclerosis with dementia. *Neurosci Lett* 1992;**139**:269–77.

315. Williams T, Day N, Ince P, Kamboj R, Shaw P. Calcium permeable a-amino-3-hydroxy-5methyl-4-isoxazole propionic acid receptors: a molecular determinant for selective vulnerability in amyotrophic lateral sclerosis. *Ann Neurol* 1997;**42**:200–7.

316. Woehrer A, Laszlo L, Finsterer J, et al. Novel crystalloid oligodendrogliopathy in hereditary spastic paraplegia. *Acta Neuropathol* 2012;**124**:583–91.

317. Workman E, Kolb S, Battle DJ. Spliceosomal small nuclear ribonucleoprotein biogenesis defects and motor neuron selectivity in spinal muscular atrophy. *Brain Res* 2012;**1462**:93–9.

318. Woulfe J. Nuclear bodies in neurodegenerative disease. *Biochim Biophys Acta* 2008;**1783**:2195–206.

319. Yamasaki M, Arita N, Hiraga S, et al. A clinical and neuroradiological study of X-linked hydrocephalus in Japan. *J Neurosurg* 1995;**83**:50–5.

320. Yang Y, Hentati A, Deng H–X, et al. The gene encoding alsin, a protein with three guanine-nucleotide exchange factor domains, is mutated in a form of recessive ALS. *Nat Genet* 2001;**29**:160–5.

321. Yokoseki A, Shiga A, Tan C, et al. TDP-43 mutation in familial amyotrophic lateral sclerosis. *Ann Neurol* 2008;**63**: 538–42.

322. Younger DS, Chou S, Hays AP, et al. Primary lateral sclerosis: a clinical diagnosis reemerges. *Arch Neurol* 1988;**45**:1304–7.

323. Zerres K, Rudnik–Schoneborn S. 93rd ENMC international workshop:non-5q-spinal muscular atrophies (SMA) – clinical picture (6–8 April 2001. Naarden, The Netherlands). *Neuromuscul Disord* 2003;**13**:179–83.

Ageing of the Brain

James Lowe

INTRODUCTION

This chapter will consider changes seen in the otherwise normal but ageing brain.

Ageing has been defined as a complex process of accumulation of molecular, cellular and organ damage leading to a loss of function and an increased vulnerability to disease and death.[37] This is an important area because many diseases affecting the brain develop in older individuals with age as the most significant risk factor. A main issue is in trying to define attributes of normal and successful brain ageing and distinguish these from changes seen in the brains of elderly people who have early forms of neurodegenerative disease. There is great interest in identifying features that are linked to successful normal brain ageing that might be adapted to therapeutic strategies. Cross-sectional studies of the past have now given way to studies using high resolution and functional imaging in longitudinal series using the same individuals with a main aim to identify biomarkers that can reliably predict cognitive decline in the elderly.[44] Epidemiology-based community cohort studies involving longitudinal cognitive assessment with neuropathological and genetic end points are also starting to be informative in relation to the significance of changes seen in the ageing brain.[11] Work in experimental systems, including fly and worm models, is providing information about signalling pathways and genes that are involved in neuronal ageing and this information is being applied to human brain studies.[8]

CLINICAL FEATURES AND CONCEPTS IN BRAIN AGEING

From a clinical perspective, ageing of the brain is usually associated with a slowly progressive decline in cognitive abilities, including memory performance, cognitive speed and flexibility. Three clinical constructs have been used to describe such age-related cognitive changes.[117]

1. Age-associated memory impairment (AAMI) applied to elderly individuals with cognitive complaints and cognitive test scores below those of young adults but not meeting criteria for dementia.
2. Age-associated cognitive decline (AACD) applied to elderly individuals with cognitive complaints and cognitive test scores greater than one standard deviation below the mean for an age-matched, symptom-free cohort but not meeting criteria for dementia.
3. Mild cognitive impairment (MCI) applied to individuals with changing cognitive abilities, objective evidence of cognitive decline but not meeting criteria for dementia.

This age-related cognitive decline is well documented in both cross-sectional and longitudinal studies.[1,15] It has long been presumed that there are underlying changes in brain structure that cause these impairments, but the exact nature of biological changes that lead to age-related cognitive decline are only just starting to be determined. Separate to cognitive impairment linked simply to ageing is the important issue that mild cognitive impairment in some elderly individuals actually represents the early stages of Alzheimer's disease. Whereas AACD implies changes due to simple brain ageing, MCI implies that changes are likely to be related to disease.[66]

Two concepts are widely presented to explain individual variation in the extent to which AACD may develop.

1. Brain reserve capacity (BRC) hypothesis emphasizes an individual's brain structure in terms of brain volume, neuronal density and synaptic density, allowing compensation for age-related effects or disease.[5]
2. The cognitive reserve (CR) hypothesis emphasizes individual differences in the flexibility and adaptability of brain networks and information processing pathways to compensate for age-related effects or disease.[114,123]

A prevailing view is that AACD is related to changes in connectivity of pathways through changes in myelinated axons, dendrites, spines and synapses.[32] This is supported by studies in animal models,[28,88] as well as in functional imaging studies comparing areas of brain activation in young and old individuals while performing the same task.[44] It is acknowledged that the evidence defining a precise neuro-anatomical substrate for AACD is weak.[98]

An important complicating factor in the interpretation of findings seen in the brains of older persons is that many diseases affecting the brain increase in frequency with age, especially AD, and also often coexist with other pathological processes; mixed pathology is common in the elderly.[133] As detailed elsewhere (Chapter 16, Dementia), Alzheimer's disease (AD) is defined clinicopathologically as a disorder in which cognitive impairment is associated with characteristic neuropathological changes in the brain. Yet, AD has a prolonged clinical prodrome, estimated to be at least 5–10 years, in which subtle memory impairments precede clinically detectable dementia. Similarly, Alzheimer-type neuropathological changes, including neurofibrillary tangles and senile plaques, are nearly universal in the brains of individuals over 60 years of age, particularly if vulnerable regions of brain are examined. The extent to which these changes represent preclinical AD or are a part of the normal physiology of ageing remains unknown, although current interpretation reflected in pathological diagnostic recommendations is the former.[73] The approach taken in this chapter on ageing is to differentiate those phenomena that are related to ageing[121] from those that appear to be prodromal or preclinical features of an age-related disease process, such as AD.

This differentiation has a major impact on interpretation and analysis of the older literature. Frequently, analyses of ageing in human brains did not take into account comorbid processes, particularly AD. Unselected cross-sectional studies will inevitably include some cases of preclinical or early AD, and this can confound attempts to study age-related changes in physical parameters, such as cortical atrophy or neuronal number. Studies in the early 1950s suggested that loss of neurons with increasing age was part of normal ageing, and the idea that humans lost neurons as they grew older was widely accepted.[12] Recent studies have challenged this idea. There is a current consensus that, for the most part, neuronal numbers remain stable in cortical and subcortical regions through advanced age, with specific exceptions (see later).

MACROSCOPIC CHANGES IN THE AGEING BRAIN

Brain Atrophy

Healthy elderly individuals tend to have a smaller volume of subcortical structures with a thinner cerebral cortex and larger cerebral ventricles compared with those who are younger.[32] Age-related decreases in brain size and weight have been well documented, but those based on cross-sectional studies need to be interpreted with caution. In particular, the possibility of secular trends in which brain size, like other measures of body mass, could have changed subtly owing to nutrition and other lifestyle changes in a modern cohort compared to a historical cohort cannot be excluded.

Second, the degree to which comorbidity of age-related diseases has been screened for and excluded varies among the studies. Unselected autopsy studies suggest relatively stable brain weight through the 50s and 60s, followed by a decline of several percent per year thereafter.[52] Imaging studies show similar findings.[33,118] It has been suggested that brain volume reductions are at a level of 0.1–0.2 per cent/year at ages of 30–50 increasing to 0.3–0.5 per cent/year over the age of 70 years.[31] Other estimates suggest that the volume of cerebral grey matter declines by about 9 per cent between the ages of 30 and 70 years, increasing to about 12 per cent by the age of 80. White matter volume increases until the mid-50s after which volume declines, such that by 70 years it is about 6 per cent and at 80 years is about 22 per cent less than volumes seen at 30 years.[2]

Although neuropathological studies examining cortical thickness have suggested only minor amounts of grey-matter alteration with normal ageing, if there is no evidence of cognitive decline,[78] longitudinal imaging studies have shown that there are significant regional differences in cortical sub-cortical and cerebellar atrophy seen in ageing with frontal regions appearing especially susceptible.[59,94] This has led to a proposed 'developmental-sensory' model of brain ageing, in which both developmental factors as well as factors specific to different cortical regions influence cortical vulnerability to atrophy.[69] High resolution MR imaging studies have been performed in populations of elderly individuals and different rates of cortical atrophy correlated with cognitive status. These studies show that progressive atrophy links with cognitive decline but that low rates of change are linked with normal cognition or some cases of stable mild age-related cognitive impairment.[25,96,102] Cerebellar atrophy is also seen in ageing and can be associated with cognitive ability, notably when significant atrophy is seen in the vermis.[71]

There has been significant interest in looking for relationships between factors known to be associated with AD and atrophy seen in normal ageing of the brain. Low levels of cerebrospinal fluid (CSF) Aβ1-42 have been linked to pathological features of AD. It has also been found that in non-demented patients, a low CSF level of Aβ1-42 is correlated with loss of brain volume in widespread areas (including those not vulnerable in AD), as well as with ventricular enlargement, suggesting that Aβ1-42 may play a role in the brain changes seen in healthy ageing. However, this is a complex area as in the same study group significant atrophy was also seen in those with high CSF levels of Aβ1-42. This implies that atrophy in healthy ageing is likely to be complex, with only some being related to Aβ1-42.[94]

White Matter Lesions in Ageing

White matter lesions are common and age-related. For example, computed tomography (CT) and magnetic resonance imaging (MRI) studies of age-related white matter changes suggest that there is frequent neuroradiological evidence of decreased white matter volume and foci of increased signal on T2-weighted MR imaging, termed white matter hyperintensities, in the elderly. The prefrontal white matter appears to be the most susceptible to the influence of age.

Two main subgroups of lesion have been described using conventional MR imaging.[101] Periventricular white matter changes appear as smooth periventricular white matter hyperintensities and histological correlation suggests that they are not of vascular origin and mainly link to disruption of the ependymal lining of the lateral ventricle with associated subependymal gliosis. Deep/subcortical white matter changes are of two main types

1. Punctate white matter hyperintensities are usually related to widening of periarteriolar spaces accompanied by rarefaction of myelin staining and loosening of texture of the neuropil around small hyalinized arteries.
2. Confluent white matter hyperintensities usually represent true ischemic lesions seen as perivascular rarefaction of myelin staining, mild to moderate axonal loss, and varying extents of astrocytic gliosis. This is presented in more detail in Chapter 16, Dementia.

Imaging using quantitative MR techniques, such as diffusion tensor imaging (DTI) and magnetization transfer imaging (MTI) is also able to show abnormalities in white matter that are not visible in conventional imaging. This is an area where correlation with histopathology is in the early stages, with studies indicating that increased fractional anisotropy and mean diffusivity derived from DTI and magnetization transfer ratio from MTI correlate with axonal loss, demyelination and dendritic injury.[101]

Studies suggest that there are trends toward higher white matter hyperintensity volume linking with cognitive decline.[48,57]

MICROSCOPIC CHANGES IN AGEING – NEURONAL, GLIAL AND VASCULAR

Histological changes associated with ageing affect meninges, neurons, glial cells and blood vessels. Changes include thickening of the arachnoid and prominence of arachnoid granulations. Age-related neuronal alterations include changes in number, size, perikaryal organelles, dendrites and synapses. Many of these neuronal changes are not associated with clinically detectable cognitive or motor signs or symptoms.

Some alterations (e.g. senile plaques, neurofibrillary tangles, granulovacuolar degeneration and Hirano bodies) are more closely associated with AD and are discussed in Chapter 16, Dementia.

Age-Related Neuronal Loss

The question of whether there is neuronal loss in ageing has been studied intensively in both man and experimental animals. The application of stereological methods has had a major impact on current thinking about neuronal loss with ageing.[131] Stereology is a counting technique developed to overcome unintended biases in density-based counts of objects. In brief, stereological methods involve the systematic random sampling of an anatomically defined region using a counting scheme, such as the optical dissector, and

mathematical rules to provide reliable and reproducible estimates of the total number of neurons, rather than density alone. The method overcomes potential difficulties in tissue fixation, such as shrinkage or swelling. Use of the optical dissector prevents the counting of an object more than once, because a neuron is counted as the user focuses through the depth of the section only if its top is included in the section. A critical requirement for the proper application of this method is the ability to define the anatomical boundary of interest. Because of difficulties of defining precise anatomical boundaries, it has been difficult to apply stereology to certain cortical and subcortical regions. It has been asserted that studies that do not meet these stereologic specifications are subject to statistical biases and, hence, misinterpretation;[130] however, in most cases there is reasonably good agreement between stereological counts and density measurements.

Overall, studies suggest significant preservation of neurons across the age span.[53] A large, unselected autopsy series of 91 brains[84] suggested a loss of 10 per cent of cortical neurons over the entire lifespan. Stereologically based studies of the superior temporal sulcus, a high-order association cortex, did not suggest loss of neurons in this region between the seventh and tenth decades.[42] Similarly, studies of the entorhinal cortex in normal individuals failed to reveal any change in neuronal number,[41,76] even when examined on a lamina-by-lamina basis. These stereological studies support previous measurements of neuronal density, which found no evidence of age-associated neuronal loss from the frontal, temporal or parietal lobes.[120] Investigations on the non-human primate visual cortex[89] showed no loss of neurons with advancing age.

Age-related loss of neurons may occur in specific subfields of the hippocampus. In two studies, a subtle atrophy and loss of neurons either in CA1 and subiculum[108] or CA1 and CA4[132] distinguished ageing from the marked neuronal losses seen in AD. In man, atrophy and neuronal loss in CA1 and the subiculum of hippocampus are difficult to interpret without a full clinical history and neuropathological study to exclude another cause of hippocampal neuronal loss, such as ischaemia or epilepsy. In the rat, there is no change in neuronal number in the hippocampus with ageing.[93,131] Similarly, in the non-human primate, there is no loss of neurons from the parahippocampal gyrus with advanced age.[70] Taken together, these studies suggest that in both human and animal models there is little or no neocortical neuronal loss with age. The same conclusion holds true for the limbic regions, with the possible exception of CA1 in the hippocampus.

Purkinje cells may show significant changes with ageing including loss of cells, cell atrophy, reduced dendritic arborization and reduced synapse density.[135]

Subcortical structures have also been examined. There is no change in the number of neurons in the putamen between the ages of 62 and 92 years.[86] Similarly, neuronal loss with increasing age has not been observed in the medial mammillary nucleus[7] or the hypothalamus.[75] The cholinergic systems including the nucleus basalis of Meynert are also felt to maintain neuronal density but may be subject to loss of connectivity,[18,49,100] in both man and non-human primates.[129] Some controversy exists about neuronal loss from the noradrenergic locus coeruleus. One analysis suggested marked neuronal loss, especially amongst three centenarians,[64] whereas other investigators did not find any change with age in the locus coeruleus.[77] Similarly, Ohm et al.[81] did not observe neuronal loss with age, but noted a nearly twofold variation between brains, indicating the difficulties in interpreting small studies.

The substantia nigra pars compacta appears to be a special case in terms of age-related neuronal loss. The consensus among available studies is that there is some degree of neuronal loss from the substantia nigra with increasing age. In monkeys, there is a 50 per cent loss of tyrosine hydroxylase-immunoreactive neurons with increasing age, and the degree of loss correlates with specific motor behaviours.[29,107] In man, Ma et al.[61] reported 10 per cent loss of nigral neurons per decade with increasing age. In contrast to these studies, Pakkenberg et al.[85] did not detect a difference between young and old rhesus monkeys in the total number of neurons in the substantia nigra.

Thus, the current literature supports the conclusion that there is little or no age-related neuronal loss in the vast majority of brain regions examined to date.[90] Nonetheless, it remains possible that specific nuclei, particularly subcortical nuclei, may be sensitive to age-related neuronal loss.[56] Moreover, recent studies suggest that the commonly held perception that neurons lack regenerative capacity in the adult brain may be incorrect. In both man and non-human primates, some neuronal neogenesis appears to occur in select brain regions, such as the dentate fascia of the hippocampus, even in the adult brain.[3,21,24,30,43] The extent to which such neogenesis might offset age-related neuronal loss is speculative,[38] because it has not been demonstrated in subcortical regions vulnerable to age-related neuronal loss.

Age-Related Dendritic, Axonal and Synaptic Changes

Ageing is associated with widespread neuronal atrophy, especially in cortical neuron.[110,113,120] There appear to be species-specific and region-specific differences in the way in which neuronal populations change with age. Dendritic structure is relatively stable with age in the entorhinal cortex in the rat,[19] but there is loss of dendritic complexity with age in monkeys in the subiculum of the hippocampus.[124] Older studies in human material have probably been complicated by the inclusion of samples from elderly subjects with neurodegenerative disease leading to conclusions that dendritic pathology develops in ageing. More recent studies suggest that small changes in both dendritic branching and spine density develop in association with ageing, but that this is region-specific.

Dendritic extent has been found to increase in dentate gyrus granule cells between the sixth and eighth decades and in contrast dendritic regression is then subsequently seen in the tenth decade.[35] Other areas of the hippocampus have shown no changes in dendritic branching with age.[34,36,50] Reduced dendritic branching has been shown in pyramidal neurons in cortical layer V, but not layer 3 of the frontal cortex.[23,125] The spine density on neocortical neurons appears to reduce with normal ageing.[125]

Studies of synaptic changes with age have often used immunohistochemical or immunochemical measures of synaptic proteins, particularly synaptophysin. These studies show a slight decrease in synaptic density with age,[67] but striking variability in old age that is not present in younger age groups suggesting the possibility that there may be contamination of the older cohorts by individuals with preclinical or subclinical disease.

These age-related structural changes have been related to physiological changes in neuronal function including changes to receptors and alterations in synaptic transmission.[27]

Swollen dystrophic axons, so-called axonal spheroids, increase with age (age-related neuroaxonal dystrophy). They are most common in the globus pallidus, pars reticularis of the substantia nigra and dorsal column nuclei of the caudal medulla. Scattered axonal spheroids are also common in the spinal cord anterior horns. They can be detected in a variety of histological methods including periodic acid–Schiff (PAS) stains. They are also immunopositive for ubiquitin. Spherical granular bodies, believed to be located within dystrophic neurites, are most numerous in the middle and upper cortical layers, especially layer II of the entorhinal cortex and the amygdala.

Neuronal Lipofuscin Pigment

Lipofuscin-filled processes of large pyramidal cells increase with age.[10] Lipofuscin is a common cytoplasmic pigment (see Chapter 1, General Pathology of the Central Nervous System). Lipofuscin granules, believed to be residual bodies derived from the lysosomal system, accumulate in variable degrees within neurons in different regions of the nervous system.[65] For example, lipofuscin is often prominent in cranial and spinal motor nuclei, the red nucleus, lateral geniculate nucleus, globus pallidus, inferior olivary nucleus and cerebellar dentate nucleus. In many sites, lipofuscin increases with age.[13] Certain nuclei (e.g. inferior olivary nucleus and anterior horn cells of the spinal cord) exhibit abundant lipofuscin granules even in very young children or infants.[13,54] Lipofuscin has a light brown to yellow granular appearance in conventional haematoxylin and eosin (H&E) preparations, and is well demonstrated with PAS or acid-fast stains. When using immunofluorescent techniques, it is important to be aware that lipofuscin has autofluorescent properties and may be mistaken for fluorochrome-labelled structures. By electron microscopy, lipofuscin has a characteristic bipartite appearance with electron-dense and electron-lucent components.

The significance of accumulation of lipofuscin has come under scrutiny with the understanding of the interdependence of cellular protein degradation pathways.[55] Lipofuscin may contribute to deranged intracellular protein elimination and its accumulation in ageing and neurodegenerative disease may be a signpost for what has been termed a 'garbage catastrophe'.[45] On the other hand, neurons that accumulate abundant lipofuscin seem to be little affected by age-related neuronal loss. It is possible that the accumulation of lipofuscin may interfere with autophagic pathways, thereby compromising other functions of this pathway including mitophagy.[116] The proteomic footprint of lipofuscin indicates that it is mainly derived from proteins of mitochondrial, cytoskeletal and cell membrane origins.[83]

Age-Related Glial Changes

Astrocytic changes may be prominent in ageing; an increase in astrocytes and in glial fibrillary acidic protein (GFAP) expression has been demonstrated in human[22,51] and animal[58] brains. Astrocytes in the aged brain may show features that have been compared to a so-called senescence-associated secretory phenotype: (1) elevated expression of glial fibrillary acidic protein and vimentin, (2) elevated expression of several cytokines and (3) elevated accumulation of proteotoxic aggregates. It has been suggested that such astrocytic changes might also contribute to age-related neuroinflammation and secondary neuronal degeneration.[97]

Corpora amylacea (see Chapter 1, General Pathology of the Central Nervous System), commonly encountered in the brains of elderly individuals, are spherical, basophilic to eosinophilic structures within astrocytic processes around blood vessels and in the subependymal and subpial spaces.[79] They are often abundant at the base of the brain, in the olfactory tract and in the spinal cord white matter. They are not detected in young children, but increase in frequency and density with age. Corpora amylacea are strongly positive with the PAS stain, reflecting their content of polyglucosan.

Thorn-shaped astrocytes, which are argyrophilic and immunoreactive for tau protein, increase in frequency with ageing, and by the eighth decade are detected in approximately 50 per cent of individuals.[103] Thorn-shaped astrocytes are most numerous at the base of the brain in subpial, perivascular and subependymal regions.

Age-related changes in myelin, formed by oligodendrocytes, are observed in ageing. Myelin sheaths show degenerative changes including the formation on myelin balloons and the development of splits containing cytoplasm. It is suggested that such degenerative changes might contribute to cognitive decline as a result of reduced conduction velocity.[87]

Microglia are normally inconspicuous in the young brain, but with ageing, microglia, particularly those in white matter, show signs of activation. In particular, expression of class II major histocompatibility antigen increases with age in man and in animal models.[105,109,128] It has been suggested that age-related changes in microglial cells may be a form of degenerative pathology[20] and cause local cytotoxicity.[115,128]

Microscopic Vascular Pathology in the Aged Brains

Cerebral atherosclerosis, small vessel disease and cerebral amyloid angiopathy are the most common arterial disorders seen in the in brain of elderly individuals,[46] however these conditions are best regarded as specific diseases of vessels affecting the brain rather than an association of normal brain ageing.

Current evidence suggests that there is decreased vascular density in ageing.[14] With age, small venules in the periventricular white matter tend to show increased deposition of collagen (so-called collagenosis)[14,74] characterized by collagen deposition in the adventitia. Other common vascular alterations in the aged brain include arteriosclerotic hyalinosis—collagen deposition in the media and adventitia of small (100 μm or less in diameter) arteries and arterioles—in

the white matter, basal ganglia and thalamus.[14] This change is particularly associated with hypertension, in which vessels may show fibrinoid changes and formation of microaneurysms—uncommon in normal ageing. Some mild dilation of perivascular Virchow–Robin spaces (so-called cribriform change) is common in association with arteriosclerotic hyalinosis, particularly in cerebral white matter and basal ganglia (see also Chapter 2, Vascular Disease, Hypoxia and Related Conditions). This change is also visible in imaging studies, and, although it may be seen in patients with normal cognition, it is more evident in the population showing cognitive dysfunction.[17,47,62,137]

An increased permeability of the blood–brain barrier is seen in ageing in healthy individuals. The consequences of this may link to development of white matter lesions, the accumulation of iron in astrocytes, and oxidative stress.[91]

MOLECULAR CHANGES IN BRAIN AGEING

Molecular changes seen in brain ageing have been reviewed together with factors that are recognized as influencing ageing in lower and higher organisms.[8,31] Several common cellular pathways are seen as being pivotal in healthy whole body ageing in many species:

- insulin/IGF-1 signalling, where reduced activity extends lifespan;
- target of rapamycin (TOR) signalling, where reduced activity extends lifespan;
- calorie-restricted diets are consistently associated with increased lifespan in many species;[112]
- modestly decreased mitochondrial function causes increased lifespan;
- sirtuins, where increased expression of sirtuins extends lifespan.

It is presently not clear which of the systems that are active in whole body ageing are also pivotal in regulating brain ageing in man. Studies of changes in gene expression in the ageing brain of many species show significant differences in the patterns of genes that are either upregulated or downregulated, which suggests that it may be difficult to extrapolate some model findings to human ageing. In addition to suggestions of involvement of the pathways and factors listed earlier, activation of cell stress and cytoprotective responses,[106] changes in calcium handling, iron accumulation and changes in protein degradation systems appear to be a feature of human brain ageing.

Protein Degradation Systems and Autophagy

The majority of cellular proteins are degraded by the ubiquitin-proteasome system (UPS) alongside involvement of the processes of autophagy. Studies have implicated autophagy as a regulator of ageing with increased autophagy linked to extended lifespan. Model systems in which genes linked to autophagy are deleted show neurodegeneration. Activation of the TOR (mammalian target of rapamycin) pathway is recognized to inhibit autophagy and in model systems inhibition of the TOR pathway extends lifespan associated with an increase in autophagy.[63]

The quality control of protein homoeostasis deteriorates with ageing, although the exact underlying mechanism is unclear. It appears that there is a decline in the activity of proteasomes, which is in part due to decreased function of proteasome in senescence and is primarily due to the reduced rates of proteasome biosynthesis and assembly.[60] There may be additional factors in neurons that affect proteasomal function. Accumulation of frameshift mutant ubiquitin is seen in aged brains and it is recognized that this inhibits proteasomal degradation of abnormal proteins. It has been suggested that this may be a mechanism that develops with ageing and that in part might cause accumulation of abnormal proteins in neurons with age.[16,119,126,127]

Ubiquitin-immunopositive structures, including granular or dot-like structures, are commonly encountered in white matter in the ageing brain.[26] With immunoelectron microscopy, ubiquitin immunoreactive foci of granular degeneration of myelin have been located within glial cells and expansion of myelin lamellae. Oligodendroglia appear to contain the brunt of the white matter ubiquitin immunoreactive material that accumulates with age. Granular degeneration of myelin is an extremely common consequence of ageing in man and animals.

Mitochondrial Factors

Several signalling pathways that are known to have an effect on lifespan (IGF-1, TOR, sirtuins) converge to inhibit oxidative stresses within mitochondria, suggesting that mitochondrial factors may be part of a final common pathway in the development of age-related changes in tissues.[92] Many studies have shown that a factor common to brain ageing in many species is reduction in mitochondrial functioning. Compared to brain from younger individuals, aged brain shows development of mtDNA defects, increased production of reactive oxygen species (ROS), and reduced mitochondrial respiratory function. It is likely that defects in electron transport lead to generation of reactive oxygen species and a downstream cascade of damage through oxidative stress. Studies have shown that pathways responding to oxidative stress are consistently activated in the ageing brain. Changes are also seen in mitochondrial dynamics (mitophagy and fission), which may be linked to changes seen in systems regulating autophagy.[80,95]

Insulin/IGF-1 Signalling

It is well recognized that inhibition of the IIS (insulin/IGF-1 (insulin-like growth factor 1)-like signalling) pathway is linked to extended lifespan in many species. Longevity increases when the transcription factor FoxO becomes activated in response to reduced insulin/IGF-1 signalling. The IIS pathway is also linked to regulation of protein homoeostasis, learning and memory and has been proposed as a key factor in normal brain ageing as well as some aspects of neurodegeneration.[82] There is speculation that some of the effects of a calorie-restricted diet are mediated by suppression of the insulin/IGF-1 and/or TOR signalling pathways. Studies in rats have not shown a uniform change in insulin signalling in different tissues in response to a calorie-restricted diet.[104]

Dysregulation of Calcium

Ca[2+] signalling is involved in a wide variety of cellular physiological processes and the calcium hypothesis of brain ageing was originally put forward in the 1980s. This has been recently reviewed and aligned with more recent concepts.[122] The central idea is that dysregulation of the triad: Ca[2+] buffering–mitochondrial function–free radical production causes a reduced mitochondrial functional reserve in neurons. This lack of functional reserve then makes neuronal populations more vulnerable to other stressors. Age-related changes in Ca[2+] homeostasis are also proposed to cause changes in neuronal excitability and affect synaptic plasticity.[122]

Accumulation of Iron in Brain Ageing

Iron accumulates in the ageing brain (reviewed in Aquino et al.[4] and Zecca et al.[134]) and can be seen in microglia and astrocytes with Perl's stain. MR imaging offers an important method for detecting iron in brain regions and has allowed cross-sectional studies to be performed that reinforce earlier studies done in post-mortem brain tissue. It has been shown that genetic variants in iron metabolism genes can influence brain iron levels in men.[6] In older adults, iron is mainly found in the globus pallidus (GP), the red nucleus, and the pars reticulata of the substantia nigra (SN), with smaller amounts seen in the dentate nucleus of the cerebellum and the neostriatum. Iron accumulation is greatest in the SN, followed by the GP, putamen and caudate nucleus. The caudate nucleus does not show iron accumulation at any age. In the putamen, iron accumulation starts in the lateral portion and eventually reaches medial portion after age 50 years (reviewed in Aquino et al.[4]). Iron accumulation in the substantia nigra has been proposed as a factor that might render this region vulnerable to additional stressors and lead to Parkinson's disease.[111] Iron accumulation has been shown in association with age-related white matter lesions with the suggestion that this might potentiate local tissue damage through generation of ROS.[40]

A variety of mechanisms has been suggested whereby local accumulation of iron leads to toxic effects, most notably through a cascade of events following the generation of reactive oxygen species.[99]

Sirtuins and Brain Ageing

Silent information regulator two proteins (sirtuins or SIRTs SIRT1-7; class III histone deactylases) are a group of histone deacetylases that act as important regulators of transcription including upregulating factors involved in metabolism and cell survival in response to metabolic changes. SIRT1 is an important regulator of transcription in response to cellular stress[72] and in part mediates the longevity effects seen in mammalian species linked to a calorie-restricted diet.[136] This factor has been found to have neuroprotective effects in models of neurodegenerative disease;[136] there are suggestions of involvement with some of the pathways[9] important in Alzheimer's disease. SIRT2 has been shown to accumulate in aged mouse brain,[68] and there are suggestions that this might inhibit autophagic and aggresomal protein degradation pathways.[39] The role of this class of proteins in normal ageing and whether changes explain in part age-related vulnerability to disease awaits further investigation.

ACKNOWLEDGEMENTS

Parts of this chapter were adapted from a section that appeared in the eighth edition of *Greenfield's Neuropathology*, co-authored by Suzanne S Mirra, Bradley T Hyman and Dennis W Dickson. Their scholarship is greatly acknowledged.

REFERENCES

1. Albert MS. The ageing brain: normal and abnormal memory. *Phil Trans R Soc Lond B Biol Sci* 1997;**352**:1703–9.

2. Allen JS, Bruss J, Brown CK, Damasio H. Normal neuroanatomical variation due to age: the major lobes and a parcellation of the temporal region. *Neurobiol Aging* 2005;**26**:1245–60; discussion 79–82.

3. Amrein I, Isler K, Lipp HP. Comparing adult hippocampal neurogenesis in mammalian species and orders: influence of chronological age and life history stage. *Eur J Neurosci* 2011;**34**:978–87.

4. Aquino D, Bizzi A, Grisoli M, *et al.* Age-related iron deposition in the basal ganglia: quantitative analysis in healthy subjects. *Radiology* 2009;**252**:165–72.

5. Bartres-Faz D, Arenaza-Urquijo EM. Structural and functional imaging correlates of cognitive and brain reserve hypotheses in healthy and pathological aging. *Brain Topogr* 2011;**24**:340–57.

6. Bartzokis G, Lu PH, Tishler TA, *et al.* Prevalent iron metabolism gene variants associated with increased brain ferritin iron in healthy older men. *J Alzheimers Dis* 2010;**20**:333–41.

7. Begega A, Cuesta M, Santin LJ, *et al.* Unbiased estimation of the total number of nervous cells and volume of medial mamillary nucleus in humans. *Exp Gerontol* 1999;**34**:771–82.

8. Bishop NA, Lu T, Yankner BA. Neural mechanisms of ageing and cognitive decline. *Nature* 2010;**464**:529–35.

9. Bonda DJ, Lee HG, Camins A, *et al.* The sirtuin pathway in ageing and Alzheimer disease: mechanistic and therapeutic considerations. *Lancet Neurol* 2011;**10**:275–9.

10. Braak H. Spindle-shaped appendages of IIIab-pyramids filled with lipofuscin: a striking pathological change of the senescent human isocortex. *Acta Neuropathol (Berl)* 1979;**46**:197–202.

11. Brayne C, Barker RA, Grupe A, *et al.* From molecule to clinic and community for neurodegeneration: research to bridge translational gaps. *J Alzheimers Dis* 2013;**33**(Suppl 1):S385–96.

12. Brody H. Organization of the cerebral cortex. III. A study of aging in the human cerebral cortex. *J Comp Neurol* 1955;**102**:511–56.

13. Brody H. The deposition of aging pigment in the human cerebral cortex. *J Gerontol* 1960;**15**:258–61.

14. Brown WR, Thore CR. Review: cerebral microvascular pathology in ageing and neurodegeneration. *Neuropathol Appl Neurobiol* 2011;**37**:56–74.

15. Buschke H, Sliwinski M, Kuslansky G, Lipton RB. Aging, encoding specificity, and memory change in the Double Memory Test. *J Int Neuropsychol Soc* 1995;**1**:483–93.

16. Chadwick L, Gentle L, Strachan J, Layfield R. Review: unchained maladie— a reassessment of the role of Ubb(+1)-capped polyubiquitin chains in Alzheimer's disease. *Neuropathol Appl Neurobiol* 2012;**38**:118–31.

17. Chen W, Song X, Zhang Y. Alzheimer's disease neuroimaging I. Assessment of the Virchow-Robin Spaces in Alzheimer disease, mild cognitive impairment, and normal aging, using high-field MR imaging. *AJNR Am J Neuroradiol* 2011;**32**:1490–95.

18. Chui HC, Bondareff W, Zarow C, Slager U. Stability of neuronal number in the human nucleus basalis of Meynert with age. *Neurobiol Aging* 1984;**5**:83–8.

19. Coleman PD, Flood DG. Net dendritic stability of layer II pyramidal neurons in F344 rat entorhinal cortex from 12 to 37 months. *Neurobiol Aging* 1991;**12**:535–41.

20. Conde JR, Streit WJ. Microglia in the aging brain. *J Neuropathol Exp Neurol* 2006;**65**:199–203.

21. Couillard-Despres S, Iglseder B, Aigner L. Neurogenesis, cellular plasticity and cognition: the impact of stem cells in the adult and aging brain: a mini-review. *Gerontology* 2011;**57**:559–64.

22. David JP, Ghozali F, Fallet-Bianco C, *et al.* Glial reaction in the hippocampal formation is highly correlated with aging in human brain. *Neurosci Lett* 1997;**235**:53–6.

23. de Brabander JM, Kramers RJ, Uylings HB. Layer-specific dendritic regression of pyramidal cells with ageing in the human prefrontal cortex. *Eur J Neurosci* 1998;**10**:1261–9.

24. Decimo I, Bifari F, Krampera M, Fumagalli G. Neural stem cell niches in health and diseases. *Curr Pharm Des* 2012;**18**:1755–83.

25. den Heijer T, van der Lijn F, Koudstaal PJ, *et al.* A 10-year follow-up of hippocampal volume on magnetic resonance imaging in early dementia and cognitive decline. *Brain* 2010;**133**(Part 4):1163–72.

26. Dickson DW, Wertkin A, Kress Y, Ksiezak-Reding H, Yen SH. Ubiquitin immunoreactive structures in normal human brains. Distribution and developmental aspects. *Lab Invest* 1990;**63**:87–99.

27. Dickstein DL, Kabaso D, Rocher AB, *et al.* Changes in the structural complexity of the aged brain. *Aging Cell* 2007;**6**:275–84.

28. Dumitriu D, Hao J, Hara Y, *et al.* Selective changes in thin spine density and morphology in monkey prefrontal cortex correlate with aging-related cognitive impairment. *J Neurosci* 2010;**30**:7507–15.

29. Emborg ME, Ma SY, Mufson EJ, *et al.* Age-related declines in nigral neuronal function correlate with motor impairments in rhesus monkeys. *J Comp Neurol* 1998;**401**:253–65.

30. Eriksson PS, Perfilieva E, Bjork-Eriksson T, *et al.* Neurogenesis in the adult human hippocampus. *Nat Med* 1998;**4**:1313–17.

31. Esiri MM. Ageing and the brain. *J Pathol* 2007;**211**:181–7.

32. Fjell AM, Walhovd KB. Structural brain changes in aging: courses, causes and cognitive consequences. *Rev Neurosci* 2010;**21**:187–221.

33. Fjell AM, Westlye LT, Amlien I, *et al.* High consistency of regional cortical thinning in aging across multiple samples. *Cereb Cortex* 2009;**19**:2001–12.

34. Flood DG. Region-specific stability of dendritic extent in normal human aging and regression in Alzheimer's disease. II. Subiculum. *Brain Res* 1991;**540**:83–95.

35. Flood DG, Buell SJ, Defiore CH, Horwitz GJ, Coleman PD. Age-related dendritic growth in dentate gyrus of human brain is followed by regression in the 'oldest old'. *Brain Res* 1985;**345**:366–8.

36. Flood DG, Guarnaccia M, Coleman PD. Dendritic extent in human CA2-3 hippocampal pyramidal neurons in normal aging and senile dementia. *Brain Res* 1987;**409**:88–96.

37. Fontana L, Partridge L, Longo VD. Extending healthy life span: from yeast to humans. *Science* 2010;**328**:321–6.

38. Fotuhi M, Do D, Jack C. Modifiable factors that alter the size of the hippocampus with ageing. *Nature Rev Neurol* 2012;**8**:189–202.

39. Gal J, Bang Y, Choi HJ. SIRT2 interferes with autophagy-mediated degradation of protein aggregates in neuronal cells under proteasome inhibition. *Neurochem Int* 2012;**61**:992–1000.

40. Gebril OH, Simpson JE, Kirby J, Brayne C, Ince PG. Brain iron dysregulation and the risk of ageing white matter lesions. *Neuromolecular Med* 2011;**13**:289–99.

41. Gomez-Isla T, Price JL, McKeel DW, Jr, *et al.* Profound loss of layer II entorhinal cortex neurons occurs in very mild Alzheimer's disease. *J Neurosci* 1996;**16**:4491–500.

42. Gomez-Isla T, Hollister R, West H, *et al.* Neuronal loss correlates with but exceeds neurofibrillary tangles in Alzheimer's disease. *Ann Neurol* 1997;**41**:17–24.

43. Gould E, Reeves AJ, Fallah M, *et al.* Hippocampal neurogenesis in adult Old World primates. *Proc Natl Acad Sci U S A* 1999;**96**:5263–7.

44. Grady C. The cognitive neuroscience of ageing. *Nat Rev Neurosci* 2012;**13**:491–505.

45. Gray DA, Woulfe J. Lipofuscin and aging: a matter of toxic waste. *Sci Aging Knowledge Environ* 2005;**5**:re1.

46. Grinberg LT, Thal DR. Vascular pathology in the aged human brain. *Acta Neuropathol* 2010;**119**:277–90.

47. Groeschel S, Chong WK, Surtees R, Hanefeld F. Virchow-Robin spaces on magnetic resonance images: normative data, their dilatation, and a review of the literature. *Neuroradiology* 2006;**48**:745–54.

48. Gunning-Dixon FM, Brickman AM, Cheng JC, Alexopoulos GS. Aging of cerebral white matter: a review of MRI findings. *Int J Geriatr Psychiatry* 2009;**24**:109–17.

49. Halliday GM, Cullen K, Cairns MJ. Quantitation and three-dimensional reconstruction of Ch4 nucleus in the human basal forebrain. *Synapse* 1993;**15**:1–16.

50. Hanks SD, Flood DG. Region-specific stability of dendritic extent in normal human aging and regression in Alzheimer's disease. I. CA1 of hippocampus. *Brain Res* 1991;**540**:63–82.

51. Hansen LA, Armstrong DM, Terry RD. An immunohistochemical quantification of fibrous astrocytes in the aging human cerebral cortex. *Neurobiol Aging* 1987;**8**:1–6.

52. Ho KC, Roessmann U, Straumfjord JV, Monroe G. Analysis of brain weight. I. Adult brain weight in relation to sex, race, and age. *Arch Pathol Lab Med* 1980;**104**:635–9.

53. Hof PR, Morrison JH. The aging brain: morphomolecular senescence of cortical circuits. *Trends Neurosci* 2004;**27**:607–13.

54. Humphrey T. Primitive neurons in the embryonic human central nervous system. *J Comp Neurol* 1944;**81**:1–45.

55. Keller JN, Dimayuga E, Chen Q, *et al.* Autophagy, proteasomes, lipofuscin, and oxidative stress in the aging brain. *Int J Biochem Cell Biol* 2004;**36**:2376–91.

56. Kemper TL, Moss MB, Rosene DL, Killiany RJ. Age-related neuronal loss in the nucleus centralis superior of the rhesus monkey. *Acta Neuropathol (Berl)* 1997;**94**:124–30.

57. Kim JH, Hwang KJ, Kim JH, *et al.* Regional white matter hyperintensities in normal aging, single domain amnestic mild cognitive impairment, and mild Alzheimer's disease. *J Clin Neurosci* 2011;**18**:1101–6.

58. Kohama SG, Goss JR, Finch CE, McNeill TH. Increases of glial fibrillary acidic protein in the aging female mouse brain. *Neurobiol Aging* 1995;**16**:59–67.

59. Lemaitre H, Goldman AL, Sambataro F, *et al.* Normal age-related brain morphometric changes: nonuniformity across cortical thickness, surface area and gray matter volume? *Neurobiol Aging* 2012;**33**:617.e1–9.

60. Low P. The role of ubiquitin-proteasome system in ageing. *Gen Comp Endocrinol* 2011;**172**:39–43.

61. Ma SY, Roytt M, Collan Y, Rinne JO. Unbiased morphometrical measurements show loss of pigmented nigral neurones with ageing. *Neuropathol Appl Neurobiol* 1999;**25**:394–9.

62. Maclullich AM, Wardlaw JM, Ferguson KJ, *et al.* Enlarged perivascular spaces are associated with cognitive function in healthy elderly men. *J Neurol Neurosurg Psychiatry* 2004;**75**:1519–23.

63. Madeo F, Tavernarakis N, Kroemer G. Can autophagy promote longevity? *Nat Cell Biol* 2010;**12**:842–6.

64. Manaye KF, McIntire DD, Mann DM, German DC. Locus coeruleus cell loss in the aging human brain: a non-random process. *J Comp Neurol* 1995;**358**:79–87.

65. Mann D, Yates P. Ageing nucleic acids and pigments. In: Cavanagh J ed. *Recent advances in neuropathology.* Edinburgh: Churchill Livingstone, 1987:109–37.

66. Mariani E, Monastero R, Mecocci P. Mild cognitive impairment: a systematic review. *J Alzheimers Dis* 2007;**12**:23–35.

67. Masliah E, Mallory M, Hansen L, DeTeresa R, Terry RD. Quantitative synaptic alterations in the human neocortex during normal aging. *Neurology* 1993;**43**:192–7.

68. Maxwell MM, Tomkinson EM, Nobles J, *et al.* The Sirtuin 2 microtubule deacetylase is an abundant neuronal protein that accumulates in the aging CNS. *Hum Mol Genet* 2011;**20**:3986–96.

69. McGinnis SM, Brickhouse M, Pascual B, Dickerson BC. Age-related changes in the thickness of cortical zones in humans. *Brain Topogr* 2011;**24**:279–91.

70. Merrill DA, Roberts JA, Tuszynski MH. Conservation of neuron number and size in entorhinal cortex layers II, III, and V/VI of aged primates. *J Comp Neurol* 2000;**422**:396–401.

71. Miller TD, Ferguson KJ, Reid LM, *et al.* Cerebellar vermis size and cognitive ability in community-dwelling elderly men. *Cerebellum* 2013;**12**:68–73.

72. Monteiro JP, Cano MI. SIRT1 deacetylase activity and the maintenance of protein homeostasis in response to stress: an overview. *Protein Pept Lett* 2011;**18**:167–73.

73. Montine TJ, Phelps CH, Beach TG, *et al.* National Institute on Aging-Alzheimer's Association guidelines for the neuropathologic assessment of Alzheimer's disease: a practical approach. *Acta Neuropathol* 2012;**123**:1–11.

74. Moody DM, Brown WR, Challa VR, Anderson RL. Periventricular venous collagenosis: association with leukoaraiosis. *Radiology* 1995;**194**:469–76.

75. Morys J, Dziewiatkowski J, Switka A, *et al.* Morphometric parameters of some hypothalamic nuclei: age-related changes. *Folia Morphol* 1994;**53**:221–9.

76. Morys J, Sadowski M, Barcikowska M, *et al.* The second layer neurones of the entorhinal cortex and the perforant path in physiological ageing and Alzheimer's disease. *Acta Neurobiol Exp* 1994;**54**:47–53.

77. Mouton PR, Pakkenberg B, Gundersen HJ, Price DL. Absolute number and size of pigmented locus coeruleus neurons in young and aged individuals. *J Chem Neuroanat* 1994;**7**:185–90.

78. Mouton PR, Martin LJ, Calhoun ME, *et al.* Cognitive decline strongly correlates with cortical atrophy in Alzheimer's dementia. *Neurobiol Aging* 1998;**19**:371–7.

79. Mrak RE, Griffin ST, Graham DI. Aging-associated changes in human brain. *J Neuropathol Exp Neurol* 1997;**56**:1269–75.

80. Muller WE, Eckert A, Kurz C, Eckert GP, Leuner K. Mitochondrial dysfunction: common final pathway in brain aging and Alzheimer's disease–therapeutic aspects. *Mol Neurobiol* 2010;**41**:159–71.

81. Ohm TG, Busch C, Bohl J. Unbiased estimation of neuronal numbers in the human nucleus coeruleus during aging. *Neurobiol Aging* 1997;**18**:393–9.

82. O'Neill C, Kiely AP, Coakley MF, Manning S, Long-Smith CM. Insulin and IGF-1 signalling: longevity, protein homoeostasis and Alzheimer's disease. *Biochem Soc Trans* 2012;**40**:721–7.

83. Ottis P, Koppe K, Onisko B, *et al.* Human and rat brain lipofuscin proteome. *Proteomics* 2012;**12**:2445–54.

84. Pakkenberg B, Gundersen HJ. Neocortical neuron number in humans: effect of sex and age. *J Comp Neurol* 1997;**384**:312–20.

85. Pakkenberg H, Andersen BB, Burns RS, Pakkenberg B. A stereological study of substantia nigra in young and old rhesus monkeys. *Brain Res* 1995;**693**:201–6.

86. Pesce C, Reale A. Aging and the nerve cell population of the putamen: a morphometric study. *Clin Neuropathol* 1987;**6**:16–18.

87. Peters A. The effects of normal aging on myelin and nerve fibers: a review. *J Neurocytol* 2002;**31**:581–93.

88. Peters A, Kemper T. A review of the structural alterations in the cerebral hemispheres of the aging rhesus monkey. *Neurobiol Aging* 2012;**33**:2357–72.

89. Peters A, Nigro NJ, McNally KJ. A further evaluation of the effect of age on striate cortex of the rhesus monkey. *Neurobiol Aging* 1997;**18**:29–36.

90. Peters A, Morrison JH, Rosene DL, Hyman BT. Are neurons lost from the primate cerebral cortex during normal aging? *Cereb Cortex* 1998;**8**:295–300.

91. Popescu BO, Toescu EC, Popescu LM, *et al.* Blood–brain barrier alterations in ageing and dementia. *J Neurol Sci* 2009;**283**:99–106.

92. Raffaello A, Rizzuto R. Mitochondrial longevity pathways. *Biochim Biophys Acta* 2011;**1813**:260–68.

93. Rapp PR, Gallagher M. Preserved neuron number in the hippocampus of aged rats with spatial learning deficits. *Proc Natl Acad Sci U S A* 1996;**93**:9926–30.

94. Raz N, Lindenberger U, Rodrigue KM, *et al.* Regional brain changes in aging healthy adults: general trends, individual differences and modifiers. *Cereb Cortex* 2005;**15**:1676–89.

95. Reddy PH, Reddy TP. Mitochondria as a therapeutic target for aging and neurodegenerative diseases. *Curr Alzheimer Res* 2011;**8**:393–409.

96. Resnick SM, Pham DL, Kraut MA, Zonderman AB, Davatzikos C. Longitudinal magnetic resonance imaging studies of older adults: a shrinking brain. *J Neurosci* 2003;**23**:3295–301.

97. Salminen A, Ojala J, Kaarniranta K, *et al.* Astrocytes in the aging brain express characteristics of senescence-associated secretory phenotype. *Eur J Neurosci* 2011;**34**:3–11.

98. Salthouse TA. Neuroanatomical substrates of age-related cognitive decline. *Psychol Bull* 2011;**137**:753–84.

99. Salvador GA, Uranga RM, Giusto NM. Iron and mechanisms of neurotoxicity. *Int J Alzheimers Dis* 2010;**2011**:720658.

100. Schliebs R, Arendt T. The cholinergic system in aging and neuronal degeneration. *Behav Brain Res* 2011;**221**:555–63.

101. Schmidt R, Schmidt H, Haybaeck J, *et al.* Heterogeneity in age-related white matter changes. *Acta Neuropathol* 2011;**122**:171–85.

102. Schuff N, Tosun D, Insel PS, et al. Nonlinear time course of brain volume

loss in cognitively normal and impaired elders. *Neurobiol Aging* 2012;33:845–55.

103. Schultz C, Ghebremedhin E, Del Tredici K, Rub U, Braak H. High prevalence of thorn-shaped astrocytes in the aged human medial temporal lobe. *Neurobiol Aging* 2004;25:397–405.

104. Sharma N, Castorena CM, Cartee GD. Tissue-specific responses of IGF-1/insulin and mTOR signaling in calorie restricted rats. *PloS One* 2012;7:e38835.

105. Sheffield LG, Berman NE. Microglial expression of MHC class II increases in normal aging of nonhuman primates. *Neurobiol Aging* 1998;19:47–55.

106. Shore DE, Carr CE, Ruvkun G. Induction of cytoprotective pathways is central to the extension of lifespan conferred by multiple longevity pathways. *PLoS Genet* 2012;8:e1002792.

107. Siddiqi Z, Kemper TL, Killiany R. Age-related neuronal loss from the substantia nigra-pars compacta and ventral tegmental area of the rhesus monkey. *J Neuropathol Exp Neurol* 1999;58:959–71.

108. Simic G, Kostovic I, Winblad B, Bogdanovic N. Volume and number of neurons of the human hippocampal formation in normal aging and Alzheimer's disease. *J Comp Neurol* 1997;379:482–94.

109. Sloane JA, Hollander W, Moss MB, et al. Increased microglial activation and protein nitration in white matter of the aging monkey. *Neurobiol Aging* 1999;20:395–405.

110. Smith DE, Roberts J, Gage FH, Tuszynski MH. Age-associated neuronal atrophy occurs in the primate brain and is reversible by growth factor gene therapy. *Proc Natl Acad Sci U S A* 1999;96:10893–8.

111. Snyder AM, Connor JR. Iron, the substantia nigra and related neurological disorders. *Biochim Biophys Acta* 2009;1790:606–14.

112. Speakman JR, Mitchell SE. Caloric restriction. *Mol Aspects Med* 2011;32:159–221.

113. Stark AK, Toft MH, Pakkenberg H, et al. The effect of age and gender on the volume and size distribution of neocortical neurons. *Neuroscience* 2007;150:121–30.

114. Steffener J, Stern Y. Exploring the neural basis of cognitive reserve in aging. *Biochim Biophys Acta* 2012;1822:467–73.

115. Streit WJ, Miller KR, Lopes KO, Njie E. Microglial degeneration in the aging brain--bad news for neurons? *Front Biosci* 2008;13:3423–38.

116. Sulzer D, Mosharov E, Talloczy Z, et al. Neuronal pigmented autophagic vacuoles: lipofuscin, neuromelanin, and ceroid as macroautophagic responses during aging and disease. *J Neurochem* 2008;106:24–36.

117. Swerdlow RH. Brain aging, Alzheimer's disease, and mitochondria. *Biochim Biophys Acta* 2011;1812:1630–39.

118. Takao H, Hayashi N, Ohtomo K. A longitudinal study of brain volume changes in normal aging. *Eur J Radiol* 2012;81:2801–4.

119. Tank EM, True HL. Disease-associated mutant ubiquitin causes proteasomal impairment and enhances the toxicity of protein aggregates. *PLoS Genet* 2009;5:e1000382.

120. Terry RD, DeTeresa R, Hansen LA. Neocortical cell counts in normal human adult aging. *Ann Neurol* 1987;21:530–39.

121. Thal DR, Del Tredici K, Braak H. Neurodegeneration in normal brain aging and disease. *Sci Aging Knowledge Environ* 2004;2004:pe26.

122. Toescu EC, Vreugdenhil M. Calcium and normal brain ageing. *Cell Calcium* 2010;47:158–64.

123. Tucker AM, Stern Y. Cognitive reserve in aging. *Curr Alzheimer Res* 2011;8:354–60.

124. Uemura E. Age-related changes in the subiculum of *Macaca mulatta*: dendritic branching pattern. *Exp Neurol* 1985;87:412–27.

125. Uylings HB, de Brabander JM. Neuronal changes in normal human aging and Alzheimer's disease. *Brain Cogn* 2002;49:268–76.

126. van Leeuwen FW, Hol EM, Fischer DF. Frameshift proteins in Alzheimer's disease and in other conformational disorders: time for the ubiquitin-proteasome system. *J Alzheimers Dis* 2006;9(3 Suppl):319–25.

127. van Leeuwen FW, Kros JM, Kamphorst W, van Schravendijk C, de Vos RA. Molecular misreading: the occurrence of frameshift proteins in different diseases. *Biochem Soc Trans* 2006;34(Part 5):738–42.

128. von Bernhardi R, Tichauer JE, Eugenin J. Aging-dependent changes of microglial cells and their relevance for neurodegenerative disorders. *J Neurochem* 2010;112:1099–114.

129. Voytko ML, Sukhov RR, Walker LC, et al. Neuronal number and size are preserved in the nucleus basalis of aged rhesus monkeys. *Dementia* 1995;6:131–41.

130. West M, Coleman P. How to count. *Neurobiol Aging* 1996;17:503.

131. West MJ, Slomianka L, Gundersen HJ. Unbiased stereological estimation of the total number of neurons in the subdivisions of the rat hippocampus using the optical fractionator. *Anat Rec* 1991;231:482–97.

132. West MJ, Coleman PD, Flood DG, Troncoso JC. Differences in the pattern of hippocampal neuronal loss in normal ageing and Alzheimer's disease. *Lancet* 1994;344:769–72.

133. Wharton SB, Brayne C, Savva GM, et al. Epidemiological neuropathology: the MRC Cognitive Function and Aging Study experience. *J Alzheimers Dis* 2011;25:359–72.

134. Zecca L, Youdim MB, Riederer P, Connor JR, Crichton RR. Iron, brain ageing and neurodegenerative disorders. *Nat Rev Neurosci* 2004;5:863–73.

135. Zhang C, Zhu Q, Hua T. Aging of cerebellar Purkinje cells. *Cell Tissue Res* 2010;341:341–7.

136. Zhang F, Wang S, Gan L, et al. Protective effects and mechanisms of sirtuins in the nervous system. *Prog Neurobiol* 2011;95:373–95.

137. Zhu YC, Tzourio C, Soumare A, et al. Severity of dilated Virchow-Robin spaces is associated with age, blood pressure, and MRI markers of small vessel disease: a population-based study. *Stroke* 2010;41:2483–90.

Dementia

James Lowe and Raj Kalaria

INTRODUCTION

The neuropathological diagnosis of disorders producing clinical dementia has become of heightened importance, in part, because of the following:

- There is increasing awareness that certain diseases are hereditary, which makes a definitive diagnosis important to families and their physicians for prognostication and counselling.
- The increasing prevalence of dementia in the ageing populations of many countries has resulted in the recognition of dementia as a priority for healthcare and social support systems.
- Clinical diagnosis does not predict pathology, but rather indicates the neuroanatomical distribution of the pathology. Only histopathological and biochemical studies provide a definitive diagnosis.
- There is increased recognition that neurodegenerative diseases previously thought to be associated only with a movement disorder have cognitive impairment or dementia at some point in their evolution. This is especially seen in disorders causing parkinsonism or motor neuron disease.
- New causes of dementia continue to be discovered at post-mortem neuropathological examination of the nervous system, and even for previously described disorders, modern diagnostic methods and molecular analyses enhance understanding of the pathogenesis.
- As novel diagnostic and therapeutic agents are developed, neuropathological diagnosis will be required to

establish the specificity and sensitivity of diagnostic tests and the effectiveness of possible interventions.

This chapter will cover the neuropathology of neurodegenerative and vascular diseases causing the clinical syndrome of dementia.

CONCEPTS IN THE NEUROPATHOLOGY OF DEMENTIA

Accumulation of Intracellular Proteins as Inclusion Bodies – Proteinopathies

The discovery that many neurodegenerative diseases are characterized by intracellular accumulation of abnormally configured proteins, with the formation of inclusion bodies (e.g. neurofibrillary tangles or Lewy bodies) has led to the concept of proteinopathies. Such conditions have been described as disorders of protein homeostasis or proteostasis. In this concept, environmental factors act in concert with genetic modifiers of the ageing process (see Chapter 15, Ageing of the Brain) and lead to disordered proteostasis with the accumulation of abnormal proteins in the cell. Once abnormal proteins start to accumulate a series of cellular responses occurs, including the formation of visible inclusion bodies within cells.[224] Certain forms of abnormal protein accumulation are believed to be toxic and lead to cell death. Accumulation of abnormal proteins in axons or dendrites may interfere with normal axoplasmic or dendritic transport, which also can compromise neuronal function.[169]

The finding that some inherited degenerative disorders are caused by mutations in genes encoding proteins that subsequently accumulate within inclusion bodies has reinforced the primary role of protein aggregates in disease. Disorders associated with abnormal accumulation of tau protein have been named 'tauopathies', those associated with accumulation of α-synuclein are referred to as 'synucleinopathies'[205] and those associated with accumulation of TDP-43 referred to as 'TDP-43 proteinopathy'.[286] Proteinopathies can be further classified as either genetic or sporadic, depending on whether there is clear genetic linkage or a known mutation associated with the disorder.

There are still uncertainties how intracellular protein accumulation produces disease. Inclusion bodies themselves may lead to cell death, but another view suggests that it is small oligomers of the abnormal protein that are toxic to cells, whereas aggregation into inclusion bodies may be protective.[224]

A common protein implicated in many studies of inclusion bodies is ubiquitin. This can direct substrates to the three major protein degradation pathways: the proteasome, the lysosome, and the autophagosome (autophagy).[145] These protein degradation systems are activated in neurodegenerative diseases associated with protein aggregation.[793]

The ubiquitin-proteasome system (UPS) is an energy-dependent system for protein degradation that conjugates the 15-kDa stress response protein ubiquitin to abnormal or misconfigured proteins through a series of enzymatic steps involving E1 ubiquitin-activating enzymes, E2 ubiquitin-conjugating enzyme and E3 ligase.[196] The multi-ubiquitin–conjugated protein is targeted to the proteasome, a large (26 S) multimeric cytoplasmic and nuclear barrel-shaped organelle that has several distinct proteolytic functions, and renders the abnormal protein into smaller peptides and free amino acid residues. Ubiquitin is recycled through action of the ubiquitin C-terminal hydrolase. The UPS is involved with elimination of abnormal intracellular proteins. Abnormal protein conformation triggers a cell stress response. Chaperone proteins, which function to restore conformation or facilitate clearance by the UPS, are implicated in inclusion body formation. Given the significant role of the UPS, it is perhaps not surprising that immunohistochemical staining for ubiquitin detects many forms of abnormal protein accumulation.[556] It is possible that ageing itself is associated with inhibition of the UPS.[63,64]

The other major cellular mechanism for clearance of abnormal proteins involves lysosomes and the process of autophagy. It is recognized that systems involved in these processes are active in neurodegenerative diseases leading to dementia.[19,133,137,406,500,618,745]

Accumulation of Extracellular Proteins as Amyloid

Several neurodegenerative diseases are characterized by the extracellular accumulation of abnormal proteins with common physicochemical properties, namely a high content of β-pleated sheet secondary structure. This structure allows binding to certain dyes, such as Congo red and thioflavine S, which intercalate between the β-pleated sheets. Birefringence on Congo red staining and the fluorescence

with thioflavine define the protein as amyloid. Disorders associated with extracellular deposits of amyloid protein are referred to as amyloidoses. The most common form of amyloid in the aged brain and in Alzheimer's disease (AD) is the extracellular accumulation of amyloid-β peptide (Aβ) within structures referred to as senile plaques. The Aβ peptide, which forms the amyloid characteristic of AD, derives from proteolytic cleavage of a widely expressed transmembrane molecule, the amyloid precursor protein (APP). The same type of amyloid protein may also accumulate within the walls of small blood vessels causing cerebral amyloid angiopathy (see Chapter 2). A detailed discussion of the biology of amyloid is found in the section on the molecular pathogenesis of AD.

Several other proteins may accumulate extracellularly in the brain as amyloid material, including ABri in familial British dementia, ADan in familial Danish dementia and PrP^{Sc} in prion diseases. Other proteins that accumulate as intracellular inclusions, including tau and α-synuclein, also have the physical properties of amyloid.

Seeding and Propagation of Misfolded Proteins in the Nervous System

It has long been accepted that the pathogenesis of prion diseases (see Chapter 18) occurs by spread of PrP^{Sc} through neuroanatomical pathways in the nervous system, leading to propagation of disease. This concept has been extended to include non-prion protein aggregates of tau, α-synuclein, Aβ, huntingtin and TDP-43 aggregates. It is proposed that these abnormal proteins move between cells and seed the misfolding of their normal conformers, thereby propagating disease.[162,177,230,613,677,855,936] A concept of 'molecular nexopathies' has been proposed in which conjunctions of pathogenic proteins and neuronal network connections define the pathological expression of disease.[940]

Neuroinflammation

Neuronal death or abnormal intracellular and extracellular proteins elicit neuroinflammation with microglial activation[3] and production of proinflammatory molecules that can lead to further neuronal damage. Microarray and immunohistochemical studies have shown that markers reflecting inflammatory pathways are upregulated in neurodegenerative diseases. In this form of neuroinflammation, classical inflammatory cell infiltrates are not seen. Rather, there is microglial activation[323] as evidenced by the expression of major histocompatibility antigens and upregulation of effectors such as cytokines and chemokines.[301,719]

Oxidative Stress

Oxidative stress is due to imbalance in the cellular redox state caused by production of reactive oxygen species or through dysfunction of antioxidant systems normally operating to limit damage. Oxidative stress is felt to play a role in a number of neurodegenerative disorders. The high oxygen consumption rate, abundant lipid and relative paucity of antioxidant enzymes render the brain particularly vulnerable to free radical damage.[581,582,679,711,843,999] Consequently, oxidative modification of proteins, lipids and nucleic acids occurs. At the cellular level oxidative stress-activated signalling pathways may lead to programmed cell death.[501]

Mitochondrial Pathology

Several neurodegenerative diseases have been linked to abnormalities in mitochondrial function, especially in Parkinson's disease (PD). It is evident that several of the systems that become abnormal in neurodegeneration may act through a final common pathway through mitochondria. Studies suggest that changes in mitochondrial biology, including fusion, fission and elimination (mitophagy), are factors leading to cell death.[418,794,814]

RNA Processing

Molecular investigation of several genetic neurodegenerative diseases has suggested that a common factor involves abnormalities in RNA processing, when mutated non-coding RNA or abnormal RNA-binding proteins cause defective RNA handling.[25,403] The term 'RNA-binding proteinopathy' has been used to describe diseases caused by mutations in RNA-binding proteins that lead to neurodegeneration (TDP-43 and fused-in-sarcoma protein [FUS]).[751] mRNAs may also be directly toxic if they contain nucleotide repeat sequences that directly bind to specific proteins and inhibit normal function.[751] A focus on involvement of stress granules in this group of diseases has been made.[69]

Neuronal Loss and Neurogenesis in Ageing

Common to all diseases leading to dementia is loss of neurons, with the particular populations of neurons vulnerable to cell death defining the clinical features. Programmed forms of cell death, including apoptosis, necroptosis and oncosis, are frequently implicated in neurodegenerative diseases.[271,435,783,921,943]

There are likely to be many triggers for neurons to enter a cell death pathway including the presence of an inclusion body or factors noted earlier. It remains unclear why some neurons appear to survive in the face of a neurodegenerative disease, and the concept of a selective vulnerability has long been discussed with the idea that understanding why some neurons survive would give a clue to developing therapies. It is likely that differences in factors such as levels of protein expression or mitochondrial distribution in different neuronal groups explains cell survival and selective vulnerability in neurodegenerative diseases.[223,225] A stressor-threshold model to explain how certain neurons and are selectively vulnerable in disease has been proposed.[806]

A controversial hypothesis related to the pathogenesis of neurodegenerative disease is that it involves the reactivation of cell division pathways, not normally activated in the adult central nervous system. In this model, pathological stimuli are mitogenic and adult neurons abortively enter the cell cycle. Cell cycle activation potentially links synaptic plasticity to cell death through disturbances of signalling pathways.[29,83,175] Aneuploid neurons can be detected in neurodegeneration and it has been proposed that these are vulnerable to cell death.[28]

Neuronal plasticity is a key process for maintaining cognition, and declines with age. It is possible that certain aspects of neurodegeneration operate through a failure of neuronal plasticity[111,638] or neurogenesis.[522,584,769]

DEMENTIA
Definition of Dementia

The American Psychiatric Association's Diagnostic and Statistical Manual of mental disorders (DSM-IV) defines dementia as 'the development of multiple cognitive deficits that include memory impairment and at least one of the following cognitive disturbances: aphasia, apraxia, agnosia or a disturbance in executive functioning. The cognitive deficits must be sufficiently severe to cause impairment in the occupational or social functioning and must represent a decline from a previously higher level of functioning.'[20] This usually occurs in the absence of delirium or depression. There is a wide range of pathologies associated with the clinical diagnosis of dementia (Table 16.1).

TABLE 16.1 Causes of dementia

Neurodegenerative disorders

Alzheimer's disease
Dementia with Lewy bodies
Parkinson's disease
Frontotemporal lobar degenerations (FTLD)
Hippocampal sclerosis
Neurodegenerative disorders of movement

Vascular disease

Large vessel disease, multiple infarcts
Small vessel disease, Binswanger's disease
Cerebral autosomal dominant arteriopathy with subcortical infarcts and leukoencephalopathy (CADASIL), other hereditary angiopathies
Familial amyloid angiopathies
Cranial arteritis and other cranial angiitis

Infectious, inflammatory and immune-mediated disorders

Granulomatous infections, e.g. tuberculous meningitis
Neurosyphilis

Viral infections

HIV infection
Progressive multifocal leukoencephalopathy
Viral encephalitis

Prion disorders

Creutzfeldt–Jakob disease
Variant Creutzfeldt–Jakob disease
Gerstmann–Sträussler–Scheinker syndrome
Kuru
Autoimmune encephalitis
Multiple sclerosis

Toxic and metabolic disorders

Alcoholism
Vitamin B_{12} deficiency
Folate deficiency
Hypothyroidism
Chronic hepatic encephalopathy
Neurometabolic diseases

Other conditions

Tumours and hydrocephalus
Traumatic brain injury

In most patients, memory impairment dominates the clinical picture of dementia; however, cognitive deficits in several neurodegenerative disorders are characterized by disturbances predominantly in other domains, such as executive function or praxis. Dementia implies a disease process that involves several brain regions. The term 'focal neurobehavioural deficit' is sometimes applied to disorders associated with isolated impairments of function, such as amnesia, aphasia or apraxia.

The term 'mild cognitive impairment' (MCI) is used to refer to the condition in which individuals have subjective memory complaints and documented cognitive problems, but no impairment in activities of daily living.[9] Individuals that come to autopsy while suffering from MCI usually have a range of Alzheimer-type pathology although other pathologic processes can also be associated with MCI.[639,714] Clinical studies suggest that MCI progresses to dementia over time at a rate of 10–15 per cent per year.[240,634] This is a significant area of interest and different investigators have used slightly different criteria and nomenclature, including 'age-associated cognitive decline', 'age-associated memory impairment', and 'cognitive impairment, no dementia'. Standardization has been proposed in guidelines from the National Institute on Aging-Alzheimer's Association work groups on diagnostic guidelines for Alzheimer's disease.[9]

Clinical Diagnosis of Dementia

In clinical practice, the diagnosis of dementia relies on tests of cognitive function that may be supplemented by wider clinical examination, laboratory investigation and neuroimaging. Tests of cognitive function range from comprehensive batteries of formal psychological tests, usually in the context of a research study, to the more simple application of bedside tests of mental state. A commonly used bedside test is the Mini Mental State Examination.[261] The severity of dementia can also be evaluated clinically using validated rating scales such as the Clinical Dementia Rating Scale, which is based upon interviews with a knowledgeable informant.

The National Institute on Aging–Alzheimer's Association work groups have provided clinical diagnostic guidelines for dementia, as follows.[611]

Dementia is clinically diagnosed when there are cognitive or behavioural symptoms that:

1. Interfere with the ability to function at work or at usual activities; and
2. Represent a decline from previous levels of functioning and performing; and
3. Are not explained by delirium or major psychiatric disorder.

Cognitive impairment is detected and diagnosed through a combination of (i) history-taking from the patient and a knowledgeable informant and (ii) an objective cognitive assessment, either a 'bedside' mental status examination or neuropsychological testing. Neuropsychological testing should be performed when the routine history and bedside mental status examination cannot provide a confident diagnosis.

The cognitive or behavioural impairment involves a minimum of two of the following domains:

1. Impaired ability to acquire and remember new information: symptoms include: repetitive questions or conversations, misplacing personal belongings, forgetting events or appointments, getting lost on a familiar route.
2. Impaired reasoning and handling of complex tasks, poor judgment: symptoms include: poor understanding of safety risks, inability to manage finances, poor decision-making ability, inability to plan complex or sequential activities.
3. Impaired visuospatial abilities: symptoms include: inability to recognize faces or common objects or to find objects in direct view despite good acuity, inability to operate simple implements, or orient clothing to the body.
4. Impaired language functions (speaking, reading, writing): symptoms include: difficulty thinking of common words while speaking, hesitations; speech, spelling, and writing errors.
5. Changes in personality, behaviour, or comportment: symptoms include: uncharacteristic mood fluctuations such as agitation, impaired motivation, initiative, apathy, loss of drive, social withdrawal, decreased interest in previous activities, loss of empathy, compulsive or obsessive behaviours, socially unacceptable behaviours.

The differentiation of dementia from MCI rests on the determination of whether or not there is significant interference in the ability to function at work or in usual daily activities.

The major dementia syndromes can be classified depending upon the predominant cognitive deficit and the anatomical region affected as follows:

- Temporal-parietal syndrome is dominated by early memory disturbances that often involve wider areas of association cortex, such as parietal and occipital lobes, as the disease progresses. The main example of a temporal-parietal syndrome is AD.
- Frontotemporal syndrome is dominated by problems in executive functioning or language disturbances with relative lack of memory disturbance early in disease. This syndrome underlies the heterogeneous group of frontotemporal dementia syndromes (FTD).
- Subcortical dementia is dominated by reduced speed and efficiency of cognition and is associated with disease of deep grey matter structures. This pattern of abnormal cognition is associated with certain degenerative parkinsonian syndromes, such as progressive supranuclear palsy (PSP) as well as vascular disease.

The term 'primary dementia' is used to refer to a degenerative disease of the brain that is not associated with additional neurological features. The pathology of a primary dementia is thought to be localized primarily to the cerebral cortex or the limbic and association regions involved in cognition.

The term 'dementia plus' refers to a disorder in which there are clinical features of a primary dementia, but also focal neurologic deficits. In this condition, there is pathology in subcortical regions (e.g. basal ganglia) or primary cortices (e.g. motor cortex) in addition to limbic and association cortices.

The term 'secondary dementia' refers to a condition in which cognitive deficits are related to potentially treatable disorders, such as mass lesions, metabolic or endocrine disturbances, vascular disease, hydrocephalus or certain infections.

Epidemiology of Dementia

Information concerning the incidence (the rate of new diagnoses within a certain time frame) and the prevalence (the proportion of people in a population affected at one point in time) of dementia worldwide may be confounded by wide methodological variation, including a lack of standard clinical and neuropathological criteria.[519] In most series of adult-onset dementia, the most common primary dementias include AD, dementia with Lewy bodies (DLB), frontotemporal dementia and vascular dementia; however, this is complicated by the frequency with which mixed pathologies occur in the elderly. There is an age-related increase in the prevalence of dementia especially seen over the age of 65 years in both male and females and in Western community populations this has been estimated as approximately <2 per cent at 65 years, around 10 per cent at 80 years and 25 per cent at 90 years.[519] The prevalence of dementia in those aged 65 years or older in the UK is estimated at around 8 per cent.[594]

A World Federation of Neurology group has estimated that in Asian and Latin American countries the age-adjusted dementia prevalence of dementia in 65 year olds is high at around 5 per cent, and low in India and sub-Saharan Africa at around 1 per cent.[463] The prevalence in China is around 2 per cent of the population at age 65, rising to around 25 per cent at age 90.[972]

There may be secular trends in dementia prevalence with older generations having a higher burden of disease than more recent generations in Western populations,[594,744,767] but this is not reported in all studies.[957]

Genetic Factors and Counselling in Dementia

Twenty-five per cent of the population have a family history of dementia involving a first-degree relative,[839] and with the increased public awareness of dementia many families express a concern that the disease 'runs in the family'. There are only a small number of penetrant genes that cause dementia. An approach to considering genetic risk in dementia as a basis for counselling and planning genetic testing has been published by Loy *et al.*[557] The authors note that for relatives of most people with dementia, their lifetime risk of dementia is around 20 per cent, compared with around 10 per cent in the general population.[557]

- The presence of a Mendelian dominantly inherited condition is usually indicated from the family tree and genetic testing can be considered in a staged manner, guided by clinical presenting features. The autosomal dominant genes involved are discussed throughout the chapter. In addition consulting online databases is recommended as the best way to review up to date information in this area:
  - The Alzheimer Disease & Frontotemporal Dementia Mutation Database http://www.molgen.ua.ac.be/ADMutations[392]
  - The ALZGENE database http://www.alzgene.org[72]
  - The Online Mendelian Inheritance in Man (OMIM) database http://www.omim.org
- It is more common for individuals to have a non-Mendelian family history of dementia. It is apparent that many neurodegenerative diseases have a genetic contribution as part of complex genetic disease in which genes make a contribution to a risk of developing disease. This applies to many of the late onset forms of dementia, for example in AD, discussed later. It is not currently advised that testing for such genes is performed because they do not have a strong predictive value in an individual patient for causing disease, and many people who carry such genes never get disease.

It has been estimated that people with a first-degree relative with AD have a 2.5 times lifetime risk of developing dementia compared with the general population. This can also be expressed as cumulative risk of developing dementia by the age of 80 years of 18 per cent. To put this in perspective, the figure is 6 per cent for the spouses of patients with dementia. This risk differs between ethnic groups and in an African-American population, the cumulative risk for people with a first-degree relative with AD of developing dementia by the age of 80 years is 30 per cent, being 13 per cent for spouses.[174,325]

ALZHEIMER'S DISEASE

In 1907, Alois Alzheimer described changes in the brain of a 55-year-old woman who died after a 4-year history of progressive dementia.[17] In his 1907 report, Alzheimer demonstrated neurofibrillary tangles by using the newly developed Bielschowsky silver impregnation method. He also observed cortical 'miliary foci' or senile plaques, described 15 years earlier by Blocq and Marinesco.[81] Decades later, sections of the brain of patient Auguste D were recovered in the Institute of Neuropathology of the University of Munich[322] and shown to have histopathological changes consistent with modern concepts of AD (Table 16.2).

TABLE 16.2 Summary of pathology of Alzheimer's disease

- Macroscopically, the cerebrum shows atrophy that is most severe in the medial temporal lobe, including the hippocampus and amygdala, and least severe in the occipital lobe.
- Extracellular amyloid is deposited as senile plaques in affected cortical and subcortical regions. The amyloid protein (Aβ) is a proteolytic product of a normal transmembrane protein, the amyloid precursor protein (APP). Amyloid is deposited in the walls of blood vessels, causing amyloid angiopathy in most cases.
- Neuronal degeneration is characterized by accumulation of cytoplasmic fibrillar aggregates of the microtubule-associated protein, tau, within neuronal perikarya (neurofibrillary tangles), within dystrophic neurites associated with senile plaques and within neuropil threads.
- Synaptic and neuronal loss, as well as astrocytosis and microglial activation, is seen in affected cortical regions.

AD is the leading cause of dementia and decreased quality of life among the elderly.[598] AD can be classified according to age of onset. About 95 per cent of patients develop disease aged >65 years (late-onset AD), and between 1 and 5 per cent have an earlier onset, usually between 45–55 years (early onset AD). AD now affects millions of individuals and greatly affects the lives of their families and caregivers. Our understanding of the pathobiology of AD has to a great extent been built upon the experience with autopsies and the observations of neuropathologists, the development of specialized AD research centres and the support of groups such as the Alzheimer's Association, the Alzheimer's Society and Alzheimer's Research UK.

Clinical Features of Alzheimer's Disease

The majority of patients with AD present with disturbances of recent memory for 1–3 years, during which time they may remain unconcerned or may present clinically and be considered to have MCI or prodromal AD. As the disease progresses, patients experience problems with visuospatial function, language and attention. The language disturbance is characterized by low verbal fluency and problems with naming. Patients become unable to evaluate situations and to make proper judgments. Although not core diagnostic features, many patients develop agitation or depression. Operational diagnostic criteria for AD have been developed.[611] Recent recommendations have been made to harmonize diagnostic terminology and agree on concepts in the diagnosis of AD that can be defined as a brain disorder, regardless of clinical status. The term symptomatic AD is used to refer to all of the clinically expressed phases of the disease, including early prodromal stages.[635] Clinical diagnostic criteria are increasingly being interpreted alongside information on biomarkers of disease, used in a research setting. Further work is needed on standardization and defining diagnostic thresholds before cerebrospinal fluid (CSF) biomarker assays or amyloid imaging of the brain can be used in the clinical diagnosis of AD.[635]

Patients with AD experience progressive decline in function with increasing frailty. They typically require residential care after about 2–4 years and typically survive for a further 3–10 years. The mean duration of survival for patients with AD in one cohort was 5.9 years and longer with earlier age at onset.[277] A main predictor of life expectancy is patient age, with a median life expectancy of around 8 years for patients diagnosed in the sixth or seventh decade and around 3 years for those diagnosed in the eighth or ninth decades.[991]

The burden of care for a significant proportion of the elderly population represents a severe economic challenge for most developed countries. With increasing frailty and immobility, patients typically die of pneumonia. An increased incidence of stroke is observed because of thromboembolism or haemorrhage related to cerebral amyloid angiopathy.[269]

Biomarkers for Alzheimer's Disease

Because the clinical diagnosis of AD depends on the inclusion and exclusion of features, a rather extensive clinical work-up is often performed including a clinical history, physical examination, haematological studies, neuroimaging, neuropsychological evaluation, and increasingly analysis of CSF. Yet, despite promise from neuroimaging studies, to date the definite diagnosis of AD depends upon neuropathological evaluation. A search for biomarkers has accordingly received a great deal of attention. A successful biomarker would help to make the diagnosis and establish a surrogate outcome for measures of drug efficacy. The best biomarker thus would be tightly linked to the pathophysiology of the disease. Moreover, biomarkers could be used for the preclinical diagnosis of AD.

There are five main established biomarkers used in AD[784] that can be divided into two groups as follows:[423]

- biomarkers of brain Aβ deposition:
  ○ amyloid PET scanning;
  ○ reduced CSF Aβ42 level;
- biomarkers of neurodegeneration:
  ○ atrophy on structural MRI scanning;
  ○ elevated levels of CSF total tau (t-tau) and phosphorylated (p-tau);
  ○ hypometabolism on FDG-PET scan.

The CSF biomarkers t-tau, p-tau and Aβ42 reflect the pathologic features of AD, as do neuronal loss, neurofibrillary tangles and Aβ plaques. Biomarker evolution is proposed to develop in a hierarchal manner.[423]

- cognitive normal:
  ○ amyloid biomarkers abnormal;
- cognitive normal:
  ○ CSF tau abnormal;
- cognitive impairment:
  ○ biomarkers of neurodegeneration abnormal (FDG-PET and atrophy on MRI).

Imaging of Aβ deposits in the brain has become clinically possible with the development of specific ligands that can be visualized by PET. The issues that need to be considered in its application and relationships to neuropathological findings have been reviewed by Jack *et al.*[424] The authors make the following observations: 30 per cent of cognitively normal and 60 per cent of patients with MCI have a positive amyloid PET scan. Cognitively normal elderly patients with MCI subjects who have a positive amyloid PET scan have been demonstrated to show faster cognitive decline than individuals who have a negative scan. Between 85–90 per cent of patients with clinically diagnosed AD have a positive amyloid PET scan. The explanation for the 10–15 per cent who have a negative scan is felt to be a consequence of the imprecision of clinical diagnostic criteria for AD.

The Alzheimer's Disease Neuroimaging Initiative is a longitudinal, multicenter study to develop clinical, imaging, genetic, and biochemical biomarkers for the detection and tracking of Alzheimer's disease. A review and summary of findings in this initiative has been published.[945]

Evaluations of a range of potential markers that reflect possible aspects of AD development such as Aβ processing, neuroinflammation and neuronal dysfunction are ongoing.[244] This includes the evaluation of proteomic techniques to establish biomarkers in the blood.[60]

Epidemiology of Alzheimer's Disease

AD remains the predominant cause of dementia in the elderly. A range of factors often linked to difficulty in precision of diagnosis limits precision in establishing incidence and prevalence. It is estimated that that around 24.3 million people have dementia with most affected individuals living in developing countries (60 per cent in 2001, rising to 71 per cent by 2040). The numbers in developed countries are forecast to increase by 100 per cent between 2001 and 2040, and by more than 300 per cent in India, China, and their south Asian and western Pacific neighbours.[255] A comprehensive review of epidemiology has summarized incidence rates for AD per hundred person years across studies between 1994 and 2005. Although there is a range between different studies, typical rates increase from 0.5 per cent at 65–69 years to 8 per cent over 85 years.[750]

Epidemiological studies have implicated a number of risk and protective factors for AD[449] (Table 16.3) but few factors are robust in that they have been replicated in multiple independent studies in different populations. The factors that are robust include age and family history. An updated online resource presenting risk factors for AD is at http://www.alzrisk.org/.

Age

Age is the primary risk factor for the development of dementia and AD. Both the incidence and the prevalence of AD rise exponentially between ages 65 and 95 years, doubling with every 5 years of age.[449,473] The incidence of dementia and AD has been found to rise exponentially up to the age of 90 years,[448] but more data are needed to determine whether or not a levelling off occurs beyond this age. Indeed, many centenarians are cognitively normal.[473]

Family History

A family history of dementia is a recognized risk factor for AD. Genetic factors contribute to risk and higher incidence of AD in individuals with a history of AD in a first-degree relative (mother, father or sibling).[174,830,920] The lifetime risk for AD is about 15 per cent at age 80 and 39 per cent at age 90 for people with a first-degree relative who had disease.[521] The MIRAGE study has produced lifetime risk estimates of dementia for first-degree relatives of those with AD by ethnicity and sex and can be used in counselling relatives about their own risk of developing AD.[325] A first-degree family history of Alzheimer's disease can be thought of as a

TABLE 16.3 Risk factors and protective factors for Alzheimer's disease

Risk factors	Protective factors
Age	Non-steroidal anti-inflammatory drugs
Family history	
Low educational attainment	Oestrogens
Cardiovascular disease	Statins
Diabetes mellitus	Nicotine
High serum homocysteine	
Head injury	

composite risk factor due to the influence of susceptibility genes as well as possible common environmental risks.[220] The genetic factors involved in AD are considered in more detail later.

Educational Attainment

Studies have shown that lifelong experiences, including educational and occupational attainment can increase this cognitive reserve and lower risk of development of dementia.[860] A systematic literature review indicated that just over half of studies have shown a relationship with education.[821] Individuals who have low education attainment have been estimated to a twofold to threefold increased risk of developing AD.[268,636,952,961] The quality of education and not the quantity is important.[139] Imaging has shown that education increases brain cortical thickness, suggesting that there is a physical basis to cognitive reserve.[546]

Diabetes Mellitus

The incidence of dementia is higher in patients with diabetes mellitus but the pathogenesis of the relationship remains uncertain.[558] Individuals with type 2 diabetes have a twofold higher risk of AD than those without diabetes.[834] Mechanisms linking diabetes with AD include a role of insulin or glucose metabolism, possibly mediated through the advance glycation end products, failure of insulin signalling, decreased inhibition of the tau-kinase GSK3β, and a role of insulin-degrading enzyme, a metalloprotease involved in degradation of Aβ. There is also the suggestion of activation of inflammatory pathways in the brain.[834]

Cardiovascular Disease

The presence of cardiovascular disease (stroke, hypertension, atheromatous disease, smoking) has been linked with an increase in risk of developing AD in some studies.[391,520,553,695,743,838]

It is likely that some of the association between cardiovascular disease and dementia is related to mixed (vascular and Alzheimer) pathology (see p. 937). The presence of mixed pathology may lower the burden of disease needed for dementia to manifest clinically, for both vascular pathology and Alzheimer-type pathology.[842] One genetic contribution to the vascular link between vascular disease and AD is mediated through apolipoprotein E (*APOE*), genetic variants of which predispose to both atheromatous vascular disease and AD.[469] It is understood that vascular contributions are important in cognitive decline with age;[316] however, studies have shown that cardiovascular risk score is no better at predicting dementia risk than age.[795]

Plasma Homocysteine, B12 and Folate

Folic acid and vitamin B12 have important roles in the brain, including the methionine-synthase mediated conversion of homocysteine to methionine.[758] Elevated levels of homocysteine have been linked to an increased risk of developing AD, although the mechanism involved remains uncertain. Patients with AD have higher homocysteine levels than controls, but a causal relationship is not supported.[245,380,673] Administration of B vitamins has, however, been shown to protect against brain atrophy. It has been proposed that B vitamins can lower homocysteine in some patients, decreasing rates of brain atrophy, and slowing cognitive decline.[222]

Head Injury

Head injury may increase the risk of AD.[447,599] It has been suggested that this is particularly true for individuals with an *APOE ε4* allele and links to an amyloid cascade model of pathogenesis of AD.[446] Those who are involved in contact sports leading to repetitive head trauma (including boxers and football players) may develop mild cognitive deficits, and those with long careers and repeated head trauma may progress to 'dementia pugilistica' or chronic traumatic encephalopathy (CTE)[79,191] (see Chapter 10). Neuropathological changes include those associated with AD, such as diffuse Aβ deposits as well as neurofibrillary tangles and neuropil threads. Additional changes include Lewy bodies and TDP-43 pathology in some cases.[79,191,604,841] Clinical and neuropathological features that might distinguish between AD changes in chronic traumatic encephalopathy and AD have been proposed.[161,827]

Other Risk Factors

Additional risk factors for AD that have been suggested, but for which evidence is weaker, include a family history of Parkinson's disease, a family history of Down's syndrome, hypothyroidism, a history of depression and maternal age. There are no conclusive epidemiological risk factors related to dietary factors in AD.[559]

Protective Factors

Protective factors that have been implicated in some studies include smoking, alcohol, education, and the use of oestrogen, non-steroidal anti-inflammatory drugs or cholesterol-lowering drugs such as statins.[228,243,473,591,748,768] Nicotine has been proposed as offering protection by preventing amyloid accumulation,[163,369] but experimental animal studies suggest that this may increase tau deposition.[681]

Macroscopic Pathology of Alzheimer's Disease

The degree of cortical atrophy noted at autopsy varies widely in AD and may correlate with cognitive decline.[637] Atrophy, exemplified by narrowing of the gyri and widening of the sulci, is seen in the frontal, temporal and parietal lobes, whereas relative sparing of the occipital cortex is common (Figure 16.1). On external examination, atrophy of the parahippocampal gyri in the medial part of the temporal lobes may be particularly prominent. Strikingly disproportionate frontal or temporal atrophy ('lobar sclerosis') may raise the possibility of Pick's disease in the differential

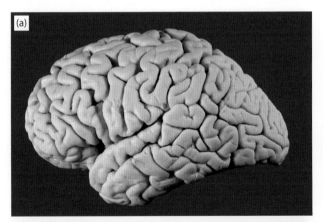
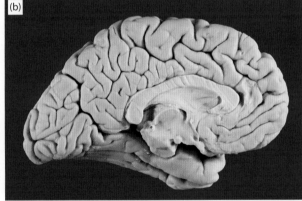
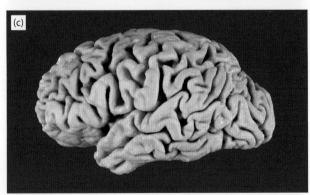
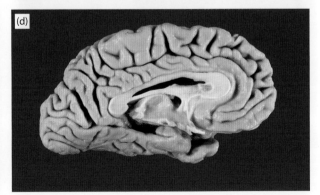

16.1 Cortical atrophy in Alzheimer's disease. (a,b) Lateral and medial views of normal brain from a 73-year-old man without neurological disease (brain weight 1371 g). **(c,d)** Atrophy of the brain of a 71-year-old man with Alzheimer's disease (brain weight 1063 g).

Photographs by Paul Reimann, kindly provided by Dr Gary Van Hoesen, University of Iowa, Iowa City, IA, USA.

diagnosis. Similarly, asymmetrical peri-Rolandic atrophy may suggest an alternative condition, such as corticobasal degeneration.

Slices through the cerebral hemispheres reveal variable narrowing of the cortical grey ribbon and widening of the sulci. The Sylvian fissure is often enlarged. The centrum semiovale may be reduced in volume, usually in proportion to the degree of cortical atrophy. Secondary hydrocephalus, exemplified by dilatation of the lateral ventricles and third ventricle and usually reflecting the degree of cortical atrophy, is common, the temporal horns of the lateral ventricles often being disproportionately enlarged (Figure 16.2). Two neighbouring structures, the amygdala (Figure 16.2) and the hippocampus (Figure 16.3), are both involved histologically in virtually all cases of AD and display atrophy in many cases. MR imaging of patients with AD demonstrates cortical and hippocampal atrophy that correlates with these post-mortem findings (Figure 16.4). 8F-Fluorodeoxyglucose positron emission tomography (FDG-PET) commonly shows reduced glucose metabolism in the posterior cingulate and precuneus and in the parietotemporal regions.[345] Imaging studies using *in vivo* amyloid imaging agents, such as the Pittsburgh compound-B, are able to detect early abnormalities in AD that correlate with macroscopic regions found to show atrophy in autopsy studies.[150,945]

Less commonly, AD pathology may be accentuated in the peri-Rolandic cortex, including the motor cortex, and may mimic corticobasal degeneration.[393] Rare cases of AD exhibiting the clinical syndrome of visual disorientation (Balint's syndrome)[616] may show more prominent involvement of the occipital and posterior parietal cortex[382,383] than in typical AD.

The gross appearance of the basal ganglia, thalamus and hypothalamus is usually unremarkable. The midbrain exhibits pallor of the substantia nigra in about one-third to

one-quarter of AD cases.[625] Pallor of the locus coeruleus in the rostral pons is common in AD, particularly in long-standing cases.[625]

Histopathology of Alzheimer's Disease

The two histological features that define AD are neurofibrillary tangles and extracellular Aβ deposits within plaques. In addition to these key diagnostic abnormalities, there is a range of other pathological features that are frequently encountered in AD but not included in current neuropathological criteria for its diagnosis. The pathology of AD includes synaptic loss, neuronal loss and reactive proliferation of astrocytes as well as microglia.[231] Neuropathological criteria have been agreed for diagnosis (see p. 886).[402,629]

Neurofibrillary Tangles

A major histopathological hallmark of AD is the neurofibrillary tangle, an intracellular filamentous lesion that often occupies the cell body and extends into the apical dendrite (Figure 16.5). Neurofibrillary tangles are not specific for AD as they are found in the brains of normal individuals as well as in a variety of other conditions.[969] Among the disorders associated with neurofibrillary tangles are PSP, dementia pugilistica, subacute sclerosing panencephalitis, Niemann-Pick disease type C, parkinsonism-dementia complex of Guam, Down's syndrome, postencephalitic parkinsonism, myotonic dystrophy, congenital muscular dystrophy (Fukuyama type), Kufs' disease, Cockayne's syndrome, Hallervorden–Spatz disease and Gerstmann–Sträussler–Scheinker syndrome.[290,552,867] Yet, despite the fact that neurofibrillary tangles occur in many other conditions, some of which are quite rare, they constitute an important correlate of cognitive dysfunction in AD. Their density parallels the duration of AD and severity of dementia.[34,73,652] Moreover, neurofibrillary tangles are a key diagnostic feature of AD.

Although some neurofibrillary tangles may be apparent in haematoxylin and eosin (H&E)-stained sections, where they have a slightly basophilic appearance (Figure16.5a), other stains are required for optimal detection. These include a variety of silver impregnation methods, such as modified

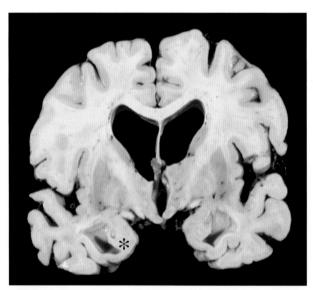

16.2 Coronal slice from a case of Alzheimer's disease. Characteristic atrophy of the entorhinal cortex is seen, with resultant dilation of the temporal horn of the lateral ventricles.

*Amygdala.

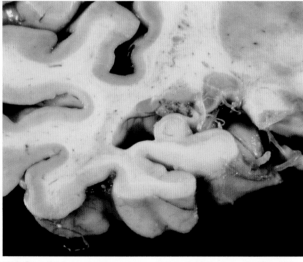

16.3 Small hippocampus in a case of Alzheimer's disease.

Bielschowsky, Gallyas (Figure 16.5b), Sevier–Munger and Bodian preparations. In addition, tangles are effectively demonstrated by immunohistochemistry with antibodies directed against tau protein, particularly antibodies to phospho-epitopes in tau, the major structural component of neurofibrillary tangles.[529] Other antigenic determinants have been identified within neurofibrillary tangles and can be used in diagnostic evaluation, e.g. immunohistochemistry for ubiquitin[554] and P62.[508]

The morphology of neurofibrillary tangles varies with the nature of the neurons in which they reside (Figure 16.6). In the pyramidal cells of the hippocampus, for example, they are usually flame shaped, whereas those in the multipolar neurons of the nucleus basalis of Meynert or locus coeruleus usually have a globose contour. In certain regions, such as the CA1 sector of the hippocampus and the pri-α layer of the entorhinal cortex, neurofibrillary tangles survive after the neurons that harbour them have died. These extracellular neurofibrillary tangles or 'ghost tangles', the tombstones of once-viable neurons, are detected on silver and other routine stains (Figure 16.7), but may also be labelled with antibodies to proteins not ordinarily detected in tangles, such as glial fibrillary acidic protein (GFAP) and Aβ. In this instance, it is believed that the extracellular neurofibrillary tangles acts as a scaffold for ingrowth of astrocytic processes or deposition of Aβ.

By electron microscopy, neurofibrillary tangles exhibit a progression of change. An early stage appears to be the

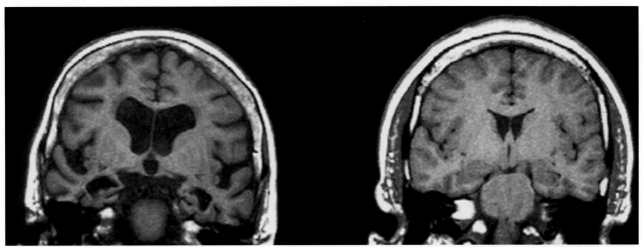

16.4 Magnetic resonance imaging of a normal control (right) and a patient with advanced Alzheimer's disease (left). Coronal sections. Note the widened sulci, enlarged ventricles and gaping temporal horns in the patient with Alzheimer's disease, reflecting cortical and hippocampal atrophy.

Courtesy of Dr J Growdon, Massachusetts General Hospital, Boston, MA, USA.

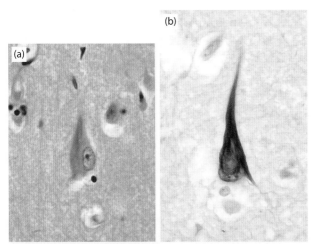

16.5 Neurofibrillary tangle. (a) In haematoxylin and eosin sections, some tangles can be seen as ill-defined basophilic structures in the cell body. A variety of silver stains show tangles as skein-like structures. This example **(b)** shows a tangle within the cell body and apical dendrite of a pyramidal cell in the hippocampus. Cytoplasmic detail and nucleus are not shown well with this technique. Gallyas.

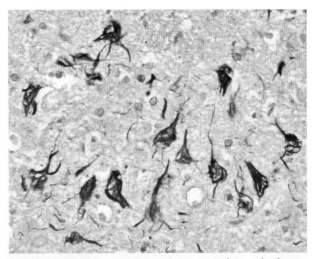

16.6 Neurofibrillary tangles adopt a variety of shapes, depending on the type of neuron in which they develop. These flame-shaped and branching tangles are from modified pyramidal neurons (stellate cells) in the entorhinal cortex. Tangles that occur in small cortical neurons adopt a compact rounded ball-like shape. Bodian silver stain.

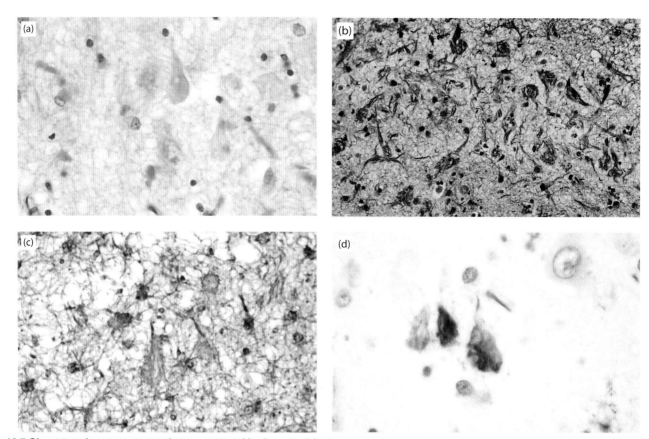

16.7 Ghost tangles are commonly encountered in the entorhinal cortex and pyramidal cells of the hippocampus. They appear as **(a)** pale tangle-shaped structures in haematoxylin and eosin-stained sections and are also detected by a variety of silver stains, **(b)** such as Bielschowsky. **(c)** Astrocytic gliosis in affected areas may be associated with ingrowth of glial processes into the extracellular tangle, as seen on glial fibrillary acidic protein (GFAP) immunostaining. **(d)** Ghost tangles may also act as a nidus for the deposition of amyloid and be detected by immunostaining for Aβ.

accumulation of perinuclear aggregates of paired helical filaments (PHFs) and other filaments, some in relation to nuclear pores (Figure 16.8). This change progresses to the formation of a densely compacted mass of PHFs within the cell body and apical dendrite impinging upon the nucleus and cytoplasmic organelles (Figure 16.9). The fine structure of filaments in neurofibrillary tangles has been extensively studied. They appear to be composed of two 10-nm filaments helically wound with regular periodic constrictions every 80 nm. Although neurofibrillary tangles were initially thought to represent twisted tubules,[880] Kidd proposed the paired helical filament double helix model, which is the term that has persisted.[477] Using tilt-stage electron microscopy with X-ray images of scale models, Wisniewski and co-workers[965] confirmed a bifilar helical structure and negative staining techniques combined with electron microscopy revealed a substructure of four protofilaments, each 3–5 nm in diameter.[966] More recent studies using atomic force microscopy suggest that PHFs are better modeled as twisted ribbons.[732] In some conditions, such as PSP and to a lesser extent AD, neurofibrillary tangles are composed predominantly of straight filaments.[825,975] Phosphorylated tau protein was shown to be a major antigenic constituent of both neurofibrillary tangles (Figure 16.10) and the dystrophic neurites associated with senile plaques.[193,493,971] Biochemical studies confirmed these observations.[170,304,963] Thread-like processes that are often numerous in the neuropil of regions vulnerable to neurofibrillary tangles, so-called neuropil threads,[94] have a similar ultrastructural and immunohistochemical profile to the tangles.

The tau protein in AD is hyperphosphorylated and is understood to be the main component of the PHFs and straight filaments that form NFTs, neuropil threads, and plaque-associated dystrophic neurites. Neurofibrillary tangles contain epitopes spanning the entire length of tau molecule, and multiple phosphorylation sites have been identified many of which are serine or threonine residues preceding a proline residue. The mechanism by which native tau converts to PHFs remains debated and is likely to be multifactorial, although hyperactivity of kinases, hypoactivity of phosphatases or an imbalance between these two processes has been suggested. In addition to abnormal phosphorylation, there are other post-translational modifications that may also play a role in tau pathology including cleavage,[278] acetylation, glycation, ubiquitylation and nitration (reviewed by Kolarova et al.[489]).

Hierarchical Distribution of Neurofibrillary Pathology

The neuroanatomical distribution of neurofibrillary pathology has been the subject of a number of investigations. Early studies have shown that neurofibrillary tangles tend to occur in the amygdala, hippocampal formation, parahippocampal gyrus and the temporal association cortex, whereas Aβ plaques are more generally distributed throughout the neocortex. Arnold and co-workers

16

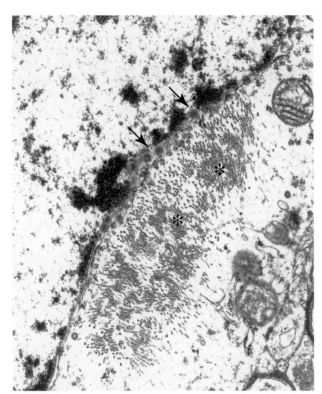

16.8 Alzheimer's disease: electron micrograph, showing an apparently early stage in the formation of a neurofibrillary tangle. Paired helical filaments (PHFs) (asterisks) are noted within the neuronal cytoplasm in close proximity to the nuclear membrane and nuclear pores (arrows).

examined 39 cytoarchitectural areas and found distinct hierarchical distributions for neurofibrillary tangles and plaques[32] (Figure 16.11). The most severe neurofibrillary degeneration occurred in the entorhinal cortex, CA1 and the subiculum of the hippocampus, the amygdala and the high-order association cortices in the temporal lobe. In general, the number of neurofibrillary tangles decreased on moving from a high-order (multimodal) association cortex (such as the superior temporal sulcus or inferior temporal gyrus) to a unimodal association cortex (that in immediate contact with sensory strips) and then to primary sensory cortices. Limbic and paralimbic areas, such as the temporal poles, the insula, the cingulate and the parahippocampal gyrus (which anatomically can be viewed as intermediate in position between the association cortices and their medial temporal lobe targets) were as affected as the high-order association cortices. In contrast, the pattern for plaques was somewhat more uniform, although the association cortices were generally more heavily involved with plaques than were primary sensory or motor areas. Thal and co-workers have proposed a staging of Aβ plaques whereby initial lesions are found in neocortical association cortices, followed by limbic areas, then basal ganglia and diencephalon and finally brain stem and cerebellum.[886] In contrast to neurofibrillary tangles, hippocampus and other medial temporal lobe structures are less severely affected by senile plaques than are neocortical areas.

Neurofibrillary tangles generally occur in a predictable laminar distribution, with some laminae being severely affected while others are spared. In the entorhinal cortex, for

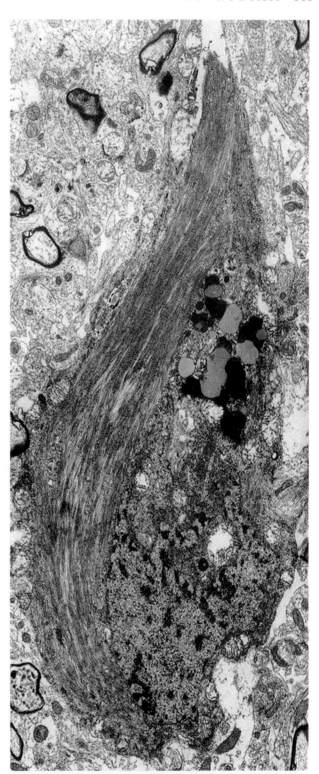

16.9 Alzheimer's disease: electron micrograph, showing advanced neurofibrillary degeneration (neurofibrillary tangle) in a cortical neuron. Note that the abnormal filaments have displaced the normal cytoplasmic organelles.

Courtesy of Dr L Carrasco, formerly of Runwell Hospital, Wickford, UK.

example, neurofibrillary tangles are virtually always present in large projection neurons of layers II and IV, whereas layers III, V and VI have relatively few[400] (Figure 16.12). In the

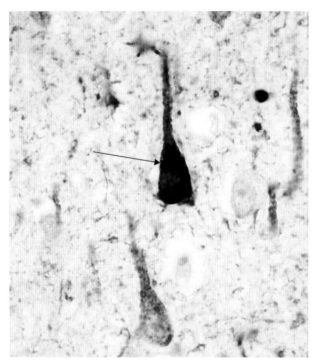

16.10 Tau immunostaining shows a pyramidal neuron containing a neurofibrillary tangle (arrow). Another neuron has a less dense and non-structured pattern of staining, representing abnormal tau protein not aggregated into paired helical filaments (PHFs) – a so-called pre-tangle. In the background are linear and granular deposits of tau, representing staining of tau in nerve cell processes, termed neuropil threads.

association cortex, the pyramidal neurons in layers II, III and V are, similarly, markedly affected. The distribution of these pathological changes has been postulated to disrupt feed-forward and feedback cortical–cortical projections as well as cortical–subcortical projections. Neurofibrillary tangles also tend to affect a selective subpopulation of neurons even within vulnerable laminae. In the visual association cortex (Brodmann area [BA] 18), a striking bilaminar distribution of neurofibrillary tangles and neuropil threads is visible with phospho-tau immunohistochemistry. Layers II, III and V are markedly affected. At the border between the association cortex and primary visual cortex (BA 17) (Figure 16.13) a pronounced change in immunostaining pattern occurs, with decreased labelling of the primary visual cortex.

In general, neurofibrillary tangle-bearing cells are large pyramidal neurons, primarily glutamatergic, and vulnerable neurons can often be labelled with antibodies to phosphorylated neurofilaments.[384] By contrast, other neuronal populations appear to be spared in AD. For example, neurons that can be labelled for calcium-binding proteins, nitric oxide synthase, γ-aminobutyric acid (GABA) and clusterin[296] are rarely affected by neurofibrillary tangles.

In 1991, Braak and Braak observed that the progression of neurofibrillary changes follows a predictable pattern[91] and subsequently confirmed these observations in a large cross-sectional autopsy study.[92] In examining autopsy brains from demented and non-demented individuals, they found a characteristic distribution of neurofibrillary tangles and neuropil threads, permitting differentiation

of six stages exemplified by progression from the transentorhinal and entorhinal layers to the neocortex. The first two stages involve neurofibrillary tangles in the entorhinal, transentorhinal (perirhinal) and CA1 and subicular portions of the hippocampal formation. Thereafter, in stages III and IV, increasing numbers of neurofibrillary tangles accumulate in the limbic system, and in stages V and VI in neocortical areas. The Braak staging of neurofibrillary degeneration is incorporated in the 2012 National Institute on Aging–Alzheimer's Association (NIA-AA) guidelines for the neuropathologic assessment of Alzheimer's disease neuropathological criteria for diagnosis of AD.[402,629] The BrainNet Europe Consortium has validated a scheme incorporating immunohistochemistry for hyperphosphorylated tau instead of silver staining for staging of neurofibrillary changes in AD.[8]

Clinicopathological studies have correlated Braak staging with clinical diagnosis and mental status.[660] Substantial involvement of the hippocampus with designation of Braak stage IV has been found to be key to the development of overt dementia. Braak staging also has provided important information on other dementing conditions, as well as in cases of combined pathologies coexisting with AD. Dementia with Lewy bodies has been found generally to be associated with lower Braak stages than are present in cases of pure AD, albeit higher than in controls.[354]

Evolution of amyloid deposition in a patterned manner similar to that of neurofibrillary degeneration is less apparent. However, Braak and colleagues examined different forms of Aβ deposition (diffuse plaques, neuritic plaques, fleecy deposits and band-like subpial deposits) in the medial temporal lobe and proposed four phases in the evolution of amyloid deposition.[885] They found that these four phases correlated significantly with the Braak staging of neurofibrillary degeneration.

Neuropil Threads

Neuropil threads[94] as their name suggests, are thread-like structures in the neuropil distributed widely in the grey matter in AD[603,710] (see Figure 16.10). Their density has been found to correlate with the severity of dementia.[603] Counts of neuropil threads correlate with numbers of neurofibrillary tangles.[590] The large portion of cortex occupied by neuropil threads and their prominent distribution within laminae II and III led Markesbery and co-workers to suggest that neuropil threads contribute to severe disorganization of intracortical and cortico-cortical connectivity.[583] Neuropil threads often have a dendritic localization[978] as evidenced by co-labelling the somatodendritic marker MAP-2.[37] A progression of cytoskeletal changes involving the dendrite and progressing to the cell body with tangle formation has been observed by Braak et al.[88] Neuropil threads are also frequently associated with extracellular neurofibrillary tangles in the hippocampus,[979] a structure that has been referred to a tangle-associated neuritic cluster.[642] Although observed near dendrites of neurofibrillary tangle-bearing neurons,[90] threads may also involve distal dendrites or axons in the amygdala or entorhinal cortex in the absence of local tangle-bearing cell bodies.[809]

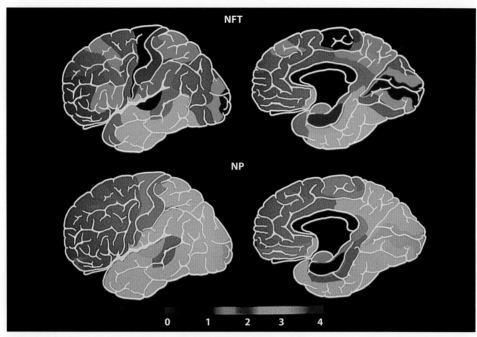

16.11 Topographical distribution of neurofibrillary tangles (NFT) and neuritic plaques (NP) in Alzheimer's disease. Thirty-nine cytoarchitectural fields were assessed on a 0–4 + scale for the presence of NFT or NP in 17 hemispheres. A consistent hierarchical distribution was observed.

Reproduced from Arnold et al.[32] By permission of Oxford University Press.

Neuropil threads are not specific to AD and may be observed in other conditions, including cortico-basal degeneration,[503] PSP,[740] Pick's disease,[149] and parkinsonism-dementia complex of Guam.[935] Their neuro-anatomical distribution is different in these other disorders. In the cortex, their density appears to be greater in AD than in other dementing disorders.[179] Immunohistochemical studies of neuropil threads have identified immunoreactivity for tau (see Figure 16.10), neurofilament protein, ubiquitin and P62.

Senile Plaques and Aβ Deposits

Immunolabelling for Aβ peptide in AD shows deposition within the neuropil, mainly as plaques (Figure 16.14 and Table 16.4). A range of morphological patterns has been described with immunohistochemistry for Aβ, some of which do not fit with older classical descriptions of plaques as assessed by silver impregnation.[967] As noted previously and described in greater detail later, Aβ (produced by pro-teolytic cleavage of APP) is not a single peptide species but rather a mixture of peptides of different length and that undergo different post-translational modifications. The major cleavage products of APP are Aβ peptide fragments of 40 or 42 amino acids in length.[142]

Focal Aβ Deposits and 'Senile' Plaques

Focal deposits of Aβ, so-called 'senile plaques', are mor-phologically diverse. The classical description of plaques is that of a spherical structure with central condensation and intense staining with amyloid stains, such as Congo red and thioflavine S. They usually have a diameter of 20–50 µm. Smaller, punctate Aβ deposits are visible with immunohistochemistry, but not usually with amyloid stains. Large diffuse Aβ deposits, sometimes over 200 µm in greatest diameter, often do not have a spherical structure or central condensation, and are weakly positive or negative with amyloid stains. Plaques are readily demonstrated with a variety of silver stains, such as the modified Bielschowsky stain. Other silver stains detect only the dystrophic neuro-nal processes associated with neuritic plaques. A low-power view of cortex in AD immunostained for Aβ reveals innumerable ('miliary') lesions (Figure 16.15). The commonly used H&E stain is not suitable for their optimal detection, showing at best only the central dense core and associated reactive glia (Figure 16.16). The reactivity of Aβ plaques with amyloid stains reflects the fact that the Aβ peptides in these lesions have a high content of β-pleated sheet second-ary structure, a feature common to all types of amyloid. At the ultrastructural level this corresponds to the pres-ence of fibrillar structures with characteristic dimensions and periodicity. The term 'fibrillized' has sometimes been used to describe Aβ with these properties, to distinguish it from Aβ deposits that are weakly positive or negative with amyloid stains. A term that is sometimes used for the latter type of Aβ deposit is 'preamyloid'.[110] It is worth noting that Aβ immunohistochemistry does not readily distinguish Aβ amyloid from preamyloid deposits. There is a wide range of focal Aβ deposits, and a variety of terms has been used to refer to these morphological variants (see Figure 16.14). Some of the more common terms and their usual meaning are summarized in Table 16.4.

Classical plaques or neuritic plaques are characterized by a dense central core of amyloid surrounded by a less com-pact peripheral halo of amyloid. The dense core and periph-eral halo are often separated by a clear zone that contains

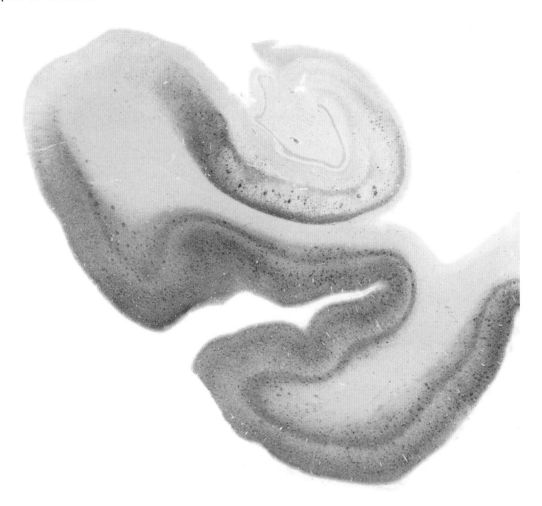

16.12 Immunohistochemical staining for phospho-tau protein, with AT8 antibody, in the temporal lobe at the level of the body of the hippocampus. Staining is most marked in the CA1 field and subiculum of the hippocampus and in layers III and V of the posterior part of the parahippocampal gyrus. Note the characteristic sparing of CA2. The punctuate labelling represents tau protein in neuritic plaques. The dark background staining represents a mixture of neurofibrillary tangles and neuropil threads.

glial cells and dystrophic neuronal processes (Figure 16.17). In AD this type of plaque is associated with morphologically and immunohistochemically heterogeneous dystrophic neurites (Figure 16.18). The presence of the abnormal neuronal cell processes defines the plaque as being neuritic, which is the type of plaque that is critical for neuropathological diagnosis of AD. Plaque-associated dystrophic neurites are abnormally distended, often radially orientated, neuronal processes that contain filamentous tau protein similar to that within neurofibrillary tangles, as well as a variable mixture of intact and degenerating cytoplasmic organelles and vesicles (see Figure 16.20). Some plaques have a perivascular orientation, usually in association with amyloid angiopathy. Microglia often abut the dense central core. Reactive astrocytes often surround the periphery of the plaque, especially those with dense amyloid cores, and increased expression of GFAP correlates significantly with neuritic plaque count.[696] Immunostaining with antibodies to specific forms of Aβ typically shows that the dense central core is enriched in Aβ40 whereas the periphery has predominantly Aβ42.[422] Neuritic plaques are present in multimodal association cortices in AD but are less common in primary motor and visual cortices. They are especially prominent in the amygdala and hippocampal subicular complex.

The fine structure of neuritic plaques highlights their complexity, as described in 1964 in the pioneering electron microscopic studies by Kidd[477] and Terry *et al.*[881] Amyloid cores with radial spikes of fibrillar amyloid may be present, surrounded by prominent astrocytic processes or abnormal neurites (Figure 16.19). Other plaques display less well-defined masses of fibrillar amyloid filling the extracellular spaces (Figure 16.20). Processes of microglia interdigitate with and surround wisps of fibrillar amyloid. Amyloid is often seen within coated pits on the surface of the cell membrane and inside coated vesicles within the cytoplasm, suggesting a process of receptor-mediated endocytosis[772] (Figure 16.21). Aβ may also be detected in the cytoplasm of plaque astrocytes[505] and can also be detected within neuronal cytoplasm.[320]

Lamellar or dense bodies are a major component of plaque neurites. Shown to contain acid phosphatase,[866] lamellar bodies are apparently lysosomal. These structures often coexist, in varying proportions, with clusters of PHFs (Figure 16.22). Our understanding of these lamellar bodies remains limited, although work by Nixon and others suggests activation of the neuronal lysosomal system in AD.[675] Processes containing only PHF are also common. Plaques in the molecular layer of the cerebellar

cortex may exhibit abnormal distended neurites that are immunoreactive for ubiquitin, but not PHF or tau.[864]

The abnormal neurites in plaques are often axonal, as is apparent by their relationship to postsynaptic densities and the presence of presynaptic vesicles. Less commonly, the abnormal neurites are dendritic. A spectrum of synaptic alterations is also appreciated within plaques.[311] Rarely, PHF-like structures are observed within myelinated axons.

Normal myelinated axons and other normal-appearing cellular processes course through the plaque (Figure 16.23) Scanning confocal microscopy has highlighted the three-dimensional structure of individual plaques.[172] Further reconstructions have revealed that the neuronal processes crossing the plaques are often morphologically different from contiguous non-plaque neurites[485] (Figure 16.24). Investigations using confocal microscopy suggest that the neuronal damage is linked to the structural characteristics of amyloid plaques, with fibrillar forms being associated with dystrophic neuritic changes. Further, the cytoskeletal changes in neuritic degeneration is different in preclinical as compared to end-stage AD, suggesting that tau accumulation is a later stage effect and more closely linked to dementia.[216]

Primitive or immature plaques are spherical deposits of amyloid that is uniformly dispersed without any core structure. They are enriched in Aβ42 and often do not have prominent dystrophic neurites. They are more abundant in superficial cortical layers, whereas plaques with dense amyloid cores are more common in deep cortical layers. Primitive plaques are also the most common type of plaque in the corpus striatum.

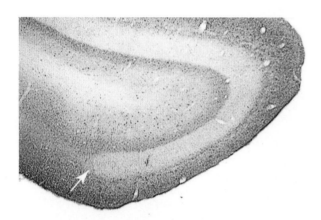

16.13 Alz50 anti-tau immunohistochemistry of the visual cortex. A clear bilaminar distribution of Alz50 immunoreactivity is observed in the visual association cortex, and there is a more diffuse, fainter pattern in the primary visual cortex. The change in staining pattern at the cytoarchitectural boundary (arrow) is dramatic.

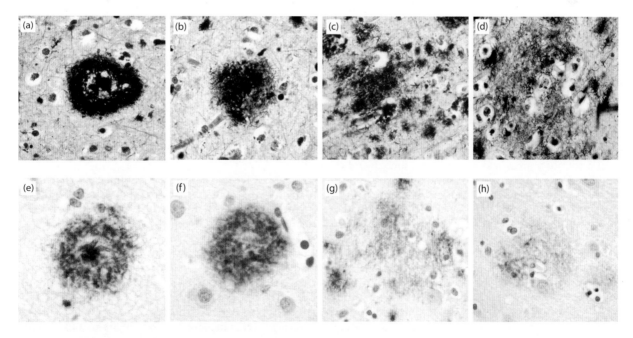

16.14 Silver staining and Aβ (a–d) immunohistochemistry, (e–h) showing a range of plaque morphologies. Focal deposits can be as cored **(a,e)** or primitive **(b,f)** plaques. Punctate condensations of amyloid may be seen within the neuropil **(c,g)**. Diffuse plaques are ill-defined and generally pale-staining **(d,h)**.

TABLE 16.4 Categories of neuropil deposits of Aβ peptide

	Amyloid	Neurites	Significance and associations
Focal Aβ deposits			
Classic or neuritic plaques	Dense central core enriched in Aβ40 and a less compact peripheral halo enriched in Aβ42	Heterogeneous neurites containing filaments and dystrophic organelles and displaying variable immunoreactivity for tau, APP and ubiquitin	Characteristic of AD and important lesions in neuropathological diagnostic criteria for AD
Primitive or immature plaques	Spherical deposits of predominantly Aβ42 dispersed within the neuropil without a dense core	Neurites either are absent or if present are not specifically increased in the vicinity of these plaques	Common in non-demented elderly people and in certain brain regions, such as the striatum
Compact or burnt-out plaques	Dense core of predominantly Aβ40	Neurites either are absent or if present are characterized by APP and ubiquitin immunoreactivity, but not tau	Common in the Purkinje cell layer of the cerebellum and globus pallidus
Cotton-wool plaques	Amyloid deposits are not compact and are composed predominantly of Aβ42	Neurites, including tau-immunopositive neurites, are variable	Have a distinctive appearance and are readily detected with routine stain. Originally described in histological methods (e.g. H&E) in familial AD with mutations in presenilin gene
Diffuse Aβ deposits			
Diffuse plaques	Similar to primitive plaques, but often have irregular contours rather than spherical shape and often encompass normal-appearing neurons and glia. In contrast to primitive plaques, they are stained weakly or negative with amyloid stains	Similar to primitive plaques, they usually have no neurites. When seen in the setting of AD, they may be permeated by tau-immunopositive neuropil threads, but neurites are not concentrated in the vicinity of diffuse plaques	Common in non-demented elderly people and also in dementia with Lewy bodies. Certain anatomical regions, such as the molecular layer of the cerebellum, typically have diffuse plaques
Fleecy, lake-like and subpial Aβ deposits	Similar to diffuse plaques with enrichment in Aβ42 and possibly amino-terminal truncated fragments of Aβ. Stained weakly or negative with amyloid stains	Similar to primitive and diffuse plaques, they usually have no neurites	Fleecy deposits are found in entorhinal layers, and lake-like deposits are typical of the parvocellular layer of the presubiculum. Subpial band-like deposits are often associated with amyloid angiopathy

AD, Alzheimer's disease; APP, amyloid precursor protein; H&E, haematoxylin and eosin.

Compact or burnt-out plaques are characterized by a dense core without a peripheral rim of non-compact amyloid and usually without much neuritic dystrophy or reactive glia. This type of plaque is most common in the globus pallidus. Dense-cored or compact plaques are also more common in primary cortices, such as the visual and motor cortices, whereas neuritic, primitive and diffuse plaques predominate in multimodal association areas and in the limbic lobe. Increased amyloid burden, as measured with amyloid imaging, is associated with changes in cortical network architecture.[738]

Cotton-wool plaques are focal deposits of Aβ that may readily be detected with routine histological methods but are poorly visualized with amyloid stains (Figure 16.25). They are spherical lesions that stand out against the pale-staining background of the neuropil and often seem to displace normal structures as would a mass lesion. They are composed predominantly of Aβ42. Cotton-wool plaques were originally described in familial AD with exon 9 deletions in the presenilin 1 (*PSEN1*) gene,[168] the most common cause of autosomal dominant AD. Subsequently, they have been described in cases with other types of *PSEN1* mutations and even

occasionally in some sporadic cases of AD.[524,989] A similar appearing plaque is also found in familial British dementia, a heritable degenerative dementia associated with plaques that do not contain Aβ, but rather a novel type of amyloid, ABri.[725]

Diffuse Aβ Deposits

Diffuse Aβ deposits are morphologically diverse. They usually are not spherical structures, but rather have irregular contours. In contrast to neuritic plaques, diffuse plaques are more amorphous and lack dystrophic neurites[407,408,977] (Figure 16.26). They may be very large, up to and exceeding 200 μm in greatest diameter. They are negative or weakly stained with amyloid stains and have been referred to as preamyloid deposits. They can be detected by some silver stains (e.g. modified Bielschowsky stain), but are best appreciated with immunohistochemistry for Aβ. They are enriched in Aβ42 and also appear to contain amino-terminally truncated or modified Aβ42: antibodies to carboxyl terminal epitopes of Aβ do not stain them, and brains with predominantly diffuse

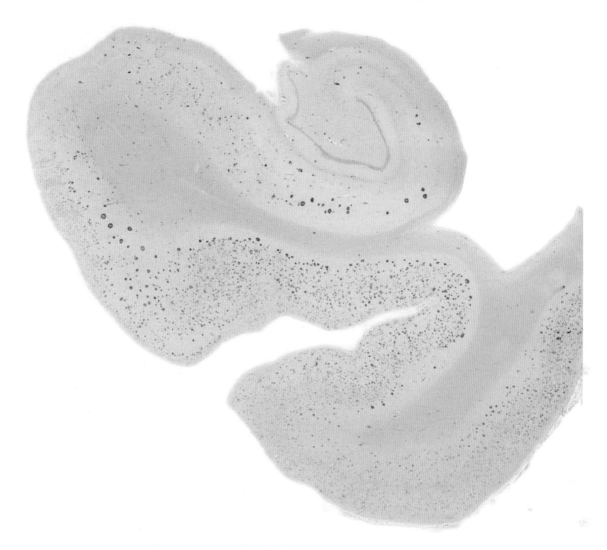

16.15 Immunostaining for Aβ peptide, the main component of plaques, in a case of Alzheimer's disease. Plaques are seen readily, even at this low magnification. This is a serial section to that shown in Figure16.12; the Aβ plaques can be mapped on to the dense punctuate staining of tau, representing the plaque neurites.

amyloid deposits are enriched in Aβ17–42 rather than full-length Aβ1-42.[321] The proportion of plaques immunopositive for Aβ40, as well as the amount of Aβ40 in the cortex, is greater in individuals carrying an *APOE* ε4 allele.[282] Ultrastructurally, diffuse plaques have little or no detectable fibrillar amyloid.[977,980]

Diffuse plaques are often numerous in the multimodal association and limbic cortices of non-demented or minimally impaired elderly individuals, and they are also the most abundant type of plaque in many cases of Lewy body dementia.[209] Diffuse plaques are the type most often found in the molecular layer of the cerebellar cortex (see Figure 16.31).

Other large diffuse Aβ deposits have been described as fleece or lake-like amyloid deposits (Figure 16.27). As in smaller diffuse plaques, these larger deposits are composed of loosely arranged Aβ deposits that lack amyloid staining properties.[885,968] Lake-like deposits are common

in the parvocellular layer of the presubiculum. Fleecy deposits and band-like subpial deposits are common in association with cerebral amyloid angiopathy. The extent of subpial amyloid deposition correlates with the severity of amyloid angiopathy.[419] The amyloid in fleecy deposits has been suggested to be amino-terminally truncated Aβ.[884]

Based upon patterns of amyloid deposition as plaques and in vessels (cerebral amyloid angiopathy [CAA]) four types of AD have been proposed.[11]

- Type 1: plaques are a dominant feature, with or without CAA within leptomeningeal vessels alone.
- Type 2: plaques are a dominant feature and CAA is present in both leptomeningeal as well as in deeper penetrating arteries.
- Type 3: capillary CAA is seen along with plaques and arterial CAA.

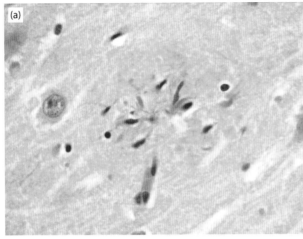

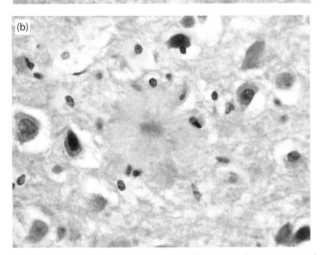

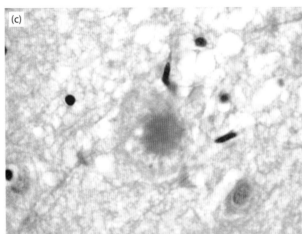

16.16 In haematoxylin and eosin-stained sections, areas of amyloid deposition may be suspected by the associated glial response, seen as radially arranged nuclei (a). Not all plaques have such a prominent glial response. Amyloid plaques may show a core-and-halo pattern within the neuropil (b). Dense amyloid cores can be seen as eosinophilic structures (c), which also stain strongly with Congo red.

- Type 4: a pattern of disease where amyloid is seen in and around blood vessels strongly associated with possession of the *APOE* ε4 allele, with typical plaques less evident.

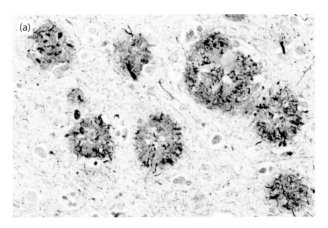

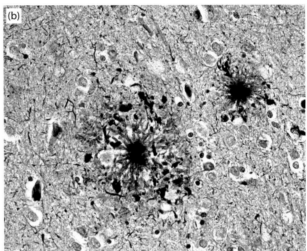

16.17 Classic or neuritic plaques stained using a silver technique that highlights the peripheral radially arranged black-stained abnormal neurites surrounding amyloid, which is seen as a region of ill-defined yellow staining in this type of preparation (Bielschowsky). The plaques in (a) do not have distinct cores. The plaques in (b) have dense central amyloid cores.

Aβ Plaque-Associated Molecules

Immunohistochemistry has revealed a wide range of molecules in Aβ plaques, either directly related to the amyloid deposits or within cells associated with plaques. Molecules associated with dystrophic neurites include tau,[971] APP,[414] growth-associated protein (GAP-43),[589] protein kinase C,[587] ubiquitin,[709] brain spectrin,[588] neurofilaments,[26] synaptophysin and chromogranin.[100] Neurotransmitters, including acetylcholine and substance P, have also been identified in neurites, indicating lack of specificity of neuritic degeneration to a particular class of neurons.[30]

Molecules associated with the amyloid deposits, *per se*, include apolipoprotein E, complement factor 1q, glycosaminoglycans, and serum amyloid protein.[203,630] Plaque-associated molecules can be grouped into six categories, which are related to their origin or class (Table 16.5). Their presence reflects deposition of amyloid in the extracellular space, products related to damage to adjacent cells, inflammatory response elements, and elements related to aggregation or dispersal of the amyloid.

TABLE 16.5 Senile plaque-associated molecules
Acute phase proteins
α1-Antichimotrypsin
α1-Antitrypsin
Serum proteins
Albumin
Serum amyloid P
Immunoglobulins
Transferrin
α2-Macroglobulin
Complement (C1q)
Apolipoproteins
ApoE
ApoJ
Extracellular matrix components
Proteoglycans (e.g. perlecan, agrin)
Acetylcholinesterase
Laminin
Midkine
Entactin
Metal ions
Zn^{2+}
Cu^{2+}
Fe^{3+}
Lipids
GM1 ganglioside
Cholesterol

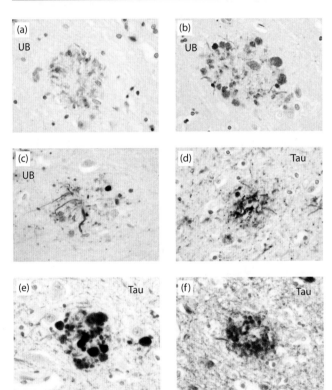

16.18 Neuritic staining in senile plaques. Anti-ubiquitin labels plaque neurites **(a–c)**. In some plaques there is fine granular staining **(a)**, while others include bulbous neurites **(b)**. These types of plaque may or may not contain tau-immunoreactive neurites. Plaques showing ubiquitin staining with both granular and curvilinear patterns **(c)** are generally associated with tau immunoreactivity. **(d–f)** Tau-immunoreactive neurites associated with plaques, reflecting the morphology seen in ubiquitin-stained sections. Note the tau-reactive neuropil threads. Although threads can be ubiquitin-immunoreactive, most cases do not show strong positivity for ubiquitin.

Cerebral Amyloid Angiopathy

Another major histopathological feature of AD is amyloid deposition within the walls of arteries and arterioles in the subarachnoid space and the cerebral and cerebellar cortex (Figure 16.28). Although CAA occurs independently, particularly in elderly individuals,[326,419] it is present to some degree in virtually all cases of AD. Capillaries and veins may also be involved but blood vessels in the white matter are generally spared. The amyloid tends to be associated with the vascular smooth muscle cells or pericytes, and Aβ deposition in periarterial interstitial fluid drainage pathways of the brain may contribute to CAA.[946] In cases of extensive amyloid deposition within cortical blood vessels, perivascular plaques may be seen in close proximity to blood vessel walls (i.e. dyshoric angiopathy). Amyloid angiopathy tends to be most severe in the occipital cortex, but its severity bears little or no relationship to the frequency with which plaques and neurofibrillary tangles are found. Vasculitis may occur with CAA, rarely in association with giant cells.[862] Other unusual associated pathologies include presentation as an inflammatory pseudotumour.[731] New data on the relationship of CAA and other forms of cerebrovascular disease (see Chapter 2), with and without dementia, have emerged and are discussed in the section on vascular dementia.

The ε4 and ε2 alleles of *APOE* have been linked to the severity of CAA.[682] Possession of the *APOE* ε4 allele favours vascular over parenchymal accumulation of Aβ in AD.[134] There is increasing evidence that patients with CAA may be predisposed to intracerebral haemorrhage as a complication of therapeutic thrombolysis.[600]

Cerebral Amyloid Angiopathy and Familial Disorders

Links between vascular disease, amyloid and dementia are exemplified by the group of familial disorders associated with cerebral amyloid angiopathy, strokes and dementia. These include hereditary cerebral haemorrhage with amyloidosis (Dutch, Icelandic and Finnish types), familial amyloid angiopathy with deafness and ocular haemorrhage (Danish type), and familial British dementia with amyloid angiopathy.[754,755]

Mann *et al.*[578] correlated amyloid angiopathy and variability in Aβ deposition in presenilin-1-linked AD to the position of the presenilin mutation. They found that cases with mutations between codons 1 and 200 exhibited mild or moderate CAA, many cortical diffuse plaques and few cored plaques, whereas cases with mutations after codon 200 had increased numbers of cored plaques, often clustered around blood vessels with severe amyloid angiopathy.

Other Microscopic Changes in Alzheimer's Disease

In addition to the two cardinal histopathological lesions that are essential for the neuropathological diagnosis of AD, there are a host of other histopathological and biochemical changes in the brain in AD. Some of these are helpful in neuropathological diagnosis and others are more of research interest.

Hirano Bodies

Hirano bodies were first described by Hirano as 'rod-like bodies' in the hippocampus of individuals with parkinsonism–dementia complex of Guam.[376] Hirano

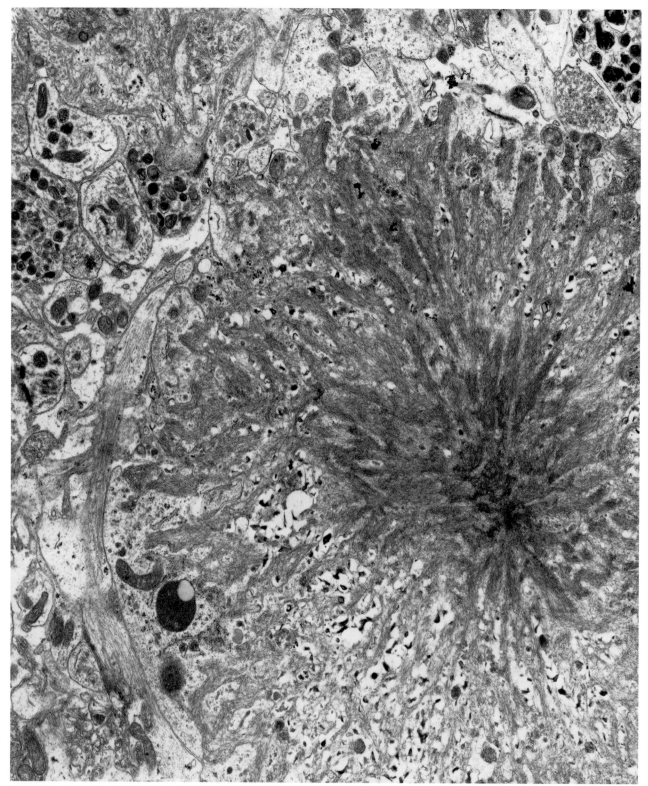

16.19 Alzheimer's disease: electron micrograph of the amyloid core of a neuritic plaque. Radiating bands of amyloid fibrils comprise the core. Note the adjacent abnormal neurites filled with dense bodies.

Reprinted from Mirra SS. Apolipoprotein E and the neuropathology of Alzheimer's disease (Editorial). Hum Pathol 1999;30:1125–7. With permission from Elsevier.

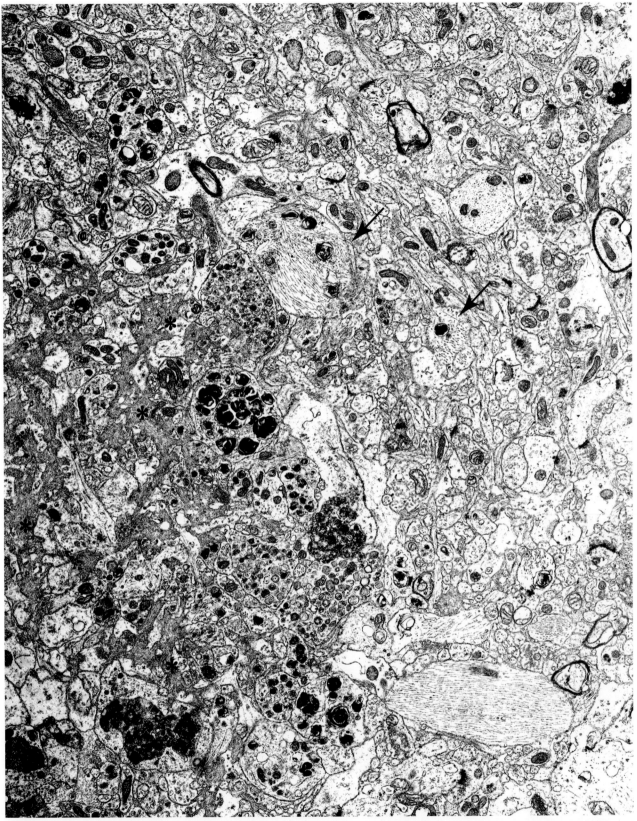

16.20 Electron micrograph of part of a neuritic plaque. Numerous neurites filled with dense bodies and paired helical filaments (arrows) along with intercellular accumulations of amyloid (asterisks) are seen. The relatively normal adjacent neuropil with many synapses is present on the right.

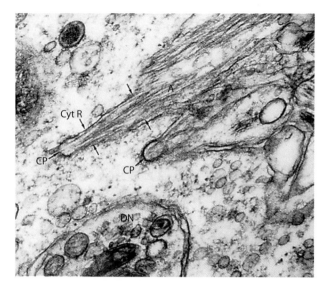

16.21 Amyloid fibrils within a neuritic plaque. Amyloid fibrils (A) are seen in close opposition to the cytoplasmic boundaries (Cyt R, arrows) of a cell within a neuritic plaque. The amyloid occupies tunnel-like invaginations of the cell surface that terminate in coated pits (CP). A dystrophic neurite (DN) filled with dense bodies is seen at the bottom left.

Courtesy of Professor EG Gray, formerly of University College London, London, UK.

bodies are refractile, brightly eosinophilic, rod-shaped, ovoid or spherical structures. By light microscopy, they appear to be located immediately adjacent to hippocampal pyramidal cells, most frequently in CA1 (Figure 16.29). In actuality, Hirano bodies are intracellular structures within neuronal processes and have a characteristic crystalloid or herringbone fine structure revealed by electron microscopy (Figure 16.30). They may occur in other regions of the central nervous system and have been reported in a Purkinje cell in AD,[981] as well as in peripheral nerves[38] and extraocular muscles.[901] Hirano bodies have also been observed in central and peripheral nerve tissues in experimental animals,[61,974] found frequently in the brains of normal elderly individuals[299] and occasionally in young adults. Hirano bodies commonly occur in AD as well as in other conditions such as Pick's disease and amyotrophic lateral sclerosis (ALS)–parkinsonism–dementia complex of Guam.[374] Often Hirano bodies coexist in the same cell with neurofibrillary tangles or granulovacuolar degeneration. The significance of Hirano bodies remains uncertain but they seem to represent a degenerative change in the cytoskeletal microfilament system.

Hirano bodies share antigenic determinants with several cytoskeletal proteins, including most prominently F-actin[309] and actin-binding proteins such as α-actinin, tropomyosin, vinculin, and cofilin. Immunolabelling has

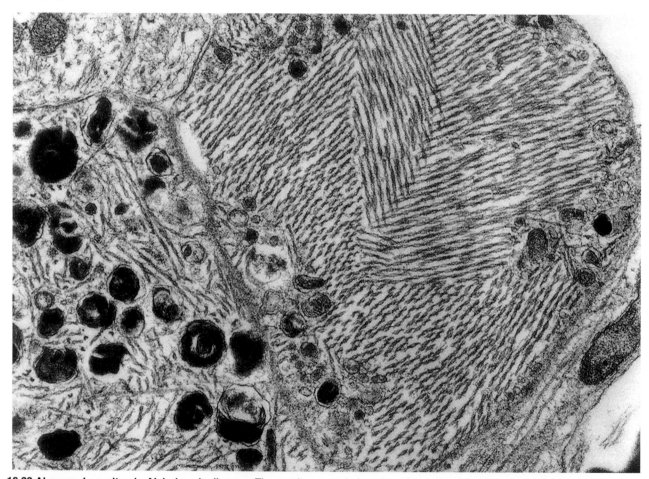

16.22 Abnormal neurites in Alzheimer's disease. The neurites contain bundles of paired helical filaments and dense bodies.

Courtesy of Dr L Carrasco, formerly of Runwell Hospital, Wickford, UK.

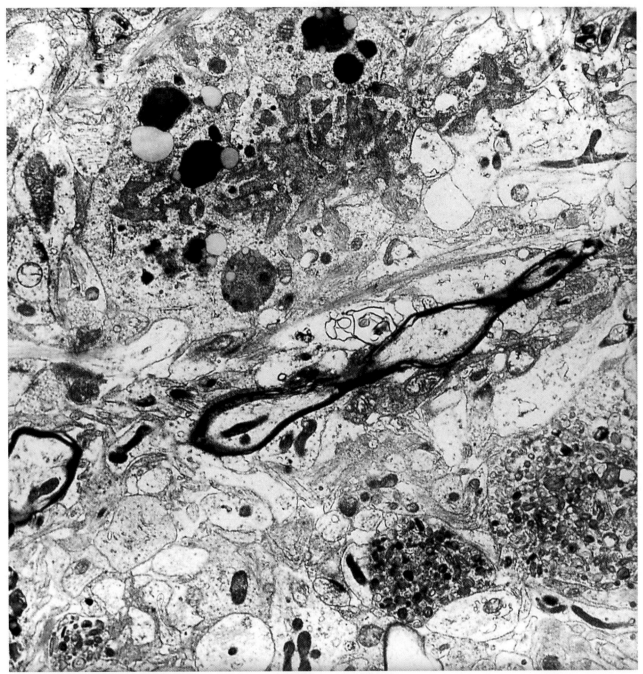

16.23 Part of a neuritic plaque containing amyloid and distended neurites. The neurites are filled with dense bodies (bottom right and left). Note an uninvolved myelinated axon coursing through the plaque (centre).

TABLE 16.6 ABC scoring scheme for AD neuropathological changes

A	Thal Plaque Phase	B	Braak NFT Stage	C	CERAD plaque score
0	0	0	None	0	None
1	1 or 2	1	I or II	1	Sparse
2	3	2	III or IV	2	Moderate
3	4 or 5	3	V or VI	3	Frequent

Modified from Hyman, BT, *et al*. National Institute on Aging-Alzheimer's Association guidelines for the neuropathologic assessment of Alzheimer's disease. *Alzheimer's Dement* 2012; Jan;8(1):1–13.

been reported for ubiquilin-1, APP-C, TGF-β3, hippocampal cholinergic neurostimulating peptide-related components SMURF1 and FAC1 (summarized by Makioka *et al.*[573] and Satoh *et al.*[805]). Model systems for the study of these inclusions have been described.[180,334,757]

Granulovacuolar Degeneration

Granulovacuolar degeneration, described by Simchowicz in 1911,[831] occurs in the cytoplasm of neurons, most commonly within the pyramidal cells of the hippocampus, predominantly in CA1 in AD (Figure 16.31) and aged brains.[48,50,973] Granulovacuolar bodies have been observed in other conditions, including Pick's disease, parkinsonism–dementia complex of Guam,[376] Down's syndrome,[49] tuberous sclerosis,[377] hypothalamic gangliocytomas[504] and PSP.[859]

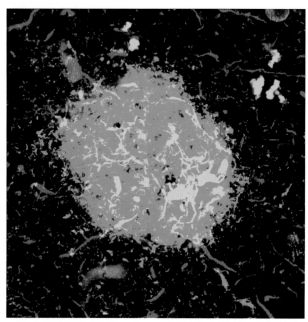

16.24 Alzheimer's disease: confocal micrograph of a senile plaque. Aβ (green) and phosphorylated tau (red) immunofluorescence is seen. Areas of overlap are yellow.

Granulovacuolar degeneration consists of single or multiple granule-containing vacuoles 3–5 μm in diameter within the neuronal cytoplasm, where they may coexist with other lesions, most often neurofibrillary tangles and Hirano bodies. Electron microscopy shows that inclusions have a double membrane and exhibit electron-dense granular cores within a translucent matrix.[376]

Granulovacuolar degeneration is readily observed on H&E staining and the granules themselves are argyrophilic and label with antibodies to phosphorylated neurofilaments,[456] tubulin,[739] phospho-tau,[84,208] ubiquitin[552] and the phosphorylated form of stress-activated protein kinase/c-Jun N-terminal kinase (p-SAPK/JNK).[509] These findings have fuelled speculation that the vacuoles are autophagic lysosomal structures in which cytoskeletal proteins are degraded. This suggestion is underpinned by studies that show localization of lysosome-associated membrane protein 1 (LAMP1) in granulovacuolar degeneration.[274] Based upon immunolocalization of the phosphorylated ribosomal protein S6 (pS6), it has been proposed that granulovacuolar degeneration relates to stress granules, and may represent a neuroprotective response.[125]

Five stages of granulovacuolar degeneration have been described based on distribution patterns in a hierarchical manner.[888]

- Stage 1: granulovacuolar degeneration seen in the hippocampal subfields CA1, CA2, and the subiculum.
- Stage 2: granulovacuolar degeneration extends to entorhinal cortex and CA4.
- Stage 3: granulovacuolar degeneration extends to temporal neocortex.
- Stage 4: granulovacuolar degeneration extends to amygdala and/or the hypothalamus.
- Stage 5: granulovacuolar degeneration reaches cingulate cortex and occasionally frontal and parietal cortices.

Perisomatic Granules

Perisomatic granules, also reported as 'non-plaque dystrophic dendrites',[33] are seen around tau-immunoreactive 'pre-tangle' pyramidal neurons in the CA1 sector of the hippocampus. They represent enlarged synaptic boutons 1–4 μ

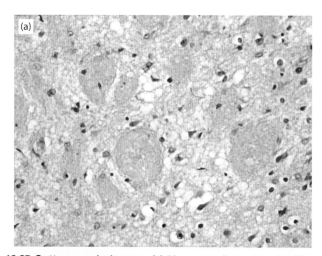
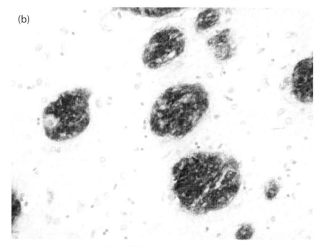

16.25 Cotton-wool plaques. (a) Haematoxylin and eosin. These have been described in Alzheimer's disease caused by specific mutations in presenilin-1 and also in some cases of sporadic Alzheimer's disease. The plaques can be stained by antibodies to Aβ peptide **(b)** and may contain variable amounts of tau protein.

16

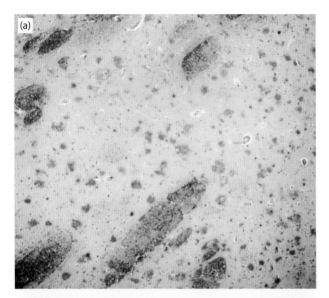

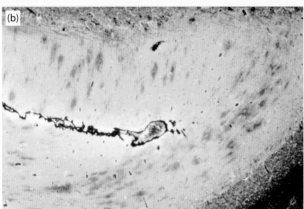

16.26 Diffuse plaques. These plaques are present in the striatum **(a)** and cerebellar cortex **(b)** in Alzheimer's disease. Bielschowsky.

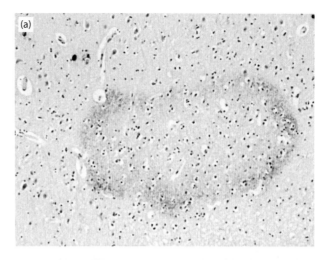

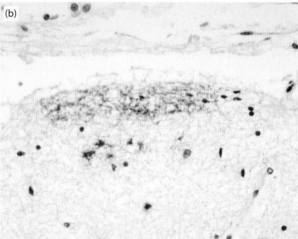

16.27 Diffuse deposits of amyloid. Lake-like deposits are seen in the transtentorial region and appear as faintly stained 'geographical' areas of staining within the neuropil **(a)**. Subpial fleecy deposits **(b)** are similarly stained only faintly. These deposits are not Congo-red-positive and are preamyloid accumulations.

in diameter containing tubulofilamentous or electron-dense floccular material. They can be detected using immunohistochemistry for ubiquitin (Figure 16.32) or the glutamate receptor GluR1-2.[33] The number of perisomatic granules has been found to increase with the severity of AD changes. They can also be seen around CA1 pyramidal cells in Pick's disease.[741,742]

Tangle-Associated Neuritic Clusters

Tangle-associated neuritic clusters represent degenerating neuronal processes associated with extracellular neurofibrillary tangles.[642] They are clusters of abnormal swollen, globular or fusiform neurites, each measuring up to 10 µ in diameter. Although they may superficially resemble a neuritic plaque, they are centred on an extracellular neurofibrillary tangle, rather than an amyloid deposit. The swollen argyrophilic neurites are similar to neurites in plaques and can be detected with immunohistochemistry for tau (Figure 16.33) or chromogranin A.[642] They are also variably ubiquitin positive. Ultrastructurally, the neurites appear as swollen nerve terminals filled with dense bodies, vesicles and degenerating organelles. These lesions are most abundant in the hippocampus, but can occasionally be found in entorhinal cortex as well.

Synaptic Dysfunction and Loss

Synaptic density, especially that in the prefrontal cortex, generally measured using assays for the presynaptic protein synaptophysin, correlates strongly with cognitive impairment in AD.[190,882] In general, synaptic proteins decline in AD as well as in other dementias.[146]

It has been suggested that a phase of reversible dysfunction may occur before the phase of irreversible degeneration and a role for diffusible oligomers of amyloid beta in synaptic dysfunction has been proposed.[27,579,674,736] The involvement of pathological tau species in the synapse has also been suggested.[855]

Spongiform Change

Spongiform change occurs in AD, similar to that in Creutzfeldt-Jakob disease (CJD), at both light and electron microscopic levels, particularly when concomitant Lewy bodies are present.[355] The spongiform change in AD usually involves the superior and inferior temporal, entorhinal and insular cortices, as well as the amygdala. In contrast, spongiform change in CJD is more widespread (although variable from case to case) with involvement of the neocortex,

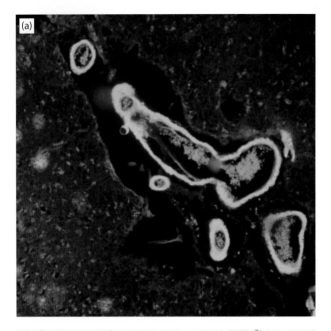

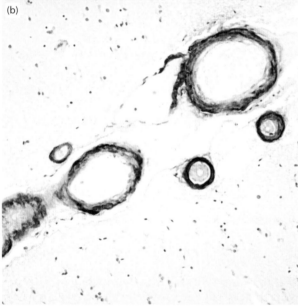

16.28 Alzheimer's disease: amyloid deposition within the walls of meningeal blood vessels. Senile plaques are also seen within the underlying cerebral cortex. **(a)** Thioflavine S, viewed under ultraviolet light. **(b)** Amyloid can also be detected by immunostaining for Aβ peptide.

striatum, thalamus and cerebellum (see Chapter 18). Spongiform change has been distinguished from the non-specific spongiosis that is seen in tissue that has been subject to severe neuronal loss and astrocytic gliosis in a wide range of neurodegenerative conditions: this latter coarse cortical microvacuolation has been termed status spongiosus.

Neurotransmitters and Neurotropic Factors

Although amyloid and tau are central to the pathogenesis of AD, a wide range of other biochemical abnormalities in AD

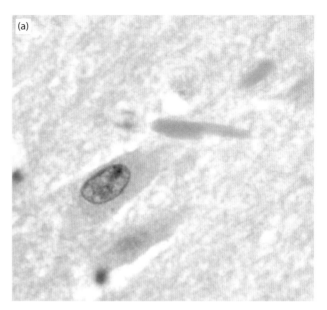

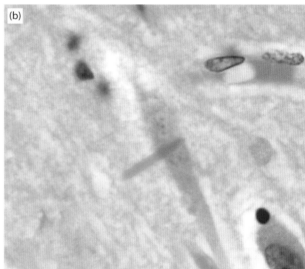

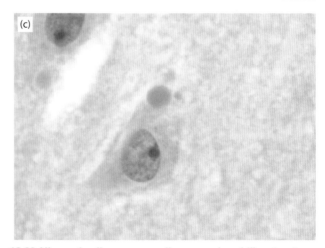

16.29 Hirano bodies seen as linear eosinophilic structures adjacent to pyramidal neurons from the hippocampus (a,b). When cut in cross-section, they appear as circular profiles **(c)**.

involves neurotransmitter systems and neurotropic factors. It is believed that these changes in neurotransmitter systems are secondary effects of the primary pathology in AD.

16.30 Electron micrograph of a Hirano body. Hirano body in a hippocampal pyramidal cell from a patient with parkinsonism–dementia complex of Guam.

Courtesy of Professor A Hirano, Montefiore Medical Center, New York, USA.[374] *With permission of John Wiley and Sons. © British Neuropathological Society.*

The cholinergic system, the most prominent neurotransmitter system affected in AD plays a major role in memory formation and consolidation (for an overview see Bentley *et al.*[68]). Loss of cholinergic neurons from the nucleus basalis of Meynert in the basal forebrain is a feature of AD.[956] Acetylcholinesterase histochemistry or ChAT immunohistochemistry and biochemical studies have revealed cholinergic loss throughout the cortex, greatest in the superficial cortical layers.[287] Addressing the cholinergic deficit has formed the basis of major therapeutic strategies to increase cholinergic transmission in AD, such as the use of cholinesterase inhibitors and receptor agonists, and acetylcholine precursors and release enhancers.[167,202]

The serotonergic system may also play a role in memory impairment (for review, see Rodriguez *et al.*[770]). The serotonergic dorsal raphe nucleus characteristically contains many neurofibrillary tangles in AD[284] and shows variable cell loss.[347]

Variable loss of the noradrenergic neurons in the locus coeruleus is common in AD. Neurofibrillary tangles are frequently seen in residual neurons of the locus coeruleus.[765,902]

Dopaminergic neurons of the substantia nigra and the ventral tegmental area show variable loss and neurofibrillary tangle formation in AD.

Defects in opioid systems have been reviewed[116] with the suggestion that these can contribute to several downstream pathways leading to clinical disease.

Neurotrophic factors that contribute to neuronal survival have a potential role in AD therapy (for review see Cattaneo *et al.*[127]). Two such factors, nerve growth factor and brain-derived neurotrophic factor,[218] promote survival

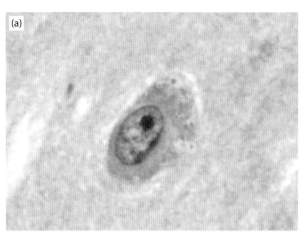

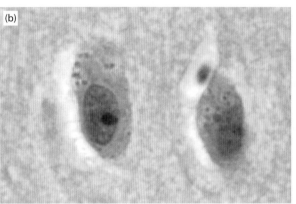

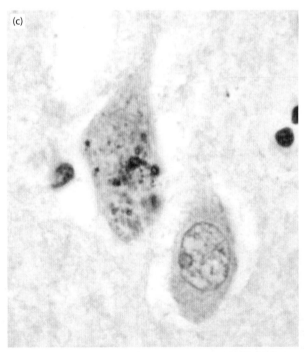

16.31 Granulovacuolar degeneration. Pyramidal neurons from CA1 of the hippocampus show prominent granulovacuolar degeneration with haematoxylin and eosin stain **(a,b)**. A proportion of granulovacuoles is immunoreactive with antibodies to phospho-tau **(c)**.

of major neuronal types affected in AD, e.g. basal forebrain, hippocampal and cortical neurons. Abnormalities in both of these factors have been observed in AD.

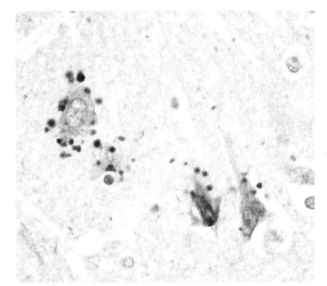

16.32 Perisomatic granules are seen adjacent to pyramidal cells in the hippocampus and are detected with anti-ubiquitin.

16.33 Tangle-associated neuritic clusters are seen in the hippocampal region and are detected by silver staining or, as here, with tau immunostaining. These clusters of abnormal neurites (arrow) are not associated with amyloid.

Brain insulin/IGF resistance has been suggested to contribute to the pathogenesis of AD.[164,184]

Neuropathological Diagnosis of Alzheimer's Disease

Despite the development of more sensitive diagnostic methods and the production of animal models recapitulating much AD pathology, neuropathological studies of human tissue remain critical for diagnosis and for furthering our understanding of the disease, and are likely to be important for the foreseeable future. Autopsies are needed not only to confirm diagnoses and define overlapping and mixed pathologies, but also to test new markers and to assess the outcomes of rapidly emerging treatment strategies. Moreover, to compare or pool data among centres and neuropathologists, standard approaches to the diagnosis and assessment of the neuropathology of AD and other dementing illnesses are needed. Several groups have addressed the need for standardization of the neuropathological evaluation of AD and other dementias.

The CERAD Neuropathology Protocol

CERAD was formed in 1986 to develop clinical, neuropsychological and neuroimaging batteries for use in the collection and pooling of information obtained from various centres. Subsequently, CERAD developed a neuropathology protocol to standardize diagnoses, promote common language, and facilitate correlation with clinical and other data.,[624,625] CERAD neuritic plaque scores (diffuse plaques are excluded) C1-4 are defined as follows:[629]

- C0 NONE diffuse plaques but not neuritic plaques;
- C1 SPARSE (1 to 5) neuritic plaques per 1 mm^2;
- C2 MODERATE (>6 but <20 neuritic plaques per 1 mm^2);
- C3 FREQUENT (>20 neuritic plaques per 1 mm^2).

The CERAD neuropathology protocol was incorporated into the recommendations of the Consortium on Dementia with Lewy Bodies,[607] as well as the NIA-Reagan Criteria for AD.

NIA–Reagan Institute of the Alzheimer's Association Criteria

A working group convened by the National Institute on Aging (NIA) and the Reagan Institute of the Alzheimer Association proposed criteria for AD that recognized it to be a heterogeneous clinicopathological entity; neither clinical history nor neuropathology could stand alone as universally predictive of its corollary.[399] The group has subsequently proposed revised criteria for the neuropathological diagnosis of AD and related disorders.[402,629]

Pathologic Staging of AD

The pathological changes in AD progress in a stepwise fashion in the majority of cases. Following systematic study using silver staining Braak and Braak proposed a staging system for AD.[91] Subsequent studies have used immunohistochemical staining methods. There are different stage descriptors for plaques and tangles.

Braak plaque stages:

- Stage A: low density of neuritic plaques in the neocortex, especially in the frontal, temporal, and occipital lobes.
- Stage B: neuritic plaques present in neocortical association areas and moderate hippocampal involvement.
- Stage C: neuritic plaques present in primary sensory and motor areas.

Braak NFT stages:

- Entorhinal stages, I and II, usually associated with normal cognition.
- Limbic stages, III and IV, may be associated with cognitive impairment.
- Neocortical stages, V and VI, usually associated with clinical dementia.

The Thal plaque phase stages the extent of involvement of the brain into five phases according to extent of plaques seen on Aβ staining:[886]

- phase 1: exclusively in neocortex;
- phase 2: additionally in allocortex;
- phase 3: extending to diencephalic nuclei, the striatum, and the cholinergic nuclei of the basal forebrain;
- phase 4: involving brain stem;
- phase 5: present in cerebellum.

The Braak staging of neurofibrillary degeneration and the Thal phases of plaques distribution have been adopted into the 2012 National Institute on Aging–Alzheimer's Association (NIA-AA) guidelines for the neuropathological assessment of AD. The BrainNet Europe Consortium has also put forward a validated system for assessment of tangles according to a staging system based on the Braak system.[8]

Sampling the Brain for Histological Evaluation

Recommendations for tissue sampling and histological evaluation in a stratified fashion have been published. Block sampling is recommended from medulla, pons (including locus coeruleus), cerebellar cortex (including dentate nucleus), thalamus and subthalamic nucleus, basal ganglia at level of anterior commissure, hippocampus and entorhinal cortex, anterior cingulate gyrus, amygdala, middle frontal gyrus, superior and middle temporal gyri, inferior parietal lobule, occipital cortex (BA 17 and 18).[629]

The BrainNet Europe Consortium has recommended the following sampling strategy that may be adapted for instances when consent is not available to retain the whole brain for diagnostic evaluation (Figure 16.34).[8,187]

Section 1: visual cortex including the calcarine fissure to include the primary visual cortex with band of Gennari (involved in stage VI) and parastriate/peristriate region (BA 18/19, the six-layered cortex (involved in stage V); section 2: middle and superior temporal gyrus (involved in stage IV); section 3: anterior hippocampus and/or amygdala at the level of uncus (involved in stage I-III); section 4: posterior hippocampus at level of lateral geniculate body (involved in stage II and III).

Neuropathological Criteria Diagnostic of AD

In the NIA-AA neuropathological diagnostic criteria it is recognized that AD is a clinicopathological entity within which there may be pathological variation.[402,629] AD lesions seen in the post-mortem brain from cognitively normal elderly people are considered pathological rather than a part of a normal ageing process. It is recommended that

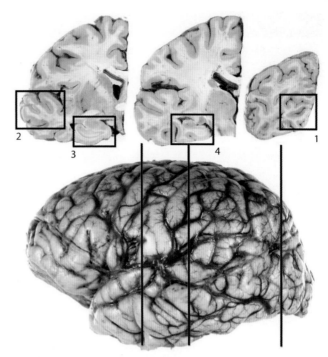

16.34 BrainNet Europe limited sampling for histological diagnosis of Alzheimer's Disease

findings of pathological changes of AD in the brain at autopsy are reported as 'AD neuropathological changes'.

In order to rate the lesions of AD a scoring system is used (see Table 16.6):

- Aβ plaques are staged according to the Thal phase scheme (A);
- NFT stage is determined according to the Braak model (B);
- Neuritic plaques are scored according to the CERAD scheme (C).

Neuropathological reports should summarize findings, as 'Alzheimer disease neuropathologic changes: A1, B0, C0' or 'Alzheimer disease neuropathologic changes: A3, B3, C3'.

Combining these three scores, termed ABC, allocates a probability that AD-associated abnormalities account for a patient's dementia in life. The likelihood that clinical dementia has been caused by AD lesions in the brain is stratified on the basis of the post-mortem neuropathological findings, as follows and Table 16.7:

- *High probability* that dementia was due to AD lesions, if the brain has both neuritic plaques and neurofibrillary tangles in the neocortex (CERAD frequent neuritic plaque score 3, Thal plaques score 3 and Braak and Braak stage V/VI score 3).
- *Intermediate probability* that dementia was due to AD lesions, if the brain has a moderate density of neocortical neuritic plaques, and neurofibrillary tangles in the limbic regions (CERAD moderate, and Braak and Braak stage III/IV).
- *Low probability* that dementia was due to AD lesions, if the brain has neuritic plaques and neurofibrillary tangles

TABLE 16.7 NIA-AA ABC scoring for Alzheimer neuropathological changes

AD neuropathologic change		B score		
A score	C score	0 or 1	2	3
0	0	Not	Not	Not
1	0 or 1	Low	Low	Low
	2 or 3	Low	Intermediate	Intermediate
2	Any C score	Low	Intermediate	Intermediate
3	0 or 1	Low	Intermediate	Intermediate
	2 or 3	Low	Intermediate	High

The probability that AD neuropathological change explains clinical dementia is assigned by applying an 'ABC' score. Aβ/amyloid plaques (A), NFT stage (B), and neuritic plaque score (C). A patient is described as having as 'Not', 'Low', 'Intermediate' or 'High' probability where 'Intermediate' or 'High' is considered sufficient explanation for dementia. Modified from Hyman BT, et al. National Institute on Aging-Alzheimer's Association guidelines for the neuropathologic assessment of Alzheimer's disease. Alzheimers Dement 2012;8(1):1-13, with permission from Elsevier. © 2012 The Alzheimer's Association.

in a more limited distribution and/or severity (CERAD infrequent, and Braak and Braak Stage I/II).

Patients who have a restricted distribution of tau-immunoreactive structures not meeting criteria for a specific tauopathy and not fitting with a Braak stage can be designated as having 'stage + neurofibrillary change'.

In applying earlier NIA-Reagan criteria it has become apparent that many laboratories have moved away from use of silver stained preparations in evaluation of AD pathology and that alternate approaches are desirable. The BrainNet Europe group has published staging criteria based on evaluation of tau-labelled histological sections from four brain regions (Figure 16.35).

Effect of Coexisting Pathologies

Coexisting pathologies are frequent in AD. A significant subset of AD cases exhibits concomitant vascular or isch-aemic pathology and may be classed as having a mixed dementia (see later)

Lewy body pathology has been reported in up to 60% of cases of AD.[352] In most studies, including studies of familial AD, Down's syndrome with AD and sporadic AD, the amygdala is particularly vulnerable.[540] In some of these cases, Lewy body pathology is confined to the amygdala,[910] whereas in others there is more typical involvement of vulnerable brainstem and basal forebrain nuclei or even neocortex. The relationship between AD with Lewy bodies confined to the amygdala and other forms of dementia associated with cortical Lewy bodies is currently unresolved.

Transactive response DNA-binding protein of 43 kDa (TDP-43) is the basis of pathological inclusions in ALS and some forms of frontotemporal lobar degeneration. TDP-43 pathology is seen in 25 per cent to 50 per cent of AD cases, linked to more severe clinical disease and more severe Alzheimer pathology. This has been proposed as a form of secondary TDP-43 proteinopathy.[960,962] It is unlikely that this represents coincident disease.[455]

Five stages of TDP-43 deposition in AD have been identified:[452]

- stage I: sparse TDP-43 in the amygdala only (17 per cent of cases);

- stage II: moderate-frequent amygdala TDP-43 with spread into entorhinal and subiculum (25 per cent of cases);
- stage III: extends into dentate gyrus and occipitotemporal cortex (31 per cent of cases);
- stage IV: extends into inferior temporal cortex (20 per cent of cases);
- stage V: involvement of frontal cortex and basal ganglia (7 per cent of cases).

TDP-positive subjects with AD pathological changes have been found to be significantly more likely to be cognitively impaired compared to those who are TDP-negative. More severe cognitive impairment and temporal atrophy are associated with the severity and extent of TDP-43 pathology.[453]

Difficulties in the Diagnosis of Alzheimer's Disease

Diagnostic neuropathology of degenerative brain disease can be challenging. It is not uncommon to be faced with diagnostic uncertainty as to the classification of a neurodegenerative disease when AD pathology is detected.[662] Some particularly difficult situations are discussed below:

- Some patients have clinical dementia and neuropathological examination reveals the presence of neurofibrillary tangles in cortical and subcortical areas in the absence of significant numbers of plaques.
  - Some of these cases have tangle-predominant dementia.[426,436] Other cases have a degenerative tauopathy, such as progressive supranuclear palsy or corticobasal degeneration. These diagnoses should be readily apparent with tau immunohistochemistry that shows presence of glial as well as neuronal lesions and extensive involvement of subcortical brain regions. Argyrophilic grain disease is another tauopathy that is usually associated with some degree of Alzheimer neurofibrillary degeneration, given that it increases in frequency with ageing. The presence of relatively selective neurofibrillary degeneration in CA2 sector of the hippocampus and swollen neurons in the limbic lobe suggest this diagnosis.[416] Additional stains, such as Gallyas silver impregnation or tau immunostains, may be needed to confirm this diagnosis.

Stage VI	Occipital NTs (++ or +++) in layer V of the striate cortex.
Stage V	Occipital NTs (++ or +++) in the superficial and deep layers of the peristriate cortex. No immunolabelling of tau in the striate cortex.
Stage IV	Middle temporal gyrus (++ or +++) in the superficial and/or deep layers. No immunolabelling of the peristriate cortex.
Stage III	In section of posterior hippocampus at the level of the lateral geniculate body, NTs (++ or +++) in the outer and inner layers of remnants of the entorhinal region, continuing into the adjacent occipito-temporal gyrus. No immunolabelling in the middle temporal gyrus.
Stage II	In section of posterior hippocampus at the level of the lateral geniculate body, NTs (++ or +++) in the outer layers of the entorhinal region and (+ or ++ or +++) in the inner layers of the entorhinal region. In section of anterior hippocampus at the level of uncus, NTs (++ or +++) in the outer layers of the entorhinal region and (+ or ++ or +++) in the inner layers of the entorhinal region. No immunolabelling in the occipito-temporal gyrus.
Stage I	In section of anterior hippocampus at level of uncus, NTs (+ or ++ or +++) limited to the transentorhinal region.
Stage +	Tangles or pre-tangles or NTs in any region with a pattern that does not fit with a defined tauopathy, including a defined AD stage (above).
Stage 0	Sampled sections all IHC tau-negative.

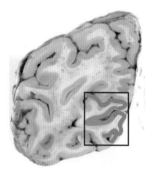
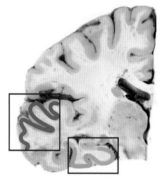
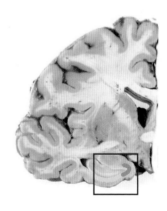

16.35 Determination of Braak stages by immunohistochemical assessment of tau pathology (BrainNet Europe).

- Some patients have an abundance of plaques but very few neurofibrillary tangles.
  - When the plaques are largely non-neuritic diffuse amyloid deposits, it is critical to exclude additional pathological processes, in particular concurrent vascular pathology or Lewy bodies. Although there are some cases of plaque-predominant AD, especially in the elderly, most cases of so-called 'plaque-only AD' have usually been found to have Lewy body disease.[356] If no cortical Lewy bodies or vascular lesions are detected, and especially in the plaques are diffuse Aβ deposits, other types of degenerative dementia (such as one of the fronto-temporal lobar degenerations) need to be excluded with appropriate methods (e.g. screening with P62 immunohistochemistry).

- Some patients have Alzheimer changes in a distribution or density that does not fit with the typical hierarchical pattern described earlier.
  - These cases may often have atypical clinical features of disease. The most common variants of AD to consider are those presenting with a posterior cortical syndrome, with severe pathology preferentially affecting the occipital lobe and the visual association cortex[85,171] and AD presenting with a frontal lobe-type dementia, or dominant associated psychosis with severe involvement of frontal cortex.[490]
- Cases with hippocampal sparing disease. Hippocampal sparing and limbic-predominant AD subtypes have been suggested to account for about 25 per cent of cases.[647]

Pathology of Mild Cognitive Impairment and Early AD

Careful longitudinal clinicopathological studies have provided insight into the pathology of early AD. Of particular interest have been neuropathological studies of individuals who at death were considered to have MCI, a transitional stage between normal cognition and frank dementia. AD-type changes have been noted in studies of MCI cases coming to autopsy; however, additional pathologic processes are common in this cohort.[442,715]

It is important to remember that MCI is a clinical syndrome with several pathological causes and that not everyone with MCI will develop AD.[639] There is great interest in identifying markers that will predict patients with MCI who are at risk of developing AD, and neuropsychological testing, imaging, and assays of potential biomarkers in blood or CSF are being evaluated.[714]

Molecular Considerations and Pathogenesis of Alzheimer's Disease

Amyloid Precursor Protein Metabolism

The APP family of proteins plays a role in synaptic development with both cell adhesion and signalling functions. Transgenic mice with reduced expression of APP have synaptic defects.[122,487,640,656]

The amyloid precursor protein, from which Aβ is cleaved, is a large, single transmembrane protein widely expressed throughout the body. There are three common isoforms, containing 695, 751 and 770 amino acid residues, derived from alternative splicing of a single gene on chromosome 21. The human APP gene contains 19 exons with Aβ derived from parts of exons 16 and 17. APP695 is the predominant form in neurons. Immunohistochemical and cell fractionation studies show a cell membrane localization as well as a synaptic and axonal localization of APP in neurons. Functional aspects of APP and the roles of its proteolytic fragments have been reviewed.[142,656] The Aβ peptide derives in part from the ectodomain and the hydrophobic transmembrane portion of the APP molecule. Three protease activities are responsible for the cleavage and processing of APP[142] (Figure 16.36).

The Non-Amyloidogenic Pathway of APP Proteolysis (α-Secretase Pathway)

The predominant cleavage pathway of APP (termed the non-amyloidogenic or secretory pathway) is brought about by the sequential action of two protease activities termed α-secretase and γ-secretase. This leads to the formation of the secreted N-terminal APP fragment sAPPα, which is implicated in neuroprotection, synaptic plasticity, neurite outgrowth and synaptogenesis (reviewed by Kogel et al.[487] and Muller et al.[640]).

The first cleavage yields an ectodomain fragment termed sAPPα and a carboxyl terminal fragment, CTFα. CTFα is then cleaved into two fragments, including a hydrophobic remnant of the Aβ domain called P3 (Figure 16.37). The cleavage site for α-secretase is in the middle of the Aβ sequence and precludes production of Aβ. The α-secretase activity probably resides in several proteases, including those of the ADAM (a disintegrin and metalloprotease) family – ADAM-10, ADAM-17 (TACE) and ADAM-9.[12,488]

The enzymatic activity described as γ-secretase is the result of a protein complex that includes four proteins, presenilin, nicastrin, Aph-1, and Pen-2 – with presenilin functioning as the catalytic subunit.[970]

Common to several of these proteolytic events is the generation of a fragment that is translocated to the nucleus to act as a transduction factor. The short CTF of APP generated by γ-secretase cleavage is referred to as APP intracellular domain and has been suggested to act as a transcriptional regulator.[703]

The Amyloidogenic Pathway of APP Proteolysis (β-Secretase Pathway)

A second and minor pathway for the proteolysis of APP (termed the amyloidogenic pathway or endosomal–lysosomal pathway) is believed to be crucial in the pathogenesis of AD. This involves the sequential action of two enzymatic activities, β-secretase and γ-secretase (Figure 16.38). The β-secretase activity has been ascribed to a novel transmembrane aspartyl protease, β-site APP cleaving enzyme (BACE), which is highly expressed in neurons.[924] There are two homologous BACE proteins (BACE1 and BACE2) derived from distinct genes. Only BACE1 is expressed in the nervous system. The action of BACE is to cleave APP to generate sAPPβ and CTFβ, which is further cleaved by the γ-secretase complex to yield Aβ and the APP intracellular domain.[923]

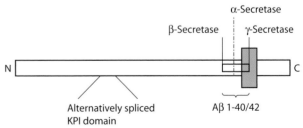

16.36 Diagrammatic representation of amyloid precursor protein (APP), indicating the secretase cleavage sites. α-Secretase cuts the Aβ sequence in half, precluding amyloid deposition. β-Secretase and γ-secretase release Aβ by cleaving Aβ at its amino terminal (β-secretase) and its carboxyl terminal (γ-secretase), the latter cleavage occurring within the cell membrane (shaded region). KPI, Kunitz protease inhibitor.

Cleavage of APP is not believed to be the main function of BACE, which has other protein substrates. Certain of the mutations in the gene for APP provide a better cleavage conformation for BACE than the native protein. BACE is clustered in cholesterol-rich membrane domains referred to as lipid rafts and APP processing is targeted to lipid rafts.

The cleavage of CTFβ by the γ-secretase complex generates a number of different Aβ peptide fragments, with the major species being 40 or 42 amino acid residues in length.[337] In AD there is an increase in the ratio of Aβ42 to Aβ40, and Aβ42 is the species that is most amyloidogenic. There is great interest in understanding factors that favour γ-secretase activity directed towards production of Aβ42, as promoting formation of Aβ40 and decreasing the ratio of Aβ42 to Aβ40 would potentially have a protective effect against amyloid deposition.

Given that the activity of BACE (and the γ-secretase complex) is required to generate Aβ, interest has been shown in developing specific inhibitors for BACE as a means of preventing Aβ production, thereby interfering with amyloid deposition. In contrast, because the γ-secretase complex has other important substrates besides the CTF of APP, the development of therapeutic inhibitors that inhibit the γ-secretase complex, while preserving other essential functions, is likely to be more problematic.

Clearance of Aβ

The formation of Aβ peptide is a physiological event and once formed it is normally cleared from the brain. Receptor-mediated transport can eliminate Aβ peptide across the blood–brain barrier by the low-density lipoprotein receptor-related protein and the receptor for advanced glycation end products.[874] Aβ peptide can also be proteolytically degraded by several peptidases and proteinases, known collectively as Aβ-degrading proteases and including insulin-degrading enzyme and neprilysin. Neprilysin is thought to be the most potent Aβ-degrading enzyme *in vivo*.[800] The full range and classes of Aβ-degrading proteases is reviewed by Saido and Leissring.[799]

There is a growing body of evidence to suggest that the drainage of CSF from the brain is a physiological pathway

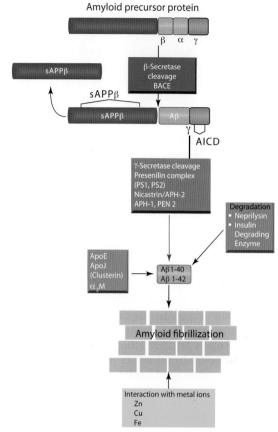

16.38 The β-secretase pathway. AICD, APP intracellular domain; ApoE, apolipoprotein E; BACE, β-site cleaving enzyme.

that can clear toxic molecules, such as Aβ, from the interstitial fluid space of the brain. With increased age there is lower CSF production, together with resistance to CSF outflow. It has been proposed that compromise of the CSF drainage pathway with age allows local concentrations of amyloidogenic peptides to increase, which may drive the development of AD.[829,946-948]

Pathological Effects of Aβ

The 'amyloid cascade hypothesis' of AD proposes that the Aβ peptide is the main driver in the pathological process leading to neurodegeneration. Despite a large body of research it remains uncertain how the pathological accumulation of Aβ peptide causes neuronal dysfunction and is related to other lesions in the brain in AD.[128,144,849,871] Recent work has concentrated on the role of small oligomeric forms of Aβ peptide as toxic intermediaries, rather than amyloid. Intracellular oligomeric forms of Aβ peptide may be locally toxic and may form small preamyloid protofilaments. Oligomers have been suggested to cause synaptic dysfunction.[674,736]

Amyloid versus Tau

For years, much debate has ensued about the relative importance of amyloid versus tau in the pathogenesis of AD. For the neuropathologist, Aβ- and tau-associated changes are so inextricably linked at a morphological level (for example in neuritic plaques both proteins are intermingled), that it is

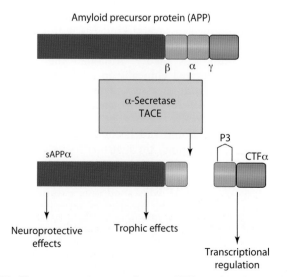

16.37 The α-secretase pathway. CTF, carboxyl terminal fragment; TACE, tumour necrosis factor α converting enzyme.

difficult to assign primacy to one over the other. The amyloid cascade hypothesis proposes a central role for Aβ amyloid in the pathogenesis of AD.[359,361] The link between Aβ generation and the formation of neurofibrillary tangles is not yet known. Several lines of evidence point to a primary role for Aβ amyloid in the pathogenesis of AD reviewed by Reitz.[749]

- Some mutations in the APP gene have been linked with rare familial forms of AD.[303] The neuropathological findings in these cases are indistinguishable from those of sporadic AD and include the presence of many neurofibrillary tangles.
- There is a multifold increase in production of Aβ42 in plasma in patients with AD caused by presenilin mutations.
- Patients with Down's syndrome are strongly predisposed to the development of AD and the Down's syndrome obligate region of chromosome 21 includes the gene for APP. There are three copies of the APP gene and they have increased Aβ42 in plasma.
- Mutant APP transgenic mice exhibit AD-like neuropathology albeit they do not develop neurofibrillary tangles.
- Diffuse plaques are detectable in the brain of patients with Down's syndrome before the development of neuritic plaques and neurofibrillary tangles.
- Mutations in the gene for tau (*MAPT*) protein have not been found to be associated with familial AD. Instead, mutations in *MAPT* are associated with frontotemporal dementia and parkinsonism with extensive tau pathology in neurons and glia, but no amyloid deposits.

Arguments against the amyloid cascade hypothesis of AD are that cognitively normal individuals can have very large numbers of neocortical plaques, although predominantly of diffuse type, and only sparse neurofibrillary tangles. Moreover, the correlation between plaque density and the severity of dementia is relatively weak. In studies that correlate severity of pathological features with clinical severity of disease, tau pathology (i.e. neuritic plaques and neurofibrillary tangles) correlates best with clinical severity of disease. Therapeutic strategies that have sought to remove amyloid have also not been effective in arresting disease progression.[749] The neuropathology of patients in a clinical trial of immunotherapy against amyloid in AD has been reviewed.[82]

Several key questions particularly cogent to neuropathologists remain. What are the steps that lead from amyloid production to tau pathology? Why does AD pathology occur or progress in some brain regions and not others? Why are some neurons vulnerable to neurofibrillary tangle formation and others largely spared? These and other questions are central to understanding AD pathogenesis and may also be critical for therapeutic approaches.

A recent appraisal of biomarkers in cohort studies has led to inconsistencies with predictions based solely on the amyloid cascade hypothesis of AD. It has been suggested that 'We are entering an era in which it is likely that the unitary view of AD as a disease with a single sequential pathological pathway, in which Aβ is considered as the only initial and causal event, will be progressively replaced with a more complex picture in which AD is considered

as a multiparameter pathology subtended by several partly independent pathological processes'.[138]

Genetic Factors in Alzheimer's Disease

Pathways to Alzheimer's Disease

Genetic factors are recognized to be very important determinants of AD and are estimated to play a role in at least 80 per cent of cases. There are two main patterns of inheritance (Figure 16.39).

- Early-onset AD (<60 years) may be associated with autosomal dominant mutations in *APP*, *PSEN1*, and *PSEN2* and represent only 5 per cent of all AD cases.
- Late-onset AD (>60 years) is associated with a growing range of gene polymorphisms that contribute to a risk of developing disease, and is thought not to be caused by the genes leading to early onset AD.[285] In this latter group, recent information from genome-wide association studies has identified several pathways to disease (see Figure 16.39).[66,331,360,614,631,873]

The identified pathways all intersect and potentially link in to the amyloid cascade hypothesis. Aβ processing takes place in lipid membranes and is likely linked to cholesterol in that location. Immune system function may relate to clearance of Aβ by microglia or to factors that lead to an inflammatory response in relation to Aβ. Despite these emerging insights a link between amyloid and tau is not forthcoming.

The Alzgene database http://www.alzgene.org/ provides an unbiased summary of genetic association studies in AD including meta-analyses for polymorphisms.[72]

Down's Syndrome and Alzheimer's Disease

Patients with Down's syndrome with trisomy of chromosome 21 inevitably develop Alzheimer-type neuropathology with increasing age. The *APP* gene is on chromosome 21,[468] and the predisposition of Down's syndrome patients to develop AD may relate to triplication of this gene. Duplication of the *APP* gene locus has been described in several kindreds with early onset AD, some of which are associated with severe CAA,[790] which supports this theory. Although the degree of dementia encountered clinically in adults with Down's syndrome is difficult to assess, dementia is believed to develop around 40–70 years of age.[511] Alzheimer's type pathology universally occurs in middle-aged adults with Down's syndrome and becomes more severe with increasing age. Although the neuropathological features of Down's syndrome recapitulate those of AD,[847] several features deserve a mention. On gross examination, about half of the cases of Down's syndrome exhibit narrowing, often bilateral, of the superior temporal gyrus. At a microscopic level, amyloid deposition is greater than that found in sporadic AD, and concomitant inheritance of an *APOE* ε4 allele is associated with still greater amyloid accumulations.[401]

Examination of the brain of younger individuals with Down's syndrome has provided an opportunity to

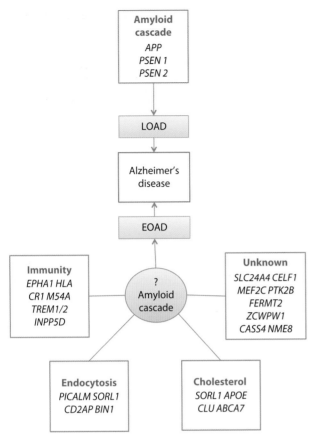

16.39 Genetic factors in Alzheimer's disease. Aβ metabolism: CLU (clusterin), ApoE, ABCA7 (ATP-binding cassette subfamily A, member 7), PICALM (phosphatidylinositol-binding clathrin assembly protein). Immune system function: CLU (clusterin), CR1 (complement receptor 1), ABCA7, MS4A cluster (membrane-spanning 4-domains subfamily A, HLA, TREM1 TREM 2 M54A and EPHA1). Cholesterol metabolism: ApoE, CLU (clusterin), ABCA7. Cell membrane/vesicle processing: PICALM, BIN1 (bridging integrator 1), CD2AP (CD2-associated protein), SORL1 and EPHA1 (ephrin receptor A1). Unknown: SLC24A4, CELF1, MEF2C, PTK2B, FERMT2, ZCWPW1, CASS4 and NME8.

study the accelerated evolution of AD pathology in these patients, including the regional development of plaques and tangles,[532,574,576,646] accrual of complement proteins in plaques,[861] cell death,[651] and amyloid deposition with respect to Aβ peptides.[421] In autopsy studies of Down's syndrome patients younger than 30 years of age, the earliest change seen was intracellular accumulation of Aβ in neurons and astrocytes. This was followed by deposition of extracellular Aβ with formation of diffuse plaques and, finally, development of neuritic plaques and neurofibrillary tangles with activation of microglia[333,850] (see also Chapter 4).

Autosomal Dominant Alzheimer's Disease

The vast majority of cases of AD are sporadic. Nonetheless, some families show an autosomal dominant mendelian pattern of inheritance. Three genes have been implicated in these rare cases: the genes for APP (*APP*), presenilin-1 (*PSEN1*) and presenilin-2 (*PSEN2*).[548]

The Alzforum maintains a database of mutations and summary of neuropathological findings at http://www.alzforum.org/mutations. Clinically, patients with *PSEN1* mutations present as early-onset AD. Many patients develop myoclonus, but it is uncertain whether this relates to any aspect of the mutation or simply to early onset disease. A distinctive phenotype of dementia associated with spastic paraparesis has been noted in some families. Patients with *PSEN2* mutations tend to have a slightly later age at onset with a longer disease duration.

A review and approach to genetic testing for counselling in dementia has been published.[557] These authors note that mendelian forms of AD are rare, with genetic mutations seen only in about 500 families with a history of AD. Eighty-six per cent of families with young-onset (<60 years) dementia in three or more generations are predicted to have a mutation in the *APP*, *PSEN1* or *PSEN2* gene. Mutation in *PSEN1* is the most common cause in around 60 per cent of families. Mutations in *APP* are seen in 15 per cent of families with duplication of the *APP* gene probably accounting for an additional 8 per cent of families. Mutations in *PSEN2* have been reported in 22 families worldwide.

Genetic investigation of autosomal dominant kindreds is likely to reveal additional genes involved in disease pathogenesis.[791]

Neuropathology of Autosomal Dominant Early-Onset Alzheimer's Disease

The neuropathology of familial AD has been reviewed.[823] In reported cases of AD due to *APP* mutations, the pathological changes have been similar to those in sporadic AD, although often very severe.[470,514,539,577] Two members of a family with the initial *APP* mutation were also noted to have Lewy bodies, an overlap feature seen in both familial and sporadic AD.[515] Individuals with *APP* mutations may have a remarkable predominance of Aβ42 within plaques.[422] Cerebral angiopathy is a typical feature of cases linked to APP mutation and in some patients is the dominant pathological finding.[823]

The neuropathological changes associated with the *PSEN1* and *PSEN2* mutations are similar to those in sporadic AD but in general are much more severe. In particular there is marked cerebellar involvement. Even in patients in whom the clinical and neuropathological phenotype is robust, there can be considerable individual variability, such that the pattern among family members carrying the same mutation may be different. This implicates a role for environment, epigenetic factors or modifying genes. On the basis of comparative studies, two main pathological types have been suggested as follows:[578]

- Type 1 pattern (generally associated with mutations before codon 200) closely resembles that in sporadic AD, being characterized by many evenly distributed diffuse and cored plaques, mainly in cortical layers II and III, with mild amyloid angiopathy confined mostly to the leptomeningeal arteries. This type 1 profile has more Aβ42 and is associated with an earlier age at onset and shorter disease duration.
- Type 2 pattern (generally associated with mutations after codon 200) also shows many diffuse and cored plaques, but these tend to be larger and clustered around arteries with amyloid angiopathy especially at

the depths of sulci. Amyloid angiopathy tends to be severe, and involves both leptomeningeal and parenchymal vessels.

Other neuropathological variants have been reported in association with *PSEN1* mutations, such as vacuolar changes[350,550] and Pick's disease.[198,348] Another presenilin mutation (N405S) has been associated with widespread cortical amyloid deposition, but limited cerebral amyloid angiopathy.[986] As noted previously, many AD cases with *PSEN1* mutations have Lewy bodies, especially in the amygdala.[540]

Several pedigrees have also been described with spastic paraparesis, with pathological evidence of wallerian degeneration of the lateral corticospinal tracts in the lower brainstem and spinal cord. The initial kindreds have functional deletions of exon 9.[168] The cortex and basal ganglia in these cases have abundant 'cotton-wool plaques' (see Figure 16.25). Another novel plaque associated with reactive microglia and astrocytes and termed an 'inflammatory plaque' has been described in *PSEN1* mutation cases. These lesions are positive with amyloid stains, such as thioflavine S, but negative with immunohistochemistry for Aβ, tau and apolipoprotein E.[824]

A detailed review of neuropathology in *PSEN2* AD suggests that affected patients present with a wide range of age of onset, often associated with seizures.[430]

There are indications that the pathogenesis of atrophy is different depending on genetic cause of disease. In one study, familial APP patients were seen to have more medial temporal lobe atrophy, whereas PSEN1 subjects showed more dominant neocortical loss.[807]

Transgenic Mouse Models of Alzheimer's Disease

The discovery of genetic risk factors for AD led to attempts to recapitulate the neuropathological changes in experimental models. Non-human primates develop amyloid deposits, including dogs[173] and bears,[879] but without any overt neurofibrillary changes. Sheep and goats spontaneously develop neurofibrillary tangles in the cerebral cortex.[661] None of these provides a model system that can be easily and rapidly modified. By contrast, transgenic mice models have been produced and, although none fully demonstrates all features of human disease, they are widely used to explore pathophysiology and develop potential therapies.[344,542]

Models have been developed by overexpressing normal or mutant proteins (e.g. APP, presenilin and tau protein) known to be involved in AD, alone or in combination. High expression levels prove to be necessary to produce a model that consistently has deposition of Aβ. Neuronal cell death occurs in some of these models, but NFTs typical of AD are hard to produce in these mice.[46] Tau-expressing models can replicate aspects of tangle pathology.[381] Knockout models, taking out key genes known to be involved in AD pathogenesis, have also been created. For example, *APOE*-null mice cross bred with APP-overexpressing mice have markedly fewer Aβ deposits than do APP-overexpressing mice.[47,390] Experiments with double transgenic animals have been informative in

showing synergistic effects. Crossing APP mice with mice that overexpress a mutant form of presenilin-1 led to a rapid acceleration of Aβ deposition, reducing the time needed to see Aβ deposits from 8–10 months to only 3–4 months (Figure 16.40).[227] A review and advice in relation to selecting a mouse model depending upon research questions to be addressed has been produced.[140]

Transgenic models have also been established in *Drosophila melanogaster* and *Caenorhabditis elegans*.[949]

DISORDERS WITH LEWY BODIES, INCLUDING PARKINSON'S DISEASE DEMENTIA

Lewy Bodies and Lewy Neurites

Lewy bodies (LBs) are concentric, hyaline cytoplasmic neuronal inclusions that were originally described in monoaminergic and cholinergic neurons of the brainstem, diencephalon and basal forebrain in Parkinson's disease, but are also known to affect neurons in the cerebral cortex (see Chapter 12). In the cortex LBs are oval to irregular, eosinophilic inclusions that displace the nucleus, but are less clearly separated from the adjacent cytoplasm than are classical Lewy bodies.[492]

Staining with α-synuclein is recommended as the best method for detecting cortical LBs,[854] but they can also be seen on P62 and ubiquitin immunolabelling, when they can be hard to distinguish from small spherical tangles in cortical neurons.

Cortical Lewy bodies are most often detected in small to medium-sized neurons in deeper cortical layers of the limbic cortex, especially in the parahippocampal and the cingulate gyri. Although Lewy bodies are often spherical, more pleomorphic inclusions or inclusions within neuronal cell processes in certain brain stem and basal forebrain regions are also common. Such 'intraneuritic' Lewy bodies are particularly common in the dorsal motor nucleus of the vagus in the medulla and the basal nucleus of Meynert. In addition to these lesions that are visible with routine histopathological methods, ubiquitin and α-synuclein immunohistochemistry also reveals widespread neuritic pathology (so-called Lewy neurites), particularly in the hippocampus[210,212] and amygdala.[95]

Lewy Bodies in Dementia Syndromes

Several neurodegenerative diseases associated with dementia are characterized by the presence of Lewy bodies. These diseases include the following.

- Some patients have a primary clinical dementia syndrome, characterized by fluctuation, early visual hallucinations and an extrapyramidal tremor, meeting the clinical criteria for DLB. These patients typically have Lewy bodies in brain stem and midbrain nuclei as well as involvement of limbic areas and neocortex by Lewy bodies. Most have associated AD-type pathology.
- Some patients with pure motor Parkinson's disease (PD) and later develop dementia (Parkinson's disease dementia [PDD]) and many of these patients have Lewy bodies

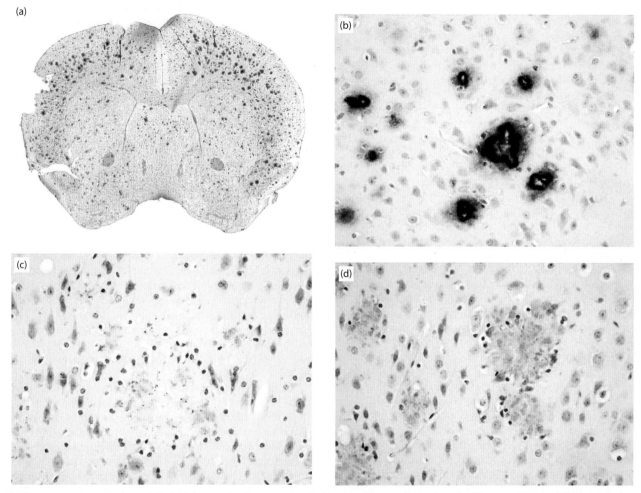

16.40 Transgenic mouse model of Alzheimer's disease, created by overexpression of mutant human APP and PS1 genes. The brain shows large numbers of amyloid plaques immunostained with Aβ peptide **(a,b)**. AT8 immunostaining shows some associated neuritic changes **(c)**. Extensive neuritic changes are demonstrated by anti-ubiquitin staining **(d)**. As in most similar models, neurofibrillary tangles are not seen.

in cortical and limbic areas of the brain, often with varying degrees of Alzheimer-type pathology. There appear to be other pathological substrates for cognitive decline in PD, but they all have at least some Lewy bodies because these are required for neuropathological diagnosis of the disease. Patients with PD may also exhibit mild cognitive impairment (PD-MCI).

- Some patients have dementia with psychosis and relatively mild Alzheimer-type pathology, but widespread cortical and limbic Lewy bodies. These patients may fit the clinical criteria for DLB. They may or may not have overt clinical parkinsonism.

A small group of patients, some with hereditary risk factors, has dementia with parkinsonism and other clinical features with widespread Lewy bodies, but no Alzheimer-type pathology. These cases represent so-called 'pure' Lewy body disease.

Many patients with AD, either familial or sporadic, have Lewy body pathology that may be relatively restricted to either the brain stem or the amygdala. Clinically normal individuals with similar restricted Lewy body pathology are considered to have 'incidental Lewy

body disease'. The Lewy bodies that occur in advanced AD may likewise be concurrent incidental Lewy body disease.

Several neuropathological terms have been used to describe the dementia disorders associated with Lewy bodies, the main ones being:

- diffuse Lewy body disease;[492]
- cortical Lewy body disease;[298]
- senile dementia of Lewy body type;[712]
- Lewy body variant of AD.[353]

The increased realization that Lewy bodies and related pathology can often be detected in brains of individuals with dementia has been largely due to the increased utilization of immunohistochemical staining for α-synuclein. As noted previously in the section on mixed pathology in AD, Lewy bodies are detected in the amygdala in many cases of advanced AD,[910] particularly those with genetic risk factors for AD.[540]

On the other hand, in retrospective studies of autopsies on clinically normal individuals, Lewy bodies are detected in about 10 per cent of those over age 60 years.[263,433]

Moreover, some neurologically unimpaired subjects have a significant burden of α-synuclein pathology.[704]

The conditions that will be discussed in this section include DLB, PDD and Alzheimer's disease with Lewy bodies. These conditions have an overlap of clinical and pathological features.

There is continued debate as to whether PDD and DLB represent the same or different entities. While there are clinical and pathological overlaps, there are distinct differences. These aspects are discussed in detail with a rehearsal of the similarities and differences by Goldman *et al*.[310]

Dementia with Lewy Bodies

Dementia with Lewy bodies is the term used to describe a clinical syndrome characterized by dementia associated with hallucinations, fluctuation in level of consciousness and spontaneous parkinsonism.[606,609] It is recognized that DLB is a common clinical pattern of dementia, mainly seen in the elderly with uncommon cases noted before the age of 65 years.[870] DLB has been estimated to accounts for about one in 25 dementia cases diagnosed in the community (4.2%) and about one in 13 cases diagnosed in secondary care (7.5%).[922]

Although the pathology in most such cases is that of widespread cortical Lewy bodies, there are exceptions, and most cases have some degree of Alzheimer-type pathology. The presence of clinical and pathological features that overlap those of both PD and AD has been difficult to reconcile. The furtherance of DLB as a distinct clinicopathological entity will remain problematic until a specific biomarker for the disease is discovered. The major reason for delineating this clinical syndrome is the belief that it has distinct clinical management issues and a different prognosis from both AD and PD. The clearest example is the need to avoid traditional neuroleptic medications – often prescribed to demented patients with psychotic features – because of the increased risk of serious sensitivity reactions in DLB. DLB patients may also be candidates for dopaminergic treatment that would not normally be considered in treating AD. In addition, DLB patients may have more favourable responses than most patients with AD to anticholinesterase therapy.[606]

Diagnostic Features of Dementia with Lewy Bodies

A distinct set of clinical features permits an ante-mortem diagnosis of DLB[609] (Table 16.8). An elevated level of α-synuclein in the CSF is being explored as a possible biomarker for this disease.[536]

The neuropathological approach to such cases has undergone evolution with guidelines proposing a means to predict the likelihood that the individual had the DLB syndrome ante mortem.[609] Guidelines for pathological evaluation have been incorporated into the recent NIAA guidelines for pathological assessment.[402]

Macroscopic Appearance

The macroscopic appearance of the brain in DLB is usually similar to that in PD, including brain weights within normal limits or only slightly lower than normal, mild cortical atrophy of the frontal lobe, and loss of neuromelanin pigment from the substantia nigra and locus coeruleus (Figure 16.41). In some cases, particularly those with a greater degree of Alzheimer-type pathology, there may be greater temporal and parietal lobe atrophy. In a few cases atrophy may be marked in medial temporal lobe and limbic structures ('lobar atrophy').[587] Whereas most cases have visible pallor in the substantia nigra and locus coeruleus, the change may be less in patients who did not have overt parkinsonism. Pallor of the locus coeruleus alone is not strongly predictive of Lewy body pathology, as it is common in most cases of AD. Although functional imaging shows decreased perfusion in the occipital lobe, there is usually no macroscopic atrophy of the visual or visual association cortices.

Histopathology

Neuropathological studies of patients with the DLB clinical syndrome usually reveal one of three main patterns: (i) widespread Lewy bodies associated with (sometimes numerous) cortical diffuse Aβ deposits and low Braak neurofibrillary tangle stages; (ii) widespread Lewy bodies with sufficient neuritic plaques and neurofibrillary tangles for an independent diagnosis of AD; and (iii) pure Lewy body disease, involving widespread cortical areas with no significant Alzheimer-type pathology.[608]

Cortical involvement varies in DLB. In some cases Lewy bodies are relatively restricted to limbic structures (so-called 'transitional Lewy body disease'), whereas in other cases Lewy bodies are widespread in cortical areas (so-called 'diffuse Lewy body disease'), most numerous in limbic and multimodal association cortex of the temporal and frontal lobe, with fewer in parietal and occipital cortex and almost none in primary sensory or motor cortex.[492]

The defining histopathological feature of DLB is the presence of Lewy bodies usually accompanied by Lewy neurites in multiple brain regions. The substantia nigra has variable degree of neuronal loss and Lewy body formation (Figure 16.42). Whereas neuronal loss in PD is often worst in the ventrolateral tier of neurons of the pars compacta, medial populations of neurons, including the ventral tegmental region, are often particularly vulnerable to Lewy bodies in cases with dementia.[763] As in PD, there is involvement of multiple regions of the central and even peripheral nervous system in DLB, including particularly the nucleus basalis of Meynert and the amygdala.

Cortical Lewy bodies are found in almost all cases of DLB and their detection is enhanced using antibodies to α-synuclein, ubiquitin or P62 (Figure 16.43). Inclusions are concentrated in small neurons of the deep cortical layers and are usually most dense in the limbic, insular and temporal cortex. The amygdala is invariably affected, where there is also frequently spongiform change and many Lewy neurites (Figure 16.44). Spongiform change is common in DLB involving the superior and inferior temporal, entorhinal and insular cortex as well as the amygdala[355] (Figure 16.45). The striatum in DLB exhibits variable reduction (more than in AD, but less than in PD) in tyrosine hydroxylase immunoreactivity and loss of dopamine markers, reflecting the degeneration of dopaminergic neurons in the substantia nigra.[723] There are also dot-like or short curvilinear α-synuclein-immunoreactive neurites in the striatum in most cases.[226]

TABLE 16.8 Clinical diagnostic criteria for dementia with Lewy bodies (DLB)

1. Dementia is the central feature of DLB
There should be progressive cognitive decline of sufficient magnitude to interfere with normal social or occupational function
Prominent or persistent memory impairment may not necessarily occur in the early stages but is usually evident with progression
Deficits on tests of attention, executive function and visuospatial ability may be especially prominent

2. Core clinical features (two for probable DLB, one for possible DLB)
Fluctuating cognition, with pronounced variations in attention and alertness
Recurrent visual hallucinations that are typically well formed and detailed
Spontaneous features of parkinsonism

3. Suggestive features (if one or more of these is present in the presence of one or more core features, then a diagnosis of probable DLB can be made; in the absence of any core features, one or more suggestive features is sufficient for possible DLB; probable DLB cannot be diagnosed on the basis of suggestive features alone):
REM sleep behaviour disorder, which may precede onset of dementia by several years
Severe neuroleptic sensitivity
Abnormal (low uptake) in basal ganglia on SPECT dopamine transporter scan

4. Supportive features (commonly present but not proven to have diagnostic specificity)
Repeated falls and syncope
Transient, unexplained loss of consciousness
Severe autonomic dysfunction, e.g. orthostatic hypotension, urinary incontinence
Systematized delusions
Hallucinations in other modalities
Depression
Relative preservation of medial temporal lobe structures on CT/MR imaging scan
Generalized low uptake on SPECT/PET perfusion scan, with reduced occipital activity
Abnormal (low uptake) MIBG myocardial scintigraphy
Prominent slow-wave activity on EEG, with temporal lobe transient sharp waves

5. Diagnosis of DLB is less likely
In the presence of cerebrovascular disease evident by clinical signs or imaging
In the presence of physical or neurological disorders sufficient to account for the syndrome
If parkinsonism appears for the first time only at a stage of severe dementia

CT, computed tomography; EEG, electroencephalography; MIBG, meta-iodobenzoguanidine; MR, magnetic resonance; PET, positron emission tomography; REM, rapid eye movement; SPECT, single-photon-emission computed tomography.

Cortical synaptic loss has been demonstrated in DLB when associated with AD pathology. Synapse loss may be less severe than in other forms of cortical dementia.[357,538] A few variably argyrophilic and α-synuclein-immunoreactive glia can also be demonstrated in some cases[934] but glial cytoplasmic inclusions of multiple system atrophy are not seen.

Alzheimer-type pathology is frequently encountered in DLB. Approximately 80 per cent of patients with DLB have numerous diffuse plaques and rest with few or no neuritic plaques. Approximately 60 per cent have neurofibrillary tangles in the entorhinal cortex in moderate to severe density and rare neocortical neurofibrillary tangles. Some of these cases meet the pathological criteria for at least possible AD. Approximately 30 per cent of patients with DLB have advanced AD-type changes, with Braak neurofibrillary tangle stages of V–VI. A small proportion of DLB patients has cortical and brainstem Lewy body pathology in the complete absence of AD changes consistent with pure diffuse Lewy body disease.

Neurochemical Pathology

In addition to nigrostriatal dopaminergic degeneration mentioned earlier, the cholinergic system is abnormal in patients with DLB. The levels of choline acetyltransferase are lower than in patients with AD with a comparable severity of dementia.[207,708,889]

Consensus Guidelines for Pathological Assessment of Dementia with Lewy Bodies

The guidelines for the pathological assessment of DLB were devised as a standardized approach to rating the severity and distribution of α-synuclein and Alzheimer-type pathology. These guidelines do not specify absolute diagnostic criteria but rather assign a probability that the observed pathologies are likely to have caused a clinical picture conforming to DLB. The approach to evaluation of AD was similar to that used in the CERAD protocol[624,625] for sampling and semiquantitative assessment of lesion density. The importance of topographical staging of the neurofibrillary pathology is also incorporated into the diagnostic scheme.

In the diagnostic criteria a semiquantitative assessment (analogous to those used in CERAD scheme for evaluation of neuritic plaques and neurofibrillary tangles) of Lewy body density based upon α-synuclein immunohistochemistry is used (Figure 16.46).[609] The severity of α-synuclein pathology is rated in five cortical regions (transentorhinal cortex [BA 29], anterior cingulate gyrus [BA 24], mid-frontal gyrus [BA 8/9], middle temporal gyrus [BA 21] and inferior parietal lobule [BA 40]). Cases are allocated to one of three categories of Lewy body type similar to those originally proposed by Kosaka, including brain stem-predominant, transitional or diffuse types.[492]

The probability that neuropathological findings are associated with a DLB clinical syndrome is directly related to the severity of Lewy body-type pathology and inversely related to the severity of Alzheimer-type pathology (Table 16.9). This is based upon prior experience that when Alzheimer-type pathology is extensive, the patient most often has a clinical syndrome indistinguishable from AD, even when Lewy body pathology is severe.

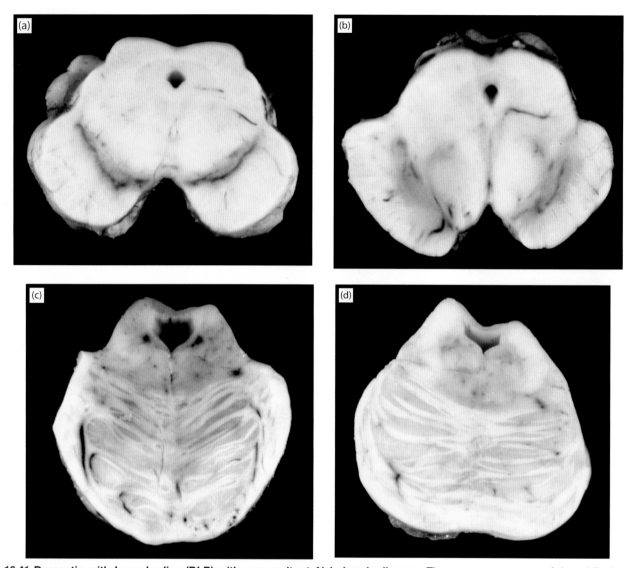

16.41 Dementia with Lewy bodies (DLB) with concomitant Alzheimer's disease. The gross appearance of the midbrain and pons can be compared with that of a normal individual. The substantia nigra shows normal pigmentation in the control subject **(a)**, whereas nigral pallor is observed in the case of DLB with concomitant Alzheimer's disease **(b)**, In contrast to the normally pigmented locus coeruleus **(c)**, pallor of this nucleus is common in longstanding cases of AD **(d)**, with or without Lewy bodies.

Reprinted with permission from Mirra et al.[625] Copyright © 1993 American Medical Association. All rights reserved.

Diagnostic Terminology in Disorders with Dementia and Lewy Bodies

The neuropathologist is often confronted with examination of brains from patients with a clinical diagnosis of DLB, a clinical diagnosis of AD, a clinical diagnosis of PDD or with minimal clinical information related to cognitive function. In each instance Lewy body pathology may be encountered, often in combination with neuropathological changes of AD. Table 16.10 summarizes commonly used clinical and neuropathological terminology in this diagnostic area.[632]

When confronted with a brain that shows neuronal loss and Lewy bodies in the substantia nigra, as well as Lewy bodies in other subcortical and cortical regions, the neuropathologist cannot determine from histopathological findings alone whether the patient had a movement disorder

or dementia or both. Consequently, some neuropathologists eschew diagnostic labels entirely, at least for research purposes, providing only descriptions to be correlated with the clinical features.

For all cases, regardless of clinical history, reporting should follow the format:[402]

- 'Lewy body disease, limbic' or
- 'Lewy body disease, amygdala predominant'

Clinicians and patients like the security of a diagnostic label. The consensus criteria for neuropathology of DLB propose an intermediate approach. Following pathological assessment, the diagnosis lists the full range of pathological findings and a pathological diagnostic statement is amended, such as 'there is a low/intermediate/high likelihood that the pathological findings are associated with

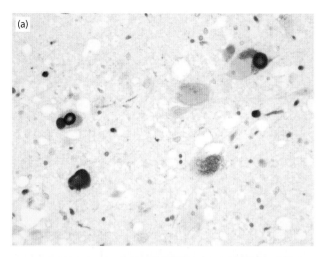

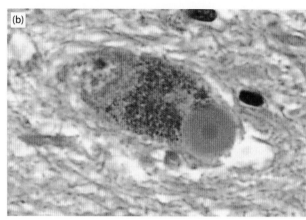

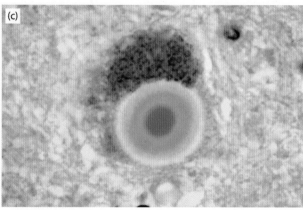

16.42 Substantia nigra in dementia with Lewy bodies. **(a)** The substantia nigra displays Lewy bodies immunostained for α-synuclein in neuromelanin-containing neurons. Labelled neurites can be seen in the neuropil. **(b,c)** Nigral Lewy bodies often exhibit peripheral halos and a concentric eosinophilic appearance. Appearances are identical to those in idiopathic Parkinson's disease.

a DLB clinical syndrome'. For individuals with cognitive impairment it has been recommended that neocortical LBs be considered an adequate explanation of cognitive impairment or dementia.[402]

As discussed earlier, in patients with clinically diagnosed AD, it is relatively common for Lewy body pathology to be found, especially if it is sought by immunostaining for α-synuclein. Although such patients can have Lewy body pathology that may extend to the neocortex, they often do not present with clinical features of DLB, especially if the Alzheimer-type pathology is advanced. By ascribing a probability that regional Lewy body pathology predicts DLB clinical features, depending on relative severity of Lewy body and AD changes, the consensus guidelines for neuropathological diagnosis addresses this issue, because such cases would be considered to have a 'low likelihood' of DLB.

Parkinson's Disease Dementia

Although James Parkinson originally suggested that the disease named after him spared the intellect, it is now realized that this is not always true. Recent long-term follow-up studies suggest that dementia affects between 40–70 per cent of patients with PD and more subtle cognitive deficits are even more common.[2,234] The pattern of dementia appears to be dominated by a dysexecutive syndrome but some patients have a clinical syndrome with many features

similar to DLB.[310,428] For a more detailed discussion of clinical and pathological aspects of PD, see Chapter 12. The International Parkinson and Movement Disorder Society (MDS) published recommended clinical diagnostic criteria for PDD requiring a diagnosis of PD and a dementia syndrome with insidious onset and slow progression.[235] It is also apparent that MCI is common in patients with PD, possibly affecting 20–30 per cent of patients.[310] The MDS has also more recently established guidelines for establishing PD with mild cognitive impairment (PD-MCI).[543,544] Longitudinal studies indicate that 50 per cent of patients with PD will develop dementia within 10 years from diagnosis.[310]

Neuropathology of Parkinson's Disease Dementia

The relative importance of Lewy- and Alzheimer-type pathologies in leading to dementia in PD has been subject to study in several series. A study of patients with PD and PDD found that a combination of Lewy- and Alzheimer-type pathologies provides the main pathological correlate of PDD.[153] The neurobiological basis for cognitive impairment in PD has recently been reviewed.[349] The pathologic basis for PDD includes LBs, AD pathology, cerebrovascular disease, and neurotransmitter changes that include loss of limbic and cortically projecting dopamine, noradrenaline, serotonin, and acetyl choline systems. It is concluded that Lewy-related pathology is the most important pathological feature

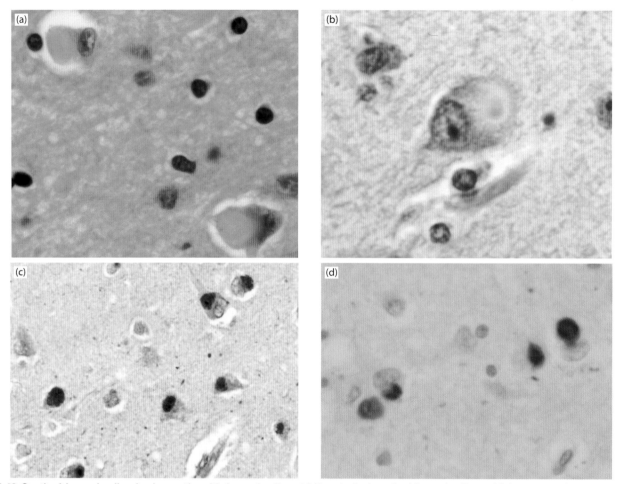

16.43 Cortical Lewy bodies in dementia with Lewy bodies. (a) Two typical cortical Lewy bodies exhibit an amorphous eosinophilic appearance. **(b)** Less commonly, cortical Lewy bodies have halos and a concentric morphology similar to nigral Lewy bodies. Detection of cortical Lewy bodies may be enhanced using antibodies to ubiquitin **(c)** and α-synuclein **(d)**.

α-Synuclein immunohistochemical preparation provided by Dr Bernardino Ghetti, Indiana University, Indianapolis, IN, USA.

TABLE 16.9 Probability of Lewy body pathology as substrate of dementia with Lewy bodies (DLB)

Lewy body pathology	Alzheimer-type pathology		
	Braak stage 0–II	Braak stage III–IV	Braak stage V–VI
Brain stem-predominant	Low	Low	Low
Limbic (transitional)	High	Intermediate	Low
Diffuse neocortical	High	High	Intermediate

that relates to the development of cognitive impairment. When the contribution of AD pathology is reviewed from published series, there are differing reports on frequency of association. Given the age of those affected by PD, concurrent AD changes are expected. The authors conclude that it is very likely that although LB pathology is the main factor in cognitive decline, there is a synergistic effect between all pathologies, a view supported by others.[410] It is also likely that the same spectrum of pathological changes is the basis for PD-MCI.[349]

Most studies suggest that DLB and PDD share common pathological features and differ primarily in mode of clinical presentation.[1,2,332,632,904] Specifically, PDD and DLB show widespread cortical and brain stem Lewy bodies with varying degrees of Alzheimer-type pathology, but usually insufficient to warrant a high likelihood diagnosis of AD. Most cases of PDD show brain stem pathology similar to that in PD. Degeneration of the substantia nigra correlates with disease duration and severity in PD. It also correlates with motor score in PD rating scales.[329] Given that parkinsonism is typically more severe in PDD than in DLB, it is not surprising that substantia nigra degeneration is typically more severe in PDD than DLB.[904] Degeneration of the basal nucleus of Meynert is common in PD with or without dementia,[655,956] and cholinergic deficits are often profound in PDD.[596,708,890]

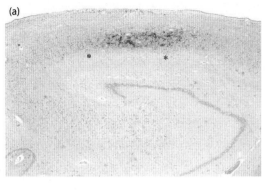

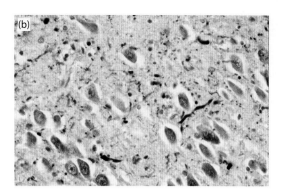

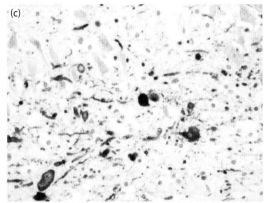

16.44 Lewy neurites in CA2–3 of the hippocampus.
(a) Clusters of ubiquitin-immunoreactive neurites (asterisks) are seen in a localized region of the hippocampus in CA2–3. **(b)** High-power appearances with **(b)** anti-ubiquitin and **(c)** anti-α-synuclein staining.

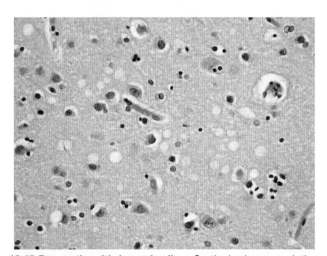

16.45 Dementia with Lewy bodies. Cortical microvacuolation resembling that seen in a prion disease may be evident in the mesial temporal lobe.

In some patients with PDD other pathological processes contribute to dementia. In some cases the changes of AD can be found to explain the dementia.[439] In others vascular lesions may lead to a diagnosis of vascular dementia or, if combined with AD pathology, mixed dementia.[395,396] Less common are cases with parkinsonism and dementia due to PSP, corticobasal degeneration (CBD), or frontotemporal dementia with parkinsonism.

Genetic Factors Leading to Cognitive Impairment in Lewy Body Disorders

Several studies have established that genetic factors play a role in the risk of development of dementia in PD. This has been reviewed by Halliday *et al.*[349]

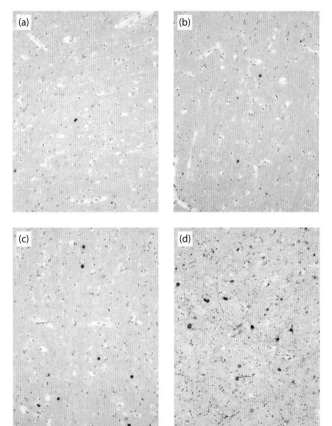

16.46 Semiquantitative rating of cortical Lewy body density. Severity of cortical Lewy body pathology using anti-α-synuclein immunostaining. **(a)** Stage 1: sparse Lewy bodies (LBs) or Lewy neurites (LNs). **(b)** Stage 2: more than one LB per high-power field and sparse LNs. **(c)** Stage 3: four or more LBs and scattered LNs in low-power field; **(d)** Stage 4: numerous LBs and LNs.

TABLE 16.10 Summary of diagnostic terminology for Lewy body disorders

Term	Notes
Lewy body disease	Neuropathological term used to include all diseases associated with Lewy body pathology
Diffuse Lewy body disease, transitional Lewy body disease, brain stem-predominant Lewy body disease	Neuropathological terms that describe the distribution of Lewy body pathology independent of clinical presentation. Some patients may have parkinsonism or dementia, but others may be clinically normal (e.g. brain stem-predominant type)
Dementia with Lewy bodies (DLB)	Clinical term for a specific syndrome as proposed by the Consortium on Dementia with Lewy Bodies. Neuropathology associated with this condition is usually diffuse or transitional Lewy body disease (see earlier). A small proportion of patients (e.g. those with progressive supranuclear palsy or vascular dementia) do not have Lewy body pathology. This merely reflects the imprecision of clinical diagnostic criteria at present
Alzheimer's disease (AD) with amygdala Lewy bodies	Diagnostic category used to describe cases of advanced AD associated with α-synuclein pathology limited largely to the amygdala. Such cases are not considered within the DLB clinical spectrum
Alzheimer's disease with incidental Lewy bodies	Clinical and neuropathological features fit with high probability those of AD, and yet Lewy bodies are detected in selected brain stem nuclei and are not accompanied by significant neuronal loss
Parkinson's disease dementia (PDD)	Clinical term describing Parkinson's disease for at least 1 year before onset of dementia. The pathology may be indistinguishable from that found in DLB, and the same diagnostic scheme is recommended. Whether it is possible to predict parkinsonism based upon severity of neuronal loss in the substantia nigra is not resolved

TABLE 16.11 Genetic factors in Parkinson's disease dementia

Gene	Familial PD: risk of cognitive impairment	Sporadic PD: risk of cognitive impairment
SNCA	Increased	
LRRK2	Decreased	
parkin	Decreased	
GBA		Increased
MAPT H1 haplotype		Increased
COMT Met/Met		Mixed findings
ApoE		Increased

Risk of cognitive impairment associated with genetic associations with Parkinson's disease.[349]

Patients with PD due to α-synuclein triplication typically have prominent early cognitive decline. Those with the E46K mutation also appear to have an increased risk for early PDD. Patients with PD associated with LRRK2 or parkin mutations are at reduced risk of PDD. This is summarized in Table 16.11.

FRONTOTEMPORAL LOBAR DEGENERATIONS, INCLUDING TAUOPATHIES

Frontotemporal lobar degeneration (FTLD) is the term that describes a clinically and pathologically heterogeneous group of non-Alzheimer neurodegenerative disorders linked by their preferential involvement of the frontal and temporal lobes and subcortical grey matter. Patients present clinically with one of the patters of cognitive disorder grouped under the term frontotemporal dementia (FTD).

The epidemiology of these diseases is difficult to evaluate because of changes in recognition and diagnostic criteria.[787] Molecular investigation has brought a more complete understanding in this group of neurodegenerative diseases, leading to recent changes in nomenclature. The interpretation of older literature for several conditions must now be done in the light of these recent findings.

Males and females appear to be equally affected with a prevalence estimated at 15–22 per 100 000, and incidence 2.7–4.1 per 100 000.[690,746] Patients diagnosed with FTD typically have a poor prognosis and die within 3 to 10 years of diagnosis. As with many neurodegenerative diseases, increasing frailty and cachexia lead to development of pneumonia as a cause of death.[678] In some of the conditions causing FTLD, neuropathology affects the brain stem and causes impairment in swallowing that would increase a risk of aspiration.

Historical Notes and Evolution of Concepts and Terminology

It has long been recognized that some patients with dementia, often associated with a disturbance of language, have neuropathology that predominantly affects the frontal and temporal lobes. The best characterized of these conditions was Pick's disease and for many years all patients falling into this clinicopathological group were classed as having Pick's disease. Pick's disease has a distinct and easily recognized set of neuropathological changes, including argyrophilic round intraneuronal inclusions referred to as Pick bodies. It subsequently became apparent that many cases clinically diagnosed as having Pick's disease did not have Pick bodies. On the basis of such findings, a classification of Pick's disease was put forward in 1974 by Constantinidis and colleagues, [154] which can now be interpreted in the light of recent understanding (Table 16.12).

In current practice the term Pick's disease is restricted to cases in which tau-immunoreactive Pick bodies are seen, associated with characteristic additional features. In the 1980s many clinical case reports based on the use of conventional neurohistology drew attention to apparently distinctive pathological changes in patients with cognitive abnormalities, often associated with an extrapyramidal movement disorder that did not fit with clinical features of AD or PD.[147]

In the 1980s and early 1990s, increasing use of immunohistochemistry to detect pathological accumulations of proteins and inclusion bodies allowed refinement of pathological diagnosis.[155] It became apparent that patients with clinical features identical to Pick's disease had novel neuropathological changes, and this led to the proposal of a whole range of clinical and neuropathological diagnostic terms. Clinical and pathological overlaps were also being described between patients with dementia and patients with parkinsonism or motor neuron disease. A group of non-Alzheimer neurodegenerations was identified that mapped to patients with cognitive dysfunction.

A variety of groups used a range of terms to describe either clinical or pathological findings in this group of disorders. These terms included FTD, frontotemporal degeneration, frontal lobe degeneration of non-Alzheimer type, FTLD, dementia lacking distinctive histopathological features, primary progressive aphasia and semantic dementia. Kertesz and colleagues proposed the concept of 'Pick complex'[474] to include a spectrum of disorders, including primary progressive aphasia, frontal lobe dementias, CBD, and motor neuron disease (MND) with dementia. This hypothesis, based on clinical symptomatology rather than neuropathology and other features, must be evaluated in view of current biochemical and genetic information that is increasingly redefining understanding of these disorders. Previous terms used for the FTLDs are as follows:

- frontotemporal lobar degeneration with motor neuron disease (FTLD-MND);
- motor neuron disease dementia or dementia with motor neuron disease (MND dementia);
- amyotrophic lateral sclerosis and dementia (excluding Guam type);
- Pick's disease combined with motor neuron disease;
- dementia lacking distinctive histology (DLDH);
- frontotemporal lobar degeneration with ubiquitin-only-immunoreactive changes (FTLD-U) ;
- neurofilament inclusion disease (NFID);
- motor neuron disease inclusion dementia (MNDID);
- frontotemporal degeneration (FTD);
- dementia with ubiquitin-reactive tau-negative inclusions;
- frontal lobe degeneration with ubiquitin neurites;
- primary aphasia with ubiquitinated neurites;
- semantic dementia with ubiquitin-positive tau-negative inclusion bodies;
- dementia with inclusions, tau and synuclein negative, ubiquitinated (ITSNU);
- frontal lobe degeneration of non-Alzheimer type.

The term FTDP-17 was defined in 1997 and it is now possible to distinguish between cases linked to MAPT gene mutations and PGR mutations, both localized to chromosome 17.

Based on an international consensus meeting, McKhann and colleagues proposed that the term frontotemporal dementia (FTD) be used to refer to the clinical syndrome, and frontotemporal lobar degenerations (FTLD) to refer to the neuropathological processes that cause the clinical syndrome.[610]

TABLE 16.12 Historical concepts of Pick's disease subtypes		
Pick's type	**Neuropathology**	**Modern interpretation**
A	Pick bodies and swollen neurons in atrophic cortex	Classical Pick's disease – now recognized as one of the tauopathies
B	Swollen neurons in affected cortical regions; cell loss from basal ganglia and substantia nigra	Cases probably represented CBD and now classed as one of the tauopathies
C	No inclusions seen in atrophic cortical regions	Now classed as one of the non-tauopathy forms of FTLD

CBD, corticobasal degeneration; FTLD, frontotemporal lobar degeneration.

Recent Evolution in Nomenclature and Mapping Morphology to Genetics

The recent evolution in terminology and classification has been documented by an international consensus group and modified to link to new findings.[120,566,567]

Once immunohistochemical studies were routinely informing diagnostic classification, two main types of FTLD became recognized: FTLD-tau (including classical Pick's disease) and a large group, characterized by the presence of neuronal inclusions immunopositive for ubiquitin, termed FTLD-MND-type or FTLD-U.[541]

In 2006 it was discovered that in around 90 per cent of the FTLD-U cases, the inclusions and related pathology were related to accumulation of transactive response DNA-binding protein 43 (TDP-43) leading to the identification of this group as FTLD-TDP.[667]

The discovery of the large group of FTLD-TDP cases led to the realization that a smaller group remained, characterized by ubiquitin-positive, TDP-43-negative inclusions, and termed atypical FTLD-U (aFTLD-U). These cases were sporadic, had a very early onset with dominant severe psychobehavioral presentations and showed ubiquitin-immunoreactive inclusions that included prominent neuronal intranuclear inclusions.[565,771] In this group the inclusions were later shown to contain FUS, leading to the delineation of FTLD-FUS.[669,912]

The remainder of the original FTLD-U group is now referred to as FTLD-ubiquitin-proteasome system (FTLD-UPS).

Alongside the evolution in nomenclature for neuropathological classification has been discovery of genetic factors linking to disease, the latest (in 2011) and most significant being the discovery of the *C9orf72* GGGGCC repeat expansion as a major contributor to FTLD as well as MND (ALS) pathogenesis.[188]

In this account the classification outlined in Figure 16.47 and Table 16.13 will be used. This emphasizes that several clinical syndromes may be caused by several distinct pathological entities. One apparently distinct clinical syndrome may have several pathological causes. Conversely, a distinct pathological process may result in several different clinical syndromes (see Table 16.13).

Pathological Types of Frontotemporal Lobar Degeneration

There are three main and one rarer type of pathology seen in frontotemporal lobar degenerations

- tauopathies (FTLD-tau) (sporadic or inherited) ~50 per cent cases;
- frontotemporal lobar degeneration with TDP-43 pathology (FTLD-TDP) (sporadic or inherited) ~45 per cent;
- frontotemporal lobar degeneration with FUS pathology (FTLD-FUS) ~5 per cent;
- frontotemporal lobar degeneration with ubiquitin pathology (FTLD-UPS) <1 per cent.

The pathological classification and nomenclature that is applied to this group of diseases was revised in 2010 and is used here.[567] An approach to diagnosis for general pathologists has been published.[74]

Clinical Features Associated with Frontotemporal Lobar Degeneration

There are four main clinical variants within FTD associated with FTLD:[317,939]

- behavioural variant frontotemporal dementia (bvFTD);
- semantic variant of progressive aphasia (svPPA);
- nonfluent variant of progressive aphasia (nfvPPA);
- logopenic variant of progressive aphasia (lvPPA);

Behavioural variant frontotemporal dementia (bvFTD) is the most common clinical presentation of FTLD, characterized by a decline in social behaviour and personal conduct, with disinhibition, apathy, loss of empathy, perseveration or stereotypic behaviours, and hyperorality. bvFTD is associated with executive deficits and there is usually sparing of memory and visuospatial functions. Imaging shows atrophy of the frontal lobes in mesiofrontal, orbitofrontal, and anterior insular regions.[761] Criteria for clinical diagnosis have been published by The International bvFTD Criteria Consortium (Table 16.14).

Primary progressive aphasia (PPA) is a group term in which there are three clinical presentations of FTD characterized by a dominant abnormality of language. The first is termed semantic variant PPA (svPPA, also called semantic dementia), the second, nonfluent variant PPA (nfvPPA also termed progressive nonfluent aphasia) and the third, most recently described, called logopenic variant of progressive aphasia (lvPPA).

- Patients with svPPA have a profound impairment of semantic memory (which relates to knowledge of the meaning of verbal or visual inputs) but have preservation of episodic memory (relating to autobiographical events); these patients often have anterior and inferior temporal atrophy, particularly asymmetrically involving the dominant hemisphere. As disease progresses, atrophy affects the frontal lobes and symptoms as seen in bvFTD may develop.
- Patients with nfvPPA have a disorder of expressive language, with problems in word retrieval yet preservation of word comprehension. This disorder is typically associated with atrophy of the peri-Sylvian regions in the dominant hemisphere. As disease progresses speech becomes sparse and patients become mute.
- Patients with logopenic variant of progressive aphasia (lvPPA) have slow speech that is not fluent and shows impaired comprehension of meaning. Imaging shows atrophy more posteriorly, with asymmetrical temporo–parietal atrophy involving the dominant hemisphere.

Patients with FTLD may also have an associated movement disorder, most commonly parkinsonism[475] or motor neuron disease. Thus, FTLD includes some patients who would be classified clinically as having parkinsonism-dementia as well as some who would be classified as having MND dementia. During progression of disease, as more cortical and subcortical regions are affected, it is possible for patients to develop overlapping clinical features.

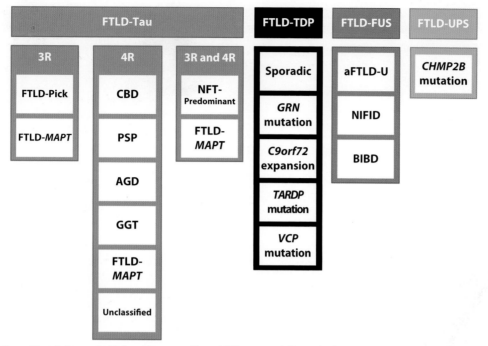

16.47 Classification of frontotemporal lobar degeneration. AGD, argyrophilic grain disease; BIBD, basophilic inclusion disease; CBD, corticobasal degeneration; *CHMP2B*, charged multivesicular protein 2B; FTLD, frontotemporal lobar degeneration; GGT, globular glial tauopathy; *GRN*, progranulin gene; *MAPT*, microtubule-associated protein tau; NFT, neurofibrillary tangle; NIFID, neuronal intermedial filament inclusion disease; PSP, progressive supranuclear palsy; *TARDP*, TAR DNA-binding protein 43; *VCP*, valosin-containing protein.

It has been suggested that FTLD in the elderly may have a different clinical profile to early onset FTLD. When developing in the elderly, FTLD may be more associated with memory loss and behavioural change rather than language and semantic dysfunction. This is reflected in the pathological findings of hippocampal sclerosis and less severe cortical atrophy than in early onset cases.[41]

Genetics of Frontotemporal Lobar Degeneration

Approximately 50 per cent of those in reported series of FTLD have a family history of dementia.[659] Within this group, 10–30 per cent of patients have an autosomal dominant pattern of inheritance.[761] The following genes have been implicated in the development of FTLD[557,828] (Table 16.15). Mouse models based on these identified genes have been produced.[764]

It is clear that a significant proportion of cases of FTLD do not have mutations in established genes. A family with ALS-dementia associated with type B TDP pathology and phosphorylated tau pathology consistent with corticobasal degeneration has been linked to chromosome 16p12.1-q12.2.[219] Genome-wide association studies are progressing in establishing additional genetic risk factors in disease. One study has suggested that immune processes (link to 6p21.3) and lysosomal or autophagy pathways (link to 11q14) are potentially involved in FTD.[250]

MAPT (Chromosome 17)

Mutations in the *MAPT* gene were discovered in 1998 and account for about 50 per cent of disease in familial cases of FTLD and about 5 per cent of all cases of FTLD. There are 42 coding-region mutations in *MAPT* and nine intronic mutations in the gene leading to changes in function of tau. Most mutations are located in exons 9–12 and the adjacent introns, and form two broad groups:[852]

- affecting protein level;
- affecting the alternative splicing of tau pre-mRNA.

Some mutations reduce the ability of tau to bind to microtubules and are believed to increase the tendency for tau to aggregate into filaments. Other mutations affect RNA level and increase the mRNA splicing of exon 10, leading to an increase in four-repeat tau compared with three-repeat tau and an increased tendency of tau to aggregate into filaments.

Online databases should be consulted for details of mutations and links to case reports of neuropathology.

GRN (Chromosome 17)

Mutations in the *GRN* (progranulin) gene were discovered in 2006.[45] They account for 3–26 per cent of familial FTLD and result in FTLD-TDP pathology (type A)[560,564,828] with several features that differ from those of other types of FTLD-TDP.[136] Progranulin is involved in a range of cell processes in tumour biology, inflammation and neuronal function. Over 70 mutations have been reported in FTLD as well as in neurodegeneration associated with other clinical patterns, for example corticobasal degeneration.[990] *GRN* mutation carriers may present with any of the patterns of FTD, including corticobasal syndrome and parkinsonism,[857] although MND is rare. Hippocampal sclerosis in AD has been linked to *GRN* mutation.[214] Imaging shows that *GRN* mutations are associated with asymmetrical inferior frontal,

TABLE 16.13 Classification of frontotemporal lobar degeneration (FTLD)*

Type	Neuropathology	Gene	Clinical disease				
			bvFTD	nfvPPA	svPPA	CBS	FTD-MND
FTLD-Tau	Pick body	*MAPT*	•	•	•		
	CBD		•	•		•	
	PSP			•		•	
	AGD		•				
	NFT dementia		•				
	MAPT mutation		•	•			
FTLD-TDP	Type A	*GRN*	•	•	•	•	•
	Type B	*C9orf72*	•	•			•
	Type C			•	•		
	Type D	*VCP*					
FTLD-FUS	aFTLD-U		•				
	NIFID		•				
	BIBD						
FTLD-UPS	FTD-3	*CHMP2B*	•				

*From Rohrer[774] and Riedl.[761]

bvFTD, behavioural variant frontotemporal dementia; CBS, corticobasal syndrome; FTD-MND, frontotemporal dementia - motor neuron disease; FTLD, frontotemporal lobar degeneration; nfvPPA, nonfluent variant of progressive aphasia; svPPA, semantic variant of progressive aphasia.

temporal and inferior parietal lobe grey matter atrophy and involvement of long intrahemispheric association white matter tracts.[774] Brain atrophy can be detected by imaging in asymptomatic mutation carriers.[722] In patients with *GRN* mutations, CSF fluid and plasma progranulin levels are low.[256] *GRN* mutations cause disease by loss of function of progranulin or haploinsufficiency.[129,291,716]

C9orf72 (Chromosome 9)

Hexanucleotide repeat expansion of this gene in patients with FTD was discovered in 2011.[188,752] The GGGGCC repeat expansion of *C9orf72* is the most common genetic cause of motor neuron disease (ALS) and FTLD.[371] The frequency of *C9orf72* expansions in Western Europe was 9.98 per cent in FTLD, with 18.52 per cent in familial and 6.26 per cent in sporadic FTLD cases.[918] It has been seen in 1 out of 7579 (0.15%) of the UK 1958 birth cohort (58BC) controls.[62] Fewer than 20 GGGGCC hexanucleotide repeats are regarded as normal, the lower limit for the disease-associated range being around 65 repeats. Intermediate expansions are regarded as risk-associated.

C9orf72-FTLD is predominantly associated with behavioural variant FTD, with psychotic features often highlighted,[201] and is linked to clinical parkinsonism and corticobasal syndrome.[156] Hippocampal sclerosis dementia has also been linked.[648,726]

The neuropathology of *C9orf72* cases is characterized by FTLD-TDP (types A and B) and associated ALS[845] (see Chapter 14). A distinctive feature is the finding of ubiquitin- and P62-positive, TDP-43-negative neuronal cytoplasmic and intranuclear inclusions in the cerebellum, hippocampal pyramidal cells and neocortical neurons

(Figure 16.48).[14,572] These inclusions contain dipeptide repeat proteins (poly-GA, -GR and -GP) derived from translation of *C9orf72* transcripts.[570,571,633] Polyclonal antibody to poly-GA was used to explore the relationship of the inclusions to TDP-43 pathology.[178]

It has been suggested that the mechanism leading to disease is related to interference with RNA processing and/or to the formation of toxic dipeptide repeat protein.[157,341] Pathology appears similar in rare patients showing homozygosity for the abnormal gene.[265]

Instances of pathology in patients with mutations in two FTD-associated genes have been reported.[914]

TDP-43 (Chromosome 1)

TAR DNA-binding protein 43 (TDP-43) is seen in inclusions and neurites in both sporadic and familial FTLD, as well as being characteristic of inclusions seen in motor neuron disease (ALS).[121] Although mutations in the gene are recognized to cause ALS they are rarely associated with FTLD.

Charged Multivesicular Body Protein 2B (Chmp2b) (Chromosome 3)

Mutation of this gene is a rare causal association of FTLD, reported in a Danish kindred and termed FTD-3.[835] Abnormalities develop in vesicular fusion events affecting the endosome-lysosome and autophagic pathways.[411] Imaging shows generalized central and cortical atrophy. Ubiquitin- and P62-positive neuronal cytoplasmic inclusions are seen in the dentate granule cell layer of the hippocampus and do not mark for other proteins characteristic of FTLD, classifying this as FTLD-UPS.[388]

TABLE 16.14 Consensus guidelines for the clinical diagnosis of bvFTLD

I.   Neurodegenerative disease
      The following symptom must be present to meet criteria for bvFTD
      A. Shows progressive deterioration of behaviour and/or cognition by observation or history (as provided by a knowledgeable informant)

II.   Possible bvFTD
      Three of the following behavioural/cognitive symptoms [A–F] must be present. Ascertainment requires that symptoms be persistent or recurrent, rather than single or rare events
      A. Early behavioural disinhibition (one of the following symptoms [A.1–A.3] must be present)
         A.1. Socially inappropriate behaviour
         A.2. Loss of manners or decorum
         A.3. Impulsive, rash, or careless actions

      B. Early apathy or inertia (one of the following symptoms [B.1–B.2] must be present)
         B.1. Apathy
         B.2. Inertia

      C. Early loss of sympathy or empathy (one of the following symptoms [C.1–C.2] must be present)
         C.1. Diminished response to other people's needs and feelings
         C.2. Diminished social interest, interrelatedness, or personal warmth

      D. Early perseverative, stereotyped, or compulsive/ritualistic behaviour (one of the following symptoms [D.1–D.3] must be present)
         D.1. Simple repetitive movements
         D.2. Complex, compulsive, or ritualistic behaviours
         D.3. Stereotypy of speech

      E. Hyperorality and dietary changes (one of the following symptoms [E.1–E.3] must be present)
         E.1. Altered food preferences
         E.2. Binge eating, increased consumption of alcohol or cigarettes
         E.3. Oral exploration or consumption of inedible objects

      F. Neuropsychological profile: executive/generation deficits with relative sparing of memory and visuospatial functions (all of the following symptoms [F.1–F.3] must be present)
         F.1. Deficits in executive tasks
         F.2. Relative sparing of episodic memory
         F.3. Relative sparing of visuospatial skills

III. Probable bvFTD
      All of the following symptoms [A–C] must be present
      A. Meets criteria for possible bvFTD
      B. Exhibits significant functional decline (by caregiver report or as evidenced by CDR or FAQ scores)
      C. Imaging results consistent with bvFTD (one of the following [C.1–C.2] must be present)
         C.1. Frontal and/or anterior temporal atrophy on MRI or CT
         C.2. Frontal hypoperfusion or hypometabolism on PET or SPECT

IV.   bvFTD with definite FTLD pathology
      Criterion A and either Criterion B or C must be present
      A. Meets criteria for possible or probable bvFTD
      B. Histopathological evidence of FTLD on biopsy or at autopsy
      C. Presence of a known pathogenic mutation

V.   Exclusionary criteria for bvFTD
      Criteria A and B must be answered negatively for any bvFTD diagnosis. Criterion C can be positive for possible bvFTD but must be negative for probable bvFTD
      A. Pattern of deficits is better accounted for by other non-degenerative nervous system or medical disorders
      B. Behavioural disturbance is better accounted for by a psychiatric diagnosis
      C. Biomarkers strongly indicative of Alzheimer's disease or other neurodegenerative process

Abbreviations: bvFTD, behavioural variant frontotemporal dementia; CDR, Clinical Dementia Rating Scale; FAQ, functional activities questionnaire scores; CT, computed tomography; FTLD, frontotemporal lobar degeneration; MRI, magnetic resonance imaging; PET, positron emission tomography; SPECT, single-photon emission computed tomography.
From Rascovsky K, Hodges JR, Knopman D, *et al*. Sensitivity of revised diagnostic criteria for the behavioural variant of frontotemporal dementia. *Brain* 2011;134(Pt 9):2456–77.

Valosin-Containing Protein (Chromosome 9)

Inclusion body myopathy associated with Paget disease of bone and frontotemporal dementia (IBMPFD) is linked to mutation in the gene for valosin-containing protein (VCP, a member of the AAA-ATPase superfamily). VCP has roles in cell cycle control, membrane fusion and the ubiquitin-proteasome degradation pathway.[942]

SQSTM1 (Chromosome 5)

Mutations in the gene for sequestosome 1 (*SQSTM1*), encoding the P62 protein, have been genetically associated with motor neuron disease (ALS) and Paget disease of bone. Coding variants have been seen in FTLD patients[523,792] with the suggestion that mutations clustering in the UBA domain of *SQSTM1* may influence disease susceptibility by doubling

TABLE 16.15 Genes associated with FTLD ranked by number of families affected

Gene	Number of mutations	Number of families reported
C9orf72	1 (expansion repeat)	336
GRN	69	231
MAPT	44	134
VCP	18	48
CHMP2B	4	5
SQSTM1	25	-

Adapted from Alzheimer Disease & Frontotemporal Dementia Mutation Database (accessed June 20th 2014) (http://goo.gl/V7KJQL).[919]

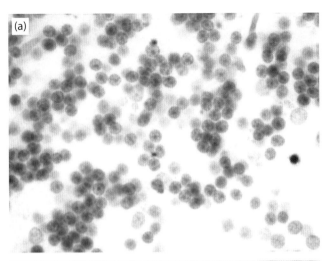

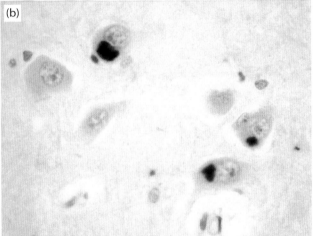

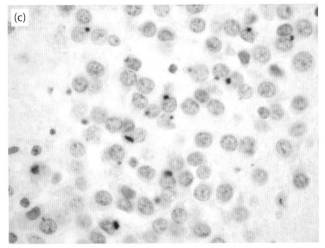

the risk for FTLD. There is limited autopsy information on such cases, but it is suggested that neuronal and glial phospho-TDP-43 is present, making this a form of FTLD-TDP.[919]

TMEM106B

The *TMEM106B* gene is a risk factor for FTLD-TDP especially in patients with *GRN* mutations. It has been proposed that *TMEM106B* can modulate levels of progranulin, possibly by modifying lysosomal degradation.[97,916]

Clinical Features Predicting Genetic Associations

Of these genes, mutations in *MAPT*, *GRN* and *C9orf72* explain 80 per cent of those cases associated with an autosomal dominant pattern of inheritance.[557] Clinical features associated with presentation can predict the likely genetic cause in autosomal dominant cases, as follows:[557]

- motor neuron disease: *C9orf72* and *GRN*;
- corticobasal syndrome: *GRN* and *MAPT*;
- psychoses: *C9orf72* and *GRN*;
- variable age of onset or reduced penetrance in family: *GRN* and *C9orf72*;
- cerebellar involvement: *C9orf72*;
- inclusion body myositis or Paget's disease of the bone: *VCP*;
- Danish ancestry: *CHMP2B*.

Imaging studies are also informing associations with underlying pathology, and indirectly to genetic causation as follows:[775]

- FTLD-tau *MAPT*: symmetrical, localized atrophy mainly in temporal lobe;
- FTLD-tau with Pick Bodies: asymmetrical distributed atrophy;
- FTLD-tau CBD: symmetrical, predominantly extratemporal atrophy;

16.48 P62 staining of inclusions in *C9orf72* FTLD. Inclusions seen in cerebellar granule cells **(a)**, pyramidal neurons in hippocampus **(b)**, and neurons in hippocampal dentate gyrus **(c)**.

- FTLD-FUS: symmetrical predominantly extratemporal atrophy;
- FTLD-TDP type C: asymmetrical localized atrophy mainly in temporal lobe.

Dementia may develop in the course of other inherited diseases and be associated with other dominant neurological presenting features. The pathology of these conditions is covered elsewhere in the book.

Frontotemporal Lobar Degenerations-Tau

The increased application of immunohistochemistry for tau in diagnostic neuropathology has led to increased recognition of the range of neurodegenerative conditions associated with tau pathology.

The tauopathies can be divided into two main groups (Table 16.16).

- The majority of patients with a tauopathy do not have a family history and have one of the conditions that are regarded as sporadic tauopathies.
- In a small number of kindreds the tauopathy is inherited, associated with mutation in the tau gene, *MAPT*.[852]

The most common of the tauopathies is AD, in which tau is an integral component of neurofibrillary tangles, dystrophic neurites in plaques, and neuropil threads. The non-Alzheimer tauopathies are characterized by pathological accumulation of tau protein in neurons or glial cells or more commonly both. Glial tau pathology is prominent in most cases and the species of tau that is seen in glia mirrors the primary cause of disease.[254]

Biology and Biochemistry of Tau

Tau protein is one of a family of microtubule-associated proteins (MAPs), which include structural MAPs and motor MAPs.[22] Tau is a phosphoprotein that is most abundant in axons, where it binds to and stabilizes microtubules in their polymerized state. Microtubules are a critical component of the cytoskeleton, important in the maintenance of cell shape, motility, transport and mitosis.[852]

Six tau isoforms exist in the adult human brain and are derived from alternative splicing of *MAPT* transcripts.[23] The shortest isoform is the only variant found in the fetal brain, with the other forms appearing shortly after birth. All six isoforms persist throughout adulthood.[305] The six tau isoforms

range from 352 to 441 amino acid residues with molecular weights of 45–65 kDa and differ by the presence or absence in the N-terminal half of 29 or 58 amino acid inserts encoded by exons 2 and 3, and an additional 31 amino acid insert in the C-terminal half of tau, encoded by exon 10. The 31 amino acid insert shares sequence homology with 30–32 amino acid repeats encoded by exons 9, 11 and 12. Tau isoforms derived from exon 10 splicing produce two major classes of tau, those with three repeats (3R tau) and those with four repeats (4R tau) in the microtubule-binding domain of tau.[107] Monoclonal antibodies that can distinguish 3R and 4R tau have been developed,[185,186] which permit assessment of distribution of the isoforms in lesions at the cellular level.

Tau protein is modified by phosphorylation, and in the tauopathies it is typically hyperphosphorylated. It also assumes an abnormal conformation, with increased β-sheet secondary structure and altered solubility. Tau protein is normally an unstructured and soluble protein, but in pathological conditions it is insoluble and forms filamentous intracellular aggregates within neurons and glia. The abnormal form of tau protein can be enriched by detergent extraction of brain homogenates, and immunoblot studies of these fractions form the basis for biochemical classification of the major tauopathies[852] (Table 16.17).

The effects of gene mutations in *MAPT* have been reviewed[852] and it is believed that disease develops through a gain of toxic function. A new consideration is whether tau protein in an aggregated form may propagate within the nervous system in a prion-like manner, seeding further aggregation and contributing to the pathogenesis of the disease.[162,852]

In establishing a diagnosis of a tauopathy a staged approach has been proposed[120] (Figure 16.49).

TABLE 16.16 Sporadic and inherited tau disorders	
Sporadic tauopathy and secondary tauopathy	Alzheimer's disease
	Progressive supranuclear palsy
	Corticobasal degeneration
	Argyrophilic grain disease
	Globular glial tauopathy
	Pick's disease
	Chronic traumatic encephalopathy (dementia pugilistica)
	Parkinsonism-dementia complex of Guam
	Gerstmann-Sträussler-Scheinker disease
	Postencephalitic parkinsonism
	Familial British and Danish dementia
	Myotonic dystrophy
	Niemann-Pick disease, type C
	Subacute sclerosing panencephalitis
Inherited tauopathy	Frontotemporal lobar degeneration with parkinsonism linked to tau (FTLD-Tau)
	Progressive supranuclear palsy pattern
	Corticobasal degeneration pattern
	Pick's disease pattern

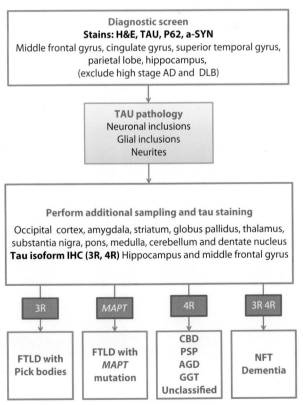

16.49 Diagnostic approach to FTLD-Tau.

3 Repeat Tauopathy

FTLD with Pick Bodies (Pick's Disease)

First described in 1892 by Pick,[720,891] the disorder that bears his name is a rare neurodegenerative disease, 50- to 100-fold less common than AD. Although Pick emphasized the language impairment and the lobar distribution of the atrophy,[788] the histopathology was detailed almost two decades later by Alzheimer[18] and subsequently by Altman,[16] who described the characteristic silver-positive cytoplasmic inclusions or Pick bodies. Pick's disease is characterized by progressive dementia and personality deterioration with signs of frontal disinhibition.[788] The onset is typically between 45 and 65 years of age. The prevalence of Pick's disease does not appear to increase with age, and it is rarely encountered after the age of 75 years. The disease duration ranges from a few years to over a decade, the average being 5–10 years. The sex incidence is similar, with inconsistent reports on male versus female predominance.[370] Although most cases are sporadic, familial cases of Pick's disease, usually with an autosomal dominant form of inheritance, have been reported; however, some of the original reports have been shown to be due to *MAPT* mutations.[721] The clinical features of frontotemporal dementia in combination with the lobar atrophy seen on CT or MRI scans (Figure 16.50) assist in differentiating Pick's disease from the more common AD. Recent review of clinical features against modern classifications in a series of 21 cases showed that 13 (62 per cent) presented with bvFTD and 8 (38 per cent) with lvFTD, with 2 patients having a mixed picture.[724] The neuropathology in a case series of Pick's disease has been reviewed by BrainNet Europe.[497]

The macroscopic appearance is that of frontotemporal atrophy with relative sparing of the posterior part of the superior temporal gyrus and the precentral and postcentral gyri (Figure 16.51a). With severe atrophy, the brain resembles a walnut and the involved cortical gyri have a so-called 'knifeblade' appearance. Asymmetry is common, and the dominant side (usually the left) tends to exhibit more atrophy (Figure 16.51b). The lateral ventricles are dilated secondary to loss of grey and white matter. The amygdala and hippocampus are usually involved and may exhibit prominent atrophy. Other regions are variably affected; atrophy and degeneration of the corpus striatum, globus pallidus

TABLE 16.17 Abnormal tau protein in the tauopathies		
Western blot of insoluble brain extracts	**Predominant tau isoform**	**Diseases**
Tau triplet: 60, 64 and 68kDa	4R and 3R	Alzheimer's disease, parkinsonism–dementia complex of Guam, postencephalitic parkinsonism, chronic traumatic encephalopathy, familial British dementia, FTDP-Tau—*MAPT* mutation
Tau doublet: 64 and 69kDa	4R	Progressive supranuclear palsy, corticobasal degeneration, argyrophilic grain disease, Globular Glial Tauopathy, FTDP-Tau—*MAPT* mutation
Tau doublet: 60 and 64kDa	3R	FTDP-Tau with Pick bodies, FTDP-Tau—*MAPT* mutation

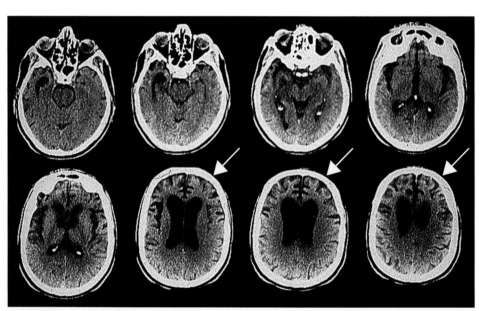

16.50 Computed tomography scans from a patient with Pick's disease. Pick's disease was subsequently proven at autopsy. Note selective atrophy of the frontal lobes (arrows).

Courtesy of Dr K Jobst, Oxford Project to Investigate Memory and Ageing, Oxford, UK.

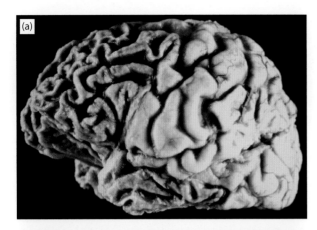

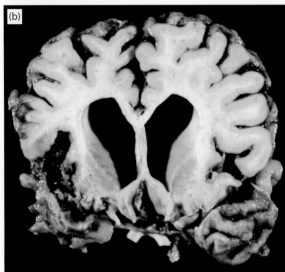

16.51 Brain in Pick's disease. (a) Atrophy of the frontal and temporal lobes is seen. Note relative preservation of the posterior portion of the superior temporal gyrus. **(b)** Coronal slice from a case of Pick's disease revealing striking and asymmetrical temporal lobe atrophy.

(a) Courtesy of Dr Chandrakant Rao, SUNY Downstate Medical Center, Brooklyn, NY, USA.

and substantia nigra are common. Some cases show such marked atrophy of the caudate nucleus that they mimic the gross appearance of Huntington's disease.

Corresponding to the regions of cortical atrophy, microscopic examination typically reveals loss of neurons, astrocytosis and extensive spongiosis of the tissue. In early disease microvacuolation may be restricted to the outer cortical layers (Figure 16.52). Neurons with swollen eosinophilic cytoplasm (so-called Pick cells) (Figure 16.53a) are seen in these areas but may be difficult to detect if the cortical structure is virtually obliterated. Pick cells are readily detected with antibodies to phosphorylated neurofilament protein,[206] low molecular weight stress protein, αB-crystallin, and less consistently with antibodies to ubiquitin and ubiquitin C-terminal hydrolase (also known as PGP9.5).[471,555] Pick cells are indistinguishable from swollen achromatic or ballooned neurons in a variety of other conditions.[206] Granulovacuolar degeneration and Hirano bodies (described in the section on Alzheimer's disease) are also commonly observed in Pick's disease.

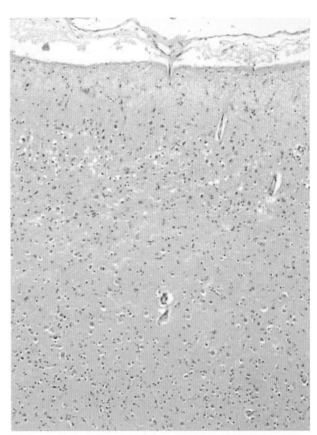

16.52 In early stages of Pick's disease neuronal loss may not be severe. Microvacuolation can be seen in the superficial cortex around layer II. This appearance is identical to that seen in several types of frontotemporal lobar degeneration.

The characteristic argyrophilic intraneuronal cytoplasmic inclusions, Pick bodies (Figure 16.53), are common in limbic and paralimbic cortices and the ventral temporal lobe. Pick bodies are often especially abundant in granule neurons of the dentate fascia and pyramidal neurons of the hippocampus. On H&E staining, they appear as round, slightly basophilic neuronal cytoplasmic inclusions. Pick bodies also occur in the deep grey matter including the basal ganglia, hypothalamus, periaqueductal grey matter, substantia nigra and locus coeruleus. Some cases show extensive loss of pigmented neurons in the substantia nigra,[909] whereas other cases of classical Pick's disease have preservation of the substantia nigra.[988]

By electron microscopy, Pick bodies exhibit loosely arranged 10–16-nm straight filaments and 22–24-nm twisted filaments with a periodicity either longer than or similar to the paired helical filaments of AD.[480,645,756,964] The major antigenic component of Pick bodies is tau,[551,733] and antibodies to 3R tau show that they are predominantly of this isoform.[185] Western blots of insoluble protein from the brain confirm that the predominant tau isoform is a doublet at 60 and 64kDa, consistent with 3R tau.[194] A distinct isoform of tau (69kDa, Tau 69) has been seen in Pick bodies in sporadic disease.[683]

Other antigenic determinants found less consistently in Pick bodies include ubiquitin,[551,554] chromogranin A and synaptophysin.[944,987] TDP-43 has also been noted in Pick's disease brain specimens.[266] Early Pick bodies may show differences from those in advanced disease, for example having less ubiquitin immunoreactivity.[620]

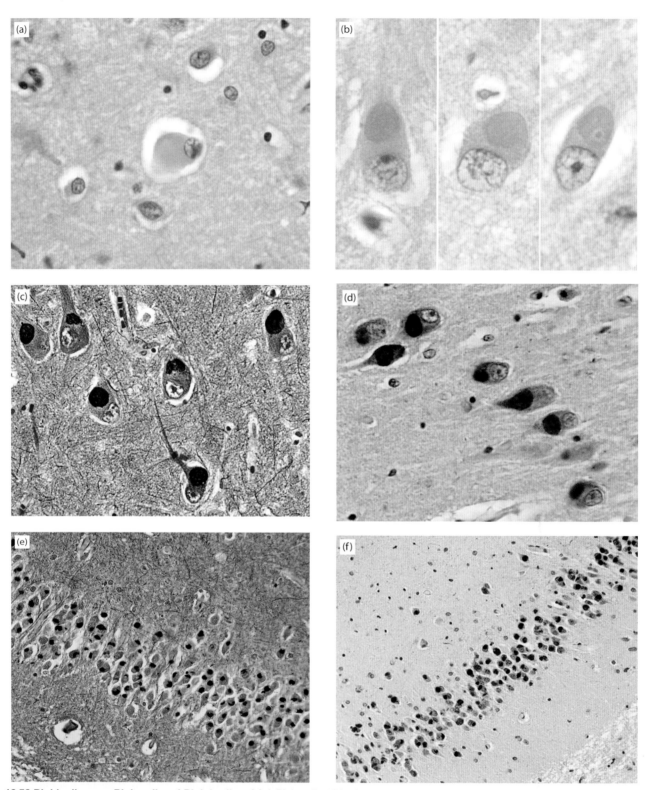

16.53 Pick's disease: Pick cell and Pick bodies. (a) A Pick cell within the cortex exhibits ballooned achromatic cytoplasm and an eccentric nucleus. **(b)** Pick bodies with a slightly basophilic appearance. Pick bodies within pyramidal **(c,d)** and dentate cells **(e,f)** of the hippocampus are readily detected with silver stain **(c,e)** and tau immunohistochemistry **(d,f)**.

Most cases of familial Pick's disease are linked to mutations in *MAPT*.[721] Familial Pick's disease associated with a *PSEN1* mutation has been reported, but the mutation does not appear to segregate with disease in this kindred and the observation remains tentative.[198]

4 Repeat Tauopathy

FTDP-Tau (Progressive Supranuclear Palsy)

Although PSP is primarily a movement disorder (described in detail in Chapter 12), cognitive impairment occurs in many cases and may even be the presenting feature.[112] Thus, any large series of cases of dementia will include cases of PSP. Cognitive impairment is usually mild, but some patients exhibit severe frontal subcortical-type dementia with a prominent dysexecutive syndrome and changes in personality.[75] In one clinicopathological study, 12 out of 13 PSP patients had cognitive impairment at some point in their clinical course.[280] Pathologically confirmed PSP has been reported to present as primary progressive aphasia[451] or with the corticobasal syndrome.[906]

The neuropathology of FTDP-Tau (PSP) has been recently reviewed.[215] The gross pathology of PSP is characterized by mild frontal atrophy that may extend to the precentral gyrus, atrophy of the midbrain often with severe atrophy of the superior cerebellar peduncle, atrophy[905] and discolouration of the subthalamic nucleus and loss of neuromelanin pigment from the substantia nigra (Figure 16.54). Microscopically, the salient feature of PSP is the argyrophilic tau-immunoreactive globose neurofibrillary tangle (Figure 16.55) found in the affected grey matter. Pre-tangles and flame-shaped tangles may also be detected by tau immunohistochemistry, which also reveals characteristic lesions in astrocytes (so-called 'tufted astrocytes') and oligodendroglia (so-called 'coiled bodies').[204] The tau immunostain also reveals thread-like processes, particularly in the basal ganglia, diencephalon and brain stem. Neocortical involvement is minimal in most cases, and often limited to the precentral gyrus.[366] In cases of PSP presenting with cortical syndromes (e.g. frontal lobe dementia, progressive aphasia or corticobasal syndrome) the tau pathology is more extensive in the cortex, where it may be associated with loss of synapses as assessed by synaptophysin immunoassays.[77] Medial temporal lobe (including hippocampus and amygdala) pathology is usually minimal, but about 20 per cent of cases have neuronal and glial pathology consistent with argyrophilic grains. These cases often have ballooned neurons in the limbic regions, as is typical in argyrophilic grain disease (AGD).

The major antigenic component of the tau pathology in PSP is 4R tau, as demonstrated by immunohistochemistry with isoform-specific tau antibodies. In Western blots, the pattern of tau protein in PSP is that of a major doublet of 64 and 69kDa, consistent with 4R tau,[820] with a variable admixture of 3R tau, depending upon the degree of concurrent Alzheimer-type pathology. Ultrastructurally, straight 15-nm filaments are detected in neuronal and glial lesions[868] (Figure 16.56).

Cases of familial PSP are considered to be part of the spectrum of disorders of FTDP-17 associated with tau mutations.[852]

FTDP-Tau (Corticobasal Degeneration)

Corticobasal degeneration is a specific tau-related disease that was originally described as a movement disorder (also see Chapter 12), but is now recognized to have a wider range of clinical presentations. Four clinical presenting forms of pathologically confirmed CBD are corticobasal syndrome, frontal behavioural-spatial syndrome (FBS), nonfluent variant of primary progressive aphasia (nfvPPA) and progressive

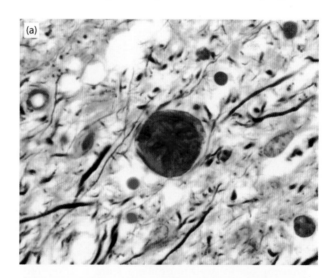

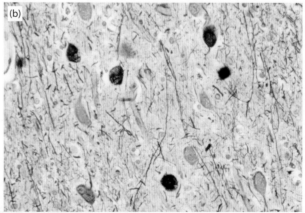

(a)

(b)

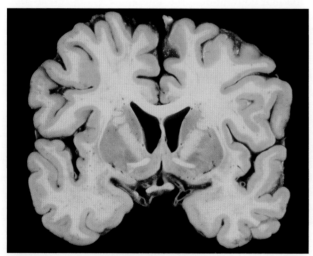

16.54 Progressive supranuclear palsy (PSP). This coronal section demonstrates bilateral atrophy and discolouration of the globus pallidus.

Reproduced by permission from Mirra S, Gearing M. The neuropathology of dementia. In: Morris J (ed). Handbook of dementing illnesses. New York: Marcel Dekker, 1994:189–226. With permission of Dekker.

16.55 Microscopic changes in progressive supranuclear palsy (PSP). Globose tangles are seen in **(a)** the globus pallidus and **(b)** the subthalamic nucleus. Sevier-Munger.

(a) Reproduced by permission from Mirra S, Gearing M. The neuropathology of dementia. In: Morris J (ed). Handbook of dementing illnesses. New York: Marcel Dekker, 1994:189–226. With permission of Dekker.

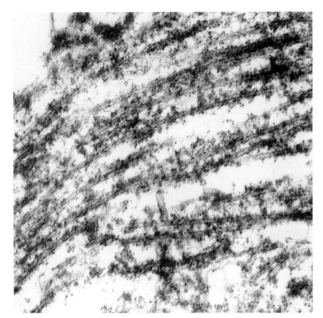

16.56 Electron microscopy of globose neurofibrillary tangle in PSP. Straight filaments 15 nm in diameter are revealed.

supranuclear palsy syndrome (PSPS).[31,112,494,528] The classic corticobasal clinical syndrome is an atypical movement disorder with cortical signs and symptoms characterized by asymmetric rigidity and apraxia, cortical sensory deficits, dystonia and myoclonus. The corticobasal clinical syndrome is not specific for tau pathology[537] and other pathological processes can produce this syndrome. These include PSP, Pick's disease, FTLD-TDP,[428,875] AD and CJD. Given a wide range of clinical syndromes associated with CBD, the neuropathologist must consider it in the differential diagnosis of dementia and motor disorders.

The pathology of FTDP-Tau(CBD) was recently reviewed.[215] Neuropathological features of corticobasal degeneration include cortical atrophy, often asymmetrical and predominantly involving the peri-Rolandic cortex but also involving other cortical regions. Cases with more pronounced cognitive problems tend to exhibit a more anterior–frontal rather than peri-Rolandic cortical atrophy and those with progressive aphasia may have prominent peri-Sylvian atrophy. The underlying white matter may display considerable loss of myelin and axons. Degeneration of the basal ganglia, substantia nigra and cerebellum varies from case to case.

Microscopically, CBD is characterized by focal cortical atrophy with superficial laminar spongiosis and astrocytic gliosis. There are usually many ballooned or achromatic neurons (Figure 16.57) in affected cortical areas, especially the cingulate gyrus and the superior frontal gyrus. The ballooned neurons are similar to those in Pick's disease and contain phosphorylated neurofilament protein and αB-crystallin. A major histopathological feature of CBD is abundant tau deposition in thread-like processes in both grey and white matter of the cortex and basal ganglia, with less involvement of the brain stem and cerebellum. Glial tau pathology including oligodendroglial coiled bodies and characteristic astrocytic lesions, so-called 'astrocytic plaques',[215,247,494] can be detected by tau immunohistochemistry or Gallyas silver impregnation (Figure 16.57). Minimal pathological diagnostic criteria for CBD are

cortical and striatal tau-positive neuronal and glial lesions, especially astrocytic plaques and thread-like lesions in both white matter and grey matter, as well as neuronal loss from the substantia nigra.[213]

In Western blots, tau protein in CBD shows a 64- and 69-kDa doublet similar to that in PSP, and 4R tau isoforms are predominant.[262] Immunohistochemistry with 4R tau antibodies shows labelling of both neuronal and glial lesions.[185] Cases of familial CBD are due to a mutation in *MAPT*.[852]

FTDP-Tau (Argyrophilic Grain Disease)

AGD or dementia with grains was first described by Braak and Braak in a demented individual,[89] but has since been shown to be a distinctive neuropathological process occurring in elderly people with or without dementia.[217,443,894,895] Grains have been described as a constant finding in brains from centenarians.[718] Younger-onset cases are described and have been associated with frontotemporal dementia;[413] however, most patients have an amnesic syndrome, sometimes associated with psychiatric features such as agitation or apathy.[893] An association was also reported with late-onset schizophrenia.[650] AGD has been estimated to affect about 5 per cent of patients with clinical dementia and may independently contribute to cognitive decline in the elderly.[887]

Neuropathologically, this condition is characterized by tau pathology that is relatively restricted to the medial temporal lobe, although there are also lesions in certain nuclei of the hypothalamus and in the nucleus acumens. The characteristic lesions are argyrophilic (especially with Gallyas silver impregnation) spindle- or comma-shaped structures occurring predominantly at branch points of apical dendrites, mostly in pyramidal neurons (Figure 16.58).[896]The grains contain hyperphosphorylated tau,[895] and isoform-specific tau antibodies as well as tau biochemistry demonstrate that the grains contain predominantly 4R tau.[892] Although conventional tau immunolabelling reveals these structures, they may be difficult to detect if there is any significant degree of Alzheimer-type neurofibrillary and neuritic pathology; however use of 4R tau-specific antibodies facilitates their detection.[272] Antibodies to the ubiquitin binding protein, P62, also detect grain-like structures.[817] Other tau-immunoreactive lesions in AGD include ramified astrocytes, most common in the amygdala, and oligodendroglial coiled bodies, most common in the temporal stem white matter.

Grains are most prominent in the anteromedial temporal lobe, including the entorhinal cortex, Ammon's horn and the amygdala.[93] Involvement of the CA2 sector of the hippocampus is frequent in AGD and other 4R tauopathies, such that when it is detected a search for grains is advisable.[416] A diffuse form of AGD has been reported and associated with frontal lobe dementia, and some *MAPT* mutation cases have diffuse grains, so it is imperative to obtain information on the family history if diffuse AGD is considered.[597]

AGD is commonly associated with ballooned neurons in limbic structures (see Figure 16.58), including medial temporal lobe, amygdala cingulate gyrus, insular cortex and claustrum.[272,896] Detection of ballooned neurons in these regions with routine histologic methods, should prompt further studies (e.g. Gallyas silver impregnation) to exclude AGD.

AD pathology is commonly associated with grains but this seems to be coincidental, because the nature of the tau within the lesions differs (3R+4R tau in AD versus 4R tau in AGD).[897] Tau in AGD is also not acetylated.[330] The grains do not clearly progress to neurofibrillary tangle formation and may be found independent of Alzheimer-type neurofibrillary pathology. Many AGD cases have a relatively low amyloid plaque load

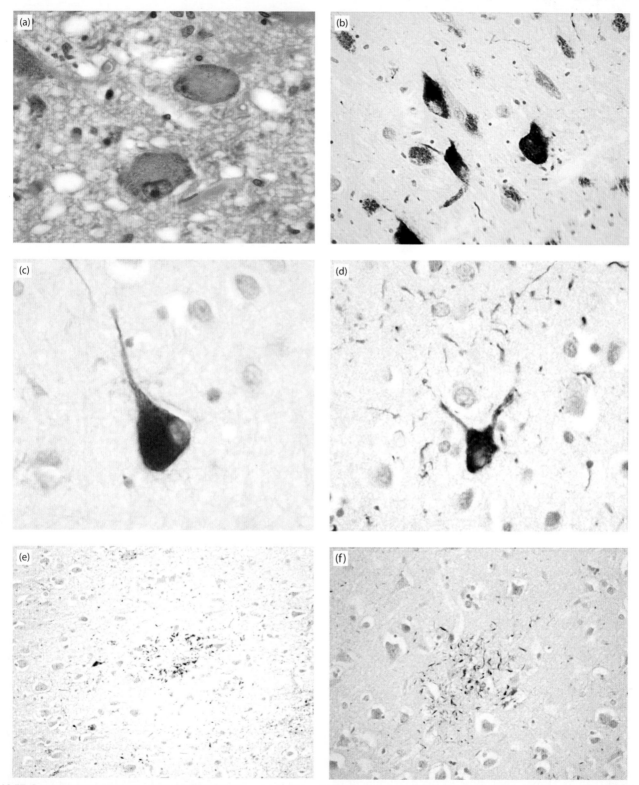

16.57 Corticobasal degeneration (CBD). (a) Neurons in the substantia nigra contain large, basophilic globose tangles. **(b)** Tau immunostaining shows both neuronal and glial accumulation in this region. In the cerebral cortex swollen neurons are a distinctive feature and can be highlighted by immunohistochemistry for αβ-crystallin **(c)**. Tau immunostaining of the cerebral cortex shows neuronal staining as well as staining of neuropil threads **(d)**. A distinctive feature is the presence of astrocytic plaques, seen on tau staining **(e,f)**. See Chapter 12, Extrapyramidal Diseases of Movement, for other details of CBD.

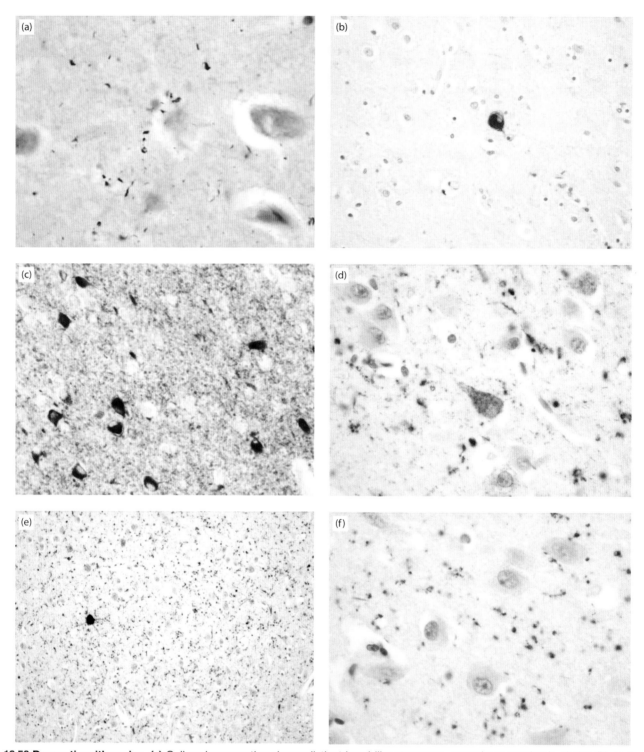

16.58 Dementia with grains. (a) Gallyas impregnation shows distinct bead-like structures, sometimes running in chains **(b)**. In the temporal lobe, swollen neurons are a common finding, here shown by αβ-crystallin staining. **(c,d)** Tau immunostaining shows an intense background neuropil pattern in affected regions together with neuronal staining composed of tangles and pre-tangles. Grain structures can be seen as dense bead-like deposits in areas with least neuropil staining. **(e,f)** Immunostaining for P62 (sequestosome 1) tends not to show background neuropil threads and better highlights the grain deposits in this condition.

with a predominance of diffuse plaques. It has been proposed that AGD may contribute to the development of dementia in AD by lowering the threshold for cognitive deficits in the presence of moderate AD-related pathology.[887] It has been difficult to determine the frequency of AGD in AD because neurofibrillary pathology masks the grains. In studies using antibodies specific to 4R tau, AGD was detected in 26 per cent of AD cases.[272] TDP-43 pathology was also reported as a frequent association with AGD pathology.[273]

Argyrophilic grains are also observed in other tau-associated disorders including PSP, Pick's disease and CBD, as well as in *MAPT* mutation cases.[585] A staging scheme for AGD pathology has been proposed.[801]

Biochemical characterization of relatively pure cases of AGD has shown 4R tau in the detergent-insoluble brain fractions.[898] Reports of association of AGD with *MAPT* haplotypes that are commonly associated with other 4R tauopathies, such as PSP and CBD, have been conflicting.[626,892] An association with *APOE* ε2 was been observed in one case series[289] but not another.[892]

Studies of genetic factors that may lead to AGD have suggested *CTNS*, the gene encoding cystinosin, a lysosomal membrane protein, as a candidate.[930]

FTLD-Tau (Globular Glial Tauopathy)

A group of sporadic 4-repeat tauopathies characterized by the presence of globular glial inclusions has been designated globular glial tauopathy (GGT). Cases had originally been described as unusual forms of tau-related disease,[71,76,252,270,295,451,495,627,735] but it was not until 2013 that a consensus statement on the clinical and pathological features of this group of diseases was published.[4] Three clinical presentations have been recognized:[4]

- type I: frontotemporal dementia;
- type II: pyramidal motor weakness;
- type III: combination of frontotemporal dementia and motor neuron disease.

Histologically two forms of glial inclusion have been defined.

- Globular oligodendroglial inclusions (GOIs) are seen in tau staining as globular structures the same size as, or slightly larger than, the nucleus (Figure 16.59a). These are Gallyas-positive.
- Globular astrocytic inclusions (GAIs) are seen in tau staining as globular or dot-like inclusions of variable size in the cytoplasm or processes of astrocytes (Figure 16.59b). They are Gallyas-negative.

Pathological correlation with the three types of clinical presentation is as follows:

- type I: predominantly frontotemporal involvement without corticospinal involvement (resembling cases originally described as sporadic MSTD);[76]
- type II: motor cortex and corticospinal tract degeneration;
- type III: frontotemporal, motor cortex and corticospinal tract involvement.

To differentiate this sporadic tauopathy from cases with similar pathology, it is recommended that suspected GGT cases should be routinely sent for *MAPT* genetic analysis to exclude the diagnosis of FTLD–*MAPT*.

At present no firmer neuropathological diagnostic criteria for GGT have been proposed; this is in order to allow for clinicopathological ascertainment of cases, following which it may be possible to devise specific criteria.[4,628]

FTLD-*MAPT* with Tau Mutation

FTLD-Tau associated with a *MAPT* mutation is mainly associated with bvFTD. However, forms of PPA are also reported. The gross features of FTLD-*MAPT* may resemble those found in other forms of FTLD, including CBD, PSP and Pick's disease, or consist simply of focal medial temporal lobe atrophy. Some cases have marked basal ganglia

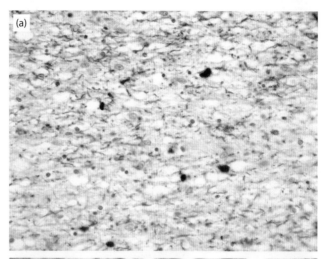

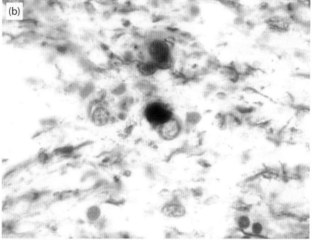

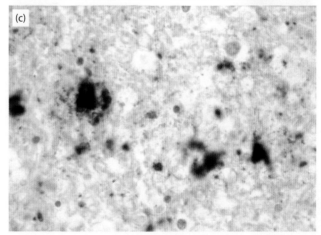

16.59 Globular glial tauopathy. **(a)** Tau staining of white matter shows threads and inclusions in glia. **(b)** Globular oligodendroglial inclusion seen on tau staining of white matter. **(c)** Globular astrocytic inclusion seen in white matter on tau staining.

atrophy, with a rust-coloured appearance of the globus pallidus because of increased iron pigment. Some have striking loss of pigment from the substantia nigra (Figure 16.60).

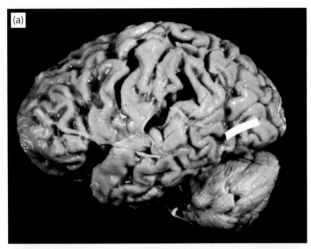

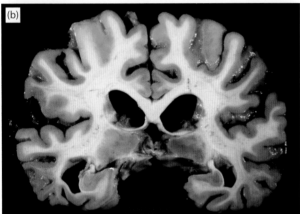

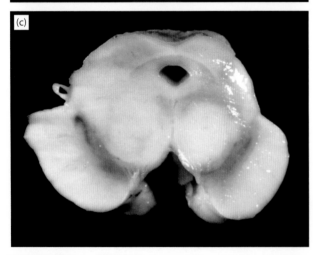

16.60 FTDP-Tau with P301L tau mutation. (a) Atrophy of frontal, temporal and parietal lobes is seen. **(b)** Coronal slice at the level of the hippocampus shows prominent cortical atrophy. **(c)** The midbrain displays pallor of the substantia nigra and bronze discolouration, reflecting iron deposition.

Diffuse tau immunoreactivity within neurons, in the form of so-called pre-tangles, is the dominant component of neuronal pathology in most cases (Figure 16.61). In some cases, neurofibrillary tangles that resemble those in AD are detected in the same vulnerable neuronal populations but not associated with Aβ deposits. In some cases globose tangles and astrocytic lesions similar to those in PSP are found (Figure 16.62). In other cases the pathology may resemble that of CBD, including the presence of many ballooned or swollen neurons. Neuropil threads are present in many cases. Pick bodies are seen in association with several mutations.

Tau-immunoreactive glial lesions are common (Figure 16.63) and include astrocytic lesions that may resemble tufted astrocytes of PSP, astrocytic plaques of CBD or ramified astrocytes of AGD depending upon the particular mutation. Oligodendroglial coiled bodies are also common. Astrocytic gliosis in the cortical and subcortical grey matter usually parallels the severity of neuronal loss. Hippocampal neuronal loss may resemble that of hippocampal sclerosis. Myelin loss and astrocytic gliosis of white matter are features in some cases.

Ultrastructural Findings In the inherited tauopathies, abnormal tau protein becomes insoluble, becomes hyperphosphorylated and aggregates into filaments. Ultrastructurally the filaments vary in structure and appearance with different mutations:[853]

- paired helical filaments, similar to those in AD, 8–20 nm wide (e.g. V337M and R406W);
- twisted ribbons, 15–27 nm wide (e.g. K257T, N279K, P301L, E10+12, E10+13, E10+16, K369I);
- straight filaments, 12–15 nm wide (e.g. R5L, P301S, P310L, S320F);
- straight tubules, 15–20 nm wide (e.g. R5H, N296H, S305N).

Genotype-Phenotype Relationships in FTLD-*MAPT* Although there is significant pathological heterogeneity both in and between different mutations, some broad correlations between different mutations and pathological findings have been suggested.[852] Online resources list all mutations in tau together with pathological associations and source publications (http://www .molgen.ua.ac.be/ADMutations/ and http://www.alzforum.org/mutations).

Missense mutations within coding regions of exons 1, 9, 11, 12 and 13 generally cause tau accumulation that is mainly neuronal. In some cases, the tau pathology resembles that in AD, whereas others resemble PSP, CBD, or have swollen neurons and rounded Pick-body-like intraneuronal inclusions. Mutations outside exon 10 are associated with tangles composed of both 4R and 3R tau. Some Pick-like pathology is associated with insoluble tau accumulation that is exclusively 3R tau. Glial pathology in this group is typically less severe than the neuronal tau pathology. Mutations that affect the splicing of exon 10, increasing the 4R:3R ratio of tau isoforms, produce neuronal pre-tangles and neurofibrillary tangles as well as many glial lesions.

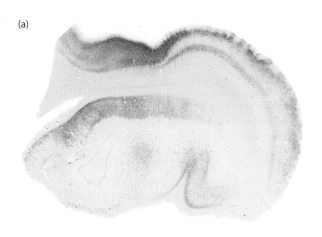

(a)

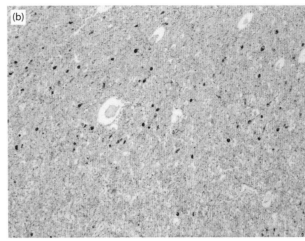

(b)

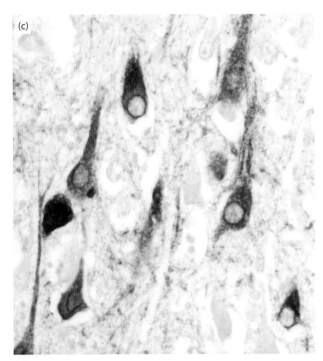

(c)

16.61 Familial FTDP-Tau. (a) Low magnification, showing distribution of tau pathology following immunostaining for phospho-tau protein in the anterior hippocampus (uncus) with adjacent entorhinal cortex, showing staining in CA1/subiculum, layers II and IV of the entorhinal cortex, and pyramidal layers of the peri-rhinal and temporal neocortex. **(b)** Tau immunostaining is widespread and dense in the cortex. Dense deposits in neurons are seen with a mass of background neuropil threads and glial staining. **(c)** Tau immunoreactivity in pyramidal neurons of the hippocampus. This is mostly pre-tangle deposition. Perinuclear staining is often reported but is not specific to familial tauopathies.

3R and 4R Tauopathy

Neurofibrillary Tangle-Predominant Dementia

In some very old individuals with dementia, the brain shows Alzheimer-type neurofibrillary tangles with either no amyloid deposits or only diffuse Aβ plaques. This has been termed tangle-only dementia, tangle-predominant dementia or senile dementia of the NFT type (SD-NFT).[976] In many cases the neurofibrillary tangles are relatively restricted to the anteromedial portion of the temporal lobe, involving the entorhinal region, Ammon's horn and the amygdala. A high proportion of the neurofibrillary tangles in these areas may be extracellular ghost tangles. Histologically, the neurofibrillary tangles are immunoreactive for both 4R and 3R tau, as in AD, and most cases have some neuropil threads and occasionally some glial pathology corresponding to the limbic stage of AD tangle pathology.[436] The *MAPT* H1 haplotype is associated with this form of disease.[804] In some series the *APOE* ε2 allele is over-represented in patients with tangle-predominant dementia.[54,55,438,976]

When this pathology is detected, it is important to exclude other tauopathies, in particular AGD, because grain pathology can easily be missed with routine histological methods. A useful clue to the diagnosis of grain dementia is the presence of swollen neurons in the mesial temporal cortex, best detected by immunostaining for αB-crystallin.

Diffuse NFT Dementia with Calcification

Diffuse neurofibrillary tangles with calcification are characterized by temporal or temporofrontal lobar atrophy, the presence of cortical neurofibrillary tangles[911] as well as brain mineralization in basal ganglia, cerebellum or cerebral cortex.[420] The condition was originally reported in a series of 16 patients with a late-onset dementia in Japan,[491] but has since been described elsewhere.[657] Glial pathology is described[363] as well as Lewy body pathology[378,907] and TDP-43 deposition.[338]

Unclassified Tauopathy

Despite advances in the characterization of diseases associated with pathological tau deposition, some remain as

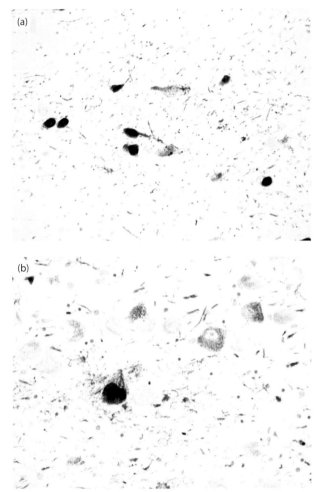

(a)

(b)

16.62 Familial FTDP-Tau. (a,b) Tau immunostaining of globose tangles, pre-tangles and neuropil threads in the substantia nigra.

unclassified according to present criteria. The group of sporadic 4R tau diseases incorporated into globular glial tauopathies is an example of how individual case reports can lead to recognition of a new subgroup of disease. A discussion of elderly patients presenting with dementia or parkinsonism and tau pathology in glial cells and neurons indicates that further groups are likely to be recognized.[496]

Rare Geographically Restricted Tauopathies

ALS-Parkinsonism-Dementia Complex

In 1945, Zimmerman documented two neurological syndromes occurring among the Chamorro population on Guam. One syndrome was characterized by ALS (named 'lytico' by the natives) and parkinsonism–dementia complex (PDC), locally known as 'bodig'.[1000] Both syndromes remain prevalent on Guam today but may be decreasing in frequency.[601] Similar disorders were described in the Kii peninsula of Japan[826] and in Papua New Guinea.[275]

Neuropathologically, PDC is characterized by extensive neurofibrillary tangle formation, the tangles showing similar morphology, immunochemical profiles and ultrastructure to those in AD. More tangles are seen in supragranular cortical layers than in AD,[385] and cases with predominant PDC have higher neurofibrillary tangle densities than cases with predominant ALS. As in AD, calcium-binding

protein-containing interneurons are not affected, but a subset of neurofilament-rich pyramidal cells involved in AD is resistant in ALS–PDC.[385] Hirano bodies, first described in this condition, and granulovacuolar degeneration are also common features of PDC.[375] The nucleus basalis of Meynert shows extensive neuronal loss.[592,654] Cases with parkinsonism show extensive loss of neurons from the substantia nigra and locus coeruleus. Argyrophilic glia have also been reported in PDC.[697] Neurofibrillary tangles in the spinal cord are immunoreactive for antibodies to tau and ubiquitin, similar to those in AD[593] (see also Chapter 14).

The cause of this disorder remains enigmatic.[527] Genetic factors have not been supported by epidemiological data showing that Chamorros living on a neighbouring island are spared.[279,984] Recent evidence suggests a genetic background to ALS in some populations.[457] Environmental factors have been investigated but no definite toxins identified, despite much speculation about aluminum, iron and other substances. A relationship with ingestion of cycad-derived foods containing the neurotoxin β-N-methylamino-l-alanine (BMAA) was postulated but not substantiated.[229,851] This hypothesis has been revisited with the possibility that the environmental concentration of substances such as BMAA may be greater than originally suspected.[53,141]

Frontotemporal Lobar Degenerations with TDP Pathology

This is the most common pathological type of frontotemporal dementia. Although most cases are not associated with a family history of disease, familial cases have been linked to mutations in *GRN* (encoding progranulin), *VCP* or *C9orf72*.

Biology and Significance of TDP-43 Pathology

Transactive response DNA-binding protein of 43 kDa (TDP-43), is a RNA- and DNA-binding protein that is normally involved in gene transcriptional regulation and RNA processing, including splicing.[819] Mutations in the gene are associated with motor neuron disease/ALS.[518] The functions of TDP-43 remain to be established fully but evidence suggests roles in RNA metabolism and protein homeostasis.[427] Biochemical investigation has revealed truncated, phosphorylated and ubiquitinated forms of TDP-43 in disease[86,668,908] Abnormalities in TDP-43 are felt to interfere with mRNAs transcribed from genes with long introns essential for neuronal integrity.[510]

Two main types of antibody have been used in investigations. Immunohistochemistry with antibodies to TDP-43 shows a widespread neuronal and glial nuclear localization of TDP-43. In pathological states, this normal diffuse nuclear immunoreactivity is lost and there is labelling of cytoplasmic inclusions or neurites, and in some instances of discrete intranuclear inclusions. Other antibodies recognize phosphorylated 43 kDa TAR DNA-binding protein (pTDP-43) and have specificity for pathological aggregates and inclusions and do not stain normal nuclei.[362]

Biochemical investigation has shown different Western blot banding patterns in TDP-43 proteinopathies, possibly representing different conformations of abnormal TDP-43 or their aggregates and linked to different patterns of

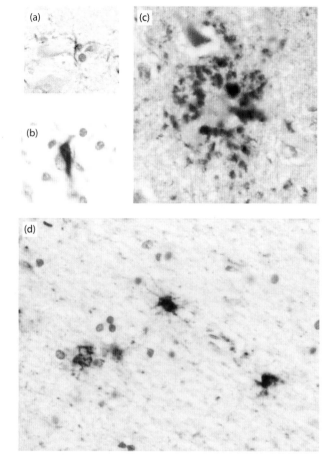

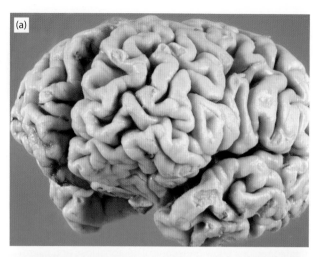

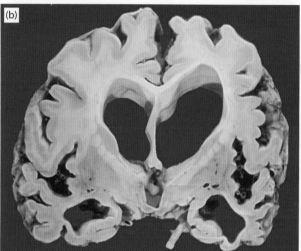

16.63 Familial tauopathy, glial tau staining. (a,b) Tau-immunoreactive inclusions in astrocytes. **(c)** Astrocytic plaque-like staining similar to that seen in corticobasal degeneration. **(d)** Tau-immunoreactive astrocytes in white matter. Tau immunohistochemistry.

(b,c) Modified from Mirra SS, Murrell JR, Gearing M, et al. Tau pathology in a family with dementia and a P301L mutation in tau. J Neuropathol Exp Neurol 1999;58:335–45. With permission from Lippincott Williams & Wilkins/Wolters Kluwers Health.

16.64 Macroscopic features of FTLD-TDP. Neuropathological assessment revealed TDP-43 pathology. **(a)** There is marked frontal lobe atrophy. **(b)** Marked ventricular enlargement, with atrophy of the temporal lobe and moderate atrophy of the basal ganglia.

disease. The most distinctive pattern of abnormal TDP-43 is seen in type C disease.[908]

Animal models have reproduced some of the features of disease.[547]

The involvement of TDP-43 and related proteins in the wider context of the FTLD group has been reviewed by Halliday *et al.*[346]

Pathological Features

Macroscopically, frontal and temporal atrophy can vary from mild or moderate to very severe (Figure 16.64). The basal ganglia may show mild to moderate atrophy, and the substantia nigra moderate pallor.

Histologically there is typically neuronal loss and astrocytic gliosis, initially limited to the superficial neocortical laminae that show microvacuolation (Figure 16.65). Later disease is characterized by transcortical loss of neurons and astrocytic gliosis. In patients with associated motor neuron disease/ALS, loss of upper or lower motor neurons can be found, together with pathology in the corticospinal tracts (see

Chapter 14). There is usually rarefaction and gliosis of white matter in the affected regions. Neuronal loss and astrocytic gliosis may involve the basal ganglia and substantia nigra.

Classification of Pathology

The extent and patterns of staining of pathological aggregates have been assessed by several groups.[563,802] In FTLD-TDP the three main abnormalities seen on immunohistochemistry for ubiquitin, P62 and TDP-43 have been reconciled into four groups, A–D, in a consensus proposal[569] (Table 16.18). The three main pathological features are as follows:

- Neuronal cytoplasmic inclusions (NCI) are seen in neocortical neurons as well as in dentate granule cells of the hippocampus. A range of morphological types of NCI has been described. Hippocampal involvement varies in the different types of FTLD-TDP (Figures 16.66 and 16.67 and Table 16.18).
- Dystrophic neurites are most evident in regions of cortical microvacuolation. There are two main types.

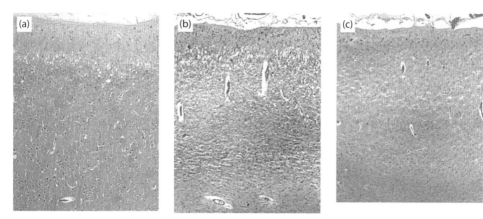

16.65 FTLD-TDP. A common feature s is cortical neuronal loss with microvacuolation and astrocytic gliosis. In early-affected regions, microvacuolation is seen in layer 2 **(a)**, As the disease progresses, there is neuronal loss from other cortical layers **(b)**, and severely affected regions show transcortical neuronal loss and microvacuolation (status spongiosus) **(c)**.

TABLE 16.18 Harmonized classification system for FTLD-TDP pathology

Type (% of cases)	NCI	DN	NII	Other	Clinical	Genetic associations
A 41%	Layer 2 Crescentic/oval and dot-like +++ Hippocampal dentate +/- Round and dot-like or granular	Layer 2 Short comma-shaped DN +++	Round or lenti-form ++/-	Hippocampal involvement less likely than in other types	bvFTD PNFA ALS/MND	50% family history *GRN* mutations Chr 9p linkage and *C9orf72* expansion
B 34%	Layer 2 Crescentic/oval and dot-like including some in deep cortical layers ++ Hippocampal dentate +/++ often granular and 'starburst'	Transcortical Fine DN +/-	+/-	Hippocampal sclerosis can be seen	bvFTD ALS/MND and FTD	30% family history *TARDP* mutations Chr 9p linkage and *C9orf72* expansion
C 25%	Neocortical +/- Hippocampal dentate +/++ dense round 'Pick-like'	Layer 2 and deeper cortex Long thick DN +++	+/-	Hippocampal sclerosis can be seen	SDbvFTD	30% family history
D Rare	Neocortical -/+ Hippocampal dentate +/-	Neocortex Short DN +++	Lentiform +++		IBMPFD	*VCP* mutations

bvFTD behavioural variant frontotemporal dementia, DN dystrophic neurites, *GRN* progranulin gene, *TARDP* TDP-43 gene, IBMPFD inclusion body myopathy with Paget's disease of bone and frontotemporal dementia, ALS/MND amyotrophic lateral sclerosis/motor neuron disease, NCI neuronal cytoplasmic inclusions, NII neuronal intranuclear inclusions, PNFA progressive nonfluent aphasia, SD semantic dementia, *VCP* valosin-containing protein gene, +/- rare or absent, + to +++ occasional to numerous.

The first are long, linear neurites mainly seen in outer cortical layers. The second are short, stubby, comma-shaped neurites that can be seen in all cortical layers. The different types of neurite are associated with different types of disease (Figures 16.68 and 16.69 and Table 16.18).

- Neuronal intranuclear inclusions are seen in some cases. These are typically elongated and lenticular in shape. There are prominent in patients with *VCP* mutations but may be seen in lower density in other types of FTLD-TDP (Figures 16.70 and 16.72 and Table 16.18).
- Hippocampal sclerosis is seen in some FTLD-TDP cases (see Table 16.18).

If immunohistochemical investigation fails to reveal TDP-43-positive inclusions yet ubiquitin or P62 immunostaining have shown NCI, then it is important to look for FUS pathology.

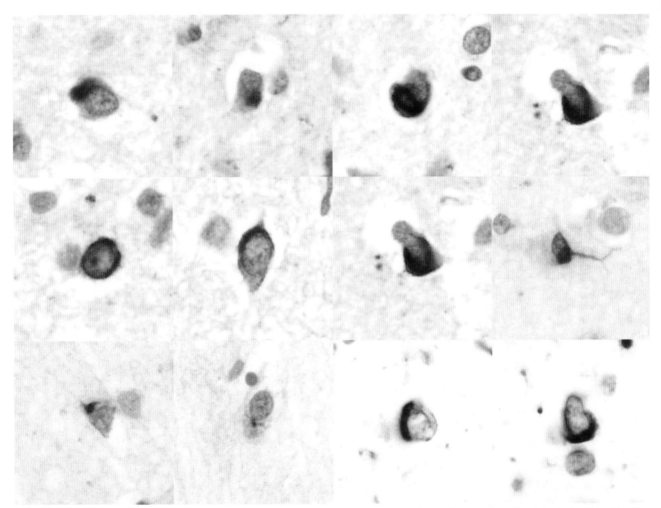

16.66 FTLD-TDP neuronal cytoplasmic inclusions. Inclusions are seen in neuronal cytoplasm and can be detected with antibodies to pTDP-43, ubiquitin or P62. They range from small paranuclear granular deposits, through crescent shaped and ring-shaped structures, to small spherical paranuclear inclusions. (P62 staining.)

Inclusion Body Myopathy Associated with Paget's Disease of Bone and Frontotemporal Dementia

Patients with autosomal dominant inclusion body myopathy associated with Paget's disease of the bone and frontotemporal dementia or IBMPFD have a combination of skeletal, muscular and neurological disease that has been related to mutations in the gene for VCP on chromosome 9p13-p12.[479] Valosin-containing protein, also known as P97, is an AAA (ATPases associated with diverse cellular activities) ATPase associated with a variety of cellular processes, including cell cycle regulation, integrity of internal membrane systems, and some aspects of function the ubiquitin-proteasome system.[619] A review of 190 members in 27 families with 10 missense mutations has established some genotype–phenotype relationships with average survival after diagnosis of myopathy and Paget's disease of around 18 years, and after dementia 6 years.[615]

This disorder is characterized neuropathologically by severe brain atrophy, especially affecting frontotemporal regions and the striatum. There is extensive neuronal loss with astrocytic gliosis in neocortical regions, mainly in layers III and V, and superficial cortical microvacuolation in areas with cortical atrophy, similar to that in most other forms of FTLD. White matter astrocytic gliosis is also common. Some remaining neurons contain intranuclear inclusions that are immunoreactive for ubiquitin and inconsistently with antibodies to valosin-containing protein.[816] This condition is associated with a type D form of TDP-43 proteinopathy (see Table 16.18).

Muscle biopsy reveals myopathic features with rimmed vacuoles similar to those in inclusion body myositis, and the bone pathology resembles typical Paget's disease of the bone. Inclusions in the muscle, brain and heart show ubiquitin, Aβ and TDP-43 immunoreactivity.[479]

Frontotemporal Lobar Degenerations with FUS Pathology

There are three conditions associated with pathological accumulation of FUS in cellular inclusions.[568]

- atypical FTLD-U (aFTLD-U);
- neuronal intermediate filament inclusion disease (NIFID);
- basophilic inclusion body disease.

In Japan, basophilic inclusion body disease is the most common subtype of FTLD-FUS and aFTLD-U is rare, a finding that contrasts with aFTLD-U being the most common form in Western countries.[486]

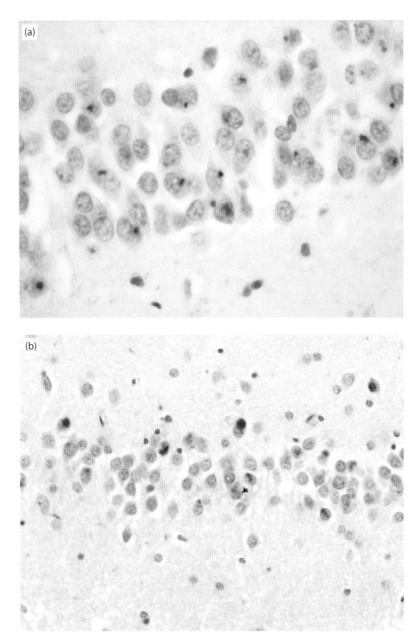

16.67 FTLD-TDP hippocampal NCI in dentate granule cells. The morphology of hippocampal neuronal cytoplasmic inclusions varies between different types of FTLD-TDP. Some are dot-like **(a)**. Others have a more dispersed granular appearance. In some cases inclusions appear as larger structures that are Pick body-like or bean-shaped **(b)**.

Molecular Pathology and Significance of Fused-in-Sarcoma

The FUS protein is encoded by the *FUS* gene on chromosome 16 and is a 526-residue protein with a molecular weight of 53 kDa. FUS is normally located in the nucleus and involved in RNA–protein interactions and transcriptional activation. It plays an important role in mRNA transport, axonal maintenance, and motor neuron development.[221,518]

Mutations in *FUS* are linked to familial ALS type 6 and juvenile onset basophil inclusion body disease[195,518] but pathogenic mutations have not been seen in patients with bvFTD.[394]

FUS, together with Ewing's sarcoma protein (EWS) and TATA-binding protein associated factor 15 (TAF15), make up the FET family of DNA/RNA-binding proteins. Other members of the FET protein family (FUS, EWS and TAF15) as well as transportin-1 have also been detected in the inclusions of FTLD-FUS, raising the possibility of a pathway defect linking similar conditions.[562]

The involvement of FUS and related proteins in the wider context of the FTLD group has been reviewed.[346] Animal models of disease have started to inform understanding of disease.[513]

Atypical FTLD with Ubiquitin-Only Immunoreactive Changes

This condition is the most common of the FTLD-FUS group and was defined as a pathological entity following the discovery that some cases formerly classified as FTLD-U did not show TDP-43 pathology. These were termed atypical FTLD

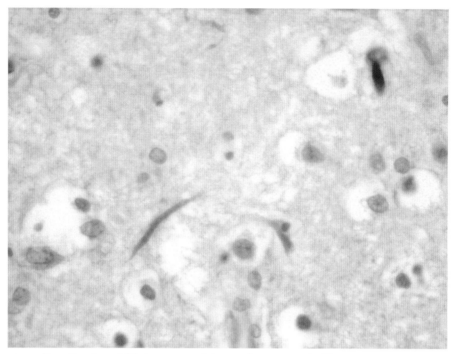

16.68 FTLD-TDP long neurites. Neurites of this type are mainly seen in outer cortical layers and usually run approximately perpendicular to the cortical surface. They can be detected with antibodies to TDP-43, P62 or ubiquitin.

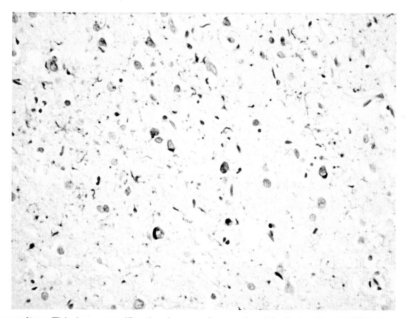

16.69 FTLD-TDP short neurites. This low magnification image of a section labelled with anti-P62 shows layer 2, which includes numerous short, twisted and comma-like DNs. Many small neurons also contain neuronal cytoplasmic inclusions.

with ubiquitin-only pathology (aFTDP-U). The inclusions in these cases are immunopositive for FUS.[669,912] Inclusions also contain the FET protein, transportin-1.[98,671,903]

Clinically, patients present with a subtype of bvFTD that has been termed the stereotypic form of FTD, dominated by obsessional, repetitive, ritualistic behaviour associated with social withdrawal and hyperorality.[517]

Macroscopically there is moderate frontal and temporal atrophy. Caudate atrophy is usually prominent and severe to a point where it has been suggested as being predictive of the aFTLD-U pattern of disease, especially if patients have

onset under 40 years. In most cases hippocampal sclerosis is also present.

Histologically there are ubiquitin- or P62-positive, tau/TDP-43-negative NCI that are strongly immunoreactive for FUS.[525] The majority of NCI appear as small, compact, round or oval inclusions, which sit next to the nucleus and are rarely larger than it. These bean-shaped inclusions are commonly seen in outer layers of the cerebral cortex, hippocampal dentate fascia (Figure 16.71) and periaqueductal grey matter. In addition to bean-shaped FUS inclusions, crescentic NCI can also be seen in the striatum. A coarse

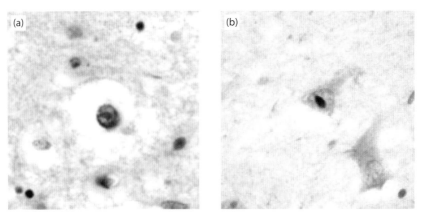

16.70 FTLD-TDP nuclear inclusions. (a,b) Inclusions can be detected by immunohistochemistry for ubiquitin P62 or TDP-43. The commonest types of inclusions are seen as lenticular structures occupying the centre of the nucleus.

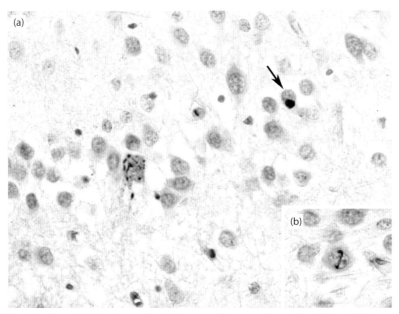

16.71 FTLD-FUS aFTLD-U cytoplasmic inclusions. Inclusions can be seen in cortical **(a)** and hippocampal and inset neuronal nuclear **(b)** neurons. FUS immunostaining.

granular pattern of FUS accumulation is demonstrable in pyramidal neurons in some cases. Neuritic pathology is not evident although rare fine FUS-positive filaments may be seen. A small proportion of lower motor neurons contain inclusions, some of which may have a skein appearance.

Transportin 1 is an abundant component of the FUS-immunopositive inclusions in aFTLD-U(FUS)[98] but is not seen in ALS-FUS inclusions, suggesting that that ALS-FUS and FTLD-FUS have a different pathogenesis.[671,903]

Another prominent abnormality is the presence of neuronal intranuclear inclusions that are long and thin, and may appear either to run straight across the nucleus as a rod or to be curved around the nuclear margin (vermiform neuronal intranuclear inclusions). These are present in the hippocampus, especially in dentate granule cells (Figure 16.72).

Neuronal Intermediate Filament Inclusion Disease

Sporadic cases of frontotemporal dementia associated with distinctive inclusions were initially characterized on the basis of the presence of intermediate filaments in the inclusions[117-119,189] but later found to be characterized by the presence of FUS in the inclusions, hence the classification of this disorder as a type of FTLD-FUS.[517,670,844]

The mean age at onset in a large review of cases was 40.8 years (range 23 to 56 years), the mean duration 4.5 years (range 2.7 to 13 years) and the mean age at death 45.3 years (range 28 to 61 years). The clinical features are those of FTD with parkinsonism. Pyramidal dysfunction has been reported in a small number of cases.[561] Rapid progression has often been noted.

Macroscopically, frontotemporal atrophy has been noted, with severe caudate atrophy in several cases. In one case there was striking putaminal atrophy resembling that of multiple system atrophy.[450] Affected regions show neuronal loss, microvacuolation and astrocytic gliosis. Swollen achromatic neurons are sometimes present. Neuronal inclusions are typically seen in the neocortex, hippocampus and basal ganglia. Inclusions can be detected with antibodies to FUS.[450]

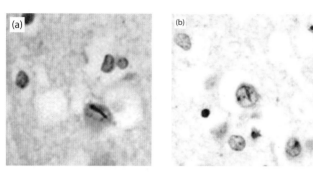

16.72 FTLD-FUS aFTLD-U nuclear inclusions. Neuronal intranuclear inclusions can be detected on immunolabelling for FUS or P62. They are typically thin rod-like bodies **(a, b)** but can appear to be curved around the nuclear membrane or to have a sinuous form (vermiform-pattern inclusions).

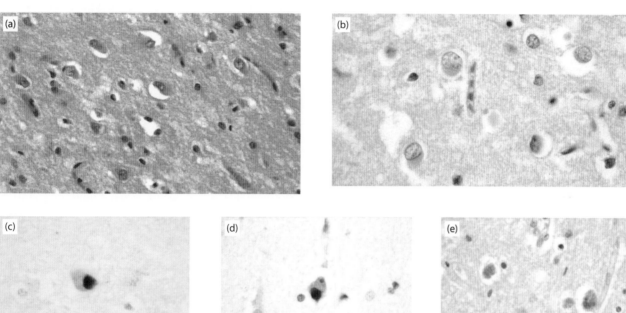

16.73 FTLD-FUS Neuronal intermediate filament inclusion disease. Ill-defined eosinophilic inclusions are seen in cortical neurons, mainly in superficial layers of the frontal and temporal cortex **(a,b)**. These inclusions are immunostained strongly with antibodies to neurofilament protein **(c,d)** and show weak immunoreactivity for ubiquitin **(e)**.

These inclusions are also variably immunoreactive for neurofilament proteins of different molecular weights, both phosphorylated and non-phosphorylated. The inclusions are typically only weakly immunoreactive for ubiquitin. They are negative for tau and α-synuclein. It has been noted that not all inclusions in NIFID show neurofilament immunoreactivity. The inclusions are also strongly immunoreactive for α-internexin.[118,119]

The neuronal cytoplasmic inclusions (Figure 16.73) mainly affect neocortical laminae II/III and V/VI and can be divided into three main types. Some inclusions resemble Pick bodies and are common in neocortical neurons, hippocampal pyramidal neurons and hippocampal dentate granule cells. Smaller numbers are noted in basal ganglia and midbrain neurons. Inclusions are not seen in the substantia nigra or cerebellum. On routine histology, these inclusions appear round or oval, usually slightly basophilic, and often displace the nucleus to one side. The inclusions are typically not stained with silver methods. They are immunopositive for class IV intermediate filaments including neurofilaments and α-internexin but are negative for tau,

TDP-43, and α-synuclein. Occasional FUS-positive oligodendroglial inclusions have been noted.

Ultrastructural examination reveals Pick body-like inclusions composed of granulofilamentous material.

Another distinct type of inclusion is the so-called hyaline conglomerate inclusion. This is less common and is largely limited to larger pyramidal neurons in the cortex. The inclusions appear as irregular lobulated areas of clearing in the neuronal cytoplasm. Some inclusions contain intensely eosinophilic punctate structures at their centre. These inclusions commonly appear argyrophilic with Bielschowsky but not Gallyas silver impregnation. They are immunopositive for class IV intermediate filaments including neurofilaments and α-internexin but are negative for tau, TDP-43, and α-synuclein. This type of inclusion shows dot-like FUS positivity.[417]

A third type of small cytoplasmic inclusion is found in the neocortex and hippocampus. Inclusions of this type appear as small-rounded, tangle-shaped, crescentic, spiculated or linear structures within the neuronal cytoplasm. Immunohistochemistry for intermediate filaments is usually not positive. Vermiform or annular intranuclear inclusions are also seen.

Basophilic Inclusion Body Disease

This condition is pathologically characterized by the presence of neuronal cytoplasmic inclusions that are well defined and basophilic on H&E staining: basophilic inclusions (Figure 16.74). Basophilic inclusion body disease may be associated with sporadic ALS/MND (including, in particular, juvenile ALS in which mutations in the *FUS* gene are found;[59] see Chapter 14), ALS/MND with dementia, and a form of pure frontotemporal dementia. Since the discovery that these inclusion bodies contain FUS as well as related proteins, basophilic inclusion body disease is classed as one of the TDP-FUS group.

On immunolabelling for FUS, cytoplasmic inclusions can be seen in neocortical neurons, hippocampal pyramidal cells and dentate granule cells, neurons in the globus pallidus and thalamus. Inclusions are also present in the midbrain, pons and medulla (including the hypoglossal nuclei) and in lower motor neurons of the spinal cord. Neuronal cytoplasmic inclusions appear as one of three main types:

- compact, round inclusions in the cerebral cortex;
- crescentic and annular inclusions in the cerebral cortex and striatum;
- large round inclusions or complex pleomorphic inclusions in brain stem and deep grey matter nuclei.

Immunohistochemically basophilic inclusions are negative for tau, TDP-43, α-synuclein, neurofilament and α-internexin, positive for P62, and variably ubiquitinated.[417,643] FUS-labelling also reveals a small number of dystrophic neurites, and glial cytoplasmic inclusions have been described. In contrast to other forms of FTLD-FUS, neuronal intranuclear inclusions are rarely seen.

Frontotemporal Lobar Degenerations with UPS Pathology (FTLD-UPS)

During the evolution of terminology used to classify FTLD cases, the main tool was use of antibodies to ubiquitin, leading to the designation of a group called FTLD-U. With discovery of TDP-43 and FUS and the application of antibodies to other components of the ubiquitin-proteasome system, especially P62, cases that show ubiquitin-only pathology are now termed FTLD-UPS. One defined set of cases in this group is linked to mutations in *CHMP2B*, the gene for charged multivesicular body protein 2b. In other cases, mutation in this gene has been excluded. It seems likely that other associated proteins will be discovered that will remove cases from this group of FTLD and define new subgroups based underlying protein abnormalities.

FTLD-UPS (CHMP2B)

A kindred with autosomal dominant frontotemporal dementia (FTD3) has been described in which disease is related to mutation of a gene in the pericentromeric region of chromosome 3 termed *CHMP2B* (charged multivesicular body protein 2B) that encodes a component of the endosomal ESCRTIII complex. Mutations in *CHMP2B* cause disruption of the endosomal pathway,[835] similar to that in several other neurodegenerative conditions.

Patients from three generations have been studied with an age at onset between 46 and 65 years. The disease presents with frontotemporal cognitive disturbances, but there is also evidence for temporoparietal syndromes. Later in the illness, individuals often develop pyramidal or extrapyramidal features. The limited available information from autopsies indicates generalized cerebral atrophy preferentially affecting the frontal lobes. Neuropathology has shown a loss of cortical neurons with microvacuolation of layer II,

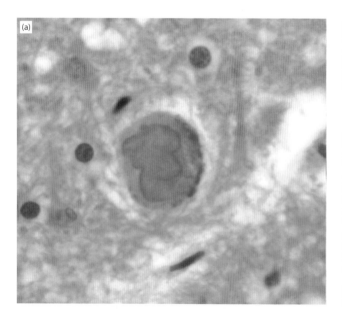
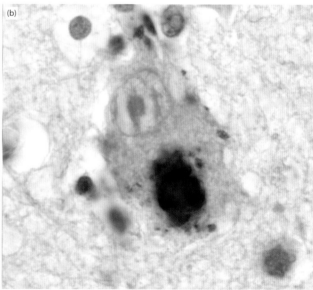

16.74 FTLD-FUS Basophilic inclusion body disease. Inclusions appear as basophilic areas in the cytoplasm, typically with crisply defined, basophilic margin on H&E stains; **(a)** inclusions can be detected using FUS staining **(b)**.

Courtesy of Dr. Olaf Ansorge, Oxford, UK.

associated with loss of myelin in deep white matter. Small, round, ubiquitin- and P62-positive neuronal cytoplasmic inclusions are seen in hippocampal dentate granule cells as well as in some neurons in frontal and temporal cortex.[387]

A mouse model of disease has suggested a toxic gain of function in the pathogenesis of FTLD-UPS.[288]

FTLD with No Inclusions (FTLD-Ni)

A very small number of cases with FTLD have no inclusions that can be detected with antibodies to tau, TDP-43, FUS or UPS. The pathological basis of such cases remains to be defined. It is important to preserve unfixed frozen material in such cases to facilitate genetic and protein-based studies.

HIPPOCAMPAL SCLEROSIS ASSOCIATED WITH AGEING

Pure hippocampal sclerosis (HS) has been defined as severe degeneration and gliosis of the CA1 sector and subiculum of the hippocampal formation in the absence of pathological changes of other significant dementing disease such as AD (Figure 16.75). It is found in 10 per cent of individuals over the age of 85 years.[663,664] This pathology is mainly encountered in an apparently pure form in elderly demented patients when it is has been termed hippocampal sclerosis dementia or HS-ageing.[7,159,211,431,663,994] This entity appears biologically different from hippocampal sclerosis (also known as medial temporal sclerosis) noted in some patients with temporal

(a)

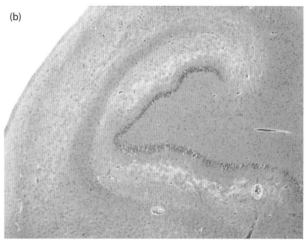

(b)

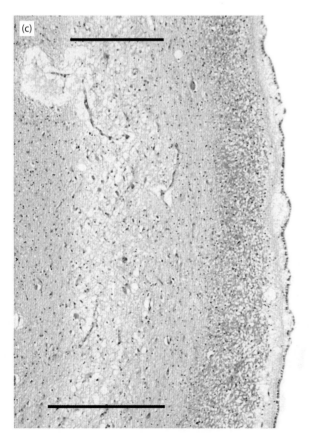

(c)

16.75 Hippocampal sclerosis. There is shrinkage of the hippocampus **(a)**, which shows pallor and microvacuolation **(b)**, associated with loss of large pyramidal neurons and astrocytic gliosis **(c)**.

lobe epilepsy (see Chapter 11), and hippocampal sclerosis recognized as a consequence of vascular disease, such as occurs after cardiac arrest or severe hypotension. Differences in hippocampal regions affected have been shown.[365]

A review of the clinical profile and significance of HS in the elderly confirmed an association with amnesic dementia but with differences from the profile in AD. Patients later shown to have HS-ageing pathology were often diagnosed clinically as having probable or possible AD.[99,663,992]

HS is generally initially diagnosed on examination of H&E-stained sections. It has been recommended that the term HS should only be applied to cases with complete/near-complete neuronal loss and gliosis in the subfields of the cornu Ammonis but not to hippocampal microinfarction.[747] If HS is identified, TDP-43 immunohistochemistry should be performed.[402] In autopsy series of patients with dementia in life 89–93 per cent of those with HS had TDP-43 pathology.[702,994] Pathology can also be seen on ubiquitin immunochemistry. A detailed autopsy study has characterized pathological and clinical features of this condition.[663] Individuals with pathological evidence of hippocampal sclerosis tended to be older than those without hippocampal sclerosis. The authors suggest that although AD is the most common disease underlying dementia and increasing age is the main risk factor, the increased risk of development of dementia in extreme old age may be linked to HS-ageing not AD.[663] Arteriolosclerosis has also been found to be associated with HS ageing.[666]

HS has been shown to have a profile of *GRN* and *TMEM106B* variants associated with frontotemporal lobar degeneration.[649] In a genome-wide association study, HS ageing was associated with polymorphism in the ATP-binding cassette, subfamily C member 9 (*ABCC9*) gene, encoding sulfonylurea receptor 2.[665]

DEMENTIA CAUSED BY CEREBROVASCULAR DISEASE AND HEREDITARY ANGIOPATHIES

Introduction

Cerebrovascular disease (CVD) is the second most common cause of age-related dementia. A wide range of vascular disorders can disrupt the supply of oxygen and vital nutrients to the brain and affect cognition and behaviour. That gradual strangulation of the blood supply to the brain could cause dementing illness was entertained by Alzheimer and Kraeplin.[70] They reasoned that progressive hardening of the arteries in old age could lead to arteriosclerotic dementia. Until the late 1960s, arteriosclerotic dementia was often diagnosed and attributed to the cerebral softening with loss of a relatively large volume, at least 50 mL but usually 100 mL or more of brain tissue.[900] It was clinically overdiagnosed in comparison to AD.[260,776] Current concepts of vascular dementia (VaD) evolved over a long period. VaD is now recognized as a cognitive disorder explained by vascular causes in the absence of other pathologies. The first clear recognition of subclasses of VaD probably should be credited to Otto Binswanger, who described subcortical arteriosclerotic encephalopathy (a type of small vessel disease-related dementia) on pathological demonstration of cerebral white matter (WM) damage in a group of patients with hypertensive disease.[70] In more recent times, based on his extensive experience C. Miller Fisher proposed that cerebrovascular dementia results from both large and small strokes and provided lucid accounts of lacunar syndromes.[260] Multiple small infarcts in association with hypertension are one of the most common causes of VaD. The recognition of subtypes of clinical VaD was an important step towards current pathological classifications. Multi-infarct dementia predominantly results from cortical infarcts attributed to large vessel disease whereas subcortical ischaemic (Binswanger-type) VaD involving subcortical WM and deep grey matter structures results from intracranial small vessel disease. In the last two decades, there have been several refinements in definitions and descriptions of cognitive impairment and dementia caused by both systemic vascular disease and CVD.

Terminology

A variety of terms are used to relate the clinical and pathological descriptions of vascular disease to cognitive impairment and dementia (Table 16.19). Cerebrovascular dementia, more commonly termed VaD, implies a clinically diagnosed dementia syndrome of ischaemic and/or haemorrhagic aetiology. The diagnosis of dementia is based on the dominant feature of abnormal executive function that is severe enough to interfere with social or occupational functioning.[782] The term vascular cognitive impairment (VCI)[87] was introduced to incorporate conditions in which impairment of function any cognitive domain has a vascular origin or results from impaired brain perfusion.[680] It is suggested that adopting this term not only encompasses more conditions, but also reflects a broader intellectual view of the potential contributions of vascular disease to impaired cognition.[316,680]

The term vascular cognitive disorder was also coined to describe any condition with cognitive impairment of vascular origin.[796] Like VCI, it encompasses cognitive disorders of vascular aetiology with diverse pathologies and clinical manifestations. Vascular cognitive disorder has been proposed as the global diagnostic category, restricting the term VCI to patients whose cognitive impairment falls short of dementia.[782] The most recent DSM-V criteria and guidelines propose that the categories of mild and major vascular cognitive disorders be recognized.[21] Major vascular cognitive disorder, as a substitute term for VaD, may be more acceptable to patients and families and is more appropriate for neurodegenerative cognitive disorders in which memory impairment is not predominant.[797]

Cognitive impairment or dementia is relatively common following stroke.[706,535] The risk of incident dementia or post-stroke dementia (PSD) is highest within the first 6 to 12 months after stroke but remains elevated for many years.[10,65,727] PSD has a complex aetiology with varying combinations of large and small vessel disease as well as non-vascular pathology. Stroke injury or CVD may unmask other preexisting disease processes such as AD, although it was recently demonstrated that most PSD is attributable to vascular pathology.[10]

Advances in neuroimaging and systematic neuropathological examination have improved our understanding of the pathological abnormalities associated with VCI.[339] Pathological assessment in VaD or VCI requires careful evaluation of potentially relevant clinical or phenotypic

TABLE 16.19 Definition of terms: cerebrovascular disease as a cause of dementia

Term	Clinical or other definition	Blood vessel type and size	Location of tissue lesions/ pathological processes
Arteriosclerotic dementia	Dementia explained by any cause of CVD	Usually carotid arteries	Ageing-related hardening or stiffening of the arteries. Diffuse WM or parenchymal ischaemic changes
Binswanger's disease	Subcortical leukoencephalopathy with cognitive impairment	Wall thickening of perforating arteries	Decreased perfusion causing strokes and micro-haemorrhages. Lacunar strokes in subcortical structures, particularly diffuse WM disease
Vascular dementia	Dementia caused by CVDs at time of event	Various sizes of vessels	Large and small infarcts throughout, caused by various atherothrombembolic events or SVD
Multi-infarct dementia	Dementia developing in stroke patients with large single or multiple infarcts	Large arteries including carotid and main cerebral arteries	Atherosclerosis or cardioembolic obstruction of usually large vessels. Multiple lesions act synergistically to compromise mental function, meeting operational criteria for dementia
Subcortical ischaemic vascular dementia or small vessel dementia*	Dementia largely caused by subcortical lesions although small cortical infarcts may be present particularly in sporadic cases	Small perforating arteries, end arteries, smaller branches of main brain arteries	Sporadic forms of SVD, include arteriolosclerosis, possible involvement of microthrombi, blood–brain barrier changes. Hereditary types include CADASIL, CARASIL, RVCL and COL4-related conditions caused by mutations in genes affecting vascular function
Strategic infarct dementia	Dementia or impairment caused by large infarcts in subcortical structures, e.g. thalamus or brain stem	Major branches of arteries	Usually lacunar infarcts in thalamus or angular gyrus
Hypoperfusive dementia	Cognitive impairment resulting from global insufficiency, e.g. cardiac arrest or autonomic dysfunction	Large vessels bring blood to the brain	Large areas of hypoperfused parenchyma; incomplete infarcts in watershed regions
CAA-related dementia*	Dementia associated with angiopathies e.g. amyloid proteins	Penetrating or perforating arteries	Small ischaemic or haemorrhagic infarcts, microinfarcts
Haemorrhagic dementia	Dementia or impairment resulting from haemorrhages caused by trauma, aneurysms or lobar bleed, e.g. in CAA	Large and small vessels	Tissue destruction resulting from bleeds
Other causes (vasculitis)	Impairment resulting from localized changes in vessels	Different types of vessels	Localized infarcts or ischaemic changes
Post-stroke dementia (PSD)	Any type of dementia caused after first stroke and includes delayed incidence	All types of vessels	Large and small infarcts caused by athero-thrombembolic events or lacunar infarcts. Impairments of cognitive function resulting from a single stroke, such as aphasia, apraxia and agnosia are usually excluded
Mixed dementia	Dementia associated with both vascular and neurodegenerative (usually AD) pathological changes	All types of vessels	Alzheimer-type lesions in limbic and neocortical areas. White matter may show severe degeneration. Cortical and subcortical vascular pathology. Both executive function and amnestic memory affected
Vascular cognitive impairment (VCI)	All causes of CVD that cause cognitive impairment; early to late and severe forms of dementia syndromes	All types of vessels	Lesions involve all regions of the brain and spinal cord
Vascular neurocognitive disorder (VCD)	Cognitive disorders of vascular aetiology as a heterogeneous group of disorders	All types of vessels	Lesions involve all regions of the brain and spinal cord

*Familial or hereditary forms of cerebral amyloid angiopathy involving ischaemic strokes and intracerebral haemorrhages may also lead to cognitive impairment and stroke.[658]
Abbreviations: CADASIL, cerebral autosomal dominant arteriopathy with subcortical infarcts and leukoencephalopathy;[983] CARASIL, cerebral autosomal recessive arteriopathy with subcortical infarcts and leukoencephalopathy;[358] CAA, cerebral amyloid angiopathy; CVD, cardiovascular disease; RVCL, autosomal dominant retinal vasculopathy with cerebral leukodystrophy;[472] SVD, small vessel disease; WM, white matter.

features with particular attention to timing of events.[251] It has been daunting to define the neuropathological substrate of VCI. This reflects the heterogeneous localization of lesions and the frequent coexistence of other pathologies including neurodegenerative changes such as those in AD.

Several factors may contribute to the overall impairment and the VaD phenotype (Tables 16.20 and 16.21). These include origin and type of vascular occlusion, presence of haemorrhage, distribution of arterial territories, and the size of vessels involved. Many brain regions including the

TABLE 16.20 Vascular lesions in neuropathology of vascular cognitive impairment or vascular dementia

Atherosclerosis
Cardiac, atherosclerotic and systemic emboli
Arteriolosclerosis
Lipohyalinosis and fibroid necrosis
Cerebral amyloid angiopathy
Vasculitis – infective and non-infective
Venous collagenosis
Arteriovenous fistulae – dural or parenchymal
Hereditary angiopathies – cerebral autosomal dominant arteriopathy with subcortical infarcts and leukoencephalopathy (CADASIL); cerebral autosomal recessive arteriopathy with subcortical infarcts and leukoencephalopathy (CARASIL); cerebroretinal vasculopathies (CRVs).
Giant cell arteritis
Berry aneurysms
Miscellaneous vasculopathies – fibromuscular dysplasia, moyamoya
Systemic microangiopathies without vascular inflammatory cell infiltrates
Cerebral venous thrombosis

TABLE 16.21 Parenchymal lesions in the neuropathology of vascular cognitive impairment or vascular dementia

Brain tissue changes of vascular aetiology[a]

Large vessel disease
  Multiple infarcts
  Single strategically placed infarct

Small vessel disease
  Multiple lacunar infarcts in white matter and deep grey matter structures
  Ischaemic white matter changes
  Dilatation of perivascular spaces
  Cortical and subcortical microinfarcts and microhaemorrhages

Brain haemorrhage
  Subarachnoid haemorrhage
  Intracerebral haemorrhage
  Perivascular haemosiderin deposits (cortical and subcortical)

Hypoperfusion
  Hippocampal sclerosis
  Laminar cortical sclerosis

[a]Microinfarcts may be localized in cortical and subcortical structures because of different aetiologies (see Cerebral Microinfarction and Cortical Microinfarcts).

territories of the anterior, posterior and middle cerebral arteries; the angular gyrus, caudate and medial thalamus in the dominant hemisphere; the amygdala and the hippocampus have been implicated in VaD.[580] Factors that define subtypes of VaD include multiplicity, size, anatomical location, laterality and age of the lesions, in addition to genetic influences and contributions of systemic cardiac and vascular disease (see Table 16.25). Subcortical ischaemic VaD is the most significant subtype of VaD.[781]

Epidemiology of Vascular Dementia

Clinical and Pathological Estimates

The prevalence of clinical VaD increases with age[549,766] from an estimated 0.4 per cent in 60–64 year olds, to 4.2 per cent in people over 85 years. In the elderly, the prevalence of VaD

doubles with every 5.3-year increase in age, compared to every 4.5 years for AD. In contrast to AD, men are more likely to develop VaD. The prevalence of VaD among the elderly varies in different regions of the world.[126,463,998] Estimates of the global prevalence of VaD range from 1.2 per cent to 4.21 per cent over the age of 65 years.[367] The age-standardized prevalence of VaD in the population-based Canadian Study of Health and Ageing was 1.5 per cent overall, ranging from 0.6 per cent in 65–74 year olds to 4.8 per cent over 85 years. The estimated annual incidence of VaD in the Canadian study was 2.5–3.8 cases per 1000.[368] Most patients in these studies had post-stroke VaD. Recent meta-analyses of major dementia subtypes in those over 65 years among East, South and Southeast Asians indicated the prevalence of VaD to vary from 0.6 per cent to 9.8 per cent.[126,463,998] In Europe, VaD accounts for 15–20 per cent of dementia and AD for 60–70 per cent.[264,549] The reported prevalence of early onset dementia VaD (before 65 years) has ranged from 3.1 per cent to 44 per cent in various clinic and population-based studies.[929]

There is variable concordance between clinical and pathological diagnosis of VaD, depending on the criteria that are used (Table 16.22). In autopsy studies of dementia, the rate of pathological diagnosis of VaD ranges widely, from as low as 0.03 per cent to as high as 58 per cent with a mean of 17 per cent.[434,580] In Western countries, the estimated rates of pathologically diagnosed VaD are between 8 per cent and 15 per cent. In studies in which the diagnosis was made using NINDS-AIREN criteria,[780] VaD is diagnosed, on average, in 7 per cent of cases. The worldwide frequency of VaD in autopsy-verified cases is estimated as 10–15 per cent, marginally lower than when clinical criteria alone are used.[56,484] In Japan, the incidence of autopsy-verified VaD was reported to be 35 per cent[818] and later 22 per cent.[5] Population-based cohorts with post-mortem assessment should provide the best estimates of the incidence of pathology verified VaD. There are relatively few such studies but they all show that microvascular lesions are more frequent than neurodegenerative lesions in elderly community-dwelling subjects with dementia.[131,812,848,955]

Risk Factors for VaD

Risk factors for VaD are essentially similar to those for CVD or stroke (Table 16.23). These may be divided into five major classes: demographic; atherosclerotic and arteriosclerotic; stroke related; poor cerebral perfusion or oxygenation; and genetic.[313,778] Clinically silent infarcts and WM lesions are also important risk factors for dementia.[409,925] In addition, other medical problems, including depressive illness may increase risk. The potential risk factors for VaD may differ substantially in different countries and population groups.[463]

The rarity with which pure VaD is observed by neuropathologists may partly reflect the relative under-representation of ethnic minorities in many post-mortem series.[463] In the USA, the proportions of elderly African-American and Hispanic people are growing at a faster rate than that of the white population, with the number of African-Americans expected to increase by 30 per cent between 1990 and 2050.[302] African-American and Japanese populations have a higher prevalence of VaD than white

TABLE 16.22 Sensitivities and specificities of clinical criteria for vascular dementia (VaD)

Clinical criteria for VaD	Pathological diagnosis of VaD of patients with dementia[a]		Detection of VaD or mixed dementia by pathology	
Criteria (category if applicable)	Sensitivity	Specificity	Sensitivity	Specificity
HIS[b]	0.43	0.88	-	-
DSM-IV	0.50	0.84	0.36	0.89
ADDTC (possible VaD)	0.70	0.78	0.52	0.87
ADDTC (probable)	0.25	0.91	0.21	0.96
ICD-10	0.20	0.94	0.19	1.00
NINDS-AIREN (possible)	0.55	0.84	0.41	0.91
NINDS-AIREN (probable)	0.20	0.93	0.19	0.98

[a] Dementia or cognitive impairment is determined by use of neuropsychometric screening instruments and batteries for cognitive function. The most widely used ones include the Mini Mental State Examination (MMSE), Cambridge Cognitive Assessment (CAMCOG), the Montreal Cognitive Assessment (MoCA) test and the cognitive screening instrument (CSI-D). The scores from these are usually compared with DSM criteria. The Honolulu-Asia Aging Study (HAAS) has consistently used the Cognitive Abilities Screening Instrument (CASI) Exam, which is a battery of the Hasegawa Dementia Screening Scale, the Mini Mental State Examination (MMSE) and the Modified Mini Mental State Examination.[276] Similar to the CAMCOG exam, the CASI includes tasks of attention, concentration, orientation, short- and long-term memory, language, visual construction, list-generating fluency, abstraction, and judgment, and has a score range of 0-100.

[b] HIS values taken from Gold et al, 1997.[306] The ADDTC and NINDS-AIREN criteria had the highest (bold figures) sensitivities and specificities for possible VaD diagnosis. These were followed by DSM-IV. The kappa statistic showed that the best agreements for criteria were between the ADDTC and NINDS-AIREN for both possible (0.81) and probable (0.89).

ADDTC, State of California Alzheimer's Disease Diagnostic and Treatment Centers; DSM-IV, the Diagnostic and Statistical Manual of Mental Disorder, Fourth Edition; HIS, Hachinski Ischaemic Score; ICD-10, International Classification of Diseases, 10th revision; NINDS-AIREN, National Institute for Neurological Disorders and Stroke with the Association Internationale pour la Recherché et l'Enseignement en Neurosciences.

Results derived from Gold et al, 2002.[307]

TABLE 16.23 Risk factors for vascular dementia

Risk category	Risk factor
Demographic	Age, male gender, lower educational level
Atherosclerotic or arteriosclerotic	Hypertension, cigarette smoking, myocardial infarction, diabetes mellitus, hyperlipidaemia
Stroke related	Volume of cerebral tissue loss, evidence of bilateral cerebral infarction, strategic infarction (e.g. thalamic, angular gyrus or subcortical frontal infarction), white matter disease
Poor cerebral perfusion or oxygenation	Obstructive sleep apnoea, congestive heart failure, cardiac arrhythmias, major surgery, orthostatic hypotension, hypoxia
Genetic	Familial vascular encephalopathies (e.g. CADASIL, CARASIL, CRV), possibly apolipoprotein E gene (*APOE*) ε4
Other biological risk factors	Silent cerebral infarcts, cerebral atrophy, ventricular size, depression

CADASIL, cerebral autosomal dominant arteriopathy with subcortical infarcts and leukoencephalopathy; CARASIL, cerebral autosomal recessive arteriopathy with subcortical infarcts and leukoencephalopathy; CRV, cerebroretinal vasculopathy.

populations.[516] Similarly, African-American and Hispanic patients had a twofold to threefold greater risk of dementia than did white patients 3 months after a stroke.[200] In a clinical study of 165 individuals aged 65 years and over, there was a higher prevalence of dementia as well as a higher frequency of CVD in African-Americans.[372] Over one-third of urban African-Americans in Atlanta, GA, were diagnosed as having mixed dementia or VaD.[40] Gorelick and colleagues, comparing demographic and medical factors among 113 African-American patients with AD and 79 African-American patients with VaD in Chicago, IL, found that the group diagnosed as having VaD had a higher frequency of vascular risk factors and focal neurological

findings.[315] In an autopsy study of 144 dementia patients in Baltimore, MD, the frequency of AD was 2.6 times higher among white patients, whereas the frequencies of multi-infarct dementia and dementia due to chronic ethanol abuse were higher among African-Americans.[628] The higher frequency of multi-infarct dementia in African-Americans has been attributed to the higher frequency of systemic hypertension.

Clinical Diagnostic Criteria

The diagnosis of dementia has long been based on fulfilment of the DSM-IV criteria after evaluation of clinical information including history, time course and findings on neuropsychometry and neuroimaging but in the absence of pathological examination (see Table 16.22). The new DSM-V criteria distinguish between minor and major neurocognitive disorders according to the severity of cognitive impairment; the designation major vascular neurocognitive disorder substitutes for VaD.[797] In DSM-IV and earlier versions of DSM, diagnosis of dementia required memory loss as a core feature, whereas most patients with VaD do not necessarily have a profound memory deficit particularly in early stages but predominantly a frontal dysexecutive syndrome.[199] The new guidelines for the diagnosis of minor and major vascular cognitive disorder focus on speed of information processing, complex attention and frontal-executive functioning.[797] Diagnostic accuracy and better recognition of subgroups of VaD with more homogeneous vascular changes are especially important for clinical trials.[238] Another advance in the diagnosis of VaD has been the use of the Montreal Cognitive Assessment (MoCA) as a brief first screen instrument. Both the full and short versions of the MoCA had excellent diagnostic accuracy in discriminating VaD patients with greater sensitivity and specificity than the Mini Mental State Examination.[267]

Historically, several sets of clinical criteria have been proposed for the diagnosis of VaD.[958,959] These have variable specificity and sensitivity, are not interchangeable, and yield substantial misclassification of dementia cases.[160,307,730] The emphasis on cognitive domains such as memory, which is primary to most forms of AD, results in relatively low sensitivity (0.20) but high specificity (0.93) for probable VaD in clinicopathological validation studies.[307] In earlier studies, the Hachinski Ischaemic Scale was used to indicate presence of multi-infarct dementia in demented patients who scored ≥7 out of 10. Subsequent developments included the Alzheimer's Disease Diagnostic and Treatment Centers (ADDTC) criteria for ischaemic VaD,[143] the National Institute for Neurological Disorders and Stroke–Association Internationale pour la Recherché et l'Enseignement en Neurosciences (NINDS-AIREN) criteria for VaD,[780] and the International Classification of Diseases (ICD-10) criteria for VaD. The ADDTC followed by the NINDS-AIREN criteria for possible (ischaemic) VaD achieve the best balance of sensitivity and specificity, and show reasonable agreement with DSM-IV criteria for possible VaD (see Table 16.22). However, none of the criteria including those of ADDTC, NINDS-AIREN and ICD-10, was consistently found to have high sensitivity for probable VaD.

Despite the deficiencies, the NINDS-AIREN criteria are widely used, particularly in research settings. The NINDS-AIREN criteria emphasize the heterogeneity of VaD syndromes and pathological subtypes (e.g. ischaemic and haemorrhagic strokes, cerebral hypoxic–ischaemic events, WM changes).[780] The three cardinal features of VaD that harmonize with NINDS-AIREN criteria for the clinical diagnosis of probable VaD include (i) acute onset of dementia, demonstrated by impairment of memory and in two other cognitive domains, such as orientation, praxis or executive dysfunction; (ii) neuroimaging evidence of cerebrovascular lesions; and (iii) evidence of a temporal relation between stroke and cognitive loss.[777] Although neuroimaging evidence of vascular lesions is required for a diagnosis of probable VaD, the NINDS-AIREN criteria do not distinguish between people with and without dementia in the context of CVD.[52] The diagnosis of 'definite' VaD requires histopathological evidence of cerebrovascular disease, an absence of neurofibrillary tangles and neuritic plaques exceeding those expected for age, and an absence of other conditions associated with dementia.[462]

Even when NINDS-AIREN and DSM-IV criteria are used, autopsy studies show limited accuracy of clinical diagnosis VaD and, invariably, the inclusion of cases with AD-type pathological changes.[237,397,460] The current clinical diagnostic criteria are reasonable predictors of vascular pathology but not necessarily 'pure' VaD.[389,484] There are currently no validated criteria for VCI or vascular cognitive disorder.[308,727,797] Unbiased criteria encompassing relevant cognitive domains for VCI still need to be evaluated fully.[197,316,339] As for AD, confirmation of VaD at autopsy requires appropriate sampling and systematic neuropathological examination (Table 16.24 and Figure 16.76) to define the nature and severity of vascular abnormalities and to rule out other significant pathological changes that might account for the cognitive impairment.[339]

More sensitive neuroimaging modalities have increased ante-mortem recognition of vascular changes in dementia patients but these have also become harder to interpret, by revealing similar lesions in non-demented individuals. As discussed earlier, diagnosis is complicated by the heterogeneity of infarcts and the poor standardization of assessment and diagnosis, especially when dealing with mixed pathologies.[308] Depending on the inclinations of the observer, cases of AD with coexistent vascular lesions such as infarcts may be classified variously as VaD, or AD with coexistent vascular pathology, or mixed dementia.[259,813] To derive more accurate prevalence or incidence estimates and consistency in pathological diagnosis, there has to be uniformity in protocols across different centres, including more extensive bilateral sampling of the brain at autopsy.[197,339,701]

Subtypes of Vascular Dementia

Based upon current clinical and neuroimaging evidence, subtypes of VaD can conveniently be divided into three major types defined by the type of vascular disease, infarct site and calibre of the abnormal vessels (Table 16.25). The type of CVD determines the specific neurological signs in VaD. Large vessel disease is more often related to lateralized sensorimotor changes and aphasia whereas small vessel disease results in more subtle bilateral neurological dysfunction, usually including extrapyramidal signs.[856] In patients with large vessel obstruction or disease, dementia may result from multiple infarcts or from infarcts that are strategically located. Cerebral small vessel disease typically produces subcortical ischaemic VaD (see Figure 16.76). However, it is recognized that overlap between these subtypes is common. Microinfarcts are seen in both large and small vessel disease and VaD is often associated

	0	1	2	3
Arteriolosclerosis				
Amyloid angiopathy				
Perivascular haemosiderin leakage				
Perivascular space dilatation				
Myelin loss				

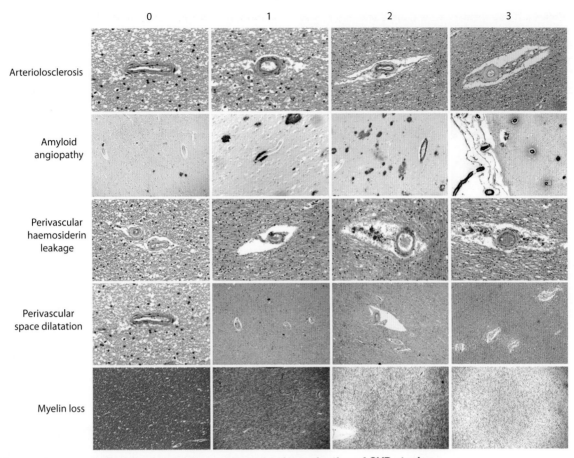

16.76 Key cerebrovascular lesions and rating templates for determination of CVD staging.

Adapted from Deramecourt et al.[197] With permission from Lippincott Williams & Wilkins/Wolters Kluwer Health.

with a combination of cortical and subcortical lesions (cortico–subcortical VaD). It is rare for vascular lesions to be exclusively cortical. Strategic infarct VaD is relatively uncommon and may result from small or large infarcts (see Table 16.25).

Multi-infarct Dementia

The term multi-infarct dementia was first coined to describe dementia resulting from multiple small and large cerebral infarcts.[340] The term is sometimes used interchangeably with cortical VaD. It results from multiple strokes of various sizes and in various locations. Compared to AD, this a relatively infrequent cause of dementia and is most often associated with hypertension and/or atherosclerosis affecting large intracranial or extracranial blood vessels, giving rise to local thromboses or emboli.[241,340] Other predisposing conditions include cardiac disorders such as atrial fibrillation and myocardial infarction that may be the source of cerebral microemboli, haematological conditions including sickle cell anaemia and other conditions predisposing to thrombosis, systemic lupus erythematosus, vasculitis and familial cerebrovascular diseases with large vessel involvement (Table 16.20).

Dementia Associated With Strategic Infarcts

Sometimes VaD is diagnosed in patients who have a lesion volume below that usually required for development of dementia but in whom the ischaemic injury is in areas critical for normal cognitive functioning (Table 16.25). Strategically located infarcts involving deep grey matter (basal ganglia, thalamus), subcortical WM or the limbic system may produce effects on cognition disproportionate to the size of the infarcts.[200,877,878] These may be single or multiple small infarcts or haemorrhages in subcortical locations or even in large vessel territories. Infarcts placed in specific frontosubcortical circuits connecting the prefrontal cortex to the basal ganglia or in non-specific thalamo-cortical projections can cause combinations of executive dysfunction, personality change or apathy, which are associated with hypoperfusion and hypometabolism predominantly in frontal cortical areas.[507] Except for lesions in the midbrain, infarcts in other infratentorial regions of the brain do not appear to be important for cognition.[57]

Subcortical Ischaemic Vascular Dementia

Subcortical ischaemic vascular dementia or dementia resulting from small vessel disease is a common cause of cognitive impairment and dementia.[699] The main vascular

TABLE 16.24 Key variables for reporting neuropathology of vascular dementia

List of items to define each case
- Identify ischaemic and/or haemorrhagic infarct(s)
- Presence of lacunes and lacunar infarcts: *état lacunaire* (grey) and *état crible* (WM)
- Location of infarcts: cortex, WM, basal ganglia, brainstem (pontine), cerebellum
- Circulation involved: arterial territories - anterior, middle or posterior
- Laterality: right or left anterior and posterior
- Sizes and number of infarcts = dimension: 0-4 mm, 5-15 mm, 16-30 mm, 31-50 mm and >50 mm; if size <5 mm determine as small or microinfarcts
- Assess presence and location of SVD: lipohyalinosis; fibroid necrosis; CAA
- Assess presence of WM disease: rarefaction or incomplete infarction
- Assess degree of microgliosis and astrocytosis: mild, moderate or severe
- Assess presence of Alzheimer pathology (including neurofibrillary tangle and neuritic plaque staging). If tangle stage exceeds III the case is mixed AD and VaD
- Assess presence of hippocampal sclerosis

For reporting purposes each of these features can be scored numerically to provide a summary.[339] For example, 0 is absent and 1 means present. Less frequent lesions include watershed infarcts and laminar necrosis. Increasing numerical value may also be assigned to the infarcts. Abbreviations: CAA, cerebral amyloid angiopathy; NFT, neurofibrillary tangles; WM, white matter.

pathology relates to sclerotic changes in intracranial arteries and arterioles (Table 16.25). There are two main patterns of subcortical ischaemic vascular dementia: multiple lacunar infarcts affecting subcortical grey matter, and diffuse or widespread incomplete infarction of WM. Patients usually exhibit motor and executive slowing, forgetfulness and dysarthria. A short-stepped gait is also common and can mimic that of parkinsonism. The dementia may be caused by disruption of pathways running from the prefrontal cortex to the basal ganglia, and of thalamocortical pathways.[13] It is now well-recognized that VaD may be exclusively due to subcortical vascular lesions, and various attempts have been made to characterize subcortical VaD.[13,15,47,699] Pure subcortical VaD with a slowly progressive course can simulate AD.[876] A special case of subcortical VaD is thalamic dementia, due to infarcts in the thalamus with relatively little involvement of other brain structures.[124]

Cerebral Haemorrhage and Cognitive Impairment

Cognitive impairment can complicate subarachnoid haemorrhage (SAH) or subdural haemorrhage. Cognitive deficits were reported in 19–62 per cent of patients following SAH,[858] the severity being related to the degree of associated brain damage.[398] The likelihood of brain damage and dementia is influenced by a range of factors including older

TABLE 16.25 Comparison of the subtypes of vascular dementia: main clinical and pathological features and mechanisms

Feature	Large vessel disease	Strategic infarcts	Small vessel disease[a]
Pathology	Combination of large infarcts involving cortex and underlying white matter, and cortical microinfarcts	Large or small arterial territory infarcts, or distal field (watershed) infarcts	Lacunar infarcts, subcortical and cortical microinfarcts
Vascular mechanisms	Atherosclerosis, thrombotic occlusion, artery-to-artery embolism, arterial dissection, dolichoectasia, coagulopathies	Emboli or thrombi, hypoperfusive injury	Arteriosclerosis, arteriolosclerosis, hyalinization and lipohyalinosis (fibroid necrosis), enlarged perivascular spaces, microangiopathy, CAA, microaneurysms, microatheroma, segmental arterial disorganization, incomplete ischaemic injury
Radiology	Large lesions, focal WM changes	Strategic infarcts or lacunes in fronto–subcortical pathways, angular gyrus, basal ganglia, dominant thalamus, pons	Lacunes, diffuse WM disease, enlarged perivascular spaces, microbleeds
Clinical onset	Abrupt	Abrupt	Majority gradual
Disease progression	Stepwise deterioration	Stepwise deterioration	Majority slow or insidious
Neurological signs/symptoms	Focal cortical signs, hemianopia, hemisensory dysfunction, reflex asymmetry	Focal cortical signs depending upon site of lesion – marked clinical heterogeneity	Sensorimotor deficits, dysarthria, dysphagia, gait disorder (parkinsonian type), extrapyramidal signs, hypokinesia, urinary disorder
Cognition	Cortical dysfunction including aphasia, alexia, agraphia, according to affected areas, impaired working memory (in later stages)	Rapid onset of cognitive dysfunction (varied dependent in site of infarct)	Cognitive slowing, inattention, apathy, speed processing, learning and verbal memory
Executive dysfunction	Common, not always explicit	Not always	Clear early signs and maintained (~100%)

[a] Several hereditary disorders such as CADASIL have pathological features similar to SVD.

CAA, cerebral amyloid angiopathy; WM, white matter.

age, the development of arterial vasospasm and delayed cerebral infarction, increased intracranial pressure, intraparenchymal and intraventricular haemorrhage, hydrocephalus, and location of the aneurysm.[858] Subdural haemorrhage is an uncommon cause of dementia. Older people with a chronic subdural haematoma may rarely have cognitive deficits,[858] sometimes progressive and not always reversible with surgical drainage.[415] Unlike subdural haemorrhage, which usually results from trauma rather than vascular pathology, SAH is a vascular disease and its association with cognitive deficits is regarded as a form of VaD. Intracerebral haemorrhages or haemorrhagic infarcts occur in 7–10 per cent of cases suspected with dementia. A common cause is sporadic CAA, but hereditary CAA[36] and other genetic disorders[983] are occasionally responsible. Neuroradiological detection of microhaemorrhages (microbleeds) is a surrogate marker of cerebral small vessel disease and associated with cognitive impairment.

Mixed Dementia

In the context of VaD, the most common form of mixed dementia comprises vascular lesions combined with Alzheimer-type pathology. This combination is particularly common in the oldest old[734] and the interaction may increase the risk of cognitive impairment.[316,996] Mixed dementia in which vascular changes, including cerebral amyloid angiopathy, are associated with Lewy body pathology or other non-AD neurodegenerative disease is relatively rare.[437] The precise clinic or population-based prevalence of mixed dementia is unknown but community studies suggest it is frequent.[131,812] Retrospective and prospective autopsy studies suggest a wide range, from 2–58 per cent, with a mean in the region of 6–12 per cent.[434] Early validation studies indicated that although mixed dementia could be distinguished clinically from AD, it could not be separated from VaD.[785] More recent studies suggest that 30–50 per cent of autopsy-verified mixed AD and VaD cases are misclassified as VaD.[306,307] Clinical and pathological evidence indicates that combined neurodegenerative and vascular pathologies quicken the progression of dementia.[242,786,846] However, this depends on vascular lesion type and location as well as the severity of concomitant AD-related pathology. Although evidence-based pathological criteria for the diagnosis of mixed dementia are still lacking, the diagnosis should be made when a primary neurodegenerative disease known to cause dementia exists with one or more of the pathological lesions defining the VaD subtypes. In a key study,[308] WM lacunes, periventricular and diffuse myelin loss, and cortical gliosis were not associated with cognition, but Braak staging, amyloid plaque burden, cortical microinfarcts and thalamic and basal ganglia lacunes predicted the presence of dementia with high sensitivity (0.80) and distinguished VaD from AD in mixed cases. The combination of autopsy-proven AD with multiple vascular or ischaemic lesions producing about 30–50 mL of infarcted/damaged brain tissue is considered mixed. Mixed dementia is also diagnosed when there are pathological changes of VaD and neurofibrillary pathology above Braak stage III is encountered.[462]

Post-Stroke Dementia

Several studies have shown that post-stroke cognitive impairment and PSD or incident dementia after the first ever stroke are relatively frequent, particularly with increasing age.[535,727] In meta-analysis,[706] the pooled prevalence estimates of PSD less than 1 year after the stroke ranged from 7.4 per cent in population-based studies of first ever stroke to 41.3 per cent in hospital-based studies of recurrent stroke. In the population studies, estimates of pre-stroke dementia were excluded whereas they were included in the hospital-based studies. In hospital-based studies, the cumulative incidence of dementia after the first year was slightly greater than expected (3.0 per cent, range 1.3–4.7) based on recurrent stroke alone.[706] Multiple lesions over time and the characteristics and complications of the stroke were found to be most strongly associated with PSD. In other cohorts of older stroke survivors, despite initial improvements in cognition following stroke, risk of progression to delayed dementia after stroke was substantial and related to the presence of vascular risk factors. During the follow-up over a mean period of 3.8 years, 24 per cent of elderly stroke survivors developed dementia and 76 per cent remained alive without dementia or had died without dementia. The incidence of delayed dementia was calculated to be 6.3 cases per 1000 person years whereas it was 8.6 for death or dementia.[10]

The risk factors for PSD are similar to those for VaD and include older age, pre-stroke cognitive deficits, history of cardiovascular disease, epilepsy, sepsis and low education.[535] Neuroimaging studies show that the volume and site of infarcts, extent and location of WM lesions as well as brain atrophy including medial temporal atrophy are important determinants of PSD.[258,535,729] Recent autopsy findings suggested that 75 per cent or more of the demented stroke survivors met the current pathological classification of VaD;[462] those who developed dementia had significantly more microinfarcts and were more likely to have additional neurodegenerative pathology.[10] However, the pathological basis of dementia in PSD may include large vessel disease, small vessel disease associated with WM damage, and lacunar infarction, as well as concurrent AD.[535]

Interaction between Vascular and Alzheimer-Type Pathologies

Multiple studies have shown that a high proportion of people fulfilling clinical criteria for diagnosis of AD also exhibit cerebrovascular pathology.[281,373,460,653,737] Variable combinations of ischaemic and haemorrhagic lesions may be present at autopsy in more than one-third of subjects with a neuropathological diagnosis of AD. Community-based studies show that older cohorts have an even greater degree of concomitant vascular pathology or mixed pathologies.[131,762,848] In one community-based sample, 38 per cent of dementia cases had mixed pathology, with both Alzheimer-type changes and vascular lesions, but 'pure AD' represented only 21–24 per cent of cases.[812] White matter ischaemic or oligaemic lesions are demonstrable in as many as 94 per cent of older community-dwelling study participants, and this change is an independent substrate for dementia.[248] In addition, atherosclerosis in cerebral arteries and the circle of Willis[773,985] is frequently present in AD. The most common overlapping vascular pathologies are smaller cerebrovascular lesions rather than large infarcts[197] and include most features of small vessel disease such as cortical infarcts, lacunes, diffuse and periventricular myelin loss, WM microvacuolation, microinfarcts, microhaemorrhages, arteriolosclerosis and focal and diffuse gliosis.[51,241,931] Conversely, AD pathology was three times more likely in VaD cases with small (<15 mL) than large infarcts.[51] These findings corroborate the critical importance of microvascular disease in VaD and in AD.

Clinicopathological studies also suggest that vascular disease not only influences the neurodegenerative disease burden[762,995] but also worsens the cognitive status of patients who may already have AD (Figure 16.77). The density of neocortical plaques was lower in AD cases with coexistent vascular lesions interpreted as contributing to the dementia.[653] In the Nun Study of elderly women in convents in the midwestern, eastern, and southern United States, participants in whom coexistent AD pathology and brain infarcts were found at autopsy had poorer cognitive function and a higher prevalence of dementia than did those without vascular change.[846] A lower burden of Alzheimer-type pathology (particularly, fewer neurofibrillary tangles) was required to reach the threshold for dementia when there were concomitant lacunar infarcts in subcortical structures including the basal ganglia, thalamus or deep WM. Similarly, in another study, after accounting for AD lesion burden, the presence of other pathologies or infarcts increased the odds of dementia over fivefold[812] and caused earlier onset of dementia.[242]

Clinicopathological correlation studies indicate that patients diagnosed as having probable AD are seldom reported by the pathologist to have a solely vascular basis for their dementia.[281,397,445,617,624,676,813] Cortical or subcortical vascular lesions comprising small old infarcts that do not involve a substantive volume of tissue loss, or WM cribriform state (état criblé) and rarefaction, or lesions in

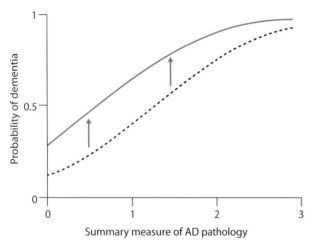

16.77 Probability of clinically diagnosed dementia proximate to death as a function of degree of AD pathology (summary measures) in persons with cerebral infarcts (solid line) and persons without infarcts (dashed line).

Modified from Schneider et al.[811] With permission from Lippincott Williams & Wilkins/Wolters Kluwer Health.

regions believed not to play an obvious role in cognitive function tend to be dismissed by neuropathologists as incidental, and such cases tend not to be distinguished from those with 'pure' AD pathology.

Epidemiological studies consistently indicate an interaction between CVD or VaD and AD.[6] This is highlighted by the association of several vascular disease risk factors, including hypertension, diabetes, high triglycerides and elevated LDL cholesterol, with cognitive impairment and AD.[459,464] Conversely, pathological evidence indicates that Alzheimer-type lesions including Aβ plaques and neurofibrillary pathology occur more often in cases of CVD or VaD than in normal ageing.[460] Aβ peptides increase intracerebrally in patients who succumb to ischaemic brain injury compared to normal elderly lacking cerebrovascular pathology.[798] Tissue accumulation of soluble Aβ in VaD subjects was equal to that in AD by the eighth decade[534] and may contribute to neurological dysfunction even if conventional histopathology does not reveal the presence of classical Alzheimer-type pathology. Experimental evidence suggests that chronic hypoperfusion upregulates expression of APP and BACE to increase Aβ accumulation[464] and ageing-related vascular changes may then impede the clearance of Aβ from the brain.[123,798] APP has been reported to accumulate along WM tracts during cerebral ischaemia in middle-aged subjects[148] and in VaD.[405]

It has also been demonstrated that increases in CSF tau are not restricted to AD but also evident in VaD.[78,644,836] The concentration of tau in lumbar CSF in patients diagnosed with probable VaD was reported comparable to that in AD and significantly increased compared to that in non-demented controls.[836] In addition, immunolabelling of synaptophysin in the cerebral cortex was decreased in VaD as well as in AD.[997] These findings provide evidence for synaptic degeneration in VaD as well as AD and support the notion that ischaemic or oligaemic events in the elderly may exacerbate Alzheimer-associated pathology. Other biological markers for AD such as decreased apolipoprotein E (ApoE) levels in CSF have also been reported in VaD.[837] It is likely that two processes, vascular and neurodegenerative, result in the final presentation of AD,[461,810] although debate continues as to the extent to which they are interactive, independent or additive.[811]

Pathological Diagnosis of Vascular Dementia

There is a lack of broad agreement as to diagnostic criteria for the neuropathological validation of VaD (see Table 16.23). Neuroimaging and clinicopathological studies suggest that VaD is related to the extent of cerebral damage. A combination of factors including origin, volume, location and number of lesions, may contribute to the development of dementia. Blessed and colleagues observed that the total volume of infarcts in demented stroke patients was usually over 50 mL and in some cases was greater than 100 mL, exceeding that in non-demented stroke patients.[80,900] Subsequent clinicopathological studies reported that only 5 of 23 patients with a pathological diagnosis of VaD had more than 50 mL of infarcted tissue, and 7 had less than 10 mL.[237] It is now clear that widespread small ischaemic lesions or multiple microinfarcts[951,953] distributed throughout the central nervous system correlate with dementia and are key predictors of cognitive impairment.[465]

Location of lesions may also be more critical than total volume.[200,239] For example, infarction in the left hemisphere disproportionately increases the risk of dementia.[130,200,314,545,728] Bilateral infarcts with more involvement of the dominant hemisphere also increase the risk of dementia after stroke.[192,237,545]

Relatively few prospective studies have validated the clinical criteria for VaD. Previous criteria for Binswanger's disease or cerebral small vessel disease (SVD)[67] proposed a clinical diagnosis of dementia accompanied by neuroimaging evidence (CT or MRI) of bilateral abnormalities and at least two out of three of the following: (i) a vascular disease risk factor or systemic vascular disease; (ii) focal cerebrovascular disease; or (iii) 'subcortical' cerebral dysfunction as evidenced by gait disorder, parkinsonism or incontinence. These criteria were validated in a prospective series of 184 patients with AD: only 1.6 per cent were diagnostically misclassified when all three clinical criteria were met.[67]

More recently, as part of the Oxford Project to Investigate Memory and Ageing (OPTIMA), researchers evaluated a simple, image-matching scoring system[840,860] to assess the relationship between SVD and cognitive function in 70 cases with insufficient pathology to meet criteria for diagnosis of AD. Cognitive scores were inversely related to the severity of SVD pathology and 43 per cent of the cases with high SVD scores had been assessed as demented. To advance standardization of neuropathological assessment and achieve better clinicopathological correlation in all subtypes of VaD including SVD, Deramecourt et al[197] proposed an algorithm for the neuropathological quantification of the CVD burden in dementia. The staging system (I–VI) needs further evaluation against cognitive function scores to determine whether this system can be used in large-scale studies to investigate clinicopathological correlation in VaD (Figure 16.78).

The neuropathological diagnosis of 'pure' VaD should be based on the exclusion of a primary neurodegenerative disease known to cause dementia and the presence of cerebrovascular pathology that defines one or more of the VaD subtypes. These would include dementia among post-stroke survivors who fulfil the NINDS-AIREN criteria[780] for probable VaD, and stroke survivors with VCI.[680] An approach to the neuropathological diagnostic subtyping of VaD was proposed by the Newcastle group that recognizes two neuropathological diagnostic categories: probable VaD, based on the exclusion of a primary neurodegenerative disease known to cause dementia plus the presence of cerebrovascular pathology that defines one or more of the VaD subtypes (Table 16.26), and possible VaD, when the brain shows vascular pathology that does not fulfil the criteria for one of the subtypes but no other explanation for dementia is found. Post-stroke cases are usually included in subtypes I–III. Cases with extensive WM disease in the absence of other significant features are included under small vessel disease. Subtype I may result from large vessel occlusion (atherothromboembolism), artery-to-artery embolism or cardioembolism. Subtype II usually involves arteriolosclerosis, lipohyalinosis, and hypertensive, arteriosclerotic, amyloid or collagen angiopathy. Subtypes I, II and V may result from aneurysms, arterial dissections,

arteriovenous malformations and various forms of arteritis (vasculitis).

Assessing the neuropathological substrates of VaD involves systematic assessment of parenchymal lesions, including microinfarcts and haemorrhages and the vascular abnormalities that may have caused them.[197,462,623,840,863] In addition, systemic factors (e.g. hypotension, hypoglycemia) may cause brain or neuronal lesions in the absence of severe vascular disease and should be taken in account when attributing causes to VaD. As discussed earlier, parenchymal abnormalities of neurodegenerative type may be present that are not obviously associated with either vascular disease or systemic factors; these include Alzheimer type or hippocampal lesions.

Types of Cerebrovascular Pathology and Dementia

Atherothromboembolism is the main cause of infarction in major arterial territories. Thromboembolic events are responsible for up to 50 per cent of all ischaemic strokes whereas intracranial SVD causes 25 per cent of these infarcts (see Chapter 2). Small vessel alterations involve arteriolosclerosis and hyalinosis and are associated with lacunar infarcts and lacunes predominantly occurring in subcortical structures. WM disease or subcortical leukoencephalopathy with incomplete infarction and SVD are common in CVD associated with dementia.[197] Other features include border zone (watershed) infarcts, laminar necrosis and amyloid angiopathy. Complicated angiopathies such as fibromuscular dysplasia, arterial dissection, granulomatous angiitis, collagen vascular disease and giant cell arteritis are much rarer causes of cerebrovascular disease and VaD.

Previous studies have recorded ischaemic, oedematous and haemorrhagic lesions induced by pathological changes in the brain circulation or perfusion in association with VaD (Figure 16.79; see also Table 16.20). In four studies,[462] only 75 per cent of VaD cases had cortical and subcortical infarcts, suggesting that other vascular pathologies involving incomplete infarction or border zone infarcts could be important factors. A quarter of the cases had cystic infarcts, whereas 50 per cent showed lacunar infarcts or microinfarcts. Severe amyloid angiopathy was present in 10 per cent of the cases. Hippocampal sclerosis, which may be caused by remote ischaemic injury, was apparent in 55 per cent of patients with a clinical diagnosis of ischaemic VaD.[931]

Large Vessel Disease and Macroinfarction

Large infarcts or macroinfarcts are defined as those visible on gross examination. Stenosis resulting from atherosclerosis within large vessels is considered the main cause of large infarcts, which may sometimes extend beyond the relevant arterial territory. In approximately 15 per cent of VaD, there is occlusion of the extracranial arteries such as the internal carotid artery and the main intracranial arteries of the circle of Willis including the middle cerebral artery, leading to multiple infarcts.[106] The relative involvement of the anterior versus posterior portions of the circle of Willis, and the left versus the right side is variable. Typical atherosclerosis or microatheromatous disease in the meningeal and smaller vessels,

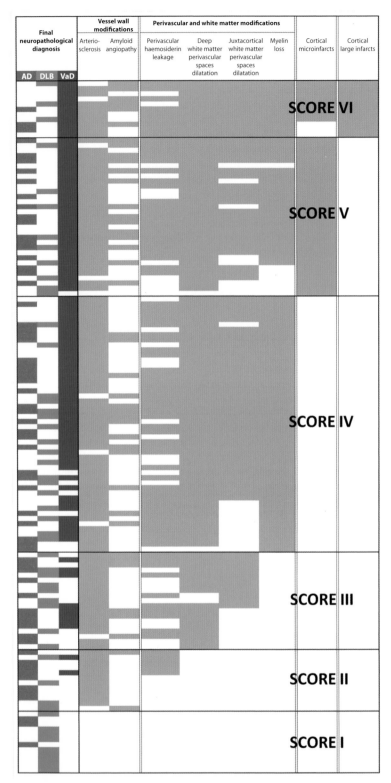

16.78 Distribution of the cerebrovascular lesions. Cases were ordered according to the severity of CVD assessment in the frontal lobe but similar hierarchical distribution was observed in the temporal lobe and the basal ganglia. Coloured columns represent cases from AD (red), DLB (green) and VaD (blue) groups. Grey-shaded areas represent moderate to severe modification or the presence of infarcts. The progressive increment of CVD was weighted into 7 scores from 0 to VI.

Adapted from Deramecourt et al.[197] With permission from Lippincott Williams & Wilkins/Wolters Kluwer Health.

beyond the circle of Willis, involving proximal segments of the middle and anterior cerebral arteries, is generally rare but may be found in the very old.[465] Dolichoectasia and fusiform aneurysms have been noted in some cases. In severe atherosclerosis, medium-sized arteries in the lepto-meninges and proximal perforating arteries are involved.

TABLE 16.26 Classification of cerebrovascular pathology and dementia

Type of lesions	Subtype[a]
Large infarct or several infarcts (>50 mL loss of tissue): multi-infarct dementia	I
Multiple small or microinfarcts (>3 with minimum diameter 5 mm); SVD in >3 coronal levels (hyalinization, CAA, lacunar infarcts, perivascular changes): small vessel vascular dementia	II
Infarcts in critical areas (e.g. thalamus, hippocampus, basal forebrain): strategic infarct dementia	III
Incomplete or diffuse infarction (hippocampal sclerosis, ischaemic–anoxic damage, cortical laminar necrosis, border zone infarcts involving three different coronal levels): vascular dementia due to cerebral hypoperfusion	IV
Multiple cerebral haemorrhages (lobar, intracerebral or subarachnoid): vascular dementia due to cerebral haemorrhages	V
Any of the above cerebrovascular changes with concurrent Alzheimer type pathology (Braak stage III): mixed dementia	VI

The proposed Newcastle categorization includes six subtypes.[462] In all of the above, the age of the vascular lesion(s) should correspond with the time when disease began. The post-stroke cases are usually included in subtypes I-III. Although these may not be different from other published subtypes[434] they are practical and simple to use. Cases with extensive WM disease in the absence of significant other features are included under SVD.

[a] Subtype I may result from large vessel occlusion (atherothromboembolism), artery-to-artery embolism or cardioembolism. Subtype II usually involves descriptions of arteriosclerosis, lipohyalinosis, hypertensive, arteriosclerotic, amyloid or collagen angiopathy. Subtypes I–II and V may result from aneurysms, arterial dissections, arteriovenous malformations and various forms of arteritis (vasculitis). For reporting purposes, each of the features can be scored numerically to provide a summary; for example, 0 is absent and 1 is present. Less frequent lesions, including watershed infarcts and laminar necrosis, may be scored similarly. Increasing numerical value may also be assigned to the infarcts.

AD, Alzheimer's disease; CAA, cerebral amyloid angiopathy; ICH, intracerebral haemorrhage; SAH, subarachnoid haemorrhage; SVD, small vessel disease.

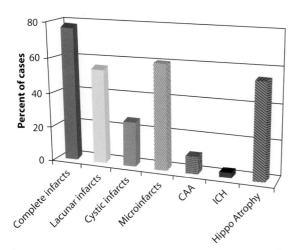

16.79 Distribution of types of vascular lesions and hippocampal changes in VaD. Histograms show the mean percent lesions reported in 2 or more studies. Data from 214 cases. Cystic infarcts (possibly also lacunar) with typically ragged edges were admixed in both cortical and subcortical structures. BG, basal ganglia; WM, white matter.

Artery-to-artery embolism involves the separation of thrombi from (often ulcerated) lesions in extracranial arteries, e.g. at the bifurcation of the common carotid artery. In addition to coagulated blood and platelets, the thrombi may contain cholesterol and calcified deposits from the underlying atheromatous plaque. Cardiogenic emboli may also find their way to the anterior or, particularly, the posterior cerebral circulation to cause infarcts in the territory of the posterior cerebral artery or superior cerebellar artery (Table 16.25).

Arterial territorial cerebral infarcts usually involve four principal regions, supplied by the major arteries: anterior cerebral artery, middle cerebral artery, posterior cerebral artery and the territory between the anterior and middle cerebral artery. The size of the infarcts is described by assessing the two largest diameters of each lesion (Chapter 2). In VaD brains subjected to global ischaemia, i.e. after transient cardiac arrest, damage may be observed in vulnerable neuronal groups within the hippocampus or neocortical laminae and degrees of gliosis or glial scars may be noted.

Cerebral Small Vessel Disease

The changes of SVD include hyalinization of the vessel wall, expansion of the perivascular space and pallor of staining adjacent perivascular myelin, with associated astrocytic gliosis (Figure 16.76). The smaller vessels of the brain including intracerebral end arteries and arterioles undergo progressive age-related changes[512] that alter perfusion and cause lacunar infarcts (cystic lesions generally <1cm) and microinfarcts. The arteriolar changes range from wall thickening by hyalinosis, reduction or increment of the intima to severe arteriolosclerosis and fibroid necrosis (Figure 16.76). Uncomplicated hyalinosis is characterized by almost complete degeneration of vascular smooth muscle cells (the tunica media becomes acellular) with concentric accumulation of extracellular matrix components and fibroblasts.[512] These changes promote thrombotic occlusion or lead to progressive stenosis with consequent acute or chronic ischaemia of the tissue beyond the vessel. Arteriolosclerotic changes are likely to cause loss of elasticity and impaired dilation and constriction in response to variation in systemic blood pressure or local auto-regulation, which interferes with maintenance of blood flow, and changes in tissue perfusion in response to metabolic demand. The deep cerebral structures and WM are most vulnerable because the vessels are end arteries, almost devoid of anastomoses. Small vessel pathology can also lead to damage of the blood–brain barrier, with chronic leakage of fluid and macromolecules into the WM.[379,937] Microvascular disease may also be associated with inflammatory changes, including the presence of lymphocytes or macrophages centred on affected blood vessels (and not necessarily part of a reaction to brain ischaemia). Particularly in older SVD cases, there is often evidence of previous microhaemorrhage, in form of perivascular haemosiderin.[197]

Lacunar Infarcts

Lacunar infarcts are cavitating lesions 1 cm or less in diameter and are frequently found in both subcortical grey and WM in VaD (Chapter 2). They represent small foci of ischaemic necrosis resulting from narrowing or occlusion of penetrating arteries that branch directly from larger cerebral arteries.[260] Lacunar infarcts are frequently multiple and bilateral and often coexist with other vascular lesions (e.g. large infarcts or diffuse WM damage). Lacunar state (état lacunaire) is the condition of multiple small lacunes in the grey matter. Lacunar infarcts should be distinguished from dilated perivascular spaces, which also produce MRI hyperintensities.[982]

Aside from critical lesions occurring most often in the internal capsule or caudate nucleus, recent meta-analyses did not identify pathological differences (in terms of lacunar lesions) between symptomatic and asymptomatic patients. Perivascular oedema and thickening, inflammation and disintegration of the arteriolar wall were common, whereas vessel occlusion was rare.[44] In neuropathological studies of elderly patients with vascular disease but without evidence of AD or other neurodegenerative pathologies, dementia was associated with severe cribriform change and associated subcortical WM damage and microinfarcts.[241] In the Honolulu-Asia Aging Study (HAAS) analysis,[953] 'microvascular ischaemic lesions' were identified as the sole or dominant abnormality in 34 per cent of the demented or definitely impaired decedents (Figure 16.80). Cases of VaD without significant AD pathology show more severe cribriform change and deep white and grey matter lacunar

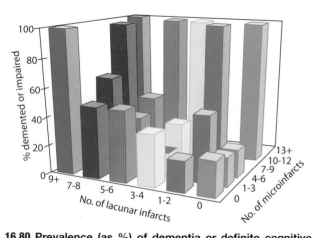

16.80 Prevalence (as %) of dementia or definite cognitive impairment related to numbers of lacunar infarcts and microinfarcts. The cases had negligible degree of other neurodegenerative lesions including AD pathology, Lewy bodies and hippocampal sclerosis. Data from 257 cases.

Courtesy of L White, University of Hawaii, Honolulu, HI, USA.

or microinfarcts than do stroke subjects with macroscopic infarcts and elderly subjects without dementia.[840] Lacunar infarcts (defined as cystic infarcts less than 1 cm in greatest dimension and visible on gross examination) and microinfarcts (infarcts only visible on microscopic examination) were the most common neuropathological features in more than 50 per cent of elderly patients with ischaemic VaD.[931] Cortical microinfarcts and subcortical lacunar lesions were also strong determinants of dementia in the Geneva brain ageing study.[297]

White Matter Disease

White matter changes predict increased risk of cerebrovascular events, stroke, cognitive impairment or dementia and death.[187,672] White matter hyperintensities on T2-weighted MRI or leukaraiosis as a decreased signal on CT is a neuroimaging construct to describe diffuse and focal WM changes. Leukoariaosis predominantly has reference to vascular disease and can encompass a range of abnormalities: WM rarefaction, incomplete infarction, lacunar strokes, enlargement of perivascular spaces, demyelination and axonal degeneration. Recent meta-analyses revealed a significant association between the burden of WM hyperintensities and occurrence of dementia of all types including incident VaD.[187] Both areas of leukoaraiosis and zones outside these lesions show decreased microvascular density indicating that leukoaraiosis is a feature of generalized CVD rather than a reflection of changes limited to the deep WM.[104] This is consistent with finding of an association between unstable carotid plaques and the number of WM lesions, suggesting a thromboembolic role in some patients with leukoaraiosis.[15] Diffusion tensor imaging[693] has demonstrated that structural changes in WM tracts are related to cognitive dysfunction. In addition to lacunar infarct count, altered diffusivity of water molecules in normal-appearing WM was an independent predictor of executive function in SVD. Altered radial diffusivity was a stronger predictor than axial diffusivity, possibly reflecting ischaemic demyelination as an important predictor of cognitive impairment in SVD.

Neuroimaging and pathological studies demonstrate that WM hyperintensities reflect degeneration of the WM that is mostly explained by SVD.[699,700] Diffuse and focal WM lesions are a hallmark of VaD[405] but also occur in approximately 30 per cent of AD and DLB cases.[236] There is some controversy as to whether deep or periventricular lesions are the more important but this depends on how the boundaries between the periventricular and deep WM are defined.[499] Lacunar infarcts are produced when the ischaemic damage is focal and of sufficient severity to result in a small area of necrosis whereas diffuse WM change is considered a form of rarefaction or incomplete infarction with selective damage to some cellular components. WM disease may comprise several patterns of alteration including pallor or swelling of myelin, loss of oligodendrocytes, axons and myelin fibres, foci of cavitation with or without presence of macrophages and areas of reactive astrogliosis.[832] Lesions in the WM also include spongiosis, i.e. vacuolation of WM and widening of perivascular spaces.[982] Affected regions do not have sharp boundaries, in contrast to the plaques of multiple sclerosis. It is argued that some of the WM damage in demented patients may simply reflect wallerian changes secondary to loss of cortical neurons. However, this is unlikely because histological changes characteristic of wallerian degeneration are not evident as WM pallor. Conversely, in AD patients with severe loss of cortical neurons similar WM lesions are often not apparent.[236] Neuroimaging investigations gauging signal enhancement after gadolinium propose that breakdown of the blood–brain barrier occurs in areas of leukoaraiosis and may mediate subsequent cellular changes.[937,938] These changes may be associated with chronic pro-thrombotic endothelial dysfunction in cerebral SVD,[364] also involving the WM.[101] The endothelial changes in SVD were shown to include upregulation of thrombomodulin.[300] Elevated immunoreactivity for hypoxia-inducible factors and neuroglobin is in keeping with a chronic hypoxic state in the deep WM,[249] which leads to upregulation of a range of genes involved in immune function, the cell cycle, proteolysis and ion transport, amongst others.[833]

Although WM changes focus on the arterial system, narrowing and, in many cases, occlusion of veins and venules by collagenous thickening of the vessel walls also occur. The thickening of the walls of periventricular veins and venules by collagen (collagenosis) increases with age, and perivenous collagenosis is increased further in brains with leukoaraiosis.[103] The presence of apoptotic cells in WM adjacent to areas of leukoaraiosis suggests that such lesions are dynamic, with progressive cell loss and expansion.[102] Vascular stenosis caused by collagenosis may induce chronic ischaemia or oedema in the deep WM leading to capillary loss and more widespread effects on the brain.[104,105]

Cerebral Microinfarction

The accumulation of small, even miniscule ischaemic lesions as an important substrate of VaD has been emphasized in recent years.[465] Microinfarcts are generally defined as small lesions visible only on microscopy (Figure 16.76 and Table 16.24). These lesions of up to about 5 mm diameter may or may not involve a small vessel at its centre but are foci with pallor, neuronal loss, axonal damage (in the WM) and gliosis. Their number in severely affected brains is estimated to reach into the thousands. Sometimes they may include regions of incomplete infarction or rarefied (subacute) change. Microinfarcts have been described as attenuated lesions of indistinct nature occurring in both cortical and subcortical regions. Such lesions or combination of these are reported as subtype II cerebrovascular pathology when they are multiple, with more than three in any region (Table 16.26). Microvascular infarcts (lacunar infarcts and microinfarcts) appear central to the most common cause of VaD (Figure 16.80) and predict poor outcome in the elderly.[51,931,954] In autopsied older Japanese–American men the importance of microvascular lesions as a likely explanation for dementia was nearly equal to that of Alzheimer lesions.[953,954] Microinfarcts in subcortical structures has been emphasized as a substrate of cognitive impairment,[35,465] and reported to correlate with increased Alzheimer type of pathology but cortical microinfarcts also appear to contribute to the progression of cognitive deficits with age.[498] In another study, a significant association was found between AD and microinfarcts in border zone (watershed) regions.[865]

Cortical Microinfarcts

In addition to microinfarcts in subcortical structures, there is increasing evidence that multiple cortical areas of microinfarction are associated with subcortical VaD or SVD (Figure 16.80). Thus these lesions should be taken into account when defining the neuropathological criteria. Cortical microinfarcts are more numerous in the presence of CAA.[685] In a recent study, cortical microinfarcts were frequently detected in AD and associated with CAA but rarely observed in subcortical VaD linked to SVD.[684,865] Microinfarcts in the cerebral cortex associated with severe CAA may be the primary pathological substrate in a significant proportion of VaD cases.[343] Cortical microinfarcts and to lesser extent periventricular demyelination were significantly associated with cognitive decline in individuals at high risk for dementia.[308] Haemodynamic changes, e.g. associated with hypotension and atherosclerosis, may play a role in the genesis of cortical watershed microinfarcts.

Cerebral Microbleeds and Haemosiderin

Cerebral microbleeds evident upon T2*-weighted MRI are small, dot-like hypointense abnormalities, thought to represent ferromagnetic haemosiderin iron derived from extravasated erythrocytes (Figure 16.76). Cerebral microbleeds detected by MRI were also associated with histopathological evidence of lipohyalinosis and CAA.[246] They are a surrogate MRI marker of small vessel disease, along with lacunes and WM changes.[917] The prevalence of radiological microbleeds in VaD ranges 35–85 per cent. Microbleeds are thought to result mainly from hypertensive vasculopathy, but their frequent co-occurrence with lobar haemorrhage suggests that CAA is also of importance, and they are also common in AD.[950] Microbleeds associated with hypertension are usually seen in the deep nuclei and brain stem and those with AD are generally lobar in location.[158,312] The relevance of this radiological construct is increasingly recognized because of their relation to clinical outcome and occurrence in anti-Aβ immunization trials.[328] The presence of multiple microbleeds in the context of VaD is related to worse performance on cognitive tests, mainly in psychomotor speed and executive functioning. Because microbleeds are common in cognitively normal older individuals, attribution of VaD to these should follow a careful exclusion of other causes of cognitive impairment and the diagnosis made only if numerous such lesions are present.

Both radiological cerebral microbleeds and foci of haemosiderin containing single crystalloids or larger perivascular aggregates are found in brains of older subjects including those diagnosed with VaD and AD but the radiological and pathological relationship between these findings has not been entirely clear. Recent evidence suggests that cerebral microbleeds detected by MR imaging are a surrogate for ischaemic SVD rather than exclusively a haemorrhagic manifestation.[425] Accumulation of haemosiderin in the putamen was significantly associated with indices of small vessel ischaemia including microinfarcts, arteriolosclerosis and enlarged perivascular spaces and with lacunes in any brain region but not large vessel disease or whole-brain measures of neurodegenerative pathology. Higher levels of putamen haemosiderin correlated with more 'microbleeds' upon MR imaging but it is possible that the imaging abnormalities may partly reflect altered brain iron homeostasis and small vessel ischaemic change rather than serving solely as a marker of minor episodes of cerebrovascular extravasation.

Hippocampal Atrophy and Sclerosis

Neuroimaging studies have shown that medial temporal lobe and hippocampal atrophy are associated with SVD[694,915] and VaD,[58,257] albeit not to the same extent as in AD.[113] In pathological studies too, ischaemic VaD and SVD have been associated with hippocampal changes and atrophy remote from ischaemic injury.[595,993] Pyramidal neurons in Sommer's sector are highly vulnerable to disturbances in the cerebral circulation or hypoperfusion caused by systemic vascular disease. Focal loss of CA1 neurons in ischaemic VaD was associated with lower hippocampal volume and memory score[993] but the degree of loss was less in VaD[502] than AD. Selective hippocampal neuronal atrophy may be seen in post-stroke dementia.[283] Thus there is a clear vascular basis for hippocampal neurodegeneration, in keeping with neuroimaging observations of hippocampal atrophy even in population-based incident VaD.[808] The simplest mechanistic explanation for the atrophy is it reflects loss of synaptic terminals or dendritic arbour and consequent loss in connectivity, which contributes to brain structural and functional changes. This is consistent with the finding that soluble synaptophysin was decreased in VaD as well as AD.[997]

Hippocampal sclerosis is probably a major contributing factor to the hippocampal atrophy. Any focal loss or patterns of hippocampal sclerosis should be graded[747] and recorded together with a note of any hippocampal microinfarcts. Sometimes the pyramidal neuron loss and gliosis in brains with AD pathology may be difficult or impossible to distinguish from ischaemic change. The aetiology of hippocampal sclerosis in dementia should be defined as in association with a neurodegenerative process, or as lacking neurodegenerative markers, in which case it is presumed to arise from ischaemic mechanisms.[747] There is a correlation between hippocampal sclerosis and arteriolosclerosis in multiple brain regions outside of the hippocampus including the frontal cortex (BA 9).[663] This is ascribed to a pathogenetic change in aged human brain arterioles that impacts multiple brain areas and contributes to hippocampal sclerosis of ageing.[666]

Border Zone (Watershed) Infarcts and Incomplete Ischaemic Injury

Border zone or watershed infarcts mostly result from haemodynamic events, usually in patients with severe internal carotid artery stenosis. They could occur bilaterally or unilaterally, and affect regions between main arterial territories in deep and superficial arterial systems. Typical borderzone infarcts are wedge-shaped regions of pallor, rarefaction and gliosis, measuring 5 mm or more across and extending through the cortex into the WM. Larger areas of incomplete infarction may extend well into the WM.[405] These are characterized by mild to moderate loss of oligodendrocytes, myelin and axons in areas where there may be hyalinized

vessels.[106] This may be accompanied by astrogliosis, some microgliosis and macrophage infiltration. The morphology of incomplete or subinfarctive changes, though suspected to be associated with impaired cognitive function, is not consistently described in VaD. It may variably manifest as tissue rarefaction assessed by conventional stains and revealed as injury responses such as microgliosis and astrocytosis, or the presence of other 'reactive' cells or surrogate markers of dendritic, synaptic or axonal damage.

Laminar Necrosis

Laminar necrosis (see Chapter 2) is most often seen in dementia after an episode of global ischaemia or hypoperfusion, e.g. after cardiac arrest. The typical topographic distribution of spongiosis and gliosis is readily apparent on histology with standard stains. If ischaemia has been incomplete, the lesions may be accentuated at, or restricted to, the arterial border zones[106,253] and may fall into the subtype IV of VaD pathology (Table 16.26).

Sporadic Cerebral Amyloid Angiopathy and Vascular Dementia

CAA is most common in AD but it often occurs in CVD in the general absence of Alzheimer pathology.[151] In sporadic cases, CAA alone is considered a substrate for cognitive impairment.[36,327,717] Tissue damage caused by CAA in the absence of intracerebral haemorrhage is independently associated with cognitive impairment.[932] The prevalence of CAA in VaD (Figure 16.79) is not known, but it is a major cause of intracerebral and lobar haemorrhages leading to profound ischaemic damage.[933] In a study of surgical biopsies exhibiting cerebral and cerebellar infarction, CAA was significantly more common in samples showing infarction than in age-matched controls with non-vascular lesions.[115] There is also an association between severe CAA and cerebrovascular lesions coexisting with AD, including lacunar infarcts, microinfarcts and haemorrhages.[233,685-688] This association apparently is not attributable to apolipoprotein E (ApoE) ε4 allele, as the vascular lesions correlated best with severity of CAA, regardless of ApoE genotype.[689] There is also some evidence to suggest CAA is related to WM changes but possibly not in the oldest of the old.[872] It appears that a first stroke-like episode in association with CAA triggers multiple cerebral bleeds, and that preceding diffuse WM changes lead to rapid decline of cognitive function.[326,327,658,931] Aβ protein accumulation within or adjacent to the vasculature may lead to the degeneration of vascular cells in both larger perforating arterial vessels as well as cerebral capillaries that represent the blood–brain barrier, leading to vascular dysfunction and causing cortical microinfarcts[342,684] and perivascular cavities similar to those associated with arteriolosclerosis in subcortical structures. Genetic factors such as the *APOE* ε4 allele (associated with increased severity of CAA), may promote or attenuate perfusion of the WM.[459] Several familial forms of CAA involving ischaemic and haemorrhagic infarcts and cerebral hypoperfusion further demonstrate the link between CAA and VaD.

Neurochemical Pathology of VaD and VCI

The neurochemical basis of cognitive decline in CVD is poorly understood. There have been few studies on the protein and lipid chemistry of VaD. Various cellular signalling and regulatory pathways including those relating to apoptosis, autophagy, oxidative stress and inflammation are altered in VaD by virtue of their involvement in cerebral ischaemia or oligaemia (see Chapter 2). In one of the first studies of its kind, Mulugeta et al.[641] reported that VaD patients exhibit a selectively attenuated neuroinflammatory response. The authors found that the monocyte chemoattractant protein-1 and interleukin-6 concentrations were significantly reduced in the frontal lobe in both VaD and mixed dementia.

The perivascular nerve plexus[351] would be expected to be vulnerable in VaD yet few neurotransmitter changes have been described in VaD. Selective changes in neurotransmitters were reported in some cases of VaD.[458] Two different groups had previously shown that choline acetyltransferase activity was reduced, albeit to a lesser degree than in AD, in the temporal cortex and hippocampus in multi-infarct dementia or VaD.[318,707] Choline acetyltransferase activity was significantly reduced (by 60–70 per cent) in frontal and temporal cortex in cerebral autosomal dominant arteriopathy with subcortical infarcts and leukoencephalopathy (CADASIL), which models SVD.[476] Choline acetyltransferase and P75 (neurotrophin receptor) immunoreactivities within cholinergic cell bodies in the basal forebrain in CADASIL were also affected, but these changes were not so pronounced. The extent of alteration in neurotransmitter levels may depend on severity of the WM degeneration.[476,575] Loss of cholinergic function is consistently greater in VaD patients with concurrent Alzheimer pathology.[779] Conversely, an unexpected increase in cholinergic activity in the frontal cortex (BA 9) was found in infarct dementia.[822] Other studies have reported deficits in monoamines including dopamine and 5-hydroxytryptamine (5-HT) in the basal ganglia and neocortex in VaD.[318] Possibly as a compensatory reaction to loss of 5-HT innervation,[232] 5-HT(1A) and 5-HT(2A) receptors were reportedly increased in the temporal cortex in multi-infarct but not subcortical VaD. Such findings, albeit fragmentary, indicate distinctions between the neurochemical pathology of VaD subtypes and suggest possibilities of pharmacological manipulation in VaD. There was also loss of glutamatergic synapses, assessed by measurement of vesicular glutamate transporter 1 concentration, in the temporal cortex in VaD[481] but preservation in the frontal cortex.

Hereditary Cerebral Amyloid Angiopathies

There are more than ten different hereditary cerebral amyloid angiopathies, caused by mutations in different genes (see Chapter 2). All these angiopathies lead to some degree of cognitive impairment or dementia. They are characterized by multiple haemorrhages and haemorrhagic or ischaemic infarcts in addition to severe amyloid deposition within walls of the meningeal and intracerebral vessels. In hereditary cerebral haemorrhage with amyloidosis of the Dutch type, dementia occurs in most patients surviving their initial stroke[336] and may occasionally be the presenting

feature.[335,941] The extensive CAA is alone sufficient to cause dementia and this has implications for CAA-related cognitive dysfunction in sporadic CAA and AD.[658] In the Icelandic type of hereditary cerebral haemorrhage with amyloidosis (HCHWA-I), which is associated with a point mutation in the gene encoding the cysteine protease inhibitor cystatin C,[533] dementia, occurring in some patients, has been attributed to the multiple vascular lesions. Individuals with gelsolin-related amyloidosis manifest facial palsy, mild peripheral neuropathy and corneal lattice dystrophy, atrophic bulbar palsy, gait ataxia and mild cognitive impairment.[483]

Familial British dementia with amyloid angiopathy is an autosomal dominant condition associated with a point mutation in the integral membrane protein 2B gene, *ITM2B*, located on chromosome 13q14.[927] The disease is characterized by dementia, progressive spastic tetraparesis and cerebellar ataxia. Onset is usually in the sixth decade.[292,612] Neuropathological features include Alzheimer-type neurofibrillary tangles and neuropil threads in the anteromedial temporal lobe (i.e. hippocampus and entorhinal cortex), predominantly non-neuritic and perivascular amyloid plaques, and widespread vascular amyloid deposits in the small cerebral and spinal arteries (Figure 16.81).[725,753] Cerebellar amyloid deposits are also common. The amyloid is composed of a highly insoluble 24 amino acid subunit (ABri) distinct from other known amyloid molecules.[293,927] A single base substitution at the stop codon of the *ITM2B* gene in affected family members leads to production of an abnormally longer ITM2B protein, the ABri precursor. Evidence suggests that the proprotein convertase, furin, mediates processing of the ABri precursor to 4-kDa ABri, which has inherent fibrillogenic properties.[478]

Familial Danish dementia (FDD), also known as heredopathia ophthalmo–oto–encephalica, is another condition with severe and widespread CAA. Familial Danish dementia is associated with a 10-nucleotide duplication between codons 265 and 266 of *ITM2B*, the same gene that causes familial British dementia.[928] The duplication results in an abnormally long precursor protein to a unique 4-kDa amyloidogenic peptide referred to as ADan. Like ABri, it is generated from the abnormal precursor protein through cleavage by furin. This autosomal dominant disorder is characterized clinically by cataracts, deafness, progressive ataxia and dementia. The neuropathological features include amyloid deposits and CAA composed of ADan, as well as Alzheimer-type neurofibrillary tangles, similar to those in familial British dementia (Fig 16.81).

Hereditary Small Vessel Disease and Dementia

Early reports suggest the existence of several familial stroke disorders unrelated to atherosclerotic disease, which lead to cognitive impairment or dementia (see Chapter 2). Most of these disorders can be classed as SVD involving small vessels in subcortical structures.[467] The characteristic pathological features and the pathogenesis of the hereditary SVDs of the brain are described in Chapter 2. CADASIL is the most common of the hereditary SVDs leading to cognitive decline and dementia.[132] Motor deficits, ataxic hemiparesis, hemianopsia and dysarthria are the most common

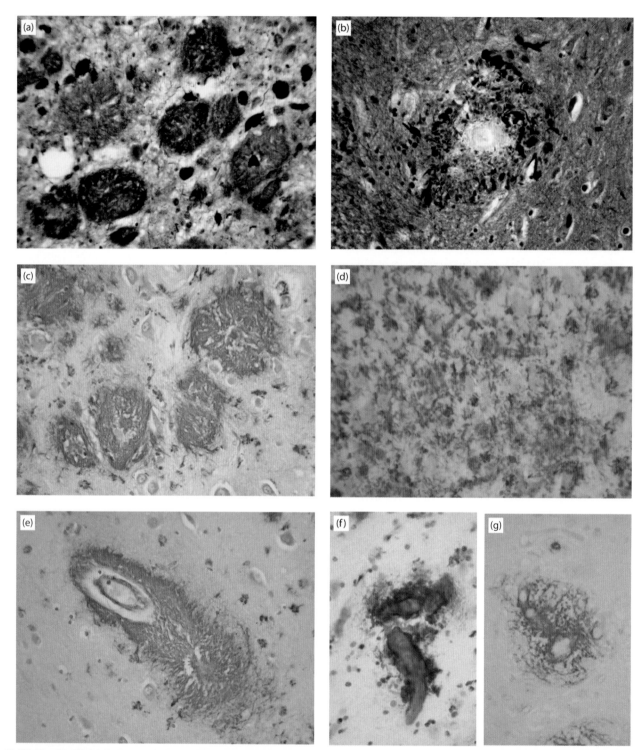

16.81 Familial British dementia (a,c,e,h) and familial Danish dementia (b,d,f,g,i). (a) Large argyrophilic plaques in the hippocampus. **(c)** An antibody recognizing ABri highlights the parenchymal amyloid plaques and **(e)** blood vessels with cerebral amyloid angiopathy and perivascular ABri deposits. **(b)** In familial Danish dementia, abnormal argyrophilic neurites cluster around blood vessels with amyloid deposition, but parenchymal plaques are not a feature. **(d)** An antibody recognizing the amyloid protein specific for the Danish mutation (ADan) demonstrates ill-defined diffuse parenchymal plaques and also **(f)** cerebral amyloid angiopathy due to ADan deposition. **(g)** In cases with familial Danish dementia, co-deposition of Aβ peptide has also been observed.

Courtesy of Professor T. Revesz, Institute of Neurology, London, UK.

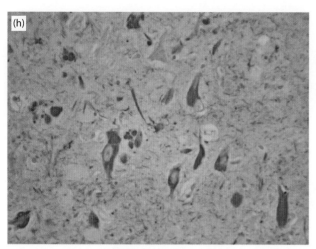
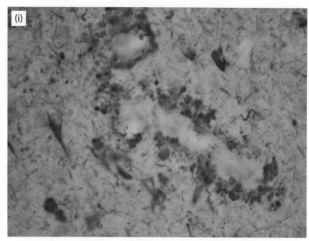

16.81 (Continued) Familial British dementia (a,c,e,h) and familial Danish dementia (b,d,f,g,i). (h) Neurofibrillary tangles, neuropil threads and amyloid plaque-associated abnormal neurites are present in the hippocampus. **(i)** Similar to familial British dementia, neurofibrillary tangles and neuropil threads are numerous in the hippocampus, but abnormal neurites are located only around blood vessels with cerebral amyloid angiopathy. (a,b Bielschowsky's silver impregnation; c,e ABri immunohistochemistry; d,f ADan immunohistochemistry; g Aβ immunohistochemistry; h,i tau immunohistochemistry.)

Courtesy of Professor T. Revesz, Institute of Neurology, London, UK.

presenting neurological features. Vascular changes, including apoptotic loss of brain vascular smooth muscle cells[324] and vessel wall thickening,[165] reduce blood flow and the vasodilatory response to cause lacunar infarcts and induce cognitive deficits, which progress to subcortical VaD.[983] Increasing age is the strongest risk factor for dementia even in CADASIL patients (Figure 16.82). Similar to that for subcortical VaD, the cognitive profile in CADASIL during the early stages primarily comprises impairments in attention, processing speed, working memory and aspects of executive function.[108,135,713] More than 75 per cent of patients, particularly men are demented by age 60, with visuospatial and reasoning abilities deteriorating soon after. Episodic memory and language are mostly preserved until the late stages. Difficulties in retrieval of memories rather than impairment of the encoding process is a distinctive feature. Deficits in verbal fluency and ideational praxis become apparent but recall, orientation and receptive language skills are largely preserved. The extensive demyelination and axonal damage in the underlying WM contributes to cortical atrophy.[166] The alterations in frontal lobe cognitive function in CADASIL are consistent with the disconnection of the fronto-subcortical circuits. Neuronal apoptosis, predominantly in neocortical layers III and V, also probably contributes to dementia in CADASIL.[324]

A few other CADASIL-like disorders lead to cognitive impairment.[983] The Maeda syndrome or cerebral autosomal recessive arteriopathy with subcortical infarcts and leukoencephalopathy (CARASIL) is an autosomal recessive disorder similar to CADASIL.[358] Patients are normotensive but exhibit severe arteriopathy, leukoencephalopathy and lacunar infarcts as well as spinal anomalies and alopecia. Strokes lead to stepwise deterioration with most subjects becoming demented in older age. Familial cerebral SVDs involving progressive visual impairment[466] cause deterioration in cognitive function. Hereditary endotheliopathy with retinopathy, nephropathy and stroke, cerebroretinal vasculopathy and hereditary vascular retinopathy were

reported independently but represent different phenotypes in the same disease spectrum.[440,691,883] These diseases, now described as autosomal dominant retinal vasculopathy with cerebral leukodystrophy, cause dementia and lead to early death.

Some rarer and less well characterized hereditary SVDs are also associated with different degrees of cognitive impairment. One of these is pseudoxanthoma elasticum, a hereditary connective tissue disorder with abnormalities in the skin and eye and multiple lacunar infarcts in deep WM and pons.[705] Others include hereditary infantile hemiparesis, retinal arteriolar tortuosity and leukoencephalopathy[319,913] and a novel autosomal dominant SVD of the brain in a large Portuguese-French family manifesting with motor hemiplegia, memory deficits, executive dysfunction and WM changes on MRI in the general absence of vascular risk. The genetic abnormality is not fully penetrant.[926] There also exists another hereditary vascular leukoencephalopathy in which main clinical manifestations include gait disturbances, transient movement disorders, stroke, and cognitive dysfunction.[369a] This disorder is characterized by a progressive, age-related hemispheric and brain stem leukoencephalopathy, the severity of which contrasts with the paucity and late onset of clinical symptoms in most cases.

Experimental Studies and Animal Models of VaD

Experimental studies in animals have enhanced our understanding of the pathogenesis of VaD. The utility of various animal models of CVD in relating the neuropathological and cognitive features has been systematically analyzed.[42,444] Intravascular surgical and emboli models have explored the molecular and cellular sequelae of focal and global ischaemia and hypoperfusive (oligaemic) injury in rats, mice and gerbils. To model SVD and white matter disease in man, rats and mice have been subjected to bilateral

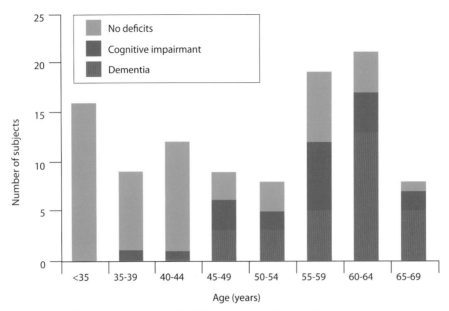

16.82 Numbers of genetically or biopsy confirmed CADASIL subjects with cognitive impairment or dementia with increasing age. Proportions of demented subjects increase with age even in autosomal dominant forms of SVD.

Courtesy of M Dichgans, The Ludwig Maximilian University of Munich, Germany.

carotid artery occlusion or stenosis.[404] Overall, these have demonstrated that chronic hypoperfusion causes blood–brain barrier disruption, glial activation, oxidative stress, oligodendrocyte loss and diffuse WM injury. To replicate more precisely gradual narrowing of the common carotid arteries, a two-vessel occlusion model in Wistar–Kyoto rats was developed using a device known as the ameroid constrictor.[482] Chronic cerebral hypoperfusion induced by the constrictor caused selective white matter changes with relatively preserved neurovascular coupling and mild to moderate metabolic and histological derangements in the grey matter including the hippocampus. The rats developed significant spatial working memory impairment, which mimics cognitive impairment associated with selective white matter damage in human VaD.

The role of vascular risk factors in VaD, particularly hypertension, diabetes and hyperhomocysteinaemia, has also been investigated in animals, including primates and stroke-prone or spontaneously hypertensive rats, to assess the relationship of pathological changes such as hippocampal neuronal loss, microvascular lesions, microhaemorrhages and diffuse white matter disease, to spatial memory and learning paradigms. Chronic hypertensive models such as the stroke-prone spontaneously hypertensive rat[43] mimic most of the features of SVD including arteriolar wall thickening, lacunar stroke, WM lesions, enlarged perivascular spaces, cortical infarcts and microhaemorrhages and endothelial changes, but do not entirely capture the sequelae of events evident in chronic hypertensive disease in man. However, one of the recently developed hypertension models, the Cyp1a1-Ren2 transgenic rat, has shown that modest, sustained hypertension is sufficient to cause cerebrovascular alterations accompanied by endothelial and inflammatory changes and that these parallel alterations in growth factor expression.[698] Consistent with the SVD phenotype, hypoperfusion induced by bilateral

carotid artery stenosis causes diffuse axonal changes and myelin pathology, which is evident on MRI by altered fractional anisotropy and magnetization transfer ratios. MRI is also a sensitive measure of diffuse, subtle WM changes in wild type mice.[386] Several transgenic mouse models have been developed to explore pathological features of VaD. The main ones include APP-overexpressing mice to study CAA,[685] and CADASIL-associated *NOTCH3* mutants[454] to evaluate the pathogenesis of CADASIL focusing on the functional significance of disease-linked mutations and the earliest pathological events that initiate brain changes. The studies have provided important insights in showing that cerebrovascular dysfunction and hypoperfusion precede WM changes and that the CADASIL phenotype likely involves a gain of novel or toxic function of NOTCH3.[454] Although there is marked heterogeneity in the phenotypic spectrum, possibly due to genetic background differences among mouse models, and obvious species differences between mouse and man,[526] the animal models have been useful in partially reproducing and allowing investigation of the pathophysiology of SVD.

NORMAL PRESSURE HYDROCEPHALUS

Clinically, idiopathic normal pressure hydrocephalus (iNPH) is characterized by the triad of gait disturbance, dementia and urinary incontinence. Imaging shows enlarged ventricles with the severity of ventricular enlargement out of proportion to the degree of cortical atrophy (distinguishing this from ventricular enlargement in a neurodegenerative disease). These changes are reflected in the gross appearance of the brain (Figure 16.83). CSF studies show normal or slightly elevated CSF pressure.

The dementia is manifested by slowing, poor concentration and reduced capacity for abstract thought, although memory may be relatively well preserved.[109] Imaging studies show that based on evaluation of ventricular size, the prevalence of probable iNPH was just under 6 per cent in those aged 80 years and older.[429] In a Japanese study, the incidence of iNPH in those over 70 years was estimated at 1.2 per 1000 persons per year. This suggests that there are preclinical states before the development of iNPH.[412] Longitudinal evaluation over 6 months has documented progression of clinical features and the recommendation that treatment should be implemented promptly.[24,899]

The major differential diagnosis of iNPH is AD. Although the iNPH condition remains a diagnostic and therapeutic challenge, some patients benefit from shunt surgery, with reversal of dementia. Deciding which patients may benefit from shunt surgery is critical. Clinical investigators have used a variety of measures to assist in predicting response.[294] CSF biomarker studies have been examined as a way of distinguishing iNPH from AD. CSF NFL neurofilament protein is elevated and APP-derived proteins and tau proteins are lower in patients with iNPH compared to normal.[441]

Neuropathological features of iNPH are poorly documented. A post-mortem study of ten patients treated for iNPH and a review of published cases showed a high frequency of vascular lesions that suggested this as an important component of pathology leading to cognitive impairment.[531] Brain and meningeal biopsies have been performed at the time of shunt insertion and the published literature on findings has been reviewed.[531] Vascular changes or findings consistent with AD are commonly observed. AD pathology has not been found to relate to survival.[530] AD is a common pathological comorbidity in patients with iNPH and this may be a factor in determining whether there is improvement in cognitive function following treatment by CSF shunting.[114]

The pathogenesis of NPH is uncertain. Recent reports have suggested the existence of a familial subgroup.[176,602,869] A recent hypothesis links age-related insufficiency of CSF production or elimination to the development of NPH. The concept of CSF circulatory failure might explain NPH whether pathological processes lead to resistance to CSF flow.[829]

ALCOHOL AND DEMENTIA

Ethanol abuse is recognized to cause cognitive impairment with impaired executive functioning and impaired memory that generally improve with abstinence. Chronic ethanol abuse may be associated with a clinical dementia syndrome. These are alcohol-related dementia and alcohol-induced persisting amnestic syndrome (Wernicke–Korsakoff syndrome). The neuropathological lesions associated with chronic ethanol abuse have been reviewed[183] and are presented in Chapter 9.

Clinical criteria have been suggested for the diagnosis of alcohol-related dementia[692] and recently reviewed.[760] In some patients with frontotemporal dementia, alcohol abuse may be symptomatic of loss of frontal lobe function, as a reflection of disinhibition. Neuropathological assessment of cases of suspected alcoholic dementia for features of a FTLD may therefore be appropriate to exclude this possibility.

CHRONIC TRAUMATIC ENCEPHALOPATHY

Dementia may be a consequence of traumatic brain injury.[827] This condition was originally termed dementia pugilistica but the term chronic traumatic encephalopathy is now used. The clinical cognitive and motor changes of chronic traumatic encephalopathy are associated with frontal and temporal lobe atrophy, with tau-related and TDP-43-related pathology.[191,605] The pathology of chronic traumatic encephalopathy is presented in Chapter 10.

RARE NEURODEGENERATIVE DISORDERS

Familial Encephalopathy with Neuroserpin Inclusion Bodies

Neuroserpin belongs to the serine protease inhibitor superfamily. It is mainly but not exclusively expressed by neurons within the central nervous system.[621]Point mutations in the gene for neuroserpin, *PI12*, on chromosome 3q26 lead to an abnormal protein conformation which results in aberrant intermolecular linkage and polymer formation.

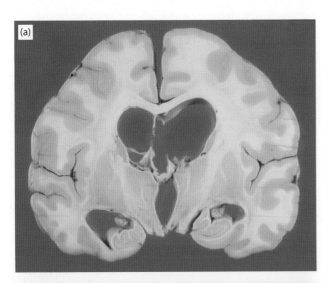

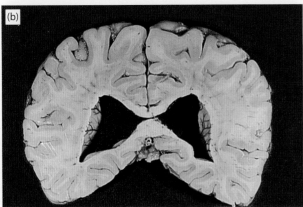

16.83 Normal pressure hydrocephalus (a) at level of third ventricle and (b) occipital lobes. Ventricles are enlarged without significant atrophy of the cortex, deep grey or white matter.

Subsequent accumulation of polymerized neuroserpin within the endoplasmic reticulum of neurons leads to neurodegenerative disease.[181,622] Cognitive changes typically develop after the age of 40 years, with a pattern of frontoparietal dysfunction and relative sparing of memory early in disease. Patients may also have myoclonic epilepsy. Clinical–genetic correlation suggests that certain mutations are associated with extensive neuronal accumulation of neuroserpin, a more severe clinical phenotype and earlier clinical onset of disease. The S49P mutation is associated with dementia after age 45 years, and neuroserpin inclusions in a small proportion of neurons. In contrast, the G392E mutation is associated with progressive myoclonus epilepsy with onset at the age of 13 years and inclusions in almost all neurons.[182]

The characteristic histological finding is the presence of numerous PAS-positive eosinophilic bodies (so-called 'Collins bodies') within neuronal cell bodies and processes (Figure 16.84); the inclusions are immunoreactive for human neuroserpin.[96] Ultrastructural studies show entangled fibrils that are similar to fibrils that occur in other serpin disorders, such as those in α1-antitrypsin deficiency.[181]

Heredodegenerative and Metabolic Causes of Early Onset Dementia

Although the clinical syndrome of dementia is usually seen in the elderly, a significant minority of cases presents before the age of 60 years. It has been estimated that in the UK, the prevalence of dementia with onset between the ages of 30 and 65 years was 54 per 100 000.[789]

This patient group is of interest as the spectrum of diseases encountered is different from that in the elderly

population.[152] There is an increased proportion of inherited causes, including familial AD and FTLD, and an increased proportion with neurometabolic disorders (Table 16.27).

An approach to diagnosis of young-onset dementia has been reviewed, including discussions of clinical features and differential diagnoses to be considered with comprehensive summary tables.[506,586,759,789,803]

DIAGNOSTIC NEUROPATHOLOGY CONSIDERATIONS

Brain Biopsy in Dementia

Brain biopsy may be considered in patients who develop clinical features of dementia. The decision to perform a biopsy must balance the clinical risk of an invasive procedure in what is usually a frail patient against the likely benefit of establishing a diagnosis that will lead to a specific treatment. Advances in diagnostic techniques, for example the ability to diagnose certain infective diseases such as Whipple's disease by PCR, have reduced the need for biopsy in certain areas of differential diagnosis. The indications for brain biopsy in dementia would be profoundly altered should any specific disease-modifying therapy be developed for one of the neurodegenerative diseases in the absence of any other diagnostic test for the condition.

In current practice, three main conditions need to be satisfied before a brain biopsy is undertaken in a patient with dementia.

i. It is clinically plausible that the patient under consideration has a disease of the brain that may respond to a specific therapy and no other diagnostic procedure is appropriate. Given that there are no specific treatments for neurodegenerative diseases, the range of diagnoses in this group mainly come down to inflammatory (especially neurosarcoidosis), immune-mediated (especially isolated vasculitis of the brain), infective or neoplastic causes of dementia.

ii. Administration of empirical treatment for a suspected diagnosis in itself might cause harm. For example,

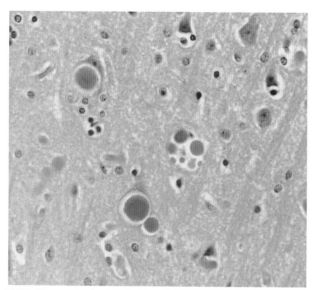

16.84 Neuroserpin mutation S52R. Numerous intraneuronal eosinophilic bodies are seen within the cerebral cortex in a patient with progressive myoclonus epilepsy, dementia and neuroserpin mutation S52R.

Case material provided by Dr Bernardino Ghetti, Indiana University, Indianapolis, IN, USA.

TABLE 16.27 Degenerative diseases that cause dementia and usually present in childhood
Lafora body disease
Mitochondrial cytopathy
Neuronal ceroid lipofuscinosis (Batten's disease)
Neuronal intranuclear inclusion disease
Wilson's disease
Adrenoleukodystrophy
Fabry's disease
Krabbe's disease
Alexander's disease
Nasu–Hakola disease
Cerebrotendinous xanthomatosis
GM1 gangliosidosis type III
GM2 gangliosidosis
Gaucher's disease
Niemann–Pick disease type C
Mucopolysaccharidosis type IIIB

giving immunosuppressive treatment when there is a possibility of an infection.

iii. The balance between the risks of an invasive procedure and the benefit of obtaining a diagnosis is in the patient's favour.

In a comprehensive recent review of the role of brain biopsy in the evaluation of patients with dementia, it was noted that in 90 cases sampled up to 2003 a specific diagnosis was achieved in about 57 per cent of brain biopsies. The main finding in non-diagnostic cases was non-specific gliosis (37per cent). Treatment was altered because of biopsy findings in 11 per cent of cases. In contrast, in biopsies taken more recently a diagnosis was achieved in 74 per cent, the most common diagnosis being CJD.[815] The precise figure one might cite in counselling a patient is difficult to determine as in many reported series some biopsies were performed to confirm a diagnosis rather than to investigate a patient when there was complete diagnostic uncertainty.

Brain biopsy in a patient with dementia should be undertaken only after close discussion between the clinical, imaging and neurosurgical teams and the neuropathologist(s). A scheme for patient investigation prior to suggesting a brain biopsy has been proposed.[815] This involves imaging, microbiological and serological testing for a range of diseases, followed by CSF sampling and studies for inherited metabolic disorders. The main categories for exclusion remain as neoplastic infiltration, inflammatory disorder or infective disorder. The type of biopsy taken may depend on site. Deep white matter lesions are virtually only accessible by stereotactic needle biopsy. Where cortical pathology is suspected, the biopsy should include meninges (with vessels) cortex and underlying white matter. A biopsy that promises a good diagnostic yield is typically about 1 cm^3 in volume.

The sample should be received fresh in the laboratory with a portion preserved for microbiological investigation or molecular investigations, or both. A small sample of cortex and white matter should be fixed in a glutaraldehyde-based fixative for electron microscopy. Formalin immersion of the main sample should allow for adequate fixation.

In view of potential risk of unsuspected prion disease in some cases, initial handling of the sample should be in accordance with local health and safety guidelines, and might include treatment of the fixed tissue with formic acid. Diagnostic evaluation of the sample should include immunohistochemical stains for Aβ and for neuronal and glial inclusions, in addition to routine histology and special stains (e.g. for microorganisms) as indicated by the initial findings and the clinical setting. Where a neurodegenerative disease is suspected on initial evaluation, following a diagnostic strategy as defined for the autopsy investigation of dementia is appropriate.

An Approach to Pathological Diagnosis

Review of the medical records of a patient who has died of dementia provides insight into the nature of clinical progression, identifies anatomical regions linked to the pattern of cognitive decline and assists in planning histological sampling. Moreover, it plays an important role in formulation of a clinicopathological summary. A number of questions should be considered, such as:

- How was the diagnosis of dementia made? Has the assessment been made by a clinician experienced in dementia? Have causes of secondary dementia been excluded? Has there been longitudinal assessment of the patient with application of bedside tests of cognitive function? It is not uncommon for a diagnosis of dementia to be applied to an elderly patient, who is cognitively impaired because of an acute problem and is therefore best classed as having an acute confusional state. Depression can also lead to poor global performance and is a recognized cause of pseudo-dementia.
- Which domains of cognition were affected first? At the end stage of disease it can be clinically difficult to discriminate between different diseases. The early clinical features obtained from medical records often give important clues to the eventual pathological diagnosis. The pathologist should try to distinguish between the various clinico–anatomic syndromes described in this chapter.
- Are there neurological features associated with the decline in cognitive function that can be a clue to the cause of dementia? (Table 16.28).
- Has any diagnostic imaging been performed to assist with diagnosis?

Strategies for the staged examination of the postmortem brain in a suspected dementia have evolved over time with the increasing use of immunohistochemical and molecular tools for diagnosis and the expanding range of diagnostic categories. A tissue sampling strategy for evaluation has been recommended and is freely available online.[402,629] The algorithm shown in Figure 16.85 is a suggested route for clinical diagnostic practice, providing a staged examination and appropriate application of investigations.

TABLE 16.28 Clinical presentations and possible diagnosis of dementia

Clinical presentation	Diagnostic considerations
Dementia with myoclonus	Prion disease Autosomal dominant AD
Dementia with ataxia	Inherited forms of ataxia including SCA2, SCA3, SCA17, DPRLA
Dementia with chorea	Huntington's disease SCA3, SCA17, DPRLA, neuroferritinopathy, neuroacanthocytosis
Dementia with dystonia	Wilson's disease Niemann–Pick disease (NPC1 and NPC2)
Dementia with progressive myoclonic epilepsy	Mitochondrial disease, Lafora body disease, Neuronal ceroid lipofuscinosis

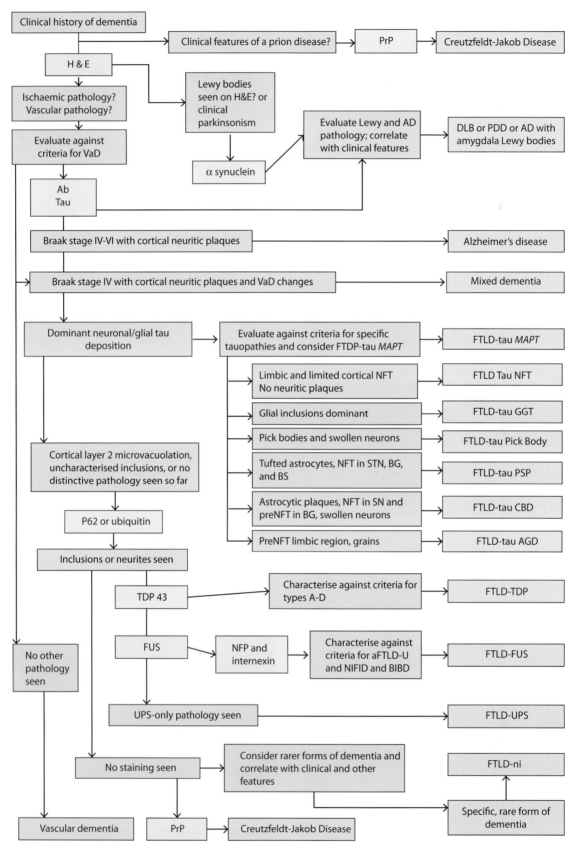

16.85 Diagnostic strategy for the pathological diagnosis of dementia. AD, Alzheimer's disease; aFTLD-U, atypical frontotemporal lobar degeneration with ubiquitin pathology; AGD, argyrophilic grain disease; CBD, corticobasal degeneration; DLB, dementia with Lewy bodies; FTLD, frontotemporal lobar degeneration; FTLD-ni, FTLD with no inclusions; FUS, fused-in-sarcoma protein; GGT, globular glial tauopathy; MAPT, microtubule-binding protein tau mutation; NFP, neurofilament protein; NFT, neurofibrillary tangles; PD, Parkinson's disease; PDD, Parkinson's disease dementia; PrP, prion protein; TDP43, TAR DNA-binding protein 43; UPS, ubiquitin-proteasome system; VaD, vascular disease.

ACKNOWLEDGMENTS

Although much of this chapter is new, the authors acknowledge previous contributors for their scholarship and generosity to permit re-use of some material and especially Bradley Hyman, Susanne Mirra and Dennis Dickson.

JL has been supported by Alzheimer's Research Trust and the Medical Research Council, UK.

RNK's work is supported by the RCUK Newcastle Centre for Brain Ageing and Vitality, Medical Research Council (UK), Alzheimer's Research UK and the Newcastle National Institute for Health Research Biomedical Research Centre in Ageing and Age-Related Diseases, Newcastle upon Tyne Hospitals National Health Service Foundation Trust.

REFERENCES

1. Aarsland D, Ballard CG, Halliday G. Are Parkinson's disease with dementia and dementia with Lewy bodies the same entity? *J Geriatr Psychiatry Neurol* 2004;**17**:137–45.

2. Aarsland D, Perry R, Brown A, Larsen JP, Ballard C. Neuropathology of dementia in Parkinson's disease: a prospective, community-based study. *Ann Neurol* 2005;**58**:773–6.

3. Aguzzi A, Barres BA, Bennett ML. Microglia: scapegoat, saboteur, or something else? *Science* 2013;**339**: 156–61.

4. Ahmed Z, Bigio EH, Budka H, Dickson DW, Ferrer I, et al. Globular glial tauopathies (GGT): consensus recommendations. *Acta Neuropathol* 2013;**126**:537–44.

5. Akatsu H, Takahashi M, Matsukawa N, Ishikawa Y, Kondo N, et al. Subtype analysis of neuropathologically diagnosed patients in a Japanese geriatric hospital. *J Neurol Sci* 2002;**196**:63–9.

6. Akinyemi RO, Mukaetova-Ladinska EB, Attems J, Ihara M, Kalaria RN. Vascular risk factors and neurodegeneration in ageing related dementias: Alzheimer's disease and vascular dementia. *Curr Alzheimer Res* 2013;**10**:642–53.

7. Ala TA, Beh GO, Frey WH, II. Pure hippocampal sclerosis: a rare cause of dementia mimicking Alzheimer's disease. *Neurology* 2000;**54**:843–8.

8. Alafuzoff I, Arzberger T, Al-Sarraj S, Bodi I, Bogdanovic N, et al. Staging of neurofibrillary pathology in Alzheimer's disease: a study of the BrainNet Europe Consortium. *Brain Pathol* 2008;**18**:484–96.

9. Albert MS, DeKosky ST, Dickson D, Dubois B, Feldman HH, et al. The diagnosis of mild cognitive impairment due to Alzheimer's disease: recommendations from the National Institute on Aging-Alzheimer's Association workgroups on diagnostic guidelines for Alzheimer's disease. *Alzheimer's Dement* 2011;**7**:270–9.

10. Allan LM, Rowan EN, Firbank MJ, Thomas AJ, Parry SW, et al. Long term incidence of dementia, predictors of mortality and pathological diagnosis in older stroke survivors. *Brain* 2012;**134**:3716–27.

11. Allen N, Robinson AC, Snowden J, Davidson YS, Mann DM. Patterns of cerebral amyloid angiopathy define histopathological phenotypes in Alzheimer's disease. *Neuropathol Appl Neurobiol* 2014;**40**:136–48.

12. Allinson TM, Parkin ET, Turner AJ, Hooper NM. ADAMs family members as amyloid precursor protein alpha-secretases. *J Neurosci Res* 2003;**74**:342–52.

13. Alonzo NC, Hyman BT, Rebeck GW, Greenberg SM. Progression of cerebral amyloid angiopathy: accumulation of amyloid-beta 40 in affected vessels. *J Neuropathol Exp Neurol* 1998;**57**: 353–9.

14. Al-Sarraj S, King A, Troakes C, Smith B, Maekawa S, et al. p62-Positive, TDP-43 negative, neuronal cytoplasmic and intranuclear inclusions in the cerebellum and hippocampus define the pathology of C9orf72-linked FTLD and MND/ALS. *Acta Neuropathol* 2011;**122**:691–702.

15. Altaf N, Daniels L, Morgan PS, Lowe J, Gladman J, et al. Cerebral white matter hyperintense lesions are associated with unstable carotid plaques. *Eur J Vasc Endovasc Surg* 2006;**31**:8–13.

16. Altman A. Uber die umschriebene Gehirnatrophie des spateren Alters. *Z Neurol Psychiatr* 1923;**83**:610–43.

17. Alzheimer A. A unique illness involving the cerebral cortex. A case report from the mental institution in Frankfurt am Main. *Allg Z Psychiatr Psych-Gerich Med* 1907;**64**:146–8.

18. Alzheimer A. Über eigenartige Krankheitsfälle des späteren Alters. *Z Gesamte Neurol Psychiatrie* 1911;**4**: 356–85.

19. Ambegaokar SS, Jackson GR. The downward spiral of tau and autolysosomes: a new hypothesis in neurodegeneration. *Autophagy* 2012;**8**:1144–5.

20. American Psychiatric Association. *Diagnostic and statistical manual of mental disorders*. 4th edn. Washington: American Psychiatric Association; 1994.

21. American Psychiatric Association. *Diagnostic and statistical manual of mental disorders*, 5th ed. Arlington: APPI; 2013.

22. Amos LA, Schlieper D. Microtubules and maps. *Adv Protein Chem* 2005;**71**: 257–98.

23. Andreadis A. Tau gene alternative splicing: expression patterns, regulation and modulation of function in normal brain and neurodegenerative diseases. *Biochim Biophys Acta* 2005;**1739**:91–103.

24. Andren K, Wikkelso C, Tisell M, Hellstrom P. Natural course of idiopathic normal pressure hydrocephalus. *J Neurol Neurosurg Psychiatry* 2014;**85**:806–10.

25. Anthony K, Gallo JM. Aberrant RNA processing events in neurological disorders. *Brain Res* 2010;**1338**:67–77.

26. Arai H, Lee VM, Otvos L, Jr, et al. Defined neurofilament, tau, and beta-amyloid precursor protein epitopes distinguish Alzheimer from non-Alzheimer senile plaques. *Proc Natl Acad Sci U S A* 1990;**87**:2249–53.

27. Arendt T. Synaptic degeneration in Alzheimer's disease. *Acta Neuropathol* 2009;**118**:167–79.

28. Arendt T. Cell cycle activation and aneuploid neurons in Alzheimer's disease. *Mol Neurobiol* 2012;**46**:125–35.

29. Arendt T, Bruckner MK. Linking cell-cycle dysfunction in Alzheimer's disease to a failure of synaptic plasticity. *Biochim Biophys Acta* 2007;**1772**:413–21.

30. Armstrong DM, LeRoy S, Shields D, Terry RD. Somatostatin-like immunoreactivity within neuritic plaques. *Brain Res* 1985;**338**:71–9.

31. Armstrong MJ, Litvan I, Lang AE, Bak TH, Bhatia KP, et al. Criteria for the diagnosis of corticobasal degeneration. *Neurology* 2013;**80**:496–503.

32. Arnold SE, Hyman BT, Flory J, et al. The topographical and neuroanatomical distribution of neurofibrillary tangles and neuritic plaques in the cerebral cortex of patients with Alzheimer's disease. *Cereb Cortex* 1991;**1**:103–16.

33. Aronica E, Dickson DW, Kress Y, Morrison JH, Zukin RS. Non-plaque dystrophic dendrites in Alzheimer hippocampus: a new pathological structure revealed by glutamate receptor immunocytochemistry. *Neuroscience* 1998;**82**:979–91.

34. Arriagada PV, Growdon JH, Hedley-Whyte ET, Hyman BT. Neurofibrillary tangles but not senile plaques parallel duration and severity of Alzheimer's disease. *Neurology* 1992;**42**:631–9.

35. Arvanitakis Z, Leurgans SE, Barnes LL, Bennett DA, Schneider JA. Microinfarct pathology, dementia, and cognitive systems. *Stroke* 2011;**42**:722–7.

36. Arvanitakis Z, Leurgans SE, Wang Z, Wilson RS, Bennett DA, Schneider JA. Cerebral amyloid angiopathy pathology and cognitive domains in older persons. *Ann Neurol* 2011;**69**:320–7.

37. Ashford JW, Soultanian NS, Zhang SX, Geddes JW. Neuropil threads are collinear with MAP2 immunostaining in neuronal dendrites of Alzheimer brain. *J Neuropathol Exp Neurol* 1998;**57**: 972–8.

38. Atsumi T, Yamamura Y, Sato T, Ikuta F. Hirano bodies in the axon of peripheral nerves in a case with progressive external ophthalmoplegia with multisystemic involvements. *Acta Neuropathol* 1980;**49**:95–100.

39. Attems J, Jellinger K, Thal DR, Van Nostrand W. Review: sporadic cerebral amyloid angiopathy. *Neuropathol Appl Neurobiol* 2011;**37**:75–93.

40. Auchus AP. Dementia in urban black outpatients: initial experience at the

Emory satellite clinics. *Gerontologist* 1997;**37**:25–9.

41. Baborie A, Griffiths TD, Jaros E, Momeni P, McKeith IG, et al. Frontotemporal dementia in elderly individuals. *Arch Neurol* 2012;**69**:1052–60.

42. Bailey EL, McCulloch J, Sudlow C, Wardlaw JM. Potential animal models of lacunar stroke: a systematic review. *Stroke* 2009;**40**:e451–8.

43. Bailey EL, Smith C, Sudlow CL, Wardlaw JM. Is the spontaneously hypertensive stroke prone rat a pertinent model of sub cortical ischemic stroke? A systematic review. *Int J Stroke* 2012;**6**:434–44.

44. Bailey EL, Smith C, Sudlow CL, Wardlaw JM. Pathology of lacunar ischemic stroke in humans – a systematic review. *Brain Pathol* 2012;**22**:583–91.

45. Baker M, Mackenzie IR, Pickering-Brown SM, Gass J, Rademakers R, et al. Mutations in progranulin cause tau-negative frontotemporal dementia linked to chromosome 17. *Nature* 2006;**442**:916–9.

46. Balducci C, Forloni G. APP transgenic mice: their use and limitations. *Neuromolecular Med* 2011;**13**:117–37.

47. Bales KR, Verina T, Dodel RC, et al. Lack of apolipoprotein E dramatically reduces amyloid beta-peptide deposition (Letter). *Nat Genet* 1997;**17**:263–4.

48. Ball MJ, Lo P. Granulovacuolar degeneration in the ageing brain and in dementia. *J Neuropathol Exp Neurol* 1977;**36**:474–87.

49. Ball MJ, Nuttall K. Topography of neurofibrillary tangles and granulovacuoles in hippocampi of patients with Down's syndrome: quantitative comparison with normal ageing and Alzheimer's disease. *Neuropathol Appl Neurobiol* 1981;**7**:13–20.

50. Ball MJ, Vis CL. Relationship of granulovacuolar degeneration in hippocampal neurones to aging and to dementia in normal-pressure hydrocephalics. *J Gerontol* 1978;**33**:815–24.

51. Ballard C, McKeith I, O'Brien J, Kalaria R, Jaros E, et al. Neuropathological substrates of dementia and depression in vascular dementia, with a particular focus on cases with small infarct volumes. *Dement Geriatr Cogn Disord* 2000;**11**:59–65.

52. Ballard CG, Burton EJ, Barber R, Stephens S, Kenny RA, et al. NINDS AIREN neuroimaging criteria do not distinguish stroke patients with and without dementia. *Neurology* 2004;**63**:983–8.

53. Banack SA, Cox PA. Biomagnification of cycad neurotoxins in flying foxes: implications for ALS-PDC in Guam. *Neurology* 2003;**61**:387–9.

54. Bancher C, Jellinger KA. Neurofibrillary tangle predominant form of senile dementia of Alzheimer type: a rare subtype in very old subjects. *Acta Neuropathol* 1994;**88**:565–70.

55. Bancher C, Egensperger R, Kosel S, Jellinger K, Graeber MB. Low prevalence of apolipoprotein E epsilon 4 allele in the neurofibrillary tangle predominant form of senile dementia. *Acta Neuropathol* 1997;**94**:403–9.

56. Barker WW, Luis CA, Kashuba A, Luis M, Harwood DG, et al. Relative frequencies of Alzheimer disease, Lewy body, vascular and frontotemporal dementia, and hippocampal sclerosis in the State of Florida Brain Bank. *Alzheimer Dis Assoc Disord* 2002;**16**:203–12.

57. Bastos Leite AJ, van der Flier WM, van Straaten EC, Scheltens P, Barkhof F. Infratentorial abnormalities in vascular dementia. *Stroke* 2006;**37**:105–10.

58. Bastos-Leite AJ, van der Flier WM, van Straaten EC, Staekenborg SS, Scheltens P, Barkhof F. The contribution of medial temporal lobe atrophy and vascular pathology to cognitive impairment in vascular dementia. *Stroke* 2007;**38**:3182–5.

59. Baumer D, Hilton D, Paine SM, Turner MR, Lowe J, et al. Juvenile ALS with basophilic inclusions is a FUS proteinopathy with FUS mutations. *Neurology* 2010;**75**:611–8.

60. Bazenet C, Lovestone S. Plasma biomarkers for Alzheimer's disease: much needed but tough to find. *Biomark Med* 2012;**6**:441–54.

61. Beal JA. Morphogenesis of the Hirano body in neurons of the squirrel monkey dorsal horn. *J Neurocytol* 1978;**7**:395–403.

62. Beck J, Poulter M, Hensman D, Rohrer JD, Mahoney CJ, et al. Large C9orf72 hexanucleotide repeat expansions are seen in multiple neurodegenerative syndromes and are more frequent than expected in the UK population. *Am J Hum Genet* 2013;**92**:345–53.

63. Bedford L, Hay D, Paine S, Rezvani N, Mee M, et al. Is malfunction of the ubiquitin proteasome system the primary cause of alpha-synucleinopathies and other chronic human neurodegenerative disease? *Biochim Biophys Acta* 2008;**1782**:683–90.

64. Bedford L, Paine S, Rezvani N, Mee M, Lowe J, Mayer RJ. The UPS and autophagy in chronic neurodegenerative disease: six of one and half a dozen of the other--or not? *Autophagy* 2009;**5**:224–7.

65. Bejot Y, Aboa-Eboule C, Durier J, Rouaud O, Jacquin A, et al. Prevalence of early dementia after first-ever stroke: a 24-year population-based study. *Stroke* 2011;**42**:607–12.

66. Benitez BA, Jin SC, Guerreiro R, Graham R, Lord J, et al. Missense variant in TREML2 protects against Alzheimer's disease. *Neurobiol Aging* 2014;**35**:1510, e19–26.

67. Bennett DA, Wilson RS, Gilley DW, Fox JH. Clinical diagnosis of Binswanger's disease. *J Neurol Neurosurg Psychiatry* 1990;**53**:961–5.

68. Bentley P, Driver J, Dolan RJ. Cholinergic modulation of cognition: insights from human pharmacological functional neuroimaging. *Prog Neurobiol* 2011;**94**:360–88.

69. Bentmann E, Haass C, Dormann D. Stress granules in neurodegeneration: lessons learnt from TAR DNA binding protein of 43 kDa and fused in sarcoma. *FEBS J* 2013;**280**:4348–70.

70. Berrios GE, Freeman HL. Alzheimer and the dementiaa. In: Berrios GE, editor. *Eponymists in medicine* series. London: Royal Society of Medicine Services; 1991; p. 69–76.

71. Berry RW, Quinn B, Johnson N, Cochran EJ, Ghoshal N, Binder LI. Pathological glial tau accumulations in neurodegenerative disease: review and case report. *Neurochem Int* 2001;**39**:469–79.

72. Bertram L, McQueen MB, Mullin K, Blacker D, Tanzi RE. Systematic meta-analyses of Alzheimer disease genetic association studies: the AlzGene database. *Nat Genet* 2007;**39**:17–23.

73. Bierer LM, Hof PR, Purohit DP, et al. Neocortical neurofibrillary tangles correlate with dementia severity in Alzheimer's disease. *Arch Neurol* 1995;**52**:81–8.

74. Bigio EH. Making the diagnosis of frontotemporal lobar degeneration. *Arch Pathol Lab Med* 2013;**137**:314–25.

75. Bigio EH, Brown DF, White CL, 3rd. Progressive supranuclear palsy with dementia: cortical pathology. *J Neuropathol Exp Neurol* 1999;**58**:359–64.

76. Bigio EH, Lipton AM, Yen SH, Hutton ML, Baker M, et al. Frontal lobe dementia with novel tauopathy: sporadic multiple system tauopathy with dementia. *J Neuropathol Exp Neurol* 2001;**60**:328–41.

77. Bigio EH, Vono MB, Satumtira S, et al. Cortical synapse loss in progressive supranuclear palsy. *J Neuropathol Exp Neurol* 2001;**60**:403–10.

78. Blennow K, Wallin A, Agren H, Spenger C, Siegfried J, Vanmechelen E. Tau protein in cerebrospinal fluid: a biochemical marker for axonal degeneration in Alzheimer disease? *Mol Chem Neuropathol* 1995;**26**:231–45.

79. Blennow K, Hardy J, Zetterberg H. The neuropathology and neurobiology of traumatic brain injury. *Neuron* 2012;**76**:886–99.

80. Blessed G, Tomlinson BE, Roth M. The association between quantitative measures of dementia and of senile change in the cerebral grey matter of elderly subjects. *Br J Psychiatry* 1968;**114**:797–811.

81. Blocq P, Marinesco G. Sur les lesions et la pathogenie de l'epilepsie dite essentielle. *Semaine Med* 1892;**12**:445–6.

82. Boche D, Denham N, Holmes C, Nicoll JA. Neuropathology after active Abeta42 immunotherapy: implications for Alzheimer's disease pathogenesis. *Acta Neuropathol* 2010;**120**:369–84.

83. Bonda DJ, Bajic VP, Spremo-Potparevic B, Casadesus G, Zhu X, et al. Review: cell cycle aberrations and neurodegeneration. *Neuropathol Appl Neurobiol* 2010;**36**:157–63.

84. Bondareff W, Wischik CM, Novak M, Roth M. Sequestration of tau by granulovacuolar degeneration in Alzheimer's disease. *Am J Pathol* 1991;**139**:641–7.

85. Borruat FX. Posterior cortical atrophy: review of the recent literature. *Curr Neurol Neurosci Rep* 2013;**13**:406.

86. Bosque PJ, Boyer PJ, Mishra P. A 43-kDa TDP-43 species is present in aggregates associated with frontotemporal lobar degeneration. *PLoS One* 2013;**8**:e62301.

87. Bowler JV, Hachinski V. Vascular cognitive impairment: a new approach to vascular dementia. *Baillieres Clin Neurol* 1995;**4**:357–76.

88. Braak H, Braak E, Mandelkow EM. A sequence of cytoskeleton changes related to the formation of neurofibrillary tangles and neuropil threads. *Acta Neuropathol* 1994;**87**:554–67.

89. Braak H, Braak E. Argyrophilic grains: characteristic pathology of cerebral cortex in cases of adult onset dementia without Alzheimer changes. *Neurosci Lett* 1987;**76**:124–7.

90. Braak H, Braak E. Neuropil threads occur in dendrites of tangle-bearing

nerve cells. *Neuropathol Appl Neurobiol* 1988;**14**:39–44.

91. Braak H, Braak E. Neuropathological stageing of Alzheimer-related changes. *Acta Neuropathol* 1991;**82**:239–59.

92. Braak H, Braak E. Diagnostic criteria for neuropathologic assessment of Alzheimer's disease. *Neurobiol Aging* 1997;**18**:S85–8.

93. Braak H, Braak E. Argyrophilic grain disease: frequency of occurrence in different age categories and neuropathological diagnostic criteria. *J Neural Transm* 1998;**105**:801–19.

94. Braak H, Braak E, Grundke-Iqbal I, Iqbal K. Occurrence of neuropil threads in the senile human brain and in Alzheimer's disease: a third location of paired helical filaments outside of neurofibrillary tangles and neuritic plaques. *Neurosci Lett* 1986;**65**:351–5.

95. Braak H, Braak E, Yilmazer D, de Vos RA, Jansen EN, et al. Amygdala pathology in Parkinson's disease. *Acta Neuropathol* 1994;**88**:493–500.

96. Bradshaw CB, Davis RL, Shrimpton AE, Holohan PD, Rea CB, et al. Cognitive deficits associated with a recently reported familial neurodegenerative disease: familial encephalopathy with neuroserpin inclusion bodies. *Arch Neurol* 2001;**58**:1429–34.

97. Brady OA, Zheng Y, Murphy K, Huang M, Hu F. The frontotemporal lobar degeneration risk factor, *TMEM106B*, regulates lysosomal morphology and function. *Hum Mol Genet* 2013; **22**:685–95.

98. Brelstaff J, Lashley T, Holton JL, Lees AJ, Rossor MN, et al. Transportin1: a marker of FTLD-FUS. *Acta Neuropathol* 2011;**122**:591–600.

99. Brenowitz WD, Monsell SE, Schmitt FA, Kukull WA, Nelson PT. Hippocampal sclerosis of aging is a key Alzheimer's disease mimic: clinical-pathologic correlations and comparisons with both Alzheimer's disease and non-tauopathic frontotemporal lobar degeneration. *J Alzheimer's Dis* 2014;**39**:691–702.

100. Brion JP, Couck AM, Bruce M, et al. Synaptophysin and chromogranin A immunoreactivities in senile plaques of Alzheimer's disease. *Brain Res* 1991;**539**:143–50.

101. Brown WR, Thore CR. Review: cerebral microvascular pathology in ageing and neurodegeneration. *Neuropathol Appl Neurobiol* 2011;**37**:56–74.

102. Brown WR, Moody DM, Challa VR, Thore CR, Anstrom JA. Apoptosis in leukoaraiosis lesions. *J Neurol Sci* 2002;**203–204**:169–71.

103. Brown WR, Moody DM, Challa VR, Thore CR, Anstrom JA. Venous collagenosis and arteriolar tortuosity in leukoaraiosis. *J Neurol Sci* 2002;**203–204**:159–63.

104. Brown WR, Moody DM, Thore CR, Challa VR, Anstrom JA. Vascular dementia in leukoaraiosis may be a consequence of capillary loss not only in the lesions, but in normal-appearing white matter and cortex as well. *J Neurol Sci* 2007;**257**:62–6.

105. Brown WR, Moody DM, Thore CR, Anstrom JA, Challa VR. Microvascular changes in the white matter in dementia. *J Neurol Sci* 2009;**283**:28–31.

106. Brun A. Pathology and pathophysiology of cerebrovascular dementia: pure subgroups

of obstructive and hypoperfusive etiology. *Dementia* 1994;**5**:145–7.

107. Buee L, Delacourte A. Comparative biochemistry of tau in progressive supranuclear palsy, corticobasal degeneration, FTDP-17 and Pick's disease. *Brain Pathol* 1999;**9**:681–93.

108. Buffon F, Porcher R, Hernandez K, Kurtz A, Pointeau S, et al. Cognitive profile in CADASIL. *J Neurol Neurosurg Psychiatry* 2006;**77**:175–80.

109. Bugalho P, Alves L, Miguel R, Ribeiro O. Profile of cognitive dysfunction and relation with gait disturbance in normal pressure hydrocephalus. *Clin Neurol Neurosurg* 2014;**118**:83–8.

110. Bugiani O, Giaccone G, Frangione B, Ghetti B, Tagliavini F. Alzheimer patients: preamyloid deposits are more widely distributed than senile plaques throughout the central nervous system. *Neurosci Lett* 1989;**103**:263–8.

111. Burke SN, Barnes CA. Neural plasticity in the ageing brain. *Nat Rev Neurosci* 2006;**7**:30–40.

112. Burrell JR, Hodges JR, Rowe JB. Cognition in corticobasal syndrome and progressive supranuclear palsy: a review. *Mov Disord* 2014;**29**:684–93.

113. Burton EJ, Barber R, Mukaetova-Ladinska EB, Robson J, Perry RH, et al. Medial temporal lobe atrophy on MRI differentiates Alzheimer's disease from dementia with Lewy bodies and vascular cognitive impairment: a prospective study with pathological verification of diagnosis. *Brain* 2009;**132**:195–203.

114. Cabral D, Beach TG, Vedders L, Sue LI, Jacobson S, et al. Frequency of Alzheimer's disease pathology at autopsy in patients with clinical normal pressure hydrocephalus. *Alzheimer's Dement* 2011;**7**:509–13.

115. Cadavid D, Mena H, Koeller K, Frommelt RA. Cerebral beta amyloid angiopathy is a risk factor for cerebral ischemic infarction. A case control study in human brain biopsies. *J Neuropathol Exp Neurol* 2000;**59**:768–73.

116. Cai Z, Ratka A. Opioid system and Alzheimer's disease. *Neuromolecular Med* 2012;**14**:91–111.

117. Cairns NJ, Grossman M, Arnold SE, Burn DJ, Jaros E, et al. Clinical and neuropathologic variation in neuronal intermediate filament inclusion disease. *Neurology* 2004;**63**:1376–84.

118. Cairns NJ, Uryu K, Bigio EH, Mackenzie IR, Gearing M, et al. Alpha-internexin aggregates are abundant in neuronal intermediate filament inclusion disease (NIFID) but rare in other neurodegenerative diseases. *Acta Neuropathol* 2004;**108**:213–23.

119. Cairns NJ, Zhukareva V, Uryu K, Zhang B, Bigio E, et al. Alpha-internexin is present in the pathological inclusions of neuronal intermediate filament inclusion disease. *Am J Pathol* 2004;**164**:2153–61.

120. Cairns NJ, Bigio EH, Mackenzie IR, Neumann M, Lee VM, et al. Neuropathologic diagnostic and nosologic criteria for frontotemporal lobar degeneration: consensus of the Consortium for Frontotemporal Lobar Degeneration. *Acta Neuropathol* 2007;**114**:5–22.

121. Cairns NJ, Neumann M, Bigio EH, Holm IE, Troost D, et al. TDP-43 in familial and sporadic frontotemporal lobar

degeneration with ubiquitin inclusions. *Am J Pathol* 2007;**171**:227–40.

122. Caldwell JH, Klevanski M, Saar M, Muller UC. Roles of the amyloid precursor protein family in the peripheral nervous system. *Mech Dev* 2013;**130**:433–46.

123. Carare RO, Hawkes CA, Jeffrey M, Kalaria RN, Weller RO. Review: cerebral amyloid angiopathy, prion angiopathy, CADASIL and the spectrum of protein elimination failure angiopathies (PEFA) in neurodegenerative disease with a focus on therapy. *Neuropathol Appl Neurobiol* 2013;**39**:593–611.

124. Carrera E, Bogousslavsky J. The thalamus and behavior: effects of anatomically distinct strokes. *Neurology* 2006;**66**:1817–23.

125. Castellani RJ, Gupta Y, Sheng B, Siedlak SL, Harris PL, et al. A novel origin for granulovacuolar degeneration in aging and Alzheimer's disease: parallels to stress granules. *Lab Invest* 2011;**91**:1777–86.

126. Catindig JA, Venketasubramanian N, Ikram MK, Chen C. Epidemiology of dementia in Asia: insights on prevalence, trends and novel risk factors. *J Neurol Sci* 2012;**321**:11–6.

127. Cattaneo A, Calissano P. Nerve growth factor and Alzheimer's disease: new facts for an old hypothesis. *Mol Neurobiol* 2012;**46**:588–604.

128. Cavallucci V, D'Amelio M, Cecconi F. Abeta toxicity in Alzheimer's disease. *Mol Neurobiol* 2012;**45**:366–78.

129. Cenik B, Sephton CF, Kutluk Cenik B, Herz J, Yu G. Progranulin: a proteolytically processed protein at the crossroads of inflammation and neurodegeneration. *J Biol Chem* 2012;**287**:32298–306.

130. Censori B, Manara O, Agostinis C, et al. Dementia after first stroke. *Stroke* 1996;**27**:1205–10.

131. CFAS. Pathological correlates of late-onset dementia in a multicentre, community-based population in England and Wales. Neuropathology Group of the Medical Research Council Cognitive Function and Ageing Study (MRC CFAS). *Lancet* 2001;**357**:169–75.

132. Chabriat H, Joutel A, Dichgans M, Tournier-Lasserve E, Bousser MG. CADASIL. *Lancet Neurol* 2009;**8**:643–53.

133. Chadwick L, Gentle L, Strachan J, Layfield R. Review: unchained maladie: a reassessment of the role of Ubb(+1)-capped polyubiquitin chains in Alzheimer's disease. *Neuropathol Appl Neurobiol* 2012;**38**:118–31.

134. Chalmers K, Wilcock GK, Love S. APOE epsilon 4 influences the pathological phenotype of Alzheimer's disease by favouring cerebrovascular over parenchymal accumulation of A beta protein. *Neuropathol Appl Neurobiol* 2003;**29**:231–8.

135. Charlton RA, Morris RG, Nitkunan A, Markus HS. The cognitive profiles of CADASIL and sporadic small vessel disease. *Neurology* 2006;**66**:1523–6.

136. Chen-Plotkin AS, Martinez-Lage M, Sleiman PM, Hu W, Greene R, et al. Genetic and clinical features of progranulin-associated frontotemporal lobar degeneration. *Arch Neurol* 2011;**68**:488–97.

137. Cherra SJ, 3rd, Dagda RK, Chu CT. Review: autophagy and neurodegeneration: survival at a cost? *Neuropathol Appl Neurobiol* 2010;**36**:125–32.

138. Chetelat G. Alzheimer disease: Abeta-independent processes-rethinking preclinical AD. *Nat Rev Neurol* 2013;9:123–4.

139. Chin AL, Negash S, Xie S, Arnold SE, Hamilton R. Quality, and not just quantity, of education accounts for differences in psychometric performance between African Americans and white non-Hispanics with Alzheimer's disease. *J Int Neuropsychol Soc* 2012;18:277–85.

140. Chin J. Selecting a mouse model of Alzheimer's disease. *Methods Mol Biol* 2011;670:169–89.

141. Chiu AS, Gehringer MM, Welch JH, Neilan BA. Does alpha-amino-beta-methylaminopropionic acid (BMAA) play a role in neurodegeneration? *Int J Environ Res Public Health* 2011;8:3728–46.

142. Chow VW, Mattson MP, Wong PC, Gleichmann M. An overview of APP processing enzymes and products. *Neuromolecular Med* 2010;12:1–12.

143. Chui HC, Victoroff JI, Margolin D, Jagust W, Shankle R, Katzman R. Criteria for the diagnosis of ischemic vascular dementia proposed by the State of California Alzheimer's Disease Diagnostic and Treatment Centers. *Neurology* 1992;42:473–80.

144. Claeysen S, Cochet M, Donneger R, Dumuis A, Bockaert J, Giannoni P. Alzheimer culprits: cellular crossroads and interplay. *Cell Signal* 2012;24:1831–40.

145. Clague MJ, Urbe S. Ubiquitin: same molecule, different degradation pathways. *Cell* 2010;143:682–5.

146. Clare R, King VG, Wirenfeldt M, Vinters HV. Synapse loss in dementias. *J Neurosci Res* 2010;88:2083–90.

147. Clark AW, White CL, III, Manz HJ, et al. Primary degenerative dementia without Alzheimer pathology. *Can J Neurol Sci* 1986;13:462–70.

148. Cochran E, Bacci B, Chen Y, Patton A, Gambetti P, Autilio-Gambetti L. Amyloid precursor protein and ubiquitin immunoreactivity in dystrophic axons is not unique to Alzheimer's disease. *Am J Pathol* 1991;139:485–9.

149. Cochran EJ, Fox JH, Mufson EJ. Severe panencephalic Pick's disease with Alzheimer's disease-like neuropil threads and synaptophysin immunoreactivity. *Acta Neuropathol* 1994;88:479–84.

150. Cohen AD, Klunk WE. Early detection of Alzheimer's disease using PiB and FDG PET. *Neurobiol Dis* 2014;72PA:117–22.

151. Cohen DL, Hedera P, Premkumar DR, Friedland RP, Kalaria RN. Amyloid-beta protein angiopathies masquerading as Alzheimer's disease? *Ann N Y Acad Sci* 1997;826:390–5.

152. Coker SB. The diagnosis of childhood neurodegenerative disorders presenting as dementia in adults. *Neurology* 1991;41:794–8.

153. Compta Y, Parkkinen L, O'Sullivan SS, Vandrovcova J, Holton JL, et al. Lewy- and Alzheimer-type pathologies in Parkinson's disease dementia: which is more important? *Brain* 2011;134:1493–505.

154. Constantinidis J, Richard J, Tissot R. Pick's disease. Histological and clinical correlations. *Eur Neurol* 1974;11:208–17.

155. Cooper PN, Jackson M, Lennox G, Lowe J, Mann DM. Tau, ubiquitin, and alpha B-crystallin immunohistochemistry define the principal causes of degenerative frontotemporal dementia. *Arch Neurol* 1995;52:1011–5.

156. Cooper-Knock J, Shaw PJ, Kirby J. The widening spectrum of *C9ORF72*-related disease; genotype/phenotype correlations and potential modifiers of clinical phenotype. *Acta Neuropathol* 2014;127:333–45.

157. Cooper-Knock J, Walsh MJ, Higginbottom A, Robin Highley J, Dickman MJ, et al. Sequestration of multiple RNA recognition motif-containing proteins by *C9orf72* repeat expansions. *Brain* 2014;137:2040–51.

158. Cordonnier C, Al-Shahi Salman R, Wardlaw J. Spontaneous brain microbleeds: systematic review, subgroup analyses and standards for study design and reporting. *Brain* 2007;130:1988–2003.

159. Corey-Bloom J, Sabbagh MN, Bondi MW, et al. Hippocampal sclerosis contributes to dementia in the elderly. *Neurology* 1997;48:154–60.

160. Cosentino SA, Jefferson AL, Carey M, Price CC, Davis-Garrett K, et al. The clinical diagnosis of vascular dementia: A comparison among four classification systems and a proposal for a new paradigm. *Clin Neuropsychol* 2004;18:6–21.

161. Costanza A, Weber K, Gandy S, Bouras C, Hof PR, et al. Review: Contact sport-related chronic traumatic encephalopathy in the elderly: clinical expression and structural substrates. *Neuropathol Appl Neurobiol* 2011;37:570–84.

162. Costanzo M, Zurzolo C. The cell biology of prion-like spread of protein aggregates: mechanisms and implication in neurodegeneration. *Biochem J* 2013;452:1–17.

163. Court JA, Johnson M, Religa D, Keverne J, Kalaria R, et al. Attenuation of Abeta deposition in the entorhinal cortex of normal elderly individuals associated with tobacco smoking. *Neuropathol Appl Neurobiol* 2005;31:522–35.

164. Craft S, Cholerton B, Baker LD. Insulin and Alzheimer's disease: untangling the web. *J Alzheimer's Dis* 2013;33:S263–75.

165. Craggs LJ, Hagel C, Kuhlenbaeumer G, Borjesson-Hanson A, Andersen O, et al. Quantitative vascular pathology and phenotyping familial and sporadic cerebral small vessel diseases. *Brain Pathol* 2013;23:547–57.

166. Craggs LJ, Yamamoto Y, Ihara M, Fenwick R, Burke M, et al. White matter pathology and disconnection in the frontal lobe in CADASIL. *Neuropathol Appl Neurobiol* 2014;40:591–602.

167. Craig LA, Hong NS, McDonald RJ. Revisiting the cholinergic hypothesis in the development of Alzheimer's disease. *Neurosci Biobehav Rev* 2011;35:1397–409.

168. Crook R, Verkkoniemi A, Perez-Tur J, et al. A variant of Alzheimer's disease with spastic paraparesis and unusual plaques due to deletion of exon 9 of presenilin 1. *Nat Med* 1998;4:452–5.

169. Crosby AH. Disruption of cellular transport: a common cause of neurodegeneration? *Lancet Neurol* 2003;2:311–6.

170. Crowther T, Goedert M, Wischik CM. The repeat region of microtubule-associated protein tau forms part of the core of the paired helical filament of Alzheimer's disease. *Ann Med* 1989;21:127–32.

171. Crutch SJ, Lehmann M, Schott JM, Rabinovici GD, Rossor MN, Fox NC. Posterior cortical atrophy. *Lancet Neurol* 2012;11:170–8.

172. Cruz L, Urbanc B, Buldyrev SV, et al. Aggregation and disaggregation of senile plaques in Alzheimer disease. *Proc Natl Acad Sci U S A* 1997;94:7612–16.

173. Cummings BJ, Head E, Afagh AJ, et al. Beta-amyloid accumulation correlates with cognitive dysfunction in the aged canine. *Neurobiol Learn Mem* 1996;66:11–23.

174. Cupples LA, Farrer LA, Sadovnick AD, Relkin N, Whitehouse P, Green RC. Estimating risk curves for first-degree relatives of patients with Alzheimer's disease: the REVEAL study. *Genet Med* 2004;6:192–6.

175. Currais A, Hortobagyi T, Soriano S. The neuronal cell cycle as a mechanism of pathogenesis in Alzheimer's disease. *Aging* 2009;1:363–71.

176. Cusimano MD, Rewilak D, Stuss DT, Barrera-Martinez JC, Salehi F, Freedman M. Normal-pressure hydrocephalus: is there a genetic predisposition? *Can J Neurol Sci* 2011;38:274–81.

177. Danzer KM, Krebs SK, Wolff M, Birk G, Hengerer B. Seeding induced by alpha-synuclein oligomers provides evidence for spreading of alpha-synuclein pathology. *J Neurochem* 2009;111:192–203.

178. Davidson YS, Barker H, Robinson AC, Thompson JC, Harris J, et al. Brain distribution of dipeptide repeat proteins in frontotemporal lobar degeneration and motor neurone disease associated with expansions in *C9ORF72*. *Acta Neuropathol Commun* 2014;2:70.

179. Davis DG, Wang HZ, Markesbery WR. Image analysis of neuropil threads in Alzheimer's, Pick's, diffuse Lewy body disease and in progressive supranuclear palsy. *J Neuropathol Exp Neurol* 1992;51:594–600.

180. Davis RC, Furukawa R, Fechheimer M. A cell culture model for investigation of Hirano bodies. *Acta Neuropathol* 2008;115:205–17.

181. Davis RL, Holohan PD, Shrimpton AE, et al. Familial encephalopathy with neuroserpin inclusion bodies. *Am J Pathol* 1999;155:1901–13.

182. Davis RL, Shrimpton AE, Carrell RW, Lomas DA, Gerhard L, et al. Association between conformational mutations in neuroserpin and onset and severity of dementia. *Lancet* 2002;359:2242–7.

183. de la Monte SM, Kril JJ. Human alcohol-related neuropathology. *Acta Neuropathol* 2014;127:71–90.

184. de la Monte SM, Tong M. Brain metabolic dysfunction at the core of Alzheimer's disease. *Biochem Pharmacol* 2014;88:548–59.

185. de Silva R, Lashley T, Gibb G, Hanger D, Hope A, et al. Pathological inclusion bodies in tauopathies contain distinct complements of tau with three or four microtubule-binding repeat domains as demonstrated by new specific monoclonal antibodies. *Neuropathol Appl Neurobiol* 2003;29:288–302.

186. de Silva R, Lashley T, Revesz T, Lees A, Powers JM. Detecting tau isoforms in archival cases. *Acta Neuropathol* 2004;107:181–2.

187. Debette S, Markus HS. The clinical importance of white matter hyperintensities on brain magnetic resonance imaging:

systematic review and meta-analysis. *BMJ* 2010;**341**:c3666.

188. DeJesus-Hernandez M, Mackenzie IR, Boeve BF, Boxer AL, Baker M, et al. Expanded GGGGCC hexanucleotide repeat in noncoding region of *C9ORF72* causes chromosome 9p-linked FTD and ALS. *Neuron* 2011;**72**:245–56.

189. DeKosky ST, Ikonomovic MD. NIFID: a new molecular pathology with a frontotemporal dementia phenotype. *Neurology* 2004;**63**:1348–9.

190. DeKosky ST, Scheff SW. Synapse loss in frontal cortex biopsies in Alzheimer's disease: correlation with cognitive severity. *Ann Neurol* 1990;**27**:457–64.

191. DeKosky ST, Blennow K, Ikonomovic MD, Gandy S. Acute and chronic traumatic encephalopathies: pathogenesis and biomarkers. *Nat Rev Neurol* 2013;**9**:192–200.

192. del Ser T, Bermejo F, Portera A, et al. Vascular dementia. A clinicopathological study. *J Neurol Sci* 1990;**96**:1–17.

193. Delacourte A, Defossez A. Alzheimer's disease: tau proteins, the promoting factors of microtubule assembly, are major components of paired helical filaments. *J Neurol Sci* 1986;**76**:173–86.

194. Delacourte A, Robitaille Y, Sergeant, et al. Specific pathological Tau protein variants characterize Pick's disease. *J Neuropathol Exp Neurol* 1996;**55**:159–68.

195. Deng H, Gao K, Jankovic J. The role of FUS gene variants in neurodegenerative diseases. *Nat Rev Neurol* 2014;**10**:337–48.

196. Dennissen FJ, Kholod N, van Leeuwen FW. The ubiquitin proteasome system in neurodegenerative diseases: culprit, accomplice or victim? *Prog Neurobiol* 2012;**96**:190–207.

197. Deramecourt V, Slade JY, Oakley AE, Perry RH, Ince PG, et al. Staging and natural history of cerebrovascular pathology in dementia. *Neurology* 2012;**78**:1043–50.

198. Dermaut B, Kumar-Singh S, Engelborghs S, Theuns J, Rademakers R, et al. A novel presenilin 1 mutation associated with Pick's disease but not beta-amyloid plaques. *Ann Neurol* 2004;**55**:617–26.

199. Desmond DW. The neuropsychology of vascular cognitive impairment: is there a specific cognitive deficit? *J Neurol Sci* 2004;**226**:3–7.

200. Desmond DW, Moroney JT, Paik MC, Sano M, Mohr JP, et al. Frequency and clinical determinants of dementia after ischemic stroke. *Neurology* 2000;**54**:1124–31.

201. Devenney E, Hornberger M, Irish M, Mioshi E, Burrell J, et al. Frontotemporal dementia associated with the *C9ORF72* mutation: a unique clinical profile. *JAMA Neurol* 2014;**71**:331–9.

202. Di Santo SG, Prinelli F, Adorni F, Caltagirone C, Musicco M. A meta-analysis of the efficacy of donepezil, rivastigmine, galantamine, and memantine in relation to severity of Alzheimer's disease. *J Alzheimer's Dis* 2013;**35**:349–61.

203. Dickson DW. The pathogenesis of senile plaques. *J Neuropathol Exp Neurol* 1997;**56**:321–39.

204. Dickson DW. Neuropathologic differentiation of progressive supranuclear palsy and corticobasal degeneration. *J Neurol* 1999;**246**:II6–15.

205. Dickson DW. Alpha-synuclein and the Lewy body disorders. *Curr Opin Neurol* 2001;**14**:423–32.

206. Dickson DW, Yen SH, Suzuki KI, et al. Ballooned neurons in select neurodegenerative diseases contain phosphorylated neurofilament epitopes. *Acta Neuropathol* 1986;**71**:216–23.

207. Dickson DW, Davies P, Mayeux R, Crystal H, Horoupian DS, et al. Diffuse Lewy body disease. Neuropathological and biochemical studies of six patients. *Acta Neuropathol* 1987;**75**:8–15.

208. Dickson DW, Ksiezak-Reding H, Davies P, Yen SH. A monoclonal antibody that recognizes a phosphorylated epitope in Alzheimer neurofibrillary tangles, neurofilaments and tau proteins immunostains granulovacuolar degeneration. *Acta Neuropathol* 1987;**73**:254–8.

209. Dickson DW, Farlo J, Davies P, et al. Alzheimer's disease. A double-labeling immunohistochemical study of senile plaques. *Am J Pathol* 1988;**132**:86–101.

210. Dickson DW, Ruan D, Crystal H, et al. Hippocampal degeneration differentiates diffuse Lewy body disease (DLBD) from Alzheimer's disease: light and electron microscopic immunocytochemistry of CA2–3 neurites specific to DLBD. *Neurology* 1991;**41**:1402–9.

211. Dickson DW, Davies P, Bevona C, et al. Hippocampal sclerosis: a common pathological feature of dementia in very old (?80 years of age) humans. *Acta Neuropathol* 1994;**88**:212–21.

212. Dickson DW, Schmidt ML, Lee VM, et al. Immunoreactivity profile of hippocampal CA2/3 neurites in diffuse Lewy body disease. *Acta Neuropathol* 1994;**87**:269–76.

213. Dickson DW, Bergeron C, Chin SS, Duyckaerts C, Horoupian D, et al. Office of Rare Diseases neuropathologic criteria for corticobasal degeneration. *J Neuropathol Exp Neurol* 2002;**61**: 935–46.

214. Dickson DW, Baker M, Rademakers R. Common variant in *GRN* is a genetic risk factor for hippocampal sclerosis in the elderly. *Neurodegener Dis* 2010;**7**:170–4.

215. Dickson DW, Kouri N, Murray ME, Josephs KA. Neuropathology of frontotemporal lobar degeneration-tau (FTLD-tau). *J Mol Neurosci* 2011;**45**:384–9.

216. Dickson TC, Vickers JC. The morphological phenotype of beta-amyloid plaques and associated neuritic changes in Alzheimer's disease. *Neuroscience* 2001;**105**:99–107.

217. Ding ZT, Wang Y, Jiang YP, Yoshida M, Mimuro M, et al. Argyrophilic grain disease: frequency and neuropathology in centenarians. *Acta Neuropathol* 2006;**111**:320–8.

218. Diniz BS, Teixeira AL. Brain-derived neurotrophic factor and Alzheimer's disease: physiopathology and beyond. *Neuromolecular Med* 2011;**13**:217–22.

219. Dobson-Stone C, Luty AA, Thompson EM, Blumbergs P, Brooks WS, et al. Frontotemporal dementia-amyotrophic lateral sclerosis syndrome locus on chromosome 16p12.1-q12.2: genetic, clinical and neuropathological analysis. *Acta Neuropathol* 2013;**125**:523–33.

220. Donix M, Small GW, Bookheimer SY. Family history and APOE-4 genetic risk in Alzheimer's disease. *Neuropsychol Rev* 2012;**22**:298–309.

221. Dormann D, Haass C. Fused in sarcoma (FUS): an oncogene goes awry in neurodegeneration. *Mol Cell Neurosci* 2013;**56**:475–86.

222. Douaud G, Refsum H, de Jager CA, Jacoby R, Nichols TE, et al. Preventing Alzheimer's disease-related gray matter atrophy by B-vitamin treatment. *Proc Natl Acad Sci U S A* 2013;**110**:9523–8.

223. Double KL, Reyes S, Werry EL, Halliday GM. Selective cell death in neurodegeneration: why are some neurons spared in vulnerable regions? *Prog Neurobiol* 2010;**92**:316–29.

224. Douglas PM, Dillin A. Protein homeostasis and aging in neurodegeneration. *J Cell Biol* 2010;**190**:719–29.

225. Dubinsky JM. Heterogeneity of nervous system mitochondria: location, location, location! *Exp Neurol* 2009;**218**:293–307.

226. Duda JE, Giasson BI, Mabon ME, Lee VM, Trojanowski JQ. Novel antibodies to synuclein show abundant striatal pathology in Lewy body diseases. *Ann Neurol* 2002;**52**:205–10.

227. Duff K, Eckman C, Zehr C, et al. Increased amyloid-?42(43) in brains of mice expressing mutant presenilin 1. *Nature* 1996;**383**:710–13.

228. Dufouil C, Richard F, Fievet N, Dartigues JF, Ritchie K, et al. APOE genotype, cholesterol level, lipid-lowering treatment, and dementia: the Three-City Study. *Neurology* 2005;**64**:1531–8.

229. Duncan MW, Kopin IJ, Garruto RM, et al. 2-Amino-3 (methylamino)-propionic acid in cycad-derived foods is an unlikely cause of amyotrophic lateral sclerosis/parkinsonism. *Lancet* 1988;**2**:631–2.

230. Duyckaerts C. Neurodegenerative lesions: seeding and spreading. *Rev Neurol (Paris)* 2013;**169**:825–33.

231. Duyckaerts C, Delatour B, Potier MC. Classification and basic pathology of Alzheimer disease. *Acta Neuropathol* 2009;**118**:5–36.

232. Elliott MS, Ballard CG, Kalaria RN, Perry R, Hortobagyi T, Francis PT. Increased binding to 5-HT1A and 5-HT2A receptors is associated with large vessel infarction and relative preservation of cognition. *Brain* 2009;**132**:1858–65.

233. Ellis RJ, Olichney JM, Thal LJ, et al. Cerebral amyloid angiopathy in the brains of patients with Alzheimer's disease: the CERAD experience, Part XV. *Neurology* 1996;**46**:1592–6.

234. Emre M. Dementia associated with Parkinson's disease. *Lancet Neurol* 2003;**2**:229–37.

235. Emre M, Aarsland D, Brown R, Burn DJ, Duyckaerts C, et al. Clinical diagnostic criteria for dementia associated with Parkinson's disease. *Mov Disord* 2007;**22**:1689–707;quiz 837.

236. Englund E. Neuropathology of white matter changes in Alzheimer's disease and vascular dementia. *Dement Geriatr Cogn Disord* 1998;**9**:6–12.

237. Erkinjuntti T, Haltia M, Palo J, Sulkava R, Paetau A. Accuracy of the clinical diagnosis of vascular dementia: a prospective clinical and post-mortem neuropathological study. *J Neurol Neurosurg Psychiatry* 1988;**51**:1037–44.

238. Erkinjuntti T, Inzitari D, Pantoni L, Wallin A, Scheltens P, et al. Limitations of clinical criteria for the diagnosis of vascular dementia in clinical trials. Is a focus on

subcortical vascular dementia a solution? *Ann N Y Acad Sci* 2000;**903**:262–72.

239. Erkinjuntti T, Inzitari D, Pantoni L, Wallin A, Scheltens P, et al. Research criteria for subcortical vascular dementia in clinical trials. *J Neural Transm Suppl* 2000;**59**:23–30.

240. Ertekin-Taner N, Ronald J, Feuk L, Prince J, Tucker M, et al. Elevated amyloid beta protein (Abeta42) and late onset alzheimer's disease are associated with single nucleotide polymorphisms in the urokinase-type plasminogen activator gene. *Hum Mol Genet* 2005;**14**:447–60.

241. Esiri MM, Wilcock GK, Morris JH. Neuropathological assessment of the lesions of significance in vascular dementia. *J Neurol Neurosurg Psychiatry* 1997;**63**:749–53.

242. Esiri MM, Nagy Z, Smith MZ, et al. Cerebrovascular disease and threshold for dementia in the early stages of Alzheimer's disease (Letter). *Lancet* 1999;**354**:919–20.

243. Etminan M, Gill S, Samii A. Effect of non-steroidal anti-inflammatory drugs on risk of Alzheimer's disease: systematic review and meta-analysis of observational studies. [see comment]. *BMJ* 2003;**327**:128.

244. Fagan AM, Perrin RJ. Upcoming candidate cerebrospinal fluid biomarkers of Alzheimer's disease. *Biomark Med* 2012;**6**:455–76.

245. Faux NG, Ellis KA, Porter L, Fowler CJ, Laws SM, et al. Homocysteine, vitamin B12, and folic acid levels in Alzheimer's disease, mild cognitive impairment, and healthy elderly: baseline characteristics in subjects of the Australian Imaging Biomarker Lifestyle study. *J Alzheimer's Dis* 2011;**27**:909–22.

246. Fazekas F, Kleinert R, Roob G, Kleinert G, Kapeller P, et al. Histopathologic analysis of foci of signal loss on gradient-echo T2*-weighted MR images in patients with spontaneous intracerebral hemorrhage: evidence of microangiopathy-related microbleeds. *AJNR Am J Neuroradiol* 1999;**20**:637–42.

247. Feany MB, Dickson DW. Widespread cytoskeletal pathology characterizes corticobasal degeneration. *Am J Pathol* 1995;**146**:1388–96.

248. Fernando MS, Ince PG, MRC Cognitive Function and Ageing Neuropathology Study Group. Vascular pathologies and cognition in a population-based cohort of elderly people. *J Neurol Sci* 2004;**226**:13–7.

249. Fernando MS, Simpson JE, Matthews F, Brayne C, Lewis CE, et al. White matter lesions in an unselected cohort of the elderly: molecular pathology suggests origin from chronic hypoperfusion injury. *Stroke* 2006;**37**:1391–8.

250. Ferrari R, Hernandez DG, Nalls MA, Rohrer JD, Ramasamy A, et al. Frontotemporal dementia and its subtypes: a genome-wide association study. *Lancet Neurol* 2014;**13**:686–99.

251. Ferrer I. Cognitive impairment of vascular origin: neuropathology of cognitive impairment of vascular origin. *J Neurol Sci* 2010;**299**:139–49.

252. Ferrer I, Hernandez I, Boada M, Llorente A, Rey MJ, et al. Primary progressive aphasia as the initial manifestation of corticobasal degeneration and unusual tauopathies. *Acta Neuropathol* 2003;**106**:419–35. Epub 2003/09/05.

253. Ferrer I, Kaste M, Kalimo H. Vascular diseases. In: Love S, Louis D, Ellison D,

editors. *Greenfield's neuropathology.* 8th ed. Oxford: Oxford University Press; 2008; p. 121–240.

254. Ferrer I, Lopez-Gonzalez I, Carmona M, Arregui L, Dalfo E, et al. Glial and neuronal tau pathology in tauopathies: characterization of disease-specific phenotypes and tau pathology progression. *J Neuropathol Exp Neurol* 2014;**73**: 81–97.

255. Ferri CP, Prince M, Brayne C, Brodaty H, Fratiglioni L, et al. Global prevalence of dementia: a Delphi consensus study. *Lancet* 2005;**366**:2112–7.

256. Finch N, Baker M, Crook R, Swanson K, Kuntz K, et al. Plasma progranulin levels predict progranulin mutation status in frontotemporal dementia patients and asymptomatic family members. *Brain* 2009;**132**:583–91.

257. Firbank MJ, He J, Blamire AM, Singh B, Danson P, et al. Cerebral blood flow by arterial spin labeling in poststroke dementia. *Neurology* 2011;**76**:1478–84.

258. Firbank MJ, Allan LM, Burton EJ, Barber R, O'Brien JT, Kalaria RN. Neuroimaging predictors of death and dementia in a cohort of older stroke survivors. *J Neurol Neurosurg Psychiatry* 2012;**83**:263–7.

259. Fischer P, Jellinger K, Gatterer G, Danielczyk W. Prospective neuropathological validation of Hachinski's Ischaemic Score in dementias. *J Neurol Neurosurg Psychiatry* 1991;**54**:580–3.

260. Fisher CM. Lacunar strokes and infarcts: a review. *Neurology* 1982;**32**:871–6.

261. Folstein MF, Folstein SE, McHugh PR. "Mini-mental state". A practical method for grading the cognitive state of patients for the clinician. *J Psychiatr Res* 1975;**12**:189–98.

262. Forman MS, Zhukareva V, Bergeron C, Chin SS, Grossman M, et al. Signature tau neuropathology in gray and white matter of corticobasal degeneration. *Am J Pathol* 2002;**160**:2045–53.

263. Forno LS. Concentric hyalin intraneuronal inclusions of Lewy type in the brains of elderly persons (50 incidental cases): relationship to parkinsonism. *J Am Geriatr Soc* 1969;**17**:557–75.

264. Fratiglioni L, Launer LJ, Andersen K, Breteler MM, Copeland JR, et al. Incidence of dementia and major subtypes in Europe: A collaborative study of population-based cohorts. Neurologic Diseases in the Elderly Research Group. *Neurology* 2000;**54**:S10–5.

265. Fratta P, Poulter M, Lashley T, Rohrer JD, Polke JM, et al. Homozygosity for the *C9orf72* GGGGCC repeat expansion in frontotemporal dementia. *Acta Neuropathol* 2013;**126**:401–9.

266. Freeman SH, Spires-Jones T, Hyman BT, Growdon JH, Frosch MP. TAR-DNA binding protein 43 in Pick disease. *J Neuropathol Exp Neurol* 2008;**67**:62–7.

267. Freitas S, Simoes MR, Alves L, Vicente M, Santana I. Montreal Cognitive Assessment (MoCA): validation study for vascular dementia. *J Int Neuropsychol Soc* 2012;**18**:1031–40.

268. Fritsch T, McClendon MJ, Smyth KA, Ogrocki PK. Effects of educational attainment and occupational status on cognitive and functional decline in persons with Alzheimer-type dementia. *Int Psychogeriatr* 2002;**14**:347–63.

269. Fu C, Chute DJ, Farag ES, Garakian J, Cummings JL, Vinters HV. Comorbidity in dementia: an autopsy study. *Arch Pathol Lab Med* 2004;**128**:32–8.

270. Fu YJ, Nishihira Y, Kuroda S, Toyoshima Y, Ishihara T, et al. Sporadic four-repeat tauopathy with frontotemporal lobar degeneration, Parkinsonism, and motor neuron disease: a distinct clinicopathological and biochemical disease entity. *Acta Neuropathol* 2010;**120**:21–32.

271. Fuchs Y, Steller H. Programmed cell death in animal development and disease. *Cell* 2011;**147**:742–58.

272. Fujino Y, Wang DS, Thomas N, Espinoza M, Davies P, Dickson DW. Increased frequency of argyrophilic grain disease in Alzheimer disease with 4R tau-specific immunohistochemistry. *J Neuropathol Exp Neurol* 2005;**64**:209–14.

273. Fujishiro H, Uchikado H, Arai T, Hasegawa M, Akiyama H, et al. Accumulation of phosphorylated TDP-43 in brains of patients with argyrophilic grain disease. *Acta Neuropathol* 2009;**117**:151–8.

274. Funk KE, Mrak RE, Kuret J. Granulovacuolar degeneration (GVD) bodies of Alzheimer's disease (AD) resemble late-stage autophagic organelles. *Neuropathol Appl Neurobiol* 2011;**37**:295–306.

275. Gajdusek DC. Foci of motor neuron disease in high incidence in isolated populations of East Asia and the Western Pacific. *Adv Neurol* 1982;**36**:363–93.

276. Galanis DJ, Petrovitch H, Launer LJ, Harris TB, Foley DJ, White LR. Smoking history in middle age and subsequent cognitive performance in elderly Japanese-American men. The Honolulu-Asia Aging Study. *Am J Epidemiol* 1997;**145**:507–15.

277. Ganguli M, Dodge HH, Shen C, Pandav RS, DeKosky ST. Alzheimer disease and mortality: a 15-year epidemiological study. *Arch Neurol* 2005;**62**:779–84.

278. Garcia-Sierra F, Mondragon-Rodriguez S, Basurto-Islas G. Truncation of tau protein and its pathological significance in Alzheimer's disease. *J Alzheimer's Dis* 2008;**14**:401–9.

279. Garruto RM, Gajdusek DC, Chen KM. Amyotrophic lateral sclerosis and parkinsonism-dementia among Filipino migrants to Guam. *Ann Neurol* 1981;**10**:341–50.

280. Gearing M, Olson DA, Watts RL, Mirra SS. Progressive supranuclear palsy: neuropathologic and clinical heterogeneity. *Neurology* 1994;**44**:1015–24.

281. Gearing M, Mirra SS, Hedreen JC, Sumi SM, Hansen LA, Heyman A. The Consortium to Establish a Registry for Alzheimer's Disease (CERAD). Part X. Neuropathology confirmation of the clinical diagnosis of Alzheimer's disease. *Neurology* 1995;**45**:461–6.

282. Gearing M, Mori H, Mirra SS. Abeta-peptide length and apolipoprotein E genotype in Alzheimer's disease. *Ann Neurol* 1996;**39**:395–9.

283. Gemmell E, Bosomworth H, Allan L, Hall R, Khundakar A, et al. Hippocampal neuronal atrophy and cognitive function in delayed poststroke and aging-related dementias. *Stroke* 2012;**43**:808–14.

284. German DC, White CL, III, Sparkman DR. Alzheimer's disease: neurofibrillary tangles in nuclei that project to the cerebral cortex. *Neuroscience* 1987;21:305–12.

285. Gerrish A, Russo G, Richards A, Moskvina V, Ivanov D, et al. The role of variation at AbetaPP, *PSEN1*, *PSEN2*, and *MAPT* in late onset Alzheimer's disease. *J Alzheimer's Dis* 2012;28:377–87.

286. Geser F, Lee VM, Trojanowski JQ. Amyotrophic lateral sclerosis and frontotemporal lobar degeneration: a spectrum of TDP-43 proteinopathies. *Neuropathology* 2010;30:103–12.

287. Geula C, Mesulam MM. Cholinergic systems in Alzheimer disease. In: Terry R, Katzman R, Bick K, Sisodia S, editors. *Alzheimer disease*. Philadelphia: Lippincott Williams & Wilkins; 1999; p. 269–92.

288. Ghazi-Noori S, Froud KE, Mizielinska S, Powell C, Smidak M, et al. Progressive neuronal inclusion formation and axonal degeneration in *CHMP2B* mutant transgenic mice. *Brain* 2012;135:819–32.

289. Ghebremedhin E, Schultz C, Botez G, Rub U, Sassin I, et al. Argyrophilic grain disease is associated with apolipoprotein E epsilon 2 allele. *Acta Neuropathol* 1998;96:222–4.

290. Ghetti B, Tagliavini F, Giaccone G, et al. Familial Gerstmann–Sträussler–Scheinker disease with neurofibrillary tangles. *Mol Neurobiol* 1994;8:41–8.

291. Ghidoni R, Paterlini A, Albertini V, Binetti G, Benussi L. Losing protein in the brain: the case of progranulin. *Brain Res* 2012;1476:172–82.

292. Ghiso J, Plant GT, Revesz T, et al. Familial cerebral amyloid angiopathy (British type) with nonneuritic amyloid plaque formation may be due to a novel amyloid protein (Letter). *J Neurol Sci* 1995;129:74–5.

293. Ghiso J, Vidal R, Rostagno A, et al. A newly formed amyloidogenic fragment due to a stop codon mutation causes familial British dementia. *Ann N Y Acad Sci* 2000;903:129–37.

294. Ghosh S, Lippa C. Diagnosis and prognosis in idiopathic normal pressure hydrocephalus. *Am J Alzheimer's Dis Other Demen* 2014;29:583–89.

295. Giaccone G, Marcon G, Mangieri M, Morbin M, Rossi G, et al. Atypical tauopathy with massive involvement of the white matter. *Neuropathol Appl Neurobiol* 2008;34:468–72.

296. Giannakopoulos P, Kovari E, French LE, et al. Possible neuroprotective role of clusterin in Alzheimer's disease: a quantitative immunocytochemical study. *Acta Neuropathol* 1998;95:387–94.

297. Giannakopoulos P, Gold G, Kovari E, von Gunten A, Imhof A, et al. Assessing the cognitive impact of Alzheimer disease pathology and vascular burden in the aging brain: the Geneva experience. *Acta Neuropathol* 2007;113:1–12.

298. Gibb WR, Esiri MM, Lees AJ. Clinical and pathological features of diffuse cortical Lewy body disease (Lewy body dementia). *Brain* 1987;110:1131–53.

299. Gibson PH, Tomlinson BE. Numbers of Hirano bodies in the hippocampus of normal and demented people with Alzheimer's disease. *J Neurol Sci* 1977;33:199–206.

300. Giwa MO, Williams J, Elderfield K, Jiwa NS, Bridges LR, et al. Neuropathologic evidence of endothelial changes in cerebral small vessel disease. *Neurology* 2012;78:167–74.

301. Glass CK, Saijo K, Winner B, Marchetto MC, Gage FH. Mechanisms underlying inflammation in neurodegeneration. *Cell* 2010;140:918–34.

302. Glymour MM, Manly JJ. Lifecourse social conditions and racial and ethnic patterns of cognitive aging. *Neuropsychol Rev* 2008;18:223–54.

303. Goate A, Chartier-Harlin M, Mullan M, et al. Segregation of a missense mutation in the amyloid precursor protein gene with familial Alzheimer's disease. *Nature* 1991;349:704–6.

304. Goedert M, Wischik CM, Crowther RA, et al. Cloning and sequencing of the cDNA encoding a core protein of the paired helical filament of Alzheimer disease: identification as the microtubule-associated protein tau. *Proc Natl Acad Sci U S A* 1988;85:4051–5.

305. Goedert M, Spillantini MG, Jakes, et al. Multiple isoforms of human microtubule-associated protein tau: sequences and localization in neurofibrillary tangles of Alzheimer's disease. *Neuron* 1989;3:519–26.

306. Gold G, Giannakopoulos P, Montes-Paixao Junior C, Herrmann FR, Mulligan R, et al. Sensitivity and specificity of newly proposed clinical criteria for possible vascular dementia. *Neurology* 1997;49:690–4.

307. Gold G, Bouras C, Canuto A, Bergallo MF, Herrmann FR, et al. Clinicopathological validation study of four sets of clinical criteria for vascular dementia. *Am J Psychiatry* 2002;159:82–7.

308. Gold G, Giannakopoulos P, Herrmann FR, Bouras C, Kovari E. Identification of Alzheimer and vascular lesion thresholds for mixed dementia. *Brain* 2007;130:2830–6.

309. Goldman JE. The association of actin with Hirano bodies. *J Neuropathol Exp Neurol* 1983;42:146–52.

310. Goldman JG, Williams-Gray C, Barker RA, Duda JE, Galvin JE. The spectrum of cognitive impairment in Lewy body diseases. *Mov Disord* 2014;29:608–21.

311. Gonatas NK, Anderson W, Evangelista I. The contribution of altered synapses in the senile plaque: an electron microscopic study in Alzheimer's dementia. *J Neuropathol Exp Neurol* 1967;26:25–39.

312. Goos JD, Kester MI, Barkhof F, Klein M, Blankenstein MA, et al. Patients with Alzheimer disease with multiple microbleeds: relation with cerebrospinal fluid biomarkers and cognition. *Stroke* 2009;40:3455–60.

313. Gorelick PB. Risk factors for vascular dementia and Alzheimer disease. *Stroke* 2004;35:2620–2.

314. Gorelick PB, Chatterjee A, Patel D, Flowerdew G, Dollear W, et al. Cranial computed tomographic observations in multi-infarct dementia. A controlled study. *Stroke* 1992;23:804–11.

315. Gorelick PB, Freels S, Harris Y, et al. Epidemiology of vascular and Alzheimer's dementia among African Americans in Chicago, IL: baseline frequency and comparison of risk factors. *Neurology* 1994;44:1391–6.

316. Gorelick PB, Scuteri A, Black SE, Decarli C, Greenberg SM, et al. Vascular contributions to cognitive impairment and dementia: a statement for healthcare professionals from the American Heart Association/American Stroke Association. *Stroke* 2011;42:2672–713.

317. Gorno-Tempini ML, Hillis AE, Weintraub S, Kertesz A, Mendez M, et al. Classification of primary progressive aphasia and its variants. *Neurology* 2011;76:1006–14.

318. Gottfries CG, Blennow K, Karlsson I, Wallin A. The neurochemistry of vascular dementia. *Dementia* 1994;5:163–7.

319. Gould DB, Phalan FC, van Mil SE, Sundberg JP, Vahedi K, et al. Role of *COL4A1* in small-vessel disease and hemorrhagic stroke. *N Engl J Med* 2006;354:1489–96.

320. Gouras GK, Almeida CG, Takahashi RH. Intraneuronal Abeta accumulation and origin of plaques in Alzheimer's disease. *Neurobiol Aging* 2005;26:1235–44.

321. Gowing E, Roher AE, Woods AS, Cotter RJ, Chaney M, et al. Chemical characterization of A beta 17-42 peptide, a component of diffuse amyloid deposits of Alzheimer disease. *J Biol Chem* 1994;269:10987–90.

322. Graeber MB, Kosel S, Grasbon-Frodl E, et al. Histopathology and APOE genotype of the first Alzheimer disease patient, Auguste D. *Neurogenetics* 1998;1:223–8.

323. Graeber MB, Li W, Rodriguez ML. Role of microglia in CNS inflammation. *FEBS Lett* 2011;585:3798–805.

324. Gray F, Polivka M, Viswanathan A, Baudrimont M, Bousser MG, Chabriat H. Apoptosis in cerebral autosomal-dominant arteriopathy with subcortical infarcts and leukoencephalopathy. *J Neuropathol Exp Neurol* 2007;66:597–607.

325. Green RC, Cupples LA, Go R, Benke KS, Edeki T, et al. Risk of dementia among white and African American relatives of patients with Alzheimer disease. *JAMA* 2002;287:329–36.

326. Greenberg SM, Vonsattel JP, Stakes JW, et al. The clinical spectrum of cerebral amyloid angiopathy: presentations without lobar hemorrhage. *Neurology* 1993;43:2073–9.

327. Greenberg SM, Gurol ME, Rosand J, Smith EE. Amyloid angiopathy-related vascular cognitive impairment. *Stroke* 2004;35:2616–9.

328. Greenberg SM, Nandigam RN, Delgado P, Betensky RA, Rosand J, et al. Microbleeds versus macrobleeds: evidence for distinct entities. *Stroke* 2009;40:2382–6.

329. Greffard S, Verny M, Bonnet AM, Beinis JY, Gallinari C, et al. Motor score of the Unified Parkinson Disease Rating Scale as a good predictor of Lewy body-associated neuronal loss in the substantia nigra. *Arch Neurol* 2006;63:584–8.

330. Grinberg LT, Wang X, Wang C, Sohn PD, Theofilas P, et al. Argyrophilic grain disease differs from other tauopathies by lacking tau acetylation. *Acta Neuropathol* 2013;125:581–93.

331. Guerreiro R, Wojtas A, Bras J, Carrasquillo M, Rogaeva E, et al. TREM2 variants in Alzheimer's disease. *N Engl J Med* 2013;368:117–27.

332. Guo L, Itaya M, Takanashi M, Mizuno Y, Mori H. Relationship between Parkinson disease with dementia and dementia with Lewy bodies. *Parkinsonism Relat Disord* 2005;11:305–9.

333. Gyure KA, Durham R, Stewart WF, et al. Intraneuronal AB-amyloid precedes development of amyloid plaques in

Down syndrome. *Arch Pathol Lab Med* 2001;**125**:489–92.

334. Ha S, Furukawa R, Stramiello M, Wagner JJ, Fechheimer M. Transgenic mouse model for the formation of Hirano bodies. *BMC Neurosci* 2011;**12**:97.

335. Haan J, Roos RA, Briet PE, et al. Hereditary cerebral hemorrhage with amyloidosis—Dutch type. Research Group Hereditary Cerebral Amyloid–Angiopathy. *Clin Neurol Neurosurg* 1989;**91**:285–90.

336. Haan J, Lanser JB, Zijderveld I, et al. Dementia in hereditary cerebral hemorrhage with amyloidosis-Dutch type. *Arch Neurol* 1990;**47**:965–7.

337. Haass C. Take five: BACE and the gamma-secretase quartet conduct Alzheimer's amyloid beta-peptide generation. *EMBO J* 2004;**23**:483–8.

338. Habuchi C, Iritani S, Sekiguchi H, Torii Y, Ishihara R, et al. Clinicopathological study of diffuse neurofibrillary tangles with calcification. With special reference to TDP-43 proteinopathy and alpha-synucleinopathy. *J Neurol Sci* 2011;**301**:77–85.

339. Hachinski V, Iadecola C, Petersen RC, Breteler MM, Nyenhuis DL, et al. National Institute of Neurological Disorders and Stroke-Canadian Stroke Network vascular cognitive impairment harmonization standards. *Stroke* 2006;**37**:2220–41.

340. Hachinski VC, Lassen NA, Marshall J. Multi-infarct dementia. A cause of mental deterioration in the elderly. *Lancet* 1974;**2**:207–10.

341. Haeusler AR, Donnelly CJ, Periz G, Simko EA, Shaw PG, et al. C9orf72 nucleotide repeat structures initiate molecular cascades of disease. *Nature* 2014;**507**:195–200.

342. Haglund M, Kalaria R, Slade JY, Englund E. Differential deposition of amyloid beta peptides in cerebral amyloid angiopathy associated with Alzheimer's disease and vascular dementia. *Acta Neuropathol* 2006;**111**:430–5.

343. Haglund M, Passant U, Sjobeck M, Ghebremedhin E, Englund E. Cerebral amyloid angiopathy and cortical microinfarcts as putative substrates of vascular dementia. *Int J Geriatr Psychiatry* 2006;**21**:681–7.

344. Hall AM, Roberson ED. Mouse models of Alzheimer's disease. *Brain Res Bull* 2012;**88**:3–12.

345. Haller S, Garibotto V, Kovari E, Bouras C, Xekardaki A, et al. Neuroimaging of dementia in 2013: what radiologists need to know. *Eur Radiol* 2013;**23**:3393–404.

346. Halliday G, Bigio EH, Cairns NJ, Neumann M, Mackenzie IR, Mann DM. Mechanisms of disease in frontotemporal lobar degeneration: gain of function versus loss of function effects. *Acta Neuropathol* 2012;**124**:373–82.

347. Halliday GM, McCann HL, Pamphlett R, et al. Brain stem serotonin-synthesizing neurons in Alzheimer's disease: a clinicopathological correlation. *Acta Neuropathol* 1992;**84**:638–50.

348. Halliday GM, Song YJ, Lepar G, Brooks WS, Kwok JB, et al. Pick bodies in a family with presenilin-1 Alzheimer's disease. *Ann Neurol* 2005;**57**:139–43.

349. Halliday GM, Leverenz JB, Schneider JS, Adler CH. The neurobiological basis of cognitive impairment in Parkinson's disease. *Mov Disord* 2014;**29**:634–50.

350. Haltia M, Viitanen M, Sulkava R, et al. Chromosome 14-encoded Alzheimer's disease: genetic and clinicopathological description. *Ann Neurol* 1994;**36**:362–7.

351. Hamel E. Perivascular nerves and the regulation of cerebrovascular tone. *J Appl Physiol* 2006;**100**:1059–64.

352. Hamilton RL. Lewy bodies in Alzheimer's disease: a neuropathological review of 145 cases using alpha-synuclein immunohistochemistry. *Brain Pathol* 2000;**10**:378–84.

353. Hansen L, Salmon, Galasko D, et al. The Lewy body variant of Alzheimer's disease: a clinical and pathologic entity. *Neurology* 1990;**40**:1–8.

354. Hansen LA, Samuel W. Criteria for Alzheimer's disease and the nosology of dementia with Lewy bodies. *Neurology* 1997;**48**:126–32.

355. Hansen LA, Masliah E, Terry RD, Mirra SS. A neuropathological subset of Alzheimer's disease with concomitant Lewy body disease and spongiform change. *Acta Neuropathol* 1989;**78**: 194–201.

356. Hansen LA, Masliah E, Galasko D, Terry RD. Plaque-only Alzheimer disease is usually the Lewy body variant, and vice versa. *J Neuropathol Exp Neurol* 1993;**52**:648–54.

357. Hansen LA, Daniel SE, Wilcock GK, Love S. Frontal cortical synaptophysin in Lewy body diseases: relation to Alzheimer's disease and dementia. *J Neurol Neurosurg Psychiatry* 1998;**64**:653–6.

358. Hara K, Shiga A, Fukutake T, Nozaki H, Miyashita A, et al. Association of HTRA1 mutations and familial ischemic cerebral small-vessel disease. *N Engl J Med* 2009;**360**:1729–39.

359. Hardy J, Allsop D. Amyloid deposition as the central event in the aetiology of Alzheimer's disease. *Trends Pharmacol Sci* 1991;**12**:383–8.

360. Hardy J, Bogdanovic N, Winblad B, Portelius E, Andreasen N, et al. Pathways to Alzheimer's disease. *J Intern Med* 2014;**275**:296–303.

361. Hardy JA, Higgins GA. Alzheimer's disease: the amyloid cascade hypothesis. *Science* 1992;**256**:184–5.

362. Hasegawa M, Arai T, Nonaka T, Kametani F, Yoshida M, et al. Phosphorylated TDP-43 in frontotemporal lobar degeneration and amyotrophic lateral sclerosis. *Ann Neurol* 2008;**64**:60–70.

363. Hashimoto N, Takeuchi T, Ishihara R, Ukai K, Kobayashi H, et al. Glial fibrillary tangles in diffuse neurofibrillary tangles with calcification. *Acta Neuropathol* 2003;**106**:150–6.

364. Hassan A, Hunt BJ, O'Sullivan M, Parmar K, Bamford JM, et al. Markers of endothelial dysfunction in lacunar infarction and ischaemic leukoaraiosis. *Brain* 2003;**126**:424–32.

365. Hatanpaa KJ, Raisanen JM, Herndon E, Burns DK, Foong C, et al. Hippocampal sclerosis in dementia, epilepsy, and ischemic injury: differential vulnerability of hippocampal subfields. *J Neuropathol Exp Neurol* 2014;**73**:136–42.

366. Hauw JJ, Verny M, Delaere P, et al. Constant neurofibrillary changes in the neocortex in progressive supranuclear palsy. Basic differences with Alzheimer's disease and aging. *Neurosci Lett* 1990;**119**:182–6.

367. Hebert R, Brayne C. Epidemiology of vascular dementia. *Neuroepidemiology* 1995;**14**:240–57.

368. Hebert R, Lindsay J, Verreault R, et al. Vascular dementia: incidence and risk factors in the Canadian study of health and aging. *Stroke* 2000;**31**:1487–93.

369. Hellstrom-Lindahl E, Mousavi M, Ravid R, Nordberg A. Reduced levels of Abeta 40 and Abeta 42 in brains of smoking controls and Alzheimer's patients. *Neurobiol Dis* 2004;**15**:351-60.

369a. Hervé D, Chabriat H, Rigal M, Dalloz MA, Kawkabani Marchini A, De Lepeleire J, et al. A novel hereditary extensive vascular leukoencephalopathy mapping to chromosome 20q13. *Neurology* 2012;**79**:2283-7.

370. Heston LL, White JA, Mastri AR. Pick's disease. Clinical genetics and natural history. *Arch Gen Psychiatry* 1987;**44**:409–11.

371. Heutink P, Jansen IE, Lynes EM. C9orf72; abnormal RNA expression is the key. *Exp Neurol* 2014;**262PB**:102–10.

372. Heyman A, Fillenbaum G, Prosnitz B, Raiford K, Burchett B, Clark C. Estimated prevalence of dementia among elderly black and white community residents. *Arch Neurol* 1991;**48**:594–8.

373. Heyman A, Fillenbaum GG, Welsh-Bohmer KA, Gearing M, Mirra SS, et al. Cerebral infarcts in patients with autopsy-proven Alzheimer's disease: CERAD, part XVIII. Consortium to Establish a Registry for Alzheimer's Disease. *Neurology* 1998;**51**:159–62.

374. Hirano A. Hirano bodies and related neuronal inclusions. *Neuropathol Appl Neurobiol* 1994;**20**:3–11.

375. Hirano A, Malamud N, Elizan TS, Kurland LT. Amyotrophic lateral sclerosis and Parkinsonism-dementia complex on Guam. Further pathologic studies. *Arch Neurol* 1966;**15**:35–51.

376. Hirano A, Dembitzer HM, Kurland LT, Zimmerman HM. The fine structure of some intraganglionic alterations. Neurofibrillary tangles, granulovacuolar bodies and 'rod-like' structures as seen in Guam amyotrophic lateral sclerosis and parkinsonism–dementia complex. *J Neuropathol Exp Neurol* 1968;**27**:167–82.

377. Hirano A, Tuazon R, Zimmerman HM. Neurofibrillary changes, granulovacuolar bodies and argentophilic globules observed in tuberous sclerosis. *Acta Neuropathol* 1968;**11**:257–61.

378. Hishikawa N, Hashizume Y, Ujihira N, Okada Y, Yoshida M, Sobue G. Alpha-synuclein-positive structures in association with diffuse neurofibrillary tangles with calcification. *Neuropathol Appl Neurobiol* 2003;**29**:280–7.

379. Ho KL, Garcia JH. Neuropathology of the small blood vessels in selected disease of the cerebral white matter. In: Pantoni L, Inzitari D, Wallin A, editors. *The matter of white matter: current issues in neurodegenerative diseases*, Vol 10. Utrecht: Academic Pharmaceutical Productions; 2000; p. 247–273.

380. Ho RC, Cheung MW, Fu E, Win HH, Zaw MH, et al. Is high homocysteine level a risk factor for cognitive decline in elderly? A systematic review, meta-analysis, and meta-regression. *Am J Geriatr Psychiatry* 2011;**19**:607–17.

381. Hochgrafe K, Sydow A, Mandelkow EM. Regulatable transgenic mouse models

16

of Alzheimer disease: onset, reversibility and spreading of Tau pathology. *FEBS J* 2013;**280**:4371–81.

382. Hof PR, Bouras C, Constantinidis J, Morrison JH. Balint's syndrome in Alzheimer's disease: specific disruption of the occipito-parietal visual pathway. *Brain Res* 1989;**493**:368–75.

383. Hof PR, Bouras C, Constantinidis J, Morrison JH. Selective disconnection of specific visual association pathways in cases of Alzheimer's disease presenting with Balint's syndrome. *J Neuropathol Exp Neurol* 1990;**49**:168–84.

384. Hof PR, Cox K, Morrison JH. Quantitative analysis of a vulnerable subset of pyramidal neurons in Alzheimer's disease: I. Superior frontal and inferior temporal cortex. *J Comp Neurol* 1990;**301**:44–54.

385. Hof PR, Nimchinsky EA, Buee-Scherrer V, et al. Amyotrophic lateral sclerosis/ parkinsonism–dementia complex of Guam: quantitative neuropathology, immunohistochemical analysis of neuronal vulnerability, and comparison with related neurodegenerative disorders. *Acta Neuropathol* 1994;**88**:397–404.

386. Holland PR, Bastin ME, Jansen MA, Merrifield GD, Coltman RB, et al. MRI is a sensitive marker of subtle white matter pathology in hypoperfused mice. *Neurobiol Aging* 2011;**32**:2325, e1–6.

387. Holm IE, Englund E, Mackenzie IR, Johannsen P, Isaacs AM. A reassessment of the neuropathology of frontotemporal dementia linked to chromosome 3. *J Neuropathol Exp Neurol* 2007;**66**:884–91.

388. Holm IE, Isaacs AM, Mackenzie IR. Absence of FUS-immunoreactive pathology in frontotemporal dementia linked to chromosome 3 (FTD-3) caused by mutation in the *CHMP2B* gene. *Acta Neuropathol* 2009;**118**:719–20.

389. Holmes C, Cairns N, Lantos P, Mann A. Validity of current clinical criteria for Alzheimer's disease, vascular dementia and dementia with Lewy bodies. *Br J Psychiatry* 1999;**174**:45–50.

390. Holtzman DM, Bales KR, Tenkova T, et al. Apolipoprotein E isoform-dependent amyloid deposition and neuritic degeneration in a mouse model of Alzheimer's disease. *Proc Natl Acad Sci U S A* 2000;**97**:2892–7.

391. Honig LS, Tang MX, Albert S, Costa R, Luchsinger J, et al. Stroke and the risk of Alzheimer disease. *Arch Neurol* 2003;**60**:1707–12.

392. Horaitis O, Talbot CC, Jr., Phommarinh M, Phillips KM, Cotton RG. A database of locus-specific databases. *Nat Genet* 2007;**39**:425.

393. Horoupian DS, Wasserstein PH. Alzheimer's disease pathology in motor cortex in dementia with Lewy bodies clinically mimicking corticobasal degeneration. *Acta Neuropathol* 1999;**98**:317–22.

394. Huey ED, Ferrari R, Moreno JH, Jensen C, Morris CM, et al. FUS and TDP43 genetic variability in FTD and CBS. *Neurobiol Aging* 2012;**33**:1016.e9-17.

395. Hughes AJ, Daniel SE, Kilford L, Lees AJ. Accuracy of clinical diagnosis of idiopathic Parkinson's disease: a clinico-pathological study of 100 cases. *J Neurol Neurosurg Psychiatry* 1992;**55**:181–4.

396. Hughes AJ, Daniel SE, Blankson S, Lees AJ. A clinicopathologic study of 100 cases of Parkinson's disease. *Arch Neurol* 1993;**50**:140–8.

397. Hulette C, Nochlin D, McKeel D, Morris JC, Mirra SS, et al. Clinical-neuropathologic findings in multi-infarct dementia: a report of six autopsied cases. *Neurology* 1997;**48**:668–72.

398. Hunt WE. Clinical assessment of SAH. *J Neurosurg* 1983;**59**:550–1.

399. Hyman BT. The neuropathological diagnosis of Alzheimer's disease: clinical–pathological studies. *Neurobiol Aging* 1997;**18**:S27–32.

400. Hyman BT, Van Hoesen GW, Damasio AR, Barnes CL. Alzheimer's disease: cell-specific pathology isolates the hippocampal formation. *Science* 1984;**225**:1168–70.

401. Hyman BT, West HL, Rebeck GW, et al. Neuropathological changes in Down's syndrome hippocampal formation. Effect of age and apolipoprotein E genotype. *Arch Neurol* 1995;**52**:373–8.

402. Hyman BT, Phelps CH, Beach TG, Bigio EH, Cairns NJ, et al. National Institute on Aging-Alzheimer's Association guidelines for the neuropathologic assessment of Alzheimer's disease. *Alzheimer's Dement* 2012;**8**:1–13.

403. Ibrahim F, Nakaya T, Mourelatos Z. RNA dysregulation in diseases of motor neurons. *Annu Rev Pathol* 2012;**7**:323–52.

404. Ihara M, Tomimoto H. Lessons from a mouse model characterizing features of vascular cognitive impairment with white matter changes. J Aging Res 2011;**2011**:978761.

405. Ihara M, Polvikoski TM, Hall R, Slade JY, Perry RH, et al. Quantification of myelin loss in frontal lobe white matter in vascular dementia, Alzheimer's disease, and dementia with Lewy bodies. *Acta Neuropathol* 2010;**119**:579–89.

406. Ihara Y, Morishima-Kawashima M, Nixon R. The ubiquitin-proteasome system and the autophagic-lysosomal system in Alzheimer's disease. *Cold Spring Harb Perspect Med* 2012;**2**:a006361.

407. Ikeda K, Haga C, Kosaka K, Oyanagi S. Senile plaque-like structures: observation of a probably unknown type of senile plaque by periodic-acid methenamine silver (PAM) electron microscopy. *Acta Neuropathol* 1989;**78**):137–42.

408. Ikeda S, Yanagisawa N, Allsop D, Glenner GG. Early senile plaques in Alzheimer's disease demonstrated by histochemistry, immunocytochemistry, and electron microscopy. *Hum Pathol* 1990;**21**:1221–6.

409. Inzitari D, Pracucci G, Poggesi A, Carlucci G, Barkhof F, et al. Changes in white matter as determinant of global functional decline in older independent outpatients: three year follow-up of LADIS (leukoaraiosis and disability) study cohort. *BMJ* 2009;**339**:b2477.

410. Irwin DJ, Lee VM, Trojanowski JQ. Parkinson's disease dementia: convergence of alpha-synuclein, tau and amyloid-beta pathologies. *Nat Rev Neurosci* 2013;**14**:626–36.

411. Isaacs AM, Johannsen P, Holm I, Nielsen JE. Frontotemporal dementia caused by *CHMP2B* mutations. *Curr Alzheimer Res* 2011;**8**:246–51.

412. Iseki C, Takahashi Y, Wada M, Kawanami T, Adachi M, Kato T. Incidence of idiopathic normal pressure hydrocephalus (iNPH): a 10-year follow-up study of a rural community in Japan. *J Neurol Sci* 2014;**339**:108–12.

413. Ishihara K, Araki S, Ihori N, Shiota J, Kawamura M, et al. Argyrophilic grain disease presenting with frontotemporal dementia: a neuropsychological and pathological study of an autopsied case with presenile onset. *Neuropathology* 2005;**25**:165–70.

414. Ishii T, Kametani F, Haga S, Sato M. The immunohistochemical demonstration of subsequences of the precursor of the amyloid A4 protein in senile plaques in Alzheimer's disease. *Neuropathol Appl Neurobiol*1989;**15**:135–47.

415. Ishikawa E, Yanaka K, Sugimoto K, Ayuzawa S, Nose T. Reversible dementia in patients with chronic subdural hematomas. *J Neurosurg* 2002;**96**:680–3.

416. Ishizawa T, Ko LW, Cookson N, Davias P, Espinoza M, Dickson DW. Selective neurofibrillary degeneration of the hippocampal CA2 sector is associated with four-repeat tauopathies. *J Neuropathol Exp Neurol* 2002;**61**:1040–7.

417. Ito H. Basophilic inclusions and neuronal intermediate filament inclusions in amyotrophic lateral sclerosis and frontotemporal lobar degeneration. *Neuropathology* 2014;**34**:589–95.

418. Itoh K, Nakamura K, Iijima M, Sesaki H. Mitochondrial dynamics in neurodegeneration. *Trends Cell Biol* 2013;**23**:64–71.

419. Itoh Y, Yamada M. Cerebral amyloid angiopathy in the elderly: the clinicopathological features, pathogenesis, and risk factors. *J Med Dent Sci* 1997;**44**:11–9.

420. Iwasaki Y, Ito M, Mori K, Deguchi A, Nagaoka M, et al. An autopsy case of diffuse neurofibrillary tangles with calcification: early stage pathologic findings. *Neuropathology* 2009;**29**:697–703.

421. Iwatsubo T, Mann DM, Odaka A, et al. Amyloid beta protein (A beta) deposition: A beta 42(43) precedes A beta 40 in Down syndrome. *Ann Neurol* 1995;**37**:294–9.

422. Iwatsubo T, Odaka A, Suzuki N, et al. Visualization of A beta 42(43) and A beta 40 in senile plaques with end-specific A beta monoclonals: evidence that an initially deposited species is A beta 42(43). *Neuron* 1994;**13**:45–53.

423. Jack CR, Jr., Holtzman DM. Biomarker modeling of Alzheimer's disease. *Neuron* 2013;**80**:1347–58.

424. Jack CR, Jr., Barrio JR, Kepe V. Cerebral amyloid PET imaging in Alzheimer's disease. *Acta Neuropathol* 2013;**126**:643–57.

425. Janaway BM, Simpson JE, Hoggard N, Highley JR, Forster G, et al. Brain haemosiderin in older people: pathological evidence for an ischaemic origin of magnetic resonance imaging (MRI) microbleeds. *Neuropathol Appl Neurobiol* 2014;**40**:258–69.

426. Janocko NJ, Brodersen KA, Soto-Ortolaza AI, Ross OA, Liesinger AM, et al. Neuropathologically defined subtypes of Alzheimer's disease differ significantly from neurofibrillary tangle-predominant dementia. *Acta Neuropathol* 2012;**124**:681–92.

427. Janssens J, Van Broeckhoven C. Pathological mechanisms underlying TDP-43 driven neurodegeneration in FTLD-

ALS spectrum disorders. *Hum Mol Genet* 2013;**22**:R77–87.

428. Janvin CC, Larsen JP, Salmon DP, Galasko D, Hugdahl K, Aarsland D. Cognitive profiles of individual patients with Parkinson's disease and dementia: comparison with dementia with Lewy bodies and Alzheimer's disease. *Mov Disord* 2006;**21**:337–42.

429. Jaraj D, Rabiei K, Marlow T, Jensen C, Skoog I, Wikkelso C. Prevalence of idiopathic normal-pressure hydrocephalus. *Neurology* 2014;**82**:1449–54.

430. Jayadev S, Leverenz JB, Steinbart E, Stahl J, Klunk W, et al. Alzheimer's disease phenotypes and genotypes associated with mutations in presenilin 2. *Brain* 2010;**133**:1143–54.

431. Jellinger KA. Hippocampal sclerosis: a common pathological feature of dementia in very old humans (Editorial). *Acta Neuropathol* 1994;**88**:599.

432. Jellinger KA. Dementia with grains. *Brain Pathol* 1998;**8**:377–86.

433. Jellinger KA. Lewy body-related alpha-synucleinopathy in the aged human brain. *J Neural Transm* 2004;**111**:1219–35.

434. Jellinger KA. The pathology of "vascular dementia": a critical update. *J Alzheimer's Dis* 2008;**14**:107–23.

435. Jellinger KA. Basic mechanisms of neurodegeneration: a critical update. *J Cell Mol Med* 2010;**14**:457–87.

436. Jellinger KA, Attems J. Neurofibrillary tangle-predominant dementia: comparison with classical Alzheimer disease. *Acta Neuropathol* 2007;**113**:107–17.

437. Jellinger KA, Attems J. Neuropathological evaluation of mixed dementia. *J Neurol Sci* 2007;**257**:80–7.

438. Jellinger KA, Bancher C. Senile dementia with tangles (tangle predominant form of senile dementia). *Brain Pathol* 1998;**8**:367–76.

439. Jellinger KA, Seppi K, Wenning GK, Poewe W. Impact of coexistent Alzheimer pathology on the natural history of Parkinson's disease. *J Neural Transm* 2002;**109**:329–39.

440. Jen J, Cohen AH, Yue Q, Stout JT, Vinters HV, et al. Hereditary endotheliopathy with retinopathy, nephropathy, and stroke (HERNS). *Neurology* 1997;**49**:1322–30.

441. Jeppsson A, Zetterberg H, Blennow K, Wikkelso C. Idiopathic normal-pressure hydrocephalus: pathophysiology and diagnosis by CSF biomarkers. *Neurology* 2013;**80**:1385–92.

442. Jicha GA, Parisi JE, Dickson DW, Johnson K, Cha R, et al. Neuropathologic outcome of mild cognitive impairment following progression to clinical dementia. *Arch Neurol* 2006;**63**:674–81.

443. Jicha GA, Petersen RC, Knopman DS, Boeve BF, Smith GE, et al. Argyrophilic grain disease in demented subjects presenting initially with amnestic mild cognitive impairment. *J Neuropathol Exp Neurol* 2006;**65**:602–9.

444. Jiwa NS, Garrard P, Hainsworth AH. Experimental models of vascular dementia and vascular cognitive impairment: a systematic review. *J Neurochem* 2010;**115**:814–28.

445. Joachim CL, Morris JH, Selkoe DJ. Clinically diagnosed Alzheimer's disease: autopsy results in 150 cases. *Ann Neurol* 1988;**24**:50–6.

446. Johnson VE, Stewart W, Smith DH. Traumatic brain injury and amyloid-beta pathology: a link to Alzheimer's disease? *Nat Rev Neurosci* 2010;**11**:361–70.

447. Jordan BD, Relkin NR, Ravdin LD, et al. Apolipoprotein E epsilon4 associated with chronic traumatic brain injury in boxing. *JAMA* 1997;**278**:136–40.

448. Jorm AF, Jolley D. The incidence of dementia: a meta-analysis. *Neurology* 1998;**51**:728–33.

449. Jorm AF, Korten AE, Henderson AS. The prevalence of dementia: a quantitative integration of the literature. *Acta Psychiatr Scand* 1987;**76**:465–79.

450. Josephs KA, Uchikado H, McComb RD, Bashir R, Wszolek Z, et al. Extending the clinicopathological spectrum of neurofilament inclusion disease. *Acta Neuropathol* 2005;**109**:427–32.

451. Josephs KA, Katsuse O, Beccano-Kelly DA, Lin WL, Uitti RJ et al. Atypical progressive supranuclear palsy with corticospinal tract degeneration. *J Neuropathol Exp Neurol* 2006;**65**:396–405.

452. Josephs KA, Murray ME, Whitwell JL, Parisi JE, Petrucelli L, et al. Staging TDP-43 pathology in Alzheimer's disease. *Acta Neuropathol* 2014;**127**:441–50.

453. Josephs KA, Whitwell JL, Weigand SD, Murray ME, Tosakulwong N, et al. TDP-43 is a key player in the clinical features associated with Alzheimer's disease. *Acta Neuropathol* 2014;published online Mar 23 (E-pub ahead of print).

454. Joutel A. Pathogenesis of CADASIL: transgenic and knock-out mice to probe function and dysfunction of the mutated gene, Notch3, in the cerebrovasculature. *Bioessays* 2010;**33**:73–80.

455. Jung Y, Dickson DW, Murray ME, Whitwell JL, Knopman DS, et al. TDP-43 in Alzheimer's disease is not associated with clinical FTLD or Parkinsonism. *J Neurol* 2014;**261**:1344–8

456. Kahn J, Anderton BH, Probst A, et al. Immunohistological study of granulovacuolar degeneration using monoclonal antibodies to neurofilaments. *J Neurol Neurosurg Psychiatry* 1985;**48**:924–6.

457. Kaji R, Izumi Y, Adachi Y, Kuzuhara S. ALS-parkinsonism-dementia complex of Kii and other related diseases in Japan. *Parkinsonism Relat Disord* 2012;**18**:S190–1.

458. Kalaria RN. The blood–brain barrier and cerebrovascular pathology in Alzheimer's disease. *Ann N Y Acad Sci* 1999;**893**:113–25.

459. Kalaria RN. The role of cerebral ischemia in Alzheimer's disease. *Neurobiol Aging* 2000;**21**:321–30.

460. Kalaria RN, Ballard C. Overlap between pathology of Alzheimer disease and vascular dementia. *Alzheimer Dis Assoc Disord* 1999;**13**:S115–23.

461. Kalaria RN, Ihara M. Dementia: Vascular and neurodegenerative pathways-will they meet? *Nat Rev Neurol* 2013;**9**:487–8.

462. Kalaria RN, Kenny RA, Ballard CG, Perry R, Ince P, Polvikoski T. Towards defining the neuropathological substrates of vascular dementia. *J Neurol Sci* 2004;**226**:75–80.

463. Kalaria RN, Maestre GE, Arizaga R, Friedland RP, Galasko D, et al. Alzheimer's disease and vascular dementia in developing countries: prevalence, management, and risk factors. *Lancet Neurol* 2008;**7**:812–26.

464. Kalaria RN, Akinyemi R, Ihara M. Does vascular pathology contribute to Alzheimer changes? *J Neurol Sci* 2012;**322**:141–7.

465. Kalaria RN, Perry RH, O'Brien J, Jaros E. Atheromatous disease in small intracerebral vessels, microinfarcts and dementia. *Neuropathol Appl Neurobiol* 2012;**38**:505–8.

466. Kalimo H, Kalaria RN. Hereditary forms of vascular dementia. In: Kalimo H, editor. *Cerebrovascular diseases, pathology & genetics*. Basel: ISN Neuropath Press; 2005; p. 324–84.

467. Kalimo H, Ruchoux MM, Viitanen M, Kalaria RN. CADASIL: a common form of hereditary arteriopathy causing brain infarcts and dementia. *Brain Pathol* 2002;**12**:371–84.

468. Kang J, Lemaire HG, Unterbeck A, et al. The precursor of Alzheimer's disease amyloid A4 protein resembles a cell-surface receptor. *Nature* 1987;**325**:733–6.

469. Kang JH, Logroscino G, De Vivo I, Hunter D, Grodstein F. Apolipoprotein E, cardiovascular disease and cognitive function in aging women. *Neurobiol Aging* 2005;**26**:475–84.

470. Karlinsky H, Vaula G, Haines JL, et al. Molecular and prospective phenotypic characterization of a pedigree with familial Alzheimer's disease and a missense mutation in codon 717 of the beta-amyloid precursor protein gene. *Neurology* 1992;**42**:1445–53.

471. Kato S, Hirano A, Umahara T, et al. Comparative immunohistochemical study on the expression of alpha B crystallin, ubiquitin and stress-response protein 27 in ballooned neurons in various disorders. *Neuropathol Appl Neurobiol* 1992;**18**:335–40.

472. Kavanagh D, Spitzer D, Kothari PH, Shaikh A, Liszewski MK, et al. New roles for the major human 3′-5′ exonuclease TREX1 in human disease. *Cell Cycle* 2008;**7**:1718–25.

473. Kawas C, Katzman R. Epidemiology of dementia and Alzheimer disease. In: Terry R, Katzman R, Bick K, Sisodia S, editors. *Alzheimer disease*. Philadelphia: Lippincott Williams & Wilkins; 1999; p. 95–116.

474. Kertesz A, Martinez-Lage P, Davidson W, Munoz DG. The corticobasal degeneration syndrome overlaps progressive aphasia and frontotemporal dementia. *Neurology* 2000;**55**:1368–75.

475. Kertesz A, McMonagle P, Jesso S. Extrapyramidal syndromes in frontotemporal degeneration. *J Mol Neurosci* 2011;**45**:336–42.

476. Keverne JS, Low WC, Ziabreva I, Court JA, Oakley AE, Kalaria RN. Cholinergic neuronal deficits in CADASIL. *Stroke* 2007;**38**:188–91.

477. Kidd M. Alzheimer's disease: an electron microscopic study. *Brain* 1964;**87**:307–20.

478. Kim SH, Wang R, Gordon DJ, et al. Furin mediates enhanced production of fibrillogenic ABri peptides in familial British dementia. *Nature Neurosci* 1999;**2**:984–8.

479. Kimonis VE, Fulchiero E, Vesa J, Watts G. VCP disease associated with myopathy, Paget disease of bone and frontotemporal dementia: review of a unique disorder. *Biochim Biophys Acta* 2008;**1782**:744–8.

480. King ME, Ghoshal N, Wall JS, et al. Structural analysis of Pick's disease-derived and in vitro-assembled tau filaments. *Am J Pathol* 2001;**158**:1481–90.

16

481. Kirvell SL, Elliott MS, Kalaria RN, Hortobagyi T, Ballard CG, Francis PT. Vesicular glutamate transporter and cognition in stroke: a case-control autopsy study. *Neurology* 2011;**75**:1803–9.

482. Kitamura A, Fujita Y, Oishi N, Kalaria RN, Washida K, et al. Selective white matter abnormalities in a novel rat model of vascular dementia. *Neurobiol Aging* 2012;**33**:1012, e25–35.

483. Kiuru S, Salonen O, Haltia M. Gelsolin-related spinal and cerebral amyloid angiopathy. *Ann Neurol* 1999;**45**:305–11.

484. Knopman DS, Parisi JE, Boeve BF, Cha RH, Apaydin H, et al. Vascular dementia in a population-based autopsy study. *Arch Neurol* 2003;**60**:569–75.

485. Knowles RB, Wyart C, Buldyrev SV, et al. Plaque-induced neurite abnormalities: implications for disruption of neural networks in Alzheimer's disease. *Proc Natl Acad Sci U S A* 1999;**96**:5274–9.

486. Kobayashi Z, Kawakami I, Arai T, Yokota O, Tsuchiya K, et al. Pathological features of FTLD-FUS in a Japanese population: analyses of nine cases. *J Neurol Sci* 2013;**335**:89–95.

487. Kogel D, Deller T, Behl C. Roles of amyloid precursor protein family members in neuroprotection, stress signaling and aging. *Exp Brain Res* 2012;**217**:471–9.

488. Kojro E, Fahrenholz F. The non-amyloidogenic pathway: structure and function of alpha-secretases. *Subcell Biochem* 2005;**38**:105–27.

489. Kolarova M, Garcia-Sierra F, Bartos A, Ricny J, Ripova D. Structure and pathology of tau protein in Alzheimer disease. *Int J Alzheimer's Dis* 2012;**2012**:731526.

490. Koppel J, Acker C, Davies P, Lopez OL, Jimenez H, et al. Psychotic Alzheimer's disease is associated with gender-specific tau phosphorylation abnormalities. *Neurobiol Aging* 2014;**35**:2021–8.

491. Kosaka K. Diffuse neurofibrillary tangles with calcification: a new presenile dementia. *J Neurol Neurosurg Psychiatry* 1994;**57**:594–6.

492. Kosaka K, Yoshimura M, Ikeda K, Budka H. Diffuse type of Lewy body disease: progressive dementia with abundant cortical Lewy bodies and senile changes of varying degree--a new disease? *Clin Neuropathol* 1984;**3**:185–92.

493. Kosik KS, Joachim CL, Selkoe DJ. Microtubule-associated protein tau (tau) is a major antigenic component of paired helical filaments in Alzheimer disease. *Proc Natl Acad Sci U S A* 1986;**83**:4044–8.

494. Kouri N, Whitwell JL, Josephs KA, Rademakers R, Dickson DW. Corticobasal degeneration: a pathologically distinct 4R tauopathy. *Nat Rev Neurol* 2011;**7**:263–72.

495. Kovacs GG, Majtenyi K, Spina S, Murrell JR, Gelpi E, et al. White matter tauopathy with globular glial inclusions: a distinct sporadic frontotemporal lobar degeneration. *J Neuropathol Exp Neurol* 2008;**67**:963–75.

496. Kovacs GG, Molnar K, Laszlo L, Strobel T, Botond G, et al. A peculiar constellation of tau pathology defines a subset of dementia in the elderly. *Acta Neuropathol* 2011;**122**:205–22.

497. Kovacs GG, Rozemuller AJ, van Swieten JC, Gelpi E, Majtenyi K, et al. Neuropathology of the hippocampus in FTLD-Tau with Pick bodies: A study of the BrainNet Europe Consortium. *Neuropathol Appl Neurobiol* 2013;**39**:166–78.

498. Kovari E, Gold G, Herrmann FR, Canuto A, Hof PR, et al. Cortical microinfarcts and demyelination significantly affect cognition in brain aging. *Stroke* 2004;**35**:410–4.

499. Kovari E, Gold G, Herrmann FR, Canuto A, Hof PR, et al. Cortical microinfarcts and demyelination affect cognition in cases at high risk for dementia. *Neurology* 2007;**68**:927–31.

500. Kragh CL, Ubhi K, Wyss-Coray T, Masliah E. Autophagy in dementias. *Brain Pathol* 2012;**22**:99–109.

501. Krantic S, Mechawar N, Reix S, Quirion R. Molecular basis of programmed cell death involved in neurodegeneration. *Trends Neurosci* 2005;**28**:670–6.

502. Kril JJ, Patel S, Harding AJ, Halliday GM. Patients with vascular dementia due to microvascular pathology have significant hippocampal neuronal loss. *J Neurol Neurosurg Psychiatry* 2002;**72**:747–51.

503. Ksiezak-Reding H, Morgan K, Mattiace LA, et al. Ultrastructure and biochemical composition of paired helical filaments in corticobasal degeneration. *Am J Pathol* 1994;**145**:1496–508.

504. Kudo M. Hypothalamic gangliocytoma. Selective appearance of neurofibrillary changes, granulovacuolar degeneration, and argentophilic bodies. *Acta Pathol Jpn* 1986**36**:1225–9.

505. Kurt MA, Davies DC, Kidd M. beta-Amyloid immunoreactivity in astrocytes in Alzheimer's disease brain biopsies: an electron microscope study. *Exp Neurol* 1999;**158**:221–8.

506. Kuruppu DK, Matthews BR. Young-onset dementia. *Semin Neurol* 2013;**33**:365–85.

507. Kurz AF. What is vascular dementia? *Int J Clin Pract Suppl* 2001;**120**:5–8.

508. Kuusisto E, Salminen A, Alafuzoff I. Early accumulation of p62 in neurofibrillary tangles in Alzheimer's disease: possible role in tangle formation. *Neuropathol Appl Neurobiol* 2002;**28**:228–37.

509. Lagalwar S, Berry RW, Binder LI. Relation of hippocampal phospho-SAPK/JNK granules in Alzheimer's disease and tauopathies to granulovacuolar degeneration bodies. *Acta Neuropathol* 2007;**113**:63–73.

510. Lagier-Tourenne C, Polymenidou M, Hutt KR, Vu AQ, Baughn M, et al. Divergent roles of ALS-linked proteins FUS/TLS and TDP-43 intersect in processing long pre-mRNAs. *Nat Neurosci* 2012;**15**:1488–97..

511. Lai F, Kammann E, Rebeck GW, et al. APOE genotype and gender effects on Alzheimer disease in 100 adults with Down syndrome. *Neurology* 1999;**53**:331–6.

512. Lammie GA. Pathology of small vessel stroke. *Br Med Bull* 2000;**56**:296–306.

513. Lanson NA, Jr., Pandey UB. FUS-related proteinopathies: lessons from animal models. *Brain Res* 2012;**1462**:44–60.

514. Lantos PL, Luthert PJ, Hanger D, et al. Familial Alzheimer's disease with the amyloid precursor protein position 717 mutation and sporadic Alzheimer's disease have the same cytoskeletal pathology. *Neurosci Lett* 1992;**137**:221–4.

515. Lantos PL, Ovenstone IM, Johnson J, et al. Lewy bodies in the brain of two members of a family with the 717 (Val to Ile) mutation of the amyloid precursor protein gene. *Neurosci Lett* 1994;**172**:77–9.

516. Larson EB. Illnesses causing dementia in the very elderly (Editorial). *N Engl J Med* 1993;**328**:203–5.

517. Lashley T, Rohrer JD, Bandopadhyay R, Fry C, Ahmed Z, et al. A comparative clinical, pathological, biochemical and genetic study of fused in sarcoma proteinopathies. *Brain* 2011;**134**:2548–64.

518. Lattante S, Rouleau GA, Kabashi E. TARDBP and FUS mutations associated with amyotrophic lateral sclerosis: summary and update. *Hum Mutat* 2013;**34**:812–26.

519. Launer LJ. Counting dementia: There is no one "best" way. *Alzheimer's Dement* 2011;**7**:10–4.

520. Launer LJ, Oudkerk M, Nilsson LG, et al. CASCADE: a European collaborative study on vascular determinants of brain lesions. Study design and objectives. *Neuroepidemiology* 2000;**19**:113–20.

521. Lautenschlager NT, Cupples LA, Rao VS, Auerbach SA, Becker R, et al. Risk of dementia among relatives of Alzheimer's disease patients in the MIRAGE study: What is in store for the oldest old? *Neurology* 1996;**46**:641–50.

522. Lazarov O, Marr RA. Neurogenesis and Alzheimer's disease: at the crossroads. *Exp Neurol* 2010;**223**:267–81.

523. Le Ber I, Camuzat A, Guerreiro R, Bouya-Ahmed K, Bras J, et al. *SQSTM1* mutations in French patients with frontotemporal dementia or frontotemporal dementia with amyotrophic lateral sclerosis. *JAMA Neurol* 2013;**70**:1403–10.

524. Le TV, Crook R, Hardy J, Dickson DW. Cotton wool plaques in non-familial late-onset alzheimer disease. *J Neuropathol Exp Neurol* 2001;**60**:1051–61.

525. Lee EB, Russ J, Jung H, Elman LB, Chahine LM, et al. Topography of FUS pathology distinguishes late-onset BIBD from aFTLD-U. *Acta Neuropathol Commun* 2013;**1**:1–11.

526. Lee JH, Bacskai BJ, Ayata C. Genetic animal models of cerebral vasculopathies. *Prog Mol Biol Transl Sci* 2012;**105**:25–55.

527. Lee SE. Guam dementia syndrome revisited in 2011. *Curr Opin Neurol*2011;**24**:517–24.

528. Lee SE, Rabinovici GD, Mayo MC, Wilson SM, Seeley WW, et al. Clinicopathological correlations in corticobasal degeneration. *Ann Neurol* 2011;**70**:327–40.

529. Lee VM, Balin BJ, Otvos L, Jr., Trojanowski JQ. A68: a major subunit of paired helical filaments and derivatized forms of normal Tau. *Science* 1991;**251**:675–8.

530. Leinonen V, Koivisto AM, Alafuzoff I, Pyykko OT, Rummukainen J, et al. Cortical brain biopsy in long-term prognostication of 468 patients with possible normal pressure hydrocephalus. *Neurodegener Dis* 2012;**10**:166–9.

531. Leinonen V, Koivisto AM, Savolainen S, Rummukainen J, Sutela A, et al. Post-mortem findings in 10 patients with presumed normal-pressure hydrocephalus and review of the literature. *Neuropathol Appl Neurobiol*2012;**38**:72–86.

532. Leverenz JB, Raskind MA. Early amyloid deposition in the medial temporal lobe of young Down syndrome patients: A regional quantitative analysis. *ExpNeurol* 1998;**50**:296–304.

533. Levy E, Lopez-Otin C, Ghiso J, et al. Stroke in Icelandic patients with hereditary amyloid angiopathy is related to a mutation in the cystatin C gene, an inhibitor of cysteine proteases. *J Exp Med* 1989;**169**:1771–8.

534. Lewis H, Beher D, Cookson N, Oakley A, Piggott M, et al. Quantification of Alzheimer pathology in ageing and dementia: age-related accumulation of amyloid-beta(42) peptide in vascular dementia. *Neuropathol Appl Neurobiol* 2006;**32**:103–18.

535. Leys D, Henon H, Mackowiak-Cordoliani MA, Pasquier F. Poststroke dementia. *Lancet Neurol* 2005;**4**:752–9.

536. Lim X, Yeo JM, Green A, Pal S. The diagnostic utility of cerebrospinal fluid alpha-synuclein analysis in dementia with Lewy bodies - a systematic review and meta-analysis. *Parkinsonism Relat Disord* 2013;**19**:851–8.

537. Ling H, O'Sullivan SS, Holton JL, Revesz T, Massey LA, et al. Does corticobasal degeneration exist? A clinicopathological re-evaluation. *Brain* 2010;**133**:2045–57.

538. Lippa CF. Synaptophysin immunoreactivity in Pick's disease: comparison with Alzheimer's disease and dementia with Lewy bodies. *Am J Alzheimer's Dis Other Demen* 2004;**19**:341–4.

539. Lippa CF, Saunders AM, Smith TW, et al. Familial and sporadic Alzheimer's disease: neuropathology cannot exclude a final common pathway. *Neurology* 1996;**46**:406–12.

540. Lippa CF, Fujiwara H, Mann DM, et al. Lewy bodies contain altered alpha-synuclein in brains of many familial Alzheimer's disease patients with mutations in presenilin and amyloid precursor protein genes. *Am J Pathol* 1998;**153**:1365–70.

541. Lipton AM, White CL 3rd, Bigio EH. Frontotemporal lobar degeneration with motor neuron disease-type inclusions predominates in 76 cases of frontotemporal degeneration. *Acta Neuropathol* 2004;**108**:379–85.

542. Lithner CU, Hedberg MM, Nordberg A. Transgenic mice as a model for Alzheimer's disease. *Curr Alzheimer Res* 2011;**8**:818–31.

543. Litvan I, Aarsland D, Adler CH, Goldman JG, Kulisevsky J, et al. MDS Task Force on mild cognitive impairment in Parkinson's disease: critical review of PD-MCI. *Mov Disord* 2011;**26**:1814–24.

544. Litvan I, Goldman JG, Troster AI, Schmand BA, Weintraub D, Petersen RC, et al. Diagnostic criteria for mild cognitive impairment in Parkinson's disease: Movement Disorder Society Task Force guidelines. *Mov Disord* 2012;**27**:349–56.

545. Liu CK, Miller BL, Cummings JL, Mehringer CM, Goldberg MA, et al. A quantitative MRI study of vascular dementia. *Neurology* 1992;**42**:138–43.

546. Liu Y, Julkunen V, Paajanen T, Westman E, Wahlund LO, et al. Education increases reserve against Alzheimer's disease--evidence from structural MRI analysis. *Neuroradiology* 2012;**54**:929–38.

547. Liu YC, Chiang PM, Tsai KJ. Disease animal models of TDP-43 proteinopathy and their pre-clinical applications. *Int J Mol Sci* 2013;**14**:20079–111.

548. Lleo A, Berezovska O, Growdon JH, Hyman BT. Clinical, pathological, and biochemical spectrum of Alzheimer disease associated with PS-1 mutations. *Am J Geriatr Psychiatry* 2004;**12**:146–56.

549. Lobo A, Launer LJ, Fratiglioni L, Andersen K, Di Carlo A, et al. Prevalence of dementia and major subtypes in Europe: A collaborative study of population-based cohorts. Neurologic Diseases in the Elderly Research Group. *Neurology* 2000;**54**:S4–9.

550. Lopera F, Ardilla A, Martinez A, et al. Clinical features of early-onset alzheimer disease in a large kindred with an E280A presenilin-1 mutation. *JAMA* 1997;**277**:793–9.

551. Love S, Saitoh T, Quijada S, et al. Alz-50: ubiquitin and tau immunoreactivity of neurofibrillary tangles, Pick bodies and Lewy bodies. *J Neuropathol Exp Neurol* 1988;**47**:393–405.

552. Love S, Bridges LR, Case CP. Neurofibrillary tangles in Niemann–Pick disease type C. *Brain* 1995;**118**:119–29.

553. Lovestone S. Diabetes and dementia: is the brain another site of end-organ damage? *Neurology* 1999;**53**:1907–9.

554. Lowe J, Blanchard A, Morrell K, et al. Ubiquitin is a common factor in intermediate filament inclusion bodies of diverse type in man, including those of Parkinson's disease, Pick's disease, and Alzheimer's disease, as well as Rosenthal fibres in cerebellar astrocytomas, cytoplasmic bodies in muscle, and mallory bodies in alcoholic liver disease. *J Pathol* 1988;**155**:9–15.

555. Lowe J, Errington DR, Lennox G, et al. Ballooned neurons in several neurodegenerative diseases and stroke contain alpha B crystallin. *Neuropathol Appl Neurobiol* 1992;**18**:341–50.

556. Lowe J, Hand N, Mayer RJ. Application of ubiquitin immunohistochemistry to the diagnosis of disease. *Methods Enzymol* 2005;**399**:86–119.

557. Loy CT, Schofield PR, Turner AM, Kwok JB. Genetics of dementia. *Lancet* 2014;**383**:828–40.

558. Luchsinger JA. Type 2 diabetes and cognitive impairment: linking mechanisms. *J Alzheimer's Dis* 2012;**30**:S185–98.

559. Luchsinger JA, Mayeux R. Dietary factors and Alzheimer's disease. *Lancet Neurol* 2004;**3**:579–87.

560. Mackenzie IR. The neuropathology and clinical phenotype of FTD with progranulin mutations. *Acta Neuropathol* 2007;**114**:49–54.

561. Mackenzie IR, Feldman H. Neurofilament inclusion body disease with early onset frontotemporal dementia and primary lateral sclerosis. *Clin Neuropathol* 2004;**23**:183–93.

562. Mackenzie IR, Neumann M. FET proteins in frontotemporal dementia and amyotrophic lateral sclerosis. *Brain Res* 2012;**1462**:40–3.

563. Mackenzie IR, Baborie A, Pickering-Brown S, Du Plessis D, Jaros E, et al. Heterogeneity of ubiquitin pathology in frontotemporal lobar degeneration: classification and relation to clinical phenotype. *Acta Neuropathol* 2006;**112**:539–49.

564. Mackenzie IR, Baker M, Pickering-Brown S, Hsiung GY, Lindholm C, et al. The neuropathology of frontotemporal lobar degeneration caused by mutations in the progranulin gene. *Brain* 2006;**129**:3081–90.

565. Mackenzie IR, Foti D, Woulfe J, Hurwitz TA. Atypical frontotemporal lobar degeneration with ubiquitin-positive, TDP-43-negative neuronal inclusions. *Brain* 2008;**131**:1282–93.

566. Mackenzie IR, Neumann M, Bigio EH, Cairns NJ, Alafuzoff I, et al. Nomenclature for neuropathologic subtypes of frontotemporal lobar degeneration: consensus recommendations. *Acta Neuropathol* 2009;**117**:15–8.

567. Mackenzie IR, Neumann M, Bigio EH, Cairns NJ, Alafuzoff I, et al. Nomenclature and nosology for neuropathologic subtypes of frontotemporal lobar degeneration: an update. *Acta Neuropathol* 2010;**119**:1–4.

568. Mackenzie IR, Munoz DG, Kusaka H, Yokota O, Ishihara K, et al. Distinct pathological subtypes of FTLD-FUS. *Acta Neuropathol* 2011;**121**:207–18.

569. Mackenzie IR, Neumann M, Baborie A, Sampathu DM, Du Plessis D, et al. A harmonized classification system for FTLD-TDP pathology. *Acta Neuropathol* 2011;**122**:111–3.

570. Mackenzie IR, Arzberger T, Kremmer E, Troost D, Lorenzl S, et al. Dipeptide repeat protein pathology in *C9ORF72* mutation cases: clinico-pathological correlations. *Acta Neuropathol* 2013;**126**:859–79.

571. Mackenzie IR, Frick P, Neumann M. The neuropathology associated with repeat expansions in the *C9ORF72* gene. *Acta Neuropathol* 2014;**127**:347–57.

572. Mahoney CJ, Beck J, Rohrer JD, Lashley T, Mok K, et al. Frontotemporal dementia with the *C9ORF72* hexanucleotide repeat expansion: clinical, neuroanatomical and neuropathological features. *Brain* 2012;**135**:736–50.

573. Makioka K, Yamazaki T, Takatama M, Ikeda M, Okamoto K. Immunolocalization of Smurf1 in Hirano bodies. *J Neurol Sci* 2014;**336**:24–8.

574. Mann DM, Esiri MM. The pattern of acquisition of plaques and tangles in the brains of patients under 50 years of age with Down's syndrome. *J Neurol Sci* 1989;**89**:169–79.

575. Mann DM, Yates PO, Marcyniuk B. The nucleus basalis of Meynert in multi-infarct (vascular) dementia. *Acta Neuropathol* 1986;**71**:332–7.

576. Mann DM, Prinja D, Davies CA, et al. Immunocytochemical profile of neurofibrillary tangles in Down's syndrome patients of different ages. *J Neurol Sci* 1989;**92**:247–60.

577. Mann DM, Jones D, Snowden JS, et al. Pathological changes in the brain of a patient with familial Alzheimer's disease having a missense mutation at codon 717 in the amyloid precursor protein gene. *Neurosci Lett* 1992;**137**:225–8.

578. Mann DM, Pickering-Brown SM, Takeuchi A, Iwatsubo T. Amyloid angiopathy and variability in amyloid beta deposition is determined by mutation position in presenilin-1-linked Alzheimer's disease. *Am J Pathol* 2001;**158**:2165–75.

579. Marcello E, Epis R, Saraceno C, Di Luca M. Synaptic dysfunction in Alzheimer's disease. *Adv Exp Med Biol* 2012;**970**:573–601.

580. Markesbery W. Vascular dementia. In: Markesbery WR, editor. *Neuropathology of dementing disorders*. London: Arnold; 1998; p. 293–311.

581. Markesbery W, Ehmann W. Oxidative stress in Alzheimer disease. In: Terry R, Katzman R, Bick K, Sisodia S, editors. *Alzheimer disease*. Philadelphia: Lippincott Williams & Wilkins; 1999; p. 401–14.

582. Markesbery WR, Carney JM. Oxidative alterations in Alzheimer's disease. *Brain Pathol* 1999;**9**:133–46.

583. Markesbery WR, Wang HZ, Kowall NW, et al. Morphometric image analysis of neuropil threads in Alzheimer's disease. *Neurobiol Aging* 1993;**14**:303–7.

584. Marlatt MW, Lucassen PJ. Neurogenesis and Alzheimer's disease: Biology and pathophysiology in mice and men. *Curr Alzheimer Res* 2010;**7**:113–25.

585. Martinez-Lage P, Munoz DG. Prevalence and disease associations of argyrophilic grains of Braak. *J Neuropathol Exp Neurol* 1997;**56**:157–64.

586. Masellis M, Sherborn K, Neto P, Sadovnick DA, Hsiung GY, et al. Early-onset dementias: diagnostic and etiological considerations. *Alzheimer's Res Ther* 2013;**5**:S7.

587. Masliah E, Cole G, Shimohama S, et al. Differential involvement of protein kinase C isozymes in Alzheimer's disease. *J Neurosci* 1990;**10**:2113–24.

588. Masliah E, Iimoto DS, Saitoh T, et al. Increased immunoreactivity of brain spectrin in Alzheimer disease: a marker for synapse loss? *Brain Res* 1990;**531**:36–44.

589. Masliah E, Mallory M, Hansen L, et al. Patterns of aberrant sprouting in Alzheimer's disease. *Neuron* 1991;**6**:729–39.

590. Masliah E, Ellisman M, Carragher B, et al. Three-dimensional analysis of the relationship between synaptic pathology and neuropil threads in Alzheimer disease. *J Neuropathol Exp Neurol* 1992;**51**:404–14.

591. Masse I, Bordet R, Deplanque D, Al Khedr A, Richard F, et al. Lipid lowering agents are associated with a slower cognitive decline in Alzheimer's disease.[see comment]. *J Neurol Neurosurg Psychiatry* 2005;**76**:1624–9.

592. Masullo C, Pocchiari M, Mariotti P, et al. The nucleus basalis of Meynert in parkinsonism-dementia of Guam: a morphometric study. *Neuropathol Appl Neurobiol* 1989;**15**:193–206.

593. Matsumoto S, Hirano A, Goto S. Spinal cord neurofibrillary tangles of Guamanian amyotrophic lateral sclerosis and parkinsonism–dementia complex: an immunohistochemical study. *Neurology* 1990;**40**:975–9.

594. Matthews FE, Arthur A, Barnes LE, Bond J, Jagger C, et al. A two-decade comparison of prevalence of dementia in individuals aged 65 years and older from three geographical areas of England: results of the Cognitive Function and Ageing Study I and II. *Lancet* 2013;**382**:1405–12

595. Matthews FE, Brayne C, Lowe J, McKeith I, Wharton SB, Ince P. Epidemiological pathology of dementia: attributable-risks at death in the Medical Research Council Cognitive Function and Ageing Study. *PLoS Med* 2009;**6**:e1000180.

596. Mattila PM, Roytta M, Lonnberg P, Marjamaki P, Helenius H, Rinne JO. Choline acetytransferase activity and striatal dopamine receptors in Parkinson's disease in relation to cognitive impairment. *Acta Neuropathol* 2001;**102**:160–6.

597. Maurage CA, Sergeant N, Schraen-Maschke S, Lebert F, Ruchoux MM, et al. Diffuse form of argyrophilic grain disease: a new variant of four-repeat tauopathy different from limbic argyrophilic grain disease. *Acta Neuropathol* 2003;**106**:575–83.

598. Mayeux R, Stern Y. Epidemiology of Alzheimer disease. *Cold Spring Harb Perspect Med* 2012;**2**:a006239.

599. Mayeux R, Ottman R, Maestre G, et al. Synergistic effects of traumatic head injury and apolipoprotein-epsilon 4 in patients with Alzheimer's disease. *Neurology* 1995;**45**:555–7.

600. McCarron MO, Nicoll JA. Cerebral amyloid angiopathy and thrombolysis-related intracerebral haemorrhage. *Lancet Neurol* 2004;**3**:484–92.

601. McGeer PL, Schwab C, McGeer EG, et al. Familial nature and continuing morbidity of the amyotrophic lateral sclerosis–parkinsonism dementia complex of Guam. *Neurology* 1997;**49**:400–9.

602. McGirr A, Cusimano MD. Familial aggregation of idiopathic normal pressure hydrocephalus: novel familial case and a family study of the NPH triad in an iNPH patient cohort. *J Neurol Sci* 2012;**321**:82–8.

603. McKee AC, Kosik KS, Kowall NW. Neuritic pathology and dementia in Alzheimer's disease. *Ann Neurol* 1991 ;**30**:156–65.

604. McKee AC, Stein TD, Nowinski CJ, Stern RA, Daneshvar DH, et al. The spectrum of disease in chronic traumatic encephalopathy. *Brain* 2013;**136**(Pt 1):43–64.

605. McKee AC, Daneshvar DH, Alvarez VE, Stein TD. The neuropathology of sport. *Acta Neuropathol* 2014;**127**:29–51.

606. McKeith I, Mintzer J, Aarsland D, Burn D, Chiu H, et al. Dementia with Lewy bodies. *Lancet Neurol* 2004;**3**:19–28.

607. McKeith IG, Galasko D, Kosaka K, et al. Consensus guidelines for the clinical and pathologic diagnosis of dementia with Lewy bodies (DLB): report of the consortium on DLB international workshop. *Neurology* 1996;**47**:1113–24.

608. McKeith IG, Ballard CG, Perry RH, et al. Prospective validation of consensus criteria for the diagnosis of dementia with Lewy bodies. *Neurology* 2000;**54**:1050–8.

609. McKeith IG, Dickson DW, Lowe J, Emre M, O'Brien JT, et al. Diagnosis and management of dementia with Lewy bodies: third report of the DLB Consortium. *Neurology* 2005;**65**:1863–72.

610. McKhann GM, Albert MS, Grossman M, Miller B, Dickson D, Trojanowski JQ. Clinical and pathological diagnosis of frontotemporal dementia: report of the Work Group on Frontotemporal Dementia and Pick's Disease. *Arch Neurol* 2001;**58**:1803–9.

611. McKhann GM, Knopman DS, Chertkow H, Hyman BT, Jack CR, Jr., et al. The diagnosis of dementia due to Alzheimer's disease: recommendations from the National Institute on Aging-Alzheimer's Association workgroups on diagnostic guidelines for Alzheimer's disease. *Alzheimer's Dement* 2011;**7**:263–9.

612. Mead S, James-Galton M, Revesz T, et al. Familial British dementia with amyloid angiopathy: early clinical, neuropsychological and imaging findings. *Brain* 2000;**123**:975–91.

613. Medina M, Avila J. The role of extracellular Tau in the spreading of neurofibrillary pathology. *Front Cell Neurosci* 2014;**8**:113.

614. Medway C, Morgan K. Review: The genetics of Alzheimer's disease; putting flesh on the bones. *Neuropathol Appl Neurobiol* 2014;**40**:97–105.

615. Mehta SG, Khare M, Ramani R, Watts GD, Simon M, et al. Genotype-phenotype studies of VCP-associated inclusion body myopathy with Paget disease of bone and/or frontotemporal dementia. *Clin Genet* 2013;**83**:422–31.

616. Mendez MF, Turner J, Gilmore GC, et al. Balint's syndrome in Alzheimer's disease: visuospatial functions. *Int Neurosci* 1990;**54**:339–46.

617. Mendez MF, Mastri AR, Sung JH, Frey WHD. Clinically diagnosed Alzheimer disease: neuropathologic findings in 650 cases. *Alzheimer Dis Assoc Disord* 1992;**6**:35–43.

618. Metcalf DJ, Garcia-Arencibia M, Hochfeld WE, Rubinsztein DC. Autophagy and misfolded proteins in neurodegeneration. *Exp Neurol* 2012;**238**:22–8.

619. Meyer H, Bug M, Bremer S. Emerging functions of the VCP/p97 AAA-ATPase in the ubiquitin system. *Nat Cell Biol* 2012;**14**:117–23.

620. Miki Y, Mori F, Tanji K, Kurotaki H, Kakita A, et al. An autopsy case of incipient Pick's disease: Immunohistochemical profile of early-stage Pick body formation. Neuropathology 2014;**34**:386–91.

621. Miranda E, Lomas DA. Neuroserpin: a serpin to think about. *Cell Mol Life Sci* 2006;**63**:709–22.

622. Miranda E, Romisch K, Lomas DA. Mutants of neuroserpin that cause dementia accumulate as polymers within the endoplasmic reticulum. *J Biol Chem* 2004;**279**:28283–91.

623. Mirra SS, Gearing M. Brain infarction and the clinical expression of Alzheimer disease (Letter). *JAMA* 1997;**278**:113; Discussion 4.

624. Mirra SS, Heyman A, McKeel D, Sumi SM, Crain BJ, et al. The Consortium to Establish a Registry for Alzheimer's Disease (CERAD). Part II. Standardization of the neuropathologic assessment of Alzheimer's disease. *Neurology* 1991;**41**:479–86.

625. Mirra SS, Hart MN, Terry RD. Making the diagnosis of Alzheimer's disease. A primer for practicing pathologists. *Arch Pathol Lab Med* 1993;**117**:132–44.

626. Miserez AR, Clavaguera F, Monsch AU, Probst A, Tolnay M. Argyrophilic grain disease: molecular genetic difference to other four-repeat tauopathies. *Acta Neuropathol* 2003;**106**:363–6.

627. Molina JA, Probst A, Villanueva C, Jimenez-Jimenez FJ, Madero S, et al. Primary progressive aphasia with glial cytoplasmic inclusions. *Eur Neurol* 1998;**40**:71–7.

628. Monte SMdl, Hutchins GM, Moore GW. Racial differences in the etiology of dementia and frequency of Alzheimer lesions in the brain. *J Nat Med Assoc* 1989;**81**:644–52.

629. Montine TJ, Phelps CH, Beach TG, Bigio EH, Cairns NJ, et al. National Institute on Aging-Alzheimer's Association guidelines for the neuropathologic assessment of Alzheimer's disease: a practical approach. *Acta Neuropathol* 2012;**123**:1–11.

630. Morgan C, Colombres M, Nunez MT, Inestrosa NC. Structure and function of amyloid in Alzheimer's disease. *Prog Neurobiol* 2004;**74**:323–49.

631. Morgan K. The three new pathways leading to Alzheimer's disease. *Neuropathol Appl Neurobiol* 2011;**37**:353–7.

632. Mori H. Pathological substrate of dementia in Parkinson's disease--its relation to DLB and DLBD. *Parkinsonism Relat Disord* 2005;**11**:S41–5.

633. Mori K, Weng SM, Arzberger T, May S, Rentzsch K, et al. The *C9orf72* GGGGCC repeat is translated into aggregating dipeptide-repeat proteins in FTLD/ALS. *Science* 2013;**339**:1335–8.

634. Morris JC, Storandt M, Miller JP, et al. Mild cognitive impairment represents early-stage Alzheimer disease. *Arch Neurol* 2001;**58**:397–405.

635. Morris JC, Blennow K, Froelich L, Nordberg A, Soininen H, et al. Harmonized diagnostic criteria for Alzheimer's disease: recommendations. *J Intern Med* 2014;**275**:204–13.

636. Mortimer JA, Snowdon DA, Markesbery WR. Head circumference, education and risk of dementia: findings from the Nun Study. *J Clin Exp Neuropsychol* 2003;**25**:671–9.

637. Mouton PR, Martin LJ, Calhoun ME, et al. Cognitive decline strongly correlates with cortical atrophy in Alzheimer's dementia. *Neurobiol Aging* 1998;**19**:371–7.

638. Mu Y, Gage FH. Adult hippocampal neurogenesis and its role in Alzheimer's disease. *Mol Neurodegener* 2011;**6**:85.

639. Mufson EJ, Binder L, Counts SE, DeKosky ST, de Toledo-Morrell L, et al. Mild cognitive impairment: pathology and mechanisms. *Acta Neuropathol* 2012;**123**:13–30.

640. Muller UC, Zheng H. Physiological functions of APP family proteins. *Cold Spring Harb Perspect Med* 2012;**2**:a006288.

641. Mulugeta E, Molina-Holgado F, Elliott MS, Hortobagyi T, Perry R, et al. Inflammatory mediators in the frontal lobe of patients with mixed and vascular dementia. *Dement Geriatr Cogn Disord* 2008;**25**:278–86..

642. Munoz DG, Wang D. Tangle-associated neuritic clusters. A new lesion in Alzheimer's disease and aging suggests that aggregates of dystrophic neurites are not necessarily associated with beta/A4. *Am J Pathol* 1992;**140**:1167–78.

643. Munoz DG, Neumann M, Kusaka H, Yokota O, Ishihara K, et al. FUS pathology in basophilic inclusion body disease. *Acta Neuropathol* 2009;**118**:617–27.

644. Munroe WA, Southwick PC, Chang L, Scharre DW, Echols CL, Jr., et al. Tau protein in cerebrospinal fluid as an aid in the diagnosis of Alzheimer's disease. *Ann Clin Lab Sci* 1995;**25**:207–17.

645. Murayama S, Mori H, Ihara Y, Tomonaga M. Immunocytochemical and ultrastructural studies of Pick's disease. *Ann Neurol* 1990;**27**:394–405.

646. Murphy GM, Jr., Ellis WG. The amygdala in Down's syndrome and familial Alzheimer's disease: four clinicopathological case reports. *Biol Psychiatry* 1991;**30**;:92–106.

647. Murray ME, Graff-Radford NR, Ross OA, Petersen RC, Duara R, Dickson DW. Neuropathologically defined subtypes of Alzheimer's disease with distinct clinical characteristics: a retrospective study. *Lancet Neurol* 2011;**10**:785–96.

648. Murray ME, Bieniek KF, Banks Greenberg M, DeJesus-Hernandez M, Rutherford NJ, et al. Progressive amnestic dementia, hippocampal sclerosis, and mutation in *C9ORF72*. *Acta Neuropathol* 2013;**126**:545–54.

649. Murray ME, Cannon A, Graff-Radford NR, Liesinger AM, Rutherford NJ, et al. Differential clinicopathologic and genetic features of late-onset amnestic dementias. *Acta Neuropathol* 2014;**128**:411–21.

650. Nagao S, Yokota O, Ikeda C, Takeda N, Ishizu H, et al. Argyrophilic grain disease as a neurodegenerative substrate in late-onset schizophrenia and delusional disorders. *Eur Arch Psychiatry Clin Neurosci* 2014;**264**:317–31.

651. Nagy Z. Mechanisms of neuronal death in Down's syndrome. *J Neural Transm Suppl* 1999;**57**:233–45.

652. Nagy Z, Esiri MM, Jobst KA, et al. Relative roles of plaques and tangles in the dementia of Alzheimer's disease: correlations using three sets of neuropathological criteria. *Dementia* 1995;**6**:21–31.

653. Nagy Z, Esiri MM, Jobst KA, Morris JH, King EM, et al. The effects of additional pathology on the cognitive deficit in Alzheimer disease. *J Neuropathol Exp Neurol* 1997;**56**:165–70.

654. Nakano I, Hirano A. Neuron loss in the nucleus basalis of Meynert in parkinsonism-dementia complex of Guam. *Ann Neurol* 1983;**13**:87–91.

655. Nakano I, Hirano A. Parkinson's disease: neuron loss in the nucleus basalis without concomitant Alzheimer's disease. *Ann Neurol* 1984;**15**:415–18.

656. Nalivaeva NN, Turner AJ. The amyloid precursor protein: a biochemical enigma in brain development, function and disease. *FEBS Lett* 2013;**587**:2046–54.

657. Nanda S, Bhatt SP, Pamula J, Woodruff WW, Fowler M, Miller D. Diffuse neurofibrillary tangles with calcification (DNTC): Kosaka-Shibayama disease in America. *Am J Alzheimer's Dis Other Demen* 2007;**22**:535–7.

658. Natte R, Maat-Schieman ML, Haan J, Bornebroek M, Roos RA, van Duinen SG. Dementia in hereditary cerebral hemorrhage with amyloidosis-Dutch type is associated with cerebral amyloid angiopathy but is independent of plaques and neurofibrillary tangles. *Ann Neurol* 2001;**50**:765–72.

659. Neary D, Snowden JS, Mann DM, et al. Frontal lobe dementia and motor neuron disease. *J Neurol Neurosurg Psychiatry* 1990;**53**:23–32.

660. Nelson PT, Alafuzoff I, Bigio EH, Bouras C, Braak H, et al. Correlation of Alzheimer disease neuropathologic changes with cognitive status: a review of the literature. *J Neuropathol Exp Neurol* 2012;**71**:362–81.

661. Nelson PT, Greenberg SG, Saper CB. Neurofibrillary tangles in the cerebral cortex of sheep. *Neurosci Lett* 1994;**170**:187–90.

662. Nelson PT, Kukull WA, Frosch MP. Thinking outside the box: Alzheimer-type neuropathology that does not map directly onto current consensus recommendations. *J Neuropathol Exp Neurol* 2010;**69**:449–54.

663. Nelson PT, Schmitt FA, Lin Y, Abner EL, Jicha GA, et al. Hippocampal sclerosis in advanced age: clinical and pathological features. *Brain* 2011;**134**:1506–18.

664. Nelson PT, Smith CD, Abner EL, Wilfred BJ, Wang WX, et al. Hippocampal sclerosis of aging, a prevalent and high-morbidity brain disease. *Acta Neuropathol* 2013;**126**:161–77.

665. Nelson PT, Estus S, Abner EL, Parikh I, Malik M, *et al. ABCC9* gene polymorphism is associated with hippocampal sclerosis of aging pathology. *Acta Neuropathol* 2014;**127**:825–43.

666. Neltner JH, Abner EL, Baker S, Schmitt FA, Kryscio RJ, et al. Arteriolosclerosis that affects multiple brain regions is linked to hippocampal sclerosis of ageing. *Brain* 2014;**137**:255–67.

667. Neumann M, Sampathu DM, Kwong LK, Truax AC, Micsenyi MC, Chou TT, et al. Ubiquitinated TDP-43 in frontotemporal lobar degeneration and amyotrophic lateral sclerosis. *Science* 2006;**314**:130–3.

668. Neumann M, Kwong LK, Lee EB, Kremmer E, Flatley A, et al. Phosphorylation of S409/410 of TDP-43 is a consistent feature in all sporadic and familial forms of TDP-43 proteinopathies. *Acta Neuropathol* 2009;**117**:137–49.

669. Neumann M, Rademakers R, Roeber S, Baker M, Kretzschmar HA, Mackenzie IR. A new subtype of frontotemporal lobar degeneration with FUS pathology. *Brain* 2009;**132**:2922–31.

670. Neumann M, Roeber S, Kretzschmar HA, Rademakers R, Baker M, Mackenzie IR. Abundant FUS-immunoreactive pathology in neuronal intermediate filament inclusion disease. *Acta Neuropathol* 2009;**118**:605–16.

671. Neumann M, Valori CF, Ansorge O, Kretzschmar HA, Munoz DG, et al. Transportin 1 accumulates specifically with FET proteins but no other transportin cargos in FTLD-FUS and is absent in FUS inclusions in ALS with FUS mutations. *Acta Neuropathol* 2012;**124**:705–16.

672. Nichtweiss M, Weidauer S, Treusch N, Hattingen E. White matter lesions and vascular cognitive impairment: part 1: typical and unusual causes. *Clin Neuroradiol* 2012;**22**:193–210.

673. Nilsson K, Gustafson L, Hultberg B. Elevated plasma homocysteine level is not primarily related to Alzheimer's disease. *Dement Geriatr Cogn Disord* 2012;**34**:121–7.

674. Nimmrich V, Ebert U. Is Alzheimer's disease a result of presynaptic failure? Synaptic dysfunctions induced by oligomeric beta-amyloid. *Rev Neurosci* 2009;**20**:1–12.

675. Nixon RA, Cataldo AM, Mathews PM. The endosomal-lysosomal system of neurons in Alzheimer's disease pathogenesis: a review. *Neurochem Res* 2000;**25**:1161–72.

676. Nolan KA, Lino MM, Seligmann AW, Blass JP. Absence of vascular dementia in an autopsy series from a dementia clinic. *J Am Geriatr Soc* 1998;**46**:597–604.

677. Nonaka T, Masuda-Suzukake M, Arai T, Hasegawa Y, Akatsu H, et al. Prion-like properties of pathological TDP-43 aggregates from diseased brains. *Cell Rep* 2013;**4**:124–34.

678. Nunnemann S, Last D, Schuster T, Forstl H, Kurz A, Diehl-Schmid J. Survival in a German population with frontotemporal lobar degeneration. *Neuroepidemiology* 2011;**37**:160–5.

679. Nunomura A, Perry G, Aliev G, et al. Oxidative damage is the earliest event in Alzheimer's disease. *J Neuropathol Exp Neurol* 2001;60:759–67.

680. O'Brien JT, Erkinjuntti T, Reisberg B, Roman G, Sawada T, et al. Vascular cognitive impairment. *Lancet Neurol* 2003;2:89–98.

681. Oddo S, Caccamo A, Green KN, Liang K, Tran L, et al. Chronic nicotine administration exacerbates tau pathology in a transgenic model of Alzheimer's disease. *Proc Natl Acad Sci U S A* 2005;102:3046–51.

682. O'Donnell HC, Rosand J, Knudsen KA, Furie KL, Segal AZ, et al. Apolipoprotein E genotype and the risk of recurrent lobar intracerebral hemorrhage. *N Engl J Med* 2000;342:240–5.

683. Ohkubo T, Sakasegawa Y, Toda H, Kishida H, Arima K, et al. Three-repeat Tau 69 is a major tau isoform in laser-microdissected Pick bodies. *Amyloid* 2006;13:1–5.

684. Okamoto Y, Ihara M, Fujita Y, Ito H, Takahashi R, Tomimoto H. Cortical microinfarcts in Alzheimer's disease and subcortical vascular dementia. *Neuroreport* 2009;20:990–6.

685. Okamoto Y, Yamamoto T, Kalaria RN, Senzaki H, Maki T, et al. Cerebral hypoperfusion accelerates cerebral amyloid angiopathy and promotes cortical microinfarcts. *Acta Neuropathol* 2012;123:381–94.

686. Olichney JM, Hansen LA, Hofstetter CR, et al. Cerebral infarction in Alzheimer's disease is associated with severe amyloid angiopathy and hypertension. *Arch Neurol* 1995;52:702–8.

687. Olichney JM, Ellis RJ, Katzman R, et al. Types of cerebrovascular lesions associated with severe cerebral amyloid angiopathy in Alzheimer's disease. *Ann N Y Acad Sci* 1997;826:493–7.

688. Olichney JM, Hansen LA, Hofstetter CR, et al. Association between severe cerebral amyloid angiopathy and cerebrovascular lesions in Alzheimer disease is not a spurious one attributable to apolipoprotein E4. *Arch Neurol* 2000;57:869–74.

689. Olichney JM, Hansen LA, Lee JH, et al. Relationship between severe amyloid angiopathy, apolipoprotein E genotype, and vascular lesions in Alzheimer's disease. *Ann N Y Acad Sci* 2000;903:138–43.

690. Onyike CU, Diehl-Schmid J. The epidemiology of frontotemporal dementia. *Int Rev Psychiatry* 2013;25:130–7.

691. Ophoff RA, DeYoung J, Service SK, Joosse M, Caffo NA, et al. Hereditary vascular retinopathy, cerebroretinal vasculopathy, and hereditary endotheliopathy with retinopathy, nephropathy, and stroke map to a single locus on chromosome 3p21.1-p21.3. *Am J Hum Genet* 2001;69:447–53.

692. Oslin D, Atkinson RM, Smith DM, Hendrie H. Alcohol related dementia: proposed clinical criteria. *Int J Geriatr Psychiatry* 1998;13:203–12.

693. O'Sullivan M. Imaging small vessel disease: lesion topography, networks, and cognitive deficits investigated with MRI. *Stroke* 2010;41:S154–8.

694. O'Sullivan M, Ngo E, Viswanathan A, Jouvent E, Gschwendtner A, et al. Hippocampal volume is an independent predictor of cognitive performance in CADASIL. *Neurobiol Aging* 2009;30:890–7.

695. Ott A, Stolk RP, van Harskamp F, et al. Diabetes mellitus and the risk of dementia: the Rotterdam Study. *Neurology* 1999;53:1937–42.

696. Overmyer M, Helisalmi S, Soininen H, et al. Astrogliosis and the ApoE genotype. An immunohistochemical study of postmortem human brain tissue. *Dement Geriatr Cogn Disord* 1999;10:252–7.

697. Oyanagi K, Makifuchi T, Ohtoh T, Chen KM, Gajdusek DC, Chase TN. Distinct pathological features of the gallyas- and tau-positive glia in the Parkinsonism-dementia complex and amyotrophic lateral sclerosis of Guam. *J Neuropathol Exp Neurol* 1997;56:308–16.

698. Pannozzo MA, Holland PR, Scullion G, Talbot R, Mullins JJ, Horsburgh K. Controlled hypertension induces cerebrovascular and gene alterations in Cyp1a1-Ren2 transgenic rats. *J Am Soc Hypertens* 2013;7:411–9.

699. Pantoni L. Cerebral small vessel disease: from pathogenesis and clinical characteristics to therapeutic challenges. *Lancet Neurol* 2010;9:689–701.

700. Pantoni L, Garcia JH. Pathogenesis of leukoaraiosis: a review. *Stroke* 1997;28:652–9.

701. Pantoni L, Sarti C, Alafuzoff I, Jellinger K, Munoz DG, et al. Postmortem examination of vascular lesions in cognitive impairment: a survey among neuropathological services. *Stroke* 2006;37:1005–9.

702. Pao WC, Dickson DW, Crook JE, Finch NA, Rademakers R, Graff-Radford NR. Hippocampal sclerosis in the elderly: genetic and pathologic findings, some mimicking Alzheimer disease clinically. *Alzheimer Dis Assoc Disord* 2011;25:364–8.

703. Pardossi-Piquard R, Checler F. The physiology of the beta-amyloid precursor protein intracellular domain AICD. *J Neurochem* 2012;120:109–24.

704. Parkkinen L, Kauppinen T, Pirttila T, Autere JM, Alafuzoff I. Alpha-synuclein pathology does not predict extrapyramidal symptoms or dementia. *Ann Neurol* 2005;57:82–91.

705. Pavlovic AM, Zidverc-Trajkovic J, Milovic MM, Pavlovic DM, Jovanovic Z, et al. Cerebral small vessel disease in pseudoxanthoma elasticum: three cases. *Can J Neurol Sci* 2005;32:115–8.

706. Pendlebury ST, Rothwell PM. Prevalence, incidence, and factors associated with pre-stroke and post-stroke dementia: a systematic review and meta-analysis. *Lancet Neurol* 2009;8:1006–18.

707. Perry EK, Gibson PH, Blessed G, Perry RH, Tomlinson BE. Neurotransmitter enzyme abnormalities in senile dementia. Choline acetyltransferase and glutamic acid decarboxylase activities in necropsy brain tissue. *J Neurol Sci* 1977;34:247–65.

708. Perry EK, Curtis M, Dick DJ, et al. Cholinergic correlates of cognitive impairment in Parkinson's disease: comparisons with Alzheimer's disease. *J Neurol Neurosurg Psychiatry* 1985;48:413–21.

709. Perry G, Friedman R, Shaw G, Chau V. Ubiquitin is detected in neurofibrillary tangles and senile plaque neurites of Alzheimer disease brains. *Proc Natl Acad Sci U S A* 1987;84:3033–6.

710. Perry G, Kawai M, Tabaton M, et al. Neuropil threads of Alzheimer's disease show a marked alteration of the normal cytoskeleton. *J Neurosci* 1991;11:1748–55.

711. Perry G, Nunomura A, Hirai K, et al. Oxidative damage in Alzheimer's disease: the metabolic dimension. *Int J Dev Neurosci* 2000;18:417–21.

712. Perry RH, Irving D, Blessed G, Perry EK, Fairbairn AF. Senile dementia of Lewy body type and spectrum of Lewy body disease. *Lancet* 1989;1:1088.

713. Peters N, Opherk C, Danek A, Ballard C, Herzog J, Dichgans M. The pattern of cognitive performance in CADASIL: a monogenic condition leading to subcortical ischemic vascular dementia. *Am J Psychiatry* 2005;162:2078–85.

714. Petersen RC. Mild cognitive impairment: current research and clinical implications. *Semin Neurol* 2007;27:22–31.

715. Petersen RC, Parisi JE, Dickson DW, Johnson KA, Knopman DS, et al. Neuropathologic features of amnestic mild cognitive impairment. *Arch Neurol* 2006;63:665–72.

716. Petkau TL, Leavitt BR. Progranulin in neurodegenerative disease. *Trends Neurosci* 2014;37:388–398.

717. Pfeifer LA, White LR, Ross GW, Petrovitch H, Launer LJ. Cerebral amyloid angiopathy and cognitive function: the HAAS autopsy study. *Neurology* 2002;58:1629–34.

718. Pham CT, de Silva R, Haik S, Verny M, Sachet A, et al. Tau-positive grains are constant in centenarians' hippocampus. *Neurobiol Aging* 2011;32:1296–303.

719. Philips T, Robberecht W. Neuroinflammation in amyotrophic lateral sclerosis: role of glial activation in motor neuron disease. *Lancet Neurol* 2011;10:253–63.

720. Pick A. Über die Beziehungen der senilen Hirnatrophie zur Aphasie. *Prager Med Wochenschr* 1892;17:165–7.

721. Pickering-Brown S, Baker M, Yen SH, et al. Pick's disease is associated with mutations in the tau gene. *Ann Neurol* 2000;48:859–67.

722. Pievani M, Paternico D, Benussi L, Binetti G, Orlandini A, et al. Pattern of structural and functional brain abnormalities in asymptomatic granulin mutation carriers. *Alzheimer's Dement* 2014;10:S354–63.

723. Piggott MA, Marshall EF, Thomas N, Lloyd S, Court JA, et al. Striatal dopaminergic markers in dementia with Lewy bodies, Alzheimer's and Parkinson's diseases: rostrocaudal distribution. *Brain* 1999;122:1449–68.

724. Piguet O, Halliday GM, Reid WG, Casey B, Carman R, et al. Clinical phenotypes in autopsy-confirmed Pick disease. *Neurology* 2011;76:253–9.

725. Plant GT, Revesz T, Barnard RO, et al. Familial cerebral amyloid angiopathy with nonneuritic amyloid plaque formation. *Brain* 1990;113:721–47.

726. Pletnikova O, Sloane KL, Renton AE, Traynor BJ, Crain BJ, et al. Hippocampal sclerosis dementia with the C9ORF72 hexanucleotide repeat expansion. *Neurobiol Aging* 2014;35:2419.e17–21.

727. Pohjasvaara T, Erkinjuntti T, Vataja R, Kaste M. Dementia three months after stroke. Baseline frequency and effect of different definitions of dementia in the Helsinki Stroke Aging Memory Study (SAM) cohort. *Stroke* 1997;28:785–92.

728. Pohjasvaara T, Erkinjuntti T, Ylikoski R, Hietanen M, Vataja R, Kaste M. Clinical

determinants of poststroke dementia. *Stroke* 1998;**29**:75–81.

729. Pohjasvaara T, Mantyla R, Salonen O, Aronen HJ, Ylikoski R, et al. MRI correlates of dementia after first clinical ischemic stroke. *J Neurol Sci* 2000;**181**:111–7.

730. Pohjasvaara T, Mantyla R, Ylikoski R, Kaste M, Erkinjuntti T. Comparison of different clinical criteria (DSM-III, ADDTC, ICD-10, NINDS-AIREN, DSM-IV) for the diagnosis of vascular dementia. National Institute of Neurological Disorders and Stroke-Association Internationale pour la Recherche et l'Enseignement en Neurosciences. *Stroke* 2000;**31**:2952–7.

731. Polivka M, Vallat AV, Woimant F, et al. Cerebral amyloid angiopathy (CAA) with presentation as a brain inflammatory pseudo-tumour. *Clin Exp Pathol* 1999;**47**:303–10.

732. Pollanen MS, Markiewicz P, Bergeron C, Goh MC. Twisted ribbon structure of paired helical filaments revealed by atomic force microscopy. *Am J Pathol* 1994;**144**:869–73.

733. Pollock NJ, Mirra SS, Binder LI, et al. Filamentous aggregates in Pick's disease, progressive supranuclear palsy, and Alzheimer's disease share antigenic determinants with microtubule-associated protein, tau [letter]. *Lancet* 1986;**2**:1211.

734. Polvikoski TM, van Straaten EC, Barkhof F, Sulkava R, Aronen HJ, et al. Frontal lobe white matter hyperintensities and neurofibrillary pathology in the oldest old. *Neurology* 2010;**75**:2071–8.

735. Powers JM, Byrne NP, Ito M, Takao M, Yankopoulou D, et al. A novel leukoencephalopathy associated with tau deposits primarily in white matter glia. *Acta Neuropathol* 2003;**106**:181–7.

736. Pozueta J, Lefort R, Shelanski ML. Synaptic changes in Alzheimer's disease and its models. *Neuroscience* 2013;**251**:51–65.

737. Premkumar DR, Cohen DL, Hedera P, Friedland RP, Kalaria RN. Apolipoprotein E-epsilon4 alleles in cerebral amyloid angiopathy and cerebrovascular pathology associated with Alzheimer's disease. *Am J Pathol* 1996;**148**:2083–95.

738. Prescott JW, Guidon A, Doraiswamy PM, Choudhury KR, Liu C, Petrella J. The Alzheimer structural connectome: changes in cortical network topology with increased amyloid plaque burden. *Radiology* 2014; published online May 22:132593 (E-pub ahead of print).

739. Price DL, Altschuler RJ, Struble RG, et al. Sequestration of tubulin in neurons in Alzheimer's disease. *Brain Res* 1986;**385**:305–10.

740. Probst A, Langui D, Lautenschlager C, et al. Progressive supranuclear palsy: extensive neuropil threads in addition to neurofibrillary tangles. Very similar antigenicity of subcortical neuronal pathology in progressive supranuclear palsy and Alzheimer's disease. *Acta Neuropathol* 1988;**77**:61–8.

741. Probst A, Herzig MC, Mistl C, Ipsen S, Tolnay M. Perisomatic granules (non-plaque dystrophic dendrites) of hippocampal CA1 neurons in Alzheimer's disease and Pick's disease: a lesion distinct from granulovacuolar degeneration. *Acta Neuropathol* 2001;**102**:636–44.

742. Probst A, Mistl C, Ipsen S, Tolnay M. Perisomatic granules of hippocampal CA1 neurons in Alzheimer's disease, pre-Alzheimer stage and Pick's disease: an overlooked pathological entity. *Adv Exp Med Biol* 2001;**487**:187–98.

743. Qiu C, Winblad B, Fratiglioni L. The age-dependent relation of blood pressure to cognitive function and dementia. *Lancet Neurol* 2005;**4**:487–99.

744. Qiu C, von Strauss E, Backman L, Winblad B, Fratiglioni L. Twenty-year changes in dementia occurrence suggest decreasing incidence in central Stockholm, Sweden. *Neurology* 2013;**80**:1888–94.

745. Rami A. Review: autophagy in neurodegeneration: firefighter and/ or incendiarist? *Neuropathol Appl Neurobiol* 2009;**35**:449–61.

746. Ratnavalli E, Brayne C, Dawson K, Hodges JR. The prevalence of frontotemporal dementia. *Neurology* 2002;**58**:1615–21.

747. Rauramaa T, Pikkarainen M, Englund E, Ince PG, Jellinger K, et al. Consensus recommendations on pathologic changes in the hippocampus: a postmortem multicenter inter-rater study. *J Neuropathol Exp Neurol* 2013;**72**:452–61.

748. Reiss AB. Cholesterol and apolipoprotein E in Alzheimer's disease. *Am J Alzheimer's Dis Other Demen* 2005;**20**:91–6.

749. Reitz C. Alzheimer's disease and the amyloid cascade hypothesis: a critical review. *Int J Alzheimer's Dis* 2012;**2012**:369808.

750. Reitz C, Brayne C, Mayeux R. Epidemiology of Alzheimer disease. *Nat Rev Neurol* 2011;**7**:137–52.

751. Renoux AJ, Todd PK. Neurodegeneration the RNA way. *Prog Neurobiol* 2012;**97**:173–89.

752. Renton AE, Majounie E, Waite A, Simon-Sanchez J, Rollinson S, et al. A hexanucleotide repeat expansion in C9ORF72 is the cause of chromosome 9p21-linked ALS-FTD. *Neuron* 2011;**72**:257–68.

753. Revesz T, Holton JL, Doshi B, et al. Cytoskeletal pathology in familial cerebral amyloid angiopathy (British type) with non-neuritic amyloid plaque formation. *Acta Neuropathol* 1999;**97**:170–6.

754. Revesz T, Ghiso J, Lashley T, Plant G, Rostagno A, et al. Cerebral amyloid angiopathies: a pathologic, biochemical, and genetic view. *J Neuropathol Exp Neurol* 2003;**62**:885–98.

755. Revesz T, Holton JL, Lashley T, Plant G, Frangione B, et al. Genetics and molecular pathogenesis of sporadic and hereditary cerebral amyloid angiopathies. *Acta Neuropathol* 2009;**118**:115–30.

756. Rewcastle NB, Ball MJ. Electron microscopic structure of the 'inclusion bodies' in Pick's disease. *Neurology* 1968;**18**:1205–13.

757. Reyes JF, Stone K, Ramos J, Maselli A. Formation of Hirano bodies after inducible expression of a modified form of an actin-cross-linking protein. *Eukaryot Cell* 2009;**8**:852–7.

758. Reynolds E. Vitamin B12, folic acid, and the nervous system. *Lancet Neurol* 2006;**5**:949–60.

759. Ridha B, Josephs KA. Young-onset dementia: a practical approach to diagnosis. *Neurologist* 2006;**12**:2–13.

760. Ridley NJ, Draper B, Withall A. Alcohol-related dementia: an update of the evidence. *Alzheimer's Res Ther* 2013;**5**:3.

761. Riedl L, Mackenzie IR, Forstl H, Kurz A, Diehl-Schmid J. Frontotemporal lobar degeneration: current perspectives. *Neuropsychiatr Dis Treat* 2014;**10**:297–310.

762. Riekse RG, Leverenz JB, McCormick W, Bowen JD, Teri L, et al. Effect of vascular lesions on cognition in Alzheimer's disease: a community-based study. *J Am Geriatr Soc* 2004;**52**:1442–8.

763. Rinne JO, Rummukainen J, Paljarvi L, Rinne UK. Dementia in Parkinson's disease is related to neuronal loss in the medial substantia nigra. *Ann Neurol* 1989;**26**:47–50.

764. Roberson ED. Mouse models of frontotemporal dementia. *Ann Neurol* 2012;**72**:837–49.

765. Robertson IH. A noradrenergic theory of cognitive reserve: implications for Alzheimer's disease. *Neurobiol Aging* 2013;**34**:298–308.

766. Rocca WA, Kokmen E. Frequency and distribution of vascular dementia. *Alzheimer Dis Assoc Disord* 1999;**13**:S9–14.

767. Rocca WA, Petersen RC, Knopman DS, Hebert LE, Evans DA, et al. Trends in the incidence and prevalence of Alzheimer's disease, dementia, and cognitive impairment in the United States. *Alzheimer's Dement* 2011;**7**:80–93.

768. Rockwood K, Darvesh S. The risk of dementia in relation to statins and other lipid lowering agents. *Neurol Res* 2003;**25**:601–4.

769. Rodriguez JJ, Verkhratsky A. Neurogenesis in Alzheimer's disease. *J Anat* 2011;**219**:78–89.

770. Rodriguez JJ, Noristani HN, Verkhratsky A. The serotonergic system in ageing and Alzheimer's disease. *Prog Neurobiol* 2012;**99**:15–41.

771. Roeber S, Mackenzie IR, Kretzschmar HA, Neumann M. TDP-43-negative FTLD-U is a significant new clinico-pathological subtype of FTLD. *Acta Neuropathol* 2008;**116**:147–57.

772. Roher A, Gray EG, Paula-Barbosa M. Alzheimer's disease: coated vesicles, coated pits and the amyloid-related cell. *Proc R Soc Lond B Biol Sci* 1988;**232**:367–73.

773. Roher AE, Esh C, Kokjohn T, Sue L, Beach T. Atherosclerosis and AD: analysis of data from the US National Alzheimer's Coordinating Center. *Neurology* 2005;**65**:974; author reply 974.

774. Rohrer JD, Ridgway GR, Modat M, Ourselin S, Mead S, et al. Distinct profiles of brain atrophy in frontotemporal lobar degeneration caused by progranulin and tau mutations. *Neuroimage* 2010;**53**:1070–6.

775. Rohrer JD, Lashley T, Schott JM, Warren JE, Mead S, et al. Clinical and neuroanatomical signatures of tissue pathology in frontotemporal lobar degeneration. *Brain* 2011;**134**:2565–81.

776. Roman GC. Vascular dementia may be the most common form of dementia in the elderly. *J Neurol Sci* 2002;**203–204**:7–10.

777. Roman GC. Vascular dementia revisited: diagnosis, pathogenesis, treatment,

and prevention. *Med Clin North Am* 2002;**86**:477–99.

778. Roman GC. Vascular dementia. Advances in nosology, diagnosis, treatment and prevention. *Panminerva Med* 2004;**46**:207–15.

779. Roman GC, Kalaria RN. Vascular determinants of cholinergic deficits in Alzheimer disease and vascular dementia. *Neurobiol Aging* 2006;**27**:1769–85.

780. Roman GC, Tatemichi TK, Erkinjuntti T, Cummings JL, Masdeu JC, et al. Vascular dementia: diagnostic criteria for research studies. Report of the NINDS-AIREN International Workshop. *Neurology* 1993;**43**:250–60.

781. Roman GC, Erkinjuntti T, Wallin A, Pantoni L, Chui HC. Subcortical ischaemic vascular dementia. *Lancet Neurol* 2002;**1**:426–36.

782. Roman GC, Sachdev P, Royall DR, Bullock RA, Orgogozo JM, et al. Vascular cognitive disorder: a new diagnostic category updating vascular cognitive impairment and vascular dementia. *J Neurol Sci* 2004;**226**:81–7.

783. Rosello A, Warnes G, Meier UC. Cell death pathways and autophagy in the central nervous system and its involvement in neurodegeneration, immunity and central nervous system infection: to die or not to die: that is the question. *Clin Exp Immunol* 2012;**168**:52–7.

784. Rosen C, Hansson O, Blennow K, Zetterberg H. Fluid biomarkers in Alzheimer's disease - current concepts. *Mol Neurodegener* 2013;**8**:20.

785. Rosen WG, Terry RD, Fuld PA, Katzman R, Peck A. Pathological verification of ischemic score in differentiation of dementias. *Ann Neurol* 1980;**7**:486–8.

786. Rossi R, Joachim C, Geroldi C, Combrinck M, Esiri MM, et al. Association between subcortical vascular disease on CT and neuropathological findings. *Int J Geriatr Psychiatry* 2004;**19**:690–5.

787. Rosso SM, Donker Kaat L, Baks T, Joosse M, de Koning I, et al. Frontotemporal dementia in The Netherlands: patient characteristics and prevalence estimates from a population-based study. *Brain* 2003;**126**:2016–22.

788. Rossor MN. Pick's disease: A clinical overview. *Neurology* 2001;**56**:S3–5.

789. Rossor MN, Fox NC, Mummery CJ, Schott JM, Warren JD. The diagnosis of young-onset dementia. *Lancet Neurol* 2010;**9**:793–806.

790. Rovelet-Lecrux A, Hannequin D, Raux G, Le Meur N, Laquerriere A, et al. APP locus duplication causes autosomal dominant early-onset alzheimer disease with cerebral amyloid angiopathy. *Nat Genet* 2006;**38**:24–6.

791. Rovelet-Lecrux A, Legallic S, Wallon D, Flaman JM, Martinaud O, et al. A genome-wide study reveals rare CNVs exclusive to extreme phenotypes of Alzheimer's disease. *Eur J Hum Genet* 2012;**20**:613–7.

792. Rubino E, Rainero I, Chio A, Rogaeva E, Galimberti D, *et al.* SQSTM1 mutations in frontotemporal lobar degeneration and amyotrophic lateral sclerosis. *Neurology* 2012;**79**:1556–62.

793. Rubinsztein DC. The roles of intracellular protein-degradation pathways in neurodegeneration. *Nature* 2006;**443**:780–6.

794. Rugarli EI, Langer T. Mitochondrial quality control: a matter of life and death for neurons. *EMBO J* 2012;**31**:1336–49.

795. Russ TC, Hamer M, Stamatakis E, Starr JM, Batty GD, Kivimaki M. Does the Framingham cardiovascular disease risk score also have predictive utility for dementia death? An individual participant meta-analysis of 11,887 men and women. *Atherosclerosis* 2013;**228**:256–8.

796. Sachdev P. Vascular cognitive disorder. *Int J Geriatr Psychiatry* 1999;**14**:402–3.

797. Sachdev P, Kalaria RN, O'Brien JT, Skoog I, Alladi S, et al. Diagnostic criteria for vascular cognitive disorders: a VASCOG statement. *Alzheimer Dis Assoc Disord* 2014;**28**:206–18.

798. Sadowski M, Pankiewicz J, Scholtzova H, Li YS, Quartermain D, et al. Links between the pathology of Alzheimer's disease and vascular dementia. *Neurochem Res* 2004;**29**:1257–66.

799. Saido T, Leissring MA. Proteolytic degradation of amyloid beta-protein. *Cold Spring Harb Perspect Med* 2012;**2**:a006379.

800. Saido TC, Iwata N. Metabolism of amyloid beta peptide and pathogenesis of Alzheimer's disease. Towards presymptomatic diagnosis, prevention and therapy. *Neurosci Res* 2006;**54**:235–53.

801. Saito Y, Ruberu NN, Sawabe M, Arai T, Tanaka N, et al. Staging of argyrophilic grains: an age-associated tauopathy. *J Neuropathol Exp Neurol* 2004;**63**:911–8.

802. Sampathu DM, Neumann M, Kwong LK, Chou TT, Micsenyi M, et al. Pathological heterogeneity of frontotemporal lobar degeneration with ubiquitin-positive inclusions delineated by ubiquitin immunohistochemistry and novel monoclonal antibodies. *Am J Pathol* 2006;**169**:1343–52.

803. Sampson EL, Warren JD, Rossor MN. Young onset dementia. *Postgrad Med J* 2004;**80**:125–39.

804. Santa-Maria I, Haggiagi A, Liu X, Wasserscheid J, Nelson PT, et al. The MAPT H1 haplotype is associated with tangle-predominant dementia. *Acta Neuropathol* 2012;**124**:693–704.

805. Satoh J, Tabunoki H, Ishida T, Saito Y, Arima K. Ubiquilin-1 immunoreactivity is concentrated on Hirano bodies and dystrophic neurites in Alzheimer's disease brains. *Neuropathol Appl Neurobiol* 2013;**39**:817–30.

806. Saxena S, Caroni P. Selective neuronal vulnerability in neurodegenerative diseases: from stressor thresholds to degeneration. *Neuron* 2011;**71**:35–48.

807. Scahill RI, Ridgway GR, Bartlett JW, Barnes J, Ryan NS, et al. Genetic influences on atrophy patterns in familial Alzheimer's disease: a comparison of APP and PSEN1 mutations. *J Alzheimer's Dis* 2013;**35**:199–212.

808. Scher AI, Xu Y, Korf ES, Hartley SW, Witter MP, et al. Hippocampal morphometry in population-based incident Alzheimer's disease and vascular dementia: the HAAS. *J Neurol Neurosurg Psychiatry* 2011;**82**:373–6.

809. Schmidt ML, Murray JM, Trojanowski JQ. Continuity of neuropil threads with tangle-bearing and tangle-free neurons in Alzheimer disease cortex. A confocal laser scanning microscopy study. *Mol Chem Neuropathol* 1993;**18**:299–312.

810. Schneider JA, Bennett DA. Where vascular meets neurodegenerative disease. *Stroke* 2010;**41**:S144–6.

811. Schneider JA, Wilson RS, Bienias JL, Evans DA, Bennett DA. Cerebral infarctions and the likelihood of dementia from Alzheimer disease pathology. *Neurology* 2004;**62**:1148–55.

812. Schneider JA, Arvanitakis Z, Bang W, Bennett DA. Mixed brain pathologies account for most dementia cases in community-dwelling older persons. *Neurology* 2007;**69**:2197–204.

813. Schneider JA, Arvanitakis Z, Leurgans SE, Bennett DA. The neuropathology of probable Alzheimer disease and mild cognitive impairment. *Ann Neurol* 2009;**66**:200–8.

814. Schon EA, Przedborski S. Mitochondria: the next (neurode)generation. *Neuron* 2011;**70**:1033–53.

815. Schott JM, Reiniger L, Thom M, Holton JL, Grieve J, et al. Brain biopsy in dementia: clinical indications and diagnostic approach. *Acta Neuropathol* 2010;**120**:327–41.

816. Schroder R, Watts GD, Mehta SG, Evert BO, Broich P, et al. Mutant valosin-containing protein causes a novel type of frontotemporal dementia. *Ann Neurol* 2005;**57**:457–61.

817. Scott IS, Lowe JS. The ubiquitin-binding protein p62 identifies argyrophilic grain pathology with greater sensitivity than conventional silver stains. *Acta Neuropathol* 2007;**113**:417–20.

818. Seno H, Ishino H, Inagaki T, Iijima M, Kaku K, Inata T. A neuropathological study of dementia in nursing homes over a 17-year period, in Shimane Prefecture, Japan. *Gerontology* 1999;**45**:44–8.

819. Sephton CF, Cenik B, Cenik BK, Herz J, Yu G. TDP-43 in central nervous system development and function: clues to TDP-43-associated neurodegeneration. *Biol Chem* 2012;**393**:589–94.

820. Sergeant N, Wattez A, Delacourte A. Neurofibrillary degeneration in progressive supranuclear palsy and corticobasal degeneration: tau pathologies with exclusively 'exon 10' isoforms. *J Neurochem* 1999;**72**:1243–9.

821. Sharp ES, Gatz M. Relationship between education and dementia: an updated systematic review. *Alzheimer Dis Assoc Disord* 2011;**25**:289–304.

822. Sharp SI, Francis PT, Elliott MS, Kalaria RN, Bajic N, et al. Choline acetyltransferase activity in vascular dementia and stroke. *Dement Geriatr Cogn Disord* 2009;**28**:233–8.

823. Shepherd C, McCann H, Halliday GM. Variations in the neuropathology of familial Alzheimer's disease. *Acta Neuropathol* 2009;**118**:37–52.

824. Shepherd CE, Gregory GC, Vickers JC, Halliday GM. Novel 'inflammatory plaque' pathology in presenilin-1 Alzheimer's disease. *Neuropathol Appl Neurobiol* 2005;**31**:503–11.

825. Shibayama H, Kitoh J. Electron microscopic structure of the Alzheimer's neurofibrillary changes in case of atypical senile dementia. *Acta Neuropathol* 1978;**41**:229–34.

826. Shiraki H, Yase Y. Amyotrophic lateral sclerosis in Japan. In: Vinken P, Bruyn G, editors. *Handbook of clinical neurology*. New York: Elsevier; 1975; p. 353–419.

827. Shively S, Scher AI, Perl DP, Diaz-Arrastia R. Dementia resulting from traumatic brain injury: what is the pathology? *Arch Neurol* 2012;**69**:1245–51.

828. Sieben A, Van Langenhove T, Engelborghs S, Martin JJ, Boon P, et al. The genetics and neuropathology of frontotemporal lobar degeneration. *Acta Neuropathol* 2012;**124**:353–72.

829. Silverberg GD, Mayo M, Saul T, Rubenstein E, McGuire D. Alzheimer's disease, normal-pressure hydrocephalus, and senescent changes in CSF circulatory physiology: a hypothesis. *Lancet Neurol* 2003;**2**:506–11.

830. Silverman JM, Raiford K, Edland S, et al. The Consortium to Establish a Registry for Alzheimer's Disease (CERAD). Part VI. Family history assessment: a multicenter study of first-degree relatives of Alzheimer's disease probands and nondemented spouse controls. *Neurology* 1994 ;**44**:1253–9.

831. Simchowicz T. Histologische Studien ueber die senile Demenz. *Histol Histopathol Arbeit Grosshirn* 1911;**4**:267–444.

832. Simpson JE, Fernando MS, Clark L, Ince PG, Matthews F, et al. White matter lesions in an unselected cohort of the elderly: astrocytic, microglial and oligodendrocyte precursor cell responses. *Neuropathol Appl Neurobiol* 2007;**33**:410–9.

833. Simpson JE, Hosny O, Wharton SB, Heath PR, Holden H, et al. Microarray RNA expression analysis of cerebral white matter lesions reveals changes in multiple functional pathways. *Stroke* 2009;**40**: 369–75.

834. Sims-Robinson C, Kim B, Rosko A, Feldman EL. How does diabetes accelerate Alzheimer disease pathology? *Nat Rev Neurol* 2010;**6**:551–9.

835. Skibinski G, Parkinson NJ, Brown JM, Chakrabarti L, Lloyd SL, et al. Mutations in the endosomal ESCRTIII-complex subunit *CHMP2B* in frontotemporal dementia. *Nat Genet* 2005;**37**:806–8.

836. Skoog I, Vanmechelen E, Andreasson LA, Palmertz B, Davidsson P, et al. A population-based study of tau protein and ubiquitin in cerebrospinal fluid in 85-year-olds: relation to severity of dementia and cerebral atrophy, but not to the apolipoprotein E4 allele. *Neurodegeneration* 1995;**4**: 433–42.

837. Skoog I, Hesse C, Fredman P, Andreasson LA, Palmertz B, Blennow K. Apolipoprotein E in cerebrospinal fluid in 85-year-old subjects. Relation to dementia, apolipoprotein E polymorphism, cerebral atrophy, and white matter lesions. *Arch Neurol* 1997;**54**:267–72.

838. Skoog I, Kalaria RN, Breteler MM. Vascular factors and Alzheimer disease. *Alzheimer Dis Assoc Disord* 1999;**13**:S106–14.

839. Slooter AJ, Cruts M, Kalmijn S, Hofman A, Breteler MM, et al. Risk estimates of dementia by apolipoprotein E genotypes from a population-based incidence study: the Rotterdam Study. *Arch Neurol* 1998;**55**:964–8.

840. Smallwood A, Oulhaj A, Joachim C, Christie S, Sloan C, et al. Cerebral subcortical small vessel disease and its relation to cognition in elderly subjects: a pathological study in the Oxford Project to Investigate Memory and Ageing (OPTIMA) cohort. *Neuropathol Appl Neurobiol* 2012;**38**:337–43.

841. Smith C. Review: the long-term consequences of microglial activation following acute traumatic brain injury.

Neuropathol Appl Neurobiol 2013; **39**:35–44.

842. Smith CD, Snowdon D, Markesbery WR. Periventricular white matter hyperintensities on MRI: correlation with neuropathologic findings. *J Neuroimaging* 2000;**10**:13–6.

843. Smith MA, Rottkamp CA, Nunomura A, et al. Oxidative stress in Alzheimer's disease. *Biochim Biophys Acta* 2000;**1502**:139–44.

844. Snowden JS, Hu Q, Rollinson S, Halliwell N, Robinson A, et al. The most common type of FTLD-FUS (aFTLD-U) is associated with a distinct clinical form of frontotemporal dementia but is not related to mutations in the FUS gene. *Acta Neuropathol* 2011;**122**:99–110.

845. Snowden JS, Rollinson S, Thompson JC, Harris JM, Stopford CL, et al. Distinct clinical and pathological characteristics of frontotemporal dementia associated with *C9ORF72* mutations. *Brain* 2012;**135**:693–708.

846. Snowdon DA, Greiner LH, Mortimer JA, Riley KP, Greiner PA, Markesbery WR. Brain infarction and the clinical expression of Alzheimer disease. The Nun Study. *JAMA* 1997;**277**:813–7.

847. Solitare GB, Lamarche JB. Alzheimer's disease and senile dementia as seen in mongoloids: neuropathological observations. *Am J Ment Defic* 1966;**70**:840–8.

848. Sonnen JA, Larson EB, Crane PK, Haneuse S, Li G, et al. Pathological correlates of dementia in a longitudinal, population-based sample of aging. *Ann Neurol* 2007;**62**:406–13.

849. Sorrentino P, Iuliano A, Polverino A, Jacini F, Sorrentino G. The dark sides of amyloid in Alzheimer's disease pathogenesis. *FEBS Lett* 2014;**588**:641–52.

850. Sparks LD, Kryscio RJ, Hunsaker JC, 3rd. Early age-related progression of AD-like neuropathology in Down's syndrome. *Am J Neurodegener Dis* 2013;**2**:121–8.

851. Spencer PS, Nunn PB, Hugon J, et al. Guam amyotrophic lateral sclerosis-parkinsonism-dementia linked to a plant excitant neurotoxin. *Science* 1987;**237**:517–22.

852. Spillantini MG, Goedert M. Tau pathology and neurodegeneration. *Lancet Neurol* 2013;**12**:609–22.

853. Spillantini MG, Bird TD, Ghetti B. Frontotemporal dementia and Parkinsonism linked to chromosome 17: a new group of tauopathies. *Brain Pathol* 1998;**8**:387–402.

854. Spillantini MG, Crowther RA, Jakes R, et al. alpha-Synuclein filamentous inclusions of Lewy bodies from Parkinson's disease and dementia with lewy bodies. *Proc Natl Acad Sci U S A* 1998;**95**:6469–73.

855. Spires-Jones TL, Hyman BT. The intersection of amyloid beta and tau at synapses in Alzheimer's disease. *Neuron* 2014;**82**:756–71.

856. Staekenborg SS, van der Flier WM, van Straaten EC, Lane R, Barkhof F, Scheltens P. Neurological signs in relation to type of cerebrovascular disease in vascular dementia. *Stroke* 2008;**39**:317–22.

857. Stamelou M, Quinn NP, Bhatia KP. "Atypical" atypical parkinsonism: new genetic conditions presenting with features of progressive supranuclear palsy,

corticobasal degeneration, or multiple system atrophy: a diagnostic guide. *Mov Disord* 2013;**28**:1184–99.

858. Starkstein SE, Jorge R, Capizzano AA. Uncommon causes of cerebrovascular dementia. *Int Psychogeriatr* 2005;**17**:S51–64.

859. Steele JC, Richardson JC, Olszewski J. Progressive supranuclear palsy: a heterogenous degeneration involving the brain stem, basal ganglia and cerebellum with vertical gaze and pseudobulbar palsy, nuchal dystonia and dementia. *Arch Neurol* 1964;**10**:333–59.

860. Stern Y. Cognitive reserve in ageing and Alzheimer's disease. *Lancet Neurol* 2012;**11**:1006–12.

861. Stoltzner SE, Grenfell TJ, Mori C, et al. Temporal accrual of complement proteins in amyloid plaques in Down's syndrome with Alzheimer's disease. *Am J Pathol* 2000;**156**:489–99.

862. Streichenberger N, Girard-Madoux P, Verejan I, et al. Giant cell angiitis of the central nervous system with amyloid angiopathy. A case report and review of the literature. *Clin Exp Pathol* 1999;**47**:311–17.

863. Strozyk D, Dickson DW, Lipton RB, Katz M, Derby CA, et al. Contribution of vascular pathology to the clinical expression of dementia. *Neurobiol Aging* 2010;**31**:1710–20.

864. Suenaga T, Hirano A, Llena JF, et al. Modified Bielschowsky and immunocytochemical studies on cerebellar plaques in Alzheimer's disease. *J Neuropathol Exp Neurol* 1990;**49**:31–40.

865. Suter OC, Sunthorn T, Kraftsik R, Straubel J, Darekar P, et al. Cerebral hypoperfusion generates cortical watershed microinfarcts in Alzheimer disease. *Stroke* 2002;**33**:1986–92.

866. Suzuki K, Terry RDT. Fine structural localization of acid phosphatase in senile plaques in Alzheimer's presenile dementia. *Acta Neuropathol* 1967;**8**:276–84.

867. Suzuki K, Parker CC, Pentchev PG, et al. Neurofibrillary tangles in Niemann-Pick disease type C. *Acta Neuropathol* 1995;**89**:227–38.

868. Takahashi M, Weidenheim KM, Dickson DW, Ksiezak-Reding H. Morphological and biochemical correlations of abnormal tau filaments in progressive supranuclear palsy. *J Neuropathol Exp Neurol* 2002;**61**:33–45.

869. Takahashi Y, Kawanami T, Nagasawa H, Iseki C, Hanyu H, Kato T. Familial normal pressure hydrocephalus (NPH) with an autosomal-dominant inheritance: a novel subgroup of NPH. *J Neurol Sci* 2011;**308**:149–51.

870. Takao M, Ghetti B, Yoshida H, Piccardo P, Narain Y, et al. Early-onset dementia with Lewy bodies. *Brain Pathol* 2004;**14**:137–47.

871. Tam JH, Pasternak SH. Amyloid and Alzheimer's disease: inside and out. *Can J Neurol Sci* 2012;**39**:286–98.

872. Tanskanen M, Kalaria RN, Notkola IL, Makela M, Polvikoski T, et al. Relationships between white matter hyperintensities, cerebral amyloid angiopathy and dementia in a population-based sample of the oldest old. *Curr Alzheimer Res* 2013;**10**:1090–7.

873. Tanzi RE. The genetics of Alzheimer disease. *Cold Spring Harb Perspect Med* 2012;**2**:a006296

874. Tanzi RE, Moir RD, Wagner SL. Clearance of Alzheimer's Abeta peptide: the many roads to perdition. *Neuron* 2004;**43**:605–8.

875. Tartaglia MC, Sidhu M, Laluz V, Racine C, Rabinovici GD, et al. Sporadic corticobasal syndrome due to FTLD-TDP. *Acta Neuropathol* 2010;**119**:365–74.

876. Tartaglia MC, Rosen HJ, Miller BL. Neuroimaging in dementia. *Neurotherapeutics* 2011;**8**:82–92.

877. Tatemichi TK, Desmond DW, Prohovnik I, et al. Confusion and memory loss from capsular genu infarction: a thalamocortical disconnection syndrome? *Neurology* 1992;**42**:1966–79.

878. Tatemichi TK, Desmond DW, Prohovnik I. Strategic infarcts in vascular dementia. A clinical and brain imaging experience. *Arzneimittelforschung* 1995;**45**:371–85.

879. Tekirian TL, Cole GM, Russell MJ, et al. Carboxy terminal of beta-amyloid deposits in aged human, canine, and polar bear brains. *Neurobiol Aging* 1996;**17**:249–57.

880. Terry RD. The fine structure of neurofibrillary tangles in Alzheimer's disease. *J Neuropathol Exp Neurol* 1963;**22**:629–42.

881. Terry RD, Gonatas NK, Weiss M. Ultrastructural studies in Alzheimer's presenile dementia. *Am J Pathol* 1964;**44**:269–97.

882. Terry RD, Masliah E, Salmon DP, et al. Physical basis of cognitive alterations in Alzheimer's disease: synapse loss is the major correlate of cognitive impairment. *Ann Neurol* 1991;**30**:572–80.

883. Terwindt GM, Haan J, Ophoff RA, Groenen SM, Storimans CW, et al. Clinical and genetic analysis of a large Dutch family with autosomal dominant vascular retinopathy, migraine and Raynaud's phenomenon. *Brain* 1998;**121**:303–16.

884. Thal DR, Sassin I, Schultz C, Haass C, Braak E, Braak H. Fleecy amyloid deposits in the internal layers of the human entorhinal cortex are comprised of N-terminal truncated fragments of Abeta. *J Neuropathol Exp Neurol* 1999;**58**:210–6.

885. Thal DR, Rub U, Schultz C, Sassin I, Ghebremedhin E, et al. Sequence of Abeta-protein deposition in the human medial temporal lobe. *J Neuropathol Exp Neurol* 2000;**59**:733–48.

886. Thal DR, Rub U, Orantes M, Braak H. Phases of A beta-deposition in the human brain and its relevance for the development of AD. *Neurology* 2002;**58**:1791–800.

887. Thal DR, Schultz C, Botez G, Del Tredici K, Mrak RE, et al. The impact of argyrophilic grain disease on the development of dementia and its relationship to concurrent Alzheimer's disease-related pathology. *Neuropathol Appl Neurobiol* 2005;**31**:270–9.

888. Thal DR, Del Tredici K, Ludolph AC, Hoozemans JJ, Rozemuller AJ, et al. Stages of granulovacuolar degeneration: their relation to Alzheimer's disease and chronic stress response. *Acta Neuropathol* 2011;**122**:577–89.

889. Tiraboschi P, Hansen LA, Alford M, et al. The decline in synapses and cholinergic activity is asynchronous in Alzheimer's disease. *Neurology* 2000;**55**:1278–83.

890. Tiraboschi P, Hansen LA, Alford M, Sabbagh MN, Schoos B, et al. Cholinergic dysfunction in diseases with Lewy bodies. *Neurology* 2000;**54**:407–11.

891. Todman D. Arnold Pick (1851–1924). *J Neurol* 2009;**256**:504–5.

892. Togo T, Sahara N, Yen SH, Cookson N, Ishizawa T, et al. Argyrophilic grain disease is a sporadic 4-repeat tauopathy. *J Neuropathol Exp Neurol* 2002;**61**:547–56.

893. Togo T, Isojima D, Akatsu H, Suzuki K, Uchikado H, et al. Clinical features of argyrophilic grain disease: a retrospective survey of cases with neuropsychiatric symptoms. *Am J Geriatr Psychiatry* 2005;**13**:1083–91.

894. Tolnay M, Schwietert M, Monsch AU, et al. Argyrophilic grain disease: distribution of grains in patients with and without dementia. *Acta Neuropathol* 1997;**94**:353–8.

895. Tolnay M, Spillantini MG, Goedert M, et al. Argyrophilic grain disease: widespread hyperphosphorylation of tau protein in limbic neurons. *Acta Neuropathol* 1997;**93**:477–84.

896. Tolnay M, Mistl C, Ipsen S, Probst A. Argyrophilic grains of Braak: occurrence in dendrites of neurons containing hyperphosphorylated tau protein. *Neuropathol Appl Neurobiol* 1998;**24**:53–9.

897. Tolnay M, Calhoun M, Pham HC, et al. Low amyloid (Ab) plaque load and relative predominance of diffuse plaques distinguish argyrophilic grain disease from Alzheimer's disease. *Neuropathol Appl Neurobiol* 1999;**25**:295–305.

898. Tolnay M, Sergeant N, Ghestem A, Chalbot S, De Vos RA, et al. Argyrophilic grain disease and Alzheimer's disease are distinguished by their different distribution of tau protein isoforms. *Acta Neuropathol* 2002;**104**:425–34.

899. Toma AK, Stapleton S, Papadopoulos MC, Kitchen ND, Watkins LD. Natural history of idiopathic normal-pressure hydrocephalus. *Neurosurg Rev* 2011;**34**:433–9.

900. Tomlinson BE, Blessed G, Roth M. Observations on the brains of demented old people. *J Neurol Sci* 1970;**11**:205–42.

901. Tomonaga M. Hirano body in extraocular muscle. *Acta Neuropathol* 1983;**60**:309–13.

902. Trillo L, Das D, Hsieh W, Medina B, Moghadam S, et al. Ascending monoaminergic systems alterations in Alzheimer's disease. translating basic science into clinical care. *Neurosci Biobehav Rev* 2013;**37**:1363–79.

903. Troakes C, Hortobagyi T, Vance C, Al-Sarraj S, Rogelj B, Shaw CE. Transportin 1 colocalization with Fused in Sarcoma (FUS) inclusions is not characteristic for amyotrophic lateral sclerosis-FUS confirming disrupted nuclear import of mutant FUS and distinguishing it from frontotemporal lobar degeneration with FUS inclusions. *Neuropathol Appl Neurobiol* 2013;**39**:553–61.

904. Tsuboi Y, Dickson DW. Dementia with Lewy bodies and Parkinson's disease with dementia: are they different? *Parkinsonism Relat Disord* 2005;**11**:S47–51.

905. Tsuboi Y, Slowinski J, Josephs KA, Honer WG, Wszolek ZK, Dickson DW. Atrophy of superior cerebellar peduncle in progressive supranuclear palsy. *Neurology* 2003;**60**:1766–9.

906. Tsuboi Y, Josephs KA, Boeve BF, Litvan I, Caselli RJ, et al. Increased tau burden in the cortices of progressive supranuclear palsy presenting with corticobasal syndrome. *Mov Disord* 2005;**20**:982–8.

907. Tsuchiya K, Nakayama H, Iritani S, Arai T, Niizato K, et al. Distribution of basal ganglia lesions in diffuse neurofibrillary tangles with calcification: a clinicopathological study of five autopsy cases. *Acta Neuropathol* 2002;**103**:555–64.

908. Tsuji H, Arai T, Kametani F, Nonaka T, Yamashita M, et al. Molecular analysis and biochemical classification of TDP-43 proteinopathy. *Brain* 2012;**135**:3380–91.

909. Uchihara T, Tsuchiya K, Kosaka K. Selective loss of nigral neurons in Pick's disease: a morphometric study. *Acta Neuropathol* 1990;**81**:155–61.

910. Uchikado H, Lin WL, DeLucia MW, Dickson DW. Alzheimer disease with amygdala Lewy bodies: a distinct form of alpha-synucleinopathy. *J Neuropathol Exp Neurol* 2006;**65**:685–97.

911. Ukai K, Shibayama H, Ishihara R, Ozaki N. Distribution of neurofibrillary tangles in diffuse neurofibrillary tangles with calcification. *Psychiatry Clin Neurosci* 2009;**63**:646–51.

912. Urwin H, Josephs KA, Rohrer JD, Mackenzie IR, Neumann M, et al. FUS pathology defines the majority of tau- and TDP-43-negative frontotemporal lobar degeneration. *Acta Neuropathol* 2010;**120**:33–41.

913. Vahedi K, Massin P, Guichard JP, Miocque S, Polivka M, et al. Hereditary infantile hemiparesis, retinal arteriolar tortuosity, and leukoencephalopathy. *Neurology* 2003;**60**:57–63.

914. van Blitterswijk M, Baker MC, DeJesus-Hernandez M, Ghidoni R, Benussi L, et al. C9ORF72 repeat expansions in cases with previously identified pathogenic mutations. *Neurology* 2013;**81**:1332–41.

915. van de Pol L, Gertz HJ, Scheltens P, Wolf H. Hippocampal atrophy in subcortical vascular dementia. *Neurodegener Dis* 2011;**8**:465–9.

916. Van Deerlin VM, Sleiman PM, Martinez-Lage M, Chen-Plotkin A, Wang LS, et al. Common variants at 7p21 are associated with frontotemporal lobar degeneration with TDP-43 inclusions. *Nat Genet* 2010;**42**:234–9.

917. Van der Flier WM, Cordonnier C. Microbleeds in vascular dementia: clinical aspects. *Exp Gerontol* 2012;**47**:853–7.

918. van der Zee J, Gijselinck I, Dillen L, Van Langenhove T, Theuns J, et al. A pan-European study of the *C9orf72* repeat associated with FTLD: geographic prevalence, genomic instability, and intermediate repeats. *Hum Mutat* 2013;**34**:363–73.

919. van der Zee J, Van Langenhove T, Kovacs GG, Dillen L, Deschamps W, et al. Rare mutations in *SQSTM1* modify susceptibility to frontotemporal lobar degeneration. *Acta Neuropathol* 2014;published online Jun 5 (E-pub ahead of print).

920. van Duijn CM, Clayton D, Chandra V, et al. Familial aggregation of Alzheimer's disease and related disorders: a collaborative re-analysis of case-control studies. EURODEM Risk Factors Research Group. *Int J Epidemiol* 1991;**20**:S13–20.

921. Vandenabeele P, Galluzzi L, Vanden Berghe T, Kroemer G. Molecular mechanisms of necroptosis: an ordered cellular explosion. *Nat Rev Mol Cell Biol* 2010;**11**:700–14.

922. Vann Jones SA, O'Brien JT. The prevalence and incidence of dementia with Lewy bodies: a systematic review of population and clinical studies. *Psychol Med* 2014;**44**:673–83.

923. Vassar R. BACE1: the beta-secretase enzyme in Alzheimer's disease. *J Mol Neurosci* 2004;**23**:105–14.

924. Vassar R, Bennett BD, Babu-Khan S, et al. Beta-secretase cleavage of Alzheimer's amyloid precursor protein by the transmembrane aspartic protease BACE. *Science* 1999;**286**:735–41.

925. Vermeer SE, Prins ND, den Heijer T, Hofman A, Koudstaal PJ, Breteler MM. Silent brain infarcts and the risk of dementia and cognitive decline. *N Engl J Med* 2003;**348**:1215–22.

926. Verreault S, Joutel A, Riant F, Neves G, Rui Silva M, et al. A novel hereditary small vessel disease of the brain. *Ann Neurol* 2006;**59**:353–7.

927. Vidal R, Frangione B, Rostagno A, et al. A stop-codon mutation in the *BRI* gene associated with familial British dementia. *Nature* 1999;**399**:776–81.

928. Vidal R, Revesz T, Rostagno A, Kim E, Holton JL, et al. A decamer duplication in the 3′ region of the *BRI* gene originates an amyloid peptide that is associated with dementia in a Danish kindred. *Proc Natl Acad Sci U S A* 2000;**97**:4920–5.

929. Vieira RT, Caixeta L, Machado S, Silva AC, Nardi AE, et al. Epidemiology of early-onset dementia: a review of the literature. *Clin Pract Epidemiol Ment Health* 2013;**9**:88–95.

930. Villela D, Kimura L, Schlesinger D, Goncalves A, Pearson PL, et al. Germline DNA copy number variation in individuals with argyrophilic grain disease reveals *CTNS* as a plausible candidate gene. *Genet Mol Biol* 2013;**36**:498–501.

931. Vinters HV, Ellis WG, Zarow C, Zaias BW, Jagust WJ, et al. Neuropathologic substrates of ischemic vascular dementia. *J Neuropathol Exp Neurol* 2000;**59**:931–45.

932. Viswanathan A, Patel P, Rahman R, Nandigam RN, Kinnecom C, et al. Tissue microstructural changes are independently associated with cognitive impairment in cerebral amyloid angiopathy. *Stroke* 2008;**39**:1988–92.

933. Vonsattel JP, Myers RH, Hedley-Whyte ET, et al. Cerebral amyloid angiopathy without and with cerebral hemorrhages: a comparative histological study. *Ann Neurol* 1991;**30**:637–49.

934. Wakabayashi K, Hayashi S, Kakita A, et al. Accumulation of alpha-synuclein/NACP is a cytopathological feature common to Lewy body disease and multiple system atrophy. *Acta Neuropathol* 1998;**96**:445–52.

935. Wakayama I, Kihira T, Yoshida S, Garruto RM. Rare neuropil threads in amyotrophic lateral sclerosis and parkinsonism–dementia on Guam and in the Kii Peninsula of Japan. *Dementia* 1993;**4**:75–80.

936. Walker LC, Diamond MI, Duff KE, Hyman BT. Mechanisms of protein seeding in neurodegenerative diseases. *JAMA Neurol* 2013;**70**:304–10.

937. Wardlaw JM. Blood-brain barrier and cerebral small vessel disease. *J Neurol Sci* 2010;**299**:66–71.

938. Wardlaw JM, Doubal F, Armitage P, Chappell F, Carpenter T, et al. Lacunar stroke is associated with diffuse blood-brain barrier dysfunction. *Ann Neurol* 2009;**65**:194–202.

939. Warren JD, Rohrer JD, Rossor MN. Clinical review. Frontotemporal dementia. *BMJ* 2013;**347**:f4827.

940. Warren JD, Rohrer JD, Schott JM, Fox NC, Hardy J, Rossor MN. Molecular nexopathies: a new paradigm of neurodegenerative disease. *Trends Neurosci* 2013;**36**:561–9.

941. Wattendorff AR, Bots GT, Went LN, Endtz LJ. Familial cerebral amyloid angiopathy presenting as recurrent cerebral haemorrhage. *J Neurol Sci* 1982;**55**:121–35.

942. Watts GD, Wymer J, Kovach MJ, Mehta SG, Mumm S, et al. Inclusion body myopathy associated with Paget disease of bone and frontotemporal dementia is caused by mutant valosin-containing protein. *Nat Genet* 2004;**36**:377–81.

943. Weerasinghe P, Buja LM. Oncosis: an important non-apoptotic mode of cell death. *Exp Mol Pathol* 2012;**93**:302–8.

944. Weiler R, Lassmann H, Fischer P, et al. A high ratio of chromogranin A to synaptin/synaptophysin is a common feature of brains in Alzheimer and Pick disease. *FEBS Lett* 1990;**263**:337–9.

945. Weiner MW, Veitch DP, Aisen PS, Beckett LA, Cairns NJ, et al. The Alzheimer's Disease Neuroimaging Initiative: a review of papers published since its inception. *Alzheimer's Dement* 2013;**9**:e111–94.

946. Weller RO, Massey A, Newman TA, et al. Cerebral amyloid angiopathy: amyloid beta accumulates in putative interstitial fluid drainage pathways in Alzheimer's disease. *Am J Pathol* 1998;**153**:725–33.

947. Weller RO, Massey A, Kuo YM, Roher AE. Cerebral amyloid angiopathy: accumulation of A beta in interstitial fluid drainage pathways in Alzheimer's disease. *Ann N Y Acad Sci* 2000;**903**:110–17.

948. Weller RO, Djuanda E, Yow HY, Carare RO. Lymphatic drainage of the brain and the pathophysiology of neurological disease. *Acta Neuropathol* 2009;**117**:1–14.

949. Wentzell J, Kretzschmar D. Alzheimer's disease and tauopathy studies in flies and worms. *Neurobiol Dis* 2010;**40**:21–8.

950. Werring DJ, Gregoire SM, Cipolotti L. Cerebral microbleeds and vascular cognitive impairment. *J Neurol Sci* 2011;**299**:131–5.

951. Westover MB, Bianchi MT, Yang C, Schneider JA, Greenberg SM. Estimating cerebral microinfarct burden from autopsy samples. *Neurology* 2013;**80**:1365–9.

952. Whalley LJ, Dick FD, McNeill G. A life-course approach to the aetiology of late-onset dementias. *Lancet Neurol* 2006;**5**:87–96.

953. White L. Brain lesions at autopsy in older Japanese-American men as related to cognitive impairment and dementia in the final years of life: a summary report from the Honolulu-Asia aging study. *J Alzheimer's Dis* 2009;**18**:713–25.

954. White L, Petrovitch H, Hardman J, Nelson J, Davis DG, et al. Cerebrovascular pathology and dementia in autopsied Honolulu-Asia Aging Study participants. *Ann N Y Acad Sci* 2002;**977**:9–23.

955. White L, Small BJ, Petrovitch H, Ross GW, Masaki K, et al. Recent clinical-pathologic research on the causes of dementia in late life: update from the Honolulu-Asia Aging Study. *J Geriatr Psychiatry Neurol* 2005;**18**:224–7..

956. Whitehouse PJ, Price DL, Struble RG, et al. Alzheimer's disease and senile dementia: loss of neurons in the basal forebrain. *Science* 1982;**215**:1237–9.

957. Wiberg P, Waern M, Billstedt E, Ostling S, Skoog I. Secular trends in the prevalence of dementia and depression in Swedish septuagenarians 1976-2006. *Psychol Med* 2013;**43**:2627–34.

958. Wiederkehr S, Simard M, Fortin C, van Reekum R. Comparability of the clinical diagnostic criteria for vascular dementia: a critical review. Part I. *J Neuropsychiatry Clin Neurosci* 2008;**20**:150–61.

959. Wiederkehr S, Simard M, Fortin C, van Reekum R. Validity of the clinical diagnostic criteria for vascular dementia: a critical review. Part II. *J Neuropsychiatry Clin Neurosci* 2008;**20**:162–77.

960. Wilson AC, Dugger BN, Dickson DW, Wang DS. TDP-43 in aging and Alzheimer's disease - a review. *Int J Clin Exp Pathol* 2011;**4**:147–55.

961. Wilson RS, Li Y, Aggarwal NT, Barnes LL, McCann JJ, et al. Education and the course of cognitive decline in Alzheimer disease. *Neurology* 2004;**63**:1198–202.

962. Wilson RS, Yu L, Trojanowski JQ, Chen EY, Boyle PA, et al. TDP-43 pathology, cognitive decline, and dementia in old age. *JAMA Neurol* 2013;**70**:1418–24.

963. Wischik CM, Novak M, Thogersen HC, et al. Isolation of a fragment of tau derived from the core of the paired helical filament of Alzheimer disease. *Proc Natl Acad Sci U S A* 1988;**85**:4506–10.

964. Wisniewski HM, Coblentz JM, Terry RD. Pick's disease. A clinical and ultrastructural study. *Arch Neurol* 1972;**26**:97–108.

965. Wisniewski HM, Narang HK, Terry RD. Neurofibrillary tangles of paired helical filaments. *J Neurol Sci* 1976;**27**:173–81.

966. Wisniewski HM, Merz PA, Iqbal K. Ultrastructure of paired helical filaments of Alzheimer's neurofibrillary tangle. *J Neuropathol Exp Neurol* 1984;**43**:643–56.

967. Wisniewski HM, Bancher C, Barcikowska M, Wen GY, Currie J. Spectrum of morphological appearance of amyloid deposits in Alzheimer's disease. *Acta Neuropathol* 1989;**78**:337–47.

968. Wisniewski HM, Sadowski M, Jakubowska-Sadowska K, Tarnawski M, Wegiel J. Diffuse, lake-like amyloid-beta deposits in the parvopyramidal layer of the presubiculum in Alzheimer disease. *J Neuropathol Exp Neurol* 1998;**57**:674–83.

969. Wisniewski K, Jervis GA, Moretz RC, Wisniewski HM. Alzheimer neurofibrillary tangles in diseases other than senile and presenile dementia. *Ann Neurol* 1979;**5**:288–94.

970. Wolfe MS. The gamma-secretase complex: membrane-embedded proteolytic ensemble. *Biochemistry* 2006;**45**:7931–9.

971. Wood JG, Mirra SS, Pollock NJ, Binder LI. Neurofibrillary tangles of Alzheimer disease share antigenic determinants with the axonal microtubule-associated protein tau (tau). *Proc Natl Acad Sci U S A* 1986;**83**:4040–3.

972. Wu YT, Lee HY, Norton S, Chen C, Chen H, et al. Prevalence studies of dementia in mainland China, Hong Kong and Taiwan: a systematic review and meta-analysis. *PLoS One* 2013;**8**:e66252.

973. Xu M, Shibayama H, Kobayashi H, et al. Granulovacuolar degeneration in the hippocampal cortex of aging and demented

patients—a quantitative study. *Acta Neuropathol* 1992;**85**:1–9.

974. Yagishita S, Itoh Y, Nakano T, et al. Crystalloid inclusions reminiscent of Hirano bodies in autolyzed peripheral nerve of normal Wistar rats. *Acta Neuropathol* 1979;**47**:231–6.

975. Yagishita S, Itoh Y, Nan W, Amano N. Reappraisal of the fine structure of Alzheimer's neurofibrillary tangles. *Acta Neuropathol* 1981;**54**:239–46.

976. Yamada M. Senile dementia of the neurofibrillary tangle type (tangle-only dementia): neuropathological criteria and clinical guidelines for diagnosis. *Neuropathology* 2003;**23**:311–7.

977. Yamaguchi H, Hirai S, Morimatsu M, et al. Diffuse type of senile plaques in the brains of Alzheimer-type dementia. *Acta Neuropathol* 1988;**77**:113–19.

978. Yamaguchi H, Nakazato Y, Shoji M, et al. Ultrastructure of the neuropil threads in the Alzheimer brain: their dendritic origin and accumulation in the senile plaques. *Acta Neuropathol* 1990;**80**:368–74.

979. Yamaguchi H, Nakazato Y, Kawarabayashi T, et al. Extracellular neurofibrillary tangles associated with degenerating neurites and neuropil threads in Alzheimer-type dementia. *Acta Neuropathol* 1991;**81**:603–9.

980. Yamaguchi H, Nakazato Y, Shoji M, et al. Ultrastructure of diffuse plaques in senile dementia of the Alzheimer type: comparison with primitive plaques. *Acta Neuropathol* 1991;**82**:13–20.

981. Yamamoto T, Hirano A. Hirano bodies in the perikaryon of the Purkinje cell in a case of Alzheimer's disease. *Acta Neuropathol* 1985;**67**:167–9.

982. Yamamoto Y, Ihara M, Tham C, Low RW, Slade JY, et al. Neuropathological correlates of temporal pole white matter hyperintensities in CADASIL. *Stroke* 2009;**40**:2004–11.

983. Yamamoto Y, Craggs L, Baumann M, Kalimo H, Kalaria RN. Molecular genetics and pathology of hereditary small vessel diseases of the brain. *Neuropathol Appl Neurobiol* 2011;**37**:94–113.

984. Yanagihara RT, Garruto RM, Gajdusek DC. Epidemiological surveillance of amyotrophic lateral sclerosis and parkinsonism-dementia in the Commonwealth of the Northern Mariana Islands. *Ann Neurol* 1983;**13**:79–86.

985. Yarchoan M, Xie SX, Kling MA, Toledo JB, Wolk DA, et al. Cerebrovascular atherosclerosis correlates with Alzheimer pathology in neurodegenerative dementias. *Brain* 2012;**135**:3749–56.

986. Yasuda M, Maeda S, Kawamata T, et al. Novel presenilin-1 mutation with widespread cortical amyloid deposition but limited cerebral amyloid angiopathy. *J Neurol Neurosurg Psychiatry* 2000;**68**:220–3.

987. Yasuhara O, Kawamata T, Aimi Y, et al. Expression of chromogranin A in lesions in the central nervous system from patients with neurological diseases. *Neurosci Lett* 1994;**170**:13–6.

988. Yokota O, Ishizu H, Terada S, Tsuchiya K, Haraguchi T, et al. Preservation of nigral neurons in Pick's disease with Pick bodies: a clinicopathological and morphometric study of five autopsy cases. *J Neurol Sci* 2002;**194**:41–8.

989. Yokota O, Terada S, Ishizu H, Ujike H, Ishihara T, et al. Variability and heterogeneity in Alzheimer's disease with cotton wool plaques: a clinicopathological study of four autopsy cases. *Acta Neuropathol* 2003;**106**:348–56.

990. Yu CE, Bird TD, Bekris LM, Montine TJ, Leverenz JB, et al. The spectrum of mutations in progranulin: a collaborative study screening 545 cases of neurodegeneration. *Arch Neurol* 2010;**67**:161–70.

991. Zanetti O, Solerte SB, Cantoni F. Life expectancy in Alzheimer's disease (AD). *Arch Gerontol Geriatr* 2009;**49**:237–43.

992. Zarow C, Sitzer TE, Chui HC. Understanding hippocampal sclerosis in the elderly: epidemiology, characterization, and diagnostic issues. *Curr Neurol Neurosci Rep* 2008;**8**:363–70.

993. Zarow C, Vinters HV, Ellis WG, Weiner MW, Mungas D, et al. Correlates of hippocampal neuron number in Alzheimer's disease and ischemic vascular dementia. *Ann Neurol* 2005;**57**:896–903.

994. Zarow C, Weiner MW, Ellis WG, Chui HC. Prevalence, laterality, and comorbidity of hippocampal sclerosis in an autopsy sample. *Brain Behav* 2012;**2**:435–42.

995. Zekry D, Duyckaerts C, Moulias R, Belmin J, Geoffre C, et al. Degenerative and vascular lesions of the brain have synergistic effects in dementia of the elderly. *Acta Neuropathol* 2002;**103**:481–7.

996. Zekry D, Hauw JJ, Gold G. Mixed dementia: epidemiology, diagnosis, and treatment. *J Am Geriatr Soc* 2002;**50**:1431–8.

997. Zhan SS, Beyreuther K, Schmitt HP. Synaptophysin immunoreactivity of the cortical neuropil in vascular dementia of Binswanger type compared with the dementia of Alzheimer type and nondemented controls. *Dementia* 1994;**5**:79–87.

998. Zhang Y, Xu Y, Nie H, Lei T, Wu Y, et al. Prevalence of dementia and major dementia subtypes in the Chinese populations: a meta-analysis of dementia prevalence surveys, 1980–2010. *J Clin Neurosci* 2012;**19**:1333–7.

999. Zhu X, Raina AK, Perry G, Smith MA. Alzheimer's disease: the two-hit hypothesis. *Lancet Neurol* 2004;**3**:219–26.

1000. Zimmerman H. *Monthly report to medical officer in command; US Naval Medical Research Unit No. 2.* Guam: Micronesian Area Research Center; 1945.

Index